CHILTON®

ASIAN DIAGNOSTIC SERVICE 2006 EDITION VOLUME II

Hyundai
Infiniti
Isuzu
Kia
Mazda
Mitsubishi
Nissan

Australia • Canada • Mexico • Singapore • Spain • United Kingdom • United States

CHILTON®
ASIAN
DIAGNOSTIC SERVICE
2006 Edition
Volume II
Hyundai, Infiniti, Isuzu, Kia Mazda, Mitsubishi, Nissan

Vice President,
Technology Professional Business Unit:
Gregory L. Clayton

Publisher,
Technology Professional Business Unit:
David Koontz

Director of Marketing:
Beth A. Lutz

Production Director:
Patty Stephan

Editorial Assistant:
Rebecca Rokitowski

Production Manager:
Andrew Crouth

Marketing Manager:
Brian McGrath

Marketing Coordinator:
Jennifer Stall

Publishing Coordinator:
Paula Baillie

Sr. Content Project Manager:
Elizabeth C. Hough

Managing Editor:
Terry Blomquist

Editors:
Terry Blomquist
Tim Crain
Nick D'Andrea
Wayne Eiffes
Holly McBain
Ryan Price

Graphical Designer:
Melinda Possinger

Printed in the United States of America
1 2 3 4 5 6 XX 10 09 08 07 06

For more information contact:
Thomson Delmar Learning
Executive Woods
5 Maxwell Drive, PO Box 8007,
Clifton Park, NY 12065-8007
Or find us on the World Wide Web at
www.delmarlearning.com

ISBN: **1-4180-2914-9**

TABLE OF CONTENTS

SECTIONS

1 INTRODUCTION TO OBD SYSTEMS

2 INDENTIFYING THE PROBLEM

3 INTRODUCTION TO OBD DIAGNOSTIC SYSTEMS

4 SYMPTOM DIAGNOSIS (NO CODES)

5 **HYUNDAI** - DIAGNOSTIC TROUBLE CODES

6 **HYUNDAI** - COMPONENT TESTING

7 **INFINITI** - DIAGNOSTIC TROUBLE CODES

8 **INFINITI** - COMPONENT TESTING

9 **ISUZU** - DIAGNOSTIC TROUBLE CODES

10 **KIA** - DIAGNOSTIC TROUBLE CODES

11 **KIA** - COMPONENT TESTING

12 **MAZDA** - DIAGNOSTIC TROUBLE CODES

13 **MAZDA** - COMPONENT TESTING

14 **MITSUBISHI** - DIAGNOSTIC TROUBLE CODES

15 **MITSUBISHI** - COMPONENT TESTING

16 **NISSAN** - DIAGNOSTIC TROUBLE CODES

17 **NISSAN** - COMPONENT TESTING

USING THIS INFORMATION

Organization

To find where a particular model section or procedure is located, look in the Table of Contents. Main topics are listed with the page number on which they may be found. Following the main topics is a listing of all of the subjects within the section and their page numbers.

Manufacturer and Model Coverage

This product covers 1996-2006 Asian models that are produced in sufficient quantities to warrant coverage, and which have technical content available from the vehicle manufacturers before our publication date. Although this information is as complete as possible at the time of publication, some manufacturers may make changes which cannot be included here. While striving for total accuracy, the publisher cannot assume responsibility for any errors, changes, or omissions that may occur in the compilation of this data.

Part Numbers & Special Tools

Part numbers and special tools are recommended by the publisher and vehicle manufacturer to perform specific jobs. Before substituting any part or tool for the one recommended, you must be completely satisfied that neither your personal safety, nor the performance of the vehicle will be endangered.

ACKNOWLEDGEMENT

The publisher would like to express appreciation to the following vehicle manufacturers for their assistance in producing this publication. No further reproduction or distribution of the material in this manual is allowed without the expressed written permission of the vehicle manufacturers and the publisher. Hyundai Group, including Hyundai and Kia Motor, Isuzu Motors America, Inc., Mazda Motor Corporation, Mitsubishi Motors North America, Inc., Nissan North America, including Infiniti and Nissan Divisions.

PRECAUTIONS

Before servicing any vehicle, please be sure to read all of the following precautions, which deal with personal safety, prevention of component
damage, and important points to take into consideration when servicing a motor vehicle:

- Always wear safety glasses or goggles when drilling, cutting, grinding or prying.
- Steel-toed work shoes should be worn when working with heavy parts. Pockets should not be used for carrying tools. A slip or fall can drive a screwdriver into your body.
- Work surfaces, including tools and the floor should be kept clean of grease, oil or other slippery material.
- When working around moving parts, don't wear loose clothing. Long hair should be tied back under a hat or cap, or in a hair net.
- Always use tools only for the purpose for which they were designed. Never pry with a screwdriver.
- Keep a fire extinguisher and first aid kit handy.
- Always properly support the vehicle with approved stands or lift.
- Always have adequate ventilation when working with chemicals or hazardous material.
- Carbon monoxide is colorless, odorless and dangerous. If it is necessary to operate the engine with vehicle in a closed area such as a garage, always use an exhaust collector to vent the exhaust gases outside the closed area.
- When draining coolant, keep in mind that small children and some pets are attracted by ethylene glycol antifreeze, and are quite likely to drink any left in an open container, or in puddles on the ground. This will prove fatal in sufficient quantity. Always drain the coolant into a sealable container.
- To avoid personal injury, do not remove the coolant pressure relief cap while the engine is operating or hot. The cooling system is under pressure; steam and hot liquid can come out forcefully when the cap is loosened slightly. Failure to follow these instructions may result in personal injury. The coolant must be recovered in a suitable, clean container for reuse. If the coolant is contaminated it must be recycled or disposed of correctly.
- When carrying out maintenance on the starting system be aware that heavy gauge leads are connected directly to the battery. Make sure the protective caps are in place when maintenance is completed. Failure to follow these instructions may result in personal injury.
- Do not remove any part of the engine emission control system. Operating the engine without the engine emission control system will reduce fuel economy and engine ventilation. This will weaken engine performance and shorten engine life. It is also a violation of Federal law.
- Due to environmental concerns, when the air conditioning system is drained, the refrigerant must be collected using refrigerant recovery/recycling equipment. Federal law requires that refrigerant be recovered into appropriate recovery equipment and the process be conducted by qualified technicians who have been certified by an approved organization, such as MACS, ASI, etc. Use of a recovery machine dedicated to the appropriate refrigerant is necessary to reduce the possibility of oil and refrigerant incompatibility concerns. Refer to the instructions provided by the equipment manufacturer when removing refrigerant from or charging the air conditioning system.

• Always disconnect the battery ground when working on or around the electrical system.

• Batteries contain sulfuric acid. Avoid contact with skin, eyes, or clothing. Also, shield your eyes when working near batteries to protect against possible splashing of the acid solution. In case of acid contact with skin or eyes, flush immediately with water for a minimum of 15 minutes and get prompt medical attention. If acid is swallowed, call a physician immediately. Failure to follow these instructions may result in personal injury.

• Batteries normally produce explosive gases. Therefore, do not allow flames, sparks or lighted substances to come near the battery. When charging or working near a battery, always shield your face and protect your eyes. Always provide ventilation. Failure to follow these instructions may result in personal injury.

• When lifting a battery, excessive pressure on the end walls could cause acid to spew through the vent caps, resulting in personal injury, damage to the vehicle or battery. Lift with a battery carrier or with your hands on opposite corners. Failure to follow these instructions may result in personal injury.

• Observe all applicable safety precautions when working around fuel. Whenever servicing the fuel system, always work in a well-ventilated area. Do not allow fuel spray or vapors to come in contact with a spark, open flame, or excessive heat (a hot drop light, for example). Keep a dry chemical fire extinguisher near the work area.
Always keep fuel in a container specifically designed for fuel storage; also, always properly seal fuel containers to avoid the possibility of fire or explosion. Do not smoke or carry lighted tobacco or open flame of any type when
working on or near any fuel related components.

• Fuel injection systems often remain pressurized, even after the engine has been turned OFF. The fuel system pressure must be relieved before disconnecting any fuel lines. Failure to do so may result in fire and/or personal injury.

• The evaporative emissions system contains fuel vapor and condensed fuel vapor. Although not present in large quantities, it still presents the danger of explosion or fire. Disconnect the battery ground cable from the battery to minimize the possibility of an electrical spark occurring, possibly causing a fire or explosion if fuel vapor or liquid fuel is present in the area. Failure to follow these instructions can result in personal injury.

• The EPA warns that prolonged contact with used engine oil may cause a number of skin disorders, including cancer! You should make every effort to minimize your exposure to used engine oil. Protective gloves should be worn when changing oil. Wash your hands and any other exposed skin
areas as soon as possible after exposure to used engine oil. Soap and water, or waterless hand cleaner should be used.

• Some vehicles are equipped with an air bag system, often referred to as a Supplemental Restraint System (SRS) or Supplemental Inflatable Restraint (SIR) system. The system must be
disabled before performing service on or around system components, steering column, instrument panel components, wiring and sensors. Failure to follow safety and disabling procedures could result in accidental air bag deployment, possible personal injury and unnecessary system repairs.

• Always wear safety goggles when working with, or around, the air bag system. When carrying a non-deployed air bag, be sure the bag and trim cover are pointed away from your body. When placing a non-deployed air bag on a work surface, always face the bag and trim cover upward, away from the
surface. This will reduce the motion of the module if it is accidentally deployed.

• Electronic modules are sensitive to electrical charges. The ABS module can be damaged if exposed to these charges.

• Brake pads and shoes may contain asbestos, which has been determined to be a cancer-causing agent. Never clean brake surfaces with compressed air. Avoid inhaling brake dust. Clean all brake surfaces with a commercially available brake cleaning fluid.

• When replacing brake pads, shoes, discs or drums, replace them as complete axle sets.

• When servicing drum brakes, disassemble and assemble one side at a time, leaving the remaining side intact for reference.

• Brake fluid often contains polyglycol ethers and polyglycols. Avoid contact with the eyes and wash your hands thoroughly after handling brake fluid. If you do get brake fluid in your eyes, flush your eyes with clean, running water for 15 minutes. If eye irritation persists, or if you have taken brake fluid internally, immediately seek medical assistance.

• Clean, high quality brake fluid from a sealed container is essential to the safe and proper operation of the brake system. You should always buy the correct type of brake fluid for your vehicle. If the brake fluid becomes contaminated, completely flush the system with new fluid. Never reuse any brake fluid. Any brake fluid that is removed from the system should be discarded. Also, do not allow any brake fluid to come in contact with a painted or plastic surface; it will damage the paint.

• Never operate the engine without the proper amount and type of engine oil; doing so will result in severe engine damage.

• Timing belt maintenance is extremely important! Many models utilize an interference-type, non-freewheeling engine. If the timing belt breaks, the valves in the cylinder head may strike the pistons, causing potentially serious (also time-consuming and expensive) engine damage.

• Disconnecting the negative battery cable on some vehicles may interfere with the functions of the on-board computer system(s) and may require the computer to undergo a relearning process once the negative battery cable is reconnected.

• Steering and suspension fasteners are critical parts because they affect performance of vital components and systems and their failure can result in major service expense. They must be replaced with the same grade or part number or an equivalent part if replacement is necessary. Do not use a replacement part of lesser quality or substitute design. Torque values must be used as specified during reassembly to ensure proper retention of these parts.

INTRODUCTION TO OBD SYSTEMS

1

Table of Contents

NOTES & CAUTIONS

Notes & Cautions 1-2

PRELIMINARY DIAGNOSTICS

History of OBD Systems 1-2
OBD I System Diagnostics 1-2
Changes In Diagnostic Routines 1-2
OBD II System Overview 1-4
Common Terminology 1-4

DIAGNOSTIC TOOLS & CIRCUIT TESTING

Hand Tools & Meter Operation 1-4
Scan Tools 1-4
Malfunction Indicator Lamp 1-4
Electronic Controls 1-4
Electricity & Electrical Circuits 1-4
Circuit Testing Tools 1-5

EFFECTIVE DIAGNOSTICS

Getting Started 1-5

INTRODUCTION TO OBD

Contents

NOTES & CAUTIONS

Notes & Cautions .... Page 1-6

PRELIMINARY DIAGNOSTICS

History of OBD Systems .... Page 1-6
OBD I System Diagnostics .... Page 1-7
Changes In Diagnostic Routines .... Page 1-7
OBD II System Overview .... Page 1-9
Common Terminology .... Page 1-9

DIAGNOSTIC TOOLS & CIRCUIT TESTING

Hand Tools & Meter Operation .... Page 1-10
Scan Tools .... Page 1-10
Malfunction Indicator Lamp .... Page 1-10
Electronic Controls .... Page 1-10
Electricity & Electrical Circuits .... Page 1-11
Circuit Testing Tools .... Page 1-11

EFFECTIVE DIAGNOSTICS

Getting Started.......................... Page 1-11

Notes & Cautions

Before servicing any vehicle, please be sure to read all of the following precautions, which deal with personal safety, prevention of component damage, and important points to take into consideration when servicing a motor vehicle:

- Observe all applicable safety precautions when working around fuel. Whenever servicing the fuel system, always work in a well-ventilated area. Do NOT allow fuel spray or vapors to come in contact with a spark, open flame, or excessive heat (a hot drop light, for example). Keep a dry chemical fire extinguisher near the work area. Always keep fuel in a container specifically designed for fuel storage; also, always properly seal fuel containers to avoid the possibility of fire or explosion. Refer to the additional fuel system precautions that follow.
- Fuel injection systems often remain pressurized, even after the engine has been turned OFF. The fuel system pressure must be relieved before disconnecting any fuel lines. Failure to do so may result in fire and/or personal injury.
- Brake fluid often contains Polyglycol Ethers and Polyglycols. Avoid contact with the eyes and wash your hands thoroughly after handling brake fluid. If you do get brake fluid in your eyes, flush your eyes with clean, running water for 15 minutes. If eye irritation persists, or if you have taken brake fluid internally, IMMEDIATELY seek medical assistance.
- The EPA warns that prolonged contact with used engine oil may cause a number of skin disorders, including cancer. You should make every effort to minimize your exposure to used engine oil. Protective gloves should be worn when changing oil. Wash your hands and any other exposed skin areas as soon as possible after exposure to used engine oil. Soap and water, or waterless hand cleaner should be used.
- The air bag system must be disabled (negative battery cable disconnected and/or air bag system main fuse removed) for at least 30 seconds before performing service on or around system components, steering column, instrument panel components, wiring and sensors. Failure to follow safety and disabling procedures could result in accidental air bag deployment, possible personal injury and unnecessary system repairs.
- Always wear safety goggles when working with, or around, the air bag system. When carrying a non-deployed air bag, be sure the bag and trim cover are pointed away from your body. When placing a non-deployed air bag on a work surface, always face the bag and trim cover upward, away from the surface. This will reduce the motion of the module if it is accidentally deployed. Refer to the additional air bag system precautions later in this section.
- Disconnecting the negative battery cable on some vehicles may interfere with the functions of the on-board computer system(s) and may require the computer to undergo a relearning process once the negative battery cable is reconnected.
- It is critically important to observe all instructions regarding ground disconnects, ignition switch positions, etc., in each diagnostic routine provided. Ignoring these instructions can result in false readings, damage to electronic components or circuits, or personal injury.

Preliminary Diagnostics

HISTORY OF OBD SYSTEMS

Starting in 1978, several vehicle manufacturers introduced a new type of control for several vehicle systems and computer control of engine management systems. These computer-controlled systems included programs to test for problems in the engine mechanical area, electrical fault identification and tests to help diagnose the computer

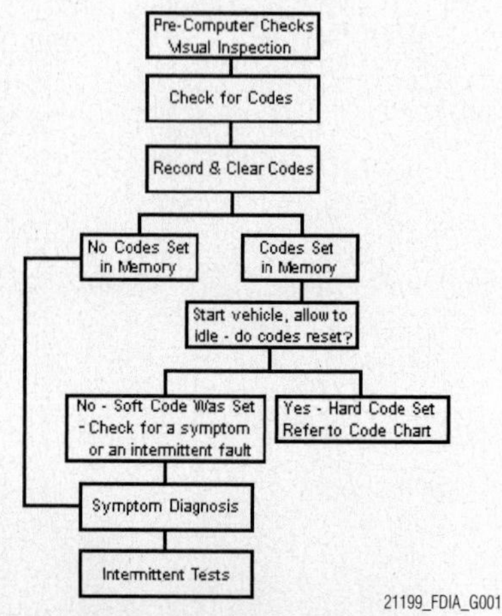

Fig. 1 OBD I diagnostic flow chart

OBD I SYSTEM DIAGNOSTICS

One of the most important things to understand about the automotive repair industry is the fact that you have to continually learn new systems and new diagnostic routines (the test procedures designed to isolate a problem on a vehicle system). For OBD I and II systems, a diagnostic routine can be defined as a procedure (a series of steps) that you follow to find the cause of a problem, make a repair and then verify the problem is fixed.

CHANGES IN DIAGNOSTIC ROUTINES

In some cases, a new Engine Control system may be similar to an earlier system, but it can have more indepth control of vehicle emissions, input and output devices and it may include a diagnostic "monitor" embedded in the engine controller designed to run a thorough set of emission control system tests.

OBD I Diagnostic Flowchart

See Figure 1.

The OBD I Diagnostic Flowchart on this page can be used to find the cause of problems related to Engine Control system trouble codes or driveability symptoms detected on OBD I systems. It includes a step-by-step procedure to use to repair these systems. Compare this flowchart with the one used on OBD II systems.

The steps in this flow chart should be followed as described (from top to bottom).

- Do the Pre-Computer Checks.
- Check for any trouble codes stored in memory.
- Read the trouble codes - If trouble codes are set, record them and then clear the codes.

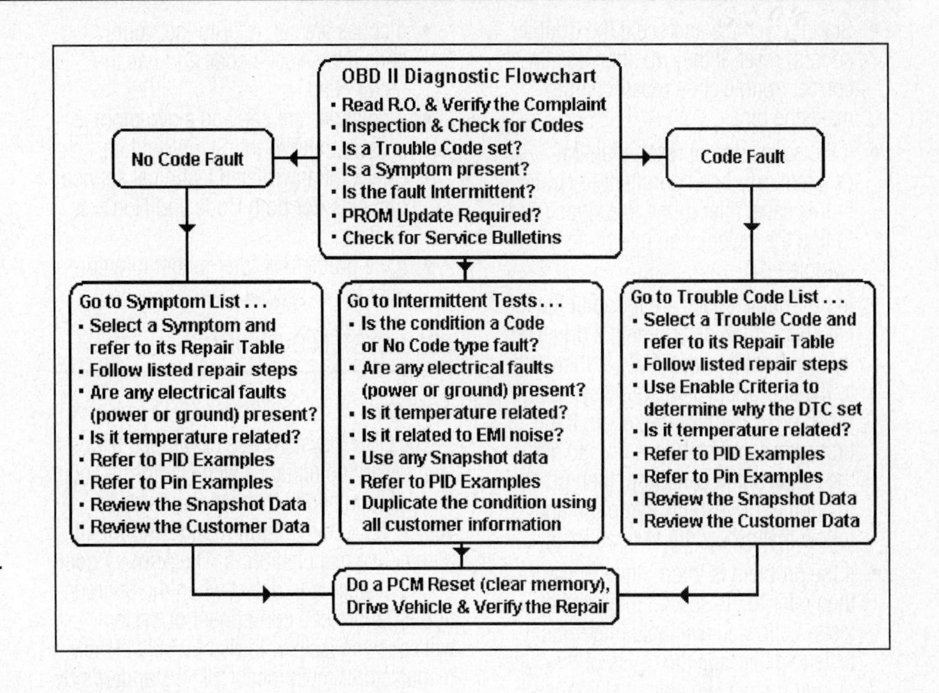

Fig. 2 OBD II diagnostic flow chart

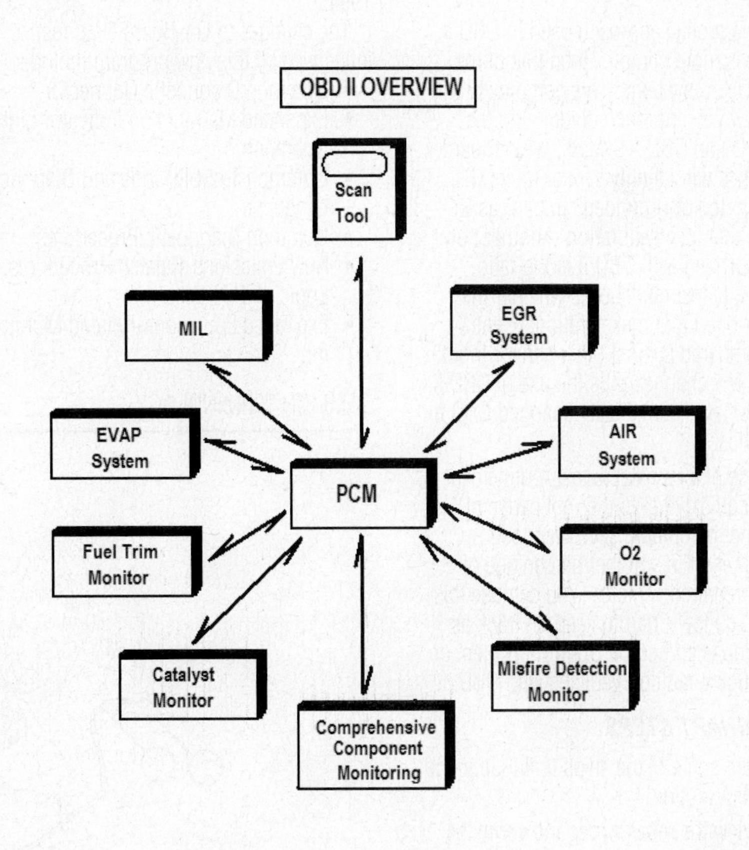

Fig. 3 PCM inputs and outputs

- Start the vehicle and see if the trouble code(s) reset. If they do, then use the correct trouble code repair chart to make the repair.
- If the codes do not reset, than the problem may be intermittent in nature. In this case, refer to the test steps used to find the cause of an intermittent fault (wiggle test).
- In no trouble codes are found at the initial check, then determine if a driveability symptom is present. If so, then refer to the approriate driveability symptom repair chart to make the repair. If the first symptom chart does not isolate the cause of the condition, then go on to another driveability symptom and follow that procedure to conclusion.
- If the problem is intermittent in nature, then refer to the special intermittent tests. Follow all available intermittent tests to determine the cause of this type of fault (usually an electrical connection problem).

OBD II System Diagnostics

See Figure 2.

The diagnostic approach used in OBD II systems is more complex than that of the one for OBD I systems. This complexity will effect how you approach diagnosing the vehicle. On an OBD II system, the onboard diagnostics will identify sensor faults (i.e., open, shorted or grounded circuits) as well as those that lose calibration. Another new test that arrived with OBD II is the rationality test (a test that checks whether the value for one input makes rational sense when compared against other sensor input values). The changes plus the use of OBD II Monitors have dramatically changed OBD II diagnostics.

The use of a repeatable test routine can help you quickly get to the root cause of a customer complaint, save diagnostic time and result in a higher percentage of properly repaired vehicles. You can use this Diagnostic Flow Chart to keep on track as you diagnose an Engine Control problem or a base engine fault on vehicles with OBD II.

FLOW CHART STEPS

Here are some of the steps included in the Diagnostic Routine:

- Review the repair order and verify the customer complaint as described
- Perform a Visual Inspection of underhood or engine related items
- If the engine will not start, refer to No Start Tests
- If codes are set, refer to the trouble code list, select a code and use the repair chart
- If no codes are set, and a symptom is present, refer to the Symptom List
- Check for any related technical service bulletins (for both Code and No Code Faults)
- If the problem is intermittent in nature, refer to the special Intermittent Tests

OBD II SYSTEM OVERVIEW

See Figure 3.

The OBD II system was developed as a step toward compliance with California and Federal regulations that set standards for vehicle emission control monitoring for all automotive manufacturers. The primary goal of this system is to detect when the degradation or failure of a component or system will cause emissions to rise by 50%. Every manufacturer must meet OBD II standards by the 1996 model year. Some manufacturers began programs that were OBD II mandated as early as 1992, but most manufacturers began an OBD II phase-in period starting in 1994.

The changes to On-Board Diagnostics influenced by this new program include:

- Common Diagnostic Connector
- Expanded Malfunction Indicator Light Operation
- Common Trouble Code and Diagnostic Language
- Common Diagnostic Procedures
- New Emissions-Related Procedures, Logic and Sensors
- Expanded Emissions-Related Monitoring

COMMON TERMINOLOGY

OBD II introduces common terms, connectors, diagnostic language and new emissions-related monitoring procedures. The most important benefit of OBD II is that all vehicles will have a common data output system with a common connector. This allows equipment Scan Tool manufacturers to read data from every vehicle and pull codes with common names and similar descriptions of fault conditions. In the future, emissions testing will require the use of an OBD II certifiable Scan Tool.

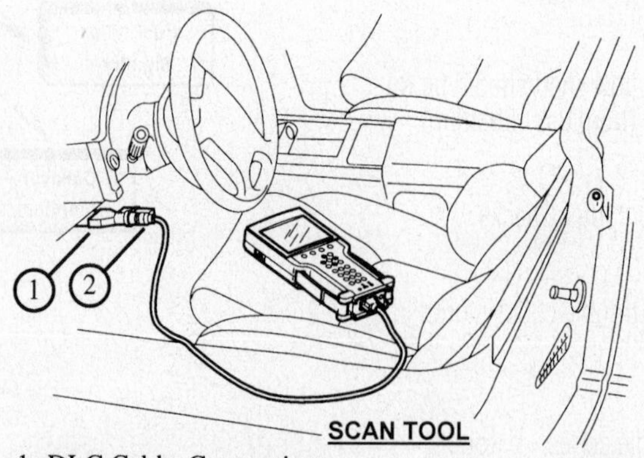

1. DLC Cable Connection
2. SAE 16/19 Pin Adapter

21199_FDIA_G004

Fig. 4 Typical scan tool hook up

Diagnostic Tools & Circuit Testing

HAND TOOLS & METER OPERATION

To effectively use this or any diagnostic information, you should have a solid understanding of how to operate required tools and test equipment.

SCAN TOOLS

See Figure 4.

Vehicle manufacturers designed their computers to have an accessible data line where a diagnostic tester could retrieve data on sensors and the status of operation for components.

These testers became known in the automotive repair industry as "Scan Tools" because they scanned the data on the computers and provided information for the technician.

The Scan Tool is your basic tool link into the on-board electronic control system of the vehicle. Scan Tools are equipped with, or have separate software cards, for each OEM needed to be diagnosed. In this case, always secure a scan tool that has the latest OEM-specific diagnostic software included. Spend some time in the scan tool user's manual to ensure you know how to properly operate the tool and how to select the necessary programs required for full and proper diagnostics.

MALFUNCTION INDICATOR LAMP

Emission regulations require that a Malfunction Indicator Lamp (MIL) be illuminated when an emissions related fault is detected and that a Diagnostic Trouble Code be stored in the vehicle controller (PCM) memory.

When the MIL is illuminated, it is an indication of a problem within one of the electronic components or circuits. When the scan tool is attached to the Data Link Connector (DLC) in the vehicle, it can access the DTCs. In some situations, without the use of a scan tool, the MIL can be activated to flash a series of long and short flashes, which correspond to the numbering of the DTC.

OBD II guidelines define when an emissions-related fault will cause the MIL to activate and set a Diagnostic Trouble Code (DTC). There are some DTCs that will not cause the MIL to illuminate. OBD II guidelines determine how quickly the onboard diagnostics must be able to identify a fault, set the trouble code in memory and activate the MIL (lamp).

ELECTRONIC CONTROLS

You should have a basic knowledge of electronic controls when performing test procedures to keep from making an incorrect

IDENTIFYING THE PROBLEM 2

Table of Contents

INTRODUCTION

System Control Modules 2-2
Powertrain Subsystems 2-2

WHERE TO BEGIN

Six-Step Procedure 2-2
Verify The Complaint & Check For TSBs 2-2
Check For Trouble Codes Or Symptoms 2-3
Problem Resolution & Repair 2-3
PCM Reset 2-3
Repair Verification 2-3
Base Engine Tests 2-3
Engine Compression Test 2-3
Engine Vacuum Test 2-4
Ignition System Tests – Distributor 2-4
Ignition System Tests – Distributorless 2-5
Symptom Diagnosis 2-5
Accessing Components & Circuits 2-5

Problem Identification

INTRODUCTION

System Control Modules

See Figures 1 and 2.

Before attempting diagnosis of the Electronic Engine Control system, familiarize yourself with the basics of how the system is designed to operate. It consists of a central processing unit: Powertrain Control Module (PCM), Engine Control Module (ECM), Transmission Control Module (TCM) and/or the Body Control Module (BCM). These units are the "heart" of the electronic control systems on the vehicle. In some cases, these units are integral with one another, and on some applications, they are separate. As you get deeper into actual diagnostic testing, you will find out which units are used on the vehicle you are testing.

The PCM is a digital computer that contains a microprocessor. The PCM receives input signals from various sensors and switches that are referred to as PCM inputs. Based on these inputs, the PCM adjusts various engine and vehicle operations through devices that are referred to as PCM outputs. Examples of the input and output devices are shown in the graphic.

Powertrain Subsystems

A key to the diagnosis of the PCM and its subsystems is to determine which subsystems are on a vehicle. Examples of typical subsystems are:

- Cranking & Charging System
- Emission Control Systems
- Engine Cooling System
- Engine Air/Fuel Controls
- Exhaust System
- Ignition System
- Speed Control System
- Transaxle Controls

WHERE TO BEGIN

See Figure 3.

Diagnosis of engine performance or drivability problems on a vehicle with an onboard computer requires that you have a logical plan on how to approach the problem. The "Six Step Test Procedure" is designed to provide a uniform approach to repair any problems that occur in one or more of the vehicle subsystems.

The diagnostic flow built into this test procedure has been field-tested for several years at dealerships - it is the starting point when a repair is required!

It should be noted that a commonly overlooked part of the "Problem Resolution" step is to check for any related Technical Service Bulletins.

Six-Step Test Procedure

The steps outlined as follows were defined to help you determine how to perform a proper diagnosis. Refer to the flow chart that outlines the Six Step Test Procedure as needed. The recommended steps include:

Verify The Complaint & Check For TSBs

To verify the customer complaint, the technician should understand the normal operation of the system. Conduct a thorough visual and operational inspection, review the service history, detect unusual sounds or odors, and gather diagnostic trouble code (DTC) information resources to achieve an effective repair.

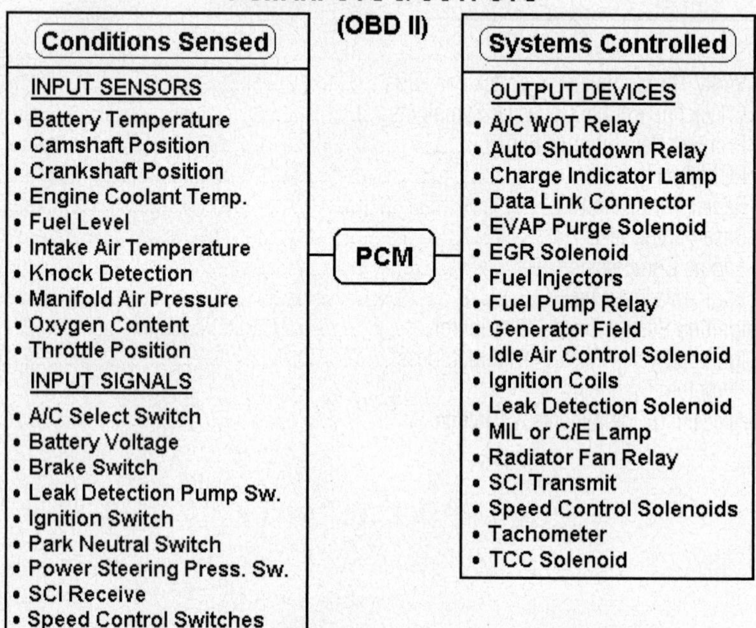

Fig. 1 An example of OBD II input and output devices 21199_FDIA_G005

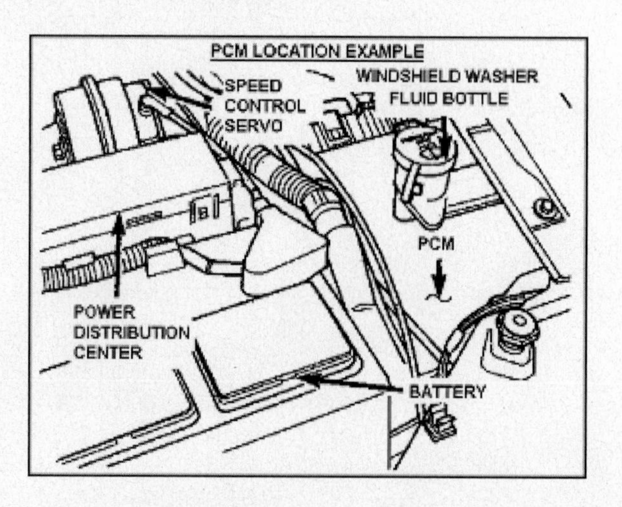

Fig. 2 Typical PCM location 21199_FDIA_G006

This check should include videos, newsletters, and any other information in the form of TSBs or Dealer Service Bulletins. Analyze the complaint and then use the recommended Six Step Test Procedure. Utilize the wiring diagrams and theory of operation articles. Combine your own knowledge with efficient use of the available service information.

Verify the cause of any related symptoms that may or may not be supported by one or more trouble codes. There are various checks that can be performed to Engine Controls that will help verify the cause of a related symptom. This step helps to lead you in an organized diagnostic approach.

Check For Trouble Codes Or Symptoms

Determine if the problem is a Code or a No Code Fault. Then refer to the appropriate published service diagnostic information to make the repair.

Problem Resolution & Repair

Once the problem component or circuit has been properly identified and verified using published diagnostic procedures, make any needed repairs or replacement to restore the vehicle to proper working order. If the condition has set a DTC, follow the designated repair chart to make an effective repair. If there is not a DTC set, but you can determine specific symptoms that are evident during the failure, select the symptom from the symptom tables and follow the diagnostic paths or suggestions to complete the repair or refer to the applicable component or system in service information.

If the vehicle does not set a DTC and has only intermittent operating failures or concerns, to resolve an intermittent fault, perform the following steps:

- Observe trouble codes, DTC modes and freeze frame data.
- Evaluate the symptoms and conditions described by the customer.
- Use a check sheet to identify the circuit or electrical system component.
- Many Aftermarket Scan Tools and Lab Scopes have data capturing features.

PCM Reset

It is a good idea, prior to tracing any faults, to clear the DTCs, attempt to replicate the condition and see if the same DTC resets. Also, once any repairs are made, it will be necessary to clear the DTC(s) - PCM Reset - to ensure the repair has totally resolved the problem. For procedures on PCM Reset, see DIAGNOSTIC TROUBLE CODES.

Repair Verification

Once a repair is completed, the next step is to verify the vehicle operates properly and that the original symptom was corrected. Verification Tests, related to specific DTC diagnostic steps, can be used to verify a repair.

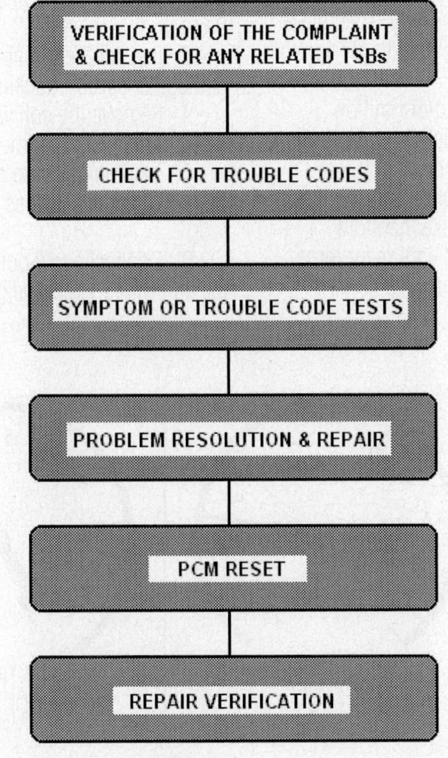

Fig. 3 Six-step diagnostic procedure

Base Engine Tests

To determine that an engine is mechanically sound, certain tests need to be performed to verify that the correct A/F mixture enters the engine, is compressed, ignited, burnt, and then discharged out of the exhaust system. These tests can be used to help determine the mechanical condition of the engine.

To diagnose an engine-related complaint, compare the results of the Compression, Cylinder Balance, Engine Cylinder Leakage (not included) and Engine Vacuum Tests.

Engine Compression Test

The Engine Compression Test is used to determine if each cylinder is contributing its equal share of power. The compression readings of all the cylinders are recorded and then compared to each other and to the manufacturer's specification (if available).

Cylinders that have low compression readings have lost their ability to seal. It this type of problem exists, the location of the compression leak must be identified. The leak can be in any of these areas: piston, head gasket, spark plugs, and exhaust or intake valves.

The results of this test can be used to determine the overall condition of the engine and to identify any problem cylinders as well as the most likely cause of the problem.

✱✱ CAUTION

Prior to starting this procedure, set the parking brake, place the gear selector in P/N and block the drive wheels for safety. The battery must be fully charged.

COMPRESSION TEST PROCEDURE

1. Allow the engine to run until it is fully warmed up.
2. Remove the spark plugs and disable the Ignition system and the Fuel system for safety. Disconnecting the CKP sensor harness connector will disable both fuel and ignition (except on NGC vehicles).
3. Carefully block the throttle to the wide-open position.
4. Insert the compression gauge into the cylinder and tighten it firmly by hand.
5. Use a remote starter switch or ignition key and crank the engine for 3-5 complete engine cycles. If the test is interrupted for any reason, release the gauge pressure and retest. Repeat this test procedure on all cylinders and record the readings.

The lowest cylinder compression reading should not be less than 70% of the highest cylinder compression reading and no cylinder should read less than 100 psi.

EVALUATING THE TEST RESULTS

To determine why an individual cylinder has a low compression reading, insert a small amount of engine oil (3 squirts) into the suspect cylinder. Reinstall the compression gauge and retest the cylinder and record the reading. Review the explanations that follow.

Reading is higher - If the reading is higher at this point, oil inserted into the cylinder helped to seal the piston rings against the cylinder walls. Look for worn piston rings.

Reading did not change - If the reading didn't change, the most likely cause of the low cylinder compression reading is the head gasket or valves.

Low readings on companion cylinders - If low compression readings were recorded from cylinders located next to each other, the most likely cause is a blown head gasket.

Readings are higher than normal - If the compression readings are higher than normal, excessive carbon may have collected on the pistons and in the exhaust areas. One way to remove the carbon is with an approved brand of "Top Engine Cleaner."

➡ **Always clean spark plug threads and seat with a spark plug thread chaser and seat cleaning tool prior to reinstallation. Use anti-seize compound on aluminum heads.**

Engine Vacuum Tests

An engine vacuum test can be used to determine if each cylinder is contributing an equal share of power. Engine vacuum, defined as any pressure lower than atmospheric pressure, is produced in each cylinder during the intake stroke. If each cylinder produces an equal amount of vacuum, the measured vacuum in the intake manifold will be even during engine cranking, at idle speed, and at off-idle speeds.

Engine vacuum is measured with a vacuum gauge calibrated to show the difference between engine vacuum (the lack of pressure in the intake manifold) and atmospheric pressure. Vacuum gauge measurements are usually shown in inches of Mercury (in. Hg).

➡ **In the tests described in this article, connect the vacuum gauge to an intake manifold vacuum source at a point below the throttle plate on the throttle body.**

ENGINE CRANKING VACUUM TEST PROCEDURE

The Engine Cranking Vacuum Test can be used to verify that low engine vacuum is not the cause of a No Start, Hard Start, Starts and Dies or Rough Idle condition (symptom).

The vacuum gauge needle fluctuations that occur during engine cranking are indications of individual cylinder problems. If a cylinder produces less than normal engine vacuum, the needle will respond by fluctuating between a steady high reading (from normal cylinders) and a lower reading (from the faulty cylinder). If more than one cylinder has a low vacuum reading, the needle will fluctuate very rapidly.

1. Prior to starting this test, set the parking brake, place the gearshift in P/N and block the drive wheels for safety. Then block the PCV valve and disable the idle air control device.
2. Disable the fuel and/or ignition system to prevent the vehicle from starting during the test (while it is cranking).
3. Close the throttle plate and connect a vacuum gauge to an intake manifold vacuum source. Crank the engine for three seconds (do this step at least twice).

The test results will vary due to engine design characteristics, the type of PCV valve and the position of the AIS or IAC motor and throttle plate. However, the engine vacuum should be steady between 1.0–4.0 in. Hg during normal cranking.

ENGINE RUNNING VACUUM TEST PROCEDURE

See Figure 4.

1. Allow the engine to run until fully warmed up. Connect a vacuum gauge to a clean intake manifold source. Connect a tachometer or Scan Tool to read engine speed.
2. Start the engine and let the idle speed stabilize. Raise the engine speed rapidly to just over 2000 rpm. Repeat the test (3) times. Compare the idle and cruise readings.

EVALUATING THE TEST RESULTS

If the engine wear is even, the gauge should read over 16 in. Hg and be steady. Test results can vary due to engine design and the altitude above or below sea level.

Ignition System Tests–Distributor

This next section provides an overview of ignition tests with examples of Engine Analyzer patterns for a Distributor Ignition System.

PRELIMINARY INSPECTION

1. Perform these checks prior to connecting the Engine Analyzer:
2. Check the battery condition (verify that it can sustain a cranking voltage of 9.6v).
3. Inspect the ignition coil for signs of damage or carbon tracking at the coil tower.
4. Remove the coil wire and check for signs of corrosion on the wire or tower.
5. Test the coil wire resistance with a DVOM (it should be less than 7 k/ohm per foot).
6. Connect a low output spark tester to the coil wire and engine ground. Verify that

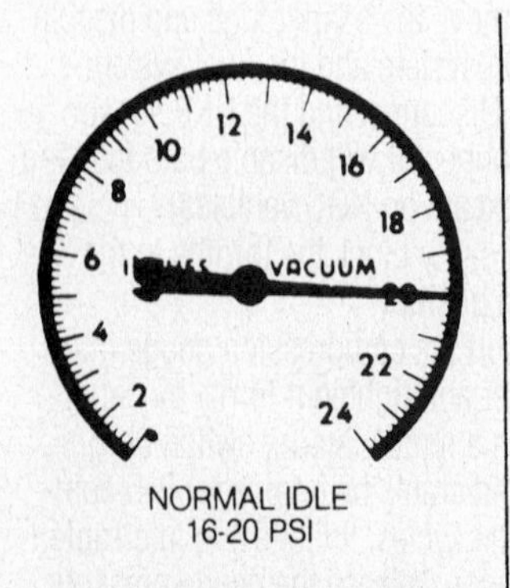

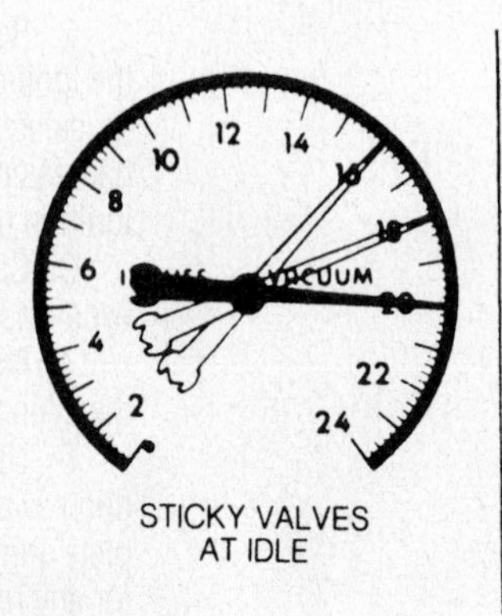

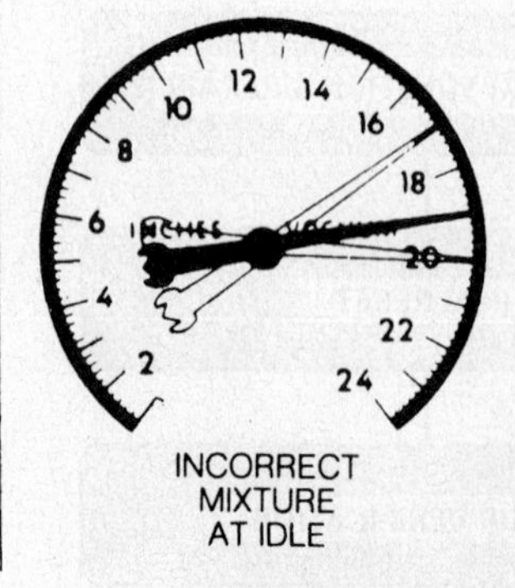

Fig. 4 Engine running vacuum test

the ignition coil can sustain adequate spark output while cranking for 3-6 seconds.

7. Connect the Engine Analyzer to the Ignition System, and choose Parade display. Run the engine at 2000 RPM, and note the display patterns, looking for any abnormalities.

Ignition System Tests–Distributorless

Perform the following checks prior to connecting the Engine Analyzer:

1. Check the battery condition (verify that it can sustain a cranking voltage of 9.6v).
2. Inspect the ignition coils for signs of damage or carbon tracking at the coil towers.
3. Remove the secondary ignition wires and check for signs of corrosion.
4. Test the plug wire resistance with a DVOM (specification varies from 15-30 k/ohm).
5. Connect a low output spark tester to a plug wire and to engine ground. Verify that the ignition coil can sustain adequate spark output for 3-6 seconds.

SECONDARY IGNITION SYSTEM SCOPE PATTERNS (V6 ENGINE)

See Figure 5.

1. Connect the Engine Analyzer to the ignition system.
2. Turn the scope selector to view the "Parade Display" of the ignition secondary.
3. Start the engine in Park or Neutral and slowly increase the engine speed from idle to 2000 rpm.
4. Compare actual display to the examples in the illustration.

Symptom Diagnosis

To determine whether vehicle problems are identified by a set Diagnostic Trouble Code, you will first have to connect a proper scan tool to the Data Link Connector and retrieve any set codes. See DIAGNOSTIC TROUBLE CODES for information on retrieving and reading codes.

If no codes are set, the problem must be diagnosed using only vehicle operating symptoms. A complete set of "No Code" symptoms is found in the SYMPTOM DIAGNOSIS (NO CODES).

DO NOT attempt to diagnose driveability symptoms without having a logical plan to use to determine which engine control system is the cause of the symptom - this plan should include a way to determine which systems do NOT have a problem! Remember, there are 2 kinds of NO CODE conditions:

- Symptom diagnosis, in which a continuous problem exists, but no DTC is set as a result. Therefore, only the operating symptoms of the vehicle can be used to pinpoint the root cause of the problem.
- Intermittent problem diagnosis, in which the problem does not occur all the time and does not set any DTCs.
- Both of these NO CODE conditions are covered in the SYMPTOM DIAGNOSIS.

Accessing Components & Circuits

See Figures 6 and 7.

Every vehicle and every diagnostic situation is different. It is a good idea to first determine the best diagnostic path to follow using flow charts, wiring diagrams, TSBs, etc. Part of choosing steps is to determine how time-consuming and effective each step will be. It may be easy to access a component or circuit in one vehicle, but difficult in

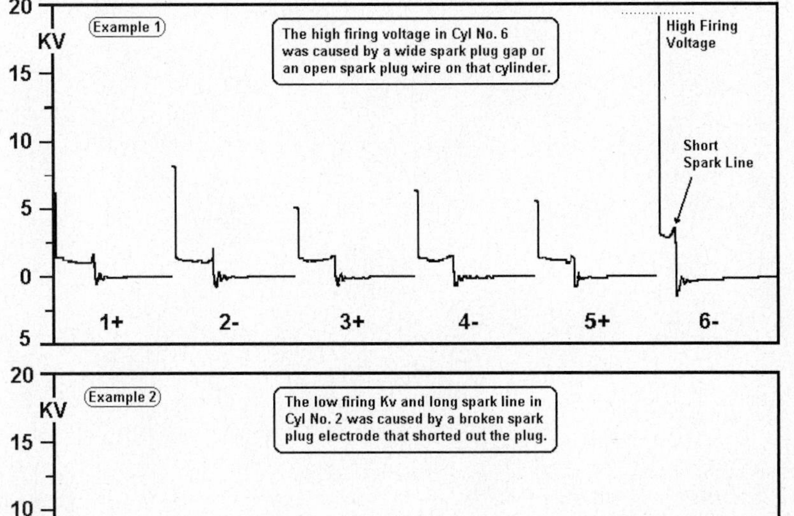

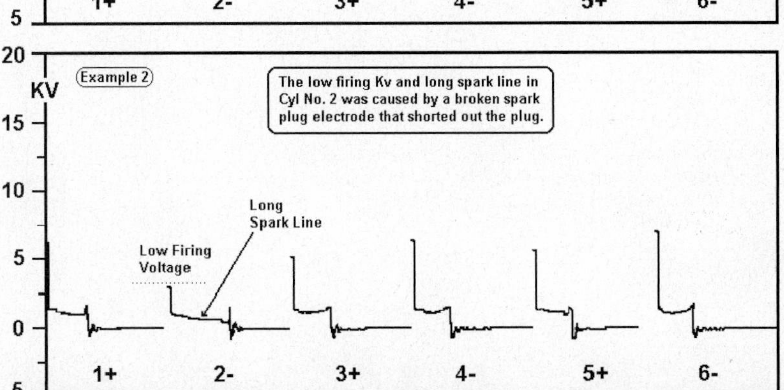

Fig. 5 Secondary ignition system (V6 engine)

21199_FDIA_G009

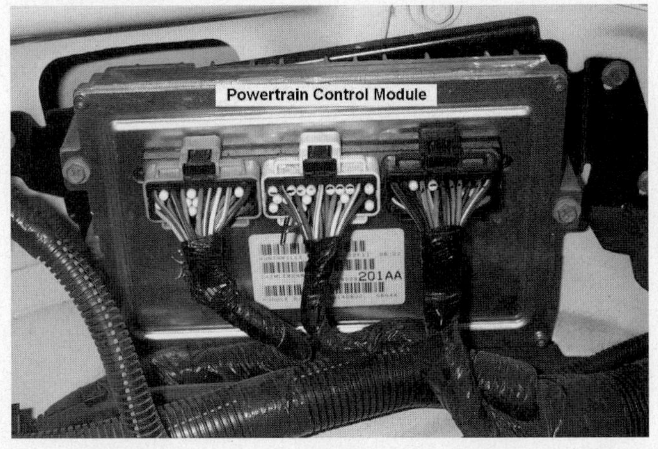

21199_FDIA_G010

Fig. 6 Circuits located at the back of the PCM connector

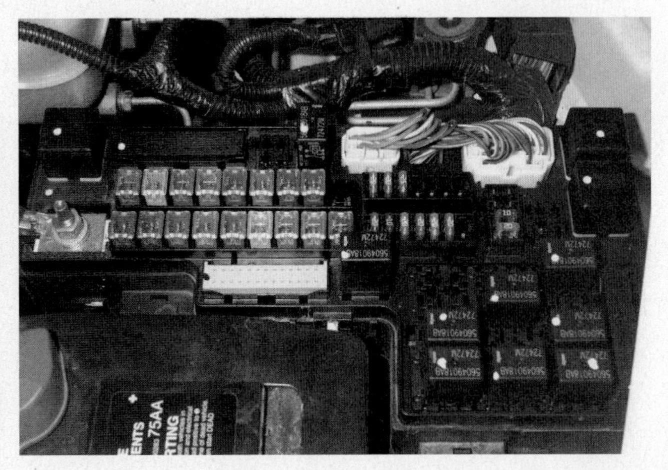

21199_FDIA_G010

Fig. 7 Typical underhood fuse block

another. Many circuits are integrated into a large harness and are difficult to test. Many components are inaccessible without disassembly of unrelated systems.

In the graphic, you will note that the protective covers have been removed from the PCM connectors, and any circuit can be easily identified and back probed. In other cases, PCM access is difficult, and it may be easier to access circuits at the component side of the harness.

Another important point to remember is that any circuit or component controlled by a relay or fused circuit can be monitored from the appropriate fuse box.

There is generally more than one of each type of relay or fuse. Therefore, swapping a suspect relay from another system may be more efficient than testing the relay itself. Relays and fuses may also be removed and replaced with fused jumper wires for testing circuits. Jumper wires can also provide a loop for inductive amperage tests.

Choosing the easiest way has its limitations, however. Remember that an appropriate signal on a PCM controlled circuit at an actuator means that the signal at the PCM is also good. However, a sensor signal at the sensor does not necessarily mean that the PCM is receiving the same signal. Think about the direction flow through a circuit, and not just what signal is appropriate, to save time without making costly assumptions.

INTRODUCTION TO OBD DIAGNOSTIC SYSTEMS

3

Table of Contents

OBD SYSTEMS

Differences Between OBD I & OBD II 3-2
Powertrain Control Module 3-2
OBD II Standardization 3-2
Changes in MIL Operation 3-2
1994 OBD II Phase-In Systems 3-2
1996 & Later OBD II Systems 3-2
Diagnostic Test Modes 3-2
Onboard Diagnostics 3-2

OBD SYSTEM TERMINOLOGY

Two-Trip Detection 3-2
Similar Conditions 3-3
OBD II Warmup Cycle 3-3
Malfunction Indicator Lamp 3-3
Freeze Frame Data 3-3
Diagnostic Trouble Codes 3-4
Data Link Connector 3-4
Standard Corporate Protocol 3-4
OBD II Monitor Software 3-4
Cylinder Bank Identification 3-4
Oxygen Sensor Identification 3-4
OBD II Monitor Test Results 3-4
Adaptive Fuel Control Strategy 3-5
Short Term Fuel Trim 3-5
Long Term Fuel Trim 3-5
Enable Criteria 3-5
Drive Cycle 3-5
OBD II Trip 3-5
OBD II Drive Cycle 3-5
Drive Cycle Procedure 3-5

OBD II SYSTEM MONITORS

Comprehensive Component Monitor 3-5
Input Strategies 3-5
Output Strategies 3-6
IAC Motor Test 3-6
Catalyst Efficiency Monitor 3-6
Catalyst Monitor Operation 3-6
EGR System Monitor 3-6
EVAP Catalyst Efficiency Monitor 3-6
On-Board Refueling Vapor Recovery System 3-7
Fuel System Monitor 3-7
Misfire Detection Monitor 3-7
Oxygen Sensor Monitor 3-9
Oxygen Sensor Heater Monitor 3-9
Air Injection System Monitor 3-9

OBD Systems

The California Air Resources Board (CARB) began regulating On-Board Diagnostic (OBD) systems for vehicles sold in California beginning with the 1988 model year. The initial requirements, known as OBD I, required the identification of the likely area of a fault with regard to the fuel metering system, EGR system, emission-related components and the PCM. Implementation of this new vehicle emission control monitoring regulation was done in several phases.

OBD I SYSTEMS

A Malfunction Indicator Lamp (MIL) labeled Check Engine Lamp or Service Engine Soon was required to illuminate and alert the driver of a fault, and the need to service the emission controls. A Diagnostic Trouble Code (DTC) was required to assist in identifying the system or component associated with the fault. If the fault that caused the MIL goes away, the MIL will go out and the code associated with the fault will disappear after a predetermined number of ignition cycles.

Following extensive research, CARB determined that by the time an Emission System component failed and caused the MIL to illuminate, that the vehicle could have emitted excess emissions over a long period of time. CARB also concluded that semi-annual or annual tailpipe tests were not catching enough of the vehicles with Emission Control systems operating at less than normal efficiency.

To take advantage of improvements in vehicle manufacturer adaptive and failsafe strategies, CARB developed new requirements designed to monitor the performance of Emission Control components, as well as to detect circuit and component hard faults. The new diagnostics were designed to operate under normal driving conditions, and the results of its tests would be viewable without any special equipment.

OBD II SYSTEMS

Beginning in the 1994 model year, both CARB and the EPA mandated Enhanced OBD systems, commonly known as OBD II. The objectives of OBD II were to improve air quality by reducing high in-use emissions caused by emission-related faults, reduce the time between the occurrence of a fault and its detection and repair, and assist in the diagnosis and repair of an emissions-related fault.

Differences Between OBD I & OBD II

As with OBD I, if an emission related problem is detected on a vehicle with OBD II, the MIL is activated and a code is set. However, that is the only real similarity between these systems. OBD II procedures that define emissions component and system tests, code clearing and drive cycles are more comprehensive than tests in the OBD I system.

Powertrain Control Module

The PCM in the OBD II system monitors almost all Emission Control systems that affect tailpipe or evaporative emissions. In most cases, the fault must be detected before tailpipe emissions exceed 1.5 times applicable 50K or 100K-mile FTP standards. If a component exceeds emission levels or fails to operate within the design specifications, the MIL is illuminated and a code is stored within two OBD II drive cycles.

The OBD II test runs continuously or once per trip (it depends on the driving mode requirement). Tests are run once per drive cycle during specific drive patterns called trips. Codes are stored in the PCM memory when a fault is first detected. In most cases, the MIL is turned on after two trips with a fault present. If the MIL is "on", it will go off after three consecutive trips if the same fault does not reappear. If the same fault is not detected after 40 engine warmup periods, the code will be erased (Fuel and Misfire faults require 80 warmup cycles).

OBD II Standardization

OBD II diagnostics require the use of a standardized Diagnostic Link Connector (DLC), standard communication protocol and messages, and standardized trouble codes and terminology. Examples of this standardization are Freeze Frame Data and I/M Readiness Monitors.

Changes in MIL Operation

An important change for OBD II involves when to activate the MIL. The MIL must be activated by at least the second trip if vehicle emissions could exceed 1.5 times the FTP standard. If any single component or system failure would allow the emissions to exceed this level, the MIL is activated and a related code is stored in the PCM.

1994 OBD II Phase-In Systems

Starting in 1994 some manufacturers began to "phase-in" the OBD II system on certain vehicles. The OBD II "phase-in" system on these vehicles included the use of a Misfire Monitor that operated with a "lower threshold" Misfire Detection system designed to monitor misfires without setting any codes. In addition, the EVAP Monitor was not operational on these vehicles.

1996 & Later OBD II Systems

By the 1996 model year, all California passenger cars and trucks up to 14,000 lb. GVWR, and all Federal passenger cars and trucks up to 8,600 lb. GWVR were required to comply with the CARB-OBD II or EPA OBD requirements. The requirements applied to diesel and gasoline vehicles, and were phased in on alternative-fuel vehicles.

Diagnostic Test Modes

The "test mode" messages available on a Scan Tool are listed below:

- Mode $01: Used to display Powertrain Data (PID data)
- Mode $02: Used to display any stored Freeze Frame data
- Mode $03: Used to request any trouble codes stored in memory
- Mode $04: Used to request that any trouble codes be cleared
- Mode $05: Used to monitor the Oxygen sensor test results
- Mode $06: Used to monitor Non-Continuous Monitor test results
- Mode $07: Used to monitor the Continuous Monitor test results
- Mode $08: Used to request control of a special test (EVAP Leak)
- Mode $09: Used to request vehicle information (INFO MENU)

Onboard Diagnostics

See Figure 1.

The Diagnostic Repair Chart should be used as follows:

- Trouble Code Diagnosis - Refer to the Code List or electronic media for a repair chart for a particular trouble code.
- Driveability Symptoms - Refer to the Driveability Symptom List in manuals or in electronic media.
- Intermittent Faults - Refer to the Intermittent Test Procedures.
- OBD II Drive Cycles - Refer to the Comprehensive Component Monitor or a Main Monitor drive cycle article.

OBD SYSTEM TERMINOLOGY

It is very important that service technicians understand terminology related to OBD II test procedures. Several of the essential OBD II terms and definitions are explained in the following text.

Two-Trip Detection

Frequently, an emission system or component must fail a Monitor test more

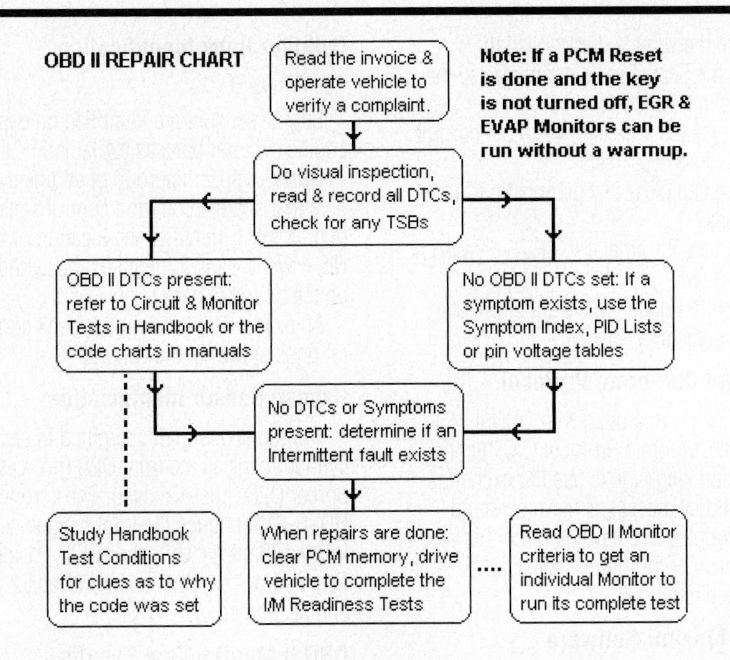

Fig. 1 OBD II repair chart

than once before the MIL is activated. In these cases, the first time an OBD II Monitor detects a fault during any drive cycle it sets a pending code in the PCM memory.

A pending code, which is read by selecting DDL from the Scan Tool menu, appears when Memory or Continuous codes are read. In order for a pending code to cause the MIL to activate, the original fault must be repeated under similar conditions.

This is a critical issue to understand as a pending code could remain in the PCM for a long time before the conditions that caused the code to set reappear. This type of OBD II trouble code logic is frequently referred to as the "Two-Trip Detection Logic".

➡ **Codes related to a Misfire fault and Fuel Trim can cause the PCM to activate the MIL after one trip because these codes are related to critical emission systems that could cause emissions to exceed the federally mandated limits.**

Similar Conditions

If a pending code is set because of a Misfire or Fuel System Monitor fault, the vehicle must meet similar conditions for a second trip before the code matures the PCM activates the MIL and stores the code in memory. Refer to prior Note for exceptions to this rule. The meaning of similar conditions is important when attempting to repair a fault detected by a Misfire or Fuel System Monitor.

To achieve similar conditions, the vehicle must reach the following engine running conditions simultaneously:

- Engine speed must be within 375 RPM of the speed when the trouble code set.
- Engine load must be within 10% of the engine load when the trouble code set.
- Engine warmup state must match a previous cold or warm state.

Summary–Similar conditions are defined as conditions that match the conditions recorded in Freeze Frame when the fault was first detected and the trouble code was set in the PCM memory.

OBD II Warmup Cycle

See Figure 2.

The meaning of the expression warmup cycle is important. Once the fault that caused an OBD II trouble code to set is gone and the MIL is turned off, the PCM will not erase that code until after 40 warmup cycles. This is the purpose of the warmup cycle: To help clear stored codes.

However, trouble codes related to a Fuel system or Misfire fault require that 80 warmup cycles occur without the fault reappearing before codes related to these monitors will be erased from the PCM memory.

➡ **A warmup cycle is defined as vehicle operation (after an engine off and cool-down period) when the engine temperature rises to at least 40°F and reaches at least 160°F.**

Malfunction Indicator Lamp

If the PCM detects an emission related component or system fault for two consecutive drive cycles on OBD II systems, the MIL is turned on and a trouble code is stored. The MIL is turned off if three consecutive drive cycles occur without the same fault being detected.

Most trouble codes related to a MIL are erased from memory after 40 warmup periods if the same fault is not repeated. The MIL can be turned off after a repair by using the Scan Tool PCM Reset function.

Freeze Frame Data

See Figure 3.

The term Freeze Frame is used to describe the engine conditions that are recorded in PCM memory at the time a Monitor detects an emissions related fault. These conditions include fuel control state, spark timing, engine speed and load.

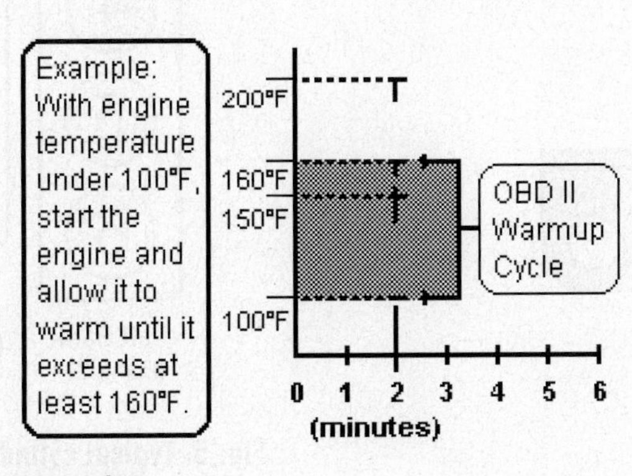

Fig. 2 OBD II warmup cycle

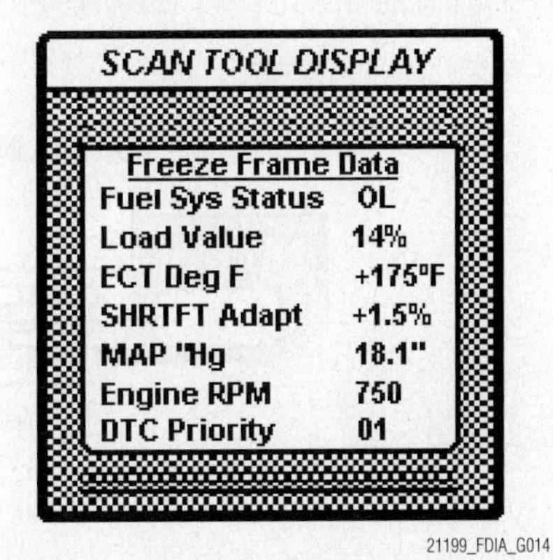

Fig. 3 Scan tool freeze frame

Freeze Frame data is recorded when a system fails the first time for two-trip type faults. The Freeze Frame Data will only be overwritten by a different fault with a "higher emission priority."

Diagnostic Trouble Codes

The OBD II system uses a Diagnostic Trouble Code (DTC) identification system established by the Society of Automotive Engineers (SAE) and the EPA. The first letter of a DTC is used to identify the type of computer system that has failed as shown below:

- The letter 'P' indicates a Powertrain related device
- The letter 'C' indicates a Chassis related device
- The letter 'B' indicates a Body related device
- The letter 'U' indicates a Data Link or Network device code.

The first DTC number indicates a generic (P0xxx) or manufacturer (P1xxx) type code. A list of trouble codes is included.

The number in the hundreds position indicates the specific vehicle system or subgroup that failed (i.e., P0300 for a Misfire code, P0400 for an emission system code, etc.).

Data Link Connector

See Figure 4.

Vehicles equipped with OBD II use a standardized Data Link Connector (DLC). It is typically located between the left end of the instrument panel and 12 inches past vehicle centerline. The connector is mounted out of sight from vehicle passengers, but should be easy to see from outside by a technician in a kneeling position (door open). However, not all of the connectors are located in this exact area.

The DLC is rectangular in design and capable of accommodating up to 16 terminals. It has keying features to allow easy connection to the Scan Tool. Both the DLC and Scan Tool have latching features used to ensure that the Scan Tool will remain connected to the vehicle during testing.

Once the Scan Tool is connected to the DLC, it can be used to:

- Display the results of the most current I/M Readiness Tests
- Read and clear any diagnostic trouble codes
- Read the Parameter ID (PID) data from the PCM
- Perform Enhanced Diagnostic Tests (manufacturer specific)

Standard Corporate Protocol

On vehicles equipped with OBD II, a Standard Corporate Protocol (SCP) communication language is used to exchange bi-directional messages between standalone modules and devices. With this type of system, two or more messages can be sent over one circuit.

OBD II Monitor Software

The Diagnostic Executive contains software designed to allow the PCM to organize and prioritize the Main Monitor tests and procedures, and to record and display test results and diagnostic trouble codes.

The functions controlled by this software include:

- To control the diagnostic system so the vehicle continues to operate in a normal manner during testing.
- To ensure the OBD II Monitors run during the first two sample periods of the Federal Test Procedure.
- To ensure that all OBD II Monitors and their related tests are sequenced so that required inputs (enable criteria) for a particular Monitor are present prior to running that particular Monitor.
- To sequence the running of the Monitors to eliminate the possibility of different Monitor tests interfering with each other or upsetting normal vehicle operation.
- To provide a Scan Tool interface by coordinating the operation of special tests or data requests.

Cylinder Bank Identification

See Figure 5.

Engine sensors are identified on each engine cylinder bank as explained next.

Bank–A specific group of engine cylinders that share a common control sensor (e.g., Bank 1 identifies the location of Cyl. No. 1 while Bank 2 identifies the cylinders on the opposite bank).

An example of the cylinder bank configuration is shown in the Graphic.

Oxygen Sensor Identification

Oxygen sensors are identified in each cylinder bank as the front O2S (pre-catalyst) or rear O2S (post-catalyst). The acronym HO2S-11 identifies the front oxygen sensor located (Bank 1) while the HO2S-21 identifies the front oxygen sensor in Bank 2 of the engine, and so on.

OBD II Monitor Test Results

Generally, when an OBD II Monitor runs and fails a particular test during a trip, a pending code is set. If the same Monitor detects a fault for two consecutive trips, the MIL is activated and a code is set in PCM memory. The results of a particular Monitor test indicate that an emission system or component failed: NOT the circuit that failed!

To determine where the fault is located; follow the correct code repair chart, symptom diagnosis or intermittent test. The code and symptom repair charts are the most efficient way to repair an OBD II system.

➡ **Two important pieces of information that can help speed up a diagnosis are code conditions (including all enable criteria), and the parameter information (PID) stored in the Freeze Frame at the time a trouble code is set and stored in memory.**

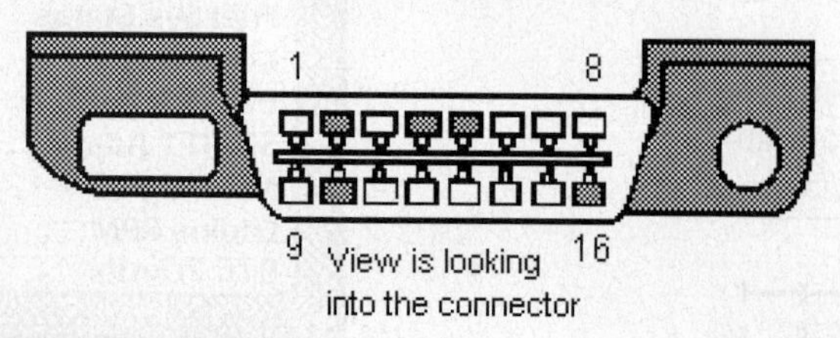

Fig. 4 Typical data link connector

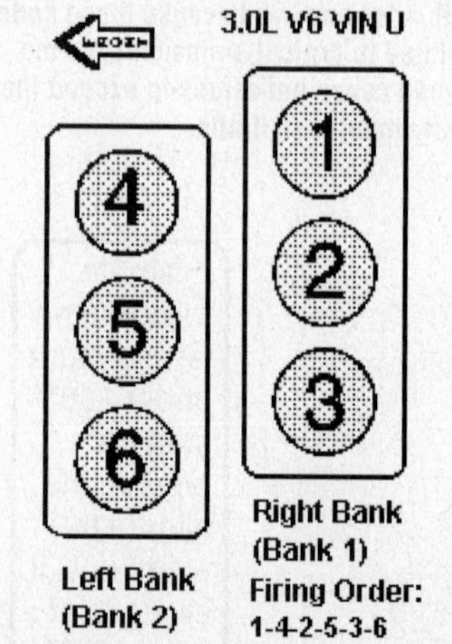

Fig. 5 Typical cylinder bank identification (V6 engine)

Adaptive Fuel Control Strategy

The PCM incorporates an Adaptive Fuel Control Strategy that includes an adaptive fuel control table stored to compensate for normal changes in fuel system devices due to age or engine wear.

During closed loop operation, the Fuel System Monitor has two methods of attempting to maintain an ideal A/F ratio of 14:7 to 1 (they are referred to as short term fuel trim and long term fuel trim).

➡ **If a fuel injector, fuel pressure regulator or oxygen sensor is replaced the, memory in the PCM should be cleared by a PCM Reset step so that the PCM will not use a previously learned strategy.**

Short Term Fuel Trim

Short term fuel trim (SHRTFT) is an engine operating parameter that indicates the amount of short term fuel adjustment made by the PCM to compensate for operating conditions that vary from the ideal A/F ratio condition. A SHRTFT number that is negative (-15%) means that the HO2S is indicating a richer than normal condition to the PCM, and that the PCM is attempting to lean the A/F mixture. If the A/F ratio conditions are near ideal, the SHRTFT number will be close to 0%.

Long Term Fuel Trim

Long term fuel trim (LONGFT) is an engine parameter that indicates the amount of long term fuel adjustment made by the PCM to correct for operating conditions that vary from ideal A/F ratios. A LONGFT number that is positive (+15%) means that the HO2S is indicating a leaner than normal condition, and that it is attempting to add more fuel to the A/F mixture. If A/F ratio conditions are near ideal, the LONGFT number will be close to 0%. The PCM adjusts the LONGFT in a range from -35 to +35%. The values are in percentage on a Scan Tool.

Enable Criteria

The term enable criteria describe the conditions necessary for any of the OBD II Monitors to run their diagnostic tests. Each Monitor has specific conditions that must be met before it will run its test.

Enable criteria information can be different for each vehicle and engine type. Examples of trouble code conditions for DTC P0460 and P1168 are shown below:

Code information includes any of the following examples:

- Air Conditioning Status
- BARO, ECT, IAT, TFT, TP and Vehicle Speed sensors
- Camshaft (CMP) and Crankshaft (CKP) sensors
- Canister Purge (duty cycle) and Ignition Control Module Signals
- Short (SHRTFT) and Long Term (LONGFT) Fuel Trim Values
- Transmission Shift Solenoid On/Off Status

Drive Cycle

The term drive cycle has been used to describe a drive pattern used to verify that a trouble code, driveability symptom or intermittent fault had been fixed. With OBD II systems, this term is used to describe a vehicle drive pattern that would allow all the OBD II Monitors to initiate and run their diagnostic tests. For OBD II purposes, a minimum drive cycle includes an engine startup with continued vehicle operation that exceeds the amount of time required to enter closed loop fuel control.

OBD II Trip

The term OBD II Trip describes a method of driving the vehicle so that one or more of the following OBD II Monitors complete their tests:

- Comprehensive Component Monitor (completes anytime in a trip)
- Fuel System Monitor (completes anytime during a trip)
- EGR System Monitor (completes after accomplishing a specific idle and acceleration period)
- Oxygen Sensor Monitor (completes after accomplishing a specific steady state cruise speed for a certain amount of time)

OBD II Drive Cycle

The ambient or inlet air temperature must be from 40-100°F to initiate the OBD II drive cycle. Allow the engine to warm to 130°F prior to starting the test.

Connect the Scan Tool prior to beginning the drive cycle. Some tools are designed to emit a three-pulse beep when all of the OBD II Monitors complete their tests.

➡ **The IAT PID must be from 50-100°F to start the drive cycle. If it is less than 50°F at any time during the highway part of the drive cycle, the EVAP Monitor may not complete. The engine should reach 130°F before starting before attempting to verify an EVAP system fault. Disengage the PTO before proceeding (PTO PID will show OFF) if applicable. For the EVAP Running Loss system, verify FLI PID is at 15-85%. Some Monitors require very specific idle and acceleration steps.**

Drive Cycle Procedure

The primary intention of the OBD II drive cycle is to clear a specific DTC. The drive cycle can also be used to assist in identifying any OBD II concerns present through total Monitor testing. Perform all of the Vehicle Preparation steps.

Connect a Scan Tool and have an assistant watch the Scan Tool I/M Readiness Status to determine when the Catalyst, EGR, EVAP, Fuel System, O2 Sensor, Secondary AIR and Misfire Monitors complete.

OBD II SYSTEM MONITORS

Comprehensive Component Monitor

OBD II regulations require that all emission related circuits and components controlled by the PCM that could affect emissions are monitored for circuit continuity and out-of-range faults. The Comprehensive Component Monitor (CCM) consists of four different monitoring strategies: two for inputs and two for output signals. The CCM is a two trip Monitor for emission faults on most vehicles.

Input Strategies

One input strategy is used to check devices with analog inputs for opens, shorts, or out-of-range values. The CCM accomplishes this task by monitoring A/D converter input voltages. The analog inputs monitored include the ECT, IAT, MAF, TP and Transmission Range Sensors signals.

DTC	Trouble Code Title & Conditions
	EVAP System Small Leak Conditions: Cold startup, engine running at off-idle conditions, then the PCM detected a small leak (a leak of more than 0.040") in the EVAP system.
	FRP Sensor in Range but Low Conditions: Engine running, then the PCM detected that the FRP sensor signal was out-of-range low. Scan Tool Tip: Monitor the FRP PID for a value below 80 psi (551 kPa).

A second input strategy is used to check devices with digital and frequency inputs by performing rationality checks. The PCM uses other sensor readings and calculations to determine if a sensor or switch reading is correct under existing conditions. Some tests run continuously, some only after actuation.

Output Strategies

An Output State Monitor in the PCM checks outputs for opens or shorts by observing the control voltage level of the related device. The control voltage is low with it on, and high with the device off.

IAC Motor Test

The PCM monitors the IAC system in order to "learn" the closed loop correlation it needs to reposition the IAC solenoid (a rationality check).

Catalyst Efficiency Monitor

The Catalyst Monitor is a PCM diagnostic run once per drive cycle that uses the downstream heated Oxygen Sensor (HO2S-12) to determine if a catalyst falls below a minimum level of effectiveness in its ability to control exhaust emissions. The PCM uses a program to determine the catalyst efficiency based on the oxygen storage capacity of the catalytic converter.

Catalyst Monitor Operation

See Figure 6.

The Catalyst Monitor is a diagnostic that tests the oxygen storage capacity of the catalyst. The PCM determines the capacity by comparing the switching frequency of the rear oxygen sensor to the switching frequency of the front oxygen sensor. If the catalyst is okay, the switching frequency of the rear oxygen sensor will be much slower than the frequency of the front oxygen sensor.

However, as the catalyst efficiency deteriorates its ability to store oxygen declines. This deterioration causes the rear oxygen sensor to switch more rapidly. If the PCM detects the switching frequency of the rear oxygen sensor is approaching the frequency of the front oxygen sensor, the test fails and a pending code is set. If the PCM detects a fault on consecutive trips (from two to six consecutive trips) the MIL is activated, and a trouble code is stored in the PCM memory.

The Catalyst Monitor runs after startup once a specified time has elapsed and the vehicle is in closed loop. The amount of time is subject to each PCM calibration. Certain inputs (enable criteria) from various engine sensors (i.e., CKP, ECT, IAT, TPS and VSS) are required before the Catalyst Monitor can run.

Once the Catalyst Monitor is activated, closed loop fuel control is temporarily transferred from the front oxygen sensor to the rear oxygen sensor. During the test, the Monitor analyzes the switching frequency of both sensors to determine if a catalyst has degraded.

Catalyst Efficiency Monitor

CATALYST TEST–STEADY STATE CATALYST EFFICIENCY TEST

The PCM transfers the input for closed loop fuel control from the front HO2S-11 to the rear HO2S-21 during this test. The PCM measures the output frequency of the rear HO2S. This "test frequency" indicates the current oxygen storage capacity of the converter. The slower the frequency of the test result, the higher the efficiency of the converter.

CATALYST TEST–CALIBRATED FREQUENCY TEST

In Part 2 of the test a second frequency is calculated based on engine speed and load. This frequency serves as a high limit threshold for the test frequency. If the PCM detects the test frequency is less than the calibrated frequency the catalyst passes the test. If the frequency is too high, the converter or system has failed (a pending code is set).

The sequence of counting the front and rear O2S switches continues until the drive cycle completes. The ratio of total HO2S-21 switches to the total of the HO2S-11 switches is calculated. If the switch ratio is over the stored threshold, the catalyst has failed and a code is set.

CATALYTIC MONITOR REPAIR VERIFICATION TRIP

See Figure 7.

Start the engine, and drive in stop and go traffic for over 20 minutes. (Ambient air temperature must be over 50°F to run this test). Drive at speeds from 25-40 mph (6 times) and then at cruise for five minutes.

POSSIBLE CAUSES OF A CATALYST EFFICIENCY FAULT

- Base Engine faults (engine mechanical)
- Exhaust leaks or contaminated fuel

EGR System Monitor

The EGR System Monitor is a PCM diagnostic run once per trip that monitors EGR system component functionality and components for faults that could cause vehicle tailpipe levels to exceed 1.5 times the FTP Standard. A series of sequenced tests is used to test the system.

HO2S-12 WAVEFORM EXAMPLES

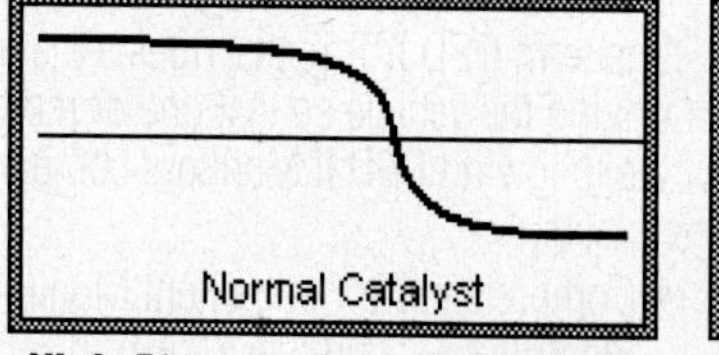

High Storage Capacity - Okay

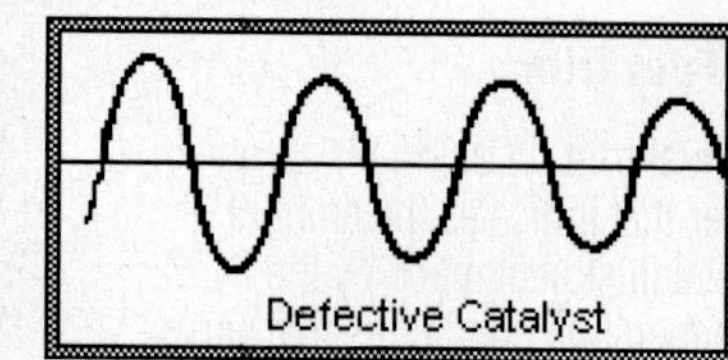

Low Storage Capacity - Not Okay

21199_FDIA_G019

Fig. 6 Typical rear oxygen sensor waveform

CATALYST MONITOR REPAIR VERIFICATION "TRIP"

Then drive on the highway for 5 minutes at a steady cruise speed.

Start engine & drive vehicle for 25 minutes - 20 minutes in stop & go traffic including 6 different steady speeds between 25 and 40 mph.

IAT & ECT Inputs

Key Off

0 20 30 35

Time (minutes)

Note: Verify the IAT PID reads 50-100°F and the ECT PID is over 130°F. Monitor the Scan Tool to verify that the EGR Monitor completes.

21199_FDIA_G020

Fig. 7 Typical catalyst monitor trip

Possible Causes of an EGR System Failure

See Figure 8.

- Leaks or disconnects in upstream or downstream vacuum hoses
- Damaged DPFE or EGR EVP sensor
- Plugged or restricted DPFE or EGR VP sensor or orifice assembly

Evap System Monitor

The EVAP System Monitor is a PCM diagnostic run once per trip that monitors the EVAP system in order to detect a loss of system integrity or leaks in the system (anywhere from 0.020" to 0.040" in diameter).

Possible Causes of an EVAP System Failure

- Cracks, leaks or disconnected hoses in the fuel vapor lines, components or plastic connectors or lines
- Backed-out or loose connectors to the Canister Purge solenoid
- Fuel filler cap (gas cap) loose or missing
- PCM has failed

On-Board Refueling Vapor Recovery System

An On-Board Refueling Vapor Recovery (ORVR) system is used on late model vehicles to recover fuel vapors during vehicle refueling.

SYSTEM OPERATION

The operation of the ORVR system during refueling is described next:

- The fuel filler pipe forms a seal to stop vapors from escaping the fuel tank while liquid is entering the tank (liquid in the 1" diameter tube blocks fuel vapor from rushing back up the fuel filler pipe).
- The fuel vapor control valve controls the flow of vapors out of the tank (it closes when the liquid level reaches a height associated with the fuel tank usable capacity). The fuel vapor control valve:

a. Limits the total amount of fuel dispensed into the fuel tank.

b. Prevents liquid gasoline from exiting the fuel tank when submerged (and also when tipped well beyond a horizontal plane as part of the vehicle rollover protection in an accident).

c. Minimizes vapor flow resistance in a refueling condition.

- Fuel vapor tubing connects the fuel vapor control valve to the EVAP canister. This routes the fuel tank vapors (that are displaced by the incoming fuel) to the canister.

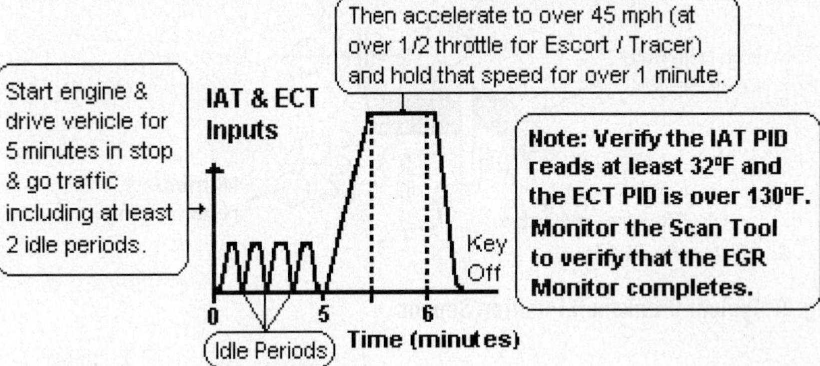

Fig. 8 Typical EGR monitor

- A check valve in the bottom of the pipe prevents any liquid from rushing back up the fuel filler pipe during liquid flow variations associated with the filler nozzle shut-off.
- Between refueling events, the charcoal canister is purged with fresh air so that it may be used again to store vapors accumulated during engine soak periods or subsequent refueling events. The vapors drawn from the canister are consumed in the engine.

Evap Monitor Test Conditions

The PCM allows canister purge to occur when the engine is warm, at wide open or part throttle (as long as the engine is not overheated). The engine can be in open or closed loop fuel control during purging.

Fuel System Monitor

The Fuel System Monitor is a PCM diagnostic that monitors the Adaptive Fuel Control system. The PCM uses adaptive fuel tables that are updated constantly and stored in long term memory (KAM) to compensate for wear and aging in the fuel system components.

FUEL SYSTEM MONITOR OPERATION

Once the PCM determines all the enable criteria has been are met (ECT, IAT and MAF PIDs in range and closed loop enabled), the PCM uses its adaptive strategy to "learn" changes needed to correct a Fuel system that is biased either rich or lean. The PCM accomplishes this task by monitoring Short Term and Long Term fuel trim in closed loop mode.

LONG AND SHORT TERM FUEL TRIM

Short Term fuel trim is a PCM parameter identification (PID) used to indicate Short Term fuel adjustments. This parameter is expressed as a percentage and its range of authority is from -10% to +10%. Once the engine enters closed loop, if the PCM receives a HO2S signal that indicates the A/F mixture is richer than desired, it moves the SHRTFT command to a more negative range to correct for the rich condition.

If the PCM detects the SHRTFT is adjusting for a rich condition for too long a time, the PCM will "learn" this fact, and move LONGFT into a negative range to compensate so that SHRTFT can return to a value close to 0%. Once a change occurs to LONGFT or SHRTFT, the PCM adds a correction factor to the injector pulsewidth calculation to adjust for variations. If the change is too large, the PCM will detect a fault.

➡ **If a fuel injector, fuel pressure regulator, etc. is replaced, clear the KAM and then drive the vehicle through the Fuel System Monitor drive pattern to reset the fuel control table in the PCM.**

Misfire Detection Monitor

The Misfire Monitor is a PCM diagnostic that continuously monitors for engine misfires under all engine positive load and speed conditions (accelerating, cruising and idling). The Misfire Monitor detects misfires caused by fuel, ignition or mechanical misfire conditions. If a misfire is detected, engine conditions present at the time of the fault are written to the Freeze Frame Data. These conditions overwrite existing data.

Misfire Monitor Operation

See Figure 9.

The Misfire Monitor is designed to measure the amount of power that each cylinder contributes to the engine. The amount of contribution is calculated based upon measurements determined by crankshaft acceleration (TDC of compression stroke to

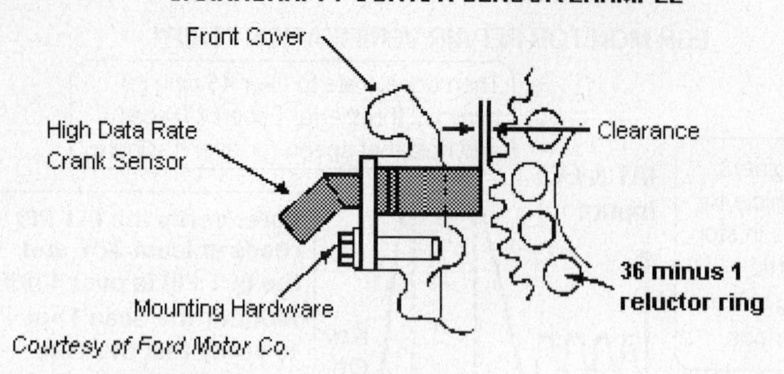

Fig. 9 Typical Crankshaft Position Sensor

BDC of the power stroke) for each cylinder. This calculation requires accurate measurement of the crankshaft angle. Crankshaft angle measurement is determined using a low data rate system on 4-Cyl engines. The high data rate system is used to determine crankshaft angle on all other engines.

Catalyst Damaging Misfire (One-Trip Detection)

If the PCM detects a Catalyst Damaging Misfire, the MIL will flash once per second within 200 engine revolutions from the point where misfire is detected. The MIL will stop flashing and remain on if the engine stops misfiring in a manner that could damage the catalyst.

High Emissions Misfire (Two-Trip Detection)

A High Emissions Misfire is set if a misfire condition is present that could cause the tailpipe emissions to exceed the FTP emissions standard by 1.5 times. If this fault is detected for two consecutive trips under similar engine speed, load and temperature conditions, the MIL is activated. It is also activated if a misfire is detected under similar conditions for two non-consecutive trips that are not 80 trips apart.

State Emissions Failure Misfire (Two-Trip Detection)

A State Emissions Failure Misfire is set if the misfire is sufficient to cause the vehicle to fail a State Inspection or Maintenance (I/M) Test. This fault is determined by identifying misfire percentages that would cause a "durability demonstration vehicle" to fail an Inspection Maintenance (I/M) Test. If the Misfire Monitor detects the fault for two consecutive trips with the engine at similar engine speed, load and temperature conditions, the MIL is activated and a code is set. The MIL is also activated if this type of misfire is detected under similar conditions for two non-consecutive trips of not more than 80 trips apart.

➡ **Some vehicles set Misfire codes because of an early version of OBD II hardware and software. If a misfire code is set and the cause of the fault is not found, clear the code and retest. Search the TSB list for possible answers or contact the dealer.**

Misfire Detection

See Figure 10.

The Misfire Monitor uses the CKP sensor signals to detect an engine misfire. The amount of contribution is calculated based upon measurements determined by crankshaft acceleration from each cylinder's power stroke.

The PCM performs various calculations to detect individual cylinder acceleration rates. If acceleration for a cylinder deviates beyond the average variation of acceleration for all cylinders, a misfire is detected.

Faults detected by the Misfire Monitor:

- Engine mechanical faults, restricted intake or exhaust system
- Dirty or faulty fuel injectors, loose or damaged injector connectors
- The vehicle has been run low on fuel or run until it ran out of fuel

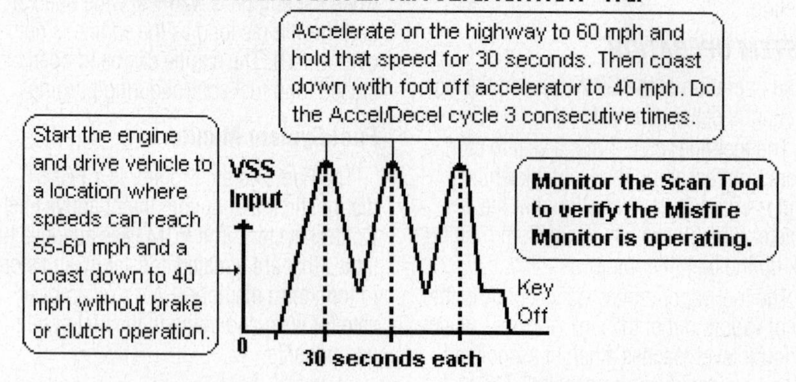

Fig. 10 Misfire Detection Monitor

Oxygen Sensor Monitor

The Oxygen Sensor Monitor is a PCM diagnostic designed to monitor the front and rear oxygen sensor for faults or deterioration that could cause tailpipe emissions to exceed 1.5 times the FTP standard. The front oxygen sensor voltage and response time are also monitored.

HO2S Monitor Operation

Fuel System and Misfire Monitors must be run and complete before the PCM will start the HO2S Monitor. Additionally, parts of the HO2S Sensor Monitor are enabled during the KOER Self-Test. The HO2S Monitor is run during each drive cycle after the CKP, ECT, IAT and MAF sensor signals are within a predetermined range.

Fixed Frequency Closed Loop Test

See Figure 11.

The HO2S Monitor constantly monitors the sensor voltage and frequency. The PCM detects a high voltage condition by comparing the HO2S signal to a preset level.

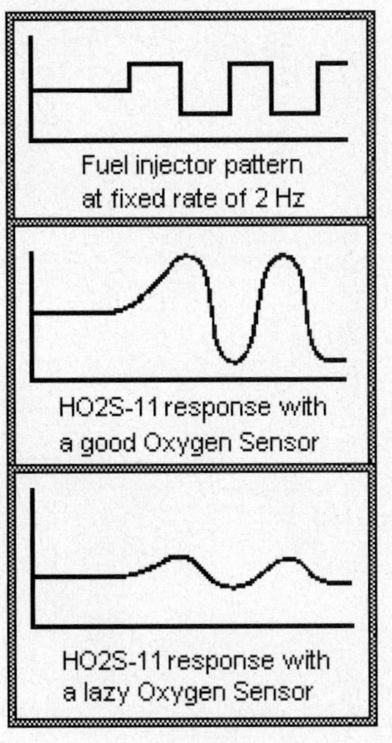

Fig. 11 Fixed Frequency Test

A Fixed Frequency Closed Loop Test is used to check the HO2S voltage and frequency. A sample of the HO2S signal is checked to determine if the sensor is capable of switching properly or has a slow response time (referred to as a lazy sensor).

Oxygen Sensor Heater Monitor

The Oxygen Sensor Heater Monitor is a PCM diagnostic designed to monitor the Oxygen Sensor Heater and its related circuits for faults.

OXYGEN SENSOR HEATER MONITOR OPERATION

The Oxygen Sensor Heater Monitor performs its task by detecting whether the proper amount of O2 sensor voltage change occurred as the HO2S Heater is turned from "on" to "off" with the engine in closed loop. The time it takes for the HO2S-11 and HO2S-12 signal to switch (the response time) is constantly monitored by the Oxygen Sensor Monitor. Once the Oxygen Sensor Heater Monitor is enabled, if the switch time for the HO2S-11 or HO2S-12 signal is too long, the PCM fails the test, the MIL is activated and a trouble code is set.

➡ **Response time is defined as the amount of time it takes for a HO2S signal to switch from Rich to Lean, and then Lean to Rich.**

FRONT AND REAR OXYGEN SENSOR HEATER OPERATION

Both upstream and downstream Oxygen sensors are used on the OBD II system. These sensors are designed with additional protection around the ceramic core to protect them from condensation that could crack them if the heater is turned on with condensation present.

The HO2S heaters are not turned on until the ECT sensor signal indicates that the engine is warm. The delay period can last for as long as 5 minutes from startup. The delay allows any condensation in the Exhaust system to evaporate.

Faults detected by the HO2S or HO2S Heater Monitor:

- A fault in the HO2S, the HO2S heater or its related circuits
- A fault in the HO2S connectors (look for moisture tracking)
- A defective Power Control Module

Air Injection System Monitor

The Air Injection System Monitor is an OBD diagnostic controlled by the PCM that monitors the Air Injection (AIR) system. The Oxygen Sensor Monitor must run and complete before the PCM will run this test. The PCM enables this test during AIR system operation after certain engine conditions are met and these enable criteria are met:

- Crankshaft Position sensor signal must be present
- ECT and IAT sensor input signals must be within limits

AIR MONITOR–ELECTRIC PUMP DESIGN

The AIR Monitor consists of these Solid State Monitor tests:

- A check of the Solid State relay for electrical faults.
- A check of the secondary side of the relay for electrical faults.
- A test to determine if the AIR system can inject additional air.

AIR MONITOR–MECHANICAL PUMP DESIGN

The AIR Monitor for the mechanical (belt-driven air pump) design uses two Output State Monitor configurations to perform two different circuit tests. One test is used to check for faults in the Secondary Air Bypass (AIRB) solenoid circuit. The normal function of the AIRB solenoid and valve assembly is to dump air into the atmosphere.

A second test is used to check for electrical faults in the Secondary Air Divert (AIRD) solenoid. The normal function of the AIRD solenoid and valve assembly is to direct the air either upstream or downstream.

FUNCTIONAL CHECK

See Figure 12.

An AIR system functional check is done at startup with the AIR pump on or during a hot idle period if the startup part of the test was not performed. A flow test is included that uses the HO2S signal to indicate the presence of extra air injected into the exhaust stream.

Diagnostic Trouble Codes

In the Diagnostic Trouble Code charts for the specific manufacturers you will see the following terms in the left column of the chart:

1. 1T–This means the code was activated when the PCM recognized the problem the first time it occurred.
2. 2T–This means the code was activated when the PCM recognized the problem and set the code after it occurred two times.
3. CCM–This means that the code and system affected is an emission related device and has a Comprehensive Component Monitor (CCM) tracking it.
4. MIL: Yes–This means that the Malfunction Indicator Light will be displayed.

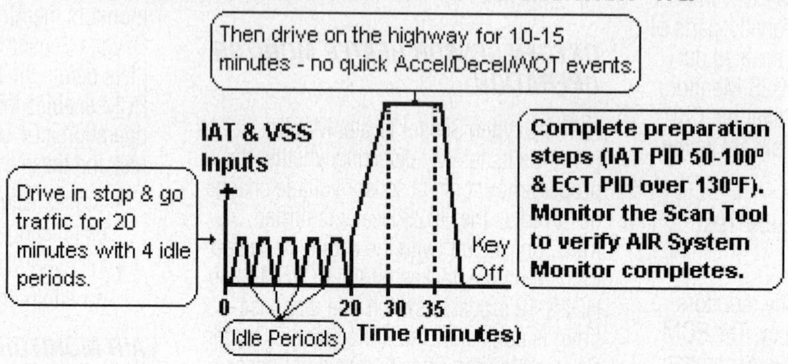
SECONDARY AIR MONITOR REPAIR VERIFICATION "TRIP"
Then drive on the highway for 10-15 minutes - no quick Accel/Decel/WOT events.
IAT & VSS Inputs
Drive in stop & go traffic for 20 minutes with 4 idle periods.
Complete preparation steps (IAT PID 50-100° & ECT PID over 130°F). Monitor the Scan Tool to verify AIR System Monitor completes.
Key Off
0
20
30
35
Idle Periods
Time (minutes)
21199_FDIA_G036

Fig. 12 Secondary AIR monitor

SYMPTOM DIAGNOSIS (NO CODES)

4

Table of Contents

WHAT TO DO WHEN THERE ARE NO DTCS

Driveability Symptom Index Table 4-2
Symptom Diagnosis - Test 1 – No Start, Hard Start Condition 4-2
Symptom Diagnosis - Test 2 – Rough, Low or High Idle Speed Condition 4-4
Symptom Diagnosis - Test 3 – Runs Rough Condition 4-6
Symptom Diagnosis - Test 4 – Cuts Out Or Misses Condition 4-7
Symptom Diagnosis - Test 5 – Surge Condition 4-8

INTERMITTENT TESTS

Test For Loose Connectors 4-9
The Wiggle Test 4-9
Vehicle Does Not Fill 4-10

What To Do When There Are No DTCs

Do not attempt to diagnose a Drivability Symptoms without having a logical plan to use to determine which Engine Control system is the cause of the symptom - this plan should include a way to determine which systems do not have a problem! Drivability symptom diagnosis is a part of an organized approach to problem solving and repair.

DRIVABILITY SYMPTOM INDEX TABLE

To use this list, locate the symptom that matches a particular problem and refer to the areas to test. The items listed under each symptom may not apply to all models, engines or vehicle systems. The repair steps indicate what vehicle component or system to test.

➡ The Drivability Symptoms in this list are intended to be generic. While they apply to most vehicles, some vehicles may not have all of the components listed. Refer to other Chilton repair information and electronic media for specific tests.

Symptom Test Table

Symptom Description	Suggested Areas to Test
Test 1 - No Start, Hard Start Condition • No Crank • Hard Start, Long Crank, Erratic Crank • Stall After Start • No Start, Normal Crank • No Start, MIL is off (if the VREF shorts to ground)	- Check battery, battery circuits to starter - Check for a damaged flywheel, engine compression, base timing and minimum air rate - Check for a failed fuel pump relay - Check for distributor rotor "punch-through" - Check for a faulty ignition control module (ICM) - Check for a VREF circuit shorted to ground - Check SKIM (security system) with a Scan Tool
Test 2 - Rough Idle or Stalls Condition • Low or slow idle speed • Fast idle speed • Hunting or rolling idle speed • Slow return to idle speed • Stalls or almost stalls	- Check for engine vacuum leaks - Check the condition of the PCV valve and lines - Check for excessive carbon buildup - Check for a restricted exhaust - Check base idle speed, check for low fuel pressure - Check the throttle linkage for sticking or binding
Test 3 - Runs Rough Condition • At idle speed • During acceleration • At cruise speed • During deceleration	- Check for engine vacuum leaks at intake manifold - Check condition of ignition secondary components - Check base timing and idle speed settings - Check for low or high fuel pressure - Check for dirty, leaking or shorted fuel injectors - Check for excessive carbon buildup on valves
Test 4 - Cuts-out, Misses Condition • At idle speed • During acceleration • At cruise speed • During deceleration	- Check for engine vacuum leaks at intake manifold - Check condition of ignition secondary components - Check that spark timing advance is available - Check for low or high fuel pressure - Check for dirty, leaking or shorted fuel injectors - Check for excessive carbon buildup on valves
Test 5 - Bucks, Jerks Condition • During acceleration • At cruise speed • During deceleration	- Check for engine vacuum leaks at intake manifold - Check condition of ignition secondary components - Check that spark timing advance is available - Check for low or high fuel pressure - Check for dirty, leaking or shorted fuel injectors - Check operation of the TCC solenoid, brake switch

Symptom Diagnosis Test 1 — No Start, Hard Start Condition

➡ If there is no spark output or fuel pressure available, check for a failed fuel pump relay, no power to the PCM, or loss of the ignition reference signal to the PCM.

PRELIMINARY CHECKS

Prior to starting this symptom test routine, inspect these underhood items:

- Check battery charge and condition, starter current draw.
- Verify the starter relay operation and that the engine cranks (turns over).
- Verify the check engine light (MIL) operation - if it does not activate, check the PCM power and ground circuits, and check for 5v supply at the MAP or TP sensor.
- Check Air Intake system for restrictions (inspect air inlet tubes, air filter for dirt, etc.).
- Check the status of the Smart Key Immobilizer System (SKIM) with the Scan Tool.

Test 1 Chart

Step	Action	Yes	No
1	**Step Description: No Start Condition Only** » Check battery cables, state of charge. » If the engine does not rotate, inspect for a locked engine (hydrostatic lockup condition). » Does the engine crank normally?	Go to Step 2.	Repair the fault in the battery, starter, or Base Engine. Retest for the symptom when all repairs are done.
2	**Step Description: Check the Fuel System** » Verify that the pump operates at key on. » Check the fuel pump relay operation. If the relay does not operate, check for blown fuse. » Inspect pump for a leak-down condition » Test fuel pressure, volume and quality. » Test the operation of the fuel regulator. » Are there any faults in the Fuel system?	Make needed repairs.	Go to Step 3.
3	**Step Description: Check the Ignition System** » Inspect ignition secondary components for damage (look for rotor "punch-through"). » Inspect the coils for signs of spark leakage at coil towers or primary connections. » Check the spark output with a spark tester. » Test Ignition system with an engine analyzer. » Are there any faults in the Ignition system?	Make repairs to the Ignition system. Then retest the symptom.	Go to Step 4.
4	**Step Description: Check the Exhaust System** » Check Exhaust system for leaks or damage. » Check the Exhaust system for a restriction using the Vacuum or Pressure Gauge Test (e.g., exhaust backpressure reading should not exceed 1.5 psi at cruise speeds). » Are there any faults in the Exhaust system?	Make repairs to the Exhaust system. Then retest the symptom.	Go to Step 5.
5	**Step Description: Check the MAP Sensor** » Disconnect the MAP sensor and attempt to start the engine. » Does the engine start and run normally?	Replace the MAP sensor. Retest for the symptom when repairs are completed.	Go to Step 6.
6	**Step Description: Check for a Hot Engine** » Check for signs of an engine overheating condition related to a Hard Start Symptom. » Does the engine appear to be overheated?	Make the repairs to correct the hot engine and then retest for the symptom when done.	Go to Step 7.
7	**Step Description: Check ECT Sensor PID** » Connect a Scan Tool and turn the key to on. » Read the ECT sensor (compare to chart). » Has the ECT sensor shifted out of range?	Replace the ECT sensor. Then retest for the symptom when all repairs are completed.	Go to Step 8.
8	**Step Description: Check the PCV System** » Inspect the PCV system components for broken parts or loose connections. » Test the operation of the PCV valve. » Are there any faults in the PCV system?	Repair the PCV system. Refer to the PCV system tests. Retest the symptom when all repairs are done.	Go to Step 9.
9	**Step Description: Check the EVAP System** » nspect for damaged or disconnected EVAP system components. » Inspect for a fuel saturated charcoal canister. » Are there any faults in the EVAP system?	Refer to the EVAP system tests. Retest for the symptom when all repairs are completed.	Go to Step 10.
10	**Step Description: Test the Base Engine** » Check the engine compression. » Test valve timing and timing chain condition. » Check for a worn camshaft or valve train. » Check for any large intake manifold leaks. » Are there any faults in the Base Engine?	Repair the Base Engine. Refer to the Base Engine Tests. Retest symptom when done.	Return to Step 2 to repeat the test steps in this series to locate and repair the "No Start, Hard Start" condition.

Symptom Diagnosis Test 2 — Rough, Low or High Idle Speed Condition

➡ **If the vehicle has a rough idle and the base timing, idle speed and the IAC (or AIS) motor operates properly, check the engine for excessive carbon buildup.**

PRELIMINARY CHECKS

Prior to starting this symptom test routine, inspect these underhood items:

- All related vacuum lines for proper routing and integrity.
- All related electrical connectors and wiring harnesses for faults (Wiggle Test).
- Check the throttle linkage for a sticking or binding condition.
- Air Intake system for restrictions (air inlet tubes, dirty air filter, etc.).
- Search for any technical service bulletins related to this symptom.
- Turn the key to off. Unplug the MAP sensor connection and restart the engine to recheck for the idle concern. If the condition is gone, replace the MAP sensor.

Test 2 Chart

Step	Action	Yes	No
1	**Step Description: Verify the rough idle or stall** » Does the engine have a warm engine rough idle, low idle or high idle condition in P or N?	Go to Step 2.	Fault is intermittent. Return to the Symptom List and select another fault.
2	**Step Description: Verify idle speed & timing** » Verify the base timing is within specifications » Verify that the base idle speed is set properly » Are the timing and idle speed set properly?	Go to Step 3.	Set the base idle speed and timing to the specifications and then retest for the symptom.
3	**Step Description: Check AIS / IAC Operation** » Check the AIS or IAC motor operation » Inspect the AIS/IAC housing in throttle body for restricted passages. Clean as needed. » Set the parking brake, block the drive wheels and turn the A/C off. Install the Scan Tool. » IAC Motor Tester - Turn the key off and then connect the IAC tester to the IAC valve. » Start the engine and use the IAC tester to extend and retract the IAC valve. » ATM Test - Start the engine. Use the tool to change the speed from min-idle to 1500 rpm. » Did the idle speed change as commanded?	Install an Aftermarket Noid light and check the operation of the PCM and AIS or IAC motor circuits. Check the motor for signs of open or shorted circuits. Replace the IAC motor or PCM as needed or make repairs to the IAC motor wiring. If all are okay, go to Step 4.	If the AIS/IAC motor passages are clean and engine speed did not change as described when the AIS/IAC motor was extended and retracted, replace the AIS/IAC motor. Then retest for the condition.
4	**Step Description: Check/compare PID values** » Connect Scan Tool & turn off all accessories. » Start the engine and allow it to fully warmup. » Monitor all related PIDs on the Scan Tool. » Verify the P/N switch input in gear and Park. » Check the O2S operation with a Lab Scope. » Are all PIDs within normal range?	Go to Step 5. Note: An IAC motor count of over 80 indicates the pintle is extended and an IAC count of (0) indicates the pintle is retracted.	One or more of the PIDs are out of range when compared to "known good" values. Make repairs to the system that is out of range, then retest for the symptom.

5	**Step Description: Check the Ignition System** » Inspect the coils for signs of spark leakage at coil towers or primary connections. » Check the spark output with a spark tester. » Test Ignition system with an engine analyzer. » Were any faults found in the Ignition system?	Make repairs as needed	Go to Step 6.
6	**Step Description: Check the Fuel System** » Inspect the Fuel delivery system for leaks. » Test the fuel pressure, quality and volume. » Test the operation of the pressure regulator. » Were any faults found in the Fuel system?	Make repairs as needed	Go to Step 7.
7	**Step Description: Check the Exhaust System** » Check Exhaust system for leaks or damage. » Check the Exhaust system for a restriction using the Vacuum or Pressure Gauge Test (e.g., exhaust backpressure reading should not exceed 1.5 psi at cruise speeds). » Were any faults found in Exhaust System?	Make repairs to the Exhaust system. Then retest the symptom.	Go to Step 8.
8	**Step Description: Check the PCV System** » Inspect the PCV system components for broken parts or loose connections. » Test the operation of the PCV valve. » Were any faults found in the PCV system?	Make repairs to the PCV system. Refer to the PCV system tests. Then retest for the condition.	Go to Step 9.
9	**Step Description: Check the EVAP System** » Inspect for damaged or disconnected EVAP system components or a saturated canister. » Were any faults found in the EVAP system?	Make repairs to EVAP system. Retest for the condition.	Go to Step 10.
10	**Step Description: Check the Base Engine** » Test the engine compression. » Test valve timing and timing chain condition. » Check for a worn camshaft or valve train. » Check for any large intake manifold leaks. » Were any faults found in the Base Engine?	Make repairs as needed to the Base Engine. Refer to the Base Engine tests. Then retest for the condition when repairs are completed.	Go to Step 2 and repeat the tests from the beginning to locate and repair the cause of the "Rough, Low or High Idle Speed" condition.

Symptom Diagnosis Test 3 — Runs Rough Condition

PRELIMINARY CHECKS

Prior to starting this symptom test routine, inspect these underhood items:

- All related vacuum lines for proper routing and integrity
- Air Intake system for restrictions (air inlet tubes, dirty air filter, etc.)
- Search for any technical service bulletins related to this symptom.

Test 3 Chart

Step	Action	Yes	No
1	**Step Description: Verify engine runs rough** » Start the engine and allow it to idle in P or N. » Does the engine run rough when warm in Park or Neutral position?	Check for any stored codes. If codes are set, repair codes and retest. If no codes are set, go to Step 3.	Go to Step 2.
2	**Step Description: Condition does not exist!** » Inspect various underhood items that could cause an intermittent Runs Rough condition (i.e., dirt in the throttle body, vacuum leaks, IAC motor connections, etc.). » Were any problems located in this step?	Correct the problems. Do a PCM reset and engine "idle relearn" procedure. Then verify the "runs rough" condition is repaired.	The problem is not present at this time. It may be an intermittent problem.
3	**Step Description: Check/compare PID values** » Connect a Scan Tool to the test connector. » Turn off all accessories. » Start the engine and allow it to fully warmup. » Monitor all related PIDs on the Scan Tool. » Were all PIDs within their normal range?	Go to Step 4. Note: The IAC motor should read from 5-50 counts. Check the LONGFT reading for a large shift into the negative range (due to a rich condition).	One or more of the PIDs are out of range when compared to "known good" values. Make repairs to the system that is out of range, then retest for the symptom.
4	**Step Description: Check the Ignition System** » Inspect the coils for signs of spark leakage at coil towers or primary connections. » Check the spark output with a spark tester. » Test Ignition system with an engine analyzer. » Were any faults found in the Ignition system?	Make repairs as needed	Go to Step 5.
5	**Step Description: Check the Fuel System** » Inspect the Fuel delivery system for leaks. » Test the fuel pressure, quality and volume. » Test the operation of the pressure regulator. » Were any faults found in the Fuel system?	Make repairs as needed	Go to Step 6.
6	**Step Description: Check the Exhaust System** » Check Exhaust system for leaks or damage. » Check the Exhaust system for a restriction using the Vacuum or Pressure Gauge Test (e.g., exhaust backpressure reading should not exceed 1.5 psi at cruise speeds). » Were any faults found in Exhaust System?	Make repairs to the Exhaust system. Then retest the symptom.	Go to Step 7.
7	**Step Description: Check the PCV System** » Inspect the PCV system components for broken parts or loose connections. » Test the operation of the PCV valve. » Were any faults found in the PCV system?	Make repairs to the PCV system. Refer to the PCV system tests. Then retest for the condition.	Go to Step 9.
8	**Step Description: Check the EVAP System** » Inspect for damaged or disconnected EVAP system components or a saturated canister. » Were any faults found in the EVAP system?	Make repairs to EVAP system. Retest for the condition.	Go to Step 10.
9	**Step Description: Check Engine Condition** » Test the engine compression. » Test valve timing and timing chain condition. » Check for a worn camshaft or valve train. » Check for any large intake manifold leaks. » Were any faults found in the Base Engine?	Make repairs as needed to the Base Engine. Refer to the Base Engine tests. Then retest for the condition when repairs are completed.	Return to Step 2 and repeat the tests from the beginning to locate and repair the cause of the "Runs Rough" condition.

Symptom Diagnosis Test 4 — Cuts-out or Misses Condition

PRELIMINARY CHECKS

Prior to starting this symptom test routine, inspect these underhood items:

- All related vacuum lines for proper routing and integrity
- Search for any technical service bulletins related to this symptom.

Test 4 Chart

Step	Action	Yes	No
1	**Step Description: Verify Cuts-out condition** » Start the engine and attempt to verify the Cuts-out or misses condition. » Does the engine have a cuts-out condition?	Check for any stored codes. If codes are set, repair codes and retest. If no codes are set, go to Step 3.	Go to Step 2.
2	**Step Description: Condition does not exist!** » Inspect various underhood items that could cause an intermittent Cuts-out condition (i.e., EVAP, Fuel or Ignition system components). » Were any problems located in this step?	Correct the problems. Do a PCM reset and "Fuel Trim Relearn" procedure. Then verify condition is repaired.	The problem is not present at this time. It may be an intermittent problem.
3	**Step Description: Check/compare PID values** » Connect a Scan Tool to the test connector. » Turn off all accessories. » Start the engine and allow it to fully warmup. » Monitor all related PIDs on the Scan Tool (i.e., ECT IAC Counts and LONGFT at idle). » Were all PIDs within their normal range?	Go to Step 4. Note: The IAC motor should be from 5-50 counts. Watch fuel trim (%) for a large shift into the negative (-) range (due to a rich condition).	One or more of the PIDs are out of range when compared to "known good" values. Make repairs to the system that is out of range, then retest for the symptom.
4	**Step Description: Check the Ignition System** » Inspect the coils for signs of spark leakage at coil towers or primary connections. » Check the spark output with a spark tester. » Test Ignition system with an engine analyzer. » Were any faults found in the Ignition system?	Make repairs as needed	Go to Step 5.
5	**Step Description: Check the Fuel System** » Inspect the Fuel delivery system for leaks. » Test the fuel pressure, quality and volume. » Test the operation of the pressure regulator. » Were any faults found in the Fuel system?	Make repairs as needed	Go to Step 6.
6	**Step Description: Check the Exhaust System** » Check Exhaust system for leaks or damage. » Check the Exhaust system for a restriction using the Vacuum or Pressure Gauge Test (e.g., exhaust backpressure reading should not exceed 1.5 psi at cruise speeds). » Were any faults found in Exhaust System?	Make repairs to the Exhaust system. Then retest the symptom.	Go to Step 7.
7	**Step Description: Check the PCV System** » Inspect the PCV system components for broken parts or loose connections. » Test the operation of the PCV valve. » Were any faults found in the PCV system?	Make repairs to the PCV system. Then retest for the condition.	Go to Step 8.
8	**Step Description: Check the EVAP System** » Inspect for damaged or disconnected EVAP system components » Check for a saturated EVAP canister. » Were any faults found in the EVAP system?	Make repairs to EVAP system. Retest for the condition.	Go to Step 9.
9	**Step Description: Check the AIR system** » Inspect AIR system for broken parts, leaking valves or disconnected hoses. » Test the operation of Secondary AIR system. » Were any faults found in the AIR system?	Make repairs as needed. Refer to the Secondary AIR system tests. Retest for the condition.	Go to Step 10.
10	**Step Description: Check Engine Condition** » Test the engine compression. » Test valve timing and timing chain condition. » Check for a worn camshaft or valve train. » Check for any large intake manifold leaks. » Were any faults found in the Base Engine?	Make repairs as needed to the Base Engine. Refer to the Base Engine tests. Then retest for the condition when repairs are completed.	Go to Step 2 and repeat the tests from the beginning to locate and repair the cause of the "Cuts Out or Misses" condition.

Symptom Diagnosis Test 5 — Surge Condition

PRELIMINARY CHECKS

1. Discuss how the operation of the torque converter clutch (TCC) or air conditioning compressor can affect the "feel" of the vehicle during normal operation. Refer to the information in the Owner's Manual to explain how these devices normally operate.
2. Search for any technical service bulletins related to this symptom.

Test 5 Chart

Step	Action	Yes	No
1	**Step Description: Verify the surge condition** » Drive the vehicle and attempt to verify that the vehicle surges at cruise speeds. » Does the engine have a surge condition?	Check for any stored codes. If codes are set, repair codes and retest. If no codes are set, go to Step 3.	Go to Step 2.
2	**Step Description: Condition does not exist!** » Inspect various underhood items that could cause an intermittent surge condition (check for leaks in the MAP sensor vacuum lines). » Were any problems located in this step?	Correct the problems. Do a PCM reset and "Fuel Trim Relearn" procedure. Then verify condition is repaired.	The problem is not present at this time. It may be an intermittent problem.
3	**Step Description: Check/compare PID values** » Connect a Scan Tool to the test connector. » IStart the engine and allow it to fully warmup. » Monitor all related PIDs on Scan Tool (HO2S switching, LONGFT, and the TCC operation) » Compare VSS PID reading to speedometer. » Were all PIDs within their normal range?	Go to Step 4. Note: Verify that the front HO2S responds quickly to throttle changes. Check for silicon contamination on the front HO2S (this can cause a rich A/F signal).	One or more of the PIDs are out of range when compared to "known good" values. Make repairs to the system that is out of range, then retest for the symptom.
4	**Step Description: Check the Ignition System** » Inspect the coils for signs of spark leakage at coil towers or primary connections. » Check the spark output with a spark tester. » Test Ignition system with an engine analyzer. » Were any faults found in the Ignition system?	Make repairs as needed	Go to Step 5.
5	**Step Description: Check the Fuel System** » Inspect the Fuel delivery system for leaks. » Test the fuel pressure, quality and volume. » Test the operation of the pressure regulator. » Were any faults found in the Fuel system?	Make repairs as needed	Go to Step 6.
6	**Step Description: Check the Exhaust System** » Check Exhaust system for leaks or damage. » Check the Exhaust system for a restriction using the Vacuum or Pressure Gauge Test (e.g., exhaust backpressure reading should not exceed 1.5 psi at cruise speeds). » Were any faults found in Exhaust System?	Make repairs to the Exhaust system. Then retest the symptom.	Return to Step 2 and repeat the tests from the beginning to locate and repair the cause of the "Surge" condition.

INTERMITTENT TESTS

Many trouble code repair charts end with a result that reads "Fault Not Present at this Time." What this expression means is that the conditions that were present when a code set or drivability symptom occurred are no longer there or were not met. In effect, the problem was present at least once, but is not present at this time. However, it is likely to return in the future, so it should be diagnosed and repaired if at all possible.

One way to find an intermittent problem is to gather the information that was present when the problem occurred. In the case of a Code Fault, this can be done in two ways: by capturing the data in Snapshot or Movie mode or by driver observations.

The PCM has to detect the fault for a specific period of time before a trouble code will set. While intermittent problems may appear to be occasional in nature, they usually occur under specific conditions. Therefore, you should identify and duplicate these conditions. Since intermittent faults are difficult to duplicate, a logical routine (checklist) must be followed when attempting to find the faulty component, system or circuit. The tests on the next page can be used to help find the cause of an intermittent fault.

Some intermittent faults occur due to a loose connection, wiring problem or warped circuit board. An intermittent fault can also be caused by poor test techniques that cause damage to the male or female ends of a connector.

Test for Loose Connectors

To test for a loose or damaged connection, take the male end of a connector from another wiring harness and carefully push it into the "suspect" female terminal to verify that the opening is tight. There should be some resistance felt as the male connector is inserted in the terminal connection.

The Wiggle Test

See Figures 1 and 2.

A wiggle test can be used to locate the cause of some intermittent faults. The sensor, switch or the PCM wiring can be backprobed, as shown, while the test is done.

During testing, move or wiggle the suspect device, connector or wiring while watching for a change.

If the DVOM has a Min/Max record mode, use this mode during the test.

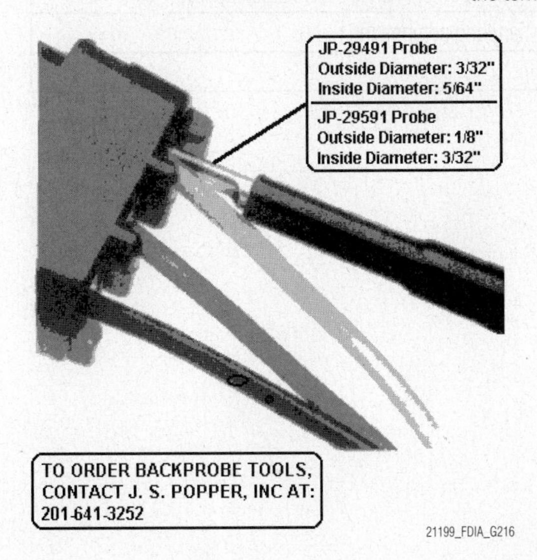

Fig. 1 Backprobing a connector

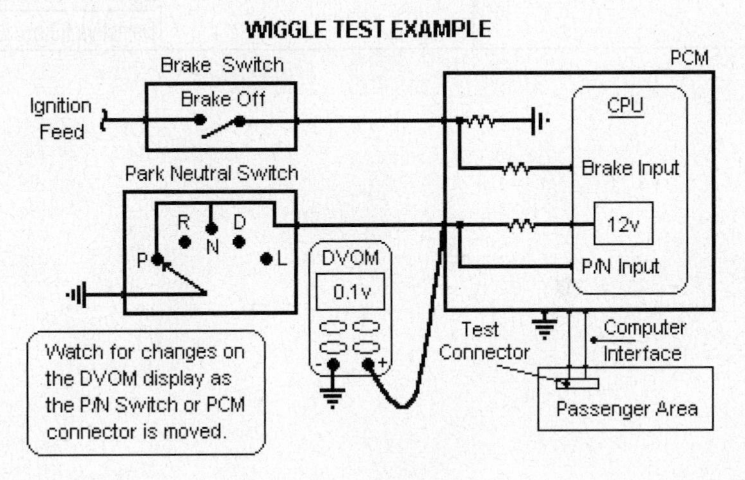

Fig. 2 Wiggle Test Example

Diagnosis And Testing - Vehicle Does Not Fill

CONDITION	POSSIBLE CAUSES	CORRECTION
Pre-Mature Nozzle Shut-Off	Defective fuel tank assembly components.	Fill tube improperly installed (sump)
		Fill tube hose pinched.
		Check valve stuck shut.
		Control valve stuck shut.
	Defective vapor/vent components.	Vent line from control valve to canister pinched.
		Vent line from canister to vent filter pinched.
		Canister vent valve failure (requires double failure, plugged to NVLD and atmosphere).
		Leak detection pump failed closed.
		Leak detection pump filter plugged.
	On-Board diagnostics evaporative system leak test just conducted.	Canister vent valve vent plugged to atmosphere.
		Engine still running when attempting to fill (System designed not to fill).
	Defective fill nozzle.	Try another nozzle.
Fuel Spits Out Of Filler Tube.	During fill.	See Pre-Mature Shut-Off.
	At conclusion of fill.	Defective fuel handling component. (Check valve stuck open).
		Defective vapor/vent handling component.
		Defective fill nozzle.

HYUNDAI
DIAGNOSTIC TROUBLE CODES

5

TABLE OF CONTENTS

VEHICLE APPLICATIONS 5-2
P0XXX 5-3
P1XXX 5-42
P2XXX 5-54

OBD II Vehicle Applications

Accent
1996-1999
1.5L I4 .... VIN K
2000-2001
1.5L I4 .... VIN B
2002-2003
1.5L I4 .... VIN G
2004-2006
1.5L I4 .... VIN B
2001
1.6L I4 .... VIN D
2002-2003
1.6L I4 .... VIN C
2004-2005
1.6L I4 .... VIN D

Elantra
1996-1998
1.8L I4 .... VIN M
1999-2000
2.0L I4 .... VIN F
2001-2006
2.0L I4 .... VIN D

Santa Fe
2001-2004
2.4L I4 .... VIN B
2001-2006
2.7L V6........ VIN D
2004-2006
3.5L V6........ VIN E

Sonata
1996-1998
2.0L I4 .... VIN F
1999-2000
2.4L I4 .... VIN D
2001-2006
2.4L I4 .... VIN S
1999-2000
2.5L I4 .... VIN E
2001-2002
2.5L I4 .... VIN V
2003-2004
2.7L V6........ VIN H
1996
3.0L V6........ VIN K
1997-98
3.0L V6........ VIN T
2006
3.3L V6........ VIN F

Tiburon
1997-2000
1.8L I4 .... VIN M
1997-2001
2.0L I4 .... VIN F
2003-2006
2.0L I4 .... VIN D
2003-2006
2.7L V6........ VIN F

Tucson
2005-2006
2.0L I4 .... VIN B
2005-2006
2.7L V6........ VIN D

XG300
2001
3.0L V6........ VIN D

XG350
2002-2005
3.5L V6........ VIN E

OBD II Trouble Code List (P0xxx Codes)

DTC	Trouble Code Title, Conditions & Possible Causes
DTC: P0010 **2T CCM, MIL: Yes** **2004** **Models:** Elantra, Tiburon **Engines:** 2.0L **Transmissions:** All	**"A" Camshaft Position Actuator Circuit Malfunction** A large deviation between camshaft position set point and actual value is detected. **Possible Causes:** • Mechanical trouble in CVVT system • ECM
DTC: P0011 **2T CCM, MIL: Yes** **2005-06** **Models:** Accent, Elantra, Sonata, Tiburon, Tucson **Engines:** 1.5L, 2.0L, 2.4L, 3.3L **Transmissions:** All	**"A" Camshaft Position Timing Over Advanced or System Performance (Bank 1)** Monitor deviation between camshaft position set point and actual value. **Possible Causes:** • Oil leakage • Faulty oil pump • Faulty intake valve control solenoid
DTC: P0012 **2T CCM, MIL: Yes** **2006** **Models:** Accent, Sonata **Engines:** 1.5L, 3.3L **Transmissions:** All	**"A" Camshaft Position- Timing Over retarded (Bank 1)** Determines if the phaser is stuck or has a steady error. **Possible Causes:** • Engine oil • OCV • CVVT stuck • Faulty PCM
DTC: P0016 **2T CCM, MIL: Yes** **2005-06** **Models:** Accent, Elantra, Santa Fe, Sonata, Tiburon, Tucson, XG350 **Engines:** 1.5L, 2.0L, 2.4L, 3.3L, 3.5L **Transmissions:** All	**Crankshaft Position- Camshaft Position Correlation (Bank 1 Sensor "A")** Monitor camshaft position in the full retard condition or during CVVT control. Camshaft switching out of 109 to 141 degrees in full retard position, 70 to 140 degrees CRK during CVVT control. **Possible Causes:** • Abnormal installation of camshaft • Abnormal installation of crankshaft • Abnormal installation of tone wheel
DTC: P0018 **2T CCM, MIL: Yes** **2006** **Models:** Sonata **Engines:** 3.3L **Transmissions:** All	**Crankshaft Position- Camshaft Position Correlation (Bank 2 Sensor A)** Determines if CAM (B2) target is aligned correctly to the crank. No active faults. **Possible Causes:** • CKPS, CMPS (B2) • CVVT • Timing misalignment • Faulty PCM
DTC: P0021 **2T CCM, MIL: Yes** **2006** **Models:** Sonata **Engines:** 3.3L **Transmissions:** All	**"A" Camshaft Position- Timing Over Advanced Or System Performance (Bank 2)** Determines if the phaser is moving at an unexpected rate. Cam off set is available. **Possible Causes:** • Excessive phasing • Binding oil pressure (blockage) • Faulty PCM
DTC: P0022 **2T CCM, MIL: Yes** **2006** **Models:** Sonata **Engines:** 3.3L **Transmissions:** All	**"A" Camshaft Position- Timing Over Retarded (Bank 2)** Determines if the phaser is stuck or has a steady state error. Off sets available. Cam velocity below threshold at 15 CAD/s. **Possible Causes:** • Engine oil • OCV • CVVT stuck • Faulty PCM
DTC: P0026 **2T CCM, MIL: Yes** **2006** **Models:** Sonata **Engines:** 3.3L **Transmissions:** All	**Intake Valve Control Solenoid Circuit Range/Performance (Bank 1)** Determines if oil control valve is stuck. Valve cleaning not in progress. Off sets available. **Possible Causes:** • Oil pressure loss • OCV seizure • Faulty PCM
DTC: P0028 **2T CCM, MIL: Yes** **2006** **Models:** Sonata **Engines:** 3.3L **Transmissions:** All	**Intake Valve Control Solenoid Circuit Range/Performance (Bank 2)** Determines if oil control valve is stuck. Valve cleaning not in progress. Off sets available. **Possible Causes:** • Oil pressure loss • OCV seizure • Faulty PCM

DTC	Trouble Code Title, Conditions & Possible Causes
DTC: P0030 **2T CCM, MIL: Yes** **2003-06** **Models:** Accent, Elantra, Santa Fe, Sonata, Tiburon, Tucson **Engines:** 1.5L, 1.6L, 2.0L, 2.4L, 2.7L, 3.3L **Transmissions:** All	**O2 Sensor Heater Control Circuit (Bank 1/Sensor 1)** Check heater current. Internal resistance above threshold temperature (exhaust temperature, heater power). **Possible Causes:** • Contaminated, deteriorated or aged sensor • Heater resistance out of range • Faulty HO2S heater • Faulty PCM • Misplaced, bent, loose or corroded terminals
DTC: P0031 **2T CCM, MIL: Yes** **2003-06** **Models:** Accent, Elantra, Santa Fe, Sonata, Tiburon, Tucson, XG350 **Engines:** 1.5L, 1.6L, 2.0L, 2.4L, 2.7L, 3.3L, 3.5L **Transmissions:** All	**O2 Sensor Heater Circuit Low (Bank 1/Sensor 1)** **Heater check, low. Open or short circuit.** **Possible Causes:** • Open in battery and control circuit • Short to ground in control circuit (pin 48 to 36) • Faulty HO2S heater • Faulty PCM
DTC: P0032 **2T CCM, MIL: Yes** **2003-06** **Models:** Accent, Elantra, Santa Fe, Sonata, Tiburon, Tucson, XG350 **Engines:** 1.5L, 1.6L, 2.0L, 2.4L, 2.7L, 3.3L, 3.5L **Transmissions:** All	**O2 Sensor Heater Circuit High (Bank 1/Sensor 1)** Heater check, high. Short circuit. **Possible Causes:** • Short to battery in control circuit • Faulty HO2S heater • Faulty PCM
DTC: P0036 **2T CCM, MIL: Yes** **2003-06** **Models:** Accent, Elantra, Santa Fe, Sonata, Tiburon, Tucson **Engines:** 1.5L, 1.6L, 2.0L, 2.4L, 2.7L, 3.3L **Transmissions:** All	**O2 Sensor Heater Control Circuit Bank 1/Sensor 2)** Check heater current. Internal resistance above threshold temperature (exhaust temperature, heater power). **Possible Causes:** • Contaminated, deteriorated or aged sensor • Heater resistance out of range • Faulty HO2S heater • Faulty PCM • Misplaced, bent, loose or corroded terminals
DTC: P0037 **2T CCM, MIL: Yes** **2003-06** **Models:** Accent, Elantra, Santa Fe, Sonata, Tiburon, Tucson, XG350 **Engines:** 1.5L, 1.6L, 2.0L, 2.4L, 2.7L, 3.3L, 3.5L **Transmissions:** All	**O2 Sensor Heater Circuit Low (Bank 1/Sensor 2)** Heater check, low. Open or short circuit. **Possible Causes:** • Open in battery and control circuit • Short to ground in control circuit (pin 48 to 36) • Faulty HO2S heater • Faulty PCM
DTC: P0038 **2T CCM, MIL: Yes** **2003-06** **Models:** Accent, Elantra, Santa Fe, Sonata, Tiburon, Tucson, XG350 **Engines:** 1.5L, 1.6L, 2.0L, 2.4L, 2.7L, 3.3L, 3.5L **Transmissions:** All	**O2 Sensor Heater Circuit High (Bank 1/Sensor 2)** Heater check, high. Short circuit. **Possible Causes:** • Short to battery in control circuit • Faulty HO2S heater • Faulty PCM
DTC: P0050 **2T CCM, MIL: Yes** **2003-06** **Models:** Santa Fe, Tiburon, Tucson **Engines:** 2.7L, 3.5L **Transmissions:** All	**OHO2S Heater Control Circuit High (Bank 2/Sensor 1)** Evaluate O2 sensor element temperature via measuring element resistance. Sensor preheating and full heating phases finished. **Possible Causes:** • Related fuse blown or missing • Heater control circuit open or short • Power supply circuit open or short • Contact resistance in connectors • Faulty HO2S

DTC	Trouble Code Title, Conditions & Possible Causes
DTC: P0051 **2T CCM, MIL: Yes** **2003-06** **Models:** Santa Fe, Sonata, Tiburon, Tucson, XG350 **Engines:** 2.7L, 3.3L, 3.5L **Transmissions:** All	**OHO2S Heater Circuit Low (Bank 2/Sensor 1)** Short circuit to ground on front HO2S heater line. Battery voltage above 10 volts. **Possible Causes:** • Related fuse blown or missing • Open or short to ground in power supply or control harness • Contact resistance in connectors • Faulty HO2S
DTC: P0052 **2T CCM, MIL: Yes** **2003-06** **Models:** Santa Fe, Sonata, Tiburon, Tucson, XG350 **Engines:** 2.7L, 3.3L, 3.5L **Transmissions:** All	**OHO2S Heater Circuit High (Bank 2/Sensor 1)** Open or short circuit to battery line on front HO2S heater line. Battery voltage above 10 volts. **Possible Causes:** • Open or short to battery in control harness • Contact resistance in connectors • Faulty HO2S
DTC: P0053 **2T CCM, MIL: Yes** **2006** **Models:** Santa Fe **Engines:** 3.5L **Transmissions:** All	**HO2S Heater Resistance (Bank 1/Sensor 1)** Heater current difference between heater "OFF" and "ON" is less than threshold value (<0.4 amp). The PCM determines that a fault exists and a DTC is stored. Performance check. **Possible Causes:** • Poor connection or damaged harness • Faulty HO2S heater • Faulty PCM
DTC: P0054 **2T CCM, MIL: Yes** **2006** **Models:** Santa Fe **Engines:** 3.5L **Transmissions:** All	**HO2S Heater Resistance (Bank 1/Sensor 2)** Heater current difference between heater "OFF" and "ON" is less than threshold value (<0.4 amp). The PCM determines that a fault exists and a DTC is stored. Performance check. **Possible Causes:** • Poor connection or damaged harness • Faulty HO2S heater • Faulty PCM
DTC: P0056 **2T CCM, MIL: Yes** **2003-06** **Models:** Santa Fe, Sonata, Tiburon, Tucson **Engines:** 2.7L, 3.5L **Transmissions:** All	**HO2S Heater Control Circuit (Bank 2/Sensor 2)** Evaluate O2 sensor element temperature via measuring element resistance. Sensor preheating and full heating phases finished. **Possible Causes:** • Related fuse blown or missing • Heater control circuit open or short • Power supply circuit open or short • Contact resistance in connections • Faulty HO2S
DTC: P0057 **2T CCM, MIL: Yes** **2003-06** **Models:** Santa Fe, Sonata, Tiburon, Tucson, XG350 **Engines:** 2.7L, 3.3L, 3.5L **Transmissions:** All	**HO2S Heater Circuit Low (Bank 2/Sensor 2)** Check short circuit to ground on rear HO2S heater line. **Possible Causes:** • Related fuse blown or missing • Open or short to ground in power supply or control harness • Contact resistance in connections • Faulty HO2S
DTC: P0058 **2T CCM, MIL: Yes** **2003-06** **Models:** Santa Fe, Sonata, Tiburon, Tucson, XG350 **Engines:** 2.7L, 3.3L, 3.5L **Transmissions:** All	**HO2S Heater Circuit High (Bank 2/Sensor 2)** Check short circuit to ground on rear HO2S heater line. **Possible Causes:** • Open or short to battery in control harness • Contact resistance in connections • Faulty HO2S
DTC: P0059 **2T CCM, MIL: Yes** **2006** **Models:** Santa Fe **Engines:** 3.5L **Transmissions:** All	**HO2S Heater Resistance (Bank 2/Sensor 1)** Heater current difference between heater "OFF" and "ON" is less than threshold value (<0.4 amp). The PCM determines that a fault exists and a DTC is stored. Performance check. **Possible Causes:** • Poor connection or damaged harness • Faulty HO2S heater • Faulty PCM

DTC	Trouble Code Title, Conditions & Possible Causes
DTC: P0060 **2T CCM, MIL: Yes** **2006** **Models:** Santa Fe **Engines:** 3.5L **Transmissions:** All	**HO2S Heater Resistance (Bank 2/Sensor 2)** Heater current difference between heater "OFF" and "ON" is less than threshold value (<0.4 amp). The PCM determines that a fault exists and a DTC is stored. Performance check. **Possible Causes:** • Poor connection or damaged harness • Faulty HO2S heater • Faulty PCM
DTC: P0068 **2T CCM, MIL: Yes** **2006** **Models:** Accent **Engines:** 1.5L **Transmissions:** All	**MAFS/MAPS- TPS Correlation** Rationality check. **Possible Causes:** • Poor connection • Faulty TPS • Faulty MAFS • Faulty ECM/PCM
DTC: P0075 **2T CCM, MIL: Yes** **2006** **Models:** Accent **Engines:** 1.5L **Transmissions:** All	**Intake Valve Control Solenoid Circuit (Bank 1)** Circuit continuity check, open. **Possible Causes:** • Poor connection • Open or short to ground in power circuit • Open in control circuit • OCV • Faulty ECM/PCM
DTC: P0076 **2T CCM, MIL: Yes** **2004-06** **Models:** Accent, Elantra, Sonata, Tiburon, Tucson **Engines:** 1.5L, 2.0L, 2.4L, 3.3L **Transmissions:** All	**Intake Valve Control Solenoid Circuit Low (Bank 1)** PCM sets the code if it detects that the intake valve control solenoid control circuit is short to ground. Electrical check. **Possible Causes:** • Short to ground in control circuit • Contact resistance in connectors • Faulty intake valve control solenoid
DTC: P0077 **2T CCM, MIL: Yes** **2004-06** **Models:** Accent, Elantra, Sonata, Tiburon, Tucson **Engines:** 1.5L, 2.0L, 2.4L, 3.3L **Transmissions:** All	**Intake Valve Control Solenoid Circuit High (Bank 1)** PCM sets the code if it detects that the OCV control circuit is open or short to battery. Electrical check. **Possible Causes:** • Open or short to battery in control circuit • Contact resistance in connectors • Faulty intake valve control solenoid
DTC: P0082 **2T CCM, MIL: Yes** **2006** **Models:** Sonata **Engines:** 3.3L **Transmissions:** All	**Intake Valve Control Solenoid Circuit Low (Bank 2)** Detects a short to ground or open circuit of VCPD bank 1 intake circuit output. No disabling faults present. Engine running. Enable time delay equal to or greater than 0.5 second. **Possible Causes:** • Poor connection • Open in power circuit • Open or short to ground in control circuit • OCV • Faulty PCM
DTC: P0083 **2T CCM, MIL: Yes** **2006** **Models:** Sonata **Engines:** 3.3L **Transmissions:** All	**Intake Valve Control Solenoid Circuit High (Bank 2)** Detects a short to battery of VCPD bank 1 intake circuit output. No disabling faults present. Engine running. Enable time delay equal to or greater than 0.5 second. **Possible Causes:** • poor connection • Short to battery in control circuit • OCV • Faulty PCM
DTC: P0100 **2T CCM, MIL: Yes** **1996-06** **Models:** Accent, Elantra, Santa Fe, Sonata, Tiburon, XG300, XG350 **Engines:** All **Transmissions:** All	**Mass or Volume Airflow Sensor Circuit Malfunction** Engine started, engine runtime over 5 seconds, and the PCM detected an unexpected voltage condition on the MAF sensor circuit during in the CCM test. **Possible Causes:** • MAF sensor signal circuit is shorted to ground • MAF sensor signal circuit is shorted to VREF or system power • MAF sensor ground circuit is open between sensor and ground • MAF sensor is damaged or has failed • PCM has failed

DTC	Trouble Code Title, Conditions & Possible Causes
DTC: P0101 **2T CCM, MIL: Yes** **1999-06** **Models:** Accent, Elantra, Santa Fe, Sonata, Tiburon, Tucson, XG300, XG350 **Engines:** All **Transmissions:** All	**Mass Airflow or Volume Airflow Sensor Performance** Engine started, engine running at idle speed, and the PCM detected the MAF sensor signal was less than 0.5 volt, or with the engine speed more than 3000 rpm, the MAF sensor signal was more than 4.5 volts. **Possible Causes:** • MAF sensor signal circuit is shorted to ground • MAF sensor signal circuit is shorted to VREF or system power • MAF sensor ground circuit is open between sensor and ground • MAF sensor is contaminated (dirty), damaged or has failed • PCM has failed
DTC: P0102 **2T CCM, MIL: Yes** **1998-06** **Models:** Accent, Elantra, Santa Fe, Sonata, Tiburon, Tucson, XG300, XG350 **Engines:** All **Transmissions:** All	**Mass or Volume Airflow Sensor Circuit Low Input** Engine started, engine runtime over 5 seconds, and the PCM detected the MAF sensor signal was less than 0.5 volt during the test. **Possible Causes:** • MAF sensor signal circuit is open or shorted to ground • MAF sensor power (VREF) circuit is open or shorted to ground • MAF sensor is damaged or has failed • PCM has failed
DTC: P0103 **2T CCM, MIL: Yes** **1996-06** **Models:** Accent, Elantra, Santa Fe, Sonata, Tiburon, Tucson, XG300, XG350 **Engines:** All **Transmissions:** All	**Volume Airflow Sensor Circuit High Input** Engine runtime over 5 seconds, and the PCM detected the MAF sensor input was out of range high. **Possible Causes:** • MAF sensor signal circuit is open between the sensor and PCM • MAF sensor signal circuit is shorted to VREF or system power • MAF sensor ground circuit is open between sensor and ground • MAF sensor is damaged or has failed • PCM has failed
DTC: P0105 **2T CCM, MIL: Yes** **1996-02** **Models:** Elantra **Engines:** All **Transmissions:** All	**Manifold Absolute Pressure Sensor Circuit Malfunction** Engine started, engine runtime over 5 seconds and the PCM detected an unexpected voltage condition (i.e., more than 4.50 volts or less than 1.95 volt) on the MAP sensor circuit for 4 seconds in the test. **Possible Causes:** • MAP sensor signal circuit is open or shorted to ground • MAP sensor ground circuit open between sensor and ground • MAP sensor power (VREF) circuit is open • MAP sensor is damaged or has failed • PCM has failed
DTC: P0106 **2T CCM, MIL: Yes** **2003-06** **Models:** Accent, Elantra, Santa Fe, Sonata, Tiburon, XG350 **Engines:** 1.5L, 1.6L, 2.0L, 2.4L, 3.3L, 3.5L **Transmissions:** All	**Manifold Absolute Pressure/Barometric Pressure Circuit Range/Performance** Map sensor output voltage is out of the threshold value. Rationality check. Coolant temperature above 19.4 degrees F. **Possible Causes:** • Poor connection • Open or short in MAP sensor circuit • Faulty MAPS • Faulty PCM
DTC: P0107 **2T CCM, MIL: Yes** **2003-06** **Models:** Accent, Elantra, Santa Fe, Sonata, Tiburon, XG350 **Engines:** 1.5L, 1.6L, 2.0L, 2.4L, 3.3L, 3.5L **Transmissions:** All	**Manifold Absolute Pressure/Barometric Pressure Circuit Low Input** Map sensor output voltage is lower than threshold value. Low voltage check. Coolant temperature above 19.4 degrees F. **Possible Causes:** • Poor connection • Open or short to ground in MAP sensor circuit • Faulty MAPS • Faulty PCM
DTC: P0108 **2T CCM, MIL: Yes** **2003-06** **Models:** Accent, Elantra, Santa Fe, Sonata, Tiburon, XG350 **Engines:** 1.5L, 1.6L, 2.0L, 2.4L, 3.3L, 3.5L **Transmissions:** All	**Manifold Absolute Pressure/Barometric Pressure Circuit High Input** Map sensor output voltage is higher than 4.9 volts. High voltage check. Coolant temperature above 19.4 degrees F. **Possible Causes:** • Poor connection • Open or short to battery in MAP sensor circuit • Faulty MAPS • Faulty PCM

DTC	Trouble Code Title, Conditions & Possible Causes
DTC: P0110 **2T CCM, MIL: Yes** **1996-06** **Models:** Accent, Elantra, Santa Fe, Sonata, Tiburon, XG300, XG350 **Engines:** All **Transmissions:** All	**Intake Air Temperature Sensor Circuit Malfunction** Key on or engine running, and the PCM detected an unexpected voltage condition on the IAT sensor signal circuit during the test. **Possible Causes:** • IAT sensor signal circuit is open or shorted to ground • IAT sensor ground circuit is open between sensor and the PCM • IAT sensor signal circuit is shorted to VREF or system power • IAT sensor is damaged or has failed • PCM has failed
DTC: P0111 **2T CCM, MIL: Yes** **2003-06** **Models:** Accent, Elantra, Santa Fe, Sonata, Tiburon, XG350 **Engines:** 1.5L, 1.6L, 2.0L, 2.4L, 2.7L, 3.5L **Transmissions:** All	**Intake Air temperature Sensor 1 Circuit Range/Performance** If the sensor is out of specification, a code is set. Output voltage is monitored. Engine coolant is above 167 degrees F. Vehicle speed is above 30MPH for more than 60 seconds. Vehicle speed is below 7mph for more than 30 seconds. **Possible Causes:** • Poor connection • Faulty IATS • Faulty PCM
DTC: P0112 **2T CCM, MIL: Yes** **1998-06** **Models:** Accent, Elantra, Santa Fe, Sonata, Tiburon, Tucson, XG300, XG350 **Engines:** All **Transmissions:** All	**Intake Air Temperature Sensor Circuit High Input** Key on or engine running, and the PCM detected the IAT sensor indicated more than 4.96 volot during the CCM test. **Possible Causes:** • IAT sensor signal circuit is open between sensor and the PCM • IAT sensor ground circuit is open between sensor and the PCM • IAT sensor signal circuit is shorted to VREF or system power • IAT sensor is damaged or has failed • PCM has failed
DTC: P0113 **2T CCM, MIL: Yes** **1998-06** **Models:** Accent, Elantra, Santa Fe, Sonata, Tiburon, Tucson, XG300, XG350 **Engines:** All **Transmissions:** All	**Intake Air Temperature Sensor Circuit Low Input** Key on or engine runtime over 5 seconds, and the PCM detected the IAT sensor indicated less than 0.20 volt during the CCM test. **Possible Causes:** • IAT sensor signal circuit is shorted to ground • IAT sensor is damaged or has failed • PCM has failed
DTC: P0115 **2T CCM, MIL: Yes** **1996-06** **Models:** Accent, Elantra, Santa Fe, Sonata, Tiburon, XG300, XG350 **Engines:** All **Transmissions:** All	**Engine Coolant Temperature Sensor Circuit Malfunction** Key on or engine running, and the PCM detected an unexpected voltage condition on the ECT sensor signal circuit during the test. **Possible Causes:** • ECT sensor signal circuit is open or shorted ground • ECT sensor ground circuit is open between sensor and PCM • ECT sensor signal circuit is shorted to VREF or system power • ECT sensor is damaged or has failed • PCM has failed
DTC: P0116 **2T CCM, MIL: Yes** **1998-06** **Models:** Accent, Elantra, Santa Fe, Sonata, Tiburon, Tucson, XG300, XG350 **Engines:** All **Transmissions:** All	**Engine Coolant Temperature Sensor Performance** Engine started, engine runtime over20 minutes, and the PCM detected the ECT sensor signal tailed more than 68ºF from the model curve stored in memory (could be an intermittent fault). **Note: Check for a possible problem related to the Cooling system.** **Possible Causes:** • ECT sensor circuit is open or shorted ground (intermittent fault) • ECT sensor ground circuit is open (an intermittent fault) • ECT sensor has drifted out of calibration or has failed • PCM has failed
DTC: P0117 **2T CCM, MIL: Yes** **1998-06** **Models:** Accent, Elantra, Santa Fe, Sonata, Tiburon, Tucson, XG300, XG350 **Engines:** All **Transmissions:** All	**Engine Coolant Temperature Sensor Low Input** Key on or engine runtime over 5 seconds, and the PCM detected that the ECT sensor input was less than 0.20 volt during the CCM test. **Possible Causes:** • ECT sensor signal circuit is shorted to ground • ECT sensor is damaged or has failed • PCM has failed

DTC	Trouble Code Title, Conditions & Possible Causes
DTC: P0118 **2T CCM, MIL: Yes** **1998-06** **Models:** Accent, Elantra, Santa Fe, Sonata, Tiburon, Tucson, XG300, XG350 **Engines:** All **Transmissions:** All	**Engine Coolant Temperature Sensor High Input** Key on or engine running, and the PCM detected the ECT sensor indicated more than 4.96 volts during the CCM test. **Possible Causes:** • ECT sensor signal circuit is open between sensor and the PCM • ECT sensor ground circuit is open between sensor and PCM • ECT sensor signal circuit is shorted to VREF or system power • ECT sensor is damaged or has failed • PCM has failed
DTC: P0119 **2T CCM, MIL: Yes** **2000-06** **Models:** Accent, Elantra, Santa Fe, Sonata, Tiburon, XG300, XG350 **Engines:** All **Transmissions:** All	**Engine Coolant Temperature Sensor Circuit Malfunction** Engine started, engine runtime over 5 seconds, and the PCM detected an intermittent loss of the ECT sensor input during the test. **Possible Causes:** • ECT sensor signal circuit is open (an intermittent fault) • ECT sensor ground circuit is open (an intermittent fault) • ECT sensor signal circuit is shorted to VREF (intermittent fault) • ECT sensor is damaged or has failed (an intermittent fault) • PCM has failed
DTC: P0120 **2T CCM, MIL: Yes** **1996-06** **Models:** Accent, Elantra, Santa Fe, Sonata **Engines:** All **Transmissions:** All	**Throttle Position Sensor Circuit Malfunction** Engine stared, engine runtime over 5 seconds, and the PCM detected the TP sensor was too high or too low for the conditions. **Possible Causes:** • TP sensor signal circuit is open or shorted to ground • TP sensor ground circuit or power (VREF) circuit is open • TP sensor is damaged or has failed • PCM has failed
DTC: P0120 **2T CCM, MIL: Yes** **2001-06** **Models:** Santa Fe, XG300, XG350 **Engines:** 2.4L, 2.7L, 3.0L, 3.5L **Transmissions:** All	**Electronic Throttle System Main Throttle Position Sensor Circuit Malfunction** Engine started, engine runtime over 5 seconds, and the PCM detected a problem in the Electronic Throttle System ETS-TP1 sensor circuit (i.e., the signal was too high or too low). **Possible Causes:** • ETS Main TP sensor signal circuit open or shorted to ground • ETS Main TP sensor ground circuit is open • ETS Main TP sensor power circuit is open (test VREF at PCM) • ETS Main TP sensor is damaged or has failed • PCM has failed
DTC: P0121 **2T CCM, MIL: Yes** **1999-06** **Models:** Accent, Elantra, Santa Fe, Sonata, Tiburon, Tucson, XG300, XG350 **Engines:** All **Transmissions:** All	**Throttle Position Sensor Range/Performance** Engine started, vehicle driven at a speed of over 20 mph, and the PCM detected the TP sensor input was incorrect when it was compared to the MAF sensor signal under these engine conditions. **Possible Causes:** • TP sensor signal circuit open or shorted to ground (intermittent) • TP sensor is loose at it mounting or the throttle is binding • Throttle body or linkage is binding or sticking • TP sensor is damaged or has failed (perform a sweep test) • PCM has failed
DTC: P0122 **2T CCM, MIL: Yes** **1998-06** **Models:** Accent, Elantra, Santa Fe, Sonata, Tiburon, Tucson, XG300, XG350 **Engines:** All **Transmissions:** All	**Throttle Position Sensor Circuit Low Input** Engine started, engine runtime over 5 seconds, and the PCM detected the TP sensor signal was 0.2 volt or lower during the test. **Possible Causes:** • TP sensor signal circuit is shorted to ground • TP sensor power circuit is open • TP sensor is damaged or has failed (perform a sweep test) • PCM has failed
DTC: P0123 **2T CCM, MIL: Yes** **1998-06** **Models:** Accent, Elantra, Santa Fe, Sonata, Tiburon, Tucson, XG300, XG350 **Engines:** All **Transmissions:** All	**Throttle Position Sensor Circuit High Input** Engine started, engine runtime over 5 seconds, and the PCM detected the TP sensor signal was 4.96 volts or higher during the test. **Possible Causes:** • TP sensor signal circuit is open or shorted to VREF • TP sensor ground circuit is open between sensor and the PCM • TP sensor is damaged or has failed (perform a sweep test) • PCM has failed

DTC	Trouble Code Title, Conditions & Possible Causes
DTC: P0124 **2T CCM, MIL: Yes** **2006** **Models:** Accent **Engines:** 1.5L **Transmissions:** All	**Throttle/Pedal Position Sensor/Switch "A" Circuit Intermittent** Rationality check. Rate of change in throttle angle 0.1221 percent. Engine speed 600 rpm. Coolant temperature 167 degrees F. **Possible Causes:** • Poor connection • TPS
DTC: P0125 **2T ECT, MIL: Yes** **1996-06** **Models:** Accent, Elantra, Santa Fe, Sonata, Tiburon, Tucson, XG300, XG350 **Engines:** All **Transmissions:** All	**Excessive Time To Enter Closed Loop** DTC P0115, P0116, P0117, P0118 and P0119 not set, engine started, engine runtime over 6 minutes, and the PCM detected the engine did not enter closed loop after an additional 30 seconds. **Possible Causes:** • Check the operation of the thermostat (it may be stuck open) • ECT sensor has failed • Inspect for low coolant level or an incorrect coolant mixture
DTC: P0128 **2T ECT, MIL: Yes** **2000-06** **Models:** Accent, Elantra, Santa Fe, Sonata, Tiburon, Tucson, XG300, XG350 **Engines:** All **Transmissions:** All	**Thermostat Malfunction Detected** ECT sensor input less than 40ºF at startup, engine started, and the PCM detected the ECT sensor did not reach 167ºF after a normal warm up period had expired during the CCM Rationality test. **Possible Causes:** • Check the operation of the thermostat (it may be stuck open) • ECT sensor is out-of-calibration or skewed • Inspect for low coolant level or for an incorrect coolant mixture • TSB 01-36-008 (5/01) contains information about this code
DTC: P0128 **2T ECT, MIL: Yes** **2000-06** **Models:** Accent, Elantra, Santa Fe, Sonata, Tiburon, Tucson, XG300, XG350 **Engines:** All **Transmissions:** All	**Thermostat Malfunction Detected** ECT sensor input less than 40ºF at startup, engine started, and the PCM detected the ECT sensor did not reach 167ºF after a normal warm up period had expired during the CCM Rationality test. **Possible Causes:** • Check the operation of the thermostat (it may be stuck open) • ECT sensor is out-of-calibration or skewed • Inspect for low coolant level or for an incorrect coolant mixture
DTC: P0130 **2T CCM, MIL: Yes** **2000-06** **Models:** Accent, Elantra, Santa Fe, Sonata, Tiburon, Tucson, XG300, XG350 **Engines:** All **Transmissions:** All	**HO2S-11 (Bank 1 Sensor 1) Circuit Malfunction** Engine started, engine running in closed loop, and the PCM detected one of the following "failure" conditions existed: - HO2S signal was too high (more than 1.0 volt at idle speed) - HO2S signal was fixed from 350-600 mv - HO2S switch time from rich-to-lean or lean-to -rich was too long - HO2S input fixed at mid-range (from 350-550 mv) **Possible Causes:** • HO2S signal circuit is open between the sensor and the PCM • HO2S signal circuit is shorted to sensor or chassis ground • HO2S signal circuit is shorted to VREF or system power (B+) • HO2S is damaged, contaminated or it has failed • PCM has failed
DTC: P0131 **2T CCM, MIL: Yes** **1998-06** **Models:** Accent, Elantra, Santa Fe, Sonata, Tiburon, Tucson, XG300, XG350 **Engines:** All **Transmissions:** All	**HO2S-11 (Bank 1 Sensor 1) Circuit Low Input** Engine started, engine running at cruise speed in closed loop for 2 minutes, and the PCM detected the HO2S-11 signal indicated less than 0.16 volt during the CCM test. **Possible Causes:** • Low fuel pressure, fuel filter restricted or fuel injectors plugged • HO2S signal circuit is shorted to ground (an intermittent fault) • HO2S may be contaminated or it has failed • HO2S heater is damaged or has failed • PCM has failed
DTC: P0132 **2T CCM, MIL: Yes** **1996-06** **Models:** Accent, Elantra, Santa Fe, Sonata, Tiburon, Tucson, XG300, XG350 **Engines:** All **Transmissions:** All	**HO2S-11 (Bank 1 Sensor 1) Circuit High Input** Engine started, engine running in closed loop at cruise speed for 2 minutes, and the PCM detected the HO2S-11 signal indicated more than 1.20 volts during the CCM test. **Possible Causes:** • Fuel pressure regulator leaking or fuel injectors leaking • HO2S signal circuit is shorted to the heater power circuit • HO2S may be contaminated or it has failed • HO2S heater is damaged or has failed • PCM has failed

DTC	Trouble Code Title, Conditions & Possible Causes
DTC: P0133 **2T O2S, MIL: Yes** **1998-06** **Models:** Accent, Elantra, Santa Fe, Sonata, Tiburon, Tucson, XG300, XG350 **Engines:** All **Transmissions:** All	**HO2S-11 (Bank 1 Sensor 1) Slow Response** DTC P0135 not set, engine started, engine running at idle speed in closed loop, and PCM detected the HO2S-11 response time to switch from rich-to-lean or from lean-to-rich was over one second. **Possible Causes:** • HO2S signal circuit is open or shorted to ground • HO2S element is contaminated or it has failed • HO2S heater is damaged or has failed • Intake air leaks, exhaust manifold leaks or PCV system leaks • MAF sensor out of calibration (it may be dirty or contaminated)
DTC: P0133 **2T O2S, MIL: Yes** **1998-06** **Models:** Accent, Elantra, Santa Fe, Sonata, Tiburon, Tucson, XG300, XG350 **Engines:** All **Transmissions:** All	**HO2S-11 (Bank 1 Sensor 1) Slow Response** DTC P0135 not set, engine started, engine running at idle speed in closed loop, and PCM detected the HO2S-11 response time to switch from rich-to-lean or from lean-to-rich was over one second. **Possible Causes:** • HO2S signal circuit is open or shorted to ground • HO2S element is contaminated or it has failed • HO2S heater is damaged or has failed • Intake air leaks, exhaust manifold leaks or PCV system leaks • MAF sensor out of calibration (it may be dirty or contaminated) • TSB 02-360010 (3/02) contains information for this code
DTC: P0134 **2T O2S, MIL: Yes** **1998-06** **Models:** Accent, Elantra, Santa Fe, Sonata, Tiburon, Tucson, XG300, XG350 **Engines:** All **Transmissions:** All	**HO2S-11 (Bank 1 Sensor 1) No Activity Detected** DTC P0135 not set, engine started, engine at idle speed and running in closed loop, and the PCM detected the HO2S-11 signal remained fixed from 400-550 mv for more than 1 minute during the CCM test. **Possible Causes:** • HO2S signal circuit is open or shorted to ground • HO2S element is contaminated or it has failed • HO2S heater is damaged or has failed • PCM has failed
DTC: P0134 **2T O2S, MIL: Yes** **1998-06** **Models:** Accent, Elantra, Santa Fe, Sonata, Tiburon, Tucson, XG300, XG350 **Engines:** All **Transmissions:** All	**HO2S-11 (Bank 1 Sensor 1) No Activity Detected** DTC P0135 not set, engine started, engine at idle speed and running in closed loop, and the PCM detected the HO2S-11 signal remained fixed from 400-550 mv for more than 1 minute during the CCM test. **Possible Causes:** • HO2S signal circuit is open or shorted to ground • HO2S element is contaminated or it has failed • HO2S heater is damaged or has failed • PCM has failed • TSB 02-360010 (3/02) contains information about this code
DTC: P0135 **2T O2S HTR1, MIL: Yes** **1996-06** **Models:** Accent, Elantra, Santa Fe, Sonata, Tiburon, XG300, XG350 **Engines:** All **Transmissions:** All	**HO2S-11 (Bank 1 Sensor 1) Heater Circuit Malfunction** Engine started, engine running in closed loop at cruise speed, and the PCM detected the HO2S-11 heater current was less than 0.2 amps, or that is was more than 3.5 amps during the test period. **Possible Causes:** • HO2S heater control circuit is open or shorted to ground • HO2S heater control circuit is shorted to power • HO2S heater power circuit is open (check fuse in Engine J/B) • HO2S heater is damaged or has failed • PCM has failed
DTC: P0136 **2T CCM, MIL: Yes** **1996-06** **Models:** Accent, Elantra, Santa Fe, Sonata, Tiburon, Tucson, XG300, XG350 **Engines:** All **Transmissions:** All	**HO2S-12 (Bank 1 Sensor 2) Circuit Malfunction** Engine started, running in closed loop, and the PCM detected an unexpected high voltage on the HO2S circuit; or the HO2S signal was fixed at mid-range (350-550 mv) or not switching properly. **Possible Causes:** • HO2S signal circuit is open between the sensor and the PCM • HO2S signal circuit is shorted to sensor or chassis ground • HO2S signal circuit is shorted to VREF or system power (B+) • HO2S is damaged, contaminated or it has failed • PCM has failed

DTC	Trouble Code Title, Conditions & Possible Causes
DTC: P0137 **2T CCM, MIL: Yes** **1998-06** **Models:** Accent, Elantra, Santa Fe, Sonata, Tiburon, Tucson, XG300, XG350 **Engines:** All **Transmissions:** All	**HO2S-12 (Bank 1 Sensor 2) Circuit Low Input** Engine started, engine running at cruise speed in closed loop for 2 minutes, and the PCM detected the HO2S-12 signal indicated less than 0.16 volt during the CCM test. **Possible Causes:** • Low fuel pressure, fuel filter restricted or fuel injectors plugged • HO2S signal circuit is shorted to ground (an intermittent fault) • HO2S may be contaminated or it has failed • HO2S heater is damaged or has failed • PCM has failed
DTC: P0138 **2T CCM, MIL: Yes** **1998-06** **Models:** Accent, Elantra, Santa Fe, Sonata, Tiburon, Tucson, XG300, XG350 **Engines:** All **Transmissions:** All	**HO2S-12 (Bank 1 Sensor 2) Circuit High Input** Engine started, engine running in closed loop at cruise speed for 2 minutes, and the PCM detected the HO2S-12 signal indicated more than 1.20 volts during the CCM test. **Possible Causes:** • Fuel pressure regulator leaking or fuel injectors leaking • HO2S signal circuit is shorted to the heater power circuit • HO2S may be contaminated or it has failed • HO2S heater is damaged or has failed • PCM has failed
DTC: P0139 **2T CCM, MIL: Yes** **1996-06** **Models:** Accent, Elantra, Santa Fe, Sonata, Tiburon, Tucson, XG300, XG350 **Engines:** All **Transmissions:** All	**HO2S-12 (Bank 1 Sensor 2) Slow Response** DTC P0141 not set, engine started, engine running at idle speed in closed loop, and PCM detected the HO2S-12 response time to switch from rich-to-lean or from lean-to-rich was over one second. **Possible Causes:** • HO2S signal circuit is open or shorted to ground • HO2S element is contaminated or it has failed • HO2S heater is damaged or has failed • Intake air leaks, exhaust manifold leaks or PCV system leaks • MAF sensor out of calibration (it may be dirty or contaminated)
DTC: P0140 **2T O2S, MIL: Yes** **1996-06** **Models:** Accent, Elantra, Santa Fe, Sonata, Tiburon, Tucson, XG300, XG350 **Engines:** All **Transmissions:** All	**HO2S-12 (Bank 1 Sensor 2) Circuit No Activity** DTC P0141 not set, engine started, engine at idle speed and running in closed loop, and the PCM detected the HO2S-12 signal remained fixed from 400-550 mv for more than 1 minute during the CCM test. **Possible Causes:** • HO2S signal circuit is open or shorted to ground • HO2S element is contaminated, damaged or has failed • PCM has failed
DTC: P0141 **2T O2S HTR1, MIL: Yes** **1996-06** **Models:** Accent, Elantra, Santa Fe, Sonata, Tiburon, XG300, XG350 **Engines:** All **Transmissions:** All	**HO2S-12 (Bank 1 Sensor 2) Heater Circuit Malfunction** Engine started, engine running in closed loop at cruise speed, and the PCM detected the HO2S-12 heater current was less than 0.2 amps, or that is was more than 3.5 amps during the test period. **Possible Causes:** • HO2S heater control circuit shorted to ground or system power • HO2S heater power circuit is open (check fuse in Engine J/B) • HO2S heater is damaged or has failed • PCM has failed
DTC: P0150 **2T CCM, MIL: Yes** **1999-06** **Models:** Santa Fe, Sonata, Tiburon, Tucson, XG300, XG350 **Engines:** 2.5 VIN V, 2.7L, 3.0L, 3.5L **Transmissions:** All	**HO2S-21 (Bank 2 Sensor 1) Circuit Malfunction** Engine started, engine running in closed loop, and the PCM detected one of the following "failure" conditions were present: - The HO2S signal was too high (more than 1.0 volt) - The HO2S signal was fixed between 350-600 mv - The HO2S switch time was too long - The HO2S signal fixed at mid-range (350-550 mv) **Possible Causes:** • HO2S signal circuit is open between the sensor and the PCM • HO2S signal circuit is shorted to sensor or chassis ground • HO2S signal circuit is shorted to VREF or system power (B+) • HO2S is damaged, contaminated or it has failed • PCM has failed

DTC	Trouble Code Title, Conditions & Possible Causes
DTC: P0152 **2T CCM, MIL: Yes** **1999-06** **Models:** Santa Fe, Sonata, Tiburon, Tucson, XG300, XG350 **Engines:** 2.5 VIN V, 2.7L, 3.0L, 3.3L **Transmissions:** All	**HO2S-21 (Bank 2 Sensor 1) Circuit High Input** Engine started, engine running in closed loop at cruise speed, and the PCM detected the HO2S-21 signal was more 1.20 volts in the test. **Possible Causes:** • Fuel pressure regulator leaking or fuel injectors leaking • HO2S signal circuit is shorted to the heater power circuit • HO2S may be contaminated or it has failed • HO2S heater is damaged or has failed • PCM has failed
DTC: P0153 **2T O2S, MIL: Yes** **2001-06** **Models:** Santa Fe, Sonata, Tiburon, Tucson, XG300, XG350 **Engines:** All **Transmissions:** All	**HO2S-21 (Bank 2 Sensor 1) Slow Response** DTC P0155 not set, engine started, engine running at idle speed in closed loop, and PCM detected the HO2S-12 response time to switch from rich-to-lean or from lean-to-rich was over one second. **Possible Causes:** • HO2S signal circuit is open or shorted to ground • HO2S element is contaminated or it has failed • HO2S heater is damaged or has failed • Intake air leaks, exhaust manifold leaks or PCV system leaks • MAF sensor out of calibration (it may be dirty or contaminated)
DTC: P0154 **2T O2S, MIL: Yes** **2001-06** **Models:** Santa Fe, Sonata, Tiburon, Tucson, XG300, XG350 **Engines:** All **Transmissions:** All	**HO2S-21 (Bank 2 Sensor 1) No Activity Detected** DTC P0141 not set, engine started, engine at idle speed and running in closed loop, and the PCM detected the HO2S-21 signal remained fixed from 400-550 mv for more than 1 minute during the CCM test. **Possible Causes:** • HO2S signal circuit is open or shorted to ground • HO2S element is contaminated or it has failed • HO2S heater is damaged or has failed • PCM has failed
DTC: P0155 **2T O2S HTR1, MIL: Yes** **1999-06** **Models:** Santa Fe, Sonata, XG300, XG350 **Engines:** 2.5 VIN V, 2.7L, 3.0L, 3.3L. **Transmissions:** All	**HO2S-21 (Bank 2 Sensor 1) Heater Circuit Malfunction** Engine started, engine running in closed loop at cruise speed, and the PCM detected the HO2S-21 heater current was less than 0.2 amps, or that is was more than 3.5 amps during the test period. **Possible Causes:** • HO2S heater control circuit is open or shorted to ground • HO2S heater control circuit is shorted to power • HO2S heater power circuit is open (check fuse in Engine J/B) • HO2S heater is damaged or has failed • PCM has failed
DTC: P0156 **2T CCM, MIL: Yes** **1999-06** **Models:** Santa Fe, Sonata, Tiburon, Tucson, XG350 **Engines:** 2.5 VIN V, 2.7L, 3.0L, 3.5L **Transmissions:** All	**HO2S-22 (Bank 2 Sensor 2) Circuit Malfunction** Engine started, engine running in closed loop, and the PCM detected one of the following "failure" conditions were present: - The HO2S signal was too high (more than 1.0 volt) - The HO2S signal was fixed between 350-600 mv - The HO2S switch time was too long - The HO2S signal fixed at mid-range (350-550 mv) **Possible Causes:** • HO2S signal circuit is open between the sensor and the PCM • HO2S signal circuit is shorted to sensor or chassis ground • HO2S signal circuit is shorted to VREF or system power (B+) • HO2S is damaged, contaminated or it has failed • PCM has failed
DTC: P0157 **2T CCM, MIL: Yes** **2003-06** **Models:** Santa Fe, Sonata, Tiburon, Tucson, XG350 **Engines:** 2.7L, 3.3L, 3.5L **Transmissions:** All	**HO2S Circuit Low Voltage (Bank 2 Sensor 2)** The signal voltage of the front or rear sensor changes the rear circuit voltage specification when air fuel ratio is rich, a DTC is set. Out of range low failure (ground short open circuit). **Possible Causes:** • Poor connection • Short to ground in HO2S circuit • Faulty HO2S • Faulty PCM
DTC: P0158 **2T CCM, MIL: Yes** **2003-06** **Models:** Santa Fe, Sonata, Tiburon, Tucson, XG350 **Engines:** 2.7L, 3.3L, 3.5L **Transmissions:** All	**HO2S Circuit High Voltage (Bank 2 Sensor 2)** **The signal voltage is higher than 1.2 volts after open in circuit. Out of range high failure.** **Possible Causes:** • Poor connection • Short to battery in HO2S circuit • Faulty HO2S • Faulty PCM

DTC	Trouble Code Title, Conditions & Possible Causes
DTC: P0159 **2T O2S, MIL: Yes** **1999-06** **Models:** Santa Fe, Sonata, Tiburon, Tucson, XG300, XG350 **Engines:** 2.5 VIN V, 2.7L, 3.0L, 3.3L, 3.5L **Transmissions:** All	**HO2S-22 (Bank 2 Sensor 2) Rationality Check** DTC P0141 not set, engine started, engine running at idle speed in closed loop, and PCM detected the HO2S-22 response time to switch from rich-to-lean or from lean-to-rich was over one second. **Possible Causes:** • HO2S signal circuit is open or shorted to ground • HO2S element is contaminated or it has failed • HO2S heater is damaged or has failed • Intake air leaks, exhaust manifold leaks or PCV system leaks • MAF sensor out of calibration (it may be dirty or contaminated)
DTC: P0160 **2T O2S, MIL: Yes** **1999-06** **Models:** Santa Fe, Sonata, Tiburon, Tucson, XG300, XG350 **Engines:** 2.5 VIN V, 2.7L, 3.0L, 3.3L, 3.5L **Transmissions:** All	**HO2S-22 (Bank 2 Sensor 2) No Activity Detected** DTC P0141 not set, engine started, engine at idle speed and running in closed loop, and the PCM detected the HO2S-22 signal remained fixed at 550 mv or less for more than 1 minute during the CCM test. **Possible Causes:** • HO2S signal circuit is open or shorted to ground • HO2S element is contaminated or it has failed • HO2S heater is damaged or has failed • PCM has failed
DTC: P0161 **2T O2S HTR1, MIL: Yes** **1999-06** **Models:** Santa Fe, Sonata, Tiburon, Tucson, XG300, XG350 **Engines:** 2.5 VIN V, 2.7L, 3.0L, 3.3L, 3.5L **Transmissions:** All	**HO2S-22 (Bank 2 Sensor 2) Heater Circuit Malfunction** Engine started, engine running in closed loop at cruise speed, and the PCM detected the HO2S-22 heater current was less than 0.2 amps, or that is was more than 3.5 amps during the test period. **Possible Causes:** • HO2S heater control circuit is open or shorted to ground • HO2S heater control circuit is shorted to power • HO2S heater power circuit is open (check fuse in Engine J/B) • HO2S heater is damaged or has failed • PCM has failed
DTC: P0170 **2T Fuel, MIL: Yes** **1996-06** **Models:** Elantra, Santa Fe, Sonata, Tiburon, Tucson **Engines:** All **Transmissions:** All	**Fuel Trim Too Rich or Too Lean (Bank 1)** Engine started, engine running at cruise speed in closed loop for 3-5 minutes, and PCM the detected the Fuel system control was too rich or too lean under these conditions in the Fuel System Monitor test. **Possible Causes:** • Air leaks after the MAF sensor, or in the EGR or PCV system • Base engine "mechanical" fault affecting one or more cylinders • Exhaust leaks located in front of the A/FS or HO2S location • Fuel control sensor is out of calibration (i.e., ECT, IAT or MAP) • Fuel delivery system supplying too little fuel during cruise or idle periods (e.g., faulty fuel pump or dirty, restricted fuel filter) • Fuel injector (one or more) dirty or pressure regulator has failed • HO2S is contaminated, deteriorated or it has failed • Vehicle driven low on fuel or until it ran out of fuel
DTC: P0171 **2T Fuel, MIL: Yes** **2001-06** **Models:** Accent, Elantra, Santa Fe, Sonata, Tiburon, Tucson, XG300, XG350 **Engines:** All **Transmissions:** All	**Fuel System Too Lean (Bank 1)** Engine started, engine running at cruise speed for 3-5 minutes in closed loop, and the PCM detected the Fuel system was too lean (i.e., it was beyond a calibrated value stored in the PCM memory). **Possible Causes:** • Air leaks after the MAF sensor, or in the EGR or PCV system • Base engine "mechanical" fault affecting one or more cylinders • Exhaust leaks located in front of the A/FS or HO2S location • Fuel control sensor is out of calibration (i.e., ECT, IAT or MAP) • Fuel delivery system supplying too little fuel during cruise or idle periods (e.g., faulty fuel pump or dirty, restricted fuel filter) • Fuel injector (one or more) dirty or pressure regulator has failed • HO2S is contaminated, deteriorated or it has failed • Vehicle driven low on fuel or until it ran out of fuel

DTC	Trouble Code Title, Conditions & Possible Causes
DTC: P0172 **2T Fuel, MIL: Yes** **2001-06** **Models:** Accent, Elantra, Santa Fe, Sonata, Tiburon, Tucson, XG300, XG350 **Engines:** All **Transmissions:** All	**Fuel System Too Rich (Bank 1)** Engine started, engine running at cruise speed for 3-5 minutes in closed loop, and the PCM detected the Fuel system was too rich (i.e., it was beyond a calibrated value stored in the PCM memory). **Possible Causes:** • Base engine "mechanical" fault affecting one or more cylinders • EVAP system component has failed or canister fuel saturated • Exhaust leaks located in front of the HO2S location • Fuel control sensor is out of calibration (i.e., ECT, IAT or MAF) • Fuel delivery system supplying too much fuel during cruise or idle periods (e.g., faulty fuel pump, or faulty pressure regulator) • Fuel injector(s) is leaking or stuck partially open (one or more) • HO2S is contaminated, deteriorated or it has failed
DTC: P0173 **2T Fuel, MIL: Yes** **1996-06** **Models:** Santa Fe, Sonata, Tiburon, Tucson **Engines:** 2.5 VIN V, 2.7L **Transmissions:** All	**Fuel Trim Too Rich or Too Lean (Bank 2)** Engine running in closed loop, and the PCM detected the Fuel system was too rich or too lean during two or more consecutive trips. **Possible Causes:** • Base engine "mechanical" fault affecting one or more cylinders • EVAP system component has failed or canister fuel saturated • Exhaust leaks located in front of the HO2S location • Fuel control sensor is out of calibration (i.e., ECT, IAT or MAF) • Fuel delivery system supplying too much fuel during cruise or idle periods (e.g., faulty fuel pump, or faulty pressure regulator) • Fuel injector(s) is leaking or stuck partially open (one or more) • HO2S is contaminated, deteriorated or it has failed
DTC: P0174 **2T Fuel, MIL: Yes** **2001-06** **Models:** Santa Fe, Sonata, Tiburon, Tucson, XG300, XG350 **Engines:** All **Transmissions:** All	**Fuel System Too Lean (Bank 2)** Engine started, engine running at cruise speed for 3-5 minutes in closed loop, and the PCM detected the Fuel system was too lean (i.e., it was beyond a calibrated value stored in the PCM memory). **Possible Causes:** • Air leaks after the MAF sensor, or in the EGR or PCV system • Base engine "mechanical" fault affecting one or more cylinders • Exhaust leaks located in front of the A/FS or HO2S location • Fuel control sensor is out of calibration (i.e., ECT, IAT or MAP) • Fuel delivery system supplying too little fuel during cruise or idle periods (e.g., faulty fuel pump or dirty, restricted fuel filter) • Fuel injector (one or more) dirty or pressure regulator has failed • HO2S is contaminated, deteriorated or it has failed • Vehicle driven low on fuel or until it ran out of fuel
DTC: P0175 **2T Fuel, MIL: Yes** **2001-06** **Models:** Santa Fe, Sonata, Tiburon, Tucson, XG300, XG350 **Engines:** All **Transmissions:** All	**Fuel System Too Rich (Bank 2)** Engine started, engine running at cruise speed for 3-5 minutes in closed loop, and the PCM detected the Fuel system was too rich (i.e., it was beyond a calibrated value stored in the PCM memory). **Possible Causes:** • Base engine "mechanical" fault affecting one or more cylinders • EVAP system component has failed or canister fuel saturated • Exhaust leaks located in front of the HO2S location • Fuel control sensor is out of calibration (i.e., ECT, IAT or MAF) • Fuel delivery system supplying too much fuel during cruise or idle periods (e.g., faulty fuel pump, or faulty pressure regulator) • Fuel injector(s) is leaking or stuck partially open (one or more) • HO2S is contaminated, deteriorated or it has failed
DTC: P0181 **2T CCM, MIL: Yes** **2003-06** **Models:** Santa Fe, Sonata, XG350 **Engines:** 2.4L, 3.5L **Transmissions:** All	**Fuel Temperature Sensor "A" Circuit Range/Performance** If the voltage difference between fuel tank temperature and engine coolant temperature at starting is greater than threshold value (above 59 degree F), a fault code exists. Rationality check. Engine coolant 14 to 122 degrees F. **Possible Causes:** • Poor connection • Faulty fuel temperature sensor • Faulty PCM
DTC: P0182 **2T CCM, MIL: Yes** **2003-06** **Models:** Santa Fe, Sonata, XG350 **Engines:** 2.4L, 3.5L **Transmissions:** All	**Fuel Temperature Sensor "A" Circuit Low Input** If the sensor signal is less than 0.1 volt after starting, a fault code exists. Output voltage (VFTS) is monitored. **Possible Causes:** • Poor connection • Short to ground in fuel temperature sensor circuit • Faulty fuel temperature sensor • Faulty PCM

DTC	Trouble Code Title, Conditions & Possible Causes
DTC: P0183 **2T CCM, MIL: Yes** **2003-06** **Models:** Santa Fe, Sonata, XG350 **Engines:** 2.4L, 3.5L **Transmissions:** All	**Fuel Temperature Sensor "A" Circuit High Input** If the sensor signal is above 4.6 volts after starting, a fault code exists. Output voltage (VFTS) is monitored. **Possible Causes:** • Poor connection • Open in fuel temperature sensor circuit • Faulty fuel temperature sensor • Faulty PCM
DTC: P0196 **2T CCM, MIL: Yes** **2004-06** **Models:** Elantra, Sonata, Tiburon, Tucson **Engines:** 2.0L, 2.4L, 3.3L **Transmissions:** All	**Engine Oil temperature Sensor Range/Performance** Stuck oil temperature sensor signal or unusual low or high signal. Condition 1 (signal high or low), engine coolant temperature more than 158 degrees F and oil temperature less than 68 degrees F. Condition 2 (signal high or low), engine coolant temperature less than 158 degrees F and oil temperature above 212 degrees F. Condition 3 (stuck signal) engine coolant temperature less than 104 degrees F. **Possible Causes:** • Contact resistance in connectors • faulty OTS
DTC: P0197 **2T CCM, MIL: Yes** **2004-06** **Models:** Elantra, Sonata, Tiburon, Tucson **Engines:** 2.0L, 2.4L, 3.3L **Transmissions:** All	**Engine Oil temperature Sensor Low Input** Signal voltage lower than the possible range of a properly operating OTS. Voltage range check. Engine coolant temperature less than 212 degrees F. Oil temperature above 309 degrees F. **Possible Causes:** • Short circuit to ground • Contact resistance in connectors • faulty OTS
DTC: P0198 **2T CCM, MIL: Yes** **2004-06** **Models:** Elantra, Sonata, Tiburon, Tucson **Engines:** 2.0L, 2.4L, 3.3L **Transmissions:** All	**Engine Oil temperature Sensor High Input** Signal voltage higher than the possible range of a properly operating OTS. Voltage range check. Five minutes after engine start if engine coolant temperature less than 14 degrees F. Oil temperature minus 33 degrees F. **Possible Causes:** • Open circuit to battery • Contact resistance in connectors • faulty OTS
DTC: P0201 **2T CCM, MIL: Yes** **1996-06** **Models:** Accent, Elantra, Santa Fe, Sonata, Tiburon, XG300, XG350 **Engines:** All **Transmissions:** All	**Fuel Injector 1 Circuit Malfunction** Engine running and the PCM detected an unexpected voltage on the Fuel Injector 1 control circuit during the Component Monitor test. **Possible Causes:** • Injector 1 control circuit open or grounded • Injector 1 control circuit shorted to system power • Injector 1 power (B+) circuit open • Injector 1 damaged or has failed • PCM has failed
DTC: P0202 **2T CCM, MIL: Yes** **1996-06** **Models:** Accent, Elantra, Santa Fe, Sonata, Tiburon, XG300, XG350 **Engines:** All **Transmissions:** All	**Fuel Injector 2 Circuit Malfunction** Engine running and the PCM detected an unexpected voltage on the Fuel Injector control circuit during the Component Monitor test. **Possible Causes:** • Injector 2 control circuit open or grounded • Injector 2 control circuit shorted to system power • Injector 2 power (B+) circuit open • Injector 2 damaged or has failed • PCM has failed
DTC: P0203 **2T CCM, MIL: Yes** **1996-06** **Models:** Accent, Elantra, Santa Fe, Sonata, Tiburon, XG300, XG350 **Engines:** All **Transmissions:** All	**Fuel Injector 3 Circuit Malfunction** Engine running and the PCM detected an unexpected voltage on the Fuel Injector control circuit during the Component Monitor test. **Possible Causes:** • Injector 3 control circuit open or grounded • Injector 3 control circuit shorted to system power • Injector 3 power (B+) circuit open • Injector 3 damaged or has failed • PCM has failed

DTC	Trouble Code Title, Conditions & Possible Causes
DTC: P0204 **2T CCM, MIL: Yes** **1996-06** **Models:** Accent, Elantra, Santa Fe, Sonata, Tiburon, XG300, XG350 **Engines:** All **Transmissions:** All	**Fuel Injector 4 Circuit Malfunction** Engine running and the PCM detected an unexpected voltage on the Fuel Injector control circuit during the Component Monitor test. **Possible Causes:** • Injector 4 control circuit open or grounded • Injector 4 control circuit shorted to system power • Injector 4 power (B+) circuit open • Injector 4 damaged or has failed • PCM has failed
DTC: P0205 **2T CCM, MIL: Yes** **1999-06** **Models:** Santa Fe, Sonata, XG300, XG350 **Engines:** 2.5 VIN V, 2.7L, 3.0L, 3.5L **Transmissions:** All	**Fuel Injector 5 Circuit Malfunction** Engine running and the PCM detected an unexpected voltage on the Fuel Injector control circuit during the Component Monitor test. **Possible Causes:** • Injector 5 control circuit open or grounded • Injector 5 control circuit shorted to system power • Injector 5 power (B+) circuit open • Injector 5 damaged or has failed • PCM has failed
DTC: P0206 **2T CCM, MIL: Yes** **1999-06** **Models:** Santa Fe, Sonata, XG300, XG350 **Engines:** 2.5 VIN V, 2.7L, 3.0L, 3.5L **Transmissions:** All	**Fuel Injector 6 Circuit Malfunction** Engine running and the PCM detected an unexpected voltage on the Fuel Injector control circuit during the Component Monitor test. **Possible Causes:** • Injector 6 control circuit open or grounded • Injector 6 control circuit shorted to system power • Injector 6 power (B+) circuit open • Injector 6 damaged or has failed • PCM has failed
DTC: P0217 **2T CCM, MIL: Yes** **2006** **Models:** Sonata **Engines:** 3.3L **Transmissions:** All	**Engine Coolant Over Temperature Condition** This diagnostic introduces a delay and also looks out for excessive engine loads. Once the delay period passes and excessive loads were not experienced, the diagnostic checks whether the undefaulted coolant temperature has exceeded a maximum threshold in order to make a pass/fail determination. No engine running status. No disabling faults present. Coolant sensor within range. Coolant temperature equal or greater than 122 degrees F. IAT equal or greater than 95 degrees F. **Possible Causes:** • Poor connection • Lack of engine coolant • Water pump problems • ECTS • Faulty PCM
DTC: P0220 **2T CCM, MIL: Yes** **2001-06** **Models:** Santa Fe, XG300, XG350 **Engines:** All **Transmissions:** A/T	**Electronic Throttle System Throttle Position Sensor Range/Performance** Key on or engine runtime over 5 seconds, and the PCM detected the Electronic Throttle System (ETS) throttle position sensor input was out of range (i.e., it was too high or too low) during the CCM test. **Possible Causes:** • APP sensor signal circuit is open or shorted to ground • APP sensor signal circuit is shorted to VREF or system power • APP sensor power circuit is open between sensor and the PCM • APP sensor is damaged or has failed • PCM has failed
DTC: P0221 **2T CCM, MIL: Yes** **2006** **Models:** Sonata **Engines:** 2.4L **Transmissions:** All	**Throttle/Pedal Position Sensor/Switch "B" Circuit Range/Performance** Plausibility check between TPS1 and TPS2. No engine start mode. No TSP adaptation request. No relevant failure. **Possible Causes:** • Poor connection or damaged harness • Air leakage in intake system • Faulty TPS2
DTC: P0222 **2T CCM, MIL: Yes** **2003-06** **Models:** Santa Fe, Sonata, XG350 **Engines:** 2.4L, 3.3L, 3.5L **Transmissions:** All	**Throttle/Pedal Position Sensor/Switch "B" Circuit Low Input** The DTC is recorded if the output voltage of the TPS 1 is lower than threshold value (Vtps1 less than or equal to 0.2 volt). TPS 1 low input. **Possible Causes:** • Poor connection • Open or short to ground in TPS circuit • Faulty TPS • Faulty PCM

DTC	Trouble Code Title, Conditions & Possible Causes
DTC: P0223 **2T CCM, MIL: Yes** **2003-06** **Models:** Santa Fe, Sonata, XG350 **Engines:** 2.4L, 3.3L, 3.5L **Transmissions:** All	**Throttle/Pedal Position Sensor/Switch "B" Circuit High Input** The DTC is recorded if the output voltage of the TPS 1 higher than threshold value (Vtps1 greater than or equal to 4.85 volts, load value, EV less than 70 percent) when TPS 2 (Vtps2 less than or equal to 2.5 volts) is normal. TPS 1 high input. **Possible Causes:** • Poor connection • Open or short to ground in TPS circuit • Faulty TPS • Faulty PCM
DTC: P0224 **2T CCM, MIL: Yes** **2003-06** **Models:** Santa Fe, XG350 **Engines:** 3.5L **Transmissions:** All	**Throttle Position Sensor/Switch "B" Linearity** **The DTC is recorded if the output voltage of the TPS 1 is too low. TPS 1 linearity.** **Possible Causes:** • Poor connection • Faulty TPS • Faulty PCM
DTC: P0230 **1T CCM, MIL: Yes** **1999-06** **Models:** Accent, Elantra, Santa Fe, Sonata, Tiburon, Tucson **Engines:** All **Transmissions:** All	**Fuel Pump Circuit Malfunction** Key on, and then the PCM detected an unexpected voltage condition on the fuel pump circuit through the fuel pump monitoring input. **Possible Causes:** • Fuel pump control circuit is open or shorted to ground • Fuel pump relay power circuit from ignition switch is open • Fuel pump relay is damaged or has failed • PCM has failed
DTC: P0231 **2T CCM, MIL: Yes** **2006** **Models:** Accent **Engines:** 1.5L **Transmissions:** All	**Electric Fuel Pump Relay Open Or Short Circuit** Circuit continuity check, high. **Possible Causes:** • Poor connection • Short to power in control circuit • Fuel pump relay • Faulty ECM/PCM
DTC: P0232 **2T CCM, MIL: Yes** **2006** **Models:** Accent **Engines:** 1.5L **Transmissions:** All	**Electric Fuel Pump Relay Short Circuit** Circuit continuity check, low. **Possible Causes:** • Poor connection • Short to ground in control circuit • Fuel pump relay • Faulty ECM/PCM
DTC: P0261 **2T CCM, MIL: Yes** **2003-06** **Models:** Accent, Elantra, Santa Fe, Sonata, Tiburon, Tucson **Engines:** 1.5L, 1.6L, 2.0L, 2.4L, 2.7L **Transmissions:** All	**Cylinder 1- Injector Circuit Low** The PCM sets the DTC if the control circuit is shorted to ground. Driver stage check. **Possible Causes:** • Open in power supply harness • Short to ground in control harness • Contact resistance in connectors • Faulty injector
DTC: P0262 **2T CCM, MIL: Yes** **2003-06** **Models:** Accent, Elantra, Santa Fe, Sonata, Tiburon, Tucson **Engines:** 1.5L, 1.6L, 2.0L, 2.4L, 2.7L **Transmissions:** All	**Cylinder 1- Injector Circuit High** The PCM sets the DTC if the control circuit is open or shorted to battery voltage. Driver stage check. **Possible Causes:** • Open or short to battery control harness • Contact resistance in connectors • Faulty injector
DTC: P0264 **2T CCM, MIL: Yes** **2003-06** **Models:** Accent, Elantra, Santa Fe, Sonata, Tiburon, Tucson **Engines:** 1.5L, 1.6L, 2.0L, 2.4L, 2.7L **Transmissions:** All	**Cylinder 2- Injector Circuit Low** The PCM sets the DTC if the control circuit is shorted to ground. Driver stage check. **Possible Causes:** • Open in power supply harness • Short to ground in control harness • Contact resistance in connectors • Faulty injector

DTC	Trouble Code Title, Conditions & Possible Causes
DTC: P0265 **2T CCM, MIL: Yes** **2003-06** **Models:** Accent, Elantra, Santa Fe, Sonata, Tiburon, Tucson **Engines:** 1.5L, 1.6L, 2.0L, 2.4L, 2.7L **Transmissions:** All	**Cylinder 2- Injector Circuit High** The PCM sets the DTC if the control circuit is open or shorted to battery voltage. Driver stage check. **Possible Causes:** • Open or short to battery control harness • Contact resistance in connectors • Faulty injector
DTC: P0267 **2T CCM, MIL: Yes** **2003-06** **Models:** Accent, Elantra, Santa Fe, Sonata, Tiburon, Tucson **Engines:** 1.5L, 1.6L, 2.0L, 2.4L, 2.7L **Transmissions:** All	**Cylinder 3- Injector Circuit Low** The PCM sets the DTC if the control circuit is shorted to ground. Driver stage check. **Possible Causes:** • Open in power supply harness • Short to ground in control harness • Contact resistance in connectors • Faulty injector
DTC: P0268 **2T CCM, MIL: Yes** **2003-06** **Models:** Accent, Elantra, Santa Fe, Sonata, Tiburon, Tucson **Engines:** 1.5L, 1.6L, 2.0L, 2.4L, 2.7L **Transmissions:** All	**Cylinder 3- Injector Circuit High** The PCM sets the DTC if the control circuit is open or shorted to battery voltage. Driver stage check. **Possible Causes:** • Open or short to battery control harness • Contact resistance in connectors • Faulty injector
DTC: P0270 **2T CCM, MIL: Yes** **2003-06** **Models:** Accent, Elantra, Santa Fe, Sonata, Tiburon, Tucson **Engines:** 1.5L, 1.6L, 2.0L, 2.4L, 2.7L **Transmissions:** All	**Cylinder 4- Injector Circuit Low** The PCM sets the DTC if the control circuit is shorted to ground. Driver stage check. **Possible Causes:** • Open in power supply harness • Short to ground in control harness • Contact resistance in connectors • Faulty injector
DTC: P0271 **2T CCM, MIL: Yes** **2003-06** **Models:** Accent, Elantra, Santa Fe, Sonata, Tiburon, Tucson **Engines:** 1.5L, 1.6L, 2.0L, 2.4L, 2.7L **Transmissions:** All	**Cylinder 4- Injector Circuit High** The PCM sets the DTC if the control circuit is open or shorted to battery voltage. Driver stage check. **Possible Causes:** • Open or short to battery control harness • Contact resistance in connectors • Faulty injector
DTC: P0273 **2T CCM, MIL: Yes** **2003-06** **Models:** Santa Fe, Sonata, Tiburon, Tucson **Engines:** 2.7L, 3.3L, 3.5L **Transmissions:** All	**Cylinder 5- Injector Circuit Low** The PCM sets the DTC if the control circuit is shorted to ground. Driver stage check. **Possible Causes:** • Open in power supply harness • Short to ground in control harness • Contact resistance in connectors • Faulty injector
DTC: P0274 **2T CCM, MIL: Yes** **2003-06** **Models:** Santa Fe, Sonata, Tiburon, Tucson **Engines:** 2.7L, 3.3L, 3.5L **Transmissions:** All	**Cylinder 5- Injector Circuit High** The PCM sets the DTC if the control circuit is open or shorted to battery voltage. Driver stage check. **Possible Causes:** • Open or short to battery control harness • Contact resistance in connectors • Faulty injector
DTC: P0276 **2T CCM, MIL: Yes** **2003-06** **Models:** Santa Fe, Sonata, Tiburon, Tucson **Engines:** 2.7L, 3.3L, 3.5L **Transmissions:** All	**Cylinder 6- Injector Circuit Low** The PCM sets the DTC if the control circuit is shorted to ground. Driver stage check. **Possible Causes:** • Open in power supply harness • Short to ground in control harness • Contact resistance in connectors • Faulty injector

DTC	Trouble Code Title, Conditions & Possible Causes
DTC: P0277 **2T CCM, MIL: Yes** **2003-06** **Models:** Santa Fe, Sonata, Tiburon, Tucson **Engines:** 2.7L, 3.3L, 3.5L **Transmissions:** All	**Cylinder 6- Injector Circuit High** The PCM sets the DTC if the control circuit is open or shorted to battery voltage. Driver stage check. **Possible Causes:** • Open or short to battery control harness • Contact resistance in connectors • Faulty injector
DTC: P0300 **2T Misfire, MIL: Yes** **1996-06** **Models:** Accent, Elantra, Santa Fe, Sonata, Tiburon, Tucson, XG300, XG350 **Engines:** All **Transmissions:** All	**Multiple Misfire Detected** Engine started, vehicle driven to over 3 mph at an engine speed of 400-3500 rpm, and the PCM detected irregular CKP sensor signals indicating a misfire condition in two or more cylinders was present during the 200-revolution or 1000-revolution Misfire Monitor test. Note: If the misfire is severe, the MIL will flash on/off on the 1st trip! **Possible Causes:** • Ignition system or fuel metering fault in 2 or more cylinders • Fuel pressure too low or too high, fuel supply contaminated • CKP/CMP signals, EVAP canister saturated, EGR valve stuck
DTC: P0301 **2T Misfire, MIL: Yes** **1996-06** **Models:** Accent, Elantra, Santa Fe, Sonata, Tiburon, Tucson, XG300, XG350 **Engines:** All **Transmissions:** All	**Cylinder 1 Misfire Detected** Engine started, vehicle driven to over 3 mph at an engine speed of 400-3500 rpm, and the PCM detected irregular CKP sensor signals indicating a misfire condition in one cylinder was present during the 200-revolution or 1000-revolution Misfire Detection Monitor test. Note: If the misfire is severe, the MIL will flash on/off on the 1st trip! **Possible Causes:** • Base engine mechanical fault that affects only one cylinder • Fuel metering fault that affects only one cylinder • EGR valve is stuck open or the PCV system has a vacuum leak • Ignition system fault (i.e., a coil) that affects only one cylinder
DTC: P0302 **2T Misfire, MIL: Yes** **1996-06** **Models:** Accent, Elantra, Santa Fe, Sonata, Tiburon, Tucson, XG300, XG350 **Engines:** All **Transmissions:** All	**Cylinder 2 Misfire Detected** Engine started, vehicle driven to over 3 mph at an engine speed of 400-3500 rpm, and the PCM detected irregular CKP sensor signals indicating a misfire condition in one cylinder was present during the 200-revolution or 1000-revolution Misfire Detection Monitor test. Note: If the misfire is severe, the MIL will flash on/off on the 1st trip! **Possible Causes:** • Base engine mechanical fault that affects only one cylinder • Fuel metering fault that affects only one cylinder • EGR valve is stuck open or the PCV system has a vacuum leak • Ignition system fault (i.e., a coil) that affects only one cylinder
DTC: P0303 **2T Misfire, MIL: Yes** **1996-06** **Models:** Accent, Elantra, Santa Fe, Sonata, Tiburon, Tucson, XG300, XG350 **Engines:** All **Transmissions:** All	**Cylinder 3 Misfire Detected** Engine started, vehicle driven to over 3 mph at an engine speed of 400-3500 rpm, and the PCM detected irregular CKP sensor signals indicating a misfire condition in one cylinder was present during the 200-revolution or 1000-revolution Misfire Detection Monitor test. Note: If the misfire is severe, the MIL will flash on/off on the 1st trip! **Possible Causes:** • Base engine mechanical fault that affects only one cylinder • Fuel metering fault that affects only one cylinder • EGR valve is stuck open or the PCV system has a vacuum leak • Ignition system fault (i.e., a coil) that affects only one cylinder
DTC: P0304 **2T Misfire, MIL: Yes** **1996-06** **Models:** Accent, Elantra, Santa Fe, Sonata, Tiburon, Tucson, XG300, XG350 **Engines:** All **Transmissions:** All	**Cylinder 4 Misfire Detected** Engine speed from 400-3500 rpm, VSS input over 3 mph, and the PCM detected irregular CKP inputs indicating a misfire condition in one cylinder during the 200-revolution or 1000-revolution test period. Note: If the misfire is severe, the MIL will flash on/off on the 1st trip! **Possible Causes:** • Base engine mechanical fault that affects only one cylinder • Fuel metering fault that affects only one cylinder • EGR valve is stuck open or the PCV system has a vacuum leak • Ignition system fault (i.e., a coil) that affects only one cylinder

DTC	Trouble Code Title, Conditions & Possible Causes
DTC: P0305 **2T Misfire, MIL: Yes** **1996-06** **Models:** Santa Fe, Sonata, Tiburon, Tucson, XG300, XG350 **Engines:** All **Transmissions:** All	**Cylinder 5 Misfire Detected** Engine started, vehicle driven to over 3 mph at an engine speed of 400-3500 rpm, and the PCM detected irregular CKP sensor signals indicating a misfire condition in one cylinder was present during the 200-revolution or 1000-revolution Misfire Detection Monitor test. Note: If the misfire is severe, the MIL will flash on/off on the 1st trip! **Possible Causes:** • Base engine mechanical fault that affects only one cylinder • Fuel metering fault that affects only one cylinder • EGR valve is stuck open or the PCV system has a vacuum leak • Ignition system fault (i.e., a coil) that affects only one cylinder
DTC: P0306 **2T Misfire, MIL: Yes** **1996-06** **Models:** Santa Fe, Sonata, Tiburon, Tucson, XG300, XG350 **Engines:** All **Transmissions:** All	**Cylinder 6 Misfire Detected** Engine started, vehicle driven to over 3 mph at an engine speed of 400-3500 rpm, and the PCM detected irregular CKP sensor signals indicating a misfire condition in one cylinder was present during the 200-revolution or 1000-revolution Misfire Detection Monitor test. Note: If the misfire is severe, the MIL will flash on/off on the 1st trip! **Possible Causes:** • Base engine mechanical fault that affects only one cylinder • Fuel metering fault that affects only one cylinder • EGR valve is stuck open or the PCV system has a vacuum leak • Ignition system fault (i.e., a coil) that affects only one cylinder
DTC: P0315 **2T CCM, MIL: Yes** **2005-06** **Models:** Elantra, Santa Fe, Sonata, Tiburon, Tucson **Engines:** 2.0L, 2.4L, 2.7L, 3.3L **Transmissions:** All	**Segment Time Acquisition Incorrect** A misfire induces a decrease in the engine speed and causes a variation in the segment period. Monitor segment time adaptation. **Possible Causes:** • Improperly installed target wheel • Contact resistance in connectors
DTC: P0320 **2T CCM, MIL: Yes** **2001-06** **Models:** Accent, Elantra, Santa Fe, Sonata, Tiburon, Tucson, XG300, XG350 **Engines:** All **Transmissions:** All	**Ignition Failure Sensor Circuit Malfunction** Engine started, engine runtime over 2 seconds, and the PCM detected a problem in the Ignition Failure Sensor or its circuit. **Possible Causes:** • Ignition failure sensor circuit is open or shorted to ground • Ignition failure sensor is damaged or has failed • PCM has failed
DTC: P0325 **2T CCM, MIL: Yes** **1998-06** **Models:** Accent, Elantra, Santa Fe, Sonata, Tiburon, Tucson, XG300, XG350 **Engines:** All **Transmissions:** All	**Knock Sensor 1 Circuit Malfunction** Engine started, engine running, and the PCM detected an unexpected voltage condition on the Knock Sensor 1 circuit. **Possible Causes:** • KS signal circuit is open, shorted to ground or to system power • KS ground circuit is open between sensor and PCM • Knock sensor is damaged or it has failed • PCM has failed
DTC: P0326 **2T CCM, MIL: Yes** **1998-06** **Models:** Accent, Elantra, Santa Fe, Sonata, Tiburon, Tucson, XG300, XG350 **Engines:** All **Transmissions:** All	**Knock Sensor 2 Range/Performance** Engine started, vehicle driven with the engine speed over 3000 rpm, ECT sensor more than 104°F, engine load over 2.5 ms, and the PCM detected an unexpected (intermittent) Knock Sensor 2 signal with the engine running as described during the CCM test. **Possible Causes:** • KS signal circuit is open, shorted to ground or to system power • Knock sensor is damaged or it has failed (an intermittent fault) • PCM has failed
DTC: P0237 **2T CCM, MIL: No** **2006** **Models:** Accent **Engines:** 1.5L **Transmissions:** All	**Knock Sensor 1 Circuit High Input** Signal check. Coolant temperature 104 degrees F. **Possible Causes:** • Poor connection • Short to power in signal circuit • Faulty knock sensor • Faulty ECM/PCM

DTC	Trouble Code Title, Conditions & Possible Causes
DTC: P0238 **2T CCM, MIL: No** **2006** **Models:** Accent **Engines:** 1.5L **Transmissions:** All	**Knock Sensor 1 Circuit Low Input** Signal check. Engine speed 2600 rpm. **Possible Causes:** • Poor connection • Open or short to ground in signal circuit • Faulty knock sensor • Faulty ECM/PCM
DTC: P0330 **2T CCM, MIL: Yes** **1996-06** **Models:** Accent, Elantra, Santa Fe, Sonata, Tiburon, Tucson, XG300, XG350 **Engines:** All **Transmissions:** All	**Knock Sensor 2 Circuit Malfunction** Engine started, engine running, and the PCM detected an unexpected voltage condition on the Knock Sensor 2 circuit. **Possible Causes:** • KS signal circuit is open, shorted to ground or to system power • KS ground circuit is open between sensor and PCM • Knock sensor is damaged or it has failed • PCM has failed
DTC: P0331 **2T CCM, MIL: Yes** **2006** **Models:** Sonata **Engines:** 3.3L **Transmissions:** All	**Knock Sensor 2 Circuit Range/Performance (Bank 2)** Signal short. Pressure in intake manifold is normal. Engine speed is equal to or less than 1600 rpm. **Possible Causes:** • Poor connection • Short in harness • Faulty knock sensor • Faulty PCM
DTC: P0335 **1T CCM, MIL: Yes** **1996-06** **Models:** Accent, Elantra, Santa Fe, Sonata, Tiburon, Tucson, XG300, XG350 **Engines:** All **Transmissions:** All	**Crankshaft Position Sensor 'A' Circuit Malfunction** Engine cranking for over 2 seconds, CMP sensor signals detected, and the PCM did not receive any CKP sensor signals during the test. **Possible Causes:** • CKP (Magnetic) sensor signal (+) or (-) circuit is open or shorted to ground between the sensor and the PCM • CKP (Magnetic) sensor signal is damaged or has failed • PCM has failed
DTC: P0336 **1T CCM, MIL: Yes** **1998-06** **Models:** Accent, Elantra, Santa Fe, Sonata, Tiburon **Engines:** All **Transmissions:** All	**Crankshaft Position Sensor Performance** Engine started, vehicle driven with the engine speed over 2000 rpm, and the PCM detected the CKP sensor inputs were either irregular, invalid or out of phase with the CMP sensor signals for 4 seconds. Note: The CKP sensor resistance at 68°F is 486-594 ohms. **Possible Causes:** • CKP sensor signal is open or shorted to ground (Intermittent) • CKP sensor is damaged or has failed • PCM has failed
DTC: P0337 **2T CCM, MIL: Yes** **2005-06** **Models:** Accent, Santa Fe, Sonata, XG350 **Engines:** 1.5L, 2.4L, 3.5L **Transmissions:** All	**Crankshaft Position Sensor "A" Circuit Low Input** If the output voltage of the CKPS remains low for more than two seconds. When the change of the CMPS output voltage is zero, the PCM determines a fault and stores a code. Change in output voltage (delta sign Vckp) is monitored. **Possible Causes:** • Poor connection • Open or short to ground in CKPS circuit • Faulty CKPS • Faulty PCM
DTC: P0338 **2T CCM, MIL: Yes** **2005-06** **Models:** Accent, Santa Fe, Sonata, XG350 **Engines:** 1.5L, 2.4L, 3.5L **Transmissions:** All	**Crankshaft Position Sensor "A" Circuit High Input** If the output voltage of the CKPS remains high for more than two seconds. When the change of the CMPS output voltage is zero, the PCM determines a fault and stores a code. Change in output voltage (delta sign Vckp) is monitored. **Possible Causes:** • Poor connection • Open or short to ground in CKPS circuit • Faulty CKPS • Faulty PCM
DTC: P0339 **2T CCM, MIL: Yes** **2006** **Models:** Accent **Engines:** 1.5L **Transmissions:** All	**Crankshaft Position Sensor "A" Circuit** Signal check. Edge counter of camshaft position sensor 8. **Possible Causes:** • Poor connection • Open or short in signal circuit • CKPS • Faulty ECM/PCM

DTC	Trouble Code Title, Conditions & Possible Causes
DTC: P0340 **2T CCM, MIL: Yes** **1996-06** **Models:** Accent, Elantra, Santa Fe, Sonata, Tiburon, Tucson, XG300, XG350 **Engines:** All **Transmissions:** All	**Camshaft Position Sensor Circuit Malfunction** Engine cranking for over 2 seconds, and the PCM detected an invalid or irregular CMP signal, or it did not detect any CMP signals. **Possible Causes:** • CMP (Resistor) sensor signal circuit open or shorted to ground • CMP (Resistor) sensor ground circuit is open • CMP (Resistor) sensor power circuit open (test power to relay) • CMP (Resistor) sensor is damaged or has failed • PCM has failed
DTC: P0341 **2T CCM, MIL: Yes** **2004-06** **Models:** Accent, Elantra, Sonata, Tiburon **Engines:** 1.5L, 2.0L, 2.4L, 3.3L **Transmissions:** All	**Camshaft Position Sensor Circuit Malfunction** No signal or no signal switching is detected. Crankshaft sensor is normal. Battery voltage is between 10 and 16 volts. **Possible Causes:** • Open or short in CMPS circuit • Faulty CMPS • Faulty PCM
DTC: P0342 **2T CCM, MIL: Yes** **1998-06** **Models:** Accent, Elantra, Santa Fe, Sonata, Tiburon, XG350 **Engines:** All **Transmissions:** All	**Camshaft Position Sensor Low Input** Engine started, engine speed over 600 rpm, and the PCM detected invalid or irregular CMP signals during the CCM test. **Possible Causes:** • CMP (Hall) sensor signal circuit is open or shorted (intermittent) • CMP (Hall) sensor is damaged or has failed • PCM has failed
DTC: P0343 **2T CCM, MIL: Yes** **1996-06** **Models:** Accent, Elantra, Santa Fe, Sonata, Tiburon, XG350 **Engines:** All **Transmissions:** All	**Camshaft Position Sensor High Input** Engine started, engine speed over 600 rpm, and the PCM did not detect any CMP signals within 200 revolutions during the CCM test. **Possible Causes:** • CMP (Hall) sensor signal circuit is open or shorted to ground • CMP (Hall) sensor ground circuit is open • CMP (Hall) sensor power circuit is open (check fuse in the I/P) • CMP (Hall) sensor is damaged or has failed • PCM has failed
DTC: P0346 **2T CCM, MIL: Yes** **2006** **Models:** Sonata **Engines:** 3.3L **Transmissions:** All	**Camshaft Position Sensor Positioning** Check if cam sensor is synchronized correctly. Engine running state. Cam tooth count #6. **Possible Causes:** • Poor connection • Open or short in harness • Electrical noise • target wheel • CMPS • Faulty PCM
DTC: P0350 **2T CCM, MIL: Yes** **1999-06** **Models:** Santa Fe, Sonata, Tiburon, Tucson, XG350 **Engines:** All **Transmissions:** All	**Ignition Coil Primary or Secondary Circuit Malfunction** Engine started, engine running, and the PCM detected an unexpected voltage condition on the Ignition Coil primary or secondary circuit during the CCM test. **Possible Causes:** • Ignition Coil 1, 2 or 3 primary/secondary circuit open or shorted • Ignition Coil power circuit is open (test power from the relay) • Ignition Coil 1, 2 or 3 is damaged or has failed • PCM has failed
DTC: P0351 **2T CCM, MIL: Yes** **1999-06** **Models:** Santa Fe, Sonata, Tiburon, Tucson **Engines:** All **Transmissions:** All	**Ignition Coil 'A' Circuit Malfunction** Engine started, engine running, and the PCM detected an unexpected voltage condition on the Ignition Coil 'A' primary circuit. **Possible Causes:** • Ignition Coil 'A' primary circuit is open or shorted to ground • Ignition Coil 'A' power circuit is open (test power from I/P fuse) • Ignition Coil 'A' is damaged or has failed • PCM has failed

DTC	Trouble Code Title, Conditions & Possible Causes
DTC: P0352 **2T CCM, MIL: Yes** **1999-06** **Models:** Santa Fe, Sonata, Tiburon, Tucson **Engines:** All **Transmissions:** All	**Ignition Coil 'B' Circuit Malfunction** Engine started, engine running, and the PCM detected an unexpected voltage condition on the Ignition Coil 'B' primary circuit. **Possible Causes:** • Ignition Coil 'B' primary circuit is open or shorted to ground • Ignition Coil 'B' power circuit is open (test power from I/P fuse) • Ignition Coil 'B' is damaged or has failed • PCM has failed
DTC: P0353 **2T CCM, MIL: Yes** **1999-06** **Models:** Santa Fe, Sonata, Tiburon, Tucson **Engines:** All **Transmissions:** All	**Ignition Coil 'C' Circuit Malfunction** Engine started, engine running, and the PCM detected an unexpected voltage condition on the Ignition Coil 'C' primary circuit. **Possible Causes:** • Ignition Coil 'C' primary circuit is open or shorted to ground • Ignition Coil 'C' power circuit is open (test power from I/P fuse) • Ignition Coil 'C' is damaged or has failed • PCM has failed
DTC: P0354 **2T CCM, MIL: Yes** **1999-06** **Models:** Santa Fe, Sonata, Tiburon, Tucson **Engines:** All **Transmissions:** All	**Ignition Coil 'D' Circuit Malfunction** Engine started, engine running, and the PCM detected an unexpected voltage condition on the Ignition Coil 'D' primary circuit. **Possible Causes:** • Ignition Coil 'D' primary circuit is open or shorted to ground • Ignition Coil 'D' power circuit is open (test power from I/P fuse) • Ignition Coil 'D' is damaged or has failed • PCM has failed
DTC: P0355 **2T CCM, MIL: Yes** **1999-06** **Models:** Santa Fe, Sonata, Tiburon, Tucson **Engines:** All **Transmissions:** All	**Ignition Coil 'E' Circuit Malfunction** Engine started, engine running, and the PCM detected an unexpected voltage condition on the Ignition Coil 'E' primary circuit. **Possible Causes:** • Ignition Coil 'E' primary circuit is open or shorted to ground • Ignition Coil 'E' power circuit is open (test power from I/P fuse) • Ignition Coil 'E' is damaged or has failed • PCM has failed
DTC: P0356 **2T CCM, MIL: Yes** **1999-06** **Models:** Santa Fe, Sonata, Tiburon, Tucson **Engines:** All **Transmissions:** All	**Ignition Coil 'F' Circuit Malfunction** Engine started, engine running, and the PCM detected an unexpected voltage condition on the Ignition Coil 'F' primary circuit. **Possible Causes:** • Ignition Coil 'F' primary circuit is open or shorted to ground • Ignition Coil 'F' power circuit is open (test power from I/P fuse) • Ignition Coil 'F' is damaged or has failed • PCM has failed
DTC: P0400 **2T EGR, MIL: Yes** **1996-99** **Models:** Sonata **Engines:** All **Transmissions:** All	**EGR System Flow Detected** Engine started, ECT sensor more than 180°F, engine speed from 900-6000 rpm, idle position switch closed, engine load under 22%, and the PCM detected too small a change in manifold air pressure (< 1.02" Hg) after the EGR valve was opened during the EGR test. **Possible Causes:** • EGR valve source vacuum supply line is open or restricted • EGR exhaust tube is clogged or restricted • EGR valve assembly or solenoid is damaged or has failed • PCM has failed
DTC: P0401 **2T EGR, MIL: Yes** **1999-06** **Models:** Santa Fe, Sonata, XG300, XG350 **Engines:** All **Transmissions:** All	**EGR System Malfunction** Engine running in closed loop with a steady throttle, VSS input from 45-60 mph for 3-5 minutes, and the PCM detected the intake manifold pressure (MAP sensor input) did not change correctly after the EGR valve was opened and closed during the EGR Monitor test. **Possible Causes:** • EGR vacuum hose to source vacuum is loose or disconnected • EGR exhaust tube is clogged or restricted • EGR valve assembly or solenoid is damaged or has failed • PCM has failed
DTC: P0403 **2T CCM, MIL: Yes** **1999-06** **Models:** Santa Fe, Sonata, XG300, XG350 **Engines:** All **Transmissions:** All	**EGR Solenoid Circuit Malfunction** Engine started, vehicle driven at a speed of from 36-55 mph in closed loop, and the PCM detected an unexpected voltage condition on the EGR solenoid circuit during the CCM test. **Possible Causes:** • EGR solenoid control circuit is open or shorted to ground • EGR solenoid control circuit is shorted to system power (B+) • EGR solenoid power circuit is open (check power to I/P fuse) • EGR solenoid is damaged or has failed

DTC	Trouble Code Title, Conditions & Possible Causes
DTC: P0420 **2T CCM, MIL: Yes** **1999-06** **Models:** Accent, Elantra, Santa Fe, Sonata, Tiburon, Tucson, XG300, XG350 **Engines:** All **Transmissions:** All	**Catalyst Efficiency Below Normal (Bank 1)** Engine started, vehicle driven at a speed of 45-60 mph for 8-10 minutes in closed loop, and the PCM detected the rear HO2S-12 switch rate was similar to the front HO2S-11 switch rate for over 3 seconds under these conditions during the Catalyst Monitor test. **Possible Causes:** • Air leaks at the exhaust manifold or in the exhaust pipes • Catalytic converter is contaminated, damaged or has failed • Front HO2S is older (aged) than the rear HO2S (HO2S is lazy) • Front HO2S or rear HO2S is contaminated with fuel or moisture
DTC: P0421 **2T CAT, MIL: Yes** **1996-06** **Models:** Accent, Santa Fe, Sonata, XG300, XG350 **Engines:** All **Transmissions:** All	**Warm up Catalyst Efficiency Below Normal (Bank 1)** Engine started, vehicle driven at a speed of 45-60 mph for 8-10 minutes in closed loop, and the PCM detected the rear HO2S-12 switch rate was similar to the front HO2S-11 switch rate for over 3 seconds under these conditions during the Catalyst Monitor test. **Possible Causes:** • Air leaks at the exhaust manifold or in the exhaust pipes • Catalytic converter is contaminated, damaged or has failed • Front HO2S is older (aged) than the rear HO2S (HO2S is lazy) • Front HO2S or rear HO2S is contaminated with fuel or moisture
DTC: P0422 **2T CAT, MIL: Yes** **1998-02** **Models:** Accent, Elantra, Tiburon **Engines:** All **Transmissions:** All	**Catalyst Efficiency Below Normal (Bank 1)** Engine started, vehicle driven at a speed of 45-60 mph for 8 minutes in closed loop, and the PCM detected the rear HO2S-12 switch rate was similar to the front HO2S-11 switch rate for over 3 seconds. **Possible Causes:** • Air leaks at the exhaust manifold or in the exhaust pipes • Catalytic converter is contaminated, damaged or has failed • Front HO2S is older (aged) than the rear HO2S (HO2S is lazy) • Front HO2S or rear HO2S is contaminated with fuel or moisture
DTC: P0430 **2T CAT, MIL: Yes** **1999-06** **Models:** Santa Fe, Sonata, Tiburon, Tucson, XG350 **Engines:** All **Transmissions:** All	**Catalyst Efficiency Below Normal (Bank 2)** Engine started, vehicle driven at a speed of 45-60 mph for 8 minutes in closed loop, and the PCM detected the rear HO2S-22 switch rate was similar to the front HO2S-21 switch rate for over 3 seconds. **Possible Causes:** • Air leaks at the exhaust manifold or in the exhaust pipes • Catalytic converter is contaminated, damaged or has failed • Front HO2S is older (aged) than the rear HO2S (HO2S is lazy) • Front HO2S or rear HO2S is contaminated with fuel or moisture
DTC: P0431 **2T CAT, MIL: Yes** **1996-06** **Models:** Santa Fe, Sonata, XG300, XG350 **Engines:** All **Transmissions:** All	**Warm up Catalyst Efficiency Below Normal (Bank 2)** Engine started, vehicle driven at a speed of 45-60 mph for 8 minutes in closed loop, and the PCM detected the rear HO2S-22 switch rate was similar to the front HO2S-21 switch rate for over 3 seconds. **Possible Causes:** • Air leaks at the exhaust manifold or in the exhaust pipes • Catalytic converter is contaminated, damaged or has failed • Front HO2S is older (aged) than the rear HO2S (HO2S is lazy) • Front HO2S or rear HO2S is contaminated with fuel or moisture
DTC: P0440 **2T Evap, MIL: Yes** **1996-02** **Models:** Accent, Elantra, Santa Fe, Sonata, Tiburon, XG300, XG350 **Engines:** All **Transmissions:** All	**EVAP System Malfunction** Cold startup requirement met (ECT sensor from 14-90ºF and IAT sensor more than 14ºF), engine started, power steering switch "off", engine running under light load conditions, ECT sensor more than 180ºF, and the PCM detected less than a 3% change in the A/F ratio after the EVAP solenoid was opened during the EVAP Monitor test. **Possible Causes:** • Charcoal canister is loaded with fuel or moisture • ECT, IAT, MAP or TP sensor signal is out-of-calibration • Fuel filler cap loose, cross-threaded, incorrect part or damaged • Fuel tank pressure sensor is damaged or has failed • Fuel tank vapor line(s) blocked, damaged or disconnected • Purge or Vent solenoid control circuit open or shorted to ground • PCM has failed

DTC	Trouble Code Title, Conditions & Possible Causes
DTC: P0441 **2T Evap, MIL: Yes** **1996-06** **Models:** Accent, Elantra, Santa Fe, Sonata, Tiburon, Tucson, XG300, XG350 **Engines:** All **Transmissions:** All	**EVAP System Incorrect Purge Flow (Stuck Open)** Engine started, vehicle driven at a speed of 35-40 for 5-10 minutes under light engine load conditions, then with the Purge solenoid commanded "on" and then "off", the PCM detected the Purge solenoid valve remained "open" during the EVAP Monitor flow test. Note: This is a "functionality" test of the EVAP system (flow test). **Possible Causes:** • Charcoal canister is damaged, clogged or restricted • Purge solenoid control circuit open, shorted to ground or power • Purge solenoid power circuit is open (check the relay or fuse) • Purge valve vacuum line is clogged or contains water or debris • Fuel filler cap loose, cross-threaded, incorrect part or damaged • Fuel tank or fuel tank sender assembly 'O' ring is leaking • Fuel tank vapor line(s) blocked, damaged or disconnected • PCM has failed
DTC: P0442 **2T Evap, MIL: Yes** **1996-06** **Models:** Accent, Elantra, Santa Fe, Sonata, Tiburon, Tucson, XG300, XG350 **Engines:** All **Transmissions:** All	**EVAP System Large Leak (0.040") Detected** Cold startup requirement met (ECT sensor from 14-90ºF and the IAT sensor more than 14ºF), engine started, vehicle driven at a speed of over 20 mph, and the PCM detected a leak (from 0.040" to 0.080") somewhere in the EVAP system during the EVAP Monitor test. **Possible Causes:** • Canister Purge valve is damaged, leaking or it has failed • Charcoal canister is loaded with fuel or moisture • Fuel filler cap loose, cross-threaded, incorrect part or damaged • Fuel tank is cracked (leaking), or a leak exists in the 'O' ring • Fuel tank pressure sensor is damaged or has failed • Fuel vapor line(s), fuel pipes or hoses damaged or leaking • Purge solenoid or vent solenoid damaged, leaking or sticking • PCM has failed
DTC: P0443 **2T CCM, MIL: Yes** **1997-06** **Models:** Accent, Elantra, Santa Fe, Sonata, Tiburon, Tucson, XG300, XG350 **Engines:** All **Transmissions:** All	**EVAP Purge Solenoid Circuit Malfunction** Engine started, engine running for 2-3 minutes, and the PCM detected an unexpected voltage condition on the Purge solenoid circuit during the CCM test. **Possible Causes:** • Purge solenoid control circuit is open or shorted to ground • Purge solenoid power circuit is open (check the power source) • Purge control solenoid is damaged or has failed • PCM has failed
DTC: P0444 **2T CCM, MIL: Yes** **1996-06** **Models:** Accent, Elantra, Santa Fe, Sonata, Tiburon, Tucson, XG300, XG350 **Engines:** All **Transmissions:** All	**EVAP Purge Solenoid Circuit Malfunction (Open)** Engine started, engine running at idle speed, and the PCM detected an unexpected "low" voltage condition on the EVAP Purge solenoid circuit as the solenoid was commanded "on" and "off" in the test. **Possible Causes:** • Purge solenoid control circuit open between solenoid and PCM • Purge solenoid power circuit is open (check the power source) • Purge control solenoid is damaged or has failed • PCM has failed
DTC: P0445 **2T CCM, MIL: Yes** **1996-06** **Models:** Accent, Elantra, Santa Fe, Sonata, Tiburon, Tucson, XG300, XG350 **Engines:** All **Transmissions:** All	**EVAP Purge Solenoid Circuit Malfunction (Shorted)** Engine started, engine running at idle speed, and the PCM detected an unexpected "high" voltage condition on the EVAP Purge solenoid circuit as the solenoid was commanded "on" and "off" in the test. **Possible Causes:** • Purge solenoid control circuit is shorted to system power • Purge control solenoid is damaged or has failed (short circuit) • PCM has failed
DTC: P0446 **2T CCM, MIL: Yes** **1998-06** **Models:** Accent, Elantra, Santa Fe, Sonata, Tiburon, Tucson, XG300, XG350 **Engines:** All **Transmissions:** All	**EVAP Vent Control Solenoid Circuit Performance (Closed)** Engine started, engine running at cruise speed for 2-3 minutes, and the PCM detected the EVAP Vent Control solenoid was in a "closed" position continuously during the CCM test. **Possible Causes:** • Vent solenoid control circuit is shorted to ground • Vent solenoid power circuit is open (check the power source) • Vent control solenoid is damaged or has failed • PCM has failed

DTC	Trouble Code Title, Conditions & Possible Causes
DTC: P0447 **2T CCM, MIL: Yes** **1996-06** **Models:** Accent, Elantra, Santa Fe, Sonata, Tiburon, Tucson, XG300, XG350 **Engines:** All **Transmissions:** All	**EVAP Vent Solenoid Circuit Short To Ground Malfunction** Engine started, engine running at cruise speed for 2-3 minutes, and the PCM detected the EVAP Vent Control solenoid control circuit was in a continuous "low" state during the CCM test. **Possible Causes:** • Purge solenoid control circuit is shorted to ground • Purge solenoid power circuit is open (check the power source) • Purge solenoid is damaged or has failed • PCM has failed
DTC: P0448 **2T CCM, MIL: Yes** **1998-06** **Models:** Accent, Elantra, Santa Fe, Sonata, Tiburon, Tucson, XG300, XG350 **Engines:** All **Transmissions:** All	**EVAP Vent Solenoid Circuit Shorted To Power Malfunction** Engine started, engine running at cruise speed for 2-3 minutes, and the PCM detected the EVAP Vent Control solenoid control circuit was in a continuous "high" state during the CCM test. **Possible Causes:** • Purge solenoid control circuit is shorted to system power • Purge solenoid is damaged or has failed • PCM has failed
DTC: P0449 **2T CCM, MIL: Yes** **2003-06** **Models:** Accent, Elantra, Santa Fe, Sonata, Tiburon, Tucson **Engines:** 1.5L, 2.0L, 2.4L, 2.7L, 3.3L **Transmissions:** All	**EVAP Emission System- Vent Valve/Solenoid Circuit** The PCM measures pressure in the fuel tank, by means of a sensor during all engine operating conditions, except start and stop. If pressure is lower than threshold (less than 1.6L volts), a DTC is set. Monitoring CCV stuck closed. **Possible Causes:** • Canister air filter contamination • Faulty CCV
DTC: P0450 **2T CCM, MIL: Yes** **1996-06** **Models:** Accent, Sonata **Engines:** All (Sonata), 1.5L (Accent) **Transmissions:** All	**EVAP Pressure Sensor Circuit Malfunction** Engine started, engine running, vehicle not moving, VSV for the EVAP Vapor sensor commanded "on", and the PCM detected an unexpected voltage condition on the EVAP Pressure sensor circuit. **Possible Causes:** • Pressure sensor signal circuit is open or shorted to ground • Pressure sensor signal circuit is shorted to VREF or power • Pressure sensor power (VREF) circuit is open • Pressure sensor is damaged or has failed • PCM has failed
DTC: P0451 **2T CCM, MIL: Yes** **1998-06** **Models:** Accent, Elantra, Santa Fe, Sonata, Tiburon, Tucson, XG300, XG350 **Engines:** All **Transmissions:** All	**EVAP Pressure Sensor Performance** Engine started, engine running with the vehicle not moving, VSV for the EVAP Vapor sensor commanded "on", and the PCM detected the EVAP pressure sensor signal was not plausible during the test. Note: This condition (code) can be due to a fuel sloshing condition. **Possible Causes:** • Pressure sensor vacuum hoses loose or damaged • Pressure sensor is damaged or out-of-calibration • VSV for the EVAP pressure sensor is damaged or has failed • PCM has failed
DTC: P0452 **2T CCM, MIL: Yes** **1998-06** **Models:** Accent, Elantra, Santa Fe, Sonata, Tiburon, Tucson, XG300, XG350 **Engines:** All **Transmissions:** All	**EVAP Pressure Sensor Circuit Low Input** Engine started, engine runtime over 5 seconds, and the PCM detected an unexpected "low" voltage condition on the EVAP Pressure sensor circuit during the CCM test. **Possible Causes:** • Pressure sensor signal circuit is shorted to ground • Pressure sensor power (VREF) circuit is open • Pressure sensor is damaged or has failed • PCM has failed
DTC: P0453 **2T CCM, MIL: Yes** **1998-06** **Models:** Accent, Elantra, Santa Fe, Sonata, Tiburon, Tucson, XG300, XG350 **Engines:** All **Transmissions:** All	**EVAP Pressure Sensor Circuit High Input** Engine started, engine runtime over 5 seconds, and the PCM detected an unexpected "low" voltage condition on the EVAP Pressure sensor circuit during the CCM test. **Possible Causes:** • Pressure sensor signal circuit is shorted to power • Pressure sensor ground circuit open between sensor and PCM • Pressure sensor is damaged or has failed • PCM has failed

DTC	Trouble Code Title, Conditions & Possible Causes
DTC: P0454 **2T CCM, MIL: Yes** **2003-06** **Models:** Elantra, Santa Fe, Sonata, Tiburon, Tucson, XG350 **Engines:** 2.0L, 2.4L, 2.7L, 3.3L, 3.5L **Transmissions:** All	**EVAP Emission System- Pressure Sensor Intermittent** The PCM measures pressure stability in the fuel tank, by means of a sensor for a predetermined duration. If fluctuation is larger than predetermined threshold a DTC is set. Sensor signal noise check. **Possible Causes:** • Contact resistance in connectors • Faulty FTPS
DTC: P0455 **2T Evap, MIL: Yes** **1998-06** **Models:** Accent, Elantra, Santa Fe, Sonata, Tiburon, Tucson, XG300, XG350 **Engines:** All **Transmissions:** All	**EVAP System Large Leak (0.080") Detected** DTC P0443 not set, engine started, ECT sensor more than 185ºF, IAT sensor from 14-122ºF, fuel level from 25-75%, engine running at cruise speed, and the PCM detected a large change in the fuel tank pressure (due to a large leak) during the EVAP Monitor leak test. **Possible Causes:** • Canister vent (CV) solenoid may be stuck in open position • EVAP canister tube, EVAP canister purge outlet tube or EVAP return tube disconnected or cracked, or canister is damaged • EVAP canister purge valve stuck closed, or canister damaged • Fuel filler cap missing, loose (not tightened) or the wrong part • Fuel vapor hoses/tubes blocked or restricted, or fuel vapor control valve tube or fuel vapor vent valve assembly blocked • Fuel tank pressure (FTP) sensor has failed (mechanical fault) • Fuel tank control valve is contaminated, damaged or has failed
DTC: P0456 **2T CCM, MIL: Yes** **2003-06** **Models:** Accent, Elantra, Santa Fe, Sonata, Tiburon, Tucson, XG350 **Engines:** 1.5L, 1.6L, 2.0L, 2.4L, 2.7L, 3.3L, 3.5L **Transmissions:** All	**EVAP Emission System- Leak Detected (very small)** Monitoring the tank pressure sensor (DTP) signal with under pressure. **Possible Causes:** • Leakage in EVAP system line • Faulty CCV, PCSV or FTPS
DTC: P0457 **2T Evap, MIL: Yes** **2001-06** **Models:** Accent, Elantra, Santa Fe, Sonata, Tiburon, XG300, XG350 **Engines:** All **Transmissions:** All	**EVAP System Leak (Fuel Tank Cap) Detected** DTC P0443 not set, engine started, ECT sensor more than 185ºF, IAT sensor from 14-122ºF, fuel level from 25-75%, and the PCM detected a large change in fuel tank pressure during the EVAP System Monitor leak test. **Possible Causes:** • Canister vent (CV) solenoid may be stuck in open position • EVAP canister tube, EVAP canister purge outlet tube or EVAP return tube disconnected or cracked, or canister is damaged • Fuel filler cap missing, loose (not tightened) or the wrong part • Fuel tank pressure (FTP) sensor has failed (mechanical fault) • Fuel tank control valve is contaminated, damaged or has failed
DTC: P0458 **2T CCM, MIL: Yes** **2006** **Models:** Accent **Engines:** 1.5L **Transmissions:** All	**Evaporative Emission System Purge Control valve Circuit Low** Circuit continuity check, low. **Possible Causes:** • Poor connection • Short to ground in control circuit • PCSV • Faulty ECM/PCM
DTC: P0459 **2T CCM, MIL: Yes** **2006** **Models:** Accent **Engines:** 1.5L **Transmissions:** All	**Evaporative Emission System Purge Control valve Circuit High** Circuit continuity check, high. **Possible Causes:** • Poor connection • Short to power in control circuit • PCSV • Faulty ECM/PCM
DTC: P0460 **2T CCM, MIL: Yes** **2003-06** **Models:** Santa Fe, Sonata, XG350 **Engines:** 2.4L, 3.5L **Transmissions:** All	**Fuel Level Sensor Circuit** Out voltage and change fuel level voltage is monitored. The PCM detects that the sum of injected fuel is more than 5 liters, but fuel level change between Max and Min is less than 3 liters. **Possible Causes:** • Faulty fuel level sensor • Faulty PCM

DTC	Trouble Code Title, Conditions & Possible Causes
DTC: P0461 **2T CCM, MIL: Yes** **2003-06** **Models:** Accent, Santa Fe, Sonata, Tiburon, Tucson, XG350 **Engines:** 1.5L, 1.6L, 2.4L, 2.7L, 3.3L, 3.5L **Transmissions:** All	**Fuel Level Sensor "A" Circuit Range/Performance** Filtered and unfiltered signal of fuel sensor are monitored. **Possible Causes:** • Poor connection • Faulty fuel level sensor • Faulty PCM
DTC: P0462 **2T CCM, MIL: No** **2003-06** **Models:** Accent, Santa Fe, Sonata, Tiburon, Tucson, XG350 **Engines:** 1.5L, 1.6L, 2.4L, 2.7L, 3.3L, 3.5L **Transmissions:** All	**Fuel Level Sensor "A" Circuit Low Input** If the sensor output voltage is higher than 4.96 volts or output voltage is less than 0.039 volt for 300 seconds, when the engine speed is not 0 miles. Output voltage (VFLS) is monitored. **Possible Causes:** • Poor connection • Faulty fuel level sensor • Short to battery in fuel level (FLS) circuit • Faulty fuel level circuit (FLS) • Faulty PCM
DTC: P0463 **2T CCM, MIL: No** **2003-06** **Models:** Accent, Santa Fe, Sonata, Tiburon, Tucson, XG350 **Engines:** 1.5L, 1.6L, 2.4L, 2.7L, 3.3L, 3.5L **Transmissions:** All	**Fuel Level Sensor "A" Circuit high Input** If the sensor output voltage is less than 0.02 volt, or output voltage is less than 0.039 volt for 300 seconds, when the engine speed is not 0 miles. Output voltage (VFLS) is monitored. **Possible Causes:** • Short to battery in fuel level (FLS) circuit • Faulty fuel level sensor • Faulty PCM
DTC: P0464 **2T CCM, MIL: No** **2003-06** **Models:** Accent, Santa Fe, Sonata, Tiburon, Tucson **Engines:** 1.5L, 1.6L, 2.4L, 2.7L, 3.3L **Transmissions:** All	**Fuel Level Sensor "A" Circuit Intermittent** Check signal for fluctuation. The ECM sets the DTC if the fuel level signal is higher than the threshold value (signal fluctuation greater than 50 percent). **Possible Causes:** • Contact resistance in connectors • Short to battery in fuel level (FLS) circuit
DTC: P0480 **2T CCM, MIL: Yes** **2006** **Models:** Sonata **Engines:** 3.3L **Transmissions:** All	**Fan 1 Control Circuit Malfunction** This will detect a short to ground, to battery or open circuit of fan relay output. Fault information provided by an output driver chip. No disabling faults present. Engine running. Enable time delay equal or greater than 0.5 seconds. **Possible Causes:** • Poor connection • Open in power circuit to cooling fan • Open or short in control circuit to PCM • Faulty fan relay • Faulty cooling fan module • Faulty PCM
DTC: P0489 **2T CCM, MIL: Yes** **2005-06** **Models:** Santa Fe, Sonata, XG350 **Engines:** 2.4L, 3.5L **Transmissions:** All	**Exhaust Gas recirculation Control Circuit Low Voltage** Surge voltage (Vps) and output level voltage is monitored. **Possible Causes:** • Poor connection • Open or short to ground in EGR solenoid valve circuit • Faulty EGR solenoid valve • Faulty PCM
DTC: P0490 **2T CCM, MIL: Yes** **2005-06** **Models:** Santa Fe, Sonata, XG350 **Engines:** 2.4L, 3.5L **Transmissions:** All	**Exhaust Gas recirculation Control Circuit High Voltage** Surge voltage (Vps) and output level voltage is monitored. Engine speed is 500 rpm. **Possible Causes:** • Poor connection • Short to battery in EGR solenoid valve circuit • Faulty EGR solenoid valve • Faulty PCM

DTC	Trouble Code Title, Conditions & Possible Causes
DTC: P0496 **2T CCM, MIL: Yes** **2006** **Models:** Accent **Engines:** 1.5L **Transmissions:** All	**Evaporative Emission System High Purge Flow** Fuel tank pressure behavior (canister purge valve stuck). Time after engine start 600 seconds. Idle speed controller activated. Mixture adaptation activated. Coolant temperature at start 11.88 degrees F. Tank ventilation must be active for 10 seconds. **Possible Causes:** • Leakage at the fuel evaporative system • PCSV • Faulty ECM/PCM
DTC: P0497 **2T CCM, MIL: Yes** **2006** **Models:** Accent **Engines:** 1.5L **Transmissions:** All	**Evaporative Emission System Low Purge Flow** Fuel tank pressure behavior (canister purge valve stuck). Time after engine start 600 seconds. Idle speed controller activated. Mixture adaptation activated. Coolant temperature at start 11.88 degrees F. Tank ventilation must be active for 10 seconds. **Possible Causes:** • Clog in the fuel evaporative system • PCSV • Faulty ECM/PCM
DTC: P0498 **2T CCM, MIL: Yes** **2006** **Models:** Accent **Engines:** 1.5L **Transmissions:** All	**Evaporative Emission System Vent Valve Control Circuit Low** Circuit continuity check, low. **Possible Causes:** • Poor connection • Short to ground in control circuit • CCV • Faulty ECM/PCM
DTC: P0499 **2T CCM, MIL: Yes** **2006** **Models:** Accent **Engines:** 1.5L **Transmissions:** All	**Evaporative Emission System Vent Valve Control Circuit High** Circuit continuity check, high. **Possible Causes:** • Poor connection • Short to power in control circuit • CCV • Faulty ECM/PCM
DTC: P0500 **2T CCM, MIL: Yes** **1999-06** **Models:** Accent, Santa Fe, Sonata, XG300, XG350 **Engines:** All **Transmissions:** All	**Vehicle Speed Sensor Performance** Engine started, vehicle driven with the engine speed over 3000 rpm at an engine load over 70%, Closed Throttle switch indicating "off", and the PCM did not detect any VSS signals for 4 seconds. **Possible Causes:** • VSS signal circuit is open or shorted to ground • VSS power or ground circuit is open • VSS is damaged or has failed • PCM has failed
DTC: P0500 **2T CCM, MIL: Yes** **1999-00** **Models:** Accent, Elantra, Sonata, Tiburon **Engines:** All **Transmissions:** All	**VSS Signal (ABS/TCS Rough Road Signal) Fault** Engine started, vehicle driven at a speed of over 1000 rpm in Drive at light to medium engine load for over 10 seconds, and the PCM did not detect any VSS signals on the "common" ABS/TCS circuit bus. **Possible Causes:** • VSS signal circuit is open, shorted to ground or system power • VSS (Hall) power circuit is open (check Fuse 5 in fuse panel) • VSS (Hall) ground circuit is open • VSS (Hall) is damaged or has failed • PCM has failed
DTC: P0501 **2T CCM, MIL: Yes** **1998-06** **Models:** Accent, Elantra, Santa Fe, Sonata, Tiburon, Tucson, XG300, XG350 **Engines:** All **Transmissions:** All	**Vehicle Speed Sensor Circuit Performance** Vehicle driven with the engine speed over 2000 rpm in Drive, engine load from 2-3 ms, and the PCM did not detect any VSS signals. **Possible Causes:** • VSS signal circuit from the sensor to the I/P Cluster to the PCM is open, shorted to ground or to system power • VSS (Magnetic) signal (+) or (-) circuit is open or shorted • VSS (Magnetic) is damaged or has failed • PCM has failed

DTC	Trouble Code Title, Conditions & Possible Causes
DTC: P0502 **2T CCM, MIL: Yes** **2005** **Models:** Accent **Engines:** 1.5L, 1.6L **Transmissions:** All	**Vehicle Speed Sensor "A" Circuit Low Input** If wheel speed sensor (WSS) signal is less than threshold value the DTC is set. Signal check. Threshold value (constant vehicle speed for more than 20 seconds). **Possible Causes:** • Open in signal circuit • Open in battery and ground circuit • Short to ground in signal circuit • Faulty VSS • Faulty PCM
DTC: P0503 **2T CCM, MIL: Yes** **2005** **Models:** Accent **Engines:** 1.5L, 1.6L **Transmissions:** All	**Vehicle Speed Sensor High Input** If wheel speed sensor (WSS) signal is less than threshold value the DTC is set. Signal check **Possible Causes:** • Open in signal circuit • Open in battery and ground circuit • Short to ground in signal circuit • Short to battery in signal circuit • Faulty VSS • Faulty PCM
DTC: P0504 **2T CCM, MIL: Yes** **2006** **Models:** Sonata **Engines:** 2.4L **Transmissions:** All	**Brake Switch "A"/"B" Correlation (1)** Comparing two brake signals during driving. Case 1: Engine works. Vehicle speed sensor is normal. Case 2: Engine works. Vehicle speed sensor is normal. Vehicle speed is over 20 kph, for at least 1 second. **Possible Causes:** • Poor connection • Open or short • Faulty PCM
DTC: P0504 **2T CCM, MIL: Yes** **2006** **Models:** Sonata **Engines:** 2.4L **Transmissions:** All	**Brake Switch "A"/"B" Correlation (2)** Plausibility check between brake light switch and brake test switch. Engine running. Time between brake light switch and brake test switch do not correlate longer than 10 seconds. **Possible Causes:** • Open or short circuit in harness • Poor connection or damaged harness • Faulty brake warning lamp or brake test switch
DTC: P0505 **2T CCM, MIL: Yes** **1996-06** **Models:** Accent Sonata **Engines:** All (Sonata), 1.5L (Accent) **Transmissions:** All	**Idle Speed Control System (Mechanical) Fault** Engine started, engine running at hot idle speed in closed loop for 30 seconds, and the PCM detected the difference between the Actual and the Target idle speed was more than 300 higher or lower than a calibrated amount in memory during the CCM Rationality test. **Possible Causes:** • Stepper motor Coil A1 or A2 circuit is open or shorted to ground • Stepper motor Coil B1 or B2 circuit is open or shorted to ground • Stepper motor coil circuit(s) shorted to system power (B+) • Stepper motor power circuit is open (check power at MFI relay) • Stepper motor is damaged or has failed • PCM has failed
DTC: P0506 **2T CCM, MIL: Yes** **1998-06** **Models:** Accent, Elantra, Santa Fe, Sonata, Tiburon, Tucson, XG300, XG350 **Engines:** All **Transmissions:** All	**Idle Speed Lower Than Expected** Engine started, engine running at idle speed under these conditions: IAT sensor less than 114°F during the last ignition cycle, Long Term fuel trim from -8% to +8%, ECT sensor more than 176°F, IAT sensor more than 14°F, system voltage over 10.0 volts, and the PCM detected the Actual idle speed was over 200 rpm lower than the Target idle speed for 10 seconds during the CCM Rationality test. Power steering pressure switch signal indicating "off", engine load less than 40%, IAT sensor more than 14°F, and the PCM detected the Actual idle speed was over 120 rpm lower than the Target idle speed for 10 seconds during the CCM Rationality test. **Possible Causes:** • ISC motor "open" circuit is open or shorted to ground • ISC motor "close" circuit is open or shorted to ground • ISC motor is damaged or has failed • PCM is damaged

DTC	Trouble Code Title, Conditions & Possible Causes
DTC: P0507 **2T CCM, MIL: Yes** **1998-06** **Models:** Accent, Elantra, Santa Fe, Sonata, Tiburon, Tucson, XG300, XG350 **Engines:** All **Transmissions:** All	**Idle Speed Higher Than Expected** Engine at idle speed in closed loop, and under these conditions: Condition 1 - IAT sensor input less than 114ºF during last drive cycle, Long Term fuel trim from -8% to +8%, ECT sensor input more than 176ºF, IAT sensor input more than 14ºF, system voltage over 10.0 volts, and the PCM detected the Actual idle speed was more than 200 rpm higher than the Target idle speed for 10 seconds. Condition 2 - Power steering pressure switch signal indicating "off", engine load less than 40%, IAT sensor input more than 14ºF, and the PCM detected the Actual idle speed was more than 120 rpm higher than the Target idle speed for 10 seconds. **Possible Causes:** • ISC motor control circuit(s) open or shorted to ground • ISC motor is damaged or has failed • PCM is damaged
DTC: P0510 **2T CCM, MIL: Yes** **1996-06** **Models:** Santa Fe, Sonata, XG350 **Engines:** All (Sonata and XG350) 3.5L (Santa Fe) **Transmissions:** All	**Closed Throttle Idle Switch Circuit Malfunction** Engine started, vehicle driven at over 30 mph and then back to a stop at least 15 times, TP sensor signal over 2.0 volts at least once, and the PCM detected the CTP switch remained "off" for over 2 seconds. **Possible Causes:** • Closed throttle position switch signal circuit is open or grounded • Closed throttle position switch signal circuit is shorted to power • Closed throttle position switch or TP sensor damaged or failed • PCM has failed
DTC: P0532 **2T CCM, MIL: No** **2006** **Models:** Accent, Sonata **Engines:** 1.5L, 2.4L, 3.3L **Transmissions:** All	**A/C Refrigerant Pressure Sensor "A" Circuit Low Input** Detects sensor signal short to low voltage. Engine works. Sensor output 0.05 volt. **Possible Causes:** • Poor connection • Open in power circuit • Open or short to ground in signal circuit • Faulty A/C pressure sensor • Faulty PCM
DTC: P0533 **2T CCM, MIL: No** **2006** **Models:** Accent, Sonata **Engines:** 1.5L, 2.4L, 3.3L **Transmissions:** All	**A/C Refrigerant Pressure Sensor "A" Circuit High Input** Detects sensor signal short to high voltage. Engine works. Sensor output 4.65 volts. **Possible Causes:** • Poor connection • Open in signal circuit open • Open in ground circuit • Faulty A/C pressure sensor • Faulty PCM
DTC: P0551 **2T CCM, MIL: Yes** **2003-06** **Models:** Santa Fe, Sonata, Tiburon, Tucson, XG350 **Engines:** 2.4L, 2.7L, 3.3L, 3.5L **Transmissions:** All	**Power Steering Pressure Sensor/Switch Circuit Range/Performance** If a power steering switch signal is ON when the engine speed is more than 2500 rpm, load value is grater than 55 percent and engine coolant temperature is above 50 degrees F, the DTC will set. Signal of power steering pressure switch is monitored. **Possible Causes:** • Poor connection • Faulty power steering switch • Open or short in power steering switch • Faulty PCM
DTC: P0552 **2T CCM, MIL: Yes** **2006** **Models:** Sonata **Engines:** 2.4L, 3.3L **Transmissions:** All	**Power Steering Pressure Sensor/Switch Circuit Low Input** Detects sensor signal short to low voltage. Engine works. Sensor output 0.25 volt. **Possible Causes:** • Poor connection • Open in power circuit • Open or short to ground in signal circuit • Faulty P/S pressure sensor • Faulty PCM
DTC: P0553 **2T CCM, MIL: Yes** **2006** **Models:** Sonata **Engines:** 2.4L, 3.3L **Transmissions:** All	**Power Steering Pressure Sensor/Switch Circuit High Input** Detects sensor signal short to low voltage. Engine works. Sensor output 4.65 volts. **Possible Causes:** • Poor connection • Short in signal circuit • Open in ground circuit • Faulty P/S pressure sensor • Faulty PCM

DTC	Trouble Code Title, Conditions & Possible Causes
DTC: P0560 **2T CCM, MIL: Yes** **1998-2000** **Models:** Elantra, Tiburon **Engines:** All **Transmissions:** All	**Battery Backup Circuit Malfunction** Key on or engine running, and the PCM detected the voltage on the Battery or Memory Backup circuit was less than 2.0 volts for 10 seconds. **Possible Causes:** • Battery backup circuit is open (check the fuse to the battery) • Battery terminals are corroded, dirty or loose • Battery has been removed or the cables were disconnected • PCM has failed • TSB 00-36-013 (12/00) contains information about this code
DTC: P0561 **2T CCM, MIL: Yes** **2006** **Models:** Accent **Engines:** 1.5L **Transmissions:** All	**System Voltage Unstable** Rationality check. **Possible Causes:** • Poor connection • Open or short to ground in control circuit • Charging system • Main relay • Faulty ECM/PCM
DTC: P0562 **2T CCM, MIL: Yes** **1998-06** **Models:** Accent, Elantra, Santa Fe, Sonata, Tiburon, Tucson **Engines:** All **Transmissions:** All	**System Voltage Low Input** Engine started engine running at idle or cruise speed, and the PCM detected an unexpected low voltage condition on the ignition circuit. **Possible Causes:** • Ignition system voltage circuit is open • Generator is damaged or has failed (generator output too low) • PCM has failed
DTC: P0563 **2T CCM, MIL: Yes** **1998-06** **Models:** Accent, Elantra, Santa Fe, Sonata, Tiburon, Tucson **Engines:** All **Transmissions:** All	**System Voltage High Input** Engine started engine running at idle or cruise speed, and the PCM detected an unexpected high voltage condition on the ignition circuit. **Possible Causes:** • Generator is damaged or has failed (generator output to high) • PCM has failed
DTC: P0564 **2T CCM, MIL: Yes** **2006** **Models:** Sonata **Engines:** 2.4L **Transmissions:** All	**Cruise Control Multifunction Input "A" Circuit** Invalid voltage range check. A DTC code is set for the following conditions. Check SET/COAST switch stuck. Check RES/ACC switch stuck. **Possible Causes:** • Open or short in harness • Poor connection or damaged harness • Faulty cruise control remote control switch
DTC: P0571 **2T CCM, MIL: Yes** **2006** **Models:** Sonata **Engines:** 3.3L **Transmissions:** All	**Brake Switch "A" Circuit** PCM detects brake light input signal when the vehicle stops. VSS is normal. Vehicle speed 0 mph, during one second or more. **Possible Causes:** • Poor connection • Open or short to ground in signal circuit • Faulty PCM
DTC: P0600 **2T CCM, MIL: Yes** **2003-06** **Models:** Elantra, Santa Fe, Sonata, Tiburon, Tucson **Engines:** 2.0L, 2.4L, 2.7L, 3.3L **Transmissions:** All	**CAN Communication Bus** **CAN message transfer incorrect?** **Possible Causes:** • Open or short in CAN line • Contact resistance in connectors • Faulty PCM
DTC: P0601 **1T PCM, MIL: Yes** **1999-06** **Models:** Accent, Elantra, Santa Fe, Sonata, Tiburon, XG300, XG350 **Engines:** All **Transmissions:** All	**PCM (Internal Controller) Checksum Error** Key on or engine running for 1 second, and the PCM detected an internal checksum data error during the initial Self-Test. **Possible Causes:** • Clear the trouble codes and retest for this trouble code. If the same trouble code resets, the PCM has failed and must be replaced to repair this problem.

DTC	Trouble Code Title, Conditions & Possible Causes
DTC: P0602 **2T CCM, MIL: Yes** **2006** **Models:** Sonata **Engines:** 3.3L **Transmissions:** All	**EEPROM Programming Error** Check internal CPU **Possible Causes:** • Faulty PCM
DTC: P0604 **2T CCM, MIL: Yes** **2006** **Models:** Sonata **Engines:** 3.3L **Transmissions:** All	**Internal Control Module Random Access Memory (RAM) Error** Check internal CPU. **Possible Causes:** • Faulty PCM
DTC: P061B **2T CCM, MIL: No** **2006** **Models:** Sonata **Engines:** 3.3L **Transmissions:** All	**Internal Control Module Torque Calculation Performance** Desired torque error. **Possible Causes:** • Faulty PCM
DTC: P0605 **1T PCM, MIL: Yes** **1999-06** **Models:** Accent, Elantra, Santa Fe, Sonata, Tiburon, Tucson, XG300, XG350 **Engines:** All **Transmissions:** All	**PCM (Internal Controller) ROM Error** Key on for 1 second, and the PCM detected an internal ROM error occurred during the initial Self-Test. **Possible Causes:** • Clear the trouble codes and retest for this trouble code. If the same trouble code resets, the PCM has failed and must be replaced to repair this problem.
DTC: P0606 **2T CCM, MIL: Yes** **2004-06** **Models:** Elantra, Sonata **Engines:** 2.0L, 3.3L **Transmissions:** All	**ECM Processor (ECU-Self Test Failed)** **Controller error. No electrical fault of the front HO2S.** **Possible Causes:** • Faulty PCM
DTC: P0624 **2T CCM, MIL: No** **2006** **Models:** Accent **Engines:** 1.5L **Transmissions:** All	**Fuel Cap Lamp Control Circuit** Circuit continuity check, (high, low, or open). **Possible Causes:** • Poor connection • Open or short • Instrument cluster • Faulty ECM/PCM
DTC: P0625 **2T CCM, MIL: No** **2006** **Models:** Sonata **Engines:** 2.4L **Transmissions:** All	**Alternator Field "F" Terminal Circuit Low** Electrical check. **Possible Causes:** • Short to battery in harness • Poor connection or damaged harness
DTC: P0626 **2T CCM, MIL: No** **2006** **Models:** Sonata **Engines:** 2.4L **Transmissions:** All	**Alternator Field "F" Terminal Circuit High** Electrical check. Time after ignition ON, 1 second. Engine speed 0. No main relay error. **Possible Causes:** • Open or short to ground in harness • Faulty charging system
DTC: P0630 **2T CCM, MIL: Yes** **2005-06** **Models:** Accent, Elantra, Santa Fe, Sonata, Tiburon, Tucson, XG350 **Engines:** 1.5L, 2.0L, 2.4L, 2.7L, 3.3L, 3.5L **Transmissions:** All	**VIN Not Programmed Or Incompatible- ECM/PCMECM** **PCM internal check. Enable condition, ignition ON. VIN does not exist in boot area.** **Possible Causes:** • PCM is new and has not yet been programmed

DTC	Trouble Code Title, Conditions & Possible Causes
DTC: P0638 **2T CCM, MIL: Yes** **2006** **Models:** Sonata **Engines:** 2.4L, 3.3L **Transmissions:** All	**Throttle Actuator Control Range/Performance** ETS position control malfunction. Battery voltage more than 5 volts. **Possible Causes:** • Throttle stuck • Open in motor circuit • Faulty motor • Faulty PCM
DTC: P0641 **2T CCM, MIL: Yes** **2006** **Models:** Sonata **Engines:** 3.3L **Transmissions:** All	**Sensor Reference Voltage "A" Circuit Open** Sensor reference voltage check. Ignition ON. **Possible Causes:** • Short in sensor power supply line • Faulty PCM
DTC: P0642 **2T CCM, MIL: No** **2006** **Models:** Sonata **Engines:** 2.4L **Transmissions:** All	**Sensor Reference Voltage "A" Circuit High** Electrical check. Ignition ON. APS2 voltage 5.5 volts, for at least 1 second. **Possible Causes:** • Open or short to battery in power circuit • Poor connection or damaged harness • Faulty ECM
DTC: P0643 **2T CCM, MIL: No** **2006** **Models:** Sonata **Engines:** 2.4L **Transmissions:** All	**Sensor Reference Voltage "A" Circuit Low** Electrical check. Ignition ON. APS2 voltage 0.7 volt, for at least 1 second. **Possible Causes:** • Open or short to ground in power circuit • Poor connection or damaged harness • Faulty ECM
DTC: P0645 **2T CCM, MIL: No** **2005-06** **Models:** Accent, Elantra **Engines:** 1.5L, 1.6L, 2.0L **Transmissions:** All	**A/C Clutch Relay Control Circuit** DTC is set if the PCM detects that the relay line is open or shorted to ground or battery line. Circuit continuity check. **Possible Causes:** • Open in battery and control circuit • Short to ground in control circuit • Short to battery in control circuit • Faulty A/C relay • Faulty PCM
DTC: P0646 **2T CCM, MIL: No** **2006** **Models:** Accent, Sonata **Engines:** 1.5L, 2.4L, 3.3L **Transmissions:** All	**A/C Clutch Relay Control Circuit Low** Detects circuit short to low voltage. No DTC exists. Engine works. After 0.5 seconds. **Possible Causes:** • Poor connection • Open or short to ground in A/C relay circuit • Faulty A/C relay • Faulty PCM
DTC: P0647 **2T CCM, MIL: No** **2006** **Models:** Accent, Sonata **Engines:** 1.5L, 2.4L, 3.3L **Transmissions:** All	**A/C Clutch Relay Control Circuit High** Detects circuit short to high voltage. No DTC exists. Engine works. After 0.5 seconds. **Possible Causes:** • Poor connection • Short to power in A/C relay circuit • Faulty A/C relay • Faulty PCM
DTC: P0650 **2T CCM, MIL: No** **2003-06** **Models:** Accent, Elantra, Santa Fe, Tiburon, Tucson **Engines:** 1.5L, 1.6L, 2.0L, 2.4L, 2.7L, 3.3L **Transmissions:** All	**Malfunction Indicator Lamp (MIL) Control Circuit** DTC is set if the PCM detects that the MIL line is open or shorted to ground or battery line. Driver stage check. **Possible Causes:** • Open or short between MIL and PCM • Contact resistance in connectors • Burned out MIL bulb
DTC: P0651 **2T CCM, MIL: Yes** **2006** **Models:** Sonata **Engines:** 3.3L **Transmissions:** All	**Sensor reference Voltage "B" Circuit Open** Sensor reference voltage check. Key ON. **Possible Causes:** • Short in sensor power supply line • Faulty PCM

DTC	Trouble Code Title, Conditions & Possible Causes
DTC: P0652 **2T CCM, MIL: Yes** **2006** **Models:** Sonata **Engines:** 2.4L **Transmissions:** All	**Sensor Reference Voltage "B" Circuit Low** Electrical check. Ignition ON. APS2 voltage 0.7 volt, for at least 0.04 second. **Possible Causes:** • Open or short to ground in power circuit • Poor connection or damaged harness • Faulty ECM
DTC: P0653 **2T CCM, MIL: Yes** **2006** **Models:** Sonata **Engines:** 2.4L **Transmissions:** All	**Sensor Reference Voltage "B" Circuit High** Electrical check. Ignition ON. TPS voltage 5.5 volts, for at least 0.04 second. **Possible Causes:** • Open or short to ground in power circuit • Poor connection or damaged harness • Faulty ECM
DTC: P0660 **2T CCM, MIL: Yes** **2006** **Models:** Sonata **Engines:** 3.3L **Transmissions:** All	**Intake Manifold Tuning Valve Control Circuit/Open (Bank 1)** Signal low, high. **Possible Causes:** • Poor connection • Open or short in VIS circuit • Faulty VIS • Faulty PCM
DTC: P0661 **2T CCM, MIL: No** **2005-06** **Models:** Sonata, Tucson **Engines:** 2.7L, 3.3L **Transmissions:** All	**Intake Manifold Tuning Valve Control Circuit Low (Bank 1) Solenoid Type** DTC is set if the ECM detects that the valve control circuit is shorted to ground. Driver stage check. **Possible Causes:** • Open in power supply harness • Short to ground in control harness • Contact resistance in connectors • Faulty valve
DTC: P0662 **2T CCM, MIL: No** **2005-06** **Models:** Sonata, Tucson **Engines:** 2.7L, 3.3L **Transmissions:** All	**Intake Manifold Tuning Valve Control Circuit High (Bank 1) Solenoid Type** DTC is set if the ECM detects that the valve control circuit is open or shorted to battery voltage. Driver stage check. **Possible Causes:** • Open or short to battery in control harness • Contact resistance in connectors • Faulty valve
DTC: P0664 **2T CCM, MIL: No** **2005-06** **Models:** Sonata, Tucson **Engines:** 2.7L, 3.3L **Transmissions:** All	**Intake Manifold Tuning Valve Control Circuit High (Bank 2) Solenoid Type** DTC is set if the ECM detects that the valve control circuit is shorted to ground. Driver stage check. **Possible Causes:** • Open in power supply harness • Short to ground in control harness • Contact resistance in connectors • Faulty valve
DTC: P0665 **2T CCM, MIL: No** **2005-06** **Models:** Sonata, Tucson **Engines:** 2.7L, 3.3L **Transmissions:** All	**Intake Manifold Tuning Valve Control Circuit High (Bank 1) Solenoid Type** DTC is set if the ECM detects that the valve control circuit is open or shorted to battery voltage. Driver stage check. **Possible Causes:** • Open or short to battery in control harness • Contact resistance in connectors • Faulty valve
DTC: P0698 **2T CCM, MIL: Yes** **2006** **Models:** Sonata **Engines:** 2.4L **Transmissions:** All	**Sensor Reference Voltage "C" Circuit Low** Electrical check. Ignition ON. APS1 voltage 0.7 volt, for at least 0.1 second. **Possible Causes:** • Open or short to ground in power circuit • Poor connection or damaged harness • Faulty ECM
DTC: P0699 **2T CCM, MIL: Yes** **2006** **Models:** Sonata **Engines:** 2.4L **Transmissions:** All	**Sensor Reference Voltage "C" Circuit High** Electrical check. Ignition ON. APS1 voltage 5.5 volts, for at least 0.01 second. **Possible Causes:** • Open or short to ground in power circuit • Poor connection or damaged harness • Faulty ECM

DTC	Trouble Code Title, Conditions & Possible Causes
DTC: P0700 **2T CCM, MIL: Yes** **1998-06** **Models:** Accent, Elantra, Santa Fe, Sonata, Tiburon, Tucson, XG300, XG350 **Engines:** All **Transmissions:** All	**Transmission Control System Signal** Key on or engine running, and the PCM received a signal from the TCM that indicating an internal problem with the TCM had occurred. **Possible Causes:** • Clear the trouble codes and retest for this trouble code. If the same trouble code resets, the TCM may have failed.
DTC: P0703 **2T CCM, MIL: Yes** **1998-02** **Models:** Accent, Elantra, Santa Fe, Sonata, Tiburon, XG300, XG350 **Engines:** All **Transmissions:** A/T	**Stop Lamp Switch Circuit Malfunction** Engine started, engine running at cruise speed, and the PCM detected the Brake Switch signal did not cycle from "high" to "low" as the brake pedal was pressed and released during the CCM test. **Possible Causes:** • Brake switch signal circuit is open or shorted to ground • Brake switch power circuit is open (check power from the relay) • Brake switch is damaged or has failed • TCM has failed
DTC: P0705 **2T CCM, MIL: Yes** **1996-02** **Models:** Accent, Elantra, Santa Fe, Sonata, Tiburon, XG300, XG350 **Engines:** All **Transmissions:** A/T	**Inhibitor (Park/Neutral) Switch Circuit Malfunction** Engine started, engine running in gear at cruise speed, and the TCM detected an unexpected voltage condition on the Park/Neutral (P/N) position circuit (of the TR switch) during the CCM test. **Possible Causes:** • P/N switch signal circuit is open or shorted to ground • P/N switch power circuit is open (check the power source) • P/N switch is out-of-adjustment, damaged or has failed • TCM has failed • TSB 01-40-020 (7/01) contains information about this code
DTC: P0707 **2T CCM, MIL: Yes** **1998-02** **Models:** Accent, Elantra, Santa Fe, Sonata, Tiburon, XG300, XG350 **Engines:** All **Transmissions:** A/T	**Transmission Range Switch Circuit Malfunction (Open)** Engine started, engine running at cruise speed with VSS signals present, and the PCM detected an unexpected "high" voltage condition on the TR switch circuit during the CCM test. **Possible Causes:** • TR sensor signal circuit is open between the switch and PCM • TR sensor is damaged or has failed • TCM has failed • TSB 01-40-020 (7/01) contains information about this code
DTC: P0708 **2T CCM, MIL: Yes** **1998-02** **Models:** Accent, Elantra, Santa Fe, Sonata, Tiburon, XG300, XG350 **Engines:** All **Transmissions:** A/T	**Transmission Range Switch Circuit Malfunction (Shorted)** Engine started, engine running at cruise speed with VSS signals present, and the PCM detected an unexpected "low" voltage condition on the TR switch circuit during the CCM test. **Possible Causes:** • TR sensor signal circuit is shorted to ground • TR sensor is damaged or has failed • TCM has failed • TSB 01-40-020 (7/01) contains information about this code
DTC: P0710 **2T CCM, MIL: Yes** **2000-02** **Models:** Sonata **Engines:** All **Transmissions:** A/T	**A/T Transmission Fluid Temperature Sensor Performance** Engine started, vehicle driven in gear for more than 10 minutes, and the PCM detected an unexpected "low" voltage condition (Scan Tool reads over 392°F) or "high" voltage condition (Scan Tool reads less more than 2.6 volts) during the CCM Rationality test. **Possible Causes:** • A/T TFT sensor signal circuit is open or shorted to ground • A/T TFT sensor is damaged or has failed • TCM has failed
DTC: P0712 **2T CCM, MIL: Yes** **1998-02** **Models:** Accent, Elantra, Santa Fe, Sonata, Tiburon, XG300, XG350 **Engines:** All **Transmissions:** A/T	**Transmission Fluid Temperature Sensor Circuit High Input** Engine started, engine running, gear selector in any position except for Neutral, and the PCM detected the TFT sensor indicated more than 4.90 volts (Scan Tool reads -40°F) during the CCM test. **Possible Causes:** • TFT sensor signal circuit is open between sensor and PCM • TFT sensor ground circuit is open between sensor and PCM • TFT sensor is damaged or has failed • PCM has failed • TSB 00-40-10 (8/00) contains information about this code

DTC	Trouble Code Title, Conditions & Possible Causes
DTC: P0713 **2T CCM, MIL: Yes** **1998-02** **Models:** Accent, Elantra, Santa Fe, Sonata, Tiburon, XG300, XG350 **Engines:** All **Transmissions:** A/T	**Transmission Fluid Temperature Sensor Circuit Low Input** Engine started, engine running, gear selector in any position except for Neutral, and the PCM detected the TFT sensor indicated less than 0.50 volt (Scan Tool reads 315ºF) during the CCM test. **Possible Causes:** • TFT sensor signal circuit is shorted to ground • TFT sensor is damaged or has failed • PCM has failed • TSB 00-40-10 (8/00) contains information about this code
DTC: P0715 **2T CCM, MIL: Yes** **1996-02** **Models:** Accent, Elantra, Santa Fe, Sonata, Tiburon, XG300, XG350 **Engines:** All **Transmissions:** A/T	**Transmission Input Speed Shaft Sensor** Engine started, vehicle driven at cruise speed for 3-5 minutes, and the TCM detected too large a change in the gear/speed ratio from the Input Speed sensor signal during the CCM Rationality test. **Possible Causes:** • Input speed sensor signal circuit is open or shorted to ground • Input speed sensor is damaged or has failed • TCM has failed • TSB 1-40-005 (3/01) contains information about this code
DTC: P0717 **2T CCM, MIL: Yes** **1996-02** **Models:** Accent, Elantra, Santa Fe, Sonata, Tiburon, XG300, XG350 **Engines:** All **Transmissions:** A/T	**A/T Pulse Generator 'A' Sensor Circuit Malfunction** Engine started, vehicle driven at a speed of over 30 mph, and the PCM detected an unexpected or intermittent "low" or "high" voltage condition on the A/T Pulse Generator 'A' circuit during the CCM test. **Possible Causes:** • A/T pulse generator 'A' (+) circuit is open or shorted to ground • A/T pulse generator 'A' (-) circuit is open or shorted to ground • A/T pulse generator 'B' is damaged or has failed • TCM has failed • TSB 01-40-006-1 (12/01) contains information about this code
DTC: P0720 **2T CCM, MIL: Yes** **1998-02** **Models:** Accent, Elantra, Santa Fe, Sonata, Tiburon, XG300, XG350 **Engines:** All **Transmissions:** A/T	**Transmission Output Speed Shaft Sensor** Engine started, vehicle driven at a speed of over 30 mph, and the PCM detected the OSS signal was less than 50% of the VSS signal for 1 second at a speed of 6 mph, and with the Stop lamp switch indicating "on" during the CCM test. **Possible Causes:** • Output speed sensor signal circuit is open or shorted to ground • Output speed sensor is damaged or has failed • TCM has failed • TSB 1-40-005 (3/01) contains information about this code
DTC: P0722 **2T CCM, MIL: Yes** **1998-02** **Models:** Accent, Elantra, Tiburon **Engines:** All **Transmissions:** A/T	**A/T Pulse Generator 'B' Sensor Circuit Malfunction** Engine started, vehicle driven at a speed of over 30 mph, and the PCM detected an unexpected or intermittent "low" or "high" voltage condition on the A/T Pulse Generator 'B' circuit during the CCM test. **Possible Causes:** • Pulse generator 'B' (+) or (-) circuit is open or shorted to ground • Pulse generator 'B' is damaged or has failed • TCM has failed • TSB 01-40-006-1 (12/01) contains information about this code
DTC: P0731 **2T CCM, MIL: Yes** **1998-02** **Models:** Accent, Elantra, Santa Fe, Sonata, Tiburon, XG300, XG350 **Engines:** All **Transmissions:** A/T	**A/T Incorrect 1st Gear Ratio** Engine started, vehicle driven at a speed of over 5 mph, then after a gear shift, the PCM detected the OSS signal (times) the gear ratio of the new gear ratio did not match the ISS signal during the CCM test. **Possible Causes:** • Solenoid or related pressure switch is damaged or has failed • Problems related to the Input Speed or Output Speed sensor • Low/Reverse clutch is damaged, leaking or has failed • Problems related to the transmission valve body
DTC: P0732 **2T CCM, MIL: Yes** **1998-02** **Models:** Accent, Elantra, Santa Fe, Sonata, Tiburon, XG300, XG350 **Engines:** All **Transmissions:** A/T	**A/T Incorrect 2nd Gear Ratio** Engine started, vehicle driven at a speed of over 5 mph, then after a gear shift, the PCM detected the OSS signal (times) the gear ratio of the new gear ratio did not match the ISS signal during the CCM test. **Possible Causes:** • Solenoid or related pressure switch is damaged or has failed • Problems related to the Input Speed or Output Speed sensor • Low/Reverse clutch is damaged, leaking or has failed • Problems related to the transmission valve body

DTC	Trouble Code Title, Conditions & Possible Causes
DTC: P0733 **2T CCM, MIL: Yes** **1998-02** **Models:** Accent, Elantra, Santa Fe, Sonata, Tiburon, XG300, XG350 **Engines:** All **Transmissions:** A/T	**A/T Incorrect 3rd Gear Ratio** Engine started, vehicle driven at a speed of over 5 mph, then after a gear shift, the PCM detected the OSS signal (times) the gear ratio of the new gear ratio did not match the ISS signal during the CCM test. **Possible Causes:** • Solenoid or related pressure switch is damaged or has failed • Problems related to the Input Speed or Output Speed sensor • Low/Reverse clutch is damaged, leaking or has failed • Problems related to the transmission valve body
DTC: P0734 **2T CCM, MIL: Yes** **1998-02** **Models:** Accent, Elantra, Santa Fe, Sonata, Tiburon, XG300, XG350 **Engines:** All **Transmissions:** A/T	**A/T Incorrect 4th Gear Ratio** Vehicle driven at a speed of over 5 mph, then after a gear shift, the PCM detected the OSS signal (times) the gear ratio of the new gear ratio did not match the ISS signal. **Possible Causes:** • Solenoid or related pressure switch is damaged or has failed • Low/Reverse clutch is damaged, leaking or has failed • Problems related to the Input Speed or Output Speed sensor • Problems related to the transmission valve body
DTC: P0735 **2T CCM, MIL: Yes** **1998-02** **Models:** Accent, Elantra, Santa Fe, Sonata, Tiburon, XG300, XG350 **Engines:** All **Transmissions:** A/T	**A/T Incorrect 5th Gear Ratio** Engine started, vehicle driven at a speed of over 5 mph, then after a gear shift, the PCM detected the OSS signal (times) the gear ratio of the new gear ratio did not match the ISS signal during the CCM test. **Possible Causes:** • 5th Gear or pressure switch is damaged or has failed • Low/Reverse clutch is damaged, leaking or has failed • Problems related to the Input Speed or Output Speed sensor • Problems related to the transmission valve body
DTC: P0736 **2T CCM, MIL: Yes** **1998-02** **Models:** Accent, Elantra, Santa Fe, Sonata, Tiburon, XG300, XG350 **Engines:** All **Transmissions:** A/T	**A/T Incorrect Reverse Gear Ratio** DTC P0500 not set, engine started, vehicle driven at a speed of over 3 mph with Reverse Gear commanded "on", and the PCM detected an incorrect Reverse Gear ratio during the CCM Rationality test. **Possible Causes:** • Reverse Gear is damaged or has failed • Low/Reverse clutch is damaged, leaking or has failed • Problems related to the Input Speed or Output Speed sensor • Problems related to the transmission valve body
DTC: P0712 **2T CCM, MIL: Yes** **1996-02** **Models:** Accent, Elantra, Tiburon **Engines:** All **Transmissions:** A/T	**A/T Damper Clutch Control System Range/Performance** Engine started, vehicle driven at a speed over 30 mph, and the PCM detected a problem in the A/T Damper Clutch System operation. **Possible Causes:** • Damper Clutch solenoid is sticking or not operating properly • Damper Clutch solenoid valve is damaged or has failed • TSB 00-40-009 (8/00) contains information about this code
DTC: P0742 **2T CCM, MIL: Yes** **1996-02** **Models:** Accent, Elantra, Tiburon **Engines:** All **Transmissions:** A/T	**A/T Damper Clutch Control System Circuit Malfunction** Engine started, vehicle driven to a speed over 30 mph, and the PCM detected an unexpected "high" voltage condition on the Damper Clutch Control circuit during the CCM test. **Possible Causes:** • Damper Clutch solenoid control circuit is open • Damper Clutch solenoid valve is damaged or has failed • TSB 00-40-009 (8/00) contains information about this code
DTC: P0743 **2T CCM, MIL: Yes** **1996-02** **Models:** Accent, Elantra, Tiburon **Engines:** All **Transmissions:** A/T	**A/T Damper Clutch Control System Circuit Malfunction** Engine started, vehicle driven to a speed over 30 mph, and the PCM detected an unexpected "low" voltage condition on the Damper Clutch Control circuit during the CCM test. **Possible Causes:** • Damper Clutch solenoid control circuit is shorted to ground • Damper Clutch solenoid valve is damaged or has failed • TSB 00-40-009 (8/00) contains information about this code
DTC: P0745 **2T CCM, MIL: Yes** **2000-02** **Models:** Accent, Elantra, Tiburon **Engines:** All **Transmissions:** A/T	**A/T Pressure Control Solenoid Circuit Malfunction** Engine started, vehicle driven to over 3 mph, and the PCM detected an unexpected high voltage condition on the PCS (solenoid) circuit. **Possible Causes:** • PCS (solenoid) control circuit is open • PCS (solenoid) is damaged or has failed • TCM is damaged or has failed • TSB 00-40-009 (8/00) contains information about this code

DTC	Trouble Code Title, Conditions & Possible Causes
DTC: P0747 **2T CCM, MIL: Yes** **1996-02** **Models:** Accent, Elantra, Tiburon **Engines:** All **Transmissions:** A/T	**A/T Pressure Control Solenoid Circuit Malfunction** Engine started, vehicle driven to over 3 mph, and the PCM detected an unexpected high voltage condition on the PCS (solenoid) circuit. **Possible Causes:** • PCS (solenoid) control circuit is open • PCS (solenoid) is damaged or has failed • TCM is damaged or has failed • TSB 00-40-009 (8/00) contains information about this code
DTC: P0748 **2T CCM, MIL: Yes** **1996-99** **Models:** Accent, Elantra, Tiburon **Engines:** All **Transmissions:** A/T	**A/T Pressure Control Solenoid Circuit Malfunction** Engine started, vehicle driven to over 3 mph, and the PCM detected an unexpected low voltage condition on the PCS (solenoid) circuit. **Possible Causes:** • PCS (solenoid) control circuit is shorted to ground • PCS (solenoid) is damaged or has failed • TCM is damaged or has failed • TSB 00-40-009 (8/00) contains information about this code
DTC: P0752 **2T CCM, MIL: Yes** **1996-99** **Models:** Accent, Elantra, Tiburon **Engines:** All **Transmissions:** A/T	**A/T Shift Solenoid 'A' Circuit Malfunction** Engine started, vehicle driven to over 3 mph, and the PCM detected an unexpected high voltage condition on the Shift solenoid 'A' circuit. **Possible Causes:** • SSA (solenoid) control circuit is open • SSA (solenoid) is damaged or has failed • TCM is damaged or has failed • TSB 00-40-009 (8/00) contains information about this code
DTC: P0753 **2T CCM, MIL: Yes** **1996-99** **Models:** Accent, Elantra, Tiburon **Engines:** All **Transmissions:** A/T	**A/T Shift Solenoid 'A' Circuit Malfunction** Engine started, vehicle driven to over 3 mph, and the PCM detected an unexpected low voltage condition on the Shift Solenoid 'A' circuit. **Possible Causes:** • SSA (solenoid) control circuit is shorted to ground • SSA (solenoid) is damaged or has failed • TCM is damaged or has failed • TSB 00-40-009 (8/00) contains information about this code
DTC: P0757 **2T CCM, MIL: Yes** **1996-99** **Models:** Accent, Elantra, Tiburon **Engines:** All **Transmissions:** A/T	**A/T Shift Solenoid 'B' Circuit Malfunction** Engine started, vehicle driven to over 3 mph, and the PCM detected an unexpected high voltage condition on the Shift solenoid 'B' circuit. **Possible Causes:** • SSA (solenoid) control circuit is open • SSB (solenoid) is damaged or has failed • TCM is damaged or has failed • TSB 00-40-009 (8/00) contains information about this code
DTC: P0758 **2T CCM, MIL: Yes** **1996-99** **Models:** Accent, Elantra, Tiburon **Engines:** All **Transmissions:** A/T	**A/T Shift Solenoid 'B' Circuit Malfunction** Engine started, vehicle driven to over 3 mph, and the PCM detected an unexpected low voltage condition on the Shift Solenoid 'B' circuit. **Possible Causes:** • SSB (solenoid) control circuit is shorted to ground • SSB (solenoid) is damaged or has failed • TCM is damaged or has failed • TSB 00-40-009 (8/00) contains information about this code
DTC: P0765 **2T CCM, MIL: Yes** **2000-02** **Models:** Accent, Elantra, Tiburon **Engines:** All **Transmissions:** A/T	**A/T Pressure Control Solenoid Circuit Malfunction** Engine started, vehicle driven to over 3 mph, and the PCM detected an unexpected low voltage condition on the PCS (solenoid) circuit. **Possible Causes:** • PCS (solenoid) control circuit is shorted to ground • PCS (solenoid) is damaged or has failed • TCM is damaged or has failed • TSB 00-40-009 (8/00) contains information about this code

DTC	Trouble Code Title, Conditions & Possible Causes
DTC: P0740 **2T CCM, MIL: Yes** **2000-02** **Models:** Accent, Elantra, Tiburon **Engines:** All **Transmissions:** A/T	**Torque Converter Clutch Solenoid Circuit Malfunction** Engine started, vehicle driven at a constant speed of over 30 mph, and the PCM detected the TCC solenoid duty cycle ratio was 100% for over 4 seconds during the CCM test. **Possible Causes:** • Transmission fluid is contaminated, or the fluid level is too low • Worn pump bushing and/or the TCC is damaged or has failed • Valve body lockup accumulator diameter is out of specification • TSB 00-40-009 (8/00) contains information about this code
DTC: P0743 **2T CCM, MIL: Yes** **2000-02** **Models:** Santa Fe, Sonata, XG300, XG350 **Engines:** All **Transmissions:** A/T	**Torque Converter Clutch Solenoid Circuit Malfunction** Engine started, engine running, relay voltage over 10 volts, and the PCM detected an unexpected voltage condition on the TCC Solenoid circuit during the CCM test. **Possible Causes:** • TCC solenoid control circuit is open or shorted to ground • TCC solenoid wiring harness connector damaged • TCC solenoid is damaged or has failed • PCM has failed • TSB 00-40-009 (8/00) contains information about this code
DTC: P0750 **2T CCM, MIL: Yes** **1998-02** **Models:** Accent, Elantra, Santa Fe, Sonata, Tiburon, XG300, XG350 **Engines:** All **Transmissions:** A/T	**Low/Reverse Solenoid Circuit Malfunction** Engine started, engine running, relay voltage over 10 volts, and the PCM detected an unexpected voltage condition the Low/Reverse solenoid circuit during the CCM test. **Possible Causes:** • L/R solenoid control circuit is open, shorted to ground or power • L/R pressure switch is open, shorted to ground or it has failed • L/R solenoid is damaged or has failed • TCM power ground circuit is open, or the TCM has failed • TSB 00-40-009 (8/00) contains information about this code
DTC: P0755 **2T CCM, MIL: Yes** **1998-02** **Models:** Accent, Elantra, Santa Fe, Sonata, Tiburon, XG300, XG350 **Engines:** All **Transmissions:** A/T	**U/D Solenoid Valve Circuit Malfunction** Engine started, engine running, relay voltage over 10 volts, and the PCM detected an unexpected voltage condition on the U/D Solenoid circuit during the CCM test. **Possible Causes:** • U/D solenoid control circuit is open, shorted to ground or power • U/D pressure switch is open, shorted to ground or it has failed • U/D solenoid is damaged or has failed • TCM has failed or the TCM power ground circuit is open • TSB 00-40-009 (8/00) contains information about this code
DTC: P0760 **2T CCM, MIL: Yes** **1998-02** **Models:** Accent, Elantra, Santa Fe, Sonata, Tiburon, XG300, XG350 **Engines:** All **Transmissions:** A/T	**A/T Second Solenoid Valve Circuit Malfunction** Engine started, vehicle driven to over 30 mph and the PCM detected an unexpected voltage condition on the 2nd solenoid circuit. **Possible Causes:** • 2nd solenoid control circuit is open or shorted to ground • 2nd solenoid is damaged or has failed • TCM has failed • TSB 00-40-009 (8/00) contains information about this code
DTC: P0765 **2T CCM, MIL: Yes** **1998-02** **Models:** Accent, Elantra, Santa Fe, Sonata, Tiburon, XG300, XG350 **Engines:** All **Transmissions:** A/T	**Overdrive (O/D) Solenoid Valve Circuit Malfunction** Engine started, engine running, relay voltage over 10 volts, and the PCM detected an unexpected voltage condition on the O/D solenoid circuit during the CCM test. **Possible Causes:** • O/D solenoid control circuit is open, shorted to ground or power • O/D pressure switch is open, shorted to ground or it has failed • O/D solenoid is damaged or has failed • TCM power ground circuit is open, or the TCM has failed • TSB 00-40-009 (8/00) contains information about this code

OBD II Trouble Code List (P1xxx Codes)

DTC	Trouble Code Title, Conditions & Possible Causes
DTC: P1100 **2T CCM, MIL: Yes** **1998-06** **Models:** Accent, Elantra, Santa Fe, Sonata, Tiburon, XG300, XG350 **Engines:** All **Transmissions:** All	**MAP Sensor Performance** Engine started, engine runtime over 60 seconds, system voltage over 8 volts, and the PCM detected the MAP sensor indicated more than 4.5 volts (Scan Tool reads 114 kPa), or it indicated less than 1.95 volts (Scan Tool reads 50 kPa) for over 4 seconds during the CCM test. **Possible Causes:** • MAP sensor signal circuit is open or shorted to ground • MAP sensor vacuum source line is clogged or contains ice • MAP sensor is damaged or has failed • PCM has failed
DTC: P1102 **2T CCM, MIL: Yes** **1998-06** **Models:** Accent, Elantra, Santa Fe, Sonata, Tiburon, XG300, XG350 **Engines:** All **Transmissions:** All	**MAP Sensor Circuit Low Input** Engine started, engine runtime over 60 seconds, system voltage over 8 volts, and the PCM detected an unexpected low voltage condition of less than 1.95 volts (Scan Tool reads 50 kPa) on the MAP sensor circuit for over 4 seconds during the CCM test. **Possible Causes:** • MAP sensor signal circuit is open or shorted to ground • MAP sensor power (VREF) circuit is open • MAP sensor is damaged or has failed • PCM has failed
DTC: P1103 **2T CCM, MIL: Yes** **1999-06** **Models:** Accent, Elantra, Santa Fe, Sonata, Tiburon, XG300, XG350 **Engines:** All **Transmissions:** All	**MAP Sensor Circuit High Input** Engine started, engine runtime over 60 seconds, system voltage over 8 volts, and the PCM detected an unexpected high voltage condition of more than 4.50 volts (Scan Tool reads 114 kPa) on the MAP sensor circuit for over 4 seconds during the CCM test. **Possible Causes:** • MAP sensor signal circuit is shorted to VREF or system power • MAP sensor ground circuit is open • MAP sensor is damaged or has failed • PCM has failed
DTC: P1106 **2T CCM, MIL: No** **2006** **Models:** Sonata **Engines:** 3.3L **Transmissions:** All	**Manifold Absolute Pressure Sensor Circuit Short- Intermittent High Input** This code detects an intermittent short to high in either the signal circuit or the MAP sensor. **Possible Causes:** • Poor connection • Short to battery in signal circuit • Open in ground circuit • Faulty MAPS • Faulty PCM
DTC: P1107 **2T CCM, MIL: No** **2006** **Models:** Sonata **Engines:** 3.3L **Transmissions:** All	**Manifold Absolute Pressure Sensor Circuit Short- Intermittent Low Input** This code detects an intermittent short to high in either the signal circuit or the MAP sensor. **Possible Causes:** • Poor connection • Open or short to ground in the power circuit • Open or short to ground in the signal circuit • Faulty MAPS • Faulty PCM
DTC: P1110 **2T CCM, MIL: Yes** **2001-06** **Models:** Santa Fe, XG300, XG350 **Engines:** All **Transmissions:** All	**Electronic Throttle System Malfunction** Key on or engine running, and the PCM detected that one or more fault conditions existed in the Electronic Throttle System (ETS). **Possible Causes:** • This code indicates the PCM has received a signal from the ETS controller indicating a problem exists with the throttle body • Refer to the additional trouble code(s) that are also set
DTC: P1111 **2T CCM, MIL: No** **2006** **Models:** Sonata **Engines:** 2.4L, 3.3L **Transmissions:** All	**Intake Air Temperature Sensor Circuit Short- Intermittent High Input** This code detects a continuous short to high in either the signal circuit or the sensor. **Possible Causes:** • Poor connection • Open or short in signal circuit • Open in ground circuit • Faulty IATS • Faulty PCM

DTC	Trouble Code Title, Conditions & Possible Causes
DTC: P1112 **2T CCM, MIL: No** **2006** **Models:** Sonata **Engines:** 3.3L **Transmissions:** All	**Intake Air Temperature Sensor Circuit Short- Intermittent Low Input** This code detects a continuous short to high in either the signal circuit or the sensor. **Possible Causes:** • Poor connection • Short to ground in the signal circuit • Open in ground circuit • Faulty IATS • Faulty PCM
DTC: P1114 **2T CCM, MIL: No** **2006** **Models:** Sonata **Engines:** 3.3L **Transmissions:** All	**Engine Coolant temperature Sensor Circuit- Intermittent Low Input** This code detects an intermittent short to ground in the signal circuit or the sensor. **Possible Causes:** • Poor connection • Short to ground in signal circuit • Open in ground circuit • Faulty ECTS • Faulty PCM
DTC: P1115 **2T CCM, MIL: No** **2006** **Models:** Sonata **Engines:** 3.3L **Transmissions:** All	**Engine Coolant temperature Sensor Circuit- Intermittent High Input** This code detects an intermittent open or short to battery in the signal circuit or the sensor. **Possible Causes:** • Poor connection • Open or short to battery in signal circuit • Open in ground circuit • Faulty ECTS • Faulty PCM
DTC: P1118 **2T CCM, MIL: Yes** **2001-06** **Models:** Santa Fe, XG300, XG350 **Engines:** All **Transmissions:** All	**Electronic Throttle System Motor Circuit Malfunction** Engine started, engine running, and the PCM detected the motor duty cycle was equal to or more than 80% with a current level less than 0.5A; or the throttle motor current level was more than 16A, or the motor current level was equal to or more than 7A for 600 ms. **Possible Causes:** • ETCS throttle motor control (+) or (-) circuit is open, shorted to ground or shorted to system power (B+) • ETCS throttle motor is damaged or has failed • PCM has failed
DTC: P1123 **2T Fuel, MIL: Yes** **1998-02** **Models:** Accent, Elantra, Santa Fe, Sonata **Engines:** All **Transmissions:** All	**Fuel Trim Adaptive Air System Too Rich** Engine started, engine speed less than 1000 rpm in closed loop, ECT sensor more than 158°F, MAF sensor indicating less than 5.5 g/sec, EVAP purge is "off", and the PCM detected the Short Term fuel trim "rich" command exceeded 10-15% for over 30 seconds. **Possible Causes:** • Base engine "mechanical" fault affecting one or more cylinders • EVAP system component has failed or canister fuel saturated • Exhaust leaks located in front of the HO2S location • Fuel control sensor is out of calibration (i.e., ECT, IAT or MAP) • Fuel delivery system supplying too much fuel during cruise or idle periods (e.g., faulty fuel pump, or faulty pressure regulator) • Fuel injector(s) is leaking or stuck partially open (one or more) • HO2S is contaminated, deteriorated or it has failed
DTC: P1124 **2T Fuel, MIL: Yes** **1999-02** **Models:** Accent, Elantra, Santa Fe, Sonata **Engines:** All **Transmissions:** All	**Fuel Trim Adaptive Air System Too Lean** Engine started, engine speed less than 1000 rpm in closed loop, ECT sensor more than 158°F, MAF sensor indicating less than 7.5 g/sec, EVAP purge is "off", and the PCM detected the Short Term fuel trim "lean" command exceeded 10-15% for over 30 seconds. **Possible Causes:** • Air leaks after the MAF sensor, or in the EGR or PCV system • Base engine "mechanical" fault affecting one or more cylinders • Exhaust leaks located in front of the HO2S location • Fuel control sensor is out of calibration (i.e., ECT, IAT or MAP) • Fuel delivery system supplying too little fuel during cruise or idle periods (e.g., faulty fuel pump or dirty, restricted fuel filter) • Fuel injector (one or more) dirty or pressure regulator has failed • HO2S is contaminated, deteriorated or it has failed • Vehicle driven low on fuel or until it ran out of fuel

DTC	Trouble Code Title, Conditions & Possible Causes
DTC: P1127 **2T Fuel, MIL: Yes** **1996-02** **Models:** Accent, Elantra, Santa Fe, Sonata **Engines:** All **Transmissions:** All	**Long Term Fuel Trim Multiplier System Too Rich** Engine speed less than 1000 rpm in closed loop, ECT sensor input more than 158°F, MAF sensor input indicating less than 5.5 g/sec, EVAP purge not enabled, engine load less than 1.8 ms, and the PCM detected the Long Term fuel trim multiplier was more than 10-15% rich for 30 seconds under these operating conditions. **Possible Causes:** • Base engine "mechanical" fault affecting one or more cylinders • EVAP system component has failed or canister fuel saturated • Exhaust leaks located in front of the HO2S location • Fuel control sensor is out of calibration (i.e., ECT, IAT or MAP) • Fuel delivery system supplying too much fuel during cruise or idle periods (e.g., faulty fuel pump, or faulty pressure regulator) • Fuel injector(s) is leaking or stuck partially open (one or more) • HO2S is contaminated, deteriorated or it has failed
DTC: P1128 **2T Fuel, MIL: Yes** **1999-02** **Models:** Accent, Elantra, Santa Fe, Sonata **Engines:** All **Transmissions:** All	**Long Term Fuel Trim Multiplier System Too Lean** Engine speed less than 1000 rpm in closed loop, ECT sensor input more than 158°F, MAF sensor input indicating less than 5.5 g/sec, EVAP purge not enabled, engine load less than 1.8 ms, and the PCM detected the Long Term fuel trim multiplier was more than 10-15% lean for 30 seconds under these operating conditions. **Possible Causes:** • Air leaks after the MAF sensor, or in the EGR or PCV system • Base engine "mechanical" fault affecting one or more cylinders • Exhaust leaks located in front of the HO2S location • Fuel control sensor is out of calibration (i.e., ECT, IAT or MAP) • Fuel delivery system supplying too little fuel during cruise or idle periods (e.g., faulty fuel pump or dirty, restricted fuel filter) • Fuel injector (one or more) dirty or pressure regulator has failed • HO2S is contaminated, deteriorated or it has failed • Vehicle driven low on fuel or until it ran out of fuel
DTC: P1134 **2T O2S, MIL: Yes** **1999-06** **Models:** Elantra, Santa Fe, Sonata, Tiburon **Engines:** All **Transmissions:** All	**HO2S-11 (Bank 1 Sensor 1) Transition Time Switch** Engine started, engine running in closed loop, and the PCM detected the average front HO2S-11 rich-to-lean or lean-to-rich switch time was more than 1.1 seconds during the CCM test. **Possible Causes:** • HO2S signal circuit is open or shorted to ground • HO2S element is contaminated or it has failed • HO2S heater is damaged or has failed • Intake air leaks, exhaust manifold leaks or PCV system leaks • MAF sensor out of calibration (it may be dirty or contaminated)
DTC: P1147 **2T CCM, MIL: Yes** **2001-02** **Models:** XG300, XG350 **Engines:** All **Transmissions:** A/T	**Electronic Throttle System Sub Accelerator Position Sensor 1 Circuit Malfunction** Key on or engine running, and the PCM detected an unexpected voltage on the Sub Accelerator Position Sensor 1 (APS1) circuit. **Possible Causes:** • APS1 signal circuit is open (>4.50 volts) or grounded (<0.20 volts) • APS1 signal circuit is shorted to VREF or system power (B+) • APS1 is damaged or has failed • PCM has failed
DTC: P1151 **2T CCM, MIL: Yes** **2001-02** **Models:** Santa Fe, XG300, XG350 **Engines:** All **Transmissions:** A/T	**Electronic Throttle System Sub Accelerator Position Sensor 2 Circuit Malfunction** Key on or engine running and the PCM detected an unexpected voltage on the Sub Accelerator Position Sensor 2 (APS2) circuit. **Possible Causes:** • APS2 signal circuit is open (>4.50 volts) or grounded (<0.20 volts) • APS2 signal circuit is shorted to VREF or system power (B+) • APS2 is damaged or has failed • PCM has failed • TSB 01-36-019 (8/01) contains information about this code
DTC: P1154 **2T O2S, MIL: Yes** **1999-06** **Models:** Santa Fe, Sonata, Tiburon **Engines:** All **Transmissions:** All	**HO2S-21 (Bank 2 Sensor 1) Transition Time Switch** Engine started, engine running in closed loop, and the PCM detected the average front HO2S-21 rich-to-lean or lean-to-rich switch time was more than 1.1 seconds during the CCM test. **Possible Causes:** • HO2S signal circuit is open or shorted to ground • HO2S element is contaminated or it has failed • HO2S heater is damaged or has failed • Intake air leaks, exhaust manifold leaks or PCV system leaks • MAF sensor out of calibration (it may be dirty or contaminated)

DTC	Trouble Code Title, Conditions & Possible Causes
DTC: P1155 **2T CCM, MIL: Yes** **2001-06** **Models:** Santa Fe, XG300, XG350 **Engines:** All **Transmissions:** All	**Electronic Throttle System Limp Home Valve Circuit Malfunction** Key on or engine running and the PCM detected an unexpected voltage on the ETS Limp Home Valve circuit. **Possible Causes:** • ETS limp home valve control circuit is open or grounded • ETS limp home valve control circuit is shorted to system power • ETS limp home valve is damaged or has failed • PCM has failed
DTC: P1166 **2T Fuel, MIL: Yes** **1999-06** **Models:** Elantra, Santa Fe, Sonata, Tiburon **Engines:** 2.5 VIN V, 2.0L, 2.7L **Transmissions:** All	**HO2S-11 Controller Adaptive Test (Bank 1)** Engine started, engine running in closed loop for 2-3 minutes, and the PCM detected the Short Term fuel trim was less than -12.5% or more than +12.5%; or it detected the Long Term fuel trim value was less than -12.5%, or more than +22.4% for 5-10 seconds in the test. **Possible Causes:** • Air leaks after the MAF sensor, or in the EGR or PCV system • Base engine "mechanical" fault affecting one or more cylinders • Exhaust leaks located in front of the A/FS or HO2S location • Fuel control sensor is out of calibration (i.e., ECT, IAT or MAP) • Fuel delivery system supplying too little fuel during cruise or idle periods (e.g., faulty fuel pump or dirty, restricted fuel filter) • Fuel injector (one or more) dirty or pressure regulator has failed • HO2S is contaminated, deteriorated or it has failed • Vehicle driven low on fuel or until it ran out of fuel
DTC: P1167 **2T Fuel, MIL: Yes** **1999-06** **Models:** Santa Fe, Sonata, Tiburon **Engines:** 2.5 VIN V, 2.7L **Transmissions:** All	**HO2S-11 Controller Adaptive Test (Bank 2)** Engine started, engine running in closed loop for 2-3 minutes, and the PCM detected the Short Term fuel trim was less than -12.5% or more than +12.5%; or it detected the Long Term fuel trim value was less than -12.5%, or more than +22.4% for 5-10 seconds in the test. **Possible Causes:** • Air leaks after the MAF sensor, or in the EGR or PCV system • Base engine "mechanical" fault affecting one or more cylinders • Exhaust leaks located in front of the A/FS or HO2S location • Fuel control sensor is out of calibration (i.e., ECT, IAT or MAP) • Fuel delivery system supplying too little fuel during cruise or idle periods (e.g., faulty fuel pump or dirty, restricted fuel filter) • Fuel injector (one or more) dirty or pressure regulator has failed • HO2S is contaminated, deteriorated or it has failed • Vehicle driven low on fuel or until it ran out of fuel
DTC: P1168 **2T CCM, MIL: Yes** **1999-00** **Models:** Sonata **Engines:** 2.5 VIN V **Transmissions:** All	**HO2S-11 (Bank 1 Sensor 2) Heater Power Incorrect** Engine started, engine runtime from 1-2 minutes, and the PCM detected an unexpected voltage condition on the front HO2S-11 Heater Power circuit during the CCM test. **Possible Causes:** • HO2S heater control circuit is open or shorted to ground • HO2S heater control circuit is shorted to system power • HO2S heater is damaged or has failed • PCM has failed
DTC: P1169 **2T CCM, MIL: Yes** **1999-00** **Models:** Sonata **Engines:** 2.5 VIN V **Transmissions:** All	**HO2S-22 (Bank 2 Sensor 2) Heater Power Incorrect** Engine started, engine runtime from 1-2 minutes, and the PCM detected an unexpected voltage condition on the rear HO2S-22 Heater Power circuit during the CCM test. **Possible Causes:** • HO2S heater control circuit is open or shorted to ground • HO2S heater control circuit is shorted to system power • HO2S heater is damaged or has failed • PCM has failed
DTC: P1171 **2T CCM, MIL: Yes** **2001-06** **Models:** Santa Fe, XG300, XG350 **Engines:** 3.0L, 3.5L **Transmissions:** A/T	**Electronic Throttle System Circuit Malfunction (Open)** Engine started, engine running, and the PCM detected the APP sensor VPA reading was equal to or less than 0.2 volt with a VPA2 reading of equal to or less than 0.5 volt; or the VPA reading was equal to or more than 4.7 volts; or the VPA reading indicated from 0.2-1.8 volts with the VPA2 reading equal to or more than 4.97 volts; or the VPA reading minus the VPA2 reading was less than 0.02 volt; or the VPA 2 reading minus the VPA reading was less than 0.02 volt for 5 seconds. **Possible Causes:** • APP sensor circuit is open or shorted to ground • APP sensor circuit is shorted to VREF • APP sensor power circuit is open between sensor and the PCM • APP sensor ground circuit is open between sensor and PCM • APP sensor is damaged or has failed • PCM has failed

DTC	Trouble Code Title, Conditions & Possible Causes
DTC: P1172 **2T CCM, MIL: Yes** **2001-06** **Models:** Santa Fe, XG300, XG350 **Engines:** 3.0L, 3.5L **Transmissions:** A/T	**Electronic Throttle System Motor Current Malfunction** Engine started, engine running, and the PCM detected the motor duty cycle was equal to or more than 80% with a current level less than 0.5 amp; or the throttle motor current level was more than 16 amp, or the motor current level was equal to or more than 7A for 600 ms. **Possible Causes:** • ETCS throttle motor control (+) or (-) circuit is open, shorted to ground or shorted to system power (B+) • ETCS throttle motor is damaged or has failed • PCM has failed
DTC: P1173 **2T CCM, MIL: Yes** **2001-06** **Models:** Santa Fe, XG300, XG350 **Engines:** 3.0L, 3.5L **Transmissions:** A/T	**Electronic Throttle System Rationality** Engine started, engine running, and the PCM detected the difference between the Electronic Throttle Control System (ETCS) APP sensor VPA and VPA2 readings was less than 0.7 volt or 1.7 volts for more than 2 seconds during the CCM test. **Possible Causes:** • APP sensor is damaged or has failed • Throttle assembly or throttle linkage is binding or has failed • PCM has failed
DTC: P1174 **2T CCM, MIL: Yes** **2001-06** **Models:** Santa Fe, XG300, XG350 **Engines:** 3.0L, 3.5L **Transmissions:** A/T	**Electronic Throttle System 1 Close Malfunction** Engine started, engine running, and the PCM detected an unexpected voltage condition on the Electronic Throttle System (ETS) No. 1 Close signal circuit during the CCM test. **Possible Causes:** • ETS 1 "close" signal circuit is open or shorted to ground • ETS 1 "close" signal circuit is shorted to system power • ETS controller is damaged or has failed • PCM has failed
DTC: P1175 **2T CCM, MIL: Yes** **2001-06** **Models:** Santa Fe, XG300, XG350 **Engines:** 3.0L, 3.5L **Transmissions:** A/T	**Electronic Throttle System No. 2 Close** Engine started, engine running, and the PCM detected an unexpected voltage condition on the Electronic Throttle System (ETS) No. 2 Close signal circuit during the CCM test. **Possible Causes:** • ETS 2 "close" signal circuit is open or shorted to ground • ETS 2 "close" signal circuit is shorted to system power • ETS controller is damaged or has failed • PCM has failed
DTC: P1176 **2T CCM, MIL: Yes** **2001-06** **Models:** Santa Fe, XG300, XG350 **Engines:** 3.0L, 3.5L **Transmissions:** A/T	**Electronic Throttle System Motor No. 1 Open or Shorted** Engine started, engine running, and the PCM detected an unexpected voltage condition on the Electronic Throttle System (ETS) Motor No. 1 signal circuit during the CCM test. **Possible Causes:** • ETS 1 Motor control circuit is open or shorted to ground • ETS 1 Motor control circuit is shorted to system power • ETS 1 Motor is damaged or has failed • ETS controller is damaged or has failed • PCM has failed
DTC: P1177 **2T CCM, MIL: Yes** **2001-06** **Models:** Santa Fe, XG300, XG350 **Engines:** 3.0L, 3.5L **Transmissions:** A/T	**Electronic Throttle System Motor No. 2 Open or Shorted** Engine started, engine running, and the PCM detected an unexpected voltage condition on the Electronic Throttle System (ETS) Motor No. 2 signal circuit during the CCM test. **Possible Causes:** • ETS 2 Motor control circuit is open or shorted to ground • ETS 2 Motor control circuit is shorted to system power • ETS 2 Motor is damaged or has failed • ETS controller is damaged or has failed • PCM has failed
DTC: P1178 **2T CCM, MIL: Yes** **2001-06** **Models:** Santa Fe, XG300, XG350 **Engines:** 3.0L, 3.5L **Transmissions:** A/T	**Electronic Throttle System Motor Battery Voltage Open** Key on, and the PCM detected an unexpected voltage condition on the Electric Throttle Control System motor battery voltage circuit. Note: The PCM shuts off power to the throttle motor and magnetic clutch (throttle valve is closed by a return spring, so the accelerator pedal can be opened with the throttle valve) when this code sets. **Possible Causes:** • Battery connections are dirty or loose • ETCS power source circuit is open (check the ETCS fuse) • PCM has failed

DTC	Trouble Code Title, Conditions & Possible Causes
DTC: P1192 **2T CCM, MIL: No** **2003-06** **Models:** Santa Fe, XG350 **Engines:** 3.5L **Transmissions:** All	**ETS Limp home Target Following Mal.** Ignition switches ON. ETS motor relay ON. Battery voltage greater than 11 volts. TPS 1 is normal. **Possible Causes:** • Poor connector • Short in ETS motor circuit • Faulty ETS motor • Faulty PCM
DTC: P1193 **2T CCM, MIL: No** **2003-06** **Models:** Santa Fe, XG350 **Engines:** 3.5L **Transmissions:** All	**ETS Limp home Low RPM** Ignition switches ON. ETS motor relay ON. Battery voltage greater than 11 volts. TPS 1 is normal. Engine speed greater than or equal to 700 rpm. **Possible Causes:** • Poor connector • Intake/exhaust system blockage • Check throttle plate for carbon deposits • Faulty ETS system • Faulty TPS • Faulty ETS motor • Faulty PCM
DTC: P1194 **2T CCM, MIL: No** **2003-06** **Models:** Santa Fe, XG350 **Engines:** 3.5L **Transmissions:** All	**ETS Limp home TPS2 Mal.** Ignition switches ON. ETS motor relay ON. Battery voltage greater than 11 volts. TPS 1 is normal. Engine coolant temperature above 158 degrees F. Vtps2 less than or equal to 0.7 volt. **Possible Causes:** • Poor connector • Short in ETS motor circuit • Faulty ETS motor • Faulty PCM
DTC: P1195 **2T CCM, MIL: Yes** **2003-06** **Models:** Santa Fe, XG350 **Engines:** 3.5L **Transmissions:** All	**ETS Limp home Target Following Delay** Ignition switches ON. ETS motor relay ON. Battery voltage greater than 11 volts. TPS 1 is normal. Engine coolant temperature above 158 degrees F. Vtps1 less than 0.2 volt. **Possible Causes:** • Poor connector • Short in ETS motor circuit • Faulty ETS motor • Faulty PCM
DTC: P1196 **2T CCM, MIL: No** **2006** **Models:** Santa Fe **Engines:** 3.5L **Transmissions:** All	**ETS Limp home Closed Throttle Stuck** Ignition switches ON. ETS motor relay ON. Battery voltage greater than 11 volts. TPS 1 is normal. **Possible Causes:** • Poor connector • Short in ETS motor circuit • Faulty ETS motor • Faulty PCM
DTC: P1295 **2T CCM, MIL: No** **2006** **Models:** Sonata **Engines:** 3.3L **Transmissions:** All	**Electronic Throttle Control (ETC) System Malfunction- Power Management** This code is set is there is a problem in the power management system. Ignition ON. **Possible Causes:** • TPS malfunction • TPS malfunction plus MAFS malfunction • MAP malfunction plus TPS malfunction • Faulty PCM
DTC: P1308 **2T CCM, MIL: Yes** **2003-2006** **Models:** Accent **Engines:** 1.5L, 1.6L **Transmissions:** All	**Acceleration Sensor Low Input** A code is set if the signal output is lower than specification. Signal check, low. Threshold value is less than 1.5L volts. **Possible Causes:** • Short to ground in signal circuit • Open in power circuit • Faulty sensor • Faulty PCM

DTC	Trouble Code Title, Conditions & Possible Causes
DTC: P1309 **2T CCM, MIL: Yes** **2003-2006** **Models:** Accent **Engines:** 1.5L, 1.6L **Transmissions:** All	**Acceleration Sensor Signal Check High** A code is set if the signal output is higher than specification. Signal check, high. Threshold value is greater than 3.5 volts. **Possible Causes:** • Open in signal circuit • Short to battery in signal circuit • Open in ground circuit • Poor connection of acceleration sensor • Short to ground in signal circuit • Faulty sensor • Faulty PCM
DTC: P1330 **2T CCM, MIL: Yes** **2001-06** **Models:** Santa Fe, Sonata, XG300, XG350 **Engines:** 2.7L (Sonata) 3.0L, 3.5L **Transmissions:** A/T	**Spark Timing Adjust Circuit Malfunction** Engine running and the PCM detected an unexpected (invalid) signal on the Spark Timing Adjust circuit during the CCM test. **Possible Causes:** • Spark timing adjust circuit is open or shorted to ground • Spark timing adjust circuit is shorted to VREF or system power • PCM has failed
DTC: P1372 **2T CCM, MIL: Yes** **1999-06** **Models:** Elantra, Santa Fe, Sonata, Tiburon, **Engines:** All **Transmissions:** All	**PCM Segment Time Acquisition Incorrect** Key on, and then the PCM detected that the internal Segment Time was incorrect. **Possible Causes:** • Target wheel in the crankshaft installed incorrectly • CKP sensor signal erratic or invalid • CKP sensor signal circuit open or grounded (intermittent fault)
DTC: P1400 **2T CCM, MIL: Yes** **1996-98** **Models:** Sonata **Engines:** All **Transmissions:** All	**Manifold Differential Pressure Sensor Circuit Malfunction** ECT sensor signal more than 65.4°F at startup, then with the engine running in closed loop at low to medium load, the PCM detected the MDP Sensor signal was more than 4.50 volts or less than 0.20 volt, either condition met for 4 seconds. **Possible Causes:** • MDP sensor signal open or shorted to ground • MDP sensor signal shorted to VREF or to system power (B+) • MDP sensor is damaged or has failed
DTC: P1440 **2T CCM, MIL: Yes** **1999-00** **Models:** Sonata **Engines:** All **Transmissions:** All	**EVAP System Vent Solenoid Circuit Malfunction** Engine running and the PCM detected an unexpected (invalid) voltage on the EVAP Vent solenoid control circuit during the test. **Possible Causes:** • Vent solenoid control circuit open or shorted to ground • Vent solenoid control circuit shorted to system power (B+) • Vent solenoid is damaged or has failed • PCM has failed
DTC: P1443 **2T CCM, MIL: Yes** **1999-00** **Models:** Sonata **Engines:** All **Transmissions:** All	**EVAP System Malfunction (Fuel Cap Missing)** Engine started, engine running at cruise speed in closed loop for 2-3 minutes, ECT sensor more than 185°F, IAT sensor from 14-122°F, fuel level from 25-75%, and the PCM detected too much change in the fuel tank pressure during the EVAP Monitor leak test. **Possible Causes:** • Canister vent (CV) solenoid may be stuck in open position • EVAP canister tube, EVAP canister purge outlet tube or EVAP return tube disconnected or cracked, or canister is damaged • Fuel filler cap missing, loose (not tightened) or the wrong part • Fuel tank pressure (FTP) sensor has failed (mechanical fault) • Fuel tank control valve is contaminated, damaged or has failed
DTC: P1502 **2T CCM, MIL: Yes** **2004** **Models:** Elantra **Engines:** 2.0L **Transmissions:** All	**Open Wire Magnetic Wheel Sensor Malfunction** Electric check error. Vehicle speed greater than 0. Battery voltage greater than 10 volts. **Possible Causes:** • Faulty wheel speed sensor (WSS) • Open or short in WSS circuit • Faulty PCM
DTC: P1505 **2T CCM, MIL: Yes** **2003-06** **Models:** Accent, Elantra, Santa Fe, Sonata, Tiburon, Tucson **Engines:** 1.5L, 1.6L, 2.0L, 2.7L **Transmissions:** All	**Idle Charge Actuator Signal Low of Coil #1** The PCM sets a DTC if the ICAV (open) control circuit is open or short to ground. Driver stage check. **Possible Causes:** • Open or short to ground in harness • Contact resistance in connectors • Faulty ICA valve

DTC	Trouble Code Title, Conditions & Possible Causes
DTC: P1506 **2T CCM, MIL: Yes** **2003-06** **Models:** Accent, Elantra, Santa Fe, Sonata, Tiburon, Tucson **Engines:** 1.5L, 1.6L, 2.0L, 2.7L **Transmissions:** All	**Idle Charge Actuator Signal High of Coil #1** The PCM sets a DTC if the ICAV (open) control circuit is short to battery. Driver stage check. **Possible Causes:** • Short to battery in harness • Contact resistance in connectors • Faulty ICA valve
DTC: P1507 **2T CCM, MIL: Yes** **2003-06** **Models:** Accent, Elantra, Santa Fe, Sonata, Tiburon, Tucson **Engines:** 1.5L, 1.6L, 2.0L, 2.7L **Transmissions:** All	**Idle Charge Actuator Signal Low of Coil #2** The PCM sets a DTC if the ICAV (open) control circuit is open or short to ground. Driver stage check. **Possible Causes:** • Open or short to ground in harness • Contact resistance in connectors • Faulty ICA valve
DTC: P1508 **2T CCM, MIL: Yes** **2003-06** **Models:** Accent, Elantra, Santa Fe, Sonata, Tiburon, Tucson **Engines:** 1.5L, 1.6L, 2.0L, 2.7L **Transmissions:** All	**Idle Charge Actuator Signal High of Coil #2** The PCM sets a DTC if the ICAV (open) control circuit is short to battery. Driver stage check. **Possible Causes:** • Short to battery in harness • Contact resistance in connectors • Faulty ICA valve
DTC: P1510 **2T CCM, MIL: Yes** **1999-02** **Models:** Sonata **Engines:** All **Transmissions:** All	**Idle Charge Actuator Command Incorrect (Coil No. 1)** Key on or engine running and the PCM detected an incorrect voltage on the Idle Charge Actuator Command circuit for Coil No. 1 during the CCM test. **Possible Causes:** • Idle charge (speed) actuator command signal for Coil No.1 open, shorted to ground, or shorted to power • Idle charge (speed) actuator valve is damaged or has failed • PCM has failed
DTC: P1510 **2T CCM, MIL: Yes** **1996-02** **Models:** Accent, Elantra, Santa Fe, Sonata, Tiburon, XG300, XG350 **Engines:** All **Transmissions:** All	**Idle Speed Control Actuator Opening Coil Circuit Shorted** Key on or engine running and the PCM detected an unexpected (invalid) voltage on the Idle Speed Control Actuator opening coil control circuit during the CCM test. **Possible Causes:** • Idle speed actuator opening coil circuit shorted to system power (B+) • Idle speed actuator is damaged or has failed • PCM has failed
DTC: P1511 **2T CCM, MIL: Yes** **1999-02** **Models:** Sonata **Engines:** All **Transmissions:** All	**Idle Charge Actuator Command Incorrect (Coil No. 2)** Key on or engine running and the PCM detected an incorrect voltage on the Idle Charge Actuator Command circuit for Coil No. 2 during the CCM test. **Possible Causes:** • Idle charge (speed) actuator command signal for Coil No.2 open, shorted to ground, or shorted to power • Idle charge (speed) actuator valve is damaged or has failed • PCM has failed
DTC: P1513 **2T CCM, MIL: Yes** **1996-02** **Models:** Accent, Elantra, Santa Fe, Sonata, Tiburon, XG300, XG350 **Engines:** All **Transmissions:** All	**Idle Speed Control Actuator Opening Coil Circuit Open** Key on or engine running and the PCM detected an unexpected (invalid) voltage on the Idle Speed Control Actuator opening coil control circuit during the CCM test. **Possible Causes:** • Idle speed actuator opening coil circuit open or shorted to ground • Idle speed actuator is damaged or has failed • PCM has failed
DTC: P1520 **2T CCM, MIL: Yes** **2000-02** **Models:** Sonata **Engines:** 2.4L VIN S **Transmissions:** All	**Generator 'FR' Terminal Circuit Malfunction** Engine running, and the PCM detected the Generator 'FR' terminal indicated a 0% duty cycle command, condition met for 20 seconds. **Possible Causes:** • Generator 'FR' signal circuit is open or shorted to ground • Generator 'FR' component (inside the unit) has failed • PCM has failed

DTC	Trouble Code Title, Conditions & Possible Causes
DTC: P1521 **2T CCM, MIL: Yes** **1999-06** **Models:** Santa Fe, Sonata, Tiburon, XG300, XG350 **Engines:** All **Transmissions:** All	**Power Steering Switch Input Circuit Malfunction** Engine speed over 2500 rpm with the engine load over 55% for 4 seconds, ECT sensor input more than 55ºF, or with the engine speed 800 rpm or less (conditions repeated at least 10 times), and the PCM detected the Power Steering switch signal remained on. **Possible Causes:** • Power steering pressure switch circuit open, shorted to ground, shorted to VREF, or shorted to system power (B+) • Power steering pressure switch is damaged or has failed • PCM has failed
DTC: P1523 **2T CCM, MIL: No** **2006** **Models:** Sonata **Engines:** 3.3L **Transmissions:** All	**Electronic Throttle Control (ETC) System Malfunction- Throttle Valve Stuck** This code is set when the throttle fails to return to unpowered default position when power to the ETC motor is turned off. Fault set for failure to return to default position within a specified time. Throttle actuation previous mode not OFF. Throttle actuator mode is OFF. ETC power control mode equal normal. TPS1 and 2 equal normal. Sensor supply voltage equals normal. **Possible Causes:** • Carbon in throttle • Broken throttle return spring • Throttle sticky • Throttle icy • Faulty PCM
DTC: P1529 **1T PCM, MIL: Yes** **1999-06** **Models:** Elantra, Santa Fe, Sonata, Tiburon **Engines:** All **Transmissions:** All	**Customer Snapshot Request Error** Engine started, engine running, and the PCM did not receive the Customer Request Snapshot Data (MIL and Freeze Frame Data) as requested from the Transmission Control Module (TCM). **Possible Causes:** • Transmission control module error • Check for trouble codes in the transmission control module • TSB 1-40-005 (3/01) contains information about this code
DTC: P1550 **2T CCM, MIL: No** **2006** **Models:** Accent **Engines:** 1.5L **Transmissions:** All	**Knock Sensor Evaluation IC** Circuit continuity check, pulse test. **Possible Causes:** • Poor connection • Open or short in control circuit • Faulty knock sensor • Faulty PCM
DTC: P1552 **2T CCM, MIL: Yes** **1998-02** **Models:** Accent, Elantra, Santa Fe, Tiburon, XG300, XG350 **Engines:** All **Transmissions:** All	**Idle Speed Control Actuator Closing Coil Circuit Shorted** Key on or engine running and the PCM detected an unexpected (invalid) voltage on the Idle Speed Control Actuator closing coil during the CCM test. **Possible Causes:** • Idle speed actuator closing coil circuit shorted to system power • Idle speed actuator is damaged or has failed • PCM has failed
DTC: P1553 **2T CCM, MIL: Yes** **1998-02** **Models:** Accent, Elantra, Santa Fe, Sonata, Tiburon, XG300, XG350 **Engines:** All **Transmissions:** All	**Idle Speed Control Actuator Closing Coil Circuit Open** Key on or engine running and the PCM detected an unexpected (invalid) voltage on the Idle Speed Control Actuator closing coil control circuit during the CCM test. **Possible Causes:** • Idle speed actuator closing coil circuit is open or shorted to ground • Idle speed actuator is damaged or has failed • PCM has failed
DTC: P1560 **2T CCM, MIL: No** **2006** **Models:** Accent **Engines:** 1.5L **Transmissions:** All	**System Voltage** SPI communication check. **Possible Causes:** • Poor connection • Faulty ECM/PCM
DTC: P1586 **1T CCM, MIL: Yes** **1996-06** **Models:** Accent, Elantra, Tiburon **Engines:** All **Transmissions:** All	**Encoding Signal Circuit Not Rational** Engine started, engine running, and the PCM determined the Encoding Signal circuit was not rational during the CCM test. **Possible Causes:** • P/N switch signal circuit is open or shorted to ground • P/N switch signal circuit is shorted to system power • P/N switch is damaged or has failed • PCM has failed

DTC	Trouble Code Title, Conditions & Possible Causes
DTC: P1602 **1T PCM, MIL: Yes** **1996-06** **Models:** Elantra, Santa Fe, Sonata, Tiburon **Engines:** All **Transmissions:** All	**Serial Communication Problem With The TCM** Key on, and the PCM detected that a serial line "timeout" condition occurred between the PCM and the TCM. **Possible Causes:** • CAN network circuit open or shorted to ground • CAN network circuit shorted to VREF or system power (B+) • TCM has failed
DTC: P1605 **2T CCM, MIL: Yes** **1998-02** **Models:** Accent, Elantra, Santa Fe, Sonata, Tiburon **Engines:** All **Transmissions:** All	**Rough Road Sensor Circuit Malfunction** Engine running with the vehicle moving, and the PCM detected an unexpected (invalid) voltage on the Rough Road Sensor circuit. Note: The normal voltage output of this sensor at idle is 2.50-2.70 volts. It will range from 0.50 volt to 4.50 volts during normal driving conditions. **Possible Causes:** • Acceleration sensor circuit is open or shorted to ground • Acceleration sensor is damaged or has failed • PCM has failed
DTC: P1606 **2T CCM, MIL: Yes** **1996-02** **Models:** Accent, Elantra, Santa Fe, Tiburon **Engines:** All **Transmissions:** All	**Rough Road Sensor Circuit Not Valid** Engine running with the vehicle moving, and the PCM detected a signal from the Rough Road sensor that was not valid. Note: The normal voltage output of this sensor at idle is 2.50-2.70 volts. It will range from 0.50 volt to 4.50 volts during normal driving conditions. **Possible Causes:** • Acceleration sensor circuit is open or shorted to ground • Acceleration sensor is damaged or has failed • PCM has failed
DTC: P1607 **2T CCM, MIL: Yes** **2001-06** **Models:** Santa Fe, XG300, XG350 **Engines:** 3.0L, 3.5L **Transmissions:** All	**Electronic Throttle System Communication Error** Key on or engine running, and then the PCM detected a communication error occurred between it and the ETS controller. **Possible Causes:** • ETS circuit to the PCM open, shorted to ground or shorted to system power • ETC controller has failed • PCM has failed
DTC: P1611 **2T CCM, MIL: Yes** **1998-02** **Models:** Accent, Elantra, Santa Fe, Sonata, Tiburon, XG300, XG350 **Engines:** All **Transmissions:** A/T	**MIL Request Signal Circuit Low Input** Engine started, engine speed less than 800 rpm, and the PCM detected an unexpected "low" voltage condition (i.e., less than 80% of system voltage) on the MIL Request circuit for 6 seconds. **Possible Causes:** • MIL request circuit between the PCM and TCM is open or shorted to ground
DTC: P1613 **2T CCM, MIL: Yes** **1996-06** **Models:** Accent, Elantra, Santa Fe, Tiburon, XG350 **Engines:** All **Transmissions:** All	**MIL Request Signal Circuit High Input** Engine started, engine speed less than 800 rpm, and the PCM detected an unexpected "high" voltage condition (i.e., more than 120% of system voltage) on the MIL Request circuit for 6 seconds. **Possible Causes:** • MIL request circuit between the PCM and TCM is shorted to system power
DTC: P1614 **2T CCM, MIL: Yes** **1998-06** **Models:** Santa Fe, Sonata, XG300, XG350 **Engines:** 3.0L, 3.5L **Transmissions:** A/T	**Electronic Throttle System Module Malfunction** Engine started, engine running, and the PCM detected the Actual throttle opening angle varied too much from the Target throttle opening angle during the CCM Rationality test. **Possible Causes:** • ETCS (system) is damaged or has failed • PCM has failed
DTC: P161B **2T CCM, MIL: Yes** **2006** **Models:** Sonata **Engines:** 3.3L **Transmissions:** All	**PCM Internal Error- Torque Calculating** This code is set if delivered torque is grossly different from the desired torque. **Possible Causes:** • Intake air leakage • Faulty ETS system • Clogged exhaust system • Faulty PCM

DTC	Trouble Code Title, Conditions & Possible Causes
DTC: P1616 **1T CCM, MIL: Yes** **1999-02** **Models:** Sonata **Engines:** All **Transmissions:** All	**Main Relay Circuit Malfunction** Key on or engine running and the PCM detected a main relay fault. **Possible Causes:** • Main relay power fuse is open • Main relay control circuit to PCM is open or grounded • Main relay is damaged or has failed • PCM has failed
DTC: P1623 **1T CCM, MIL: Yes** **1999-02** **Models:** Sonata **Engines:** All **Transmissions:** All	**Diagnostic Lamp (MIL) Power Stage Malfunction** Key on or engine running and the PCM detected an unexpected voltage condition on the Diagnostic Lamp Power Stage circuit. **Possible Causes:** • MIL control circuit is open or shorted to ground • MIL control circuit is shorted to system power (B+) • MIL lamp is missing or has failed • PCM has failed
DTC: P1624 **1T CCM, MIL: Yes** **1999-06** **Models: Accent, Santa Fe, Sonata, Tiburon, XG300** **Engines:** All **Transmissions:** All	**Cooling Fan Relay Low Circuit** Key on or engine running and the PCM detected an unexpected (invalid) voltage on the Cooling Fan Relay "low" circuit. **Possible Causes:** • Cooling Fan Relay control circuit open or shorted to ground • Cooling Fan Relay power circuit from Main Relay is open • Cooling Fan Relay is damaged or has failed • PCM has failed • TSB 00-36-013 (9/00) contains information about this code
DTC: P1625 **1T CCM, MIL: Yes** **1999-06** **Models:** Accent, Santa Fe, Sonata, Tiburon **Engines:** All **Transmissions:** All	**Cooling Fan Relay High Circuit Malfunction** Key on or engine running, and the PCM detected an unexpected (invalid) voltage on the Cooling Fan Relay "high" circuit. **Possible Causes:** • Cooling Fan Relay control circuit shorted to system power (B+) • Cooling Fan Relay is damaged or has failed • PCM has failed
DTC: P1665 **2T CCM, MIL: Yes** **1996-99** **Models:** Accent **Engines:** All **Transmissions:** A/T	**Power Stage Group 'A' Circuit Malfunction** Key on, and the PCM detected a fault in Power Stage Group 'A' related to the transmission controller functions. **Possible Causes:** • Power Stage 'A' control circuit open or shorted to ground • TCM has failed
DTC: P1670 **2T CCM, MIL: Yes** **1996-99** **Models:** Accent **Engines:** All **Transmissions:** A/T	**Power Stage Group 'B' Circuit Malfunction** Key on, and the PCM detected a fault in Power Stage Group 'B' related to the transmission controller functions. **Possible Causes:** • Power Stage 'B' control circuit open or shorted to ground • TCM has failed
DTC: P1690 **2T CCM, MIL: No** **2003-04** **Models:** Accent, Elantra, Tiburon **Engines:** 1.5L, 1.6L, 2.0L **Transmissions:** All	**Smartra Error** No answer from SMARTRA. Invalid message from SMARTRA to ECM. **Possible Causes:** • Open or short in antenna or SMARTRA circuit • Antenna • SMARTRA • Faulty transponder • Faulty ECM
DTC: P1691 **2T CCM, MIL: No** **2003-04** **Models:** Accent, Elantra, Tiburon **Engines:** 1.5L, 1.6L, 2.0L **Transmissions:** All	**Antenna Error** Antenna error. **Possible Causes:** • Open or short in antenna or SMARTRA circuit • Antenna • SMARTRA • Faulty transponder • Faulty ECM

DTC	Trouble Code Title, Conditions & Possible Causes
DTC: P1693 **2T CCM, MIL: No** **2003-04** **Models:** Accent, Elantra, Tiburon **Engines:** 1.5L, 1.6L, 2.0L **Transmissions:** All	**Antenna Error** Passive mode invalid. Programming error. **Possible Causes:** • Open or short in antenna or SMARTRA circuit • Antenna • SMARTRA • Faulty transponder • Faulty ECM
DTC: P1694 **2T CCM, MIL: No** **2003-04** **Models:** Accent, Elantra, Tiburon **Engines:** 1.5L, 1.6L, 2.0L **Transmissions:** All	**ECM Signal Error** Invalid request from ECM or corrupted data **Possible Causes:** • Open or short in antenna or SMARTRA circuit • Antenna • SMARTRA • Faulty transponder • Faulty ECM
DTC: P1695 **2T CCM, MIL: No** **2003-04** **Models:** Accent, Elantra, Tiburon **Engines:** 1.5L, 1.6L, 2.0L **Transmissions:** All	**EEPROM Error** Inconsistent data from EEPROM. Invalid write operation from EEPROM. Not plausible immobilizer indicator store in ECM. No valid data from SMARTRA after three attempts from the ECM. Invalid tester message or unexpected request from tester. **Possible Causes:** • Open or short in antenna or SMARTRA circuit • Antenna • SMARTRA • Faulty transponder • Faulty ECM
DTC: P1709 **2T CCM, MIL: Yes** **1996-00** **Models:** Accent, Elantra, Tiburon **Engines:** All **Transmissions:** A/T	**A/T Kick Down Servo Switch Circuit Malfunction** Engine started, vehicle driven in 1st or 3rd gear position for over 5 seconds, Transfer Shaft speed over 900 rpm, ECT sensor more than 140ºF, and the TCM detected the Kick Down Servo Switch indicated "off" during the CCM test. **Possible Causes:** • Kick Down servo switch circuit open between switch and PCM • Kick Down servo switch circuit is shorted to ground • Kick Down servo switch is damaged or has failed • TCM has failed
DTC: P1715 **2T CCM, MIL: Yes** **1996-98** **Models:** Sonata **Engines:** All **Transmissions:** A/T	**A/T Pulse Generator 'A' Circuit Malfunction** Engine runtime over 15 seconds, and the PCM detected an unexpected (invalid) voltage on the Pulse Generator 'A' signal from the automatic transaxle. **Possible Causes:** • Pulse Generator 'A' signal circuit is open, shorted to ground or shorted to system power (B+) • Pulse Generator 'A' is damaged or has failed • PCM has failed
DTC: P1750 **2T CCM, MIL: Yes** **1996-98** **Models:** Sonata **Engines:** All **Transmissions:** A/T	**A/T Shift, Pressure Clutch Solenoid Circuit Malfunction** Engine started, engine runtime over 15 seconds, and the PCM detected an unexpected voltage condition on the Shift Control Solenoid "A", on Shift Solenoid 'B', or on the Pressure Control Solenoid Valve circuit, or it detected an invalid voltage drop on the Damper Clutch Control Solenoid Valve circuit during the CCM test. **Possible Causes:** • Shift control signal circuit is open, shorted to ground or shorted to system power (B+), or the Shift Control solenoid has failed • Pressure Control circuit is open, shorted to ground or shorted to system power (B+), or the Shift Control solenoid has failed • PCM has failed

OBD II Trouble Code List (P2xxx Codes)

DTC	Trouble Code Title, Conditions & Possible Causes
DTC: P2015 **2T CCM, MIL: No** **2005-06** **Models:** Santa Fe, XG350 **Engines:** 3.5L **Transmissions:** All	**Intake Manifold Runner Position Sensor/Switch Circuit Range/Performance (Bank 1)** The DTC is set if the VIS system could not approach a target position. Engine speed equal to or greater than 3750 rpm. **Possible Causes:** • Poor connection • Open or short to battery in harness • Faulty VIS motor or VIS motor rotation sensor • Faulty PCM
DTC: P2016 **2T CCM, MIL: No** **2005** **Models:** XG350 **Engines:** 3.5L **Transmissions:** All	**Limp Home Valve Mal.** **ON/OFF malfunction. Low voltage is detected when the LHV solenoid is OFF.** **Possible Causes:** • Poor connection • Faulty limp home value (LHV) • Open or short in limp home value (LHV) circuit • Faulty PCM
DTC: P2096 **2T CCM, MIL: Yes** **2004-06** **Models:** Accent, Elantra, Sonata **Engines:** 1.5L, 1.6L, 2.0L, 2.4L, 3.3L **Transmissions:** All	**Post Catalyst Fuel Trim System Too Lean (Bank 1)** Case 1: Monitoring deviation of fuel trim control (long term). No relevant failure. Long term fuel trim active. Case 2: Monitoring deviation of fuel trim control (short term). No relevant failure. Short term fuel trim active. Current engine speed less than 500 rpm. Current mass air flow less than 400mg/rev. Current lambda correction mean value less than 4 percent. **Possible Causes:** • Three way catalytic converter (TWC) • Rear HO2S
DTC: P2097 **2T CCM, MIL: Yes** **2004-06** **Models:** Accent, Elantra, Sonata **Engines:** 1.5L, 1.6L, 2.0L, 2.4L, 3.3L **Transmissions:** All	**Post Catalyst Fuel Trim System Too Rich (Bank 1)** Case 1: Monitoring deviation of fuel trim control (long term). No relevant failure. Long term fuel trim active. Case 2: Monitoring deviation of fuel trim control (short term). No relevant failure. Short term fuel trim active. Current engine speed less than 500 rpm. Current mass air flow less than 400mg/rev. Current lambda correction mean value less than 4 percent. **Possible Causes:** • Three way catalytic converter (TWC) • Rear HO2S
DTC: P2100 **2T CCM, MIL: Yes** **2005-06** **Models:** Santa Fe, XG350 **Engines:** 3.5L **Transmissions:** All	**Throttle Actuator Control Motor Current Range/Performance/Throttle Actuator Control Motor Circuit Open** Vb open. Motor relay ON. Voltage to detect circuit open less than or equal to 4.0 volts. **Possible Causes:** • Poor connection • Open in ETS relay circuit • Faulty ETS relay/fuse • Faulty PCM
DTC: P2101 **2T CCM, MIL: No** **2006** **Models:** Sonata **Engines:** 2.4L **Transmissions:** All	**Throttle Actuator Control Motor Circuit** Hardware check. Battery voltage 9 volts. ECU power stage error. **Possible Causes:** • Poor connection or damaged harness • Faulty ETC motor
DTC: P2102 **2T CCM, MIL: Yes** **2005-06** **Models:** Santa Fe, XG350 **Engines:** 3.5L **Transmissions:** All	**Throttle Actuator Control Motor Circuit Range/Performance/Throttle Actuator Control Motor Circuit Low** Motor circuit low. Ignition switch ON. **Possible Causes:** • Poor connection • Short to ground in ETS motor circuit • Faulty ETS motor • Faulty PCM
DTC: P2103 **2T CCM, MIL: Yes** **2005-06** **Models:** Santa Fe, XG350 **Engines:** 3.5L **Transmissions:** All	**Throttle Actuator Control Motor Circuit Range/Performance/Throttle Actuator Control Motor Circuit High** Motor circuit High. Ignition switch ON. **Possible Causes:** • Poor connection • Short to battery in ETS motor circuit • Faulty ETS motor • Faulty PCM

DTC	Trouble Code Title, Conditions & Possible Causes
DTC: P2104 **2T CCM, MIL: Yes** **2006** **Models:** Sonata **Engines:** 2.4L, 3.3L **Transmissions:** All	**Electronic Throttle Control (ETC) System Malfunction- Forced Idle** This code is set if the system is in forced idle mode. Ignition ON. **Possible Causes:** • Faulty AFS • Faulty AFS plus brake • Faulty AFS plus vehicle speed sensor • Faulty AFS plus brake plus vehicle speed sensor • Faulty PCM
DTC: P2105 **2T CCM, MIL: Yes** **2006** **Models:** Sonata **Engines:** 2.4L, 3.3L **Transmissions:** All	**Electronic Throttle Control (ETC) System Malfunction- Forced Engine Shutdown** This code is set if the system is in forced engine shutdown mode. Ignition ON. **Possible Causes:** • Faulty AFS plus MAPS plus ETS • Faulty PCM
DTC: P2106 **2T CCM, MIL: Yes** **2005-06** **Models:** Santa Fe, Sonata, XG350 **Engines:** 2.4L, 3.3L, 3.5L **Transmissions:** All	**Limp Home Value Mal.** Case 1: ON/OFF malfunction. Low voltage detected. Case 2: Valve stuck open. Engine coolant temperature 167 degrees F. Vehicle speed 0 mph. Threshold value of TPS2 less than 1 volt and engine speed greater than or equal to 2000 rpm. **Possible Causes:** • Poor connection • Faulty limp home valve (LHV) • Open or short in limp home valve (LHV) • Faulty PCM
DTC: P2107 **2T CCM, MIL: Yes** **2005-06** **Models:** Santa Fe, XG350 **Engines:** 3.5L **Transmissions:** All	**ETS-ECM Malfunction (EEPROM R/W)** The DTC will set if the PCM can't read or write on the EEPROM. Check reading and writing. Ignition switch ON. Reading or writing error will occur. **Possible Causes:** • Poor connection • Faulty PCM
DTC: P2108 **2T CCM, MIL: Yes** **2005-06** **Models:** Santa Fe, XG350 **Engines:** 3.5L **Transmissions:** All	**ETS-ECM Malfunction (EEPROM R/W)** The DTC will set if the PCM detects an error to itself. Case 1: PCM/ETS error. Ignition switch ON. Something wrong with the communication PCM to ETS. Case 2: ETS/PCM error. Ignition switch ON. Something wrong with the communication ETS to PCM. Case 3: Computer malfunction. Ignition switch ON. **Possible Causes:** • Faulty PCM
DTC: P2110 **2T CCM, MIL: Yes** **2005-06** **Models:** Santa Fe, Sonata, XG350 **Engines:** 2.4L, 3.5L **Transmissions:** All	**Throttle Actuator Control Module Performance/Throttle Actuator Control System Stuck Closed (IG OFF)** Valve stuck closed (#1). Ignition switch OFF. TPS output as throttle valve is closed less than 0.025 volt. **Possible Causes:** • Poor connector • Faulty throttle valve • Faulty ETS motor • Faulty PCM
DTC: P2111 **2T CCM, MIL: Yes** **2005-06** **Models:** Santa Fe, XG350 **Engines:** 3.5L **Transmissions:** All	**Throttle Actuator Control Module Performance/Throttle Actuator Control System Stuck Open** Valve stuck open. Ignition switch ON. Motor relay ON. **Possible Causes:** • Poor connector • Faulty throttle valve • Faulty ETS motor • Faulty PCM
DTC: P2112 **2T CCM, MIL: Yes** **2005-06** **Models:** Santa Fe, XG350 **Engines:** 3.5L **Transmissions:** All	**Throttle Actuator Control Module Performance/Throttle Actuator Control System Stuck Closed (IG OFF)** Valve close stuck #2. Ignition switch OFF. TPS output as throttle valve is closed less than 0.06 volt. **Possible Causes:** • Poor connector • Faulty throttle valve • Faulty ETS motor • Faulty PCM

DTC	Trouble Code Title, Conditions & Possible Causes
DTC: P2118 **2T CCM, MIL: Yes** **2005-06** **Models:** Santa Fe, Sonata, XG350 **Engines:** 2.4L, 3.3L, 3.5L **Transmissions:** All	**Throttle Actuator Control Motor Current Range/Performance/Throttle Actuator Control Motor Circuit Open** Vb open. Motor relay ON. Voltage to detect circuit open less than or equal to 4.0 volts. **Possible Causes:** • Poor connection • Open in ETS relay circuit • Faulty ETS relay/fuse • Faulty PCM
DTC: P2118 **2T CCM, MIL: Yes** **2005-06** **Models:** Santa Fe, Sonata, XG350 **Engines:** 2.4L, 3.3L, 3.5L **Transmissions:** All	**Throttle Actuator Control Motor Circuit Range/Performance/Throttle Actuator Control Motor Circuit Low** **Motor circuit low. Ignition switch ON.** **Possible Causes:** • Poor connection • Short to ground in ETS motor circuit • Faulty ETS motor • Faulty PCM
DTC: P2118 **2T CCM, MIL: Yes** **2005-06** **Models:** Santa Fe, Sonata, XG350 **Engines:** 2.4L, 3.3L, 3.5L **Transmissions:** All	**Throttle Actuator Control Motor Circuit Range/Performance/Throttle Actuator Control Motor Circuit High** Motor circuit High. Ignition switch ON. **Possible Causes:** • Poor connection • Short to battery in ETS motor circuit • Faulty ETS motor • Faulty PCM
DTC: P2119 **2T CCM, MIL: Yes** **2005-06** **Models:** Santa Fe, Sonata, XG350 **Engines:** 2.4L, 3.3L, 3.5L **Transmissions:** All	**Throttle Actuator Control Module Performance/Throttle Actuator Control System Stuck Closed (IG OFF)** Valve stuck closed (#1). Ignition switch OFF. TPS output as throttle valve is closed less than 0.025 volt. **Possible Causes:** • Poor connector • Faulty throttle valve • Faulty ETS motor • Faulty PCM
DTC: P2122 **2T CCM, MIL: Yes** **2005-06** **Models:** Santa Fe, Sonata, XG350 **Engines:** 2.4L, 3.3L, 3.5L **Transmissions:** All	**Throttle/Pedal Position Sensor/Switch "D" Circuit Low Input** Accelerator position sensor (APS1) low input. ETS/PCM communication is normal. Output voltage of APS1 is less than 0.2 volt. **Possible Causes:** • Poor connector • Faulty APS1 • Open or short in APS1 circuit • Faulty PCM
DTC: P2123 **2T CCM, MIL: Yes** **2005-06** **Models:** Santa Fe, Sonata, XG350 **Engines:** 2.4L, 3.3L, 3.5L **Transmissions:** All	**Throttle/Pedal Position Sensor/Switch "D" Circuit High Input** Accelerator position sensor (APS1) high input. ETS/PCM communication is normal. Output voltage of APS1 is equal to or greater than 4.9 volts. Output voltage of APS2 is less than 4.1 volts. **Possible Causes:** • Poor connector • Faulty APS1 • Open or short in APS1 circuit • Faulty PCM
DTC: P2125 **2T CCM, MIL: Yes** **2005-06** **Models:** Santa Fe, XG350 **Engines:** 3.5L **Transmissions:** All	**Throttle/Pedal Position Sensor/Switch "E" Circuit** Accelerator position sensor (APS2) circuit. ETS/PCM communication is normal. Vaps1 and Vaps2: 0.2 to 4.9 volts. Idle switch ON. **Possible Causes:** • Poor connector • Faulty APS1 • Open or short in APS1 circuit • Faulty PCM
DTC: P2127 **2T CCM, MIL: Yes** **2005-06** **Models:** Santa Fe, XG350 **Engines:** 3.5L **Transmissions:** All	**Throttle/Pedal Position Sensor/Switch "E" Circuit Low Input** Accelerator position sensor (APS2) low input. ETS/PCM communication is normal. Output voltage of APS2 is less than 0.2 volt. **Possible Causes:** • Poor connection • Faulty APS2 • Open or short in APS2 circuit • Faulty PCM

DTC	Trouble Code Title, Conditions & Possible Causes
DTC: P2128 **2T CCM, MIL: Yes** **2005-06** **Models:** Santa Fe, XG350 **Engines:** 3.5L **Transmissions:** All	**Throttle/Pedal Position Sensor/Switch "E" Circuit High Input** Accelerator position sensor (APS2) high input. ETS/PCM communication is normal. Output voltage of APS2 is greater than or equal to 4.9 volts. Output voltage of ASP1 is less than 4.1 volts. **Possible Causes:** • Poor connection • Faulty APS2 • Open or short in APS2 circuit • Faulty PCM
DTC: P2135 **2T CCM, MIL: Yes** **2005-06** **Models:** Santa Fe, Sonata, XG350 **Engines:** 3.3L, 3.5L **Transmissions:** All	**Accelerator Position Sensor 2 Range/Performance** Monitoring abnormal TPS. Ignition switch ON. Vtps1, Vtps2: 0.2 to 4.85 volts. **Possible Causes:** • Poor connection • Faulty TPS • Faulty PCM
DTC: P2138 **2T CCM, MIL: Yes** **2005-06** **Models:** Santa Fe, Sonata, XG350 **Engines:** 2.4L, 3.3L, 3.5L **Transmissions:** All	**Throttle/Pedal Position Sensor/Switch "D/E" Voltage Correlation** Monitoring abnormal APS. Output voltage of APS1: 0.2 to 4.9 volts. Output voltage of APS2: 0.2 to 4.9 volts. Ignition switch ON. **Possible Causes:** • Poor connection • Faulty APS • Faulty PCM
DTC: P2173 **2T CCM, MIL: Yes** **2006** **Models:** Sonata **Engines:** 3.3L **Transmissions:** All	**Electronic Throttle Control (ETC) System Malfunction- High Air Flow Detected** The engine airflow measurements are not based on throttle position. They are compared with throttle position based on estimated air flow. If measured air flow is much higher, the throttle body may not be throttling the engine. Engine running. Throttle actuation mode is not off. MAP sensor is not failed. MAF sensor is not failed. IAT sensor is not failed. **Possible Causes:** • Air leakage between TPS and MAFS • Faulty throttle body • Faulty PCM
DTC: P2176 **2T CCM, MIL: Yes** **2005-06** **Models:** Santa Fe **Engines:** 3.5L **Transmissions:** All	**Throttle Actuator Control Module Performance/Throttle Actuator Control System Stuck Closed (IG OFF)** Valve stuck closed (#1). Ignition switch OFF. TPS output as throttle valve is closed less than 0.025 volt. **Possible Causes:** • Poor connector • Faulty throttle valve • Faulty ETS motor • Faulty PCM
DTC: P2187 **2T CCM, MIL: Yes** **2004-06** **Models:** Accent, Sonata **Engines:** 1.5L, 1.6L, 2.4L, 3.3L **Transmissions:** All	**Fuel Trim Malfunction System Too Lean At Idle (Additive)** Fuel trim limit. Coolant temperature greater than 158 degrees F. Intake air temperature less than 176 degrees F. Throttle angle less than 60 percent. Integrated air mass greater than 10 grams. Closed loop control enabled. No transient control phase. No canister purge phase. Air mass less than 24 kg/h. **Possible Causes:** • Faulty ignition system • EVAP PCSV malfunction • Faulty fuel injectors • Leak in exhaust system • Faulty MAP, TPS, ECTS • Faulty front HO2S • Faulty PCM
DTC: P2188 **2T CCM, MIL: Yes** **2004-06** **Models:** Accent, Sonata **Engines:** 1.5L, 1.6L, 2.4L, 3.3L **Transmissions:** All	**System Too Rich At Idle (Bank 1)** Fuel trim limit. Coolant temperature greater than 158 degrees F. Intake air temperature less than 176 degrees F. Throttle angle less than 60 percent. Integrated air mass greater than 10 grams. Closed loop control enabled. No transient control phase. No canister purge phase. Engine speed less than 920 rpm. **Possible Causes:** • Faulty ignition system • EVAP PCSV malfunction • Faulty fuel injectors • Leak in exhaust system • Faulty MAP, TPS, ECTS • Faulty front HO2S • Faulty PCM

DTC	Trouble Code Title, Conditions & Possible Causes
DTC: P2189 **2T CCM, MIL: Yes** **2006** **Models:** Sonata **Engines:** 3.3L **Transmissions:** All	**System Too Lean At Idle (additive) Bank 2** Fuel trim idle condition (option limits exceeded). **Possible Causes:** • Sensors related to fuel trim • Intake system • Fuel pressure • Faulty PCM
DTC: P2190 **2T CCM, MIL: Yes** **2006** **Models:** Sonata **Engines:** 3.3L **Transmissions:** All	**System Too Rich At Idle (additive) Bank 2** Fuel trim idle condition (option limits exceeded). **Possible Causes:** • Sensors related to fuel trim • Intake system • Fuel pressure • Faulty PCM
DTC: P2191 **2T CCM, MIL: Yes** **2004-06** **Models:** Accent, Sonata **Engines:** 1.5L, 1.6L, 2.4L, 3.3L **Transmissions:** All	**System Too Lean At Higher Load (Multiple) (Bank 1)** Fuel trim limit. Coolant temperature greater than 158 degrees F. Intake air temperature less than 176 degrees F. Throttle angle less than 60 percent. Integrated air mass greater than 10 grams. Closed loop control enabled. No transient control phase. No canister purge phase. Air mass1 40 to 80 kg/h. Air mass2 greater than 100 kg/h. **Possible Causes:** • Faulty ignition system • EVAP PCSV malfunction • Faulty fuel injectors • Leak in exhaust system • Faulty MAP, TPS, ECTS • Faulty front HO2S • Faulty PCM
DTC: P2192 **2T CCM, MIL: Yes** **2004-06** **Models:** Accent, Sonata **Engines:** 1.5L, 1.6L, 3.3L **Transmissions:** All	**System Too Rich At Higher Load (Bank 1)** Fuel trim limit. Coolant temperature greater than 158 degrees F. Intake air temperature less than 176 degrees F. Throttle angle less than 60 percent. Integrated air mass greater than 10 grams. Closed loop control enabled. No transient control phase. No canister purge phase. Engine load1 30 to 55 percent. Engine load2 greater than 70 percent. **Possible Causes:** • Faulty ignition system • EVAP PCSV malfunction • Faulty fuel injectors • Leak in exhaust system • Faulty MAP, TPS, ECTS • Faulty front HO2S • Faulty PCM
DTC: P2195 **2T CCM, MIL: Yes** **2004-06** **Models:** Elantra, Sonata **Engines:** 2.0L, 3.3L **Transmissions:** All	**HO2S Signal Stuck Lean (Bank 1 Sensor 1)** Sensor characteristic line shifted to lean. No relevant failure. No misfire detected. Fuel trim control active. **Possible Causes:** • Contact resistance in connectors • Faulty HO2S
DTC: P2196 **2T CCM, MIL: Yes** **2004-06** **Models:** Elantra, Sonata **Engines:** 2.0L, 3.3L **Transmissions:** All	**HO2S Signal Stuck Rich (Bank 1 Sensor 1)** Sensor characteristic line shifted to lean. No relevant failure. No misfire detected. Fuel trim control active. **Possible Causes:** • Contact resistance in connectors • Faulty HO2S
DTC: P2197 **2T CCM, MIL: Yes** **2006** **Models:** Sonata **Engines:** 3.3L **Transmissions:** All	**HO2S Signal Stuck Lean (Bank 2 Sensor 1)** Determines if O2 sensor indicates lean exhaust while in power enrichment. Sensor not in cooled status flag. Not in transient conditions status flag. Device control not active. Engine running. Minimum air flow present is equal or greater than 2 g/s. Engine coolant warm (140 degrees F. Above conditions met for at least 1.5L seconds. **Possible Causes:** • Poor connection • Faulty HO2S • Faulty PCM

DTC	Trouble Code Title, Conditions & Possible Causes
DTC: P2198 **2T CCM, MIL: Yes** **2006** **Models:** Sonata **Engines:** 3.3L **Transmissions:** All	**HO2S Signal Stuck Rich (Bank 2 Sensor 1)** Determines if O2 sensor indicates rich exhaust while in decal fuel cut off (DFCO). Sensor not in cooled status flag. Not in transient conditions status flag. Device control not active. Engine running. Minimum air flow present is equal or greater than 2 g/s. Ignition voltage equal to or greater than 10 volts. Fuel reduction not active. Engine running long enough (more than 60 seconds). Engine coolant warm (140 degrees F. Above conditions met for at least 1.5L seconds. **Possible Causes:** • Poor connection • Faulty HO2S • Faulty PCM
DTC: P2226 **2T CCM, MIL: Yes** **2006** **Models:** Accent **Engines:** 1.5L **Transmissions:** All	**Barometric Pressure Circuit** Rationality check. **Possible Causes:** • Clog at sensing hole • Faulty ECM
DTC: P2227 **2T CCM, MIL: Yes** **2006** **Models:** Accent **Engines:** 1.5L **Transmissions:** All	**Barometric Pressure Circuit Range/Performance** Rationality check. **Possible Causes:** • Clog at sensing hole • Faulty ECM
DTC: P2228 **2T CCM, MIL: Yes** **2006** **Models:** Accent **Engines:** 1.5L **Transmissions:** All	**Barometric Pressure Circuit Low Input** Signal check low. **Possible Causes:** • Faulty ECM
DTC: P2229 **2T CCM, MIL: Yes** **2006** **Models:** Accent **Engines:** 1.5L **Transmissions:** All	**Barometric Pressure Circuit High Input** Signal check, high. **Possible Causes:** • Faulty ECM
DTC: P2231 **2T CCM, MIL: Yes** **2004-06** **Models:** Elantra **Engines:** 2.0L **Transmissions:** All	**HO2S Signal Circuit Shorted To Heater Circuit (Bank 1 Sensor 1)** Front HO2S signal monitoring. Exhaust temperature greater than 752 degrees F. No relevant failure. Amplitude of forced lambda simulation less than 0.05. Period time of forced lambda simulation less than 2.55 seconds. Current engine speed less than 500 rpm. Current mass air flow less than 400 mg/rev. **Possible Causes:** • Contact resistance in connectors • Interference in HO2S
DTC: P2232 **2T CCM, MIL: Yes** **2006** **Models:** Accent **Engines:** 1.5L **Transmissions:** All	**HO2S Signal Circuit Shorted To Heater Circuit (Bank 1 Sensor 2)** Rationality check. **Possible Causes:** • Poor connection • Short to power in signal circuit • B1S2 • Faulty ECM
DTC: P2237 **2T CCM, MIL: Yes** **2004-06** **Models:** Elantra **Engines:** 2.0L **Transmissions:** All	**HO2S Pumping Current Circuit/Open Bank 1, Sensor 1** Open circuit of front HO2S circuit. No relevant failure. **Possible Causes:** • Contact resistance in connectors • Open or short to ground in HO2S circuit • Front HO2S sensor
DTC: P2243 **2T CCM, MIL: Yes** **2004-06** **Models:** Elantra **Engines:** 2.0L **Transmissions:** All	**HO2S Reference Voltage Circuit/Open Bank 1, Sensor 1** Open circuit of front HO2S circuit. No relevant failure. **Possible Causes:** • Contact resistance in connectors • Open or short to ground in HO2S circuit • Front HO2S sensor

DTC	Trouble Code Title, Conditions & Possible Causes
DTC: P2251 **2T CCM, MIL: Yes** **2004-06** **Models:** Elantra **Engines:** 2.0L **Transmissions:** All	**HO2S Reference Ground Circuit/Open Bank 1, Sensor 1** Open circuit of front HO2S circuit. No pump current malfunction. **Possible Causes:** • Contact resistance in connectors • Open or short to ground in HO2S circuit • Front HO2S sensor
DTC: P2252 **2T CCM, MIL: Yes** **2006** **Models:** Santa Fe **Engines:** 3.5L **Transmissions:** All	**HO2S Reference Ground Circuit Low** Electrical check. Time after start greater than two seconds. Voltage is less than 0.4 volt. **Possible Causes:** • Poor connection or damaged harness • Open or short to ground in reference ground circuit • Faulty HO2S sensor
DTC: P2253 **2T CCM, MIL: Yes** **2006** **Models:** Santa Fe **Engines:** 3.5L **Transmissions:** All	**HO2S Reference Ground Circuit High** Electrical check. Time after start greater than two seconds. Voltage is greater than 3.7 volts. **Possible Causes:** • Poor connection or damaged harness • Short to battery in reference ground circuit • Faulty HO2S sensor
DTC: P2270 **2T CCM, MIL: Yes** **2004-06** **Models:** Elantra, Sonata **Engines:** 2.0L, 3.3L **Transmissions:** All	**HO2S Signal Stuck Rich (Bank 1 Sensor 2)** Plausibility check during shift of lambda set point to rich from lean. No fuel cut off. No full load phase. No fuel trim error detected. Delay time to start diagnosis: 13 to 30 seconds. No relevant failure. **Possible Causes:** • Three way catalytic converter (TWC) • Air leakage in exhaust system • Faulty rear HO2S sensor
DTC: P2271 **2T CCM, MIL: Yes** **2004-06** **Models:** Elantra, Sonata **Engines:** 2.0L, 3.3L **Transmissions:** All	**O2 Signal Stuck Rich (Bank 1/2 Sensor 1)** Plausibility check during shift of lambda set point to rich from lean. No fuel cut off. No full load phase. No fuel trim error detected. Delay time to start diagnosis: 13 to 30 seconds. No relevant failure. **Possible Causes:** • Three way catalytic converter (TWC) • Air leakage in exhaust system • Faulty rear HO2S sensor
DTC: P2272 **2T CCM, MIL: Yes** **2006** **Models:** Sonata **Engines:** 3.3L **Transmissions:** All	**HO2S Signal Stuck Lean (Bank 2 Sensor 2)** Determines if O2 sensor indicates lean exhaust while in power enrichment mode. Sensor not in cooled status flag. Not in transient conditions status flag. Device control not active. Engine running. Minimum air flow present is equal or greater than 2 g/s. Ignition voltage equal to or greater than 10 volts. Fuel reduction not active. Engine running long enough (more than 60 seconds). Engine coolant warm (140 degrees F. Above conditions met for at least 2.5 seconds. **Possible Causes:** • Poor connection • Faulty HO2S • Faulty PCM
DTC: P2273 **2T CCM, MIL: Yes** **2006** **Models:** Sonata **Engines:** 3.3L **Transmissions:** All	**HO2S Signal Stuck Rich (Bank 2 Sensor 2)** Determines if O2 sensor indicates rich exhaust while in decal fuel cut off (DFCO). Sensor not in cooled status flag. Not in transient conditions status flag. Device control not active. Engine running. Minimum air flow present is equal or greater than 2 g/s. Ignition voltage equal to or greater than 10 volts. Fuel reduction not active. Engine running long enough (more than 60 seconds). Engine coolant warm (140 degrees F. Above conditions met for at least 2.0L seconds. **Possible Causes:** • Poor connection • Faulty HO2S • Faulty PCM
DTC: P2400 **2T CCM, MIL: Yes** **2004-05** **Models:** Accent **Engines:** 1.5L, 1.6L **Transmissions:** All	**Evaporative Emission System Leak Detection Pump Control Circuit/Open** Pump motor circuit continuity check. **Possible Causes:** • Open in control circuit • Faulty DMTL • Faulty PCM

DTC	Trouble Code Title, Conditions & Possible Causes
DTC: P2401 **2T CCM, MIL: Yes** **2004-05** **Models:** Accent **Engines:** 1.5L, 1.6L **Transmissions:** All	**Evaporative Emission System Leak Detection Pump Control Circuit Low** Pump motor circuit continuity check. Shorted to ground. **Possible Causes:** • Open in control circuit • Short to ground in control circuit • Faulty DMTL • Faulty PCM
DTC: P2402 **2T CCM, MIL: Yes** **2004-05** **Models:** Accent **Engines:** 1.5L, 1.6L **Transmissions:** All	**Evaporative Emission System Leak Detection Pump Control Circuit High** Pump motor circuit continuity check. Shorted to battery voltage. **Possible Causes:** • Short to battery in control circuit • Faulty DMTL • Faulty PCM
DTC: P2404 **2T CCM, MIL: Yes** **2004-05** **Models:** Accent **Engines:** 1.5L, 1.6L **Transmissions:** All	**Evaporative Emission System Leak Detection Pump Sensor Circuit Range/Performance** DM-TL: pump motor current. Current at valve check is greater than reference current-2 mA. **Possible Causes:** • Leakage on EVAP system • Faulty DMTL • Faulty PCM
DTC: P2405 **2T CCM, MIL: Yes** **2004-05** **Models:** Accent **Engines:** 1.5L, 1.6L **Transmissions:** All	**Evaporative Emission System Leak Detection Pump Sensor Circuit Low** DM-TL: pump motor current. Current at reference mode phase is less than 12 Ma. **Possible Causes:** • Leakage on EVAP system • Faulty DMTL • Faulty PCM
DTC: P2406 **2T CCM, MIL: Yes** **2004-05** **Models:** Accent **Engines:** 1.5L, 1.6L **Transmissions:** All	**Evaporative Emission System Leak Detection Pump Sensor Circuit High** DM-TL: pump motor current. Current at reference mode phase is greater than 40 Ma. **Possible Causes:** • Leakage on EVAP system • Faulty DMTL • Faulty PCM
DTC: P2407 **2T CCM, MIL: Yes** **2004-05** **Models:** Accent **Engines:** 1.5L, 1.6L **Transmissions:** All	**Evaporative Emission System Leak Detection Pump Sensor Circuit Intermittent/Erratic** DM-TL: pump motor current. **Possible Causes:** • Leakage on EVAP system • Faulty DMTL • Faulty PCM
DTC: P2414 **2T CCM, MIL: Yes** **2004-06** **Models:** Elantra **Engines:** 2.0L **Transmissions:** All	**HO2S Exhaust Sample Error Bank 1 Sensor 1** Sensor not mounted. Plausibility check in part load or full load conditions. Sensor tip temperature 1202 degrees F. Part load or full load. No relevant failure. **Possible Causes:** • Incorrect installation of HO2S sensor • Contact resistance in connectors
DTC: P2418 **2T CCM, MIL: Yes** **2004-05** **Models:** Accent **Engines:** 1.5L, 1.6L **Transmissions:** All	**Evaporative Emission System Switching Valve Control Circuit Open** Circuit continuity check (change over valve). **Possible Causes:** • Open in control circuit • Faulty DMTL • Faulty PCM
DTC: P2419 **2T CCM, MIL: Yes** **2004-05** **Models:** Accent **Engines:** 1.5L, 1.6L **Transmissions:** All	**Evaporative Emission System Switching Valve Control Circuit Low** Circuit continuity check (change over valve). **Possible Causes:** • Open in control circuit • Faulty DMTL • Faulty PCM
DTC: P2420 **2T CCM, MIL: Yes** **2004-05** **Models:** Accent **Engines:** 1.5L, 1.6L **Transmissions:** All	**Evaporative Emission System Switching Valve Control Circuit High** Circuit continuity check (change over valve). Shorted to battery voltage. **Possible Causes:** • Short to battery in control unit • Faulty DMTL • Faulty PCM

DTC	Trouble Code Title, Conditions & Possible Causes
DTC: P2422 **2T CCM, MIL: Yes** **2005-06** **Models:** Santa Fe, Sonata, XG350 **Engines:** 2.4L, 3.3L, 3.5L **Transmissions:** All	**Evaporative Emission System Canister Clogging** Clogging is monitored. Purge duty greater than or equal to 60 percent. Canister close valve OFF. Canister close valve greater than 1600 rpm. Load value 20 to 85 percent. **Possible Causes:** • Blockage of canister filter • Faulty CCV • Faulty canister • Faulty fuel cut valve
DTC: P2610 **2T CCM, MIL: Yes** **2006** **Models:** Sonata **Engines:** 3.3L **Transmissions:** All	**ECM/PCM Internal Engine Off Timer Performance** The LPC SPI diagnostic allows the low power counter to count down and simultaneously enables a test timer to run for a calibrated length of time and then compares the lapsed time recorded by the counter to make a pass/fail determination. Engine running. Enough time (10 seconds). Battery voltage 8 volts. No memory failure. **Possible Causes:** • Faulty PCM
DTC: P2A00 **2T CCM, MIL: Yes** **2006** **Models:** Sonata **Engines:** 3.3L **Transmissions:** All	**O2 Sensor Not ready (Bank 1 Sensor 1)** Detects loss of O2 ready status, which would lead to open loop fueling operation, a default mode. Engine running. Ignition ON. DFCO not present too long (less than 15 seconds). No disabling faults present. All of the above for at least 20 seconds. **Possible Causes:** • Poor connection • Faulty HO2S • Faulty PCM
DTC: P2A03 **2T CCM, MIL: Yes** **2006** **Models:** Sonata **Engines:** 3.3L **Transmissions:** All	**O2 Sensor Not ready (Bank 1 Sensor 2)** Detects loss of O2 ready status, which would lead to open loop fueling operation, a default mode. Engine running. Ignition ON. DFCO not present too long (less than 15 seconds). No disabling faults present. All of the above for at least 20 seconds. **Possible Causes:** • Poor connection • Faulty HO2S • Faulty PCM
DTC: P2626 **2T CCM, MIL: Yes** **2004-06** **Models:** Elantra **Engines:** 2.0L **Transmissions:** All	**HO2S Pumping Current Trim Circuit Open Bank 1 Sensor 1** Check open circuit of front HO2S sensor. **Possible Causes:** • Contact resistance in connectors • Open or short to ground in HO2S circuit • Faulty canister • Faulty front HO2S sensor

HYUNDAI
COMPONENT TESTING

6

TABLE OF CONTENTS

Component Testing 6-2
- Accent 6-2
- Elantra 6-8
- Santa Fe 6-14
- Sonata 6-21
- Tiburon 6-33
- Tucson 6-41
- XG300 6-46
- XG350 6-49

COMPONENT TESTING

ACCENT

See Figures 1 through 5.

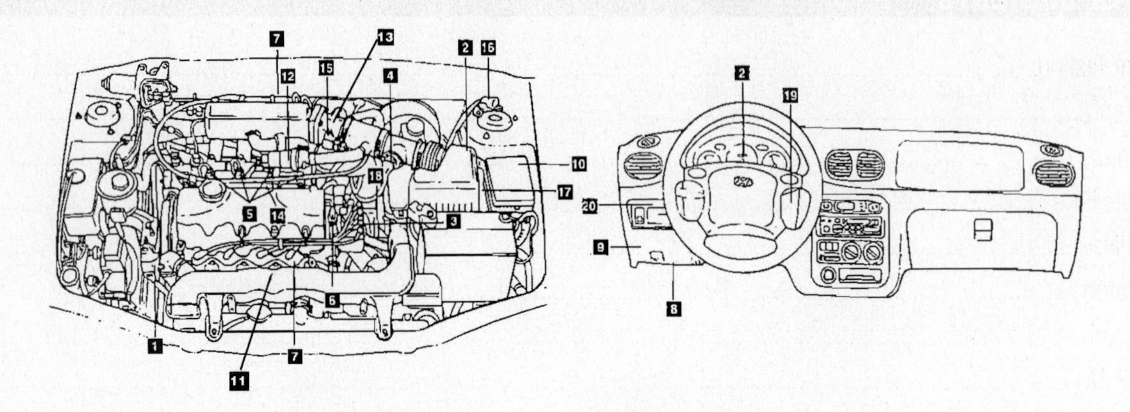

1. Engine Coolant Temperature (ECT) sensor
2. Vehicle Speed Sensor (VSS)
3. Ignition coil
4. Mass Air Flow (MAF) sensor
5. Injector
6. Camshaft Position (CMP) sensor
7. Crankshaft Position (CKP) sensor
8. Engine Control Module (ECM)
9. MFI control relay
10. Air coditioning relay
11. Heated Oxygen sensor (HO2S)
12. Idle Speed Control (ISC) Actuator
13. Intake Air Temperature (IAT) sensor
14. Knock sensor
15. Throttle Position Sensor (TPS)
16. Acceleration sensor
17. Evaporative emission canister purge solenoid valve
18. Transaxle Range (TR) switch
19. Ignition switch
20. Data link connector

Fig. 1 Underhood sensor locations—Accent 1996–1999 1.5L SOHC engine

29130_HYUN_G0001

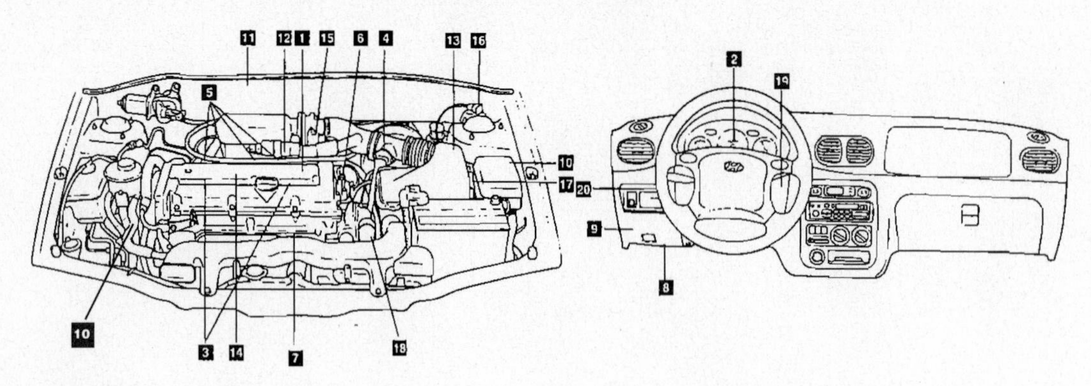

1. Engine Coolant Temperature (ECT) sensor
2. Vehicle Speed Sensor (VSS)
3. Ignition coil
4. Mass Air Flow (MAF) sensor
5. Injector
6. Camshaft Position (CMP) sensor
7. Crankshaft Position (CKP) sensor
8. Engine Control Module (ECM)
9. MFI control relay
10. Air coditioning relay
11. Heated Oxygen sensor (HO2S)
12. Idle Speed Control (ISC) Actuator
13. Intake Air Temperature (IAT) sensor
14. Knock sensor
15. Throttle Position Sensor (TPS)
16. Acceleration sensor
17. Evaporative emission canister purge solenoid valve
18. Transale Range (TR) switch
19. Ignition switch
20. Data link connector

Fig. 2 Underhood sensor locations—Accent 1996–1997 1.5L DOHC engine

29157_TOYO_G0004

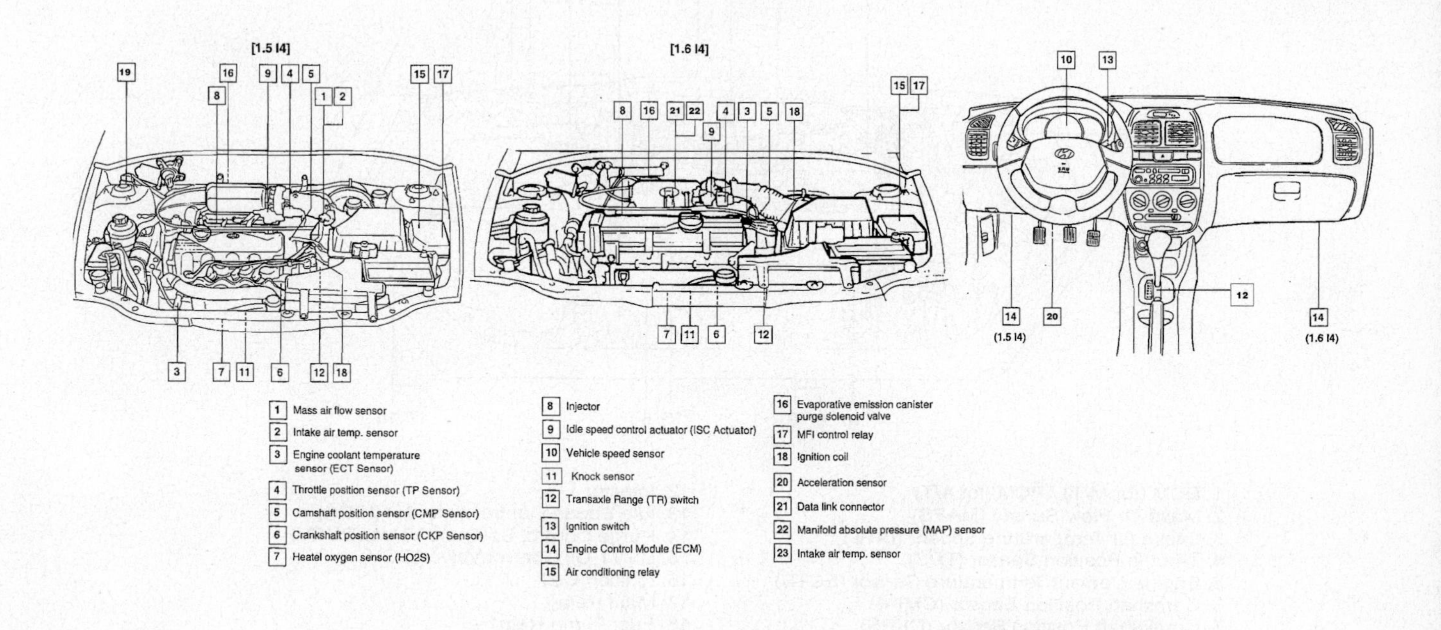

Fig. 3 Underhood sensor locations—Accent 2000–2001 1.5L and 1.6L engines

29130_HYUN_G0003

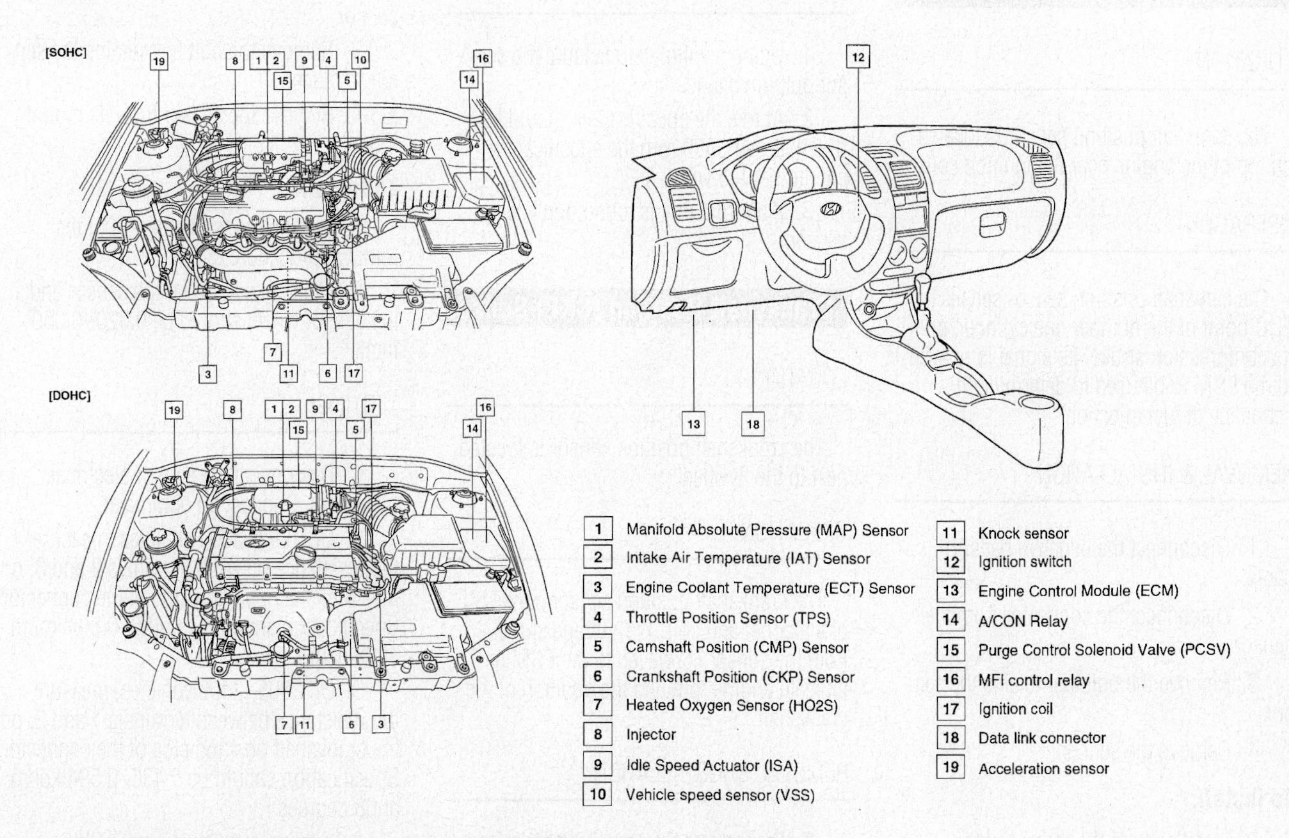

Fig. 4 Underhood sensor locations—Accent 2002–2005 1.5L and 1.6L engines

29130_HYUN_G0004

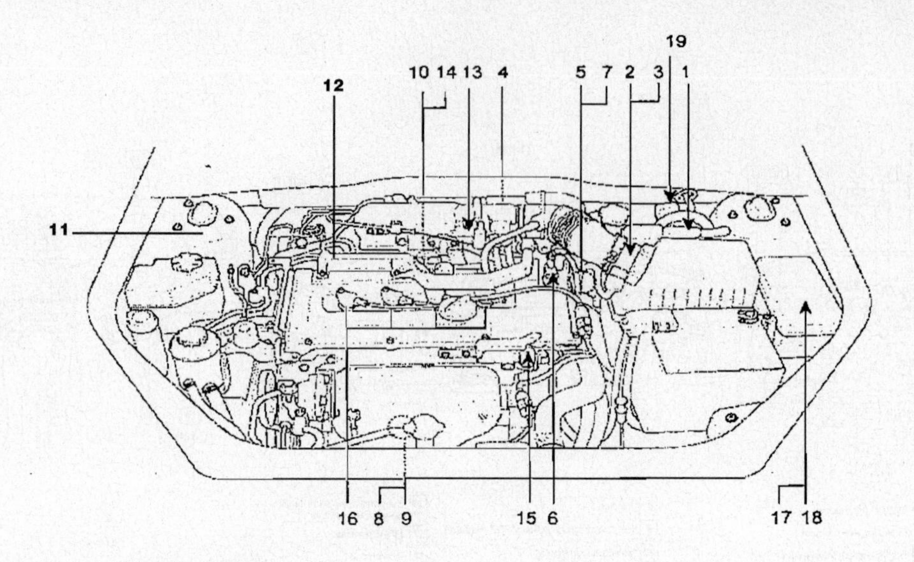

1. ECM (for M/T) / PCM (for A/T)
2. Mass Air Flow Sensor (MAFS)
3. Intake Air Temperature Sensor (IATS)
4. Throttle Position Sensor (TPS)
5. Engine Coolant Temperature Sensor (ECTS)
6. Camshaft Position Sensor (CMPS)
7. Crankshaft Position Sensor (CKPS)
8. Heated Oxygen Sensor (HO2S) [Bank 1/Sensor 1]
9. Heated Oxygen Sensor (HO2S) [Bank 1/Sensor 2]
10. Knock Sensor (KS)
11. Wheel Speed Sensor (WSS)
12. Injector
13. Idle Speed Control Actuator (ISCA)
14. Purge Control Solenoid Valve (PCSV)
15. CVVT Oil Control Valve (OCV)
16. Ignition Coil
17. Main Relay
18. Fuel Pump Relay
19. Multi Purpose Check Connector (20 pin)

Fig. 5 Underhood sensor locations—Accent 2006 1.5L engine

29130_HYUN_G0005

Camshaft Position (CMP) Sensor

LOCATION

The camshaft position sensor is located on top of the engine near the ignition coils.

OPERATION

The camshaft position sensor senses the TDC point of the number one cylinder, on its compression stroke. Its signal is relayed to the ECM to be used to determine the sequence of fuel injection.

REMOVAL & INSTALLATION

1. Disconnect the negative battery cable.
2. Disconnect the connector from the sensor.
3. Remove the bolt that retains the sensor.
4. Remove the sensor.

To install:

5. Installation is the reverse of the removal procedure.

TESTING

1. Using a voltmeter measure the sensor output voltage.
2. At idle the specification should be 0–5 volts. At 3000 rpm the specification should be 0–5 volts.
3. If abnormality is found, replace the sensor.

Crankshaft Position (CKP) Sensor

LOCATION

The crankshaft position sensor is located next to the flywheel.

OPERATION

The crankshaft position sensor consists of a magnet and coil. The voltage signal from the sensor is relayed to the ECM to indicate engine rpm and the position of the crankshaft.

REMOVAL & INSTALLATION

1. Disconnect the negative battery cable.
2. Disconnect the connector from the sensor.
3. Remove the bolt that retains the sensor in place.
4. Remove the sensor from its mounting.

To install:

5. Installation is the reverse of the removal procedure.
6. Clearance between the sensor and the sensor wheel should be 0.020–0.059 inch.

TESTING

1. Disconnect the sensor electrical connector. Connect an ohmmeter.
2. On 1997–2004 vehicles, measure the resistance between terminals 2 and 3, on the crankshaft position side of the connector. Specification should be 0.486–0.594 kohm at 68 degrees F.
3. On 2005–2006 vehicles, measure the resistance between terminals 1 and 2, on the crankshaft position side of the connector. Specification should be 0.486–0.594 kohm at 68 degrees F.
4. If measured value is not within specification, replace the sensor.

Electronic Control Module (ECM)

OPERATION

The ECM controls the vehicle engine operating system.

REMOVAL & INSTALLATION

1. Disconnect the negative battery cable.
2. Remove the lower inner trim..
3. As required, detach the floor mat. As required, remove the protective cover.
4. Remove the ECM bracket retaining nuts. Remove the clip from the bracket.
5. Disconnect the connectors.
6. Remove the ECM from the vehicle.

To install:

7. Installation is the reverse of the removal procedure.

➡ **When replacing the ECM, be careful not to use the wrong part number, as damage to the injection system could occur.**

Engine Coolant Temperature (ECT) Sensor

LOCATION

The ECT is installed in the engine coolant passage of the cylinder head.

OPERATION

This component detects the temperature of the engine coolant and relays the information to the electronic control assembly. This component employs a thermistor which is sensitive to temperature changes. The electric resistance of the thermistor decreases in response to temperature rise. The ECM judges coolant temperature by the sensor output voltage and provides optimum fuel enrichment when the engine is cold.

REMOVAL & INSTALLATION

1. Disconnect the negative battery cable.
2. Disconnect the connector from the sensor.
3. Drain the cooling system, as required.
4. Remove the sensor from its mounting.

To install:

5. Installation is the reverse of the removal procedure.

TESTING

1. To measure the power supply voltage, disconnect the connector. Turn the ignition switch to the ON position.
2. On 1996–1999 vehicles measure between 2 of the harness side and ground. The voltage should be 4.6–4.8 volts.
3. On 2000–2006 vehicles measure between 1 of the harness side and ground. The voltage should be 4.8–5.2 volts.
4. Using an ohmmeter, check for continuity of the ground circuit.
5. On 1996–1999 vehicles measure between 1 of the harness side and ground.
6. On 2000–2006 vehicles equipped with the SOHC engine, measure between terminal 3 of the harness side and ground.
7. On 2000–2006 vehicles equipped with the DOHC engine, measure between terminal 2 of the harness side and ground.

Heated Oxygen (HO2S) Sensor

LOCATION

The sensors are located in the exhaust system. On some vehicles one sensor is located up at the exhaust manifold and the other sensor is located down at the catalytic converter. On other vehicles both sensors are located down at the catalytic converter.

OPERATION

The sensor senses the oxygen concentration in the exhaust gas then converts it into a voltage and sends it on to the ECM. The sensor gives an output of about 800mV, when the air fuel ratio is richer than the theoretical ratio and output of about 100mv when the ratio is leaner. The ECM controls fuel injection based on the signal so that the air fuel ratio is maintained at the theoretical ration.

REMOVAL & INSTALLATION

1. Disconnect the electrical connector from the sensor.
2. Remove the oxygen sensor.

To install:

3. Installation is the reverse of the removal procedure.

➡ **Apply anti-seize compound to the threaded portion of the sensor, prior to installation. Never apply anti-seize compound to the protector of the sensor.**

4. Tighten the sensor to 37–44 ft. lbs.

TESTING

Perform a visual inspection of the sensor as follows:

1. If the sensor tip has a black/sooty deposit, this may indicate a rich fuel mixture.
2. If the sensor tip has a white, gritty deposit, this may indicate an internal coolant leak.
3. If the sensor tip has a brown deposit, this could indicate excessive oil consumption.
4. Warm the engine until the coolant temperature reaches operating temperature.
5. Accelerate the engine to 4000 rpm. When decelerating suddenly from 4000 rpm, the voltmeter reading should be 200mV or lower.
6. When the engine is suddenly raced, the voltmeter reading should be 600–1000mV.
7. If measured value is not within specification, replace the sensor.

Idle Speed Control Actuator (ISCA)

OPERATION

The idle speed control actuator sensor is a double coil type driven by separate driver stages in the ECM. Depending on the pulse duty factor, the equilibrium of the magnetic forces of the two coils will result in different angles of the motor. A bypass hose line is positioned, parallel to the throttle valve where the idle speed actuator is inserted.

REMOVAL & INSTALLATION

1. Disconnect the negative battery cable.
2. Disconnect the connector from the sensor.
3. Remove the sensor retaining screws.
4. Remove the sensor from its mounting.

To install:

5. Installation is the reverse of the removal procedure.

TESTING

1. Disconnect the connector at the sensor. Connect an ohmmeter.

2. On 1997–2003 vehicles, measure the resistance between terminal 1 and 2, specification should be 10.5–14 ohms. Measure the resistance between terminal 2 and 3, specification should be 10–12.5 ohms at 68 degrees F.

3. On 2004–2006 vehicles, measure the resistance between terminal 1 and 2, specification should be 16.6–18.6 ohms. Measure the resistance between terminal 2 and 3, specification should be 14.5–16.5 ohms at 68 degrees F.

4. If the measured value is not within specification, replace the sensor.

Intake Air Temperature (IAT) Sensor

LOCATION

The IAT sensor is located against the firewall to the left side of the engine. It is combined with the Manifold Absolute Pressure (MAP) sensor.

OPERATION

This sensor is a resistor based sensor which detects the intake air temperature. According to the intake air temperature reading the ECM will control the necessary amount of fuel injection.

REMOVAL & INSTALLATION

1. Disconnect the negative battery cable.

2. Disconnect the connector from the sensor.

3. Remove the sensor retaining screws.

4. Remove the sensor from its mounting.

To install:

5. Installation is the reverse of the removal procedure.

TESTING

1. Turn the ignition switch to the ON position.

2. Using a multimeter, measure the sensor resistance between terminals 3 and 4.

3. At 32 degrees F the resistance should be 4.5–7.5 kohm. At 68 degrees F the resistance should be 2.0–3.0 kohm. At 104 degrees F the resistance should be 0.7–1.6 kohm. At 166 degrees F the resistance should be 0.2–0.4 kohm.

4. If the measured value is not within specification, replace the sensor.

Knock Sensor (KS)

LOCATION

The knock sensor is located in the side of the cylinder block.

OPERATION

The knock sensor is used to detect engine vibrations caused by preignition or detonation and provides information to the ECM, which then retards the timing to eliminate detonation.

REMOVAL & INSTALLATION

1. Disconnect the negative battery cable.

2. Disconnect the sensor connector.

3. Remove the sensor from its mounting.

To install:

4. Installation is the reverse of the removal procedure.

5. Tighten the sensor to 12–18 ft. lbs.

TESTING

1. Disconnect the sensor electrical connector. Connect an ohmmeter.

2. On 1997–2000 vehicles measure the resistance between terminals 2 and 3. Specification should be 5mohm at 68 degrees F.

3. On 2001–2006 vehicles measure the resistance between terminals 1 and 2. Specification should be 5mohm at 68 degrees F.

4. If not within specification, replace the sensor.

Manifold Absolute Pressure (MAP) Sensor

LOCATION

The MAP sensor is located against the firewall to the left side of the engine.

OPERATION

This sensor is a pressure sensitive variable resistor. It measures the changes in the intake manifold pressure which result from engine load and speed changes, and converts to a voltage output. This sensor is used to measure the barometric pressure at start up, and under certain conditions, allows the ECM to automatically adjust for different altitudes. The ECM supplies 5 volts to the sensor and monitors the voltage on a signal line. The sensor provides a path to ground through its variable resistor. The sensor input affects fuel delivery and ignition timing controls in the ECM.

REMOVAL & INSTALLATION

1. Disconnect the negative battery cable.

2. Disconnect the connector from the sensor.

3. Remove the sensor retaining screws.

4. Remove the sensor from its mounting.

To install:

5. Installation is the reverse of the removal procedure.

TESTING

1. Turn the ignition switch ON.

2. Measure the voltage between terminal 1 (sensor output) and terminal 4 (sensor ground).

3. Specification should be 4–5 volts.

4. If not within specification, replace the sensor.

Mass Air Flow (MAF) Sensor

LOCATION

The MAF is mounted in the intake air hose of the air cleaner assembly.

OPERATION

The MAF is measured by detection of heat transfer from a hot film probe because the change of the mass air flow rate causes change in the amount of heat being transferred from the hot film probe surface to the air flow. The air flow sensor generates a pulse so it repeatedly opens and closes between the voltage (5V) supplied from the ECM. This results in a change of the temperature of the hot film probe and in the change of resistance.

REMOVAL & INSTALLATION

1. Disconnect the negative battery cable.
2. Disconnect the connector from the sensor.
3. Remove the sensor retaining screws.
4. Remove the sensor from its mounting.

To install:

5. Installation is the reverse of the removal procedure.

TESTING

1. To measure the power supply voltage, disconnect the connector. Turn the ignition switch to the ON position. Specification is battery voltage. Measure between terminal 2 of the harness side and ground.
2. To check for an open circuit, or a short circuit to ground between the ECM and the sensor, disconnect the ECM connector. Measure between terminal 1 of the harness side and ground.
3. To check for an open circuit, or a short circuit to ground between the ECM and the sensor, disconnect the MAF sensor connector. Measure between terminal 1 of the harness side and ground.
4. To check for continuity of the ground circuit, disconnect the connector. Measure between terminal 3 of the harness side and ground.
5. To check for continuity of the ground circuit, disconnect the connector. Measure between terminal 4 of the harness side and ground.

Throttle Position Sensor (TPS)

OPERATION

The throttle position sensor is a rotating type variable resistor that rotates with the throttle body's throttle shaft to sense the throttle valve angle. As the throttle shaft rotates, the throttle angle of the sensor changes and the ECM detects the throttle valve opening based on the TPS output voltage.

REMOVAL & INSTALLATION

1. Disconnect the negative battery cable.
2. Disconnect the sensor connector.
3. Remove the sensor retaining screws. Remove the sensor from its mounting.

To install:

4. Installation is the reverse of the removal procedure.

TESTING

1. Disconnect the connector from the sensor.
2. Measure the resistance between terminals 2 (sensor ground) and 3 (sensor power).
3. Specification should be 0.7–3.0 kohm.
4. Connect an analog ohmmeter between terminals 2 (sensor ground) and 1 (sensor output).
5. Operate the throttle valve slowly from the idle position to the full open position, and check that the resistance changes smoothly in proportion with the throttle valve opening angle.
6. If the resistance is out of specification, or fails to change smoothly, replace the sensor.

Vehicle Speed Sensor (VSS)

LOCATION

On 1997–2003 vehicles, the vehicle speed sensor is built into the speedometer. On 2004–2006 vehicles, the vehicle speed sensor is attached to the output shaft of the transaxle.

OPERATION

On 1997–2003 vehicles, the sensor is a reed switch type. On 2004–2006 vehicles, the sensor is a hall effect sensor. The sensor converts the transaxle gear revolutions into pulse signals, which are sent to the ECM.

REMOVAL & INSTALLATION

2004–2006

1. Raise and support the vehicle safely.
2. Place a drip pan below the speed sensor to catch any spilled fluid.
3. Disconnect the connector.
4. Remove the sensor from its mounting.

To install:

5. Installation is the reverse of the removal procedure.
6. Replace any lost fluid.

TESTING

1997–2003

1. To check the sensor output circuit for continuity, disconnect the ECM connector. Move the vehicle or turn the speedometer cable.
2. To measure the power supply voltage, disconnect the connector. Turn the ignition switch to the ON position.
3. The voltage should be 4.5–4.9 volts, between the connector terminals.
4. Using an ohmmeter, check for continuity of the ground circuit.

2004–2006

1. To measure the power supply voltage, disconnect the connector. Turn the ignition switch to the ON position. Specification is battery voltage (between ground and terminal 2 of the harness connector).
2. To check for an open circuit, or a short circuit to ground between the ECM and the sensor, disconnect the ECM connector. Between ground and terminal 3 of the harness connector.
3. To check for an open circuit, or a short circuit to ground between the ECM and the sensor, disconnect the vehicle speed sensor connector. Between ground and terminal 1 of the harness connector).
4. To check for an open circuit between the vehicle speed sensor and ground, disconnect the vehicle speed sensor connector.

ELANTRA

See Figures 6 through 9.

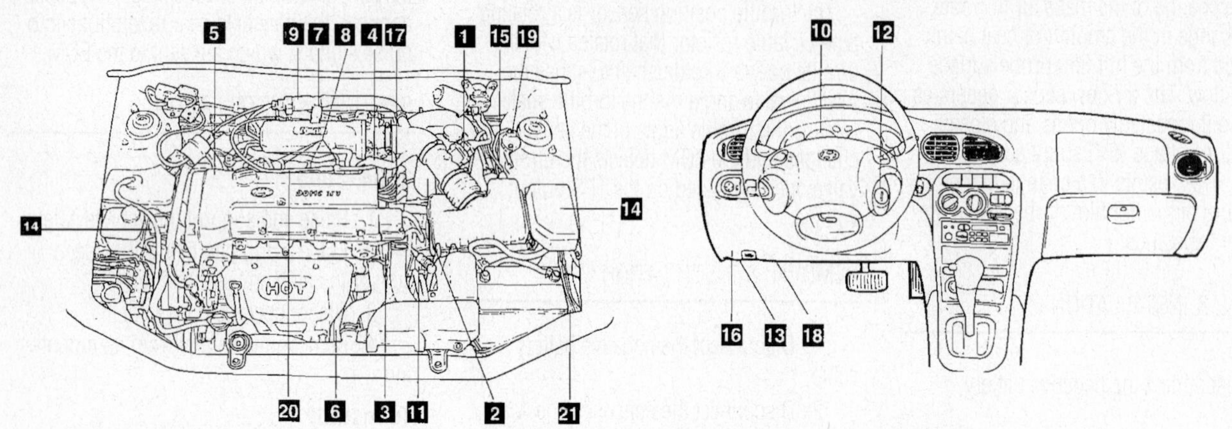

1. MAF sensor
2. Intake air temp. sensor (IAT Sensor)
3. Engine coolant temperature sensor (ECT Sensor)
4. Throttle position sensor (TP Sensor)
5. Camshaft position sensor (CMP Sensor)
6. Crankshaft position sensor (CKP Sensor)
7. Heated oxygen sensor (H2OS)
8. Injector
9. Idle speed control actuator (ISC Actuator)
10. Vehicle speed sensor
11. Transaxle Range (TR) switch
12. Ignition switch
13. Engine Control Module (ECM)
14. Air conditioning relay
15. Evaporative emission canister purge solenoid valve
16. MFI control relay
17. Ignition coil
18. Data link connector
19. Acceleration sensor
20. Knock sensor
21. Canister close valve

Fig. 6 Underhood sensor locations—Elantra 1996 1.8L engine

29130_HYUN_G0006

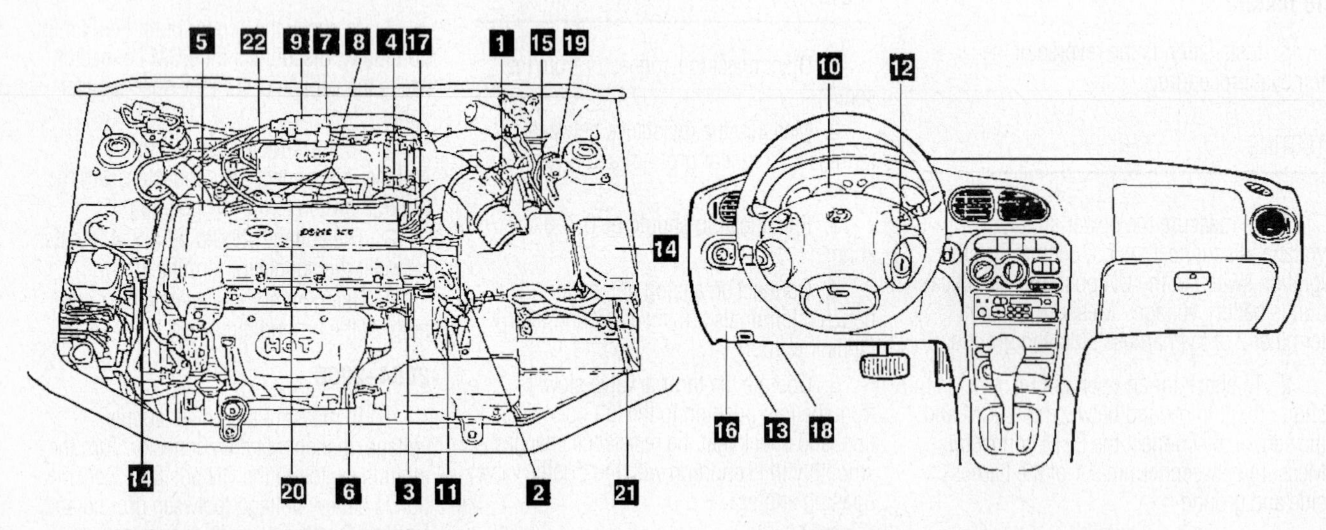

1. MAF (Mass Air Flow) sensor
2. Intake Air Temperature (IAT) sensor
3. Engine Coolant Temperature (ECT) sensor
4. Throttle Position Sensor (TPS)
5. Camshaft Position (CMP) sensor
6. Crankshaft Position (CKP) sensor
7. Heated Oxygen Sensor (HO2S)
8. Injectors
9. Idle Speed Control (ISC) actuator
10. Vehicle Speed Sensor (VSS)
11. Transaxle Range (TR) switch
12. Ignition switch
13. Engine Control Module (ECM)
14. Air conditioning relay
15. EVAP. canister purge solenoid valve
16. MFI control relay
17. Ignition coil
18. Data Link Connector (DLC)
19. Acceleration sensor
20. Knock sensor
21. Canister Close Valve (CCV)
22. Fuel Pump Check Connector

Fig. 7 Underhood sensor locations—Elantra 1997–1998 1.8L engine and 1999–2000 2.0L engine

29157_TOYO_G0007

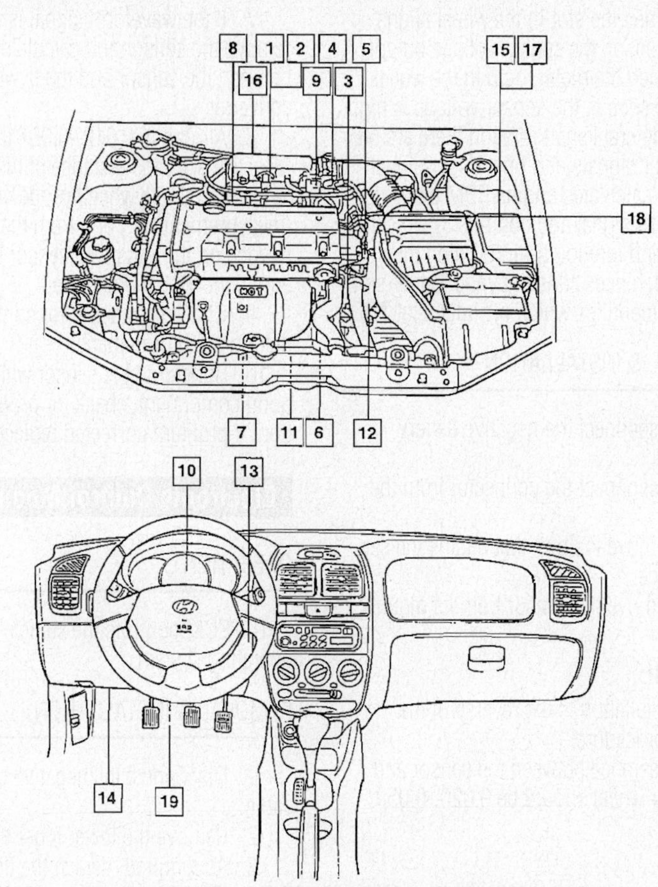

1. Manifold Absolute Pressure (MAP) Sensor
2. Intake Air Temp. (IAT) Sensor
3. Engine Coolant Temp. (ECT) Sensor
4. Throttle Position Sensor (TPS)
5. Camshaft Position Sensor (CMP)
6. Crankshaft Position Sensor (CKP)
7. Heated Oxygen Sensor
8. Injector
9. Idle Speed Actuator (ISA)
10. Vehicle Speed Sensor (VSS)
11. Knock Sensor
12. Inhibitor Switch
13. Ignition Switch
14. ECM
15. Air Conditioner Relay
16. Purge Control Solenoid Valve (PCSV)
17. Control Relay
18. Ignition Coil
19. Data Link Connector (DLC)

Fig. 8 Underhood sensor locations—Elantra 2001–2003 2.0L engine

29130_HYUN_G0008

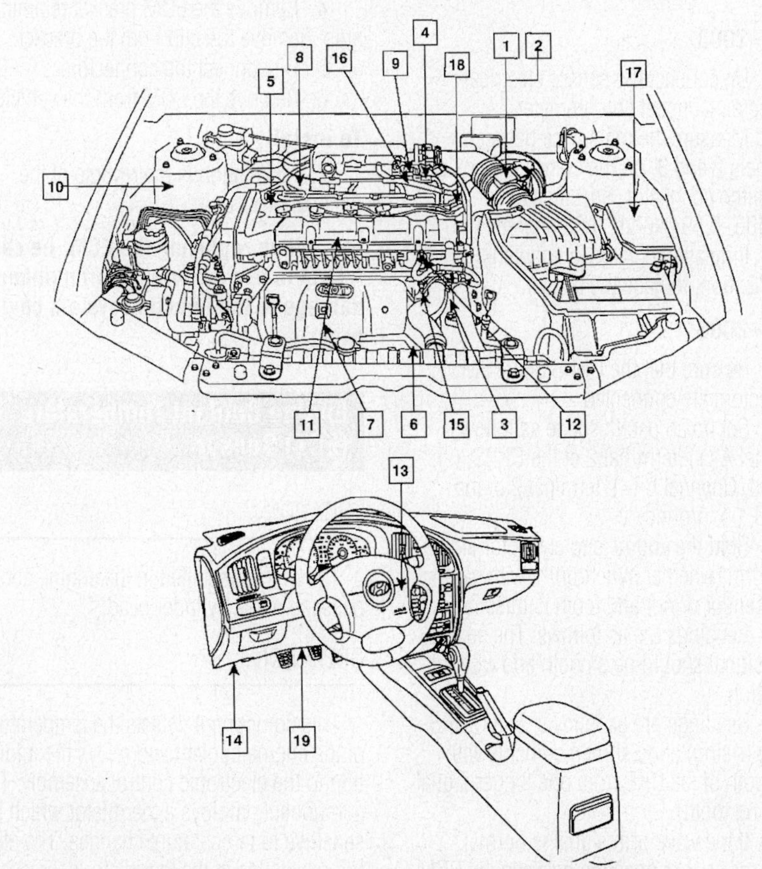

1. Mass Air Flow Sensor (MAFS)
2. Intake Air Temperature Sensor (IATS)
3. Engine Coolant Temperature Sensor (ECTS)
4. Throttle Position Sensor (TPS)
5. Camshaft Position Sensor (CMPS)
6. Crankshaft Position Sensor (CKPS)
7. Heated Oxygen Sensor (HO2S, Sensor 1)
8. Injector
9. Idle Speed Control Actuator (ISCA)
10. Vehicle Speed Sensor (VSS)
11. Knock Sensor
12. CVVT Oil Control Valve (OCV)
13. Ignition Switch
14. ECM
15. CVVT Oil Temperature Sensor (OTS)
16. Purge Control Solenoid Valve (PCSV)
17. Main Relay
18. Ignition Coil
19. DLC (Diagnostic Link Cable)

Fig. 9 Underhood sensor locations—Elantra 2004–2006 2.0L engine

29130_HYUN_G0009

Camshaft Position (CMP) Sensor

LOCATION

The camshaft position sensor is located on top of the engine near the ignition coils.

OPERATION

The camshaft position sensor senses the TDC point of the number one cylinder, on its compression stroke. Its signal is relayed to the ECM to be used to determine the sequence of fuel injection.

REMOVAL & INSTALLATION

1. Disconnect the negative battery cable.
2. Disconnect the connector from the sensor.
3. Remove the bolt that retains the sensor.
4. Remove the sensor.

To install:

5. Installation is the reverse of the removal procedure.

TESTING

1. Using a voltmeter measure the sensor output voltage.
2. At idle the specification should be 0–5 volts. At 3000 rpm the specification should be 0–5 volts.
3. If abnormality is found, replace the sensor.

Crankshaft Position (CKP) Sensor

LOCATION

The crankshaft position sensor is located next to the flywheel/torque converter.

OPERATION

1996–2003

The crankshaft position sensor consists of a magnet and coil. The voltage signal from the sensor is relayed to the ECM to indicate engine rpm and the position of the crankshaft.

2004–2006

The crankshaft position sensor is a Hall Effect type sensor that generates voltage using a sensor and a target wheel mounted on the crankshaft. There are 58 slots in the target wheel where one is longer than the others. When the slot in the wheel aligns with the sensor, the sensor voltage output is low. When the metal tooth in the wheel aligns the sensor, the sensor voltage is high. During one crankshaft rotation there are 58 rectangular signals and one longer signal. The PCM calculates engine RPM by using the sensor's signal and controls the injection duration and ignition timing. Using the signal differences caused by the longer slot, the PCM identifies which cylinder is at TDC.

REMOVAL & INSTALLATION

1. Disconnect the negative battery cable.
2. Disconnect the connector from the sensor.
3. Remove the bolt that retains the sensor in place.
4. Remove the sensor from its mounting.

To install:

5. Installation is the reverse of the removal procedure.
6. Clearance between the sensor and the sensor wheel should be 0.020–0.059 inch.

TESTING

1996–2003

1. Disconnect the sensor electrical connector. Connect an ohmmeter.
2. Measure the resistance between terminals 2 and 3, on the crankshaft position side of the connector. Specification should be 0.486–0.594 kohm at 68 degrees F.
3. If measured value is not within specification, replace the sensor.

2004–2006

1. Be sure that the CKPS and PCM connectors are connected.
2. Set up an oscilloscope as follows: Channel A (+): terminal 2 of the CKPS, (-): ground. Channel B (+): terminal 2 of the CMPS, (-): ground.
3. Start the engine and check for signal waveform (whether synchronize with camshaft sensor or not and tooth is missing).
4. Readings are as follows: The square wave signal should be smooth and without distortion.
5. Readings are as follows: The CMPS falling (rising) edge should coincide with 3–5 tooth of the CKP from one longer signal (missing tooth).
6. If the waveform signal is normal, check for poor connection between the PCM and the components.
7. If the waveform signal is not normal, remove the sensor and calculate the air gap between the sensor and the flywheel/torque converter.
8. Air gap is 0.012–0.067 inch. Measure from the distance of the housing to teeth on the flywheel/torque converter (measurement "A") and from the mounting surface on the sensor to sensor tip (measurement "B"), then subtract "B" from "A".
9. Check the sensor for contamination, deterioration or damage.
10. Substitute the sensor with a known good component, check for proper operation. If problem corrected replace the sensor.

Electronic Control Module (ECM)

OPERATION

The ECM controls the vehicle engine operating system.

REMOVAL & INSTALLATION

1. Disconnect the negative battery cable.
2. Remove the lower inner trim..
3. As required, detach the floor mat. As required, remove the protective cover.
4. Remove the ECM bracket retaining nuts. Remove the clip from the bracket.
5. Disconnect the connectors.
6. Remove the ECM from the vehicle.

To install:

7. Installation is the reverse of the removal procedure.

➡ When replacing the ECM, be careful not to use the wrong part number, as damage to the injection system could occur.

Engine Coolant Temperature (ECT) Sensor

LOCATION

The ECT is installed in the engine coolant passage of the cylinder head.

OPERATION

This component detects the temperature of the engine coolant and relays the information to the electronic control assembly. This component employs a thermistor which is sensitive to temperature changes. The electric resistance of the thermistor decreases in response to temperature rise. The ECM judges coolant temperature by the sensor output voltage and provides optimum fuel enrichment when the engine is cold.

REMOVAL & INSTALLATION

1. Disconnect the negative battery cable.
2. Disconnect the connector from the sensor.
3. Drain the cooling system, as required.
4. Remove the sensor from its mounting.

To install:

5. Installation is the reverse of the removal procedure.

TESTING

1996–2003

1. Remove the sensor from the engine.
2. With the sensing portion of the sensor immersed in hot water, check the resistance.
3. Specification should be 0.538–0.650 kohm at 140 degrees F.
4. Replace the sensor, as required.

2004

1. Turn the ignition switch to the OFF position.
2. Check the resistance of the sensor connector between terminals 1 and 3.
3. Specification should be 2.27–2.73 kohm at 68 degrees F. and 0.30–0.32 kohm at 176 degrees F.
4. Replace the sensor, as required.

2005–2006

1. Turn the ignition switch to the OFF position.
2. Check the resistance of the sensor connector between terminals 1 and ground.
3. Specification should be 2.31–2.59 kohm at 68 degrees F. and 0.32 kohm at 176 degrees F.
4. Replace the sensor, as required.

Heated Oxygen (HO2S) Sensor

LOCATION

The sensors are located in the exhaust system. On some vehicles one sensor is located up at the exhaust manifold and the other sensor is located down at the catalytic converter. On other vehicles both sensors are located down at the catalytic converter.

OPERATION

The exhaust gas oxygen sensor supplies the electronic control assembly with a signal which indicates either a rich or lean mixture condition, during the engine operation.

REMOVAL & INSTALLATION

1. Disconnect the electrical connector from the sensor.
2. Remove the oxygen sensor.

To install:

3. Installation is the reverse of the removal procedure.

➡ **Apply anti-seize compound to the threaded portion of the sensor, prior to installation. Never apply anti-seize compound to the protector of the sensor.**

TESTING

1996–2003

Perform a visual inspection of the sensor as follows:

1. If the sensor tip has a black/sooty deposit, this may indicate a rich fuel mixture.
2. If the sensor tip has a white, gritty deposit, this may indicate an internal coolant leak.
3. If the sensor tip has a brown deposit, this could indicate excessive oil consumption.
4. Warm the engine until the coolant temperature reaches operating temperature.
5. Accelerate the engine to 4000 rpm. When decelerating suddenly from 4000 rpm, the voltmeter reading should be 200mV or lower.
6. When the engine is suddenly raced, the voltmeter reading should be 600–1000mV.
7. If measured value is not within specification, replace the sensor.

2004

Perform a visual inspection of the sensor as follows:

1. If the sensor tip has a black/sooty deposit, this may indicate a rich fuel mixture.
2. If the sensor tip has a white, gritty deposit, this may indicate an internal coolant leak.
3. If the sensor tip has a brown deposit, this could indicate excessive oil consumption.
4. Turn the ignition switch to the OFF position.
5. Disconnect the sensor connector.
6. Measure the resistance between terminals 3 and 4, of the connector.
7. Specification should be 9 ohms at 68–75.2 degrees F.
8. If not, replace the sensor.

2005–2006

Perform a visual inspection of the sensor as follows:

1. If the sensor tip has a black/sooty deposit, this may indicate a rich fuel mixture.
2. If the sensor tip has a white, gritty deposit, this may indicate an internal coolant leak.
3. If the sensor tip has a brown deposit, this could indicate excessive oil consumption.
4. Disconnect the sensor connector.
5. Measure the resistance between terminals 5 and 6 (component side).
6. Specification should be 9 ohms at 20–24 degrees C.
7. If within specification, check for poor connection between the PCM and the component.
8. If not, substitute the sensor with a known good component, check for proper operation. If problem corrected replace the sensor.

Idle Speed Control Actuator (ISCA)

OPERATION

The idle speed control actuator sensor is a double coil type driven by separate driver stages in the ECM. Depending on the pulse duty factor, the equilibrium of the magnetic forces of the two coils will result in different angles of the motor. A bypass hose line is positioned, parallel to the throttle valve where the idle speed actuator is inserted.

REMOVAL & INSTALLATION

1. Disconnect the negative battery cable.
2. Disconnect the connector from the sensor.
3. Remove the sensor retaining screws.
4. Remove the sensor from its mounting.

To install:

5. Installation is the reverse of the removal procedure.

TESTING

1. Disconnect the connector at the sensor. Connect an ohmmeter.

2. Measure the resistance between terminal 1 and 2, specification should be 10.5–14 ohms. Measure the resistance between terminal 2 and 3, specification should be 10–12.5 ohms at 68 degrees F.

3. If the measured value is not within specification, replace the sensor.

Intake Air Temperature (IAT) Sensor

LOCATION

This sensor is located in the air intake plenum assembly. On some vehicles this sensor is combined with the MAP/MAF sensor.

OPERATION

This sensor is a resistor based sensor which detects the intake air temperature. According to the intake air temperature reading the ECM will control the necessary amount of fuel injection.

REMOVAL & INSTALLATION

1. Disconnect the negative battery cable.
2. Disconnect the connector from the sensor.
3. Remove the sensor retaining screws, as required.
4. Remove the air cleaner and air intake assembly, as required.
5. Remove the sensor from its mounting.

To install:

6. Installation is the reverse of the removal procedure.

TESTING

1996–2003

1. Using a multimeter, measure the sensor resistance between terminals 3 and 4.
2. At 32 degrees F the resistance should be 4.5–7.5 kohm. At 68 degrees F the resistance should be 2.0–3.0 kohm. At 104 degrees F the resistance should be 0.7–1.6 kohm. At 166 degrees F the resistance should be 0.2–0.4 kohm.
3. If the measured value is not within specification, replace the sensor.

2004

1. Turn the ignition switch to the OFF position.
2. Disconnect the sensor connector.
3. Measure the resistance between terminals 5 and 2, of the connector.
4. Specification should be 2–3 kohms at 68 degrees F.
5. If not, replace the sensor.

2005–2006

1. Turn the ignition switch OFF.
2. Disconnect the sensor connector.
3. Measure the resistance between terminals 1 and 5 (component side).
4. Specification should be 2.35–3.54 kohms at 68 degrees F.
5. If within specification, check for poor connection between the PCM and the component.
6. If not, substitute the sensor with a known good component, check for proper operation. If problem corrected replace the sensor.

Knock Sensor (KS)

LOCATION

The knock sensor is located in the side of the cylinder block.

OPERATION

The knock sensor is used to detect engine vibrations caused by preignition or detonation and provides information to the ECM, which then retards the timing to eliminate detonation.

REMOVAL & INSTALLATION

1. Disconnect the negative battery cable.
2. Disconnect the sensor connector.
3. Remove the sensor from its mounting.

To install:

4. Installation is the reverse of the removal procedure.
5. Tighten the sensor to 12–18 ft. lbs.

TESTING

1. Disconnect the sensor electrical connector. Connect an ohmmeter.
2. Measure the resistance between terminals 1 and 2. Specification should be 5mohm at 68 degrees F.
3. If not within specification, replace the sensor.

Manifold Absolute Pressure (MAP) Sensor

LOCATION

The MAP sensor is located against the firewall to the left side of the engine.

OPERATION

This sensor is a pressure sensitive variable resistor. It measures the changes in the intake manifold pressure which result from engine load and speed changes, and converts to a voltage output. This sensor is used to measure the barometric pressure at start up, and under certain conditions, allows the ECM to automatically adjust for different altitudes. The ECM supplies 5 volts to the sensor and monitors the voltage on a signal line. The sensor provides a path to ground through its variable resistor. The sensor input affects fuel delivery and ignition timing controls in the ECM.

REMOVAL & INSTALLATION

1. Disconnect the negative battery cable.
2. Disconnect the connector from the sensor.
3. Remove the sensor retaining screws.
4. Remove the sensor from its mounting.

To install:

5. Installation is the reverse of the removal procedure.

TESTING

1. Turn the ignition switch ON.
2. Measure the voltage between terminal 1 (sensor output) and terminal 4 (sensor ground).
3. Specification should be 4–5 volts.
4. If not within specification, replace the sensor.

Mass Air Flow (MAF) Sensor

LOCATION

The MAF is mounted in the intake air hose of the air cleaner assembly.

OPERATION

The MAF is measured by detection of heat transfer from a hot film probe because the change of the mass air flow rate causes change in the amount of heat being transferred from the hot film probe surface to the air flow. The air flow sensor generates a pulse so it repeatedly opens and closes between the voltage (5V) supplied from the ECM. This results in a change of the temperature of the hot film probe and in the change of resistance.

REMOVAL & INSTALLATION

1. Disconnect the negative battery cable.
2. Disconnect the connector from the sensor.
3. Remove the air cleaner and air intake assembly, as required.
4. Remove the sensor from its mounting.

To install:

5. Installation is the reverse of the removal procedure.

TESTING

2004

1. Be sure that the ECM and the MAFS connectors are connected.
2. Connect the Hi-Scan tool to the data link connector.
3. Start the engine.
4. Monitor the MAPS signals.
5. If not within specification, replace the sensor.
6. Specification should be 0.0–5.0 volts.
7. Replace the sensor, as required.

2005–2006

1. Turn the engine ON.
2. Install the scan tool and monitor the MAPS signal.
3. Specification should be approximately 10–20 kg/h at idle and no load.
4. If within specification, check for poor connection between the PCM and the component.
5. If not, substitute the sensor with a known good component, check for proper operation. If problem corrected replace the sensor.

Throttle Position Sensor (TPS)

OPERATION

The throttle position sensor is a rotating type variable resistor that rotates with the throttle body's throttle shaft to sense the throttle valve angle. As the throttle shaft rotates, the throttle angle of the sensor changes and the ECM detects the throttle valve opening based on the TPS output voltage.

REMOVAL & INSTALLATION

1. Disconnect the negative battery cable.
2. Disconnect the sensor connector.
3. Remove the sensor retaining screws. Remove the sensor from its mounting.

To install:

4. Installation is the reverse of the removal procedure.

TESTING

1996–2003

1. Disconnect the connector from the sensor.
2. Measure the resistance between terminals 2 (sensor ground) and 3 (sensor power).
3. Specification should be 0.7–3.0 kohm.
4. Connect an analog ohmmeter between terminals 3 (sensor ground) and 1 (sensor output), on 1996–1997 vehicles and between terminals 2 (sensor ground) and 1 (sensor output), on 1998–2003.
5. Operate the throttle valve slowly from the idle position to the full open position, and check that the resistance changes smoothly in proportion with the throttle valve opening angle.
6. If the resistance is out of specification, or fails to change smoothly, replace the sensor.

2004

1. Turn the ignition switch to the OFF position.
2. Measure the resistance between terminals 1 and 2 of the TPS.
3. Measure the resistance between terminals 2 and 3 of the TPS.
4. Specification should be 0.7–3.0 kohms.
5. If not within specification, replace the sensor.

2005–2006

1. Turn the ignition switch to the OFF position.
2. Disconnect the TPS connector.
3. Measure the resistance between terminals 2 and 3 of the TPS.
4. Specification should be 1.6–2.4 kohms at all throttle position..
5. With the connector still disconnected, measure the resistance between terminals 1 and 2.
6. Operate the throttle valve slowly from the idle position to the full open position and check that the resistance changes smoothly in proportion with the throttle valve opening angle.
7. Specification should be 0.71–1.38 kohm at closed throttle valve and 2.2–3.4 kohm at wide open throttle.
8. If within specification, check for poor connection between the PCM and the component.
9. If not, substitute the sensor with a known good component, check for proper operation. If problem corrected replace the sensor.

Vehicle Speed Sensor (VSS)

LOCATION

On 1996–2003 vehicles, the vehicle speed sensor is built into the speedometer. On 2004–2006 vehicles, the vehicle speed sensor is attached to the output shaft of the transaxle.

OPERATION

On 1997–2003 vehicles, the sensor is a reed switch type. On 2004–2006 vehicles, the sensor is a hall effect sensor. The sensor converts the transaxle gear revolutions into pulse signals, which are sent to the ECM.

REMOVAL & INSTALLATION

2004–2006

1. Raise and support the vehicle safely.
2. Place a drip pan below the speed sensor to catch any spilled fluid.
3. Disconnect the connector.
4. Remove the sensor from its mounting.

To install:

5. Installation is the reverse of the removal procedure.
6. Replace any lost fluid.

TESTING

1996–2003

1. To check the sensor output circuit for continuity, disconnect the ECM connector. Move the vehicle or turn the speedometer cable.

2. To measure the power supply voltage, disconnect the connector. Turn the ignition switch to the ON position.
3. The voltage should be 4.5–4.9 volts.
4. Using an ohmmeter, check for continuity of the ground circuit.

2004–2006

1. To measure the power supply voltage, disconnect the connector. Turn the ignition switch to the ON position. Specification is battery voltage (between ground and terminal 2 of the harness connector).
2. To check for an open circuit, or a short circuit to ground between the ECM and the sensor, disconnect the ECM connector. Between ground and terminal 3 of the harness connector.
3. To check for an open circuit, or a short circuit to ground between the ECM and the sensor, disconnect the vehicle speed sensor connector. Between ground and terminal 1 of the harness connector).
4. To check for an open circuit between the vehicle speed sensor and ground, disconnect the vehicle speed sensor connector.

SANTA FE

See Figures 10 through 13.

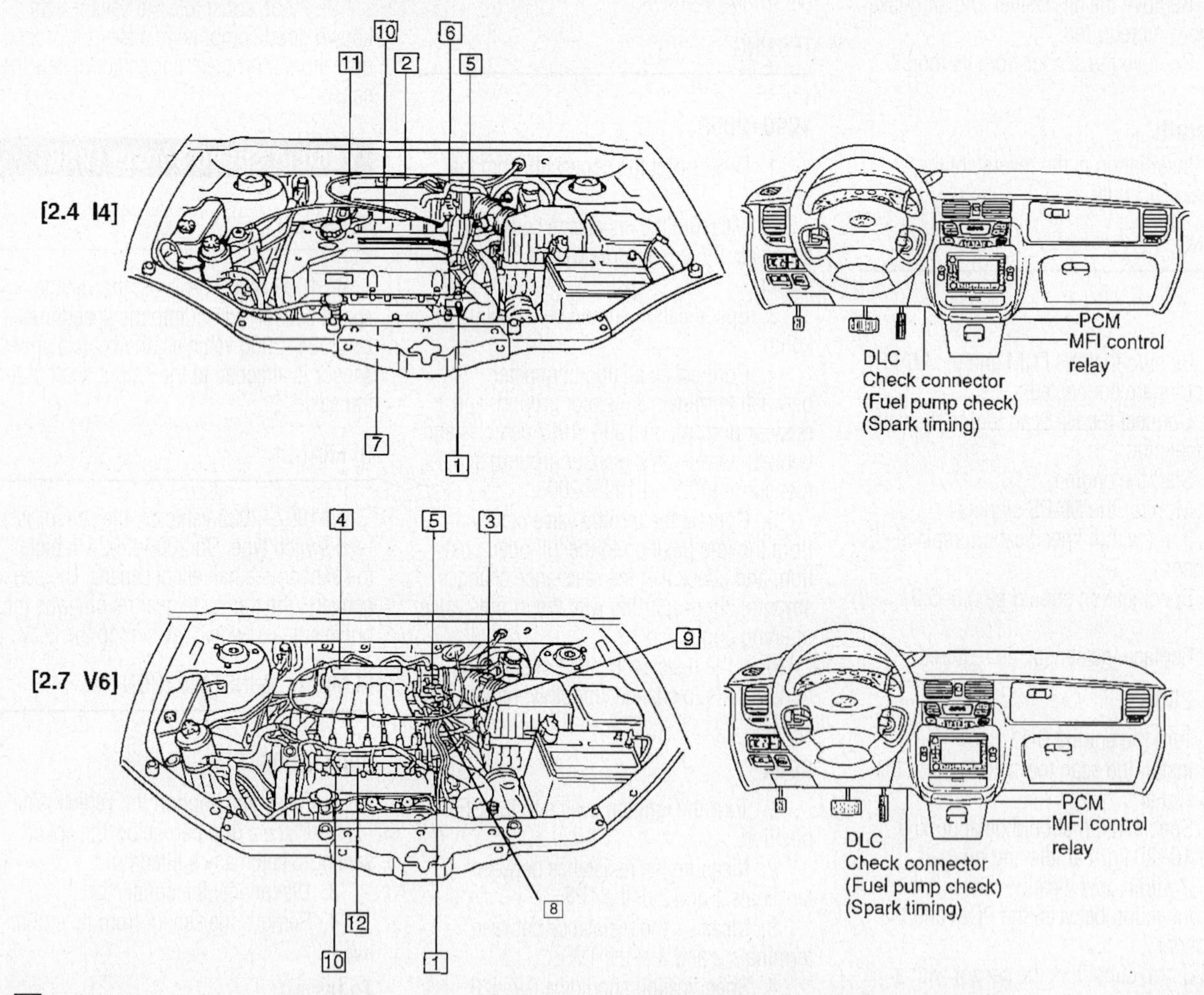

Fig. 10 Underhood sensor locations—Santa Fe 2001–2004 2.4L and 2.7L engines

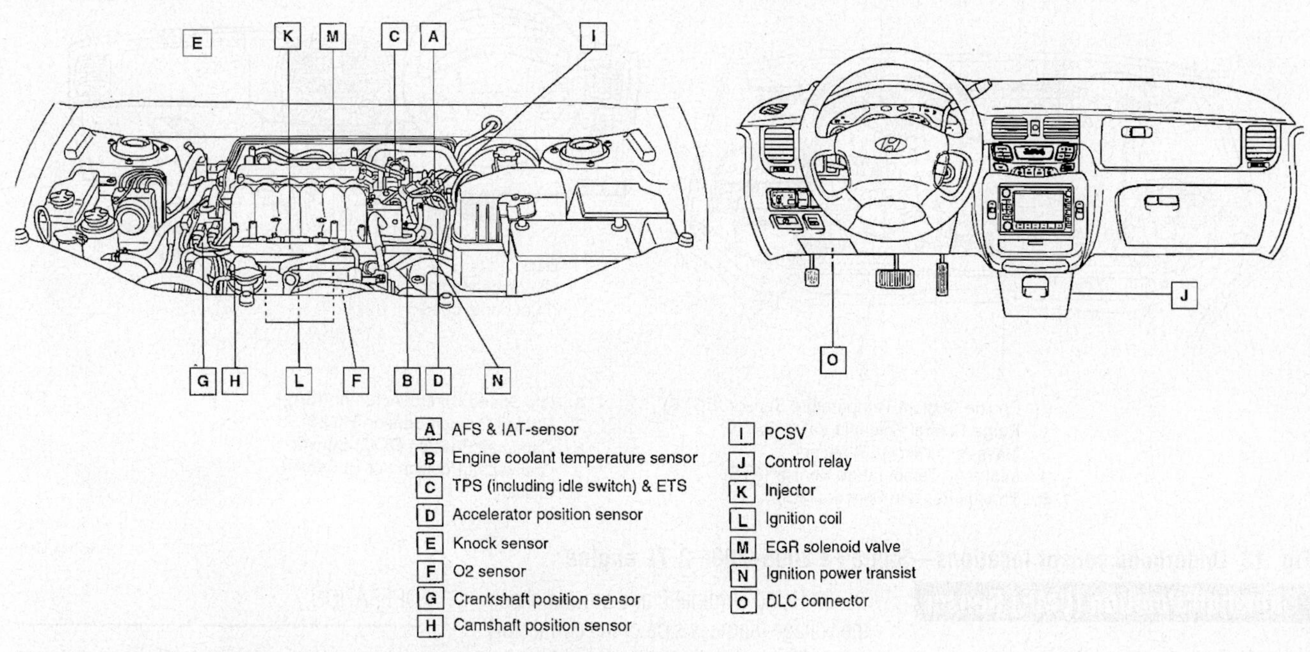

Fig. 11 Underhood sensor locations—Santa Fe 2004 3.5L engine

29130_HYUN_G0011

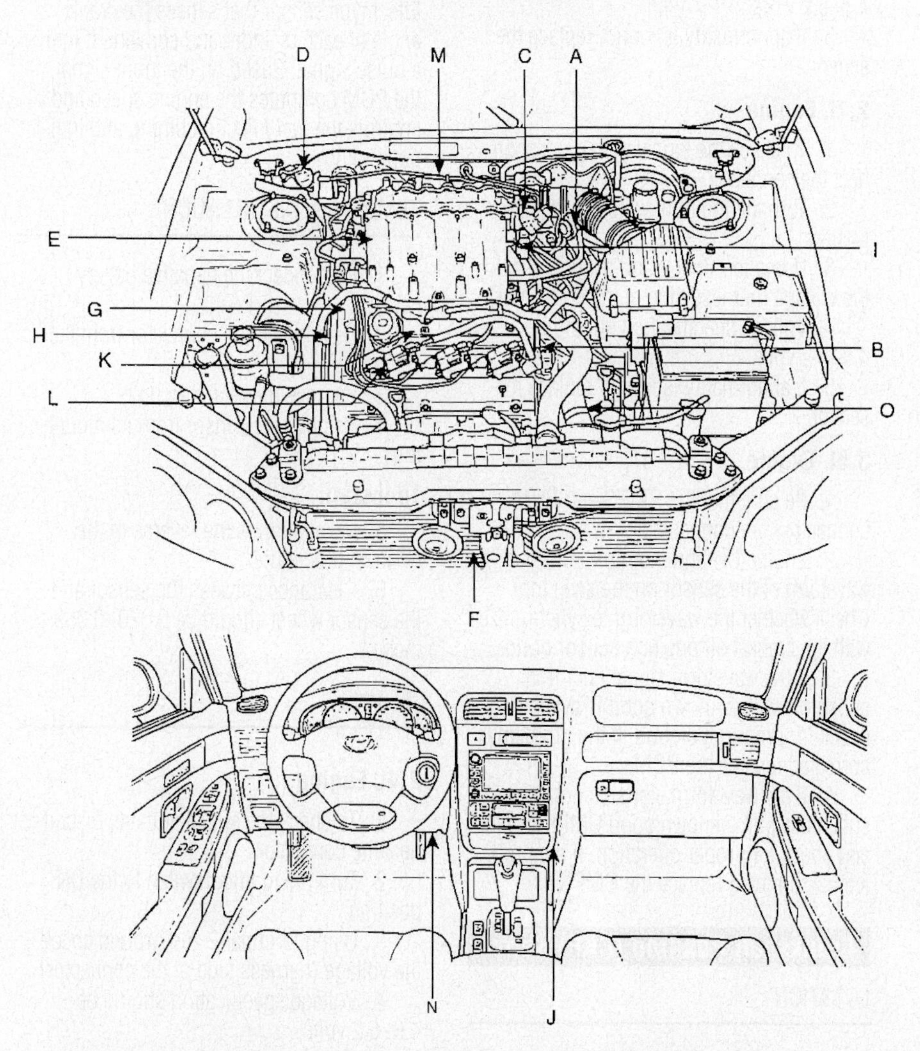

A. AFS & IAT-sensor
B. Engine coolant temperature sensor
C. TPS (including idle switch) & ETS
D. Accelerator position sensor
E. Knock sensor
F. O2 sensor
G. Crankshaft position sensor
H. Camshaft position sensor
I. PCSV
J. Control relay
K. Injector
L. Ignition coil
M. EGR solenoid valve
N. DLC connector

Fig. 12 Underhood sensor locations—Santa Fe 2005–2006 3.5L engine

29130_HYUN_G0012

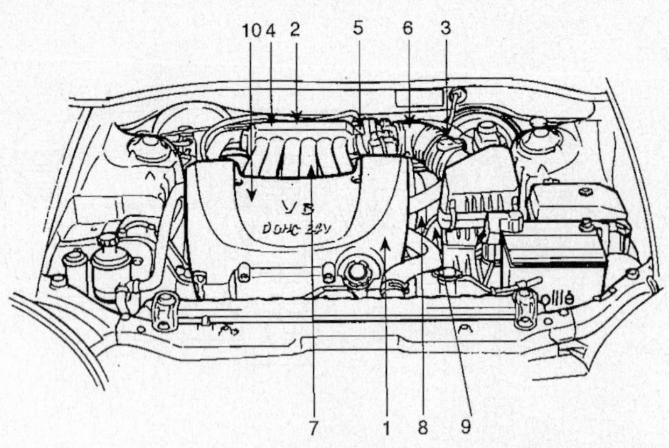

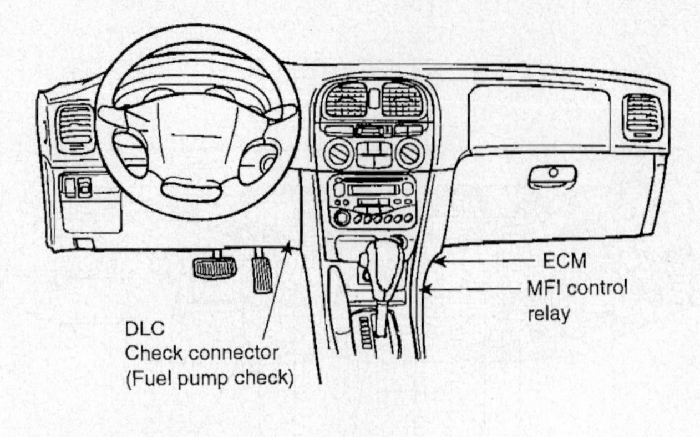

1. Engine Coolant Temperature Sensor (ECTS)
2. Purge Control Solenoid Valve (PCSV)
3. Mass Air Flow (MAF) sensor
4. Intake Air Temperature Sensor (ATS)
5. Throttle Position Sensor (TPS)
6. Idle Speed Control Actuator (ISA)
7. Heated Oxygen Sensor (HO2S)
8. Camshaft Position (TDC) Sensor
9. Crankshaft Position (CKP) Sensor
10. Injector

Fig. 13 Underhood sensor locations—Santa Fe 2005–2006 2.7L engine 29130_HYUN_G0013

Camshaft Position (CMP) Sensor

LOCATION

On the 2.4L and 2.7L engines the camshaft position sensor is located on top of the engine near the ignition coils. On the 3.5L engine the sensor is located at the front of the engine near the camshaft gear.

OPERATION

The camshaft position sensor senses the TDC point of the number one cylinder, on its compression stroke. Its signal is relayed to the PCM to be used to determine the sequence of fuel injection.

REMOVAL & INSTALLATION

1. Disconnect the negative battery cable.
2. Disconnect the connector from the sensor.
3. Remove the bolt that retains the sensor.
4. Remove the sensor.

To install:

5. Installation is the reverse of the removal procedure.

TESTING

2.4L Engine

1. To check the signal voltage, disconnect the connector.
2. Turn the ignition switch to the ON position.
3. Using terminal 2 and ground check the voltage (harness side of the connector).
4. Voltage specification should be 4.8–5.2 volts.
5. If abnormality is found, replace the sensor.

2.7L Engine

1. To check the signal voltage, disconnect the connector.
2. Turn the ignition switch to the ON position.
3. Using terminal 2 and ground check the voltage (harness side of the connector).
4. Voltage specification should be 4.8–5.2 volts.
5. If abnormality is found, replace the sensor.

3.5L Engine

1. Be sure that the CMPS and PCM connectors are connected.
2. Engine ON and monitor the signal waveform of the sensor on the scan tool. Check whether the waveform is synchronized with the crankshaft position sensor or not.
3. If the waveform signal is normal, substitute with a known good PCM and check for proper operation. If the problem is corrected replace the PCM.
4. If the waveform signal is not normal, substitute with a known good CMPS sensor and check for proper operation. If the problem is corrected replace the CMPS.

Crankshaft Position (CKP) Sensor

LOCATION

The crankshaft position sensor is located at the front of the engine near the timing belt.

OPERATION

The crankshaft position sensor is a Hall Effect type sensor that senses the crank angle of each cylinder and converts it into a pulse signal. Based on the input signal, the PCM computes the engine speed and controls the fuel injection timing and ignition timing.

REMOVAL & INSTALLATION

1. Disconnect the negative battery cable.
2. Disconnect the connector from the sensor.
3. Remove the retaining bolts.
4. Remove the sensor from its mounting.

To install:

5. Installation is the reverse of the removal procedure.
6. Clearance between the sensor and the sensor wheel should be 0.020–0.059 inch.

TESTING

2.4L Engine

1. To check the signal voltage, disconnect the connector.
2. Turn the ignition switch to the ON position.
3. Using terminal 2 and ground check the voltage (harness side of the connector).
4. Voltage specification should be 4.8–5.2 volts.
5. If abnormality is found, replace the sensor.

2.7L Engine

1. To check the signal voltage, disconnect the connector.
2. Turn the ignition switch to the ON position.
3. Using terminal 2 and ground check the voltage (harness side of the connector).
4. Voltage specification should be 4.8–5.2 volts.
5. If abnormality is found, replace the sensor.

3.5L Engine

1. Be sure that the CKPS and PCM connectors are connected.
2. Engine ON and monitor the signal waveform of the sensor on the scan tool. Check whether the waveform is synchronized with the camshaft position sensor or not.
3. If the waveform signal is normal, check the PCM.
4. If the waveform signal is not normal, substitute with a known good CKPS sensor and check for proper operation. If the problem is corrected replace the CMPS.

Electronic Control Module (ECM)

OPERATION

The ECM controls the vehicle engine operating system.

REMOVAL & INSTALLATION

1. Disconnect the negative battery cable.
2. Remove the lower inner trim..
3. As required, detach the floor mat. As required, remove the protective cover.
4. Remove the ECM bracket retaining nuts. Remove the clip from the bracket.
5. Disconnect the connectors.
6. Remove the ECM from the vehicle.

To install:

7. Installation is the reverse of the removal procedure.

➡ **When replacing the ECM, be careful not to use the wrong part number, as damage to the injection system could occur.**

Engine Coolant Temperature (ECT) Sensor

LOCATION

The ECT is installed in the engine coolant passage of the cylinder head.

OPERATION

This component detects the temperature of the engine coolant and relays the information to the PCM. This component employs a thermistor which is sensitive to temperature changes. The electric resistance of the thermistor decreases in response to temperature rise. The ECM/PCM judges coolant temperature by the sensor output voltage and provides optimum fuel enrichment when the engine is cold.

REMOVAL & INSTALLATION

1. Disconnect the negative battery cable.
2. Disconnect the connector from the sensor.
3. Drain the cooling system, as required.
4. Remove the sensor from its mounting.

To install:

5. Installation is the reverse of the removal procedure.

TESTING

1. Remove the sensor from the engine.
2. With the sensing portion of the sensor immersed in hot water, check the resistance.
3. Specification should be 0.3 kohm at 176 degrees F.
4. Replace the sensor, as required.

Heated Oxygen (HO2S) Sensor

LOCATION

The sensors are located in the exhaust system. On some vehicles one sensor is located up at the exhaust manifold(s) and the other sensor is located down at the catalytic converter. On other vehicles both sensors are located down at the catalytic converter.

OPERATION

The exhaust gas oxygen sensor supplies the electronic control assembly with a signal which indicates either a rich or lean mixture condition, during the engine operation.

REMOVAL & INSTALLATION

1. Disconnect the electrical connector from the sensor.
2. Remove the oxygen sensor.

To install:

3. Installation is the reverse of the removal procedure.

➡ **Apply anti-seize compound to the threaded portion of the sensor, prior to installation. Never apply anti-seize compound to the protector of the sensor.**

TESTING

2.4L Engine

Perform a visual inspection of the sensor as follows:

1. If the sensor tip has a black/sooty deposit, this may indicate a rich fuel mixture.
2. If the sensor tip has a white, gritty deposit, this may indicate an internal coolant leak.
3. If the sensor tip has a brown deposit, this could indicate excessive oil consumption.
4. Warm the engine until the coolant temperature reaches operating temperature.
5. Disconnect the sensor electrical connector and measure the resistance between terminal 3 and 4.
6. Specification should be 30 ohms, or more.
7. Replace the sensor, if there is a malfunction.
8. Apply battery voltage directly between terminal 3 and 4.

➡ **Be careful when applying the voltage. Damage will result if terminals 1 and 2 are connected to any voltage.**

9. Connect a voltmeter between terminal 1 and 2.
10. While racing the engine repeatedly, measure the output voltage.
11. Specification should be 0.6 volt (minimum) and resistance should be 30 ohm or more.
12. Replace the sensor, as required.

2.7L Engine

Perform a visual inspection of the sensor as follows:

1. If the sensor tip has a black/sooty deposit, this may indicate a rich fuel mixture.
2. If the sensor tip has a white, gritty deposit, this may indicate an internal coolant leak.
3. If the sensor tip has a brown deposit, this could indicate excessive oil consumption.
4. Warm the engine until the coolant temperature reaches operating temperature.

5. Disconnect the sensor electrical connector and measure the resistance between terminal 3 and 4.

6. Specification should be 4.0–5.2 ohms at 23 degrees C.

7. Apply battery voltage directly between terminal 3 and 4.

➡ **Be careful when applying the voltage. Damage will result if the terminals are incorrect or are short circuited.**

8. Connect a voltmeter between terminal 1 and 2.

9. While racing the engine repeatedly, measure the output voltage.

10. Specification should be 4000–4500mV.

11. Replace the sensor, as required.

3.5L Engine

Perform a visual inspection of the sensor as follows:

1. If the sensor tip has a black/sooty deposit, this may indicate a rich fuel mixture.

2. If the sensor tip has a white, gritty deposit, this may indicate an internal coolant leak.

3. If the sensor tip has a brown deposit, this could indicate excessive oil consumption.

4. Warm the engine until the coolant temperature reaches operating temperature.

5. Disconnect the sensor electrical connector and measure the resistance between terminal 3 and 4.

6. Specification should be 30 ohms, or more.

7. Replace the sensor, if there is a malfunction.

8. Apply battery voltage directly between terminal 3 and 4.

➡ **Be careful when applying the voltage. Damage will result if terminals 1 and 2 are connected to any voltage.**

9. Connect a voltmeter between terminal 1 and 2.

10. While racing the engine repeatedly, measure the output voltage.

11. Specification should be 0.6 volt (minimum) and resistance should be 30 ohm or more.

12. Replace the sensor, as required.

Idle Speed Control Actuator (ISCA)

OPERATION

The idle speed control actuator sensor is a double coil type driven by separate driver stages in the ECM. Depending on the pulse duty factor, the equilibrium of the magnetic forces of the two coils will result in different angles of the motor. A bypass line is positioned, parallel to the throttle valve where the idle speed actuator is inserted.

REMOVAL & INSTALLATION

1. Disconnect the negative battery cable.

2. Disconnect the connector from the sensor.

3. Remove the sensor retaining screws.

4. Remove the sensor from its mounting.

To install:

5. Installation is the reverse of the removal procedure.

TESTING

2.4L Engine

1. Disconnect the connector at the sensor.

2. Measure the resistance between terminal 3 and 2, specification should be 10.5–14 ohms. Measure the resistance between terminal 1 and 3, specification should be 10–12.5 ohms at 68 degrees F.

3. If the measured value is not within specification, replace the sensor.

2.7L Engine

2001–2003

1. Disconnect the connector at the sensor. Connect an ohmmeter.

2. Measure the resistance between terminal 3 and 2, specification should be 10.5–14 ohms. Measure the resistance between terminal 1 and 3, specification should be 10–12.5 ohms at 68 degrees F.

3. If the measured value is not within specification, replace the sensor.

2004–2006

1. Disconnect the connector at the sensor. Connect an ohmmeter.

2. Measure the resistance between the terminals.

3. Specification should be 14.5–16.5 ohms at terminal 1 and 2 (open).

4. Specification should be 16.6–18.6 ohms at 68 degrees F. at terminals 3 and 2 (closing).

5. If the measured value is not within specification, replace the sensor.

Intake Air Temperature (IAT) Sensor

LOCATION

On some vehicles this sensor is located in the air intake plenum assembly. On some vehicles this sensor is combined with the MAF sensor. On some vehicles, this sensor is located on the intake manifold.

OPERATION

This sensor is a resistor based sensor which detects the intake air temperature. According to the intake air temperature reading the ECM/PCM will control the necessary amount of fuel injection.

REMOVAL & INSTALLATION

1. Disconnect the negative battery cable.

2. Disconnect the connector from the sensor.

3. Remove the sensor retaining screws, as required.

4. Remove the air cleaner and air intake assembly, as required.

5. Remove the sensor from its mounting.

To install:

6. Installation is the reverse of the removal procedure.

TESTING

2.4L Engine

1. Allow the engine coolant to reach 176–198 degrees F.

2. Using a voltmeter check the sensor output voltage.

3. Specification should be 0.5 volt at idle, 1.0 volt at 2000 rpm.

4. If the measured value is not within specification, replace the sensor.

2.7L Engine

2001–2003

1. Remove the sensor connector.

2. Turn the ignition switch to the ON position.

3. Measure the voltage between the sensor terminal 1 and 2.

4. Specification should be 4.3 volts at 32 degrees F. and 3.44 volts at 68 degrees F.

5. Replace the sensor, as required.

2004–2006

1. Remove the sensor connector.

2. Turn the ignition switch to the ON position.
3. Measure the voltage between the sensor terminal 1 and 2.
4. Specification should be 3.52–3.90 volts at 32 degrees F. and 2.63–2.92 volts at 68 degrees F.
5. Replace the sensor, as required.

3.5L Engine

1. Be sure the engine is at operating temperature.
2. Remove the sensor connector.
3. Measure the output voltage.
4. Specification should be 1.1–1.8 volt at idle and 3.0–3.5 volts at 2100 rpm at WOT (Drive).
5. Replace the sensor, as required.

Knock Sensor (KS)

LOCATION

The knock sensor is located in the side of the cylinder block.

OPERATION

The knock sensor is used to detect engine vibrations caused by preignition or detonation and provides information to the ECM, which then retards the timing to eliminate detonation.

REMOVAL & INSTALLATION

1. Disconnect the negative battery cable.
2. Disconnect the sensor connector.
3. Remove the sensor from its mounting.

To install:

4. Installation is the reverse of the removal procedure.
5. Tighten the sensor to 12–18 ft. lbs.

TESTING

2.4L Engine

1. Disconnect the sensor electrical connector. Connect an ohmmeter.
2. Measure the resistance between terminals 2 and 3. Specification should be 5mohm at 68 degrees F.
3. If resistance is zero, replace the sensor.
4. Measure the capacitance between terminal 2 and 3. Specification should be 800–1600 pF.
5. If not within specification, replace the sensor.

2.7L Engine

2001–2003

1. Disconnect the sensor electrical connector. Connect an ohmmeter.
2. Measure the resistance between terminals 2 and 3. Specification should be 5mohm at 68 degrees F.
3. If resistance is zero, replace the sensor.
4. Measure the capacitance between terminal 2 and 3. Specification should be 800–1600 pF.
5. If not within specification, replace the sensor.

2004–2006

➡ **If the sensor is suspected of being defective, it should be replaced with a known good component for testing purposes.**

1. Check the sensor torque. It should be 12–18 ft. lbs.
2. If the sensor is still not functioning, replace it with a known good component.
3. Recheck the sensor.

3.5L Engine

1. Disconnect the sensor electrical connector. Connect an ohmmeter.
2. Measure the resistance between terminals 2 and 3. Specification should be 5mohm at 68 degrees F.
3. If resistance is zero, replace the sensor.
4. Measure the capacitance between terminal 2 and 3. Specification should be 800–1600 pF.
5. If not within specification, replace the sensor.

Manifold Absolute Pressure

LOCATION

The MAP sensor is located against the firewall to the left side of the engine.

OPERATION

This sensor is a pressure sensitive variable resistor. It measures the changes in the intake manifold pressure which result from engine load and speed changes, and converts to a voltage output. This sensor is used to measure the barometric pressure at start up, and under certain conditions, allows the ECM to automatically adjust for different altitudes. The ECM supplies 5 volts to the sensor and monitors the voltage on a signal line. The sensor provides a path to ground through its variable resistor. The sensor input affects fuel delivery and ignition timing controls in the ECM.

REMOVAL & INSTALLATION

1. Disconnect the negative battery cable.
2. Disconnect the connector from the sensor.
3. Remove the sensor retaining screws.
4. Remove the sensor from its mounting.

To install:

5. Installation is the reverse of the removal procedure.

TESTING

1. Turn the ignition switch ON.
2. Measure the voltage between terminal 1 (sensor output) and terminal 4 (sensor ground).
3. Specification should be 4–5 volts.
4. If not within specification, replace the sensor.

Mass Air Flow (MAF) Sensor

LOCATION

The MAF is mounted in the intake air hose of the air cleaner assembly. On some vehicles it is combined with the IAT sensor.

OPERATION

The MAF is measured by detection of heat transfer from a hot film probe because the change of the mass air flow rate causes change in the amount of heat being transferred from the hot film probe surface to the air flow. The air flow sensor generates a pulse so it repeatedly opens and closes between the voltage (5V) supplied from the PCM. This results in a change of the temperature of the hot film probe and in the change of resistance.

REMOVAL & INSTALLATION

1. Disconnect the negative battery cable.
2. Disconnect the connector from the sensor.
3. Remove the air cleaner and air intake assembly, as required.
4. Remove the sensor from its mounting.

To install:

5. Installation is the reverse of the removal procedure.

TESTING

2.7L Engine

1. Allow the engine coolant to reach 176–198 degrees F.
2. Using a voltmeter check the sensor output voltage.
3. Specification should be 0.5 volt at idle, 1.0 volt at 2000 rpm.
4. If the measured value is not within specification, replace the sensor.

Throttle Position Sensor (TPS)

OPERATION

The throttle position sensor is a rotating type variable resistor that rotates with the throttle body's throttle shaft to sense the throttle valve angle. As the throttle shaft rotates, the throttle angle of the sensor changes and the ECM/PCM detects the throttle valve opening based on the TPS output voltage.

REMOVAL & INSTALLATION

1. Disconnect the negative battery cable.
2. Disconnect the sensor connector.
3. Remove the sensor retaining screws. Remove the sensor from its mounting.

To install:

4. Installation is the reverse of the removal procedure.

TESTING

2.4L Engine

1. Disconnect the sensor connector.
2. Measure the resistance between terminal 1 (sensor ground) and terminal 2 (sensor power).
3. Specification should be 3.5–6.5 kohm.
4. Connect and analog ohmmeter between terminal 1 (sensor ground) and terminal 3 (sensor output).
5. Operate the throttle valve slowly from the idle position to the full open position, and check that the resistance changes smoothly in proportion with the throttle valve opening angle.
6. If the resistance is out of specification, or fails to change smoothly, replace the sensor.

2.7L Engine

2001–2003

1. Disconnect the sensor connector.
2. Measure the resistance between terminal 2 (sensor ground) and terminal 1 (sensor power).
3. Specification should be 3.5–6.5 kohm.
4. Connect and analog ohmmeter between terminal 2 (sensor ground) and terminal 3 (sensor output).
5. Operate the throttle valve slowly from the idle position to the full open position, and check that the resistance changes smoothly in proportion with the throttle valve opening angle.
6. If the resistance is out of specification, or fails to change smoothly, replace the sensor.

2004–2006

1. Disconnect the sensor connector.
2. Measure the resistance between terminal 2 (sensor ground) and terminal 1 (sensor power).
3. Specification should be 1.6–2.4 kohm.
4. Connect and analog ohmmeter between terminal 2 (sensor ground) and terminal 3 (sensor output).
5. Operate the throttle valve slowly from the idle position to the full open position, and check that the resistance changes smoothly in proportion with the throttle valve opening angle.
6. If the resistance is out of specification, or fails to change smoothly, replace the sensor.

3.5L Engine

1. Disconnect the sensor connector.
2. Measure the resistance between terminal 3 (sensor ground) and terminal 1 (sensor power).
3. Specification should be 3.5–6.5 kohm.
4. Connect and analog ohmmeter between terminal 3 (sensor ground) and terminal 2 (sensor output).
5. Operate the throttle valve slowly from the idle position to the full open position, and check that the resistance changes smoothly in proportion with the throttle valve opening angle.
6. If the resistance is out of specification, or fails to change smoothly, replace the sensor.

SONATA

See Figures 14 through 22.

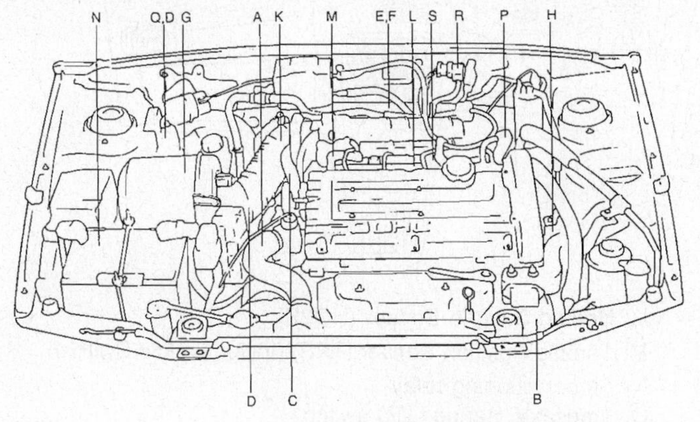

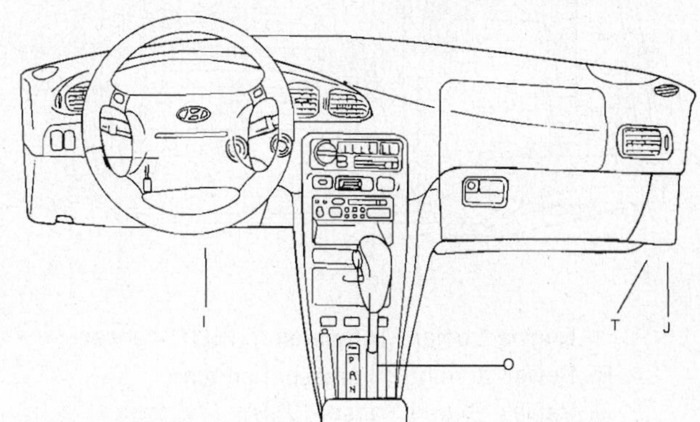

A. Engine coolant temperature sensor (ECT Sensor)
B. Power steering oil pressure switch
C. Vehicle speed sensor
D. Ignition timing adjustment terminal
E. EVAP Canister Purge Solenoid Valve
F. EGR solenoid valve
G. Volume air flow sensor (Including intake air temp. sensor and barometric pressure sensor)
H. Crankshaft position (CKP) sensor and camshaft position (CMP) sensor connector
I. Data link connector
J. MFI control relay
K. ISC motor (Stepper motor)
L. Heated oxygen sensor (HO2S)
M. Throttle position sensor (Including idle switch)
N. Air conditioning relay
O. Transaxle range (TR) switch
P. Ignition Power transistor
Q. Fuel pump
R. MDP sensor
T. Engine control module (ECM)
S. Injectors

29130_HYUN_G0014

Fig. 14 Underhood sensor locations—Sonata 1996 2.0L engine

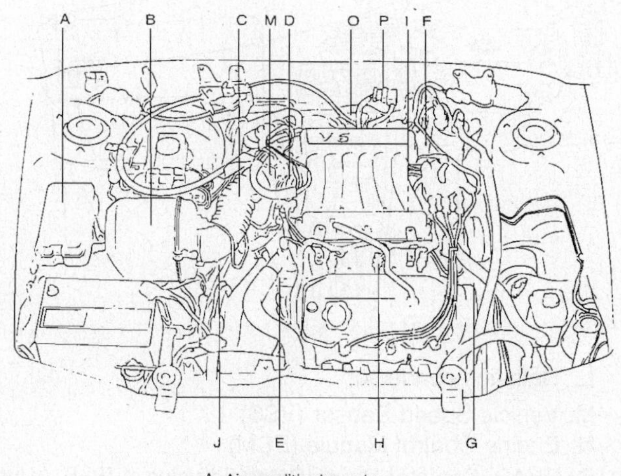

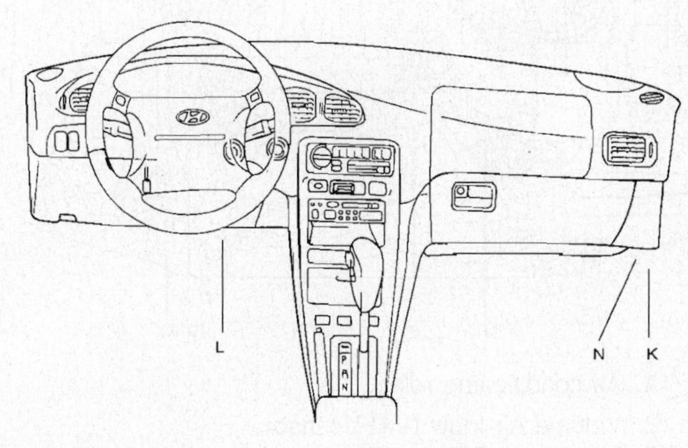

A. Air conditioning relay
B. VAF Sensor
C. ISC (Idle Speed Control) motor
D. Throttle position sensor (TP Sensor) (Including idle position switch
E. Engine coolant temperature sensor (ECT Sensor)
F. Ignition Power transistor
G. Crankshaft position (CKP) sensor & CMP sensor connector
H. Injectors
I. Heated oxygen sensor (HO2S)
J. Transaxle range (TR) switch
K. MFI control relay
L. Data link connector
M. Vehicle speed sensor
N. Engine control module (ECM)
O. EVAP Canister Purge Solenoid valve & EGR valve
P. MDP sensor

29130_HYUN_G0015

Fig. 15 Underhood sensor locations—Sonata 1996 3.0L engine

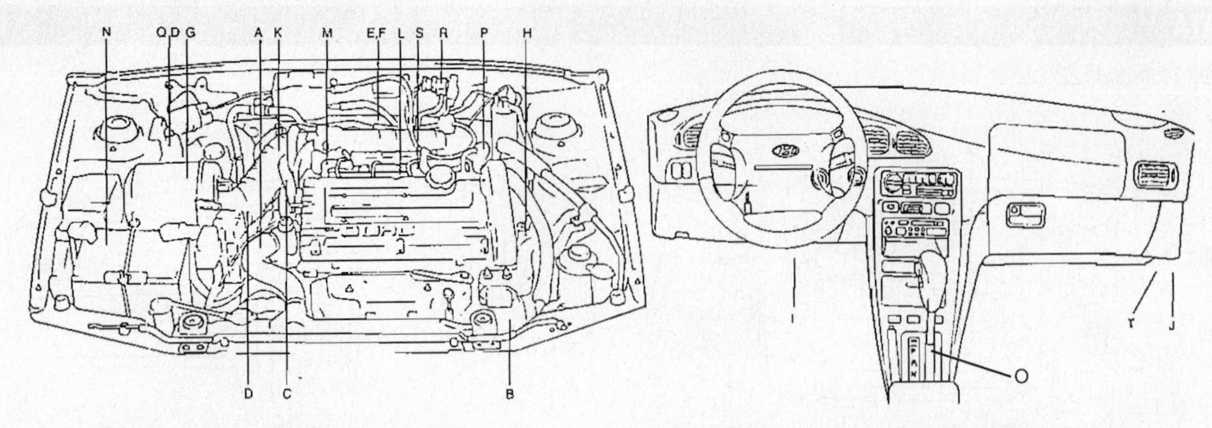

A. Engine Coolant Temperature (ECT) sensor
B. Power steering oil pressure switch
C. Vehicle Speed Sensor (VSS)
D. Ignition timing adjustment terminal
E. EVAP. Canister Purge Solenoid Valve
F. EGR solenoid valve (Federal only)
G. Volume Air Flow (VAF) sensor (Including IAT sensor and BPS)
H. Crankshaft Position (CKP) and Camshaft Position (CMP) sensor
I. Data Link Connector (DLC)
J. MFI control relay
K. Idle Speed Control (ISC) actuator
L. Heated Oxygen Sensor (HO2S)
M. Throttle Position Sensor (TPS) (Including Idle Switch)
N. Air conditioning relay
O. Transaxle Range (TR) switch
P. Ignition power transistor
Q. Fuel Pump
R. Manifold Differential Pressure (MDP) sensor
S. Injectors
T. Engine Control Module (ECM)

29130_HYUN_G0016

Fig. 16 Underhood sensor locations—Sonata 1997–1998 2.0L engine

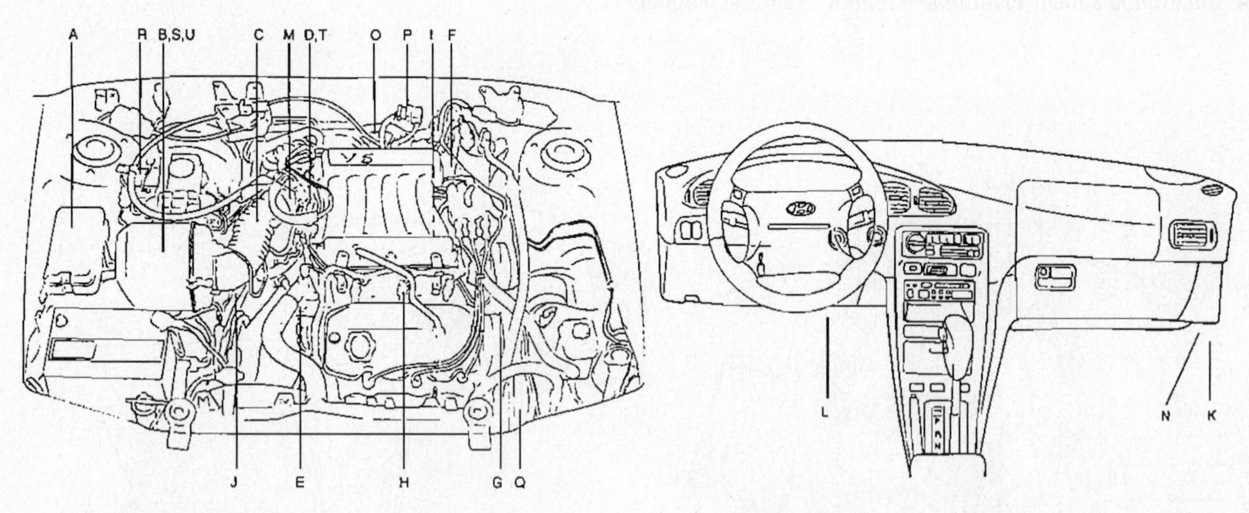

A. Air conditioning relay
B. Volume Air Flow (VAF) sensor
C. Idle Speed Control (ISC) actuator
D. Throttle Position Sensor (TPS)
E. Engine Coolant Temperature (ECT) sensor
F. Ignition Power Transistor
G. Crankshaft Position (CKP) & Camshaft Position (CMP sensor
H. Injectors
I. Heated Oxygen Sensor (HO2S)
J. Transaxle Range (TR) switch
K. MFI control relay
L. Data link connector
M. Vehicle Speed Sensor (VSS)
N. Engine Control Module (ECM)
O. EVAP. Canister Purge Solenoid valve & EGR valve
P. Manifold Differential Pressure (MDP) sensor
Q. Power steering oil pressure switch
R. Fuel pump
S. Barometric Pressure Sensor (BPS)
T. IDLE position switch
U. Intake air temperature sensor

29130_HYUN_G0017

Fig. 17 Underhood sensor locations—Sonata 1997–1998 3.0L engine

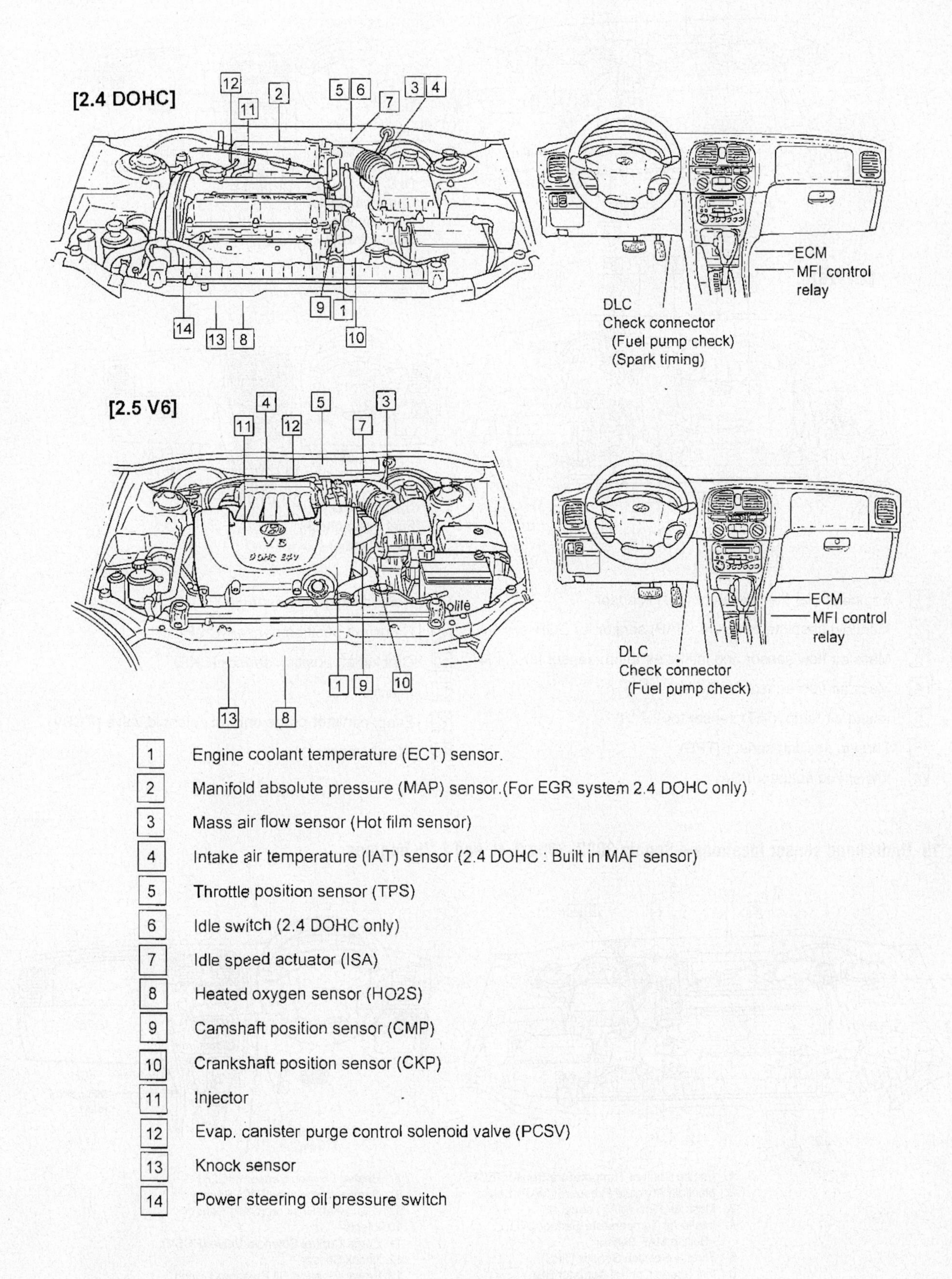

Fig. 18 Underhood sensor locations—Sonata 1999–2001 2.4L and 2.5L engines

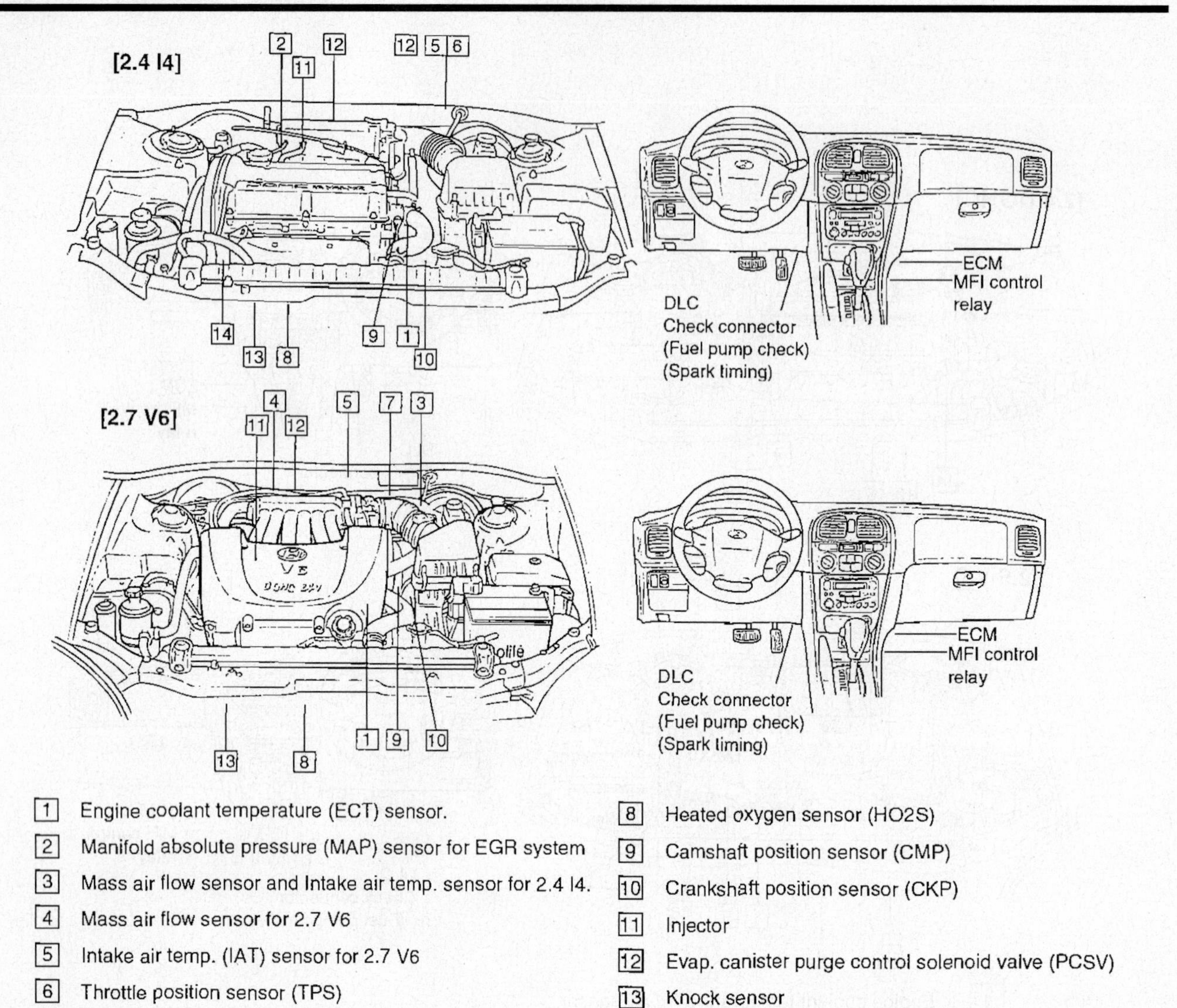

29130_HYUN_G0019

Fig. 19 Underhood sensor locations—Sonata 2002–2004 2.4L and 2.7L engines

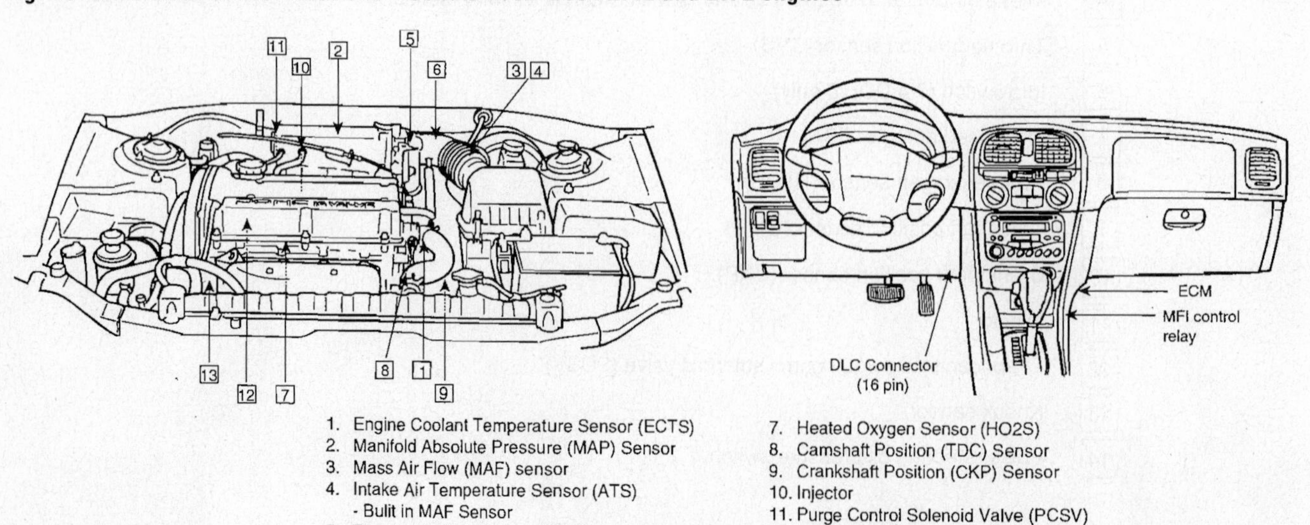

29130_HYUN_G0020

Fig. 20 Underhood sensor locations—Sonata 2005 2.4L engine

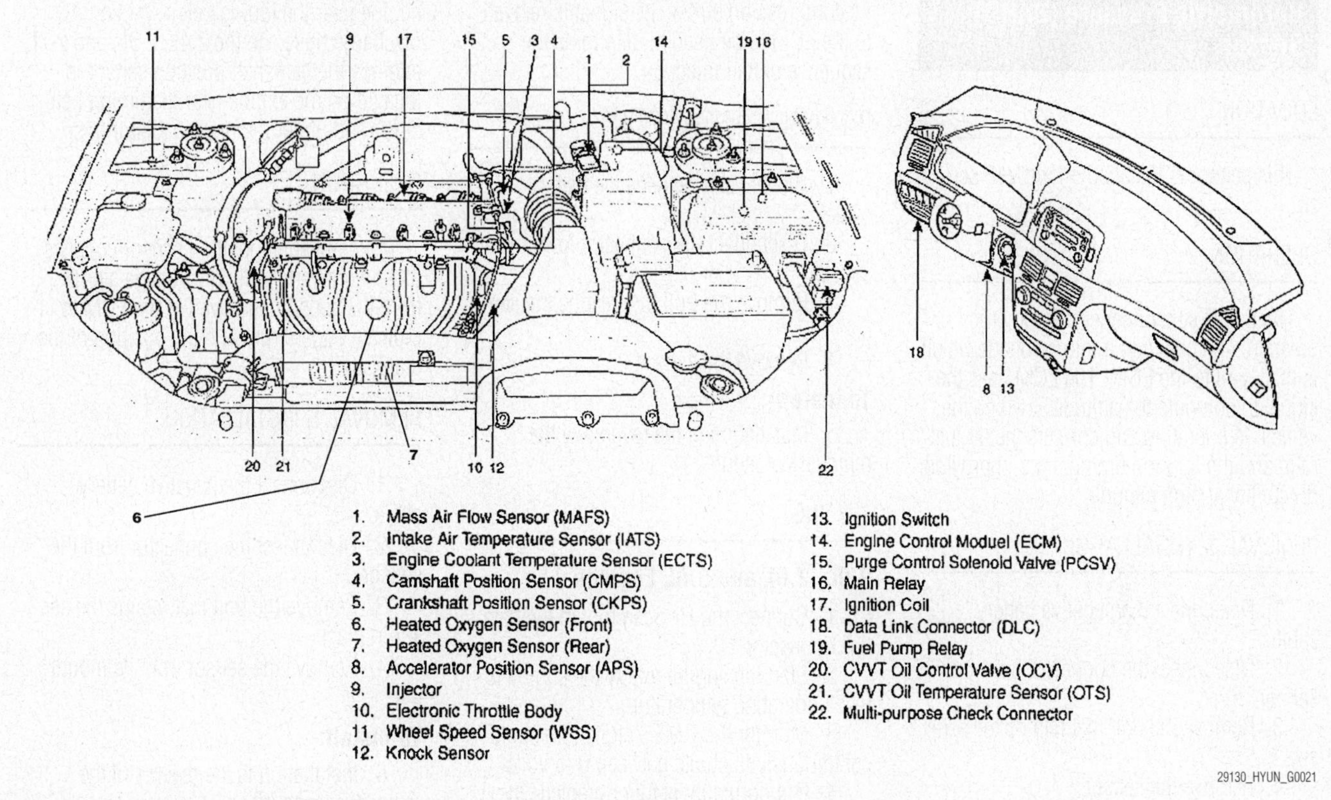

Fig. 21 Underhood sensor locations—Sonata 2006 2.4L engine

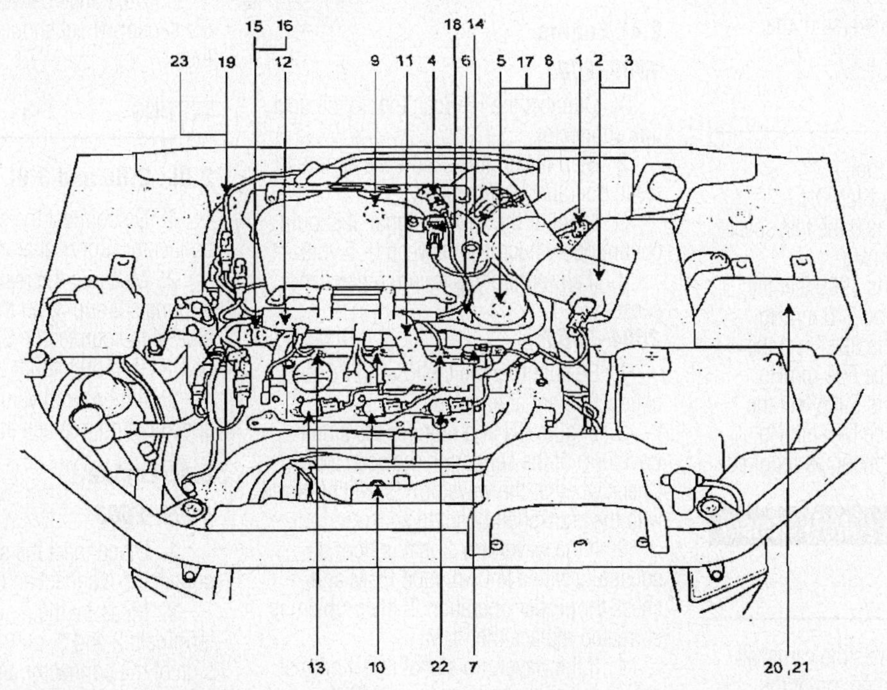

Fig. 22 Underhood sensor locations—Sonata 2006 3.3L engine

Barometric Pressure (BARO) Sensor

LOCATION

This sensor is installed on the VAF sensor.

OPERATION

The BARO sensor senses barometric pressure and converts it into a voltage which is than sent to the ECM. The ECM uses the signal to compute the altitude at which the vehicle is operating and corrects the air fuel ratio and the ignition timing, thus improving drivability at high altitude.

REMOVAL & INSTALLATION

1. Disconnect the negative battery cable.
2. Disconnect the connector from the sensor.
3. Remove the bolt that retains the sensor.
4. Remove the sensor.

To install:

5. Installation is the reverse of the removal procedure.

TESTING

1. Connect the scan tool.
2. Turn the ignition switch ON.
3. When the altitude is 0 the test specification should be 760 mmHg.
4. When the altitude is 1969 feet the test specification should be 710 mmHg.
5. When the altitude is 3937 feet the test specification should be 660 mmHg.
6. When the altitude is 5906 feet the test specification should be 610 mmHg.
7. Replace the component, as required.

Camshaft Position (CMP) Sensor

LOCATION

On 2.0L and 3.0L engines the camshaft position sensor is located on top of the engine near the timing belt cover and AC discharge hose. On the 2.4L, 2.5L and 2.7L engines the camshaft position sensor is located on the engine near the timing belt cover and under the upper radiator hose.

OPERATION

The camshaft position sensor senses the TDC point of the number one cylinder, on its compression stroke. Its signal is relayed to the ECM to be used to determine the sequence of fuel injection.

REMOVAL & INSTALLATION

1. Disconnect the negative battery cable.
2. Disconnect the connector from the sensor.
3. Remove the bolt that retains the sensor.
4. Remove the sensor.

To install:

5. Installation is the reverse of the removal procedure.

TESTING

2.0L, 2.5L and 3.0L Engines

1. Connect the Hi-Scan tool to the data link connector.
2. Start the engine. Allow the engine to reach operating temperature.
3. Monitor the CMPS signal. It should continuously fluctuate between 0–5 volts.
4. If abnormality is found, replace the sensor.

2.4L Engine

1999–2003

1. Connect the Hi-Scan tool to the data link connector.
2. Start the engine. Allow the engine to reach operating temperature.
3. Monitor the CMPS signal. It should continuously fluctuate between 0–5 volts.
4. If abnormality is found, replace the sensor.

2004–2006

1. Be sure that the CMPS and PCM connectors are connected.
2. Engine ON and monitor the signal waveform of the sensor on the scan tool. Check whether the waveform is synchronized with the crankshaft position sensor or not.
3. If the waveform signal is normal, substitute with a known good PCM and check for proper operation. If the problem is corrected replace the PCM.
4. If the waveform signal is not normal, substitute with a known good CMPS sensor and check for proper operation. If the problem is corrected replace the CMPS.

Crankshaft Position (CKP) Sensor

LOCATION

On 2.0L and 3.0L engines the camshaft position sensor is located on top of the engine near the timing belt cover and AC discharge hose. On the 2.4L, 2.5L, and 2.7L engines the camshaft position sensor is located on the engine near the timing belt cover and under the upper radiator hose.

OPERATION

The crankshaft position sensor consists of a magnet and coil. The voltage signal from the sensor is relayed to the ECM to indicate engine rpm and the position of the crankshaft.

REMOVAL & INSTALLATION

1. Disconnect the negative battery cable.
2. Disconnect the connector from the sensor.
3. Remove the bolt that retains the sensor in place.
4. Remove the sensor from its mounting.

To install:

5. Installation is the reverse of the removal procedure.
6. Clearance between the sensor and the sensor wheel should be 0.020–0.059 inch.

TESTING

2.0L, 2.5L and 3.0L Engines

1. Disconnect the sensor electrical connector. Connect an ohmmeter.
2. Measure the resistance between terminals 2 and 3, on the crankshaft position side of the connector. Specification should be 0.486–0.594 kohm at 68 degrees F.
3. If measured value is not within specification, replace the sensor.

2.4L Engine

1999–2003

1. Disconnect the sensor electrical connector. Connect an ohmmeter.
2. Measure the resistance between terminals 2 and 3, on the crankshaft position side of the connector. Specification should be 0.486–0.594 kohm at 68 degrees F.
3. If measured value is not within specification, replace the sensor.

2004–2006

1. Be sure that the CKPS and PCM connectors are connected.
2. Engine ON and monitor the signal waveform of the sensor on the scan tool. Check whether the waveform is synchronized with the crankshaft position sensor or not.

3. If the waveform signal is normal, substitute with a known good PCM and check for proper operation. If the problem is corrected replace the PCM.

4. If the waveform signal is not normal, substitute with a known good CMPS sensor and check for proper operation. If the problem is corrected replace the CMPS.

Electronic Control Module (ECM)

OPERATION

The ECM controls the vehicle engine operating system.

REMOVAL & INSTALLATION

1. Disconnect the negative battery cable.
2. Remove the lower inner trim..
3. As required, detach the floor mat. As required, remove the protective cover.
4. Remove the ECM bracket retaining nuts. Remove the clip from the bracket.
5. Disconnect the connectors.
6. Remove the ECM from the vehicle.

To install:

7. Installation is the reverse of the removal procedure.

➡ **When replacing the ECM, be careful not to use the wrong part number, as damage to the injection system could occur.**

Engine Coolant Temperature (ECT) Sensor

LOCATION

The ECT is installed in the engine coolant passage of the cylinder head.

OPERATION

This component detects the temperature of the engine coolant and relays the information to the electronic control assembly. This component employs a thermistor which is sensitive to temperature changes. The electric resistance of the thermistor decreases in response to temperature rise. The ECM judges coolant temperature by the sensor output voltage and provides optimum fuel enrichment when the engine is cold.

REMOVAL & INSTALLATION

1. Disconnect the negative battery cable.
2. Disconnect the connector from the sensor.
3. Drain the cooling system, as required.
4. Remove the sensor from its mounting.

To install:

5. Installation is the reverse of the removal procedure.

TESTING

1. Remove the sensor.
2. With the temperature sensing portion of the sensor immersed in hot engine coolant, check the resistance.
3. Specification should be 5.9 kohms at 32 degrees F., 2.5 kohms at 68 degrees F., 1.1 kohms at 104 degrees F., 0.3 kohms at 176 degrees F. for all engines except 2004–2006 2.4L engine.
4. Specification should be 5.79 kohms at 32 degrees F., 2.31–2.59 kohms at 68 degrees F., 1.15kohms at 104 degrees F., 0.32 kohms at 176 degrees F. for 2004–2006 2.4L engine.
5. If not within specification, replace the sensor.

Heated Oxygen (HO2S) Sensor

LOCATION

The sensors are located in the exhaust system. On some vehicles one sensor is located up at the exhaust manifold(s) and the other sensor is located down at the catalytic converter. On other vehicles both sensors are located down at the catalytic converter.

OPERATION

The exhaust gas oxygen sensor supplies the electronic control assembly with a signal which indicates either a rich or lean mixture condition, during the engine operation.

REMOVAL & INSTALLATION

1. Disconnect the electrical connector from the sensor.
2. Remove the oxygen sensor.

To install:

3. Installation is the reverse of the removal procedure.

➡ **Apply anti-seize compound to the threaded portion of the sensor, prior to installation. Never apply anti-seize compound to the protector of the sensor.**

TESTING

2.0L and 3.0L Engines

Perform a visual inspection of the sensor as follows:

1. If the sensor tip has a black/sooty deposit, this may indicate a rich fuel mixture.
2. If the sensor tip has a white, gritty deposit, this may indicate an internal coolant leak.
3. If the sensor tip has a brown deposit, this could indicate excessive oil consumption.
4. Warm the engine until the coolant temperature reaches 176–205 degrees F.
5. When decelerating suddenly from 4000 rpm, the voltmeter reading should be 200mV or lower.
6. When the engine is suddenly raced, the voltmeter reading should be 600–1000mV.
7. If measured value is not within specification, replace the sensor.
8. Turn the ignition switch to the OFF position.
9. Disconnect the sensor connector.
10. Measure the resistance between terminals 3 and 4 (component side).
11. Specification should be 30 ohms or more at 752 degrees F.
12. If not, replace the sensor.
13. Apply battery voltage directly between terminal 3 and 4.

➡ **Take care when applying the voltage. Damage will result if the terminals are incorrect or short circuited.**

14. Connect a digital voltmeter between terminal 1 and 2.
15. While racing the engine, measure the sensor output voltage. Specification should be 0.6 volt (minimum).
16. If not, replace the sensor.

2.4L Engine

1999–2003

Perform a visual inspection of the sensor as follows:

1. If the sensor tip has a black/sooty deposit, this may indicate a rich fuel mixture.
2. If the sensor tip has a white, gritty deposit, this may indicate an internal coolant leak.
3. If the sensor tip has a brown deposit, this could indicate excessive oil consumption.
4. Warm the engine until the coolant temperature reaches 176–205 degrees F.
5. Accelerate the engine to 4000 rpm. When decelerating suddenly from 4000 rpm, the voltmeter reading should be 200mV or lower.

6. When the engine is suddenly raced, the voltmeter reading should be 600–1000mV.

7. If measured value is not within specification, replace the sensor.

8. Turn the ignition switch to the OFF position.

9. Disconnect the sensor connector.

10. Measure the resistance between terminals 3 and 4 (component side).

11. Specification should be 30 ohms or more at 752 degrees F.

12. If not, replace the sensor.

13. Apply battery voltage directly between terminal 3 and 4.

➡ **Take care when applying the voltage. Damage will result if the terminals are incorrect or short circuited.**

14. Connect a digital voltmeter between terminal 1 and 2.

15. While racing the engine, measure the sensor output voltage. Specification should be 0.6 volt (minimum).

16. If not, replace the sensor.

2004–2006

Perform a visual inspection of the sensor as follows:

1. If the sensor tip has a black/sooty deposit, this may indicate a rich fuel mixture.

2. If the sensor tip has a white, gritty deposit, this may indicate an internal coolant leak.

3. If the sensor tip has a brown deposit, this could indicate excessive oil consumption.

4. Warm the engine until the coolant temperature reaches 176–205 degrees F.

5. Accelerate the engine to 4000 rpm. When decelerating suddenly from 4000 rpm, the voltmeter reading should be 200mV or lower.

6. When the engine is suddenly raced, the voltmeter reading should be 600–1000mV.

7. If measured value is not within specification, replace the sensor.

8. Disconnect the sensor connector.

9. Measure the resistance between terminals 3 and 4 (component side).

10. Specification should be 5.0–7.0 ohms.

11. If not, replace the sensor.

12. Apply battery voltage directly between terminal 3 and 4.

➡ **Take care when applying the voltage. Damage will result if the terminals are incorrect or short circuited.**

13. Connect a digital voltmeter between terminal 1 and 2.

14. While racing the engine, measure the sensor output voltage. Specification should be 0.6 volt (minimum).

15. If not, replace the sensor.

2.5L and 2.7L Engines

Perform a visual inspection of the sensor as follows:

1. If the sensor tip has a black/sooty deposit, this may indicate a rich fuel mixture.

2. If the sensor tip has a white, gritty deposit, this may indicate an internal coolant leak.

3. If the sensor tip has a brown deposit, this could indicate excessive oil consumption.

4. Warm the engine until the coolant temperature reaches 176–205 degrees F.

5. Accelerate the engine to 4000 rpm. When decelerating suddenly from 4000 rpm, the voltmeter reading should be 200mV or lower.

6. When the engine is suddenly raced, the voltmeter reading should be 600–1000mV.

7. If measured value is not within specification, replace the sensor.

8. Apply battery voltage directly between terminal 3 and 4.

➡ **Take care when applying the voltage. Damage will result if the terminals are incorrect or short circuited.**

9. Connect a digital voltmeter between terminal 1 and 2.

10. While racing the engine, measure the sensor output voltage. Specification should be 4.5 volt for 2.5L engine and 4000–4500 mV for 2.7l engine.

11. If not, replace the sensor.

Idle Speed Control Actuator (ISCA)

OPERATION

The idle speed control actuator sensor is a double coil type driven by separate driver stages in the ECM. Depending on the pulse duty factor, the equilibrium of the magnetic forces of the two coils will result in different angles of the motor. A bypass hose line is positioned, parallel to the throttle valve where the idle speed actuator is inserted.

REMOVAL & INSTALLATION

1. Disconnect the negative battery cable.

2. Disconnect the connector from the sensor.

3. Remove the sensor retaining screws.

4. Remove the sensor from its mounting.

To install:

5. Installation is the reverse of the removal procedure.

TESTING

2.4L Engine

1999–2000

1. Disconnect the connector at the sensor. Connect an ohmmeter.

2. Measure the resistance between terminal 1 and 2, specification should be 10.5–14 ohms.

3. Measure the resistance between terminal 2 and 3, specification should be 10–12.5 ohms at 68 degrees F.

4. If the measured value is not within specification, replace the sensor.

2001–2002

1. Disconnect the connector at the sensor. Connect an ohmmeter.

2. Measure the resistance between terminal 3 and 2, specification should be 10.5–14 ohms.

3. Measure the resistance between terminal 1 and 3, specification should be 10–12.5 ohms at 68 degrees F.

4. If the measured value is not within specification, replace the sensor.

2003

1. Disconnect the connector at the sensor. Connect an ohmmeter.

2. Measure the resistance between terminal 1 and 2, specification should be 17.0–18.2 ohms.

3. Measure the resistance between terminal 2 and 3, specification should be 15.0–16.0 ohms.

4. If the measured value is not within specification, replace the sensor.

2004–2006

1. Disconnect the connector at the sensor. Connect an ohmmeter.

2. Measure the resistance between terminal 3 and 2 (open), specification should be 14.5–16.5 ohms.

3. Measure the resistance between terminal 1 and 2 (closing), specification should be 16.6–18.6 ohms at 68 degrees F.

4. If the measured value is not within specification, replace the sensor.

2.5L Engine

1999–2000

1. Disconnect the connector at the sensor. Connect an ohmmeter.

2. Measure the resistance between terminal 1 and 2, specification should be 10.5–14 ohms.

3. Measure the resistance between terminal 2 and 3, specification should be 10–12.5 ohms at 68 degrees F.

4. If the measured value is not within specification, replace the sensor.

2001–2002

1. Disconnect the connector at the sensor. Connect an ohmmeter.

2. Measure the resistance between terminal 3 and 2, specification should be 10.5–14 ohms.

3. Measure the resistance between terminal 1 and 3, specification should be 10–12.5 ohms at 68 degrees F.

4. If the measured value is not within specification, replace the sensor.

2.7L Engine

1. Disconnect the connector at the sensor. Connect an ohmmeter.

2. Measure the resistance between terminal 1 and 2, specification should be 17.0–18.2 ohms.

3. Measure the resistance between terminal 2 and 3, specification should be 15.0–16.0 ohms.

4. If the measured value is not within specification, replace the sensor.

Idle Speed Control Actuator (ISC)

OPERATION

The intake air volume at idle is controlled by opening or closing the motor valve provided in the air path that bypasses the throttle valve.

REMOVAL & INSTALLATION

1. Disconnect the negative battery cable.

2. Disconnect the connector from the sensor.

3. Remove the sensor retaining screws.

4. Remove the sensor from its mounting.

To install:

5. Installation is the reverse of the removal procedure.

TESTING

1. Check for the sound of the stepper motor after the ignition is switched ON (but without starting the motor).

2. If the operation sound cannot be heard, check the stepper motor's circuit.

3. If the circuit is normal, measure the resistance between terminals 2–3 and/or 1, specification should be 28–33 ohms at 68 degrees F.

4. Measure the resistance between terminals 5–4 and/or 6, specification should be 28–33 ohms at 68 degrees F.

5. Replace the component, as required.

Intake Air Temperature (IAT) Sensor

LOCATION

On the 2.0L and 3.0L engines, this sensor is located in the air intake plenum assembly. It is built into the volume air flow sensor (VAF). On all other engines this sensor is located in the air plenum assembly. On some vehicles it is combined with the MAF sensor.

OPERATION

This sensor is a resistor based sensor which detects the intake air temperature. According to the intake air temperature reading the ECM will control the necessary amount of fuel injection.

REMOVAL & INSTALLATION

1. Disconnect the negative battery cable.

2. Disconnect the connector from the sensor.

3. Remove the sensor retaining screws, as required.

4. Remove the air cleaner and air intake assembly, as required.

5. Remove the sensor from its mounting.

To install:

6. Installation is the reverse of the removal procedure.

TESTING

2.0L and 3.0L Engines

1. Disconnect the VAF connectors.

2. Measure the resistance between terminals 4 and 6.

3. Specification should be 6.0 kohm at 32 degrees F, 2.7 kohm at 68 degrees F, 0.4 kohm at 176 degrees F.

4. Measure the resistance while heating the sensor using a hair dryer.

5. Specification should be 6.0 kohm at 32 degrees F, 2.7 kohm at 68 degrees F, 0.4 kohm at 176 degrees F.

6. Replace the VAF assembly, as required.

2.4L Engine

1999–2002

1. Be sure the ignition switch in the ON position.

2. Measure the voltage between the sensor terminal 1 and 2.

3. Specification should be 3.3–3.7 volts at 32 degrees F., 2.4–2.8 volts at 68 degrees F., 1.6–2.0 volts at 104 degrees F., 0.5–0.9 volts at 176 degrees F.

4. Replace the sensor, as required.

2003

1. Be sure the ignition switch in the ON position.

2. Measure the voltage between the sensor terminal 1 and 2.

3. Specification should be 2.22–2.82 volts at 68 degrees F.

4. Replace the sensor, as required.

2004–2006

1. Be sure the ignition switch in the ON position, or running.

2. Measure the resistance, using an ohmmeter.

3. Specification should be 33.85–61.20 kohms at -40 degrees F., 2.33–2.97 kohms at 68 degrees F., 0.31–0.43 kohms at 176 degrees F.,

4. Replace the sensor, as required.

2.5L Engine

1. Be sure the ignition switch in the ON position.

2. Measure the voltage between the sensor terminal 1 and 2.

3. Specification should be 3.3–3.7 volts at 32 degrees F., 2.4–2.8 volts at 68 degrees F., 1.6–2.0 volts at 104 degrees F., 0.5–0.9 volts at 176 degrees F.

4. Replace the sensor, as required.

2.7L Engine

1. Be sure the ignition switch in the ON position.

2. Measure the voltage between the sensor terminal 1 and 2.

3. Specification should be 2.22–2.82 volts at 68 degrees F.

4. Replace the sensor, as required.

Knock Sensor (KS)

LOCATION

The knock sensor is located in the side of the cylinder block.

OPERATION

The knock sensor is used to detect engine vibrations caused by preignition or detonation and provides information to the ECM, which then retards the timing to eliminate detonation.

REMOVAL & INSTALLATION

1. Disconnect the negative battery cable.
2. Disconnect the sensor connector.
3. Remove the sensor from its mounting.

To install:

4. Installation is the reverse of the removal procedure.
5. Tighten the sensor to 12–18 ft. lbs.

TESTING

2.4L Engine

1999–2003

1. Disconnect the sensor electrical connector. Connect an ohmmeter
2. Measure the resistance between terminals 2 and 3. Specification should be 5mohm at 68 degrees F.
3. If resistance is zero, replace the sensor.
4. Measure the capacitance between terminal 2 and 3. Specification should be 800–1600 pF.
5. If not within specification, replace the sensor.

2004–2006

➡ If the sensor is suspected of being defective, it should be replaced with a known good component for testing purposes.

1. Check the sensor torque. It should be 12–18 ft. lbs.
2. If the sensor is still not functioning, replace it with a known good component.
3. Recheck the sensor.

2.5L and 2.7L Engines

1. Disconnect the sensor electrical connector. Connect an ohmmeter
2. Measure the resistance between terminals 2 and 3. Specification should be 5mohm at 68 degrees F.
3. If resistance is zero, replace the sensor.
4. Measure the capacitance between terminal 2 and 3. Specification should be 800–1600 pF.
5. If not within specification, replace the sensor.

Mass Air Flow (MAF) Sensor

LOCATION

The MAF is mounted in the intake air hose of the air cleaner assembly. On some vehicles this sensor is combined wit the ATS sensor.

OPERATION

The MAF is measured by detection of heat transfer from a hot film probe because the change of the mass air flow rate causes change in the amount of heat being transferred from the hot film probe surface to the air flow. The air flow sensor generates a pulse so it repeatedly opens and closes between the voltage (5V) supplied from the ECM. This results in a change of the temperature of the hot film probe and in the change of resistance.

REMOVAL & INSTALLATION

1. Disconnect the negative battery cable.
2. Disconnect the connector from the sensor.
3. Remove the air cleaner and air intake assembly, as required.
4. Remove the sensor from its mounting.

To install:

5. Installation is the reverse of the removal procedure.

TESTING

1. Using a voltmeter check the output voltage of the sensor.
2. Specification should be 0.5 volt at idle and 1.0 volt at 2000 rpm.
3. Replace the sensor, as required.

Manifold Absolute Pressure (MAP) Sensor

OPERATION

This sensor converts the pressure in the intake manifold to voltage signal The ECM judges the condition of the EGR by this signal.

REMOVAL & INSTALLATION

1. Disconnect the negative battery cable.
2. Disconnect the connector from the sensor.
3. Remove the sensor retaining screws.
4. Remove the sensor from its mounting.

To install:

5. Installation is the reverse of the removal procedure.

TESTING

2.4L Engine

1999–2003

1. Measure the voltage between terminal 1 (sensor ground) and terminal 4 (sensor output), 1999–2000 vehicles.
2. Measure the voltage between terminal 4 (sensor ground) and terminal 1 (sensor output), 2001–2003 vehicles.
3. Specification should be 4–5 volts, with the ignition switch ON.
4. Specification should be 0.8–2.4 volts at idle.
5. If not within specification, replace the sensor.

2004–2006

1. Measure the voltage between terminal 4 (sensor ground) and terminal 1 (sensor output).
2. Specification should be 4–5 volts, with the ignition switch ON.
3. Specification should be 0.8–2.4 volts at idle.
4. If not within specification, replace the sensor.

2.5L Engine

1. Measure the voltage between terminal 1 (sensor ground) and terminal 4 (sensor output), 1999–2000 vehicles.
2. Measure the voltage between terminal 4 (sensor ground) and terminal 1 (sensor output), 2001–2003 vehicles.
3. Specification should be 4–5 volts, with the ignition switch ON.
4. Specification should be 0.8–2.4 volts at idle.
5. If not within specification, replace the sensor.

2.7L Engine

1. Measure the voltage between terminal 4 (sensor ground) and terminal 1 (sensor output).
2. Specification should be 4–5 volts, with the ignition switch ON.
3. Specification should be 0.8–2.4 volts at idle.
4. If not within specification, replace the sensor.

Manifold Differential Pressure (MDP) Sensor

LOCATION

The MDP sensor is located against the firewall near the windshield wiper motor.

OPERATION

This sensor converts the pressure in the intake manifold to voltage and inputs it to the ECM. The ECM judges the condition of the EGR by this signal. If there is abnormal condition the engine warning light is turned on to notify the driver.

REMOVAL & INSTALLATION

1. Disconnect the negative battery cable.
2. Disconnect the connector from the sensor.
3. Remove the sensor retaining screws.
4. Remove the sensor from its mounting.

To install:

5. Installation is the reverse of the removal procedure.

Throttle Position Sensor (TPS)

OPERATION

The throttle position sensor is a rotating type variable resistor that rotates with the throttle body's throttle shaft to sense the throttle valve angle. As the throttle shaft rotates, the throttle angle of the sensor changes and the ECM detects the throttle valve opening based on the TPS output voltage.

REMOVAL & INSTALLATION

1. Disconnect the negative battery cable.
2. Disconnect the sensor connector.
3. Remove the sensor retaining screws. Remove the sensor from its mounting.

To install:

4. Installation is the reverse of the removal procedure.

TESTING

2.0L and 3.0L Engines

1. Disconnect the connector from the sensor.
2. Measure the resistance between terminals 1 (sensor ground) and 4 (sensor power).
3. Specification should be 3.5–6.5 kohm.
4. Connect an analog ohmmeter between terminals 1 (sensor ground) and 3 (sensor output.
5. Operate the throttle valve slowly from the idle position to the full open position, and check that the resistance changes smoothly in proportion with the throttle valve opening angle.
6. If the resistance is out of specification, or fails to change smoothly, replace the sensor.

2.4L Engine

1999–2000

1. Disconnect the connector from the sensor.
2. Measure the resistance between terminals 1 (sensor ground) and 4 (sensor power) on 2.4L engine and between terminals 2 (sensor ground) and 3 (sensor power) on 2.5L engine.
3. Specification should be 3.5–6.5 kohm.
4. Connect an analog ohmmeter between terminals 1 (sensor ground) and 3 (sensor output), on 2.4L engine and between terminals 2 (sensor ground) and 1 (sensor output), on 2.5L engine.
5. Operate the throttle valve slowly from the idle position to the full open position, and check that the resistance changes smoothly in proportion with the throttle valve opening angle.
6. If the resistance is out of specification, or fails to change smoothly, replace the sensor.

2001–2003

1. Disconnect the connector from the sensor.
2. Measure the resistance between terminals 4 (sensor ground) and 1 (sensor power) on 2.4L engine and between terminals 2 (sensor ground) and 1 (sensor power) on 2.5L and 2.7L engines.
3. Specification should be 3.5–6.5 kohm.
4. Connect an analog ohmmeter between terminals 4 (sensor ground) and 2 (sensor output), on 2.4L engine and between terminals 2 (sensor ground) and 3 (sensor output), on 2.5L and 2.7L engines.
5. Operate the throttle valve slowly from the idle position to the full open position, and check that the resistance changes smoothly in proportion with the throttle valve opening angle.
6. If the resistance is out of specification, or fails to change smoothly, replace the sensor.

2004–2006

1. Disconnect the connector from the sensor.
2. Measure the resistance between terminals 1 (sensor ground) and 2 (sensor power).
3. Specification should be 1.6–2.4 kohm.
4. Connect an analog ohmmeter between terminals 1 (sensor ground) and 2 (sensor output).
5. Operate the throttle valve slowly from the idle position to the full open position, and check that the resistance changes smoothly in proportion with the throttle valve opening angle.
6. If the resistance is out of specification, or fails to change smoothly, replace the sensor.

2.5L Engine

1999–2000

1. Disconnect the connector from the sensor.
2. Measure the resistance between terminals 1 (sensor ground) and 4 (sensor power) on 2.4L engine and between terminals 2 (sensor ground) and 3 (sensor power) on 2.5L engine.
3. Specification should be 3.5–6.5 kohm.
4. Connect an analog ohmmeter between terminals 1 (sensor ground) and 3 (sensor output), on 2.4L engine and between terminals 2 (sensor ground) and 1 (sensor output), on 2.5L engine.
5. Operate the throttle valve slowly from the idle position to the full open position, and check that the resistance changes smoothly in proportion with the throttle valve opening angle.
6. If the resistance is out of specification, or fails to change smoothly, replace the sensor.

2001–2002

1. Disconnect the connector from the sensor.
2. Measure the resistance between terminals 4 (sensor ground) and 1 (sensor power) on 2.4L engine and between terminals 2 (sensor ground) and 1 (sensor power) on 2.5L and 2.7L engines.
3. Specification should be 3.5–6.5 kohm.
4. Connect an analog ohmmeter between terminals 4 (sensor ground) and 2 (sensor output), on 2.4L engine and between

terminals 2 (sensor ground) and 3 (sensor output), on 2.5L and 2.7L engines.

5. Operate the throttle valve slowly from the idle position to the full open position, and check that the resistance changes smoothly in proportion with the throttle valve opening angle.

6. If the resistance is out of specification, or fails to change smoothly, replace the sensor.

2.7L Engine

1. Disconnect the connector from the sensor.

2. Measure the resistance between terminals 4 (sensor ground) and 1 (sensor power) on 2.4L engine and between terminals 2 (sensor ground) and 1 (sensor power) on 2.5L and 2.7L engines.

3. Specification should be 3.5–6.5 kohm.

4. Connect an analog ohmmeter between terminals 4 (sensor ground) and 2 (sensor output), on 2.4L engine and between terminals 2 (sensor ground) and 3 (sensor output), on 2.5L and 2.7L engines.

5. Operate the throttle valve slowly from the idle position to the full open position, and check that the resistance changes smoothly in proportion with the throttle valve opening angle.

6. If the resistance is out of specification, or fails to change smoothly, replace the sensor.

Volume Air Flow (VAF) Sensor

LOCATION

This sensor is located in the air intake plenum assembly.

OPERATION

This sensor measures the intake air volume. It makes use of a Karman vortex to detect the sir flow rate and sends it to the ECM as the intake air volume signal. The ECM uses this intake air volume signal to decide the basic fuel injection duration.

REMOVAL & INSTALLATION

1. Disconnect the negative battery cable.

2. Disconnect the sensor connector.

3. Remove the air filter cover.

4. Remove the sensor retaining screws. Remove the sensor from its mounting.

To install:

5. Installation is the reverse of the removal procedure.

TESTING

1. Connect the scan tool.

2. Allow the engine to reach operating temperature.

3. Be sure all accessories are off. Be sure the transaxle is in Park for automatic transaxle and Neutral for manual transaxle.

4. On 2.0L engine specification should be 25–50Hz at idle (750 rpm) and 70–90Hz at 2000 rpm.

5. On 3.0L engine specification should be 30–45Hz at idle (750 rpm) and 85–105Hz at 2000 rpm.

6. Replace the component, as required.

Vehicle Speed Sensor (VSS)

OPERATION

This sensor is a hall effect sensor. The sensor converts the transaxle gear revolutions into pulse signals, which are sent to the ECM.

REMOVAL & INSTALLATION

2.0L and 3.0L Engines

1. Disconnect the connector.

2. Remove the sensor from its mounting.

To install:

3. Installation is the reverse of the removal procedure.

TESTING

2.0L and 3.0L Engines

1. Remove the sensor.

2. Connect the positive lead from the battery cable to terminal 2 and the negative lead to terminal 1.

3. Connect the positive lead from the tester to terminal 1 and the negative lead to terminal 3.

4. Turn the speedometer shaft, using a suitable tool.

5. Check for a voltage change from about 0–11 volts or more between terminals 1 and 3.

6. The voltage change should be four times per each revolution of the shaft.

7. Replace the component, as required.

TIBURON

See Figures 23 through 27.

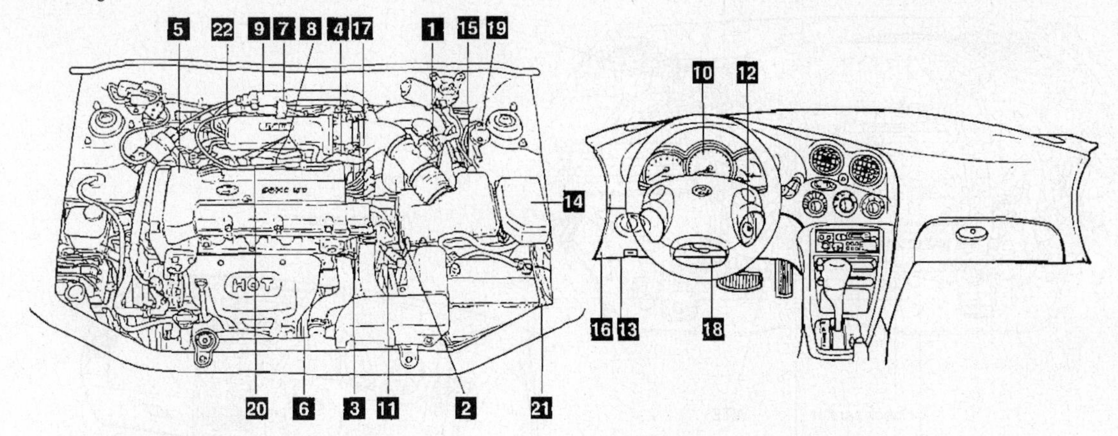

1. MAF (Mass Air Flow) sensor
2. Intake Air Temperature (IAT) sensor
3. Engine Coolant Temperature (ECT) sensor
4. Throttle Position Sensor (TPS)
5. Camshaft Position (CMP) sensor
6. Crankshaft Position (CKP) sensor
7. Heated Oxygen Sensor (HO2S)
8. Injectors
9. Idle Speed Control (ISC) actuator
10. Vehicle Speed Sensor (VSS)
11. Transaxle Range (TR) switch
12. Ignition switch
13. Engine Control Module (ECM)
14. Air conditioning relay
15. EVAP. canister purge solenoid valve
16. MFI control relay
17. Ignition coil
18. Data Link Connector (DLC)
19. Accleration sensor
20. Knock sensor
21. Canister Close Valve (CCV)
22. Fuel Pump Check Connector

Fig. 23 Underhood sensor locations—Tiburon 1997–2000 1.8L engine and 1997–2001 2.0L engine 29130_HYUN_G0023

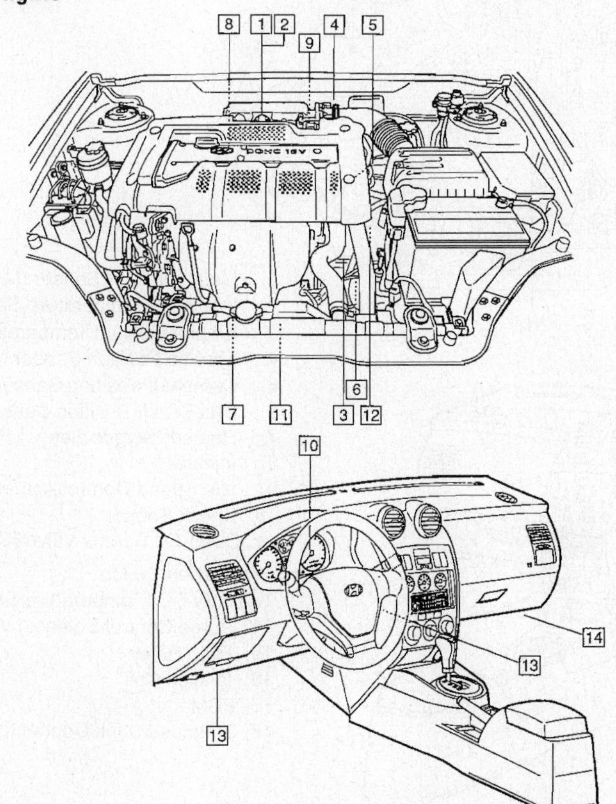

1 Manifold Absolute Pressure (MAP) Sensor
2 Intake Air Temp. (IAT) Sensor
3 Engine Coolant Temp. (ECT) Sensor
4 Throttle Position Sensor (TPS)
5 Camshaft Position Sensor (CMP)
6 Crankshaft Position Sensor (CKP)
7 Heated Oxygen Sensor
8 Injector
9 Idle Speed Actuator (ISA)
10 Vehicle Speed Sensor (VSS)
11 Knock Sensor
12 Inhibitor Switch
13 Ignition Switch
14 ECM
13 Data Link Connector (DLC)

Fig. 24 Underhood sensor locations—Tiburon 2003–2004 2.0L engine 29130_HYUN_G0024

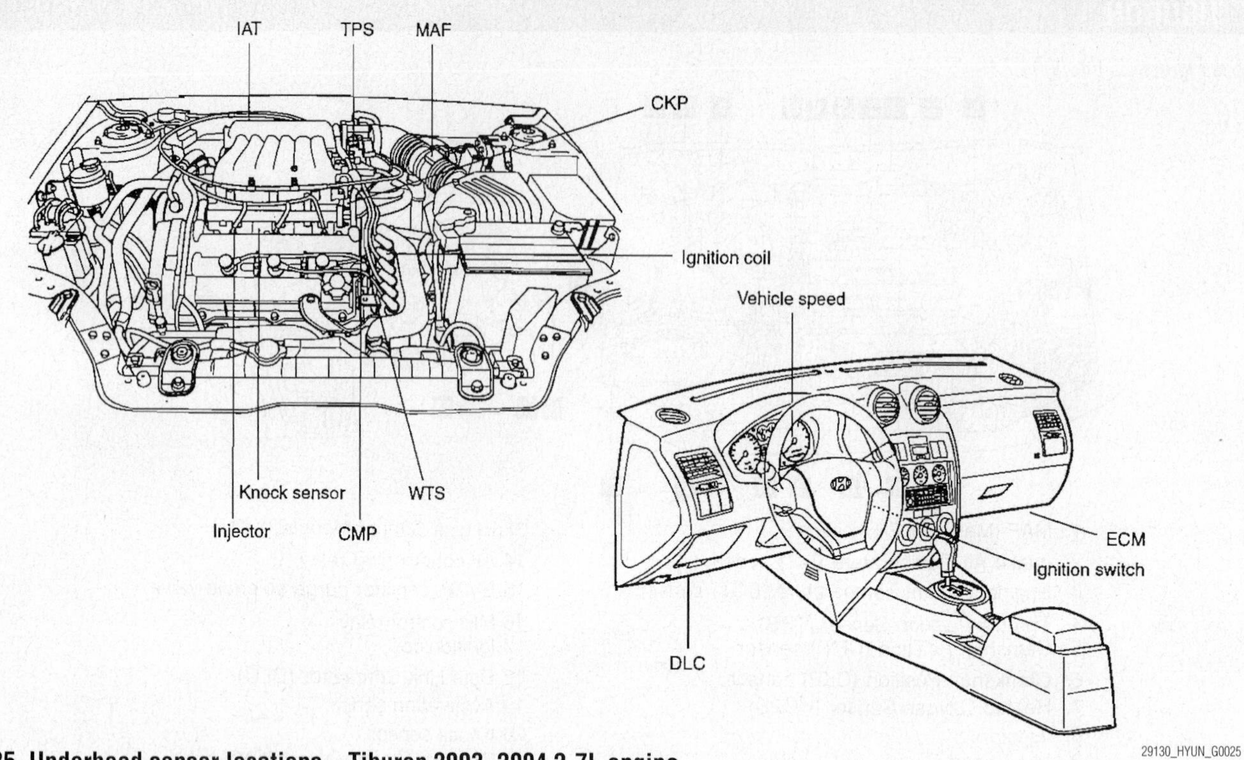

Fig. 25 Underhood sensor locations—Tiburon 2003–2004 2.7L engine

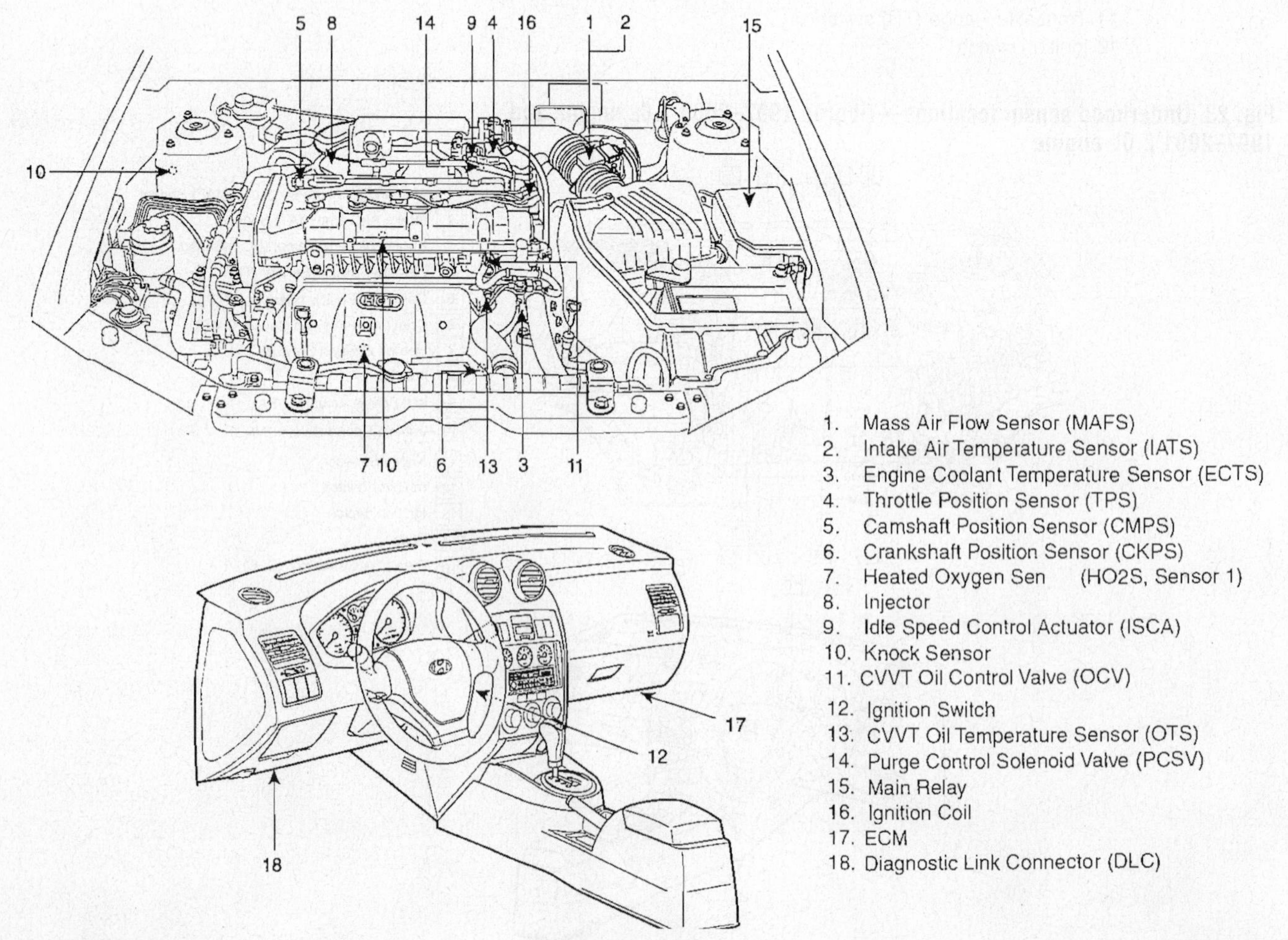

Fig. 26 Underhood sensor locations—Tiburon 2005–2006 2.0L engine

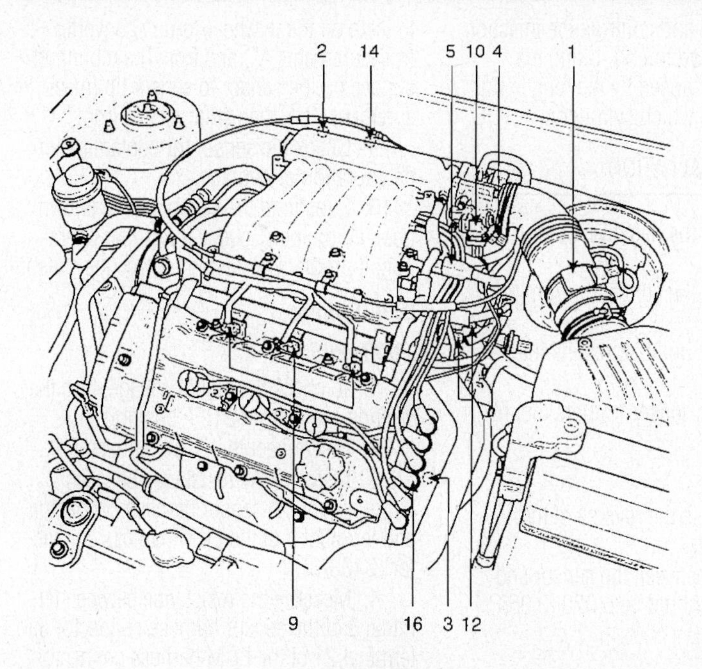

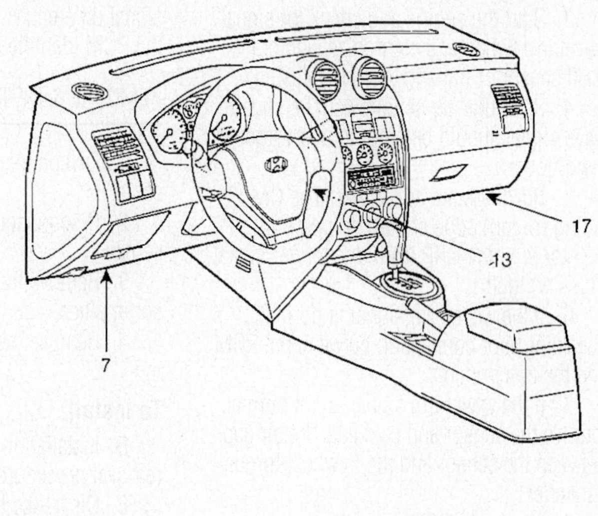

1. Mass Air Flow Sensor (MAFS)
2. Intake Air Temperature Sensor (IATS)
3. Engine Coolant Temperature Sensor (ECTS)
4. Throttle Position Sensor (TPS)
5. Camshaft Position Sensor (CMPS)
6. Crankshaft Position Sensor (CKPS)
7. Diagnostic Link Connector (DLC)
8. Injector
9. Idle Speed Control Actuator (ISCA)
10. Knock Sensor
11. Ignition Switch
12. Purge Control Solenoid Valve (PCSV)
13. Ignition Coil
14. ECM

29130_HYUN_G0027

Fig. 27 Underhood sensor locations—Tiburon 2005–2006 2.7L engine

Camshaft Position (CMP) Sensor

LOCATION

The camshaft position sensor is either located near the top of the engine, passenger's side, or on the right side of the engine near the ignition coils.

OPERATION

The camshaft position sensor senses the TDC point of the number one cylinder, on its compression stroke. Its signal is relayed to the ECM to be used to determine the sequence of fuel injection.

REMOVAL & INSTALLATION

1. Disconnect the negative battery cable.
2. Disconnect the connector from the sensor.
3. Remove the bolt that retains the sensor.
4. Remove the sensor.

To install:

5. Installation is the reverse of the removal procedure.

TESTING

Except 2.7L Engine

1997–2003

1. Using a voltmeter measure the sensor output voltage.
2. At idle (800 rpm) the specification should be 0–5 volts. At 3000 rpm the specification should be 0–5 volts.
3. If abnormality is found, replace the sensor.

2004–2006

1. Be sure that the CMPS and PCM connectors are connected.
2. Set up an oscilloscope as follows: Channel A (+): terminal 2 of the CKPS, (-): ground. Channel B (+): terminal 2 of the CMPS, (-): ground.
3. Start the engine and check for signal waveform (whether synchronize with crankshaft sensor or not and tooth is missing).
4. Readings are as follows: The square wave signal should be smooth and without distortion.
5. Readings are as follows: The CMPS falling (rising) edge should coincide with 3–5 tooth of the CKP from one longer signal (missing tooth).
6. If the waveform signal is normal, check for poor connection between the PCM and the components.
7. If the waveform signal is not normal, remove the sensor and calculate the air gap between the sensor and the flywheel/torque converter.
8. Air gap is 0.07 inch. Measure from the distance of the housing to teeth on the flywheel/torque converter (measurement "A") and from the mounting surface on the sensor to sensor tip (measurement "B"), then subtract "B" from "A".
9. Check the sensor for contamination, deterioration or damage.
10. Substitute the sensor with a known good component, check for proper operation. If problem corrected replace the sensor.

2.7L Engine

2003–2004

1. Connect the Hi-Scan tool to the data link connector.
2. Start the engine. Allow the engine to reach operating temperature.
3. Monitor the CMPS signal. It should continuously fluctuate between 0–5 volts.
4. If abnormality is found, replace the sensor.

2005–2006

1. Be sure that the CMPS and PCM connectors are connected.
2. Set up an oscilloscope as follows: Channel A (+): terminal 2 of the CKPS, (-):

ground. Channel B (+): terminal 2 of the CMPS, (-): ground.

3. Start the engine and check for signal waveform (whether synchronize with crankshaft sensor or not and tooth is missing).

4. Readings are as follows: The square wave signal should be smooth and without distortion.

5. Readings are as follows: The CMPS falling (rising) edge should coincide with 3–5 tooth of the CKP from one longer signal (missing tooth).

6. If the waveform signal is normal, check for poor connection between the PCM and the components.

7. If the waveform signal is not normal, remove the sensor and calculate the air gap between the sensor and the flywheel/torque converter.

8. Air gap is 0.07 inch. Measure from the distance of the housing to teeth on the flywheel/torque converter (measurement "A") and from the mounting surface on the sensor to sensor tip (measurement "B"), then subtract "B" from "A".

9. Check the sensor for contamination, deterioration or damage.

10. Substitute the sensor with a known good component, check for proper operation. If problem corrected replace the sensor.

Crankshaft Position (CKP) Sensor

LOCATION

The crankshaft position sensor is located on the driver's side of the vehicle, down by the lower radiator hose.

OPERATION

1997–2003

The crankshaft position sensor consists of a magnet and coil. The voltage signal from the sensor is relayed to the ECM to indicate engine rpm and the position of the crankshaft.

2004–2006

The crankshaft position sensor is a Hall Effect type sensor that generates voltage using a sensor and a target wheel mounted on the crankshaft. There are 58 slots in the target wheel where one is longer than the others. When the slot in the wheel aligns with the sensor, the sensor voltage output is low. When the metal tooth in the wheel aligns the sensor, the sensor voltage is high. During one crankshaft rotation there are 58 rectangular signals and one longer signal. The PCM calculates engine RPM by using the sensor's signal and controls the injection duration and ignition timing. Using the signal differences caused by the longer slot, the PCM identifies which cylinder is at TDC.

REMOVAL & INSTALLATION

1. Disconnect the negative battery cable.

2. Disconnect the connector from the sensor.

3. Remove the bolt that retains the sensor in place.

4. Remove the sensor from its mounting.

To install:

5. Installation is the reverse of the removal procedure.

6. Clearance between the sensor and the sensor wheel should be 0.020–0.059 inch.

TESTING

Except 2.7L Engine

1997–2003

1. Disconnect the sensor electrical connector. Connect an ohmmeter

2. Measure the resistance between terminals 2 and 3, on the crankshaft position side of the connector. Specification should be 0.486–0.594 kohm at 68 degrees F.

3. If measured value is not within specification, replace the sensor.

2004–2006

1. Be sure that the CKPS and PCM connectors are connected.

2. Set up an oscilloscope as follows: Channel A (+): terminal 2 of the CKPS, (-): ground. Channel B (+): terminal 2 of the CMPS, (-): ground.

3. Start the engine and check for signal waveform (whether synchronize with camshaft sensor or not and tooth is missing).

4. Readings are as follows: The square wave signal should be smooth and without distortion.

5. Readings are as follows: The CMPS falling (rising) edge should coincide with 3–5 tooth of the CKP from one longer signal (missing tooth).

6. If the waveform signal is normal, check for poor connection between the PCM and the components.

7. If the waveform signal is not normal, remove the sensor and calculate the air gap between the sensor and the flywheel/torque converter.

8. Air gap is 0.012–0.067 inch. Measure from the distance of the housing to teeth on the flywheel/torque converter (measurement "A") and from the mounting surface on the sensor to sensor tip (measurement "B"), then subtract "B" from "A".

9. Check the sensor for contamination, deterioration or damage.

10. Substitute the sensor with a known good component, check for proper operation. If problem corrected replace the sensor.

2.7L Engine

2003–2004

1. To check the signal voltage, turn the ignition switch to the OFF position.

2. Disconnect the ECM connector.

3. Measure the resistance between terminal 2 of the sensor harness connector and terminal 8 of the ECM harness connector (C133-3).

4. Measure the resistance between terminal 3 of the sensor harness connector and terminal 21 of the ECM harness connector (C133-3).

5. Specification should be below 1 ohm.

6. If not, replace the sensor.

2005–2006

1. Be sure that the CKPS and PCM connectors are connected.

2. Set up an oscilloscope as follows: Channel A (+): terminal 2 of the CKPS, (-): ground. Channel B (+): terminal 2 of the CMPS, (-): ground.

3. Start the engine and check for signal waveform (whether synchronize with camshaft sensor or not and tooth is missing).

4. Readings are as follows: The square wave signal should be smooth and without distortion.

5. Readings are as follows: The CMPS falling (rising) edge should coincide with 3–5 tooth of the CKP from one longer signal (missing tooth).

6. If the waveform signal is normal, check for poor connection between the PCM and the components.

7. If the waveform signal is not normal, remove the sensor and calculate the air gap between the sensor and the flywheel/torque converter.

8. Air gap is 0.012–0.067 inch. Measure from the distance of the housing to teeth on the flywheel/torque converter (measurement "A") and from the mounting surface on the sensor to sensor tip (measurement "B"), then subtract "B" from "A".

9. Check the sensor for contamination, deterioration or damage.

10. Substitute the sensor with a known good component, check for proper operation. If problem corrected replace the sensor.

Electronic Control Module (ECM)

OPERATION

The ECM controls the vehicle engine operating system.

REMOVAL & INSTALLATION

1. Disconnect the negative battery cable.
2. Remove the lower inner trim..
3. As required, detach the floor mat. As required, remove the protective cover.
4. Remove the ECM bracket retaining nuts. Remove the clip from the bracket.
5. Disconnect the connectors.
6. Remove the ECM from the vehicle.

To install:

7. Installation is the reverse of the removal procedure.

➡ **When replacing the ECM, be careful not to use the wrong part number, as damage to the injection system could occur.**

Engine Coolant Temperature (ECT) Sensor

LOCATION

The ECT is installed in the engine coolant passage of the cylinder head.

OPERATION

This component detects the temperature of the engine coolant and relays the information to the electronic control assembly. This component employs a thermistor which is sensitive to temperature changes. The electric resistance of the thermistor decreases in response to temperature rise. The ECM judges coolant temperature by the sensor output voltage and provides optimum fuel enrichment when the engine is cold.

REMOVAL & INSTALLATION

1. Disconnect the negative battery cable.
2. Disconnect the connector from the sensor.
3. Drain the cooling system, as required.
4. Remove the sensor from its mounting.

To install:

5. Installation is the reverse of the removal procedure.

TESTING

Except 2.7L Engine

1997–2003

1. Remove the sensor from the engine.
2. With the sensing portion of the sensor immersed in hot water, check the resistance.
3. Specification should be 0.538–0.650 kohm at 140 degrees F.
4. Replace the sensor, as required.

2004–2006

1. Turn the ignition switch to the OFF position.
2. Check the resistance of the sensor connector between terminals 1 and 3 (component side).
3. Specification should be 2.27–2.73 kohm at 68 degrees F. for 2004 vehicles and 2.31–2.59 kohm at 68 degrees F. for 2005–2006 vehicles.
4. Replace the sensor, as required.

2.7L Engine

2003

1. Remove the sensor from the engine.
2. With the sensing portion of the sensor immersed in hot water, check the resistance.
3. Specification should be 2.5 kohm at 68 degrees F.
4. Replace the sensor, as required.

2004–2006

1. Turn the ignition switch to the OFF position.
2. Disconnect the ECTS connector.
3. Check the resistance of the sensor connector between terminals 1 and 3 (component side).
4. Specification should be 2.31–2.59 kohm at 68 degrees F.
5. Replace the sensor, as required.

Heated Oxygen (HO2S) Sensor

LOCATION

The sensors are located in the exhaust system. On some vehicles one sensor is located up at the exhaust manifold(s) and the other sensor is located down at the catalytic converter. On other vehicles both sensors are located down at the catalytic converter.

OPERATION

The exhaust gas oxygen sensor supplies the electronic control assembly with a signal which indicates either a rich or lean mixture condition, during the engine operation.

REMOVAL & INSTALLATION

1. Disconnect the electrical connector from the sensor.
2. Remove the oxygen sensor.

To install:

3. Installation is the reverse of the removal procedure.

➡ **Apply anti-seize compound to the threaded portion of the sensor, prior to installation. Never apply anti-seize compound to the protector of the sensor.**

TESTING

Except 2.7L Engine

1997–2003

Perform a visual inspection of the sensor as follows:

1. If the sensor tip has a black/sooty deposit, this may indicate a rich fuel mixture.
2. If the sensor tip has a white, gritty deposit, this may indicate an internal coolant leak.
3. If the sensor tip has a brown deposit, this could indicate excessive oil consumption.
4. Warm the engine until the coolant temperature reaches operating temperature.
5. Accelerate the engine to 4000 rpm. When decelerating suddenly from 4000 rpm, the voltmeter reading should be 200mV or lower.
6. When the engine is suddenly raced, the voltmeter reading should be 600–1000mV.
7. If measured value is not within specification, replace the sensor.

2004–2006

Perform a visual inspection of the sensor as follows:

1. If the sensor tip has a black/sooty deposit, this may indicate a rich fuel mixture.
2. If the sensor tip has a white, gritty deposit, this may indicate an internal coolant leak.
3. If the sensor tip has a brown deposit, this could indicate excessive oil consumption.
4. Turn the ignition switch to the OFF position.
5. Disconnect the sensor connector.
6. Measure the resistance between terminals 3 and 4 (component side).
7. Specification should be 9 ohms at 68–75.2 degrees F.
8. If not, replace the sensor.

2.7L Engine

2003–2004

Perform a visual inspection of the sensor as follows:

1. If the sensor tip has a black/sooty deposit, this may indicate a rich fuel mixture.
2. If the sensor tip has a white, gritty deposit, this may indicate an internal coolant leak.
3. If the sensor tip has a brown deposit, this could indicate excessive oil consumption.
4. Warm the engine until the coolant temperature reaches operating temperature.
5. Accelerate the engine to 4000 rpm. When decelerating suddenly from 4000 rpm, the voltmeter reading should be 200mV or lower.
6. When the engine is suddenly raced, the voltmeter reading should be 600–1000mV.
7. If measured value is not within specification, replace the sensor.

2004–2006 (SENSOR ONE)

Perform a visual inspection of the sensor as follows:

1. If the sensor tip has a black/sooty deposit, this may indicate a rich fuel mixture.
2. If the sensor tip has a white, gritty deposit, this may indicate an internal coolant leak.
3. If the sensor tip has a brown deposit, this could indicate excessive oil consumption.
4. Warm the engine until the engine reaches operating temperature.
5. Connect the scan tool and monitor sensor operation.
6. Verify that the signal is switching from rich (above 0.45 volt) to lean (below 0.45 volt) a minimum of 3 times in 10 seconds (voltage will vary between 0.1–0.9 volt).
7. If the sensor is operating properly, check for poor connection between the ECM and the component.
8. If not, check the sensor for contamination, deterioration or damage.
9. Substitute the sensor with a known good component, check for proper operation. If problem corrected replace the sensor.

2004–2006 (SENSOR TWO)

Perform a visual inspection of the sensor as follows:

1. If the sensor tip has a black/sooty deposit, this may indicate a rich fuel mixture.
2. If the sensor tip has a white, gritty deposit, this may indicate an internal coolant leak.
3. If the sensor tip has a brown deposit, this could indicate excessive oil consumption.
4. Warm the engine until the engine reaches operating temperature.
5. Measure the resistance between terminals 3 and 4 (component side).
6. Specification should be 3–4 ohms at 68–75.2 degrees F.
7. If not, replace the sensor.

Idle Speed Control Actuator (ISCA)

OPERATION

The idle speed control actuator sensor is a double coil type driven by separate driver stages in the ECM. Depending on the pulse duty factor, the equilibrium of the magnetic forces of the two coils will result in different angles of the motor. A bypass hose line is positioned, parallel to the throttle valve where the idle speed actuator is inserted.

REMOVAL & INSTALLATION

1. Disconnect the negative battery cable.
2. Disconnect the connector from the sensor.
3. Remove the sensor retaining screws.
4. Remove the sensor from its mounting.

To install:

5. Installation is the reverse of the removal procedure.

TESTING

Except 2.7L Engine

1. Disconnect the connector at the sensor. Connect an ohmmeter.
2. Measure the resistance between terminal 1 and 2, specification should be 10.5–14 ohms. Measure the resistance between terminal 2 and 3, specification should be 10–12.5 ohms at 68 degrees F.
3. If the measured value is not within specification, replace the sensor.

2.7L Engine

2003–2004

1. Disconnect the connector at the sensor. Connect an ohmmeter.
2. Measure the resistance between terminal 3 and 2, specification should be 10.5–14 ohms. Measure the resistance between terminal 1 and 3, specification should be 10–12.5 ohms at 68 degrees F.
3. If the measured value is not within specification, replace the sensor.

2005–2006

1. Turn the ignition switch to the OFF position.
2. Remove the valve and check the throttle bore, throttle plate and valve passages for damage and chocking. Repair/clean as required.
3. Reinstall the valve.
4. With the engine off and the ignition switch in the ON position, connect the scan tool.
5. Select "IDLE SPEED ACTUATOR" parameter on the "ACTUATION TEST" mode.
6. Activate the ISCA valve, by pressing the "STAT" key.
7. Check the ISCA for a clicking sound and visually verify that the valve opens and closes. Repeat numerous times to ensure a working valve.
8. If the valve is operating properly, check for poor connection between the ECM and the component.
9. If not, check the valve for contamination, deterioration or damage.
10. Substitute the valve with a known good component, check for proper operation. If problem corrected replace the sensor.

Intake Air Temperature (IAT) Sensor

LOCATION

This sensor is located in the air intake plenum assembly. On some vehicles this sensor is combined with the MAP/MAF sensor.

OPERATION

This sensor is a resistor based sensor which detects the intake air temperature. According to the intake air temperature reading the ECM will control the necessary amount of fuel injection.

REMOVAL & INSTALLATION

1. Disconnect the negative battery cable.
2. Disconnect the connector from the sensor.
3. Remove the sensor retaining screws, as required.
4. Remove the air cleaner and air intake assembly, as required.
5. Remove the sensor from its mounting.

To install:

6. Installation is the reverse of the removal procedure.

TESTING

Except 2.7L Engine

1997–2001

1. Using a multimeter, measure the sensor resistance between terminals 1 and 2.
2. At 68 degrees F the resistance should be 2.4–2.8 kohm.
3. If the measured value is not within specification, replace the sensor.

2003–2004

1. Turn the ignition switch to the OFF position.
2. Disconnect the sensor connector.
3. Measure the resistance between terminals 5 and 2, of the connector.
4. Specification should be 2.0–3.0 kohms at 68 degrees F.
5. If not, replace the sensor.

2005–2006

1. Turn the ignition switch OFF.
2. Disconnect the sensor connector.
3. Measure the resistance between terminals 1 and 5 (component side).
4. Specification should be 2.35–3.54 kohms at 68 degrees F.
5. If within specification, check for poor connection between the PCM and the component.
6. If not, substitute the sensor with a known good component, check for proper operation. If problem corrected replace the sensor.

2.7L Engine

2003–2004

1. Remove the sensor connector.
2. Measure the voltage between the sensor terminal 1 and 2.
3. Specification should be 3.3–3.7 volts at 32 degrees F. and 2.4–2.8 volts at 68 degrees F.
4. Replace the sensor, as required.

2005–2006

1. Remove the sensor connector.
2. Measure the voltage between the sensor terminal 1 and 2.
3. Specification should be 2.22–2.82 volts at 68 degrees F.
4. Replace the sensor, as required.

Knock Sensor (KS)

LOCATION

The knock sensor is located in the side of the cylinder block.

OPERATION

The knock sensor is used to detect engine vibrations caused by preignition or detonation and provides information to the ECM, which then retards the timing to eliminate detonation.

REMOVAL & INSTALLATION

1. Disconnect the negative battery cable.
2. Disconnect the sensor connector.
3. Remove the sensor from its mounting.

To install:

4. Installation is the reverse of the removal procedure.
5. Tighten the sensor to 12–18 ft. lbs.

TESTING

Except 2.7L Engine

1997–2001

1. Disconnect the sensor electrical connector. Connect an ohmmeter.
2. Measure the resistance between terminals 2 and 3. Specification should be 5mohm at 68 degrees F.
3. If resistance is zero, replace the sensor.
4. Measure the capacitance between terminal 2 and 3. Specification should be 800–1600 pF.
5. If not within specification, replace the sensor.

2003–2006

1. Disconnect the sensor electrical connector. Connect an ohmmeter.
2. Measure the resistance between terminals 1 and 2, (component side).
3. Specification should be 5mohm at 68 degrees F.
4. Replace the sensor, as required.

2.7L Engine

1. Disconnect the sensor electrical connector. Connect an ohmmeter.
2. Measure the resistance between terminals 2 and 3. Specification should be 5mohm at 68 degrees F.
3. If resistance is zero, replace the sensor.
4. Measure the capacitance between terminal 2 and 3. Specification should be 800–1600 pF.
5. If not within specification, replace the sensor.

Manifold Absolute Pressure (MAP) Sensor

LOCATION

The MAP sensor is located against the firewall to the left side of the engine.

OPERATION

This sensor is a pressure sensitive variable resistor. It measures the changes in the intake manifold pressure which result from engine load and speed changes, and converts to a voltage output. This sensor is used to measure the barometric pressure at start up, and under certain conditions, allows the ECM to automatically adjust for different altitudes. The ECM supplies 5 volts to the sensor and monitors the voltage on a signal line. The sensor provides a path to ground through its variable resistor. The sensor input affects fuel delivery and ignition timing controls in the ECM.

REMOVAL & INSTALLATION

1. Disconnect the negative battery cable.
2. Disconnect the connector from the sensor.
3. Remove the sensor retaining screws.
4. Remove the sensor from its mounting.

To install:

5. Installation is the reverse of the removal procedure.

TESTING

1. Turn the ignition switch ON.
2. Measure the voltage between terminal 1 (sensor output) and terminal 4 (sensor ground).
3. Specification should be 4–5 volts.
4. If not within specification, replace the sensor.

Mass Air Flow (MAF) Sensor

LOCATION

The MAF is mounted in the intake air hose of the air cleaner assembly.

OPERATION

The MAF is measured by detection of heat transfer from a hot film probe because the change of the mass air flow rate causes change in the amount of heat being transferred from the hot film probe surface to the air flow. The air flow sensor generates a pulse so it repeatedly opens and closes between the voltage (5V) supplied from the ECM. This results in a change of the temperature of the hot film probe and in the change of resistance.

REMOVAL & INSTALLATION

1. Disconnect the negative battery cable.
2. Disconnect the connector from the sensor.
3. Remove the air cleaner and air intake assembly, as required.
4. Remove the sensor from its mounting.

To install:

5. Installation is the reverse of the removal procedure.

TESTING

Except 2.7L Engine

1997–2003

1. Using a voltmeter check the output voltage of the sensor.
2. Use MAF sensor side connector terminal 1 or ECM harness side connector terminal 41.
3. Specification should be 0.7–1.1 volt at idle (800 rpm) and 1.3–2.0 volts at 3000 rpm.
4. Replace the sensor, as required.

2004

1. Be sure that the ECM and the MAFS connectors are connected.
2. Connect the Hi-Scan tool to the data link connector.
3. Start the engine.
4. Monitor the MAPS signals.
5. If not within specification, replace the sensor.
6. Specification should be 0.0–5.0 volts.
7. Replace the sensor, as required.

2005–2006

1. Turn the engine ON.
2. Install the scan tool and monitor the MAPS signal.
3. Specification should be approximately 10–20 kg/h at idle and no load.
4. If within specification, check for poor connection between the PCM and the component.
5. If not, substitute the sensor with a known good component, check for proper operation. If problem corrected replace the sensor.

2.7L Engine

2003–2004

1. Allow the engine coolant to reach 176–198 degrees F.
2. Using a voltmeter check the sensor output voltage.
3. Specification should be 0.5 volt at idle, 1.0 volt at 2000 rpm.
4. If the measured value is not within specification, replace the sensor.

2005–2006

1. Connect the scan tool and monitor the "MASS AIR FLOW (V)" parameter on the scan tool data list.
2. Monitor the "MASS AIR FLOW (V)" parameter on the scan tool.
3. Specification is approximately 0.6–1.0 volt at idle (no load) and approximately 1.0–1.3 volt at idle (air conditioning ON).
4. If within specification, check for poor connection between the PCM and the component.
5. If not, substitute the sensor with a known good component, check for proper operation. If problem corrected replace the sensor.

Throttle Position Sensor (TPS)

OPERATION

The throttle position sensor is a rotating type variable resistor that rotates with the throttle body's throttle shaft to sense the throttle valve angle. As the throttle shaft rotates, the throttle angle of the sensor changes and the ECM detects the throttle valve opening based on the TPS output voltage.

REMOVAL & INSTALLATION

1. Disconnect the negative battery cable.
2. Disconnect the sensor connector.
3. Remove the sensor retaining screws. Remove the sensor from its mounting.

To install:

4. Installation is the reverse of the removal procedure.

TESTING

Except 2.7L Engine

1997–2003

1. Disconnect the connector from the sensor.
2. Measure the resistance between terminals 2 (sensor ground) and 3 (sensor power).
3. Specification should be 0.7–3.0 kohm.
4. Connect an analog ohmmeter between terminals 3 (sensor ground) and 1 (sensor output), on 1997–2001 vehicles and between terminals 2 (sensor ground) and 1 (sensor output), on 2003 vehicles.
5. Operate the throttle valve slowly from the idle position to the full open position, and check that the resistance changes smoothly in proportion with the throttle valve opening angle.
6. If the resistance is out of specification, or fails to change smoothly, replace the sensor.

2004

1. Turn the ignition switch to the OFF position.
2. Measure the resistance between terminals 1 and 2 of the TPS.
3. Measure the resistance between terminals 2 and 3 of the TPS.
4. Specification should be 0.7–3.0 kohms.
5. If not within specification, replace the sensor.

2005–2006

1. Turn the ignition switch to the OFF position.
2. Disconnect the TPS connector.
3. Measure the resistance between terminals 2 and 3 of the TPS.
4. Specification should be 1.6–2.4 kohms at all throttle position..
5. With the connector still disconnected, measure the resistance between terminals 1 and 2.
6. Operate the throttle valve slowly from the idle position to the full open position and check that the resistance changes smoothly in proportion with the throttle valve opening angle.
7. Specification should be 0.71–1.38 kohm at closed throttle valve and 2.2–3.4 kohm at wide open throttle.
8. If within specification, check for poor connection between the PCM and the component.
9. If not, substitute the sensor with a known good component, check for proper operation. If problem corrected replace the sensor.

2.7L Engine

2003

1. Disconnect the sensor connector.
2. Measure the resistance between terminal 2 (sensor ground) and terminal 1 (sensor power).
3. Specification should be 3.5–6.5 kohm.

4. Connect and analog ohmmeter between terminal 2 (sensor ground) and terminal 3 (sensor output).

5. Operate the throttle valve slowly from the idle position to the full open position, and check that the resistance changes smoothly in proportion with the throttle valve opening angle.

6. If the resistance is out of specification, or fails to change smoothly, replace the sensor.

2004–2006

1. Disconnect the sensor connector.
2. Measure the resistance between terminal 1 and terminal 2, (component side).
3. Specification should be 1.6–2.4 kohm, at all throttle positions.
4. Measure the resistance between terminal 2 and terminal 3, (component side).
5. Specification should be 0.71–1.38 kohm, at closed throttle and 2.2–3.4 kohm at wide open throttle.
6. If within specification, check for poor connection between the PCM and the component.
7. If not, substitute the sensor with a known good component, check for proper operation. If problem corrected replace the sensor.

Vehicle Speed Sensor (VSS)

LOCATION

On 1997–2003 vehicles, the vehicle speed sensor is built into the speedometer. On 2004–2006 vehicles, the vehicle speed sensor is attached to the output shaft of the transaxle.

OPERATION

On 1997–2003 vehicles, the sensor is a reed switch type. On 2004–2006 vehicles, the sensor is a hall effect sensor. The sensor converts the transaxle gear revolutions into pulse signals, which are sent to the ECM.

REMOVAL & INSTALLATION

Except 2.7L Engine

2004–2006

1. Raise and support the vehicle safely.
2. Place a drip pan below the speed sensor to catch any spilled fluid.
3. Disconnect the connector.
4. Remove the sensor from its mounting.

To install:

5. Installation is the reverse of the removal procedure.
6. Replace any lost fluid.

TESTING

Except 2.7L Engine

1997–2003

1. To check the sensor output circuit for continuity, disconnect the ECM connector. Move the vehicle or turn the speedometer cable.

2. To measure the power supply voltage, disconnect the connector. Turn the ignition switch to the ON position.

3. The voltage should be 4.5–4.9 volts.

4. Using an ohmmeter, check for continuity of the ground circuit.

2004–2006

2004–2006

1. To measure the power supply voltage, disconnect the connector. Turn the ignition switch to the ON position. Specification is battery voltage (between ground and terminal 2 of the harness connector).

2. To check for an open circuit, or a short circuit to ground between the ECM and the sensor, disconnect the ECM connector. Between ground and terminal 3 of the harness connector.

3. To check for an open circuit, or a short circuit to ground between the ECM and the sensor, disconnect the vehicle speed sensor connector. Between ground and terminal 1 of the harness connector).

4. To check for an open circuit between the vehicle speed sensor and ground, disconnect the vehicle speed sensor connector.

TUCSON

See Figures 28 and 29.

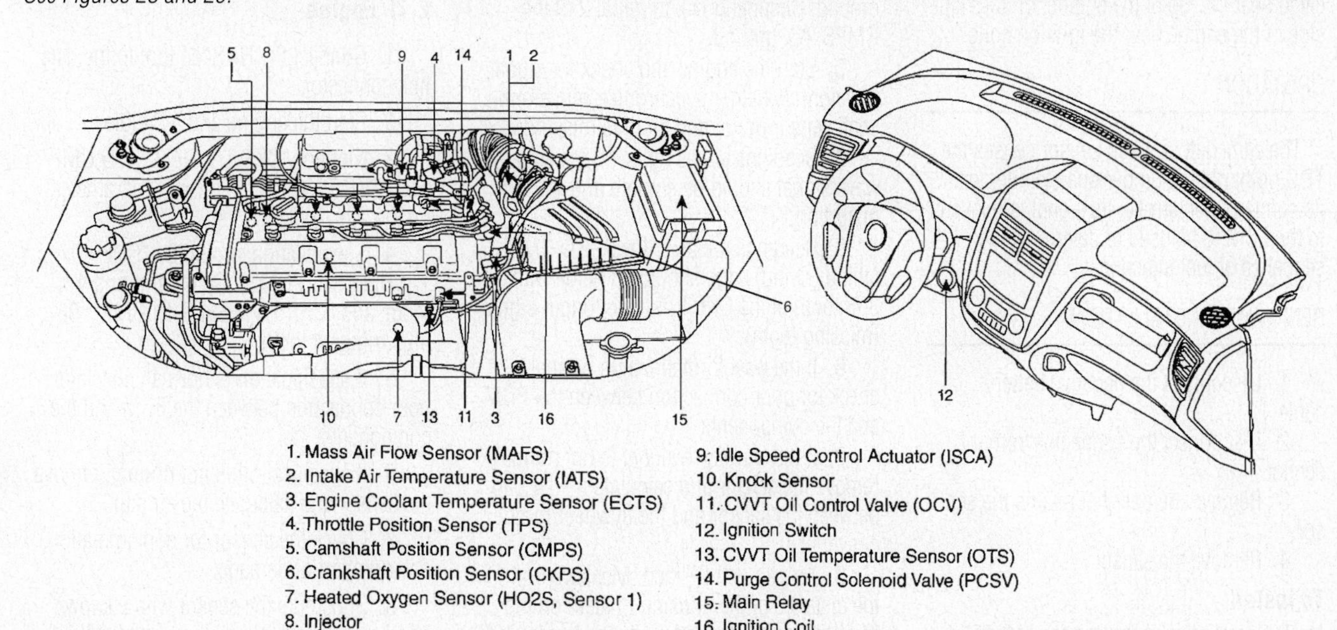

Fig. 28 Underhood sensor locations—Tucson 2005–2006 2.0L engine

29130_HYUN_G0028

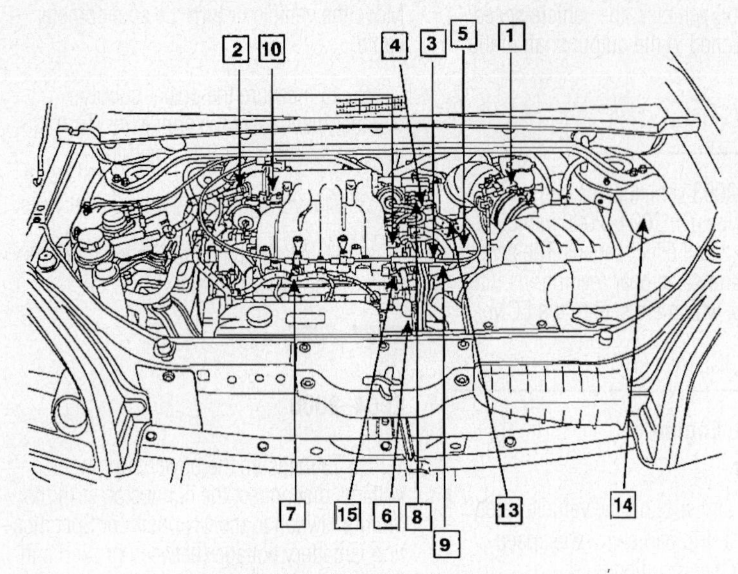

12
16
11

1. Mass Air Flow Sensor (MAFS) [With CVVT]
2. Intake Air Temperature Sensor (IATS)
3. Engine Coolant Temperature Sensor (ECTS)
4. Throttle Position Sensor (TPS)
5. Camshaft Position Sensor (CMPS)
6. Crankshaft Position Sensor (CKPS)
7. Injector
8. Idle Speed Control Actuator (ISCA)
9. Knock Sensor
10. VIS Control solenoid valve
11. Ignition Switch
12. ECM
13. Purge Control Solenoid Valve (PCSV)
14. Main Relay
15. Ignition Coil
16. DLC (Diagnostic Link Connector)

Fig. 29 Underhood sensor locations—Tucson 2005–2006 2.7L engine

29130_HYUN_G0029

Camshaft Position (CMP) Sensor

LOCATION

On the 2.0L engine the camshaft position sensor is located near the top of the engine, on the left side of the engine. On the 2.7L engine the camshaft position sensor is located near the top of the engine, on the right side of the engine near the ignition coils.

OPERATION

The camshaft position sensor senses the TDC point of the number one cylinder, on its compression stroke. Its signal is relayed to the ECM to be used to determine the sequence of fuel injection.

REMOVAL & INSTALLATION

1. Disconnect the negative battery cable.
2. Disconnect the connector from the sensor.
3. Remove the bolt that retains the sensor.
4. Remove the sensor.

To install:

5. Installation is the reverse of the removal procedure.

TESTING

2.0L Engine

1. Be sure that the CMPS and PCM connectors are connected.
2. Set up an oscilloscope as follows: Channel A (+): terminal 2 of the CKPS, (-): ground. Channel B (+): terminal 2 of the CMPS, (-): ground.
3. Start the engine and check for signal waveform (whether synchronize with crankshaft sensor or not and tooth is missing).
4. Readings are as follows: The square wave signal should be smooth and without distortion.
5. Readings are as follows: The CMPS falling (rising) edge should coincide with 3–5 tooth of the CKP from one longer signal (missing tooth).
6. If the waveform signal is normal, check for poor connection between the PCM and the components.
7. If the waveform signal is not normal, remove the sensor and calculate the air gap between the sensor and the flywheel/torque converter.
8. Air gap is 0.07 inch. Measure from the distance of the housing to teeth on the flywheel/torque converter (measurement "A") and from the mounting surface on the sensor to sensor tip (measurement "B"), then subtract "B" from "A".
9. Check the sensor for contamination, deterioration or damage.
10. Substitute the sensor with a known good component, check for proper operation. If problem corrected replace the sensor.

2.7L Engine

1. Connect the Hi-Scan tool to the data link connector.
2. Start the engine and let it idle.
3. Monitor "CKP T/WHEELS-LO CMP" and "CKP T/WHEELS-HI CMP" parameters on the scan tool data list.
4. Specification should be "CKP T/WHEELS-LO CMP": 38 plus/minus 4 tooth, and "CKP T/WHEELS-HI CMP": 98 plus/minus 4 tooth.
5. If specification is normal, check for poor connection between the ECM and the components.
6. If specification is not normal, remove the sensor and calculate the air gap.
7. Check the sensor for contamination, deterioration or damage.
8. Substitute the sensor with a known good component, check for proper operation. If problem corrected replace the sensor.

Crankshaft Position (CKP) Sensor

LOCATION

The crankshaft position sensor is located on the driver's side of the vehicle, down by the lower radiator hose.

OPERATION

The crankshaft position sensor is a Hall Effect type sensor that generates voltage using a sensor and a target wheel mounted on the crankshaft. There are 58 slots in the target wheel where one is longer than the others. When the slot in the wheel aligns with the sensor, the sensor voltage output is low. When the metal tooth in the wheel aligns the sensor, the sensor voltage is high. During one crankshaft rotation there are 58 rectangular signals and one longer signal. The PCM calculates engine RPM by using the sensor's signal and controls the injection duration and ignition timing. Using the signal differences caused by the longer slot, the PCM identifies which cylinder is at TDC.

REMOVAL & INSTALLATION

1. Disconnect the negative battery cable.
2. Disconnect the connector from the sensor.
3. Remove the bolt that retains the sensor in place.
4. Remove the sensor from its mounting.

To install:

5. Installation is the reverse of the removal procedure.
6. Clearance between the sensor and the sensor wheel should be 0.020–0.059 inch.

TESTING

2.0L Engine

1. Be sure that the CKPS and PCM connectors are connected.
2. Set up an oscilloscope as follows: Channel A (+): terminal 2 of the CKPS, (-): ground. Channel B (+): terminal 2 of the CMPS, (-): ground.
3. Start the engine and check for signal waveform (whether synchronize with camshaft sensor or not and tooth is missing).
4. Readings are as follows: The square wave signal should be smooth and without distortion.
5. Readings are as follows: The CMPS falling (rising) edge should coincide with 3–5 tooth of the CKP from one longer signal (missing tooth).
6. If the waveform signal is normal, check for poor connection between the PCM and the components.
7. If the waveform signal is not normal, remove the sensor and calculate the air gap between the sensor and the flywheel/torque converter.
8. Air gap is 0.012–0.067 inch. Measure from the distance of the housing to teeth on the flywheel/torque converter (measurement "A") and from the mounting surface on the sensor to sensor tip (measurement "B"), then subtract "B" from "A".
9. Check the sensor for contamination, deterioration or damage.
10. Substitute the sensor with a known good component, check for proper operation. If problem corrected replace the sensor.

2.7L Engine

1. Connect the Hi-Scan tool to the data link connector.
2. Start the engine and let it idle.
3. Monitor "CKP T/WHEELS-LO CMP" and "CKP T/WHEELS-HI CMP" parameters on the scan tool data list.
4. Specification should be "CKP T/WHEELS-LO CMP": 38 plus/minus 4 tooth, and "CKP T/WHEELS-HI CMP": 98 plus/minus 4 tooth.
5. If specification is normal, check for poor connection between the ECM and the components.
6. If specification is not normal, remove the sensor and calculate the air gap between the sensor and the flywheel/torque converter.
7. Air gap is 0.012–0.067 inch. Measure from the distance of the housing to teeth on the flywheel/torque converter (measurement "A") and from the mounting surface on the sensor to sensor tip (measurement "B"), then subtract "B" from "A".
8. Check the sensor for contamination, deterioration or damage.
9. Substitute the sensor with a known good component, check for proper operation. If problem corrected replace the sensor.

Electronic Control Module (ECM)

OPERATION

The ECM controls the vehicle engine operating system.

REMOVAL & INSTALLATION

1. Disconnect the negative battery cable.
2. Remove the lower inner trim..
3. As required, detach the floor mat. As required, remove the protective cover.
4. Remove the ECM bracket retaining nuts. Remove the clip from the bracket.
5. Disconnect the connectors.
6. Remove the ECM from the vehicle.

To install:

7. Installation is the reverse of the removal procedure.

➡ **When replacing the ECM, be careful not to use the wrong part number, as damage to the injection system could occur.**

Engine Coolant Temperature (ECT) Sensor

LOCATION

The ECT is installed in the engine coolant passage of the cylinder head.

OPERATION

This component detects the temperature of the engine coolant and relays the information to the electronic control assembly. This component employs a thermistor which is sensitive to temperature changes. The electric resistance of the thermistor decreases in response to temperature rise. The ECM judges coolant temperature by the sensor output voltage and provides optimum fuel enrichment when the engine is cold.

REMOVAL & INSTALLATION

1. Disconnect the negative battery cable.
2. Disconnect the connector from the sensor.
3. Drain the cooling system, as required.
4. Remove the sensor from its mounting.

To install:

5. Installation is the reverse of the removal procedure.

TESTING

1. Turn the ignition switch to the OFF position.
2. Disconnect the ECTS connector.
3. Check the resistance of the sensor connector between terminals 1 and 3 (component side).
4. Specification should be 2.31–2.59 kohm at 68 degrees F.
5. Replace the sensor, as required.

Heated Oxygen (HO2S) Sensor

LOCATION

The sensors are located in the exhaust system. On some vehicles one sensor is located up at the exhaust manifold(s) and the other sensor is located down at the catalytic converter.

OPERATION

The exhaust gas oxygen sensor supplies the electronic control assembly with a signal which indicates either a rich or lean mixture condition, during the engine operation.

REMOVAL & INSTALLATION

1. Disconnect the electrical connector from the sensor.
2. Remove the oxygen sensor.

To install:

3. Installation is the reverse of the removal procedure.

➡ **Apply anti-seize compound to the threaded portion of the sensor, prior to installation. Never apply anti-seize compound to the protector of the sensor.**

TESTING

2.0L Engine

SENSOR ONE

Perform a visual inspection of the sensor as follows:

1. If the sensor tip has a black/sooty deposit, this may indicate a rich fuel mixture.
2. If the sensor tip has a white, gritty deposit, this may indicate an internal coolant leak.
3. If the sensor tip has a brown deposit, this could indicate excessive oil consumption.
4. Warm the engine until the engine reaches operating temperature.
5. Connect the scan tool and monitor sensor operation.
6. Verify that the signal is switching from rich (above 0.45 volt) to lean (below 0.45 volt) a minimum of 3 times in 10 seconds (voltage will vary between 0.1–0.9 volt).
7. If the sensor is operating properly, check for poor connection between the ECM and the component.
8. If not, check the sensor for contamination, deterioration or damage.
9. Substitute the sensor with a known good component, check for proper operation. If problem corrected replace the sensor.

SENSOR TWO

Perform a visual inspection of the sensor as follows:

1. If the sensor tip has a black/sooty deposit, this may indicate a rich fuel mixture.
2. If the sensor tip has a white, gritty deposit, this may indicate an internal coolant leak.
3. If the sensor tip has a brown deposit, this could indicate excessive oil consumption.
4. Warm the engine until the engine reaches operating temperature. Check that the HO2S signal is active.
5. Connect the scan tool and monitor the "O2 SNSR VOLT." parameter on the scan tool data list.
6. Test condition should be Engine ON and in idle at closed loop condition. Specification should be above 0.6 volt.
7. If the sensor is operating properly, check for poor connection between the PCM and the component.
8. If not, check the sensor for contamination, deterioration or damage.
9. Substitute the sensor with a known good component, check for proper operation. If problem corrected replace the sensor.

2.7L Engine

SENSOR ONE

Perform a visual inspection of the sensor as follows:

1. If the sensor tip has a black/sooty deposit, this may indicate a rich fuel mixture.
2. If the sensor tip has a white, gritty deposit, this may indicate an internal coolant leak.
3. If the sensor tip has a brown deposit, this could indicate excessive oil consumption.
4. Warm the engine until the engine reaches operating temperature.
5. Connect the scan tool and monitor sensor operation.
6. Verify that the signal is switching from rich (above 0.45 volt) to lean (below 0.45 volt) a minimum of 3 times in 10 seconds (voltage will vary between 0.1–0.9 volt).
7. If the sensor is operating properly, check for poor connection between the ECM and the component.
8. If not, check the sensor for contamination, deterioration or damage.
9. Substitute the sensor with a known good component, check for proper operation. If problem corrected replace the sensor.

SENSOR TWO

Perform a visual inspection of the sensor as follows:

1. If the sensor tip has a black/sooty deposit, this may indicate a rich fuel mixture.
2. If the sensor tip has a white, gritty deposit, this may indicate an internal coolant leak.
3. If the sensor tip has a brown deposit, this could indicate excessive oil consumption.
4. Warm the engine until the engine reaches operating temperature.
5. Measure the resistance between terminals 3 and 4 (component side).
6. Specification should be 3–4 ohms at 68–75.2 degrees F.
7. If not, replace the sensor.

Idle Speed Control Actuator (ISCA)

LOCATION

The ISCA is installed on the intake manifold.

OPERATION

The idle speed control actuator maintains idle speed according to various engine loads and conditions. This component also provides additional air during startup. It consists of an opening coil, a closing coil and a permanent magnet. According to the control signals from the ECM, the valve rotor rotates to control the bypass air flow into the engine.

REMOVAL & INSTALLATION

1. Disconnect the negative battery cable.
2. Disconnect the connector from the component.
3. Remove the retaining screws.
4. Remove the component from its mounting.

To install:

5. Installation is the reverse of the removal procedure.

TESTING

2.0L Engine

1. Disconnect the connector.
2. Measure the resistance between terminal 1 and 2 (component side); specification should be 11.1–12.7 ohms.

3. Measure the resistance between terminal 2 and 3 (component side); specification should be 14.5–16.1 ohms.
4. If the component is operating properly, check for poor connection, damage or deterioration.
5. If not, substitute the sensor with a known good component, check for proper operation. If problem corrected replace the component.

2.7L Engine

1. Disconnect the connector.
2. Measure the resistance between terminal 1 and 2 (component side); specification should be 15.0–16.2 ohms.
3. Measure the resistance between terminal 2 and 3 (component side); specification should be 17.0–18.2 ohms.
4. If the component is operating properly, check for poor connection, damage or deterioration.
5. If not, substitute the sensor with a known good component, check for proper operation. If problem corrected replace the component.

Intake Air Temperature (IAT) Sensor

LOCATION

This sensor is located in the air intake plenum assembly. This sensor is combined with the MAF sensor.

OPERATION

This sensor is a resistor based sensor which detects the intake air temperature. According to the intake air temperature reading the ECM will control the necessary amount of fuel injection.

REMOVAL & INSTALLATION

1. Disconnect the negative battery cable.
2. Disconnect the connector from the sensor.
3. Remove the sensor retaining screws, as required.
4. Remove the air cleaner and air intake assembly, as required.
5. Remove the sensor from its mounting.

To install:

6. Installation is the reverse of the removal procedure.

TESTING

2.0L Engine

1. Turn the ignition switch OFF.
2. Disconnect the sensor connector.
3. Measure the resistance between terminals 1 and 5 (component side).
4. Specification should be 2.35–3.54 kohms at 68 degrees F.
5. If within specification, check for poor connection between the PCM and the component.
6. If not, substitute the sensor with a known good component, check for proper operation. If problem corrected replace the sensor.

2.7L Engine

1. Remove the sensor connector.
2. Measure the voltage between the sensor terminal 1 and 2.
3. Specification should be 2.35–2.54 volts at 68 degrees F.
4. Replace the sensor, as required.

Knock Sensor (KS)

LOCATION

The knock sensor is located in the side of the cylinder block.

OPERATION

The knock sensor is used to detect engine vibrations caused by preignition or detonation and provides information to the ECM, which then retards the timing to eliminate detonation.

REMOVAL & INSTALLATION

1. Disconnect the negative battery cable.
2. Disconnect the sensor connector.
3. Remove the sensor from its mounting.

To install:

4. Installation is the reverse of the removal procedure.
5. Tighten the sensor to 12–18 ft. lbs.

TESTING

1. Disconnect the sensor electrical connector. Connect an ohmmeter.
2. Measure the resistance between terminals 1 and 2 (component side). Specification should be 5mohm at 68 degrees F.
3. Replace the sensor, as required.

Mass Air Flow (MAF) Sensor

LOCATION

This sensor is located in the air intake plenum assembly, between the air cleaner assembly and the throttle body.

OPERATION

The MAF is measured by detection of heat transfer from a hot film probe because the change of the mass air flow rate causes change in the amount of heat being transferred from the hot film probe surface to the air flow. The air flow sensor generates a pulse so it repeatedly opens and closes between the voltage (5V) supplied from the ECM. This results in a change of the temperature of the hot film probe and in the change of resistance.

REMOVAL & INSTALLATION

1. Disconnect the negative battery cable.
2. Disconnect the connector from the sensor.
3. Remove the air cleaner and air intake assembly, as required.
4. Remove the sensor from its mounting.

To install:

5. Installation is the reverse of the removal procedure.

TESTING

2.0L Engine

1. Turn the engine ON.
2. Install the scan tool and monitor the "MASS AIR FLOW" parameter on the scan tool data list.
3. Specification should be approximately 10–20 kg/h at idle and no load.
4. If within specification, check for poor connection between the PCM and the component.
5. If not, substitute the sensor with a known good component, check for proper operation. If problem corrected replace the sensor.

2.7L Engine

1. Connect the scan tool and monitor the "MASS AIR FLOW (V)" parameter on the scan tool data list.
2. Monitor the "MASS AIR FLOW (V)" parameter on the scan tool.
3. Specification is approximately 0.6–1.0 volt at idle (no load) and approxi-

mately 1.0–1.3 volt at idle (air conditioning ON).

4. If within specification, check for poor connection between the ECM and the component.

5. If not, substitute the sensor with a known good component, check for proper operation. If problem corrected replace the sensor.

Throttle Position Sensor (TPS)

OPERATION

The TPS has a variable resistor whose characteristic is the resistance changing according to throttle angle. The ECM supplies a reference of 5 volts to the TPS and the output voltage increases directly with the opening of the throttle valve. The TPS voltage will vary from 0.2–0.8 volt at closed throttle and 4.3–4.8 volts at wide open throttle.

REMOVAL & INSTALLATION

1. Disconnect the negative battery cable.

2. Disconnect the sensor connector.

3. Remove the sensor retaining screws. Remove the sensor from its mounting.

To install:

4. Installation is the reverse of the removal procedure.

TESTING

2.0L Engine

1. Turn the ignition switch to the OFF position.

2. Disconnect the TPS connector.

3. Measure the resistance between terminals 2 and 3 of the TPS.

4. Specification should be 1.6–2.4 kohms at all throttle position.

5. With the connector still disconnected, measure the resistance between terminals 1 and 2.

6. Operate the throttle valve slowly from the idle position to the full open position and check that the resistance changes smoothly in proportion with the throttle valve opening angle.

7. Specification should be 0.71–1.38 kohm at closed throttle valve and 2.2–3.4 kohm at wide open throttle.

8. If within specification, check for poor connection between the PCM and the component.

9. If not, substitute the sensor with a known good component, check for proper operation. If problem corrected replace the sensor.

2.7L Engine

1. Disconnect the sensor connector.

2. Measure the resistance between terminal 1 and terminal 2, (component side).

3. Specification should be 1.6–2.4 kohm, at all throttle positions.

4. Measure the resistance between terminal 2 and terminal 3, (component side).

5. Specification should be 0.71–1.38 kohm, at closed throttle and 2.2–3.4 kohm at wide open throttle.

6. If within specification, check for poor connection between the ECM and the component.

7. If not, substitute the sensor with a known good component, check for proper operation. If problem corrected replace the sensor.

XG300

See Figure 30.

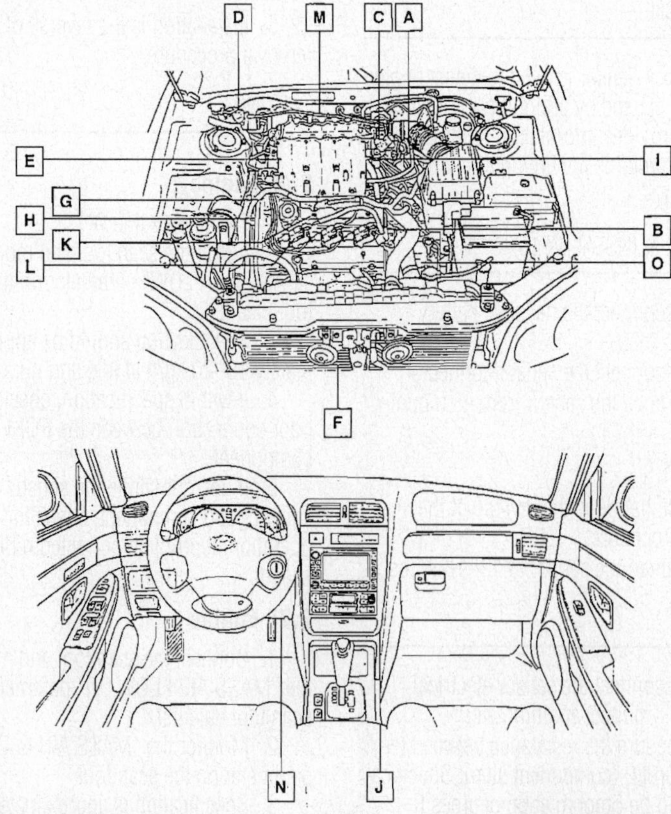

Fig. 30 Underhood sensor locations—XG300

29130_HYUN_G0030

Camshaft Position (CMP) Sensor

LOCATION

The camshaft position sensor is located on the engine near the camshaft pulley.

OPERATION

The camshaft position sensor senses the TDC point of the number one cylinder, on its compression stroke. Its signal is relayed to the ECM to be used to determine the sequence of fuel injection.

REMOVAL & INSTALLATION

1. Disconnect the negative battery cable.
2. Disconnect the connector from the sensor.
3. Remove the bolt that retains the sensor.
4. Remove the sensor.

To install:

5. Installation is the reverse of the removal procedure.

TESTING

1. Using a voltmeter measure the sensor output voltage.
2. At idle the specification should be 0–5 volts. At 3000 rpm the specification should be 0–5 volts.
3. If abnormality is found, replace the sensor.

Crankshaft Position (CKP) Sensor

LOCATION

The crankshaft position sensor is located on the engine near the crankshaft pulley.

OPERATION

The crankshaft position sensor consists of a magnet and coil. The voltage signal from the sensor is relayed to the ECM to indicate engine rpm and the position of the crankshaft.

REMOVAL & INSTALLATION

1. Disconnect the negative battery cable.
2. Disconnect the connector from the sensor.
3. Remove the bolt that retains the sensor in place.
4. Remove the sensor from its mounting.

To install:

5. Installation is the reverse of the removal procedure.

TESTING

1. Turn the ignition ON.
2. Disconnect the sensor electrical connector. Be sure the PCM connector is connected.
3. Measure the voltage between ground and terminal 2 of the sensor harness connector.
4. Specification should be 5 volts.
5. If so repair or replace the sensor.

Electronic Control Module (ECM)

OPERATION

The ECM controls the vehicle engine operating system.

REMOVAL & INSTALLATION

1. Disconnect the negative battery cable.
2. Remove the lower inner trim..
3. As required, detach the floor mat. As required, remove the protective cover.
4. Remove the ECM bracket retaining nuts. Remove the clip from the bracket.
5. Disconnect the connectors.
6. Remove the ECM from the vehicle.

To install:

7. Installation is the reverse of the removal procedure.

➡ **When replacing the ECM, be careful not to use the wrong part number, as damage to the injection system could occur.**

Engine Coolant Temperature (ECT) Sensor

LOCATION

The ECT is installed in the engine coolant passage of the cylinder head.

OPERATION

This component detects the temperature of the engine coolant and relays the information to the electronic control assembly. This component employs a thermistor which is sensitive to temperature changes. The electric resistance of the thermistor decreases in response to temperature rise. The ECM judges coolant temperature by the sensor output voltage and provides optimum fuel enrichment when the engine is cold.

REMOVAL & INSTALLATION

1. Disconnect the negative battery cable.
2. Disconnect the connector from the sensor.
3. Drain the cooling system, as required.
4. Remove the sensor from its mounting.

To install:

5. Installation is the reverse of the removal procedure.

TESTING

1. Remove the sensor.
2. With the temperature sensing portion of the sensor immersed in hot engine coolant, check the resistance.
3. Specification should be 5.9 kohms at 32 degrees F., 2.5 kohms at 68 degrees F., 1.1 kohms at 104 degrees F., 0.3 kohms at 176 degrees F.
4. If not within specification, replace the sensor.

Heated Oxygen (HO2S) Sensor

LOCATION

The sensors are located in the exhaust system. On some vehicles one sensor is located up at the exhaust manifold and the other sensor is located down at the catalytic converter. On other vehicles both sensors are located down at the catalytic converter.

OPERATION

The sensor senses the oxygen concentration in the exhaust gas then converts it into a voltage and sends it on to the ECM. The sensor gives an output of about 800mV, when the air fuel ratio is richer than the theoretical ratio and output of about 100mv when the ratio is leaner. The ECM controls fuel injection based on the signal so that the air fuel ratio is maintained at the theoretical ratio.

REMOVAL & INSTALLATION

1. Disconnect the electrical connector from the sensor.

2. Remove the oxygen sensor.

To install:

3. Installation is the reverse of the removal procedure.

➡ **Apply anti-seize compound to the threaded portion of the sensor, prior to installation. Never apply anti-seize compound to the protector of the sensor.**

4. Tighten the sensor to 37–44 ft. lbs.

TESTING

Perform a visual inspection of the sensor as follows:

1. If the sensor tip has a black/sooty deposit, this may indicate a rich fuel mixture.
2. If the sensor tip has a white, gritty deposit, this may indicate an internal coolant leak.
3. If the sensor tip has a brown deposit, this could indicate excessive oil consumption.
4. Warm the engine until the coolant temperature reaches operating temperature.
5. Accelerate the engine to 4000 rpm. When decelerating suddenly from 4000 rpm, the voltmeter reading should be 200mV or lower.
6. When the engine is suddenly raced, the voltmeter reading should be 600–1000mV.
7. If measured value is not within specification, replace the sensor.

Intake Air Temperature (IAT) Sensor

LOCATION

This sensor is located in the air intake plenum assembly. It is combined with the Manifold Absolute Pressure (AFS) sensor.

OPERATION

This sensor is a resistor based sensor which detects the intake air temperature. According to the intake air temperature reading the ECM will control the necessary amount of fuel injection.

REMOVAL & INSTALLATION

1. Disconnect the negative battery cable.
2. Disconnect the connector from the sensor.
3. Remove the sensor retaining screws.
4. Remove the sensor from its mounting.

To install:

5. Installation is the reverse of the removal procedure.

TESTING

1. Allow the engine coolant to reach 176–198 degrees F.
2. Using a voltmeter check the sensor output voltage.
3. Specification should be 0.5 volt at idle, 1.0 volt at 2000 rpm.
4. If the measured value is not within specification, replace the sensor.

Knock Sensor (KS)

LOCATION

The knock sensor is located in the side of the cylinder block.

OPERATION

The knock sensor is used to detect engine vibrations caused by preignition or detonation and provides information to the ECM, which then retards the timing to eliminate detonation.

REMOVAL & INSTALLATION

1. Disconnect the negative battery cable.
2. Disconnect the sensor connector.
3. Remove the sensor from its mounting.

To install:

4. Installation is the reverse of the removal procedure.
5. Tighten the sensor to 12–18 ft. lbs.

TESTING

1. Disconnect the sensor electrical connector. Connect an ohmmeter.
2. Measure the resistance between terminals 2 and 3. Specification should be 5mohm at 68 degrees F.
3. If the resistance is zero, replace the sensor.
4. Measure the capacitance between terminal 2 and 3.
5. Specification should be 800–1600 pF.

Mass Air Flow (MAF) Sensor

LOCATION

This sensor is located in the air intake plenum assembly. It is combined with the IAT sensor.

OPERATION

The MAF is measured by detection of heat transfer from a hot film probe because the change of the mass air flow rate causes change in the amount of heat being transferred from the hot film probe surface to the air flow. The air flow sensor generates a pulse so it repeatedly opens and closes between the voltage (5V) supplied from the ECM. This results in a change of the temperature of the hot film probe and in the change of resistance.

REMOVAL & INSTALLATION

1. Disconnect the negative battery cable.
2. Disconnect the connector from the sensor.
3. Remove the sensor retaining screws.
4. Remove the sensor from its mounting.

To install:

5. Installation is the reverse of the removal procedure.

TESTING

1. Allow the engine coolant to reach 176–198 degrees F.
2. Using a voltmeter check the sensor output voltage.
3. Specification should be 0.5 volt at idle, 1.0 volt at 2000 rpm.
4. If the measured value is not within specification, replace the sensor.

Throttle Position Sensor (TPS)

OPERATION

The throttle position sensor is a rotating type variable resistor that rotates with the throttle body's throttle shaft to sense the throttle valve angle. As the throttle shaft rotates, the throttle angle of the sensor changes and the ECM detects the throttle valve opening based on the TPS output voltage.

REMOVAL & INSTALLATION

1. Disconnect the negative battery cable.
2. Disconnect the sensor connector.
3. Remove the sensor retaining screws. Remove the sensor from its mounting.

To install:

4. Installation is the reverse of the removal procedure.

TESTING

1. Disconnect the connector from the sensor.
2. Measure the resistance between terminals 1 (sensor ground) and 4 (sensor power).
3. Specification should be 3.5–6.5 kohm.
4. Connect an analog ohmmeter between terminals 1 (sensor ground) and 3 (sensor output).
5. Operate the throttle valve slowly from the idle position to the full open position, and check that the resistance changes smoothly in proportion with the throttle valve opening angle.
6. If the resistance is out of specification, or fails to change smoothly, replace the sensor.

XG350

See Figure 31.

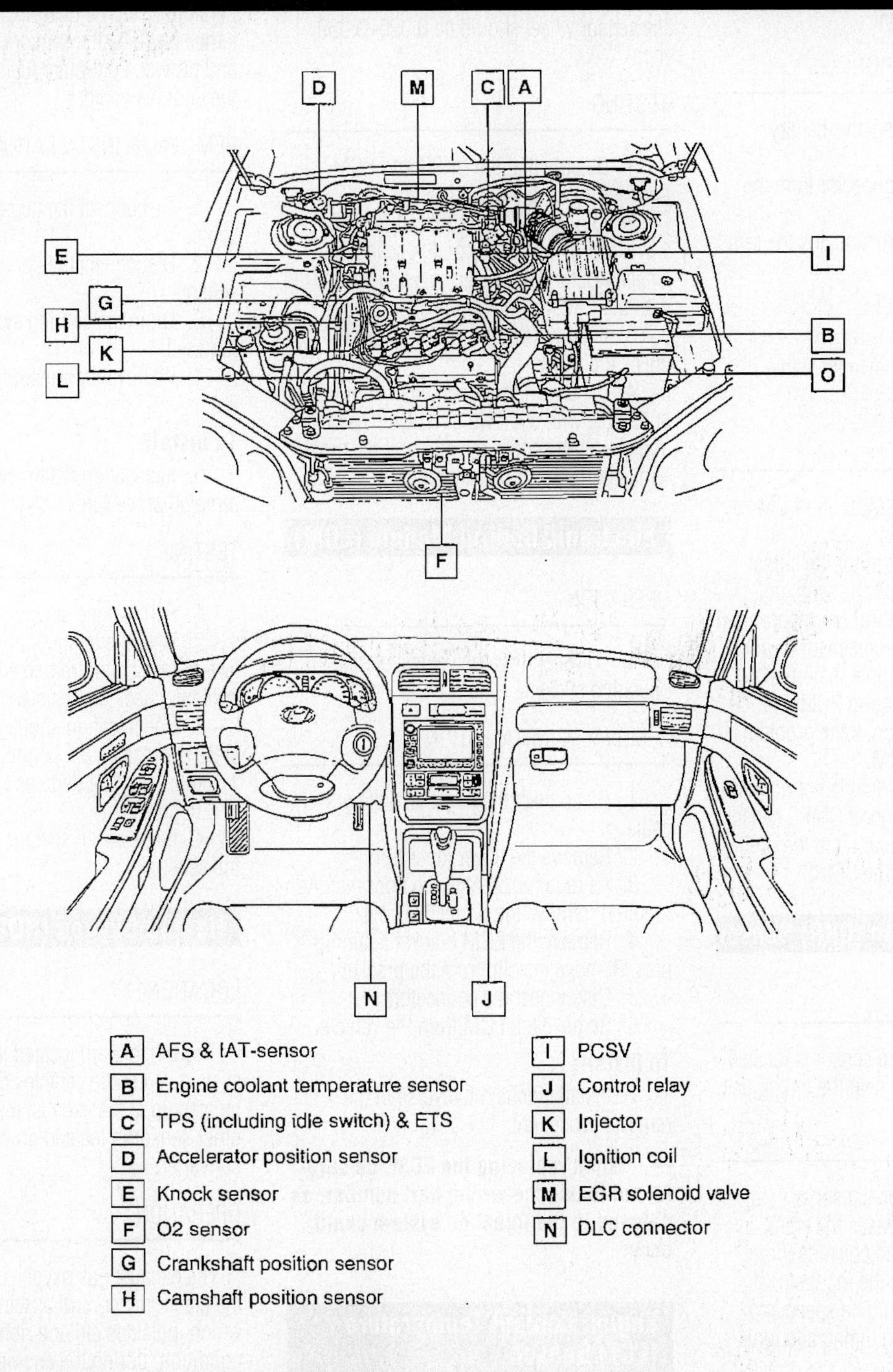

Fig. 31 Underhood sensor locations—XG350

Camshaft Position (CMP) Sensor

LOCATION

This sensor is located at the front of the engine near the camshaft gear.

OPERATION

The camshaft position sensor senses the TDC point of the number one cylinder, on its compression stroke. Its signal is relayed to the PCM to be used to determine the sequence of fuel injection.

REMOVAL & INSTALLATION

1. Disconnect the negative battery cable.
2. Disconnect the connector from the sensor.
3. Remove the bolt that retains the sensor.
4. Remove the sensor.

To install:

5. Installation is the reverse of the removal procedure.

TESTING

1. Be sure that the CMPS and PCM connectors are connected.
2. Engine ON and monitor the signal waveform of the sensor on the scan tool. Check whether the waveform is synchronized with the crankshaft position sensor or not.
3. If the waveform signal is normal, substitute with a known good PCM and check for proper operation. If the problem is corrected replace the PCM.
4. If the waveform signal is not normal, substitute with a known good CMPS sensor and check for proper operation. If the problem is corrected replace the CMPS.

Crankshaft Position (CKP) Sensor

LOCATION

The crankshaft position sensor is located at the front of the engine near the timing belt.

OPERATION

The crankshaft position sensor is a Hall Effect type sensor that senses the crank angle of each cylinder and converts it into a pulse signal. Based on the input signal, the PCM computes the engine speed and controls the fuel injection timing and ignition timing.

REMOVAL & INSTALLATION

1. Disconnect the negative battery cable.
2. Disconnect the connector from the sensor.
3. Remove the retaining bolts.
4. Remove the sensor from its mounting.

To install:

5. Installation is the reverse of the removal procedure.
6. Clearance between the sensor and the sensor wheel should be 0.020–0.059 inch.

TESTING

1. Be sure that the CKPS and PCM connectors are connected.
2. Engine ON and monitor the signal waveform of the sensor on the scan tool. Check whether the waveform is synchronized with the camshaft position sensor or not.
3. If the waveform signal is normal, check the PCM.
4. If the waveform signal is not normal, substitute with a known good CKPS sensor and check for proper operation. If the problem is corrected replace the CMPS.

Electronic Control Module (ECM)

OPERATION

The ECM controls the vehicle engine operating system.

REMOVAL & INSTALLATION

1. Disconnect the negative battery cable.
2. Remove the lower inner trim..
3. As required, detach the floor mat. As required, remove the protective cover.
4. Remove the ECM bracket retaining nuts. Remove the clip from the bracket.
5. Disconnect the connectors.
6. Remove the ECM from the vehicle.

To install:

7. Installation is the reverse of the removal procedure.

➡ **When replacing the ECM, be careful not to use the wrong part number, as damage to the injection system could occur.**

Engine Coolant Temperature (ECT) Sensor

LOCATION

The ECT is installed in the engine coolant passage of the cylinder head.

OPERATION

This component detects the temperature of the engine coolant and relays the information to the PCM. This component employs a thermistor which is sensitive to temperature changes. The electric resistance of the thermistor decreases in response to temperature rise. The ECM/PCM judges coolant temperature by the sensor output voltage and provides optimum fuel enrichment when the engine is cold.

REMOVAL & INSTALLATION

1. Disconnect the negative battery cable.
2. Disconnect the connector from the sensor.
3. Drain the cooling system, as required.
4. Remove the sensor from its mounting.

To install:

5. Installation is the reverse of the removal procedure.

TESTING

1. Remove the sensor from the engine.
2. With the temperature sensing portion of the sensor immersed in hot engine coolant, check the resistance.
3. Specification should be 5.9 kohms at 32 degrees F., 2.5 kohms at 68 degrees F., 1.1 kohms at 104 degrees F., 0.3 kohms at 176 degrees F.
4. If not within specification, replace the sensor.

Heated Oxygen (HO2S) Sensor

LOCATION

The sensors are located in the exhaust system. On some vehicles one sensor is located up at the exhaust manifold(s) and the other sensor is located down at the catalytic converter.

OPERATION

The exhaust gas oxygen sensor supplies the electronic control assembly with a signal which indicates either a rich or lean mixture condition, during the engine operation.

REMOVAL & INSTALLATION

1. Disconnect the electrical connector from the sensor.
2. Remove the oxygen sensor.

To install:

3. Installation is the reverse of the removal procedure.

➡ **Apply anti-seize compound to the threaded portion of the sensor, prior to installation. Never apply anti-seize compound to the protector of the sensor.**

TESTING

Perform a visual inspection of the sensor as follows:

1. If the sensor tip has a black/sooty deposit, this may indicate a rich fuel mixture.
2. If the sensor tip has a white, gritty deposit, this may indicate an internal coolant leak.
3. If the sensor tip has a brown deposit, this could indicate excessive oil consumption.
4. Warm the engine until the coolant temperature reaches operating temperature.
5. Disconnect the sensor electrical connector and measure the resistance between terminal 3 and 4.
6. Specification should be 7–40 ohms.
7. If within specification, substitute with a known good PCM and check for proper operation. If the problem is corrected replace the PCM.
8. If not within specification, substitute with a known good sensor and check for proper operation. If the problem is corrected replace the sensor.

Intake Air Temperature (IAT) Sensor

LOCATION

This sensor is located in the air intake plenum assembly. It is combined with the Manifold Absolute Pressure (AFS) sensor.

OPERATION

This sensor is a resistor based sensor which detects the intake air temperature. According to the intake air temperature reading the ECM will control the necessary amount of fuel injection.

REMOVAL & INSTALLATION

1. Disconnect the negative battery cable.
2. Disconnect the connector from the sensor.
3. Remove the sensor retaining screws.
4. Remove the sensor from its mounting.

To install:

5. Installation is the reverse of the removal procedure.

TESTING

1. Be sure the engine is at operating temperature.
2. Remove the sensor connector.
3. Measure the output voltage.
4. Specification should be 1.1–1.8 volt at idle and 3.0–3.5 volts at 2100 rpm at WOT (Drive).
5. Replace the sensor, as required.

Knock Sensor (KS)

LOCATION

The knock sensor is located in the side of the cylinder block.

OPERATION

The knock sensor is used to detect engine vibrations caused by preignition or detonation and provides information to the ECM, which then retards the timing to eliminate detonation.

REMOVAL & INSTALLATION

1. Disconnect the negative battery cable.
2. Disconnect the sensor connector.
3. Remove the sensor from its mounting.

To install:

4. Installation is the reverse of the removal procedure.
5. Tighten the sensor to 12–18 ft. lbs.

TESTING

1. Disconnect the sensor electrical connector. Connect an ohmmeter.
2. Measure the resistance between terminals 2 and 3. Specification should be 5mohm at 68 degrees F.
3. If resistance is zero, replace the sensor.
4. Measure the capacitance between terminal 2 and 3. Specification should be 800–1600 pF.
5. If not within specification, replace the sensor.

Mass Air Flow (MAF) Sensor

LOCATION

The MAF is mounted in the intake air hose of the air cleaner assembly. On some vehicles it is combined with the IAT sensor.

OPERATION

The MAF is measured by detection of heat transfer from a hot film probe because the change of the mass air flow rate causes change in the amount of heat being transferred from the hot film probe surface to the air flow. The air flow sensor generates a pulse so it repeatedly opens and closes between the voltage (5V) supplied from the PCM. This results in a change of the temperature of the hot film probe and in the change of resistance.

REMOVAL & INSTALLATION

1. Disconnect the negative battery cable.
2. Disconnect the connector from the sensor.
3. Remove the air cleaner and air intake assembly, as required.
4. Remove the sensor from its mounting.

To install:

5. Installation is the reverse of the removal procedure.

TESTING

1. Be sure the engine is at operating temperature.
2. Remove the sensor connector.
3. Measure the output voltage.
4. Specification should be 1.1–1.8 volt at idle and 3.0–3.5 volts at 2100 rpm at WOT (Drive).
5. Replace the sensor, as required.

Throttle Position Sensor (TPS)

OPERATION

The throttle position sensor is a rotating type variable resistor that rotates with the throttle body's throttle shaft to sense the throttle valve angle. As the throttle shaft rotates, the throttle angle of the sensor changes and the ECM/PCM detects the throttle valve opening based on the TPS output voltage.

REMOVAL & INSTALLATION

1. Disconnect the negative battery cable.
2. Disconnect the sensor connector.
3. Remove the sensor retaining screws. Remove the sensor from its mounting.

To install:

4. Installation is the reverse of the removal procedure.

TESTING

1. Disconnect the sensor connector.
2. Measure the resistance between terminal 3 (sensor ground) and terminal 1 (sensor power).
3. Specification should be 3.5–6.5 kohm.
4. Connect and analog ohmmeter between terminal 3 (sensor ground) and terminal 2 (sensor output).
5. Operate the throttle valve slowly from the idle position to the full open position, and check that the resistance changes smoothly in proportion with the throttle valve opening angle.
6. If the resistance is out of specification, or fails to change smoothly, replace the sensor.

INFINITI

DIAGNOSTIC TROUBLE CODES

7

TABLE OF CONTENTS

VEHICLE APPLICATIONS 7-2
P0XXX 7-3
P1XXX 7-29
P2XXX 7-45
U1XXX 7-46

OBD II VEHICLE APPLICATIONS

INFINITI

FX35 FX45
2003–2006
3.5L V6 VQ35DE.......VIN C
4.5L V8 VK45DE.......A

G20
1995–2002
2.0 L4 SR20DE.......VIN C

G35
2003–2006
3.5L V6 VQ35DE.......VIN C

I30
1996–2001
3.0L V6 VQ30DE.......VIN C

I35
2002–2004
3.0L V6 VQ35DE.......VIN D

J30
1995–1997
3.0L V6 VG30DE.......VIN Y

M35
2006
3.5L V6 VQ35DE.......VIN A

M45
2003–2006
4.5L V8 VK45DE.......VIN B

Q45
1995–2006
4.1L V8 VH41DE.......VIN B
4.5L V8 VH45DE.......VIN N
4.5L V8 VK45DE.......VIN B

QX4
1997–2003
3.3L V6.......VIN A
3.5L V6.......VIN D

QX56
2004–2006
5.6L V8.......VIN A

OBD II Trouble Code List (P0XXX Codes)

DTC	Trouble Code Title, Conditions & Possible Causes
DTC: P0011 **2T PCM, MIL: Yes** **2002-06** **Models:** FX35, FX45, G35, I35, M35, M45, QX4, Q45 **Engines:** All **Transmissions:** All	**Intake Valve Timing Control Range/Performance (Bank 1):** Engine started, engine running at idle speed, followed by a quick and steady acceleration in 1st gear to over 2000 rpm, and the PCM detected a problem in the operation of the IVT control during testing. **Possible Causes:** • Check for an accumulation of debris on the CMP sensor pickup • CKP POS sensor or CMP Phase sensor signal is erratic • IVT connector is damaged or loose (an intermittent fault) • IVT solenoid control circuit is open or shorted to ground • IVT solenoid is damaged or has failed • PCM has failed
DTC: P0021 **2T PCM, MIL: Yes** **2002-06** **Models:** FX35, FX45, G35, I35, M35, M45, QX4, Q45 **Engines:** All **Transmissions:** All	**Intake Valve Timing Control Performance (Bank 2):** Engine started, engine running at idle speed, followed by a quick and steady acceleration in 1st gear to over 2000 rpm, and the PCM detected a problem in the operation of the IVT control during testing. **Possible Causes:** • Check for an accumulation of debris on the CMP sensor pickup • CKP POS sensor or CMP Phase sensor signal is erratic • IVT connector is damaged or loose (an intermittent fault) • IVT solenoid control circuit is open or shorted to ground • IVT solenoid is damaged or has failed • PCM has failed
DTC: P0031 **2T CCM, MIL: Yes** **2002-06** **Models:** FX35, FX45, G35, I35, M35, M45, QX4, Q45 **Engines:** All **Transmissions:** All	**Heated Oxygen Sensor (Bank 1 Sensor 1) Heater Control Circuit Low:** Engine started, system voltage from 10.5-16.0v, and the PCM detected the HO2S-11 heater control circuit input (voltage) was excessively low (i.e., it was operating out of its normal range). **Possible Causes:** • HO2S heater control connector is damaged or loose (open) • HO2S heater power circuit is open (check the No. 8 fuse 15A) • HO2S heater assembly is damaged or has failed • PCM has failed
DTC: P0032 **2T CCM, MIL: Yes** **2002-06** **Models:** FX35, FX45, G35, I35, M35, M45, QX4, Q45 **Engines:** All **Transmissions:** All	**HO2S-11 (Bank 1 Sensor 1) Heater Circuit High Input:** Engine started, system voltage from 10.5-16.0v, and the PCM detected the HO2S-11 heater control circuit input (voltage) was excessively high (i.e., it was operating out of its normal range). **Possible Causes:** • HO2S heater control connector is damaged or loose (shorted) • HO2S heater control circuit is shorted to ground • HO2S heater assembly is damaged or has failed • PCM has failed
DTC: P0037 **2T CCM, MIL: Yes** **2002-06** **Models:** FX35, FX45, G35, I35, M35, M45, QX4 **Engines:** All **Transmissions:** All	**HO2S-12 (Bank 1 Sensor 2) Heater Circuit Low Input:** Engine started, system voltage from 10.5-16.0v, and the PCM detected the HO2S-12 heater control circuit input (voltage) was excessively low (i.e., it was operating out of its normal range). **Possible Causes:** • HO2S heater control connector is damaged or loose (open) • HO2S heater power circuit is open (check the No. 8 fuse 15A) • HO2S heater assembly is damaged or has failed • PCM has failed
DTC: P0038 **2T CCM, MIL: Yes** **2002-06** **Models:** FX35, FX45, G35, I35, M35, M45, QX4, Q45 **Engines:** All **Transmissions:** All	**HO2S-12 (Bank 1 Sensor 2) Heater Circuit High Input:** Engine started, system voltage from 10.5-16.0v, and the PCM detected the HO2S-12 heater control circuit input (voltage) was excessively high (i.e., it was operating out of its normal range). **Possible Causes:** • HO2S heater control connector is damaged or loose (shorted) • HO2S heater control circuit is shorted to ground • HO2S heater assembly is damaged or has failed • PCM has failed
DTC: P0051 **2T CCM, MIL: Yes** **2002-06** **Models:** FX35, FX45, G35, I35, M35, M45, QX4, Q45 **Engines:** All **Transmissions:** All	**HO2S-21 (Bank 2 Sensor 1) Heater Circuit Low Input:** Engine started, system voltage from 10.5-16.0v, and the PCM detected the HO2S-21 heater control circuit input (voltage) was excessively low (i.e., it was operating out of its normal range). **Possible Causes:** • HO2S heater control connector is damaged or loose (open) • HO2S heater power circuit is open (check the No. 8 fuse 15A) • HO2S heater assembly is damaged or has failed • PCM has failed

DTC	Trouble Code Title, Conditions & Possible Causes
DTC: P0052 **2T PCM, MIL: Yes** **2002-06** **Models:** FX35, FX45, G35, I35, M35, M45, QX4, Q45 **Engines:** All **Transmissions:** All	**HO2S-21 (Bank 2 Sensor 1) Heater Control Circuit High Input:** Engine started, system voltage from 10.5-16.0v, and the PCM detected the HO2S-21 heater control circuit input (voltage) was excessively high (i.e., it was operating out of its normal range). **Possible Causes:** • HO2S heater control connector is damaged or loose (shorted) • HO2S heater control circuit is shorted to ground • HO2S heater assembly is damaged or has failed • PCM has failed
DTC: P0057 **2T CCM, MIL: Yes** **2002-06** **Models:** FX35, FX45, G35, I35, M35, M45, QX4, Q45 **Engines:** All **Transmissions:** All	**HO2S-22 (Bank 2 Sensor 2) Heater Circuit Low Input:** Engine started, system voltage from 10.5-16.0v, and the PCM detected the HO2S-22 heater control circuit input (voltage) was excessively low (i.e., it was operating out of its normal range). **Possible Causes:** • HO2S heater control connector is damaged or loose (open) • HO2S heater power circuit is open (check the No. 8 fuse 15A) • HO2S heater assembly is damaged or has failed • PCM has failed
DTC: P0058 **2T PCM, MIL: Yes** **2002-06** **Models:** FX35, FX45, G35, I35, M35, M45, QX4, Q45 **Engines:** All **Transmissions:** All	**HO2S-22 (Bank 2 Sensor 2) Heater Circuit High Input:** Engine started, system voltage from 10.5-16.0v, and the PCM detected the HO2S-22 heater control circuit input (voltage) was excessively high (i.e., it was operating out of its normal range). **Possible Causes:** • HO2S heater control connector is damaged or loose (shorted) • HO2S heater control circuit is shorted to ground • HO2S heater assembly is damaged or has failed • PCM has failed
DTC: P0100 **2T CCM, MIL: Yes** **1996-2003** **Models:** G20, I30, M45, Q45 **Engines:** All **Transmissions:** All	**Mass Airflow Sensor Circuit Malfunction:** Key on, and the PCM detected the an excessively high MAF sensor signal, or with that the MAF sensor signal was excessively low with the engine running, or the MAF signal was high at light engine load, or the MAF signal was low at high engine load conditions, or that the MAF signal remained below 1.0v during any operating conditions. **Possible Causes:** • Intake air leaks after the MAF sensor location • MAF sensor signal circuit is open or shorted to ground • MAF sensor signal circuit is shorted to VREF or system power • MAF sensor has failed • TSB 01-029A (9/01) contains information related to this code
DTC: P0101 **1T CCM, MIL: Yes** **2002-06** **Models:** FX35, FX45, G35, I35, M35, M45, QX4, Q45 **Engines:** All **Transmissions:** All	**Mass Airflow Sensor Signal Range/Performance:** Engine started, engine running at light engine load, and the PCM detected the MAF sensor signal was excessively high; or with the engine running under high engine load, the PCM detected the MAF sensor signal was excessively low under these operating conditions. **Possible Causes:** • Harness or connectors (The sensor circuit is open or shorted.) • Intake air temperature sensor • Mass air flow sensor • Intake air leaks
DTC: P0102 **1T CCM, MIL: Yes** **2002-06** **Models:** FX35, FX45, G35, I35, QX4, Q45 **Engines:** All **Transmissions:** All	**Mass Air Flow Sensor Circuit Low Input:** Key on or engine running; and the PCM detected an unexpected low voltage condition on the MAF sensor signal circuit in the CCM test. **Possible Causes:** • MAF sensor connector is damaged, loose or shorted • MAF sensor signal circuit is shorted to ground • MAF sensor power circuit is open (check for power from relay) • MAF sensor is damaged or has failed • PCM is damaged
DTC: P0103 **1T CCM, MIL: Yes** **2002-06** **Models:** FX35, FX45, G35, I35, M35, M45, QX4, Q45 **Engines:** All **Transmissions:** All	**Mass Air Flow Sensor Circuit High Input:** Key on or engine running; and PCM detected an unexpected high voltage condition on the MAF sensor signal circuit in the CCM test. **Possible Causes:** • MAF sensor signal circuit is open (check for loose connector) • MAF sensor is shorted to VREF or system power • MAF sensor is damaged or has failed • PCM is damaged

DTC	Trouble Code Title, Conditions & Possible Causes
DTC: P0105 **2T CCM, MIL: Yes** **1998-2003** **Models:** G20, I30, J30, M45, Q45 **Engines:** All **Transmissions:** All	**Absolute Pressure Sensor Circuit Malfunction:** Key on or engine running; and the PCM detected an unexpected voltage on the Absolute Pressure signal during the CCM test. **Note: This signal is used during several of the Main Monitor tests.** **Possible Causes:** • Absolute pressure sensor signal circuit open, shorted to ground • Absolute pressure sensor signal circuit is shorted to VREF • Absolute pressure sensor is damaged or has failed • PCM has failed
DTC: P0107 **2T CCM, MIL: Yes** **2002-04** **Models:** FX35, FX45, I35, QX4 **Engines:** All **Transmissions:** All	**Absolute Pressure Sensor Circuit Low Input:** Key on or engine running; and the PCM detected an unexpected low voltage on the Absolute Pressure signal circuit during the CCM test. **Note: This signal is used during several of the Main Monitor tests.** **Possible Causes:** • Absolute pressure sensor signal circuit is shorted to ground • Absolute pressure sensor is damaged or has failed • PCM has failed
DTC: P0108 **2T CCM, MIL: Yes** **2002-04** **Models:** FX35, FX45, I35, QX4 **Engines:** All **Transmissions:** All	**Absolute Pressure Sensor Circuit High Input:** Key on or engine running; and the PCM detected an unexpected high voltage on the Absolute Pressure signal circuit in the CCM test. **Note: This signal is used during several of the Main Monitor tests.** **Possible Causes:** • Absolute pressure sensor connector is damaged (open circuit) • Absolute pressure sensor signal circuit open, shorted to VREF • Absolute pressure sensor ground circuit is open • Absolute pressure sensor is damaged or has failed • PCM has failed
DTC: P0110 **2T CCM, MIL: Yes** **1996-2004** **Models:** G20, I30, I35, J30, M45, Q45 **Engines:** All Transmission: All	**Intake Air Temperature Sensor Signal Range/Performance:** Engine started, and the PCM detected an unexpected low or high voltage on the IAT sensor circuit, or the IAT sensor signal was not plausible when compared to the ECT signal. **Possible Causes:** • IAT sensor signal circuit is open, shorted to ground or to VREF • IAT sensor is damaged or has failed (out of calibration) • PCM has failed
DTC: P0112 **2T CCM, MIL: Yes** **2002-06** **Models:** FX35, FX45, G35, I35, M35, M45, QX4, Q45 **Engines:** All **Transmissions:** All	**Intake Air Temperature Sensor Circuit Low Input:** Key on or engine running; and the PCM detected an unexpected low voltage condition on the IAT sensor signal circuit in the CCM test. **Possible Causes:** • IAT sensor connector is damaged (it may be shorted) • IAT sensor signal circuit is shorted to ground • IAT sensor is damaged or has failed • PCM has failed
DTC: P0113 **2T CCM, MIL: Yes** **2002-04** **Models:** FX35, FX45, G35, I35, M35, M45, QX4 **Engines:** All **Transmissions:** All	**Intake Air Temperature Sensor Circuit High Input:** Key on or engine running; and PCM detected an unexpected high voltage condition on the IAT sensor signal circuit in the CCM test. **Possible Causes:** • IAT sensor signal circuit is open or shorted to VREF • IAT sensor ground circuit is open • IAT sensor is damaged or has failed • PCM has failed
DTC: P0115 **2T CCM, MIL: Yes** **1996-2003** **Models:** G20, I30, J30, M45, Q45 **Engines:** All **Transmissions:** All	**Engine Coolant Temperature Sensor Circuit Malfunction:** Engine started, and the PCM detected the ECT sensor input was either too high or too low during the CCM test. **Possible Causes:** • ECT sensor connector is damaged or loose (intermittent fault) • ECT sensor signal circuit open, shorted to ground or to VREF • ECT sensor is damaged or has failed (out of calibration) • PCM has failed

DTC	Trouble Code Title, Conditions & Possible Causes
DTC: P0117 **1T CCM, MIL: Yes** **2002-06** **Models:** FX35, FX45, G35, I35, M35, M45, QX4, Q45 **Engines:** All **Transmissions:** All	**Engine Coolant Temperature Sensor Circuit Low Input:** Key on or engine running; and the PCM detected an unexpected low voltage condition on the ECT sensor signal circuit in the CCM test. **Possible Causes:** • ECT sensor signal circuit is shorted to ground • ECT sensor is damaged or has failed • PCM has failed
DTC: P0118 **1T CCM, MIL: Yes** **2002-06** **Models:** FX35, FX45, G35, I35, M35, M45, QX4, Q45 **Engines:** All **Transmissions:** All	**Engine Coolant Temperature Sensor Circuit High Input:** Key on or engine running; and PCM detected an unexpected high voltage condition on the ECT sensor signal circuit in the CCM test. **Possible Causes:** • ECT sensor signal circuit is open (the connector may be loose) • ECT sensor ground circuit is open (connector may be loose) • ECT sensor signal circuit is shorted to sensor VREF • ECT sensor is damaged or has failed • PCM has failed
DTC: P0120 **2T CCM, MIL: Yes** **1996-02** **Models:** G20, I30, J30, M45, Q45 **Engines:** All **Transmissions:** All	**Throttle Position Sensor Circuit Malfunction** Key on for at least 5 seconds, and the PCM detected an unexpected TP sensor signal, or with the engine running, that the TP signal was not plausible when it was compared to the IACV-AAC valve, CMP and MAF sensor signals. **Possible Causes:** • TP sensor signal circuit open or shorted to ground • TP sensor signal circuit shorted to VREF or to system power • TP sensor is damaged or has failed (perform a sweep test) • PCM has failed
DTC: P0121 **2T CCM, MIL: Yes** **2002-03** **Models:** QX4 **Engines:** All **Transmissions:** All	**Throttle Position Sensor Signal Range/Performance** DTC P0510 not set, engine started, and the PCM detected a high voltage condition on the TP sensor signal circuit under low engine load conditions, or the PCM detected a low voltage condition on the TP sensor signal circuit under high engine load conditions. **Possible Causes:** • TP sensor connector is damaged or loose (intermittent fault) • TP sensor signal circuit is open, shorted to ground or to VREF • TP sensor is damaged • PCM has failed
DTC: P0122 **1T CCM, MIL: Yes** **2002** **Models:** QX4 **Engines:** All **Transmissions:** All	**Throttle Position Sensor Circuit Low Input:** DTC P0510 not set, engine started, system voltage over 10.0v, and the PCM detected an unexpected low voltage condition on the TP sensor signal during the CCM test. **Possible Causes:** • TP sensor connector is damaged (possible intermittent short) • TP sensor signal circuit is shorted to ground • TP sensor is damaged • PCM has failed
DTC: P0122 **1T CCM, MIL: Yes** **2003-06** **Models:** FX35, FX45, G35, I35, M35, M45, Q45, QX56 **Engines:** All **Transmissions:** All	**Throttle Position Sensor 2 Circuit Low Input** Key on or engine running; and the PCM detected an unexpected low voltage condition on the TP sensor 2 signal during the CCM test. The ETC Actuator consists of a throttle control motor and TP sensors. The TP sensors respond to movement of the throttle valve. The PCM judges the actual opening angle of the throttle valve from the signals from the TP1 and TP2 sensors during engine operation. **Possible Causes:** • TP sensor 2 connector is damaged or loose (intermittent fault) • TP sensor 2 signal circuit is shorted to ground • TP sensor 2 is damaged, or the ETC Actuator has failed
DTC: P0123 **1T CCM, MIL: Yes** **2002** **Models:** QX4 **Engines:** All **Transmissions:** All	**Throttle Position Sensor Circuit High Input** DTC P0510 not set, engine started, system voltage over 10.0v, and the PCM detected an unexpected high voltage condition on the TP sensor signal during the CCM test. **Possible Causes:** • TP sensor connector is damaged or loose (intermittent open) • TP sensor signal circuit is open or shorted to VREF • TP sensor is damaged • PCM has failed

DTC	Trouble Code Title, Conditions & Possible Causes
DTC: P0123 **1T CCM, MIL: Yes** **2002-06** **Models:** FX35, FX45, G35, I35, M35, M45, Q45, QX56 **Engines:** All **Transmissions:** All	**Throttle Position Sensor 2 Circuit High Input:** DTC P0510 not set, engine started, system voltage over 10.0v, and the PCM detected an unexpected high voltage condition on the TP sensor signal during the CCM test. **Possible Causes:** • TP sensor 2 connector is damaged or loose (intermittent fault) • TP sensor 2 signal circuit is open or shorted to VREF • TP sensor 2 is damaged, or the ETC Actuator has failed
DTC: P0125 **1T ECT, MIL: Yes** **1996-2006** **Models:** FX35, FX45, G20, G35, 30, I30, I35, J30, M45, Q45, QX4, QX56 **Engines:** All **Transmissions:** All	**ECT Sensor Excessive Time To Enter Closed Loop:** DTC P0115 not set, engine started, engine runtime over 6 minutes, ECT sensor more than 176°F, and the PCM detected that the engine did not achieve closed loop operation during the test period. **Note: If DTC P0125 is set with DTC P0115, repair DTC P0115 first!** **Possible Causes:** • Check the operation of the thermostat (it may be stuck open) • ECT sensor signal circuit has a high resistance condition • ECT sensor is damaged or has failed (out of calibration)
DTC: P0127 **2T ECT, MIL: Yes** **2003-06** **Models:** FX35, FX45, G35, I35, M35, M45, QX4, Q45 **Engines:** All **Transmissions:** All	**Intake Air Temperature Sensor Range/Performance:** Engine started cold, vehicle driven at a speed of over 45 mph for 10 minutes, and the PCM detected the IAT sensor input was not valid when compared to the change in the ECT sensor signal in the test. **Possible Causes:** • IAT sensor connector is damaged or loose (intermittent fault) • IAT sensor is damaged or has failed • PCM has failed
DTC: P0128 **2T ECT, MIL: Yes** **2003-06** **Models:** FX35, FX45, G35, I35, M35, M45, QX4, Q45 **Engines:** All **Transmissions:** All	**Thermostat function:** The engine coolant temperature does not reach to specified temperature even though the engine has run long enough. **Possible Causes:** • Check the operation of the thermostat (it may be stuck open) • Leakage from sealing portion of thermostat • Engine coolant temperature sensor
DTC: P0130 **2T CCM, MIL: Yes** **1996-2002** **Models:** G20, I30, J30, Q45 **Engines:** All **Transmissions:** All	**HO2S-11 (Bank 1 Sensor 1) Closed Loop Fuel Control:** Engine started, and after the vehicle was driven at a speed of 30-55 mph for 2 minutes, the PCM detected the HO2S signal voltage was excessively high or low, or it detected the HO2S signal did not reach the maximum or minimum voltages, or it took too long for the HO2S signal to switch from rich-to-lean or from lean-to-rich during the test. **Possible Causes:** • HO2S signal circuit is open or shorted to ground • HO2S is damaged or it has failed • PCM has failed
DTC: P0130 **2T CCM, MIL: Yes** **1998-2001** **Models:** QX4 **Engines:** All **Transmissions:** All	**HO2S-11 (Bank 1 Sensor 1) Closed Loop Fuel Control:** Engine started, and after the vehicle was driven at a speed of 30-55 mph for 2 minutes, the PCM detected the HO2S signal voltage was excessively high or low, or it detected the HO2S signal did not reach the maximum or minimum voltages, or it took too long for the HO2S signal to switch from rich-to-lean or from lean-to-rich during the test. **Possible Causes:** • HO2S signal circuit is open or shorted to ground • HO2S is damaged or it has failed • PCM has failed
DTC: P0130 **2T CCM, MIL: Yes** **2006** **Models:** FX35, FX45, M35, M45, QX56 **Engines:** All **Transmissions:** All	**HO2S-11 (Bank 1 Sensor 1) Closed Loop Fuel Control:** Engine started, and after the vehicle was driven at a speed of 30-55 mph for 2 minutes, the PCM detected the HO2S signal voltage was excessively high or low, or it detected the HO2S signal did not reach the maximum or minimum voltages, or it took too long for the HO2S signal to switch from rich-to-lean or from lean-to-rich during the test. **Possible Causes:** • HO2S signal circuit is open or shorted to ground • HO2S is damaged or it has failed • PCM has failed

DTC	Trouble Code Title, Conditions & Possible Causes
DTC: P0131 **2T CCM, MIL: Yes** **1996-2003** **Models:** G20, I30, J30, M45, Q45 **Engines:** All **Transmissions:** All	**HO2S-11 (Bank 1 Sensor 1) Circuit Low Input (Lean Shift):** Engine running in closed loop at a speed of over 20 mph for at least 20 seconds, IAT sensor signal from 14-122ºF, fuel level over 25%, and the PCM detected the HO2S-11 input did not reach a maximum voltage level of 0.60v, or it remained fixed at approximately 300 mv. **Possible Causes:** • Low fuel pressure, fuel filter restricted or fuel injectors plugged • HO2S heater is damaged or has failed • HO2S is contaminated or it has failed • PCM has failed
DTC: P0131 **2T CCM, MIL: Yes** **1998-2001** **Models:** QX4 **Engines:** All **Transmissions:** All	**HO2S-11 (Bank 1 Sensor 1) Circuit Low Input (Lean Shift):** Engine running in closed loop at a speed of over 20 mph for at least 20 seconds, IAT sensor signal from 14-122ºF, fuel level over 25%, and the PCM detected the HO2S-11 input did not reach a maximum voltage level of 0.60v, or it remained fixed at approximately 300 mv. **Possible Causes:** • Low fuel pressure, fuel filter restricted or fuel injectors plugged • HO2S heater is damaged or has failed • HO2S is contaminated or it has failed • PCM has failed
DTC: P0131 **2T CCM, MIL: Yes** **2006** **Models:** FX35, FX45, M35, M45, QX56 **Engines:** All **Transmissions:** All	**HO2S-11 (Bank 1 Sensor 1) Circuit Low Input (Lean Shift):** Engine running in closed loop at a speed of over 20 mph for at least 20 seconds, IAT sensor signal from 14-122ºF, fuel level over 25%, and the PCM detected the HO2S-11 input did not reach a maximum voltage level of 0.60v, or it remained fixed at approximately 300 mv. **Possible Causes:** • Low fuel pressure, fuel filter restricted or fuel injectors plugged • HO2S heater is damaged or has failed • HO2S is contaminated or it has failed • PCM has failed
DTC: P0132 **2T CCM, MIL: Yes** **1998-2006** **Models:** FX35, FX45, G20, G35, I30, I35, J30, M35, M45, Q45, QX4 **Engines:** All **Transmissions:** All	**HO2S-11 (Bank 1 Sensor 1) Circuit High Input (Rich Shift):** Engine started, IAT sensor signal from 14-122ºF, fuel level over 25%, vehicle driven at over 20 mph while in closed loop, and the PCM detected the minimum HO2S-11signal was more than 600 mv. **Possible Causes:** • Fuel pressure regulator leaking or fuel injectors leaking • HO2S connector is damaged or loose (intermittent open circuit) • HO2S signal circuit is open or shorted to power • HO2S is contaminated or it has failed
DTC: P0133 **2T O2S, MIL: Yes** **1998-2006** **Models:** FX35, FX45, G20, G35, I30, I35, J30, M35, M45, Q45, QX4, QX56 **Engines:** All **Transmissions:** All	**HO2S-11 (Bank 1 Sensor 1) Slow Response:** Engine started, vehicle driven in Drive to over 50 mph at 1200-3100 rpm, IAT sensor signal from 14-122ºF, Fuel Schedule from 2.5-12.0 ms, and the PCM detected the average HO2S-11 response time indicated more than 1 second. **Possible Causes:** • HO2S signal circuit is open or shorted to ground • HO2S element is contaminated, or HO2S heater has failed • Intake air leaks, exhaust manifold leaks or PCV system leaks • MAF sensor out of calibration (it may be dirty or contaminated)
DTC: P0134 **2T O2S, MIL: Yes** **1998-2005** **Models:** G20, G35, I30, I35, J30, M45, Q45, QX4 **Engines:** All **Transmissions:** All	**HO2S-11 (Bank 1 Sensor 1) Insufficient Activity Detected :** Engine started, vehicle driven at over 20 mph while in closed loop, IAT sensor signal from 14-122ºF, and the PCM detected the HO2S signal was fixed under 300 mv, or the HO2S signal was fixed over 600 mv for 10 seconds, or it detected that the HO2S-11 input switched from rich to lean less than 5 times in a 10 second period. **Possible Causes:** • HO2S heater is damaged or has failed • HO2S signal circuit is open or shorted to ground • HO2S is damaged, has fuel contamination or has failed • TSB 02-024 (5/02) contains information related to this code
DTC: P0135 **2T O2S, MIL: Yes** **1996-2003** **Models:** G20, I30, J30, M45, Q45 **Engines:** All **Transmissions:** All	**HO2S-11 (Bank 1 Sensor 1) Heater Circuit Malfunction:** Engine started, engine running in closed loop at under 3000 rpm, and the PCM detected an unexpected voltage condition (either too high or too low a voltage) on the HO2S-11 heater circuit. **Possible Causes:** • HO2S heater control circuit is open or shorted to ground • HO2S heater control circuit is shorted to power • HO2S heater is damaged or has failed • PCM has failed

DTC	Trouble Code Title, Conditions & Possible Causes
DTC: P0135 **2T O2S, MIL: Yes** **1998-2001** **Models:** QX4 **Engines:** All **Transmissions:** All	**HO2S-11 (Bank 1 Sensor 1) Heater Circuit Malfunction:** Engine started, engine running in closed loop at under 3000 rpm, and the PCM detected an unexpected voltage condition (either too high or too low a voltage) on the HO2S-11 heater circuit. **Possible Causes:** • HO2S heater control circuit is open or shorted to ground • HO2S heater control circuit is shorted to power • HO2S heater is damaged or has failed • PCM has failed
DTC: P0136 **2T CCM, MIL: Yes** **1996-2003** **Models:** I30, J30, M45, Q45 **Engines:** All **Transmissions:** All	**HO2S-12 (Bank 1 Sensor 2) Closed Loop Fuel Control:** Engine started, vehicle driven at 30-55 mph for 2 minutes, and the PCM detected the HO2S signal was too high or low, or it detected the HO2S signal did not reach the maximum or minimum voltages, or it took too long for the HO2S signal to switch from rich to lean. **Possible Causes:** • HO2S signal circuit is open or shorted to ground • HO2S is deteriorated or has a fuel contamination condition • HO2S is damaged or it has failed
DTC: P0136 **2T CCM, MIL: Yes** **1996-97** **Models:** G20, QX4 **Engines:** All **Transmissions:** All	**HO2S-12 (Bank 1 Sensor 2) Closed Loop Fuel Control:** Engine started, vehicle driven at 30-55 mph for 2 minutes, and the PCM detected the HO2S signal was too high or low, or it detected the HO2S signal did not reach the maximum or minimum voltages, or it took too long for the HO2S signal to switch from rich to lean. **Possible Causes:** • HO2S signal circuit is open or shorted to ground • HO2S is deteriorated or has a fuel contamination condition • HO2S is damaged or it has failed
DTC: P0137 **2T CCM, MIL: Yes** **1998-2006** **Models:** FX35, FX45, G20, I30, J30, M35, M45, Q45, QX56 **Engines:** All **Transmissions:** All	**HO2S-12 (Bank 1 Sensor 2) Circuit Low Input (Lean Shift):** Engine running in closed loop at a speed of over 20 mph for at least 20 seconds, IAT sensor signal from 14-122°F, fuel level over 25%, and the PCM detected the HO2S signal did not reach a maximum voltage level of 0.60v, or it remained fixed at approximately 300 mv. **Possible Causes:** • Low fuel pressure, fuel filter restricted or fuel injectors plugged • HO2S may be contaminated or it has failed • HO2S heater is damaged or has failed
DTC: P0137 **2T CCM, MIL: Yes** **1998-2001** **Models:** QX4 **Engines:** All **Transmissions:** All	**HO2S-12 (Bank 1 Sensor 2) Circuit Low Input (Lean Shift):** Engine running in closed loop at a speed of over 20 mph for at least 20 seconds, IAT sensor signal from 14-122°F, fuel level over 25%, and the PCM detected the HO2S signal did not reach a maximum voltage level of 0.60v, or it remained fixed at approximately 300 mv. **Possible Causes:** • Low fuel pressure, fuel filter restricted or fuel injectors plugged • HO2S may be contaminated or it has failed • HO2S heater is damaged or has failed
DTC: P0138 **2T CCM, MIL: Yes** **1998-2006** **Models:** G20, I30, J30, M35, M45, Q45 **Engines:** All **Transmissions:** All	**HO2S-12 (Bank 1 Sensor 2) Circuit High Input (Rich Shift):** Engine running in closed loop at a speed of over 20 mph for at least 20 seconds, IAT sensor signal from 14-122°F, fuel level over 25%, and the PCM detected the HO2S-12 input did not reach a maximum voltage level of 0.60v, or it remained fixed at approximately 300 mv. **Possible Causes:** • Fuel pressure regulator leaking or fuel injectors leaking • HO2S connector is damaged or loose (an open circuit) • HO2S signal circuit is open or shorted to power • HO2S assembly is contaminated or it has failed • TSB 02-022 (4/02) contains information related to this code
DTC: P0138 **2T CCM, MIL: Yes** **1998-2003** **Models:** QX4 **Engines:** All **Transmissions:** All	**HO2S-12 (Bank 1 Sensor 2) Circuit High Input (Rich Shift):** Engine started, IAT sensor signal from 14-122°F, fuel level over 25%, vehicle driven at over 20 mph for 2-3 minutes, and the PCM detected the HO2S-12 input did not reach a maximum voltage level of 0.60v, or the signal remained fixed at approximately 300 mv. **Possible Causes:** • Fuel pressure regulator leaking or fuel injectors leaking • HO2S connector is damaged or loose (an open circuit) • HO2S signal circuit is open or shorted to power • HO2S assembly is contaminated or it has failed

DTC	Trouble Code Title, Conditions & Possible Causes
DTC: P0138 **2T CCM, MIL: Yes** **2003-06** **Models:** FX35, FX45, G35, I35, QX56 **Engines:** All **Transmissions:** All	**HO2S-12 (Bank 1 Sensor 2) Circuit High Input (Rich Shift):** Engine started, IAT sensor signal from 14-122ºF, fuel level over 25%, vehicle driven at over 20 mph for 2-3 minutes, and the PCM detected the HO2S-12 input did not reach a maximum voltage level of 0.60v, or the signal remained fixed at approximately 300 mv. **Possible Causes:** • Fuel pressure regulator leaking or fuel injectors leaking • HO2S connector is damaged or loose (an open circuit) • HO2S signal circuit is open or shorted to power • HO2S assembly is contaminated or it has failed
DTC: P0139 **2T O2S, MIL: Yes** **1998-2006** **Models:** G20, I30, J30, M35, M45, Q45 **Engines:** All **Transmissions:** All	**HO2S-12 (Bank 1 Sensor 2) Slow Response** Engine started, IAT sensor input from 14-122ºF, fuel level over 25%, vehicle driven at over 20 mph for 20 seconds, and the PCM detected the average HO2S-12 response time indicated more than 1 second. **Possible Causes:** • HO2S signal circuit is open or shorted to ground • HO2S element is contaminated, or HO2S heater has failed • Intake air leaks, exhaust manifold leaks or PCV system leaks • MAF sensor out of calibration (it may be dirty or contaminated) • TSB 02-022 (4/02) contains information related to this code
DTC: P0139 **2T O2S, MIL: Yes** **1998-2003** **Models:** QX4 **Engines:** All **Transmissions:** All	**HO2S-12 (Bank 1 Sensor 2) Slow Response** Engine started, IAT sensor input from 14-122ºF, fuel level over 25%, vehicle driven at over 20 mph for 20 seconds, and the PCM detected the average HO2S-12 response time indicated more than 1 second. **Possible Causes:** • HO2S signal circuit is open or shorted to ground • HO2S element is contaminated, or HO2S heater has failed • Intake air leaks, exhaust manifold leaks or PCV system leaks • MAF sensor out of calibration (it may be dirty or contaminated)
DTC P0139 **2T O2S, MIL: Yes** **2002-06** **Models:** FX35, FX45, G35, I35, QX56 **Engines:** All **Transmissions:** All	**HO2S-12 (Bank 1 Sensor 2) Slow Response** Engine started, IAT sensor input from 14-122ºF, fuel level over 25%, vehicle driven at over 20 mph for 20 seconds, and the PCM detected the average HO2S-12 response time indicated more than 1 second. **Possible Causes:** • HO2S signal circuit is open or shorted to ground • HO2S element is contaminated, or HO2S heater has failed • Intake air leaks, exhaust manifold leaks or PCV system leaks • MAF sensor out of calibration (it may be dirty or contaminated)
DTC: P0140 **2T O2S, MIL: Yes** **1998-2003** **Models:** G20, I30, J30, M45, Q45 **Engines:** All **Transmissions:** All	**HO2S-12 (Bank 1 Sensor 2) Insufficient Activity Detected:** Engine started, IAT sensor signal from 14-122ºF, fuel level over 25%, engine running at over 20 mph in closed loop, and the PCM detected the HO2S signal was fixed under 300 mv or over 600 mv, or it switched from rich to lean less than 5 times in 10 seconds. **Possible Causes:** • HO2S heater is damaged or has failed • HO2S signal circuit is open or shorted to ground • HO2S element is contaminated, damaged or it has failed • TSB 02-022 (4/02) contains information related to this code
DTC: P0140 **2T O2S, MIL: Yes** **1998-2001** **Models:** QX4 **Engines:** All **Transmissions:** All	**HO2S-12 (Bank 1 Sensor 2) Insufficient Activity Detected:** Engine started, IAT sensor signal from 14-122ºF, fuel level over 25%, engine running at over 20 mph in closed loop, and the PCM detected the HO2S signal was fixed under 300 mv or over 600 mv, or it switched from rich to lean less than 5 times in 10 seconds. **Possible Causes:** • HO2S heater is damaged or has failed • HO2S signal circuit is open or shorted to ground • HO2S element is contaminated, damaged or it has failed • TSB 02-022 (4/02) contains information related to this code
DTC: P0141 **2T O2S, MIL: Yes** **1996-2003** **Models:** G20, I30, J30, M45, Q45 **Engines:** All **Transmissions:** All	**HO2S-12 (Bank 1 Sensor 2) Heater Circuit Malfunction:** Engine running in closed loop at less than 3000 rpm, and the PCM detected an unexpected voltage on the HO2S heater circuit. **Note: The current level of the HO2S circuit was too high or too low.** **Possible Causes:** • HO2S heater control circuit is open or shorted to ground • HO2S heater control circuit is shorted to power • HO2S heater is contaminated, damaged or it has failed • PCM has failed

DTC	Trouble Code Title, Conditions & Possible Causes
DTC: P0141 **2T O2S, MIL: Yes** **1998-2001** **Models:** QX4 **Engines:** All **Transmissions:** All	**HO2S-12 (Bank 1 Sensor 2) Heater Circuit Malfunction:** Engine running in closed loop at less than 3000 rpm, and the PCM detected an unexpected voltage on the HO2S heater circuit. **Note: The current level of the HO2S circuit was too high or too low.** **Possible Causes:** • HO2S heater control circuit is open or shorted to ground • HO2S heater control circuit is shorted to power • HO2S heater is contaminated, damaged or it has failed • PCM has failed
DTC: P0150 **2T CCM, MIL: Yes** **1997-2001** **Models:** QX4 **Engines:** All **Transmissions:** All	**HO2S-21 (Bank 2 Sensor 1) Closed Loop Fuel Control:** Engine started, vehicle driven at a speed of 30-55 mph for 2 minutes, and the PCM detected the HO2S signal was excessively high or low, or it detected the HO2S signal did not reach the maximum or minimum voltages, or it took too long for the HO2S signal to switch from rich to lean. **Possible Causes:** • HO2S signal circuit is open or shorted to ground • HO2S is damaged or it has failed
DTC: P0150 **2T CCM, MIL: Yes** **2006** **Models:** FX35, FX45, M35, M45, QX56 **Engines:** All **Transmissions:** All	**HO2S-21 (Bank 2 Sensor 1) Circuit Low Input (Lean Shift):** Engine running in closed loop at a speed of over 20 mph for at least 20 seconds, IAT sensor signal from 14-122°F, fuel level over 25%, and the PCM detected the HO2S signal did not reach a maximum voltage level of 0.60v, or it remained fixed at approximately 300 mv. **Possible Causes:** • Low fuel pressure, fuel filter restricted or fuel injectors plugged • HO2S may be contaminated or it has failed • HO2S heater is damaged or has failed
DTC: P0151 **2T CCM, MIL: Yes** **2006** **Models:** FX35, FX45, M35, M45, QX56 **Engines:** All **Transmissions:** All	**HO2S-21 (Bank 2 Sensor 1) Circuit Low Input (Lean Shift):** Engine running in closed loop at a speed of over 20 mph for at least 20 seconds, IAT sensor signal from 14-122°F, fuel level over 25%, and the PCM detected the HO2S signal did not reach a maximum voltage level of 0.60v, or it remained fixed at approximately 300 mv. **Possible Causes:** • Low fuel pressure, fuel filter restricted or fuel injectors plugged • HO2S may be contaminated or it has failed • HO2S heater is damaged or has failed
DTC: P0151 **2T CCM, MIL: Yes** **1998-2003** **Models:** I30, J30, M45, Q45 **Engines:** All **Transmissions:** All	**HO2S-21 (Bank 2 Sensor 1) Circuit Low Input (Lean Shift):** Engine running in closed loop at a speed of over 20 mph for at least 20 seconds, IAT sensor signal from 14-122°F, fuel level over 25%, and the PCM detected the HO2S signal did not reach a maximum voltage level of 0.60v, or it remained fixed at approximately 300 mv. **Possible Causes:** • Low fuel pressure, fuel filter restricted or fuel injectors plugged • HO2S may be contaminated or it has failed • HO2S heater is damaged or has failed
DTC: P0152 **2T CCM, MIL: Yes** **1998-2006** **Models:** I30, J30, M35, M45, Q45, QX4 **Engines:** All **Transmissions:** All	**HO2S-21 (Bank 2 Sensor 1) Circuit High Input (Rich Shift):** Engine started, IAT sensor input from 14-122°F, fuel level over 25%, vehicle driven at over 20 mph while in closed loop, and the PCM detected the minimum HO2S-21signal indicated more than 600 mv. **Possible Causes:** • Fuel pressure regulator leaking or fuel injectors leaking • HO2S connector is damaged or loose (an open circuit) • HO2S signal circuit is open or shorted to power • HO2S is contaminated or it has failed
DTC: P0152 **2T CCM, MIL: Yes** **2002-06** **Models:** FX35, FX45, G35, I35, QX56 **Engines:** All **Transmissions:** All	**HO2S-21 (Bank 2 Sensor 1) Circuit High Input (Rich Shift):** Engine started, IAT sensor input from 14-122°F, fuel level over 25%, vehicle driven at over 20 mph while in closed loop, and the PCM detected the minimum HO2S-21signal indicated more than 600 mv. **Possible Causes:** • Fuel pressure regulator leaking or fuel injectors leaking • HO2S connector is damaged or loose (an open circuit) • HO2S signal circuit is open or shorted to power • HO2S is contaminated or it has failed
DTC: P0153 **2T O2S, MIL: Yes** **1998-2006** **Models:** I30, J30, M35, M45, Q45, QX4 **Engines:** All **Transmissions:** All	**HO2S-21 (Bank 2 Sensor 1) Slow Response:** Engine started, IAT sensor input from 14-122°F, fuel level over 25%, vehicle driven at over 20 mph while in closed loop, and the PCM detected the average HO2S response time was more than 1 second. **Possible Causes:** • HO2S signal circuit is open or shorted to ground • HO2S element is contaminated, or HO2S heater has failed • Intake air leaks, exhaust manifold leaks or PCV system leaks • MAF sensor out of calibration (it may be dirty or contaminated)

DTC	Trouble Code Title, Conditions & Possible Causes
DTC P0153 **2T O2S, MIL: Yes** **2002-06** **Models:** FX35, FX45, G35, I35, QX56 **Engines:** All **Transmissions:** All	**HO2S-21 (Bank 2 Sensor 1) Slow Response:** Engine started, IAT sensor input from 14-122ºF, fuel level over 25%, vehicle driven at over 20 mph while in closed loop, and the PCM detected the average HO2S response time was more than 1 second. **Possible Causes:** • HO2S signal circuit is open or shorted to ground • HO2S element is contaminated, or HO2S heater has failed • Intake air leaks, exhaust manifold leaks or PCV system leaks • MAF sensor out of calibration (it may be dirty or contaminated)
DTC: P0154 **2T O2S, MIL: Yes** **1998-2005** **Models:** I30, J30, M45, Q45, QX4 **Engines:** All **Transmissions:** All	**HO2S-21 (Bank 2 Sensor 1) Insufficient Activity Detected:** Engine started, IAT sensor input from 14-122ºF, fuel level over 25%, vehicle driven at over 20 mph while in closed loop, and the PCM detected the HO2S-21 signal was fixed under 300 mv, or the HO2S signal was fixed over 600 mv for 10 seconds, or the HO2S signal switched from rich to lean less than 5 times in 10 seconds. **Possible Causes:** • HO2S element is contaminated • HO2S heater is damaged or has failed • HO2S signal circuit is open or shorted to ground • HO2S is damaged or it has failed
DTC: P0154 **2T O2S, MIL: Yes** **2002-05** **Models:** FX35, FX45, G35, I35 **Engines:** All **Transmissions:** All	**HO2S-21 (Bank 2 Sensor 1) Insufficient Activity Detected:** Engine started, IAT sensor input from 14-122ºF, fuel level over 25%, vehicle driven at over 20 mph while in closed loop, and the PCM detected the HO2S-21 signal was fixed under 300 mv, or the HO2S signal was fixed over 600 mv for 10 seconds, or the HO2S signal switched from rich to lean less than 5 times in 10 seconds. **Possible Causes:** • HO2S element is contaminated • HO2S heater is damaged or has failed • HO2S signal circuit is open or shorted to ground • HO2S is damaged or it has failed
DTC: P0155 **2T O2S, MIL: Yes** **1996-2003** **Models:** I30, J30, M45, Q45 **Engines:** All **Transmissions:** All	**HO2S-21 (Bank 2 Sensor 1) Heater Circuit Malfunction:** Engine started, engine speed less than 3000 rpm, and the PCM detected an unexpected voltage on the HO2S heater circuit. **Note: The current level of the HO2S circuit was too high or too low.** **Possible Causes:** • HO2S heater control circuit is open or shorted to ground • HO2S heater control circuit is shorted to power • HO2S heater is damaged or has failed • PCM has failed
DTC: P0155 **2T O2S, MIL: Yes** **1997-2001** **Models:** QX4 **Engines:** All **Transmissions:** All	**HO2S-21 (Bank 2 Sensor 1) Heater Circuit Malfunction:** Engine started, engine speed less than 3000 rpm, and the PCM detected an unexpected voltage on the HO2S heater circuit. **Note: The current level of the HO2S circuit was too high or too low.** **Possible Causes:** • HO2S heater control circuit is open or shorted to ground • HO2S heater control circuit is shorted to power • HO2S heater is damaged or has failed • PCM has failed
DTC: P0156 **2T CCM, MIL: Yes** **1996-2003** **Models:** I30, J30, M45, Q45 **Engines:** All **Transmissions:** All	**HO2S-22 (Bank 2 Sensor 2) Closed Loop Fuel Control:** Engine started, IAT sensor input from 14-122ºF, fuel level over 25%, vehicle driven at over 20 mph while in closed loop, and the PCM detected the HO2S signal was excessively high or low, or it detected the HO2S signal did not reach the maximum or minimum voltages, or it took too long for the HO2S signal to switch from rich to lean. **Possible Causes:** • HO2S connector is damaged or loose • HO2S signal circuit is open or shorted to ground • HO2S is damaged or it has failed
DTC: P0156 **2T CCM, MIL: Yes** **1997** **Models:** QX4 **Engines:** All **Transmissions:** All	**HO2S-22 (Bank 2 Sensor 2) Closed Loop Fuel Control:** Engine started, IAT sensor input from 14-122ºF, fuel level over 25%, vehicle driven at over 20 mph while in closed loop, and the PCM detected the HO2S signal was excessively high or low, or it detected the HO2S signal did not reach the maximum or minimum voltages, or it took too long for the HO2S signal to switch from rich to lean. **Possible Causes:** • HO2S connector is damaged or loose • HO2S signal circuit is open or shorted to ground • HO2S is damaged or it has failed

DTC	Trouble Code Title, Conditions & Possible Causes
DTC: P0157 **2T CCM, MIL: Yes** **1998-2003** **Models:** I30, J30, M45, Q45 **Engines:** All **Transmissions:** All	**HO2S-22 (Bank 2 Sensor 2) Circuit Low Input (Lean Shift):** Engine started, IAT sensor input from 14-122°F, fuel level over 25%, vehicle driven at over 20 mph while in closed loop, and the PCM detected the HO2S signal did not reach a maximum voltage level of 0.60v, or it remained fixed at approximately 300 mv. **Possible Causes:** • Low fuel pressure, fuel filter restricted or fuel injectors plugged • HO2S may be contaminated or it has failed • HO2S heater is damaged or has failed
DTC: P0157 **2T CCM, MIL: Yes** **1998-2001** **Models:** QX4 **Engines:** All **Transmissions:** All	**HO2S-22 (Bank 2 Sensor 2) Circuit Low Input (Lean Shift):** Engine started, IAT sensor input from 14-122°F, fuel level over 25%, vehicle driven at over 20 mph while in closed loop, and the PCM detected the HO2S signal did not reach a maximum voltage level of 0.60v, or it remained fixed at approximately 300 mv. **Possible Causes:** • Low fuel pressure, fuel filter restricted or fuel injectors plugged • HO2S may be contaminated or it has failed • HO2S heater is damaged or has failed
DTC: P0157 **2T CCM, MIL: Yes** **2006** **Models:** FX35, FX45, M35, M45, QX56 **Engines:** All **Transmissions:** All	**HO2S-22 (Bank 2 Sensor 2) Circuit Low Input (Lean Shift):** Engine started, IAT sensor input from 14-122°F, fuel level over 25%, vehicle driven at over 20 mph while in closed loop, and the PCM detected the HO2S signal did not reach a maximum voltage level of 0.60v, or it remained fixed at approximately 300 mv. **Possible Causes:** • Low fuel pressure, fuel filter restricted or fuel injectors plugged • HO2S may be contaminated or it has failed • HO2S heater is damaged or has failed
DTC: P0158 **2T CCM, MIL: Yes** **1998-2006** **Models:** I30, J30, M35, M45, Q45, QX4, QX56 **Engines:** All **Transmissions:** All	**HO2S-22 (Bank 2 Sensor 2) Circuit High Input (Rich Shift):** Engine started, IAT sensor input from 14-122°F, fuel level over 25%, vehicle driven at over 20 mph while in closed loop, and the PCM detected the HO2S-22 input did not reach a maximum voltage level of 0.60v, or it remained fixed at approximately 300 mv. **Possible Causes:** • Low fuel pressure, fuel filter restricted or fuel injectors plugged • HO2S may be contaminated or it has failed • HO2S heater is damaged or has failed • TSB 02-008 (5/02) contains information related to this code
DTC: P0158 **2T CCM, MIL: Yes** **2002-06** **Models:** FX35, FX45, G35, I35 **Engines:** All **Transmissions:** All	**HO2S-22 (Bank 2 Sensor 2) Circuit High Input (Rich Shift):** Engine started, IAT sensor input from 14-122°F, fuel level over 25%, vehicle driven at over 20 mph while in closed loop, and the PCM detected the HO2S-22 input did not reach a maximum voltage level of 0.60v, or it remained fixed at approximately 300 mv. **Possible Causes:** • Low fuel pressure, fuel filter restricted or fuel injectors plugged • HO2S may be contaminated or it has failed • HO2S heater is damaged or has failed
DTC: P0159 **2T O2S, MIL: Yes** **1998-2006** **Models:** I30, J30, M35, M45, Q45 **Engines:** All **Transmissions:** All	**HO2S-22 (Bank 2 Sensor 2) Slow Response:** Engine started, IAT sensor input from 14-122°F, fuel level over 25%, vehicle driven at over 20 mph while in closed loop, and the PCM detected the average HO2S response time was more than 1 second. **Possible Causes:** • HO2S signal circuit is open or shorted to ground • HO2S element is contaminated, or HO2S heater has failed • Intake air leaks, exhaust manifold leaks or PCV system leaks • MAF sensor out of calibration (it may be dirty or contaminated) • TSB 02-008 (5/02) contains information related to this code
DTC: P0159 **2T O2S, MIL: Yes** **1998-2001** **Models:** QX4 **Engines:** All **Transmissions:** All	**HO2S-22 (Bank 2 Sensor 2) Slow Response:** Engine started, IAT sensor input from 14-122°F, fuel level over 25%, vehicle driven at over 20 mph while in closed loop, and the PCM detected the average HO2S response time was more than 1 second. **Possible Causes:** • HO2S signal circuit is open or shorted to ground • HO2S element is contaminated, or HO2S heater has failed • Intake air leaks, exhaust manifold leaks or PCV system leaks • MAF sensor out of calibration (it may be dirty or contaminated) • TSB 02-008 (5/02) contains information related to this code

DTC	Trouble Code Title, Conditions & Possible Causes
DTC: P0159 **2T O2S, MIL: Yes** **2002-06** **Models:** FX35, FX45, G35, I35, QX56 **Engines:** All **Transmissions:** All	**HO2S-22 (Bank 2 Sensor 2) Slow Response:** Engine started, IAT sensor input from 14-122ºF, fuel level over 25%, vehicle driven at over 20 mph while in closed loop, and the PCM detected the average HO2S response time was more than 1 second. **Possible Causes:** • HO2S signal circuit is open or shorted to ground • HO2S element is contaminated, or HO2S heater has failed • Intake air leaks, exhaust manifold leaks or PCV system leaks • MAF sensor out of calibration (it may be dirty or contaminated)
DTC: P0160 **2T O2S, MIL: Yes** **1998-2003** **Models:** I30, J30, M45, Q45, QX4 **Engines:** All **Transmissions:** All	**HO2S-22 (Bank 2 Sensor 2) Insufficient Activity Detected:** Engine started, IAT sensor input from 14-122ºF, fuel level over 25%, vehicle driven at over 20 mph while in closed loop, and the PCM detected the HO2S signal was fixed under 300 mv, or the HO2S signal was fixed over 600 mv for 10 seconds, or the HO2S signal switched from rich to lean less than 5 times in 10 seconds. **Possible Causes:** • HO2S element is contaminated • HO2S heater is damaged or has failed • HO2S signal circuit is open or shorted to ground • HO2S is damaged or it has failed • TSB 02-008 (5/02) contains information related to this code
DTC: P0161 **2T O2S, MIL: Yes** **1996-2003** **Models:** I30, J30, M45, Q45 **Engines:** All **Transmissions:** All	**HO2S-22 (Bank 2 Sensor 2) Heater Circuit Malfunction:** Engine started, engine speed less than 3000 rpm, and the PCM detected an unexpected voltage on the HO2S heater circuit. **Note: The current level of the HO2S circuit was too high or too low.** **Possible Causes:** • HO2S heater control circuit is open or shorted to ground • HO2S heater control circuit is shorted to power • HO2S heater is damaged or has failed • PCM has failed
DTC: P0161 **2T O2S, MIL: Yes** **1997-2001** **Models:** QX4 **Engines:** All **Transmissions:** All	**HO2S-22 (Bank 2 Sensor 2) Heater Circuit Malfunction:** Engine started, engine speed less than 3000 rpm, and the PCM detected an unexpected voltage on the HO2S heater circuit. **Note: The current level of the HO2S circuit was too high or too low.** **Possible Causes:** • HO2S heater control circuit is open or shorted to ground • HO2S heater control circuit is shorted to power • HO2S heater is damaged or has failed • PCM has failed
DTC: P0171 **2T FUEL, MIL: Yes** **1996-2006** **Models:** G20, I30, J30, M35, M45, Q45 **Engines:** All **Transmissions:** All	**Fuel Trim Lean (Bank 1):** Engine started, engine running at cruise speed in closed loop, and the PCM detected the Bank 1 fuel system indicated a lean condition. **Possible Causes:** • Air leaks after the MAF sensor, EGR, Intake Manifold or PCV • Exhaust system leaks before or near the front HO2S location • Fuel system not supplying enough fuel under high fuel demand conditions (i.e., fuel pump weak, fuel filter or fuel injectors dirty) • Vehicle running out of fuel, or driven until it runs out of fuel •TSB 02-024 (5/02) contains information related to this code
DTC: P0171 **2T FUEL, MIL: Yes** **2002-06** **Models:** FX35, FX45, G35, I35, QX4, QX56 **Engines:** All **Transmissions:** All	**Fuel Trim Lean (Bank 1):** Engine started, engine running at cruise speed in closed loop, and the PCM detected the Bank 1 fuel system indicated a lean condition. **Possible Causes:** • Air leaks after the MAF sensor, EGR, Intake Manifold or PCV • Exhaust system leaks before or near the front HO2S location • Fuel system not supplying enough fuel under high fuel demand conditions (i.e., fuel pump weak, fuel filter or fuel injectors dirty) • Vehicle running out of fuel, or driven until it runs out of fuel • TSB 02-024 (5/02) contains information related to this code

DTC	Trouble Code Title, Conditions & Possible Causes
DTC: P0172 **2T FUEL, MIL: Yes** **1996 2006** **Models:** G20, I30, J30, M35, M45, Q45 **Engines:** All **Transmissions:** All	**Fuel Trim Rich (Bank 1):** Engine started, engine running at cruise speed in closed loop, and the PCM detected the Bank 1 fuel system indicated a rich condition. **Possible Causes:** • Engine oil overfill condition, or camshaft timing is incorrect • EVAP recovery system failure (pulling in excessive fuel vapors) • HO2S element may be contaminated with water or alcohol • Leaking fuel pressure regulator or leaking fuel injector(s) • MAF sensor signals invalid (it may be dirty or out of calibration)
DTC P0172 **2T FUEL, MIL: Yes** **2002-06** **Models:** FX35, FX45, G35, I35, QX4, QX56 **Engines:** All **Transmissions:** All	**Fuel Trim Rich (Bank 1):** Engine started, engine running at cruise speed in closed loop, and the PCM detected the Bank 1 fuel system indicated a rich condition. **Possible Causes:** • Engine oil overfill condition, or camshaft timing is incorrect • EVAP recovery system failure (pulling in excessive fuel vapors) • HO2S element may be contaminated with water or alcohol • Leaking fuel pressure regulator or leaking fuel injector(s) • MAF sensor signals invalid (it may be dirty or out of calibration)
DTC: P0174 **2T FUEL, MIL: Yes** **1996-2006** **Models:** I30, J30, M35, M45, Q45 **Engines:** All **Transmissions:** All	**Fuel Trim Lean (Bank 2):** Engine started, engine running at cruise speed in closed loop, and the PCM detected the Bank 2 fuel system indicated a lean condition. **Possible Causes:** • Air leaks after the MAF sensor, EGR, Intake Manifold or PCV • Exhaust system leaks before or near the front HO2S location • Fuel system not supplying enough fuel under high fuel demand conditions (i.e., fuel pump weak, fuel filter or fuel injectors dirty) • Vehicle running out of fuel, or driven until it runs out of fuel
DTC: P0174 **2T FUEL, MIL: Yes** **2002-06** **Models:** FX35, FX45, G35, I35, QX4, QX56 **Engines:** All **Transmissions:** All	**Fuel Trim Lean (Bank 2):** Engine started, engine running at cruise speed in closed loop, and the PCM detected the Bank 2 fuel system indicated a lean condition. **Possible Causes:** • Air leaks after the MAF sensor, EGR, Intake Manifold or PCV • Exhaust system leaks before or near the front HO2S location • Fuel system not supplying enough fuel under high fuel demand conditions (i.e., fuel pump weak, fuel filter or fuel injectors dirty) • Vehicle running out of fuel, or driven until it runs out of fuel
DTC: P0175 **2T FUEL, MIL: Yes** **1996-2006** **Models:** I30, J30, M35, M45, Q45 **Engines:** All **Transmissions:** All	**Fuel Trim Rich (Bank 2):** Engine started, engine running at cruise speed in closed loop, and the PCM detected the Bank 2 fuel system indicated a lean condition. **Possible Causes:** • Engine oil overfill condition, or camshaft timing is incorrect • EVAP recovery system failure (pulling in excessive fuel vapors) • HO2S element may be contaminated with water or alcohol • Leaking or contaminated fuel injector(s) • Leaking fuel pressure regulator • MAF sensor signals invalid (it may be dirty or out of calibration)
DTC: P0175 **2T FUEL, MIL: Yes** **2002-06** **Models:** FX35, FX45, G35, I35, QX4, QX56 **Engines:** All **Transmissions:** All	**Fuel Trim Rich (Bank 2):** Engine started, engine running at cruise speed in closed loop, and the PCM detected the Bank 2 fuel system indicated a lean condition. **Possible Causes:** • Engine oil overfill condition, or camshaft timing is incorrect • EVAP recovery system failure (pulling in excessive fuel vapors) • HO2S element may be contaminated with water or alcohol • Leaking or contaminated fuel injector(s) • Leaking fuel pressure regulator • MAF sensor signals invalid (it may be dirty or out of calibration)

DTC	Trouble Code Title, Conditions & Possible Causes
DTC: P0180 **2T CCM, MIL: Yes** **1996-2003** **Models:** G20, I30, J30, M45 **Engines:** All **Transmissions:** All	**Fuel Tank Temperature Sensor Circuit Malfunction:** Key on for over 5 seconds, and the PCM detected the Fuel Tank Temperature signal was too high or tool low, or that the value was not plausible when compared to the ECT and IAT sensor signals. **Note: The fuel tank temperature sensor should read 3.5v at 68ºF.** **Possible Causes:** • FTT sensor connector is damaged or loose (open or shorted) • FTT sensor signal circuit is open or shorted to ground • FTT sensor ground circuit is open • Fuel tank temperature sensor is damaged or has failed • PCM has failed
DTC: P0181 **2T CCM, MIL: Yes** **2002-06** **Models:** FX35, FX45, G35, I35, M35, M45, Q45, QX4, QX56 **Engines:** All **Transmissions:** All	**Fuel Tank Temperature Sensor Range/Performance:** Engine started, engine running for 5-10 minutes, and the PCM detected the Fuel Tank Temperature (FTT) sensor signal was not plausible when compared to the ECT sensor and IAT sensor signals. **Note: The fuel tank temperature sensor should read 3.5v at 68ºF.** **Possible Causes:** • Fuel tank temperature sensor connector is loose (open circuit) • Fuel tank temperature sensor signal open or shorted to ground • Fuel tank temperature sensor is damaged or has failed
DTC: P0182 **2T CCM, MIL: Yes** **2002-06** **Models:** FX35, FX45, G35, I35, M35, M45, Q45, QX4, QX56 **Engines:** All **Transmissions:** All	**Fuel Tank Temperature Sensor Circuit Low Input:** Key on or engine running; and the PCM detected an unexpected low voltage (less than 0.20v) on the FTT sensor circuit in the CCM test. **Note: The fuel tank temperature sensor should read 3.5v at 68ºF.** **Possible Causes:** • FTT sensor signal circuit is shorted to ground • FTT sensor connector is damaged (causing a short circuit) • FTT sensor is damaged or has failed
DTC: P0183 **2T CCM, MIL: Yes** **2002-06** **Models:** FX35, FX45, G35, I35, M35, M45, Q45, QX4, QX56 **Engines:** All **Transmissions:** All	**Fuel Tank Temperature Sensor Circuit High Input:** Key on or engine running; and PCM detected an unexpected high voltage (more than 4.90v) on the FTT sensor circuit in the CCM test. **Note: The fuel tank temperature sensor should read 3.5v at 68ºF.** **Possible Causes:** • FTT sensor connector is damaged (causing an open circuit) • FTT sensor signal circuit is open or shorted to VREF • FTT sensor ground circuit is open • FTT sensor is damaged or has failed
DTC P0217 **1T CCM, MIL: Yes** **2000-03** **Models:** G20, I30, QX4 **Engines:** All **Transmissions:** All	**Engine Over-Temperature Condition Detected:** Engine started, engine running in closed loop, IAT sensor from 14-122ºF, and the PCM detected the ECT sensor signal indicated that the engine temperature was high under "low" engine load conditions. **Possible Causes:** • Blocked air passages at front of the vehicle (recent damage?) • Blocked or restricted radiator passages • Cooling fan control circuit open or shorted to ground • Cooling system problems (low coolant, thermostat closed)
DTC: P0222 **1T CCM, MIL: Yes** **2002-06** **Models:** FX35, FX45, G35, I35 **Engines:** All **Transmissions:** All	**ETC Throttle Position Sensor 1 Signal Low Input:** Key on or engine running; and the PCM detected an unexpected low voltage on the ETC TP Sensor 1 signal circuit. The ETC Actuator consists of a throttle control motor and TP sensors. The TP sensors respond to movement of the throttle valve. The PCM judges the actual opening angle of the throttle valve from the signals from the TP1 and TP2 sensors during engine operation. **Possible Causes:** • ETC TP1 sensor connector is damaged (it may be shorted) • ETC TP1 sensor signal circuit is shorted to ground • ETC TP1 sensor is damaged or it failed (ETC actuator failure)
DTC: P0222 **1T CCM, MIL: Yes** **2003-06** **Models:** M35, M45, Q45, QX56 **Engines:** All **Transmissions:** All	**ETC Throttle Position Sensor 1 Signal Low Input:** Key on or engine running; and the PCM detected an unexpected low voltage on the ETC TP Sensor 1 signal circuit. The ETC Actuator consists of a throttle control motor and TP sensors. The TP sensors respond to movement of the throttle valve. The PCM judges the actual opening angle of the throttle valve from the signals from the TP1 and TP2 sensors during engine operation. **Possible Causes:** • ETC TP1 sensor connector is damaged (it may be shorted) • ETC TP1 sensor signal circuit is shorted to ground • ETC TP1 sensor is damaged or it failed (ETC actuator failure)

DTC	Trouble Code Title, Conditions & Possible Causes
DTC: P0223 **1T CCM, MIL: Yes** **2002-06** **Models:** FX35, FX45, G35, I35 **Engines:** All **Transmissions:** All	**ETC Throttle Position Sensor 1 Signal High Input:** Key on or engine running; and the PCM detected an unexpected high voltage on the ETC TP Sensor 1 signal circuit. The ETC Actuator consists of a throttle control motor and TP sensors. The TP sensors respond to movement of the throttle valve. The PCM judges the actual opening angle of the throttle valve from the signals from the TP1 and TP2 sensors during engine operation. **Possible Causes:** • ETC TP1 sensor connector is damaged (it may be open) • ETC TP1 sensor signal circuit is open between TP1 and PCM • ETC TP1 sensor is damaged or it failed (ETC actuator failure)
DTC: P0223 **1T CCM, MIL: Yes** **2003-06** **Models:** M35, M45, Q45, QX56 **Engines:** All **Transmissions:** All	**ETC Throttle Position Sensor 1 Signal High Input:** Key on or engine running; and the PCM detected an unexpected high voltage on the ETC TP Sensor 1 signal circuit. The ETC Actuator consists of a throttle control motor and TP sensors. The TP sensors respond to movement of the throttle valve. The PCM judges the actual opening angle of the throttle valve from the signals from the TP1 and TP2 sensors during engine operation. **Possible Causes:** • ETC TP1 sensor connector is damaged (it may be open) • ETC TP1 sensor signal circuit is open between TP1 and PCM • ETC TP1 sensor is damaged or it failed (ETC actuator failure)
DTC: P0226 **1T CCM, MIL: Yes** **2002-03** **Models:** FX35, FX45, I35 **Engines:** All **Transmissions:** All	**Accelerator Pedal Position Sensor 1 Signal Range/Performance:** Engine started, and the PCM detected a signal from the APP1 sensor that was not plausible when compared to the engine speed and engine load signals during the test period. **Possible Causes:** • APP1 sensor connector is damaged (may be open or shorted) • APP1 sensor signal circuit is damaged or has failed • PCM is damaged or has failed
DTC: P0227 **1T CCM, MIL: Yes** **2002-03** **Models:** FX35, FX45, I35 **Engines:** All **Transmissions:** All	**Accelerator Pedal Position Sensor 1 Signal Low Input:** Key on or engine running; and the PCM detected an unexpected low voltage condition the APP1 sensor signal circuit. **Possible Causes:** • APP1 sensor connector is damaged (it may be shorted) • APP1 sensor signal circuit is shorted to ground • APP1 sensor signal circuit is damaged or has failed • PCM is damaged or has failed
DTC: P0228 **1T CCM, MIL: Yes** **2002-03** **Models:** FX35, FX45, I35 **Engines:** All **Transmissions:** All	**Accelerator Pedal Position Sensor 1 Signal High Input:** Key on or engine running; and the PCM detected an unexpected high voltage condition the APP1 sensor signal circuit. **Possible Causes:** • APP1 sensor connector is damaged (it may be open) • APP1 sensor signal circuit is shorted to VREF or system power • APP1 sensor signal circuit is damaged or has failed • TSB 01-064 (10/19/01) contains information related to this code
DTC: P0300 **2T MISFIRE, MIL: Yes** **1996-2006** **Models:** FX35, FX45, G20, G35, I30, I35, J30, M35, M45, Q45, QX4, QX56 **Engines:** All **Transmissions:** All	**Multiple Misfire Detected:** Engine started, engine speed from 400-3500 rpm, VSS more than 3 mph, and the PCM detected an irregular CKP sensor signal that indicated a multiple misfire condition existed in two or more cylinders during the 200 or 1000 revolution period of the Misfire Monitor test. **Possible Causes:** • Base engine problem that affects two or more cylinders • CMP or CKP sensor signals erratic or out of phase • EGR valve stuck open, or EVAP purge system has failed • Fuel delivery component fault that affects two or more cylinders (e.g., a contaminated, dirty or sticking fuel injector) • Ignition system problem that affects two or more cylinders • Vehicle driven while quite low on fuel (less than 1/8 of a tank)
DTC: P0301 **2T MISFIRE, MIL: Yes** **1996-2006** **Models:** FX35, FX45, G20, G35, I30, I35, J30, M35, M45, Q45, QX4, QX56 **Engines:** All **Transmissions:** All	**Cylinder 1 Misfire Detected:** Engine started, engine speed from 400-3500 rpm, VSS over 3 mph, and the PCM detected an irregular CKP sensor signal indicating a misfire present in Cylinder 1 during the 200 or 1000 revolution test. **Possible Causes:** • Base engine mechanical problem that affects only Cylinder 1 • EGR valve stuck open, or EVAP purge system has failed • Fuel delivery component fault that affects only Cylinder 1 (e.g., a contaminated, dirty or sticking fuel injector) • Ignition system fault (Coil or Plug) that affects only Cylinder 1

DTC	Trouble Code Title, Conditions & Possible Causes
DTC: P0302 **2T MISFIRE, MIL: Yes** **1996-2006** **Models:** FX35, FX45, G20, G35, I30, I35, J30, M35, M45, Q45, QX4, QX56 **Engines:** All **Transmissions:** All	**Cylinder 2 Misfire Detected:** Engine started, engine speed from 400-3500 rpm, VSS over 3 mph, and the PCM detected an irregular CKP sensor signal indicating a misfire present in Cylinder 2 during the 200 or 1000 revolution test. **Possible Causes:** • Base engine mechanical problem that affects only Cylinder 2 • EGR valve stuck open, or EVAP purge system has failed • Fuel delivery component fault that affects only Cylinder 2 (e.g., a contaminated, dirty or sticking fuel injector) • Ignition system fault (Coil or Plug) that affects only Cylinder 2
DTC: P0303 **2T MISFIRE, MIL: Yes** **1996-2006** **Models:** FX35, FX45, G20, G35, I30, I35, J30, M35, M45, Q45, QX4, QX56 **Engines:** All **Transmissions:** All	**Cylinder 3 Misfire Detected:** Engine started, engine speed from 400-3500 rpm, VSS over 3 mph, and the PCM detected an irregular CKP sensor signal indicating a misfire present in Cylinder 3 during the 200 or 1000 revolution test. **Possible Causes:** • Base engine mechanical problem that affects only Cylinder 3 • EGR valve stuck open, or EVAP purge system has failed • Fuel delivery component fault that affects only Cylinder 3 (e.g., a contaminated, dirty or sticking fuel injector) • Ignition system fault (Coil or Plug) that affects only Cylinder 3
DTC: P0304 **2T MISFIRE, MIL: Yes** **1996-2006** **Models:** FX35, FX45, G20, G35, I30, I35, J30, M35, M45, Q45, QX4, QX56 **Engines:** All **Transmissions:** All	**Cylinder 4 Misfire Detected:** Engine started, engine speed from 400-3500 rpm, VSS over 3 mph, and the PCM detected an irregular CKP sensor signal indicating a misfire present in Cylinder 4 during the 200 or 1000 revolution test. **Possible Causes:** • Base engine mechanical problem that affects only Cylinder4 • EGR valve stuck open, or EVAP purge system has failed • Fuel delivery component fault that affects only Cylinder 4 (e.g., a contaminated, dirty or sticking fuel injector) • Ignition system fault (Coil or Plug) that affects only Cylinder 4
DTC: P0305 **2T MISFIRE, MIL: Yes** **1996-2006** **Models:** FX35, FX45, G20, G35, I30, I35, J30, M35, M45, Q45, QX4, QX56 **Engines:** All **Transmissions:** All	**Cylinder 5 Misfire Detected:** Engine started, engine speed from 400-3500 rpm, VSS over 3 mph, and the PCM detected an irregular CKP sensor signal indicating a misfire present in Cylinder 5 during the 200 or 1000 revolution test. **Possible Causes:** • Base engine mechanical problem that affects only Cylinder 5 • EGR valve stuck open, or EVAP purge system has failed • Fuel delivery component fault that affects only Cylinder 5 (e.g., a contaminated, dirty or sticking fuel injector) • Ignition system fault (Coil or Plug) that affects only Cylinder 5
DTC: P0306 **2T MISFIRE, MIL: Yes** **1996-2006** **Models:** FX35, FX45, G20, G35, I30, I35, J30, M35, M45, Q45, QX4, QX56 **Engines:** All **Transmissions:** All	**Cylinder 6 Misfire Detected:** Engine started, engine speed from 400-3500 rpm, VSS over 3 mph, and the PCM detected an irregular CKP sensor signal indicating a misfire present in Cylinder 6 during the 200 or 1000 revolution test. **Possible Causes:** • Base engine mechanical problem that affects only Cylinder 6 • EGR valve stuck open, or EVAP purge system has failed • Fuel delivery component fault that affects only Cylinder 6 (e.g., a contaminated, dirty or sticking fuel injector) • Ignition system fault (Coil or Plug) that affects only Cylinder 6
DTC: P0307 **2T MISFIRE, MIL: Yes** **1996-2006** **Models:** FX45, M45, Q45, QX56 **Engines:** All **Transmissions:** All	**Cylinder 7 Misfire Detected:** Engine started, engine speed from 400-3500 rpm, VSS over 3 mph, and the PCM detected an irregular CKP sensor signal indicating a misfire present in Cylinder 7 during the 200 or 1000 revolution test. **Possible Causes:** • Base engine mechanical problem that affects only Cylinder 7 • EGR valve stuck open, or EVAP purge system has failed • Fuel delivery component fault that affects only Cylinder 7 (e.g., a contaminated, dirty or sticking fuel injector) • Ignition system fault (Coil or Plug) that affects only Cylinder 7
DTC: P0308 **2T MISFIRE, MIL: Yes** **1996-2006** **Models:** FX45, M45, Q45, QX56 **Engines:** All **Transmissions:** All	**Cylinder 8 Misfire Detected:** Engine started, engine speed from 400-3500 rpm, VSS over 3 mph, and the PCM detected an irregular CKP sensor signal indicating a misfire present in Cylinder 8 during the 200 or 1000 revolution test. **Possible Causes:** • Base engine mechanical problem that affects only Cylinder 8 • EGR valve stuck open, or EVAP purge system has failed • Fuel delivery component fault that affects only Cylinder 8 (e.g., a contaminated, dirty or sticking fuel injector) • Ignition system fault (Coil or Plug) that affects only Cylinder 8

DTC	Trouble Code Title, Conditions & Possible Causes
DTC: P0325 **2T CCM, MIL: No** **1996-2003** **Models:** G20, I30, J30, M45, Q45 **Engines:** All **Transmissions:** All	**Knock Sensor Circuit Malfunction (Bank 1):** Key on or engine running; system voltage over 10.0v, and the PCM detected the Knock sensor signal was too high or too low in the test. **Note: The Knock sensor signal will read 2.5v at idle if no fault exists.** **Possible Causes:** • Knock sensor signal circuit is open or shorted to ground • Knock sensor signal circuit is shorted to VREF or system power • Knock sensor is damaged or has failed • PCM has failed
DTC: P0327 **2T CCM, MIL: No** **2002-06** **Models:** FX35, FX45, G35, I35, M35, M45, Q45, QX4, QX56 **Engines:** All **Transmissions:** All	**Knock Sensor Circuit Low Input (Bank 1):** Engine started, system voltage over 10.0v, and the PCM detected an excessively low voltage on the Bank 1 Knock sensor circuit. **Note: The Knock sensor signal will read 2.5v at idle if no fault exists.** **Possible Causes:** • Knock sensor signal circuit is shorted to ground • Knock sensor is damaged or has failed • PCM has failed
DTC: P0328 **2T CCM, MIL: No** **2002-06** **Models:** FX35, FX45, G35, I35, M35, M45, Q45, QX4, QX56 **Engines:** All **Transmissions:** All	**Knock Sensor Circuit High Input (Bank 1):** Engine started, system voltage over 10.0v, and the PCM detected an excessively high voltage on the Bank 1 Knock sensor circuit. **Note: The Knock sensor signal will read 2.5v at idle if no fault exists.** **Possible Causes:** • Knock sensor signal circuit is open • Knock sensor signal circuit is shorted to VREF or system power • Knock sensor is damaged or has failed • PCM has failed
DTC: P0330 **1T CCM, MIL: No** **1996-2003** **Models:** FX45, M45, Q45 **Engines:** All **Transmissions:** A/T	**Knock Sensor Circuit Malfunction (Bank 2):** Engine started, system voltage over 10.0v, and the PCM detected the Knock sensor signal was either too high or too low. **Note: The Knock sensor signal will read 2.5v at idle if no fault exists.** **Possible Causes:** • Knock sensor signal circuit is open or shorted to ground • Knock sensor signal circuit is shorted to VREF or system power • Knock sensor is damaged or has failed • PCM has failed
DTC: P0332 **2T CCM, MIL: No** **2003-06** **Models:** Q45, QX56 **Engines:** All **Transmissions:** All	**Knock Sensor Circuit Low Input (Bank 2):** **Engine started, system voltage over 10.0v, and the PCM detected an excessively low voltage on the Bank 2 Knock sensor circuit. Note: The Knock sensor signal will read 2.5v at idle if no fault exists.** **Possible Causes:** • Knock sensor signal circuit is shorted to ground • Knock sensor is damaged or has failed • PCM has failed
DTC: P0333 **2T CCM, MIL: No** **2003-06** **Models:** Q45, QX56 **Engines:** All **Transmissions:** All	**Knock Sensor Circuit High Input (Bank 2):** Engine started, system voltage over 10.0v, and the PCM detected an excessively high voltage on the Bank 2 Knock sensor circuit. **Note: The Knock sensor signal will read 2.5v at idle if no fault exists.** **Possible Causes:** • Knock sensor signal circuit is open • Knock sensor signal circuit is shorted to VREF or system power • Knock sensor is damaged or has failed • PCM has failed
DTC: P0335 **2T CCM, MIL: Yes** **1996-2006** **Models:** FX35, FX45, G20, G35, I30, I35, J30, M35, M45, Q45, QX4, QX56 **Engines:** All **Transmissions:** All	**Crankshaft Position Sensor Circuit Malfunction:** Engine cranking for 2 seconds, and the PCM did not detect a proper CKP sensor (1°) signal, or with engine running, it did not detect a normal pattern of CKP sensor signals. **Possible Causes:** • CKP sensor signal circuit is open or shorted to ground • CKP sensor signal is shorted to VREF or system power • CKP sensor is damaged or has failed • PCM has failed

DTC	Trouble Code Title, Conditions & Possible Causes
DTC: P0340 **2T CCM, MIL: Yes** **1996-2002** **Models:** G20, I30, J30, M45, Q45, QX4 **Engines:** All **Transmissions:** All	**Camshaft Position Sensor Circuit Malfunction:** Engine cranking for over 2 seconds, and the PCM did not detect any CMP sensor signals, or with engine running, the PCM did not detect a normal pattern of CMP signals during the CCM test. **Possible Causes:** • CMP sensor signal circuit is open or shorted to ground • CMP sensor signal is shorted to VREF or system power • CMP sensor is damaged or has failed • PCM has failed
DTC: P0340 **2T CCM, MIL: Yes** **2003-06** **Models:** FX35, FX45, G35, I35, M35, M45, Q45, QX56 **Engines:** All **Transmissions:** All	**Camshaft Position Sensor Circuit Malfunction (Bank 1):** Engine cranking for over 2 seconds, and the PCM did not detect any CMP sensor signals, or with engine running, the PCM did not detect any CMP signals, or it detected an abnormal CMP sensor pattern. **Possible Causes:** • Camshaft (intake) is damaged or has failed (Bank 1) • CMP Phase sensor signal circuit is open or shorted to ground • CMP Phase sensor is damaged or has failed • Low battery voltage condition present during engine cranking • Starter motor circuit problem during engine cranking • PCM has failed
DTC: P0345 **2T CCM, MIL: Yes** **2002-06** **Models:** FX35, FX45, G35, I35, M35, M45 **Engines:** All **Transmissions:** All	**Camshaft Position Sensor Circuit Malfunction (Bank 2):** Engine cranking for over 2 seconds, and the PCM did not detect any CMP sensor signals, or with engine running, the PCM did not detect any CMP signals, or it detected an abnormal CMP sensor pattern. **Possible Causes:** • Camshaft (intake) is damaged or has failed (Bank 2) • CMP Phase sensor signal circuit is open or shorted to ground • CMP Phase sensor is damaged or has failed • Low battery voltage condition present during engine cranking • Starter motor circuit problem during engine cranking • PCM has failed
DTC: P0400 **2T EGR, MIL: Yes** **1996-2002** **Models:** G20, I30, J30, M45, Q45 **Engines:** All **Transmissions:** All	**EGR System Recirculation Flow:** Engine started, vehicle driven to a speed of 19-35 mph at an engine speed of 1952-2400 rpm in closed loop, ECT sensor more than 158ºF, and the PCM detected the EGR Temperature sensor indicated too little or too much EGR flow with the Dual EGR/EVAP solenoid switched "on" and "off" during the EGR Monitor test. **Possible Causes:** • EGR gas temperature sensor is damaged or has failed • EGR valve is stuck partially open or closed • Dual EGR/EVAP solenoid is damaged or has failed • Exhaust system is damaged or has collapsed
DTC: P0400 **2T EGR, MIL: Yes** **1997-2000** **Models:** QX4 **Engines:** All **Transmissions:** All	**EGR System Recirculation Flow:** Engine started, vehicle driven to a speed of 19-35 mph at an engine speed of 1952-2400 rpm in closed loop, ECT sensor more than 158ºF, and the PCM detected the EGR Temperature sensor indicated too little or too much EGR flow with the Dual EGR/EVAP solenoid switched "on" and "off" during the EGR Monitor test. **Possible Causes:** • EGR gas temperature sensor is damaged or has failed • EGR valve is stuck partially open or closed • Dual EGR/EVAP solenoid is damaged or has failed • Exhaust system is damaged or has collapsed
DTC: P0402 **2T CCM, MIL: Yes** **1996-2003** **Models:** G20, I30, J30, M45, Q45 **Engines:** All **Transmissions:** All	**EGRC-BPT Valve Function Conditions:** Engine started, engine speed at 1952-2400 rpm in closed loop, ECT sensor input more than 158ºF, VSS over 19-35 mph, and the PCM detected an unexpected voltage on the EGRC-BPT valve circuit. **Possible Causes:** • EGRC-BPT solenoid control circuit is open or shorted to ground • EGRC-BPT valve vacuum hose is clogged or disconnected • Orifice missing in vacuum hose between EGRC-BPT valve and the EGRC solenoid valve • Exhaust system is damaged or has collapsed
DTC: P0402 **2T CCM, MIL: Yes** **1997-2000** **Models:** QX4 **Engines:** All **Transmissions:** All	**EGRC-BPT Valve Function Conditions:** Engine started, engine speed at 1952-2400 rpm in closed loop, ECT sensor input more than 158ºF, VSS over 19-35 mph, and the PCM detected an unexpected voltage on the EGRC-BPT valve circuit. **Possible Causes:** • EGRC-BPT solenoid control circuit is open or shorted to ground • EGRC-BPT valve vacuum hose is clogged or disconnected • Orifice missing in vacuum hose between EGRC-BPT valve and the EGRC solenoid valve • Exhaust system is damaged or has collapsed

DTC	Trouble Code Title, Conditions & Possible Causes
DTC: P0403 **2T CCM, MIL: Yes** **2000-02** **Models:** G20, I30, I35 **Engines:** All **Transmissions:** All	**EGR Volume Control Valve Circuit Malfunction:** Key on or engine running; system voltage over 10.0v, and the PCM detected an unexpected voltage condition on the EGR Volume Control valve circuit during the CCM test. **Possible Causes:** • EGR volume control valve circuit open or shorted to ground • EGR volume control valve circuit shorted to system power (B+) • EGR volume control valve is damaged or has failed • PCM has failed
DTC: P0420 **1T CAT, MIL: Yes** **1996-2002** **Models:** G20, I30, J30, M45, Q45, QX4 **Engines:** All **Transmissions:** All	**Catalyst Efficiency Below Normal (Bank 1):** Engine started, ECT sensor signal over 158ºF, vehicle driven at 53-60 mph for 5-6 minutes, and the PCM detected the rear HO2S-22 switch rate was similar to the front HO2S switch rate for 3 seconds. **Possible Causes:** • Air leaks at the exhaust manifold or in the exhaust pipes • Catalytic converter is contaminated, damaged or has failed • Continuous engine misfire conditions, or low ignition coil output • Front HO2S is older (aged) than the rear HO2S (HO2S is lazy) • Front HO2S or rear HO2S is contaminated with fuel or moisture • Rear HO2S is loose in the mounting hole (check it for a leak)
DTC: P0420 **1T CAT, MIL: Yes** **2002-06** **Models:** FX35, FX45, G35, I35, M35, M45, Q45, QX56 **Engines:** All **Transmissions:** All	**Catalyst Efficiency Below Normal (Bank 1):** Engine started, ECT sensor signal over 158ºF, vehicle driven at 53-60 mph for 5-6 minutes, and the PCM detected the rear HO2S-22 switch rate was similar to the front HO2S switch rate for 3 seconds. **Possible Causes:** • Air leaks at the exhaust manifold or in the exhaust pipes • Catalytic converter is contaminated, damaged or has failed • Continuous engine misfire conditions, or low ignition coil output • Front HO2S is older (aged) than the rear HO2S (HO2S is lazy) • Front HO2S or rear HO2S is contaminated with fuel or moisture • Rear HO2S is loose in the mounting hole (check it for a leak)
DTC: P0430 **1T CAT, MIL: Yes** **1996-2003** **Models:** I30, I35, J30, M45, Q45, QX4 **Engines:** All **Transmissions:** All	**Catalyst Efficiency Below Normal (Bank 2):** Engine started, ECT sensor signal over 158ºF, vehicle driven at 53-60 mph for 5-6 minutes, and the PCM detected the rear HO2S-22 switch rate was similar to the front HO2S switch rate for 3 seconds. **Possible Causes:** • Air leaks at the exhaust manifold or in the exhaust pipes • Catalytic converter is contaminated, damaged or has failed • Continuous engine misfire conditions, or low ignition coil output • Front HO2S is older (aged) than the rear HO2S (HO2S is lazy) • Front HO2S or rear HO2S is contaminated with fuel or moisture • Rear HO2S is loose in the mounting hole (check it for a leak)
DTC: P0430 **1T CAT, MIL: Yes** **2003-06** **Models:** FX35, FX45, G35, I35, M35, M45, Q45, QX56 **Engines:** All **Transmissions:** All	**Catalyst Efficiency Below Normal (Bank 2):** Engine started, ECT sensor signal over 158ºF, vehicle driven at 53-60 mph for 5-6 minutes, and the PCM detected the rear HO2S-22 switch rate was similar to the front HO2S switch rate for 3 seconds. **Possible Causes:** • Air leaks at the exhaust manifold or in the exhaust pipes • Catalytic converter is contaminated, damaged or has failed • Continuous engine misfire conditions, or low ignition coil output • Front HO2S is older (aged) than the rear HO2S (HO2S is lazy) • Front HO2S or rear HO2S is contaminated with fuel or moisture • Rear HO2S is loose in the mounting hole (check it for a leak)
DTC: P0440 **2T EVAP, MIL: Yes** **1996-2003** **Models:** G20, I30, J30, M45, Q45, QX4 **Engines:** All **Transmissions:** All	**EVAP System Small Leak (0.040") Detected:** DTC P1440 and P1448 not set, engine started, IAT sensor from 32-86ºF, ECT sensor from 32-158ºF, Fuel Level from 25-75%, EVAP purge and vent solenoids both enabled, and the PCM detected the presence of a leak in the EVAP system during the EVAP leak test. **Note: If DTC P1448 is set, repair the cause of this trouble code first.** **Possible Causes:** • EVAP purge solenoid is damaged or has failed • EVAP emission canister clogged or restricted • Fuel tank cap loose, damaged or missing, or vacuum line is off • TSB 00-046 (7/00) contains information related to this code

DTC	Trouble Code Title, Conditions & Possible Causes
DTC: P0441 **2T EVAP, MIL: Yes** **2002-06** **Models:** FX35, FX45, G35, I35, M35, M45, Q45, QX56 **Engines:** All **Transmissions:** All	**EVAP System Incorrect Purge Flow Detected:** DTC P0226, P0227, P0228, P1227 and P1228 not set, IAT sensor signal from 32-86ºF, ECT sensor signal from 32-158ºF, engine running at cruise, then with the EVAP purge solenoid open, the PCM detected an incorrect amount of pressure drop, or the presence of a leak in the EVAP system during the EVAP Monitor test period. **Possible Causes:** • EVAP canister purge volume control solenoid is stuck closed • EVAP canister purge volume control solenoid circuit is shorted • EVAP control system pressure sensor is damaged or has failed • EVAP emission canister or rubber tube is cracked or restricted • EVAP purge port is blocked or restricted • EVAP canister vent control valve is damaged or has failed
DTC: P0442 **2T EVAP, MIL: Yes** **2002-06** **Models:** FX35, FX45, G35, I35, M35, M45, Q45, QX56 **Engines:** All **Transmissions:** All	**EVAP System Small Leak (0.040") Detected:** DTC P1448 not set, IAT sensor signal from 32-86ºF, ECT sensor signal from 32-158ºF, fuel level from 25-75%, engine started, then after the EVAP cut-valve was opened (to clear the line between the tank and EVAP purge valve), the PCM closed the Vent control valve, opened the Purge Volume Control valve to depressurize the system, and after a period of time, it closed the Purge Volume Control valve, and then the PCM determined a small leak present in the system. **Possible Causes:** • EVAP control system pressure sensor is damaged or has failed • EVAP purge solenoid or canister vacuum line loose or off • EVAP purge solenoid is damaged or has failed • EVAP emission canister is dirty, restricted or full of water • EVAP canister vent control valve is damaged or has failed • Fuel tank cap loose, missing or it is the wrong part number • Fuel tank temperature sensor is damaged or has failed • Fuel tank vacuum relief valve is damaged or has failed • Small leak in the EVAP canister or in the fuel tank area • ORVR system is damaged or has a leak in a component • Refueling control valve is damaged or has failed • Vacuum hose leaking between intake manifold and Purge valve
DTC: P0443 **2T CCM, MIL: Yes** **1996-2003** **Models:** G20, I30, J30, M45, Q45, QX4 **Engines:** All **Transmissions:** All	**EVAP Canister Purge Solenoid Circuit Malfunction:** Engine running at cruise speed with light engine load, and the PCM detected an unexpected voltage condition on the Purge solenoid circuit, or it detected an invalid EVAP signal present with the Purge solenoid commanded "on" and "off" during the CCM test period. **Possible Causes:** • Purge solenoid connector is damaged/loose (intermittent fault) • Purge solenoid control circuit open, shorted to ground or power • Purge solenoid is damaged or has failed • PCM has failed • TSB 00-046 (7/00) contains information related to this code
DTC: P0443 **2T CCM, MIL: Yes** **2006** **Models:** FX35, FX45, M35, M45, QX56 **Engines:** All **Transmissions:** All	**EVAP Canister Purge Solenoid Circuit Malfunction:** Engine running at cruise speed with light engine load, and the PCM detected an unexpected voltage condition on the Purge solenoid circuit, or it detected an invalid EVAP signal present with the Purge solenoid commanded "on" and "off" during the CCM test period. **Possible Causes:** • Purge solenoid connector is damaged/loose (intermittent fault) • Purge solenoid control circuit open, shorted to ground or power • Purge solenoid is damaged or has failed • PCM has failed • TSB 00-046 (7/00) contains information related to this code
DTC: P0444 **2T CCM, MIL: Yes** **2002-06** **Models:** FX35, FX45, G35, I35, M35, M45, Q45, QX4, QX56 **Engines:** All **Transmissions:** All	**EVAP Canister Purge Volume Control Solenoid Circuit Low Input:** Key on or engine running; and the PCM detected an unexpected low voltage condition on the EVAP Canister Purge Volume Control Solenoid control circuit during the CCM test period. **Possible Causes:** • Canister purge volume control solenoid connector is damaged • Canister purge volume control solenoid circuit is shorted • Canister purge volume control solenoid is damaged or failed • PCM is damaged or has failed
DTC: P0445 **2T CCM, MIL: Yes** **2002** **Models:** I35 **Engines:** All **Transmissions:** All	**EVAP Canister Purge Volume Control Solenoid Circuit High Input:** Key on or engine running; and the PCM detected an unexpected high voltage condition on the EVAP Canister Purge Volume Control Solenoid control circuit during the CCM test period. **Possible Causes:** • Canister purge volume control solenoid connector is damaged • Canister purge volume control solenoid circuit is open • Canister purge volume control solenoid is damaged or failed • PCM is damaged or has failed

DTC	Trouble Code Title, Conditions & Possible Causes
DTC: P0550 **2T CCM, MIL: Yes** **2002-06** **Models:** FX35, FX45, G35, I35, M35, M45, Q45, QX56 **Engines:** All **Transmissions:** All	**Power Steering Pressure Switch Circuit Malfunction:** Key on or engine running; and the PCM detected an unexpected low or high voltage on the Power Steering Pressure switch circuit. **Possible Causes:** • PSPS connector is damaged or loose (intermittent fault) • PSPS signal circuit is open or shorted to ground • PSPS assembly is damaged or has failed • PCM is damaged or has failed
DTC: P0600 **2T PCM, MIL: No** **1996-2002** **Models:** G20, I30, I35, J30, M45, Q45, QX4 **Engines:** All **Transmissions:** A/T	**A/T Communication Line Circuit Malfunction:** Engine started, system voltage over 10.0v, and the PCM detected an unexpected (continuous) voltage on the communication data circuit between the TCM and the PCM during the CCM test. **Possible Causes:** • TCM communication line circuit is open or shorted to ground • TCM communication line circuit shorted to VREF or power (B+) • Clear the codes and retest for DTC P0600. If the same code resets, substitute a known good TCM or PCM and retest. If the code does not reset, the original TCM PCM has failed
TC: P0605 **T PCM, MIL: Yes** **996-2006** **odels:** FX35, FX45, G20, G35, 0, I35, J30, M35, M45, Q45, QX4, X56 **gines:** All **nsmissions:** All	**PCM Internal Fault Detected:** Engine runtime over 30 seconds, and the PCM detected an internal calculation function problem or a problem in the self-shutoff function. **Possible Causes:** • Clear the codes and retest for DTC P0605. If the same code resets, substitute a known good PCM and retest. If the code does not reset, the original PCM has failed.
C: P0650 **PCM, MIL: Yes** **02-03** **dels:** FX35, FX45, I35 **ines:** All **nsmissions:** All	**Malfunction Indicator Lamp Circuit Malfunction:** Engine started, and the PCM detected an unexpected high voltage condition on the MIL control circuit with the MIL activated, or it detected a low voltage on the MIL circuit with the MIL not activated. **Possible Causes:** • MIL control circuit is open (high voltage condition) • MIL control circuit is shorted to ground (high voltage condition) • PCM is damaged or has failed.
: P0705 **CM, MIL: Yes** **-2006** **els:** FX35, FX45, G20, G35, 35, J30, M35, M45, Q45, QX4, **nes:** All **smissions:** A/T	**Park Neutral Position Switch Circuit Malfunction:** Engine started, and the PCM detected multiple P/N switch inputs, or it did not a change in the P/N switch inputs with the vehicle moving. **Possible Causes:** • P/N switch connector is damaged or loose • P/N switch signal circuit is open or shorted to ground • P/N switch signal circuit is shorted to VREF or system power • P/N switch is damaged or has failed • PCM has failed
P0710 **M, MIL: Yes** **2006** **s:** FX35, FX45, G20, G35, , J30, M35, M45, Q45, QX4, **s:** All **nissions:** A/T	**Transmission Fluid Temperature Sensor Circuit Malfunction:** Engine started, vehicle driven at a speed over 6 mph in Drive with the TP signal over 1.20v, and the PCM detected the TFT sensor signal from the TCM was either too high or low during the CCM test. **Possible Causes:** • TFT sensor signal circuit is open or shorted to ground • TFT sensor signal circuit is shorted to VREF or system power • TFT sensor is damaged or has failed • PCM has failed
0720 **, MIL: Yes** **006** **:** FX35, FX45, G20, G35, J30, M35, M45, Q45, QX4, **:** All **ssions:** A/T	**TCM Revolution Sensor Circuit Malfunction:** Engine started, vehicle driven at a speed of 19-35 mph in Drive with the TP signal over 1.20v for 30 seconds, and the TCM did not detect any signals from the Revolution sensor during the CCM test. **Possible Causes:** • Revolution sensor connector is damage, open or shorted • Revolution sensor signal circuit is open or shorted to ground • Revolution sensor signal circuit shorted to VREF or power (B+) • Revolution sensor is damaged or has failed • PCM has failed

DTC	Trouble Code Title, Conditions & Possible Causes
DTC: P0445 **2T CCM, MIL: Yes** **2003-06** **Models:** FX35, FX45, G35, I35, M35, M45, Q45, QX56 **Engines:** All **Transmissions:** All	**EVAP Canister Purge Volume Control Solenoid Circuit High Input:** Key on or engine running; and the PCM detected an unexpected high voltage condition on the EVAP Canister Purge Volume Control Solenoid control circuit during the CCM test period. **Possible Causes:** • Canister purge volume control solenoid connector is damaged • Canister purge volume control solenoid circuit is open • Canister purge volume control solenoid is damaged or failed • PCM is damaged or has failed
DTC: P0446 **2T CCM, MIL: Yes** **1998-2002** **Models:** G20, Q45, QX4 **Engines:** All **Transmissions:** All	**EVAP Canister Vent Control Solenoid Circuit Malfunction:** Key on or engine running; and the PCM detected an unexpected voltage condition on the EVAP Canister Vent solenoid control circuit. **Possible Causes:** • Canister vent control solenoid connector is damaged or loose • Canister vent control solenoid circuit is open (high voltage) • Canister vent control solenoid is damaged or failed • PCM is damaged or has failed
DTC: P0447 **2T CCM, MIL: Yes** **2002-06** **Models:** FX35, FX45, G35, I35, M35, M45, Q45, QX56 **Engines:** All **Transmissions:** All	**EVAP Canister Vent Control Solenoid Circuit Malfunction:** Key on or engine running; and the PCM detected an unexpected voltage condition on the EVAP Canister Vent solenoid control circuit. **Possible Causes:** • Canister vent control solenoid connector is damaged or loose • Canister vent control solenoid circuit is open (high voltage) • Canister vent control solenoid is damaged or failed • PCM is damaged or has failed
DTC: P0450 **2T CCM, MIL: Yes** **1998-2003** **Models:** G20, I30, J30, M45, Q45, QX4 **Engines:** All **Transmissions:** All	**EVAP Pressure Sensor Circuit Malfunction:** Engine started engine running at idle speed, VSS at 0 mph, EVAP purge commanded "on", and the PCM detected an unexpected condition on the EVAP Pressure sensor signal during the CCM test. **Possible Causes:** • EVAP pressure sensor circuit is open, shorted to ground or B+ • EVAP pressure sensor is damaged or has failed • PCM is damaged • TSB 00-046 (7/00) contains information related to this code
DTC: P0451 **2T CCM, MIL: Yes** **2002-06** **Models:** FX35, FX45, G35, I35, M35, M45, QX56 **Engines:** All **Transmissions:** All	**EVAP System Pressure Sensor Range/Performance:** Engine started, vehicle driven at cruise speed for 2-3 minutes, and the PCM detected an EVAP Pressure sensor signal that was not plausible when compared to various other sensor inputs. **Possible Causes:** • EVAP pressure sensor circuit connector is damaged or loose • EVAP pressure sensor circuit open, shorted to ground or VREF • EVAP pressure sensor is damaged or has failed • PCM is damaged
DTC: P0452 **2T CCM, MIL: Yes** **2002-06** **Models:** FX35, FX45, G35, I35, M35, M45, Q45, QX56 **Engines:** All **Transmissions:** All	**EVAP System Pressure Sensor Circuit Low Input:** Key on or engine running; and the PCM detected an unexpected low voltage condition on the EVAP Pressure sensor signal circuit. **Possible Causes:** • EVAP pressure sensor circuit connector is damaged (loose) • EVAP pressure sensor circuit is shorted to ground • EVAP pressure sensor is damaged or has failed • PCM is damaged
DTC: P0453 **2T CCM, MIL: Yes** **2002-06** **Models:** FX35, FX45, G35, I35, M35, M45, Q45, QX56 **Engines:** All **Transmissions:** All	**EVAP System Pressure Sensor Circuit High Input:** Key on or engine running; and the PCM detected an unexpected high voltage condition on the EVAP Pressure sensor signal circuit. **Possible Causes:** • EVAP pressure sensor circuit connector is damaged (shorted) • EVAP pressure sensor circuit is open or shorted to VREF • EVAP pressure sensor is damaged or has failed • PCM is damaged

DTC	Trouble Code Title, Conditions & Possible Causes
DTC: P0455 **2T EVAP, MIL: Yes** **2000-06** **Models:** FX35, FX45, G20, G35, I30, I35, M35, M45, Q45, QX4, QX56 **Engines:** All **Transmissions:** All	**EVAP System Gross Leak (0.080") Detected:** DTC P1448 not set, vehicle driven at 53-60 mph for 6-8 minutes, ECT sensor signal from 32-158ºF, IAT sensor signal from 32-86ºF, fuel level from 25-75%, and the PCM detected a large leak (larger than 0.040") somewhere between the fuel tank and EVAP canister purge volume control solenoid during the EVAP Monitor leak test. **Possible Causes:** • EVAP control system pressure sensor is damaged or has failed • EVAP purge solenoid or canister vacuum line loose or off • EVAP purge solenoid is damaged or has failed • EVAP emission canister is dirty, restricted or full of water • EVAP canister vent control valve is damaged or has failed • Fuel tank cap loose, missing or it is the wrong part number • Fuel tank temperature sensor is damaged or has failed • Fuel tank vacuum relief valve is damaged or has failed • Large leak in the EVAP canister or in the fuel tank area • ORVR system is damaged or has a large leak in a component • Refueling control valve is damaged or has failed • Vacuum hose leaking between intake manifold and Purge valve • TSB 00-046 (7/00) contains information related to this code
DTC: P0456 **2T EVAP, MIL: Yes** **1996-2006** **Models:** FX35, FX45, G20, G35, I30, I35, J30, M35, M45, Q45, QX4, QX56 **Engines:** All **Transmissions:** All	**EVAP System Very Small Leak (0.020") Detected:** IAT sensor signal from 32-86ºF, ECT sensor signal from 32-158ºF, fuel level from 25-75%, engine started, vehicle driven with the EVAP purge and vent solenoids both enabled, and the PCM detected the presence of a very small leak in the EVAP system in the leak test. **Possible Causes:** • EVAP control system pressure sensor is damaged or has failed • EVAP purge solenoid is damaged or has failed • EVAP canister vent control valve is damaged or has failed • Fuel tank cap not tightened properly or the wrong part number • Fuel tank temperature sensor is damaged or has failed • Fuel tank vacuum relief valve is damaged or has failed • Very small leak in the EVAP canister or in the fuel tank area • ORVR system is damaged or has a small leak in a component • Refueling control valve is damaged or has failed • TSB 00-046 (7/00) contains information related to this code
DTC: P0460 **2T CCM, MIL: Yes** **2000-03** **Models:** FX35, FX45 G20, G35, I30, I35, M35, M45, Q45, QX4, QX56 **Engines:** All **Transmissions:** All	**Fuel Level Sensor Slosh Or Noise Detected:** Engine started, engine running with gear selector in Park (vehicle not moving), and the PCM detected too much variation in the Fuel Level sensor input (i.e., a "noise" condition with the vehicle stopped). **Possible Causes:** • Fuel level sensor connector damaged or loose (disconnected) • Fuel level sensor signal open or shorted to ground (intermittent) • Fuel level sensor is damaged or has failed (intermittent fault)
DTC: P0461 **2T CCM, MIL: Yes** **2000-06** **Models:** FX35, FX45, G20, G35, I30, I35, M35, M45, Q45, QX4, QX56 **Engines:** All **Transmissions:** All	**Fuel Level Sensor Range/Performance:** Engine started, vehicle driven a distance of more than 30 miles, the PCM did not detect any change in the Fuel Level sensor input (i.e., there was no change after driving the vehicle for several miles). **Possible Causes:** • Fuel level sensor signal open or shorted to ground • Fuel level sensor is damaged (stuck) or has failed • PCM has failed
DTC: P0462 **2T CCM, MIL: Yes** **2002-06** **Models:** FX35, FX45, G35, I35, M35, M45, Q45, QX56 **Engines:** All **Transmissions:** All	**Fuel Level Sensor Circuit Low Input:** Key on or engine running; and the PCM detected an unexpected low voltage condition on the Fuel Level sensor circuit during testing. The fuel level sensor is mounted in the fuel level sensor unit in the fuel tank. This sensor sends a signal to the PCM indicating the fuel level. **Possible Causes:** • Fuel level sensor connector is damaged or loose (intermittent) • Fuel level sensor signal is shorted to ground • Fuel level sensor is damaged (stuck) or has failed

DTC	Trouble Code Title, Conditions & Possible Causes
DTC: P0463 **2T CCM, MIL: Yes** **2000-06** **Models:** FX35, FX45, G20, G35, I30, I35, M35, M45, Q45, QX4, QX56 **Engines:** All **Transmissions:** All	**Fuel Level Sensor Circuit High Input:** Key on or engine running; and PCM detected an unexpected high voltage condition on the Fuel Level sensor input during testing fuel level sensor is mounted in the fuel level sensor unit in the fuel tank. This sensor sends a signal to the PCM indicating th level. **Possible Causes:** • Fuel level sensor connector is damaged or loose (intermittent) • Fuel level sensor signal shorted to ground • Fuel level sensor is damaged or has failed
DTC: P0464 **2T CCM, MIL: Yes** **2000-03** **Models:** G20, I30, M45, Q45, QX4 **Engines:** All **Transmissions:** All	**Fuel Level Sensor Circuit Malfunction:** Engine running for 5 seconds, and the PCM detected an unexpected high or low voltage condition on the Fuel Level sensor **Possible Causes:** • Fuel level sensor connector is damaged or loose (intermittent) • Fuel level sensor signal circuit is open or shorted to ground • Fuel level sensor signal circuit shorted to VREF or power (B+) • Fuel level sensor is damaged or has failed
DTC: P0500 **2T CCM, MIL: Yes** **1996-2006** **Models:** FX35, FX45, G20, G35, I30, I35, J30, M35, M45, Q45, QX4, QX56 **Engines:** All **Transmissions:** All	**Vehicle Speed Sensor Circuit Malfunction:** Engine started, vehicle driven at over 1500 rpm at light engine load for over 1 minute, and the PCM did not detect any V **Possible Causes:** • ABS/TCS control unit is damaged or has failed • Combination meter is damaged or has failed • VSS signal circuit is open, shorted to ground or shorted to B+ • VSS is damaged or has failed • PCM has failed
DTC: P0505 **2T CCM, MIL: Yes** **1996-2002** **Models:** G20, I30, J30, M45, Q45, QX4 **Engines:** All **Transmissions:** All	**Idle Air Control, Auxiliary Air Control Valve:** Engine started, engine at hot idle speed for 30 seconds, and the PCM detected the IAC control volume was incorre valve was commanded open and closed, it detected the change in air volume did not correlate to the engine air volu comparing the air volume to the volume of the MAF sensor signal). **Possible Causes:** • IACV-AAC valve control circuit is open or shorted to ground • IACV-AAC valve control circuit is shorted to system power (B+) • IACV-AAC valve is damaged or has failed • PCM is damaged or has failed
DTC: P0506 **2T CCM, MIL: Yes** **2002-06** **Models:** FX35, FX45, G35, I35, M35, M45, Q45, QX56 **Engines:** All **Transmissions:** All	**Idle Speed Control System RPM Lower Than Expected:** Engine started, engine running at hot idle speed for 30 seconds, and the PCM detected the Idle Speed Control was more than 100 rpm lower than the desired (control) idle speed. **Possible Causes:** • Air inlet duct is collapsed, loose or air filter element is clogged • Base engine problem (i.e., a compression or misfire problem) • Idle air inlet passage or throttle bore is dirty or full of deposits • Electronic throttle control actuator connector is open • Electronic throttle control actuator circuit is open • Electronic throttle control actuator is damaged or has failed
DTC: P0507 **2T CCM, MIL: Yes** **2002-06** **Models:** FX35, FX45, G35, I35, M35, M45, Q45, QX56 **Engines:** All **Transmissions:** All	**Idle Speed Control System RPM Higher Than Expected:** Engine started, engine running at hot idle speed for 30 seconds, and the PCM detected the Idle Speed Co was more than 100 rpm higher than the desired (control) idle speed. **Possible Causes:** • Engine vacuum leaks, PCM valve is leaking or the wrong valve • Idle air inlet passage or throttle bore is dirty or full of deposits • Electronic throttle control actuator connector is shorted • Electronic throttle control actuator circuit is shorted • Electronic throttle control actuator is damaged or has failed
DTC: P0510 **2T CCM, MIL: Yes** **1996-2003** **Models:** G20, I30, J30, M45, Q45, QX4 **Engines:** All **Transmissions:** All	**Closed Throttle Position Switch Circuit Malfunction:** Engine started, vehicle driven at a speed of 5-20 mph, and the PCM did not detect any change of status switch circuit during the CCM test period. **Possible Causes:** • Closed throttle position switch connector is damaged or loose • Closed throttle position switch signal circuit is open or grounded • Closed throttle position switch signal circuit is shorted to power • Closed throttle position switch or TP sensor is damaged/failed

DTC	Trouble Code Title, Conditions & Possible Causes
DTC: P0725 **2T CCM, MIL: Yes** **1996-2003** **Models:** FX35, FX45, G20, G35, I30, I35, J30, M45, Q45, QX4 **Engines:** All **Transmissions:** A/T	**TCM Engine Speed Signal Circuit Malfunction:** Engine started, vehicle driven at 19-35 mph in Drive with the TP signal over 1.20v for 30 seconds, and the TCM did not detect any Engine Speed signals during the CCM continuous test. **Possible Causes:** • Engine speed signal is open or shorted to ground • Engine speed sensor signal shorted to VREF or power (B+) • Engine speed sensor is damaged or has failed • PCM has failed
DTC: P0725 **2T CCM, MIL: Yes** **2006** M35, M45 **Engines:** All **Transmissions:** A/T	**TCM Engine Speed Signal Circuit Malfunction:** Engine started, vehicle driven at 19-35 mph in Drive with the TP signal over 1.20v for 30 seconds, and the TCM did not detect any Engine Speed signals during the CCM continuous test. **Possible Causes:** • Engine speed signal is open or shorted to ground • Engine speed sensor signal shorted to VREF or power (B+) • Engine speed sensor is damaged or has failed • PCM has failed
DTC: P0731 **2T CCM, MIL: Yes** **1996-2003** **Models:** FX35, FX45, G20, G35, I30, I35, J30, M45, Q45, QX4 **Engines:** All **Transmissions:** A/T	**A/T First Gear Circuit Malfunction:** Engine started, vehicle driven at a steady cruise speed, and the TCM detected an incorrect voltage condition on the 1st Gear circuit. **Note: During this test, the CCM monitors the Actual gear position by checking the torque converter slip ratio. The slip ratio is calculated as this equation: A X C/B (where 'A' is the output shaft revolution signal, 'B' is the engine speed signal from the PCM, and 'C' is the gear ratio inferred TCM from other inputs. If the Actual gear ratio is higher than the gear position (1st2nd) inferred by the TCM, the slip ratio will be too high. If the slip ratio exceeds a certain value, the TCM determines that a fault exists and signals the PCM.** **Possible Causes:** • Shift Solenoid 'A' is stuck in open position • Shift Solenoid 'B' is stuck in open position
DTC: P0732 **2T CCM, MIL: Yes** **1996-2003** **Models:** FX35, FX45, G20, G35, I30, I35, J30, M45, Q45, QX4 **Engines:** All **Transmissions:** A/T	**A/T Second Gear Circuit Malfunction:** Engine started, vehicle driven at a steady cruise speed, and the TCM detected an incorrect voltage condition on the 2nd Gear circuit. **Note: During this test, the CCM monitors the Actual gear position by checking the torque converter slip ratio. The slip ratio is calculated as this equation: A X C/B (where 'A' is the output shaft revolution signal, 'B' is the engine speed signal from the PCM, and 'C' is the gear ratio inferred TCM from other inputs. If the Actual gear ratio is higher than the gear position (2nd) inferred by the TCM, the slip ratio will be too high. If the slip ratio exceeds a certain value, the TCM determines that a fault exists and signals the PCM to set this code.** **Possible Causes:** • Control valve sticking or binding • Solenoid valve is damaged or not operating
DTC: P0733 **2T CCM, MIL: Yes** **1996-2003** **Models:** FX35, FX45, G20, G35, I30, I35, J30, M45, Q45, QX4 **Engines:** All **Transmissions:** A/T	**A/T Third Gear Circuit Malfunction:** Engine started, vehicle driven at a steady cruise speed, and the TCM detected an incorrect voltage condition on the 3rd Gear circuit. **Note: During this test, the CCM monitors the Actual gear position by checking the torque converter slip ratio. The slip ratio is calculated as this equation: A X C/B (where 'A' is the output shaft revolution signal, 'B' is the engine speed signal from the PCM, and 'C' is the gear ratio inferred TCM from other inputs. If the Actual gear ratio is higher than the gear position (3rd) inferred by the TCM, the slip ratio will be too high. If the slip ratio exceeds a certain value, the TCM determines that a fault exists and signals the PCM.** **Possible Causes:** • Control valve sticking, solenoid valve damaged or not operating • Servo piston or brake band is damaged or is not operating
DTC: P0734 **2T CCM, MIL: Yes** **1996-2003** **Models:** FX35, FX45, G20, G35, I30, I35, J30, M45, Q45, QX4 **Engines:** All **Transmissions:** A/T	**A/T Fourth Gear Circuit Malfunction:** Engine started, vehicle driven at a steady cruise speed, and the TCM detected an incorrect voltage condition on the 4th Gear circuit. **Note: During this test, the CCM monitors the Actual gear position by checking the torque converter slip ratio. The slip ratio is calculated as this equation: A X C/B (where 'A' is the output shaft revolution signal, 'B' is the engine speed signal from the PCM, and 'C' is the gear ratio inferred TCM from other inputs. If the Actual gear ratio is higher than the gear position (4th) inferred by the TCM, the slip ratio will be too high. If the slip ratio exceeds a certain value, the TCM determines that a fault exists and signals the PCM.** **Possible Causes:** • Oil pump is damaged, or the TCC is not operating • Shift Solenoid 'B' may be stuck closed

DTC	Trouble Code Title, Conditions & Possible Causes
DTC: P0740 **2T CCM, MIL: Yes** **1996-2006** **Models:** FX35, FX45, G20, G35, I30, I35, J30, M35, M45, Q45, QX4, QX56 **Engines:** All **Transmissions:** A/T	**A/T TCC Solenoid Circuit Malfunction:** Engine started, vehicle driven at a steady cruise speed, and the TCM detected an unexpected voltage condition on the TCC solenoid control circuit during the CCM continuous test. **Possible Causes:** • TCC solenoid connector is damaged, open or shorted • TCC solenoid control circuit open or shorted to ground • TCC solenoid control circuit shorted to system power (B+) • TCC solenoid is damaged or has failed
DTC: P0744 **2T CCM, MIL: Yes** **1996-2006** **Models:** FX35, FX45, G20, G35, I30, I35, J30, M35, M45, Q45, QX4, QX56 **Engines:** All **Transmissions:** A/T	**A/T Servo Valve Solenoid Circuit Malfunction:** Engine started, vehicle driven at a steady cruise speed, and the TCM detected an unexpected voltage condition on the Servo Valve solenoid control circuit during the CCM continuous test. **Possible Causes:** • Servo valve solenoid connector is damaged, open or shorted • Servo valve solenoid control circuit open or shorted to ground • Servo valve solenoid control circuit shorted to system power • Servo valve solenoid is damaged or has failed • TSB 03-034 (5/03) contains information related to this code
DTC: P0745 **2T CCM, MIL: Yes** **1996-2006** **Models:** FX35, FX45, G20, G35, I30, I35, J30, M35, M45, Q45, QX4, QX56 **Engines:** All **Transmissions:** A/T	**A/T Low Pressure Solenoid Circuit Malfunction:** Engine started, vehicle driven at a steady cruise speed, and the TCM detected an unexpected voltage condition on the Low Pressure solenoid control circuit during the CCM test. **Possible Causes:** • Low pressure solenoid connector is damaged, open or shorted • Low pressure solenoid control circuit open or shorted to ground • Low pressure solenoid control circuit shorted to system power • Low pressure solenoid is damaged or has failed • TSB 03-034 (5/03) contains information related to this code
DTC: P0750 **2T CCM, MIL: No** **1996-2003** **Models:** FX35, FX45, G20, G35, I30, I35, J30, M45, Q45, QX4 **Engines:** All **Transmissions:** A/T	**A/T Shift Solenoid 'A' Circuit Malfunction:** Engine started, vehicle driven at a steady cruise speed, and the TCM detected an unexpected voltage condition on the Shift Solenoid 'A' control circuit during the CCM test. **Possible Causes:** • Shift Solenoid 'A' connector is damaged, open or shorted • Shift Solenoid 'A' control circuit open or shorted to ground • Shift Solenoid 'A' control circuit shorted to VREF or power (B+) • Shift Solenoid 'A' is damaged or has failed • PCM or TCM has failed
DTC: P0755 **2T CCM, MIL: No** **1996-2003** **Models:** FX35, FX45, G20, G35, I30, I35, J30, M45, Q45, QX4 **Engines:** All **Transmissions:** A/T	**A/T Shift Solenoid 'B' Circuit Malfunction:** Engine started, vehicle driven at a steady cruise speed, and the TCM detected an unexpected voltage condition on the Shift Solenoid 'B' control circuit during the CCM test. **Possible Causes:** • Shift Solenoid 'B' connector is damaged, open or shorted • Shift Solenoid 'B' control circuit open or shorted to ground • Shift Solenoid 'B' control circuit shorted to VREF or power (B+) • Shift Solenoid 'B' is damaged or has failed • PCM or TCM has failed

OBD II Trouble Code List (P1XXX Codes)

DTC	Trouble Code Title, Conditions & Possible Causes
DTC: P1065 **2T PCM, MIL: Yes** **2002-05** **Models:** FX35, FX45, G35, I35, M45, Q45, QX56 **Engines:** All **Transmissions:** All	**PCM Backup Circuit Malfunction:** Key on or engine running; and the PCM detected an internal problem in its ECM Backup circuit during the CCM test. **Possible Causes:** • PCM Backup circuit (to battery) is open or shorted to ground • Clear the codes and retest for DTC P1065. If the same code resets, substitute a known good PCM and retest. If the code does not reset, the original PCM has failed
DTC: P1102 **2T PCM, MIL: Yes** **2002-03** **Models:** FX35, I35, QX4 **Engines:** All **Transmissions:** All	**Mass Airflow Sensor Range/Performance:** Engine started, engine running at idle and cruise, and the PCM detected the MAF sensor signal remained near 1.0v at all times. **Possible Causes:** • MAF sensor connector is damaged (it may be shorted) • MAF sensor signal circuit is shorted • MAF sensor is damaged or has failed • PCM is damaged or has failed
DTC: P1105 **2T CCM, MIL: Yes** **1996-2003** **Models:** FX45, I30, J30, M45, Q45 **Engines:** All **Transmissions:** All	**MAP/BARO Switch Solenoid Circuit Malfunction:** Engine started, then after the PCM cycled the MAP/BARO Switch Solenoid (to change the MAP input signal to read the atmospheric pressure instead of manifold pressure), the PCM detected the MAP sensor input did not change to reflect the atmospheric pressure. **Possible Causes:** • Absolute pressure sensor is damaged or out of calibration • MAP/BARO switch solenoid control circuit is open or grounded • MAP/BARO switch solenoid control circuit is shorted to power • MAP/BARO switch solenoid is damaged or has failed • MAP/BARO switch solenoid vacuum hoses loose or restricted • PCM has failed
DTC: P1105 **2T CCM, MIL: Yes** **1997-2000** **Models:** QX4 **Engines:** All **Transmissions:** All	**MAP/BARO Switch Solenoid Circuit Malfunction:** Engine started, then after the PCM cycled the MAP/BARO Switch Solenoid (to change the MAP input signal to read the atmospheric pressure instead of manifold pressure), the PCM detected the MAP sensor input did not change to reflect the atmospheric pressure. **Possible Causes:** • Absolute pressure sensor is damaged or out of calibration • MAP/BARO switch solenoid control circuit is open or grounded • MAP/BARO switch solenoid control circuit is shorted to power • MAP/BARO switch solenoid is damaged or has failed • MAP/BARO switch solenoid vacuum hoses loose or restricted • PCM has failed
DTC: P1110 **2T CCM, MIL: Yes** **1998-2003** **Models:** FX45, M45, Q45, QX4 **Engines:** All **Transmissions:** All	**Intake Valve Timing Control System Performance:** Engine started, ECT sensor input at 59-230°F, vehicle driven to over 4 mph at an engine speed of 1100-4600 rpm at high load conditions, the PCM detected the difference between the Intake Valve Timing solenoid "on" position did not correlate to the position wit it "off". **Possible Causes:** • Intake valve timing position control sensor circuit is open • Intake valve timing position control solenoid damaged or failed • Signal pickup portion of camshaft is contaminated with debris
DTC: P1111 **2T CCM, MIL: Yes** **2001-05** **Models:** FX35, G35, QX4, Q45 **Engines:** All **Transmissions:** All	**Intake Valve Timing Control Circuit Malfunction (Bank 1):** Engine started, and the PCM detected an unexpected voltage condition in the Intake Valve Timing Control solenoid circuit for Cylinder Bank 1 during the CCM test. **Possible Causes:** • Intake valve timing control solenoid circuit is open • Intake valve timing position control solenoid damaged or failed • PCM has failed
DTC: P1120 **2T CCM, MIL: Yes** **1996-2003** **Models:** FX45, M45, Q45 **Engines:** All **Transmissions:** A/T	**Secondary Throttle Position Sensor Circuit Malfunction:** Key on or engine running; and the PCM detected an out of range high or low signal from the Secondary Throttle Position sensor, or with the engine running, it detected a signal that was not practical when compared to the IACV-AAC valve, CMP and MAP sensors. **Possible Causes:** • Secondary throttle position sensor circuit is open or shorted • Secondary throttle position sensor is damaged or has failed • PCM has failed

DTC	Trouble Code Title, Conditions & Possible Causes
DTC: P1121 **1T CCM, MIL: Yes** **2001-05** **Models:** FX35, G35, I35, Q45, QX56 **Engines:** All **Transmissions:** A/T	**Electronic Throttle Control Actuator Malfunction:** Key on or engine running; and the PCM detected a problem in the operation of the Electronic Throttle Control Actuator operation (two-trip failure detection), or that the throttle was stuck "open" due to a return spring mechanical problem (one-trip failure detection). **Possible Causes:** • Electronic throttle control actuator is damaged or has failed • Throttle return spring is damaged or has failed
DTC: P1122 **1T CCM, MIL: Yes** **2002-05** **Models:** FX35, I35, Q45, QX56 **Engines:** All **Transmissions:** A/T	**Electronic Throttle Control Actuator Malfunction:** DTC P1121 and P1126 not set; key on or engine running; and the PCM detected a problem in the operation of the Electronic Throttle Control Actuator operation during the CCM test period. **Possible Causes:** • Electronic throttle control actuator connector is damaged/loose • Electronic throttle control actuator circuit is open or shorted • Electronic throttle control actuator is damaged or has failed
DTC: P1124 **1T CCM, MIL: Yes** **2002-05** **Models:** FX35, G35, I35, Q45, QX56 **Engines:** All **Transmissions:** A/T	**Electronic Throttle Control Actuator Relay High Input:** Key on or engine running; and the PCM detected an unexpected high voltage on the Electronic Throttle Control Actuator Relay power circuit after the key was turned "off" during the CCM shutdown test. **Possible Causes:** • ETC Actuator relay is damaged or failed (stuck in "on" position) • ETC Actuator relay control circuit is shorted to ground • ETC Actuator is damaged or has failed
DTC: P1125 **1T CCM, MIL: Yes** **1996-2003** **Models:** FX45, M45, Q45 **Engines:** All **Transmissions:** A/T	**Tandem Throttle Position Sensor Circuit Malfunction:** Engine started, engine runtime over 15 seconds, and the PCM detected the Tandem Throttle Position sensor signal was not practical when compared to signals from the IACV-AAC, CMP and MAF sensors. **Possible Causes:** • Secondary throttle position sensor is damaged or has failed • Tandem throttle position sensor circuit is open or shorted • Tandem throttle position sensor is damaged or has failed • Throttle actuator control (TAC) module has failed
DTC: P1126 **1T CCM, MIL: Yes** **2002-05** **Models:** FX35, G35, I35, Q45, QX56 **Engines:** All **Transmissions:** A/T	**Electronic Throttle Control Actuator Relay Low Input:** Key on or engine running; and the PCM detected an unexpected low voltage on the Electronic Throttle Control Actuator Relay power circuit during the CCM test period. **Possible Causes:** • ETC Actuator relay is damaged or failed (stuck in "off" position) • ETC Actuator relay control circuit is open • ETC Actuator is damaged or has failed
DTC: P1126 **1T ECT, MIL: Yes** **1996-2002** **Models:** G20, I30, J30 **Engines:** All **Transmissions:** All	**Thermostat Malfunction (Stuck Open):** DTC P0115 not set, IAT sensor signal more than 14ºF, engine runtime over 10 minutes, and the PCM detected that the engine temperature did not reach at least 176ºF under these conditions. **Possible Causes:** • Check the operation of the thermostat (it may be stuck open) • Inspect for low coolant level or an incorrect coolant mixture
DTC: P1128 **1T CCM, MIL: Yes** **2002-05** **Models:** FX35, G35, I35, Q45, QX56 **Engines:** All **Transmissions:** A/T	**Electronic Throttle Control Actuator Motor Circuit Malfunction:** Key on or engine running; and the PCM detected an unexpected low voltage condition both of the Electronic Throttle Control Actuator motor circuits during the CCM test period. **Possible Causes:** • ETC Actuator motor is damaged or has failed • ETC Actuator motor circuits are shorted to ground (low voltage)

DTC	Trouble Code Title, Conditions & Possible Causes
DTC: P1130 **2T CCM, MIL: Yes** **1999-2003** **Models:** I30, I35, QX4 **Engines:** All **Transmissions:** All	**Swirl Control Solenoid Valve Performance:** DTC P1165 not set, engine started, ECT sensor input at 50-131°F, and the PCM detected an unexpected voltage on the Swirl Control valve circuit, or the vacuum signal was not sent to the Swirl Control valve with the valve commanded "on", or that the vacuum signal was not sent to the Swirl Control valve with the valve commanded "off". **Note: If DTC P1165 is also set, repair that trouble code first.** **Possible Causes:** • Swirl Control valve control circuit is open or shorted to ground • Swirl Control valve is damaged or has failed • Swirl Control valve vacuum hoses are loose, restricted or dirty
DTC: P1130 **2T CCM, MIL: Yes** **2002-03** **Models:** I35, QX4 **Engines:** All **Transmissions:** All	**Swirl Control Solenoid Valve Circuit Malfunction:** Key on or engine running; and the PCM detected an unexpected voltage condition on the Swirl Control Valve control circuit. The Swirl Control Solenoid Valve is controlled by signals from the PCM. When the solenoid control signal is grounded ("on"), the solenoid valve is bypassed to apply intake manifold vacuum to the swirl control valve actuator. When the solenoid valve is turned "off", the vacuum signal to the valve is "cut", and the swirl valve opens. **Possible Causes:** • Swirl control valve connector is damaged or loose • Swirl control valve control circuit is open or shorted to ground • Swirl control valve is damaged or has failed • PCM has failed
DTC: P1135 **2T CCM, MIL: Yes** **1998-2003** **Models:** FX45, M45, Q45, QX4 **Engines:** All **Transmissions:** All	**Intake Valve Timing Control Circuit Malfunction (Bank 2):** Engine started, and the PCM detected an unexpected voltage condition in the Intake Valve Timing Control solenoid circuit for Cylinder Bank 2 during the CCM test. **Possible Causes:** • Intake valve timing control solenoid circuit is open • Intake valve timing position control solenoid is damaged/failed • PCM has failed
DTC: P1136 **2T CCM, MIL: Yes** **2001-05** **Models:** FX35, G35, QX4, Q45 **Engines:** All **Transmissions:** All	**Intake Valve Timing Control Circuit Malfunction (Bank 2):** Engine started, and the PCM detected an unexpected voltage condition in the Intake Valve Timing Control solenoid circuit for Cylinder Bank 2 during the CCM test. The Intake Valve Timing Control Solenoid Valve is controlled by a duty cycle signal from the PCM. This valve changes the amount of oil (and direction of the oil flow) through the Intake Valve Timing Control unit, or stops the flow. **Possible Causes:** • Intake valve timing control solenoid circuit is open or shorted • Intake valve timing position control solenoid is damaged/failed • PCM has failed
DTC: P1137 **2T CCM, MIL: Yes** **1999-2003** **Models:** I30, I35 **Engines:** All **Transmissions:** All	**Swirl Control Valve Control Position Sensor Circuit Malfunction:** Engine running at idle speed, and the PCM detected an excessively high or low signal from the SCV Control Position sensor circuit. **Possible Causes:** • Swirl Control valve control position sensor open or shorted • Swirl Control valve control position sensor is damaged • Swirl Control valve control position sensor has failed • PCM has failed
DTC: P1138 **2T CCM, MIL: Yes** **1999-2002** **Models:** I30, I35 **Engines:** All **Transmissions:** All	**Swirl Control Valve Control Position Sensor Performance:** Engine running at idle speed, and the PCM detected the Target opening angle and Actual signal from the sensor did not correlate. **Possible Causes:** • Swirl Control valve control position sensor is damaged • Swirl Control valve control position sensor has failed • PCM has failed
DTC: P1140 **2T CCM, MIL: Yes** **1998-2005** **Models:** FX45, M45, Q45, QX4 **Engines:** All **Transmissions:** All	**Intake Valve Timing Control Position Sensor Circuit Malfunction (Bank 1):** Engine started, and the PCM detected an unexpected voltage on the Bank 1 Intake Valve Timing Control Position sensor circuit. **Possible Causes:** • Intake valve timing control position sensor open or shorted • Intake valve timing control position sensor is damaged • Intake valve timing control position sensor has failed • PCM has failed

DTC	Trouble Code Title, Conditions & Possible Causes
DTC: P1143 **2T CCM, MIL: Yes** **2001-05** **Models:** FX35, G35, I35, QX4, Q45 **Engines:** All **Transmissions:** All	**HO2S-11 (Bank 1 Sensor 1) Lean Shift Monitoring (Bank 1):** Engine started, vehicle driven at over 48 mph for 2 minutes, and the PCM detected the minimum and maximum voltages of the HO2S signal circuit were not reached during the CCM test period. **Possible Causes:** • Air leaks in the intake manifold, throttle body or PCV system • Fuel pressure is excessively low (lean air fuel) • Fuel injectors restricted (one or more has a mechanical fault) • HO2S assembly is damaged or has failed • HO2S heater assembly is damaged or has failed
DTC: P1144 **2T CCM, MIL: Yes** **2001-05** **Models:** G35, I35, QX4, Q45 **Engines:** All **Transmissions:** All	**HO2S-11 (Bank 1 Sensor 1) Rich Shift Monitoring (Bank 1):** Engine started, vehicle driven at over 48 mph for 2 minutes, and the PCM detected the minimum and maximum voltages of the HO2S signal circuit were not reached during the CCM test period. **Possible Causes:** • Fuel pressure is excessively high (rich air fuel) • Fuel injectors sticking open (one or more has a mechanical fault) • HO2S assembly is damaged or has failed • HO2S heater assembly is damaged or has failed
DTC: P1145 **2T CCM, MIL: Yes** **1998-2005** **Models:** FX45, M45, Q45, QX4 **Engines:** All **Transmissions:** All	**Intake Valve Timing Control Position Sensor Circuit Malfunction (Bank 2):** Engine started, vehicle driven at over 48 mph for 2 minutes, and the PCM detected an unexpected voltage on the Bank 2 Intake Valve Timing Control Position sensor circuit during the CCM test. **Possible Causes:** • Intake valve timing control position sensor open or shorted • Intake valve timing control position sensor is damaged • Intake valve timing control position sensor has failed • PCM has failed
DTC P1146 **2T CCM, MIL: Yes** **2002-05** **Models:** FX35, G35, I35, QX4, Q45, QX56 **Engines:** All **Transmissions:** All	**HO2S-12 (Bank 1 Sensor 2) Lean Shift Monitoring (Bank 1):** Engine started, vehicle driven at over 48 mph for 2 minutes, and the PCM detected the minimum and maximum voltages of the HO2S-12 signal circuit were not reached during the CCM test period. **Possible Causes:** • Air leaks in the intake manifold, throttle body or PCV system • Fuel pressure is excessively low (lean air fuel) • Fuel injectors restricted (one or more has a mechanical fault) • HO2S assembly is damaged or has failed • HO2S heater assembly is damaged or has failed
DTC: P1147 **2T CCM, MIL: Yes** **2002-05** **Models:** FX35, G35, I35, QX4, Q45, QX56 **Engines:** All **Transmissions:** All	**HO2S-12 (Bank 1 Sensor 2) Rich Shift Monitoring (Bank 1):** Engine started, vehicle driven at over 48 mph for 2 minutes, and the PCM detected the minimum and maximum voltages of the HO2S-12 signal circuit were not reached during the CCM test period. **Possible Causes:** • Fuel pressure is excessively high (rich air fuel) • Fuel injectors sticking open (one or more has a mechanical fault) • HO2S assembly is damaged or has failed • HO2S heater assembly is damaged or has failed
DTC: P1148 **2T CCM, MIL: Yes** **1998-2006** **Models:** G20, I30, J30, FX45, M45, Q45, QX4, QX56 **Engines:** All **Transmissions:** All	**Closed Loop Malfunction Detected (Bank 1):** Engine started, engine running in closed loop for over 2 minutes, and the PCM determined the engine was not operating in closed loop mode after the additional driving period had elapsed. **Possible Causes:** • HO2S signal circuit is open or shorted to ground • HO2S is damaged, contaminated or has failed • HO2S heater is damaged or has failed
DTC: P1148 **1T CCM, MIL: Yes** **2002-06** **Models:** FX35, G35, I35, M35 **Engines:** All **Transmissions:** All	**Closed Loop Malfunction Detected (Bank 1):** Engine started, engine running in closed loop for over 2 minutes, and the PCM determined the engine was not operating in closed loop mode after the additional driving period had elapsed. **Possible Causes:** • HO2S signal circuit is open or shorted to ground • HO2S is damaged, contaminated or has failed • HO2S heater is damaged or has failed

DTC	Trouble Code Title, Conditions & Possible Causes
DTC: P1163 **2T CCM, MIL: Yes** **2002-05** **Models:** FX35, G35, I35, QX4, Q45 **Engines:** All **Transmissions:** All	**HO2S-12 (Bank 1 Sensor 2) Lean Shift Monitoring (Bank 2):** Engine started, vehicle driven at over 48 mph for 2 minutes, and the PCM detected the minimum and maximum voltages of the HO2S-12 signal circuit were not reached during the CCM test period. **Possible Causes:** • Air leaks in the intake manifold, throttle body or PCV system • Fuel pressure is excessively low (lean air fuel) • Fuel injectors restricted (one or more has a mechanical fault) • HO2S assembly is damaged or has failed • HO2S heater assembly is damaged or has failed
DTC: P1164 **2T CCM, MIL: Yes** **2002-05** **Models:** FX35, G35, I35, QX4, Q45 **Engines:** All **Transmissions:** All	**HO2S-12 (Bank 1 Sensor 2) Rich Shift Monitoring (Bank 2):** Engine started, vehicle driven at over 48 mph for 2 minutes, and the PCM detected the minimum and maximum voltages of the HO2S-12 signal circuit were not reached during the CCM test period. **Possible Causes:** • Fuel pressure is excessively high (rich air fuel) • Fuel injectors sticking open (one or more mechanical faults) • HO2S assembly is damaged or has failed • HO2S heater assembly is damaged or has failed
DTC: P1165 **2T CCM, MIL: Yes** **1999-2003** **Models:** I30, QX4 **Engines:** All **Transmissions:** All	**Swirl Control Valve Control Vacuum Check Switch:** Engine started, ECT sensor from 59-122ºF, and the PCM detected an unexpected voltage condition on the Swirl Control circuit during the CCM test. **Possible Causes:** • SCV vacuum hoses are switched or disconnected • SCV control vacuum switch circuit is opened or shorted • SCV control vacuum switch is damaged or has failed • SCV solenoid is damaged or has failed
DTC: P1166 **2T CCM, MIL: Yes** **2002-05** **Models:** FX35, G35, I35, QX4, QX56, Q45 **Engines:** All **Transmissions:** All	**HO2S-22 (Bank 2 Sensor 2) Lean Shift Monitoring (Bank 2):** Engine started, vehicle driven at over 48 mph for 2 minutes, and the PCM detected the minimum and maximum voltages of the HO2S-22 signal circuit were not reached during the CCM test period. **Possible Causes:** • Air leaks in the intake manifold, throttle body or PCV system • Fuel pressure is excessively low (lean air fuel) • Fuel injectors restricted (one or more has a mechanical fault) • HO2S assembly is damaged or has failed • HO2S heater assembly is damaged or has failed
DTC: P1167 **2T CCM, MIL: Yes** **2002-05** **Models:** FX35, G35, I35, QX4, QX56, Q45 **Engines:** All **Transmissions:** All	**HO2S-22 (Bank 2 Sensor 2) Rich Shift Monitoring (Bank 2):** Engine started, vehicle driven at over 48 mph for 2 minutes, and the PCM detected the minimum and maximum voltages of the HO2S-22 signal circuit were not reached during the CCM test period. **Possible Causes:** • Fuel pressure is excessively high (rich air fuel) • Fuel injectors sticking open (one or more with mechanical fault) • HO2S assembly is damaged or has failed • HO2S heater assembly is damaged or has failed
DTC: P1168 **1T CCM, MIL: Yes** **1998-2006** **Models:** G20, I30, J30, FX45, M45, Q45, QX4, QX56 **Engines:** All **Transmissions:** All	**Closed Loop Malfunction Detected (Bank 2):** Engine started, engine running in closed loop for over 2 minutes, and the PCM determined the engine was not operating in closed loop mode after the additional driving period had elapsed. **Possible Causes:** • HO2S signal circuit is open or shorted to ground • HO2S is damaged, contaminated or has failed • HO2S heater is damaged or has failed
DTC: P1168 **2T CCM, MIL: Yes** **2002-06** **Models:** FX35, G35, I35, M35 **Engines:** All **Transmissions:** All	**Closed Loop Malfunction Detected (Bank 2):** Engine started, engine running in closed loop for over 2 minutes, and the PCM determined the engine was not operating in closed loop mode after the additional driving period had elapsed. **Possible Causes:** • HO2S signal circuit is open or shorted to ground • HO2S is damaged, contaminated or has failed • HO2S heater is damaged or has failed

DTC	Trouble Code Title, Conditions & Possible Causes
DTC: P1210 **2T CCM, MIL: No** **1996-2001** **Models:** G20, I30, J30, Q45 **Engines:** All **Transmissions:** All	**Traction Control System Signal Missing:** Engine started, engine runtime over 3 seconds, and the PCM detected an excessively high or excessively low Traction Control System (TCS) signal from the Throttle Actuator Control module. **Possible Causes:** • Traction Control communication circuit is open or shorted • Throttle Actuator Control (TAC) module has failed
DTC: P1211 **2T CCM, MIL: No** **2000-06** **Models:** FX35, G35, I30, I35, M35, M45, Q45, QX56 **Engines:** All **Transmissions:** All	**VDC/TCS/ABS Control Unit Malfunction:** Key on or engine running; and the PCM detected an unexpected signal from the VDC/TCSABS control unit, or it detected the Traction Control System fuel-cut operation continued for too long of a time. **Possible Causes:** • Traction Control System (TCS) related parts have failed • VDCTCS/ABS controller is damaged or has failed
DTC: P1212 **2T CCM, MIL: No** **2000-06** **Models:** FX35, G35, I30, I35, M35, M45, Q45, QX56 **Engines:** All **Transmissions:** All	**VDC/TCS/ABS Communication Line Malfunction:** Key on or engine running; and the PCM detected a continuous incorrect signal (digital pulse signal) from the VDC/TCS/ABS unit. **Possible Causes:** • VDCTCS/ABS communication line is open or shorted to ground • VDCTCS/ABS controller has failed • System voltage is too low (a severely discharged battery)
DTC: P1217 **1T CCM, MIL: Yes** **2000-06** **Models:** FX35, G35, I30, I35, M35, M45, Q45, QX56 **Engines:** All **Transmissions:** All	**Engine Over-Temperature Condition:** Engine running in closed loop, and the PCM detected an engine overheated (engine over temperature) condition for too long a time. **Possible Causes:** • Check cooling system components (i.e., hoses, cap, coolant) • Check the thermostat operation (it may be stuck partly closed) • Engine cooling fan circuit(s) open or shorted to ground • Engine cooling fan is damaged or has failed
DTC: P1217 **1T CCM, MIL: Yes** **2002-03** **Models:** QX4 **Engines:** All **Transmissions:** All	**Engine Over-Temperature Condition:** Engine running in closed loop, and the PCM detected an engine overheated (engine over temperature) condition for too long a time. **Possible Causes:** • Check cooling system components (i.e., hoses, cap, coolant) • Check the thermostat operation (it may be stuck partly closed) • Engine cooling fan circuit(s) open or shorted to ground • Engine cooling fan is damaged or has failed
DTC: P1220 **2T CCM, MIL: Yes** **1996-2005** **Models:** FX45, M45, Q45 **Engines:** All **Transmissions:** All	**Fuel Pump Control Module Circuit Malfunction:** Engine runtime over 16 seconds, and the PCM detected an incorrect signal from the Fuel Pump Control Module during the CCM test. **Possible Causes:** • FPCM signal circuit to the PCM open or shorted to ground • FPCM signal circuit to the PCM shorted to system power (B+) • FPCM signal circuit dropping resistor is damaged or has failed • FPCM is damaged or has failed
DTC: P1223 **2T CCM, MIL: No** **2002-03** **Models:** I35 **Engines:** All **Transmissions:** All	**Throttle Position Sensor 2 Circuit Low Input:** Key on or engine running; and the PCM detected an unexpected low voltage condition on the Throttle Position Sensor 2 signal circuit. The ETC Actuator consists of the ETC motor, TP1 and TP2 sensors. The sensors send a throttle position signal to the PCM. **Possible Causes:** • TP2 sensor connector is damaged or loose • TP2 sensor signal circuit is open or shorted to ground • TP2 sensor power circuit is open • TP2 sensor is damaged or has failed
DTC: P1224 **2T CCM, MIL: No** **2002-03** **Models:** I35 **Engines:** All **Transmissions:** All	**Throttle Position Sensor 2 Circuit High Input:** Key on or engine running; and the PCM detected an unexpected high voltage condition on the Throttle Position Sensor 2 signal circuit. The ETC Actuator consists of the ETC motor, TP1 and TP2 sensors. The sensors send a throttle position signal to the PCM. **Possible Causes:** • TP2 sensor connector is damaged or loose • TP2 sensor signal circuit is shorted to VREF or system power • TP2 sensor is damaged or has failed

DTC	Trouble Code Title, Conditions & Possible Causes
DTC: P1225 **2T CCM, MIL: No** **2002-06** **Models:** FX35, G35, I35, M35, M45, Q45, QX56 **Engines:** All **Transmissions:** All	**Closed Throttle Learning Position Range/Performance:** Engine started, system voltage over 10.0v, and the PCM detected the Closed Throttle Learning Position voltage was excessively low in the test. The ETC Actuator consists of the ETC motor, TP1 and TP2 sensors. These sensors send a throttle position signal to the PCM. **Possible Causes:** • Electronic Throttle Control Actuator is damaged or has failed (i.e., the TP1 and TP2 sensor assemblies in the ETCA failed).
DTC: P1226 **2T CCM, MIL: No** **2002-06** **Models:** FX35, G35, I35, M35, M45, Q45, QX56 **Engines:** All **Transmissions:** All	**Closed Throttle Learning Position Range/Performance:** Key on or engine running; system voltage over 10.0v, and the PCM determined that the Closed Throttle Learning Position step was not performed properly several times during the CCM test period. **Possible Causes:** • Electronic Throttle Control Actuator is damaged or has failed (i.e., the TP1 and TP2 sensor assemblies in the ETCA failed).
DTC: P1227 **2T CCM, MIL: No** **2002-03** **Models:** I35 **Engines:** All **Transmissions:** All	**Accelerator Pedal Position Sensor 2 Circuit Low Input:** Engine started, system voltage over 10.0v, and the PCM detected an unexpected low voltage condition on the Accelerator Pedal Position (APP) Sensor 2 signal circuit during the CCM test. The APP2 sensor is installed on the upper end of the throttle unit. The sensor detects the accelerator position and sends it to the PCM. **Possible Causes:** • APP2 sensor connector is damaged or disconnected (open) • APP2 sensor signal circuit is open • APP2 sensor is damaged or has failed
DTC: P1228 **2T CCM, MIL: No** **2002-03** **Models:** I35 **Engines:** All **Transmissions:** All	**Accelerator Pedal Position Sensor 2 Circuit High Input:** Engine started, system voltage over 10.0v, and the PCM detected an unexpected high voltage condition on the Accelerator Pedal Position (APP) Sensor 2 signal circuit during the CCM test. The APP2 sensor is installed on the upper end of the throttle unit. The sensor detects the accelerator position and sends it to the PCM. **Possible Causes:** • APP2 sensor connector is damaged (shorted to VREF) • APP2 sensor signal circuit is shorted to VREF • APP2 sensor is damaged or has failed
DTC: P1229 **1T CCM, MIL: Yes** **2002-05** **Models:** FX35, G35, I35, Q45, QX56 **Engines:** All **Transmissions:** All	**Accelerator Pedal Position Sensor 2 Power Circuit Short:** Engine started, system voltage over 10.0v, and the PCM detected an unexpected high or low voltage condition on the Accelerator Pedal Position (APP) Sensor 2 power source (VREF) circuit. The APP2 sensor is installed on the upper end of the throttle unit. The sensor detects the accelerator position and sends it to the PCM. **Possible Causes:** • AAP1 and/or APP2 sensor circuits are shorted to ground • APP2 sensor connector is damaged (shorted to ground) • ETC Actuator assembly is internally shorted to power or ground • TP1 and TP2 sensor circuits are shorted together or to ground • PCM terminal to the VREF power circuit is open • VREF power circuit to the APP1, APP2, MAF, EVAP pressure sensor, PSPS or refrigerant pressure sensor shorted to ground
DTC: P1320 **2T CCM, MIL: Yes** **1996-2003** **Models:** G20, I30, I35, J30, M45, Q45, QX4 **Engines:** All **Transmissions:** All	**Ignition Control Signal Circuit Malfunction:** Engine started, engine runtime over 20 seconds, and the PCM did not detect any IC signals from the power transistor during the test. **Possible Causes:** • Ignition coil primary circuit is open or shorted • Ignition coil condenser is shorted • CKP sensor signal is erratic, missing or the sensor has failed • CMP sensor signal is erratic, missing or the sensor has failed • TSB 01-052 (9/01) contains information related to this code
DTC: P1335 **2T CCM, MIL: Yes** **1996-2003** **Models:** FX45, G20, I30, I35, J30, M45, Q45, QX4 **Engines:** All **Transmissions:** All	**Crankshaft Position Reference Sensor Circuit Malfunction:** Engine cranking; and the PCM did not receive the CKP sensor 120° reference signals, or with the engine running, it did not detect any 120° reference signals, or it detected the signals changed too much. **Possible Causes:** • CKP sensor connector is damaged, loose or shorted • CKP reference signal circuit is open or shorted to ground • CKP reference sensor is damaged or has failed • Starter motor or starter motor circuit is damaged (not working)

DTC	Trouble Code Title, Conditions & Possible Causes
DTC: P1336 **2T CCM, MIL: Yes** **1996-2003** **Models:** FX35, G20, G35, I30, I35, J30, M45, Q45, QX4 **Engines:** All **Transmissions:** All	**Crankshaft Position Sensor Circuit Malfunction:** Engine started, engine runtime more than 5 seconds, and the PCM detected that the crankshaft position (CKP POS) signals were erratic or missing during the CCM continuous test. **Possible Causes:** • CKP sensor connector is damaged, loose or shorted • CKP POS signal circuit is open or shorted to ground • Crankshaft position (POS) sensor is damaged or has failed • Engine flywheel or drive plate gear tooth may be chipped or damaged (causing erratic CKP signals - check with Lab Scope)
DTC: P1400 **2T CCM, MIL: Yes** **1996-97** **Models:** G20, I30, J30, Q45, QX4 **Engines:** All **Transmissions:** All	**EGR/EVAP Control Solenoid Circuit Malfunction:** Engine started, engine running at cruise speed under light engine load, and the PCM detected an unexpected voltage on the EGR/EVAP solenoid control circuit during the CCM test. **Possible Causes:** • EGR/EVAP solenoid connector is damaged, open or shorted • EGR/EVAP solenoid control circuit open or shorted to ground • EGR/EVAP solenoid is damaged or has failed • PCM has failed
DTC: P1400 **2T CCM, MIL: Yes** **1998-2003** **Models:** FX45, G20, I30, J30, M45, Q45, QX4 **Engines:** All **Transmissions:** All	**EGR Control Solenoid Circuit Malfunction:** Engine started, engine running at cruise speed under light engine load, and the PCM detected an improper voltage signal from the EGR Control solenoid while the solenoid was being operated. **Possible Causes:** • EGRC solenoid connector is damaged, open or shorted • EGRC solenoid control circuit is open or shorted to ground • EGRC solenoid control circuit is shorted to system power (B+) • EGRC solenoid is damaged or has failed • PCM has failed
DTC: P1401 **2T CCM, MIL: Yes** **1996-2003** **Models:** FX45, G20, I30, J30, M45, Q45, QX4 **Engines:** All **Transmissions:** All	**EGR Temperature Sensor Circuit Malfunction:** Engine started, engine running at light load in closed loop at over 5 mph, and the PCM detected the EGR Temperature sensor input was too low or too high when compared to the ECT sensor input. **Possible Causes:** • EGR temperature sensor connector is damaged or shorted • EGR temperature sensor signal circuit open/shorted to ground • EGR temperature sensor signal circuit shorted to VREF • EGR temperature sensor is damaged or has failed
DTC: P1402 **2T EGR, MIL: Yes** **1998-2003** **Models:** FX45, G20, I30, J30, M45, Q45, QX4 **Engines:** All **Transmissions:** All	**EGR System Malfunction (Open) Detected:** Engine started, engine running at over 5 mph at light engine conditions, and the PCM detected excessive EGR gas flow. **Possible Causes:** • EGR valve seat leaking or EGR valve stuck in open position • EGRC solenoid is damaged or has failed • EGR temperature sensor is damaged or has failed • EGRC-BPT valve is damaged or has failed
DTC: P1440 **2T EVAP, MIL: Yes** **1998-2001** **Models:** G20, I30, J30, M45, Q45, QX4 **Engines:** All **Transmissions:** All	**EVAP Canister System Small Leak (0.040") Detected:** Engine started, vehicle driven at cruise speed at light engine load, and the PCM detected a small leak present in the EVAP system with the Purge and Canister Control Vent solenoids "on" in the Leak test. **Possible Causes:** • Fuel filler cap loose, damaged or incorrect application • EVAP canister vent valve is damaged or contaminated • EVAP system leaks (in canister or related vapor lines) • EVAP control system vapor pressure sensor line is blocked • MAP/BARO switch solenoid valve is damaged or has failed • MAP/BARO switch solenoid vacuum line is blocked or bent • Fuel tank temperature sensor is damaged or has failed • Absolute pressure sensor is damaged or has failed

DTC	Trouble Code Title, Conditions & Possible Causes
DTC: P1440 **2T EVAP, MIL: Yes** **2002-03** **Models:** G20, Q45 **Engines:** All **Transmissions:** All	**EVAP Canister System Small Leak (0.040") Detected:** Engine started, vehicle driven at cruise speed at light engine load, and the PCM detected a small leak present in the EVAP system with the Purge and Canister Control Vent solenoids "on" in the Leak test. **Possible Causes:** • Fuel filler cap loose, damaged or incorrect application • EVAP canister vent valve is damaged or contaminated • EVAP system leaks (in canister or related vapor lines) • EVAP control system vapor pressure sensor line is blocked • MAP/BARO switch solenoid valve is damaged or has failed • MAP/BARO switch solenoid vacuum line is blocked or bent • Fuel tank temperature sensor is damaged or has failed • Absolute pressure sensor is damaged or has failed
DTC: P1441 **2T CCM, MIL: Yes** **1998-2002** **Models:** G20, I30, Q45, QX4 **Engines:** All **Transmissions:** All	**Vacuum Cut Valve Bypass Valve Circuit Malfunction:** Engine started, vehicle driven at cruise speed at light engine load, and the PCM detected an unexpected voltage on the EVAP Vacuum Cut Valve Bypass circuit during the CCM test. **Possible Causes:** • EVAP vacuum cut valve bypass valve circuit is open • EVAP vacuum cut valve bypass valve circuit shorted to ground • EVAP vacuum cut valve bypass valve is damaged or has failed • PCM has failed
DTC: P1443 **2T CCM, MIL: Yes** **1996-2002** **Models:** G20, I30, J30, Q45, QX4 **Engines:** All **Transmissions:** All	**Canister Control Vacuum Switch Circuit Malfunction:** Engine started, vehicle driven at cruise speed at light engine load, and the PCM detected the EVAP Canister Control Vacuum Switch indicated off even without any vacuum supplied to the Purge valve. **Possible Causes:** • EVAP canister control vacuum switch circuit is open • EVAP canister control vacuum switch is damaged or has failed • PCM has failed
DTC: P1444 **2T CCM, MIL: Yes** **1998-2005** **Models:** FX45, G20, I30, M45, Q45, QX4, QX56 **Engines:** All **Transmissions:** All	**Canister Purge Volume Control Valve Circuit Malfunction:** Engine started, vehicle driven at cruise speed at light engine load, and the PCM detected purge flow through the canister with the Purge Volume Control valve commanded closed during the test. **Possible Causes:** • EVAP control system pressure sensor is damaged or has failed • EVAP purge volume control valve is stuck open • EVAP canister vent control valve is damaged or has failed • EVAP canister is damaged, or hoses are connected incorrectly
DTC: P1444 **2T CCM, MIL: Yes** **2002-05** **Models:** FX35, G35, I35 **Engines:** All **Transmissions:** All	**Canister Purge Volume Control Valve Circuit Malfunction:** Engine started, vehicle driven at cruise speed at light engine load, and the PCM detected purge flow through the canister with the Purge Volume Control valve commanded closed during the test. **Possible Causes:** • EVAP control system pressure sensor is damaged or has failed • EVAP purge volume control valve is stuck open • EVAP canister vent control valve is damaged or has failed • EVAP canister is damaged, or hoses are connected incorrectly
DTC: P1445 **2T CCM, MIL: Yes** **1996-97** **Models:** I30 **Engines:** All **Transmissions:** All	**EVAP Purge Volume Control Valve Range/Performance:** Engine started, vehicle driven at cruise speed at light engine load, and the PCM detected an unexpected voltage on the Purge Volume Control valve circuit, or it detected canister purge flow in the system with the EVAP Canister Purge Volume Control valve turned "off". **Possible Causes:** • EVAP canister purge volume control valve is stuck open • EVAP canister purge control valve is damaged or has failed • EVAP control system pressure sensor is damaged or has failed • EVAP canister is damaged, or hoses are connected incorrectly
DTC: P1446 **2T EVAP, MIL: Yes** **1998-2005** **Models:** FX45, G20, I30, M45, Q45, QX4, QX56 **Engines:** All **Transmissions:** All	**EVAP Canister Vent Valve Closed Malfunction:** Engine started, vehicle driven at cruise speed for 1-2 minutes, and the PCM detected the EVAP Canister Vent Valve remained closed. **Possible Causes:** • EVAP canister vent control valve is damaged or has failed • EVAP canister vent control valve rubber tube is blocked • EVAP control system pressure sensor is damaged or has failed • EVAP water separator is damaged or has failed • EVAP canister is saturated with water

DTC	Trouble Code Title, Conditions & Possible Causes
DTC: P1446 **2T EVAP, MIL: Yes** **2002-05** **Models:** FX35, G35, I35 **Engines:** All **Transmissions:** All	**EVAP Canister Vent Valve Closed Malfunction:** Engine started, vehicle driven at cruise speed for 1-2 minutes, and the PCM detected the EVAP Canister Vent Valve remained closed. **Possible Causes:** • EVAP canister vent control valve is damaged or has failed • EVAP canister vent control valve rubber tube is blocked • EVAP control system pressure sensor is damaged or has failed • EVAP water separator is damaged or has failed • EVAP canister is saturated with water
DTC: P1447 **2T EVAP, MIL: Yes** **1998-2003** **Models:** FX45, G20, I30, M45, Q45, QX4 **Engines:** All **Transmissions:** All	**EVAP Control System Purge Flow Malfunction:** Engine started, vehicle driven at cruise for 1 minute, and the PCM detected a leak between the intake manifold and the EVAP pressure sensor, or that the EVAP control system did not purge properly. **Possible Causes:** • EVAP canister purge volume control valve stuck closed • EVAP canister purge control valve stuck closed • EVAP control pressure sensor is damaged or has failed • EVAP canister vent control valve is damaged or has failed • EVAP rubber tube is blocked, disconnected or bent • EVAP canister purge control solenoid is damaged or has failed • MAP/BARO switch solenoid vacuum line is blocked or bent • EVAP canister is leaking (it may be cracked) • EVAP 1-way valve connected improperly or purge port blocked
DTC: P1448 **2T EVAP, MIL: Yes** **1998-2005** **Models:** FX45, G20, I30, M45, Q45, QX4 **Engines:** All **Transmissions:** All	**EVAP Canister Vent Control Valve Malfunction (Open):** DTC P0442 not set, engine started, vehicle driven at cruise speed at light engine load, and the PCM detected the EVAP Canister Vent Control Valve remained open after the PCM commanded it closed. **Possible Causes:** • EVAP canister vent control valve is damaged or has failed • EVAP control pressure sensor is damaged or has failed • EVAP canister vent control valve rubber tube is blocked • EVAP water separator has failed or canister is full of water • EVAP vacuum cut valve is damaged or has failed • TSB 00-046 (7/00) contains information related to this code
DTC: P1448 **2T EVAP, MIL: Yes** **2002-04** **Models:** I35 **Engines:** All **Transmissions:** All	**EVAP Canister Vent Control Valve Malfunction (Open):** DTC P0442 not set, engine started, vehicle driven at cruise speed at light engine load, and the PCM detected the EVAP Canister Vent Control Valve remained open after the PCM commanded it closed. **Possible Causes:** • EVAP canister vent control valve is damaged or has failed • EVAP control pressure sensor is damaged or has failed • EVAP canister vent control valve rubber tube is blocked • EVAP water separator has failed or canister is full of water • EVAP vacuum cut valve is damaged or has failed
DTC: P1456 **2T EVAP, MIL: Yes** **2002-04** **Models:** I35 **Engines:** All **Transmissions:** All	**EVAP System Very Small Leak (0.020") Detected:** IAT sensor signal from 32-86ºF, ECT sensor signal from 32-158ºF, fuel level from 25-75%, engine started, vehicle driven with the EVAP purge and vent solenoids both enabled, and the PCM detected the presence of a very small leak in the EVAP system in the leak test. **Possible Causes:** • EVAP control system pressure sensor is damaged or has failed • EVAP purge solenoid is damaged or has failed • EVAP canister vent control valve is damaged or has failed • Fuel tank cap not tightened properly or the wrong part number • Fuel tank temperature sensor is damaged or has failed • Fuel tank vacuum relief valve is damaged or has failed • Very small leak in the EVAP canister or in the fuel tank area • ORVR system is damaged or has a small leak in a component • Refueling control valve is damaged or has failed • TSB 00-046 (7/00) contains information related to this code

DTC	Trouble Code Title, Conditions & Possible Causes
DTC: P1456 **2T EVAP, MIL: Yes** **2003-05** **Models:** Q45 **Engines:** All **Transmissions:** All	**EVAP System Very Small Leak (0.020") Detected:** IAT sensor signal from 32-86ºF, ECT sensor signal from 32-158ºF, fuel level from 25-75%, engine started, vehicle driven with the EVAP purge and vent solenoids both enabled, and the PCM detected the presence of a very small leak in the EVAP system in the leak test. **Possible Causes:** • EVAP control system pressure sensor is damaged or has failed • EVAP purge solenoid is damaged or has failed • EVAP canister vent control valve is damaged or has failed • Fuel tank cap not tightened properly or the wrong part number • Fuel tank temperature sensor is damaged or has failed • Fuel tank vacuum relief valve is damaged or has failed • Very small leak in the EVAP canister or in the fuel tank area • ORVR system is damaged or has a small leak in a component • Refueling control valve is damaged or has failed • TSB 00-046 (7/00) contains information related to this code
DTC: P1464 **2T CCM, MIL: Yes** **2000-03** FX45, G20, I30, I35, M45, Q45, **Models:** QX4 **Engines:** All **Transmissions:** All	**Fuel Level Sensor Ground Circuit Malfunction:** Key on or engine running; and the PCM detected an unexpected high voltage signal on the Fuel Level sensor ground circuit during the CCM test. This circuit normally reads near 0.00v with the key on. **Possible Causes:** • Fuel level sensor signal circuit is open or grounded • Fuel level sensor circuit is shorted to VREF • Fuel level sensor is damaged or has failed
DTC: P1490 **2T CCM, MIL: Yes** **1998-2005** **Models:** FX45, G20, I30, I35, M45, Q45, QX4 **Engines:** All **Transmissions:** All	**EVAP Vacuum Cut Valve Bypass Valve Circuit Malfunction:** Key on or engine running; and the PCM detected an unexpected voltage condition on the EVAP Vacuum Cut Valve Bypass Valve circuit during the CCM test. **Possible Causes:** • EVAP vacuum cut valve bypass valve connector is damaged • EVAP vacuum cut valve bypass valve control circuit is open • EVAP vacuum cut valve bypass valve control circuit is shorted • EVAP vacuum cut valve bypass valve is damaged or has failed • PCM is damaged
DTC: P1491 **2T CCM, MIL: Yes** **1998-2005** **Models:** FX45, G20, I30, I35, M45, Q45, QX4 **Engines:** All **Transmissions:** All	**EVAP Vacuum Cut Valve/Bypass Range/Performance:** Engine started, ECT sensor signal from 41-86ºF, engine at idle speed or 70 seconds, and the PCM detected the Vacuum Vent Cut Valve/Bypass Valve did not operate properly during the CCM test. **Possible Causes:** • EVAP vacuum cut bypass valve is damaged or has failed • EVAP vacuum cut valve is damaged or has failed • EVAP control system pressure sensor damaged or has failed • Vapor hose between fuel tank and cut valve is blocked or bent • Vapor hose between canister and vacuum cut valve is plugged • EVAP canister is damaged or EVAP purge port at tank is dirty • TSB 02-015 (3/02) contains information that relates to this code
DTC: P1492 **2T CCM, MIL: Yes** **1998-2003** **Models:** FX45, G20, I30, M45, Q45, QX4 **Engines:** All **Transmissions:** All	**EVAP Canister Purge Control Solenoid Circuit Malfunction:** Key on or engine running; and the PCM detected an unexpected voltage on the Canister Purge Control Valve/Solenoid Valve circuit. **Possible Causes:** • EVAP canister purge control solenoid control circuit is open • EVAP canister purge control solenoid control circuit is shorted • EVAP canister purge control solenoid is damaged or has failed • PCM has failed
DTC: P1493 **2T CCM, MIL: Yes** **1998-2003** **Models:** FX45, G20, I30, M45, Q45, QX4 **Engines:** All **Transmissions:** All	**EVAP Canister Purge Control Solenoid Performance (Stuck Open):** Engine running in closed loop under light engine load conditions, and the PCM detected that the EVAP Canister Purge Control Valve did not operate correctly (it may be stuck in open position). **Possible Causes:** • EVAP canister purge control valve is damaged or has failed • EVAP canister purge control solenoid valve is damaged/failed • EVAP vacuum hoses are clogged or disconnected • EVAP canister vent control valve is damaged or has failed • EVAP canister is full of water, or the vapor separator has failed

DTC	Trouble Code Title, Conditions & Possible Causes
DTC: P1564 **1T CCM, MIL: No** **2002-06** **Models:** FX35, FX45, G35, I35, M35, M45, Q45, QX56 **Engines:** All **Transmissions:** All	**ASCD Steering Switch Circuit Malfunction:** Engine started, and PCM detected an unexpected high voltage condition on the ASCD Steering switch circuit, or the ASCD Steering switch signal was out of its specified range, or the Steering switch signal indicated it was always stuck "on". The status of the ASCD is displayed in the Combination Meter (Cruise or Set). **Possible Causes:** • ASCD steering switch connector is damaged or loose • ASCD steering switch signal circuit is open • ASCD steering switch signal circuit is shorted to ground • ASCD steering switch is damaged or has failed
DTC: P1568 **1T CCM, MIL: No** **2002-06** **Models:** FX35, FX45, G35, I35, M35, M45, Q45, QX56 **Engines:** All **Transmissions:** All	**ASCD Command Valve Circuit Malfunction:** If DTC P1568 is displayed with DTC U1000 and/or U1001, diagnose the cause of the DTC U1000 and/or U1001 first. If DTC P1568 is displayed with DTC P0605, diagnose the cause of DTC P0605 first. Engine started, and PCM detected that the signals from the command valve unit (ICC unit) were out of range. **Possible Causes:** • CAN communication line is open, shorted to ground or power • ICC unit is damaged or has failed • PCM has failed
DTC: P1572 **1T CCM, MIL: No** **2002-06** **Models:** FX35, FX45, G35, I35, M35, M45, Q45, QX56 **Engines:** All **Transmissions:** All	**ASCD Brake Switch Circuit Malfunction:** DTC P0605, P1706 and P1805 not set, engine started, vehicle driven at over 19 mph, and the PCM detected an "on" signal from both the ASCD and brake switches at the same time. The status of the ASCD is displayed in the Combination Meter (Cruise or Set). **Possible Causes:** • ASCD switch connector is damaged, loose or shorted • ASCD or stop lamp switch is incorrectly installed or positioned • ASCD or stop lamp switch circuit is open or shorted to ground • ASCD or stop lamp switch is damaged or has failed
DTC: P1574 **1T CCM, MIL: No** **2002-06** **Models:** FX35, FX45, G35, I35, M35, M45, Q45, QX56 **Engines:** All **Transmissions:** All	**ASCD Vehicle Speed Sensor Signal Malfunction:** DTC P0500, P0605, U1000 and U1001 not set, engine started, vehicle driven at over 25 mph, and the PCM detected a difference in the VSS signals from the Combination Meter and TCM. The status of the ASCD is displayed in the Combination Meter (Cruise or Set). **Possible Causes:** • CAN communication connector is damaged or loose (open) • CAN communication line is open, shorted to power or to ground • Combination Meter VSS circuit is open or shorted to ground • Combination Meter is damaged or has failed • Vehicle Speed sensor is damaged or has failed
DTC: P1605 **1T CCM, MIL: Yes** **1996-2003** **Models:** G20, I30, J30, M45, Q45, QX4 **Engines:** All **Transmissions:** A/T	**TCM A/T Diagnosis Communication Line Malfunction:** Engine runtime over 30 seconds, system voltage more than 10.5v, and the PCM detected an incorrect or no signal on the A/T DLC line. **Possible Causes:** • TCM communication line circuit is open or shorted to ground • TCM communication line circuit shorted to VREF or power (B+) • TCM or PCM has failed • TSB 01-002 (1/01) contains information related to this code
DTC: P1610 **2T CCM, MIL: No** **PATS: Yes** **2000-06** **Models:** FX35, FX45, G20, G35, I30, I35, M35, M45, Q45, QX4, QX56 **Engines:** All **Transmissions:** A/T	**NATS Initialization Failure:** Key on or engine cranking, system voltage over 10.5v, and the PCM detected a problem in the Nissan Antitheft System (NATS). **Possible Causes:** • Refer to the information that explains how to diagnose a problem in the NATS (system) - use a compatible Scan Tool.
DTC: P1611 **2T CCM, MIL: No** **PATS: Yes** **2000-06** **Models:** FX35, FX45, G20, G35, I30, I35, M35, M45, Q45, QX4, QX56 **Engines:** All **Transmissions:** A/T	**NATS ID DISCORD, IMM-ECM:** Key on or engine cranking, system voltage over 10.5v, and the PCM detected a problem in the Nissan Antitheft System (NATS). **Possible Causes:** • Refer to the information that explains how to diagnose a problem in the NATS (system) - use a compatible Scan Tool.

DTC	Trouble Code Title, Conditions & Possible Causes
DTC: P1612 **2T CCM, MIL: No** **PATS: Yes** **2000-06** **Models:** FX35, FX45, G20, G35, I30, I35, M35, M45, Q45, QX4, QX56 **Engines:** All **Transmissions:** A/T	**NATS CHAIN OF IMMU-ECM:** Key on or engine cranking, system voltage over 10.5v, and the PCM detected a problem in the Nissan Antitheft System (NATS). **Possible Causes:** • Refer to the information that explains how to diagnose a problem in the NATS (system) - use a compatible Scan Tool.
DTC: P1613 **2T CCM, MIL: No** **PATS: Yes** **2000-06** **Models:** FX35, FX45, G20, G35, I30, I35, M35, M45, Q45, QX4, QX56 **Engines:** All **Transmissions:** A/T	**NATS ECM INT CIRC-IMMU:** Key on or engine cranking, system voltage over 10.5v, and the PCM detected a problem in the Nissan Antitheft System (NATS). **Possible Causes:** • Refer to the information that explains how to diagnose a problem in the NATS (system) - use a compatible Scan Tool.
DTC: P1614 **2T CCM, MIL: Yes** **PATS: Yes** **2000-06** **Models:** FX35, FX45, G20, G35, I30, I35, M35, M45, Q45, QX4, QX56 **Engines:** All **Transmissions:** A/T	**NATS CHAIN OF IMMU-KEY:** Key on or engine cranking, system voltage over 10.5v, and the PCM detected a problem in the Nissan Antitheft System (NATS). **Possible Causes:** • Refer to the information that explains how to diagnose a problem in the NATS (system) - use a compatible Scan Tool.
DTC: P1615 **2T CCM, MIL: Yes** **PATS: Yes** **2000-06** **Models:** FX35, FX45, G20, G35, I30, I35, M35, M45, Q45, QX4, QX56 **Engines:** All **Transmissions:** A/T	**NATS INDIFFERENCE OF KEY:** Key on or engine cranking, system voltage over 10.5v, and the PCM detected a problem in the Nissan Antitheft System (NATS). **Possible Causes:** • Refer to the information that explains how to diagnose a problem in the NATS (system) - use a compatible Scan Tool.
DTC: P1705 **1T CCM, MIL: Yes** **1996-2003** **Models:** FX45, G20, I30, J30, M45, Q45, QX4 **Engines:** All **Transmissions:** A/T	**Throttle Position Sensor Circuit Malfunction:** Engine runtime over 30 seconds, and the TCM detected the TP Switch input was too high or too low regardless of whether the Idle Switch was on or was off (accelerator pedal is depressed = off). **Possible Causes:** • TP sensor signal circuit is open between the sensor and TCM • TP sensor signal circuit is shorted between sensor and TCM • TP sensor is damaged or has failed • TCM has failed
DTC: P1705 **1T CCM, MIL: Yes** **2002-03** **Models:** FX35, G35, I35 **Engines:** All **Transmissions:** A/T	**Throttle Position Sensor Circuit Malfunction:** Engine runtime over 30 seconds, and the TCM detected the TP Switch input was too high or too low regardless of whether the Idle Switch was on or was off (accelerator pedal is depressed = off). **Possible Causes:** • TP sensor signal circuit is open between the sensor and TCM • TP sensor signal circuit is shorted between sensor and TCM • TP sensor is damaged or has failed • TSB 03-034 (5/03) contains information related to this code
DTC: P1706 **2T CCM, MIL: Yes** **1998-2005** **Models:** FX35, FX45, G20, G35, I30, I35, M45, Q45, QX4 **Engines:** All **Transmissions:** A/T	**A/T Park Neutral Position Switch Circuit Malfunction:** Engine started, vehicle driven to a speed of over 5 mph, and the TCM determined the PNP switch input was not plausible with the vehicle moving in a forward gear, or the signal did not change during the start to run transition. **Possible Causes:** • PNP switch signal circuit is open or shorted to ground • PNP switch is shorted to VREF or system power (B+) • PNP switch or the PNP relay is damaged or has failed • PCM has failed

DTC	Trouble Code Title, Conditions & Possible Causes
DTC: P1716 **2T CCM, MIL: Yes** **2003-03** **Models:** FX35, G35 **Engines:** All **Transmissions:** A/T	**A/T Turbine Revolution Sensor Circuit Malfunction:** Engine started, vehicle driven to a speed of over 5 mph, and the PCM detected an unexpected voltage condition on the A/T Turbine Revolution sensor circuit during the CCM test period. **Possible Causes:** • Turbine revolution sensor connector is damaged or loose • Turbine revolution signal circuit is open or shorted to ground • Turbine revolution sensor is damaged or has failed • TSB 02-074A (1/03) contains information related to this code
DTC: P1716 **2T CCM, MIL: Yes** **2004-05** **Models:** Q45, QX56 **Engines:** All **Transmissions:** A/T	**A/T Turbine Revolution Sensor Circuit Malfunction:** Engine started, vehicle driven to a speed of over 5 mph, and the PCM detected an unexpected voltage condition on the A/T Turbine Revolution sensor circuit during the CCM test period. **Possible Causes:** • Turbine revolution sensor connector is damaged or loose • Turbine revolution signal circuit is open or shorted to ground • Turbine revolution sensor is damaged or has failed • TSB 02-074A (1/03) contains information related to this code
DTC: P1730 **1T CCM, MIL: Yes** **2003-06** **Models:** FX35, G35, M35, M45, Q45, QX56 **Engines:** All **Transmissions:** A/T	**A/T Interlock System Circuit Malfunction:** Key on or engine cranking; and the PCM detected an unexpected voltage condition on the A/T Interlock System circuit during testing. **Possible Causes:** • A/T Interlock system connector is damaged or loose • A/T Interlock system circuit is open or shorted to ground • A/T Interlock system is damaged or has failed
DTC 1752 **1T CCM, MIL: Yes** **2003-06** **Models:** FX35, G35, M35, M45, Q45, QX56 **Engines:** All **Transmissions:** A/T	**A/T Turbine Revolution Sensor Circuit Malfunction:** Engine started, vehicle driven to a speed of over 5 mph, and the PCM detected an unexpected voltage condition on the A/T Turbine Revolution sensor circuit during the CCM test period. **Possible Causes:** • Turbine revolution sensor connector is damaged or loose • Turbine revolution signal circuit is open or shorted to ground • Turbine revolution sensor is damaged or has failed
DTC: P1754 **1T CCM, MIL: Yes** **1996-2002** **Models:** G20, I30, J30, M45, Q45 **Engines:** All **Transmissions:** A/T	**A/T Overrun Clutch Solenoid Valve Circuit Malfunction:** Engine started, engine runtime over 30 seconds, and the TCM detected an incorrect voltage condition while operating the Overrun Clutch Solenoid valve. **Possible Causes:** • Overrun clutch solenoid control circuit is open • Overrun clutch solenoid control circuit is shorted to ground • Overrun clutch solenoid is damaged or has failed • TCM has failed
DTC: P1754 **1T CCM, MIL: Yes** **2002-06** **Models:** FX35, FX45 G35, I35, M35, M45, Q45, QX56 **Engines:** All **Transmissions:** A/T	**A/T I/C Solenoid Valve Circuit Malfunction:** Engine started, engine runtime over 30 seconds, and PCM detected an unexpected voltage condition during the operation of the A/T I/C Solenoid valve circuit. **Possible Causes:** • I/C solenoid connector is damaged or has failed • I/C solenoid control circuit is open or shorted to ground • I/C solenoid is damaged or has failed • TCM or PCM has failed
DTC: P1757 **1T CCM, MIL: Yes** **2002-06** **Models:** FX35, FX45, G35, I35, M35, M45, Q45, QX56 **Engines:** All **Transmissions:** A/T	**A/T FR/B Solenoid Valve Circuit Malfunction:** Engine started, engine runtime over 30 seconds, and PCM detected an unexpected voltage condition on the A/T FR/B solenoid circuit. **Possible Causes:** • FR/B solenoid connector is damaged or has failed • FR/B solenoid control circuit is open or shorted to ground • FR/B solenoid is damaged or has failed • TCM or PCM has failed

DTC	Trouble Code Title, Conditions & Possible Causes
DTC: P1759 **1T CCM, MIL: Yes** **2002-06** **Models:** FX35, FX45, G35, I35, M35, M45, Q45, QX56 **Engines:** All **Transmissions:** A/T	**A/T FR/B Solenoid Valve Performance:** Engine started, engine runtime over 30 seconds, and PCM detected incorrect operation of the A/T FR/B solenoid circuit during the test. **Possible Causes:** • FR/B solenoid connector is damaged or has failed • FR/B solenoid control circuit is open or shorted to ground • FR/B solenoid is damaged or has failed • TCM or PCM has failed
DTC: P1760 **1T CCM, MIL: Yes** **1997-2003** **Models:** QX4 **Engines:** All **Transmissions:** A/T	**A/T Overrun Clutch Solenoid Valve Circuit Malfunction:** Engine started, vehicle driven at 55-65 mph and then back to idle, and the PCM detected an unexpected voltage condition on the A/T Overrun Clutch solenoid control circuit during the CCM test. **Possible Causes:** • O/R clutch solenoid connector is damaged or has failed • O/R clutch solenoid control circuit is open or shorted to ground • O/R clutch solenoid is damaged or has failed • TCM or PCM has failed
DTC: P1762 **1T CCM, MIL: Yes** **2002-06** **Models:** FX35, FX45, G35, I35, M35, M45, Q45, QX56 **Engines:** All **Transmissions:** A/T	**A/T D/C Solenoid Valve Circuit Malfunction:** Engine started, engine runtime over 30 seconds, and PCM detected an unexpected voltage condition on the A/T D/C solenoid circuit. **Possible Causes:** • D/C solenoid connector is damaged or has failed • D/C solenoid control circuit is open or shorted to ground • D/C solenoid is damaged or has failed • TCM or PCM has failed
DTC: P1764 **1T CCM, MIL: Yes** **2002-06** **Models:** FX35, FX45, G35, I35, M35, M45, Q45, QX56 **Engines:** All **Transmissions:** A/T	**A/T D/C Solenoid Valve Circuit Malfunction:** Engine started, engine runtime over 30 seconds, and PCM detected incorrect operation of the A/T D/C solenoid circuit during the test. **Possible Causes:** • D/C solenoid connector is damaged or has failed • D/C solenoid control circuit is open or shorted to ground • D/C solenoid is damaged or has failed • TCM or PCM has failed
DTC: P1767 **1T CCM, MIL: Yes** **2002-06** **Models:** FX35, FX45, G35, I35, M35, M45, Q45, QX56 **Engines:** All **Transmissions:** A/T	**A/T HLR/C Solenoid Valve Circuit Malfunction:** Engine started, engine runtime over 30 seconds, and PCM detected an unexpected voltage condition on the A/T HLR/C solenoid circuit. **Possible Causes:** • HLR/C solenoid connector is damaged or has failed • HLR/C solenoid control circuit is open or shorted to ground • HLR/C solenoid is damaged or has failed • TCM or PCM has failed
DTC: P1769 **1T CCM, MIL: Yes** **2002-06** **Models:** FX35, FX45, G35, I35, M35, M45, Q45, QX56 **Engines:** All **Transmissions:** A/T	**A/T HLR/C Solenoid Valve Circuit Malfunction:** Engine started, engine runtime over 30 seconds, and PCM detected incorrect operation of the A/T HLR/C solenoid circuit during the test. **Possible Causes:** • HLR/C solenoid connector is damaged or has failed • HLR/C solenoid control circuit is open or shorted to ground • HLR/C solenoid is damaged or has failed • TCM or PCM has failed
DTC: P1772 **1T CCM, MIL: Yes** **2002-06** **Models:** FX35, FX45, G35, I35, M35, M45, Q45, QX56 **Engines:** All **Transmissions:** A/T	**A/T LC/B Solenoid Valve Circuit Malfunction:** Engine started, engine runtime over 30 seconds, and PCM detected an unexpected voltage condition on the A/T HLR/C solenoid circuit. **Possible Causes:** • LC/B solenoid connector is damaged or has failed • LC/B solenoid control circuit is open or shorted to ground • LC/B solenoid is damaged or has failed • TCM or PCM has failed

DTC	Trouble Code Title, Conditions & Possible Causes
DTC: P1774 **1T CCM, MIL: Yes** **2002-06** **Models:** FX35, FX45, G35, I35, M35, M45, Q45, QX56 **Engines:** All **Transmissions:** A/T	**A/T LC/B Solenoid Valve Circuit Malfunction:** Engine started, engine runtime over 30 seconds, and PCM detected incorrect operation of the A/T HLR/C solenoid circuit during the test. **Possible Causes:** • LC/B solenoid connector is damaged or has failed • LC/B solenoid control circuit is open or shorted to ground • LC/B solenoid is damaged or has failed • TCM or PCM has failed
DTC: P1800 **2T CCM, MIL: Yes** **2002-04** **Models:** I35 **Engines:** All **Transmissions:** All	**VIAS Control Solenoid Circuit Low Input (Open):** Engine started, engine running at cruise speed, system voltage over 11.0v, and the PCM detected an unexpected low voltage condition on the VIAS Control solenoid circuit during the CCM test. The VIAS control solenoid valve cuts the intake manifold vacuum signal for power valve control. It is controlled by On/Off signals from the PCM. **Possible Causes:** • VIAS solenoid connector is damaged or loose (open circuit) • VIAS solenoid control circuit is open or shorted to ground • VIAS solenoid is damaged or has failed
DTC: P1805 **2T CCM, MIL: Yes** **2002-06** **Models:** FX35, FX45, G35, I35, M35, M45, Q45, QX56 **Engines:** All **Transmissions:** All	**A/T Brake Switch Circuit Malfunction:** Engine started, engine running at cruise speed for several minutes, system voltage over 11.0v, and the PCM detected a continuous "on" signal (12v) on the Brake Switch circuit. The Brake Switch signal is used to decrease the engine speed while the vehicle is moving. **Possible Causes:** • Stop lamp switch connector is damaged or loose (open circuit) • Stop lamp switch power circuit is open (test the No. 2 15A fuse) • Stop lamp switch is damaged or has failed (it may be shorted)
DTC: P1900 **2T CCM, MIL: Yes** **1996-97** **Models:** G20, I30, J30, Q45 **Engines:** All **Transmissions:** All	**Cooling Fan Control Circuit Open Malfunction:** DTC P0115 not set, engine run time over 5 seconds, hot engine condition present, and the PCM detected the Cooling Fan did not operate, or the Cooling Fan system did not operate. **Note: This trouble code can be set even with the coolant level okay.** **Possible Causes:** • Cooling fan control circuit is open or shorted to ground • Cooling fan is damaged or has failed • Cooling system component failure (i.e., radiator hose, radiator cap, water pump or thermostat may be stuck partially closed

OBD II Trouble Code List (P2XXX Codes)

DTC	Trouble Code Title, Conditions & Possible Causes
DTC: P2122 **1T CCM, MIL: Yes** **2003-06** **Models:** FX35, FX45 G35, M35, M45, Q45, QX56 **Engines:** All **Transmissions:** All	**Accelerator Pedal Position Sensor 1 Signal Low Input:** Key on or engine running; and the PCM detected an unexpected low voltage condition the APP1 sensor signal circuit during testing. **Possible Causes:** • APP1 sensor connector is damaged (it may be shorted) • APP1 sensor signal circuit is shorted to ground • APP1 sensor signal circuit is damaged or has failed • PCM is damaged or has failed
DTC: P2123 **1T CCM, MIL: Yes** **2003-06** **Models:** FX35, FX45 G35, M35, M45, Q45, QX56 **Engines:** All **Transmissions:** All	**Accelerator Pedal Position Sensor 1 Signal High Input:** Key on or engine running; and the PCM detected an unexpected high voltage condition the APP1 sensor signal circuit during testing. **Possible Causes:** • APP1 sensor connector is damaged (it may be open) • APP1 sensor signal circuit is shorted to VREF or system power • APP1 sensor signal circuit is damaged or has failed • TSB 01-064 (10/19/01) contains information related to this code
DTC: P2127 **1T CCM, MIL: Yes** **2003-06** **Models:** FX35,FX45 G35, M35, M45, Q45, QX56 **Engines:** All **Transmissions:** All	**Accelerator Pedal Position Sensor 2 Signal Low Input:** Key on or engine running; and the PCM detected an unexpected low voltage condition the APP2 sensor signal circuit during testing. **Possible Causes:** • APP2 sensor connector is damaged (it may be shorted) • APP2 sensor signal circuit is shorted to ground • APP2 sensor signal circuit is damaged or has failed • PCM is damaged or has failed
DTC: P2128 **1T CCM, MIL: Yes** **2003-06** **Models:** FX35,FX45 G35, M35, M45, Q45, QX56 **Engines:** All **Transmissions:** All	**Accelerator Pedal Position Sensor 2 Signal High Input:** Key on or engine running; and the PCM detected an unexpected high voltage condition the APP2 sensor signal circuit during testing. **Possible Causes:** • APP2 sensor connector is damaged (it may be open) • APP2 sensor signal circuit is shorted to VREF or system power • APP2 sensor signal circuit is damaged or has failed
DTC: P2135 **1T CCM, MIL: Yes** **2003** **2003-06** **Models:** FX35,FX45 G35, M35, M45, Q45, QX56 **Engines:** All **Transmissions:** All	**Throttle Position Sensor Range/Performance:** Engine started, and the PCM detected a signal that was not plausible when compared to the signals from the TP1 and TP2 sensors during the CCM test period. The ETC Actuator consists of a throttle control motor and a pair of throttle position sensors. The TP sensors respond to movement of the throttle valve. The PCM judges the actual opening angle of the throttle valve from the signals from the TP1 and TP2 sensors during engine operation. **Possible Causes:** • TP1 or TP2 sensor connector is damaged (open or shorted) • TP1 sensor signal circuit is open or shorted to ground • TP1 sensor signal circuit is damaged or has failed • TP2 sensor signal circuit is open or shorted to ground • TP2 sensor signal circuit is damaged or has failed • PCM is damaged or has failed
DTC: P2138 **1T CCM, MIL: Yes** **2003-06** **Models:** FX35,FX45 G35, M35, M45, Q45, QX56 **Engines:** All **Transmissions:** All	**Accelerator Pedal Position Sensor Range/Performance:** Engine started, and the PCM detected a signal that was not plausible when compared to the signals from the APP1 and APP2 sensors during the CCM test period. **Possible Causes:** • APP1 or APP2 sensor connector is damaged (open or shorted) • APP1 sensor signal circuit is open or shorted to ground • APP1 sensor signal circuit is damaged or has failed • APP2 sensor signal circuit is open or shorted to ground • APP2 sensor signal circuit is damaged or has failed • PCM is damaged or has failed

OBD II Trouble Code List (U1XXX Codes)

DTC	Trouble Code Title, Conditions & Possible Causes
DTC: U1000 **1T CCM, MIL: Yes** **2002-06** **Models:** FX35, FX45, G35, I35, M35, M45, QX4, Q45, QX56 **Engines:** All **Transmissions:** All	**Controller Area Network Line Malfunction:** Key on or engine running; and one or more of the controllers on the Controller Area Network (CAN) determined that it could not communicate with one or more of the other modules. The Controller Area Network (CAN) is a serial communication line for real-time application. It is an on-vehicle multiplex communication line with high data communication speed and excellent error detection ability. A vehicle can have several Electronic Control Modules, and each ECM shares information and links with other control units during vehicle operation (not independently). In CAN communication, control units are connected via two (2) communication lines (a CAN 'H' line and a CAN 'L' line) that allow for a high rate of information transmission with less wiring. Each ECM transmits and receives data, but still selectively reads only data that it requires. **Possible Causes:** • CAN connector is damaged or is disconnected • CAN communication line is open, shorted to ground or to power • One of the modules is shorted and pulling the CAN voltage low
DTC: U1001 **1T CCM, MIL: No** **2002-06** **Models:** FX35, FX45, G35, I35, M35, M45, QX4, Q45, QX56 **Engines:** All **Transmissions:** All	**Controller Area Network Line Malfunction:** Key on or engine running; and one or more of the controllers on the Controller Area Network (CAN) determined that it could not communicate for more than a specified time period. The Controller Area Network (CAN) is a serial communication line for real-time application. It is an on-vehicle multiplex communication line with high data communication speed and excellent error detection ability. A vehicle can have several Electronic Control Modules, and each ECM shares information and links with other control units during vehicle operation (not independently). In CAN communication, control units are connected via two (2) communication lines (a CAN 'H' line and a CAN 'L' line) that allow for a high rate of information transmission with less wiring. Each ECM transmits and receives data, but still selectively reads only data that it requires. **Possible Causes:** • CAN connector is damaged or is disconnected • CAN communication line is open, shorted to ground or to power • One of the modules is shorted and pulling the CAN voltage low

INFINITI
COMPONENT TESTING

8

TABLE OF CONTENTS

Component Testing 8-2
- Camshaft Position Sensor 8-2
- Crankshaft Position Sensor 8-2
- EGR Temperature Sensor 8-3
- Engine Control Module 8-3
- Engine Coolant Temperature Sensor 8-3
- Fuel Level Sensor 8-4
- Fuel Temperature Sensor 8-5
- Heated Oxygen Sensor 8-5
- Intake Air Temperature Sensor 8-6
- Knock Sensor 8-6
- Mass Air Flow Sensor 8-7
- Manifold Absolute Pressure Sensor 8-8
- Throttle Position Sensor 8-8
- Vehicle Speed Sensor 8-9

COMPONENT TESTING

Camshaft Position (CMP) Sensor

OPERATION

G20, I30, J30, Q45 and Q56

ALL ENGINES

See Figure 1.

The camshaft position sensor monitors the engine speed and position of the pistons. It sends signals to the ECM to control fuel injection, ignition timing and other functions. The camshaft position sensor has a rotor plate and a wave-forming circuit. The rotor plate has 360 slits for one degree signal and eight slits for 90 degree signal. Light Emitting Diodes (LED) and photo diodes are built into the wave-forming circuit. When the rotor plate passes between the LED and the photo diode, the slits in the rotor plate continually cut the transmitted light to the photo diode from the LED. This generates rough-shaped pulses converted into on-off pulses by the wave forming circuit sent to the ECM. For some models, the distributor is not repairable and must be replaced as an assembly except the cap and rotor head.

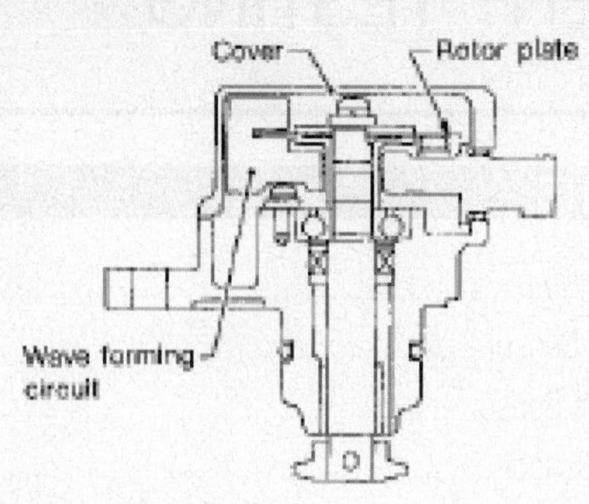

Fig. 1 The camshaft position sensor

TESTING

G20

2.0L SR20DE ENGINE

1. Start the engine and warm it up to normal operating temperature.
2. Check the voltage between terminals 75, 85 and engine ground.
3. At idle to 2000 rpm, the voltage between terminal 75 and ground should be 0.1–0.4 volts.
4. At idle, the voltage between terminal 85 and ground should be 2.5 volts.
5. At 2000 rpm, the voltage between terminal 85 and ground should be 2.4 volts.
6. If these readings are not met, replace the camshaft position sensor.

I30, I35 and J30

ALL ENGINES

1. Start the engine.
2. Check the voltage between terminals 3 and 4 and ground.
3. At idle, the voltage between terminal 3 and ground should be approximately 2.5 volts.
4. At idle, the voltage between terminal 4 and ground should be approximately 2.1 volts.
5. If these readings are not met, replace the camshaft position sensor.

Q45

4.1L VH41DE ENGINE

1. Start the engine and warm it up to the normal operating temperature.
2. Check the voltage between the ECM terminals 59 and 58 (or 55) and ground.
3. With the engine at idle, the voltage between terminal 59 and ground should be 2.5 volts.
4. With the engine at idle, the voltage between terminals 58 (or 55) and ground should be 0.7– 1.2 volts.
5. If these readings are not met, replace the camshaft position sensor.

4.5L VH45DE ENGINE

1. Start the engine.
2. Check the voltage between terminals 3 and 4 and ground.
3. At idle, the voltage between terminal 3 and ground should be approximately 2.5 volts.
4. At idle, the voltage between terminal 4 and ground should be approximately 1.8 volts.
5. If these readings are not met, replace the camshaft position sensor.

Crankshaft Position (CKP) Sensor

LOCATION

G20, Q45 and Q56

ALL ENGINES

The crankshaft position sensor is located on the transaxle housing, facing the gear teeth of the drive plate.

OPERATION

G20, G35, I30, J30, Q45 and Q56

ALL ENGINES

See Figures 2 and 3.

The crankshaft position sensor detects the fluctuation of the engine revolution. The sensor consists of a permanent magnet, a core and a coil. When the engine is running, the high and low parts of the teeth cause the gap with the sensor to change. The changing gap causes the magnetic field near the sensor to

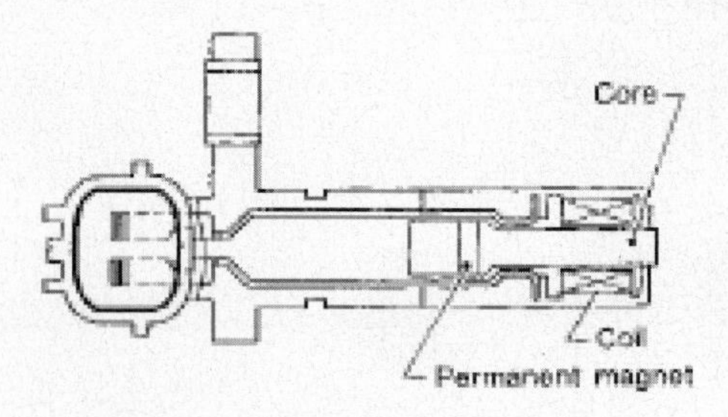

Fig. 2 The crankshaft position sensor

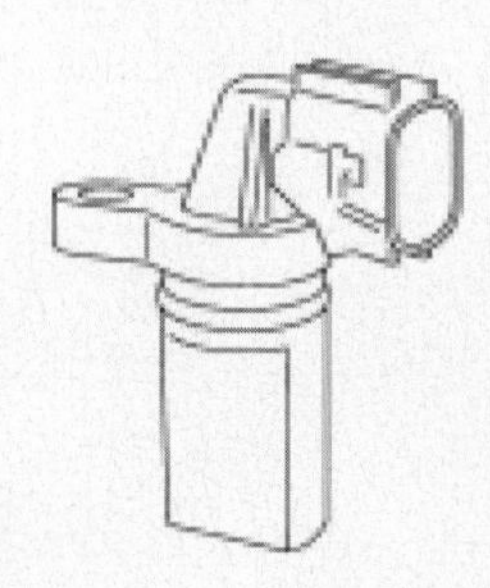

Fig. 3 The crankshaft position sensor–G35

change. Due to the changing magnetic field, the voltage from the sensor changes and the ECM receives the voltage signal and detects the fluctuation of the engine revolution. The sensor is not directly used to control the engine system, only for the on-board diagnosis of potential misfire.

TESTING

G20, I30, J30, Q45 and Q56

ALL ENGINES

1. Disconnect the crankshaft position sensor harness connector.
2. Loosen the mounting bolt of the sensor.
3. Remove the sensor and visually check the sensor for chipping or damage.
4. Check the resistance of the sensor. Resistance should be 166.5–203.5 ohms at 77°F (at 68–F for 4.1L and 2.0L engines).

FX35, FX45 and G35

ALL ENGINES

1. Disconnect the crankshaft position sensor harness connector.
2. Loosen the mounting bolt of the sensor.
3. Remove the sensor and visually check the sensor for chipping or damage.
4. Check the resistance of the sensor. The results should be neither zero nor infinity.

EGR Temperature Sensor

OPERATION

Q45 and Q56

ALL ENGINES

See Figure 4.

The EGR temperature sensor detects temperature changes in the EGR passageway. When the EGR valve opens, hot exhaust gases flow, and the temperature in the passageway changes. The EGR temperature sensor is a thermistor that modifies a voltage signal sent from the ECM. This modified signal than returns to the ECM as an input signal. As the temperature increases, the EGR temperature sensor resistance decreases. This sensor is not directly used to control the engine system, as it is used for the on-board diagnosis.

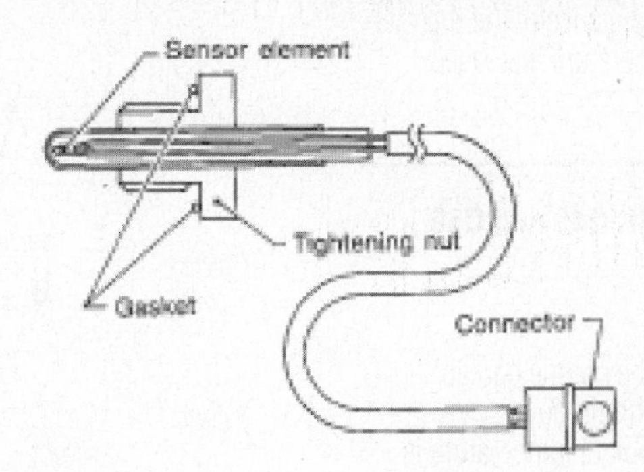

Fig. 4 The EGR temperature sensor

TESTING

Q45 and Q56

ALL ENGINES

The resistance of the EGR temperature sensor should be:

- If the temperature of the EGR is 32°F the voltage should be 4.51 and the resistance should be 0.68–1.11 megohms.
- If the temperature of the EGR is 122°F the voltage should be 2.53 and the resistance should be 0.092–0.12 megohms.
- If the temperature of the EGR is 212°F the voltage should be 0.87 and the resistance should be 0.017–0.024 megohms.

Engine Control Module (ECM)

OPERATION

G20, G35, I30, J30, Q45 and Q56

ALL ENGINES

See Figures 5, 6 and 7.

The ECM consists of a microcomputer, diagnostic test mode selector and connectors for signal input and output and for the power supply. The unit controls the engine.

Engine Coolant Temperature (ECT) Sensor

OPERATION

G20, G35, I30, J30, Q45 and Q56

ALL ENGINES

See Figure 8.

The engine coolant temperature sensor is used to detect the engine coolant temperature. The sensor modifies a voltage signal from the ECM. The modified signal returns to the ECM as the engine coolant temperature input. The sensor uses a thermistor which is sensitive to the change in tempera-

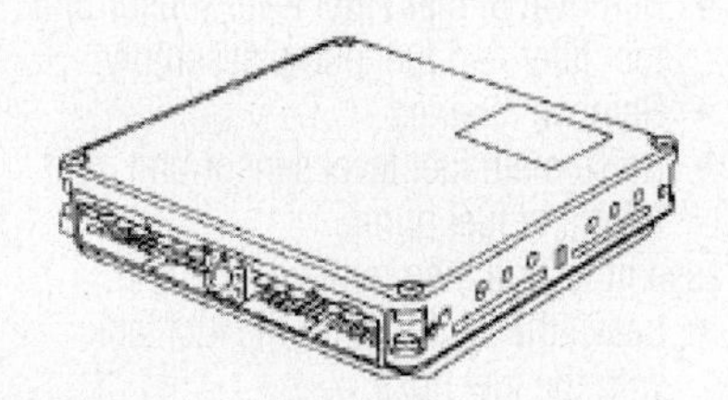

Fig. 5 The engine control module

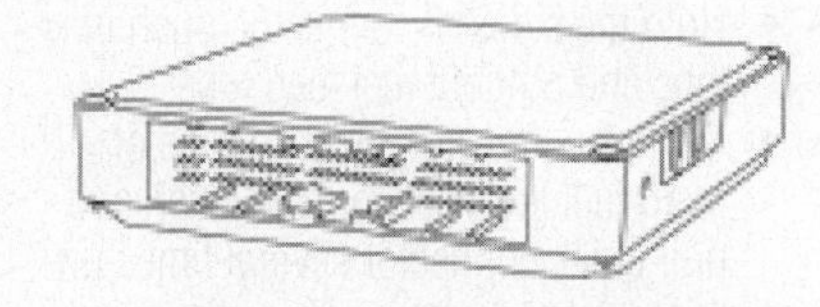

Fig. 6 The engine control module–2.0L engine

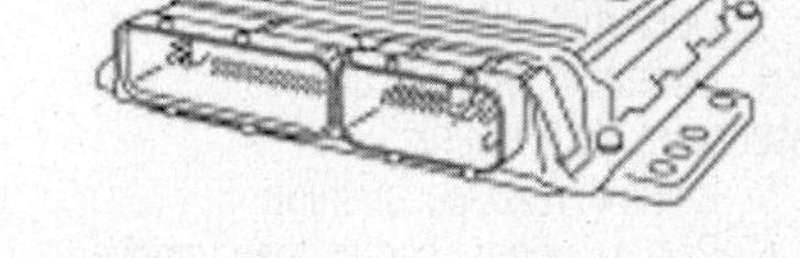

Fig. 7 The engine control module–3.5L engine (G35)

ture. The electrical resistance of the thermistor decreases as temperature increases.

TESTING

G20, G35, I30, J30, Q45 and Q56

ALL ENGINES

See Figure 9.

The resistance of the engine coolant temperature sensor should be:

- If the engine coolant temperature is 68°F, resistance should be 2.1–2.9 kohms.
- If the engine coolant temperature is 122°F, resistance should be 0.68–1.00 kohms.
- If the engine coolant temperature is 194°F, resistance should be 0.236–0.260 kohms.

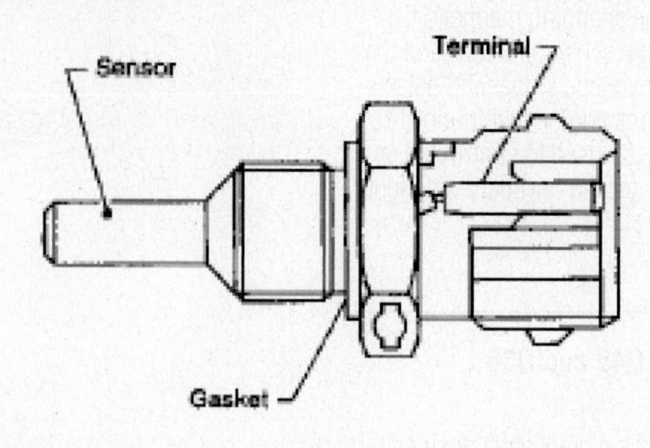

Fig. 8 The engine coolant temperature sensor

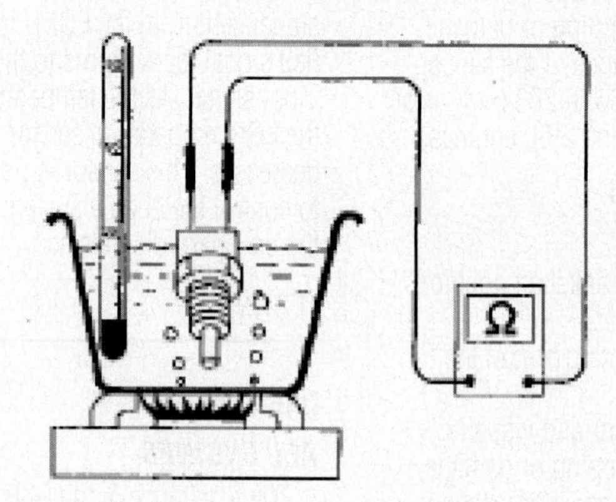

Fig. 9 The engine coolant temperature sensor

Fuel Level Sensor

OPERATION

FX35, FX45 and G35

ALL ENGINES

See Figure 10.

The fuel level sensor is mounted in the fuel level sensor unit. The sensor detects a fuel level in the fuel tank and transmits a signal to the combination meter. The combination meter sends the fuel level sensor signal to the ECM through CAN communication line. It consists of two parts, one is a mechanical float and the other is a variable resistor. Fuel level sensor output voltage changes depending on the movement of the fuel mechanical float.

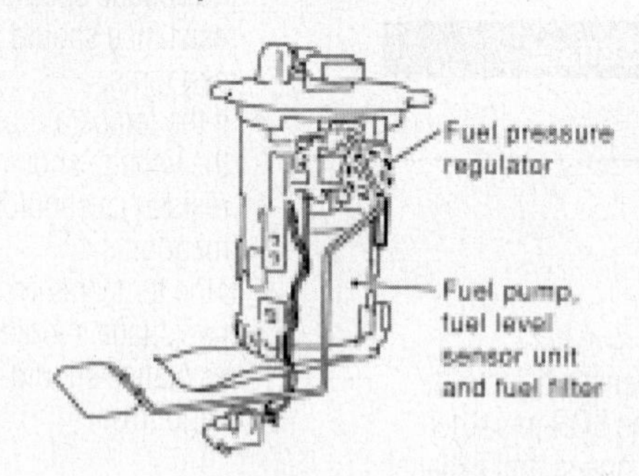

Fig. 10 The fuel level sensor

REMOVAL & INSTALLATION

FX35, FX45 and G35

ALL ENGINES

1. Check fuel level on fuel gauge. If fuel gauge indicates full or almost full, drain fuel from fuel tank until fuel gauge indicates nearly three-quarters full.
2. Release the fuel pressure from the fuel lines.
3. Open fuel filler lid.
4. Open fuel filler cap and release the pressure inside fuel tank.
5. Remove rear seat cushion.
6. Peel away floor carpet, then remove the inspection hole cover for the main and sub fuel level sensor units by turning clips clockwise by 90°.
7. Disconnect the harness connector and the fuel feed tube.
8. Disconnect quick connector as follows:
   - Hold the sides of connector, push in tabs and pull out fuel feed tube.
   - If quick connector sticks to tube of main fuel level sensor unit, push and pull quick connector several times until they start to move. Then disconnect them by pulling.
9. Remove the main fuel level sensor unit, the fuel filter and the fuel pump assembly, and the sub fuel level sensor unit as follows:
   - Removal of main fuel level sensor unit, fuel filter and fuel pump assembly:
   - Remove retainer.
   - Raise main fuel level sensor unit, fuel filter and fuel pump assembly, and then remove jet pump.
   - Leave the jet pump on the fuel tank with the fuel hose.
   - Removal of sub fuel level sensor unit:
   - Remove retainer.
   - Raise and release sub fuel level sensor unit to remove.

To install:

10. Install the sensor in reverse order of removal with the following notes:
    - When installing the jet pump, insert them fully until a clicking sound of the full stopper engagement is heard.
    - Face the main and the sub fuel level sensor units and install it with the dowel pin on back aligned with the pin hole on the fuel tank.
    - Install the retainer so that its notch becomes parallel with the notch on the fuel tank.
    - Tighten the retainer mounting bolts evenly.

Fuel Temperature Sensor

OPERATION

FX35, FX45 and G35

ALL ENGINES

See Figure 11.

The fuel temperature sensor is used to detect the fuel temperature inside the fuel tank. The sensor modifies a voltage signal from the ECM. The modified signal returns to the ECM as the fuel temperature input. The sensor uses a thermistor which is sensitive to the change in temperature. The electrical resistance of the thermistor decreases as temperature increases.

Q45

4.1L VH41DE ENGINE

See Figure 12.

The fuel temperature sensor is used to detect the fuel temperature inside the fuel tank. The sensor modifies a voltage signal from the ECM. The modified signal returns to the ECM as the fuel temperature input. The sensor uses a thermistor which is sensitive to the change in temperature. The electrical resistance of the thermistor decreases as temperature increases.

TESTING

FX35, FX45 and G35

ALL ENGINES

1. Remove the fuel temperature sensor.
2. Check the resistance of the fuel temperature sensor by immersing the sensor in hot water (or using a hot gun) and measuring the results.
3. At 68°F, resistance should be 2.3–2.7 kohms.
4. At 122°F, resistance should be 0.79–0.90 kohms.
5. If neither of these results are found, replace the fuel tank temperature sensor

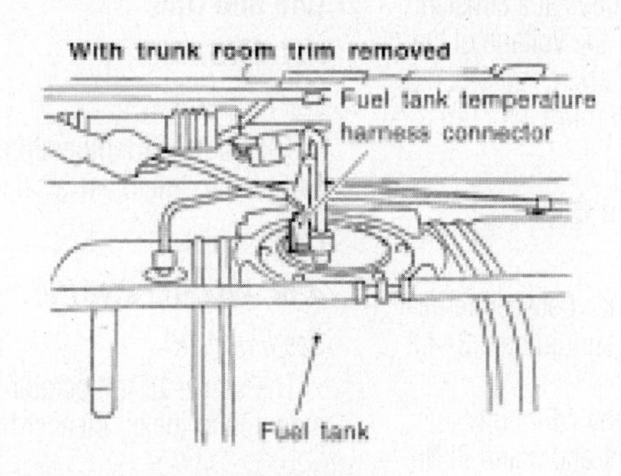

Fig. 11 The fuel tank temperature sensor

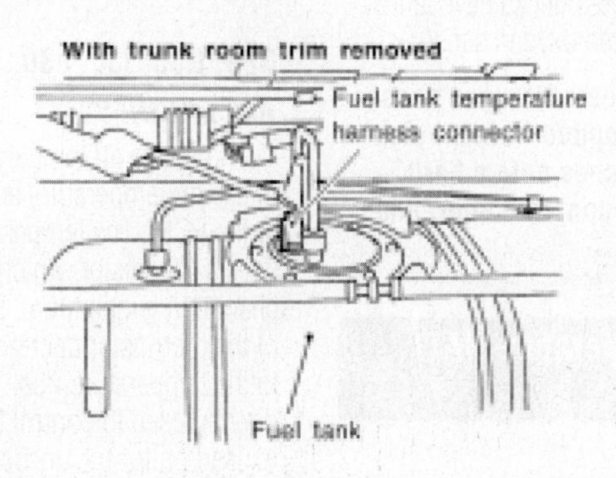

Fig. 12 The fuel tank temperature sensor

Q45 and Q56

ALL ENGINES

1. Check the resistance of the fuel temperature sensor by immersing the sensor in hot water and measuring the results.
2. At 68°F, resistance should be 2.3–2.7 kohms.
3. At 122°F, resistance should be 0.79–0.90 kohms.
4. If neither of these results are found, replace the fuel tank temperature sensor

Heated Oxygen (HO2S) Sensor

LOCATION

G20

2.0L SR20DE

The heated oxygen sensor is located in the exhaust manifold.

OPERATION

G20, I30, J30, Q45 and Q56

ALL ENGINES

See Figure 13.

The heated oxygen sensor detects the amount of oxygen in the exhaust gas compared to the outside air. The front heated oxygen sensor (left bank) has a closed-end tube made of ceramic zirconia. The zirconia generates voltage from approximately one volt in richer conditions to zero volts in leaner conditions. The front heated oxygen sensor (left bank) signal is sent to the ECM. The ECM adjusts the injection pulse duration to achieve the ideal air/fuel ratio, which occurs near the radical change from one to zero volts.

TESTING

G20

2.0L SR20DE

1. Start the engine and let it warm up to normal operating temperatures.
2. Set the voltmeter probes between the ECM terminal 62 and the engine ground.

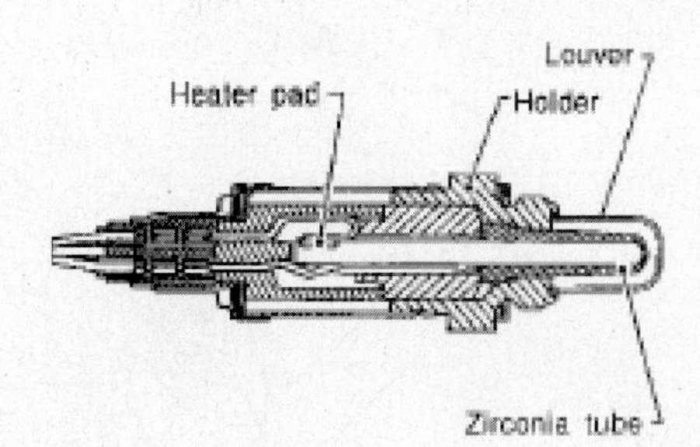

Fig. 13 The heated oxygen sensor (left bank)

3. With the engine speed at a constant 2000 rpm under no load, the voltage of the sensor should fluctuate between zero to 0.3 volts and 0.6 and 1.0 volt more than five times within 10 seconds.

I30, I35, J30, Q45 and Q56

ALL ENGINES

1. Check the resistance between terminals 3 and 1. Resistance should be 2.3–4.3 ohms at 77°F.
2. Check if there is any continuity between terminals 2 and 1 and 3 and 2. No continuity should exist.
3. If either of these checks have failed, replace the front heated oxygen sensor.

➡ **Discard any heated oxygen sensor which has been dropped from a height of more than 20 inches onto a hard surface, such as a concrete floor, and use a new one.**

Intake Air Temperature (IAT) Sensor

LOCATION

G20

2.0L SR20DE

The intake air temperature sensor is mounted to the intake air duct housing.

FX35, FX45 and G35

ALL ENGINES

The intake air temperature sensor is built into the mass air flow sensor.

I30, I35 and J30

ALL ENGINES

The intake air temperature sensor is mounted to the intake air duct behind the left-hand headlamp.

Q45 and Q56

ALL ENGINES

See Figure 14.

The intake air temperature sensor is mounted to the intake air duct.

Q45

4.5L VH45DE ENGINE

See Figure 15.

The intake air temperature sensor is mounted to the air cleaner housing.

OPERATION

G20, G35, I30, J30, Q45 and Q56

ALL ENGINES

The intake air temperature sensor detects intake air temperature and transmits a signal to the ECM. The temperature sensing unit uses a thermistor, which is sensitive to the change in temperature. Electrical resistance of the thermistor decreases in response to the temperature rise. This sensor is not directly used to control the engine system. It is used on for the on-board diagnostics.

TESTING

G20, I30, J30, Q45 and Q56

ALL ENGINES

See Figure 16.

The resistance of the intake air temperature sensor should be:

- If the intake air temperature is 68 degrees F, than the resistance should be 2.1–2.9 kohms.
- If the intake air temperature is 176 degrees F, than the resistance should be 0.27–0.38 kohms.

**** CAUTION**

Do not use the ECM terminals when measuring the voltage as you may damage the ECM. Use a ground other than the ECM terminals, such as the body ground.

FX35, FX45 and G35

ALL ENGINES

The resistance of the intake air temperature sensor between terminals 5 and 6 should be 1.8–2.2 kohms at 77°F.

Knock Sensor (KS)

LOCATION

G20, G35, Q45 and Q56

ALL ENGINES

The knock sensor is attached to the cylinder block.

OPERATION

G20

2.0L SR20DE

The knock sensor senses engine knocking using a piezoelectric element. A knocking vibration from the cylinder block is sensed as a vibration pressure. This pressure is converted into a voltage signal and sent to the ECM. Freeze frame data will not be stored in the ECM for the knock sensor. The MIL will not light for a knock sensor malfunction, but the sensor has one trip detection logic.

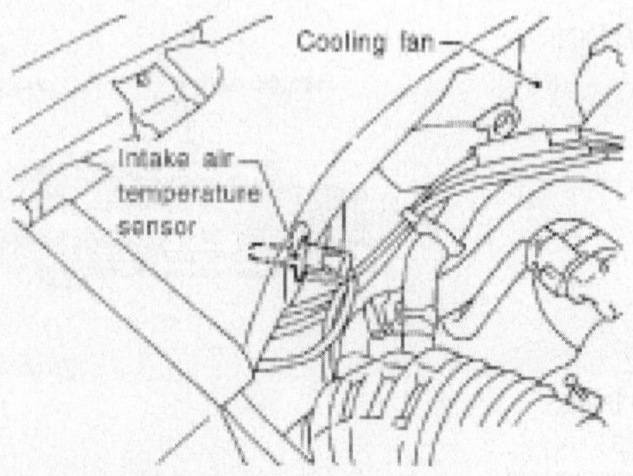

Fig. 14 The intake air temperature sensor–4.1L engine

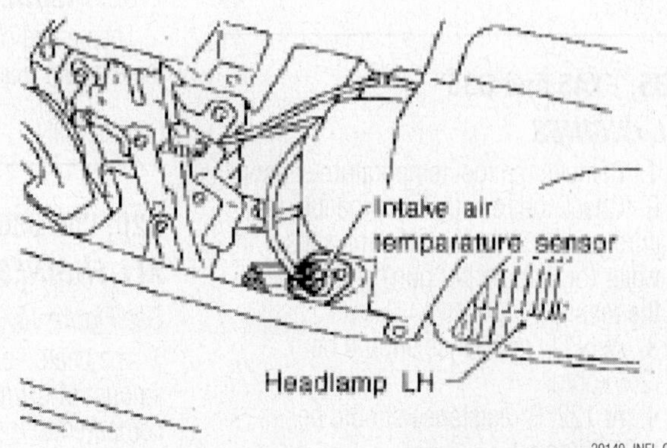

Fig. 15 The intake air temperature sensor–4.5L engine

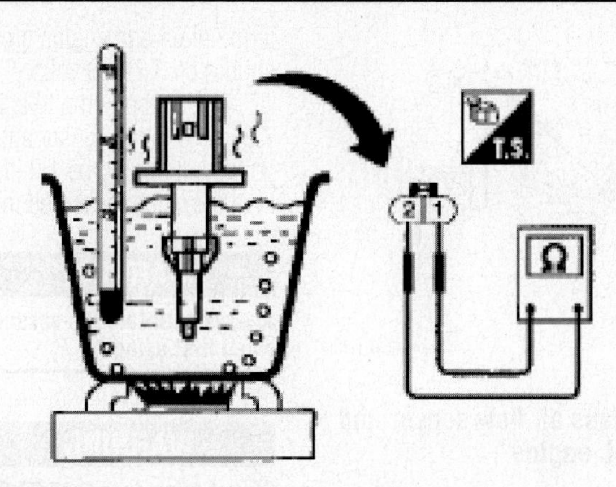

29149_INFI_G0003

Fig. 16 The intake air temperature sensor

FX35, FX45, G35, Q45 and Q56

ALL ENGINES

See Figure 17.

The knock sensor senses engine knocking using a plezoelectric element. A knocking vibration from the cylinder block is sensed as a vibration pressure. This pressure is converted into a voltage signal and sent to the ECM.

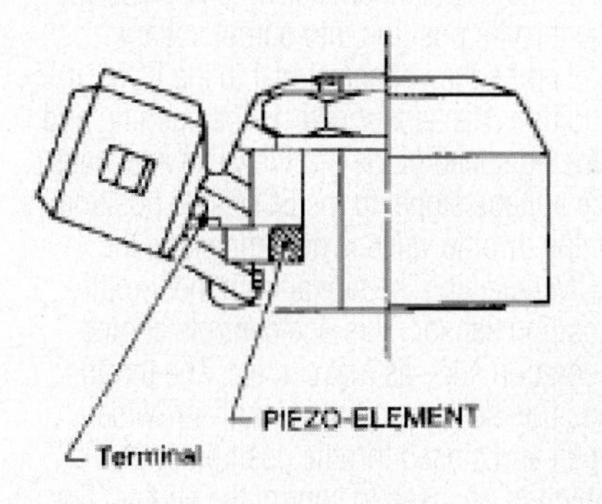

29149_INFI_G0008

Fig. 17 The knock sensor

TESTING

G20, Q45 and Q56

ALL ENGINES

See Figure 18.

1. Disconnect the knock sensor harness connector.
2. Check the resistance between terminal 2 and ground. Resistance should read approximately 500–620 kohms at 77°F.

➡ **Do not use any knock sensors that have been dropped or physically damaged.**

FX35, FX45 and G35

ALL ENGINES

1. Disconnect the knock sensor harness connector.
2. Check the resistance between terminal 1 and ground. Resistance should read approximately 532–588 kohms at 68°F.

➡ **Do not use any knock sensors that have been dropped or physically damaged.**

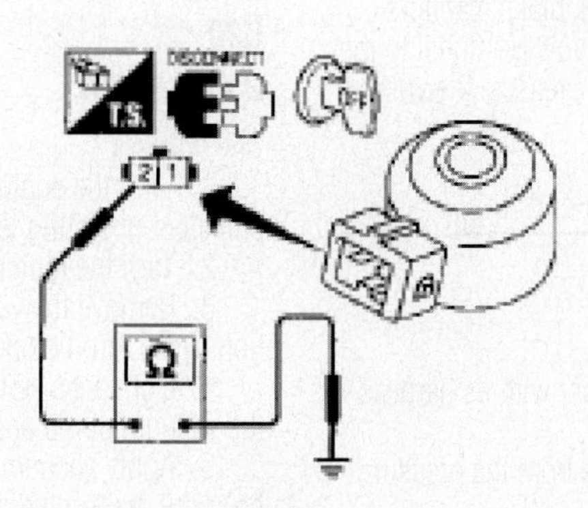

29149_INFI_G0009

Fig. 18 Testing the knock sensor

Mass Air Flow (MAF) Sensor

LOCATION

G20, G35, I30, J30, Q45 and Q56

ALL ENGINES

The mass air flow sensor is located in the intake air duct.

OPERATION

G20, G35, I30, J30, Q45 and Q56

ALL ENGINES

The mass air flow sensor measures the intake flow rate by measuring a part of the entire intake flow. It consists of a hot film that is supplied with electric current from the ECM. The temperature of the hot film is controlled by the ECM a certain amount. The heat generated by the hot film is reduced as the intake air flows around it. The more air, the greater the heat loss. Therefore, the ECM must supply more electric current to the hot film as the air flow increases. This maintains the temperature of the hot film. The ECM detects the air flow by means of this current change.

TESTING

G20

2.0L SR20DE

1. Turn the ignition switch to the on position.
2. Start the engine and warm it up to the normal operating temperature.
3. Check the voltage between terminal 4 and ground.
   - With the ignition switch on, the voltage should read less than 1.0 volts.
   - At idle, the voltage should be 1.3–1.7 volts.

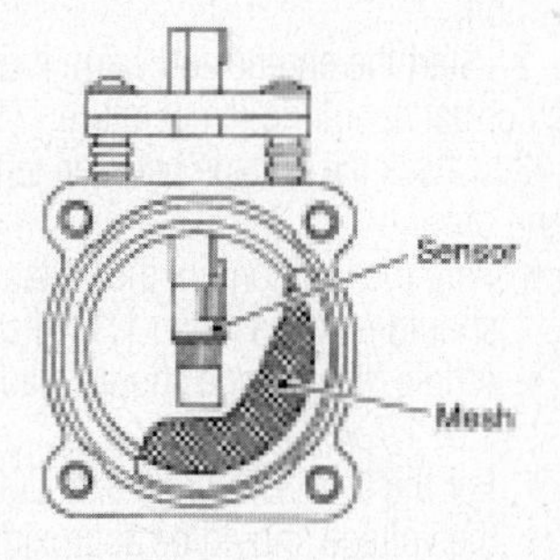

29149_INFI_G0019

Fig. 19 The mass air flow sensor–2.0L engine

- At 2500rpm, the voltage should be 1.8–2.4 volts.
- From idle to about 4000 rpm, the voltage should be 1.3–1.7 to approximately 4.0 volts.

4. If the voltage is out of specification, disconnect the mass air flow sensor harness connector and connect it again. Repeat the above check.

5. If the results aren't obtained, remove the mass air flow sensor from the air duct and check the hot film for damage or dust.

FX35, FX45 and G35

ALL ENGINES

1. Start the engine and warm it up to the normal operating temperature.
2. Check the voltage between the ECM terminal 51 (sensor signal) and ground.
   - If the ignition switch is on but the engine is stopped, the voltage should read approximately 0.4 volts.
   - At idle, the voltage should be 0.9–1.2 volts.
   - At 2500rpm, the voltage should be 1.5–1.9 volts.
   - From idle to about 4000 rpm, the voltage should be 0.9–1.2 to approximately 2.4 volts.
3. If the voltage is out of specification, disconnect the mass air flow sensor harness connector and connect it again.
4. Also check for crushed air ducts, malfunctioning seals on the cleaner element, uneven dirt on the element, or other improper specifications of the intake air system.
5. If everything checks out to be in proper working order, replace the mass air flow sensor.

I30, I35, J30, Q45 and Q56

ALL ENGINES

See Figure 20.

1. Turn the ignition switch to the on position.
2. Start the engine and warm it up to the normal operating temperature.
3. Check the voltage between terminal 4 and ground.
   - With the ignition on, the voltage should read less than 1.0 volts.
   - At idle, the voltage should read 1.0–1.7 volts.
   - For the Q45 4.1L engine, at 2000 rpm, the voltage should be approximately 2.1 volts.
   - Between idle and 4000 rpm, the voltage should be 1.0–1.7 and approximately 4.0 volts. Check for a linear voltage rise in response to an increase in engine speed to about 4000 rpm.

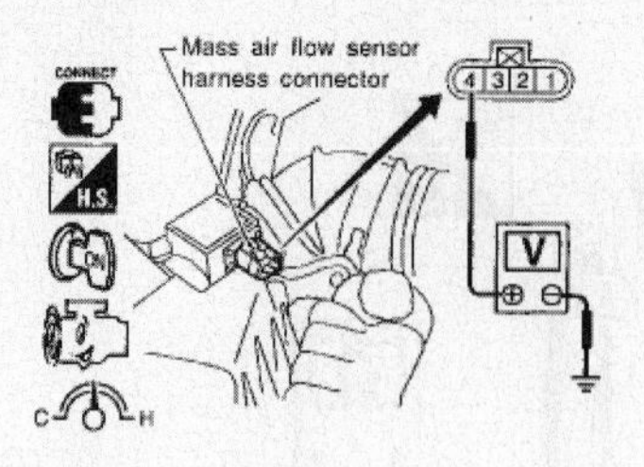

Fig. 20 The mass air flow sensor and connector–4.5L engine

4. If these voltages are not read under those conditions, remove the mass air flow sensor from the air duct and check the hot film for damage or dust. Replace it and conduct the above test again.

Manifold Absolute Pressure (MAP) Sensor

LOCATION

G20

2.0L SR20DE

The manifold absolute pressure sensor is built into the ECM.

OPERATION

G20

2.0L SR20DE

The sensor detects ambient barometric pressure and intake manifold pressure and sends the voltage signal to the ECM. As the pressure increases, the voltage rises.

Q45 and Q56

ALL ENGINES

The absolute pressure sensor is connected to the MAP/BARO switch solenoid valve by a hose. The sensor detects ambient barometric pressure and intake manifold pressure and sends the voltage signal to the ECM. As the pressure increases, the voltage rises.

TESTING

Q45 and Q56

ALL ENGINES

1. Remove the sensor with its harness connector connected.
2. Remove the hose from the pressure sensor.
3. Turn the ignition switch on and check the output voltage between the terminal 64 and engine ground. The voltage should be 3.2–4.8 volts.
4. Use the pump to apply vacuum of 7.8 in. hg to the sensor and check the output voltage. It should be 1.0–1.4 volts lower than the value measured in the above step.

**** WARNING**

Always calibrate the vacuum pump gauge when first using it.

**** WARNING**

Do not apply less than 27.5 in. hg or more than 29.9 in. hg of pressure.

5. If these results are yielded, replace the manifold absolute pressure sensor.

Throttle Position Sensor (TPS)

OPERATION

G20, G35, I30, J30, Q45 and Q56

ALL ENGINES

The throttle position sensor responds to the accelerator pedal movement. This sensor is a kind of potentiometer which transforms the throttle position into output voltage and emits the voltage signal to the ECM. In addition, the sensor detects the opening and closing speed of the throttle valve and feeds the voltage signal to the ECM. Idle position of the throttle valve is determined by the ECM receiving the signal from the throttle position sensor. This one controls engine operation such as fuel cut, etc. The throttle position sensor unit has a built-in "wide open and closed throttle position switch," which is not used to control the engine. For the G35, the throttle position sensor is part of the throttle control actuator.

TESTING

G20

2.0L SR20DE

See Figure 21.

1. Start the engine and warm it up to a sufficient operating temperature.
2. Turn the ignition switch off.
3. Remove the vacuum hose connection to the throttle opener.
4. Connect a suitable vacuum hose to the vacuum pump and the opener.
5. Apply vacuum (more than 11.8 in. hg) until the throttle drum becomes free from the rod of the throttle opener.
6. Turn ignition switch on.

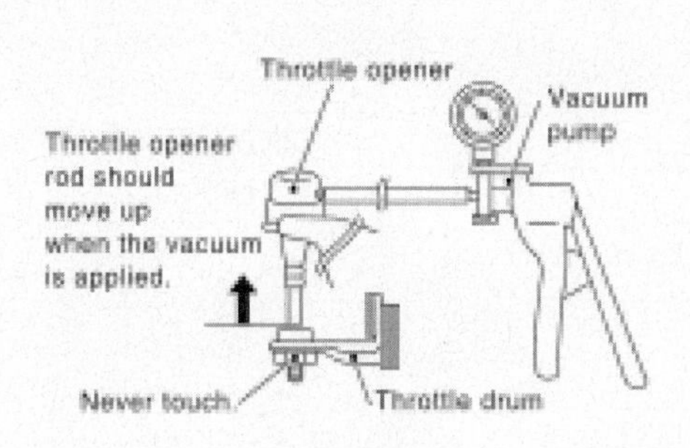

29149_INFI_G0020

Fig. 21 Testing the throttle position sensor–2.0L engine

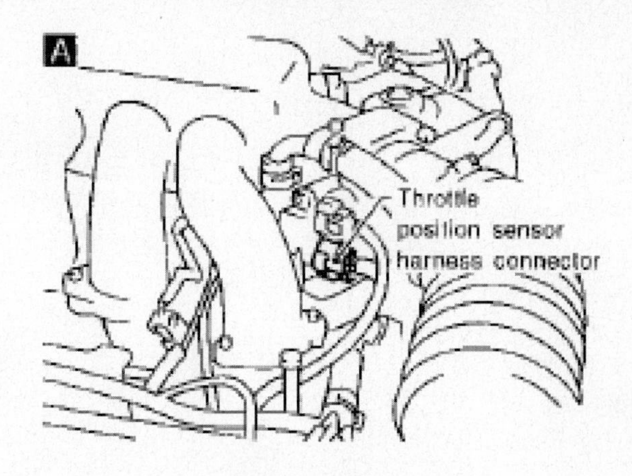

29149_INFI_G0006

Fig. 22 The throttle position sensor

7. Check the voltage between the ECM terminal 92 (the throttle position sensor signal) and ground. The voltage measurements must be made with the throttle position sensor installed in the vehicle.

- When completely closed, the voltage should be 0.15–0.85 volts.
- When partially open, the voltage should be 0.15–4.7 volts.
- When completely closed, the voltage should be 3.5–4.7 volts.

FX35, FX45 and G35

ALL ENGINES

1. Turn on the ignition and set the shift lever to D (for A/T) or first (for M/T) position.
2. Check the voltage between the ECM terminals 50, 69 and ground.

- For terminal 50 (throttle position sensor 1), the voltage should read more than 0.35 volts if the accelerator pedal is fully released and less than 4.75 volts if the pedal is full depressed.
- For terminal 69 (throttle position sensor 2), the voltage should read more than 4.75 volts if the accelerator pedal is fully released and less than 0.35 volts if the pedal is full depressed.

3. If these results are found, replace the electric throttle control actuator.

I30, I35, J30, Q45 and Q56

ALL ENGINES

See Figure 22.

1. Start the engine and warm it up to a sufficient operating temperature.
2. Turn the ignition switch off.
3. Disconnect the throttle position sensor harness connector.
4. Make sure that the resistance between terminals 2 and 3 changes when opening the throttle valve manually.

- When the throttle position sensor is completely closed, the resistance at 77°F should be approximately 0.6 kohms.
- When the throttle position sensor is partially open, the resistance at 77°F should be 0.6–4.0 kohms.
- When the throttle position sensor is completely open, the resistance at 77°F should be approximately 5 kohms (4 kohms for the J30).

Vehicle Speed Sensor (VSS)

LOCATION

G20, I30, J30, Q45 and Q56

ALL ENGINES

See Figure 23.

The vehicle speed sensor is installed on the transmission.

OPERATION

G20, I30, J30, Q45 and Q56

ALL ENGINES

See Figure 24.

The vehicle speed sensor contains a pulse generator which provides a vehicle speed signal to the speedometer. The speedometer then sends a signal to the ECM.

FX35, FX45 and G35

ALL ENGINES

The vehicle speed signal is sent to the combination meter from the VDC/TCS/ABS control unit by CAN communication line. The combination meter then sends the signal to the ECM by CAN communication line.

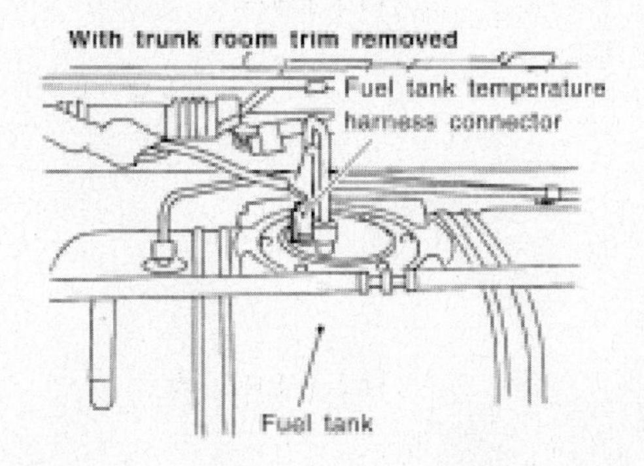

29149_INFI_G0015

Fig. 23 The vehicle speed sensor–4.1L engine

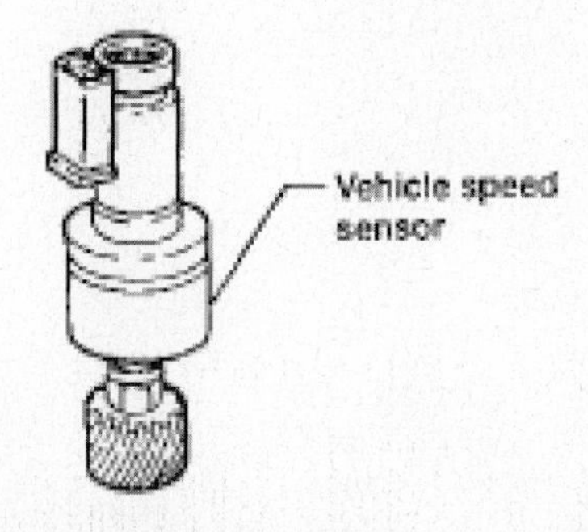

29149_INFI_G0012

Fig. 24 The vehicle speed sensor

ISUZU
DIAGNOSTIC TROUBLE CODES

9

TABLE OF CONTENTS

VEHICLE APPLICATIONS 9-2
P0XXX 9-3
P1XXX 9-85
P2XXX 9-117

OBD II Vehicle Applications

ISUZU

Ascender

2001
4.2L I6 .... VIN S
5.3L V8.......... VIN P

Amigo

1998-2000
2.2L I4 .... VIN D
1998-2000
3.2L V6... .... VIN W

Axiom

2002-
3.5L V6.......... VIN X

Hombre

1996-2000
2.2L I4 .... VIN 4
1996-1999
4.3L V6.......... VIN X
2000
4.3L V6.......... VIN W

Oasis

1996-1998
2.2L I4 .... VIN 4
1998-1999
2.3L I4 .... VIN 7

Rodeo

1996-1997
2.6L I4 .... VIN E
3.2L V6.......... VIN V
1998-2002
2.2L I4 .... VIN D
1998-2002
3.2L V6.......... VIN W

Rodeo Sport

2001-2002
2.2L I4 .... VIN D
3.2L V6.......... VIN W

Trooper

1996-1997
3.2L V6.......... VIN V
1998-2002
3.5L V6.......... VIN X

VehiCROSS

1999-2001
3.5L V6.......... VIN X

Gas Engine OBD II Trouble Code List (P0xxx Codes)

DTC	Trouble Code Title
DTC: P0101 **1T CCM, MIL: YES** **1998, 1999, 2000** **Models:** Amigo **Engines:** 3.2L VIN W **Transmissions:** A/T, M/T	**Mass Airflow Sensor Circuit Range/Performance** DTC P0102, P0103, P0122 and P0123 not set, system voltage at 11-16v, engine speed stable, throttle angle less than 90% and stable, Purge duty cycle less than 99%, EGR duty cycle less than 90%, EGR pintle position less than 90%, conditions stable for 2 seconds, and the PCM detected the Actual MAF sensor value was not equal to the Calculated MAF sensor value for 5 seconds. **Note: The Calculated MAF sensor value is determined from RPM, BARO pressure and the MAP sensor signal at the initial key "on".** **Possible Causes:** • Air leaks after the MAF sensor, or in the EGR or PCV system • Engine oil cap missing, engine oil dipstick not fully seated • MAF sensor is contaminated, dirty or out-of-calibration • MAF sensor ground circuit has high resistance • MAF minimum airflow rate to low at idle or during deceleration • MAP or TP sensor signal skewed, stuck or out of calibration • High signal interference (i.e., electrical noise from the ignition) • PCM has failed
DTC: P0101 **1T CCM, MIL: YES** **1997, 1998, 1999, 2000** **Models:** Hombre **Engines:** 4.3L VIN X **Transmissions:** A/T, M/T	**Mass Airflow Sensor Circuit Range/Performance** No CKP, EGR (DTC P0401 not active), EVAP, MAF, MAP or TP sensor codes set, system voltage from 11-16v, engine running at a steady throttle of under 90%, throttle angle less than 89.8% with any change less than 3.9%, Purge duty cycle less than 99.6%, EGR duty cycle less than 89.8%, EGR pintle position less than 89.8%, engine vacuum less than or equal to 75 kPa, conditions met for 2 seconds, and the PCM detected the Actual MAF sensor value was not equal to the Calculated MAF sensor value for 5 seconds during the test. **Note: The Calculated MAF sensor value is determined from RPM, BARO pressure and the MAP sensor signal at the initial key "on".** **Possible Causes:** • Air leaks after the MAF sensor, or in the EGR or PCV system • MAF sensor is contaminated, dirty or out-of-calibration • MAF sensor ground circuit has high resistance • MAF minimum airflow rate to low at idle or during deceleration • MAP or TP sensor signal skewed, stuck or out of calibration • High signal interference (i.e., electrical noise from the ignition) • PCM has failed
DTC: P0101 **2T CCM, MIL: YES** **1996, 1997, 1998, 1999, 2000, 2001, 2002** **Models:** Amigo, Axiom, Rodeo, Rodeo Sport, Trooper, VehiCROSS **Engines:** 3.2L VIN V, 3.2L VIN W, 3.5L VIN X **Transmissions:** A/T, M/T	**Mass Airflow Sensor Circuit Range/Performance** DTC P0106, P0107, P0108, P0121, P0122 and P0123 not set, system voltage at 11-16v, engine running at a stable idle speed, throttle angle stable (±1%), Calculated airflow from 25-40 g/sec, conditions met for 1 second, and the PCM detected a MAF sensor frequency that was significantly higher or lower than a "predicted" MAF airflow based on throttle position and engine speed for 12.5 seconds over a 25 second period during the CCM Rationality test. **Possible Causes:** • Air leaks after the MAF sensor, or in the EGR or PCV system • Engine oil cap missing, engine oil dipstick not fully seated • MAF sensor is contaminated, dirty or out-of-calibration • MAF sensor ground circuit has high resistance • MAF minimum airflow rate to low at idle or during deceleration • MAP or TP sensor signal skewed, stuck or out of calibration • High signal interference (i.e., electrical noise from the ignition) • PCM has failed
DTC: P0102 **1T CCM, MIL: YES** **1998, 1999, 2000** **Models:** Amigo **Engines:** 3.2L VIN W **Transmissions:** A/T, M/T	**MAF Sensor Circuit Low Frequency** Engine started, engine speed over 500 rpm for 10 seconds, system voltage over 11.5v, the PCM detected the MAF sensor frequency was equivalent to less than 1.6 g/sec in 50% of the last 100 samples measured during the CCM Rationality test. **Possible Causes:** • MAF sensor signal is shorted to ground • MAF sensor power circuit is open • MAF sensor is contaminated, dirty or is damaged • PCM has failed
DTC: P0102 **1T CCM, MIL: YES** **1997, 1998, 1999, 2000** **Models:** Hombre **Engines:** 4.3L VIN X **Transmissions:** A/T, M/T	**MAF Sensor Circuit Low Frequency** Engine started, engine runtime over 300 ms, system voltage over 9v, the PCM detected the MAF sensor frequency was less than 10 Hz for 1 second during the CCM test. **Possible Causes:** • MAF sensor signal circuit is shorted to ground • MAF sensor power circuit is open • MAF sensor is contaminated, dirty or is damaged • PCM has failed

DTC	Trouble Code Title, Conditions & Possible Causes
DTC: P0102 **1T CCM, MIL: YES** **1996, 1997, 1998, 1999, 2000, 2001, 2002** **Models:** Amigo, Axiom, Rodeo, Rodeo Sport, Trooper, VehiCROSS **Engines:** 3.2L VIN V, 3.2L VIN W, 3.5L VIN X **Transmissions:** A/T, M/T	**MAF Sensor Circuit Low Frequency** Engine started, engine speed over 500 rpm for 10 seconds, system voltage over 11.5v, the PCM detected the MAF sensor frequency was less than 1000 Hz for a total of 50% of the last 100 samples in the CCM Rationality test (a sample is taken every cylinder event). **Possible Causes:** • MAF sensor signal is shorted to ground • MAF sensor power circuit is open • MAF sensor is contaminated, dirty or is damaged • PCM has failed
DTC: P0103 **1T CCM, MIL: YES** **1998, 1999, 2000** **Models:** Amigo **Engines:** 3.2L VIN W **Transmissions:** A/T, M/T	**MAF Sensor Circuit High Frequency** Engine started, engine speed over 500 rpm for 10 seconds, system voltage over 11.5v, the PCM detected the MAF sensor frequency was equivalent to more than 40 g/sec in 50% of the last 200 samples measured during the CCM Rationality test. **Possible Causes:** • RFI or EMI interference from the Generator or Ignition system • RFI or EMI interference from an Ignition system component • MAF sensor is contaminated, dirty or is damaged • PCM has failed
DTC: P0103 **1T CCM, MIL: YES** **1997, 1998, 1999, 2000** **Models:** Hombre **Engines:** 4.3L VIN X **Transmissions:** A/T, M/T	**MAF Sensor Circuit High Frequency** Engine started, engine runtime over 300 ms, system voltage over 9v, the PCM detected the MAF sensor frequency was more than 1100 Hz for 1 second during the CCM test. **Possible Causes:** • RFI or EMI interference from the Generator or Ignition system • RFI or EMI interference from an Ignition system component • MAF sensor is contaminated, dirty or is damaged • PCM has failed
DTC: P0103 **1T CCM, MIL: YES** **1996, 1997, 1998, 1999, 2000, 2001, 2002** **Models:** Amigo, Axiom, Rodeo, Rodeo Sport, Trooper, VehiCROSS **Engines:** 3.2L VIN V, 3.2L VIN W, 3.5L VIN X **Transmissions:** A/T, M/T	**MAF Sensor Circuit High Frequency** Engine started, engine speed over 500 rpm for 10 seconds, system voltage over 11.5v, the PCM detected the MAF sensor frequency was more than 10,000 Hz for a total of 50% of the last 200 samples in the CCM Rationality test (a sample is taken every cylinder event). **Possible Causes:** • RFI or EMI interference from the Generator or Ignition system • RFI or EMI interference from an Ignition system component • MAF sensor is contaminated, dirty or is damaged • PCM has failed
DTC: P0105 **2T CCM, MIL: YES** **1999, 2000** **Models:** Hombre **Engines:** 2.2L VIN 4 **Transmissions:** A/T, M/T	**MAP Sensor Circuit Range/Performance** No CKP, ECT, EVAP, IAC, Knock Sensor, MAP, Fuel Trim, Injector, Misfire, TP or VSS codes set, engine runtime over 40 seconds, engine speed from 900-6375 rpm, engine speed stable within 50 rpm, TCC operation stable within 2.5%, IAC valve position stable within 5 counts, TP sensor stable within 2%, conditions met for 1.5 seconds, and the PCM detected the MAP sensor signal was out of its expected range for 14 of 16 seconds during the CCM test. **Possible Causes:** • MAP sensor source vacuum line is leaking or restricted • MAP sensor source vacuum line is plugged at intake manifold • MAP sensor is damaged, out-of-calibration or has failed • PCM has failed
DTC: P0106 **2T CCM, MIL: YES** **1998, 1999, 2000, 2001, 2002** **Models:** Amigo, Rodeo, Rodeo Sport **Engines:** 2.2L VIN D **Transmissions:** A/T, M/T	**MAP Sensor Signal Range/Performance** No CKP, ECT, EVAP, IAC, Knock Sensor, MAP, Fuel Trim, Injector, Misfire, TP or VSS codes set, engine speed from 1000-4000 rpm and steady within ±100 rpm, throttle angle change less than 5%, EGR flow steady with change less than 2%, IAC counts steady with change less than 3 counts, no change in PSPS, Brake switch, A/C clutch, and TCC status, conditions met for 1.5 seconds, and the PCM detected the Actual MAP sensor value varied over 10 kPa from the Expected MAP value for 10 seconds over a 20 second period. **Possible Causes:** • MAP sensor circuit open or shorted to ground (intermittent) • MAP sensor source vacuum line is leaking or restricted • MAP sensor source vacuum line is plugged at intake manifold • MAP sensor is damaged, out-of-calibration or has failed • PCM has failed

DTC	Trouble Code Title, Conditions & Possible Causes
DTC: P0106 **2T CCM, MIL: YES** **1996, 1997, 1998, 1999, 2000, 2001, 2002** **Models:** Amigo, Axiom, Rodeo, Rodeo Sport, Trooper, VehiCROSS **Engines:** 3.2L VIN V, 3.2L VIN W, 3.5L VIN X **Transmissions:** A/T, M/T	**MAP Sensor Signal Range/Performance** DTC P0121, P0122 and P0123 not set, engine speed stable (±100 rpm), throttle angle stable (±1%), IAC counts steady (±10 counts), EGR flow stable (±4%), no change in the A/C clutch, PSPS, Brake switch or TCC status, conditions met for 1 second, and the PCM detected the Actual MAP sensor value varied more than 10 kPa from the Expected MAP value for 10 seconds over a 20 second period. **Possible Causes:** • MAP sensor circuit open or shorted to ground (intermittent) • MAP sensor source vacuum line is leaking or restricted • MAP sensor source vacuum line is plugged at intake manifold • MAP sensor is damaged, out-of-calibration or has failed • PCM has failed
DTC: P0106 **2T CCM, MIL: YES** **1997, 1998, 1999, 2000** **Models:** Hombre **Engines:** 4.3L VIN X **Transmissions:** A/T	**MAP Sensor Signal Range/Performance** No EGR or TP sensor codes set, engine speed from 400-5000 rpm and steady within ±75 rpm, throttle angle change less than 1.5%, IAC counts steady with change less than 4 counts, EGR flow steady with change less than 2%, no change in Brake switch and A/C clutch status for 2 seconds, and the PCM detected the MAP sensor signal was out of the acceptable limits for 24 seconds. Possible Causes: • MAP sensor circuit open or shorted to ground (intermittent) • MAP sensor source vacuum line is leaking or restricted • MAP sensor source vacuum line is plugged at intake manifold • MAP sensor is damaged, out-of-calibration or has failed • PCM has failed
DTC: P0106 **2T CCM, MIL: YES** **1997, 1998, 1999, 2000** **Models:** Hombre **Engines:** 4.3L VIN X **Transmissions:** M/T	**MAP Sensor Signal Range/Performance** DTC P0121, P0122, P0123, P0401, P0404 and P0405 not set, engine speed from 400-5000 rpm and steady within ±75 rpm, throttle angle change less than 1.5%, IAC counts steady with change less than 4 counts, EGR flow steady with change less than 2%, no change in the Clutch position, Brake switch or A/C clutch status for 2 seconds, and the PCM detected the MAP sensor signal was out of the acceptable limits for 24 seconds. **Possible Causes:** • MAP sensor circuit open or shorted to ground (intermittent) • MAP sensor source vacuum line is leaking or restricted • MAP sensor source vacuum line is plugged at intake manifold • MAP sensor is damaged, out-of-calibration or has failed • PCM has failed
DTC: P0106 **2T CCM, MIL: YES** **1996, 1997, 1998** **Models:** Hombre **Engines:** 2.2L VIN 4 **Transmissions:** A/T, M/T	**MAP Sensor Signal Range/Performance** DTC P0107, P0108, P0117, P0118, P0121, P0122, P0123, P0131, P0132, P0171, P0172, P0200, P0300, P0301-P0304, P0325, P0341, P0342, P0404, P0405, P0440, P0442, P0446, P0452, P0453, P0502, P0506, P0507, P0601, P0602 and P1441 not set. Idle Speed Test • Engine speed over 800 rpm, throttle angle less than 50% with any change less than 12%, MAP sensor more than 60 kPa, and the PCM detected the MAP sensor signal did not change as expected when compared to the change in the throttle position. • Cruse and Idle Speed Test • Engine speed over 800 rpm, throttle angle less than 50% with any change less than 12%, initial MAP sensor signal less than 65 kPa, and the PCM detected the MAP sensor did not change as expected when compared to the amount of change in the throttle angle. **Possible Causes:** • MAP sensor circuit open or shorted to ground (intermittent) • MAP sensor source vacuum line is leaking or restricted • MAP sensor source vacuum line is plugged at intake manifold • MAP sensor is damaged, out-of-calibration or has failed • PCM has failed
DTC: P0106 **2T CCM, MIL: YES** **1996, 1997, 1998, 1999** **Models:** Oasis **Engines:** All **Transmissions:** A/T, M/T	**Manifold Air Pressure Sensor Range/Performance** Engine runtime over 1 second, and the PCM detected the MAP sensor signal was more 11.8" Hg during the test period. **Possible Causes:** • MAP sensor source vacuum line is leaking or disconnected • MAP sensor source vacuum line is plugged at intake manifold • MAP sensor is damaged, out-of-calibration or has failed • PCM has failed

DTC	Trouble Code Title, Conditions & Possible Causes
DTC: P0106 **2T CCM, MIL: YES** **1996, 1997** **Models:** Rodeo **Engines:** 2.6L VIN E **Transmissions:** M/T	**MAP Sensor Signal Range/Performance** DTC P0121, P0122 and P0123 not set, engine running with the engine speed stable ±100 rpm, throttle angle stable with any change less than 1%, IAC counts stable with change less than 10 counts, EGR flow stable with any change less than 4%, no change in the A/C clutch, Brake switch, Power Steering switch and TCC status for 1 second, and the PCM detected the MAP sensor signal varied by more than 10 kPa for 10 seconds over a 20 second time period. Possible Causes: • MAP sensor circuit open or shorted to ground (intermittent) • MAP sensor source vacuum line is leaking or restricted • MAP sensor source vacuum line is plugged at intake manifold • MAP sensor is damaged, out-of-calibration or has failed • PCM has failed
DTC: P0107 **2T CCM, MIL: YES** **1998, 1999, 2000, 2001, 2002** **Models:** Amigo, Rodeo, Rodeo Sport **Engines:** 2.2L VIN D **Transmissions:** A/T, M/T	**MAP Sensor Circuit Low Input** Engine started, then with the engine speed below 1000 rpm and the throttle angle over 1%, or with the throttle angle over 2%, and the engine speed over 1000 rpm, the PCM detected the MAP sensor signal indicated less than 11 kPa during the CCM test. **Possible Causes:** • MAP sensor circuit shorted to ground between sensor and PCM • MAP sensor power circuit is open or shorted to ground • MAP sensor is damaged or has failed • PCM has failed
DTC: P0107 **1T CCM, MIL: YES** **1996, 1997, 1998, 1999, 2000, 2001, 2002** **Models:** Amigo, Axiom, Rodeo, Rodeo Sport, Trooper, VehiCROSS **Engines:** 3.2L VIN V, 3.2L VIN W, 3.5L VIN X **Transmissions:** A/T, M/T	**MAP Sensor Circuit Low Input** DTC P0121, P0122 and P0123 not set, engine started, system voltage from 11-16v, then with the engine speed below 1000 rpm and throttle angle over 1%, or with the engine speed over 1000 rpm and throttle angle over 2%, the PCM detected the MAP sensor was less than 0.04v (11 kPa) for 10 seconds over a 20 second period. **Possible Causes:** • MAP sensor circuit shorted to ground between sensor and PCM • MAP sensor power circuit is open or shorted to ground • MAP sensor is damaged or has failed • PCM has failed
DTC: P0107 **2T CCM, MIL: YES** **1996, 1997, 1998, 1999, 2000** **Models:** Hombre **Engines:** 2.2L VIN 4 **Transmissions:** A/T, M/T	**MAP Sensor Circuit Low Input** DTC P0122 and P0123 not set, engine started, then with the engine speed below 1000 rpm, or with the engine speed above 1000 rpm, and the throttle angle over 15%, the PCM detected the MAP sensor signal indicated less than 0.08v (11.8 kPa) for 6.25 seconds. **Possible Causes:** • MAP sensor circuit shorted to ground between sensor and PCM • MAP sensor power circuit is open or shorted to ground • MAP sensor is damaged or has failed • PCM has failed
DTC: P0107 **2T CCM, MIL: YES** **1997, 1998, 1999, 2000** **Models:** Hombre **Engines:** 4.3L VIN X **Transmissions:** A/T, M/T	**MAP Sensor Circuit Low Input** DTC P0122 and P0123 not set, engine started, then with the engine speed below 800 rpm and the throttle angle over 0%, or with the engine speed above 800 rpm and the throttle angle over 12.5%, the PCM detected MAP sensor signal was less than 0.04v for 1 second. Possible Causes: • MAP sensor circuit shorted to ground between sensor and PCM • MAP sensor power circuit is open or shorted to ground • MAP sensor is damaged or has failed • PCM has failed
DTC: P0107 **1T CCM, MIL: YES** **1996, 1997, 1998, 1999** **Models:** Oasis **Engines:** All **Transmissions:** A/T, M/T	**Manifold Air Pressure Sensor Circuit Low Input** Engine running in closed loop conditions, and the PCM detected the MAP sensor signal was near 0.0" Hg during the CCM test. **Note: The key on, engine off MAP sensor input should be near 2.9v.** **Possible Causes:** • MAP sensor 5-volt power circuit open or shorted to ground • MAP Sensor signal circuit is shorted to ground • MAP Sensor is damaged or has failed • PCM has failed

DTC	Trouble Code Title, Conditions & Possible Causes
DTC: P0107 **2T CCM, MIL: YES** **1996, 1997** **Models:** Rodeo **Engines:** 2.6L VIN E **Transmissions:** M/T	**MAP Sensor Circuit Low Input** DTC P0121, P0122 and P0123 not set, engine started, then with the engine speed below 1000 rpm and throttle angle over 1%, or with the engine speed above 1000 rpm and throttle angle over 2%, the PCM detected MAP sensor was under 0.04v (11 kPa) for 10 seconds over a 16 second time period during the CCM test. **Possible Causes:** • MAP sensor circuit shorted to ground between sensor and PCM • MAP sensor power circuit is open or shorted to ground • MAP sensor is damaged or has failed • PCM has failed
DTC: P0108 **2T CCM, MIL: YES** **1998, 1999, 2000, 2001, 2002** **Models:** Amigo, Rodeo, Rodeo Sport **Engines:** 2.2L VIN D **Transmissions:** A/T, M/T	**MAP Sensor Circuit High Input** Engine started, then with the engine speed less than 1000 rpm and the throttle angle under 3%, or with the engine speed over 1000 rpm and the throttle angle under 10%, the PCM detected the MAP sensor signal indicated more than 90 kPa during the CCM test. **Possible Causes:** • MAP sensor circuit is open between the sensor and the PCM • MAP sensor signal circuit is shorted to VREF or system power • MAP sensor ground circuit is open between sensor and PCM • MAP sensor is damaged or has failed • PCM has failed
DTC: P0108 **1T CCM, MIL: YES** **1996, 1997, 1998, 1999, 2000, 2001, 2002** **Models:** Amigo, Axiom, Rodeo, Rodeo Sport, Trooper, VehiCROSS Engines: 3.2L VIN V, 3.2L VIN W, 3.5L VIN X **Transmissions:** A/T, M/T	**MAP Sensor Circuit High Input** DTC P0121, P0122 and P0123 not set, engine started, engine speed less than 1000 rpm and the throttle angle less than 3%, or with engine speed more than 1000 rpm and the throttle angle less than 10%, the PCM detected the MAP sensor was more than 4.40v (90 kPa) for 10 seconds over a 16 second period during the test. **Possible Causes:** • MAP sensor circuit is open between the sensor and the PCM • MAP sensor signal circuit is shorted to VREF or system power • MAP sensor ground circuit is open between sensor and PCM • MAP sensor is damaged or has failed • PCM has failed
DTC: P0108 **2T CCM, MIL: YES** **1996, 1997, 1998, 1999, 2000** **Models:** Hombre **Engines:** 2.2L VIN 4 **Transmissions:** A/T, M/T	**MAP Sensor Circuit High Input** DTC P0122 and P0123 not set, engine runtime from 20-40 seconds, throttle angle less than 12%, vehicle speed less than 1 mph, and the PCM detected the MAP sensor signal was more than 3.80v (82 kPa) for over 1.25 seconds during the CCM test. **Possible Causes:** • MAP sensor circuit is open between the sensor and the PCM • MAP sensor signal circuit is shorted to VREF or system power • MAP sensor ground circuit is open between sensor and PCM • MAP sensor is damaged or has failed • PCM has failed
DTC: P0108 **2T CCM, MIL: YES** **1997, 1998, 1999, 2000** **Models:** Hombre **Engines:** 4.3L VIN X **Transmissions:** A/T, M/T	**MAP Sensor Circuit High Input** DTC P0122 and P0123 not set, engine speed less than 1200 rpm with the throttle angle below 0.4%, or with engine speed more than 1200 rpm and the throttle angle below 20%, the PCM detected the MAP sensor indicated more than 4.40v for 1 second during the test. **Possible Causes:** • MAP sensor circuit is open between the sensor and the PCM • MAP sensor signal circuit is shorted to VREF or system power • MAP sensor ground circuit is open between sensor and PCM • MAP sensor is damaged or has failed • PCM has failed
DTC: P0108 **1T CCM, MIL: YES** **1996, 1997, 1998, 1999** **Models:** Oasis **Engines:** All **Transmissions:** A/T, M/T	**Manifold Air Pressure Sensor Circuit High Input** Engine running in closed loop conditions, and the PCM detected the MAP sensor signal was near 29.9" Hg during the test period. **Note: The key on, engine off MAP sensor input should be near 2.9v.** **Possible Causes:** • MAP sensor signal circuit is open, or the ground circuit is open • MAP sensor signal circuit shorted to 5v VREF or system power • MAP sensor is damaged (due to an open circuit) or has failed • PCM has failed

DTC	Trouble Code Title, Conditions & Possible Causes
DTC: P0108 **2T CCM, MIL: YES** **1996, 1997** **Models:** Rodeo **Engines:** 2.6L VIN E **Transmissions:** M/T	**MAP Sensor Circuit High Input** DTC P0121, P0122 and P0123 not set, engine started, engine speed less than 1000 rpm and the throttle angle less than 3%, or with engine speed more than 1000 rpm and the throttle angle less than 10%, the PCM detected the MAP sensor signal was more than 4.40v (90 kPa) for 10 seconds over a 16 second time period. **Possible Causes:** • MAP sensor circuit is open between the sensor and the PCM • MAP sensor signal circuit is shorted to VREF or system power • MAP sensor ground circuit is open between sensor and PCM • MAP sensor is damaged or has failed • PCM has failed
DTC: P0111 **2T CCM, MIL: YES** **1996, 1997, 1998, 1999** **Models:** Oasis **Engines:** All **Transmissions:** A/T, M/T	**Intake Air Temperature Sensor Range/Performance** Engine running for over 10 minutes, and the PCM detected the Intake Air Temperature (IAT) sensor signal changed too much in too short a period of time. **Possible Causes:** • IAT sensor signal value less than 1.0v • IAT sensor ground circuit has high resistance • IAT sensor signal circuit has high resistance • IAT sensor is damaged or has failed • PCM has failed
DTC: P0112 **1T CCM, MIL: YES** **1998, 1999, 2000, 2001, 2002** **Models:** Amigo, Rodeo, Rodeo Sport **Engines:** 2.2L VIN D **Transmissions:** A/T, M/T	**Intake Air Temperature Sensor Circuit Low Input** DTC P0502 not set, engine started, engine runtime over 2 minutes, vehicle speed over 30 mph, and the PCM detected the IAT sensor indicated less than 0.10v (Scan Tool reads 298°F) for 12.5 seconds over a 25 second period during the test. Possible Causes: • IAT sensor circuit shorted to ground between sensor and PCM • IAT sensor is damaged, out-of-calibration or has failed • PCM has failed
DTC: P0112 **1T CCM, MIL: YES** **1996, 1997, 1998, 1999, 2000, 2001, 2002** **Models:** Amigo, Axiom, Rodeo, Rodeo Sport, Trooper, VehiCROSS Engines: 3.2L VIN V, 3.2L VIN W, 3.5L VIN X **Transmissions:** A/T, M/T	**Intake Air Temperature Sensor Circuit Low Input** DTC P0502 not set, engine started, engine runtime over 2 minutes, vehicle speed over 30 mph, and the PCM detected the IAT sensor indicated less than 0.10v (Scan Tool reads 298°F) for 12.5 seconds over a 20 second period during the CCM test. **Possible Causes:** • IAT sensor circuit shorted to ground between sensor and PCM • IAT sensor is damaged, out-of-calibration or has failed • PCM has failed
DTC: P0112 **1T CCM, MIL: YES** **1996, 1997, 1998, 1999, 2000** **Models:** Hombre **Engines:** 2.2L VIN 4 **Transmissions:** A/T, M/T	**Intake Air Temperature Sensor Circuit Low Input** DTC P0502 not set, engine started, engine runtime over 320 seconds, vehicle speed over 15 mph, the PCM detected the IAT sensor indicated less than 0.08v (Scan Tool reads 262°F) for 3.125 seconds during the CCM test. **Possible Causes:** • IAT sensor circuit shorted to ground between sensor and PCM • IAT sensor is damaged, out-of-calibration or has failed • PCM has failed
DTC: P0112 **2T CCM, MIL: YES** **1997, 1998, 1999, 2000** **Models:** Hombre **Engines:** 4.3L VIN X **Transmissions:** A/T, M/T	**Intake Air Temperature Sensor Circuit Low Input** DTC P0502 and P0503 not set, engine started, engine runtime over 100 seconds, vehicle speed over 2 mph, and the PCM detected the IAT sensor indicated less than 0.08v (Scan Tool reads 262°F) for 5 seconds during the CCM test. **Possible Causes:** • IAT sensor circuit shorted to ground between sensor and PCM • IAT sensor is damaged, out-of-calibration or has failed • PCM has failed
DTC: P0112 **1T CCM, MIL: YES** **1996, 1997, 1998, 1999** **Models:** Oasis **Engines:** All **Transmissions:** A/T, M/T	**Intake Air Temperature Sensor Circuit Low Input** Key on or engine running, and the PCM detected the Intake Air Temperature (IAT) sensor signal indicated less than 0.1v (302°F). **Possible Causes:** • IAT sensor signal shorted to chassis ground • IAT sensor signal shorted to sensor ground circuit • IAT sensor has an internal failure (it is shorted) or has failed • PCM has failed

DTC	Trouble Code Title, Conditions & Possible Causes
DTC: P0112 **1T CCM, MIL: YES** **1996, 1997** **Models:** Rodeo **Engines:** 2.6L VIN E **Transmissions:** M/T	**Intake Air Temperature Sensor Circuit Low Input** DTC P0502 not set, engine started, engine runtime over 2 minutes, vehicle speed over 30 mph, and the PCM detected the IAT sensor indicated less than 0.10v (Scan Tool reads 298°F) for 12.5 seconds over a 20 second period during the CCM test. **Possible Causes:** • IAT sensor circuit shorted to ground between sensor and PCM • IAT sensor is damaged, out-of-calibration or has failed • PCM has failed
DTC: P0113 **1T CCM, MIL: YES** **1998, 1999, 2000, 2001, 2002** **Models:** Amigo, Rodeo, Rodeo Sport **Engines:** 2.2L VIN D **Transmissions:** A/T, M/T	**Intake Air Temperature Sensor Circuit High Input** Engine started, then with the engine runtime over 2 minutes, ECT sensor more than 140°F, MAF sensor than 20 g/sec, VSS indicating under 1 mph, the PCM detected the IAT sensor indicated more than 4.90v (Scan Tool reads -39°F) for 12.5 seconds of a 25-second period during the CCM test. **Possible Causes:** • IAT sensor signal circuit open between the sensor and PCM • IAT sensor signal is shorted to VREF or system power (B+) • IAT sensor is damaged, out-of-calibration or has failed • PCM has failed
DTC: P0113 **2T CCM, MIL: YES** **1996, 1997, 1998, 1999, 2000** **Models:** Hombre **Engines:** 2.2L VIN 4 **Transmissions:** A/T, M/T	**Intake Air Temperature Sensor Circuit High Input** DTC P0117, P0118, P0502 and P0503 not set, engine started, engine runtime over 320 seconds, the PCM detected the IAT sensor indicated more than 4.90v (Scan Tool reads -39°F) for 3.125 seconds during the CCM test. **Possible Causes:** • IAT sensor signal circuit open between the sensor and PCM • IAT sensor signal is shorted to VREF or system power (B+) • IAT sensor is damaged, out-of-calibration or has failed • PCM has failed
DTC: P0113 **1T CCM, MIL: YES** **1996, 1997, 1998, 1999, 2000, 2001, 2002** **Models:** Amigo, Axiom, Rodeo, Rodeo Sport, Trooper, VehiCROSS **Engines:** 3.2L VIN V, 3.2L VIN W, 3.5L VIN X **Transmissions:** A/T, M/T	**Intake Air Temperature Sensor Circuit High Input** DTC P0502 not set, engine runtime 4 minutes, vehicle speed over 20 mph, ECT sensor more than 140°F, MAF sensor less than 20 g/sec, and the PCM detected the IAT sensor indicated more than 4.90v (Scan Tool reads -38°F) for 12.5 second over a 25 second period during the CCM test. **Possible Causes:** • IAT sensor signal circuit open between the sensor and PCM • IAT sensor signal is shorted to VREF or system power (B+) • IAT sensor is damaged, out-of-calibration or has failed • PCM has failed
DTC: P0113 **2T CCM, MIL: YES** **1997, 1998, 1999, 2000** **Models:** Hombre **Engines:** 4.3L VIN X **Transmissions:** A/T, M/T	**Intake Air Temperature Sensor Circuit High Input** DTC P0102, P0103, P0117, P0118, P0502 and P0503 not set, engine started, engine runtime over 100 seconds, vehicle speed over 2 mph, ECT sensor more than 185.5°F, MAF sensor less than 20 g/sec, and the PCM detected the IAT sensor indicated more than 4.90v (Scan Tool reads -38°F) for 5 seconds during the CCM test. **Possible Causes:** • IAT sensor signal circuit open between the sensor and PCM • IAT sensor signal is shorted to VREF or system power (B+) • IAT sensor is damaged, out-of-calibration or has failed • PCM has failed
DTC: P0113 **1T CCM, MIL: YES** **1996, 1997, 1998, 1999** **Models:** Oasis **Engines:** All **Transmissions:** A/T, M/T	**Intake Air Temperature Sensor Circuit High Input** Key on or engine running, and the PCM detected the Intake Air Temperature (IAT) sensor signal indicated more than 4.90 (-4°F). **Possible Causes:** • IAT sensor signal shorted to VREF or system power • IAT sensor signal circuit is open • IAT sensor ground circuit is open • Sensor has an internal failure (it is open) • PCM has failed
DTC: P0113 **1T CCM, MIL: YES** **1996, 1997** **Models:** Rodeo **Engines:** 2.6L VIN E **Transmissions:** M/T	**Intake Air Temperature Sensor Circuit High Input** DTC P0502 not set, engine started, engine runtime over 4 minutes, vehicle speed over 20 mph, ECT sensor more than 140°F, MAF sensor less than 20 g/sec, and the PCM detected the IAT sensor indicated more than 4.90v (Scan Tool reads -38°F) for 12.5 second over a 25 second period during the CCM test. **Possible Causes:** • IAT sensor signal circuit open between the sensor and PCM • IAT sensor signal is shorted to VREF or system power (B+) • IAT sensor is damaged, out-of-calibration or has failed • PCM has failed

DTC	Trouble Code Title, Conditions & Possible Causes
DTC: P0116 **1T CCM, MIL: YES** **1996, 1997, 1998, 1999** **Models:** Oasis **Engines:** All **Transmissions:** A/T, M/T	**Engine Coolant Temperature Sensor Range/Performance** Engine started, engine running, and the PCM detected the ECT sensor signal changed too much in too short a period of time. **Note: The ECT sensor should read 0.47v-0.78v at hot idle speed.** **Possible Causes:** • ECT sensor ground circuit is open (an intermittent fault) • ECT sensor signal circuit is open (an intermittent fault) • ECT sensor is damaged or has failed (an intermittent fault) • PCM has failed
DTC: P0117 **1T CCM, MIL: YES** **1998, 1999, 2000, 2001, 2002** **Models:** Amigo, Rodeo, Rodeo Sport **Engines:** 2.2L VIN D **Transmissions:** A/T, M/T	**Engine Coolant Temperature Sensor Circuit Low Input** Engine started, engine runtime over 1 minute, and the PCM detected the ECT sensor indicated less than 0.10v (Scan Tool reads 302°F) for 50 seconds over a 100-second period in the CCM test. **Possible Causes:** • ECT sensor circuit shorted to ground between sensor and PCM • ECT sensor is damaged, out-of-calibration or has failed • PCM has failed
DTC: P0117 **1T CCM, MIL: YES** **1996, 1997, 1998, 1999, 2000, 2001, 2002** **Models:** Amigo, Axiom, Rodeo, Rodeo Sport, Trooper, VehiCROSS **Engines:** 3.2L VIN V, 3.2L VIN W, 3.5L VIN X **Transmissions:** A/T, M/T	**Engine Coolant Temperature Sensor Circuit Low Input** Engine started, engine runtime over 1 minute, and the PCM detected the ECT sensor indicated less than 0.10v (Scan Tool reads 302°F) for 6.25 seconds for 50 seconds over a 100 second period during the CCM test. **Possible Causes:** • ECT sensor circuit shorted to ground between sensor and PCM • ECT sensor is damaged, out-of-calibration or has failed • PCM has failed
DTC: P0117 **1T CCM, MIL: YES** **1996, 1997, 1998, 1999, 2000** **Models:** Hombre **Engines:** 2.2L VIN 4 **Transmissions:** A/T, M/T	**Engine Coolant Temperature Sensor Circuit Low Input** Engine started, engine runtime over 128 seconds, and the PCM detected the ECT sensor indicated less than 0.14v (Scan Tool reads 280°F) for 6.25 seconds during the CCM test. **Possible Causes:** • ECT sensor circuit shorted to ground between sensor and PCM • ECT sensor is damaged, out-of-calibration or has failed • PCM has failed
DTC: P0117 **1T CCM, MIL: YES** **1997, 1998, 1999, 2000** **Models:** Hombre **Engines:** 4.3L VIN X **Transmissions**: A/T, M/T	**Engine Coolant Temperature Sensor Circuit Low Input** Engine started, engine runtime over 5 seconds, and the PCM detected the ECT sensor indicated less than 0.14v (Scan Tool reads 302°F) for 6.25 seconds during the CCM test. **Possible Causes:** • ECT sensor circuit shorted to ground between sensor and PCM • ECT sensor is damaged, out-of-calibration or has failed • PCM has failed
DTC: P0117 **1T CCM, MIL: YES** **1996, 1997, 1998, 1999** **Models:** Oasis **Engines:** All **Transmissions**: A/T, M/T	**Engine Coolant Temperature Sensor Circuit Low Input** Key on or engine running, and the PCM detected the Engine Coolant Temperature (ECT) sensor signal indicated more than 302°F (0.1v). **Note: The normal range of the ECT sensor is from 0.47v to 0.78v.** **Possible Causes:** • ECT sensor signal shorted to chassis ground • ECT sensor signal shorted to sensor ground circuit • ECT sensor has an internal failure (it is shorted) or has failed • PCM has failed
DTC: P0117 **1T CCM, MIL: YES** **1996, 1997** **Models:** Rodeo **Engines:** 2.6L VIN E **Transmissions:** M/T	**Engine Coolant Temperature Sensor Circuit Low Input** Engine started, engine runtime over 1 minute, and the PCM detected the ECT sensor indicated less than 0.10v (Scan Tool reads 302°F) for 6.25 seconds for 50 seconds over a 100 second period during the CCM test. **Possible Causes:** • ECT sensor circuit shorted to ground between sensor and PCM • ECT sensor is damaged, out-of-calibration or has failed • PCM has failed

DTC	Trouble Code Title, Conditions & Possible Causes
DTC: P0118 **1T CCM, MIL: YES** **1998, 1999, 2000, 2001, 2002** **Models:** Amigo, Rodeo, Rodeo Sport **Engines:** 2.2L VIN D **Transmissions:** A/T, M/T	**Intake Air Temperature Sensor Circuit High Input** Engine started, engine runtime over 2 minutes, and the PCM detected the ECT sensor indicated more than 4.90v (Scan Tool reads -38ºF) for 50 seconds of a 100-second period during the test. **Possible Causes:** • ECT sensor signal circuit is open • ECT sensor signal circuit is shorted to VREF or system power • ECT sensor is damaged or has failed • PCM has failed
DTC: P0118 **1T CCM, MIL: YES** **1996, 1997, 1998, 1999, 2000, 2001, 2002** **Models:** Amigo, Axiom, Rodeo, Rodeo Sport, Trooper, VehiCROSS **Engines:** 3.2L VIN V, 3.2L VIN W, 3.5L VIN X **Transmissions:** A/T, M/T	**Engine Coolant Temperature Sensor Circuit High Input** Engine started, engine runtime over 90 seconds, and the PCM detected the ECT sensor was more than 4.90v (Scan Tool reads -38ºF) for 50 seconds over a 100 second period during the CCM test. **Possible Causes:** • ECT sensor signal circuit is open • ECT sensor signal circuit is shorted to VREF or system power • ECT sensor is damaged or has failed • PCM has failed
DTC: P0118 **1T CCM, MIL: YES** **1996, 1997, 1998, 1999, 2000** **Models:** Hombre **Engines:** 2.2L VIN 4 **Transmissions:** A/T, M/T	**Engine Coolant Temperature Sensor Circuit High Input** Engine started, engine runtime over 128 seconds, and the PCM detected the ECT sensor indicated more than 4.90v (Scan Tool reads -39ºF) for 8.25 seconds during the CCM test. **Possible Causes:** • ECT sensor signal circuit is open • ECT sensor signal circuit is shorted to VREF or system power • ECT sensor is damaged or has failed • PCM has failed
DTC: P0118 **1T CCM, MIL: YES** **1997, 1998, 1999, 2000** **Models:** Hombre **Engines:** 4.3L VIN X **Transmissions:** A/T, M/T	**Engine Coolant Temperature Sensor Circuit High Input** Engine started, engine runtime over 5 seconds, and the PCM detected the ECT sensor indicated more than 4.90v (Scan Tool reads -38ºF) for 6.25 seconds during the CCM test. **Possible Causes:** • ECT sensor signal circuit is open • ECT sensor signal circuit is shorted to VREF or system power • ECT sensor is damaged or has failed • PCM has failed
DTC: P0118 **1T CCM, MIL: YES** **1996, 1997, 1998, 1999** **Models:** Oasis **Engines:** All **Transmissions:** A/T, M/T	**Engine Coolant Temperature Sensor Circuit High Input** Key on or engine running, and the PCM detected the Engine Coolant Temperature (ECT) sensor signal indicated less than -4ºF (4.9v). **Note: The normal range of the ECT sensor is from 0.47v to 0.78v.** **Possible Causes:** • IAT sensor signal shorted to VREF or system power • IAT sensor signal circuit is open • IAT sensor ground circuit is open • Sensor has an internal failure (it is open) • PCM has failed
DTC: P0118 **1T CCM, MIL: YES** **1996, 1997** **Models:** Rodeo **Engines:** 2.6L VIN E **Transmissions:** M/T	**Engine Coolant Temperature Sensor Circuit High Input** Engine started, engine runtime over 90 seconds, and the PCM detected the ECT sensor was more than 4.90v (Scan Tool reads -38ºF) for 50 seconds over a 100 second period during the CCM test. **Possible Causes:** • ECT sensor signal circuit is open • ECT sensor signal circuit is shorted to VREF or system power • ECT sensor is damaged or has failed • PCM has failed

DTC	Trouble Code Title, Conditions & Possible Causes
DTC: P0121 **2T CCM, MIL: YES** **1998, 1999, 2000** **Models:** Amigo Engines: 2.2L VIN D **Transmissions:** A/T, M/T	**Throttle Position Sensor Range/Performance** No CKP, ECT, EGR, EVAP, MAP or TP codes set, engine started, engine at idle speed with IAC command from 10-160 counts, ECT sensor signal over 14°F MAP sensor signal stable within less than a 2% change, and the PCM detected the MAP sensor signal remained over 55 kPa (stuck high) or it remained below 50 kPa. **Note: In both cases, the PCM estimates the correct throttle angle based on engine speed (RPM) and manifold pressure (MAP signal).** **Possible Causes:** • TP sensor signal circuit is open to the PCM (intermittent fault) • TP sensor ground circuit is open (an intermittent fault) • MAP sensor is out of calibration • Throttle body is damaged or throttle linkage is bent or binding • TP sensor is damaged or has failed
DTC: P0121 **2T CCM, MIL: YES** **1998, 1999** **Models:** Amigo **Engines:** 3.2L VIN W **Transmissions:** A/T, M/T	**Throttle Position Sensor Range/Performance** No CKP, ECT, EGR, EVAP, MAP or TP codes set, engine started, engine at idle speed with IAC command from 10-160 counts, ECT sensor signal over 14°F MAP sensor signal stable within less than a 2% change, and the PCM detected the MAP sensor signal remained over 55 kPa (stuck high) or it remained below 50 kPa. **Note: In both cases, the PCM estimates the correct throttle angle based on engine speed (RPM) and manifold pressure (MAP signal).** **Possible Causes:** • TP sensor signal circuit is open to the PCM (intermittent fault) • TP sensor ground circuit is open (an intermittent fault) • MAP sensor is out of calibration • Throttle body is damaged or throttle linkage is bent or binding • TP sensor is damaged or has failed
DTC: P0121 **2T CCM, MIL: YES** **1996, 1997, 1998** **Models:** Hombre **Engines:** 2.2L VIN 4 **Transmissions:** A/T, M/T	**Throttle Position Sensor Range/Performance** DTC P0106, P0107, P0108, P0171, P0172, P0200, P0300, P0301-P0304, P0325, P0335, P0341, P0342, P0404, P0405, P0440, P0442, P0446, P0452, P0453, P0502, P0506, P0507, P0601, P0602 and P1441 not set, engine started, engine at idle speed, throttle angle steady (± 2%), ECT sensor signal more than 68°F, MAP sensor signal less than 45 kPa, and the PCM detected the throttle angle indicated one of the following values for 6 seconds: TP sensor signal more than 2% at zero (0) rpm TP sensor signal more than 10% at 800 rpm TP sensor signal more than 20% at 1600 rpm TP sensor signal more than 25% at 2400 rpm TP sensor signal more than 30% at 3200 rpm TP sensor signal more than 35% at 4000 to 4500 rpm TP sensor signal more than 40% at 5500 to 6500 rpm **Possible Causes:** • TP sensor signal circuit is open to the PCM (intermittent fault) • TP sensor ground circuit is open (an intermittent fault) • MAP sensor damaged or out-of-calibration • Throttle body is damaged or throttle linkage is bent or binding • TP sensor is damaged or has failed
DTC: P0121 **2T CCM, MIL: YES** **1997, 1998, 1999, 2000** **Models**: Hombre **Engines:** 4.3L VIN X **Transmissions:** A/T, M/T	**Throttle Position Sensor Circuit Range/Performance** DTC P0106, P0107, P0108, P0122, P0123, P0506 and P0507 not set, engine started, BARO sensor signal not in default, throttle angle steady, and the PCM detected the Actual throttle angle was more than the Calculated throttle angle (Stuck High Test), or the Actual throttle angle was less than the Calculated throttle angle (Stuck Low Test), condition met for 5 seconds during the CCM Rationality test. **Possible Causes:** • TP sensor signal circuit is open to the PCM (intermittent fault) • TP sensor ground circuit is open (an intermittent fault) • MAP sensor damaged or out-of-calibration • Throttle body is damaged or throttle linkage is bent or binding • TP sensor is damaged or has failed
DTC: P0121 **2T CCM, MIL: YES** **1999, 2000, 2001** **Models:** VehiCROSS **Engines:** 3.5L VIN X **Transmissions:** A/T, M/T	**Throttle Position Sensor Circuit Range/Performance** DTC P0121, P0122, P0123 and P1122 not set, engine started, MAP sensor less than 55 kPa, throttle angle stable (±1%), and the PCM detected the Actual throttle angle was not close to the Predicted throttle angle for 12.5 seconds over a 25 second period in the test. **Possible Causes:** • TP sensor signal circuit is open to the PCM (intermittent fault) • TP sensor ground circuit is open (an intermittent fault) • MAP sensor damaged or out-of-calibration • Throttle body is damaged or throttle linkage is bent or binding • TP sensor is damaged or has failed

DTC	Trouble Code Title, Conditions & Possible Causes
DTC: P0121 **2T CCM, MIL: YES** **1996, 1997** **Models:** Rodeo **Engines:** 2.6L VIN E Transmissions: M/T	**Throttle Position Sensor Circuit Range/Performance** DTC P0106, P0107, P0108, P0122 and P0123 not set, engine started, MAP sensor signal less than 55 kPa, throttle angle stable with any change less than 1%, and the PCM detected the Actual throttle angle was not close to the Predicted throttle angle for 12.5 seconds over a 25 second period during the CCM Rationality test. **Possible Causes:** • TP sensor signal circuit is open to the PCM (intermittent fault) • TP sensor ground circuit is open (an intermittent fault) • MAP sensor damaged or out-of-calibration • Throttle body is damaged or throttle linkage is bent or binding • TP sensor is damaged or has failed
DTC: P0121 **2T CCM, MIL: YES** **1996, 1997, 1998, 1999, 2000, 2001, 2002** **Models:** Rodeo, Rodeo Sport, Trooper Engines: 3.2L VIN V, 3.2L VIN W, 3.5L VIN X Transmissions: A/T, M/T	**Throttle Position Sensor Circuit Range/Performance** DTC P0121, P0122, P0123 and P1122 not set, engine started, MAP sensor less than 55 kPa, throttle angle stable (±1%), and the PCM detected the Actual throttle angle was not close to the Predicted throttle angle for 12.5 seconds over a 25 second period in the test. **Possible Causes:** • TP sensor signal circuit is open to the PCM (intermittent fault) • TP sensor ground circuit is open (an intermittent fault) • MAP sensor damaged or out-of-calibration • Throttle body is damaged or throttle linkage is bent or binding • TP sensor is damaged or has failed
DTC: P0122 **1T CCM, MIL: YES** **1998, 1999, 2000, 2001, 2002** **Models:** Amigo, Rodeo, Rodeo Sport **Engines:** 2.2L VIN D **Transmissions:** A/T, M/T	**Throttle Position Sensor Circuit Low Input** Key on or engine running, and the PCM detected the TP sensor indicated less than 0.20v for 0.78 seconds over a 1.5 second period during the CCM test. **Possible Causes:** • TP sensor signal circuit is grounded between sensor and PCM • Throttle body is damaged • Throttle linkage is bent or binding (sticking) • TP sensor is damaged or has failed
DTC: P0122 **1T CCM, MIL: YES** **1998, 1999** **Models:** Amigo **Engines:** 3.2L VIN W **Transmissions:** A/T, M/T	**Throttle Position Sensor Circuit Low Input** Key on or engine running, and the PCM detected the TP sensor indicated less than 0.20v for 0.78 seconds over a 1.5 second period during the CCM test. **Possible Causes:** • TP sensor signal circuit is grounded between sensor and PCM • Throttle body is damaged • Throttle linkage is bent or binding (sticking) • TP sensor is damaged or has failed
DTC: P0122 **1T CCM, MIL: YES** **1996, 1997, 1998, 1999, 2000** **Models:** Hombre **Engines:** 2.2L VIN 4 **Transmissions:** A/T, M/T	**Throttle Position Sensor Circuit Low Input** Engine started, engine running, and the PCM detected the TP sensor signal indicated less than 0.20v for 6.25 seconds in the test. **Possible Causes:** • TP sensor signal circuit is grounded between sensor and PCM • Throttle body is damaged • Throttle linkage is bent or binding (sticking) • TP sensor is damaged or has failed
DTC: P0122 **1T CCM, MIL: YES** **1997, 1998, 1999, 2000** **Models:** Hombre **Engines:** 4.3L VIN X **Transmissions:** A/T, M/T	**Throttle Position Sensor Circuit Low Input** Key on or engine running, and the PCM detected the TP sensor signal indicated less than 0.25v for 1 second during the test. **Possible Causes:** • TP sensor signal circuit is grounded between sensor and PCM • Throttle body is damaged • Throttle linkage is bent or binding (sticking) • TP sensor is damaged or has failed
DTC: P0122 **1T CCM, MIL: YES** **1996, 1997, 1998, 1999** **Models:** Oasis **Engines:** All **Transmissions:** A/T, M/T	**Throttle Position Sensor Circuit Low Input** Engine started, engine running at hot idle, and the PCM detected the closed throttle TP signal was less than 0.16v (less than 10% open). **Possible Causes:** • TP sensor signal circuit is shorted to ground • TP sensor VREF circuit is open or shorted to ground • TP sensor is damaged (it may be shorted internally) • PCM has failed

DTC	Trouble Code Title, Conditions & Possible Causes
DTC: P0122 **1T CCM, MIL: YES** **1996, 1997** **Models:** Rodeo **Engines:** 2.6L VIN E **Transmissions:** M/T I4 VIN E	**Throttle Position Sensor Circuit Low Input** Key on or engine running, and the PCM detected the TP sensor indicated less than 0.22v for 0.78 seconds over a 1.5 second period. **Possible Causes:** • TP sensor signal circuit is grounded between sensor and PCM • Throttle body is damaged • Throttle linkage is bent or binding (sticking) • TP sensor is damaged or has failed
DTC: P0122 **1T CCM, MIL: YES** **1996, 1997, 1998, 1999, 2000, 2001, 2002** **Models:** Rodeo, Rodeo Sport, Trooper **Engines:** 3.2L VIN V, 3.2L VIN W, 3.5L VIN X **Transmissions:** A/T, M/T	**Throttle Position Sensor Circuit Low Input** Key on or engine running, and the PCM detected the TP sensor indicated less than 0.22v for 0.78 seconds over a 1.5 second period. **Possible Causes:** • TP sensor signal circuit is grounded between sensor and PCM • Throttle body is damaged • Throttle linkage is bent or binding (sticking) • TP sensor is damaged or has failed
DTC: P0122 **2T CCM, MIL: YES** **1999, 2000, 2001** **Models:** VehiCROSS **Engines:** 3.5L VIN X **Transmissions:** A/T, M/T	**Throttle Position Sensor Circuit Low Input** Key on or engine running, and the PCM detected the TP sensor indicated less than 0.20v for 0.78 seconds over a 1.5 second period during the CCM test. **Possible Causes:** • TP sensor signal circuit is grounded between sensor and PCM • Throttle body is damaged • Throttle linkage is bent or binding (sticking) • TP sensor is damaged or has failed
DTC: P0123 **1T CCM, MIL: YES** **1998, 1999, 2000, 2001, 2002** **Models:** Amigo, Rodeo, Rodeo Sport **Engines:** 2.2L VIN D **Transmissions:** A/T, M/T	**Throttle Position Sensor Circuit High Input** Key on or engine running, and the PCM detected the TP sensor was more than 4.88v for 0.78 seconds over a 1.5 second period during the CCM test. **Possible Causes:** • TP sensor signal circuit or ground circuit is open • TP sensor signal circuit is shorted to VREF or system power • TP sensor is damaged or has failed • PCM has failed
DTC: P0123 **1T CCM, MIL: YES** **1998, 1999** **Models:** Amigo **Engines:** 3.2L VIN W **Transmissions:** A/T, M/T	**Throttle Position Sensor Circuit High Input** Key on or engine running, and the PCM detected the TP sensor was more than 4.88v for 0.78 seconds over a 1.5 second period during the CCM test. **Possible Causes:** • TP sensor signal circuit or ground circuit is open • TP sensor signal circuit is shorted to VREF or system power • TP sensor is damaged or has failed • PCM has failed
DTC: P0123 **1T CCM, MIL: YES** **1996, 1997, 1998, 1999, 2000** **Models:** Hombre **Engines:** 2.2L VIN 4 Transmissions: A/T, M/T	**Throttle Position Sensor Circuit High Input** Key on or engine running, and the PCM detected the TP sensor signal indicated more than 4.88v for 0.78 seconds out of a 1.5 second period during the CCM test. **Possible Causes:** • TP sensor signal circuit or ground circuit is open • TP sensor signal circuit is shorted to VREF or system power • TP sensor is damaged or has failed • PCM has failed
DTC: P0123 **1T CCM, MIL: YES** **1997, 1998, 1999, 2000** **Models:** Hombre Engines: 4.3L VIN X **Transmissions:** A/T, M/T	**Throttle Position Sensor Circuit High Input** Key on or engine running, and the PCM detected the TP sensor signal was more than 4.70v for 1 second during the CCM test. **Possible Causes:** • TP sensor signal circuit is open • TP sensor ground circuit is open • TP sensor signal circuit is shorted to VREF or system power • TP sensor is damaged or has failed • PCM has failed

DTC	Trouble Code Title, Conditions & Possible Causes
DTC: P0123 **1T CCM, MIL: YES** **1996, 1997, 1998, 1999** **Models:** Oasis **Engines:** All **Transmissions:** A/T, M/T	**Throttle Position Sensor Circuit High Input** Engine started, engine running at hot idle, and the PCM detected the wide-open-throttle TP signal was more than 4.60v (more than 90%). **Possible Causes:** • TP sensor signal circuit is open • TP sensor signal circuit is shorted to VREF • TP sensor ground circuit is open between sensor and the PCM • TP sensor is damaged or has failed • PCM has failed
DTC: P0123 **1T CCM, MIL: YES** **1996, 1997** **Models:** Rodeo **Engines:** 2.6L VIN E **Transmissions:** M/T	**Throttle Position Sensor Circuit High Input** Key on or engine running, and the PCM detected the TP sensor signal was more than 4.78v for 0.78 seconds over a 1.5 second period during the CCM test. **Possible Causes:** • TP sensor signal circuit is open • TP sensor ground circuit is open • TP sensor signal circuit is shorted to VREF or system power • TP sensor is damaged or has failed • PCM has failed
DTC: P0123 **2T CCM, MIL: YES** **1996, 1997, 1998, 1999, 2000, 2001, 2002** **Models:** Rodeo, Rodeo Sport, Trooper **Engines:** 3.2L VIN V, 3.2L VIN W, 3.5L VIN X **Transmissions:** A/T, M/T	**Throttle Position Sensor Circuit High Input** Key on or engine running, and the PCM detected the Throttle Position (TP) sensor was more than 4.88v for 0.78 seconds out of a 1.5 second period during the CCM test. **Possible Causes:** • TP sensor signal circuit or ground circuit is open • TP sensor signal circuit is shorted to VREF or system power • TP sensor is damaged or has failed • PCM has failed
DTC: P0123 **2T CCM, MIL: YES** **1999, 2000, 2001** **Models:** VehiCROSS **Engines:** 3.5L VIN X **Transmissions:** A/T, M/T	**Throttle Position Sensor Circuit High Input** Key on or engine running, and the PCM detected the Throttle Position (TP) sensor was more than 4.88v for 0.78 seconds out of a 1.5 second period during the CCM test. **Possible Causes:** • TP sensor signal circuit or ground circuit is open • TP sensor signal circuit is shorted to VREF or system power • TP sensor is damaged or has failed • PCM has failed
DTC: P0125 **2T CCM, MIL: YES** **1998, 1999, 2000, 2001, 2002** **Models:** Amigo, Rodeo, Rodeo Sport **Engines:** 2.2L VIN D **Transmissions:** A/T, M/T	**Insufficient Coolant Temperature For Closed Loop** No ECT, IAT, MAP, Misfire, TP or VSS codes set, then under: Cold Case Startup Conditions • IAT sensor signal from 20-50°F, engine runtime under 225 seconds, accumulated airflow over 2000 grams, and the PCM detected it took over 300 seconds for the coolant to reach “closed loop” conditions. • Warm Case Startup Conditions • ECT sensor less than 84°F, IAT sensor signal more than 50°F, engine runtime over 90 seconds, accumulated airflow over 1500 grams, and the PCM detected it took over 120 seconds for the coolant to reach “closed loop” conditions. • Other Case Startup Conditions • IAT sensor signal from -22°F to 20°F, engine runtime under 450 seconds, accumulated airflow over 3600 grams, and the PCM detected it took over 600 seconds for the coolant to reach “closed loop” conditions **Possible Causes:** • Inspect for low coolant level or an incorrect coolant mixture • Check the operation of the thermostat (it may be stuck open) • ECT sensor is damaged or out-of-calibration (it is “skewed”) • ECT sensor signal circuit has high resistance • ECT sensor has failed • PCM has failed

DTC	Trouble Code Title, Conditions & Possible Causes
DTC: P0125 **2T CCM, MIL: YES** **1996, 1997, 1998, 1999, 2000, 2001, 2002** **Models:** Amigo, Axiom, Rodeo, Rodeo Sport, Trooper, VehiCROSS **Engines:** 3.2L VIN V, 3.2L VIN W, 3.5L VIN X **Transmissions:** A/T, M/T	**Insufficient Coolant Temperature For Closed Loop** DTC P0112, P0113, P0117, P0118, P1111, P1112, P1114 and P1115 not set, engine started, ECT and IAT sensors from 14-82°F at startup, then with the IAT sensor from 17-50°F at startup, the PCM detected the ECT signal did not reach 84°F after 20 minutes; or with the IAT sensor more than 50°F at startup, the PCM detected the ECT sensor signal was less than 84°F after 2 minutes, condition met at least 20 times during the CCM Rationality test. **Possible Causes:** • Inspect for low coolant level or an incorrect coolant mixture • Check the operation of the thermostat (it may be stuck open) • ECT sensor is damaged or out-of-calibration (it is "skewed") • ECT sensor signal circuit has high resistance • ECT sensor has failed • PCM has failed
DTC: P0125 **2T CCM, MIL: YES** **1996, 1997, 1998, 1999, 2000** **Models:** Hombre **Engines:** 2.2L VIN 4 **Transmissions:** A/T, M/T	**Insufficient Coolant Temperature For Closed Loop** Region 1: Engine started, engine runtime over 120 seconds, maximum idle time less than 90 seconds, ECT sensor and IAT sensor signals more than 50°F. Region 2: Engine started, engine runtime over 300 seconds, maximum idle time less than 225 seconds, ECT sensor and IAT sensor signals more than 45°F. Region 3: Engine started, engine runtime over 1350 seconds, maximum idle time less than 1015 seconds, ECT sensor more than 45°F and IAT sensor signals less than 45°F. Then with DTC P0112, P0113, P0117 and P0118 not set, engine at idle speed with the minimum Calculated airflow less than 10 g/sec, the PCM detected the ECT sensor signal was less than 113°F. **Possible Causes:** • Inspect for low coolant level or an incorrect coolant mixture • Check the operation of the thermostat (it may be stuck open) • ECT sensor is damaged or out-of-calibration (it is "skewed") • ECT sensor signal circuit has high resistance • ECT sensor has failed • PCM has failed
DTC: P0125 **2T CCM, MIL: YES** **1997, 1998, 1999, 2000** **Models:** Hombre **Engines:** 4.3L VIN X **Transmissions:** A/T, M/T	**Insufficient Coolant Temperature For Closed Loop** DTC P0112, P0113, P0117 and P0118 not set, then with the ECT and IAT sensors both less than -40°F at startup, accumulated idle time under 6 minutes and airflow since startup more than 7500 grams, the PCM detected the ECT sensor signal was less than 68°F after 8 minutes; or with the ECT and IAT sensors both more than 20°F at startup, accumulated idle time under 225 seconds and airflow since startup more than 4500 grams, the ECT sensor signal was less than 68°F after 5 minutes; or with the ECT and IAT sensors both more than 50°F at startup, accumulated idle time under 90 seconds and airflow since startup more than 1600 grams, the ECT sensor signal was less than 68°F after 2 minutes in the CCM test. **Possible Causes:** • Inspect for low coolant level or an incorrect coolant mixture • Check the operation of the thermostat (it may be stuck open) • ECT sensor is damaged or out-of-calibration (it is "skewed") • ECT sensor signal circuit has high resistance • ECT sensor has failed • PCM has failed
DTC: P0125 **2T CCM, MIL: YES** **1996, 1997** **Models:** Rodeo **Engines:** 2.6L VIN E **Transmissions:** M/T	**Insufficient Coolant Temperature For Closed Loop** DTC P0112, P0113, P0117, P0118, P1111, P1112, P1114 and P1115 not set, engine started, ECT sensor less than -20°F and the IAT sensor less than 50°F at startup, the PCM detected the ECT sensor signal was less than 84°F after 5 minutes; or with the ECT and IAT sensor both more than 50°F at startup, the PCM detected the ECT sensor signal was less than 84°F after 2 minutes, condition met 20 times during the CCM Rationality test. **Possible Causes:** • Inspect for low coolant level or an incorrect coolant mixture • Check the operation of the thermostat (it may be stuck open) • ECT sensor is damaged or out-of-calibration (it is "skewed") • ECT sensor signal circuit has high resistance • ECT sensor has failed • PCM has failed

DTC	Trouble Code Title, Conditions & Possible Causes
DTC: P0128 **2T CCM, MIL: YES** **2000, 2001, 2002** **Models:** Amigo, Axiom, Rodeo, Rodeo Sport, Trooper, VehiCROSS **Engines:** All **Transmissions:** A/T, M/T	**Thermostat Malfunction** DTC P0101, P0102, P0103, P0112, P0113, P0117, P0118 and P0502 not set, and the PCM detected under one of these cases: Cold Case Startup Conditions • IAT sensor from 20-50ºF, and the PCM detected it took over 263 seconds to reach a stabilized thermostat regulated temperature. • Warm Case Startup Conditions • IAT sensor from 50-128ºF, and the PCM detected it took over 239 seconds to reach a stabilized thermostat regulated temperature. **Possible Causes:** • Check the operation of the thermostat (it may be stuck open) • ECT sensor is damaged or out-of-calibration (it is "skewed") • PCM has failed
DTC: P0131 **1T CCM, MIL: YES** **1998, 1999, 2000, 2001, 2002** **Models:** Amigo, Rodeo, Rodeo Sport **Engines:** 2.2L VIN D **Transmissions:** A/T, M/T	**HO2S-11 (Bank 1 Sensor 1) Circuit Low Input** DTC P0106, P0107, P0108, P0112, P0113, P0117, P0118, P0121, P0122, P0123, P0171, P0172, P0300, P0301-P0304, P0404, P0405, P0506, P0507 and P0601 not set, ECT sensor more than 140ºF, engine running in closed loop with A/F ratio from 14.5-14.8:1, throttle angle from 3-19%, and the PCM detected the HO2S-11 signal was less than 22 mv for 77 seconds over a 90 second period. **Possible Causes:** • HO2S signal circuit is open or shorted to ground • HO2S is water or fuel contaminated, or it has failed • PCM has failed
DTC: P0131 **1T CCM, MIL: YES** **1996, 1997, 1998, 1999, 2000, 2001, 2002** **Models:** Amigo, Axiom, Rodeo, Rodeo Sport, Trooper, VehiCROSS **Engines:** 3.2L VIN V, 3.2L VIN W, 3.5L VIN X **Transmissions:** A/T, M/T	**HO2S-11 (Bank 1 Sensor 1) Circuit Low Input** DTC P0106, P0107, P0108, P0112, P0113, P0117, P0118, P0121, P0122, P0123, P0171, P0172 and P0300, P0301-P0306 not set, engine running in closed loop with the A/F ratio from 14.5-14.8:1, ECT sensor more than 140ºF, throttle angle from 3-19% for 5 seconds, the PCM detected the HO2S-11 signal was less than 26 mv for 77 seconds over a 90 second period during the test. **Possible Cause:** • HO2S signal circuit is open or shorted to ground • HO2S is water or fuel contaminated, or it has failed • PCM has failed
DTC: P0131 **1T CCM, MIL: YES** **1996, 1997, 1998, 1999, 2000** **Models:** Hombre **Engines:** 2.2L VIN 4 **Transmissions:** A/T, M/T	**HO2S-11 (Bank 1 Sensor 1) Circuit Low Input** DTC P0106, P0107, P0108, P0112, P0113, P0117, P0118, P0121, P0122, P0123, P0171, P0172, P0200, P0300, P0301-P0304, P0325, P0335, P0341, P0342, P0404, P0405, P0506, P0507, P0601 and P0602 not set, ECT sensor signal more than 158ºF, Calculated airflow more than 3 g/sec, engine running in closed loop, throttle angle from 5-50%, conditions met for 20 seconds, and the PCM detected the front HO2S-11 signal was less than 44 mv for 125 seconds during the CCM test. **Possible Causes:** • HO2S signal circuit is open or shorted to ground • HO2S is water or fuel contaminated, or it has failed • PCM has failed
DTC: P0131 **1T CCM, MIL: YES** **1997, 1998, 1999, 2000** **Models:** Hombre **Engines:** 4.3L VIN X **Transmissions:** A/T, M/T	**HO2S-11 (Bank 1 Sensor 1) Circuit Low Input** DTC P0106, P0107, P0108, P0112, P0113, P0117, P0118, P0121, P0122, P0123, P0171, P0172, P0300, P0301-P0304, P0440, P0442 and P0446 not set, no Intrusive or Device Control tests active, then under Lean Test conditions (engine running in closed loop, throttle angle from 3.5-19% for 5 seconds), the PCM detected the HO2S-11 signal was less than 86 mv for 50 seconds; or with Power Enrichment conditions (engine running in closed loop, Power Enrichment Lean Test mode active and High Speed Fuel Cutoff mode inactive, elapsed time since the last test enabled over 1 second), the PCM detected the HO2S-11 signal was less than 598 mv for 30 seconds during the CCM test. **Possible Causes:** • HO2S signal circuit is open or shorted to ground • HO2S is water or fuel contaminated, or it has failed • PCM has failed
DTC: P0131 **1T CCM, MIL: YES** **1996, 1997, 1998, 1999** **Models:** Oasis **Engines:** All **Transmissions:** A/T	**HO2S-11 (Bank 1 Sensor 1) Circuit Low Input** Engine running in closed loop in D4 position at cruise speed, and the PCM detected the HO2S signal was fixed at less than 0.50v. **Note: The actual value to set the code is stored in the PCM memory.** **Possible Causes:** • HO2S signal circuit is open or it is shorted to ground • HO2S may be contaminated or may have failed • Fuel supply system is too lean (fuel filter is clogged or dirty) • PCM has failed

DTC	Trouble Code Title, Conditions & Possible Causes
DTC: P0131 **1T CCM, MIL: YES** **1996, 1997** **Models:** Rodeo **Engines:** 2.6L VIN E **Transmissions:** M/T	**HO2S-11 (Bank 1 Sensor 1) Circuit Low Input** DTC P0106, P0107, P0108, P0112, P0113, P0117, P0118, P0121, P0122, P0123, P0171, P0172 and P0300 not set, no Intrusive or Device Control tests active, engine started, engine running in closed loop with the A/F ratio command at 14.5-14.8:1, throttle angle from 3-19% for 5 seconds, the PCM detected the HO2S-11 signal was less than 26 mv for 77 seconds over a 90 second period in the test. **Possible Causes:** • HO2S signal circuit is open or shorted to ground • HO2S is water or fuel contaminated, or it has failed • PCM has failed
DTC: P0132 **1T CCM, MIL: YES** **1998, 1999, 2000, 2001, 2002** **Models:** Amigo, Rodeo, Rodeo Sport **Engines:** 2.2L VIN D **Transmissions:** A/T, M/T	**HO2S-11 (Bank 1 Sensor 1) Circuit High Input** No HO2S codes set, ECT sensor signal more than 140ºF, engine running in closed loop with the A/F ratio at 14.5-14.8:1, throttle angle from 3-19%, conditions met for 300 ms, and the PCM detected the front HO2S-11 signal was more than 952 mv for 76.5 seconds out of a 900 second period of time, or it was over 500 mv in DFCO mode. **Possible Causes:** • HO2S signal circuit shorted to system power (oil in connector) • HO2S is water or fuel contaminated • HO2S is damaged or has failed • PCM has failed
DTC: P0132 **1T CCM, MIL: YES** **1996, 1997, 1998, 1999, 2000, 2001, 2002** **Models:** Amigo, Axiom, Rodeo, Rodeo Sport, Trooper, VehiCROSS **Engines:** 3.2L VIN V, 3.2L VIN W, 3.5L VIN X **Transmissions:** A/T, M/T	**HO2S-11 (Bank 1 Sensor 1) Circuit High Input** DTC P0106, P0107, P0108, P0112, P0113, P0117, P0118, P0121, P0122, P0123, P0171, P0172 and P0300, P0301-P0306 not set, engine running in closed loop with the A/F ratio from 14.5-14.8:1, throttle angle from 3-19% for 5 seconds, the PCM detected the HO2S-11 signal was less than 952 mv for 77 seconds over a 90 second period; or the HO2S-11 signal was more than 500 mv during Decel Fuel Cutoff mode for 3 seconds during the CCM test. **Possible Causes:** • HO2S signal circuit shorted to system power (oil in connector) • HO2S is water or fuel contaminated • HO2S is damaged or has failed • PCM has failed
DTC: P0132 **1T CCM, MIL: YES** **1996, 1997, 1998, 1999, 2000** **Models:** Hombre **Engines:** 2.2L VIN 4 **Transmissions:** A/T, M/T	**HO2S-11 (Bank 1 Sensor 1) Circuit High Input** DTC P0106, P0107, P0108, P0112, P0113, P0117, P0118, P0121, P0122, P0123, P0171, P0172, P0200, P0300, P0301-P0304, P0325, P0335, P0341, P0342, P0404, P0405, P0506, P0507, P0601 and P0602 not set, ECT sensor signal more than 158ºF, Calculated airflow more than 3 g/sec, engine running in closed loop, throttle angle from 5-50%, conditions met for 20 seconds, and the PCM detected the front HO2S-11 signal was more than 946 mv for 50 seconds, or it was more than 1042 mv for 50 seconds while in Deceleration Fuel Cutoff (DFCO) mode during the CCM test. **Possible Causes:** • HO2S signal circuit shorted to system power (oil in connector) • HO2S is water or fuel contaminated • HO2S is damaged or has failed • PCM has failed
DTC: P0132 **1T CCM, MIL: YES** **1997, 1998, 1999, 2000** **Models:** Hombre **Engines:** 4.3L VIN X **Transmissions:** A/T, M/T	**HO2S-11 (Bank 1 Sensor 1) Circuit High Input** DTC P0106, P0107, P0108, P0112, P0113, P0117, P0118, P0121, P0122, P0123, P0171, P0172, P0200, P0300, P0301-P0304, P0440, P0442 and P0446 not set, no Intrusive or Device Control tests active, then under Rich Test conditions (running in closed loop, A/F ratio from 14.5-14.8:1, throttle angle from 0-50% for 5 seconds), the PCM detected the HO2S-11 signal was less than 976 mv for 40 seconds; or under Decel Fuel Cutoff conditions (running in closed loop, Decel Fuel Cutoff mode active, elapsed time since the last test enabled over 2 seconds), the PCM detected the HO2S-11 signal was more than 468 mv for 30 seconds during the CCM test. **Possible Causes:** • HO2S signal circuit shorted to system power (oil in connector) • HO2S is water or fuel contaminated • HO2S is damaged or has failed • PCM has failed
DTC: P0132 **1T CCM, MIL: YES** **1996, 1997, 1998, 1999** **Models:** Oasis **Engines:** All **Transmissions:** A/T	**HO2S-11 (Bank 1 Sensor 1) Circuit High Input** Engine running in closed loop in D4 position at cruise speed, and the PCM detected the HO2S signal was fixed at more than 0.90v **Note: The actual value to set the code is stored in the PCM memory.** **Possible Causes:** • HO2S signal circuit is shorted to system power • HO2S signal tracking (wet/oily) in connector causing a short between the signal circuit and heater power circuit • PCM has failed

DTC	Trouble Code Title, Conditions & Possible Causes
DTC: P0132 **1T CCM, MIL: YES** **1996, 1997** **Models**: Rodeo **Engines:** 2.6L VIN E **Transmissions:** M/T	**HO2S-11 (Bank 1 Sensor 1) Circuit High Input** DTC P0106, P0107, P0108, P0112, P0113, P0117, P0118, P0121, P0122, P0123, P0171, P0172 and P0300 not set, no Device Control or Intrusive tests active, engine running in closed loop with the A/F ratio at 14.5-14.8:1, throttle angle from 3-19% for 5 seconds, and the PCM detected the HO2S-11 signal was less than 952 mv for 77 seconds over a 90 second period; or the HO2S-11 signal was more than 500 mv during Decel Fuel Cutoff mode for 3 seconds. **Possible Causes:** • HO2S signal circuit shorted to system power (oil in connector) • HO2S is water or fuel contaminated • HO2S is damaged or has failed • PCM has failed
DTC: P0133 **2T OBD/O2S, MIL: YES** **1998, 1999, 2000, 2001, 2002** **Models:** Amigo, Rodeo, Rodeo Sport **Engines:** 2.2L VIN D Transmissions: A/T, M/T	**HO2S-11 (Bank 1 Sensor 1) Slow Response** No HO2S-11 codes set, engine runtime 2 minutes, engine speed from 1500-3000 rpm in closed loop for 90 seconds, ECT sensor signal over 140ºF, Purge duty cycle over 2%, calculated airflow from 17-32 g/sec, system voltage more than 10v, and PCM detected the front HO2S-11 lean-to-rich response time was over 100 ms, or the rich-to-lean time was over 150 ms while in the 300-600 mv range. **Possible Causes:** • Exhaust leak present in the exhaust manifold or exhaust pipes • HO2S element fuel contamination • HO2S element has deteriorated • PCM has failed
DTC: P0133 **2T OBD/O2S, MIL: YES** **1996, 1997, 1998, 1999, 2000, 2001, 2002** **Models:** Amigo, Axiom, Rodeo, Rodeo Sport, Trooper, VehiCROSS Engines: 3.2L VIN V, 3.2L VIN W, 3.5L VIN X **Transmissions:** A/T, M/T	**HO2S-11 (Bank 1 Sensor 1) Slow Response** DTC P0106, P0107, P0108, P0112, P0113, P0117, P0118, P0121, P0122, P0123, P0171, P0172 and P0300, P0301-P0306 not set, engine runtime 1 minute in closed loop, ECT sensor more than 122ºF, engine speed from 1500-3000 rpm, MAF sensor from 9-42 g/sec, Purge duty cycle over 1%, conditions met for 3 seconds, then 90 seconds after entering closed loop, the PCM detected the HO2S-11 lean-to-rich average transition time was over 94 ms, or the rich-to-lean average transition time was over 105 ms during the test. **Possible Causes:** • Exhaust leak present in the exhaust manifold or exhaust pipes • HO2S element fuel contamination • HO2S element has deteriorated • PCM has failed
DTC: P0133 **2T OBD/O2S, MIL: YES** **1996, 1997, 1998, 1999, 2000** **Models:** Hombre Engines: 2.2L VIN 4 **Transmissions:** A/T, M/T	**HO2S-11 (Bank 1 Sensor 1) Slow Response** DTC P0106, P0107, P0108, P0112, P0113, P0117, P0118, P0121, P0122, P0123, P0171, P0172, P0200, P0300, P0301-P0304, P0341, P0404, P0405, P0506, P0507, P0601 and P0602 not set, system voltage from 10-16v, ECT sensor over 140ºF, engine speed from 1600-2400 rpm in closed loop, Purge duty cycle over 60%, Purge learn memory over 191, conditions met for 100 seconds, and the PCM detected the HO2S-11 lean-to-rich or rich-to-lean response time was over 249 ms, or the ratio of response times was more than 3.75, or the ratio of response times was less than 0.4 in the test. **Possible Causes:** • Exhaust leak present in the exhaust manifold or exhaust pipes • HO2S element fuel contamination • HO2S element has deteriorated • PCM has failed
DTC: P0133 **2T OBD/O2S, MIL: YES** **1997, 1998, 1999, 2000** **Models:** Hombre **Engines:** 4.3L VIN X **Transmissions:** A/T, M/T	**HO2S-11 (Bank 1 Sensor 1) Slow Response** DTC P0101, P0102, P0103, P0106, P0107, P0108, P0112, P0113, P0117, P0118, P0121, P0122, P0123, P0131, P0132, P0134, P0135, P0300, P0301-P0304, P0341, P0440, P0442 and P0446 not set, No Intrusive or Device Control tests enabled, system voltage from 11-16v, ECT sensor more than 135ºF, engine speed from 1100-3000 rpm in closed loop for 100 seconds, MAF sensor signal from 15-55 g/sec, Purge duty cycle enabled, and the PCM detected the HO2S-11 average lean-to-rich or the rich-to-lean response time was over 125 ms during the Oxygen Sensor Monitor test. **Possible Causes:** • Exhaust leak present in the exhaust manifold or exhaust pipes • HO2S element fuel contamination • HO2S element has deteriorated • PCM has failed
DTC: P0133 **1T OBD/O2S1, MIL: YES** **1996, 1997, 1998, 1999** **Models:** Oasis **Engines:** All **Transmissions:** A/T	**HO2S-11 (Bank 1 Sensor 1) Circuit Slow Response** Engine running in closed loop in D4 position at over 55 mph at steady speed, and the PCM detected the HO2S response time to switch between 300-600 mv was too slow, or that the rich to lean or lean to rich switch time was too slow. **Possible Causes:** • Exhaust leak present in the exhaust manifold or exhaust pipes • O2S element fuel contamination • O2S element has deteriorated • PCM has failed

DTC	Trouble Code Title, Conditions & Possible Causes
DTC: P0133 **2T OBD/O2S, MIL: YES** **1996, 1997** **Models:** Rodeo **Engines:** 2.6L VIN E **Transmissions:** A/T, M/T	**HO2S-11 (Bank 1 Sensor 1) Slow Response** DTC P0106, P0107, P0108, P0112, P0113, P0117, P0118, P0121, P0122, P0123, P0171, P0172 and P0300 not set, engine runtime over 1 minute, ECT sensor more than 122ºF, engine speed from 1500-3000 rpm, Calculated airflow from 9-42 g/sec, conditions met for 3 seconds, then 90 seconds after entering closed loop, the PCM detected the HO2S-11 lean-to-rich average transition time was over 85 ms, or the rich-to-lean average transition time was 100 ms during the Oxygen Sensor Monitor test. **Possible Causes:** • Exhaust leak present in the exhaust manifold or exhaust pipes • HO2S element fuel contamination • HO2S element has deteriorated • PCM has failed
DTC: P0134 **2T OBD/O2S, MIL: YES** **1998, 1999, 2000, 2001, 2002** **Models:** Amigo, Rodeo, Rodeo Sport **Engines:** 2.2L VIN D **Transmissions:** A/T, M/T	**HO2S-11 (Bank 1 Sensor 1) Insufficient Activity Detected** No related codes set, system voltage from 11-16v, engine runtime 120 seconds, front HO2S heater test passed and the HO2S signal varying, and the PCM detected the front HO2S-11 signal was fixed between 400-500 mv for 77 seconds over a 90 second period of time in the Oxygen Sensor Monitor test. **Possible Causes:** • Exhaust leak present in exhaust manifold or exhaust pipes • HO2S element fuel contamination or has deteriorated • HO2S signal circuit or the ground circuit has high resistance • HO2S heater element has failed, or the heater circuit is open • PCM has failed
DTC: P0134 **1T OBD/O2S, MIL: YES** **1996, 1997, 1998, 1999, 2000, 2001, 2002** **Models:** Amigo, Axiom, Rodeo, Rodeo Sport, Trooper, VehiCROSS **Engines:** 3.2L VIN V, 3.2L VIN W, 3.5L VIN X **Transmissions:** A/T, M/T	**HO2S-11 (Bank 1 Sensor 1) Insufficient Activity Detected** DTC P0106, P0107, P0108, P0112, P0113, P0117, P0118, P0121, P0122, P0123, P0171, P0172 and P0300, P0301-306 not set, system voltage from 11-16v, engine runtime over 40 seconds, then after the PCM determined the Oxygen Sensor Heater test passed, it detected the HO2S-11 signal remained from 400-500 mv for 77 seconds over a 90 second period in the Oxygen Sensor Monitor test. **Possible Causes:** • Exhaust leak present in exhaust manifold or exhaust pipes • HO2S element fuel contamination or has deteriorated • HO2S signal circuit or the ground circuit has high resistance • HO2S heater element has failed, or the heater circuit is open • PCM has failed
DTC: P0134 **2T OBD/O2S, MIL: YES** **1996, 1997, 1998, 1999, 2000** **Models:** Hombre **Engines:** 2.2L VIN 4 **Transmissions:** A/T, M/T	**HO2S-11 (Bank 1 Sensor 1) Insufficient Activity Detected** DTC P0106, P0107, P0108, P0112, P0113, P0117, P0118, P0121, P0122, P0123, P0171, P0172, P0200, P0300, P0301-P0304, P0341, P0404, P0506, P0507, P0601 and P0602 not set, system voltage from 11-16v, engine runtime over 30 seconds, Calculated airflow over 3 g/sec, ECT sensor signal more than 158ºF, throttle angle from 4-56% for 20 seconds, and the PCM detected the front HO2S-11 signal was fixed between 399-499 mv for 125 seconds during the Oxygen Sensor Monitor test. **Possible Causes:** • Exhaust leak present in exhaust manifold or exhaust pipes • HO2S element fuel contamination or has deteriorated • HO2S signal circuit or the ground circuit has high resistance • HO2S heater element has failed, or the heater circuit is open • PCM has failed
DTC: P0134 **1T OBD/O2S, MIL: YES** **1996, 1997** **Models:** Rodeo **Engines:** 2.6L VIN E **Transmissions:** M/T Trouble Code ID: P0134	**HO2S-11 (Bank 1 Sensor 1) Insufficient Activity Detected** DTC P0106, P0107, P0108, P0112, P0113, P0117, P0118, P0121, P0122, P0123, P0171, P0172 and P0300 not set, system voltage from 10-16v, engine runtime over 90 seconds, then after the PCM determined the Oxygen Sensor Heater test passed, it detected the HO2S-11 signal remained from 400-500 mv for 77 seconds over a 90 second period during the Oxygen Sensor Monitor test. **Possible Causes:** • Exhaust leak present in exhaust manifold or exhaust pipes • HO2S element fuel contamination or has deteriorated • HO2S signal circuit or the ground circuit has high resistance • HO2S heater element has failed, or the heater circuit is open • PCM has failed

DTC	Trouble Code Title, Conditions & Possible Causes
DTC: P0134 **2T OBD/O2S, MIL: YES** **1997, 1998, 1999, 2000** **Models:** Hombre **Engines:** 4.3L VIN X **Transmissions:** A/T, M/T	**HO2S-11 (Bank 1 Sensor 1) Insufficient Activity Detected** DTC P0101, P0102, P0103, P0106, P0107, P0108, P0112, P0113, P0117, P0118, P0121, P0122, P0123, P0135, P0300, P0301-P0304, P0440, P0442 and P0446 not set, no Intrusive or Device Control tests active, system voltage from 11-16v, engine runtime over 2 minutes, then during the Oxygen Sensor Temperature Test (ECT sensor signal more than 137ºF, MAF sensor signal more than 13 g/sec, and Decel Fuel Cutoff not active); or with the Oxygen Sensor Open Test enabled (DTC P0135 not active, engine running in closed loop), the PCM detected the HO2S-11 signal was fixed at 300-600 mv for 80 seconds during the Oxygen Sensor Monitor test. **Possible Causes:** • Exhaust leak present in exhaust manifold or exhaust pipes • HO2S element fuel contamination or has deteriorated • HO2S signal circuit or the ground circuit has high resistance • HO2S heater element has failed, or the heater circuit is open • PCM has failed
DTC: P0135 **2T OBD/O2S HTR1, MIL: YES** **1998, 1999, 2000, 2001, 2002** **Models:** Amigo, Rodeo, Rodeo Sport **Engines:** 2.2L VIN D **Transmissions:** A/T, M/T	**HO2S-11 (Bank 1 Sensor 1) Heater Circuit Malfunction** No HO2S-11 codes set, ECT and IAT sensor signals less than 90ºF, and within 14ºF of each other at startup, system voltage at 11-16v, throttle angle under 40%, average calculated airflow less than 18 g/sec in the sample period, and the PCM detected the HO2S-11 signal did not vary more than 150 mv from the bias voltage of 400 to 500 mv for a long period of time (maximum time is 240 seconds). **Possible Causes:** • HO2S heater power circuit is open (from the O2S heater fuse) • HO2S heater ground circuit is open • HO2S heater element has high resistance • HO2S heater element has failed (open or shorted) • PCM has failed
DTC: P0135 **2T OBD/O2S HTR1, MIL: YES** **1996, 1997, 1998, 1999, 2000, 2001, 2002** **Models:** Amigo, Axiom, Rodeo, Rodeo Sport, Trooper, VehiCROSS **Engines:** 3.2L VIN V, 3.2L VIN W, 3.5L VIN X **Transmissions:** A/T, M/T	**HO2S-11 (Bank 1 Sensor 1) Heater Circuit Malfunction** No HO2S-11 codes set, ECT and IAT sensors less than 90ºF and within 14ºF at startup, engine running, system voltage from 11-16v, then with the average Calculated airflow less than 15 g/sec during the test period, the PCM detected the HO2S-11 signal did not vary more than 150 mv from the bias voltage of 400 to 500 mv for up to 150 seconds during the Oxygen Sensor Heater Monitor test. **Possible Causes:** • HO2S heater power circuit is open (check the 20A heater fuse) • HO2S heater ground circuit is open • HO2S heater element has high resistance or has failed • PCM has failed
DTC: P0135 **2T OBD/O2S HTR1, MIL: YES** **1997, 1998, 1999, 2000** **Models:** Hombre **Engines:** 4.3L VIN X **Transmissions:** A/T, M/T	**HO2S-11 (Bank 1 Sensor 1) Heater Circuit Malfunction** No HO2S-11 codes set, ECT and IAT sensor signals less than 91ºF and within 9ºF of each other at startup, system voltage from 11-16v, engine runtime over 2 seconds, MAF sensor signal less than 27 g/sec, and the PCM detected the HO2S-11 signal did not vary more than 150 mv from the bias voltage of 400 to 500 mv for 130 seconds under these conditions in the Oxygen Sensor Heater Monitor test. **Possible Causes:** • HO2S heater power circuit is open (check the O2S heater fuse) • HO2S heater ground circuit is open • HO2S heater element has high resistance • HO2S heater element has failed (open or shorted) • PCM has failed
DTC: P0135 **1T OBD/O2S HTR1, MIL: YES** **1996, 1997, 1998, 1999** **Models:** Oasis **Engines:** All **Transmissions:** A/T, M/T	**HO2S-11 (Bank 1 Sensor 1) Heater Circuit Malfunction** Engine runtime over 80 seconds, and the PCM detected an incorrect signal value at the HO2S heater circuit during the test period. **Possible Causes:** • Main relay output (power) circuit to the heater is open • O2S heater ground circuit is open • O2S heater element has high resistance • O2S heater element has an open condition • O2S heater element has a shorted condition • PCM has failed
DTC: P0135 **2T OBD/O2S HTR1, MIL: YES** **1996, 1997** **Models:** Rodeo **Engines:** 2.6L VIN E **Transmissions:** M/T	**HO2S-11 (Bank 1 Sensor 1) Heater Circuit Malfunction** No HO2S-11 codes set, ECT and IAT sensors less than 90ºF and within 14ºF at startup, engine running, system voltage from 10-16v, then with the average Calculated airflow less than 15 g/sec during the test period, the PCM detected the HO2S-11 signal did not vary more than 150 mv from the bias voltage of 400 to 500 mv after 150 seconds during the Oxygen Sensor Heater Monitor test. **Possible Causes:** • HO2S heater power circuit is open (check the 20A heater fuse) • HO2S heater ground circuit is open • HO2S heater element has high resistance or has failed • PCM has failed

DTC	Trouble Code Title, Conditions & Possible Causes
DTC: P0137 **1T CCM, MIL: YES** **1998, 1999, 2000, 2001, 2002** **Models:** Amigo, Rodeo, Rodeo Sport **Engines:** 2.2L VIN D **Transmissions:** A/T, M/T	**HO2S-12 (Bank 1 Sensor 2) Circuit Low Input** DTC P0106, P0107, P0108, P0112, P0113, P0117, P0118, P0121, P0122, P0123, P0171, P0172, P0300, P0301-P0304, P0341, P0404, P0506, P0507 and P0601 not set, ECT sensor more than 140ºF, engine running in closed loop with A/F ratio at 14.5-14.8:1, throttle angle from 3-19%, and the PCM detected the HO2S-12 signal was less than 22 mv for 77 seconds over a 90 second period. **Possible Causes:** • HO2S signal circuit is open or shorted to ground • HO2S is water or fuel contaminated • HO2S is damaged or has failed • PCM has failed
DTC: P0137 **1T CCM, MIL: YES** **1996, 1997, 1998, 1999, 2000, 2001, 2002** **Models:** Amigo, Axiom, Rodeo, Rodeo Sport, Trooper, VehiCROSS **Engines:** 3.2L VIN V, 3.2L VIN W, 3.5L VIN X **Transmissions:** A/T, M/T	**HO2S-12 (Bank 1 Sensor 2) Circuit Low Input** DTC P0106, P0107, P0108, P0112, P0113, P0117, P0118, P0121, P0122, P0123, P0171, P0172 and P0300, P0301-P0306 not set, engine running in closed loop with the A/F ratio from 14.5-14.8:1, ECT sensor more than 140ºF, throttle angle from 3-19% for 5 seconds, the PCM detected the HO2S-12 signal was less than 26 mv for 106 seconds over a 125 second period during the CCM test. **Possible Causes:** • HO2S signal circuit is open or shorted to ground • HO2S is water or fuel contaminated, or it has failed • PCM has failed
DTC: P0137 **1T CCM, MIL: YES** **1996, 1997, 1998, 1999, 2000** **Models:** Hombre **Engines:** 2.2L VIN 4 **Transmissions:** A/T, M/T	**HO2S-12 (Bank 1 Sensor 2) Circuit Low Input** DTC P0106, P0107, P0108, P0112, P0113, P0117, P0118, P0121, P0122, P0123, P0171, P0172, P0200, P0300, P0301-P0304, P0341, P0404, P0506, P0507, P0601 and P0602 not set, system voltage from 11-16v, ECT sensor signal more than 104ºF, Calculated airflow more than 5.5 g/sec, throttle angle from 5-50%, and the PCM detected the front HO2S-12 signal was less than 22 mv for 150 seconds during the CCM test. **Possible Causes:** • HO2S signal circuit is open or shorted to ground • HO2S is water or fuel contaminated • HO2S is damaged or has failed • PCM has failed
DTC: P0137 **1T CCM, MIL: YES** **1996, 1997, 1998, 1999** **Models:** Oasis **Engines:** All **Transmissions:** A/T	**HO2S-12 (Bank 1 Sensor 2) Circuit Low Input** Engine running in closed loop in D4 position at cruise speed, and the PCM detected the HO2S signal was fixed at less than 0.30v. **Note: The actual value where the code sets is in the PCM memory.** **Possible Causes:** • HO2S signal circuit is open • HO2S signal circuit is shorted to ground • HO2S ground circuit is open • HO2S may be contaminated or may have failed • PCM has failed
DTC: P0137 **1T CCM, MIL: YES** **1996, 1997** **Models:** Rodeo **Engines:** 2.6L VIN E **Transmissions:** M/T	**HO2S-12 (Bank 1 Sensor 2) Circuit Low Input** DTC P0106, P0107, P0108, P0112, P0113, P0117, P0118, P0121, P0122, P0123, P0171, P0172 and P0300 not set, no Intrusive or Device Control tests active, engine started, engine running in closed loop with the A/F ratio command at 14.5-14.8:1, throttle angle from 3-19% for 5 seconds, the PCM detected the HO2S-12 signal was less than 26 mv for 106 seconds over a 125 second period, or it detected the HO2S-12 signal remained below 400 mv during Power Enrichment mode for 5 seconds during the CCM Rationality test. **Possible Causes:** • HO2S signal circuit is open or shorted to ground • HO2S is water or fuel contaminated, or it has failed • PCM has failed
DTC: P0138 **1T CCM, MIL: YES** **1998, 1999, 2000, 2001, 2002** **Models:** Amigo, Rodeo, Rodeo Sport **Engines:** 2.2L VIN D **Transmissions:** A/T, M/T	**HO2S-12 (Bank 1 Sensor 2) Circuit High Input** No HO2S-12 codes set, ECT sensor signal more than 140ºF, engine running in closed loop with the A/F ratio at 14.5-14.8:1, throttle angle from 3-19%, and PCM detected the front HO2S-11 signal was more than 952 mv for 102 seconds of a 125 second period of time, or it was over 500 mv for over 3 seconds while operating in DFCO mode. **Possible Causes:** • HO2S signal circuit shorted to system power (oil in connector) • HO2S is water or fuel contaminated, damaged or it has failed • PCM has failed

DTC	Trouble Code Title, Conditions & Possible Causes
DTC: P0138 **1T CCM, MIL: YES** **1996, 1997, 1998, 1999, 2000, 2001, 2002** **Models:** Amigo, Axiom, Rodeo, Rodeo Sport, Trooper, VehiCROSS **Engines:** 3.2L VIN V, 3.2L VIN W, 3.5L VIN X Transmissions: A/T, M/T	**HO2S-12 (Bank 1 Sensor 2) Circuit High Input** DTC P0106, P0107, P0108, P0112, P0113, P0117, P0118, P0121, P0122, P0123, P0171, P0172 and P0300, P0301-P0306 not set, engine running in closed loop with the A/F ratio command at 14.5-14.8:1, throttle angle from 3-19% for 5 seconds, the PCM detected the HO2S-12 signal was less than 952 mv for 106 seconds over a 125 second period; or the HO2S-12 signal was more than 500 mv during Decel Fuel Cutoff mode for 3 seconds during the CCM test. **Possible Causes:** • HO2S signal circuit shorted to system power (oil in connector) • HO2S is water or fuel contaminated • HO2S is damaged or has failed • PCM has failed
DTC: P0138 **1T CCM, MIL: YES** **1996, 1997, 1998, 1999, 2000** **Models:** Hombre **Engines:** 2.2L VIN 4 **Transmissions:** A/T, M/T	**HO2S-12 (Bank 1 Sensor 2) Circuit High Input** DTC P0106, P0107, P0108, P0112, P0113, P0117, P0118, P0121, P0122, P0123, P0171, P0172, P0200, P0300, P0301-P0304, P0341, P0404, P0506, P0507, P0601 and P0602 not set, system voltage from 11-16v, ECT sensor signal more than 104ºF, Calculated airflow more than 5.5 g/sec, throttle angle from 5-50%, conditions met for 300 ms, and the PCM detected the front HO2S-12 signal was more than 1042 mv for 50-75 seconds during the test. **Possible Causes:** • HO2S signal circuit shorted to system power (oil in connector) • HO2S is water or fuel contaminated, damaged or it has failed • PCM has failed
DTC: P0138 **1T CCM, MIL: YES** **1996, 1997, 1998, 1999** **Models:** Oasis **Engines:** All **Transmissions:** A/T	**HO2S-12 (Bank 1 Sensor 2) Circuit High Input** Engine running in closed loop in D4 position at cruise speed, and the PCM detected the HO2S signal was fixed at more than 0.60v **Note: The actual value where the code sets is in the PCM memory.** **Possible Causes:** • HO2S signal circuit is shorted to system power • HO2S signal tracking (wet/oily) in connector causing a short between the signal circuit and heater power circuit • PCM has failed
DTC: P0138 **1T CCM, MIL: YES** **1996, 1997** **Models:** Rodeo **Engines:** 2.6L VIN E **Transmissions:** M/T	**HO2S-12 (Bank 1 Sensor 2) Circuit High Input** DTC P0106, P0107, P0108, P0112, P0113, P0117, P0118, P0121, P0122, P0123, P0171, P0172 and P0300 not set, no Intrusive or Device Control tests active, engine running in closed loop with the A/F ratio command at 14.5-14.8:1, throttle angle from 3-19% for 5 seconds, and the PCM detected the HO2S-12 signal was less than 952 mv for 106 seconds over a 125 second period, or it detected the HO2S-12 signal was above 500 mv in Decel Fuel Cutoff (DFCO) mode for up to 5 seconds during the CCM Rationality test. **Possible Causes:** • HO2S signal circuit shorted to system power (oil in connector) • HO2S is water or fuel contaminated, damaged or it has failed • PCM has failed
DTC: P0139 **2T OBD/O2S1, MIL: YES** **1996, 1997, 1998, 1999** **Models:** Oasis Engines: All **Transmissions:** A/T	**HO2S-12 (Bank 1 Sensor 2) Circuit Slow Response** Engine running in closed loop in D4 position at over 55 mph at steady speed, and the PCM detected the HO2S response time to switch between 300-600 mv was too slow, or that the rich to lean or lean to rich switch time was too slow. **Possible Causes:** • Exhaust leak present in the exhaust manifold or exhaust pipes • HO2S element fuel contamination • HO2S element has deteriorated • PCM has failed
DTC: P0140 **1T OBD/O2S, MIL: YES** **1998, 1999, 2000, 2001, 2002** **Models:** Amigo, Rodeo, Rodeo Sport **Engines:** 2.2L VIN D **Transmissions:** A/T, M/T	**HO2S-12 (Bank 1 Sensor 2) Insufficient Activity Detected** No HO2S-12 codes set, system voltage over 10v, engine runtime over 2 minutes, HO2S-12 heater test passed, HO2S-12 signal is varying, and the PCM detected the HO2S-12 signal was fixed between 426-474 mv for 106 seconds over a 125 second period of time during the Oxygen Sensor Monitor test. **Possible Causes:** • Exhaust leak present in exhaust manifold or exhaust pipes • O2S element fuel contamination or has deteriorated • O2S signal circuit or the ground circuit has high resistance • PCM has failed
DTC: P0140 **1T OBD/O2S, MIL: YES** **1996, 1997, 1998, 1999, 2000, 2001, 2002** **Models:** Amigo, Axiom, Rodeo, Rodeo Sport, Trooper, VehiCROSS **Engines:** 3.2L VIN V, 3.2L VIN W, 3.5L VIN X **Transmissions:** A/T, M/T	**HO2S-12 (Bank 1 Sensor 2) Insufficient Activity Detected** DTC P0106, P0107, P0108, P0112, P0113, P0117, P0118, P0121, P0122, P0123, P0171, P0172 and P0300, P0301-306 not set, system voltage from 11-16v, engine runtime over 40 seconds, then after the PCM determined the Oxygen Sensor Heater test passed, it detected the HO2S-12 signal remained from 426-474 mv for 105 seconds of a 125 second period in the Oxygen Sensor Monitor test. **Possible Causes:** • Exhaust leak present in exhaust manifold or exhaust pipes • HO2S element fuel contamination or has deteriorated • HO2S signal circuit or the ground circuit has high resistance • HO2S heater element has failed, or the heater circuit is open • PCM has failed

DTC	Trouble Code Title, Conditions & Possible Causes
DTC: P0140 **2T OBD/O2S, MIL: YES** **1996, 1997, 1998, 1999, 2000** **Models:** Hombre **Engines:** 2.2L VIN 4 **Transmissions:** A/T, M/T	**HO2S-12 (Bank 1 Sensor 2) Insufficient Activity Detected** DTC P0106, P0107, P0108, P0112, P0113, P0117, P0118, P0121, P0122, P0123, P0171, P0172, P0200, P0300, P0301-P0304, P0341, P0404, P0506, P0507, P0601 and P0602 not set, system voltage from 11-16v, engine runtime over 30 seconds, Calculated airflow over 5.5 g/sec, ECT sensor signal more than 104ºF, throttle angle from 5-50%, and the PCM detected the front HO2S-12 signal was fixed between 425-460 mv for 125 seconds during the Oxygen Sensor Monitor test. **Possible Causes:** • Exhaust leak present in exhaust manifold or exhaust pipes • HO2S element fuel contamination or has deteriorated • HO2S signal circuit or the ground circuit has high resistance • PCM has failed
DTC: P0140 **1T OBD/O2S, MIL: YES** **1996, 1997** **Models:** Rodeo **Engines:** 2.6L VIN E Transmissions: M/T	**HO2S-12 (Bank 1 Sensor 2) Insufficient Activity Detected** DTC P0106, P0107, P0108, P0112, P0113, P0117, P0118, P0121, P0122, P0123, P0171, P0172 and P0300 not set, system voltage from 10-16v, engine runtime over 90 seconds, then after the PCM determined the Oxygen Sensor Heater test passed, it detected the HO2S-12 signal remained from 426-474 mv for 105 seconds over a 125 second period during the Oxygen Sensor Monitor test. **Possible Causes**: • Exhaust leak present in exhaust manifold or exhaust pipes • HO2S element fuel contamination or has deteriorated • HO2S signal circuit or the ground circuit has high resistance • HO2S heater element has failed, or the heater circuit is open • PCM has failed
DTC: P0141 **2T OBD/O2S HTR1, MIL: YES** **1998, 1999, 2000, 2001, 2002** **Models:** Amigo, Rodeo, Rodeo Sport **Engines:** 2.2L VIN D **Transmissions:** A/T, M/T	**HO2S-12 (Bank 1 Sensor 2) Heater Circuit Malfunction** No related codes set, ECT and IAT sensors less than 90ºF, and within 14ºF at startup, system voltage from 11-16v, throttle angle under 40%, average calculated airflow less than 18 g/sec in the sample period, and the PCM detected the HO2S-12 signal did not vary more than 150 mv from the bias voltage (400-500 mv) for too long a period of time (maximum time is 120 seconds). **Possible Causes:** • HO2S power circuit is open (from the O2S heater fuse) • HO2S heater ground circuit is open • HO2S heater element has high resistance or has failed • PCM has failed
DTC: P0141 **1T OBD/O2S HTR1, MIL: YES** **1996, 1997, 1998, 1999** **Models:** Oasis **Engines:** All **Transmissions:** A/T	**HO2S-12 (Bank 1 Sensor 2) Heater Circuit Malfunction** Engine runtime over 80 seconds, and the PCM detected an incorrect signal value at the HO2S heater circuit during the test period. **Possible Causes:** • Main relay output (power) circuit to the heater is open • HO2S heater ground circuit is open • HO2S heater element has high resistance • HO2S heater element has an open condition • HO2S heater element has a shorted condition • PCM has failed
DTC: P0141 **2T OBD/O2S HTR1, MIL: YES** **1996, 1997** **Models:** Rodeo **Engines:** 2.6L VIN E **Transmissions:** M/T	**HO2S-12 (Bank 1 Sensor 2) Heater Circuit Malfunction** No HO2S-12 codes set, ECT and IAT sensors less than 90ºF and within 14ºF at startup, engine running, system voltage from 10-16v, then with the average Calculated airflow less than 15 g/sec during the test period, the PCM detected the HO2S-11 signal did not vary more than 150 mv from the bias voltage of 400 to 500 mv for up to 150 seconds during the Oxygen Sensor Heater Monitor test. **Possible Causes:** • HO2S heater power circuit is open (check the 20A heater fuse) • HO2S heater ground circuit is open • HO2S heater element has high resistance or has failed • PCM has failed
DTC: P0141 **2T OBD/O2S HTR1, MIL: YES** **1996, 1997, 1998, 1999, 2000, 2001, 2002** **Models:** Amigo, Axiom, Rodeo, Rodeo Sport, Trooper, VehiCROSS **Engines:** 3.2L VIN V, 3.2L VIN W, 3.5L VIN X **Transmissions:** A/T, M/T	**HO2S-12 (Bank 1 Sensor 2) Heater Circuit Malfunction** No HO2S-12 codes set, ECT and IAT sensors less than 90ºF and within 11ºF at startup, engine running, system voltage from 10-16v, then with the average Calculated airflow less than 23 g/sec during the test period, the PCM detected the HO2S-12 signal did not vary more than 150 mv from the bias voltage of 400 to 500 mv for up to 300 seconds during the Oxygen Sensor Heater Monitor test. **Possible Causes:** • HO2S heater power circuit is open (check the 20A heater fuse) • HO2S heater ground circuit is open • HO2S heater element has high resistance or has failed • PCM has failed

DTC	Trouble Code Title, Conditions & Possible Causes
DTC: P0141 **2T OBD/O2S HTR1, MIL: YES** **1996, 1997, 1998, 1999, 2000** **Models:** Hombre **Engines:** 2.2L VIN 4 **Transmissions:** A/T, M/T	**HO2S-12 (Bank 1 Sensor 2) Heater Circuit Malfunction** No HO2S-12 codes set, ECT and IAT sensor signals less than 104°F, and within 45°F at startup, system voltage from 11-16v, throttle angle less than 18%, and the PCM detected the HO2S-12 signal changed less than 150 mv over a 100-150 second period of time during the Oxygen Sensor Heater Monitor test. **Possible Causes:** • HO2S power circuit is open (from the O2S heater fuse) • HO2S heater ground circuit is open • HO2S heater element has high resistance or has failed • PCM has failed
DTC: P0143 **2T CCM, MIL: YES** **1997, 1998, 1999, 2000** **Models:** Hombre **Engines:** 4.3L VIN X **Transmissions:** A/T, M/T	**HO2S-13 (Bank 1 Sensor 3) Circuit Low Input** No ECT, IAT, EVAP, MAF, MAP, Misfire, or TP codes set, system voltage from 11-16v, Catalyst or EGR Intrusive tests not active, no Device Control tests active, then during one of these test conditions: Lean Test – Engine running in closed loop, A/F ratio at 14.5-14.8:1, throttle angle from 3-19% for 2 seconds, and the PCM detected the HO2S-13 signal was less than 26 mv for more than 110 seconds. Power Enrichment Lean Test – Engine running in closed loop, Power Enrichment Mode "on", High Speed Fuel Cutoff "off", and the PCM detected the HO2S-13 signal was less than 399 mv for 40 seconds. **Possible Causes:** • HO2S signal circuit is open or shorted to ground • HO2S heater ground circuit is open • HO2S is water or fuel contaminated • HO2S is damaged or has failed • PCM has failed
DTC: P0143 **1T CCM, MIL: YES** **1996, 1997** **Models:** Rodeo **Engines:** 3.2L VIN V **Transmissions:** M/T	**HO2S-13 (Bank 1 Sensor 3) Circuit Low Input** DTC P0106, P0107, P0108, P0112, P0113, P0117, P0118, P0121, P0122, P0123, P0171, P0172 and P0300, P0301-P0306 not set, engine running in closed loop with the A/F ratio from 14.5-14.8:1, ECT sensor more than 140°F, throttle angle from 3-19% for 5 seconds, the PCM detected the HO2S-13 signal was less than 22 mv for 106 seconds over a 125 second period during the CCM test. **Possible Causes:** • HO2S signal circuit is open or shorted to ground • HO2S is water or fuel contaminated, or it has failed • PCM has failed
DTC: P0143 **1T CCM, MIL: YES** **1996, 1997, 1998, 1999** **Models:** Trooper **Engines:** 3.2L VIN V, 3.5L VIN X **Transmissions:** M/T	**HO2S-13 (Bank 1 Sensor 3) Circuit Low Input** DTC P0106, P0107, P0108, P0112, P0113, P0117, P0118, P0121, P0122, P0123, P0171, P0172 and P0300, P0301-P0306 not set, engine running in closed loop with the A/F ratio from 14.5-14.8:1, ECT sensor more than 140°F, throttle angle from 3-19% for 5 seconds, the PCM detected the HO2S-13 signal was less than 22 mv for 106 seconds over a 125 second period during the CCM test. **Possible Causes:** • HO2S signal circuit is open or shorted to ground • HO2S is water or fuel contaminated, or it has failed • PCM has failed
DTC: P0144 **2T CCM, MIL: YES** **1997, 1998, 1999, 2000** **Models:** Hombre **Engines:** 4.3L VIN X **Transmissions:** A/T, M/T	**HO2S-13 (Bank 1 Sensor 3) Circuit High Input** No ECT, IAT, EVAP, MAF, MAP, Misfire, or TP codes set, system voltage from 11-17v, no device controls active, Catalyst and EGR Intrusive tests not active, then during one of these test conditions: Rich Test – Engine running in closed loop, A/F ratio at 14.5-14.8:1, throttle angle from 0-50% for over 5 seconds, and the PCM detected the middle HO2S-13 signal was more than 994 mv for 110 seconds. Decel Fuel Cutoff Rich Test – Engine running in closed loop, Decel Fuel Cutoff Mode active for over 2 seconds, and the PCM detected the HO2S-13 signal was more than 469 mv for over 40 seconds. **Possible Causes:** • HO2S signal is shorted to power (intermittent fault) • HO2S signal circuit shorted to system power (oil in connector) • HO2S is water or fuel contaminated, damaged or it has failed • PCM has failed
DTC: P0144 **1T CCM, MIL: YES** **1996, 1997** **Models:** Rodeo **Engines:** 3.2L VIN V **Transmissions:** M/T	**HO2S-13 (Bank 1 Sensor 3) Circuit High Input** DTC P0106, P0107, P0108, P0112, P0113, P0117, P0118, P0121, P0122, P0123, P0171, P0172 and P0300, P0301-P0306 not set, engine running in closed loop with the A/F ratio from 14.5-14.8:1, ECT sensor more than 140°F, throttle angle from 3-19% for 5 seconds, the PCM detected the HO2S-13 signal was more than 952 mv for 106 seconds over a 125 second period during the CCM test. **Possible Causes:** • HO2S signal is open or shorted to power (intermittent fault) • HO2S signal circuit shorted to system power (oil in connector) • HO2S is water or fuel contaminated, damaged or it has failed • PCM has failed

DTC	Trouble Code Title, Conditions & Possible Causes
DTC: P0144 **1T CCM, MIL: YES** **1996, 1997, 1998, 1999** **Models:** Trooper **Engines:** 3.2L VIN V, 3.5L VIN X **Transmissions:** M/T	**HO2S-13 (Bank 1 Sensor 3) Circuit High Input** DTC P0106, P0107, P0108, P0112, P0113, P0117, P0118, P0121, P0122, P0123, P0171, P0172 and P0300, P0301-P0306 not set, engine running in closed loop with the A/F ratio from 14.5-14.8:1, ECT sensor more than 140ºF, throttle angle from 3-19% for 5 seconds, the PCM detected the HO2S-13 signal was more than 952 mv for 106 seconds over a 125 second period during the CCM test. **Possible Causes:** • HO2S signal is shorted to power (intermittent fault) • HO2S signal circuit shorted to system power (oil in connector) • HO2S is water or fuel contaminated, damaged or it has failed • PCM has failed
DTC: P0146 **1T OBD/O2S, MIL: YES** **1997, 1998, 1999, 2000** **Models:** Hombre **Engines:** 4.3L VIN W **Transmissions:** A/T, M/T	**HO2S-13 (Bank 1 Sensor 3) Insufficient Activity Detected** No ECT, IAT, EVAP, MAF, MAP, Misfire, or TP codes set, system voltage from 11-17v, no device controls active, Catalyst and EGR Intrusive tests not active, engine runtime over 2 minutes, and then during these test conditions: Oxygen Sensor Temperature Test • ECT sensor signal more than 137ºF, MAF sensor signal more than 13 g/sec, and with the Oxygen Sensor Open Test enabled (the Oxygen Sensor Temperature Test "equals" true and DTC P0147 not active while running in closed loop mode, the PCM detected the middle HO2S-13 signal was fixed between 399-473 mv for more than 100 seconds during the Oxygen Sensor Monitor test. **Possible Causes:** • HO2S signal circuit is open or shorted to system power • HO2S ground circuit is open or shorted to system power • HO2S heater is damaged, or the heater power circuit is open • PCM has failed
DTC: P0146 **1T OBD/O2S, MIL: YES** **1996, 1997** **Models:** Rodeo **Engines:** 3.2L VIN V **Transmissions:** M/T	**HO2S-13 (Bank 1 Sensor 3) Insufficient Activity Detected** DTC P0106, P0107, P0108, P0112, P0113, P0117, P0118, P0121, P0122, P0123, P0171, P0172 and P0300, P0301-306 not set, system voltage from 11-16v, engine runtime over 40 seconds, then after the PCM determined the Oxygen Sensor Heater test passed, it detected the HO2S-13 signal remained from 426-474 mv for 105 seconds of a 125 second period in the Oxygen Sensor Monitor test. **Possible Causes:** • Exhaust leak present in exhaust manifold or exhaust pipes • HO2S element fuel contamination or has deteriorated • HO2S signal circuit or the ground circuit has high resistance • HO2S heater element has failed, or the heater circuit is open • PCM has failed
DTC: P0146 **1T CCM, MIL: YES** **1996, 1997, 1998, 1999** **Models:** Trooper **Engines:** 3.2L VIN V, 3.5L VIN X **Transmissions:** M/T	**HO2S-13 (Bank 1 Sensor 3) Insufficient Activity Detected** DTC P0106, P0107, P0108, P0112, P0113, P0117, P0118, P0121, P0122, P0123, P0171, P0172 and P0300, P0301-306 not set, system voltage from 11-16v, engine runtime over 40 seconds, then after the PCM determined the Oxygen Sensor Heater test passed, it detected the HO2S-13 signal remained from 426-474 mv for 105 seconds of a 125 second period in the Oxygen Sensor Monitor test. **Possible Causes:** • Exhaust leak present in exhaust manifold or exhaust pipes • HO2S element fuel contamination or has deteriorated • HO2S signal circuit or the ground circuit has high resistance • HO2S heater element has failed, or the heater circuit is open • PCM has failed
DTC: P0147 **1T OBD/O2S, MIL: YES** **1997, 1998, 1999, 2000** **Models:** Hombre **Engines:** 4.3L VIN W **Transmissions:** A/T, M/T	**HO2S-13 (Bank 1 Sensor 3) Insufficient Activity Detected** No HO2S codes set, ECT and IAT sensor signals less than 91ºF and within 9ºF of each other at startup, system voltage from 11-17v and stable, MAF sensor signal less than 64 g/sec, and the PCM detected the middle HO2S-13 signal remained within 150 mv of its startup voltage for 245 seconds after a "cold" engine startup period. **Possible Causes:** • HO2S heater power circuit open between the heater and fuse • HO2S ground circuit is open between the heater and the PCM • HO2S heater element is damaged or has failed • PCM has failed

DTC	Trouble Code Title, Conditions & Possible Causes
DTC: P0147 **2T OBD/O2S HTR1, MIL: YES** **1997, 1998, 1999, 2000** **Models:** Hombre **Engines:** 4.3L VIN X **Transmissions:** A/T, M/T	**HO2S-13 (Bank 1 Sensor 3) Heater Circuit Malfunction** No HO2S-13 codes set, ECT and IAT sensor signals less than 91°F and within 9°F of each other at startup, system voltage from 11-16v, engine runtime over 2 seconds, MAF sensor signal less than 27 g/sec, and the PCM detected the HO2S-13 signal did not vary more than 150 mv from the bias voltage of 400 to 500 mv for 130 seconds after a cold startup during the Oxygen Sensor Heater Monitor test. **Possible Causes:** • HO2S power circuit is open (check the O2S heater fuse) • HO2S heater ground circuit is open • HO2S heater element has high resistance • HO2S heater element has failed (open or shorted) • PCM has failed
DTC: P0147 **2T OBD/O2S HTR1, MIL: YES** **1996, 1997** **Models:** Rodeo **Engines:** 3.2L VIN V **Transmissions:** M/T	**HO2S-13 (Bank 1 Sensor 3) Heater Circuit Malfunction** No HO2S-13 codes set, ECT and IAT sensors less than 90°F and within 11°F at startup, engine running, system voltage from 11-16v, then with the average Calculated airflow less than 23 g/sec during the test period, the PCM detected the HO2S-13 signal did not vary more than 150 mv from the bias voltage of 400 to 500 mv for up to 300 seconds during the Oxygen Sensor Heater Monitor test. **Possible Causes:** • HO2S power circuit is open (check the O2S heater fuse) • HO2S heater ground circuit is open • HO2S heater element has high resistance • HO2S heater element has failed (open or shorted) • PCM has failed
DTC: P0147 **1T CCM, MIL: YES** **1996, 1997, 1998, 1999** **Models:** Trooper **Engines:** 3.2L VIN V, 3.5L VIN X **Transmissions:** M/T	**HO2S-13 (Bank 1 Sensor 3) Heater Circuit Malfunction** No HO2S-13 codes set, ECT and IAT sensors less than 90°F and within 11°F at startup, engine running, system voltage from 11-16v, then with the average Calculated airflow less than 23 g/sec during the test period, the PCM detected the HO2S-13 signal did not vary more than 150 mv from the bias voltage of 400 to 500 mv for up to 300 seconds during the Oxygen Sensor Heater Monitor test. **Possible Causes:** • HO2S power circuit is open (check the O2S heater fuse) • HO2S heater ground circuit is open • HO2S heater element has high resistance • HO2S heater element has failed (open or shorted) • PCM has failed
DTC: P0151 **1T CCM, MIL: YES** **1996, 1997, 1998, 1999, 2000, 2001, 2002** **Models:** Amigo, Axiom, Rodeo, Rodeo Sport, Trooper, VehiCROSS **Engines:** 3.2L VIN V, 3.2L VIN W, 3.5L VIN X **Transmissions:** A/T, M/T	**HO2S-21 (Bank 2 Sensor 1) Circuit Low Input** DTC P0106, P0107, P0108, P0112, P0113, P0117, P0118, P0121, P0122, P0123, P0171, P0172 and P0300, P0301-P0306 not set, engine running in closed loop with the A/F ratio from 14.5-14.8:1, ECT sensor more than 140°F, throttle angle from 3-19% for 5 seconds, the PCM detected the HO2S-21 signal was less than 22 mv for 77 seconds over a 90 second period during the CCM test. **Possible Causes:** • HO2S signal circuit is open or shorted to ground • HO2S is water or fuel contaminated • HO2S is damaged or has failed • PCM has failed
DTC: P0151 **2T CCM, MIL: YES** **1997, 1998, 1999, 2000** **Models:** Hombre **Engines:** 4.3L VIN X **Transmissions:** A/T, M/T	**HO2S-21 (Bank 2 Sensor 1) Circuit Low Input** DTC P0106, P0107, P0108, P0112, P0113, P0117, P0118, P0121, P0122, P0123, P0171, P0172, P0300, P0301-P0304, P0440, P0442 and P0446 not set, no Intrusive or Device Control tests active, then under Lean Test conditions (engine running in closed loop, throttle angle from 3.5-99% for 5 seconds), the PCM detected the HO2S-21 signal was less than 86 mv for 50 seconds; or with Power Enrichment conditions (engine running in closed loop, Power Enrichment Lean Test mode active and High Speed Fuel Cutoff mode inactive, elapsed time since the last test enabled over 1 second), the PCM detected the HO2S-21 signal was less than 598 mv for 30 seconds during the CCM test. **Possible Causes:** • HO2S signal circuit is open or shorted to ground • HO2S is water or fuel contaminated, or it has failed • PCM has failed
DTC: P0152 **1T CCM, MIL: YES** **1996, 1997, 1998, 1999, 2000, 2001, 2002** **Models:** Amigo, Axiom, Rodeo, Rodeo Sport, Trooper, VehiCROSS **Engines:** 3.2L VIN V, 3.2L VIN W, 3.5L VIN X **Transmissions:** A/T, M/T	**HO2S-21 (Bank 2 Sensor 1) Circuit High Input** DTC P0106, P0107, P0108, P0112, P0113, P0117, P0118, P0121, P0122, P0123, P0171, P0172 and P0300, P0301-P0306 not set, engine running in closed loop with the A/F ratio command at 14.5-14.8:1, throttle angle from 3-19% for 5 seconds, the PCM detected the HO2S-21 signal was less than 952 mv for 77 seconds over a 90 second period; or the HO2S-21 signal was more than 500 mv during Decel Fuel Cutoff mode for 3 seconds during the CCM test. **Possible Causes:** • HO2S signal circuit shorted to system power (oil in connector) • HO2S is water or fuel contaminated • HO2S is damaged or has failed • PCM has failed

DTC	Trouble Code Title, Conditions & Possible Causes
DTC: P0152 **2T CCM, MIL: YES** **1997, 1998, 1999, 2000** **Models:** Hombre **Engines:** 4.3L VIN X **Transmissions:** A/T, M/T	**HO2S-21 (Bank 2 Sensor 1) Circuit High Input** DTC P0106, P0107, P0108, P0112, P0113, P0117, P0118, P0121, P0122, P0123, P0171, P0172, P0200, P0300, P0301-P0304, P0440, P0442 and P0446 not set, no Intrusive or Device Control tests active, then under Rich Test conditions (running in closed loop, A/F ratio from 14.5-14.8:1, throttle angle from 0-50% for 5 seconds), the PCM detected the HO2S-21 signal was less than 976 mv for 40 seconds; or under Decel Fuel Cutoff conditions (running in closed loop, Decel Fuel Cutoff mode active, elapsed time since the last test enabled over 2 seconds), the PCM detected the HO2S-21 signal was more than 468 mv for 30 seconds during the CCM test. **Possible Causes:** • HO2S signal circuit shorted to system power (oil in connector) • HO2S is water or fuel contaminated • HO2S is damaged or has failed • PCM has failed
DTC: P0153 **2T OBD/O2S, MIL: YES** **1996, 1997, 1998, 1999, 2000, 2001, 2002** **Models:** Amigo, Axiom, Rodeo, Rodeo Sport, Trooper, VehiCROSS **Engines:** 3.2L VIN V, 3.2L VIN W, 3.5L VIN X **Transmissions:** A/T, M/T	**HO2S-21 (Bank 2 Sensor 1) Slow Response** DTC P0106, P0107, P0108, P0112, P0113, P0117, P0118, P0121, P0122, P0123, P0171, P0172 and P0300, P0301-P0306 not set, engine runtime 1 minute in closed loop, ECT sensor more than 122ºF, engine speed from 1500-3000 rpm, MAF sensor from 9-42 g/sec, Purge duty cycle over 1%, conditions met for 3 seconds, then 90 seconds after entering closed loop, the PCM detected the HO2S-21 lean-to-rich average transition time was over 94 ms, or the rich-to-lean average transition time was over 105 ms during the Oxygen Sensor Monitor test. **Possible Causes:** • Exhaust leak present in the exhaust manifold or exhaust pipes • HO2S element fuel contamination • HO2S element has deteriorated • PCM has failed
DTC: P0153 **2T OBD/O2S, MIL: YES** **1997, 1998, 1999, 2000** **Models:** Hombre **Engines:** 4.3L VIN X **Transmissions:** A/T, M/T	**HO2S-21 (Bank 2 Sensor 1) Slow Response** DTC P0101, P0102, P0103, P0106, P0107, P0108, P0112, P0113, P0117, P0118, P0121, P0122, P0123, P0131, P0132, P0134, P0135, P0300, P0301-P0304, P0341, P0440, P0442 and P0446 not set, No Intrusive or Device Control tests enabled, system voltage from 11-16v, ECT sensor signal over 135ºF, engine speed from 1100-3000 rpm in closed loop for 100 seconds, MAF sensor signal from 15-55 g/sec, Purge duty cycle enabled, and the PCM detected the HO2S-21 average lean-to-rich or the rich-to-lean response time was over 125 ms during the Oxygen Sensor Monitor test. **Possible Causes:** • Exhaust leak present in the exhaust manifold or exhaust pipes • HO2S element fuel contamination • HO2S element has deteriorated • PCM has failed
DTC: P0154 **1T OBD/O2S, MIL: YES** **1996, 1997, 1998, 1999, 2000, 2001, 2002** **Models:** Amigo, Axiom, Rodeo, Rodeo Sport, Trooper, VehiCROSS **Engines:** 3.2L VIN V, 3.2L VIN W, 3.5L VIN X Transmissions: A/T, M/T	**HO2S-21 (Bank 2 Sensor 1) Insufficient Activity Detected** DTC P0106, P0107, P0108, P0112, P0113, P0117, P0118, P0121, P0122, P0123, P0171, P0172 and P0300, P0301-306 not set, system voltage from 11-16v, engine runtime over 40 seconds, then after the PCM determined the Oxygen Sensor Heater test passed, it detected the HO2S-21 signal remained from 400-500 mv for 77 seconds over a 90 second period in the Oxygen Sensor Monitor test. **Possible Causes:** • Exhaust leak present in exhaust manifold or exhaust pipes • HO2S element fuel contamination or has deteriorated • HO2S signal circuit or the ground circuit has high resistance • HO2S heater element has failed, or the heater circuit is open • PCM has failed
DTC: P0154 **2T OBD/O2S, MIL: YES** **1997, 1998, 1999, 2000** **Models:** Hombre **Engines:** 4.3L VIN X **Transmissions:** A/T, M/T	**HO2S-21 (Bank 2 Sensor 1) Insufficient Activity Detected** DTC P0101, P0102, P0103, P0106, P0107, P0108, P0112, P0113, P0117, P0118, P0121, P0122, P0123, P0135, P0300, P0301-P0304, P0440, P0442 and P0446 not set, no Intrusive or Device Control tests active, system voltage from 11-16v, engine runtime over 2 minutes, then during the Oxygen Sensor Temperature Test (ECT sensor signal more than 137ºF, MAF sensor signal more than 13 g/sec, and Decel Fuel Cutoff not active); or with the Oxygen Sensor Open Test enabled (DTC P0135 not active, engine running in closed loop), the PCM detected the HO2S-21 signal was fixed at 300-600 mv for 80 seconds during the Oxygen Sensor Monitor test. **Possible Causes:** • Exhaust leak present in exhaust manifold or exhaust pipes • HO2S element fuel contamination or has deteriorated • HO2S signal circuit or the ground circuit has high resistance • HO2S heater element has failed, or the heater circuit is open • PCM has failed

DTC	Trouble Code Title, Conditions & Possible Causes
DTC: P0155 **2T OBD/O2S, MIL: YES** **1996, 1997, 1998, 1999, 2000, 2001, 2002** **Models:** Amigo, Axiom, Rodeo, Rodeo Sport, Trooper, VehiCROSS **Engines:** 3.2L VIN V, 3.2L VIN W, 3.5L VIN X **Transmissions:** A/T, M/T	**HO2S-21 (Bank 2 Sensor 1) Heater Circuit Malfunction** DTC P0151, P0152, P0153 and P0154 not set, ECT and IAT sensor signals less than 90ºF, and within 11ºF at startup, system voltage from 11-16v, throttle angle under 40%, average Calculated airflow less than 18 g/sec in the sample period, and the PCM detected the HO2S-21 signal did not vary more than 150 mv from the bias voltage (400-500 mv) for too long a period (maximum time is 120 seconds). **Possible Causes:** • HO2S power circuit is open (from the O2S heater fuse) • HO2S heater ground circuit is open • HO2S heater element has high resistance • HO2S heater element has failed (open or shorted) • PCM has failed
DTC: P0155 **2T OBD/O2S HTR1, MIL: YES** **1997, 1998, 1999, 2000** **Models:** Hombre **Engines:** 4.3L VIN X **Transmissions:** A/T, M/T	**HO2S-21 (Bank 2 Sensor 1) Heater Circuit Malfunction** DTC P0151, P0152, P0153 and P0154 not set, ECT and IAT sensor signals less than 91ºF and within 9ºF at startup, system voltage from 11-16v, engine running, average Calculated airflow less than 21 g/sec, and the PCM detected the HO2S-21 signal did not vary more than 150 mv from bias voltage of 400 to 500 mv for 150 seconds. **Possible Causes:** • HO2S power circuit is open (check the O2S heater fuse) • HO2S heater ground circuit is open • HO2S heater element has high resistance • HO2S heater element has failed (open or shorted) • PCM has failed
DTC: P0157 **1T CCM, MIL: YES** **1996, 1997, 1998, 1999, 2000, 2001, 2002** **Models:** Amigo, Axiom, Rodeo, Rodeo Sport, Trooper, VehiCROSS **Engines:** 3.2L VIN V, 3.2L VIN W, 3.5L VIN X **Transmissions:** A/T, M/T	**HO2S-22 (Bank 2 Sensor 2) Circuit Low Input** No related codes set, ECT sensor signal more than 140ºF, engine running in closed loop with the A/F ratio at 14.5-14.8:1, throttle angle from 3-19%, and the PCM detected the HO2S-22 signal was less than 26 mv for 106 seconds out of a 125 second period; or that it was more than 400 mv in Power Enrichment Mode during the test. **Possible Causes:** • HO2S signal circuit is open or shorted to ground • HO2S is water or fuel contaminated • HO2S is damaged or has failed • PCM has failed
DTC: P0158 **1T CCM, MIL: YES** **1996, 1997, 1998, 1999, 2000, 2001, 2002** **Models:** Amigo, Axiom, Rodeo, Rodeo Sport, Trooper, VehiCROSS **Engines:** 3.2L VIN V, 3.2L VIN W, 3.5L VIN X **Transmissions:** A/T, M/T	**HO2S-22 (Bank 2 Sensor 2) Circuit High Input** DTC P0106, P0107, P0108, P0112, P0113, P0117, P0118, P0121, P0122, P0123, P0171, P0172 and P0300, P0301-P0306 not set, engine running in closed loop with the A/F ratio command at 14.5-14.8:1, throttle angle from 3-19% for 5 seconds, the PCM detected the HO2S-22 signal was less than 952 mv for 106 seconds over a 125 second period; or the HO2S-12 signal was more than 500 mv during Decel Fuel Cutoff mode for 3 seconds during the CCM test. **Possible Causes:** • HO2S signal circuit shorted to system power (oil in connector) • HO2S is water or fuel contaminated • HO2S is damaged or has failed • PCM has failed
DTC: P0160 **1T OBD/O2S, MIL: YES** **1996, 1997, 1998, 1999, 2000, 2001, 2002** **Models:** Amigo, Axiom, Rodeo, Rodeo Sport, Trooper, VehiCROSS **Engines:** 3.2L VIN V, 3.2L VIN W, 3.5L VIN X **Transmissions:** A/T, M/T	**HO2S-22 (Bank 2 Sensor 2) Insufficient Activity Detected** DTC P0106, P0107, P0108, P0112, P0113, P0117, P0118, P0121, P0122, P0123, P0171, P0172 and P0300, P0301-P0306 not set, system voltage from 11-16v, engine runtime over 40 seconds, then after the PCM determined the Oxygen Sensor Heater test passed, it detected the HO2S-22 signal remained from 426-474 mv for 105 seconds of a 125 second period in the Oxygen Sensor Monitor test. **Possible Causes:** • Exhaust leak present in exhaust manifold or exhaust pipes • HO2S element has fuel contamination or has deteriorated • HO2S signal circuit or the ground circuit has high resistance • PCM has failed
DTC: P0161 **2T OBD/O2S HTR1, MIL: YES** **1996, 1997, 1998, 1999, 2000, 2001, 2002** **Models:** Amigo, Axiom, Rodeo, Rodeo Sport, Trooper, VehiCROSS **Engines:** 3.2L VIN V, 3.2L VIN W, 3.5L VIN X **Transmissions:** A/T, M/T	**HO2S-22 (Bank 2 Sensor 2) Heater Circuit Malfunction** No HO2S-22 codes set, ECT and IAT sensor less than 90ºF, and within 11ºF at startup, system voltage at 11-16v, average Calculated airflow during the test period less than 23 g/sec, and the PCM detected the HO2S-22 signal changed less than 150 mv from the bias voltage of 400-500 mv for up to 300 seconds during the test. **Possible Causes:** • HO2S power circuit is open (from the O2S heater fuse) • HO2S heater element has failed (it may be open or shorted) • PCM has failed

DTC	Trouble Code Title, Conditions & Possible Causes
DTC: P0171 **2T OBD/FUEL, MIL: YES** **1998, 1999, 2000, 2001, 2002** **Models:** Amigo, Rodeo, Rodeo Sport **Engines:** 2.2L VIN D **Transmissions:** A/T, M/T	**Fuel System Too Lean (Bank 1)** No EGR, EVAP, HO2S, MAP or TPS codes set, system voltage over 10v, ECT sensor signal from 149-219ºF, IAT sensor signal from -40ºF to 248ºF, no Scan Tool tests "active", MAP sensor signal from 24-99 kPa, BARO sensor signal over 72.3 kPa, engine speed from 400-6000 rpm while operating in closed loop mode, and the PCM detected the average Short Term fuel trim values were over 0.97, or the average of the adaptive index multiplier samples was over 1.21. **Possible Causes:** • Air leaks after the MAF sensor, or in the EGR or PCV system • Base engine "mechanical" fault affecting one or more cylinders • Exhaust leaks before or near where the front HO2S is mounted • Fuel control sensor is out of calibration (i.e., ECT, IAT or MAP) • Fuel delivery system supplying too little fuel during cruise or idle periods (e.g., faulty fuel pump or dirty, restricted fuel filter) • Fuel injector (one or more) dirty or pressure regulator has failed • HO2S is contaminated, deteriorated or it has failed • Vehicle driven low on fuel or until it ran out of fuel
DTC: P0171 **2T OBD/FUEL, MIL: YES** **1996, 1997, 1998, 1999, 2000, 2001, 2002** **Models:** Amigo, Axiom, Rodeo, Rodeo Sport, Trooper, VehiCROSS **Engines:** 3.2L VIN V, 3.2L VIN W, 3.5L VIN X **Transmissions:** A/T, M/T	**Fuel System Too Lean (Bank 1)** DTC P0106, P0107, P0108, P0112, P0113, P0117, P0118, P0121, P0122, P0123, P0131, P0132, P0133, P0134, P0135, P0137, P0138, P0201-206, P0300, P0301=P0306, P0401, P0502, P0503, P0506, P0507, P1406 and P1441 not set, engine running in closed loop, system voltage from 11-16v, BARO sensor over 72.5 kPa, ECT sensor from 77-212ºF, IAT sensor from -40 to 248ºF, MAP sensor from 24-99 kPa, throttle angle less than 95%, VSS under 85 mph, engine speed from 400-6000 rpm, MAF sensor from 2-20 g/sec, Purge duty cycle over 0%, and the PCM detected the average of the Long Term fuel trim values was more than +20%. **Possible Causes:** • Air leaks after the MAF sensor, or in the EGR or PCV system • Base engine "mechanical" fault affecting one or more cylinders • Exhaust leaks before or near where the front HO2S is mounted • Fuel control sensor is out of calibration (i.e., ECT, IAT or MAP) • Fuel delivery system supplying too little fuel during cruise or idle periods (e.g., faulty fuel pump or dirty, restricted fuel filter) • Fuel injector (one or more) dirty or pressure regulator has failed • HO2S is contaminated, deteriorated or it has failed • Vehicle driven low on fuel or until it ran out of fuel
DTC: P0171 **2T OBD/FUEL, MIL: YES** **1996, 1997, 1998, 1999, 2000** **Models:** Hombre **Engines:** 2.2L VIN 4 **Transmissions:** A/T	**Fuel System Too Lean (Bank 1)** DTC P0106, P0107, P0108, P0112, P0113, P0117, P0118, P0121, P0122, P0123, P0131, P0132, P0133, P0134, P0200, P0300, P0301-P0304, P0325, P0335, P0341, P0342, P0401, P0404, P0405, P0502, P0503, P0506, P0507, P0601, P0602, P1404, P1133 and P1441 not set, ECT sensor signal from 140-239ºF, IAT sensor signal from -13ºF to 175ºF, BARO sensor signal over 72.5 kPa, MAP sensor signal over 26 kPa, throttle angle steady below 95%, VSS under 77 mph, Purge command "off", engine speed from 550-3400 rpm while operating in closed loop, and the PCM detected the Short Term fuel trim index was over 170 (+33%) for 4 seconds. **Possible Causes:** • Air leaks after the MAF sensor, or in the EGR or PCV system • Base engine "mechanical" fault affecting one or more cylinders • Exhaust leaks before or near where the front HO2S is mounted • Fuel control sensor is out of calibration (i.e., ECT, IAT or MAP) • Fuel delivery system supplying too little fuel during cruise or idle periods (e.g., faulty fuel pump or dirty, restricted fuel filter) • Fuel injector (one or more) dirty or pressure regulator has failed • HO2S is contaminated, deteriorated or it has failed • Vehicle driven low on fuel or until it ran out of fuel
DTC: P0171 **2T OBD/FUEL, MIL: YES** **1996, 1997, 1998, 1999, 2000** **Models:** Hombre **Engines:** 2.2L VIN 4 **Transmissions:** M/T	**Fuel System Too Lean (Bank 1)** DTC P0106, P0107, P0108, P0112, P0113, P0117, P0118, P0121, P0122, P0123, P0131, P0132, P0133, P0134, P0200, P0300, P0301-P0304, P0325, P0335, P0341, P0342, P0401, P0404, P0405, P0502, P0503, P0506, P0507, P0601, P0602, P1404, P1133 and P1441 not set, ECT sensor signal from 140-239ºF, IAT sensor signal from -13ºF to 175ºF, BARO sensor signal over 72.5 kPa, MAP sensor signal over 26 kPa, throttle angle steady below 95%, VSS under 77 mph, Purge command "off", engine speed from 850-3400 rpm while operating in closed loop, and the PCM detected the Short Term fuel trim index was over 170 (+33%) for 4 seconds. **Possible Causes:** • Air leaks after the MAF sensor, or in the EGR or PCV system • Base engine "mechanical" fault affecting one or more cylinders • Exhaust leaks before or near where the front HO2S is mounted • Fuel control sensor is out of calibration (i.e., ECT, IAT or MAP) • Fuel delivery system supplying too little fuel during cruise or idle periods (e.g., faulty fuel pump or dirty, restricted fuel filter) • Fuel injector (one or more) dirty or pressure regulator has failed • HO2S is contaminated, deteriorated or it has failed • Vehicle driven low on fuel or until it ran out of fuel

DTC	Trouble Code Title, Conditions & Possible Causes
DTC: P0171 **2T OBD/FUEL, MIL: YES** **1997, 1998, 1999, 2000** **Models:** Hombre **Engines:** 4.3L VIN X **Transmissions:** A/T, M/T	**Fuel System Too Lean (Bank 1)** DTC P0106, P0107, P0108, P0112, P0113, P0117, P0118, P0121, P0122, P0123, P0131, P0132, P0133, P0134, P0200, P0300, P0301-P0304, P0325, P0336, P0341, P0342, P0401, P0404, P0405, P0502, P0503, P0506, P0507, P0601, P0602, P1404, P1133 and P1441 not set, engine running in closed loop, ECT sensor from 167-237ºF, IAT sensor from 46-169ºF, BARO sensor over 70 kPa, MAF signal from 3-85 g/sec, MAP sensor from 22-85 kPa, throttle angle less than 70%, VSS under 85 mph, engine speed from 575-4000 rpm, and the PCM detected the average of the Short Term fuel trim values indicated a Lean A/F condition. **Possible Causes:** • Air leaks after the MAF sensor, or in the EGR or PCV system • Base engine "mechanical" fault affecting one or more cylinders • Exhaust leaks before or near where the front HO2S is mounted • Fuel control sensor is out of calibration (i.e., ECT, IAT or MAP) • Fuel delivery system supplying too little fuel during cruise or idle periods (e.g., faulty fuel pump or dirty, restricted fuel filter) • Fuel injector (one or more) dirty or pressure regulator has failed • HO2S is contaminated, deteriorated or it has failed • Vehicle driven low on fuel or until it ran out of fuel
DTC: P0171 **2T CCM, MIL: YES** **1996, 1997, 1998, 1999** **Models:** Oasis **Engines:** All **Transmissions:** A/T, M/T	**Fuel System Too Lean (Bank 1)** DTC P0107, P0108, P0135, P0137, P0138, P0141, P1128, P1129 and P1259 not set, engine running in closed loop, and the PCM detected the LONGFT value exceeded the calibrated lean limit value **Possible Causes:** • Air leaks in intake manifold, exhaust pipes or exhaust manifold • One or more injectors restricted or pressure regulator has failed • Air is being drawn in from leaks in gaskets or other seals • O2S element is deteriorated or has failed • A "fuel control" sensor is out of calibration (ECT, IAT or MAP) • PCM has failed
DTC: P0171 **2T OBD/FUEL, MIL: YES** **1996, 1997** **Models:** Rodeo **Engines:** 2.6L VIN E **Transmissions:** M/T	**Fuel System Too Lean (Bank 1)** DTC P0106, P0107, P0108, P0112, P0113, P0117, P0118, P0121, P0122, P0123, P0131, P0132, P0133, P0134, P0135, P0137, P0138, P0201-204, P0300, P0325, P0336, P0341, P0342, P0401, P0502, P0503, P0506, P0507, P1406 and P1441 not set, engine running in closed loop, BARO sensor over 72.5 kPa, ECT sensor signal from 77-212ºF, IAT sensor signal from -40 to 248ºF, MAP sensor signal at 24-99 kPa, throttle angle less than 95%, VSS under 85 mph, engine speed from 400-6000 rpm, MAF sensor signal from 2-20 g/sec, Purge duty cycle over 0% if "on", and the PCM detected the average of the Long Term fuel trim values was more than +20%. **Possible Causes:** • Air leaks after the MAF sensor, or in the EGR or PCV system • Base engine "mechanical" fault affecting one or more cylinders • Exhaust leaks before or near where the front HO2S is mounted • Fuel control sensor is out of calibration (i.e., ECT, IAT or MAP) • Fuel delivery system supplying too little fuel during cruise or idle periods (e.g., faulty fuel pump or dirty, restricted fuel filter) • Fuel injector (one or more) dirty or pressure regulator has failed • HO2S is contaminated, deteriorated or it has failed • Vehicle driven low on fuel or until it ran out of fuel
DTC: P0172 **2T OBD/FUEL, MIL: YES** **1998, 1999, 2000, 2001, 2002** **Models:** Amigo, Rodeo, Rodeo Sport **Engines:** 2.2L VIN D **Transmissions:** A/T, M/T	**Fuel System Too Rich (Bank 1)** No EGR, EVAP, HO2S, MAP or TPS codes set, system voltage over 10v, ECT sensor signal from 149-219ºF, IAT sensor signal from -40ºF to 248ºF, no Scan Tool tests "active", MAP sensor signal from 24-99 kPa, BARO sensor signal over 72.3 kPa, engine speed from 400-6000 rpm while operating in closed loop mode, and the PCM detected the average Short Term fuel trim values was more than 0.97, or the average of the adaptive index multiplier samples was more than or equal to 1.21 during the Fuel System Monitor test. **Possible Causes:** • Air leak at the exhaust pipe or manifold, or at air injection pipes • Base engine "mechanical" fault affecting one or more cylinders • EVAP system component has failed or canister fuel saturated • Fuel control sensor is out of calibration (i.e., ECT, IAT or MAP) • Fuel delivery system supplying too much fuel during cruise or idle periods (e.g., faulty fuel pump, or faulty pressure regulator) • Fuel injector(s) is leaking or stuck partially open (one or more) • HO2S is contaminated, deteriorated or it has failed

DTC	Trouble Code Title, Conditions & Possible Causes
DTC: P0172 **2T OBD/FUEL, MIL: YES** **1996, 1997, 1998, 1999, 2000, 2001, 2002** **Models:** Amigo, Axiom, Rodeo, Rodeo Sport, Trooper, VehiCROSS **Engines:** 3.2L VIN V, 3.2L VIN W, 3.5L VIN X **Transmissions:** A/T, M/T	**Fuel System Too Rich (Bank 1)** DTC P0106, P0107, P0108, P0112, P0113, P0117, P0118, P0121, P0122, P0123, P0131, P0132, P0133, P0134, P0135, P0137, P0138, P0201-206, P0300, P0301=P0306, P0401, P0502, P0503, P0506, P0507, P1406 and P1441 not set, engine running in closed loop, BARO sensor over 72.5 kPa, ECT sensor from 77-212°F, IAT sensor from -40 to 248°F, MAP sensor from 24-99 kPa, throttle angle less than 95%, VSS under 85 mph, engine speed from 400-6000 rpm, MAF sensor from 2-20 g/sec, Purge duty cycle over 0%, and the PCM detected the average of the Long Term fuel trim values was more than -14% during the Fuel System Monitor test. **Possible Causes:** • Air leak at the exhaust pipe or manifold, or at air injection pipes • Base engine "mechanical" fault affecting one or more cylinders • EVAP system component has failed or canister fuel saturated • Fuel control sensor is out of calibration (i.e., ECT, IAT or MAP) • Fuel delivery system supplying too much fuel during cruise or idle periods (e.g., faulty fuel pump, or faulty pressure regulator) • Fuel injector(s) is leaking or stuck partially open (one or more) • HO2S is contaminated, deteriorated or it has failed
DTC: P01072 **2T OBD/FUEL, MIL: YES** **1996, 1997, 1998, 1999, 2000** **Models:** Hombre **Engines:** 2.2L VIN 4 **Transmissions:** A/T	**Fuel System Too Rich (Bank 1)** DTC P0106, P0107, P0108, P0112, P0113, P0117, P0118, P0121, P0122, P0123, P0131, P0132, P0133, P0134, P0200, P0300, P0301-P0304, P0325, P0335, P0341, P0342, P0401, P0404, P0405, P0502, P0503, P0506, P0507, P0601, P0602, P1404, P1133 and P1441 not set, ECT sensor signal from 140-239°F, IAT sensor signal from -13°F to 175°F, BARO sensor signal over 72.5 kPa, MAP sensor signal over 26 kPa, throttle angle steady below 95%, VSS under 77 mph, Purge command "off", engine speed from 550-3400 rpm while operating in closed loop, and the PCM detected the Short Term fuel trim index was under 75 (-40%) for 4 seconds. **Possible Causes:** • Air leak at the exhaust pipe or manifold, or at air injection pipes • Base engine "mechanical" fault affecting one or more cylinders • EVAP system component has failed or canister fuel saturated • Fuel control sensor is out of calibration (i.e., ECT, IAT or MAP) • Fuel delivery system supplying too much fuel during cruise or idle periods (e.g., faulty fuel pump, or faulty pressure regulator) • Fuel injector(s) is leaking or stuck partially open (one or more) • HO2S is contaminated, deteriorated or it has failed
DTC: P0172 **2T OBD/FUEL, MIL: YES** **1996, 1997, 1998, 1999, 2000** **Models:** Hombre **Engines:** 2.2L VIN 4 **Transmissions:** M/T	**Fuel System Too Rich (Bank 1)** DTC P0106, P0107, P0108, P0112, P0113, P0117, P0118, P0121, P0122, P0123, P0131, P0132, P0133, P0134, P0200, P0300, P0301-P0304, P0325, P0335, P0341, P0342, P0401, P0404, P0405, P0502, P0503, P0506, P0507, P0601, P0602, P1404, P1133 and P1441 not set, ECT sensor signal from 140-239°F, IAT sensor signal from -13°F to 175°F, BARO sensor signal over 72.5 kPa, MAP sensor signal over 26 kPa, throttle angle steady below 95%, VSS under 77 mph, Purge command "off", engine speed from 850-3400 rpm while operating in closed loop, and the PCM detected the Short Term fuel trim index was under 75 (-42%) for 4 seconds. **Possible Causes:** • Air leak at the exhaust pipe or manifold, or at air injection pipes • Base engine "mechanical" fault affecting one or more cylinders • EVAP system component has failed or canister fuel saturated • Fuel control sensor is out of calibration (i.e., ECT, IAT or MAP) • Fuel delivery system supplying too much fuel during cruise or idle periods (e.g., faulty fuel pump, or faulty pressure regulator) • Fuel injector(s) is leaking or stuck partially open (one or more) • HO2S is contaminated, deteriorated or it has failed
DTC: P0172 **2T OBD/FUEL, MIL: YES** **1997, 1998, 1999, 2000** **Models:** Hombre **Engines:** 4.3L VIN X **Transmissions:** A/T, M/T	**Fuel System Too Rich (Bank 1)** DTC P0106, P0107, P0108, P0112, P0113, P0117, P0118, P0121, P0122, P0123, P0131, P0132, P0133, P0134, P0200, P0300, P0301-P0304, P0325, P0336, P0341, P0342, P0401, P0404, P0405, P0502, P0503, P0506, P0507, P0601, P0602, P1404, P1133 and P1441 not set, ECT sensor signal from 167-237°F, IAT sensor signal from 46-169°F, BARO sensor signal over 70 kPa, MAF sensor signal at 3-85 g/sec, MAP sensor signal at 22-85 kPa, throttle angle under 70%, VSS under 85 mph, engine speed from 575-4000 rpm in closed loop, and the PCM detected the average of the Short Term fuel trim values for 120 seconds indicated a rich A/F condition. **Possible Causes:** • Air leak at the exhaust pipe or manifold, or at air injection pipes • Base engine "mechanical" fault affecting one or more cylinders • EVAP system component has failed or canister fuel saturated • Fuel control sensor is out of calibration (i.e., ECT, IAT or MAP) • Fuel delivery system supplying too much fuel during cruise or idle periods (e.g., faulty fuel pump, or faulty pressure regulator) • Fuel injector(s) is leaking or stuck partially open (one or more) • HO2S is contaminated, deteriorated or it has failed

DTC	Trouble Code Title, Conditions & Possible Causes
DTC: P0172 **2T CCM, MIL: YES** **1996, 1997, 1998, 1999** **Models:** Oasis **Engines:** All **Transmissions:** A/T, M/T	**Fuel System Too Rich (Bank 1)** DTC P0107, P0108, P0135, P0137, P0138, P0141, P1128, P1129 and P1259 not set, engine running in closed loop, and the PCM detected the LONGFT value exceeded the calibrated rich limit. **Note: A high MAP sensor signal at idle can cause this code to set.** **Possible Causes:** • Leaking/contaminated fuel injector(s) or fuel pressure regulator • HO2S element may be contaminated with water or alcohol • EVAP vapor recovery system has failed (pulling vacuum) • Base engine fault (i.e., cam timing incorrect, engine oil too high
DTC: P0172 **2T OBD/FUEL, MIL: YES** **1996, 1997** **Models:** Rodeo **Engines:** 2.6L VIN E **Transmissions:** M/T	**Fuel System Too Rich (Bank 1)** DTC P0106, P0107, P0108, P0112, P0113, P0117, P0118, P0121, P0122, P0123, P0131, P0132, P0133, P0134, P0135, P0137, P0138, P0201-204, P0300, P0325, P0336, P0341, P0342, P0401, P0502, P0503, P0506, P0507, P1406 and P1441 not set, engine running in closed loop, BARO sensor over 72.5 kPa, ECT sensor from 77-212ºF, IAT sensor from -40 to 248ºF, MAP sensor from 24-99 kPa, throttle angle less than 95%, VSS under 85 mph, engine speed from 400-6000 rpm, MAF sensor signal from 2-20 g/sec, Purge duty cycle over 0% if "on", and the PCM detected the average of the Long Term fuel trim values was more than -14%. **Possible Causes:** • Air leak at the exhaust pipe or manifold, or at air injection pipes • Base engine "mechanical" fault affecting one or more cylinders • EVAP system component has failed or canister fuel saturated • Fuel control sensor is out of calibration (i.e., ECT, IAT or MAP) • Fuel delivery system supplying too much fuel during cruise or idle periods (e.g., faulty fuel pump, or faulty pressure regulator) • Fuel injector(s) is leaking or stuck partially open (one or more) • HO2S is contaminated, deteriorated or it has failed
DTC: P0174 **2T OBD/FUEL, MIL: YES** **1996, 1997, 1998, 1999, 2000, 2001, 2002** **Models:** Amigo, Axiom, Rodeo, Rodeo Sport, Trooper, VehiCROSS **Engines:** 3.2L VIN V, 3.2L VIN W, 3.5L VIN X **Transmissions:** A/T, M/T	**Fuel System Too Lean (Bank 2)** DTC P0106, P0107, P0108, P0112, P0113, P0117, P0118, P0121, P0122, P0123, P0131, P0132, P0133, P0134, P0135, P0137, P0138, P0201-206, P0300, P0301=P0306, P0401, P0502, P0503, P0506, P0507, P1406 and P1441 not set, engine running in closed loop, system voltage from 11-16v, BARO sensor over 72.5 kPa, ECT sensor from 77-212ºF, IAT sensor from -40 to 248ºF, MAP sensor from 24-99 kPa, throttle angle steady at less than 95%, VSS under 85 mph, engine speed from 400-6000 rpm, MAF sensor from 2-20 g/sec, Purge duty cycle over 0%, and the PCM detected the average of the Long Term fuel trim values was more than +20%. **Possible Causes:** • Air leaks after the MAF sensor, or in the EGR or PCV system • Base engine "mechanical" fault affecting one or more cylinders • Exhaust leaks before or near where the front HO2S is mounted • Fuel control sensor is out of calibration (i.e., ECT, IAT or MAP) • Fuel delivery system supplying too little fuel during cruise or idle periods (e.g., faulty fuel pump or dirty, restricted fuel filter) • Fuel injector (one or more) dirty or pressure regulator has failed • HO2S is contaminated, deteriorated or it has failed • Vehicle driven low on fuel or until it ran out of fuel
DTC: P0174 **2T OBD/FUEL, MIL: YES** **1997, 1998, 1999, 2000** **Models:** Hombre **Engines:** 4.3L VIN X **Transmissions:** A/T, M/T	**Fuel System Too Lean (Bank 2)** DTC P0106, P0107, P0108, P0112, P0113, P0117, P0118, P0121, P0122, P0123, P0131, P0132, P0133, P0134, P0200, P0300, P0301-P0304, P0325, P0336, P0341, P0342, P0401, P0404, P0405, P0502, P0503, P0506, P0507, P0601, P0602, P1404, P1133 and P1441 not set, ECT sensor signal from 167-237ºF, IAT sensor signal from 46-169ºF, BARO sensor signal over 70 kPa, MAF sensor signal at 3-85 g/sec, MAP sensor signal at 22-85 kPa, throttle angle less than 70%, VSS under 85 mph, engine speed from 575-4000 rpm in closed loop, and the PCM detected the average of the Short Term fuel trim values indicated a Lean A/F condition. **Possible Causes:** • Air leaks after the MAF sensor, or in the EGR or PCV system • Base engine "mechanical" fault affecting one or more cylinders • Exhaust leaks before or near where the front HO2S is mounted • Fuel control sensor is out of calibration (i.e., ECT, IAT or MAP) • Fuel delivery system supplying too little fuel during cruise or idle periods (e.g., faulty fuel pump or dirty, restricted fuel filter) • Fuel injector (one or more) dirty or pressure regulator has failed • HO2S is contaminated, deteriorated or it has failed • Vehicle driven low on fuel or until it ran out of fuel

DTC	Trouble Code Title, Conditions & Possible Causes
DTC: P0175 **2T OBD/FUEL, MIL: YES** **1996, 1997, 1998, 1999, 2000, 2001, 2002** **Models:** Amigo, Axiom, Rodeo, Rodeo Sport, Trooper, VehiCROSS **Engines:** 3.2L VIN V, 3.2L VIN W, 3.5L VIN X **Transmissions:** A/T, M/T	**Fuel System Too Rich (Bank 2)** DTC P0106, P0107, P0108, P0112, P0113, P0117, P0118, P0121, P0122, P0123, P0131, P0132, P0133, P0134, P0135, P0137, P0138, P0201-206, P0300, P0301=P0306, P0401, P0502, P0503, P0506, P0507, P1406 and P1441 not set, engine running in closed loop, BARO sensor over 72.5 kPa, ECT sensor from 77-212°F, IAT sensor from -40 to 248°F, MAP sensor from 24-99 kPa, throttle angle less than 95%, VSS under 85 mph, engine speed from 400-6000 rpm, MAF sensor from 2-20 g/sec, Purge duty cycle over 0%, and the PCM detected the average of the Long Term fuel trim values was more than -14% during the Fuel System Monitor test. **Possible Causes:** • Air leak at the exhaust pipe or manifold, or at air injection pipes • Base engine "mechanical" fault affecting one or more cylinders • EVAP system component has failed or canister fuel saturated • Fuel control sensor is out of calibration (i.e., ECT, IAT or MAP) • Fuel delivery system supplying too much fuel during cruise or idle periods (e.g., faulty fuel pump, or faulty pressure regulator) • Fuel injector(s) is leaking or stuck partially open (one or more) • HO2S is contaminated, deteriorated or it has failed
DTC: P0175 **2T OBD/FUEL, MIL: YES** **1997, 1998, 1999, 2000** **Models:** Hombre **Engines:** 4.3L VIN X **Transmissions:** A/T, M/T	**Fuel System Too Rich (Bank 2)** DTC P0106, P0107, P0108, P0112, P0113, P0117, P0118, P0121, P0122, P0123, P0131, P0132, P0133, P0134, P0200, P0300, P0301-P0304, P0325, P0336, P0341, P0342, P0401, P0404, P0405, P0502, P0503, P0506, P0507, P0601, P0602, P1404, P1133 and P1441 not set, ECT sensor signal from 167-237°F, IAT sensor signal from 46-169°F, BARO sensor signal over 70 kPa, MAF sensor signal at 3-85 g/sec, MAP sensor signal at 22-85 kPa, throttle angle under 70%, VSS under 85 mph, engine speed from 575-4000 rpm in closed loop, and the PCM detected the average of the Short Term fuel trim values for 120 seconds indicated a rich A/F condition. **Possible Causes:** • Air leak at the exhaust pipe or manifold, or at air injection pipes • Base engine "mechanical" fault affecting one or more cylinders • EVAP system component has failed or canister fuel saturated • Fuel control sensor is out of calibration (i.e., ECT, IAT or MAP) • Fuel delivery system supplying too much fuel during cruise or idle periods (e.g., faulty fuel pump, or faulty pressure regulator) • Fuel injector(s) is leaking or stuck partially open (one or more) • HO2S is contaminated, deteriorated or it has failed
DTC: P0200 **2T CCM, MIL: YES** **1998, 1999, 2000** **Models**: Hombre **Engines:** 2.2L VIN 4 **Transmissions:** A/T, M/T	**Fuel Injector Control Circuit Malfunction** Engine running, system voltage more than 9.0v, and the PCM detected the injector current was less than 1 amp, or the injector current was continuously high, either condition met for 7 seconds. **Possible Causes:** • Fuel injector control circuit is open or shorted to ground • Fuel injector control circuit is shorted to system power • Fuel injector power circuit is open between injector and relay • Fuel Injector has failed • PCM has failed (injector driver circuit may be open or shorted)
DTC: P0201 **1T CCM, MIL: YES** **1998, 1999, 2000, 2001, 2002** **Models:** Amigo, Rodeo, Rodeo Sport **Engines:** 2.2L VIN D **Transmissions:** A/T, M/T	**Injector Circuit Malfunction - Cylinder 1** Engine running, system voltage over 9v, and PCM detected the injector voltage for Cylinder 1 did not equal the ignition voltage with the injector commanded "off", or that the injector voltage did not equal zero (0) volts with the injector commanded "on". **Possible Causes:** • Fuel injector control circuit is open or shorted to ground • Fuel injector power circuit is open between injector and relay • Fuel Injector has failed • PCM has failed (injector driver circuit may be open or shorted)
DTC: P0202 **1T CCM, MIL: YES** **1998, 1999, 2000, 2001, 2002** **Models:** Amigo, Rodeo, Rodeo Sport **Engines:** 2.2L VIN D **Transmissions:** A/T, M/T	**Injector Circuit Malfunction - Cylinder 2** Engine running, system voltage over 9v, and PCM detected the injector voltage for Cylinder 2 did not equal the ignition voltage with the injector commanded "off", or that the injector voltage did not equal zero (0) volts with the injector commanded "on". **Possible Causes:** • Fuel injector control circuit is open or shorted to ground • Fuel injector power circuit open between injector and ECM fuse • Fuel Injector has failed • PCM has failed (injector driver circuit may be open or shorted)

DTC	Trouble Code Title, Conditions & Possible Causes
DTC: P0203 **1T CCM, MIL: YES** **1998, 1999, 2000, 2001, 2002** **Models:** Amigo, Rodeo, Rodeo Sport **Engines:** 2.2L VIN D **Transmissions:** A/T, M/T	**Injector Circuit Malfunction - Cylinder 3** Engine running, system voltage over 9v, and PCM detected the injector voltage for Cylinder 3 did not equal the ignition voltage with the injector commanded "off", or that the injector voltage did not equal zero (0) volts with the injector commanded "on". **Possible Causes:** • Fuel injector control circuit is open or shorted to ground • Fuel injector power circuit is open between injector and relay • Fuel Injector has failed • PCM has failed (injector driver circuit may be open or shorted)
DTC: P0204 **1T CCM, MIL: YES** **1998, 1999, 2000, 2001, 2002** **Models:** Amigo, Rodeo, Rodeo Sport **Engines:** 2.2L VIN D **Transmissions:** A/T, M/T	**Injector Circuit Malfunction - Cylinder 4** Engine running, system voltage over 9v, and PCM detected the injector voltage for Cylinder 4 did not equal the ignition voltage with the injector commanded "off", or that the injector voltage did not equal zero (0) volts with the injector commanded "on". **Possible Causes:** • Fuel injector control circuit is open or shorted to ground • Fuel injector power circuit is open between injector and relay • Fuel Injector has failed • PCM has failed (injector driver circuit may be open or shorted)
DTC: P0201 **1T CCM, MIL: YES** **1996, 1997, 1998, 1999, 2000, 2001, 2002** **Models:** Amigo, Axiom, Rodeo, Rodeo Sport, Trooper, VehiCROSS **Engines:** 3.2L VIN V, 3.2L VIN W, 3.5L VIN X **Transmissions:** A/T, M/T	**Injector Circuit Malfunction – Cylinder 1** Engine started, engine running, system voltage over 9v, and PCM detected the injector voltage for Cylinder 1 did not equal the system voltage with the injector commanded "off", or that the injector voltage did not equal zero (0) volts with the injector commanded "on". **Possible Causes:** • Fuel injector control circuit is open or shorted to ground • Fuel injector power circuit is open between injector and relay • Fuel Injector has failed • PCM has failed (injector driver circuit may be open or shorted)
DTC: P0202 **1T CCM, MIL: YES** **1996, 1997, 1998, 1999, 2000, 2001, 2002** **Models:** Amigo, Axiom, Rodeo, Rodeo Sport, Trooper, VehiCROSS **Engines:** 3.2L VIN V, 3.2L VIN W, 3.5L VIN X **Transmissions**: A/T, M/T	**Injector Circuit Malfunction – Cylinder 2** Engine started, engine running, system voltage over 9v, and PCM detected the injector voltage for Cylinder 2 did not equal the system voltage with the injector commanded "off", or that the injector voltage did not equal zero (0) volts with the injector commanded "on". **Possible Causes:** • Fuel injector control circuit is open or shorted to ground • Fuel injector power circuit open between injector and ECM fuse • Fuel Injector has failed • PCM has failed (injector driver circuit may be open or shorted)
DTC: P0203 **1T CCM, MIL: YES** **1996, 1997, 1998, 1999, 2000, 2001, 2002** **Models:** Amigo, Axiom, Rodeo, Rodeo Sport, Trooper, VehiCROSS **Engines:** 3.2L VIN V, 3.2L VIN W, 3.5L VIN X **Transmissions:** A/T, M/T	**Injector Circuit Malfunction – Cylinder 3** Engine started, engine running, system voltage over 9v, and PCM detected the injector voltage for Cylinder 3 did not equal the system voltage with the injector commanded "off", or that the injector voltage did not equal zero (0) volts with the injector commanded "on". **Possible Causes:** • Fuel injector control circuit is open or shorted to ground • Fuel injector power circuit open between injector and ECM fuse • Fuel Injector has failed • PCM has failed (injector driver circuit may be open or shorted)
DTC: P0204 **1T CCM, MIL: YES** **1996, 1997, 1998, 1999, 2000, 2001, 2002** **Models:** Amigo, Axiom, Rodeo, Rodeo Sport, Trooper, VehiCROSS **Engines:** 3.2L VIN V, 3.2L VIN W, 3.5L VIN X **Transmissions:** A/T, M/T	**Injector Circuit Malfunction – Cylinder 4** Engine started, engine running, system voltage over 9v, and PCM detected the injector voltage for Cylinder 4 did not equal the system voltage with the injector commanded "off", or that the injector voltage did not equal zero (0) volts with the injector commanded "on". **Possible Causes:** • Fuel injector control circuit is open or shorted to ground • Fuel injector power circuit open between injector and ECM fuse • Fuel Injector has failed • PCM has failed (injector driver circuit may be open or shorted)

DTC	Trouble Code Title, Conditions & Possible Causes
DTC: P0205 **1T CCM, MIL: YES** **1996, 1997, 1998, 1999, 2000, 2001, 2002** **Models:** Amigo, Axiom, Rodeo, Rodeo Sport, Trooper, VehiCROSS **Engines:** 3.2L VIN V, 3.2L VIN W, 3.5L VIN X **Transmissions:** A/T, M/T	**Injector Circuit Malfunction – Cylinder 5** Engine started, engine running, system voltage over 9v, and PCM detected the injector voltage for Cylinder 5 did not equal the system voltage with the injector commanded "off", or that the injector voltage did not equal zero (0) volts with the injector commanded "on". **Possible Causes:** • Fuel injector control circuit is open or shorted to ground • Fuel injector power circuit open between injector and ECM fuse • Fuel Injector has failed • PCM has failed (injector driver circuit may be open or shorted)
DTC: P0206 **1T CCM, MIL: YES** **1996, 1997, 1998, 1999, 2000, 2001, 2002** **Models:** Amigo, Axiom, Rodeo, Rodeo Sport, Trooper, VehiCROSS **Engines:** 3.2L VIN V, 3.2L VIN W, 3.5L VIN X **Transmissions:** A/T, M/T	**Injector Circuit Malfunction – Cylinder 6** Engine started, engine running, system voltage over 9v, and PCM detected the injector voltage for Cylinder 6 did not equal the system voltage with the injector commanded "off", or that the injector voltage did not equal zero (0) volts with the injector commanded "on". **Possible Causes:** • Fuel injector control circuit is open or shorted to ground • Fuel injector power circuit open between injector and ECM fuse • Fuel Injector has failed • PCM has failed (injector driver circuit may be open or shorted)
DTC: P0201 **1T CCM, MIL: YES** **1996, 1997** **Models:** Rodeo **Engines:** 2.6L VIN E **Transmissions:** M/T Trouble Code ID: P0201	**Injector Circuit Malfunction – Cylinder 1** Engine running, system voltage over 9v, and PCM detected the injector voltage for Cylinder 1 did not equal the ignition voltage with the injector commanded "off", or that the injector voltage did not equal zero (0) volts with the injector commanded "on". **Possible Causes:** • Fuel injector control circuit is open or shorted to ground • Fuel injector power circuit is open between injector and relay • Fuel Injector has failed • PCM has failed (injector driver circuit may be open or shorted)
DTC: P0202 **1T CCM, MIL: YES** **1996, 1997** **Models:** Rodeo **Engines:** 2.6L VIN E **Transmissions:** M/T Trouble Code ID: P0202	**Injector Circuit Malfunction – Cylinder 2** Engine running, system voltage over 9v, and PCM detected the injector voltage for Cylinder 2 did not equal the ignition voltage with the injector commanded "off", or that the injector voltage did not equal zero (0) volts with the injector commanded "on". **Possible Causes:** • Fuel injector control circuit is open or shorted to ground • Fuel injector power circuit open between injector and ECM fuse • Fuel Injector has failed • PCM has failed (injector driver circuit may be open or shorted)
DTC: P0203 **1T CCM, MIL: YES** **1996, 1997** **Models:** Rodeo **Engines:** 2.6L VIN E **Transmissions:** M/T	**Injector Circuit Malfunction – Cylinder 3** Engine running, system voltage over 9v, and PCM detected the injector voltage for Cylinder 3 did not equal the ignition voltage with the injector commanded "off", or that the injector voltage did not equal zero (0) volts with the injector commanded "on". **Possible Causes:** • Fuel injector control circuit is open or shorted to ground • Fuel injector power circuit open between injector and ECM fuse • Fuel Injector has failed • PCM has failed (injector driver circuit may be open or shorted)
DTC: P0204 **1T CCM, MIL: YES** **1996, 1997** **Models:** Rodeo **Engines:** 2.6L VIN E Transmissions: M/T Trouble Code ID: P0204	**Injector Circuit Malfunction – Cylinder 4** Engine running, system voltage over 9v, and PCM detected the injector voltage for Cylinder 4 did not equal the ignition voltage with the injector commanded "off", or that the injector voltage did not equal zero (0) volts with the injector commanded "on". **Possible Causes:** • Fuel injector control circuit is open or shorted to ground • Fuel injector power circuit open between injector and ECM fuse • Fuel Injector has failed • PCM has failed (injector driver circuit may be open or shorted)
DTC: P0218 **1T CCM, MIL: NO** **1998, 1999, 2000, 2001, 2002** **Models:** Amigo, Rodeo, Rodeo Sport **Engines:** All Transmissions: A/T	**Transmission Fluid Over-Temperature Malfunction** DTC P0712 and P0713 not set, and the PCM detected the Transmission Fluid Temperature (TFT) sensor signal indicated more than 275°F, condition met for 21 seconds during the CCM test. **Note: The CHECK TRANS lamp is "on" if the TFT sensor signal exceeds 293°F.** **Possible Causes:** • TFT sensor is out-of-calibration (skewed), or it has failed • Torque converter stator is damaged or has failed • PCM has failed

DTC	Trouble Code Title, Conditions & Possible Causes
DTC: P0101 **1T CCM, MIL: NO** **1997, 1998, 1999, 2000** **Models:** Hombre Engines: All **Transmissions:** A/T	**Transmission Fluid Over-Temperature Malfunction** DTC P0711, P0712 and P0713 not set, key on for 5 seconds, and the PCM detected the Transmission Fluid Temperature (TFT) sensor signal indicated more than 266°F for over 10 minutes. **Note: For additional help with this code, view the Failure Records.** **Possible Causes:** • TFT sensor is out-of-calibration (skewed), or it has failed • Torque converter stator is damaged or has failed • PCM has failed
DTC: P0218 **1T CCM, MIL: YES** **1996, 1997, 1998, 1999, 2000, 2001, 2002** **Models:** Axiom, Trooper, VehiCROSS **Engines:** All **Transmissions:** A/T	**Transmission Fluid Over-Temperature Malfunction** DTC P0712 and P0713 not set, and the PCM detected the Transmission Fluid Temperature (TFT) sensor signal indicated more than 284°F, condition met for 21 seconds. **Note: The ATF LAMP is "on" if the TFT sensor signal exceeds 293°F.** **Possible Causes:** • TFT sensor signal circuit is shorted to ground • TFT sensor is out-of-calibration (skewed), or it has failed • Torque converter stator is damaged or has failed • PCM has failed
DTC: P0300 **2T CCM, MIL: YES** **1998, 1999, 2000, 2001, 2002** **Models:** Amigo, Rodeo, Rodeo Sport Engines: 2.2L VIN D **Transmissions:** A/T, M/T	**Multiple Cylinder Misfire Detected** No CKP, CMP, ECT, MAP, TP or VSS codes set, ECT sensor signal from 20-240°F, system voltage from 11-16v, engine started and driven to a speed of over 20 mph, engine speed of 600-6250 rpm with the engine under load, throttle angle change less than ±2.73% and steady, TP sensor signal less than 3.125% in a 100 ms period of time, and the PCM detected a crankshaft variation that indicated a multiple misfire condition in more than one cylinder during the test. **Note: If the misfire is severe, the MIL will flash on/off on the 1st trip!** **Possible Causes:** • Base engine mechanical fault that affects one or more cylinders • CMP sensor is damaged or failed (problem may be intermittent) • Fuel metering fault that affects more than one cylinder • Fuel pressure too low or too high, fuel supply contaminated • EVAP system problem or the EVAP canister is fuel saturated • EGR valve is stuck open or the PCV system has a vacuum leak • Ignition system fault (a coil) that affects more than one cylinder • TSB 01-04-S004 (3/01) contains information related to the code
DTC: P0301 **2T CCM, MIL: YES** **1998, 1999, 2000, 2001, 2002** **Models:** Amigo, Rodeo, Rodeo Sport **Engines:** 2.2L VIN D **Transmissions:** A/T, M/T	**Misfire Detected - Cylinder 1** No CKP, CMP, ECT, MAP, TP or VSS codes set, ECT sensor signal from 20-240°F, system voltage from 11-16v, engine started and driven to a speed of over 20 mph, engine speed of 600-6250 rpm with the engine under load, throttle angle change less than ±2.73% and steady, TP sensor signal less than 3.125% in a 100 ms period of time, and the PCM detected a crankshaft variation that indicated a multiple misfire condition in more than one cylinder during the test. **Note: If the misfire is severe, the MIL will flash on/off on the 1st trip!** **Possible Causes:** • Base engine mechanical fault that affects only one cylinder • Fuel metering fault that affects only one cylinder • EGR valve is stuck open or the PCV system has a vacuum leak • Ignition system fault (i.e., a coil) that affects only one cylinder
DTC: P0302 **2T CCM, MIL: YES** **1998, 1999, 2000, 2001, 2002** **Models:** Amigo, Rodeo, Rodeo Sport **Engines:** 2.2L VIN D **Transmissions:** A/T, M/T	**Misfire Detected - Cylinder 2** No CKP, CMP, ECT, MAP, TP or VSS codes set, ECT sensor signal from 20-240°F, system voltage from 11-16v, engine started and driven to a speed of over 20 mph, engine speed of 600-6250 rpm with the engine under load, throttle angle change less than ±2.73% and steady, TP sensor signal less than 3.125% in a 100 ms period of time, and the PCM detected a crankshaft variation that indicated a multiple misfire condition in more than one cylinder during the test. **Note: If the misfire is severe, the MIL will flash on/off on the 1st trip!** **Possible Causes:** • Base engine mechanical fault that affects only one cylinder • Fuel metering fault that affects only one cylinder • EGR valve is stuck open or the PCV system has a vacuum leak • Ignition system fault (i.e., a coil) that affects only one cylinder

DTC	Trouble Code Title, Conditions & Possible Causes
DTC: P0303 **2T CCM, MIL: YES** **1998, 1999, 2000, 2001, 2002** **Models:** Amigo, Rodeo, Rodeo Sport **Engines:** 2.2L VIN D **Transmissions:** A/T, M/T	**Misfire Detected - Cylinder 3** No CKP, CMP, ECT, MAP, TP or VSS codes set, ECT sensor signal from 20-240ºF, system voltage from 11-16v, engine started and driven to a speed of over 20 mph, engine speed of 600-6250 rpm with the engine under load, throttle angle change less than ±2.73% and steady, TP sensor signal less than 3.125% in a 100 ms period of time, and the PCM detected a crankshaft variation that indicated a multiple misfire condition in more than one cylinder during the test. **Note: If the misfire is severe, the MIL will flash on/off on the 1st trip!** **Possible Causes:** • Base engine mechanical fault that affects only one cylinder • Fuel metering fault that affects only one cylinder • EGR valve is stuck open or the PCV system has a vacuum leak • Ignition system fault (i.e., a coil) that affects only one cylinder
DTC: P0304 **2T CCM, MIL: YES** **1998, 1999, 2000, 2001, 2002** **Models:** Amigo, Rodeo, Rodeo Sport **Engines:** 2.2L VIN D **Transmissions:** A/T, M/T	**Misfire Detected - Cylinder 4** No CKP, CMP, ECT, MAP, TP or VSS codes set, ECT sensor signal from 20-240ºF, system voltage from 11-16v, engine started and driven to a speed of over 20 mph, engine speed of 600-6250 rpm with the engine under load, throttle angle change less than ±2.73% and steady, TP sensor signal less than 3.125% in a 100 ms period of time, and the PCM detected a crankshaft variation that indicated a multiple misfire condition in more than one cylinder during the test. **Note: If the misfire is severe, the MIL will flash on/off on the 1st trip!** **Possible Causes:** • Base engine mechanical fault that affects only one cylinder • Fuel metering fault that affects only one cylinder • EGR valve is stuck open or the PCV system has a vacuum leak • Ignition system fault (i.e., a coil) that affects only one cylinder
DTC: P0300 **2T CCM, MIL: YES** **1996, 1997, 1998, 1999, 2000, 2001, 2002** **Models:** Amigo, Axiom, Rodeo, Rodeo Sport, Trooper, VehiCROSS **Engines:** 3.2L VIN V, 3.2L VIN W, 3.5L VIN X **Transmissions:** A/T, M/T	**Multiple Cylinder Misfire Detected** DTC P0101, P0102, P0103, P0106, P0107, P0108, P0117, P0118, P0121, P0122, P0123, P0336, P0341, P0342, P0502 and P0503 not set, ECT sensor from 20-248ºF, system voltage from 11-16v, engine speed from 800-5500 rpm, throttle angle stable (± 3%), and the PCM detected a crankshaft speed variation in one or more cylinders characteristic of a misfire condition during the Misfire Monitor test. **Note: If the misfire is severe, the MIL will flash on/off on the 1st trip!** **Possible Causes:** • Base engine mechanical fault that affects one or more cylinders • Fuel metering fault that affects more than one cylinder • Fuel pressure too low or too high, fuel supply contaminated • EVAP system problem or the EVAP canister is fuel saturated • EGR valve is stuck open or the PCV system has a vacuum leak • IC control circuit is shorted to ground (an intermittent fault) • Ignition system fault (a coil) that affects more than one cylinder • MAF sensor contamination (it can cause a very lean condition)
DTC: P0301 **2T CCM, MIL: YES** **1996, 1997, 1998, 1999, 2000, 2001, 2002** **Models:** Amigo, Axiom, Rodeo, Rodeo Sport, Trooper, VehiCROSS **Engines:** 3.2L VIN V, 3.2L VIN W, 3.5L VIN X **Transmissions:** A/T, M/T	**Misfire Detected - Cylinder 1** DTC P0101, P0102, P0103, P0106, P0107, P0108, P0117, P0118, P0121, P0122, P0123, P0336, P0341, P0342, P0502 and P0503 not set, ECT sensor from 20-248ºF, system voltage from 11-16v, engine speed from 800-5500 rpm, throttle angle stable (± 3%), and the PCM detected a crankshaft speed variation in one cylinder characteristic of a misfire condition during the Misfire Diagnostic Monitor test. **Note: If the misfire is severe, the MIL will flash on/off on the 1st trip!** **Possible Causes:** • Base engine mechanical fault that affects only one cylinder • Fuel metering fault that affects only one cylinder • EGR valve is stuck open or the PCV system has a vacuum leak • Ignition system fault (i.e., a coil) that affects only one cylinder
DTC: P0302 **2T CCM, MIL: YES** **1996, 1997, 1998, 1999, 2000, 2001, 2002** **Models:** Amigo, Axiom, Rodeo, Rodeo Sport, Trooper, VehiCROSS **Engines:** 3.2L VIN V, 3.2L VIN W, 3.5L VIN X **Transmissions:** A/T, M/T	**Misfire Detected - Cylinder 2** DTC P0101, P0102, P0103, P0106, P0107, P0108, P0117, P0118, P0121, P0122, P0123, P0336, P0341, P0342, P0502 and P0503 not set, ECT sensor from 20-248ºF, system voltage from 11-16v, engine speed from 800-5500 rpm, throttle angle stable (± 3%), and the PCM detected a crankshaft speed variation in one cylinder characteristic of a misfire condition during the Misfire Diagnostic Monitor test. **Note: If the misfire is severe, the MIL will flash on/off on the 1st trip!** **Possible Causes:** • Base engine mechanical fault that affects only one cylinder • Fuel metering fault that affects only one cylinder • EGR valve is stuck open or the PCV system has a vacuum leak • Ignition system fault (i.e., a coil) that affects only one cylinder

DTC	Trouble Code Title, Conditions & Possible Causes
DTC: P0303 **2T CCM, MIL: YES** **1996, 1997, 1998, 1999, 2000, 2001, 2002** **Models:** Amigo, Axiom, Rodeo, Rodeo Sport, Trooper, VehiCROSS **Engines:** 3.2L VIN V, 3.2L VIN W, 3.5L VIN X **Transmissions:** A/T, M/T	**Misfire Detected - Cylinder 3** DTC P0101, P0102, P0103, P0106, P0107, P0108, P0117, P0118, P0121, P0122, P0123, P0336, P0341, P0342, P0502 and P0503 not set, ECT sensor from 20-248°F, system voltage from 11-16v, engine speed from 800-5500 rpm, throttle angle stable (± 3%), and the PCM detected a crankshaft speed variation in one cylinder characteristic of a misfire condition during the Misfire Diagnostic Monitor test. **Note: If the misfire is severe, the MIL will flash on/off on the 1st trip!** **Possible Causes:** • Base engine mechanical fault that affects only one cylinder • Fuel metering fault that affects only one cylinder • EGR valve is stuck open or the PCV system has a vacuum leak • Ignition system fault (i.e., a coil) that affects only one cylinder
DTC: P0304 **2T CCM, MIL: YES** **1996, 1997, 1998, 1999, 2000, 2001, 2002** **Models:** Amigo, Axiom, Rodeo, Rodeo Sport, Trooper, VehiCROSS **Engines:** 3.2L VIN V, 3.2L VIN W, 3.5L VIN X **Transmissions:** A/T, M/T	**Misfire Detected - Cylinder 4** DTC P0101, P0102, P0103, P0106, P0107, P0108, P0117, P0118, P0121, P0122, P0123, P0336, P0341, P0342, P0502 and P0503 not set, ECT sensor from 20-248°F, system voltage from 11-16v, engine speed from 800-5500 rpm, throttle angle stable (± 3%), and the PCM detected a crankshaft speed variation in one cylinder characteristic of a misfire condition during the Misfire Diagnostic Monitor test. **Note: If the misfire is severe, the MIL will flash on/off on the 1st trip!** **Possible Causes:** • Base engine mechanical fault that affects only one cylinder • Fuel metering fault that affects only one cylinder • EGR valve is stuck open or the PCV system has a vacuum leak • Ignition system fault (i.e., a coil) that affects only one cylinder
DTC: P0305 **2T CCM, MIL: YES** **1996, 1997, 1998, 1999, 2000, 2001, 2002** **Models:** Amigo, Axiom, Rodeo, Rodeo Sport, Trooper, VehiCROSS **Engines:** 3.2L VIN V, 3.2L VIN W, 3.5L VIN X **Transmissions:** A/T, M/T	**Misfire Detected - Cylinder 5** DTC P0101, P0102, P0103, P0106, P0107, P0108, P0117, P0118, P0121, P0122, P0123, P0336, P0341, P0342, P0502 and P0503 not set, ECT sensor from 20-248°F, system voltage from 11-16v, engine speed from 800-5500 rpm, throttle angle stable (± 3%), and the PCM detected a crankshaft speed variation in one cylinder characteristic of a misfire condition during the Misfire Diagnostic Monitor test. **Note: If the misfire is severe, the MIL will flash on/off on the 1st trip!** **Possible Causes:** • Base engine mechanical fault that affects only one cylinder • Fuel metering fault that affects only one cylinder • EGR valve is stuck open or the PCV system has a vacuum leak • Ignition system fault (i.e., a coil) that affects only one cylinder
DTC: P0306 **2T CCM, MIL: YES** **1996, 1997, 1998, 1999, 2000, 2001, 2002** **Models:** Amigo, Axiom, Rodeo, Rodeo Sport, Trooper, VehiCROSS **Engines:** 3.2L VIN V, 3.2L VIN W, 3.5L VIN X **Transmissions:** A/T, M/T	**Misfire Detected - Cylinder 6** DTC P0101, P0102, P0103, P0106, P0107, P0108, P0117, P0118, P0121, P0122, P0123, P0336, P0341, P0342, P0502 and P0503 not set, ECT sensor from 20-248°F, system voltage from 11-16v, engine speed from 800-5500 rpm, throttle angle stable (± 3%), and the PCM detected a crankshaft speed variation in one cylinder characteristic of a misfire condition during the Misfire Diagnostic Monitor test. **Note: If the misfire is severe, the MIL will flash on/off on the 1st trip!** **Possible Causes:** • Base engine mechanical fault that affects only one cylinder • Fuel metering fault that affects only one cylinder • EGR valve is stuck open or the PCV system has a vacuum leak • Ignition system fault (i.e., a coil) that affects only one cylinder
DTC: P0300 **2T CCM, MIL: YES** **1996, 1997, 1998, 1999, 2000** **Models:** Hombre **Engines:** 2.2L VIN 4 **Transmissions:** A/T, M/T	**Multiple Cylinder Misfire Detected** DTC P0106, P0107, P0108, P0112, P0113, P0117, P0118, P0121, P0122, P0123, P0131, P0132, P0133, P0134, P0200, P0325, P0335, P0341, P0342, P0401, P0404, P0405, P0502, P0503, P0506, P0507, P0601, P0740, P0742, P1133, P1404 and P1621 not set, ECT sensor signal from 20-254°F, system voltage from 11-16v, engine runtime over 15 seconds, engine speed from 469-6000 rpm, throttle angle steady and any change less than 8% in a 1 second, and the PCM detected the "misfire total" value was more than 12 counts in one or more cylinders in the Misfire Monitor Detection test. **Note: If the misfire is severe, the MIL will flash on/off on the 1st trip!** **Possible Causes:** • Base engine mechanical fault that affects one or more cylinders • CMP sensor is damaged or failed (problem may be intermittent) • Fuel metering fault that affects more than one cylinder • Fuel pressure too low or too high, fuel supply contaminated • EVAP system problem or the EVAP canister is fuel saturated • EGR valve is stuck open or the PCV system has a vacuum leak • Ignition system fault (a coil) that affects more than one cylinder • MAF sensor is contaminated (this can cause a lean condition)

DTC	Trouble Code Title, Conditions & Possible Causes
DTC: P0301 **2T CCM, MIL: YES** **1996, 1997, 1998, 1999, 2000** **Models:** Hombre **Engines:** 2.2L VIN 4 **Transmissions:** A/T, M/T	**Misfire Detected - Cylinder 1** DTC P0106, P0107, P0108, P0112, P0113, P0117, P0118, P0121, P0122, P0123, P0131, P0132, P0133, P0134, P0200, P0325, P0335, P0341, P0342, P0401, P0404, P0405, P0502, P0503, P0506, P0507, P0601, P0740, P0742, P1133, P1404 and P1621 not set, ECT sensor signal from 20-254°F, system voltage from 11-16v, engine runtime over 15 seconds, engine speed from 469-6000 rpm, throttle angle steady and any change less than 8% in 1 second, and the PCM detected the misfire total value was more than 12 counts in one cylinder during the Misfire Detection test. **Note: If the misfire is severe, the MIL will flash on/off on the 1st trip!** **Possible Causes:** • Base engine mechanical fault that affects only one cylinder • Fuel metering fault that affects only one cylinder • EGR valve is stuck open or the PCV system has a vacuum leak • Ignition system fault (i.e., a coil) that affects only one cylinder
DTC: P0302 **2T CCM, MIL: YES** **1996, 1997, 1998, 1999, 2000** **Models:** Hombre Engines: 2.2L VIN 4 Transmissions: A/T, M/T	**Misfire Detected - Cylinder 2** DTC P0106, P0107, P0108, P0112, P0113, P0117, P0118, P0121, P0122, P0123, P0131, P0132, P0133, P0134, P0200, P0325, P0335, P0341, P0342, P0401, P0404, P0405, P0502, P0503, P0506, P0507, P0601, P0740, P0742, P1133, P1404 and P1621 not set, ECT sensor signal from 20-254°F, system voltage from 11-16v, engine runtime over 15 seconds, engine speed from 469-6000 rpm, throttle angle steady and any change less than 8% in 1 second, and the PCM detected the misfire total value was more than 12 counts in one cylinder during the Misfire Detection test. **Note: If the misfire is severe, the MIL will flash on/off on the 1st trip!** **Possible Causes:** • Base engine mechanical fault that affects only one cylinder • Fuel metering fault that affects only one cylinder • EGR valve is stuck open or the PCV system has a vacuum leak • Ignition system fault (i.e., a coil) that affects only one cylinder
DTC: P0303 **2T CCM, MIL: YES** **1996, 1997, 1998, 1999, 2000** **Models:** Hombre **Engines**: 2.2L VIN 4 **Transmissions:** A/T, M/T	**Misfire Detected - Cylinder 3** DTC P0106, P0107, P0108, P0112, P0113, P0117, P0118, P0121, P0122, P0123, P0131, P0132, P0133, P0134, P0200, P0325, P0335, P0341, P0342, P0401, P0404, P0405, P0502, P0503, P0506, P0507, P0601, P0740, P0742, P1133, P1404 and P1621 not set, ECT sensor signal from 20-254°F, system voltage from 11-16v, engine runtime over 15 seconds, engine speed from 469-6000 rpm, throttle angle steady with any change less than 8% in 1 second, and the PCM detected the misfire total value was more than 12 counts in one cylinder during the Misfire Detection test. **Note: If the misfire is severe, the MIL will flash on/off on the 1st trip!** **Possible Causes:** • Base engine mechanical fault that affects only one cylinder • Fuel metering fault that affects only one cylinder • EGR valve is stuck open or the PCV system has a vacuum leak • Ignition system fault (i.e., a coil) that affects only one cylinder
DTC: P0304 **2T CCM, MIL: YES** **1996, 1997, 1998, 1999, 2000** **Models:** Hombre **Engines:** 2.2L VIN 4 **Transmissions:** A/T, M/T	**Misfire Detected - Cylinder 4** DTC P0106, P0107, P0108, P0112, P0113, P0117, P0118, P0121, P0122, P0123, P0131, P0132, P0133, P0134, P0200, P0325, P0335, P0341, P0342, P0401, P0404, P0405, P0502, P0503, P0506, P0507, P0601, P0740, P0742, P1133, P1404 and P1621 not set, ECT sensor signal from 20-254°F, system voltage from 11-16v, engine runtime over 15 seconds, engine speed from 469-6000 rpm, throttle angle steady with any change less than 8% in 1 second, and the PCM detected the misfire total value was more than 12 counts in one cylinder during the Misfire Detection test. **Note: If the misfire is severe, the MIL will flash on/off on the 1st trip!** **Possible Causes:** • Base engine mechanical fault that affects only one cylinder • Fuel metering fault that affects only one cylinder • EGR valve is stuck open or the PCV system has a vacuum leak • Ignition system fault (i.e., a coil) that affects only one cylinder

DTC	Trouble Code Title, Conditions & Possible Causes
DTC: P0300 **2T CCM, MIL: YES** **1997, 1998, 1999, 2000** **Models:** Hombre **Engines:** 4.3L VIN X **Transmissions:** A/T, M/T	**Multiple Cylinder Misfire Detected** DTC P0101, P0102, P0103, P0121, P0122, P0123, P0336, P0341, P0342, P0502 and P0503 not set, ECT sensor signal more than 70°F (if the ECT signal is less than 20°F, the test is delayed until it reaches 70°F, and if the ECT signal is over 20°F, the test is delayed for 5 seconds), system voltage from 11-16v, engine runtime over 15 seconds, fuel level over 10%, engine speed from 450-5000 rpm, throttle angle steady within 2% for 100 ms, and the PCM detected a deceleration in crankshaft speed in one or more cylinders characteristic of a misfire during the Misfire Monitor Detection test. **Note: If the misfire is severe, the MIL will flash on/off on the 1st trip!** **Possible Causes:** • Base engine mechanical fault that affects one or more cylinders • Fuel metering fault that affects more than one cylinder • Fuel pressure too low or too high, fuel supply contaminated • EVAP system problem or the EVAP canister is fuel saturated • EGR valve is stuck open or the PCV system has a vacuum leak • IC control circuit is shorted to ground (an intermittent fault) • Ignition system fault (a coil) that affects more than one cylinder • MAF sensor contamination (it can cause a very lean condition)
DTC: P0301 **2T CCM, MIL: YES** **1997, 1998, 1999, 2000** **Models:** Hombre **Engines:** 4.3L VIN X **Transmissions:** A/T, M/T	**Misfire Detected - Cylinder 1** DTC P0101, P0102, P0103, P0121, P0122, P0123, P0336, P0341, P0342, P0502 and P0503 not set, ECT sensor signal more than 70°F (if the ECT signal is less than 20°F, the test is delayed until it reaches 70°F, and if the ECT signal is over 20°F, the test is delayed for 5 seconds), system voltage from 11-16v, engine runtime over 15 seconds, fuel level over 10%, engine speed from 450-5000 rpm, throttle angle steady within 2% for 100 ms, and the PCM detected a deceleration in crankshaft speed in one cylinder characteristic of a misfire during the Misfire Monitor Detection test. **Note: If the misfire is severe, the MIL will flash on/off on the 1st trip!** **Possible Causes:** • Base engine mechanical fault that affects only one cylinder • Fuel metering fault that affects only one cylinder • EGR valve is stuck open or the PCV system has a vacuum leak • Ignition system fault (i.e., a coil) that affects only one cylinder
DTC: P0302 **2T CCM, MIL: YES** **1997, 1998, 1999, 2000** **Models:** Hombre **Engines:** 4.3L VIN X **Transmissions:** A/T, M/T	**Misfire Detected - Cylinder 2** DTC P0101, P0102, P0103, P0121, P0122, P0123, P0336, P0341, P0342, P0502 and P0503 not set, ECT sensor signal more than 70°F (if the ECT signal is less than 20°F, the test is delayed until it reaches 70°F, and if the ECT signal is over 20°F, the test is delayed for 5 seconds), system voltage from 11-16v, engine runtime over 15 seconds, fuel level over 10%, engine speed from 450-5000 rpm, throttle angle steady within 2% for 100 ms, and the PCM detected a deceleration in crankshaft speed in one cylinder characteristic of a misfire during the Misfire Monitor Detection test. **Note: If the misfire is severe, the MIL will flash on/off on the 1st trip!** **Possible Causes:** • Base engine mechanical fault that affects only one cylinder • Fuel metering fault that affects only one cylinder • EGR valve is stuck open or the PCV system has a vacuum leak • Ignition system fault (i.e., a coil) that affects only one cylinder
DTC: P0303 **1T CCM, MIL: YES** **1997, 1998, 1999, 2000** **Models:** Hombre **Engines:** 4.3L VIN X **Transmissions:** A/T, M/T	**Misfire Detected - Cylinder 3** DTC P0101, P0102, P0103, P0121, P0122, P0123, P0336, P0341, P0342, P0502 and P0503 not set, ECT sensor signal more than 70°F (if the ECT signal is less than 20°F, the test is delayed until it reaches 70°F, and if the ECT signal is over 20°F, the test is delayed for 5 seconds), system voltage from 11-16v, engine runtime over 15 seconds, fuel level over 10%, engine speed from 450-5000 rpm, throttle angle steady within 2% for 100 ms, and the PCM detected a deceleration in crankshaft speed in one cylinder characteristic of a misfire during the Misfire Monitor Detection test. **Note: If the misfire is severe, the MIL will flash on/off on the 1st trip!** **Possible Causes:** • Base engine mechanical fault that affects only one cylinder • Fuel metering fault that affects only one cylinder • EGR valve is stuck open or the PCV system has a vacuum leak • Ignition system fault (i.e., a coil) that affects only one cylinder

DTC	Trouble Code Title, Conditions & Possible Causes
DTC: P0304 **2T CCM, MIL: YES** **1997, 1998, 1999, 2000** **Models:** Hombre **Engines:** 4.3L VIN X **Transmissions:** A/T, M/T	**Misfire Detected - Cylinder 4** DTC P0101, P0102, P0103, P0121, P0122, P0123, P0336, P0341, P0342, P0502 and P0503 not set, ECT sensor signal more than 70ºF (if the ECT signal is less than 20ºF, the test is delayed until it reaches 70ºF, and if the ECT signal is over 20ºF, the test is delayed for 5 seconds), system voltage from 11-16v, engine runtime over 15 seconds, fuel level over 10%, engine speed from 450-5000 rpm, throttle angle steady within 2% for 100 ms, and the PCM detected a deceleration in crankshaft speed in one cylinder characteristic of a misfire during the Misfire Monitor Detection test. **Note: If the misfire is severe, the MIL will flash on/off on the 1st trip!** **Possible Causes:** • Base engine mechanical fault that affects only one cylinder • Fuel metering fault that affects only one cylinder • EGR valve is stuck open or the PCV system has a vacuum leak • Ignition system fault (i.e., a coil) that affects only one cylinder
DTC: P0305 **2T CCM, MIL: YES** **1997, 1998, 1999, 2000** **Models:** Hombre **Engines:** 4.3L VIN X **Transmissions:** A/T, M/T	**Misfire Detected - Cylinder 5** DTC P0101, P0102, P0103, P0121, P0122, P0123, P0336, P0341, P0342, P0502 and P0503 not set, ECT sensor signal more than 70ºF (if the ECT signal is less than 20ºF, the test is delayed until it reaches 70ºF, and if the ECT signal is over 20ºF, the test is delayed for 5 seconds), system voltage from 11-16v, engine runtime over 15 seconds, fuel level over 10%, engine speed from 450-5000 rpm, throttle angle steady within 2% for 100 ms, and the PCM detected a deceleration in crankshaft speed in one cylinder characteristic of a misfire during the Misfire Monitor Detection test. **Note: If the misfire is severe, the MIL will flash on/off on the 1st trip!** **Possible Causes:** • Base engine mechanical fault that affects only one cylinder • Fuel metering fault that affects only one cylinder • EGR valve is stuck open or the PCV system has a vacuum leak • Ignition system fault (i.e., a coil) that affects only one cylinder
DTC: P0306 **2T CCM, MIL: YES** **1997, 1998, 1999, 2000** **Models:** Hombre **Engines:** 4.3L VIN X Transmissions: A/T, M/T	**Misfire Detected - Cylinder 6** DTC P0101, P0102, P0103, P0121, P0122, P0123, P0336, P0341, P0342, P0502 and P0503 not set, ECT sensor signal more than 70ºF (if the ECT signal is less than 20ºF, the test is delayed until it reaches 70ºF, and if the ECT signal is over 20ºF, the test is delayed for 5 seconds), system voltage from 11-16v, engine runtime over 15 seconds, fuel level over 10%, engine speed from 450-5000 rpm, throttle angle steady within 2% for 100 ms, and the PCM detected a deceleration in crankshaft speed in one cylinder characteristic of a misfire during the Misfire Monitor Detection test. **Note: If the misfire is severe, the MIL will flash on/off on the 1st trip!** **Possible Causes:** • Base engine mechanical fault that affects only one cylinder • Fuel metering fault that affects only one cylinder • EGR valve is stuck open or the PCV system has a vacuum leak • Ignition system fault (i.e., a coil) that affects only one cylinder
DTC: P0300 **2T CCM, MIL: YES** **1996, 1997, 1998, 1999** **Models:** Oasis **Engines:** All **Transmissions:** A/T, M/T	**Multiple Misfire Detected** DTC P0107, P0108, P0131, P0132, P0171, P0172, P1128, P0335, P0336, P0505, P1128, P1129, P1259, P1361, P1362, P1366, P1367 and P1519 not set, engine running under positive torque conditions, and the PCM detected a misfire in 2 or more cylinders. **Note: If the misfire is severe, the MIL will flash on/off on the 1st trip!** **Possible Causes:** • CKP or CMP sensor problem affecting more than one cylinder • Fuel system problem affecting more than one cylinder • Ignition system problem affecting more than one cylinder • Base engine mechanical fault affecting more than 1 cylinder
DTC: P0301 **2T CCM, MIL: YES** **1996, 1997, 1998, 1999** **Models:** Oasis **Engines:** All **Transmissions:** A/T, M/T	**Cylinder 1 Misfire Detected** DTC P0107, P0108, P0131, P0132, P0171, P0172, P1128, P0335, P0336, P0505, P1128, P1129, P1259, P1361, P1362, P1366, P1367 and P1519 not set, engine running under positive torque conditions, and the PCM detected a misfire condition in one cylinder. **Note: If the misfire is severe, the MIL will flash on/off on the 1st trip!** **Possible Causes:** • Fuel system problem affecting only Cylinder 1 • Ignition system problem affecting Cylinder 1 • Base engine (mechanical) problem affecting only Cylinder 1
DTC: P0302 **2T CCM, MIL: YES** **1996, 1997, 1998, 1999** **Models:** Oasis **Engines:** All **Transmissions:** A/T, M/T	**Cylinder 2 Misfire Detected** DTC P0107, P0108, P0131, P0132, P0171, P0172, P1128, P0335, P0336, P0505, P1128, P1129, P1259, P1361, P1362, P1366, P1367 and P1519 not set, engine running under positive torque conditions, and the PCM detected a misfire condition in one cylinder. **Note: If the misfire is severe, the MIL will flash on/off on the 1st trip!** **Possible Causes:** • Fuel system problem affecting only Cylinder 2 • Ignition system problem affecting Cylinder 2 • Base engine (mechanical) problem affecting only Cylinder 2

DTC	Trouble Code Title, Conditions & Possible Causes
DTC: P0303 **2T CCM, MIL: YES** **1996, 1997, 1998, 1999** **Models:** Oasis **Engines:** All **Transmissions:** A/T, M/T	**Cylinder 3 Misfire Detected** DTC P0107, P0108, P0131, P0132, P0171, P0172, P1128, P0335, P0336, P0505, P1128, P1129, P1259, P1361, P1362, P1366, P1367 and P1519 not set, engine running under positive torque conditions, and the PCM detected a misfire condition in one cylinder. **Note: If the misfire is severe, the MIL will flash on/off on the 1st trip!** **Possible Causes:** • Fuel system problem affecting only Cylinder 3 • Ignition system problem affecting Cylinder 3 • Base engine (mechanical) problem affecting only Cylinder 3
DTC: P0304 **2T CCM, MIL: YES** **1996, 1997, 1998, 1999** **Models:** Oasis **Engines:** All **Transmissions:** A/T, M/T	**Cylinder 4 Misfire Detected** DTC P0107, P0108, P0131, P0132, P0171, P0172, P1128, P0335, P0336, P0505, P1128, P1129, P1259, P1361, P1362, P1366, P1367 and P1519 not set, engine running under positive torque conditions, and the PCM detected a misfire condition in one cylinder. **Note: If the misfire is severe, the MIL will flash on/off on the 1st trip!** **Possible Causes:** • Fuel system problem affecting only Cylinder 4 • Ignition system problem affecting Cylinder 4 • Base engine (mechanical) problem affecting only Cylinder 4
DTC: P0300 **2T CCM, MIL: YES** **1996, 1997** **Models:** Rodeo **Engines:** 2.6L VIN E **Transmissions:** M/T	**Multiple Cylinder Misfire Detected** DTC P0106, P0107, P0108, P0117, P0118, P0121, P0122, P0123, P0336, P0341, P0342, P0502, P0503, P1390, P1391, P392 and P1393 not set, ECT sensor from 20-248°F, system voltage from 11-16v, engine speed from 800-5500 rpm, throttle angle stable with any change less than 3% within 125 ms, and the PCM detected a deceleration in crankshaft speed in more than one cylinder characteristic of a misfire during the Misfire Monitor Detection test. **Note: If the misfire is severe, the MIL will flash on/off on the 1st trip!** **Possible Causes:** • Base engine mechanical fault that affects one or more cylinders • Fuel metering fault that affects more than one cylinder • Fuel pressure too low or too high, fuel supply contaminated • EVAP system problem or the EVAP canister is fuel saturated • EGR valve is stuck open or the PCV system has a vacuum leak • IC control circuit is shorted to ground (an intermittent fault) • Ignition system fault (a coil) that affects more than one cylinder • MAF sensor contamination (it can cause a very lean condition)
DTC: P0325 **2T CCM, MIL: YES** **1998, 1999, 2000, 2001, 2002** **Models:** Amigo, Rodeo, Rodeo Sport **Engines:** 2.2L VIN D **Transmissions:** A/T, M/T	**Knock Sensor Circuit Malfunction** DTC P0327 not set, engine started, engine runtime over 10 seconds, engine speed over 2500 rpm, and the PCM detected an unexpected voltage condition (KS voltage signal of over 1.5625 or an instantaneous voltage signal of below 0.0195) on the Knock Sensor (KS) signal circuit for 8.75 seconds of a 10 second period of time during the CCM Rationality test. **Possible Causes:** • KS signal circuit is open or shorted to ground • KS signal circuit is shorted to VREF or system power (B+) • KS Module is damaged or has failed • PCM has failed
DTC: P0325 **1T CCM, MIL: YES** **2000, 2001, 2002** **Models:** Amigo, Rodeo, Rodeo Sport **Engines**: 3.2L VIN W **Transmissions:** A/T, M/T	**ION Sensing Module or ION Sensor Module Knock Intensity Circuit Malfunction** DTC P0336, P0337, P1311 and P1323 not set, engine started, system voltage from 11-16v, and the PCM detected an unexpected voltage condition on the ION Module Knock Intensity (KI) signal circuit during the test. **Possible Causes:** • KI signal circuit is open or shorted to ground • KI signal circuit is shorted to VREF or system power (B+) • ION Sensing Module is damaged or has failed • PCM has failed
DTC: P0325 **2T CCM, MIL: YES** **1999, 2000, 2001, 2002** **Models:** Axiom, VehiCROSS **Engines:** 3.5L VIN X **Transmissions:** A/T, M/T	**Knock Sensor Module Circuit Performance** DTC P0327 not set, engine started, system voltage from 11-16v, engine runtime over 120 seconds, and the PCM detected the Knock Sensor (KS) signal was present for over 5 seconds in the CCM test. **Possible Causes:** • KS signal circuit is open or shorted to ground • KS signal circuit is shorted to VREF or system power (B+) • Knock Sensor is damaged or has failed • PCM has failed

DTC	Trouble Code Title, Conditions & Possible Causes
DTC: P0325 **1T CCM, MIL: YES** **1996, 1997, 1998, 1999, 2000** **Models:** Hombre **Engines:** 2.2L VIN 4 **Transmissions:** A/T, M/T	**Knock Sensor Circuit Malfunction** Engine started, engine runtime over 20 seconds, engine speed over 1600 rpm, ECT sensor signal more than 131°F, MAP sensor signal more than 60 kPa, calculated engine vacuum less than 33 kPa, and the PCM detected the Knock Sensor (KS) variation was not within its expected operating range for over 60 seconds during the CCM test. **Possible Causes:** • KS signal circuit is open or shorted to ground • KS signal circuit is shorted to VREF or system power (B+) • Knock Sensor is damaged or has failed • PCM has failed
DTC: P0325 **2T CCM, MIL: YES** **1997, 1998, 1999, 2000** **Models:** Hombre **Engines:** 4.3L VIN X **Transmissions:** A/T, M/T	**Knock Sensor Module Range/Performance** DTC P0327 not set, engine started, system voltage from 11-16v, engine runtime over 120 seconds, and the PCM detected the Knock Sensor (KS) signal was present for more than 5 seconds during the CCM test. **Possible Causes:** • KS signal circuit is open or shorted to ground • KS signal circuit is shorted to VREF or system power (B+) • Knock Sensor is damaged or has failed • PCM has failed
DTC: P0325 **2T CCM, MIL: YES** **1998, 1999** **Models:** Oasis **Engines:** All **Transmissions:** A/T, M/T	**Knock Sensor Circuit Malfunction** Engine started, engine runtime over 1 minute, and the PCM detected an unexpected voltage condition on the Knock Sensor (KS) circuit during the CCM test. **Possible Causes:** • Knock sensor signal circuit is open between sensor and PCM • Knock sensor signal circuit is shorted to ground • Knock sensor is damaged or has failed (it may have an internal open or shorted condition) • PCM has failed
DTC: P0325 **2T CCM, MIL: YES** **1996, 1997, 1998, 1999** **Models:** Rodeo, Trooper **Engines:** 3.2L VIN V, 3.2L VIN W **Transmissions:** A/T, M/T	**Knock Sensor Module Circuit Performance** DTC P0327 not set, engine started, system voltage from 11-16v, engine runtime over 120 seconds, and the PCM detected the Knock Sensor (KS) signal was present for over 5 seconds in the CCM test. **Possible Causes:** • KS signal circuit is open or shorted to ground • KS signal circuit is shorted to VREF or system power (B+) • Knock Sensor is damaged or has failed • PCM has failed
DTC: P0327 **2T CCM, MIL: YES** **1998, 1999, 2000, 2001, 2002** **Models:** Amigo, Rodeo, Rodeo Sport **Engines:** 2.2L VIN D **Transmissions:** A/T, M/T	**Knock Sensor Circuit Low Input** Engine started, engine speed over 2000 rpm, and the PCM detected a Knock Sensor (KS) signal less than 0.0977 for 8.75 seconds over a 10 second period during the CCM test. **Possible Causes:** • KS signal circuit is open or shorted to ground • KS signal circuit is shorted to VREF or system power (B+) • KS Module is damaged or has failed • PCM has failed
DTC: P0327 **2T CCM, MIL: YES** **1999, 2000, 2001** **Models:** VehiCROSS **Engines:** 3.5L VIN X **Transmissions:** A/T, M/T	**Knock Sensor Circuit Low Input** Engine started, engine runtime over 10 seconds, system voltage from 11-16v, ECT sensor more than 140°F, engine speed from 2000-4000 rpm, throttle angle over 5%, and the PCM detected the Knock Sensor (KS) signal indicated less than 0.20v, or indicated ore than 4.8v for over 15 seconds during the CCM test. **Possible Causes:** • KS signal circuit is open or shorted to ground • KS signal circuit is shorted to VREF or system power (B+) • Knock Sensor is damaged or has failed • PCM has failed
DTC: P0327 **2T CCM, MIL: YES** **1997, 1998, 1999, 2000** **Models:** Hombre **Engines:** 4.3L VIN X **Transmissions:** A/T, M/T	**Knock Sensor Circuit Low Input** DTC P0117, P0118, P0121, P0122 or P0123 not set, engine started, system voltage from 11-16v, ECT sensor signal more than 140°F, engine runtime over 2 minutes, engine speed from 2000-3000 rpm, spark timing retard at 0 degrees or less, Knock sensor "noise" less than 3 counts, and the PCM detected the difference between the Actual Knock sensor noise value and learned value was less than 50 counts, or it was more than 200 counts during the CCM test. **Possible Causes:** • KS signal circuit is open or shorted to ground • KS signal circuit is shorted to VREF or system power (B+) • Knock Sensor is damaged or has failed • PCM has failed

DTC	Trouble Code Title, Conditions & Possible Causes
DTC: P0335 **2T CCM, MIL: YES** **1996, 1997, 1998, 1999, 2000** **Models:** Hombre **Engines:** 2.2L VIN 4 **Transmissions:** A/T, M/T	**Crankshaft Position Sensor Circuit Malfunction** Engine started, engine running, and the PCM detected the CKP Resync counter indicated more than 15 counts within a 4 minute, 15 second period of time during the CCM Rationality test. **Possible Causes:** • CKP sensor signal (+) circuit is open or shorted to ground • CKP sensor signal (-) circuit is open or shorted to ground • CKP sensor is damaged, or the reluctor wheel is damaged • ICM is damaged or has failed • PCM has failed
DTC: P0327 **2T CCM, MIL: YES** **1996, 1997, 1998, 1999** **Models:** Rodeo, Trooper **Engines:** 3.2L VIN V, 3.2L VIN W **Transmissions:** A/T, M/T	**Knock Sensor Circuit Low Input** Engine started, engine runtime over 10 seconds, system voltage from 11-16v, ECT sensor more than 140°F, engine speed from 2000-4000 rpm, throttle angle over 5%, and the PCM detected the Knock Sensor (KS) signal indicated less than 0.20v, or indicated ore than 4.8v for over 15 seconds during the CCM test. **Possible Causes:** • KS signal circuit is open or shorted to ground • KS signal circuit is shorted to VREF or system power (B+) • Knock Sensor is damaged or has failed • PCM has failed
DTC: P0335 **2T CCM, MIL: YES** **1996, 1997, 1998, 1999** **Models:** Oasis **Engines:** All **Transmissions:** A/T, M/T	**CKP Sensor 'A' Circuit Malfunction (No Signal)** Engine running, and the PCM did not detect any signals from the Crankshaft Position (CKP) Sensor 'A' during the test period. **Note: The engine will crank for a longer period of time, may buck or jerk, but it will start and run without the CKP sensor signal present.** **Possible Causes:** • CKP Sensor 'A' signal circuit is open or shorted to ground • CKP Sensor 'A' signal circuit shorted to VREF or system power • CKP Sensor 'A' is damaged or has failed • PCM has failed
DTC: P0336 **2T CCM, MIL: YES** **1998, 1999, 2000, 2001, 2002** **Models:** Amigo, Rodeo, Rodeo Sport **Engines:** 2.2L VIN D **Transmissions:** A/T, M/T	**Crankshaft Position Sensor 58X Circuit Performance** Engine running, and the PCM detected extra or missing CKP 58X pulses occurred between consecutive CKP 58X pulses during 10 out of 100 crankshaft rotations during the CCM Rationality test. **Possible Causes:** • CKP sensor signal "high" circuit is open or shorted to ground • CKP sensor signal "low" circuit is open or shorted to ground • CKP sensor ground circuit is open • CKP sensor is damaged, or the reluctor wheel is damaged • PCM has failed (the Ignition module function is inside the PCM)
DTC: P0336 **2T CCM, MIL: YES** **1996, 1997, 1998, 1999, 2000, 2001, 2002** **Models:** Amigo, Axiom, Rodeo, Rodeo Sport, Trooper, VehiCROSS **Engines:** 3.2L VIN V, 3.2L VIN W, 3.5L VIN X **Transmissions:** A/T, M/T	**Crankshaft Position 58X Sensor Circuit Malfunction** Engine started, and the PCM detected extra or missing pulses between consecutive Crankshaft Position (CKP) 58X sensor signals during 10 out of 100 revolutions during the CCM test. **Possible Causes:** • CKP sensor 58X signal circuit is open or shorted to ground • CKP sensor ground circuit is open or has high resistance • CKP sensor power circuit is open between the sensor and PCM • CKP sensor is damaged, or the reluctor wheel is damaged • PCM has failed (the Ignition module function is inside the PCM)
DTC: P0336 **2T CCM, MIL: YES** **1997, 1998, 1999, 2000** **Models:** Hombre **Engines**: 4.3L VIN X **Transmissions:** A/T, M/T	**Crankshaft Position 3X Sensor Circuit Performance** Engine cranking, four or more Camshaft Position (CMP) sensor signals detected, and the PCM did not detect any Crankshaft Position (CKP) sensor signals for 500 ms during the CCM test. **Possible Causes:** • CKP sensor 3X signal circuit is open or shorted to ground • CKP sensor ground circuit is open or has high resistance • CKP sensor power circuit is open (check the ECM 1 fuse) • CKP sensor is damaged, or the reluctor wheel is damaged • PCM has failed (the Ignition module function is inside the PCM)

DTC	Trouble Code Title, Conditions & Possible Causes
DTC: P0336 **2T CCM, MIL: YES** **1996, 1997, 1998, 1999** **Models:** Oasis **Engines:** All **Transmissions:** A/T, M/T	**CKP Sensor 'A' Circuit Range/Performance** Engine running, and the PCM detected the Crankshaft Position (CKP) Sensor 'A' signal was missing for a short period of time. **Note: This trouble code is usually caused by an intermittent fault.** **Possible Causes:** • CKP Sensor 'A' signal circuit is open or shorted to ground • CKP Sensor 'A' signal circuit shorted to VREF or system power • CKP Sensor 'A' is damaged or has failed • PCM has failed
DTC: P0336 **2T CCM, MIL: YES** **1996, 1997** **Models:** Rodeo **Engines:** 2.6L VIN E **Transmissions:** M/T	**Crankshaft Position 58X Sensor Circuit Malfunction** Engine started, and the PCM detected extra or missing pulses between consecutive Crankshaft Position (CKP) 58X sensor signals during 10 out of 100 revolutions during the CCM test. **Possible Causes:** • CKP sensor 58X signal circuit is open or shorted to ground • CKP sensor ground circuit is open or has high resistance • CKP sensor power circuit is open between the sensor and PCM • CKP sensor is damaged, or the reluctor wheel is damaged • PCM has failed (the Ignition module function is inside the PCM)
DTC: P0337 **2T CCM, MIL: YES** **1998, 1999, 2000, 2001, 2002** **Models:** Amigo, Rodeo, Rodeo Sport **Engines:** 2.2L VIN D **Transmissions:** A/T, M/T	**CKP Sensor Circuit Low Frequency** No CMP codes set, engine cranking, and the PCM did not detect any CKP pulses present between two CMP pulses, or it did not detect any CKP pulses within 8 CMP pulses during the CCM test. **Possible Causes:** • CKP sensor signal "high" or "low circuit open • CKP sensor signal "high" or "low circuit is shorted to ground • CKP sensor ground circuit is open (an intermittent fault) • CKP sensor is damaged, or the reluctor wheel is damaged • PCM has failed (the Ignition module function is inside the PCM)
DTC: P0337 **2T CCM, MIL: YES** **1996, 1997, 1998, 1999, 2000, 2001, 2002** **Models:** Amigo, Axiom, Rodeo, Rodeo Sport, Trooper, VehiCROSS **Engines:** 3.2L VIN V, 3.2L VIN W, 3.5L VIN X **Transmissions:** A/T, M/T	**Crankshaft Position 58X Sensor Circuit Low Input** DTC P0341 and P0342 not set, engine started, and the PCM did not detect any Crankshaft Position (CKP) 58X sensor pulses present between two (2) CMP sensor pulses, or it did not detect any CKP sensor pulses within 8 CMP sensor pulses during the CCM test. **Possible Causes:** • CKP sensor 58X signal circuit is open or shorted to ground • CKP sensor ground circuit is open or has high resistance • CKP sensor power circuit is open between the sensor and PCM • CKP sensor is damaged, or the reluctor wheel is damaged • PCM has failed (the Ignition module function is inside the PCM)
DTC: P0337 **2T CCM, MIL: YES** **1997, 1998, 1999, 2000** **Models:** Hombre **Engines:** 4.3L VIN X **Transmissions:** A/T, M/T	**Crankshaft Position Sensor Circuit Low Input** Engine started, engine speed less than 4000 rpm, MAF sensor signal over 5 g/sec, and the PCM detected the Crankshaft Position (CKP) sensor duty cycle was less than a calibrated amount. **Possible Causes:** • CKP sensor reluctor wheel is damaged or chipped • CKP sensor reluctor wheel is the wrong part • CKP sensor is not aligned correctly to the reluctor wheel • Excessive crankshaft end-play exists • PCM has failed
DTC: P0337 **2T CCM, MIL: YES** **1996, 1997** **Models:** Rodeo **Engines:** 2.6L VIN E **Transmissions:** M/T	**Crankshaft Position 58X Sensor Circuit Low Input** DTC P0341 and P0342 not set, engine started, and the PCM did not detect any Crankshaft Position (CKP) 58X sensor signals present between two (2) CMP sensor pulses, or it did not detect any CKP sensor signals over a 24 second period during the CCM test. **Possible Causes:** • CKP sensor 58X signal circuit is open or shorted to ground • CKP sensor ground circuit is open or has high resistance • CKP sensor power circuit is open between the sensor and PCM • CKP sensor is damaged, or the reluctor wheel is damaged • PCM has failed (the Ignition module function is inside the PCM)

DTC	Trouble Code Title, Conditions & Possible Causes
DTC: P0338 **2T CCM, MIL: YES** **1997, 1998, 1999, 2000** **Models:** Hombre **Engines:** 4.3L VIN X **Transmissions:** A/T, M/T	**Crankshaft Position Sensor Circuit High Input** Engine started, engine speed less than 4000 rpm, MAF sensor signal over 5 g/sec, and the PCM detected the Crankshaft Position (CKP) sensor duty cycle was more than a calibrated amount. **Possible Causes:** • CKP sensor reluctor wheel is damaged or chipped • CKP sensor reluctor wheel is the wrong part • CKP sensor is not aligned correctly to the reluctor wheel • Excessive crankshaft end-play exists • PCM has failed
DTC: P0339 **2T CCM, MIL: YES** **1997, 1998, 1999, 2000** **Models:** Hombre **Engines:** 4.3L VIN X **Transmissions:** A/T, M/T	**Crankshaft Position Sensor Circuit Malfunction** Engine started, engine running, MAF sensor signal over 5 g/sec, and the PCM detected the measured change in engine speed was more than 1000 rpm in a 125 ms period, or the measure change in engine speed was zero (0) rpm with 4 or more CMP sensor signals present. **Possible Causes:** • CKP sensor tip contains metal shavings • CKP sensor reluctor wheel is damaged or chipped • CKP sensor reluctor wheel is the wrong part • CKP sensor is not aligned correctly to the reluctor wheel • Excessive crankshaft end-play exists • PCM has failed
DTC: P0340 **2T CCM, MIL: YES** **1997, 1998, 1999, 2000** **Models:** Hombre **Engines:** 4.3L VIN X **Transmissions:** A/T, M/T	**Camshaft Position Sensor Circuit Malfunction** Engine started, engine running, and the PCM did not detect any Camshaft Position (CMP) sensor signals after detecting that two complete crankshaft revolutions had occurred during the CCM test. **Possible Causes:** • CMP sensor circuit is open or shorted to ground (intermittent) • CMP sensor ground circuit is open (an intermittent fault) • CMP sensor is damaged or has failed • EMI or RFI interference or "noise" is present on the circuit
DTC: P0341 **1T CCM, MIL: YES** **1998, 1999, 2000, 2001, 2002** **Models:** Amigo, Rodeo, Rodeo Sport **Engines:** 2.2L VIN D **Transmissions:** A/T, M/T	**Camshaft Position Sensor Range/Performance** Engine started, engine running with CMP sensor pulses received, and the PCM detected an incorrect number of CMP signals were received during 10 tests over a 100-test sample period (that lasts 15.6 ms) during the CCM Rationality test. **Note: If a CKP sensor code is also set, check the common ground circuit between the CKP and CMP sensors for an open condition. If a fuel injector code is also set, check the power feed circuit as it connects to the fuel injectors and to the CMP sensor.** **Possible Causes:** • CMP sensor circuit is open or shorted to ground (intermittent) • CMP sensor ground circuit is open (an intermittent fault) • CMP sensor is damaged or has failed
DTC: P0341 **2T CCM, MIL: YES** **1996, 1997, 1998, 1999, 2000** **Models:** Amigo, Rodeo, Rodeo Sport, Trooper **Engines:** 3.2L VIN V, 3.2L VIN W **Transmissions:** A/T, M/T	**Camshaft Position Sensor Range/Performance** Engine started, engine running with CMP sensor pulses received, and the PCM detected an incorrect number of CMP signals were received during 10 tests over a 100-test sample period (that lasts 15.6 ms) during the CCM Rationality test. **Note: If a CKP sensor code is also set, check the common ground circuit between the CKP and CMP sensors for an open condition. If a fuel injector code is also set, check the power feed circuit as it connects to the fuel injectors and to the CMP sensor.** **Possible Causes:** • CMP sensor circuit is open or shorted to ground (intermittent) • CMP sensor ground circuit is open (an intermittent fault) • CMP sensor is damaged or has failed
DTC: P0341 **2T CCM, MIL: YES** **1996, 1997, 1998, 1999, 2000** **Models:** Hombre Engines: 2.2L VIN 4 **Transmissions:** A/T, M/T	**Camshaft Position Sensor Range/Performance** Engine started, engine running, and the PCM did not detected that the CKP sensor "Resync" counter indicated more than 15 counts in a period of 256 seconds (4 minutes, 16 seconds) in the CCM test. **Possible Causes:** • CMP sensor signal circuit is or shorted to ground (intermittent) • CMP sensor ground or power circuit is open (intermittent fault) • CMP sensor power circuit is open (an intermittent fault) • PCM has failed (the Ignition module function is inside the PCM)

DTC	Trouble Code Title, Conditions & Possible Causes
DTC: P0341 **2T CCM, MIL: YES** **1997, 1998, 1999, 2000** **Models:** Hombre **Engines:** 4.3L VIN X **Transmissions:** A/T, M/T	**Camshaft Position Sensor Range/Performance** Engine started, engine running, and the PCM the Camshaft Position (CMP) sensor signals were not detected at the correct time interval after detecting two complete crankshaft revolutions during the test. **Possible Causes:** • CMP sensor circuit is open or shorted to ground (intermittent) • CMP sensor ground circuit is open (an intermittent fault) • CMP sensor is damaged or has failed • PCM has failed
DTC: P0341 **2T CCM, MIL: YES** **1996, 1997** **Models:** Rodeo **Engines:** 2.6L VIN E **Transmissions:** M/T	**Camshaft Position Sensor Range/Performance** Engine started, engine running, and the PCM the Camshaft Position (CMP) sensor signals were not detected at the correct time interval after detecting for 100 occurrences in a 200-test sample in the test. **Possible Causes:** • CMP sensor circuit is open or shorted to ground (intermittent) • CMP sensor ground circuit is open (an intermittent fault) • CMP sensor is damaged or has failed • PCM has failed
DTC: P0341 **2T CCM, MIL: YES** **1999, 2000, 2001, 2002** **Models:** Axiom, VehiCROSS **Engines:** 3.2L VIN V, 3.2L VIN W **Transmissions:** A/T, M/T	**Camshaft Position Sensor Circuit Range/Performance** Engine started, engine running with CMP sensor pulses received, and the PCM detected an incorrect number of CMP signals were received during 10 tests over a 100-test sample period (that lasts 15.6 ms) during the CCM Rationality test. **Note: If a CKP sensor code is also set, check the common ground circuit between the CKP and CMP sensors for an open condition. If a fuel injector code is also set, check the power feed circuit as it connects to the fuel injectors and to the CMP sensor.** **Possible Causes:** • CMP sensor circuit is open or shorted to ground (intermittent) • CMP sensor ground circuit is open (an intermittent fault) • CMP sensor is damaged or has failed
DTC: P0342 **2T CCM, MIL: YES** **1998, 1999, 2000, 2001, 2002** **Models:** Amigo, Rodeo, Rodeo Sport **Engines:** 2.2L VIN D **Transmissions:** A/T, M/T	**Camshaft Position Sensor Circuit Low Input** Engine started, engine running, and PCM detected any CMP sensor signal once for every four (4) rotations of the crankshaft in a 10 second time period. **Possible Causes:** • CMP sensor signal circuit is open • CMP sensor signal circuit is shorted to ground (intermittent) • CMP sensor power circuit is open (check the Ignition fuse) • CMP sensor ground circuit is open (an intermittent fault) • CMP sensor is damaged or has failed
DTC: P0342 **2T CCM, MIL: YES** **1996, 1997, 1998, 1999, 2000** **Models:** Amigo, Rodeo, Rodeo Sport, Trooper **Engines:** 3.2L VIN V, 3.2L VIN W **Transmissions:** A/T, M/T	**Camshaft Position Sensor Circuit Low Input** Engine started, engine running, and PCM did not detect a Camshaft Position (CMP) sensor pulse at least once for every six (6) rotations of the crankshaft in a 10 second period during the CCM test. **Possible Causes:** • CMP sensor signal circuit is open • CMP sensor signal circuit is shorted to ground • CMP sensor power circuit open between the sensor and PCM • CMP sensor ground circuit is open • CMP sensor is damaged or has failed
DTC: P0342 **2T CCM, MIL: YES** **1999, 2000, 2001, 2002** **Models:** Axiom, VehiCROSS **Engines:** 3.2L VIN V, 3.2L VIN W **Transmissions:** A/T, M/T	**Camshaft Position Sensor Circuit Low Input** Engine started, engine running, and PCM did not detect a Camshaft Position (CMP) sensor pulse at least once for every six (6) rotations of the crankshaft in a 10 second period during the CCM test. **Possible Causes:** • CMP sensor signal circuit is open • CMP sensor signal circuit is shorted to ground • CMP sensor power circuit open between the sensor and PCM • CMP sensor ground circuit is open • CMP sensor is damaged or has failed
DTC: P0342 **1T CCM, MIL: YES** **1997, 1998, 1999, 2000** **Models:** Hombre **Engines:** 2.2L VIN 4 **Transmissions:** A/T, M/T	**Camshaft Position Sensor Circuit Low Input** Engine started, engine running, and PCM detected the CMP Active counter did not count up or did not increment during the CCM test. **Possible Causes:** • CMP sensor signal circuit is open • CMP sensor signal circuit is shorted to ground • CMP sensor power circuit is open (check the Ignition fuse) • CMP sensor ground circuit is open • CMP sensor is damaged or has failed

DTC	Trouble Code Title, Conditions & Possible Causes
DTC: P0342 **2T CCM, MIL: YES** **1996, 1997** **Models:** Rodeo **Engines:** 2.6L VIN E **Transmissions:** M/T	**Camshaft Position Sensor Circuit Low Input** Engine started, engine running, and PCM did not detect a Camshaft Position (CMP) sensor pulse at least once for every four (4) rotations of the crankshaft in a 10 second period during the CCM test. **Possible Causes:** • CMP sensor signal circuit is open • CMP sensor signal circuit is shorted to ground • CMP sensor power circuit open between the sensor and PCM • CMP sensor ground circuit is open • CMP sensor is damaged or has failed
DTC: P0351 **1T CCM, MIL: YES** **1998, 1999, 2000, 2001, 2002** **Models:** Amigo, Rodeo, Rodeo Sport **Engines:** 2.2L VIN D **Transmissions:** A/T, M/T	**Ignition Coil 'A' Primary/Secondary Circuit Malfunction** Key on, and the PCM detected the IC output signal did not equal 5v with the output commanded "on", or that it did not equal 0v with the output commanded "off" in 20 test faults over a 40 sample period. **Possible Causes:** • Ignition Coil 'A' control circuit is open • Ignition Coil 'A' control circuit is shorted to ground • Ignition Coil 'A' control circuit is shorted to system power (B+) • Ignition Module 'A' ground circuit is open • Ignition Coil 'A' power circuit is open (check the IGN COIL fuse) • Ignition control module is damaged or has failed
DTC: P0352 **1T CCM, MIL: YES** **1998, 1999, 2000, 2001, 2002** **Models:** Amigo, Rodeo, Rodeo Sport Engines: 2.2L VIN D **Transmissions:** A/T, M/T	**Ignition Coil 'B' Primary/Secondary Circuit Malfunction** Key on, and the PCM detected the IC output signal did not equal 5v with the output commanded "on", or that it did not equal 0v with the output commanded "off" in 20 test faults over a 40 sample period. **Possible Causes:** • Ignition Coil 'B' control circuit is open • Ignition Coil 'B' control circuit is shorted to ground • Ignition Coil 'B' control circuit is shorted to system power (B+) • Ignition Coil 'B' ground circuit is open • Ignition Coil 'B' power circuit is open (check the IGN COIL fuse) • Ignition control module is damaged or has failed
DTC: P0351 **2T CCM, MIL: YES** **1996, 1997** **Models:** Rodeo **Engines:** 2.6L VIN E **Transmissions:** M/T	**Ignition Coil Primary Circuit Malfunction** Engine running, and the PCM detected the IC output signal did not equal 5v with the output commanded "on", or it did not equal 0v with the output commanded "off" in 20 tests over a 40 sample period. **Possible Causes:** • Ignition Coil control circuit is open or shorted to ground • Ignition Coil control circuit is shorted to system power (B+) • Ignition Module ground circuit is open • Ignition Coil power circuit is open (check the COIL 15A fuse) • Ignition control module is damaged or has failed
DTC: P0351 **1T CCM, MIL: YES** **1996, 1997, 1998, 1999** **Models:** Amigo, Axiom, Rodeo, Rodeo Sport, Trooper, VehiCROSS **Engines:** 3.2L VIN V, 3.2L VIN W, 3.5L VIN X **Transmissions:** A/T, M/T	**Ignition Control Module Circuit 1 Malfunction** Engine running with CKP 58X signals received, and the PCM detected the IC output signal did not equal 5v with the output commanded "on", or it did not equal 0v with the output commanded "off" in 20 tests over a 40 sample period during the CCM test. **Possible Causes:** • EST signal circuit is open between module and coil or the PCM • EST signal circuit shorted between module and coil or the PCM • Ignition Control module is damaged or has failed • PCM has failed
DTC: P0351 **2T CCM, MIL: YES** **2000, 2001, 2002** **Models:** Amigo, Axiom, Rodeo, Rodeo Sport, Trooper, VehiCROSS **Engines:** 3.2L VIN V, 3.2L VIN W, 3.5L VIN X **Transmissions:** A/T, M/T	**Ignition Control Module Circuit 1 Malfunction** Engine running with CKP 58X signals received, and the PCM detected the IC output signal did not equal 5v with the output commanded "on", or it did not equal 0v with the output commanded "off" in 20 tests over a 40 sample period during the CCM test. **Possible Causes:** • EST signal circuit is open between module and coil or the PCM • EST signal circuit shorted between module and coil or the PCM • ION module is damaged or has failed • PCM has failed

DTC	Trouble Code Title, Conditions & Possible Causes
DTC: P0352 **1T CCM, MIL: YES** **1996, 1997, 1998, 1999** **Models:** Amigo, Axiom, Rodeo, Rodeo Sport, Trooper, VehiCROSS **Engines:** 3.2L VIN V, 3.2L VIN W, 3.5L VIN X **Transmissions:** A/T, M/T	**Ignition Control Module Circuit 2 Malfunction** Engine running with CKP 58X signals received, and the PCM detected the IC output signal did not equal 5v with the output commanded "on", or it did not equal 0v with the output commanded "off" in 20 tests over a 40 sample period during the CCM test. **Possible Causes:** • EST signal circuit is open between module and coil or the PCM • EST signal circuit shorted between module and coil or the PCM • Ignition Control module is damaged or has failed • PCM has failed
DTC: P0352 **1T CCM, MIL: YES** **2000, 2001, 2002** **Models:** Amigo, Axiom, Rodeo, Rodeo Sport, Trooper, VehiCROSS **Engines:** 3.2L VIN V, 3.2L VIN W, 3.5L VIN X **Transmissions:** A/T, M/T	**Ignition Control Module Circuit 2 Malfunction** Engine running with CKP 58X signals received, and the PCM detected the IC output signal did not equal 5v with the output commanded "on", or it did not equal 0v with the output commanded "off" in 20 tests over a 40 sample period during the CCM test. **Possible Causes:** • EST signal circuit is open between module and coil or the PCM • EST signal circuit shorted between module and coil or the PCM • ION module is damaged or has failed • PCM has failed
DTC: P0353 **1T CCM, MIL: YES** **1996, 1997, 1998, 1999** **Models:** Amigo, Axiom, Rodeo, Rodeo Sport, Trooper, VehiCROSS **Engines:** 3.2L VIN V, 3.2L VIN W, 3.5L VIN X **Transmissions:** A/T, M/T	**Ignition Control Module Circuit 3 Malfunction** Engine running with CKP 58X signals received, and the PCM detected the IC output signal did not equal 5v with the output commanded "on", or it did not equal 0v with the output commanded "off" in 20 tests over a 40 sample period during the CCM test. **Possible Causes:** • EST signal circuit is open between module and coil or the PCM • EST signal circuit shorted between module and coil or the PCM • Ignition Control module is damaged or has failed • PCM has failed
DTC: P0353 **1T CCM, MIL: YES** **2000, 2001, 2002** **Models:** Amigo, Axiom, Rodeo, Rodeo Sport, Trooper, VehiCROSS **Engines:** 3.2L VIN V, 3.2L VIN W, 3.5L VIN X **Transmissions:** A/T, M/T	**Ignition Control Module Circuit 3 Malfunction** Engine running with CKP 58X signals received, and the PCM detected the IC output signal did not equal 5v with the output commanded "on", or it did not equal 0v with the output commanded "off" in 20 tests over a 40 sample period during the CCM test. **Possible Causes:** • EST signal circuit is open between module and coil or the PCM • EST signal circuit shorted between module and coil or the PCM • ION module is damaged or has failed • PCM has failed
DTC: P0354 **1T CCM, MIL: YES** **1996, 1997, 1998, 1999** **Models:** Amigo, Rodeo, Rodeo Sport **Engines:** 3.2L VIN V, 3.2L VIN W **Transmissions:** A/T, M/T	**Ignition Control Module Circuit 4 Malfunction** Engine running with CKP 58X signals received, and the PCM detected the IC output signal did not equal 5v with the output commanded "on", or it did not equal 0v with the output commanded "off" in 20 tests over a 40 sample period during the CCM test. **Possible Causes:** • EST signal circuit is open between module and coil or the PCM • EST signal circuit shorted between module and coil or the PCM • Ignition Control module is damaged or has failed • PCM has failed
DTC: P0354 **1T CCM, MIL: YES** **2000, 2001, 2002** **Models:** Amigo, Rodeo, Rodeo Sport **Engines:** 3.2L VIN V, 3.2L VIN W **Transmissions:** A/T, M/T	**Ignition Control Module Circuit 4 Malfunction** Engine running with CKP 58X signals received, and the PCM detected the IC output signal did not equal 5v with the output commanded "on", or it did not equal 0v with the output commanded "off" in 20 tests over a 40 sample period during the CCM test. **Possible Causes:** • EST signal circuit is open between module and coil or the PCM • EST signal circuit shorted between module and coil or the PCM • ION module is damaged or has failed • PCM has failed
DTC: P0355 **1T CCM, MIL: YES** **1996, 1997, 1998, 1999** **Models:** Amigo, Axiom, Rodeo, Rodeo Sport, Trooper, VehiCROSS **Engines:** 3.2L VIN V, 3.2L VIN W, 3.5L VIN X Transmissions: A/T, M/T	**Ignition Control Module Circuit 5 Malfunction** Engine running with CKP 58X signals received, and the PCM detected the IC output signal did not equal 5v with the output commanded "on", or it did not equal 0v with the output commanded "off" in 20 tests over a 40 sample period during the CCM test. **Possible Causes:** • EST signal circuit is open between module and coil or the PCM • EST signal circuit shorted between module and coil or the PCM • Ignition Control module is damaged or has failed • PCM has failed

DTC	Trouble Code Title, Conditions & Possible Causes
DTC: P0355 **1T CCM, MIL: YES** **2000, 2001, 2002** **Models:** Amigo, Axiom, Rodeo, Rodeo Sport, Trooper, VehiCROSS **Engines:** 3.2L VIN V, 3.2L VIN W, 3.5L VIN X **Transmissions:** A/T, M/T	**Ignition Control Module Circuit 5 Malfunction** Engine running with CKP 58X signals received, and the PCM detected the IC output signal did not equal 5v with the output commanded "on", or it did not equal 0v with the output commanded "off" in 20 tests over a 40 sample period during the CCM test. **Possible Causes:** • EST signal circuit is open between module and coil or the PCM • EST signal circuit shorted between module and coil or the PCM • ION module is damaged or has failed • PCM has failed
DTC: P0356 **1T CCM, MIL: YES** **1996, 1997, 1998, 1999** **Models:** Amigo, Axiom, Rodeo, Rodeo Sport, Trooper, VehiCROSS **Engines:** 3.2L VIN V, 3.2L VIN W, 3.5L VIN X **Transmissions:** A/T, M/T	**Ignition Control Module Circuit 6 Malfunction** Engine running with CKP 58X signals received, and the PCM detected the IC output signal did not equal 5v with the output commanded "on", or it did not equal 0v with the output commanded "off" in 20 tests over a 40 sample period during the CCM test. **Possible Causes:** • EST signal circuit is open between module and coil or the PCM • EST signal circuit shorted between module and coil or the PCM • Ignition Control module is damaged or has failed • PCM has failed
DTC: P0356 **1T CCM, MIL: YES** **2000, 2001, 2002** **Models:** Amigo, Axiom, Rodeo, Rodeo Sport, Trooper, VehiCROSS **Engines:** 3.2L VIN V, 3.2L VIN W, 3.5L VIN X **Transmissions:** A/T, M/T	**Ignition Control Module Circuit 6 Malfunction** Engine running with CKP 58X signals received, and the PCM detected the IC output signal did not equal 5v with the output commanded "on", or it did not equal 0v with the output commanded "off" in 20 tests over a 40 sample period during the CCM test. **Possible Causes:** • EST signal circuit is open between module and coil or the PCM • EST signal circuit shorted between module and coil or the PCM • ION module is damaged or has failed • PCM has failed
DTC: P0401 **1T CCM, MIL: YES** **1998, 1999, 2000, 2001, 2002** **Models:** Amigo, Rodeo, Rodeo Sport **Engines:** 2.2L VIN D **Transmissions:** A/T, M/T	**Insufficient EGR System Flow Detected** No ECT, EGR Pintle Position, EVAP, IAC, IAT, MAP, Misfire, TP or VSS codes set, system voltage from 11-16v, ECT sensor signal more than 140°F, BARO sensor signal more than 72 kPa, IAC position steady at ±5 counts, A/C Clutch and TCC status unchanged, and VSS over 14 mph, then with the throttle angle under 0.8%, EGR duty cycle under 1%, MAP sensor signal steady with any change less than 1 kPa, engine speed from 1200-2000 rpm, the PCM detected the compensated MAP sensor signal indicated a value from 10.3-49.8 kPa during the test. **Note: To determine the amount of change in the MAP signal during the test, the PCM records the Delta MAP sensor value with the EGR valve "open", and this value is ramped over a preset time interval. This test is aborted if the vehicle speed changes by over 10 mph, or if the engine speed (RPM) changes by more than 100 rpm, or if the EGR valve is opened less than 95% during the test sequence.** **Possible Causes:** • Linear EGR valve "low" circuit is open or shorted to ground • Linear EGR valve "low" circuit is shorted to system power (B+) • EGR valve VREF (5-volt) is open between sensor and the PCM • EGR valve feedback circuit is open or shorted to ground • EGR valve is stuck closed, or partially open during the test • EGR exhaust flow path may be restricted • EGR valve is damaged, or has failed • PCM has failed
DTC: P0401 **2T CCM, MIL: YES** **1996, 1997, 1998, 1999** **Models:** Oasis **Engines:** All **Transmissions:** A/T, M/T	**EGR System Insufficient Flow Detected** Cold engine startup requirement met (ECT sensor less than 76°F at startup), engine running in closed loop at 40-55 mph for 2 minutes in Drive (D4), followed by a deceleration period back to 35 mph with the throttle closed, and the PCM detected a signal from the EGR position sensor that indicated insufficient EGR flow during the test. **Possible Causes:** • EGR valve source vacuum supply line open or restricted • EGR intake or exhaust manifold passages are restricted • EGR valve assembly or solenoid valve damaged or has failed • EGR constant vacuum control (CVC) valve is dirty or damaged • PCM has failed
DTC: P0402 **2T CCM, MIL: YES** **1998, 1999, 2000, 2001, 2002** **Models:** Amigo, Rodeo, Rodeo Sport **Engines:** 2.2L VIN D **Transmissions:** A/T, M/T	**Excessive EGR System Flow Detected** Engine started, ECT sensor signal more than 41°F at startup, engine speed over 500 rpm, system voltage at 11-16v, and the PCM detected the EGR position sensor signal indicated more than 55 counts during the EGR System flow test at engine startup. **Possible Causes:** • Linear EGR valve control circuit is shorted to ground • EGR valve is stuck partially open during the initial startup test • EGR valve is damaged, or has excessive carbon buildup • PCM has failed

DTC	Trouble Code Title, Conditions & Possible Causes
DTC: P0404 **2T CCM, MIL: YES** **1998, 1999, 2000, 2001, 2002** **Models:** Amigo, Rodeo, Rodeo Sport **Engines:** 2.2L VIN D **Transmissions:** A/T, M/T	**EGR Position Sensor Circuit Range/Performance** Engine started, then with the engine at idle speed and the EGR valve commanded "on", the PCM detected the difference between the Actual EGR position and the Desired EGR Position was more than 15% for a period of 5 seconds during the CCM Rationality test. **Possible Causes:** • Linear EGR valve has moisture on it, and is "frozen" in position • EGR valve is stuck partially open during the test sequence • EGR valve is damaged, or has excessive carbon buildup • PCM has failed
DTC: P0405 **1T CCM, MIL: YES** **1998, 1999, 2000, 2001, 2002** **Models:** Amigo, Rodeo, Rodeo Sport **Engines:** 2.2L VIN D **Transmissions:** A/T, M/T	**EGR Position Sensor Circuit Low Input** Engine started, engine running, IAT sensor signal more than 41ºF, and the PCM detected the EGR feedback signal was less than 0.10v, condition met for 10 seconds during the CCM test. **Possible Causes:** • EGR feedback signal circuit is shorted to ground • EGR sensor is damaged or has failed • PCM has failed
DTC: P0406 **1T CCM, MIL: YES** **1998, 1999, 2000, 2001, 2002** **Models:** Amigo, Rodeo, Rodeo Sport **Engines:** 2.2L VIN D **Transmissions:** A/T, M/T	**EGR Position Sensor Circuit High Input** Engine started, engine running, IAT sensor signal more than 41ºF, and the PCM detected the EGR feedback signal was more than 4.80v, condition met for 10 seconds during the CCM test. **Possible Causes:** • EGR sensor signal circuit is open between sensor and PCM • EGR sensor ground circuit is open between sensor and PCM • EGR sensor is damaged or has failed • PCM has failed
DTC: P0401 **1T CCM, MIL: YES** **1996, 1997, 1998, 1999, 2000, 2001, 2002** **Models:** Amigo, Axiom, Rodeo, Rodeo Sport, Trooper, VehiCROSS **Engines:** 3.2L VIN V, 3.2L VIN W, 3.5L VIN X **Transmissions:** A/T, M/T	**Insufficient EGR System Flow Detected** No ECT, EGR Pintle Position, EVAP, IAC, IAT, MAP, Misfire, TP or VSS codes set, system voltage from 11-16v, ECT sensor more than 140ºF, BARO sensor over 75 kPa, IAC position stable (±10 counts), A/C Clutch and TCC status unchanged, and VSS over 15 mph, then with the throttle closed (TP angle under 1%), EGR duty cycle under 1%, MAP sensor from 10-40 kPa (±2 kPa), engine speed from 1100-2000 rpm, the PCM detected the compensated MAP sensor signal indicated a value from 10.3-49.8 kPa during the EGR System test. **Possible Causes:** • Linear EGR valve "low" circuit is open or shorted to ground • Linear EGR valve "low" circuit is shorted to system power (B+) • EGR valve VREF (5-volt) is open between sensor and the PCM • EGR valve feedback circuit is open or shorted to ground • EGR valve is stuck closed, or partially open during the test • EGR exhaust flow path may be restricted • EGR valve is damaged, or has failed • PCM has failed
DTC: P0402 **2T CCM, MIL: YES** **1998, 1999, 2000, 2001, 2002** **Models:** Amigo, Axiom, Rodeo, Rodeo Sport, Trooper, VehiCROSS **Engines:** 3.2L VIN W, 3.5L VIN X **Transmissions:** A/T, M/T	**Excessive EGR System Excessive Flow Detected** Engine started, IAT sensor more than 38ºF, engine running, system voltage at 11-16v, and the PCM detected the EGR position sensor signal indicated more than 21% over a 625 ms period during the EGR System flow test right after engine startup. **Possible Causes:** • Linear EGR valve control circuit is shorted to ground • EGR valve is stuck partially open during the initial startup test • EGR valve is damaged, or has excessive carbon buildup • PCM has failed
DTC: P0404 **2T CCM, MIL: YES** **1998, 1999, 2000, 2001, 2002** **Models:** Amigo, Axiom, Rodeo, Rodeo Sport, Trooper, VehiCROSS **Engines:** 3.2L VIN W, 3.5L VIN X **Transmissions:** A/T, M/T	**EGR Pintle Position Sensor Circuit Range/Performance** Engine started, IAT sensor more than 38ºF, engine speed less than 600 rpm, system voltage at 11-16v, then with the Desired EGR position at over 0%, the PCM detected the difference between the Actual and Desired EGR position was more than 15% for over 15 seconds. This fault must occur 3 times in a single trip to set a code. **Possible Causes:** • EGR valve is stuck partially open during the initial startup test • EGR valve is damaged, or has excessive carbon buildup • PCM has failed

DTC	Trouble Code Title, Conditions & Possible Causes
DTC: P0405 **2T CCM, MIL: YES** **1998, 1999, 2000, 2001, 2002** **Models:** Amigo, Axiom, Rodeo, Rodeo Sport, Trooper, VehiCROSS **Engines:** 3.2L VIN W, 3.5L VIN X **Transmissions:** A/T, M/T	**EGR Pintle Position Sensor Circuit Low Input** Key on or engine running, system voltage from 11-16v, IAT sensor more than 140°F, and the PCM detected the EGR position sensor indicated less than 0.10v for 10 seconds during the CCM test. **Possible Causes:** • Linear EGR valve control circuit is shorted to ground • EGR valve is stuck partially open during the initial startup test • EGR valve is damaged, or has excessive carbon buildup • PCM has failed
DTC: P0406 **1T CCM, MIL: YES** **1998, 1999, 2000, 2001, 2002** **Models:** Amigo, Axiom, Rodeo, Rodeo Sport, Trooper, VehiCROSS **Engines:** 3.2L VIN W, 3.5L VIN X **Transmissions:** A/T, M/T	**EGR Pintle Position Sensor Circuit High Input** Engine started, engine running, system voltage from 11-16v, IAT sensor more than 41°F, and the PCM detected the EGR position sensor indicated more than 4.80v for 10 seconds during the test. **Possible Causes:** • EGR sensor ground circuit is open between sensor and PCM • EGR sensor is damaged or has failed • PCM has failed
DTC: P0401 **1T CCM, MIL: YES** **1996, 1997, 1998, 1999, 2000** **Models**: Hombre **Engines:** 2.2L VIN 4 **Transmissions:** A/T, M/T	**Insufficient EGR System Flow Detected** DTC P0106, P0107, P0108, P0112, P0113, P0117, P0118, P0121, P0122, P0123, P0200, P0300, P0301-P0304, P0335, P0502, P0506, P0507 and P1441 not set, ECT sensor signal more than 167°F, BARO sensor signal more than 72.5 kPa, throttle angle less than 1%, vehicle speed over 20 mph, and the PCM detected the EGR system flow was restricted in the EGR System Monitor test. **Possible Causes:** • EGR valve control circuit is shorted to chassis ground • EGR valve is stuck partially open during the initial startup test • EGR valve is damaged, or has excessive carbon buildup • PCM has failed
DTC: P0404 **2T CCM, MIL: YES** **1997** **Models:** Hombre **Engines:** 2.2L VIN 4 **Transmissions:** A/T, M/T	**EGR Position Sensor Circuit Range/Performance** DTC P0106, P0107, P0108, P0112, P0113, P0117, P0118, P0121, P0122, P0123, P0200, P0300, P0301-P0304, P0335, P0502, P0506, P0507 and P1441 not set, system voltage more than 11.7v, engine started, engine running, EGR commanded "on" with the Desired EGR Position indicating over 0%, and the PCM detected the Actual EGR Position did not agree with the Desired EGR Position by more than 9% for 11 seconds during the CCM test. **Possible Causes:** • EGR valve has moisture on it, and is "frozen" in position • EGR valve is stuck partially open during the test sequence • EGR valve is damaged, or has excessive carbon buildup • PCM has failed
DTC: P0405 **2T CCM, MIL: YES** **1997** **Models:** Hombre **Engines:** 2.2L VIN 4 **Transmissions:** A/T, M/T	**EGR Position Sensor Circuit Low Input** DTC P0106, P0107, P0108, P0112, P0113, P0117, P0118, P0121, P0122, P0123, P0200, P0300, P0301-P0304, P0335, P0502, P0506, P0507 and P1441 not set, system voltage more than 11.7v, engine started, engine running, and the PCM detected the EGR sensor signal was less than 0.11v during the CCM test. **Possible Causes:** • EGR sensor signal circuit is shorted to ground • EGR sensor VREF circuit is open or shorted to ground • EGR sensor is damaged or has failed • PCM has failed
DTC: P0401 **1T CCM, MIL: YES** **1997, 1998, 1999, 2000** **Models:** Hombre **Engines**: 4.3L VIN X **Transmissions:** A/T	**Insufficient EGR System Flow Detected** DTC P0106, P0107, P0108, P0112, P0113, P0117, P0118, P0121, P0122, P0123, P0300, P0301-P0306, P0404, P0405, P0500, P0506 and P0507 not set, ECT sensor signal more than 158°F, BARO sensor signal over 70 kPa, throttle angle under 1%, IAC position less than 8 counts, vehicle speed over 43 mph, EGR position under 1%, engine speed from 800-1600 rpm, any change in the MAP sensor signal less than 0.5 kPa, DFCO not active, Transmission and A/C clutch status unchanged, and the PCM detected the change in the MAP sensor signal during the EGR System Monitor test was less than a calibrated amount for 2 seconds. **Possible Causes:** • EGR valve control circuit is shorted to chassis ground • EGR valve is stuck partially open during the initial startup test • EGR valve is damaged, or has excessive carbon buildup • PCM has failed

DTC	Trouble Code Title, Conditions & Possible Causes
DTC: P0401 **1T CCM, MIL: YES** **1997, 1998, 1999, 2000** **Models:** Hombre **Engines:** 4.3L VIN X **Transmissions:** M/T Trouble Code ID: P0401	**Insufficient EGR System Flow Detected** DTC P0106, P0107, P0108, P0112, P0113, P0117, P0118, P0121, P0122, P0123, P0300, P0301-P0306, P0404, P0405, P0500, P0506 and P0507 not set, ECT sensor signal more than 158°F, BARO sensor signal over 70 kPa, throttle angle under 1%, IAC position less than 8 counts, vehicle speed over 43 mph, EGR position under 1%, engine speed from 1000-1900 rpm, any change in the MAP sensor signal less than 0.5 kPa, DFCO not active, Transmission and A/C clutch status unchanged, and the PCM detected the change in the MAP sensor signal was less than a calibrated amount for 2 seconds. **Possible Causes:** • EGR valve control circuit is shorted to chassis ground • EGR valve is stuck partially open during the initial startup test • EGR valve is damaged, or has excessive carbon buildup • PCM has failed
DTC: P0404 **1T CCM, MIL: YES** **1997, 1998, 1999, 2000** **Models:** Hombre **Engines:** 4.3L VIN X **Transmissions:** M/T	**EGR Position Sensor Circuit Range/Performance** Engine started, engine running, system voltage over 10v, and the PCM detected the difference between the Actual and the Desired EGR position was over 10% for 10 seconds during the CCM test. **Possible Causes:** • EGR valve has moisture on it, and is "frozen" in position • EGR valve is stuck partially open during the test sequence • EGR valve is damaged, or has excessive carbon buildup • PCM has failed
DTC: P0405 **1T CCM, MIL: YES** **1997, 1998, 1999, 2000** **Models:** Hombre **Engines:** 4.3L VIN X **Transmissions:** M/T	**EGR Position Sensor Circuit Range/Performance** Engine started, engine running, system voltage over 10v, and the PCM detected the difference between the Actual and the Desired EGR position was less than 0.12v for 10 seconds during the test. **Possible Causes:** • EGR feedback signal circuit is shorted to ground • EGR sensor is damaged or has failed • PCM has failed
DTC: P0401 **2T CCM, MIL: YES** **1996, 1997** **Models:** Rodeo **Engines:** 2.6L VIN E **Transmissions:** M/T	**Insufficient EGR System Flow Detected** No ECT, EGR Pintle Position sensor, EVAP, IAC, IAT, MAP, Misfire, TP or VSS codes set, engine running, system voltage from 11-16v, ECT sensor more than 140°F, BARO sensor more than 75 kPa, VSS more than 15 mph, IAC position stable with any change less than 10 counts, engine speed from 1100-2000 rpm, A/C clutch status stable, then with the throttle angle under 1%, EGR pintle position less than 1%, and during a gradual deceleration period with the EGR valve command over 95%, any engine speed change less than 100 rpm and any vehicle speed change less than 5 mph, the PCM detected the compensated MAP sensor signal indicated a value 10-40 kPa. **Possible Causes:** • EGR valve "low" circuit is open or shorted to ground • EGR valve "low" circuit is shorted to system power (B+) • EGR valve VREF (5-volt) is open between sensor and the PCM • EGR valve feedback circuit is open or shorted to ground • EGR valve is stuck closed, or partially open during the test • EGR exhaust flow path may be restricted • EGR valve is damaged, or has failed • PCM has failed
DTC: P0410 **2T CCM, MIL: YES** **1997, 1998, 1999, 2000** **Models:** Hombre **Engines:** 4.3L VIN X **Transmissions:** A/T, M/T	**Secondary Air Injection System Performance** No ECT, EVAP, Fuel Trim, HO2S, IAC, IAT, MAF, MAP, Misfire or TP codes set, system voltage over 11.7v, engine running in closed loop at over 550 rpm with the commanded A/F ratio at 14.7:1 for 20 seconds, ECT sensor signal from 176-230°F, MAF sensor signal less than 25 gm/s, engine load under 34%, SHRTFT between 124-132 counts, Decel Fuel Cutoff, Power Enrichment, and Catalyst Over-Temperature Mode all "off", and the PCM detected the HO2S signal was less than 222 mv for 1.3 seconds, or the SHRTFT change exceeded 20% with the Air Pump "on" while in closed loop. **Possible Causes:** • AIR system hose is pinched, kinked, disconnected or leaking • AIR system hose "melted" in front of the check valve • AIR solenoid control circuit open or shorted (intermittent fault) • PCM has failed

DTC	Trouble Code Title, Conditions & Possible Causes
DTC: P0420 **1T OBD CAT1, MIL: YES** **1998, 1999, 2000, 2001, 2002** **Models:** Amigo, Rodeo, Rodeo Sport **Engines:** 2.2L VIN D **Transmissions:** A/T, M/T	**Catalyst Efficiency Below Normal (Bank 1)** DTC P0106, P0107, P0108, P0112, P0113, P0117, P0118, P0121, P0122, P0123, P0131-P0134, P0137, P0138, P0140, P0141, P0171, P0172, P0300, P0301-P0304, P0336, P0341, P0342, P0401, P0502, P0506 and P0507 not set, ECT sensor more than 140°F, engine running in closed loop at less than 3500 rpm, engine load less than 99% with any change in engine load under 3.91%, vehicle speed from 15-75 mph, Calculated airflow from 10-32 g/sec, calculated Catalyst temperature over 750°F, and the PCM detected the Catalyst efficiency was less than the acceptable threshold. **Possible Causes:** • Air leaks at the exhaust manifold or in the exhaust pipes • Front HO2S or rear HO2S is contaminated with fuel or moisture • Front HO2S older (aged) than the rear HO2S (HO2S is lazy) • Front HO2S and/or the rear HO2S is loose in the mounting hole • Catalytic converter is damaged or has failed
DTC: P0420 **1T OBD/CAT1, MIL: YES** **1996, 1997, 1998, 1999, 2000, 2001, 2002** **Models:** Amigo, Axiom, Rodeo, Rodeo Sport, Trooper, VehiCROSS **Engines:** 3.2L VIN V, 3.2L VIN W, 3.5L VIN X **Transmissions:** A/T, M/T	**Catalyst Efficiency Below Normal (Bank 1)** DTC P0106, P0107, P0108, P0112, P0113, P0117, P0118, P0121, P0122, P0123, P0131, P0132, P0133, P0134, P0137, P0138, P0140, P0141, P0171, P0172, P0300, P0301-P0306, P0336, P0341, P0342, P0401, P0502, P0506 and P0507 not set, engine speed less than 3500 rpm in closed loop, ECT sensor more than 140°F, MAF sensor from 8-50 g/sec, engine load less than 99% (±8%), predicted Catalyst temperature over 750°F, vehicle speed from 16-75 mph, and the PCM detected the catalyst oxygen storage capacity was below an acceptable threshold during the Catalyst test. **Possible Causes:** • Air leaks at the exhaust manifold or in the exhaust pipes • Front HO2S or rear HO2S is contaminated with fuel or moisture • Front HO2S older (aged) than the rear HO2S (HO2S is lazy) • Front HO2S and/or the rear HO2S is loose in the mounting hole • Catalytic converter is damaged or has failed
DTC: P0430 **1T OBD/CAT2, MIL: YES** **1996, 1997, 1998, 1999, 2000, 2001, 2002** **Models:** Amigo, Axiom, Rodeo, Rodeo Sport, Trooper, VehiCROSS **Engines:** 3.2L VIN V, 3.2L VIN W, 3.5L VIN X **Transmissions:** A/T, M/T	**Catalyst Efficiency Below Normal (Bank 2)** DTC P0106, P0107, P0108, P0112, P0113, P0117, P0118, P0121, P0122, P0123, P0131, P0132, P0133, P0134, P0137, P0138, P0140, P0141, P0171, P0172, P0300, P0301-P0306, P0336, P0341, P0342, P0401, P0502, P0506 and P0507 not set, engine speed less than 3500 rpm in closed loop, ECT sensor more than 140°F, MAF sensor from 8-50 g/sec, engine load less than 99% (±8%), predicted Catalyst temperature over 750°F, vehicle speed from 16-75 mph, and the PCM detected the catalyst oxygen storage capacity was below an acceptable threshold during the Catalyst test. **Possible Causes:** • Air leaks at the exhaust manifold or in the exhaust pipes • Front HO2S or rear HO2S is contaminated with fuel or moisture • Front HO2S older (aged) than the rear HO2S (HO2S is lazy) • Front HO2S and/or the rear HO2S is loose in the mounting hole • Catalytic converter is damaged or has failed
DTC: P0420 **1T OBD/CAT2, MIL: YES** **1996, 1997, 1998, 1999, 2000** **Models:** Hombre **Engines:** 2.2L VIN 4 **Transmissions:** A/T, M/T	**Catalyst Efficiency Below Normal (Bank 1)** DTC P0106, P0107, P0108, P0112, P0113, P0117, P0118, P0121, P0122, P0123, P0131, P0132, P0133, P0134, P0137, P0138, P0140, P0141, P0171, P0172, P0200, P0300, P0301-P0304, P0502, P0506, P0507, P0562, P0563, P0801, P1133, P1171, P1336 and P1441 not set, ECT sensor signal from 158-239°F, BARO sensor signal more than 72 kPa, system voltage over 10v, engine runtime over 510 seconds in closed loop, vehicle stopped at idle speed in Drive or Neutral, throttle angle at 0%, ECT sensor signal from 167-257°F, IAT sensor signal from -4°F to 176°F, Short Term fuel trim from -28% to +28%, catalytic converter temperature more than 653°F, and the PCM detected the rear HO2S-12 response time was too quick indicating a low efficiency catalyst. **Possible Causes:** • Air leaks at the exhaust manifold or in the exhaust pipes • Front HO2S or rear HO2S is contaminated with fuel or moisture • Front HO2S older (aged) than the rear HO2S (HO2S is lazy) • Front HO2S and/or the rear HO2S is loose in the mounting hole • Catalytic converter is damaged or has failed
DTC: P0420 **1T OBD/CAT2, MIL: YES** **1997, 1998, 1999, 2000** **Models:** Hombre **Engines:** 4.3L VIN X **Transmissions:** A/T, M/T	**Catalyst Efficiency Below Normal (Bank 1)** DTC P0101, P0102, P0103, P0106, P0107, P0108, P0112, P0113, P0117, P0118, P0121, P0122, P0123, P0131, P0132, P0133, P0134, P0137, P0138, P0140, P0141, P0171, P0172, P0300, P0301-P0306, P0336, P0341, P0342, P0401, P0404, P0405, P0440, P0442, P0446, P0500, P0506, P0507 and P0705 not set, ECT sensor from 158-239°F, engine running at idle speed for 35 seconds since last idle speed, engine runtime over 346 seconds in closed loop with Long Term fuel trim stable, BARO sensor over 72 kPa, predicted catalyst temperature over 997°F, IAT sensor from 20°F to 167°F, idle speed within 100 rpm of the Desired idle speed for 60 seconds, and the PCM detected the Catalyst oxygen storage was below an acceptable threshold during the Catalyst Monitor test. **Possible Causes:** • Air leaks at the exhaust manifold or in the exhaust pipes • Front HO2S or rear HO2S is contaminated with fuel or moisture • Front HO2S older (aged) than the rear HO2S (HO2S is lazy) • Front HO2S and/or the rear HO2S is loose in the mounting hole • Catalytic converter is damaged or has failed

DTC	Trouble Code Title, Conditions & Possible Causes
DTC: P0420 **2T CCM, MIL: YES** **1996, 1997, 1998, 1999** **Models:** Oasis **Engines:** All **Transmissions:** A/T, M/T	**Catalyst Efficiency Below Threshold (Bank 1)** DTC P0137, P0138 and P0141 not set, engine running in closed loop at 40-55 mph for 2 minutes, followed by a deceleration period to 35 mph at closed throttle, and the PCM detected excessive activity in the Catalyst oxygen sensor (rear HO2S) during the test. **Possible Causes:** • Air leaks in at the exhaust manifold or exhaust pipes • Catalytic converter damaged or has failed (deteriorated) • Front HO2S older (aged) than the rear HO2S (HO2S is lazy) • PCM has failed
DTC: P0420 **1T OBD/CAT2, MIL: YES** **1996, 1997** **Models:** Rodeo **Engines:** 2.6L VIN E **Transmissions:** M/T	**Catalyst Efficiency Below Normal (Bank 1)** DTC P0106, P0107, P0108, P0112, P0113, P0117, P0118, P0121, P0122, P0123, P0131, P0132, P0133, P0134, P0137, P0138, P0140, P0141, P0171, P0172, P0300, P0336, P0341, P0342, P0401, P0502, P0506 and P0507 not set, ECT sensor more than 140°F, engine speed less than 3500 rpm in closed loop, Calculated airflow from 7-41 g/sec, any change in engine load less than 8%, predicted Catalyst temperature over 750°F, vehicle speed from 16-75 mph, and the PCM detected the catalyst oxygen storage capacity was below an acceptable threshold during the Catalyst Monitor test. **Possible Causes:** • Air leaks at the exhaust manifold or in the exhaust pipes • Front HO2S or rear HO2S is contaminated with fuel or moisture • Front HO2S older (aged) than the rear HO2S (HO2S is lazy) • Front HO2S and/or the rear HO2S is loose in the mounting hole • Catalytic converter is damaged or has failed
DTC: P0440 **1T CCM, MIL: YES** **1998, 1999, 2000, 2001, 2002** **Models:** Amigo, Rodeo, Rodeo Sport **Engines:** 2.2L VIN D **Transmissions:** A/T, M/T	**EVAP System Performance** DTC P0106, P0107, P0108, P0112, P0113, P0117, P0118, P0121, P0122, P0123, P0125, P0131, P0132, P0122, P0134 and P0135 not set, ECT and IAT sensor signals from 39-90°F and the difference between the ECT and IAT sensor values less than 12.2°F, system voltage from 11-16v, BARO sensor signal more than 72.3 kPa, fuel level from 10-90%, fuel level counts vary by less than 15 counts in a 125 ms period of time with the maximum engine runtime less than 540 seconds, and the PCM detected the difference in Expected and Actual fuel tank pressure exceeded a calculated value in memory. **Possible Causes:** • Charcoal canister is loaded with fuel or moisture • ECT, IAT, MAP, VSS or TP sensor signals out-of-calibration • Fuel filler cap loose, cross-threaded, incorrect part or damaged • Fuel tank pressure sensor is damaged or has failed • Fuel tank or fuel tank sender assembly 'O' ring is leaking • Fuel tank vapor line(s) block, damaged or disconnected • Purge or Vent solenoid control circuit open or shorted to ground • Purge or Vent solenoid power circuit is open (check the fuse) • TSB 02-02-S001 (4/02) contain information related to this code
DTC: P0440 **2T CCM, MIL: YES** **1996, 1997** **Models:** Rodeo, Trooper **Engines:** 3.2L VIN V **Transmissions:** A/T, M/T	**EVAP System Performance** DTC P0106, P0107, P0108, P0112, P0113, P0117, P0118, P0121, P0122, P0123, P1640 ad P1650 not set, ECT and IAT sensors less than 90°F and within 13°F at startup, ECT sensor over 39°F at startup, IAT sensor over 4°F at startup, system voltage from 11-16v, BARO sensor over 75 kPa, throttle angle from 7-30%, fuel level from 15-85% with minimum fuel slosh, vehicle speed under 75 mph, and the PCM determined it was unable to achieve or maintain vacuum in the system for 60-180 seconds during the EVAP Monitor leak test. **Possible Causes:** • Charcoal canister is loaded with fuel or moisture • ECT, IAT, MAP, VSS or TP sensor signals out-of-calibration • Fuel filler cap loose, cross-threaded, incorrect part or damaged • Fuel tank pressure sensor is damaged or has failed • Fuel tank or fuel tank sender assembly 'O' ring is leaking • Fuel tank vapor line(s) block, damaged or disconnected • Purge or Vent solenoid control circuit open or shorted to ground • Purge or Vent solenoid power circuit is open (check the fuse)

DTC	Trouble Code Title, Conditions & Possible Causes
DTC: P0440 **2T CCM, MIL: YES** **1998, 1999, 2000, 2001, 2002** **Models:** Amigo, Axiom, Rodeo, Rodeo Sport, Trooper, VehiCROSS **Engines:** 3.2L VIN W, 3.5L VIN X **Transmissions:** A/T, M/T	**EVAP System Performance** DTC P0106, P0107, P0108, P0112, P0113, P0117, P0118, P0121, P0122, P0123, P1640 ad P1650 not set, ECT and IAT sensors less than 90ºF and within 13ºF at startup, ECT sensor over 39ºF at startup, IAT sensor over 4ºF at startup, system voltage from 11-16v, BARO sensor over 75 kPa, throttle angle from 7-30%, fuel level from 15-85% with minimum fuel slosh, vehicle speed under 75 mph, and the PCM determined it was unable to achieve or maintain vacuum in the system for 60-180 seconds during the EVAP Monitor leak test. **Possible Causes:** • Charcoal canister is loaded with fuel or moisture • ECT, IAT, MAP, VSS or TP sensor signals out-of-calibration • Fuel filler cap loose, cross-threaded, incorrect part or damaged • Fuel tank pressure sensor is damaged or has failed • Fuel tank or fuel tank sender assembly 'O' ring is leaking • Fuel tank vapor line(s) block, damaged or disconnected • Purge or Vent solenoid control circuit open or shorted to ground • Purge or Vent solenoid power circuit is open (check the fuse)
DTC: P0440 **2T CCM, MIL: YES** **1996, 1997, 1998, 1999, 2000** **Models:** Hombre **Engines:** 2.2L VIN 4 **Transmissions:** A/T, M/T	**EVAP System Performance** DTC P0106, P0107, P0108, P0112, P0113, P0117, P0118, P0121, P0122, P0123, P0125, P0131, P0132, P0133, P0134 and P1133 not set, ECT and IAT sensor signals from 41-84ºF at startup, ECT signal within 12ºF of the IAT signal at startup, and the IAT signal within 2ºF of the ECT signal at startup, system voltage from 11-16v, BARO sensor signal more than 75 kPa, fuel level from 26-74%, throttle angle from 7-35%, Purge duty cycle at 50% within 65 seconds of the engine runtime, and the PCM determined it was unable to achieve or maintain vacuum in the system during the EVAP Monitor leak test. **Possible Causes:** • Charcoal canister is damaged, clogged or restricted • Purge solenoid circuit is open or shored to ground (intermittent) • Purge valve vacuum line is clogged, restricted or disconnected • Purge vacuum switch or fuel tank pressure sensor is damaged • PCM has failed
DTC: P0440 **2T CCM, MIL: YES** **1997, 1998, 1999, 2000** **Models:** Hombre **Engines:** 4.3L VIN X **Transmissions:** A/T, M/T	**EVAP System Performance** DTC P0106, P0107, P0108, P0112, P0113, P0117, P0118, P0121, P0122, P0123, P0125, P0131, P0132, P0133, P0134 and P1133 not set, ECT and IAT sensor signals from 39-86ºF at startup, ECT signal within 12ºF of the IAT signal and the IAT signal within 35ºF of the ECT signal at startup, system voltage from 11-16v, BARO sensor signal more than 73 kPa, fuel level from 26-74%, Purge duty cycle over 50%, and the PCM determined it was unable to achieve or maintain a vacuum in the system during the EVAP Monitor leak test. **Possible Causes:** • Charcoal canister is damaged, clogged or restricted • Purge solenoid circuit is open or shored to ground (intermittent) • Purge valve vacuum line is clogged, restricted or disconnected • Purge vacuum switch or fuel tank pressure sensor is damaged • PCM has failed
DTC: P0441 **2T CCM, MIL: YES** **1996, 1997, 1998, 1999** **Models:** Oasis **Engines:** All **Transmissions:** A/T, M/T	**EVAP System Incorrect Purge Flow** Cold engine startup ECT sensor less than 154ºF and IAT sensor more than 14ºF at startup), engine runtime from 3-5 minutes, followed by an acceleration period to 50-60 mph at steady throttle, and then the PCM did not detect enough purge flow through the EVAP system during the EVAP purge flow test. **Possible Causes:** • EVAP purge control diaphragm valve hose loose/disconnected • EVAP purge cutoff or purge control solenoid valve is damaged • EVAP purge control diaphragm valve is damaged or has failed • PCM has failed
DTC: P0441 **2T CCM, MIL: YES** **1996, 1997** **Models:** Rodeo, Trooper **Engines:** 2.6L VIN E, 3.5L VIN X **Transmissions:** A/T, M/T	**EVAP System No Flow During Purge Detected** DTC P0106, P0107, P0108, P0112, P0113, P0117, P0118, P0121, P0122, P0123 and P1442 not set, ECT and IAT sensors more than 41ºF and within 45ºF at startup, ECT sensor less than 158ºF during testing, BARO sensor over 85 kPa, Calculated manifold pressure over 10 kPa, throttle angle over 14%, engine speed from 800-6000 rpm, system voltage from 11-16v, Purge duty cycle over 95%, and the PCM detected the EVAP vacuum switch indicated a "closed" position (with Purge enabled) for 3 seconds in the EVAP Purge test. **Possible Causes:** • Charcoal canister is damaged, clogged or restricted • Purge solenoid circuit is open or shored to ground (intermittent) • Purge valve vacuum line is clogged, restricted or disconnected • Purge vacuum switch is damaged or has failed • PCM has failed

DTC	Trouble Code Title, Conditions & Possible Causes
DTC: P0442 **1T CCM, MIL: YES** **1998, 1999, 2000, 2001, 2002** **Models:** Amigo, Rodeo, Rodeo Sport **Engines:** 2.2L VIN D **Transmissions:** A/T, M/T	**EVAP System Small Leak Detected** DTC P0106, P0107, P0108, P0112, P0113, P0117, P0118, P0121, P0122, P0123, P0562, P0563, P0440 and P1442 not set, ECT sensor and IAT sensors more than 41ºF and within 45ºF at startup, ECT and IAT sensor signals from 38-90ºF and within 12.2ºF at startup, system voltage at 11-16v, BARO sensor more than 72.3 kPa, throttle angle over 75%, engine runtime over 540 seconds, vehicle speed under 60 mph, fuel tank level sensor signal from 10-90%, fuel level change under 8 counts over a 125 ms period of time, and the PCM detected a vacuum decaying condition was present (indicating a small leak) during the EVAP Leak test. **Possible Causes:** • Canister Purge valve is damaged, leaking or has failed • Charcoal canister is loaded with fuel or moisture • ECT, IAT, MAF, VSS or TP sensor signals out-of-calibration • Fuel filler cap loose, cross-threaded, incorrect part or damaged • Fuel tank is cracked (leaking), or a leak exists in the 'O' ring • Fuel tank pressure sensor is damaged or has failed • Fuel vapor line(s), fuel pipes or hoses damaged or leaking • PCM has failed • TSB 02-02-S001 (4/02) contain information related to this code
DTC: P0442 **2T CCM, MIL: YES** **1996, 1997, 1998, 1999, 2000, 2001, 2002** **Models:** Amigo, Rodeo, Rodeo Sport, Trooper **Engines:** 3.2L VIN V, 3.2L VIN W **Transmissions:** A/T, M/T	**EVAP System Small Leak (0.040") Detected** DTC P0106, P0107, P0108, P0112, P0113, P0117, P0118, P0121, P0122, P0123, P0440, P0446, P0562, P0563, P1640 and P1650 not set, ECT and IAT sensors less than 90ºF and within 13ºF at startup, ECT sensor over 39ºF at startup, IAT sensor over 4ºF at startup, system voltage from 11-16v, BARO sensor over 75 kPa, throttle angle from 7-30%, fuel level from 15-85% with minimum fuel slosh, vehicle speed under 75 mph, and the PCM detected a vacuum decaying condition characteristic of a small leak (0.040") in the system during the EVAP Monitor leak test. **Possible Causes:** • Charcoal canister is loaded with fuel or moisture • ECT, IAT, MAP, VSS or TP sensor signals out-of-calibration • Fuel filler cap loose, cross-threaded, incorrect part or damaged • Fuel tank pressure sensor is damaged or has failed • Fuel tank or fuel tank sender assembly 'O' ring is leaking • Fuel tank vapor line(s) block, damaged or disconnected • Purge or Vent solenoid control circuit open or shorted to ground • Purge or Vent solenoid power circuit is open (check the fuse)
DTC: P0442 **2T CCM, MIL: YES** **1999, 2000, 2001, 2002** **Models:** Axiom, Trooper, VehiCROSS **Engines:** 3.5L VIN X **Transmissions:** A/T, M/T	**EVAP System Small Leak (0.040") Detected** DTC P0106, P0107, P0108, P0112, P0113, P0121, P0122, P0123, P0440, P1650 and P1650 not set, ECT and IAT sensors less than 90ºF and within 13ºF at startup, ECT sensor over 39ºF at startup, IAT sensor over 4ºF at startup, system voltage from 11-16v, BARO sensor over 75 kPa, throttle angle from 7-30%, fuel level from 15-85% with minimum fuel slosh, vehicle speed under 75 mph, and the PCM detected a vacuum decaying condition characteristic of a small leak (0.040") in the system during the EVAP Monitor leak test. **Possible Causes:** • Charcoal canister is loaded with fuel or moisture • ECT, IAT, MAP, VSS or TP sensor signals out-of-calibration • Fuel filler cap loose, cross-threaded, incorrect part or damaged • Fuel tank pressure sensor is damaged or has failed • Fuel tank or fuel tank sender assembly 'O' ring is leaking • Fuel tank vapor line(s) block, damaged or disconnected • Purge or Vent solenoid control circuit open or shorted to ground • Purge or Vent solenoid power circuit is open (check the fuse)
DTC: P0442 **2T CCM, MIL: YES** **1996, 1997, 1998, 1999, 2000** **Models:** Hombre **Engines:** 2.2L VIN 4 **Transmissions:** A/T, M/T	**EVAP System Small Leak (0.040") Detected** DTC P0106, P0107, P0108, P0112, P0113, P0117, P0118, P0121, P0122, P0123, P0125, P0131, P0132, P0133, P0134 and P1133 not set, ECT and IAT sensor signals from 41-84ºF at startup, ECT signal within 12ºF of the IAT signal at startup, and the IAT signal within 2ºF of the ECT signal at startup, system voltage from 11-16v, BARO sensor signal more than 75 kPa, fuel level from 26-74%, throttle angle from 7-35%, Purge duty cycle at 50% within 65 seconds of the engine runtime, and the PCM detected a small leak (0.040") was present in the EVAP system during the EVAP Monitor leak test. **Possible Causes:** • Canister Purge valve is damaged, leaking or has failed • Fuel filler cap loose, cross-threaded, incorrect part or damaged • Fuel tank is cracked (leaking), or a leak exists in the 'O' ring • Fuel vacuum switch or fuel tank pressure sensor is damaged • Fuel vapor line(s), fuel pipes or hoses damaged or leaking • PCM has failed

DTC	Trouble Code Title, Conditions & Possible Causes
DTC: P0442 **2T CCM, MIL: YES** **1997, 1998, 1999, 2000** **Models:** Hombre **Engines:** 4.3L VIN X **Transmissions:** A/T, M/T	**EVAP System Small Leak (0.040") Detected** DTC P0106, P0107, P0108, P0112, P0113, P0117, P0118, P0121, P0122, P0123, P0125, P0131, P0132, P0133, P0134, P0137, P0138, P0141 and P0500 not set, ECT and IAT sensor signals from 41-84°F at startup, ECT signal within 12°F of the IAT signal at startup, and the IAT signal within 2°F of the ECT signal at startup, system voltage from 11-16v, BARO sensor signal more than 75 kPa, fuel level from 26-74%, throttle angle under 75%, vehicle speed less than 65 mph, any change in fuel tank vacuum less than 0.8" H2O or any change in the fuel level less than 8% and the PCM detected it was unable to hold a specified vacuum level for a certain number of seconds in the EVAP Monitor leak test due to a small leak (0.040"). **Possible Causes:** • Canister Purge valve is damaged, leaking or has failed • Fuel filler cap loose, cross-threaded, incorrect part or damaged • Fuel tank is cracked (leaking), or a leak exists in the 'O' ring • Fuel vacuum switch or fuel tank pressure sensor is damaged • Fuel vapor line(s), fuel pipes or hoses damaged or leaking • PCM has failed
DTC: P0443 **1T CCM, MIL: YES** **1998, 1999, 2000, 2001, 2002** **Models:** Amigo, Rodeo, Rodeo Sport **Engines:** 2.2L VIN D **Transmissions:** A/T, M/T	**EVAP Purge Control Solenoid Circuit Malfunction** Engine started, engine runtime over 32 seconds, system voltage from 11-16v, and the PCM detected the Purge solenoid control circuit voltage was "high" with the solenoid commanded "on", or it detected the voltage was "low" with the solenoid commanded "off". **Possible Causes:** • Purge solenoid control circuit is open or shorted to ground • Purge solenoid power circuit is open (check the fuse) • Purge solenoid is damaged or has failed • PCM has failed
DTC: P0444 **1T CCM, MIL: YES** **2001, 2002** **Models:** Axiom, Rodeo, Rodeo Sport, Trooper, VehiCROSS **Engines:** 3.2L VIN W, 3.5L VIN X **Transmissions:** A/T, M/T	**EVAP Purge Solenoid Circuit Malfunction (Open)** Engine started, engine running, system voltage from 11-16v, and the PCM detected an unexpected "high" voltage condition on the EVAP Purge solenoid control circuit with the solenoid commanded "on". **Possible Causes:** • Purge solenoid control circuit is open between device and PCM • Purge solenoid control circuit is shorted to system power (B+) • Purge solenoid is damaged or has failed • PCM has failed (the solenoid driver circuit is damaged)
DTC: P0445 **1T CCM, MIL: YES** **2001, 2002** **Models:** Axiom, Rodeo, Rodeo Sport, Trooper, VehiCROSS **Engines:** 3.2L VIN W, 3.5L VIN X **Transmissions:** A/T, M/T	**EVAP Purge Solenoid Circuit Malfunction (Shorted)** Engine started, engine running, system voltage from 11-16v, and the PCM detected an unexpected "low" voltage condition on the EVAP Purge solenoid control circuit with the solenoid commanded "off". **Possible Causes:** • Purge solenoid control circuit is shorted to chassis ground • Purge solenoid power circuit is open (check the ENGINE fuse) • Purge solenoid is damaged or has failed • PCM has failed (the solenoid driver circuit is damaged)
DTC: P0446 **1T CCM, MIL: YES** **1998, 1999, 2000, 2001, 2002** **Models:** Amigo, Rodeo, Rodeo Sport **Engines:** 2.2L VIN D **Transmissions:** A/T, M/T	**EVAP Vent Control System Performance** DTC P0106, P0107, P0108, P0112, P0113, P0121, P0122, P0123, P0562, P0563, P1640 and P1650 not set, ECT and IAT inputs from 39-86°F and within 12.2°F of each other at startup, system voltage from 11-16v, BARO sensor signal more than 72 kPa, fuel tank level from 12-87%, and the PCM detected the difference between the Actual fuel tank pressure and the Expected fuel tank pressure was less than a value in memory. **Possible Causes:** • Charcoal canister is clogged, plugged or restricted • EVAP vent control solenoid control circuit is shorted to ground • EVAP vent control solenoid hose is bent, kinked or plugged • EVAP vent control solenoid is damaged or has failed • PCM has failed (the solenoid driver circuit may be shorted)
DTC: P0446 **1T CCM, MIL: YES** **1996, 1997, 1998, 1999, 2000, 2001, 2002** **Models:** Amigo, Rodeo, Rodeo Sport **Engines:** 3.2L VIN V, 3.2L VIN W **Transmissions:** A/T, M/T	**EVAP Vent Control System Performance** DTC P0106, P0107, P0108, P0112, P0113, P0121, P0122, P0123, P1640 and P1650 not set, ECT and IAT sensors from 39-86°F at startup, ECT signal within 12°F of the IAT signal at startup, and the IAT signal within 2°F of the ECT signal at startup, system voltage from 11-16v, BARO sensor more than 72 kPa, fuel tank level from 12-87%, Purge duty cycle over 50%, and the PCM detected the FTP sensor did no indicate close to -10" H2O under normal purge conditions with the Canister Vent solenoid "open", or it detected the FTP sensor did not indicate about -1.5 to +1.5" H2O at key "on". **Possible Causes:** • Charcoal canister is clogged, plugged or restricted • EVAP vent control solenoid control circuit is shorted to ground • EVAP vent control solenoid hose is bent, kinked or plugged • EVAP vent control solenoid is damaged or has failed • PCM has failed (the solenoid driver circuit may be shorted)

DTC	Trouble Code Title, Conditions & Possible Causes
DTC: P0446 **1T CCM, MIL: YES** **1999, 2000, 2001, 2002** **Models:** Axiom, Trooper, VehiCROSS **Engines:** 3.5L VIN X **Transmissions:** A/T, M/T	**EVAP Vent Control System Performance** DTC P0106, P0107, P0108, P0112, P0113, P0121, P0122, P0123, P1640 and P1650 not set, ECT and IAT sensors from 39-86°F at startup, ECT signal within 12°F of the IAT signal at startup, and the IAT signal within 2°F of the ECT signal at startup, system voltage from 11-16v, BARO sensor more than 72 kPa, fuel tank level from 12-87%, Purge duty cycle over 50%, and the PCM detected the FTP sensor did no indicate close to -10" H2O under normal purge conditions with the Canister Vent solenoid "open", or it detected the FTP sensor did not indicate about -1.5 to +1.5" H2O at key "on". **Possible Causes:** • Charcoal canister is clogged, plugged or restricted • EVAP vent control solenoid control circuit is shorted to ground • EVAP vent control solenoid hose is bent, kinked or plugged • EVAP vent control solenoid is damaged or has failed • PCM has failed (the solenoid driver circuit may be shorted)
DTC: P0446 **1T CCM, MIL: YES** **1996, 1997, 1998, 1999, 2000** **Models:** Hombre **Engines:** 2.2L VIN 4 **Transmissions:** A/T, M/T	**EVAP Vent Control System Performance** DTC P0106, P0107, P0108, P0112, P0113, P0117, P0118, P0121, P0122, P0123, P0125, P0131, P0132, P0133, P0134 and P1133 not set, ECT and IAT sensor signals from 41-84°F at startup, ECT signal within 12°F of the IAT signal at startup, and the IAT signal within 2°F of the ECT signal at startup, system voltage from 11-16v, BARO sensor signal more than 75 kPa, fuel level from 26-74%, throttle angle from 7-35%, Purge duty cycle at 50% within 65 seconds of the engine runtime, and the PCM determined it was unable to achieve vacuum in the EVAP system during the EVAP Monitor leak test. **Possible Causes:** • Charcoal canister is clogged, plugged or restricted • EVAP vent control solenoid control circuit is shorted to ground • EVAP vent control solenoid hose is bent, kinked or plugged • EVAP vent control solenoid is damaged or has failed • PCM has failed (the solenoid driver circuit may be shorted)
DTC: P0446 **1T CCM, MIL: YES** **1997, 1998, 1999, 2000** **Models:** Hombre **Engines:** 4.3L VIN X **Transmissions:** A/T, M/T	**EVAP Vent Control System Performance** DTC P0106, P0107, P0108, P0112, P0113, P0117, P0118, P0121, P0122, P0123, P0125, P0131, P0132, P0133, P0134, P0137, P0138, P0141 and P0500 not set, ECT and IAT sensor signals from 39-86°F at startup, ECT signal within 12°F of the IAT signal at startup, and the IAT signal within 2°F of the ECT signal at startup, system voltage from 11-16v, BARO sensor signal over 72 kPa, fuel level from 12-87%, Purge duty cycle over 50%, and the PCM detected the vacuum level in the EVAP system did not drop enough over a calibrated amount of time with the Vent solenoid "open". **Possible Causes:** • Charcoal canister is clogged, plugged or restricted • EVAP vent control solenoid control circuit is shorted to ground • EVAP vent control solenoid hose is bent, kinked or plugged • EVAP vent control solenoid is damaged or has failed • PCM has failed (the solenoid driver circuit may be shorted)
DTC: P0447 **1T CCM, MIL: YES** **2001, 2002** **Models:** Axiom, Rodeo, Trooper, VehiCROSS **Engines:** 3.2L VIN W, 3.5L VIN X **Transmissions:** A/T, M/T	**EVAP Vent Solenoid Circuit Malfunction (Open)** Engine started, engine running, system voltage from 11-16v, and the PCM detected an unexpected "high" voltage condition on the EVAP Vent solenoid control circuit with the solenoid commanded "on". **Possible Causes:** • Vent solenoid control circuit is open between device and PCM • Vent solenoid control circuit is shorted to system power (B+) • Vent solenoid is damaged or has failed • PCM has failed (the solenoid driver circuit is damaged)
DTC: P0448 **1T CCM, MIL: YES** **2001, 2002** **Models:** Axiom, Rodeo, Trooper, VehiCROSS **Engines:** 3.2L VIN W, 3.5L VIN X **Transmissions:** A/T, M/T	**EVAP Vent Solenoid Circuit Malfunction (Shorted)** Engine started, engine running, system voltage from 11-16v, and the PCM detected an unexpected "low" voltage condition on the EVAP Vent solenoid control circuit with the solenoid commanded "off". **Possible Causes:** • Vent solenoid control circuit is shorted to chassis ground • Vent solenoid power circuit is open (check the ENGINE fuse) • Vent solenoid is damaged or has failed • PCM has failed (the solenoid driver circuit is damaged)
DTC: P0449 **1T CCM, MIL: YES** **1998, 1999, 2000, 2001, 2002** **Models:** Amigo, Rodeo, Rodeo Sport **Engines:** 2.2L VIN D Transmissions: A/T, M/T	**EVAP Vent Control Solenoid Circuit Malfunction** Engine started, engine runtime over 32 seconds, system voltage over 10v, and the PCM detected the EVAP Vent solenoid control circuit was "high" with the solenoid commanded "on", or that it was "low" with the solenoid commanded "off" during the CCM test. **Possible Causes:** • EVAP vent control solenoid control circuit is open • EVAP vent control solenoid control circuit is shorted to ground • EVAP vent control solenoid is damaged or has failed • PCM has failed (the solenoid driver circuit is open or shorted)

DTC	Trouble Code Title, Conditions & Possible Causes
DTC: P0452 **1T CCM, MIL: YES** **1998** **Models:** Amigo, Rodeo Engines: All **Transmissions:** A/T, M/T	**Fuel Tank Pressure Sensor Circuit Low Input** Key on or engine running, and the PCM detected the Fuel Tank Pressure (FTP) sensor signal was less than 0.20v. The test failed 100 times out of 200 test events during the CCM test. **Possible Causes:** • FTP sensor signal circuit is shorted to ground • FTP sensor power circuit is open between sensor and the PCM • FTP sensor is damaged or has failed • PCM has failed • TSB 01-04-S001 (7/01) contain information related to this code
DTC: P0453 **1T CCM, MIL: YES** **1998, 1999, 2000, 2001, 2002** **Models:** Amigo, Rodeo, Rodeo Sport **Engines:** All **Transmissions:** A/T, M/T	**Fuel Tank Pressure Sensor Circuit High Input** Key on or engine running, and the PCM detected the Fuel Tank Pressure (FTP) sensor signal was more than 4.98v. The test failed 100 times out of 200 test events during the CCM test. **Possible Causes:** • FTP sensor signal circuit is open or shorted to VREF • FTP sensor ground circuit open between sensor and the PCM • FTP sensor is damaged or has failed • PCM has failed
DTC: P0452 **1T CCM, MIL: YES** **1997, 1998, 1999, 2000** **Models:** Hombre **Engines:** All **Transmissions:** A/T, M/T	**Fuel Tank Pressure Sensor Circuit Low Input** Key on or engine running, and the PCM detected the Fuel Tank Pressure (FTP) sensor signal was less than 0.10v for 5 seconds during the CCM test. **Possible Causes:** • FTP sensor signal circuit is shorted to ground • FTP sensor power circuit is open between sensor and the PCM • FTP sensor is damaged or has failed • PCM has failed
DTC: P0453 **1T CCM, MIL: YES** **1997, 1998, 1999, 2000** **Models:** Hombre **Engines:** All **Transmissions:** A/T, M/T	**Fuel Tank Pressure Sensor Circuit High Input** Key on or engine running, and the PCM detected the Fuel Tank Pressure (FTP) sensor signal was more than 4.98v for 5 seconds during the CCM test. **Possible Causes:** • FTP sensor signal circuit is open or shorted to VREF • FTP sensor ground circuit open between sensor and the PCM • FTP sensor is damaged or has failed • PCM has failed
DTC: P0452 **1T CCM, MIL: YES** **1998, 1999, 2000, 2001, 2002** **Models:** Axiom, Trooper, VehiCROSS **Engines:** All **Transmissions:** A/T, M/T	**Fuel Tank Pressure Sensor Circuit Low Input** Key on or engine running, and the PCM detected the Fuel Tank Pressure (FTP) sensor signal was less than 0.20v for 12.5 seconds with 100 test failures occurring in 200 test samples in the CCM test. **Possible Causes:** • FTP sensor signal circuit is shorted to ground • FTP sensor power circuit is open between sensor and the PCM • FTP sensor is damaged or has failed • PCM has failed
DTC: P0453 **1T CCM, MIL: YES** **1998, 1999, 2000, 2001, 2002** **Models:** Axiom, Trooper, VehiCROSS **Engines:** All **Transmissions:** A/T, M/T , VehiCROSS	**Fuel Tank Pressure Sensor Circuit High Input** Key on or engine running, and the PCM detected the Fuel Tank Pressure (FTP) sensor signal was more than 4.90v for 12.5 seconds with 100 test failures occurring in 200 test samples in the CCM test. **Possible Causes:** • FTP sensor signal circuit is open or shorted to VREF • FTP sensor ground circuit open between sensor and the PCM • FTP sensor is damaged or has failed • PCM has failed
DTC: P0452 **1T CCM, MIL: YES** **1998, 1999** **Models:** Oasis **Engines:** All **Transmissions:** A/T, M/T	**Fuel Tank Pressure Sensor Circuit Low Input** Key on or engine running, and the PCM detected the fuel tank pressure (FTP) sensor signal was less than 0.16v during the test. **Note: The FTP sensor PID should be near 2.5v with the fuel cap off.** **Possible Causes:** • FTP sensor signal circuit is shorted to ground • Fuel tank pressure sensor is damaged or has failed • PCM has failed

DTC	Trouble Code Title, Conditions & Possible Causes
DTC: P0453 **1T CCM, MIL: YES** **1998, 1999** **Models:** Oasis **Engines:** All **Transmissions:** A/T, M/T	**Fuel Tank Pressure Sensor Circuit High Input** Key on or engine running, and the PCM detected the fuel tank pressure (FTP) sensor signal was more than 4.90v during the test. **Note: The FTP sensor PID should be near 2.5v with the fuel cap off.** **Possible Causes:** • FTP sensor signal circuit is shorted to VREF or power (B+) • FTP sensor ground circuit is open • FTP sensor vacuum lines loose, damaged or disconnected • Fuel tank pressure sensor is damaged or has failed • PCM has failed
DTC: P0456 **2T CCM, MIL: YES** **2000, 2001, 2002** **Models:** Amigo, Axiom, Rodeo, Rodeo, Sport, Trooper, VehiCROSS **Engines:** 3.2L VIN W, 3.5L VIN X **Transmissions:** A/T, M/T	**EVAP System Very Small Leak (0.020") Detected** DTC P0106, P0107, P0108, P0112, P0113, P0117, P0118, P0121, P0122, P0123, P0452 and P0453 codes set, BARO sensor more than 70 kPa, system voltage at 11-16v, engine speed under 1200 rpm, VSS indicating less than 5 mph, fuel tank level from 40-80%, throttle angle below 1.1%, VSS over 65 mph, and the PCM detected a vacuum decay condition that indicated the presence of a very small leak (under 0.020" in diameter) somewhere in the system. **Possible Causes:** • Canister Purge solenoid is damaged, leaking or has failed • EVAP purge or vent solenoid is damaged or has failed • Fuel filler cap loose, cross-threaded, incorrect part or damaged • Fuel tank is cracked (leaking), or a leak exists in the 'O' ring • Fuel tank pressure sensor is damaged or has failed • Fuel vapor line(s), fuel pipes or hoses damaged or leaking • PCM has failed
DTC: P0461 **1T CCM, MIL: NO** **1998, 1999, 2000, 2001, 2002** **Models:** Amigo, Rodeo, Rodeo Sport **Engines:** 2.2L VIN D **Transmissions:** A/T, M/T	**Fuel Level Sensor Circuit Range/Performance** Engine started, engine running, fuel tank level from 15-85%, Fuel Tank Level Slosh Test completed, Fuel Level Main Test completed, and the PCM detected the fuel level signal changed by less than 0.14v (7 counts) over a distance of 116 miles (240 km) in the test. **Possible Causes:** • Fuel level signal circuit is open, shorted to ground or to power • Fuel tank empty or overfull (fuel sender is stuck mechanically) • Wrong fuel gauge is installed, or instrument panel is damaged • Fuel gauge sender unit is damaged or has failed • PCM has failed
DTC: P0462 **1T CCM, MIL: NO** **1998, 1999, 2000, 2001, 2002** **Models:** Amigo, Rodeo, Rodeo Sport **Engines:** 2.2L VIN D **Transmissions:** A/T, M/T	**Fuel Level Sensor Circuit Low Input** Engine started, engine running, fuel tank level from 15-85%, Fuel Tank Level Slosh Test completed, Fuel Level Main Test completed, and the PCM detected the fuel level signal was less than 0.06v, condition met for 20 seconds during the CCM test. **Possible Causes:** • Fuel level signal circuit is open between the sender and gauge • Wrong fuel gauge is installed, or instrument panel is damaged • Fuel gauge sender unit is damaged or has failed • PCM has failed
DTC: P0463 **1T CCM, MIL: NO** **1998, 1999, 2000, 2001, 2002** **Models:** Amigo, Rodeo, Rodeo Sport **Engines:** 2.2L VIN D **Transmissions:** A/T, M/T	**Fuel Level Sensor Circuit Low Input** Engine started, engine running, fuel tank level from 15-85%, Fuel Tank Level Slosh Test completed, Fuel Level Main Test completed, and the PCM detected the fuel level signal was more than 4.90v, condition met for 20 seconds during the CCM test. **Possible Causes:** • Fuel level signal circuit is open between the sender and gauge • Fuel level signal circuit is shorted to VREF or system power • Wrong fuel gauge is installed, or instrument panel is damaged • Fuel gauge sender unit is damaged or has failed • PCM has failed
DTC: P0462 **1T CCM, MIL: NO** **2000, 2001, 2002** **Models:** Axiom, Trooper, VehiCROSS **Engines:** 3.5L VIN X **Transmissions:** A/T, M/T	**Fuel Level Sensor Circuit Low Input** Engine started, engine running, fuel tank level from 15-85%, Fuel Tank Level Slosh Test completed, Fuel Level Main Test completed, Fuel Tank Level data is valid, and the PCM detected the fuel level was less than 0.06v for 100 test failures over a 200 test sample. **Possible Causes:** • Fuel level signal circuit is open between the sender and gauge • Wrong fuel gauge is installed, or instrument panel is damaged • Fuel gauge sender unit is damaged or has failed • PCM has failed

DTC	Trouble Code Title, Conditions & Possible Causes
DTC: P0463 **1T CCM, MIL: NO** **2000, 2001, 2002** **Models:** Axiom, Trooper, VehiCROSS **Engines:** 3.5L VIN X **Transmissions:** A/T, M/T	**Fuel Level Sensor Circuit Low Input** Engine started, engine running, fuel tank level from 15-85%, Fuel Tank Level Slosh Test completed, Fuel Level Main Test completed, Fuel Tank Level data is valid, and the PCM detected the fuel level was more than 4.90v for 100 test failures over a 200 test sample. **Possible Causes:** • Fuel level signal circuit is open between the sender and gauge • Fuel level signal circuit is shorted to VREF or system power • Wrong fuel gauge is installed, or instrument panel is damaged • Fuel gauge sender unit is damaged or has failed • PCM has failed
DTC: P0464 **1T CCM, MIL: NO** **2002** **Models:** Axiom **Engines:** 3.5L VIN X **Transmissions:** A/T, M/T	**Fuel Level Sensor Circuit Malfunction (Noisy)** DTC P0461 not set, fuel level over 3%, IAT sensor more than 40°F, engine running, and the PCM detected an unexpected "noisy" condition on the Fuel level sensor circuit during the CCM test. **Possible Causes:** • Fuel level sensor signal circuit open (an intermittent fault) • Fuel level sensor signal circuit shorted to power (intermittent) • Fuel level sensor damaged or has failed (an intermittent fault) • PCM has failed
DTC: P0460 **1T CCM, MIL: NO** **1996, 1997, 1998** **Models:** Hombre **Engines:** 2.2L VIN 4 **Transmissions:** A/T, M/T	**Fuel Level Sensor Circuit Low Input** Engine started, vehicle driven for 120 miles, and the PCM detected the fuel level did not change more than 1.6% (4 counts) during the CCM Rationality test. **Note: For additional help with this code, view the Failure Records.** **Possible Causes:** • Fuel level signal circuit is open between the sender and gauge • Fuel level signal circuit is shorted to VREF or system power • Wrong fuel gauge is installed, or instrument panel is damaged • Fuel gauge sender unit is damaged or has failed • PCM has failed
DTC: P0461 **1T CCM, MIL: NO** **1999, 2000** **Models:** Hombre **Engines:** 2.2L VIN 4 **Transmissions:** A/T, M/T	**Fuel Level Sensor Circuit Range/Performance** Engine started, engine running, fuel tank level from 15-85%, Fuel Tank Level Slosh Test completed, Fuel Level Main Test completed, and the PCM detected the fuel level signal changed by less than 0.14v (7 counts) over a distance of 116 miles (240 km) in the test. **Possible Causes:** • Fuel level signal circuit is open, shorted to ground or to power • Fuel tank empty or overfull (fuel sender is stuck mechanically) • Wrong fuel gauge is installed, or instrument panel is damaged • Fuel gauge sender unit is damaged or has failed • PCM has failed
DTC: P0461 **1T CCM, MIL: NO** **1997, 1998, 1999, 2000** **Models:** Hombre **Engines:** 4.3L VIN X **Transmissions:** A/T, M/T	**Fuel Level Sensor Circuit Range/Performance** Engine started, engine running, fuel tank level from 15-85%, vehicle driven over 200 miles (320 Km), and the PCM detected the fuel level signal changed by less than 0.14v (7 counts) during the CCM test. **Possible Causes:** • Fuel level signal circuit is open, shorted to ground or to power • Fuel tank empty or overfull (fuel sender is stuck mechanically) • Wrong fuel gauge is installed, or instrument panel is damaged • Fuel gauge sender unit is damaged or has failed • PCM has failed
DTC: P0462 **1T CCM, MIL: NO** **1996, 1997, 1998, 1999, 2000** **Models:** Hombre **Engines:** All **Transmissions:** A/T, M/T	**Fuel Level Sensor Circuit Low Input** Key on or engine running, and the PCM detected the Fuel Level sensor signal was less than 0.39v for 20 seconds during the test. **Note: For additional help with this code, view the Failure Records.** **Possible Causes:** • Fuel level signal circuit is shorted to ground • Wrong fuel gauge is installed, or instrument panel is damaged • Fuel gauge sender unit is damaged or has failed • PCM has failed
DTC: P0463 **1T CCM, MIL: NO** **1996, 1997, 1998, 1999, 2000** **Models:** Hombre **Engines:** All **Transmissions:** A/T, M/T	**Fuel Level Sensor Circuit High Input** Key on or engine running, and the PCM detected the Fuel Level sensor signal was more than 4.90v for 20 seconds during the test. **Note: For additional help with this code, view the Failure Records.** **Possible Causes:** • Fuel level signal circuit is open or shorted to system power (B+) • Wrong fuel gauge is installed, or instrument panel is damaged • Fuel gauge sender unit is damaged or has failed • PCM has failed

DTC	Trouble Code Title, Conditions & Possible Causes
DTC: P0452 **1T CCM, MIL: NO** **1998, 1999, 2000, 2001, 2002** **Models:** Amigo, Rodeo, Rodeo Sport **Engines:** 3.2L VIN V, 3.2L VIN W **Transmissions:** A/T, M/T	**Fuel Tank Pressure Sensor Circuit Low Input** Key on or engine running, and the PCM detected the Fuel Tank Pressure (FTP) sensor signal was less than 0.20v for 12.5 seconds, 100 test failures over a 200 test period during the CCM test. **Possible Causes:** • FTP sensor signal circuit is shorted to ground • FTP sensor power circuit is open between sensor and the PCM • FTP sensor is damaged or has failed • PCM has failed
DTC: P0453 **1T CCM, MIL: NO** **1998, 1999, 2000, 2001, 2002** **Models:** Amigo, Rodeo, Rodeo Sport **Engines:** 3.2L VIN V, 3.2L VIN W **Transmissions:** A/T, M/T	**Fuel Tank Pressure Sensor Circuit High Input** Key on or engine running, and the PCM detected the Fuel Tank Pressure (FTP) sensor signal was more than 4.98v for 12.5 seconds, 100 test failures over a 200 test period in the CCM test. **Possible Causes:** • FTP sensor signal circuit is shorted to ground • FTP sensor power circuit is open between sensor and the PCM • FTP sensor is damaged or has failed • PCM has failed
DTC: P0461 **1T CCM, MIL: YES** **1996, 1997, 1998, 1999, 2000, 2001, 2002** **Models:** Amigo, Rodeo, Rodeo Sport Engines: 3.2L VIN V, 3.2L VIN W **Transmissions:** A/T, M/T	**Fuel Level Sensor Circuit Range/Performance** Engine started, engine running, fuel tank level from 15-85%, Fuel Tank Level Slosh Test and Fuel Level Main Test both completed, fuel tank data is valid, and the PCM detected the fuel level did not change over a distance of 62.2 miles during the CCM test. **Note: For additional help with this code, view the Failure Records.** **Possible Causes:** • Fuel level signal circuit is open, shorted to ground or to power • Fuel tank empty or overfull (fuel sender is stuck mechanically) • Wrong fuel gauge is installed, or instrument panel is damaged • Fuel gauge sender unit is damaged or has failed • PCM has failed
DTC: P0462 **1T CCM, MIL: YES** **1996, 1997, 1998, 1999, 2000, 2002** **Models:** Amigo, Rodeo, Rodeo Sport **Engines:** 3.2L VIN V, 3.2L VIN W **Transmissions:** A/T, M/T	**Fuel Level Sensor Circuit Low input** Key on or engine running, and the PCM detected the Fuel Level sensor signal was less than 0.39v for 20 seconds during the test. **Note: For additional help with this code, view the Failure Records.** **Possible Causes:** • Fuel level signal circuit is shorted to ground • Wrong fuel gauge is installed, or instrument panel is damaged • Fuel gauge sender unit is damaged or has failed • PCM has failed
DTC: P0463 **1T CCM, MIL: YES** **1996, 1997, 1998, 1999, 2000, 2001, 2002** **Models:** Amigo, Rodeo, Rodeo Sport **Engines:** 3.2L VIN V, 3.2L VIN W **Transmissions:** A/T, M/T	**Fuel Level Sensor Circuit High input** Key on or engine running, and the PCM detected the Fuel Level sensor signal was less than 2.90v for 20 seconds during the test. **Note: For additional help with this code, view the Failure Records.** **Possible Causes:** • Fuel level signal circuit is open or shorted to system power (B+) • Wrong fuel gauge is installed, or instrument panel is damaged • Fuel gauge sender unit is damaged or has failed • PCM has failed
DTC: P0464 **1T CCM, MIL: YES** **2001, 2002** **Models:** Rodeo **Engines:** All **Transmissions:** A/T, M/T	**Fuel Level Sensor Circuit Malfunction (Noisy)** DTC P0461, P0462 and P0463 not set, fuel level over 3%, IAT sensor signal more than 40ºF, engine running, and the PCM detected an unexpected "noisy" condition on the Fuel level sensor circuit during the CCM Rationality test. **Possible Causes:** • Fuel level sensor signal circuit open (an intermittent fault) • Fuel level sensor signal circuit shorted to power (intermittent) • Fuel level sensor damaged or has failed (an intermittent fault) • PCM has failed
DTC: P0480 **1T CCM, MIL: NO** **1998, 1999, 2000, 2001, 2002** **Models:** Amigo, Rodeo, Rodeo Sport **Engines:** 2.2L VIN D **Transmissions:** A/T, M/T	**Cooling Fan 1 Control Circuit Malfunction** Engine started, engine runtime over 32 seconds, system voltage over 10v, and the PCM detected the Cooling Fan 1 control circuit was "high" with the Fan 1commanded "on", or that it was "low" with the Fan 1 commanded "off" for 25 seconds of the 50 second test. **Possible Causes:** • Cooling Fan 1 control circuit open (high voltage condition) • Cooling Fan 1 control circuit is shorted to ground (low voltage) • Cooling Fan 1 relay power circuit is open (check the fuse) • Cooling Fan 1 is damaged or has failed • PCM has failed

DTC	Trouble Code Title, Conditions & Possible Causes
DTC: P0481 **1T CCM, MIL: YES** **1998, 1999, 2000, 2001, 2002** **Models:** Amigo, Rodeo, Rodeo Sport **Engines:** 2.2L VIN D **Transmissions:** A/T, M/T	**Cooling Fan 2 Control Circuit Malfunction** Engine started, engine runtime over 32 seconds, system voltage over 10v, and the PCM detected the Cooling Fan 2 control circuit was "high" with the Fan 2 commanded "on", or that it was "low" with the Fan 2 commanded "off" for 25 seconds of the 50 second test. **Possible Causes:** • Cooling Fan 2 control circuit open (high voltage condition) • Cooling Fan 2 control circuit is shorted to ground (low voltage) • Cooling Fan 2 relay power circuit is open (check the fuse) • Cooling Fan 2 is damaged or has failed • PCM has failed
DTC: P0503 **1T CCM, MIL: YES** **1996** **Models:** Hombre **Engines:** 2.2L VIN 4 **Transmissions:** A/T, M/T	**Vehicle Speed Sensor Circuit Low Input** DTC P0107, P0108, P0122 an P0123 not set, engine speed from 1700-3600, throttle angle from 0-1%, engine vacuum from 72-80 kPa, conditions met for 5 seconds, and the PCM detected the vehicle speed sensor signal was less than 2 mph in the CCM test. **Possible Causes:** • VSS (+) signal circuit is open or shorted to ground • VSS (-) signal circuit is open or shorted to ground • VSS is damaged or has failed, or the VSS rotor is cracked • PCM has failed
DTC: P0503 **1T CCM, MIL: YES** **1997, 1998, 1999, 2000** **Models:** Hombre **Engines:** All **Transmissions:** A/T, M/T	**Vehicle Speed Sensor Circuit Malfunction** DTC P1810 not set, engine speed over 450 for 5 seconds, not in Fuel Cutoff mode, time since last gear change over 6 seconds, and the PCM detected the transmission output speed did not rise 600 rpm, or it dropped by over 1300 rpm for 3 seconds with the gear selector not in Park or Neutral during the CCM Rationality test. **Possible Causes:** • VSS (+) signal circuit is open or shorted to ground (intermittent) • VSS (-) signal circuit is open or shorted to ground (intermittent) • VSS is damaged or has failed, or the VSS rotor is cracked • PCM has failed
DTC: P0500 **1T CCM, MIL: YES** **1997, 1998, 1999, 2000** **Models:** Hombre **Engines:** 4.3L VIN X **Transmissions:** A/T, M/T	**Vehicle Speed Sensor Circuit Malfunction** DTC P0107 and P0108 not set, ECT sensor signal more than 140ºF or higher, MAP sensor signal less than 20 kPa, engine speed from 1400-4000, throttle angle less than 3.125%, and the PCM detected the vehicle speed sensor indicated less than 1 mph for 5 seconds. **Possible Causes:** • Inspect the VSS rotor for cracks or signs of metal shavings • VSS (+) signal circuit is open or shorted to ground • VSS (-) signal circuit is open or shorted to ground • VSS is damaged or has failed • PCM has failed
DTC: P0500 **1T CCM, MIL: YES** **1996, 1997, 1998, 1999** **Models:** Oasis **Engines:** All **Transmissions:** A/T, M/T	**Vehicle Speed Sensor Circuit Low Input** Engine running, then the vehicle was accelerated to 4000 rpm in 2nd gear, followed by a deceleration period to 1500 rpm at closed throttle and the PCM did not detect any VSS signal during the CCM test. **Note: The VSS signal should pulse from 0-5v as the vehicle moves.** **Possible Causes:** • VSS signal circuit is open • VSS signal circuit is shorted to ground • VSS signal circuit is shorted to VREF or system power (B+) • VSS is damaged or has failed
DTC: P0501 **2T CCM, MIL: YES** **1996, 1997, 1998, 1999** **Models:** Oasis **Engines:** All **Transmissions:** A/T, M/T	**Vehicle Speed Sensor Circuit Performance** Engine running at Cruise speed under road load conditions, and the PCM detected the VSS signal was erratic or too low during the test. **Note: The VSS signal should pulse from 0-5v as the vehicle moves.** **Possible Causes:** • VSS signal circuit is open • VSS signal circuit is shorted to ground • VSS signal circuit is shorted to VREF or system power (B+) • VSS is damaged or has failed

DTC	Trouble Code Title, Conditions & Possible Causes
DTC: P0502 **2T CCM, MIL: YES** **1997, 1998, 1999, 2000** **Models:** Hombre **Engines:** 2.2L VIN 4 **Transmissions:** A/T, M/T	**Vehicle Speed Sensor Circuit Low Input** DTC P0107, P0108, P0122 and P0123 not set, engine speed from 1700-3600, throttle angle from 0-1%, engine vacuum from 0-105 kPa, engine torque 40-400 lbs, gear selector not in Park or Neutral, and the PCM detected the transmission output speed was less than 150 mph for 2.5 seconds during the CCM Rationality test. **Possible Causes:** • Inspect the VSS rotor for cracks or signs of metal shavings • VSS (+) signal circuit is open or shorted to ground • VSS (-) signal circuit is open or shorted to ground • VSS is damaged or has failed • PCM has failed
DTC: P0502 **1T CCM, MIL: YES** **1997, 1998, 1999, 2000** **Models:** Hombre **Engines:** 4.3L VIN X **Transmissions:** A/T, M/T	**Vehicle Speed Sensor Circuit Low Input** DTC P0107, P0108, P0122, P0123 and P1810 not set, engine speed over 3000, throttle angle over 20%, engine vacuum from 70-80 kPa, conditions met for 5 seconds, and the PCM detected the vehicle speed sensor signal was less than 2 mph in the CCM test. **Possible Causes:** • VSS (+) signal circuit is open or shorted to ground • VSS (-) signal circuit is open or shorted to ground • VSS is damaged or has failed, or the VSS rotor is cracked • PCM has failed
DTC: P0502 **1T CCM, MIL: YES** **1996, 1997** **Models:** Rodeo **Engines:** 2.6L VIN E **Transmissions:** A/T, M/T	**Vehicle Speed Sensor Circuit Low Input** Engine started, engine speed from 1800-2500, system voltage from 10-16v, ECT sensor more than 140ºF, throttle angle from 10-40%, engine load over 40%, conditions met for 5 seconds, and the PCM did not detect any vehicle speed sensor signals during the CCM test. **Possible Causes:** • VSS signal circuit is open or shorted to ground • VSS power circuit is open (check the Meter 15A fuse) • VSS ground circuit is open between the sensor and ground • VSS is damaged or has failed, or the VSS rotor is cracked • PCM has failed
DTC: P0502 **1T CCM, MIL: YES** **1998, 1999, 2000, 2001, 2002** **Models:** Amigo, Rodeo, Rodeo Sport **Engines:** 2.2L VIN D **Transmissions:** A/T, M/T	**Vehicle Speed Sensor Circuit Low Input** Engine started, engine running, system voltage from 11-16v, ECT sensor more than 140ºF, then during the Decel Test with the engine speed from 1500-3500, throttle angle less than 8%, MAP sensor less than 35%, and the PCM detected the VSS indicated less than 5 mph, or in the Power Test with the engine speed from 2700-4400 rpm, throttle angle from 25-70%, MAP sensor over 50 kPa, the PCM detected the VSS indicated less than 5 mph during the CCM test. **Possible Causes:** • VSS signal circuit is open or shorted to ground • VSS power circuit is open (check the Engine 15A fuse) • VSS ground circuit is open between the sensor and ground • VSS is damaged or has failed, or the VSS rotor is cracked • PCM has failed • TSB 00-04-S008 (12/00) contains information related to this code
DTC: P0502 **1T CCM, MIL: YES** **1996, 1997, 1998, 1999, 2000, 2001, 2002** **Models:** Axiom, Trooper, VehiCROSS **Engines:** 3.2L VIN V, 3.2L VIN W, 3.5L VIN X **Transmissions:** A/T, M/T	**Vehicle Speed Sensor Circuit Low Input** Engine started, engine speed from 1800-2500, system voltage from 11-16v, ECT sensor more than 140ºF, throttle angle from 10-40%, engine load over 50%, and the PCM did not detect any VSS signals for 12.5 seconds over a 15 second period during the CCM test. **Possible Causes:** • VSS signal circuit is open or shorted to ground • VSS power circuit is open (check the Meter 10A fuse) • VSS ground circuit is open between the sensor and ground • VSS is damaged or has failed, or the VSS rotor is cracked • PCM has failed
DTC: P0505 **1T CCM, MIL: YES** **1996, 1997, 1998, 1999** **Models:** Oasis **Engines:** All **Transmissions:** A/T, M/T	**Idle Speed Control System** DTC P1519 not set, engine running at hot idle speed, and the PCM detected the Actual and Target idle speed values were too far apart. **Possible Causes:** • IAC valve circuit open, shorted to ground or to power (B+) • IAC valve is damaged or has failed • Fast idle thermo valve is damaged or has failed (some models) • Throttle body is dirty or full of sludge • PCM has failed

DTC	Trouble Code Title, Conditions & Possible Causes
DTC: P0506 **2T CCM, MIL: YES** **1998, 1999, 2000, 2001, 2002** **Models:** Amigo, Rodeo, Rodeo Sport **Engines:** 2.2L VIN D **Transmissions:** A/T, M/T	**Idle Air Control System Low RPM** DTC P0106, P0107, P0108, P0112, P0113, P0117, P0118, P0121, P0122, P0123, P0125, P0131, P0132, P0133, P0134, P0200, P0300, P0301, P0302, P0303, P0304, P0335, P0341, P0342, P0404, P0405, P0440, P0442, P0446, P0452, P0453, P0502, P0507, P0601, P0602, P0705, P1133, P1404 and P1441 not set, system voltage from 11-16v, engine running with throttle closed, ECT sensor signal over 122ºF, IAT sensor signal more than -40ºF, BARO sensor signal over 72.7 kPa, MAP sensor signal over 60 kPa, Purge duty cycle over 0%, IAC command over 145 counts, and the PCM detected the Actual idle speed was over 100 rpm below the Desired idle speed for 10 seconds during the CCM Rationality test. **Possible Causes:** • High resistance between the main relay and IAC valve • High resistance between PCM and IAC valve control circuits • IAC valve is damaged, dirty, sticking or has failed • The throttle plate is carbon fouled (it may need to be cleaned)
DTC: P0507 **2T CCM, MIL: YES** **1998, 1999, 2000, 2001, 2002** **Models:** Amigo, Rodeo, Rodeo Sport **Engines:** 2.2L VIN D **Transmissions:** A/T, M/T	**Idle Air Control System High RPM** DTC P0106, P0107, P0108, P0112, P0113, P0117, P0118, P0121, P0122, P0123, P0125, P0131, P0132, P0133, P0134, P0200, P0300, P0301, P0302, P0303, P0304, P0335, P0341, P0342, P0404, P0405, P0440, P0442, P0446, P0452, P0453, P0502, P0507, P0601, P0602, P0705, P1133, P1404 and P1441 not set, system voltage from 11-16v, engine running with throttle closed, ECT sensor signal over 122ºF, IAT sensor signal more than -40ºF, BARO sensor signal over 72.7 kPa, MAP sensor signal over 60 kPa, Purge duty cycle over 0%, IAC command over 145 counts, and the PCM detected the Actual idle speed was over 200 rpm above the Desired idle speed for 10 seconds during the CCM Rationality test. **Possible Causes:** • High resistance between the main relay and IAC valve • High resistance between PCM and IAC valve control circuits • IAC valve is damaged, dirty, sticking or has failed • The throttle plate is carbon fouled (it may need to be cleaned)
DTC: P0506 **1T CCM, MIL: YES** **1996, 1997, 1998, 1999, 2000, 2001, 2002** **Models:** Amigo, Axiom, Rodeo, Rodeo Sport, Trooper, VehiCROSS **Engines:** 3.2L VIN V, 3.2L VIN W, 3.5L VIN X **Transmissions:** A/T, M/T	**Idle Air Control System Low RPM** DTC P0106, P0107, P0108, P0112, P0113, P0117, P0118, P0121, P0122, P0123, P0125, P0131, P0132, P0133, P0134, P0200-206, P0300, P0301-P306, P0335, P0341, P0342, P0404, P0405, P0440, P0442, P0446, P0452, P0453, P0502, P0507, P0601, P0602, P0705, P1133, P1404 and P1441 not set, engine runtime over 125 seconds, system voltage from 11-16v, ECT sensor more than 122ºF, IAT sensor more than -40ºF, MAP sensor less than 40 kPa, Purge duty cycle over 10%, BARO sensor over 75 kPa, vehicle speed less than 2 mph with the throttle closed, and the PCM detected the Actual speed was 100-200 rpm below the Desired idle speed for 10 seconds based on the current engine coolant temperature. **Possible Causes:** • High resistance between the IAC 'A' high or low control circuits • Short to ground between the IAC 'B' high or low control circuits • IAC valve is damaged, dirty, sticking or has failed • The throttle plate is carbon fouled (it may need to be cleaned)
DTC: P0507 **1T CCM, MIL: YES** **1996, 1997, 1998, 1999, 2000, 2001, 2002** **Models:** Amigo, Axiom, Rodeo, Rodeo Sport, Trooper, VehiCROSS **Engines:** 3.2L VIN V, 3.2L VIN W, 3.5L VIN X **Transmissions:** A/T, M/T	**Idle Air Control System High RPM** DTC P0106, P0107, P0108, P0112, P0113, P0117, P0118, P0121, P0122, P0123, P0125, P0131, P0132, P0133, P0134, P0201-206, P0300, P0301-P0306, P0335, P0341, P0342, P0404, P0405, P0440, P0442, P0446, P0452, P0453, P0502, P0507, P0601, P0602, P0705, P1133, P1404 and P1441 not set, engine runtime over 125 seconds, system voltage from 11-16v, ECT sensor more than 122ºF, IAT sensor more than -40ºF, MAP sensor less than 40 kPa, Purge duty cycle over 10%, BARO sensor over 75 kPa, vehicle speed under 2 mph with the throttle closed, and the PCM detected the Actual speed was 100-200 rpm above the Desired idle speed for 10 seconds based on the current engine coolant temperature. **Possible Causes:** • High resistance between the IAC 'A' high or low control circuits • Short to ground between the IAC 'B' high or low control circuits • IAC valve is damaged, dirty, sticking or has failed • The throttle plate is carbon fouled (it may need to be cleaned)
DTC: P0506 **1T CCM, MIL: YES** **1996, 1997** **Models:** Rodeo **Engines:** 2.6L VIN E **Transmissions:** A/T, M/T	**Idle Air Control System Low RPM** DTC P0106, P0107, P0108, P112, P113, P0117, P0118, P0121, P0122, P0123, P0171, P0172, P0300, P0441, P0502, P0562 and P0563 not set, engine running at idle speed with the throttle closed, system voltage from 10-16v, ECT sensor more than 118ºF, IAT sensor more than -40ºF, MAP sensor less than 40 kPa, Purge duty cycle over 10%, BARO sensor over 75kPa, vehicle speed less than 2 mph, then if the non-intrusive test fails, the PCM runs the Intrusive test, and the PCM detected the Actual idle speed was 100-200 rpm below the Desired idle speed based on the coolant temperature. **Possible Causes:** • High resistance between the IAC 'A' high or low control circuits • Short to ground between the IAC 'B' high or low control circuits • IAC valve is damaged, dirty, sticking or has failed • The throttle plate is carbon fouled (it may need to be cleaned)

DTC	Trouble Code Title, Conditions & Possible Causes
DTC: P0507 **2T CCM, MIL: YES** **1996, 1997** **Models:** Rodeo **Engines:** 2.6L VIN E **Transmissions:** A/T, M/T	**Idle Air Control System High RPM** DTC P0106, P0107, P0108, P112, P113, P0117, P0118, P0121, P0122, P0123, P0171, P0172, P0300, P0441, P0502, P0562 and P0563 not set, engine running at idle speed with the throttle closed, system voltage from 10-16v, ECT sensor more than 118ºF, IAT sensor more than -40ºF, MAP sensor less than 40 kPa, Purge duty cycle over 10%, BARO sensor over 75kPa, vehicle speed less than 2 mph, then if the non-intrusive test fails, the PCM runs the Intrusive test, and the PCM detected the Actual idle speed was 100-200 rpm above the Desired idle speed based on the coolant temperature. **Possible Causes:** • High resistance between the IAC 'A' high or low control circuits • Short to ground between the IAC 'B' high or low control circuits • IAC valve is damaged, dirty, sticking or has failed • The throttle plate is carbon fouled (it may need to be cleaned)
DTC: P0506 **2T CCM, MIL: YES** **1996, 1997, 1998, 1999, 2000** **Models:** Hombre **Engines:** 2.2L VIN 4 **Transmissions:** A/T, M/T	**Idle Air Control System Low RPM** DTC P0106, P0107, P0108, P0112, P0113, P0117, P0118, P0121, P0122, P0123, P0125, P0131, P0132, P0133, P0134, P0200, P0300, P0301, P0302, P0303, P0304, P0335, P0341, P0342, P0404, P0405, P0440, P0442, P0446, P0452, P0453, P0502, P0507, P0601, P0602, P0705, P1133, P1404 and P1441 not set, system voltage from 11-16v, engine running with throttle closed, BARO sensor signal over 72 kPa, Purge duty cycle over 10%, ECT sensor signal over 104ºF, IAC command over 145 counts, and the PCM detected the Actual idle speed was more than 100 rpm below the Desired idle speed for 5 seconds in the CCM Rationality test. **Possible Causes:** • High resistance between the main relay and IAC valve • High resistance between PCM and IAC valve control circuits • IAC valve is damaged, dirty, sticking or has failed • The throttle plate is carbon fouled (it may need to be cleaned)
DTC: P0507 **2T CCM, MIL: YES** **1996, 1997, 1998, 1999, 2000** **Models:** Hombre **Engines:** 2.2L VIN 4 **Transmissions:** A/T, M/T	**Idle Air Control System High RPM** DTC P0106, P0107, P0108, P0112, P0113, P0117, P0118, P0121, P0122, P0123, P0125, P0131, P0132, P0133, P0134, P0200, P0300, P0301, P0302, P0303, P0304, P0335, P0341, P0342, P0404, P0405, P0440, P0442, P0446, P0452, P0453, P0502, P0507, P0601, P0602, P0705, P1133, P1404 and P1441 not set, system voltage from 11-16v, engine running with throttle closed, BARO sensor signal over 72 kPa, Purge duty cycle over 10%, ECT sensor signal over 104ºF, IAC command under 2 counts, and the PCM detected the Actual idle speed was more than 60 rpm above the Desired idle speed for 5 seconds in the CCM Rationality test. **Possible Causes:** • High resistance between the main relay and IAC valve • High resistance between PCM and IAC valve control circuits • IAC valve is damaged, dirty, sticking or has failed • The throttle plate is carbon fouled (it may need to be cleaned)
DTC: P0506 **2T CCM, MIL: YES** **1997, 1998, 1999, 2000** **Models:** Hombre **Engines:** 4.3L VIN X **Transmissions:** A/T, M/T	**Idle Air Control System Low RPM** DTC P0106, P0107, P0108, P0117, P0118, P0121, P0122, P0123 and P0502 not set, system voltage from 11-16v, ECT sensor signal over 122ºF, IAT sensor signal more than -13ºF, engine runtime 30 seconds with the throttle closed, BARO sensor signal over 70 kPa, vehicle speed less than 2 mph, then if the non-intrusive test fails, the PCM runs the Intrusive test, vehicle speed from 25-75 mph, and with the calculated airflow from 17.5-50 g/sec, throttle angle under 1%, and the IAC motor commanded to 10% for 100 ms, it detected the change in idle speed was less than 50 rpm during the CCM test. **Possible Causes:** • High resistance between the IAC No. 1, 2, 3 or 4 control circuits • Short to ground between the IAC No. 1, 2, 3 or 4 control circuits • IAC valve is damaged, dirty, sticking or has failed • The throttle plate is carbon fouled (it may need to be cleaned)
DTC: P0507 **2T CCM, MIL: YES** **1997, 1998, 1999, 2000** **Models:** Hombre **Engines:** 4.3L VIN X **Transmissions:** A/T, M/T	**Idle Air Control System High RPM** DTC P0106, P0107, P0108, P0117, P0118, P0121, P0122, P0123 and P0502 not set, system voltage from 11-16v, ECT sensor signal over 122ºF, IAT sensor signal more than -13ºF, engine runtime 30 seconds with the throttle closed, BARO sensor signal over 70 kPa, vehicle speed less than 2 mph, then if the non-intrusive test fails, the PCM runs the Intrusive test, vehicle speed from 25-75 mph, and with the calculated airflow from 17.5-50 g/sec, throttle angle under 1%, and the IAC motor commanded to 10% for 100 ms, it detected the change in idle speed was more than 50 rpm during the CCM test. **Possible Causes:** • High resistance between the IAC No. 1, 2, 3 or 4 control circuits • Short to ground between the IAC No. 1, 2, 3 or 4 control circuits • IAC valve is damaged, dirty, sticking or has failed • The throttle plate is carbon fouled (it may need to be cleaned)

DTC	Trouble Code Title, Conditions & Possible Causes
DTC: P0530 **1T CCM, MIL: YES** **1996, 1997, 1998, 1999, 2000** **Models:** Hombre **Engines:** 2.2L VIN 4 **Transmissions:** A/T, M/T	**A/C Refrigerant Pressure Sensor Circuit Malfunction** DTC P0112 and P0113 not set, Key on or engine running, IAT sensor signal more than 32°F, and the PCM detected the A/C Refrigerant Pressure sensor signal was less than 0.10v (32°F), or it was more than 4.0v (363°F) with the A/C "off", or it was more than 4.90v (453°F) with the A/C requested during the CCM test. **Note: For additional help with this code, view the Failure Records.** **Possible Causes:** • A/C refrigerant pressure sensor signal circuit is open • A/C refrigerant pressure sensor signal circuit shorted to ground • A/C refrigerant pressure sensor is damaged or has failed • PCM has failed
DTC: P0532 **1T CCM, MIL: YES** **1998, 1999, 2000, 2001, 2002** **Models:** Amigo, Rodeo, Rodeo Sport **Engines:** 2.2L VIN D Transmissions: A/T, M/T	**A/C Refrigerant Pressure Sensor Circuit Low Input** Engine started, engine running, and the PCM detected the A/C Refrigerant Pressure sensor signal was less than 0.10v (5 counts), condition met for 125 seconds over a 250 sample period. **Note: For additional help with this code, view the Failure Records.** **Possible Causes:** • A/C refrigerant pressure sensor signal circuit shorted to ground • A/C refrigerant pressure sensor is damaged or has failed • PCM has failed
DTC: P0533 **1T CCM, MIL: YES** **1998, 1999, 2000, 2001, 2002** **Models:** Amigo, Rodeo, Rodeo Sport **Engines:** 2.2L VIN D **Transmissions:** A/T, M/T	**A/C Refrigerant Pressure Sensor Circuit High Input** Engine started, engine running, and the PCM detected the A/C Refrigerant Pressure sensor signal was more than 4.88v (5 counts), condition met for 125 seconds over a 250 sample period. **Note: For additional help with this code, view the Failure Records.** **Possible Causes:** • A/C refrigerant pressure sensor signal circuit shorted to ground • A/C refrigerant pressure sensor is damaged or has failed • PCM has failed
DTC: P0560 **1T CCM, MIL: YES** **1996, 1997, 1998** **Models:** Trooper **Engines:** 3.2L VIN V, 3.5L VIN X **Transmissions:** A/T, M/T	**System Voltage Malfunction** Engine started, engine speed over 1000 rpm, TFT sensor more than 302°F, and the PCM detected the system voltage was under 10v, or with the TFT sensor signal less than -40°F, the PCM detected the system voltage was less than 7.3v, or it detected the system voltage was over 16v for 2 seconds under any operating conditions. **Possible Causes:** • Check the drive belt for excessive wear and the proper tension • Check for high resistance at the battery connections or at the starter solenoid connection that connects to PCM power circuit • Check the generator output and the battery condition
DTC: P0562 **1T CCM, MIL: YES** **1999, 2000, 2001, 2002** **Models:** Axiom, Trooper, VehiCROSS **Engines:** All **Transmissions:** A/T, M/T	**System Voltage Low Input** Engine started, engine running, and the PCM detected the system voltage was less than 11.5v for 15 minutes during the CCM test. **Note: For additional help with this code, view the Failure Records.** **Possible Causes:** • Check the drive belt for excessive wear and the proper tension • Check for high resistance at the battery connections or at the starter solenoid connection that connects to PCM power circuit • Check the generator output and the battery condition
DTC: P0563 **1T CCM, MIL: YES** **1999, 2000, 2001, 2002** **Models:** Axiom, Trooper, VehiCROSS **Engines:** All **Transmissions:** A/T, M/T	**System Voltage High Input** Engine started, engine running, and the PCM detected the system voltage was more than 16v for 15 minutes during the CCM test. **Note: For additional help with this code, view the Failure Records.** **Possible Causes:** • Check for high resistance at the battery connections or at the starter solenoid connection that connects to PCM power circuit • Check the generator output and the battery condition • Check for a problem in an accessory (maybe causing spiking)
DTC: P0562 **1T CCM, MIL: YES** **1998, 1999, 2000, 2001, 2002** **Models:** Amigo, Rodeo, Rodeo Sport **Engines:** All **Transmissions:** A/T, M/T	**System Voltage Low Input** Engine started, engine running, and the PCM detected the system voltage was less than 11.5v for over 4 minutes during the CCM test. **Note: For additional help with this code, view the Failure Records.** **Possible Causes:** • Check the drive belt for excessive wear and the proper tension • Check for high resistance at the battery connections or at the starter solenoid connection that connects to PCM power circuit • Check the generator output and the battery condition

DTC	Trouble Code Title, Conditions & Possible Causes
DTC: P0562 **1T CCM, MIL: NO** **1998, 1999, 2000, 2001, 2002** **Models:** Amigo, Rodeo, Rodeo Sport **Engines:** All **Transmissions:** A/T, M/T	**System Voltage High Input** Engine started, engine running, and the PCM detected the system voltage was more than 16v for over 4 minutes during the CCM test. **Note: For additional help with this code, view the Failure Records.** **Possible Causes:** • Check for high resistance at the battery connections or at the starter solenoid connection that connects to PCM power circuit • Check the generator output and the battery condition • Check for a problem in an accessory (maybe causing spiking)
DTC: P0562 **1T CCM, MIL: NO** **1996, 1997, 1998, 1999, 2000** **Models:** Hombre **Engines:** 2.2L VIN 4 **Transmissions:** A/T, M/T	**System Voltage Low Input** Engine started, engine speed over 1300 rpm, and the PCM detected the system voltage was less than 10v for 4 minutes during the test. **Note: For additional help with this code, view the Failure Records.** **Possible Causes:** • Check the drive belt for excessive wear and the proper tension • Check for high resistance at the battery connections or at the starter solenoid connection that connects to PCM power circuit • Check the generator output and the battery condition
DTC: P0563 **1T CCM, MIL: YES** **1996, 1997, 1998, 1999, 2000** **Models:** Hombre **Engines:** 2.2L VIN 4 **Transmissions:** A/T, M/T	**System Voltage High Input** Engine started, engine speed over 1300 rpm, and the PCM detected the system voltage was more than 17v for 4 minutes during the test. **Note: For additional help with this code, view the Failure Records.** **Possible Causes:** • Check for high resistance at the battery connections or at the starter solenoid connection that connects to PCM power circuit • Check the generator output and the battery condition • Check for a problem in an accessory (maybe causing spiking)
DTC: P0562 **2T CCM, MIL: YES** **1996, 1997, 1998, 1999, 2000, 2001, 2002** **Models:** Rodeo, Rodeo Sport **Engines**: All **Transmissions:** A/T, M/T	**System Voltage Low Input** Engine started, engine runtime over 15 minutes, and the PCM detected the system voltage was less than 11.5v in the CCM test. **Note: For additional help with this code, view the Failure Records.** **Possible Causes:** • Check the drive belt for excessive wear and the proper tension • Check for high resistance at the battery connections or at the starter solenoid connection that connects to PCM power circuit • Check the generator output and the battery condition
DTC: P0563 **2T CCM, MIL: YES** **1996, 1997, 1998, 1999, 2000, 2001, 2002** **Models:** Rodeo, Rodeo Sport **Engines:** All **Transmissions:** A/T, M/T	**System Voltage High Input** Engine started, engine runtime over 15 minutes, and the PCM detected the system voltage was more than 17.0v in the CCM test. **Note: For additional help with this code, view the Failure Records.** **Possible Causes:** • Check for high resistance at the battery connections or at the starter solenoid connection that connects to PCM power circuit • Check the generator output and the battery condition • Check for a problem in an accessory (maybe causing spiking)
DTC: P0565 **1T CCM, MIL: NO** **2000, 2001, 2002** **Models:** Axiom, Rodeo, Rodeo Sport, Trooper **Engines**: 2.2L VIN D, 3.2L VIN W, 3.5L VIN X **Transmissions:** A/T, M/T	**Cruise Control Main Switch Circuit Malfunction** Engine started, engine running, system voltage from 11-16v, and the PCM detected noise from Cruise Control (C/C) main switch contacts 60 times within one second, or the C/C main switch remained "on" for 15 seconds during the CCM Rationality test. **Possible Causes:** • C/C main switch signal circuit is shorted to ground • C/C main switch signal circuit is shorted to another circuit • C/C main switch is damaged or has failed
DTC: P0566 **1T CCM, MIL: NO** **2000, 2001, 2002** **Models:** Axiom, Rodeo, Rodeo Sport, Trooper **Engines:** 2.2L VIN D, 3.2L VIN W, 3.5L VIN X **Transmissions:** A/T, M/T	**Cruise Control Cancel Switch Circuit Malfunction** Engine started, engine running, system voltage from 11-16v, and the PCM detected noise from Cruise Control (C/C) cancel switch contacts 100 times within 1.6 seconds, or the C/C cancel switch remained "on" for 40 seconds during the CCM Rationality test. **Possible Causes:** • C/C cancel switch signal circuit is shorted to ground • C/C cancel switch signal circuit is shorted to another circuit • C/C cancel switch is damaged or has failed

DTC	Trouble Code Title, Conditions & Possible Causes
DTC: P0567 **1T CCM, MIL: NO** **2000, 2001, 2002** **Models:** Axiom, Rodeo, Rodeo Sport, Trooper **Engines:** 2.2L VIN D, 3.2L VIN W, 3.5L VIN X **Transmissions:** A/T, M/T	**Cruise Control Resume Switch Circuit Malfunction** Engine started, engine running, system voltage from 11-16v, and the PCM detected noise from Cruise Control (C/C) resume switch contacts 100 times within 1.6 seconds, or the C/C resume switch remained "on" for 50 seconds during the CCM Rationality test. **Possible Causes:** • C/C resume switch signal circuit is shorted to ground • C/C resume switch signal circuit is shorted to another circuit • C/C resume switch is damaged or has failed
DTC: P0568 **1T CCM, MIL: NO** **2000, 2001, 2002** **Models:** Axiom, Rodeo, Rodeo Sport, Trooper Engines: 2.2L VIN D, 3.2L VIN W, 3.5L VIN X **Transmissions:** A/T, M/T	**Cruise Control Set Switch Circuit Malfunction** Engine started, engine running, system voltage from 11-16v, and the PCM detected noise from Cruise Control (C/C) set switch contacts 100 times within 1.6 seconds, or the C/C set switch remained "on" for 120 seconds during the CCM Rationality test. **Possible Causes:** • C/C set switch signal circuit is shorted to ground • C/C set switch signal circuit is shorted to another circuit • C/C set switch is damaged or has failed
DTC: P0571 **1T CCM, MIL: NO** **1998, 1999, 2000, 2001, 2002** **Models:** Amigo, Axiom, Rodeo, Rodeo Sport **Engines:** 3.2L VIN W **Transmissions:** A/T, M/T	**Brake Light Switch Circuit Malfunction** DTC P0502 not set, engine started, then driven to a speed of over 12.5 mph, and the PCM detected that two (2) brake switch signals did not agree with the brake switch status during the CCM test. **Possible Causes:** • Brake switch signal circuit is open between switch and the PCM • Brake switch signal circuit is shorted to ground • Brake switch power circuit open between switch and 15A fuse • Check the battery condition (it may have failed)
DTC: P0601 **1T CCM, MIL: NO** **1996, 1997, 1998, 1999, 2000, 2001, 2002** **Models:** Amigo, Rodeo, Rodeo Sport, Trooper **Engines:** All Transmissions: A/T, M/T	**PCM Internal Check Sum Error** Key on, and the PCM detected a check sum error had occurred during its initial self-test. **Note: For additional help with this code, view the Failure Records.** **Possible Causes:** • The contents of the EEPROM have changed • PCM needs to be replaced and reprogrammed to repair this trouble code
DTC: P0602 **1T CCM, MIL: NO** **1998, 1999, 2000, 2001, 2002** **Models:** Amigo **Engines:** 2.2L VIN D, 3.2L VIN W **Transmissions:** A/T, M/T	**PCM Programming Error** Key on or engine cranking, and the PCM detected that it was not programmed properly. **Note: A failure record is stored when this trouble code is set.** **Possible Causes:** • PCM must be replaced and then reprogrammed to repair this trouble code
DTC: P0604 **1T CCM, MIL: NO** **1998, 1999, 2000, 2001, 2002** **Models:** Amigo **Engines:** 2.2L VIN D, 3.2L VIN W **Transmissions:** A/T, M/T	**PCM Random Access Memory Error** Key on, and the PCM detected inconsistencies between the Main CPU and the Watchdog CPU software calibration **Note: A failure record is stored when this trouble code is set.** **Possible Causes:** • PCM must be replaced and then reprogrammed to repair this trouble code
DTC: P0606 **1T CCM, MIL: NO** **1998, 1999, 2000, 2001, 2002** **Models:** Amigo **Engines:** 2.2L VIN D, 3.2L VIN W **Transmissions:** A/T, M/T	**PCM Internal Performance Error** Key on, and the PCM detected inconsistencies between the Main CPU and the Watchdog CPU software calibration **Note: A failure record is stored when this trouble code is set.** **Possible Causes:** • PCM must be replaced and then reprogrammed to repair this trouble code

DTC	Trouble Code Title, Conditions & Possible Causes
DTC: P0600 **1T CCM, MIL: YES** **1996** **Models:** Hombre **Engines:** 2.2L VIN 4 **Transmissions:** A/T, M/T	**PCM Serial Communication Link Lost** Key on or engine cranking, and the PCM detected that it could not communicate between its 2 sides on the serial communication link. **Note: For additional help with this code, view the Failure Records.** **Possible Causes:** • PCM must be replaced and then reprogrammed to repair this trouble code
DTC: P0601 **1T CCM, MIL: NO** **1996, 1997, 1998, 1999, 2000** **Models:** Hombre **Engines:** All **Transmissions:** A/T, M/T	**PCM Memory Error** Key on or engine cranking, and the PCM detected more than three (3) check sum errors occurred during its initial self-test. **Note: For additional help with this code, view the Failure Records.** **Possible Causes:** • The contents of the module flash memory have changed • PCM must be replaced to repair this trouble code
DTC: P0602 **1T CCM, MIL: YES** **1996, 1997, 1998, 1999, 2000** **Models:** Hombre **Engines:** All **Transmissions:** A/T, M/T	**PCM Programming Error** Key on or engine cranking, and the PCM detected that it was not programmed properly. **Note: For additional help with this code, view the Failure Records.** **Possible Causes:** • Attempt to reprogram the PCM (at least twice) • If this code continues to reset, the PCM must be replaced and then reprogrammed to repair the trouble code.
DTC: P0603 **1T CCM, MIL: YES** **1997, 1998, 1999, 2000** **Models:** Hombre **Engines:** 4.3L VIN X **Transmissions:** A/T, M/T	**PCM Memory Reset Error** Key on or engine cranking, and the PCM detected a difference in the data stored at key "off" versus the data retrieved in KAM at key "on". **Note: For additional help with this code, view the Failure Records.** **Possible Causes:** • PCM power circuit is open (an intermittent fault) • PCM main ground circuit is open (an intermittent fault) • PCM keep alive battery circuit is opened (battery disconnected) • PCM must be reprogrammed to repair this trouble code
DTC: P0604 **1T CCM, MIL: YES** **1997, 1998, 1999, 2000** **Models:** Hombre **Engines**: 4.3L VIN X **Transmissions:** A/T, M/T	**PCM Random Access Memory Error** Key on, and the PCM detected a difference in the data read from a memory location and the data written to that location during the test. **Note: For additional help with this code, view the Failure Records.** **Possible Causes:** • PCM must be replaced and then reprogrammed to repair this trouble code
DTC: P0605 **1T CCM, MIL: YES** **1997, 1998, 1999, 2000** **Models:** Hombre **Engines:** 4.3L VIN X **Transmissions:** A/T, M/T	**PCM Programming Read Only Memory Error** Key on, and the PCM determined the data checksum did not match the expected value during an internal RAM check. **Note: A failure record is stored when this trouble code is set.** **Possible Causes:** • PCM must be replaced and then reprogrammed to repair this trouble code
DTC: P0601 **1T CCM, MIL: NO** **2000, 2001, 2002** **Models:** Axiom, VehiCROSS **Engines:** All **Transmissions:** A/T, M/T	**PCM Internal Check Sum Error** Key on, and the PCM detected a check sum error had occurred. **Note: For additional help with this code, view the Failure Records.** **Possible Causes:** • The contents of the EEPROM have changed • PCM must be replaced to repair this trouble code
DTC: P0602 **1T CCM, MIL: NO** **2000, 2001, 2002** **Models:** Axiom, Trooper, VehiCROSS **Engines:** All **Transmissions:** A/T, M/T	**PCM Not Programmed** Key on, and the PCM detected inconsistencies between the Main CPU and the Watchdog CPU software calibration. **Note: For additional help with this code, view the Failure Records.** **Possible Causes:** • PCM needs to be replace and then reprogrammed to repair this trouble code

DTC	Trouble Code Title, Conditions & Possible Causes
DTC: P0604 **1T CCM, MIL: NO** **2000, 2001, 2002** **Models:** Axiom, Rodeo, Rodeo Sport, Trooper, VehiCROSS **Engines:** All **Transmissions**: A/T, M/T	**PCM Random Access Memory Error** Key on, and the PCM detected a difference between the data read to a location and the data written to that location. **Note: For additional help with this code, view the Failure Records.** **Possible Causes:** • The contents of the EEPROM have changed • PCM must be replaced to repair this trouble code
DTC: P0606 **1T CCM, MIL: NO** **2000, 2001, 2002** **Models**: Axiom, Rodeo, Rodeo Sport, Trooper, VehiCROSS **Engines:** All **Transmissions:** A/T, M/T	**PCM Internal Performance Error** Key on, and the PCM detected inconsistencies between the Main CPU and the Watchdog CPU software calibration **Note: A failure record is stored when this trouble code is set.** **Possible Causes:** • PCM must be replaced and then reprogrammed to repair this trouble code
DTC: P0700 **1T CCM, MIL: YES** **1996, 1997, 1998, 1999** **Models:** Oasis **Engines:** All **Transmissions:** A/T	**Automatic Transaxle** Engine running and the PCM detected an Automatic Transaxle fault. **Note: DTC P0700 sets along with several other TCM trouble codes.** **Possible Causes:** • Refer to the repair instructions in a transmission repair manual or the information in other electronic media to repair this code.
DTC: P0704 **1T CCM, MIL: YES** **1997, 1998, 1999, 2000** **Models**: Hombre **Engines:** 4.3L VIN X **Transmissions**: M/T	**Clutch Pedal Switch Circuit Malfunction** No VSS codes set, engine started, then driven to a speed of over 40 mph, and the PCM detected the VSS signal indicated 0 mph without detecting any change in the Clutch Switch status (the clutch was not depressed and released) during the CCM Rationality test. **Possible Causes:** • Clutch switch signal circuit is open or shorted to ground • Clutch switch power circuit is open (check the ABS 10A fuse) • Clutch switch is damaged or has failed • PCM has failed
DTC: P0705 **1T CCM, MIL: NO** **1998, 1999, 2000, 2001, 2002** **Models:** Amigo, Rodeo, Rodeo Sport **Engines:** All **Transmissions**: A/T	**Transmission Range Switch Illegal Position Malfunction** Engine started, then driven to a speed of over 8 mph, and the PCM detected "illegal" TR Range or Mode switch signals for 5 seconds. **Note: For additional help with this code, view the Failure Records.** **Possible Causes:** • TR range switch signal is open • TR range switch signal shorted to another switch position signal • TR range switch is damaged or has failed • PCM has failed
DTC: P0705 **1T CCM, MIL: NO** **1996, 1997, 1998, 1999, 2000** **Models:** Hombre **Engines:** 2.2L VIN 4 **Transmissions:** A/T	**Transmission Range Switch Circuit Malfunction** Engine started, then driven to a speed of over 5 mph, and the PCM detected an invalid PRNDL parameter (signal) for 5 seconds. **Note: For additional help with this code, view the Failure Records.** **Possible Causes:** • PRNDL switch signal is open • PRNDL switch signal shorted to another switch position signal • PRNDL switch is damaged or has failed • PCM has failed
DTC: P0705 **1T CCM, MIL: NO** **1996, 1997, 1998, 1999, 2000, 2001, 2002** **Models:** Axiom, Trooper, VehiCROSS **Engines:** All **Transmissions:** A/T	**Transmission Range Switch Illegal Position Malfunction** Engine started, then driven to a speed of over 8 mph, and the PCM detected "illegal" TR Range or Mode switch signals for 5 seconds. **Note: For additional help with this code, view the Failure Records.** **Possible Causes:** • TR range switch signal is open • TR range switch signal shorted to another switch position signal • TR range switch is damaged or has failed • PCM has failed

DTC	Trouble Code Title, Conditions & Possible Causes
DTC: P0706 **1T CCM, MIL: NO** **1998, 1999, 2000, 2001, 2002** **Models:** Amigo, Rodeo, Rodeo Sport **Engines:** All **Transmissions:** A/T	**Transmission Range Switch Circuit Performance** DTC P0122, P0123, P0722 and P0723 not set, engine started, then driven with the output speed over 3200 rpm, and the PCM detected the TR Switch indicated Reverse position, or with the Output speed under 3000 rpm and the throttle angle over 20%, it detected the TR switch indicated Park or Neutral position for 4 seconds in the test. **Note: For additional help with this code, view the Failure Records.** **Possible Causes:** • TR switch signal is open • TR switch signal shorted to another switch position signal • TR switch is damaged or has failed • PCM has failed
DTC: P0706 **1T CCM, MIL: NO** **1996, 1997, 1998, 1999, 2000, 2001, 2002** **Models:** Axiom, Trooper, VehiCROSS **Engines:** All **Transmissions:** A/T	**Transmission Range Switch Circuit Performance** DTC P0122, P0123, P0722 and P0723 not set, engine started, then driven with the output speed over 3200 rpm, and the PCM detected the TR Switch indicated Reverse position, or with the Output speed under 3000 rpm and the throttle angle over 20%, it detected the TR switch indicated Park or Neutral position for 4 seconds in the test. **Note: For additional help with this code, view the Failure Records.** **Possible Causes:** • TR switch signal is open • TR switch signal shorted to another switch position signal • TR switch is damaged or has failed • PCM has failed
DTC: P0711 **1T CCM, MIL: NO** **1998, 1999, 2000, 2001, 2002** **Models:** Amigo, Axiom, Rodeo, Rodeo Sport, Trooper, VehiCROSS **Engines:** All Transmissions: A/T	**Transmission Fluid Temperature Sensor Performance** DTC P0722, P0723 and P1870 not set, engine started, system voltage from 11-16v, TFT sensor from -40°F to 69.8°F at startup, ECT sensor more than 150°F and has changed more than 90°F since startup, vehicle speed over 5 mph with the TCC slip speed over 120 rpm for 410 seconds, and the PCM detected the TFT sensor changed less than 2 counts since startup, or that its delta change was over 36°F at least 14 times during a 7 second period. **Possible Causes**: • TFT signal or ground circuit has a high resistance condition • TFT sensor is out-of-calibration (it may be skewed) • TFT sensor is damaged or has failed • PCM has failed
DTC: P0711 **1T CCM, MIL: NO** **1997, 1998, 1999, 2000** **Models:** Hombre **Engines:** All **Transmissions:** A/T	**Transmission Fluid Temperature Sensor Performance** DTC P0502, P0503 and P1870 not set, system voltage from 11-16v, engine runtime over 409 seconds, TFT sensor signal from -40°F to +70°F at startup, and TFT signal from -36 to 304°F during the test, ECT sensor signal more than 158°F and has changed over 90°F since startup, vehicle speed over 5 mph with the TCC slip speed over 120 rpm for 409 seconds cumulative, and the PCM detected the TFT sensor signal changed less than 2.7°F since startup, or it changed more than 36°F within 200 ms 14 times in 7 seconds. **Possible Causes:** • TFT signal or ground circuit has a high resistance condition • TFT sensor is out-of-calibration (it may be skewed) • TFT sensor is damaged or has failed • PCM has failed
DTC: P0712 **1T CCM, MIL: NO** **1998, 1999, 2000, 2001, 2002** **Models**: Amigo, Rodeo, Rodeo Sport **Engines:** All **Transmissions**: A/T	**Transmission Fluid Temperature Sensor Low Input** Engine started, system voltage from 11-16v, and the PCM detected the TFT sensor was less than 0.40v for 20 seconds during the test. **Note: For additional help with this code, view the Failure Records.** **Possible Causes:** • TFT sensor signal circuit is shorted to sensor ground • TFT sensor signal circuit is shorted to chassis ground • TFT sensor is damaged (it may be shorted internally) • PCM has failed
DTC: P0712 **1T CCM, MIL: NO** **1997, 1998, 1999, 2000** **Models:** Hombre **Engines:** All Transmissions: A/T	**Transmission Fluid Temperature Sensor Low Input** Engine started, system voltage from 11-16v, and the PCM detected the TFT sensor signal was less than 0.20v for 10 seconds. **Possible Causes:** • TFT sensor signal circuit is shorted to sensor ground • TFT sensor signal circuit is shorted to chassis ground • TFT sensor is damaged (it may be shorted internally) • PCM has failed

DTC	Trouble Code Title, Conditions & Possible Causes
DTC: P0712 **1T CCM, MIL: NO** **1996, 1997, 1998, 1999, 2000, 2001, 2002** **Models**: Axiom, Trooper, VehiCROSS **Engines:** All **Transmissions:** A/T	**Transmission Fluid Temperature Sensor Low Input** Engine started, system voltage from 11-16v, and the PCM detected the TFT sensor was less than 0.40v for 20 seconds during the test. **Note: For additional help with this code, view the Failure Records.** **Possible Causes:** • TFT sensor signal circuit is shorted to sensor ground • TFT sensor signal circuit is shorted to chassis ground • TFT sensor is damaged (it may be shorted internally) • PCM has failed
DTC: P0713 **1T CCM, MIL: NO** **1998, 1999, 2000, 2001, 2002** **Models:** Amigo, Rodeo, Rodeo Sport **Engines:** All **Transmissions:** A/T	**Transmission Fluid Temperature Sensor High Input** Engine started, system voltage from 11-16v, and the PCM detected the TFT sensor was more than 4.86v for 20 seconds during the test. **Note: For additional help with this code, view the Failure Records.** **Possible Causes:** • TFT sensor signal circuit is open between the sensor and PCM • TFT sensor ground circuit is open between sensor and ground • TFT sensor signal circuit is shorted to VREF or system power • TFT sensor is damaged (it may be open internally) • PCM has failed
DTC: P0713 **1T CCM, MIL: NO** **1997, 1998, 1999, 2000** **Models:** Hombre **Engines:** All **Transmissions:** A/T	**Transmission Fluid Temperature Sensor High Input** Engine started, system voltage from 11-16v, and the PCM detected the TFT sensor signal was more than 4.92v for 409 seconds. **Possible Causes:** • TFT sensor signal circuit is open between the sensor and PCM • TFT sensor ground circuit is open between sensor and ground • TFT sensor signal circuit is shorted to VREF or system power • TFT sensor is damaged (it may be open internally) • PCM has failed
DTC: P0713 **1T CCM, MIL: NO** **1996, 1997, 1998, 1999, 2000, 2001, 2002** **Models:** Axiom, Trooper, VehiCROSS **Engines:** All Transmissions: A/T	**Transmission Fluid Temperature Sensor High Input** Engine started, system voltage from 11-16v, and the PCM detected the TFT sensor signal was more than 4.92v for 409 seconds. **Possible Causes:** • TFT sensor signal circuit is open between the sensor and PCM • TFT sensor ground circuit is open between sensor and ground • TFT sensor signal circuit is shorted to VREF or system power • TFT sensor is damaged (it may be open internally) • PCM has failed
DTC: P0715 **1T CCM, MIL: YES** **1996, 1997, 1998, 1999** **Models:** Oasis **Engines**: All **Transmissions:** A/T	**TCM A/T Mainshaft Speed Sensor Circuit Malfunction** Engine running with VSS inputs received, and the PCM detected an unexpected voltage condition on the Mainshaft speed sensor circuit. **Possible Causes:** • Mainshaft speed sensor circuit is open or shorted to ground • Mainshaft speed sensor circuit is shorted to VREF or power • Mainshaft speed sensor is damaged or has failed • PCM has failed
DTC: P0719 **1T CCM, MIL: NO** **1998, 1999, 2000, 2001, 2002** **Models:** Amigo, Rodeo, Rodeo Sport **Engines:** All **Transmissions:** A/T	**TCC Brake Switch Circuit Low Input** DTC P0722 and P0723 not set, engine started, then driven to a speed over 5 mph, then driven to over 20 mph for 5 seconds, then back to a speed of 5-20 mph for 4 seconds, and the PCM detected an "open" Brake switch condition for 15 minutes with no change in its status, condition occurred at least 7 times during the CCM test. **Possible Causes:** • TCC brake switch signal circuit is open or shorted to ground • TCC brake switch is damaged (it may be open internally) • PCM has failed
DTC: P0719 **1T CCM, MIL: NO** **1996, 1997, 1998, 1999, 2000, 2001, 2002** **Models:** Axiom, Trooper, VehiCROSS **Engines:** All **Transmissions:** A/T	**TTCC Brake Switch Circuit Low input** DDTC P0502 and P0503 not set, engine started, then driven to a speed under 5 mph, then driven from 5-20 mph for 4 seconds, then back to a speed over 20 mph for 6 seconds, and the PCM detected an "open" Brake switch condition for 15 minutes without it changing its status, conditions occurred at least 7 times during the CCM test. **Possible Causes:** • TCC brake switch signal circuit is open or shorted to ground • TCC brake switch is damaged (it may be open internally) • PCM has failed

DTC	Trouble Code Title, Conditions & Possible Causes
DTC: P0719 **1T CCM, MIL: NO** **1997, 1998, 1999, 2000** **Models:** Hombre **Engines:** All **Transmissions:** A/T	**TCC Brake Switch Circuit Low Input** DTC P0502 and P0503 not set, engine started, then driven to a speed under 5 mph, then driven from 5-20 mph for 4 seconds, then back to a speed over 20 mph for 6 seconds, and the PCM detected an "open" Brake switch condition for 15 minutes without it changing its status, conditions occurred at least 7 times during the CCM test. **Possible Causes:** • TCC brake switch signal circuit is open or shorted to ground • TCC brake switch is damaged (it may be open internally) • PCM has failed
DTC: P0720 **1T CCM, MIL: YES** **1996, 1997, 1998, 1999** **Models:** Oasis **Engines:** All **Transmissions:** A/T	**TCM A/T Countershaft Speed Sensor Circuit Malfunction** Engine running with VSS inputs received, and the PCM detected an unexpected voltage on the Countershaft Speed Sensor circuit. **Possible Causes:** • Countershaft speed sensor circuit is open or shorted to ground • Countershaft speed sensor circuit is shorted to VREF or power • Countershaft speed sensor is damaged or has failed • PCM has failed
DTC: P0722 **2T CCM, MIL: YES** **1998, 1999, 2000, 2001, 2002** **Models**: Amigo, Rodeo, Rodeo Sport **Engines:** All **Transmissions:** A/T	**Output Speed Sensor Low Input** DTC P0106, P0107, P0108, P0122, P0123, P1106 and P1107 not set, TR switch indicating other than Park or Neutral position, throttle angle over 10%, engine vacuum from 0-70 kPa, engine speed from 3000-5000 rpm, and the PCM detected the Output Speed Sensor (OSS) signal indicated zero (0) rpm, condition met for 5 seconds. **Possible Causes:** • OSS (+) signal circuit is open or shorted to ground • OSS (-) signal circuit is open or shorted to ground • OSS is damaged or has failed • PCM has failed
DTC: P0722 **1T CCM, MIL: NO** **1996, 1997, 1998, 1999, 2000, 2001, 2002** **Models:** Axiom, Trooper, VehiCROSS **Engines**: All **Transmissions:** A/T	**Output Speed Sensor Low Input** DTC P0106, P0107, P0108, P0122, P0123, P1106 and P1107 not set, TR switch indicating other than Park or Neutral position, throttle angle over 10%, engine vacuum from 0-70 kPa, engine speed from 3000-5000 rpm, and the PCM detected the Output Speed Sensor (OSS) signal indicated zero (0) rpm, condition met for 5 seconds. **Possible Causes:** • OSS (+) signal circuit is open or shorted to ground • OSS (-) signal circuit is open or shorted to ground • OSS is damaged or has failed • PCM has failed
DTC: P0723 **2T CCM, MIL: YES** **1998, 1999, 2000, 2001, 2002** **Models:** Amigo, Rodeo, Rodeo Sport **Engines:** All **Transmissions:** A/T	**Output Speed Sensor Signal Malfunction (Intermittent)** Engine started, engine running, TR switch indicating not in Park or Neutral, OSS signal over 1000 rpm for 2 seconds, then OSS signal over 512 rpm for 2 seconds, engine vacuum less than 20 kPa, then OSS signal over 1380 for 1 second, then the NORAW-NOLAST was less than 200 rpm for 2-6 seconds, and the PCM determined the OSS signal did not vary during the CCM Rationality test. **Note: NORAW indicates the latest OSS raw data and NOLAST indicated the filtered previous data from the OSS.** **Possible Causes:** • OSS (+) signal circuit is open or shorted to ground (intermittent) • OSS (-) signal circuit is open or shorted to ground (intermittent) • OSS is damaged or has failed (an intermittent fault) • PCM has failed
DTC: P0723 **1T CCM, MIL: NO** **1996, 1997, 1998, 1999, 2000, 2001, 2002** **Models:** Axiom, Trooper, VehiCROSS **Engines:** All **Transmissions**: A/T	**Output Speed Sensor Signal Malfunction (Intermittent)** Engine started, engine runtime over 5 seconds, OSS signal more than 1300 rpm for 2 seconds, then the NORAW-NORAWLAST less than 200 rpm for 2 seconds, and the PCM detected a transmission negative OSS signal change of more than 1300 rpm for 3 seconds. **Possible Causes:** • OSS (+) signal circuit is open or shorted to ground (intermittent) • OSS (-) signal circuit is open or shorted to ground (intermittent) • OSS is damaged or has failed • PCM has failed
DTC: P0724 **1T CCM, MIL: NO** **1998, 1999, 2000, 2001, 2002** **Models:** Amigo, Rodeo, Rodeo Sport **Engines:** All **Transmissions:** A/T	**TCC Brake Switch Circuit High Input (Stuck Closed)** DTC P0722 and P0723 not set, engine started, then driven to a speed of over 5 mph, then driven to over 20 mph for 5 seconds, then back to a speed of 5-20 mph for 4 seconds, and the PCM detected a "closed" Brake switch condition for 15 minutes without it changing its status, conditions occurred at least 7 times during the CCM test. **Possible Causes:** • TCC brake switch signal circuit is shorted to system power (B+) • TCC brake switch is damaged (it may be shorted internally) • PCM has failed

DTC	Trouble Code Title, Conditions & Possible Causes
DTC: P0724 **1T CCM, MIL: NO** **1996, 1997, 1998, 1999, 2000, 2001, 2002** **Models:** Axiom, Trooper, VehiCROSS **Engines:** All **Transmissions:** A/T	**TCC Brake Switch Circuit High Input (Stuck Closed)** DTC P0722 and P0723 not set, engine started, then driven to a speed of over 5 mph, then driven to over 20 mph for 5 seconds, then back to a speed of 5-20 mph for 4 seconds, and the PCM detected a "closed" Brake switch condition for 15 minutes without it changing its status, conditions occurred at least 7 times during the CCM test. **Possible Causes:** • TCC brake switch signal circuit is shorted to system power (B+) • TCC brake switch is damaged (it may be shorted internally) • PCM has failed
DTC: P0724 **1T CCM, MIL: NO** **1997, 1998, 1999, 2000** **Models:** Hombre **Engines:** All **Transmissions:** A/T	**TCC Brake Switch Circuit High Input (Stuck Closed)** DTC P0502 and P0503 not set, engine started, then driven to a speed of over 20 mph for 6 seconds, then driven to over 5-20 mph for 4 seconds, then to a speed of 5 mph, and the PCM detected a "closed" Brake switch condition (12v all the time) without it changing its status, conditions occurred at least 7 times during the CCM test. **Possible Causes:** • TCC brake switch signal circuit is shorted to system power (B+) • TCC brake switch is damaged (it may be shorted internally) • PCM has failed
DTC: P0725 **1T CCM, MIL: YES** **1996, 1997, 1998, 1999** **Models:** Oasis **Engines:** All **Transmissions:** A/T	**Automatic Transaxle** Engine running and the PCM detected an Automatic Transaxle fault. **Note: This trouble code sets along with several other Automatic Transaxle related trouble codes.** **Possible Causes:** • Refer to the repair instructions in a transmission repair manual or the information in other electronic media to repair this code.
DTC: P0730 **1T CCM, MIL: NO** **1998, 1999, 2000** **Models:** Amigo **Engines:** 3.2L VIN W **Transmissions:** A/T	**Transmission Incorrect Gear Ratio** DTC P0722 and P0723 not set, gear selector not in Park, Neutral or Reverse position, engine speed more than 3500 rpm, 3 seconds have passed since an Upshift event, and the PCM detected a slip value over 720 rpm in 1st gear, a slip value over 680 rpm in 2nd gear, a slip value over 660 rpm in 3rd gear, or a slip value over 650 rpm in 4th gear, condition met for 5.5 seconds during the CCM test. **Possible Causes:** • ATF fluid level to low, too high, or burnt or contaminated • OSS signal is open or shorted to ground (an intermittent fault) • OSS is out-of-calibration (i.e., the tire size or rear axle ratio) • OSS is damaged or has failed • PCM has failed
DTC: P0730 **1T CCM, MIL: NO** **1998, 1999** **Models:** Rodeo **Engines:** 3.2L VIN W **Transmissions:** A/T	**Transmission Incorrect Gear Ratio** DTC P0722 and P0723 not set, gear selector not in Park, Neutral or Reverse position, engine speed more than 3500 rpm, 3 seconds have passed since an Upshift event, and the PCM detected a slip value over 753 rpm in 1st gear, a slip value over 713 rpm in 2nd gear, a slip value over 694 rpm in 3rd gear, or a slip value over 685 rpm in 4th gear, condition met for 5.5 seconds during the CCM test. **Possible Causes:** • ATF fluid level to low, too high, or burnt or contaminated • OSS signal is open or shorted to ground (an intermittent fault) • OSS is out-of-calibration (i.e., the tire size or rear axle ratio) • OSS is damaged or has failed • PCM has failed
DTC: P0730 **1T CCM, MIL: NO** **2000, 2001, 2002** **Models:** Rodeo, Rodeo Sport **Engines:** All **Transmissions:** A/T	**Transmission Gear Error Without Input Speed** DTC P0705, P0706, P0722 and P0723 not set, gear selector not in Park, Neutral or Reverse position, engine speed more than 3500 rpm, 3 seconds have passed since an Upshift event, and the PCM detected a slip value over 720 rpm in 1st gear, a slip value over 680 rpm in 2nd gear, a slip value over 660 rpm in 3rd gear, or a slip value over 650 rpm in 4th gear for 5.5 seconds during the CCM test. **Possible Causes:** • ATF fluid level to low, too high, or burnt or contaminated • OSS signal is open or shorted to ground (an intermittent fault) • OSS is out-of-calibration (i.e., the tire size or rear axle ratio) • OSS is damaged or has failed • PCM has failed

DTC	Trouble Code Title, Conditions & Possible Causes
DTC: P0730 **1T CCM, MIL: NO** **1996, 1997, 1998, 1999, 2000, 2001, 2002** **Models:** Axiom, Trooper, VehiCROSS **Engines:** All **Transmissions:** A/T	**Transmission Incorrect Gear Ratio** DTC P0722 and P0723 not set, gear selector not in Park, Neutral or Reverse position, engine speed more than 3500 rpm, 3 seconds have passed since an Upshift event, and the PCM detected a slip value over 753 rpm in 1st gear, a slip value over 713 rpm in 2nd gear, a slip value over 694 rpm in 3rd gear, or a slip value over 685 rpm in 4th gear, condition met for 5.5 seconds during the CCM test. **Possible Causes:** • ATF fluid level to low, too high, or burnt or contaminated • OSS signal is open or shorted to ground (an intermittent fault) • OSS is out-of-calibration (i.e., the tire size or rear axle ratio) • OSS is damaged or has failed • PCM has failed
DTC: P0730 **1T CCM, MIL: YES** **1996, 1997, 1998, 1999** **Models:** Oasis **Engines:** All **Transmissions:** A/T	**TCM A/T Shift Control System** No other A/T trouble codes set, engine running at cruise speed with VSS inputs received, and the PCM detected the lockup clutch did not lock or unlock correctly. **Possible Causes:** • Refer to the repair instructions in a transmission repair manual or the information in other electronic media to repair this code.
DTC: P0740 **1T CCM, MIL: YES** **1997, 1998, 1999, 2000** **Models:** Hombre **Engines:** All **Transmissions:** A/T	**Torque Converter Clutch Circuit Malfunction** Engine started, engine speed over 450 rpm for 5 seconds, DFCO not "on", and the PCM detected an unexpected high condition on the TCC control circuit with the TCC commanded "on", or a low condition with the TCC commanded "off" during the CCM test. **Possible Causes:** • TCC control circuit is open or shorted to ground • TCC control circuit is shorted to system power (B+) • TCC power circuit is open (check the Cluster 10A fuse) • Internal transmission concerns exist
DTC: P0740 **1T CCM, MIL: YES** **1996, 1997, 1998, 1999** **Models:** Oasis **Engines:** All **Transmissions:** A/T	**TCM A/T Lockup Clutch System** No other A/T trouble codes set, engine running at cruise speed with VSS inputs received, and the PCM detected the lockup clutch did not engage or disengage correctly. **Possible Causes:** • Refer to the repair instructions in a transmission repair manual or the information in other electronic media to repair this code.
DTC: P0742 **2T CCM, MIL: YES** **2000, 2001, 2002** **Models:** Amigo, Rodeo, Rodeo Sport **Engines:** 2.2L VIN D **Transmissions:** A/T	**Torque Converter Clutch Circuit Malfunction (Stuck On)** DTC P0122, P0123, P0722, P0723 and P1860 not set, engine speed from 500-3000 rpm, throttle angle over 20%, engine vacuum from 0-105 kPa, commanded gear not in 1st, TCC is "off", gear range is D4, TCC commanded "off", speed ratio between 0.9-1.8, vehicle speed from 15-75 mph, and the PCM detected the TCC slip speed was from -20 to -40 rpm for 5 seconds during the CCM test. **Possible Causes:** • TCC assembly has failed (it may be mechanically stuck "on") • Internal transmission concerns exist
DTC: P0742 **2T CCM, MIL: YES** **1998, 1999, 2000, 2001, 2002** **Models:** Amigo, Rodeo, Rodeo Sport **Engines:** 3.2L VIN W **Transmissions:** A/T	**Torque Converter Clutch Circuit Malfunction (Stuck On)** DTC P0107, P0108, P0122, P0123, P0502, P0503, P0740, P1810 and P1860 not set, engine started engine speed over 450 rpm for 5 seconds, throttle angle from 15-60%, not in Fuel Cutoff mode, engine speed from 1000-3000 rpm, engine vacuum from 0-105 kPa, engine torque from 40-400 lbs, speed ratio from 0.65-1.25, VSS indicating 20-65 mph, commanded gear not in 1st, TCC is "off", gear range is D4 with no change for 6 seconds, and the PCM detected the TCC slip speed was from -40 to +30 rpm for 5 seconds. **Possible Causes:** • TCC assembly has failed (it may be mechanically stuck "on") • Internal transmission concerns exist
DTC: P0742 **2T CCM, MIL: YES** **1997, 1998, 1999, 2000** **Models**: Hombre **Engines:** All **Transmissions:** A/T	**Torque Converter Clutch Circuit Malfunction (Stuck On)** DTC P0107, P0108, 122, P0123, P0502, P0503, P0722, P0723, P0740, P1810, P1860 and P1870 not set, vehicle driven to a speed of 15-50 mph, throttle angle from 17-45%, engine speed from 500-3000 rpm, engine vacuum from 0-105 kPa, not in 1st Gear, gear selector in D4 range, speed ratio from 0.64-1.8, TCC commanded "off", and the PCM detected the TCC slip speed was from -20 to +20 for 5 seconds during the CCM Rationality test. **Possible Causes:** • TCC assembly has failed (it may be mechanically stuck "on") • Internal transmission concerns exist

DTC	Trouble Code Title, Conditions & Possible Causes
DTC: P0742 **1T CCM, MIL: NO** **1996, 1997, 1998, 1999, 2000, 2001, 2002** **Models:** Axiom, Trooper, VehiCROSS **Engines:** All **Transmissions:** A/T	**Torque Converter Clutch Circuit Malfunction (Stuck On)** DTC P0107, P0108, P0122, P0123, P0502, P0503, P0740, P1810 and P1860 not set, engine started engine speed over 450 rpm for 5 seconds, throttle angle from 15-60%, not in Fuel Cutoff mode, engine speed from 1000-3000 rpm, engine vacuum from 0-105 kPa, engine torque from 40-400 lbs, speed ratio from 0.65-1.25, VSS indicating 20-65 mph, commanded gear not in 1st, TCC is "off", gear range is D4 with no change for 6 seconds, and the PCM detected the TCC slip speed was from -40 to +30 rpm for 5 seconds. **Possible Causes:** • TCC assembly has failed (it may be mechanically stuck "on") • Internal transmission concerns exist
DTC: P0748 **2T CCM, MIL: NO** **2000, 2001, 2002** **Models:** Amigo, Rodeo, Rodeo Sport **Engines:** 2.2L VIN D **Transmissions:** A/T	**A/T Pressure Control Solenoid Circuit Malfunction** Engine started, engine speed over 300 rpm, system voltage from 11-16v, and the PCM detected the difference between the Actual and Commanded Pressure Control Solenoid (PCS) current level was more than 200 mA for over 1 second during the CCM test. **Possible Causes:** • PCS control circuit is open or shorted to ground • PCS power circuit is open (check the fuse) • Pressure control solenoid is damaged or has failed • PCM has failed
DTC: P0748 **2T CCM, MIL: NO** **1998, 1999, 2000, 2001, 2002** **Models**: Amigo, Rodeo, Rodeo Sport **Engines:** 3.2L VIN W **Transmissions:** A/T	**A/T Pressure Control Solenoid Circuit Malfunction** Engine started, engine running, and the PCM detected the difference between the Actual and Commanded Pressure Control solenoid (PCS) current level was more than 200 mA during the CCM test. **Note: For additional help with this code, view the Failure Records.** **Possible Causes:** • Pressure control solenoid control circuit is shorted to ground • Pressure control solenoid is damaged or has failed • PCM has failed
DTC: P0748 **2T CCM, MIL: NO** **1997, 1998, 1999, 2000** **Models:** Hombre **Engines:** All **Transmissions:** A/T	**A/T Pressure Control Solenoid Circuit Malfunction** Engine started, engine running, system voltage from 10-16v, and the PCM detected the Pressure Control solenoid (PCS) command duty cycle reached its "high" (95%) or "low" (0%) limit during the test. **Possible Causes:** • PCS control circuit is open or shorted to ground • PCS power circuit is open (check the fuse) • Pressure control solenoid is damaged or has failed • PCM has failed
DTC: P0748 **1T CCM, MIL: NO** **1996, 1997, 1998, 1999, 2000, 2001, 2002** **Models:** Axiom, Trooper, VehiCROSS **Engines:** All **Transmissions:** A/T	**A/T Pressure Control Solenoid Circuit Malfunction** Engine started, engine running, and the PCM detected the difference between the Actual and Commanded Pressure Control solenoid (PCS) current level was more than 200 mA during the CCM test. **Note: For additional help with this code, view the Failure Records.** **Possible Causes:** • Pressure control solenoid control circuit is shorted to ground • Pressure control solenoid is damaged or has failed • PCM has failed
DTC: P0751 **2T CCM, MIL: YES** **2000, 2001, 2002** **Models:** Amigo, Rodeo, Rodeo Sport **Engines:** 2.2L VIN D **Transmissions:** A/T	**A/T Pressure Control Solenoid Circuit Malfunction** Engine started, engine speed over 300 rpm, system voltage from 11-16v, and the PCM detected the difference between the Actual and Commanded Pressure Control Solenoid (PCS) current level was more than 200 mA for over 1 second during the CCM test. **Possible Causes:** • PCS control circuit is open or shorted to ground • PCS power circuit is open (check the fuse) • Pressure control solenoid is damaged or has failed • PCM has failed
DTC: P0751 **2T CCM, MIL: YES** **2000, 2001, 2002** **Models:** Amigo, Rodeo, Rodeo Sport **Engines:** 2.2L VIN D **Transmissions:** A/T	**Shift Solenoid 'A' Performance (Stuck On)** DTC P0705, P0706, P0722, P0723, P0742, P0753, P0758, P1860 and P1870 not set, engine started, then driven in D4 with the Output speed over 375 rpm, TFT sensor signal from 68-266°F, and with 2nd Gear commanded "on" for 1 second, the PCM detected: 40 less than engine torque less than 400 Nm, speed ratio more than 0.6, throttle angle more than 10%, 800 less than TCC slip less than 4000 rpm, 2.75 less than modeled ratio less than 3.2 for 1 second; or with 3rd Gear commanded "on" for 1 second, the PCM detected: 40 less than engine torque less than 400 Nm, 800 less than TCC slip less than 8000 rpm, speed ratio more than 0.45, throttle angle more than 10%, 0.62 less than modeled ratio less than 0.95 for 3 seconds. **Note: This fault must occur twice in one trip to set this trouble code.** **Possible Causes:** • Shift solenoid 'A' is damaged or has failed mechanically (on) • Other internal transmission concerns can cause this fault

DTC	Trouble Code Title, Conditions & Possible Causes
DTC: P0751 **2T CCM, MIL: YES** **1998, 1999, 2000, 2001, 2002** **Models:** Amigo, Rodeo, Rodeo Sport **Engines:** 3.2L VIN W **Transmissions:** A/T	**Shift Solenoid 'A' Performance Without Input Speed** DTC P0122, P0123, P0722, P0723, P0742, P0753, P0758 and P1860 not set, vehicle driven in D4 Gear at over 6.25 mph, TFT sensor signal at 68-257°F, then during a 1-2 Shift, TP angle at 10-60% (± 3%), VSS at 11-31 mph, the PCM detected the engine speed in 2nd Gear was 100 rpm more than it was in 1st Gear (1); or during a 2-3 Shift, TP angle at 13-60% (± 5%), VSS at 20-45 mph, the engine speed in 3rd Gear was 64 rpm less than it was in 2nd Gear (2); or during a 3-4 Shift, TP at 7-60% (± 5%), VSS at 25-87 mph, the engine speed in 4th gear was 60 rpm more than it was in 3rd Gear (3); or while in 4th Gear, TP angle at 13-60% (± 5%), speed ratio at 0.85 to 1.2, the TCC slip speed was 100-2000 rpm for 3 seconds (4); or while in 4th Gear with TCC "on", speed ratio at 0.5-0.85, the TCC slip speed was -50 to +500 for 3 seconds (5). **Note: This code is set if the conditions in (1), (2), (3) or (4) are met, or if the conditions in (1), (2), (3) or (5) are met twice in a row.** **Possible Causes:** • Shift solenoid 'A' is damaged or has failed mechanically (on) • Other internal transmission concerns can cause this problem
DTC: P0751 **1T CCM, MIL: NO** **1996, 1997, 1998, 1999, 2000** **Models:** Hombre **Engines:** 2.2L VIN 4 **Transmissions:** A/T	**1-2 Shift Solenoid Valve Range/Performance** DTC P0122, P0123, P0502, P0503, P0742, P0753, P0758, P0785, P1810, and P1860 not set, vehicle driven in D4 to over 5 mph, Fuel Cutoff Mode "off", TFT sensor signal at 68-266°F, then during a 1-2 Shift, TP angle at 10-50% (± 6%), VSS at 5-35 mph, the PCM detected the 2nd gear engine speed was 80 rpm more than it was in 1st gear (1); or during a 2-3 Shift, TP angle at 10-50% (± 7%), VSS at 15-60 mph, the 3rd gear speed was 100 rpm more than it was in 2nd gear (2); or during a 3-4 Shift, TP angle at 10-50% (± 7%), VSS at 30-65 mph, the 4th gear speed was 100 rpm more than it was in 3rd gear (3); or while in 4th Gear with TCC "on", TP angle at 10-50%, speed ratio at 0.95-1.25, the TCC slip speed was 400-1200 rpm (4); or while in 4th Gear with TCC "on", TP angle at 10-50%, speed ratio at 0.65-0.80, the TCC slip speed was -20 to +50 rpm (5). **Note: This code is set if the conditions in (1), (2), (3) or (4) are met, or if the conditions in (1), (2), (3) or (5) are met three times.** **Possible Causes:** • 1-2 Shift solenoid is damaged or has failed mechanically (on) • Other internal transmission concerns can cause this fault
DTC: P0751 **1T CCM, MIL: NO** **1997, 1998, 1999, 2000** **Models:** Hombre **Engines:** 4.3L VIN X **Transmissions:** A/T	**1-2 Shift Solenoid Valve Range/Performance** DTC P0122, P0123, P0502, P0503, P0740, P0742, P0753, P0758, P0785, P1810, and P1860 not set, vehicle driven in D4 to over 5 mph, Fuel Cutoff Mode "off", TFT sensor signal at 68-266°F, then during a 1-2 Shift, TP angle at 10-50% (± 6%), VSS at 5-35 mph, the PCM detected the 2nd gear engine speed was 80 rpm more than it was in 1st gear and the PCM detected the speed ratio was 0.85-1.2 or the TCC slip speed was 200-1000 rpm for 4 seconds; or during a 2-3 Shift, VSS at 15-60 mph, the 3rd gear speed was 100 rpm more than it was in 2nd gear; or during a 3-4 Shift, VSS at 30-65 mph, the 4th gear speed was 100 rpm more than it was in 3rd gear and the PCM detected speed ratio was 0.6-0.8, or that the TCC slip speed was from -20 to +50 rpm for 4 seconds during the CCM test. **Possible Causes:** • 1-2 Shift solenoid is damaged or has failed mechanically (on) • Other internal transmission concerns can cause this fault
DTC: P0751 **1T CCM, MIL: NO** **1996, 1997, 1998, 1999, 2000, 2001, 2002** **Models:** Axiom, Trooper, VehiCROSS **Engines:** All **Transmissions:** A/T	**Shift Solenoid 'A' Performance Without Input Speed** DTC P0122, P0123, P0722, P0723, P0742, P0753, P0758 and P1860 not set, vehicle driven in D4 Gear at over 6.25 mph, TFT sensor signal at 68-257°F, then during a 1-2 Shift, TP angle at 10-60% (± 3%), VSS at 11-31 mph, the PCM detected the engine speed in 2nd Gear was 100 rpm more than it was in 1st Gear (1); or during a 2-3 Shift, TP angle at 13-60% (± 5%), VSS at 20-45 mph, the engine speed in 3rd Gear was 64 rpm less than it was in 2nd Gear (2); or during a 3-4 Shift, TP at 7-60% (± 5%), VSS at 25-87 mph, the engine speed in 4th gear was 60 rpm more than it was in 3rd Gear (3); or while in 4th Gear, TP angle at 13-60% (± 5%), speed ratio at 0.85 to 1.2, the TCC slip speed was 100-2000 rpm for 3 seconds (4); or while in 4th Gear with TCC "on", speed ratio at 0.5-0.85, the TCC slip speed was -50 to +500 for 3 seconds (5). **Note: This code is set if the conditions in (1), (2), (3) or (4) are met, or if the conditions in (1), (2), (3) or (5) are met twice in a row.** **Possible Causes:** • Shift solenoid 'A' is damaged or has failed mechanically (on) • Other internal transmission concerns can cause this problem
DTC: P0752 **2T CCM, MIL: YES** **2000, 2001, 2002** **Models:** Amigo, Axiom, Rodeo, Rodeo Sport **Engines:** 2.2L VIN D, 3.2L VIN W, 3.5L VIN X **Transmissions:** A/T	**Shift Solenoid 'A' Performance (Stuck Off)** DTC P0705, P0706, P0722, P0723, P0742, P0753, P0758, P1860 and P1870 not set, engine started, then driven in D4 with the output speed over 375 rpm, TFT sensor signal from 68-266°F, and with 1st Gear commanded "on" for 1 second, the PCM detected: 40 less than engine torque less than 400 Nm, speed ratio more than 0.3, throttle angle more than 10%, 800 less than TCC slip less than 8000 rpm, transmission output speed more than 375 rpm, 0.62 less than modeled ratio less than 2.4 for 0.687 seconds; or with 4th Gear commanded "on" for 1 second, the PCM detected: 40 less than engine torque less than 400 Nm, throttle angle more than 10%, 800 less than TCC slip less than 8000 rpm, speed ratio more than 0.6, 0.92 less than modeled ratio less than 1.5 for 7 seconds.. **Note: This fault must occur twice in one trip to set this trouble code.** **Possible Causes:** • Shift solenoid 'A' is damaged or has failed mechanically (off) • Other internal transmission concerns can cause this fault

DTC	Trouble Code Title, Conditions & Possible Causes
DTC: P0753 **2T CCM, MIL: YES** **2000, 2001, 2002** **Models:** Amigo, Rodeo, Rodeo Sport **Engines:** 2.2L VIN D **Transmissions:** A/T	**Shift Solenoid 'A' Circuit Malfunction** Engine started, engine running, then with Shift Solenoid 'A' (SSA) commanded "on", and PCM detected the solenoid control signal was 12v, or with SSA commanded "off", or the solenoid control signal was near 0v, condition met for 0.84-1.0 second in the CCM test. **Possible Causes:** • SSA control circuit is open or shorted to ground • SSA control circuit is shorted to system power (B+) • SSA power circuit (from the PCM) is open • SSA is damaged or has failed • PCM has failed
DTC: P0753 **2T CCM, MIL: YES** **1998, 1999, 2000, 2001, 2002** **Models:** Amigo, Rodeo, Rodeo Sport **Engines:** 3.2L VIN W **Transmissions:** A/T	**Shift Solenoid 'A' Circuit Malfunction** Engine started, engine running, then with Shift Solenoid 'A' (SSA) commanded "on", and PCM detected the solenoid control signal was 12v, or with SSA commanded "off", the solenoid control signal was near 0v, either condition met for 0.84 to 1.0 seconds during the test. **Possible Causes:** • SSA control circuit is open or shorted to ground • SSA control circuit is shorted to system power (B+) • SSA power circuit (from the PCM) is open • SSA is damaged or has failed • PCM has failed
DTC: P0753 **1T CCM, MIL: NO** **1996, 1997, 1998, 1999, 2000** **Models:** Hombre **Engines:** 2.2L VIN 4 **Transmissions:** A/T	**1-2 Shift Solenoid Circuit Malfunction** Engine started, engine speed over 450 rpm for 5 seconds, not in Fuel Cutoff mode, then with the 1-2 Shift Solenoid commanded "on", it detected a "high" voltage condition (12v) on the solenoid control circuit, or with the solenoid commanded "off", it detected a "low" voltage condition (0v) during the CCM Rationality test. **Possible Causes:** • 1-2 shift solenoid control circuit is open or shorted to ground • 1-2 shift solenoid control circuit is shorted to power • 1-2 shift solenoid power circuit is open (check the fuse) • 1-2 shift solenoid is damaged or has failed • PCM has failed
DTC: P0753 **1T CCM, MIL: YES** **1997, 1998, 1999, 2000** **Models:** Hombre **Engines:** 4.3L VIN X **Transmissions:** A/T	**1-2 Shift Solenoid Circuit Malfunction** Engine started, engine speed over 450 rpm for 5 seconds, not in Fuel Cutoff mode, then with the 1-2 Shift Solenoid commanded "on", the PCM detected a "high" voltage condition (12v) on the solenoid control circuit, or with the solenoid commanded "off", it detected a "low" voltage condition (0v) during the CCM Rationality test. **Possible Causes:** • 1-2 shift solenoid control circuit is open or shorted to ground • 1-2 shift solenoid control circuit is shorted to power • 1-2 shift solenoid power circuit is open (check the fuse) • 1-2 shift solenoid is damaged or has failed • PCM has failed
DTC: P0753 **1T CCM, MIL: NO** **1996, 1997, 1998, 1999, 2000, 2001, 2002** **Models:** Axiom, Trooper, VehiCROSS **Engines:** All **Transmissions:** A/T	**Shift Solenoid 'A' Circuit Malfunction** Engine started, engine running, then with Shift Solenoid 'A' (SSA) commanded "on", and PCM detected the solenoid control signal was 12v, or with SSA commanded "off", the solenoid control signal was near 0v, either condition met for 0.84 to 1.0 seconds during the test. **Possible Causes:** • SSA control circuit is open or shorted to ground • SSA control circuit is shorted to system power (B+) • SSA power circuit (from the PCM) is open • SSA is damaged or has failed • PCM has failed
DTC: P0753 **1T CCM, MIL: YES** **1996, 1997, 1998, 1999** **Models:** Oasis **Engines:** All **Transmissions:** A/T	**TCM A/T Lockup Solenoid 'A' Circuit Malfunction** Engine running with VSS inputs, and the PCM detected an unexpected voltage condition on the Solenoid Valve 'A' circuit. **Possible Causes:** • TCM Solenoid 'A' control circuit is open or shorted to ground • TCM Solenoid 'A' control circuit is shorted to system power • TCM Solenoid 'A' is damaged or has failed • TCM or PCM has failed

DTC	Trouble Code Title, Conditions & Possible Causes
DTC: P0756 **1T CCM, MIL: YES** **2000, 2001, 2002** **Models:** Amigo, Rodeo, Rodeo Sport **Engines:** 2.2L VIN D **Transmissions:** A/T	**Shift Solenoid 'B' Circuit Malfunction** DTC P0705, P0706, P0722, P0723, P0742, P0753, P0758, P1860 and P1870 not set, engine started, then driven in D4 with the output speed over 375 rpm, TFT sensor signal from 68-266°F, and with 1st Gear commanded "on" for 1 second, the PCM detected: 40 less than engine torque less than 400 Nm, speed ratio more than 0.3, transmission output speed more than 400 rpm, throttle angle more than 10%, 0.60 less than modeled ratio less than 1.49 for 1 second; or with 2nd Gear commanded "on" for 1 second, the PCM detected: 40 less than engine torque less than 400 Nm, throttle angle more than 10%, -8000 less than TCC slip less than 8000 rpm, speed ratio more than 0.6, 0.92 less than modeled ratio less than 1.5 for 0.687 seconds during the CCM Rationality test. **Note: This fault must occur twice in one trip to set this trouble code.** **Possible Causes:** • Shift solenoid 'A' is damaged or has failed mechanically (off) • Other internal transmission concerns can cause this fault
DTC: P0756 **2T CCM, MIL: YES** **1998, 1999, 2000, 2001, 2002** **Models**: Amigo, Rodeo, Rodeo Sport **Engines:** 3.2L VIN W **Transmissions:** A/T	**Shift Solenoid 'B' Performance Without Input Speed** DTC P0122, P0123, P705, P706, P0722, P0723, P0742, P0753, P0758, P1106, P1107 and P1860 not set, vehicle driven in D4 Gear to over 6.25 mph at under 8000 rpm, MAP at 0-70 kPa, TFT signal at 68-257°F, the TCC "off", TP angle over 4%, 1st Gear "on", speed ratio at 0.5-2.65, TSS signal at 320-2000 rpm, the TCC slip speed was -200 to -4000 rpm for 1.8 seconds (1), or during a 2-3 Shift, TP angle at 10-60% (± 5%), VSS at 20-45 mph, the 3rd Gear speed was 64 rpm less than it was in 2nd gear (2), or during a 3-4 shift, TP angle at 7-60% (± 5%), VSS at 25-87 mph, the 4th Gear speed was 60 rpm less than it was in 3rd Gear (3); or in 4th Gear, TP angle at 13-60, speed ratio at 0.5-1.20, the TCC slip speed was 100-2000 rpm (4), or in 4th Gear, TP angle at 13-60, speed ratio at 0.5-0.85, the TCC slip speed was -50 to -500 rpm (5). **Note: This code is set if the conditions in (1), (3) or (4) are met twice (stuck on), or if the conditions in (1) and (3) are met twice (stuck off).** **Possible Causes:** • Shift solenoid 'B' is damaged or has failed mechanically (on) • Other internal transmission concerns can cause this fault
DTC: P0756 **1T CCM, MIL: YES** **1996, 1997, 1998, 1999, 2000** **Models:** Hombre **Engines:** 2.2L VIN 4 **Transmissions:** A/T	**2-3 Shift Solenoid Valve Performance** DTC P0122, P0123, P0502, P0503, P0740, P0742, P0753, P0758, P0785, P1810, and P1860 not set, vehicle driven in D4 to over 5 mph, Fuel Cutoff mode "off", TFT signal at 68-266°F, 1st Gear "on" for 2 seconds, engine torque at 4-400 lbs, speed ratio is 0.7-3.0, TSS signal from 400-1500 rpm, the PCM detected the TCC slip speed was -100 to -3000 rpm for 1.5 seconds (1); or 3rd Gear "on" for 2-6 seconds, TP angle at 10-50% (± 7%), engine torque at 40-400 lbs, speed ratio in 3rd gear within 0.35 of the last speed ratio in 2nd gear, the TCC slip speed in 3rd gear was 300 rpm above TCC slip speed in 2nd gear for 1.5 seconds (2); or 4th gear "on" for 1 second, engine torque at 40-400 lbs, speed ratio at 1.6-3.5, the TCC slip speed was 1000-4000 rpm with the TSS speed from 1400-3000 rpm for 1 second (3). **Note: This code sets if the conditions in (1) and (2), or the conditions in (2) and (3) occur at least three times in a row.** **Possible Causes:** • 2-3 Shift solenoid is damaged or has failed mechanically (on) • Other internal transmission concerns can cause this fault
DTC: P0756 **1T CCM, MIL: YES** **1996, 1997, 1998, 1999, 2000** **Models**: Hombre **Engines:** 4.3L VIN X **Transmissions:** A/T	**2-3 Shift Solenoid Valve Performance** DTC P0122, P0123, P0502, P0503, P711, P12, P713, P0740, P0742, P0753, P0758, P0785, P1810, and P1860 not set, vehicle driven in D4 to over 5 mph, not in Fuel Cutoff mode, TFT signal at 68-266°F, engine torque at 5-450 lbs, then with 1st Gear "on", speed ratio is 0.5-3.0, TSS signal from 400-1500 rpm, the PCM detected the TCC slip speed was -100 to -3000 rpm for 1.5 seconds; or with 3rd Gear "on" for 2-4 seconds, it detected the speed ratio in 3rd Gear does not drop by more than 0.3 from the last speed ratio in 2nd gear; or with 4th Gear "on" for one second, it detected the TCC slip speed in 3rd gear remains 300 rpm higher than the last TCC slip speed in 2nd gear; or with 4th gear "on", or the transmission output speed was 1400-2500 rpm or the speed ratio was 2.05 to 0.8. **Possible Causes:** • 2-3 Shift solenoid is damaged or has failed mechanically (on) • Other internal transmission concerns can cause this fault
DTC: P0756 **1T CCM, MIL: NO** **1996, 1997, 1998, 1999, 2000, 2001, 2002** **Models:** Axiom, Trooper, VehiCROSS **Engines:** All **Transmissions:** A/T	**Shift Solenoid 'B' Performance Without Input Speed** DTC P0122, P0123, P705, P706, P0722, P0723, P0742, P0753, P0758, P1106, P1107 and P1860 not set, vehicle driven in D4 Gear to over 6.25 mph at under 8000 rpm, MAP at 0-70 kPa, TFT signal at 68-257°F, the TCC "off", TP angle over 4%, 1st Gear "on", speed ratio at 0.5-2.65, TSS signal at 320-2000 rpm, the TCC slip speed was -200 to -4000 rpm for 1.8 seconds (1), or during a 2-3 Shift, TP angle at 10-60% (± 5%), VSS at 20-45 mph, the 3rd Gear speed was 64 rpm less than it was in 2nd gear (2), or during a 3-4 shift, TP angle at 7-60% (± 5%), VSS at 25-87 mph, the 4th Gear speed was 60 rpm less than it was in 3rd Gear (3); or in 4th Gear, TP angle at 13-60, speed ratio at 0.5-1.20, the TCC slip speed was 100-2000 rpm (4), or in 4th Gear, TP angle at 13-60, speed ratio at 0.5-0.85, the TCC slip speed was -50 to -500 rpm (5). **Note: This code is set if the conditions in (1), (3) or (4) are met twice (stuck on), or if the conditions in (1) and (3) are met twice (stuck off).** **Possible Causes:** • Shift solenoid 'B' is damaged or has failed mechanically (on) • Other internal transmission concerns can cause this fault

DTC	Trouble Code Title, Conditions & Possible Causes
DTC: P0757 **2T CCM, MIL: YES** **2000, 2001, 2002** **Models:** Amigo, Axiom, Rodeo, Rodeo Sport **Engines:** 2.2L VIN D, 3.2L VIN W, 3.5L VIN X **Transmissions:** A/T	**Shift Solenoid 'B' Performance (Stuck On)** DTC P0705, P0706, P0722, P0723, P0742, P0753, P0758, P1860 and P1870 not set, engine started, then driven in D4 with the output speed over 375 rpm, TFT sensor signal from 68-266°F, and with 1st Gear commanded "on" for 1 second, the PCM detected: 40 less than engine torque less than 400 Nm, speed ratio more than 0.6, transmission output speed more than 375 rpm, throttle angle more than 10%, 1.44 less than modeled ratio less than 2.4.49 for 41 seconds; or with 4th Gear commanded "on" for 1 second, the PCM detected: 15 less than engine torque less than 400 Nm, throttle angle more than 10%, -8000 less than TCC slip less than 8000 rpm, speed ratio more than 0.6, 2.75 less than modeled ratio less than 3.2 for 2 seconds during the CCM Rationality test. **Note: This fault must occur twice in one trip to set this trouble code.** **Possible Causes:** • Shift solenoid 'A' is damaged or has failed mechanically (on) • Other internal transmission concerns can cause this fault
DTC: P0758 **2T CCM, MIL: YES** **2000, 2001, 2002** **Models:** Amigo, Rodeo, Rodeo Sport **Engines:** 2.2L VIN D **Transmissions:** A/T	**Shift Solenoid 'B' Circuit Malfunction** Engine started, engine running, then with Shift Solenoid 'B' (SSB) commanded "on", and PCM detected the solenoid control signal was 12v, or with SSB commanded "off", the solenoid control signal was near 0v, either condition met for 0.84 to 1.0 seconds during the test. **Possible Causes:** • SSB control circuit is open or shorted to ground • SSB control circuit is shorted to system power (B+) • SSB power circuit (from the PCM) is open • SSB is damaged or has failed • PCM has failed
DTC: P0758 **2T CCM, MIL: YES** **1998, 1999, 2000, 2001, 2002** **Models:** Amigo, Rodeo, Rodeo Sport **Engines:** 3.2L VIN W **Transmissions:** A/T	**Shift Solenoid 'B' Circuit Malfunction** Engine started, engine running, then with Shift Solenoid 'B' (SSB) commanded "on", and PCM detected the solenoid control signal was 12v, or with SSB commanded "off", the solenoid control signal was near 0v, either condition met for 0.84 to 1.0 seconds during the test. **Possible Causes:** • SSB control circuit is open or shorted to ground • SSB control circuit is shorted to system power (B+) • SSB power circuit (from the PCM) is open • SSB is damaged or has failed • PCM has failed
DTC: P0758 **1T CCM, MIL: NO** **1997, 1998, 1999, 2000** **Models:** Hombre **Engines:** All **Transmissions:** A/T	**2-3 Shift Solenoid Circuit Malfunction** Engine started, engine speed over 450 rpm for 5 seconds, not in Fuel Cutoff mode, then with the 2-3 Shift Solenoid commanded "on", it detected a "high voltage condition (12v) on the solenoid control circuit, or with the 2-3 Shift Solenoid commanded "off", it detected a "low" voltage condition (0v) during the CCM Rationality test. Possible Causes: • 2-3 shift solenoid control circuit is open or shorted to ground • 2-3 shift solenoid control circuit is shorted to power • 2-3 shift solenoid power circuit is open (check the fuse) • 2-3 shift solenoid is damaged or has failed • PCM has failed
DTC: P0758 **1T CCM, MIL: NO** **1996, 1997, 1998, 1999, 2000, 2001, 2002** **Models:** Axiom, Trooper, VehiCROSS **Engines:** All **Transmissions:** A/T	**Shift Solenoid 'B' Circuit Malfunction** Engine started, engine running, then with Shift Solenoid 'B' (SSB) commanded "on", and PCM detected the solenoid control signal was 12v, or with SSB commanded "off", the solenoid control signal was near 0v, either condition met for 0.84 to 1.0 seconds during the test. **Possible Causes:** • SSB control circuit is open or shorted to ground • SSB control circuit is shorted to system power (B+) • SSB power circuit (from the PCM) is open • SSB is damaged or has failed • PCM has failed
DTC: P0758 **1T CCM, MIL: YES** **1996, 1997, 1998, 1999** **Models:** Oasis **Engines:** All **Transmissions:** A/T	**TCM A/T Lockup Solenoid 'B' Circuit Malfunction** Engine running with VSS inputs, and the PCM detected an unexpected voltage condition on the Solenoid Valve 'B' circuit. **Possible Causes:** • TCM Solenoid 'B' control circuit is open or shorted to ground • TCM Solenoid 'B' control circuit is shorted to system power • TCM Solenoid 'B' is damaged or has failed • TCM or PCM has failed

DTC	Trouble Code Title, Conditions & Possible Causes
DTC: P0763 **1T CCM, MIL: YES** **1996, 1997, 1998, 1999** Models: Oasis **Engines: All** **Transmissions:** A/T	**TCM A/T Control Unit or Related Circuit Malfunction** Engine running with VSS inputs received, and the PCM detected a fault in the TCM A/T Control Unit or one of its related circuits. **Possible Causes:** • Refer to the repair instructions in a transmission repair manual or the information in other electronic media to repair this code.
DTC: P0780 **1T CCM, MIL: YES** **1996, 1997, 1998, 1999** **Models: Oasis** **Engines:** All **Transmissions:** A/T	**Automatic Transaxle Malfunction** Engine running and the PCM detected an Automatic Transaxle fault. **Note: This trouble code (P0780) sets with along with several TCM related trouble codes.** **Possible Causes:** • Refer to the repair instructions in a transmission repair manual or the information in other electronic media to repair this code.
DTC: P0785 **1T CCM, MIL: NO** **2000** **Models:** Hombre **Engines:** All **Transmissions:** A/T	**3-2 Shift Solenoid Circuit Malfunction** Engine started, engine speed over 450 rpm for 5 seconds, not in Fuel Cutoff mode, then with the 3-2 Shift Solenoid commanded "on", it detected a "high voltage condition (12v) on the solenoid control circuit, or with the solenoid commanded "off", it detected a "low" voltage condition (0v) during the CCM Rationality test. **Possible Causes:** • 3-2 shift solenoid control circuit is open or shorted to ground • 3-2 shift solenoid control circuit is shorted to power • 3-2 shift solenoid power circuit is open (check the fuse) • 3-2 shift solenoid is damaged or has failed • PCM has failed

Gas Engine OBD II Trouble Code List (P1xxx Codes)

DTC	Trouble Code Title
DTC: P1106 **1T CCM, MIL: NO** **1998, 1999, 2000, 2001, 2002** **Models:** Amigo, Rodeo, Rodeo Sport **Engines:** 2.2L VIN D **Transmissions:** A/T, M/T	**MAP Sensor Circuit High Input (Intermittent)** DTC P0121, P0122 and P0123 not set, engine speed less than 1000 rpm with the throttle angle at 2.7% or less, or with the engine speed over 1000 rpm with the throttle angle at 10% or less, the PCM detected the MAP sensor signal was interrupted (e.g., the MAP value indicated over 90 kPa for 5 seconds over a 16 second period). **Note: For additional help with this code, view the Failure Records.** **Possible Causes:** • MAP sensor signal circuit is open (an intermittent fault) • MAP sensor ground circuit is open (an intermittent fault) • MAP sensor signal circuit is shorted to VREF or system power • MAP sensor is damaged or has failed
DTC: P1106 **1T CCM, MIL: NO** **1998, 1999, 2000** **Models:** Hombre **Engines:** 4.3L VIN X **Transmissions:** A/T, M/T	**MAP Sensor Circuit High Input (Intermittent)** DTC P0121, P0122 and P0123 not set, engine speed less than 1200 rpm with the throttle angle at 4% or less, or with the engine speed over 1200 rpm with the throttle angle at 20% or less, and the PCM detected the MAP sensor signal was interrupted (e.g., the MAP sensor indicated over 4.4v for 5 seconds over a 16 second period). **Note: For additional help with this code, view the Failure Records.** **Possible Causes:** • MAP sensor signal circuit is open (an intermittent fault) • MAP sensor ground circuit is open (an intermittent fault) • MAP sensor signal circuit is shorted to VREF or system power • MAP sensor is damaged or has failed
DTC: P1106 **1T CCM, MIL: YES** **1996, 1997, 1998, 1999** **Models:** Oasis **Engines:** All **Transmissions:** A/T, M/T	**BARO Pressure Sensor Performance** Engine running in 4th gear, then accelerated to WOT, and the PCM detected the BARO sensor input did not change sufficiently within a specified period of time. **Possible Causes:** • BARO sensor signal circuit is open or shorted to ground • BARO sensor ground circuit has high resistance • BARO sensor is damaged or it may be out of calibration • PCM has failed
DTC: P1107 **1T CCM, MIL: YES** **1996, 1997, 1998, 1999** **Models:** Oasis **Engines:** All **Transmissions:** A/T, M/T	**BARO Pressure Sensor Circuit Low Input** Key on or engine running, and the PCM detected the BARO sensor signal was less than a value in stored in backup memory. **Possible Causes:** • BARO sensor signal circuit is shorted to signal ground • BARO sensor signal circuit is shorted to chassis ground • BARO sensor is damaged (it may be shorted internally) • BARO sensor signal circuit to the TCM is open or grounded • TCM or the PCM has failed
DTC: P1108 **1T CCM, MIL: YES** **1996, 1997, 1998, 1999** **Models:** Oasis **Engines:** All **Transmissions:** A/T, M/T	**BARO Pressure Sensor Circuit High Input** Key on or engine running, and the PCM detected the BARO sensor signal was more than a value in stored in backup memory. **Possible Causes:** • BARO sensor signal circuit shorted to VREF • BARO sensor signal circuit is shorted to system power (B+) • BARO sensor is damaged (it may be open internally) • BARO sensor signal circuit to the TCM is shorted to power • TCM or the PCM has failed
DTC: P1106 **1T CCM, MIL: YES** **1996, 1997, 1998, 1999** **Models:** Oasis **Engines:** All **Transmissions:** A/T, M/T	**BARO Pressure Sensor Performance** Engine running in 4th gear, then accelerated to WOT, and the PCM detected the BARO sensor input did not change sufficiently within a specified period of time. **Possible Causes:** • BARO sensor signal circuit is open or shorted to ground • BARO sensor ground circuit has high resistance • BARO sensor is damaged or it may be out of calibration • PCM has failed

DTC	Trouble Code Title, Conditions & Possible Causes
DTC: P1106 **1T CCM, MIL: NO** **1996, 1997** **Models:** Rodeo **Engines:** 2.6L VIN E **Transmissions:** A/T, M/T	**MAP Sensor Circuit High Input (Intermittent)** DTC P0121, P0122 and P0123 not set, engine speed less than 1000 rpm with the throttle angle less than 3%, or with the engine speed over 1000 rpm with the throttle angle less than 10%, and the PCM detected the MAP sensor circuit was interrupted (e.g., the MAP input indicated over 80 kPa for 5 seconds over a 16 second period). **Note: For additional help with this code, view the Failure Records.** **Possible Causes:** • MAP sensor signal circuit is open (an intermittent fault) • MAP sensor ground circuit is open (an intermittent fault) • MAP sensor signal circuit is shorted to VREF or system power • MAP sensor is damaged or has failed
DTC: P1106 **1T CCM, MIL: NO** **1996, 1997, 1998, 1999, 2000, 2001, 2002** **Models:** Amigo, Axiom, Rodeo, Rodeo Sport, Trooper, VehiCROSS **Engines:** 3.2L VIN V, 3.2L VIN W, 3.5L VIN X **Transmissions:** A/T, M/T	**MAP Sensor Circuit High Input (Intermittent)** DTC P0121, P0122 and P0123 not set, engine runtime 10 seconds, engine speed less than 1000 rpm with throttle angle less than 3%, or the engine speed is more than 1000 rpm with the throttle angle less than 10%, and the PCM detected an unexpected high value (over 80 kPa) on the MAP sensor circuit for 5 seconds of a 16 second period. **Note: For additional help with this code, view the Failure Records.** **Possible Causes:** • MAP sensor signal circuit is open (an intermittent fault) • MAP sensor ground circuit is open (an intermittent fault) • MAP sensor signal circuit is shorted to VREF or system power • MAP sensor is damaged or has failed
DTC: P1107 **1T CCM, MIL: NO** **1998, 1999, 2000, 2001, 2002** **Models**: Amigo, Rodeo, Rodeo Sport **Engines:** 2.2L VIN D **Transmissions:** A/T, M/T	**MAP Sensor Circuit Low Input (Intermittent)** DTC P0121, P0122 and P0123 not set, engine speed less than 1300 rpm with the throttle angle at 0%, or with the engine speed over 1300 rpm with the throttle angle at 5% or more, the PCM detected the MAP sensor signal was interrupted (e.g., the MAP value was less than 0.04v (11 kPa) for 5 seconds over a 16 second period). **Note: For additional help with this code, view the Failure Records.** **Possible Causes:** • MAP sensor signal circuit shorted to ground (intermittent fault) • MAP VREF circuit open or shorted to ground (intermittent fault) • MAP sensor is damaged or has failed
DTC: P1107 **1T CCM, MIL: YES** **1998, 1999, 2000** **Models:** Hombre **Engines:** 4.3L VIN X **Transmissions:** A/T, M/T	**MAP Sensor Circuit Low Input (Intermittent)** DTC P0121, P0122 and P0123 not set, engine started, engine speed less than 800 rpm with the throttle angle at 0%, or with the engine speed over 800 rpm and the throttle angle at 12.5% or more, the PCM detected a sudden low voltage condition (less than 0.04v) on the MAP sensor circuit for 5 seconds over a 16 second period. **Note: For additional help with this code, view the Failure Records.** **Possible Causes:** • MAP sensor signal circuit shorted to ground (intermittent fault) • MAP VREF circuit open or shorted to ground (intermittent fault) • MAP sensor is damaged or has failed
DTC: P1107 **1T CCM, MIL: NO** **1996, 1997** **Models:** Rodeo **Engines:** 2.6L VIN E **Transmissions:** A/T, M/T	**MAP Sensor Circuit Low Input (Intermittent)** DTC P0121, P0122 and P0123 not set, engine started, engine speed less than 1000 rpm with the throttle angle less than 3%, or with the engine speed over 1000 rpm and the throttle angle over 2%, the PCM detected a sudden low voltage condition on the MAP sensor circuit on the MAP sensor circuit (i.e., the MAP sensor signal was less 11 kPa for 5 seconds over a 16 second period). **Note: For additional help with this code, view the Failure Records.** **Possible Causes:** • MAP sensor signal circuit shorted to ground (intermittent fault) • MAP VREF circuit open or shorted to ground (intermittent fault) • MAP sensor is damaged or has failed
DTC: P1107 **1T CCM, MIL: NO** **1996, 1997, 1998, 1999, 2000, 2001, 2002** **Models:** Amigo, Axiom, Rodeo, Rodeo Sport, Trooper, VehiCROSS **Engines:** 3.2L VIN V, 3.2L VIN W, 3.5L VIN X **Transmissions:** A/T, M/T	**MAP Sensor Circuit Low Input (Intermittent)** DTC P0121, P0122, P0123 not set, engine running, engine speed less than 1000 rpm and throttle angle over 1%, or the engine speed more than 1000 rpm and throttle angle more than 2%, and the PCM detected an unexpected low value (below 11 kPa) on the MAP sensor circuit for 5 seconds of a 16 second period. **Note: For additional help with this code, view the Failure Records.** **Possible Causes:** • MAP sensor signal circuit shorted to ground (intermittent fault) • MAP VREF circuit open or shorted to ground (intermittent fault) • MAP sensor is damaged or has failed

DTC	Trouble Code Title, Conditions & Possible Causes
DTC: P1107 **1T CCM, MIL: YES** **1996, 1997, 1998, 1999** **Models:** Oasis **Engines:** All **Transmissions:** A/T, M/T	**BARO Pressure Sensor Circuit Low Input** Key on or engine running, and the PCM detected the BARO sensor signal was less than a value in stored in backup memory. **Possible Causes:** • BARO sensor signal circuit is shorted to signal ground • BARO sensor signal circuit is shorted to chassis ground • BARO sensor is damaged (it may be shorted internally) • BARO sensor signal circuit to the TCM is open or grounded • TCM or the PCM has failed
DTC: P1108 **1T CCM, MIL: YES** **1996, 1997, 1998, 1999** **Models:** Oasis **Engines:** All **Transmissions:** A/T, M/T	**BARO Pressure Sensor Circuit High Input** Key on or engine running, and the PCM detected the BARO sensor signal was more than a value in stored in backup memory. **Possible Causes:** • BARO sensor signal circuit shorted to VREF • BARO sensor signal circuit is shorted to system power (B+) • BARO sensor is damaged (it may be open internally) • BARO sensor signal circuit to the TCM is shorted to power • TCM or the PCM has failed
DTC: P1111 **1T CCM, MIL: NO** **1998, 1999, 2000, 2001, 2002** **Models:** Amigo, Rodeo, Rodeo Sport **Engines:** 2.2L VIN D **Transmissions:** A/T, M/T	**IAT Sensor Circuit High Input (Intermittent)** Engine started, engine runtime over 4 minutes, ECT sensor more than 140ºF, vehicle driven to a speed of under 20 mph, MAF sensor less than 20 g/sec, and the PCM detected an unexpected high voltage signal of over 4.94v (Scan Tool reads -38ºF) on the IAT sensor circuit for 2.5 seconds over a 25 second period. **Note: For additional help with this code, view the Failure Records.** **Possible Causes:** • IAT sensor signal circuit is open (an intermittent fault) • IAT sensor is damaged or has failed (an intermittent fault) • PCM has failed
DTC: P1111 **1T CCM, MIL: NO** **1997, 1998, 1999, 2000** **Models:** Hombre **Engines:** 4.3L VIN X **Transmissions:** A/T, M/T	**IAT Sensor Circuit High Input (Intermittent)** DTC P0101, P0102, P0103, P0117, P0118, P0121, P0122 and P0123 not set, engine started, engine runtime over 100 seconds, ECT sensor over 184ºF, VSS less than 2 mph, MAF sensor below 250 g/sec, and the PCM detected an unexpected high signal of over 4.94v [Scan Tool reads -38ºF] on the IAT sensor circuit for 1 second. **Note: For additional help with this code, view the Failure Records.** **Possible Causes:** • IAT sensor signal circuit is open (an intermittent fault) • IAT sensor is damaged or has failed (an intermittent fault) • PCM has failed
DTC: P1111 **1T CCM, MIL: NO** **1996, 1997** **Models:** Rodeo **Engines:** 2.6L VIN E **Transmissions:** A/T, M/T	**IAT Sensor Circuit High Input (Intermittent)** Engine started, engine runtime over 4 minutes, ECT sensor more than 140ºF, vehicle driven to a speed of over 20 mph, Calculated airflow less than 20 g/sec, and the PCM detected an unexpected high voltage condition (over 4.90v) on the IAT sensor circuit (Scan Tool reads -38ºF) for 2.5 seconds over a 25 second period. **Note: For additional help with this code, view the Failure Records.** **Possible Causes:** • IAT sensor signal circuit is open (an intermittent fault) • IAT sensor is damaged or has failed (an intermittent fault) • PCM has failed
DTC: P1111 **1T CCM, MIL: NO** **1996, 1997, 1998, 1999, 2000, 2001, 2002** **Models:** Amigo, Axiom, Rodeo, Rodeo Sport, Trooper, VehiCROSS **Engines:** 3.2L VIN V, 3.2L VIN W, 3.5L VIN X **Transmissions:** A/T, M/T	**IAT Sensor Circuit High Input (Intermittent)** Engine started, engine runtime over 4 minutes, ECT sensor more than 140ºF, vehicle speed under 20 mph, MAF sensor less than 20 g/sec, and the PCM detected an unexpected high signal of over 4.90v (Scan Tool reads -38ºF) on the IAT sensor circuit for 2.5 seconds over a 25 second period during the CCM test. **Note: For additional help with this code, view the Failure Records.** **Possible Causes:** • IAT sensor signal circuit is open (an intermittent fault) • IAT sensor is damaged or has failed (an intermittent fault) • PCM has failed

DTC	Trouble Code Title, Conditions & Possible Causes
DTC: P1112 **1T CCM, MIL: NO** **1998, 1999, 2000, 2001, 2002** **Models:** Amigo, Rodeo, Rodeo Sport **Engines:** 2.2L VIN D **Transmissions:** A/T, M/T	**IAT Sensor Circuit Low Input (Intermittent)** Engine started, engine runtime over 2 minutes, ECT sensor more than 140ºF, vehicle driven to a speed of more than 30 mph, and the PCM detected an unexpected low voltage of less than 0.10v (Scan Tool reads 298ºF) on the IAT sensor signal circuit for 2.5 seconds over a 25 second period during the CCM test. **Note: For additional help with this code, view the Failure Records.** **Possible Causes:** • IAT sensor signal circuit is shorted to ground (intermittent fault) • IAT sensor is damaged or has failed (an intermittent fault) • PCM has failed
DTC: P1112 **1T CCM, MIL: NO** **1997, 1998, 1999, 2000** **Models:** Hombre **Engines:** 4.3L VIN X **Transmissions:** A/T, M/T	**IAT Sensor Circuit Low Input (Intermittent)** DTC P0101, P0102, P0103, P0117, P0118, P0121, P0122 and P0123 not set, engine started, engine runtime over 100 seconds, ECT sensor signal over 184ºF, VSS less than 2 mph, MAF sensor signal less than 250 g/sec, and the PCM detected an unexpected low voltage of less than 0.10v (Scan Tool reads 298ºF) on the IAT sensor circuit for 1 second during the CCM test. **Note: For additional help with this code, view the Failure Records.** **Possible Causes:** • IAT sensor signal circuit is shorted to ground (intermittent fault) • IAT sensor is damaged or has failed (an intermittent fault) • PCM has failed
DTC: P1112 **1T CCM, MIL: NO** **1996, 1997** **Models:** Rodeo **Engines:** 2.6L VIN E **Transmissions:** A/T, M/T	**IAT Sensor Circuit Intermittent Low Input Conditions:** Engine started, engine runtime over 2 minutes, ECT sensor more than 140ºF, vehicle speed more than 30 mph, MAF sensor more than 20 g/sec, and the PCM detected an unexpected low signal of under 0.10v [Scan Tool reads 298ºF]) on the IAT sensor circuit for 2.5 seconds over a 25 second time period. **Note: For additional help with this code, view the Failure Records.** **Possible Causes:** • IAT sensor signal circuit is shorted to ground (intermittent fault) • IAT sensor is damaged or has failed (an intermittent fault) • PCM has failed
DTC: P1112 **1T CCM, MIL: NO** **1996, 1997, 1998, 1999, 2000, 2001, 2002** **Models:** Amigo, Axiom, Rodeo, Rodeo Sport, Trooper, VehiCROSS **Engines:** 3.2L VIN V, 3.2L VIN W, 3.5L VIN X **Transmissions:** A/T, M/T	**IAT Sensor Circuit Low Input (Intermittent)** Engine started, engine runtime over 4 minutes, ECT sensor more than 140ºF, vehicle driven to a speed of over 20 mph, Calculated airflow less than 20 g/sec, and the PCM detected an unexpected low voltage condition of less than 0.10v (Scan Tool reads 298ºF) on the IAT sensor circuit for 2.5 seconds over a 25 second period. **Note: For additional help with this code, view the Failure Records.** **Possible Causes:** • IAT sensor signal circuit is shorted to ground (intermittent fault) • IAT sensor is damaged or has failed (an intermittent fault) • PCM has failed
DTC: P1114 **1T CCM, MIL: NO** **1998, 1999, 2000, 2001, 2002** **Models:** Amigo, Rodeo, Rodeo Sport **Engines:** 2.2L VIN D **Transmissions:** A/T, M/T	**ECT Sensor Circuit Low Input (Intermittent)** Engine started, engine runtime over 2 minutes, and the PCM detected an unexpected low voltage condition of less than 0.10v (Scan Tool reads 302ºF) on the ECT sensor circuit for 10 seconds of a 100 second period. **Note: For additional help with this code, view the Failure Records.** **Possible Causes:** • ECT sensor signal circuit shorted to ground (intermittent fault) • ECT sensor is damaged or has failed • PCM has failed
DTC: P1114 **1T CCM, MIL: NO** **1997, 1998, 1999, 2000** **Models:** Hombre **Engines:** 4.3L VIN X **Transmissions:** A/T, M/T	**ECT Sensor Circuit Low Input (Intermittent)** Engine started, engine runtime 5 seconds, and the PCM detected an unexpected low voltage condition (under 0.10v) on the ECT sensor circuit (Scan Tool reads 302ºF) for 1 second during the CCM test. **Note: For additional help with this code, view the Failure Records.** **Possible Causes:** • ECT sensor signal circuit shorted to ground (intermittent fault) • ECT sensor is damaged or has failed • PCM has failed
DTC: P1114 **1T CCM, MIL: NO** **1996, 1997** **Models:** Rodeo **Engines:** 2.6L VIN E **Transmissions:** A/T, M/T	**ECT Sensor Circuit Low Input (Intermittent)** Engine started, engine runtime more than 1 minute, and the PCM detected an unexpected low voltage condition of under 0.10v (Scan Tool reads 302ºF) on the ECT sensor circuit for 10 seconds of a 100 second period. **Note: For additional help with this code, view the Failure Records.** **Possible Causes:** • ECT sensor signal circuit shorted to ground (intermittent fault) • ECT sensor is damaged or has failed • PCM has failed

DTC	Trouble Code Title, Conditions & Possible Causes
DTC: P1114 **1T CCM, MIL: NO** **1996, 1997, 1998, 1999, 2000, 2001, 2002** **Models:** Amigo, Axiom, Rodeo, Rodeo Sport, Trooper, VehiCROSS **Engines:** 3.2L VIN V, 3.2L VIN W, 3.5L VIN X **Transmissions:** A/T, M/T	**ECT Sensor Circuit Low Input (Intermittent)** Engine started, engine runtime more than 1 minute, and the PCM detected an unexpected low voltage condition of less than 0.10v (Scan Tool read 302ºF) on the ECT sensor circuit for 10 seconds of a 100 second period. **Note: For additional help with this code, view the Failure Records.** **Possible Causes:** • ECT sensor signal circuit shorted to ground (intermittent fault) • ECT sensor is damaged or has failed • PCM has failed
DTC: P1115 **1T CCM, MIL: NO** **1998, 1999, 2000, 2001, 2002** **Models:** Amigo, Rodeo, Rodeo Sport **Engines:** 2.2L VIN D **Transmissions:** A/T, M/T	**ECT Sensor Circuit High Input (Intermittent)** Engine started, engine runtime more than 3 minutes, and the PCM detected an unexpected high voltage condition of more than 4.94v (Scan Tool reads -38ºF) on the ECT sensor for 10 seconds of a 100 second period during the CCM test. **Note: For additional help with this code, view the Failure Records.** **Possible Causes:** • ECT sensor signal circuit is open (an intermittent fault) • ECT sensor is damaged or has failed (an intermittent fault) • PCM has failed
DTC: P1115 **1T CCM, MIL: NO** **1997, 1998, 1999, 2000** **Models**: Hombre **Engines**: 4.3L VIN X **Transmissions:** A/T, M/T	**ECT Sensor Circuit High Input (Intermittent)** Engine started, engine runtime more than 5 seconds, and the PCM detected an unexpected high voltage condition of more than 4.90v (Scan Tool reads -38ºF) on the ECT sensor circuit for one second during the CCM test. **Note: For additional help with this code, view the Failure Records.** **Possible Causes:** • ECT sensor signal circuit is open (an intermittent fault) • ECT sensor is damaged or has failed (an intermittent fault) • PCM has failed
DTC: P1115 **1T CCM, MIL: NO** **1996, 1997** **Models**: Rodeo **Engines:** 2.6L VIN E **Transmissions:** A/T, M/T	**ECT Sensor Circuit High Input (Intermittent)** Engine started, engine running for 1 minute, and the PCM detected an unexpected high voltage of over 4.90v (Scan Tool reads -38ºF) on the ECT sensor circuit for 10 seconds of a 100 second period. **Note: For additional help with this code, view the Failure Records.** **Possible Causes:** • ECT sensor signal circuit is open (an intermittent fault) • ECT sensor is damaged or has failed (an intermittent fault) • PCM has failed
DTC: P1115 **1T CCM, MIL: NO** **1996, 1997, 1998, 1999, 2000, 2001, 2002** **Models:** Amigo, Axiom, Rodeo, Rodeo Sport, Trooper, VehiCROSS **Engines:** 3.2L VIN V, 3.2L VIN W, 3.5L VIN X **Transmissions:** A/T, M/T	**ECT Sensor Circuit High Input (Intermittent)** Engine started, engine running for 1 minute, and the PCM detected an unexpected high voltage of over 4.90v (Scan Tool reads -38ºF)] on the ECT sensor circuit for 10 seconds of a 100 second period. **Note: For additional help with this code, view the Failure Records.** **Possible Causes:** • ECT sensor signal circuit is open (an intermittent fault) • ECT sensor is damaged or has failed (an intermittent fault) • PCM has failed
DTC: P1120 **2T CCM, MIL: YES** **2000, 2001, 2002** **Models:** Amigo, Rodeo, Rodeo Sport **Engines**: 3.2L VIN W **Transmissions:** A/T, M/T	**Throttle Position Sensor 1 Circuit Malfunction** Key on or engine running, and the PCM detected the TP Sensor 1 (TP1) circuit indicated less than 2.5%, or more than 97.5% for 93.6 ms or for 18 tests out of 500 test samples during the CCM test. **Possible Causes:** • TP1 sensor signal circuit is open between the sensor and PCM • TP1 sensor signal circuit is grounded between sensor and PCM • TP1 sensor VREF circuit is open between sensor and the PCM • TP1 sensor is damaged or has failed • TSB 01-04-S001 (7/01) contain information related to this code
DTC: P1120 **2T CCM, MIL: YES** **2000, 2001, 2002** **Models**: Axiom, Trooper **Engines**: 3.5L VIN X **Transmissions:** A/T, M/T	**Throttle Position Sensor 1 Circuit Malfunction** Key on or engine running, and the PCM detected the TP Sensor 1 (TP1) circuit indicated less than 2.5% or more than 97.5% of the TP sensor 5v VREF value for 93.6 ms during the CCM Rationality test. **Possible Causes:** • TP1 sensor signal circuit is open between the sensor and PCM • TP1 sensor signal circuit is grounded between sensor and PCM • TP1 sensor VREF circuit is open between sensor and the PCM • TP1 sensor is damaged or has failed • TSB 01-04-S001 (7/01) contain information related to this code

DTC	Trouble Code Title, Conditions & Possible Causes
DTC: P1121 **1T CCM, MIL: NO** **1998, 1999, 2000, 2001, 2002** **Models:** Amigo, Rodeo, Rodeo Sport **Engines:** 2.2L VIN D **Transmissions:** A/T, M/T	**Throttle Position Sensor Intermittent High Input** Engine started, engine running for 5 seconds, and the PCM detected an unexpected high signal of over 4.88v on the TP sensor circuit for 150 ms of a 1.5 second period (check the Failure records first!). **Possible Causes:** • TP sensor signal circuit is open between the sensor and PCM • TP sensor ground circuit is open between the sensor and PCM • TP sensor signal circuit is shorted to VREF or system power • TP sensor is damaged or has failed
DTC: P1121 **1T CCM, MIL: NO** **1996, 1997, 1998, 1999** **Models:** Amigo, Rodeo, Rodeo Sport **Engines:** 3.2L VIN V, 3.2L VIN W **Transmissions:** A/T, M/T	**Throttle Position Sensor Intermittent High Input** Engine started, engine running for 5 seconds, and the PCM detected the TP sensor was over 4.90v for 0.15 seconds of a 1.5 second period during the CCM test (check the Failure Records first!). **Possible Causes:** • TP sensor signal circuit is open between the sensor and PCM • TP sensor ground circuit is open between the sensor and PCM • TP sensor signal circuit is shorted to VREF or system power • TP sensor is damaged or has failed
DTC: P1121 **1T CCM, MIL: NO** **1999, 2000, 2001, 2002** **Models:** VehiCROSS **Engines:** 3.5L VIN X **Transmissions:** A/T, M/T	**Throttle Position Sensor Intermittent High Input** Engine running, and the PCM detected an unexpected high signal of over 4.90v on the TP sensor circuit for 150 ms over a 1.5 second period during the CCM test (check the Failure Records first!). **Possible Causes:** • TP sensor signal circuit is open between the sensor and PCM • TP sensor ground circuit is open between the sensor and PCM • TP sensor signal circuit is shorted to VREF or system power • TP sensor is damaged or has failed
DTC: P1121 **1T CCM, MIL: NO** **1997, 1998, 1999, 2000** **Models**: Hombre **Engines:** 4.3L VIN X **Transmissions:** A/T, M/T	**Throttle Position Sensor Intermittent High Input** Engine running, and the PCM detected an unexpected high signal of over 4.70v on the TP sensor circuit for one second in the CCM test. **Note: For additional help with this code, view the Failure Records.** **Possible Causes:** • TP sensor signal circuit is open between the sensor and PCM • TP sensor ground circuit is open between the sensor and PCM • TP sensor signal circuit is shorted to VREF or system power • TP sensor is damaged or has failed
DTC: P1121 **1T CCM, MIL: YES** **1996, 1997, 1998, 1999** **Models:** Oasis **Engines:** All **Transmissions:** A/T, M/T	**TP Sensor Input Lower Than Expected** Engine running, and the PCM detected the TP sensor input was lower than an expected value with the throttle wide open (<13.7%). **Note: This trouble code sets if this circuit fails the rationality test.** **Possible Causes:** • Throttle plate is dirty, clogged, or it is binding • TP sensor circuit open or shorted to ground between the PCM and the TCM • TP sensor is damaged or has failed • PCM has failed
DTC: P1121 **1T CCM, MIL: NO** **1996, 1997** **Models:** Rodeo **Engines:** 2.6L VIN E **Transmissions:** A/T, M/T	**Throttle Position Sensor Intermittent High Input** Engine running, and the PCM detected an unexpected high signal of over 4.90v on the TP sensor circuit for 0.15 seconds over a 1.5 second period during the CCM test. **Note: For additional help with this code, view the Failure Records.** **Possible Causes:** • TP sensor signal circuit is open between the sensor and PCM • TP sensor ground circuit is open between the sensor and PCM • TP sensor signal circuit is shorted to VREF or system power • TP sensor is damaged or has failed
DTC: P1122 **1T CCM, MIL: NO** **1998, 1999, 2000, 2001, 2002** **Models:** Amigo, Rodeo, Rodeo Sport **Engines:** 2.2L VIN D **Transmissions:** A/T, M/T	**Throttle Position Sensor Intermittent Low Input** Engine running, and the PCM detected an unexpected low signal of less than 0.10v on the TP sensor circuit for 150 ms over a 1.5 second time period during the CCM test. **Note: For additional help with this code, view the Failure Records.** **Possible Causes:** • TP sensor signal circuit is shorted to ground • TP sensor VREF circuit shorted to ground (test other sensors) • TP sensor is damaged or has failed

DTC	Trouble Code Title, Conditions & Possible Causes
DTC: P1122 **1T CCM, MIL: NO** **1996, 1997, 1998, 1999** **Models:** Amigo, Rodeo, Rodeo Sport **Engines:** 3.2L VIN V, 3.2L VIN W **Transmissions:** A/T, M/T	**Throttle Position Sensor Intermittent Low Input** Engine running, and the PCM detected an unexpected low signal of less than 0.22v on the TP sensor circuit for 0.15 seconds over a 1.5 second period during the CCM test. **Note: For additional help with this code, view the Failure Records.** **Possible Causes:** • TP sensor signal circuit is shorted to ground • TP sensor VREF circuit shorted to ground (test other sensors) • TP sensor is damaged or has failed
DTC: P1122 **1T CCM, MIL: NO** **1999, 2000, 2001, 2002** **Models**: VehiCROSS **Engines:** 3.5L VIN X **Transmissions:** A/T, M/T	**Throttle Position Sensor Intermittent Low Input** Engine running, and the PCM detected an unexpected low signal of less than 0.10v on the TP sensor circuit for 150 ms over a 1.5 second time period during the CCM test. **Note: For additional help with this code, view the Failure Records.** **Possible Causes:** • TP sensor signal circuit is shorted to ground • TP sensor VREF circuit shorted to ground (test other sensors) • TP sensor is damaged or has failed
DTC: P1122 **1T CCM, MIL: NO** **1997, 1998, 1999, 2000** **Models:** Hombre **Engines:** 4.3L VIN X **Transmissions:** A/T, M/T	**Throttle Position Sensor Intermittent Low Input** Engine running, and the PCM detected an unexpected low signal of less than 0.25v on the TP sensor circuit for over one second during the CCM test. **Note: For additional help with this code, view the Failure Records.** **Possible Causes:** • TP sensor signal circuit is shorted to ground • TP sensor VREF circuit shorted to ground (test other sensors) • TP sensor is damaged or has failed
DTC: P1122 **1T CCM, MIL: YES** **1996, 1997, 1998, 1999** **Models:** Oasis **Engines:** All **Transmissions:** A/T, M/T	**TP Sensor Input Higher Than Expected** Engine running, and the PCM detected that the TP sensor input higher than the expected value with the throttle closed (>16.9%). **Note: This trouble code sets if this circuit fails the rationality test.** **Possible Causes:** • Throttle plate is dirty, clogged, or it is binding • TP sensor signal circuit shorted to VREF or it is open between the PCM and the TCM • TP sensor is damaged or has failed • PCM has failed
DTC: P1122 **1T CCM, MIL: NO** **1996, 1997** **Models**: Rodeo **Engines:** 2.6L VIN E **Transmissions:** A/T, M/T	**Throttle Position Sensor Intermittent Low Input** Engine running, and the PCM detected an unexpected low signal of less than 0.22v on the TP sensor circuit for 0.15 seconds over a 1.5 second period during the CCM test. **Note: For additional help with this code, view the Failure Records.** **Possible Causes:** • TP sensor signal circuit is shorted to ground • TP sensor VREF circuit shorted to ground (test other sensors) • TP sensor is damaged or has failed
DTC: P1125 **1T CCM, MIL: YES** **2000, 2001, 2002** **Models:** Amigo, Rodeo, Rodeo Sport **Engines:** 3.2L VIN W **Transmissions:** A/T, M/T	**Electronic Throttle Control Limit Performance Mode** Key on or engine running, and the PCM detected the Electronic Throttle Control (ETC) system was in Limited Performance Mode. **Possible Causes:** • The ETC controller has detected the throttle angle was not within a range of 13% (idle) to 87% (wide open throttle). • Limited Performance and Multiple DTC Mode is "active" • ETC module is damaged or has failed • TSB 01-04-S001 (7/01) contain information related to this code
DTC: P1125 **1T CCM, MIL: YES** **2000, 2001, 2002** **Models:** Axiom, Trooper **Engines:** 3.5L VIN X **Transmissions:** A/T, M/T	**Electronic Throttle Control Limit Performance Mode** Key on or engine running, and the PCM detected the Electronic Throttle Control (ETC) system was in Limited Performance Mode. **Possible Causes:** • The ETC controller has detected the throttle angle was not within a range of 13% (idle) to 87% (wide open throttle). • Limited Performance and Multiple DTC Mode is "active" • ETC module is damaged or has failed • TSB 01-04-S001 (7/01) contain information related to this code

DTC	Trouble Code Title, Conditions & Possible Causes
DTC: P1128 **1T CCM, MIL: YES** **1996, 1997, 1998, 1999** **Models:** Oasis **Engines:** All **Transmissions:** A/T, M/T	**MAP Sensor Value Less Than Expected** Engine running at cruise speed, then back to idle speed, and the PCM detected a MAP sensor signal lower than the expected value. **Note: This trouble code sets if this circuit fails the rationality test.** **Possible Causes:** • MAP sensor signal circuit shorted to ground (intermittent fault) • MAP sensor vacuum line bent or plugged at intake manifold • MAP sensor is damaged or it is out-of-calibration • PCM has failed
DTC: P1129 **1T CCM, MIL: YES** **1996, 1997, 1998, 1999** **Models:** Oasis **Engines:** All **Transmissions:** A/T, M/T	**MAP Sensor Value Higher Than Expected** Engine running at cruise speed, then back to idle speed, and the PCM detected a MAP sensor signal higher than the expected value. **Note: This trouble code sets if this circuit fails the rationality test.** **Possible Causes:** • MAP sensor signal circuit shorted to VREF (intermittent fault) • MAP sensor ground circuit has high resistance • MAP sensor is damaged or it is out-of-calibration • PCM has failed
DTC: P1133 **2T OBD/O2S1, MIL: YES** **1998, 1999, 2000, 2001, 2002** **Models:** Amigo, Rodeo, Rodeo Sport **Engines:** 2.2L VIN D **Transmissions:** A/T, M/T	**HO2S-11 (Bank 1 Sensor 1) Insufficient Switching** Engine started, engine runtime over 2 minutes, engine speed from 1500-3500 rpm in closed loop for 90 seconds, ECT sensor input over 140°F, Purge duty cycle over 2%, Calculated airflow from 17-32 g/sec, conditions met for 1 second, and the PCM detected less than 12 rich-to-lean or less than 12 lean-to-rich switches during the test. **Possible Causes:** • Air leaks at the exhaust manifold or exhaust pipes • HO2S signal circuit is open or shorted to ground (intermittent) • HO2S heater power circuit is open, or the heater has failed • HO2S contaminated with wrong fuel, has deteriorated or failed
DTC: P1134 **2T OBD/O2S1, MIL: YES** **1998, 1999, 2000, 2001, 2002** **Models:** Amigo, Rodeo, Rodeo Sport **Engines:** 2.2L VIN D **Transmissions:** A/T, M/T	**HO2S-11 (Bank 1 Sensor 1) Transition Time Ratio Error** Engine started, engine runtime over 2 minutes, engine speed from 1500-3500 rpm in closed loop for 90 seconds, ECT sensor signal over 140°F, Purge duty cycle over 2%, Calculated airflow from 17-32 g/sec, conditions met for 1 second, and the PCM detected the transition time ratio between lean-to-rich and rich-to-lean was less than 0.4, or it was more than 3.8 in the Oxygen Sensor Monitor Test. **Possible Causes:** • Air leaks at the exhaust manifold or exhaust pipes • HO2S signal circuit is open or shorted to ground (intermittent) • HO2S heater power circuit is open, or the heater has failed • HO2S contaminated with wrong fuel, has deteriorated or failed
DTC: P1133 **2T OBD/O2S1, MIL: YES** **1996, 1997** **Models:** Rodeo **Engines:** 2.6L VIN E **Transmissions:** M/T	**HO2S-11 (Bank 1 Sensor 1) Insufficient Switching** DTC P0106, P0107, P0108, P0112, P0113, P0117, P0118, P0121, P0122, P0123, P0131, P0132, P0133, P0134, P0135, P0300, P0441 and P1441 not set, system voltage from 10-16v, engine speed from 1500-3000 rpm in closed loop for 1 minute, ECT sensor more than 122°F, Calculated airflow from 9-42 g/sec, Purge duty cycle over 2%, conditions met for 3 seconds, then 90 seconds after entering closed loop control, the PCM detected less than 18 rich-to-lean or lean-to-rich switches from the front HO2S-11 during the Oxygen Sensor Monitor test. **Possible Causes:** • Air leaks after the MAF sensor, or air leaks in the PCV system • Air leaks at the EGR gasket, or at the EGR valve diaphragm • Exhaust leaks before or near where the front HO2S is mounted • Fuel control sensor is out of calibration (i.e., ECT, IAT or MAP) • Fuel delivery system supplying too much or too little fuel during cruise or idle periods (e.g., faulty fuel pump, or dirty fuel filter) • Fuel injector (one or more) dirty, leaking or sticking • Fuel pressure regulator leaking, damaged or has failed • HO2S is contaminated, deteriorated or it has failed
DTC: P1134 **2T CCM, MIL: YES** **1996, 1997** **Models:** Rodeo **Engines:** 2.6L VIN E **Transmissions:** M/T	**HO2S-11 (Bank 1 Sensor 1) Transition Time Ratio Error** DTC P0106, P0107, P0108, P0112, P0113, P0117, P0118, P0121, P0122, P0123, P0131, P0132, P0133, P0134, P0135, P0300, P0441 and P1441 not set, system voltage from 10-16v, engine speed from 1500-3000 rpm in closed loop for 1 minute, ECT sensor more than 122°F, Calculated airflow from 9-42 g/sec, Purge duty cycle over 2%, conditions met for 3 seconds, then 90 seconds after entering closed loop control, the PCM detected the transition time ratio to switch from lean-to-rich or rich to lean from the HO2S-11 was below 0.8 or over 3.8 during the Oxygen Sensor Monitor test. **Possible Causes:** • Air leaks at the exhaust manifold or exhaust pipes • HO2S signal circuit is open or shorted to ground (intermittent) • HO2S heater power circuit is open, or the heater has failed • HO2S contaminated with wrong fuel, has deteriorated or failed

DTC	Trouble Code Title, Conditions & Possible Causes
DTC: P1133 **2T OBD/O2S1, MIL: YES** **1996, 1997, 1998, 1999, 2000, 2001, 2002** **Models:** Amigo, Axiom, Rodeo, Rodeo Sport, Trooper, VehiCROSS **Engines:** 3.2L VIN V, 3.2L VIN W, 3.5L VIN X **Transmissions:** A/T, M/T	**HO2S-11 (Bank 1 Sensor 1) Insufficient Switching** DTC P0101, P0102, P0103, P0106, P0107, P0108, P0117, P0118, P0121, P0122, P0123, P0131, P0132, P0133, P0134, P0135, P0300, P0301-P0306, P0441 and P1441 not set, engine speed from 1500-3000 rpm in closed loop for 1 minute, system voltage from 11-16v, ECT sensor more than 122°F, MAF sensor from 9-42 g/sec, Purge duty cycle over 2%, conditions met for 3 seconds, and the PCM detected less than 23 rich-to-lean or lean-to-rich switches on the HO2S-11 signal circuit. **Possible Causes:** • Air leaks after the MAF sensor, or in the EGR or PCV system • Exhaust leaks before or near where the front HO2S is mounted • Fuel control sensor is out of calibration (i.e., ECT, IAT or MAP) • Fuel delivery system supplying too much or too little fuel during cruise or idle periods (e.g., faulty fuel pump, or dirty fuel filter) • Fuel injector (one or more) dirty, leaking or sticking • Fuel pressure regulator leaking, damaged or has failed • HO2S is contaminated, deteriorated or it has failed
DTC: P1134 **2T OBD/O2S1, MIL: YES** **1996, 1997, 1998, 1999, 2000, 2001, 2002** **Models:** Amigo, Axiom, Rodeo, Rodeo Sport, Trooper, VehiCROSS **Engines:** 3.2L VIN V, 3.2L VIN W, 3.5L VIN X **Transmissions:** A/T, M/T	**HO2S-11 (Bank 1 Sensor 1) Transition Time Ratio Error** DTC P0101, P0102, P0103, P0106, P0107, P0108, P0117, P0118, P0121, P0122, P0123, P0131, P0132, P0133, P0134, P0135, P0300, P0301-P0306, P0441 and P1441 not set, engine speed from 1500-3000 rpm in closed loop for 1 minute, system voltage from 11-16v, ECT sensor more than 122-167°F, MAF sensor at 18-42 g/sec, Purge duty cycle over 2%, conditions met for 3 seconds, and the PCM detected the HO2S-11 transition ratio from lean-to-rich and rich-to-lean was less than 0.44 or more than 3.8 during the test. **Possible Causes:** • Air leaks after the MAF sensor, or in the EGR or PCV system • Exhaust leaks before or near where the front HO2S is mounted • Fuel control sensor is out of calibration (i.e., ECT, IAT or MAP) • Fuel delivery system supplying too much or too little fuel during cruise or idle periods (e.g., faulty fuel pump, or dirty fuel filter) • Fuel injector (one or more) dirty, leaking or sticking • Fuel pressure regulator leaking, damaged or has failed • HO2S is contaminated, deteriorated or it has failed
DTC: P1153 **2T CCM, MIL: YES** **1996, 1997, 1998, 1999, 2000, 2001, 2002** **Models:** Amigo, Axiom, Rodeo, Rodeo Sport, Trooper, VehiCROSS **Engines:** 3.2L VIN V, 3.2L VIN W, 3.5L VIN X **Transmissions:** A/T, M/T	**HO2S-21 (Bank 2 Sensor 1) Insufficient Switching** DTC P0101, P0102, P0103, P0106, P0107, P0108, P0117, P0118, P0121, P0122, P0123, P0131, P0132, P0133, P0134, P0135, P0300, P0301-P0306, P0441 and P1441 not set, system voltage from 11-16v, engine speed from 1500-3000 rpm in closed loop for 1 minute, ECT sensor more than 122-167°F, MAF sensor from 9-42 g/sec, Purge duty cycle over 2%, conditions met for 3 seconds, and the PCM detected less than 27 rich-to-lean or lean-to-rich switches on the HO2S-21 signal circuit in the Oxygen Sensor Monitor test. **Possible Causes:** • Air leaks after the MAF sensor, or in the EGR or PCV system • Exhaust leaks before or near where the front HO2S is mounted • Fuel control sensor is out of calibration (i.e., ECT, IAT or MAP) • Fuel delivery system supplying too much or too little fuel during cruise or idle periods (e.g., faulty fuel pump, or dirty fuel filter) • Fuel injector (one or more) dirty, leaking or sticking • Fuel pressure regulator leaking, damaged or has failed • HO2S is contaminated, deteriorated or it has failed
DTC: P1154 **2T OBD/O2S1, MIL: YES** **1996, 1997, 1998, 1999, 2000, 2001, 2002** **Models:** Amigo, Axiom, Rodeo, Rodeo Sport, Trooper, VehiCROSS **Engines:** 3.2L VIN V, 3.2L VIN W, 3.5L VIN X **Transmissions:** A/T, M/T	**HO2S-21 (Bank 2 Sensor 1) Transition Time Ratio** DTC P0101, P0102, P0103, P0106, P0107, P0108, P0117, P0118, P0121, P0122, P0123, P0131, P0132, P0133, P0134, P0135, P0300, P0301-P0306, P0441 and P1441 not set, system voltage from 10-16v, engine speed from 1500-3000 rpm in closed loop for 1 minute, ECT sensor more than 122-167°F, MAF sensor from 9-42 g/sec, Purge duty cycle over 2%, conditions met for 3 seconds, then 90 seconds after entering closed loop control, the PCM detected the transition time ratio to switch from lean-to-rich or rich to lean from the HO2S-21 was less than 0.44 or more than 3.8 during the test. **Possible Causes:** • Air leaks at the exhaust manifold or exhaust pipes • HO2S signal circuit is open or shorted to ground (intermittent) • HO2S heater power circuit is open, or the heater has failed • HO2S contaminated with wrong fuel, has deteriorated or failed

DTC	Trouble Code Title, Conditions & Possible Causes
DTC: P1133 **2T OBD/O2S1, MIL: YES** **1996, 1997, 1998, 1999, 2000** **Models:** Hombre **Engines:** 2.2L VIN 4 **Transmissions:** A/T, M/T	**HO2S-11 (Bank 1 Sensor 1) Insufficient Switching** DTC P0105, P0106, P0107, P0108, P0112, P0113, P0117, P0118, P0121, P0122, P0123, P0171, P0200, P0300, P0301-304, P0341, P0404, P0506, P0507 and P0601 not set, engine speed from 1600-2600 rpm in closed loop, throttle angle from 9-20%, Purge duty cycle over 80%, conditions met for 2 minutes, and the PCM detected less than 10 rich-to-lean switches, or less than 15 lean-to-rich switches. **Possible Causes:** • Air leaks after the MAF sensor, or in the EGR or PCV system • Exhaust leaks before or near where the front HO2S is mounted • Fuel control sensor is out of calibration (i.e., ECT, IAT or MAP) • Fuel delivery system supplying too much or too little fuel during cruise or idle periods (e.g., faulty fuel pump, or dirty fuel filter) • Fuel injector (one or more) dirty, leaking or sticking • Fuel pressure regulator leaking, damaged or has failed • HO2S is contaminated, deteriorated or it has failed
DTC: P1133 **2T OBD/O2S1, MIL: YES** **1997, 1998, 1999, 2000** **Models:** Hombre **Engines:** 4.3L VIN X **Transmissions:** A/T, M/T	**HO2S-11 (Bank 1 Sensor 1) Insufficient Switching** DTC P0101, P0102, P0103, P0106, P0107, P0108, P0112, P0113, P0117, P0118, P0121, P0122, P0123, P0131, P0132, P0133, P0134, P0135, P0300, P0301-P0306, P0440, P0442 and P0446 not set, engine speed from 1100-3000 rpm in closed loop for 5 seconds, system voltage from 11-16v, throttle angle from 9-20%, Purge duty cycle "on", conditions met for 2 seconds, and the PCM detected the number of HO2S-11 rich-to-lean or lean-to-rich switches during a 100 second sample period was less than a calibrated amount. **Possible Causes:** • Air leaks after the MAF sensor, or in the EGR or PCV system • Exhaust leaks before or near where the front HO2S is mounted • Fuel control sensor is out of calibration (i.e., ECT, IAT or MAP) • Fuel delivery system supplying too much or too little fuel during cruise or idle periods (e.g., faulty fuel pump, or dirty fuel filter) • Fuel injector (one or more) dirty, leaking or sticking • Fuel pressure regulator leaking, damaged or has failed • HO2S is contaminated, deteriorated or it has failed
DTC: P1134 **2T OBD/O2S1, MIL: YES** **1997, 1998, 1999, 2000** **Models:** Hombre **Engines:** 4.3L VIN X **Transmissions:** A/T, M/T	**HO2S-11 (Bank 1 Sensor 1) Transition Time Ratio Error** DTC P0101, P0102, P0103, P0106, P0107, P0108, P0112, P0113, P0117, P0118, P0121, P0122, P0123, P0131, P0132, P0133, P0134, P0135, P0300, P0301-306, P0341, P0440, P0442 and P0446 not set, no Intrusive Tests active, system voltage from 11-16v, engine speed from 1100-3000 rpm in closed loop, engine runtime over 75 seconds, ECT signal more than 135°F, MAF sensor signal from 15-55 g/sec, Purge duty cycle "on", conditions met for 2 seconds, and the PCM detected the HO2S-11 average transition time during a 100 second period was not within the specified range. **Possible Causes:** • Air leaks at the exhaust manifold or exhaust pipes • HO2S signal circuit is open or shorted to ground (intermittent) • HO2S heater power circuit is open, or the heater has failed • HO2S contaminated with wrong fuel, has deteriorated or failed
DTC: P1153 **2T OBD/O2S1, MIL: YES** **1997, 1998, 1999, 2000** **Models:** Hombre **Engines:** 4.3L VIN X **Transmissions:** A/T, M/T	**HO2S-21 (Bank 2 Sensor 1) Insufficient Switching** DTC P0101, P0102, P0103, P0106, P0107, P0108, P0112, P0113, P0117, P0118, P0121, P0122, P0123, P0131, P0132, P0133, P0134, P0135, P0300, P0301-P0306, P0440, P0442 and P0446 not set, engine speed from 1100-3000 rpm in closed loop for 5 seconds, system voltage from 11-16v, throttle angle from 9-20%, Purge duty cycle "on", conditions met for 2 seconds, and the PCM detected the number of HO2S-21 rich-to-lean or lean-to-rich switches during a 100 second sample period was less than a calibrated amount. **Possible Causes:** • Air leaks after the MAF sensor, or in the EGR or PCV system • Exhaust leaks before or near where the front HO2S is mounted • Fuel control sensor is out of calibration (i.e., ECT, IAT or MAP) • Fuel delivery system supplying too much or too little fuel during cruise or idle periods (e.g., faulty fuel pump, or dirty fuel filter) • Fuel injector (one or more) dirty, leaking or sticking • Fuel pressure regulator leaking, damaged or has failed • HO2S is contaminated, deteriorated or it has failed

DTC	Trouble Code Title, Conditions & Possible Causes
DTC: P1154 **2T OBD/O2S1, MIL: YES** **1997, 1998, 1999, 2000** **Models:** Hombre **Engines:** 4.3L VIN X **Transmissions**: A/T, M/T	**HO2S-21 (Bank 2 Sensor 1) Transition Time Ratio Error** DTC P0101, P0102, P0103, P0106, P0107, P0108, P0112, P0113, P0117, P0118, P0121, P0122, P0123, P0131, P0132, P0133, P0134, P0135, P0300, P0301-306, P0341, P0440, P0442 and P0446 not set, no Intrusive Tests active, system voltage from 11-16v, engine speed from 1100-3000 rpm in closed loop, engine runtime over 75 seconds, ECT signal more than 135ºF, MAF sensor signal from 15-55 g/sec, Purge duty cycle is "on", conditions met for 2 seconds, and the PCM detected the HO2S-21 average transition time during a 100 second period was not within the specified range. **Possible Causes:** • Air leaks at the exhaust manifold or exhaust pipes • HO2S signal circuit is open or shorted to ground (intermittent) • HO2S heater power circuit is open, or the heater has failed • HO2S contaminated with wrong fuel, has deteriorated or failed
DTC: P1167 **1T CCM, MIL: YES** **2000, 2001, 2002** **Models:** Amigo, Axiom, Rodeo, Rodeo Sport **Engines:** 3.2L VIN W, 3.5L VIN X **Transmissions:** A/T, M/T	**Fuel System Rich During Decel Fuel Cutoff (Bank 1)** No related codes set, ECT sensor signal more than 140ºF, engine running in Power Enrichment mode in closed loop for 3 seconds, and the PCM detected the HO2S-11 signal was more than 600 mv in a Decel Fuel Cutoff period during the Fuel System Monitor test. **Possible Causes:** • Base engine "mechanical" fault affecting one or more cylinders • EVAP system component has failed or canister fuel saturated • Exhaust leaks before or near where the front HO2S is mounted • Fuel control sensor is out of calibration (i.e., ECT, IAT or MAP) • Fuel delivery system supplying too much fuel during cruise or idle periods (e.g., faulty fuel pump, or faulty pressure regulator) • Fuel injectors (one or more) leaking, pressure regulator leaking • HO2S is contaminated, deteriorated or it has failed
DTC: P1169 **1T CCM, MIL: NO** **2000, 2001, 2002** **Models:** Amigo, Axiom, Rodeo, Rodeo Sport **Engines:** 3.2L VIN W, 3.5L VIN X **Transmissions:** A/T, M/T	**Fuel System Rich During Decel Fuel Cutoff (Bank 2)** No related codes set, ECT sensor signal more than 140ºF, engine running in closed loop, Power Enrichment mode "on" for 3 seconds, and the PCM detected the Bank 2 HO2S-21 signal remained above 600 mv in a Decel Fuel Cutoff period during the Fuel System Test. **Possible Causes:** • Base engine "mechanical" fault affecting one or more cylinders • EVAP system component has failed or canister fuel saturated • Exhaust leaks before or near where the front HO2S is mounted • Fuel control sensor is out of calibration (i.e., ECT, IAT or MAP) • Fuel delivery system supplying too much fuel during cruise or idle periods (e.g., faulty fuel pump, or faulty pressure regulator) • Fuel injectors (one or more) leaking, pressure regulator leaking • HO2S is contaminated, deteriorated or it has failed
DTC: P1171 **1T CCM, MIL: YES** **1996, 1997, 1998, 1999, 2000, 2001, 2002** **Models:** Amigo, Axiom, Rodeo, Rodeo Sport, Trooper, VehiCROSS **Engines:** 3.2L VIN V, 3.2L VIN W, 3.5L VIN X **Transmissions:** A/T	**Fuel System Lean During Acceleration Detected** DTC P0131, P0132, P0133, P0134 and P1133 not set, ECT sensor more than 140ºF, engine running in Power Enrichment mode in closed loop, and the PCM detected the HO2S-11 signal indicated less than 400 mv for 5 seconds during the Fuel System Monitor test. **Possible Causes:** • Air intake leaks in the engine, or in the PCV system (valve) • Air leaks at the EGR gasket, or at the EGR valve diaphragm • Base engine "mechanical" fault affecting one or more cylinders • Exhaust leaks before or near where the front HO2S is mounted • Fuel injectors (one or more) restricted (allowing too little fuel) • Fuel delivery system supplying too little fuel during acceleration periods (e.g., faulty fuel pump, dirty or restricted fuel filter) • Fuel control sensor out of calibration (i.e., IAT, MAF or MAP) • HO2S is contaminated, deteriorated or it has failed • Vehicle driven low on fuel or until it ran out of fuel

DTC	Trouble Code Title, Conditions & Possible Causes
DTC: P1171 **1T CCM, MIL: YES** **1996, 1997** **Models:** Rodeo **Engines:** 2.6L VIN E **Transmissions:** A/T, M/T	**Fuel System Lean During Acceleration Detected** DTC P0131, P0132, P0133, P0134 and P1133 not set, ECT sensor signal more than 140°F, engine running in closed loop mode, Power Enrichment mode enabled, and the PCM detected the HO2S-11 signal indicated less than 400 mv for 5 seconds during the test. **Possible Causes:** • Air intake leaks in the engine, or in the PCV system (valve) • Air leaks at the EGR gasket, or at the EGR valve diaphragm • Base engine "mechanical" fault affecting one or more cylinders • Exhaust leaks before or near where the front HO2S is mounted • Fuel injectors (one or more) restricted (allowing too little fuel) • Fuel delivery system supplying too little fuel during acceleration periods (e.g., faulty fuel pump, dirty or restricted fuel filter) • Fuel control sensor out of calibration (i.e., IAT, MAF or MAP) • HO2S is contaminated, deteriorated or it has failed • Vehicle driven low on fuel or until it ran out of fuel
DTC: P1171 **1T CCM, MIL: YES** **1996, 1997, 1998, 1999, 2000** **Models:** Hombre **Engines:** 2.2L VIN 4 **Transmissions:** A/T, M/T	**Fuel System Lean During Acceleration Detected** DTC P0131, P0132, P0133, P0134 and P1133 not set, ECT sensor signal more than 140°F, engine runtime over 20 seconds while in closed loop, Power Enrichment mode enabled, fuel injector base pulsewidth more than 0 ms, and the PCM detected the HO2S signal indicated less than 300 mv for 5 seconds during the test. **Possible Causes:** • Air intake leaks in the engine, or in the PCV system (valve) • Air leaks at the EGR gasket, or at the EGR valve diaphragm • Base engine "mechanical" fault affecting one or more cylinders • Exhaust leaks before or near where the front HO2S is mounted • Fuel injectors (one or more) restricted (allowing too little fuel) • Fuel delivery system supplying too little fuel during acceleration periods (e.g., faulty fuel pump, dirty or restricted fuel filter) • Fuel control sensor out of calibration (i.e., IAT, MAF or MAP) • HO2S is contaminated, deteriorated or it has failed • Vehicle driven low on fuel or until it ran out of fuel
DTC: P1220 **1T CCM, MIL: YES** **2000, 2001, 2002** **Models:** Amigo, Axiom, Rodeo, Rodeo Sport, Trooper **Engines:** 3.2L VIN W, 3.5L VIN X **Transmissions:** A/T, M/T	**Throttle Position Sensor 2 Circuit Malfunction** Key on or engine running, and the PCM detected the TP Sensor 2 (TP2) circuit was less than 2.5% or over 97.5% of VREF (5v), condition met for 18 counts within 500 test samples over 15.6 ms. **Possible Causes**: • TP2 sensor signal circuit is open or shorted to ground • TP2 sensor signal circuit is shorted to VREF or system power • TP2 sensor power circuit is open between sensor and the PCM • TP2 sensor is damaged or has failed • PCM has failed • TSB 01-04-S001 (7/01) contain information related to this code
DTC: P1221 **1T CCM, MIL: YES** **2000, 2001, 2002** **Models:** Amigo, Axiom, Rodeo, Rodeo Sport, Trooper **Engines:** 3.2L VIN W, 3.5L VIN X **Transmissions:** A/T, M/T	**Throttle Position Sensor 1-2 Circuit Performance** Key on or engine running, and the PCM detected the TP Sensor 2 (TP2) signal did not correlate with the TP Sensor 1 (TP1) signal. **Possible Causes:** • Check for excessive deposits in the ETC passage or the spring • Check for excessive deposits in throttle bore and throttle valve • Check for objects blocking the DC motor or throttle bore • ETC DC motor circuit(s) has high resistance • TP2 sensor signal circuit has high resistance • TSB 01-04-S001 (7/01) contain information related to this code
DTC: P1259 **1T CCM, MIL: YES** **1998, 1999** **Models:** Oasis **Engines:** All **Transmissions:** A/T, M/T	**VTEC System Malfunction** Engine running in closed loop, then accelerated in 1st gear to over 6000 rpm for 2 seconds, and the PCM detected a fault in the VTEC solenoid or the VTEC switch. **Possible Causes:** • VTEC solenoid is damaged or has failed • VTEC switch is damaged or has failed • PCM has failed

DTC	Trouble Code Title, Conditions & Possible Causes
DTC: P1271 **1T CCM, MIL: YES** **2000, 2001, 2002** **Models:** Amigo, Axiom, Rodeo, Rodeo Sport, Trooper **Engines:** 3.2L VIN W, 3.5L VIN X **Transmissions**: A/T, M/T	**Throttle Position Sensor 1-2 Correlation Error** Key on or engine running, and the PCM detected the difference in angle of the accelerator pedal for ASP1 and ASP2 was less than 4.5% for 50 counts within 50 test samples over a 15.6 time period. **Possible Causes:** • Check for excessive deposits in the ETC passage or the spring • Check for excessive deposits in throttle bore and throttle valve • Check for objects blocking the DC motor or throttle bore • ETC DC motor circuit(s) has high resistance • TP2 sensor signal circuit has high resistance • TSB 01-04-S001 (7/01) contain information related to this code
DTC: P1272 **1T CCM, MIL: YES** **2000, 2001, 2002** **Models:** Amigo, Axiom, Rodeo, Rodeo Sport, Trooper **Engines:** 3.2L VIN W, 3.5L VIN X **Transmissions:** A/T, M/T	**Acceleration Position Sensor 2-3 Correlation Error** Key on or engine running, and the PCM detected the difference in angle of the accelerator pedal for ASP2 and ASP3 was less than 4.5% for 50 counts within 50 test samples over a 15.6 time period. **Possible Causes:** • Check for excessive deposits in the ETC passage or the spring • Check for excessive deposits in throttle bore and throttle valve • Check for objects blocking the DC motor or throttle bore • ETC DC motor circuit(s) has high resistance • TP2 sensor signal circuit has high resistance • TSB 01-04-S001 (7/01) contain information related to this code
DTC: P1273 **1T CCM, MIL: YES** **2000, 2001, 2002** **Models:** Amigo, Axiom, Rodeo, Rodeo Sport, Trooper **Engines:** 3.2L VIN W, 3.5L VIN X **Transmissions:** A/T, M/T	**Acceleration Position Sensor 1-3 Correlation Error** Key on or engine running, and the PCM detected the difference in angle of the accelerator pedal for ASP1 and ASP3 was less than 4.5% for 50 counts within 50 test samples over a 15.6 time period. **Possible Causes:** • Check for excessive deposits in the ETC passage or the spring • Check for excessive deposits in throttle bore and throttle valve • Check for objects blocking the DC motor or throttle bore • ETC DC motor circuit(s) has high resistance • TP2 sensor signal circuit has high resistance • TSB 01-04-S001 (7/01) contain information related to this code
DTC: P1275 **1T CCM, MIL: YES** **2000, 2001, 2002** **Models**: Amigo, Axiom, Rodeo, Rodeo Sport, Trooper **Engines:** 3.2L VIN W, 3.5L VIN X **Transmissions:** A/T, M/T	**Acceleration Position Sensor 1 Circuit Malfunction** Key on or engine running, and the PCM detected the Acceleration Position Sensor 1 (APS1) circuit was less than 2.5% or over 97.5% of VREF (5v) for 12 counts within 500 test samples over 15.6 ms. **Possible Causes:** • APS1 sensor signal circuit is open or shorted to ground • APS1 sensor signal circuit is shorted to VREF or system power • APS1 sensor power circuit is open between sensor and PCM • APS1 sensor is damaged or has failed • PCM has failed • TSB 01-04-S001 (7/01) contain information related to this code
DTC: P1280 **1T CCM, MIL: YES** **2000, 2001, 2002** **Models:** Amigo, Axiom, Rodeo, Rodeo Sport, Trooper **Engines:** 3.2L VIN W, 3.5L VIN X **Transmissions:** A/T, M/T	**Acceleration Position Sensor 2 Circuit Malfunction** Key on or engine running, and the PCM detected the Acceleration Position Sensor 2 (APS2) circuit was less than 2.5% or over 97.5% of VREF (5v) for 12 counts within 500 test samples over 15.6 ms. **Possible Causes:** • APS2 sensor signal circuit is open or shorted to ground • APS2 sensor signal circuit is shorted to VREF or system power • APS2 sensor power circuit is open between sensor and PCM • APS2 sensor is damaged or has failed • PCM has failed • TSB 01-04-S001 (7/01) contain information related to this code
DTC: P1285 **1T CCM, MIL: YES** **2000, 2001, 2002** **Models:** Amigo, Axiom, Rodeo, Rodeo Sport, Trooper **Engines:** 3.2L VIN W, 3.5L VIN X **Transmissions:** A/T, M/T	**Acceleration Position Sensor 3 Circuit Malfunction** Key on or engine running, and the PCM detected the Acceleration Position Sensor 3 (APS3) circuit was less than 2.5% or over 97.5% of VREF (5v) for 12 counts within 500 test samples over 15.6 ms. **Possible Causes:** • APS3 sensor signal circuit is open or shorted to ground • APS3 sensor signal circuit is shorted to VREF or system power • APS3 sensor power circuit is open between sensor and PCM • APS3 sensor is damaged or has failed • PCM has failed • TSB 01-04-S001 (7/01) contain information related to this code

DTC	Trouble Code Title, Conditions & Possible Causes
DTC: P1290 **1T CCM, MIL: YES** **2000, 2001, 2002** **Models:** Amigo, Axiom, Rodeo, Rodeo Sport, Trooper **Engines:** 3.2L VIN W, 3.5L VIN X **Transmissions:** A/T, M/T	**Electronic Throttle Control Forced Idle Mode** Key on or engine running, and the PCM detected the Electronic Throttle Control (ETC) system had entered Forced Idle Mode. **Possible Causes:** • ETC module has detected a problem in the system • Forced Idle Mode is "active" • ETC module is damaged or has failed
DTC: P1295 **1T CCM, MIL: YES** **2000, 2001, 2002** **Models**: Amigo, Axiom, Rodeo, Rodeo Sport, Trooper **Engines:** 3.2L VIN W, 3.5L VIN X **Transmissions:** A/T, M/T	**Electronic Throttle Control Power Management Mode** Key on or engine running, and the PCM detected the Electronic Throttle Control (ETC) system had entered Power Management Mode and is operating under "failsafe" mode conditions. **Possible Causes:** • ETC module has detected a problem in the system • Power Management Mode is "active" • ETC module is damaged or has failed
DTC: P1297 **1T CCM, MIL: YES** **1996, 1997, 1998, 1999** **Models:** Oasis **Engines:** All **Transmissions:** A/T, M/T	**Electrical Load Detector Circuit Low Input** Engine running at hot idle speed or at cruise speed, headlights "on", and the PCM detected the ELD signal was less than a stored value. **Possible Causes:** • ELD sensor signal circuit is open or shorted to ground • ELD sensor power circuit is open or shorted to ground • ELD sensor is damaged or has failed • PCM has failed
DTC: P1298 **1T CCM, MIL: YES** **1996, 1997, 1998, 1999** **Models:** Oasis **Engines:** All **Transmissions:** A/T, M/T	**Electrical Load Detector Circuit High Input** Engine running at hot idle speed or at cruise speed, headlights "on", and the PCM detected the ELD signal was more than a stored value. **Possible Causes:** • ELD sensor signal circuit is shorted to VREF • ELD sensor signal circuit is shorted to system power (B+) • ELD sensor is damaged or has failed • PCM has failed
DTC: P1299 **1T CCM, MIL: YES** **2000, 2001, 2002** **Models:** Amigo, Rodeo, Rodeo Sport **Engines:** 3.2L VIN W **Transmissions:** A/T, M/T	**Electronic Throttle Control Forced Engine Shutdown Mode** Key on or engine running, and the PCM detected the Electronic Throttle Control (ETC) system had entered Forced Engine Shutdown Mode and is operating under "failsafe" mode conditions. **Possible Causes:** • The ETC controller has detected the throttle angle was not within a range of 8% (idle) to 92% (wide open throttle). • Engine Forced Shutdown Mode is "active" • ETC module is damaged or has failed
DTC: P1310 **1T CCM, MIL: YES** **2000, 2001, 2002** **Models:** Amigo, Axiom, Rodeo, Rodeo Sport, Trooper **Engines:** 3.2L VIN W, 3.5L VIN X **Transmissions:** A/T, M/T	**ION Sensing Module Diagnosis** No CKP or System Voltage codes set, engine started, engine speed from 650-6500 rpm, system voltage from 10-16v, MAF sensor signal from 26-100 g/sec, fuel level indicating more than 10%, and the PCM detected "missing" Combustion Quality (CQ) pulses, multiple CQ pulses or CQ pulsewidth calculation errors in the CQ quality test. **Possible Causes:** • Ignition coil is open, shorted or damaged • ION sensing module circuit is open or shorted to ground • ION sensing module ground circuit open or has high resistance • ION sensor module is damaged or has failed
DTC: P1311 **1T CCM, MIL: YES** **2000, 2001, 2002** **Models:** Amigo, Axiom, Rodeo, Rodeo Sport, Trooper **Engines:** 3.2L VIN W, 3.5L VIN X **Transmissions**: A/T, M/T	**ION Sensing Module Secondary Line 1 Circuit Malfunction** Engine started, engine speed from 650-6500 rpm, system voltage from 10-16v, MAF sensor signal from 26-100 g/sec, fuel level indicating more than 10%, and the PCM detected an unexpected voltage condition on the Secondary Line 1 circuit to the ION Module. **Possible Causes:** • Secondary Line 1 circuit is open or shorted to ground • Secondary Line 1 circuit is shorted to system power (B+) • ION sensing module is damaged or has failed

DTC	Trouble Code Title, Conditions & Possible Causes
DTC: P1312 **1T CCM, MIL: YES** **2000, 2001, 2002** **Models:** Amigo, Axiom, Rodeo, Rodeo Sport, Trooper **Engines:** 3.2L VIN W, 3.5L VIN X **Transmissions:** A/T, M/T	**ION Sensing Module Secondary Line 2 Circuit Malfunction** Engine started, engine speed from 650-6500 rpm, system voltage from 10-16v, MAF sensor signal from 26-100 g/sec, fuel level indicating more than 10%, and the PCM detected an unexpected voltage condition on the Secondary Line 2 circuit to the ION Module. **Possible Causes:** • Secondary Line 2 circuit is open or shorted to ground • Secondary Line 2 circuit is shorted to system power (B+) • ION sensing module is damaged or has failed
DTC: P1326 **1T CCM, MIL: YES** **2000, 2001, 2002** **Models:** Amigo, Axiom, Rodeo, Rodeo Sport, Trooper **Engines:** 3.2L VIN W, 3.5L VIN X **Transmissions:** A/T, M/T	**ION Sensing Module Combustion Quality** No CKP or CMP Sensor codes set, engine started, system voltage from 10-16v, and the PCM detected an unexpected voltage condition on the Combustion Quality (CQ) line circuit, faults in the ION Module or one or more faults in analog input signals to the PCM. **Possible Causes:** • CQ line circuit is open or shorted to ground • One or more analog inputs to the PCM has failed or is missing • ION Sensing Module is damaged or has failed
DTC: P1336 **1T CCM, MIL: YES** **1999, 2000, 2001, 2002** **Models:** Amigo, Rodeo, Rodeo Sport **Engines:** 2.2L VIN D **Transmissions:** A/T, M/T	**Crankshaft Position System Not Learned** No CKP, CMP, ECT, Injector or Knock Sensor codes set, engine started, engine running, and the PCM detected it had not learned the crankshaft position (Tooth Error Correction) sensor variation after three (3) attempts, or after traveling a distance of 5 miles. **Note: The CKP "relearn" position step only needs to be done once for the life cycle of the vehicle. This data is in Keep Alive Memory. This procedure should be done after replacement of the crankshaft, crankshaft balancer, the PCM or after an engine replacement.** **Possible Causes:** • CKP sensor or related hardware was removed or replaced • Crankshaft Position Tooth Error Correction step not performed
DTC: P1336 **1T CCM, MIL: YES** **1997, 1998, 1999, 2000** **Models: Hombre** **Engines: 2.2L VIN 4** **Transmissions: A/T, M/T**	**Crankshaft Position System Not Learned** DTC P0335, P0340 and P0341 not set, engine started, and the PCM detected it had not learned the crankshaft position (Tooth Error Correction) sensor variation after three (3) unsuccessful attempts. **Note: The CKP "relearn" position step only needs to be done once for the life cycle of the vehicle. This data is in Keep Alive Memory. This procedure should be done after replacement of the crankshaft, crankshaft balancer, the PCM or after an engine replacement.** **Possible Causes:** • CKP sensor or related hardware was removed or replaced • Crankshaft Position Tooth Error Correction step not performed
DTC: P1336 **1T CCM, MIL: YES** **1997, 1998, 1999, 2000** **Models:** Hombre **Engines:** 4.3L VIN X **Transmissions:** A/T, M/T	**Crankshaft Position System Not Learned** DTC P0336, P0340 and P0341 not set, engine started, and the PCM detected it had not learned the crankshaft position (Tooth Error Correction) sensor variation after three (3) unsuccessful attempts. **Note: The CKP "relearn" position step only needs to be done once for the life cycle of the vehicle. This data is in Keep Alive Memory. This procedure should be done after replacement of the crankshaft, crankshaft balancer, the PCM or after an engine replacement.** **Possible Causes:** • CKP sensor or related hardware was removed or replaced • Crankshaft Position Tooth Error Correction step not performed
DTC: P1340 **1T CCM, MIL: YES** **2000** **Models:** Amigo **Engines:** 3.2L VIN W **Transmissions:** A/T, M/T	**ION Sensing Module Cylinder ID Malfunction** No ECT, Fuel Trim, Injector, Misfire or System Voltage codes set, engine started, system voltage from 10-16v, and the PCM detected that the "cylinder synchronization" routine had not been completed after a predetermined number of events occurred after startup. **Possible Causes:** • Knock sensor and Combustion Line connectors are swapped • One or more spark plugs (or boots) are shorted to ground • ION Sensing Module is damaged or has failed • PCM related hardware is damaged or has failed • TSB 01-04-S001 (2/01) contains information about this code
DTC: P1340 **1T CCM, MIL: YES** **2000, 2001, 2002** **Models:** Axiom, Rodeo, Rodeo Sport, Trooper **Engines:** 3.2L VIN W, 3.5L VIN X **Transmissions:** A/T, M/T	**ION Sensing Module Cylinder ID Malfunction** No ECT, Fuel Trim, Injector, Misfire or System Voltage codes set, engine started, system voltage from 10-16v, and the PCM detected that the "cylinder synchronization" routine had not been completed after a predetermined number of events occurred after startup. **Possible Causes:** • Knock sensor and Combustion Line connectors are swapped • One or more spark plugs (or boots) are shorted to ground • ION Sensing Module is damaged or has failed • PCM related hardware is damaged or has failed • TSB 01-04-S001 (2/01) contains information about this code

DTC	Trouble Code Title, Conditions & Possible Causes
DTC: P1345 **1T CCM, MIL: YES** **1997, 1998, 1999, 2000** **Models:** Hombre **Engines:** 4.3L VIN X **Transmissions:** A/T, M/T	**Crankshaft Position or Camshaft Position Correlation** Engine started, engine running, and the PCM detected the CMP sensor pulses were not at the right position relative to the CKP sensor pulses during the CCM Rationality test. **Possible Causes:** • CMP sensor is loose at its mounting location • Excessive free-play in the timing chain and gear assembly • Distributor installed one tooth out of phase (advance or retard) • Distributor not tightened (loose), or the distributor rotor is loose
DTC: P1358 **2T CCM, MIL: YES** **Models:** Hombre **Engines:** 4.3L VIN X **Transmissions:** A/T, M/T	**Ignition Coil Control Circuit High Input** Engine started, ignition control enabled with the engine speed less than 250 rpm, and the PCM detected an unexpected high voltage (over 4.9v) on the Ignition Coil Control circuit during the CCM test. **Possible Causes:** • IC timing signal circuit is open • IC module is damaged or has failed • PCM has failed
DTC: P1359 **1T CCM, MIL: YES** **1996, 1997, 1998, 1999** **Models:** Oasis **Engines:** All **Transmissions:** A/T, M/T string	**CKP/TDC Sensor Circuit Malfunction** Engine running, and the PCM detected an unexpected voltage condition on the CKP/TDC sensor circuit during the CCM test. **Possible Causes:** • TDC signal circuit is open or shorted to ground • TDC signal circuit is shorted to VREF or system power (B+) • TDC pickup assembly or its pulse rotor is damaged • TDC is damaged or has failed • PCM has failed
DTC: P1361 **1T CCM, MIL: YES** **1997, 1998, 1999, 2000** **Models:** Hombre **Engines:** 4.3L VIN X **Transmissions:** A/T, M/T	**Ignition Coil Control Circuit Low Input** Engine started, ignition control enabled with the engine speed less than 250 rpm, and the PCM detected an unexpected high voltage (under 0.04v) on the Ignition Coil Control circuit during the CCM test. **Possible Causes:** • IC timing signal circuit is shorted to ground • IC module is damaged or has failed • PCM has failed
DTC: P1361 **1T CCM, MIL: YES** **1996, 1997, 1998, 1999** **Models:** Oasis **Engines:** All **Transmissions:** A/T, M/T	**Top Dead Center Sensor 1 Circuit Intermittent Signal** Engine running, and the PCM detected an unexpected or intermittent interruption of the Top Dead Center 1 (TDC1) sensor signal. **Possible Causes:** • TDC1 signal circuit is open or shorted to ground • TDC1 signal circuit is shorted to VREF or system power • TDC1 pickup assembly or its pulse rotor is damaged • TDC1 is damaged or has failed • PCM has failed
DTC: P1362 **1T CCM, MIL: YES** **1996, 1997, 1998, 1999** **Models:** Oasis **Engines:** All **Transmissions:** A/T, M/T	**Top Dead Center Sensor 1 No Signal** Engine cranking or running, and the PCM did not receive any signals from the Top Dead Center 1 (TDC1) sensor during the CCM test. **Note: The engine will start and run without the TDC sensor 1 signal.** **Possible Causes:** • TDC1 signal circuit is open or shorted to ground • TDC1 pickup assembly or its pulse rotor is damaged • TDC1 is damaged or has failed • PCM has failed
DTC: P1380 **1T CCM, MIL: NO** **1998, 1999** **Models:** Amigo, Rodeo **Engines:** 2.2L VIN D **Transmissions:** A/T, M/T	**Antilock Brake System Rough Road System Malfunction** Engine started, then driven at a speed of over 1 mph with the engine speed at 5800 rpm or less, engine load at 89% or less, and the PCM detected 20 unusable ABS data values within a 50 value sample due to a problem with the ABS rough road data with an ABS code set. **Note: For additional help with this code, view the Failure Records.** **Possible Causes:** • Class 2 data line open or shorted to ground (check for loose connections or frayed wiring as the fault may be intermittent) • ABS module or the PCM has failed

DTC	Trouble Code Title, Conditions & Possible Causes
DTC: P1381 **1T CCM, MIL: NO** **1998, 1999** **Models:** Amigo, Rodeo **Engines:** 2.2L VIN D **Transmissions:** A/T, M/T	**ABS Module Rough Road/Class 2 Serial Data Malfunction** DTC P0300, P0301, P0302, P0303 and P0304 not set, then driven at a speed of over 1 mph with the engine speed less than 3406 rpm, and the PCM determined it did not receive any ABS Rough Road information for over 2.5 seconds. **Note: For additional help with this code, view the Failure Records.** **Possible Causes:** • Class 2 data line open or shorted to ground (check for loose connections or frayed wiring as the fault may be intermittent) • Check for a legitimate misfire condition in the engine that is not related to the ABS Rough Road information
DTC: P1380 **1T CCM, MIL: NO** **1996, 1997, 1998, 1999, 2000** **Models:** Hombre **Engines:** 2.2L VIN 4 **Transmissions:** A/T, M/T	**Antilock Brake System Rough Road System Malfunction** DTC P0300, P0301-304 has been set, engine started, then driven at a speed of over 1 mph with the engine speed less than 3400 rpm, MAP sensor signal less than 99 kPa, and the PCM detected a problem with the ABS rough road information with an ABS code set. **Note: For additional help with this code, view the Failure Records.** **Possible Causes:** • Class 2 data line open or shorted to ground (check for loose connections or frayed wiring as the fault may be intermittent) • ABS module or the PCM has failed
DTC: P1380 **1T CCM, MIL: NO** **1997, 1998, 1999, 2000** **Models:** Hombre **Engines:** 4.3L VIN X **Transmissions:** A/T, M/T	**Antilock Brake System Rough Road System Malfunction** DTC P0300, P0301-306 has been set, engine started, then driven at a speed of over 1 mph with the engine speed less than 5800 rpm, engine load less than 89%, and the PCM detected a problem with the ABS rough road information with an ABS code set. **Note: For additional help with this code, view the Failure Records.** **Possible Causes:** • Class 2 data line open or shorted to ground (check for loose connections or frayed wiring as the fault may be intermittent) • ABS module or the PCM has failed
DTC: P1381 **1T CCM, MIL: NO** **1996, 1997, 1998, 1999, 2000** **Models:** Hombre **Engines:** 2.2L VIN 4 **Transmissions:** A/T, M/T	**ABS Module Rough Road/Class 2 Serial Data Malfunction** DTC P0300, P0301, P0302, P0303 or P0304 set, engine started, then driven at a speed of over 1 mph with the engine speed less than 3406 rpm, and the PCM determined it did not receive any ABS Rough Road information for over 2.5 seconds. **Note: For additional help with this code, view the Failure Records.** **Possible Causes:** • Class 2 data line open or shorted to ground (check for loose connections or frayed wiring as the fault may be intermittent) • Check for a legitimate misfire condition in the engine that is not related to the ABS Rough Road information
DTC: P1381 **1T CCM, MIL: NO** **1997, 1998, 1999, 2000** **Models:** Hombre **Engines:** 4.3L VIN X **Transmissions:** A/T, M/T	**ABS Module Rough Road/Class 2 Serial Data Malfunction** DTC P0300, P0301-P0306 is set, engine started, then driven at a speed of over 1 mph and the PCM determined it did not receive any ABS Rough Road information through the serial data bus circuit. **Note: For additional help with this code, view the Failure Records.** **Possible Causes:** • Class 2 data line open or shorted to ground (check for loose connections or frayed wiring as the fault may be intermittent) • Check for a legitimate misfire condition in the engine that is not related to the ABS Rough Road information
DTC: P1380 **1T CCM, MIL: YES** **1998, 1999** **Models:** Amigo, Rodeo **Engines:** 3.2L VIN W **Transmissions:** A/T, M/T	**Antilock Brake System Rough Road System Malfunction** A DTC P0300 or P0301-306 is set, engine started, then driven at a speed of over 5 mph, engine speed less than 6250 rpm, engine load less than 99%, and the PCM received an ABS fault signal from the ABS module, 100 test failures in 120 test samples. **Note: For additional help with this code, view the Failure Records.** **Possible Causes:** • Class 2 data line open or shorted to ground (check for loose connections or frayed wiring as the fault may be intermittent) • ABS module or the PCM has failed
DTC: P1381 **1T CCM, MIL: NO** **1998, 1999** **Models:** Rodeo **Engines:** 3.2L VIN W **Transmissions:** A/T, M/T	**ABS Module Rough Road/Class 2 Serial Data Malfunction** A DTC P0300, P0301-P0306 is set, engine started, then driven to a speed over 1 mph, engine speed below 6250 rpm for 2.5 seconds, engine load less than 99%, and the PCM did not receive any ABS signals from the ABS module, 100 test failures in 120 test samples. **Note: For additional help with this code, view the Failure Records.** **Possible Causes:** • Class 2 data line open or shorted to ground (check for loose connections or frayed wiring as the fault may be intermittent) • Check for a legitimate misfire condition in the engine that is not related to the ABS Rough Road information
DTC: P1380 **1T CCM, MIL: NO** **2000, 2001, 2002** **Models:** Axiom, Trooper, VehiCROSS **Engines:** 3.5L VIN X **Transmissions:** A/T, M/T	**Antilock Brake System Rough Road System Malfunction** A DTC P0300 or P0301-306 is set, engine started, then driven at a speed of over 5 mph, engine speed less than 6250 rpm, engine load less than 99%, and the PCM received an ABS fault signal from the ABS module, 100 test failures in 120 test samples. **Note: For additional help with this code, view the Failure Records.** **Possible Causes:** • Class 2 data line open or shorted to ground (check for loose connections or frayed wiring as the fault may be intermittent) • ABS module or the PCM has failed

DTC	Trouble Code Title, Conditions & Possible Causes
DTC: P1381 **1T CCM, MIL: NO** **2000, 2001, 2002** **Models:** Axiom, Trooper, VehiCROSS **Engines:** 3.5L VIN X **Transmissions:** A/T, M/T	**ABS Module Rough Road/Class 2 Serial Data Malfunction** A DTC P0300, P0301-P0306 is set, engine started, then driven to a speed over 1 mph, engine speed below 6250 rpm for 2.5 seconds, engine load less than 99%, and the PCM did not receive any ABS signals from the ABS module, 100 test failures in 120 test samples. **Note: For additional help with this code, view the Failure Records.** **Possible Causes:** • Class 2 data line open or shorted to ground (check for loose connections or frayed wiring as the fault may be intermittent) • Check for a legitimate misfire condition in the engine that is not related to the ABS Rough Road information
DTC: P1381 **1T CCM, MIL: YES** **1996, 1997, 1998, 1999** **Models:** Oasis **Engines:** All **Transmissions**: A/T, M/T string	**Camshaft Position Sensor 1 Circuit Malfunction** Engine running and the PCM detected an unexpected or intermittent interruption of the Camshaft Position (CMP) sensor 1 signal. **Possible Causes:** • CMP signal circuit is open or shorted to ground • CMP signal circuit is shorted to VREF or system power • CMP pickup assembly or CMP sensor is damaged or has failed • PCM has failed
DTC: P1382 **1T CCM, MIL: YES** **1996, 1997, 1998, 1999, 2000** **Models:** Oasis **Engines:** All **Transmissions:** A/T, M/T	**Camshaft Position Sensor 1 No Signal** Engine cranking or running, and the PCM did not detect any signals from the Camshaft Position (CMP) sensor 1 during the CCM test. **Note: The engine will start and run without the CMP sensor 1 signal.** **Possible Causes:** • CMP signal circuit is open or shorted to ground • CMP pickup assembly or CMP sensor is damaged or has failed • PCM has failed
DTC: P1390 **1T CCM, MIL: NO** **1996, 1997** **Models:** Rodeo **Engines:** 2.6L VIN E, 3.2L VIN V **Transmissions:** A/T, M/T	**Acceleration 'G' Sensor Intermittent Low Input** Engine started, engine running, and the PCM detected an unexpected low signal of less than 0.5v on the 'G' Sensor circuit over a 6.25 second period. The 'G' sensor is used to sense vertical acceleration due to road vibration during Misfire diagnostics. **Note: For additional help with this code, view the Failure Records.** **Possible Causes:** • 'G' sensor signal circuit is shorted to ground • 'G' sensor is damaged or has failed • PCM has failed
DTC: P1391 **1T CCM, MIL: NO** **1996, 1997** **Models:** Rodeo **Engines:** 2.6L VIN E, 3.2L VIN V **Transmissions:** A/T, M/T	**Acceleration 'G' Sensor Range/Performance** Engine started, engine running, vehicle speed indicating (0) mph, and the PCM detected the G-Sensor signal was more than 2.5v, or it was less than 1.5v for 1 minute, or with vehicle speed at 30-80 mph, the G-Sensor signal indicated a change of 0.0002v each 10 ms. **Note: For additional help with this code, view the Failure Records.** **Possible Causes:** • 'G' sensor signal circuit is shorted to ground (intermittent fault) • 'G' sensor VREF circuit is open or shorted to ground (this circuit is also connected to the MAP sensor - check the MAP sensor) • 'G' sensor ground circuit is open (an intermittent fault) • 'G' sensor is damaged or has failed • PCM has failed
DTC: P1392 **1T CCM, MIL: YES** **1996, 1997** **Models:** Rodeo **Engines:** 2.6L VIN E, 3.2L VIN V **Transmissions:** A/T, M/T	**Acceleration 'G' Sensor Circuit Low Input** Engine started, engine running, and the PCM detected the 'G' Sensor signal indicated less than 0.5v for 12.5 seconds over a 25 second period of time during the CCM test. **Possible Causes:** • 'G' sensor signal circuit is shorted to ground • 'G' sensor VREF circuit is open or shorted to ground (this circuit is also connected to the MAP sensor - check the MAP sensor) • 'G' sensor is damaged or has failed • PCM has failed
DTC: P1393 **1T CCM, MIL: YES** **1996, 1997** **Models:** Rodeo **Engines:** 2.6L VIN E, 3.2L VIN V **Transmissions:** A/T, M/T	**Acceleration 'G' Sensor Circuit High Input** Engine started, engine running, and the PCM detected the 'G' Sensor signal indicated more than 4.5v for 12.5 seconds over a 25 second period of time during the CCM test. **Possible Causes:** • 'G' sensor signal circuit is shorted to VREF or system power • 'G' sensor ground circuit is open between sensor and ground • 'G' sensor is damaged or has failed • PCM has failed

DTC	Trouble Code Title, Conditions & Possible Causes
DTC: P1394 **1T CCM, MIL: NO** **1996, 1997** **Models:** Rodeo **Engines:** 2.6L VIN E, 3.2L VIN V **Transmissions:** A/T, M/T	**Acceleration 'G' Sensor Intermittent High Input** Engine started, engine running, and the PCM detected the 'G' Sensor signal indicated an intermittent high signal (4.8v) for 12.5 seconds over a 25 second time period. This sensor is used to sense vertical acceleration due to road vibration during Misfire diagnostics. **Possible Causes:** • 'G' sensor signal circuit is shorted to VREF (intermittent fault) • 'G' sensor ground circuit is open (an intermittent fault) • 'G' sensor is damaged or has failed (an intermittent fault) • PCM has failed
DTC: P1404 **2T CCM, MIL: YES** **1998, 1999, 2000, 2001, 2002** **Models:** Amigo, Rodeo, Rodeo Sport **Engines:** 2.2L VIN D **Transmissions:** A/T, M/T	**EGR Valve Stuck Closed Malfunction** Engine started, engine running, IAT sensor signal more than 14°F, and the PCM detected the Actual EGR pintle position was 16 counts below the Desired EGR position threshold for over 6.3 seconds. **Possible Causes:** • EGR valve sticking or binding (check for deposits on the valve) • EGR valve is damaged or has failed (if the valve shows signs of excessive heat, check the converter and pipes for a restriction) • PCM has failed
DTC: P1404 **2T CCM, MIL: YES** **1998, 1999, 2000, 2001, 2002** **Models:** Amigo, Rodeo, Rodeo Sport **Engines:** 3.2L VIN W **Transmissions:** A/T, M/T	**EGR Valve Stuck Closed Malfunction** Engine started, engine running, system voltage from 11-16v, ECT sensor from 176-248, IAT sensor less than 212°F, Desired EGR valve position is 0%, and the PCM detected the difference between the Actual and Desired EGR pintle position was over 30% for 5 seconds. This test must fail three times in one trip to set this code. **Possible Causes:** • EGR valve sticking or binding (check for deposits on the valve) • EGR valve is damaged or has failed (if the valve shows signs of excessive heat, check the converter and pipes for a restriction) • PCM has failed
DTC: P1404 **2T CCM, MIL: YES** **1997, 1998, 1999, 2000** **Models:** Hombre **Engines:** 2.2L VIN 4 **Transmissions:** A/T, M/T	**EGR Valve Stuck Closed Malfunction** DTC P0106, P0107, P0108, P0112, P0113, P0117, P0118, P0121, P0122, P0123, P0125, P0300, P0301-P0304, P0335, P0440, P0442, P0446, P0502, P0506, P0507, P1336 and P1441 not set, engine running, system voltage at 11-16v, EGR valve "off", and the PCM detected the EGR pintle position was 15% for 20 seconds. **Possible Causes:** • EGR valve sticking or binding (check for deposits on the valve) • EGR valve is damaged or has failed (if the valve shows signs of excessive heat, check the converter and pipes for a restriction) • PCM has failed
DTC: P1404 **2T CCM, MIL: YES** **1997, 1998, 1999, 2000** **Models:** Hombre **Engines:** 4.3L VIN X **Transmissions:** A/T, M/T	**EGR Valve Stuck Closed Malfunction** Engine started, engine running, system voltage from 11-16v, ECT sensor from 176-248°F, IAT sensor less than 212°F, Desired EGR position is 0%, and the PCM detected the difference between the Actual and Desired EGR pintle position was more than 40%. **Possible Causes:** • EGR valve sticking or binding (check for deposits on the valve) • EGR valve is damaged or has failed (if the valve shows signs of excessive heat, check the converter and pipes for a restriction) • PCM has failed
DTC: P1404 **2T CCM, MIL: YES** **1998, 1999, 2000, 2001, 2002** **Models:** Axiom, Trooper, VehiCROSS **Engines:** 3.5L VIN X **Transmissions:** A/T, M/T	**EGR Valve Stuck Closed Malfunction** Engine started, engine running, system voltage from 11-16v, ECT sensor from 176-248, IAT sensor less than 212°F, Desired EGR valve position is 0%, and the PCM detected the difference between the Actual and Desired EGR pintle position was more than 30% for 5 seconds. The test must fail three times in one trip to set this code. **Possible Causes:** • EGR valve sticking or binding (check for deposits on the valve) • EGR valve is damaged or has failed (if the valve shows signs of excessive heat, check the converter and pipes for a restriction) • PCM has failed

DTC	Trouble Code Title, Conditions & Possible Causes
DTC: P1406 **2T CCM, MIL: YES** **1996** **Models:** Hombre **Engines:** 2.2L VIN 4 **Transmissions:** A/T, M/T	**EGR Valve Pintle Position Sensor Circuit Malfunction** DTC P0106, P0107, P0108, P0112, P0113, P0117, P0118, P0121, P0122, P0123, P0125, P0200, P0300, P0301, P0302, P0303, P0304, P0335, P0440, P0442, P0446, P0502, P0503, P0506, P0507 and P441 not set, system voltage over 12v, engine running with the EGR valve commanded "on" (Desired EGR position is 0%), and the PCM detected the EGR position was more than 3% for 25 seconds or the Actual EGR position differed from the Desired EGR position by more than 6% for 20 seconds, or with the EGR valve commanded "off" (Desired EGR position is 0%) it detected the Actual EGR position differed from the Desired EGR position by more than 9% for 18 seconds with the Desired EGR position less than 99% or the Actual EGR position differed from the Desired EGR position by more than 20% for 18 seconds with the Desired EGR position more than 99% during the EGR System Monitor Test. **Possible Causes:** • EGR position sensor signal circuit is open (intermittent fault) • EGR valve sticking or binding (check for moisture or deposits on the EGR valve or on the valve seat - the valve may lock up) • EGR valve is damaged or has failed (if the valve shows signs of excessive heat, check the converter and pipes for a restriction) • PCM has failed
DTC: P1406 **2T CCM, MIL: YES** **1996, 1997** **Models:** Rodeo, Trooper **Engines:** 2.6L VIN E, 3.2L VIN V **Transmissions:** A/T, M/T	**EGR Valve Pintle Position Sensor Circuit Malfunction** Engine started, system voltage over 12v, engine running with the EGR valve commanded to 0%, and the PCM detected the Actual EGR position was 0.20v more or less than the EGR closed valve position for 5 seconds; or with the EGR valve commanded to more than 0%, the PCM detected the Actual EGR position was more than 15% greater than, or 15% less than the Desired EGR position for 5 seconds; or the PCM detected the Actual EGR position was less than 0.10v for over 5 seconds at any time during the CCM test. **Possible Causes:** • EGR position sensor signal circuit is open (intermittent fault) • EGR valve sticking or binding (check for moisture or deposits on the EGR valve or on the valve seat - the valve may lock up) • EGR valve is damaged or has failed (if the valve shows signs of excessive heat, check the converter and pipes for a restriction) • PCM has failed
DTC: P1415 **2T CCM, MIL: YES** **1999, 2000** **Models:** Hombre **Engines:** 4.3L VIN X **Transmissions:** A/T, M/T	**Secondary Air Injection System Malfunction (Bank 1)** No CKP, ECT, Fuel Trim, HO2S, IAT, MAF, MAP, Misfire, TP or VSS sensor codes set, system voltage over 11.7v, engine speed over 550 rpm in closed loop with the A/F Ratio at 14.7:1, ECT sensor from 140-244°F, MAF sensor less than 25 g/sec, engine load less than 34%, SHRTFT counts from 124-132, VSS at 16-75 mph, engine load under 34%, Decel Fuel Cutoff, Power Enrichment and Catalyst Over-Temperature Mode all inactive, then in the Passive Test, the PCM detected the HO2S-11 signal did not go below 370 mv in open loop or in the Intrusive Test, the PCM detected the HO2S-21 did not go below 222 mv within 1 second with air injected. **Possible Causes:** • Check for pinched, kinked, leaking or loose Air System hoses • A restriction in the Air Pump Inlet valve, or a failed Check valve
DTC: P1416 **2T CCM, MIL: YES** **1999, 2000** **Models:** Hombre **Engines:** 4.3L VIN X **Transmissions:** A/T, M/T	**Secondary Air Injection System Malfunction (Bank 2)** No CKP, ECT, Fuel Trim, HO2S, IAT, MAF, MAP, Misfire, TP or VSS sensor codes set, system voltage over 11.7v, engine speed over 550 rpm in closed loop with the A/F Ratio at 14.7:1, ECT sensor from 140-244°F, MAF sensor less than 25 g/sec, engine load less than 34%, SHRTFT counts from 124-132, VSS at 16-75 mph, engine load under 34%, Decel Fuel Cutoff, Power Enrichment and Catalyst Over-Temperature Mode all inactive, then in the Passive Test, the PCM detected the HO2S-11 signal did not go below 370 mv in open loop or in the Intrusive Test, the PCM detected the HO2S-21 did not go below 222 mv within 1 second with air injected. **Possible Causes:** • Check for pinched, kinked, leaking or loose Air System hoses • A restriction in the Air Pump Inlet valve, or a failed Check valve
DTC: P1441 **2T CCM, MIL: YES** **1998, 1999, 2000, 2001, 2002** **Models:** Amigo, Rodeo, Rodeo Sport **Engines:** 2.2L VIN D **Transmissions:** A/T, M/T	**EVAP System Flow During Non-Purge Error** No ECT, IAT, MAP, Vacuum Switch, TP or VSS codes set, BARO sensor signal more than 72.3 kPa, ECT and IAT inputs from 38-90°F and within 12.2°F of each other at startup, fuel tank level at 10-90%, engine started, then driven to a speed of less than 60 mph, and the PCM detected the fuel tank vacuum level was more than 6" of water for 500 ms during the EVAP Monitor flow test. **Possible Causes:** • EVAP vacuum switch signal circuit is open or shorted to ground • Fuel tank pressure sensor is damaged or has failed • Purge solenoid control circuit shorted to ground • PCM has failed

DTC	Trouble Code Title, Conditions & Possible Causes
DTC: P1441 **2T CCM, MIL: YES** **1996, 1997, 1998, 1999, 2000, 2001, 2002** **Models:** Amigo, Axiom, Rodeo, Rodeo Sport, **Trooper, VehiCROSS** **Engines:** 3.2L VIN V, 3.2L VIN W, 3.5L VIN X **Transmissions:** A/T, M/T	**EVAP Vacuum Switch Circuit High Input** DTC P0106, P0107, P0108, P0112, P0113, P0121, P0122, P0123, P1640 and P1650 not set, system voltage at 11-16v, IAT sensor more than 32ºF, fuel level from 15-85%, and the PCM detected a continuous "open" purge condition during the EVAP Monitor test. **Possible Causes:** • Vacuum switch signal circuit is open between switch and PCM • Vacuum switch ground circuit open between switch and ground • Vacuum switch is damaged or has failed • PCM has failed
DTC: P1441 **1T CCM, MIL: YES** **1996, 1997, 1998, 1999, 2000** **Models:** Hombre **Engines:** 2.2L VIN 4 **Transmissions:** A/T, M/T	**EVAP System Flow During Non-Purge Error** DTC P0106, P0107, P0108, P0112, P0113, P0117, P0118, P0121, P0122, P0123, P0125, P0131, P0132, P0133, P0134 and P1133 not set, BARO sensor signal more than 75 kPa, ECT and IAT inputs from 41-84ºF and within 12ºF of each other at startup, fuel tank level from 26-74%, engine started, engine running, EVAP solenoid duty cycle command at 0% within 65 seconds of the engine runtime, and the PCM detected the presence of vacuum in the EVAP system. **Possible Causes:** • Purge solenoid control circuit is shorted to ground • Purge solenoid is leaking, or stuck partially open • Purge and engine vacuum lines may be switched at the connections to the EVAP purge valve assembly • PCM has failed
DTC: P1441 **1T CCM, MIL: YES** **1997, 1998, 1999, 2000** **Models:** Hombre **Engines:** 4.3L VIN X **Transmissions:** A/T, M/T	**EVAP System Flow During Non-Purge Error** DTC P0106, P0107, P0108, P0112, P0113, P0117, P0118, P0121, P0122, P0123, P0125, P0131, P0132, P0133, P0134 and P1133 not set, BARO sensor signal over 72 kPa, ECT and IAT sensor signals at 41-84ºF and within 12ºF of each other at startup, throttle angle under 75%, fuel tank level from 12-88%, engine running, EVAP Vent solenoid commanded open, vehicle speed under 75 mph, fuel tank vacuum indicating less than 7" Hg, and the PCM detected a vacuum in the system over a calibrated amount with Purge commanded off. **Possible Causes:** • Purge solenoid control circuit is shorted to ground • Purge solenoid is leaking, or stuck partially open • Purge and engine vacuum lines may be switched at the connections to the EVAP purge valve assembly • PCM has failed
DTC: P1441 **2T CCM, MIL: YES** **1996, 1997** **Models:** Rodeo **Engines:** 2.6L VIN E **Transmissions:** A/T, M/T	**EVAP Vacuum Switch Circuit High Input** No IAT, MAP or TP sensor codes set, system voltage at 11-16v, ECT and IAT sensor signals more than 41ºF, startup ECT sensor signal less than 122ºF, key on (prior to engine cranking), and the PCM detected the EVAP Purge Switch indicated 12v (open circuit) at least 4 out of the last 8 times the ignition switch was turned "on". **Possible Causes:** • Vacuum switch signal circuit is open between switch and PCM • Vacuum switch ground circuit open between switch and ground • Vacuum switch is damaged or has failed • PCM has failed
DTC: P1442 **2T CCM, MIL: YES** **1996, 1997** **Models:** Rodeo **Engines:** 2.6L VIN E **Transmissions:** A/T, M/T	**EVAP Vacuum Switch Circuit High Input During Key On** DTC P0106, P0107, P0108, P0112, P0113, P0117, P0118, P0121, P0122 or P0123 not set, ECT and IAT sensors more than 41ºF at startup, and the difference between the startup ECT sensor and IAT sensor more than 45ºF, system voltage at 11-16v, engine speed from 800-6000 rpm, throttle angle over 14%, Purge duty cycle below 1%, and the PCM detected the EVAP Purge Switch indicated 12v (the switch is open) for 5 seconds during the CCM Rationality test. **Possible Causes:** • Vacuum switch signal circuit is open between switch and PCM • Vacuum switch ground circuit open between switch and ground • Vacuum switch is damaged or has failed • PCM has failed
DTC: P1442 **2T CCM, MIL: YES** **1996, 1997, 1998** **Models:** Trooper **Engines:** 3.2L VIN V, 3.5L VIN X **Transmissions:** A/T, M/T	**EVAP Vacuum Switch Circuit High Input During Key On** DTC P0106, P0107, P0108, P0112, P0113, P0117, P0118, P0121, P0122 or P0123 not set, ECT and IAT sensors more than 41ºF at startup, and the difference between the startup ECT sensor and IAT sensor more than 45ºF, system voltage at 11-16v, engine speed from 800-6000 rpm, throttle angle over 14%, Purge duty cycle below 1%, and the PCM detected the EVAP Purge Switch indicated 12v (the switch is open) for 5 seconds during the CCM Rationality test. **Possible Causes:** • Vacuum switch signal circuit is open between switch and PCM • Vacuum switch ground circuit open between switch and ground • Vacuum switch is damaged or has failed • PCM has failed

DTC	Trouble Code Title, Conditions & Possible Causes
DTC: P1456 **2T CCM, MIL: YES** **1998, 1999** **Models:** Oasis **Engines:** All **Transmissions:** A/T, M/T	**EVAP System Leak Detected (Fuel Tank Area)** Cold startup completed (IAT sensor signal from 32-86ºF at engine startup), vehicle driven at over 5 mph for over 2 minutes, then with ECT sensor signal more than 154ºF and the EVAP Control and Vent solenoids enabled, the PCM detected the fuel tank pressure was incorrect due to a leak in the fuel tank area during the Leak Test. **Possible Causes:** • Fuel tank cap damaged, loose or the wrong part number • Fuel tank leaks at the fuel fill pipe or at the fuel tank seals • Fuel vapor control valve is damaged or has failed • Fuel tank vapor recirculation valve or vapor tube is damaged • Fuel tank vapor control vent tube is damaged or has failed
DTC: P1457 **2T CCM, MIL: YES** **1998, 1999** **Models:** Oasis **Engines:** All **Transmissions:** A/T, M/T	**EVAP System Leak Detected (Canister Area)** Cold startup completed (IAT sensor signal from 32-86ºF at engine startup), vehicle driven at over 5 mph for over 2 minutes, then with ECT sensor signal more than 154ºF and the EVAP Control and Vent solenoids enabled, the PCM detected the fuel tank pressure was incorrect due to a leak in the canister area during the Leak Test. **Possible Causes:** • EVAP canister is leaking, damaged or full of water • EVAP canister purge line is loose, damaged or blocked • EVAP two-way valve or ORVR vent shut valve is damaged • EVAP fuel tank vapor control valve is damaged or has failed • PCM has failed
DTC: P1491 **2T CCM, MIL: YES** **1996, 1997, 1998, 1999** **Models:** Oasis **Engines:** All **Transmissions:** A/T, M/T	**EGR Valve Lift Sensor Insufficient Flow Detected** Vehicle driven in closed loop at 1700-2500 rpm for over 10 minutes, and the PCM detected the EGR valve lift sensor (EGRV) signal indicated insufficient EGR flow during the EGR Monitor test. **Possible Causes:** • EGR valve lift sensor is stuck, damaged or has failed • EGR control solenoid circuit is open or shorted to ground • EGR control solenoid valve is damaged or has failed • PCM has failed
DTC: P1498 **1T CCM, MIL: YES** **1996, 1997, 1998, 1999** **Models:** Oasis **Engines:** All **Transmissions:** A/T, M/T	**EGR Valve Lift Sensor High Input** Key on or engine running, and the PCM detected the EGR Valve Lift sensor signal was more than an allowable range stored in memory. **Possible Causes:** • EGR valve lift sensor circuit is open or shorted to power • EGR valve lift sensor is shorted to VREF or system power (B+) • EGR valve lift sensor is stuck, damaged or has failed • PCM has failed
DTC: P1508 **2T CCM, MIL: YES** **1998, 1999, 2000** **Models:** Amigo **Engines:** 3.2L VIN W **Transmissions:** A/T, M/T	**Idle Speed Control System Low RPM** No ECT, EGR, Fuel Trim, IAT, IC Control, Injector, MAF, MAP, EVAP Purge, TP or VSS codes set, system voltage from 10-16v, ECT sensor signal over 120ºF, BARO sensor signal over 75 kPa, engine runtime over 125 seconds, throttle closed, vehicle speed less than 1 mph, no Scan Tool tests "active", and the PCM detected the Actual idle speed was 100-200 rpm less than the Desired Idle speed based upon the current engine coolant temperature (ECT signal). **Possible Causes:** • Air intake system leaks in the EGR or PCV system (PCV valve) • Fuel delivery system is too lean or too rich during the test • Inspect the engine mounts for damage to the mounts • Inspect the throttle linkage adjustment and tension • Inspect the throttle body bore for dirt or foreign material • IAC valve may be damaged or have failed • IAC motor control circuit is open or shorted to ground • Perform an IAC Reset function with the Scan Tool

DTC	Trouble Code Title, Conditions & Possible Causes
DTC: P1509 **2T CCM, MIL: YES** **1998, 1999, 2000** **Models:** Amigo **Engines:** 3.2L VIN W **Transmissions:** A/T, M/T	**Idle Speed Control System High RPM** No ECT, EGR, Fuel Trim, IAT, IC Control, Injector, MAF, MAP, EVAP Purge, TP or VSS codes set, system voltage from 10-16v, ECT sensor signal over 120ºF, BARO sensor signal over 75 kPa, engine runtime over 125 seconds, throttle closed, vehicle speed less than 1 mph, no Scan Tool tests "active", and the PCM detected the Actual idle speed was 100-200 rpm more than Desired Idle speed based upon the current engine coolant temperature (ECT signal). **Possible Causes:** • Air intake system leaks in the EGR or PCV system (PCV valve) • Fuel delivery system is too lean or too rich during the test • Inspect the engine mounts for damage to the mounts • Inspect the throttle linkage adjustment and tension • Inspect the throttle body bore for dirt or foreign material • IAC valve may be damaged or have failed • IAC motor control circuit is open or shorted to ground • Perform an IAC Reset function with the Scan Tool
DTC: P1508 **2T CCM, MIL: YES** **1997, 1998, 1999, 2000** **Models:** Hombre **Engines:** 4.3L VIN X **Transmissions:** A/T, M/T	**Idle Speed Control System Low RPM** No ECT, MAP, TP or VSS codes set, system voltage from 10-16v, ECT sensor signal over 122ºF, IAT sensor signal over -13ºF, BARO sensor signal over 70 kPa, engine runtime over 30 seconds, then after the Non-Intrusive Test "passed", and during the Intrusive Test with the vehicle driven to a speed of 25-75 mph, MAF sensor signal from 17-50 g/sec, change in throttle position less than 1%, with the IAC valve commanded to move a specified number of steps and the engine speed changed less than 50 rpm, the PCM detected less than a 3 g/sec change in the MAF signal in the CCM Rationality test. **Possible Causes:** • Fuel delivery system is too lean or too rich during the test • Inspect the engine mounts for damage to the mounts • Inspect the throttle linkage adjustment and tension • Inspect the throttle body bore for dirt or foreign material • Inspect for any air intake system leaks in the engine or hoses • IAC valve may be damaged or have failed • IAC motor control circuit is open or shorted to ground • Perform an IAC Reset function with the Scan Tool
DTC: P1509 **2T CCM, MIL: YES** **1997, 1998, 1999, 2000** **Models:** Hombre **Engines:** 4.3L VIN X **Transmissions:** A/T, M/T	**Idle Speed Control System High RPM** No ECT, MAP, TP or VSS codes set, system voltage from 10-16v, ECT sensor signal over 122ºF, IAT sensor signal over -13ºF, BARO sensor signal over 70 kPa, engine runtime over 30 seconds, then after the Non-Intrusive Test "passed", and during the Intrusive Test with the vehicle driven to a speed of 25-75 mph, MAF sensor signal from 17-50 g/sec, change in throttle position less than 1%, with the IAC valve commanded to move a specified number of steps and the engine speed changed less than 50 rpm, the PCM detected more than a 3 g/sec change in the MAF signal in the CCM Rationality test. **Possible Causes:** • Fuel delivery system is too lean or too rich during the test • Inspect the engine mounts for damage to the mounts • Inspect the throttle linkage adjustment and tension • Inspect the throttle body bore for dirt or foreign material • Inspect for any air intake system leaks in the engine or hoses • IAC valve may be damaged or have failed • IAC motor control circuit is open or shorted to ground • Perform an IAC Reset function with the Scan Tool
DTC: P1508 **2T CCM, MIL: YES** **1996, 1997, 1998, 1999, 2000, 2001, 2002** **Models:** Rodeo, VehiCROSS **Engines:** 3.2L VIN V, 3.2L VIN W, 3.5L VIN X **Transmissions:** A/T, M/T	**Idle Speed Control System Low RPM** DTC P0101, P0102, P0103, P0106, P0107, P0108, P0112, P0113, P0117, P0118, P0121, P0122, P0123, P0171, P0172, P0201-P0206, P0351-P0356, P0401, P0440, engine runtime 120 seconds, system voltage from 11-16v, ECT sensor over 120ºF, BARO sensor over 75 kPa, Scan Tool tests all "off", vehicle speed under 1 mph with the throttle closed, and the PCM detected the Actual engine speed was 100-200 rpm less than the Desired idle speed for 5 seconds (Desired idle speed is based on ECT signal during the test). **Possible Causes:** • Fuel delivery system is too lean or too rich during the test • Inspect the engine mounts for damage to the mounts • Inspect the throttle linkage adjustment and tension • Inspect the throttle body bore for dirt or foreign material • Inspect for any air intake system leaks in the engine or hoses • IAC valve may be damaged or have failed • IAC motor control circuit is open or shorted to ground • Perform an IAC Reset function with the Scan Tool

DTC	Trouble Code Title, Conditions & Possible Causes
DTC: P1509 **2T CCM, MIL: YES** **1996, 1997, 1998, 1999, 2000, 2001, 2002** **Models:** Rodeo, VehiCROSS **Engines:** 3.2L VIN V, 3.2L VIN W, 3.5L VIN X **Transmissions:** A/T, M/T	**Idle Speed Control System High RPM** DTC P0101, P0102, P0103, P0106, P0107, P0108, P0112, P0113, P0117, P0118, P0121, P0122, P0123, P0171, P0172, P0201-P0206, P0351-P0356, P0401, P0440, engine runtime 120 seconds, system voltage from 11-16v, ECT sensor over 120ºF, BARO sensor over 75 kPa, Scan Tool tests all "off", vehicle speed under 1 mph with the throttle closed, and the PCM detected the Actual engine speed was 100-200 rpm more than the Desired idle speed for 5 seconds (Desired idle speed is based on ECT signal during the test). **Possible Causes:** • Fuel delivery system is too lean or too rich during the test • Inspect the engine mounts for damage to the mounts • Inspect the throttle linkage adjustment and tension • Inspect the throttle body bore for dirt or foreign material • Inspect for any air intake system leaks in the engine or hoses • IAC valve may be damaged or have failed • IAC motor control circuit is open or shorted to ground • Perform an IAC Reset function with the Scan Tool
DTC: P1508 **2T CCM, MIL: YES** **1996, 1997, 1998, 1999** **Models:** Trooper **Engines:** 3.2L VIN V, 3.5L VIN X **Transmissions:** A/T, M/T	**Idle Speed Control System Low RPM** DTC P0101, P0102, P0103, P0106, P0107, P0108, P0112, P0113, P0117, P0118, P0121, P0122, P0123, P0171, P0172, P0201-P0206, P0351-P0356, P0401, P0440, engine runtime 120 seconds, system voltage from 11-16v, ECT sensor over 120ºF, BARO sensor over 75 kPa, Scan Tool tests all "off", vehicle speed under 1 mph with the throttle closed, and the PCM detected the Actual engine speed was 100-200 rpm less than the Desired idle speed for 5 seconds (Desired idle speed is based on ECT signal during the test). **Possible Causes:** • Fuel delivery system is too lean or too rich during the test • Inspect the engine mounts for damage to the mounts • Inspect the throttle linkage adjustment and tension • Inspect the throttle body bore for dirt or foreign material • Inspect for any air intake system leaks in the engine or hoses • IAC valve may be damaged or have failed • IAC motor control circuit is open or shorted to ground • Perform an IAC Reset function with the Scan Tool
DTC: P1509 **2T CCM, MIL: YES** **1996, 1997, 1998, 1999** **Models:** Trooper **Engines:** 3.2L VIN V, 3.5L VIN X **Transmissions:** A/T, M/T	**Idle Speed Control System High RPM** DTC P0101, P0102, P0103, P0106, P0107, P0108, P0112, P0113, P0117, P0118, P0121, P0122, P0123, P0171, P0172, P0201-P0206, P0351-P0356, P0401, P0440, engine runtime 120 seconds, system voltage from 11-16v, ECT sensor over 120ºF, BARO sensor over 75 kPa, Scan Tool tests all "off", vehicle speed under 1 mph with the throttle closed, and the PCM detected the Actual engine speed was 100-200 rpm more than the Desired idle speed for 5 seconds (Desired idle speed is based on ECT signal during the test). **Possible Causes:** • Fuel delivery system is too lean or too rich during the test • Inspect the engine mounts for damage to the mounts • Inspect the throttle linkage adjustment and tension • Inspect the throttle body bore for dirt or foreign material • Inspect for any air intake system leaks in the engine or hoses • IAC valve may be damaged or have failed • IAC motor control circuit is open or shorted to ground • Perform an IAC Reset function with the Scan Tool
DTC: P1514 **1T CCM, MIL: YES** **2000, 2001, 2002** **Models:** Amigo, Axiom, Rodeo, Rodeo Sport, Trooper **Engines:** 3.2L VIN W, 3.5L VIN X **Transmissions:** A/T, M/T	**TP Sensor To MAF Sensor Correlation** No MAF sensor codes set, engine running with throttle actuated mode "on", and after the Electronic Throttle Control (ETC) estimated MAF sensor reading was less than 40 g/sec with 250 test failures within 1000 test samples during the CCM Rationality test. **Possible Causes:** • Air intake system is plugged or restricted (air filter clogged) • MAF or TP sensor signal circuit open or shorted (intermittent) • ETC module is damaged or has failed
DTC: P1515 **1T CCM, MIL: YES** **2000, 2001, 2002** **Models:** Amigo, Axiom, Rodeo, Rodeo Sport, Trooper **Engines:** 3.2L VIN W, 3.5L VIN X **Transmissions:** A/T, M/T	**TP Sensor To MAF Sensor Correlation** Engine started, engine running with throttle actuator in normal mode, and the PCM detected the Command Throttle position to the Actual Throttle position was over 5%, or the Actual Throttle position was less than 40% and the Command Throttle position to Actual Throttle Position was more than -5%, or the Command Throttle Position to Actual Throttle position was more than -20%, condition met for 150 test failures within 1000 test samples for 15.6 ms in the CCM test. **Possible Causes:** • TP1 sensor circuit open or shorted to ground (intermittent fault) • TP2 sensor circuit open or shorted to ground (intermittent fault) • TP1, TP2 VREF circuit open or shorted to ground (intermittent) • TP1, TP2 ground circuit open or shorted to ground (intermittent) • ETC module is damaged or has failed

DTC	Trouble Code Title, Conditions & Possible Causes
DTC: P1516 **1T CCM, MIL: YES** **2000, 2001, 2002** **Models:** Amigo, Axiom, Rodeo, Rodeo Sport, Trooper **Engines:** 3.2L VIN W, 3.5L VIN X **Transmissions:** A/T, M/T	**ETC Command Throttle To TPS Correlation Error** Engine started, engine running with throttle actuator in normal mode, and the PCM detected the Command Throttle position to the Actual Throttle position was under 8%, with the Desired TP sensor signal steady with 0.5% for 30 seconds during the CCM Rationality test. **Possible Causes:** • TP1 sensor circuit open or shorted to ground (intermittent fault) • TP2 sensor circuit open or shorted to ground (intermittent fault) • TP1, TP2 VREF circuit open or shorted to ground (intermittent) • TP1, TP2 ground circuit open or shorted to ground (intermittent) • ETC module is damaged or has failed
DTC: P1519 **1T CCM, MIL: YES** **1998, 1999** **Models:** Oasis **Engines:** All **Transmissions:** A/T, M/T	**Idle Air Control Valve Circuit Malfunction** Key on or engine running, and the PCM detected an unexpected voltage condition on the Idle Air Control (IAC) valve control circuit. **Possible Causes:** • IAC valve control circuit is open or shorted to power • IAC valve control circuit is shorted to system power (B+) • IAC valve power circuit is open or shorted to ground • IAC valve is damaged or has failed • PCM is damaged
DTC: P1523 **1T CCM, MIL: YES** **2000, 2001, 2002** **Models**: Amigo, Axiom, Rodeo, Rodeo Sport, Trooper **Engines:** 3.2L VIN W, 3.5L VIN X **Transmissions:** A/T, M/T	**Throttle Actuator Control Return Performance** Key on, and the PCM detected the Actual TP sensor value near 0% while the Normalized TP sensor value was over 25% during the test. **Possible Causes:** • TP1 sensor circuit open or shorted to ground (intermittent fault) • TP2 sensor circuit open or shorted to ground (intermittent fault) • TP1, TP2 VREF circuit open or shorted to ground (intermittent) • TP1, TP2 ground circuit open or shorted to ground (intermittent) • ETC module is damaged or has failed
DTC: P1546 **1T CCM, MIL: NO** **1998, 1999, 2000, 2001, 2002** **Models:** Amigo **Engines:** All **Transmissions:** A/T, M/T	**A/C Compressor Clutch Output Circuit Malfunction** Engine started, engine runtime over 32 seconds, system voltage over 10v, and the PCM detected the A/C Clutch Output circuit was high with the A/C compressor clutch commanded "on", or it was low with the clutch commanded "off" for 25 seconds over a 50 second test sample during the CCM Rationality test. **Possible Causes:** • A/C clutch signal circuit is open or shorted to ground • A/C clutch signal circuit is shorted to system power (B+) • A/C clutch power circuit is open (check the A/C 10A fuse) • A/C compressor relay is damaged or has failed • PCM has failed
DTC: P1571 **1T CCM, MIL: NO** **2001, 2002** **Models:** Rodeo, Rodeo Sport **Engines:** 3.2L VIN W **Transmissions:** A/T, M/T	**Brake Switch Circuit - No Operation Detected** DTC P0502 not set, engine started, then after an acceleration and deceleration period, the PCM did not detect two changes in the Brake switch signal status after the signal changed one time. **Possible Causes:** • TP1 sensor circuit open or shorted to ground (intermittent fault) • TP2 sensor circuit open or shorted to ground (intermittent fault) • TP1, TP2 VREF circuit open or shorted to ground (intermittent) • TP1, TP2 ground circuit open or shorted to ground (intermittent) • ETC module is damaged or has failed
DTC: P1574 **1T CCM, MIL: YES** **2000** **Models:** Amigo, Rodeo **Engines:** 3.2L VIN W **Transmissions:** A/T, M/T	**Brake Light Switch Circuit Malfunction** DTC P0502 not set, engine started, vehicle driven to a speed of over 15 mph for 3 seconds, then back to a stop for 5 seconds, and the PCM did not detect any change in the Brake Light switch status during the CCM Rationality test. **Possible Causes:** • Brake light switch signal circuit is open or shorted to ground • Brake light switch power circuit is open (check the STOP fuse) • Brake light switch is damaged or has failed • PCM has failed
DTC: P1607 **1T CCM, MIL: YES** **1996, 1997, 1998, 1999** **Models:** Oasis **Engines**: All **Transmissions:** A/T, M/T	**PCM Internal Circuit 'A' Malfunction** Key on, and the PCM detected an Internal Fault 'A' condition. **Note: This trouble code indicates an internal failure in the PCM. The OEM repair procedure recommends replacing the original PCM with a "known good" PCM and then verify the code does not reset.** **Possible Causes:** • PCM is damaged or has failed

DTC	Trouble Code Title, Conditions & Possible Causes
DTC: P1618 **1T CCM, MIL: NO** **1999, 2000, 2001** **Models**: Axiom, VehiCROSS **Engines:** All **Transmissions:** A/T, M/T	**Serial Peripheral Interface Communication** Key on for 2 seconds, system voltage over 9v, and the PCM detected an internal program fault (the check sum of the data communications error) for 3 out of 6 seconds with no TCM resets during the test period of 2 seconds. **Note: For additional help with this code, view the Failure Records.** **Possible Causes:** • Check the PCM calibration to verify it is the latest calibration • Recalibrate the PCM as required • PCM may need to be replaced and reprogrammed
DTC: P1618 **1T CCM, MIL: NO** **1998, 1999, 2000** **Models:** Amigo **Engines:** 3.2L VIN W **Transmissions:** A/T, M/T	**Serial Peripheral Interface Communication** Key on for 2 seconds, system voltage over 9v, and the PCM detected an internal program fault (the check sum of the data communications error) for 3 out of 6 seconds with no TCM resets during the test period of 2 seconds. **Note: For additional help with this code, view the Failure Records.** **Possible Causes:** • Check the PCM calibration to verify it is the latest calibration • Recalibrate the PCM as required • PCM may need to be replaced and reprogrammed
DTC: P1618 **1T CCM, MIL: NO** **1996, 1997, 1998, 1999, 2000, 2001, 2002** **Models**: Rodeo, Rodeo Sport, Trooper **Engines:** 3.2L VIN V, 3.2L VIN W **Transmissions:** A/T, M/T	**Serial Peripheral Interface Communication** Key on for 2 seconds, system voltage over 9v, and the PCM detected an internal program fault (the check sum of the data communications error) for 3 out of 6 seconds with no TCM resets during the test period of 2 seconds. **Note: For additional help with this code, view the Failure Records.** **Possible Causes:** • Check the PCM calibration to verify it is the latest calibration • Recalibrate the PCM as required • PCM may need to be replaced and reprogrammed
DTC: P1621 **1T CCM, MIL: YES** **1997, 1998, 1999, 2000** **Models:** Hombre **Engines:** All **Transmissions:** A/T, M/T	**PCM Long Term Memory Performance** Key on or engine cranking, and PCM detected it was not able to communicate internally (i.e., to read data correctly from the EEPROM memory) during the initial self-test. **Note: For additional help with this code, view the Failure Records.** **Possible Causes:** • PCM needs to be replaced and reprogrammed
DTC: P1625 **1T CCM, MIL: NO** **1998, 1999, 2000, 2001, 2002** **Models:** Amigo, Axiom, Rodeo, Rodeo Sport, Trooper, VehiCROSS **Engines:** All **Transmissions:** A/T, M/T	**PCM Unexpected Reset Occurred** Key on, and the PCM detected a Clock or Computer Operating Properly (COP) reset or illegal software code interrupt occurred. **Note: For additional help with this code, view the Failure Records.** **Possible Causes:** • Perform a PCM Reset and retrieve the trouble codes
DTC: P1626 **1T CCM, MIL: NO** **1997, 1998, 1999, 2000** **Models:** Hombre **Engines:** All **Transmissions:** A/T, M/T	**Vehicle Theft Deterrent Fuel Enable Input Not Detected** Engine cranking, VTD system enabled, fuel decision point reached, and the PCM could not communicate with the VTD (Passlock) module, or it did not receive a valid password before reaching the fuel decision point. **Note: For additional help with this code, view the Failure Records.** **Possible Causes:** • Class 2 serial data line is open or shorted to ground • Passlock module is damaged or has failed • PCM is damaged or has failed
DTC: P1627 **1T PCM, MIL: YES** **1998, 1999, 2000, 2001, 2002** **Models:** Amigo, Rodeo, Rodeo Sport **Engines:** 2.2L VIN D **Transmissions:** A/T, M/T	**PCM Analog To Digital Conversion Error** Key on, and the PCM detected an Analog to Digital (A/D) Conversion error had occurred (an internal problem was detected). **Note: For additional help with this code, view the Failure Records.** **Possible Causes:** • PCM needs to be replaced and reprogrammed

DTC	Trouble Code Title, Conditions & Possible Causes
DTC: P1631 **1T PCM, MIL: NO** **1997, 1998, 1999, 2000** **Models:** Hombre **Engines**: All **Transmissions:** A/T, M/T	**Theft Deterrent Start Enable Input Incorrect** DTC P1626 not set, engine cranking, VCM not in password "learn" mode, VTD System enabled, fuel decision point has been reached, and the PCM received a number of invalid passwords before the fuel decision point was reached. If DTC B2960 is set with this code, then there is a fault in the Passlock components or the related circuits. **Note: For additional help with this code, view the Failure Records.** **Possible Causes:** • PCM or VCM has been replaced (PCM/VCM must relearn a valid password and then relearn the correct crankshaft variation • VTD controller has been replaced • Wait at least 10 minutes and then try to restart the vehicle
DTC: P1632 **1T PCM, MIL: NO** **1997, 1998, 1999, 2000** **Models:** Hombre **Engines:** All **Transmissions:** A/T, M/T	**Theft Deterrent Fuel Disable Input Received** Key on, and the PCM received a Fuel Disable signal from the BCM, or it received an undecided password from the BCM. **Note: For additional help with this code, view the Failure Records.** **Possible Causes:** • If the password is not recognized, the engine will not start • If no password or an incorrect password is received, the engine will start and immediately stall (the Theft System Telltale lamp in the I/P Cluster will flash for 4 seconds. • If over (3) invalid passwords are received the engine is disabled for 10 minutes. The Telltale lamp will come on for 3 seconds and then flash for 10 minutes (while the fuel system is disabled)
DTC: P1635 **1T CCM, MIL: YES** **2000, 2001, 2002** **Models:** Amigo, Rodeo, Rodeo Sport **Engines:** All **Transmissions:** A/T, M/T	**5-Volt Reference Voltage 'A' Circuit Malfunction** Engine started, engine running, system voltage more than 6.3v, and the PCM detected the 5v VREF 'A' was less than 4.88v or more than 5.12v for 5 seconds within a 10 second sample during the CCM test. **Possible Causes:** • VREF circuit is open (an internal fault inside the PCM) • VREF circuit is shorted to ground (an internal fault in the PCM) • VREF circuit shorted to ground between a sensor and the PCM • PCM has failed • TSB 01-04-S001 (7/01) contain information related to this code
DTC: P1639 **1T CCM, MIL: YES** **2000, 2001, 2002** **Models:** Amigo, Rodeo, Rodeo Sport **Engines:** All **Transmissions:** A/T, M/T	**5-Volt Reference Voltage 'B' Circuit Malfunction** Engine started, engine running, system voltage more than 6.3v, and the PCM detected the 5v VREF 'B' was less than 4.88v or more than 5.12v for 5 seconds within a 10 second sample during the CCM test. **Possible Causes:** • VREF circuit is open (an internal fault inside the PCM) • VREF circuit is shorted to ground (an internal fault in the PCM) • VREF circuit shorted to ground between a sensor and the PCM • PCM has failed
DTC: P1640 **1T CCM, MIL: NO** **1998, 1999, 2000, 2001, 2002** **Models:** Amigo, Rodeo, Rodeo Sport **Engines:** 2.2L VIN D **Transmissions:** A/T, M/T	**Output Driver Module 'A' Circuit Malfunction** Engine started, engine running, and the PCM detected an unexpected voltage condition on the Output Driver Module 'A' circuit (A/C Clutch Relay and Purge Solenoid) for more than 2.5 seconds during the CCM Rationality test. **Note: For additional help with this code, view the Failure Records.** **Possible Causes:** • One or more output device driver circuits connected to ODM 'A' has an open circuit condition • One or more output device driver circuits connected to ODM 'A' has a short-to-voltage condition • Check for an open power circuit to the related output devices • Disconnect the A/C clutch relay and Purge solenoid to find fault
DTC: P1640 **1T CCM, MIL: NO** **1998, 1999, 2000, 2001, 2002** **Models:** Amigo, Axiom, Rodeo, Rodeo Sport, Trooper, VehiCROSS **Engines**: 3.2L VIN W, 3.5L VIN X **Transmissions:** A/T, M/T	**Output Driver Module 'A' Circuit Malfunction** DTC P1618 not set, engine started, engine running, system voltage over 13.2v for 4 seconds, and the PCM detected an open circuit condition and an unexpected high voltage condition on the Output Driver Module circuit (A/C Clutch Relay or Purge Solenoid) with the device "on" for 2.5 seconds during the CCM Rationality test. **Note: For additional help with this code, view the Failure Records.** **Possible Causes:** • One or more output device driver circuits connected to ODM 'A' has an open circuit condition • One or more output device driver circuits connected to ODM 'A' has a short-to-voltage condition • Check for an open power circuit to the related output devices • Disconnect the A/C clutch relay and Purge solenoid to find fault

DTC	Trouble Code Title, Conditions & Possible Causes
DTC: P1640 **1T CCM, MIL: NO** **1996, 1997** **Models:** Rodeo **Engines:** 2.6L VIN E **Transmissions:** M/T	**Output Driver Module Circuit Malfunction** DTC P1618 not set, engine started, engine running, system voltage over 13.2v for 4 seconds, and the PCM detected a "high" voltage condition on the Output Driver Module (ODM) circuit with the device "on", or a "low" voltage condition on the ODM circuit with the device "off", condition met for 1 second during the CCM Rationality test. **Note: For additional help with this code, view the Failure Records.** **Possible Causes:** • One or more output device driver circuits connected to ODM has an open circuit condition • One or more output device driver circuits connected to ODM has a short-to-voltage condition • Check for an open power circuit to the related output devices • Disconnect the A/C clutch relay and Purge solenoid to find fault
DTC: P1650 **1T CCM, MIL: NO** **1998, 1999, 2000, 2001, 2002** **Models:** Amigo, Axiom, Rodeo, Rodeo Sport, Trooper, VehiCROSS **Engines:** 3.2L VIN W, 3.5L VIN X **Transmissions:** A/T, M/T	**Output Driver Module Circuit Malfunction** DTC P1618 not set, engine started, engine running, system voltage over 13.2v for 4 seconds, and the PCM detected the voltage on the Output Driver Module (ODM) circuit did not indicate less than 1.0 volt with the device commanded "on" for 0.5 seconds in the CCM test. **Note: For additional help with this code, view the Failure Records.** **Possible Causes:** • One or more output device driver circuits connected to ODM has an open circuit condition or has a short-to-voltage condition • Check for an open power circuit to the related output devices • Disconnect the A/C clutch relay and Purge solenoid to find fault
DTC: P1650 **1T CCM, MIL: NO** **1998, 1999, 2000** **Models:** Amigo **Engines:** All **Transmissions:** A/T, M/T	**Output Driver Module Circuit Malfunction** DTC P1618 not set, engine started and the PCM detected an unexpected voltage condition on the Output Driver Module circuit (purge solenoid or VIM) with the device "on" for more than 2.5 seconds during the CCM Rationality test. **Note: For additional help with this code, view the Failure Records.** **Possible Causes:** • One or more output device driver circuits connected to ODM have an open circuit condition • One or more output device driver circuits connected to ODM have a short-to-voltage condition • Check for an open power circuit to the related output devices • Disconnect the Purge solenoid and VIM to find the fault
DTC: P1705 **1T CCM, MIL: YES** **1996, 1997, 1998, 1999** **Models:** Oasis **Engines:** All **Transmissions:** A/T	**TCM A/T Gear Position Switch Low Input** Engine running, and the PCM detected a "low input" condition in the Gear Position Switch. **Note: The transaxle has no lockup function when this code is set.** **Possible Causes:** • A/T gear position switch signal circuit is shorted to ground • A/T gear position switch signal circuit is shorted to another wire • A/T gear position switch is damaged or has failed
DTC: P1706 **1T CCM, MIL: YES** **1996, 1997, 1998, 1999** **Models:** Oasis **Engines:** All **Transmissions:** A/T	**TCM A/T Gear Position Switch High Input** Key on or engine running, and the PCM detected a "high input" condition in the Gear Position Switch. **Note: The transaxle has no lockup function when this code is set.** **Possible Causes:** • A/T gear position switch signal circuit is open • A/T gear position switch is damaged or has failed
DTC: P1733 **1T CCM, MIL: YES** **1996, 1997, 1998, 1999** **Models:** Oasis **Engines:** All **Transmissions:** A/T	**TCM A/T Clutch Pressure Solenoid 'B' Circuit Malfunction** Vehicle driven in 1st, 2nd, 3rd and 4th gears, and the PCM detected an unexpected voltage condition on the Clutch Pressure Solenoid 'B' during the CCM test. **Note: The D4 lamp on the dash will blink when this code is set.** **Possible Causes:** • Clutch pressure solenoid 'B' circuit is open or shorted to ground • Clutch solenoid 'B' control circuit is shorted to system power • Clutch solenoid valve 'B' is damaged or has failed • TCM or PCM has failed
DTC: P1738 **1T CCM, MIL: NO** **1996, 1997, 1998, 1999** **Models:** Oasis **Engines:** All **Transmissions:** A/T	**TCM A/T 2nd Pressure Switch Circuit Malfunction** Engine running in gear, and the PCM detected an unexpected voltage condition on the 2nd pressure switch circuit during the test. **Possible Causes:** • A/T 2nd pressure switch signal circuit is open • A/T 2nd pressure switch signal circuit is shorted to ground • A/T 2nd pressure switch signal circuit shorted to system power • A/T 2nd pressure switch is damaged or has failed • TCM or PCM has failed

DTC	Trouble Code Title, Conditions & Possible Causes
DTC: P1739 **1T CCM, MIL: NO** **1996, 1997, 1998, 1999** **Models:** Oasis **Engines:** All **Transmissions:** A/T	**TCM A/T 3rd Pressure Switch Circuit Malfunction** Engine running in gear, and the PCM detected an unexpected voltage condition on the 3rd pressure switch circuit during the test. **Possible Causes**: • A/T 3rd pressure switch signal circuit is open • A/T 3rd pressure switch signal circuit is shorted to ground • A/T 3rd pressure switch signal circuit shorted to system power • A/T 3rd pressure switch is damaged or has failed • TCM or PCM has failed
DTC: P1753 **1T CCM, MIL: YES** **1996, 1997, 1998, 1999** **Models:** Oasis **Engines:** All **Transmissions:** A/T	**TCM A/T Lockup Solenoid Valve 'A' Circuit Malfunction** Vehicle driven in 1st, 2nd, 3rd and 4th gears, and the PCM detected an unexpected voltage condition on the Solenoid 'A' circuit during the CCM test. **Note: The D4 lamp on the dash will blink when this code is set.** **Possible Causes:** • Lockup solenoid 'A' control circuit is open or shorted to ground • Lockup solenoid 'A' circuit is shorted to system power • Lockup solenoid 'A' is damaged or has failed • TCM or PCM has failed
DTC: P1768 **1T CCM, MIL: YES** **1996, 1997, 1998, 1999** **Models:** Oasis **Engines**: All **Transmissions:** A/T	**TCM A/T Clutch Pressure Solenoid 'A' Circuit Malfunction** Vehicle driven in 1st, 2nd, 3rd and 4th gears, and the PCM detected an unexpected voltage condition on the Clutch Pressure Solenoid 'A' during the CCM test. **Note: The D4 lamp on the dash will blink when this code is set.** **Possible Causes:** • Clutch pressure solenoid 'A' circuit is open or shorted to ground • Clutch solenoid 'A' control circuit is shorted to system power • Clutch solenoid valve 'A' is damaged or has failed • TCM or PCM has failed
DTC: P1790 **2T PCM, MIL: YES** **2001, 2002** **Models:** Rodeo, Rodeo Sport **Engines:** 2.2L VIN D **Transmissions:** A/T	**PCM ROM (Transmission Side) Check Sum Error** Key on, and the PCM detected a ROM Check Sum Error occurred in the Transmission Side of the controller for 1 second. **Note: For additional help with this code, view the Failure Records.** **Possible Causes:** • Reprogram the Transmission EEPROM • Recheck for the trouble code, and if it resets, the PCM may need to be replaced and reprogrammed
DTC: P1790 **2T PCM, MIL: YES** **1998, 1999, 2000, 2001, 2002** **Models:** Amigo, Rodeo, Rodeo Sport **Engines:** 3.2L VIN W **Transmissions:** A/T	**PCM ROM (Transmission Side) Check Sum Error** Key on, and the PCM detected a ROM Check Sum Error occurred in the Transmission Side of the controller for 1 second. **Note: For additional help with this code, view the Failure Records.** **Possible Causes:** • Reprogram the Transmission EEPROM • Recheck for the trouble code, and if it resets, the PCM may need to be replaced and reprogrammed
DTC: P1790 **2T PCM, MIL: YES** **1996, 1997, 1998, 1999** **Models:** Trooper, VehiCROSS **Engines:** All **Transmissions:** A/T	**PCM ROM (Transmission Side) Check Sum Error** Key on, and the PCM detected a ROM Check Sum Error occurred in the Transmission Side of the controller for 1 second. **Note: For additional help with this code, view the Failure Records.** **Possible Causes:** • Reprogram the Transmission EEPROM • Recheck for the trouble code, and if it resets, the PCM may need to be replaced and reprogrammed
DTC: P1792 **2T CCM, MIL: YES** **1998, 1999, 2000, 2001, 2002** **Models:** Amigo, Rodeo, Rodeo Sport **Engines:** 3.2L VIN W **Transmissions:** A/T	**VCM EEPROM (Transmission Side) Check Sum Error** Key on, and the PCM detected an EEPROM Check Sum Error occurred in the Transmission Side of the controller for 1 second. **Note: For additional help with this code, view the Failure Records.** **Possible Causes:** • Reprogram the Transmission EEPROM • Recheck for the trouble code, and if it resets, the PCM may need to be replaced and reprogrammed
DTC: P1792 **2T CCM, MIL: YES** **2001, 2002** **Models**: Rodeo, Rodeo Sport **Engines:** 2.2L VIN D **Transmissions:** A/T	**VCM EEPROM (Transmission Side) Check Sum Error** Key on, and the PCM detected an EEPROM Check Sum Error occurred in the Transmission Side of the controller for 1 second. **Note: For additional help with this code, view the Failure Records.** **Possible Causes:** • Reprogram the Transmission EEPROM • Recheck for the trouble code, and if it resets, the PCM may need to be replaced and reprogrammed

DTC	Trouble Code Title, Conditions & Possible Causes
DTC: P1792 **2T PCM, MIL: YES** **1996, 1997, 1998, 1999** **Models:** Trooper, VehiCROSS **Engines:** All **Transmissions:** A/T	**VCM EEPROM (Transmission Side) Check Sum Error** Key on, and the PCM detected an EEPROM Check Sum Error occurred in the Transmission Side of the controller for 1 second. **Note: For additional help with this code, view the Failure Records.** **Possible Causes:** • Reprogram the Transmission EEPROM • Recheck for the trouble code, and if it resets, the PCM may need to be replaced and reprogrammed
DTC: P1810 **2T CCM, MIL: NO** **1996, 1997, 1998, 1999, 2000** **Models:** Hombre **Engines:** All **Transmissions:** A/T	**TFP Manual Valve Position Switch Circuit Malfunction** DTC P0502 and P0503 not set, engine started, engine speed over 450 rpm for 5 seconds, system voltage from 10-16v, Fuel Cutoff mode "off", and the PCM detected an illegal TFP Valve Position switch state for 60 seconds; or with the engine speed less than 80 rpm for 1 second, then from 80-600 rpm for 800 ms, then over 600 rpm, VSS indicating over 2 mph, the detected gear range was D2, D4 or Reverse for 5 seconds at startup; or while in D4 Gear range, TP angle at 10-50%, engine torque 40-450 lbs, TCC is "on", speed ratio at 0.65-0.75, the detected gear range was Park/Neutral position for 10 seconds when the gear selector should indicate the D4 range. **Note: The Speed Ratio is calculated by dividing the engine speed value by the transmission output speed value.** **Possible Causes:** • TFP position switch signal circuit is open (one or more circuits) • TFP position switch signal circuit is shorted to ground • TFP position switch is damaged, dirty or out-of-adjustment • PCM has failed
DTC: P1835 **1T CCM, MIL: NO** **2000, 2001, 2002** **Models:** Amigo, Rodeo, Rodeo Sport **Engines:** 2.2L VIN D **Transmissions:** A/T	**Kick Down Switch Circuit Low Input** DTC P0122 and P0123 not set, engine started, engine running, throttle position less than 70%, and the PCM detected the Kick Down Switch remained in "on" position during the CCM test. **Note: For additional help with this code, view the Failure Records.** **Possible Causes:** • Kick down switch signal circuit is shorted to ground • Kick down switch ground circuit is open (from switch to ground) • Kick down switch is damaged, out-of-adjustment or has failed • PCM has failed
TC: P1850 **1T CCM, MIL: NO** **1998, 1999, 2000, 2001, 2002** **Models:** Amigo, Axiom, Rodeo, Rodeo Sport, Trooper, VehiCROSS **Engines:** 3.2L VIN W, 3.5L VIN X **Transmissions:** A/T	**Kick Down Switch Circuit Low Input** DTC P0122 and P0123 not set, engine started, engine running, throttle position less than 70%, and the PCM detected the Kick Down Switch remained in "on" position during the CCM test. **Note: For additional help with this code, view the Failure Records.** **Possible Causes:** • Kick down switch signal circuit is shorted to ground • Kick down switch ground circuit open between switch and ground • Kick down switch is damaged, out-of-adjustment or has failed • PCM has failed
DTC: P1850 **1T CCM, MIL: NO** **1998, 1999, 2000, 2001, 2002** **Models:** Amigo, Axiom, Rodeo, Rodeo Sport, Trooper, VehiCROSS **Engines:** 3.2L VIN W, 3.5L VIN X **Transmissions:** A/T	**A/T Brake Band Apply Solenoid Circuit Malfunction** Engine started, engine running, then with A/T Brake Band Apply solenoid commanded "on", and PCM detected the solenoid control signal was 12v, or with A/T Brake Band Apply solenoid commanded "off", the solenoid control signal was 1.34 to 1.56 seconds. **Note: The PCM controls this solenoid with a pulsewidth modulated (PWM) control signal.** **Possible Causes:** • A/T Brake band solenoid circuit is open or shorted to ground • A/T Brake band solenoid High circuit open or shorted to ground • A/T brake band apply solenoid is damaged or has failed • PCM has failed
DTC: P1850 **1T CCM, MIL: NO** **2000, 2001, 2002** **Models:** Amigo, Rodeo, Rodeo Sport **Engines:** 2.2L VIN D **Transmissions:** A/T	**A/T Brake Band Apply Solenoid Circuit Malfunction** Engine started, engine running, then with A/T Brake Band Apply solenoid commanded "on", and PCM detected the solenoid control signal was 12v, or with A/T Brake Band Apply solenoid commanded "off", the solenoid control signal was 1.34 to 1.56 seconds. **Note: The PCM controls this solenoid with a pulsewidth modulated (PWM) control signal.** **Possible Causes:** • A/T Brake band solenoid circuit is open or shorted to ground • A/T Brake band solenoid High circuit open or shorted to ground • A/T brake band apply solenoid is damaged or has failed • PCM has failed

DTC	Trouble Code Title, Conditions & Possible Causes
DTC: P1860 **2T CCM, MIL: YES** **2000, 2001, 2002** **Models:** Amigo, Rodeo, Rodeo Sport **Engines:** 2.2L VIN D **Transmissions:** A/T	**Torque Converter Clutch PWM Solenoid Circuit Malfunction** DTC P0751, P0752, P0753, P0756, P0757 and P0758 not set, engine started, engine running, then with the TCC solenoid commanded "on", the PCM detected the solenoid control circuit signal was 12v, or with the TCC solenoid commanded "off", it detected the TCC control circuit signal was 0v for 1.25 seconds. **Possible Causes:** • TCC PWM solenoid control circuit open or shorted to ground • TCC PWM solenoid control circuit is shorted to power • TCC PWM solenoid is damaged or has failed • PCM has failed
DTC: P1860 **2T CCM, MIL: YES** **1998, 1999, 2000, 2001, 2002** **Models:** Amigo, Axiom, Rodeo, Rodeo Sport, Trooper, VehiCROSS **Engines:** 3.2L VIN W, 3.5L VIN X **Transmissions:** A/T	**Torque Converter Clutch PWM Solenoid Circuit Malfunction** DTC P0751, P0752, P0753, P0756, P0757 and P0758 not set, engine started, engine running, then with the TCC solenoid commanded "on", the PCM detected the solenoid control circuit signal was 12v, or with the TCC solenoid commanded "off", it detected the TCC control circuit signal was 0v for 1.25 seconds. **Possible Causes:** • TCC PWM solenoid control circuit open or shorted to ground • TCC PWM solenoid control circuit is shorted to power • TCC PWM solenoid is damaged or has failed • PCM has failed
DTC: P1860 **1T CCM, MIL: YES** **1996, 1997, 1998, 1999, 2000** **Models**: Hombre **Engines:** All **Transmissions:** A/T	**TCC PWM Solenoid Circuit Malfunction** Engine started, 1st Gear commanded "on", and the PCM detected the TCC solenoid control circuit voltage indicated "high" with the TCC solenoid command at 90%, or it detected the TCC solenoid control circuit indicated "low" with the TCC solenoid command 0%. **Possible Causes:** • TCC PWM solenoid control circuit open or shorted to ground • TCC PWM solenoid control circuit is shorted to power • TCC PWM solenoid is damaged or has failed • PCM has failed
DTC: P1870 **2T CCM, MIL: YES** **2000, 2001, 2002** **Models:** Amigo, Rodeo, Rodeo Sport **Engines:** 2.2L VIN 4 **Transmissions:** A/T	**Transmission Component Slipping Malfunction** DTC P0722, P0723, P0742, P0751, P0752, P0753, P0756, P0757, P0758, P1860 and P1870 not set, engine started, then driven to a speed of 15-58 mph, engine speed from 1000-3500 rpm, TP sensor signal from 15-99%, MAP sensor signal from 0-70 kPa, 50 less than Engine Torque less than 300 Nm, gear selector in D4, TFT sensor signal from 68-302ºF, speed ratio at 0.6-0.95, and the PCM detected the TCC slip speed was 250-800 rpm (event occurred 3 times within 7 seconds). **Possible Causes:** • Engine speed signal circuit open or shorted (intermittent fault) • Internal transmission component problem • TCC PWM or Shift Solenoids have failed (mechanical fault) • TR switch is damaged, out-of-adjustment or has failed
DTC: P1870 **2T CCM, MIL: YES** **1998, 1999, 2000, 2001, 2002** **Models:** Amigo, Axiom, Rodeo, Rodeo Sport, Trooper, VehiCROSS **Engines:** 3.2L VIN W, 3.5L VIN X **Transmissions:** A/T	**Transmission Component Slipping Malfunction** DTC P0722, P0723, P0742, P0751, P0752, P0753, P0756, P0757, P0758, P1860 and P1870 not set, engine started, then driven to a speed of 15-58 mph, engine speed from 1000-3500 rpm, TP sensor signal from 15-99%, MAP sensor signal from 0-70 kPa, 50 < Engine Torque < 300 Nm, gear selector in D4, TFT sensor signal from 68-302ºF, speed ratio at 0.6-0.95, and the PCM detected the TCC slip speed was 250-800 rpm (event occurred 3 times within 7 seconds). **Possible Causes:** • Engine speed signal circuit open or shorted (intermittent fault) • Internal transmission component problem • TCC PWM or Shift Solenoids have failed (mechanical fault) • TR switch is damaged, out-of-adjustment or has failed
DTC: P1870 **2T CCM, MIL: YES** **1996, 1997, 1998, 1999, 2000** **Models:** Hombre **Engines:** 2.2L VIN 4 **Transmissions:** A/T	**Transmission Component Slipping Malfunction** DTC P0P0122, P0123, P0502, P0503, P0740, P0753, P0758, P0785, P1810 and P1860 not set, engine started, then driven to a speed of 35-65 mph in D4 range (not in 1st Gear), engine speed from 1500-3500 rpm, Fuel Cutoff mode "off", TP sensor signal from 15-99%, engine torque from 40-450 lbs, speed ratio from 0.6-0.95, TFT sensor signal from 68-266ºF, TCC commanded "on" at maximum duty cycle for 5 seconds, shift solenoid performance diagnostic counters are at zero (0), and the PCM detected the TCC slip speed was 300-1000 rpm, conditions met for 7 seconds for two consecutive times. **Possible Causes:** • Internal transmission component problem • TCC PWM or Shift Solenoids have failed (mechanical fault)

DTC	Trouble Code Title, Conditions & Possible Causes
DTC: P1870 **2T CCM, MIL: YES** **1996, 1997, 1998, 1999, 2000** **Models:** Hombre **Engines:** 4.3L VIN X **Transmissions:** A/T	**Transmission Component Slipping Malfunction** DTC P0P0122, P0123, P0502, P0503, P0711, P0712, P0713, P0740, P0753, P0758, P0785, P1810 and P1860 not set, engine started, then driven to a speed of 35-65 mph in D4 range (not in 1st Gear), engine speed from 1500-3500 rpm, Fuel Cutoff mode "off", TP sensor signal from 15-99%, engine torque from 50-400 lbs, speed ratio from 0.69-0.88, TFT sensor signal from 68-266°F, Shift Solenoid performance diagnostic counters are at zero (0), and the PCM detected with the TCC commanded "on" at a 95% duty cycle for 5 seconds, that the TCC slip speed was 130-800 rpm, conditions occurred during three (3) different TCC cycles during the CCM test. **Possible Causes:** • Internal transmission component problem • TCC PWM or Shift Solenoids have failed (mechanical fault)
DTC: P1875 **2T CCM, MIL: YES** **2000** **Models:** Hombre **Engines:** 2.2L VIN 4 **Transmissions:** A/T	**4-Wheel Drive Low Switch Circuit Malfunction** DTC P0P0122, P0123, P0502, P0503, P0740, P0751, P0753, P0756, P0758, P1810 and P1860 not set, engine started, engine speed over 450 rpm for 5 seconds, Fuel Cutoff not "on", TP angle from 17-50%, engine vacuum from 0-105 kPa, engine torque at 50-400 lbs, TFT sensor signal from 68-266°F, gear selector in D4 Range, Shift Solenoid Performance diagnostic counters at zero (0), TCC slip speed from 300-1000 rpm, then with the transfer case ratio at 0.8-1.2, the PCM detected the 4WD Low Switch was in 4WD low with the Transfer Case not in 4WD for 5 seconds; or with the TCC "on" and TCC slip speed at 100-3000 rpm, transfer case ratio at 2.5-2.9, the PCM detected the Transfer Case was in 4WD low while the 4WD low switch did not indicate it was in 4WD low. **Possible Causes:** • 4WD low switch signal circuit is open • 4WD low switch signal circuit is shorted to ground • 4WD low switch is damaged or has failed
DTC: P1875 **2T CCM, MIL: YES** **2000** **Models:** Hombre **Engines:** 4.3L VIN X **Transmissions:** A/T	**4-Wheel Drive Low Switch Circuit Malfunction** DTC P0P0122, P0123, P0502, P0503, P0740, P0751, P0753, P0756, P0758, P1810 and P1860 not set, engine started, engine speed over 450 rpm for 5 seconds, Fuel Cutoff not "on", throttle angle from 17-50%, engine vacuum from 0-105 kPa, engine torque at 50-400 lbs, TFT sensor signal from 68-266°F, gear in D4 Range, Shift Solenoid Performance diagnostic counters at zero; and the PCM detected the 4WD switch was in 4WD low with the transfer case not in 4WD low, and the TCC slip speed from -3000 to -50 rpm or the transfer case ratio was 0.8-1.2; or the PCM detected the 4WD Low Switch was not in 4WD low with the Transfer Case in 4WD, TCC commanded "on", and the TCC slip speed was 100-3000 rpm, or the transfer case ratio was 2.5-2.9, condition met for 10 seconds. **Possible Causes:** • 4WD low switch signal circuit is open • 4WD low switch signal circuit is shorted to ground • 4WD low switch is damaged or has failed

Gas Engine OBD II Trouble Code List (U1xxx Codes)

DTC	Trouble Code Title
DTC: U1000 **1T CCM, MIL: YES** **1997, 1998, 1999, 2000, 2001, 2002** **Models:** Amigo, Hombre Engines: All **Transmissions:** A/T, M/T	**No Class 2 Communication (ID Not Learned)** Key on, then the PCM determined that it could not communicate with the EBCM or the Truck Body Controller Module (TBM) for 15 seconds after the ignition key was turned "on". **Note: If electrical interference is suspected, install an additional ground wire from the IPC to the battery negative terminal as a fix.** **Possible Causes:** • Class 2 serial data bus circuit is open or shorted to ground • One or more modules on the data bus are shorted • EBCM or TBM has failed
DTC: U1026 **1T CCM, MIL: YES** **1998, 1999, 2000** **Models**: Amigo, Hombre **Engines**: All **Transmissions:** A/T, M/T	**Loss of Electronic Brake Controller Communication** Key on or engine running, and after the PCM/VCM established communications and received NODE Alive/SOH messages during this ignition cycle from the Active Transfer Case (ATC) module, the PCM/VCM did not detect communications, and no NODE Alive/SOH messages were received from the ATC module for 5 seconds. **Possible Causes:** • Class 2 serial data bus circuit is open or shorted to ground • One or more modules on the data bus are shorted • ATC module has failed
DTC: U1041 **1T CCM, MIL: YES** **1998, 1999, 2000** **Models:** Amigo, Hombre **Engines:** All **Transmissions:** A/T, M/T	**Loss of Electronic Brake Controller Communication** Key on, then the PCM determined that it could not communicate with the Electronic Brake Controller Module (EBCM) for 15 seconds after it had already established communication with the EBCM (at the initial key on period). **Note: If electrical interference is suspected, install an additional ground wire from the IPC to the battery negative terminal as a fix.** **Possible Causes:** • Class 2 serial data bus circuit is open or shorted to ground • One or more modules on the data bus are shorted • EBCM has failed
DTC: U1064 **1T CCM, MIL: YES** **1998, 1999, 2000** **Models:** Amigo, Hombre **Engines:** All **Transmissions:** A/T, M/T	**Loss of Truck Body Controller Communication** Key on, then the PCM determined that it could not communicate with the Truck Body Controller Module (TBM) for 15 seconds after it had already established communication with the TBM (at the initial key on period). **Possible Causes:** • Class 2 serial data bus circuit is open or shorted to ground • One or more modules on the data bus are shorted • TBM has failed

KIA
DIAGNOSTIC TROUBLE CODES

10

TABLE OF CONTENTS

VEHICLE APPLICATIONS 10-2
P0XXX 10-3
P1XXX 10-41
P2XXX 10-59

DIAGNOSTIC TROUBLE CODES

OBD II VEHICLE APPLICATIONS

KIA

Amanti

2004-2006
3.5L .... VIN 4

Optima

2001-2004
2.4L .... VIN S

2005
2.4L .... VIN 6

2006
2.4L .... VIN 3

2001
2.5L .... VIN 4

2002-2005
2.7L .... VIN 8

2006
2.7L .... VIN 4

Rio

2001-2002
2.5L .... VIN 3

2003-2005
1.6L .... VIN 5

2006
1.6L .... VIN 3

Sedona

2002-2005
3.5L .... VIN 1

2006
3.8L .... VIN 3

Sephia

1996-1997
1.6L .... VIN 3

1996-1997
1.6L .... VIN 4

1996-1998
1.8L .... VIN 5

1998-2001
1.8L .... VIN 1

Sorento

2003-2006
3.5L .... VIN 3

Spectra

2000-2003
1.8L .... VIN 1

2004-2006
2.0L .... VIN 1

2004-2006
2.0L .... VIN 2

Sportage

1996-2002
2.0L .... VIN 3

2005-2006
2.0L .... VIN 2

2005-2006
2.0L .... VIN 4

2005-2006
2.7L .... VIN 3

Gas Engine OBD II Trouble Code List (P0xxx Codes)

DTC	Trouble Code Title, Conditions & Possible Causes
DTC: P0011 **2T CCM, MIL: Yes** **2004-06** **Models:** Optima, Rio, Sedona, Spectra, Sportage **Engines:** 1.6L, 2.0L, 2.4L, 2.7L, 3.8L **Transmissions:** All	**"A" Camshaft Position Timing Over Advanced or System Performance (Bank 1)** Monitor deviation between camshaft position set point and actual value. **Possible Causes:** • Oil leakage • Faulty oil pump • Faulty intake valve control solenoid
DTC: P0012 **2T CCM, MIL: Yes** **2006** **Models:** Optima, Rio, Sedona **Engines:** 1.6L, 2.7L, 3.8L **Transmissions:** All	**"A" Camshaft Position- Timing Over retarded (Bank 1)** Determines if the phaser is stuck or has a steady error. **Possible Causes:** • Engine oil • OCV • CVVT stuck • Faulty PCM
DTC: P0016 **2T CCM, MIL: Yes** **2004-06** **Models:** Amanti, Optima, Rio, Sedona, Spectra, Sportage **Engines:** 1.6L, 2.0L, 2.4L, 2.7L, 3.5L, 3.8L **Transmissions:** All	**Crankshaft Position- Camshaft Position Correlation (Bank 1 Sensor "A")** Monitor camshaft position in the full retard condition or during CVVT control. Camshaft switching out of 109 to 141 degrees in full retard position, 70 to 140 degrees CRK during CVVT control. **Possible Causes:** • Abnormal installation of camshaft • Abnormal installation of crankshaft • Abnormal installation of tone wheel
DTC: P0018 **2T CCM, MIL: Yes** **2006** **Models:** Optima, Sedona **Engines:** 2.7L, 3.8L **Transmissions:** All	**Crankshaft Position- Camshaft Position Correlation (Bank 2 Sensor A)** Determines if CAM (B2) target is aligned correctly to the crank. No active faults. **Possible Causes:** • CKPS, CMPS (B2) • CVVT • Timing misalignment • Faulty PCM
DTC: P0021 **2T CCM, MIL: Yes** **2006** **Models:** Optima, Sedona **Engines:** 2.7L, 3.8L **Transmissions:** All	**"A" Camshaft Position- Timing Over Advanced Or System Performance (Bank 2)** Determines if the phaser is moving at an unexpected rate. Cam off set is available. **Possible Causes:** • Excessive phasing • Binding oil pressure (blockage) • Faulty PCM
DTC: P0022 **2T CCM, MIL: Yes** **2006** **Models:** Optima, Sedona **Engines:** 2.7L, 3.8L **Transmissions:** All	**"A" Camshaft Position- Timing Over Retarded (Bank 2)** Determines if the phaser is stuck or has a steady state error. Off sets available. Cam velocity below threshold at 15 CAD/s. **Possible Causes:** • Engine oil • OCV • CVVT stuck • Faulty PCM
DTC: P0026 **2T CCM, MIL: Yes** **2006** **Models:** Optima, Sedona **Engines:** 2.7L, 3.8L **Transmissions:** All	**Intake Valve Control Solenoid Circuit Range/Performance (Bank 1)** Determines if oil control valve is stuck. Valve cleaning not in progress. Off sets available. **Possible Causes:** • Oil pressure loss • OCV seizure • Faulty PCM
DTC: P0028 **2T CCM, MIL: Yes** **2006** **Models:** Optima, Sedona **Engines:** 2.7L, 3.8L **Transmissions:** All	**Intake Valve Control Solenoid Circuit Range/Performance (Bank 2)** Determines if oil control valve is stuck. Valve cleaning not in progress. Off sets available. **Possible Causes:** • Oil pressure loss • OCV seizure • Faulty PCM

DTC	Trouble Code Title, Conditions & Possible Causes
DTC: P0030 **2T CCM, MIL: Yes** **2001-06** **Models:** Optima, Rio, Spectra, Sportage **Engines:** All **Transmissions:** All	**HO2S-11 Heater Circuit Malfunction** Engine started, engine runtime over 3 minutes, and the PCM determined the resistance of the HO2S heater was more than a calculated amount. **Possible Causes:** • HO2S heater control circuit is open or shorted to ground • HO2S heater control circuit is shorted to system power (B+) • HO2S heater is damaged or has failed • PCM has failed
DTC: P0031 **2T CCM, MIL: Yes** **2003-06** **Models:** Amanti, Optima, Rio, Sedona, Spectra, Sportage **Engines:** 1.6L, 1.8, 2.0L, 2.4L, 2.7L, 3.5L, 3.8L **Transmissions:** All	**O2 Sensor Heater Circuit Low (Bank 1/Sensor 1)** Heater check, low. Open or short circuit. **Possible Causes:** • Open in battery and control circuit • Short to ground in control circuit (pin 48 to 36) • Faulty HO2S heater • Faulty PCM
DTC: P0032 **2T CCM, MIL: Yes** **2003-06** **Models:** Amanti, Optima, Rio, Sedona, Spectra, Sportage **Engines:** 1.6L, 1.8, 2.0L, 2.4L, 2.7L, 3.5L, 3.8L **Transmissions:** All	**O2 Sensor Heater Circuit High (Bank 1/Sensor 1)** **Heater check, high. Short circuit.** **Possible Causes:** • Short to battery in control circuit • Faulty HO2S heater • Faulty PCM
DTC: P0075 **2T CCM, MIL: Yes** **2006** **Models:** Rio **Engines:** 1.6L **Transmissions:** All	**Intake Valve Control Solenoid Circuit (Bank 1)** Circuit continuity check, open. **Possible Causes:** • Poor connection • Open or short to ground in power circuit • Open in control circuit • OCV • Faulty ECM/PCM
DTC: P0076 **2T CCM, MIL: Yes** **2004-06** **Models:** Optima, Rio, Sedona, Spectra, Sportage **Engines:** 1.6L, 2.0L, 2.4L, 2.7L, 3.8L **Transmissions:** All	**Intake Valve Control Solenoid Circuit Low (Bank 1)** PCM sets the code if it detects that the intake valve control solenoid control circuit is short to ground. Electrical check. **Possible Causes:** • Faulty ECM/PCM • Short to ground in control circuit • Contact resistance in connectors • Faulty intake valve control solenoid
DTC: P0077 **2T CCM, MIL: Yes** **2004-06** **Models:** Optima, Rio, Sedona, Spectra, Sportage **Engines:** 1.6L, 2.0L, 2.4L, 2.7L, 3.8L **Transmissions:** All	**Intake Valve Control Solenoid Circuit High (Bank 1)** PCM sets the code if it detects that the OCV control circuit is open or short to battery. Electrical check. **Possible Causes:** • Open or short to battery in control circuit • Contact resistance in connectors • Faulty intake valve control solenoid • Faulty PCM
DTC: P0082 **2T CCM, MIL: Yes** **2006** **Models:** Optima, Sedona **Engines:** 2.7L, 3.8L **Transmissions:** All	**Intake Valve Control Solenoid Circuit Low (Bank 2)** Detects a short to ground or open circuit of VCPD bank 1 intake circuit output. No disabling faults present. Engine running. Enable time delay equal to or greater than 0.5 second. **Possible Causes:** • Poor connection • Open in power circuit • Open or short to ground in control circuit • OCV • Faulty PCM

DTC	Trouble Code Title, Conditions & Possible Causes
DTC: P0159 **2T O2S2, MIL: Yes** **2002-06** **Models:** Amanti, Optima, Sedona, Sorento, Sportage **Engines:** All **Transmissions:** All	**HO2S-22 (Bank 2 Sensor 2) Slow Response** Engine started, engine running in closed loop at a speed over 3 mph for more than 2 minutes, and the PCM detected the average ratio between the HO2S Actual and maximum allowed frequency during 100 Lambda cycles was more than the Threshold value (e.g., 0.66 Hz) during the Oxygen Sensor Monitor test. **Possible Causes:** • Exhaust leak present in the exhaust manifold or exhaust pipes • Front HO2S failed, or front & rear HO2S connections reversed • HO2S has deteriorated, is contaminated or has failed
DTC: P0160 **2T CCM, MIL: Yes** **2002-06** **Models:** Amanti, Optima, Sedona, Sorento, Sportage **Engines:** All **Transmissions:** All	**HO2S-22 (Bank 2 Sensor 2) Circuit Malfunction** Engine started, engine running in closed loop at a speed over 3 mph, and the PCM detected an unexpected high voltage condition on the HO2S circuit during the CCM test. **Possible Causes:** • HO2S signal circuit is shorted to VREF or to the Heater power • HO2S signal circuit is shorted to system power (B+) • HO2S is damaged or has failed • PCM has failed
DTC: P0161 **2T O2S2 HTR2, MIL: Yes** **2002-06** **Models:** Amanti, Optima, Sedona, Sorento **Engines:** All **Transmissions:** All	**HO2S-22 (Bank 2 Sensor 2) Heater Circuit Malfunction** Engine started, engine running and PCM detected the HO2S heater current was more than 2 amps, or it was less than 0.25 amps. **Possible Causes:** • HO2S power feed circuit from the Main Relay is open • HO2S heater control circuit is open • HO2S heater element has high resistance, is shorted or open • HO2S heater is damaged or has failed • PCM has failed
DTC: P0170 **2T FUEL, MIL: Yes** **1996-06** **Models:** Optima, Rio, Sedona, Sephia, Spectra, Sportage **Engines:** All **Transmissions:** All	**Fuel System Too Rich or Too Lean (Bank 1)** DTC P0171and P0172 not set, engine running in closed loop for over 2 minutes, and the PCM detected the amount of rich or lean Fuel Trim correction exceeded the Threshold maximum. **Possible Causes:** • Air leaks present in the exhaust manifold or exhaust pipes • Air being drawn in from leaks in engine gaskets or other seals • Incorrect fuel pressure, or one or more fuel injectors has failed • Front HO2S element is contaminated or has failed • A "fuel control" sensor is out of calibration (BARO, ECT or IAT)
DTC: P0171 **2T FUEL, MIL: Yes** **1998-06** **Models:** Amanti, Optima, Rio, Sedona, Sorento, Sephia, Spectra, Sportage **Engines:** All **Transmissions:** All	**Fuel System Too Lean (Bank 1)** Engine running in closed loop at a speed of over 5 mph for 2-3 minutes, and the PCM detected the Lambda correction value exceeded the "high" Threshold limit, condition met for 200 seconds. **Possible Causes:** • Air leaks in intake manifold, exhaust pipes or exhaust manifold • One or more injectors restricted or pressure regulator has failed • Air is being drawn in from leaks in gaskets or other seals • O2S element is deteriorated or has failed • A "fuel control" sensor is out of calibration (ECT, IAT or MAP)
DTC: P0172 **2T FUEL, MIL: Yes** **1998-06** **Models:** Amanti, Optima, Rio, Sedona, Sorento, Sephia, Spectra, Sportage **Engines:** All **Transmissions:** All	**Fuel System Too Rich (Bank 1)** Engine running in closed loop at a speed of over 5 mph for 2-3 minutes, and the PCM detected the Lambda correction value exceeded the "low" Threshold limit, condition met for 240 seconds. **Possible Causes:** • One or more injectors leaking or pressure regulator is leaking • O2S element is deteriorated or has failed • EVAP vapor recovery system has failed (canister full of fuel) • Base engine fault (i.e., cam timing incorrect, engine oil too high • A "fuel control" sensor is out of calibration (ECT, IAT or MAP)
DTC: P0173 **2T Fuel, MIL: Yes** **2005-06** **Models:** Sportage **Engines:** 2.7L **Transmissions:** All	**Fuel Trim Too Rich or Too Lean (Bank 2)** Engine running in closed loop, and the PCM detected the Fuel system was too rich or too lean during two or more consecutive trips. **Possible Causes:** • Base engine "mechanical" fault affecting one or more cylinders • EVAP system component has failed or canister fuel saturated • Exhaust leaks located in front of the HO2S location • Fuel control sensor is out of calibration (i.e., ECT, IAT or MAF) • Fuel delivery system supplying too much fuel during cruise or idle periods (e.g., faulty fuel pump, or faulty pressure regulator) • Fuel injector(s) is leaking or stuck partially open (one or more) • HO2S is contaminated, deteriorated or it has failed

DTC	Trouble Code Title, Conditions & Possible Causes
DTC: P0174 **2T FUEL, MIL: Yes** **2001-06** **Models:** Amanti, Optima, Sedona, Sorento, Sportage **Engines:** All **Transmissions:** All	**Fuel System Too Lean (Bank 2)** Engine running in closed loop at a speed of over 5 mph for 2-3 minutes, and the PCM detected the Lambda correction value exceeded the "high" Threshold limit, condition met for 200 seconds. **Possible Causes:** • Air leaks in intake manifold, exhaust pipes or exhaust manifold • One or more injectors restricted or pressure regulator has failed • Air is being drawn in from leaks in gaskets or other seals • O2S element is deteriorated or has failed • A "fuel control" sensor is out of calibration (ECT, IAT or MAP)
DTC: P0175 **2T FUEL, MIL: Yes** **2001-06** **Models:** Amanti, Optima, Sedona, Sorento, Sportage **Engines:** All **Transmissions:** All	**Fuel System Too Rich (Bank 1)** Engine running in closed loop at a speed of over 5 mph for 2-3 minutes, and the PCM detected the Lambda correction value exceeded the "low" Threshold limit, condition met for 240 seconds. **Possible Causes:** • One or more injectors leaking or pressure regulator is leaking • O2S element is deteriorated or has failed • EVAP vapor recovery system has failed (canister full of fuel) • Base engine fault (i.e., cam timing incorrect, engine oil too high • A "fuel control" sensor is out of calibration (ECT, IAT or MAP)
DTC: P0180 **2T CCM, MIL: Yes** **2002-02** **Models:** Sedona **Engines:** All **Transmissions:** All	**Fuel Temperature Sensor Circuit Malfunction** Key on or engine running and the PCM detected an unexpected voltage condition on the Fuel Temperature sensor circuit. **Possible Causes:** • Fuel temperature sensor signal circuit open, shorted to ground • Fuel temperature sensor circuit shorted to VREF or power (B+) • Fuel temperature sensor is damaged or has failed • PCM has failed
DTC: P0181 **2T CCM, MIL: Yes** **2003-06** **Models:** Amanti, Optima, Sedona, Sorento **Engines:** 2.4L, 3.5L **Transmissions:** All	**Fuel Temperature Sensor "A" Circuit Range/Performance** If the voltage difference between fuel tank temperature and Engine coolant temperature at starting is greater than threshold value (above 59 degree F), a fault code exists. Rationality check. Engine coolant 14 to 122 degrees F. **Possible Causes:** • Poor connection • Faulty fuel temperature sensor • Faulty PCM
DTC: P0182 **2T CCM, MIL: Yes** **2003-06** **Models:** Amanti, Optima, Sedona, Sorento **Engines:** 2.4L, 3.5L **Transmissions:** All	**Fuel Temperature Sensor "A" Circuit Low Input** **If the sensor signal is less than 0.1 volt after starting, a fault code exists. Output voltage (VFTS) is monitored.** **Possible Causes:** • Poor connection • Short to ground in fuel temperature sensor circuit • Faulty fuel temperature sensor • Faulty PCM
DTC: P0183 **2T CCM, MIL: Yes** **2003-06** **Models:** Amanti, Optima, Sedona, Sorento **Engines:** 2.4L, 3.5L **Transmissions:** All	**Fuel Temperature Sensor "A" Circuit High Input** **If the sensor signal is above 4.6 volts after starting, a fault code exists. Output voltage (VFTS) is monitored.** **Possible Causes:** • Poor connection • Open in fuel temperature sensor circuit • Faulty fuel temperature sensor • Faulty PCM
DTC: P0196 **2T CCM, MIL: Yes** **2004-06** **Models:** Optima, Sedona, Spectra, Sporatge **Engines:** 2.0L, 2.4L, 2.7L, 3.8L **Transmissions:** All	**Engine Oil temperature Sensor Range/Performance** Stuck oil temperature sensor signal or unusual low or high signal. Condition 1 (signal high or low), Engine coolant temperature more than 158 degrees F and oil temperature less than 68 degrees F. Condition 2 (signal high or low), Engine coolant temperature less than 158 degrees F and oil temperature above 212 degrees F. Condition 3 (stuck signal) Engine coolant temperature less than 104 degrees F. **Possible Causes:** • Contact resistance in connectors • faulty OTS
DTC: P0197 **2T CCM, MIL: Yes** **2004-06** **Models:** Optima, Sedona, Spectra, Sporatge **Engines:** 2.0L, 2.4L, 2.7L, 3.8L **Transmissions:** All	**Engine Oil temperature Sensor Low Input** Signal voltage lower than the possible range of a properly operating OTS. Voltage range check. Engine coolant temperature less than 212 degrees F. Oil temperature above 309 degrees F. **Possible Causes:** • Short circuit to ground • Contact resistance in connectors • faulty OTS

DTC	Trouble Code Title, Conditions & Possible Causes
DTC: P0198 **2T CCM, MIL: Yes** **2004-06** **Models:** Optima, Sedona, Spectra, Sporatge **Engines:** 2.0L, 2.4L, 2.7L, 3.8L **Transmissions:** All	**Engine Oil temperature Sensor High Input** Signal voltage higher than the possible range of a properly operating OTS. Voltage range check. Five minutes after engine start if Engine coolant temperature less than 14 degrees F. Oil temperature minus 33 degrees F. **Possible Causes:** • Open circuit to battery • Contact resistance in connectors • faulty OTS
DTC: P0201 **2T CCM, MIL: Yes** **2000-06** **Models:** Amanti, Optima, Rio, Sedona, Sorento, Sephia, Spectra, Sportage **Engines:** All **Transmissions:** All	**Cylinder 1 Injector Circuit Malfunction** Engine running and the PCM detected the identified fuel injector control circuit signal was more than the upper limit, or that it was less than the lower limit, or that no control signal was present. **Possible Causes:** • Main relay power supply circuit to the injector is open • Fuel injector 1 control circuit is open or shorted to ground • Fuel injector 1 is damaged or has failed • Injector "driver" circuit in the PCM is damaged or has failed
DTC: P0202 **2T CCM, MIL: Yes** **2000-06** **Models:** Amanti, Optima, Rio, Sedona, Sorento, Sephia, Spectra, Sportage **Engines:** All **Transmissions:** All	**Cylinder 2 Injector Circuit Malfunction** Engine running and the PCM detected the identified fuel injector control circuit signal was more than the upper limit, or that it was less than the lower limit, or that no control signal was present. **Possible Causes:** • Main relay power supply circuit to the injector is open • Fuel injector 2 control circuit is open or shorted to ground • Fuel injector 2 is damaged or has failed • Injector "driver" circuit in the PCM is damaged or has failed
DTC: P0203 **2T CCM, MIL: Yes** **2000-06** **Models:** Amanti, Optima, Rio, Sedona, Sorento, Sephia, Spectra, Sportage **Engines:** All **Transmissions:** All	**Cylinder 3 Injector Circuit Malfunction** Engine running and the PCM detected the identified fuel injector control circuit signal was more than the upper limit, or that it was less than the lower limit, or that no control signal was present. **Possible Causes:** • Main relay power supply circuit to the injector is open • Fuel injector 3 control circuit is open or shorted to ground • Fuel injector 3 is damaged or has failed • Injector "driver" circuit in the PCM is damaged or has failed
DTC: P0204 **2T CCM, MIL: Yes** **2000-06** **Models:** Amanti, Optima, Rio, Sedona, Sorento, Sephia, Spectra, Sportage **Engines:** All **Transmissions:** All	**Cylinder 4 Injector Circuit Malfunction** Engine running and the PCM detected the identified fuel injector control circuit signal was more than the upper limit, or that it was less than the lower limit, or that no control signal was present. **Possible Causes:** • Main relay power supply circuit to the injector is open • Fuel injector 4 control circuit is open or shorted to ground • Fuel injector 4 is damaged or has failed • Injector "driver" circuit in the PCM is damaged or has failed
DTC: P0205 **2T CCM, MIL: Yes** **2001-06** **Models:** Amanti, Optima, Sedona, Sorento **Engines:** All **Transmissions:** All	**Cylinder 5 Injector Circuit Malfunction** Engine running and the PCM detected the identified fuel injector control circuit signal was more than the upper limit, or that it was less than the lower limit, or that no control signal was present. **Possible Causes:** • Main relay power supply circuit to the injector is open • Fuel injector 5 control circuit is open or shorted to ground • Fuel injector 5 is damaged or has failed • Injector "driver" circuit in the PCM is damaged or has failed
DTC: P0206 **2T CCM, MIL: Yes** **2001-06** **Models:** Amanti, Optima, Sedona, Sorento **Engines:** All **Transmissions:** All	**Cylinder 6 Injector Circuit Malfunction** Engine running and the PCM detected the identified fuel injector control circuit signal was more than the upper limit, or that it was less than the lower limit, or that no control signal was present. **Possible Causes:** • Main relay power supply circuit to the injector is open • Fuel injector 6 control circuit is open or shorted to ground • Fuel injector 6 is damaged or has failed • Injector "driver" circuit in the PCM is damaged or has failed

DTC	Trouble Code Title, Conditions & Possible Causes
DTC: P0217 **2T CCM, MIL: Yes** **2006** **Models:** Amanti, Optima, Sedona **Engines:** 2.7L, 3.5L, 3.8L **Transmissions:** All	**Engine Coolant Over Temperature Condition** Engine running and no disabling faults present. Coolant sensor within range. **Possible Causes:** • Poor connection • Lack of engine coolant • Faulty water pump • ECTS • Faulty PCM
DTC: P0220 **1T CCM, MIL: Yes** **2004-06** **Models:** Amanti **Engines:** 3.5L **Transmissions:** All	**Throttle/Pedal Position Sensor/Switch "B" Circuit** The DTC is recorded if the deviation between TPS target value and TPS output is out of threshold. Ignition switch is ON. **Possible Causes:** • Poor connection • Open or short to ground in TPS circuit • Faulty PCM
DTC: P0221 **2T CCM, MIL: Yes** **2006** **Models:** Optima **Engines:** 2.4L **Transmissions:** All	**Throttle/Pedal Position Sensor/Switch "B" Circuit Range/Performance** The ECM compares the TPS1 and TPS2 signal, and sets a code as required. Enabling conditions are as follows; no engine stop and engine start, no TPS adaptation request, no relevant failure. **Possible Causes:** • Poor connection or damaged harness • Air leakage in intake system • Faulty TPS2
DTC: P0222 **2T CCM, MIL: Yes** **2004-06** **Models:** Amanti, Optima, Sedona **Engines:** 2.4L, 2.7L, 3.5L, 3.8L **Transmissions:** All	**Throttle/Pedal Position Sensor/Switch "B" Circuit Low Input** The DTC is recorded If the output voltage of the TPS 1 is lower than threshold value (Vtps1 less than or equal to 0.2 volt). TPS 1 low input. **Possible Causes:** • Poor connection • Open or short to ground in TPS circuit • Faulty TPS • Faulty PCM
DTC: P0223 **2T CCM, MIL: Yes** **2004-06** **Models:** Amanti, Optima, Sedona **Engines:** 2.4L, 2.7L, 3.5L, 3.8L **Transmissions:** All	**Throttle/Pedal Position Sensor/Switch "B" Circuit High Input** The DTC is recorded If the output voltage of the TPS 1 higher than threshold value (Vtps1 greater than or equal to 4.85 volts, load value, EV less than 70 percent) when TPS 2 (Vtps2 less than or equal to 2.5 volts) is normal. TPS 1 high input. **Possible Causes:** • Poor connection • Open or short to ground in TPS circuit • Faulty TPS • Faulty PCM
DTC: P0224 **2T CCM, MIL: Yes** **2004-06** **Models:** Amanti **Engines:** 3.5L **Transmissions:** All	**Throttle Position Sensor/Switch "B" Linearity** The DTC is recorded If the output voltage of the TPS 1 is too low. TPS 1 linearity. **Possible Causes:** • Poor connection • Faulty TPS • Faulty PCM
DTC: P0225 **2T CCM, MIL: Yes** **2005** **Models:** Amanti **Engines:** 3.5L **Transmissions:** ALL	Accelerator position sensor (APS) Property Malfunction Disconnect sensor connector. Ignition switch ON. **Possible Causes:** • Open or short to ground • Faulty sensor
DTC: P0227 **2T CCM, MIL: Yes** **2005** **Models:** Amanti **Engines:** 3.5L **Transmissions:** ALL	Accelerator position sensor (APS #1 sub) Circuit Low Voltage Disconnect sensor connector. Ignition switch ON. **Possible Causes:** • Open or short to ground • Faulty sensor

DTC	Trouble Code Title, Conditions & Possible Causes
DTC: P0228 **2T CCM, MIL: Yes** **2005** **Models:** Amanti **Engines:** 3.5L **Transmissions:** ALL	Accelerator position sensor (APS #1 sub) Circuit High Voltage Disconnect sensor connector. Ignition switch ON. **Possible Causes:** • Open or short to ground • Faulty sensor
DTC: P0230 **1T CCM, MIL: Yes** **2003-06** **Models:** Optima, Rio, Sedona, Spectra, Sportage **Engines:** 1.6L, 2.0L, 2.4L, 2.7L, 3.8L **Transmissions:** All	**Fuel Pump Circuit Malfunction** Key on, and then the PCM detected an unexpected voltage condition on the fuel pump circuit through the fuel pump monitoring input. **Possible Causes:** • Fuel pump control circuit is open or shorted to ground • Fuel pump relay power circuit from ignition switch is open • Fuel pump relay is damaged or has failed • PCM has failed
DTC: P0231 **2T CCM, MIL: Yes** **2006** **Models:** Rio **Engines:** 1.6L **Transmissions:** All	**Electric Fuel Pump Relay Open Or Short Circuit** Circuit continuity check, high. **Possible Causes:** • Poor connection • Short to power in control circuit • Fuel pump relay • Faulty ECM/PCM
DTC: P0232 **2T CCM, MIL: Yes** **2006** **Models:** Rio **Engines:** 1.6L **Transmissions:** All	**Electric Fuel Pump Relay Short Circuit** Circuit continuity check, low. **Possible Causes:** • Poor connection • Short to ground in control circuit • Fuel pump relay • Faulty ECM/PCM
DTC: P0261 **2T CCM, MIL: Yes** **1998-06** **Models:** Optima, Rio, Sedona, Spectra, Sportage **Engines:** All **Transmissions:** All	**Fuel Injector 1 Circuit Low Input** Engine running and the PCM detected the Injector 1 signal was in a low signal state (0 volt) with the injector commanded off in the test. **Possible Causes:** • Main relay power supply circuit to the injector is open • Fuel injector 1 control circuit is shorted to ground • Fuel injector 1 is damaged or has failed • Injector "driver" circuit in the PCM is damaged or has failed
DTC: P0262 **2T CCM, MIL: Yes** **1998-06** **Models:** Optima, Rio, Sedona, Spectra, Sportage **Engines:** All **Transmissions:** All	**Fuel Injector 1 Circuit High Input** Engine running and the PCM detected the Injector 1 signal was in a high signal state (12 volts) with the injector commanded off in the test. **Possible Causes:** • Fuel injector 1 control circuit is shorted to system power (B+) • Fuel injector 1 is damaged or has failed • Injector 1 "driver" circuit in the PCM is damaged or has failed
DTC: P0264 **2T CCM, MIL: Yes** **1998-06** **Models:** Optima, Rio, Sedona, Spectra, Sportage **Engines:** All **Transmissions:** All	**Fuel Injector 2 Circuit Low Input** Engine running and the PCM detected the Injector 2 signal was in a low signal state (0 volt) with the injector commanded off in the test. **Possible Causes:** • Main relay power supply circuit to the injector is open • Fuel injector 2 control circuit is shorted to ground • Fuel injector 2 is damaged or has failed • Injector 2 "driver" circuit in the PCM is damaged or has failed
DTC: P0265 **2T CCM, MIL: Yes** **1998-06** **Models:** Optima, Rio, Sedona, Spectra, Sportage **Engines:** All **Transmissions:** All	**Fuel Injector 2 Circuit High Input** Engine running and the PCM detected the Injector 2 signal was in a high signal state (12 volts) with the injector commanded off in the test. **Possible Causes:** • Fuel injector 2 control circuit is shorted to system power (B+) • Fuel injector 2 is damaged or has failed • Injector 2 "driver" circuit in the PCM is damaged or has failed

DTC	Trouble Code Title, Conditions & Possible Causes
DTC: P0267 **2T CCM, MIL: Yes** **1998-06** **Models:** Optima, Rio, Sedona, Spectra, Sportage **Engines:** All **Transmissions:** All	**Fuel Injector 3 Circuit Low Input** Engine running and the PCM detected the Injector 3 signal was in a low signal state (0 volt) with the injector commanded off in the test. **Possible Causes:** • Main relay power supply circuit to the injector is open • Fuel injector 3 control circuit is shorted to ground • Fuel injector 3 is damaged or has failed • Injector 3 "driver" circuit in the PCM is damaged or has failed
DTC: P0268 **2T CCM, MIL: Yes** **1998-06** **Models:** Optima, Rio, Sedona, Spectra, Sportage **Engines:** All **Transmissions:** All	**Fuel Injector 3 Circuit High Input** Engine running and the PCM detected the Injector 3 signal was in a high signal state (12 volts) with the injector commanded off in the test. **Possible Causes:** • Fuel injector 3 control circuit is shorted to system power (B+) • Fuel injector 3 is damaged or has failed • Injector 3 "driver" circuit in the PCM is damaged or has failed
DTC: P0270 **2T CCM, MIL: Yes** **1998-06** **Models:** Optima, Rio, Sedona, Spectra, Sportage **Engines:** All **Transmissions:** All	**Fuel Injector 4 Circuit Low Input** Engine running and the PCM detected the Injector 4 signal was in a low signal state (0 volt) with the injector commanded off in the test. **Possible Causes:** • Fuel injector 4 control circuit is shorted to system power (B+) • Fuel injector 4 is damaged or has failed • Injector 4 "driver" circuit in the PCM is damaged or has failed
DTC: P0271 **2T CCM, MIL: Yes** **1998-06** **Models:** Optima, Rio, Sedona, Spectra, Sportage **Engines:** All **Transmissions:** All	**Fuel Injector 4 Circuit High Input** Engine running and the PCM detected the Injector 4 signal was in a high signal state (12 volts) with the injector commanded off in the test. **Possible Causes:** • Fuel injector 4 control circuit is shorted to system power (B+) • Fuel injector 4 is damaged or has failed • Injector 4 "driver" circuit in the PCM is damaged or has failed
DTC: P0273 **2T CCM, MIL: Yes** **2003-06** **Models:** Optima, Sedona, Sportage **Engines:** 2.7L, 3.8L **Transmissions:** All	**Cylinder 5- Injector Circuit Low** The PCM sets the DTC if the control circuit is shorted to ground. Driver stage check. **Possible Causes:** • Open in power supply harness • Short to ground in control harness • Contact resistance in connectors • Faulty injector
DTC: P0274 **2T CCM, MIL: Yes** **2003-06** **Models:** Optima, Sedona, Sportage **Engines:** 2.7L, 3.8L **Transmissions:** All	**Cylinder 5- Injector Circuit High** The PCM sets the DTC if the control circuit is open or shorted to battery voltage. Driver stage check. **Possible Causes:** • Open or short to battery control harness • Contact resistance in connectors • Faulty injector
DTC: P0276 **2T CCM, MIL: Yes** **2004-06** **Models:** Sedona, Sportage **Engines:** 2.7L, 3.8L **Transmissions:** All	**Cylinder 6- Injector Circuit Low** The PCM sets the DTC if the control circuit is shorted to ground. Driver stage check. **Possible Causes:** • Open in power supply harness • Short to ground in control harness • Contact resistance in connectors • Faulty injector
DTC: P0277 **2T CCM, MIL: Yes** **2004-06** **Models:** Sedona, Sportage **Engines:** 2.7L, 3.8L **Transmissions:** All	**Cylinder 6- Injector Circuit High** The PCM sets the DTC if the control circuit is open or shorted to battery voltage. Driver stage check. **Possible Causes:** • Open or short to battery control harness • Contact resistance in connectors • Faulty injector

DTC	Trouble Code Title, Conditions & Possible Causes
DTC: P0300 **2T MISFIRE, MIL: Yes** **1996-06** **Models:** Amanti, Optima, Rio, Sedona, Sorento, Sephia, Spectra, Sportage **Engines:** All **Transmissions:** All	**Random Cylinder Misfire Detected** Engine runtime 3 seconds, engine speed change under 1200 rpm, and the PCM detected a random misfire condition in more than one cylinder during the 200 revolution or the 1000 revolution test range. **Note: If the misfire is severe, the MIL will flash on/off on the 1st trip!** **Possible Causes:** • Vehicle driven under low fuel condition (less than 1/8 of a tank) • CKP or CMP sensor signal erratic or out of phase • Base engine problem affecting two or more engine cylinders • Ignition system problem affecting two or more engine cylinders
DTC: P0301 **2T MISFIRE, MIL: Yes** **1996-06** **Models:** Amanti, Optima, Rio, Sedona, Sorento, Sephia, Spectra, Sportage **Engines:** All **Transmissions:** All	**Cylinder 1 Misfire Detected** Engine runtime 3 seconds, engine speed change under 1200 rpm, and the PCM detected a misfire condition present in one cylinder during the 200 (Catalyst) or 1000 revolution (Emission) test range. **Note: If the misfire is severe, the MIL will flash on/off on the 1st trip!** **Possible Causes:** • Fuel metering (fuel injector dirty) problem affecting Cylinder 1 • Base engine (compression) problem affecting Cylinder 1 • Ignition system (spark plug or plug wire) problem on Cylinder 1
DTC: P0302 **2T MISFIRE, MIL: Yes** **1996-06** **Models:** Amanti, Optima, Rio, Sedona, Sorento, Sephia, Spectra, Sportage **Engines:** All **Transmissions:** All	**Cylinder 2 Misfire Detected** Engine runtime 3 seconds, engine speed change under 1200 rpm, and the PCM detected a misfire condition present in one cylinder during the 200 (Catalyst) or 1000 revolution (Emission) test range. **Note: If the misfire is severe, the MIL will flash on/off on the 1st trip!** **Possible Causes:** • Fuel metering (fuel injector dirty) problem affecting Cylinder 2 • Base engine (compression) problem affecting Cylinder 2 • Ignition system (spark plug or plug wire) problem on Cylinder 2
DTC: P0303 **2T MISFIRE, MIL: Yes** **1996-06** **Models:** Amanti, Optima, Rio, Sedona, Sorento, Sephia, Spectra, Sportage **Engines:** All **Transmissions:** All	**Cylinder 3 Misfire Detected** Engine runtime 3 seconds, engine speed change under 1200 rpm, and the PCM detected a misfire condition present in one cylinder during the 200 (Catalyst) or 1000 revolution (Emission) test range. **Note: If the misfire is severe, the MIL will flash on/off on the 1st trip!** **Possible Causes:** • Fuel metering (fuel injector dirty) problem affecting Cylinder 3 • Base engine (compression) problem affecting Cylinder 3 • Ignition system (spark plug or plug wire) problem on Cylinder 3
DTC: P0304 **2T MISFIRE, MIL: Yes** **1996-06** **Models:** Amanti, Optima, Rio, Sedona, Sorento, Sephia, Spectra, Sportage **Engines:** All **Transmissions:** All	**Cylinder 4 Misfire Detected** Engine runtime 3 seconds, engine speed change under 1200 rpm, and the PCM detected a misfire condition present in one cylinder during the 200 (Catalyst) or 1000 revolution (Emission) test range. **Note: If the misfire is severe, the MIL will flash on/off on the 1st trip!** **Possible Causes:** • Fuel metering (fuel injector dirty) problem affecting Cylinder 4 • Base engine (compression) problem affecting Cylinder 4 • Ignition system (spark plug or plug wire) problem on Cylinder 4
DTC: P0305 **2T MISFIRE, MIL: Yes** **2001-06** **Models:** Amanti, Optima, Sedona, Sorento, Sportage Sedona **Engines:** All **Transmissions:** All	**Cylinder 5 Misfire Detected** Engine runtime 3 seconds, engine speed change under 1200 rpm, and the PCM detected a misfire condition present in one cylinder during the 200 (Catalyst) or 1000 revolution (Emission) test range. **Note: If the misfire is severe, the MIL will flash on/off on the 1st trip!** **Possible Causes:** • Fuel metering (fuel injector dirty) problem affecting Cylinder 5 • Base engine (compression) problem affecting Cylinder 5 • Ignition system (spark plug or plug wire) problem on Cylinder 5
DTC: P0306 **2T MISFIRE, MIL: Yes** **2001-06** **Models:** Amanti, Optima, Sedona, Sorento, Sportage **Engines:** All **Transmissions:** All	**Cylinder 6 Misfire Detected** Engine runtime 3 seconds, engine speed change under 1200 rpm, and the PCM detected a misfire condition present in one cylinder during the 200 (Catalyst) or 1000 revolution (Emission) test range. **Note: If the misfire is severe, the MIL will flash on/off on the 1st trip!** **Possible Causes:** • Fuel metering (fuel injector dirty) problem affecting Cylinder 6 • Base engine (compression) problem affecting Cylinder 6 • Ignition system (spark plug or plug wire) problem on Cylinder 6

DTC	Trouble Code Title, Conditions & Possible Causes
DTC: P0315 **2T CCM, MIL: Yes** **2005-06** **Models:** Optima, Sedona, Spectra, Sportage **Engines:** 2.0L, 2.4L, 2.7L, 3.8L **Transmissions:** All	**Segment Time Acquisition Incorrect** A misfire induces a decrease in the engine speed and causes a variation in the segment period. Monitor segment time adaptation. **Possible Causes:** • Improperly installed target wheel • Contact resistance in connectors • Faulty PCM • Faulty CKPS
DTC: P0320 **2T CCM, MIL: Yes** **2000-06** **Models:** Amanti, Optima, Sedona, Sorento **Engines:** All **Transmissions:** All	**Ignition Failure System Circuit Malfunction** Engine running and the PCM detected an unexpected circuit condition on the Ignition Failure System circuit to the ignition coil. **Possible Causes:** • IFS circuit to the PCM is open or shorted to ground • IFS ground circuit is open • IFS is damaged or has failed • PCM has failed
DTC: P0325 **2T CCM, MIL: Yes** **2000-06** **Models:** Amanti, Optima, Rio, Sedona, Sorento, Spectra, Sportage **Engines:** All **Transmissions:** All	**Knock Sensor Circuit Malfunction (Bank 1)** Engine running at over 1000 rpm for 5 seconds, and the PCM did not detect enough variation in the KS signals (e.g., 0.049 volt). **Possible Causes:** • Knock sensor signal circuit open or shorted to ground • Knock sensor signal circuit shorted to VREF or system power • Knock sensor is damaged or has failed • PCM has failed
DTC: P0326 **2T CCM, MIL: Yes** **1996-06** **Models:** Rio, Sedona, Sportage **Engines:** All **Transmissions:** All	**Knock Sensor Circuit Malfunction (Bank 1)** Engine speed from 1000-2200 rpm, ECT sensor signal more than 104°F, engine load more than 2 ms, and the PCM detected the Knock sensor signal was out of range at a calculated engine speed. **Possible Causes:** • Knock sensor signal circuit open or shorted to ground • Knock sensor signal circuit shorted to VREF or system power • Knock sensor is damaged or has failed • PCM has failed
DTC: P0327 **2T CCM, MIL: Yes** **2006** **Models:** Rio **Engines:** 1.6L **Transmissions:** All	**Knock Sensor 1 Circuit Low Input** Engine speed greater than 2600 rpm. **Possible Causes:** • Poor connection • Open or short to ground in signal circuit • Knock sensor is damaged or has failed • PCM/ECM has failed
DTC: P0328 **2T CCM, MIL: Yes** **2006** **Models:** Rio **Engines:** 1.6L **Transmissions:** All	**Knock Sensor 1 Circuit High Input** Coolant temperature more than 104 degrees F. **Possible Causes:** • Poor connection • Short to power in signal circuit • Knock sensor is damaged or has failed • PCM/ECM has failed
DTC: P0330 **2T CCM, MIL: Yes** **2000-06** **Models:** Optima, Sedona, Sportage **Engines:** All **Transmissions:** All	**Knock Sensor Circuit Malfunction (Bank 2)** Engine running at over 1000 rpm for 5 seconds, and the PCM did not detect enough variation in the KS signals (e.g., 0.049 volt). **Possible Causes:** • Knock sensor signal circuit open or shorted to ground • Knock sensor signal circuit shorted to VREF or system power • Knock sensor is damaged or has failed • PCM has failed
DTC: P0331 **2T CCM, MIL: Yes** **2006** **Models:** Sedona **Engines:** 3.8L **Transmissions:** All	**Knock Sensor 2 Circuit Range/Performance (Bank 2)** Signal short. Pressure in intake manifold is normal. Engine speed is equal to or less than 1600 rpm. **Possible Causes:** • Poor connection • Short in harness • Faulty knock sensor • Faulty PCM

DTC	Trouble Code Title, Conditions & Possible Causes
DTC: P0335 **1T CCM, MIL: Yes** **1996-06** **Models:** Amanti, Optima, Rio, Sedona, Sorento Sephia, Spectra, Sportage **Engines:** All **Transmissions:** All	**Crankshaft Position Sensor Circuit Malfunction** Engine cranking for 5 seconds, and the PCM did not detect any CKP sensor signals, or the vehicle was driven to a speed over 15.5 mph, and the PCM did not detect any CKP sensor signals for 5 seconds. **Possible Causes:** • CKP sensor signal circuit open or shorted to ground • CKP sensor signal circuit shorted to VREF or system power • CKP sensor is damaged or has failed • PCM has failed
DTC: P0336 **1T CCM, MIL: Yes** **1996-06** **Models:** Optima, Rio, Sedona, Sephia, Spectra, Sportage **Engines:** All **Transmissions:** All	**Crankshaft Position Sensor Performance** Engine running and the PCM detected the number of CKP sensor signals counted (between the reference mark gap) did not equal the Actual number of available teeth (i.e., the CKP signals were out or the normal window" of operation with the CMP sensor signals okay). **Possible Causes:** • CKP sensor signal circuit connections loose (intermittent fault) • CKP sensor wiring harness has a connection fault (intermittent) • CKP to Target Wheel "air gap" is incorrect • CKP sensor is damaged or has failed
DTC: P0337 **2T CCM, MIL: Yes** **2005-06** **Models:** Amanti, Rio **Engines:** 1.6L, 3.5L **Transmissions:** All	**Crankshaft Position Sensor "A" Circuit Low Input** If the output voltage of the CKPS remains low for more than two seconds. When the change of the CMPS output voltage is zero, the PCM determines a fault and stores a code. Change in output voltage (delta sign Vckp) is monitored. **Possible Causes:** • Poor connection • Open or short to ground in CKPS circuit • Faulty CKPS • Faulty PCM
DTC: P0338 **2T CCM, MIL: Yes** **2005-06** **Models:** Amanti, Rio **Engines:** 1.6L, 3.5L **Transmissions:** All	**Crankshaft Position Sensor "A" Circuit High Input** If the output voltage of the CKPS remains high for more than two seconds. When the change of the CMPS output voltage is zero, the PCM determines a fault and stores a code. Change in output voltage (delta sign Vckp) is monitored. **Possible Causes:** • Poor connection • Open or short to ground in CKPS circuit • Faulty CKPS • Faulty PCM
DTC: P0339 **2T CCM, MIL: Yes** **2006** **Models:** Rio **Engines:** 1.6L **Transmissions:** All	**Crankshaft Position Sensor "A" Circuit** Signal check. Edge counter of camshaft position sensor 8. **Possible Causes:** • Poor connection • Open or short in signal circuit • CKPS • Faulty ECM/PCM
DTC: P0340 **2T CCM, MIL: Yes** **1996-06** **Models:** Amanti, Optima, Rio, Sedona, Sorento, Sephia, Spectra, Sportage **Engines:** All **Transmissions:** All	**Camshaft Position Sensor Circuit Malfunction** Engine speed over 600 rpm, and the PCM detected less than one CMP sensor signal was present, condition met for 1.5 seconds. **Possible Causes:** • CMP sensor signal circuit is open or shorted to ground • CMP sensor signal circuit is shorted to VREF or system power • CMP sensor is damaged or has failed • PCM has failed
DTC: P0341 **2T CCM, MIL: Yes** **2006** **Models:** Optima, Rio, Sedona **Engines:** 1.6L, 2.4L, 3.8L **Transmissions:** All	**Camshaft Position Sensor Circuit Malfunction** No signal or no signal switching is detected. Crankshaft sensor is normal. Battery voltage is between 10 and 16 volts. **Possible Causes:** • Open or short in CMPS circuit • Faulty CMPS • Faulty PCM
DTC: P0342 **2T CCM, MIL: Yes** **1996-06** **Models:** Amanti, Optima, Rio, Sedona, Sephia, Spectra, Sportage **Engines:** All **Transmissions:** All	**Camshaft Position Sensor Low Input** Engine speed over 600 rpm, and the PCM detected an unexpected low voltage condition on the CMP sensor signal circuit. **Possible Causes:** • CMP sensor signal circuit is open or shorted to ground • CMP sensor ground circuit is open • CMP sensor is damaged or has failed • PCM has failed

DTC	Trouble Code Title, Conditions & Possible Causes
DTC: P0343 **2T CCM, MIL: Yes** **1996-97** **Models:** Sportage **Engines:** All **Transmissions:** All	**Camshaft Position Sensor High Input** Engine speed over 600 rpm, and the PCM detected an unexpected high voltage condition on the CMP sensor signal circuit. **Possible Causes:** • CMP sensor signal circuit is shorted to VREF or system power • CMP sensor is damaged or has failed • PCM has failed
DTC: P0343 **2T CCM, MIL: Yes** **1998-06** **Models:** Amanti, Optima, Rio, Sedona, Sephia, Spectra, Sportage **Engines:** All **Transmissions:** All	**Camshaft Position Sensor High Input** Engine cranking and the PCM detected the CMP sensor signal was above a threshold value stored in memory during the CCM test. **Possible Causes:** • CMP sensor signal circuit is shorted to VREF or system power • CMP sensor is damaged or has failed • PCM has failed
DTC: P0346 **2T CCM, MIL: Yes** **2006** **Models:** Sedona **Engines:** 3.8L **Transmissions:** All	**Camshaft Position Sensor "A" Circuit Range/Performance (Bank 2)** Engine running and the PCM detected the CMP sensor signal was above a threshold value stored in memory during the CCM test. **Possible Causes:** • Poor connection • Open or Short in harness • Electrical noise • Target wheel • CMPS • PCM has failed
DTC: P0350 **2T CCM, MIL: Yes** **2002-06** **Models:** Amanti, Optima, Sorento, Sedona, Sportage **Engines:** All **Transmissions:** All	**Ignition Coil Primary/Secondary Circuit Malfunction** Engine started, engine running for 4 seconds, and the PCM detected an unexpected voltage condition either the ignition coil primary or on the ignition coil secondary circuit during the CCM test. **Possible Causes:** • Ignition coil primary circuit open or shorted together • Ignition coil secondary components (coil or plug wires) arching • Ignition coil is damaged or has failed • PCM has failed
DTC: P0351 **2T CCM, MIL: Yes** **2003-06** **Models:** Optima, Sedona, Sportage **Engines:** 2.0L, 2.7L, 3.5L **Transmissions:** All	**Ignition Coil 'A' Circuit Malfunction** Engine started, engine running, and the PCM detected an unexpected voltage condition on the Ignition Coil 'A' primary circuit. **Possible Causes:** • Ignition Coil 'A' primary circuit is open or shorted to ground • Ignition Coil 'A' power circuit is open (test power from I/P fuse) • Ignition Coil 'A' is damaged or has failed • PCM has failed
DTC: P0352 **2T CCM, MIL: Yes** **2003-06** **Models:** Optima, Sedona, Sportage **Engines:** 2.0L, 2.7L, 3.5L **Engines:** All **Transmissions:** All	**Ignition Coil 'B' Circuit Malfunction** Engine started, engine running, and the PCM detected an unexpected voltage condition on the Ignition Coil 'B' primary circuit. **Possible Causes:** • Ignition Coil 'B' primary circuit is open or shorted to ground • Ignition Coil 'B' power circuit is open (test power from I/P fuse) • Ignition Coil 'B' is damaged or has failed • PCM has failed
DTC: P0353 **2T CCM, MIL: Yes** **2003-06** **Models:** Optima, Sedona, Sportage **Engines:** 2.0L, 2.7L, 3.5L **Transmissions:** All	**Ignition Coil 'C' Circuit Malfunction** Engine started, engine running, and the PCM detected an unexpected voltage condition on the Ignition Coil 'C' primary circuit. **Possible Causes:** • Ignition Coil 'C' primary circuit is open or shorted to ground • Ignition Coil 'C' power circuit is open (test power from I/P fuse) • Ignition Coil 'C' is damaged or has failed • PCM has failed
DTC: P0354 **2T CCM, MIL: Yes** **2003-06** **Models:** Optima, Sedona, Sportage **Engines:** 2.0L, 2.7L, 3.5L **Transmissions:** All	**Ignition Coil 'D' Circuit Malfunction** Engine started, engine running, and the PCM detected an unexpected voltage condition on the Ignition Coil 'D' primary circuit. **Possible Causes:** • Ignition Coil 'D' primary circuit is open or shorted to ground • Ignition Coil 'D' power circuit is open (test power from I/P fuse) • Ignition Coil 'D' is damaged or has failed • PCM has failed

DTC	Trouble Code Title, Conditions & Possible Causes
DTC: P0355 **2T CCM, MIL: Yes** **2003-06** **Models:** Optima, Sedona, Sportage **Engines:** 2.0L, 2.7L, 3.5L **Transmissions:** All	**Ignition Coil 'E' Circuit Malfunction** Engine started, engine running, and the PCM detected an unexpected voltage condition on the Ignition Coil 'E' primary circuit. **Possible Causes:** • Ignition Coil 'E' primary circuit is open or shorted to ground • Ignition Coil 'E' power circuit is open (test power from I/P fuse) • Ignition Coil 'E' is damaged or has failed • PCM has failed
DTC: P0356 **2T CCM, MIL: Yes** **2003-06** **Models:** Optima, Sedona, Sportage **Engines:** 2.0L, 2.7L, 3.5L **Transmissions:** All	**Ignition Coil 'F' Circuit Malfunction** Engine started, engine running, and the PCM detected an unexpected voltage condition on the Ignition Coil 'F' primary circuit. **Possible Causes:** • Ignition Coil 'F' primary circuit is open or shorted to ground • Ignition Coil 'F' power circuit is open (test power from I/P fuse) • Ignition Coil 'F' is damaged or has failed • PCM has failed
DTC: P0400 **2T EGR, MIL: Yes** **1996-1997** **Models:** Sephia, Sportage **Engines:** All **Transmissions:** All	**Insufficient EGR System Flow Detected** Engine speed over 1000 rpm at a vehicle speed of over 6.2 mph, ECT sensor signal more than 176°F, and the PCM detected too small a change in the EGR differential pressure sensor after the EGR valve was cycled from "on" to "off" in the EGR Monitor test. **Possible Causes:** • Air cleaner housing or element is restricted • EGR control valve or EGR solenoid valve is leaking • EGR valve or valve seat is leaking • EGR valve or EGR differential pressure sensor is damaged • EGR valve position sensor is damaged or has failed • EGR solenoid valve is damaged or has failed • Exhaust manifold or intake manifold areas is restricted
DTC: P0401 **2T EGR, MIL: Yes** **2001-06** **Models:** Amanti, Optima **Engines:** All **Transmissions:** All	**Insufficient EGR System Flow Detected** Engine started, engine speed from 1000-2000 rpm, ECT sensor more than 176°F, closed throttle position switch indicating "on", EGR solenoid commanded open for 2 seconds, and the PCM detected too small an amount of change in EGR flow during the EGR test. **Possible Causes:** • EGR valve or EGR differential pressure sensor is damaged • EGR solenoid valve is damaged or has failed • MAP sensor is damaged or out of calibration • Leaks or restrictions in these EGR components:
DTC: P0403 **2T EGR, MIL: Yes** **2001-06** **Models:** Amanti, Optima **Engines:** All **Transmissions:** All	**EGR Solenoid Circuit Malfunction** Engine started, system voltage more than 10.0 volts, and the PCM did not detect any solenoid surge voltage (system voltage +2 volts) when the EGR solenoid was cycled from "on" to "off" during the CCM test. **Possible Causes:** • EGR solenoid control circuit open or shorted to ground • EGR solenoid control circuit shorted to VREF or system power • EGR solenoid is damaged or has failed • PCM has failed (the EGR solenoid driver may be open/shorted)
DTC: P0420 **2T CAT1, MIL: Yes** **1996-97** **Models:** Sephia **Engines:** All **Transmissions:** All	**Catalyst Efficiency Below Normal (Bank 1)** Engine started, vehicle driven at a speed of 45-60 mph in closed loop for over 1 minute, and the PCM detected the rear HO2S and front HO2S signal voltage amplitudes were too similar. **Possible Causes:** • Air leaks at the exhaust manifold or in the exhaust pipes • Catalytic converter is damaged or has failed • Front HO2S or rear HO2S is contaminated with fuel or moisture • Front HO2S or rear HO2S heater is damaged or has failed • Front HO2S or rear HO2S is contaminated with fuel or moisture
DTC: P0420 **2T CAT1, MIL: Yes** **2001-06** **Models:** Amanti, Optima, Rio, Sedona, Spectra, Sportage **Engines:** 1.6L, 2.0L, 2.4L,2.5L (vin 4), 2.7L, 3.5L, 3.8L **Transmissions:** All	**Catalyst Efficiency Below Normal (Bank 1)** DTC P0130, P0133, P0134, P0135, P0136, P0139, P0140 and P0141 not set, engine started, engine running in closed loop at a speed of 45-60 mph for 2-3 minutes, and the PCM detected that the rear HO2S and front HO2S voltage amplitudes were too similar. **Possible Causes:** • Air leaks at the exhaust manifold or in the exhaust pipes • Catalytic converter is damaged or has failed • Front HO2S or rear HO2S is contaminated with fuel or moisture • Front HO2S or rear HO2S is contaminated with fuel or moisture

DTC	Trouble Code Title, Conditions & Possible Causes
DTC: P0421 **2T CAT1, MIL: Yes** **2001** **Models:** Optima **Engines:** 2.4L (vin 3) **Transmissions:** All	**Catalyst Efficiency Below Normal (Bank 1)** DTC P0100, P0102, P0103, P0136, P0139, P0140, P0171 and P0172 not set, engine started, engine running in closed loop at 45-60 mph for 2-3 minutes, and the PCM detected the rear HO2S and front HO2S voltage amplitudes were too similar during the test. **Possible Causes:** • Air leaks at the exhaust manifold or in the exhaust pipes • Catalytic converter is damaged or has failed • Front HO2S or rear HO2S is contaminated with fuel or moisture • Front HO2S or rear HO2S is contaminated with fuel or moisture
DTC: P0421 **2T CAT1, MIL: Yes** **2002-06** **Models:** Amanti, Optima, Rio, Sorento **Engines:** All **Transmissions:** All	**Main Catalyst Efficiency Deterioration (Bank 1)** DTC P0136, P0139 and P0140 not set, engine running at 3000 rpm or high in closed loop at 45-60 mph for 140 seconds, and the PCM detected the rear HO2S and front HO2S signals were too similar. **Possible Causes:** • Air leaks at the exhaust manifold or in the exhaust pipes • Catalytic converter is damaged or has failed • Front HO2S or rear HO2S is contaminated with fuel or moisture • Front HO2S or rear HO2S heater is damaged or has failed • Front HO2S or rear HO2S is contaminated with fuel or moisture
DTC: P0421 **2T CAT1, MIL: Yes** **2002-06** **Models:** Sedona **Engines:** All **Transmissions:** All	**Main Catalyst Efficiency Deterioration (Bank 1)** DTC P0136, P0139 and P0140 not set, engine running at 3000 rpm or high in closed loop at 45-60 mph for 140 seconds, and the PCM detected the rear HO2S and front HO2S signals were too similar. **Possible Causes:** • Air leaks at the exhaust manifold or in the exhaust pipes • Catalytic converter is damaged or has failed • Front HO2S or rear HO2S is contaminated with fuel or moisture • Front HO2S or rear HO2S heater is damaged or has failed • Front HO2S or rear HO2S is contaminated with fuel or moisture
DTC: P0422 **2T CCM, MIL: Yes** **1998-02** **Models:** Rio, Sedona, Sephia, Spectra, Sportage **Engines:** All **Transmissions:** All	**Catalyst Efficiency Below Normal (Bank 1)** DTC P0100, P0102, P0103, P0136, P0139, P0140, P0171 and P0172 not set, engine running in closed loop at 720-2520 rpm, vehicle speed from 45-60 mph for 80 seconds, and the PCM detected the averaged Catalyst malfunction index was higher than the Threshold value for 3 seconds. **Possible Causes:** • Air leaks at the exhaust manifold or in the exhaust pipes • Catalytic converter is damaged or has failed • Front HO2S or rear HO2S is contaminated with fuel or moisture • Front HO2S or rear HO2S heater is damaged or has failed • Front HO2S or rear HO2S is contaminated with fuel or moisture
DTC: P0430 **2T CAT1, MIL: Yes** **2001-06** **Models:** Amanti, Optima, Sedona, Sportage **Engines:** 2.5L (vin 4), 2.7L, 3.5L, 3.8L **Transmissions:** All	**Catalyst Efficiency Below Normal (Bank 2)** DTC P0150, P0153, P0154, P0155, P0156, P0160 and P0161 not set, engine started, engine running in closed loop at 45-60 mph for 2-3 minutes, and the PCM detected the rear HO2S and front HO2S voltage amplitudes were too similar during the Catalyst Monitor test. **Possible Causes:** • Air leaks at the exhaust manifold or in the exhaust pipes • Catalytic converter is damaged or has failed • Front HO2S or rear HO2S is contaminated with fuel or moisture • Front HO2S or rear HO2S heater is damaged or has failed • Front HO2S or rear HO2S is contaminated with fuel or moisture
DTC: P0431 **2T CAT1, MIL: Yes** **2002-2006** **Models:** Amanti, Optima, Sedona, Sorento **Engines:** All **Transmissions:** All	**Main Catalyst Efficiency Deterioration (Bank 2)** DTC P0150, P0154, P0159, P0158, P0160 and P0161 not set, engine started, engine running at 3000 rpm or high in closed loop at 45-60 mph for 140 seconds, and the PCM detected the rear HO2S and front HO2S signals were too similar in the Catalyst Monitor test. **Possible Causes:** • Air leaks at the exhaust manifold or in the exhaust pipes • Catalytic converter is damaged or has failed • Front HO2S or rear HO2S is contaminated with fuel or moisture • Front HO2S or rear HO2S heater is damaged or has failed • Front HO2S or rear HO2S is contaminated with fuel or moisture

DTC	Trouble Code Title, Conditions & Possible Causes
DTC: P0431 **2T CAT1, MIL: Yes** **2002-2006** **Models:** Amanti, Optima, Sedona, Sorento **Engines:** All **Transmissions:** All	**Main Catalyst Efficiency Deterioration (Bank 2)** DTC P0150, P0154, P0159, P0158, P0160 and P0161 not set, engine started, engine running at 3000 rpm or high in closed loop at 45-60 mph for 140 seconds, and the PCM detected the rear HO2S and front HO2S signals were too similar in the Catalyst Monitor test. **Possible Causes:** • Air leaks at the exhaust manifold or in the exhaust pipes • Catalytic converter is damaged or has failed • Front HO2S or rear HO2S is contaminated with fuel or moisture • Front HO2S or rear HO2S heater is damaged or has failed • Front HO2S or rear HO2S is contaminated with fuel or moisture
DTC: P0440 **2T EVAP, MIL: Yes** **1996** **Models:** Sephia **Engines:** All **Transmissions:** A/T	**EVAP System Malfunction** ECT sensor input from 38-95°F at startup, vehicle driven at 50-65 mph with gear selector in Overdrive for 3 minutes, followed by a deceleration period to 35-45 mph, then driven at 35-45 mph for 4 minutes, and returned to idle speed, then back to 35-45 mph, and the PCM detected the FTP sensor indicated less than 2.3 kPa. **Possible Causes:** • Fuel filler cap damaged, cross-threaded or loosely installed • Small hoses or cuts present in the EVAP vapor hoses/lines • EVAP purge valve is damaged or has failed • PCM has failed
DTC: P0440 **2T EVAP, MIL: Yes** **1996** **Models:** Sephia **Engines:** All **Transmissions:** M/T	**EVAP System Malfunction** ECT sensor input from 38-95°F at startup, vehicle driven at 50-65 mph with gear selector in 5th Gear for 3 minutes, followed by a deceleration period to 35-45 mph, then driven at 35-45 mph for 4 minutes, and returned to idle speed, then back to 35-45 mph, and the PCM detected the FTP sensor indicated less than 2.3 kPa. **Possible Causes:** • Fuel filler cap damaged, cross-threaded or loosely installed • Small hoses or cuts present in the EVAP vapor hoses/lines • EVAP purge valve is damaged or has failed • PCM has failed
DTC: P0440 **2T EVAP, MIL: Yes** **1997-02** **Models:** Optima, Rio, Sedona, Sephia, Spectra, Sportage **Engines:** All **Transmissions:** All	**EVAP System Malfunction** Engine started, ECT sensor signal less than158°F at startup, IAT sensor signal more than 9.5°F, system voltage over 10.0 volts engine runtime 15-20 minutes at cruise speed, then returned to idle speed, VSS indicating 0 mph, load value 2.2 ms, canister load factor less than 4.0, fuel tank pressure less than 0.5" Hg, then after the Idle Control system and Fuel Trim had stabilized, the PCM detected a fuel vapor leak (as small as 0.040") in the EVAP system during the EVAP Leak Test. **Possible Causes:** • Fuel filler cap damaged, cross-threaded or loosely installed • Small hoses or cuts present in the EVAP vapor hoses/lines • EVAP purge valve is damaged or has failed
DTC: P0441 **2T EVAP, MIL: Yes** **1996** **Models:** Sportage **Engines:** All **Transmissions:** All	**EVAP System Incorrect Purge Flow** Engine started, ECT sensor signal from 38-95°F at startup, vehicle driven to a speed over 6.2 mph at a steady throttle in closed loop for 2 minutes, and the PCM detected that insufficient Purge of the charcoal canister occurred during the EVAP System Monitor test. **Possible Causes:** • Fuel filler cap damaged, cross-threaded or loosely installed • Small hoses or cuts present in the EVAP vapor hoses/lines • EVAP purge valve is damaged or has failed
DTC: P0441 **2T EVAP, MIL: Yes** **1997-06** **Models:** Amanti, Optima, Rio, Sedona, Sorento, Sephia, Spectra, Sportage **Engines:** All **Transmissions:** All	**EVAP System Malfunction** ECT sensor signal less than158°F at startup, IAT sensor signal more than 9.05°F, system voltage more than 10.9 volts engine runtime 15-20 minutes at cruise speed, then returned to idle speed, VSS indicating 0 mph, load value 2.2 ms, canister load factor less than 4.0, fuel tank pressure less than 0.5" Hg, then after the Idle Control system and Fuel Trim stabilized, the PCM detected a continuous purge condition in the EVAP system during the Purge flow test. **Possible Causes:** • Small hoses or cuts present in the EVAP vapor hoses/lines • EVAP canister purge solenoid is damaged or is stuck open • PCM has failed
DTC: P0442 **2T EVAP, MIL: Yes** **1996-06** **Models:** Amanti, Optima, Rio, Sedona, Sorento, Sephia, Spectra, Sportage **Engines:** All **Transmissions:** All	**EVAP System Small Leak (0.040") Detected** ECT sensor signal less than158°F at startup, IAT sensor signal more than 9.05°F, system voltage more than 10.9 volts engine runtime 15-20 minutes at cruise speed, then returned to idle speed, VSS indicating 0 mph, load value 2.2 ms, canister load factor less than 4.0, fuel tank pressure less than 0.5" Hg, then after the Idle Control system and Fuel Trim had stabilized, the PCM detected a fuel vapor leak (as small as 0.040") in the EVAP system during the EVAP Leak Test. **Possible Causes:** • Fuel filler cap damaged, cross-threaded or loosely installed • Small leaks or cuts present in the EVAP vapor hoses/lines • EVAP purge valve is damaged or has failed • PCM has failed

DTC	Trouble Code Title, Conditions & Possible Causes
DTC: P0443 **2T CCM, MIL: Yes** **1996-05** **Models:** Amanti, Optima, Rio, Sedona, Sorento, Sephia, Spectra, Sportage **Engines:** All **Transmissions:** All	**EVAP Purge Control Valve Circuit Malfunction** Key on or engine running and the PCM detected an unexpected voltage condition on the EVAP Purge Control solenoid circuit. **Possible Causes:** • Injector Fuse (B+) open to the Purge solenoid power circuit • Purge control solenoid control circuit is open or grounded • Purge control solenoid circuit is shorted to system power (B+) • Purge control solenoid is damaged or has failed • PCM has failed (Purge solenoid driver may be open or shorted)
DTC: P0444 **2T CCM, MIL: Yes** **2003-06** **Models:** Amanti, Optima, Rio, Sedona, Spectra, Sportage **Engines:** 1.6L, 2.0L, 2.4L, 2.7L, 3.5L, 3.8L **Transmissions:** All	**EVAP Emission System- Purge Control Valve Circuit Open** Engine running. Checking output signals from PCSV every 10 seconds, under detecting condition. **Possible Causes:** • Poor connection • Open or short to ground in harness • PCVS • PCM
DTC: P0445 **2T CCM, MIL: Yes** **2003-06** **Models:** Amanti, Optima, Rio, Sedona, Spectra, Sportage **Engines:** 1.6L, 2.0L, 2.4L, 2.7L, 3.5L, 3.8L **Transmissions:** All	**EVAP Emission System- Purge Control Valve Circuit Shorted** Engine running. Checking output signals from PCSV every 10 seconds, under detecting condition. **Possible Causes:** • Poor connection • Short to battery in harness • PCVS • PCM
DTC: P0445 **2T CCM, MIL: Yes** **2002-06** **Models:** Amanti, Optima, Rio, Sedona, Spectra, Sportage **Engines:** All **Transmissions:** All	**EVAP Vent Control Circuit Malfunction** Key on or engine running and the PCM detected an unexpected voltage condition on the EVAP Vent Control solenoid circuit. **Possible Causes:** • Vent control solenoid power circuit from Main Relay is open • Vent control solenoid control circuit is open or grounded • Vent control solenoid circuit is shorted to system power (B+) • Vent control solenoid is damaged or has failed • PCM has failed (Vent solenoid driver may be open or shorted)
DTC: P0446 **2T CCM, MIL: Yes** **1996-02** **Models:** Optima, Rio, Sedona, Sephia, Spectra, Sportage **Engines:** All **Transmissions:** All	**EVAP Vent Control Circuit Malfunction** Key on or engine running and the PCM detected an unexpected voltage condition on the EVAP Vent Control solenoid circuit. **Possible Causes:** • Injector Fuse (B+) open to the Vent solenoid power circuit • Vent control solenoid control circuit is open or grounded • Vent control solenoid circuit is shorted to system power (B+) • Vent control solenoid is damaged or has failed • PCM has failed (Vent solenoid driver may be open or shorted)
DTC: P0446 **2T CCM, MIL: Yes** **2003-06** **Models:** Amanti, Optima, Rio, Sedona, Sorento **Engines:** 1.6L, 2.4L, 3.5L **Transmissions:** All	**EVAP Emission System- Vent Control Circuit** CCV stuck open. Time after engine start greater than 600 seconds. Idle speed controller activated. Coolant temperature less than 12 degrees F. **Possible Causes:** • Poor connection • CCV • ECM/PCM
DTC: P0447 **2T CCM, MIL: Yes** **2003-06** **Models:** Amanti, Optima, Rio, Sedona, Spectra, Sportage **Engines:** 1.6L, 2.0L,2.4L, 2.7L,3.5L, 3.8L **Transmissions:** All	**EVAP Emission System- Vent Control Circuit Open** Detects a short to ground or open circuit on vent valve output circuit. No disabling faults present. Engine running. **Possible Causes:** • Poor connection • Open or short in power circuit • Open or short in control circuit • CCV • ECM/PCM

DTC	Trouble Code Title, Conditions & Possible Causes
DTC: P0448 **2T CCM, MIL: Yes** **2003-06** **Models:** Amanti, Optima, Rio, Sedona, Spectra, Sportage **Engines:** 1.6L, 2.0L,2.4L, 2.7L,3.5L, 3.8L **Transmissions:** All	**EVAP Emission System- Vent Control Circuit Shorted** Detects a short to battery on vent valve output circuit. No disabling faults present. Engine running. **Possible Causes:** • Poor connection • Short to battery in CCV circuit • CCV • ECM/PCM
DTC: P0449 **2T CCM, MIL: Yes** **2003-06** **Models:** Optima, Rio, Spectra, Sportage **Engines:** 1.6L, 2.0L,2.4L, 2.7L **Transmissions:** All	**EVAP Emission System- Vent Valve/Solenoid Circuit** Circuit continuity check open. **Possible Causes:** • Poor connection • Open or short to ground in power circuit • CCV • ECM/PCM
DTC: P0450 **2T CCM, MIL: Yes** **1997-06** **Models:** Rio, Sephia **Engines:** All **Transmissions:** All	**EVAP Pressure Sensor Circuit Malfunction** Key on or engine running and the PCM detected the Fuel Tank Pressure sensor signal was more than 4.9 volts or less than 0.14 volt. **Possible Causes:** • FTP sensor signal circuit open or shorted to ground • FTP sensor signal circuit shorted to VREF or system power • FTP sensor is damaged or has failed • PCM has failed
DTC: P0451 **2T CCM, MIL: Yes** **1998-06** **Models:** Amanti, Optima, Rio, Sedona, Sorento, Sephia, Spectra, Sportage **Engines:** All **Transmissions:** All	**EVAP Pressure Sensor Performance** Engine at idle speed with the vehicle speed indicating 0 mph, then with the EVAP Vent Control solenoid commanded "on", the PCM detected the FTP sensor signal variation was less than 15 mv. **Possible Causes:** • FTP sensor signal or ground circuit has high resistance • FTP sensor is damaged or out of calibration • EVAP canister close valve (CCV) is stuck closed • PCM has failed
DTC: P0452 **2T CCM, MIL: Yes** **1996-06** **Models:** Amanti, Optima, Rio, Sedona, Sorento, Sephia, Spectra, Sportage **Engines:** All **Transmissions:** All	**EVAP Pressure Sensor Circuit Low Input** Key on or engine running and the PCM detected the Fuel Tank Pressure (FTP) sensor signal was less than 0.14 volt during the test. **Possible Causes:** • FTP sensor signal circuit is shorted to ground • FTP sensor is damaged or has failed • PCM has failed
DTC: P0453 **2T CCM, MIL: Yes** **1996-06** **Models:** Amanti, Optima, Rio, Sedona, Sorento, Sephia, Spectra, Sportage **Engines:** All **Transmissions:** All	**EVAP Pressure Sensor Circuit High Input** Key on or engine running and the PCM detected the Fuel Tank Pressure (FTP) sensor signal was more than 4.90 volts during the test. **Possible Causes:** • FTP sensor signal circuit is open • FTP sensor ground circuit is open • FTP sensor is damaged or has failed • PCM has failed
DTC: P0454 **2T CCM, MIL: Yes** **2003-06** **Models:** Amanti, Optima, Sedona, Sportage **Engines:** 2.0L, 2.4L, 2.7L, 3.5L, 3.8L **Transmissions:** All	**EVAP Emission System- Pressure Sensor Intermittent** The PCM measures pressure stability in the fuel tank, by means of a sensor for a predetermined duration. If fluctuation is larger than predetermined threshold a DTC is set. Sensor signal noise check. **Possible Causes:** • Contact resistance in connectors • Faulty FTPS • Faulty ECM/PCM • Faulty FTPS
DTC: P0455 **2T EVAP, MIL: Yes** **1996-06** **Models:** Amanti, Optima, Rio, Sedona, Sorento, Sephia, Spectra, Sportage **Engines:** All **Transmissions:** All	**EVAP System Large Leak Detected** ECT input at 38-95°F at startup, engine running at a steady throttle at over 6.2 mph for over 2 minutes, and the PCM detected the FTP sensor signal indicated more than -15 kPa in the EVAP Leak test. **Possible Causes:** • Fuel filler cap damaged, cross-threaded or loosely installed • Small leaks or cuts present in the EVAP vapor hoses/lines • EVAP purge valve is damaged or has failed • PCM has failed

DTC	Trouble Code Title, Conditions & Possible Causes
DTC: P0456 **2T EVAP, MIL: Yes** **2001-06** **Models:** Amanti, Optima, Rio, Sedona, Sorento, Spectra, Sportage **Engines:** All **Transmissions:** All	**EVAP System Very Small Leak (0.020") Detected** ECT input at 38-95°F at startup, vehicle driven at a steady speed of over 6.2 mph for 2 minutes, and then the PCM detected a very small leak (less than 0.020") in the EVAP system during the Leak test. **Possible Causes:** • Fuel filler is damaged, cross-threaded, loose or missing • Fuel filler pipe is damaged, or a fuel vapor hose is leaking • Rollover valve or ORVR (valve) had failed allowing fuel in lines • Canister close valve clogged or stuck in open or closed position • Purge solenoid valve is damaged or installed improperly • FTP sensor is damaged or has failed • Leaks in the charcoal canister, or at the fuel tank seals
DTC: P0457 **2T CCM, MIL: Yes** **2006** **Models:** Rio **Engines:** 1.6L **Transmissions:** All	**Evaporative Emission System- Leak Detected (Tank Cap Loose/Off)** Large leak caused by fuel cap loosened. **Possible Causes:** • Fuel cap • Faulty ECM/PCM
DTC: P0458 **2T CCM, MIL: Yes** **2006** **Models:** Rio **Engines:** 1.6L **Transmissions:** All	**Evaporative Emission System Purge Control valve Circuit Low** Circuit continuity check, low. **Possible Causes:** • Poor connection • Short to ground in control circuit • PCSV • Faulty ECM/PCM
DTC: P0459 **2T CCM, MIL: Yes** **2006** **Models:** Rio **Engines:** 1.6L **Transmissions:** All	**Evaporative Emission System Purge Control valve Circuit High** Circuit continuity check, high. **Possible Causes:** • Poor connection • Short to power in control circuit • PCSV • Faulty ECM/PCM
DTC: P0460 **2T CCM, MIL: Yes** **2002-06** **Models:** Amanti, Optima, Sedona, Sorento **Engines:** All **Transmissions:** All	**Fuel Level Sensor Circuit Malfunction** Engine running and the PCM detected the Fuel Level sensor input did not match the fuel level during the CCM test. **Possible Causes:** • Empty fuel tank, or the fuel level indicator mechanism is stuck • Fuel gauge incorrectly installed • Fuel level indicator signal circuit is open or shorted to ground • Fuel tank has been overfilled, or fuel level indicator is damaged • PCM has failed
DTC: P0461 **2T CCM, MIL: Yes** **2003-06** **Models:** Amanti, Optima, Rio, Sedona, Sportage **Engines:** 1.6L, 2.0L, 2.4L, 2.7L, 3.5L, 3.8L **Transmissions:** All	**Fuel Level Sensor "A" Circuit Range/Performance** **Filtered and unfiltered signal of fuel sensor are monitored.** **Possible Causes:** • Poor connection • Faulty fuel level sensor • Faulty PCM
DTC: P0462 **2T CCM, MIL: Yes** **2002-06** **Models:** Amanti, Optima, Rio, Sedona, Sportage **Engines:** All **Transmissions:** All	**Fuel Level Sensor Input Low (Sticking)** Key on or engine running system voltage more than 10 volts, and the PCM detected the fuel level sensing unit signal was less than 0.2 volt. **Possible Causes:** • Fuel level sending unit signal circuit shorted to VREF • Fuel level sending unit signal circuit shorted to system power • Fuel level sensing unit is damaged or the fuel tank is damaged • BCM or PCM has failed

DTC	Trouble Code Title, Conditions & Possible Causes
DTC: P0463 **2T CCM, MIL: Yes** **2002-06** **Models:** Amanti, Optima, Rio, Sedona, Sorento, Spectra, Sportage **Engines:** All **Transmissions:** All	**Fuel Level Sensor Input High (Sticking)** Key on or engine running system voltage more than 10 volts, and the PCM detected the fuel level sensing unit signal was more than 4.5 volts. **Possible Causes:** • Fuel level sending unit signal circuit shorted to VREF • Fuel level sending unit signal circuit shorted to system power • Fuel level sensing unit is damaged or the fuel tank is damaged • BCM or PCM has failed
DTC: P0464 **2T CCM, MIL: No** **2003-06** **Models:** Optima, Sedona, Sportage **Engines:** 2.0L, 2.4L, 2.7L, 3.8L **Transmissions:** All	**Fuel Level Sensor "A" Circuit Intermittent** Check signal for fluctuation. The ECM sets the DTC if the fuel level signal is higher than the threshold value (signal fluctuation greater than 50 percent). **Possible Causes:** • Contact resistance in connectors • Short to battery in fuel level (FLS) circuit • Faulty ECM
DTC: P0480 **2T CCM, MIL: Yes** **2006** **Models:** Optima, Sedona **Engines:** 2.7L, 3.8L **Transmissions:** All	**Fan 1 Control Circuit Malfunction** This will detect a short to ground, to battery or open circuit of fan relay output. Fault information provided by an output driver chip. No disabling faults present. Engine running. Enable time delay equal or greater than 0.5 seconds. **Possible Causes:** • Poor connection • Open in power circuit to cooling fan • Open or short in control circuit to PCM • Faulty fan relay • Faulty cooling fan module • Faulty PCM
DTC: P0481 **2T CCM, MIL: Yes** **2006** **Models:** Optima **Engines:** 2.7L **Transmissions:** All	**Fan 2 Control Circuit Malfunction** This will detect a short to ground, to battery or open circuit of fan relay output. Fault information provided by an output driver chip. No disabling faults present. Engine running. Enable time delay equal or greater than 0.5 seconds. **Possible Causes:** • Poor connection • Open in power circuit to cooling fan • Open or short in control circuit to PCM • Faulty fan relay
DTC: P0489 **2T CCM, MIL: Yes** **2005-06** **Models:** Amanti **Engines:** 3.5L **Transmissions:** All	**Exhaust Gas recirculation Control Circuit Low Voltage** Surge voltage (Vps) and output level voltage is monitored. **Possible Causes:** • Poor connection • Open or short to ground in EGR solenoid valve circuit • Faulty EGR solenoid valve • Faulty PCM
DTC: P0490 **2T CCM, MIL: Yes** **2005-06** **Models:** Amanti **Engines:** 3.5L **Transmissions:** All	**Exhaust Gas recirculation Control Circuit High Voltage** Surge voltage (Vps) and output level voltage is monitored. Engine speed is 500 rpm. **Possible Causes:** • Poor connection • Short to battery in EGR solenoid valve circuit • Faulty EGR solenoid valve • Faulty PCM
DTC: P0496 **2T CCM, MIL: Yes** **2006** **Models:** Rio **Engines:** 1.6L **Transmissions:** All	**Evaporative Emission System High Purge Flow** Fuel tank pressure behavior (canister purge valve stuck). Time after engine start 600 seconds. Idle speed controller activated. Mixture adaptation activated. Coolant temperature at start 11.88 degrees F. Tank ventilation must be active for 10 seconds. **Possible Causes:** • Leakage at the fuel evaporative system • PCSV • Faulty ECM/PCM

DTC	Trouble Code Title, Conditions & Possible Causes
DTC: P0497 **2T CCM, MIL: Yes** **2006** **Models:** Rio **Engines:** 1.6L **Transmissions:** All	**Evaporative Emission System Low Purge Flow** Fuel tank pressure behavior (canister purge valve stuck). Time after engine start 600 seconds. Idle speed controller activated. Mixture adaptation activated. Coolant temperature at start 11.88 degrees F. Tank ventilation must be active for 10 seconds. **Possible Causes:** • Clog in the fuel evaporative system • PCSV • Faulty ECM/PCM
DTC: P0498 **2T CCM, MIL: Yes** **2006** **Models:** Rio **Engines:** 1.6L **Transmissions:** All	**Evaporative Emission System Vent Valve Control Circuit Low** Circuit continuity check, low. **Possible Causes:** • Poor connection • Short to ground in control circuit • CCV • Faulty ECM/PCM
DTC: P0499 **2T CCM, MIL: Yes** **2006** **Models:** Rio **Engines:** 1.6L **Transmissions:** All	**Evaporative Emission System Vent Valve Control Circuit High** Circuit continuity check, high. **Possible Causes:** • Poor connection • Short to power in control circuit • CCV • Faulty ECM/PCM
DTC: P0500 **2T CCM, MIL: Yes** **1996-97** **Models:** Sephia **Engines:** All **Transmissions:** All	**Vehicle Speed Sensor Circuit Malfunction** Engine running in gear at light load for over 1 second, and the PCM did not detect any VSS signals during the CCM test. **Possible Causes:** • VSS signal circuit is open or shorted to ground • VSS signal circuit is shorted to VREF or to system power (B+) • VSS is damaged or has failed • PCM has failed
DTC: P0500 **2T CCM, MIL: Yes** **2001-06** **Models:** Optima, Sorento **Engines:** All **Transmissions:** All	**Vehicle Speed Sensor Circuit Malfunction** Engine speed over 3000 rpm with Closed Throttle switch indicating "off", engine load over 70%, and the PCM detected the VSS signal did not change voltage or was missing for 30 seconds. **Possible Causes:** • VSS signal circuit is open or shorted to ground • VSS signal circuit is shorted to VREF or to system power (B+) • VSS is damaged or has failed • PCM has failed
DTC: P0501 **2T CCM, MIL: Yes** **1996-97** **Models:** Sportage **Engines:** All **Transmissions:** All	**Vehicle Speed Sensor Range/Performance** Engine running in gear at high speed and for over 1 second, and the PCM did not detect any VSS signals. **Possible Causes:** • VSS signal circuit is open or shorted to ground • VSS signal circuit is shorted to VREF or to system power (B+) • VSS is damaged or has failed • PCM has failed
DTC: P0501 **2T CCM, MIL: Yes** **1998-06** **Models:** Optima, Rio, Sedona, Sephia, Spectra, Sportage **Engines:** All **Transmissions:** All	**Vehicle Speed Sensor Performance** Engine running in gear at high speed and lover for over 1 second, and the PCM did not detect any VSS signals. **Possible Causes:** • VSS signal circuit is open or shorted to ground • VSS signal circuit is shorted to VREF or to system power (B+) • VSS is damaged or has failed • PCM has failed
DTC: P0504 **2T CCM, MIL: Yes** **2006** **Models:** Optima, Sedona **Engines:** 2.4L, 2.7L, 3.8L **Transmissions:** All	**Brake Switch "A"/"B" Correlation (1)** Comparing two brake signals during driving. Case 1: Engine works. Vehicle speed sensor is normal. Case 2: Engine works. Vehicle speed sensor is normal. Vehicle speed is over 20 kph, for at least 1 second. **Possible Causes:** • Poor connection • Open or short • Faulty PCM

DTC	Trouble Code Title, Conditions & Possible Causes
DTC: P0504 **2T CCM, MIL: Yes** **2006** **Models:** Optima, Sedona **Engines:** 2.4L, 2.7L, 3.8L **Transmissions:** All	**Brake Switch "A"/"B" Correlation (2)** Plausibility check between brake light switch and brake test switch. Engine running. Time between brake light switch and brake test switch do not correlate longer than 10 seconds. **Possible Causes:** • Open or short circuit in harness • Poor connection or damaged harness • Faulty brake warning lamp or brake test switch
DTC: P0505 **2T CCM, MIL: Yes** **1996-06** **Models:** Rio, Sephia **Engines:** All **Transmissions:** All	**Idle Speed System Malfunction** Engine running at idle speed while in closed loop, and the PCM detected the Actual idle speed more than 100-200 rpm above or below the Target idle speed during the test. **Possible Causes:** • High resistance between the main relay and IAC valve • High resistance between PCM and IAC valve control circuits • IAC valve is damaged or has failed • The throttle plate is carbon fouled (it may need to be cleaned)
DTC: P0506 **2T CCM, MIL: Yes** **1996-97** **Models:** Sportage **Engines:** All **Transmissions:** All	**Idle Speed Lower Than Expected** Engine running at idle speed while in closed loop, and the PCM detected the Actual idle speed was more than 100 rpm below the Target idle speed during the CCM test. **Possible Causes:** • High resistance between the main relay and IAC valve • High resistance between PCM and IAC valve control circuits • IAC valve is damaged or has failed • The throttle plate is carbon fouled (it may need to be cleaned)
DTC: P0506 **2T CCM, MIL: Yes** **1998-06** **Models:** Amanti, Optima, Rio, Sedona, Sorento, Sephia, Spectra, Sportage **Engines:** All **Transmissions:** All	**Idle Speed Lower Than Expected** Engine running at idle speed while in closed loop, and the PCM detected the Actual idle speed was more than 100 rpm below the Target idle speed during the CCM test. **Possible Causes:** • High resistance between the main relay and IAC valve • High resistance between PCM and IAC valve control circuits • IAC valve is damaged or has failed • The throttle plate is carbon fouled (it may need to be cleaned)
DTC: P0507 **2T CCM, MIL: Yes** **1996-06** **Models:** Amanti, Optima, Rio, Sedona, Sorento, Sephia, Spectra, Sportage **Engines:** All **Transmissions:** All	**Idle Speed Higher Than Expected** Engine running at idle speed while in closed loop, and the PCM detected the Actual idle speed was more than 200 rpm above the Target idle speed during the CCM test. **Possible Causes:** • High resistance between the main relay and IAC valve • High resistance between PCM and IAC valve control circuits • IAC valve is damaged or has failed • Intake air leak located below the throttle plate assembly
DTC: P0510 **2T CCM, MIL: Yes** **1996-98** **Models:** Sephia **Engines:** 1.6L (vin 4) **Transmissions:** All	**Closed Throttle Position Switch Circuit Malfunction** Engine running and the PCM detected the Closed Throttle Position switch status did not change from "off" to "on" during normal driving. **Possible Causes:** • Closed throttle switch signal circuit open or shorted to ground • Closed throttle switch signal circuit shorted to VREF or power • Closed throttle switch is damaged or has failed • PCM has failed
DTC: P0510 **2T CCM, MIL: Yes** **2001-06** **Models:** Amanti, Optima, Sedona, Sorento **Engines:** All **Transmissions:** All	**Closed Throttle Position Switch Circuit Malfunction** Engine running and the PCM detected the Closed Throttle Position switch status did not change from "off" to "on" during normal driving. **Possible Causes:** • Closed throttle switch signal circuit open or shorted to ground • Closed throttle switch signal circuit shorted to VREF or power • Closed throttle switch is damaged or has failed • PCM has failed

DTC	Trouble Code Title, Conditions & Possible Causes
DTC: P0532 **2T CCM, MIL: No** **2006** **Models:** Optima, Rio, Sedona **Engines:** 1.6L, 2.4L, 2.7L, 3.8L **Transmissions:** All	**A/C Refrigerant Pressure Sensor "A" Circuit Low Input** Detects sensor signal short to low voltage. Engine works. Sensor output 0.05 volt. **Possible Causes:** • Poor connection • Open in power circuit • Open or short to ground in signal circuit • Faulty A/C pressure sensor • Faulty PCM
DTC: P0533 **2T CCM, MIL: No** **2006** **Models:** Optima, Rio, Sedona **Engines:** 1.6L, 2.4L, 2.7L, 3.8L **Transmissions:** All	**A/C Refrigerant Pressure Sensor "A" Circuit High Input** Detects sensor signal short to high voltage. Engine works. Sensor output 4.65 volts. **Possible Causes:** • Poor connection • Open in signal circuit open • Open in ground circuit • Faulty A/C pressure sensor • Faulty PCM
DTC: P0551 **2T CCM, MIL: Yes** **2003-06** **Models:** Amanti, Optima, Sedona, Sorento, **Engines:** 2.0L, 2.4L, 2.7L, 3.5L **Transmissions:** All	**Power Steering Pressure Sensor/Switch Circuit Range/Performance** If a power steering switch signal is ON when the engine speed is more than 2500 rpm, load value is grater than 55 percent and Engine coolant temperature is above 50 degrees F, the DTC will set. Signal of power steering pressure switch is monitored. **Possible Causes:** • Poor connection • Faulty power steering switch • Open or short in power steering switch • Faulty PCM
DTC: P0560 **2T CCM, MIL: Yes** **2001-06** **Models:** Amanti, Optima, Rio, Sedona, Sorento, Spectra, Sportage **Engines:** All **Transmissions:** All	**Battery Backup Line Circuit Malfunction** Engine runtime over 4 minutes and the PCM did not detect any system voltage on the Battery Backup circuit for 5 seconds. **Possible Causes:** • Battery backup circuit to the PCM is open • Battery backup fuse to the PCM is open or missing • Battery backup circuit to the PCM has high resistance • PCM has failed
DTC: P0561 **2T CCM, MIL: No** **1998-02** **Models:** Sephia **Engines:** All **Transmissions:** All	**System Voltage Unstable** Engine runtime over 4 minutes, and the PCM detected the system voltage rapidly changed its value by more than 3 volts. **Note: If the Battery Backup circuit is open, the vehicle will not run.** **Possible Causes:** • Charging system problem (charging voltage interrupted) • Backup voltage circuit to the PCM open (intermittent fault) • PCM has failed
DTC: P0561 **2T CCM, MIL: No** **1996-06** **Models:** Rio, Sedona, Spectra, Sportage **Engines:** All **Transmissions:** All	**System Voltage Unstable** Engine runtime over 4 minutes, and the PCM detected the system voltage rapidly changed its value by more than 3 volts. **Note: If the Battery Backup circuit is open, the vehicle will not run.** **Possible Causes:** • Charging system problem (charging voltage interrupted) • Backup voltage circuit to the PCM open (intermittent fault) • PCM has failed
DTC: P0562 **2T CCM, MIL: No** **1996-06** **Models:** Optima, Rio, Sedona, Spectra, Sportage **Engines:** All **Transmissions:** All	**System Voltage Low Input** Engine runtime over 4 minutes, and the PCM detected the system voltage was less than 8.0 volts, condition met for 5 seconds. **Note: If the Battery Backup circuit is open, the vehicle will not run.** **Possible Causes:** • Charging system problem (charging voltage too low) • Battery backup circuit to the PCM has high resistance • Backup voltage circuit to the PCM open (intermittent fault) • PCM has failed
DTC: P0563 **2T CCM, MIL: No** **1996-06** **Models:** Optima, Rio, Sedona, Sephia, Spectra, Sportage **Engines:** All **Transmissions:** All	**System Voltage High Input** Engine runtime over 4 minutes, and the PCM detected the system voltage was more than 17.0 volts, condition met for 5 seconds. **Note: If the Battery Backup circuit is open, the vehicle will not run.** **Possible Causes:** • Charging system problem (charging voltage too high) • Backup voltage circuit to the PCM open (intermittent fault) • PCM has failed

DTC	Trouble Code Title, Conditions & Possible Causes
DTC: P0564 **2T CCM, MIL: Yes** **2006** **Models:** Amanti **Engines:** 3.5L **Transmissions:** All	**Cruise Control Multifunction Input "A" Circuit** Invalid voltage range check. A DTC code is set for the following conditions. Check SET/COAST switch stuck. Check RES/ACC switch stuck. **Possible Causes:** • Open or short in harness • Poor connection or damaged harness • Faulty cruise control remote control switch
DTC: P0571 **2T CCM, MIL: Yes** **2006** **Models:** Amanti, Optima, Sedona **Engines:** 2.7L, 3.5L, 3.8L **Transmissions:** All	**Brake Switch "A" Circuit** PCM detects brake light input signal when the vehicle stops. VSS is normal. Vehicle speed 0 mph, during one second or more. **Possible Causes:** • Poor connection • Open or short to ground in signal circuit • Faulty PCM
DTC: P0600 **2T CCM, MIL: Yes** **2004-06** **Models:** Amanti, Optima, Spectra, Sportage **Engines:** 2.0L, 2.4L, 3.5L **Transmissions:** All	**CAN Communication Bus** **CAN message transfer incorrect?** **Possible Causes:** • Open or short in CAN line • Contact resistance in connectors • Faulty PCM
DTC: P0601 **2T PCM, MIL: Yes** **1996-06** **Models:** Optima, Rio, Sedona, Sephia, Spectra, Sportage **Engines:** All **Transmissions:** All	**PCM or TCM Internal Random Check Sum Error** Key on or engine running and the PCM or the TCM detected that a Random Check Sum Error was present. **Possible Causes:** • Poor terminal contact at the ECM ISC Backup Voltage circuit • PCM or TCM has an internal problem or has failed
DTC: P0601 **1T PCM, MIL: Yes** **2006** **Models:** Optima, Sedona **Engines:** 2.7L, 3.8L **Transmissions:** All	**PCM (Internal Controller) Checksum Error** Key on or engine running for 1 second, and the PCM detected an internal checksum data error during the initial Self-Test. **Possible Causes:** • Clear the trouble codes and retest for this trouble code. If the same trouble code resets, the PCM has failed and must be replaced to repair this problem.
DTC: P0602 **2T CCM, MIL: Yes** **2006** **Models:** Optima, Sedona **Engines:** 2.7L, 3.8L **Transmissions:** All	**EEPROM Programming Error** Check internal CPU **Possible Causes:** • Faulty PCM
DTC: P0604 **2T PCM, MIL: Yes** **1996-06** **Models:** Optima, Sedona, Sephia, Spectra, Sportage **Engines:** All **Transmissions:** All	**PCM or TCM Internal Random Access Memory Error** Key on or engine running and the PCM or TCM detected an Internal Random Access Memory (RAM) error was present. **Possible Causes:** • Poor terminal contact at the ECM Backup Voltage circuit • PCM or TCM has an internal problem or has failed
DTC: P0605 **1T PCM, MIL: Yes** **2003-06** **Models:** Optima, Rio, Sedona, Spectra, Sportage **Engines:** 1.6L, 2.0L, 2.4L, 2.7L **Transmissions:** All	**PCM (Internal Controller) ROM Error** Key on for 1 second, and the PCM detected an internal ROM error occurred during the initial Self-Test. **Possible Causes:** • Clear the trouble codes and retest for this trouble code. If the same trouble code resets, the PCM has failed and must be replaced to repair this problem.
DTC: P0606 **2T CCM, MIL: Yes** **2003-06** **Models:** Sedona, Spectra **Engines:** 2.0L, 3.8L **Transmissions:** All	**ECM Processor (ECU-Self Test Failed)** Controller error. No electrical fault of the front HO2S. **Possible Causes:** • Faulty PCM

DTC	Trouble Code Title, Conditions & Possible Causes
DTC: P061B **2T CCM, MIL: No** **2006** **Models:** Optima, Sedona **Engines:** 2.7L, 3.8L **Transmissions:** All	**Internal Control Module Torque Calculation Performance** Desired torque error. **Possible Causes:** • Faulty PCM
DTC: P0624 **2T CCM, MIL: No** **2006** **Models:** Rio **Engines:** 1.6L **Transmissions:** All	**Fuel Cap Lamp Control Circuit** Circuit continuity check, (high, low, or open). **Possible Causes:** • Poor connection • Open or short • Instrument cluster • Faulty ECM/PCM
DTC: P0625 **2T CCM, MIL: No** **2006** **Models:** Optima **Engines:** 2.4L **Transmissions:** All	**Alternator Field "F" Terminal Circuit Low** Electrical check. **Possible Causes:** • Short to battery in harness • Poor connection or damaged harness
DTC: P0626 **2T CCM, MIL: No** **2006** **Models:** Optima **Engines:** 2.4L **Transmissions:** All	**Alternator Field "F" Terminal Circuit High** Electrical check. Time after ignition ON, 1 second. Engine speed 0. No main relay error. **Possible Causes:** • Open or short to ground in harness • Faulty charging system
DTC: P0630 **2T CCM, MIL: Yes** **2005-06** **Models:** Amanti, Optima, Rio, Sedona, Spectra, Sportage **Engines:** 1.6L, 2.0L, 2.4L, 2.7L, 3.5L, 3.8L **Transmissions:** All	**VIN Not Programmed Or Incompatible- ECM/PCMECM** PCM internal check. Enable condition, ignition ON. VIN does not exist in boot area. **Possible Causes:** • PCM is new and has not yet been programmed • Faulty PCM
DTC: P0638 **2T CCM, MIL: Yes** **2006** **Models:** Sedona, Optima **Engines:** 2.7L, 3.8L **Transmissions:** All	**Throttle Actuator Control Range/Performance** ETS position control malfunction. Battery voltage more than 5 volts. **Possible Causes:** • Throttle stuck • Open in motor circuit • Faulty motor • Faulty PCM
DTC: P0641 **2T CCM, MIL: Yes** **2006** **Models:** Optima, Sedona **Engines:** 2.7L, 3.8L **Transmissions:** All	**Sensor Reference Voltage "A" Circuit Open** Sensor reference voltage check. Ignition ON. **Possible Causes:** • Short in sensor power supply line • Faulty PCM
DTC: P0642 **2T CCM, MIL: Yes** **2006** **Models:** Optima, Rio **Engines:** 1.6L, 2.4L **Transmissions:** All	**Sensor Reference Voltage "A" Circuit Low** Sensor reference voltage check. Battery voltage 11-16 volts. **Possible Causes:** • Spoor connection • Short to ground in 5V, voltage circuit. • ECM/PCM
DTC: P0643 **2T CCM, MIL: Yes** **2006** **Models:** Optima, Rio **Engines:** 1.6L, 2.4L **Transmissions:** All	**Sensor Reference Voltage "A" Circuit High** Sensor reference voltage check. Battery voltage 11-16 volts. **Possible Causes:** • Spoor connection • Short to power in 5V, voltage circuit. • ECM/PCM

DTC	Trouble Code Title, Conditions & Possible Causes
DTC: P0645 **2T CCM, MIL: No** **2006** **Models:** Rio **Engines:** 1.6L **Transmissions:** All	**A/C Clutch Relay Control Circuit** DTC is set if the PCM detects that the relay line is open or shorted to ground or battery line. Circuit continuity check. **Possible Causes:** • Open in battery and control circuit • Short to ground in control circuit • Short to battery in control circuit • Faulty A/C relay • Faulty PCM
DTC: P0646 **2T CCM, MIL: No** **2006** **Models:** Optima, Rio, Sedona **Engines:** 1.6L, 2.4L, 2.7L, 3.8L **Transmissions:** All	**A/C Clutch Relay Control Circuit Low** Detects circuit short to low voltage. No DTC exists. Engine works. After 0.5 seconds. **Possible Causes:** • Poor connection • Open or short to ground in A/C relay circuit • Faulty A/C relay • Faulty PCM
DTC: P0647 **2T CCM, MIL: No** **2006** **Models:** Optima, Rio, Sedona **Engines:** 1.6L, 2.4L, 2.7L, 3.8L **Transmissions:** All	**A/C Clutch Relay Control Circuit High** Detects circuit short to high voltage. No DTC exists. Engine works. After 0.5 seconds. **Possible Causes:** • Poor connection • Short to power in A/C relay circuit • Faulty A/C relay • Faulty PCM
DTC: P0650 **2T CCM, MIL: No** **2001-06** **Models:** Optima, Rio, Sedona, Sephia, Spectra, Sportage **Engines:** All **Transmissions:** All	**Malfunction Indicator Lamp Circuit Malfunction** Key on or engine running and the PCM detected an unexpected voltage condition on the Malfunction Indicator Lamp (MIL) circuit. **Possible Causes:** • MIL control circuit open • MIL control circuit shorted to ground • MIL "bulb" is damaged or missing • PCM has failed (MIL control "driver" may be open or shorted)
DTC: P0651 **2T CCM, MIL: Yes** **2006** **Models:** Optima, Sedona **Engines:** 2.7L, 3.8L **Transmissions:** All	**Sensor reference Voltage "B" Circuit Open** Sensor reference voltage check. Key ON. **Possible Causes:** • Short in sensor power supply line • Faulty PCM
DTC: P0652 **2T CCM, MIL: Yes** **2006** **Models:** Optima **Engines:** 2.4L **Transmissions:** All	**Sensor Reference Voltage "B" Circuit Low** Electrical check. Ignition ON. APS2 voltage 0.7 volt, for at least 0.04 second. **Possible Causes:** • Open or short to ground in power circuit • Poor connection or damaged harness • Faulty ECM
DTC: P0653 **2T CCM, MIL: Yes** **2006** **Models:** Optima **Engines:** 2.4L **Transmissions:** All	**Sensor Reference Voltage "B" Circuit High** Electrical check. Ignition ON. TPS voltage 5.5 volts, for at least 0.04 second. **Possible Causes:** • Open or short to ground in power circuit • Poor connection or damaged harness • Faulty ECM
DTC: P0660 **2T CCM, MIL: Yes** **2006** **Models:** Optima, Sedona **Engines:** 2.7L, 3.8L **Transmissions:** All	**Intake Manifold Tuning Valve Control Circuit/Open (Bank 1)** Signal low, high. **Possible Causes:** • Poor connection • Open or short in VIS circuit • Faulty VIS • Faulty PCM
DTC: P0661 **2T CCM, MIL: No** **2005-06** **Models:** Sportage **Engines:** 2.0L, 2.7L **Transmissions:** All	**Intake Manifold Tuning Valve Control Circuit Low (Bank 1) Solenoid Type** DTC is set if the ECM detects that the valve control circuit is shorted to ground. Driver stage check. **Possible Causes:** • Open in power supply harness • Short to ground in control harness • Contact resistance in connectors • Faulty valve

DTC	Trouble Code Title, Conditions & Possible Causes
DTC: P0662 **2T CCM, MIL: No** **2005-06** **Models:** Sportage **Engines:** 2.0L, 2.7L **Transmissions:** All	**Intake Manifold Tuning Valve Control Circuit High (Bank 1) Solenoid Type** DTC is set if the ECM detects that the valve control circuit is open or shorted to battery voltage. Driver stage check. **Possible Causes:** • Open or short to battery in control harness • Contact resistance in connectors • Faulty valve
DTC: P0663 **2T CCM, MIL: Yes** **2006** **Models:** Optima **Engines:** 2.7L **Transmissions:** All	**Intake Manifold Tuning Valve Control Circuit/Open (Bank 2)** Signal low, high. **Possible Causes:** • Poor connection • Open or short in VIS #2 circuit • Faulty VIS #2 • Faulty PCM
DTC: P0664 **2T CCM, MIL: No** **2005-06** **Models:** Sportage **Engines:** 2.7L **Transmissions:** All	**Intake Manifold Tuning Valve Control Circuit High (Bank 2) Solenoid Type** DTC is set if the ECM detects that the valve control circuit is shorted to ground. Driver stage check. **Possible Causes:** • Open in power supply harness • Short to ground in control harness • Contact resistance in connectors • Faulty valve
DTC: P0665 **2T CCM, MIL: No** **2005-06** **Models:** Sportage **Engines:** 2.7L **Transmissions:** All	**Intake Manifold Tuning Valve Control Circuit Low (Bank 1) Solenoid Type** DTC is set if the ECM detects that the valve control circuit is open or shorted to battery voltage. Driver stage check. **Possible Causes:** • Open or short to battery in control harness • Contact resistance in connectors • Faulty valve
DTC: P0685 **2T CCM, MIL: No** **2006** **Models:** Optima, Sedona **Engines:** 2.7L, 3.8L **Transmissions:** All	**ECM/PCM Power Relay Control Circuit/Open** Engine running. Ignition voltage less than or equal to 11 volts. **Possible Causes:** • poor connection • Open or short to in control circuit • Main relay • PCM
DTC: P0698 **2T CCM, MIL: Yes** **2006** **Models:** Optima **Engines:** 2.4L **Transmissions:** All	**Sensor Reference Voltage "C" Circuit Low** Electrical check. Ignition ON. ASP1 voltage less than 0.7 volt, for at least 0.1 second. **Possible Causes:** • Open or short to ground in power circuit • Poor connection or damaged harness • Faulty ECM
DTC: P0699 **2T CCM, MIL: Yes** **2006** **Models:** Optima **Engines:** 2.4L **Transmissions:** All	**Sensor Reference Voltage "C" Circuit High** Electrical check. Ignition ON. APS1 voltage 5.5 volts, for at least 0.1 second. **Possible Causes:** • Open or short to ground in power circuit • Poor connection or damaged harness • Faulty ECM
DTC: P0700 **2T CCM, MIL: Yes** **2005-06** **Models:** Optima, Rio, Sorento, Spectra, Sportage **Engines:** 1.6L, 2.0L, 2.4L, 2.7L, 3.5L **Transmissions:** A/T	**TCU Request For MIL "ON"** Engine at normal operating temperature. Check for additional DTC's. **Possible Causes:** • poor connection • TCM • PCM/ECM

DTC	Trouble Code Title, Conditions & Possible Causes
DTC: P0703 **2T CCM, MIL: Yes** **1996-97** **Models:** Sephia **Engines:** All **Transmissions:** A/T	**TCC/Brake Switch Input Circuit Malfunction** Engine running with VSS inputs received, and the PCM did not detect a change in the Brake Switch status with the brake "on" for 33 seconds, or with the brake pedal continuously "off" for 15 seconds. **Possible Causes:** • Brake pedal switch signal circuit is open or shorted to ground • Brake pedal switch signal circuit is shorted to VREF or power • Brake pedal switch is damaged or has failed • PCM has failed
DTC: P0705 **2T CCM, MIL: Yes** **1996-02** **Models:** Optima, Rio, Sedona, Sephia, Spectra, Sportage **Engines:** All **Transmissions:** A/T	**Transmission Range Switch Circuit Malfunction** Engine running with VSS inputs received, and the PCM detected invalid signals or multiple TR switch inputs with the gearshift in drive. **Note: If DTC P1500 is also set, check the Meter Fuse and circuit.** **Possible Causes:** • TR switch signal circuit is open or shorted to ground • TR switch signal circuit is shorted to VREF or system power • TR switch is damaged or has failed • PCM has failed • TSB 2TD007 (12/01) contains information related to this code
DTC: P0707 **2T CCM, MIL: Yes** **2002** **Models:** Sedona **Engines:** All **Transmissions:** A/T	**Transmission Range Switch Circuit Malfunction** Engine running with VSS inputs received, and the PCM did not detect any TR switch signal for 10 seconds during the CCM test. **Possible Causes:** • TR switch signal circuit is open or shorted to ground • TR switch signal circuit is shorted to VREF or system power • TR switch is damaged or has failed • PCM has failed
DTC: P0708 **2T CCM, MIL: Yes** **2002** **Models:** Sedona **Engines:** All **Transmissions:** A/T	**Transmission Range Switch Circuit Performance** Engine running with VSS inputs received, and the PCM detected two (2) or more TR switch signals simultaneously for over 30 seconds. **Possible Causes:** • TR switch signal circuit is open or shorted to ground • TR switch signal circuit is shorted to VREF or system power • TR switch is damaged or has failed • PCM has failed
DTC: P0710 **2T CCM, MIL: Yes** **1996-97** **Models:** Sephia **Engines:** All **Transmissions:** A/T	**Transmission Fluid Temperature Sensor Performance** Engine runtime more than 10 minutes, and the PCM detected the TFT sensor signal indicated less than -40°F, or it detected that it indicated more than 300°F during the CCM rationality test period. **Possible Causes:** • TFT sensor signal circuit open or shorted to ground • TFT sensor signal circuit is shorted to VREF or system power • TFT sensor is damaged or has failed • PCM is damaged
DTC: P0710 **2T CCM, MIL: Yes** **2002** **Models:** Sedona **Engines:** All **Transmissions:** A/T	**Transmission Fluid Temperature Sensor Performance** Engine runtime more than 10 minutes, and the PCM detected the TFT sensor signal indicated less than -40°F, or it detected that it indicated more than 300°F during the CCM rationality test period. **Possible Causes:** • TFT sensor signal circuit open or shorted to ground • TFT sensor signal circuit is shorted to VREF or system power • TFT sensor is damaged or has failed • PCM is damaged
DTC: P0712 **2T CCM, MIL: Yes** **1998-02** **Models:** Optima, Rio, Sedona, Sephia, Spectra **Engines:** All **Transmissions:** A/T	**Transmission Fluid Temperature Low Input** Key on or engine running and the PCM detected the TFT sensor signal indicated less than 0.49 volt 300°F) for more than 1 second. **Note: The TFT sensor signal at 68°F is 4.0 volts, and at 266°F it is 1.5 volts.** **Possible Causes:** • TFT sensor signal circuit shorted to ground (sensor to TCM) • TFT sensor signal circuit shorted to ground (TCM to PCM) • TFT sensor is damaged or has failed • PCM is damaged

DTC	Trouble Code Title, Conditions & Possible Causes
DTC: P0713 **2T CCM, MIL: Yes** **1998-02** **Models:** Optima, Rio, Sedona, Sephia, Spectra **Engines:** All **Transmissions:** A/T	**Transmission Fluid Temperature High Input** Key on or engine running and the PCM detected the TFT sensor signal indicated more than 4.57 volts (-40°F) during the CCM test period. **Note: The TFT sensor signal at 68°F is 4.0 volts, and at 266°F it is 1.5 volts.** **Possible Causes:** • TFT sensor signal circuit is open (sensor circuit to the TCM) • TFT sensor signal circuit is open (TCM circuit to the PCM) • TFT sensor is damaged or has failed • PCM is damaged
DTC: P0715 **2T CCM, MIL: Yes** **1996-02** **Models:** Optima, Sedona, Sephia, Spectra **Engines:** All **Transmissions:** A/T	**Input/Turbine Speed Sensor Circuit Malfunction** Vehicle driven to a speed of over 15.5 mph with the engine speed over 1500 rpm in 3rd or 4th gear, and the PCM did not detect any Input/Turbine speed sensor signals for 5 seconds. **Possible Causes:** • Input/Turbine speed sensor signal circuit open or shorted • Input/Turbine speed sensor is damaged or has failed • PCM has failed
DTC: P0716 **2T CCM, MIL: Yes** **1998-02** **Models:** Optima, Rio, Sephia, Spectra, Sportage **Engines:** All **Transmissions:** A/T	**Input/Turbine Speed Sensor Performance** Vehicle drive to a speed of over 25 mph, gear ratio indicating the vehicle is in Drive, 2nd or 1st gear, and the PCM did not detect any Input/Turbine Speed sensor signals, or that the Input/Turbine sensor signals indicated less than 98 rpm during the CCM test. **Possible Causes:** • Input/Turbine speed sensor signal circuit open or shorted • Input/Turbine speed sensor is damaged or has failed • PCM has failed
DTC: P0717 **2T CCM, MIL: Yes** **1998-02** **Models:** Optima, Rio, Sephia, Spectra, Sportage **Engines:** All **Transmissions:** A/T	**Input/Turbine Speed Sensor No Signal** Vehicle drive to a speed of over 25 mph, gear ratio indicating the vehicle is in Drive, 2nd or 1st gear, and the PCM detected that the Input/Turbine Speed sensor signal indicated less than 98 rpm. **Possible Causes:** • Input/Turbine speed sensor signal circuit open or shorted • Input/Turbine speed sensor is damaged or has failed • PCM has failed
DTC: P0720 **2T CCM, MIL: Yes** **2002** **Models:** Sedona **Engines:** All **Transmissions:** A/T	**Output Shaft Speed Sensor Circuit Malfunction** Vehicle driven in 3rd or 4th gear at a speed of over 20 mph, and the PCM detected an unexpected voltage condition on the OSS sensor circuit. If this code is generated four (4) times or more, the PCM will lock the transmission into either 2nd or 3rd gear for safety reasons. **Possible Causes:** • Output shaft speed sensor signal circuit open or shorted • Output shaft speed sensor is damaged or has failed • PCM has failed
DTC: P0722 **2T CCM, MIL: Yes** **1996-02** **Models:** Sportage **Engines:** All **Transmissions:** A/T	**Output Shaft Speed Sensor No Signal** Vehicle driven in Drive, 2nd or 1st gear, gear ratio indicating the vehicle is in Drive, 2nd or 1st gear, Input shaft speed over 775 rpm, and the PCM did not detect any OSS sensor signals during the test. **Possible Causes:** • Output shaft speed sensor signal circuit open or shorted • Output shaft speed sensor is damaged or has failed • PCM has failed
DTC: P0726 **2T CCM, MIL: Yes** **1996-02** **Models:** Rio, Sportage **Engines:** All **Transmissions:** A/T	**Engine Speed Signal Performance** Engine running, throttle position angle from 5-94%, and the TCM detected the Engine Speed signal indicated more than 7500 rpm. **Possible Causes:** • Engine Speed Signal circuit open (PCM to the TCM) • Engine Speed Signal circuit shorted to ground (PCM to TCM) • PCM has failed • TCM has failed
DTC: P0726 **2T CCM, MIL: Yes** **1998-02** **Models:** Sephia **Engines:** All **Transmissions:** A/T	**Engine Speed Signal High Input** Engine running, throttle position angle from 5-94%, and the TCM detected the Engine Speed signal indicated more than 7500 rpm. **Possible Causes:** • Engine Speed Signal circuit open (PCM to the TCM) • Engine Speed Signal circuit shorted to ground (PCM to TCM) • PCM has failed • TCM has failed

DTC	Trouble Code Title, Conditions & Possible Causes
DTC: P0727 **2T CCM, MIL: Yes** **1996-02** **Models:** Rio, Sportage **Engines:** All **Transmissions:** A/T	**Engine Speed Signal Low Input** Engine running, throttle position angle from 5-94%, gear selector not indicating 'P' or 'N' position, and the TCM detected the Engine Speed signal indicated less than 96 rpm during the CCM test. **Possible Causes:** • Engine Speed Signal circuit open (PCM to the TCM) • Engine Speed Signal circuit shorted to ground (PCM to TCM) • PCM has failed • TCM has failed
DTC: P0727 **2T CCM, MIL: Yes** **1998-02** **Models:** Sephia **Engines:** All **Transmissions:** A/T	**Engine Speed Signal Low Input** Engine running, throttle position angle from 5-94%, gear selector not indicating 'P' or 'N' position, and the TCM detected the Engine Speed signal indicated less than 96 rpm during the CCM test. **Possible Causes:** • Engine Speed Signal circuit open (PCM to the TCM) • Engine Speed Signal circuit shorted to ground (PCM to TCM) • PCM has failed • TCM has failed
DTC: P0731 **2T CCM, MIL: Yes** **1996-02** **Models:** Optima, Rio, Sedona, Sephia, Spectra, Sportage **Engines:** All **Transmissions:** A/T	**TCM Incorrect First Gear Ratio** Vehicle driven at 12-32 mph in 3rd gear, shift solenoids 'A', 'B' and 'C', input/turbine speed sensor and TFT sensor inputs all indicating okay, and the PCM detected the 1st gear ratio was too high. **Possible Causes:** • ATF fluid level too low or line pressure low • Control valve stuck or solenoid valve is damaged or has failed • Forward clutch, 3-4 brake band or 1-way clutch No. 1 slippage • PCM has failed • TSB 2TD007 (12/01) contains information related to this code
DTC: P0732 **2T CCM, MIL: Yes** **1996-02** **Models:** Optima, Rio, Sedona, Sephia, Spectra, Sportage **Engines:** All **Transmissions:** A/T	**TCM Incorrect Second Gear Ratio** Vehicle driven at 17-60 mph in 2nd gear, shift solenoids 'A', 'B' and 'C', input/turbine speed sensor and TFT sensor inputs all indicating okay, and the PCM detected the 2nd gear ratio was too high. **Possible Causes:** • ATF fluid level too low or line pressure low • Control valve stuck or solenoid valve is damaged or has failed • Forward clutch, 2-4 brake band or 1-way clutch No. 1 slippage • PCM has failed • TSB 2TD004 (8/00) contains information related to this code • TSB 2TD007 (12/01) contains information related to this code • TSB 3TD008 (3/02) contains information related to this code
DTC: P0733 **2T CCM, MIL: Yes** **1996-02** **Models:** Optima, Rio, Sedona, Sephia, Spectra, Sportage **Engines:** All **Transmissions:** A/T	**TCM Third Gear Incorrect Ratio** Vehicle driven at 19-32 mph in 3rd gear, shift solenoids 'A', 'B' and 'C', input/turbine speed sensor and TFT sensor inputs all indicating okay, and the PCM detected the 3rd gear ratio was too high. **Possible Causes:** • ATF fluid level too low or line pressure low • Control valve stuck or solenoid valve is damaged or has failed • Forward clutch, 3-4 brake band or 1-way clutch No. 1 slippage • PCM has failed • TSB 2TD004 (8/00) contains information related to this code • TSB 2TD007 (12/01) contains information related to this code • TSB 3TD008 (3/02) contains information related to this code
DTC: P0734 **2T CCM, MIL: Yes** **1996-02** **Models:** Optima, Sedona, Sephia, Spectra, Sportage **Engines:** All **Transmissions:** A/T	**TCM Fourth Gear Incorrect Ratio** Vehicle driven at 44-65 mph in 4th gear, shift solenoids 'A', 'B' and 'C', input/turbine speed sensor and TFT sensor inputs all indicating okay, and the PCM detected the 4th gear ratio was too high. **Possible Causes:** • ATF fluid level too low or line pressure low • Control valve stuck or solenoid valve is damaged or has failed • 2-4 brake band or 3-4 clutch slippage • PCM has failed • TSB 2TD004 (8/00) contains information related to this code • TSB 2TD007 (12/01) contains information related to this code • TSB 3TD008 (3/02) contains information related to this code

DTC	Trouble Code Title, Conditions & Possible Causes
DTC: P0735 **2T CCM, MIL: Yes** **2002** **Models:** Sedona **Engines:** All **Transmissions:** A/T	**TCM Fifth Gear Incorrect Ratio** Vehicle driven at 44-65 mph in 5th gear, shift solenoids 'A', 'B' and 'C', input/turbine speed sensor and TFT sensor inputs all indicating okay, and the PCM detected the 5th gear ratio was too high. If this code sets 4 times or more, the PCM will lock the gear into 3rd gear. **Possible Causes:** • ATF fluid level too low or line pressure low • Control valve stuck or solenoid valve is damaged or has failed • 2-4 brake band or 3-4 clutch slippage • PCM has failed
DTC: P0736 **2T CCM, MIL: Yes** **2002** **Models:** Sedona **Engines:** All **Transmissions:** A/T	**TCM Reverse Gear Incorrect Ratio** Vehicle driven in Reverse gear, and the PCM detected the output speed sensor signal did not match the input speed sensor signal. If this code sets 4 times, the PCM will lock the gear into 3rd gear. **Possible Causes:** • ATF fluid level too low or line pressure low • Input speed sensor or output speed sensor damaged or failed • UD clutch retainer has failed or DIR planetary carrier has failed • Reverse clutch failure or LR brake line or RED brake line failure
DTC: P0740 **2T CCM, MIL: Yes** **1996-02** **Models:** Optima, Rio, Sedona, Sephia, Spectra, Sportage **Engines:** All **Transmissions:** A/T	**TCC System Malfunction** Vehicle driven at 44-65 mph in 4th gear in TCC, shift solenoids 'A', 'B' and 'C', input/turbine speed sensor and TFT sensor inputs all okay, and the PCM detected the difference in the number of engine revolutions and reverse/forward drum revolutions was over 100 rpm. **Possible Causes:** • ATF fluid level too low or line pressure low • Torque converter clutch slippage excessive • Control valve stuck or TCC solenoid is damaged or has failed • PCM has failed • TSB 2TD004 (8/00) contains information related to this code • TSB 2TD005 (11/00) contains information related to this code • TSB 2TD007 (12/01) contains information related to this code • TSB 3TD008 (3/02) contains information related to this code
DTC: P0743 **2T CCM, MIL: Yes** **1996-02** **Models:** Optima, Rio, Sephia, Spectra, Sportage **Engines:** All **Transmissions:** A/T	**TCC Solenoid Circuit Malfunction** Engine running in gear with VSS inputs received, and the PCM detected an unexpected voltage condition on the TCC circuit. **Possible Causes:** • TCC solenoid control circuit is open or shorted to ground • TCC solenoid control circuit is shorted to system power (B+) • TCC solenoid is damaged or has failed • PCM or TCM has failed
DTC: P0748 **2T CCM, MIL: Yes** **1998-02** **Models:** Optima, Rio, Sephia, Spectra **Engines:** All **Transmissions:** A/T	**TCM Linear Solenoid Circuit Malfunction** Engine running in gear with VSS inputs received, and the PCM detected an unexpected voltage on the Linear solenoid circuit. **Possible Causes:** • Linear solenoid control circuit open or shorted to ground • Linear solenoid control circuit shorted to system power (B+) • Linear solenoid is damaged or has failed • PCM or TCM has failed
DTC: P0750 **2T CCM, MIL: Yes** **1996-97** **Models:** Sephia, Sportage **Engines:** All **Transmissions:** A/T	**TCM Shift Solenoid 'A' Circuit Malfunction** Engine running in gear with VSS inputs received, and the PCM detected that the TCC solenoid was always "on" or always "off". **Possible Causes:** • SSA control circuit is open or shorted to ground • SSA is damaged or has failed • PCM or TCM has failed
DTC: P0750 **2T CCM, MIL: Yes** **2002** **Models:** Sedona **Engines:** All **Transmissions:** A/T	**TCM Shift Solenoid 'A' Circuit Malfunction** Engine running in gear with VSS inputs received, and the PCM detected that the TCC solenoid was always "on" or always "off". **Possible Causes:** • SSA control circuit is open or shorted to ground • SSA is damaged or has failed • PCM or TCM has failed

DTC	Trouble Code Title, Conditions & Possible Causes
DTC: P0753 **2T CCM, MIL: Yes** **1998-02** **Models:** Optima, Rio, Sephia, Spectra, Sportage **Engines:** All **Transmissions:** A/T	**TCM Shift Solenoid 'A' Circuit Malfunction** Engine running in gear, VSS inputs received, and PCM detected an unexpected voltage condition on the SSA circuit during the test. **Possible Causes:** • SSA control circuit open or shorted to ground • SSA control circuit shorted to system power (B+) • SSA is damaged or has failed • PCM or TCM has failed • TSB 3TD008 (3/02) contains information related to this code
DTC: P0755 **2T CCM, MIL: Yes** **1996-97** **Models:** Sephia **Engines:** All **Transmissions:** A/T	**TCM Shift Solenoid 'B' Circuit Malfunction** Engine running in gear with VSS inputs received, and the PCM detected that the SSB was always "on" or always "off". **Possible Causes:** • SSB is damaged or has failed • PCM or TCM has failed
DTC: P0755 **2T CCM, MIL: Yes** **1996-02** **Models:** Sedona, Sportage **Engines:** All **Transmissions:** A/T	**TCM Shift Solenoid 'B' Circuit Malfunction** Engine running in gear with VSS inputs received, and the PCM detected that the SSB was always "on" or always "off". **Possible Causes:** • SSB is damaged or has failed • PCM or TCM has failed
DTC: P0758 **2T CCM, MIL: Yes** **1998-02** **Models:** Optima, Rio, Sephia, Spectra, Sportage **Engines:** All **Transmissions:** A/T	**TCM Shift Solenoid 'B' Circuit Malfunction** Engine running in gear, VSS inputs received, and PCM detected an open or shorted condition in the SSB circuit during the CCM test. **Possible Causes:** • SSB control circuit open or shorted to ground • SSB control circuit shorted to system power (B+) • SSB is damaged or has failed • PCM or TCM has failed • TSB 3TD008 (3/02) contains information related to this code
DTC: P0758 **2T CCM, MIL: Yes** **1996-97** **Models:** Sportage **Engines:** All **Transmissions:** A/T	**TCM Shift Solenoid 'B' Circuit Malfunction** Engine running in gear, VSS inputs received, and PCM detected an open or shorted condition in the SSB circuit during the CCM test. **Possible Causes:** • SSB control circuit open or shorted to ground • SSB control circuit shorted to system power (B+) • SSB is damaged or has failed • PCM or TCM has failed
DTC: P0760 **2T CCM, MIL: Yes** **1996-97** **Models:** Sephia **Engines:** All **Transmissions:** A/T	**TCM Shift Solenoid 'C' Circuit Malfunction** Engine running in gear, VSS inputs received, and PCM detected an unexpected voltage condition on the SSC circuit during the test. **Possible Causes:** • SSC control circuit open or shorted to ground • SSC control circuit shorted to system power (B+) • SSC is damaged or has failed • PCM or TCM has failed
DTC: P0760 **2T CCM, MIL: Yes** **2002** **Models:** Sedona **Engines:** All **Transmissions:** A/T	**TCM Shift Solenoid 'C' Circuit Malfunction** Engine running in gear, VSS inputs received, and PCM detected an unexpected voltage condition on the SSC circuit during the test. **Possible Causes:** • SSC control circuit open or shorted to ground • SSC control circuit shorted to system power (B+) • SSC is damaged or has failed • PCM or TCM has failed
DTC: P0760 **2T CCM, MIL: Yes** **2000** **Models:** Sedona **Engines:** All **Transmissions:** A/T	**TCM Shift Solenoid 'C' Circuit Malfunction** Engine running in gear, VSS inputs received, and PCM detected an unexpected voltage condition on the SSC circuit during the test. **Possible Causes:** • SSC control circuit open or shorted to ground • SSC control circuit shorted to system power (B+) • SSC is damaged or has failed • PCM or TCM has failed

DTC	Trouble Code Title, Conditions & Possible Causes
DTC: P0765 **2T CCM, MIL: Yes** **1998-02** **Models:** Optima, Rio, Sephia, Spectra **Engines:** All **Transmissions:** A/T	**TCM Overdrive Solenoid Circuit Malfunction** Engine running in gear, VSS inputs received, and PCM detected an unexpected voltage condition on the O/D valve circuit during the test. **Possible Causes:** • O/D solenoid control circuit open or shorted to ground • O/D solenoid control circuit shorted to system power (B+) • O/D solenoid valve is damaged or has failed • PCM or TCM has failed
DTC: P0770 **2T CCM, MIL: Yes** **2002** **Models:** Sedona **Engines:** All **Transmissions:** A/T	**TCM RED Solenoid Circuit Malfunction** Engine running in gear, VSS inputs received, and PCM detected an unexpected voltage condition on the RED solenoid circuit in the test. **Possible Causes:** • RED control circuit open or shorted to ground • RED control circuit shorted to system power (B+) • RED is damaged or has failed • PCM or TCM has failed

Gas Engine OBD II Trouble Code List (P1xxx Codes)

DTC	Trouble Code Title, Conditions & Possible Causes
DTC: P1100 **2T CCM, MIL: Yes** **2002** **Models:** Optima **Engines:** 2.4L (vin 3) **Transmissions:** All	**MAP Sensor Circuit Malfunction** ECT sensor signal less than 32°F at startup, engine runtime over 8 minutes, engine load from 30-55%, ECT sensor signal 113°F or more, IAT sensor signal 41°F or more, the PCM detected a MAP sensor signal of more than 4.60 volts or less than 0.1 volt for 4 seconds. **Possible Causes:** • MAP sensor signal circuit open or shorted to ground • MAP sensor signal circuit shorted to VREF or system power • MAP sensor is damaged or has failed • PCM is damaged
DTC: P1102 **2T OS2 HTR2, MIL: Yes** **1996** **Models:** Sportage **Engines:** All **Transmissions:** All	**HO2S-11 (Bank 1 Sensor 1) Heater Circuit High Input** Engine running, and PCM detected an unexpected voltage condition on the front HO2S heater control circuit. **Note: Inspect the HO2S connector for oil and water contamination.** **Possible Causes:** • HO2S heater control circuit is shorted to system power • HO2S heater control circuit is shorted to heater power feed • HO2S heater is damaged or has failed
DTC: P1102 **2T CCM, MIL: Yes** **2002** **Models:** Optima **Engines:** 2.4L (vin 3) **Transmissions:** All	**MAP Sensor Low Input** ECT sensor signal less than 32°F at startup, engine runtime over 8 minutes, engine load from 30-55%, ECT sensor signal 113°F or more, IAT sensor signal 41°F or more, and the PCM detected a MAP sensor signal was less than 0.1 volt for 4 seconds. **Possible Causes:** • MAP sensor signal circuit open or shorted to ground • MAP sensor is damaged or has failed • PCM is damaged
DTC: P1103 **2T CCM, MIL: Yes** **2002** **Models:** Optima **Engines:** 2.4L (vin 3) **Transmissions:** All	**MAP Sensor High Input** ECT sensor signal less than 32°F at startup, engine runtime over 8 minutes, engine load from 30-55%, ECT sensor signal 113°F or more, IAT sensor signal 41°F or more, and the PCM detected the MAP sensor signal was more than 4.6 volts for 4 seconds. **Possible Causes:** • MAP sensor signal circuit open or shorted to ground • MAP sensor is damaged or has failed • PCM is damaged
DTC: P1105 **2T 2 HTR2, MIL: Yes** **1996** **Models:** Sportage **Engines:** All **Transmissions:** All	**HO2S-12 (Bank 1 Sensor 2) Heater Circuit High Input** Engine running, and PCM detected an unexpected voltage condition on the rear HO2S heater control circuit. **Note: Inspect the HO2S connector for oil and water contamination.** **Possible Causes:** • HO2S heater control circuit is shorted to system power • HO2S heater control circuit is shorted to heater power feed • HO2S heater is damaged or has failed • PCM has failed
DTC: P1106 **2T CCM, MIL: No** **2006** **Models:** Optima, Sedona **Engines:** 2.7L 3.8L **Transmissions:** All	**Manifold Absolute Pressure Sensor Circuit Short- Intermittent High Input** This code detects an intermittent short to high in either the signal circuit or the MAP sensor. **Possible Causes:** • Poor connection • Short to battery in signal circuit • Open in ground circuit • Faulty MAPS • Faulty PCM
DTC: P1107 **2T CCM, MIL: No** **2006** **Models:** Optima, Sedona **Engines:** 2.7L 3.8L **Transmissions:** All	**Manifold Absolute Pressure Sensor Circuit Short- Intermittent Low Input** This code detects an intermittent short to high in either the signal circuit or the MAP sensor. **Possible Causes:** • Poor connection • Open or short to ground in the power circuit • Open or short to ground in the signal circuit • Faulty MAPS • Faulty PCM

DTC	Trouble Code Title, Conditions & Possible Causes
DTC: P1111 **2T CCM, MIL: No** **2006** **Models:** Optima, Sedona **Engines:** 2.7L 3.8L **Transmissions:** All	**Intake Air Temperature Sensor Circuit Short- Intermittent High Input** This code detects a continuous short to high in either the signal circuit or the sensor. **Possible Causes:** • Poor connection • Open or short in signal circuit • Open in ground circuit • Faulty IATS • Faulty PCM
DTC: P1112 **2T CCM, MIL: No** **2006** **Models:** Optima, Sedona **Engines:** 2.7L 3.8L **Transmissions:** All	**Intake Air Temperature Sensor Circuit Short- Intermittent Low Input** This code detects a continuous short to high in either the signal circuit or the sensor. **Possible Causes:** • Poor connection • Short to ground in the signal circuit • Open in ground circuit • Faulty IATS • Faulty PCM
DTC: P1114 **2T CCM, MIL: No** **2006** **Models:** Optima, Sedona **Engines:** 2.7L 3.8L **Transmissions:** All	Engine coolant temperature Sensor Circuit- Intermittent Low Input This code detects an intermittent short to ground in the signal circuit or the sensor. **Possible Causes:** • Poor connection • Short to ground in signal circuit • Open in ground circuit • Faulty ECTS • Faulty PCM
DTC: P1115 **2T 2 HTR2, MIL: Yes** **1996-06** **Models:** Optima, Sedona, Sportage **Engines:** All **Transmissions:** All	**HO2S-12 (Bank 1 Sensor 2) Heater Circuit Low Input** Engine running, and PCM detected an unexpected voltage condition on the rear HO2S heater control circuit. **Note: Inspect the HO2S connector for oil and water contamination.** **Possible Causes:** • HO2S heater control circuit open between HO2S and the PCM • HO2S heater control circuit is shorted to ground • HO2S heater is damaged or has failed • PCM has failed
DTC: P1115 **2T CCM, MIL: Yes** **1998-02** **Models:** Sportage **Engines:** All **Transmissions:** A/T	**ECT Water Temperature Signal Circuit Malfunction** Engine running, and PCM detected an unexpected voltage condition on the ECT signal circuit (PCM circuit to the TCM). **Possible Causes:** • ECT sensor circuit open between the PCM and the TCM • ECT sensor circuit shorted to power between PCM and TCM • ECT sensor circuit shorted to ground between PCM and TCM
DTC: P1117 **2T 2 HTR2, MIL: Yes** **1996** **Models:** Sportage **Engines:** All **Transmissions:** A/T	**HO2S-11 (Bank 1 Sensor 1) Heater Circuit Low Input** Engine running, and PCM detected an unexpected voltage condition on the rear HO2S heater control circuit. **Note: Inspect the HO2S connector for oil and water contamination.** **Possible Causes:** • HO2S heater control circuit open between HO2S and the PCM • HO2S heater control circuit is shorted to ground • HO2S heater is damaged or has failed • PCM has failed
DTC: P1121 **2T CCM, MIL: Yes** **1998-02** **Models:** Sportage **Engines:** All **Transmissions:** A/T	**Throttle Position Signal Circuit Malfunction** Engine runtime over 4 seconds, and the PCM detected an unexpected voltage on the TPS signal circuit (PCM circuit to TCM). **Note: If DTC P0121 or P0123 are set, repair these codes first!** **Possible Causes:** • TP sensor circuit open between the PCM and the TCM • TP sensor circuit shorted to power between PCM and TCM • TP sensor circuit shorted to ground between PCM and TCM

DTC	Trouble Code Title, Conditions & Possible Causes
DTC: P1123 **2T FUEL, MIL: Yes** **1996** **Models:** Sportage **Engines:** All **Transmissions:** All	**Long Term F/T Adaptive Air System Low** Engine running in closed loop at road load for 2 minutes, and the PCM detected a Long Term fuel trim correction that indicated a very "rich" condition existed. **Possible Causes:** • One or more injectors leaking or pressure regulator is leaking • O2S element is deteriorated or has failed • EVAP vapor recovery system has failed (canister full of fuel) • Base engine fault (i.e., cam timing incorrect, engine oil too high • A "fuel control" sensor is out of calibration (ECT, IAT or MAF)
DTC: P1124 **2T FUEL, MIL: Yes** **1996** **Models:** Sportage **Engines:** All **Transmissions:** All	**Long Term F/T Adaptive Air System High** Engine running in closed loop at road load for 2 minutes, and the PCM detected a Long Term fuel trim correction that indicated a very "lean" condition existed. **Possible Causes:** • One or more injectors is dirty or restricted • Fuel filter is severely restricted or the fuel pressure is too low • Air leaks exist in the engine (i.e., the intake manifold or PCV) • Base engine fault (i.e., cam timing incorrect, engine oil too high • A "fuel control" sensor is out of calibration (ECT, IAT or MAF)
DTC: P1127 **2T FUEL, MIL: Yes** **1996** **Models:** Sportage **Engines:** All **Transmissions:** All	**Long Term F/T Multiplier Air System Low** Engine runtime 2 minutes in closed loop, and the PCM detected the Long Term Fuel Trim Multiplicative correction value indicated a rich A/F ratio condition existed that was too rich to correct. **Possible Causes:** • One or more injectors leaking • Fuel pressure regulator is leaking or fuel pressure is too high • EVAP vapor recovery system has failed (canister full of fuel) • A "fuel control" sensor is out of calibration (ECT, IAT or MAF)
DTC: P1128 **2T FUEL, MIL: Yes** **1996** **Models:** Optima, Rio, Sedona, Sephia, Spectra, Sportage **Engines:** All **Transmissions:** All	**Long Term F/T Multiplier Air System High** Engine runtime 2 minutes in closed loop, and the PCM detected the Long Term Fuel Trim Multiplicative correction value indicated a rich A/F ratio condition existed that was too lean to correct. **Possible Causes:** • One or more injectors is dirty or restricted • Fuel filter is severely restricted or the fuel pressure is too low • Air leaks exist in the engine (i.e., the intake manifold or PCV) • A "fuel control" sensor is out of calibration (ECT, IAT or MAF)
DTC: P1134 **2T O2S, MIL: Yes** **2003-05** **Models:** Optima **Engines:** 2.7L **Transmissions:** All	**HO2S-11 (Bank 1 Sensor 1) Transition Time Switch** Engine started, engine running in closed loop, and the PCM detected the average front HO2S-11 rich-to-lean or lean-to-rich switch time was more than 1.1 seconds during the CCM test. **Possible Causes:** • HO2S signal circuit is open or shorted to ground • HO2S element is contaminated or it has failed • HO2S heater is damaged or has failed • Intake air leaks, exhaust manifold leaks or PCV system leaks • MAF sensor out of calibration (it may be dirty or contaminated)
DTC: P1140 **2T CCM, MIL: Yes** **1996** **Models:** Sportage **Engines:** All **Transmissions:** All	**Load Calculation Cross Check** Engine running and the PCM detected the Load Calculation Cross Check function was incorrect due to input signals that did not match. **Possible Causes:** • MAF sensor signal is invalid (it may be out of calibration) • TP sensor signal is invalid (it may be erratic or broken) • PCM has failed
DTC: P1151 **2T CCM, MIL: Yes** **2005** **Models:** Amanti **Engines:** 3.5L **Transmissions:** A/T	Electronic Throttle System Sub Accelerator position sensor 2 Circuit Malfunction Key on or engine running and the PCM detected an unexpected voltage on the Sub Accelerator position sensor 2 (APS2) circuit. **Possible Causes:** • APS2 signal circuit is open (>4.50 volts) or grounded (<0.20 volts) • APS2 signal circuit is shorted to VREF or system power (B+) • APS2 is damaged or has failed • PCM has failed • TSB 01-36-019 (8/01) contains information about this code

DTC	Trouble Code Title, Conditions & Possible Causes
DTC: P1152 **2T CCM, MIL: Yes** **2004** **Models:** Amanti **Engines:** 3.5L **Transmissions:** A/T	Accelerator position sensor (APS #2 Main) Circuit Low Voltage Turn ignition switch OFF. **Possible Causes:** • Poor connection • Defective sensor
DTC: P1154 **2T O2S, MIL: Yes** **2005** **Models:** Optima **Engines:** 2.7L **Transmissions:** All	**HO2S-21 (Bank 2 Sensor 1) Transition Time Switch** Engine started, engine running in closed loop, and the PCM detected the average front HO2S-21 rich-to-lean or lean-to-rich switch time was more than 1.1 seconds during the CCM test. **Possible Causes:** • HO2S signal circuit is open or shorted to ground • HO2S element is contaminated or it has failed • HO2S heater is damaged or has failed • Intake air leaks, exhaust manifold leaks or PCV system leaks • MAF sensor out of calibration (it may be dirty or contaminated)
DTC: P1159 **2T COM, MIL: Yes** **2003-05** **Models:** Amanti, Sedona **Engines:** 3.5L **Transmissions:** All	**VICS Valve Circuit Malfunction** Engine started. **Possible Causes:** • Open or short to chassis ground between VICS valve and ECM • Faulty VICS valve
DTC: P1166 **2T Fuel, MIL: Yes** **2003-06** **Models:** Optima, Rio, Spectra **Engines:** 1.6L, 2.0L, 2.7L **Transmissions:** All	**HO2S-11 Controller Adaptive Test (Bank 1)** Engine started, engine running in closed loop for 2-3 minutes, and the PCM detected the Short Term fuel trim was less than -12.5% or more than +12.5%; or it detected the Long Term fuel trim value was less than -12.5%, or more than +22.4% for 5-10 seconds in the test. **Possible Causes:** • Air leaks after the MAF sensor, or in the EGR or PCV system • Base engine "mechanical" fault affecting one or more cylinders • Exhaust leaks located in front of the A/FS or HO2S location • Fuel control sensor is out of calibration (i.e., ECT, IAT or MAP) • Fuel delivery system supplying too little fuel during cruise or idle periods (e.g., faulty fuel pump or dirty, restricted fuel filter) • Fuel injector (one or more) dirty or pressure regulator has failed • HO2S is contaminated, deteriorated or it has failed • Vehicle driven low on fuel or until it ran out of fuel
DTC: P1167 **2T Fuel, MIL: Yes** **2003-05** **Models:** Optima **Engines:** 2.7L **Transmissions:** All	**HO2S-11 Controller Adaptive Test (Bank 2)** Engine started, engine running in closed loop for 2-3 minutes, and the PCM detected the Short Term fuel trim was less than -12.5% or more than +12.5%; or it detected the Long Term fuel trim value was less than -12.5%, or more than +22.4% for 5-10 seconds in the test. **Possible Causes:** • Air leaks after the MAF sensor, or in the EGR or PCV system • Base engine "mechanical" fault affecting one or more cylinders • Exhaust leaks located in front of the A/FS or HO2S location • Fuel control sensor is out of calibration (i.e., ECT, IAT or MAP) • Fuel delivery system supplying too little fuel during cruise or idle periods (e.g., faulty fuel pump or dirty, restricted fuel filter) • Fuel injector (one or more) dirty or pressure regulator has failed • HO2S is contaminated, deteriorated or it has failed • Vehicle driven low on fuel or until it ran out of fuel
DTC: P1170 **2T CCM, MIL: Yes** **1996-97** **Models:** Sephia **Engines:** All **Transmissions:** All	**HO2S-11 (Bank 1 Sensor 1) Circuit Stuck at Mid-Range** Engine running in closed loop, and the PCM detected the front HO2S signal was fixed at mid-range (300-500 mv) for 20 seconds. **Possible Causes:** • HO2S signal circuit is open or shorted to ground • Air leaks exist in the engine (i.e., the intake manifold or PCV) • Low fuel pressure or restricted fuel injectors (misfire present) • Vehicle driven while low on fuel (less than 1/8 of a tank)

DTC	Trouble Code Title, Conditions & Possible Causes
DTC: P1192 **2T CCM, MIL: No** **2005-06** **Models:** Amanti **Engines:** 3.5L **Transmissions:** All	**ETS Limp home Target Following Mal.** Ignition switches ON. ETS motor relay ON. Battery voltage greater than 11 volts. TPS 1 is normal. **Possible Causes:** • Poor connector • Short in ETS motor circuit • Faulty ETS motor • Faulty PCM
DTC: P1193 **2T CCM, MIL: No** **2005-06** **Models:** Amanti **Engines:** 3.5L **Transmissions:** All	**ETS Limp home Low RPM** Ignition switches ON. ETS motor relay ON. Battery voltage greater than 11 volts. TPS 1 is normal. Engine speed greater than or equal to 700 rpm. **Possible Causes:** • Poor connector • Intake/exhaust system blockage • Check throttle plate for carbon deposits • Faulty ETS system • Faulty TPS • Faulty ETS motor • Faulty PCM
DTC: P1194 **2T CCM, MIL: No** **2005-06** **Models:** Amanti **Engines:** 3.5L **Transmissions:** All	**ETS Limp home TPS2 Mal.** Ignition switches ON. ETS motor relay ON. Battery voltage greater than 11 volts. TPS 1 is normal. Engine coolant temperature above 158 degrees F. Vtps2 less than or equal to 0.7 volt. **Possible Causes:** • Poor connector • Short in ETS motor circuit • Faulty ETS motor • Faulty PCM
DTC: P1195 **2T CCM, MIL: Yes** **2006** **Models:** Optima, Sedona **Engines:** 2.7L, 3.8L **Transmissions:** All	**EGR Boost Or Pressure Sensor Circuit Malfunction** Engine running in closed loop for 2 minutes, and the PCM detected the EGR Boost sensor (BP DOHC) or Pressure Sensor (B6 DOHC) signal was less than 0.2 volt, or that it was more than 4.5 volts for 1 second. **Possible Causes:** • EGR Boost sensor or Pressure sensor circuit is open • EGR Boost sensor or Pressure sensor circuit shorted to ground • EGR Boost sensor or Pressure sensor circuit is damaged • PCM has failed
DTC: P1196 **2T CCM, MIL: Yes** **1996-06** **Models:** Amanti, Sephia **Engines:** 1.6L (vin 4), 3.5L **Transmissions:** All	**Ignition Switch Start Circuit Malfunction** Engine cranking and the PCM detected the Starter Switch signal was present for over 20 seconds after the engine was running. **Possible Causes:** • Starter switch signal circuit is shorted to power • Starter switch is damaged or has failed • PCM has failed
DTC: P1213 **1T CCM, MIL: Yes** **1996** **Models:** Sportage **Engines:** All **Transmissions:** All	**Fuel Injector 1 Circuit High Input** Engine running at over 500 rpm, and the PCM detected the fuel injector 1 control circuit was continuously high for 5 seconds. **Possible Causes:** • Fuel injector 1 control circuit shorted to system power (B+) • Fuel injector 1 is damaged or has failed • PCM has failed
DTC: P1214 **1T CCM, MIL: Yes** **1996** **Models:** Sportage **Engines:** All **Transmissions:** All	**Fuel Injector 2 Circuit High Input** Engine running at over 500 rpm, and the PCM detected the fuel injector 2 control circuit was continuously high for 5 seconds. **Possible Causes:** • Fuel injector 2 control circuit shorted to system power (B+) • Fuel injector 2 is damaged or has failed • PCM has failed
DTC: P1215 **1T CCM, MIL: Yes** **1996** **Models:** Sportage **Engines:** All **Transmissions:** All	**Fuel Injector 3 Circuit High Input** Engine running at over 500 rpm, and the PCM detected the fuel injector 1 control circuit was continuously high for 5 seconds. **Possible Causes:** • Fuel injector 3 control circuit shorted to system power (B+) • Fuel injector 3 is damaged or has failed • PCM has failed

DTC	Trouble Code Title, Conditions & Possible Causes
DTC: P1216 **1T CCM, MIL: Yes** **1996** **Models:** Sportage **Engines:** All **Transmissions:** All	**Fuel Injector 4 Circuit High Input** Engine running at over 500 rpm, and the PCM detected the fuel injector 4 control circuit was continuously high for 5 seconds. **Possible Causes:** • Fuel injector 4 control circuit shorted to system power (B+) • Fuel injector 4 is damaged or has failed • PCM has failed
DTC: P1225 **1T CCM, MIL: Yes** **1996** **Models:** Sportage **Engines:** All **Transmissions:** All	**Fuel Injector 1 Circuit Low Input** Engine running at over 500 rpm, and the PCM detected the fuel injector 1 control circuit was continuously low for 5 seconds. **Possible Causes:** • Fuel injector 1 control circuit shorted to ground • Fuel injector 1 power circuit is open between injector and relay • Fuel injector 1 is damaged or has failed • PCM has failed
DTC: P1226 **1T CCM, MIL: Yes** **1996** **Models:** Sportage **Engines:** All **Transmissions:** All	**Fuel Injector 2 Circuit Low Input** Engine running at over 500 rpm, and the PCM detected the fuel injector 2 control circuit was continuously low for 5 seconds. **Possible Causes:** • Fuel injector 2 control circuit shorted to ground • Fuel injector 2 power circuit is open between injector and relay • Fuel injector 2 is damaged or has failed • PCM has failed
DTC: P1227 **1T CCM, MIL: Yes** **1996** **Models:** Sportage **Engines:** All **Transmissions:** All	**Fuel Injector 3 Circuit Low Input** Engine running at over 500 rpm, and the PCM detected the fuel injector 1 control circuit was continuously low for 5 seconds. **Possible Causes:** • Fuel injector 3 control circuit shorted to ground • Fuel injector 3 power circuit is open between injector and relay • Fuel injector 3 is damaged or has failed • PCM has failed
DTC: P1228 **1T CCM, MIL: Yes** **1996** **Models:** Sportage **Engines:** All **Transmissions:** All	**Fuel Injector 4 Circuit Low Input** Engine running at over 500 rpm, and the PCM detected the fuel injector 4 control circuit was continuously low for 5 seconds. **Possible Causes:** • Fuel injector 4 control circuit shorted to ground • Fuel injector 4 power circuit is open between injector and relay • Fuel injector 4 is damaged or has failed • PCM has failed
DTC: P1229 **2T CCM, MIL: Yes** **1996-97** **Models:** Sephia **Engines:** All **Transmissions:** All	**Pressure Regulator Control Solenoid Circuit Malfunction** Engine runtime over 1 minute and the PCM detected an unexpected voltage condition in the PRC solenoid control circuit during the test. **Possible Causes:** • Pressure regulator control solenoid circuit is open • Pressure regulator control solenoid circuit is shorted to ground • Pressure regulator control solenoid is damaged or has failed • PCM has failed
DTC: P1295 **2T CCM, MIL: No** **2006** **Models:** Optima, Sedona **Engines:** 2.7L, 3.8L **Transmissions:** All	**Electronic Throttle Control (ETC) System Malfunction- Power Management** This code is set is there is a problem in the power management system. Ignition ON. **Possible Causes:** • TPS malfunction • TPS malfunction plus MAFS malfunction • MAP malfunction plus TPS malfunction • Faulty PCM
DTC: P1307 **2T CCM, MIL: Yes** **1997-03** **Models:** Sephia, Spectra, Sportage **Engines:** All **Transmissions:** All	**Chassis Acceleration Sensor Range/Performance** Vehicle driven to a speed of over 10 mph and then returned to a stop and allowed to idle for 15 seconds, and the PCM detected the Chassis Acceleration Sensor input was not plausible during the test. **Note: If DTC P0501 or P1500 is set, repair these trouble codes first.** **Possible Causes:** • Chassis Acceleration sensor signal circuit is open • Chassis Acceleration sensor signal circuit is shorted to ground • Chassis Acceleration sensor power circuit is open • Chassis Acceleration sensor is damaged or has failed • PCM is damaged

DTC	Trouble Code Title, Conditions & Possible Causes
DTC: P1308 **2T CCM, MIL: Yes** **1997-03** **Models:** Sephia, Spectra, Sportage **Engines:** All **Transmissions:** All	**Chassis Acceleration Sensor Low Input** Vehicle driven to a speed of over 10 mph, and the PCM detected the Chassis Acceleration Sensor signal was too low during the test. **Note: If DTC P0501 or P1500 is set, repair these trouble codes first.** **Possible Causes:** • Chassis Acceleration sensor signal circuit is shorted to ground • Chassis Acceleration sensor power circuit is open • Chassis Acceleration sensor is damaged or has failed • PCM is damaged
DTC: P1309 **2T CCM, MIL: Yes** **1997-03** **Models:** Sephia, Spectra, Sportage **Engines:** All **Transmissions:** All	**Chassis Acceleration Sensor High Input** Vehicle driven to a speed of over 10 mph, and the PCM detected the Chassis Acceleration Sensor signal was too high during the test. **Note: If DTC P0501 or P1500 is set, repair these trouble codes first.** **Possible Causes:** • Chassis Acceleration sensor signal circuit is shorted to power • Chassis Acceleration sensor ground circuit is open • Chassis Acceleration sensor is damaged or has failed • PCM is damaged
DTC: P1330 **2T CCM, MIL: Yes** **2001-06** **Models:** Amanti, Optima, Sedona, Sorento **Engines:** 2.4L (vin 3), 2.4L, 3.5L **Transmissions:** All	**Spark Timing Adjustment Circuit Malfunction** Key on, and the PCM detected an unexpected voltage condition between the ROM Change Tool circuit and the PCM. **Possible Causes:** • Check the RCT circuit for short to ground • PCM has failed
DTC: P1345 **2T CCM, MIL: Yes** **1996-97** **Models:** Sephia **Engines:** 1.6L (vin 4) **Transmissions:** All	**No SGC (CMP) Signal To PCM** Engine cranking for 6 cycles, and the PCM did not detect any CMP signals, or no signals were detected with engine running. **Possible Causes:** • CMP sensor signal circuit is open or shorted to ground • CMP sensor signal power circuit is open • CMP sensor is damaged or has failed • PCM has failed
DTC: P1372 **2T CCM, MIL: Yes** **2001-06** **Models:** Optima, Spectra **Engines:** 2.0L, 2.5L (vin 4), 2.7L **Transmissions:** All	**Segment Time Acquisition Incorrect** Vehicle driven to a speed of over 10 mph, and the PCM detected an incorrect Segment Time Acquisition signal was received. **Note: If DTC P519 is set, repair this trouble code first.** **Possible Causes:** • Target wheel improperly installed or damaged • Wheel speed sensor circuits to the ECM are open or grounded • RH front wheel speed sensor or rotor is damaged or has failed • PCM has failed
DTC: P1386 **2T CCM, MIL: Yes** **2000-02** **Models:** Sephia, Spectra **Engines:** All **Transmissions:** All	**Knock Sensor Control Zero Test** Engine speed over 2200 rpm, ECT sensor signal more than 104°F, engine load more than 3 ms, and the PCM detected an unexpected voltage condition on the Knock Sensor 1 circuit during the test. **Possible Causes:** • Knock sensor 1 circuit is open • Knock sensor 1 circuit is shorted to ground • Knock sensor 1 is damaged or has failed • PCM has failed
DTC: P1386 **2T CCM, MIL: Yes** **1996-02** **Models:** Sportage **Engines:** All **Transmissions:** All	**Knock Sensor Control Zero Test** Engine speed over 2200 rpm, ECT sensor signal more than 104°F, engine load more than 3 ms, and the PCM detected an unexpected voltage condition on the Knock Sensor 1 or 2 circuit during the test. **Possible Causes:** • Knock sensor 1 or 2 circuit is open • Knock sensor 1 or 2 circuit is shorted to ground • Knock sensor 1 or knock sensor 2 is damaged or has failed • PCM has failed
DTC: P1401 **2T CCM, MIL: Yes** **1996** **Models:** Sportage **Engines:** All **Transmissions:** All	**EGR Control Solenoid Circuit Signal Low** Engine running for 5 seconds, and the PCM detected the EGR Solenoid control circuit was in a continuous low state during the test. **Possible Causes:** • EGR control solenoid control circuit open or shorted to ground • EGR control solenoid power circuit is open or shorted to ground • EGR control solenoid is damaged or has failed • PCM has failed (EGR solenoid driver circuit may be shorted)

DTC	Trouble Code Title, Conditions & Possible Causes
DTC: P1402 **2T CCM, MIL: Yes** **1996** **Models:** Sportage **Engines:** All **Transmissions:** All	**EGR Control Solenoid Circuit Signal High** Engine running for 5 seconds, and the PCM detected the EGR Solenoid control circuit was in a continuous high state during the test. **Possible Causes:** • EGR control solenoid control circuit shorted to VREF • EGR control solenoid power circuit shorted to system power • EGR control solenoid is damaged or has failed • PCM has failed (EGR solenoid driver circuit may be open)
DTC: P1402 **2T CCM, MIL: Yes** **1996-97** **Models:** Sephia **Engines:** All **Transmissions:** All	**EGR Valve Position Sensor Circuit Malfunction** Engine running, and the PCM detected the EGR Valve Position signal was less than 0.2 volt, or that it was more than 4.5 volts for 1 second. **Possible Causes:** • EGR valve position sensor signal circuit is open • EGR valve position sensor signal circuit is grounded • EGR valve position sensor is damaged or has failed • PCM is damaged
DTC: P1402 **2T CCM, MIL: Yes** **2003** **Models:** Spectra **Engines:** 1.8L **Transmissions:** All	**Leak Detection Module Pump Circuit Malfunction** Ignition OFF. **Possible Causes:** • Open or short to ground between leakage detection pump and ECM • Short to battery between leakage detection pump and ECM • Faulty or damaged leakage detection pump in diagnostic module tank leakage(DMTL)
DTC: P1403 **2T CCM, MIL: Yes** **2003** **Models:** Spectra **Engines:** 1.8L **Transmissions:** All	**Leak Detection Module Solenoid Valve Circuit Malfunction** Ignition OFF. **Possible Causes:** • Open or short to ground between leakage detection solenoid valve and ECM • Short to battery between leakage detection solenoid valve and ECM • Faulty or damaged leakage solenoid valve in diagnostic module tank leakage(DMTL)
DTC: P1404 **2T CCM, MIL: Yes** **2003** **Models:** Spectra **Engines:** 1.8L **Transmissions:** All	**Leak Detection Module Heater Malfunction** Ignition OFF. **Possible Causes:** • Open or short to ground between leakage detection heater and ECM • Short to battery between leakage detection heater and ECM • Faulty or damaged leakage detection heater in diagnostic module tank leakage(DMTL)
DTC: P1410 **2T CCM, MIL: Yes** **1996** **Models:** Sportage **Engines:** All **Transmissions:** All	**EVAP Purge Solenoid Circuit High Input** Engine running and the PCM detected the EVAP Purge solenoid control signal was in a continuous high state during the CCM test. **Possible Causes:** • EVAP purge solenoid control circuit shorted to system power • EVAP purge solenoid is damaged or has failed • PCM is damaged
DTC: P1412 **2T CCM, MIL: Yes** **1996** **Models:** Sportage **Engines:** All **Transmissions:** All	**EGR Differential Pressure Sensor Low Input Condition:** Engine running for 2 minutes, and the PCM detected the EGR Pressure sensor signal circuit was in a continuously low state. **Possible Causes:** • EGR differential pressure sensor signal circuit is grounded • EGR differential pressure sensor ground circuit open • EGR differential pressure sensor is damaged or has failed • PCM has failed
DTC: P1413 **2T CCM, MIL: Yes** **1996** **Models:** Sportage **Engines:** All **Transmissions:** All	**EGR Differential Pressure Sensor High Input** Engine running, and the PCM detected the EGR Pressure sensor signal circuit was in a continuous high state. **Possible Causes:** • EGR differential pressure sensor signal circuit is open • EGR differential pressure sensor signal circuit shorted to power • EGR differential pressure sensor is damaged or has failed • PCM has failed

DTC	Trouble Code Title, Conditions & Possible Causes
DTC: P1425 **2T CCM, MIL: Yes** **1996-97** **Models:** Sportage **Engines:** All **Transmissions:** All	**EVAP Purge Solenoid Circuit Low Input** Engine running and the PCM detected the EVAP Purge solenoid control circuit was in a continuous low state. **Possible Causes:** • Purge solenoid control circuit is open • Purge solenoid control circuit is shorted to ground • Purge solenoid power circuit from Main Relay is open • Purge solenoid is damaged or has failed • PCM has failed (purge solenoid driver circuit is open or shorted)
DTC: P1440 **2T CCM, MIL: Yes** **2001-02** **Models:** Optima **Engines:** All **Transmissions:** All	**EVAP Canister Control System Vent Circuit Malfunction** Engine running and the PCM detected an unexpected voltage condition on the Canister Control Vent solenoid during the CCM test. **Possible Causes:** • Canister control vent valve stuck in closed position • Canister control vent solenoid is open or shorted to ground • Canister control vent solenoid is shorted to system power (B+) • Canister control vent solenoid is damaged or has failed • PCM has failed (vent solenoid driver circuit is open or shorted)
DTC: P1440 **2T CCM, MIL: Yes** **2002** **Models:** Sedona **Engines:** All **Transmissions:** All	**Power Steering Switch Circuit Malfunction** Engine running and the PCM detected an unexpected condition on the Power Steering Pressure Switch circuit during the CCM test. **Possible Causes:** • PSP switch signal circuit is open • PSP switch signal circuit is shorted to ground • PSP switch signal circuit is shorted to VREF or system power • PSP switch is damaged or has failed • PCM has failed
DTC: P1446 **2T CCM, MIL: Yes** **2000-02** **Models:** Sephia, Spectra **Engines:** All **Transmissions:** All	**Leak Detection Pump Solenoid Circuit Malfunction** Key on or engine running and the PCM detected an unexpected voltage condition on the Leak Detection Pump circuit (PCM pin 19). **Possible Causes:** • LDP power circuit open to the Injector Fuse • LDP control solenoid circuit open or shorted to ground • LDP control solenoid circuit shorted to system power • LDP is damaged or has failed • PCM has failed (pump solenoid driver circuit is open or shorted)
DTC: P1447 **2T CCM, MIL: Yes** **2000-02** **Models:** Sephia, Spectra **Engines:** All **Transmissions:** All	**Leak Detection Pump Control Circuit Malfunction** Key on or engine running and the PCM detected an unexpected voltage condition on the Leak Detection Pump circuit (PCM pin 20). **Possible Causes:** • LDP pump power circuit open to the Injector Fuse • LDP motor circuit open or shorted to ground • LDP motor circuit is shorted to system power • LDP is damaged or has failed • PCM has failed (pump motor driver circuit is open or shorted)
DTC: P1448 **2T CCM, MIL: Yes** **2000-02** **Models:** Sephia, Spectra **Engines:** All **Transmissions:** All	**Leak Detection Module Malfunction** Key on or engine running and the PCM detected an unexpected voltage condition on the Leak Detection Heater circuit to the module. **Possible Causes:** • Leak detection heater circuit is open or shorted to ground • Leak detection heater circuit is shorted to system power (B+) • Leak detection heater is damaged or has failed • PCM has failed
DTC: P1449 **2T CCM, MIL: Yes** **1996-97** **Models:** Sephia **Engines:** 1.8L (vin 5) **Transmissions:** All	**Canister Drain Cut Valve Solenoid Circuit Malfunction** Engine running, and then the PCM detected an unexpected voltage condition in the Canister Drain Cut Valve circuit during the CCM test. **Possible Causes:** • Canister drain cut solenoid control circuit is open • Canister drain cut solenoid control circuit shorted to ground • Canister drain cut solenoid control circuit shorted to power • Canister drain cut solenoid is damaged or has failed • PCM has failed (canister drain cut solenoid circuit is open)

DTC	Trouble Code Title, Conditions & Possible Causes
DTC: P1450 **2T EVAP, MIL: Yes** **1996-97** **Models:** Sephia **Engines:** All **Transmissions:** All	**Excessive EVAP System Vacuum Leak** Cold startup requirement met (ECT sensor signal less than 76°F), vehicle driven at a cruise speed of less than 62 mph, and the PCM detected the fuel tank pressure was less than -3.92 kPa for over 10 seconds in a period of time from 3.1 to 400 seconds after startup. **Possible Causes:** • Canister drain cut valve is stuck • EVAP hoses and pipes between fuel tank, EVAP canister and air filter clogged, bent or restricted
DTC: P1455 **2T CCM, MIL: Yes** **1996-97** **Models:** Sephia **Engines:** All **Transmissions:** All	**Fuel Tank Sending Unit Circuit Malfunction** Engine running and the PCM detected an unexpected voltage condition on the Fuel Tank Sending Unit circuit during the CCM test. **Possible Causes:** • Fuel tank sending unit signal circuit open or shorted to ground • Fuel tank sending unit ground circuit is open • Fuel tank sensor unit power circuit open (check 15a meter fuse) • Fuel tank sending unit is damaged or has failed • PCM has failed
DTC: P1458 **2T CCM, MIL: Yes** **1996-02** **Models:** Sportage **Engines:** All **Transmissions:** All	**Air Conditioning Compressor Clutch Circuit Malfunction** Engine running and PCM detected an unexpected voltage condition on the A/C Compressor Clutch signal circuit during the test. **Possible Causes:** • A/C compressor clutch circuit is open • A/C compressor clutch circuit is shorted to ground • A/C compressor clutch circuit is shorted to system power • PCM has failed
DTC: P1458 **2T CCM, MIL: Yes** **1998-02** **Models:** Sephia, Spectra, **Engines:** All **Transmissions:** All	**Air Conditioning Compressor Clutch Circuit Malfunction** Engine running and PCM detected an unexpected voltage condition on the A/C Compressor Clutch signal circuit during the test. **Possible Causes:** • A/C compressor clutch circuit is open • A/C compressor clutch circuit is shorted to ground • A/C compressor clutch circuit is shorted to system power • PCM has failed
DTC: P1485 **2T CCM, MIL: Yes** **1996-97** **Models:** Sephia **Engines:** 1.6L (vin 4) **Transmissions:** All	**EGR Solenoid Valve (Vacuum) Circuit Malfunction** Key on or engine running and PCM detected an unexpected voltage condition on the EGR Vent Solenoid circuit during the CCM test. **Possible Causes:** • EGR solenoid valve control circuit is open (vacuum side) • EGR solenoid valve control circuit is shorted to ground • EGR solenoid valve is damaged or has failed • PCM has failed (vent control solenoid "driver" circuit is open)
DTC: P1486 **2T CCM, MIL: Yes** **1996-97** **Models:** Sephia **Engines:** 1.6L (vin 4) **Transmissions:** All	**EGR Solenoid Valve (Vent) Circuit Malfunction** Key on or engine running and PCM detected an unexpected voltage condition on the EGR Vent Solenoid circuit during the CCM test. **Possible Causes:** • EGR solenoid valve control circuit is open (vent side) • EGR solenoid valve control circuit is shorted to ground • EGR solenoid valve is damaged or has failed • PCM has failed (vent control solenoid "driver" circuit is open)
DTC: P1487 **2T CCM, MIL: Yes** **1996-97** **Models:** Sephia **Engines:** 1.8L (vin 5) **Transmissions:** All	**EGR Boost Sensor Solenoid Circuit Malfunction** Key on or engine running and PCM detected an unexpected voltage condition on the EGR Boost Sensor solenoid circuit during the test. **Possible Causes:** • EGR boost solenoid control circuit is open (vent side) • EGR boost solenoid control circuit is shorted to ground • EGR boost solenoid is damaged or has failed • PCM has failed (boost sensor solenoid "driver" circuit is open)
DTC: P1496 **2T CCM, MIL: Yes** **1996-97** **Models:** Sephia **Engines:** 1.8L (vin 5) **Transmissions:** All	**EGR Stepper Motor 1 Circuit Malfunction** Key on or engine running and PCM detected an unexpected voltage condition on the EGR Stepper Motor 1 circuit during the CCM test. **Possible Causes:** • EGR stepper motor 1 control circuit open or shorted to ground • EGR stepper motor 1 power circuit is open to the Main Relay • EGR stepper motor 1 is damaged or has failed • PCM has failed (EGR stepper motor 1 "driver" circuit is open)

DTC	Trouble Code Title, Conditions & Possible Causes
DTC: P1497 **2T CCM, MIL: Yes** **1996-97** **Models:** Sephia **Engines:** 1.8L (vin 5) **Transmissions:** All	**EGR Stepper Motor 2 Circuit Malfunction** Key on or engine running and PCM detected an unexpected voltage condition on the EGR Stepper Motor 2 circuit during the CCM test. **Possible Causes:** • EGR stepper motor 2 control circuit open or shorted to ground • EGR stepper motor 2 power circuit is open to the Main Relay • EGR stepper motor 2 is damaged or has failed • PCM has failed (EGR stepper motor 2 "driver" circuit is open)
DTC: P1498 **2T CCM, MIL: Yes** **1996-97** **Models:** Sephia **Engines:** 1.8L (vin 5) **Transmissions:** All	**EGR Stepper Motor 3 Circuit Malfunction** Key on or engine running and PCM detected an unexpected voltage condition on the EGR Stepper Motor 3 circuit during the CCM test. **Possible Causes:** • EGR stepper motor 3 control circuit open or shorted to ground • EGR stepper motor 3 power circuit is open to the Main Relay • EGR stepper motor 3 is damaged or has failed • PCM has failed (EGR stepper motor 3 "driver" circuit is open)
DTC: P1499 **2T CCM, MIL: Yes** **1996-97** **Models:** Sephia **Engines:** 1.8L (vin 5) **Transmissions:** All	**EGR Stepper Motor 4 Circuit Malfunction** Key on or engine running and PCM detected an unexpected voltage condition on the EGR Stepper Motor 4 circuit during the CCM test. **Possible Causes:** • EGR stepper motor 4 control circuit open or shorted to ground • EGR stepper motor 4 power circuit is open to the Main Relay • EGR stepper motor 4 is damaged or has failed • PCM has failed (EGR stepper motor 4 "driver" circuit is open)
DTC: P1500 **2T CCM, MIL: Yes** **1998-02** **Models:** Optima, Rio, Sedona, Sephia, Spectra, Sportage **Engines:** All **Transmissions:** All	**Vehicle Speed Sensor Performance** Engine running OSS signal over 775 rpm, L4 switch indicating "off", and the PCM did not detect any VSS signals during the CCM test. **Possible Causes:** • Vehicle speed sensor circuit to the Meter Fuse is open • VSS signal ground circuit is open • VSS signal circuit is open or shorted to ground • VSS signal circuit is shorted to VREF or system power (B+) • VSS is damaged or failed, or the speedometer is damaged • TSB 2TD004 (8/00) contains information related to this code
DTC: P1505 **2T CCM, MIL: Yes** **1997-06** **Models:** Optima, Rio, Sephia, Spectra, Sportage **Engines:** All **Transmissions:** All	**IAC Valve Opening Coil Signal Low** Engine runtime over 5 seconds, and the PCM detected the IAC Valve Opening Coil signal remained in a low state during the test. **Possible Causes:** • IAC valve control signal is open • IAC valve control signal is shorted to ground • IAC valve is damaged or has failed • PCM has failed (IAC "driver" circuit may be open in the PCM)
DTC: P1506 **2T CCM, MIL: Yes** **1997-06** **Models:** Optima, Rio, Sephia, Spectra, Sportage **Engines:** All **Transmissions:** All	**IAC Valve Opening Coil Signal High** Engine runtime over 5 seconds, and the PCM detected the IAC Valve Opening Coil signal remained in a high state during the test. **Possible Causes:** • IAC valve control signal is shorted to system power (B+) • IAC valve is damaged or has failed • PCM has failed (IAC "driver" circuit may be shorted in the PCM)
DTC: P1507 **2T CCM, MIL: Yes** **1997-06** **Models:** Optima, Rio, Sephia, Spectra, Sportage **Engines:** All **Transmissions:** All	**IAC Valve Closing Coil Signal Low** Engine runtime over 5 seconds, and the PCM detected the IAC Valve Closing Coil signal remained in a low state during the test. **Possible Causes:** • IAC valve control signal is open • IAC valve control signal is shorted to ground • IAC valve is damaged or has failed • PCM has failed (IAC "driver" circuit may be open in the PCM)

DTC	Trouble Code Title, Conditions & Possible Causes
DTC: P1508 **2T CCM, MIL: Yes** **1997-06** **Models:** Optima, Rio, Sephia, Spectra, Sportage **Engines:** All **Transmissions:** All	**IAC Valve Closing Coil Signal High** Engine runtime over 5 seconds, and the PCM detected the IAC Valve Closing Coil signal remained in a high state during the test. **Possible Causes:** • IAC valve control signal is shorted to system power (B+) • IAC valve is damaged or has failed • PCM has failed (IAC "driver" circuit may be shorted in the PCM)
DTC: P1510 **2T CCM, MIL: Yes** **2001-02** **Models:** Optima **Engines:** 2.5L (vin 4) **Transmissions:** All	**Idle Charge Actuator Signal Incorrect (Coil 1)** Engine running and the PCM detected an unexpected voltage condition on the Idle Charge Actuator control circuit during the test. **Possible Causes:** • Idle charge actuator control signal is open • Idle charge actuator control signal is shorted to ground • Idle charge actuator control signal is shorted to system power • Idle charge actuator is damaged or has failed • PCM has failed (the "driver" circuit may be open in the PCM)
DTC: P1510 **2T CCM, MIL: Yes** **1996** **Models:** Sportage **Engines:** All **Transmissions:** All	**Idle Air Control Valve Opening Coil Signal High** Engine runtime over 5 seconds, and the PCM detected the IAC Valve Opening Coil signal remained in a high state during the test. **Possible Causes:** • IAC valve control signal is shorted to system power (B+) • IAC valve is damaged or has failed • PCM has failed (IAC "driver" circuit may be shorted in the PCM)
DTC: P1511 **2T CCM, MIL: Yes** **2001-02** **Models:** Optima **Engines:** 2.5L (vin 4) **Transmissions:** All	**Idle Charge Actuator Signal Incorrect (Coil 2)** Engine running and the PCM detected an unexpected voltage condition on the IAC Valve Closing Coil circuit during the test. **Possible Causes:** • Idle charge actuator control signal is open • Idle charge actuator control signal is shorted to ground • Idle charge actuator control signal is shorted to system power • Idle charge actuator is damaged or has failed • PCM has failed (the "driver" circuit may be open in the PCM)
DTC: P1513 **2T CCM, MIL: Yes** **1996** **Models:** Sportage **Engines:** All **Transmissions:** All	**Idle Air Control Valve Opening Coil Signal Low** Engine runtime over 5 seconds, and the PCM detected the IAC Valve Opening Coil signal remained in a low state during the test. **Possible Causes:** • IAC valve control signal is open • IAC valve control signal is shorted to ground • IAC valve is damaged or has failed • PCM has failed (IAC "driver" circuit may be open in the PCM)
DTC: P1515 **2T CCM, MIL: Yes** **1996-97** **Models:** Sportage **Engines:** All **Transmissions:** All	**A/T To M/T Codification** Engine speed over 2200 rpm, and the PCM detected the A/T to M/T Codification input was "off" with the gear switch "on", or that the A/T to M/T Codification input was "on" with the gear switch input "off". **Possible Causes:** • A/T to M/T codification input circuit is open • A/T to M/T codification input circuit is shorted to ground • A/T to M/T codification input circuit is VREF or system power • PCM has failed
DTC: P1523 **2T CCM, MIL: Yes** **1996-96** **Models:** Optima, Sedona, Sephia **Engines:** All **Transmissions:** All	**VICS Solenoid Valve Circuit Malfunction** Key on or engine running and PCM detected an unexpected voltage condition on the VICS Solenoid valve circuit during the CCM test. **Possible Causes:** • VICS solenoid control circuit is open • VICS solenoid control circuit is shorted to ground • VICS solenoid is damaged or has failed • PCM has failed
DTC: P1529 **1T PCM, MIL: Yes** **2003-06** **Models:** Optima, Rio, Spectra **Engines:** 1.6L, 1.8L, 2.0L, 2.7L **Transmissions:** All	**Customer Snapshot Request Error** Engine started, engine running, and the PCM did not receive the Customer Request Snapshot Data (MIL and Freeze Frame Data) as requested from the Transmission Control Module (TCM). **Possible Causes:** • Transmission control module error • Check for trouble codes in the transmission control module • TSB 1-40-005 (3/01) contains information about this code

DTC	Trouble Code Title, Conditions & Possible Causes
DTC: P1550 **2T CCM, MIL: No** **2006** **Models:** Rio **Engines:** 1.6L **Transmissions:** All	**Knock Sensor Evaluation IC** Circuit continuity check, pulse test. **Possible Causes:** • Poor connection • Open or short in control circuit • Faulty knock sensor • Faulty PCM
DTC: P1552 **2T CCM, MIL: Yes** **1996** **Models:** Sportage **Engines:** All **Transmissions:** All	**Idle Air Control Valve Closing Coil Signal Low** Engine runtime over 5 seconds, and the PCM detected the IAC Valve Closing Coil signal remained in a low state during the test. **Possible Causes:** • IAC valve control signal is open • IAC valve control signal is shorted to ground • IAC valve is damaged or has failed • PCM has failed (IAC "driver" circuit may be open in the PCM)
DTC: P1553 **2T CCM, MIL: Yes** **1996** **Models:** Sportage **Engines:** All **Transmissions:** All	**Idle Air Control Valve Closing Coil Signal High** Engine runtime over 5 seconds, and the PCM detected the IAC Valve Closing Coil signal remained in a high state during the test. **Possible Causes:** • IAC valve control signal is shorted to system power (B+) • IAC valve is damaged or has failed • PCM has failed (IAC "driver" circuit may be shorted in the PCM)
DTC: P1560 **2T CCM, MIL: No** **2006** **Models:** Rio **Engines:** 1.6L **Transmissions:** All	**System Voltage** SPI communication check. **Possible Causes:** • Poor connection • Faulty ECM/PCM
DTC: P161B **2T CCM, MIL: Yes** **2006** **Models:** Optima, Sedona **Engines:** 2.7L, 3.8L **Transmissions:** All	**PCM Internal Error- Torque Calculating** This code is set if delivered torque is grossly different from the desired torque. **Possible Causes:** • Intake air leakage • Faulty ETS system • Clogged exhaust system • Faulty PCM
DTC: P1586 **2T CCM, MIL: Yes** **1998-05** **Models:** Rio, Spectra, Sportage **Engines:** All **Transmissions:** All	**A/T To M/T Codification** Engine speed over 2200 rpm, and the PCM detected the A/T to M/T Codification input was "off" with the gear switch "on", or that the A/T to M/T Codification input was "on" with the gear switch input "off". **Possible Causes:** • A/T to M/T codification input circuit is open • A/T to M/T codification input circuit is shorted to ground • A/T to M/T codification input circuit is VREF or system power • PCM has failed
DTC: P1602 **2T PCM, MIL: Yes** **2001-02** **Models:** Optima **Engines:** 2.5L (vin 4) **Transmissions:** All	**Serial Communication Problem With TCM** Key on or engine running, and the PCM detected problem when it tried to communicate with the TCM over the CAN network (time out). **Possible Causes:** • PCM to TCM serial communication circuit is open • PCM to TCM serial communication circuit is shorted to ground • PCM to TCM serial communication circuit is shorted to power • TCM has failed
DTC: P1606 **2T CCM, MIL: Yes** **1996-05** **Models:** Optima, Spectra, Sportage **Engines:** All **Transmissions:** All	**Chassis Accelerator Sensor Signal Performance** Engine running and the PCM detected an unexpected voltage condition on the Chassis Accelerator sensor circuit during the test. **Possible Causes:** • Chassis acceleration sensor signal circuit is open • Chassis acceleration sensor signal circuit is shorted to ground • Chassis acceleration sensor signal circuit is shorted to power • Chassis acceleration sensor is damaged or has failed • PCM has failed

DTC	Trouble Code Title, Conditions & Possible Causes
DTC: P1608 **2T PCM, MIL: Yes** **1996-97** **Models:** Sephia **Engines:** All **Transmissions:** All	**PCM Internal Fault** Key on or engine running, and the PCM detected that it could not read trouble codes from the output devices. **Possible Causes:** • Output device signal circuit is open • Output device signal circuit is shorted to ground • PCM has failed
DTC: P1611 **2T CCM, MIL: Yes** **1998-02** **Models:** Sephia, Spectra **Engines:** All **Transmissions:** All	**MIL Request Circuit Low Input** Key on or engine running, and the PCM detected the MIL Request circuit remained in a low state during the CCM test. **Possible Causes:** • MIL request control circuit is shorted to sensor ground • MIL request control circuit is shorted to chassis ground • PCM is damaged
DTC: P1611 **2T CCM, MIL: Yes** **1996-02** **Models:** Sportage **Engines:** All **Transmissions:** All	**MIL Request Circuit Low Input** Key on or engine running, and the PCM detected the MIL Request circuit remained in a low state during the CCM test. **Possible Causes:** • MIL request control circuit is shorted to sensor ground • MIL request control circuit is shorted to chassis ground • PCM is damaged
DTC: P1613 **2T PCM, MIL: Yes** **2001-02** **Models:** Optima **Engines:** 2.5L (vin 4) **Transmissions:** All	**Engine Control Unit Self-Test Failed** Key on, and the PCM detected that it failed an internal self-test. **Possible Causes:** • PCM (ECU) has failed (it must be replaced to repair this fault)
DTC: P1614 **2T CCM, MIL: Yes** **1998-02** **Models:** Sephia, Spectra **Engines:** All **Transmissions:** All	**MIL Request Circuit High Input** Key on or engine running, and the PCM detected the MIL Request circuit remained in a high state during the CCM test. **Possible Causes:** • MIL request control circuit is shorted to system power (B+) • MIL "bulb" circuit is open or missing • PCM is damaged
DTC: P1614 **2T CCM, MIL: Yes** **1996-02** **Models:** Sportage **Engines:** All **Transmissions:** All	**MIL Request Circuit High Input** Key on or engine running, and the PCM detected the MIL Request circuit remained in a high state during the CCM test. **Possible Causes:** • MIL request control circuit is shorted to system power (B+) • MIL "bulb" circuit is open or missing • PCM is damaged
DTC: P1616 **2T CCM, MIL: Yes** **1996** **Models:** Sportage **Engines:** All **Transmissions:** All	**Chassis Accelerator Sensor Signal Low Input** Engine running and the PCM detected an unexpected voltage condition on the Chassis Accelerator sensor circuit during the test. **Possible Causes:** • Chassis acceleration sensor signal circuit is shorted to sensor ground or to chassis ground • Chassis acceleration sensor is damaged or has failed • PCM has failed
DTC: P1616 **2T CCM, MIL: Yes** **2001-02** **Models:** Optima **Engines:** All **Transmissions:** All	**Main Relay Malfunction** Key on, and the PCM detected an unexpected voltage condition on the Main Relay control circuit during the CCM test. **Possible Causes:** • Main relay control circuit is open or shorted to ground • Main relay power circuit from ignition is open (check the fuse) • Main relay is damaged or has failed • PCM has failed
DTC: P1617 **2T CCM, MIL: Yes** **1996** **Models:** Sportage **Engines:** All **Transmissions:** All	**Chassis Accelerator Sensor Signal High Input** Engine running and the PCM detected an unexpected voltage condition on the Chassis Accelerator sensor circuit during the test. **Possible Causes:** • Chassis acceleration sensor signal circuit is open • Chassis acceleration sensor signal circuit is shorted to VREF • Chassis acceleration sensor signal circuit is shorted to power • Chassis acceleration sensor is damaged or has failed • PCM has failed

DTC	Trouble Code Title, Conditions & Possible Causes
DTC: P1624 **2T CCM, MIL: Yes** **1997-02** **Models:** Sephia, Spectra, Sportage **Engines:** All **Transmissions:** All	**MIL Request Circuit (TCM To PCM) Circuit Malfunction** Key on or engine running and PCM detected a signal from the TCM indicating it had detected a problem in the MIL Request circuit. **Possible Causes:** • Perform the related TCM diagnostics • TCM may have failed
DTC: P1624 **2T CCM, MIL: Yes** **2001-05** **Models:** Optima **Engines:** All **Transmissions:** All	**Cooling Fan Low Relay Circuit Malfunction** Key on or engine running and the PCM detected the Cooling Fan Low Relay control circuit was in a low state during the CCM test. **Possible Causes:** • Cooling fan low relay control circuit is open • Cooling fan low relay control circuit is shorted to ground • Cooling fan low relay is damaged or has failed • PCM has failed
DTC: P1625 **2T CCM, MIL: Yes** **2001-05** **Models:** Optima **Engines:** All **Transmissions:** All	**Cooling Fan High Relay Circuit Malfunction** Key on or engine running, and the PCM detected the Cooling Fan High Relay 1 or 2 Control circuit was in a low state during the test. **Possible Causes:** • Cooling fan high relay 1 or 2 control circuit is open • Cooling fan high relay 1 or 2 control circuit is shorted to ground • Cooling fan high relay 1 or 2 is damaged or has failed • PCM has failed
DTC: P1631 **2T CCM, MIL: Yes** **1996-97** **Models:** Sephia **Engines:** All **Transmissions:** All	**Generator Voltage Detection Circuit Malfunction** Engine running, and the PCM detected an unexpected voltage condition on the Generator 'T' circuit during the CCM test. **Possible Causes:** • Circuit C249-21 open between the Generator and the PCM • Circuit C249-22 open between the Generator and the PCM • Generator is damaged or has failed • PCM has failed
DTC: P1632 **2T CCM, MIL: Yes** **1996-06** **Models:** Amanti, Sephia **Engines:** All **Transmissions:** All	**Generator Battery Detection Circuit Malfunction** Engine running, and the PCM detected the Battery Detection circuit voltage was less than 8 volts during the CCM test. **Possible Causes:** • Circuit C249-9 open between the Generator and the PCM • Generator is damaged or has failed • PCM has failed
DTC: P1633 **2T CCM, MIL: Yes** **1996-97** **Models:** Sephia **Engines:** All **Transmissions:** All	**Battery Overcharge Detected** Engine running, and the PCM detected an unexpected voltage condition on the Generator 'T' circuit during the CCM test. **Possible Causes:** • Generator (momentary) voltage spike occurred • Generator is damaged or has failed • PCM has failed
DTC: P1634 **2T CCM, MIL: Yes** **1996-97** **Models:** Sephia **Engines:** 1.8L (vin 5) **Transmissions:** All	**Generator 'B' Circuit Open Malfunction** Engine running and the PCM detected the Generator 'B' circuit voltage was over 18 volts while the system voltage was less than 8 volts. **Possible Causes:** • Generator 'B' circuit is open between the Generator and Battery • Generator is damaged or has failed • Battery has failed
DTC: P1640 **2T CCM, MIL: Yes** **2001-02** **Models:** Rio **Engines:** All **Transmissions:** All	**Main Relay Malfunction** Key on or engine running, and the PCM detected the Main relay voltage was less than 6 volts (low), or that it was more than 6 volts (too high). **Possible Causes:** • Charging system voltage is too low or too high • Main relay power circuit to the PCM has high resistance
DTC: P1642 **12T PCM, MIL: No** **2001-06** **Models:** Optima, Sportage **Engines:** All **Transmissions:** All	**Non-Immobilizer EMS Connected to Immobilizer** Key on, and the PCM detected the signal from the Immobilizer Control Unit (ICU) was inconsistent during the self-test period. **Possible Causes:** • Check the ICU (controller) for any related trouble codes

DTC	Trouble Code Title, Conditions & Possible Causes
DTC: P1655 **2T CCM, MIL: Yes** **1996** **Models:** Sportage **Engines:** All **Transmissions:** All	**Unused Power Stage 'B' Circuit Malfunction** Key on, and the PCM detected an unexpected voltage condition on the Power Stage 'B' circuit (EGR solenoid and IAC valve circuits). **Possible Causes:** • Power Stage 'B' circuit is open • PCM has failed
DTC: P1660 **2T CCM, MIL: Yes** **1996** **Models:** Sportage **Engines:** All **Transmissions:** All	**Unused Power Stage 'A' Circuit Malfunction** Key on, and the PCM detected an unexpected voltage condition on the Power Stage 'A' circuit (EGR solenoid and IAC valve circuits). **Possible Causes:** • Power Stage 'A' circuit is open • PCM has failed
DTC: P1665 **2T CCM, MIL: Yes** **1996** **Models:** Sportage **Engines:** All **Transmissions:** All	**Power Group Stage 'A' Circuit Malfunction** Key on, and the PCM detected an unexpected voltage condition on the Power Group Stage 'A' circuit (fuel injectors and purge valve circuits). **Possible Causes:** • Power Group Stage 'A' circuit is open • PCM has failed
DTC: P1670 **2T CCM, MIL: Yes** **1996** **Models:** Sportage **Engines:** All **Transmissions:** All	**Power Group Stage 'B' Circuit Malfunction** Key on, and the PCM detected an unexpected voltage condition on the Power Group Stage 'B' circuit (EGR solenoid and IAC valve circuits). **Possible Causes:** • Power Group Stage 'B' circuit is open • PCM has failed
DTC: P1673 **2T CCM, MIL: Yes** **1998-00** **Models:** Sephia **Engines:** All **Transmissions:** All	**Cooling Fan Circuit Malfunction** Key on or the engine running and the PCM detected an unexpected voltage condition on the Fan control circuit during the CCM test. **Note: The cooling fan relay pull-in coil resistance is 80 ohms at 68°F.** **Possible Causes:** • Cooling fan relay control circuit is open • Cooling fan relay control circuit is shorted to ground • Cooling fan relay is damaged or has failed • PCM has failed
DTC: P1690 **2T CCM, MIL: No** **2004-05** **Models:** Spectra **Engines:** 2.0L **Transmissions:** All	**Smartra Error** **No answer from SMARTRA. Invalid message from SMARTRA to ECM.** **Possible Causes:** • Open or short in antenna or SMARTRA circuit • Antenna • SMARTRA • Faulty transponder • Faulty ECM
DTC: P1691 **2T CCM, MIL: No** **2004-05** **Models:** Spectra, Sporatge **Engines:** 2.0L **Transmissions:** All	**Antenna Error** **Antenna error.** **Possible Causes:** • Open or short in antenna or SMARTRA circuit • Antenna • SMARTRA • Faulty transponder • Faulty ECM
DTC: P1693 **2T CCM, MIL: No** **2003-05** **Models:** Rio, Spectra, Sportage **Engines:** 1.6L, 1.8L, 2.0L **Transmissions:** All	**Antenna Error** **Passive mode invalid. Programming error.** **Possible Causes:** • Open or short in antenna or SMARTRA circuit • Antenna • SMARTRA • Faulty transponder • Faulty ECM

DTC	Trouble Code Title, Conditions & Possible Causes
DTC: P1693 **2T CCM, MIL: Yes** **1998-02** **Models:** Rio, Sephia, Sportage **Engines:** All **Transmissions:** All	**MIL Circuit Malfunction** Key on, and the PCM detected an unexpected voltage condition on the TCM to MIL control circuit during the CCM test. **Possible Causes:** • MIL control circuit is open • MIL control circuit is shorted to ground • MIL control circuit is shorted to system power (B+) • MIL "bulb" is open or missing • PCM has failed
DTC: P1694 **2T CCM, MIL: No** **2004-05** **Models:** Spectra, Sporatge **Engines:** 2.0L **Transmissions:** All	**ECM Signal Error** **Invalid request from ECM or corrupted data** **Possible Causes:** • Open or short in antenna or SMARTRA circuit • Antenna • SMARTRA • Faulty transponder • Faulty ECM
DTC: P1695 **2T CCM, MIL: No** **2004-05** **Models:** Spectra, Sporatge **Engines:** 2.0L **Transmissions:** All	**EEPROM Error** **Inconsistent data from EEPROM. Invalid write operation from EEPROM. Not plausible immobilizer indicator store in ECM. No valid data from SMARTRA after three attempts from the ECM. Invalid tester message or unexpected request from tester.** **Possible Causes:** • Open or short in antenna or SMARTRA circuit • Antenna • SMARTRA • Faulty transponder • Faulty ECM
DTC: P1700 **2T CCM, MIL: Yes** **1998-02** **Models:** Rio, Sephia, Spectra **Engines:** All **Transmissions:** A/T	**Overdrive (O/D) Off Lamp Circuit Malfunction** Engine running and the PCM detected an unexpected voltage condition on the Overdrive Off Lamp circuit during the CCM test. **Possible Causes:** • O/D "Off" lamp circuit is open between PCM and the TCM • O/D "Off" lamp circuit is shorted to ground between the TCM and the PCM • O/D "Off" lam circuit is shorted to system power between the TCM and the PCM
DTC: P1723 **2T CCM, MIL: Yes** **2002** **Models:** Sedona **Engines:** All **Transmissions:** A/T	**TCM A/T Control Relay Line Circuit Malfunction** Key on or engine running and the PCM detected the A/T control relay voltage as less than 7 volts. The PCM locks the gear selection into 3rd gear for safety when this trouble code sets. **Possible Causes:** • O/D "Off" lamp circuit is open between PCM and the TCM • O/D "Off" lamp circuit is shorted to ground between the TCM and the PCM • O/D "Off" lam circuit is shorted to system power between the TCM and the PCM
DTC: P1743 **2T CCM, MIL: Yes** **1996-97** **Models:** Sephia **Engines:** All **Transmissions:** A/T	**Torque Converter Clutch Solenoid Circuit Malfunction** Vehicle driven in Drive to a speed of over 30 mph, and the PCM detected an unexpected voltage condition on the TCC Solenoid circuit. **Possible Causes:** • TCC solenoid control circuit is open or shorted to ground • TCC solenoid control circuit is shorted to power • TCC solenoid is damaged or has failed • PCM has failed (TCC solenoid "driver" circuit may be open)
DTC: P1765 **2T CCM, MIL: No** **2003** **Models:** Spectra **Engines:** 2.0L **Transmissions:** A/T	**Transmission Control Spark Advance** Engine running and the PCM detected an unexpected circuit condition on the Transmission Controlled Spark Advance circuit. **Possible Causes:** • Open or short to ground between TCM and ECM • TCM has failed
DTC: P1780 **2T CCM, MIL: No** **1998-01** **Models:** Sephia **Engines:** All **Transmissions:** A/T	**Transmission Control Spark Advance Circuit Malfunction** Engine running and the PCM detected an unexpected circuit condition on the Transmission Controlled Spark Advance circuit. **Possible Causes:** • Spark advance circuit is open between the TCM and the PCM • Spark advance circuit shorted to ground between the TCM and the PCM • TCM or the PCM has failed

DTC	Trouble Code Title, Conditions & Possible Causes
DTC: P1794 **2T CCM, MIL: Yes** **1996-97** **Models:** Sephia **Engines:** All **Transmissions:** All	**Direct Battery Circuit Open Malfunction** Key on or the engine running and the PCM detected an open condition in the battery or backup power circuit. **Possible Causes:** • Battery direct circuit between the PCM and the battery is open • Battery direct circuit between the PCM and battery is shorted • Battery direct circuit fuse is open or missing • PCM has failed
DTC: P1795 **2T CCM, MIL: Yes** **1996-02** **Models:** Sportage **Engines:** All **Transmissions:** All	**4x4 Low Switch Signal Circuit Malfunction** Vehicle driven to a speed over 25 mph, TR switch indicating Drive, TP angle over 2%, and the PCM detected the VSS ratio indicated over 1.5 with the 4x4 switch "off", or less than 1.5 with switch "on". **Possible Causes:** • 4x4 low switch circuit is open • 4x4 low switch circuit is shorted to ground • 4x4 low switch circuit is damaged or has failed • PCM has failed
DTC: P1797 **2T CCM, MIL: Yes** **1996-97** **Models:** Sephia **Engines:** All **Transmissions:** All	**A/T Park Neutral Range Switch Circuit Malfunction** Engine running and the PCM did not detect a change in the A/T Park Neutral (P/N) position signal for 33 seconds after the gearshift selector was moved. **Possible Causes:** • A/T gear position switch signal circuit is open • A/T gear position switch signal circuit is shorted to ground • A/T gear position switch signal circuit is shorted to power • A/T gear position switch is damaged or has failed • PCM has failed
DTC: P1797 **2T CCM, MIL: Yes** **1998-02** **Models:** Sephia **Engines:** All **Transmissions:** All	**Clutch Pedal Switch Clutch Malfunction** Engine running, and the PCM did not detect a change in the Clutch switch signal for 33 seconds after the clutch was engaged and released. **Possible Causes:** • Clutch pedal position switch signal circuit is open • Clutch pedal position switch signal circuit is shorted to ground • Clutch pedal position switch signal circuit is shorted to power • Clutch pedal position switch is damaged or has failed • PCM has failed
DTC: P1800 **2T CCM, MIL: Yes** **1998-02** **Models:** Rio, Sephia, Spectra **Engines:** All **Transmissions:** A/T	**Engine Torque Circuit Malfunction** Engine running, and the PCM detected an unexpected voltage condition on the Engine Torque Signal circuit during the CCM test. **Possible Causes:** • Engine torque signal circuit is open between PCM and the TCM • Engine torque signal circuit is shorted to ground between the TCM and the PCM • Engine torque signal circuit is shorted to system power between the TCM and the PCM

Gas Engine OBD II Trouble Code List (P2xxx Codes)

DTC	Trouble Code Title, Conditions & Possible Causes
DTC: P2015 **2T CCM, MIL: No** **2005-06** **Models:** Amanti **Engines:** 3.5L **Transmissions:** All	**Intake Manifold Runner Position Sensor/Switch Circuit Range/Performance (Bank 1)** The DTC is set if the VIS system could not approach a target position. Engine speed equal to or greater than 3750 rpm. **Possible Causes:** • Poor connection • Open or short to battery in harness • Faulty VIS motor or VIS motor rotation sensor • Faulty PCM
DTC: P2096 **2T CCM, MIL: Yes** **2004-06** **Models:** Optima, Rio, Spectra **Engines:** 1.6L, 2.0L, 2.4L **Transmissions:** All	**Post Catalyst Fuel Trim System Too Lean (Bank 1)** Case 1: Monitoring deviation of fuel trim control (long term). No relevant failure. Long term fuel trim active. Case 2: Monitoring deviation of fuel trim control (short term). No relevant failure. Short term fuel trim active. Current engine speed less than 500 rpm. Current mass air flow less than 400mg/rev. Current lambda correction mean value less than 4 percent. **Possible Causes:** • Three way catalytic converter (TWC) • Rear HO2S
DTC: P2097 **2T CCM, MIL: Yes** **2004-06** **Models:** Optima, Rio, Spectra **Engines:** 1.6L, 2.0L, 2.4L **Transmissions:** All	**Post Catalyst Fuel Trim System Too Rich (Bank 1)** Case 1: Monitoring deviation of fuel trim control (long term). No relevant failure. Long term fuel trim active. Case 2: Monitoring deviation of fuel trim control (short term). No relevant failure. Short term fuel trim active. Current engine speed less than 500 rpm. Current mass air flow less than 400mg/rev. Current lambda correction mean value less than 4 percent. **Possible Causes:** • Three way catalytic converter (TWC) • Rear HO2S
DTC: P2101 **2T CCM, MIL: No** **2006** **Models:** Optima **Engines:** 2.4L **Transmissions:** All	**Throttle Actuator Control Motor Circuit** Hardware check. Battery voltage 9 volts. ECU power stage error. **Possible Causes:** • Poor connection or damaged harness • Faulty ETC motor
DTC: P2104 **2T CCM, MIL: Yes** **2006** **Models:** Optima, Sedons **Engines:** 2.4L, 2.7L, 3.8L **Transmissions:** All	**Electronic Throttle Control (ETC) System Malfunction- Forced Idle** This code is set if the system is in forced idle mode. Ignition ON. **Possible Causes:** • Faulty AFS • Faulty AFS plus brake • Faulty AFS plus vehicle speed sensor • Faulty AFS plus brake plus vehicle speed sensor • Faulty PCM
DTC: P2105 **2T CCM, MIL: Yes** **2006** **Models:** Optima, Sedona **Engines:** 2.4L, 2.7L, 3.8L **Transmissions:** All	**Electronic Throttle Control (ETC) System Malfunction- Forced Engine Shutdown** This code is set if the system is in forced engine shutdown mode. Ignition ON. **Possible Causes:** • Faulty AFS plus MAPS plus ETS • Faulty PCM
DTC: P2106 **2T CCM, MIL: Yes** **2006** **Models:** Amanti, Optima, Sedona **Engines:** 2.4L, 2.7L, 3.5L, 3.8L **Transmissions:** All	**Electronic Throttle Control (ETC) System Malfunction- Forced Limited Power** This code is set if the system is in forced limited power mode. Ignition ON. **Possible Causes:** • Faulty APS • Faulty APS + Brake • Faulty APS + vehicle speed sensor • Faulty APS + vehicle speed sensor + brake • Faulty PCM
DTC: P2107 **2T CCM, MIL: Yes** **2005-06** **Models:** Amanti **Engines:** 3.5L **Transmissions:** All	**ETS-ECM Malfunction (EEPROM R/W)** The DTC will set if the PCM can't read or write on the EEPROM. Check reading and writing. Ignition switch ON. Reading or writing error will occur. **Possible Causes:** • Poor connection • Faulty PCM

DTC	Trouble Code Title, Conditions & Possible Causes
DTC: P2108 **2T CCM, MIL: Yes** **2005-06** **Models:** Amanti **Engines:** 3.5L **Transmissions:** All	**ETS-ECM Malfunction (EEPROM R/W)** The DTC will set if the PCM detects an error to itself. Case 1: PCM/ETS error. Ignition switch ON. Something wrong with the communication PCM to ETS. Case 2: ETS/PCM error. Ignition switch ON. Something wrong with the communication ETS to PCM. Case 3: Computer malfunction. Ignition switch ON. **Possible Causes:** • Faulty PCM
DTC: P2110 **2T CCM, MIL: Yes** **2005-06** **Models:** Amanti, Optima **Engines:** 2.4L, 3.5L **Transmissions:** All	**Throttle Actuator Control Module Performance/Throttle Actuator Control System Stuck Closed (IG OFF)** Valve stuck closed (#1). Ignition switch OFF. TPS output as throttle valve is closed less than 0.025 volt. **Possible Causes:** • Poor connector • Faulty throttle valve • Faulty ETS motor • Faulty PCM
DTC: P2111 **2T CCM, MIL: Yes** **2005-06** **Models:** Amanti **Engines:** 3.5L **Transmissions:** All	**Throttle Actuator Control Module Performance/Throttle Actuator Control System Stuck Open** Valve stuck open. Ignition switch ON. Motor relay ON. **Possible Causes:** • Poor connector • Faulty throttle valve • Faulty ETS motor • Faulty PCM
DTC: P2112 **2T CCM, MIL: Yes** **2005-06** **Models:** Amanti **Engines:** 3.5L **Transmissions:** All	**Throttle Actuator Control Module Performance/Throttle Actuator Control System Stuck Closed (IG OFF)** Valve close stuck #2. Ignition switch OFF. TPS output as throttle valve is closed less than 0.06 volt. **Possible Causes:** • Poor connector • Faulty throttle valve • Faulty ETS motor • Faulty PCM
DTC: P2118 **2T CCM, MIL: Yes** **2005-06** **Models:** Amanti, Optima **Engines:** 2.4L, 3.5L **Transmissions:** All	**Throttle Actuator Control Motor Current Range/Performance/Throttle Actuator Control Motor Circuit Open** Vb open. Motor relay ON. Voltage to detect circuit open less than or equal to 4.0 volts. **Possible Causes:** • Poor connection • Open in ETS relay circuit • Faulty ETS relay/fuse • Faulty PCM
DTC: P2118 **2T CCM, MIL: Yes** **2005-06** **Models:** Amanti, Optima **Engines:** 2.4L, 3.5L **Transmissions:** All	**Throttle Actuator Control Motor Circuit Range/Performance/Throttle Actuator Control Motor Circuit Low** Motor circuit low. Ignition switch ON. **Possible Causes:** • Poor connection • Short to ground in ETS motor circuit • Faulty ETS motor • Faulty PCM
DTC: P2118 **2T CCM, MIL: Yes** **2005-06** **Models:** Amanti, Optima **Engines:** 2.4L, 3.5L **Transmissions:** All	**Throttle Actuator Control Motor Circuit Range/Performance/Throttle Actuator Control Motor Circuit High** Motor circuit High. Ignition switch ON. **Possible Causes:** • Poor connection • Short to battery in ETS motor circuit • Faulty ETS motor • Faulty PCM
DTC: P2119 **2T CCM, MIL: Yes** **2006** **Models:** Optima **Engines:** 2.4L **Transmissions:** All	**Throttle Actuator Control Module Performance/Throttle Actuator Control System Stuck Closed (IG OFF)** Valve stuck closed (#1). Ignition switch OFF. TPS output as throttle valve is closed less than 0.025 volt. **Possible Causes:** • Poor connector • Faulty throttle valve • Faulty ETS motor • Faulty PCM

DTC	Trouble Code Title, Conditions & Possible Causes
DTC: P2122 **2T CCM, MIL: Yes** **2005-06** **Models:** Amanti, Optima, Sedona **Engines:** 2.4L, 2.7L, 3.5L, 3.8L **Transmissions:** All	**Throttle/Pedal Position Sensor/Switch "D" Circuit Low Input** Accelerator position sensor (APS1) low input. ETS/PCM communication is normal. Output voltage of APS1 is less than 0.2 volt. **Possible Causes:** • Poor connector • Faulty APS1 • Open or short in APS1 circuit • Faulty PCM
DTC: P2123 **2T CCM, MIL: Yes** **2005-06** **Models:** Amanti, Optima, Sedona **Engines:** 2.4L, 2.7L, 3.5L, 3.8L **Transmissions:** All	**Throttle/Pedal Position Sensor/Switch "D" Circuit High Input** Accelerator position sensor (APS1) high input. ETS/PCM communication is normal. Output voltage of APS1 is equal to or greater than 4.9 volts. Output voltage of APS2 is less than 4.1 volts. **Possible Causes:** • Poor connector • Faulty APS1 • Open or short in APS1 circuit • Faulty PCM
DTC: P2125 **2T CCM, MIL: Yes** **2005-06** **Models:** Amanti **Engines:** 3.5L **Transmissions:** All	**Throttle/Pedal Position Sensor/Switch "E" Circuit** Accelerator position sensor (APS2) circuit. ETS/PCM communication is normal. Vaps1 and Vaps2: 0.2 to 4.9 volts. Idle switch ON. **Possible Causes:** • Poor connector • Faulty APS1 • Open or short in APS1 circuit • Faulty PCM
DTC: P2127 **2T CCM, MIL: Yes** **2005-06** **Models:** Amanti, Optima, Sedona **Engines:** 2.4L, 2.7L, 3.5L, 3.8L **Transmissions:** All	**Throttle/Pedal Position Sensor/Switch "E" Circuit Low Input** Accelerator position sensor (APS2) low input. ETS/PCM communication is normal. Output voltage of APS2 is less than 0.2 volt. **Possible Causes:** • Poor connection • Faulty APS2 • Open or short in APS2 circuit • Faulty PCM
DTC: P2128 **2T CCM, MIL: Yes** **2005-06** **Models:** Amanti, Optima, Sedona **Engines:** 2.4L, 2.7L, 3.5L, 3.8L **Transmissions:** All	**Throttle/Pedal Position Sensor/Switch "E" Circuit High Input** Accelerator position sensor (APS2) high input. ETS/PCM communication is normal. Output voltage of APS2 is greater than or equal to 4.9 volts. Output voltage of ASP1 is less than 4.1 volts. **Possible Causes:** • Poor connection • Faulty APS2 • Open or short in APS2 circuit • Faulty PCM
DTC: P2135 **2T CCM, MIL: Yes** **2005-06** **Models:** Amanti, Optima, Sedona **Engines:** 2.7L, 3.5L, 3.8L **Transmissions:** All	**Throttle/Pedal Position Sensor/Switch "A"/"B" Voltage Correlation** Determines if TPS #1 disagrees with TPS #2. Ignition "ON". **Possible Causes:** • Poor connection • Open or short in TPS circuit • Faulty TPS • Faulty PCM
DTC: P2138 **2T CCM, MIL: Yes** **2005-06** **Models:** Amanti, Optima, Sedona **Engines:** 2.4L, 2.7L, 3.5L, 3.8L **Transmissions:** All	**Throttle/Pedal Position Sensor/Switch "D/E" Voltage Correlation** Monitoring abnormal APS. Output voltage of APS1: 0.2 to 4.9 volts. Output voltage of APS2: 0.2 to 4.9 volts. Ignition switch ON. **Possible Causes:** • Poor connection • Faulty APS • Faulty PCM
DTC: P2159 **2T CCM, MIL: Yes** **2006** **Models:** Optima **Engines:** 2.4L **Transmissions:** All	**Vehicle Speed Sensor "B" Range/Performance** Plausibility check. Enabling conditions are as follows: engine speed greater than 2100 rpm, air mass flow greater than 0.44 g/rev, no fuel injection shut off, coolant temperature 140 degrees F. **Possible Causes:** • Open or short in harness • Poor connection or damaged harness • VSS

DTC	Trouble Code Title, Conditions & Possible Causes
DTC: P2173 **2T CCM, MIL: Yes** **2006** **Models:** Optima, Sedona **Engines:** 2.7L, 3.8L **Transmissions:** All	**Electronic Throttle Control (ETC) System Malfunction- High Air Flow Detected** The engine airflow measurements are not based on throttle position. They are compared with throttle position based on estimated air flow. If measured air flow is much higher, the throttle body may not be throttling the engine. Engine running. Throttle actuation mode is not off. MAP sensor is not failed. MAF sensor is not failed. IAT sensor is not failed. **Possible Causes:** • Air leakage between TPS and MAFS • Faulty throttle body • Faulty PCM
DTC: P2176 **2T CCM, MIL: Yes** **2005-06** **Models:** Amanti **Engines:** 3.5L **Transmissions:** All	**Throttle Actuator Control Module Performance/Throttle Actuator Control System Stuck Closed (IG OFF)** Valve stuck closed (#1). Ignition switch OFF. TPS output as throttle valve is closed less than 0.025 volt. **Possible Causes:** • Poor connector • Faulty throttle valve • Faulty ETS motor • Faulty PCM
DTC: P2187 **2T CCM, MIL: Yes** **2006** **Models:** Optima, Sedona **Engines:** 2.4L, 2.7L, 3.8L **Transmissions:** All	**System Too Lean At Idle (Additive) (Bank 1)** Engine coolant temperature 140 degrees F. Intake air temperature 140 degrees F. System voltage greater than 11 volts. Closed loop active. **Possible Causes:** • Sensors related to fuel trim • Intake system • Fuel pressure • Faulty PCM
DTC: P2188 **2T CCM, MIL: Yes** **2006** **Models:** Optima, Sedona **Engines:** 2.4L, 2.7L, 3.8L **Transmissions:** All	**System Too Lean At Idle (Additive) (Bank 2)** Engine coolant temperature 140 degrees F. Intake air temperature 140 degrees F. System voltage greater than 11 volts. Closed loop active. **Possible Causes:** • Sensors related to fuel trim • Intake system • Fuel pressure • Faulty PCM
DTC: P2188 **2T CCM, MIL: Yes** **2006** **Models:** Optima, Sedona **Engines:** 2.7L, 3.8L **Transmissions:** All	**System Too Rich At Idle (Additive) (Bank 2)** Engine coolant temperature 140 degrees F. Intake air temperature 140 degrees F. System voltage greater than 11 volts. Closed loop active. **Possible Causes:** • Sensors related to fuel trim • Intake system • Fuel pressure • Faulty PCM
DTC: P2189 **2T CCM, MIL: Yes** **2006** **Models:** Optima **Engines:** 2.7L **Transmissions:** All	**System Too Lean At Idle (Additive) (Bank 2)** Engine coolant temperature 140 degrees F. Intake air temperature 140 degrees F. System voltage greater than 11 volts. Closed loop active. **Possible Causes:** • Sensors related to fuel trim • Intake system • Fuel pressure • Faulty PCM
DTC: P2190 **2T CCM, MIL: Yes** **2006** **Models:** Optima, Sedona **Engines:** 2.7L, 3.8L **Transmissions:** All	**System Too Rich At Idle (Additive) (Bank 2)** Engine coolant temperature 140 degrees F. Intake air temperature 140 degrees F. System voltage greater than 11 volts. Closed loop active. **Possible Causes:** • Sensors related to fuel trim • Intake system • Fuel pressure • Faulty PCM

DTC	Trouble Code Title, Conditions & Possible Causes
DTC: P2191 **2T CCM, MIL: Yes** **2006** **Models:** Optima **Engines:** 2.4L **Transmissions:** All	**System Too Lean At Higher Load (Multiple) (Bank 1)** Fuel trim limit. Coolant temperature greater than 158 degrees F. Intake air temperature less than 176 degrees F. Throttle angle less than 60 percent. Integrated air mass greater than 10 grams. Closed loop control enabled. No transient control phase. No canister purge phase. Air mass1 40 to 80 kg/h. Air mass2 greater than 100 kg/h. **Possible Causes:** • Faulty ignition system • EVAP PCSV malfunction • Faulty fuel injectors • Leak in exhaust system • Faulty MAP, TPS, ECTS • Faulty front HO2S • Faulty PCM
DTC: P2192 **2T CCM, MIL: Yes** **2006** **Models:** Optima **Engines:** 2.4L **Transmissions:** All	**System Too Rich At Higher Load (Bank 1)** Fuel trim limit. Coolant temperature greater than 158 degrees F. Intake air temperature less than 176 degrees F. Throttle angle less than 60 percent. Integrated air mass greater than 10 grams. Closed loop control enabled. No transient control phase. No canister purge phase. Engine load1 30 to 55 percent. Engine load2 greater than 70 percent. **Possible Causes:** • Faulty ignition system • EVAP PCSV malfunction • Faulty fuel injectors • Leak in exhaust system • Faulty MAP, TPS, ECTS • Faulty front HO2S • Faulty PCM
DTC: P2195 **2T CCM, MIL: Yes** **2004-06** **Models:** Optima, Sedona, Spectra **Engines:** 2.0L, 2.7L, 3.8L **Transmissions:** All	**HO2S Signal Stuck Lean (Bank 1 Sensor 1)** **Sensor characteristic line shifted to lean. No relevant failure. No misfire detected. Fuel trim control active.** **Possible Causes:** • Contact resistance in connectors • Faulty HO2S • Faulty PCM
DTC: P2196 **2004-06** **Models:** Optima, Sedona, Spectra **Engines:** 2.0L, 2.7L, 3.8L **Transmissions:** All	**HO2S Signal Stuck Rich (Bank 1 Sensor 1)** **Sensor characteristic line shifted to lean. No relevant failure. No misfire detected. Fuel trim control active.** **Possible Causes:** • Contact resistance in connectors • Faulty HO2S • Faulty PCM
DTC: P2197 **2T CCM, MIL: Yes** **2006** **Models:** Optima, Sedona **Engines:** 2.7L, 3.8L **Transmissions:** All	**HO2S Signal Stuck Lean (Bank 2 Sensor 1)** Determines if O2 sensor indicates lean exhaust while in power enrichment. Sensor not in cooled status flag. Not in transient conditions status flag. Device control not active. Engine running. Minimum air flow present is equal or greater than 2 g/s. Engine coolant warm (140 degrees F. Above conditions met for at least 1.5L seconds. **Possible Causes:** • Poor connection • Faulty HO2S • Faulty PCM
DTC: P2198 **2T CCM, MIL: Yes** **2006** **Models:** Optima, Sedona **Engines:** 2.7L, 3.8L **Transmissions:** All	**HO2S Signal Stuck Rich (Bank 2 Sensor 1)** Determines if O2 sensor indicates rich exhaust while in decal fuel cut off (DFCO). Sensor not in cooled status flag. Not in transient conditions status flag. Device control not active. Engine running. Minimum air flow present is equal or greater than 2 g/s. Ignition voltage equal to or greater than 10 volts. Fuel reduction not active. Engine running long enough (more than 60 seconds). Engine coolant warm (140 degrees F. Above conditions met for at least 1.5L seconds. **Possible Causes:** • Poor connection • Faulty HO2S • Faulty PCM
DTC: P2226 **2T CCM, MIL: Yes** **2006** **Models:** Rio **Engines:** 1.6L **Transmissions:** All	**Barometric Pressure Circuit** Rationality check. **Possible Causes:** • Clog at sensing hole • Faulty ECM

DTC	Trouble Code Title, Conditions & Possible Causes
DTC: P2227 **2T CCM, MIL: Yes** **2006** **Models:** Rio **Engines:** 1.6L **Transmissions:** All	**Barometric Pressure Circuit Range/Performance** Rationality check. **Possible Causes:** • Clog at sensing hole • Faulty ECM
DTC: P2228 **2T CCM, MIL: Yes** **2006** **Models:** Rio **Engines:** 1.6L **Transmissions:** All	**Barometric Pressure Circuit Low Input** Signal check low. **Possible Causes:** • Faulty ECM
DTC: P2229 **2T CCM, MIL: Yes** **2006** **Models:** Rio **Engines:** 1.6L **Transmissions:** All	**Barometric Pressure Circuit High Input** Signal check, high. **Possible Causes:** • Faulty ECM
DTC: P2231 **2T CCM, MIL: Yes** **2004-06** **Models:** Spectra **Engines:** 2.0L **Transmissions:** All	**HO2S Signal Circuit Shorted To Heater Circuit (Bank 1 Sensor 1)** Front HO2S signal monitoring. Exhaust temperature greater than 752 degrees F. No relevant failure. Amplitude of forced lambda simulation less than 0.05. Period time of forced lambda simulation less than 2.55 seconds. Current engine speed less than 500 rpm. Current mass air flow less than 400 mg/rev. **Possible Causes:** • Contact resistance in connectors • Interference in HO2S
DTC: P2232 **2T CCM, MIL: Yes** **2006** **Models:** Rio **Engines:** 1.6L **Transmissions:** All	**HO2S Signal Circuit Shorted To Heater Circuit (Bank 1 Sensor 2)** Rationality check. **Possible Causes:** • Poor connection • Short to power in signal circuit • B1S2 • Faulty ECM
DTC: P2237 **2T CCM, MIL: Yes** **2004-06** **Models:** Spectra **Engines:** 2.0L **Transmissions:** All	**HO2S Pumping Current Circuit/Open Bank 1, Sensor 1** Open circuit of front HO2S circuit. No relevant failure. **Possible Causes:** • Contact resistance in connectors • Open or short to ground in HO2S circuit • Front HO2S sensor
DTC: P2243 **2T CCM, MIL: Yes** **2004-06** **Models:** Spectra **Engines:** 2.0L **Transmissions:** All	**HO2S Reference Voltage Circuit/Open Bank 1, Sensor 1** Open circuit of front HO2S circuit. No relevant failure. **Possible Causes:** • Contact resistance in connectors • Open or short to ground in HO2S circuit • Front HO2S sensor
DTC: P2251 **2T CCM, MIL: Yes** **2004-06** **Models:** Spectra **Engines:** 2.0L **Transmissions:** All	**HO2S Reference Ground Circuit/Open Bank 1, Sensor 1** Open circuit of front HO2S circuit. No pump current malfunction. **Possible Causes:** • Contact resistance in connectors • Open or short to ground in HO2S circuit • Front HO2S sensor
DTC: P2270 **2T CCM, MIL: Yes** **2004-06** **Models:** Optima, Sedona, Spectra **Engines:** 2.0L, 2.7L, 3.8L **Transmissions:** All	**HO2S Signal Stuck Rich (Bank 1 Sensor 2)** Plausibility check during shift of lambda set point to rich from lean. No fuel cut off. No full load phase. No fuel trim error detected. Delay time to start diagnosis: 13 to 30 seconds. No relevant failure. **Possible Causes:** • Three way catalytic converter (TWC) • Air leakage in exhaust system • Faulty rear HO2S sensor

DTC	Trouble Code Title, Conditions & Possible Causes
DTC: P2271 **2T CCM, MIL: Yes** **2004-06** **Models:** Optima, Sedona, Spectra **Engines:** 2.0L, 2.7L, 3.8L **Transmissions:** All	**O2 Signal Stuck Rich (Bank 1/2 Sensor 1)** Plausibility check during shift of lambda set point to rich from lean. No fuel cut off. No full load phase. No fuel trim error detected. Delay time to start diagnosis: 13 to 30 seconds. No relevant failure. **Possible Causes:** • Three way catalytic converter (TWC) • Air leakage in exhaust system • Faulty rear HO2S sensor
DTC: P2272 **2T CCM, MIL: Yes** **2006** **Models:** Optima, Sedona **Engines:** 2.7L, 3.8L **Transmissions:** All	**HO2S Signal Stuck Lean (Bank 2 Sensor 2)** Determines if O2 sensor indicates lean exhaust while in power enrichment mode. Sensor not in cooled status flag. Not in transient conditions status flag. Device control not active. Engine running. Minimum air flow present is equal or greater than 2 g/s. Ignition voltage equal to or greater than 10 volts. Fuel reduction not active. Engine running long enough (more than 60 seconds). Engine coolant warm (140 degrees F. Above conditions met for at least 2.5 seconds. **Possible Causes:** • Poor connection • Faulty HO2S • Faulty PCM
DTC: P2273 **2T CCM, MIL: Yes** **2006** **Models:** Optima, Sedona **Engines:** 2.7L, 3.8L **Transmissions:** All	**HO2S Signal Stuck Rich (Bank 2 Sensor 2)** Determines if O2 sensor indicates rich exhaust while in decal fuel cut off (DFCO). Sensor not in cooled status flag. Not in transient conditions status flag. Device control not active. Engine running. Minimum air flow present is equal or greater than 2 g/s. Ignition voltage equal to or greater than 10 volts. Fuel reduction not active. Engine running long enough (more than 60 seconds). Engine coolant warm (140 degrees F. Above conditions met for at least 2.0L seconds. **Possible Causes:** • Poor connection • Faulty HO2S • Faulty PCM
DTC: P2422 **2T CCM, MIL: Yes** **2005-06** **Models:** Amanti, Optima, Sedona **Engines:** 2.7L, 3.5L, 3.8L **Transmissions:** All	**Evaporative Emission System- Canister Clogging** Ignition voltage 10-16 volts. Barometric pressure 72kpa. Engine run time, one second. **Possible Causes:** • Faulty canister close valve • Clogging of canister air filter • Open in ground harness of FTPS • Faulty PCM
DTC: P2414 **2T CCM, MIL: Yes** **2004-06** **Models:** Spectra **Engines:** 2.0L **Transmissions:** All	**HO2S Exhaust Sample Error Bank 1 Sensor 1** Sensor not mounted. Plausibility check in part load or full load conditions. Sensor tip temperature 1202 degrees F. Part load or full load. No relevant failure. **Possible Causes:** • Incorrect installation of HO2S sensor • Contact resistance in connectors
DTC: P2610 **2T CCM, MIL: Yes** **2006** **Models:** Optima, Sedona **Engines:** 2.7L, 3.8L **Transmissions:** All	**ECM/PCM Internal Engine Off Timer Performance** The LPC SPI diagnostic allows the low power counter to count down and simultaneously enables a test timer to run for a calibrated length of time and then compares the lapsed time recorded by the counter to make a pass/fail determination. Engine running. Enough time (10 seconds). Battery voltage 8 volts. No memory failure. **Possible Causes:** • Faulty PCM
DTC: P2A00 **2T CCM, MIL: Yes** **2006** **Models:** Optima, Sedona **Engines:** 2.7L, 3.8L **Transmissions:** All	**O2 Sensor Not ready (Bank 1 Sensor 1)** Detects loss of O2 ready status, which would lead to open loop fueling operation, a default mode. Engine running. Ignition ON. DFCO not present too long (less than 15 seconds). No disabling faults present. All of the above for at least 20 seconds. **Possible Causes:** • Poor connection • Faulty HO2S • Faulty PCM
DTC: P2A03 **2T CCM, MIL: Yes** **2006** **Models:** Optima, Sedona **Engines:** 2.7L, 3.8L **Transmissions:** All	**O2 Sensor Not ready (Bank 1 Sensor 2)** Detects loss of O2 ready status, which would lead to open loop fueling operation, a default mode. Engine running. Ignition ON. DFCO not present too long (less than 15 seconds). No disabling faults present. All of the above for at least 20 seconds. **Possible Causes:** • Poor connection • Faulty HO2S • Faulty PCM

DTC	Trouble Code Title, Conditions & Possible Causes
DTC: P2626 **2T CCM, MIL: Yes** **2004-06** **Models:** Spectra **Engines:** 2.0L **Transmissions:** All	**HO2S Pumping Current Trim Circuit Open Bank 1 Sensor 1** Check open circuit of front HO2S sensor. **Possible Causes:** • Contact resistance in connectors • Open or short to ground in HO2S circuit • Faulty canister • Faulty front HO2S sensor

KIA
COMPONENT TESTING

11

TABLE OF CONTENTS

Component Testing 11-2
- Amanti 11-2
- Optima 11-7
- Rio 11-12
- Sedona 11-17
- Sephia 11-22
- Sorento 11-27
- Spectra 11-30
- Sportage 11-35

COMPONENT TESTING

AMANTI

See Figure 1.

Camshaft Position (CMP) Sensor

LOCATION

The camshaft position sensor is mounted on the engine near the camshaft gear by the ignition coils.

OPERATION

The camshaft position sensor senses the TDC point of the number one cylinder, on its compression stroke. Its signal is relayed to the ECM to be used to determine the sequence of fuel injection.

REMOVAL & INSTALLATION

1. Disconnect the negative battery cable.
2. Disconnect the connector from the sensor.
3. Remove the bolt that retains the sensor.
4. Remove the sensor.

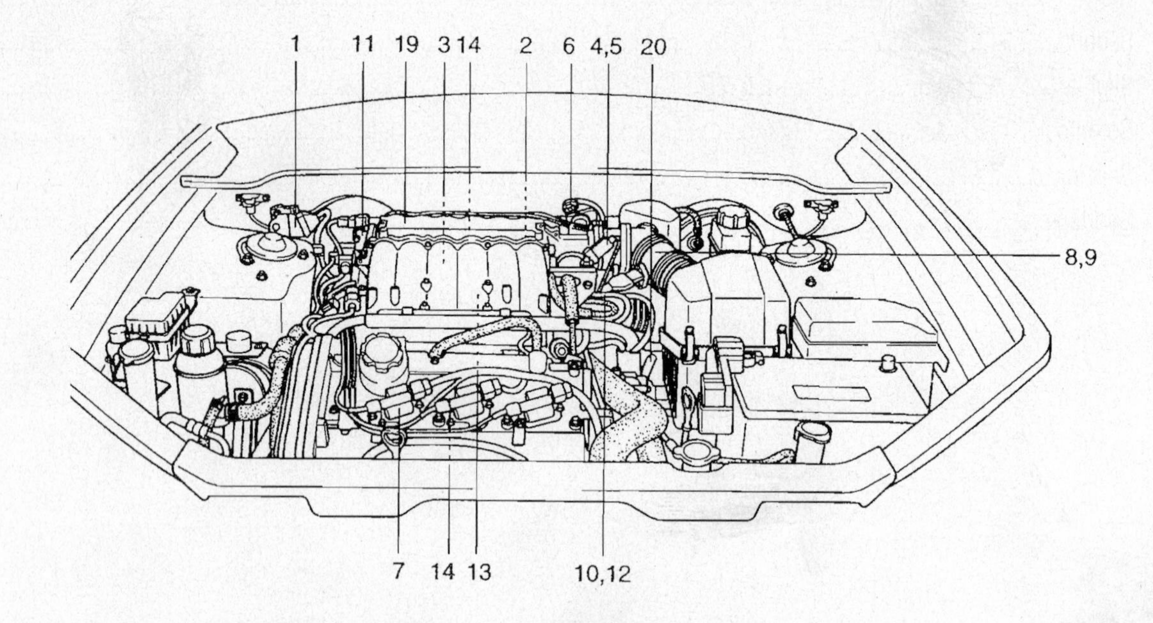

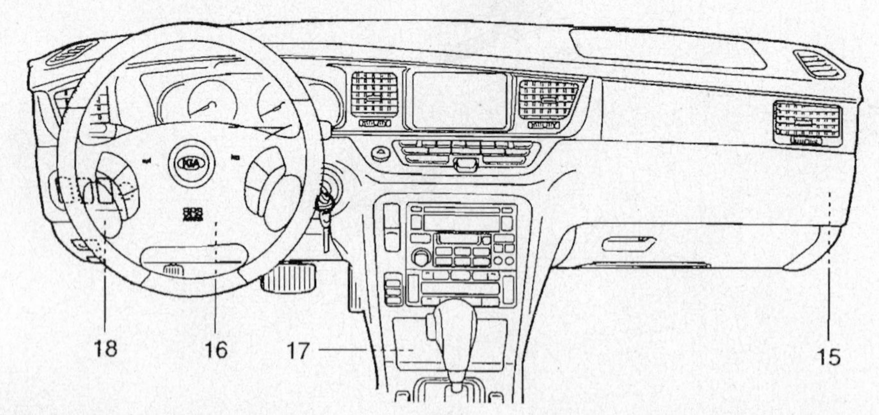

1. Accelerator Position Sensor (APS)
2. EGR Valve
3. Knock Sensor
4. Throttle Position Sensor (TPS)
5. Limp-Home Valve
6. ETS Motor
7. Ignition Coil
8. Mass Air Flow Sensor (MAF) - With ATS
9. Intake Air Temperature Sensor (ATS)
10. Purge Control Solenoid Valve (PCSV)
11. Variable Intake Manifold Control Motor (V.I.Motor)
12. Water Temperature Sensor (WTS)
13. Injector
14. Heated Oxygen Sensor (HO2S)
15. ETS (Electronic Throttle System) Control Unit
16. OBD Connector (16 pin)
17. ECU
18. Check Connector (10 pin)
19. Manifold Absolute Pressure (MAP) Sensor
20. Ignition Failure Sensor (IFS)

29130_KIA_G0001

Fig. 1 Underhood sensor locations—Amanti

To install:

5. Installation is the reverse of the removal procedure.

TESTING

1. Be sure that the CMPS and PCM connectors are connected.
2. Engine ON and monitor the signal waveform of the sensor on the scan tool. Check whether the waveform is synchronized with the crankshaft position sensor or not.
3. If the waveform signal is normal, substitute with a known good PCM and check for proper operation. If the problem is corrected replace the PCM.
4. If the waveform signal is not normal, substitute with a known good sensor and check for proper operation. If the problem is corrected replace the sensor.

Crankshaft Position (CKP) Sensor

LOCATION

The crankshaft position sensor is mounted on the engine near the crankshaft.

OPERATION

The crankshaft position sensor is a Hall-Effect sensor that senses the crank angle (piston position) of each cylinder and converts it into a pulse signal. The signal from the sensor is relayed to the ECM to indicate engine rpm and the position of the crankshaft.

REMOVAL & INSTALLATION

1. Disconnect the negative battery cable.
2. Disconnect the connector from the sensor.
3. Remove the bolt that retains the sensor in place.
4. Remove the sensor from its mounting.

To install:

5. Installation is the reverse of the removal procedure.
6. Clearance between the sensor and the sensor wheel should be 0.020–0.059 inch.

TESTING

2004

1. Properly connect the scan tool.
2. Start the engine and allow it to idle.
3. Be sure the idle position switch is ON.
4. Specifications are as follows:
   - Temperature - 4 degrees F, test specification 1500–1700 RPM
   - Temperature 32 degrees F, test specification 1350–1550 RPM
   - Temperature 68 degrees F, test specification 1200–1400 RPM
   - Temperature 104 degrees F, test specification 1000–1200 RPM
   - Temperature 176 degrees F, test specification 650–850 RPM
5. Replace sensor, as required.

2005–2006

1. Be sure that the CKPS and PCM connectors are connected.
2. Engine ON and monitor the signal waveform of the sensor on the scan tool. Check whether the waveform is synchronized with the camshaft position sensor or not.
3. If the waveform signal is normal, substitute with a known good PCM and check for proper operation. If the problem is corrected replace the PCM.
4. If the waveform signal is not normal, substitute with a known good sensor and check for proper operation. If the problem is corrected replace the sensor.

Electronic Control Module (ECM)

OPERATION

The ECM controls the vehicle engine operating system.

REMOVAL & INSTALLATION

1. Disconnect the negative battery cable.
2. Remove the lower inner trim.
3. As required, detach the floor mat. As required, remove the protective cover.
4. Remove the ECM bracket retaining nuts. Remove the clip from the bracket.
5. Disconnect the connectors.
6. Remove the ECM from the vehicle.

To install:

7. Installation is the reverse of the removal procedure.

➡ **When replacing the ECM, be careful not to use the wrong part number, as damage to the injection system could occur.**

Electronic Throttle System (ETS)

OPERATION

On 2004 vehicles the ETS consists of inputs from the Accelerator Position Sensor (APS), Throttle Position Sensor (TPS), ignition switch and serial communication line, ETS motor and Limp/Home value unit. These components are used to control idle speed, throttle valve and TCS using only one motor. In the ECU and ETS the signals from the APS and TPS are most important. They have two circuits in the sensor module, one is flowed to the ECU and the other is flowed to the ETS. The ECU calculates the throttle target value with the signals from APS#2 (main) and TPS#2 (sub) and the ETS control module controls the ETS motor with the signals from the APS#1 (sub) and TPS#1 (main).

On 2005–2006 vehicles the electronic throttle system module is integrated with the PCM module. The ETS is the throttle body, whose throttle angle is controlled by the control module. It consists of and ETS motor, Accelerator Position Sensor (APS), Throttle Position Sensor (TPS), and Limp/Home value unit. The APS consists of APS#1 and APS#2 and an idle switch which detects the closed throttle position. When the driver steps on the accelerator pedal, the output signal of APS#1 is used for the sub signal and that of APS#2 is used for the main signal. If the PCM receives these signals and other input values, is calculates throttle valve target opening value and then controls the ETS motor to meet the throttle valve target opening value.

REMOVAL & INSTALLATION

1. Disconnect the negative battery cable.
2. Disconnect the sensor(s) connector(s).
3. Remove the sensor retaining screws. Remove the sensor(s) from its mounting.

To install:

4. Installation is the reverse of the removal procedure.

TESTING

Throttle Position Sensor

2004

1. Disconnect the sensor connector.
2. Using an ohmmeter, measure the resistance between terminal 3 (sensor ground) and terminal 1 (sensor power).
3. Specification should be 1.5–3.5 kohm.
4. Connect the ohmmeter between terminal 3 (sensor ground) and terminal 2 (sensor output).
5. Inspect the sensor signal while rotating the throttle valve from closed throttle to wide open throttle. At this time the sensor signal varies proportional to throttle angle.

6. If the resistance is out of specification, or is not directly proportional to throttle angle, replace the sensor.

2005–2006

1. Disconnect the sensor connector.
2. Using an ohmmeter, measure the resistance between terminal 1 and 3 of the connector.
3. Specification should be 3.5–6.5 kohm.
4. Connect the ohmmeter between terminal 2 and 3 of the TPS#1 connector and measure the resistance (resistance should change smoothly in proportion with the throttle valve opening angle).
5. Connect the ohmmeter between terminal 3 and 4 of the TPS#2 connector and measure the resistance (resistance should change smoothly in inverse proportion with the throttle valve opening angle).
6. If within specification, substitute the PCM with a known good one. Recheck for proper operation. Replace the PCM, as required.
7. If not within specification, substitute the TPS with a known good sensor. Recheck for proper operation. Replace the sensor, as required.

Accelerator Pedal Sensor

2004

1. Disconnect the sensor connector.
2. Using an ohmmeter, measure the resistance between ground and sensor supply (APS#1: terminal 7 and 8. APS#2: terminal 1 and 2).
3. Specification should be 3.5–6.5 kohm.
4. Using an ohmmeter, measure the resistance between ground and sensor signal output (APS#1: terminal 6 and 7. APS#2: terminal 3 and 1).
5. Inspect the sensor signal while pressing the accelerator pedal, at this time the sensor signal will vary proportional to the accelerator pedal angle.
6. If the resistance is out of specification, or is not directly proportional to the accelerator pedal angle, replace the sensor.

2005–2006

1. Connect the scan tool.
2. With the ignition OFF, disconnect the connector.
3. Measure APS#1 and APS#2 signal waveform when stepping on the accelerator pedal.
4. Specification should be approximately 0.3–1.0 volt (when not stepping on the accelerator pedal) and approximately 4.0–5.5 volts (when stepping on the accelerator pedal).
5. If within specification, substitute the PCM with a known good one. Recheck for proper operation. Replace the PCM, as required.
6. If not within specification, substitute the TPS with a known good sensor. Recheck for proper operation. Replace the sensor, as required.

Engine Coolant Temperature (ECT) Sensor

LOCATION

The ECT is installed in the engine coolant passage of the cylinder head.

OPERATION

This component detects the temperature of the engine coolant and relays the information to the electronic control assembly. This component employs a thermistor which is sensitive to temperature changes. The electric resistance of the thermistor decreases in response to temperature rise. The ECM judges coolant temperature by the sensor output voltage and provides optimum fuel enrichment when the engine is cold.

REMOVAL & INSTALLATION

1. Disconnect the negative battery cable.
2. Disconnect the connector from the sensor.
3. Drain the cooling system, as required.
4. Remove the sensor from its mounting.

To install:

5. Installation is the reverse of the removal procedure.

TESTING

2004

1. Remove the sensor from the engine.
2. With the sensing portion of the sensor immersed in hot water, check the resistance.
3. Specification should be 0.32 kohm at 176 degrees F.
4. Replace the sensor, as required.

2005–2006

1. Be sure the ignition switch is in the OFF position.
2. Disconnect the sensor electrical connector.
3. Using an ohmmeter measure the resistance between terminals 1 and 3 of the sensor connector.
4. Specifications are as follows (error range is 2–3 percent):
   - Temperature 32 degrees F, resistance 2.45 kohms
   - Temperature 68 degrees F, resistance 1.15 kohms
   - Temperature 176 degrees F, resistance 0.587 kohms
   - Temperature 212 degrees F, resistance 0.322 kohms
   - Temperature 230 degrees F, resistance 0.188 kohms
   - Temperature 248 degrees F, resistance 0.147 kohms
5. If within specification, check the PCM as follows, ignition OFF. Connect the scan tool. Turn the ignition ON and engine OFF. Select simulation function on the scan tool. Simulate voltage at terminal 3 of the sensor connector.
6. If the signal voltage value does not change, substitute a known good PCM and repeat the test. Replace the PCM as required. If the signal voltage value does change, check connectors for looseness, poor connection, binding etc. Correct and retest.
7. If the temperature and resistance check is not within specification, check for deterioration and damage to the sensor.
8. Retest, if still not within specification substitute a known good sensor and retest. Replace the sensor, as required.

Heated Oxygen (HO2S) Sensor

LOCATION

The sensors are located in the exhaust system. On some vehicles one sensor is located up at the exhaust manifold(s) and the other sensor is located down at the catalytic converter. On other vehicles both sensors are located down at the catalytic converter.

OPERATION

The exhaust gas oxygen sensor supplies the electronic control assembly with a signal which indicates either a rich or lean mixture condition, during the engine operation.

REMOVAL & INSTALLATION

1. Disconnect the electrical connector from the sensor.
2. Remove the oxygen sensor.

To install:

3. Installation is the reverse of the removal procedure.

➡ **Apply anti-seize compound to the threaded portion of the sensor, prior to installation. Never apply anti-seize compound to the protector of the sensor.**

TESTING

2004

Perform a visual inspection of the sensor as follows:

1. If the sensor tip has a black/sooty deposit, this may indicate a rich fuel mixture.
2. If the sensor tip has a white, gritty deposit, this may indicate an internal coolant leak.
3. If the sensor tip has a brown deposit, this could indicate excessive oil consumption.
4. Warm the engine until the coolant temperature reaches operating temperature.
5. Disconnect the sensor electrical connector and measure the resistance between terminal 3 and 4.
6. Specification should be as follows:
   - Bank 1 Sensor 1, resistance 3.3 kohms
   - Bank 2 Sensor 1, resistance 3.3 kohms
   - Bank 1 Sensor 2, resistance 6.0 kohms
   - Bank 2 Sensor 2, resistance 6.0 kohms
7. Replace the sensor, if there is a malfunction.
8. Apply battery voltage directly between terminal 3 and 4.

➡ **Be careful when applying the voltage. Damage will result if terminals 1 and 2 are connected to any voltage.**

9. Connect a voltmeter between terminal 1 and 2.
10. While racing the engine repeatedly, measure the output voltage.
11. Specification should be 0.6 volt (minimum).
12. Replace the sensor, as required.

2005–2006

SENSOR ONE

Perform a visual inspection of the sensor as follows:

1. If the sensor tip has a black/sooty deposit, this may indicate a rich fuel mixture.
2. If the sensor tip has a white, gritty deposit, this may indicate an internal coolant leak.
3. If the sensor tip has a brown deposit, this could indicate excessive oil consumption.
4. Warm the engine until the engine reaches operating temperature.
5. Connect the scan tool and monitor sensor operation.
6. Verify that the signal is switching from rich (above 0.45 volt) to lean (below 0.45 volt) a minimum of 3 times in 10 seconds (voltage will vary between 0.1–0.9 volt).
7. If within specification, check the PCM as follows, ignition OFF. Connect the scan tool. Turn the ignition ON. Select simulation function on the scan tool. Simulate voltage at terminal 1 of the sensor connector.

➡ **Never simulate voltage over 1 volt, because output voltage range of the sensor is 0.1–0.9 volt.**

8. If the signal voltage value does not change, substitute a known good PCM and repeat the test. Replace the PCM as required. If the signal voltage value does change, check connectors for looseness, poor connection, binding etc. Correct and retest.
9. If the signal readings are not within specification, check the sensor for contamination, deterioration or damage.
10. Substitute the sensor with a known good component, check for proper operation. If problem corrected replace the sensor.

SENSOR TWO

Perform a visual inspection of the sensor as follows:

1. If the sensor tip has a black/sooty deposit, this may indicate a rich fuel mixture.
2. If the sensor tip has a white, gritty deposit, this may indicate an internal coolant leak.
3. If the sensor tip has a brown deposit, this could indicate excessive oil consumption.
4. If the sensor is physically okay, check the PCM as follows, ignition OFF. Connect the scan tool. Turn the ignition ON. Select simulation function on the scan tool. Simulate voltage at terminal 1 of the sensor connector.

➡ **Never simulate voltage over 1 volt, because output voltage range of the sensor is 0.1–0.9 volt.**

5. If the signal voltage value does not change, substitute a known good PCM and repeat the test. Replace the PCM as required. If the signal voltage value does change, check connectors for looseness, poor connection, binding etc. Correct and retest.
6. Using an ohmmeter, measure the resistance between terminals 3 and 4 (component side).
7. Specification should be 7–40 ohms at 68 degrees F.
8. If not, replace the sensor.

Intake Air Temperature (IAT) Sensor

LOCATION

This sensor is located in the air intake plenum assembly. This sensor is combined with the MAF sensor.

OPERATION

This sensor is a resistor based sensor which detects the intake air temperature. According to the intake air temperature reading the ECM/PCM will control the necessary amount of fuel injection.

REMOVAL & INSTALLATION

1. Disconnect the negative battery cable.
2. Disconnect the connector from the sensor.
3. Remove the sensor retaining screws, as required.
4. Remove the air cleaner and air intake assembly, as required.
5. Remove the sensor from its mounting.

To install:

6. Installation is the reverse of the removal procedure.

TESTING

1. Be sure the ignition is OFF.
2. Remove the sensor connector.
3. Using an ohmmeter measure the resistance between terminal 5 and 3 (component side).
4. On 2004 vehicles specification should be as follows:
   - Intake air temperature -40 degrees, resistance 33.85–61.20 kohms
   - Intake air temperature 68 degrees, resistance 2.33–2.97 kohms
   - Intake air temperature 176 degrees, resistance 0.31–0.43 kohms
5. On 2005–2006 vehicles specification should be as follows:
   - Intake air temperature -40 degrees, resistance 45.3 kohms
   - Intake air temperature 68 degrees, resistance 2.50 kohms
   - Intake air temperature 176 degrees, resistance 0.33 kohms
6. Replace the sensor, as required.

Knock Sensor (KS)

OPERATION

The knock sensor is used to detect engine vibrations caused by preignition or detonation and provides information to the ECM, which then retards the timing to eliminate detonation.

REMOVAL & INSTALLATION

1. Disconnect the negative battery cable.
2. Remove the necessary components to gain access to the sensor.
3. Disconnect the sensor connector.
4. Remove the sensor from its mounting.

To install:

5. Installation is the reverse of the removal procedure.
6. Tighten the sensor to 11–18 ft. lbs.

TESTING

➡ **If the sensor is suspected of being defective, it should be replaced with a known good component for testing purposes.**

1. Check the sensor torque. It should be 11–18 ft. lbs.
2. If the sensor is still not functioning, replace it with a known good component.
3. Recheck the sensor.

Manifold Absolute Pressure (MAP) Sensor

OPERATION

This sensor measures the change of pressure in the intake manifold. This sensor measures the changes in the intake manifold pressure which result from engine load and speed changes, and converts to a voltage output that is monitored by the PCM.

REMOVAL & INSTALLATION

1. Disconnect the negative battery cable.
2. Disconnect the connector from the sensor.
3. Remove the sensor retaining screws.
4. Remove the sensor from its mounting.

To install:

5. Installation is the reverse of the removal procedure.

TESTING

2004

1. Turn the ignition switch ON.
2. Using a voltmeter, measure the voltage between terminal 1 (sensor output) and terminal 4 (sensor ground).
3. Specification should be 4–5 volts. At idle specification should be 0.8–2.4 volts.
4. If not within specification, replace the sensor.

2005–2006

1. Connect the scan tool.
2. Be sure the ignition is OFF.
3. Remove the MAP sensor and connect MAP sensor connector.
4. Turn the ignition ON, engine OFF.
5. Using a hand held vacuum pump, apply vacuum and measure the "MAP SENSOR" parameter on the scan tool.
6. Specification should be as follows:
   - Applied vacuum 1.45 psi, output voltage 0.06 volt
   - Applied vacuum 2.18 psi, output voltage 0.25 volt
   - Applied vacuum 5.8 psi, output voltage 1.3 volt
   - Applied vacuum 11.6 psi, output voltage 3.0 volts
   - Applied vacuum 17.4 psi, output voltage 4.8 volts
7. If not within specification substitute a known good sensor and recheck for proper operation. Replace defective components as required.
8. If within specification, check the PCM as follows, ignition OFF. Disconnect the sensor connector. Connect the scan tool. Turn the ignition ON and engine OFF. Select simulation function on the scan tool. Simulate voltage at terminal 1 of the sensor connector.
9. If the signal voltage value does not change, substitute a known good PCM and repeat the test. Replace the PCM as required. If the signal voltage value does change, check connectors for looseness, poor connection, binding etc. Correct and retest.

Mass Air Flow (MAF) Sensor

LOCATION

This sensor is located in the air intake plenum assembly. This sensor is combined with the IAT sensor.

OPERATION

The MAF is measured by detection of heat transfer from a hot film probe because the change of the mass air flow rate causes change in the amount of heat being transferred from the hot film probe surface to the air flow. The air flow sensor generates a pulse so it repeatedly opens and closes between the voltages supplied from the PCM. This results in a change of the temperature of the hot film probe and in the change of resistance.

REMOVAL & INSTALLATION

1. Disconnect the negative battery cable.
2. Disconnect the connector from the sensor.
3. Remove the air cleaner and air intake assembly, as required.
4. Remove the sensor from its mounting.

To install:

5. Installation is the reverse of the removal procedure.

TESTING

2004

1. Using a voltmeter check the sensor output voltage.
2. Specification should be as follows:
   - Intake air quantity 15 kg/h, output voltage 1.34 volts
   - Intake air quantity 30 kg/h, output voltage 1.64 volts
   - Intake air quantity 60 kg/h, output voltage 2.07 volts
   - Intake air quantity 120 kg/h, output voltage 2.61 volts
   - Intake air quantity 250 kg/h, output voltage 3.28 volts
   - Intake air quantity 370 kg/h, output voltage 3.68 volts
   - Intake air quantity 480 kg/h, output voltage 3.96 volts
   - Intake air quantity 640 kg/h, output voltage 4.28 volts
3. If not within specification, replace the sensor.
4. If within specification, erase the DTC code. If code is reset replace the ECM with a known good component and recheck. Replace the ECM, as required.

2005–2006

1. Connect the scan tool.
2. Select "MAF SENSOR" parameter on the CURRENT DATA.

3. Monitor the signal waveform of the "MAF SENSOR" on the scan tool.

4. Specification, check that the value is increased correspondently with TPS when pressing the accelerator pedal.

5. If not within specification substitute a known good sensor and recheck for proper operation. Replace defective components as required.

6. If within specification, check the PCM as follows, ignition OFF. Connect the scan tool. Turn the ignition ON and engine OFF. Select simulation function on the scan tool. Simulate voltage at terminal 1 of the sensor connector.

7. If the signal voltage value does not change, substitute a known good PCM and repeat the test. Replace the PCM as required. If the signal voltage value does change, check connectors for looseness, poor connection, binding etc. Correct and retest.

OPTIMA

See Figures 2 and 3.

Camshaft Position (CMP) Sensor

OPERATION

The camshaft position sensor senses the TDC point of the number one cylinder, on its compression stroke. Its signal is relayed to the ECM to be used to determine the sequence of fuel injection.

REMOVAL & INSTALLATION

1. Disconnect the negative battery cable.
2. Disconnect the connector from the sensor.
3. Remove the bolt that retains the sensor.
4. Remove the sensor.

To install:

5. Installation is the reverse of the removal procedure.

TESTING

2001–2003

1. Connect the scan tool to the data link connector.
2. Start the engine. Allow the engine to reach operating temperature.
3. Monitor the CMPS signal. It should continuously fluctuate between 0–5 volts.
4. If abnormality is found, replace the sensor.

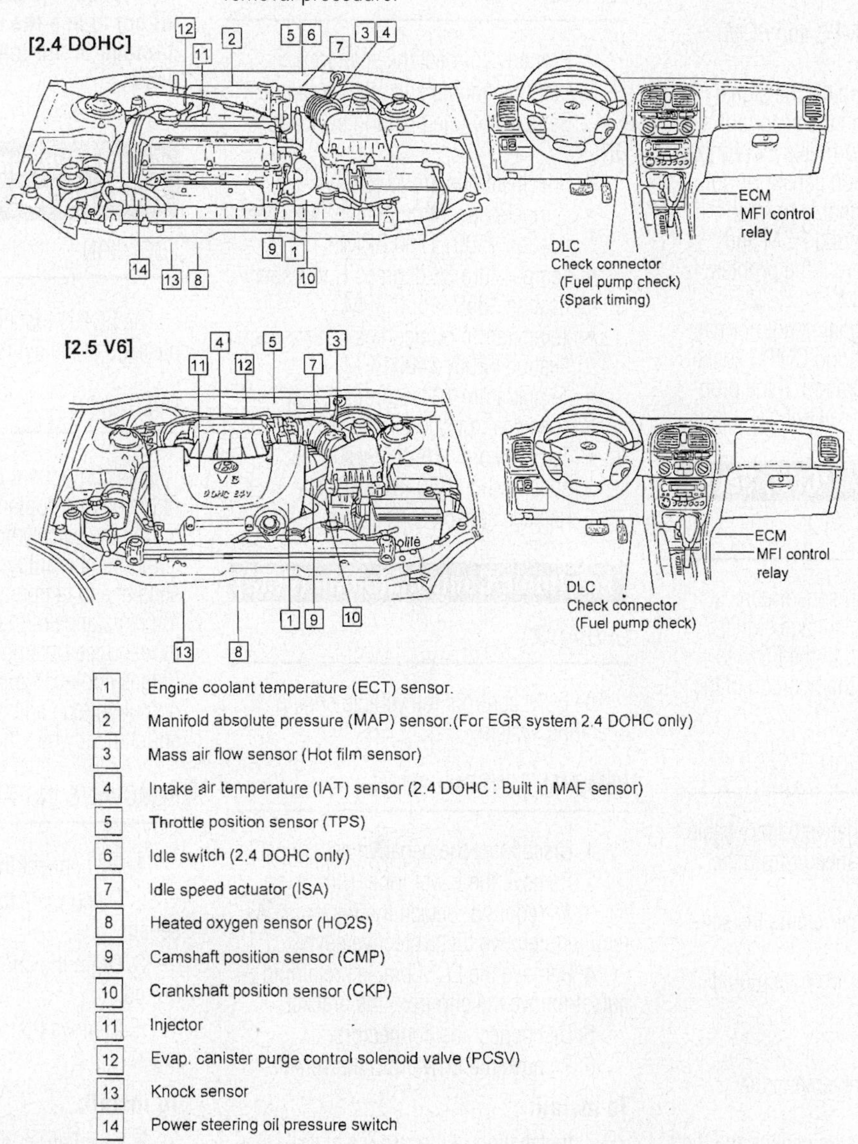

Fig. 2 Underhood sensor locations—Optima 2001–2006 2.4L (MS) engine and 2001 2.5L engine

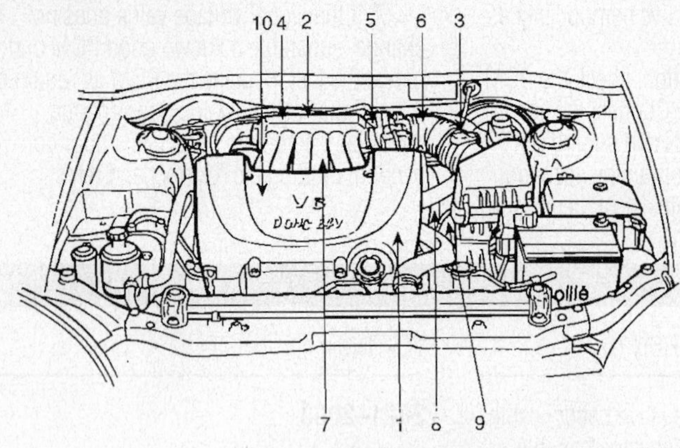

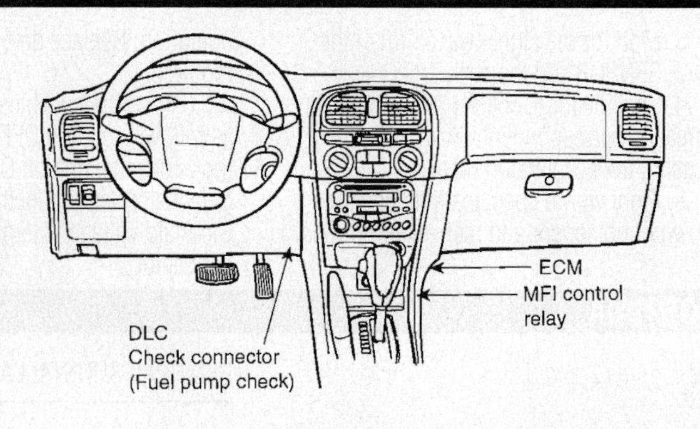

1. Engine Coolant Temperature Sensor (ECTS)
2. Purge Control Solenoid Valve (PCSV)
3. Mass Air Flow (MAF) sensor
4. Intake Air Temperature Sensor (ATS)
5. Throttle Position Sensor (TPS)
6. Idle Speed Control Actuator (ISA)
7. Heated Oxygen Sensor (HO2S)
8. Camshaft Position (TDC) Sensor
9. Crankshaft Position (CKP) Sensor
10. Injector

29130_HYUN_G0013

Fig. 3 Underhood sensor locations—Optima 2002–2006 2.7L (MS) engine

2004–2006

1. Be sure that the CMPS and PCM connectors are connected.
2. Engine ON and monitor the signal waveform of the sensor on the scan tool. Check whether the waveform is synchronized with the crankshaft position sensor or not.
3. If the waveform signal is normal, substitute with a known good PCM and check for proper operation. If the problem is corrected replace the PCM.
4. If the waveform signal is not normal, substitute with a known good CMPS sensor and check for proper operation. If the problem is corrected replace the CMPS.

Crankshaft Position (CKP) Sensor

OPERATION

The crankshaft position sensor consists of a magnet and coil. The voltage signal from the sensor is relayed to the ECM to indicate engine rpm and the position of the crankshaft.

REMOVAL & INSTALLATION

1. Disconnect the negative battery cable.
2. Disconnect the connector from the sensor.
3. Remove the bolt that retains the sensor in place.
4. Remove the sensor from its mounting.

To install:

5. Installation is the reverse of the removal procedure.
6. Clearance between the sensor and the sensor wheel should be 0.020–0.059 inch.

TESTING

1. Properly connect the scan tool.
2. Start the engine and allow it to idle.
3. Be sure the idle position switch is ON.
4. Specifications are as follows:
   - Temperature - 4 degrees F, test specification 1500–1700 RPM
   - Temperature 32 degrees F, test specification 1350–1550 RPM
   - Temperature 68 degrees F, test specification 1200–1400 RPM
   - Temperature 104 degrees F, test specification 1000–1200 RPM
   - Temperature 176 degrees F, test specification 650–850 RPM
5. Replace sensor, as required.

Electronic Control Module (ECM)

OPERATION

The ECM controls the vehicle engine operating system.

REMOVAL & INSTALLATION

1. Disconnect the negative battery cable.
2. Remove the lower inner trim.
3. As required, detach the floor mat. As required, remove the protective cover.
4. Remove the ECM bracket retaining nuts. Remove the clip from the bracket.
5. Disconnect the connectors.
6. Remove the ECM from the vehicle.

To install:

7. Installation is the reverse of the removal procedure.

➡ When replacing the ECM, be careful not to use the wrong part number, as damage to the injection system could occur.

Engine Coolant Temperature (ECT) Sensor

LOCATION

The ECT is installed in the engine coolant passage of the cylinder head.

OPERATION

This component detects the temperature of the engine coolant and relays the information to the electronic control assembly. This component employs a thermistor which is sensitive to temperature changes. The electric resistance of the thermistor decreases in response to temperature rise. The ECM judges coolant temperature by the sensor output voltage and provides optimum fuel enrichment when the engine is cold.

REMOVAL & INSTALLATION

1. Disconnect the negative battery cable.
2. Disconnect the connector from the sensor.
3. Drain the cooling system, as required.
4. Remove the sensor from its mounting.

To install:

5. Installation is the reverse of the removal procedure.

TESTING

1. Remove the sensor.

2. With the temperature sensing portion of the sensor immersed in hot engine coolant, check the resistance, using an ohmmeter.

3. On 2001–2003 vehicles specification should be as follows:

- Temperature 68 degrees F, resistance 2.31–2.59 kohms
- Temperature 176 degrees F, resistance 0.31–0.34 kohms

4. On 2004–2006 vehicles specification should be as follows:

- Temperature 68 degrees F, resistance 2.31–2.59 kohms
- Temperature 176 degrees F, resistance 0.32 kohms

Heated Oxygen (HO2S) Sensor

LOCATION

The sensors are located in the exhaust system. On some vehicles one sensor is located up at the exhaust manifold(s) and the other sensor is located down at the catalytic converter. On other vehicles both sensors are located down at the catalytic converter.

OPERATION

The exhaust gas oxygen sensor supplies the electronic control assembly with a signal which indicates either a rich or lean mixture condition, during the engine operation.

REMOVAL & INSTALLATION

1. Disconnect the electrical connector from the sensor.

2. Remove the oxygen sensor.

To install:

3. Installation is the reverse of the removal procedure.

➡ **Apply anti-seize compound to the threaded portion of the sensor, prior to installation. Never apply anti-seize compound to the protector of the sensor.**

TESTING

2001–2003

Perform a visual inspection of the sensor as follows:

1. If the sensor tip has a black/sooty deposit, this may indicate a rich fuel mixture.

2. If the sensor tip has a white, gritty deposit, this may indicate an internal coolant leak.

3. If the sensor tip has a brown deposit, this could indicate excessive oil consumption.

4. Warm the engine until the coolant temperature reaches 176–205 degrees F.

5. Accelerate the engine to 4000 rpm. When decelerating suddenly from 4000 rpm, the voltmeter reading should be 200mV or lower.

6. When the engine is suddenly raced, the voltmeter reading should be 600–1000mV.

7. If measured value is not within specification, replace the sensor.

8. Turn the ignition switch to the OFF position.

9. Disconnect the sensor connector.

10. Measure the resistance between terminals 3 and 4 (component side).

11. On 2001 vehicles specification should be 30 ohms or more at 752 degrees F.

12. On 2002 vehicles, specification should be 3.0–6.2 ohms at 68 degrees F.

13. On 2003 vehicles, specification should be 3.0–6.2 ohms at 68 degrees F. if equipped with a 2.4L engine and 4.0–5.2 ohms if equipped with a 2.7L engine.

14. If not, replace the sensor.

15. Apply battery voltage directly between terminal 3 and 4.

➡ **Take care when applying the voltage. Damage will result if the terminals are incorrect or short circuited.**

16. Connect a digital voltmeter between terminal 1 and 2.

17. While racing the engine, measure the sensor output voltage. Specification should be 0.6 volt (minimum) on 2.4L engine and 4.5 volts on 2.5L and 2.7L engines.

18. If not, replace the sensor.

2004–2006

Perform a visual inspection of the sensor as follows:

1. If the sensor tip has a black/sooty deposit, this may indicate a rich fuel mixture.

2. If the sensor tip has a white, gritty deposit, this may indicate an internal coolant leak.

3. If the sensor tip has a brown deposit, this could indicate excessive oil consumption.

4. Warm the engine until the coolant temperature reaches 176–205 degrees F.

5. Accelerate the engine to 4000 rpm. When decelerating suddenly from 4000 rpm, the voltmeter reading should be 200mV or lower.

6. When the engine is suddenly raced, the voltmeter reading should be 600–1000mV.

7. If measured value is not within specification, replace the sensor.

8. Disconnect the sensor connector.

9. Measure the resistance between terminals 3 and 4 (component side).

10. On the 2.4L engine, specification should be 5.0–7.0 ohms.

11. On the 2.7L engine, specification should be 4.0–5.2 ohms.

12. If not, replace the sensor.

13. Apply battery voltage directly between terminal 3 and 4.

➡ **Take care when applying the voltage. Damage will result if the terminals are incorrect or short circuited.**

14. Connect a digital voltmeter between terminal 1 and 2.

15. While racing the engine, measure the sensor output voltage. Specification should be 0.6 volt (minimum).

16. If not, replace the sensor.

Idle Speed Control Actuator (ISCA)

OPERATION

The idle speed control actuator sensor is a double coil type driven by separate driver stages in the ECM. Depending on the pulse duty factor, the equilibrium of the magnetic forces of the two coils will result in different angles of the motor. A bypass hose line is positioned, parallel to the throttle valve where the idle speed actuator is inserted.

REMOVAL & INSTALLATION

1. Disconnect the negative battery cable.

2. Disconnect the connector from the sensor.

3. Remove the sensor retaining screws.

4. Remove the sensor from its mounting.

To install:

5. Installation is the reverse of the removal procedure.

TESTING

2001

1. Disconnect the connector at the sensor. Connect an ohmmeter.

2. Measure the resistance between terminal 1 and 2, specification should be 10.5–14 ohms at 68 degrees F.

3. Measure the resistance between terminal 3 and 2, specification should be 17.0–18.2 ohms.
4. If the measured value is not within specification, replace the sensor.

2002–2003

1. Disconnect the connector at the sensor. Connect an ohmmeter.
2. Measure the resistance between terminal 1 and 2, specification should be 17.0–18.2 ohms at 68 degrees F.
3. Measure the resistance between terminal 2 and 3, specification should be 15.0–16.0 ohms at 68 degrees F.
4. If the measured value is not within specification, replace the sensor.

2004–2006

1. Disconnect the connector at the sensor. Connect an ohmmeter.
2. Measure the resistance between terminal 3 and 2 (open), specification should be 14.5–16.5 ohms.
3. Measure the resistance between terminal 1 and 2 (closing), specification should be 16.6–18.6 ohms at 68 degrees F.
4. If the measured value is not within specification, replace the sensor.

Intake Air Temperature (IAT) Sensor

LOCATION

The MAF is mounted in the intake air hose of the air cleaner assembly. On some vehicles this sensor is combined with the IAT sensor.

OPERATION

This sensor is a resistor based sensor which detects the intake air temperature. According to the intake air temperature reading the ECM will control the necessary amount of fuel injection.

REMOVAL & INSTALLATION

1. Disconnect the negative battery cable.
2. Disconnect the connector from the sensor.
3. Remove the sensor retaining screws, as required.
4. Remove the air cleaner and air intake assembly, as required.
5. Remove the sensor from its mounting.

To install:

6. Installation is the reverse of the removal procedure.

TESTING

1. Connect a multimeter.
2. Disconnect the connector.
3. Be sure the ignition switch is in the ON position.
4. Measure the voltage between the sensor terminal 1 and 2.
5. On 2001–2003 vehicles, specification should be 2.22–2.82 kohms at 68 degrees F. and 0.29–0.36 kohms at 176 degrees F.
6. On 2004–2006 vehicles equipped with the 2.4L engine, specification should be 2.33–2.97 kohms at 68 degrees F. and 0.31–0.43 kohms at 176 degrees F.
7. On 2004–2006 vehicles equipped with the 2.7L engine, specification should be 2.22–2.82 kohms at 68 degrees F. and 0.30–0.36 kohms at 176 degrees F.
8. Replace the sensor, as required.

Knock Sensor (KS)

OPERATION

The knock sensor is used to detect engine vibrations caused by preignition or detonation and provides information to the ECM, which then retards the timing to eliminate detonation.

REMOVAL & INSTALLATION

1. Disconnect the negative battery cable.
2. Disconnect the sensor connector.
3. Remove the sensor from its mounting.

To install:

4. Installation is the reverse of the removal procedure.
5. Tighten the sensor to 12–18 ft. lbs.

TESTING

➡ **If the sensor is suspected of being defective, it should be replaced with a known good component for testing purposes.**

1. Check the sensor torque. It should be 12–18 ft. lbs.
2. If the sensor is still not functioning, replace it with a known good component.
3. Recheck the sensor.

Mass Air Flow (MAF) Sensor

LOCATION

The MAF is mounted in the intake air hose of the air cleaner assembly. On some vehicles this sensor is combined with the IAT sensor.

OPERATION

The MAF is measured by detection of heat transfer from a hot film probe because the change of the mass air flow rate causes change in the amount of heat being transferred from the hot film probe surface to the air flow. The air flow sensor generates a pulse so it repeatedly opens and closes between the voltage (5V) supplied from the ECM. This results in a change of the temperature of the hot film probe and in the change of resistance.

REMOVAL & INSTALLATION

1. Disconnect the negative battery cable.
2. Disconnect the connector from the sensor.
3. Remove the air cleaner and air intake assembly, as required.
4. Remove the sensor from its mounting.

To install:

5. Installation is the reverse of the removal procedure.

TESTING

2001–2003

1. Using a voltmeter check the output voltage of the sensor.
2. On 2001 2.4L and 2.5L engines specification should be 0.5 volt at idle and 1.0 volt at 2000 rpm.
3. On 2002–2003 vehicles with 2.4L engine specification should be 1.2–1.6 volts at idle.
4. On 2002–2003 vehicles with 274L engine specification should be 0.6–1.0 volt at idle.
5. Replace the sensor, as required.

2004–2006

1. Using a voltmeter check the sensor output voltage.
2. On 2.4L engine specification should be as follows:
   - Intake air quantity 15 kg/h, output voltage 1.40 volts
   - Intake air quantity 30 kg/h, output voltage 1.77 volts
   - Intake air quantity 60 kg/h, output voltage 2.24 volts
   - Intake air quantity 120 kg/h, output voltage 2.82 volts
   - Intake air quantity 250 kg/h, output voltage 3.54 volts
   - Intake air quantity 370 kg/h, output voltage 3.97 volts
   - Intake air quantity 480 kg/h, output voltage 4.28 volts

3. On 2.7L engine specification should be as follows:

- Intake air quantity 7.34 kg/h, output voltage 0.3 volt
- Intake air quantity 10.06 kg/h, output voltage 0.5 volt
- Intake air quantity 19.85 kg/h, output voltage 1.0 volt
- Intake air quantity 35.58 kg/h, output voltage 1.50 volts
- Intake air quantity 58.79 kg/h, output voltage 2.0 volts
- Intake air quantity 94.70 kg/h, output voltage 2.5 volts
- Intake air quantity 149.07 kg/h, output voltage 3.0 volts
- Intake air quantity 226.66 kg/h, output voltage 3.5 volts
- Intake air quantity 335.66 kg/h, output voltage 4.0 volts
- Intake air quantity 500.16 kg/h, output voltage 4.5 volts
- Intake air quantity 614.61 kg/h, output voltage 4.76 volts
- Intake air quantity 730.01 kg/h, output voltage 5.0 volts

4. If not within specification, replace the sensor.

5. If within specification, erase the DTC code. If code is reset replace the ECM with a known good component and recheck. Replace the ECM, as required.

Manifold Absolute Pressure (MAP) Sensor

OPERATION

This sensor converts the pressure in the intake manifold to voltage signal The ECM judges the condition of the EGR by this signal.

REMOVAL & INSTALLATION

1. Disconnect the negative battery cable.
2. Disconnect the connector from the sensor.
3. Remove the sensor retaining screws.
4. Remove the sensor from its mounting.

To install:

5. Installation is the reverse of the removal procedure.

TESTING

2.4L and 2.7L Engines

1. Using a voltmeter, measure the voltage between terminal 1 (sensor ground) and terminal 4 (sensor output), 2002–2003 vehicles.
2. Using a voltmeter, measure the voltage between terminal 4 (sensor ground) and terminal 1 (sensor output), 2001 vehicles and 2004–2006 vehicles.
3. Specification should be 4–5 volts, with the ignition switch ON.
4. Specification should be 0.8–2.4 volts at idle.
5. If not within specification, replace the sensor.

2.5L Engine

1. Using a voltmeter, measure the voltage between terminal 4 (sensor ground) and terminal 1 (sensor output).
2. Specification should be 4–5 volts, with the ignition switch ON.
3. Specification should be 0.8–2.4 volts at idle.
4. If not within specification, replace the sensor.

Throttle Position Sensor (TPS)

OPERATION

The throttle position sensor is a rotating type variable resistor that rotates with the throttle body's throttle shaft to sense the throttle valve angle. As the throttle shaft rotates, the throttle angle of the sensor changes and the ECM detects the throttle valve opening based on the TPS output voltage.

REMOVAL & INSTALLATION

1. Disconnect the negative battery cable.
2. Disconnect the sensor connector.
3. Remove the sensor retaining screws. Remove the sensor from its mounting.

To install:

4. Installation is the reverse of the removal procedure.

TESTING

2001

1. Disconnect the connector from the sensor. Connect an ohmmeter.
2. Measure the resistance between terminals 1 (sensor ground) and 2 (sensor power)
3. Specification should be 3.5–6.5 kohm for the 2.4L engine and 1.6–2.4 kohm for the 2.5L engine.
4. Connect an analog ohmmeter between terminals 1 (sensor ground) and 3 (sensor output), on 2.4L engine and between terminals 2 (sensor ground) and 3 (sensor output), on 2.5L engine.
5. Operate the throttle valve slowly from the idle position to the full open position, and check that the resistance changes smoothly in proportion with the throttle valve opening angle.
6. If the resistance is out of specification, or fails to change smoothly, replace the sensor.

2002–2003

1. Disconnect the connector from the sensor. Connect an ohmmeter.
2. Measure the resistance between terminals 3 (sensor ground) and 2 (sensor power) on 2.4L engine and between terminals 2 (sensor ground) and 3 (sensor power) on 2.7L engine.
3. Specification should be 3.5–6.5 kohm for the 2.4L engine and 1.6–2.4 kohm for the 2.7L engine.
4. Connect an analog ohmmeter between terminals 3 (sensor ground) and 1 (sensor output), on 2.4L engine and between terminals 2 (sensor ground) and 1 (sensor output), on 2.7L engine.
5. Operate the throttle valve slowly from the idle position to the full open position, and check that the resistance changes smoothly in proportion with the throttle valve opening angle.
6. If the resistance is out of specification, or fails to change smoothly, replace the sensor.

2004–2006

1. Disconnect the connector from the sensor. Connect an ohmmeter.
2. Measure the resistance between terminals 1 (sensor ground) and 2 (sensor power) on 2.4L engine and between terminals 2 (sensor ground) and 3 (sensor power) on 2.7L engine.
3. Specification should be 1.6–2.4 kohm.
4. Connect an analog ohmmeter between terminals 1 (sensor ground) and 2 (sensor output), on 2.4L engine and between terminals 2 (sensor ground) and 1 (sensor output), on 2.7L engine.
5. Operate the throttle valve slowly from the idle position to the full open position, and check that the resistance changes smoothly in proportion with the throttle valve opening angle.
6. If the resistance is out of specification, or fails to change smoothly, replace the sensor.

RIO

See Figure 4.

Camshaft Position (CMP) Sensor

LOCATION

On 2001–2005 vehicles the camshaft position sensor is mounted at the rear of the cylinder head. On 2006 vehicles, the camshaft position sensor is located on top of the engine, on the engine head cover.

OPERATION

On 2001–2005 vehicles, the sensor sends a signal to the ECM when the number one cylinder is at TDC. The sensor consists of a Hall Effect device and magnet. When the magnetic field of the Hall Effect device is interrupted by the half moon shape of the camshaft target wheel, the sensor voltage outputs 0 volts. If not, the sensor voltage outputs 5 volts. On 2006 vehicles the camshaft position sensor senses the TDC point of the number one cylinder, on its compression stroke. Its signal is relayed to the ECM to be used to determine the sequence of fuel injection.

REMOVAL & INSTALLATION

1. Disconnect the negative battery cable.
2. Disconnect the connector from the sensor.
3. Remove the bolt that retains the sensor.
4. Remove the sensor.

To install:

5. Installation is the reverse of the removal procedure.

TESTING

2001–2005

1. Disconnect the sensor and check half-moon installation.
2. If the half-moon wheel is installed properly, install a known good sensor and recheck. Replace the sensor, as required.
3. If the half-moon wheel is not installed properly, remove the camshaft and install the half-moon wheel correctly.

2006

1. Connect a scan tool.
2. Be sure the ignition is ON. Do not disconnect the sensors.
3. Select "VEHICLE SCOPEMETER" in the menu, and connect channel A of the scan tool with terminal 2 of the sensor harness connector.

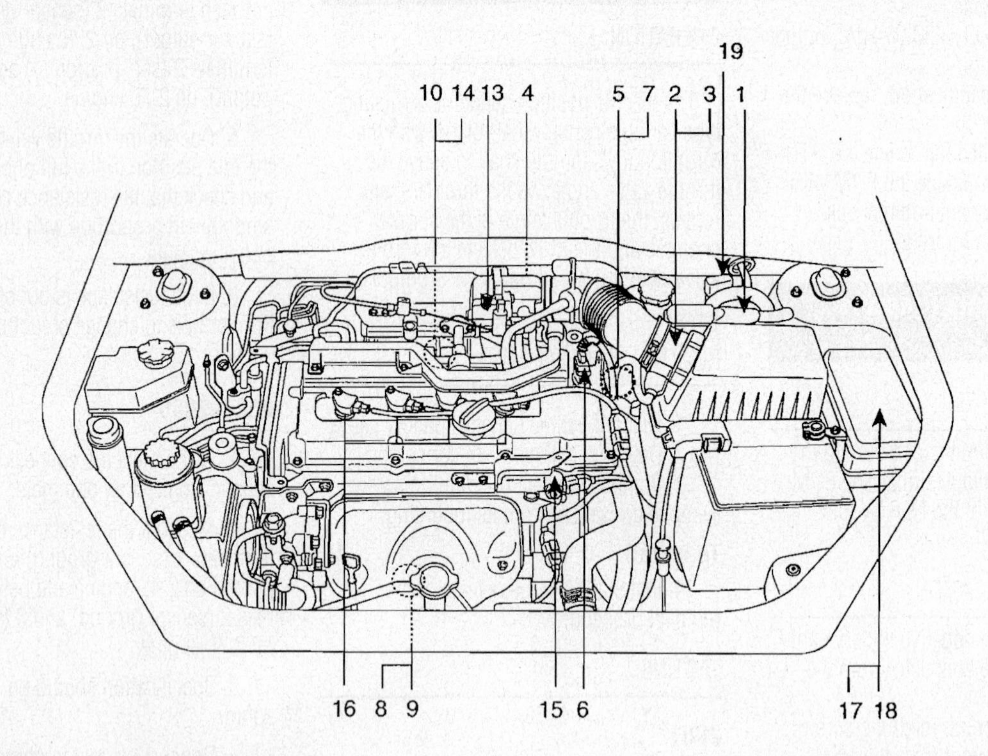

1. ECM (for M/T) / PCM (for A/T)
2. Mass Air Flow Sensor (MAFS)
3. Intake Air Temperature Sensor (IATS)
4. Throttle Position Sensor (TPS)
5. Engine Coolant Temperature Sensor (ECTS)
6. Camshaft Position Sensor (CMPS)
7. Crankshaft Position Sensor (CKPS)
8. Heated Oxygen Sensor (HO2S) [Bank 1/Sensor 1]
9. Heated Oxygen Sensor (HO2S) [Bank 1/Sensor 2]
10. Knock Sensor (KS)
13. Idle Speed Control Actuator (ISCA)
14. Purge Control Solenoid Valve (PCSV)
15. CVVT Oil Control Valve (OCV)
16. Ignition Coil
17. Main Relay
18. Fuel Pump Relay
19. Multi Purpose Check Connector (20 pin)

Fig. 4 Underhood sensor locations—Rio 2006

29130_KIA_G0002

4. Start the engine. Check the signal waveforms.

5. If they are within specification, substitute a known good ECM/PCM and recheck. Replace the component as required.

6. If not within specification, substitute a known good sensor and recheck. Replace the component as required.

Crankshaft Position (CKP) Sensor

LOCATION

On 2001–2005 vehicles the crankshaft sensor is mounted on the front of the transmission case adjacent to the flywheel. On 2006 vehicles the crankshaft position sensor is located next to the radiator hose inlet and outlet assembly.

OPERATION

On 2001–2005 vehicles the sensor is a Hall Effect type sensor that generates voltage using a sensor and a target wheel mounted on the crankshaft. There are 58 slots and one longer slot on the target wheel. When the slot in the wheel aligns with the sensor, the sensor voltage output is low (approximately 0 volts). It is high (approximately 5 volts) when a slot is not aligned with the sensor. The ECM calculates the frequency to compute engine rpm to determine the injection duration and ignition timing. On 2006 vehicles the crankshaft position sensor consists of a magnet and coil. The voltage signal from the sensor is relayed to the ECM to indicate engine rpm and the position of the crankshaft.

REMOVAL & INSTALLATION

1. Disconnect the negative battery cable.
2. Disconnect the connector from the sensor.
3. Remove the bolt that retains the sensor in place.
4. Remove the sensor from its mounting.

To install:

5. Installation is the reverse of the removal procedure.

TESTING

2001–2005

1. Measure the gap between the sensor and the target wheel.
2. Specification should be 0.012–0.067 inch.
3. If the gap is within specification, install a known good sensor and recheck. Replace the sensor, as required.
4. If the gap is not within specification, check the target wheel for proper installation. Adjust the target wheel gap.

2006

1. Connect a scan tool.
2. Be sure the ignition is ON. Do not disconnect the sensors.
3. Select "VEHICLE SCOPEMETER" in the menu, and connect channel A of the scan tool with terminal 1 of the sensor harness connector.
4. Start the engine. Check the signal waveforms.
5. If they are within specification, substitute a known good ECM/PCM and recheck. Replace the component as required.
6. If not within specification, substitute a known good sensor and recheck. Replace the component as required.

Electronic Control Module (ECM)

OPERATION

The ECM controls the vehicle engine operating system.

REMOVAL & INSTALLATION

1. Disconnect the negative battery cable.
2. Remove the lower inner trim..
3. As required, detach the floor mat. As required, remove the protective cover.
4. Remove the ECM bracket retaining nuts. Remove the clip from the bracket.
5. Disconnect the connectors.
6. Remove the ECM from the vehicle.

To install:

7. Installation is the reverse of the removal procedure.

➡ **When replacing the ECM, be careful not to use the wrong part number, as damage to the injection system could occur.**

Engine Coolant Temperature (ECT) Sensor

LOCATION

On 2001–2005 vehicles, the ECT is installed in an engine coolant passage, usually in the cylinder head. On 2006 vehicles this sensor is located near the thermostat housing of the cylinder head.

OPERATION

This component detects the temperature of the engine coolant and relays the information to the electronic control assembly. The ECM judges coolant temperature by the sensor output voltage and provides optimum fuel enrichment when the engine is cold.

REMOVAL & INSTALLATION

1. Disconnect the negative battery cable.
2. Disconnect the connector from the sensor.
3. Drain the cooling system, as required.
4. Remove the sensor from its mounting.

To install:

5. Installation is the reverse of the removal procedure.

TESTING

2001–2002

1. Connect an ohmmeter.
2. Measure the resistance between sensor connector B11, terminal 1 and 2.
3. Specifications are as follows:
   - Temperature - 4 degrees F, resistance 15.04 (plus 1.79, minus 1.60) kohm
   - Temperature 68 degrees F, resistance 2.45 (plus 0.19, minus 0.18) kohm
   - Temperature 176 degrees F, resistance 0.318 (plus 0.011, minus 0.011) kohm
4. If within specification, check poor terminal contact, bent or misplaced terminal. Correct as required.
5. If not within specification, install a known good sensor and recheck. Replace the sensor, as required.

2003–2005

1. Turn the ignition ON.
2. Using a voltmeter, measure the voltage of the ECT sensor signal circuit between the ECT sensor harness connector and ground.
3. Specification should be below 0.5 volt.
4. If within specification, install a known good sensor and recheck. Replace the sensor, as required.
5. If not within specification, check for short circuit between ETC sensor harness connector and ECM harness connector. Repair as required.

2006

1. Be sure the ignition switch is OFF.
2. Disconnect the sensor electrical connector.
3. Measure the resistance between terminal 1 and 3 of the harness connector (component side).
4. Specifications are as follows:
   - Temperature - 4 degrees F, resistance 14.13–16.83 kohm
   - Temperature 32 degrees F, resistance 5.79 kohm
   - Temperature 68 degrees F, resistance 2.31–2.59 kohm
   - Temperature 176 degrees F, resistance 0.32 kohm
5. If not within specification, install a known good sensor and recheck. Replace the sensor, as required.
6. If within specification, install a known good ECM/PCM and recheck. Replace the component, as required.

Heated Oxygen (HO2S) Sensor

LOCATION

The sensors are located in the exhaust system. On some vehicles one sensor is located up at the exhaust manifold and the other sensor is located down at the catalytic converter. On other vehicles both sensors are located down at the catalytic converter.

OPERATION

The exhaust gas oxygen sensor supplies the electronic control assembly with a signal which indicates either a rich or lean mixture condition, during the engine operation.

REMOVAL & INSTALLATION

1. Disconnect the electrical connector from the sensor.
2. Remove the oxygen sensor.

To install:

3. Installation is the reverse of the removal procedure.

➡ Apply anti-seize compound to the threaded portion of the sensor, prior to installation. Never apply anti-seize compound to the protector of the sensor.

4. Tighten the sensor to 22–36 ft. lbs for 2001–2005 vehicles and to 37–44 ft. lbs. for 2006 vehicles.

TESTING

2001–2005

Perform a visual inspection of the sensor as follows:

1. If the sensor tip has a black/sooty deposit, this may indicate a rich fuel mixture.
2. If the sensor tip has a white, gritty deposit, this may indicate an internal coolant leak.
3. If the sensor tip has a brown deposit, this could indicate excessive oil consumption.
4. Allow the engine to reach normal operating temperature.
5. Run the engine at idle speed.
6. Connect a voltmeter between terminal 1 (LG/R) and ground.
7. Increase and decrease the engine speed quickly several times.
8. Verify that the meter reading varies between 0–1 volt.

➡ The rear sensor voltage does not fluctuate as quickly as the front sensor.

9. If not a specified, inspect the on board diagnostic system, intake manifold vacuum, fuel line pressure and the exhaust system.
10. If all systems are normal, replace the sensor.
11. To inspect the sensor heater, be sure the ignition switch is OFF.
12. Disconnect the sensor electrical connector, if not already doe.
13. Connect an ohmmeter between terminals 1 and 3.
14. Specification should be 3–7 ohm at 68 degrees F.
15. If not within specification, replace the sensor.

2006

Perform a visual inspection of the sensor as follows:

1. If the sensor tip has a black/sooty deposit, this may indicate a rich fuel mixture.
2. If the sensor tip has a white, gritty deposit, this may indicate an internal coolant leak.
3. If the sensor tip has a brown deposit, this could indicate excessive oil consumption.
4. Warm the engine until the coolant temperature reaches operating temperature.
5. Be sure the ignition switch is OFF.
6. Disconnect the sensor connector.
7. Using an ohmmeter, measure the resistance between terminal 3 and 4 of the sensor (component side).
8. Specification should be 9.0 ohm at 68 degrees F.
9. If not within specification, install a known good sensor and recheck. Replace the sensor, as required.
10. If within specification, install a known good ECM/PCM and recheck. Replace the component, as required.

Idle Speed Control Actuator (ISCA)

OPERATION

This sensor is designed to maintain a steady desired idle speed. Idle air flow is adjusted through the idle air actuator in order to maintain the desired idle speed under various load conditions.

REMOVAL & INSTALLATION

1. Disconnect the negative battery cable.
2. Disconnect the connector from the sensor.
3. Remove the sensor retaining screws.
4. Remove the sensor from its mounting.

To install:

5. Installation is the reverse of the removal procedure.

TESTING

1. Be sure the ignition is OFF.
2. Disassemble and check the valve for damage and contamination. Reassemble the valve.
3. If it is not functioning properly, replace it with a known good component. Recheck and replace the component as required.
4. If it is still not functioning properly, disassemble the valve and using an ohmmeter, measure the resistance between terminal 1 and 2 of the harness connector (component side).
5. Specification should be 14.6–16.2 ohms (closing coil resistance) and 11.1–12.7 ohms (opening coil resistance) at 68–95 degrees F.
6. If not within specification, install a known good sensor and recheck. Replace the sensor, as required.
7. If within specification, install a known good ECM/PCM and recheck. Replace the component, as required.

Intake Air Temperature (IAT) Sensor

OPERATION

This sensor is a resistor based sensor which detects the intake air temperature. According to the intake air temperature reading the ECM will control the necessary amount of fuel injection.

REMOVAL & INSTALLATION

1. Disconnect the negative battery cable.
2. Disconnect the connector from the sensor.
3. Remove the sensor retaining screws.
4. Remove the sensor from its mounting.

To install:

5. Installation is the reverse of the removal procedure.

TESTING

2001–2002

1. Connect an ohmmeter.
2. Measure the resistance between sensor connector B11, terminal 1 and 2.
3. Specifications are as follows:
   - Temperature - 4 degrees F, resistance 15.04 (plus 1.79, minus 1.60) kohm
   - Temperature 68 degrees F, resistance 2.45 (plus 0.19, minus 0.18) kohm
   - Temperature 176 degrees F, resistance 0.318 (plus 0.011, minus 0.011) kohm
4. If within specification, check poor terminal contact, bent or misplaced terminal. Correct as required.
5. If not within specification, install a known good sensor and recheck. Replace the sensor, as required.

2003–2005

1. Be sure the ignition switch is OFF.
2. Disconnect the IAT connector. Disconnect the ECM connector.
3. Using an ohmmeter measure the resistance between the IAT sensor signal circuit and the ECTS ground circuit. Specification is infinite.
4. Using an ohmmeter measure the resistance between the IAT sensor harness connector and chassis ground at the IAT sensor signal circuit. Specification is infinite.
5. If within specification, install a known good sensor and recheck. Replace the sensor, as required.
6. If not within specification, check for short circuit between IAT sensor signal circuit and ground circuit. Check for short to ground between IAT sensor harness connector and ECM harness connector. Repair as required.

2006

1. Be sure the ignition switch is OFF.
2. Disconnect the sensor electrical connector.
3. Measure the resistance between terminal 1 and 5 of the connector (component side).
4. Specifications are as follows:
   - Temperature - 4 degrees F, resistance 14.26–16.02 kohm
   - Temperature 32 degrees F, resistance 5.50–6.05 kohm
   - Temperature 68 degrees F, resistance 2.35–2.54 kohm
   - Temperature 176 degrees F, resistance 0.31–0.32 kohm
5. If not within specification, install a known good sensor and recheck. Replace the sensor, as required.
6. If within specification, install a known good ECM/PCM and recheck. Replace the component, as required.

Knock Sensor (KS)

OPERATION

The knock sensor is used to detect engine vibrations caused by preignition or detonation and provides information to the ECM, which then retards the timing to eliminate detonation.

REMOVAL & INSTALLATION

2001–2005

1. Disconnect the negative battery cable.
2. Remove the intake manifold support bracket.
3. Disconnect the sensor connector.
4. Remove the sensor from its mounting.

To install:

5. Installation is the reverse of the removal procedure.
6. Tighten the sensor to 12–18 ft. lbs.

2006

1. Disconnect the negative battery cable.
2. Disconnect the sensor connector.
3. Remove the sensor from its mounting.

To install:

4. Installation is the reverse of the removal procedure.
5. Tighten the sensor to 12–18 ft. lbs.

TESTING

2001–2005

1. Disconnect the sensor electrical connector.
2. Remove the sensor from the engine.
3. Mount the sensor in the jaws of a bench vise.
4. Connect a voltmeter between terminal 1 and 2.
5. Wrap the bench vise sharply with a hammer and note the voltmeter reading.
6. Verify that the voltage spike is less than 1 volt.
7. If no voltage spike is observed, replace the sensor.

2006

➡ **If the sensor is suspected of being defective, it should be replaced with a known good component for testing purposes.**

1. Replace the sensor with a known good component.
2. If the signal is normal, replace the sensor.
3. If the signal is still not normal, replace the ECM/PCM with a known good component and recheck. Replace the component as required.

Mass Air Flow (MAF) Sensor

LOCATION

This sensor is located in the air intake plenum assembly.

OPERATION

On 2001–2005 vehicles, this sensor is the most direct method of measuring engine load because it measures the mass of air intake. The MAF sensor provides a signal to the ECM for incoming air temperature. This information is used to adjust the injector pulse width and in turn the air/fuel ratio. On 2006 vehicles this sensor is an air mass flowmeter, which operates on the principle of hot film anemometry. A heated element is placed within the air stream, and maintained at a constant temperature above the air temperature. The amount of electrical power required to maintain the heated element at the proper temperature is a direct function of the flow rate of the air mass past the element. The ECM/PCM uses this information to determine the injection duration and ignition timing of the desired air/fuel ratio.

REMOVAL & INSTALLATION

2001–2005

➡ **Do not drop or subject the sensor to shock. Do not put objects inside the sensor.**

1. Disconnect the sensor electrical connector.
2. Loosen the air intake hose retaining clamps on both sides of the sensor.
3. Disconnect the air intake hose from the sensor.
4. Remove the two bolts attaching the sensor to the mounting bracket.
5. Remove the sensor.

To install:

6. Installation is the reverse of the removal procedure.
7. Torque the retaining screws to 69–96 inch lbs.

2006

1. Disconnect the negative battery cable.
2. Disconnect the connector from the sensor.
3. Remove the sensor retaining screws.
4. Remove the sensor from its mounting.

To install:

5. Installation is the reverse of the removal procedure.

TESTING

2001–2005

1. Allow the engine to reach operating temperature.
2. Connect a voltmeter between terminal 2 (G/L) and ground.
3. Verify that the voltage varies between 0.6–4.0 volts.
4. If not within specification, replace the sensor.

Throttle Position Sensor (TPS)

OPERATION

The throttle position sensor is a rotating type variable resistor that rotates with the throttle body's throttle shaft to sense the throttle valve angle. As the throttle shaft rotates, the throttle angle of the sensor changes and the ECM detects the throttle valve opening based on the TPS output voltage.

REMOVAL & INSTALLATION

1. Disconnect the negative battery cable.
2. Disconnect the sensor connector.
3. Remove the sensor retaining screws. Remove the sensor from its mounting.

To install:

4. Installation is the reverse of the removal procedure.

TESTING

2001–2005

1. Disconnect the sensor electrical connector.
2. Connect an ohmmeter between sensor terminals 1 and 2.
3. Verify that the resistance increases linearly according to throttle angle.
4. Specification should be 1.6–2.4 ohm (throttle valve closed).
5. If resistance does not meet specification, replace the sensor.
6. Verify that the throttle is at the closed throttle position.
7. Turn the ignition switch ON (engine OFF).
8. Connect a voltmeter between terminal 2 (L/G) and terminal 3 (L/O) on the TPS connector.
9. Measure the voltage at the full open and full closed positions.
10. Specification is 0.2–0.8 volt (full closed position) and 4.0–4.8 volts (full open position).
11. If not within specification, replace the sensor.

2006

1. Disconnect the connector from the sensor.
2. Using an ohmmeter, measure the resistance between terminals 2 and 3 (component side).
3. Specification should be 1.6–2.4 kohm at 68 degrees F.
4. If not within specification, install a known good sensor and recheck. Replace the sensor, as required.
5. If within specification, install a known good ECM/PCM and recheck. Replace the component, as required.

Vehicle Speed Sensor (VSS)

OPERATION

The vehicle speed sensor sends a signal to the ECM. The ECM uses this information to control transmission shift patterns.

TESTING

2001–2002

1. Using an ohmmeter, check for continuity between connector B02-18 and the vehicle speed sensor B18-2.
2. If there is continuity, install a known good sensor and recheck. Replace the sensor, as required.
3. If there is no continuity, check for open circuit between connector B18-2 and ECM connector B02-18. Repair as required.

2003–2005

1. Turn the ignition switch ON.
2. Using a voltmeter measure the voltage between the wheel speed sensor signal terminal and ground.
3. Specification should be B+ at ignition ON and 0–B+ at driving.
4. If the voltage is not within specification, install a known good sensor and recheck. Replace the sensor, as required.
5. If the voltage is within specification, check for poor terminal contacts. If okay, check for open circuit or short circuit between VSS and ECM. Repair as required.

SEDONA

See Figures 5 and 6.

Camshaft Position (CMP) Sensor

LOCATION

The camshaft position sensor is mounted on the engine near the camshaft gear by the ignition coils.

OPERATION

The camshaft position sensor senses the TDC point of the number one cylinder, on its compression stroke. Its signal is relayed to the ECM to be used to determine the sequence of fuel injection.

REMOVAL & INSTALLATION

1. Disconnect the negative battery cable.
2. Disconnect the connector from the sensor.
3. Remove the bolt that retains the sensor.
4. Remove the sensor.

To install:

5. Installation is the reverse of the removal procedure.

TESTING

2002–2005

1. Be sure that the CMPS and PCM connectors are connected.
2. Engine ON and monitor the signal waveform of the sensor on the scan tool. Check whether the waveform is synchronized with the crankshaft position sensor or not.
3. If the waveform signal is normal, substitute with a known good PCM and check for proper operation. If the problem is corrected replace the PCM.
4. If the waveform signal is not normal, substitute with a known good sensor and check for proper operation. If the problem is corrected replace the sensor.

2006

1. Be sure that the ignition switch is OFF. Connect the scan tool.
2. Engine ON and monitor the signal waveform at terminal 3 of the sensor.

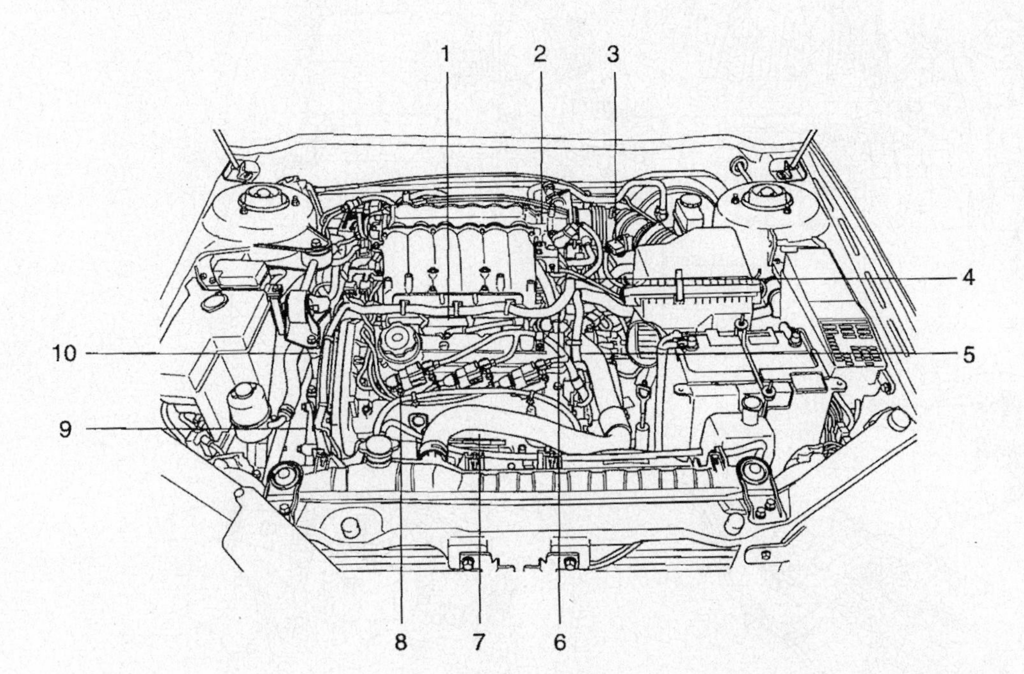

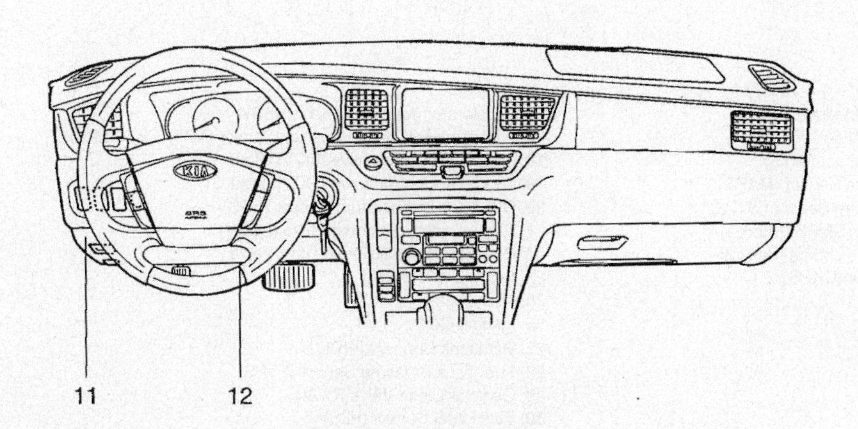

Fig. 5 Underhood sensor locations—Sedona 2002–2005

3. If the waveform signal is normal, substitute with a known good PCM and check for proper operation. If the problem is corrected replace the PCM.

4. If the waveform signal is not normal, substitute with a known good sensor and check for proper operation. If the problem is corrected replace the sensor.

Crankshaft Position (CKP) Sensor

LOCATION

The crankshaft position sensor is mounted on the engine near the crankshaft.

OPERATION

The crankshaft position sensor is a Hall-Effect sensor that senses the crank angle (piston position) of each cylinder and converts it into a pulse signal. The signal from the sensor is relayed to the ECM to indicate engine rpm and the position of the crankshaft.

REMOVAL & INSTALLATION

1. Disconnect the negative battery cable.
2. Disconnect the connector from the sensor.
3. Remove the bolt that retains the sensor in place.
4. Remove the sensor from its mounting.

To install:

5. Installation is the reverse of the removal procedure.
6. Clearance between the sensor and the sensor wheel should be 0.020–0.059 inch.

TESTING

2002–2005

1. Properly connect the scan tool.
2. Start the engine and allow it to idle.

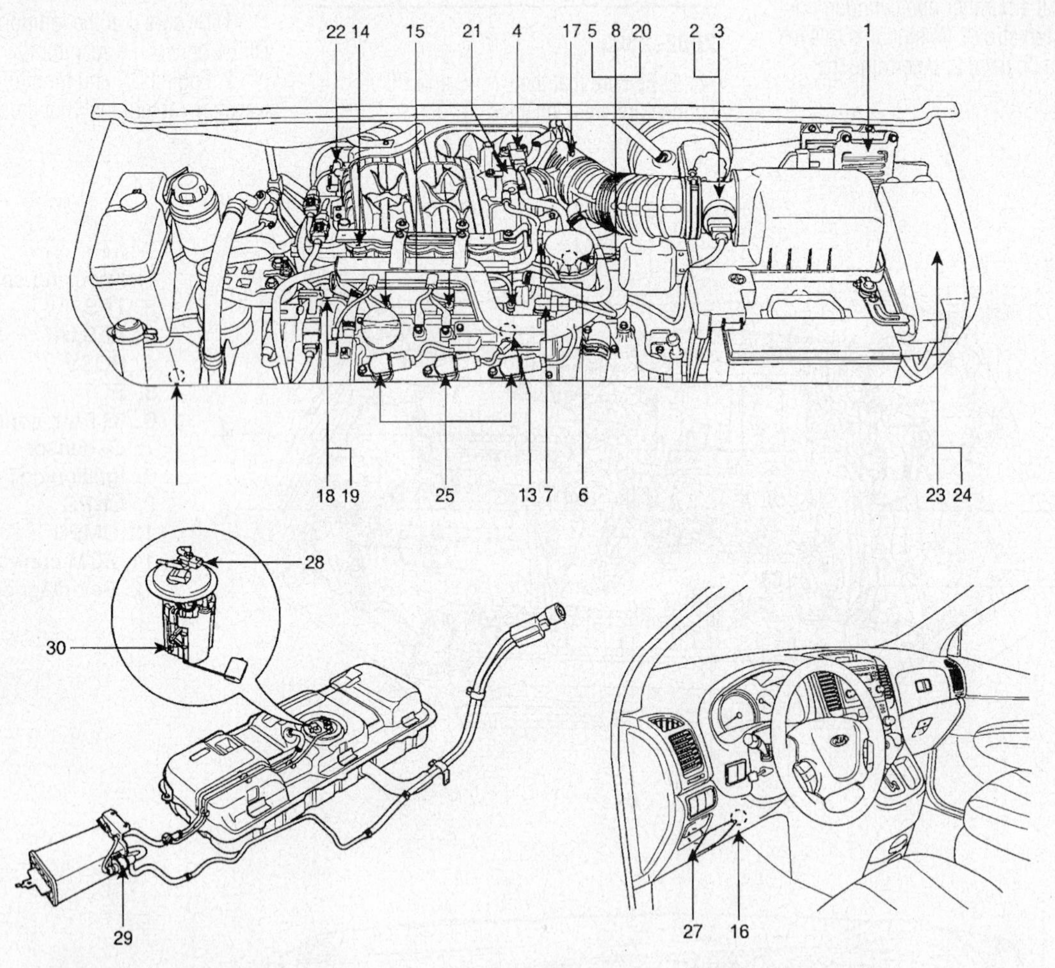

1. PCM (Powertrain Control Module)
2. Mass Air Flow Sensor (MAFS)
3. Intake Air Temperature Sensor (IATS)
4. Manifold Absolute Pressure Sensor (MAPS)
5. Engine Coolant Temperature Sensor (ECTS)
6. Camshaft Position Sensor (CMPS) [Bank 1]
7. Camshaft Position Sensor (CMPS) [Bank 2]
8. Crankshaft Position Sensor (CKPS)
13. Knock Sensor (KS) #1
14. Knock Sensor (KS) #2
15. Injector
16. Accelerator Position Sensor (APS)
17. ETC Module [Throttle Position Sensor (TPS) + ETC Motor]
18. CVVT Oil Control Valve (OCV) [Bank 1]
19. CVVT Oil Control Valve (OCV) [Bank 2]
20. CVVT Oil Temperature Sensor (OTS)
21. Purge Control Solenoid Valve (PCSV)
22. Variable Intake Solenoid (VIS) Valve
23. Fuel Pump Relay
24. Main Relay
25. Ignition Coil
27. Data Link Connector (DLC)
28. Fuel Tank Pressure Sensor (FTPS)
29. Canister Close Valve (CCV)
30. Fuel Level Sensor (FLS)

29130_KIA_G0004

Fig. 6 Underhood sensor locations—Sedona 2006

3. Be sure the idle position switch is ON.

4. Specifications are as follows:

- Temperature - 4 degrees F, test specification 1500–1700 RPM
- Temperature 32 degrees F, test specification 1350–1550 RPM
- Temperature 68 degrees F, test specification 1200–1400 RPM
- Temperature 104 degrees F, test specification 1000–1200 RPM
- Temperature 176 degrees F, test specification 650–850 RPM

5. Replace sensor, as required.

2006

1. Be sure the ignition switch is OFF.

2. Disconnect the sensor connector.

3. Using an ohmmeter, measure the resistance between terminal 1 and 2 of the connector (component side).

4. Specification should be 630<en dash770 ohms.

5. If not within specification substitute a known good sensor and recheck for proper operation. Replace the sensor, if required.

6. If within specification, check the waveform signal as follows.

7. Be sure the ignition is OFF. Install the scan tool. With the ignition switch in the ON position, measure the waveform signal at terminal 1 or 2 of the sensor.

8. If the signal is normal check the connectors for looseness, poor connection, bending, corrosion etc.

9. If the signal is not normal, substitute a known good PCM and recheck for proper operation. Replace the PCM, if required.

Electronic Control Module (ECM)

OPERATION

The ECM controls the vehicle engine operating system.

REMOVAL & INSTALLATION

1. Disconnect the negative battery cable.

2. Remove the lower inner trim.

3. As required, detach the floor mat. As required, remove the protective cover.

4. Remove the ECM bracket retaining nuts. Remove the clip from the bracket.

5. Disconnect the connectors.

6. Remove the ECM from the vehicle.

To install:

7. Installation is the reverse of the removal procedure.

➡ **When replacing the ECM, be careful not to use the wrong part number, as damage to the injection system could occur.**

Engine Coolant Temperature (ECT) Sensor

LOCATION

The ECT is installed in the engine coolant passage of the cylinder head.

OPERATION

This component detects the temperature of the engine coolant and relays the information to the electronic control assembly. This component employs a thermistor which is sensitive to temperature changes. The electric resistance of the thermistor decreases in response to temperature rise. The ECM judges coolant temperature by the sensor output voltage and provides optimum fuel enrichment when the engine is cold.

REMOVAL & INSTALLATION

1. Disconnect the negative battery cable.

2. Disconnect the connector from the sensor.

3. Drain the cooling system, as required.

4. Remove the sensor from its mounting.

To install:

5. Installation is the reverse of the removal procedure.

TESTING

2002–2005

1. Remove the sensor from the engine.

2. With the sensing portion of the sensor immersed in hot water, check the resistance using an ohmmeter.

3. Specifications are as follows:

- Temperature 32 degrees F, resistance 5.9 kohms
- Temperature 68 degrees F, resistance 2.31–2.59 kohms
- Temperature 104 degrees F, resistance 1.1 kohms
- Temperature 176 degrees F, resistance 0.310–0.331 kohms

4. Replace the sensor, as required.

2006

1. Be sure the ignition switch is OFF.

2. Disconnect the sensor connector.

3. Using an ohmmeter, measure the resistance between terminal 1 and 3 of the connector (component side).

4. Specifications are as follows:

- Temperature 32 degrees F, resistance 5.79 kohms
- Temperature 68 degrees F, resistance 2.31–2.59 kohms
- Temperature 104 degrees F, resistance 1.15 kohms
- Temperature 176 degrees F, resistance 0.32 kohms

5. If not within specification substitute a known good sensor and recheck for proper operation. Replace the sensor, if required.

6. If within specification, check the PCM as follows, ignition OFF. Connect the scan tool. Turn the ignition ON and engine OFF. Select simulation function on the scan tool. Simulate voltage at terminal 3 of the sensor harness connector.

7. If the signal voltage value does not change, substitute a known good PCM and repeat the test. Replace the PCM as required. If the signal voltage value does change, check connectors for looseness, poor connection, binding etc. Correct and retest.

Heated Oxygen (HO2S) Sensor

LOCATION

The sensors are located in the exhaust system. On some vehicles one sensor is located up at the exhaust manifold(s) and the other sensor is located down at the catalytic converter. On other vehicles both sensors are located down at the catalytic converter.

OPERATION

The exhaust gas oxygen sensor supplies the electronic control assembly with a signal which indicates either a rich or lean mixture condition, during the engine operation.

REMOVAL & INSTALLATION

1. Disconnect the electrical connector from the sensor.

2. Remove the oxygen sensor.

To install:

3. Installation is the reverse of the removal procedure.

➡ **Apply anti-seize compound to the threaded portion of the sensor, prior to installation. Never apply anti-seize compound to the protector of the sensor.**

TESTING

2002–2005

Perform a visual inspection of the sensor as follows:

1. If the sensor tip has a black/sooty deposit, this may indicate a rich fuel mixture.
2. If the sensor tip has a white, gritty deposit, this may indicate an internal coolant leak.
3. If the sensor tip has a brown deposit, this could indicate excessive oil consumption.
4. Warm the engine until the coolant temperature reaches operating temperature.
5. Disconnect the sensor electrical connector, and using a DVOM mete, measure the resistance between terminal 4 and 3.
6. Specification should be 30 ohms, or more.
7. Replace the sensor, if there is a malfunction.
8. Apply battery voltage directly between terminal 4 (power) and 3 (ground).

➡ **Be careful when applying the voltage. Damage will result if terminals 1 and 2 are connected to any voltage.**

9. Connect a voltmeter between terminal 2 (sensor output)and 1 (sensor ground).
10. While racing the engine repeatedly, measure the output voltage.
11. Specification should be 1 volt.
12. Replace the sensor, as required.

2006

SENSOR ONE

Perform a visual inspection of the sensor as follows:

1. If the sensor tip has a black/sooty deposit, this may indicate a rich fuel mixture.
2. If the sensor tip has a white, gritty deposit, this may indicate an internal coolant leak.
3. If the sensor tip has a brown deposit, this could indicate excessive oil consumption.
4. Warm the engine until the engine reaches operating temperature.
5. Be sure the ignition is OFF. Disconnect the sensor connector.
6. Check the sensor for damage or contamination caused by foreign substance.
7. If the sensor is not normal, replace the component with a known good one and recheck. Replace the sensor as required.
8. If the sensor is normal, substitute a known good PCM and repeat the test. Replace the PCM as required.

SENSOR TWO

Perform a visual inspection of the sensor as follows:

1. If the sensor tip has a black/sooty deposit, this may indicate a rich fuel mixture.
2. If the sensor tip has a white, gritty deposit, this may indicate an internal coolant leak.
3. If the sensor tip has a brown deposit, this could indicate excessive oil consumption.
4. Warm the engine until the engine reaches operating temperature.
5. Be sure the ignition is OFF. Disconnect the sensor connector.
6. Using an ohmmeter, measure the resistance between terminals 1 and 2 (component side).
7. Specification should be 9.6 plus/minus 1.5 ohms at 69.8 degrees F.
8. If not, replace the sensor.

Intake Air Temperature (IAT) Sensor

LOCATION

This sensor is located in the air intake plenum assembly. This sensor is combined with the MAF sensor.

OPERATION

This sensor is a resistor based sensor which detects the intake air temperature. According to the intake air temperature reading the ECM/PCM will control the necessary amount of fuel injection.

REMOVAL & INSTALLATION

1. Disconnect the negative battery cable.
2. Disconnect the connector from the sensor.
3. Remove the sensor retaining screws, as required.
4. Remove the air cleaner and air intake assembly, as required.
5. Remove the sensor from its mounting.

To install:

6. Installation is the reverse of the removal procedure.

TESTING

2002–2005

1. Be sure the ignition is ON.
2. Remove the sensor connector.
3. Using an ohmmeter measure the resistance between terminal 5 and 3 (component side).
4. Specification should be as follows:
   - Intake air temperature -40 degrees, resistance 33.85–61.20 kohms
   - Intake air temperature 68 degrees, resistance 2.22–2.82 kohms
   - Intake air temperature 176 degrees, resistance 0.299–0.375 kohms
5. Replace the sensor, as required.

2006

1. Be sure the ignition is OFF.
2. Remove the sensor connector.
3. Using an ohmmeter measure the resistance between terminal 4 and 5 (component side).
4. Specification should be as follows:
   - Intake air temperature -40 degrees, resistance 95.95–105.78 kohms
   - Intake air temperature 32 degrees, resistance 9.08–9.72 kohms
   - Intake air temperature 68 degrees, resistance 3.42–3.61 kohms
   - Intake air temperature 176 degrees, resistance 0.33–0.34 kohms
5. Replace the sensor, as required.

Knock Sensor (KS)

OPERATION

The knock sensor is used to detect engine vibrations caused by preignition or detonation and provides information to the ECM, which then retards the timing to eliminate detonation.

REMOVAL & INSTALLATION

1. Disconnect the negative battery cable.
2. Remove the necessary components to gain access to the sensor.
3. Disconnect the sensor connector.
4. Remove the sensor from its mounting.

To install:

5. Installation is the reverse of the removal procedure.
6. Tighten the sensor to 15–18 ft. lbs.

TESTING

2002–2005

1. Be sure the ignition switch is OFF. Disconnect the sensor electrical connector and the ECM connector. Connect an ohmmeter.

2. Measure the resistance between the sensor harness and ground at the sensor signal circuit.

3. Measure the resistance between the sensor signal circuit and the ground circuit.

4. Specification is infinite.

5. If resistance indicates an open circuit, install a known good sensor. Recheck, if the problem is corrected replace the sensor

2006

➡ **If the sensor is suspected of being defective, it should be replaced with a known good component for testing purposes.**

1. Check the sensor torque. It should be 11–18 ft. lbs.

2. If the sensor is still not functioning, replace it with a known good component.

3. Recheck the sensor.

Manifold Absolute Pressure (MAP) Sensor

OPERATION

This sensor measures the change of pressure in the intake manifold. This sensor measures the changes in the intake manifold pressure which result from engine load and speed changes, and converts to a voltage output that is monitored by the PCM.

REMOVAL & INSTALLATION

1. Disconnect the negative battery cable.

2. Disconnect the connector from the sensor.

3. Remove the sensor retaining screws.

4. Remove the sensor from its mounting.

To install:

5. Installation is the reverse of the removal procedure.

TESTING

2002–2005

1. Turn the ignition switch ON.

2. Using a voltmeter, measure the voltage between terminal 1 (sensor output) and terminal 4 (sensor ground).

3. Specification should be more than 4 volts. At idle specification should be less than 2.5 volts.

4. If not within specification, replace the sensor.

2006

1. Be sure the ignition is OFF. Connect the scan tool.

2. Connect a probe to the MAPS and TPS to check the signal waveform, by using oscilloscope function.

3. Engine ON, and monitor signal waveform during acceleration and deceleration.

4. Specification should be as follows:

- Pressure 20 kpa, voltage 0.789
- Pressure 35 kpa, voltage 1.382
- Pressure 60 kpa, voltage 2.369
- Pressure 95 kpa, voltage 3.750
- Pressure 101.32 kpa, voltage 4.000

5. If not within specification substitute a known good sensor and recheck for proper operation. Replace the sensor, if required.

6. If within specification, check the PCM as follows, ignition OFF. Connect the scan tool. Turn the ignition ON and engine OFF. Select simulation function on the scan tool. Simulate voltage at terminal 1 of the sensor harness connector.

7. If the signal voltage value does not change, substitute a known good PCM and repeat the test. Replace the PCM as required. If the signal voltage value does change, check connectors for looseness, poor connection, binding etc. Correct and retest.

Mass Air Flow (MAF) Sensor

LOCATION

This sensor is located in the air intake plenum assembly. This sensor is combined with the IAT sensor.

OPERATION

The MAF is measured by detection of heat transfer from a hot film probe because the change of the mass air flow rate causes change in the amount of heat being transferred from the hot film probe surface to the air flow. The air flow sensor generates a pulse so it repeatedly opens and closes between the voltages supplied from the PCM. This results in a change of the temperature of the hot film probe and in the change of resistance.

REMOVAL & INSTALLATION

1. Disconnect the negative battery cable.

2. Disconnect the connector from the sensor.

3. Remove the air cleaner and air intake assembly, as required.

4. Remove the sensor from its mounting.

To install:

5. Installation is the reverse of the removal procedure.

TESTING

2002–2005

1. Run the engine until it reaches operating temperature.

2. Using a voltmeter check the sensor output voltage.

3. Specification should be as follows:

- Air mass 7.30 kg/h, output voltage 0.3 volt
- Air mass 10.06 kg/h, output voltage 0.5 volt
- Air mass 19.85 kg/h, output voltage 1.0 volt
- Air mass 35.58 kg/h, output voltage 1.50 volts
- Air mass 58.79 kg/h, output voltage 2.0 volts
- Air mass 94.70 kg/h, output voltage 2.5 volts
- Air mass 149.07 kg/h, output voltage 3.0 volts
- Air mass 226.66 kg/h, output voltage 3.5 volts
- Air mass 335.66 kg/h, output voltage 4.0 volts
- Air mass 500.16 kg/h, output voltage 4.5 volts
- Air mass 614.61 kg/h, output voltage 4.76 volts
- Air mass 730.01 kg/h, output voltage 5.0 volts

4. If not within specification, replace the sensor.

2006

1. Be sure the ignition switch is OFF.

2. Connect the scan tool.

3. Engine ON, monitor "MAFS" on the service data.

4. Monitor signal waveform at terminal 1 of the sensor with the scan tool.

5. Be aware that the signal of the sensor is not a voltage display but a frequency display.

6. If the signal pattern is normal, substitute a known good PCM and repeat the test. Replace the PCM as required.

7. If the sensor is not normal, replace the component with a known good one and recheck. Replace the sensor as required.

Throttle Position Sensor (TPS)

OPERATION

The throttle position sensor is a rotating type variable resistor that rotates with the throttle body's throttle shaft to sense the throttle valve angle. As the throttle shaft rotates, the throttle angle of the sensor changes and the ECM/PCM detects the throttle valve opening based on the TPS output voltage.

REMOVAL & INSTALLATION

1. Disconnect the negative battery cable.
2. Disconnect the sensor connector.
3. Remove the sensor retaining screws. Remove the sensor from its mounting.

To install:

4. Installation is the reverse of the removal procedure.

TESTING

2002–2005

1. Disconnect the sensor connector.
2. Measure the resistance between terminal 1 (sensor ground) and terminal 4 (sensor power).
3. Specification should be 3.5–6.5 kohm.
4. Connect and analog ohmmeter between terminal 1 (sensor ground) and terminal 3 (sensor output).
5. Operate the throttle valve slowly from the idle position to the full open position, and check that the resistance changes smoothly in proportion with the throttle valve opening angle.
6. If the resistance is out of specification, or fails to change smoothly, replace the sensor.

SEPHIA

See Figures 7 and 8.

Camshaft Position (CMP) Sensor

OPERATION

The camshaft position sensor senses the TDC point of the number one cylinder, on its compression stroke. Its signal is relayed to the ECM to be used to determine the sequence of fuel injection.

REMOVAL & INSTALLATION

1998–2001

1. Disconnect the negative battery cable.
2. Disconnect the connector from the sensor.
3. Remove the mounting bolts.
4. Remove the sensor, from the cylinder head.

To install:

5. Installation is the reverse of the removal procedure.

TESTING

1996–1997

1.6L ENGINE

1. Disconnect the negative battery cable.
2. Remove the distributor.
3. Disconnect the distributor's 7 pin connector.

➡ Turning the ignition switch ON with the fuel injector connector still connected will actuate the fuel injector. Turning the ignition switch ON with the distributor 3 pin connector still attached will generate sparks, which can cause electrical shock. Disconnect this connector and prevent it from grounding to the vehicle body.

4. Turn the ignition switch ON.
5. Using a voltmeter measure the voltage at PCM terminals C249-16 and C249-4.
6. Rotate the distributor drive, by hand, and check the output signal.
7. Specification should be approximately 5 volts (4 pulses/rev) connector C249-16.
8. If not as specified inspect the wiring harness and connector between the distributor and the PCM terminals.

1.8L ENGINE

1. Disconnect the negative battery cable.
2. Remove the distributor.
3. Disconnect the distributor's 6 pin connector.

➡ Turning the ignition switch ON with the fuel injector connector still connected will actuate the fuel injector. Turning the ignition switch ON with the distributor 3 pin connector still attached will generate sparks, which can cause electrical shock. Disconnect this connector and prevent it from grounding to the vehicle body.

4. Turn the ignition switch ON.
5. Using a voltmeter measure the voltage at PCM terminals C249-16 and C249-4.
6. Rotate the distributor drive, by hand, and check the output signal.
7. Specification should be approximately 5 volts (4 pulses/rev) connector C249-16, and approximately 5 volts (1 pulse/rev) connector C249-4.
8. If not as specified inspect the wiring harness and connector between the distributor and the PCM terminals.

Crankshaft Position (CKP) Sensor

OPERATION

The CKP is used to inform the ECM when the No.1 piston is at top dead center. This information is used for controlling and adjusting ignition and fuel injector timing.

REMOVAL & INSTALLATION

1998–2001

1. Disconnect the negative battery cable.
2. Disconnect the connector from the sensor.
3. Remove the bolt that retains the sensor in place.
4. Remove the sensor from the transmission cover

To install:

5. Installation is the reverse of the removal procedure.

TESTING

1996–1997

1. Disconnect the 3 pin sensor connector.
2. Measure the resistance between terminals 1 and 2, using an ohmmeter.
3. Specification should be 500–600 at 68 degrees F.
4. If not within specification, replace the sensor.
5. Measure the air gap between the crankshaft pulley and the sensor, using a feeler gauge.
6. Specification should be 0.020–0.059 inch.
7. If not within specification, replace the sensor.

EGR Boost Sensor
Heated Oxygen Sensor (Rear)
Crankshaft Position Sensor
Powertrain Control Module (PCM)
Muffler
Throttle Position Sensor
Clutch Pedal Position Switch (M/T)
EGR Control Valve
Air Intake Hose
Main Relay
'ower Steering
ressure Switch
Heated Oxygen Sensor (Front)
Neutral Position Switch (M/T)
Engine Coolant Temperature Sensor
MAF Sensor (Includes Intake Air Temperature Sensor)
Camshaft Position Sensor

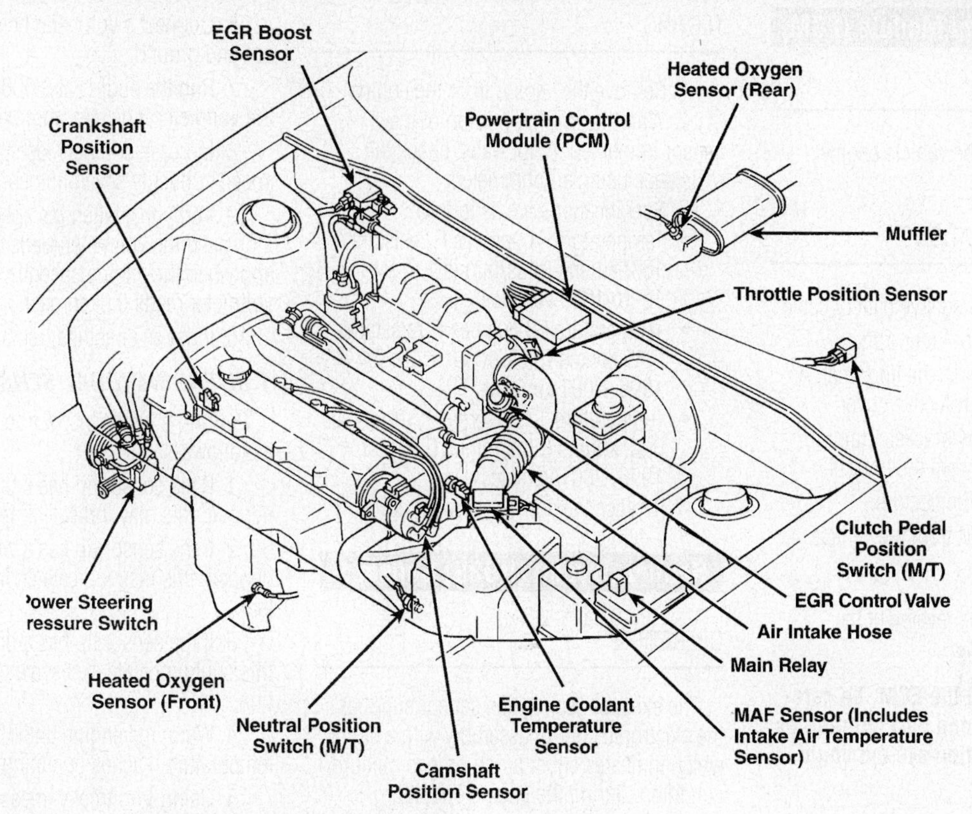

Fig. 7 Underhood sensor locations—Sephia 1996–1997

29130_KIA_G0005

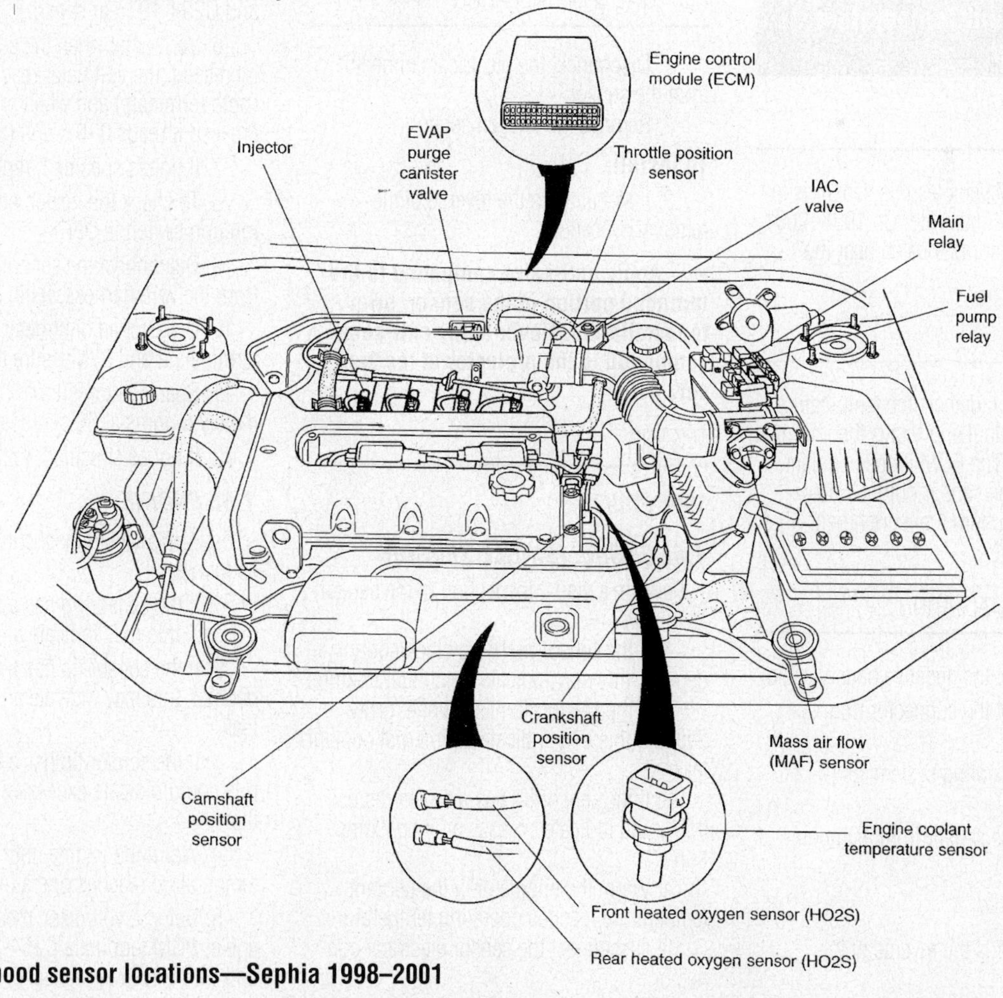

Engine control module (ECM)
Injector
EVAP purge canister valve
Throttle position sensor
IAC valve
Main relay
Fuel pump relay
Crankshaft position sensor
Mass air flow (MAF) sensor
Camshaft position sensor
Engine coolant temperature sensor
Front heated oxygen sensor (HO2S)
Rear heated oxygen sensor (HO2S)

Fig. 8 Underhood sensor locations—Sephia 1998–2001

29130_KIA_G0006

Electronic Control Module (ECM)

OPERATION

The ECM controls the vehicle engine operating system.

REMOVAL & INSTALLATION

1. Disconnect the negative battery cable.
2. Remove the lower inner trim.
3. As required, detach the floor mat. As required, remove the protective cover.
4. Remove the ECM bracket retaining nuts. Remove the clip from the bracket.
5. Disconnect the connectors.
6. Remove the ECM from the vehicle.

To install:

7. Installation is the reverse of the removal procedure.

➡ **When replacing the ECM, be careful not to use the wrong part number, as damage to the injection system could occur.**

Engine Coolant Temperature (ECT) Sensor

LOCATION

On 1996–1997 vehicles this sensor is located in the intake manifold. On 1998–2001 vehicles, this sensor is located near the thermostat.

OPERATION

This component detects the temperature of the engine coolant and relays the information to the ECM. The ECM judges coolant temperature by the sensor output voltage and provides optimum fuel enrichment when the engine is cold.

REMOVAL & INSTALLATION

1. Disconnect the negative battery cable.
2. Disconnect the connector from the sensor.
3. Drain the cooling system, as required.
4. Remove the sensor from its mounting.

To install:

5. Installation is the reverse of the removal procedure.

TESTING

1. Remove the sensor from the engine.
2. With the sensing portion of the sensor immersed in hot water, check the resistance using an ohmmeter.
3. Specifications are as follows:
   - Temperature -4 degrees F, resistance 16.2 kohms (plus/minus 1.6) for 1996–1997 vehicles
   - Temperature 68 degrees F, resistance 2.45 kohms (plus/minus 0.25) for 1996–2001 vehicles
   - Temperature 176 degrees F, resistance 0.32 kohms (plus/minus 0.03) for 1996–2001 vehicles
4. Replace the sensor, as required.

Heated Oxygen (HO2S) Sensor

OPERATION

The exhaust gas oxygen sensor supplies the electronic control assembly with a signal which indicates either a rich or lean mixture condition, during the engine operation.

REMOVAL & INSTALLATION

1. Disconnect the electrical connector from the sensor.
2. Remove the oxygen sensor.

To install:

3. Installation is the reverse of the removal procedure.

➡ **Apply anti-seize compound to the threaded portion of the sensor, prior to installation. Never apply anti-seize compound to the protector of the sensor.**

TESTING

1996–1997

1.6L ENGINE (SINGLE SENSOR)

Perform a visual inspection of the sensor as follows:

1. If the sensor tip has a black/sooty deposit, this may indicate a rich fuel mixture.
2. If the sensor tip has a white, gritty deposit, this may indicate an internal coolant leak.
3. If the sensor tip has a brown deposit, this could indicate excessive oil consumption.
4. Warm the engine until the coolant temperature reaches operating temperature.
5. Disconnect the sensor electrical connector.
6. Connect a voltmeter between the sensor and ground.
7. Run the engine at 3,000 rpm's, until the voltmeter indicates approximately 0.55V.
8. Increase and decrease the engine speed suddenly several times.
9. Verify that when the speed is increased the voltmeter reads 0.5–1.0 volt, and when the engine speed is decreased the voltmeter reads 0.0<en dash0.4 volt.
10. If not as specified, replace the sensor.

1.6L ENGINE (DUAL SENSORS)

Perform a visual inspection of the sensor as follows:

1. If the sensor tip has a black/sooty deposit, this may indicate a rich fuel mixture.
2. If the sensor tip has a white, gritty deposit, this may indicate an internal coolant leak.
3. If the sensor tip has a brown deposit, this could indicate excessive oil consumption.
4. Warm the engine until the coolant temperature reaches operating temperature.
5. Using a voltmeter, measure the voltage at PCM terminals C254-2 (front sensor) and C254-10 (rear sensor).
6. Verify that when the engine speed is increased, the voltmeter reads 0.5–1.0 volt (both terminals) and when the speed is decreased it reads 0–0.5 volt (both terminals).
7. If not as specified, replace the sensor.
8. To check the heater, be sure that the ignition switch is OFF.
9. Disconnect the sensor connector from the wire harness, if not already done.
10. Connect an ohmmeter between terminals 2 and 3. Measure the resistance.
11. Specification should be approximately 6 ohms at 68 degrees F.
12. If not as specified, replace the sensor.

1.8L ENGINE

Perform a visual inspection of the sensor as follows:

1. If the sensor tip has a black/sooty deposit, this may indicate a rich fuel mixture.
2. If the sensor tip has a white, gritty deposit, this may indicate an internal coolant leak.
3. If the sensor tip has a brown deposit, this could indicate excessive oil consumption.
4. Warm the engine until the coolant temperature reaches operating temperature.
5. Using a voltmeter, measure the voltage at PCM terminals C254-2 (front sensor) and C254-10 (rear sensor).

6. Verify that when the engine speed is increased, the voltmeter reads 0.5–1.0 volt (both terminals) and when the speed is decreased it reads 0–0.5 volt (both terminals).
7. If not as specified, replace the sensor.
8. To check the heater, be sure that the ignition switch is OFF.
9. Disconnect the sensor connector from the wire harness, if not already done.
10. Connect an ohmmeter between terminals 2 and 3. Measure the resistance.
11. Specification should be approximately 13ohms at 68 degrees F.
12. If not as specified, replace the sensor.

1998–2000

Perform a visual inspection of the sensor as follows:

1. If the sensor tip has a black/sooty deposit, this may indicate a rich fuel mixture.
2. If the sensor tip has a white, gritty deposit, this may indicate an internal coolant leak.
3. If the sensor tip has a brown deposit, this could indicate excessive oil consumption.
4. Warm the engine until the coolant temperature reaches operating temperature.
5. Run the engine at idle speed.
6. Connect a voltmeter between terminal 4 (blk) and ground.
7. Increase and decrease the engine speed quickly several times.
8. Specification should be 0–1.0 volt.

➡ **The rear sensor voltage does not fluctuate as quickly as the front sensor.**

9. If not as specified, inspect the on-board diagnostic system, exhaust system, intake manifold system and fuel line pressure.
10. If all systems are normal, replace the sensor.
11. To check the heater, be sure that the ignition switch is OFF.
12. Disconnect the sensor connector from the wire harness, if not already done.
13. Connect an ohmmeter between terminals 2 and 4. Measure the resistance.
14. Specification should be approximately 6 ohms at 68 degrees F.
15. If not as specified, replace the sensor.

2001

Perform a visual inspection of the sensor as follows:

1. If the sensor tip has a black/sooty deposit, this may indicate a rich fuel mixture.
2. If the sensor tip has a white, gritty deposit, this may indicate an internal coolant leak.
3. If the sensor tip has a brown deposit, this could indicate excessive oil consumption.
4. Warm the engine until the coolant temperature reaches operating temperature.
5. Run the engine at idle speed.
6. Connect a voltmeter between terminal 4 (grn/brn) and ground.
7. Increase and decrease the engine speed quickly several times.
8. Specification should be 0–1.0 volt.

➡ **The rear sensor voltage does not fluctuate as quickly as the front sensor.**

9. If not as specified, inspect the on-board diagnostic system, exhaust system, intake manifold system and fuel line pressure.
10. If all systems are normal, replace the sensor.
11. To check the heater, be sure that the ignition switch is OFF.
12. Disconnect the sensor connector from the wire harness, if not already done.
13. Connect an ohmmeter between terminals 1 and 3. Measure the resistance.
14. Specification should be approximately 2–4 ohms at 68 degrees F.
15. If not as specified, replace the sensor.

Intake Air Temperature (IAT) Sensor

OPERATION

This sensor is a resistor based sensor which detects the intake air temperature. According to the intake air temperature reading the ECM/PCM will control the necessary amount of fuel injection.

REMOVAL & INSTALLATION

1. Disconnect the negative battery cable.
2. Disconnect the connector from the sensor.
3. Remove the sensor retaining screws, as required.
4. Remove the air cleaner and air intake assembly, as required.
5. Remove the sensor from its mounting.

To install:

6. Installation is the reverse of the removal procedure.

TESTING

1996–1997

1. Turn the ignition switch to the OFF position.
2. Remove the sensor connector.
3. Using an ohmmeter measure the resistance between terminal 3 and 4 (component side).
4. Specification should be 2.21–2.69 ohms at 68 degrees F.
5. Replace the sensor, as required.

Knock Sensor (KS)

OPERATION

The knock sensor is used to detect engine vibrations caused by preignition or detonation and provides information to the ECM, which then retards the timing to eliminate detonation.

REMOVAL & INSTALLATION

1. Disconnect the negative battery cable.
2. Remove the intake manifold support bracket.
3. Disconnect the sensor connector.
4. Remove the sensor from its mounting.

To install:

5. Installation is the reverse of the removal procedure.

TESTING

1. Disconnect the sensor electrical connector.
2. Remove the sensor from the engine.
3. Mount the sensor in the jaws of a bench vise.
4. Connect a voltmeter between terminal 1 and 2.
5. Wrap the bench vise sharply with a hammer and note the voltmeter reading.
6. Verify that the voltage spike is less than 1 volt.
7. If no voltage spike is observed, replace the sensor.

Mass Air Flow (MAF) Sensor

OPERATION

The MAF provides a signal to the ECM for incoming air temperature. This information is used to adjust the injector pulse width and in turn the air/fuel ratio.

REMOVAL & INSTALLATION

1. Disconnect the negative battery cable.
2. Disconnect the connector from the sensor.
3. Remove the air cleaner and air intake assembly, as required.

4. Remove the sensor from its mounting.

To install:

5. Installation is the reverse of the removal procedure.

TESTING

1996–1997

1. Using a voltmeter, back probe the MAF sensor connector terminals 2 and 5.
2. Specification at terminal 2 should be below 1.0 volt with the ignition switch ON and 1.0–2.0 volts at idle.
3. Specification at terminal 5 should be below 1.0 volt with the ignition switch ON and/or at idle.
4. If not within specification replace the sensor.

1998–2001

1. Allow the engine to reach operating temperature.
2. Let the engine idle.
3. Connect a voltmeter between terminal 4 (red/grn on 1998 vehicles and red/wht on 1999–2001 vehicles).
4. Verify that the voltage varies between 0.8–1.2 volts.
5. Rev the engine and verify that the voltage varies between 3.0–4.0 volts for 1998 vehicles and 3.5–4.0 volts for 1999–2001 vehicles.
6. If not within specification, replace the sensor.

Throttle Position Sensor (TPS)

OPERATION

The throttle position sensor provides a signal to the ECM that is related to the relative throttle plate position. As the throttle plate moves in relation to driving conditions, a signal is sent to the control unit which adjusts the injector pulse width and air/fuel ratio. As the throttle plate is opened further, more air is taken into the combustion chambers, and as a result the relative fuel demand of the engine changes. The throttle position sensor relays this information to the ECM which in alters the fuel amount.

REMOVAL & INSTALLATION

1996–1997

1.6L ENGINE

1. Disconnect the negative battery cable.
2. Disconnect the sensor connector.
3. Remove the sensor retaining screws. Remove the sensor from its mounting.

To install:

4. Installation is the reverse of the removal procedure.

1.8L ENGINE

1. Disconnect the electrical connector.
2. Remove the attaching screws.
3. Remove the sensor.

To install:

4. Verify that the throttle valve is fully closed.
5. Open the throttle valve slightly and catch the tang of the throttle body on the new throttle position sensor plastic tabs. Align the tan on the throttle body with the tab on the sensor.

➡ **Tangs on the throttle body mate with the tab on the sensor, on the edge of the tab without a lot of pressure.**

6. Position the sensor on the throttle body so that the mounting holes align.
7. Install and hand tighten the attaching screws.
8. Release the throttle.
9. Adjust the sensor output voltage and close the switch.

1998–2001

1. Be sure the engine is in the OFF position.
2. Disconnect the sensor connector.
3. Remove the sensor retaining bolts.
4. Remove the sensor.

To install:

5. Installation is the reverse of the removal procedure.

TESTING

1996–1997

1.6L ENGINE (MANUAL TRANSMISSION)

1. Disconnect the connector from the sensor.
2. Connect an ohmmeter to the sensor.
3. Insert a feeler gauge between the throttle stop screw and the stop lever and check the continuity between the terminals.
4. Specification should be as follows:
   - gauge reading 0.004 inch, IDL/E yes, POW/E no
   - gauge reading 0.039 inch, IDL/E no, POW/E no
   - gauge reading WOT, IDL/E no, POW/E yes
5. Replace as required.

1.6L ENGINE (AUTOMATIC TRANSMISSION)

1. Disconnect the connector from the sensor.
2. Connect an ohmmeter to the sensor, between terminal 1 (E) and 2 (IDL).
3. Insert a feeler gauge between the throttle stop screw and the stop lever and check the continuity between the terminals.
4. Specification should be as follows:
   - gauge reading 0.004 inch, continuity yes
   - gauge reading 0.024 inch, continuity no
5. Connect an ohmmeter to the sensor, between terminal 3 (VT) and 1 (E).
6. Verify that the resistance increases according to throttle valve opening angle.
7. Specification should be as follows:
   - throttle valve fully closed, resistance below 1 kohm
   - throttle valve fully open, resistance approximately 5 kohms
8. Replace as required.

1.8L ENGINE

1. Be sure that the engine is at operating temperature.
2. Turn the ignition switch to the ON position.
3. Check the voltage output at the signal return voltage terminal (lt grn/wht), using a voltmeter.
4. Specification should be 0.4–0.6 volt with the throttle valve fully closed.
5. If not within specification replace the sensor.
6. Verify that the sensor signal return voltage is within specification, using a voltmeter.
7. Slowly open the throttle valve and verify that the voltage increases are proportionate to the throttle valve opening angle.
8. Specification should be approximately 3.5–4.3 volts at WOT.
9. If not within specification, replace the sensor.

1998–2001

1. Disconnect the connector from the sensor.
2. Connect an ohmmeter between sensor terminals 2 and 3 on 1998–2000 vehicles and terminals 1 and 3 on 2001 vehicles.
3. Verify that the resistance increases according to throttle angle.
4. Specification should be 2 kohms (plus or minus 0.4 kohm) with the throttle valve closed.
5. If not within specification, replace the sensor.

SORENTO

See Figure 9.

Camshaft Position (CMP) Sensor

LOCATION

The camshaft position sensor is mounted on the engine near the camshaft gear by the ignition coils.

OPERATION

The camshaft position sensor senses the TDC point of the number one cylinder, on its compression stroke. Its signal is relayed to the ECM to be used to determine the sequence of fuel injection.

REMOVAL & INSTALLATION

1. Disconnect the negative battery cable.
2. Disconnect the connector from the sensor.
3. Remove the bolt that retains the sensor.
4. Remove the sensor.

To install:

5. Installation is the reverse of the removal procedure.

TESTING

1. Be sure that the CMPS and PCM connectors are connected.
2. Engine ON and monitor the signal waveform of the sensor on the scan tool. Check whether the waveform is synchronized with the crankshaft position sensor or not.
3. If the waveform signal is normal, substitute with a known good PCM and check for proper operation. If the problem is corrected replace the PCM.
4. If the waveform signal is not normal, substitute with a known good sensor and check for proper operation. If the problem is corrected replace the sensor.

Crankshaft Position (CKP) Sensor

LOCATION

The crankshaft position sensor is mounted on the engine near the crankshaft.

OPERATION

The crankshaft position sensor is a Hall-Effect sensor that senses the crank angle (piston position) of each cylinder and converts it into a pulse signal. The signal from the sensor is relayed to the ECM to indicate engine rpm and the position of the crankshaft.

REMOVAL & INSTALLATION

1. Disconnect the negative battery cable.
2. Disconnect the connector from the sensor.
3. Remove the bolt that retains the sensor in place.

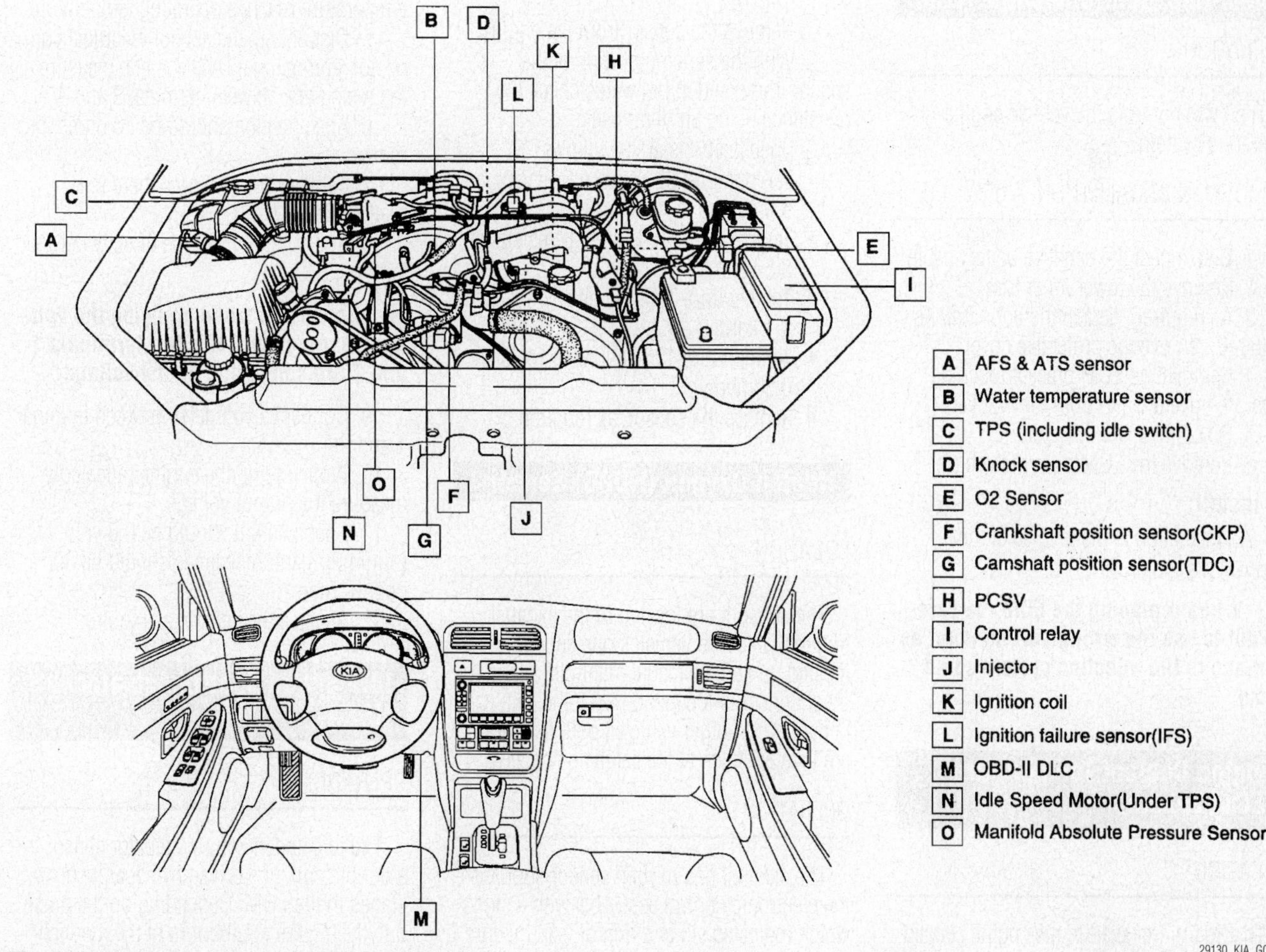

Fig. 9 Underhood sensor locations—Sorento

4. Remove the sensor from its mounting.

To install:

5. Installation is the reverse of the removal procedure.

6. Clearance between the sensor and the sensor wheel should be 0.020–0.059 inch.

TESTING

1. Properly connect the scan tool.
2. Start the engine and allow it to idle.
3. Be sure the idle position switch is ON.
4. Specifications are as follows:

- Temperature - 4 degrees F, test specification 1500–1700 RPM
- Temperature 32 degrees F, test specification 1350–1550 RPM
- Temperature 68 degrees F, test specification 1200–1400 RPM
- Temperature 104 degrees F, test specification 1000–1200 RPM
- Temperature 176 degrees F, test specification 650–850 RPM

5. Replace sensor, as required.

Electronic Control Module (ECM)

OPERATION

The ECM controls the vehicle engine operating system.

REMOVAL & INSTALLATION

1. Disconnect the negative battery cable.
2. Remove the lower inner trim.
3. As required, detach the floor mat. As required, remove the protective cover.
4. Remove the ECM bracket retaining nuts. Remove the clip from the bracket.
5. Disconnect the connectors.
6. Remove the ECM from the vehicle.

To install:

7. Installation is the reverse of the removal procedure.

➡ **When replacing the ECM, be careful not to use the wrong part number, as damage to the injection system could occur.**

Engine Coolant Temperature (ECT) Sensor

LOCATION

The ECT is installed in the engine coolant passage of the cylinder head.

OPERATION

This component detects the temperature of the engine coolant and relays the information to the electronic control assembly. This component employs a thermistor which is sensitive to temperature changes. The electric resistance of the thermistor decreases in response to temperature rise. The ECM judges coolant temperature by the sensor output voltage and provides optimum fuel enrichment when the engine is cold.

REMOVAL & INSTALLATION

1. Disconnect the negative battery cable.
2. Disconnect the connector from the sensor.
3. Drain the cooling system, as required.
4. Remove the sensor from its mounting.

To install:

5. Installation is the reverse of the removal procedure.

TESTING

1. Remove the sensor from the engine.
2. With the sensing portion of the sensor immersed in hot water, check the resistance using an ohmmeter.
3. Specifications are as follows:

- Temperature 32 degrees F, resistance 5.9 kohms
- Temperature 68 degrees F, resistance 2.5 kohms
- Temperature 104 degrees F, resistance 1.1 kohms
- Temperature 176 degrees F, resistance 0.3 kohms

4. Replace the sensor, as required.

Heated Oxygen (HO2S) Sensor

LOCATION

The sensors are located in the exhaust system. On some vehicles one sensor is located up at the exhaust manifold(s) and the other sensor is located down at the catalytic converter. On other vehicles both sensors are located down at the catalytic converter.

OPERATION

The exhaust gas oxygen sensor supplies the electronic control assembly with a signal which indicates either a rich or lean mixture condition, during the engine operation.

REMOVAL & INSTALLATION

1. Disconnect the electrical connector from the sensor.
2. Remove the oxygen sensor.

To install:

3. Installation is the reverse of the removal procedure.

➡ **Apply anti-seize compound to the threaded portion of the sensor, prior to installation. Never apply anti-seize compound to the protector of the sensor.**

TESTING

Perform a visual inspection of the sensor as follows:

1. If the sensor tip has a black/sooty deposit, this may indicate a rich fuel mixture.
2. If the sensor tip has a white, gritty deposit, this may indicate an internal coolant leak.
3. If the sensor tip has a brown deposit, this could indicate excessive oil consumption.
4. Warm the engine until the coolant temperature reaches operating temperature.
5. Disconnect the sensor electrical connector, and using a DVOM mete, measure the resistance between terminal 3 and 4.
6. Specification should be 30 ohms, or more.
7. Replace the sensor, if there is a malfunction.
8. Apply battery voltage directly between terminal 3 and 4.

➡ **Be careful when applying the voltage. Damage will result if terminals 1 and 2 are connected to any voltage.**

9. Connect a voltmeter between terminal 1 and 2.
10. While racing the engine repeatedly, measure the output voltage.
11. Specification should be 0.6 volt (minimum) and resistance should be 30 ohm or more.
12. Replace the sensor, as required.

Idle Speed Control Actuator (ISCA)

OPERATION

The idle speed control actuator sensor is a double coil type driven by separate driver stages in the ECM. Depending on the pulse duty factor, the equilibrium of the magnetic forces of the two coils will result in different angles of the motor. A bypass line is positioned, parallel to the throttle valve where the idle speed actuator is inserted.

REMOVAL & INSTALLATION

1. Disconnect the negative battery cable.
2. Disconnect the connector from the sensor.
3. Remove the sensor retaining screws.
4. Remove the sensor from its mounting.

To install:

5. Installation is the reverse of the removal procedure.

TESTING

1. Disconnect the connector at the sensor. Connect an ohmmeter.
2. Measure the resistance between terminal 1 and 3, specification should be 36.5–39.5 ohms, at 68 degrees F.
3. Measure the resistance between terminal 4 and 6, specification should be 36.5–39.5 ohms, at 68 degrees F.
4. If the measured value is not within specification, replace the sensor.

Intake Air Temperature (IAT)

LOCATION

This sensor is located in the air intake plenum assembly. This sensor is combined with the MAF sensor.

OPERATION

This sensor is a resistor based sensor which detects the intake air temperature. According to the intake air temperature reading the ECM/PCM will control the necessary amount of fuel injection.

REMOVAL & INSTALLATION

1. Disconnect the negative battery cable.
2. Disconnect the connector from the sensor.
3. Remove the sensor retaining screws, as required.
4. Remove the air cleaner and air intake assembly, as required.
5. Remove the sensor from its mounting.

To install:

6. Installation is the reverse of the removal procedure.

TESTING

1. Be sure the engine is at operating temperature.
2. Remove the sensor connector.
3. Using a voltmeter, measure the output voltage.
4. Specification should be 0.5 volt at idle and 1.0 volt at 2000 rpm.
5. Replace the sensor, as required.

Knock Sensor (KS)

OPERATION

The knock sensor is used to detect engine vibrations caused by preignition or detonation and provides information to the ECM, which then retards the timing to eliminate detonation.

REMOVAL & INSTALLATION

1. Disconnect the negative battery cable.
2. Remove the necessary components to gain access to the sensor.
3. Disconnect the sensor connector.
4. Remove the sensor from its mounting.

To install:

5. Installation is the reverse of the removal procedure.
6. Tighten the sensor to 15–18 ft. lbs.

TESTING

1. Disconnect the sensor electrical connector. Connect an ohmmeter.
2. Measure the resistance between terminals 1 and 2. Specification should be 5mohm at 68 degrees F.
3. If resistance is zero, replace the sensor.
4. Measure the capacitance between terminal 1 and 2. Specification should be 800–1600 pF.
5. If not within specification, replace the sensor.

Manifold Absolute Pressure (MAP) Sensor

OPERATION

This sensor measures the change of pressure in the intake manifold. This sensor measures the changes in the intake manifold pressure which result from engine load and speed changes, and converts to a voltage output that is monitored by the PCM.

REMOVAL & INSTALLATION

1. Disconnect the negative battery cable.
2. Disconnect the connector from the sensor.
3. Remove the sensor retaining screws.
4. Remove the sensor from its mounting.

To install:

5. Installation is the reverse of the removal procedure.

TESTING

1. Turn the ignition switch ON.
2. Using a voltmeter, measure the voltage between terminal 1 (sensor output) and terminal 4 (sensor ground).
3. Specification should be more than 4 volts. At idle specification should be less than 2.5 volts.
4. If not within specification, replace the sensor.

Mass Air Flow (MAF) Sensor

LOCATION

This sensor is located in the air intake plenum assembly. This sensor is combined with the IAT sensor.

OPERATION

The MAF is measured by detection of heat transfer from a hot film probe because the change of the mass air flow rate causes change in the amount of heat being transferred from the hot film probe surface to the air flow. The air flow sensor generates a pulse so it repeatedly opens and closes between the voltages supplied from the PCM. This results in a change of the temperature of the hot film probe and in the change of resistance.

REMOVAL & INSTALLATION

1. Disconnect the negative battery cable.
2. Disconnect the connector from the sensor.
3. Remove the air cleaner and air intake assembly, as required.
4. Remove the sensor from its mounting.

To install:

5. Installation is the reverse of the removal procedure.

TESTING

1. Be sure the engine is at operating temperature.
2. Remove the sensor connector.
3. Using a voltmeter, measure the output voltage.
4. Specification should be 0.5 volt at idle and 1.0 volt at 2000 rpm.
5. Replace the sensor, as required.

Throttle Position Sensor (TPS)

OPERATION

The throttle position sensor is a rotating type variable resistor that rotates with the throttle body's throttle shaft to sense the throttle valve angle. As the throttle shaft rotates, the throttle angle of the sensor changes and the ECM/PCM detects the throttle valve opening based on the TPS output voltage.

REMOVAL & INSTALLATION

1. Disconnect the negative battery cable.
2. Disconnect the sensor connector.
3. Remove the sensor retaining screws. Remove the sensor from its mounting.

To install:

4. Installation is the reverse of the removal procedure.

TESTING

1. Disconnect the sensor connector.
2. Measure the resistance between terminal 1 (sensor ground) and terminal 4 (sensor power).
3. Specification should be 3.5–6.5 kohm.
4. Connect and analog ohmmeter between terminal 1 (sensor ground) and terminal 3 (sensor output).
5. Operate the throttle valve slowly from the idle position to the full open position, and check that the resistance changes smoothly in proportion with the throttle valve opening angle.
6. If the resistance is out of specification, or fails to change smoothly, replace the sensor.

SPECTRA

See Figures 10 and 11.

Camshaft Position (CMP) Sensor

OPERATION

The camshaft position sensor senses the TDC point of the number one cylinder, on its compression stroke. Its signal is relayed to the ECM to be used to determine the sequence of fuel injection.

REMOVAL & INSTALLATION

1. Disconnect the negative battery cable.
2. Disconnect the connector from the sensor.
3. Remove the mounting bolts.
4. Remove the sensor, from the cylinder head.

To install:

5. Installation is the reverse of the removal procedure.

TESTING

2000–2003

1. Be sure the ignition switch is in the OFF position.
2. Disconnect the sensor connector.
3. Check for proper sensor installation.
4. If the sensor is installed properly, install a known good sensor and recheck. Replace the sensor, as required.
5. If the sensor is not installed properly, remove the sensor and reinstall it. Check for proper operation.

2004–2005

1. Be sure the ignition switch is in the OFF position.
2. Disconnect the sensor connector.
3. Remove the sensor from the engine.
4. Apply battery voltage to terminal 1 and 5 volts to terminal 2. Ground terminal 3 on the CMPS.
5. Install an LED between plus 5 volt power and CMPS terminal 2 and a steel wheel-tooth wheel (or anything made of steel) at the CMPS tip.
6. Rotate the steel wheel slowly and check if the LED light flashes.
7. If the LED flashes, the sensor is working.
8. Replace the sensor, as required.

2006

1. Be sure that the CMPS and PCM connectors are connected.
2. Set up an oscilloscope as follows. Channel "A" (+): terminal 2 of the CKPS, (-): ground. Channel "B" (+): terminal 2 of the CMPS, (-): ground.
3. Start the engine and check for a signal waveform and whether synchronization with CKPS has occurred or not and a tooth is missing.
4. Specification should be as follows. The square waveform signal should be smooth and without distortion. The CMPS falling (rising) edge is coincided with 3–5 tooth of the CKP from one longer signal (missing tooth).
5. If the waveform pattern is normal check for proper connection between the PCM and the component.
6. If the waveform signal is not normal, remove the sensor and calculate the air gap it should be 0.07 inch. Recheck for proper operation. If the problem is corrected, replace the sensor.

Crankshaft Position (CKP) Sensor

OPERATION

The CKP is used to inform the ECM when the No.1 piston is at top dead center. This information is used for controlling and adjusting ignition and fuel injector timing.

REMOVAL & INSTALLATION

1. Disconnect the negative battery cable.
2. Disconnect the connector from the sensor.
3. Remove the bolt that retains the sensor in place.
4. Remove the sensor from the transmission cover

To install:

5. Installation is the reverse of the removal procedure.

TESTING

2000–2003

1. Measure the gap between the sensor and the target wheel.
2. Specification should be 0.037–0.067 inch.
3. If the gap is within specification, install a known good sensor and recheck. Replace the sensor, as required.
4. If the gap is not within specification, check the target wheel for proper installation. Adjust the target wheel gap.

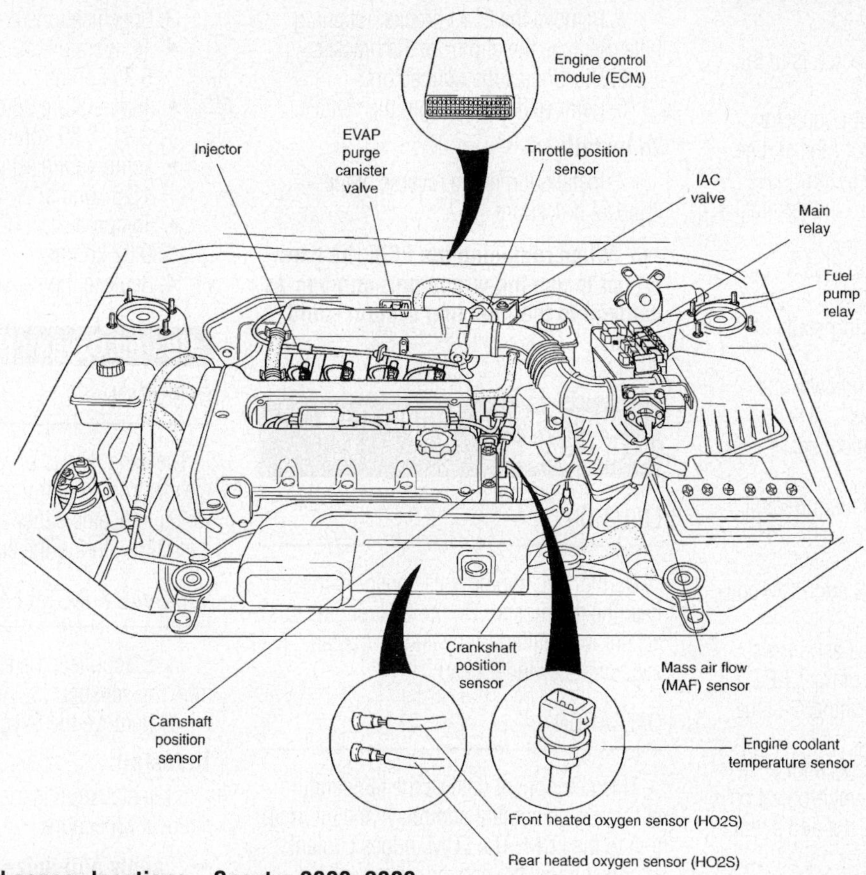

Fig. 10 Underhood sensor locations—Spectra 2000–2003

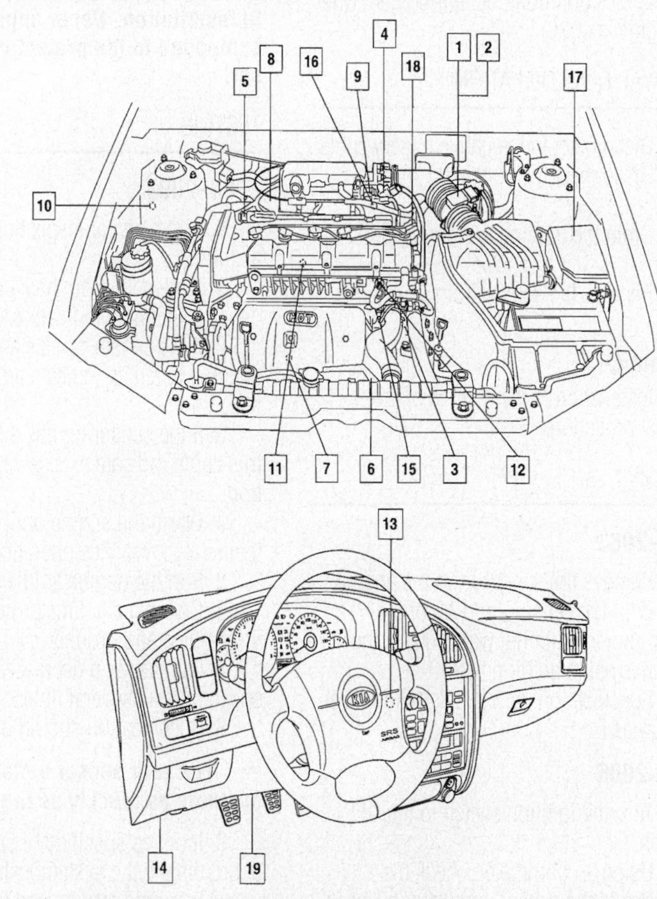

1. Mass Air Flow Sensor (MAFS)
2. Intake Air Temperature Sensor (IATS)
3. Engine Coolant Temperature Sensor (ECTS)
4. Throttle Position Sensor (TPS)
5. Camshaft Position Sensor (CMPS)
6. Crankshaft Position Sensor (CKPS)
7. Heated Oxygen Sensor (HO2S, Sensor 1)
8. Injector
9. Idle Speed Control Actuator (ISCA)
10. Vehicle Speed Sensor (VSS)
11. Knock Sensor
12. CVVT Oil Control Valve (OCV)
13. Ignition Switch
14. ECM
15. CVVT Oil Temperature Sensor (OTS)
16. Purge Control Solenoid Valve (PCSV)
17. Main Relay
18. Ignition Coil
19. DLC (Diagnostic Link Cable)

29130_KIA_G0009

Fig. 11 Underhood sensor locations—Spectra 2004–2006

2004–2005

1. Be sure the ignition switch is in the OFF position.
2. Disconnect the sensor connector.
3. Remove the sensor from the engine.
4. Apply battery voltage to terminal 1 and 5 volts to terminal 2. Ground terminal 3 of the CKPS.
5. Install an LED between plus 5 volt power and CMPS terminal 2 and a steel wheel-tooth wheel (or anything made of steel) at the CMPS tip.
6. Rotate the steel wheel slowly and check if the LED light flashes.
7. If the LED flashes, the sensor is working.
8. Replace the sensor, as required.

2006

1. Be sure that the CKPS and PCM connectors are connected.
2. Set up an oscilloscope as follows. Channel "A" (+): terminal 2 of the CKPS, (-): ground. Channel "B" (+): terminal 2 of the CMPS, (-): ground.
3. Start the engine and check for a signal waveform and whether synchronization with CMPS has occurred or not and a tooth is missing.
4. Specification should be as follows. The square waveform signal should be smooth and without distortion. The CMPS falling (rising) edge is coincided with 3–5 tooth of the CKP from one longer signal (missing tooth).
5. If the waveform pattern is normal check for proper connection between the PCM and the component.
6. If the waveform signal is not normal, remove the sensor and measure the distance from the housing to teeth on the flywheel/ torque converter (measurement "A") and from the mounting surface on the sensor to the sensor tip (measurement "B"). Subtract "B" from "A", specification should be 0.012–0.067 inch.
7. Recheck for proper operation. If the problem is corrected, replace the sensor.

Electronic Control Module (ECM)

OPERATION

The ECM controls the vehicle engine operating system.

REMOVAL & INSTALLATION

1. Disconnect the negative battery cable.
2. Remove the lower inner trim.
3. As required, detach the floor mat. As required, remove the protective cover.
4. Remove the ECM bracket retaining nuts. Remove the clip from the bracket.
5. Disconnect the connectors.
6. Remove the ECM from the vehicle.

To install:

7. Installation is the reverse of the removal procedure.

➡ **When replacing the ECM, be careful not to use the wrong part number, as damage to the injection system could occur.**

Engine Coolant Temperature (ECT) Sensor

LOCATION

On the 1.8L engine, this sensor is located near the thermostat. On the 2.0L engine this sensor is installed in the engine coolant passage of the cylinder head

OPERATION

This component detects the temperature of the engine coolant and relays the information to the ECM. The ECM judges coolant temperature by the sensor output voltage and provides optimum fuel enrichment when the engine is cold.

REMOVAL & INSTALLATION

1. Disconnect the negative battery cable.
2. Disconnect the connector from the sensor.
3. Drain the cooling system, as required.
4. Remove the sensor from its mounting.

To install:

5. Installation is the reverse of the removal procedure.

TESTING

2000–2003

1. Remove the sensor from the engine.
2. With the sensing portion of the sensor immersed in hot water, check the resistance using an ohmmeter.
3. Specification is 2.31–2.59 kohms at 68 degrees F.

2004–2006

1. Turn the ignition switch to the OFF position.
2. Using an ohmmeter, check the resistance of the sensor connector between terminals 1 and 3 (component side).
3. Specifications are as follows:
   - Temperature 32 degrees F, resistance 5.79 kohms
   - Temperature 68 degrees F, resistance 2.31–2.59 kohms
   - Temperature 104 degrees F, resistance 1.15 kohms
   - Temperature 176 degrees F, resistance 0.32 kohms
4. Replace the sensor, as required.

Heated Oxygen (HO2S) Sensor

OPERATION

The exhaust gas oxygen sensor supplies the electronic control assembly with a signal which indicates either a rich or lean mixture condition, during the engine operation.

REMOVAL & INSTALLATION

1. Disconnect the electrical connector from the sensor.
2. Remove the oxygen sensor.

To install:

3. Installation is the reverse of the removal procedure.

➡ **Apply anti-seize compound to the threaded portion of the sensor, prior to installation. Never apply anti-seize compound to the protector of the sensor.**

TESTING

2000–2003

Perform a visual inspection of the sensor as follows:

1. If the sensor tip has a black/sooty deposit, this may indicate a rich fuel mixture.
2. If the sensor tip has a white, gritty deposit, this may indicate an internal coolant leak.
3. If the sensor tip has a brown deposit, this could indicate excessive oil consumption.
4. Warm the engine until the coolant temperature reaches operating temperature.
5. Run the engine at idle speed.
6. Connect a voltmeter between terminal 4 (grn/brn) and ground.
7. Increase and decrease the engine speed quickly several times.
8. Specification should be 0–1.0 volt.

➡ **The rear sensor voltage does not fluctuate as quickly as the front sensor.**

9. If not as specified, inspect the onboard diagnostic system, exhaust system, intake manifold system and fuel line pressure.

10. If all systems are normal, replace the sensor.

11. To check the heater, be sure that the ignition switch is OFF.

12. Disconnect the sensor connector from the wire harness, if not already done.

13. Connect an ohmmeter between terminals 1 and 3. Measure the resistance.

14. Specification should be approximately 2–4 ohms at 68 degrees F.

15. If not as specified, replace the sensor.

2004–2006

SENSOR ONE

Perform a visual inspection of the sensor as follows:

1. If the sensor tip has a black/sooty deposit, this may indicate a rich fuel mixture.
2. If the sensor tip has a white, gritty deposit, this may indicate an internal coolant leak.
3. If the sensor tip has a brown deposit, this could indicate excessive oil consumption.
4. Turn the ignition switch to the OFF position. Disconnect the sensor connector.
5. Using an ohmmeter measure the resistance between terminal 5 and 6 of the sensor connector.
6. Specifications are as follows:
   - Temperature 68 degrees F, resistance 2.4 ohms
   - Temperature 212 degrees F, resistance 3.3 ohms
   - Temperature 392 degrees F, resistance 4.0 ohms
   - Temperature 572 degrees F, resistance 4.6 ohms
   - Temperature 752 degrees F, resistance 5.3 ohms
   - Temperature 932 degrees F, resistance 5.8 ohms
   - Temperature 1112 degrees F, resistance 6.4 ohms
   - Temperature 1292 degrees F, resistance 6.9 ohms
7. Replace the sensor if not within specification.

SENSOR TWO

Perform a visual inspection of the sensor as follows:

1. If the sensor tip has a black/sooty deposit, this may indicate a rich fuel mixture.
2. If the sensor tip has a white, gritty deposit, this may indicate an internal coolant leak.
3. If the sensor tip has a brown deposit, this could indicate excessive oil consumption.
4. Turn the ignition switch to the OFF position. Disconnect the sensor connector.
5. Using an ohmmeter measure the resistance between terminal 3 and 4 of the sensor connector.
6. Specifications are as follows:
   - Temperature 68 degrees F, resistance 9.2 ohms
   - Temperature 212 degrees F, resistance 10.7 ohms
   - Temperature 392 degrees F, resistance 13.1 ohms
   - Temperature 572 degrees F, resistance 14.6 ohms
   - Temperature 752 degrees F, resistance 17.7 ohms
   - Temperature 932 degrees F, resistance 19.2 ohms
   - Temperature 1112 degrees F, resistance 20.7 ohms
   - Temperature 1292 degrees F, resistance 22.5 ohms
7. Replace the sensor if not within specification.

Idle Speed Control Actuator (ISCA)

OPERATION

The idle speed control actuator sensor is installed on the intake manifold. This sensor controls the intake airflow that is bypassed around the throttle plate to keep constant engine speed when the throttle valve is closed. According to signals from the ECM, the valve rotor rotates to control the by pass airflow into the engine.

REMOVAL & INSTALLATION

1. Disconnect the negative battery cable.
2. Disconnect the connector from the sensor.
3. Remove the sensor retaining screws.
4. Remove the sensor from its mounting.

To install:

5. Installation is the reverse of the removal procedure.

TESTING

2004–2005

1. Turn the ignition switch to the OFF position.
2. Disconnect the sensor connector.
3. Using an ohmmeter, measure the resistance between terminals 2 and 3, of the closing coil.
4. Specifications are as follows:
   - Temperature -4 degrees F, resistance 9.2–10.8 ohms
   - Temperature 32 degrees F, resistance 10.2–11.8 ohms
   - Temperature 68 degrees F, resistance 11.1–12.7 ohms
   - Temperature 104 degrees F, resistance 12.0–13.6 ohms
   - Temperature 140 degrees F, resistance 12.9–14.5 ohms
   - Temperature 176 degrees F, resistance 13.8–15.4 ohms
5. Using a scan tool inspect the signal waveform of the closing coil.
6. If not within specification, replace the sensor.

2006

1. Turn the ignition switch to the OFF position.
2. Disconnect the sensor connector.
3. Using a multimeter, measure the resistance between terminals 1 and 2, of the valve connector (component side).
4. Specifications are as follows:
   - ISCA normal parameter: opening coil 11.1–12.7 ohms, closing coil 14.6–16.2 ohms
   - TPS normal parameter: idle 0.2–0.8 volt, WOT 4.3–4.8 volts
5. If not within specification, check for contamination, deterioration or damage. Substitute a known good component. Recheck, and replace the sensor as required.

Intake Air Temperature (IAT) Sensor

OPERATION

This sensor is a resistor based sensor which detects the intake air temperature. According to the intake air temperature reading the ECM/PCM will control the necessary amount of fuel injection.

REMOVAL & INSTALLATION

1. Disconnect the negative battery cable.
2. Disconnect the connector from the sensor.
3. Remove the sensor retaining screws, as required.
4. Remove the air cleaner and air intake assembly, as required.
5. Remove the sensor from its mounting.

To install:

6. Installation is the reverse of the removal procedure.

TESTING

2004–2005

1. Turn the ignition switch to the OFF position.
2. Disconnect the sensor connector.
3. Using an ohmmeter, measure the resistance between terminals 1 and 5, of the connector.
4. Specifications are as follows:
- Temperature -4 degrees F, resistance 14.26–16.02 ohms
- Temperature 32 degrees F, resistance 5.50–6.05 ohms
- Temperature 68 degrees F, resistance 2.35–2.54 ohms
- Temperature 104 degrees F, resistance 1.11–1.19 ohms
- Temperature 140 degrees F, resistance 0.57–0.60 ohms
- Temperature 176 degrees F, resistance 0.31–0.32 ohms
5. If not, replace the sensor.

2006

1. Turn the ignition switch to the OFF position.
2. Disconnect the sensor connector.
3. Using an ohmmeter, measure the resistance between terminals 1 and 5, of the connector.
4. Specifications are as follows:
- Temperature -4 degrees F, resistance 14.26–16.02 ohms
- Temperature 32 degrees F, resistance 5.50–6.05 ohms
- Temperature 68 degrees F, resistance 2.35–2.54 ohms
- Temperature 104 degrees F, resistance 1.11–1.19 ohms
- Temperature 140 degrees F, resistance 0.57–0.60 ohms
- Temperature 176 degrees F, resistance 0.31–0.32 ohms
5. If not within specification, check for contamination, deterioration or damage. Substitute a known good component. Recheck, and replace the sensor as required.
6. If within specification, check for poor connection between the PCM and the component.

Knock Sensor (KS)

OPERATION

The knock sensor is used to detect engine vibrations caused by preignition or detonation and provides information to the ECM, which then retards the timing to eliminate detonation.

REMOVAL & INSTALLATION

1. Disconnect the negative battery cable.
2. Remove the intake manifold support bracket.
3. Disconnect the sensor connector.
4. Remove the sensor from its mounting.

To install:

5. Installation is the reverse of the removal procedure.

TESTING

2000–2003

1. Disconnect the sensor electrical connector.
2. Remove the sensor from the engine.
3. Mount the sensor in the jaws of a bench vise.
4. Connect a voltmeter between terminal 1 and 2.
5. Wrap the bench vise sharply with a hammer and note the voltmeter reading.
6. Verify that the voltage spike is less than 1 volt.
7. If no voltage spike is observed, replace the sensor.

2004–2005

1. Be sure that the ECM and sensor connectors are connected.
2. Connect the scan tool.
3. Start the engine and monitor the knock sensor signal at idle.
4. If the signal is not normal, replace it.
5. Turn the ignition switch OFF.
6. Disconnect the sensor electrical connector.
7. Measure the capacitance between terminal 1 and 2 of the sensor connector.
8. Specification should be 800–1600 pF.
9. If not within specification, replace the sensor.

2006

1. Turn the ignition switch OFF.
2. Disconnect the sensor electrical connector. Connect an ohmmeter.
3. Measure the resistance between terminals 1 and 2 of the sensor connector (component side).
4. Specification should be 5mohm at 68 degrees F.
5. If not within specification, replace the sensor.
6. To check the output signal, remove the sensor from its mounting.
7. Mount the sensor in the jaws of a bench vise.
8. Connect a voltmeter between terminal 1 and 2.
9. Wrap the bench vise sharply with a hammer and note the voltmeter reading.
10. Verify that the voltage spike is less than 1 volt.
11. If no voltage spike is observed, replace the sensor.

Mass Air Flow (MAF) Sensor

OPERATION

The MAF provides a signal to the ECM for incoming air temperature. This information is used to adjust the injector pulse width and in turn the air/fuel ratio.

REMOVAL & INSTALLATION

1. Disconnect the negative battery cable.
2. Disconnect the connector from the sensor.
3. Remove the air cleaner and air intake assembly, as required.
4. Remove the sensor from its mounting.

To install:

5. Installation is the reverse of the removal procedure.

TESTING

2000–2003

1. Allow the engine to reach operating temperature.
2. Let the engine idle.
3. Connect a voltmeter between terminal 4 (red//wht).
4. Verify that the voltage varies between 0.8–1.2 volts.
5. Rev the engine and verify that the voltage varies between 3.5–4.0 volts.

If not within specification, replace the sensor.

2004–2005

1. Be sure that the ECM and the MAFS connectors are connected.
2. Connect the scan tool to the data link connector.
3. Start the engine.
4. Monitor the MAPS signals.
5. Specification should be 0.6–1.0 volt at idle and 1.7–2.0 volts at 3000 rpm.
6. Replace the sensor, as required.

2006

1. Connect the scan tool. Turn the engine ON.

2. Install the scan tool and monitor the "MASS AIR FLOW" parameter on the scan tool data list.

3. Specification should be 0.6–1.0 volt at idle and no load.

4. If within specification, check for poor connection between the PCM and the component.

5. If not, substitute the sensor with a known good component, check for proper operation. If problem corrected replace the sensor.

Throttle Position Sensor (TPS)

OPERATION

On the 1.8L engine, the throttle position sensor provides a signal to the ECM that is related to the relative throttle plate position. As the throttle plate moves in relation to driving conditions, a signal is sent to the control unit which adjusts the injector pulse width and air/fuel ratio. As the throttle plate is opened further, more air is taken into the combustion chambers, and as a result the relative fuel demand of the engine changes. The throttle position sensor relays this information to the ECM which in alters the fuel amount. On the 2.0L engine, the throttle position sensor is a rotating type variable resistor that rotates with the throttle body's throttle shaft to sense the throttle valve angle. As the throttle shaft rotates, the throttle angle of the sensor changes and the ECM detects the throttle valve opening based on the TPS output voltage.

REMOVAL & INSTALLATION

1. Be sure the engine is in the OFF position.

2. Disconnect the sensor connector.

3. Remove the sensor retaining bolts.

4. Remove the sensor.

To install:

5. Installation is the reverse of the removal procedure.

TESTING

2000–2003

1. Disconnect the connector from the sensor.

2. Connect an ohmmeter between sensor terminals 1 and 3.

3. Verify that the resistance increases according to throttle angle.

4. Specification should be 2 kohms (plus or minus 0.4 kohm) with the throttle valve closed.

5. If not within specification, replace the sensor.

2004–2005

1. Turn the ignition switch to the OFF position.

2. Disconnect the TPS connector.

3. Using an ohmmeter, measure the resistance between terminals 1 and 2 of the TPS connector.

4. Using an ohmmeter, measure the resistance between terminals 2 and 3 of the TPS connector.

5. Specification at terminal 1 and 2 should be 0.71–1.38 kohms at closed throttle and 2.7 kohms at WOT.

6. Specification at terminal 2 and 3 should be 1.6–2.4 kohms at all throttle positions.

7. If not within specification, replace the sensor.

2006

1. Turn the ignition switch to the OFF position.

2. Disconnect the TPS connector.

3. Using an ohmmeter, measure the resistance between terminals 2 and 3 of the TPS.

4. Specification should be 1.6–2.4 kohms at all throttle positions.

5. With the connector still disconnected, measure the resistance between terminals 1 and 2.

6. Operate the throttle valve slowly from the idle position to the full open position and check that the resistance changes smoothly in proportion with the throttle valve opening angle.

7. Specification should be 0.71–1.38 kohm at closed throttle valve and 2.7 kohm at wide open throttle.

8. If within specification, check for poor connection between the PCM and the component.

9. If not, substitute the sensor with a known good component, check for proper operation. If problem corrected replace the sensor.

SPORTAGE

See Figures 12 through 15.

Camshaft Position (CMP) Sensor

LOCATION

On the 2005–2006 2.0L engine the camshaft position sensor is located near the top of the engine, on the left side of the engine. On the 2.7L engine the camshaft position sensor is located near the top of the engine, on the right side of the engine near the ignition coils.

OPERATION

The camshaft position sensor senses the TDC point of the number one cylinder, on its compression stroke. Its signal is relayed to the ECM to be used to determine the sequence of fuel injection.

REMOVAL & INSTALLATION

1998–2002

1. Disconnect the negative battery cable.

2. Disconnect the connector from the sensor.

3. Remove the mounting bolts.

4. Remove the sensor, from the cylinder head.

To install:

5. Installation is the reverse of the removal procedure.

2005–2006

1. Disconnect the negative battery cable.

2. Disconnect the connector from the sensor.

3. Remove the mounting bolts.

4. Remove the sensor, from the cylinder head.

To install:

5. Installation is the reverse of the removal procedure.

TESTING

1996–1997

1. Disconnect the sensor electrical connector.

2. Using a multimeter, check for continuity between terminal 3 and ground of the harness side connector.

3. If continuity exists, replace the sensor.

4. If continuity does not exist repair or replace the wiring harness.

1998–2002

1. Be sure the ignition switch is OFF.

2. Remove the sensor from its mounting.

3. Install a known good component.

4. If the sensor is functioning properly, replace it.

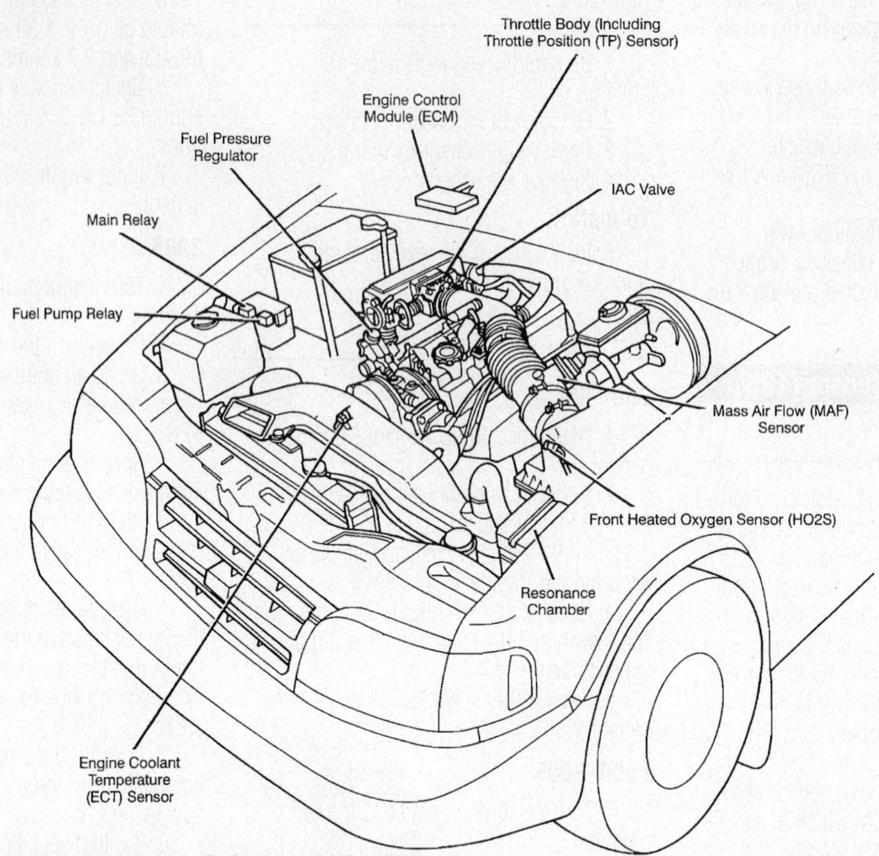

Fig. 12 Underhood sensor locations—Sportage 1996–1997

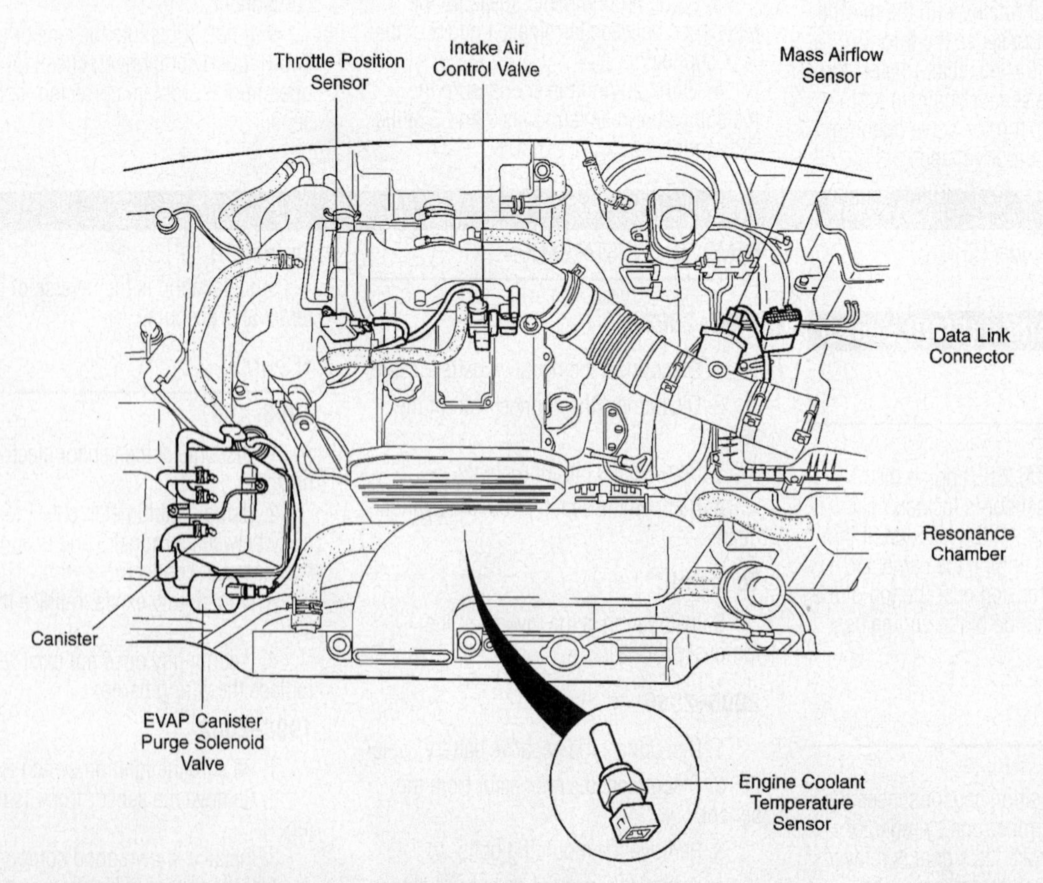

Fig. 13 Underhood sensor locations—Sportage 1998–2002

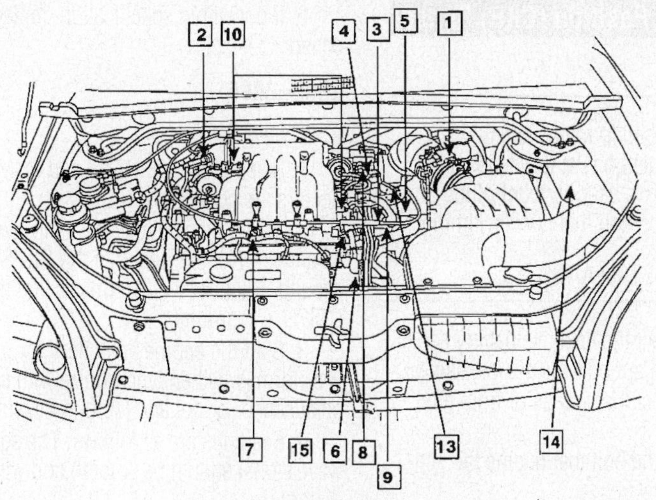

1. Mass Air Flow Sensor (MAFS) [With CVVT]
2. Intake Air Temperature Sensor (IATS)
3. Engine Coolant Temperature Sensor (ECTS)
4. Throttle Position Sensor (TPS)
5. Camshaft Position Sensor (CMPS)
6. Crankshaft Position Sensor (CKPS)
7. Injector
8. Idle Speed Control Actuator (ISCA)
9. Knock Sensor
10. VIS Control solenoid valve
11. Ignition Switch
12. ECM
13. Purge Control Solenoid Valve (PCSV)
14. Main Relay
15. Ignition Coil
16. DLC (Diagnostic Link Connector)

29130_HYUN_29

Fig. 14 Underhood sensor locations—Sportage 2005–2006 2.0L engine

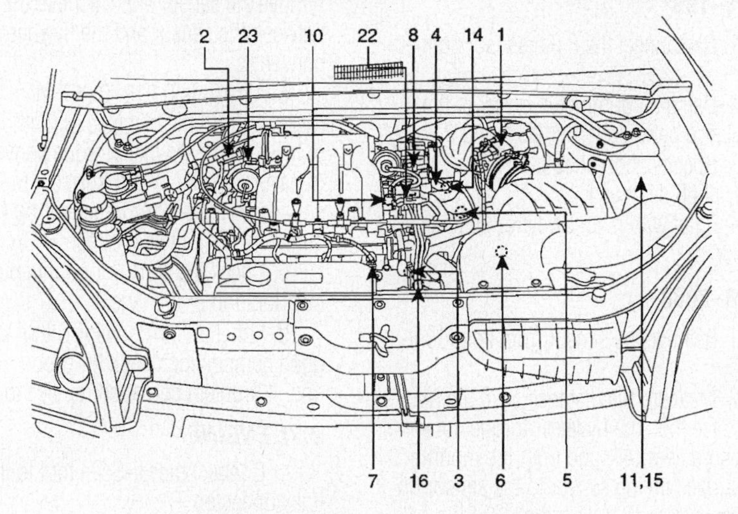

1. Mass Air Flow Sensor (MAFS)
2. Intake Air Temperature Sensor (IATS)
3. Engine Coolant Temperature Sensor (ECTS)
4. Throttle Position Sensor (TPS)
5. Camshaft Position Sensor (CMPS)
6. Crankshaft Position Sensor (CKPS)
7. Injector
8. Idle Speed Control Actuator (ISCA)
10. Knock Sensor
11. Fuel Pump Relay
12. Ignition Switch
13. ECM
14. Purge Control Solenoid Valve (PCSV)
15. Main Relay
16. Ignition Coil
17. DLC (Diagnostic Link Connector)
23. Intake Manifold Tuning Valve #1 (Intake Manifold Side)

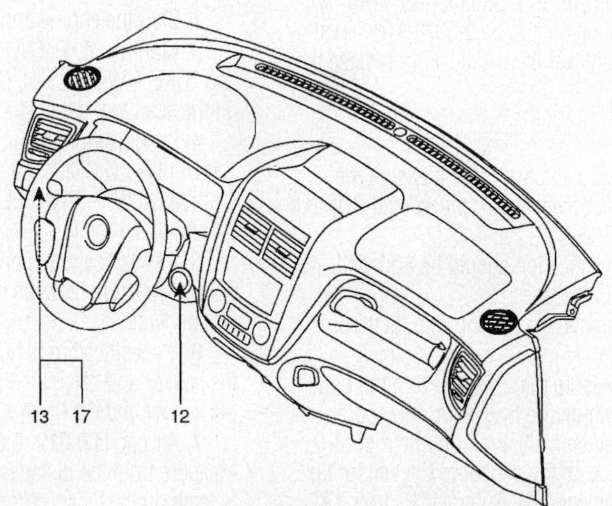

29130_KIA_G0012

Fig. 15 Underhood sensor locations—Sportage 2005–2006 2.7L engine

2005–2006

2.0L ENGINE

1. Be sure that the CMPS and PCM connectors are connected.

2. Set up an oscilloscope as follows: Channel A (+): terminal 2 of the CKPS, (-): ground. Channel B (+): terminal 2 of the CMPS, (-): ground.

3. Start the engine and check for signal waveform (whether synchronize with crankshaft sensor or not and tooth is missing).

4. Readings are as follows: The square wave signal should be smooth and without distortion.

5. Readings are as follows: The CMPS falling (rising) edge should coincide with 3–5 tooth of the CKP from one longer signal (missing tooth).

6. If the waveform signal is normal, check for poor connection between the PCM and the components.

7. If the waveform signal is not normal, remove the sensor and calculate the air gap between the sensor and the flywheel/torque converter.

8. Air gap is 0.07 inch. Measure from the distance of the housing to teeth on the flywheel/torque converter (measurement "A") and from the mounting surface on the sensor to sensor tip (measurement "B"), then subtract "B" from "A".

9. Check the sensor for contamination, deterioration or damage.

10. Substitute the sensor with a known good component, check for proper operation. If problem corrected replace the sensor.

2.7L ENGINE

1. Connect the scan tool to the data link connector.

2. Start the engine and let it idle.

3. Monitor "CKP T/WHEELS-LO CMP" and "CKP T/WHEELS-HI CMP" parameters on the scan tool data list.

4. Specification should be "CKP T/WHEELS-LO CMP": 38 plus/minus 4 tooth, and "CKP T/WHEELS-HI CMP": 98 plus/minus 4 tooth.

5. If specification is normal, check for poor connection between the ECM and the components.

6. If specification is not normal, remove the sensor and calculate the air gap.

7. Check the sensor for contamination, deterioration or damage.

8. Substitute the sensor with a known good component, check for proper operation. If problem corrected replace the sensor.

Crankshaft Position (CKP) Sensor

OPERATION

The CKP is used to inform the ECM when the No.1 piston is at top dead center. This information is used for controlling and adjusting ignition and fuel injector timing.

REMOVAL & INSTALLATION

1. Disconnect the negative battery cable.
2. Raise and support the vehicle safely.
3. Disconnect the connector from the sensor.
4. Remove the bolt that retains the sensor in place.
5. Remove the sensor from the transmission cover

To install:

6. Installation is the reverse of the removal procedure.

TESTING

1996–1997

1. Disconnect the 3 pin sensor connector.
2. Using an ohmmeter, measure the resistance between terminals 1 and 2.
3. Specification should be 500–600 ohms at 68 degrees F.
4. If not within specification, replace the sensor.

1998–2000

1. Remove the sensor from its mounting.
2. Measure the distance from the housing to teeth on the flywheel/torque converter (measurement "A") and from the mounting surface on the sensor to the sensor tip (measurement "B"). Subtract "B" from "A", specification should be 0.037–0.067 inch.
3. If not within specification, replace the sensor.

2001

1. Using an ohmmeter, measure the resistance between BOP pin 16 and BOB pin 43.
2. Specification should be 800–900 ohms at 68 degrees F.
3. Remove the sensor from its mounting.
4. Measure the distance from the housing to teeth on the flywheel/torque converter (measurement "A") and from the mounting surface on the sensor to the sensor tip (measurement "B"). Subtract "B" from "A", specification should be 0.037–0.067 inch.
5. If not within specification, replace the sensor.

2005–2006

2.0L ENGINE

1. Be sure that the CKPS and PCM connectors are connected.

2. Set up an oscilloscope as follows: Channel A (+): terminal 2 of the CKPS, (-): ground. Channel B (+): terminal 2 of the CMPS, (-): ground.

3. Start the engine and check for signal waveform (whether synchronize with camshaft sensor or not and tooth is missing).

4. Readings are as follows: The square wave signal should be smooth and without distortion.

5. Readings are as follows: The CMPS falling (rising) edge should coincide with 3–5 tooth of the CKP from one longer signal (missing tooth).

6. If the waveform signal is normal, check for poor connection between the PCM and the components.

7. If the waveform signal is not normal, remove the sensor and calculate the air gap between the sensor and the flywheel/torque converter.

8. Air gap is 0.012–0.067 inch. Measure from the distance of the housing to teeth on the flywheel/torque converter (measurement "A") and from the mounting surface on the sensor to sensor tip (measurement "B"), then subtract "B" from "A".

9. Check the sensor for contamination, deterioration or damage.

10. Substitute the sensor with a known good component, check for proper operation. If problem corrected replace the sensor.

2.7L ENGINE

1. Connect the Hi-Scan tool to the data link connector.

2. Start the engine and let it idle.

3. Monitor "CKP T/WHEELS-LO CMP" and "CKP T/WHEELS-HI CMP" parameters on the scan tool data list.

4. Specification should be "CKP T/WHEELS-LO CMP": 38 plus/minus 4 tooth, and "CKP T/WHEELS-HI CMP": 98 plus/minus 4 tooth.

5. If specification is normal, check for poor connection between the ECM and the components.

6. If specification is not normal, remove the sensor and calculate the air gap between the sensor and the flywheel/torque converter.

7. Air gap is 0.012–0.067 inch. Measure from the distance of the housing to teeth on the flywheel/torque converter (measurement "A") and from the mounting

surface on the sensor to sensor tip (measurement "B"), then subtract "B" from "A".

8. Check the sensor for contamination, deterioration or damage.

9. Substitute the sensor with a known good component, check for proper operation. If problem corrected replace the sensor.

Electronic Control Module (ECM)

OPERATION

The ECM controls the vehicle engine operating system.

REMOVAL & INSTALLATION

1. Disconnect the negative battery cable.
2. Remove the lower inner trim.
3. As required, detach the floor mat. As required, remove the protective cover.
4. Remove the ECM bracket retaining nuts. Remove the clip from the bracket.
5. Disconnect the connectors.
6. Remove the ECM from the vehicle.

To install:

7. Installation is the reverse of the removal procedure.

➡ **When replacing the ECM, be careful not to use the wrong part number, as damage to the injection system could occur.**

Engine Coolant Temperature (ECT) Sensor

OPERATION

This component detects the temperature of the engine coolant and relays the information to the ECM. The ECM judges coolant temperature by the sensor output voltage and provides optimum fuel enrichment when the engine is cold.

REMOVAL & INSTALLATION

1. Disconnect the negative battery cable.
2. Disconnect the connector from the sensor.
3. Drain the cooling system, as required.
4. Remove the sensor from its mounting.

To install:

5. Installation is the reverse of the removal procedure.

TESTING

1996–2002

1. Remove the sensor from the engine.
2. With the sensing portion of the sensor immersed in hot water, check the resistance using an ohmmeter.
3. Specifications are as follows:
   - Temperature 68 degrees F, resistance 2.45 kohms (plus/minus 0.25)
   - Temperature 176 degrees F, resistance 0.322 kohms (plus/minus 0.032)
4. Replace the sensor, as required.

2005–2006

1. Turn the ignition switch to the OFF position.
2. Disconnect the ECTS connector.
3. Check the resistance of the sensor connector between terminals 1 and 3 (component side).
4. Specifications should be as follows.
   - Temperature 32 degrees F, resistance 5.79 kohms (2.0L engine)
   - Temperature 68 degrees F, resistance 2.31–2.59 kohms (2.0L and 2.7L engines)
   - Temperature 176 degrees F, resistance 0.32 kohms (2.7L engine)
5. If specification is normal, check for poor connection between the ECM and the components.
6. If not within specification, substitute the sensor with a known good component, check for proper operation. If problem corrected replace the sensor.

Heated Oxygen (HO2S) Sensor

OPERATION

The exhaust gas oxygen sensor supplies the electronic control assembly with a signal which indicates either a rich or lean mixture condition, during the engine operation.

REMOVAL & INSTALLATION

1. Disconnect the electrical connector from the sensor.
2. Remove the oxygen sensor.

To install:

3. Installation is the reverse of the removal procedure.

➡ **Apply anti-seize compound to the threaded portion of the sensor, prior to installation. Never apply anti-seize compound to the protector of the sensor.**

TESTING

1996–2002

Perform a visual inspection of the sensor as follows:

1. If the sensor tip has a black/sooty deposit, this may indicate a rich fuel mixture.
2. If the sensor tip has a white, gritty deposit, this may indicate an internal coolant leak.
3. If the sensor tip has a brown deposit, this could indicate excessive oil consumption.
4. Warm the engine until the coolant temperature reaches operating temperature.
5. Run the engine at idle speed.
6. Connect a voltmeter between terminal 4 (blk) and ground.
7. Increase and decrease the engine speed quickly several times.
8. Specification should be 0–1.0 volt.

➡ **The rear sensor voltage does not fluctuate as quickly as the front sensor.**

9. If not as specified, inspect the on-board diagnostic system, exhaust system, intake manifold system and fuel line pressure.
10. If all systems are normal, replace the sensor.
11. To check the heater, be sure that the ignition switch is OFF.
12. Disconnect the sensor connector from the wire harness, if not already done.
13. On 1996–1997 vehicles connect an ohmmeter between terminals 1 and 3. Measure the resistance. Specification should be approximately 6 ohms at 68 degrees F.
14. On 1998–2002 vehicles connect an ohmmeter between terminals 2 and 4. Measure the resistance. Specification should be approximately 5 ohms at 68 degrees F.
15. If not as specified, replace the sensor.

2005–2006

2.0L ENGINE (SENSOR ONE)

Perform a visual inspection of the sensor as follows:

1. If the sensor tip has a black/sooty deposit, this may indicate a rich fuel mixture.
2. If the sensor tip has a white, gritty deposit, this may indicate an internal coolant leak.
3. If the sensor tip has a brown deposit, this could indicate excessive oil consumption.
4. Warm the engine until the engine reaches operating temperature.
5. Connect the scan tool and monitor sensor operation.

6. Verify that the signal is switching from rich (above 0.45 volt) to lean (below 0.45 volt) a minimum of 3 times in 10 seconds (voltage will vary between 0.1–0.9 volt).

7. If the sensor is operating properly, check for poor connection between the ECM and the component.

8. If not, check the sensor for contamination, deterioration or damage.

9. Substitute the sensor with a known good component, check for proper operation. If problem corrected replace the sensor.

2.0L ENGINE (SENSOR TWO)

Perform a visual inspection of the sensor as follows:

1. If the sensor tip has a black/sooty deposit, this may indicate a rich fuel mixture.

2. If the sensor tip has a white, gritty deposit, this may indicate an internal coolant leak.

3. If the sensor tip has a brown deposit, this could indicate excessive oil consumption.

4. Warm the engine until the engine reaches operating temperature. Check that the HO2S signal is active.

5. Connect the scan tool and monitor the "O2 SNSR VOLT." parameter on the scan tool data list.

6. Test condition should be Engine ON and in idle at closed loop condition. Specification should be above 0.6 volt.

7. If the sensor is operating properly, check for poor connection between the PCM and the component.

8. If not, check the sensor for contamination, deterioration or damage.

9. Substitute the sensor with a known good component, check for proper operation. If problem corrected replace the sensor.

2.7L ENGINE (SENSOR ONE)

Perform a visual inspection of the sensor as follows:

1. If the sensor tip has a black/sooty deposit, this may indicate a rich fuel mixture.

2. If the sensor tip has a white, gritty deposit, this may indicate an internal coolant leak.

3. If the sensor tip has a brown deposit, this could indicate excessive oil consumption.

4. Warm the engine until the engine reaches operating temperature.

5. Connect the scan tool and monitor sensor operation.

6. Verify that the signal is switching from rich (above 0.45 volt) to lean (below 0.45 volt) a minimum of 3 times in 10 seconds (voltage will vary between 0.1–0.9 volt).

7. If the sensor is operating properly, check for poor connection between the ECM and the component.

8. If not, check the sensor for contamination, deterioration or damage.

9. Substitute the sensor with a known good component, check for proper operation. If problem corrected replace the sensor.

2.7L ENGINE (SENSOR TWO)

Perform a visual inspection of the sensor as follows:

1. If the sensor tip has a black/sooty deposit, this may indicate a rich fuel mixture.

2. If the sensor tip has a white, gritty deposit, this may indicate an internal coolant leak.

3. If the sensor tip has a brown deposit, this could indicate excessive oil consumption.

4. Warm the engine until the engine reaches operating temperature.

5. Connect the scan tool and monitor sensor operation.

6. Test condition: engine ON and in idle (closed loop) condition. Specification should be above 0.6 volt.

7. If the sensor is operating properly, check for poor connection between the ECM and the component.

8. If not, check the sensor for contamination, deterioration or damage.

9. Substitute the sensor with a known good component, check for proper operation. If problem corrected replace the sensor.

Idle Speed Control Actuator (ISCA)

LOCATION

The ISCA is installed on the intake manifold.

OPERATION

The idle speed control actuator maintains idle speed according to various engine loads and conditions. This component also provides additional air during startup. It consists of an opening coil, a closing coil and a permanent magnet. According to the control signals from the ECM, the valve rotor rotates to control the bypass air flow into the engine.

REMOVAL & INSTALLATION

1. Disconnect the negative battery cable.

2. Disconnect the connector from the component.

3. Remove the retaining screws.

4. Remove the component from its mounting.

To install:

5. Installation is the reverse of the removal procedure.

TESTING

1. Be sure the ignition is OFF

2. Remove the valve from the throttle body. Check the throttle bore, throttle plate and the ISCA passages for chocking and for foreign objects. Repair as required.

3. Install the valve.

4. Be sure the ignition is ON and the engine is OFF.

5. Install the scan tool. Select the "IDLE SPEED ACTUATOR" parameter on the "Actuation Test" mode.

6. Activate the valve by pressing the "STAT" key.

7. Check the valve for a clicking sound and visually verify that the valve opens and closes.

➡ **Repeat numerous times to ensure reliability.**

8. If the component is operating properly, check for poor connection between ECM and component. Correct, as required.

9. If not, substitute the sensor with a known good component, check for proper operation. If problem corrected replace the component.

Intake Air Temperature (IAT) Sensor

OPERATION

This sensor is a resistor based sensor which detects the intake air temperature. According to the intake air temperature reading the ECM/PCM will control the necessary amount of fuel injection.

REMOVAL & INSTALLATION

1. Disconnect the negative battery cable.

2. Disconnect the connector from the sensor.

3. Remove the sensor retaining screws, as required.

4. Remove the air cleaner and air intake assembly, as required.

5. Remove the sensor from its mounting.

To install:

6. Installation is the reverse of the removal procedure.

TESTING

1998–2002

1. Turn the ignition switch to the OFF position.
2. Disconnect the sensor electrical connector.
3. Using an ohmmeter measure the resistance between terminal 1 and 2 (component side).
4. Specification should be 2.21–2.69 ohms at 68 degrees F.
5. Replace the sensor, as required.

2005–2006

2.0L ENGINE

1. Turn the ignition switch OFF.
2. Disconnect the sensor connector.
3. Using and ohmmeter, measure the resistance between terminals 1 and 5 (component side).
4. Specifications should be as follows.
   - Temperature 32 degrees F, resistance 5.50–6.05 kohms
   - Temperature 68 degrees F, resistance 2.35–2.54 kohms
   - Temperature 104 degrees F, resistance 1.11–1.19 kohms
   - Temperature 176 degrees F, resistance 0.31–0.32 kohms
5. If within specification, check for poor connection between the ECM and the component.
6. If not, substitute the sensor with a known good component, check for proper operation. If problem corrected replace the sensor.

2.7L ENGINE

1. Disconnect the sensor connector.
2. Using and ohmmeter, measure the resistance between terminals 1 and 2 (component side).
3. Specifications should be as follows.
   - Temperature -4 degrees F, resistance 14.26–16.02 kohms
   - Temperature 68 degrees F, resistance 2.35–2.54 kohms
   - Temperature 176 degrees F, resistance 0.31–0.32 kohms
4. If within specification, check for poor connection between the ECM and the component.
5. If not, substitute the sensor with a known good component, check for proper operation. If problem corrected replace the sensor.

Knock Sensor (KS)

OPERATION

The knock sensor is used to detect engine vibrations caused by preignition or detonation and provides information to the ECM, which then retards the timing to eliminate detonation.

REMOVAL & INSTALLATION

1. Disconnect the negative battery cable.
2. Remove the intake manifold support bracket.
3. Disconnect the sensor connector.
4. Remove the sensor from its mounting.

To install:

5. Installation is the reverse of the removal procedure.

TESTING

1996–2002

1. With the sensor still connected, install a voltmeter between terminal 3 (yel/grn) and ground.
2. Turn the ignition switch ON.
3. Tap the engine crossmember bracket, using a wrench.
4. Verify that a voltage spike (less than 1 volt) is output from the sensor.
5. If no voltage spike is observed, replace the sensor.

2005–2006

1. Be sure the ignition switch is OFF.
2. Disconnect the sensor electrical connector. Connect an ohmmeter.
3. Measure the resistance between terminals 1 and 2 (component side). Specification should be 5mohm at 68 degrees F.
4. Replace the sensor, as required.
5. To check the output signal, remove the sensor from its mounting.
6. Mount the sensor in the jaws of a bench vise.
7. Connect a voltmeter between terminal 1 and 2.
8. Wrap the bench vise sharply with a hammer and note the voltmeter reading.
9. Verify that the voltage spike is less than 1 volt.
10. If no voltage spike is observed, replace the sensor.

Mass Air Flow (MAF) Sensor

OPERATION

The MAF provides a signal to the ECM for incoming air temperature. This information is used to adjust the injector pulse width and in turn the air/fuel ratio.

REMOVAL & INSTALLATION

1. Disconnect the negative battery cable.
2. Disconnect the connector from the sensor.
3. Remove the air cleaner and air intake assembly, as required.
4. Remove the sensor from its mounting.

To install:

5. Installation is the reverse of the removal procedure.

TESTING

1996–2002

1. Allow the engine to reach operating temperature.
2. Let the engine idle.
3. Connect a voltmeter between terminal 4 (red/grn) and ground.
4. Verify that the voltage varies between 0.8–1.2 volts.
5. Rev the engine and verify that the voltage varies between 3.5–4.0 volts.
6. If not within specification, replace the sensor.

2005–2006

2.0L ENGINE

1. Turn the engine ON.
2. Install the scan tool and monitor the "MASS AIR FLOW" parameter on the scan tool data list.
3. Specification should be approximately 9.8–15.2 kg/h at idle and no load.
4. If within specification, check for poor connection between the PCM and the component.
5. If not, substitute the sensor with a known good component, check for proper operation. If problem corrected replace the sensor.

2.7L ENGINE

1. Connect the scan tool and monitor the "MASS AIR FLOW (V)" parameter on the scan tool data list.
2. Monitor the "MASS AIR FLOW (V)" parameter on the scan tool.

3. Specification is approximately 0.6–1.0 volt at idle (no load) and approximately 1.0–1.3 volt at idle (air conditioning ON).

4. If within specification, check for poor connection between the ECM and the component.

5. If not, substitute the sensor with a known good component, check for proper operation. If problem corrected replace the sensor.

Manifold Absolute Pressure (MAP) Sensor

OPERATION

This sensor is a pressure sensitive variable resistor. It measures the changes in the intake manifold pressure which result from engine load and speed changes, and converts to a voltage output. This sensor is used to measure the barometric pressure at start up, and under certain conditions, allows the ECM to automatically adjust for different altitudes.

REMOVAL & INSTALLATION

1. Disconnect the negative battery cable.
2. Disconnect the connector from the sensor.
3. Remove the sensor retaining screws.
4. Remove the sensor from its mounting.

To install:

5. Installation is the reverse of the removal procedure.

TESTING

1996–2001

1. Apply vacuum to the EGR differential pressure sensor, using a vacuum pump.
2. Turn the ignition switch ON.
3. With the EGR differential pressure sensor still connected, measure the voltage at the ERGR differential pressure sensor connector (pin 3 gry).
4. Specifications are as follows:
   - 6.0 inch Hg approximately 3.6 volts
   - 12.0 inch Hg approximately 2.6 volts
   - 16.0 inch Hg approximately 1.9 volts
   - 25.0 inch Hg approximately 0.4 volt
5. If not within specification, replace the sensor.

Throttle Position Sensor (TPS)

OPERATION

On 1996–2002 vehicles, the throttle position sensor provides a signal to the ECM that is related to the relative throttle plate position. As the throttle plate moves in relation to driving conditions, a signal is sent to the control unit which adjusts the injector pulse width and air/fuel ratio. As the throttle plate is opened further, more air is taken into the combustion chambers, and as a result the relative fuel demand of the engine changes. The throttle position sensor relays this information to the ECM which in alters the fuel amount. On 2005–2006 vehicles, the TPS has a variable resistor whose characteristic is the resistance changing according to throttle angle. The ECM supplies a reference of 5 volts to the TPS and the output voltage increases directly with the opening of the throttle valve. The TPS voltage will vary from 0.2–0.8 volt at closed throttle and 4.3–4.8 volts at wide open throttle.

REMOVAL & INSTALLATION

1. Disconnect the negative battery cable.
2. Be sure that the ignition switch is in the OFF position.
3. Disconnect the sensor connector.
4. Remove the sensor retaining screws. Remove the sensor from its mounting.

To install:

5. Installation is the reverse of the removal procedure.

TESTING

1996–2002

1. Disconnect the connector from the sensor.
2. Connect an ohmmeter between sensor terminals 2 and 3.
3. Verify that the resistance increases according to throttle angle.
4. Specifications are as follows:
   - On 1996–1997 vehicles, measuring condition: fully closed resistance approximately 1 kohm
   - On 1996–1997 vehicles, measuring condition: fully open resistance approximately 2.4 kohm
   - On 1998–2002 vehicles, measuring condition: fully closed resistance approximately 2.4 kohm
   - On 1998–2002 vehicles, measuring condition: fully open resistance approximately 1 kohm
5. If not within specification, replace the sensor.
6. Verify that the throttle is at the closed throttle position.
7. Turn the ignition switch to the ON position.
8. Connect a voltmeter between terminal 3 (yel/blk) and terminal 2 (grn/yel) on the TPS connector.
9. Once again, verify that the throttle is at the closed throttle position.
10. Fully open the throttle valve and verify that the voltage at terminal 3 is within specification.
11. Specifications are as follows:
    - Measuring condition: fully closed, 0.5 volt
    - Measuring condition: fully open, 4.1 volts
12. If not within specification, replace the sensor.

2005–2006

1. Turn the ignition switch to the OFF position.
2. Disconnect the TPS connector.
3. Using an ohmmeter, measure the resistance between terminals 2 and 3 of the TPS.
4. Specification should be 1.6–2.4 kohms at all throttle position.
5. With the connector still disconnected, measure the resistance between terminals 1 and 2.
6. Operate the throttle valve slowly from the idle position to the full open position and check that the resistance changes smoothly in proportion with the throttle valve opening angle.
7. Specification should be 0.71–1.38 kohm at closed throttle valve and 2.7 kohm at wide open throttle.
8. If within specification, check for poor connection between the ECM and the component.
9. If not, substitute the sensor with a known good component, check for proper operation. If problem corrected replace the sensor.

MAZDA

DIAGNOSTIC TROUBLE CODES

12

TABLE OF CONTENTS

VEHICLE APPLICATIONS.......12-2
P0XXX.......12-3
P1XXX.......12-126
P2XXX.......12-210
U0XXX-U2XXX.......12-222

DIAGNOSTIC TROUBLE CODES

OBD II Vehicle Applications

MAZDA

626
1996-2002
2.0L I4 .... VIN C
1996-2002
2.5L V6.......... VIN D

B2300
1996-1997
2.3L I4 .... VIN A
2001
2.3L I4 .... VIN D

B2500
1998-2001
2.5L I4 .... VIN C

B3000
1996-1999
3.0L V6.......... VIN U
1999-2000
3.0L V6.......... VIN V
2001
3.0L V6.......... VIN U
2002-2003
3.0L V6.......... VIN V

B4000
1996-2000
4.0L V6.......... VIN X
2001
4.0L V6.......... VIN E

CX-7
1997-2001
3.5L V6.......... VIN B

Mazda3
2004
2.0L I4 .... CODE LF
2004
2.3L I4 .... CODE L3

Mazda5
1998
2.4L I4 .... VIN D
1999
3.3L V6.......... VIN E
2001
3.3L V6.......... VIN M

Mazda6
2003
2.3L I4 .... VIN C
2003
3.0L V6.......... VIN D

MazdaSpeed6
2006
3.5L V6.......... VIN A

Miata MX-5
1996
1.8L I4 .... CODE BP
2001-
3.5L V6.......... VIN D

Millenia
1996
2.3L V6.......... VIN 2
1996
2.5L V6.......... VIN 1

MPV
1996-1998
3.0L V6.......... VIN 2
1996-1998
3.0L V6.......... VIN 3
2000-2001
2.5L V6.......... VIN G
2002
3.0L V6.......... VIN 1

MX-3
1996
1.6L I4 .... VIN A

MX-6
1996-1997
2.0L I4 .... VIN C
1996-1997
2.5L V6.......... VIN D

Protégé
1996-1998
1.5L I4 .... VIN 1
1996-1998
1.8L I4 .... VIN 2
1999-2002
1.6L I4 .... VIN 1
1999-2000
1.8L I4 .... VIN 1
2001-2003
2.0L I4 .... VIN 5
2003
2.0L I4 .... VIN 7

Protege5
2002-2003
2.0L I4 .... VIN 5

RX-8
2000
2.4L I4 .... VIN D
2004
2.4L I4 .... VIN A
2004
2.4L I4 .... VIN B

Tribute
2001
2.0L I4 .... VIN B
2001
3.0L V6.......... VIN 1

Gas Engine OBD II Trouble Code List (P0xxx Codes)

DTC	Trouble Code Title
DTC: P0010 **1T CCM, MIL: YES** **2001, 2002, 2003, 2004, 2005** **Models:** MX-5 Miata **Engines:** All **Transmissions:** A/T, M/T	**Camshaft Position Actuator Circuit Malfunction** Engine running, and the PCM detected the Oil Control Valve (OCV) current was too high or too low (as calculated from system voltage). **Possible Causes:** • OCV solenoid control circuit is open or shorted to ground • OCV solenoid power circuit is open • OCV solenoid ground circuit is open • OCV is damaged or has failed • PCM has failed
DTC: P0011 **1T CCM, MIL: YES** **2001, 2002, 2003, 2004, 2005** **Models:** MX-5 Miata **Engines:** All **Transmissions:** A/T, M/T	**Camshaft Position Timing Over-Advanced** Engine running, and the PCM detected the Actual valve timing was more than 10 degrees advanced for 1 second from the Target valve timing with the engine running at maximum valve timing retard. **Possible Causes:** • OCV valve is damaged or has failed • OCV spool valve is stuck in the advanced position • Variable valve timing actuator is stuck in the retard position • Oil runners between oil pressure switch and the OCV, or between the OCV and the VVT actuator are dirty or clogged
DTC: P0011 **1T CCM, MIL: YES** **2003, 2004, 2005** **Models:** Mazda3, Mazda6 **Engines:** All **Transmissions:** A/T, M/T	**Camshaft Position Timing Over-Advanced** Engine running, and the PCM detected the Actual valve timing was more than 10 degrees advanced for 1 second from the Target valve timing with the engine running at maximum valve timing retard. **Possible Causes:** • OCV valve is damaged or has failed • OCV spool valve is stuck in the advanced position • Variable valve timing actuator is stuck in the retard position • Oil runners between oil pressure switch and the OCV, or between the OCV and the VVT actuator are dirty or clogged
DTC: P0012 **1T CCM, MIL: YES** **2001, 2002, 2003, 2004, 2005** **Models:** MX-5 Miata **Engines:** All : A/T, M/T	**Camshaft Position Timing Over-Retarded** Engine running, and the PCM detected the Actual valve timing was more than 15 degrees retarded for 5 seconds from the Target valve timing with the Oil Control Valve system within feedback range. **Possible Causes:** • OCV valve is damaged or has failed • Low engine oil pressure condition • OCV spool valve is stuck in the retard position • Variable valve timing actuator is stuck in the advanced position • Timing belt is loose or improper valve timing due to a loose belt • PCM has failed
DTC: P0012 **1T CCM, MIL: YES** **2003, 2004, 2005** **Models:** Mazda3, Mazda6 **Engines:** All **Transmissions:** A/T, M/T	**Camshaft Position Timing Over-Retarded** Engine running, and the PCM detected the Actual valve timing was more than 15 degrees retarded for 5 seconds from the Target valve timing with the Oil Control Valve system within feedback range. **Possible Causes:** • OCV valve is damaged or has failed • Low engine oil pressure condition • OCV spool valve is stuck in the retard position • Variable valve timing actuator is stuck in the advanced position • Timing belt is loose or improper valve timing due to a loose belt • PCM has failed
DTC: P0016 **2T CCM, MIL: YES** **2004, 2005** **Models:** Mazda3 **Engines:** All **Transmissions:** A/T, M/T	**CKP-CPM correlation** The PCM monitors the input pulses from the CKP sensor and CMP sensor. If the input pulse pick-up timing does not match each other, the PCM determines that the camshaft position does not coincide with the crankshaft position. **Possible Causes:** • Poor connection of connector • CMP sensor malfunction • CKP sensor malfunction • Damaged or foreign material on CKP or CMP sensor • Improper valve timing

DTC	Trouble Code Title, Conditions & Possible Causes
DTC: P0030 **2T CCM, MIL: YES** **2004, 2005** **Models:** Mazda6, RX-8 **Engines:** All **Transmissions:** A/T, M/T	**HO2S Heater control circuit (Bank1, Sensor 1)** The PCM monitors the front HO2S impedance when under the front HO2S heater control for 200 s. If the impedance is more than 44 ohms, the PCM determines that there is a front HO2S heater control circuit problem. **Possible Causes:** • Front HO2S heater malfunction • Connector or terminal damage • PCM has failed
DTC: P0031 **2T CCM, MIL: YES** **2001, 2002, 2003, 2004, 2005** **Models:** MX-5 Miata, MPV, Protégé, Protégé5, Tribute **Engines:** All **Transmissions:** A/T, M/T	**HO2S-11 (Bank 1 Sensor 1) Heater Circuit Low Input** Engine running, and after the front HO2S Heater Control duty cycle signal was commanded "off", the PCM detected a low input condition on the heater control circuit. **Possible Causes:** • HO2S heater control circuit is shorted to chassis ground • HO2S heater is damaged or has failed • PCM has failed
DTC: P0031 **2T CCM, MIL: YES** **2003, 2004, 2005** **Models:** Mazda3, Mazda6, RX-8 **Engines:** All **Transmissions:** A/T, M/T	**HO2S-11 (Bank 1 Sensor 1) Heater Circuit Low Input** Engine running, and after the front HO2S Heater Control duty cycle signal was commanded "off", the PCM detected a low input condition on the heater control circuit. **Possible Causes:** • HO2S heater control circuit is shorted to chassis ground • HO2S heater is damaged or has failed • PCM has failed
DTC: P0032 **2T CCM, MIL: YES** **2001, 2002, 2003, 2004, 2005** **Models:** MX-5 Miata, MPV, Protégé, Protégé5, Tribute **Engines:** All **Transmissions:** A/T, M/T	**HO2S-11 (Bank 1 Sensor 1) Heater Circuit High Input** Engine running, and after the front HO2S Heater Control duty cycle signal was commanded "on", the PCM detected a high input condition on the heater control circuit. **Possible Causes:** • HO2S heater control circuit is shorted to system power (B+) • HO2S heater is damaged or has failed • PCM has failed
DTC: P0032 **2T CCM, MIL: YES** **2003, 2004, 2005** **Models:** Mazda3, Mazda6 **Engines:** All **Transmissions:** A/T, M/T	**HO2S-11 (Bank 1 Sensor 1) Heater Circuit High Input** Engine running, and after the front HO2S Heater Control duty cycle signal was commanded "on", the PCM detected a high input condition on the heater control circuit. **Possible Causes:** • HO2S heater control circuit is shorted to system power (B+) • HO2S heater is damaged or has failed • PCM has failed
DTC: P0037 **2T CCM, MIL: YES** **2001, 2002, 2003, 2004, 2005** **Models:** MX-5 Miata, MPV, Protégé, Protégé5, Tribute **Engines:** All **Transmissions:** A/T, M/T	**HO2S-12 (Bank 1 Sensor 2) Heater Circuit Low Input** Engine running, and after the rear HO2S Heater Control duty cycle signal was commanded "off", the PCM detected a low input condition on the heater control circuit. **Possible Causes:** • HO2S heater control circuit is shorted to chassis ground • HO2S heater is damaged or has failed • PCM has failed
DTC: P0037 **2T CCM, MIL: YES** **2003, 2004, 2005** **Models:** Mazda3, Mazda6, RX-8 **Engines:** All **Transmissions:** A/T, M/T	**HO2S-12 (Bank 1 Sensor 2) Heater Circuit Low Input** Engine running, and after the rear HO2S Heater Control duty cycle signal was commanded "off", the PCM detected a low input condition on the heater control circuit. **Possible Causes:** • HO2S heater control circuit is shorted to chassis ground • HO2S heater is damaged or has failed • PCM has failed
DTC: P0038 **2T CCM, MIL: YES** **2001, 2002, 2003, 2004, 2005** **Models:** MX-5 Miata, MPV, Protégé, Protégé5, Tribute **Engines:** All **Transmissions:** A/T, M/T	**HO2S-12 (Bank 1 Sensor 2) Heater Circuit High Input** Engine running, and after the rear HO2S Heater Control duty cycle signal was commanded "on", the PCM detected a high input condition on the heater control circuit. **Possible Causes:** • HO2S heater control circuit is shorted to system power (B+) • HO2S heater is damaged or has failed • PCM has failed

DTC	Trouble Code Title, Conditions & Possible Causes
DTC: P0038 **2T CCM, MIL: YES** **2003, 2004, 2005** **Models:** Mazda3, Mazda6 **Engines:** All **Transmissions:** A/T, M/T	**HO2S-12 (Bank 1 Sensor 2) Heater Circuit High Input** Engine running, and after the rear HO2S Heater Control duty cycle signal was commanded "on", the PCM detected a high input condition on the heater control circuit. **Possible Causes:** • HO2S heater control circuit is shorted to system power (B+) • HO2S heater is damaged or has failed • PCM has failed
DTC: P0040 **2T CCM, MIL: NO** **2004, 2005** **Models:** Tribute **Engines:** 3.0L only **Transmissions:** A/T, M/T	**HO2S-12 (Bank 1 Sensor 2) Heater Circuit High Input** Engine running, and after the rear HO2S Heater Control duty cycle signal was commanded "on", the PCM detected a high input condition on the heater control circuit. **Possible Causes:** • HO2S heater control circuit is shorted to system power (B+) • HO2S heater is damaged or has failed • PCM has failed
DTC: P0041 **2T CCM, MIL: YES** **2004, 2005** **Models:** Tribute **Engines:** All **Transmissions:** A/T, M/T	**HO2S-12 (Bank 1 Sensor 2) Heater Circuit High Input** Engine running, and after the rear HO2S Heater Control duty cycle signal was commanded "on", the PCM detected a high input condition on the heater control circuit. **Possible Causes:** • HO2S heater control circuit is shorted to system power (B+) • HO2S heater is damaged or has failed • PCM has failed
DTC: P0043 **2T CCM, MIL: YES** **2004, 2005** **Models:** Mazda3, Mazda6, Tribute **Engines:** All **Transmissions:** A/T, M/T	**HO2S-12 (Bank 1 Sensor 2) Heater Circuit High Input** Engine running, and after the rear HO2S Heater Control duty cycle signal was commanded "on", the PCM detected a high input condition on the heater control circuit. **Possible Causes:** • HO2S heater control circuit is shorted to system power (B+) • HO2S heater is damaged or has failed • PCM has failed
DTC: P0044 **2T CCM, MIL: YES** **2004, 2005** **Models:** Mazda3, Mazda6, Tribute **Engines:** All **Transmissions:** A/T, M/T	**HO2S-12 (Bank 1 Sensor 2) Heater Circuit High Input** Engine running, and after the rear HO2S Heater Control duty cycle signal was commanded "on", the PCM detected a high input condition on the heater control circuit. **Possible Causes:** • HO2S heater control circuit is shorted to system power (B+) • HO2S heater is damaged or has failed • PCM has failed
DTC: P0051 **2T CCM, MIL: YES** **2001, 2002, 2003, 2004, 2005** **Models:** MPV, Tribute **Engines:** All **Transmissions:** A/T, M/T	**HO2S-11 (Bank 1 Sensor 1) Heater Circuit Low Input** Engine running, and after the front HO2S Heater Control duty cycle signal was commanded "off", the PCM detected a low input condition on the heater control circuit. **Possible Causes:** • HO2S heater control circuit is shorted to chassis ground • HO2S heater is damaged or has failed • PCM has failed
DTC: P0052 **2T CCM, MIL: YES** **2001, 2002, 2003, 2004, 2005** **Models:** MPV, Tribute **Engines:** All **Transmissions:** A/T, M/T	**HO2S-21 (Bank 2 Sensor 1) Heater Circuit High Input** Engine running, and after the front HO2S Heater Control duty cycle signal was commanded "on", the PCM detected a high input condition on the heater control circuit. **Possible Causes:** • HO2S heater control circuit is shorted to system power (B+) • HO2S heater is damaged or has failed • PCM has failed
DTC: P0053 **2T CCM, MIL: YES** **2004, 2005** **Models:** Tribute **Engines:** All **Transmissions:** A/T, M/T	**HO2S-12 (Bank 1 Sensor 2) Heater Circuit High Input** Engine running, and after the rear HO2S Heater Control duty cycle signal was commanded "on", the PCM detected a high input condition on the heater control circuit. **Possible Causes:** • HO2S heater control circuit is shorted to system power (B+) • HO2S heater is damaged or has failed • PCM has failed

DTC	Trouble Code Title, Conditions & Possible Causes
DTC: P0054 **2T CCM, MIL: YES** **2004, 2005** **Models:** Tribute **Engines:** All **Transmissions:** A/T, M/T	**HO2S-12 (Bank 1 Sensor 2) Heater Circuit High Input** Engine running, and after the rear HO2S Heater Control duty cycle signal was commanded "on", the PCM detected a high input condition on the heater control circuit. **Possible Causes:** • HO2S heater control circuit is shorted to system power (B+) • HO2S heater is damaged or has failed • PCM has failed
DTC: P00057 **2T CCM, MIL: YES** **2001, 2002, 2003, 2004, 2005** **Models:** MPV **Engines:** All **Transmissions:** A/T, M/T	**HO2S-22 (Bank 2 Sensor 2) Heater Circuit Low Input** Engine running, and after the rear HO2S Heater Control duty cycle signal was commanded "off", the PCM detected a low input condition on the heater control circuit. **Possible Causes:** • HO2S heater control circuit is shorted to chassis ground • HO2S heater is damaged or has failed • PCM has failed
DTC: P0058 **2T CCM, MIL: YES** **2001, 2002, 2003, 2004, 2005** **Models:** MPV **Engines:** All **Transmissions:** A/T, M/T	**HO2S-22 (Bank 2 Sensor 2) Heater Circuit High Input** Engine running, and after the rear HO2S Heater Control duty cycle signal was commanded "on", the PCM detected a high input condition on the heater control circuit. **Possible Causes:** • HO2S heater control circuit is shorted to system power (B+) • HO2S heater is damaged or has failed • PCM has failed
DTC: P0059 **2T CCM, MIL: YES** **2005** **Models:** Tribute **Engines:** 3.0L **Transmissions:** A/T, M/T	**HO2S-12 (Bank 1 Sensor 2) Heater Circuit High Input** Engine running, and after the rear HO2S Heater Control duty cycle signal was commanded "on", the PCM detected a high input condition on the heater control circuit. **Possible Causes:** • HO2S heater control circuit is shorted to system power (B+) • HO2S heater is damaged or has failed • PCM has failed
DTC: P0060 **2T CCM, MIL: YES** **2005** **Models:** Tribute **Engines:** All **Transmissions:** A/T, M/T	**HO2S-12 (Bank 1 Sensor 2) Heater Circuit High Input** Engine running, and after the rear HO2S Heater Control duty cycle signal was commanded "on", the PCM detected a high input condition on the heater control circuit. **Possible Causes:** • HO2S heater control circuit is shorted to system power (B+) • HO2S heater is damaged or has failed • PCM has failed
DTC: P0068 **2T CCM, MIL: YES** **2005** **Models:** Tribute **Engines:** All **Transmissions:** A/T, M/T	**PCM – Mass Air Flow (MAF/IAT) Sensor – Throttle Position (TP) Sensor** TP signal went below 0.24 volts with a load greater than 55%, or TP signal went above 2.44 volts with a load less than 30%. **Possible Causes:** • TP Sensor has failed • PCM has failed
DTC: P0076 **2T CCM, MIL: NO** **2005** **Models:** RX-8 **Engines:** All **Transmissions:** A/T, M/T	**VDI solenoid valve control circuit low** The PCM monitors the VDI solenoid valve control voltage when the PCM turns off the VDI solenoid valve. If the control voltage is low, the PCM determines that the VDI solenoid control circuit voltage is low. **Possible Causes:** • VDI solenoid valve malfunction • Connector or terminal malfunction • PCM has failed
DTC: P0077 **2T CCM, MIL: NO** **2005** **Models:** RX-8 **Engines:** All **Transmissions:** A/T, M/T	**VDI solenoid valve control circuit high** The PCM monitors the VDI solenoid valve control voltage when the PCM turns off the VDI solenoid valve. If the control voltage is high, the PCM determines that the VDI solenoid control circuit voltage is high. **Possible Causes:** • VDI solenoid valve malfunction • Connector or terminal malfunction • PCM has failed

DTC	Trouble Code Title, Conditions & Possible Causes
DTC: P0096 **2T CCM, MIL: YES** **2004, 2005** **Models:** MX-5 Miata **Engines:** Engine Code BP with turbocharger **Transmissions:** A/T, M/T	**IAT No. 2 circuit performance problem** If air intake temperature is lower than engine coolant temperature by -9.4 degrees F (-23 degrees C) for 1.2 seconds with the ignition switch ON, the PCM determines that there is a IAT No. 2 performance problem. **Possible Causes:** • IAT No. 2 sensor malfunction • ECT sensor malfunction • PCM has failed
DTC: P0097 **1T CCM, MIL: YES** **2004, 2005** **Models:** MX-5 Miata **Engines:** Engine Code BP with turbocharger **Transmissions:** A/T, M/T	**IAT No. 2 circuit low input** The PCM monitors the IAT No. 2 sensor at PCM terminal 4R. If the voltage at PCM terminal 4R is below 0.1 v, the PCM determines that the IAT NO. 2 sensor has malfunctioned. **Possible Causes:** • IAT No. 2 sensor malfunction • Short to the ground circuit between IAT No. 2 sensor and PCM terminal connection 4R • PCM has failed
DTC: P0098 **1T CCM, MIL: YES** **2004, 2005** **Models:** MX-5 Miata **Engines:** Engine Code BP with turbocharger **Transmissions:** A/T, M/T	**IAT No. 2 circuit high input** The PCM monitors the IAT No. 2 sensor at PCM terminal 4R. If the voltage at PCM terminal 4R is above 4.9 v, the PCM determines that the IAT NO. 2 sensor has malfunctioned. **Possible Causes:** • IAT No. 2 sensor malfunction • Short to the ground circuit between IAT No. 2 sensor and PCM terminal connection 4R • PCM has failed
DTC: P0100 **1T CCM, MIL: YES** **1996, 1997, 1998, 1999, 2000, 2001, 2002, 2003, 2004, 2005** **Models:** All **Engines:** All **Transmissions:** A/T, M/T	**MAF or VAF Sensor Circuit Malfunction** Key on or engine running and the PCM detected the MAF or VAF sensor signal was less than 0.20v, or that it was more than 4.90v at any time during the CCM test. **Possible Causes:** • MAF sensor signal circuit is open or shorted to ground • MAF sensor power circuit is open or the ground circuit is open • MAF sensor is damaged or has failed • PCM has failed • TSB SSP056 (2/02) contains a MAF sensor warranty (Protégé)
DTC: P0101 **2T CCM, MIL: YES** **2003, 2004, 2005** **Models:** Mazda3, MPV, Mazda6, MX-5 Miata **Engines:** All **Transmissions:** A/T, M/T	**MAF Sensor inconsistent with TP Sensor** The PCM compares the actual input signal from MAF sensor with expected input signal from MAF sensor which PCM calculates by input voltage from TP sensor or engine speed. With engine running and throttle opening angle at 50% for 5 seconds, if MAF amount is less than 8.93 g/sec., PCM determines that detected MAF amount is too low. With engine running at 2,000 rpm for 5 seconds, if MAF amount is over 103 g/sec., PSM determines that detected MAF amount is too high. **Possible Causes:** • MAF sensor malfunction • TP sensor malfunction • Electric corrosion in MAF signal circuit or MAF return circuit • Voltage drops in MAF signal circuit • Voltage drops in ground circuit
DTC: P0101 **2T CCM, MIL: YES** **2004, 2005** **Models:** RX-8 **Engines:** All **Transmissions:** A/T, M/T	**MAF Sensor inconsistent with TP Sensor** The PCM compares the actual input signal from MAF sensor with expected input signal from MAF sensor which PCM calculates by input voltage from TP sensor or engine speed. With engine running and throttle opening angle at 50% for 5 seconds, if MAF amount is less than 8.93 g/sec., PCM determines that detected MAF amount is too low. With engine running at 2,000 rpm for 5 seconds, if MAF amount is over 103 g/sec., PSM determines that detected MAF amount is too high. **Possible Causes:** • MAF sensor malfunction • TP sensor malfunction • Electric corrosion in MAF signal circuit or MAF return circuit • Voltage drops in MAF signal circuit • Voltage drops in ground circuit

DTC	Trouble Code Title, Conditions & Possible Causes
DTC: P0102 **1T CCM, MIL: YES** **1996, 1997, 1998, 1999, 2000, 2001, 2002, 2003, 2004, 2005** **Models:** All **Engines:** All **Transmissions:** A/T, M/T	**MAF Sensor Circuit Low Input** Key on or engine running and the PCM detected the MAF sensor signal was less than 0.36v at any time during the CCM test. **Possible Causes:** • MAF sensor signal circuit is open • MAF sensor signal is shorted to ground • MAF sensor power circuit is open • MAF sensor is damaged or has failed • PCM has failed
DTC: P0103 **1T CCM, MIL: YES** **1996, 1997, 1998, 1999, 2000, 2001, 2002, 2003, 2004, 2005** **Models:** All **Engines:** All **Transmissions:** A/T, M/T	**MAF Sensor Circuit High Input** Key on or engine running and the PCM detected the MAF sensor signal was more than 4.97v at any time during the CCM test. **Possible Causes:** • MAF sensor ground circuit is open • MAF sensor signal is shorted to VREF or system power (B+) • MAF sensor is damaged or has failed • PCM has failed
DTC: P0105 **1T CCM, MIL: YES** **1996, 1997, 1998, 1999, 2000, 2001, 2002** **Models:** Millenia **Engines:** 2.3L VIN 2 **Transmissions:** A/T, M/T	**MAP/BP Sensor Circuit Malfunction** Key on, and the PCM detected the MAP/BP sensor signal was less than 0.20v, or that it was more than 4.90v, or with the ECT sensor signal more than 176°F, the engine running and the MAP sensor solenoid enabled, the difference between the atmospheric pressure and intake manifold pressure was less than 6.43 kPa (1.90" Hg). **Possible Causes:** • MAP sensor signal circuit is open or shorted to ground • MAP sensor power circuit is open or the ground circuit is open • MAP sensor is damaged or has failed • PCM has failed
DTC: P0106 **2T CCM, MIL: YES** **1996, 1997, 1998, 1999, 2000, 2001, 2002** **Models:** 626 **Engines:** All **Transmissions:** A/T, M/T	**BARO Sensor Circuit Performance** Key on or engine running, and the PCM detected the BARO sensor signal was less than 0.22v, or that it was more than 4.97v during the CCM test. **Possible Causes:** • BARO sensor signal circuit is open or shorted to ground • BARO sensor signal is shorted to VREF or system power • BARO sensor is damaged or has failed • PCM has failed
DTC: P0106 **2T CCM, MIL: YES** **1996, 1997, 1998, 1999, 2000, 2001, 2002, 2003, 2004, 2005** **Models:** MX-5 Miata, MPV, Protégé, Protégé5 **Engines:** All **Transmissions:** A/T, M/T	**BARO Sensor Circuit Performance** Engine running, and the PCM detected the BARO sensor signal was less than 0.22v, or that it was more than 4.97v during the CCM test. **Possible Causes:** • BARO sensor signal circuit is open or shorted to ground • BARO sensor signal is shorted to VREF or system power • BARO sensor is damaged or has failed • PCM has failed
DTC: P0106 **2T CCM, MIL: YES** **2003, 2004, 2005** **Models:** Mazda3, Mazda6 **Engines:** All **Transmissions:** A/T, M/T	**BARO Sensor Circuit Performance** Engine running, and the PCM detected the BARO sensor signal was less than 0.22v, or that it was more than 4.97v during the CCM test. **Possible Causes:** • BARO sensor signal circuit is open or shorted to ground • BARO sensor signal is shorted to VREF or system power • BARO sensor is damaged or has failed • PCM has failed
DTC: P0107 **1T CCM, MIL: YES** **1996, 1997, 1998, 1999, 2000, 2001, 2002** **Models:** 626 **Engines:** All **Transmissions:** A/T, M/T	**BARO Sensor Circuit Low Input** Engine running and the PCM detected the BARO sensor signal was less than 0.01v at any time during the CCM test. **Possible Causes:** • BARO sensor signal circuit is shorted to ground • BARO sensor VREF circuit is open or shorted to ground • BARO sensor is damaged or has failed • PCM has failed

DTC	Trouble Code Title, Conditions & Possible Causes
DTC: P0107 **1T CCM, MIL: YES** **1996, 1997, 1998, 1999, 2000, 2001, 2002, 2003, 2004, 2005** **Models:** MX-5 Miata, MPV, Protégé, Protégé5 **Engines:** All **Transmissions:** A/T, M/T	**BARO Sensor Circuit Low Input** Engine running, EGR Boost Sensor Solenoid commanded "off" so that BARO pressure is applied to the sensor), IAT sensor signal more than 50°F, and the PCM detected the BARO sensor signal indicated less than 0.21v during the CCM test. **Possible Causes:** • EGR Boost sensor signal circuit is shorted to ground • EGR boost sensor VREF circuit is open or shorted to ground • EGR Boost sensor is damaged or has failed • PCM has failed
DTC: P0107 **1T CCM, MIL: YES** **2003, 2004, 2005** **Models:** Mazda3, Mazda6, RX-8 **Engines:** All **Transmissions:** A/T, M/T	**BARO Sensor Circuit Low Input** Engine running, EGR Boost Sensor Solenoid commanded "off" so that BARO pressure is applied to the sensor), IAT sensor signal more than 50°F, and the PCM detected the BARO sensor signal indicated less than 0.21v during the CCM test. **Possible Causes:** • EGR Boost sensor signal circuit is shorted to ground • EGR boost sensor VREF circuit is open or shorted to ground • EGR Boost sensor is damaged or has failed • PCM has failed
DTC: P0107 **1T CCM, MIL: YES** **2004, 2005** **Models:** B2300, B3000, B4000 **Engines:** All **Transmissions:** A/T, M/T	**BARO Sensor Circuit Low Input** Engine running, EGR Boost Sensor Solenoid commanded "off" so that BARO pressure is applied to the sensor), IAT sensor signal more than 50°F, and the PCM detected the BARO sensor signal indicated less than 0.21v during the CCM test. **Possible Causes:** • EGR Boost sensor signal circuit is shorted to ground • EGR boost sensor VREF circuit is open or shorted to ground • EGR Boost sensor is damaged or has failed • PCM has failed
DTC: P0107 **1T CCM, MIL: YES** **2005** **Models:** Tribute **Engines:** All **Transmissions:** A/T, M/T	**BARO Sensor Circuit Low Input** Engine running, EGR Boost Sensor Solenoid commanded "off" so that BARO pressure is applied to the sensor), IAT sensor signal more than 50°F, and the PCM detected the BARO sensor signal indicated less than 0.21v during the CCM test. **Possible Causes:** • EGR Boost sensor signal circuit is shorted to ground • EGR boost sensor VREF circuit is open or shorted to ground • EGR Boost sensor is damaged or has failed • PCM has failed
DTC: P0108 **1T CCM, MIL: YES** **1996, 1997, 1998, 1999, 2000, 2001, 2002** **Models:** 626 **Engines:** All **Transmissions:** A/T, M/T	**BARO Sensor Circuit High Input** Engine running and the PCM detected the BARO sensor signal was more than 4.99v at any time during the CCM test. **Possible Causes:** • EGR boost sensor signal circuit open from sensor to the PCM • EGR boost sensor ground circuit open from sensor to the PCM • EGR Boost sensor signal circuit is shorted to VREF or power • EGR Boost sensor is damaged or has failed • PCM has failed
DTC: P0108 **1T CCM, MIL: YES** **1996, 1997, 1998, 1999, 2000, 2001, 2002, 2003, 2004, 2005** **Models:** MX-5 Miata, MPV, Protégé, Protégé5 **Engines:** All **Transmissions:** A/T, M/T	**BARO Sensor Circuit High Input** Engine running, EGR Boost Sensor Solenoid commanded "off" so that BARO pressure is applied to the sensor), IAT sensor signal more than 50°F, and the PCM detected the BARO sensor signal indicated more than 4.80v during the CCM test. **Possible Causes:** • EGR boost sensor signal circuit open from sensor to the PCM • EGR boost sensor ground circuit open from sensor to the PCM • EGR Boost sensor signal circuit is shorted to VREF or power • EGR Boost sensor is damaged or has failed • PCM has failed
DTC: P0108 **1T CCM, MIL: YES** **2003, 2004, 2005** **Models:** Mazda3, Mazda6, RX-8 **Engines:** All **Transmissions:** A/T, M/T	**BARO Sensor Circuit High Input** Engine running, EGR Boost Sensor Solenoid commanded "off" so that BARO pressure is applied to the sensor), IAT sensor signal more than 50°F, and the PCM detected the BARO sensor signal indicated more than 4.80v during the CCM test. **Possible Causes:** • EGR boost sensor signal circuit open from sensor to the PCM • EGR boost sensor ground circuit open from sensor to the PCM • EGR Boost sensor signal circuit is shorted to VREF or power • EGR Boost sensor is damaged or has failed • PCM has failed

DTC	Trouble Code Title, Conditions & Possible Causes
DTC: P0108 **1T CCM, MIL: YES** **2004, 2005** **Models:** B2300, B3000, B4000 **Engines:** All **Transmissions:** A/T, M/T	**BARO Sensor Circuit High Input** Engine running, EGR Boost Sensor Solenoid commanded "off" so that BARO pressure is applied to the sensor), IAT sensor signal more than 50°F, and the PCM detected the BARO sensor signal indicated more than 4.80v during the CCM test. **Possible Causes:** • EGR boost sensor signal circuit open from sensor to the PCM • EGR boost sensor ground circuit open from sensor to the PCM • EGR Boost sensor signal circuit is shorted to VREF or power • EGR Boost sensor is damaged or has failed • PCM has failed
DTC: P0108 **1T CCM, MIL: YES** **2005** **Models:** Tribute **Engines:** All **Transmissions:** A/T, M/T	**BARO Sensor Circuit High Input** Engine running, EGR Boost Sensor Solenoid commanded "off" so that BARO pressure is applied to the sensor), IAT sensor signal more than 50°F, and the PCM detected the BARO sensor signal indicated more than 4.80v during the CCM test. **Possible Causes:** • EGR boost sensor signal circuit open from sensor to the PCM • EGR boost sensor ground circuit open from sensor to the PCM • EGR Boost sensor signal circuit is shorted to VREF or power • EGR Boost sensor is damaged or has failed • PCM has failed
DTC: P0109 **1T CCM, MIL: YES** **2004, 2005** **Models:** B2300, B3000, B4000 **Engines:** All **Transmissions:** A/T, M/T	**Manifold Absolute Pressure (MAP) Sensor** MAP sensor signal to the PCM is failing intermittently. **Possible Causes:** • MAP sensor is damaged or has failed • PCM has failed
DTC: P0109 **1T CCM, MIL: YES** **2005** **Models:** Tribute **Engines:** All **Transmissions:** A/T, M/T	**BARO Sensor Circuit High Input** MAP sensor signal to the PCM is failing intermittently. **Possible Causes:** • MAP sensor is damaged or has failed • PCM has failed
DTC: P0110 **1T CCM, MIL: YES** **1996, 1997, 1998, 1999, 2000, 2001, 2002** **Models:** Millenia **Engines:** All **Transmissions:** A/T, M/T	**Intake Air Temperature Sensor Circuit Malfunction** Key on or engine running, and the PCM detected the IAT sensor signal indicated less than 0.10v, or it detected the signal was more than 4.80v at any time during the CCM test period. **Possible Causes:** • IAT sensor signal circuit is open or shorted to ground • IAT sensor is damaged or has failed • PCM has failed
DTC: P0110 **1T CCM, MIL: YES** **1996, 1997** **Models:** 626 **Engines:** All **Transmissions:** A/T, M/T	**Intake Air Temperature Sensor Circuit Malfunction** Key on or engine running, and the PCM detected the IAT sensor signal indicated less than 0.10v, or it detected the signal was more than 4.80v at any time during the CCM test period. **Possible Causes:** • IAT sensor signal circuit is open • IAT sensor signal circuit is shorted to ground • IAT sensor is damaged or has failed • PCM has failed
DTC: P0110 **1T CCM, MIL: YES** **1996, 1997, 1998** **Models:** MX-5 Miata, Protégé **Engines:** All **Transmissions:** A/T, M/T	**Intake Air Temperature Sensor Circuit Malfunction** Key on or engine running, and the PCM detected the IAT sensor signal indicated less than 0.10v, or it detected the signal was more than 4.80v at any time during the CCM test period. **Possible Causes:** • IAT sensor signal circuit is open or shorted to ground • IAT sensor is damaged or has failed • PCM has failed

DTC	Trouble Code Title, Conditions & Possible Causes
DTC: P0110 **1T CCM, MIL: YES** **2004** **Models:** Tribute **Engines:** All **Transmissions:** A/T, M/T	**Intake Air Temperature Sensor Circuit Malfunction** Key on or engine running, and the PCM detected the IAT sensor signal indicated less than 0.10v, or it detected the signal was more than 4.80v at any time during the CCM test period. **Possible Causes:** • IAT sensor signal circuit is open or shorted to ground • IAT sensor is damaged or has failed • PCM has failed
DTC: P0111 **2T CCM, MIL: YES** **1996, 1997, 1998, 1999, 2000, 2001, 2002, 2003, 2004, 2005** **Models:** MX-5 Miata, MPV **Engines:** All **Transmissions:** A/T, M/T	**Intake Air Temperature Sensor Circuit Range/Performance** Key on or engine running and the PCM detected the IAT sensor signal was more than 104ºF higher than the ECT sensor signal. **Possible Causes:** • IAT sensor signal circuit has high resistance • IAT sensor has drifted out of calibration • IAT sensor is damaged or has failed • PCM has failed
DTC: P0111 **2T CCM, MIL: YES** **1999, 2000, 2001, 2002, 2003** **Models:** Protégé, Protégé5 **Engines:** All **Transmissions:** A/T, M/T	**Intake Air Temperature Sensor Circuit Range/Performance** Key on or engine running, and the PCM detected the IAT sensor signal was more than 104ºF higher than the ECT sensor signal. **Possible Causes:** • IAT sensor signal circuit has high resistance • IAT sensor has drifted out of calibration • IAT sensor is damaged or has failed • PCM has failed
DTC: P0111 **2T CCM, MIL: YES** **2003, 2004, 2005** **Models:** Mazda3, Mazda6, RX-8 **Engines:** All **Transmissions:** A/T, M/T	**Intake Air Temperature Sensor Circuit Range/Performance** Key on or engine running, and the PCM detected the IAT sensor signal was more than 104ºF higher than the ECT sensor signal. **Possible Causes:** • IAT sensor signal circuit has high resistance • IAT sensor has drifted out of calibration • IAT sensor is damaged or has failed • PCM has failed
DTC: P0112 **1T CCM, MIL: YES** **1996, 1997, 1998, 1999, 2000, 2001, 2002, 2003, 2004, 2005** **Models:** B2300, B2500, B3000, B4000, MX6, Millenia, MPV, Tribute **Engines:** All **Transmissions:** A/T, M/T	**Intake Air Temperature Sensor Circuit Low Input** Key on or engine running, and the PCM detected the IAT sensor signal indicated less than 0.20v (a Scan Tool PID near 250°F) at any time during the CCM test period. **Possible Causes:** • IAT sensor signal circuit is shorted to ground • IAT sensor is damaged or has failed • PCM has failed
DTC: P0112 **1T CCM, MIL: YES** **1999, 2000, 2001, 2002, 2003, 2004, 2005** **Models:** MX-5 Miata, Protégé, Protégé5 **Engines:** All **Transmissions:** A/T, M/T	**Intake Air Temperature Sensor Circuit Low Input** Key on or engine running, and the PCM detected the IAT sensor signal indicated less than 0.20v (a Scan Tool PID near 250°F) at any time during the CCM test period. **Possible Causes:** • IAT sensor signal circuit is shorted to ground • IAT sensor is damaged or has failed • PCM has failed
DTC: P0112 **1T CCM, MIL: YES** **1998, 1999, 2000, 2001, 2002** **Models:** 626 **Engines:** All **Transmissions:** A/T, M/T	**Intake Air Temperature Sensor Circuit Low Input** Key on or engine running, and the PCM detected the IAT sensor signal indicated less than 0.20v (a Scan Tool PID near 250°F) at any time during the CCM test period. **Possible Causes:** • IAT sensor signal circuit is shorted to ground • IAT sensor is damaged or has failed • PCM has failed

DTC	Trouble Code Title, Conditions & Possible Causes
DTC: P0112 **1T CCM, MIL: YES** **2003, 2004, 2005** **Models:** Mazda3, Mazda6, RX-8 **Engines:** All **Transmissions:** A/T, M/T	**Intake Air Temperature Sensor Circuit Low Input** Key on or engine running, and the PCM detected the IAT sensor signal indicated less than 0.20v (a Scan Tool PID near 250°F) at any time during the CCM test period. **Possible Causes:** • IAT sensor signal circuit is shorted to ground • IAT sensor is damaged or has failed • PCM has failed
DTC: P0113 **1T CCM, MIL: YES** **1996, 1997, 1998, 1999, 2000, 2001, 2002, 2003, 2004, 2005** **Models:** B2300, B2500, B3000, B4000, MX6, Millenia, MPV, Tribute **Engines:** All **Transmissions:** A/T, M/T	**Intake Air Temperature Sensor Circuit High Input** Key on or engine running, and the PCM detected the IAT sensor signal indicated more than 4.60v (a Scan Tool PID near -46°F) at any time during the CCM test period. **Possible Causes:** • IAT sensor signal circuit is open • IAT sensor signal circuit is shorted to VREF or system power • IAT sensor is damaged or has failed • PCM has failed
DTC: P0113 **1T CCM, MIL: YES** **1998, 1999, 2000, 2001, 2002** **Models:** 626 **Engines:** All **Transmissions:** A/T, M/T	**Intake Air Temperature Sensor Circuit High Input** Key on or engine running, and the PCM detected the IAT sensor signal indicated more than 4.60v (a Scan Tool PID near -46°F) at any time during the CCM test period. **Possible Causes:** • IAT sensor signal circuit is open • IAT sensor signal circuit is shorted to VREF or system power • IAT sensor is damaged or has failed • PCM has failed
DTC: P0113 **1T CCM, MIL: YES** **1999, 2000, 2001, 2002, 2003, 2004, 2005** **Models:** MX-5 Miata, Protégé, Protégé5 **Engines:** All **Transmissions:** A/T, M/T	**Intake Air Temperature Sensor Circuit High Input** Key on or engine running, and the PCM detected the IAT sensor signal indicated more than 4.60v (a Scan Tool PID near -46°F) at any time during the CCM test period. **Possible Causes:** • IAT sensor signal circuit is open • IAT sensor signal circuit is shorted to VREF or system power • IAT sensor is damaged or has failed • PCM has failed
DTC: P0113 **1T CCM, MIL: YES** **2003, 2004, 2005** **Models:** Mazda3, Mazda6, RX-8 **Engines:** All **Transmissions:** A/T, M/T	**Intake Air Temperature Sensor Circuit High Input** Key on or engine running, and the PCM detected the IAT sensor signal indicated more than 4.60v (a Scan Tool PID near -46°F) at any time during the CCM test period. **Possible Causes:** • IAT sensor signal circuit is open • IAT sensor signal circuit is shorted to VREF or system power • IAT sensor is damaged or has failed • PCM has failed
DTC: P0113 **1T CCM, MIL: YES** **2004, 2005** **Models:** B2300, B3000, B4000 **Engines:** All **Transmissions:** A/T, M/T	**Intake Air Temperature Sensor Circuit High Input** Key on or engine running, and the PCM detected the IAT sensor signal indicated more than 4.60v (a Scan Tool PID near -46°F) at any time during the CCM test period. **Possible Causes:** • IAT sensor signal circuit is open • IAT sensor signal circuit is shorted to VREF or system power • IAT sensor is damaged or has failed • PCM has failed
DTC: P0114 **1T CCM, MIL: YES** **2005** **Models:** Tribute **Engines:** All **Transmissions:** A/T, M/T	**PCM - Intake Air Temperature (IAT) Sensor** IAT sensor signal erratic. **Possible Causes:** • IAT sensor signal circuit is open • IAT sensor signal circuit is shorted to VREF or system power • IAT sensor is damaged or has failed • PCM has failed

DTC	Trouble Code Title, Conditions & Possible Causes
DTC: P0115 **1T CCM, MIL: YES** **1996, 1997, 1998, 1999, 2000, 2001, 2002** **Models:** Millenia **Engines:** All **Transmissions:** A/T, M/T	**Engine Coolant Temperature Sensor Circuit Malfunction** Key on or engine running, and the PCM detected the ECT sensor signal indicated less than 0.10v, or it detected the signal was more than 4.80v at any time during the CCM test period. **Possible Causes:** • ECT sensor signal circuit is open • ECT sensor signal circuit is shorted to ground • ECT sensor is damaged or has failed • PCM has failed
DTC: P0115 **1T CCM, MIL: YES** **1996, 1997** **Models:** 626 **Engines:** All **Transmissions:** A/T, M/T	**Intake Air Temperature Sensor Circuit Malfunction** Key on or engine running, and the PCM detected the IAT sensor signal indicated less than 0.10v, or it detected the signal was more than 4.80v at any time during the CCM test period. **Possible Causes:** • IAT sensor signal circuit is open • IAT sensor signal circuit is shorted to ground • IAT sensor is damaged or has failed • PCM has failed
DTC: P0115 **1T CCM, MIL: YES** **1996, 1997, 1998** **Models:** MX-5 Miata, Protégé **Engines:** All **Transmissions:** A/T, M/T	**Intake Air Temperature Sensor Circuit Malfunction** Key on or engine running, and the PCM detected the IAT sensor signal indicated less than 0.10v, or it detected the signal was more than 4.80v at any time during the CCM test period. **Possible Causes:** • IAT sensor signal circuit is open or shorted to ground • IAT sensor is damaged or has failed • PCM has failed
DTC: P0115 **1T CCM, MIL: YES** **2004** **Models:** Tribute **Engines:** All **Transmissions:** A/T, M/T	**Intake Air Temperature Sensor Circuit Malfunction** Key on or engine running, and the PCM detected the IAT sensor signal indicated less than 0.10v, or it detected the signal was more than 4.80v at any time during the CCM test period. **Possible Causes:** • IAT sensor signal circuit is open or shorted to ground • IAT sensor is damaged or has failed • PCM has failed
DTC: P0116 **2T CCM, MIL: YES** **2003, 2004, 2005** **Models:** MX-5 Miata, MPV, Tribute **Engines:** All **Transmissions:** A/T, M/T	**Engine Coolant Temperature Sensor Circuit Performance Problem** Key on or engine running, and the PCM detected the ECT sensor signal maximum value and minimum value is below 5.6 degrees C (4.2 degrees F), the PCM determines that ECT signal circuit has malfunctioned. **Possible Causes:** • ECT sensor malfunction • Poor connection at ECT sensor or PCM connector • PCM malfunction
DTC: P0117 **1T CCM, MIL: YES** **1996, 1997, 1998, 1999, 2000, 2001, 2002, 2003, 2004, 2005** **Models:** B2300, B2500, B3000, B4000, MX6, Millenia, MPV, Tribute **Engines:** All **Transmissions:** A/T, M/T	**Engine Coolant Temperature Sensor Circuit Low Input** Key on or engine running, and the PCM detected the ECT sensor signal indicated less than 0.20v (a Scan Tool PID near 250°F) at any time during the CCM test period. **Possible Causes:** • ECT sensor signal circuit is shorted to ground • ECT sensor is damaged or has failed • PCM has failed
DTC: P0117 **1T CCM, MIL: YES** **1999, 2000, 2001, 2002, 2003, 2004, 2005** **Models:** MX-5 Miata, Protégé, Protégé5 **Engines:** All **Transmissions:** A/T, M/T	**Engine Coolant Temperature Sensor Circuit Low Input** Key on or engine running, and the PCM detected the ECT sensor signal indicated less than 0.20v (a Scan Tool PID near 250°F) at any time during the CCM test period. **Possible Causes:** • ECT sensor signal circuit is shorted to ground • ECT sensor is damaged or has failed • PCM has failed

DTC	Trouble Code Title, Conditions & Possible Causes
DTC: P0117 **1T CCM, MIL: YES** **1998, 1999, 2000, 2001, 2002** **Models:** 626 **Engines:** All **Transmissions:** A/T, M/T	**Engine Coolant Temperature Sensor Circuit Low Input** Key on or engine running, and the PCM detected the ECT sensor signal indicated less than 0.20v (a Scan Tool PID near 250°F) at any time during the CCM test period. **Possible Causes:** • ECT sensor signal circuit is shorted to ground • ECT sensor is damaged or has failed • PCM has failed
DTC: P0117 **1T CCM, MIL: YES** **2003, 2004, 2005** **Models:** Mazda3, Mazda6, RX-8 **Engines:** All **Transmissions:** A/T, M/T	**Engine Coolant Temperature Sensor Circuit Low Input** Key on or engine running, and the PCM detected the ECT sensor signal indicated less than 0.20v (a Scan Tool PID near 250°F) at any time during the CCM test period. **Possible Causes:** • ECT sensor signal circuit is shorted to ground • ECT sensor is damaged or has failed • PCM has failed
DTC: P0118 **1T CCM, MIL: YES** **1996, 1997, 1998, 1999, 2000, 2001, 2002, 2003, 2004, 2005** **Models:** B2300, B2500, B3000, B4000, MX6, Millenia, MPV, Tribute **Engines:** All **Transmissions:** A/T, M/T	**Engine Coolant Temperature Sensor Circuit High Input** Key on or engine running, and the PCM detected the ECT sensor signal indicated more than 4.60v (a Scan Tool PID near -46°F) at any time during the CCM test period. **Possible Causes:** • ECT sensor signal circuit is open • ECT sensor signal circuit is shorted to VREF or system power • ECT sensor is damaged or has failed • PCM has failed
DTC: P0118 **1T CCM, MIL: YES** **1999, 2000, 2001, 2002, 2003, 2004, 2005** **Models:** MX-5 Miata, Protégé, Protégé5 **Engines:** All **Transmissions:** A/T, M/T	**Engine Coolant Temperature Sensor Circuit High Input** Key on or engine running, and the PCM detected the ECT sensor signal indicated more than 4.60v (a Scan Tool PID near -46°F) at any time during the CCM test period. **Possible Causes:** • ECT sensor signal circuit is open • ECT sensor signal circuit is shorted to VREF or system power • ECT sensor is damaged or has failed • PCM has failed
DTC: P0118 **1T CCM, MIL: YES** **1998, 1999, 2000, 2001, 2002** **Models:** 626 **Engines:** All **Transmissions:** A/T, M/T	**Engine Coolant Temperature Sensor Circuit High Input** Key on or engine running, and the PCM detected the ECT sensor signal indicated more than 4.60v (a Scan Tool PID near -46°F) at any time during the CCM test period. **Possible Causes:** • ECT sensor signal circuit is open • ECT sensor signal circuit is shorted to VREF or system power • ECT sensor is damaged or has failed • PCM has failed
DTC: P0118 **1T CCM, MIL: YES** **2003, 2004, 2005** **Models:** Mazda3, Mazda6, RX-8 **Engines:** All **Transmissions:** A/T, M/T	**Engine Coolant Temperature Sensor Circuit High Input** Key on or engine running, and the PCM detected the ECT sensor signal indicated more than 4.60v (a Scan Tool PID near -46°F) at any time during the CCM test period. **Possible Causes:** • ECT sensor signal circuit is open • ECT sensor signal circuit is shorted to VREF or system power • ECT sensor is damaged or has failed • PCM has failed
DTC: P0119 **1T CCM, MIL: YES** **2005** **Models:** Tribute **Engines:** All **Transmissions:** A/T, M/T	**PCM - Engine Coolant Temperature (ECT) Sensor/Coolant Head Temperature (CHT) Sensor** ECT / CHT sensor signal is erratic. **Possible Causes:** • ECT / CHT sensor signal circuit is open • ECT / CHT sensor signal circuit is shorted to VREF or system power • ECT / CHT sensor is damaged or has failed • PCM has failed

DTC	Trouble Code Title, Conditions & Possible Causes
DTC: P0120 **1T CCM, MIL: YES** **1996, 1997, 1998, 1999, 2000, 2001, 2002** **Models:** Millenia **Engines:** All **Transmissions:** A/T, M/T	**TP Sensor Circuit Malfunction** Key on or engine running, and the PCM detected the TP sensor signal was less than 0.10v, or that it was more than 4.70v. **Possible Causes:** • TP sensor signal circuit is open or shorted to ground • TP sensor VREF circuit is open or shorted to ground • TP sensor signal circuit is shorted to VREF or system power • TP sensor is damaged or has failed • PCM has failed
DTC: P0120 **1T CCM, MIL: YES** **1996, 1997, 1998** **Models:** Protégé **Engines:** All **Transmissions:** A/T, M/T	**TP Sensor Circuit Malfunction** Key on or engine running, and the PCM detected the TP sensor signal was less than 0.10v, or that it was more than 4.70v. **Possible Causes:** • TP sensor signal circuit is open or shorted to ground • TP sensor VREF circuit is open or shorted to ground • TP sensor signal circuit is shorted to VREF or system power • TP sensor is damaged or has failed • PCM has failed
DTC: P0120 **1T CCM, MIL: YES** **1996, 1997** **Models:** 626, MX-6 **Engines:** 2.0L VIN C **Transmissions:** M/T	**Intake Air Temperature Sensor Circuit Malfunction** Key on or engine running, and the PCM detected the IAT sensor signal indicated less than 0.10v, or it detected the signal was more than 4.80v at any time during the CCM test period. **Possible Causes:** • IAT sensor signal circuit is open • IAT sensor signal circuit is shorted to ground • IAT sensor is damaged or has failed • PCM has failed
DTC: P0120 **1T CCM, MIL: YES** **1996, 1997** **Models:** 626, MX-6 **Engines:** 2.5L VIN D **Transmissions:** A/T, M/T	**Intake Air Temperature Sensor Circuit Malfunction** Key on or engine running, and the PCM detected the IAT sensor signal indicated less than 0.10v, or it detected the signal was more than 4.80v at any time during the CCM test period. **Possible Causes:** • IAT sensor signal circuit is open • IAT sensor signal circuit is shorted to ground • IAT sensor is damaged or has failed • PCM has failed
DTC: P0120 **1T CCM, MIL: YES** **2004** **Models:** Tribute **Engines:** 2.5L VIN D **Transmissions:** A/T, M/T	**Intake Air Temperature Sensor Circuit Malfunction** Key on or engine running, and the PCM detected the IAT sensor signal indicated less than 0.10v, or it detected the signal was more than 4.80v at any time during the CCM test period. **Possible Causes:** • IAT sensor signal circuit is open • IAT sensor signal circuit is shorted to ground • IAT sensor is damaged or has failed • PCM has failed
DTC: P0121 **2T CCM, MIL: YES** **1996, 1997** **Models:** MX-6 **Engines:** All **Transmissions:** A/T	**TP Sensor In-Range Operating Circuit Malfunction** Trouble Code Conditions: Engine running at cruise speed and then back to idle speed, and the PCM detected the TP sensor signal was less than 0.17v (3.42%) or more than 4.60v (93%). This code can set if the TP sensor signal does not correlate when compared to the MAF sensor signal. **Possible Causes:** • TP sensor VREF circuit open or shorted to ground (intermittent) • MAF sensor or TP sensor has drifted out of calibration • TP sensor is damaged or has failed
DTC: P0121 **2T CCM, MIL: YES** **1996, 1997, 1998, 1999, 2000, 2001, 2002, 2003, 2004, 2005** **Models:** B2300, B2500, B3000, B4000, MPV, Tribute **Engines:** All **Transmissions:** A/T, M/T	**TP Sensor In-Range Operating Circuit Malfunction** Vehicle driven at light engine load at over 20 mph, and the PCM detected the TP sensor signal did not correlate when it was compared to the MAF sensor signal during the CCM test. **Possible Causes:** • TP sensor signal circuit is open to the PCM (intermittent fault) • TP sensor ground circuit is open • Throttle body is damaged or throttle linkage is bent or binding • TP sensor is damaged or has failed

DTC	Trouble Code Title, Conditions & Possible Causes
DTC: P0122 **1T CCM, MIL: YES** **1996, 1997, 1998, 1999, 2000, 2001, 2002, 2003, 2004, 2005** **Models:** 626, B2300, B2500, B3000, B4000, MX6, MPV, MX-5 Miata, Tribute **Engines:** All **Transmissions:** A/T, M/T	**TP Sensor Circuit Low Input** Key on or engine running, and the PCM detected the TP sensor signal indicated less than 0.17v (a throttle opening of 3.43%) during the CCM test. **Possible Causes:** • TP sensor signal circuit is shorted to ground • TP sensor VREF circuit is open or shorted to ground • Throttle body is damaged • Throttle linkage is bent or binding (sticking) • TP sensor is damaged or has failed
DTC: P0122 **1T CCM, MIL: YES** **1999, 2000, 2001, 2002, 2003** **Models:** Protégé, Protégé5 **Engines:** All **Transmissions:** A/T, M/T	**TP Sensor Circuit Low Input** Key on or engine running, and the PCM detected the TP sensor signal indicated less than 0.17v (a throttle opening of 3.43%) during the CCM test. **Possible Causes:** • TP sensor signal circuit is shorted to ground • TP sensor VREF circuit is open or shorted to ground • Throttle body is damaged • Throttle linkage is bent or binding (sticking) • TP sensor is damaged or has failed
DTC: P0122 **1T CCM, MIL: YES** **2003, 2004, 2005** **Models:** Mazda3, Mazda6, MX-5 Miata, RX-8 **Engines:** All **Transmissions:** A/T, M/T	**TP Sensor Circuit Low Input** Key on or engine running, and the PCM detected the TP sensor signal indicated less than 0.17v (a throttle opening of 3.43%) during the CCM test. **Possible Causes:** • TP sensor signal circuit is shorted to ground • TP sensor VREF circuit is open or shorted to ground • Throttle body is damaged • Throttle linkage is bent or binding (sticking) • TP sensor is damaged or has failed
DTC: P0123 **1T CCM, MIL: YES** **1996, 1997, 1998, 1999, 2000, 2001, 2002, 2003, 2004, 2005** **Models:** 626, B2300, B2500, B3000, B4000, MX6, MPV, MX-5 Miata, Tribute **Engines:** All **Transmissions:** A/T, M/T	**TP Sensor Circuit High Input** Key on or engine running, and the PCM detected the TP sensor signal indicated more than 4.60v (a throttle opening of 92.27%). **Note: This trouble code may set due to an intermittent fault.** **Possible Causes:** • TP sensor signal return circuit is open • TP sensor signal circuit is shorted to VREF or system power • TP sensor not mounted properly to the throttle body • TP sensor is damaged or has failed • PCM has failed
DTC: P0123 **1T CCM, MIL: YES** **1999, 2000, 2001, 2002, 2003** **Models:** Protégé, Protégé5 **Engines:** All **Transmissions:** A/T, M/T	**TP Sensor Circuit High Input** Key on or engine running, and the PCM detected the TP sensor signal indicated more than 4.60v (a throttle opening of 92.27%). **Note: This trouble code may set due to an intermittent fault.** **Possible Causes:** • TP sensor signal return circuit is open • TP sensor signal circuit is shorted to VREF or system power • TP sensor not mounted properly to the throttle body • TP sensor is damaged or has failed • PCM has failed
DTC: P0123 **1T CCM, MIL: YES** **2003, 2004, 2005** **Models:** Mazda3, Mazda6, RX-8 **Engines:** All **Transmissions:** A/T, M/T	**TP Sensor Circuit High Input** Key on or engine running, and the PCM detected the TP sensor signal indicated more than 4.60v (a throttle opening of 92.27%). **Note: This trouble code may set due to an intermittent fault.** **Possible Causes:** • TP sensor signal return circuit is open • TP sensor signal circuit is shorted to VREF or system power • TP sensor not mounted properly to the throttle body • TP sensor is damaged or has failed • PCM has failed

DTC	Trouble Code Title, Conditions & Possible Causes
DTC: P0125 **2T CCM, MIL: YES** **1996, 1997, 1998, 1999, 2000, 2001, 2002, 2003, 2004, 2005** **Models:** All **Engines:** All **Transmissions:** A/T, M/T	**Excessive Time To Enter Closed Loop** DTC P0115, P0117 and P0118 not set, cold startup requirement met (ECT input less than 140°F, IAT sensor more than 50°Fat startup), engine runtime over 10 minutes, and the PCM detected the ECT sensor signal did not reach 78°F after the engine warm-up period. **Possible Causes:** • Inspect for low coolant level or an incorrect coolant mixture • Check the operation of the thermostat (it may be stuck open) • ECT sensor signal circuit has high resistance • ECT sensor has failed
DTC: P0126 **2T CCM, MIL: YES** **2000, 2001, 2002, 2003, 2004, 2005** **Models:** 626, MX-5 Miata, MPV, Protégé, Protégé5 **Engines:** All **Transmissions:** A/T, M/T	**Thermostat Malfunction** Cold startup requirement met (IAT sensor more than 14°F and the difference between ECT and IAT sensor signals less than 43°F), engine runtime over 10 minutes, and the PCM detected the ECT sensor did not reach 160°F after the engine warm-up period. **Possible Causes:** • Inspect for low coolant level or an incorrect coolant mixture • Check the operation of the thermostat (it may be stuck open) • ECT sensor has drifted out of calibration or it has failed • PCM has failed
DTC: P0126 **2T CCM, MIL: YES** **2003, 2004, 2005** **Models:** Mazda3, Mazda6, MX-5 Miata, RX-8 **Engines:** All **Transmissions:** A/T, M/T	**Thermostat Malfunction** Cold startup requirement met (IAT sensor more than 14°F and the difference between ECT and IAT sensor signals less than 43°F), engine runtime over 10 minutes, and the PCM detected the ECT sensor did not reach 160°F after the engine warm-up period. **Possible Causes:** • Inspect for low coolant level or an incorrect coolant mixture • Check the operation of the thermostat (it may be stuck open) • ECT sensor has drifted out of calibration or it has failed • PCM has failed
DTC: P0128 **2T CCM, MIL: YES** **2001, 2002, 2003** **Models:** 626, Protégé, Protégé5 **Engines:** All **Transmissions:** A/T, M/T	**Thermostat Malfunction** Cold startup requirement met (ECT input less than 95°F, IAT sensor more than 14°F and the difference between the ECT and IAT sensor signals less than 43°F), VSS input over 15 mph and the MAF, IAT and ECT sensor signals all within normal range, the PCM detected the radiator heat ratio exceeded its threshold after a warm-up period. **Possible Causes:** • Check the operation of the thermostat (it may be stuck open) • ECT sensor has failed • PCM has failed
DTC: P0128 **1T CCM, MIL: YES** **2000, 2001, 2002, 2003, 2004, 2005** **Models:** MPV **Engines:** All **Transmissions:** A/T, M/T	**Thermostat Malfunction** Cold startup requirement met (ECT input less than 95°F, IAT sensor more than 14°F and the difference between the ECT and IAT sensor signals less than 43°F), VSS input over 15 mph and the MAF, IAT and ECT sensor signals all within normal range, the PCM detected the radiator heat ratio exceeded its threshold after a warm-up period. **Possible Causes:** • Check the operation of the thermostat (it may be stuck open) • ECT sensor has failed • PCM has failed
DTC: P0128 **1T CCM, MIL: YES** **2003, 2004, 2005** **Models:** Mazda3, Mazda6, MX-5 Miata **Engines:** All **Transmissions:** A/T, M/T	**Thermostat Malfunction** Cold startup requirement met (ECT input less than 95°F, IAT sensor more than 14°F and the difference between the ECT and IAT sensor signals less than 43°F), VSS input over 15 mph and the MAF, IAT and ECT sensor signals all within normal range, the PCM detected the radiator heat ratio exceeded its threshold after a warm-up period. **Possible Causes:** • Check the operation of the thermostat (it may be stuck open) • ECT sensor has failed • PCM has failed
DTC: P0128 **1T CCM, MIL: YES** **2004, 2005** **Models:** Tribute **Engines:** All **Transmissions:** A/T, M/T	**Thermostat Malfunction** Cold startup requirement met (ECT input less than 95°F, IAT sensor more than 14°F and the difference between the ECT and IAT sensor signals less than 43°F), VSS input over 15 mph and the MAF, IAT and ECT sensor signals all within normal range, the PCM detected the radiator heat ratio exceeded its threshold after a warm-up period. **Possible Causes:** • Check the operation of the thermostat (it may be stuck open) • ECT sensor has failed • PCM has failed

DTC	Trouble Code Title, Conditions & Possible Causes
DTC: P0130 **2T OBD/O2S1, MIL: YES** **1996, 1997, 1998, 1999, 2000, 2001, 2002, 2003** **Models:** Miata, Millenia, Protégé, Protégé5, MPV **Engines:** All **Transmissions:** A/T, M/T	**HO2S-11 (Bank 1 Sensor 1) Circuit Malfunction** Engine speed from 1500-3000 rpm while in closed loop, ECT sensor signal more than 14ºF, engine load from 28-59%, vehicle speed over 3.5 mph, and after the PCM monitored the HO2S inversion cycle period, lean to rich response and rich to lean response time under these conditions, it determined that the HO2S failed one of the tests. **Possible Causes:** • Air leaks in intake manifold, exhaust pipes or exhaust manifold • Fuel delivery system component has failed (i.e., a clogged fuel filter, dirty or restricted fuel injectors, low or high fuel pressure) • Base engine "mechanical" problem (low cylinder compression, incorrect camshaft timing, intake or exhaust manifold leaks) • HO2S is damaged or has failed • HO2S heater element has deteriorated or failed • PRC solenoid or Purge solenoid is damaged or has failed • PCM has failed
DTC: P0130 **2T OBD/O2S1, MIL: YES** **2004** **Models:** Tribute **Engines:** All **Transmissions:** A/T, M/T	**HO2S-11 (Bank 1 Sensor 1) Circuit Malfunction** Engine speed from 1500-3000 rpm while in closed loop, ECT sensor signal more than 14ºF, engine load from 28-59%, vehicle speed over 3.5 mph, and after the PCM monitored the HO2S inversion cycle period, lean to rich response and rich to lean response time under these conditions, it determined that the HO2S failed one of the tests. **Possible Causes:** • Air leaks in intake manifold, exhaust pipes or exhaust manifold • Fuel delivery system component has failed (i.e., a clogged fuel filter, dirty or restricted fuel injectors, low or high fuel pressure) • Base engine "mechanical" problem (low cylinder compression, incorrect camshaft timing, intake or exhaust manifold leaks) • HO2S is damaged or has failed • HO2S heater element has deteriorated or failed • PRC solenoid or Purge solenoid is damaged or has failed • PCM has failed
DTC: P0130 **2T OBD/O2S1, MIL: YES** **2004, 2005** **Models:** Mazda3, Mazda6, RX-8 **Engines:** All **Transmissions:** A/T, M/T	**HO2S-11 (Bank 1 Sensor 1) Circuit Malfunction** Engine speed from 1500-3000 rpm while in closed loop, ECT sensor signal more than 14ºF, engine load from 28-59%, vehicle speed over 3.5 mph, and after the PCM monitored the HO2S inversion cycle period, lean to rich response and rich to lean response time under these conditions, it determined that the HO2S failed one of the tests. **Possible Causes:** • Air leaks in intake manifold, exhaust pipes or exhaust manifold • Fuel delivery system component has failed (i.e., a clogged fuel filter, dirty or restricted fuel injectors, low or high fuel pressure) • Base engine "mechanical" problem (low cylinder compression, incorrect camshaft timing, intake or exhaust manifold leaks) • HO2S is damaged or has failed • HO2S heater element has deteriorated or failed • PRC solenoid or Purge solenoid is damaged or has failed • PCM has failed
DTC: P0131 **2T CCM, MIL: YES** **1996, 1997, 1998, 1999, 2000, 2001, 2002, 2003, 2004, 2005** **Models:** B2300, B2500, B3000, B4000, MPV, Tribute **Engines:** All **Transmissions:** A/T, M/T	**HO2S-11 (Bank 1 Sensor 1) Circuit Malfunction** Vehicle driven in closed loop at cruise speed and back to idle speed, and the PCM detected a negative voltage on the HO2S circuit. **Possible Causes:** • HO2S is water or fuel contaminated • HO2S is damaged or has failed • PCM has failed
DTC: P0131 **2T CCM, MIL: YES** **2003, 2004, 2005** **Models:** Mazda3, Mazda6, RX-8 **Engines:** All **Transmissions:** A/T, M/T	**HO2S-11 (Bank 1 Sensor 1) Circuit Malfunction** Vehicle driven in closed loop at cruise speed and back to idle speed, and the PCM detected a negative voltage on the HO2S circuit. **Possible Causes:** • HO2S is water or fuel contaminated • HO2S is damaged or has failed • PCM has failed

DTC	Trouble Code Title, Conditions & Possible Causes
DTC: P0132 **2T CCM, MIL: YES** **2003, 2004, 2005** **Models:** Mazda3, Mazda6, MX-5 Miata, MPV, Tribute **Engines:** All **Transmissions:** A/T, M/T	**HO2S-11 (Bank 1 Sensor 1) Circuit Malfunction** Vehicle driven in closed loop at cruise speed and back to idle speed, and the PCM detected a negative voltage on the HO2S circuit. **Possible Causes:** • HO2S is water or fuel contaminated • HO2S is damaged or has failed • PCM has failed
DTC: P0132 **2T CCM, MIL: YES** **2003, 2004, 2005** **Models:** Mazda3, Mazda6, RX-8 **Engines:** All **Transmissions:** A/T, M/T	**HO2S-11 (Bank 1 Sensor 1) Circuit Malfunction** Vehicle driven in closed loop at cruise speed and back to idle speed, and the PCM detected a negative voltage on the HO2S circuit. **Possible Causes:** • HO2S is water or fuel contaminated • HO2S is damaged or has failed • PCM has failed
DTC: P0133 **2T OBD/O2S1, MIL: YES** **1996, 1997, 1998, 1999, 2000, 2001, 2002, 2003, 2004, 2005** **Models:** B2300, B2500, B3000, B4000, Tribute, 626 **Engines:** All **Transmissions:** A/T, M/T	**HO2S-11 (Bank 1 Sensor 1) Circuit Slow Response** Vehicle driven in closed loop at cruise speed and back to idle speed, and the PCM detected the frequency and the amplitude response rate of the front HO2S was less than a calibrated window in memory. **Possible Causes:** • Air leaks in the intake manifold or exhaust manifold or pipes • IAT sensor or MAF sensor has deteriorated (out of calibration) • HO2S signal circuit is open or shorted to ground (intermittent) • HO2S contaminated with wrong fuel, has deteriorated or failed • PCM has failed
DTC: P0133 **2T OBD/O2S1, MIL: YES** **1996, 1997, 1998, 1999, 2000, 2001, 2002, 2003, 2004, 2005** **Models:** B2300, B2500, B3000, B4000, Tribute, 626 **Engines:** All **Transmissions:** A/T, M/T	**HO2S-11 (Bank 1 Sensor 1) Circuit Slow Response** Vehicle driven in closed loop at cruise speed and back to idle speed, and the PCM detected the frequency and the amplitude response rate of the front HO2S was less than a calibrated window in memory. **Possible Causes:** • Air leaks in the intake manifold or exhaust manifold or pipes • IAT sensor or MAF sensor has deteriorated (out of calibration) • HO2S signal circuit is open or shorted to ground (intermittent) • HO2S contaminated with wrong fuel, has deteriorated or failed • PCM has failed
DTC: P0133 **2T OBD/O2S1, MIL: YES** **2003, 2004, 2005** **Models:** Mazda3, Mazda6, MX-5 Miata, MPV, RX-8 **Engines:** All **Transmissions:** A/T, M/T	**HO2S-11 (Bank 1 Sensor 1) Circuit Slow Response** Vehicle driven in closed loop at cruise speed and back to idle speed, and the PCM detected the frequency and the amplitude response rate of the front HO2S was less than a calibrated window in memory. **Possible Causes:** • Air leaks in the intake manifold or exhaust manifold or pipes • IAT sensor or MAF sensor has deteriorated (out of calibration) • HO2S signal circuit is open or shorted to ground (intermittent) • HO2S contaminated with wrong fuel, has deteriorated or failed • PCM has failed
DTC: P0134 **2T OBD/O2S1, MIL: YES** **1996, 1997, 1998, 1999, 2000, 2001, 2002, 2003, 2004, 2005** **Models:** Mazda3, Mazda6, MX-5 Miata, Millenia, Protégé, Protégé5, MPV, Tribute **Engines:** All **Transmissions:** A/T, M/T	**HO2S-11 (Bank 1 Sensor 1) Circuit No Activity Detected** Engine speed from 1500-3000 rpm while in closed loop, ECT sensor signal more than 176ºF, and the PCM detected the front HO2S signal did not exceed 550 mv for over 54.2 seconds during the test. **Possible Causes:** • Air leaks in the intake manifold or exhaust manifold or pipes • IAT sensor or MAF sensor has deteriorated (out of calibration) • HO2S signal circuit is open or shorted to ground (intermittent) • HO2S contaminated with wrong fuel, has deteriorated or failed • PCM has failed
DTC: P0135 **2T OBD/O2S2, MIL: YES** **1996, 1997, 1998, 1999, 2000, 2001, 2002, 2003, 2004, 2005** **Models:** B2300, B2500, B3000, B4000, Tribute, MPV **Engines:** All **Transmissions:** A/T, M/T	**HO2S-11 (Bank 1 Sensor 1) Heater Circuit Malfunction** Engine started, engine running, and the PCM detected the HO2S heater circuit was more than 11.5v with the heater turned "on", or it was less than 5.8v with the heater turned "off" for 331-361 seconds. **Possible Causes:** • HO2S heater control circuit is open or shorted to ground • HO2S heater control circuit is shorted to system power (B+) • HO2S heater power circuit open between the heater and PCM • HO2S heater is damaged or has failed • PCM has failed (it controls the heater with a duty cycle signal)

DTC	Trouble Code Title, Conditions & Possible Causes
DTC: P01035 **2T OBD/O2S2, MIL: YES** **1996, 1997, 1998, 1999, 2000, 2001, 2002** **Models:** 626 **Engines:** All **Transmissions:** A/T, M/T	**HO2S-11 (Bank 1 Sensor 1) Heater Circuit Malfunction** Engine started, engine runtime over 254 seconds, HO2S heater turned "on", and the PCM detected the HO2S current was more than 3.0 amps, or that it was less than 0.14 amps during the CCM test. **Possible Causes:** • HO2S heater control circuit is open or shorted to ground • HO2S heater power circuit open between the heater and PCM • HO2S heater is damaged or has failed • PCM has failed (it controls the heater with a duty cycle signal)
DTC: P0135 **2T OBD/O2S2, MIL: YES** **1996, 1997, 1998, 1999, 2000, 2001, 2002** **Models:** Millenia **Engines:** All **Transmissions:** A/T, M/T	**HO2S-11 (Bank 1 Sensor 1) Heater Circuit Malfunction** Engine started, engine running, and the PCM detected the HO2S heater circuit indicated more than 11.5v with the heater turned "on", or it indicated less than 5.8v with the heater turned "off" for 331-361 seconds during the CCM test. **Possible Causes:** • HO2S heater control circuit is open or shorted to ground • HO2S heater control circuit is shorted to system power (B+) • HO2S heater power circuit open between the heater and PCM • HO2S heater is damaged or has failed • PCM has failed (it controls the heater with a duty cycle signal)
DTC: P0135 **2T OBD/O2S2, MIL: YES** **1996, 1997, 1998** **Models:** Protégé **Engines:** 1.8L VIN 2 **Transmissions:** A/T, M/T	**HO2S-11 (Bank 1 Sensor 1) Heater Circuit Malfunction** Engine started, engine running, and the PCM detected the HO2S heater circuit indicated more than 11.5v with the heater turned "on", or it indicated less than 5.8v with the heater turned "off" for 331-361 seconds during the CCM test. **Possible Causes:** • HO2S heater control circuit is open or shorted to ground • HO2S heater control circuit is shorted to system power (B+) • HO2S heater power circuit open between the heater and PCM • HO2S heater is damaged or has failed • PCM has failed (it controls the heater with a duty cycle signal)
DTC: P0135 **2T OBD/O2S2, MIL: YES** **1996, 1997, 1998** **Models:** Protégé **Engines:** 1.5L VIN 1 **Transmissions:** A/T, M/T	**HO2S-11 (Bank 1 Sensor 1) Heater Circuit Malfunction** Engine started, engine runtime over 60 seconds, system voltage at 10-16v, and the PCM detected the HO2S heater current was more than 6.0 amps or less than 0.14 amps with the heater turned "on". **Possible Causes:** • HO2S heater control circuit is open or shorted to ground • HO2S heater power circuit open between the heater and PCM • HO2S heater is damaged or has failed • PCM has failed (it controls the heater with a duty cycle signal)
DTC: P0136 **2T OBD/O2S1, MIL: YES** **1996, 1997, 1998, 1999, 2000, 2001, 2002, 2003, 2004, 2005** **Models:** B2300, B2500, B3000, B4000, Tribute **Engines:** All **Transmissions:** A/T, M/T	**HO2S-12 (Bank 1 Sensor 2) No Activity Detected** Engine running in closed loop for 5 minutes, and the PCM detected the middle HO2S signal was less than a calibrated value in memory. **Possible Causes:** • Air leaks in the intake manifold or exhaust manifold or pipes • IAT sensor or MAF sensor has deteriorated (out of calibration) • HO2S signal circuit is open or shorted to ground • HO2S contaminated with wrong fuel, has deteriorated or failed • PCM has failed
DTC: P0136 **2T CCM, MIL: YES** **1996, 1997** **Models:** 626, MX-6 **Engines:** 2.5L VIN D **Transmissions:** A/T	**HO2S-12 (Bank 1 Sensor 2) No Activity Detected** Vehicle driven in closed loop at cruise speed for 5 minutes, then back to idle speed, and the PCM detected the HO2S signal was less than a calibrated value stored in memory during the CCM test. **Possible Causes:** • HO2S wiring crossed at the connector or shorted in the harness • Air leaks at the exhaust manifold or exhaust pipes • HO2S signal circuit is open or shorted to ground • HO2S contaminated with wrong fuel, has deteriorated or failed • PCM has failed

DTC	Trouble Code Title, Conditions & Possible Causes
DTC: P0136 **2T CCM, MIL: YES** **1998, 1999, 2000, 2001, 2002** **Models:** 626 **Engines:** All **Transmissions:** A/T, M/T	**HO2S-12 (Bank 1 Sensor 2) No Activity Detected** Engine started, vehicle driven at cruise at low engine load for over 5 minutes, and the PCM detected the middle HO2S signal remained too high, the signal was fixed from 350-550 mv, or the HO2S signal switch time was too long during the CCM test. **Possible Causes:** • Air leaks at the exhaust manifold or exhaust pipes • HO2S wiring crossed at the connector or shorted in the harness • HO2S signal circuit is open or shorted to ground • HO2S contaminated with wrong fuel, has deteriorated or failed
DTC: P0137 **2T CCM, MIL: YES** **2001, 2002** **Models:** 626 **Engines:** All **Transmissions:** A/T, M/T	**HO2S-12 (Bank 1 Sensor 2) Lack of Switching (Too Lean)** Vehicle driven in closed loop at cruise speed at low engine load for over 5 minutes, and the PCM detected a lack of switching on the HO2S signal with the circuit indicating a lean status during the test. **Possible Causes:** • HO2S signal circuit is open or shorted to ground • Air leaks at the exhaust manifold or exhaust pipes • HO2S contaminated with wrong fuel, has deteriorated or failed • PCM has failed
DTC: P0137 **2T CCM, MIL: YES** 1997, 1997 **Models:** B2300, B2500, B3000, B4000 **Engines:** All **Transmissions:** A/T, M/T	**HO2S-12 (Bank 1 Sensor 2) Circuit Low Input** Vehicle driven in closed loop at cruise speed at low engine load for over 5 minutes, and the PCM detected the HO2S signal remained at less than 500 mv during the CCM test for 55 seconds. **Possible Causes:** • HO2S signal circuit is open or shorted to ground • Air leaks at the exhaust manifold or exhaust pipes • HO2S contaminated with wrong fuel, has deteriorated or failed • PCM has failed
DTC: P0137 **2T CCM, MIL: YES** **2003** **Models:** Mazda6 **Engines:** All **Transmissions:** A/T, M/T	**HO2S-12 (Bank 1 Sensor 2) Circuit Low Input** Vehicle driven in closed loop at cruise speed at low engine load for over 5 minutes, and the PCM detected the HO2S signal remained at less than 500 mv during the CCM test for 55 seconds. **Possible Causes:** • HO2S signal circuit is open or shorted to ground • Air leaks at the exhaust manifold or exhaust pipes • HO2S contaminated with wrong fuel, has deteriorated or failed • PCM has failed
DTC: P0137 **2T CCM, MIL: YES** **2004** **Models:** Tribute **Engines:** All **Transmissions:** A/T, M/T	**HO2S-12 (Bank 1 Sensor 2) Circuit Low Input** Vehicle driven in closed loop at cruise speed at low engine load for over 5 minutes, and the PCM detected the HO2S signal remained at less than 500 mv during the CCM test for 55 seconds. **Possible Causes:** • HO2S signal circuit is open or shorted to ground • Air leaks at the exhaust manifold or exhaust pipes • HO2S contaminated with wrong fuel, has deteriorated or failed • PCM has failed
DTC: P0138 **2T CCM, MIL: YES** **2001, 2002, 2003** **Models:** 626, Protégé, Protégé5 **Engines:** All **Transmissions:** A/T, M/T	**HO2S-12 (Bank 1 Sensor 2) Lack of Switching (Too Rich)** Vehicle driven in closed loop at cruise speed at low engine load for over 5 minutes, and the PCM detected a lack of switching on the HO2S signal with the circuit indicating a rich status during the test. **Possible Causes:** • HO2S signal tracking (wet/oily) in connector causing a short between the signal circuit and heater power circuit • HO2S signal circuit is shorted to system power • HO2S contaminated with wrong fuel, has deteriorated or failed • PCM has failed
DTC: P0138 **2T CCM, MIL: YES** 1997, 1997 **Models:** B2300, B2500, B3000, B4000 **Engines:** All **Transmissions:** A/T, M/T	**HO2S-12 (Bank 1 Sensor 2) Circuit High Input** Vehicle driven in closed loop at cruise speed at low engine load for over 5 minutes, and the PCM detected the HO2S signal remained at more than 500 mv during the CCM test for 55 seconds. **Possible Causes:** • HO2S signal tracking (wet/oily) in connector causing a short between the signal circuit and heater power circuit • HO2S signal circuit is shorted to system power HO2S contaminated with wrong fuel, has deteriorated or failed • PCM has failed

DTC	Trouble Code Title, Conditions & Possible Causes
DTC: P0138 **2T CCM, MIL: YES** **2003, 2004, 2005** **Models:** Mazda3, Mazda6, MX-5 Miata, MPV **Engines:** All **Transmissions:** A/T, M/T	**HO2S-12 (Bank 1 Sensor 2) Circuit High Input** Vehicle driven in closed loop at cruise speed at low engine load for over 5 minutes, and the PCM detected the HO2S signal remained at more than 500 mv during the CCM test for 55 seconds. **Possible Causes:** • HO2S signal tracking (wet/oily) in connector causing a short between the signal circuit and heater power circuit • HO2S signal circuit is shorted to system power HO2S contaminated with wrong fuel, has deteriorated or failed • PCM has failed
DTC: P0138 **2T CCM, MIL: YES** **2005** **Models:** RX-8, Tribute **Engines:** All **Transmissions:** A/T, M/T	**HO2S-12 (Bank 1 Sensor 2) Circuit High Input** Vehicle driven in closed loop at cruise speed at low engine load for over 5 minutes, and the PCM detected the HO2S signal remained at more than 500 mv during the CCM test for 55 seconds. **Possible Causes:** • HO2S signal tracking (wet/oily) in connector causing a short between the signal circuit and heater power circuit • HO2S signal circuit is shorted to system power HO2S contaminated with wrong fuel, has deteriorated or failed • PCM has failed
DTC: P0139 **2T CCM, MIL: YES** **2004, 2005** **Models:** Mazda3, MX-5 Miata, RX-8 **Engines:** All **Transmissions:** A/T, M/T	**Middle HO2S Circuit problem** The PCM monitors inversion cycle period, middle HO2S output voltage inclination. The PCM detects that the voltage inclinations are below threshold consecutive 5 times when following conditions are met, the PCM determines that circuit has malfunction. Under the following monitoring conditions, if 0.3 V or more is detected three times even if fuel cut is performed for 3 seconds or more, a circuit malfunction is determined. **Possible Causes:** • Middle HO2S circuit deterioration • Middle HO2S circuit malfunction • PCM has failed
DTC: P0140 **2T OBD/O2S1, MIL: YES** **1996, 1997, 1998, 1999, 2000, 2001, 2002, 2003, 2004, 2005** **Models:** Mazda3, Mazda6, MX-5 Miata, Millenia, Protégé, Protégé5, MPV **Engines:** All **Transmissions:** A/T, M/T	**HO2S-12 (Bank 1 Sensor 2) Circuit No Activity Detected** Engine speed from 1500-3000 rpm while in closed loop, ECT sensor signal more than 176°F, and the PCM detected the HO2S signal voltage did not exceed 550 mv for over 54.2 seconds during the test. **Possible Causes:** • Air leaks in the intake manifold or exhaust manifold or pipes • IAT sensor or MAF sensor has deteriorated (out of calibration) • HO2S signal circuit is open or shorted to ground • HO2S contaminated with wrong fuel, has deteriorated or failed • TSB 0100700 (3/00) contains information for this code (MPV)
DTC: P0140 **2T OBD/O2S1, MIL: YES** **2004** **Models:** Tribute **Engines:** All **Transmissions:** A/T, M/T	**HO2S-12 (Bank 1 Sensor 2) Circuit No Activity Detected** Engine speed from 1500-3000 rpm while in closed loop, ECT sensor signal more than 176°F, and the PCM detected the HO2S signal voltage did not exceed 550 mv for over 54.2 seconds during the test. **Possible Causes:** • Air leaks in the intake manifold or exhaust manifold or pipes • IAT sensor or MAF sensor has deteriorated (out of calibration) • HO2S signal circuit is open or shorted to ground • HO2S contaminated with wrong fuel, has deteriorated or failed • TSB 0100700 (3/00) contains information for this code (MPV)
DTC: P0141 **2T OBD/O2S, MIL: YES** **1996, 1997, 1998, 1999, 2000, 2001, 2002, 2003, 2004, 2005** **Models:** B2300, B2500, B3000, B4000, Tribute, MPV **Engines:** All **Transmissions:** A/T, M/T	**HO2S-12 (Bank 1 Sensor 2) Heater Circuit Malfunction** Engine started, engine running, and the PCM detected the HO2S heater circuit indicated more than 11.5v with the heater turned "on", or it indicated less than 5.8v with the heater turned "off" for 331-361 seconds during the CCM test. **Possible Causes:** • HO2S heater control circuit is open or shorted to ground • HO2S heater control circuit is shorted to system power (B+) • HO2S heater power circuit open between the heater and PCM • HO2S heater is damaged or has failed • PCM has failed (it controls the heater with a duty cycle signal)
DTC: P0141 **2T OBD/O2S, MIL: YES** **1996, 1997, 1998, 1999, 2000, 2001, 2002** **Models:** 626 **Engines:** All **Transmissions:** A/T, M/T	**HO2S-21 (Bank 1 Sensor 1) Heater Circuit Malfunction** Engine started, engine runtime over 254 seconds, HO2S heater turned "on", and the PCM detected the HO2S current was more than 3.0 amps, or that it was less than 0.14 amps during the CCM test. **Possible Causes:** • HO2S heater control circuit is open or shorted to ground • HO2S heater power circuit open between the heater and PCM • HO2S heater is damaged or has failed • PCM has failed (it controls the heater with a duty cycle signal)

DTC	Trouble Code Title, Conditions & Possible Causes
DTC: P0141 **2T OBD/O2S, MIL: YES** **1996, 1997, 1998, 1999, 2000, 2001, 2002** **Models:** Millenia **Engines:** All **Transmissions:** A/T, M/T	**HO2S-21 (Bank 2 Sensor 1) Heater Circuit Malfunction** Engine started, engine running, and the PCM detected the HO2S heater circuit indicated more than 11.5v with the heater turned "on", or it indicated less than 5.8v with the heater turned "off" for 331-361 seconds during the CCM test. **Possible Causes:** • HO2S heater control circuit is open or shorted to ground • HO2S heater control circuit is shorted to system power (B+) • HO2S heater power circuit open between the heater and PCM • HO2S heater is damaged or has failed • PCM has failed (it controls the heater with a duty cycle signal)
DTC: P0141 **2T OBD/O2S, MIL: YES** **1996, 1997, 1998** **Models:** Protégé **Engines:** 1.5L VIN 1 **Transmissions:** A/T, M/T	**HO2S-12 (Bank 1 Sensor 2) Heater Circuit Malfunction** Engine started, engine runtime over 60 seconds, system voltage at 10-16v, and the PCM detected the HO2S heater current was more than 6.0 amps or less than 0.14 amps with the heater turned "on". **Possible Causes:** • HO2S heater control circuit is open or shorted to ground • HO2S heater power circuit open between the heater and PCM • HO2S heater is damaged or has failed • PCM has failed (it controls the heater with a duty cycle signal)
DTC: P0141 **2T OBD/O2S, MIL: YES** **1996, 1997, 1998** **Models:** Protégé **Engines:** 1.8L VIN 2 **Transmissions:** A/T, M/T	**HO2S-12 (Bank 1 Sensor 2) Heater Circuit Malfunction** Engine started, engine running, and the PCM detected the HO2S heater circuit indicated more than 11.5v with the heater turned "on", or it indicated less than 5.8v with the heater turned "off" for 331-361 seconds during the CCM test. **Possible Causes:** • HO2S heater control circuit is open or shorted to ground • HO2S heater control circuit is shorted to system power (B+) • HO2S heater power circuit open between the heater and PCM • HO2S heater is damaged or has failed • PCM has failed (it controls the heater with a duty cycle signal)
DTC: P0142 **2T CCM, MIL: YES** **2001, 2002** **Models:** 626 **Engines:** All **Transmissions:** A/T, M/T	**HO2S-12 (Bank 1 Sensor 2) Circuit Malfunction** Vehicle driven in closed loop at cruise speed at low engine load for over 5 minutes, and the PCM detected an unexpected voltage condition on the HO2S circuit during the CCM test. **Possible Causes:** • Air leaks at the exhaust manifold or exhaust pipes • HO2S contaminated due to wrong fuel, or it has deteriorated • PCM has failed
DTC: P0142 **2T CCM, MIL: YES** **1998, 1999, 2000, 2001, 2002** **Models:** 626 **Engines:** 2.5L VIN D	**HO2S-13 (Bank 1 Sensor 3) Circuit Malfunction** Vehicle driven in closed loop at cruise speed at low engine load for over 5 minutes, and the PCM detected the HO2S signal was outside of its functional calibrated window during the CCM test. **Possible Causes:** • Air leaks at the exhaust manifold or exhaust pipes • HO2S signal and grounded wire shorted inside the harness • HO2S contaminated due to wrong fuel, or it has deteriorated • PCM has failed
DTC: P0142 **2T CCM, MIL: YES** **2004** **Models:** Tribute **Engines:** 2.5L VIN D	**HO2S-13 (Bank 1 Sensor 3) Circuit Malfunction** Vehicle driven in closed loop at cruise speed at low engine load for over 5 minutes, and the PCM detected the HO2S signal was outside of its functional calibrated window during the CCM test. **Possible Causes:** • Air leaks at the exhaust manifold or exhaust pipes • HO2S signal and grounded wire shorted inside the harness • HO2S contaminated due to wrong fuel, or it has deteriorated • PCM has failed
DTC: P0144 **2T CCM, MIL: YES** **1999, 2000** **Models:** Millenia, MPV **Engines:** 2.5L VIN 1, 2.5L VIN G **Transmissions:** A/T, M/T	**HO2S-13 (Bank 1 Sensor 3) Circuit Malfunction** Vehicle driven in closed loop at cruise speed for 5 minutes, followed by a deceleration period, and the PCM detected the HO2S signal remained above 0.45v for 5 seconds under these conditions. **Possible Causes:** • HO2S signal circuit is shorted to system power (heater circuit) at the connector or somewhere in the wiring harness • HO2S contaminated due to wrong fuel, or it has deteriorated • PCM has failed

DTC	Trouble Code Title, Conditions & Possible Causes
DTC: P0144 **2T CCM, MIL: YES** **2004, 2005** **Models:** Tribute **Engines:** 2.5L VIN 1, 2.5L VIN G **Transmissions:** A/T, M/T	**HO2S-13 (Bank 1 Sensor 3) Circuit Malfunction** Vehicle driven in closed loop at cruise speed for 5 minutes, followed by a deceleration period, and the PCM detected the HO2S signal remained above 0.45v for 5 seconds under these conditions. **Possible Causes:** • HO2S signal circuit is shorted to system power (heater circuit) at the connector or somewhere in the wiring harness • HO2S contaminated due to wrong fuel, or it has deteriorated • PCM has failed
DTC: P0144 **2T CCM, MIL: YES** **2004, 2005** **Models:** Mazda3, Mazda6 **Engines:** All **Transmissions:** A/T, M/T	**HO2S-13 (Bank 1 Sensor 3) Circuit Malfunction** Vehicle driven in closed loop at cruise speed for 5 minutes, followed by a deceleration period, and the PCM detected the HO2S signal remained above 0.45v for 5 seconds under these conditions. **Possible Causes:** • HO2S signal circuit is shorted to system power (heater circuit) at the connector or somewhere in the wiring harness • HO2S contaminated due to wrong fuel, or it has deteriorated • PCM has failed
DTC: P0145 **2T CCM, MIL: YES** **2004** **Models:** Mazda3 **Engines:** All **Transmissions:** A/T, M/T	**Rear HO2S Circuit problem** The PCM monitors inversion cycle period, Rear HO2S output voltage inclination. The PCM detects that the voltage inclinations are below threshold consecutive 5 times when following conditions are met, or isn't below 0.3 V for 2 seconds after fuel cut, The PCM determines that circuit has malfunction. **Possible Causes:** • Rear HO2S malfunction • Rear HO2S contaminated due to wrong fuel, or it has deteriorated • PCM has failed
DTC: P0146 **2T CCM, MIL: YES** **1998, 1999, 2000** **Models:** Millenia, MPV **Engines:** 2.5L VIN 1, 2.5L VIN G **Transmissions:** A/T, M/T	**HO2S-13 (Bank 1 Sensor 3) Circuit No Activity Detected** Vehicle driven in closed loop at over 1150 rpm for 5 minutes and with the HO2S heater commanded "on", the PCM detected the HO2S signal did not exceed 600 mv for 80 seconds during the test. **Possible Causes:** • Air leaks in the intake manifold or exhaust manifold or pipes • IAT sensor or MAF sensor has deteriorated (out of calibration) • Fuel injector leaking or restricted, or pressure regulator leaking • HO2S signal circuit is open or shorted to ground • HO2S contaminated with wrong fuel, has deteriorated or failed • Ignition system component has failed or is damaged
DTC: P0146 **2T CCM, MIL: YES** **2004** **Models:** Tribute **Engines:** 2.5L VIN 1, 2.5L VIN G **Transmissions:** A/T, M/T	**HO2S-13 (Bank 1 Sensor 3) Circuit No Activity Detected** Vehicle driven in closed loop at over 1150 rpm for 5 minutes and with the HO2S heater commanded "on", the PCM detected the HO2S signal did not exceed 600 mv for 80 seconds during the test. **Possible Causes:** • Air leaks in the intake manifold or exhaust manifold or pipes • IAT sensor or MAF sensor has deteriorated (out of calibration) • Fuel injector leaking or restricted, or pressure regulator leaking • HO2S signal circuit is open or shorted to ground • HO2S contaminated with wrong fuel, has deteriorated or failed • Ignition system component has failed or is damaged
DTC: P0147 **2T OBD/O2S, MIL: YES** **1998, 1999, 2000, 2001, 2002** **Models:** 626 **Engines:** 2.5L VIN D **Transmissions:** A/T, M/T	**HO2S-13 (Bank 1 Sensor 3) Heater Circuit Malfunction** Engine started, engine runtime over 254 seconds, HO2S heater turned "on", and the PCM detected the HO2S current was more than 3.0 amps, or that it was less than 0.14 amps during the CCM test. **Possible Causes:** • HO2S heater control circuit is open or shorted to ground • HO2S heater power circuit open between the heater and PCM • HO2S heater is damaged or has failed • PCM has failed (it controls the heater with a duty cycle signal)
DTC: P0147 **2T OBD/O2S, MIL: YES** **1998, 1999, 2000** **Models:** 626, Millenia **Engines:** 2.5L VIN 1 **Transmissions:** A/T, M/T	**HO2S-13 (Bank 1 Sensor 3) Heater Circuit Malfunction** Engine started, engine running, and the PCM detected the HO2S heater circuit indicated more than 11.5v with the heater turned "on", or it indicated less than 5.8v with the heater turned "off" for 331-361 seconds during the CCM test. **Possible Causes:** • HO2S heater control circuit is open or shorted to ground • HO2S heater control circuit is shorted to system power (B+) • HO2S heater power circuit open between the heater and PCM • HO2S heater is damaged or has failed • PCM has failed (it controls the heater with a duty cycle signal)

DTC	Trouble Code Title, Conditions & Possible Causes
DTC: P0147 **2T OBD/O2S, MIL: YES** **2004, 2005** **Models:** Tribute **Engines:** 2.5L VIN 1 **Transmissions:** A/T, M/T	**HO2S-13 (Bank 1 Sensor 3) Heater Circuit Malfunction** Engine started, engine running, and the PCM detected the HO2S heater circuit indicated more than 11.5v with the heater turned "on", or it indicated less than 5.8v with the heater turned "off" for 331-361 seconds during the CCM test. **Possible Causes:** • HO2S heater control circuit is open or shorted to ground • HO2S heater control circuit is shorted to system power (B+) • HO2S heater power circuit open between the heater and PCM • HO2S heater is damaged or has failed • PCM has failed (it controls the heater with a duty cycle signal)
DTC: P0148 **2T OBD/O2S, MIL: YES** **2004, 2005** **Models:** Tribute **Engines:** 2.5L VIN 1 **Transmissions:** A/T, M/T	**HO2S-13 (Bank 1 Sensor 3) Heater Circuit Malfunction** At least one bank lean at wide open throttle. **Possible Causes:** • HO2S heater control circuit is open or shorted to ground • HO2S heater control circuit is shorted to system power (B+) • HO2S heater power circuit open between the heater and PCM • HO2S heater is damaged or has failed • PCM has failed (it controls the heater with a duty cycle signal)
DTC: P0150 **2T CCM, MIL: YES** **1996, 1997, 1998, 1999, 2000, 2001, 2002** **Models:** Millenia **Engines:** 2.3L VIN 2 **Transmissions:** A/T, M/T	**HO2S-21 (Bank 2 Sensor 1) Circuit Malfunction** Engine started, vehicle driven to over 3.5 mph at 1500-3000 rpm in closed loop, ECT sensor more than 14°F, engine load from 28-59%, and after the PCM monitored the HO2S inversion cycle period, lean to rich response and rich to lean response times under these conditions, it detected the HO2S failed one of the tests. **Possible Causes:** • Air leaks in intake manifold, exhaust pipes or exhaust manifold • Fuel delivery system component has failed (i.e., a clogged fuel filter, dirty or restricted fuel injectors, low or high fuel pressure) • Base engine "mechanical" problem (low cylinder compression, incorrect camshaft timing, intake or exhaust manifold leaks) • HO2S is damaged or has failed • HO2S heater element has deteriorated or failed • PRC solenoid or Purge solenoid is damaged or has failed • PCM has failed
DTC: P0150 **2T CCM, MIL: YES** **1999, 2000, 2001, 2002** **Models:** Millenia **Engines:** 2.5L VIN 1 **Transmissions:** A/T, M/T	**HO2S-21 (Bank 2 Sensor 1) Circuit Malfunction** Engine speed from 1500-3000 rpm while in closed loop, ECT sensor signal more than 14°F, engine load from 28-59%, vehicle speed over 3.5 mph, and after the PCM monitored the HO2S inversion cycle period, lean to rich response and rich to lean response time under these conditions, it determined that the HO2S failed one of the tests. **Possible Causes:** • Air leaks in intake manifold, exhaust pipes or exhaust manifold • Fuel delivery system component has failed (i.e., a clogged fuel filter, dirty or restricted fuel injectors, low or high fuel pressure) • Base engine "mechanical" problem (low cylinder compression, incorrect camshaft timing, intake or exhaust manifold leaks) • HO2S is damaged or has failed • HO2S heater element has deteriorated or failed • PRC solenoid or Purge solenoid is damaged or has failed • PCM has failed
DTC: P0150 **2T CCM, MIL: YES** **2000, 2001, 2002** **Models:** MPV **Engines:** All **Transmissions:** A/T, M/T	**HO2S-21 (Bank 2 Sensor 1) Circuit Malfunction** Engine started, vehicle driven at over 3.5 mph at 1500-3000 rpm in closed loop, ECT sensor more than 14°F, engine load from 28-59%, and after the PCM monitored the HO2S inversion cycle period, lean to rich response and rich to lean response times under these conditions, it determined the HO2S-21 failed one of these tests. **Possible Causes:** • Air leaks in intake manifold, exhaust pipes or exhaust manifold • Fuel delivery system component has failed (i.e., a clogged fuel filter, dirty or restricted fuel injectors, low or high fuel pressure) • Base engine "mechanical" problem (low cylinder compression, incorrect camshaft timing, intake or exhaust manifold leaks) • HO2S is damaged or has failed • HO2S heater element has deteriorated or failed • PRC solenoid or Purge solenoid is damaged or has failed • PCM has failed

DTC	Trouble Code Title, Conditions & Possible Causes
DTC: P0150 **2T CCM, MIL: YES** **2004** **Models:** Tribute **Engines:** All **Transmissions:** A/T, M/T	**HO2S-21 (Bank 2 Sensor 1) Circuit Malfunction** Engine started, vehicle driven at over 3.5 mph at 1500-3000 rpm in closed loop, ECT sensor more than 14ºF, engine load from 28-59%, and after the PCM monitored the HO2S inversion cycle period, lean to rich response and rich to lean response times under these conditions, it determined the HO2S-21 failed one of these tests. **Possible Causes:** • Air leaks in intake manifold, exhaust pipes or exhaust manifold • Fuel delivery system component has failed (i.e., a clogged fuel filter, dirty or restricted fuel injectors, low or high fuel pressure) • Base engine "mechanical" problem (low cylinder compression, incorrect camshaft timing, intake or exhaust manifold leaks) • HO2S is damaged or has failed • HO2S heater element has deteriorated or failed • PRC solenoid or Purge solenoid is damaged or has failed • PCM has failed
DTC: P0151 **2T CCM, MIL: YES** **1996, 1997, 1998, 1999, 2000, 2001, 2002, 2003, 2004, 2005** **Models:** B3000, B4000, Tribute **Engines:** All **Transmissions:** A/T, M/T	**HO2S-21 (Bank 2 Sensor 1) Circuit Malfunction** Engine started, vehicle driven in closed loop at cruise speed and then back to idle speed, and the PCM detected a negative voltage on the HO2S circuit during the CCM test. **Possible Causes:** • HO2S is water or fuel contaminated • HO2S is damaged or has failed • PCM has failed
DTC: P0151 **2T CCM, MIL: YES** **2003** **Models:** MPV **Engines:** All **Transmissions:** A/T, M/T	**HO2S-21 (Bank 2 Sensor 1) Circuit Malfunction** Engine started, vehicle driven in closed loop at cruise speed and then back to idle speed, and the PCM detected a negative voltage on the HO2S circuit during the CCM test. **Possible Causes:** • HO2S is water or fuel contaminated • HO2S is damaged or has failed • PCM has failed
DTC: P0152 **2T CCM, MIL: YES** **2003, 2004, 2005** **Models:** MPV, Tribute **Engines:** All **Transmissions:** A/T, M/T	**HO2S-21 (Bank 2 Sensor 1) Circuit Malfunction** Engine started, vehicle driven in closed loop at cruise speed and then back to idle speed, and the PCM detected a negative voltage on the HO2S circuit during the CCM test. **Possible Causes:** • HO2S is water or fuel contaminated • HO2S is damaged or has failed • PCM has failed
DTC: P0153 **2T CCM, MIL: YES** **1996, 1997, 1998, 1999, 2000, 2001, 2002, 2003, 2004, 2005** **Models:** B3000, B4000, Tribute, 626 **Engines:** All **Transmissions:** A/T, M/T	**HO2S-21 (Bank 2 Sensor 1) Circuit Slow Response** Vehicle driven in closed loop at cruise speed and back to idle speed, and the PCM detected the frequency and the amplitude response rate of the front HO2S was less than a calibrated window in memory. **Possible Causes:** • Air leaks in the intake manifold or exhaust manifold or pipes • IAT sensor or MAF sensor has deteriorated (out of calibration) • HO2S signal circuit is open or shorted to ground (intermittent) • HO2S contaminated with wrong fuel, has deteriorated or failed
DTC: P0153 **2T CCM, MIL: YES** **2003** **Models:** MPV **Engines:** All **Transmissions:** A/T, M/T	**HO2S-21 (Bank 2 Sensor 1) Circuit Slow Response** Vehicle driven in closed loop at cruise speed and back to idle speed, and the PCM detected the frequency and the amplitude response rate of the front HO2S was less than a calibrated window in memory. **Possible Causes:** • Air leaks in the intake manifold or exhaust manifold or pipes • IAT sensor or MAF sensor has deteriorated (out of calibration) • HO2S signal circuit is open or shorted to ground (intermittent) • HO2S contaminated with wrong fuel, has deteriorated or failed
DTC: P0154 **2T OBD/O2S1, MIL: YES** **1996, 1997, 1998, 1999, 2000, 2001, 2002, 2003** **Models:** Millenia, MPV **Engines:** All **Transmissions:** A/T, M/T	**HO2S-21 (Bank 2 Sensor 1) Circuit No Activity Detected** Engine speed from 1500-3000 rpm while in closed loop, ECT sensor signal more than 176ºF, and the PCM detected the front HO2S signal did not exceed 0.55v for over 54.2 seconds during the test. **Possible Causes:** • Air leaks in the intake manifold or exhaust manifold or pipes • IAT sensor or MAF sensor has deteriorated (out of calibration) • HO2S signal circuit is open or shorted to ground • HO2S contaminated with wrong fuel, has deteriorated or failed

DTC	Trouble Code Title, Conditions & Possible Causes
DTC: P0154 **2T OBD/O2S1, MIL: YES** **2004** **Models:** Tribute **Engines:** All **Transmissions:** A/T, M/T	**HO2S-21 (Bank 2 Sensor 1) Circuit No Activity Detected** Engine speed from 1500-3000 rpm while in closed loop, ECT sensor signal more than 176°F, and the PCM detected the front HO2S signal did not exceed 0.55v for over 54.2 seconds during the test. **Possible Causes:** • Air leaks in the intake manifold or exhaust manifold or pipes • IAT sensor or MAF sensor has deteriorated (out of calibration) • HO2S signal circuit is open or shorted to ground • HO2S contaminated with wrong fuel, has deteriorated or failed
DTC: P0155 **2T OBD/O2S, MIL: YES** **1996, 1997, 1998, 1999, 2000, 2001, 2002, 2003, 2004, 2005** **Models:** B3000, B4000, Tribute, MPV **Engines:** All **Transmissions:** A/T, M/T	**HO2S-21 (B2 S1) Heater Circuit Conditions** Engine started, engine running, and the PCM detected the HO2S heater circuit indicated more than 11.5v with the heater turned "on", or it indicated less than 5.8v with the heater turned "off" for 331-361 seconds during the CCM test. **Possible Causes:** • HO2S heater control circuit is open or shorted to ground • HO2S heater control circuit is shorted to system power (B+) • HO2S heater power circuit open between the heater and PCM • HO2S heater is damaged or has failed • PCM has failed (it controls the heater with a duty cycle signal)
DTC: P0155 **2T OBD/O2S, MIL: YES** **1996, 1997, 1998, 1999, 2000, 2001, 2002** **Models:** 626 **Engines:** All **Transmissions:** A/T, M/T	**HO2S-21 (Bank 2 Sensor 1) Heater Circuit Malfunction** Engine started, engine runtime over 254 seconds, HO2S heater turned "on", and the PCM detected the HO2S current was more than 3.0 amps, or that it was less than 0.14 amps during the CCM test. **Possible Causes:** • HO2S heater control circuit is open or shorted to ground • HO2S heater power circuit open between the heater and PCM • HO2S heater is damaged or has failed • PCM has failed (it controls the heater with a duty cycle signal)
DTC: P0155 **2T OBD/O2S, MIL: YES** **1996, 1997, 1998, 1999, 2000, 2001, 2002** **Models:** Millenia **Engines:** All **Transmissions:** A/T, M/T	**HO2S-21 (Bank 2 Sensor 1) Heater Circuit Malfunction** Engine started, engine running, and the PCM detected the HO2S heater circuit indicated more than 11.5v with the heater turned "on", or it indicated less than 5.8v with the heater turned "off" for 331-361 seconds during the CCM test. **Possible Causes:** • HO2S heater control circuit is open or shorted to ground • HO2S heater control circuit is shorted to system power (B+) • HO2S heater power circuit open between the heater and PCM • HO2S heater is damaged or has failed • PCM has failed (it controls the heater with a duty cycle signal)
DTC: P0156 **2T OBD/O2S1, MIL: YES** **1999, 2000** **Models:** B3000, B4000 **Engines:** All **Transmissions:** A/T, M/T	**HO2S-22 (Bank 2 Sensor 2) Circuit High Input** Engine started, vehicle driven at cruise at low engine load for over 5 minutes, and the PCM detected the rear HO2S signal remained too high, the signal was fixed from 350-550 mv, or the HO2S signal switch time was too long during the CCM test. **Possible Causes:** • Air leaks at the exhaust manifold or exhaust pipes • HO2S signal tracking (wet/oily) in connector causing a short between the signal circuit and heater power circuit • HO2S contaminated with wrong fuel, has deteriorated or failed • PCM has failed
DTC: P0156 **2T OBD/O2S1, MIL: YES** **2003, 2004, 2005** **Models:** Tribute **Engines:** All **Transmissions:** A/T, M/T	**HO2S-22 (Bank 2 Sensor 2) Circuit High Input** Engine started, vehicle driven at cruise at low engine load for over 5 minutes, and the PCM detected the rear HO2S signal remained too high, the signal was fixed from 350-550 mv, or the HO2S signal switch time was too long during the CCM test. **Possible Causes:** • Air leaks at the exhaust manifold or exhaust pipes • HO2S signal tracking (wet/oily) in connector causing a short between the signal circuit and heater power circuit • HO2S contaminated with wrong fuel, has deteriorated or failed • PCM has failed
DTC: P0157 **2T OBD/O2S1, MIL: YES** **2004** **Models:** Tribute **Engines:** All **Transmissions:** A/T, M/T	**HO2S-22 (Bank 2 Sensor 2) Circuit High Input** Engine started, vehicle driven at cruise at low engine load for over 5 minutes, and the PCM detected the rear HO2S signal remained too high, the signal was fixed from 350-550 mv, or the HO2S signal switch time was too long during the CCM test. **Possible Causes:** • Air leaks at the exhaust manifold or exhaust pipes • HO2S signal tracking (wet/oily) in connector causing a short between the signal circuit and heater power circuit • HO2S contaminated with wrong fuel, has deteriorated or failed • PCM has failed

DTC	Trouble Code Title, Conditions & Possible Causes
DTC: P0158 **2T OBD/O2S1, MIL: YES** **2004, 2005** **Models:** Tribute, MPV **Engines:** All **Transmissions:** A/T, M/T	**HO2S-22 (Bank 2 Sensor 2) Circuit High Input** Engine started, vehicle driven at cruise at low engine load for over 5 minutes, and the PCM detected the rear HO2S signal remained too high, the signal was fixed from 350-550 mv, or the HO2S signal switch time was too long during the CCM test. **Possible Causes:** • Air leaks at the exhaust manifold or exhaust pipes • HO2S signal tracking (wet/oily) in connector causing a short between the signal circuit and heater power circuit • HO2S contaminated with wrong fuel, has deteriorated or failed • PCM has failed
DTC: P0158 **2T OBD/O2S1, MIL: YES** **2004, 2005** **Models:** MPV **Engines:** All **Transmissions:** A/T, M/T	**HO2S-22 (Bank 2 Sensor 2) Circuit High Input** Engine started, vehicle driven at cruise at low engine load for over 5 minutes, and the PCM detected the rear HO2S signal remained too high, the signal was fixed from 350-550 mv, or the HO2S signal switch time was too long during the CCM test. **Possible Causes:** • Air leaks at the exhaust manifold or exhaust pipes • HO2S signal tracking (wet/oily) in connector causing a short between the signal circuit and heater power circuit • HO2S contaminated with wrong fuel, has deteriorated or failed • PCM has failed
DTC: P0160 **2T OBD/O2S1, MIL: YES** **1999, 2000** **Models:** B3000, B4000 **Engines:** All **Transmissions:** A/T, M/T	**HO2S-22 (Bank 2 Sensor 2) Circuit No Activity Detected** Engine started, engine speed from 1500-3000 rpm in closed loop, ECT sensor more than 180ºF, and the PCM detected the rear HO2S signal did not exceed 0.55v for 30-54 seconds during the CCM test. **Possible Causes:** • Air leaks in the intake manifold or exhaust manifold or pipes • IAT sensor or MAF sensor has deteriorated (out of calibration) • HO2S signal circuit is open or shorted to ground • HO2S contaminated with wrong fuel, has deteriorated or failed • TSB 0100700 (3/00) contains information for this code (MPV)
DTC: P0160 **2T OBD/O2S1, MIL: YES** **1996** **Models:** 626, MX-6 **Engines:** 2.5L VIN D **Transmissions:** A/T, M/T	**HO2S-22 (Bank 2 Sensor 2) No Activity Detected** Engine running, and the PCM detected the HO2S signal remained under 500 mv, or with the ECT input over 180ºF and engine speed over 1500 rpm for 2 minutes, it detected the rear HO2S-22 signal remained at less than 500 mv during the CCM test. **Possible Causes:** • Air leaks in the intake manifold or exhaust manifold or pipes • IAT sensor or MAF sensor has deteriorated (out of calibration) • HO2S signal circuit is open or shorted to ground • HO2S contaminated with wrong fuel, has deteriorated or failed
DTC: P0160 **2T OBD/O2S1, MIL: YES** **1996, 1997, 1998, 1999, 2000, 2001, 2002, 2003, 2004, 2005** **Models:** Millenia, MPV **Engines:** 2.3L VIN 2 **Transmissions:** A/T, M/T	**HO2S-22 (Bank 2 Sensor 2) No Activity Detected** Engine started, engine running in closed loop, and the PCM detected the HO2S signal remained under 500 mv, or with the ECT input over 154ºF and engine speed over 1500 rpm for 2 minutes, it detected the HO2S-22 signal did not exceed 500 mv during the test. **Possible Causes:** • Air leaks in the intake manifold or exhaust manifold or pipes • IAT sensor or MAF sensor has deteriorated (out of calibration) • HO2S signal circuit is open or shorted to ground • HO2S contaminated with wrong fuel, has deteriorated or failed
DTC: P0160 **2T OBD/O2S1, MIL: YES** **2004** **Models:** Tribute **Engines:** 2.3L VIN 2 **Transmissions:** A/T, M/T	**HO2S-22 (Bank 2 Sensor 2) No Activity Detected** Engine started, engine running in closed loop, and the PCM detected the HO2S signal remained under 500 mv, or with the ECT input over 154ºF and engine speed over 1500 rpm for 2 minutes, it detected the HO2S-22 signal did not exceed 500 mv during the test. **Possible Causes:** • Air leaks in the intake manifold or exhaust manifold or pipes • IAT sensor or MAF sensor has deteriorated (out of calibration) • HO2S signal circuit is open or shorted to ground • HO2S contaminated with wrong fuel, has deteriorated or failed
DTC: P0161 **2T OBD/O2S, MIL: YES** **1996, 1997, 1998, 1999, 2000, 2001, 2002** **Models:** Millenia **Engines:** 2.3L VIN 2 **Transmissions:** A/T, M/T	**HO2S-22 (Bank 2 Sensor 2) Heater Circuit Malfunction** Engine started, engine running, and the PCM detected the HO2S heater circuit indicated more than 11.5v with the heater turned "on", or it indicated less than 5.8v with the heater turned "off" for 331-361 seconds during the CCM test. **Possible Causes:** • HO2S heater control circuit is open or shorted to ground • HO2S heater control circuit is shorted to system power (B+) • HO2S heater power circuit open between the heater and PCM • HO2S heater is damaged or has failed • PCM has failed (it controls the heater with a duty cycle signal)

DTC	Trouble Code Title, Conditions & Possible Causes
DTC: P0161 **2T OBD/O2S, MIL: YES** **2003, 2004, 2005** **Models:** Tribute **Engines:** All **Transmissions:** A/T, M/T	**HO2S-22 (Bank 2 Sensor 2) Heater Circuit Malfunction** Engine started, engine running, and the PCM detected the HO2S heater circuit indicated more than 11.5v with the heater turned "on", or it indicated less than 5.8v with the heater turned "off" for 331-361 seconds during the CCM test. **Possible Causes:** • HO2S heater control circuit is open or shorted to ground • HO2S heater control circuit is shorted to system power (B+) • HO2S heater power circuit open between the heater and PCM • HO2S heater is damaged or has failed • PCM has failed (it controls the heater with a duty cycle signal)
DTC: P0170 **2T CCM, MIL: YES** **1996, 1997** **Models:** MX-6, 626 **Engines:** 2.0L VIN C **Transmissions:** M/T !	**Adaptive Fuel Trim Too Rich or Too Lean (Bank 1)** Engine running in closed loop at cruise speed for 2 minutes, and the PCM detected the A/F ratio remained richer or leaner than the fuel correction limit for 10 seconds during the Fuel System Monitor test. **Possible Causes:** • Vehicle driven low on fuel or driven until it ran out of fuel • One or more injectors restricted or pressure regulator has failed • Fuel delivery system supplying too much or too little fuel during cruise or idle speed (weak fuel pump, dirty/leaking fuel injector) • Air leaks after the MAF sensor, or air leaks in the PCV system • Air leaks at the EGR gasket, or at the EGR valve diaphragm • Exhaust leaks before or near where the front HO2S is mounted • HO2S is contaminated, deteriorated or it has failed • Fuel control sensor is out of calibration (i.e., ECT, IAT or MAP) • Base engine "mechanical" fault affecting one or more cylinders
DTC: P0170 **2T CCM, MIL: YES** **2004** **Models:** Tribute **Engines:** 2.5L VIN D **Transmissions:** A/T, M/T	**Adaptive Fuel Trim Too Rich or Too Lean (Bank 1)** Engine running in closed loop at cruise speed for 2 minutes, and the PCM detected the A/F ratio remained richer or leaner than the fuel correction limit for 10 seconds during the Fuel System Monitor test. **Possible Causes:** • Air leaks after the MAF sensor, or air leaks in the PCV system • Air leaks at the EGR gasket, or at the EGR valve diaphragm • Base engine "mechanical" fault affecting one or more cylinders • Exhaust leaks before or near where the front HO2S is mounted • Fuel control sensor is out of calibration (i.e., ECT, IAT or MAP) • Fuel delivery system supplying too much or too little fuel during idle or cruise speed (weak fuel pump, weak pressure regulator) • Fuel Injectors (one or more) is dirty, restricted or leaking fuel • HO2S is contaminated, deteriorated or it has failed • Vehicle driven low on fuel or driven until it ran out of fuel
DTC: P0171 **2T CCM, MIL: YES** **2004, 2005** **Models:** Mazda3, RX-8 **Engines:** All **Transmissions:** A/T, M/T	**Adaptive Fuel Trim Too Lean (Bank 1)** Engine running in closed loop at cruise speed for 2 minutes, and the PCM detected the A/F ratio remained leaner than the fuel correction limit in memory for 10 seconds during the Fuel System Monitor test. **Possible Causes:** • Air leaks after the MAF sensor, or in the PCV or EGR system • Base engine "mechanical" fault affecting one or more cylinders • Exhaust leaks before or near where the front HO2S is mounted • Fuel control sensor out of calibration (i.e., ECT, IAT or MAF) • Fuel delivery system supplying too much or too little fuel during idle or cruise speed (weak fuel pump, weak pressure regulator) • Fuel Injectors (one or more) is dirty, restricted or leaking fuel • HO2S is contaminated, deteriorated or it has failed • Vehicle driven low on fuel or driven until it ran out of fuel
DTC: P0172 **2T CCM, MIL: YES** **2004, 2005** **Models:** Mazda3, RX-8 **Engines:** All **Transmissions:** A/T, M/T	**Adaptive Fuel Trim Too Rich (Bank 1)** Engine running in closed loop at cruise speed for 2 minutes, and the PCM detected the A/F ratio remained richer than the fuel correction limit in memory for 10 seconds during the Fuel System Monitor test. **Possible Causes:** • Base engine "mechanical" fault affecting one or more cylinders • EVAP system component has failed or canister fuel saturated • Exhaust leaks before or near where the front HO2S is mounted • Fuel control sensor out of calibration (i.e., IAT, MAP or MAF) • Fuel injectors (one or more) sticking or leaking fuel • Fuel delivery system supplying too much fuel during cruise or idle periods (e.g., faulty fuel pump, or faulty pressure regulator) • HO2S is contaminated, deteriorated or it has failed

DTC	Trouble Code Title, Conditions & Possible Causes
DTC: P0173 **2T CCM, MIL: YES** **1996, 1997** **Models:** MX-6, 626 **Engines:** 2.5L VIN D **Transmissions:** A/T, M/T	**Adaptive Fuel Trim Too Rich or Too Lean (Bank 2)** Engine running in closed loop at cruise speed for 2 minutes, and the PCM detected the A/F ratio remained leaner or leaner than the fuel correction limit for 10 seconds during the Fuel System Monitor test. **Possible Causes:** • Air leaks after the MAF sensor, or air leaks in the PCV system • Air leaks at the EGR gasket, or at the EGR valve diaphragm • Base engine "mechanical" fault affecting one or more cylinders • Exhaust leaks before or near where the front HO2S is mounted • Fuel control sensor is out of calibration (i.e., ECT, IAT or MAP) • Fuel delivery system supplying too much or too little fuel during idle or cruise speed (weak fuel pump, weak pressure regulator) • Fuel Injectors (one or more) is dirty, restricted or leaking fuel • HO2S is contaminated, deteriorated or it has failed • Vehicle driven low on fuel or driven until it ran out of fuel
DTC: P0173 **2T CCM, MIL: YES** **2004** **Models:** Tribute **Engines:** 2.5L VIN D **Transmissions:** A/T, M/T	**Adaptive Fuel Trim Too Rich or Too Lean (Bank 2)** Engine running in closed loop at cruise speed for 2 minutes, and the PCM detected the A/F ratio remained leaner or leaner than the fuel correction limit for 10 seconds during the Fuel System Monitor test. **Possible Causes:** • Air leaks after the MAF sensor, or air leaks in the PCV system • Air leaks at the EGR gasket, or at the EGR valve diaphragm • Base engine "mechanical" fault affecting one or more cylinders • Exhaust leaks before or near where the front HO2S is mounted • Fuel control sensor is out of calibration (i.e., ECT, IAT or MAP) • Fuel delivery system supplying too much or too little fuel during idle or cruise speed (weak fuel pump, weak pressure regulator) • Fuel Injectors (one or more) is dirty, restricted or leaking fuel • HO2S is contaminated, deteriorated or it has failed • Vehicle driven low on fuel or driven until it ran out of fuel
DTC: P0171 **2T CCM, MIL: YES** **1996, 1997** **Models:** MX-6, 626 **Engines:** 2.0L VIN C **Transmissions:** A/T	**Adaptive Fuel Trim Too Lean (Bank 1)** Engine running in closed loop at cruise speed for 2 minutes, and the PCM detected the A/F ratio remained leaner than the fuel correction limit in memory for 10 seconds during the Fuel System Monitor test. **Possible Causes:** • Air leaks after the MAF sensor, or in the PCV or EGR system • Base engine "mechanical" fault affecting one or more cylinders • Exhaust leaks before or near where the front HO2S is mounted • Fuel control sensor out of calibration (i.e., ECT, IAT or MAF) • Fuel delivery system supplying too much or too little fuel during idle or cruise speed (weak fuel pump, weak pressure regulator) • Fuel Injectors (one or more) is dirty, restricted or leaking fuel • HO2S is contaminated, deteriorated or it has failed • Vehicle driven low on fuel or driven until it ran out of fuel
DTC: P0171 **2T CCM, MIL: YES** **2003, 2004, 2005** **Models:** All **Engines:** All **Transmissions:** A/T	**Adaptive Fuel Trim Too Lean (Bank 1)** Engine running in closed loop at cruise speed for 2 minutes, and the PCM detected the A/F ratio remained leaner than the fuel correction limit in memory for 10 seconds during the Fuel System Monitor test. **Possible Causes:** • Air leaks after the MAF sensor, or in the PCV or EGR system • Base engine "mechanical" fault affecting one or more cylinders • Exhaust leaks before or near where the front HO2S is mounted • Fuel control sensor out of calibration (i.e., ECT, IAT or MAF) • Fuel delivery system supplying too much or too little fuel during idle or cruise speed (weak fuel pump, weak pressure regulator) • Fuel Injectors (one or more) is dirty, restricted or leaking fuel • HO2S is contaminated, deteriorated or it has failed • Vehicle driven low on fuel or driven until it ran out of fuel

DTC	Trouble Code Title, Conditions & Possible Causes
DTC: P0172 **2T CCM, MIL: YES** **1996, 1997** **Models:** MX-6, 626 **Engines:** 2.0L VIN C **Transmissions:** A/T	**Adaptive Fuel Trim Too Rich (Bank 1)** Engine running in closed loop at cruise speed for 2 minutes, and the PCM detected the A/F ratio remained richer than the fuel correction limit in memory for 10 seconds during the Fuel System Monitor test. **Possible Causes:** • Base engine "mechanical" fault affecting one or more cylinders • EVAP system component has failed or canister fuel saturated • Exhaust leaks before or near where the front HO2S is mounted • Fuel control sensor out of calibration (i.e., IAT, MAP or MAF) • Fuel injectors (one or more) sticking or leaking fuel • Fuel delivery system supplying too much fuel during cruise or idle periods (e.g., faulty fuel pump, or faulty pressure regulator) • HO2S is contaminated, deteriorated or it has failed
DTC: P0171 **2T CCM, MIL: YES** **1998, 1999, 2000, 2001, 2002** **Models:** MX-6, 626 **Engines:** All **Transmissions:** A/T, M/T	**Adaptive Fuel Trim Too Lean (Bank 1)** Engine running in closed loop at cruise speed for 2 minutes, and the PCM detected the A/F ratio remained leaner than the fuel correction limit in memory for 10 seconds during the Fuel System Monitor test. **Possible Causes:** • Air leaks after the MAF sensor, or in the PCV or EGR system • Base engine "mechanical" fault affecting one or more cylinders • Exhaust leaks before or near where the front HO2S is mounted • Fuel control sensor out of calibration (i.e., ECT, IAT or MAF) • Fuel delivery system supplying too little fuel at idle or cruise • Fuel Injectors (one or more) is dirty, restricted or leaking fuel • HO2S is contaminated, deteriorated or it has failed • Vehicle driven low on fuel or driven until it ran out of fuel
DTC: P0172 **2T CCM, MIL: YES** **2004, 2005** **Models:** Tribute **Engines:** All **Transmissions:** A/T, M/T	**Adaptive Fuel Trim Too Rich (Bank 1)** Engine running in closed loop at cruise speed for 2 minutes, and the PCM detected the A/F ratio remained richer than the fuel correction limit in memory for 10 seconds during the Fuel System Monitor test. **Possible Causes:** • Base engine "mechanical" fault affecting one or more cylinders • EVAP system component has failed or canister fuel saturated • Exhaust leaks before or near where the front HO2S is mounted • Fuel control sensor out of calibration (i.e., IAT, MAP or MAF) • Fuel injectors (one or more) sticking or leaking fuel • Fuel delivery system supplying too much fuel during cruise or idle periods (e.g., faulty fuel pump, or faulty pressure regulator) • HO2S is contaminated, deteriorated or it has failed
DTC: P0172 **2T CCM, MIL: YES** **2003, 2004, 2005** **Models:** All **Engines:** All **Transmissions:** A/T, M/T	**Adaptive Fuel Trim Too Rich (Bank 1)** Engine running in closed loop at cruise speed for 2 minutes, and the PCM detected the A/F ratio remained richer than the fuel correction limit in memory for 10 seconds during the Fuel System Monitor test. **Possible Causes:** • Base engine "mechanical" fault affecting one or more cylinders • EVAP system component has failed or canister fuel saturated • Exhaust leaks before or near where the front HO2S is mounted • Fuel control sensor out of calibration (i.e., IAT, MAP or MAF) • Fuel injectors (one or more) sticking or leaking fuel • Fuel delivery system supplying too much fuel during cruise or idle periods (e.g., faulty fuel pump, or faulty pressure regulator) • HO2S is contaminated, deteriorated or it has failed
DTC: P0174 **2T CCM, MIL: YES** **1998, 1999, 2000, 2001, 2002** **Models:** MX-6, 626 **Engines:** 2.5L VIN D **Transmissions:** A/T, M/T	**Adaptive Fuel Trim Too Lean (Bank 1)** Engine running in closed loop at cruise speed for 2 minutes, and the PCM detected the A/F ratio remained leaner than the fuel correction limit in memory for 10 seconds during the Fuel System Monitor test. **Possible Causes:** • Air leaks after the MAF sensor, or in the PCV or EGR system • Base engine "mechanical" fault affecting one or more cylinders • Exhaust leaks before or near where the front HO2S is mounted • Fuel control sensor out of calibration (i.e., ECT, IAT or MAF) • Fuel delivery system supplying too little fuel at idle or cruise • Fuel Injectors (one or more) is dirty, restricted or leaking fuel • HO2S is contaminated, deteriorated or it has failed • Vehicle driven low on fuel or driven until it ran out of fuel

DTC	Trouble Code Title, Conditions & Possible Causes
DTC: P0174 **2T CCM, MIL: YES** **2004, 2005** **Models:** Tribute **Engines:** 2.5L VIN D **Transmissions:** A/T, M/T	**Adaptive Fuel Trim Too Lean (Bank 1)** Engine running in closed loop at cruise speed for 2 minutes, and the PCM detected the A/F ratio remained leaner than the fuel correction limit in memory for 10 seconds during the Fuel System Monitor test. **Possible Causes:** • Air leaks after the MAF sensor, or in the PCV or EGR system • Base engine "mechanical" fault affecting one or more cylinders • Exhaust leaks before or near where the front HO2S is mounted • Fuel control sensor out of calibration (i.e., ECT, IAT or MAF) • Fuel delivery system supplying too little fuel at idle or cruise • Fuel Injectors (one or more) is dirty, restricted or leaking fuel • HO2S is contaminated, deteriorated or it has failed • Vehicle driven low on fuel or driven until it ran out of fuel
DTC: P0175 **2T CCM, MIL: YES** **1998, 1999, 2000, 2001, 2002** **Models:** MX-6, 626 **Engines:** All **Transmissions:** A/T, M/T	**Adaptive Fuel Trim Too Rich (Bank 2)** Engine running in closed loop at cruise speed for 2 minutes, and the PCM detected the A/F ratio remained richer than the fuel correction limit in memory for 10 seconds during the Fuel System Monitor test. **Possible Causes:** • One or more injectors leaking or pressure regulator is leaking • Fuel delivery system supplying too much fuel during cruise or idle periods (e.g., faulty fuel pump, or faulty pressure regulator) • Exhaust leaks before or near where the front HO2S is mounted • HO2S is contaminated, deteriorated or it has failed • EVAP system component has failed or canister fuel saturated • Fuel control sensor is out of calibration (i.e., ECT, IAT or MAP) • Base engine "mechanical" fault affecting one or more cylinders
DTC: P0175 **2T CCM, MIL: YES** **2004, 2005** **Models:** Tribute **Engines:** All **Transmissions:** A/T, M/T	**Adaptive Fuel Trim Too Rich (Bank 2)** Engine running in closed loop at cruise speed for 2 minutes, and the PCM detected the A/F ratio remained richer than the fuel correction limit in memory for 10 seconds during the Fuel System Monitor test. **Possible Causes:** • One or more injectors leaking or pressure regulator is leaking • Fuel delivery system supplying too much fuel during cruise or idle periods (e.g., faulty fuel pump, or faulty pressure regulator) • Exhaust leaks before or near where the front HO2S is mounted • HO2S is contaminated, deteriorated or it has failed • EVAP system component has failed or canister fuel saturated • Fuel control sensor is out of calibration (i.e., ECT, IAT or MAP) • Base engine "mechanical" fault affecting one or more cylinders
DTC: P0150 **2T CCM, MIL: YES** **1996, 1997, 1998** **Models:** Millenia **Engines:** All **Transmissions:** A/T, M/T	**Adaptive Fuel Trim Too Rich or Too Lean (Bank 1)** Engine running in closed loop at cruise speed for 2 minutes, and the PCM detected the A/F ratio remained richer or leaner than the fuel correction limit for 10 seconds during the Fuel System Monitor test. **Possible Causes:** • Air leaks after the MAF sensor, or air leaks in the PCV system • Air leaks at the EGR gasket, or at the EGR valve diaphragm • Base engine "mechanical" fault affecting one or more cylinders • Exhaust leaks before or near where the front HO2S is mounted • Fuel control sensor out of calibration (i.e., ECT, IAT or MAF) • Fuel delivery system supplying too much or too little fuel during cruise or idle speed (weak fuel pump, weak pressure regulator) • Fuel injectors (one or more) dirty, leaking or restricted • HO2S is contaminated, deteriorated or it has failed • Vehicle driven low on fuel or driven until it ran out of fuel
DTC: P0173 **2T CCM, MIL: YES** **1996, 1997, 1998** **Models:** Millenia **Engines:** All **Transmissions:** A/T, M/T	**Adaptive Fuel Trim Too Rich or Too Lean (Bank 2)** Engine running in closed loop at cruise speed for 2 minutes, and the PCM detected the A/F ratio remained richer or leaner than the fuel correction limit for 10 seconds during the Fuel System Monitor test. **Possible Causes:** • Air leaks after the MAF sensor, or air leaks in the PCV system • Air leaks at the EGR gasket, or at the EGR valve diaphragm • Base engine "mechanical" fault affecting one or more cylinders • Exhaust leaks before or near where the front HO2S is mounted • Fuel control sensor out of calibration (i.e., ECT, IAT or MAF) • Fuel delivery system supplying too much or too little fuel during cruise or idle speed (weak fuel pump, weak pressure regulator) • Fuel injectors (one or more) dirty, leaking or restricted • HO2S is contaminated, deteriorated or it has failed • Vehicle driven low on fuel or driven until it ran out of fuel

DTC	Trouble Code Title, Conditions & Possible Causes
DTC: P0171 **2T CCM, MIL: YES** **1999, 2000, 2001, 2002** **Models:** Millenia **Engines:** All **Transmissions:** A/T, M/T	**Adaptive Fuel Trim Too Lean (Bank 1)** Engine running in closed loop at cruise speed for 2 minutes, and the PCM detected the A/F ratio remained leaner than the fuel correction limit in memory for 10 seconds during the Fuel System Monitor test. **Possible Causes:** • Air leaks after the MAF sensor, or air leaks in the PCV system • Air leaks at the EGR gasket, or at the EGR valve diaphragm • Base engine "mechanical" fault affecting one or more cylinders • Fuel control sensor out of calibration (i.e., ECT, IAT or MAF) • Fuel delivery system supplying too little fuel during cruise or idle speed (weak fuel pump, dirty/leaking fuel injector) • Exhaust leaks before or near where the front HO2S is mounted • HO2S is contaminated, deteriorated or it has failed • Vehicle driven low on fuel or driven until it ran out of fuel
DTC: P0172 **2T CCM, MIL: YES** **1999, 2000, 2001, 2002** **Models:** Millenia **Engines:** All **Transmissions:** A/T, M/T	**Adaptive Fuel Trim Too Rich (Bank 1)** Engine running in closed loop at cruise speed for 2 minutes, and the PCM detected the A/F ratio remained richer than the fuel correction limit in memory for 10 seconds during the Fuel System Monitor test. **Possible Causes:** • Base engine "mechanical" fault affecting one or more cylinders • EVAP system component has failed or canister fuel saturated • Exhaust leaks before or near where the front HO2S is mounted • Fuel control sensor out of calibration (i.e., IAT, MAP or MAF) • Fuel injectors (one or more) sticking or leaking fuel • Fuel delivery system supplying too much fuel during cruise or idle periods (e.g., faulty fuel pump, or faulty pressure regulator) • HO2S is contaminated, deteriorated or it has failed
DTC: P0174 **2T CCM, MIL: YES** **1999, 2000, 2001, 2002** **Models:** Millenia **Engines:** All **Transmissions:** A/T, M/T	**Adaptive Fuel Trim Too Lean (Bank 2)** Engine running in closed loop at cruise speed for 2 minutes, and the PCM detected the A/F ratio remained richer than the fuel correction limit in memory for 10 seconds during the Fuel System Monitor test. **Possible Causes:** • Air leaks after the MAF sensor, or air leaks in the PCV system • Air leaks at the EGR gasket, or at the EGR valve diaphragm • Base engine "mechanical" fault affecting one or more cylinders • Fuel delivery system supplying too little fuel during cruise or idle speed (weak fuel pump, dirty/leaking fuel injector) • Exhaust leaks before or near where the front HO2S is mounted • HO2S is contaminated, deteriorated or it has failed • Vehicle driven low on fuel or driven until it ran out of fuel
DTC: P0175 **2T CCM, MIL: YES** **1999, 2000, 2001, 2002** **Models:** Millenia **Engines:** All **Transmissions:** A/T, M/T	**Adaptive Fuel Trim Too Rich (Bank 2)** Engine running in closed loop at cruise speed for 2 minutes, and the PCM detected the A/F ratio remained richer than the fuel correction limit in memory for 10 seconds during the Fuel System Monitor test. **Possible Causes:** • One or more injectors leaking or pressure regulator is leaking • Fuel delivery system supplying too much fuel during cruise or idle periods (e.g., faulty fuel pump, or faulty pressure regulator) • Exhaust leaks before or near where the front HO2S is mounted • HO2S is contaminated, deteriorated or it has failed • EVAP system component has failed or canister fuel saturated • Fuel control sensor is out of calibration (i.e., ECT, IAT or MAP) • Base engine "mechanical" fault affecting one or more cylinders
DTC: P0170 **2T CCM, MIL: YES** **1996, 1997, 1998** **Models:** MX-5 Miata **Engines:** All **Transmissions:** A/T, M/T	**Adaptive Fuel Trim Too Rich or Too Lean (Bank 1)** Engine running in closed loop at cruise speed for 2 minutes, and the PCM detected the A/F ratio remained richer or leaner than the fuel correction limit for 10 seconds during the Fuel System Monitor test. **Possible Causes:** • Vehicle driven low on fuel or driven until it ran out of fuel • One or more injectors restricted or pressure regulator has failed • Fuel delivery system supplying too much or too little fuel during cruise or idle speed (weak fuel pump, dirty/leaking fuel injector) • Air leaks after the MAF sensor, or air leaks in the PCV system • Air leaks at the EGR gasket, or at the EGR valve diaphragm • Exhaust leaks before or near where the front HO2S is mounted • HO2S is contaminated, deteriorated or it has failed • Fuel control sensor is out of calibration (i.e., ECT, IAT or MAP) • Base engine "mechanical" fault affecting one or more cylinders

DTC	Trouble Code Title, Conditions & Possible Causes
DTC: P0171 **2T CCM, MIL: YES** **1999, 2000, 2001, 2002** **Models:** MX-5 Miata **Engines:** All **Transmissions:** A/T, M/T	**Adaptive Fuel Trim Too Lean (Bank 1)** Engine running in closed loop at cruise speed for 2 minutes, and the PCM detected the A/F ratio remained leaner than the fuel correction limit in memory for 10 seconds during the Fuel System Monitor test. **Possible Causes:** • Vehicle driven low on fuel or until it ran out of fuel • One or more injectors restricted or pressure regulator has failed • Fuel delivery system supplying too much or too little fuel during cruise or idle periods (e.g., faulty fuel pump, or dirty fuel filter) • Air leaks after the MAF sensor, or air leaks in the PCV system • Air leaks at the EGR gasket, or at the EGR valve diaphragm • Exhaust leaks before or near where the front HO2S is mounted • HO2S is contaminated, deteriorated or it has failed • Fuel control sensor is out of calibration (i.e., ECT, IAT or MAP) • Base engine "mechanical" fault affecting one or more cylinders
DTC: P0172 **2T CCM, MIL: YES** **1999, 2000, 2001, 2002** **Models:** MX-5 Miata **Engines:** All **Transmissions:** A/T, M/T	**Adaptive Fuel Trim Too Rich (Bank 1)** Engine running in closed loop at cruise speed for 2 minutes, and the PCM detected the A/F ratio remained richer than the fuel correction limit in memory for 10 seconds during the Fuel System Monitor test. **Possible Causes:** • One or more injectors leaking or pressure regulator is leaking • Fuel delivery system supplying too much fuel during cruise or idle periods (e.g., faulty fuel pump, or faulty pressure regulator) • Exhaust leaks before or near where the front HO2S is mounted • HO2S is contaminated, deteriorated or it has failed • EVAP system component has failed or canister fuel saturated • Fuel control sensor is out of calibration (i.e., ECT, IAT or MAP) • Base engine "mechanical" fault affecting one or more cylinders
DTC: P0170 **2T CCM, MIL: YES** **1996, 1997, 1998** **Models:** Protégé **Engines:** All **Transmissions:** A/T, M/T	**Adaptive Fuel Trim Too Rich or Too Lean (Bank 1)** Engine running in closed loop at cruise speed for 2 minutes, and the PCM detected the A/F ratio remained richer or leaner than the fuel correction limit for 10 seconds during the Fuel System Monitor test. **Possible Causes:** • Vehicle driven low on fuel or driven until it ran out of fuel • One or more injectors restricted or pressure regulator has failed • Fuel delivery system supplying too much or too little fuel during cruise or idle speed (weak fuel pump, dirty/leaking fuel injector) • Air leaks after the MAF sensor, or air leaks in the PCV system • Air leaks at the EGR gasket, or at the EGR valve diaphragm • Exhaust leaks before or near where the front HO2S is mounted • HO2S is contaminated, deteriorated or it has failed • Fuel control sensor is out of calibration (i.e., ECT, IAT or MAP) • Base engine "mechanical" fault affecting one or more cylinders
DTC: P0171 **2T CCM, MIL: YES** **1999, 2000, 2001, 2002** **Models:** Protégé, Protégé5 **Engines:** All **Transmissions:** A/T, M/T	**Adaptive Fuel Trim Too Lean (Bank 1)** Engine running in closed loop at cruise speed for 2 minutes, and the PCM detected the A/F ratio remained leaner than the fuel correction limit in memory for 10 seconds during the Fuel System Monitor test. **Possible Causes:** • Vehicle driven low on fuel or until it ran out of fuel • One or more injectors restricted or pressure regulator has failed • Fuel delivery system supplying too much or too little fuel during cruise or idle periods (e.g., faulty fuel pump, or dirty fuel filter) • Air leaks after the MAF sensor, or air leaks in the PCV system • Air leaks at the EGR gasket, or at the EGR valve diaphragm • Exhaust leaks before or near where the front HO2S is mounted • HO2S is contaminated, deteriorated or it has failed • Fuel control sensor is out of calibration (i.e., ECT, IAT or MAP) • Base engine "mechanical" fault affecting one or more cylinders

DTC	Trouble Code Title, Conditions & Possible Causes
DTC: P0172 **2T CCM, MIL: YES** **1999, 2000, 2001, 2002** **Models:** Protégé, Protégé5 **Engines:** All **Transmissions:** A/T, M/T	**Adaptive Fuel Trim Too Rich (Bank 1)** Engine running in closed loop at cruise speed for 2 minutes, and the PCM detected the A/F ratio remained richer than the fuel correction limit in memory for 10 seconds during the Fuel System Monitor test. **Possible Causes:** • Base engine "mechanical" fault affecting one or more cylinders • EVAP system component has failed or canister fuel saturated • Exhaust leaks before or near where the front HO2S is mounted • Fuel control sensor out of calibration (i.e., IAT, MAP or MAF) • Fuel injectors (one or more) sticking or leaking fuel • Fuel delivery system supplying too much fuel during cruise or idle periods (e.g., faulty fuel pump, or faulty pressure regulator) • HO2S is contaminated, deteriorated or it has failed
DTC: P0170 **2T CCM, MIL: YES** **1996, 1997, 1998** **Models:** MPV **Engines:** All **Transmissions:** A/T, M/T	**Adaptive Fuel Trim Too Rich or Too Lean (Bank 1)** Engine running in closed loop at cruise speed for 2 minutes, and the PCM detected the A/F ratio remained richer or leaner than the fuel correction limit for 10 seconds during the Fuel System Monitor test. **Possible Causes:** • Vehicle driven low on fuel or driven until it ran out of fuel • One or more injectors restricted or pressure regulator has failed • Fuel delivery system supplying too much or too little fuel during cruise or idle speed (weak fuel pump, dirty/leaking fuel injector) • Air leaks after the MAF sensor, in the PCV or EGR system • Exhaust leaks before or near where the front HO2S is mounted • HO2S is contaminated, deteriorated or it has failed • Fuel control sensor is out of calibration (i.e., ECT, IAT or MAP) • Base engine "mechanical" fault affecting one or more cylinders
DTC: P0173 **2T CCM, MIL: YES** **1996, 1997, 1998** **Models:** MPV **Engines:** All **Transmissions:** A/T, M/T	**Adaptive Fuel Trim Too Rich or Too Lean (Bank 2)** Engine running in closed loop at cruise speed for 2 minutes, and the PCM detected the A/F ratio remained richer or leaner than the fuel correction limit for 10 seconds during the Fuel System Monitor test. **Possible Causes:** • Air leaks after the MAF sensor, in the PCV or EGR system • Base engine "mechanical" fault affecting one or more cylinders • One or more injectors is restricted, is sticking or leaking fuel • Fuel delivery system supplying too much or too little fuel during cruise or idle speed (weak fuel pump, weak pressure regulator) • Exhaust leaks before or near where the front HO2S is mounted • HO2S is contaminated, deteriorated or it has failed • Fuel control sensor is out of calibration (i.e., ECT, IAT or MAP) • Vehicle driven low on fuel or driven until it ran out of fuel
DTC: P0171 **2T CCM, MIL: YES** **2000, 2001, 2002** **Models:** MPV **Engines:** All **Transmissions:** A/T, M/T	**Adaptive Fuel Trim Too Lean (Bank 1)** Engine running in closed loop at cruise speed for 2 minutes, and the PCM detected the A/F ratio remained leaner than the fuel correction limit in memory for 10 seconds during the Fuel System Monitor test. **Possible Causes:** • Air leaks after the MAF sensor, or air leaks in the PCV system • Air leaks at the EGR gasket, or at the EGR valve diaphragm • Base engine "mechanical" fault affecting one or more cylinders • Fuel delivery system supplying too little fuel during cruise or idle speed (weak fuel pump, dirty/leaking fuel injector) • Exhaust leaks before or near where the front HO2S is mounted • HO2S is contaminated, deteriorated or it has failed • Vehicle driven low on fuel or driven until it ran out of fuel
DTC: P0172 **2T CCM, MIL: YES** **2000, 2001, 2002** **Models:** MPV **Engines:** All **Transmissions:** A/T, M/T	**Adaptive Fuel Trim Too Rich (Bank 1)** Engine running in closed loop at cruise speed for 2 minutes, and the PCM detected the A/F ratio remained richer than the fuel correction limit in memory for 10 seconds during the Fuel System Monitor test. **Possible Causes:** • Base engine "mechanical" fault affecting one or more cylinders • EVAP system component has failed or canister fuel saturated • Exhaust leaks before or near where the front HO2S is mounted • Fuel control sensor out of calibration (i.e., IAT, MAP or MAF) • Fuel injectors (one or more) sticking or leaking fuel • Fuel delivery system supplying too much fuel at idle or cruise • HO2S is contaminated, deteriorated or it has failed

DTC	Trouble Code Title, Conditions & Possible Causes
DTC: P0174 **2T CCM, MIL: YES** **2000, 2001, 2002, 2003** **Models:** MPV **Engines:** All **Transmissions:** A/T, M/T	**Adaptive Fuel Trim Too Lean (Bank 2)** Engine running in closed loop at cruise speed for 2 minutes, and the PCM detected the A/F ratio remained leaner than the fuel correction limit in memory for 10 seconds during the Fuel System Monitor test. **Possible Causes:** • Vehicle driven low on fuel or until it ran out of fuel • One or more injectors restricted or pressure regulator has failed • Fuel delivery system supplying too much or too little fuel during cruise or idle periods (e.g., faulty fuel pump, or dirty fuel filter) • Air leaks after the MAF sensor, or air leaks in the PCV system • Air leaks at the EGR gasket, or at the EGR valve diaphragm • Exhaust leaks before or near where the front HO2S is mounted • HO2S is contaminated, deteriorated or it has failed • Fuel control sensor is out of calibration (i.e., ECT, IAT or MAP) • Base engine "mechanical" fault affecting one or more cylinders
DTC: P0175 **2T CCM, MIL: YES** **2000, 2001, 2002, 2003** **Models:** MPV **Engines:** All **Transmissions:** A/T, M/T	**Adaptive Fuel Trim Too Rich (Bank 2)** Engine running in closed loop at cruise speed for 2 minutes, and the PCM detected the A/F ratio remained richer than the fuel correction limit in memory for 10 seconds during the Fuel System Monitor test. **Possible Causes:** • One or more injectors leaking or pressure regulator is leaking • Fuel delivery system supplying too much fuel during cruise or idle periods (e.g., faulty fuel pump, or faulty pressure regulator) • Exhaust leaks before or near where the front HO2S is mounted • HO2S is contaminated, deteriorated or it has failed • EVAP system component has failed or canister fuel saturated • Fuel control sensor is out of calibration (i.e., ECT, IAT or MAP) • Base engine "mechanical" fault affecting one or more cylinders
DTC: P0171 **2T CCM, MIL: YES** **1996, 1997, 1998, 1999, 2000, 2001, 2002, 2003, 2004, 2005** **Models:** B2300, B2500, B3000, B4000, Tribute **Engines:** All **Transmissions:** A/T, M/T	**Adaptive Fuel Trim Too Lean (Bank 1)** Engine running in closed loop at cruise speed for 2 minutes, and the PCM detected the A/F ratio remained leaner than the fuel correction limit in memory for 10 seconds during the Fuel System Monitor test. **Possible Causes:** • Vehicle driven low on fuel or until it ran out of fuel • One or more injectors restricted or pressure regulator has failed • Fuel delivery system supplying too much or too little fuel during cruise or idle periods (e.g., faulty fuel pump, or dirty fuel filter) • Air leaks after the MAF sensor, or air leaks in the PCV system • Air leaks at the EGR gasket, or at the EGR valve diaphragm • Exhaust leaks before or near where the front HO2S is mounted • HO2S is contaminated, deteriorated or it has failed • Fuel control sensor is out of calibration (i.e., ECT, IAT or MAP) • Base engine "mechanical" fault affecting one or more cylinders
DTC: P0172 **2T CCM, MIL: YES** **1996, 1997, 1998, 1999, 2000, 2001, 2002, 2003, 2004, 2005** **Models:** B2300, B2500, B3000, B4000, Tribute **Engines:** All **Transmissions:** A/T, M/T	**Adaptive Fuel Trim Too Rich (Bank 1)** Engine running in closed loop at cruise speed for 2 minutes, and the PCM detected the A/F ratio remained richer than the fuel correction limit in memory for 10 seconds during the Fuel System Monitor test. **Possible Causes:** • One or more injectors leaking or pressure regulator is leaking • Fuel delivery system supplying too much fuel during cruise or idle periods (e.g., faulty fuel pump, or faulty pressure regulator) • Exhaust leaks before or near where the front HO2S is mounted • HO2S is contaminated, deteriorated or it has failed • EVAP system component has failed or canister fuel saturated • Fuel control sensor is out of calibration (i.e., ECT, IAT or MAP) • Base engine "mechanical" fault affecting one or more cylinders

DTC	Trouble Code Title, Conditions & Possible Causes
DTC: P0174 **2T CCM, MIL: YES** **1996, 1997, 1998, 1999, 2000, 2001, 2002, 2003, 2004, 2005** **Models:** B2300, B3000, B4000, Tribute **Engines:** All **Transmissions:** A/T, M/T	**Adaptive Fuel Trim Too Lean (Bank 2)** Engine running in closed loop at cruise speed for 2 minutes, and the PCM detected the A/F ratio remained leaner than the fuel correction limit in memory for 10 seconds during the Fuel System Monitor test. **Possible Causes:** • Vehicle driven low on fuel or until it ran out of fuel • One or more injectors restricted or pressure regulator has failed • Fuel delivery system supplying too much or too little fuel during cruise or idle periods (e.g., faulty fuel pump, or dirty fuel filter) • Air leaks after the MAF sensor, or air leaks in the PCV system • Air leaks at the EGR gasket, or at the EGR valve diaphragm • Exhaust leaks before or near where the front HO2S is mounted • HO2S is contaminated, deteriorated or it has failed • Fuel control sensor is out of calibration (i.e., ECT, IAT or MAP) • Base engine "mechanical" fault affecting one or more cylinders
DTC: P0175 **2T CCM, MIL: YES** **1996, 1997, 1998, 1999, 2000, 2001, 2002, 2003, 2004, 2005** **Models:** B2300, B3000, B4000, Tribute **Engines:** All **Transmissions:** A/T, M/T	**Adaptive Fuel Trim Too Rich (Bank 2)** Engine running in closed loop at cruise speed for 2 minutes, and the PCM detected the A/F ratio remained richer than the fuel correction limit in memory for 10 seconds during the Fuel System Monitor test. **Possible Causes:** • One or more injectors leaking or pressure regulator is leaking • Fuel delivery system supplying too much fuel during cruise or idle periods (e.g., faulty fuel pump, or faulty pressure regulator) • Exhaust leaks before or near where the front HO2S is mounted • HO2S is contaminated, deteriorated or it has failed • EVAP system component has failed or canister fuel saturated • Fuel control sensor is out of calibration (i.e., ECT, IAT or MAP) • Base engine "mechanical" fault affecting one or more cylinders
DTC: P0176 **1T CCM, MIL: YES** **1999, 2000** **Models:** B3000 **Engines:** 3.0L VIN V **Transmissions:** A/T, M/T	**Flexible Fuel Sensor Circuit Malfunction** Key on or engine running and the PCM detected an unexpected voltage condition on the Flexible Fuel Sensor circuit during the test. **Possible Causes:** • Flexible fuel sensor signal circuit is open or shorted to ground • Flexible fuel sensor ground circuit is open • Flexible fuel sensor power circuit is open or shorted to ground • Flexible fuel sensor is damaged or has failed • PCM has failed
DTC: P0176 **1T CCM, MIL: YES** **2004** **Models:** B2300, B3000, B4000 **Engines:** All **Transmissions:** A/T, M/T	**Flexible Fuel Sensor Circuit Malfunction** Key on or engine running and the PCM detected an unexpected voltage condition on the Flexible Fuel Sensor circuit during the test. **Possible Causes:** • Flexible fuel sensor signal circuit is open or shorted to ground • Flexible fuel sensor ground circuit is open • Flexible fuel sensor power circuit is open or shorted to ground • Flexible fuel sensor is damaged or has failed • PCM has failed
DTC: P0180 **1T CCM, MIL: YES** **2005** **Models:** Tribute **Engines:** 3.0L VIN V **Transmissions:** A/T, M/T	**Flexible Fuel Sensor Circuit Malfunction** Key on or engine running and the PCM detected an unexpected voltage condition on the Flexible Fuel Sensor circuit during the test. **Possible Causes:** • Flexible fuel sensor signal circuit is open or shorted to ground • Flexible fuel sensor ground circuit is open • Flexible fuel sensor power circuit is open or shorted to ground • Flexible fuel sensor is damaged or has failed • PCM has failed
DTC: P0181 **1T CCM, MIL: YES** **2005** **Models:** Tribute **Engines:** 3.0L VIN V **Transmissions:** A/T, M/T	**Flexible Fuel Sensor Circuit Malfunction** Key on or engine running and the PCM detected an unexpected voltage condition on the Flexible Fuel Sensor circuit during the test. **Possible Causes:** • Flexible fuel sensor signal circuit is open or shorted to ground • Flexible fuel sensor ground circuit is open • Flexible fuel sensor power circuit is open or shorted to ground • Flexible fuel sensor is damaged or has failed • PCM has failed

DTC	Trouble Code Title, Conditions & Possible Causes
DTC: P0182 **1T CCM, MIL: YES** **2005** **Models:** Tribute **Engines:** 3.0L VIN V **Transmissions:** A/T, M/T	**Flexible Fuel Sensor Circuit Malfunction** Key on or engine running and the PCM detected an unexpected voltage condition on the Flexible Fuel Sensor circuit during the test. **Possible Causes:** • Flexible fuel sensor signal circuit is open or shorted to ground • Flexible fuel sensor ground circuit is open • Flexible fuel sensor power circuit is open or shorted to ground • Flexible fuel sensor is damaged or has failed • PCM has failed
DTC: P0183 **1T CCM, MIL: YES** **2005** **Models:** Tribute **Engines:** 3.0L VIN V **Transmissions:** A/T, M/T	**Flexible Fuel Sensor Circuit Malfunction** Key on or engine running and the PCM detected an unexpected voltage condition on the Flexible Fuel Sensor circuit during the test. **Possible Causes:** • Flexible fuel sensor signal circuit is open or shorted to ground • Flexible fuel sensor ground circuit is open • Flexible fuel sensor power circuit is open or shorted to ground • Flexible fuel sensor is damaged or has failed • PCM has failed
DTC: P0190 **1T CCM, MIL: YES** **2005** **Models:** Tribute **Engines:** 3.0L VIN V **Transmissions:** A/T, M/T	**Flexible Fuel Sensor Circuit Malfunction** Key on or engine running and the PCM detected an unexpected voltage condition on the Flexible Fuel Sensor circuit during the test. **Possible Causes:** • Flexible fuel sensor signal circuit is open or shorted to ground • Flexible fuel sensor ground circuit is open • Flexible fuel sensor power circuit is open or shorted to ground • Flexible fuel sensor is damaged or has failed • PCM has failed
DTC: P0191 **1T CCM, MIL: YES** **2005** **Models:** Tribute **Engines:** 3.0L VIN V **Transmissions:** A/T, M/T	**Flexible Fuel Sensor Circuit Malfunction** Key on or engine running and the PCM detected an unexpected voltage condition on the Flexible Fuel Sensor circuit during the test. **Possible Causes:** • Flexible fuel sensor signal circuit is open or shorted to ground • Flexible fuel sensor ground circuit is open • Flexible fuel sensor power circuit is open or shorted to ground • Flexible fuel sensor is damaged or has failed • PCM has failed
DTC: P0192 **1T CCM, MIL: YES** **2005** **Models:** Tribute **Engines:** 3.0L VIN V **Transmissions:** A/T, M/T	**Flexible Fuel Sensor Circuit Malfunction** Key on or engine running and the PCM detected an unexpected voltage condition on the Flexible Fuel Sensor circuit during the test. **Possible Causes:** • Flexible fuel sensor signal circuit is open or shorted to ground • Flexible fuel sensor ground circuit is open • Flexible fuel sensor power circuit is open or shorted to ground • Flexible fuel sensor is damaged or has failed • PCM has failed
DTC: P0193 **1T CCM, MIL: YES** **2005** **Models:** Tribute **Engines:** 3.0L VIN V **Transmissions:** A/T, M/T	**Flexible Fuel Sensor Circuit Malfunction** Key on or engine running and the PCM detected an unexpected voltage condition on the Flexible Fuel Sensor circuit during the test. **Possible Causes:** • Flexible fuel sensor signal circuit is open or shorted to ground • Flexible fuel sensor ground circuit is open • Flexible fuel sensor power circuit is open or shorted to ground • Flexible fuel sensor is damaged or has failed • PCM has failed
DTC: P0201 **1T CCM, MIL: YES** **2000, 2001, 2002, 2003, 2004, 2005** **Models:** B2300, B2500, B3000, B4000, Tribute **Engines:** All **Transmissions:** A/T, M/T	**Fuel Injector 1 Control Circuit Malfunction** Engine running and the PCM detected an unexpected voltage condition on the fuel injector control circuit during the CCM test. **Possible Causes:** • Fuel injector 1 control circuit is open or grounded • Fuel injector 1 power circuit open between injector and VPWR • Fuel injector 1 has failed • PCM has failed (i.e., the PCM driver for Injector 1 has failed)

DTC	Trouble Code Title, Conditions & Possible Causes
DTC: P0202 **1T CCM, MIL: YES** **2000, 2001, 2002, 2003, 2004, 2005** **Models:** B2300, B2500, B3000, B4000, Tribute **Engines:** All **Transmissions:** A/T, M/T	**Fuel Injector 2 Control Circuit Malfunction** Engine running and the PCM detected an unexpected voltage condition on the fuel injector control circuit during the CCM test. **Possible Causes:** • Fuel injector 2 control circuit is open or grounded • Fuel injector 2 power circuit open between injector and VPWR • Fuel injector 2 has failed • PCM has failed (i.e., the PCM driver for Injector 2 has failed)
DTC: P0203 **1T CCM, MIL: YES** **2000, 2001, 2002, 2003, 2004, 2005** **Models:** B2300, B2500, B3000, B4000, Tribute **Engines:** All **Transmissions:** A/T, M/T	**Fuel Injector 3 Control Circuit Malfunction** Engine running and the PCM detected an unexpected voltage condition on the fuel injector control circuit during the CCM test. **Possible Causes:** • Fuel injector 3 control circuit is open or grounded • Fuel injector 3 power circuit open between injector and VPWR • Fuel injector 3 has failed • PCM has failed (i.e., the PCM driver for Injector 3 has failed)
DTC: P0204 **1T CCM, MIL: YES** **2000, 2001, 2002, 2003, 2004, 2005** **Models:** B2300, B2500, B3000, B4000, Tribute **Engines:** All **Transmissions:** A/T, M/T	**Fuel Injector 4 Control Circuit Malfunction** Engine running and the PCM detected an unexpected voltage condition on the fuel injector control circuit during the CCM test. **Possible Causes:** • Fuel injector 4 control circuit is open or grounded • Fuel injector 4 power circuit open between injector and VPWR • Fuel injector 4 has failed • PCM has failed (i.e., the PCM driver for Injector 4 has failed)
DTC: P0205 **1T CCM, MIL: YES** **2000, 2001, 2002, 2003, 2004, 2005** **Models:** B3000, B4000, Tribute **Engines:** 3.0L VIN V, 3.0L VIN 1, 4.0L VIN X, 4.0L VIN E **Transmissions:** A/T, M/T	**Fuel Injector 5 Control Circuit Malfunction** Engine running and the PCM detected an unexpected voltage condition on the fuel injector control circuit during the CCM test. **Possible Causes:** • Fuel injector 5 control circuit is open or grounded • Fuel injector 5 power circuit open between injector and VPWR • Fuel injector 5 has failed • PCM has failed (i.e., the PCM driver for Injector 5 has failed)
DTC: P0206 **1T CCM, MIL: YES** **2000, 2001, 2002, 2003, 2004, 2005** **Models:** B3000, B4000, Tribute **Engines:** 3.0L VIN V, 3.0L VIN 1, 4.0L VIN X, 4.0L VIN E **Transmissions:** A/T, M/T	**Fuel Injector 6 Control Circuit Malfunction** Engine running and the PCM detected an unexpected voltage condition on the fuel injector control circuit during the CCM test. **Possible Causes:** • Fuel injector 6 control circuit is open or grounded • Fuel injector 6 power circuit open between injector and VPWR • Fuel injector 6 has failed • PCM has failed (i.e., the PCM driver for Injector 6 has failed)
DTC: P0219 **1T CCM, MIL: YES** **2004, 2005** **Models:** Tribute **Engines:** All **Transmissions:** A/T, M/T	**PCM – Engine RPM Limiter, PCM software** Indicates the vehicle has been operated in a manner, which caused the engine speed to exceed a calibration limit. The engine speed is continuously monitored and evaluated by The PCM. The DTC is set when the rpm exceeds the calibrated limit set with in The PCM. **Possible Causes:** • TP Sensor malfunction • Connector or Terminal Malfunction • Open circuit between throttle body terminal A and PCM terminal 3M • Open circuit between throttle body terminal E and PCM terminal 3N
DTC: P0222 **1T CCM, MIL: YES** **2003, 2004, 2005** **Models:** Mazda3, Mazda6, MX-5 Miata, RX-8 **Engines:** All **Transmissions:** A/T, M/T	**TP Sensor No. 2 circuit high input** Key on or engine running, and the PCM detects TP Sensor No. 2 voltage at PCM terminal 3J is below 0.2 v after ignition key to ON, PCM determines that TP circuit has a malfunction. **Possible Causes:** • TP Sensor malfunction • Connector or Terminal Malfunction • Open circuit between throttle body terminal A and PCM terminal 3M • Open circuit between throttle body terminal E and PCM terminal 3N

DTC	Trouble Code Title, Conditions & Possible Causes
DTC: P0223 **1T CCM, MIL: YES** **2003, 2004, 2005** **Models:** Mazda3, Mazda6. MX-5 Miata, RX-8 **Engines:** All **Transmissions:** A/T, M/T	**TP Sensor No. 2 circuit low input** Key on or engine running, and the PCM detects TP Sensor No. 2 voltage at PCM terminal 3J is above 4.85 v after ignition key to ON, PCM determines that TP circuit has a malfunction. **Possible Causes:** • TP Sensor malfunction • Connector or Terminal Malfunction • Open circuit between throttle body terminal D and PCM terminal 3K • Open circuit between throttle body terminal C and PCM terminal 3J
DTC: P0230 **1T CCM, MIL: YES** **1996, 1997, 1998, 1999, 2000, 2001, 2002** **Models:** 626 **Engines:** All **Transmissions:** A/T, M/T	**Fuel Pump Primary Circuit Malfunction** Key on or engine running and the PCM detected an unexpected voltage condition on the fuel pump primary circuit during the test. **Note: This trouble code may set due an intermittent fault!** **Possible Causes:** • Fuel pump primary circuit is open or shorted to ground • Fuel pump primary circuit is shorted to system power (VREF) • Fuel pump relay is damaged or has failed • PCM has failed
DTC: P0230 **1T CCM, MIL: YES** **1996, 1997, 1998, 1999, 2000, 2001, 2002, 2003, 2004, 2005** **Models:** B2300, B2500, B3000, B4000, Tribute **Engines:** All **Transmissions:** A/T, M/T	**Fuel Pump Primary or Secondary Circuit Malfunction** Key on or engine running and the PCM detected an unexpected voltage condition on the fuel pump primary circuit during the test. **Note: This trouble code may set due an intermittent fault!** **Possible Causes:** • Fuel pump primary circuit is open or shorted to ground • Fuel pump primary circuit is shorted to system power (VREF) • Fuel pump relay is damaged or has failed • PCM has failed
DTC: P0231 **1T CCM, MIL: YES** **1996, 1997, 1998, 1999, 2000, 2001, 2002, 2003, 2004, 2005** **Models:** B2300, B2500, B3000, B4000, Tribute **Engines:** All **Transmissions:** A/T, M/T	**Fuel Pump Primary Circuit or Secondary Malfunction** Key on, then with the Fuel Pump commanded "on", the PCM did not detect system voltage (B+) on the fuel pump monitor circuit. **Possible Causes:** • Fuel pump feed circuit open between feed circuit and the pump • Fuel pump relay contacts "open" that provide B+ to the pump • FP PWR circuit open between relay and connection to the FPM • Fuel pump relay is damaged or has failed • PCM has failed (the engine will not start if the PCM circuit fails)
DTC: P0232 **1T CCM, MIL: YES** **1996, 1997, 1998, 1999, 2000, 2001, 2002, 2003, 2004, 2005** **Models:** B2300, B2500, B3000, B4000, Tribute **Engines:** All **Transmissions:** A/T, M/T	**Fuel Pump Primary Circuit or Secondary Malfunction** Key on, then with the Fuel Pump commanded "on", the PCM did not detect system voltage (B+) on the fuel pump monitor circuit. **Note: the Fuel Pump Driver Module (FPM) modulates the voltage to the fuel pump to achieve the correct fuel pressure. Power to the fuel pump is supplied by the power relay or FPDM power supply relay.** **Possible Causes (No Start Condition)** Inertia switch needs resetting or its internal contacts are open • Fuel pump circuit open between the FPM connection and the FP PWWR circuit • Fuel pump ground circuit is open or has high resistance • Fuel pump relay is damaged or has failed • PCM has failed (the engine will not start and run) • Possible Causes (Engine Starts Condition) • Fuel pump secondary circuit is shorted to system power (B+) • Fuel pump relay contacts closed all of the time • Fuel pump circuit open between the PCM connection and the FP PWR circuit • PCM has failed (the engine will start and run)
DTC: P0245 **1T CCM, MIL: YES** **2004, 2005** **Models:** MX-5 Miata **Engines:** Engine Code BP with turbocharger **Transmissions:** A/T, M/T	**Turbo Charger wastegate regulating valve control circuit low** The PCM monitors the turbocharger wastegate regulating valve control voltage when the PCM turns the turbocharger wastegate regulating valve on. If the control voltage is low, the PCM determines that the turbocharger wastegate regulating valve control circuit voltage is low. **Possible Causes** • Turbo charger wastegate malfunctioning • Connector or terminal malfunction • Short to ground in wiring harness between turbo charger wastegate regulating valve terminal A and PCM terminal 4D • Short to ground in wiring harness between turbo charger wastegate regulating valve terminal A and PCM terminal 4E • PCM has failed

DTC	Trouble Code Title, Conditions & Possible Causes
DTC: P0246 **1T CCM, MIL: YES** **2004, 2005** **Models:** MX-5 Miata **Engines:** Engine Code BP with turbocharger **Transmissions:** A/T, M/T	**Turbo Charger wastegate regulating valve control circuit high** The PCM monitors the turbocharger wastegate regulating valve control voltage when the PCM turns the turbocharger wastegate regulating valve on. If the control voltage is high, the PCM determines that the turbocharger wastegate regulating valve control circuit voltage is high. **Possible Causes** • Turbo charger wastegate malfunctioning • Connector or terminal malfunction • Short to ground in wiring harness between turbo charger wastegate regulating valve terminal A and PCM terminal 4D • Short to ground in wiring harness between turbo charger wastegate regulating valve terminal A and PCM terminal 4E • PCM has failed
DTC: P0298 **1T CCM, MIL: YES** **2004, 2005** **Models:** Tribute **Engines:** All **Transmissions:** A/T, M/T	**Engine Oil Over Temperature Condition** Oil protection strategy in the PCM has been activated. PCM uses an oil algorithm to infer actual oil temperature. **Possible Causes** • Overheating condition • Basic engine concerns • Engine cooling concerns • Open circuit or short in ECT harness • PCM has failed
DTC: P0300 **2T CCM, MIL: YES** **1996, 1997, 1998, 1999, 2000, 2001, 2002, 2003** **Models:** Mazda6, 626, MX-6 **Engines:** All **Transmissions:** A/T, M/T	**Random Misfire Detected** Engine started, vehicle speed over 3 mph at 400-4000 rpm, and the PCM detected a random Misfire condition in two or more cylinders during the 200 or 1000-revolution Misfire test, or the PCM could not identify the misfiring cylinder due to a problem in the CMP sensor. **Note: If the misfire is severe, the MIL will flash on/off on the 1st trip!** **Possible Causes:** • Base engine mechanical fault affecting one or more cylinders • CMP sensor is damaged or failed (problem may be intermittent) • Fuel metering fault that affects more than one cylinder • Fuel pressure too low or too high, fuel supply contaminated • EVAP system problem or the EVAP canister is fuel saturated • EGR valve is stuck open or the PCV system has a vacuum leak • Ignition system fault that affects more than one cylinder • MAF sensor is contaminated (this can cause a lean condition)
DTC: P0300 **2T CCM, MIL: YES** **2003, 2004, 2005** **Models:** All **Engines:** All **Transmissions:** A/T, M/T	**Random Misfire Detected** Engine started, vehicle speed over 3 mph at 400-4000 rpm, and the PCM detected a random Misfire condition in two or more cylinders during the 200 or 1000-revolution Misfire test, or the PCM could not identify the misfiring cylinder due to a problem in the CMP sensor. **Note: If the misfire is severe, the MIL will flash on/off on the 1st trip!** **Possible Causes:** • Base engine mechanical fault affecting one or more cylinders • CMP sensor is damaged or failed (problem may be intermittent) • Fuel metering fault that affects more than one cylinder • Fuel pressure too low or too high, fuel supply contaminated • EVAP system problem or the EVAP canister is fuel saturated • EGR valve is stuck open or the PCV system has a vacuum leak • Ignition system fault that affects more than one cylinder • MAF sensor is contaminated (this can cause a lean condition)
DTC: P0301 **2T CCM, MIL: YES** **1996, 1997, 1998, 1999, 2000, 2001, 2002** **Models:** 626, MX-6 **Engines:** All **Transmissions:** A/T, M/T	**Cylinder 1 Misfire Detected** Engine started, vehicle speed over 3 mph at 400-4000 rpm, and the PCM detected a Misfire condition in a single cylinder during the 200 or 1000-revolution Misfire test under positive engine load conditions. **Note: If the misfire is severe, the MIL will flash on/off on the 1st trip!** **Possible Causes:** • Base engine mechanical fault affecting only Cylinder 1 • CKP sensor is damaged or failed (problem may be intermittent) • Fuel metering fault that affects only Cylinder 1 • Ignition system fault that affects only Cylinder 1

<table>
<tr><th>DTC</th><th>Trouble Code Title, Conditions & Possible Causes</th></tr>
<tr><td>DTC: P0301
2T CCM, MIL: YES
2003, 2004, 2005
Models: All
Engines: All
Transmissions: A/T, M/T</td><td>Cylinder 1 Misfire Detected
Engine started, vehicle speed over 3 mph at 400-4000 rpm, and the PCM detected a Misfire condition in a single cylinder during the 200 or 1000-revolution Misfire test under positive engine load conditions.
Note: If the misfire is severe, the MIL will flash on/off on the 1st trip!
Possible Causes:
• Base engine mechanical fault affecting only Cylinder 1
• CKP sensor is damaged or failed (problem may be intermittent)
• Fuel metering fault that affects only Cylinder 1
• Ignition system fault that affects only Cylinder 1</td></tr>
<tr><td>DTC: P0302
2T CCM, MIL: YES
2003, 2004, 2005
Models: All
Engines: All
Transmissions: A/T, M/T</td><td>Cylinder 2 Misfire Detected
Engine started, vehicle speed over 3 mph at 400-4000 rpm, and the PCM detected a Misfire condition in a single cylinder during the 200 or 1000-revolution Misfire test under positive engine load conditions.
Note: If the misfire is severe, the MIL will flash on/off on the 1st trip!
Possible Causes:
• Base engine mechanical fault affecting only Cylinder 2
• CKP sensor is damaged or failed (problem may be intermittent)
• Fuel metering fault that affects only Cylinder 2
• Ignition system fault that affects only Cylinder 2</td></tr>
<tr><td>DTC: P0302
2T CCM, MIL: YES
1996, 1997, 1998, 1999, 2000, 2001, 2002
Models: 626, MX-6
Engines: All
Transmissions: A/T, M/T</td><td>Cylinder 2 Misfire Detected
Engine started, vehicle speed over 3 mph at 400-4000 rpm, and the PCM detected a Misfire condition in a single cylinder during the 200 or 1000-revolution Misfire test under positive engine load conditions.
Note: If the misfire is severe, the MIL will flash on/off on the 1st trip!
Possible Causes:
• Base engine mechanical fault affecting only Cylinder 2
• CKP sensor is damaged or failed (problem may be intermittent)
• Fuel metering fault that affects only Cylinder 2
• Ignition system fault that affects only Cylinder 2</td></tr>
<tr><td>DTC: P0303
2T CCM, MIL: YES
1996, 1997, 1998, 1999, 2000, 2001, 2002
Models: 626, MX-6
Engines: All
Transmissions: A/T, M/T</td><td>Cylinder 3 Misfire Detected
Engine started, vehicle speed over 3 mph at 400-4000 rpm, and the PCM detected a Misfire condition in a single cylinder during the 200 or 1000-revolution Misfire test under positive engine load conditions.
Note: If the misfire is severe, the MIL will flash on/off on the 1st trip!
Possible Causes:
• Base engine mechanical fault affecting only Cylinder 3
• CKP sensor is damaged or failed (problem may be intermittent)
• Fuel metering fault that affects only Cylinder 3
• Ignition system fault that affects only Cylinder 3</td></tr>
<tr><td>DTC: P0303
2T CCM, MIL: YES
2003, 2004, 2005
Models: All
Engines: All
Transmissions: A/T, M/T</td><td>Cylinder 3 Misfire Detected
Engine started, vehicle speed over 3 mph at 400-4000 rpm, and the PCM detected a Misfire condition in a single cylinder during the 200 or 1000-revolution Misfire test under positive engine load conditions.
Note: If the misfire is severe, the MIL will flash on/off on the 1st trip!
Possible Causes:
• Base engine mechanical fault affecting only Cylinder 3
• CKP sensor is damaged or failed (problem may be intermittent)
• Fuel metering fault that affects only Cylinder 3
• Ignition system fault that affects only Cylinder 3</td></tr>
<tr><td>DTC: P0304
2T CCM, MIL: YES
1996, 1997, 1998, 1999, 2000, 2001, 2002
Models: 626, MX-6
Engines: All
Transmissions: A/T, M/T
Trouble Code ID: P0304</td><td>Cylinder 4 Misfire Detected
Engine started, vehicle speed over 3 mph at 400-4000 rpm, and the PCM detected a Misfire condition in a single cylinder during the 200 or 1000-revolution Misfire test under positive engine load conditions.
Note: If the misfire is severe, the MIL will flash on/off on the 1st trip!
Possible Causes:
• Base engine mechanical fault affecting only Cylinder 4
• CKP sensor is damaged or failed (problem may be intermittent)
• Fuel metering fault that affects only Cylinder 4
• Ignition system fault that affects only Cylinder 4</td></tr>
</table>

DTC	Trouble Code Title, Conditions & Possible Causes
DTC: P0304 **2T CCM, MIL: YES** **2003, 2004, 2005** **Models:** All **Engines:** All **Transmissions:** A/T, M/T Trouble Code ID: P0304	**Cylinder 4 Misfire Detected** Engine started, vehicle speed over 3 mph at 400-4000 rpm, and the PCM detected a Misfire condition in a single cylinder during the 200 or 1000-revolution Misfire test under positive engine load conditions. **Note: If the misfire is severe, the MIL will flash on/off on the 1st trip!** **Possible Causes:** • Base engine mechanical fault affecting only Cylinder 4 • CKP sensor is damaged or failed (problem may be intermittent) • Fuel metering fault that affects only Cylinder 4 • Ignition system fault that affects only Cylinder 4
DTC: P0305 **2T CCM, MIL: YES** **2003, 2004, 2005** **Models:** All **Engines:** All **Transmissions:** A/T, M/T	**Cylinder 5 Misfire Detected** Engine started, vehicle speed over 3 mph at 400-4000 rpm, and the PCM detected a Misfire condition in a single cylinder during the 200 or 1000-revolution Misfire test under positive engine load conditions. **Note: If the misfire is severe, the MIL will flash on/off on the 1st trip!** **Possible Causes:** • Base engine mechanical fault affecting only Cylinder 5 • CKP sensor is damaged or failed (problem may be intermittent) • Fuel metering fault that affects only Cylinder 5 • Ignition system fault that affects only Cylinder 5
DTC: P0305 **2T CCM, MIL: YES** **1996, 1997, 1998, 1999, 2000, 2001, 2002** **Models:** 626, MX-6 **Engines:** 2.5L VIN D **Transmissions:** A/T, M/T	**Cylinder 5 Misfire Detected** Engine started, vehicle speed over 3 mph at 400-4000 rpm, and the PCM detected a Misfire condition in a single cylinder during the 200 or 1000-revolution Misfire test under positive engine load conditions. **Note: If the misfire is severe, the MIL will flash on/off on the 1st trip!** **Possible Causes:** • Base engine mechanical fault affecting only Cylinder 5 • CKP sensor is damaged or failed (problem may be intermittent) • Fuel metering fault that affects only Cylinder 5 • Ignition system fault that affects only Cylinder 5
DTC: P0306 **2T CCM, MIL: YES** **1996, 1997, 1998, 1999, 2000, 2001, 2002** **Models:** 626, MX-6 **Engines:** 2.5L VIN D **Transmissions:** A/T, M/T	**Cylinder 6 Misfire Detected** Engine started, vehicle speed over 3 mph at 400-4000 rpm, and the PCM detected a Misfire condition in a single cylinder during the 200 or 1000-revolution Misfire test under positive engine load conditions. **Note: If the misfire is severe, the MIL will flash on/off on the 1st trip!** **Possible Causes:** • Base engine mechanical fault affecting only Cylinder 6 • CKP sensor is damaged or failed (problem may be intermittent) • Fuel metering fault that affects only Cylinder 6 • Ignition system fault that affects only Cylinder 6
DTC: P0300 **2T CCM, MIL: YES** **1996, 1997, 1998, 1999, 2000, 2001, 2002, 2003, 2004, 2005** **Models:** B2300, B2500, B3000, B4000, Millenia, MX-5 Miata, MPV, Protégé, Protégé5, Tribute **Engines:** All **Transmissions:** A/T, M/T	**Multiple Misfire Detected** Engine started, vehicle speed over 3 mph at 400-4000 rpm, then the PCM detected irregular CKP signals indicating a random Misfire in two or more cylinders during the 200 or 1000-revolution Misfire test. **Note: If the misfire is severe, the MIL will flash on/off on the 1st trip!** **Possible Causes:** • Base engine mechanical fault affecting one or more cylinders • CMP sensor is damaged or failed (problem may be intermittent) • Fuel metering fault that affects more than one cylinder • Fuel pressure too low or too high, fuel supply contaminated • EVAP system problem or the EVAP canister is fuel saturated • EGR valve is stuck open or the PCV system has a vacuum leak • Ignition system fault that affects more than one cylinder • MAF sensor is contaminated (this can cause a lean condition)
DTC: P0301 **2T CCM, MIL: YES** **1996, 1997, 1998, 1999, 2000, 2001, 2002, 2003, 2004, 2005** **Models:** B2300, B2500, B3000, B4000, Tribute, Millenia, MX-5 Miata, MPV, Protégé, Protégé5 **Engines:** All **Transmissions:** A/T, M/T	**Cylinder 1 Misfire Detected** Engine started, vehicle speed over 3 mph at 400-4000 rpm, and the PCM detected irregular CKP signals indicating a Misfire condition present in one cylinder during the 200 or 1000-revolution Misfire test. **Note: If the misfire is severe, the MIL will flash on/off on the 1st trip!** **Possible Causes:** • Base engine mechanical fault affecting only Cylinder 1 • CKP sensor is damaged or failed (problem may be intermittent) • Fuel metering fault that affects only Cylinder 1 • Ignition system fault that affects only Cylinder 1

DTC	Trouble Code Title, Conditions & Possible Causes
DTC: P0302 **2T CCM, MIL: YES** **1996, 1997, 1998, 1999, 2000, 2001, 2002, 2003, 2004, 2005** **Models:** B2300, B2500, B3000, B4000, Tribute, Millenia, MX-5 Miata, MPV, Protégé, Protégé5 **Engines:** All **Transmissions:** A/T, M/T	**Cylinder 2 Misfire Detected** Engine started, vehicle speed over 3 mph at 400-4000 rpm, and the PCM detected irregular CKP signals indicating a Misfire condition present in one cylinder during the 200 or 1000-revolution Misfire test. **Note: If the misfire is severe, the MIL will flash on/off on the 1st trip!** **Possible Causes:** • Base engine mechanical fault affecting only Cylinder 2 • CKP sensor is damaged or failed (problem may be intermittent) • Fuel metering fault that affects only Cylinder 2 • Ignition system fault that affects only Cylinder 2
DTC: P0303 **2T CCM, MIL: YES** **1996, 1997, 1998, 1999, 2000, 2001, 2002, 2003, 2004, 2005** **Models:** B2300, B2500, B3000, B4000, Tribute, Millenia, MX-5 Miata, MPV, Protégé, Protégé5 **Engines:** All **Transmissions:** A/T, M/T	**Cylinder 3 Misfire Detected** Engine started, vehicle speed over 3 mph at 400-4000 rpm, and the PCM detected irregular CKP signals indicating a Misfire condition present in one cylinder during the 200 or 1000-revolution Misfire test. **Note: If the misfire is severe, the MIL will flash on/off on the 1st trip!** **Possible Causes:** • Base engine mechanical fault affecting only Cylinder 3 • CKP sensor is damaged or failed (problem may be intermittent) • Fuel metering fault that affects only Cylinder 3 • Ignition system fault that affects only Cylinder 3
DTC: P0304 **2T CCM, MIL: YES** **1996, 1997, 1998, 1999, 2000, 2001, 2002, 2003, 2004, 2005** **Models:** B2300, B2500, B3000, B4000, Tribute, Millenia, MX-5 Miata, MPV, Protégé, Protégé5 **Engines:** All **Transmissions:** A/T, M/T	**Cylinder 4 Misfire Detected** Engine started, vehicle speed over 3 mph at 400-4000 rpm, and the PCM detected irregular CKP signals indicating a Misfire condition present in one cylinder during the 200 or 1000-revolution Misfire test. **Note: If the misfire is severe, the MIL will flash on/off on the 1st trip!** **Possible Causes:** • Base engine mechanical fault affecting only Cylinder 4 • CKP sensor is damaged or failed (problem may be intermittent) • Fuel metering fault that affects only Cylinder 4 • Ignition system fault that affects only Cylinder 4
DTC: P0305 **2T CCM, MIL: YES** **1996, 1997, 1998, 1999, 2000, 2001, 2002, 2003, 2004, 2005** **Models:** B2300, B3000, B4000, Millenia, MPV **Engines:** All **Transmissions:** A/T, M/T	**Cylinder 5 Misfire Detected** Engine started, vehicle speed over 3 mph at 400-4000 rpm, and the PCM detected irregular CKP signals indicating a Misfire condition present in one cylinder during the 200 or 1000-revolution Misfire test. **Note: If the misfire is severe, the MIL will flash on/off on the 1st trip!** **Possible Causes:** • Base engine mechanical fault affecting only Cylinder 5 • CKP sensor is damaged or failed (problem may be intermittent) • Fuel metering fault that affects only Cylinder 5 • Ignition system fault that affects only Cylinder 5
DTC: P0305 **2T CCM, MIL: YES** **2001, 2002, 2003** **Models:** Tribute **Engines:** 3.0L VIN 1 **Transmissions:** A/T, M/T	**Cylinder 5 Misfire Detected** Engine started, vehicle speed over 3 mph at 400-4000 rpm, and the PCM detected irregular CKP signals indicating a Misfire condition present in one cylinder during the 200 or 1000-revolution Misfire test. **Note: If the misfire is severe, the MIL will flash on/off on the 1st trip!** **Possible Causes:** • Base engine mechanical fault affecting only Cylinder 5 • CKP sensor is damaged or failed (problem may be intermittent) • Fuel metering fault that affects only Cylinder 5 • Ignition system fault that affects only Cylinder 5
DTC: P0306 **2T CCM, MIL: YES** **1996, 1997, 1998, 1999, 2000, 2001, 2002, 2003, 2004, 2005** **Models:** B2300, B3000, B4000, Millenia, MPV **Engines:** All **Transmissions:** A/T, M/T	**Cylinder 6 Misfire Detected** Engine started, vehicle speed over 3 mph at 400-4000 rpm, and the PCM detected irregular CKP signals indicating a Misfire condition present in one cylinder during the 200 or 1000-revolution Misfire test. **Note: If the misfire is severe, the MIL will flash on/off on the 1st trip!** **Possible Causes:** • Base engine mechanical fault affecting only Cylinder 6 • CKP sensor is damaged or failed (problem may be intermittent) • Fuel metering fault that affects only Cylinder 6 • Ignition system fault that affects only Cylinder 6

DTC	Trouble Code Title, Conditions & Possible Causes
DTC: P0306 **2T CCM, MIL: YES** **2001, 2002, 2003, 2004, 2005** **Models:** Tribute **Engines:** 3.0L VIN 1 **Transmissions:** A/T, M/T	**Cylinder 6 Misfire Detected** Engine started, vehicle speed over 3 mph at 400-4000 rpm, and the PCM detected irregular CKP signals indicating a Misfire condition present in one cylinder during the 200 or 1000-revolution Misfire test. **Note: If the misfire is severe, the MIL will flash on/off on the 1st trip!** **Possible Causes:** • Base engine mechanical fault affecting only Cylinder 6 • CKP sensor is damaged or failed (problem may be intermittent) • Fuel metering fault that affects only Cylinder 6 • Ignition system fault that affects only Cylinder 6
DTC: P0315 **2T CCM, MIL: YES** **2001, 2002, 2003, 2004, 2005** **Models:** Tribute **Engines:** All **Transmissions:** A/T, M/T	**Misfire Detected** Engine started, vehicle speed over 3 mph at 400-4000 rpm, and the PCM detected irregular CKP signals indicating a Misfire condition present in one cylinder during the 200 or 1000-revolution Misfire test. **Note: If the misfire is severe, the MIL will flash on/off on the 1st trip!** **Possible Causes:** • Base engine mechanical fault • CKP sensor is damaged or failed (problem may be intermittent) • Fuel metering fault • Ignition system fault
DTC: P0316 **2T CCM, MIL: YES** **2001, 2002, 2003, 2004, 2005** **Models:** Tribute **Engines:** All **Transmissions:** A/T, M/T	**Misfire Detected** Engine started, vehicle speed over 3 mph at 400-4000 rpm, and the PCM detected irregular CKP signals indicating a Misfire condition present in one cylinder during the 200 or 1000-revolution Misfire test. **Note: If the misfire is severe, the MIL will flash on/off on the 1st trip!** **Possible Causes:** • Base engine mechanical fault • CKP sensor is damaged or failed (problem may be intermittent) • Fuel metering fault • Ignition system fault
DTC: P0316 **2T CCM, MIL: YES** **2004, 2005** **Models:** B2300, B3000, B4000 **Engines:** All **Transmissions:** A/T, M/T	**Misfire Detected** Engine started, vehicle speed over 3 mph at 400-4000 rpm, and the PCM detected irregular CKP signals indicating a Misfire condition present in one cylinder during the 200 or 1000-revolution Misfire test. **Note: If the misfire is severe, the MIL will flash on/off on the 1st trip!** **Possible Causes:** • Base engine mechanical fault • CKP sensor is damaged or failed (problem may be intermittent) • Fuel metering fault • Ignition system fault
DTC: P0320 **1T CCM, MIL: YES** **1996, 1997, 1998, 1999, 2000, 2001, 2002** **Models:** 626, MX-6 **Engines:** All **Transmissions:** A/T, M/T	**Ignition Engine Speed (PIP) Signal Error** Engine running, and the PCM did not detect any engine speed (PIP) signals, or it detected that erratic speed signals were present. **Possible Causes:** • Engine speed or PIP signal circuit is open (an intermittent fault) • Engine speed signal circuit shorted to ground (intermittent) • Ignition system components arcing (ignition coils or wires) • Vehicle onboard transmitter interruption (i.e., 2-way radio)
DTC: P0320 **1T CCM, MIL: YES** **1996, 1997, 1998, 1999, 2000, 2001, 2002, 2003, 2004, 2005** **Models:** B2300, B2500, B3000, B4000, Tribute **Engines:** All **Transmissions:** A/T, M/T	**Ignition Engine Speed (PIP) Signal Error** Engine running, and the PCM did not detect any engine speed (PIP) signals, or it detected that erratic speed signals were present. **Possible Causes:** • Engine speed or PIP signal circuit is open (an intermittent fault) • Engine speed signal circuit shorted to ground (intermittent) • Ignition system components arcing (ignition coils or wires) • Vehicle onboard transmitter interruption (i.e., 2-way radio)

DTC	Trouble Code Title, Conditions & Possible Causes
DTC: P0325 **1T CCM, MIL: YES** **1996, 1997** **Models:** 626, MX-6 **Engines:** 2.5L VIN D **Transmissions:** A/T, M/T	**Knock Sensor Circuit Malfunction** Engine running at idle speed, and the PCM detected an unexpected voltage condition on the Knock Sensor circuit during the CCM test. **Note: Check the Knock Sensor installation torque and connection.** **Possible Causes:** • Knock sensor signal circuit is open • Knock sensor signal circuit is shorted to ground • Knock sensor is damaged or has failed • PCM is damaged
DTC: P0325 **1T CCM, MIL: YES** **1998, 1999, 2000, 2001, 2002** **Models:** 626 **Engines:** 2.5L VIN D **Transmissions:** A/T, M/T	**Knock Sensor Circuit Malfunction** Engine running at idle speed, and the PCM detected an unexpected voltage condition on the Knock Sensor circuit during the CCM test. **Note: Check the Knock Sensor installation torque and connection.** **Possible Causes:** • Knock sensor signal circuit is open • Knock sensor signal circuit is shorted to ground • Knock sensor is damaged or has failed • PCM is damaged
DTC: P0325 **1T CCM, MIL: YES** **1996, 1997, 1998, 1999, 2000, 2001, 2002** **Models:** Millenia **Engines:** All **Transmissions:** A/T, M/T	**Knock Sensor Circuit Malfunction** Engine running at idle speed, and the PCM detected an unexpected voltage condition on the Knock Sensor circuit during the CCM test. **Note: Check the Knock Sensor installation torque and connection.** **Possible Causes:** • Knock sensor signal circuit is open • Knock sensor signal circuit is shorted to ground • Knock sensor is damaged or has failed • PCM is damaged
DTC: P0325 **1T CCM, MIL: YES** **1999, 2000, 2001, 2002, 2003** **Models:** Protégé, Protégé5 **Engines:** 1.8L VIN 1, 2.0L VIN C **Transmissions:** A/T, M/T	**Knock Sensor Circuit Malfunction** Engine running at idle speed, and the PCM detected an unexpected voltage condition on the Knock Sensor circuit during the CCM test. **Note: Check the Knock Sensor installation torque and connection.** **Possible Causes:** • Knock sensor signal circuit is open • Knock sensor signal circuit is shorted to ground • Knock sensor is damaged or has failed • PCM is damaged
DTC: P0325 **1T CCM, MIL: YES** **1996, 1997, 1998** **Models:** MPV **Engines:** All **Transmissions:** A/T, M/T	**Knock Sensor Circuit Malfunction** Engine running at idle speed, and the PCM detected an unexpected voltage condition on the Knock Sensor circuit during the CCM test. **Note: Check the Knock Sensor installation torque and connection.** **Possible Causes:** • Knock sensor signal circuit is open • Knock sensor signal circuit is shorted to ground • Knock sensor is damaged or has failed • PCM is damaged
DTC: P0325 **1T CCM, MIL: YES** **2004, 2005** **Models:** Tribute **Engines:** All **Transmissions:** A/T, M/T	**Knock Sensor Circuit Malfunction** Engine running at idle speed, and the PCM detected an unexpected voltage condition on the Knock Sensor circuit during the CCM test. **Note: Check the Knock Sensor installation torque and connection.** **Possible Causes:** • Knock sensor signal circuit is open • Knock sensor signal circuit is shorted to ground • Knock sensor is damaged or has failed • PCM is damaged

DTC	Trouble Code Title, Conditions & Possible Causes
DTC: P0325 **1T CCM, MIL: YES** **2004, 2005** **Models:** B2300, B3000, B4000 **Engines:** All **Transmissions:** A/T, M/T	**Knock Sensor Circuit Malfunction** Engine running at idle speed, and the PCM detected an unexpected voltage condition on the Knock Sensor circuit during the CCM test. **Note: Check the Knock Sensor installation torque and connection.** **Possible Causes:** • Knock sensor signal circuit is open • Knock sensor signal circuit is shorted to ground • Knock sensor is damaged or has failed • PCM is damaged
DTC: P0326 **1T CCM, MIL: YES** **1996, 1997, 1998** **Models:** MPV **Engines:** All **Transmissions:** A/T, M/T	**Knock Sensor Circuit Range/Performance** Engine running at idle speed, and the PCM detected the knock sensor (KS1) signal was more than the calibrated value in memory. **Possible Causes:** • Knock sensor signal circuit is open (intermittent fault) • Knock sensor signal circuit shorted to ground (intermittent fault) • Knock sensor is damaged or has failed • PCM is damaged
DTC: P0325 **1T CCM, MIL: YES** **2000, 2001, 2002** **Models:** MPV **Engines:** All **Transmissions:** A/T, M/T	**Knock Sensor 1 Circuit Malfunction** Engine running at idle speed, and the PCM detected an unexpected voltage condition on the Knock Sensor circuit during the CCM test. **Note: Check the Knock Sensor installation torque and connection.** **Possible Causes:** • Knock sensor signal circuit is open • Knock sensor signal circuit is shorted to ground • Knock sensor is damaged or has failed • PCM is damaged
DTC: P0326 **1T CCM, MIL: YES** **1996, 1997, 1998** **Models:** MPV **Engines:** All **Transmissions:** A/T, M/T	**Knock Sensor Circuit Range/Performance** Engine running at idle speed, and the PCM detected the knock sensor (KS1) signals during engine acceleration and deceleration changes that were outside of a calibrated value stored in memory. **Possible Causes:** • Knock sensor signal circuit is open (intermittent fault) • Knock sensor signal circuit shorted to ground (intermittent fault) • Knock sensor is damaged or has failed • PCM is damaged
DTC: P0325 **1T CCM, MIL: YES** **2001, 2002, 2003, 2004, 2005** **Models:** B2300, B2500, B3000, B4000, Tribute **Engines:** All **Transmissions:** A/T, M/T	**Knock Sensor 1 Circuit Malfunction** Engine running at idle speed, and the PCM detected an unexpected voltage condition on the Knock Sensor circuit during the CCM test. **Note: Check the Knock Sensor installation torque and connection.** **Possible Causes:** • Knock sensor signal circuit is open • Knock sensor signal circuit is shorted to ground • Knock sensor is damaged or has failed • PCM is damaged
DTC: P0326 **1T CCM, MIL: YES** **2001, 2002, 2003, 2004, 2005** **Models:** B2300, B2500, B3000, B4000, Tribute **Engines:** All **Transmissions:** A/T, M/T	**Knock Sensor Circuit Range/Performance** Engine running at idle speed, and the PCM detected the knock sensor (KS1) signals during engine acceleration and deceleration changes that were outside of a calibrated value stored in memory. **Possible Causes:** • Knock sensor signal circuit is open (intermittent fault) • Knock sensor signal circuit shorted to ground (intermittent fault) • Knock sensor is damaged or has failed • PCM is damaged
DTC: P0327 **1T CCM, MIL: YES** **2001, 2002, 2003, 2004, 2005** **Models:** Mazda3, Mazda6, MX-5 Miata, RX-8 **Engines:** All **Transmissions:** A/T, M/T	**Knock Sensor Circuit Low Input** Key on or engine running and the PCM detected the knock sensor (KS1) signal indicated less than 1.25v during the CCM test. **Possible Causes:** • Knock sensor signal circuit is shorted to ground • Knock sensor is damaged or has failed • PCM is damaged

DTC	Trouble Code Title, Conditions & Possible Causes
DTC: P0327 **1T CCM, MIL: YES** **2004** **Models:** Tribute **Engines:** All **Transmissions:** A/T, M/T	**Knock Sensor Circuit Low Input** Key on or engine running and the PCM detected the knock sensor (KS1) signal indicated less than 1.25v during the CCM test. **Possible Causes:** • Knock sensor signal circuit is shorted to ground • Knock sensor is damaged or has failed • PCM is damaged
DTC: P0328 **1T CCM, MIL: YES** **2001, 2002, 2003, 2004, 2005** **Models:** Mazda3, Mazda6, MX-5 Miata, RX-8 **Engines:** All **Transmissions:** A/T, M/T	**Knock Sensor Circuit High Input** Key on or engine running and the PCM detected the knock sensor (KS1) signal indicated more than 3.75v during the CCM test. **Possible Causes:** • Knock sensor signal circuit is open • Knock sensor signal circuit is shorted to VREF or system power • Knock sensor is damaged or has failed • PCM is damaged
DTC: P0328 **1T CCM, MIL: YES** **2004** **Models:** Tribute **Engines:** All **Transmissions:** A/T, M/T	**Knock Sensor Circuit High Input** Key on or engine running and the PCM detected the knock sensor (KS1) signal indicated more than 3.75v during the CCM test. **Possible Causes:** • Knock sensor signal circuit is open • Knock sensor signal circuit is shorted to VREF or system power • Knock sensor is damaged or has failed • PCM is damaged
DTC: P0330 **1T CCM, MIL: YES** **2004, 2005** **Models:** Tribute **Engines:** 2.3L only **Transmissions:** A/T, M/T	**Knock Sensor Circuit Malfunction** Key on or engine running and the PCM detected the knock sensor (KS1) signal indicated a malfunction **Possible Causes:** • Knock sensor signal circuit is open • Knock sensor signal circuit is shorted to VREF or system power • Knock sensor is damaged or has failed • PCM is damaged
DTC: P0335 **1T CCM, MIL: YES** **1996, 1997** **Models:** 626, MX-6 **Engines:** 2.0L VIN C **Transmissions:** M/T	**Crankshaft Position (NE) Sensor Circuit Malfunction** Engine at idle speed, then over 1500 rpm, and the PCM did not detect any CKP (NE) signals for 1.4 seconds during the CCM test. **Possible Causes:** • CKP sensor signal is open or shorted to ground • CKP sensor signal is shorted to VREF or system power (B+) • CKP sensor is damaged or has failed • Trigger wheel or tone wheel is damaged • PCM has failed
DTC: P0335 **1T CCM, MIL: YES** **1996, 1997, 1998, 1999, 2000, 2001, 2002, 2003, 2004, 2005** **Models:** Millenia, MX-5 Miata, RX-8 **Engines:** All **Transmissions:** A/T, M/T	**Crankshaft Position (NE) Sensor Circuit Malfunction** Engine cranking for 1.4 seconds, or engine running, and the PCM did not detect any CKP (NE) signals for 1.1 seconds during the test. **Possible Causes:** • CKP sensor signal is open or shorted to ground • CKP sensor signal is shorted to VREF or system power (B+) • CKP sensor is damaged or has failed • Trigger wheel or tone wheel is damaged • PCM has failed
DTC: P0335 **1T CCM, MIL: YES** **1996, 1997, 1998, 1999, 2000, 2001, 2002, 2003** **Models:** Protégé, Protégé5 **Engines:** All **Transmissions:** A/T, M/T	**Crankshaft Position (NE) Sensor Circuit Malfunction** Engine running, MAF sensor signal indicating over 2.43 g/sec, and the PCM did not detect any CKP (NE) signals for 4.3 seconds. **Possible Causes:** • CKP sensor signal is open or shorted to ground • CKP sensor signal is shorted to VREF or system power (B+) • CKP sensor is damaged or has failed • Trigger wheel or tone wheel is damaged • PCM has failed

DTC	Trouble Code Title, Conditions & Possible Causes
DTC: P0335 **1T CCM, MIL: YES** **1996, 1997, 1998, 2000, 2001, 2002** **Models:** MPV **Engines:** All **Transmissions:** A/T, M/T	**Crankshaft Position (NE) Sensor Circuit Malfunction** Engine running, MAF sensor signal indicating over 2.43 g/sec, and the PCM did not detect any CKP (NE) signals for 4.2 seconds. **Possible Causes:** • CKP sensor signal is open or shorted to ground • CKP sensor signal is shorted to VREF or system power (B+) • CKP sensor is damaged or has failed • Trigger wheel or tone wheel is damaged • PCM has failed
DTC: P0335 **1T CCM, MIL: YES** **2003, 2004, 2005** **Models:** Mazda3, Mazda6 **Engines:** All **Transmissions:** A/T, M/T	**Crankshaft Position (NE) Sensor Circuit Malfunction** Engine running, MAF sensor signal indicating over 2.43 g/sec, and the PCM did not detect any CKP (NE) signals for 4.2 seconds. **Possible Causes:** • CKP sensor signal is open or shorted to ground • CKP sensor signal is shorted to VREF or system power (B+) • CKP sensor is damaged or has failed • Trigger wheel or tone wheel is damaged • PCM has failed
DTC: P0335 **1T CCM, MIL: YES** **2004** **Models:** Tribute **Engines:** All **Transmissions:** A/T, M/T	**Crankshaft Position (NE) Sensor Circuit Malfunction** Engine running, MAF sensor signal indicating over 2.43 g/sec, and the PCM did not detect any CKP (NE) signals for 4.2 seconds. **Possible Causes:** • CKP sensor signal is open or shorted to ground • CKP sensor signal is shorted to VREF or system power (B+) • CKP sensor is damaged or has failed • Trigger wheel or tone wheel is damaged • PCM has failed
DTC: P0336 **1T CCM, MIL: YES** **2004, 2005** **Models:** RX-8, Tribute **Engines:** All **Transmissions:** A/T, M/T	**Crankshaft Position (CKP) Sensor Circuit Malfunction** Engine running, MAF sensor signal indicating over 2.43 g/sec, and the PCM did not detect any CKP (NE) signals for 4.2 seconds. **Possible Causes:** • CKP sensor signal is open or shorted to ground • CKP sensor signal is shorted to VREF or system power (B+) • CKP sensor is damaged or has failed • Trigger wheel or tone wheel is damaged • PCM has failed
DTC: P0339 **1T CCM, MIL: YES** **1999, 2000, 2001, 2002, 2003** **Models:** MX-5 Miata **Engines:** All **Transmissions:** A/T, M/T	**Crankshaft Position Sensor Circuit Malfunction** Engine running and the PCM detected less than eight (8) pulses on the CKP sensor circuit during 2 crankshaft revolutions in the test. **Possible Causes:** • CKP sensor signal is open (fault may be intermittent) • CKP sensor signal shorted to ground (fault may be intermittent) • CKP sensor is damaged or has failed • Crankshaft pulley may be damaged • PCM has failed
DTC: P0340 **2T CCM, MIL: YES** **1996, 1997** **Models:** 626, MX-6, MX-5 Miata **Engines:** 2.0L VIN C **Transmissions:** A/T, M/T	**Camshaft Position Sensor Circuit Malfunction** Engine running and the PCM did not detect any CMP sensor signals for over 3 seconds during the CCM test. **Possible Causes:** • CMP sensor signal is open or shorted to ground • CMP sensor signal is shorted to VREF or system power (B+) • CMP sensor is damaged, or the sensor signal shielding is open • PCM has failed
DTC: P0340 **2T CCM, MIL: YES** **1998, 1999, 2000, 2001, 2002** **Models:** 626 **Engines:** All **Transmissions:** A/T, M/T	**Camshaft Position Sensor Circuit Malfunction** Engine running and the PCM did not detect any CMP sensor signals for 40 seconds during the CCM test. **Possible Causes:** • CMP sensor signal is open or shorted to ground • CMP sensor signal is shorted to VREF or system power (B+) • CMP sensor is damaged or has failed • CMP sensor shielding may be open or damaged • PCM has failed

DTC	Trouble Code Title, Conditions & Possible Causes
DTC: P0340 **2T CCM, MIL: YES** **1996, 1997, 1998** **Models:** Protégé **Engines:** All **Transmissions:** A/T, M/T	**Camshaft Position Sensor Circuit Malfunction** Engine running, MAF sensor signal over 4.2 g/sec, and the PCM did not detect any CMP (SGT) sensor signals for 4.2 seconds during the CCM test. **Possible Causes:** • CMP sensor signal is open or shorted to ground • CMP sensor signal is shorted to VREF or system power (B+) • CMP sensor is damaged or has failed • CMP sensor is damaged, or the sensor signal shielding is open • PCM has failed
DTC: P0340 **1T CCM, MIL: YES** **1996, 1997, 1998, 1999, 2000, 2001, 2002, 2003, 2004, 2005** **Models:** B2300, B2500, B3000, B4000, Tribute **Engines:** All **Transmissions:** A/T, M/T	**Camshaft Position Sensor Circuit Malfunction** Engine running, MAF sensor signal over 4.2 g/sec, and the PCM did not detect any CMP (SGT) sensor signals for 4.2 seconds during the CCM test. **Possible Causes:** • CMP sensor signal is open or shorted to ground • CMP sensor signal is shorted to VREF or system power (B+) • CMP sensor is damaged, or the sensor signal shielding is open • PCM has failed
DTC: P0340 **1T CCM, MIL: YES** **2003, 2004, 2005** **Models:** Mazda3, Mazda6, MX-5 Miata **Engines:** All **Transmissions:** A/T, M/T	**Camshaft Position Sensor Circuit Malfunction** Engine running, MAF sensor signal over 4.2 g/sec, and the PCM did not detect any CMP (SGT) sensor signals for 4.2 seconds during the CCM test. **Possible Causes:** • CMP sensor signal is open or shorted to ground • CMP sensor signal is shorted to VREF or system power (B+) • CMP sensor is damaged, or the sensor signal shielding is open • PCM has failed
DTC: P0341 **1T CCM, MIL: YES** **2004** **Models:** Tribute **Engines:** All **Transmissions:** A/T, M/T	**Camshaft Position Sensor Circuit Malfunction** Engine running, MAF sensor signal over 4.2 g/sec, and the PCM did not detect any CMP (SGT) sensor signals for 4.2 seconds during the CCM test. **Possible Causes:** • CMP sensor signal is open or shorted to ground • CMP sensor signal is shorted to VREF or system power (B+) • CMP sensor is damaged, or the sensor signal shielding is open • PCM has failed
DTC: P0350 **2T CCM, MIL: YES** **1996, 1997, 1998** **Models:** B2300, B2500 **Engines:** All **Transmissions:** A/T, M/T	**Ignition Coil Primary Circuit Malfunction** Engine running and the PCM detected an unexpected voltage condition on the ignition coil primary circuit during the CCM test. **Possible Causes:** • Ignition coil primary circuit is open or shorted to ground • Ignition coil primary circuit is shorted to system power (B+) • Ignition coil power circuit is open between coil and Start circuit • PCM has failed
DTC: P0350 **2T CCM, MIL: YES** **2004, 2005** **Models:** B2300, B3000, B4000 **Engines:** All **Transmissions:** A/T, M/T	**Ignition Coil Primary Circuit Malfunction** Engine running and the PCM detected an unexpected voltage condition on the ignition coil primary circuit during the CCM test. **Possible Causes:** • Ignition coil primary circuit is open or shorted to ground • Ignition coil primary circuit is shorted to system power (B+) • Ignition coil power circuit is open between coil and Start circuit • PCM has failed
DTC: P0350 **2T CCM, MIL: YES** **2004, 2005** **Models:** Tribute **Engines:** All **Transmissions:** A/T, M/T	**Ignition Coil Primary Circuit Malfunction** Engine running and the PCM detected an unexpected voltage condition on the ignition coil primary circuit during the CCM test. **Possible Causes:** • Ignition coil primary circuit is open or shorted to ground • Ignition coil primary circuit is shorted to system power (B+) • Ignition coil power circuit is open between coil and Start circuit • PCM has failed

<table>
<tr><th>DTC</th><th>Trouble Code Title, Conditions & Possible Causes</th></tr>
<tr><td>DTC: P0351
2T CCM, MIL: YES
1996, 1997, 1998, 1999, 2000, 2001, 2002, 2003, 2004, 2005
Models: B2300, B2500
Engines: All
Transmissions: A/T, M/T</td><td>Ignition Coil 1 Primary Circuit Malfunction
Engine running and the PCM detected an unexpected voltage condition on the Ignition Coil 1 primary circuit during the CCM test.
Note: DTC P0351 is related to the ignition coil on cylinder 1.
Possible Causes:
• Ignition coil primary circuit is open, shorted to ground or power
• Ignition coil power circuit is open between coil and Start circuit
• Ignition coil has failed (arcing between primary and secondary)
• PCM has failed</td></tr>
<tr><td>DTC: P0351
2T CCM, MIL: YES
2004, 2005
Models: Tribute
Engines: All
Transmissions: A/T, M/T</td><td>Ignition Coil 1 Primary Circuit Malfunction
Engine running and the PCM detected an unexpected voltage condition on the Ignition Coil 1 primary circuit during the CCM test.
Note: DTC P0351 is related to the ignition coil on cylinder 1.
Possible Causes:
• Ignition coil primary circuit is open, shorted to ground or power
• Ignition coil power circuit is open between coil and Start circuit
• Ignition coil has failed (arcing between primary and secondary)
• PCM has failed</td></tr>
<tr><td>DTC: P0352
2T CCM, MIL: YES
1996, 1997, 1998, 1999, 2000, 2001, 2002, 2003, 2004, 2005
Models: B2300, B2500
Engines: All
Transmissions: A/T, M/T</td><td>Ignition Coil 2 Primary Circuit Malfunction
Engine running and the PCM detected an unexpected voltage condition on the Ignition Coil 2 primary circuit during the CCM test.
Note: DTC P0352 is related to the ignition coil on cylinder 2.
Possible Causes:
• Ignition coil primary circuit is open, shorted to ground or power
• Ignition coil power circuit is open between coil and Start circuit
• Ignition coil has failed (arcing between primary and secondary)
• PCM has failed</td></tr>
<tr><td>DTC: P0352
2T CCM, MIL: YES
2004, 2005
Models: Tribute
Engines: All
Transmissions: A/T, M/T</td><td>Ignition Coil 2 Primary Circuit Malfunction
Engine running and the PCM detected an unexpected voltage condition on the Ignition Coil 2 primary circuit during the CCM test.
Note: DTC P0352 is related to the ignition coil on cylinder 2.
Possible Causes:
• Ignition coil primary circuit is open, shorted to ground or power
• Ignition coil power circuit is open between coil and Start circuit
• Ignition coil has failed (arcing between primary and secondary)
• PCM has failed</td></tr>
<tr><td>DTC: P0353
2T CCM, MIL: YES
1996, 1997, 1998, 1999, 2000, 2001, 2002, 2003, 2004, 2005
Models: B2300, B2500
Engines: All
Transmissions: A/T, M/T</td><td>Ignition Coil 3 Primary Circuit Malfunction
Engine running and the PCM detected an unexpected voltage condition on the Ignition Coil 3 primary circuit during the CCM test.
Note: DTC P0353 is related to the ignition coil on cylinder 3.
Possible Causes:
• Ignition coil primary circuit is open, shorted to ground or power
• Ignition coil power circuit is open between coil and Start circuit
• Ignition coil has failed (arcing between primary and secondary)
• PCM has failed</td></tr>
<tr><td>DTC: P0353
2T CCM, MIL: YES
2004, 2005
Models: Tribute
Engines: All
Transmissions: A/T, M/T</td><td>Ignition Coil 3 Primary Circuit Malfunction
Engine running and the PCM detected an unexpected voltage condition on the Ignition Coil 3 primary circuit during the CCM test.
Note: DTC P0353 is related to the ignition coil on cylinder 3.
Possible Causes:
• Ignition coil primary circuit is open, shorted to ground or power
• Ignition coil power circuit is open between coil and Start circuit
• Ignition coil has failed (arcing between primary and secondary)
• PCM has failed</td></tr>
<tr><td>DTC: P0354
2T CCM, MIL: YES
1996, 1997, 1998, 1999, 2000, 2001, 2002, 2003, 2004, 2005
Models: B2300, B2500
Engines: All
Transmissions: A/T, M/T</td><td>Ignition Coil 4 Primary Circuit Malfunction
Engine running and the PCM detected an unexpected voltage condition on the Ignition Coil 4 primary circuit during the CCM test.
Note: DTC P0354 is related to the ignition coil on cylinder 4.
Possible Causes:
• Ignition coil primary circuit is open, shorted to ground or power
• Ignition coil power circuit is open between coil and Start circuit
• Ignition coil has failed (arcing between primary and secondary)
• PCM has failed</td></tr>
</table>

DTC	Trouble Code Title, Conditions & Possible Causes
DTC: P0354 **2T CCM, MIL: YES** **2004, 2005** **Models:** Tribute **Engines:** All **Transmissions:** A/T, M/T	**Ignition Coil 4 Primary Circuit Malfunction** Engine running and the PCM detected an unexpected voltage condition on the Ignition Coil 4 primary circuit during the CCM test. **Note: DTC P0354 is related to the ignition coil on cylinder 4.** **Possible Causes:** • Ignition coil primary circuit is open, shorted to ground or power • Ignition coil power circuit is open between coil and Start circuit • Ignition coil has failed (arcing between primary and secondary) • PCM has failed
DTC: P0350 **2T CCM, MIL: YES** **1996, 1997, 1998** **Models:** B3000, B4000 **Engines:** All **Transmissions:** A/T, M/T	**Ignition Coil Primary Circuit Malfunction** Engine running and the PCM detected an unexpected voltage condition on the ignition coil primary circuit during the CCM test. **Possible Causes:** • Ignition coil primary circuit is open or shorted to ground • Ignition coil primary circuit is shorted to system power (B+) • Ignition coil power circuit is open between coil and Start circuit • PCM has failed
DTC: P0351 **2T CCM, MIL: YES** **1996, 1997, 1998, 1999, 2000, 2001, 2002, 2003, 2004, 2005** **Models:** B3000, B4000 **Engines:** All **Transmissions:** A/T, M/T	**Ignition Coil 1 Primary Circuit Malfunction** Engine running and the PCM detected an unexpected voltage condition on the Ignition Coil 1 primary circuit during the CCM test. **Note: DTC P0351 is related to the ignition coil for cylinders 1 and 5.** **Possible Causes:** • Ignition coil primary circuit is open, shorted to ground or power • Ignition coil power circuit is open between coil and Start circuit • Ignition coil has failed (arcing between primary and secondary) • PCM has failed
DTC: P0352 **2T CCM, MIL: YES** **1996, 1997, 1998, 1999, 2000, 2001, 2002, 2003** **Models:** B3000, B4000 **Engines:** All **Transmissions:** A/T, M/T	**Ignition Coil 2 Primary Circuit Malfunction** Engine running and the PCM detected an unexpected voltage condition on the Ignition Coil 2 primary circuit during the CCM test. **Note: DTC P0351 is related to the ignition coil for cylinders 4 and 3.** **Possible Causes:** • Ignition coil primary circuit is open, shorted to ground or power • Ignition coil power circuit is open between coil and Start circuit • Ignition coil has failed (arcing between primary and secondary) • PCM has failed
DTC: P0353 **2T CCM, MIL: YES** **1996, 1997, 1998, 1999, 2000, 2001, 2002, 2003** **Models:** B3000, B4000 **Engines:** All **Transmissions:** A/T, M/T	**Ignition Coil 3 Primary Circuit Malfunction** Engine running and the PCM detected an unexpected voltage condition on the Ignition Coil 3 primary circuit during the CCM test. **Note: DTC P0351 is related to the ignition coil for cylinders 2 and 6.** **Possible Causes:** • Ignition coil primary circuit is open, shorted to ground or power • Ignition coil power circuit is open between coil and Start circuit • Ignition coil has failed (arcing between primary and secondary) • PCM has failed
DTC: P0350 **2T CCM, MIL: YES** **2001, 2002, 2003** **Models:** Tribute **Engines:** 2.0L VIN B **Transmissions:** A/T, M/T	**Ignition Coil Primary Circuit Malfunction** Engine running and the PCM detected an unexpected voltage condition on the ignition coil primary circuit during the CCM test. **Possible Causes:** • Ignition coil primary circuit is open or shorted to ground • Ignition coil primary circuit is shorted to system power (B+) • Ignition coil power circuit is open between coil and Start circuit • PCM has failed
DTC: P0351 **2T CCM, MIL: YES** **2001, 2002** **Models: Tribute, 2003** **Engines:** 2.0L VIN B **Transmissions:** A/T, M/T	**Ignition Coil 1 Primary Circuit Malfunction** Engine running and the PCM detected an unexpected voltage condition on the Ignition Coil 1 primary circuit during the CCM test. **Note: DTC P0351 is related to the ignition coil on cylinders 1 and 4.** **Possible Causes:** • Ignition coil primary circuit is open, shorted to ground or power • Ignition coil power circuit is open between coil and Start circuit • Ignition coil has failed (arcing between primary and secondary) • PCM has failed

DTC	Trouble Code Title, Conditions & Possible Causes
DTC: P0403 **2T CCM, MIL: YES** **2004, 2005** **Models:** Tribute **Engines:** All **Transmissions:** A/T, M/T	**EGR Valve (stepper motor) malfunction** PCM monitors input voltage from EGR valve. If voltage at PCM terminals 4E, 4H, 4K and/or 4N remain low or high, PCM determines that ERG valve circuit has a malfunction. **Possible Causes:** • EGR valve malfunction • Connector or terminal malfunction • Short to power circuit in wiring to EGR valve terminals and PCM terminals • Open circuit in wiring between EGR valve terminals and PCM terminals • PCM has failed
DTC: P0404 **2T CCM, MIL: YES** **1998, 1999, 2000, 2001, 2002, 2003** **Models:** Millenia, Protégé, Protégé5 **Engines:** All **Transmissions:** A/T, M/T	**EGR Valve Position Sensor Circuit Range/Performance** Engine speed from 1810-2190, TP angle from 3-13.4%, engine load 25-50%, and the PCM detected that the EGR solenoid accumulated on-time period exceeded a threshold with the EGR system enabled, or it detected an EGR vent solenoid problem existed during the test. **Possible Causes:** • EGR boost solenoid control circuit is open or shorted to ground • EGR boost solenoid power circuit is open to the Main Relay • EGR boost solenoid is damaged or has failed • EGR boost sensor is damaged or has failed • EGR valve position sensor is damaged, stuck or has failed • EGR valve assembly is leaking, damaged or has failed • EGR valve vacuum hose(s) lose, damaged or disconnected • Exhaust pipe to EGR boost sensor or solenoid leaking/plugged • PCM has failed
DTC: P0405 **2T CCM, MIL: YES** **2004, 2005** **Models:** B2300, B3000, B4000, Tribute **Engines:** All **Transmissions:** A/T, M/T	**EGR Valve Position Sensor Circuit Range/Performance** Engine speed from 1810-2190, TP angle from 3-13.4%, engine load 25-50%, and the PCM detected that the EGR solenoid accumulated on-time period exceeded a threshold with the EGR system enabled, or it detected an EGR vent solenoid problem existed during the test. **Possible Causes:** • EGR boost solenoid control circuit is open or shorted to ground • EGR boost solenoid power circuit is open to the Main Relay • EGR boost solenoid is damaged or has failed • EGR boost sensor is damaged or has failed • EGR valve position sensor is damaged, stuck or has failed • EGR valve assembly is leaking, damaged or has failed • EGR valve vacuum hose(s) lose, damaged or disconnected • Exhaust pipe to EGR boost sensor or solenoid leaking/plugged • PCM has failed
DTC: P0406 **2T CCM, MIL: YES** **2004, 2005** **Models:** B2300, B3000, B4000, Tribute **Engines:** All **Transmissions:** A/T, M/T	**EGR Valve Position Sensor Circuit Range/Performance** Engine speed from 1810-2190, TP angle from 3-13.4%, engine load 25-50%, and the PCM detected that the EGR solenoid accumulated on-time period exceeded a threshold with the EGR system enabled, or it detected an EGR vent solenoid problem existed during the test. **Possible Causes:** • EGR boost solenoid control circuit is open or shorted to ground • EGR boost solenoid power circuit is open to the Main Relay • EGR boost solenoid is damaged or has failed • EGR boost sensor is damaged or has failed • EGR valve position sensor is damaged, stuck or has failed • EGR valve assembly is leaking, damaged or has failed • EGR valve vacuum hose(s) lose, damaged or disconnected • Exhaust pipe to EGR boost sensor or solenoid leaking/plugged • PCM has failed

DTC	Trouble Code Title, Conditions & Possible Causes
DTC: P0410 **2T CCM, MIL: YES** **2005** **Models:** RX-8 **Engines:** All **Transmissions:** A/T, M/T	**EGR Valve Position Sensor Circuit Range/Performance** Engine speed from 1810-2190, TP angle from 3-13.4%, engine load 25-50%, and the PCM detected that the EGR solenoid accumulated on-time period exceeded a threshold with the EGR system enabled, or it detected an EGR vent solenoid problem existed during the test. **Possible Causes:** • EGR boost solenoid control circuit is open or shorted to ground • EGR boost solenoid power circuit is open to the Main Relay • EGR boost solenoid is damaged or has failed • EGR boost sensor is damaged or has failed • EGR valve position sensor is damaged, stuck or has failed • EGR valve assembly is leaking, damaged or has failed • EGR valve vacuum hose(s) lose, damaged or disconnected • Exhaust pipe to EGR boost sensor or solenoid leaking/plugged • PCM has failed
DTC: P0420 **2T OBD/CAT1, MIL: YES** **1996, 1997, 1998, 1999, 2000** **Models:** 626, MX-6 **Engines:** All **Transmissions:** A/T, M/T	**Catalyst Efficiency Below Normal (Bank 1)** Vehicle driven at a speed of 16-64 mph at 1090-3090 rpm with the calculated engine load from 16-55% for 2-3 minutes, and the PCM detected the inversion ratio of the rear HO2S and front HO2S was less than a stored threshold during the Catalyst Monitor test. **Possible Causes:** • Air leaks at the exhaust manifold or in the exhaust pipes • Catalytic converter is contaminated, damaged or has failed • Front HO2S and/or the rear HO2S is loose in the mounting hole • Front HO2S older (aged) than the rear HO2S (HO2S is lazy) • Front HO2S or rear HO2S is contaminated with fuel or moisture
DTC: P0420 **2T OBD/CAT1, MIL: YES** **2003, 2004, 2005** **Models:** Mazda6, MX-5 Miata **Engines:** All **Transmissions:** A/T, M/T	**Catalyst Efficiency Below Normal (Bank 1)** Vehicle driven at a speed of 16-64 mph at 1090-3090 rpm with the calculated engine load from 16-55% for 2-3 minutes, and the PCM detected the inversion ratio of the rear HO2S and front HO2S was less than a stored threshold during the Catalyst Monitor test. **Possible Causes:** • Air leaks at the exhaust manifold or in the exhaust pipes • Catalytic converter is contaminated, damaged or has failed • Front HO2S and/or the rear HO2S is loose in the mounting hole • Front HO2S older (aged) than the rear HO2S (HO2S is lazy) • Front HO2S or rear HO2S is contaminated with fuel or moisture
DTC: P0420 **2T OBD/CAT1, MIL: YES** **2004, 2005** **Models:** Tribute **Engines:** All **Transmissions:** A/T, M/T	**Catalyst Efficiency Below Normal (Bank 1)** Vehicle driven at a speed of 16-64 mph at 1090-3090 rpm with the calculated engine load from 16-55% for 2-3 minutes, and the PCM detected the inversion ratio of the rear HO2S and front HO2S was less than a stored threshold during the Catalyst Monitor test. **Possible Causes:** • Air leaks at the exhaust manifold or in the exhaust pipes • Catalytic converter is contaminated, damaged or has failed • Front HO2S and/or the rear HO2S is loose in the mounting hole • Front HO2S older (aged) than the rear HO2S (HO2S is lazy) • Front HO2S or rear HO2S is contaminated with fuel or moisture
DTC: P0420 **2T OBD/CAT1, MIL: YES** **2004, 2005** **Models:** B2300, B3000, B4000 **Engines:** All **Transmissions:** A/T, M/T	**Catalyst Efficiency Below Normal (Bank 1)** Vehicle driven at a speed of 16-64 mph at 1090-3090 rpm with the calculated engine load from 16-55% for 2-3 minutes, and the PCM detected the inversion ratio of the rear HO2S and front HO2S was less than a stored threshold during the Catalyst Monitor test. **Possible Causes:** • Air leaks at the exhaust manifold or in the exhaust pipes • Catalytic converter is contaminated, damaged or has failed • Front HO2S and/or the rear HO2S is loose in the mounting hole • Front HO2S older (aged) than the rear HO2S (HO2S is lazy) • Front HO2S or rear HO2S is contaminated with fuel or moisture
DTC: P0430 **2T OBD/CAT3, MIL: YES** **1996, 1997, 1998, 1999, 2000** **Models:** 626, MX-6 **Engines:** 2.5L VIN D **Transmissions:** A/T, M/T	**Catalyst Efficiency Below Normal (Bank 2)** Vehicle driven at a speed of 16-64 mph at 1090-3090 rpm with the calculated engine load from 16-55% for 2-3 minutes, and the PCM detected the inversion ratio of the rear HO2S and front HO2S was less than a stored threshold during the Catalyst Monitor test. **Possible Causes:** • Air leaks at the exhaust manifold or in the exhaust pipes • Catalytic converter is contaminated, damaged or has failed • Front HO2S and/or the rear HO2S is loose in the mounting hole • Front HO2S older (aged) than the rear HO2S (HO2S is lazy) • Front HO2S or rear HO2S is contaminated with fuel or moisture

DTC	Trouble Code Title, Conditions & Possible Causes
DTC: P0420 **2T OBD/CAT1, MIL: YES** **1996, 1997, 1998, 1999, 2000** **Models:** Millenia **Engines:** All **Transmissions:** A/T, M/T	**Catalyst Efficiency Below Normal (Bank 1)** Vehicle driven at a speed of 16-64 mph at 1090-3090 rpm with the calculated engine load from 16-55% for 2-3 minutes, and the PCM detected the inversion ratio of the rear HO2S and front HO2S was less than a stored threshold during the Catalyst Monitor test. **Possible Causes:** • Air leaks at the exhaust manifold or in the exhaust pipes • Catalytic converter is contaminated, damaged or has failed • Front HO2S and/or the rear HO2S is loose in the mounting hole • Front HO2S older (aged) than the rear HO2S (HO2S is lazy) • Front HO2S or rear HO2S is contaminated with fuel or moisture
DTC: P0430 **2T OBD/CAT3, MIL: YES** **1996, 1997, 1998** **Models:** Millenia **Engines:** 2.3L VIN 2 **Transmissions:** A/T, M/T	**Catalyst Efficiency Below Normal (Bank 2)** Vehicle driven at a speed of 16-64 mph at 1090-3090 rpm with the calculated engine load from 16-55% for 2-3 minutes, and the PCM detected the inversion ratio of the rear HO2S and front HO2S was less than a stored threshold during the Catalyst Monitor test. **Possible Causes:** • Air leaks at the exhaust manifold or in the exhaust pipes • Catalytic converter is contaminated, damaged or has failed • Front HO2S and/or the rear HO2S is loose in the mounting hole • Front HO2S older (aged) than the rear HO2S (HO2S is lazy) • Front HO2S or rear HO2S is contaminated with fuel or moisture
DTC: P0430 **2T OBD/CAT3, MIL: YES** **2004, 2005** **Models:** Tribute **Engines:** All **Transmissions:** A/T, M/T	**Catalyst Efficiency Below Normal (Bank 2)** Vehicle driven at a speed of 16-64 mph at 1090-3090 rpm with the calculated engine load from 16-55% for 2-3 minutes, and the PCM detected the inversion ratio of the rear HO2S and front HO2S was less than a stored threshold during the Catalyst Monitor test. **Possible Causes:** • Air leaks at the exhaust manifold or in the exhaust pipes • Catalytic converter is contaminated, damaged or has failed • Front HO2S and/or the rear HO2S is loose in the mounting hole • Front HO2S older (aged) than the rear HO2S (HO2S is lazy) • Front HO2S or rear HO2S is contaminated with fuel or moisture
DTC: P0430 **2T OBD/CAT3, MIL: YES** **2004, 2005** **Models:** B2300, B3000, B4000 **Engines:** All **Transmissions:** A/T, M/T	**Catalyst Efficiency Below Normal (Bank 2)** Vehicle driven at a speed of 16-64 mph at 1090-3090 rpm with the calculated engine load from 16-55% for 2-3 minutes, and the PCM detected the inversion ratio of the rear HO2S and front HO2S was less than a stored threshold during the Catalyst Monitor test. **Possible Causes:** • Air leaks at the exhaust manifold or in the exhaust pipes • Catalytic converter is contaminated, damaged or has failed • Front HO2S and/or the rear HO2S is loose in the mounting hole • Front HO2S older (aged) than the rear HO2S (HO2S is lazy) • Front HO2S or rear HO2S is contaminated with fuel or moisture
DTC: P0421 **2T OBD/CAT2, MIL: YES** **1999, 2000, 2001, 2002** **Models:** Millenia **Engines:** 2.3L VIN 2 **Transmissions:** A/T, M/T	**Catalyst Efficiency Below Normal (Bank 1)** Vehicle driven at a speed of 22-75 mph at 1250-2094 rpm with the calculated engine load from 20-50% for 2-3 minutes, and the PCM detected the inversion ratio of the right bank rear HO2S and front HO2S was less than a stored threshold in the Catalyst Monitor test. **Possible Causes:** • Air leaks at the exhaust manifold or in the exhaust pipes • Catalytic converter is contaminated, damaged or has failed • Front HO2S and/or the rear HO2S is loose in the mounting hole • Front HO2S older (aged) than the rear HO2S (HO2S is lazy) • Front HO2S or rear HO2S is contaminated with fuel or moisture
DTC: P0421 **2T OBD/CAT2, MIL: YES** **2004, 2005** **Models:** Tribute **Engines:** 2.3L VIN 2 **Transmissions:** A/T, M/T	**Catalyst Efficiency Below Normal (Bank 1)** Vehicle driven at a speed of 22-75 mph at 1250-2094 rpm with the calculated engine load from 20-50% for 2-3 minutes, and the PCM detected the inversion ratio of the right bank rear HO2S and front HO2S was less than a stored threshold in the Catalyst Monitor test. **Possible Causes:** • Air leaks at the exhaust manifold or in the exhaust pipes • Catalytic converter is contaminated, damaged or has failed • Front HO2S and/or the rear HO2S is loose in the mounting hole • Front HO2S older (aged) than the rear HO2S (HO2S is lazy) • Front HO2S or rear HO2S is contaminated with fuel or moisture

DTC	Trouble Code Title, Conditions & Possible Causes
DTC: P0431 **2T OBD/CAT4, MIL: YES** **1996, 1997, 1998, 1999, 2000, 2001, 2002** **Models:** Millenia **Engines:** 2.3L VIN 2 **Transmissions:** A/T, M/T	**Warm-up Catalyst Efficiency Below Normal (Bank 2)** Vehicle driven at a speed of 22-75 mph at 1250-2094 rpm with the calculated engine load from 20-50% for 2-3 minutes, and the PCM detected the inversion ratio of the right bank rear HO2S and front HO2S was less than a stored threshold in the Catalyst Monitor test. **Possible Causes:** • Air leaks at the exhaust manifold or in the exhaust pipes • Catalytic converter is contaminated, damaged or has failed • Front HO2S and/or the rear HO2S is loose in the mounting hole • Front HO2S older (aged) than the rear HO2S (HO2S is lazy) • Front HO2S or rear HO2S is contaminated with fuel or moisture
DTC: P0431 **2T OBD/CAT4, MIL: YES** **2004** **Models:** Tribute **Engines:** 2.3L VIN 2 **Transmissions:** A/T, M/T	**Warm-up Catalyst Efficiency Below Normal (Bank 2)** Vehicle driven at a speed of 22-75 mph at 1250-2094 rpm with the calculated engine load from 20-50% for 2-3 minutes, and the PCM detected the inversion ratio of the right bank rear HO2S and front HO2S was less than a stored threshold in the Catalyst Monitor test. **Possible Causes:** • Air leaks at the exhaust manifold or in the exhaust pipes • Catalytic converter is contaminated, damaged or has failed • Front HO2S and/or the rear HO2S is loose in the mounting hole • Front HO2S older (aged) than the rear HO2S (HO2S is lazy) • Front HO2S or rear HO2S is contaminated with fuel or moisture
DTC: P0421 **2T OBD/CAT2, MIL: YES** **2001, 2002** **Models:** Millenia **Engines:** 2.5L VIN 1 **Transmissions:** A/T, M/T	**Catalyst Efficiency Below Normal (Bank 1)** Vehicle driven at a speed of 24-75 mph at 1250-2500 rpm with the calculated engine load from 16-50% for 2-3 minutes, and the PCM detected the inversion ratio of the left bank rear HO2S and front HO2S was less than a stored threshold in the Catalyst Monitor test. **Possible Causes:** • Air leaks at the exhaust manifold or in the exhaust pipes • Catalytic converter is contaminated, damaged or has failed • Front HO2S and/or the rear HO2S is loose in the mounting hole • Front HO2S older (aged) than the rear HO2S (HO2S is lazy) • Front HO2S or rear HO2S is contaminated with fuel or moisture
DTC: P0431 **2T OBD/CAT4, MIL: YES** **2001, 2002** **Models:** Millenia **Engines:** 2.5L VIN 1 **Transmissions:** A/T, M/T	**Catalyst Efficiency Below Normal (Bank 2)** Vehicle driven at a speed of 24-75 mph at 1250-2500 rpm with the calculated engine load from 16-50% for 2-3 minutes, and the PCM detected the inversion ratio of the right bank rear HO2S and front HO2S was less than a stored threshold in the Catalyst Monitor test. **Possible Causes:** • Air leaks at the exhaust manifold or in the exhaust pipes • Catalytic converter is contaminated, damaged or has failed • Front HO2S and/or the rear HO2S is loose in the mounting hole • Front HO2S older (aged) than the rear HO2S (HO2S is lazy) • Front HO2S or rear HO2S is contaminated with fuel or moisture
DTC: P0420 **2T OBD/CAT1, MIL: YES** **1996, 1997, 1999, 2000, 2001, 2002, 2003, 2004, 2005** **Models:** MX-5 Miata **Engines:** All **Transmissions:** A/T, M/T	**Catalyst Efficiency Below Normal (Bank 1)** Vehicle driven at a speed of 16-64 mph at 1090-3090 rpm with the calculated engine load from 16-55% for 2-3 minutes, and the PCM detected the inversion ratio of the rear HO2S and front HO2S was less than a stored threshold during the Catalyst Monitor test. **Possible Causes:** • Air leaks at the exhaust manifold or in the exhaust pipes • Catalytic converter is contaminated, damaged or has failed • Front HO2S and/or the rear HO2S is loose in the mounting hole • Front HO2S older (aged) than the rear HO2S (HO2S is lazy) • Front HO2S or rear HO2S is contaminated with fuel or moisture
DTC: P0420 **2T OBD/CAT1, MIL: YES** **1996, 1997, 1998** **Models:** Protégé **Engines:** 1.8L VIN 2 **Transmissions:** A/T, M/T	**Catalyst Efficiency Below Normal (Bank 1)** Vehicle driven at a speed of 24-60 mph at 1000-3000 rpm with the calculated engine load from 22-45% for 2-3 minutes, and the PCM detected the inversion ratio of the rear HO2S was close to the inversion ratio of the front HO2S during the Catalyst Monitor test. **Possible Causes:** • Air leaks at the exhaust manifold or in the exhaust pipes • Catalytic converter is contaminated, damaged or has failed • Front HO2S and/or the rear HO2S is loose in the mounting hole • Front HO2S older (aged) than the rear HO2S (HO2S is lazy) • Front HO2S or rear HO2S is contaminated with fuel or moisture

DTC	Trouble Code Title, Conditions & Possible Causes
DTC: P0420 **2T OBD/CAT1, MIL: YES** **1999, 2000** **Models:** Protégé **Engines:** All **Transmissions:** A/T, M/T	**Catalyst Efficiency Below Normal (Bank 1)** Vehicle driven at a speed of 17-74 mph at 1500-3000 rpm with the calculated engine load from 20-37% for 2-3 minutes, and the PCM detected the inversion ratio of the rear HO2S was close to the inversion ratio of the front HO2S during the Catalyst Monitor test. **Possible Causes:** • Air leaks at the exhaust manifold or in the exhaust pipes • Catalytic converter is contaminated, damaged or has failed • Front HO2S and/or the rear HO2S is loose in the mounting hole • Front HO2S older (aged) than the rear HO2S (HO2S is lazy) • Front HO2S or rear HO2S is contaminated with fuel or moisture
DTC: P0421 **2T OBD/CAT2, MIL: YES** **1999, 2000** **Models:** Protégé **Engines:** All **Transmissions:** A/T, M/T	**Warm-up Catalyst Efficiency Below Normal (Bank 1)** Vehicle driven at a speed of 17-74 mph at 1500-3000 rpm with the calculated engine load from 20-37% for 2-3 minutes, and the PCM detected the inversion ratio of the rear HO2S was close to the inversion ratio of the front HO2S during the Catalyst Monitor test. **Possible Causes:** • Air leaks at the exhaust manifold or in the exhaust pipes • Catalytic converter is contaminated, damaged or has failed • Front HO2S and/or the rear HO2S is loose in the mounting hole • Front HO2S older (aged) than the rear HO2S (HO2S is lazy) • Front HO2S or rear HO2S is contaminated with fuel or moisture
DTC: P0421 **2T OBD/CAT2, MIL: YES** **1996, 1997, 1998** **Models:** Protégé **Engines:** 1.5L VIN 1 **Transmissions:** A/T, M/T	**Warm-up Catalyst Efficiency Below Normal (Bank 1)** Vehicle driven at a speed of 24-60 mph at 1000-3000 rpm with the calculated engine load from 22-45% for 2-3 minutes, and the PCM detected the inversion ratio of the rear HO2S was close to the inversion ratio of the front HO2S during the Catalyst Monitor test. **Possible Causes:** • Air leaks at the exhaust manifold or in the exhaust pipes • Catalytic converter is contaminated, damaged or has failed • Front HO2S and/or the rear HO2S is loose in the mounting hole • Front HO2S older (aged) than the rear HO2S (HO2S is lazy) • Front HO2S or rear HO2S is contaminated with fuel or moisture
DTC: P0421 **2T OBD/CAT2, MIL: YES** **2001, 2002, 2003, 2004, 2005** **Models:** Mazda3, MX-5 Miata, Protégé, Protégé5 **Engines:** All **Transmissions:** A/T, M/T	**Warm-up Catalyst Efficiency Below Normal (Bank 1)** Vehicle driven at a speed of 17-74 mph at 1500-3000 rpm with the calculated engine load from 15-48% for 2-3 minutes, and the PCM detected the inversion ratio of the rear HO2S was close to the inversion ratio of the front HO2S during the Catalyst Monitor test. **Possible Causes:** • Air leaks at the exhaust manifold or in the exhaust pipes • Catalytic converter is contaminated, damaged or has failed • Front HO2S and/or the rear HO2S is loose in the mounting hole • Front HO2S older (aged) than the rear HO2S (HO2S is lazy) • Front HO2S or rear HO2S is contaminated with fuel or moisture
DTC: P0421 **2T OBD/CAT2, MIL: YES** **2004, 2005** **Models:** Mazda6 **Engines:** All **Transmissions:** A/T, M/T	**Warm-up Catalyst Efficiency Below Normal (Bank 1)** Vehicle driven at a speed of 17-74 mph at 1500-3000 rpm with the calculated engine load from 15-48% for 2-3 minutes, and the PCM detected the inversion ratio of the rear HO2S was close to the inversion ratio of the front HO2S during the Catalyst Monitor test. **Possible Causes:** • Air leaks at the exhaust manifold or in the exhaust pipes • Catalytic converter is contaminated, damaged or has failed • Front HO2S and/or the rear HO2S is loose in the mounting hole • Front HO2S older (aged) than the rear HO2S (HO2S is lazy) • Front HO2S or rear HO2S is contaminated with fuel or moisture
DTC: P0420 **2T OBD/CAT1, MIL: YES** **1996, 1997, 1998** **Models:** MPV **Engines:** All **Transmissions:** A/T, M/T	**Catalyst Efficiency Below Normal (Bank 1)** Vehicle driven at a speed of 16-64 mph at 1090-3090 rpm with the calculated engine load from 16-55% for 2-3 minutes, and the PCM detected the inversion ratio of the rear HO2S and front HO2S was less than a stored threshold during the Catalyst Monitor test. **Possible Causes:** • Air leaks at the exhaust manifold or in the exhaust pipes • Catalytic converter is contaminated, damaged or has failed • Front HO2S and/or the rear HO2S is loose in the mounting hole • Front HO2S older (aged) than the rear HO2S (HO2S is lazy) • Front HO2S or rear HO2S is contaminated with fuel or moisture

DTC	Trouble Code Title, Conditions & Possible Causes
DTC: P0430 **2T OBD/CAT3, MIL: YES** 1998 **Models:** MPV **Engines:** All **Transmissions:** A/T, M/T	**Catalyst Efficiency Below Normal (Bank 2)** Vehicle driven at a speed of 16-64 mph at 1090-3090 rpm with the calculated engine load from 16-55% for 2-3 minutes, and the PCM detected the inversion ratio of the rear HO2S and front HO2S was less than a stored threshold during the Catalyst Monitor test. **Possible Causes:** • Air leaks at the exhaust manifold or in the exhaust pipes • Catalytic converter is contaminated, damaged or has failed • Front HO2S and/or the rear HO2S is loose in the mounting hole • Front HO2S older (aged) than the rear HO2S (HO2S is lazy) • Front HO2S or rear HO2S is contaminated with fuel or moisture
DTC: P0421 **2T OBD/CAT2, MIL: YES** **2000, 2001, 2002** **Models:** 626, MPV **Engines:** 2.0L VIN C, 2.5L VIN D, 2.5L VIN G **Transmissions:** A/T, M/T	**Catalyst Efficiency Below Normal (Bank 1)** Vehicle driven at a speed of 24-62 mph at 1250-2500 rpm with the calculated engine load from 16-60% for 2-3 minutes, and the PCM detected the inversion ratio of the rear HO2S and front HO2S was less than a stored threshold during the Catalyst Monitor test. **Possible Causes:** • Air leaks at the exhaust manifold or in the exhaust pipes • Catalytic converter is contaminated, damaged or has failed • Front HO2S and/or the rear HO2S is loose in the mounting hole • Front HO2S older (aged) than the rear HO2S (HO2S is lazy) • Front HO2S or rear HO2S is contaminated with fuel or moisture
DTC: P0431 **2T OBD/CAT4, MIL: YES** **2000, 2001, 2002, 2003, 2004, 2005** **Models:** 626, MPV **Engines:** 2.5L VIN D, 2.5L VIN G **Transmissions:** A/T, M/T	**Warm-up Catalyst Efficiency Below Normal (Bank 2)** Vehicle driven at a speed of 24-60 mph at 1000-3000 rpm with the calculated engine load from 22-45% for 2-3 minutes, and the PCM detected the inversion ratio of the rear HO2S and front HO2S was less than a stored threshold during the Catalyst Monitor test. **Possible Causes:** • Air leaks at the exhaust manifold or in the exhaust pipes • Catalytic converter is contaminated, damaged or has failed • Front HO2S and/or the rear HO2S is loose in the mounting hole • Front HO2S older (aged) than the rear HO2S (HO2S is lazy) • Front HO2S or rear HO2S is contaminated with fuel or moisture
DTC: P0420 **2T OBD/CAT1, MIL: YES** **1996, 1997, 1998, 1999, 2000, 2001, 2002, 2003, 2004, 2005** **Models:** B2300, B2500, B3000, B4000, Tribute **Engines:** All **Transmissions:** A/T, M/T	**Catalyst Efficiency Below Normal (Bank 1)** Vehicle driven at a speed of 16-64 mph at 1090-3090 rpm with the calculated engine load from 16-55% for 2-3 minutes, and the PCM detected the switch rate of the rear HO2S was close to the switch rate of the front HO2S-11 during the Catalyst Monitor test. **Possible Causes:** • Air leaks at the exhaust manifold or in the exhaust pipes • Catalytic converter is contaminated, damaged or has failed • Front HO2S and/or the rear HO2S is loose in the mounting hole • Front HO2S older (aged) than the rear HO2S (HO2S is lazy) • Front HO2S or rear HO2S is contaminated with fuel or moisture
DTC: P0430 **2T OBD/CAT3, MIL: YES** **1996, 1997, 1998, 1999, 2000, 2001, 2002, 2003, 2004, 2005** **Models:** Tribute, B300, B4000 **Engines:** 3.0L VIN 1, 3.0L VIN V, 4.0L VIN X, 4.0L VIN E **Transmissions:** A/T, M/T	**Catalyst Efficiency Below Normal (Bank 2)** Vehicle driven at a speed of 16-64 mph at 1090-3090 rpm with the calculated engine load from 16-55% for 2-3 minutes, and the PCM detected the switch rate of the rear HO2S was close to the switch rate of the front HO2S-11 during the Catalyst Monitor test. **Possible Causes:** • Air leaks at the exhaust manifold or in the exhaust pipes • Catalytic converter is contaminated, damaged or has failed • Front HO2S and/or the rear HO2S is loose in the mounting hole • Front HO2S older (aged) than the rear HO2S (HO2S is lazy) • Front HO2S or rear HO2S is contaminated with fuel or moisture
DTC: P0440 **2T CCM, MIL: YES** **1996** **Models:** MPV **Engines:** All **Transmissions:** A/T, M/T	**EVAP System Malfunction** DTC P0443 not set, ECT sensor signal from 14-90°F and IAT sensor signal less than 14°F, engine running in closed loop at a cruise speed over 15 mph, then with the purge solenoid commanded "on", the PCM detected a fault in the EVAP Monitor purge flow test. **Possible Causes:** • EVAP purge solenoid is damaged or has failed • Canister drain cut valve is damaged or has failed • Fuel gauge sender unit is damaged or has failed • EVAP cut valve, catch tank or liquid separator has failed • Fuel tank pressure sensor circuit is open or shorted to ground • Fuel tank pressure sensor is damaged or has failed • EVAP charcoal canister is loaded with fuel or moisture • Vapor line between the purge solenoid and the intake manifold, tank pressure control valve, cut valve, separator, FTP sensor, fuel tank, canister and drain valve damaged, loose or restricted • PCM has failed

DTC	Trouble Code Title, Conditions & Possible Causes
DTC: P0442 **2T CCM, MIL: YES** 1997, 1998 **Models:** MPV **Engines:** All **Transmissions:** A/T, M/T	**EVAP System Small Leak (0.040") Detected** DTC P0443 not set, ECT sensor signal from 14-90ºF and IAT sensor signal less than 14ºF at startup, vehicle driven at cruise speed, then with the purge solenoid enabled, the PCM detected excessive fuel pressure variation with the canister drain cut valve in closed position. **Possible Causes:** • Fuel filler cap loose, cross-threaded, incorrect part or damaged • EVAP purge solenoid is damaged or has failed • EVAP charcoal canister is loaded with fuel or moisture • EVAP vent solenoid is damaged (it may be stuck partially open) • EVAP check valve or vapor valve is damaged or leaking • Vapor line between the purge solenoid and the intake manifold, EVAP purge solenoid valve and the canister, or between the canister, check valve and fuel vapor valve damaged or leaking • PCM has failed
DTC: P0442 **2T CCM, MIL: YES** **1999, 2000, 2001, 2002, 2003, 2004, 2005** **Models:** MPV **Engines:** All **Transmissions:** A/T, M/T	**EVAP System Small Leak (0.040") Detected** DTC P0443 not set, ECT sensor signal from 14-90ºF and IAT sensor signal less than 14ºF at startup, fuel level over 75%, vehicle driven at cruise speed, then with the purge solenoid commanded "on", the PCM detected a leak as small as 0.040" in the EVAP system during a period with less than 2.5" H2O bleed-up over a 15 second period. **Possible Causes:** • Fuel filler cap loose, cross-threaded, incorrect part or damaged • EVAP purge solenoid is damaged or has failed • EVAP charcoal canister is loaded with fuel or moisture • EVAP vent solenoid is damaged (it may be stuck partially open) • EVAP check valve or vapor valve is damaged or leaking • Vapor lines damaged or leaking between the purge solenoid and intake manifold, purge solenoid and canister, or between the canister vent solenoid, check valve and fuel vapor valve • PCM has failed
DTC: P0440 **2T CCM, MIL: YES** **1996** **Models:** MX-5 Miata, MX-6, 626, Millenia, Protégé **Engines:** All **Transmissions:** A/T, M/T	**EVAP System Malfunction** DTC P0443 not set, ECT sensor signal from 14-90ºF and IAT sensor signal less than 14ºF, engine running in closed loop at a cruise speed over 15 mph, then with the purge solenoid commanded "on", the PCM detected a fault in the EVAP Monitor purge flow test. **Possible Causes:** • EVAP purge solenoid is damaged or has failed • EVAP charcoal canister is loaded with fuel or moisture • Vapor line between the purge solenoid and the intake manifold reservoir, or the vapor line between the EVAP purge solenoid, check valve and canister is leaking, restricted or disconnected • PCM has failed
DTC: P0440 **2T CCM, MIL: YES** **2004** **Models:** Tribute **Engines:** All **Transmissions:** A/T, M/T	**EVAP System Malfunction** DTC P0443 not set, ECT sensor signal from 14-90ºF and IAT sensor signal less than 14ºF, engine running in closed loop at a cruise speed over 15 mph, then with the purge solenoid commanded "on", the PCM detected a fault in the EVAP Monitor purge flow test. **Possible Causes:** • EVAP purge solenoid is damaged or has failed • EVAP charcoal canister is loaded with fuel or moisture • Vapor line between the purge solenoid and the intake manifold reservoir, or the vapor line between the EVAP purge solenoid, check valve and canister is leaking, restricted or disconnected • PCM has failed
DTC: P0441 **2T CCM, MIL: YES** 1997 **Models:** MX-5 Miata, MX-6, 626 **Engines:** All **Transmissions:** A/T, M/T	**EVAP System Malfunction** DTC P0443 not set, ECT sensor signal from 14-90ºF and IAT sensor signal less than 14ºF, engine running in closed loop at a cruise speed over 15 mph, then with the purge solenoid commanded "on", the PCM detected the fuel control "feedback" signal was less than a value stored in memory with "purge" enabled during the EVAP test. **Possible Causes:** • EVAP purge solenoid is damaged or has failed • EVAP charcoal canister is loaded with fuel or moisture • Vapor line is damaged or restricted between the purge solenoid and intake manifold, or between check valve and vapor valve • PCM has failed

DTC	Trouble Code Title, Conditions & Possible Causes
DTC: P0441 **2T CCM, MIL: YES** **2004, 2005** **Models:** Mazda3, Tribute **Engines:** All **Transmissions:** A/T, M/T	**EVAP System Malfunction** DTC P0443 not set, ECT sensor signal from 14-90ºF and IAT sensor signal less than 14ºF, engine running in closed loop at a cruise speed over 15 mph, then with the purge solenoid commanded "on", the PCM detected the fuel control "feedback" signal was less than a value stored in memory with "purge" enabled during the EVAP test. **Possible Causes:** • EVAP purge solenoid is damaged or has failed • EVAP charcoal canister is loaded with fuel or moisture • Vapor line is damaged or restricted between the purge solenoid and intake manifold, or between check valve and vapor valve • PCM has failed
DTC: P0441 **2T CCM, MIL: YES** **2003, 2004, 2005** **Models:** Mazda6, MX-5 Miata, RX-8 **Engines:** All **Transmissions:** A/T, M/T	**EVAP System Malfunction** DTC P0443 not set, ECT sensor signal from 14-90ºF and IAT sensor signal less than 14ºF, engine running in closed loop at a cruise speed over 15 mph, then with the purge solenoid commanded "on", the PCM detected the fuel control "feedback" signal was less than a value stored in memory with "purge" enabled during the EVAP test. **Possible Causes:** • EVAP purge solenoid is damaged or has failed • EVAP charcoal canister is loaded with fuel or moisture • Vapor line is damaged or restricted between the purge solenoid and intake manifold, or between check valve and vapor valve • PCM has failed
DTC: P0442 **2T CCM, MIL: YES** 1997, 1998 **Models:** Millenia, Protégé **Engines:** All **Transmissions:** A/T, M/T	**EVAP System Small Leak (0.040") Detected** DTC P0443 not set, ECT sensor signal from 14-90ºF and IAT sensor signal less than 14ºF at startup, vehicle driven at cruise speed, then with the purge solenoid enabled, the PCM detected excessive fuel pressure variation with the canister drain cut valve in closed position. **Possible Causes:** • Fuel filler cap loose, cross-threaded, incorrect part or damaged • EVAP purge solenoid is damaged or has failed • EVAP charcoal canister is loaded with fuel or moisture • EVAP canister drain cut valve is damaged or leaking • EVAP check valve or vapor valve is damaged or leaking • Vapor line between the purge solenoid and the intake manifold, EVAP purge solenoid valve and the canister, or between the canister, check valve and fuel vapor valve damaged or leaking • PCM has failed
DTC: P0442 **2T CCM, MIL: YES** **1999, 2000, 2001, 2002, 2003** **Models:** Protégé, Protégé5 **Engines:** All **Transmissions:** A/T, M/T	**EVAP System Small Leak (0.040") Detected** DTC P0443 not set, ECT sensor signal from 14-90ºF, IAT sensor signal more than 14ºF at startup, BARO sensor signal over 72 kPa, IAT sensor signal from 14-140ºF, fuel level from 15-85%, vehicle driven at 24-75 mph at an engine speed of 1000-4000 rpm, engine load from 9-65% with the throttle opening from 3.1-12.5%, then with the purge and vent solenoids both closed, the PCM detected a leak as small as 0.040" in the system during the EVAP Leak Monitor test. **Note: The threshold FTP sensor value for this test is 1.17-3.91 kPa (the actual threshold value depends upon the ECT sensor signal).** **Possible Causes:** • Fuel filler cap loose, cross-threaded, incorrect part or damaged • EVAP purge solenoid is damaged or has failed • EVAP charcoal canister is loaded with fuel or moisture • EVAP canister drain cut valve is damaged or has failed • EVAP rollover valve or catch tank valve is damaged or leaking • EGR boost sensor is damaged or has failed • Fuel level sensor is damaged or has failed • Fuel tank pressure sensor is damaged or has failed • ECT, IAT, MAF, VSS or TP sensor signals out-of-calibration • Vapor line(s) damaged or leaking between the purge solenoid and the intake manifold, EVAP purge solenoid valve and the canister, canister drain cut valve and the rollover valve • PCM has failed

DTC	Trouble Code Title, Conditions & Possible Causes
DTC: P0442 **2T CCM, MIL: YES** **1999, 2000, 2001, 2002** **Models:** Millenia **Engines:** 2.5L VIN 1 **Transmissions:** A/T, M/T	**EVAP System Small Leak (0.040") Detected** DTC P0443 not set, ECT sensor signal from 14-90ºF, IAT sensor signal more than 14ºF at startup, BARO sensor signal over 72 kPa, IAT sensor signal from 14-140ºF, fuel level from 15-85%, vehicle driven at 25-65.3 mph at an engine speed of 1000-3000 rpm, engine load from 9-70% with the throttle opening from 3.0-44.0%, then with the purge and vent solenoids both closed, the PCM detected a leak as small as 0.040" in the system during the EVAP Leak Monitor test. **Note: The target pressure for this test is minus (-) 2.65 kPa.** **Possible Causes:** • Fuel filler cap loose, cross-threaded, incorrect part or damaged • Purge solenoid valve is damaged or has failed • Charcoal canister is loaded with fuel or moisture • Tank pressure control valve (TPCV) is damaged or has failed • Canister drain cut valve (CDCV) is damaged or has failed • Rollover valve or catch tank valve is damaged or leaking • Fuel tank is cracked (leaking), or a leak exists in the tank seal • Fuel tank pressure sensor is damaged or has failed • ECT, IAT, MAF, VSS or TP sensor signals out-of-calibration • Vapor line(s) damaged or leaking between the purge solenoid and the intake manifold, EVAP purge solenoid valve and the canister, canister drain cut valve and the rollover valve • PCM has failed
DTC: P0442 **2T CCM, MIL: YES** **1999, 2000, 2001, 2002** **Models:** Millenia **Engines:** 2.3L VIN 2 **Transmissions:** A/T, M/T	**EVAP System Small Leak (0.040") Detected** DTC P0443 not set, ECT sensor signal from 14-90ºF, IAT sensor signal more than 14ºF at startup, BARO sensor signal over 72 kPa, IAT sensor signal from 14-140ºF, fuel level from 15-85%, vehicle driven at 24-65 mph at an engine speed of 1000-4400 rpm, engine load from 9-65% with the throttle opening from 3.1-31.1%, then with the purge and vent solenoids both closed, the PCM detected a leak as small as 0.040" in the system during the EVAP Leak Monitor test. **Note: The fuel tank target pressure for this test is minus (-) 2.65 kPa.** **Possible Causes:** • Fuel filler cap loose, cross-threaded, incorrect part or damaged • Purge solenoid valve is damaged or has failed • Charcoal canister is loaded with fuel or moisture • Tank pressure control valve (TPCV) is damaged or has failed • Canister drain cut valve (CDCV) is damaged or has failed • Rollover valve or catch tank valve is damaged or leaking • Fuel tank is cracked (leaking), or a leak exists in the tank seal • Fuel tank pressure sensor is damaged or has failed • ECT, IAT, MAF, VSS or TP sensor signals out-of-calibration • Vapor line(s) damaged or leaking between the purge solenoid and the intake manifold, EVAP purge solenoid valve and the canister, canister drain cut valve and the rollover valve • PCM has failed
DTC: P0442 **2T CCM, MIL: YES** **1999, 2000, 2001, 2002, 2003, 2004, 2005** **Models:** Mazda3, Mazda6, MX-5 Miata, RX-8 **Engines:** All **Transmissions:** A/T, M/T	**EVAP System Small Leak (0.040") Detected** DTC P0443 not set, ECT sensor signal from 14-90ºF, IAT sensor signal more than 14ºF at startup, BARO sensor signal over 72 kPa, IAT sensor signal from 14-140ºF and ECT sensor signal from 158-212ºF during the EVAP leak test, fuel level from 15-85%, vehicle driven at 24-65 mph at an engine speed of 1000-4000 rpm, engine load from 9-65% with the throttle opening from 3.1-31.6%, then with the purge and CDCV valves both closed, the PCM detected a leak as small as 0.040" in the system during the EVAP Leak Monitor test. **Note: The fuel tank target pressure for this test is plus (+) 127 kPa.** **Possible Causes:** • Fuel filler cap loose, cross-threaded, incorrect part or damaged • Purge solenoid valve is damaged or has failed • Charcoal canister is loaded with fuel or moisture • Canister drain cut valve (CDCV) is damaged or has failed • Rollover valve, catch tank valve or fuel tank damaged/ leaking • Fuel tank pressure sensor is damaged or has failed • ECT, IAT, MAF, VSS or TP sensor signals out-of-calibration • Vapor line(s) damaged or leaking between the purge solenoid and the intake manifold, EVAP purge solenoid valve and the canister, canister drain cut valve and the rollover valve • PCM has failed

DTC	Trouble Code Title, Conditions & Possible Causes
DTC: P0442 **2T CCM, MIL: YES** **1998, 1999, 2000, 2001, 2002** **Models:** 626 **Engines:** All **Transmissions:** A/T, M/T	**EVAP System Small Leak (0.040") Detected** DTC P0443 not set, ECT sensor signal from 14-90ºF and IAT sensor signal less than 14ºF at startup, vehicle driven at cruise speed, then with the purge solenoid enabled, the PCM detected excessive fuel pressure variation with the canister drain cut valve in closed position. **Possible Causes:** • Fuel filler cap loose, cross-threaded, incorrect part or damaged • Purge solenoid valve is damaged or has failed • Charcoal canister is loaded with fuel or moisture • Tank pressure control valve (TPCV) is damaged or has failed • Canister drain cut valve (CDCV) is damaged or has failed • Rollover valve or catch tank valve is damaged or leaking • Fuel tank is cracked (leaking), or a leak exists in the tank seal • Fuel tank pressure sensor is damaged or has failed • ECT, IAT, MAF, VSS or TP sensor signals out-of-calibration • Vapor line(s) damaged or leaking between the purge solenoid and the intake manifold, EVAP purge solenoid valve and the canister, canister drain cut valve and the rollover valve • PCM has failed
DTC: P0442 **2T CCM, MIL: YES** **1999, 2000, 2001, 2002, 2003, 2004, 2005** **Models:** B2300, B2500, B3000, B4000, Tribute **Engines:** All **Transmissions:** A/T, M/T	**EVAP System Small Leak (0.040") Detected** DTC P0443 not set, ECT sensor from 14-90ºF and IAT sensor less than 14ºF at startup, vehicle driven at cruise speed, then with the purge solenoid enabled, the PCM detected a leak as small as 0.040" in the system during the EVAP Running Loss Monitor test. **Note: This code sets when there is less than 2.5" H2O bleed-up over a 15 second period with the fuel tank level more than 75% full. The bleed-up and evaluation time vary as a function of fuel level. Vapor generation is more than 2.5" H2O over a 120-second period of time.** **Possible Causes:** • Fuel filler cap loose, cross-threaded, incorrect part or damaged • EVAP purge solenoid valve is damaged or has failed • EVAP vent control solenoid is damaged or has failed • EVAP charcoal canister is loaded with fuel or moisture • Fuel tank pressure sensor damaged or has failed • Vapor line(s) damaged or leaking between the purge solenoid and the intake manifold, purge solenoid valve and the canister, fuel vapor control valve tube or fuel vapor vent valve assembly • PCM has failed
DTC: P0442 **2T CCM, MIL: YES** 1998 **Models:** B2500, B3000, B4000 **Engines:** All **Transmissions:** A/T, M/T	**EVAP System Small Leak (0.040") Detected** DTC P0443 not set, ECT sensor signal from 14-90ºF and IAT sensor signal more than 14ºF at startup, fuel level from 15-85%, BARO sensor signal more than 72 kPa, engine speed at 1000-3000 rpm, VSS input from 25-63 mph, calculated load at 9-70%, TP angle from 3-44%, then with the VMV solenoid commanded "on", the PCM detected a leak as small as 0.040" in the system in the Leak test. **Possible Causes:** • Fuel filler cap loose, cross-threaded, incorrect part or damaged • EVAP VMV solenoid is damaged or has failed • EVAP charcoal canister is loaded with fuel or moisture • Fuel tank pressure sensor damaged or has failed • Vapor line(s) damaged or leaking between the VMV solenoid and the intake manifold, or the VMV solenoid and the canister • PCM has failed
DTC: P0443 **1T CCM, MIL: YES** **1996, 1997, 1998, 1999, 2000, 2001, 2002, 2003, 2004, 2005** **Models:** Mazda6, MX-5 Miata, MX-6, 626, MPV, Millenia, Protégé, Protégé5, RX-8 **Engines:** All **Transmissions:** A/T, M/T	**EVAP Canister Purge Solenoid Circuit Malfunction** Key on or engine running and the PCM detected an unexpected voltage condition on the EVAP Purge solenoid circuit during the test. **Note: This is a diagnostic support code only – not stored in the PCM!** **Possible Causes:** • Purge solenoid control circuit is open or shorted to ground • Purge solenoid control circuit is shorted to system power (B+) • Purge solenoid power circuit is open or shorted to ground • Purge solenoid is damaged or has failed • PCM has failed (the purge solenoid driver may be damaged)

DTC	Trouble Code Title, Conditions & Possible Causes
DTC: P0443 **1T CCM, MIL: YES** **1996, 1997, 1998** **Models:** B2300, B2500, B3000, B4000 **Engines:** All **Transmissions:** A/T, M/T	**EVAP Vapor Management Valve Circuit Malfunction** Key on or engine running and the PCM detected an unexpected voltage condition on the EVAP VMV solenoid circuit during the test. **Note: This is a diagnostic support code only – not stored in the PCM!** **Possible Causes:** • VMV solenoid control circuit is open or shorted to ground • VMV solenoid control circuit is shorted to system power (B+) • VMV solenoid power circuit is open or shorted to ground • VMV solenoid is damaged or has failed • PCM has failed (the VMV solenoid driver may be damaged)
DTC: P0443 **1T CCM, MIL: YES** **1999, 2000, 2001, 2002, 2003, 2004, 2005** **Models:** B2300, B2500, B3000, B4000, Tribute **Engines:** All **Transmissions:** A/T, M/T	**EVAP Purge Solenoid Circuit Malfunction** Key on or engine running and the PCM detected an unexpected voltage condition on the EVAP purge solenoid circuit during the test. **Note: This is a diagnostic support code only – not stored in the PCM!** **Possible Causes:** • VMV solenoid control circuit is open or shorted to ground • VMV solenoid control circuit is shorted to system power (B+) • VMV solenoid power circuit is open or shorted to ground • VMV solenoid is damaged or has failed • PCM has failed (the VMV solenoid driver may be damaged)
DTC: P0443 **1T CCM, MIL: YES** **2004, 2005** **Models:** Mazda3 **Engines:** All **Transmissions:** A/T, M/T	**EVAP Purge Solenoid Circuit Malfunction** Key on or engine running and the PCM detected an unexpected voltage condition on the EVAP purge solenoid circuit during the test. **Note: This is a diagnostic support code only – not stored in the PCM!** **Possible Causes:** • VMV solenoid control circuit is open or shorted to ground • VMV solenoid control circuit is shorted to system power (B+) • VMV solenoid power circuit is open or shorted to ground • VMV solenoid is damaged or has failed • PCM has failed (the VMV solenoid driver may be damaged)
DTC: P0444 **1T CCM, MIL: YES** **2004** **Models:** Tribute **Engines:** All **Transmissions:** A/T, M/T	**EVAP Purge Solenoid Circuit Malfunction** Key on or engine running and the PCM detected an unexpected voltage condition on the EVAP purge solenoid circuit during the test. **Note: This is a diagnostic support code only – not stored in the PCM!** **Possible Causes:** • VMV solenoid control circuit is open or shorted to ground • VMV solenoid control circuit is shorted to system power (B+) • VMV solenoid power circuit is open or shorted to ground • VMV solenoid is damaged or has failed • PCM has failed (the VMV solenoid driver may be damaged)
DTC: P0445 **1T CCM, MIL: YES** **2004** **Models:** Tribute **Engines:** All **Transmissions:** A/T, M/T	**EVAP Purge Solenoid Circuit Malfunction** Key on or engine running and the PCM detected an unexpected voltage condition on the EVAP purge solenoid circuit during the test. **Note: This is a diagnostic support code only – not stored in the PCM!** **Possible Causes:** • VMV solenoid control circuit is open or shorted to ground • VMV solenoid control circuit is shorted to system power (B+) • VMV solenoid power circuit is open or shorted to ground • VMV solenoid is damaged or has failed • PCM has failed (the VMV solenoid driver may be damaged)
DTC: P0446 **1T CCM, MIL: YES** 1998 **Models:** MX-5 Miata, Millenia **Engines:** All **Transmissions:** A/T, M/T	**EVAP System Vent Control Malfunction** DTC P0443 not set, ECT sensor signal from 14-90ºF and IAT sensor signal more than 14ºF at startup, engine speed at 1000-3000 rpm, VSS input from 25-63 mph, calculated load at 9-70%, TP angle from 3-44%, Constant Drain Cut Valve closed and then opened, and the PCM detected the change in fuel tank pressure was too small. **Possible Causes:** • CDCV is damaged, sticking or has failed • Canister Vent side filter is damaged or missing • Charcoal canister is loaded with fuel or moisture • FTP sensor signal circuit is open, shorted to ground or to power • FTP sensor power circuit is open or shorted to ground • FTP sensor is damaged or has failed • BARO, ECT, IAT, MAF, VSS or TP sensor is out-of-calibration • Fuel tank level sensor is damaged or out-of-calibration • Fuel vapor storage canister vent is plugged or restricted • Fuel vapor line(s) kinked or blocked between the CDVC solenoid and the intake manifold

DTC	Trouble Code Title, Conditions & Possible Causes
DTC: P0446 **1T CCM, MIL: YES** **1998, 1999, 2000** **Models:** 626 **Engines:** All **Transmissions:** A/T, M/T	**EVAP System Vent Control Malfunction** DTC P0443 not set, ECT sensor signal from 14-90ºF and IAT sensor signal more than 14ºF at startup, engine speed at 1000-3000 rpm, VSS input from 25-63 mph, calculated load at 9-70%, TP angle from 3-44%, Constant Drain Cut Valve closed and then opened, and the PCM detected the change in fuel tank pressure was too small. **Possible Causes:** • CDCV is damaged, sticking or has failed • Canister Vent side filter is damaged or missing • Charcoal canister is loaded with fuel or moisture • FTP sensor signal circuit is open, shorted to ground or to power • FTP sensor power circuit is open or shorted to ground • FTP sensor is damaged or has failed • BARO, ECT, IAT, MAF, VSS or TP sensor is out-of-calibration • Fuel tank level sensor is damaged or out-of-calibration • Fuel vapor storage canister vent is plugged or restricted • Fuel vapor line(s) kinked or blocked between the CDVC solenoid and the intake manifold
DTC: P0446 **1T CCM, MIL: YES** 1998 **Models:** B2500 **Engines:** All **Transmissions:** A/T, M/T	**EVAP System Vent Control Malfunction** DTC P0443 not set, ECT sensor signal from 14-90ºF and IAT sensor signal more than 14ºF at startup, engine speed at 1000-3000 rpm, VSS input from 25-63 mph, calculated load at 9-70%, TP angle from 3-44%, VMV and CV solenoid closed and then opened, and the PCM detected the change in fuel tank pressure was too small. **Possible Causes:** • VMV or CV solenoid is damaged or has failed • Canister Vent (CV) filter is damaged or missing • Charcoal canister fuel vapor vent is dirty or restricted • FTP sensor signal circuit is open, shorted to ground or to power • FTP sensor power circuit is open or shorted to ground • FTP sensor is damaged or has failed • Fuel vapor line(s) kinked or blocked between the fuel tank, VMV and the canister(s), or between the storage canister(s) and the CV solenoid and atmosphere
DTC: P0446 **1T CCM, MIL: YES** 1998 **Models:** B3000, B4000 **Engines:** All **Transmissions:** A/T, M/T	**EVAP System Vent Control Malfunction** DTC P0443 not set, ECT sensor signal from 14-90ºF and IAT sensor signal more than 14ºF at startup, engine speed at 1000-3000 rpm, VSS input from 25-63 mph, calculated load at 9-70%, TP angle from 3-44%, then with the CDCV closed and then reopened, the PCM detected the change in fuel tank pressure was too small. **Possible Causes:** • Canister drain cut valve (CDCV) is damaged, sticking or failed • Tank pressure control valve (TPCV) is damaged or has failed • Charcoal canister is loaded with fuel or moisture • Air filter is severely restricted, or 2-way check valve is clogged • FTP sensor signal circuit is open, shorted to ground or to power • FTP sensor power circuit is open or shorted to ground • FTP sensor is damaged or has failed • BARO, ECT, IAT, MAF, VSS or TP sensor is out-of-calibration • Fuel tank level sensor is damaged or out-of-calibration • Fuel vapor line(s) kinked or blocked between the CDVC valve and intake manifold, or between the TPCV and the canister
DTC: P0446 **1T CCM, MIL: YES** **2003, 2004, 2005** **Models:** Mazda6, MX-5 Miata, MPV, RX-8 **Engines:** All **Transmissions:** A/T, M/T	**EVAP System Vent Control Malfunction** DTC P0443 not set, ECT sensor signal from 14-90ºF and IAT sensor signal more than 14ºF at startup, engine speed at 1000-3000 rpm, VSS input from 25-63 mph, calculated load at 9-70%, TP angle from 3-44%, then with the CDCV closed and then reopened, the PCM detected the change in fuel tank pressure was too small. **Possible Causes:** • Canister drain cut valve (CDCV) is damaged, sticking or failed • Tank pressure control valve (TPCV) is damaged or has failed • Charcoal canister is loaded with fuel or moisture • Air filter is severely restricted, or 2-way check valve is clogged • FTP sensor signal circuit is open, shorted to ground or to power • FTP sensor power circuit is open or shorted to ground • FTP sensor is damaged or has failed • BARO, ECT, IAT, MAF, VSS or TP sensor is out-of-calibration • Fuel tank level sensor is damaged or out-of-calibration • Fuel vapor line(s) kinked or blocked between the CDVC valve and intake manifold, or between the TPCV and the canister

DTC	Trouble Code Title, Conditions & Possible Causes
DTC: P0446 **1T CCM, MIL: YES** **2004, 2005** **Models:** Mazda3, Tribute **Engines:** All **Transmissions:** A/T, M/T	**EVAP System Vent Control Malfunction** DTC P0443 not set, ECT sensor signal from 14-90°F and IAT sensor signal more than 14°F at startup, engine speed at 1000-3000 rpm, VSS input from 25-63 mph, calculated load at 9-70%, TP angle from 3-44%, then with the CDCV closed and then reopened, the PCM detected the change in fuel tank pressure was too small. **Possible Causes:** • Canister drain cut valve (CDCV) is damaged, sticking or failed • Tank pressure control valve (TPCV) is damaged or has failed • Charcoal canister is loaded with fuel or moisture • Air filter is severely restricted, or 2-way check valve is clogged • FTP sensor signal circuit is open, shorted to ground or to power • FTP sensor power circuit is open or shorted to ground • FTP sensor is damaged or has failed • BARO, ECT, IAT, MAF, VSS or TP sensor is out-of-calibration • Fuel tank level sensor is damaged or out-of-calibration • Fuel vapor line(s) kinked or blocked between the CDVC valve and intake manifold, or between the TPCV and the canister
DTC: P0450 **2T CCM, MIL: YES** **1997, 1998, 1999, 2000, 2001, 2002** **Models:** Millenia **Engines:** All **Transmissions:** A/T, M/T	**EVAP Pressure Sensor Circuit Malfunction** Engine running, ECT sensor signal less than 176°F, and the PCM detected the Fuel Tank Pressure (FTP) sensor signal was less than 0.20v, or that it was more than 4.80v during the CCM test. **Possible Causes:** • FTP sensor signal circuit is open (4.80v reading) • FTP sensor signal circuit is shorted to ground (0.20v reading) • FTP sensor power circuit is open between sensor and the PCM • FTP sensor ground circuit is open between sensor and ground • FTP sensor signal circuit is shorted to VREF or system power • FTP sensor is damaged or has failed • PCM has failed
DTC: P0450 **2T CCM, MIL: YES** **2004** **Models:** Tribute **Engines:** All **Transmissions:** A/T, M/T	**EVAP Pressure Sensor Circuit Malfunction** Engine running, ECT sensor signal less than 176°F, and the PCM detected the Fuel Tank Pressure (FTP) sensor signal was less than 0.20v, or that it was more than 4.80v during the CCM test. **Possible Causes:** • FTP sensor signal circuit is open (4.80v reading) • FTP sensor signal circuit is shorted to ground (0.20v reading) • FTP sensor power circuit is open between sensor and the PCM • FTP sensor ground circuit is open between sensor and ground • FTP sensor signal circuit is shorted to VREF or system power • FTP sensor is damaged or has failed • PCM has failed
DTC: P0451 **2T CCM, MIL: YES** **1996, 1997, 1998, 1999, 2000, 2001, 2002, 2003** **Models:** Protégé, Protégé5 **Engines:** All **Transmissions:** A/T, M/T	**EVAP Pressure Sensor Circuit Range/Performance** DTC P0443 not set, ECT sensor signal from 14-90°F, IAT sensor signal more than 14°F at startup, BARO sensor more than 72 kPa, fuel level from 15-85%, IAT sensor signal 14-131°F during testing, vehicle driven at 24-65 mph at an engine speed of 1000-4000 rpm, calculated engine load of 9-65% and a throttle angle from 3-31.6%, then with the CDCV (valve) closed, the PCM detected the fuel tank pressure (FTP) sensor indicated too small or too large a variation. **Possible Causes:** • Air filter element or charcoal canister is very dirty or clogged • CDCV (valve) is damaged or has failed • Pressure control valve is damaged or has failed • Connection problems at the CDCV FTP sensor or at the PCM • CDCV (valve) or FTP sensor wiring harness is open or shorted • PCM has failed
DTC: P0451 **2T CCM, MIL: YES** **1999, 2000, 2001, 2002, 2003** **Models:** MX-5 Miata, Millenia, MPV **Engines:** All **Transmissions:** A/T, M/T	**EVAP Pressure Sensor Circuit Range/Performance** DTC P0443 not set, ECT sensor signal from 14-90°F, IAT sensor signal more than 14°F at startup, BARO sensor more than 72 kPa, fuel level from 15-85%, IAT sensor signal 14-131°F during testing, vehicle driven at 24-65 mph at an engine speed of 1000-4000 rpm, calculated engine load of 9-65% and a throttle angle from 3-31.6%, then with the CDCV (valve) closed, the PCM detected the fuel tank pressure (FTP) sensor indicated too small or too large a variation. **Possible Causes:** • Air filter element or charcoal canister is very dirty or clogged • CDCV (valve) is damaged or has failed • Purge solenoid valve is damaged or has failed • Connection problems at the CDCV FTP sensor or at the PCM • CDCV (valve) or FTP sensor open or shorted in wiring harness • PCM has failed

DTC	Trouble Code Title, Conditions & Possible Causes
DTC: P0451 **2T CCM, MIL: YES** **2001, 2002, 2003, 2004, 2005** **Models:** B2300, B2500, B3000, B4000, Tribute **Engines:** All **Transmissions:** A/T, M/T	**EVAP Pressure Sensor Circuit Range/Performance** Engine running, and the PCM detected the Fuel Tank Pressure (FTP) sensor changed more than 14" H2O within a 10 second period (indicating the FTP sensor circuit is noisy). **Possible Causes:** • FTP sensor signal circuit is open (intermittent fault) • FTP sensor signal circuit is shorted to ground (intermittent fault) • FTP sensor is damaged or has failed • PCM has failed
DTC: P0452 **2T CCM, MIL: YES** **1999, 2000, 2001, 2002, 2003** **Models:** MX-5 Miata, Protégé, Protégé5 **Engines:** All **Transmissions:** A/T, M/T	**EVAP Pressure Sensor Circuit Low Input** Engine running, ECT sensor signal less than 176ºF, and the PCM detected the Fuel Tank Pressure (FTP) sensor signal indicated less than 0.22v for two (2) seconds during the CCM test. **Note: The FTP sensor PID should be 2.5v at key on, engine off.** **Possible Causes:** • FTP sensor is completely submerged in fuel inside the fuel tank • FTP sensor signal circuit is shorted to sensor ground • FTP sensor signal circuit is shorted to chassis ground • FTP sensor is damaged or has failed • PCM has failed
DTC: P0452 **2T CCM, MIL: YES** **1998, 1999, 2000, 2001, 2002** **Models:** 626 **Engines:** All **Transmissions:** A/T, M/T	**EVAP Pressure Sensor Circuit Low Input** Engine running ECT sensor signal less than 176ºF, and the PCM detected the Fuel Tank Pressure (FTP) sensor signal indicated less than 0.22v for two (2) seconds during the CCM test. **Note: The FTP sensor PID should be 2.5v at key on, engine off.** **Possible Causes:** • FTP sensor is completely submerged in fuel inside the fuel tank • FTP sensor signal circuit is shorted to sensor ground • FTP sensor signal circuit is shorted to chassis ground • FTP sensor is damaged or has failed • PCM has failed
DTC: P0452 **2T CCM, MIL: YES** **1998, 1999, 2000, 2001, 2002, 2003, 2004, 2005** **Models:** B2300, B2500, B3000, B4000, MPV, Tribute **Engines:** All **Transmissions:** A/T, M/T	**EVAP Pressure Sensor Circuit Low Input** Engine running and the PCM detected the Fuel Tank Pressure (FTP) sensor signal indicated less than 0.22v during the CCM test. **Possible Causes:** • FTP sensor signal circuit is shorted to sensor ground • FTP sensor signal circuit is shorted to chassis ground • FTP sensor signal is shorted in the connector due to moisture • FTP sensor is damaged or has failed • PCM has failed
DTC: P0453 **2T CCM, MIL: YES** **1999, 2000, 2001, 2002, 2003** **Models:** MX-5 Miata, Protégé, Protégé5 **Engines:** All **Transmissions:** A/T, M/T	**EVAP Pressure Sensor Circuit High Input** Engine running, ECT sensor signal less than 176ºF, and the PCM detected the Fuel Tank Pressure (FTP) sensor signal indicated more than 4.80v for two (2) seconds during the CCM test. **Possible Causes:** • FTP sensor signal circuit is shorted to VREF or power (B+) • FTP sensor ground circuit is open • FTP sensor is damaged or has failed • PCM has failed
DTC: P0453 **2T CCM, MIL: YES** **1998, 1999, 2000, 2001, 2002** **Models:** 626 **Engines:** All **Transmissions:** A/T, M/T	**EVAP Pressure Sensor Circuit High Input** Engine running ECT sensor signal less than 176ºF, and the PCM detected the Fuel Tank Pressure (FTP) sensor signal indicated more than 4.80v for two (2) seconds during the CCM test. **Possible Causes:** • FTP sensor signal circuit is shorted to VREF or power (B+) • FTP sensor ground circuit is open • FTP sensor is damaged or has failed • PCM has failed
DTC: P0453 **2T CCM, MIL: YES** **1998, 1999, 2000, 2001, 2002, 2003, 2004, 2005** **Models:** B2300, B2500, B3000, B4000, MPV, Tribute **Engines:** All **Transmissions:** A/T, M/T	**EVAP Pressure Sensor Circuit High Input** Engine running and the PCM detected the Fuel Tank Pressure (FTP) sensor signal indicated less than 0.22v during the CCM test. **Possible Causes:** • FTP sensor signal circuit is shorted to VREF or power (B+) • FTP sensor ground circuit is open • FTP sensor is damaged or has failed • PCM has failed

DTC	Trouble Code Title, Conditions & Possible Causes
DTC: P0455 **2T CCM, MIL: YES** **1999, 2000** **Models:** MX-5 Miata **Engines:** All **Transmissions:** A/T, M/T	**EVAP System Large Leak (0.080") Detected** DTC P0443 not set, ECT sensor signal from 14-90°F, IAT sensor signal more than 14°F at startup, BARO sensor more than 72 kPa, fuel level from 15-85%, IAT sensor signal 14-131°F during testing, vehicle driven at 24-65 mph at an engine speed of 1000-4000 rpm, calculated engine load of 9-65% and a throttle angle from 3-31.6%, then with the CDCV (valve) closed, the PCM detected the fuel tank pressure was less than a threshold due to a blockage or a large leak (0.080") somewhere in the system during the EVAP Monitor test. **Possible Causes:** • Fuel filler cap loose, cross-threaded, incorrect part or damaged • Purge solenoid valve is damaged or has failed • Charcoal canister is loaded with fuel or moisture, or has failed • Canister drain cut valve (CDCV) is damaged or leaking • Tank pressure control valve (TPCV) is damaged or has failed • Vent cut valve is damaged, leaking or has failed • Catch tank or rollover valve is damaged, or fuel tank is leaking • Boost, ECT, MAP, MAF, VSS or TP sensor is out-of-calibration • Fuel tank level sensor is damaged or out-of-calibration • Fuel tank pressure sensor is damaged or out-of-calibration • Vapor line(s) damaged or leaking between the purge solenoid and intake manifold, CDCV and the canister, or between the TPCV and the tank, or the check valve and the rollover valve • PCM has failed
DTC: P0455 **2T CCM, MIL: YES** **2001, 2002, 2003, 2004, 2005** **Models:** Mazda3, Mazda6, MX-5 Miata **Engines:** All **Transmissions:** A/T, M/T	**EVAP System Large Leak (0.080") Detected** DTC P0443 not set, ECT sensor signal from 14-90°F, IAT sensor signal more than 14°F at startup, BARO sensor more than 72 kPa, fuel level from 15-85%, ECT sensor signal from 158-212°F and IAT sensor signal from 14-131°F during testing, vehicle driven at 24-65 mph at an engine speed of 1000-4000 rpm, calculated engine load of 9-65% and a throttle angle from 3-31.6%, then with the CDCV (valve) closed, the PCM detected the fuel tank pressure was less than a threshold due to a blockage or a large leak (0.080") somewhere in the system during the EVAP Monitor test. **Note: The target fuel tank pressure for this test is minus (-) 0.99 kPa.** **Possible Causes:** • Fuel filler cap loose, cross-threaded, incorrect part or damaged • Purge solenoid valve is damaged or has failed • Charcoal canister is loaded with fuel or moisture, or has failed • Canister drain cut valve (CDCV) is damaged or leaking • Fuel tank pressure (FTP) sensor is damaged or has failed • Catch tank or rollover valve is damaged, or fuel tank is leaking • Vapor line(s) damaged or leaking between the purge solenoid and intake manifold, purge solenoid and canister, or between the pressure control valve, check valve and rollover valve • PCM has failed
DTC: P0455 **2T CCM, MIL: YES** 1998, 1999 **Models:** 626, Millenia **Engines:** All **Transmissions:** A/T, M/T	**EVAP System Large Leak (0.080") Detected** DTC P0443 not set, ECT sensor signal from 14-90°F, IAT sensor signal more than 14°F at startup, BARO sensor more than 72 kPa, fuel level from 15-85%, ECT sensor signal from 158-212°F and IAT sensor signal from 14-140°F during testing, vehicle driven at 25-65 mph at an engine speed of 1000-4000 rpm, calculated engine load of 9-65% and a throttle angle from 3-12.5%, then with the purge solenoid opened, the PCM detected the fuel tank pressure was too high right after the CDCV was closed due to a large leak (0.080") or blockage somewhere in the system during the EVAP Monitor test. **Possible Causes:** • Fuel filler cap loose, cross-threaded, incorrect part or damaged • Charcoal canister is loaded with fuel or moisture, or has failed • Canister drain cut valve (CDCV) is damaged or leaking • Fuel tank pressure (FTP) sensor is damaged or has failed • Fuel level sensor is damaged or out-of-calibration • Purge solenoid valve is damaged or has failed • Vent cut valve is damaged, is leaking or has failed • IAT, ECT, MAF, MAP, VSS or TP sensor is out-of-calibration • Catch tank or rollover valve is damaged, or fuel tank is leaking • Vapor line(s) damaged or leaking between the purge solenoid and intake manifold, purge solenoid and canister, or between the CDCV and vent cut valve, check valve and rollover valve • PCM has failed

DTC	Trouble Code Title, Conditions & Possible Causes
DTC: P0455 **2T CCM, MIL: YES** **2000, 2001, 2002, 2003** **Models:** Millenia **Engines:** All **Transmissions:** A/T, M/T	**EVAP System Large Leak (0.080") Detected** DTC P0443 not set, ECT sensor signal from 14-90ºF, IAT sensor signal more than 14ºF at startup, BARO sensor more than 72 kPa, fuel level from 15-85%, ECT sensor signal from 158-212ºF and IAT sensor signal from 14-131ºF during testing, vehicle driven at 25-65 mph at an engine speed of 1000-3000 rpm, calculated engine load of 9-70% and a throttle angle from 2.9-44%, then with the CDCV (valve) closed, the PCM detected the fuel tank pressure was less than a threshold due to a blockage or a large leak (0.080") somewhere in the system during the EVAP Monitor test. **Note: The fuel tank pressure for this test is minus (-) 2.16 kPa.** **Possible Causes:** • Fuel filler cap loose, cross-threaded, incorrect part or damaged • Charcoal canister is loaded with fuel or moisture, or has failed • Canister drain cut valve (CDCV) is damaged or leaking • Pressure control valve is damaged or leaking • Purge solenoid valve is damaged, leaking or has failed • Fuel tank pressure (FTP) sensor is damaged or has failed • Catch tank or rollover valve is damaged, or fuel tank is leaking • EGR Boost, ECT, MAF, VSS or TP sensor is out-of-calibration • Vapor line(s) damaged or leaking between the purge solenoid, intake manifold, tank pressure control valve and rollover valve • PCM has failed
DTC: P0455 **2T CCM, MIL: YES** **1998, 1999, 2000** **Models:** Protégé **Engines:** All **Transmissions:** A/T, M/T	**EVAP System Large Leak (0.080") Detected** DTC P0443 not set, ECT sensor signal from 14-90ºF, IAT sensor signal more than 14ºF at startup, BARO sensor more than 72 kPa, fuel level from 15-85%, ECT sensor signal from 158-212ºF and IAT sensor signal from 14-140ºF during testing, vehicle driven at 25-65 mph at an engine speed of 1000-4000 rpm, calculated engine load of 9-65% and a throttle angle from 3-12.5%, then with the purge solenoid opened, the PCM detected the fuel tank pressure was too high right after the CDCV was closed due to a large leak (0.080") or blockage somewhere in the system during the EVAP Monitor test. **Note: The target fuel tank pressure for this test is minus (-) 1.81 kPa.** **Possible Causes:** • Fuel filler cap loose, cross-threaded, incorrect part or damaged • Charcoal canister is loaded with fuel or moisture, or has failed • Canister drain cut valve (CDCV) is damaged or leaking • Fuel tank pressure (FTP) sensor is damaged or has failed • Purge solenoid valve is damaged or has failed • Tank pressure control valve (TPCV) is damaged or has failed • Catch tank or rollover valve is damaged, or fuel tank is leaking • Vapor line(s) damaged or leaking between the purge solenoid and intake manifold, purge solenoid and canister, or between the pressure control valve, check valve and rollover valve • PCM has failed
DTC: P0455 **2T CCM, MIL: YES** **2000, 2001, 2002, 2003** **Models:** Protégé, Protégé5 **Engines:** All **Transmissions:** A/T, M/T	**EVAP System Large Leak (0.080") Detected** DTC P0443 not set, ECT sensor signal from 14-90ºF, IAT sensor signal more than 14ºF at startup, BARO sensor more than 72 kPa, fuel level from 15-85%, ECT sensor signal from 158-212ºF and IAT sensor signal from 14-131ºF during testing, vehicle driven at 24-75 mph at an engine speed of 1000-4000 rpm, calculated engine load of 9-65% and a throttle angle from 3-12.5%, then with the CDCV (valve) closed, the PCM detected the fuel tank pressure was less than a threshold due to a blockage or a large leak (0.080") somewhere in the system during the EVAP Monitor test. **Note: The fuel tank pressure for this test is minus (-) 1.3 – 1.95 kPa.** **Possible Causes:** • Fuel filler cap loose, cross-threaded, incorrect part or damaged • Charcoal canister is loaded with fuel or moisture, or has failed • Canister drain cut valve (CDCV) is damaged or leaking • Purge solenoid valve is damaged, leaking or has failed • Fuel tank pressure (FTP) sensor is damaged or has failed • Catch tank or rollover valve is damaged, or fuel tank is leaking • EGR Boost, ECT, MAF, VSS or TP sensor is out-of-calibration • Vapor line(s) damaged or leaking between the purge solenoid, intake manifold, tank pressure control valve and rollover valve • PCM has failed

DTC	Trouble Code Title, Conditions & Possible Causes
DTC: P0455 **2T CCM, MIL: YES** **1999, 2000** **Models:** MPV **Engines:** All **Transmissions:** A/T, M/T	**EVAP System Large Leak (0.080") Detected** DTC P0443 not set, ECT sensor signal from 14-90°F, IAT sensor signal more than 14°F at startup, BARO sensor more than 72 kPa, fuel level from 15-85%, IAT sensor signal 14-131°F during testing, vehicle driven at 24-65 mph at an engine speed of 1000-4000 rpm, calculated engine load of 9-65% and a throttle angle from 3-31.6%, then with the CDCV (valve) closed, the PCM detected the fuel tank pressure was less than a threshold due to a blockage or a large leak (0.080") somewhere in the system during the EVAP Monitor test. **Possible Causes:** • Fuel filler cap loose, cross-threaded, incorrect part or damaged • Purge solenoid valve is damaged or has failed • Charcoal canister is loaded with fuel or moisture, or has failed • Canister drain cut valve (CDCV) is damaged or leaking • Tank pressure control valve (TPCV) is damaged or has failed • Vent cut valve is damaged, leaking or has failed • Catch tank or rollover valve is damaged, or fuel tank is leaking • Boost, ECT, MAP, MAF, VSS or TP sensor is out-of-calibration • Fuel tank level sensor is damaged or out-of-calibration • Fuel tank pressure sensor is damaged or out-of-calibration • Vapor line(s) damaged or leaking between the purge solenoid and intake manifold, CDCV and the canister, or between the TPCV and the tank, or the check valve and the rollover valve • PCM has failed
DTC: P0455 **2T CCM, MIL: YES** **2001, 2002, 2003** **Models:** MPV **Engines:** All **Transmissions:** A/T, M/T	**EVAP System Large Leak (0.080") Detected** DTC P0443 not set, ECT sensor signal from 14-90°F and IAT sensor signal more than 14°F at startup, BARO sensor more than 72 kPa, fuel level from 15-85%, IAT sensor signal 14-131°F during testing, vehicle driven at 25-80 mph at an engine speed of 1100-3400 rpm, calculated engine load at 7-80% and throttle angle from 3-12.5%, then with the purge solenoid closed, the PCM detected the fuel tank pressure was less than a threshold due to a blockage or a large leak (0.080") somewhere in the system during the EVAP Monitor test. **Note: The target fuel tank pressure for this test is minus (-) 2.16 kPa.** **Possible Causes:** • Fuel filler cap loose, cross-threaded, incorrect part or damaged • EVAP purge solenoid is damaged or has failed • EVAP charcoal canister is loaded with fuel or moisture • Canister drain cut valve (CDCV) is damaged or leaking • EVAP catch tank or rollover valve is damaged or leaking • Vapor line(s) damaged or leaking between the purge solenoid and intake manifold, purge solenoid and canister, or between the pressure control valve, check valve and rollover valve • PCM has failed
DTC: P0455 **2T CCM, MIL: YES** **2005** **Models:** RX-8 **Engines:** All **Transmissions:** A/T, M/T	**EVAP System Large Leak (0.080") Detected** DTC P0443 not set, ECT sensor signal from 14-90°F, IAT sensor signal more than 14°F at startup, BARO sensor more than 75 kPa, fuel level from 15-85%, IAT sensor signal 14-131°F during testing, vehicle driven at 25-80 mph at an engine speed of 1100-3400 rpm, calculated engine load of 7-80% and a throttle angle from 3-12.5%, and with the purge and vent solenoids closed, the PCM detected the FTP value was minus (-) 7.0" H2O due to a leak (0.080") or blockage in the system for 30 seconds in the EVAP Flow/Running Loss test. **Note: The target fuel tank pressure for this test is minus (-) 1.74 kPa.** **Possible Causes:** • Fuel filler cap loose, cross-threaded, incorrect part or damaged • Charcoal canister is loaded with fuel or moisture • Purge solenoid may be stuck mechanically in "closed" position • Canister vent (CV) solenoid is stuck open or may be damaged • Fuel vapor control valve tube assembly is restricted or blocked • Fuel vapor vent valve assembly is restricted or blocked • Fuel pressure sensor has failed (i.e., it failed mechanically) • Vapor tube(s) loose, disconnected or blocked to the canister tube, EVAP canister purge outlet tube or the EVAP return tube • PCM has failed

DTC	Trouble Code Title, Conditions & Possible Causes
DTC: P0455 **2T CCM, MIL: YES** **1999, 2000, 2001, 2002, 2003, 2004, 2005** **Models:** B2300, B2500, B3000, B4000, Tribute **Engines:** All **Transmissions:** A/T, M/T	**EVAP System Large Leak (0.080") Detected** DTC P0443 not set, ECT sensor signal from 14-90°F, IAT sensor signal more than 14°F at startup, BARO sensor more than 75 kPa, fuel level from 15-85%, IAT sensor signal 14-131°F during testing, vehicle driven at 25-80 mph at an engine speed of 1100-3400 rpm, calculated engine load of 7-80% and a throttle angle from 3-12.5%, and with the purge and vent solenoids closed, the PCM detected the FTP value was minus (-) 7.0" H2O due to a leak (0.080") or blockage in the system for 30 seconds in the EVAP Flow/Running Loss test. **Note: The target fuel tank pressure for this test is minus (-) 1.74 kPa.** **Possible Causes:** • Fuel filler cap loose, cross-threaded, incorrect part or damaged • Charcoal canister is loaded with fuel or moisture • Purge solenoid may be stuck mechanically in "closed" position • Canister vent (CV) solenoid is stuck open or may be damaged • Fuel vapor control valve tube assembly is restricted or blocked • Fuel vapor vent valve assembly is restricted or blocked • Fuel pressure sensor has failed (i.e., it failed mechanically) • Vapor tube(s) loose, disconnected or blocked to the canister tube, EVAP canister purge outlet tube or the EVAP return tube • PCM has failed
DTC: P0456 **2T CCM, MIL: YES** **2000, 2001, 2002, 2003, 2004, 2005** **Models:** B2500, B3000, B4000, MPV, RX-8 **Engines:** All **Transmissions:** A/T, M/T	**EVAP System Small Leak (0.020") Detected** DTC P0443 not set, ECT sensor signal from 14-90°F, IAT sensor signal more than 14°F at startup, engine speed at 1000-3000 rpm, VSS input from 25-63 mph, calculated load at 9-70%, fuel level over 75%, TP angle from 3-44%, and the PCM detected less than 2.5" H2O bleed-up over a 15 second period during the EVAP leak test. **Note: The vapor generation limit is over 2.5" H2O in 120 seconds.** **Possible Causes:** • Fuel filler cap loose, cross-threaded, incorrect part or damaged • Small holes or cuts in any of the vapor hoses and/or tubes • CV solenoid stuck part-way open while it is commanded closed • EVAP system component seals leaking at the purge valve, FTP sensor, CV solenoid, fuel vapor control valve or vapor vent tube • Fuel vapor hose or tube connections loose at the component • PCM has failed
DTC: P0456 **2T CCM, MIL: YES** **Mazda3, 2004, 2005** **Models:** Tribute **Engines:** All **Transmissions:** A/T, M/T	**EVAP System Small Leak (0.020") Detected** DTC P0443 not set, ECT sensor signal from 14-90°F, IAT sensor signal more than 14°F at startup, engine speed at 1000-3000 rpm, VSS input from 25-63 mph, calculated load at 9-70%, fuel level over 75%, TP angle from 3-44%, and the PCM detected less than 2.5" H2O bleed-up over a 15 second period during the EVAP leak test. **Note: The vapor generation limit is over 2.5" H2O in 120 seconds.** **Possible Causes:** • Fuel filler cap loose, cross-threaded, incorrect part or damaged • Small holes or cuts in any of the vapor hoses and/or tubes • CV solenoid stuck part-way open while it is commanded closed • EVAP system component seals leaking at the purge valve, FTP sensor, CV solenoid, fuel vapor control valve or vapor vent tube • Fuel vapor hose or tube connections loose at the component • PCM has failed
DTC: P0456 **2T CCM, MIL: YES** **2004, 2005** **Models:** Mazda3, Mazda6, MX-5 Miata **Engines:** All **Transmissions:** A/T, M/T	**EVAP System Small Leak (0.020") Detected** DTC P0443 not set, ECT sensor signal from 14-90°F, IAT sensor signal more than 14°F at startup, engine speed at 1000-3000 rpm, VSS input from 25-63 mph, calculated load at 9-70%, fuel level over 75%, TP angle from 3-44%, and the PCM detected less than 2.5" H2O bleed-up over a 15 second period during the EVAP leak test. **Note: The vapor generation limit is over 2.5" H2O in 120 seconds.** **Possible Causes:** • Fuel filler cap loose, cross-threaded, incorrect part or damaged • Small holes or cuts in any of the vapor hoses and/or tubes • CV solenoid stuck part-way open while it is commanded closed • EVAP system component seals leaking at the purge valve, FTP sensor, CV solenoid, fuel vapor control valve or vapor vent tube • Fuel vapor hose or tube connections loose at the component • PCM has failed

DTC	Trouble Code Title, Conditions & Possible Causes
DTC: P0457 **2T CCM, MIL: YES** **2000, 2001, 2002, 2003, 2004, 2005** **Models:** B2300, B2500, B3000, B4000, Tribute **Engines:** All **Transmissions:** A/T, M/T	**EVAP System Gross Leak Detected** Engine running, and immediately after a vehicle "refueling" event, the PCM detected it could not achieve any initial vacuum in the EVAP system with excessive vapor flow present (Gross EVAP leak). **Possible Causes:** • Fuel filler cap does not fit properly (it is the wrong part number) • Fuel filler cap is missing
DTC: P0460 **2T CCM, MIL: NO** **1998, 1999, 2000, 2001, 2002** **Models:** 626 **Engines:** All **Transmissions:** A/T, M/T	**Fuel Level Indicator Signal Circuit Malfunction** Engine running, and the PCM detected the Fuel Level Indicator (FLI) signal indicated less than 0.1v, or that it indicated more than 3.48v during the CCM test period. **Possible Causes:** • FLI signal circuit is open, shorted to ground or to power (B+) • Fuel tank empty or overfull (FP module is stuck mechanically) • Wrong fuel gauge is installed, or instrument panel is damaged • Fuel gauge sender unit is damaged or has failed • PCM has failed
DTC: P0460 **2T CCM, MIL: YES** **1999, 2000, 2001, 2002, 2003, 2004, 2005** **Models:** B2300, B2500, B3000, B4000, Tribute **Engines:** All **Transmissions:** A/T, M/T	**Fuel Level Indicator Signal Circuit Malfunction** Engine running, and the PCM detected the Fuel Level Indicator (FLI) signal indicated less than 0.1v, or that it indicated more than 3.48v during the CCM test period. **Possible Causes:** • FLI signal circuit is open, shorted to ground or to power (B+) • Fuel tank empty or overfull (FP module is stuck mechanically) • Wrong fuel gauge is installed, or instrument panel is damaged • Fuel gauge sender unit is damaged or has failed • PCM has failed
DTC: P0461 **2T CCM, MIL: YES** **2000, 2001, 2002, 2003, 2004, 2005** **Models:** B2300, B2500, B3000, B4000, Tribute **Engines:** All **Transmissions:** A/T, M/T	**Fuel Level Indicator Signal Circuit Range/Performance** Engine running, and the PCM detected the Fuel Level Indicator (FLI) signal indicated the circuit was noisy. **Possible Causes:** • FLI signal circuit is open or shorted to ground (intermittent fault) • FLI assembly is damaged or has failed • PCM has failed
DTC: P0461 **2T CCM, MIL: YES** **2000, 2001, 2002** **Models:** 626 **Engines:** All **Transmissions:** A/T, M/T	**Fuel Level Indicator Signal Circuit Range/Performance** Engine running and the PCM detected the Fuel Level Indicator (FLI) signal by more than 0.8v for 25 seconds during the CCM test period. **Note: The PCM monitors the FLI signal once per second.** **Possible Causes:** • FLI signal circuit is open, shorted to ground or to power (B+) • Fuel tank empty or overfull (FP module is stuck mechanically) • Wrong fuel gauge is installed, or instrument panel is damaged • Fuel gauge sender unit is damaged or has failed • PCM has failed • TSB 0101000 (6/01) contains information related to this code
DTC: P0461 **2T CCM, MIL: YES** **1999, 2000, 2001, 2002, 2003, 2004, 2005** **Models:** Mada3, Mazda6, MX-5 Miata, RX-8 **Engines:** All **Transmissions:** A/T, M/T	**Fuel Tank Level Sensor Circuit Range/Performance** Engine running and the PCM determined the fuel gauge sender unit signal was operating in too narrow a range during the CCM test. **Possible Causes:** • Fuel gauge sending unit is damaged or has failed • Fuel gauge sending unit signal is open or shorted to ground • Fuel gauge sending unit power circuit is open • Instrument cluster is damaged or has failed • PCM has failed

DTC	Trouble Code Title, Conditions & Possible Causes
DTC: P0461 **2T CCM, MIL: YES** **2000, 2001, 2002** **Models:** MPV **Engines:** All **Transmissions:** A/T, M/T	**Fuel Level Indicator Signal Circuit Range/Performance** Engine running for several minutes, and the PCM determined the FLI sensor signal reflected less 5% of what the PCM calculated after it had determined that the consumption volume reached 17.5 liters. **Possible Causes:** • Fuel gauge sending unit is damaged or has failed • Fuel gauge sending unit signal is open or shorted to ground • Fuel gauge sending unit power circuit is open • Fuel gauge sender signal is shorted to fuel pump terminal • PCM has failed
DTC: P0461 **2T CCM, MIL: YES** **1999, 2000, 2001, 2002, 2003** **Models:** Protégé, Protégé5 **Engines:** All **Transmissions:** A/T, M/T	**Fuel Tank Level Sensor Circuit Malfunction** Engine running for several minutes, and the PCM determined the FLI sensor signal reflected less than the amount the PCM calculated after it had determined that fuel consumption reached 17.5 liters. **Possible Causes:** • Fuel gauge sending unit is damaged or has failed • Fuel gauge sending unit signal is open or shorted to ground • Fuel gauge sending unit power circuit is open • Fuel gauge sender signal is shorted to fuel pump terminal • PCM has failed
DTC: P0461 **2T CCM, MIL: YES** **2004, 2005** **Models:** Mazda6 **Engines:** All **Transmissions:** A/T, M/T	**Fuel Tank Level Sensor Circuit Malfunction** Engine running for several minutes, and the PCM determined the FLI sensor signal reflected less than the amount the PCM calculated after it had determined that fuel consumption reached 17.5 liters. **Possible Causes:** • Fuel gauge sending unit is damaged or has failed • Fuel gauge sending unit signal is open or shorted to ground • Fuel gauge sending unit power circuit is open • Fuel gauge sender signal is shorted to fuel pump terminal • PCM has failed
DTC: P0462 **2T CCM, MIL: YES** **1999, 2000, 2001, 2002, 2003, 2004, 2005** **Models:** Mazda3, Mazda6, MX-5 Miata, MPV, Protégé, Protégé5, RX-8 **Engines:** All **Transmissions:** A/T, M/T	**Fuel Level Sensor Circuit Low Input** Engine running, system voltage from 11-16v, and the PCM detected the fuel level sensor signal was less than 0.08v during the CCM test. **Possible Causes:** • Fuel gauge sending unit signal is shorted to sensor ground • Fuel gauge sending unit signal is shorted to chassis ground • Fuel gauge sending unit is damaged or has failed • PCM has failed
DTC: P0462 **2T CCM, MIL: YES** **2005** **Models:** Tribute **Engines:** All **Transmissions:** A/T, M/T	**Fuel Level Sensor Circuit Low Input** Engine running, system voltage from 11-16v, and the PCM detected the fuel level sensor signal was less than 0.08v during the CCM test. **Possible Causes:** • Fuel gauge sending unit signal is shorted to sensor ground • Fuel gauge sending unit signal is shorted to chassis ground • Fuel gauge sending unit is damaged or has failed • PCM has failed
DTC: P0463 **2T CCM, MIL: YES** **1999, 2000, 2001, 2002, 2003, 2004, 2005** **Models:** Mazda3, Mazda6, MX-5 Miata, MPV, Protégé, Protégé5 **Engines:** All **Transmissions:** A/T, M/T	**Fuel Level Sensor Circuit High Input** Engine running, system voltage from 11-16v, and the PCM detected the fuel level sensor signal was more than 4.92v during the CCM test. **Possible Causes:** • Fuel gauge sending unit signal shorted to VREF or power • Fuel gauge sending unit ground circuit is open • Fuel gauge sender signal is damaged or has failed • PCM has failed
DTC: P0463 **2T CCM, MIL: YES** **2005** **Models:** RX-8, Tribute **Engines:** All **Transmissions:** A/T, M/T	**Fuel Level Sensor Circuit High Input** Engine running, system voltage from 11-16v, and the PCM detected the fuel level sensor signal was more than 4.92v during the CCM test. **Possible Causes:** • Fuel gauge sending unit signal shorted to VREF or power • Fuel gauge sending unit ground circuit is open • Fuel gauge sender signal is damaged or has failed • PCM has failed

DTC	Trouble Code Title, Conditions & Possible Causes
DTC: P0464 **2T CCM, MIL: YES** **1999, 2000, 2001, 2002, 2003** **Models:** Protégé, Protégé5 **Engines:** All **Transmissions:** A/T, M/T	**Fuel Level Indicator Sensor Circuit Malfunction** Engine running and the PCM detected the minimum and maximum Fuel Level Indicator (FLI) signal varied by more than 30 immediately after engine startup. **Possible Causes:** • Fuel gauge sending unit is damaged or has failed • Fuel gauge sending unit signal is open or shorted to ground • PCM has failed
DTC: P0464 **2T CCM, MIL: YES** **1999, 2000, 2001, 2002, 2003** **Models:** Millenia, MPV **Engines:** All **Transmissions:** A/T, M/T	**Fuel Tank Level Sensor Circuit Malfunction** DTC P1445 not set, engine runtime over 5 minutes, system voltage is more than 10v, vehicle is stopped, and the PCM detected the Fuel Level sensor signal did not flatten out under these condition for 14 seconds (this test is part of the Fuel Slosh Check for the EVAP test). **Possible Causes:** • Fuel gauge sending unit is damaged or has failed • Fuel gauge sending unit signal is open or shorted to ground • PCM has failed
DTC: P0470 **2T CCM, MIL: YES** **1996, 1997, 1998** **Models:** MX-5 Miata **Engines:** All **Transmissions:** A/T, M/T	**Exhaust Pressure Sensor Circuit Malfunction** Engine running, IAT sensor signal more than 50°F, Idle Air Control system operating in the feedback zone with the difference in the intake manifold pressure and the barometric pressure less than 4.45 kPa (33.4 mmHg or 1.31" Hg), and the PCM detected the EGR Boost sensor signal was less than 0.98v, or that it was more than 4.88 during the CCM test. **Possible Causes:** • EGR boost sensor vacuum hose is loose, damage or clogged • EGR boost sensor signal circuit is open or shorted to ground • EGR boost sensor signal circuit is shorted to VREF • EGR boost sensor is damaged or has failed • PCM has failed
DTC: P0480 **2T CCM, MIL: NO** **2000, 2001** **Models:** Millenia **Engines:** All **Transmissions:** A/T, M/T	**Condenser Fan Relay 1 Control Circuit Malfunction** Key on or engine running and the PCM detected an unexpected voltage condition on the Condenser Fan Relay 1 control circuit with the Air Conditioning turned "on" or turned "off" during the CCM test. **Possible Causes:** • Condenser fan relay 1 control circuit open or shorted to ground • Condenser fan relay 1 control power circuit is open • Condenser fan relay 1 is damaged or has failed • PCM has failed
DTC: P0480 **2T CCM, MIL: NO** **2000, 2001, 2002, 2003, 2004, 2005** **Models:** Mazda3, Mazda6, MX-5 Miata, Protégé, Protégé5, RX-8 **Engines:** All **Transmissions:** A/T, M/T	**Condenser Fan Relay Control Circuit Malfunction** Key on or engine running and the PCM detected an unexpected high or low voltage condition on the Condenser Fan Relay control circuit during the CCM test. **Possible Causes:** • Condenser fan relay control circuit open or shorted to ground • Condenser fan relay control power circuit is open • Condenser fan relay is damaged or has failed • PCM has failed
DTC: P0481 **2T CCM, MIL: NO** **2000, 2001** **Models:** Millenia, MX-5, Miata **Engines:** All **Transmissions:** A/T, M/T	**Cooling Fan Relay 1 and Condenser Fan Relay 1 Control Circuit Malfunction** Key on, and the PCM detected an unexpected high or low voltage on the Cooling Fan Relay 1 or Condenser Fan Relay 1 control circuit. **Possible Causes:** • Cooling or condenser fan relay 1 circuit open, shorted to ground • Cooling or condenser fan relay 1 control power circuit is open • Cooling or condenser fan relay 1 is damaged or has failed • PCM has failed
DTC: P0481 **2T CCM, MIL: NO** **2003, 2004, 2005** **Models:** Mazda6 **Engines:** All **Transmissions:** A/T, M/T	**Cooling Fan Relay 1 and Condenser Fan Relay 1 Control Circuit Malfunction** Key on, and the PCM detected an unexpected high or low voltage on the Cooling Fan Relay 1 or Condenser Fan Relay 1 control circuit. **Possible Causes:** • Cooling or condenser fan relay 1 circuit open, shorted to ground • Cooling or condenser fan relay 1 control power circuit is open • Cooling or condenser fan relay 1 is damaged or has failed • PCM has failed

DTC	Trouble Code Title, Conditions & Possible Causes
DTC: P0481 **2T CCM, MIL: NO** **2004, 2005** **Models:** RX-8, Tribute **Engines:** All **Transmissions:** A/T, M/T	**Cooling Fan Relay 1 and Condenser Fan Relay 1 Control Circuit Malfunction** Key on, and the PCM detected an unexpected high or low voltage on the Cooling Fan Relay 1 or Condenser Fan Relay 1 control circuit. **Possible Causes:** • Cooling or condenser fan relay 1 circuit open, shorted to ground • Cooling or condenser fan relay 1 control power circuit is open • Cooling or condenser fan relay 1 is damaged or has failed • PCM has failed
DTC: P0482 **2T CCM, MIL: NO** **2000, 2001** **Models:** Millenia, MX-5 Miata **Engines:** All **Transmissions:** A/T, M/T	**Cooling Fan Relay and Condenser Fan Relay 2 Control Circuit Malfunction** Key on, and the PCM detected an unexpected high or low voltage on the Cooling Fan Relay or Condenser Fan Relay 2 control circuit. **Possible Causes:** • Cooling or condenser fan relay 2 circuit open, shorted to ground • Cooling or condenser fan relay 2 control power circuit is open • Cooling or condenser fan relay 1 is damaged or has failed • PCM has failed
DTC: P0482 **2T CCM, MIL: NO** **2003, 2004, 2005** **Models:** Mazda6 **Engines:** All **Transmissions:** A/T, M/T	**Cooling Fan Relay and Condenser Fan Relay 2 Control Circuit Malfunction** Key on, and the PCM detected an unexpected high or low voltage on the Cooling Fan Relay or Condenser Fan Relay 2 control circuit. **Possible Causes:** • Cooling or condenser fan relay 2 circuit open, shorted to ground • Cooling or condenser fan relay 2 control power circuit is open • Cooling or condenser fan relay 1 is damaged or has failed • PCM has failed
DTC: P0482 **2T CCM, MIL: NO** **2005** **Models:** Tribute **Engines:** All **Transmissions:** A/T, M/T	**Cooling Fan Relay and Condenser Fan Relay 2 Control Circuit Malfunction** Key on, and the PCM detected an unexpected high or low voltage on the Cooling Fan Relay or Condenser Fan Relay 2 control circuit. **Possible Causes:** • Cooling or condenser fan relay 2 circuit open, shorted to ground • Cooling or condenser fan relay 2 control power circuit is open • Cooling or condenser fan relay 1 is damaged or has failed • PCM has failed
DTC: P0500 **2T CCM, MIL: YES** **1996, 1997, 1998, 1999, 2000** **Models:** 626 **Engines:** All **Transmissions:** A/T	**Vehicle Speed Sensor Circuit Malfunction** Vehicle driven in Drive, 2nd or Low gear position, engine speed over 2000 rpm with the charging efficiency over 40%, and the PCM detected the VSS input indicated less than 2.34 mph for 33 seconds. **Possible Causes:** • VSS positive (+) signal circuit is open or shorted to ground • VSS negative (-) signal circuit is open or shorted to ground • VSS is damaged or has failed • PCM has failed
DTC: P0500 **2T CCM, MIL: YES** **1996, 1997, 1998, 1999, 2000** **Models:** 626 **Engines:** All **Transmissions:** M/T	**Vehicle Speed Sensor Circuit Malfunction** Vehicle driven in gear with the clutch released, engine speed over 2000 rpm with the charging efficiency over 20%, and the PCM detected the VSS input indicated less than 2.34 mph for 33 seconds. **Possible Causes:** • VSS signal circuit is open, shorted to ground or to power • VSS power circuit is open or shorted to ground • VSS is damaged or has failed • PCM has failed
DTC: P0500 **2T CCM, MIL: YES** **1997, 1998, 1999, 2000, 2001, 2002** **Models:** Millenia **Engines:** All **Transmissions:** A/T, M/T	**Vehicle Speed Sensor Circuit Malfunction** Vehicle driven with the Drive, 2nd or Low Transaxle Range switch indicating "on", ECT sensor signal more than 140°F, engine speed over 1810 rpm, and the PCM did not detect any VSS signals. **Possible Causes:** • VSS signal circuit is open, shorted to ground or to power • VSS power circuit is open or shorted to ground • VSS is damaged or has failed • PCM has failed

DTC	Trouble Code Title, Conditions & Possible Causes
DTC: P0500 **2T CCM, MIL: YES** **1996, 1997, 1998, 1999, 2000, 2001, 2002, 2003, 2004, 2005** **Models:** Mazda3, Mazda6, MX-5 Miata, RX-8 **Engines:** All **Transmissions:** A/T	**Vehicle Speed Sensor Circuit Malfunction** Vehicle driven in Drive, 2nd or Low gear position, engine speed over 2000 rpm with the charging efficiency over 40%, and the PCM detected the VSS input indicated less than 2.34 mph for 33 seconds. **Possible Causes:** • VSS signal circuit is open, shorted to ground or to power • VSS power circuit is open or shorted to ground • VSS is damaged or has failed • PCM has failed
DTC: P0500 **2T CCM, MIL: YES** **1996, 1997, 1998, 1999, 2000, 2001, 2002, 2003** **Models:** MX-5 Miata **Engines:** All **Transmissions:** M/T	**Vehicle Speed Sensor Circuit Malfunction** Vehicle driven in gear with the clutch released, engine speed over 2000 rpm with the charging efficiency over 20%, and the PCM detected the VSS input indicated less than 2.34 mph for 33 seconds. **Possible Causes:** • VSS signal circuit is open, shorted to ground or to power • VSS power circuit is open or shorted to ground • VSS is damaged or has failed • PCM has failed
DTC: P0500 **2T CCM, MIL: YES** **1996, 1997, 1998, 1999** **Models:** Protégé **Engines:** All **Transmissions:** A/T	**Vehicle Speed Sensor Circuit Malfunction** Engine speed over 2000 rpm with the charging efficiency over 40%, and the PCM detected a VSS signal that indicated less than 2.34 mph for a period of 33 seconds during the CCM test. **Possible Causes:** • VSS signal circuit is open, shorted to ground or to power • VSS power circuit is open or shorted to ground • VSS is damaged or has failed • PCM has failed
DTC: P0500 **2T CCM, MIL: YES** **1996, 1997, 1998, 1999** **Models:** Protégé **Engines:** All **Transmissions:** M/T	**Vehicle Speed Sensor Circuit Malfunction** Engine speed over 2000 rpm with the charging efficiency over 20%, and the PCM detected a VSS signal that indicated less than 2.34 mph for a period of 33 seconds during the CCM test. **Possible Causes:** • VSS signal circuit is open, shorted to ground or to power • VSS power circuit is open or shorted to ground • VSS is damaged or has failed • PCM has failed
DTC: P0500 **2T CCM, MIL: YES** **2004, 2005** **Models:** Tribute **Engines:** All **Transmissions:** A/T	**Vehicle Speed Sensor Circuit Malfunction** Vehicle driven with the Drive, 1st or 2nd Transaxle Range switch indicating "on" with the Neutral Range switch indicating "off", ECT sensor signal more than 140°F, Turbine Shaft Speed sensor more than 1500 rpm, and the PCM did not detect any VSS signals. **Possible Causes:** • VSS signal circuit is open, shorted to ground or to power • VSS power circuit is open or shorted to ground • VSS is damaged or has failed • PCM has failed
DTC: P0500 **2T CCM, MIL: YES** **2000, 2001, 2002, 2003** **Models:** Protégé, Protégé5 **Engines:** All **Transmissions:** M/T	**Vehicle Speed Sensor Circuit Malfunction** Vehicle driven with Gear Position switch indicating a position other than Neutral position, engine load more than 40%, engine speed over 2000 rpm, and the PCM did not detect any VSS signals. **Possible Causes:** • VSS signal circuit is open, shorted to ground or to power • VSS power circuit is open or shorted to ground • VSS is damaged or has failed • PCM has failed
DTC: P0500 **2T CCM, MIL: YES** **1996, 1997, 1998, 2000, 2001, 2002, 2003** **Models:** MPV **Engines:** All **Transmissions:** A/T	**Vehicle Speed Sensor Circuit Malfunction** Vehicle driven with the Drive, 1st or 2nd Transaxle Range switch indicating "on" with the Neutral Range switch indicating "off", ECT sensor signal more than 140°F, Turbine Shaft Speed sensor more than 1500 rpm, and the PCM did not detect any VSS signals. **Possible Causes:** • VSS signal circuit is open, shorted to ground or to power • VSS power circuit is open or shorted to ground • VSS is damaged or has failed • PCM has failed

DTC	Trouble Code Title, Conditions & Possible Causes
DTC: P0500 **2T CCM, MIL: YES** **1996, 1997** **Models:** B2300, B3000, B4000 **Engines:** All **Transmissions:** A/T	**Vehicle Speed Sensor Circuit Malfunction** Vehicle driven in gear at more than 1000 rpm, ECT sensor signal more than 140°F, and the PCM detected an unexpected voltage condition (always high, low or missing) on the VSS circuit. **Possible Causes:** • VSS signal circuit is open, shorted to ground or to power • VSS power circuit is open or shorted to ground • VSS is damaged or has failed • PCM has failed
DTC: P0500 **2T CCM, MIL: YES** **1998, 1999, 2000** **Models:** B2300, B2500, B3000, B4000 **Engines:** All **Transmissions:** A/T	**Vehicle Speed Sensor Circuit Malfunction** Vehicle driven in gear at more than 1000 rpm, ECT sensor signal more than 140°F, and the PCM detected an intermittent VSS signal received from the ABS, CTM or GEM module during the CCM test. **Possible Causes:** • ABS wheel speed sensor circuit open or shorted to ground • ABS wheel speed sensor circuit shorted to VREF or power • ABS, CTM or GEM control module has failed • VSS positive (+) signal circuit is open or shorted to ground • VSS negative (-) signal circuit is open or shorted to ground • PCM has failed
DTC: P0501 **2T CCM, MIL: NO** **1998, 1999, 2000** **Models:** B2300, B2500, B3000, B4000 **Engines:** All **Transmissions:** A/T	**Vehicle Speed Sensor Circuit Malfunction** Vehicle driven in gear at more than 1000 rpm, ECT sensor signal more than 140°F, and the PCM detected an invalid VSS signal from the ABS, CTM or GEM module during the CCM test. **Possible Causes:** • ABS wheel speed sensor circuit open or shorted to ground • ABS wheel speed sensor circuit shorted to VREF or power • ABS, CTM or GEM control module has failed • VSS positive (+) signal circuit is open or shorted to ground • VSS negative (-) signal circuit is open or shorted to ground • PCM has failed
DTC: P0501 **2T CCM, MIL: NO** **2004, 2005** **Models:** Tribute **Engines:** All **Transmissions:** A/T	**Vehicle Speed Sensor Circuit Malfunction** Vehicle driven in gear at more than 1000 rpm, ECT sensor signal more than 140°F, and the PCM detected an invalid VSS signal from the ABS, CTM or GEM module during the CCM test. **Possible Causes:** • ABS wheel speed sensor circuit open or shorted to ground • ABS wheel speed sensor circuit shorted to VREF or power • ABS, CTM or GEM control module has failed • VSS positive (+) signal circuit is open or shorted to ground • VSS negative (-) signal circuit is open or shorted to ground • PCM has failed
DTC: P0503 **2T CCM, MIL: YES** **1996, 1997** **Models:** B2300, B3000, B4000 **Engines:** All **Transmissions:** A/T	**Vehicle Speed Sensor Circuit Malfunction** Vehicle driven in gear at more than 1000 rpm, ECT sensor signal more than 140°F, and the PCM detected an unexpected voltage condition (the signal was intermittent) on the VSS circuit. **Possible Causes:** • VSS signal circuit is open, shorted to ground or to power • VSS power circuit is open or shorted to ground • VSS is damaged or has failed • PCM has failed
DTC: P0503 **2T CCM, MIL: YES** **1998, 1999, 2000** **Models:** B2300, B2500, B3000, B4000 **Engines:** All **Transmissions:** A/T	**Vehicle Speed Sensor Circuit Malfunction** Vehicle driven in gear at more than 1000 rpm, ECT sensor signal more than 140°F, and the PCM detected a noisy or poor VSS signal received from the ABS, CTM or GEM module during the CCM test. **Possible Causes:** • ABS wheel speed sensor circuit open or shorted to ground • ABS wheel speed sensor circuit shorted to VREF or power • ABS, CTM or GEM control module has failed • VSS positive (+) signal circuit is open or shorted to ground • VSS negative (-) signal circuit is open or shorted to ground • PCM has failed

DTC	Trouble Code Title, Conditions & Possible Causes
DTC: P0620 **1T CCM, MIL: YES** **2004, 2005** **Models:** Tribute **Engines:** All **Transmissions:** A/T, M/T	**Regulator / Generator malfunction** PCM internal CPU malfunction. **Possible Causes:** • PCM internal malfunction • Configuration has not been properly completed.
DTC: P0622 **1T CCM, MIL: YES** **2004, 2005** **Models:** Tribute **Engines:** All **Transmissions:** A/T, M/T	**Control Module performance** PCM internal CPU malfunction. **Possible Causes:** • PCM internal malfunction • Configuration has not been properly completed.
DTC: P0638 **1T CCM, MIL: YES** **2003, 2004, 2005** **Models:** Mazda6, RX-8 **Engines:** All **Transmissions:** A/T, M/T	**Throttle Actuator control range performance** If the PCM detects that actual throttle angle opening is smaller or larger than the target throttle opening angle, the PCM determines that the throttle actuator control system has a malfunction **Possible Causes:** • PCM internal malfunction • Throttle Body malfunction
DTC: P0645 **1T CCM, MIL: YES** **2004, 2005** **Models:** Tribute **Engines:** All **Transmissions:** A/T, M/T	**A/C clutch relay (ACCR) malfunction** If the PCM detects excessive current draw when PCM grounds the circuit or voltage is not detected on the ACCR circuit when it is not grounded by the PCM, the PCM determines that the ACCR has a malfunction **Possible Causes:** • PCM internal malfunction • ACCR malfunction
DTC: P0660 **2T CCM, MIL: YES** **2001, 2002, 2003** **Models:** Protégé, Protégé5 **Engines:** All **Transmissions:** A/T, M/T	**Variable Inertia Charging System Circuit Malfunction** Key on or engine running and the PCM detected an unexpected voltage condition on the Variable Inertia Charging System Solenoid (VICS) circuit during the CCM test period. **Possible Causes:** • VICS control circuit is open or shorted to ground • VICS power circuit is open (check the power from the relay) • VICS valve is damaged or has failed • PCM has failed
DTC: P0660 **2T CCM, MIL: YES** **2004, 2005** **Models:** MPV, Tribute **Engines:** All **Transmissions:** A/T, M/T	**Variable Inertia Charging System Circuit Malfunction** Key on or engine running and the PCM detected an unexpected voltage condition on the Variable Inertia Charging System Solenoid (VICS) circuit during the CCM test period. **Possible Causes:** • VICS control circuit is open or shorted to ground • VICS power circuit is open (check the power from the relay) • VICS valve is damaged or has failed • PCM has failed
DTC: P0661 **2T CCM, MIL: NO** **2003, 2004, 2005** **Models:** Mazda6, MPV, RX-8 **Engines:** All **Transmissions:** A/T, M/T	**Variable intake-air system (VIS) control solenoid valve circuit low input** PCM monitors the VIS control solenoid valve signal at PCM terminal 4R. If PCM turns VIS control solenoid valve OFF but voltage at PCM terminal 4R remains low, PCM determines that VIS control solenoid valve circuit has a malfunction. **Possible Causes:** • VIS control valve malfunction • Connector or terminal malfunction • Open circuit between VIS and PSM • PCM has failed
DTC: P0662 **2T CCM, MIL: NO** **2003, 2004, 2005** **Models:** Mazda6, MPV, RX-8 **Engines:** All **Transmissions:** A/T, M/T	**Variable intake-air system (VIS) control solenoid valve circuit high input** PCM monitors the VIS control solenoid valve signal at PCM terminal 4R. If PCM turns VIS control solenoid valve OFF but voltage at PCM terminal 4R remains high, PCM determines that VIS control solenoid valve circuit has a malfunction. **Possible Causes:** • VIS control valve malfunction • Connector or terminal malfunction • Open circuit between VIS and PSM • PCM has failed

DTC	Trouble Code Title, Conditions & Possible Causes
DTC: P0703 **2T CCM, MIL: NO** **1996, 1997, 1998, 1999, 2000, 2001, 2002, 2003, 2004, 2005** **Models:** Mazda6, MX-6, 626 **Engines:** 2.0L VIN C **Transmissions:** A/T	**Brake On/Off Switch Circuit Malfunction (Self-Test)** KOEO Self-Test: Key on, and the PCM did not detect any change in the Brake Switch status after the brake pedal was pressed and released during the self-test. KOER Self-Test: and the PCM did not detect any change in the Brake Switch status after the brake pedal was pressed and released during the self-test. **Possible Causes:** • Brake switch signal circuit is open or shorted ground • Brake switch power circuit (B+) open between switch and PCM • Brake switch is damaged, misadjusted or installed improperly • PCM has failed
DTC: P0703 **2T CCM, MIL: NO** **1996, 1997, 1998** **Models:** MX-6, 626 **Engines:** 2.5L VIN D **Transmissions:** A/T	**Brake On/Off Switch Circuit Malfunction (Self-Test)** KOEO Self-Test: Key on, and the PCM did not detect any change in the Brake Switch status after the brake pedal was pressed and released during the self-test. KOER Self-Test: and the PCM did not detect any change in the Brake Switch status after the brake pedal was pressed and released during the self-test. **Possible Causes:** • Brake switch signal circuit is open or shorted ground • Brake switch power circuit open between switch and battery (+) • Brake switch is damaged, misadjusted or installed improperly • PCM has failed
DTC: P0703 **2T CCM, MIL: NO** **1999, 2000** **Models:** 626 **Engines:** 2.5L VIN D **Transmissions:** A/T	**Brake On/Off Switch Circuit Malfunction** Engine started, then after the vehicle accelerated to over 30 mph, and back to idle speed over 10 times, the PCM detected the Brake Switch signal did not change status (i.e., did not cycle "on" to "off"). **Possible Causes:** • Brake switch signal circuit is open or shorted ground • Brake switch power circuit open between switch and battery (+) • Brake switch is damaged, misadjusted or installed improperly, or noise filter has failed • PCM has failed
DTC: P0703 **2T CCM, MIL: NO** **1996, 1997, 1998, 1999, 2000, 2001, 2002, 2003** **Models:** MX-5 Miata **Engines:** All **Transmissions:** A/T	**Brake On/Off Switch Circuit Malfunction** Engine running in gear or in deceleration mode, and the PCM did not detect a TCC/Brake switch signal with the brake pedal depressed, or it detected a TCC/Brake switch signal when the brake pedal was not depressed, condition met for 33 seconds. **Possible Causes:** • Brake switch signal circuit is open or shorted ground • Brake switch power circuit (B+) open between switch and PCM • Brake switch is damaged, misadjusted or installed improperly • PCM has failed
DTC: P0703 **2T CCM, MIL: YES** **1996, 1997, 1998, 1999, 2000, 2001, 2002, 2003** **Models:** Millenia, MPV, Protégé, Protégé5, RX-8 **Engines:** All **Transmissions:** A/T	**Brake On/Off Switch Circuit Malfunction** Engine started, then the vehicle was accelerated from 0-19 mph and back to idle speed at least 10 times, and the PCM did not detect and change in the Brake Switch status (i.e., it did not cycle "on" to "off"). **Possible Causes:** • Brake switch signal circuit is open or shorted ground • Brake switch power circuit open between switch and battery (+) • Brake switch is damaged, misadjusted or installed improperly, or noise filter has failed • PCM has failed
DTC: P0703 **2T CCM, MIL: NO** **1996, 1997, 1998, 1999, 2000, 2001, 2002, 2003, 2004, 2005** **Models:** B2300, B2500, B3000, B4000 **Engines:** All **Transmissions:** A/T	**Brake On/Off Switch Circuit Malfunction (Self-Test)** **KOEO Self-Test:** Key on, and the PCM did not detect any change in the Brake Switch status after the brake pedal was pressed and released during the self-test. **KOER Self-Test:** and the PCM did not detect any change in the Brake Switch status after the brake pedal was pressed and released during the self-test. **Possible Causes:** • Brake switch signal circuit is open or shorted ground • Brake switch power circuit (B+) is open or shorted to ground • Brake switch is damaged, misadjusted or installed improperly • PCM has failed

DTC	Trouble Code Title, Conditions & Possible Causes
DTC: P0703 **2T CCM, MIL: NO** **2001, 2002, 2003, 2004, 2005** **Models:** B2300, B2500, B3000, B4000, Tribute **Engines:** All **Transmissions:** A/T	**Brake On/Off Switch Circuit Malfunction (Self-Test)** **KOEO Self-Test:** Key on, and the PCM did not detect any change in the Brake Switch status after the brake pedal was pressed and released during the self-test. **KOER Self-Test:** and the PCM did not detect any change in the Brake Switch status after the brake pedal was pressed and released during the self-test. **Possible Causes:** • Brake switch signal circuit is open or shorted ground • Brake switch power circuit (B+) is open or shorted to ground • Brake switch is damaged, misadjusted or installed improperly • PCM has failed
DTC: P0704 **2T CCM, MIL: NO** **1998, 1999, 2000, 2001, 2002** **Models:** 626 **Engines:** All **Transmissions:** M/T	**Clutch Pedal Switch Circuit Malfunction** Engine started, the after the vehicle was accelerated from 0-16 mph, the PCM detected the Clutch Pedal Position (CPP) switch input did not change status correctly under these conditions in the CCM test. **Possible Causes:** • Clutch switch signal circuit is open or shorted ground • Clutch switch power circuit (B+) open between switch and PCM • Clutch switch is damaged, misadjusted or installed improperly • PCM has failed
DTC: P0704 **2T CCM, MIL: NO** **2003, 2004, 2005** **Models:** Mazda6, RX-8 **Engines:** All **Transmissions:** M/T	**Clutch Pedal Switch Circuit Malfunction** Engine started, the after the vehicle was accelerated from 0-16 mph, the PCM detected the Clutch Pedal Position (CPP) switch input did not change status correctly under these conditions in the CCM test. **Possible Causes:** • Clutch switch signal circuit is open or shorted ground • Clutch switch power circuit (B+) open between switch and PCM • Clutch switch is damaged, misadjusted or installed improperly • PCM has failed
DTC: P0704 **2T CCM, MIL: NO** **1999, 2000, 2001, 2002, 2003** **Models:** Protégé, Protégé5 **Engines:** All **Transmissions:** M/T	**Clutch Pedal Switch Circuit Malfunction** Engine started, the after the vehicle was accelerated from 0-16 mph, the PCM detected the Clutch Pedal Position (CPP) switch input did not change status correctly during the CCM test. **Possible Causes:** • Clutch switch signal circuit is open or shorted ground • Clutch switch power circuit (B+) open between switch and PCM • Clutch switch is damaged, misadjusted or installed improperly • PCM has failed
DTC: P0704 **2T CCM, MIL: NO** **1996, 1997, 1998, 1999, 2000, 2001, 2002, 2003** **Models:** B2300, B2500, B3000, B4000, Tribute **Engines:** All **Transmissions:** M/T	**Clutch Pedal Switch Circuit Malfunction (Self-Test)** Engine started, the after the vehicle was accelerated from 0-16 mph, the PCM detected the Clutch Pedal Position (CPP) switch input did not change status correctly, or it did not change during the KOEO Self-Test procedure (the clutch may not have been depressed). **Possible Causes:** • Starter relay disconnected during the Quick Test Procedure • Clutch switch signal circuit is open or shorted ground • Clutch switch power circuit (B+) open between switch and PCM • Clutch switch signal circuit is shorted to VREF or system power • Clutch switch is damaged, misadjusted or installed improperly • PCM has failed
DTC: P0705 **2T CCM, MIL: YES** **1996, 1997, 1998, 1999, 2000, 2001, 2002** **Models:** 626, MX-6 **Engines:** 2.5L VIN D **Transmissions:** A/T	**Transmission Range Switch Circuit Malfunction** Key on or engine running, and the PCM did not detect any TR switch inputs, or it detected 2 switch inputs simultaneously during the test. **Possible Causes:** • TR signal 'P' circuit shorted to the PCM power circuit (12v) • TR signal 'D' circuit shorted to the PCM power circuit (12v) • TR signal 'S' circuit shorted to the PCM power circuit (12v) • TR signal 'L' circuit shorted to the PCM power circuit (12v) • TR switch or its connector is damaged, or the switch has failed • PCM has failed

DTC	Trouble Code Title, Conditions & Possible Causes
DTC: P0705 **2T CCM, MIL: YES** **204, 2005** **Models:** Tribute **Engines:** All **Transmissions:** A/T	**Transmission Range Switch Circuit Malfunction** Key on or engine running, and the PCM did not detect any TR switch inputs, or it detected 2 switch inputs simultaneously during the test. **Possible Causes:** • TR signal 'P' circuit shorted to the PCM power circuit (12v) • TR signal 'D' circuit shorted to the PCM power circuit (12v) • TR signal 'S' circuit shorted to the PCM power circuit (12v) • TR signal 'L' circuit shorted to the PCM power circuit (12v) • TR switch or its connector is damaged, or the switch has failed • PCM has failed
DTC: P0706 **2T CCM, MIL: YES** **1996, 1997, 1998, 1999, 2000, 2001, 2002** **Models:** 626, MX-6 **Engines:** All **Transmissions:** A/T	**Transmission Range Switch Circuit Malfunction** Key on or engine running, and the PCM did not detect any TR switch inputs, or it detected 2 switch inputs simultaneously during the test. **Possible Causes:** • TR signal 'P' circuit open between the switch and the PCM (0v) • TR signal 'D' circuit open between the switch and the PCM (0v) • TR signal 'S' circuit open between the switch and the PCM (0v) • TR signal 'L' circuit open between the switch and the PCM (0v) • TR switch or its connector is damaged, or the switch has failed • PCM has failed
DTC: P0706 **2T CCM, MIL: YES** **2004, 2005** **Models:** Mazda3, MX-5 Miata, Tribute **Engines:** All **Transmissions:** A/T	**Transmission Range Switch Circuit Malfunction** Key on or engine running, and the PCM did not detect any TR switch inputs, or it detected 2 switch inputs simultaneously during the test. **Possible Causes:** • TR signal 'P' circuit open between the switch and the PCM (0v) • TR signal 'D' circuit open between the switch and the PCM (0v) • TR signal 'S' circuit open between the switch and the PCM (0v) • TR signal 'L' circuit open between the switch and the PCM (0v) • TR switch or its connector is damaged, or the switch has failed • PCM has failed
DTC: P0707 **2T CCM, MIL: YES** **1996, 1997, 1998, 1999, 2000, 2001, 2002** **Models:** 626, MX-6 **Engines:** 2.0L VIN C **Transmissions:** A/T	**Transmission Range Switch Range/Performance** Engine running, and the PCM detected multiple T/R switch inputs, or the TR Switch signal did not change with the vehicle moving. **Possible Causes:** • TR switch signal circuit is shorted to another switch signal • TR switch signal circuit is open (problem may be intermittent) • TR switch is damaged, out of adjustment or it has failed • PCM has failed
DTC: P0707 **2T CCM, MIL: YES** **2004, 2005** **Models:** Mazda3, Tribute **Engines:** All **Transmissions:** A/T	**Transmission Range Switch Range/Performance** Engine running, and the PCM detected multiple T/R switch inputs, or the TR Switch signal did not change with the vehicle moving. **Possible Causes:** • TR switch signal circuit is shorted to another switch signal • TR switch signal circuit is open (problem may be intermittent) • TR switch is damaged, out of adjustment or it has failed • PCM has failed
DTC: P0708 **2T CCM, MIL: YES** **1996, 1997, 1998, 1999, 2000, 2001, 2002** **Models:** 626, MX-6 **Engines:** 2.0L VIN C **Transmissions:** A/T	**Transmission Range Switch Circuit Malfunction** Key on or engine running and the PCM detected an unexpected high or low voltage condition on the TR switch circuit during the test. **Possible Causes:** • TR switch signal circuit is open between the switch and PCM • TR switch signal circuit is shorted between switch and PCM • TR switch signal circuit is shorted to VREF or system power • TR switch ground circuit is open between switch and ground • TR switch or its connector is damaged, or the switch has failed • PCM has failed

DTC	Trouble Code Title, Conditions & Possible Causes
DTC: P0708 **2T CCM, MIL: YES** **2004, 2005** **Models:** Mazda3, Tribute **Engines:** All **Transmissions:** A/T	**Transmission Range Switch Circuit Malfunction** Key on or engine running and the PCM detected an unexpected high or low voltage condition on the TR switch circuit during the test. **Possible Causes:** • TR switch signal circuit is open between the switch and PCM • TR switch signal circuit is shorted between switch and PCM • TR switch signal circuit is shorted to VREF or system power • TR switch ground circuit is open between switch and ground • TR switch or its connector is damaged, or the switch has failed • PCM has failed
DTC: P0705 **1T CCM, MIL: YES** **1996, 1997, 1998, 1999, 2000, 2001, 2002** **Models:** Millenia **Engines:** All **Transmissions:** M/T	**Transmission Range Switch Circuit Malfunction** Engine speed over 530 rpm, P/N or Reverse signal indicating "on", and the PCM detected a 'D', 'S' or 'L' input from the TR switch. **Possible Causes:** • TR signal 'D' circuit shorted between switch and the PCM (12v) • TR signal 'S' circuit shorted between switch and the PCM (12v) • TR signal 'L' circuit shorted between switch and the PCM (12v) • TR switch or its connector is damaged, or the switch has failed • PCM has failed
DTC: P0705 **1T CCM, MIL: YES** **1996, 1997, 1998, 1999, 2000, 2001, 2002, 2003** **Models:** MX-5 Miata **Engines:** All **Transmissions:** M/T	**Neutral Switch Circuit Malfunction** Engine running, vehicle accelerated to over 18.6 mph and back to idle speed with the Clutch Switch input changing status 14 times, and the PCM did not detect any change in the Neutral Switch signal. **Possible Causes:** • Neutral switch signal circuit open between the switch and PCM • Neutral switch signal circuit shorted between switch and PCM • Neutral switch signal circuit shorted to the switch VREF (12v) • Neutral switch ground circuit open between switch and ground • Neutral switch or its connector is damaged, or switch has failed • PCM has failed
DTC: P0705 **1T CCM, MIL: YES** **1999, 2000, 2001, 2002, 2003** **Models:** Protégé, Protégé5 **Engines:** All **Transmissions:** M/T	**Neutral Switch Circuit Malfunction** Engine started, the after the vehicle was accelerated from 0-16 mph and back to idle with a change in the Clutch Pedal Position switch signal, the PCM did not detect a change in the Neutral Switch status. **Possible Causes:** • Neutral switch signal circuit is open or shorted ground • Neutral switch power circuit open between switch and PCM • Neutral switch ground circuit open between switch and ground • Neutral switch is damaged, misadjusted or installed improperly • PCM has failed
DTC: P0705 **1T CCM, MIL: YES** **1996, 1997, 1998, 1999, 2000, 2001, 2002, 2003** **Models:** MX-5 Miata, MPV, Protégé, Protégé5 **Engines:** All **Transmissions:** A/T	**Transmission Range Switch Circuit Malfunction** Engine running, P/N or 'R' switch input indicating "on", and the PCM detected a Transmission Range (TR) sensor Drive, Second or Low signal indicating "on" or it did not detect any TR switch signal. **Possible Causes:** • TR signal 'P' circuit shorted between switch and the PCM (12v) • TR signal 'N' circuit shorted between switch and the PCM (12v) • TR signal 'D' circuit shorted between switch and the PCM (12v) • TR signal '2' circuit shorted between switch and the PCM (12v) • TR signal '1' circuit shorted between switch and the PCM (12v) • TR switch or its connector is damaged, or the switch has failed • PCM has failed
DTC: P0706 **2T CCM, MIL: YES** **1996, 1997, 1998, 1999, 2000, 2001, 2002** **Models:** Millenia **Engines:** All **Transmissions:** M/T	**Transmission Range Switch Circuit Malfunction** Engine speed over 530 rpm, and the PCM did not receive any valid 'P', 'R', 'N', 'D', 'S' or 'L' inputs from the TR switch during the test. **Possible Causes:** • TR signal 'P' circuit open between the switch and the PCM (0v) • TR signal 'R' circuit open between the switch and the PCM (0v) • TR signal 'N' circuit open between the switch and the PCM (0v) • TR signal 'D' circuit open between the switch and the PCM (0v) • TR signal 'L' circuit open between the switch and the PCM (0v) • TR switch or its connector is damaged, or the switch has failed • PCM has failed

DTC	Trouble Code Title, Conditions & Possible Causes
DTC: P0706 **2T CCM, MIL: YES** **1996, 1997, 1998, 1999, 2000, 2001, 2002, 2003, 2004, 2005** **Models:** MX-5 Miata, MPV, Protégé, Protégé5 **Engines:** All **Transmissions:** A/T	**Transmission Range Switch Circuit Malfunction** Engine started, vehicle driven to a speed of over 37 mph, and the PCM did not detect any change in the TR Switch P-N-D-2-1 status. **Possible Causes:** • TR signal 'P' circuit open between the switch and the PCM (0v) • TR signal 'D' circuit open between the switch and the PCM (0v) • TR signal 'N' circuit open between the switch and the PCM (0v) • TR signal '2' circuit open between the switch and the PCM (0v) • TR signal '1' circuit open between the switch and the PCM (0v) • TR switch or its connector is damaged, or the switch has failed • PCM has failed
DTC: P0705 **1T CCM, MIL: YES** **1996, 1997** **Models:** B2300, B2500, B3000, B4000, Tribute **Engines:** All **Transmissions:** A/T	**Transmission Range Switch Circuit Malfunction** Key on or engine running and the PCM detected an incorrect TR switch signal, or it detected two signals at the same time in the test. **Possible Causes:** • TR signal circuit is open between the switch and the PCM • TR signal circuit is shorted to sensor ground or chassis ground • TR signal is shorted to VREF or system power (B+) • TR switch or its connector is damaged, or the switch has failed • PCM has failed
DTC: P0705 **1T CCM, MIL: YES** **1998, 1999, 2000, 2001, 2002, 2003, 2004, 2005** **Models:** B2300, B2500, B3000, B4000, Tribute **Engines:** All **Transmissions:** A/T	**Digital Transmission Range Switch Circuit Malfunction** Key on or engine running and the PCM detected an unexpected low or high voltage condition on one of the Digital TR sensor circuits. **Possible Causes:** • TR4 signal circuit is open or shorted to ground (0v or 12v) • TR3A signal circuit is open or shorted to ground (0v or 12v) • TR2 signal circuit is open or shorted to ground (0v or 12v) • TR1 signal circuit is open or shorted to ground (0v or 12v) • DTR switch or its connector is damaged, or switch has failed • PCM has failed
DTC: P0707 **1T CCM, MIL: YES** **1996, 1997** **Models:** B2300, B2500, B3000, B4000, Tribute **Engines:** All **Transmissions:** A/T	**Transmission Range Switch Circuit Low Input** Key on or engine running and the PCM detected an unexpected "low" voltage condition on the TR switch circuit during the test. **Possible Causes:** • TR switch signal shorted to the switch ground circuit • TR switch signal shorted to chassis ground • TR switch or its connector is damaged, or the switch has failed • PCM has failed
DTC: P0708 **1T CCM, MIL: YES** **1996, 1997** **Models:** B2300, B2500, B3000, B4000, Tribute **Engines:** All **Transmissions:** A/T	**Transmission Range Switch Circuit High Input** Key on or engine running and the PCM detected an unexpected "high" voltage condition on the TR switch circuit during the test. **Possible Causes:** • TR switch signal shorted to another switch signal circuit • TR switch signal shorted to VREF or system power • TR switch or its connector is damaged, or the switch has failed • PCM has failed
DTC: P0708 **2T CCM, MIL: YES** **1998, 1999, 2000, 2001, 2002, 2003, 2004, 2005** **Models:** B2300, B2500, B3000, B4000, Tribute **Engines:** All **Transmissions:** A/T	**Digital Transmission Range Switch Circuit Malfunction** Key on or engine running and the PCM detected an unexpected "low" voltage condition on the TR switch circuit during the CCM test. **Possible Causes:** • TR switch signal circuit open between switch and the PCM (0v) • DTR switch or its connector is damaged, or switch has failed • PCM has failed
DTC: P0707 **1T CCM, MIL: YES** **2001, 2002, 2003, 2004, 2005** **Models:** Tribute **Engines:** All **Transmissions:** A/T	**Transmission Range Switch Circuit Malfunction** Key on or engine running and the PCM detected an unexpected "low" voltage condition on the TR sensor circuit during the CCM test. **Possible Causes:** • TR switch circuit is shorted to ground switch ground (0v) • TR switch circuit is shorted to chassis ground (0v) • TR switch or its connector is damaged, or the switch has failed • PCM has failed

DTC	Trouble Code Title, Conditions & Possible Causes
DTC: P0708 **2T CCM, MIL: YES** **2001, 2002, 2003, 2004, 2005** **Models:** Tribute **Engines:** All **Transmissions:** A/T	**Digital Transmission Range Switch Circuit Malfunction** Key on or engine running and the PCM detected an unexpected high voltage condition on the TR switch circuit during the CCM test. **Possible Causes:** • TR switch circuit is open between the switch and PCM (0v) • TR switch or its connector is damaged, or the switch has failed • PCM has failed
DTC: P0707 **2T CCM, MIL: YES** **1996, 1997, 1998, 1999, 2000, 2001, 2002** **Models:** 626, MX-6 **Engines:** 2.0L VIN C **Transmissions:** A/T	**Transaxle Range Switch Circuit Malfunction** Engine started, and driven to over 18 mph and back to idle and the PCM detected the TR switch signal status remained low all the time. **Possible Causes:** • TR switch signal circuit is shorted to the switch ground • TR switch signal circuit is shorted to chassis ground • TR switch is damaged, out of adjustment or has failed • PCM has failed
DTC: P0708 **2T CCM, MIL: YES** **1996, 1997, 1998, 1999, 2000, 2001, 2002** **Models:** 626, MX-6 **Engines:** 2.0L VIN C **Transmissions:** A/T	**Transaxle Range Switch Circuit Malfunction** Engine started, and driven to over 18 mph and back to idle, and the PCM detected the TR switch signal remained in a high voltage state. **Possible Causes:** • TR switch signal circuit is open • TR switch ground circuit is open • TR switch is damaged, out of adjustment or has failed • PCM has failed
DTC: P0710 **1T CCM, MIL: YES** **1996, 1997, 1998, 1999, 2000, 2001, 2002** **Models:** 626, MX-6 **Engines:** 2.5L VIN D **Transmissions:** A/T	**Transmission Fluid Temperature Sensor Circuit Malfunction** Engine started, vehicle driven to a speed of over 12 mph, and the PCM detected the TFT sensor signal was less than 0.10v, or it was more than 4.90v, condition met for 100 seconds. **Possible Causes:** • TFT sensor signal circuit is open or shorted to ground • TFT sensor ground circuit is open between sensor and PCM • TFT sensor signal circuit shorted to VREF or system power • TFT sensor is damaged or has failed • PCM has failed
DTC: P0710 **1T CCM, MIL: YES** **1996, 1997, 1998, 1999, 2000, 2001, 2002, 2003** **Models:** Millenia, MPV, Protégé, Protégé5 **Engines:** All **Transmissions:** A/T	**Transmission Fluid Temperature Sensor Circuit Malfunction** Engine started, vehicle driven to a speed of over 12 mph, and the PCM detected the TFT sensor signal was less than 0.10v, or it was more than 4.90v, condition met for 100 seconds. **Possible Causes:** • TFT sensor signal circuit is open or shorted to ground • TFT sensor ground circuit is open between sensor and PCM • TFT sensor signal circuit shorted to VREF or system power • TFT sensor is damaged or has failed • PCM has failed
DTC: P0710 **1T CCM, MIL: YES** **1996, 1997** **Models:** MX-5 Miata **Engines:** All **Transmissions:** A/T	**Transmission Fluid Temperature Sensor Circuit Malfunction** Engine started, vehicle driven to a speed of over 12 mph, and the PCM detected the TFT sensor signal was less than 0.10v, or it was more than 4.90v, condition met for 100 seconds. **Possible Causes:** • TFT sensor signal circuit is open or shorted to ground • TFT sensor ground circuit is open between sensor and PCM • TFT sensor signal circuit shorted to VREF or system power • TFT sensor is damaged or has failed • PCM has failed
DTC: P0710 **1T CCM, MIL: YES** **2004, 2005** **Models:** Tribute **Engines:** All **Transmissions:** A/T	**Transmission Fluid Temperature Sensor Circuit Malfunction** Engine started, vehicle driven to a speed of over 12 mph, and the PCM detected the TFT sensor signal was less than 0.10v, or it was more than 4.90v, condition met for 100 seconds. **Possible Causes:** • TFT sensor signal circuit is open or shorted to ground • TFT sensor ground circuit is open between sensor and PCM • TFT sensor signal circuit shorted to VREF or system power • TFT sensor is damaged or has failed • PCM has failed

DTC	Trouble Code Title, Conditions & Possible Causes
DTC: P0711 **2T CCM, MIL: YES** **1996, 1997, 1998, 1999, 2000, 2001, 2002** **Models:** 626, MX-6 **Engines:** 2.5L VIN D **Transmissions:** A/T	**Transmission Fluid Temperature Sensor Circuit Range/Performance** DTC P0710 not set, engine started, vehicle driven to a speed of over 37 mph for 430 seconds, and the PCM detected the TFT sensor signal remained at 3.70v (68ºF) or more in the CCM rationality test. **Possible Causes:** • TFT sensor signal circuit has a high resistance condition • TFT sensor ground circuit has a high resistance condition • TFT sensor is damaged or has failed • PCM has failed
DTC: P0711 **2T CCM, MIL: YES** **1996, 1997, 1998, 1999, 2000, 2001, 2002, 2003** **Models:** Millenia, MPV **Engines:** All **Transmissions:** A/T	**Transmission Fluid Temperature Sensor Circuit Range/Performance** DTC P0710 not set, engine started, vehicle driven to a speed of over 37 mph for 430 seconds, and the PCM detected the TFT sensor signal remained at 3.70v (68ºF) or more in the CCM rationality test. **Possible Causes:** • TFT sensor signal circuit has a high resistance condition • TFT sensor ground circuit has a high resistance condition • TFT sensor is damaged or has failed • PCM has failed
DTC: P0711 **2T CCM, MIL: YES** **1996, 1997** **Models:** MX-5 Miata **Engines:** All **Transmissions:** A/T	**Transmission Fluid Temperature Sensor Circuit Range/Performance** DTC P0710 not set, engine started, vehicle driven to a speed of over 37 mph for 430 seconds, and the PCM detected the TFT sensor signal remained at 3.70v (68ºF) or more in the CCM rationality test. **Possible Causes:** • TFT sensor signal circuit has a high resistance condition • TFT sensor ground circuit has a high resistance condition • TFT sensor is damaged or has failed • PCM has failed
DTC: P0711 **2T CCM, MIL: YES** **1998, 1999, 2000, 2001, 2002, 2003** **Models:** Protégé, Protégé5 **Engines:** All **Transmissions:** A/T	**Transmission Fluid Temperature Sensor Circuit Range/Performance** DTC P0710 not set, engine started, vehicle driven to a speed of over 37 mph for 430 seconds, and the PCM detected the TFT sensor signal was less than 0.09v, or it was more than 4.99v in the test. **Possible Causes:** • TFT sensor signal circuit is open between sensor and the PCM • TFT sensor signal circuit is shorted to ground • TFT sensor is damaged or has failed • PCM has failed
DTC: P0711 **2T CCM, MIL: YES** **2004, 2005** **Models:** Mazda3, Tribute **Engines:** All **Transmissions:** A/T	**Transmission Fluid Temperature Sensor Circuit Range/Performance** DTC P0710 not set, engine started, vehicle driven to a speed of over 37 mph for 430 seconds, and the PCM detected the TFT sensor signal was less than 0.09v, or it was more than 4.99v in the test. **Possible Causes:** • TFT sensor signal circuit is open between sensor and the PCM • TFT sensor signal circuit is shorted to ground • TFT sensor is damaged or has failed • PCM has failed
DTC: P0712 **1T CCM, MIL: NO** **1996, 1997, 1998, 1999, 2000, 2001, 2002, 2003, 2004, 2005** **Models:** B2300, B2500, B3000, B4000, Tribute **Engines:** All **Transmissions:** A/T	**Transmission Fluid Temperature Sensor Low Input (High Temperature)** Key on or engine running and the PCM detected the Transmission Fluid Temperature (TFT) sensor signal indicated more than 315ºF during the CCM test. **Possible Causes:** • TFT sensor signal circuit has a short to ground condition • TFT sensor is damaged or has failed (it may be shorted) • PCM has failed
DTC: P0712 **1T CCM, MIL: NO** **1996, 1997, 1998, 1999, 2000, 2001, 2002** **Models:** 626, MX-6 **Engines:** 2.0L VIN C **Transmissions:** A/T	**Transaxle Fluid Temperature Sensor Low Input (High Temperature)** Key on or engine running and the PCM detected the Transaxle Fluid Temperature (TFT) sensor signal indicated less than the minimum acceptable voltage during the CCM test. **Possible Causes:** • TFT sensor signal circuit has a short toe ground condition • TFT sensor is damaged or has failed (it may be shorted) • PCM has failed

DTC	Trouble Code Title, Conditions & Possible Causes
DTC: P0712 **1T CCM, MIL: NO** **1996, 1997, 1998, 1999, 2000, 2001, 2002** **Models:** 626, MX-6 **Engines:** 2.5L VIN D **Transmissions:** A/T	**Transmission Fluid Temperature Sensor Low Input (High Temperature)** DTC P0710 not set, engine started, vehicle driven to a speed of over 37 mph for 430 seconds, and the PCM detected the TFT sensor indicated less than 0.20v (Scan Tool reads more than 315ºF) during the CCM rationality test. **Possible Causes:** • TFT sensor signal circuit has a short toe ground condition • TFT sensor is damaged or has failed (it may be shorted) • PCM has failed
DTC: P0712 **1T CCM, MIL: NO** **2004, 2005** **Models:** Mazda3 **Engines:** All **Transmissions:** A/T	**Transmission Fluid Temperature Sensor Low Input (High Temperature)** DTC P0710 not set, engine started, vehicle driven to a speed of over 37 mph for 430 seconds, and the PCM detected the TFT sensor indicated less than 0.20v (Scan Tool reads more than 315ºF) during the CCM rationality test. **Possible Causes:** • TFT sensor signal circuit has a short toe ground condition • TFT sensor is damaged or has failed (it may be shorted) • PCM has failed
DTC: P0713 **1T CCM, MIL: NO** **1996, 1997, 1998, 1999, 2000, 2001, 2002, 2003, 2004, 2005** **Models:** B2300, B2500, B3000, B4000, Tribute **Engines:** All **Transmissions:** A/T	**Transmission Fluid Temperature Sensor High Input (Low Temperature)** Key on or engine running and the PCM detected an unexpected "high" voltage condition on the Transaxle Fluid Temperature (TFT) sensor circuit during the CCM test. **Possible Causes:** • TFT sensor signal circuit is open between sensor and the PCM • TFT sensor signal circuit has a short to power condition • TFT sensor is damaged or has failed (it may be open) • PCM has failed
DTC: P0713 **1T CCM, MIL: NO** **1996, 1997, 1998, 1999, 2000, 2001, 2002** **Models:** 626, MX-6 **Engines:** 2.0L VIN C **Transmissions:** A/T	**Transmission Fluid Temperature Sensor High Input (Low Temperature)** Key on or engine running and the PCM detected an unexpected "high" voltage condition on the Transmission Fluid Temperature (TFT) sensor circuit (Scan Tool read less than -40ºF) during the test. **Possible Causes:** • TFT sensor signal circuit is open between sensor and the PCM • TFT sensor signal circuit has a short to power condition • TFT sensor is damaged or has failed (it may be open) • PCM has failed
DTC: P0713 **1T CCM, MIL: NO** **1996, 1997, 1998, 1999, 2000, 2001, 2002** **Models:** 626, MX-6 **Engines:** 2.5L VIN D **Transmissions:** A/T	**Transmission Fluid Temperature Sensor High Input (Low Temperature)** DTC P0710 not set, engine started, vehicle driven to a speed of over 37 mph for 430 seconds, and the PCM detected the TFT sensor indicated more than 4.96v (Scan Tool reads less than -40ºF) during the CCM rationality test. **Possible Causes:** • TFT sensor signal circuit is open between sensor and the PCM • TFT sensor signal circuit has a short to power condition • TFT sensor is damaged or has failed (it may be open) • PCM has failed
DTC: P0713 **1T CCM, MIL: NO** **2004, 2005** **Models:** Mazda3 **Engines:** All **Transmissions:** A/T	**Transmission Fluid Temperature Sensor High Input (Low Temperature)** DTC P0710 not set, engine started, vehicle driven to a speed of over 37 mph for 430 seconds, and the PCM detected the TFT sensor indicated more than 4.96v (Scan Tool reads less than -40ºF) during the CCM rationality test. **Possible Causes:** • TFT sensor signal circuit is open between sensor and the PCM • TFT sensor signal circuit has a short to power condition • TFT sensor is damaged or has failed (it may be open) • PCM has failed
DTC: P0715 **1T CCM, MIL: YES** **1996, 1997, 1998, 1999, 2000, 2001, 2002, 2003, 2004, 2005** **Models:** 626, MX-6, Millenia, MX-5 Miata **Engines:** All **Transmissions:** A/T	**Input Shaft or Turbine Shaft Sensor Circuit Malfunction** DTC P0710 not set, engine started, vehicle driven to a speed of over 25 mph for 430 seconds, and the PCM detected the ISS/TSS signal dropped out for more than 1 second during the CCM test. **Possible Causes:** • ISS/TSS (+) signal circuit is open (an intermittent fault) • ISS/TSS (+) signal circuit shorted to ground (intermittent fault) • ISS/TSS (-) signal circuit is open (an intermittent fault) • ISS/TSS (-) signal circuit shorted to ground (intermittent fault) • ISS/TSS is damaged or has failed • PCM has failed

DTC	Trouble Code Title, Conditions & Possible Causes
DTC: P0715 **1T CCM, MIL: YES** **1996, 1997, 1998, 1999, 2000, 2001, 2002, 2003** **Models:** Protégé, Protégé5 **Engines:** All **Transmissions:** A/T	**Input Shaft Sensor Circuit Malfunction** DTC P0710 not set, engine started, vehicle driven to a speed of over 25 mph for 430 seconds, and the PCM detected the ISS signal dropped out for more than 1 second during the CCM test. **Possible Causes:** • ISS (+) signal circuit is open (an intermittent fault) • ISS (+) signal circuit shorted to ground (intermittent fault) • ISS (-) signal circuit is open (an intermittent fault) • ISS (-) signal circuit shorted to ground (intermittent fault) • ISS is damaged or has failed • PCM has failed
DTC: P0715 **1T CCM, MIL: YES** **2000, 2001, 2002, 2003** **Models:** MPV **Engines:** All **Transmissions:** A/T	**Input Shaft or Turbine Shaft Sensor Circuit Malfunction** DTC P0710 not set, engine started, vehicle driven to a speed of over 25 mph, and the PCM detected the ISS/TSS signal dropped out for more than 1 second during the CCM test. **Possible Causes:** • ISS/TSS (+) signal circuit is open (an intermittent fault) • ISS/TSS (+) signal circuit shorted to ground (intermittent fault) • ISS/TSS (-) signal circuit is open (an intermittent fault) • ISS/TSS (-) signal circuit shorted to ground (intermittent fault) • ISS/TSS is damaged or has failed • PCM has failed
DTC: P0715 **1T CCM, MIL: YES** **1996, 1997, 1998, 1999, 2000, 2001, 2002, 2003, 2004, 2005** **Models:** B2300, B2500, Tribute **Engines:** All **Transmissions:** A/T	**Turbine Shaft Sensor Circuit Malfunction** Engine started, vehicle driven to a speed of 31 mph, and the PCM detected the ISS/TSS signal dropped out for more than 1 second. **Possible Causes:** • TSS (+) signal circuit is open (an intermittent fault) • TSS (+) signal circuit shorted to ground (intermittent fault) • TSS (-) signal circuit is open (an intermittent fault) • TSS (-) signal circuit shorted to ground (intermittent fault) • TSS is damaged or has failed • PCM has failed
DTC: P0715 **1T CCM, MIL: YES** **2004, 2005** **Models:** Mazda3 **Engines:** All **Transmissions:** A/T	**Turbine Shaft Sensor Circuit Malfunction** Engine started, vehicle driven to a speed of 31 mph, and the PCM detected the ISS/TSS signal dropped out for more than 1 second. **Possible Causes:** • TSS (+) signal circuit is open (an intermittent fault) • TSS (+) signal circuit shorted to ground (intermittent fault) • TSS (-) signal circuit is open (an intermittent fault) • TSS (-) signal circuit shorted to ground (intermittent fault) • TSS is damaged or has failed • PCM has failed
DTC: P0717 **1T CCM, MIL: YES** **2001, 2002, 2003, 2004, 2005** **Models:** Tribute **Engines:** All **Transmissions:** A/T	**Turbine Shaft Sensor Circuit Malfunction** Engine started, vehicle driven to a speed of 31 mph, and the PCM did not detect any TSS signals during the CCM test. **Possible Causes:** • TSS (+) signal circuit is open or shorted to ground • TSS (-) signal circuit is open or shorted to ground • TSS is damaged or has failed • PCM has failed
DTC: P0718 **1T CCM, MIL: YES** **2001, 2002, 2003, 2004, 2005** **Models:** Tribute **Engines:** All **Transmissions:** A/T	**Turbine Shaft Sensor Circuit Malfunction** Engine started, vehicle driven to a speed of 31 mph, and the PCM detected an erratic or noisy TSS signal during the CCM test. **Possible Causes:** • TSS (+) signal circuit is open or shorted to ground • TSS (-) signal circuit is open or shorted to ground • TSS reluctor has metal chips or is damaged, or has failed • PCM has failed
DTC: P0720 **1T CCM, MIL: YES** **1999, 2000, 2001, 2002, 2003, 2004, 2005** **Models:** MX-5 Miata **Engines:** All **Transmissions:** A/T	**Output Shaft Sensor Circuit Malfunction** Engine started, vehicle driven to a speed of over 37 mph, and the PCM detected the OSS signal had dropped out (the signal was lost) for more than 85 seconds during the CCM rationality test. **Possible Causes:** • OSS (+) signal circuit is open (fault may be intermittent) • OSS (+) signal circuit shorted to ground (an intermittent fault) • OSS (-) signal circuit is open (fault may be intermittent) • OSS (-) signal circuit shorted to ground (an intermittent fault) • OSS is damaged or has failed • PCM has failed

DTC	Trouble Code Title, Conditions & Possible Causes
DTC: P0731 **2T CCM, MIL: YES** **1996, 1997, 1998, 1999, 2000, 2001, 2002** **Models:** 626, MX-6 **Engines:** All **Transmissions:** A/T	**Transmission 1st Gear Ratio Incorrect** DTC P0500, P0710, P0755 and P0760 not set, engine started and the vehicle driven to a speed of 4-37 mph at more than 500 rpm, TSS signal more than 75 rpm, gear selector in 1GR (Drive range), TFT sensor more than 68ºF, brake switch indicating "off", TP angle over 6.25%, and the PCM detected difference between the turbine speed and the output shaft speed sensor signals was higher than or lower than the preprogrammed difference (ratio) stored in memory. **Possible Causes:** • Engine performance conditions are erratic or unstable • Engine or transmission temperature is too high • OSS or TSS is damaged or has failed • An internal transaxle component is damaged or has failed • PCM has failed
DTC: P0731 **2T CCM, MIL: YES** **1996, 1997, 1998, 1999, 2000, 2001, 2002** **Models:** Millenia **Engines:** 2.5L VIN 1 **Transmissions:** A/T	**Transmission 1st Gear Ratio Incorrect** DTC P0500, P0710, P0755 and P0760 not set, engine started and the vehicle driven to a speed of 4-37 mph at more than 500 rpm, TSS signal more than 75 rpm, gear selector in 1GR (Drive range), TFT sensor more than 68ºF, brake switch indicating "off", TP angle over 6.25%, and the PCM detected difference between the turbine speed and the output shaft speed sensor signals was higher than or lower than the preprogrammed difference (ratio) stored in memory. **Possible Causes:** • Engine performance conditions are erratic or unstable • Engine or transmission temperature is too high • OSS or TSS is damaged or has failed • An internal transaxle component is damaged or has failed • PCM has failed
DTC: P0731 **2T CCM, MIL: YES** **1996, 1997, 1998** **Models:** MX-5 Miata, Millenia **Engines:** 1.8L VIN 3, 2.3L VIN 2 **Transmissions:** A/T	**Transmission 1st Gear Ratio Incorrect** DTC P0500, P0710, P0755 and P0760 not set, engine started and the vehicle driven to a speed of 4-37 mph at more than 500 rpm, TSS signal more than 75 rpm, gear selector in 1GR (Drive range), TFT sensor more than 68ºF, brake switch indicating "off", TP angle over 6.25%, and the PCM detected difference between the turbine speed and the output shaft speed sensor signals was higher than or lower than the preprogrammed difference (ratio) stored in memory. **Possible Causes:** • Engine performance conditions are erratic or unstable • Engine or transmission temperature is too high • OSS or TSS is damaged or has failed • An internal transaxle component is damaged or has failed • PCM has failed
DTC: P0731 **1T CCM, MIL: NO** **1996, 1997, 1998, 1999, 2000, 2001, 2002, 2003** **Models:** Protégé, Protégé5 **Engines:** All **Transmissions:** A/T	**Transmission 1st Gear Ratio Incorrect** DTC P0500, P0705, P0706, P0710, P0715, P0751, P0752, P0753, P0756, P0757, P0758, P0761, P0762, P0763, P0766, P0767, P0768, P0771, P0772 and P0773 not set, engine running in Drive in 1st Gear (1GR), ATF temperature more than 68ºF, throttle opening over 3.13% on FS engine or over 3.91 on ZM engine, Turbine speed at 4988 rpm (± 225), differential gear output speed over 35 mph, and the PCM detected the revolution ratio of the forward clutch drum revolution to the differential gear case revolution was less than 2.157 during the CCM rationality test. **Possible Causes:** • ATF level is too low, or the ATF is badly deteriorated • A/T line pressure is low, or the oil pump has failed • A/T control valve is stuck (a mechanical fault) • A/T Pressure control solenoid is stuck (a mechanical fault) • A/T shift solenoid 'A' is stuck (a mechanical fault) • A/T one-way clutch is slipping, or the forward clutch is slipping • PCM has failed
DTC: P0731 **2T CCM, MIL: YES** **1996, 1997, 1998, 1999, 2000, 2001, 2002, 2003, 2004, 2005** **Models:** B3000, B4000, Tribute, MPV **Engines:** All **Transmissions:** A/T	**Transmission 1st Gear Ratio Incorrect** DTC P0500, P0710, P0715, P0755 and P0760 not set, engine started, then driven in Drive in 1st Gear (1GR) to a speed from 4-37 mph, engine speed over 500 rpm, TSS signal more than 75 rpm, TP angle over 6.25%, ATF sensor signal more than 68ºF, brake switch indicating "off", and the PCM detected the ratio of the Turbine speed to the Vehicle speed was less than a preset value stored in memory. **Possible Causes:** • ATF level is too low, or the ATF is badly deteriorated • A/T line pressure is low, or the oil pump has failed • SSA, SSB, SSC or the PCS is stuck (mechanical fault) • A/T 1-2 clutch is slipping (mechanical fault) • A/T 1-2 shift valve is stuck (mechanical fault) • A/T Pressure regulator or the pressure modifier valve is stuck • A/T solenoid reducing valve is stuck (mechanical fault) • PCM has failed

DTC	Trouble Code Title, Conditions & Possible Causes
DTC: P0731 **2T CCM, MIL: YES** **2004, 2005** **Models:** Mazda3 **Engines:** All **Transmissions:** A/T	**Transmission 1st Gear Ratio Incorrect** DTC P0500, P0710, P0715, P0755 and P0760 not set, engine started, then driven in Drive in 1st Gear (1GR) to a speed from 4-37 mph, engine speed over 500 rpm, TSS signal more than 75 rpm, TP angle over 6.25%, ATF sensor signal more than 68ºF, brake switch indicating "off", and the PCM detected the ratio of the Turbine speed to the Vehicle speed was less than a preset value stored in memory. **Possible Causes:** • ATF level is too low, or the ATF is badly deteriorated • A/T line pressure is low, or the oil pump has failed • SSA, SSB, SSC or the PCS is stuck (mechanical fault) • A/T 1-2 clutch is slipping (mechanical fault) • A/T 1-2 shift valve is stuck (mechanical fault) • A/T Pressure regulator or the pressure modifier valve is stuck • A/T solenoid reducing valve is stuck (mechanical fault) • PCM has failed
DTC: P0732 **1T CCM, MIL: YES** **1996, 1997, 1998, 1999, 2000, 2001, 2002** **Models:** 626, MX-6 **Engines:** All **Transmissions:** A/T	**Transmission 2nd Gear Ratio Incorrect** DTC P0500, P0710, P0755 and P0760 not set, engine started and the vehicle driven to a speed of 6-66 mph at more than 500 rpm, TSS signal more than 75 rpm, gear selector in 2GR (Drive range), TFT sensor more than 68ºF, brake switch indicating "off", TP angle over 6.25%, and the PCM detected difference between the turbine speed and the output shaft speed sensor signals was higher than or lower than the preprogrammed difference (ratio) stored in memory. **Possible Causes:** • Engine performance conditions are erratic or unstable • Engine or transmission temperature is too high • OSS or TSS is damaged or has failed • An internal transaxle component is damaged or has failed • PCM has failed
DTC: P0732 **1T CCM, MIL: YES** **1996, 1997, 1998, 1999, 2000, 2001, 2002** **Models:** Millenia **Engines:** 2.5L VIN 1 **Transmissions:** A/T	**Transmission 2nd Gear Ratio Incorrect** DTC P0500, P0710, P0755 and P0760 not set, engine started and the vehicle driven to a speed of 6-66 mph at more than 500 rpm, TSS signal more than 75 rpm, gear selector in 2GR (Drive range), TFT sensor more than 68ºF, brake switch indicating "off", TP angle over 6.25%, and the PCM detected difference between the turbine speed and the output shaft speed sensor signals was higher than or lower than the preprogrammed difference (ratio) stored in memory. **Possible Causes:** • Engine performance conditions are erratic or unstable • Engine or transmission temperature is too high • OSS or TSS is damaged or has failed • An internal transaxle component is damaged or has failed • PCM has failed
DTC: P0732 **1T CCM, MIL: YES** **1996, 1997, 1998** **Models:** MX-5 Miata, Millenia **Engines:** 1.8L VIN 3, 2.3L VIN 2 **Transmissions:** A/T	**Transmission 2nd Gear Ratio Incorrect** DTC P0500, P0710, P0755 and P0760 not set, engine started and the vehicle driven to a speed of 6-66 mph at more than 500 rpm, TSS signal more than 75 rpm, gear selector in 2GR (Drive range), TFT sensor more than 68ºF, brake switch indicating "off", TP angle over 6.25%, and the PCM detected difference between the turbine speed and the output shaft speed sensor signals was higher than or lower than the preprogrammed difference (ratio) stored in memory. **Possible Causes:** • Engine performance conditions are erratic or unstable • Engine or transmission temperature is too high • OSS or TSS is damaged or has failed • An internal transaxle component is damaged or has failed • PCM has failed
DTC: P0732 **1T CCM, MIL: NO** **1996, 1997, 1998, 1999, 2000, 2001, 2002, 2003** **Models:** Protégé, Protégé5 **Engines:** All **Transmissions:** A/T	**Transmission 2nd Gear Ratio Incorrect** DTC P0500, P0705, P0706, P0710, P0715, P0751, P0752, P0753, P0756, P0757, P0758, P0761, P0762, P0763, P0766, P0767, P0768, P0771, P0772 and P0773 not set, engine running in Drive in 2nd gear (2GR), ATF temperature more than 68ºF, Turbine speed at 4988 rpm (± 225), differential gear output speed over 35 mph, and the PCM detected the revolution ratio of the forward clutch drum revolution to the differential gear case revolution was less than 1.249, or it was more than 2.157 during the CCM rationality test. **Possible Causes:** • ATF level is too low, or the ATF is badly deteriorated • A/T line pressure is low, or the oil pump has failed • A/T control valve is stuck (a mechanical fault) • A/T Pressure control solenoid is stuck (a mechanical fault) • A/T SSA, SSB or SSC is stuck (a mechanical fault) • A/T 2-4 band is slipping, or the forward clutch is slipping • PCM has failed

DTC	Trouble Code Title, Conditions & Possible Causes
DTC: P0732 **2T CCM, MIL: YES** **1996, 1997, 1998, 1999, 2000, 2001, 2002, 2003, 2004, 2005** **Models:** B3000, B4000, Tribute, MPV **Engines:** All **Transmissions:** A/T	**Transmission 2nd Gear Ratio Incorrect** DTC P0500, P0710, P0715, P0755 and P0760 not set, engine started, then driven in Drive in 2nd Gear (2GR) to a speed of 66 mph (± 6 mph), engine speed over 500 rpm, throttle opening over 6.25%, TSS signal more than 75 rpm, ATF sensor signal more than 68ºF, brake switch indicating "off", and the PCM detected the ratio of the Turbine speed to the Vehicle speed was less than a preset value stored in memory. **Possible Causes:** • ATF level is too low, or the ATF is badly deteriorated • A/T line pressure is low, or the oil pump has failed • SSA, SSB, SSC or the PCS is stuck (mechanical fault) • A/T 2-3 clutch is slipping (mechanical fault) • A/T 1-2 or the 2-3 shift valve is stuck (mechanical fault) • A/T Pressure regulator or the pressure modifier valve is stuck • A/T solenoid reducing valve is stuck (mechanical fault) • PCM has failed
DTC: P0732 **2T CCM, MIL: YES** **2004, 2005** **Models:** Mazda3 **Engines:** All **Transmissions:** A/T	**Transmission 2nd Gear Ratio Incorrect** DTC P0500, P0710, P0715, P0755 and P0760 not set, engine started, then driven in Drive in 2nd Gear (2GR) to a speed of 66 mph (± 6 mph), engine speed over 500 rpm, throttle opening over 6.25%, TSS signal more than 75 rpm, ATF sensor signal more than 68ºF, brake switch indicating "off", and the PCM detected the ratio of the Turbine speed to the Vehicle speed was less than a preset value stored in memory. **Possible Causes:** • ATF level is too low, or the ATF is badly deteriorated • A/T line pressure is low, or the oil pump has failed • SSA, SSB, SSC or the PCS is stuck (mechanical fault) • A/T 2-3 clutch is slipping (mechanical fault) • A/T 1-2 or the 2-3 shift valve is stuck (mechanical fault) • A/T Pressure regulator or the pressure modifier valve is stuck • A/T solenoid reducing valve is stuck (mechanical fault) • PCM has failed
DTC: P0733 **2T CCM, MIL: YES** **1996, 1997, 1998, 1999, 2000, 2001, 2002** **Models:** 626, MX-6 **Engines:** All **Transmissions:** A/T	**Transmission 3rd Gear Ratio Incorrect** DTC P0500, P0710, P0755 and P0760 not set, engine started and the vehicle driven to a speed of 19-30 mph at more than 500 rpm, TSS signal more than 75 rpm, gear selector in 3GR (Drive range), TFT sensor more than 68ºF, brake switch indicating "off", TP angle over 6.25%, and the PCM detected difference between the turbine speed and the output shaft speed sensor signals was higher than or lower than the preprogrammed difference (ratio) stored in memory. **Possible Causes:** • ATF level is too low, or the ATF is badly deteriorated • A/T SSB is stuck "on", or it is damaged (a mechanical fault) • A/T control valve is stuck, or it is damaged (a mechanical fault) • PCM has failed
DTC: P0733 **2T CCM, MIL: YES** **1996, 1997, 1998, 1999, 2000, 2001, 2002** **Models:** Millenia **Engines:** 2.5L VIN 1 **Transmissions:** A/T	**Transmission 3rd Gear Ratio Incorrect** DTC P0500, P0710, P0755 and P0760 not set, engine started and the vehicle driven to a speed of 19-30 mph at more than 500 rpm, TSS signal more than 75 rpm, gear selector in 3GR (Drive range), TFT sensor more than 68ºF, brake switch indicating "off", TP angle over 6.25%, and the PCM detected difference between the turbine speed and the output shaft speed sensor signals was higher than or lower than the preprogrammed difference (ratio) stored in memory. **Possible Causes:** • Engine performance conditions are erratic or unstable • Engine or transmission temperature is too high • OSS or TSS is damaged or has failed • An internal transaxle component is damaged or has failed • PCM has failed
DTC: P0733 **2T CCM, MIL: YES** **1996, 1997, 1998** **Models:** MX-5 Miata, Millenia **Engines:** 1.8L VIN 3, 2.3L VIN 2 **Transmissions:** A/T	**Transmission 3rd Gear Ratio Incorrect** DTC P0500, P0710, P0755 and P0760 not set, engine started and the vehicle driven to a speed of 19-30 mph at more than 500 rpm, TSS signal more than 75 rpm, gear selector in 3GR (Drive range), TFT sensor more than 68ºF, brake switch indicating "off", TP angle over 6.25%, and the PCM detected difference between the turbine speed and the output shaft speed sensor signals was higher than or lower than the preprogrammed difference (ratio) stored in memory. **Possible Causes:** • Engine performance conditions are erratic or unstable • Engine or transmission temperature is too high • OSS or TSS is damaged or has failed • An internal transaxle component is damaged or has failed • PCM has failed

DTC	Trouble Code Title, Conditions & Possible Causes
DTC: P0733 **1T CCM, MIL: NO** **1996, 1997, 1998, 1999, 2000, 2001, 2002, 2003** **Models:** Protégé, Protégé5 **Engines:** All **Transmissions:** A/T	**Transmission 3rd Gear Ratio Incorrect** DTC P0500, P0705, P0706, P0710, P0715, P0751, P0752, P0753, P0756, P0757, P0758, P0761, P0762, P0763, P0766, P0767, P0768, P0771, P0772 and P0773 not set, engine running in Drive in 3rd Gear (3GR), ATF temperature more than 68ºF, Turbine speed at 4988 rpm (± 225), differential gear output speed over 35 mph, and the PCM detected the revolution ratio of the forward clutch drum revolution to the differential gear case revolution was less than 0.863, or it was more than 1.249 during the CCM rationality test. **Possible Causes:** • ATF level is too low, or the ATF is badly deteriorated • A/T line pressure is low, or the oil pump has failed • A/T control valve is stuck (i.e., Bypass, TCC or 3-4 valve shift) • A/T Pressure control solenoid is stuck (a mechanical fault) • A/T shift solenoids 'A' or 'C' are stuck (a mechanical fault) • A/T 3-4 clutch is slipping, or the forward clutch is slipping • PCM has failed
DTC: P0733 **2T CCM, MIL: YES** **1996, 1997, 1998, 1999, 2000, 2001, 2002, 2003, 2004, 2005** **Models:** B3000, B4000, Tribute, MPV **Engines:** All **Transmissions:** A/T	**Transmission 3rd Gear Ratio Incorrect** DTC P0500, P0710, P0715, P0755 and P0760 not set, engine started, then driven in Drive in 3rd Gear (3GR) to a speed of 19-30 mph, engine speed over 500 rpm, TSS signal more than 75 rpm, ATF sensor signal more than 68ºF, brake switch indicating "off", and the PCM detected the ratio of the Turbine speed to the Vehicle speed was less than a preset value stored in memory. **Possible Causes:** • ATF level is too low, or the ATF is badly deteriorated • A/T line pressure is low, or the oil pump has failed • SSA, SSB, SSC or the PCS is stuck (mechanical fault) • A/T 3-4 clutch is slipping (mechanical fault) • A/T 2-3 or the 3-4 shift valve is stuck (mechanical fault) • A/T Pressure regulator or the pressure modifier valve is stuck • A/T solenoid reducing valve is stuck (mechanical fault) • PCM has failed
DTC: P0733 **2T CCM, MIL: YES** **2004, 2005** **Models:** Mazda3 **Engines:** All **Transmissions:** A/T	**Transmission 3rd Gear Ratio Incorrect** DTC P0500, P0710, P0715, P0755 and P0760 not set, engine started, then driven in Drive in 3rd Gear (3GR) to a speed of 19-30 mph, engine speed over 500 rpm, TSS signal more than 75 rpm, ATF sensor signal more than 68ºF, brake switch indicating "off", and the PCM detected the ratio of the Turbine speed to the Vehicle speed was less than a preset value stored in memory. **Possible Causes:** • ATF level is too low, or the ATF is badly deteriorated • A/T line pressure is low, or the oil pump has failed • SSA, SSB, SSC or the PCS is stuck (mechanical fault) • A/T 3-4 clutch is slipping (mechanical fault) • A/T 2-3 or the 3-4 shift valve is stuck (mechanical fault) • A/T Pressure regulator or the pressure modifier valve is stuck • A/T solenoid reducing valve is stuck (mechanical fault) • PCM has failed
DTC: P0734 **2T CCM, MIL: YES** **1996, 1997, 1998, 1999, 2000, 2001, 2002** **Models:** 626, MX-6 **Engines:** All **Transmissions:** A/T	**Transmission 4th Gear Ratio Incorrect** DTC P0500, P0710, P0755 and P0760 not set, engine started and the vehicle driven to a speed of 47-64 mph at more than 500 rpm, TSS signal more than 75 rpm, gear selector in 4GR (Drive range), TFT sensor more than 68ºF, brake switch indicating "off", TP angle over 6.25%, and the PCM detected difference between the turbine speed and the output shaft speed sensor signals was higher than or lower than the preprogrammed difference (ratio) stored in memory. **Possible Causes:** • Engine performance conditions are erratic or unstable • Engine or transmission temperature is too high • OSS or TSS is damaged or has failed • An internal transaxle component is damaged or has failed • PCM has failed
DTC: P0734 **2T CCM, MIL: YES** **1996, 1997, 1998, 1999, 2000, 2001, 2002** **Models:** Millenia **Engines:** 2.5L VIN 1 **Transmissions:** A/T	**Transmission 4th Gear Ratio Incorrect** DTC P0500, P0710, P0755 and P0760 not set, engine started and the vehicle driven to a speed of 47-64 mph at more than 500 rpm, TSS signal more than 75 rpm, gear selector in 4GR (Drive range), TFT sensor more than 68ºF, brake switch indicating "off", TP angle over 6.25%, and the PCM detected difference between the turbine speed and the output shaft speed sensor signals was higher than or lower than the preprogrammed difference (ratio) stored in memory. **Possible Causes:** • Engine performance conditions are erratic or unstable • Engine or transmission temperature is too high • OSS or TSS is damaged or has failed • An internal transaxle component is damaged or has failed • PCM has failed

DTC	Trouble Code Title, Conditions & Possible Causes
DTC: P0734 **2T CCM, MIL: YES** **1996, 1997, 1998** **Models:** MX-5 Miata, Millenia **Engines:** 1.8L VIN 3, 2.3L VIN 2 **Transmissions:** A/T	**Transmission 4th Gear Ratio Incorrect** DTC P0500, P0710, P0755 and P0760 not set, engine started and the vehicle driven to a speed of 47-64 mph at more than 500 rpm, TSS signal more than 75 rpm, gear selector in 4GR (Drive range), TFT sensor than 68ºF, brake switch indicating "off", TP angle over 6.25%, and the PCM detected difference between the turbine speed and the output shaft speed sensor signals was higher than or lower than the preprogrammed difference (ratio) stored in memory. **Possible Causes:** • Engine performance conditions are erratic or unstable • Engine or transmission temperature is too high • OSS or TSS is damaged or has failed • An internal transaxle component is damaged or has failed • PCM has failed
DTC: P0734 **1T CCM, MIL: NO** **1996, 1997, 1998, 1999, 2000, 2001, 2002, 2003** **Models:** Protégé, Protégé5 **Engines:** All **Transmissions:** A/T	**Transmission 4th Gear Ratio Incorrect** DTC P0500, P0705, P0706, P0710, P0715, P0751, P0752, P0753, P0756, P0757, P0758, P0761, P0762, P0763, P0766, P0767, P0768, P0771, P0772 and P0773 not set, engine running in Drive in 4th Gear (4GR) a speed of over 31 mph with the throttle closed, ATF temperature more than 68ºF, Turbine speed at 4988 rpm (± 225), differential gear output speed over 35 mph, and the PCM detected the revolution ratio of the forward clutch drum revolution to the differential gear case revolution was less than 0.60, or it was more than 1.249 during the CCM rationality test. **Possible Causes:** • ATF level is too low, or the ATF is badly deteriorated • A/T line pressure is low, or the oil pump has failed • A/T control valve is stuck (i.e., Bypass, or the 3-4 valve shift) • A/T shift solenoids 'A', 'B' or 'C' are stuck (a mechanical fault) • A/T Pressure control solenoid is stuck (a mechanical fault) • A/T 2-4 brake band is slipping (a mechanical fault) • A/T 3-4 clutch is slipping, or the forward clutch is slipping • PCM has failed
DTC: P0734 **2T CCM, MIL: YES** **1996, 1997, 1998, 1999, 2000, 2001, 2002, 2003, 2004, 2005** **Models:** B3000, B4000, Tribute, MPV **Engines:** All **Transmissions:** A/T	**Transmission 4th Gear Ratio Incorrect** DTC P0500, P0710, P0715, P0755 and P0760 not set, engine started, then driven in Drive in 4th Gear (4GR) to a speed of 47-64 mph, engine speed over 500 rpm, TSS signal more than 75 rpm, ATF sensor signal more than 68ºF, brake switch indicating "off", and the PCM detected the ratio of the Turbine speed to the Vehicle speed was less than a preset value stored in memory. **Possible Causes:** • ATF level is too low, or the ATF is badly deteriorated • A/T line pressure is low, or the oil pump has failed • SSA, SSB, SSC or the PCS is stuck (mechanical fault) • A/T 2-4 brake bank or 3-4 clutch is slipping (mechanical fault) • A/T 1-2, 2-3 or 3-4 shift valve is stuck (mechanical fault) • A/T Pressure regulator, pressure modifier or solenoid reducing valve is stuck • PCM has failed
DTC: P0734 **2T CCM, MIL: YES** **2004, 2005** **Models:** Mazda3 **Engines:** All **Transmissions:** A/T	**Transmission 4th Gear Ratio Incorrect** DTC P0500, P0710, P0715, P0755 and P0760 not set, engine started, then driven in Drive in 4th Gear (4GR) to a speed of 47-64 mph, engine speed over 500 rpm, TSS signal more than 75 rpm, ATF sensor signal more than 68ºF, brake switch indicating "off", and the PCM detected the ratio of the Turbine speed to the Vehicle speed was less than a preset value stored in memory. **Possible Causes:** • ATF level is too low, or the ATF is badly deteriorated • A/T line pressure is low, or the oil pump has failed • SSA, SSB, SSC or the PCS is stuck (mechanical fault) • A/T 2-4 brake bank or 3-4 clutch is slipping (mechanical fault) • A/T 1-2, 2-3 or 3-4 shift valve is stuck (mechanical fault) • A/T Pressure regulator, pressure modifier or solenoid reducing valve is stuck • PCM has failed
DTC: P0735 **1T CCM, MIL: YES** **1998, 1999, 2000, 2001, 2002, 2003, 2004, 2005** **Models:** B2300, B3000, B4000 **Engines:** All **Transmissions:** A/T	**Output Shaft Sensor Circuit Erratic Malfunction** Engine started, vehicle driven to a speed of over 12 mph, and the PCM detected the OSS signal was erratic during the CCM test. **Possible Causes:** • OSS (+) signal circuit is open or shorted to ground (intermittent) • OSS (-) signal circuit is open or shorted to ground (intermittent) • OSS reluctor contains metal chips or is damaged • OSS is damaged or has failed • PCM has failed

DTC	Trouble Code Title, Conditions & Possible Causes
DTC: P0703 **1T CCM, MIL: YES** **1996, 1997** **Models:** B2300, B3000, B4000 **Engines:** All **Transmissions:** A/T	**Transmission Reverse Gear Incorrect Ratio** Engine started, and then driven in Reverse gear at over 1 mph, and the PCM and the PCM detected the difference between the reverse speed and the output shaft speed sensor signals were higher than or lower than the preprogrammed difference (ratio) stored in memory. **Possible Causes:** Engine performance conditions are erratic or unstable • Engine or transmission temperature is too high • An internal transaxle component is damaged or has failed • PCM has failed
DTC: P0740 **2T CCM, MIL: YES** **1996, 1997, 1998, 1999, 2000, 2001, 2002** **Models:** 626, MX-6 **Engines:** 2.5L VIN D **Transmissions:** A/T	**Torque Converter Clutch System Malfunction** Engine started, then the vehicle was driven to a speed of 30-40 mph, and the PCM commanded the TCC solenoid to engage, then the PCM detected the TCC system did not operate correctly. **Possible Causes:** • ATF level is too low • Transmission line pressure is too low • TCC slippage is occurring • A/T control valve has failed (it may be stuck) • TCC assembly is damaged or has failed • PCM has failed
DTC: P0740 **2T CCM, MIL: YES** **1996, 1997, 1998, 1999, 2000, 2001, 2002** **Models:** Millenia **Engines:** All **Transmissions:** A/T	**Torque Converter Clutch System Malfunction** Engine started, vehicle driven to a speed of 30-40 mph, and after the PCM commanded the TCC solenoid to engage, it detected the TCC system did not function correctly. **Possible Causes:** • ATF level is too low, or the fluid is burnt or contaminated • A/T control valve has failed (it may be stuck) • TCC control solenoid valve is damaged or has failed • TCC slippage is occurring • Transmission line pressure is too low • TCM has failed
DTC: P0740 **2T CCM, MIL: YES** **1996, 1997** **Models:** MX-5 Miata **Engines:** All **Transmissions:** A/T	**Torque Converter Clutch System Malfunction** Engine started, vehicle driven to a speed of 30-40 mph, and after the PCM commanded the TCC solenoid to engage, it detected the TCC system did not function correctly. **Possible Causes:** • ATF level is too low, or the fluid is burnt or contaminated • A/T control valve has failed (it may be stuck) • TCC control solenoid valve is damaged or has failed • TCC slippage is occurring • Transmission line pressure is too low • TCM has failed
DTC: P0740 **2T CCM, MIL: YES** **1996, 1997, 1998** **Models:** Protégé **Engines:** All **Transmissions:** A/T	**Torque Converter Clutch System Malfunction** Engine started, vehicle driven to a speed of 30-40 mph, and after the PCM commanded the TCC solenoid to engage, it detected the TCC system did not function correctly. **Possible Causes:** • ATF level is too low, or the ATF is badly deteriorated • A/T line pressure is low, or the oil pump has failed • A/T control valve has failed (it may be stuck) • TCC is slipping, is damaged or has failed (mechanical fault) • PCM has failed

DTC	Trouble Code Title, Conditions & Possible Causes
DTC: P0740 **2T CCM, MIL: YES** **1996, 1997, 1998, 1999, 2000, 2001,2002, 2003** **Models:** MPV **Engines:** All **Transmissions:** A/T	**Torque Converter Clutch System Malfunction** DTC P0500, P0710, P0715, P0755 and P0760 not set, engine started, then driven in Drive in 4th Gear (4GR) to a speed of 47-64 mph with the throttle open, engine speed over 500 rpm, ATF sensor signal more than 68ºF, Turbine speed over 75 rpm, brake switch indicating "off", TCC system operating, O/D OFF Switch is "off", and the PCM detected the difference between the engine speed and the Turbine speed was more than a preset value stored in memory. **Possible Causes:** • ATF level is too low, or the ATF is badly deteriorated • A/T line pressure is low, or the oil pump has failed • A/T control valve has failed (it may be stuck) • TCC solenoid valve or pressure control valve is stuck • TCC in the torque converter is slipping (mechanical fault) • TCC shift valve or converter relief valve is stuck • Pressure modifier or the pressure regulator valve is stuck • Solenoid reducing valve is stuck (mechanical fault) • PCM has failed
DTC: P0740 **2T CCM, MIL: YES** **2004, 2005** **Models:** Tribute **Engines:** All **Transmissions:** A/T	**Torque Converter Clutch System Malfunction** DTC P0500, P0710, P0715, P0755 and P0760 not set, engine started, then driven in Drive in 4th Gear (4GR) to a speed of 47-64 mph with the throttle open, engine speed over 500 rpm, ATF sensor signal more than 68ºF, Turbine speed over 75 rpm, brake switch indicating "off", TCC system operating, O/D OFF Switch is "off", and the PCM detected the difference between the engine speed and the Turbine speed was more than a preset value stored in memory. **Possible Causes:** • ATF level is too low, or the ATF is badly deteriorated • A/T line pressure is low, or the oil pump has failed • A/T control valve has failed (it may be stuck) • TCC solenoid valve or pressure control valve is stuck • TCC in the torque converter is slipping (mechanical fault) • TCC shift valve or converter relief valve is stuck • Pressure modifier or the pressure regulator valve is stuck • Solenoid reducing valve is stuck (mechanical fault) • PCM has failed
DTC: P0741 **1T CCM, MIL: NO** **1996, 1997, 1998, 1999** **Models:** 626, MX-6 **Engines:** 2.0L VIN C **Transmissions:** A/T	**Torque Converter Clutch Control Engagement Error** Engine started, and then driven to a speed of over 30 mph, and the PCM detected the TCC system did not operate (due to a mechanical fault. **Possible Causes:** • Brake switch is damaged, out of adjustment or has failed • Engine performance conditions are erratic or unstable • Engine or transmission temperature is too high • TCC control valve is stuck in the "off" position • An internal transaxle component is damaged or has failed • PCM has failed
DTC: P0741 **2T CCM, MIL: YES** **1999, 2000, 2001, 2002, 2003, 2004, 2005** **Models:** MX-5 Miata **Engines:** All **Transmissions:** A/T, M/T	**Torque Converter Clutch Performance (Mechanical)** DTC P0500, P0705, P0706, P0715, P0720, P1740, P1742, P1751, P1752, P1756, P1757, P1771 and P1772 not set, engine started, vehicle driven in Drive at 60 mph (± 3 mph) for 20 seconds, engine speed over 600 rpm, ECT sensor signal more than 113ºF, brake pedal switch indicating "off", throttle angle over 8%, and the PCM detected the difference between the engine speed and Turbine speed was more than a predetermined value. **Possible Causes:** • ATF level is too low, or the ATF is badly deteriorated • A/T line pressure is low, or the oil pump has failed • A/T control valve is stuck, or it is damaged (a mechanical fault) • TCC solenoid is stuck "off", or is damaged (a mechanical fault) • TCC system has failed • TCM has failed

DTC	Trouble Code Title, Conditions & Possible Causes
DTC: P0741 **1T CCM, MIL: NO** **1999, 2000, 2001, 2002, 2003** **Models:** Protégé, Protégé5 **Engines:** All **Transmissions:** A/T	**Torque Converter Clutch Performance (Mechanical)** DTC P0500, P0705, P0706, P0710, P0715, P0751, P0752, P0753, P0756, P0757, P0758, P0761, P0762, P0763, P0766, P0767, P0768, P0771, P0772 and P0773 not set, engine started, then driven at 37-62 mph in Drive in 4th Gear (4GR), Turbine speed at 4988 rpm (± 225), engine speed over 450 rpm, ATF temperature over 68ºF, TCC operation enabled while in normal or power mode, Shift Solenoid 'A' commanded to more than a 99% duty cycle, and the PCM detected the difference between the engine speed and Turbine speed was more than 100 rpm in the CCM rationality test. **Possible Causes:** • ATF level is too low, or the ATF is badly deteriorated • A/T line pressure is low, or the oil pump has failed • A/T control valve is stuck, or it is damaged (a mechanical fault) • SSA, SSB, SSC, SSD, SSE or the PCS is stuck "off" • A/T 2-4 brake band is slipping (a mechanical fault) • A/T 3-4 clutch is slipping (a mechanical fault) • TCM has failed
DTC: P0741 **1T CCM, MIL: NO** **1996, 1997** **Models:** B2300, B3000, B4000 **Engines:** All **Transmissions:** A/T	**Torque Converter Clutch Control Engagement Error** Engine started, then driven to a speed of over 30 mph, and the PCM detected the TCC system did not operate (due to a mechanical fault. **Possible Causes:** • Brake switch is damaged, out of adjustment or has failed • Engine performance conditions are erratic or unstable • Engine or transmission temperature is too high • TCC control valve is stuck in the "off" position • An internal transaxle component is damaged or has failed • PCM has failed
DTC: P0741 **1T CCM, MIL: NO** **2004, 2005** **Models:** Mazda3, Tribute **Engines:** All **Transmissions:** A/T	**Torque Converter Clutch Control Engagement Error** Engine started, then driven to a speed of over 30 mph, and the PCM detected the TCC system did not operate (due to a mechanical fault. **Possible Causes:** • Brake switch is damaged, out of adjustment or has failed • Engine performance conditions are erratic or unstable • Engine or transmission temperature is too high • TCC control valve is stuck in the "off" position • An internal transaxle component is damaged or has failed • PCM has failed
DTC: P0742 **2T CCM, MIL: YES** **1999, 2000, 2001, 2002, 2003, 2004, 2005** **Models:** MX-5 Miata **Engines:** All **Transmissions:** A/T, M/T	**Torque Converter Clutch Performance (Mechanical)** DTC P0705, P0706, P0715, P0720, P1740, P1742, P1751, P1752, P1756, P1757, P1771 and P1772 not set, engine started, vehicle driven in Drive at 60 mph (± 3 mph) for 20 seconds, engine speed over 600 rpm, ECT sensor signal more than 113ºF, brake pedal switch indicating "off", throttle angle over 8%, and the PCM detected the difference between the engine speed and Turbine speed was more than a predetermined value. **Possible Causes:** • ATF level is too low, or the ATF is badly deteriorated • A/T line pressure is low, or the oil pump has failed • A/T control valve is stuck, or it is damaged (a mechanical fault) • TCC solenoid is stuck "on", or is damaged (a mechanical fault) • TCC system has failed • TCM has failed
DTC: P0742 **1T CCM, MIL: NO** **1999, 2000, 2001, 2002, 2003** **Models:** Protégé, Protégé5 **Engines:** All **Transmissions:** A/T	**Torque Converter Clutch Performance (Mechanical)** P0734 not set, engine started, then driven at below 45 mph in Drive in 4th Gear (4GR), Turbine speed at 4988 rpm (± 225), engine speed over 450 rpm, ATF temperature over 68ºF, TCC operation not enabled, and the PCM detected the difference between the engine speed and Turbine speed was less than 100 rpm in the CCM test. **Possible Causes:** • ATF level is too low, or the ATF is badly deteriorated • A/T line pressure is low, or the oil pump has failed • A/T control valve is stuck, or it is damaged (a mechanical fault) • SSA, SSB, SSC, SSD, SSE or the PCS is stuck "on" • A/T 2-4 brake band is slipping (a mechanical fault) • A/T 3-4 clutch is slipping (a mechanical fault) • TCM has failed

DTC	Trouble Code Title, Conditions & Possible Causes
DTC: P0743 **1T CCM, MIL: YES** **1996, 1997, 1998, 1999, 2000, 2001, 2002** **Models:** 626, MX-6 **Engines:** All **Transmissions:** A/T	**Torque Converter Clutch Solenoid Circuit Malfunction (Self-Test)** KOER Self Test Enabled: Engine running and the PCM detected an unexpected voltage condition on the TCC solenoid control circuit during the CCM test. **Possible Causes:** • TCC control circuit is open (continuous high signal) • TCC control circuit is shorted to ground (continuous low signal) • TCC control circuit is shorted to system power (B+) • TCC solenoid is damaged or has failed • PCM has failed
DTC: P0743 **1T CCM, MIL: YES** **2004, 2005** **Models:** Mazda3 **Engines:** All **Transmissions:** A/T	**Torque Converter Clutch Solenoid Circuit Malfunction (Self-Test)** KOER Self Test Enabled: Engine running and the PCM detected an unexpected voltage condition on the TCC solenoid control circuit during the CCM test. **Possible Causes:** • TCC control circuit is open (continuous high signal) • TCC control circuit is shorted to ground (continuous low signal) • TCC control circuit is shorted to system power (B+) • TCC solenoid is damaged or has failed • PCM has failed
DTC: P0743 **1T CCM, MIL: YES** **2000, 2001, 2002, 2003, 2004, 2005** **Models:** MPV, MX-5 Miata, Tribute **Engines:** All **Transmissions:** A/T	**Torque Converter Clutch Solenoid Circuit Malfunction** Engine running and the PCM detected an unexpected voltage condition on the Torque Converter Clutch (TCC) solenoid control circuit during the CCM test. **Possible Causes:** • TCC control circuit is open (continuous high signal) • TCC control circuit is shorted to ground (continuous low signal) • TCC control circuit is shorted to system power (B+) • TCC solenoid is damaged or has failed • PCM has failed
DTC: P0743 **1T CCM, MIL: YES** **1996, 1997, 1998, 1999, 2000, 2001, 2002, 2003** **Models:** B3000, B4000 **Engines:** All **Transmissions:** A/T	**Torque Converter Clutch Solenoid Circuit Malfunction** Engine running and the PCM detected an unexpected voltage condition on the Torque Converter Clutch (TCC) solenoid control circuit during the CCM test. **Possible Causes:** • TCC control circuit is open (continuous high signal) • TCC control circuit is shorted to ground (continuous low signal) • TCC control circuit is shorted to system power (B+) • TCC solenoid is damaged or has failed • PCM has failed
DTC: P0745 **1T CCM, MIL: NO** **1996, 1997, 1998, 1999, 2000, 2001, 2002** **Models:** 626, MX-6 **Engines:** 2.5L VIN D **Transmissions:** A/T	**Pressure Control Solenoid Circuit Malfunction** Engine running and the PCM detected an unexpected voltage condition on the Pressure Control Solenoid (PCS) control circuit during the CCM test. **Possible Causes:** • PCS control circuit is open (continuous high signal) • PCS control circuit is shorted to ground (continuous low signal) • PCS control circuit is shorted to system power (B+) • PCS is damaged or has failed • PCM has failed
DTC: P0745 **1T CCM, MIL: NO** **1996, 1997, 1998, 1999, 2000, 2001, 2002** **Models:** Millenia **Engines:** All **Transmissions:** A/T	**Pressure Control Solenoid Circuit Malfunction** Engine running, vehicle driven to a speed of over 37 mph, and the PCM detected an unexpected voltage condition on the Pressure Control Solenoid (PCS) control circuit during the CCM test. **Possible Causes:** • PCS control circuit is open (continuous high signal) • PCS circuit is shorted to ground (continuous low signal) • PCS control circuit is shorted to system power (B+) • PCS is damaged or has failed • PCM has failed

DTC	Trouble Code Title, Conditions & Possible Causes
DTC: P0745 **1T CCM, MIL: NO** **1996, 1997, 1998, 1999, 2000, 2001, 2002, 2003** **Models:** Protégé, Protégé5 **Engines:** All **Transmissions:** A/T	**Pressure Control Solenoid Circuit Malfunction** Engine running and the PCM detected an unexpected voltage condition on the Pressure Control Solenoid (PCS) circuit in the CCM test. **Possible Causes:** • PCS control circuit is open (continuous high signal) • PCS circuit is shorted to ground (continuous low signal) • PCS control circuit is shorted to system power (B+) • PCS is damaged or has failed • PCM has failed
DTC: P0745 **1T CCM, MIL: YES** **1996, 1997, 1998, 1999, 2000, 2001, 2002, 2003** **Models:** MPV **Engines:** All **Transmissions:** A/T	**Pressure Control Solenoid Circuit Malfunction** Engine running and the PCM detected an unexpected voltage condition on the Pressure Control Solenoid (PCS) control circuit during the CCM test. **Possible Causes:** • PCS control circuit is open (continuous high signal) • PCS control circuit is shorted to ground (continuous low signal) • PCS control circuit is shorted to system power (B+) • PCS is damaged or has failed • PCM has failed
DTC: P0745 **1T CCM, MIL: NO** **1996, 1997** **Models:** B2300, B3000, B4000 **Engines:** 2.5L VIN D **Transmissions:** A/T	**Electronic Pressure Control Solenoid Circuit Malfunction** Engine running vehicle driven to a speed of over 37 mph, and the PCM detected an unexpected voltage condition on the Electronic Pressure Control (EPC) solenoid control circuit during the CCM test. **Possible Causes:** • PCS control circuit is open (continuous high signal) • PCS control circuit is shorted to ground (continuous low signal) • PCS control circuit is shorted to system power (B+) • PCS is damaged or has failed • PCM has failed
DTC: P0745 **1T CCM, MIL: NO** **2004, 2005** **Models:** Mazda3, Tribute **Engines:** All **Transmissions:** A/T	**Electronic Pressure Control Solenoid Circuit Malfunction** Engine running vehicle driven to a speed of over 37 mph, and the PCM detected an unexpected voltage condition on the Electronic Pressure Control (EPC) solenoid control circuit during the CCM test. **Possible Causes:** • PCS control circuit is open (continuous high signal) • PCS control circuit is shorted to ground (continuous low signal) • PCS control circuit is shorted to system power (B+) • PCS is damaged or has failed • PCM has failed
DTC: P0748 **1T CCM, MIL: NO** **2005** **Models:** Tribute **Engines:** All **Transmissions:** A/T	**Electronic Pressure Control Solenoid Circuit Malfunction** Engine running vehicle driven to a speed of over 37 mph, and the PCM detected an unexpected voltage condition on the Electronic Pressure Control (EPC) solenoid control circuit during the CCM test. **Possible Causes:** • PCS control circuit is open (continuous high signal) • PCS control circuit is shorted to ground (continuous low signal) • PCS control circuit is shorted to system power (B+) • PCS is damaged or has failed • PCM has failed
DTC: P0750 **2T CCM, MIL: YES** **1996, 1997, 1998, 1999, 2000, 2001, 2002** **Models:** 626, MX-6, Millenia **Engines:** All **Transmissions:** A/T	**A/T Shift Solenoid 'A' Circuit Malfunction** Engine started, then driven to a speed of over 37 mph, and the PCM detected an unexpected voltage condition on the Shift Solenoid 'A' (SSA) control circuit during the CCM test. **Possible Causes:** • SSA control circuit is open (continuous high signal) • SSA control circuit is shorted to ground (continuous low signal) • SSA control circuit is shorted to system power (B+) • SSA is damaged or has failed • PCM has failed

DTC	Trouble Code Title, Conditions & Possible Causes
DTC: P0750 **2T CCM, MIL: YES** **1996, 1997** **Models:** MX-5 Miata **Engines:** All **Transmissions:** A/T	**A/T Shift Solenoid 'A' Circuit Malfunction** Engine started, then driven to a speed of over 37 mph, and the PCM detected an unexpected voltage condition on the Shift Solenoid 'A' (SSA) control circuit during the CCM test. **Possible Causes:** • SSA control circuit is open (continuous high signal) • SSA control circuit is shorted to ground (continuous low signal) • SSA control circuit is shorted to system power (B+) • SSA is damaged or has failed • PCM has failed
DTC: P0750 **1T CCM, MIL: NO** **1996, 1997, 1998** **Models:** Protégé **Engines:** All **Transmissions:** A/T	**A/T Shift Solenoid 'A' Circuit Malfunction** Engine started, and the PCM detected an unexpected voltage condition on the Shift Solenoid 'A' (SSA) circuit during the CCM test. **Possible Causes:** • SSA control circuit is open (continuous high signal) • SSA control circuit is shorted to ground (continuous low signal) • SSA control circuit is shorted to system power (B+) • SSA is damaged or has failed • PCM has failed
DTC: P0750 **1T CCM, MIL: YES** **1996, 1997, 1998, 1999, 2000, 2001, 2002, 2003, 2004, 2005** **Models:** B3000, B4000, MPV, Tribute **Engines:** All **Transmissions:** A/T	**A/T Shift Solenoid 'A' Circuit Malfunction** Engine started, and the PCM detected an unexpected voltage condition on the Shift Solenoid 'A' (SSA) circuit during the CCM test. **Possible Causes:** • SSA control circuit is open (continuous high signal) • SSA control circuit is shorted to ground (continuous low signal) • SSA control circuit is shorted to system power (B+) • SSA is damaged or has failed • PCM has failed
DTC: P0751 **2T CCM, MIL: YES** **1996, 1997, 1998, 1999, 2000, 2001, 2002** **Models:** 626, MX-6 **Engines:** 2.0L VIN C **Transmissions:** A/T	**A/T Shift Solenoid 'A' Performance (Mechanical)** Engine started, then driven to a speed of over 37 mph, and the PCM detected a mechanical shift failure in the Shift Solenoid 'A' (SSA) during the CCM test. **Possible Causes:** • SSA is stuck in "off" position (mechanical problem) • SSA is damaged is damaged or has failed (mechanical fault) • SSA has a hydraulic problem • PCM has failed
DTC: P0751 **2T CCM, MIL: YES** **1999, 2000, 2001, 2002, 2003, 2004, 2005** **Models:** MX-5 Miata **Engines:** All **Transmissions:** A/T	**A/T Shift Solenoid 'A' Performance (Mechanical)** DTC P0705, P0706, P0715, P0720, P1740, P1742, P1751, P1752, P1756, P1757, P1771 and P1772 not set, engine started, vehicle driven in Drive at 60 mph (± 3 mph) for 20 seconds, engine speed over 600 rpm, ECT sensor signal more than 113°F, brake pedal switch indicating "off", throttle angle over 8%, and the PCM detected the difference between the engine speed and Turbine speed was more than a predetermined value. **Possible Causes:** • ATF level is too low, or the ATF is badly deteriorated • A/T line pressure is low, or the oil pump has failed • A/T control valve is stuck, or it is damaged (a mechanical fault) • SSA is stuck "off", or is damaged (a mechanical fault) • Transmission is damaged or has failed • TCM has failed
DTC: P0751 **2T CCM, MIL: YES** **1999, 2000, 2001, 2002, 2003** **Models:** Protégé, Protégé5 **Engines:** All **Transmissions:** A/T	**A/T Shift Solenoid 'A' Performance (Mechanical)** DTC P0731, P0732 and P0733 not set, engine started, then driven in Drive in 4th Gear (4GR), Turbine speed at 4988 rpm (± 225), ATF temperature over 68°F, engine speed over 450 rpm, differential gear case revolution speed over 35 mph, TCC not enabled, and the PCM detected the revolution ratio of the forward clutch drum revolution to the differential gear revolution was with a range of from 0.91-1.09. **Possible Causes:** • ATF level is too low, or the ATF is badly deteriorated • A/T line pressure is low, or the oil pump has failed • A/T control valve is stuck, or it is damaged (a mechanical fault) • SSA is stuck "off", or is damaged (a mechanical fault) • Transmission is damaged or has failed • TCM has failed

DTC	Trouble Code Title, Conditions & Possible Causes
DTC: P0751 **2T CCM, MIL: YES** **1996, 1997, 1998, 1999, 2000, 2001, 2002, 2003, 2004, 2005** **Models:** B3000, B4000, Tribute **Engines:** All **Transmissions:** A/T	**A/T Shift Solenoid 'A' Performance (Mechanical)** Engine started, then driven to a speed of over 37 mph, and the PCM detected a mechanical shift failure in the Shift Solenoid 'A' (SSA) during the CCM test. **Possible Causes:** • SSA is stuck in "off" position (mechanical problem) • SSA is damaged is damaged or has failed (mechanical fault) • SSA has a hydraulic problem • PCM has failed
DTC: P0751 **2T CCM, MIL: YES** **2004, 2005** **Models:** Mazda3 **Engines:** All **Transmissions:** A/T	**A/T Shift Solenoid 'A' Performance (Mechanical)** Engine started, then driven to a speed of over 37 mph, and the PCM detected a mechanical shift failure in the Shift Solenoid 'A' (SSA) during the CCM test. **Possible Causes:** • SSA is stuck in "off" position (mechanical problem) • SSA is damaged is damaged or has failed (mechanical fault) • SSA has a hydraulic problem • PCM has failed
DTC: P0752 **2T CCM, MIL: YES** **1999, 2000, 2001, 2002, 2003, 2004, 2005** **Models:** MX-5 Miata **Engines:** All **Transmissions:** A/T	**A/T Shift Solenoid 'A' Performance (Mechanical)** DTC P0705, P0706, P0715, P0720, P1740, P1742, P1751, P1752, P1756, P1757, P1771 and P1772 not set, engine started, vehicle driven in Drive at 60 mph (± 3 mph) for 20 seconds, engine speed over 600 rpm, ECT sensor signal more than 113°F, brake pedal switch indicating "off", throttle angle over 8%, and the PCM detected the difference between the engine speed and Turbine speed was more than a predetermined value. **Possible Causes:** • ATF level is too low, or the ATF is badly deteriorated • A/T line pressure is low, or the oil pump has failed • A/T control valve is stuck, or it is damaged (a mechanical fault) • SSA is stuck "on", or is damaged (a mechanical fault) • Transmission is damaged or has failed • TCM has failed
DTC: P0752 **2T CCM, MIL: YES** **1999, 2000, 2001, 2002, 2003** **Models:** Protégé, Protégé5 **Engines:** All **Transmissions:** A/T	**A/T Shift Solenoid 'A' Performance (Mechanical)** DTC P0500, P0705, P0706, P0710, P0715, P0734, P0751, P0753, P0756, P0758, P0761, P0762, P0763, P0766, P0767, P0768, P0771, P0772 and P0773 not set, engine started, then driven in Drive in 1st Gear (1GR) or 2nd Gear (2GR), Turbine speed at 4988 rpm (± 225), ATF temperature over 68°F, engine speed over 450 rpm, throttle position indicating closed, brake pedal depressed, VSS at 0 mph, and the PCM detected the Input/Turbine speed sensor signal was 187.5 or higher. **Possible Causes:** • ATF level is too low, or the ATF is badly deteriorated • A/T line pressure is low, or the oil pump has failed • A/T control valve is stuck, or it is damaged (a mechanical fault) • SSA is stuck "on", or is damaged (a mechanical fault) • Transmission is damaged or has failed • TCM has failed
DTC: P0752 **2T CCM, MIL: YES** **2004, 2005** **Models:** Mazda3, Tribute **Engines:** All **Transmissions:** A/T	**A/T Shift Solenoid 'A' Performance (Mechanical)** DTC P0500, P0705, P0706, P0710, P0715, P0734, P0751, P0753, P0756, P0758, P0761, P0762, P0763, P0766, P0767, P0768, P0771, P0772 and P0773 not set, engine started, then driven in Drive in 1st Gear (1GR) or 2nd Gear (2GR), Turbine speed at 4988 rpm (± 225), ATF temperature over 68°F, engine speed over 450 rpm, throttle position indicating closed, brake pedal depressed, VSS at 0 mph, and the PCM detected the Input/Turbine speed sensor signal was 187.5 or higher. **Possible Causes:** • ATF level is too low, or the ATF is badly deteriorated • A/T line pressure is low, or the oil pump has failed • A/T control valve is stuck, or it is damaged (a mechanical fault) • SSA is stuck "on", or is damaged (a mechanical fault) • Transmission is damaged or has failed • TCM has failed
DTC: P0753 **1T CCM, MIL: YES** **1999, 2000, 2001, 2002, 2003** **Models:** Protégé, Protégé5 **Engines:** All **Transmissions:** A/T	**A/T Shift Solenoid 'B' Circuit Malfunction** Engine started, then driven to a speed of over 37 mph, and the PCM detected an unexpected voltage condition on the Shift Solenoid 'B' (SSB) control circuit during the CCM test. **Possible Causes:** • SSB control circuit is open (continuous high signal) • SSB control circuit is shorted to ground (continuous low signal) • SSB control circuit is shorted to system power (B+) • SSB is damaged or has failed • PCM has failed

DTC	Trouble Code Title, Conditions & Possible Causes
DTC: P0753 **1T CCM, MIL: YES** **2004, 2005** **Models:** Mazda3, Tribute **Engines:** All **Transmissions:** A/T	**A/T Shift Solenoid 'B' Circuit Malfunction** Engine started, then driven to a speed of over 37 mph, and the PCM detected an unexpected voltage condition on the Shift Solenoid 'B' (SSB) control circuit during the CCM test. **Possible Causes:** • SSB control circuit is open (continuous high signal) • SSB control circuit is shorted to ground (continuous low signal) • SSB control circuit is shorted to system power (B+) • SSB is damaged or has failed • PCM has failed
DTC: P0755 **1T CCM, MIL: YES** **1996, 1997, 1998, 1999, 2000, 2001, 2002** **Models:** 626, MX-6, Millenia **Engines:** All **Transmissions:** A/T	**A/T Shift Solenoid 'B' Circuit Malfunction** Engine started, then driven to a speed of over 37 mph, and the PCM detected an unexpected voltage condition on the Shift Solenoid 'B' (SSB) control circuit during the CCM test. **Possible Causes:** • SSB control circuit is open (continuous high signal) • SSB control circuit is shorted to ground (continuous low signal) • SSB control circuit is shorted to system power (B+) • SSB is damaged or has failed • PCM has failed
DTC: P0755 **2T CCM, MIL: YES** **1996, 1997** **Models:** MX-5 Miata **Engines:** All **Transmissions:** A/T	**A/T Shift Solenoid 'B' Circuit Malfunction** Engine started, then driven to a speed of over 37 mph, and the PCM detected an unexpected voltage condition on the Shift Solenoid 'B' (SSB) control circuit during the CCM test. **Possible Causes:** • SSB control circuit is open (continuous high signal) • SSB control circuit is shorted to ground (continuous low signal) • SSB control circuit is shorted to system power (B+) • SSB is damaged or has failed • PCM has failed
DTC: P0755 **1T CCM, MIL: NO** **1996, 1997, 1998** **Models:** Protégé **Engines:** All **Transmissions:** A/T	**A/T Shift Solenoid 'B' Circuit Malfunction** Engine started, then driven to a speed of over 37 mph, and the PCM detected an unexpected voltage condition on the Shift Solenoid 'B' (SSB) control circuit during the CCM test. **Possible Causes:** • SSB control circuit is open (continuous high signal) • SSB control circuit is shorted to ground (continuous low signal) • SSB control circuit is shorted to system power (B+) • SSB is damaged or has failed • PCM has failed
DTC: P0755 **1T CCM, MIL: YES** **1996, 1997, 1998, 1999, 2000, 2001, 2002, 2003, 2004, 2005** **Models:** B2300, B3000, B4000, MPV, Tribute **Engines:** All **Transmissions:** A/T	**A/T Shift Solenoid 'B' Circuit Malfunction** Engine started, and the PCM detected an unexpected voltage condition on the Shift Solenoid 'B' (SSB) circuit during the CCM test. **Possible Causes:** • SSB control circuit is open (continuous high signal) • SSB control circuit is shorted to ground (continuous low signal) • SSB control circuit is shorted to system power (B+) • SSB is damaged or has failed • PCM has failed
DTC: P0756 **2T CCM, MIL: YES** **1996, 1997, 1998, 1999, 2000, 2001, 2002** **Models:** 626, MX-6 **Engines:** 2.0L VIN C **Transmissions:** A/T	**A/T Shift Solenoid 'B' Performance (Mechanical)** Engine started, then driven to a speed of over 37 mph, and the PCM detected a mechanical shift failure in the Shift Solenoid 'B' (SSB) during the CCM test. **Possible Causes:** • SSB is stuck in "on" position (mechanical problem) • SSB is damaged is damaged or has failed (mechanical fault) • SSB has a hydraulic problem • PCM has failed

DTC	Trouble Code Title, Conditions & Possible Causes
DTC: P0756 **2T CCM, MIL: YES** **1999, 2000, 2001, 2002, 2003** **Models:** Protégé, Protégé5 **Engines:** All **Transmissions:** A/T	**A/T Shift Solenoid 'B' Performance (Mechanical)** 1st Gear Test DTC P0732, P0733 and P0734 not set, engine started, then driven in Drive in 1st Gear (1GR), Turbine speed at 4988 rpm (± 225), ATF temperature at 68°F or higher, engine speed over 450 rpm, throttle angle over 6.25% on FP engine or over 3.91% on ZM engine, differential gear case output speed over 35 mph, and the PCM detected the revolution ratio of the forward clutch drum revolution to the differential gear case revolution was less than 2.157. **Possible Causes:** • ATF level is too low, or the ATF is badly deteriorated • A/T SSB is stuck "off", or it is damaged (a mechanical fault) • A/T control valve is stuck, or it is damaged (a mechanical fault) • PCM has failed
DTC: P0756 **2T CCM, MIL: YES** **1996, 1997, 1998, 1999, 2000, 2001, 2002, 2003, 2004, 2005** **Models:** MX-5 Miata **Engines:** All **Transmissions:** A/T	**A/T Shift Solenoid 'B' Performance (Mechanical)** DTC P0705, P0706, P0715, P0720, P1740, P1742, P1751, P1752, P1756, P1757, P1771 and P1772 not set, engine started, vehicle driven in Drive at 60 mph (± 3 mph) for 20 seconds, engine speed over 600 rpm, ECT sensor signal more than 113°F, brake pedal switch indicating "off", throttle angle over 8%, and the PCM detected the difference between the engine speed and Turbine speed was more than a predetermined value. **Possible Causes:** • ATF level is too low, or the ATF is badly deteriorated • A/T line pressure is low, or the oil pump has failed • A/T control valve is stuck, or it is damaged (a mechanical fault) • SSB is stuck "off", or is damaged (a mechanical fault) • Transmission is damaged or has failed • TCM has failed
DTC: P0756 **2T CCM, MIL: YES** **1996, 1997, 1998, 1999, 2000, 2001, 2002, 2003, 2004, 2005** **Models:** B3000, B4000, Tribute **Engines:** All **Transmissions:** A/T	**A/T Shift Solenoid 'B' Performance (Mechanical)** Engine started, then driven to a speed of over 37 mph, and the PCM detected a mechanical shift failure in the Shift Solenoid 'B' (SSB) during the CCM test. **Possible Causes:** • SSB is stuck in "on" position (a mechanical fault) • SSB is damaged is damaged or has failed (a mechanical fault) • SSB has a hydraulic problem • PCM has failed
DTC: P0756 **2T CCM, MIL: YES** **2004, 2005** **Models:** Mazda3 **Engines:** All **Transmissions:** A/T	**A/T Shift Solenoid 'B' Performance (Mechanical)** Engine started, then driven to a speed of over 37 mph, and the PCM detected a mechanical shift failure in the Shift Solenoid 'B' (SSB) during the CCM test. **Possible Causes:** • SSB is stuck in "on" position (a mechanical fault) • SSB is damaged is damaged or has failed (a mechanical fault) • SSB has a hydraulic problem • PCM has failed
DTC: P0757 **2T CCM, MIL: YES** **1996, 1997, 1998, 1999, 2000, 2001, 2002, 2003, 2004, 2005** **Models:** MX-5 Miata **Engines:** All **Transmissions:** A/T	**A/T Shift Solenoid 'B' Performance (Mechanical)** DTC P0705, P0706, P0715, P0720, P1740, P1742, P1751, P1752, P1756, P1757, P1771 and P1772 not set, engine started, vehicle driven in Drive at 60 mph (± 3 mph) for 20 seconds, engine speed over 600 rpm, ECT sensor signal more than 113°F, brake pedal switch indicating "off", throttle angle over 8%, and the PCM detected the difference between the engine speed and Turbine speed was more than a predetermined value. **Possible Causes:** • ATF level is too low, or the ATF is badly deteriorated • A/T line pressure is low, or the oil pump has failed • A/T control valve is stuck, or it is damaged (a mechanical fault) • SSB is stuck "on", or is damaged (a mechanical fault) • Transmission is damaged or has failed • TCM has failed

DTC	Trouble Code Title, Conditions & Possible Causes
DTC: P0757 **2T CCM, MIL: YES** **1999, 2000, 2001, 2002, 2003** **Models:** Protégé, Protégé5 **Engines:** All **Transmissions:** A/T	**A/T Shift Solenoid 'B' Performance (Mechanical)** 2nd Gear Test DTC P0731 and P0733 not set, engine started, then driven in Drive in 2nd Gear (2GR), Turbine speed at 4988 rpm (± 225), engine speed over 450 rpm, ATF temperature over 68°F, differential gear output speed over 35 mph, and the PCM detected the revolution ratio of the forward clutch drum revolution to the differential gear case revolution was less than 1.249, or more than 2.157. 4th Gear Test With the conditions listed above for the 2nd Gear Test met, with the throttle closed, the PCM detected the revolution ratio of the forward clutch drum revolution to differential gear case revolution was less than 0.60, or it was 1.249 or higher while driving in 4th gear (4GR). **Possible Causes:** • ATF level is too low, or the ATF is badly deteriorated • A/T SSB is stuck "on", or it is damaged (a mechanical fault) • A/T control valve is stuck, or it is damaged (a mechanical fault) • PCM has failed
DTC: P0757 **2T CCM, MIL: YES** **1999, 2000, 2001, 2004, 2005** **Models:** Mazda3, Tribute **Engines:** All **Transmissions:** A/T	**A/T Shift Solenoid 'B' Performance (Mechanical)** 2nd Gear Test DTC P0731 and P0733 not set, engine started, then driven in Drive in 2nd Gear (2GR), Turbine speed at 4988 rpm (± 225), engine speed over 450 rpm, ATF temperature over 68°F, differential gear output speed over 35 mph, and the PCM detected the revolution ratio of the forward clutch drum revolution to the differential gear case revolution was less than 1.249, or more than 2.157. 4th Gear Test With the conditions listed above for the 2nd Gear Test met, with the throttle closed, the PCM detected the revolution ratio of the forward clutch drum revolution to differential gear case revolution was less than 0.60, or it was 1.249 or higher while driving in 4th gear (4GR). **Possible Causes:** • ATF level is too low, or the ATF is badly deteriorated • A/T SSB is stuck "on", or it is damaged (a mechanical fault) • A/T control valve is stuck, or it is damaged (a mechanical fault) • PCM has failed
DTC: P0758 **1T CCM, MIL: YES** **1999, 2000, 2001, 2002, 2003** **Models:** Protégé, Protégé5 **Engines:** All **Transmissions:** A/T	**A/T Shift Solenoid 'B' Circuit Malfunction** Engine started, and the PCM detected an unexpected voltage condition on the Shift Solenoid 'B' (SSB) control circuit during the CCM test. **Possible Causes:** • SSB control circuit is open (continuous high signal) • SSB control circuit is shorted to ground (continuous low signal) • SSB control circuit is shorted to system power (B+) • SSB is damaged or has failed • PCM has failed
DTC: P0758 **1T CCM, MIL: YES** **1999, 2000, 2001, 2002, 2003, 2004, 2005** **Models:** Tribute **Engines:** All **Transmissions:** A/T	**A/T Shift Solenoid 'B' Circuit Malfunction** Engine started, and the PCM detected an unexpected voltage condition on the Shift Solenoid 'B' (SSB) control circuit during the CCM test. **Possible Causes:** • SSB control circuit is open (continuous high signal) • SSB control circuit is shorted to ground (continuous low signal) • SSB control circuit is shorted to system power (B+) • SSB is damaged or has failed • PCM has failed
DTC: P0758 **1T CCM, MIL: YES** **2004, 2005** **Models:** Mazda3 **Engines:** All **Transmissions:** A/T	**A/T Shift Solenoid 'B' Circuit Malfunction** Engine started, and the PCM detected an unexpected voltage condition on the Shift Solenoid 'B' (SSB) control circuit during the CCM test. **Possible Causes:** • SSB control circuit is open (continuous high signal) • SSB control circuit is shorted to ground (continuous low signal) • SSB control circuit is shorted to system power (B+) • SSB is damaged or has failed • PCM has failed

DTC	Trouble Code Title, Conditions & Possible Causes
DTC: P0760 **1T CCM, MIL: YES** **1996, 1996, 1998, 1999, 2000, 2001, 2002** **Models:** 626, MX-6 **Engines:** All **Transmissions:** A/T	**A/T Shift Solenoid 'C' Circuit Malfunction** Engine started, then driven to over 37 mph, and the PCM detected an unexpected voltage condition on the Shift Solenoid 'C' (SSC) circuit during the CCM test. **Possible Causes:** • SSC control circuit is open (continuous high signal) • SSC control circuit is shorted to ground (continuous low signal) • SSC control circuit is shorted to system power (B+) • SSC is damaged or has failed • PCM has failed
DTC: P0760 **1T CCM, MIL: YES** **1996, 1996, 1998, 1999, 2000, 2001, 2002** **Models:** Millenia **Engines:** 2.5L VIN 1 **Transmissions:** A/T	**A/T Shift Solenoid 'C' Circuit Malfunction** Engine started, then driven to over 37 mph, and the PCM detected an unexpected voltage condition on the Shift Solenoid 'C' (SSC) control circuit during the CCM test. **Possible Causes:** • SSC control circuit is open (continuous high signal) • SSC control circuit is shorted to ground (continuous low signal) • SSC control circuit is shorted to system power (B+) • SSC is damaged or has failed • PCM has failed
DTC: P0760 **2T CCM, MIL: YES** **1996, 1997** **Models:** MX-5 Miata **Engines:** All **Transmissions:** A/T	**A/T Shift Solenoid 'C' Circuit Malfunction** Engine started, then driven to over 37 mph, and the PCM detected an unexpected voltage condition on the Shift Solenoid 'C' (SSC) control circuit during the CCM test. **Possible Causes:** • SSC control circuit is open (continuous high signal) • SSC control circuit is shorted to ground (continuous low signal) • SSC control circuit is shorted to system power (B+) • SSC is damaged or has failed • PCM has failed
DTC: P0760 **1T CCM, MIL: NO** **1996, 1997, 1998** **Models:** Protégé **Engines:** All **Transmissions:** A/T	**A/T Shift Solenoid 'C' Circuit Malfunction** Engine started, then driven to a speed of over 37 mph, and the PCM detected an unexpected voltage condition on the Shift Solenoid 'C' (SSC) control circuit during the CCM test. **Possible Causes:** • SSC control circuit is open (continuous high signal) • SSC control circuit is shorted to ground (continuous low signal) • SSC control circuit is shorted to system power (B+) • SSC is damaged or has failed • PCM has failed
DTC: P0760 **1T CCM, MIL: YES** **2000, 2001, 2002, 2003** **Models:** MPV **Engines:** All **Transmissions:** A/T	**A/T Shift Solenoid 'C' Circuit Malfunction** Engine started, then driven to a speed of over 37 mph, and the PCM detected an unexpected voltage condition on the Shift Solenoid 'C' (SSC) circuit during the CCM test. **Possible Causes:** • SSC control circuit is open (continuous high signal) • SSC control circuit is shorted to ground (continuous low signal) • SSC control circuit is shorted to system power (B+) • SSC is damaged or has failed • PCM has failed
DTC: P0760 **1T CCM, MIL: YES** **1996, 1997, 1998, 1999, 2000, 2001, 2002, 2003, 2004, 2005** **Models:** B2300, B3000, B4000 **Engines:** All **Transmissions:** A/T	**A/T Shift Solenoid 'C' Circuit Malfunction** Engine started, then driven to a speed of over 37 mph, and the PCM detected an unexpected voltage condition on the Shift Solenoid 'C' (SSC) circuit during the CCM test. **Possible Causes:** • SSC control circuit is open (continuous high signal) • SSC control circuit is shorted to ground (continuous low signal) • SSC control circuit is shorted to system power (B+) • SSC is damaged or has failed • PCM has failed

DTC	Trouble Code Title, Conditions & Possible Causes
DTC: P0761 **2T CCM, MIL: YES** **1999, 2000, 2001, 2002, 2003** **Models:** Protégé, Protégé5 **Engines:** All **Transmissions:** A/T	**A/T Shift Solenoid 'C' Performance (Mechanical)** 1st Gear Test DTC P0733 and P0734 not set, engine started, and then driven in Drive in 1st Gear (1GR), Turbine speed at 4988 rpm (± 225), ATF temperature at 68ºF or higher, engine speed over 450 rpm, differential gear case output speed over 35 mph, and the PCM detected the revolution ratio of the forward clutch drum revolution to the differential gear case revolution was less than 2.157. 2nd Gear Test With the conditions listed above for the 1st Gear Test met, the PCM detected the revolution ratio of the forward clutch drum revolution to differential gear case revolution was less than 1.249, or it was 2.157 or higher while driving in 2nd gear (2GR) during the CCM test. **Possible Causes:** • ATF level is too low, or the ATF is badly deteriorated • A/T SSC is stuck "off", or it is damaged (a mechanical fault) • A/T control valve is stuck, or it is damaged (a mechanical fault) • PCM has failed
DTC: P0761 **1T CCM, MIL: YES** **1996, 1997, 1998, 1999, 2000, 2001, 2002, 2003, 2004, 2005** **Models:** B2300, B3000, B4000 **Engines:** All **Transmissions:** A/T	**A/T Shift Solenoid 'C' Performance (Mechanical)** 1st Gear Test DTC P0733 and P0734 not set, engine started, and then driven in Drive in 1st Gear (1GR), Turbine speed at 4988 rpm (± 225), ATF temperature at 68ºF or higher, engine speed over 450 rpm, differential gear case output speed over 35 mph, and the PCM detected the revolution ratio of the forward clutch drum revolution to the differential gear case revolution was less than 2.157. **Possible Causes:** • ATF level is too low, or the ATF is badly deteriorated • A/T SSC is stuck "off", or it is damaged (a mechanical fault) • A/T control valve is stuck, or it is damaged (a mechanical fault) • PCM has failed
DTC: P0761 **1T CCM, MIL: YES** **2004, 2005** **Models:** Mazda3 **Engines:** All **Transmissions:** A/T	**A/T Shift Solenoid 'C' Performance (Mechanical)** 1st Gear Test DTC P0733 and P0734 not set, engine started, and then driven in Drive in 1st Gear (1GR), Turbine speed at 4988 rpm (± 225), ATF temperature at 68ºF or higher, engine speed over 450 rpm, differential gear case output speed over 35 mph, and the PCM detected the revolution ratio of the forward clutch drum revolution to the differential gear case revolution was less than 2.157. **Possible Causes:** • ATF level is too low, or the ATF is badly deteriorated • A/T SSC is stuck "off", or it is damaged (a mechanical fault) • A/T control valve is stuck, or it is damaged (a mechanical fault) • PCM has failed
DTC: P0762 **2T CCM, MIL: YES** **1999, 2000, 2001, 2002, 2003** **Models:** Protégé, Protégé5 **Engines:** 2.0L VIN C **Transmissions:** A/T	**A/T Shift Solenoid 'C' Performance (Mechanical)** 3rd Gear Test DTC P0731 and P0732 not set, engine started, and then driven in Drive in 3rd Gear (3GR), Turbine speed at 4988 rpm (± 225), ATF temperature at 68ºF or higher, engine speed over 450 rpm, differential gear case output speed over 35 mph, and the PCM detected the revolution ratio of the forward clutch drum revolution to the differential gear case revolution was less than 0.863, or it was more than 1.249 during the CCM rationality test. 4th Gear Test With the conditions listed above for the 3rd Gear Test met, the PCM detected the revolution ratio of the forward clutch drum revolution to differential gear case revolution was less than 0.6, or it was 1.249 or higher while driving in 4th gear (4GR) in the CCM rationality test. **Possible Causes:** • ATF level is too low, or the ATF is badly deteriorated • A/T SSC is stuck "on", or it is damaged (a mechanical fault) • A/T control valve stuck "on", or it is damaged (mechanical fault) • PCM has failed

DTC	Trouble Code Title, Conditions & Possible Causes
DTC: P0762 **2T CCM, MIL: YES** **2004, 2005** **Models:** Mazda3 **Engines:** All **Transmissions:** A/T	**A/T Shift Solenoid 'C' Performance (Mechanical)** 3rd Gear Test DTC P0731 and P0732 not set, engine started, and then driven in Drive in 3rd Gear (3GR), Turbine speed at 4988 rpm (± 225), ATF temperature at 68°F or higher, engine speed over 450 rpm, differential gear case output speed over 35 mph, and the PCM detected the revolution ratio of the forward clutch drum revolution to the differential gear case revolution was less than 0.863, or it was more than 1.249 during the CCM rationality test. 4th Gear Test With the conditions listed above for the 3rd Gear Test met, the PCM detected the revolution ratio of the forward clutch drum revolution to differential gear case revolution was less than 0.6, or it was 1.249 or higher while driving in 4th gear (4GR) in the CCM rationality test. **Possible Causes:** • ATF level is too low, or the ATF is badly deteriorated • A/T SSC is stuck "on", or it is damaged (a mechanical fault) • A/T control valve stuck "on", or it is damaged (mechanical fault) • PCM has failed
DTC: P0763 **1T CCM, MIL: YES** **1999, 2000, 2001, 2002, 2003** **Models:** Protégé, Protégé5 **Engines:** All **Transmissions:** A/T	**A/T Shift Solenoid 'C' Circuit Malfunction** Engine started, then driven in Drive in 4th Gear and the PCM detected an unexpected voltage condition on the Shift Solenoid 'C' (SSC) circuit during the CCM test. **Possible Causes:** • SSC control circuit is open (continuous high signal) • SSC control circuit is shorted to ground (continuous low signal) • SSC control circuit is shorted to system power (B+) • SSC is damaged or has failed • PCM has failed
DTC: P0763 **1T CCM, MIL: YES** **2004, 2005** **Models:** Mazda3 **Engines:** All **Transmissions:** A/T	**A/T Shift Solenoid 'C' Circuit Malfunction** Engine started, then driven in Drive in 4th Gear and the PCM detected an unexpected voltage condition on the Shift Solenoid 'C' (SSC) circuit during the CCM test. **Possible Causes:** • SSC control circuit is open (continuous high signal) • SSC control circuit is shorted to ground (continuous low signal) • SSC control circuit is shorted to system power (B+) • SSC is damaged or has failed • PCM has failed
DTC: P0765 **1T CCM, MIL: YES** **1996, 1997, 1998, 1999, 2000, 2001, 2002, 2003, 2004, 2005** **Models:** B2300, B3000, B4000 **Engines:** All **Transmissions:** A/T	**A/T Shift Solenoid and TCC Solenoid Performance** Engine started, then driven in Drive to over 35 mph, and the PCM detected a problem in one or more the Shift solenoids or the TCC solenoid during the CCM rationality test. **Possible Causes:** • ATF level is too low, or the ATF is badly deteriorated • A/T SSA, SSB, SSC, SSD or TCC is stuck "off", or it is damaged (a mechanical fault) • A/T control valve is stuck, or it is damaged (a mechanical fault) • PCM has failed
DTC: P0766 **2T CCM, MIL: YES** **1999, 2000, 2001, 2002, 2003** **Models:** Protégé, Protégé5 **Engines:** All **Transmissions:** A/T	**A/T Shift Solenoid 'D' Performance (Mechanical)** 4th Gear Test DTC P0731, P0132 and P0733 not set, engine started, then driven in Drive in 4th Gear (4GR), Turbine speed at 4988 rpm (± 225), ATF temperature over 68°F, engine speed over 450 rpm, differential gear case output speed over 35 mph, and the PCM detected the revolution ratio of the forward clutch drum revolution to the differential gear case revolution was less than 0.60, or it was 1.249 or higher during the CCM rationality test. **Possible Causes:** • ATF level is too low, or the ATF is badly deteriorated • A/T SSD is stuck "off", or it is damaged (a mechanical fault) • A/T control valve is stuck, or it is damaged (a mechanical fault) • PCM has failed

DTC	Trouble Code Title, Conditions & Possible Causes
DTC: P0766 **2T CCM, MIL: YES** **2004, 2005** **Models:** Mazda3 **Engines:** All **Transmissions:** A/T	**A/T Shift Solenoid 'D' Performance (Mechanical)** 4th Gear Test DTC P0731, P0132 and P0733 not set, engine started, then driven in Drive in 4th Gear (4GR), Turbine speed at 4988 rpm (± 225), ATF temperature over 68ºF, engine speed over 450 rpm, differential gear case output speed over 35 mph, and the PCM detected the revolution ratio of the forward clutch drum revolution to the differential gear case revolution was less than 0.60, or it was 1.249 or higher during the CCM rationality test. **Possible Causes:** • ATF level is too low, or the ATF is badly deteriorated • A/T SSD is stuck "off", or it is damaged (a mechanical fault) • A/T control valve is stuck, or it is damaged (a mechanical fault) • PCM has failed
DTC: P0767 **2T CCM, MIL: YES** **1999, 2000, 2001, 2002, 2003** **Models:** Protégé, Protégé5 **Engines:** All **Transmissions:** A/T	**A/T Shift Solenoid 'D' Performance (Mechanical)** 3rd Gear Test DTC P0731, P0132, P0734 and P0741 not set, engine started, then driven in Drive in 3rd Gear (3GR) Turbine speed at 4988 rpm (± 225), engine speed over 450 rpm, differential gear case output speed over 35 mph, ATF temperature over 68ºF, and the PCM detected the revolution ratio of the forward clutch drum revolution to the differential gear case revolution was less than 0.863, or it was 1.249 or higher during the CCM rationality test. **Possible Causes:** • ATF level is too low, or the ATF is badly deteriorated • A/T SSD is stuck "on", or it is damaged (a mechanical fault) • A/T control valve is stuck, or it is damaged (a mechanical fault) • PCM has failed
DTC: P0767 **2T CCM, MIL: YES** **2004, 2005** **Models:** Mazda3 **Engines:** All **Transmissions:** A/T	**A/T Shift Solenoid 'D' Performance (Mechanical)** 3rd Gear Test DTC P0731, P0132, P0734 and P0741 not set, engine started, then driven in Drive in 3rd Gear (3GR) Turbine speed at 4988 rpm (± 225), engine speed over 450 rpm, differential gear case output speed over 35 mph, ATF temperature over 68ºF, and the PCM detected the revolution ratio of the forward clutch drum revolution to the differential gear case revolution was less than 0.863, or it was 1.249 or higher during the CCM rationality test. **Possible Causes:** • ATF level is too low, or the ATF is badly deteriorated • A/T SSD is stuck "on", or it is damaged (a mechanical fault) • A/T control valve is stuck, or it is damaged (a mechanical fault) • PCM has failed
DTC: P0768 **1T CCM, MIL: YES** **1999, 2000, 2001, 2002, 2003** **Models:** Protégé, Protégé5 **Engines:** All **Transmissions:** A/T	**A/T Shift Solenoid 'D' Circuit Malfunction** Engine started, then driven in Drive in 4th Gear, and the PCM detected an unexpected voltage condition on the Shift Solenoid 'D' (SSD) control circuit during the CCM test. **Possible Causes:** • SSD control circuit is open (continuous high signal) • SSD control circuit is shorted to ground (continuous low signal) • SSD control circuit is shorted to system power (B+) • SSD is damaged or has failed • PCM has failed
DTC: P0768 **1T CCM, MIL: YES** **2004, 2005** **Models:** Mazda3 **Engines:** All **Transmissions:** A/T	**A/T Shift Solenoid 'D' Circuit Malfunction** Engine started, then driven in Drive in 4th Gear, and the PCM detected an unexpected voltage condition on the Shift Solenoid 'D' (SSD) control circuit during the CCM test. **Possible Causes:** • SSD control circuit is open (continuous high signal) • SSD control circuit is shorted to ground (continuous low signal) • SSD control circuit is shorted to system power (B+) • SSD is damaged or has failed • PCM has failed
DTC: P0771 **2T CCM, MIL: YES** **1999, 2000, 2001, 2002, 2003** **Models:** Protégé, Protégé5 **Engines:** All **Transmissions:** A/T	**A/T Shift Solenoid 'E' Performance (Mechanical)** 4th Gear Test DTC P0731, P0132 and P0733 not set, engine started, then driven in Drive in 4th Gear (4GR) at 37-62 mph, Turbine speed at 4988 rpm (± 225), engine speed over 450 rpm, TCC operating in normal or power mode, SSA commanded to over 99%, ATF temperature at 68ºF or higher, and the PCM detected the difference between the engine speed and turbine speed was more than 100 rpm. **Possible Causes:** • ATF level is too low, or the ATF is badly deteriorated • A/T SSE is stuck "off", or it is damaged (a mechanical fault) • A/T control valve is stuck, or it is damaged (a mechanical fault) • PCM has failed

DTC	Trouble Code Title, Conditions & Possible Causes
DTC: P0771 **2T CCM, MIL: YES** **2004, 2005** **Models:** Mazda3 **Engines:** All **Transmissions:** A/T	**A/T Shift Solenoid 'E' Performance (Mechanical)** 4th Gear Test DTC P0731, P0132 and P0733 not set, engine started, then driven in Drive in 4th Gear (4GR) at 37-62 mph, Turbine speed at 4988 rpm (± 225), engine speed over 450 rpm, TCC operating in normal or power mode, SSA commanded to over 99%, ATF temperature at 68°F or higher, and the PCM detected the difference between the engine speed and turbine speed was more than 100 rpm. **Possible Causes:** • ATF level is too low, or the ATF is badly deteriorated • A/T SSE is stuck "off", or it is damaged (a mechanical fault) • A/T control valve is stuck, or it is damaged (a mechanical fault) • PCM has failed
DTC: P0772 **2T CCM, MIL: YES** **1999, 2000, 2001, 2002, 2003** **Models:** Protégé, Protégé5 **Engines:** All **Transmissions:** A/T	**A/T Shift Solenoid 'E' Performance (Mechanical)** 4th Gear Test DTC P0734 not set, engine started, then driven in Drive in 4th Gear (4GR) at a speed of less than 43 mph, Turbine speed at 4988 rpm (± 225), engine speed over 450 rpm, ATF temperature over 68°F, TCC not enabled, throttle angle over 6.25% for 10 seconds, or throttle angle closed for 10 seconds, or throttle angle from 3.125-6.25% for at least 3 seconds, and the PCM detected the difference between the engine speed and turbine speed was less than 50 rpm. **Possible Causes:** • ATF level is too low, or the ATF is badly deteriorated • A/T SSE is stuck "on", or it is damaged (a mechanical fault) • A/T control valve is stuck, or it is damaged (a mechanical fault) • PCM has failed
DTC: P0772 **2T CCM, MIL: YES** **2004, 2005** **Models:** Mazda3 **Engines:** All **Transmissions:** A/T	**A/T Shift Solenoid 'E' Performance (Mechanical)** 4th Gear Test DTC P0734 not set, engine started, then driven in Drive in 4th Gear (4GR) at a speed of less than 43 mph, Turbine speed at 4988 rpm (± 225), engine speed over 450 rpm, ATF temperature over 68°F, TCC not enabled, throttle angle over 6.25% for 10 seconds, or throttle angle closed for 10 seconds, or throttle angle from 3.125-6.25% for at least 3 seconds, and the PCM detected the difference between the engine speed and turbine speed was less than 50 rpm. **Possible Causes:** • ATF level is too low, or the ATF is badly deteriorated • A/T SSE is stuck "on", or it is damaged (a mechanical fault) • A/T control valve is stuck, or it is damaged (a mechanical fault) • PCM has failed
DTC: P0773 **1T CCM, MIL: YES** **1999, 2000, 2001, 2002, 2003** **Models:** Protégé, Protégé5 **Engines:** All **Transmissions:** A/T	**A/T Shift Solenoid 'E' Circuit Malfunction** Engine started, vehicle driven in Drive (4th Gear) with the TCC system engaged, and the PCM detected an unexpected voltage condition on the SSE circuit during the CCM test. **Possible Causes:** • SSE control circuit is open (continuous high signal) • SSE control circuit is shorted to ground (continuous low signal) • SSE control circuit is shorted to system power (B+) • SSE is damaged or has failed • PCM has failed
DTC: P0773 **1T CCM, MIL: YES** **2004, 2005** **Models:** Mazda3 **Engines:** All **Transmissions:** A/T	**A/T Shift Solenoid 'E' Circuit Malfunction** Engine started, vehicle driven in Drive (4th Gear) with the TCC system engaged, and the PCM detected an unexpected voltage condition on the SSE circuit during the CCM test. **Possible Causes:** • SSE control circuit is open (continuous high signal) • SSE control circuit is shorted to ground (continuous low signal) • SSE control circuit is shorted to system power (B+) • SSE is damaged or has failed • PCM has failed
DTC: P0778 **1T CCM, MIL: YES** **2005** **Models:** Tribute **Engines:** All **Transmissions:** A/T	**A/T Shift Solenoid 'E' Circuit Malfunction** Engine started, vehicle driven in Drive (4th[h] Gear) with the TCC system engaged, and the PCM detected an unexpected voltage condition on the SSE circuit during the CCM test. **Possible Causes:** • SSE control circuit is open (continuous high signal) • SSE control circuit is shorted to ground (continuous low signal) • SSE control circuit is shorted to system power (B+) • SSE is damaged or has failed • PCM has failed

DTC	Trouble Code Title, Conditions & Possible Causes
DTC: P0781 **2T CCM, MIL: YES** **1996, 1997** **Models:** B2300, B3000, B4000 **Engines:** All **Transmissions:** A/T	**Transmission 1-2 Shift Circuit Malfunction** Engine running with VSS inputs received, and the PCM detected an unexpected voltage condition on the Transmission 1-2 Shift circuit. **Possible Causes:** • A/T 1-2 shift circuit is open • A/T 1-2 shift circuit is shorted to ground • A/T 1-2 shift circuit is shorted to system power (B+) • A/T 1-2 shift component is damaged or has failed • PCM has failed
DTC: P0782 **2T CCM, MIL: YES** **1996, 1997** **Models:** B2300, B3000, B4000 **Engines:** All **Transmissions:** A/T	**Transmission 2-3 Shift Circuit Malfunction** Engine running with VSS inputs received, and the PCM detected an unexpected voltage condition on the Transmission 2-3 Shift circuit. **Possible Causes:** • A/T 2-3 shift circuit is open • A/T 2-3 shift circuit is shorted to ground • A/T 2-3 shift circuit is shorted to system power (B+) • A/T 2-3 shift component is damaged or has failed • PCM has failed
DTC: P0783 **2T CCM, MIL: YES** **1996, 1997** **Models:** B2300, B3000, B4000 **Engines:** All **Transmissions:** A/T	**Transmission 3-4 Shift Circuit Malfunction** Engine running with VSS inputs received, and the PCM detected an unexpected voltage condition on the Transmission 3-4 Shift circuit. **Possible Causes:** • A/T 3-4 shift circuit is open • A/T 3-4 shift circuit is shorted to ground • A/T 3-4 shift circuit is shorted to system power (B+) • A/T 3-4 shift component is damaged or has failed • PCM has failed
DTC: P0784 **2T CCM, MIL: YES** **1996, 1997** **Models:** B2300, B3000, B4000 **Engines:** All **Transmissions:** A/T	**Transmission 4-5 Shift Circuit Malfunction** Engine running with VSS inputs received, and the PCM detected an unexpected voltage condition on the Transmission 4-5 Shift circuit. **Possible Causes:** • A/T 4-5 shift circuit is open • A/T 4-5 shift circuit is shorted to ground • A/T 4-5 shift circuit is shorted to system power (B+) • A/T 4-5 shift component is damaged or has failed • PCM has failed
DTC: P0791 **2T CCM, MIL: YES** **2004, 2005** **Models:** B2300, B3000, B4000 **Engines:** All **Transmissions:** A/T	**Input Shaft Speed (ISS) Sensor** PCM detected ISS sensor signal failure. **Possible Causes:** • Input Shaft Speed Sensor circuit is open • ISS circuit is shorted to ground • ISS circuit is shorted to system power (B+) • ISS Sensor is damaged or has failed • PCM has failed
DTC: P0794 **2T CCM, MIL: YES** **2004, 2005** **Models:** B2300, B3000, B4000 **Engines:** All **Transmissions:** A/T	**Input Shaft Speed (ISS) Sensor** PCM detected and intermittent failure with the ISS sensor signal. **Possible Causes:** • ISS circuit is open • ISS circuit is shorted to ground • ISS circuit is shorted to system power (B+) • ISS sensor is damaged or has failed • PCM has failed
DTC: P0812 **2T CCM, MIL: YES** **2005** **Models:** Tribute **Engines:** All **Transmissions:** A/T	**Reverse Input circuit malfunction.** **Possible Causes:** • Reverse Input circuit is open • Reverse Input circuit is shorted to ground • Reverse Input circuit is shorted to system power (B+) • Reverse Input circuit is damaged or has failed • PCM has failed

DTC	Trouble Code Title, Conditions & Possible Causes
DTC: P0841 **2T CCM, MIL: YES** **2004, 2005** **Models:** Mazda3 **Engines:** All **Transmissions:** A/T	**Neutral switch input circuit problem** The PCM monitors changes in input voltage from the neutral switch. If the PCM does not detect PCM terminal 1S voltage changes while running vehicle with vehicle speed above 30 km/h {19 mph} and clutch pedal turns press and depress 10 times repeatedly, the PCM determines that the neutral switch circuit has malfunction. **Possible Causes:** • Neutral switch malfunction • Poor connection of neutral switch connector or PCM connector • Short to ground in wiring harness between neutral switch terminal A and PCM terminal 1S • Open circuit in wiring harness between ground and neutral switch terminal B • PCM has failed
DTC: P0850 **2T CCM, MIL: YES** **2004, 2005** **Models:** Mazda3, Mazda6, MX-5 Miata, RX-8 **Engines:** All **Transmissions:** A/T	**Neutral switch input circuit problem** The PCM monitors changes in input voltage from the neutral switch. If the PCM does not detect PCM terminal 1S voltage changes while running vehicle with vehicle speed above 30 km/h {19 mph} and clutch pedal turns press and depress 10 times repeatedly, the PCM determines that the neutral switch circuit has malfunction. **Possible Causes:** • Neutral switch malfunction • Poor connection of neutral switch connector or PCM connector • Short to ground in wiring harness between neutral switch terminal A and PCM terminal 1S • Open circuit in wiring harness between ground and neutral switch terminal B • PCM has failed
DTC: P0928 **2T CCM, MIL: YES** **2005** **Models:** Tribute **Engines:** All **Transmissions:** A/T	**Gear Shift Lock solenoid** **PCM detects a problem with the Gear Shift Lock Solenoid.** **Possible Causes:** • Gear shift lock solenoid circuit is open • Gear shift lock solenoid is damaged or has failed • PCM has failed
DTC: P0930 **2T CCM, MIL: YES** **2005** **Models:** Tribute **Engines:** All **Transmissions:** A/T	**Gear Shift Lock solenoid** PCM detects a problem with the Gear Shift Lock Solenoid. **Possible Causes:** • Gear shift lock solenoid circuit is low • Gear shift lock solenoid is damaged or has failed • PCM has failed
DTC: P0960 **2T CCM, MIL: YES** **2005** **Models:** Tribute **Engines:** All **Transmissions:** A/T	**Electronic Pressure Control (EPC)** **EPC solenoid circuit failure, open circuit.** **Possible Causes:** • EPC solenoid circuit is open • PCM has failed
DTC: P0962 **2T CCM, MIL: YES** **2005** **Models:** Tribute **Engines:** All **Transmissions:** A/T	**Electronic Pressure Control (EPC)** EPC solenoid circuit failure, shorted to ground. **Possible Causes:** • EPC circuit is shorted to ground • PCM has failed
DTC: P0963 **2T CCM, MIL: YES** **2005** **Models:** Tribute **Engines:** All **Transmissions:** A/T	**Electronic Pressure Control** EPC solenoid circuit failure, short circuit to VBAT. **Possible Causes:** • EPC circuit is shorted to VBAT • PCM has failed
DTC: P0964 **2T CCM, MIL: YES** **2005** **Models:** Tribute **Engines:** All **Transmissions:** A/T	**Timing / Coast Clutch Solenoid (2-3T/CCS)** PCM detects 3-2T/CCS solenoid failure. **Possible Causes:** • 3-2T/CCS solenoid circuit is open • 3-2T/CCS solenoid is damaged or has failed • PCM has failed

DTC	Trouble Code Title, Conditions & Possible Causes
DTC: P0966 **2T CCM, MIL: YES** **2005** **Models:** Tribute **Engines:** All **Transmissions:** A/T	**Timing / Coast Clutch Solenoid (2-3T/CCS)** PCM detects 3-2T/CCS solenoid failure. **Possible Causes:** • 3-2T/CCS solenoid circuit is shorted to ground • 3-2T/CCS solenoid is damaged or has failed • PCM has failed
DTC: P0967 **2T CCM, MIL: YES** **2005** **Models:** Tribute **Engines:** All **Transmissions:** A/T	**Timing / Coast Clutch Solenoid** PCM detects 3-2T/CCS solenoid failure. **Possible Causes:** • 3-2T/CCS solenoid circuit is shorted to VBAT • 3-2T/CCS solenoid is damaged or has failed • PCM has failed

Gas Engine OBD II Trouble Code List (P1xxx Codes)

DTC: P1000 **1T CCM, MIL: YES** **1996, 1997, 1998, 1999, 2000, 2001, 2002, 2003, 2004, 2005** **Models:** MX-5 Miata, MX-6, 626, Protégé, Protege5, MPV, Millenia, B2300, B2500, B3000, B4000, Tribute **Engines:** All **Transmissions:** A/T, M/T	**OBD II Monitor Testing No Completed** Key on or engine running, and the PCM detected that DTC P1000 was set because one of the OBD II Main Monitors did not complete the I/M Readiness Test sequence. The recommended next step is to drive vehicle through the complete Mazda Drive Cycle pattern. **Note: This trouble code is deleted if the MIL is activated.** **Possible Causes:** • A PCM Reset step was performed with an OBD II Scan Tool • Battery keep alive power (KAPWR) was removed to the PCM • One or more OBD II Monitors did not complete during an official OBD II Drive Cycle
DTC: P1001 **1T CCM, MIL: NO** **1996, 1997, 1998, 1999, 2000, 2001, 2002, 2003, 2004, 2005** **Models:** MX-5 Miata, MX-6, 626, Protégé, Protege5, MPV, Millenia, B2300, B2500, B3000, B4000, Tribute **Engines:** All **Transmissions:** A/T, M/T	**Data Link Connector Circuit Malfunction** Engine running and the PCM detected an unexpected voltage condition on the SCP communication circuit to the PCM. **Note: This code indicates the Scan Tool could not "talk" to the PCM.** **Possible Causes:** • DLC connector or the related terminals/pins are damaged • DLC Bus (+) circuit is open or shorted to ground • DLC Bus (+) circuit is shorted at an associated module • DLC ground circuit is open or has a high resistance condition • PCM power relay is damaged or has failed (no power to DLC) • Incorrect Self-Test procedure followed, or idle speed to low • PCM has failed (e.g., the PCM VREF may be out-of-range)
DTC: P1100 **2T CCM, MIL: YES** **1996, 1997, 1998, 1999, 2000, 2001, 2002** **Models:** 626, MX-6 **Engines:** All **Transmissions:** A/T, M/T	**Mass Airflow Sensor Circuit Malfunction (Intermittent)** Engine running, and the PCM detected the MAF signal was less than 0.23v, or it was more than 4.60v in the last 40 warm-up cycles. **Note: This code can set due to an interruption of the MAF signal.** **Possible Causes:** • MAF sensor continuity problems at the connector or harness • MAF sensor circuit is open (the fault may be intermittent) • MAF sensor is damaged or has failed • PCM has failed
DTC: P1100 **2T CCM, MIL: YES** **1996, 1997, 1998, 1999, 2000, 2001, 2002, 2003, 2004, 2005** **Models:** B2300, B2500, B3000, B4000 **Engines:** All **Transmissions:** A/T, M/T	**Mass Airflow Sensor Circuit Malfunction (Intermittent)** Engine running and the PCM detected the MAF signal was less than 0.39v, or it was more than 3.90v in the last 40 warm-up cycles. **Note: This code can set due to an interruption of the MAF signal.** **Possible Causes:** • MAF sensor continuity problems at the connector or harness • MAF sensor circuit is open (the fault may be intermittent) • MAF sensor is damaged or has failed • PCM has failed
DTC: P1100 **2T CCM, MIL: YES** **2001, 2002, 2003, 2004, 2005** **Models:** Tribute **Engines:** All **Transmissions:** A/T, M/T	**Mass Airflow Sensor Circuit Malfunction (Intermittent)** Engine running, and the PCM detected the MAF signal was less than 0.21v, or it was more than 4.60v in the last 40 warmup cycles. **Note: This code can set due to an interruption of the MAF signal.** **Possible Causes:** • MAF sensor continuity problems at the connector or harness • MAF sensor circuit is open (the fault may be intermittent) • MAF sensor is damaged or has failed • PCM has failed
DTC: P1101 **2T CCM, MIL: YES** **1996, 1997, 1998, 1999, 2000, 2001, 2002, 2003** **Models:** 626, MX-6 **Engines:** All **Transmissions:** A/T, M/T	**Mass Airflow Sensor Malfunction (Self-Test)** KOEO Self-Test Enabled: Key on, and the PCM detected the MAF signal was more than 0.27v. KOER Self-Test Enabled: Engine running and the PCM detected the MAF signal was less than 0.46v, or that it was more than 2.44v during the CCM test. **Possible Causes:** • MAF sensor is damaged, dirty or has failed • MAF sensor (inlet) screen is dirty or blocked • An intake air leak is present near the MAF sensor location • PCM has failed

DTC	Trouble Code Title, Conditions & Possible Causes
DTC: P1101 **2T CCM, MIL: YES** **1996, 1997, 1998, 1999, 2000, 2001, 2002, 2003, 2004, 2005** **Models:** B2300, B2500, B3000, B4000, Tribute **Engines:** All **Transmissions:** A/T, M/T	**Mass Airflow Sensor Malfunction (Self-Test)** KOEO Self-Test Enabled: Key on, and the PCM detected the MAF signal was more than 0.27v. KOER Self-Test Enabled: Engine running and the PCM detected the MAF signal was not in the normal operating range of 0.46-2.44v during the CCM test. **Possible Causes:** • MAF sensor is damaged, dirty or has failed • MAF sensor (inlet) screen is dirty or blocked • An intake air leak is present near the MAF sensor location • PCM has failed
DTC: P1101 **2T CCM, MIL: YES** **2001, 2002, 2003, 2004, 2005** **Models:** Tribute **Engines:** All **Transmissions:** A/T, M/T	**Mass Airflow Sensor Malfunction (Self-Test)** KOEO Self-Test Enabled: Key on, and the PCM detected the MAF signal was more than 0.27v. KOER Self-Test Enabled: Engine running and the PCM detected the MAF signal was not in the normal operating range of 0.46-2.44v during the CCM test. **Possible Causes:** • MAF sensor is damaged, dirty or has failed • MAF sensor (inlet) screen is dirty or blocked • An intake air leak is present near the MAF sensor location • PCM has failed
DTC: P1102 **2T CCM, MIL: YES** **1999, 2000, 2001, 2002, 2003** **Models:** MX-5 Miata **Engines:** All **Transmissions:** A/T, M/T	**MAF Signal Inconsistent with TP Sensor (Too Low)** Engine started, engine speed over 500 rpm, throttle angle more than 50% (open), and the PCM detected the MAF sensor indicated the airflow was less than 5.3 g/sec for 5 seconds during the CCM test. **Possible Causes:** • MAF sensor signal circuit is open or grounded (intermittent) • MAF sensor is damaged, dirty or has failed • MAF sensor (inlet) screen is dirty or blocked • TP sensor is damaged or erratic in operation • PCM has failed
DTC: P1102 **2T CCM, MIL: YES** **1999, 2000, 2001, 2002, 2003** **Models:** Protégé, Protege5 **Engines:** All **Transmissions:** A/T, M/T	**MAF Signal Inconsistent With TP Sensor (Too Low)** Engine started, engine speed over 500 rpm, throttle angle more than 50% (open), and the PCM detected the MAF sensor indicated the airflow was less than 4.8 g/sec for 5 seconds during the CCM test. **Possible Causes:** • MAF sensor signal circuit is open or grounded (intermittent) • MAF sensor is damaged, dirty or has failed • MAF sensor (inlet) screen is dirty or blocked • TP sensor is damaged or erratic in operation • PCM has failed
DTC: P1102 **2T CCM, MIL: YES** **2000, 2001, 2002, 2003** **Models:** MPV **Engines:** All **Transmissions:** A/T, M/T	**MAF Signal Inconsistent With TP Sensor (Too Low)** Engine started, engine speed over 500 rpm, throttle angle more than 50% (open), and the PCM detected the MAF sensor indicated the airflow was less than 8.25 g/sec for 5 seconds during the CCM test. **Possible Causes:** • MAF sensor signal circuit is open or grounded (intermittent) • MAF sensor is damaged, dirty or has failed • MAF sensor (inlet) screen is dirty or blocked • TP sensor is damaged or erratic in operation • PCM has failed
DTC: P1102 **2T CCM, MIL: YES** **2002** **Models:** Tribute **Engines:** All **Transmissions:** A/T, M/T	**MAF Signal Inconsistent With TP Sensor (Too Low)** Engine started, engine speed over 500 rpm, throttle angle more than 50% (open), and the PCM detected the MAF sensor indicated the airflow was less than 8.25 g/sec for 5 seconds during the CCM test. **Possible Causes:** • MAF sensor signal circuit is open or grounded (intermittent) • MAF sensor is damaged, dirty or has failed • MAF sensor (inlet) screen is dirty or blocked • TP sensor is damaged or erratic in operation • PCM has failed

DTC	Trouble Code Title, Conditions & Possible Causes
DTC: P1103 **2T CCM, MIL: YES** **1999, 2000, 2001, 2002, 2003** **Models:** MX-5 Miata **Engines:** All **Transmissions:** A/T, M/T	**MAF Signal Inconsistent with Engine Speed (Too High)** Engine started, engine speed less than 2000 rpm, and the PCM detected the MAF signal indicated the airflow was more than 74.7 g/sec for 5 seconds during the CCM rationality test. **Possible Causes:** • MAF sensor signal circuit is open or grounded (intermittent) • MAF sensor is damaged, dirty or has failed • MAF sensor (inlet) screen is dirty or blocked • An intake air leak is present near the MAF sensor location • PCM has failed
DTC: P1103 **2T CCM, MIL: YES** **1999, 2000, 2001, 2002, 2003** **Models:** Protégé, Protege5 **Engines:** All **Transmissions:** A/T, M/T	**MAF Signal Inconsistent with Engine Speed (Too High)** Engine started, engine speed less than 2000 rpm, and the PCM detected the MAF signal indicated the airflow was more than 66.6 g/sec for 5 seconds during the CCM rationality test. **Possible Causes:** • MAF sensor signal circuit is open or grounded (intermittent) • MAF sensor is damaged, dirty or has failed • MAF sensor (inlet) screen is dirty or blocked • An intake air leak is present near the MAF sensor location • PCM has failed
DTC: P1103 **2T CCM, MIL: YES** **2000, 2001, 2002, 2003** **Models:** MPV **Engines:** All **Transmissions:** A/T, M/T	**MAF Signal Inconsistent with Engine Speed (Too High)** Engine started, engine speed less than 2000 rpm, and the PCM detected the MAF signal indicated the airflow was more than 103 g/sec for 5 seconds during the CCM rationality test. **Possible Causes:** • MAF sensor signal circuit is open or grounded (intermittent) • MAF sensor is damaged, dirty or has failed • MAF sensor (inlet) screen is dirty or blocked • An intake air leak is present near the MAF sensor location • PCM has failed
DTC: P1103 **2T CCM, MIL: YES** **2004** **Models:** Tribute **Engines:** All **Transmissions:** A/T, M/T	**MAF Signal Inconsistent with Engine Speed (Too High)** Engine started, engine speed less than 2000 rpm, and the PCM detected the MAF signal indicated the airflow was more than 103 g/sec for 5 seconds during the CCM rationality test. **Possible Causes:** • MAF sensor signal circuit is open or grounded (intermittent) • MAF sensor is damaged, dirty or has failed • MAF sensor (inlet) screen is dirty or blocked • An intake air leak is present near the MAF sensor location • PCM has failed
DTC: P1110 **2T CCM, MIL: YES** **1996, 1997, 1998, 1999, 2000, 2001, 2002** **Models:** Millenia **Engines:** 2.3L VIN 2 **Transmissions:** A/T, M/T	**Intake Air Temperature Sensor Circuit Malfunction (Dynamic Chamber)** Key on or engine running, IAT sensor signal more than 14ºF, and the PCM detected the IAT Dynamic Chamber signal was less than 0.10v, or that it was more than 4.92v, or that the IAT sensor signal was higher than the ECT sensor signal by 104ºF with the ECT sensor signal more than 158ºF during the CCM Rationality test. **Possible Causes:** • IAT sensor signal circuit open or shorted to ground • IAT sensor ground circuit open between sensor and the PCM • IAT sensor is damaged or has failed • PCM has failed
DTC: P1110 **1T CCM, MIL: YES** **1996, 1997, 1998, 1999, 2000, 2001, 2002, 2003** **Models:** MPV **Engines:** All **Transmissions:** A/T, M/T	**Intake Air Temperature Sensor Circuit Malfunction (Dynamic Chamber)** Key on or engine running, IAT sensor signal more than 14ºF, and the PCM detected the IAT Dynamic Chamber signal was less than 0.10v, or that it was more than 4.92v, or that the IAT sensor signal was higher than the ECT sensor signal by 104ºF with the ECT sensor signal more than 158ºF during the CCM Rationality test. **Possible Causes:** • IAT sensor signal circuit open or shorted to ground • IAT sensor ground circuit open between sensor and the PCM • IAT sensor is damaged or has failed • PCM has failed

DTC	Trouble Code Title, Conditions & Possible Causes
DTC: P1110 **1T CCM, MIL: YES** **2004** **Models:** Tribute **Engines:** All **Transmissions:** A/T, M/T	**Intake Air Temperature Sensor Circuit Malfunction (Dynamic Chamber)** Key on or engine running, IAT sensor signal more than 14°F, and the PCM detected the IAT Dynamic Chamber signal was less than 0.10v, or that it was more than 4.92v, or that the IAT sensor signal was higher than the ECT sensor signal by 104°F with the ECT sensor signal more than 158°F during the CCM Rationality test. **Possible Causes:** • IAT sensor signal circuit open or shorted to ground • IAT sensor ground circuit open between sensor and the PCM • IAT sensor is damaged or has failed • PCM has failed
DTC: P1112 **1T CCM, MIL: NO** **1996, 1997, 1998, 1999, 2000, 2001, 2002** **Models:** 626, MX-6 **Engines:** All **Transmissions:** A/T, M/T	**Intake Air Temperature Sensor Signal Circuit Malfunction** Key on or engine running, and the PCM detected the IAT sensor signal was interrupted (it went low or high) during the CCM test. **Possible Causes:** • IAT sensor signal circuit is open (intermittent fault) • IAT sensor signal circuit is shorted to ground (intermittent fault) • IAT sensor ground circuit is open (intermittent fault) • IAT sensor is damaged or has failed • PCM has failed
DTC: P1112 **1T CCM, MIL: NO** **1996, 1997, 1998, 1999, 2000, 2001, 2002, 2003, 2004, 2005** **Models:** B2300, B2500, B3000, B4000, Tribute **Engines:** All **Transmissions:** A/T, M/T	**Intake Air Temperature Sensor Signal Circuit Malfunction** Key on or engine running, IAT sensor signal more than 50°F, and the PCM detected the IAT sensor signal was interrupted (dropped-out) during the CCM test. **Possible Causes:** • IAT sensor signal circuit is open (an intermittent fault) • IAT sensor signal circuit is shorted to ground (intermittent fault) • IAT sensor is damaged or has failed (an intermittent fault) • PCM has failed
DTC: P1113 **1T CCM, MIL: YES** **1996, 1997, 1998, 1999, 2000, 2001, 2002** **Models:** Millenia **Engines:** 2.3L VIN 2 **Transmissions:** A/T, M/T	**Intake Air Temperature Sensor Circuit Malfunction** Engine running, IAT sensor signal over 0°F, and the PCM detected the IAT Dynamic Chamber signal was more than 4.90v, or that the IAT signal was more than 3.48v higher than the ECT sensor signal with the ECT input more than 176°F with the IAT sensor (air cleaner) signal more than 14°F, or after the PCM reduced the charge pressure, the IAT sensor signal did not drop to less than 302°F. **Note: If DTC P1450 is set, repair the cause of this trouble code first.** **Possible Causes:** • IAT sensor signal circuit open or shorted to ground • IAT sensor ground circuit open between sensor and the PCM • IAT sensor is damaged or has failed • PCM has failed
DTC: P1113 **1T CCM, MIL: YES** **2004** **Models:** Tribute **Engines:** All **Transmissions:** A/T, M/T	**Intake Air Temperature Sensor Circuit Malfunction** Engine running, IAT sensor signal over 0°F, and the PCM detected the IAT Dynamic Chamber signal was more than 4.90v, or that the IAT signal was more than 3.48v higher than the ECT sensor signal with the ECT input more than 176°F with the IAT sensor (air cleaner) signal more than 14°F, or after the PCM reduced the charge pressure, the IAT sensor signal did not drop to less than 302°F. **Note: If DTC P1450 is set, repair the cause of this trouble code first.** **Possible Causes:** • IAT sensor signal circuit open or shorted to ground • IAT sensor ground circuit open between sensor and the PCM • IAT sensor is damaged or has failed • PCM has failed
DTC: P1114 **1T CCM, MIL: YES** **1999, 2000, 2001, 2002, 2003** **Models:** B2300, B2500, B3000, B4000 **Engines:** All **Transmissions:** A/T, M/T	**Intake Air Temperature Sensor Circuit Low Input** Key on or engine running, and the PCM detected the IAT sensor signal was less than 0.2v (Scan Tool PID indicates 250°F). **Possible Causes:** • IAT sensor signal circuit shorted to sensor ground • IAT sensor signal circuit shorted to chassis ground • IAT sensor is damaged (it may be shorted internally) • PCM has failed

DTC	Trouble Code Title, Conditions & Possible Causes
DTC: P1115 **1T CCM, MIL: YES** **1999, 2000, 2001, 2002, 2003** **Models:** B2300, B2500, B3000, B4000 **Engines:** All **Transmissions:** A/T, M/T	**Intake Air Temperature Sensor Circuit High Input** Key on or engine running, and the PCM detected the IAT sensor signal was more than 4.60v (Scan Tool PID indicates -46ºF). **Possible Causes:** • IAT sensor signal circuit open between sensor and the PCM • IAT sensor ground circuit open between sensor and the PCM • IAT sensor is damaged (it may be open internally) • PCM has failed
DTC: P1116 **1T CCM, MIL: NO** **1996, 1997, 1998, 1999, 2000, 2001, 2002** **Models:** 626, MX-6 **Engines:** All **Transmissions:** A/T, M/T	**Engine Coolant Temperature Sensor Circuit Malfunction (Self-Test)** KOER Self-Test Enabled: Engine running, IAT sensor signal more than 50°F, and the PCM detected the ECT signal was less than 0.30v, or that it was more than 3.70v during the CCM test. **Possible Causes:** • ECT sensor signal circuit open or shorted to ground • ECT sensor signal shorted to VREF or system power • ECT sensor
DTC: P1116 **1T CCM, MIL: YES** **1996, 1997, 1998, 1999, 2000, 2001, 2002, 2003, 2004, 2005** **Models:** B2300, B2500, B3000, B4000, Tribute **Engines:** All **Transmissions:** A/T, M/T	**Engine Coolant Temperature Sensor Circuit Malfunction (Self-Test)** KOER Self-Test Enabled: Engine running, IAT sensor signal more than 50°F, and the PCM detected the ECT signal was less than 0.30v, or that it was more than 3.70v during the CCM test. **Possible Causes:** • ECT sensor signal circuit open or shorted to ground • ECT sensor signal shorted to VREF or system power • ECT sensor
DTC: P1117 **2T CCM, MIL: YES** **1996, 1997, 1998, 1999, 2000, 2001, 2002** **Models:** 626, MX-6 **Engines:** All **Transmissions:** A/T, M/T	**Engine Coolant Temperature Sensor Circuit Malfunction** Key on or engine running and the PCM detected an expected "low" or "high" voltage condition on the ECT sensor signal during the test. **Possible Causes:** • ECT sensor signal is open (fault may be intermittent) • ECT sensor signal shorted to ground (fault may be intermittent) • ECT sensor signal shorted to VREF (fault may be intermittent) • ECT sensor ground circuit is open (fault may be intermittent) • ECT sensor is damaged or has failed • PCM has failed
DTC: P1117 **2T CCM, MIL: YES** **1996, 1997, 1998, 1999, 2000, 2001, 2002, 2003, 2004, 2005** **Models:** B2300, B2500, B3000, B4000, Tribute **Engines:** All **Transmissions:** A/T, M/T	**Engine Coolant Temperature Sensor Circuit Malfunction** Engine running, and the PCM detected the ECT sensor signal was interrupted (dropped-out) at some point during the CCM test. **Possible Causes:** • ECT sensor signal is open (an intermittent fault) • ECT sensor signal shorted to ground (an intermittent fault) • ECT sensor is damaged or has failed (an intermittent fault) • PCM has failed
DTC: P1120 **1T CCM, MIL: YES** **1996, 1997, 1998, 1999, 2000, 2001, 2002** **Models:** 626, MX-6 **Engines:** All **Transmissions:** A/T, M/T	**TP Sensor Signal Out of Range Low** Key on or engine running at idle speed, and the PCM detected the TP sensor signal was less than 0.17 (Scan Tool indicates 3.43%), or that it was more than 0.49v (Scan Tool indicates 9.80%) in the test. **Possible Causes:** • TP sensor contacts are loose, broken or damaged • TP sensor pins are loose in the connector • TP sensor signal circuit shorted to ground (an intermittent fault) • TP sensor is damaged or has failed • PCM has failed
DTC: P1120 **1T CCM, MIL: YES** **1996, 1997, 1998, 1999, 2000, 2001, 2002, 2003, 2004, 2005** **Models:** B2300, B2500, B3000, B4000, Tribute **Engines:** All **Transmissions:** A/T, M/T	**TP Sensor Signal Out of Range Low** Engine running at idle speed, and the PCM detected the TP sensor signal was less minimum operating range of 0.17-0.49v (Scan Tool indicates 3.43-98%) during the CCM Rationality test. **Possible Causes:** • TP sensor contacts are loose, broken or damaged • TP sensor pins are loose in the connector • TP sensor signal circuit shorted to ground (an intermittent fault) • TP sensor is damaged or has failed • PCM has failed

DTC	Trouble Code Title, Conditions & Possible Causes
DTC: P11121 **1T CCM, MIL: YES** **1996, 1997, 1998, 1999, 2000, 2001, 2002** **Models:** 626, MX-6 **Engines:** All **Transmissions:** A/T, M/T	**Throttle Position Signal Not Consistent With MAF Signal** Engine running, and the PCM detected the TP sensor and MAF sensor signals were not consistent with the calibrated load values stored in the PCM memory (i.e., the PCM detected the mass intake airflow was too low). **Possible Causes:** • MAF sensor is damaged or has failed • TP sensor is damaged or has failed • Air leaks between the MAF sensor and the throttle body • PCM has failed
DTC: P1121 **1T CCM, MIL: YES** **1996, 1997, 1998, 1999, 2000, 2001, 2002, 2003, 2004, 2005** **Models:** B2300, B2500, B3000, B4000, Tribute **Engines:** All **Transmissions:** A/T, M/T	**Throttle Position Signal Not Consistent With MAF Signal** Engine running, and the PCM detected the TP sensor and MAF sensor signals were not consistent with the calibrated load values in memory (i.e., the PCM detected the mass airflow was too low). **Possible Causes:** • MAF sensor is damaged or has failed • TP sensor is damaged or has failed • Air leaks between the MAF sensor and the throttle body • PCM has failed
DTC: P1122 **2T CCM, MIL: YES** **1999, 2000, 2001, 2002, 2003** **Models:** MX-5 Miata **Engines:** All **Transmissions:** A/T, M/T	**Throttle Position Signal Low Input (Stuck Closed)** Engine running, ECT sensor signal more than 176°F, MAF sensor signal over 63.2 g/sec, and the PCM detected the throttle valve opening angle was less than 12.5% for 5 seconds in the CCM test. **Possible Causes:** • TP sensor is broken damaged or has failed • TP sensor signal circuit is corroded or shorted in the connector • High resistance in the VREF circuit or intermittent loss of VREF • PCM has failed
DTC: P1122 **2T CCM, MIL: YES** **1999, 2000, 2001, 2002, 2003** **Models:** Protégé, Protege5 **Engines:** All **Transmissions:** A/T, M/T	**Throttle Position Signal Low Input (Stuck Closed)** Engine running, ECT sensor signal more than 176°F, MAF sensor signal over 58.3 g/sec, and the PCM detected the throttle valve opening angle was less than 12.5% for 5 seconds in the CCM test. **Possible Causes:** • TP sensor is broken damaged or has failed • TP sensor signal circuit is corroded or shorted in the connector • High resistance in the VREF circuit or intermittent loss of VREF • PCM has failed
DTC: P1122 **1T CCM, MIL: YES** **2000, 2001, 2002, 2003** **Models:** MPV **Engines:** All **Transmissions:** A/T, M/T	**Throttle Position Signal Low Input (Stuck Closed)** Engine running, ECT sensor signal more than 158°F, MAF sensor signal over 88.3 g/sec, and the PCM detected the throttle valve opening angle was less than 12.5% for 5 seconds in the CCM test. **Possible Causes:** • TP sensor is broken damaged or has failed • TP sensor signal circuit is corroded or shorted in the connector • High resistance in the VREF circuit or intermittent loss of VREF • PCM has failed
DTC: P1122 **1T CCM, MIL: YES** **2004** **Models:** Tribute **Engines:** All **Transmissions:** A/T, M/T	**Throttle Position Signal Low Input (Stuck Closed)** Engine running, ECT sensor signal more than 158°F, MAF sensor signal over 88.3 g/sec, and the PCM detected the throttle valve opening angle was less than 12.5% for 5 seconds in the CCM test. **Possible Causes:** • TP sensor is broken damaged or has failed • TP sensor signal circuit is corroded or shorted in the connector • High resistance in the VREF circuit or intermittent loss of VREF • PCM has failed
DTC: P1123 **2T CCM, MIL: YES** **1999, 2000, 2001, 2002, 2003** **Models:** Protégé, Protege5 **Engines:** All **Transmissions:** A/T, M/T	**Throttle Position Signal High Input (Stuck Open)** Engine started, then with the engine speed over 500 rpm, MAF sensor signal under 5.3 g/sec, the PCM detected the throttle valve opening angle indicated over 50% for 5 seconds in the CCM test. **Possible Causes:** • TP sensor is broken damaged or has failed • MAF sensor has drifted out-of-calibration, or it is contaminated • TP sensor signal circuit is corroded or shorted in the connector • High resistance in the TP sensor ground circuit • PCM has failed

DTC	Trouble Code Title, Conditions & Possible Causes
DTC: P1123 **1T CCM, MIL: YES** **2000, 2001, 2002, 2003** **Models:** MPV **Engines:** All **Transmissions:** A/T, M/T	**Throttle Position Signal High Input (Stuck Open)** Engine started, then with the engine speed over 500 rpm, MAF sensor signal under 8.25 g/sec, the PCM detected the throttle valve opening angle was more than 50% for 5 seconds in the CCM test. **Possible Causes:** • TP sensor is broken damaged or has failed • MAF sensor has drifted out-of-calibration, or it is contaminated • TP sensor signal circuit is corroded or shorted in the connector • High resistance in the TP sensor ground circuit • PCM has failed
DTC: P1123 **1T CCM, MIL: YES** **2004** **Models:** Tribute **Engines:** All **Transmissions:** A/T, M/T	**Throttle Position Signal High Input (Stuck Open)** Engine started, then with the engine speed over 500 rpm, MAF sensor signal under 8.25 g/sec, the PCM detected the throttle valve opening angle was more than 50% for 5 seconds in the CCM test. **Possible Causes:** • TP sensor is broken damaged or has failed • MAF sensor has drifted out-of-calibration, or it is contaminated • TP sensor signal circuit is corroded or shorted in the connector • High resistance in the TP sensor ground circuit • PCM has failed
DTC: P1124 **1T CCM, MIL: NO** **1996, 1997, 1998, 1999, 2000, 2001, 2002** **Models:** 626, MX-6 **Engines:** All **Transmissions:** A/T, M/T	**Throttle Position Sensor Signal Out Of Range (Self-Test)** Key on or engine running, and the PCM detected the TP sensor signal (the rotational setting) was not in the range of 0.66-1.20v (Scan Tool indicates 13.23-24.02%) during the KOER Self-Test. **Possible Causes:** • Throttle linkage is bent or binding • TP sensor is not seated properly • Throttle plate below closed throttle position, or throttle plate screw misadjusted • TP sensor is damaged • PCM has failed
DTC: P1124 **1T CCM, MIL: NO** **1996, 1997, 1998, 1999, 2000, 2001, 2002, 2003, 2004, 2005** **Models:** B2300, B2500, B3000, B4000 **Engines:** All **Transmissions:** A/T, M/T	**Throttle Position Sensor Signal Out Of Range (Self-Test)** KOER Self-Test Enabled: Engine running, and the PCM detected the TP sensor signal (the rotational setting) was not in the range of 0.66-1.20v (Scan Tool indicates 13.23-24.02%) during the CCM Rationality test. **Possible Causes:** • Throttle linkage is bent or binding • TP sensor is not seated properly • Throttle plate below closed throttle position, or throttle plate screw misadjusted • TP sensor is damaged • PCM has failed
DTC: P1124 **1T CCM, MIL: NO** **2001, 2002, 2003, 2004, 2005** **Models:** Tribute **Engines:** All **Transmissions:** A/T, M/T	**Throttle Position Sensor Signal Out Of Range (Self-Test)** KOER Self-Test Enabled: Engine running and the PCM detected the TP sensor signal (the rotational setting) was not in the range of 0.66-1.17v (Scan Tool indicates 13.27-23.52%) during the CCM Rationality test. **Possible Causes:** • Throttle linkage is bent or binding • TP sensor is not seated properly • Throttle plate below closed throttle position, or throttle plate screw misadjusted • TP sensor is damaged • PCM has failed
DTC: P1125 **1T CCM, MIL: NO** **1996, 1997, 1998, 1999, 2000, 2001, 2002** **Models:** 626, MX-6 **Engines:** All **Transmissions:** A/T, M/T	**Throttle Position Sensor Signal Circuit Malfunction** Engine running and the PCM detected the TP sensor signal was less than 0.49v or more than 4.60v (due to an interruption of the signal) at some point within the last 80 warm-up cycles. **Possible Causes:** • TP sensor wiring harness is open (an intermittent fault) • TP sensor connector or pins are open (an intermittent fault) • TP sensor signal circuit is open (an intermittent fault) • TP sensor signal circuit is grounded (an intermittent fault)

DTC	Trouble Code Title, Conditions & Possible Causes
DTC: P1125 **1T CCM, MIL: NO** **1996, 1997, 1998, 1999, 2000, 2001, 2002, 2003, 2004, 2005** **Models:** B2300, B2500, B3000, B4000, Tribute **Engines:** All **Transmissions:** A/T, M/T	**Throttle Position Sensor Signal Circuit Malfunction** Engine running and the PCM detected the TP sensor signal was less than 0.49v (9.8%) or more than 4.65v (due to an interruption of the signal) at some point during the last 80 warm-up cycles. **Possible Causes:** • TP sensor wiring harness is open (an intermittent fault) • TP sensor connector or pins are open (an intermittent fault) • TP sensor signal circuit is open (an intermittent fault) • TP sensor signal circuit is grounded (an intermittent fault)
DTC: P1127 **1T CCM, MIL: NO** **1996, 1997, 1998, 1999, 2000, 2001, 2002** **Models:** 626, MX-6 **Engines:** All **Transmissions:** A/T, M/T	**HO2S-12 Heater Not Enabled (Self-Test)** KOER Self-Test Enabled: Engine running, and the PCM detected the HO2S-12 heater was not hot enough for testing during the Self-Test. **Possible Causes:** • Engine not operated long enough prior to beginning the KOER Self-Test • Exhaust system temperature is too cold to run the test
DTC: P1127 **1T CCM, MIL: NO** **1996, 1997, 1998, 1999, 2000, 2001, 2002, 2003, 2004, 2005** **Models:** B2300, B2500, B3000, 4000, Tribute **Engines:** All **Transmissions:** A/T, M/T	**HO2S-12 (Bank 1 Sensor 2) Heater Not Enabled (Self-Test)** KOER Self-Test Enabled: Engine running, and the PCM detected the HO2S-12 heater was not hot enough for testing during the Self-Test. **Possible Causes:** • Engine not operated long enough prior to starting the KOER Self-Test • Exhaust system temperature is too cold to run the test
DTC: P1128 **1T CCM, MIL: NO** **1996, 1997, 1998, 1999, 2000, 2001, 2002, 2003, 2004, 2005** **Models:** B2300, B3000, 4000 **Engines:** All **Transmissions:** A/T, M/T	**HO2S-11 (Bank 1 Sensor 1) Signals Swapped (Self-Test)** KOER Self-Test Enabled: Engine running and the PCM detected the front HO2S-11 signals indicated the wrong cylinder bank responded to a shift in fuel ratio. **Possible Causes:** • HO2S harness connectors are crossed • HO2S wiring is crossed at the harness connector • HO2S wiring is crossed at the PCM connector • PCM has failed
DTC: P1128 **2T OBD/O2S2, MIL: NO** **2001, 2002, 2003, 2004, 2005** **Models:** Tribute **Engines:** 3.0L VIN 1 **Transmissions:** A/T, M/T	**HO2S-11 (Bank 1 Sensor 1) Signals Swapped (Self-Test)** KOER Self-Test Enabled: Engine running and the PCM detected the front HO2S-11 signals indicated the wrong cylinder bank responded to a shift in fuel ratio. **Possible Causes:** • HO2S harness connectors are crossed • HO2S wiring is crossed at the harness connector • HO2S wiring is crossed at the PCM connector • PCM has failed
DTC: P1129 **1T OBD/O2S2, MIL: NO** **1999, 2000, 2001, 2002, 2003** **Models:** B3000, 4000 **Engines:** All **Transmissions:** A/T, M/T	**HO2S-12 (Bank 1 Sensor 2) Signals Swapped (Self-Test)** KOER Self-Test Enabled: Engine running and the PCM detected the front HO2S-12 signals indicated the wrong cylinder bank responded to a shift in fuel ratio. **Possible Causes:** • HO2S harness connectors are crossed • HO2S wiring is crossed at the harness connector • HO2S wiring is crossed at the PCM connector • PCM has failed
DTC: P1129 **1T OBD/O2S2, MIL: NO** **2001, 2002, 2003, 2004, 2005** **Models:** Tribute **Engines:** 3.0L VIN 1 **Transmissions:** A/T, M/T	**HO2S-12 (Bank 1 Sensor 2) Signals Swapped (Self-Test)** KOER Self-Test Enabled: Engine running and the PCM detected the front HO2S-12 signals indicated the wrong cylinder bank responded to a shift in fuel ratio. **Possible Causes:** • HO2S harness connectors are crossed • HO2S wiring is crossed at the harness connector • HO2S wiring is crossed at the PCM connector • PCM has failed

DTC	Trouble Code Title, Conditions & Possible Causes
DTC: P1130 **2T CCM, MIL: YES** **1996, 1997** **Models:** 626, MX-6 **Engines:** 2.0L VIN C **Transmissions:** A/T	**HO2S-11 Fuel Control Limit Reached** Engine running in closed loop at cruise speed, and the PCM detected the front HO2S-11 signal indicated the fuel control had reached its maximum lean or rich limit (HO2S-11 is not switching). **Possible Causes:** • HO2S is damaged or has failed • HO2S circuits wet or oily, corroded, or poor terminal contact • HO2S signal circuit open, shorted to ground, short to power • Low fuel pressure or vehicle driven until it was out of fuel • Fuel pressure regulator leaking, or fuel injectors leaking • EVAP vapor recovery system has failed • Air leaks located after the MAF sensor • Exhaust leaks before or near the front HO2S • EGR valve stuck, EGR diaphragm leaking, or gasket leaking • Oil dipstick not seated or engine oil level too high (overfilled)
DTC: P1130 **2T CCM, MIL: YES** **1998, 1999, 2000, 2001, 2002** **Models:** 626 **Engines:** All **Transmissions:** A/T, M/T	**HO2S-11 (Bank 1 Sensor 1) Fuel Control Limit Reached** Engine running in closed loop at cruise speed, and the PCM detected the front HO2S-11 signal indicated the fuel control had reached its maximum lean or rich limit (HO2S-11 is not switching). **Possible Causes:** • HO2S is damaged or has failed • HO2S circuits wet or oily, corroded, or poor terminal contact • HO2S signal circuit open, shorted to ground, short to power • Low fuel pressure or vehicle driven until it was out of fuel • Fuel pressure regulator leaking, or fuel injectors leaking • EVAP vapor recovery system has failed • Air leaks located after the MAF sensor • Exhaust leaks before or near the front HO2S • EGR valve stuck, EGR diaphragm leaking, or gasket leaking • Oil dipstick not seated or engine oil level too high (overfilled)
DTC: P1130 **2T CCM, MIL: YES** **1996, 1997, 1998, 1999, 2000, 2001, 2002, 2003, 2004, 2005** **Models:** B2300, B2500, B3000, B4000, Tribute **Engines:** All **Transmissions:** A/T, M/T	**HO2S-11 (Bank 1 Sensor 1) Fuel Control Limit Reached** Engine running in closed loop at cruise speed, and the PCM detected the front HO2S-11 signal indicated the fuel control had reached its maximum lean or rich limit (HO2S-11 is not switching). **Possible Causes:** • HO2S is damaged or has failed • HO2S circuits wet or oily, corroded, or poor terminal contact • HO2S signal circuit open, shorted to ground, short to power • Low fuel pressure or vehicle driven until it was out of fuel • Fuel pressure regulator leaking, or fuel injectors leaking • EVAP vapor recovery system has failed • Air leaks located after the MAF sensor • Exhaust leaks before or near the front HO2S • EGR valve stuck, EGR diaphragm leaking, or gasket leaking • Oil dipstick not seated or engine oil level too high (overfilled)
DTC: P1131 **2T CCM, MIL: YES** **1996, 1997** **Models:** 626, MX-6 **Engines:** 2.0L VIN C **Transmissions:** A/T	**HO2S-11 (Bank 1 Sensor 1) Indicates A/F Ratio Too Lean** Engine running in closed loop at cruise speed, and the PCM detected the front HO2S-11 signal indicated the Air Fuel (A/F) ratio was too lean (i.e., the PCM is correcting for an over-lean condition). **Possible Causes:** • Air leaks located after the MAF sensor • Base Engine: Check for vacuum leaks with a Smoke Tester • Intake air system is leaking, obstructed and damaged • Low fuel pressure or vehicle driven until it was out of fuel • PCV system has air leaks at the valve or related hoses
DTC: P11131 **2T CCM, MIL: YES** **1998, 1999, 2000, 2001, 2002** **Models:** 626 **Engines:** All **Transmissions:** A/T, M/T	**HO2S-11 (Bank 1 Sensor 1) Indicates A/F Ratio Too Lean** Engine running in closed loop at cruise speed, and the PCM detected the front HO2S-11 signal indicated the Air Fuel (A/F) ratio was too lean (i.e., the PCM is correcting for an over-lean condition). **Possible Causes:** • Air leaks located after the MAF sensor • Base Engine: Check for vacuum leaks with a Smoke Tester • Intake air system is leaking, obstructed and damaged • Low fuel pressure or vehicle driven until it was out of fuel • PCV system has air leaks at the valve or related hoses

DTC	Trouble Code Title, Conditions & Possible Causes
DTC: P1131 **2T CCM, MIL: YES** **1996, 1997, 1998, 1999, 2000, 2001, 2002, 2003, 2004, 2005** **Models:** B2300, B2500, B3000, B4000, Tribute **Engines:** All **Transmissions:** A/T, M/T	**HO2S-11 (Bank 1 Sensor 1) Indicates A/F Ratio Too Lean** Engine running in closed loop at cruise speed, and the PCM detected the front HO2S-11 signal indicated the Air Fuel (A/F) ratio was too lean (i.e., the PCM is correcting for an over-lean condition). **Possible Causes:** • Air leaks located after the MAF sensor • Base Engine: Check for vacuum leaks with a Smoke Tester • Intake air system is leaking, obstructed and damaged • Low fuel pressure or vehicle driven until it was out of fuel • PCV system has air leaks at the valve or related hoses
DTC: P1132 **2T CCM, MIL: YES** **1996, 1997** **Models:** 626, MX-6 **Engines:** 2.0L VIN C **Transmissions:** A/T	**HO2S-11 (Bank 1 Sensor 1) Indicates A/F Ratio Too Rich** Engine running in closed loop at cruise speed, and the PCM detected the front HO2S-11 signal indicated the Air Fuel (A/F) ratio was too rich (i.e., the PCM is correcting for an over-rich condition). **Possible Causes:** • Check air cleaner element, air cleaner housing for blockage • Fuel pressure regulator is leaking, or fuel pressure is too high • Fuel injectors leaking or contaminated (they may be sticking) • EVAP vapor recovery system has failed (canister full of fuel)
DTC: P1132 **2T CCM, MIL: YES** **1996, 1997, 1998, 1999, 2000, 2001, 2002** **Models:** 626 **Engines:** All **Transmissions:** A/T, M/T	**HO2S-11 (Bank 1 Sensor 1) Indicates A/F Ratio Too Rich** Engine running in closed loop at cruise speed, and the PCM detected the front HO2S-11 signal indicated the Air Fuel (A/F) ratio was too rich (i.e., the PCM is correcting for an over-rich condition). **Possible Causes:** • Check air cleaner element, air cleaner housing for blockage • Fuel pressure regulator is leaking, or fuel pressure is too high • Fuel injectors leaking or contaminated (they may be sticking) • EVAP vapor recovery system has failed (canister full of fuel)
DTC: P1132 **2T CCM, MIL: YES** **1996, 1997, 1998, 1999, 2000, 2001, 2002, 2003, 2004, 2005** **Models:** B2300, B2500, B3000, B4000, Tribute **Engines:** All **Transmissions:** A/T, M/T	**HO2S-11 (Bank 1 Sensor 1) Indicates A/F Ratio Too Lean** Engine running in closed loop at cruise speed, and the PCM detected the front HO2S-11 signal indicated the Air Fuel (A/F) ratio was too lean (i.e., the PCM is correcting for an over-lean condition). **Possible Causes:** • Air leaks located after the MAF sensor • Base Engine: Check for vacuum leaks with a Smoke Tester • Intake air system is leaking, obstructed and damaged • Low fuel pressure or vehicle driven until it was out of fuel • PCV system has air leaks at the valve or related hoses
DTC: P1133 **2T CCM, MIL: YES** **2004** **Models:** Tribute **Engines:** All **Transmissions:** A/T, M/T	**HO2S-11 (Bank 1 Sensor 1) Indicates A/F Ratio Too Lean** Engine running in closed loop at cruise speed, and the PCM detected the front HO2S-11 signal indicated the Air Fuel (A/F) ratio was too lean (i.e., the PCM is correcting for an over-lean condition). **Possible Causes:** • Air leaks located after the MAF sensor • Base Engine: Check for vacuum leaks with a Smoke Tester • Intake air system is leaking, obstructed and damaged • Low fuel pressure or vehicle driven until it was out of fuel • PCV system has air leaks at the valve or related hoses
DTC: P1134 **2T CCM, MIL: YES** **2004** **Models:** Tribute **Engines:** All **Transmissions:** A/T, M/T	**HO2S-11 (Bank 1 Sensor 1) Indicates A/F Ratio Too Lean** Engine running in closed loop at cruise speed, and the PCM detected the front HO2S-11 signal indicated the Air Fuel (A/F) ratio was too lean (i.e., the PCM is correcting for an over-lean condition). **Possible Causes:** • Air leaks located after the MAF sensor • Base Engine: Check for vacuum leaks with a Smoke Tester • Intake air system is leaking, obstructed and damaged • Low fuel pressure or vehicle driven until it was out of fuel • PCV system has air leaks at the valve or related hoses
DTC: P1135 **2T CCM, MIL: YES** **1999, 2000, 2003** **Models:** MX-5 Miata **Engines:** All **Transmissions:** A/T, M/T	**HO2S-11 (Bank 1 Sensor 1) Heater Circuit Low Input** Engine started, and with the front HO2S-11 commanded "off", the PCM detected an unexpected "low" voltage condition (less than 5.8v for 322-327 seconds) on the HO2S-11 heater control circuit after startup during the CCM test. **Possible Causes:** • HO2S heater control circuit is open or shorted to ground • HO2S heater power circuit open between sensor and ignition • HO2S heater is damaged or has failed • PCM has failed

DTC	Trouble Code Title, Conditions & Possible Causes
DTC: P1135 **2T CCM, MIL: YES** **1999, 2000** **Models:** Protégé **Engines:** All **Transmissions:** A/T, M/T	**HO2S-11 (Bank 1 Sensor 1) Heater Circuit Low Input** Engine started, and with the front HO2S-11 commanded "off", the PCM detected an unexpected "low" voltage condition on the front HO2S-11 heater control circuit after startup during the CCM test. **Possible Causes:** • HO2S heater control circuit is open or shorted to ground • HO2S heater power circuit open between sensor and ignition • HO2S heater is damaged or has failed • PCM has failed
DTC: P1135 **2T CCM, MIL: YES** **2004** **Models:** Tribute **Engines:** All **Transmissions:** A/T, M/T	**HO2S-11 (Bank 1 Sensor 1) Heater Circuit Low Input** Engine started, and with the front HO2S-11 commanded "off", the PCM detected an unexpected "low" voltage condition on the front HO2S-11 heater control circuit after startup during the CCM test. **Possible Causes:** • HO2S heater control circuit is open or shorted to ground • HO2S heater power circuit open between sensor and ignition • HO2S heater is damaged or has failed • PCM has failed
DTC: P1136 **2T CCM, MIL: YES** **1999, 2000** **Models:** MX-5 Miata **Engines:** All **Transmissions:** A/T, M/T	**HO2S-11 (Bank 1 Sensor 1) Heater Circuit High Input** Engine started, and with the front HO2S-11 commanded "on", the PCM detected an unexpected voltage condition (more than 11.5v) on the HO2S-11 heater control circuit after startup in the CCM test. **Possible Causes:** • HO2S heater control circuit is shorted to power • HO2S heater is damaged or has failed • PCM has failed
DTC: P1136 **2T CCM, MIL: YES** **1999, 2000** **Models:** Protégé **Engines:** All **Transmissions:** A/T, M/T	**HO2S-11 (Bank 1 Sensor 1) Heater Circuit High Input** Engine started, and with the front HO2S-11 commanded "on", the PCM detected an unexpected "high" voltage condition on the front HO2S-11 heater control circuit after startup in the CCM test. **Possible Causes:** • HO2S heater control circuit is shorted to power • HO2S heater is damaged or has failed • PCM has failed
DTC: P1136 **2T CCM, MIL: YES** **2004** **Models:** Tribute **Engines:** All **Transmissions:** A/T, M/T	**HO2S-11 (Bank 1 Sensor 1) Heater Circuit High Input** Engine started, and with the front HO2S-11 commanded "on", the PCM detected an unexpected "high" voltage condition on the front HO2S-11 heater control circuit after startup in the CCM test. **Possible Causes:** • HO2S heater control circuit is shorted to power • HO2S heater is damaged or has failed • PCM has failed
DTC: P1137 **1T CCM, MIL: NO** **1996, 1997** **Models:** 626, MX-6 **Engines:** 2.0L VIN C **Transmissions:** A/T	**HO2S-12 Lack of Switching, Indicates Lean (Self-Test)** KOER Self-Test Enabled: Engine running in closed loop, and the PCM detected the rear HO2S-12 signal indicated the fuel control had reached its maximum lean limit (i.e., the right bank HO2S-12 is not switching properly). **Possible Causes:** • Air leaks located after the MAF sensor • Base Engine: Check for vacuum leaks with a Smoke Tester • HO2S wiring may be crossed at the connector • Check intake air system for leaks, obstructions and damage • EGR system malfunction (the EGR valve may be stuck open) • PCV system has air leaks at the valve or related hoses • Fuel pressure is too low (fuel filter is dirty, or weak fuel pump) • Vehicle driven until it was out of fuel

DTC	Trouble Code Title, Conditions & Possible Causes
DTC: P1157 **1T CCM, MIL: NO** **2000, 2001, 2002** **Models:** 626 **Engines:** 2.5L VIN D **Transmissions:** A/T, M/T	**HO2S-22 Lack of Switching, Indicates Lean (Self-Test)** KOER Self-Test Enabled: Engine running in closed loop, and the PCM detected the rear HO2S-22 signal indicated the fuel control had reached its maximum lean limit (i.e., the right bank HO2S-22 is not switching properly). **Possible Causes:** • An air leak that is located after the MAF sensor • Base Engine: Check for vacuum leaks with a Smoke Tester • HO2S wiring may be crossed at the connector • Check intake air system for leaks, obstructions and damage • EGR system malfunction (the EGR valve may be stuck open) • PCV system has air leaks at the valve or related hoses • Fuel pressure is too low (fuel filter is dirty, or weak fuel pump) • Vehicle driven until it was out of fuel
DTC: P1157 **2T CCM, MIL: YES** **2001, 2002, 2003, 2004, 2005** **Models:** Tribute **Engines:** 3.0L VIN 1 **Transmissions:** A/T, M/T	**HO2S-22 Lack of Switching, Indicates Lean (Self-Test)** KOER Self-Test Enabled: Engine running in closed loop, and the PCM detected the rear HO2S-22 signal indicated the fuel control had reached its maximum lean limit (i.e., the right bank HO2S-22 is not switching properly). **Possible Causes:** • Air leaks located after the MAF sensor • Base Engine: Check for vacuum leaks with a Smoke Tester • HO2S wiring may be crossed at the connector • Check intake air system for leaks, obstructions and damage • EGR system malfunction (the EGR valve may be stuck open) • PCV system has air leaks at the valve or related hoses • Fuel pressure is too low (fuel filter is dirty, or weak fuel pump) • Vehicle driven until it was out of fuel
DTC: P1158 **1T CCM, MIL: NO** **2000, 2001, 2002** **Models:** 626 **Engines:** 2.5L VIN D **Transmissions:** A/T, M/T	**HO2S-22 Lack of Switching, Indicates Rich (Self-Test)** KOER Self-Test Enabled: Engine running in closed loop, and the PCM detected the rear HO2S-22 signal indicated the fuel control had reached its maximum rich limit (i.e., the right bank HO2S-22 is not switching properly). **Possible Causes:** • Check air cleaner element, air cleaner housing for blockage • Fuel pressure regulator is leaking, or fuel pressure is too high • HO2S wiring may be crossed at the connector • Fuel injectors leaking or contaminated (they may be sticking) • EVAP vapor recovery system has failed (canister full of fuel)
DTC: P1158 **2T CCM, MIL: YES** **2001, 2002, 2003, 2004, 2005** **Models:** Tribute **Engines:** 3.0L VIN 1 **Transmissions:** A/T, M/T	**HO2S-22 Lack of Switching, Indicates Rich (Self-Test)** KOER Self-Test Enabled: Engine running in closed loop, and the PCM detected the rear HO2S-22 signal indicated the fuel control had reached its maximum rich limit (i.e., the right bank HO2S-22 is not switching properly). **Possible Causes:** • Check air cleaner element, air cleaner housing for blockage • Fuel pressure regulator is leaking, or fuel pressure is too high • HO2S wiring may be crossed at the connector • Fuel injectors leaking or contaminated (they may be sticking) • EVAP vapor recovery system has failed (canister full of fuel)
DTC: P1168 **2T CCM, MIL: YES** **2005** **Models:** Tribute **Engines:** 3.0L VIN 1 **Transmissions:** A/T, M/T	**HO2S-22 Lack of Switching, Indicates Rich (Self-Test)** KOER Self-Test Enabled: Engine running in closed loop, and the PCM detected the rear HO2S-22 signal indicated the fuel control had reached its maximum rich limit (i.e., the right bank HO2S-22 is not switching properly). **Possible Causes:** • Check air cleaner element, air cleaner housing for blockage • Fuel pressure regulator is leaking, or fuel pressure is too high • HO2S wiring may be crossed at the connector • Fuel injectors leaking or contaminated (they may be sticking) • EVAP vapor recovery system has failed (canister full of fuel)

DTC	Trouble Code Title, Conditions & Possible Causes
DTC: P1169 **2T CCM, MIL: YES** **1996, 1997** **Models:** 626, MX-6 **Engines:** 2.5L VIN D **Transmissions:** A/T, M/T	**HO2S-12 (Bank 1 Sensor 2) Circuit No Inversion** Engine started, vehicle driven at cruise speed at over 1150 rpm, ECT sensor more than 176ºF, and the PCM detected the HO2S-12 signal was fixed above or below 0.45v for over 42 seconds. **Note: This trouble code indicates an inversion problem is present.** **Possible Causes:** • An air leak that is located after the MAF sensor • Base engine mechanical fault affecting one or more cylinders • EVAP Purge valve stuck or the hoses to the valve are reversed • Fuel metering fault (fuel injector sticking open or leaking) • Fuel pressure too low or too high, fuel supply contaminated • HO2S element is contaminated or the HO2S heater has failed • PCV system has air leaks at the valve or related hoses
DTC: P1169 **2T CCM, MIL: YES** **1996, 1997, 1998, 1999, 2000** **Models:** Millenia **Engines:** 2.5L VIN 1 **Transmissions:** A/T, M/T	**HO2S-12 (Bank 1 Sensor 2) Circuit No Inversion** Engine started, vehicle driven at cruise speed at over 1150 rpm, ECT sensor more than 176ºF, and the PCM detected the HO2S-12 signal was fixed above or below 0.45v for over 42 seconds. **Note: This trouble code indicates an inversion problem is present.** **Possible Causes:** • An air leak that is located after the MAF sensor • Base engine mechanical fault affecting one or more cylinders • EVAP Purge valve stuck or the hoses to the valve are reversed • Fuel metering fault (fuel injector sticking open or leaking) • Fuel pressure too low or too high, fuel supply contaminated • HO2S element is contaminated or the HO2S heater has failed • PCV system has air leaks at the valve or related hoses
DTC: P1169 **2T CCM, MIL: YES** **2000** **Models:** MPV **Engines:** All **Transmissions:** A/T, M/T	**HO2S-12 (Bank 1 Sensor 2) Circuit No Inversion** Engine started, then driven at cruise speed at over 1500 rpm, ECT sensor signal more than 158ºF, and the PCM detected the rear HO2S-12 signal remained above or below 0.45v for 42 seconds. **Note: This trouble code indicates an inversion problem is present.** **Possible Causes:** • An air leak that is located after the MAF sensor • Base engine mechanical fault affecting one or more cylinders • EVAP Purge valve stuck or the hoses to the valve are reversed • Fuel metering fault (fuel injector sticking open or leaking) • Fuel pressure too low or too high, fuel supply contaminated • HO2S element is contaminated or the HO2S heater has failed • PCV system has air leaks at the valve or related hoses
DTC: P1169 **2T CCM, MIL: YES** **2004, 2005** **Models:** Tribute **Engines:** All **Transmissions:** A/T, M/T	**HO2S-12 (Bank 1 Sensor 2) Circuit No Inversion** Engine started, then driven at cruise speed at over 1500 rpm, ECT sensor signal more than 158ºF, and the PCM detected the rear HO2S-12 signal remained above or below 0.45v for 42 seconds. **Note: This trouble code indicates an inversion problem is present.** **Possible Causes:** • An air leak that is located after the MAF sensor • Base engine mechanical fault affecting one or more cylinders • EVAP Purge valve stuck or the hoses to the valve are reversed • Fuel metering fault (fuel injector sticking open or leaking) • Fuel pressure too low or too high, fuel supply contaminated • HO2S element is contaminated or the HO2S heater has failed • PCV system has air leaks at the valve or related hoses
DTC: P1170 **2T CCM, MIL: YES** **1996, 1997** **Models:** 626, MX-6 **Engines:** 2.5L VIN D **Transmissions:** A/T, M/T	**HO2S-11 (Bank 1 Sensor 1) Circuit No Inversion** Engine started, then driven at cruise speed at over 1150 rpm, ECT sensor signal more than 176ºF, and the PCM detected the front HO2S-11 signal remained above 0.45v for over 42 seconds. **Note: This trouble code indicates an inversion problem is present.** **Possible Causes:** • An air leak that is located after the MAF sensor • Base engine mechanical fault affecting one or more cylinders • EVAP Purge valve stuck or the hoses to the valve are reversed • Fuel metering fault (fuel injector sticking open or leaking) • Fuel pressure too low or too high, fuel supply contaminated • HO2S element is contaminated or the HO2S heater has failed • PCV system has air leaks at the valve or related hoses

DTC	Trouble Code Title, Conditions & Possible Causes
DTC: P1170 **2T CCM, MIL: YES** **1996, 1997** **Models:** 626, MX-6 **Engines:** 2.0L VIN C **Transmissions:** M/T	**HO2S-11 (Bank 1 Sensor 1) Circuit No Inversion** Engine started, then driven at cruise speed at over 1500 rpm, ECT sensor signal more than 176ºF, and the PCM detected the front HO2S-11 signal remained above 0.45v for over 37 seconds. **Note: This trouble code indicates an inversion problem is present.** **Possible Causes:** • An air leak that is located after the MAF sensor • Base engine mechanical fault affecting one or more cylinders • EVAP Purge valve stuck or the hoses to the valve are reversed • Fuel metering fault (fuel injector sticking open or leaking) • Fuel pressure too low or too high, fuel supply contaminated • HO2S element is contaminated or the HO2S heater has failed • PCV system has air leaks at the valve or related hoses
DTC: P1170 **2T CCM, MIL: YES** **1996, 1997, 1999, 2000, 2001, 2002, 2003** **Models:** MX-5 Miata **Engines:** All **Transmissions:** A/T, M/T	**HO2S-11 (Bank 1 Sensor 1) Circuit No Inversion** DTC P0102, P0103, P0117, P0118, P0122, P0123, P0443, P0500, P1102, P1103, P1122, P1123, P1496, P1498 and P1499 not set; engine started, vehicle driven at cruise speed at over 1500 rpm with the closed throttle switch indicating "off", ECT sensor signal more than 176ºF, and the PCM detected the front HO2S-11 signal remained above 0.45v for 42 seconds during the HO2S Monitor test. **Note: This trouble code indicates an inversion problem is present.** **Possible Causes:** • An air leak that is located after the MAF sensor • Base engine mechanical fault affecting one or more cylinders • EVAP Purge valve stuck or the hoses to the valve are reversed • Fuel metering fault (fuel injector sticking open or leaking) • Fuel pressure too low or too high, fuel supply contaminated • HO2S element is contaminated or the HO2S heater has failed • PCV system has air leaks at the valve or related hoses
DTC: P1170 **2T CCM, MIL: YES** **1996, 1997, 1998, 1999, 2000, 2001, 2002** **Models:** Millenia **Engines:** All **Transmissions:** A/T, M/T	**HO2S-11 (Bank 1 Sensor 1) Circuit No Inversion** Engine started, then driven at cruise speed at over 1150 rpm, ECT sensor signal more than 176ºF, and the PCM detected the rear HO2S-11 signal remained above or below 0.45v for 42 seconds. **Possible Causes:** • An air leak that is located after the MAF sensor • Base Engine: engine compression and/or valve train problems • Check intake air system for leaks, obstructions and damage • Fuel pressure is too low (fuel filter is dirty, or weak fuel pump) • Ignition system component problems (plugs, plug wires or coil) • PCV valve or related vacuum hose leaking, or engine vacuum leaks (check for signs of vacuum leaks with a Smoke Tester)
DTC: P1170 **2T CCM, MIL: YES** **1996, 1997, 1998, 1999, 2000, 2001, 2002, 2003** **Models:** Protégé, Protege5 **Engines:** All **Transmissions:** A/T, M/T	**HO2S-11 (Bank 1 Sensor 1) Circuit No Inversion** Engine started, then driven at cruise speed at over 1500 rpm, ECT sensor signal more than 176ºF, and the PCM detected the rear HO2S-11 signal remained above or below 0.55v for 42 seconds. **Possible Causes:** • An air leak that is located after the MAF sensor • Base engine mechanical fault affecting one or more cylinders • EVAP Purge valve stuck or the hoses to the valve are reversed • Fuel metering fault (fuel injector sticking open or leaking) • Fuel pressure too low or too high, fuel supply contaminated • HO2S element is contaminated or the HO2S heater has failed • PCV system has air leaks at the valve or related hoses
DTC: P1170 **2T CCM, MIL: YES** **1996, 1997, 1998, 1999, 2000, 2001, 2002, 2003** **Models:** MPV **Engines:** All **Transmissions:** A/T, M/T	**HO2S-11 (Bank 1 Sensor 1) Circuit No Inversion** Engine started, then driven at cruise speed at over 1500 rpm, ECT sensor signal more than 158ºF, and the PCM detected the rear HO2S-11 signal remained above or below 0.45v for 42 seconds. **Possible Causes:** • Base engine mechanical fault affecting one or more cylinders • EVAP Purge valve stuck or the hoses to the valve are reversed • Fuel metering fault (fuel injector sticking open or leaking) • Fuel pressure too low or too high, fuel supply contaminated • HO2S element is contaminated or the HO2S heater has failed • PCV system has air leaks at the valve or related hoses

DTC	Trouble Code Title, Conditions & Possible Causes
DTC: P1173 **2T CCM, MIL: YES** **1996, 1997** **Models:** 626, MX-6 **Engines:** 2.5L VIN D **Transmissions:** A/T, M/T	**HO2S-21 (Bank 2 Sensor 1) Circuit No Inversion** Engine started, then driven at cruise speed at over 1150 rpm, ECT sensor signal more than 176°F, and the PCM detected the front HO2S-21 signal remained above or below 0.45v for 42 seconds. **Note: This trouble code indicates an inversion problem is present.** **Possible Causes:** • Base engine mechanical fault affecting one or more cylinders • EVAP Purge valve stuck or the hoses to the valve are reversed • Fuel metering fault (fuel injector sticking open or leaking) • Fuel pressure too low or too high, fuel supply contaminated • HO2S element is contaminated or the HO2S heater has failed • PCV system has air leaks at the valve or related hoses
DTC: P1173 **2T CCM, MIL: YES** **1998, 1999, 2000, 2001, 2002** **Models:** Millenia **Engines:** All **Transmissions:** A/T, M/T	**HO2S-21 (Bank 2 Sensor 1) Circuit Malfunction** Engine started, then driven at cruise speed at over 1150 rpm, ECT sensor signal more than 176°F, and the PCM detected the rear HO2S-21 signal remained above or below 0.45v for 42 seconds. **Possible Causes:** • Base engine mechanical fault affecting one or more cylinders • EVAP Purge valve stuck or the hoses to the valve are reversed • Fuel metering fault (fuel injector sticking open or leaking) • Fuel pressure too low or too high, fuel supply contaminated • HO2S element is contaminated or the HO2S heater has failed • PCV system has air leaks at the valve or related hoses
DTC: P1173 **2T CCM, MIL: YES** **2000, 2001, 2002, 2003** **Models:** MPV **Engines:** All **Transmissions:** A/T, M/T	**HO2S-21 (Bank 2 Sensor 1) Circuit Malfunction** Engine started, then driven at cruise speed at over 1150 rpm, ECT sensor signal more than 176°F, and the PCM detected the rear HO2S-21 signal remained above or below **0.45v for 42 seconds.** **Possible Causes:** • Base engine mechanical fault affecting one or more cylinders • EVAP Purge valve stuck or the hoses to the valve are reversed • Fuel metering fault (fuel injector sticking open or leaking) • Fuel pressure too low or too high, fuel supply contaminated • HO2S element is contaminated or the HO2S heater has failed • PCV system has air leaks at the valve or related hoses
DTC: P1173 **2T CCM, MIL: YES** **2004** **Models:** Tribute **Engines:** All **Transmissions:** A/T, M/T	**HO2S-21 (Bank 2 Sensor 1) Circuit Malfunction** Engine started, then driven at cruise speed at over 1150 rpm, ECT sensor signal more than 176°F, and the PCM detected the rear HO2S-21 signal remained above or below **0.45v for 42 seconds.** **Possible Causes:** • Base engine mechanical fault affecting one or more cylinders • EVAP Purge valve stuck or the hoses to the valve are reversed • Fuel metering fault (fuel injector sticking open or leaking) • Fuel pressure too low or too high, fuel supply contaminated • HO2S element is contaminated or the HO2S heater has failed • PCV system has air leaks at the valve or related hoses
DTC: P1195 **1T CCM, MIL: YES** **1996, 1997** **Models:** 626, MX-6 **Engines:** 2.0L VIN C **Transmissions:** M/T a M/T	**EGR Boost Sensor Circuit Malfunction** Key on or engine running and the PCM detected the EGR Boost Sensor (EGRBS) or BARO Pressure signal was less than 0.21v, or that it indicated more than 4.90v during the CCM test. **Possible Causes:** • EGRBS signal circuit is open (4.90v) or shorted to ground • EGRBS ground circuit is open (4.90v indicated) • EGRBS vacuum hose is loose, damaged or clogged • EGRBS is damaged or has failed • PCM has failed
DTC: P1195 **1T CCM, MIL: YES** **1996, 1997** **Models:** 626, MX-6 **Engines:** 2.5L VIN D **Transmissions:** M/T	**EGR Boost Sensor Circuit Malfunction** Key on or engine running and the PCM detected the EGR Boost Sensor (EGRBS) or BARO Pressure signal was less than 0.21v, or that it indicated more than 4.84v or the difference in the intake manifold pressure before and after the EGR Boost Sensor valve was switched was less than 6.4 kPa (1.90" Hg) during the CCM test. **Possible Causes:** • EGRBS signal circuit is open (4.90v) or shorted to ground • EGRBS ground circuit is open (4.90v indicated) • EGRBS vacuum hose is loose, damaged or clogged • EGRBS is damaged or has failed • PCM has failed

DTC	Trouble Code Title, Conditions & Possible Causes
DTC: P1195 **1T CCM, MIL: YES** **1996, 1997** **Models:** MX-5 Miata **Engines:** All **Transmissions:** A/T, M/T	**EGR Boost Sensor Circuit Malfunction** Key on or engine running and the PCM detected the EGR Boost Sensor (EGRBS) signal was less than 1.39v, or that it indicated more than 4.61v, or the difference in the intake manifold pressure before and after the EGR Boost Sensor solenoid valve was switched was less than 0.49 kPa (0.146" Hg) during the CCM Rationality test. **Possible Causes:** • EGRBS signal circuit is open (4.90v) or shorted to ground • EGRBS ground circuit is open (4.90v indicated) • EGRBS vacuum hose is loose, damaged or clogged • EGRBS is damaged or has failed • PCM has failed
DTC: P1195 **1T CCM, MIL: YES** **1996, 1997, 1998** **Models:** Millenia **Engines:** 2.3L VIN 2 **Transmissions:** A/T, M/T	**EGR Boost Sensor Circuit Malfunction** Key on or engine running and the PCM detected the EGR Boost Sensor (EGRBS) signal was less than 0.35v, or that it indicated more than 4.49v, or the difference in the intake manifold pressure before and after the EGR Boost Sensor solenoid valve was switched was less than 6.40 kPa (0.146" Hg) during the CCM Rationality test. **Possible Causes:** • EGRBS signal circuit is open (4.90v) or shorted to ground • EGRBS ground circuit is open (4.90v indicated) • EGRBS vacuum hose is loose, damaged or clogged • EGRBS is damaged or has failed • PCM has failed
DTC: P1195 **1T CCM, MIL: YES** **1996, 1997, 1998** **Models:** MPV **Engines:** All **Transmissions:** A/T, M/T	**EGR Boost Sensor Circuit Malfunction** Engine running and the PCM detected the EGR Boost Sensor (EGRBS) signal indicated less than 1.39v, or more than 4.61v, or that the difference in the intake manifold pressure before and after the EGR Boost Sensor solenoid valve was switched was less than 0.49 kPa (0.146" Hg) during the CCM Rationality test. **Possible Causes:** • EGRBS is damaged or has failed • PCM has failed
DTC: P1195 **2T CCM, MIL: YES** **1996, 1997, 1998** **Models:** Protégé **Engines:** All **Transmissions:** A/T, M/T	**EGR Boost Sensor Circuit Malfunction** Key on or engine running and the PCM detected the EGR Boost Sensor (EGRBS) signal was less than 0.20v, or that it indicated more than 4.79v, or the difference in the intake manifold pressure before and after the EGR Boost Sensor solenoid valve was switched was less than 6.40 kPa (0.190" Hg) during the CCM Rationality test. **Possible Causes:** • EGRBS signal circuit is open (4.90v indicated) • EGRBS signal circuit is shorted to ground (0.10v indicated) • EGRBS ground circuit is open (4.90v indicated) • EGRBS vacuum hose is loose, damaged or clogged • EGRBS is damaged or has failed • PCM has failed
DTC: P1196 **1T CCM, MIL: YES** **1996, 1997** **Models:** 626, MX-6 **Engines:** 2.5L VIN D **Transmissions:** A/T, M/T	**Ignition Switch Start Circuit Malfunction** Engine started, and then engine running at over 2000 rpm, and the PCM detected an unexpected voltage condition on the Ignition Start signal (system voltage present) for 20 seconds during the CCM test. **Possible Causes:** • Ignition Start terminal circuit shorted to system power (B+) • Starter is damaged or has failed
DTC: P1196 **1T CCM, MIL: YES** **1996, 1997, 1998, 1999, 2000, 2001, 2002** **Models:** Millenia, MPV **Engines:** All **Transmissions:** A/T, M/T	**Ignition Switch Start Circuit Malfunction** Engine started, then engine running at over 2000 rpm, and the PCM detected an unexpected voltage condition on the Ignition Start signal (system voltage present) for 20 seconds during the CCM test. **Possible Causes:** • Ignition Start terminal circuit shorted to system power (B+) • Starter is damaged or has failed
DTC: P1233 **1T CCM, MIL: YES** **2005** **Models:** Tribute **Engines:** All **Transmissions:** A/T, M/T	**Fuel Pump Control Circuit Malfunction** Key on, and then the PCM detected an unexpected voltage condition on the Fuel Pump relay circuit during the CCM test. **Possible Causes:** • Fuel pump relay control circuit is open • Fuel pump relay control circuit is shorted to ground • Fuel pump relay power circuit is open • PCM has failed

DTC	Trouble Code Title, Conditions & Possible Causes
DTC: P1235 **1T CCM, MIL: YES** **1996** **Models:** 626, MX-6 **Engines:** 2.0L VIN C **Transmissions:** A/T, M/T	**Fuel Pump Control Circuit Malfunction** Key on, and then the PCM detected an unexpected voltage condition on the Fuel Pump relay circuit during the CCM test. **Possible Causes:** • Fuel pump relay control circuit is open • Fuel pump relay control circuit is shorted to ground • Fuel pump relay power circuit is open • PCM has failed
DTC: P1235 **1T CCM, MIL: YES** **2005** **Models:** Tribute **Engines:** All **Transmissions:** A/T, M/T	**Fuel Pump Control Circuit Malfunction** Key on, and then the PCM detected an unexpected voltage condition on the Fuel Pump relay circuit during the CCM test. **Possible Causes:** • Fuel pump relay control circuit is open • Fuel pump relay control circuit is shorted to ground • Fuel pump relay power circuit is open • PCM has failed
DTC: P1236 **1T CCM, MIL: YES** **1996** **Models:** 626, MX-6 **Engines:** 2.0L VIN C **Transmissions:** A/T, M/T	**Fuel Pump Control Circuit Range/Performance** Engine running and the PCM detected the Fuel Pump relay voltage was out-of-range during the CCM test. **Possible Causes:** • Fuel pump relay control circuit is shorted to system power (B+) • Fuel pump relay power circuit is open • PCM has failed
DTC: P1237 **1T CCM, MIL: YES** **2005** **Models:** Tribute **Engines:** All **Transmissions:** A/T, M/T	**Fuel Pump Control Circuit Range/Performance** Engine running and the PCM detected the Fuel Pump relay voltage was out-of-range during the CCM test. **Possible Causes:** • Fuel pump relay control circuit is shorted to system power (B+) • Fuel pump relay power circuit is open • PCM has failed
DTC: P1244 **1T CCM, MIL: YES** **2004, 2005** **Models:** Tribute **Engines:** All **Transmissions:** A/T, M/T	**Generator / Regulator System** Engine running and the PCM detected the Generator Load Input (GLI) signal was not within a calibrated range for a period of time. **Possible Causes:** • GLI circuit is open or shorted to ground • GLI circuit is shorted to system power (B+) • Generator or voltage regulator is damaged or has failed • PCM has failed
DTC: P1245 **1T CCM, MIL: YES** **2004, 2005** **Models:** Tribute **Engines:** All **Transmissions:** A/T, M/T	**Generator / Regulator System** Engine running and the PCM detected the Generator Load Input (GLI) signal was not within a calibrated range for a period of time. **Possible Causes:** • GLI circuit is open or shorted to ground • GLI circuit is shorted to system power (B+) • Generator or voltage regulator is damaged or has failed • PCM has failed
DTC: P1246 **1T CCM, MIL: YES** **2001, 2002, 2003, 2004, 2005** **Models:** Tribute **Engines:** All **Transmissions:** A/T, M/T	**Fuel Pump Control Circuit Range/Performance** Engine running and the PCM detected the Generator Load Input (GLI) signal was not within a calibrated range for a period of time. **Possible Causes:** • GLI circuit is open or shorted to ground • GLI circuit is shorted to system power (B+) • Generator or voltage regulator is damaged or has failed • PCM has failed
DTC: P1250 **1T CCM, MIL: NO** **1996, 1997, 1998, 1999, 2000, 2001, 2002** **Models:** 626, MX-6 **Engines:** 2.0L VIN C **Transmissions:** A/T, M/T	**Pressure Regulator Control Solenoid Circuit Malfunction (Self-Test)** KOEO Self-Test enabled: Key on or engine running and the PCM detected an unexpected "low" or "high" voltage condition on the Pressure Regulator Control (PRC) solenoid circuit. KOER Self-Test enabled: Engine running and the PCM detected an unexpected "low" voltage condition on the Pressure Regulator Control (PRC) solenoid circuit. **Possible Causes:** • PRC control solenoid circuit is open or shorted to ground • Fuel pump relay power circuit is open • PCM has failed

DTC	Trouble Code Title, Conditions & Possible Causes
DTC: P1250 **1T CCM, MIL: NO** **1996, 1997, 1999, 2000, 2001, 2002, 2003** **Models:** MX-5 Miata **Engines:** All **Transmissions:** A/T, M/T	**Pressure Regulator Control Solenoid Circuit Malfunction** Key on or engine running and the PCM detected an unexpected "low" or "high" voltage condition on the Pressure Regulator Control (PRC) solenoid circuit. **Note: This is a diagnostic support code - the PCM monitors this circuit once per key cycle, but does not set a pending code.** **Possible Causes:** • PRC solenoid control circuit is open • PRC solenoid control circuit is shorted to ground • Fuel pump relay power circuit is open • PCM has failed
DTC: P1250 **1T CCM, MIL: NO** **1996, 1997, 1998, 1999, 2000, 2001, 2002** **Models:** Millenia **Engines:** All **Transmissions:** A/T, M/T	**Pressure Regulator Control Solenoid Circuit Malfunction** Key on or engine running and the PCM detected an unexpected "low" or "high" voltage condition on the Pressure Regulator Control (PRC) solenoid circuit during the CCM test. **Note: This is a diagnostic support code - the PCM monitors this circuit once per key cycle, but does not set a pending code.** **Possible Causes:** • PRC solenoid circuit is open or shorted to ground • Fuel pump relay power circuit is open • PCM has failed
DTC: P1250 **1T CCM, MIL: NO** **1996, 1997, 1998, 1999, 2000, 2001, 2002, 2003** **Models:** Protégé, Protege5 **Engines:** All **Transmissions:** A/T, M/T	**Pressure Regulator Control Solenoid Circuit Malfunction** Key on or engine running and the PCM detected an unexpected "low" voltage condition on the Pressure Regulator Control (PRC) solenoid circuit during the CCM test. **Note: This is a diagnostic support code - the PCM monitors this circuit once per key cycle, but does not set a pending code.** **Possible Causes:** • PRC solenoid control circuit is open • PRC solenoid control circuit is shorted to ground • Fuel pump relay power circuit is open • PCM has failed
DTC: P1250 **1T CCM, MIL: NO** **1996, 1997, 1998, 1999, 2000, 2001, 2002** **Models:** MPV **Engines:** All **Transmissions:** A/T, M/T	**Pressure Regulator Control Solenoid Circuit Malfunction** Key on or engine running and the PCM detected an unexpected "low" or "high" voltage condition on the Pressure Regulator Control (PRC) solenoid circuit during the CCM test. **Note: This is a diagnostic support code - the PCM monitors this circuit once per key cycle, but does not set a pending code.** **Possible Causes:** • PRC solenoid circuit is open or shorted to ground • Fuel pump relay power circuit is open • PCM has failed
DTC: P1251 **1T CCM, MIL: NO** **2004** **Models:** Tribute **Engines:** All **Transmissions:** A/T, M/T	**Pressure Regulator Control Solenoid Circuit Malfunction** Key on or engine running and the PCM detected an unexpected "low" or "high" voltage condition on the Pressure Regulator Control (PRC) solenoid circuit during the CCM test. **Note: This is a diagnostic support code - the PCM monitors this circuit once per key cycle, but does not set a pending code.** **Possible Causes:** • PRC solenoid circuit is open or shorted to ground • Fuel pump relay power circuit is open • PCM has failed
DTC: P1252 **2T CCM, MIL: YES** **1996, 1997, 1998, 1999, 2000, 2001, 2002** **Models:** MPV **Engines:** All **Transmissions:** A/T, M/T	**Pressure Regulator Control Solenoid 2 Circuit Malfunction** Engine running and the PCM detected an unexpected voltage condition on the Pressure Regulator Control solenoid 2' circuit during the CCM test. **Possible Causes:** • PRC solenoid control circuit is open • PRC solenoid control circuit is shorted to ground • Fuel pump relay power circuit is open • PCM has failed

DTC	Trouble Code Title, Conditions & Possible Causes
DTC: P1260 **1T CCM, MIL: YES** **1996, 1997, 1998, 1999, 2000, 2001, 2002, 2003, 2004, 2005** **Models:** B2300, B2500, B3000, B4000, RX-8, Tribute **Engines:** All **Transmissions:** A/T, M/T	**Anti-Theft System Signal Detected, Engine Disabled** Key on, and the PCM received a signal from the Anti-Theft System that a theft condition had occurred. The theft indicator on the dash will flash rapidly or remain on "solid" with the ignition switch in the "on" position. The engine may "start and stall", or may not crank if the vehicle is equipped with the PATS starter "disable" feature. **Possible Causes:** • Previous theft condition has occurred • Anti-Theft System is damaged or has failed
DTC: P1260 **1T CCM, MIL: YES** **2004, 2005** **Models:** Mazda3 **Engines:** All **Transmissions:** A/T, M/T	**Anti-Theft System Signal Detected, Engine Disabled** Key on, and the PCM received a signal from the Anti-Theft System that a theft condition had occurred. The theft indicator on the dash will flash rapidly or remain on "solid" with the ignition switch in the "on" position. The engine may "start and stall", or may not crank if the vehicle is equipped with the PATS starter "disable" feature. **Possible Causes:** • Previous theft condition has occurred • Anti-Theft System is damaged or has failed
DTC: P1270 **1T CCM, MIL: YES** **1996** **Models:** 626, MX-6 **Engines:** 2.0L VIN C **Transmissions:** A/T, M/T	**Engine RPM or Vehicle Speed Limit Reached** Engine running and the PCM detected that the vehicle had been driven in a manner where the engine speed or the vehicle speed had exceeded a calibrated limit stored in memory. **Possible Causes:** • Excessive wheel slippage (due to water, ice, mud or snow) • Engine over-revved with the gear selector in Neutral position • Vehicle was driven at a very high rate of speed
DTC: P1270 **1T CCM, MIL: YES** **1996, 1997, 1998, 1999, 2000, 2001, 2002, 2003, 2004, 2005** **Models:** B2300, B2500, B3000, B4000, Tribute **Engines:** All **Transmissions:** A/T, M/T	**Engine RPM or Vehicle Speed Limit Reached** Engine running and the PCM detected that the vehicle had been driven in a manner where the engine speed or the vehicle speed had exceeded a calibrated limit stored in memory. **Possible Causes:** • Excessive wheel slippage (due to water, ice, mud or snow) • Engine over-revved with the gear selector in Neutral position • Vehicle was driven at a very high rate of speed
DTC: P1285 **1T CCM, MIL: YES** **2001, 2002, 2003, 2004, 2005** **Models:** Tribute **Engines:** All **Transmissions:** A/T, M/T	**Cylinder Head Temperature Sensor - Engine Overheated** Engine running and the PCM detected a signal from the Cylinder Head Temperature (CHT) sensor that indicated an engine overheat condition existed during the CCM Rationality test. **Possible Causes:** • Engine coolant level is low, or wrong coolant mixture • Engine overheat condition exists • CHT sensor is out-of-calibration or damaged • PCM has failed
DTC: P1285 **1T CCM, MIL: YES** **2004, 2005** **Models:** B2300, B3000, B4000 **Engines:** All **Transmissions:** A/T, M/T	**Cylinder Head Temperature Sensor - Engine Overheated** Engine running and the PCM detected a signal from the Cylinder Head Temperature (CHT) sensor that indicated an engine overheat condition existed during the CCM Rationality test. **Possible Causes:** • Engine coolant level is low, or wrong coolant mixture • Engine overheat condition exists • CHT sensor is out-of-calibration or damaged • PCM has failed
DTC: P1288 **1T CCM, MIL: YES** **2001, 2002, 2003, 2004, 2005** **Models:** Tribute **Engines:** All **Transmissions:** A/T, M/T	**Cylinder Head Temperature Sensor Circuit Malfunction (Self-Test)** KOER Self-Test Enabled: Engine running IAT sensor signal more than 50°F, and the PCM detected a signal from the Cylinder Head Temperature (CHT) sensor that indicated an engine was not warm enough during the CCM test. **Possible Causes:** • Engine not warm enough to begin the KOER Self-Test • CHT sensor is out-of-calibration or damaged • PCM has failed

DTC	Trouble Code Title, Conditions & Possible Causes
DTC: P1288 **1T CCM, MIL: YES** **2004, 2005** **Models:** B2300, B3000, B4000 **Engines:** All **Transmissions:** A/T, M/T	**Cylinder Head Temperature Sensor Circuit Malfunction (Self-Test)** KOER Self-Test Enabled: Engine running IAT sensor signal more than 50°F, and the PCM detected a signal from the Cylinder Head Temperature (CHT) sensor that indicated an engine was not warm enough during the CCM test. **Possible Causes:** • Engine not warm enough to begin the KOER Self-Test • CHT sensor is out-of-calibration or damaged • PCM has failed
DTC: P1289 **1T CCM, MIL: YES** **2001, 2002, 2003, 2004, 2005** **Models:** Tribute **Engines:** All **Transmissions:** A/T, M/T	**Cylinder Head Temperature Sensor Overheat Condition** Engine started, engine running and the PCM detected the Cylinder Head Temperature (CHT) sensor indicated an overheated engine condition, and the Fail-Safe Cooling (FMEM) strategy was enabled. **Possible Causes:** • Serious engine overheat condition exists • CHT sensor is out-of-calibration or damaged • PCM has failed
DTC: P1289 **1T CCM, MIL: YES** **2004, 2005** **Models:** B2300, B3000, B4000 **Engines:** All **Transmissions:** A/T, M/T	**Cylinder Head Temperature Sensor Overheat Condition** Engine started, engine running and the PCM detected the Cylinder Head Temperature (CHT) sensor indicated an overheated engine condition, and the Fail-Safe Cooling (FMEM) strategy was enabled. **Possible Causes:** • Serious engine overheat condition exists • CHT sensor is out-of-calibration or damaged • PCM has failed
DTC: P1290 **1T CCM, MIL: YES** **2001, 2002, 2003, 2004, 2005** **Models:** Tribute **Engines:** All **Transmissions:** A/T, M/T	**Cylinder Head Temperature Sensor Low Input** Engine running and the PCM detected an unexpected "low" voltage condition on the Cylinder Head Temperature (CHT) sensor circuit. **Possible Causes:** • CHT sensor signal circuit is shorted to ground • CHT sensor is damaged or has failed • PCM has failed
DTC: P1290 **1T CCM, MIL: YES** **2004, 2005** **Models:** B2300, B3000, B4000 **Engines:** All **Transmissions:** A/T, M/T	**Cylinder Head Temperature Sensor Low Input** Engine running and the PCM detected an unexpected "low" voltage condition on the Cylinder Head Temperature (CHT) sensor circuit. **Possible Causes:** • CHT sensor signal circuit is shorted to ground • CHT sensor is damaged or has failed • PCM has failed
DTC: P1299 **1T CCM, MIL: YES** **2001, 2002, 2003, 2004, 2005** **Models:** Tribute **Engines:** All **Transmissions:** A/T, M/T	**Cylinder Head Temperature Sensor High Input** Engine running and the PCM detected an unexpected "high" voltage condition on the Cylinder Head Temperature (CHT) sensor circuit. **Possible Causes:** • CHT sensor signal circuit is open • CHT sensor signal circuit is shorted to VRED • CHT sensor ground circuit is open • CHT sensor is damaged or has failed • PCM has failed
DTC: P1299 **1T CCM, MIL: YES** **2004, 2005** **Models:** B2300, B3000, B4000 **Engines:** All **Transmissions:** A/T, M/T	**Cylinder Head Temperature Sensor High Input** Engine running and the PCM detected an unexpected "high" voltage condition on the Cylinder Head Temperature (CHT) sensor circuit. **Possible Causes:** • CHT sensor signal circuit is open • CHT sensor signal circuit is shorted to VRED • CHT sensor ground circuit is open • CHT sensor is damaged or has failed • PCM has failed

DTC	Trouble Code Title, Conditions & Possible Causes
DTC: P1309 **2T CCM, MIL: YES** **1998, 1999, 2000, 2001, 2002, 2003, 2004, 2005** **Models:** B2300, B2500, B3000, B4000, MPV, Tribute **Engines:** All **Transmissions:** A/T, M/T	**Misfire Detection Monitor** Engine running for more from 2-6 complete engine cycles, and the PCM detected that the Misfire Monitor was not "enabled" due to an internal calculation error or problem. **Possible Causes:** • CMP sensor is not properly synchronized in the engine • CMP sensor is damaged or has failed • PCM has failed
DTC: P1309 **2T CCM, MIL: YES** **2000, 2001, 2002, 2003** **Models:** MPV **Engines:** All **Transmissions:** A/T, M/T	**Misfire Detection Monitor** Engine running for more from 2-6 complete engine cycles, and the PCM detected that the Misfire Monitor was not "enabled" due to an internal calculation error or problem. **Possible Causes:** • CMP sensor is not properly synchronized in the engine • CMP sensor is damaged or has failed • PCM has failed
DTC: P1336 **2T CCM, MIL: YES** **2004, 2005** **Models:** Tribute **Engines:** All **Transmissions:** A/T, M/T	**Camshaft Position (SGC) Sensor Circuit Malfunction** Engine cranking with 4 to 5 crankshaft rotations occurring and the PCM did not detect any CMP (SGC) sensor signals during the test. **Possible Causes:** • CMP sensor signal circuit is open or shorted to ground • CMP sensor signal circuit is shorted to VREF or system power • CMP sensor power circuit open between sensor and main relay • CMP sensor is damaged or has failed • PCM has failed
DTC: P1345 **1T CCM, MIL: YES** **1996, 1997, 1999, 2000, 2001, 2002, 2003** **Models:** 626, MX-6, MX-5 Miata **Engines:** All **Transmissions:** A/T, M/T	**Camshaft Position (SGC) Sensor Circuit Malfunction** Engine cranking with 4 to 5 crankshaft rotations occurring and the PCM did not detect any CMP (SGC) sensor signals during the test. **Possible Causes:** • CMP sensor signal circuit is open or shorted to ground • CMP sensor signal circuit is shorted to VREF or system power • CMP sensor power circuit open between sensor and main relay • CMP sensor is damaged or has failed • PCM has failed
DTC: P1345 **1T CCM, MIL: YES** **1996, 1997, 1998, 1999, 2000, 2001, 2002** **Models:** Millenia **Engines:** All **Transmissions:** A/T, M/T	**Camshaft Position (SGC) Sensor Circuit Malfunction** Engine cranking with 4 to 5 crankshaft rotations occurring and the PCM did not detect any CMP (SGC) sensor signals during the test. **Possible Causes:** • CMP sensor signal circuit is open or shorted to ground • CMP sensor signal circuit is shorted to VREF or system power • CMP sensor power circuit open between sensor and main relay • CMP sensor is damaged or has failed • PCM has failed
DTC: P1345 **1T CCM, MIL: YES** **1996, 1997, 1998, 1999, 2000, 2001, 2002** **Models:** MPV **Engines:** All **Transmissions:** A/T, M/T	**Camshaft Position (SGC) Sensor Circuit Malfunction** Engine running, MAF sensor signal more than 2.43 g/sec, and the PCM did not detect any CMP (SGC) sensor signals during the test. **Possible Causes:** • CMP sensor signal circuit is open or shorted to ground • CMP sensor signal circuit is shorted to VREF or system power • CMP sensor power circuit open between sensor and main relay • CMP sensor is damaged or has failed • PCM has failed
DTC: P1345 **2T CCM, MIL: YES** **1996, 1997, 1998, 1999, 2000, 2001, 2002, 2003** **Models:** Protégé, Protege5 **Engines:** 1.5L VIN 2 **Transmissions:** A/T, M/T	**Camshaft Position (SGC) Sensor Circuit Malfunction** Engine cranking, and then after detecting (5) crankshaft rotations, the PCM did not detect any CMP (SGC) sensor signals during the test. **Possible Causes:** • CMP sensor signal circuit is open or shorted to ground • CMP sensor signal circuit is shorted to VREF or system power • CMP sensor power circuit open between sensor and main relay • CMP sensor is damaged or has failed • PCM has failed

DTC	Trouble Code Title, Conditions & Possible Causes
DTC: P1345 **2T CCM, MIL: YES** **2004** **Models:** Tribute **Engines:** All **Transmissions:** A/T, M/T	**Camshaft Position (SGC) Sensor Circuit Malfunction** Engine cranking, and then after detecting (5) crankshaft rotations, the PCM did not detect any CMP (SGC) sensor signals during the test. **Possible Causes:** • CMP sensor signal circuit is open or shorted to ground • CMP sensor signal circuit is shorted to VREF or system power • CMP sensor power circuit open between sensor and main relay • CMP sensor is damaged or has failed • PCM has failed
DTC: P1351 **2T CCM, MIL: YES** **1996** **Models:** B2300, B3000, B4000 **Engines:** All **Transmissions:** A/T, M/T	**IDM Signal Lost To PCM or Out Of Range** Engine running and the PCM detected the Ignition Diagnostic Monitor (IDM) signal was missing or that the IDM signal dropped out. **Note: This problem can cause a No Start vehicle condition!** **Possible Causes:** • IDM signal circuit is open or shorted to ground • Ignition Module is damaged or has failed • PCM has failed
DTC: P1352 **2T CCM, MIL: YES** **1996** **Models:** B2300, B3000, B4000 **Engines:** All **Transmissions:** A/T, M/T	**Ignition Coil 'A' Primary Circuit Malfunction** Engine running and the PCM detected an interruption of the signal from the Ignition Coil 'A' primary circuits during the CCM test. **Possible Causes:** • Ignition Coil 'A' primary coil circuit is open (an intermittent fault) • Ignition Coil 'A' primary coil circuit is shorted to ground • Ignition Coil 'A' is damaged or has failed • PCM has failed
DTC: P1353 **1T CCM, MIL: YES** **1996** **Models:** B2300, B3000, B4000 **Engines:** All **Transmissions:** A/T, M/T	**Ignition Coil 'B' Primary Circuit Malfunction** Engine running, and the PCM detected an interruption of the signal from the Ignition Coil 'B' primary circuits during the CCM test. **Possible Causes:** • Ignition Coil 'B' primary coil circuit is open (an intermittent fault) • Ignition Coil 'B' primary coil circuit is shorted to ground • Ignition Coil 'B' is damaged or has failed • PCM has failed
DTC: P1354 **2T CCM, MIL: YES** **1996** **Models:** B3000, B4000 **Engines:** All **Transmissions:** A/T, M/T	**Ignition Coil 'C' Primary Circuit Malfunction** Engine running and the PCM detected an interruption of the signal from the Ignition Coil 'C' primary circuits during the CCM test. **Possible Causes:** • Ignition Coil 'C' primary coil circuit is open (an intermittent fault) • Ignition Coil 'C' primary coil circuit is shorted to ground • Ignition Coil 'C' is damaged or has failed • PCM has failed
DTC: P1358 **2T CCM, MIL: YES** **1996** **Models:** B2300, B3000, B4000 **Engines:** All **Transmissions:** A/T, M/T	**IDM Signal Circuit Malfunction** KOER Self-Test Enabled: Engine running and the PCM detected the Ignition Diagnostic Monitor signal was interrupted (dropped out or missing) in the test. **Possible Causes:** • IDM signal circuit is open or shorted to ground • Ignition Module is damaged or has failed • PCM has failed
DTC: P1358 **2T CCM, MIL: YES** **1996** **Models:** B2300, B2500, B3000, B4000 **Engines:** All **Transmissions:** A/T, M/T	**SPOUT Signal Circuit Malfunction** Engine running and the PCM detected the Spark Output Signal (SPOUT) was interrupted (dropped out or missing) during the test. **Possible Causes:** • IDM signal circuit is open or shorted to ground • Ignition Module is damaged or has failed • PCM has failed
DTC: P1360 **2T CCM, MIL: YES** **1996** **Models:** B2300, B3000, B4000 **Engines:** All **Transmissions:** A/T, M/T	**Ignition Coil 'A' Secondary Circuit Malfunction** Engine running and the PCM detected an intermittent signal from the Ignition Coil 'A' secondary circuit during the CCM test. **Possible Causes:** • Ignition Coil 'A' secondary coil circuit is open • Ignition Coil 'A' secondary coil circuit is shorted to ground • Ignition Coil 'A' is damaged or has failed • PCM has failed

DTC	Trouble Code Title, Conditions & Possible Causes
DTC: P1361 **2T CCM, MIL: YES** **1996** **Models:** B2300, B3000, B4000 **Engines:** All **Transmissions:** A/T, M/T	**Ignition Coil 'B' Secondary Circuit Malfunction** Engine running, and the PCM detected an intermittent signal from the Ignition Coil 'B' secondary circuit during the CCM test. **Possible Causes:** • Ignition Coil 'B' secondary coil circuit is open • Ignition Coil 'B' secondary coil circuit is shorted to ground • Ignition Coil 'B' is damaged or has failed • PCM has failed
DTC: P1362 **2T CCM, MIL: YES** **1996** **Models:** B3000, B4000 **Engines:** All **Transmissions:** A/T, M/T	**Ignition Coil 'C' Secondary Circuit Malfunction** Engine running and the PCM detected an intermittent signal from the Ignition Coil 'C' secondary circuit during the CCM test. **Possible Causes:** • Ignition Coil 'C' secondary coil circuit is open • Ignition Coil 'C' secondary coil circuit is shorted to ground • Ignition Coil 'C' is damaged or has failed • PCM has failed
DTC: P1364 **2T CCM, MIL: YES** **1996** **Models:** B2300, B3000, B4000 **Engines:** All **Transmissions:** A/T, M/T	**Ignition Coil Primary Circuit Malfunction** Engine running and the PCM detected the IGF signals were erratic or interrupted due to a fault in the Ignition Coil primary circuit. **Possible Causes:** • Ignition coil primary circuit is open • Ignition coil primary circuit is shorted to ground • Ignition coil primary is damaged or has failed • PCM has failed
DTC: P1364 **2T CCM, MIL: YES** **2004** **Models:** Tribute **Engines:** All **Transmissions:** A/T, M/T	**Ignition Coil Primary Circuit Malfunction** Engine running and the PCM detected the IGF signals were erratic or interrupted due to a fault in the Ignition Coil primary circuit. **Possible Causes:** • Ignition coil primary circuit is open • Ignition coil primary circuit is shorted to ground • Ignition coil primary is damaged or has failed • PCM has failed
DTC: P1365 **2T CCM, MIL: YES** **1996** **Models:** B2300, B3000, B4000 **Engines:** All **Transmissions:** A/T, M/T	**Ignition Coil Secondary Circuit Malfunction** Engine running and the PCM detected the IGF signals were erratic or interrupted due to a fault in the Ignition Coil secondary circuit. **Possible Causes:** • Ignition coil secondary circuit is open • Ignition coil secondary circuit is shorted to ground • Ignition coil secondary is damaged or has failed • PCM has failed
DTC: P1382 **2T CCM, MIL: YES** **2004** **Models:** Tribute **Engines:** All **Transmissions:** A/T, M/T	**Camshaft position (CMP) sensor** PCM detects a malfunction in the CMP sensor. **Possible Causes:** • CMP sensor circuit is open • CMP sensor circuit is shorted to ground • CMP sensor is damaged or has failed • PCM has failed
DTC: P1387 **2T CCM, MIL: YES** **2004** **Models:** Tribute **Engines:** All **Transmissions:** A/T, M/T	**Camshaft positions (CMP) sensor** PCM detects a malfunction in the CMP sensor. **Possible Causes:** • CMP sensor circuit is open • CMP sensor circuit is shorted to ground • CMP sensor is damaged or has failed • PCM has failed
DTC: P1390 **1T CCM, MIL: NO** **1996, 1997** **Models:** B2300, B3000, B4000 **Engines:** All **Transmissions:** A/T, M/T	**Octane Adjust Shorting Bar Circuit Open** Key on or engine running and the PCM detected the Octane Adjust circuit was high (open circuit) during the CCM test. **Possible Causes:** • Octane adjust circuit is open • Octane adjust shorting bar is missing or has been removed • PCM has failed

DTC	Trouble Code Title, Conditions & Possible Causes
DTC: P1400 **1T CCM, MIL: NO** **1996, 1997** **Models:** 626, MX-6 **Engines:** 2.0L VIN C **Transmissions:** A/T, M/T	**EGR Valve Position Sensor Circuit Low Input (Self-Test)** KOEO Self-Test enabled: Key on, and the PCM detected the EGRP sensor signal was less than 0.20v during the CCM test. **Possible Causes:** • EGRP sensor signal circuit is shorted to ground • EGRP sensor power circuit is open • EGRP sensor is damaged or has failed • PCM has failed
DTC: P1400 **1T CCM, MIL: NO** **1996, 1997, 1998, 1999, 2000, 2001, 2002, 2003, 2004, 2005** **Models:** B2300, B2500, B3000, B4000, Tribute **Engines:** All **Transmissions:** A/T, M/T	**EGR Valve Position Sensor Circuit Low Input (Self-Test)** KOER Self-Test enabled: Key on, and the PCM detected the EGRP sensor signal was less than 0.20v during the CCM Rationality test. **Possible Causes:** • EGRP sensor signal circuit is shorted to ground • EGRP sensor power circuit is open • EGRP sensor is damaged or has failed • PCM has failed
DTC: P1401 **1T CCM, MIL: NO** **1996, 1997** **Models:** 626, MX-6 **Engines:** 2.0L VIN C **Transmissions:** A/T, M/T	**EGR Valve Position Sensor Circuit High Input (Self-Test)** KOER Self-Test enabled: Key on, and the PCM detected the EGRP sensor signal was more than 4.01v during the CCM test. **Possible Causes:** • EGRP sensor signal circuit is open • EGRP sensor signal circuit shorted to VREF or system power • EGRP sensor ground circuit is open • EGRP sensor is damaged or has failed • PCM has failed
DTC: P1401 **1T CCM, MIL: NO** **1996, 1997, 1998, 1999, 2000, 2001, 2002, 2003, 2004, 2005** **Models:** B2300, B2500, B3000, B4000, Tribute **Engines:** All **Transmissions:** A/T, M/T	**EGR Valve Position Sensor Circuit High Input (Self-Test)** KOER Self-Test enabled: Key on, and the PCM detected the EGRP sensor signal was more than 4.50v during the CCM test. **Possible Causes:** • EGRP sensor signal circuit is open • EGRP sensor ground circuit is open • EGRP sensor is damaged or has failed • PCM has failed
DTC: P1402 **1T CCM, MIL: YES** **1996, 1997** **Models:** 626, MX-6 **Engines:** All **Transmissions:** A/T, M/T	**EGR Valve Position Sensor Circuit Malfunction** Engine running, and the PCM detected the EGR Valve Position (EVP) sensor signal was less than 0.20v, or it was more than 4.90v. **Possible Causes:** • EGRP sensor signal circuit is open or shorted to ground • EGRP sensor signal circuit shorted to VREF or system power • EGRP sensor ground circuit or the power circuit is open • EGRP sensor is damaged or has failed • PCM has failed
DTC: P1402 **1T CCM, MIL: YES** **1996, 1997** **Models:** MX-5 Miata **Engines:** All **Transmissions:** A/T, M/T	**EGR Valve Position Sensor Circuit Malfunction** Engine running, and the PCM detected the EGR Valve Position (EVP) sensor signal was less than 0.20v, or it was more than 4.90v. **Possible Causes:** • EGRP sensor signal circuit is open or shorted to ground • EGRP sensor signal circuit shorted to VREF or system power • EGRP sensor ground circuit or the power circuit is open • EGRP sensor is damaged or has failed • PCM has failed
DTC: P1402 **1T CCM, MIL: YES** **1996, 1997, 1998, 1999, 2000, 2001, 2002** **Models:** Millenia **Engines:** All **Transmissions:** A/T, M/T	**EGR Valve Position Sensor Circuit Malfunction** Engine running, and the PCM detected the EGR Valve Position (EVP) sensor signal was less than 0.20v, or it was more than 4.80v. **Possible Causes:** • EGRP sensor signal circuit is open or shorted to ground • EGRP sensor signal circuit shorted to VREF or system power • EGRP sensor ground circuit or power circuit is open • EGRP sensor is damaged or has failed • PCM has failed

DTC	Trouble Code Title, Conditions & Possible Causes
DTC: P1402 **1T CCM, MIL: YES** **2004** **Models:** Tribute **Engines:** All **Transmissions:** A/T, M/T	**EGR Valve Position Sensor Circuit Malfunction** Engine running, and the PCM detected the EGR Valve Position (EVP) sensor signal was less than 0.20v, or it was more than 4.80v. **Possible Causes:** • EGRP sensor signal circuit is open or shorted to ground • EGRP sensor signal circuit shorted to VREF or system power • EGRP sensor ground circuit or power circuit is open • EGRP sensor is damaged or has failed • PCM has failed
DTC: P1405 **2T CCM, MIL: NO** **1996, 1997, 1998, 1999, 2000, 2001, 2002, 2003, 2004, 2005** **Models:** B2300, B2500, B3000, B4000, Tribute **Engines:** All **Transmissions:** A/T, M/T	**DPFE Sensor Upstream Hose Off or Plugged** KOER Self-Test Enabled: Engine running and the PCM detected the DPFE sensor upstream hose was off or plugged during the CCM test. **Possible Causes:** • DPF EGR sensor upstream hose is disconnected • DPF EGR sensor upstream hose is plugged (ice) • EGR orifice tube is plugged or damaged
DTC: P1406 **2T CCM, MIL: YES** **1996, 1997, 1998, 1999, 2000, 2001, 2002, 2003** **Models:** B2300, B2500, B3000, B4000, Tribute **Engines:** All **Transmissions:** A/T, M/T	**DPFE Sensor Downstream Hose Off or Plugged** KOER Self-Test Enabled: Engine running and the PCM detected the DPFE sensor upstream hose was off or plugged during the CCM test. **Possible Causes:** • DPF EGR sensor downstream hose is disconnected • DPF EGR sensor downstream hose is plugged (ice) • EGR orifice tube is plugged or damaged
DTC: P1407 **2T CCM, MIL: YES** **1996, 1997** **Models:** 626, MX-6 **Engines:** 2.0L VIN C **Transmissions:** A/T, M/T	**No EGR Flow Detected** Vehicle driven to a speed of over 7 mph, then back to idle speed, ECT sensor signal more than 131°F, and the PCM detected no change in the EGR Back Pressure with the EGR solenoid commanded "on" and then "off" during the EGR Monitor test. **Possible Causes:** • EGRA solenoid is damaged or has failed • EGRC solenoid is damaged or has failed • EGRV solenoid is damaged or has failed • EGRBP sensor is damaged or has failed • EGR valve assembly is leaking, damaged or has failed • EGRA, EGRC, EGRV or EGRBP harness/connector damaged • Exhaust system is partially blocked or restricted • PCM has failed
DTC: P1407 **2T CCM, MIL: YES** **2004** **Models:** Tribute **Engines:** All **Transmissions:** A/T, M/T	**No EGR Flow Detected** Vehicle driven to a speed of over 7 mph, then back to idle speed, ECT sensor signal more than 131°F, and the PCM detected no change in the EGR Back Pressure with the EGR solenoid commanded "on" and then "off" during the EGR Monitor test. **Possible Causes:** • EGRA solenoid is damaged or has failed • EGRC solenoid is damaged or has failed • EGRV solenoid is damaged or has failed • EGRBP sensor is damaged or has failed • EGR valve assembly is leaking, damaged or has failed • EGRA, EGRC, EGRV or EGRBP harness/connector damaged • Exhaust system is partially blocked or restricted • PCM has failed
DTC: P1408 **1T CCM, MIL: NO** **1996, 1997, 1998, 1999, 2000, 2001, 2002, 2003, 2004, 2005** **Models:** B2300, B2500, B3000, B4000, Tribute **Engines:** All **Transmissions:** A/T, M/T	**EGR System Flow Out Of Range (Self-Test)** KOER Self-Test enabled: Vehicle driven to a speed of over 7 mph, then back to idle speed, ECT sensor signal more than 131°F, and the PCM detected little or no change in the EGR Differential Pressure sensor (DPFE) with the EGR vent commanded "on" and then "off" in the EGR Monitor test. **Possible Causes:** • EGR valve stuck closed or iced up, or the flow path is restricted • EGR valve diaphragm leaking, hose is off, plugged or leaking • EGR system vacuum lines are loose or damaged • EGRA solenoid is damaged or has failed • EGRV solenoid is damaged or has failed • PCM has failed

DTC	Trouble Code Title, Conditions & Possible Causes
DTC: P1409 **2T CCM, MIL: NO** **1996, 1997** **Models:** 626, MX-6 **Engines:** 2.0L VIN C **Transmissions:** A/T	**EGR Vacuum Regulator Solenoid Circuit Malfunction** Engine running and the PCM detected an unexpected voltage condition on the EGR VR valve (stepper motor) during the test. **Possible Causes:** • EGR VR solenoid control circuit is open • EGR VR solenoid control circuit is shorted to ground • EGR VR solenoid control circuit is shorted to system power • EGR VR solenoid is damaged or has failed • PCM has failed
DTC: P1409 **2T CCM, MIL: NO** **1998, 1999, 2000, 2001, 2002** **Models:** 626 **Engines:** All **Transmissions:** A/T	**EGR Vacuum Regulator Solenoid Circuit Malfunction** Engine running and the PCM detected an unexpected voltage condition on the EGR VR valve (stepper motor) during the test. **Possible Causes:** • EGR VR solenoid control circuit is open • EGR VR solenoid control circuit is shorted to ground • EGR VR solenoid control circuit is shorted to system power • EGR VR solenoid is damaged or has failed • PCM has failed
DTC: P1409 **2T CCM, MIL: NO** **1996, 1997,** **1998, 1999, 2000, 2001, 2002,** **2003** **Models:** B2300, B2500, B3000, B4000, Tribute **Engines:** All **Transmissions:** A/T, M/T	**EGR Vacuum Regulator Solenoid Circuit Malfunction** Engine running and the PCM detected an unexpected voltage condition on the EGR VR valve (stepper motor) during the test. **Possible Causes:** • EGR VR solenoid control circuit is open • EGR VR solenoid control circuit is shorted to ground • EGR VR solenoid control circuit is shorted to system power • EGR VR solenoid is damaged or has failed • PCM has failed
DTC: P1410 **1T CCM, MIL: YES** **1996, 1997, 1998, 1999, 2000,** **2001, 2002** **Models:** 626, MX-6 **Engines:** 2.0L VIN C **Transmissions:** A/T, M/T	**EGR Boost Sensor Solenoid Valve Stuck** Engine running and the PCM detected the EGR Pressure continued to indicate Barometric pressure during the EGR Monitor test (after the EGRC solenoid was cycled "on" and "off" during the EGR test). **Possible Causes:** • Air leaks located after the MAF sensor assembly • EGR/BARO sensor is damaged or has failed • EGRC solenoid valve is damaged or has failed • EGR sensor vacuum hose to EGR Boost solenoid is loose • EGR sensor vacuum hose to the intake manifold is loose • EGRC or EGR/BARO sensor harness/connector damaged • PCM has failed
DTC: P1410 **1T CCM, MIL: YES** **2003** **Models:** Mazda6 **Engines:** All **Transmissions:** A/T, M/T	**EGR Boost Sensor Solenoid Valve Stuck** Engine running, and the PCM detected the EGR Pressure continued to indicate Barometric pressure during the EGR Monitor test (after the EGRC solenoid was cycled "on" and "off" during the EGR test). **Possible Causes:** • Air leaks located after the MAF sensor assembly • EGR/BARO sensor is damaged or has failed • EGRC solenoid valve is damaged or has failed • EGR sensor vacuum hose to EGR Boost solenoid is loose • EGR sensor vacuum hose to the intake manifold is loose • EGRC or EGR/BARO sensor harness/connector damaged • PCM has failed
DTC: P1412 **1T CCM, MIL: YES** **2004** **Models:** Tribute **Engines:** All **Transmissions:** A/T, M/T	**EGR Boost Sensor Solenoid Valve Stuck** Engine running, and the PCM detected the EGR Pressure continued to indicate Barometric pressure during the EGR Monitor test (after the EGRC solenoid was cycled "on" and "off" during the EGR test). **Possible Causes:** • Air leaks located after the MAF sensor assembly • EGR/BARO sensor is damaged or has failed • EGRC solenoid valve is damaged or has failed • EGR sensor vacuum hose to EGR Boost solenoid is loose • EGR sensor vacuum hose to the intake manifold is loose • EGRC or EGR/BARO sensor harness/connector damaged • PCM has failed

DTC	Trouble Code Title, Conditions & Possible Causes
DTC: P1432 **1T CCM, MIL: YES** **2004** **Models:** B2300, B3000, B4000 **Engines:** All **Transmissions:** A/T, M/T	**Thermostat Heater Control (THTRC)** **PCM detected the voltage of THTRC circuit fell below a calibrated limit for a calibration amount of time.** **Possible Causes:** • THTRC circuit is damaged or has failed • THTRC circuit harness/connector damaged • PCM has failed
DTC: P1443 **2T CCM, MIL: YES** **1996, 1997, 1998, 1999, 2000, 2001, 2002, 2003, 2004, 2005** **Models:** B2300, B2500, B3000, B4000 **Engines:** All **Transmissions:** A/T, M/T	**EVAP System Purge Flow Malfunction** Engine running in closed loop at cruise speed, TP Angle steady at medium load, VSS at 35-55 mph, and the PCM detected a leak or blockage between the intake manifold, canister purge valve and the charcoal canister during the EVAP Monitor test. **Possible Causes:** • Fuel vapor hose blocked between EVAP purge valve and PF sensor, or blocked between purge valve and intake manifold, or vacuum hose blocked between purge valve and intake manifold • EVAP canister purge solenoid stuck closed (mechanically)
DTC: P1443 **2T CCM, MIL: YES** **2001, 2002, 2003, 2004, 2005** **Models:** Tribute **Engines:** All **Transmissions:** A/T, M/T	**EVAP System Purge Flow Malfunction** Engine running in closed loop at cruise speed, TP Angle steady at medium load, VSS at 35-55 mph, and the PCM detected a large leak and no purge flow (less than -1.74 kPa) in the system during the EVAP Monitor Leak Test. **Possible Causes:** • Fuel vapor hose blocked between EVAP purge valve and FTP sensor, or blocked between purge valve and intake manifold, or vacuum hose blocked between purge valve and intake manifold • EVAP canister purge solenoid stuck closed (mechanically)
DTC: P1444 **2T CCM, MIL: YES** **1996, 1997** **Models:** B2300, B3000, B4000 **Engines:** All **Transmissions:** A/T, M/T	**EVAP Purge Flow Sensor Circuit Low Input** Key on or engine running and the PCM detected the EVAP Purge Flow (PF) sensor signal was less than 0.20v during the CCM test. **Possible Causes:** • Purge flow sensor signal circuit is shorted to ground • Purge flow sensor power circuit is open • Purge flow sensor is damaged or has failed • PCM has failed
DTC: P1445 **2T CCM, MIL: YES** **1996, 1997** **Models:** B2300, B3000, B4000 **Engines:** All **Transmissions:** A/T, M/T	**EVAP Purge Flow Sensor Circuit Low Input** Key on or engine running and the PCM detected the EVAP Purge Flow (PF) sensor signal was more than 4.89v during the CCM test. **Possible Causes:** • Purge flow sensor signal circuit is open • Purge flow sensor ground circuit is open • Purge flow sensor signal circuit is shorted to VREF or power • Purge flow sensor is damaged or has failed • PCM has failed
DTC: P1446 **2T CCM, MIL: YES** **2004** **Models:** Tribute **Engines:** All **Transmissions:** A/T, M/T	**EVAP Purge Flow Sensor Circuit Low Input** Key on or engine running and the PCM detected the EVAP Purge Flow (PF) sensor signal was more than 4.89v during the CCM test. **Possible Causes:** • Purge flow sensor signal circuit is open • Purge flow sensor ground circuit is open • Purge flow sensor signal circuit is shorted to VREF or power • Purge flow sensor is damaged or has failed • PCM has failed
DTC: P1449 **2T CCM, MIL: NO** **1999, 2000, 2001, 2002, 2003** **Models:** MX-5 Miata **Engines:** All **Transmissions:** A/T, M/T	**CDCV Circuit Malfunction** Key on, and the PCM detected an unexpected low voltage condition on the Canister Drain Cut Valve (CDCV) control circuit during the CCM test at initial key "on". **Note: This is a diagnostic support code - the PCM monitors this circuit once per key cycle, but does not set a pending code.** **Possible Causes:** • CDCV control circuit is open or shorted to ground • CDCV control circuit is shorted to power • CDCV power circuit is open between the valve and main relay • CDCV is damaged or has failed • PCM has failed

DTC	Trouble Code Title, Conditions & Possible Causes
DTC: P1485 **1T CCM, MIL: NO** **1996, 1997** **Models:** MX-5 Miata **Engines:** All **Transmissions:** A/T, M/T	**EGR Vacuum Solenoid Circuit Malfunction** Key on or engine running and the PCM detected an unexpected voltage condition on the EGR Vacuum solenoid in the CCM test. **Possible Causes:** • EGR vacuum solenoid control circuit open or shorted to ground • EGR vacuum solenoid control circuit shorted to system power • EGR vacuum solenoid power circuit open or shorted to ground • EGR vacuum solenoid is damaged or has failed • PCM has failed
DTC: P1485 **1T CCM, MIL: NO** **1996, 1997, 1998** **Models:** Millenia **Engines:** 2.3L VIN 2 **Transmissions:** A/T, M/T	**EGR Vacuum Solenoid Circuit Malfunction** Key on or engine running and the PCM detected an unexpected voltage condition on the EGR Vacuum solenoid in the CCM test. **Possible Causes:** • EGR vacuum solenoid control circuit open or shorted to ground • EGR vacuum solenoid control circuit shorted to system power • EGR vacuum solenoid power circuit open or shorted to ground • EGR vacuum solenoid is damaged or has failed • PCM has failed
DTC: P1485 **1T CCM, MIL: NO** **2004** **Models:** Tribute **Engines:** 2.3L VIN 2 **Transmissions:** A/T, M/T	**EGR Vacuum Solenoid Circuit Malfunction** Key on or engine running and the PCM detected an unexpected voltage condition on the EGR Vacuum solenoid in the CCM test. **Possible Causes:** • EGR vacuum solenoid control circuit open or shorted to ground • EGR vacuum solenoid control circuit shorted to system power • EGR vacuum solenoid power circuit open or shorted to ground • EGR vacuum solenoid is damaged or has failed • PCM has failed
DTC: P1485 **2T CCM, MIL: YES** **1996** **Models:** Protégé **Engines:** All **Transmissions:** A/T, M/T	**EGR Vacuum Solenoid Circuit Malfunction** Key on or engine running and the PCM detected an unexpected voltage condition on the EGR Vacuum solenoid in the CCM test. **Possible Causes:** • EGR vacuum solenoid control circuit open or shorted to ground • EGR vacuum solenoid control circuit shorted to system power • EGR vacuum solenoid power circuit open or shorted to ground • EGR vacuum solenoid is damaged or has failed • PCM has failed
DTC: P1486 **1T CCM, MIL: NO** **1996, 1997** **Models:** 626, MX-6 **Engines:** 2.0L VIN C **Transmissions:** M/T	**EGR Vent Solenoid Circuit Malfunction** Key on or engine running and the PCM detected an unexpected voltage condition on the EGR Vent Solenoid during the CCM test. **Possible Causes:** • EGR vent solenoid control circuit open or shorted to ground • EGR vent solenoid control circuit shorted to system power • EGR vent solenoid power circuit open or shorted to ground • EGR vent solenoid is damaged or has failed • PCM has failed
DTC: P1486 **1T CCM, MIL: NO** **1996, 1997** **Models:** MX-5 Miata **Engines:** All **Transmissions:** A/T, M/T	**EGR Vent Solenoid Circuit Malfunction** Key on or engine running and the PCM detected an unexpected voltage condition on the EGR Vent Solenoid during the CCM test. **Possible Causes:** • EGR vent solenoid control circuit open or shorted to ground • EGR vent solenoid control circuit shorted to system power • EGR vent solenoid power circuit open or shorted to ground • EGR vent solenoid is damaged or has failed • PCM has failed
DTC: P1486 **1T CCM, MIL: NO** **1996, 1997, 1998** **Models:** Millenia **Engines:** All **Transmissions:** A/T, M/T	**EGR Vent Solenoid Circuit Malfunction** Key on or engine running and the PCM detected an unexpected voltage condition on the EGR Vent Solenoid in the CCM test. **Possible Causes:** • EGR vent solenoid control circuit open or shorted to ground • EGR vent solenoid control circuit shorted to system power • EGR vent solenoid power circuit open or shorted to ground • EGR vent solenoid is damaged or has failed • PCM has failed

DTC	Trouble Code Title, Conditions & Possible Causes
DTC: P1486 **2T CCM, MIL: YES** **1996** **Models:** Protégé **Engines:** All **Transmissions:** A/T, M/T	**EGR Vent Solenoid Circuit Malfunction** Key on or engine running and the PCM detected an unexpected voltage condition on the EGR Vent Solenoid in the CCM test. **Possible Causes:** • EGR vent solenoid control circuit open or shorted to ground • EGR vent solenoid control circuit shorted to system power • EGR vent solenoid power circuit open or shorted to ground • EGR vent solenoid is damaged or has failed • PCM has failed
DTC: P1486 **2T CCM, MIL: YES** **2004** **Models:** Tribute **Engines:** All **Transmissions:** A/T, M/T	**EGR Vent Solenoid Circuit Malfunction** Key on or engine running and the PCM detected an unexpected voltage condition on the EGR Vent Solenoid in the CCM test. **Possible Causes:** • EGR vent solenoid control circuit open or shorted to ground • EGR vent solenoid control circuit shorted to system power • EGR vent solenoid power circuit open or shorted to ground • EGR vent solenoid is damaged or has failed • PCM has failed
DTC: P1487 **1T CCM, MIL: NO** **1996, 1997** **Models:** 626, MX-6 **Engines:** 2.0L VIN C **Transmissions:** M/T	**EGR Boost Sensor Solenoid Circuit Malfunction** Key on or engine running and the PCM detected an unexpected voltage condition on the EGR Boost Sensor Check solenoid during the CCM test. **Possible Causes:** • EGR-CHK solenoid control circuit is open or shorted to ground • EGR-CHK solenoid power circuit is open or shorted to ground • EGR-CHK solenoid is damaged or has failed • PCM has failed
DTC: P1487 **1T CCM, MIL: YES** **1998, 1999, 2000, 2001, 2002** **Models:** 626, MX-6 **Engines:** All **Transmissions:** A/T, M/T	**EGR Boost Sensor Solenoid Circuit Malfunction** Key on, and the PCM detected an unexpected voltage condition on the EGR Boost Sensor solenoid control circuit during the CCM test. **Possible Causes:** • EGR boost sensor solenoid circuit is open or shorted to ground • EGR boost sensor solenoid power circuit is open • EGR boost sensor solenoid is damaged or has failed • PCM has failed
DTC: P1487 **1T CCM, MIL: YES** **2004** **Models:** Tribute **Engines:** All **Transmissions:** A/T, M/T	**EGR Boost Sensor Solenoid Circuit Malfunction** Key on, and the PCM detected an unexpected voltage condition on the EGR Boost Sensor Solenoid control circuit during the CCM test. **Possible Causes:** • EGR boost sensor solenoid circuit is open or shorted to ground • EGR boost sensor solenoid circuit is shorted to system power • EGR boost sensor solenoid power circuit is open • EGR boost sensor solenoid is damaged or has failed • PCM has failed
DTC: P1487 **1T CCM, MIL: YES** **1996, 1997, 1998, 1999, 2000, 2001, 2002** **Models:** Millenia **Engines:** 2.5L VIN 2 **Transmissions:** A/T, M/T	**EGR Boost Sensor Solenoid Circuit Malfunction** Key on, and the PCM detected an unexpected voltage condition on the EGR Boost Sensor Solenoid control circuit during the CCM test. **Possible Causes:** • EGR boost sensor solenoid circuit is open or shorted to ground • EGR boost sensor solenoid circuit is shorted to system power • EGR boost sensor solenoid power circuit is open • EGR boost sensor solenoid is damaged or has failed • PCM has failed
DTC: P1487 **1T CCM, MIL: YES** **1996, 1997, 1998** **Models:** Millenia **Engines:** 2.3L VIN 1 **Transmissions:** A/T, M/T	**EGR Boost Sensor Solenoid Circuit Malfunction** Key on, and the PCM detected an unexpected voltage condition on the EGR Boost Sensor Solenoid control circuit during the CCM test. **Possible Causes:** • EGR boost sensor solenoid circuit is open or shorted to ground • EGR boost sensor solenoid circuit is shorted to system power • EGR boost sensor solenoid power circuit is open • EGR boost sensor solenoid is damaged or has failed • PCM has failed

DTC	Trouble Code Title, Conditions & Possible Causes
DTC: P1487 **1T CCM, MIL: YES** **1999, 2000, 2001, 2002** **Models:** Millenia **Engines:** 2.3L VIN 1 **Transmissions:** A/T, M/T	**MAP Sensor Solenoid Circuit Malfunction** Key on, and the PCM detected an unexpected voltage condition on the MAP Sensor Solenoid control circuit during the CCM test. **Possible Causes:** • MAP sensor solenoid circuit is open or shorted to ground • MAP sensor solenoid circuit is shorted to system power • MAP sensor solenoid power circuit is open • MAP sensor solenoid is damaged or has failed • PCM has failed
DTC: P1487 **1T CCM, MIL: NO** **1996, 1997, 1998, 1999, 2000, 2001, 2002, 2003** **Models:** Protégé, Protege5 **Engines:** All **Transmissions:** A/T, M/T	**EGR Boost Sensor Solenoid Valve Circuit Malfunction** Key on, and the PCM detected an unexpected voltage condition on the EGR Boost Sensor Check Solenoid valve during the CCM test. **Note: This is a diagnostic support code - the PCM monitors this circuit once per key cycle, but does not set a pending code.** **Possible Causes:** • EGR boost sensor solenoid circuit is open or shorted to ground • EGR boost sensor solenoid circuit is shorted to system power • EGR boost sensor solenoid circuit is open or shorted to ground • EGR boost sensor solenoid is damaged or has failed • PCM has failed
DTC: P1487 **1T CCM, MIL: YES** **2000, 2001, 2002, 2003, 2004** **Models:** MPV **Engines:** All **Transmissions:** A/T, M/T	**EGR Boost Sensor Solenoid Valve Circuit Malfunction** Key on, and the PCM detected an unexpected voltage condition on the EGR Boost Sensor Check Solenoid valve during the CCM test. **Note: This is a diagnostic support code - the PCM monitors this circuit once per key cycle, but does not set a pending code.** **Possible Causes:** • EGR boost sensor solenoid circuit is open or shorted to ground • EGR boost sensor solenoid circuit is shorted to system power • EGR boost sensor solenoid circuit is open or shorted to ground • EGR boost sensor solenoid is damaged or has failed • PCM has failed
DTC: P1487 **1T CCM, MIL: YES** **2004, 2005** **Models:** MX-5 Miata **Engines:** All **Transmissions:** A/T, M/T	**EGR Boost Sensor Solenoid Valve Circuit Malfunction** Key on, and the PCM detected an unexpected voltage condition on the EGR Boost Sensor Check Solenoid valve during the CCM test. **Note: This is a diagnostic support code - the PCM monitors this circuit once per key cycle, but does not set a pending code.** **Possible Causes:** • EGR boost sensor solenoid circuit is open or shorted to ground • EGR boost sensor solenoid circuit is shorted to system power • EGR boost sensor solenoid circuit is open or shorted to ground • EGR boost sensor solenoid is damaged or has failed • PCM has failed
DTC: P1496 **1T CCM, MIL: NO** **1999, 2000, 2001, 2002, 2003** **Models:** MX-5 Miata **Engines:** All **Transmissions:** A/T, M/T	**EGR Valve Stepper Motor Coil 1 Circuit Malfunction** Key on, and the PCM detected an unexpected low voltage condition (less than 2.7v) on the EGR Valve Stepper Motor Coil 1 control circuit with the valve commanded "off" during the CCM test. **Note: This is a diagnostic support code - the PCM monitors this circuit once per key cycle, but does not set a pending code.** **Possible Causes:** • EGR Coil 1 control circuit is open or shorted to ground • EGR Coil 1 control circuit is shorted to power • EGR Coil 1 power circuit open between valve and main relay • EGR Coil 1 is damaged or has failed • PCM has failed
DTC: P1496 **1T CCM, MIL: YES** **1997, 1998, 1999, 2000, 2001, 2002, 2003** **Models:** Protégé, Protege5 **Engines:** All **Transmissions:** A/T, M/T	**EGR Valve Stepper Motor Coil 1 Circuit Malfunction** Key on, and the PCM detected an unexpected voltage condition on the EGR valve Stepper Motor Coil 1 control circuit in the CCM test. **Note: This is a diagnostic support code - the PCM monitors this circuit once per key cycle, but does not set a pending code.** **Possible Causes:** • EGR Coil 1 control circuit is open or shorted to ground • EGR Coil 1 control circuit is shorted to power • EGR Coil 1 power circuit open between valve and main relay • EGR Coil 1 is damaged or has failed • PCM has failed

DTC	Trouble Code Title, Conditions & Possible Causes
DTC: P1496 **1T CCM, MIL: NO** **2000, 2001, 2002** **Models:** MPV **Engines:** All **Transmissions:** A/T, M/T	**EGR Valve Stepper Motor Coil 1 Circuit Malfunction** Key on, and the PCM detected an unexpected low voltage condition (less than 2.7v) on the EGR Valve Stepper Motor Coil 1 control circuit with the valve commanded "off" during the CCM test. **Note: This is a diagnostic support code - the PCM monitors this circuit once per key cycle, but does not set a pending code.** **Possible Causes:** • EGR Coil 1 control circuit is open or shorted to ground • EGR Coil 1 control circuit is shorted to power • EGR Coil 1 power circuit open between valve and main relay • EGR Coil 1 is damaged or has failed • PCM has failed
DTC: P1496 **1T CCM, MIL: NO** **2004** **Models:** Tribute **Engines:** All **Transmissions:** A/T, M/T	**EGR Valve Stepper Motor Coil 1 Circuit Malfunction** Key on, and the PCM detected an unexpected low voltage condition (less than 2.7v) on the EGR Valve Stepper Motor Coil 1 control circuit with the valve commanded "off" during the CCM test. **Note: This is a diagnostic support code - the PCM monitors this circuit once per key cycle, but does not set a pending code.** **Possible Causes:** • EGR Coil 1 control circuit is open or shorted to ground • EGR Coil 1 control circuit is shorted to power • EGR Coil 1 power circuit open between valve and main relay • EGR Coil 1 is damaged or has failed • PCM has failed
DTC: P1497 **1T CCM, MIL: NO** **1999, 2000, 2001, 2002, 2003** **Models:** MX-5 Miata **Engines:** All **Transmissions:** A/T, M/T	**EGR Valve Stepper Motor Coil 2 Circuit Malfunction** Key on, and the PCM detected an unexpected low voltage condition (less than 0.6v) on the EGR Valve Stepper Motor Coil 2 control circuit with the valve commanded "off" during the CCM test. **Note: This is a diagnostic support code - the PCM monitors this circuit once per key cycle, but does not set a pending code.** **Possible Causes:** • EGR Coil 2 control circuit is open or shorted to ground • EGR Coil 2 control circuit is shorted to power • EGR Coil 2 power circuit open between valve and main relay • EGR Coil 2 is damaged or has failed • PCM has failed
DTC: P1497 **1T CCM, MIL: YES** **1997, 1998, 1999, 2000, 2001, 2002, 2003** **Models:** Protégé, Protege5 **Engines:** All **Transmissions:** A/T, M/T	**EGR Valve Stepper Motor Coil 2 Circuit Malfunction** Key on, and the PCM detected an unexpected voltage condition on the EGR valve Stepper Motor Coil 2 control circuit in the CCM test. **Note: This is a diagnostic support code - the PCM monitors this circuit once per key cycle, but does not set a pending code.** **Possible Causes:** • EGR Coil 2 control circuit is open or shorted to ground • EGR Coil 2 control circuit is shorted to power • EGR Coil 2 power circuit open between valve and main relay • EGR Coil 2 is damaged or has failed • PCM has failed
DTC: P1497 **1T CCM, MIL: NO** **2000, 2001, 2002** **Models:** MPV **Engines:** All **Transmissions:** A/T, M/T	**EGR Valve Stepper Motor Coil 2 Circuit Malfunction** Key on, and the PCM detected an unexpected low voltage condition (less than 0.6v) on the EGR Valve Stepper Motor Coil 2 control circuit with the valve commanded "off" during the CCM test. **Note: This is a diagnostic support code - the PCM monitors this circuit once per key cycle, but does not set a pending code.** **Possible Causes:** • EGR Coil 2 control circuit is open or shorted to ground • EGR Coil 2 control circuit is shorted to power • EGR Coil 2 power circuit open between valve and main relay • EGR Coil 2 is damaged or has failed • PCM has failed

DTC	Trouble Code Title, Conditions & Possible Causes
DTC: P1497 **1T CCM, MIL: NO** **2004** **Models:** Tribute **Engines:** All **Transmissions:** A/T, M/T	**EGR Valve Stepper Motor Coil 2 Circuit Malfunction** Key on, and the PCM detected an unexpected low voltage condition (less than 0.6v) on the EGR Valve Stepper Motor Coil 2 control circuit with the valve commanded "off" during the CCM test. **Note: This is a diagnostic support code - the PCM monitors this circuit once per key cycle, but does not set a pending code.** **Possible Causes:** • EGR Coil 2 control circuit is open or shorted to ground • EGR Coil 2 control circuit is shorted to power • EGR Coil 2 power circuit open between valve and main relay • EGR Coil 2 is damaged or has failed • PCM has failed
DTC: P1498 **1T CCM, MIL: NO** **1999, 2000, 2001, 2002, 2003** **Models:** MX-5 Miata **Engines:** All **Transmissions:** A/T, M/T	**EGR Valve Stepper Motor Coil 3 Circuit Malfunction** Key on, and the PCM detected an unexpected low voltage condition (less than 0.6v) on the EGR Valve Coil 3 control circuit with the valve command "off" during the CCM test at startup. **Note: This is a diagnostic support code - the PCM monitors this circuit once per key cycle, but does not set a pending code.** **Possible Causes:** • EGR Coil 3 control circuit is open or shorted to ground • EGR Coil 3 control circuit is shorted to power • EGR Coil 3 power circuit open between valve and main relay • EGR Coil 3 is damaged or has failed • PCM has failed
DTC: P1498 **1T CCM, MIL: NO** **1997, 1998, 1999, 2000, 2001, 2002, 2003** **Models:** Protégé, Protege5 **Engines:** All **Transmissions:** A/T, M/T	**EGR Valve Stepper Motor Coil 3 Circuit Malfunction** Key on, and the PCM detected an unexpected voltage condition on the EGR valve Stepper Motor Coil 3 control circuit in the CCM test. **Note: This is a diagnostic support code - the PCM monitors this circuit once per key cycle, but does not set a pending code.** **Possible Causes:** • EGR Coil 3 control circuit is open or shorted to ground • EGR Coil 3 control circuit is shorted to power • EGR Coil 3 power circuit open between valve and main relay • EGR Coil 3 is damaged or has failed • PCM has failed
DTC: P1498 **1T CCM, MIL: NO** **2000, 2001, 2002** **Models:** MPV **Engines:** All **Transmissions:** A/T, M/T	**EGR Valve Stepper Motor Coil 3 Circuit Malfunction** Key on, and the PCM detected an unexpected low voltage condition (less than 0.6v) on the EGR Valve Coil 3 control circuit with the valve command "off" during the CCM test at startup. **Note: This is a diagnostic support code - the PCM monitors this circuit once per key cycle, but does not set a pending code.** **Possible Causes:** • EGR Coil 3 control circuit is open or shorted to ground • EGR Coil 3 control circuit is shorted to power • EGR Coil 3 power circuit open between valve and main relay • EGR Coil 3 is damaged or has failed • PCM has failed
DTC: P1498 **1T CCM, MIL: NO** **2004** **Models:** Tribute **Engines:** All **Transmissions:** A/T, M/T	**EGR Valve Stepper Motor Coil 3 Circuit Malfunction** Key on, and the PCM detected an unexpected low voltage condition (less than 0.6v) on the EGR Valve Coil 3 control circuit with the valve command "off" during the CCM test at startup. **Note: This is a diagnostic support code - the PCM monitors this circuit once per key cycle, but does not set a pending code.** **Possible Causes:** • EGR Coil 3 control circuit is open or shorted to ground • EGR Coil 3 control circuit is shorted to power • EGR Coil 3 power circuit open between valve and main relay • EGR Coil 3 is damaged or has failed • PCM has failed

DTC	Trouble Code Title, Conditions & Possible Causes
DTC: P1499 **1T CCM, MIL: NO** **1999, 2000, 2001, 2002, 2003** **Models:** MX-5 Miata **Engines:** All **Transmissions:** A/T, M/T	**EGR Valve Stepper Motor Coil 4 Circuit Malfunction** Key on, and the PCM detected an unexpected low voltage condition (less than 2.7v) on the EGR Valve Coil 4 control circuit with the valve command "off" during the CCM test at startup. **Note: This is a diagnostic support code - the PCM monitors this circuit once per key cycle, but does not set a pending code.** **Possible Causes:** • EGR Coil 4 control circuit is open or shorted to ground • EGR Coil 4 control circuit is shorted to power • EGR Coil 4 power circuit open between valve and main relay • EGR Coil 4 is damaged or has failed • PCM has failed
DTC: P1499 **1T CCM, MIL: NO** **1997, 1998, 1999, 2000, 2001, 2002, 2003** **Models:** Protégé, Protege5 **Engines:** All **Transmissions:** A/T, M/T	**EGR Valve Stepper Motor Coil 4 Circuit Malfunction** Key on, and the PCM detected an unexpected voltage condition on the EGR valve Stepper Motor Coil 4 control circuit in the CCM test. **Note: This is a diagnostic support code - the PCM monitors this circuit once per key cycle, but does not set a pending code.** **Possible Causes:** • EGR Coil 4 control circuit is open or shorted to ground • EGR Coil 4 control circuit is shorted to power • EGR Coil 4 power circuit open between valve and main relay • EGR Coil 4 is damaged or has failed • PCM has failed
DTC: P1499 **1T CCM, MIL: NO** **2000, 2001, 2002** **Models:** MPV **Engines:** All **Transmissions:** A/T, M/T	**EGR Valve Stepper Motor Coil 4 Circuit Malfunction** Key on, and the PCM detected an unexpected low voltage condition (less than 2.7v) on the EGR Valve Coil 4 control circuit with the valve command "off" during the CCM test at startup. **Note: This is a diagnostic support code - the PCM monitors this circuit once per key cycle, but does not set a pending code.** **Possible Causes:** • EGR Coil 4 control circuit is open or shorted to ground • EGR Coil 4 control circuit is shorted to power • EGR Coil 4 power circuit open between valve and main relay • EGR Coil 4 is damaged or has failed • PCM has failed
DTC: P1499 **1T CCM, MIL: NO** **2004** **Models:** Tribute **Engines:** All **Transmissions:** A/T, M/T	**EGR Valve Stepper Motor Coil 4 Circuit Malfunction** Key on, and the PCM detected an unexpected low voltage condition (less than 2.7v) on the EGR Valve Coil 4 control circuit with the valve command "off" during the CCM test at startup. **Note: This is a diagnostic support code - the PCM monitors this circuit once per key cycle, but does not set a pending code.** **Possible Causes:** • EGR Coil 4 control circuit is open or shorted to ground • EGR Coil 4 control circuit is shorted to power • EGR Coil 4 power circuit open between valve and main relay • EGR Coil 4 is damaged or has failed • PCM has failed
DTC: P1500 **1T CCM, MIL: YES** **1996, 1997** **Models:** 626, MX-6 **Engines:** 2.0L VIN C **Transmissions:** A/T, M/T	**Vehicle Speed Sensor Intermittent Signal** Engine running in gear, VSS inputs indicate the vehicle is moving, and the PCM detected an erratic or noisy VSS signal. **Possible Causes:** • VSS signal circuit is open (an intermittent fault) • VSS signal circuit is shorted to ground (an intermittent fault) • VSS signal circuit is shorted to VREF (an intermittent fault) • VSS is damaged or has failed • PCM has failed
DTC: P1500 **1T CCM, MIL: YES** **1996, 1997, 1998, 1999, 2000, 2001, 2002, 2003, 2004, 2005** **Models:** B2300, B2500, B3000, B4000 **Engines:** All **Transmissions:** A/T, M/T	**Vehicle Speed Sensor Intermittent Signal** Engine running in gear, VSS inputs indicate the vehicle is moving, and the PCM detected an erratic, intermittent or noisy VSS signal. **Possible Causes:** • VSS signal circuit is open (an intermittent fault) • VSS signal circuit is shorted to ground (an intermittent fault) • VSS signal circuit is shorted to VREF (an intermittent fault) • VSS is damaged or has failed • PCM has failed

DTC	Trouble Code Title, Conditions & Possible Causes
DTC: P1500 **1T CCM, MIL: YES** **2001, 2002, 2003, 2004, 2005** **Models:** Tribute **Engines:** All **Transmissions:** A/T, M/T	**Vehicle Speed Sensor Intermittent Signal** Engine running in gear, VSS inputs indicate the vehicle is moving, and the PCM detected the VSS signal was interrupted in the test. **Possible Causes:** • VSS signal circuit is open (an intermittent fault) • VSS signal circuit is shorted to ground (an intermittent fault) • VSS signal circuit is shorted to VREF (an intermittent fault) • VSS is damaged or has failed • PCM has failed
DTC: P1501 **1T CCM, MIL: NO** **1998, 1999, 2000, 2001, 2002** **Models:** 626 **Engines:** All **Transmissions:** A/T, M/T	**Vehicle Speed Sensor Circuit Malfunction (Self-Test)** KOER Self-Test Enabled: Engine running, gear selector in Park or Neutral, and the PCM detected a VSS signal with the vehicle not moving during the test. **Possible Causes:** • Charging system problem (check alternator for AC leakage) • Ignition system component has failed (i.e., check for signs of primary or secondary voltage leaking from the coil, coil wire, spark plug wires or boots, or the spark plug insulators • PCM has failed
DTC: P1501 **1T CCM, MIL: YES** **1996, 1997, 1998, 1999, 2000** **Models:** B2300, B2500, B3000, B4000 **Engines:** All **Transmissions:** A/T, M/T	**Vehicle Speed Sensor Circuit Malfunction (Self-Test)** KOER Self-Test Enabled: Engine running, gear selector in Park or Neutral, and the PCM detected a VSS signal with the vehicle not moving during the test. **Possible Causes:** • Charging system problem (check alternator for AC leakage) • Ignition system component has failed (i.e., check for signs of primary or secondary voltage leaking from the coil, coil wire, spark plug wires or boots, or the spark plug insulators • PCM has failed
DTC: P1501 **1T CCM, MIL: YES** **2001, 2002, 2003, 2004, 2005** **Models:** Tribute **Engines:** 2.0L VIN B **Transmissions:** A/T, M/T	**Vehicle Speed Sensor Circuit Malfunction (Self-Test)** KOER Self-Test Enabled: Engine running, gear selector in Park or Neutral, and the PCM detected a VSS signal with the vehicle not moving during the test. **Possible Causes:** • Charging system problem (check alternator for AC leakage) • Ignition system component has failed (i.e., check for signs of primary or secondary voltage leaking from the coil, coil wire, spark plug wires or boots, or the spark plug insulators • PCM has failed
DTC: P1502 **2T CCM, MIL: NO** **1999, 2000** **Models:** B2300, B2500, B3000, B4000, Tribute **Engines:** All **Transmissions:** A/T	**Vehicle Speed Sensor Circuit Malfunction** Engine running at cruise speed and load, and the PCM detected an error in the VSS data from ABS controller, Generic Electronic Module (GEM) or the Central Timer Module (CTM) during the test. **Possible Causes:** • VSS circuit open or grounded between PCM and other modules • VSS (+) or VSS (-) circuit is open, shorted to ground or power • ABS wheel speed sensors (one or more) is damaged or failed • ABS control module wiring harness is damaged or shorted • ABS control module is damaged or has failed
DTC: P1504 **2T CCM, MIL: YES** **1996, 1997** **Models:** 626, MX-6 **Engines:** 2.0L VIN C **Transmissions:** A/T	**Idle Air Control Solenoid Circuit Malfunction** Engine running and the PCM detected an unexpected voltage condition in the IAC solenoid circuit during the CCM test. **Note: This solenoid is controlled by an On or Off signal from the PCM. The resistance (range) of the IAC solenoid is 7.7-9.3 ohms.** **Possible Causes:** • IAC solenoid control circuit is open (an intermittent fault) • IAC solenoid is shorted to ground (an intermittent fault) • IAC solenoid power circuit open between solenoid and the PCM • IAC solenoid is damaged or has failed (an intermittent fault) • PCM has failed
DTC: P1504 **1T CCM, MIL: YES** 1998, 1999 **Models:** 626 **Engines:** All **Transmissions:** A/T, M/T	**Idle Air Control Solenoid Circuit Malfunction** Engine running and the PCM detected an unexpected voltage condition in the IAC solenoid circuit during the CCM test. **Note: This solenoid is controlled by an On or Off signal from the PCM. The resistance (range) of the IAC solenoid is 7.7-9.3 ohms.** **Possible Causes:** • IAC solenoid control circuit is open or shorted to ground • IAC solenoid power circuit is open between the valve and PCM • IAC solenoid is damaged or has failed • PCM has failed

DTC	Trouble Code Title, Conditions & Possible Causes
DTC: P1504 **1T CCM, MIL: YES** **2000, 2001, 2002** **Models:** 626 **Engines:** All **Transmissions:** A/T, M/T	**Idle Air Control Solenoid Circuit Malfunction** Engine running and the PCM detected an unexpected voltage over-current condition on the IAC solenoid circuit during the CCM test. **Note: This solenoid is controlled by a duty cycle signal from the PCM. The resistance (range) of the IAC solenoid is 7.7-9.3 ohms.** **Possible Causes:** • IAC solenoid No. 1 or No. 2 control circuit is open • IAC solenoid No. 1 or No. 2 control circuit is shorted to ground • IAC solenoid is damaged or has failed (an intermittent fault) • PCM has failed
DTC: P1504 **2T CCM, MIL: YES** **1999, 2000, 2001, 2002, 2003** **Models:** MX-5 Miata **Engines:** All **Transmissions:** A/T, M/T	**IAC Solenoid Circuit Malfunction** Engine running and the PCM detected an unexpected "low" voltage condition for 1 second on the IAC solenoid circuit in the CCM test. **Note: This solenoid is controlled by a duty cycle signal from the PCM. The resistance (range) of the IAC solenoid is 7.7-9.3 ohms.** **Possible Causes:** • IAC solenoid No. 1 or No. 2 control circuit is open • IAC solenoid No. 1 or No. 2 control circuit is shorted to ground • IAC solenoid No. 1 or No. 2 control circuit is shorted to power • IAC solenoid is damaged or has failed (an intermittent fault) • PCM has failed
DTC: P1504 **1T CCM, MIL: YES** **1999, 2000, 2001, 2002, 2003** **Models:** Protégé, Protege5 **Engines:** All **Transmissions:** A/T, M/T	**Idle Air Control Solenoid Circuit Malfunction** Engine running ECT sensor signal more than 77°F, then after the PCM commanded the IAC duty cycle to a range of 18-70%, it detected the IAC solenoid current was less than 400 milliamps, or that it was more than 4.60 amps for 1 second during the CCM test. **Note: This solenoid is controlled by a duty cycle signal from the PCM. The resistance (range) of the IAC solenoid is 7.7-9.3 ohms.** **Possible Causes:** • IAC solenoid No. 1 or No. 2 control circuit is open • IAC solenoid No. 1 or No. 2 control circuit is shorted to ground • IAC solenoid is damaged or has failed (an intermittent fault) • PCM has failed
DTC: P1504 **1T CCM, MIL: YES** **2000, 2001, 2002** **Models:** MPV **Engines:** All **Transmissions:** A/T, M/T	**Idle Air Control Solenoid Circuit Malfunction** Engine running, ECT sensor signal more than 77°F, then after the PCM commanded the IAC duty cycle to a range of 18-70%, it detected the IAC solenoid current was less than 400 milliamps, or that it was more than 4.60 amps for 1 second during the CCM test. **Note: This solenoid is controlled by a duty cycle signal from the PCM. The resistance (range) of the IAC solenoid is 7.7-9.3 ohms.** **Possible Causes:** • IAC solenoid No. 1 or No. 2 control circuit is open • IAC solenoid No. 1 or No. 2 control circuit is shorted to ground • IAC solenoid is damaged or has failed (an intermittent fault) • PCM has failed
DTC: P1504 **1T CCM, MIL: YES** **1996, 1997, 1998, 1999, 2000, 2001, 2002, 2003, 2004, 2005** **Models:** B2300, B2500, B3000, B4000, Tribute **Engines:** All **Transmissions:** A/T, M/T	**Idle Air Control Solenoid Circuit Malfunction** Engine running and the PCM detected an electrical "load" failure on the IAC solenoid circuit during the CCM Rationality test. **Note: This solenoid is controlled by an On or Off signal from the PCM. The resistance (range) of the IAC solenoid is 7.7-9.3 ohms.** **Possible Causes:** • IAC solenoid control circuit shorted to ground • IAC solenoid is damaged or has failed • PCM has failed
DTC: P1505 **2T CCM, MIL: YES** **1996** **Models:** 626, MX-6 **Engines:** 2.0L VIN C **Transmissions:** A/T	**Idle Air Control System at Adaptive Clip** Engine running and the PCM detected the IAC System command indicated the valve had reached its Adaptive Clip during the test. **Possible Causes:** • IAC valve is stuck in one position (it may be dirty or has failed) • An intake vacuum leak may be present, or PCV valve is leaking • IAC solenoid or valve is damaged or has failed • Throttle body is dirty, damaged or contaminated • PCM has failed

DTC	Trouble Code Title, Conditions & Possible Causes
DTC: P1505 **2T CCM, MIL: YES** **1996, 1997, 1998** **Models:** B2300, B2500, B3000, B4000 **Engines:** All **Transmissions:** A/T, M/T	**Idle Air Control System at Adaptive Clip** Engine running and the PCM detected the IAC System command indicated the valve had reached its Adaptive Clip during the test. **Possible Causes:** • IAC valve is stuck in one position (it may be dirty or has failed) • An intake vacuum leak may be present, or PCV valve is leaking • IAC solenoid or valve is damaged or has failed • Throttle body is dirty, damaged or contaminated • PCM has failed
DTC: P1506 **2T CCM, MIL: YES** **1996, 1997** **Models:** 626, MX-6 **Engines:** 2.0L VIN C **Transmissions:** A/T	**Idle Air Control System RPM Higher Than Expected** Engine running at hot idle in closed loop, and the PCM detected an IAC System Overspeed condition during the CCM test. **Possible Causes:** • EGR valve is stuck open or leaking, or the gasket is leaking • EVAP purge valve is leaking or stuck in open position • IAC solenoid control circuit is shorted to ground (intermittent) • IAC valve is stuck open (it may be dirty or have failed) • PCV valve is leaking, or a leak in the intake manifold exists • PCM has failed
DTC: P1506 **2T CCM, MIL: YES** 1998, 1999 **Models:** 626 **Engines:** All **Transmissions:** A/T, M/T	**Idle Air Control System RPM Higher Than Expected** Engine running at hot idle in closed loop, and the PCM detected an IAC System Overspeed condition during the CCM test. **Possible Causes:** • EGR valve is stuck open or leaking, or the gasket is leaking • EVAP purge valve is leaking or stuck in open position • IAC solenoid control circuit is shorted to ground (intermittent) • IAC valve is stuck open (it may be dirty or have failed) • PCV valve is leaking, or a leak in the intake manifold exists • PCM has failed
DTC: P1506 **2T CCM, MIL: YES** **2002, 2001, 2002** **Models:** 626 **Engines:** All **Transmissions:** A/T, M/T	**Idle Air Control System RPM Higher Than Expected** Engine running at hot idle in closed loop, BARO sensor signal over 72 kPa, IAT sensor signal more than 14°F, brake pedal indicating "on" (brake pedal depressed), power steering switch indicating "off" (steering wheel straight ahead), and the PCM detected the Actual idle speed was more than 200 rpm above the Desired idle speed for 14 seconds during the CCM Rationality test **Possible Causes:** • EGR valve is stuck open or leaking, or the gasket is leaking • EVAP purge valve is leaking or stuck in open position • IAC solenoid control circuit is shorted to ground (intermittent) • IAC valve is stuck open (it may be dirty or have failed) • PCV valve is leaking, or a leak in the intake manifold exists • PCM has failed
DTC: P1506 **2T CCM, MIL: YES** **1996, 1997, 1998, 1999, 2000, 2001, 2002, 2003, 2004, 2005** **Models:** B2300, B2500, B3000, B4000, Tribute **Engines:** All **Transmissions:** A/T, M/T	**Idle Air Control System RPM Higher Than Expected** Engine running at hot idle in closed loop, and the PCM detected the Actual idle speed was more than the Desired idle speed in the test. **Possible Causes:** • EGR valve is stuck open or leaking, or the gasket is leaking • EVAP purge valve is leaking or stuck in open position • IAC solenoid control circuit is shorted to ground (intermittent) • IAC valve is stuck open (it may be dirty or have failed) • PCV valve is leaking, or a leak in the intake manifold exists • PCM has failed
DTC: P1507 **2T CCM, MIL: YES** **1996, 1997** **Models:** 626, MX-6 **Engines:** 2.0L VIN C **Transmissions:** A/T	**Idle Air Control System RPM Lower Than Expected** Engine running at hot idle in closed loop, and the PCM detected an IAC System Underspeed condition during the CCM test. **Possible Causes:** • Air filter element is severely restricted, or air inlet is plugged • Base engine problems: cylinder compression or valve timing • IAC valve is stuck in one position (it may be dirty or has failed) • IAC solenoid or valve is damaged or has failed • Throttle body is dirty, damaged or contaminated • PCM has failed

DTC	Trouble Code Title, Conditions & Possible Causes
DTC: P1507 **2T CCM, MIL: YES** 1998, 1999 **Models:** 626 **Engines:** All **Transmissions:** A/T, M/T	**Idle Air Control System RPM Lower Than Expected** Engine running at hot idle in closed loop, and the PCM detected an IAC System Underspeed condition during the CCM test. **Possible Causes:** • Air filter element is severely restricted, or air inlet is plugged • Base engine problems: cylinder compression or valve timing • IAC valve is stuck in one position (it may be dirty or has failed) • IAC solenoid or valve is damaged or has failed • Throttle body is dirty, damaged or contaminated • PCM has failed
DTC: P1507 **2T CCM, MIL: YES** **2000, 2001, 2002** **Models:** 626 **Engines:** 2.0L VIN C **Transmissions:** A/T	**Idle Air Control System RPM Lower Than Expected** Engine running at hot idle in closed loop, BARO sensor signal over 72 kPa, IAT sensor signal more than 14ºF, brake pedal indicating "on" (brake pedal depressed), power steering switch indicating "off" (steering wheel straight ahead), and The PCM detected the Actual idle speed was more than 100 rpm below the Desired idle speed for 14 seconds during the CCM Rationality test. **Possible Causes:** • EGR valve is stuck open or leaking, or the gasket is leaking • EVAP purge valve is leaking or stuck in open position • IAC solenoid control circuit is shorted to ground (intermittent) • IAC valve is stuck open (it may be dirty or have failed) • PCV valve is leaking, or a leak in the intake manifold exists • PCM has failed
DTC: P1507 **2T CCM, MIL: YES** **1996, 1997, 1998, 1999, 2000, 2001, 2002, 2003, 2004, 2005** **Models:** B2300, B2500, B3000, B4000, Tribute **Engines:** All **Transmissions:** A/T, M/T	**Idle Air Control System RPM Lower Than Expected** Engine running at hot idle in closed loop, and the PCM detected the Actual idle speed was less than the Desired idle speed in the test. **Possible Causes:** • Air filter element is severely restricted, or air inlet is plugged • Base engine problems: cylinder compression or valve timing • IAC valve is stuck in one position (it may be dirty or has failed) • IAC solenoid or valve is damaged or has failed • Throttle body is dirty, damaged or contaminated • PCM has failed
DTC: P1508 **2T CCM, MIL: YES** **1996, 1997, 1998, 1999, 2000, 2001, 2002** **Models:** Millenia **Engines:** 2.3L VIN 2 **Transmissions:** A/T, M/T	**Bypass Air Solenoid 1 Circuit Malfunction** Key on or engine running and the PCM detected an unexpected voltage condition on the Bypass Air Solenoid 1 circuit during the test. **Possible Causes:** • Bypass air solenoid control circuit is open or shorted to ground • Bypass air solenoid control circuit is shorted to system power • Bypass air solenoid is damaged or has failed • PCM has failed
DTC: P1508 **2T CCM, MIL: YES** **2004** **Models:** Tribute **Engines:** All **Transmissions:** A/T, M/T	**Bypass Air Solenoid 1 Circuit Malfunction** Key on or engine running and the PCM detected an unexpected voltage condition on the Bypass Air Solenoid 1 circuit during the test. **Possible Causes:** • Bypass air solenoid control circuit is open or shorted to ground • Bypass air solenoid control circuit is shorted to system power • Bypass air solenoid is damaged or has failed • PCM has failed
DTC: P1509 **2T CCM, MIL: YES** **1996, 1997, 1998, 1999, 2000, 2001, 2002** **Models:** Millenia **Engines:** 2.3L VIN 2 **Transmissions:** A/T, M/T	**Bypass Air Solenoid 2 Circuit Malfunction** Key on or engine running and the PCM detected an unexpected voltage condition on the Bypass Air Solenoid 2 circuit during the test. **Possible Causes:** • Bypass air solenoid control circuit is open or shorted to ground • Bypass air solenoid control circuit is shorted to system power • Bypass air solenoid is damaged or has failed • PCM has failed
DTC: P1509 **2T CCM, MIL: YES** **2004** **Models:** Tribute **Engines:** All **Transmissions:** A/T, M/T	**Bypass Air Solenoid 2 Circuit Malfunction** Key on or engine running and the PCM detected an unexpected voltage condition on the Bypass Air Solenoid 2 circuit during the test. **Possible Causes:** • Bypass air solenoid control circuit is open or shorted to ground • Bypass air solenoid control circuit is shorted to system power • Bypass air solenoid is damaged or has failed • PCM has failed

DTC	Trouble Code Title, Conditions & Possible Causes
DTC: P1510 **2T CCM, MIL: YES** **2004** **Models:** Tribute **Engines:** All **Transmissions:** A/T, M/T	**Bypass Air Solenoid 2 Circuit Malfunction** Key on or engine running and the PCM detected an unexpected voltage condition on the Bypass Air Solenoid 2 circuit during the test. **Possible Causes:** • Bypass air solenoid control circuit is open or shorted to ground • Bypass air solenoid control circuit is shorted to system power • Bypass air solenoid is damaged or has failed • PCM has failed
DTC: P1511 **2T CCM, MIL: YES** **2004** **Models:** Tribute **Engines:** All **Transmissions:** A/T, M/T	**Bypass Air Solenoid 2 Circuit Malfunction** Key on or engine running and the PCM detected an unexpected voltage condition on the Bypass Air Solenoid 2 circuit during the test. **Possible Causes:** • Bypass air solenoid control circuit is open or shorted to ground • Bypass air solenoid control circuit is shorted to system power • Bypass air solenoid is damaged or has failed • PCM has failed
DTC: P1512 **2T CCM, MIL: YES** **2000, 2001, 2002** **Models:** 626 **Engines:** 2.0L VIN C **Transmissions:** A/T, M/T	**Variable Timing Control System Malfunction** Engine running, engine speed less than 3000 rpm, ECT sensor signal less than 149°F, and the PCM detected a problem in the operation of the VTCS valve during the CCM test. **Note: The PCM turns the VTCS solenoid "off" for 150 ms at startup.** **Possible Causes:** • VTCS actuator control circuit is shorted to ground • VTCS shutter valve actuator is stuck closed • VTCS vacuum hose is loose or disconnected • VTCS valve is damaged or has failed • PCM has failed
DTC: P1512 **2T CCM, MIL: YES** **1999, 2000, 2001, 2002, 2003** **Models:** MX-5 Miata **Engines:** All **Transmissions:** A/T, M/T	**Variable Timing Control System Malfunction (Stuck Closed)** Engine running, engine speed less than 3000 rpm, ECT sensor signal less than 149°F, and the PCM detected the VTS switch signal indicated low with the VTS valve commanded "off" during the test. **Note: The PCM turns the VTCS solenoid "off" for 150 ms at startup.** **Possible Causes:** • Fuel control sensor is out-of-calibration (ECT, IAT, MAF or TP) • VTCS actuator control circuit is shorted to ground • VTCS shutter valve actuator is stuck closed • VTCS vacuum hose is loose or disconnected • PCM has failed
DTC: P1512 **2T CCM, MIL: YES** **1999, 2000, 2001, 2002, 2003** **Models:** Protégé, Protege5 **Engines:** All **Transmissions:** A/T, M/T	**Variable Timing Control System Shutter Valve (Stuck Closed)** Engine started, then with the engine speed under 4000 rpm, ECT sensor signal more than 176°F, BARO sensor signal over 102 kPa, IAT sensor signal more than 68°F, and the throttle valve angle over 75%, the PCM detected the engine airflow amount was more than 37 g/sec (4.9 lb/min.) during the CCM Rationality test. **Possible Causes:** • CKP sensor is damaged or has failed • EGR boost sensor is out of calibration • Fuel control sensor is out-of-calibration (ECT, IAT, MAF or TP) • VTCS control circuit shorted to ground or to system power (B+) • VTCS shutter valve actuator is stuck closed • VTCS vacuum hose is loose or disconnected • PCM has failed
DTC: P1512 **2T CCM, MIL: YES** **2000, 2001, 2002** **Models:** MPV **Engines:** All **Transmissions:** A/T, M/T	**Intake Manifold Runner Control Shutter Valve (Stuck Closed)** Engine started, and with the Intake Manifold Runner Control (IMRC) solenoid in open position, the PCM detected the IMRC control circuit indicated from 1.6-4.9v for 5 seconds during the CCM test. **Possible Causes:** • IMRC control circuit shorted to ground or to system power (B+) • IMRC shutter valve actuator is stuck closed • IMRC vacuum hose is loose or disconnected • PCM has failed

DTC	Trouble Code Title, Conditions & Possible Causes
DTC: P1512 **2T CCM, MIL: YES** **2005** **Models:** Tribute **Engines:** All **Transmissions:** A/T, M/T	**Intake Manifold Runner Control Shutter Valve (Stuck Closed)** Engine started, and with the Intake Manifold Runner Control (IMRC) solenoid in open position, the PCM detected the IMRC control circuit indicated from 1.6-4.9v for 5 seconds during the CCM test. **Possible Causes:** • IMRC control circuit shorted to ground or to system power (B+) • IMRC shutter valve actuator is stuck closed • IMRC vacuum hose is loose or disconnected • PCM has failed
DTC: P1516 **2T CCM, MIL: YES** **2000, 2001, 2002** **Models:** 626 **Engines:** All **Transmissions:** M/T	**Neutral Switch Circuit Malfunction** Engine started, then driven to a speed of 16 mph, and the PCM did not detect a change in the status of the Neutral switch in the test. **Possible Causes:** • Neutral switch signal circuit is open or shorted to ground • Neutral switch power circuit is open or shorted to ground • Neutral switch is damaged or has failed • PCM has failed
DTC: P1516 **2T CCM, MIL: YES** **2005** **Models:** Tribute **Engines:** All **Transmissions:** M/T	**Neutral Switch Circuit Malfunction** Engine started, then driven to a speed of 16 mph, and the PCM did not detect a change in the status of the Neutral switch in the test. **Possible Causes:** • Neutral switch signal circuit is open or shorted to ground • Neutral switch power circuit is open or shorted to ground • Neutral switch is damaged or has failed • PCM has failed
DTC: P1518 **2T CCM, MIL: YES** **1999, 2000, 2001, 2002, 2003** **Models:** MX-5 Miata **Engines:** All **Transmissions:** A/T, M/T	**Variable Timing Control System Malfunction (Stuck Open)** Engine running, engine speed less than 3000 rpm, ECT sensor signal less than 149ºF, and the PCM detected the VTS switch signal indicated high with the VTS valve commanded "on" during the test. **Note: The PCM turns the VTCS solenoid "off" for 150 ms at startup.** **Possible Causes:** • VTCS actuator control circuit is shorted to ground • VTCS shutter valve actuator is stuck closed • VTCS vacuum hose is loose or disconnected • VTCS valve is damaged or has failed • PCM has failed
DTC: P1518 **2T CCM, MIL: YES** **2000, 2001, 2002** **Models:** MPV **Engines:** All **Transmissions:** A/T, M/T	**Intake Manifold Runner Control Shutter Valve (Stuck Open)** Engine started, and with the Intake Manifold Runner Control (IMRC) solenoid commanded to close, the PCM detected the IMRC control circuit indicated less than 1.58v for 3 seconds during the CCM test. **Possible Causes:** • IMRC control circuit shorted to ground or to system power (B+) • IMRC shutter valve actuator is stuck closed • IMRC vacuum hose is loose or disconnected • PCM has failed
DTC: P1518 **2T CCM, MIL: YES** **2004** **Models:** B2300, B3000, B4000 **Engines:** All **Transmissions:** A/T, M/T	**Intake Manifold Runner Control Shutter Valve (Stuck Open)** Engine started, and with the Intake Manifold Runner Control (IMRC) solenoid commanded to close, the PCM detected the IMRC control circuit indicated less than 1.58v for 3 seconds during the CCM test. **Possible Causes:** • IMRC control circuit shorted to ground or to system power (B+) • IMRC shutter valve actuator is stuck closed • IMRC vacuum hose is loose or disconnected • PCM has failed
DTC: P1518 **2T CCM, MIL: YES** **2005** **Models:** Tribute **Engines:** All **Transmissions:** A/T, M/T	**Intake Manifold Runner Control Shutter Valve (Stuck Open)** Engine started, and with the Intake Manifold Runner Control (IMRC) solenoid commanded to close, the PCM detected the IMRC control circuit indicated less than 1.58v for 3 seconds during the CCM test. **Possible Causes:** • IMRC control circuit shorted to ground or to system power (B+) • IMRC shutter valve actuator is stuck closed • IMRC vacuum hose is loose or disconnected • PCM has failed

DTC	Trouble Code Title, Conditions & Possible Causes
DTC: P1519 **2T CCM, MIL: YES** **2004** **Models:** B2300, B3000, B4000 **Engines:** All **Transmissions:** A/T, M/T	**Intake Manifold Runner Control Shutter Valve (Stuck Open)** Engine started, and with the Intake Manifold Runner Control (IMRC) solenoid commanded to close, the PCM detected the IMRC control circuit indicated less than 1.58v for 3 seconds during the CCM test. **Possible Causes:** • IMRC control circuit shorted to ground or to system power (B+) • IMRC shutter valve actuator is stuck closed • IMRC vacuum hose is loose or disconnected • PCM has failed
DTC: P1519 **2T CCM, MIL: YES** **2005** **Models:** Tribute **Engines:** All **Transmissions:** A/T, M/T	**Intake Manifold Runner Control Shutter Valve (Stuck Open)** Engine started, and with the Intake Manifold Runner Control (IMRC) solenoid commanded to close, the PCM detected the IMRC control circuit indicated less than 1.58v for 3 seconds during the CCM test. **Possible Causes:** • IMRC control circuit shorted to ground or to system power (B+) • IMRC shutter valve actuator is stuck closed • IMRC vacuum hose is loose or disconnected • PCM has failed
DTC: P1520 **2T CCM, MIL: YES** **2000, 2001, 2002** **Models:** MPV **Engines:** All **Transmissions:** A/T, M/T	**Intake Manifold Runner Control Circuit Malfunction** Engine started, and with the Intake Manifold Runner Control (IMRC) solenoid in open position, the PCM detected an unexpected "low" voltage condition on the IMRC control circuit during the CCM test. **Possible Causes:** • IMRC control circuit is open • IMRC control circuit is shorted to ground • IMRC power circuit open or shorted to ground (check the fuse) • PCM has failed
DTC: P1520 **2T CCM, MIL: YES** **2004** **Models:** Tribute **Engines:** All **Transmissions:** A/T, M/T	**Intake Manifold Runner Control Circuit Malfunction** Engine started, and with the Intake Manifold Runner Control (IMRC) solenoid in open position, the PCM detected an unexpected "low" voltage condition on the IMRC control circuit during the CCM test. **Possible Causes:** • IMRC control circuit is open • IMRC control circuit is shorted to ground • IMRC power circuit open or shorted to ground (check the fuse) • PCM has failed
DTC: P1520 **2T CCM, MIL: YES** **2004** **Models:** B2300, B3000, B4000 **Engines:** All **Transmissions:** A/T, M/T	**Intake Manifold Runner Control Circuit Malfunction** Engine started, and with the Intake Manifold Runner Control (IMRC) solenoid in open position, the PCM detected an unexpected "low" voltage condition on the IMRC control circuit during the CCM test. **Possible Causes:** • IMRC control circuit is open • IMRC control circuit is shorted to ground • IMRC power circuit open or shorted to ground (check the fuse) • PCM has failed
DTC: P1521 **1T CCM, MIL: NO** **1996, 1997, 1998, 1999, 2000, 2001, 2002** **Models:** Millenia **Engines:** 2.5L VIN 1 **Transmissions:** A/T, M/T	**VRIS Solenoid 1 Circuit Malfunction** Key on or engine running and the PCM detected an unexpected voltage condition on the Variable Reduction Intake Solenoid (VRIS) 1 circuit during the CCM test period. **Possible Causes:** • VRIS control circuit is open or shorted to ground • VRIS power circuit is open (the fuse may be blow) • VRIS is damaged or has failed • PCM has failed
DTC: P1521 **1T CCM, MIL: NO** **1996, 1997, 1998, 1999, 2000, 2001, 2002** **Models:** MPV **Engines:** All **Transmissions:** A/T, M/T	**VRIS Solenoid 1 Circuit Malfunction** Key on or engine running and the PCM detected an unexpected voltage condition on the Variable Reduction Intake Solenoid (VRIS) 1 circuit during the CCM test period. **Possible Causes:** • VRIS control circuit is open or shorted to ground • VRIS power circuit is open (the fuse may be blow) • VRIS is damaged or has failed • PCM has failed

DTC	Trouble Code Title, Conditions & Possible Causes
DTC: P1522 **1T CCM, MIL: NO** **1996, 1997, 1998, 1999, 2000, 2001, 2002** **Models:** Millenia **Engines:** 2.5L VIN 1 **Transmissions:** A/T, M/T	**VRIS Solenoid 2 Circuit Malfunction** Key on or engine running and the PCM detected an unexpected voltage condition on the Variable Reduction Intake Solenoid (VRIS) 2 circuit during the CCM test period. **Possible Causes:** • VRIS control circuit is open or shorted to ground • VRIS power circuit is open (the fuse may be blow) • VRIS is damaged or has failed • PCM has failed
DTC: P1523 **1T CCM, MIL: YES** **1999, 2000** **Models:** MX-5 Miata **Engines:** All **Transmissions:** A/T, M/T	**Variable Inertia Charging System Solenoid Circuit Malfunction** Key on or engine running and the PCM detected an unexpected voltage condition on the Variable Inertia Charging System Solenoid (VICS) circuit during the CCM test period. **Possible Causes:** • VICS control circuit is open • VICS control circuit is shorted to ground • VICS valve is damaged or has failed • PCM has failed
DTC: P1523 **1T CCM, MIL: YES** **1997, 1998, 1999, 2000** **Models:** Protégé **Engines:** 1.8L VIN 1 **Transmissions:** A/T, M/T	**Variable Inertia Charging System Solenoid Circuit Malfunction** Key on, and the PCM detected an unexpected "low" voltage condition on the Variable Inertia Charging System Solenoid (VICS) circuit during the CCM test period. **Possible Causes:** • VICS control circuit is open or shorted to ground • VICS valve is damaged or has failed • PCM has failed
DTC: P1524 **1T CCM, MIL: NO** **1996, 1997, 1998, 1999, 2000, 2001, 2002** **Models:** Millenia **Engines:** 2.3L VIN 2 **Transmissions:** A/T, M/T	**Charge Air Cooler Bypass Solenoid Circuit Malfunction** Key on or engine running and the PCM detected an unexpected voltage condition on the Charge Air Cooler Bypass Solenoid Valve circuit during the CCM test period. **Possible Causes:** • CAC bypass solenoid control circuit open or shorted to ground • CAC bypass solenoid control circuit is shorted to system power • CAC bypass solenoid is damaged or has failed • PCM has failed
DTC: P1525 **2T CCM, MIL: YES** **1996, 1997, 1998, 1999, 2000, 2001, 2002** **Models:** Millenia **Engines:** 2.3L VIN 2 **Transmissions:** A/T, M/T	**ABV Vacuum Solenoid Circuit Malfunction** Key on or engine running and the PCM detected an unexpected voltage condition on the Air Bypass Valve Vacuum solenoid circuit during the CCM test. **Possible Causes:** • ABV vacuum solenoid control circuit open or shorted to ground • ABV vacuum solenoid control circuit is shorted to system power • ABV vacuum solenoid is damaged or has failed • PCM has failed
DTC: P1525 **2T CCM, MIL: YES** **2004** **Models:** Tribute **Engines:** All **Transmissions:** A/T, M/T	**ABV Vacuum Solenoid Circuit Malfunction** Key on or engine running and the PCM detected an unexpected voltage condition on the Air Bypass Valve Vacuum solenoid circuit during the CCM test. **Possible Causes:** • ABV vacuum solenoid control circuit open or shorted to ground • ABV vacuum solenoid control circuit is shorted to system power • ABV vacuum solenoid is damaged or has failed • PCM has failed
DTC: P1526 **2T CCM, MIL: YES** **1996, 1997, 1998, 1999, 2000, 2001, 2002** **Models:** Millenia **Engines:** 2.3L VIN 2 **Transmissions:** A/T, M/T	**ABV Vent Solenoid Circuit Malfunction** Key on or engine running and the PCM detected an unexpected voltage condition on the Air Bypass Valve Vent solenoid circuit during the CCM test. **Possible Causes:** • ABV vent solenoid control circuit open or shorted to ground • ABV vent solenoid control circuit is shorted to system power • ABV vent solenoid is damaged or has failed • PCM has failed

DTC	Trouble Code Title, Conditions & Possible Causes
DTC: P1526 **2T CCM, MIL: YES** **2004** **Models:** Tribute **Engines:** All **Transmissions:** A/T, M/T	**ABV Vent Solenoid Circuit Malfunction** Key on or engine running and the PCM detected an unexpected voltage condition on the Air Bypass Valve Vent solenoid circuit during the CCM test. **Possible Causes:** • ABV vent solenoid control circuit open or shorted to ground • ABV vent solenoid control circuit is shorted to system power • ABV vent solenoid is damaged or has failed • PCM has failed
DTC: P1527 **2T CCM, MIL: YES** **1996, 1997, 1998, 1999, 2000, 2001, 2002** **Models:** MPV **Engines:** All **Transmissions:** A/T, M/T	**Bypass Air Solenoid Circuit Malfunction** Engine running, system voltage over 11.5v, and PCM detected the Bypass Air solenoid signal was less than 5.8v with the solenoid "off", or the difference in Intake Airflow was less than 2.3 g/sec before and after the Bypass Air Solenoid valve was turned "on" during the test. **Possible Causes:** • Bypass air solenoid control circuit is open or shorted to ground • Bypass air solenoid power circuit is open • Bypass air solenoid is damaged or has failed • PCM has failed
DTC: P1527 **2T CCM, MIL: YES** **2004** **Models:** Tribute **Engines:** All **Transmissions:** A/T, M/T	**Bypass Air Solenoid Circuit Malfunction** Engine running, system voltage over 11.5v, and PCM detected the Bypass Air solenoid signal was less than 5.8v with the solenoid "off", or the difference in Intake Airflow was less than 2.3 g/sec before and after the Bypass Air Solenoid valve was turned "on" during the test. **Possible Causes:** • Bypass air solenoid control circuit is open or shorted to ground • Bypass air solenoid power circuit is open • Bypass air solenoid is damaged or has failed • PCM has failed
DTC: P1537 **2T CCM, MIL: YES** **2004** **Models:** B2300, B3000, B4000 **Engines:** All **Transmissions:** A/T, M/T	**Bypass Air Solenoid Circuit Malfunction** Engine running, system voltage over 11.5v, and PCM detected the Bypass Air solenoid signal was less than 5.8v with the solenoid "off", or the difference in Intake Airflow was less than 2.3 g/sec before and after the Bypass Air Solenoid valve was turned "on" during the test. **Possible Causes:** • Bypass air solenoid control circuit is open or shorted to ground • Bypass air solenoid power circuit is open • Bypass air solenoid is damaged or has failed • PCM has failed
DTC: P1537 **2T CCM, MIL: YES** **2005** **Models:** Tribute **Engines:** All **Transmissions:** A/T, M/T	**Bypass Air Solenoid Circuit Malfunction** Engine running, system voltage over 11.5v, and PCM detected the Bypass Air solenoid signal was less than 5.8v with the solenoid "off", or the difference in Intake Airflow was less than 2.3 g/sec before and after the Bypass Air Solenoid valve was turned "on" during the test. **Possible Causes:** • Bypass air solenoid control circuit is open or shorted to ground • Bypass air solenoid power circuit is open • Bypass air solenoid is damaged or has failed • PCM has failed
DTC: P1538 **2T CCM, MIL: YES** **2004** **Models:** B2300, B3000, B4000 **Engines:** All **Transmissions:** A/T, M/T	**Bypass Air Solenoid Circuit Malfunction** Engine running, system voltage over 11.5v, and PCM detected the Bypass Air solenoid signal was less than 5.8v with the solenoid "off", or the difference in Intake Airflow was less than 2.3 g/sec before and after the Bypass Air Solenoid valve was turned "on" during the test. **Possible Causes:** • Bypass air solenoid control circuit is open or shorted to ground • Bypass air solenoid power circuit is open • Bypass air solenoid is damaged or has failed • PCM has failed
DTC: P1540 **2T CCM, MIL: YES** **1996, 1997, 1998, 1999, 2000, 2001, 2002** **Models:** Millenia **Engines:** 2.3L VIN 2 **Transmissions:** A/T, M/T	**Air Bypass Valve System** Key on or engine running and the PCM detected a problem in the Air Bypass Valve system during normal operation in the CCM test. **Possible Causes:** • Engine control sensor is out-of-range (i.e., ECT, IAT or MAF) • ABV vacuum or vent solenoid is damaged or has failed • Air intake leak in the engine or related vacuum hose or pipes • Air intake restriction in a component related to the ABV system • PCM has failed

DTC	Trouble Code Title, Conditions & Possible Causes
DTC: P1540 **2T CCM, MIL: YES** **2004** **Models:** Tribute **Engines:** All **Transmissions:** A/T, M/T	**Air Bypass Valve System** Key on or engine running and the PCM detected a problem in the Air Bypass Valve system during normal operation in the CCM test. **Possible Causes:** • Engine control sensor is out-of-range (i.e., ECT, IAT or MAF) • ABV vacuum or vent solenoid is damaged or has failed • Air intake leak in the engine or related vacuum hose or pipes • Air intake restriction in a component related to the ABV system • PCM has failed
DTC: P1549 **2T CCM, MIL: YES** **2005** **Models:** Tribute **Engines:** All **Transmissions:** A/T, M/T	**Air Bypass Valve System** Key on or engine running and the PCM detected a problem in the Air Bypass Valve system during normal operation in the CCM test. **Possible Causes:** • Engine control sensor is out-of-range (i.e., ECT, IAT or MAF) • ABV vacuum or vent solenoid is damaged or has failed • Air intake leak in the engine or related vacuum hose or pipes • Air intake restriction in a component related to the ABV system • PCM has failed
DTC: P1550 **2T CCM, MIL: YES** **2004, 2005** **Models:** Tribute **Engines:** All **Transmissions:** A/T, M/T	**Air Bypass Valve System** Key on or engine running and the PCM detected a problem in the Air Bypass Valve system during normal operation in the CCM test. **Possible Causes:** • Engine control sensor is out-of-range (i.e., ECT, IAT or MAF) • ABV vacuum or vent solenoid is damaged or has failed • Air intake leak in the engine or related vacuum hose or pipes • Air intake restriction in a component related to the ABV system • PCM has failed
DTC: P1562 **1T CCM, MIL: YES** **1999, 2000, 2001, 2002, 2003** **Models:** MX-5 Miata **Engines:** All **Transmissions:** A/T, M/T	**PCM +BB Voltage Low** Key on, engine cranking, and the PCM detected the +BB circuit voltage was too low (less than 1.4v during the CCM test period. **Note: The +BB circuit is the backup or direct voltage circuit.** **Possible Causes:** • +BB circuit is open • +BB circuit is shorted to ground (check for a blown fuse) • +BB circuit has high resistance (check the battery connections) • PCM has failed
DTC: P1562 **1T CCM, MIL: YES** **1999, 2000, 2001, 2002, 2003** **Models:** Protégé, Protege5 **Engines:** All **Transmissions:** A/T, M/T	**PCM +BB Voltage Low** Key on, engine cranking, and the PCM detected the +BB circuit voltage was too low (less than 1.4v) during the CCM test period. **Note: The +BB circuit is the backup or direct voltage circuit.** **Possible Causes:** • +BB circuit is open • +BB circuit is shorted to ground (check for a blown fuse) • +BB circuit has high resistance (check the battery connections) • PCM has failed
DTC: P1562 **1T CCM, MIL: YES** **2004** **Models:** Tribute **Engines:** All **Transmissions:** A/T, M/T	**PCM +BB Voltage Low** Key on, engine cranking, and the PCM detected the +BB circuit voltage was too low (less than 2.5v) during the CCM test period. **Note: The +BB circuit is the backup or direct voltage circuit.** **Possible Causes:** • +BB circuit is open • +BB circuit is shorted to ground (check for a blown fuse) • +BB circuit has high resistance (check the battery connections) • PCM has failed
DTC: P1562 **1T CCM, MIL: YES** **2000, 2001, 2002, 2003, 2004** **Models:** Mazda6, MPV **Engines:** All **Transmissions:** A/T, M/T	**PCM +BB Voltage Low** Key on, engine cranking, and the PCM detected the +BB circuit voltage was too low (less than 2.5v) during the CCM test period. **Note: The +BB circuit is the backup or direct voltage circuit.** **Possible Causes:** • +BB circuit is open • +BB circuit is shorted to ground (check for a blown fuse) • +BB circuit has high resistance (check the battery connections) • PCM has failed

DTC	Trouble Code Title, Conditions & Possible Causes
DTC: P1565 **1T CCM, MIL: YES** **2004, 2005** **Models:** Tribute **Engines:** All **Transmissions:** A/T, M/T	**PCM +BB Voltage Low** Key on, engine cranking, and the PCM detected the +BB circuit voltage was too low (less than 2.5v) during the CCM test period. **Note: The +BB circuit is the backup or direct voltage circuit.** **Possible Causes:** • +BB circuit is open • +BB circuit is shorted to ground (check for a blown fuse) • +BB circuit has high resistance (check the battery connections) • PCM has failed
DTC: P1566 **1T CCM, MIL: YES** **2000, 2001, 2002** **Models:** Millenia **Engines:** All **Transmissions:** A/T, M/T	**TCM +BB Voltage Low** Key on, engine cranking, and the TCM detected the +BB circuit voltage was too low (less than 2.5v) for 2 seconds during the test. **Note: The +BB circuit is the backup or direct voltage circuit.** **Possible Causes:** • +BB circuit is open • +BB circuit is shorted to ground (check for a blown fuse) • +BB circuit has high resistance (check the battery connections) • TCM has failed
DTC: P1566 **1T CCM, MIL: YES** **2004, 2005** **Models:** Tribute **Engines:** All **Transmissions:** A/T, M/T	**TCM +BB Voltage Low** Key on, engine cranking, and the TCM detected the +BB circuit voltage was too low (less than 2.5v) for 2 seconds during the test. **Note: The +BB circuit is the backup or direct voltage circuit.** **Possible Causes:** • +BB circuit is open • +BB circuit is shorted to ground (check for a blown fuse) • +BB circuit has high resistance (check the battery connections) • TCM has failed
DTC: P1567 **1T CCM, MIL: YES** **2004, 2005** **Models:** Tribute **Engines:** All **Transmissions:** A/T, M/T	**TCM +BB Voltage Low** Key on, engine cranking, and the TCM detected the +BB circuit voltage was too low (less than 2.5v) for 2 seconds during the test. **Note: The +BB circuit is the backup or direct voltage circuit.** **Possible Causes:** • +BB circuit is open • +BB circuit is shorted to ground (check for a blown fuse) • +BB circuit has high resistance (check the battery connections) • TCM has failed
DTC: P1568 **1T CCM, MIL: YES** **2004, 2005** **Models:** Tribute **Engines:** All **Transmissions:** A/T, M/T	**TCM +BB Voltage Low** Key on, engine cranking, and the TCM detected the +BB circuit voltage was too low (less than 2.5v) for 2 seconds during the test. **Note: The +BB circuit is the backup or direct voltage circuit.** **Possible Causes:** • +BB circuit is open • +BB circuit is shorted to ground (check for a blown fuse) • +BB circuit has high resistance (check the battery connections) • TCM has failed
DTC: P1569 **2T CCM, MIL: YES** **2000, 2001, 2002** **Models:** 626 **Engines:** 2.0L VIN C **Transmissions:** A/T, M/T	**Variable Timing Control System Circuit Low Input** Engine running, engine speed less than 3000 rpm, ECT sensor signal less than 149ºF, and the PCM detected an unexpected "low" voltage condition on the VTCS valve during the CCM test. **Note: The PCM turns the VTCS valve "off" for 150 ms at startup.** **Possible Causes:** • VTCS control circuit is shorted to ground • VTCS solenoid valve is damaged or has failed • PCM is damaged
DTC: P1569 **2 CCM, MIL: YES** **2001, 2002, 2003** **Models:** MX-5 Miata **Engines:** All **Transmissions:** A/T, M/T	**Variable Timing Control System Circuit Low Input** Engine running, engine speed less than 3500 rpm, ECT sensor signal less than 140ºF, and the PCM detected an unexpected "low" voltage condition on the VTCS valve during the CCM test. **Note: The PCM turns the VTCS valve "off" for 150 ms at startup.** **Possible Causes:** • VTCS control circuit is open or shorted to ground • VTCS power circuit open between the solenoid and main relay • VTCS solenoid valve is damaged or has failed • PCM is damaged

DTC	Trouble Code Title, Conditions & Possible Causes
DTC: P1569 **2T CCM, MIL: YES** **1999, 2000, 2001, 2002, 2003** **Models:** Protégé, Protege5 **Engines:** 1.6L VIN 1 **Transmissions:** A/T, M/T	**Variable Timing Control System Circuit Low Input** Key on, and the PCM detected an unexpected "low" voltage condition on the VTCS valve circuit during the CCM test. **Note: The PCM turns the VTCS valve "off" for 150 ms at startup.** **Possible Causes:** • VTCS control circuit is open or shorted to ground • VTCS power circuit open between the solenoid and main relay • VTCS solenoid valve is damaged or has failed • PCM is damaged
DTC: P1570 **2T CCM, MIL: YES** **2001, 2002, 2003** **Models:** MX-5 Miata **Engines:** All **Transmissions:** A/T, M/T	**Variable Timing Control System Circuit High Input** Engine running, engine speed less than 3500 rpm, ECT sensor signal less than 140ºF, and the PCM detected an unexpected "high" voltage condition on the VTCS valve during the CCM test. **Note: The PCM turns the VTCS valve "off" for 150 ms at startup.** **Possible Causes:** • VTCS control circuit is shorted to system power (B+) • VTCS solenoid valve is damaged or has failed • PCM is damaged
DTC: P1570 **2T CCM, MIL: YES** **1999, 2000, 2001, 2002, 2003** **Models:** MX-5 Miata **Engines:** All **Transmissions:** A/T, M/T	**Variable Timing Control System Circuit High Input** Engine running, engine speed less than 3500 rpm, ECT sensor signal less than 140ºF, and the PCM detected an unexpected "high" voltage condition on the VTCS valve during the CCM test. **Note: The PCM turns the VTCS valve "off" for 150 ms at startup.** **Possible Causes:** • VTCS control circuit is shorted to system power (B+) • VTCS solenoid valve is damaged or has failed • PCM is damaged
DTC: P1570 **2T CCM, MIL: YES** **1999, 2000, 2001, 2002, 2003** **Models:** Protégé, Protege5 **Engines:** 1.6L VIN 1 **Transmissions:** A/T, M/T	**Variable Timing Control System Circuit High Input** Engine started, engine speed over 2500 rpm with throttle angle over 10%, ECT sensor signal more than 176ºF, and the PCM detected an unexpected "high" voltage condition on the VTCS circuit in the test. **Note: The PCM turns the VTCS valve "off" for 150 ms at startup.** **Possible Causes:** • VTCS control circuit is shorted to system power (B+) • VTCS solenoid valve is damaged or has failed • PCM is damaged
DTC: P1572 **2T CCM, MIL: YES** **2004, 2005** **Models:** Tribute **Engines:** All **Transmissions:** A/T, M/T	**Variable Timing Control System Circuit High Input** Engine started, engine speed over 2500 rpm with throttle angle over 10%, ECT sensor signal more than 176ºF, and the PCM detected an unexpected "high" voltage condition on the VTCS circuit in the test. **Note: The PCM turns the VTCS valve "off" for 150 ms at startup.** **Possible Causes:** • VTCS control circuit is shorted to system power (B+) • VTCS solenoid valve is damaged or has failed • PCM is damaged
DTC: P1574 **2T CCM, MIL: YES** **2004, 2005** **Models:** RX-8 **Engines:** All **Transmissions:** A/T, M/T	**TP sensor Output Incongruent** The PCM compares the TP from TP sensor No.1 with the TP from TP sensor No.2 when the engine is running. If the difference is more than the specification, The PCM determines that the TP sensor outputs are incongruent. **Possible Causes:** • TP sensor No. 1 malfunction • TP sensor No. 2 is damaged or has failed • PCM is damaged
DTC: P1577 **2T CCM, MIL: YES** **2004, 2005** **Models:** RX-8 **Engines:** All **Transmissions:** A/T, M/T	**APP sensor Output Incongruent** The PCM compares the APP from APP sensor No.1 with the APP from APP sensor No.2 when the engine is running. If the difference is more than the specification, The PCM determines that the APP sensor outputs are incongruent. **Possible Causes:** • APP sensor No. 1 malfunction • APP sensor No. 2 malfunction • PCM is damaged

DTC	Trouble Code Title, Conditions & Possible Causes
DTC: P1600 **2T CCM, MIL: YES** **2004** **Models:** Tribute **Engines:** All **Transmissions:** A/T, M/T	**PCM Malfunction** Key on, and the PCM detected an unexpected voltage condition on the TCM communication line during the CCM communication test. **Possible Causes:** • PCM link to the TCM is open • PCM link to the TCM is shorted to ground • PCM link to the TCM is shored to power • PCM or the TCM has failed
DTC: P1601 **1T PCM, MIL: NO** **1996, 1997, 1999, 2000, 2001, 2002, 2003** **Models:** MX-5 Miata **Engines:** All **Transmissions:** A/T, M/T	**PCM Communication Line to TCM Error** Key on, and the PCM detected an unexpected voltage condition on the TCM communication line during the CCM communication test. **Possible Causes:** • PCM link to the TCM is open • PCM link to the TCM is shorted to ground • PCM link to the TCM is shored to power • PCM or the TCM has failed
DTC: P1601 **1T PCM, MIL: NO** **1996, 1997, 1998, 1999, 2000, 2001, 2002** **Models:** Millenia **Engines:** 2.3L VIN 2 **Transmissions:** A/T, M/T	**PCM Communication Line to TCM Error** Key on, and the PCM detected an unexpected voltage condition on the TCM communication line during the CCM communication test. **Possible Causes:** • PCM link to the TCM is open • PCM link to the TCM is shorted to ground • PCM link to the TCM is shored to power • PCM or the TCM has failed
DTC: P1601 **1T PCM, MIL: NO** **1996, 1997, 1998, 1999, 2000, 2001, 2002** **Models:** MPV **Engines:** All **Transmissions:** A/T, M/T	**PCM Communication Line to TCM Error** Key on, and the PCM detected an unexpected voltage condition on the TCM communication line during the CCM communication test. **Possible Causes:** • PCM link to the TCM is open • PCM link to the TCM is shorted to ground • PCM link to the TCM is shored to power • PCM or the TCM has failed
DTC: P1601 **1T PCM, MIL: NO** **2004** **Models:** Tribute **Engines:** All **Transmissions:** A/T, M/T	**PCM Communication Line to TCM Error** Key on, and the PCM detected an unexpected voltage condition on the TCM communication line during the CCM communication test. **Possible Causes:** • PCM link to the TCM is open • PCM link to the TCM is shorted to ground • PCM link to the TCM is shored to power • PCM or the TCM has failed
DTC: P1602 **1T PCM, MIL: NO** **1998, 1999, 2000, 2001, 2002, 2003, 2004, 2005** **Models:** 626, MX-5 Miata **Engines:** All **Transmissions:** A/T, M/T	**PCM Communication Line to TCM Error** Key on, and the PCM detected the command transmission to the Immobilizer unit was too long, or it did not receive any response. **Possible Causes:** • Key (the transponder) is damaged or has failed • PCM communication line to Immobilizer is open, shorted to ground or shorted to system power • Transponder "coil" is damaged or has failed • PCM or the Immobilizer unit has failed
DTC: P1602 **1T PCM, MIL: NO** **2000, 2001, 2002, 2003, 2004, 2005** **Models:** MPV **Engines:** All **Transmissions:** A/T, M/T	**PCM Communication Line to TCM Error** Key on, and the PCM detected the command transmission to the Immobilizer unit was too long, or it did not receive any response. **Possible Causes:** • Key (the transponder) is damaged or has failed • PCM communication line to Immobilizer is open, shorted to ground or shorted to system power • Transponder "coil" is damaged or has failed • PCM or the Immobilizer unit has failed

DTC	Trouble Code Title, Conditions & Possible Causes
DTC: P1602 **1T PCM, MIL: NO** **2004** **Models:** Tribute **Engines:** All **Transmissions:** A/T, M/T	**PCM Communication Line to TCM Error** Key on, and the PCM detected the command transmission to the Immobilizer unit was too long, or it did not receive any response. **Possible Causes:** • Key (the transponder) is damaged or has failed • PCM communication line to Immobilizer is open, shorted to ground or shorted to system power • Transponder "coil" is damaged or has failed • PCM or the Immobilizer unit has failed
DTC: P1603 **1T PCM, MIL: NO** **1998, 1999, 2000, 2001, 2002, 2003, 2004, 2005** **Models:** 626, MX-5 Miata **Engines:** All **Transmissions:** A/T, M/T	Code Word Not Registered In The PCM Key on, and the PCM detected the Immobilizer procedure "code word" was not performed after the PCM was replaced. **Possible Causes:** • Check for any related Technical Service Bulletins • Perform the Immobilizer "code word" reprogramming procedure
DTC: P1603 **1T PCM, MIL: NO** **2000, 2001, 2002, 2003, 2004, 2005** **Models:** MPV **Engines:** All **Transmissions:** A/T, M/T	Code Word Not Registered In The PCM Key on, and the PCM detected the Immobilizer procedure "code word" was not performed after the PCM was replaced. **Possible Causes:** • Check for any related Technical Service Bulletins • Perform the Immobilizer "code word" reprogramming procedure
DTC: P1603 **1T PCM, MIL: NO** **2004** **Models:** Tribute **Engines:** All **Transmissions:** A/T, M/T	Code Word Not Registered In The PCM Key on, and the PCM detected the Immobilizer procedure "code word" was not performed after the PCM was replaced. **Possible Causes:** • Check for any related Technical Service Bulletins • Perform the Immobilizer "code word" reprogramming procedure
DTC: P1604 **1T PM, MIL: NO** **1998, 1999, 2000, 2001, 2002, 2003, 2004, 2005** **Models:** 626, MX-5 Miata **Engines:** All **Transmissions:** A/T, M/T	Key Identification Numbers Not Registered In The PCM Key on, and the PCM detected the Key Identification (ID) numbers were not registered in the PCM was it was replaced. **Possible Causes:** • Check for any related Technical Service Bulletins • Perform the Immobilizer "Key ID" reprogramming procedure
DTC: P1604 **1T PCM, MIL: NO** **2000, 2001, 2002, 2003, 2004, 2005** **Models:** MPV **Engines:** All **Transmissions:** A/T, M/T	Key Identification Numbers Not Registered In The PCM Key on, and the PCM detected the Key Identification (ID) numbers were not registered in the PCM was it was replaced. **Possible Causes:** • Check for any related Technical Service Bulletins • Perform the Immobilizer "Key ID" reprogramming procedure
DTC: P1604 **1T PCM, MIL: NO** **2004** **Models:** Tribute **Engines:** All **Transmissions:** A/T, M/T	Key Identification Numbers Not Registered In The PCM Key on, and the PCM detected the Key Identification (ID) numbers were not registered in the PCM was it was replaced. **Possible Causes:** • Check for any related Technical Service Bulletins • Perform the Immobilizer "Key ID" reprogramming procedure
DTC: P1605 **1T PCM, MIL: NO** **Yrs. : 1996** **Models:** 626, MX-6 **Engines:** 2.0L VIN C **Transmissions:** A/T	**PCM Keep Alive Memory Test Error** Key on or engine running and the PCM detected an interruption in the Keep Alive Memory power circuit. **Possible Causes:** • Battery cables are loose, or the connections are corroded • Keep Alive Power circuit to the PCM is shorted to ground • Keep Alive Power circuit to the PCM is open (intermittent fault) • PCM has failed

DTC	Trouble Code Title, Conditions & Possible Causes
DTC: P1605 **1T PCM, MIL: NO** **1996, 1997, 1998, 1999, 2000, 2001, 2002, 2003, 2004** **Models:** B2300, B2500, B3000, B4000, Tribute **Engines:** All **Transmissions:** A/T	**PCM Keep Alive Memory or Read Only Memory Test Error** Key on or engine running and the PCM detected an interruption in the Keep Alive Memory circuit, or a Read Only Memory test error. **Possible Causes:** • Battery cables are loose, or the connections are corroded • Keep Alive Power circuit to the PCM is shorted to ground • Keep Alive Power circuit to the PCM is open (intermittent fault) • PCM has failed
DTC: P1605 **1T PCM, MIL: NO** **2004** **Models:** Tribute **Engines:** All **Transmissions:** A/T	**PCM Keep Alive Memory or Read Only Memory Test Error** Key on or engine running and the PCM detected an interruption in the Keep Alive Memory circuit, or a Read Only Memory test error. **Possible Causes:** • Battery cables are loose, or the connections are corroded • Keep Alive Power circuit to the PCM is shorted to ground • Keep Alive Power circuit to the PCM is open (intermittent fault) • PCM has failed
DTC: P1608 **1T PCM, MIL: NO** **1996, 1997** **Models:** 626, MX-6 **Engines:** 2.0L VIN C **Transmissions:** M/T	**PCM Output Device Circuit Malfunction** Key on or engine running and the PCM detected an unexpected voltage condition on one or more of the output device circuits. The PCM cannot diagnose the devices properly under these conditions. **Possible Causes:** • Output device (relay or solenoid) control circuit(s) shorted to ground between the device and the PCM • If the output device circuits are okay, the PCM has failed
DTC: P1608 **1T PCM, MIL: NO** **1996, 1997, 1999, 2000, 2001, 2002, 2003, 2004, 2005** **Models:** MX-5 Miata **Engines:** All **Transmissions:** A/T, M/T	**PCM Output Device Circuit Malfunction** Key on or engine running and the PCM detected an unexpected voltage condition on one or more of the output device circuits. The PCM cannot diagnose the devices properly under these conditions. **Possible Causes:** • Output device (relay or solenoid) control circuit(s) shorted to ground between the device and the PCM • If the output device circuits are okay, the PCM has failed
DTC: P1608 **1T PCM, MIL: NO** **1998, 1999, 2000, 2001, 2002** **Models:** Millenia, MPV **Engines:** All **Transmissions:** A/T, M/T	**PCM Output Device Circuit Malfunction** Key on or engine running and the PCM detected an unexpected voltage condition on one or more of the output device circuits. The PCM cannot diagnose the devices properly under these conditions. **Possible Causes:** • Output device (relay or solenoid) control circuit(s) shorted to ground between the device and the PCM • If none of the output device control circuits are shorted to ground (key on, voltage drop check), the PCM has failed
DTC: P1608 **1T PCM, MIL: NO** **1996, 1997, 1998, 1999, 2000** **Models:** Protégé **Engines:** All **Transmissions:** A/T, M/T	**PCM Output Device Circuit Malfunction** Key on or engine running and the PCM detected an unexpected voltage condition on one or more of the output device circuits. The PCM cannot diagnose the devices properly under these conditions. **Possible Causes:** • Output device (relay or solenoid) control circuit(s) shorted to ground between the device and the PCM • If none of the output device control circuits are shorted to ground (key on, voltage drop check), the PCM has failed
DTC: P1608 **1T PCM, MIL: NO** **2004** **Models:** Tribute **Engines:** All **Transmissions:** A/T, M/T	**PCM Output Device Circuit Malfunction** Key on or engine running and the PCM detected an unexpected voltage condition on one or more of the output device circuits. The PCM cannot diagnose the devices properly under these conditions. **Possible Causes:** • Output device (relay or solenoid) control circuit(s) shorted to ground between the device and the PCM • If none of the output device control circuits are shorted to ground (key on, voltage drop check), the PCM has failed
DTC: P1609 **1T CCM, MIL: NO** **1996, 1997** **Models:** 626, MX-6 **Engines:** 2.5L VIN D **Transmissions:** A/T, M/T	**Knock Sensor Integrated Circuit Malfunction** Engine running and the PCM detected an unexpected voltage signal on the Knock sensor integrated circuit during the CCM test. **Possible Causes:** • Perform a PCM "Reset" function • Then turn the key to on, engine off. If this trouble code resets, the PCM circuitry for the Knock sensor is damaged, and the PCM must be replaced to correct the condition.

DTC	Trouble Code Title, Conditions & Possible Causes
DTC: P1609 **1T CCM, MIL: NO** **1999, 2000** **Models:** MX-5 Miata **Engines:** All **Transmissions:** A/T, M/T	**Knock Sensor Integrated Circuit Malfunction** Engine running and the PCM detected an unexpected voltage signal on the Knock sensor integrated circuit during the CCM test. **Possible Causes:** • Perform a PCM "Reset" function • Then turn the key to on, engine off. If this trouble code resets, the PCM circuitry for the Knock sensor is damaged, and the PCM must be replaced to correct the condition.
DTC: P1609 **1T PCM, MIL: NO** **1996, 1997, 1998, 1999, 2000, 2001, 2002** **Models:** Millenia **Engines:** All **Transmissions:** A/T, M/T	**Knock Sensor Integrated Circuit Malfunction** Engine running and the PCM detected an unexpected voltage signal on the Knock sensor integrated circuit during the CCM test. **Possible Causes:** • Perform a PCM "Reset" function • Then turn the key to on, engine off. If this trouble code resets, the PCM circuitry for the Knock sensor is damaged, and the PCM must be replaced to correct the condition.
DTC: P1609 **1T PCM, MIL: NO** **2004** **Models:** Tribute **Engines:** All **Transmissions:** A/T, M/T	**Knock Sensor Integrated Circuit Malfunction** Engine running and the PCM detected an unexpected voltage signal on the Knock sensor integrated circuit during the CCM test. **Possible Causes:** • Perform a PCM "Reset" function • Then turn the key to on, engine off. If this trouble code resets, the PCM circuitry for the Knock sensor is damaged, and the PCM must be replaced to correct the condition.
DTC: P1621 **1T PCM, MIL: NO** **1998, 1999, 2000, 2001, 2002, 2003, 2004, 2005** **Models:** 626, MX-5 Miata **Engines:** All **Transmissions:** A/T, M/T	Key Identification Number Not Registered In The PCM Engine cranking and the PCM detected the "code word" in the PCM and in the Immobilizer did not match. **Possible Causes:** • Immobilizer problem during transformation of the "code word" • PCM problem during transformation of the "code word" • Immobilizer Unit has failed • PCM has failed
DTC: P1621 **1T PCM, MIL: NO** **2000, 2001, 2002, 2003, 2004, 2005** **Models:** MPV **Engines:** All **Transmissions:** A/T, M/T	Key Identification Number Not Registered In The PCM Engine cranking and the PCM detected the "code word" in the PCM and in the Immobilizer did not match. **Possible Causes:** • Immobilizer problem during transformation of the "code word" • PCM problem during transformation of the "code word" • Immobilizer Unit has failed • PCM has failed
DTC: P1621 **1T PCM, MIL: NO** **2004** **Models:** Tribute **Engines:** All **Transmissions:** A/T, M/T	Key Identification Number Not Registered In The PCM Engine cranking and the PCM detected the "code word" in the PCM and in the Immobilizer did not match. **Possible Causes:** • Immobilizer problem during transformation of the "code word" • PCM problem during transformation of the "code word" • Immobilizer Unit has failed • PCM has failed
DTC: P1622 **1T PCM, MIL: NO** **1998, 1999, 2000, 2001, 2002, 2003, 2004, 2005** **Models:** 626, MX-5 Miata **Engines:** All **Transmissions:** A/T, M/T	Key Identification Number Does Not Match In The PCM Engine cranking, and the PCM detected the "code word" in the PCM and the Immobilizer did not match after the Immobilizer Unit was replaced. **Possible Causes:** • An "unregistered" key was used during Step 3 of the Immobilizer Unit replacement procedure • A problem occurred during transformation of the Key ID number stored in the PCM
DTC: P1622 **1T PCM, MIL: NO** **2000, 2001, 2002, 2003, 2004, 2005** **Models:** MPV **Engines:** All **Transmissions:** A/T, M/T	Key Identification Number Does Not Match In The PCM Engine cranking, and the PCM detected the "code word" in the PCM and the Immobilizer did not match after the Immobilizer Unit was replaced. **Possible Causes:** • An "unregistered" key was used during Step 3 of the Immobilizer Unit replacement procedure • A problem occurred during transformation of the Key ID number stored in the PCM

DTC	Trouble Code Title, Conditions & Possible Causes
DTC: P1650 **1T CCM, MIL: NO** **1996, 1997** **Models:** 626, MX-6 **Engines:** 2.0L VIN C **Transmissions:** A/T	**Power Steering Pressure Switch Circuit Malfunction (Self-Test)** KOER Self-Test Enabled: Engine running and the PCM detected an unexpected voltage condition on the Power Steering Pressure Switch (PSPS) circuit. **Note: If the steering wheel is not turned during at the correct time during the Self-Test step, the PCM will set this trouble code.** **Possible Causes:** • PSP signal circuit open, shorted to ground or shorted to power • PSPS ground circuit is open (at the switch mounting point) • PSPS is damaged or has failed • PCM has failed
DTC: P1650 **1T CCM, MIL: NO** **1998, 1999, 2000, 2001, 2002** **Models:** 626 **Engines:** All **Transmissions:** A/T, M/T	**Power Steering Pressure Switch Circuit Malfunction (Self-Test)** KOER Self-Test Enabled: Engine running and the PCM detected an unexpected voltage condition on the Power Steering Pressure Switch (PSPS) circuit. **Note: If the steering wheel is not turned during at the correct time during the Self-Test step, the PCM will set this trouble code.** **Possible Causes:** • PSP signal circuit is open or shorted to ground • PSPS ground circuit is open (at the switch mounting point) • PSPS is damaged or has failed • PCM has failed
DTC: P1650 **1T CCM, MIL: NO** **1996, 1997, 1998, 1999, 2000, 2001, 2002, 2003, 2004, 2005** **Models:** B2300, B2500, B3000, B4000 **Engines:** All **Transmissions:** A/T, M/T	**Power Steering Pressure Switch Circuit Malfunction (Self-Test)** KOER Self-Test Enabled: Engine running and the PCM detected an unexpected voltage condition on the Power Steering Pressure Switch (PSPS) circuit. **Note: If the steering wheel is not turned during at the correct time during the Self-Test step, the PCM will set this trouble code.** **Possible Causes:** • PSP signal circuit is open or shorted to ground • PSPS ground circuit is open (at the switch mounting point) • PSPS is damaged or has failed • PCM has failed
DTC: P1650 **1T CCM, MIL: NO** **2001, 2002, 2003. 2004, 2005** **Models:** Tribute **Engines:** 2.0L VIN B **Transmissions:** A/T, M/T	**Power Steering Pressure Switch Circuit Malfunction (Self-Test)** KOEO or KOER Self-Test Enabled: Key on or engine running and the PCM detected an unexpected voltage condition on the Power Steering Pressure Switch circuit. **Note: If the steering wheel is not turned during at the correct time during the Self-Test step, the PCM will set this trouble code.** **Possible Causes:** • PSP signal circuit is open or shorted to ground • PSPS ground circuit is open (at the switch mounting point) • PSPS is damaged or has failed • PCM has failed
DTC: P1651 **1T CCM, MIL: NO** **1996, 1997** **Models:** 626, MX-6 **Engines:** 2.0L VIN C **Transmissions:** A/T	**Power Steering Pressure Switch Circuit Malfunction** Engine started, then vehicle driven to a speed over 37.4 mph for 1 minute, ECT sensor signal more than 140°F, and the PCM did not detect any change in the Power Steering Pressure Switch (PSPS) signal under these conditions during the CCM Rationality test. **Possible Causes:** • PSP signal circuit is open or shorted to ground • PSPS ground circuit is open (at the switch mounting point) • PSPS is damaged or has failed • PCM has failed
DTC: P1651 **1T CCM, MIL: NO** **1996, 1997, 1998, 1999, 2000, 2001, 2002, 2003, 2004, 2005** **Models:** B2300, B2500 **Engines:** 2.0L VIN C **Transmissions:** A/T	**Power Steering Pressure Switch Circuit Malfunction** Engine started, then vehicle driven to a speed over 37.4 mph for 1 minute, ECT sensor signal more than 140°F, and the PCM did not detect any change in the Power Steering Pressure Switch (PSPS) signal under these conditions during the CCM Rationality test. **Possible Causes:** • PSP signal circuit is open or shorted to ground • PSPS ground circuit is open (at the switch mounting point) • PSPS is damaged or has failed • PCM has failed

DTC	Trouble Code Title, Conditions & Possible Causes
DTC: P1651 **1T CCM, MIL: NO** **2001, 2002, 2003, 2004, 2005** **Models:** Tribute **Engines:** 2.0L VIN B **Transmissions:** A/T	**Power Steering Pressure Switch Circuit Malfunction** Engine started, then vehicle driven to a speed over 37.4 mph for 1 minute, ECT sensor signal more than 140ºF, and the PCM did not detect any change in the Power Steering Pressure Switch (PSPS) signal under these conditions during the CCM Rationality test. **Possible Causes:** • PSP signal circuit is open or shorted to ground • PSPS ground circuit is open (at the switch mounting point) • PSPS is damaged or has failed • PCM has failed
DTC: P1652 **2T CCM, MIL: NO** **1997, 1998, 1999, 2000, 2001, 2002** **Models:** 626 **Engines:** All **Transmissions:** A/T, M/T	**Power Steering Pressure Switch Circuit Malfunction** Engine started, then vehicle driven to a speed over 37.4 mph for 1 minute, ECT sensor signal more than 140ºF, and the PCM did not detect any change in the Power Steering Pressure Switch (PSPS) signal under these conditions during the CCM Rationality test. **Possible Causes:** • PSP signal circuit is open or shorted to ground • PSPS ground circuit is open (at the switch mounting point) • PSPS is damaged or has failed • PCM has failed
DTC: P1652 **2T CCM, MIL: NO** **2004** **Models:** Tribute **Engines:** All **Transmissions:** A/T, M/T	**Power Steering Pressure Switch Circuit Malfunction** Engine started, then vehicle driven to a speed over 37.4 mph for 1 minute, ECT sensor signal more than 140ºF, and the PCM did not detect any change in the Power Steering Pressure Switch (PSPS) signal under these conditions during the CCM Rationality test. **Possible Causes:** • PSP signal circuit is open or shorted to ground • PSPS ground circuit is open (at the switch mounting point) • PSPS is damaged or has failed • PCM has failed
DTC: P1686 **2T CCM, MIL: NO** **2004, 2005** **Models:** RX-8 **Engines:** All **Transmissions:** A/T, M/T	**Metering oil pump control circuit low flow side problem** The PCM monitors the input signal from the metering oil pump switch when the metering oil pump stepping motor is more than the standard step. If the input signal is off, the PCM determines that the metering oil pump control circuit has a problem on the low flow side. **Possible Causes:** • Metering oil pump malfunction • Metering oil pump switch malfunction • Connector or terminal malfunction, short or open circuit • PCM has failed
DTC: P1687 **2T CCM, MIL: NO** **2004, 2005** **Models:** RX-8 **Engines:** All **Transmissions:** A/T, M/T	**Metering oil pump control circuit high flow side problem** The PCM monitors the input signal from the metering oil pump switch when the metering oil pump stepping motor is less than the standard step. If the input signal is on, the PCM determines that the metering oil pump control circuit has a problem on the high flow side. **Possible Causes:** • Metering oil pump malfunction • Metering oil pump switch malfunction • Connector or terminal malfunction, short or open circuit • PCM has failed
DTC: P1688 **2T CCM, MIL: NO** **2004, 2005** **Models:** RX-8 **Engines:** All **Transmissions:** A/T, M/T	**Metering oil pump control circuit initial check problem** The PCM monitors the input signal from the metering oil pump switch when the metering oil pump stepping motor initial check is operating. If the input signal is on, the PCM determines that there is a metering oil pump control circuit initial check problem. **Possible Causes:** • Metering oil pump malfunction • Metering oil pump switch malfunction • Connector or terminal malfunction, short or open circuit • PCM has failed
DTC: P1700 **2T CCM, MIL: NO** **2004, 2005** **Models:** Tribute **Engines:** All **Transmissions:** A/T, M/T	**Transaxle failure** Internal component failure. Direct one-way clutch failure. Failed a neutral condition. FMEM becomes active - engine rpm limited to 4,000 rpms. **Possible Causes:** • Transaxle malfunction • PCM has failed

DTC	Trouble Code Title, Conditions & Possible Causes
DTC: P1701 **2T CCM, MIL: YES** 1998, 1999 **Models:** 626 **Engines:** 2.0L VIN C **Transmissions:** A/T	**Turbine Shaft Sensor Circuit Malfunction** Engine started, vehicle driven to a speed over 31 mph, and the PCM detected an unexpected voltage condition on the A/T Turbine Shaft Sensor (TSS) circuit during the CCM test. **Possible Causes:** • TSS signal circuit is open (an intermittent fault) • TSS signal circuit is shorted to ground (an intermittent fault) • TSS is damaged or has failed (an intermittent fault) • PCM has failed
DTC: P1701 **2T CCM, MIL: YES** **1996, 1997, 1998, 1999, 2000** **Models:** B2300, B2500, B3000, B4000 **Engines:** All **Transmissions:** A/T	**Turbine Speed Sensor Circuit Malfunction** Engine started, vehicle driven to a speed over 31 mph, and the PCM detected an unexpected voltage condition on the Transmission Speed Sensor (TSS) circuit during the CCM test. **Possible Causes:** • TSS signal circuit is open (an intermittent fault) • TSS signal circuit is shorted to ground (an intermittent fault) • TSS is damaged or has failed (an intermittent fault) • PCM has failed
DTC: P1701 **1T CCM, MIL: YES** 1998, 1999 **Models:** 626 **Engines:** 2.0L VIN C **Transmissions:** A/T	**Transmission Range Sensor Circuit Malfunction** Engine started, vehicle driven to a speed over 31 mph, and the PCM did not detect any change in the Transmission Range (TR) sensor signal during the CCM test. **Possible Causes:** • TR sensor signal circuit is open (an intermittent fault) • TR sensor signal circuit is shorted to ground (intermittent fault) • TR sensor signal circuit is shorted to VREF (intermittent fault) • TR sensor is damaged or has failed (an intermittent fault) • PCM has failed
DTC: P1702 **1T CCM, MIL: YES** **1999, 2000, 2001, 2002, 2003, 2004, 2005** **Models:** B2300, B2500, B3000, B4000, Tribute **Engines:** All **Transmissions:** A/T	**Transmission Range Sensor Circuit Malfunction** Engine started, vehicle driven to a speed over 31 mph, and the PCM detected an interruption inn the Transmission Range (TR) sensor signal during the CCM test (DTC P0707 or P0708 may also be set). **Possible Causes:** • TR sensor signal circuit is open (an intermittent fault) • TR sensor signal circuit is shorted to ground (intermittent fault) • TR sensor signal circuit is shorted to VREF (intermittent fault) • TR sensor is damaged or has failed (an intermittent fault) • PCM has failed
DTC: P1703 **1T CCM, MIL: NO** **1996, 1997** **Models:** 626, MX-6 **Engines:** All **Transmissions:** A/T	**Brake Switch Circuit Malfunction (Self-Test)** KOEO Self-Test Enabled: Key on, and the PCM detected an unexpected voltage condition on the Brake Switch circuit during the CCM test. **Possible Causes:** • Brake was not depressed at the proper time during KOEO test • Brake was depressed all of the time during the KOEO test • Brake switch signal circuit shorted to system power (B+) • Brake switch is damaged or has failed, or is out of adjustment • PCM has failed
DTC: P1703 **1T CCM, MIL: NO** **1998, 1999, 2000, 2001, 2002** **Models:** 626 **Engines:** All **Transmissions:** A/T	**Brake Switch Circuit Malfunction (Self-Test)** KOEO Self-Test Enabled: Key on, and the PCM detected an unexpected voltage condition on the Brake Switch circuit during the CCM test. **Possible Causes:** • Brake was not depressed at the proper time during KOEO test • Brake was depressed all of the time during the KOEO test • Brake switch signal circuit shorted to system power (B+) • Brake switch is damaged or has failed, or is out of adjustment • PCM has failed

DTC	Trouble Code Title, Conditions & Possible Causes
DTC: P1703 **1T CCM, MIL: NO** **1996, 1997, 1998, 1999, 2000, 2001, 2002, 2003, 2004, 2005** **Models:** B2300, B2500, B3000, B4000, Tribute **Engines:** All **Transmissions:** A/T	**Brake Switch Circuit Malfunction (Self-Test)** KOEO Self-Test Enabled: Key on, and the PCM detected an unexpected "high" voltage condition on the Brake Switch circuit during the CCM test. KOER Self-Test Enabled: Engine running and the PCM detected an unexpected voltage "low" or "high" voltage condition on the Brake Switch circuit in the test. **Possible Causes:** • Brake not depressed at the proper time during KOEO Self-Test. • Brake was depressed at all times during the KOEO Self-Test. • Brake was not cycled "on" and "off" during the KOER Self-Test. • Brake switch signal circuit shorted to system power (B+) • Brake switch is damaged or has failed, or is out of adjustment • PCM has failed
DTC: P1704 **1T CCM, MIL: NO** **1999, 2000, 2001, 2002, 2003, 2004, 2005** **Models:** B2300, B2500, B3000, B4000 **Engines:** All **Transmissions:** A/T	**Transmission Range Sensor Circuit Malfunction (Self-Test)** KOEO Self-Test Enabled: Key on, and the PCM detected an invalid TR switch signal occurred. KOER Self-Test Enabled: Engine running, and the PCM detected an invalid TR switch signal occurred during the CCM test. **Possible Causes:** • TR switch signal circuit is open or shorted to ground • TR switch is damaged, has failed or is out of adjustment • Transmission shift cable is damaged or out-of-adjustment • PCM had failed
DTC: P1705 **1T CCM, MIL: NO** **1996, 1997** **Models:** 626, MX-6 **Engines:** 2.0L VIN C **Transmissions:** A/T	**Transaxle Range Switch Circuit Malfunction (Self-Test)** KOER Self-Test Enabled: Key on, and the PCM detected a TR switch signal that indicated the gear selector was not in Park during the CCM test. **Possible Causes:** • The gear selector was not In Park during the KOER Self-Test • TR switch is damaged, has failed or is out of adjustment • Starter or starter relay is damaged or has failed • Backup lamp circuit is open or shorted • PCM had failed
DTC: P1705 **1T CCM, MIL: NO** **1998, 1999, 2000, 2001, 2002** **Models:** 626 **Engines:** All **Transmissions:** A/T	**Transaxle Range Switch Circuit Malfunction (Self-Test)** **KOEO Self-Test Enabled:** Key on, and the PCM detected a TR switch signal that indicated the gear selector was not in Park during the CCM test. KOER Self-Test Enabled: Engine running and the PCM detected a TR switch signal that indicated the gear selector was not in Park during the CCM test. **Possible Causes:** • The gear selector was not In Park during the Self-Test • TR switch is damaged, has failed or is out of adjustment • Starter or starter relay is damaged or has failed • Backup lamp circuit is open or shorted • PCM had failed
DTC: P1705 **1T CCM, MIL: NO** **1996, 1997, 1998, 1999, 2000, 2001, 2002, 2003, 2004, 2005** **Models:** B2300, B2500, B3000, B4000, Tribute **Engines:** All **Transmissions:** A/T	**Digital Transmission Range Sensor Circuit Malfunction (Self-Test)** KOEO Self-Test Enabled: Key on, and the PCM detected a TR switch signal that indicated the gear selector was not in Park during the CCM test. **Possible Causes:** • The gear selector was not In Park during the Self-Test • TR switch is damaged, has failed or is out of adjustment • Starter or starter relay is damaged or has failed • Backup lamp circuit is open or shorted • PCM had failed
DTC: P1707 **1T CCM, MIL: NO** **1998, 1999, 2000, 2001, 2002** **Models:** 626 **Engines:** All **Transmissions:** A/T	**Transaxle Range Switch Circuit Malfunction** Key on or engine running, and the PCM did not detect any TR switch inputs, or it detected multiple inputs simultaneously during the test. **Possible Causes:** • TR signal circuit shorted between one or more positions • TR switch connector is damaged • TR switch is damaged or has failed • PCM has failed

DTC	Trouble Code Title, Conditions & Possible Causes
DTC: P1879 **1T CCM, MIL: YES** **2005** **Models:** Tribute **Engines:** All **Transmissions:** M/T	**PCM – Timing/Coast Clutch (2-3T/CCS) solenoid** PCM detected a malfunction with the 3-2T/CCS solenoid. **Possible Causes:** • 3-2T/CCS solenoid signal is open or shorted to ground • 3-2T/CCS solenoid is damaged or has failed • PCM has failed
DTC: P1900 **2T CCM, MIL: NO** **1999, 2000, 2001, 2002, 2003, 2004, 2005** **Models:** B2300, B3000, B4000 **Engines:** All **Transmissions:** A/T	**Output Speed Sensor Circuit Malfunction** Engine started, vehicle driven at cruise speed, and the PCM detected an interruption (or drop out) of the Output Speed Sensor (OSS) signal during the CCM Rationality test. **Possible Causes:** • OSS signal circuit is open (an intermittent fault) • OSS signal circuit is shorted to ground (an intermittent fault) • OSS signal circuit is shorted to system power (intermittent fault) • OSS is damaged or has failed (an intermittent fault) • PCM has failed
DTC: P1900 **2T CCM, MIL: NO** **2005** **Models:** Tribute **Engines:** All **Transmissions:** A/T	**Output Speed Sensor Circuit Malfunction** Engine started, vehicle driven at cruise speed, and the PCM detected an interruption (or drop out) of the Output Speed Sensor (OSS) signal during the CCM Rationality test. **Possible Causes:** • OSS signal circuit is open (an intermittent fault) • OSS signal circuit is shorted to ground (an intermittent fault) • OSS signal circuit is shorted to system power (intermittent fault) • OSS is damaged or has failed (an intermittent fault) • PCM has failed
DTC: P1901 **2T CCM, MIL: NO** **1999, 2000, 2001, 2002** **Models:** 626 **Engines:** 2.0L VIN C **Transmissions:** A/T	**Turbine Speed Sensor Circuit Malfunction** Engine running at cruise speed at over 35 mph, and the PCM detected an interruption (drop out) of the Turbine Speed Sensor (TSS) signal during the CCM Rationality test. **Possible Causes:** • TSS signal circuit is open (an intermittent fault) • TSS signal circuit is shorted to ground (an intermittent fault) • TSS signal circuit is shorted to system power (intermittent fault) • TSS is damaged or has failed • PCM has failed
DTC: P1901 **2T CCM, MIL: NO** **1999, 2000** **Models:** B3000, B4000 **Engines:** All **Transmissions:** A/T	**Turbine Speed Sensor Circuit Malfunction** Engine running at cruise speed at over 35 mph, and the PCM detected an interruption (drop out) of the Turbine Speed Sensor (TSS) signal during the CCM Rationality test. **Possible Causes:** • TSS signal circuit is open (an intermittent fault) • TSS signal circuit is shorted to ground (an intermittent fault) • TSS signal circuit is shorted to system power (intermittent fault) • TSS is damaged or has failed • PCM has failed
DTC: P1901 **2T CCM, MIL: NO** **2005** **Models:** Tribute **Engines:** All **Transmissions:** A/T	**Turbine Speed Sensor Circuit Malfunction** Engine running at cruise speed at over 35 mph, and the PCM detected an interruption (drop out) of the Turbine Speed Sensor (TSS) signal during the CCM Rationality test. **Possible Causes:** • TSS signal circuit is open (an intermittent fault) • TSS signal circuit is shorted to ground (an intermittent fault) • TSS signal circuit is shorted to system power (intermittent fault) • TSS is damaged or has failed • PCM has failed

Gas Engine OBD II Trouble Code List (P2xxx Codes)

DTC: P2004 **2T CCM, MIL: YES** **2003, 2004, 2005** **Models:** Mazda3, Mazda6, MX-5 Miata, MPV, RX-8 **Engines:** All **Transmissions:** A/T, M/T	**Variable Tumble Control System (VTCS) shutter valve stuck open** PCM monitors mass VTCS shutter valve position using VTCS position sensor. If PCM turns VTCS solenoid valve ON but the VTCS position still remains open (VTCS shutter valve switch output: approx. 5.0 V), PCM determines VTCS shutter valve has been stuck open. **Possible Causes:** • VTCS shutter valve malfunction (stuck open) • Misconnecting or pull out the vacuum hose • Variable tumble control valve malfunction • PCM has failed
DTC: P2004 **2T CCM, MIL: YES** **2005** **Models:** Tribute **Engines:** All **Transmissions:** A/T, M/T	**Variable Tumble Control System (VTCS) shutter valve stuck open** PCM monitors mass VTCS shutter valve position using VTCS position sensor. If PCM turns VTCS solenoid valve ON but the VTCS position still remains open (VTCS shutter valve switch output: approx. 5.0 V), PCM determines VTCS shutter valve has been stuck open. **Possible Causes:** • VTCS shutter valve malfunction (stuck open) • Misconnecting or pull out the vacuum hose • Variable tumble control valve malfunction • PCM has failed
DTC: P2006 **2T CCM, MIL: YES** **2003, 2004, 2005** **Models:** Mazda3, Mazda6, MX-5 Miata, MPV, RX-8 **Engines:** All **Transmissions:** A/T, M/T	**Variable Tumble Control System (VTCS) shutter valve stuck closed** PCM monitors mass VTCS shutter valve position using VTCS position sensor. If PCM turns VTCS solenoid valve ON but the VTCS position still remains closed (VTCS shutter valve switch output: approx. 5.0 V), PCM determines VTCS shutter valve has been stuck closed. **Possible Causes:** • VTCS shutter valve malfunction (stuck closed) • Misconnecting or pull out the vacuum hose • Variable tumble control valve malfunction • PCM has failed
DTC: P2006 **2T CCM, MIL: YES** **2005** **Models:** Tribute **Engines:** All **Transmissions:** A/T, M/T	**Variable Tumble Control System (VTCS) shutter valve stuck closed** PCM monitors mass VTCS shutter valve position using VTCS position sensor. If PCM turns VTCS solenoid valve ON but the VTCS position still remains closed (VTCS shutter valve switch output: approx. 5.0 V), PCM determines VTCS shutter valve has been stuck closed. **Possible Causes:** • VTCS shutter valve malfunction (stuck closed) • Misconnecting or pull out the vacuum hose • Variable tumble control valve malfunction • PCM has failed
DTC: P2008 **2T CCM, MIL: YES** **2004, 2005** **Models:** RX-8, Tribute **Engines:** All **Transmissions:** A/T, M/T	**Variable Tumble Control System (VTCS) shutter valve stuck closed** PCM monitors mass VTCS shutter valve position using VTCS position sensor. If PCM turns VTCS solenoid valve ON but the VTCS position still remains closed (VTCS shutter valve switch output: approx. 5.0 V), PCM determines VTCS shutter valve has been stuck closed. **Possible Causes:** • VTCS shutter valve malfunction (stuck closed) • Misconnecting or pull out the vacuum hose • Variable tumble control valve malfunction • PCM has failed
DTC: P2009 **2T CCM, MIL: YES** **2003, 2004, 2005** **Models:** Mazda3, Mazda6, MX-5 Miata, MPV, RX-8 **Engines:** All **Transmissions:** A/T, M/T	**Variable Tumble Control Solenoid Valve circuit input low** PCM monitors Variable Tumble Control Solenoid Valve signal at PCM terminal 4T. If PCM turns variable tumble control solenoid valve OFF but voltage at PCM terminal 4T remains low, PCM determines VTCS solenoid valve circuit has a malfunction. **Possible Causes:** • Poor connection of connectors for PCM and/or VTCS • Open circuit or short between VTCS terminals and PCM terminals • Variable tumble control valve malfunction • PCM has failed

DTC	Trouble Code Title, Conditions & Possible Causes
DTC: P2010 **2T CCM, MIL: YES** **2003, 2004, 2005** **Models:** Mazda3, Mazda6, MX-5 Miata, MPV, RX-8 **Engines:** All **Transmissions:** A/T, M/T	**Variable Tumble Control Solenoid Valve circuit input high** PCM monitors Variable Tumble Control Solenoid Valve signal at PCM terminal 4T. If PCM turns variable tumble control solenoid valve OFF but voltage at PCM terminal 4T remains high, PCM determines VTCS solenoid valve circuit has a malfunction. **Possible Causes:** • Poor connection of connectors for PCM and/or VTCS • Open circuit or short between VTCS terminals and PCM terminals • Variable tumble control valve malfunction • PCM has failed
DTC: P2014 **2T CCM, MIL: YES** **2005** **Models:** Tribute **Engines:** All **Transmissions:** A/T, M/T	**SSV malfunction** The PCM monitors the input signal from the SSV switch when The PCM turns the SSV solenoid valve off. If the input signal is on, The PCM determines that the SSV is stuck open. **Possible Causes:** • SSV stuck open • SSV control malfunction • SSV actuator malfunction • PCM has failed
DTC: P2016 **2T CCM, MIL: YES** **2005** **Models:** RX-8 **Engines:** All **Transmissions:** A/T, M/T	**AVP position sensor circuit low input** The PCM monitors the input voltage from the APV position sensor when the engine is running. If the input voltage is less than 0.2 V, The PCM determines that the APV position sensor circuit input voltage is low. **Possible Causes:** • AVP position sensor malfunction • Open circuit or short between AVP terminals and PCM terminals • PCM has failed
DTC: P2017 **2T CCM, MIL: YES** **2004, 2005** **Models:** RX-8 **Engines:** All **Transmissions:** A/T, M/T	**AVP position sensor circuit high input** The PCM monitors the input voltage from the APV position sensor when the engine is running. If the input voltage is greater than 4.2 V, The PCM determines that the APV position sensor circuit input voltage is high. **Possible Causes:** • AVP position sensor malfunction • Open circuit or short between AVP terminals and PCM terminals • PCM has failed
DTC: P2067 **2T CCM, MIL: YES** **2005** **Models:** RX-8 **Engines:** All **Transmissions:** A/T, M/T	**Fuel gauge sender unit (sub) circuit low input** The PCM monitors the fuel tank level and input voltage from the fuel gauge sender unit (sub) when the engine is running. If the input voltage is less than 0.78 V and fuel tank level is full, the PCM determines that the fuel gauge sender unit (sub) circuit input voltage is low. **Possible Causes:** • Fuel gauge sender unit (sub) malfunction • Instrument cluster malfunction • PCM has failed
DTC: P2068 **2T CCM, MIL: YES** **2005** **Models:** RX-8 **Engines:** All **Transmissions:** A/T, M/T	**Fuel gauge sender unit (sub) circuit high input** The PCM monitors the fuel tank level and input voltage from the fuel gauge sender unit (sub) when the engine is running. If the input voltage is greater than 4.9 V and fuel tank level is full, the PCM determines that the fuel gauge sender unit (sub) circuit input voltage is hlow. **Possible Causes:** • Fuel gauge sender unit (sub) malfunction • Instrument cluster malfunction • PCM has failed
DTC: P2070 **2T CCM, MIL: YES** **2004, 2005** **Models:** RX-8, Tribute **Engines:** All **Transmissions:** A/T, M/T	**SSV malfunction** The PCM monitors the input signal from the SSV switch when The PCM turns the SSV solenoid valve off. If the input signal is on, The PCM determines that the SSV is stuck open. **Possible Causes:** • SSV stuck open • SSV control malfunction • SSV actuator malfunction • PCM has failed

DTC	Trouble Code Title, Conditions & Possible Causes
DTC: P2071 **2T CCM, MIL: YES** **2005** **Models:** Tribute **Engines:** All **Transmissions:** A/T, M/T	**SSV malfunction** The PCM monitors the input signal from the SSV switch when the PCM turns the SSV solenoid valve off. If the input signal is on, the PCM determines that the SSV is stuck open. **Possible Causes:** • SSV stuck open • SSV control malfunction • SSV actuator malfunction • PCM has failed
DTC: P2088 **2T CCM, MIL: YES** **2003, 2004, 2005** **Models:** Mazda3, Mazda6, MX-5 Miata, MPV **Engines:** All **Transmissions:** A/T, M/T	**CMP actuator circuit low** PCM monitors OCV voltage. If PCM detects OCV voltage (calculated from OCV) is below the threshold voltage (calculated from battery positive voltage), PCM determines that OCV circuit has a malfunction. **Possible Causes:** • Poor connection of connectors for PCM and/or OCV • Open circuit or short between OCV terminals and PCM terminals • OCV malfunction • PCM has failed
DTC: P2089 **2T CCM, MIL: YES** **2003, 2004, 2005** **Models:** Mazda3, Mazda6, MX-5 Miata, MPV **Engines:** All **Transmissions:** A/T, M/T	**CMP actuator circuit high** PCM monitors OCV voltage. If PCM detects OCV voltage (calculated from OCV) is above the threshold voltage (calculated from battery positive voltage), PCM determines that OCV circuit has a malfunction. **Possible Causes:** • Poor connection of connectors for PCM and/or OCV • Open circuit or short between OCV terminals and PCM terminals • OCV malfunction • PCM has failed
DTC: P2096 **2T CCM, MIL: YES** **2003, 2004, 2005** **Models:** Mazda3, Mazda6, MPV, RX-8 **Engines:** All **Transmissions:** A/T, M/T	**Target A/F feedback system too lean (right bank)** The PCM monitors the target A/F fuel trim when under the target A/F feedback control. If the fuel trim is more than the specification, the PCM determines that the target A/F feedback system too lean. **Possible Causes:** • Leaking exhaust gas • HO2S (RR, LR) malfunction • MAF malfunction • PCM has failed
DTC: P2097 **2T CCM, MIL: YES** **2003, 2004, 2005** **Models:** Mazda3, Mazda6, MPV, RX-8 **Engines:** All **Transmissions:** A/T, M/T	**Target A/F feedback system too rich (right bank)** The PCM monitors the target A/F fuel trim when under the target A/F feedback control. If the fuel trim is more than the specification, the PCM determines that the target A/F feedback system too rich. **Possible Causes:** • Leaking exhaust gas • HO2S (RR, LR) malfunction • MAF malfunction • PCM has failed
DTC: P2098 **2T CCM, MIL: YES** **2004, 2005** **Models:** MPV **Engines:** All **Transmissions:** A/T, M/T	**Target A/F feedback system too lean (left bank)** The PCM monitors the target A/F fuel trim when under the target A/F feedback control. If the fuel trim is more than the specification, the PCM determines that the target A/F feedback system too lean. **Possible Causes:** • Leaking exhaust gas • HO2S (RR, LR) malfunction • MAF malfunction • PCM has failed
DTC: P2099 **2T CCM, MIL: YES** **2004, 2005** **Models:** MPV **Engines:** All **Transmissions:** A/T, M/T	**Target A/F feedback system too rich (left bank)** The PCM monitors the target A/F fuel trim when under the target A/F feedback control. If the fuel trim is more than the specification, The PCM determines that the target A/F feedback system too rich. **Possible Causes:** • Leaking exhaust gas • HO2S (RR, LR) malfunction • MAF malfunction • PCM has failed

DTC	Trouble Code Title, Conditions & Possible Causes
DTC: P2100 **1T CCM, MIL: YES** **2003, 2004, 2005** **Models:** Mazda3, Mazda6, MX-5 Miata, MPV **Engines:** All **Transmissions:** A/T, M/T	**Throttle Actuator circuit open** PCM monitors electronic throttle valve motor current. If PCM detects that the electronic throttle valve motor current is below the threshold current, PCM determines that the electronic throttle valve motor current has a malfunction. **Possible Causes:** • Poor connection of connectors for PCM and/or OCV • Open circuit or short between OCV terminals and PCM terminals • OCV malfunction • PCM has failed
DTC: P2101 **1T CCM, MIL: YES** **2003, 2004, 2005** **Models:** Mazda3, Mazda6, MX-5 Miata, MPV, RX-8 **Engines:** All **Transmissions:** A/T, M/T	**Throttle Actuator circuit range/performance** If the PCM detects any of the following conditions, the PCM determines that the electronic throttle valve motor current has a malfunction: Default throttle angle that PCM memorized and the throttle angle with ET control relay OFF is not much; Voltage from ET relay is too high or too low; PCM detects a big voltage difference between the ET control relay and from tha main relay; or PCM internal malfunction. **Possible Causes:** • ET control relay and related circuit malfunction • Main relay and related circuit malfunction • PCM has failed
DTC: P2102 **1T CCM, MIL: YES** **2003, 2004, 2005** **Models:** Mazda3, Mazda6, MX-5 Miata, MPV, RX-8 **Engines:** All **Transmissions:** A/T, M/T	**Throttle Actuator circuit low input** The PCM monitors the throttle actuator circuit current. If PCM detects throttle actuator circuit current is excessively low, the PCM determines the throttle actuator circuit has a malfunction. **Possible Causes:** • Open circuits between throttle body terminals and PCM terminals • Short circuit between throttle body terminals and PCM terminals • Poor connection of throttle body or PCM connector • Throttle Valve motor malfunction • PCM has failed
DTC: P2103 **1T CCM, MIL: YES** **2003, 2004, 2005** **Models:** Mazda3, Mazda6, MX-5 Miata, MPV, RX-8 **Engines:** All **Transmissions:** A/T, M/T	**Throttle Actuator circuit high input** The PCM monitors the throttle actuator circuit current. If PCM detects throttle actuator circuit current is excessively high, the PCM determines the throttle actuator circuit has a malfunction. **Possible Causes:** • Open circuits between throttle body terminals and PCM terminals • Short circuits between throttle body terminals and PCM terminals • Poor connection of throttle body or PCM connector • Throttle Valve motor malfunction • PCM has failed
DTC: P2106 **1T CCM, MIL: YES** **2004, 2005** **Models:** RX-8 **Engines:** All **Transmissions:** A/T, M/T	**Throttle Actuator control system-forced limited power** The PCM monitors the throttle actuator control current when the ignition switch is on. If the control current is less than 8 A or more than 11 A, The PCM determines that the throttle actuator control system is under forced limited power. **Possible Causes:** • Throttle actuator malfunction • Poor connection of throttle body or PCM connector • PCM has failed
DTC: P2107 **1T CCM, MIL: YES** **2003, 2004, 2005** **Models:** Mazda3, Mazda6, MX-5 Miata, MPV, RX-8 **Engines:** All **Transmissions:** A/T, M/T	**Throttle Actuator control module processor** If the PCM detects that either the electronic control module has malfunctioned or the target throttle opening angle is more than the actual throttle opening angle, the PCM determines that the throttle actuator control module has a malfunction. **Possible Causes:** • PCM has failed
DTC: P2108 **1T CCM, MIL: YES** **2003, 2004, 2005** **Models:** Mazda3, Mazda6, MX-5 Miata, MPV, RX-8 **Engines:** All **Transmissions:** A/T, M/T	**Throttle Actuator control module performance** If the PCM detects any of the following conditions, the PCM determines that the throttle actuator control system has a malfunction: TP sensor power supply voltage below 4.4 V; TP sensor No. 1 output voltage is below 0.20 V or above 4.85 V; TP sensor No. 2 output voltage is below 0.20 V or above 4.85 V; PCM internal circuit for TP sensor No. 1 input circuit malfunction; or the wrong communication between main CPU and throttle control system CPU in PCM internal. **Possible Causes:** • Open circuits between throttle body terminals and PCM terminals • Short circuits between throttle body terminals and PCM terminals • Poor connection of throttle body or PCM connector • TP sensor No. 1 or TP sensor No. 2 malfunction • PCM has failed

DTC	Trouble Code Title, Conditions & Possible Causes
DTC: P2109 **1T CCM, MIL: YES** **2004, 2005** **Models:** Mazda3, RX-8 **Engines:** All **Transmissions:** A/T, M/T	**TP sensor minimum stop range/performance problem** The PCM monitors the minimum TP when the closed TP learning is completed. If the TP is less than 11.5% or more than 24.3%, The PCM determines that there is a TP sensor minimum stop range/performance problem. **Possible Causes:** • Drive-by-wire control system malfunction • Throttle actuator malfunction • Throttle valve malfunction • PCM has failed
DTC: P2112 **1T CCM, MIL: YES** **2003, 2004, 2005** **Models:** Mazda3, Mazda6, MX-5 Miata, MPV, RX-8 **Engines:** All **Transmissions:** A/T, M/T	**TP sensor minimum stop range/performance problem** The PCM monitors the minimum TP when the closed TP learning is completed. If the TP is less than 11.5% or more than 24.3%, The PCM determines that there is a TP sensor minimum stop range/performance problem. **Possible Causes:** • Drive-by-wire control system malfunction • Throttle actuator malfunction • Throttle valve malfunction • PCM has failed
DTC: P2119 **1T CCM, MIL: YES** **2003, 2004, 2005** **Models:** Mazda3, Mazda6, MX-5 Miata, MPV, RX-8 **Engines:** All **Transmissions:** A/T, M/T	**TP sensor minimum stop range/performance problem** The PCM monitors the minimum TP when the closed TP learning is completed. If the TP is less than 11.5% or more than 24.3%, The PCM determines that there is a TP sensor minimum stop range/performance problem. **Possible Causes:** • Drive-by-wire control system malfunction • Throttle actuator malfunction • Throttle valve malfunction • PCM has failed
DTC: P2122 **1T CCM, MIL: YES** **2003, 2004, 2005** **Models:** Mazda3, Mazda6, MX-5 Miata, MPV, RX-8 **Engines:** All **Transmissions:** A/T, M/T	**APP sensor No. 1 circuit low input** The PCM monitors the input voltage from the APP sensor No.1 when the engine is running. If the input voltage is less than 0.35 V, The PCM determines that the APP sensor No.1 circuit input voltage is low. **Possible Causes:** • Open circuits between APP sensor terminals and PCM terminals • Short circuits between APP sensor terminals and PCM terminals • Poor connection of APP sensor or PCM connector • APP sensor No. 1 malfunction • PCM has failed
DTC: P2123 **1T CCM, MIL: YES** **2003, 2004, 2005** **Models:** Mazda3, Mazda6, MX-5 Miata, MPV, RX-8 **Engines:** All **Transmissions:** A/T, M/T	**APP sensor No. 1 circuit high input** The PCM monitors the input voltage from the APP sensor No.1 when the engine is running. If the input voltage is more than 4.8 V, The PCM determines that the APP sensor No.1 circuit input voltage is high. **Possible Causes:** • Open circuits between APP sensor terminals and PCM terminals • Short circuits between APP sensor terminals and PCM terminals • Poor connection of APP sensor or PCM connector • APP sensor No. 1 malfunction • PCM has failed
DTC: P2126 **1T CCM, MIL: YES** **2004, 2005** **Models:** Mazda3 **Engines:** All **Transmissions:** A/T, M/T	**Throttle Actuator control module performance** If the PCM detects any of the following conditions, the PCM determines that the throttle actuator control system has a malfunction: TP sensor power supply voltage below 4.4 V; TP sensor No. 1 output voltage is below 0.20 V or above 4.85 V; TP sensor No. 2 output voltage is below 0.20 V or above 4.85 V; PCM internal circuit for TP sensor No. 1 input circuit malfunction; or the wrong communication between main CPU and throttle control system CPU in PCM internal. **Possible Causes:** • Open circuits between throttle body terminals and PCM terminals • Short circuits between throttle body terminals and PCM terminals • Poor connection of throttle body or PCM connector • TP sensor No. 1 or TP sensor No. 2 malfunction • PCM has failed

DTC	Trouble Code Title, Conditions & Possible Causes
DTC: P2127 **1T CCM, MIL: YES** **2003, 2004, 2005** **Models:** Mazda3, Mazda6, MX-5 Miata, MPV, RX-8 **Engines:** All **Transmissions:** A/T, M/T	**APP sensor No. 2 circuit low input** The PCM monitors the input voltage from the APP sensor No.1 when the engine is running. If the input voltage is less than 0.35 V, The PCM determines that the APP sensor No.1 circuit input voltage is low. **Possible Causes:** • Open circuits between APP sensor terminals and PCM terminals • Short circuits between APP sensor terminals and PCM terminals • Poor connection of APP sensor or PCM connector • APP sensor No. 1 malfunction • PCM has failed
DTC: P2128 **1T CCM, MIL: YES** **2003, 2004, 2005** **Models:** Mazda3, Mazda6, MX-5 Miata, MPV, RX-8 **Engines:** All **Transmissions:** A/T, M/T	**APP sensor No. 2 circuit high input** The PCM monitors the input voltage from the APP sensor No.1 when the engine is running. If the input voltage is more than 4.8 V, The PCM determines that the APP sensor No.1 circuit input voltage is high. **Possible Causes:** • Open circuits between APP sensor terminals and PCM terminals • Short circuits between APP sensor terminals and PCM terminals • Poor connection of APP sensor or PCM connector • APP sensor No. 1 malfunction • PCM has failed
DTC: P2135 **1T CCM, MIL: YES** **2003, 2004, 2005** **Models:** Mazda3, Mazda6, MX-5 Miata, MPV, RX-8 **Engines:** All **Transmissions:** A/T, M/T	**TP sensor No. 1/No. 2 voltage correlation problem** The PCM compares the input voltage from TP sensor No.1 with the input voltage from TP sensor No.2 when the engine is running. If the difference is more than the specification, The PCM determines that there is a TP sensor No.1/No.2 voltage correlation problem. **Possible Causes:** • TP sensor No. 1 malfunction • TP sensor No. 2 malfunction • Poor connection of throttle body or PCM connector • PCM has failed
DTC: P2136 **1T CCM, MIL: YES** **2004, 2005** **Models:** RX-8 **Engines:** All **Transmissions:** A/T, M/T	**TP sensor No. 1/No. 3 voltage correlation problem** The PCM compares the input voltage from TP sensor No.1 with the input voltage from TP sensor No.3 when the engine is running. If the difference is more than the specification, The PCM determines that there is a TP sensor No.1/No.3 voltage correlation problem. **Possible Causes:** • TP sensor No. 1 malfunction • TP sensor No. 3 malfunction • Poor connection of throttle body or PCM connector • PCM has failed
DTC: P2138 **1T CCM, MIL: YES** **2003, 2004, 2005** **Models:** Mazda3, Mazda6, MX-5 Miata, MPV, RX-8 **Engines:** All **Transmissions:** A/T, M/T	**TP sensor No. 3/No. 4 voltage correlation problem** The PCM compares the input voltage from TP sensor No.3 with the input voltage from TP sensor No.4 when the engine is running. If the difference is more than the specification, The PCM determines that there is a TP sensor No.3/No.4 voltage correlation problem. **Possible Causes:** • TP sensor No. 3 malfunction • TP sensor No. 4 malfunction • Poor connection of throttle body or PCM connector • PCM has failed
DTC: P2144 **1T CCM, MIL: YES** **2004, 2005** **Models:** MPV **Engines:** All **Transmissions:** A/T, M/T	**EGR boost sensor solenoid valve circuit low input** The PCM monitors the EGR boost sensor solenoid valve control signal. If PCM turns the EGR boost sensor solenoid valve off but voltage still remains low, The PCM determines that the EGR boost sensor solenoid valve circuit has malfunction. **Possible Causes:** • EGR boost sensor solenoid malfunction • Connector or terminal malfunction • PCM has failed
DTC: P2145 **1T CCM, MIL: YES** **2004, 2005** **Models:** MPV **Engines:** All **Transmissions:** A/T, M/T	**EGR boost sensor solenoid valve circuit low input** The PCM monitors the EGR boost sensor solenoid valve control signal. If PCM turns the EGR boost sensor solenoid valve off but voltage still remains low, The PCM determines that the EGR boost sensor solenoid valve circuit has malfunction. **Possible Causes:** • EGR boost sensor solenoid malfunction • Connector or terminal malfunction • PCM has failed

DTC	Trouble Code Title, Conditions & Possible Causes
DTC: P2177 **1T CCM, MIL: YES** **2003, 2004, 2005** **Models:** Mazda6, MX-5 Miata, MPV **Engines:** All **Transmissions:** A/T, M/T	**Fuel system too lean at off idle (right bank)** PCM monitors short term fuel trim (SHRTFT), long term fuel trim (LONGFT) during closed loop fuel control at off-idle. If the LONGFT and the sum total of these fuel trims exceed preprogrammed criteria. PCM determines that fuel system is too lean at off-idle. **Possible Causes:** • Misfire • Exhaust system leak • Fuel pump malfunction • Fuel filter clogged • PCM has failed
DTC: P2178 **1T CCM, MIL: YES** **2003, 2004, 2005** **Models:** Mazda6, MX-5 Miata, MPV **Engines:** All **Transmissions:** A/T, M/T	**Fuel system too rich at off idle (right bank)** PCM monitors short term fuel trim (SHRTFT), long term fuel trim (LONGFT) during closed loop fuel control at off-idle. If the LONGFT and the sum total of these fuel trims exceed preprogrammed criteria. PCM determines that fuel system is too rich at off-idle. **Possible Causes:** • Misfire • Exhaust system leak • Fuel pump malfunction • Fuel filter clogged • PCM has failed
DTC: P2179 **1T CCM, MIL: YES** **2004, 2005** **Models:** MPV **Engines:** All **Transmissions:** A/T, M/T	**Fuel system too lean at off idle (left bank)** PCM monitors short term fuel trim (SHRTFT), long term fuel trim (LONGFT) during closed loop fuel control at off-idle. If the LONGFT and the sum total of these fuel trims exceed preprogrammed criteria. PCM determines that fuel system is too lean at off-idle. **Possible Causes:** • Misfire • Exhaust system leak • Fuel pump malfunction • Fuel filter clogged • PCM has failed
DTC: P2180 **1T CCM, MIL: YES** **2004, 2005** **Models:** MPV **Engines:** All **Transmissions:** A/T, M/T	**Fuel system too rich at off idle (left bank)** PCM monitors short term fuel trim (SHRTFT), long term fuel trim (LONGFT) during closed loop fuel control at off-idle. If the LONGFT and the sum total of these fuel trims exceed preprogrammed criteria. PCM determines that fuel system is too rich at off-idle. **Possible Causes:** • Misfire • Exhaust system leak • Fuel pump malfunction • Fuel filter clogged • PCM has failed
DTC: P2187 **1T CCM, MIL: YES** **2003, 2004, 2005** **Models:** Mazda6, MX-5 Miata, MPV **Engines:** All **Transmissions:** A/T, M/T	**Fuel system too lean at off idle (right bank)** PCM monitors short term fuel trim (SHRTFT), long term fuel trim (LONGFT) during closed loop fuel control at off-idle. If the LONGFT and the sum total of these fuel trims exceed preprogrammed criteria. PCM determines that fuel system is too lean at off-idle. **Possible Causes:** • Misfire • Exhaust system leak • Fuel pump malfunction • Fuel filter clogged • PCM has failed
DTC: P2188 **1T CCM, MIL: YES** **2003, 2004, 2005** **Models:** Mazda6, MX-5 Miata, MPV **Engines:** All **Transmissions:** A/T, M/T	**Fuel system too lean at off idle (left bank)** PCM monitors short term fuel trim (SHRTFT), long term fuel trim (LONGFT) during closed loop fuel control at off-idle. If the LONGFT and the sum total of these fuel trims exceed preprogrammed criteria. PCM determines that fuel system is too lean at off-idle. **Possible Causes:** • Misfire • Exhaust system leak • Fuel pump malfunction • Fuel filter clogged • PCM has failed

DTC	Trouble Code Title, Conditions & Possible Causes
DTC: P2189 **1T CCM, MIL: YES** **2004, 2005** **Models:** MPV **Engines:** All **Transmissions:** A/T, M/T	**Fuel system too lean at off idle (left bank)** PCM monitors short term fuel trim (SHRTFT), long term fuel trim (LONGFT) during closed loop fuel control at off-idle. If the LONGFT and the sum total of these fuel trims exceed preprogrammed criteria. PCM determines that fuel system is too lean at off-idle. **Possible Causes:** • Misfire • Exhaust system leak • Fuel pump malfunction • Fuel filter clogged • PCM has failed
DTC: P2190 **1T CCM, MIL: YES** **2004, 2005** **Models:** MPV **Engines:** All **Transmissions:** A/T, M/T	**Fuel system too rich at off idle (left bank)** PCM monitors short term fuel trim (SHRTFT), long term fuel trim (LONGFT) during closed loop fuel control at off-idle. If the LONGFT and the sum total of these fuel trims exceed preprogrammed criteria. PCM determines that fuel system is too rich at off-idle. **Possible Causes:** • Misfire • Exhaust system leak • Fuel pump malfunction • Fuel filter clogged • PCM has failed
DTC: P2195 **1T CCM, MIL: YES** **2004, 2005** **Models:** Mazda6, MX-5 Miata, MPV, RX-8, Tribute **Engines:** All **Transmissions:** A/T, M/T	**HO2S (RF) signal stuck lean** The PCM monitors the HO2S (RF, LF) output voltage when the following conditions are met. If output voltage is less than 0.45 V for 25.6 s, The PCM determines that the HO2S (RF, LF) signal remains lean. **Possible Causes:** • HO2S (RF, LF) malfunction • Exhaust gas leaking • MAF sensor malfunction • ECT sensor malfunction • PCM has failed
DTC: P2196 **1T CCM, MIL: YES** **2004, 2005** **Models:** Mazda6, MPV, RX-8, Tribute **Engines:** All **Transmissions:** A/T, M/T	**HO2S (RF) signal stuck rich** The PCM monitors the HO2S (RF, LF) output voltage when the following conditions are met. If output voltage is more than 0.45 V for 25.6 s, The PCM determines that the HO2S (RF, LF) signal remains rich. **Possible Causes:** • HO2S (RF, LF) malfunction • Exhaust gas leaking • MAF sensor malfunction • ECT sensor malfunction • PCM has failed
DTC: P2197 **1T CCM, MIL: YES** **2004, 2005** **Models:** MPV, Tribute **Engines:** All **Transmissions:** A/T, M/T	**HO2S (LF) signal stuck lean** The PCM monitors the HO2S (RF, LF) output voltage when the following conditions are met. If output voltage is less than 0.45 V for 25.6 s, The PCM determines that the HO2S (RF, LF) signal remains lean. **Possible Causes:** • HO2S (RF, LF) malfunction • Exhaust gas leaking • MAF sensor malfunction • ECT sensor malfunction • PCM has failed
DTC: P2198 **1T CCM, MIL: YES** **2004, 2005** **Models:** MPV, Tribute **Engines:** All **Transmissions:** A/T, M/T	**HO2S (LF) signal stuck rich** The PCM monitors the HO2S (RF, LF) output voltage when the following conditions are met. If output voltage is more than 0.45 V for 25.6 s, The PCM determines that the HO2S (RF, LF) signal remains rich. **Possible Causes:** • HO2S (RF, LF) malfunction • Exhaust gas leaking • MAF sensor malfunction • ECT sensor malfunction • PCM has failed

DTC	Trouble Code Title, Conditions & Possible Causes
DTC: P2227 **1T CCM, MIL: YES** **2004, 2005** **Models:** MPV **Engines:** All **Transmissions:** A/T, M/T	**EGR boost sensor circuit performance problem** The PCM monitors the EGR boost sensor solenoid valve control signal. If PCM turns the EGR boost sensor solenoid valve off but voltage still remains low, The PCM determines that the EGR boost sensor solenoid valve circuit has malfunction. **Possible Causes:** • EGR boost sensor solenoid malfunction • Connector or terminal malfunction • PCM has failed
DTC: P2228 **1T CCM, MIL: YES** **2004, 2005** **Models:** Mazda3, Mazda6, MPV **Engines:** All **Transmissions:** A/T, M/T	**EGR boost sensor solenoid valve circuit low input** The PCM monitors the EGR boost sensor solenoid valve control signal. If PCM turns the EGR boost sensor solenoid valve off but voltage still remains low, The PCM determines that the EGR boost sensor solenoid valve circuit has malfunction. **Possible Causes:** • EGR boost sensor solenoid malfunction • Connector or terminal malfunction • PCM has failed
DTC: P2229 **1T CCM, MIL: YES** **2004, 2005** **Models:** Mazda3, Mazda6, MPV **Engines:** All **Transmissions:** A/T, M/T	**EGR boost sensor solenoid valve circuit high input** The PCM monitors the EGR boost sensor solenoid valve control signal. If PCM turns the EGR boost sensor solenoid valve off but voltage still remains high, The PCM determines that the EGR boost sensor solenoid valve circuit has malfunction. **Possible Causes:** • EGR boost sensor solenoid malfunction • Connector or terminal malfunction • PCM has failed
DTC: P2257 **1T CCM, MIL: YES** **2004, 2005** **Models:** RX-8 **Engines:** All **Transmissions:** A/T, M/T	**AIR pump relay circuit low** The PCM monitors the AIR pump relay control voltage when the AIR pump is not operating. If the control voltage is less than 5.8 V, The PCM determines that the AIR pump relay control circuit voltage is low. **Possible Causes:** • AIR pump relay malfunction • Connector or terminal malfunction • PCM has failed
DTC: P2258 **1T CCM, MIL: YES** **2004, 2005** **Models:** RX-8 **Engines:** All **Transmissions:** A/T, M/T	**AIR pump relay circuit high** The PCM monitors the AIR pump relay control voltage when the AIR pump is not operating. If the control voltage is more than 11.5 V, The PCM determines that the AIR pump relay control circuit voltage is low. **Possible Causes:** • AIR pump relay malfunction • Connector or terminal malfunction • PCM has failed
DTC: P2259 **1T CCM, MIL: YES** **2004, 2005** **Models:** RX-8 **Engines:** All **Transmissions:** A/T, M/T	**AIR solenoid valve control circuit low** The PCM monitors the AIR solenoid valve control voltage when the AIR pump is not operating. If the control voltage is less than 5.8 V, The PCM determines that the AIR solenoid valve control circuit voltage is low. **Possible Causes:** • AIR solenoid valve malfunction • Connector or terminal malfunction • PCM has failed
DTC: P2260 **1T CCM, MIL: YES** **2004, 2005** **Models:** RX-8 **Engines:** All **Transmissions:** A/T, M/T	**AIR solenoid valve control circuit low** The PCM monitors the AIR solenoid valve control voltage when the AIR pump is not operating. If the control voltage is more than 11.5 V, The PCM determines that the AIR solenoid valve control circuit voltage is low. **Possible Causes:** • AIR solenoid valve malfunction • Connector or terminal malfunction • PCM has failed
DTC: P2270 **1T CCM, MIL: YES** **2004, 2005** **Models:** Mazda3, Mazda6, RX-8, Tribute **Engines:** All **Transmissions:** A/T, M/T	**Rear HO2S signal stuck lean** The PCM monitors the input voltage from the rear HO2S when the following conditions are met. If the input voltage is less than 0.4 V for 40 seconds, The PCM determines that the rear HO2S signal remains lean. **Possible Causes:** • Front or Rear HO2S malfunction • Front or Rear HO2S heater malfunction • Fuel pressure malfunction • Fuel injector malfunction • PCM has failed

DTC	Trouble Code Title, Conditions & Possible Causes
DTC: P2271 **1T CCM, MIL: YES** **2004, 2005** **Models:** Mazda3, Mazda6, RX-8, Tribute **Engines:** All **Transmissions:** A/T, M/T	**Rear HO2S signal stuck rich** The PCM monitors the input voltage from the rear HO2S when the following conditions are met. If the input voltage is more than 0.85 V for 40 seconds, The PCM determines that the rear HO2S signal remains lean. **Possible Causes:** • Front or Rear HO2S malfunction • Front or Rear HO2S heater malfunction • Fuel pressure malfunction • Fuel injector malfunction • PCM has failed
DTC: P2272 **1T CCM, MIL: YES** **2004, 2005** **Models:** Mazda6, Tribute **Engines:** All **Transmissions:** A/T, M/T	**Rear HO2S signal stuck lean** The PCM monitors the input voltage from the rear HO2S when the following conditions are met. If the input voltage is less than 0.4 V for 40 seconds, The PCM determines that the rear HO2S signal remains lean. **Possible Causes:** • Front or Rear HO2S malfunction • Front or Rear HO2S heater malfunction • Fuel pressure malfunction • Fuel injector malfunction • PCM has failed
DTC: P2273 **1T CCM, MIL: YES** **2004, 2005** **Models:** Tribute **Engines:** All **Transmissions:** A/T, M/T	**Rear HO2S signal stuck rich** The PCM monitors the input voltage from the rear HO2S when the following conditions are met. If the input voltage is more than 0.85 V for 40 seconds, The PCM determines that the rear HO2S signal remains lean. **Possible Causes:** • Front or Rear HO2S malfunction • Front or Rear HO2S heater malfunction • Fuel pressure malfunction • Fuel injector malfunction • PCM has failed
DTC: P2274 **1T CCM, MIL: YES** **2004, 2005** **Models:** Mazda3, Mazda6 **Engines:** All **Transmissions:** A/T, M/T	**Rear HO2S signal stuck lean** The PCM monitors the input voltage from the rear HO2S when the following conditions are met. If the input voltage is less than 0.4 V for 40 seconds, The PCM determines that the rear HO2S signal remains lean. **Possible Causes:** • Front or Rear HO2S malfunction • Front or Rear HO2S heater malfunction • Fuel pressure malfunction • Fuel injector malfunction • PCM has failed
DTC: P2275 **1T CCM, MIL: YES** **2004, 2005** **Models:** Mazda3, Mazda6 **Engines:** All **Transmissions:** A/T, M/T	**Rear HO2S signal stuck rich** The PCM monitors the input voltage from the rear HO2S when the following conditions are met. If the input voltage is more than 0.85 V for 40 seconds, The PCM determines that the rear HO2S signal remains lean. **Possible Causes:** • Front or Rear HO2S malfunction • Front or Rear HO2S heater malfunction • Fuel pressure malfunction • Fuel injector malfunction • PCM has failed
DTC: P2401 **1T CCM, MIL: YES** **2004, 2005** **Models:** Mazda3, Mazda6, MX-5 Miata, MPV, RX-8 **Engines:** All **Transmissions:** A/T, M/T	**EVAP system leak detection pump motor circuit low** The PCM monitors pump load current (EVAP line pressure), while evaporative leak monitor is operating. If the pump load current is lower than specified, The PCM determines EVAP system leak detection pump motor circuit has a malfunction. **Possible Causes:** • EVAP system leak detection pump malfunction • Poor connection of EVAP system or PCM connector • PCM has failed
DTC: P2402 **1T CCM, MIL: YES** **2004, 2005** **Models:** Mazda3, Mazda6, MX-5 Miata, MPV, RX-8 **Engines:** All **Transmissions:** A/T, M/T	**EVAP system leak detection pump motor circuit low** The PCM monitors pump load current (EVAP line pressure), while evaporative leak monitor is operating. If the pump load current is higher than specified, The PCM determines EVAP system leak detection pump motor circuit has a malfunction. **Possible Causes:** • EVAP system leak detection pump malfunction • Poor connection of EVAP system or PCM connector • PCM has failed

DTC	Trouble Code Title, Conditions & Possible Causes
DTC: P2404 **1T CCM, MIL: YES** **2004, 2005** **Models:** Mazda3, Mazda6, MX-5 Miata, MPV, RX-8 **Engines:** All **Transmissions:** A/T, M/T	**EVAP system leak detection pump sense circuit problem** The PCM monitors pump load current (EVAP line pressure), while evaporative leak monitor is operating. After obtaining the reference current value, if the time in which the pump load current reaches the reference current value is less than the specification, The PCM determines air filter has a malfunction. **Possible Causes:** • Air filter clogging • EVAP hose bending • PCM has failed
DTC: P2405 **1T CCM, MIL: YES** **2004, 2005** **Models:** Mazda6, MX-5 Miata, MPV, RX-8 **Engines:** All **Transmissions:** A/T, M/T	**EVAP system leak detection pump sense circuit low input** The PCM monitors pump load current (EVAP line pressure), while evaporative leak monitor is operating. If the current is lower than the specification while The PCM obtains the reference current value, The PCM determines EVAP system leak detection pump orifice has a malfunction. **Possible Causes:** • EVAP system leak detection orifice has fallen off • EVAP system leak detection pump motor malfunction • PCM has failed
DTC: P2407 **1T CCM, MIL: YES** **2004, 2005** **Models:** Mazda3. Mazda6, MX-5 Miata, MPV, RX-8 **Engines:** All **Transmissions:** A/T, M/T	**EVAP system leak detection pump sense circuit intermittent** The PCM monitors pump load current (EVAP line pressure), while evaporative leak monitor is operating. When either of the following is detected 6 times or more successively, the PCM determines EVAP system leak detection pump heater has a malfunction. **Possible Causes:** • EVAP system leak detection pump heater malfunction • PCM has failed
DTC: P2502 **1T CCM, MIL: YES** **2004, 2005** **Models:** Mazda6, MX-5 Miata, MPV, RX-8 **Engines:** All **Transmissions:** A/T, M/T	**Charging system voltage problem** **PCM judges' generator output voltage is above 17 V or battery voltage is below 11 V during engine running.** **Possible Causes:** • Open circuits between battery terminals • Battery is malfunctioning • Poor connection of PCM connectors • Generator malfunction • PCM has failed
DTC: P2503 **1T CCM, MIL: YES** **2004, 2005** **Models:** Mazda6, MX-5 Miata, MPV, RX-8 **Engines:** All **Transmissions:** A/T, M/T	**Charging system voltage low** **PCM needs more than 20 A from generator, and judges' generator output voltage to be below 8.5 V during engine running.** **Possible Causes:** • Open circuits between battery terminals • Battery is malfunctioning • Poor connection of PCM connectors • Generator malfunction • PCM has failed
DTC: P2504 **1T CCM, MIL: YES** **2004, 2005** **Models:** Mazda3, Mazda6, MX-5 Miata, MPV, RX-8 **Engines:** All **Transmissions:** A/T, M/T	**Charging system voltage high** **PCM judges generator output voltage is above 18.5 V or battery voltage is above 16.0 V during engine running.** **Possible Causes:** • Short to power circuit between generator and PCM • Generator malfunction • PCM has failed
DTC: P2507 **1T CCM, MIL: YES** **2004, 2005** **Models:** Mazda3, Mazda6, MX-5 Miata, MPV **Engines:** All **Transmissions:** A/T, M/T	**PCM B+ voltage low** The PCM monitors the voltage of back-up battery positive terminal at PCM terminal 4AG. If The PCM detected battery positive terminal voltage below 2.5 V for 2 seconds, The PCM determines that the backup voltage circuit has malfunction. **Possible Causes:** • Melt down fuse • Poor connection of PCM • PCM has failed

DTC	Trouble Code Title, Conditions & Possible Causes
DTC: P2510 **1T CCM, MIL: YES** **2004, 2005** **Models:** MX-5 Miata **Engines:** All **Transmissions:** A/T, M/T	**Timer error in PCM** **PCM internal timer is damaged.** **Possible Causes:** • PCM internal timer is damaged
DTC: P2676 **1T CCM, MIL: YES** **2004, 2005** **Models:** Mazda6 **Engines:** All **Transmissions:** A/T, M/T	**Variable Air Duct (VAD) control solenoid valve circuit low input** The PCM monitors VAD solenoid valve control signal at PCM terminal 68. If The PCM turns VAD solenoid valve off but voltage at PCM terminal 68 still remains low, The PCM determines that VAD solenoid valve circuit has malfunction. **Possible Causes:** • Open circuits between main relay and VAD terminals • Short circuits between VAD terminals and PCM terminals • Poor connection of PCM connector • VAD solenoid valve malfunction • PCM has failed
DTC: P2677 **1T CCM, MIL: YES** **2004, 2005** **Models:** Mazda6 **Engines:** All **Transmissions:** A/T, M/T	**Variable Air Duct (VAD) control solenoid valve circuit high input** The PCM monitors VAD solenoid valve control signal at PCM terminal 68. If The PCM turns VAD solenoid valve off but voltage at PCM terminal 68 still remains high, The PCM determines that VAD solenoid valve circuit has malfunction. **Possible Causes:** • Open circuits between main relay and VAD terminals • Short circuits between VAD terminals and PCM terminals • Poor connection of PCM connector • VAD solenoid valve malfunction • PCM has failed

Gas Engine OBD II Trouble Code List (U1xxx Codes)

DTC: U0073 **1T CCM, MIL: NO** **1998, 1999, 2000, 2001, 2002, 2003, 2004, 2005** **Models:** All **Engines:** All **Transmissions:** A/T	**Data Circuit Malfunction** Key on, and the PCM detected the SCP was invalid, or that it was missing data from the A/C System during the CCM test. **Possible Causes:** • SCP data bus to other controller(s) is open or shorted to ground • A/C System controller is damaged or has failed • PCM has failed
DTC: U0101 **1T CCM, MIL: NO** **1998, 1999, 2000, 2001, 2002, 2003, 2004, 2005** **Models:** All **Engines:** All **Transmissions:** A/T	**Data Circuit Malfunction** Key on, and the PCM detected the SCP was invalid, or that it was missing data from the A/C System during the CCM test. **Possible Causes:** • SCP data bus to other controller(s) is open or shorted to ground • A/C System controller is damaged or has failed • PCM has failed
DTC: U0121 **1T CCM, MIL: NO** **1998, 1999, 2000, 2001, 2002, 2003, 2004, 2005** **Models:** All **Engines:** All **Transmissions:** A/T	**Data Circuit Malfunction** Key on, and the PCM detected the SCP was invalid, or that it was missing data from the A/C System during the CCM test. **Possible Causes:** • SCP data bus to other controller(s) is open or shorted to ground • A/C System controller is damaged or has failed • PCM has failed
DTC: U0155 **1T CCM, MIL: NO** **1998, 1999, 2000, 2001, 2002, 2003, 2004, 2005** **Models:** All **Engines:** All **Transmissions:** A/T	**Data Circuit Malfunction** Key on, and the PCM detected the SCP was invalid, or that it was missing data from the A/C System during the CCM test. **Possible Causes:** • SCP data bus to other controller(s) is open or shorted to ground • A/C System controller is damaged or has failed • PCM has failed
DTC: U0167 **1T CCM, MIL: NO** **1998, 1999, 2000, 2001, 2002, 2003, 2004, 2005** **Models:** All **Engines:** All **Transmissions:** A/T	**Data Circuit Malfunction** Key on, and the PCM detected the SCP was invalid, or that it was missing data from the A/C System during the CCM test. **Possible Causes:** • SCP data bus to other controller(s) is open or shorted to ground • A/C System controller is damaged or has failed • PCM has failed
DTC: U0302 **1T CCM, MIL: NO** **1998, 1999, 2000, 2001, 2002, 2003, 2004, 2005** **Models:** All **Engines:** All **Transmissions:** A/T	**Data Circuit Malfunction** Key on, and the PCM detected the SCP was invalid, or that it was missing data from the A/C System during the CCM test. **Possible Causes:** • SCP data bus to other controller(s) is open or shorted to ground • A/C System controller is damaged or has failed • PCM has failed
DTC: U1020 **1T CCM, MIL: NO** **1998, 1999, 2000, 2001, 2002, 2003, 2004, 2005** **Models:** All **Engines:** All **Transmissions:** A/T	**Data Circuit Malfunction** Key on, and the PCM detected the SCP was invalid, or that it was missing data from the A/C System during the CCM test. **Possible Causes:** • SCP data bus to other controller(s) is open or shorted to ground • A/C System controller is damaged or has failed • PCM has failed

DTC	Trouble Code Title, Conditions & Possible Causes
DTC: U1039 **1T CCM, MIL: NO** **1998, 1999, 2000, 2001, 2002, 2003, 2004, 2005** **Models:** All **Engines:** All **Transmissions:** A/T	**Data Circuit Malfunction** Key on, and the PCM detected the SCP was invalid, or that it was missing data from the Vehicle Speed System during the CCM test. **Possible Causes:** • SCP data bus to other controller(s) is open or shorted to ground • Vehicle Speed System controller is damaged or has failed • PCM has failed
DTC: U1040 **1T CCM, MIL: NO** **1998, 1999, 2000, 2001, 2002, 2003, 2004, 2005** **Models:** All **Engines:** All **Transmissions:** A/T	**Data Circuit Malfunction** Key on, and the PCM detected the SCP was invalid, or that it was missing data from the Vehicle Speed System during the CCM test. **Possible Causes:** • SCP data bus to other controller(s) is open or shorted to ground • Vehicle Speed System controller is damaged or has failed • PCM has failed
DTC: U1051 **1T CCM, MIL: NO** **1998, 1999, 2000, 2001, 2002, 2003, 2004, 2005** **Models:** All **Engines:** All **Transmissions:** A/T	**Data Circuit Malfunction** Key on, and the PCM detected the SCP was invalid, or that it was missing data from the Antilock Brake System during the CCM test. **Possible Causes:** • SCP data bus to other controller(s) is open or shorted to ground • ABS System controller is damaged or has failed • PCM has failed
DTC: U1131 **1T CCM, MIL: NO** **1998, 1999, 2000, 2001, 2002, 2003, 2004, 2005** **Models:** All **Engines:** All **Transmissions:** A/T	**Data Circuit Malfunction** Key on, and the PCM detected the SCP was invalid, or that it was missing data from the Fuel System during the CCM test. **Possible Causes:** • SCP data bus to other controller(s) is open or shorted to ground • Fuel System controller is damaged or has failed • PCM has failed
DTC: U1147 **1T CCM, MIL: NO** **1998, 1999, 2000, 2001, 2002, 2003, 2004, 2005** **Models:** All **Engines:** All **Transmissions:** A/T	**Data Circuit Malfunction** Key on, and the PCM detected the SCP was invalid, or that it was missing data from the Vehicle Security System during the CCM test. **Possible Causes:** • SCP data bus to other controller(s) is open or shorted to ground • Vehicle Security System controller is damaged or has failed • PCM has failed
DTC: U1451 **1T CCM, MIL: NO** **1998, 1999, 2000, 2001, 2002, 2003, 2004, 2005** **Models:** All **Engines:** All **Transmissions:** A/T	**Data Circuit Malfunction** Key on, and the PCM detected the SCP was invalid, or that it was missing data from the Anti-Theft System during the CCM test. **Possible Causes:** • SCP data bus to other controller(s) is open or shorted to ground • Anti-Theft System controller is damaged or has failed • PCM has failed
DTC: U1262 **1T CCM, MIL: NO** **1998, 1999, 2000, 2001, 2002, 2003, 2004, 2005** **Models:** All **Engines:** All **Transmissions:** A/T	**Data Circuit Malfunction** Key on, and the PCM detected a problem in the SCP communications data bus circuit during the CCM test. **Note: Perform a network communications test of the whole system.** **Possible Causes:** • SCP data bus to other controller(s) is open or shorted to ground • Anti-Theft System controller is damaged or has failed • PCM has failed

DTC	Trouble Code Title, Conditions & Possible Causes
DTC: U1900 **1T CCM, MIL: NO** **1998, 1999, 2000, 2001, 2002, 2003, 2004, 2005** **Models:** All **Engines:** All **Transmissions:** A/T	**Data Circuit Malfunction** Key on, and the PCM detected the SCP was invalid, or that it was missing data from the Exterior Environment System in the CCM test. **Possible Causes:** • SCP data bus to other controller(s) is open or shorted to ground • Vehicle Security System controller is damaged or has failed • PCM has failed
DTC: U2013 **1T CCM, MIL: NO** **1998, 1999, 2000, 2001, 2002, 2003, 2004, 2005** **Models:** All **Engines:** All **Transmissions:** A/T	**Data Circuit Malfunction** Key on, and the PCM detected the SCP was invalid, or that it was missing data from the Exterior Environment System in the CCM test. **Possible Causes:** • SCP data bus to other controller(s) is open or shorted to ground • Vehicle Security System controller is damaged or has failed • PCM has failed
DTC: U2023 **1T CCM, MIL: NO** **1998, 1999, 2000, 2001, 2002, 2003, 2004, 2005** **Models:** All **Engines:** All **Transmissions:** A/T	**Data Circuit Malfunction** Key on, and the PCM detected the SCP was invalid, or that it was missing data from the Exterior Environment System in the CCM test. **Possible Causes:** • SCP data bus to other controller(s) is open or shorted to ground • Vehicle Security System controller is damaged or has failed • PCM has failed
DTC: U2050 **1T CCM, MIL: NO** **1998, 1999, 2000, 2001, 2002, 2003, 2004, 2005** **Models:** All **Engines:** All **Transmissions:** A/T	**Data Circuit Malfunction** Key on, and the PCM detected the SCP was invalid, or that it was missing data from the Exterior Environment System in the CCM test. **Possible Causes:** • SCP data bus to other controller(s) is open or shorted to ground • Vehicle Security System controller is damaged or has failed • PCM has failed
DTC: U2051 **1T CCM, MIL: NO** **1998, 1999, 2000, 2001, 2002, 2003, 2004, 2005** **Models:** All **Engines:** All **Transmissions:** A/T	**Data Circuit Malfunction** Key on, and the PCM detected the SCP was invalid, or that it was missing data from the Exterior Environment System in the CCM test. **Possible Causes:** • SCP data bus to other controller(s) is open or shorted to ground • Vehicle Security System controller is damaged or has failed • PCM has failed
DTC: U2243 **1T CCM, MIL: NO** **1998, 1999, 2000, 2001, 2002, 2003, 2004, 2005** **Models:** All **Engines:** All **Transmissions:** A/T	**Data Circuit Malfunction** Key on, and the PCM detected the SCP was invalid, or that it was missing data from the Exterior Environment System in the CCM test. **Possible Causes:** • SCP data bus to other controller(s) is open or shorted to ground • Vehicle Security System controller is damaged or has failed • PCM has failed

MAZDA
COMPONENT TESTING

13

TABLE OF CONTENTS

Component Locations ... 113-2
Accelerator Pedal Position Sensor ... 13-22
Camshaft Position Sensor ... 13-22
Crankshaft Position Sensor ... 13-24
Engine Coolant Temperature Sensor ... 13-26
Heated Oxygen Sensor ... 13-26
Idle Air Control Valve ... 13-27
Intake Air Temperature Sensor ... 13-28
Knock Sensor ... 13-28
Mass Air Flow Sensor ... 13-30
Manifold Absolute Pressure Sensor ... 13-32
Powertrain Control Module ... 13-32
Throttle Position Sensor ... 13-34
Vehicle Speed Sensor ... 13-36

Component Locations

1995–2002 626 L4

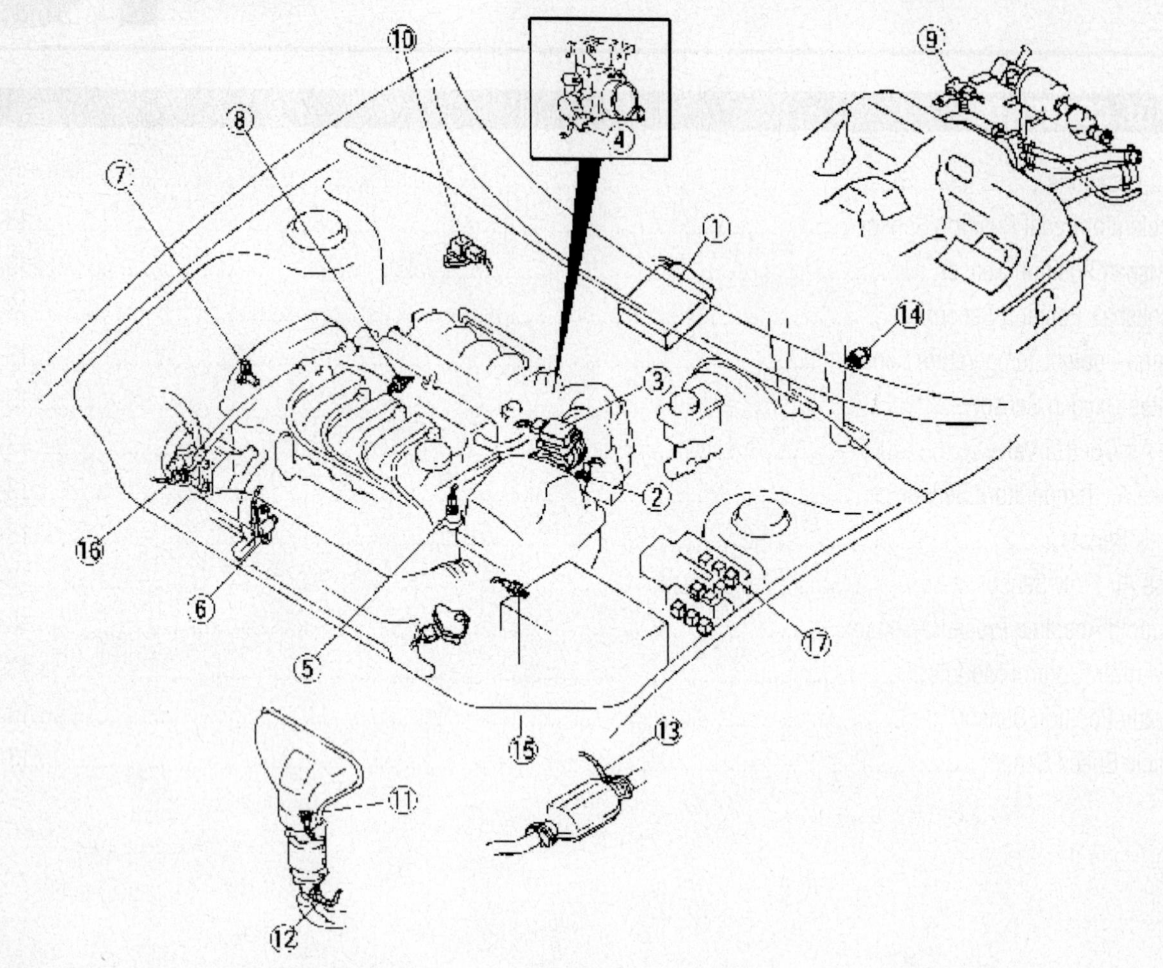

Component Location – 1995-2002 626 L4

1	PCM
2	Intake air temperature sensor
3	Mass air flow sensor
4	Throttle position sensor
5	Engine coolant temperature sensor
6	Crankshaft position sensor
7	Camshaft position sensor
8	Knock sensor
9	Fuel tank pressure sensor
10	EGR boost sensor
11	Heated oxygen sensor (Front: California)
12	Heated oxygen sensor (Rear: California), (Front: Except California)
13	Heated oxygen sensor (Rear: Except California)
14	Clutch switch
15	Neutral switch
16	Power steering pressure switch
17	Main relay

1995–2002 626 V6

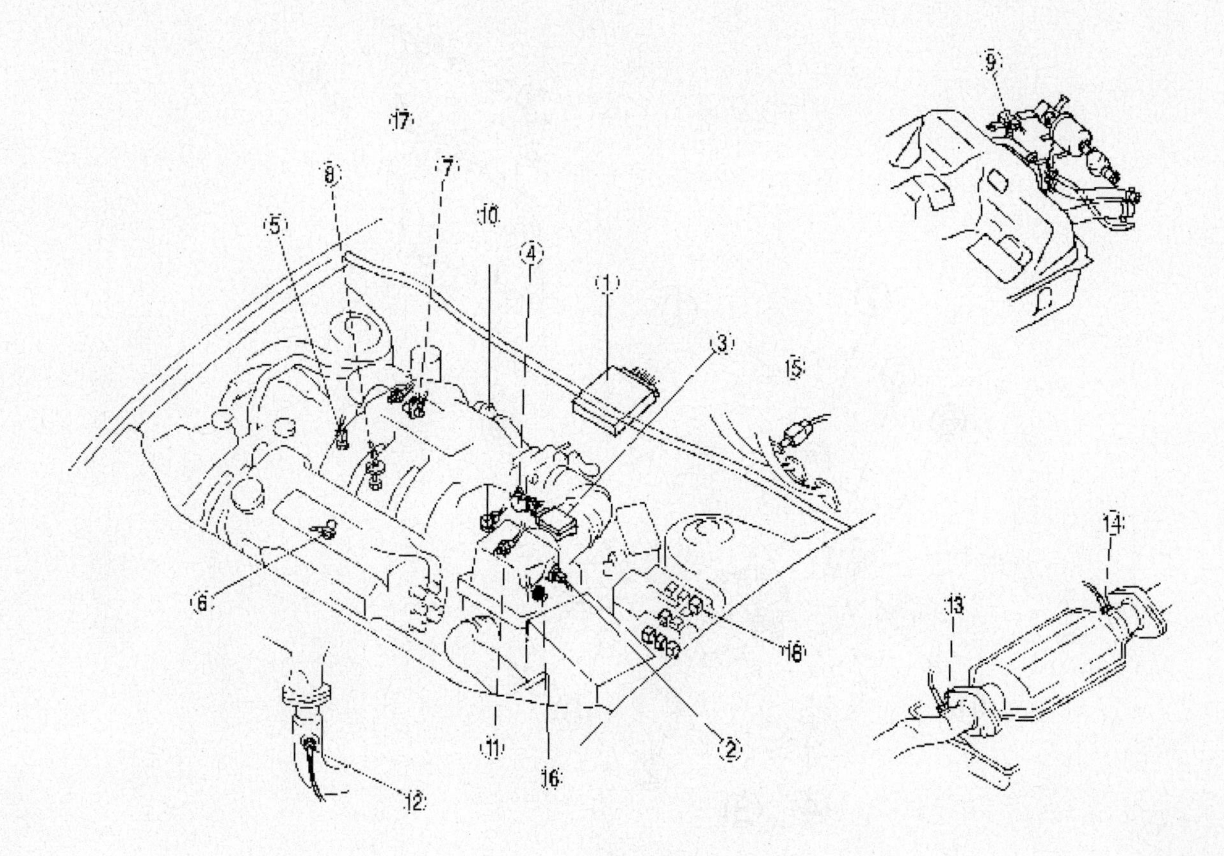

Component Location – 1995-2002 626 V6

1	PCM
2	Intake air temperature sensor
3	Mass air flow sensor
4	Throttle position sensor
5	Engine coolant temperature sensor
6	Crankshaft position sensor
7	Camshaft position sensor
8	Knock sensor
9	Fuel tank pressure sensor
10	EGR boost sensor
11	Heated oxygen sensor (Front RH)
12	Heated oxygen sensor (Front LH)
13	Heated oxygen sensor (Middle)
14	Heated oxygen sensor (Rear)
15	Clutch switch (MTX)
16	Neutral switch (MTX)
17	Power steering pressure switch
18	Main relay

1995–1998 MPV

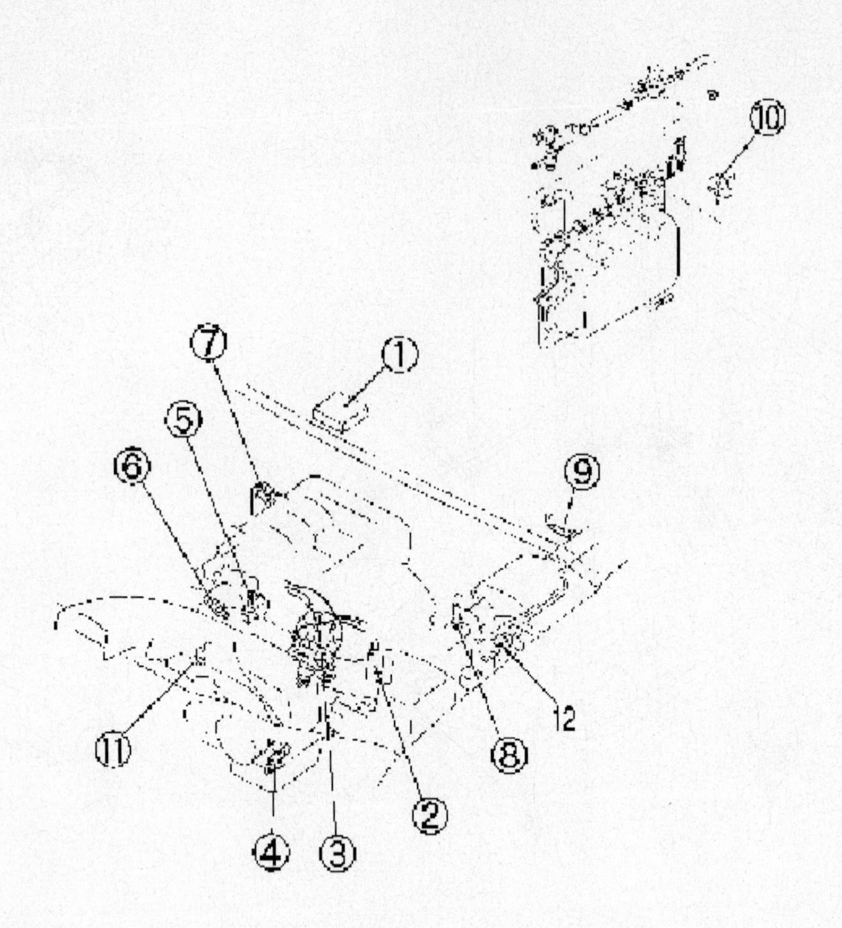

Component Location – 1995-98 MPV

1	Powertrain control module (PCM)
2	Mass air flow sensor
3	Camshaft position sensor (built-in distributor)
4	Crankshaft position sensor
5	Throttle position sensor
6	Engine coolant temperature sensor
7	Intake air temperature sensor (Dynamic chamber)
8	Heated oxygen sensor (Front)
9	Heated oxygen sensor (Rear)
10	Fuel tank pressure sensor
11	Power steering pressure switch
12	Main relay

1995–1999 Millenia

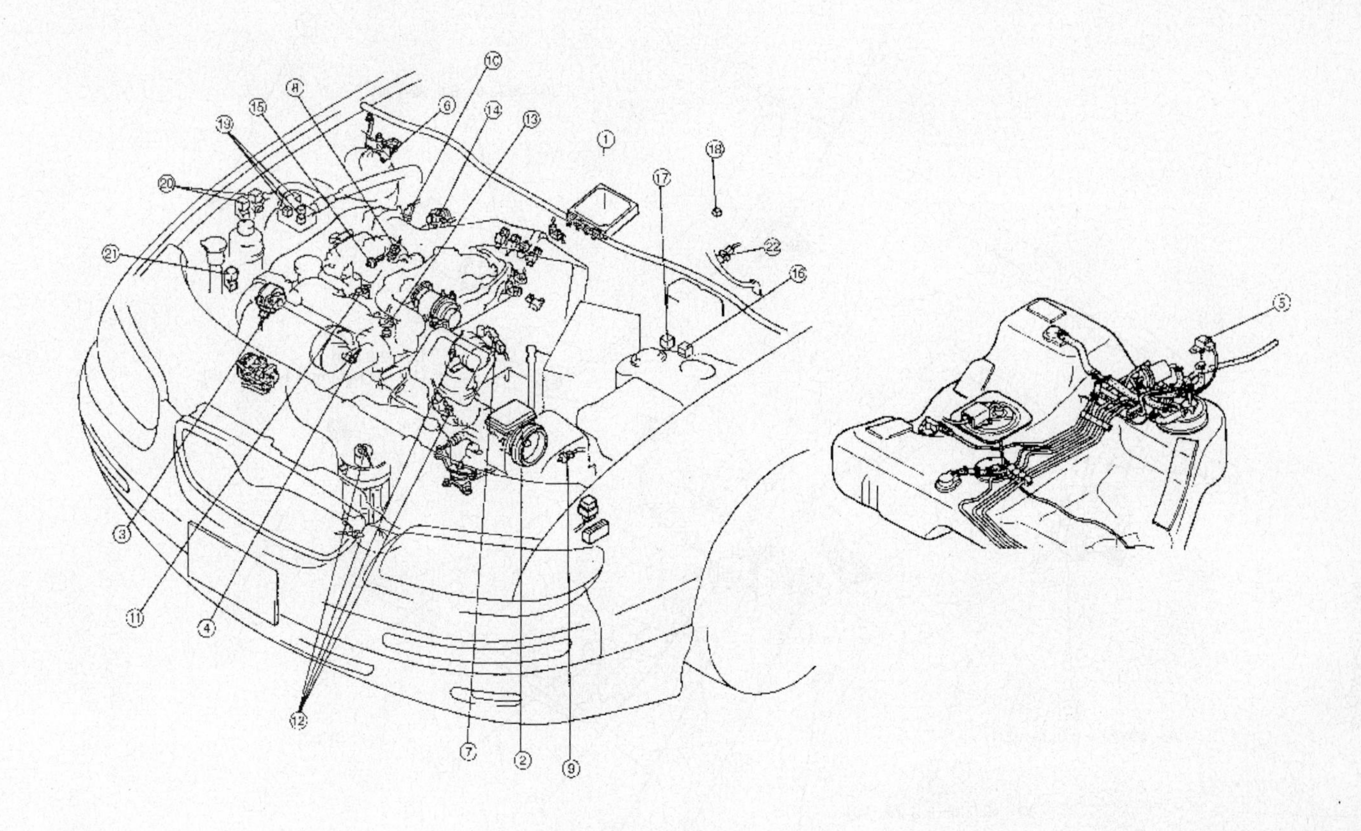

29149_MAZD_G0004

Component Location – 1995-99 Millenia

1	Powertrain control module (PCM)
2	Mass air flow sensor
3	Camshaft position sensor
4	Crankshaft position sensor
5	Fuel tank pressure sensor
6	Manifold absolute pressure sensor
7	Throttle position sensor
8	Engine coolant temperature sensor
9	Intake air temperature sensor (Air cleaner)
10	Intake air temperature sensor (Dynamic chamber)
11	Intake air temperature sensor (L/C)
12	Heated oxygen sensor (front, rear)
13	Knock sensor
14	EGR valve position sensor
15	Power steering pressure switch
16	Main relay
17	Fuel pump relay
18	Vehicle speed sensor
19	Cooling fan relay
20	Condenser fan relay
21	A/C relay
22	Brake switch

1995–1998 Protege

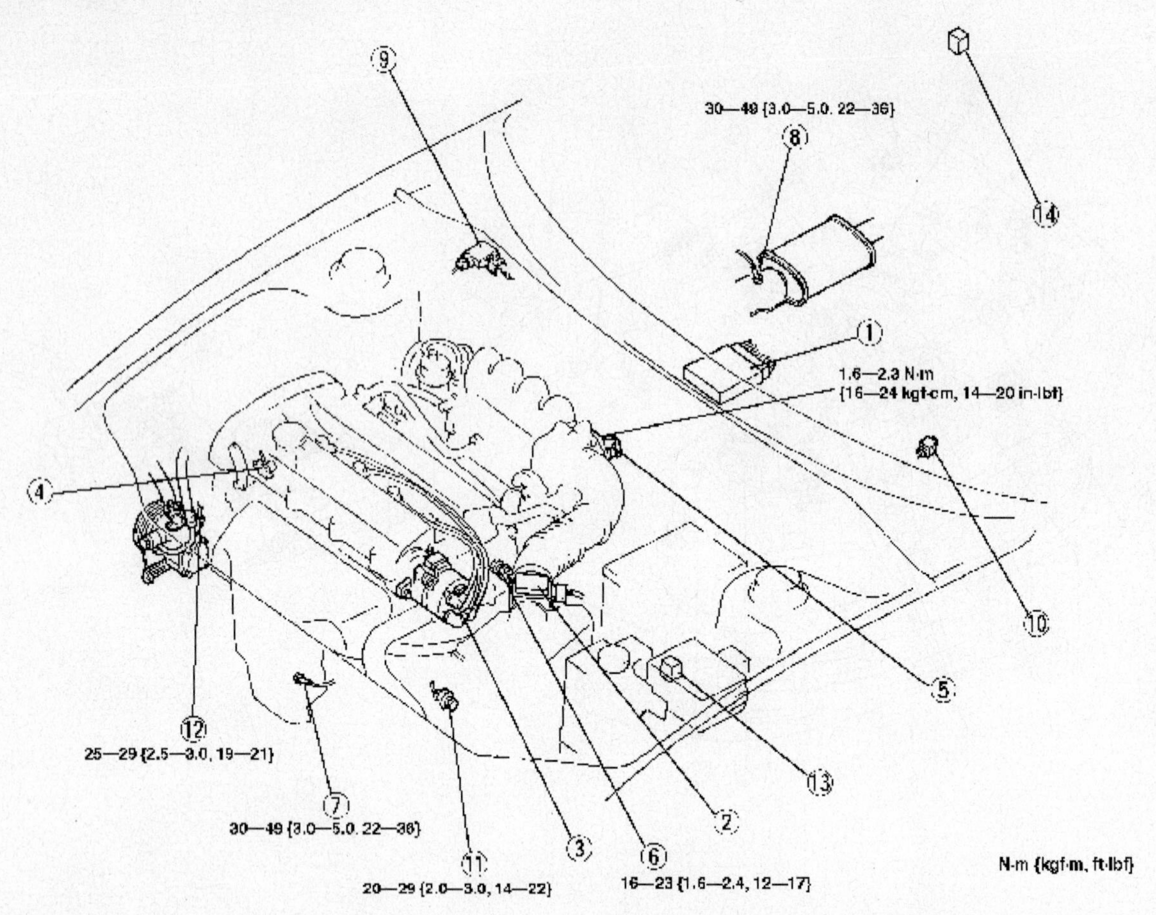

Component Location – 1995–1998 Protege

1 Powertrain control module (PCM)
2 Mass air flow sensor (Include intake air temperature sensor)
3 Camshaft position sensor
4 Crankshaft position sensor
5 Throttle position sensor
6 Engine coolant temperature sensor
7 Heated oxygen sensor (Front)
8 Heated oxygen sensor (Rear)
9 EGR boost sensor
10 Clutch switch
11 Neutral switch
12 Power steering pressure switch
13 Main relay
14 Fuel tank pressure sensor

1999–2000 Miata

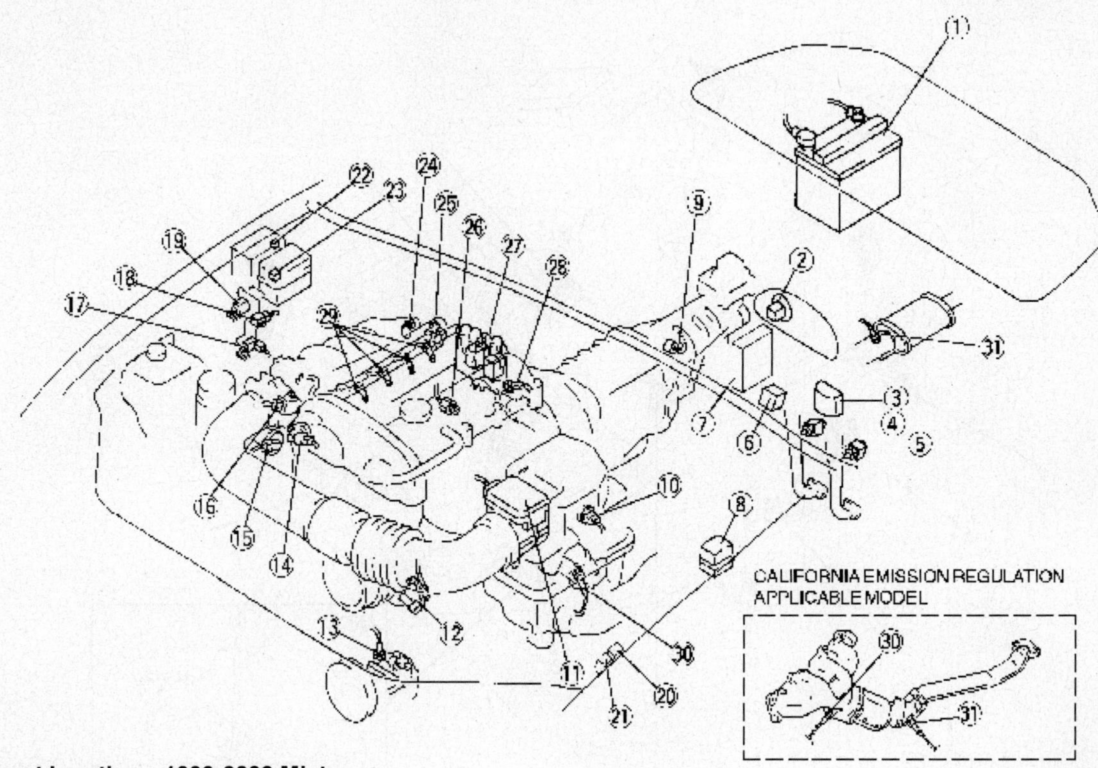

Component Location – 1999-2000 Miata

1	Battery
2	Vehicle speed sensor
3	DLC-2
4	Brake switch
5	Clutch switch (MT)
6	Fuel pump relay
7	PCM
8	DLC
9	Neutral switch (MT)
10	IAT sensor
11	MAF sensor
12	Crankshaft position sensor
13	PSP switch
14	Camshaft position sensor
15	IAC valve
16	TP sensor
17	Purge solenoid valve
18	EGR boost sensor
19	EGR boost sensor solenoid valve
20	Condenser fan relay
21	A/C relay
22	Cooling fan relay
23	Main relay
24	VICS solenoid valve
25	EGR valve
26	Knock sensor
27	Ignition coil
28	ECT sensor
29	Fuel injectors
30	HO2S (Front)
31	HO2S (Rear)

1999–2001 Protege

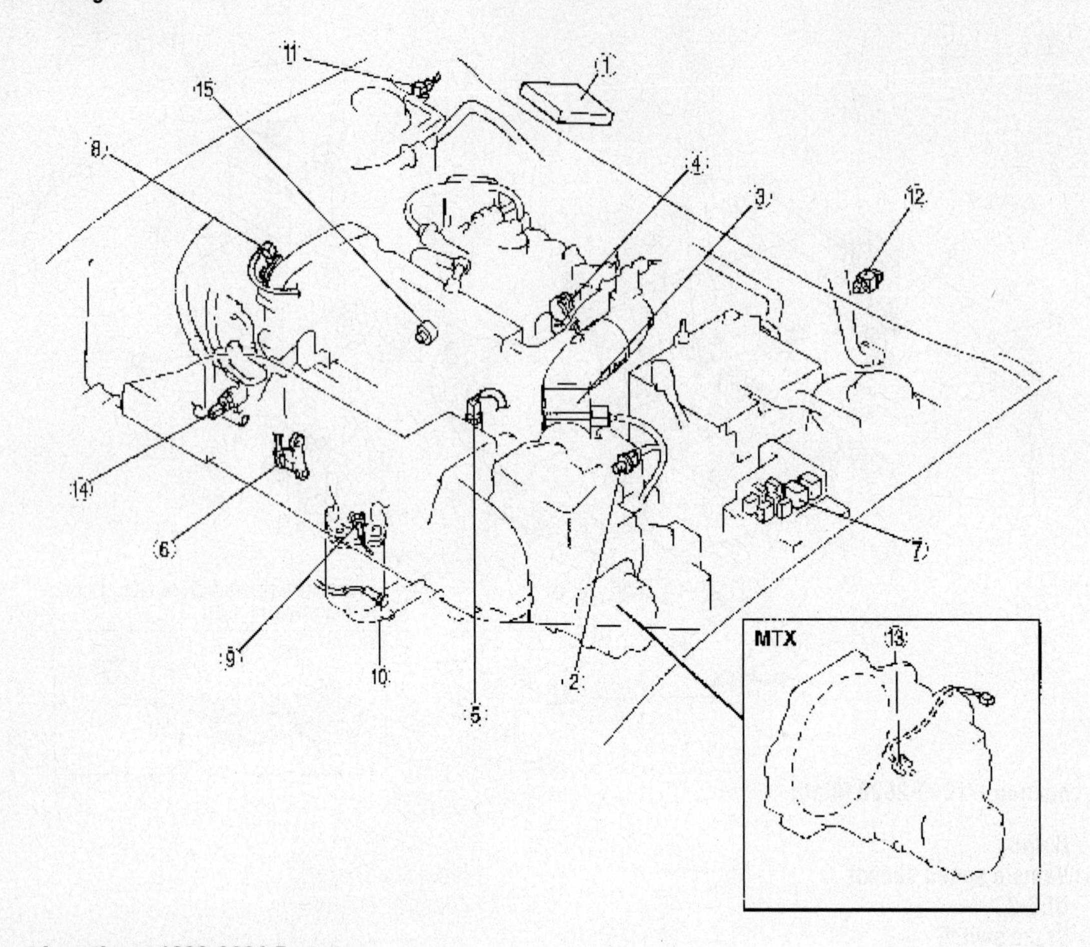

Component Location – 1999-2001 Protege

1	PCM
2	Intake air temperature (IAT) sensor
3	Mass air flow (MAF) sensor
4	Throttle position (TP) sensor
5	Engine coolant temperature (ECT) sensor
6	Crankshaft position (CKP) sensor
7	Main relay
8	Camshaft position (CMP) sensor
9	Heated oxygen sensor (front)
10	Heated oxygen sensor (rear)
11	EGR boost sensor
12	Clutch switch
13	Neutral switch
14	Power steering pressure (PSP) switch
15	Knock sensor

2000 Millenia

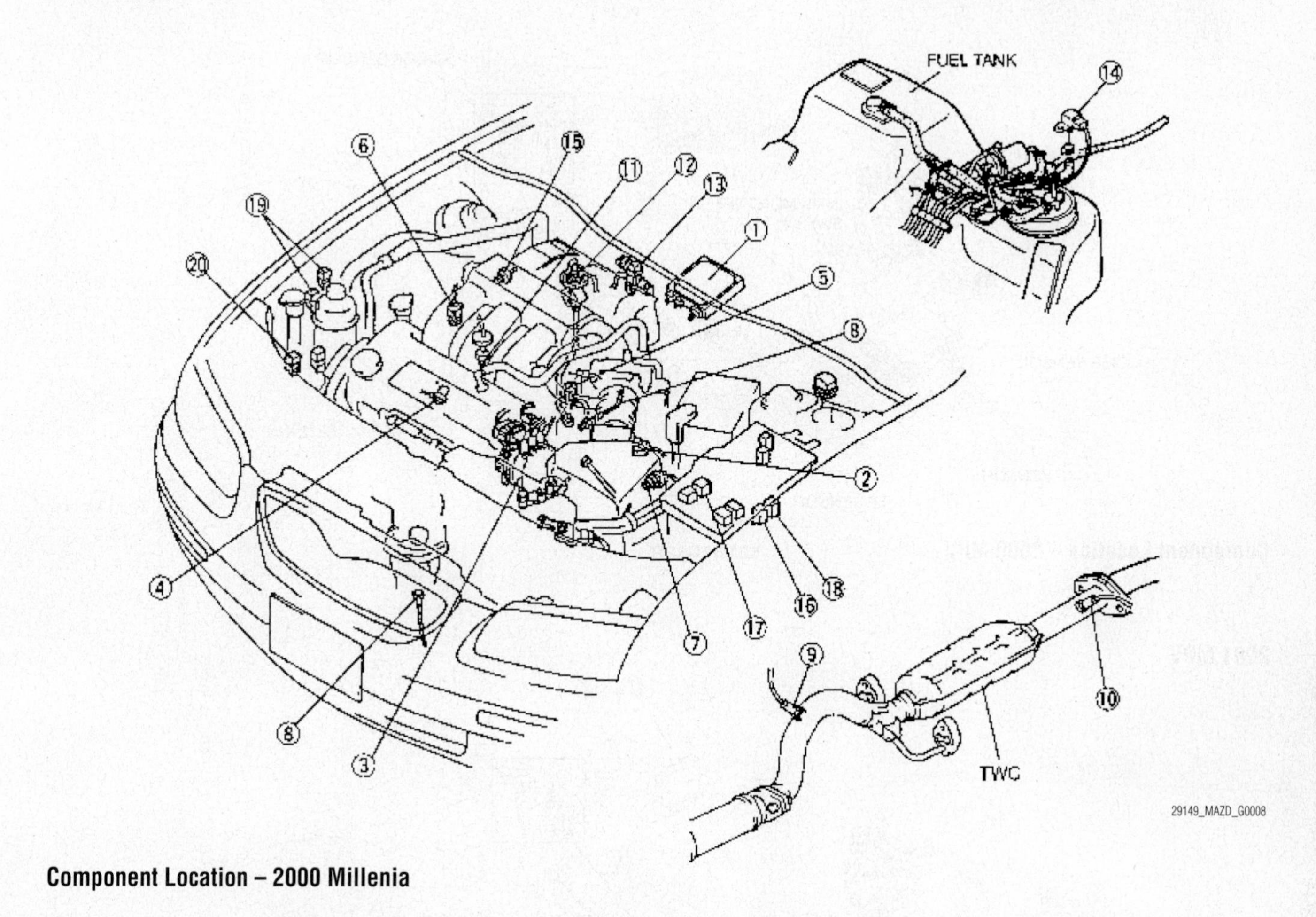

Component Location – 2000 Millenia

1	Powertrain control module (PCM)
2	Mass air flow (MAF) sensor
3	Camshaft position (CMP) sensor
4	Crankshaft position (CKP) sensor
5	Throttle position (TP) sensor
6	Engine coolant temperature (ECT) sensor
7	Intake air temperature (IAT) sensor
8	Heated oxygen sensor (HO2S) (front)
9	Heated oxygen sensor (HO2S) (middle)
10	Heated oxygen sensor (HO2S) (rear)
11	Knock sensor (KS)
12	EGR valve position sensor
13	EGR boost sensor
14	Fuel tank pressure sensor
15	Power steering pressure (PSP) switch
16	Main relay
17	Fuel pump relay
18	Cooling fan relay
19	Condenser fan relay
20	A/C relay

2000 MPV

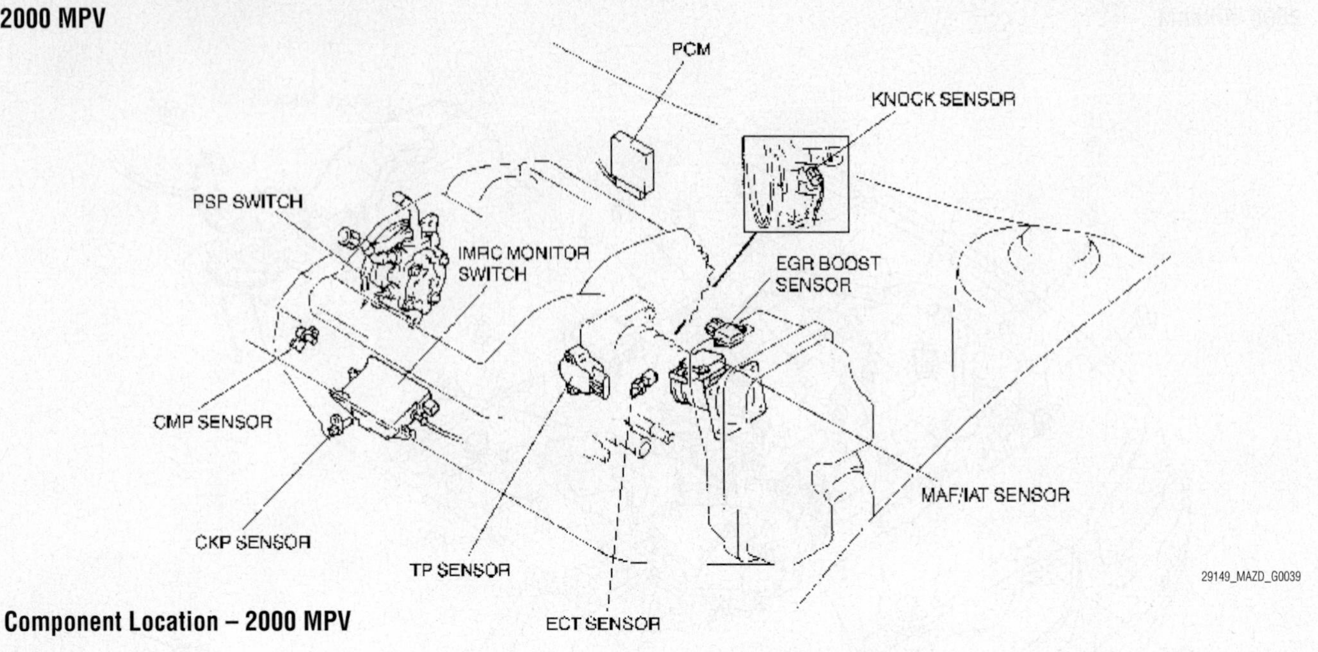

Component Location – 2000 MPV

2001 MPV

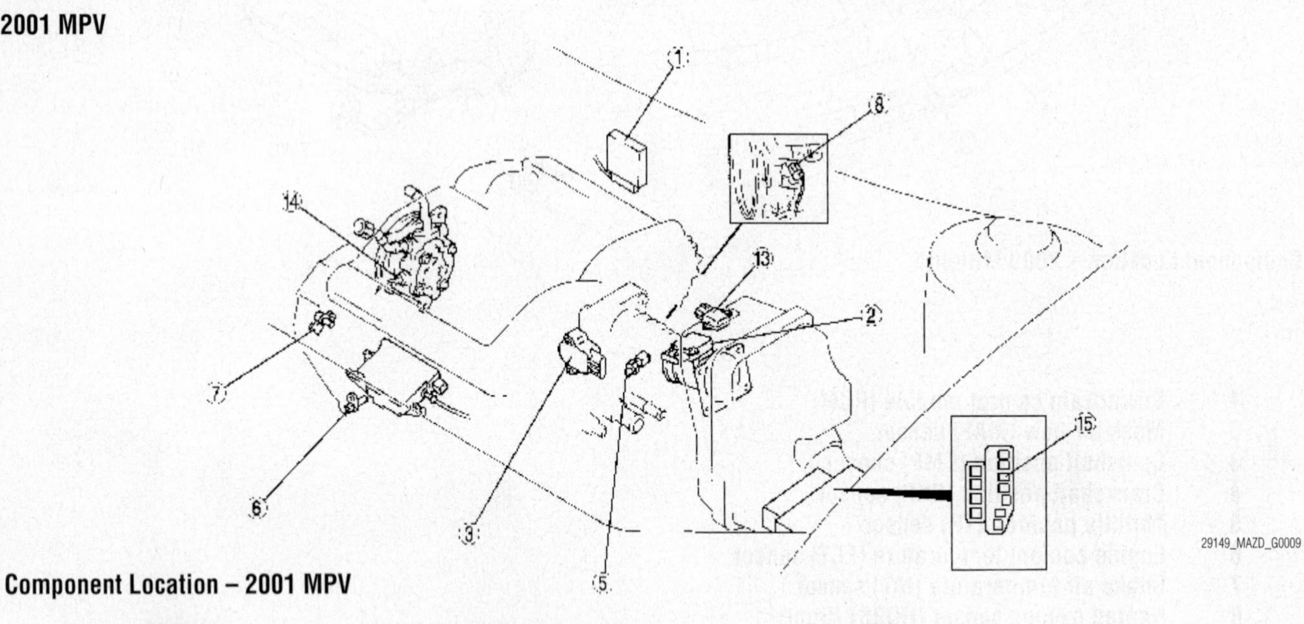

Component Location – 2001 MPV

1. PCM
2. Mass air flow (MAF)/intake air temperature (IAT) sensor
3. Throttle position (TP) sensor
4. Fuel tank pressure sensor
5. Engine coolant temperature (ECT) sensor
6. Crankshaft position (CKP) sensor
7. Camshaft position (CMP) sensor
8. Knock sensor
9. Heated oxygen sensor (HO2S) (front, LH)
10. Heated oxygen sensor (HO2S) (front, RH)
11. Heated oxygen sensor (HO2S) (rear, LH)
12. Heated oxygen sensor (HO2S) (rear, RH)
13. EGR boost sensor
14. Power steering pressure (PSP) switch
15. Main relay

2001–02 Millenia

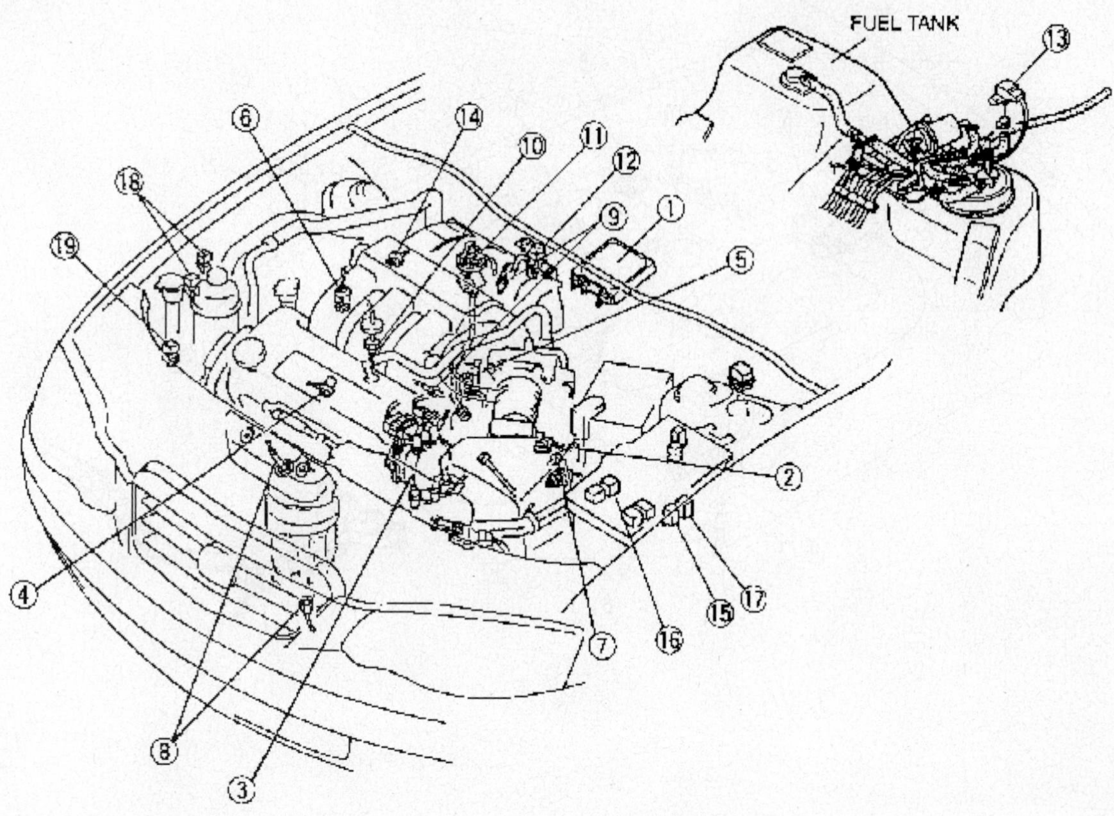

Component Location – 2001-02 Millenia

1 Powertrain control module (PCM)
2 Mass air flow (MAF) sensor
3 Camshaft position (CMP) sensor
4 Crankshaft position (CKP) sensor
5 Throttle position (TP) sensor
6 Engine coolant temperature (ECT) sensor
7 Intake air temperature (IAT) sensor
8 Heated oxygen sensor (HO2S) (LH)
9 Heated oxygen sensor (HO2S) (RH)
10 Knock sensor
11 EGR valve position sensor
12 EGR boost sensor
13 Fuel tank pressure sensor
14 Power steering pressure (PSP) switch
15 Main relay
16 Fuel pump relay
17 Cooling fan relay
18 Condenser fan relay
19 A/C relay

2001–04 Miata

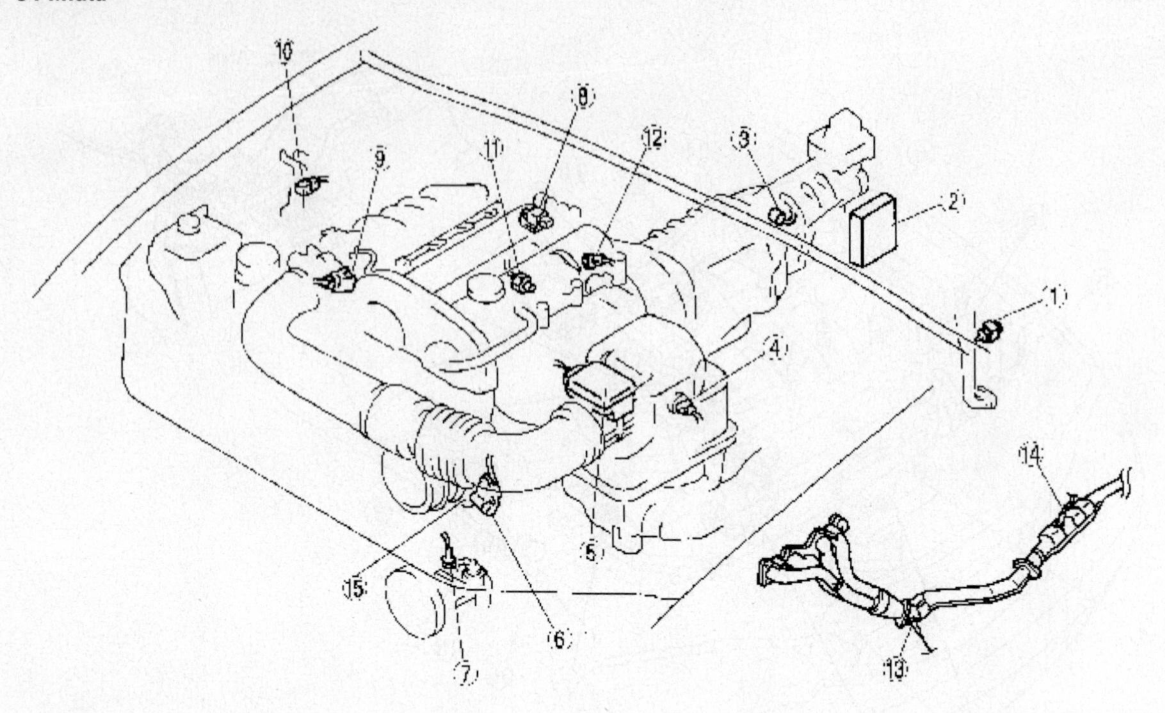

Component Location – 2001-04 Miata

1	Clutch switch (MT)
2	PCM
3	Neutral switch (MT)
4	Intake air temperature (IAT) sensor
5	Mass air flow (MAF) sensor
6	Crankshaft position (CKP) sensor
7	Power steering position (PSP) switch
8	Camshaft position (CMP) sensor
9	Throttle position (TP) sensor
10	EGR boost sensor
11	Knock sensor (KS)
12	Engine coolant temperature (ECT) sensor
13	Heated oxygen sensor (HO2S)(Front)
14	Heated oxygen sensor (HO2S)(Rear)
15	Plate

2002–06 MPV

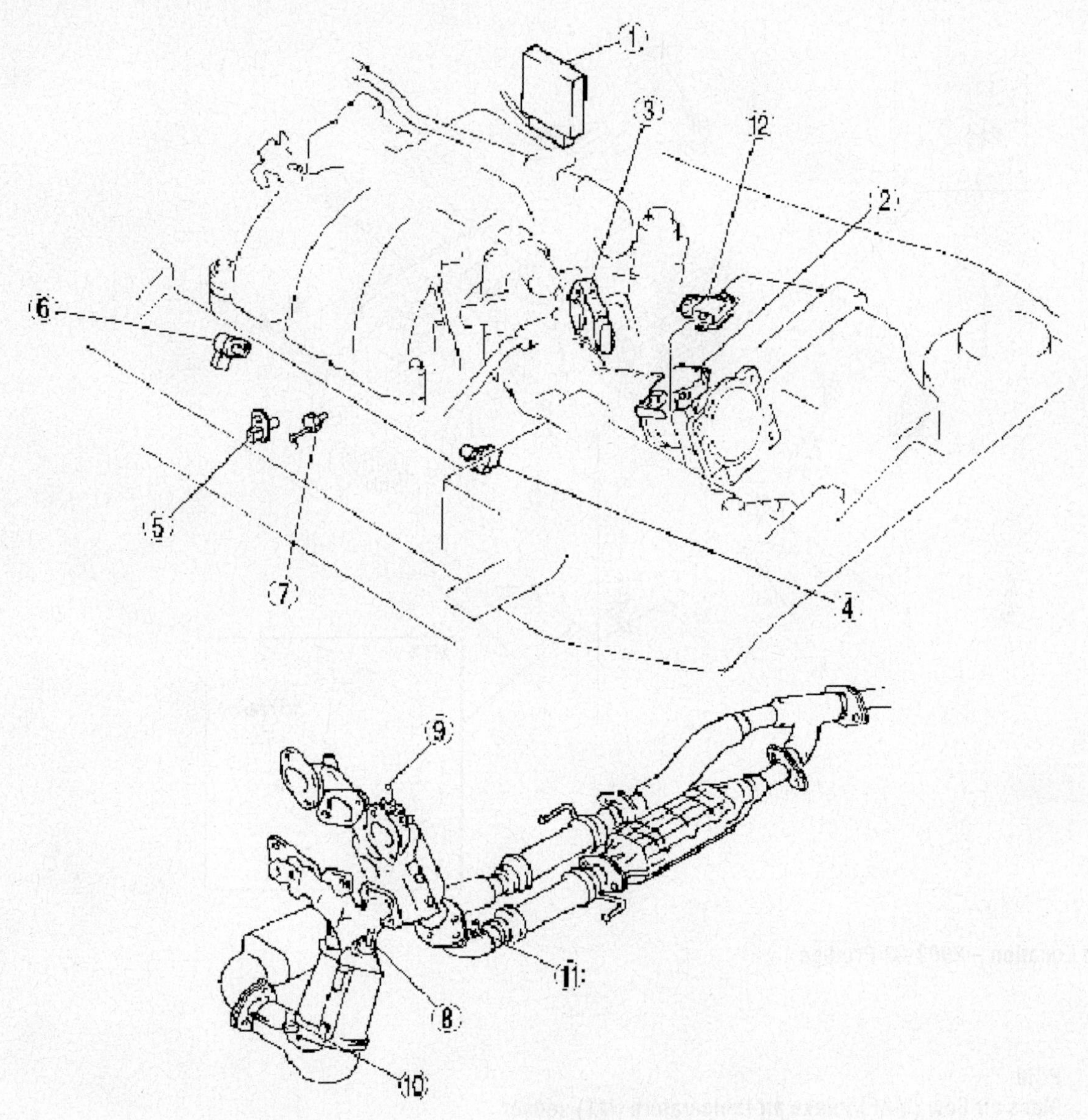

Component Location – 2002-06 MPV

1	PCM
2	Mass air flow (MAF)/Intake air temperature (IAT) sensor
3	Throttle position (TP) sensor
4	Engine coolant temperature (ECT) sensor
5	Crankshaft position (CKP) sensor
6	Camshaft position (CMP) sensor
7	Knock sensor
8	Heated oxygen sensor (HO2S) (front, LH)
9	Heated oxygen sensor (HO2S) (front, RH)
10	Heated oxygen sensor (HO2S) (rear, LH)*
11	Heated oxygen sensor (HO2S) (rear, RH)*
12	EGR boost sensor

2002–03 Protege

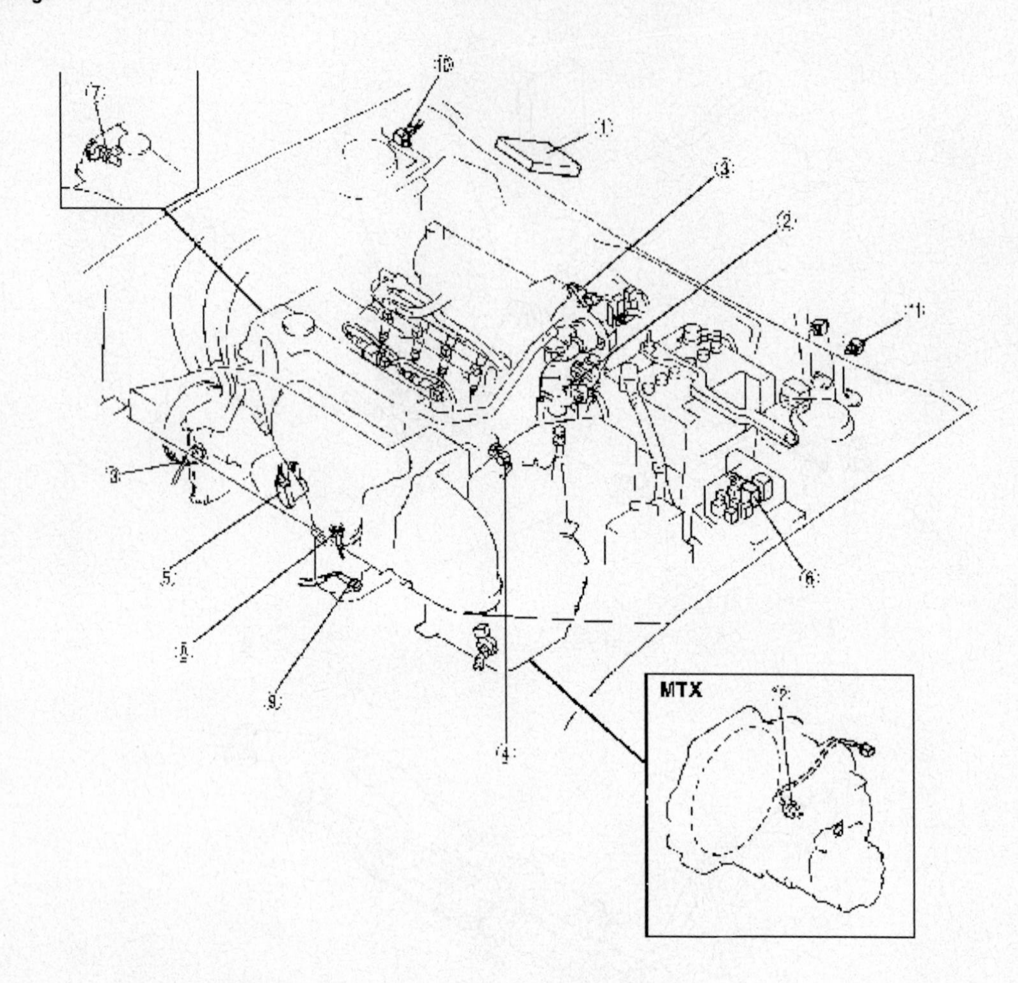

Component Location – 2002-03 Protege

1 PCM
2 Mass air flow (MAF)/intake air temperature (IAT) sensor
3 Throttle position (TP) sensor
4 Engine coolant temperature (ECT) sensor
5 Crankshaft position (CKP) sensor
6 Main relay
7 Camshaft position (CMP) sensor
8 Heated oxygen sensor (front)
9 Heated oxygen sensor (rear)
10 EGR boost sensor
11 Clutch switch
12 Neutral switch
13 Power steering pressure (PSP) switch

2003–06 Mazda 6

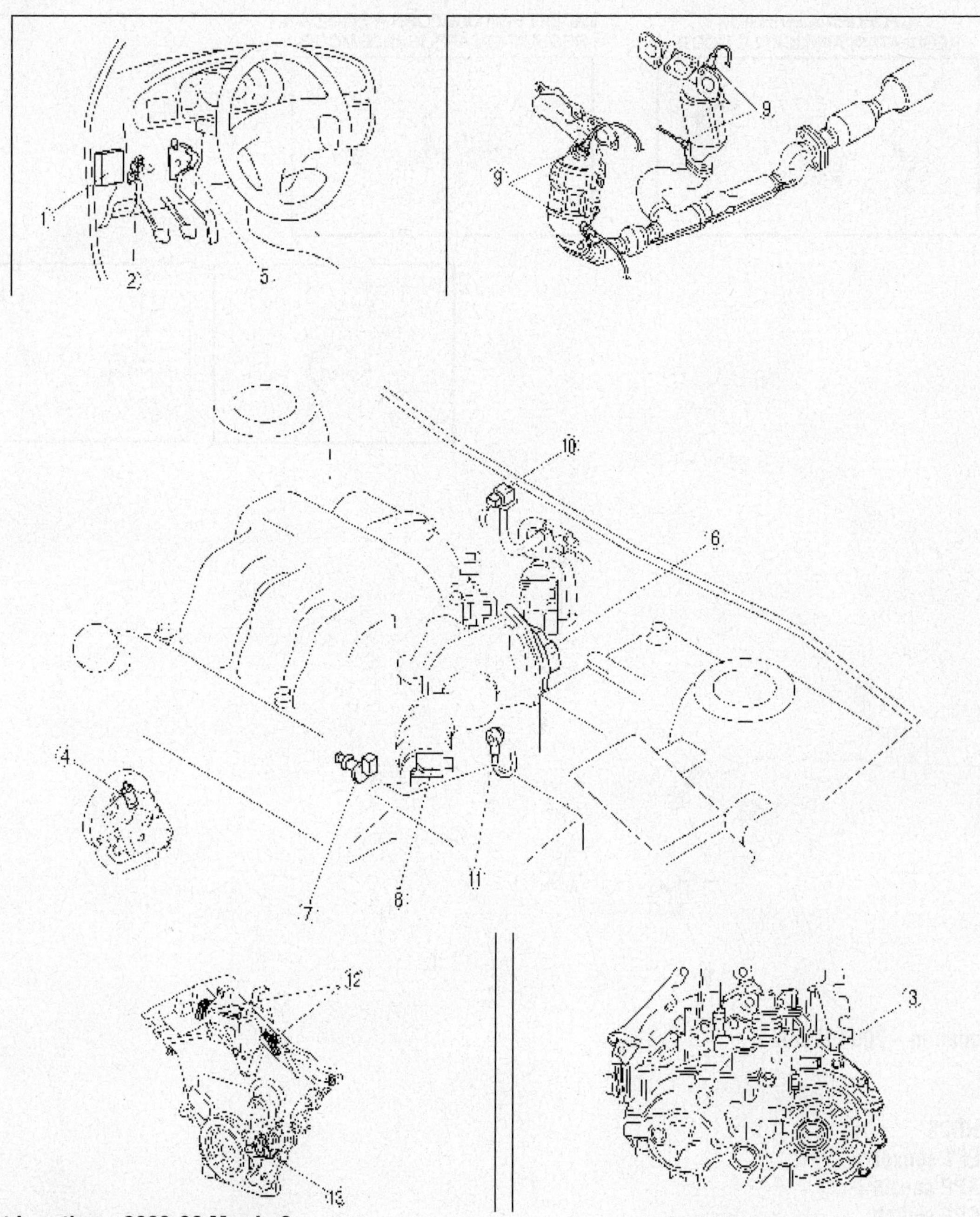

Component Location – 2003-06 Mazda 6

1	PCM
2	Clutch switch (MTX)
3	Neutral switch (MTX)
4	PSP switch
5	APP sensor
6	TP sensor and throttle actuator
7	ECT sensor
8	MAF/IAT sensor
9	HO2S
10	EGR boost sensor
11	KS
12	CMP sensor
13	CKP sensor

2004 Mazda 3

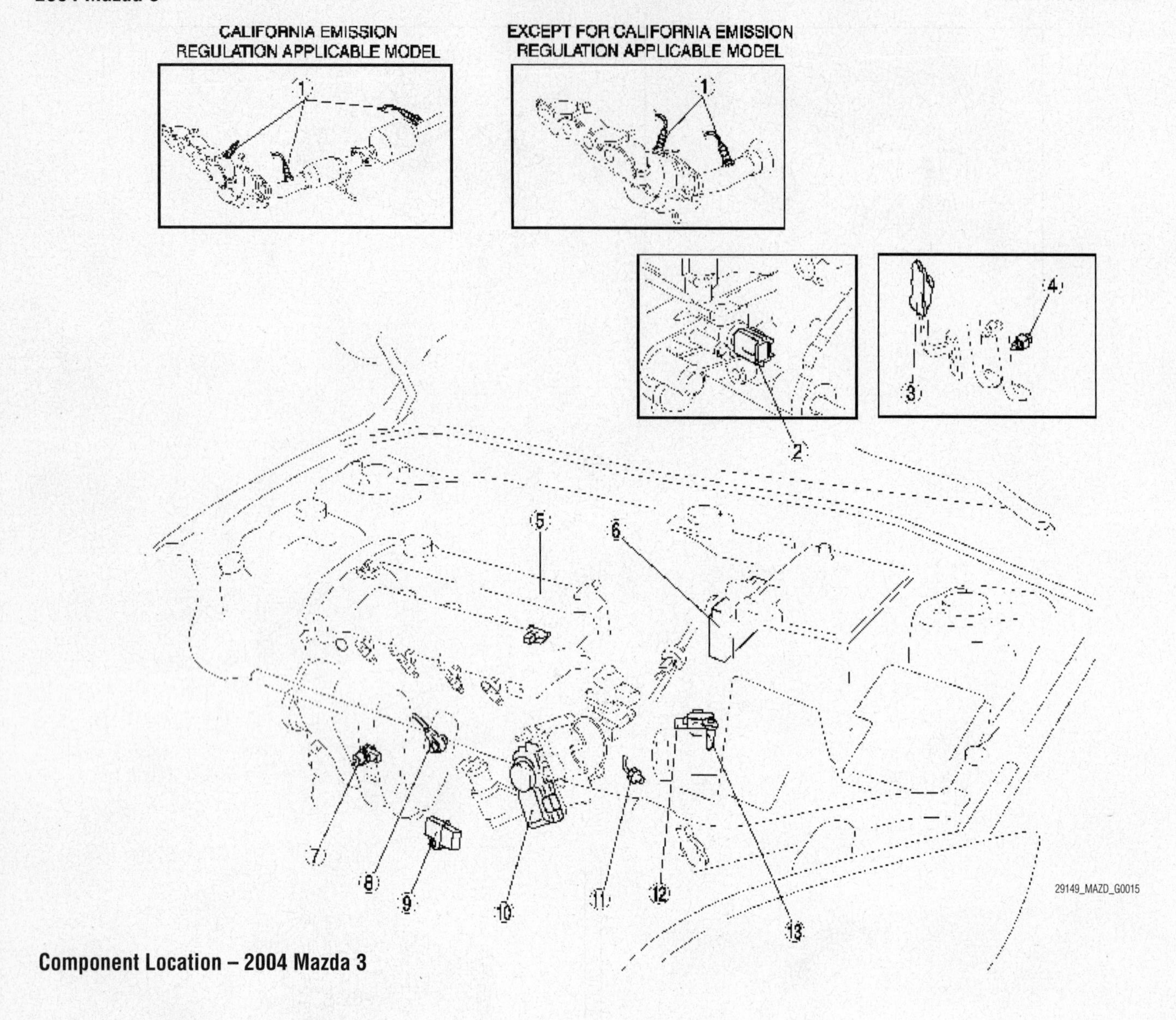

Component Location – 2004 Mazda 3

1	HO2S
2	ECT sensor
3	APP sensor
4	CPP switch
5	CMP sensor
6	PCM
7	CKP sensor
8	KS
9	MAP sensor
10	TP sensor
11	Neutral switch
12	BARO sensor
13	MAF/IAT sensor

2005 Miata

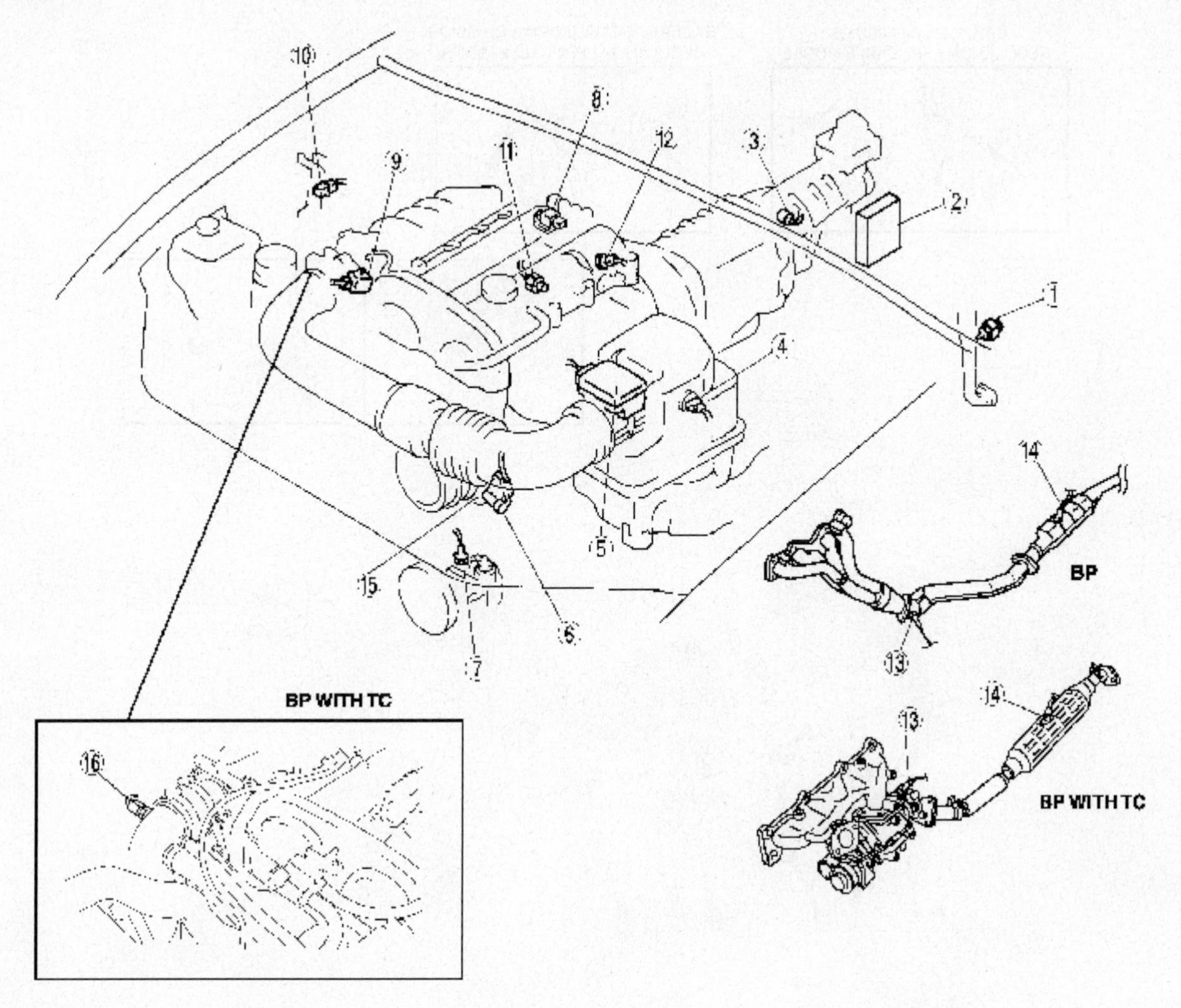

Component Location – 2005 Miata

1	Clutch switch (MT)
2	PCM
3	Neutral switch (MT)
4	Intake air temperature (IAT) sensor (BP) Intake air temperature (IAT) sensor No.1 (BP WITH TC)
5	Mass air flow (MAF) sensor
6	Crankshaft position (CKP) sensor
7	Power steering position (PSP) switch
8	Camshaft position (CMP) sensor
9	Throttle position (TP) sensor
10	EGR boost sensor (BP) Barometric pressure (BARO)/manifold absolute pressure (MAP) sensor (BP WITH TC)
11	Knock sensor (KS)
12	Engine coolant temperature (ECT) sensor
13	Heated oxygen sensor (HO2S) (Front)
14	Heated oxygen sensor (HO2S) (Rear)
15	Plate
16	Intake air temperature (IAT) sensor No.2

2005–06 Mazda 3

CALIFORNIA EMISSION REGULATION APPLICABLE MODEL

EXCEPT FOR CALIFORNIA EMISSION REGULATION APPLICABLE MODEL

29149_MAZD_G0017

Component Location – 2005-06 Mazda 3

1	HO2S
2	ECT sensor
3	APP sensor
4	CPP switch (MTX)
5	CMP sensor
6	PCM (built-in BARO sensor)
7	CKP sensor
8	KS
9	MAP sensor
10	TP sensor
11	Neutral switch (MTX)
12	MAF/IAT sensor

2006 Mazda 5

29149_MAZD_G0018

Component Location – 2006 Mazda 5

1	HO2S
2	ECT sensor
3	APP sensor
4	CPP switch (MTX)
5	CMP sensor
6	PCM
7	CKP sensor
8	KS
9	MAP sensor
10	TP sensor
11	Neutral switch (MTX)
12	MAF/IAT sensor
13	EGR valve

2006 Miata

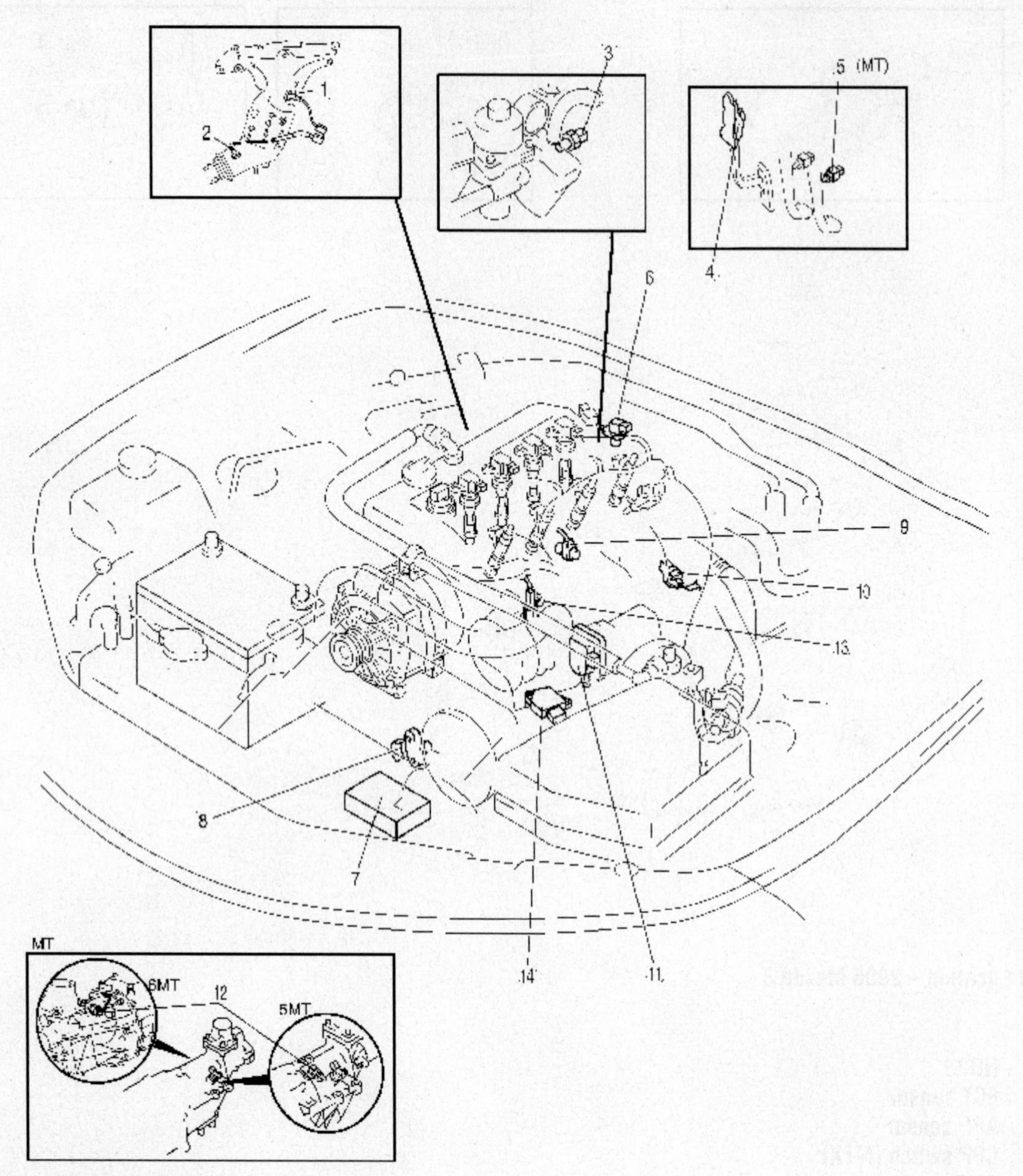

Component Location – 2006 Miata

1	Front HO2S
2	Rear HO2S
3	ECT sensor
4	APP sensor
5	CPP switch
6	CMP sensor
7	PCM (built into BARO sensor)
8	CKP sensor
9	KS
10	MAP sensor
11	TP sensor
12	Neutral switch
13	PSP switch
14	MAF/IAT sensor

2006 RX8

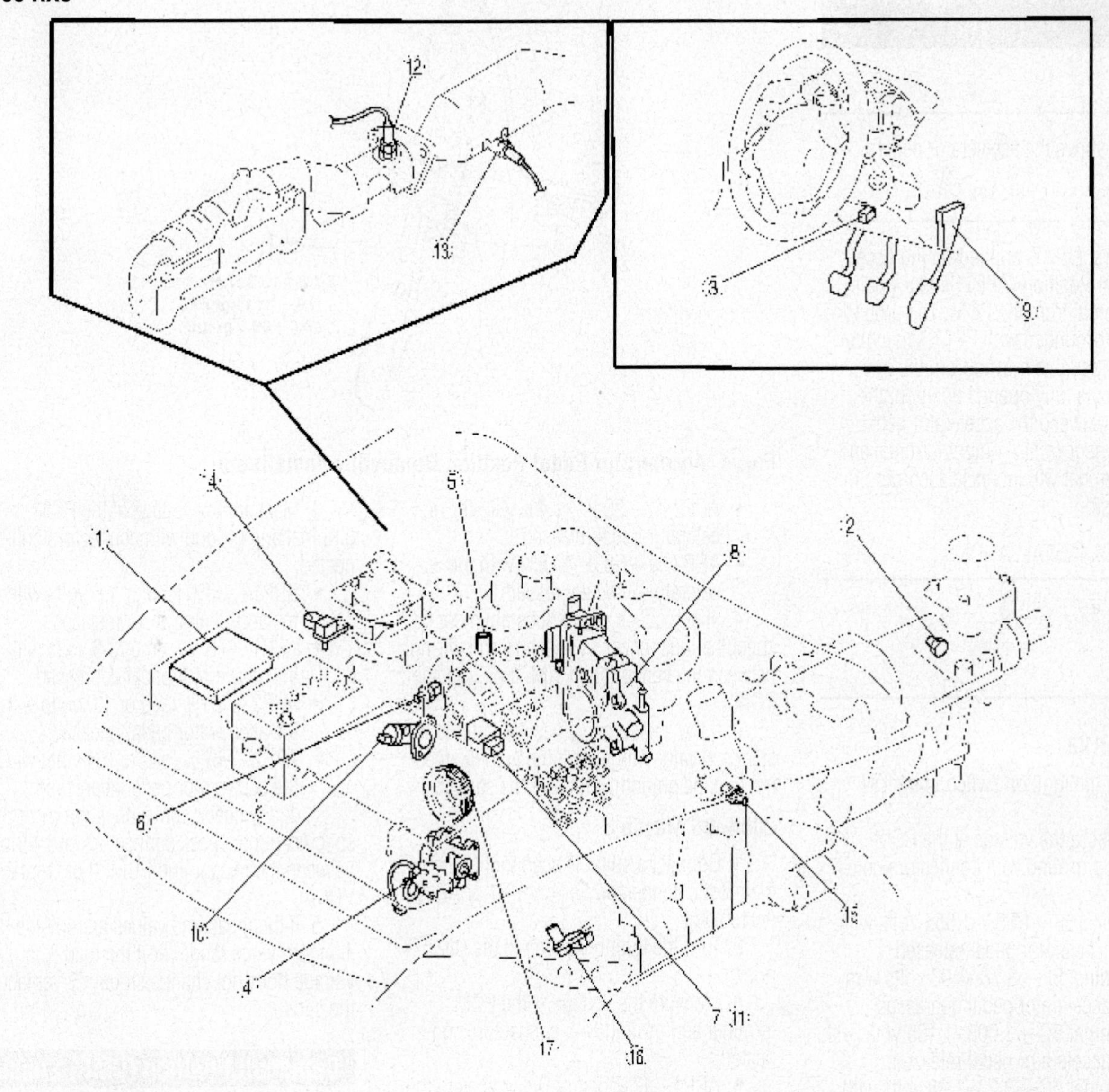

Component Location – 2006 RX8

1	PCM
2	Neutral switch (MT)
3	CPP switch (MT)
4	Metering oil pump switch
5	SSV switch
6	ECT sensor
7	IAT sensor
8	TP sensor
9	APP sensor
10	APV position sensor
11	MAF sensor.
12	Front HO2S
13	Rear HO2S
14	BARO sensor
15	KS
16	Eccentric shaft position sensor
17	Eccentric shaft position plate

Accelerator Pedal Position (APP) Sensor

LOCATION

Located above the accelerator pedal.

OPERATION

A power supply is applied on the Accelerator Pedal Position (APP) Sensor (main) power terminal from the PCM. The ground terminal is grounded with PCM. When the accelerator pedal is moved from the idle position to the fully opened position, the resistance between the accelerator pedal position sensor (main) output terminal and ground terminal will increase according to the depression.

REMOVAL & INSTALLATION

See Figure 1.

TESTING

2004–06 RX8

1. Turn the ignition switch to the ON position.
2. Measure the voltage at the PCM terminal and ground with connectors connected:
   - Terminal 5F—1.555–1.655 volts with the accelerator pedal released
   - Terminal 5F—3.78–3.93 volts with the accelerator pedal depressed
   - Terminal 5C—1.005–1.105 volts with the accelerator pedal released
   - Terminal 5C—3.23–3.38 volts with the accelerator pedal depressed
3. If any check above does not meet the specifications, check connectors and wiring between the sensor and PCM. If ok, replace PCM.
4. If the measured values are outside the standard value range, or if the voltage does not change smoothly, replace the sensor.

2003–06 Mazda 6

1. Connect a suitable scan tool to the diagnostic connector. Select APP1 or APP2 percentage.
2. Turn the ignition switch to the ON position.
3. Measure the voltage at the PCM terminal and ground with connectors connected:
   - APP1 %—31.0–32.4% with the accelerator pedal released
   - APP1 %—69.8–81.8% with the accelerator pedal depressed

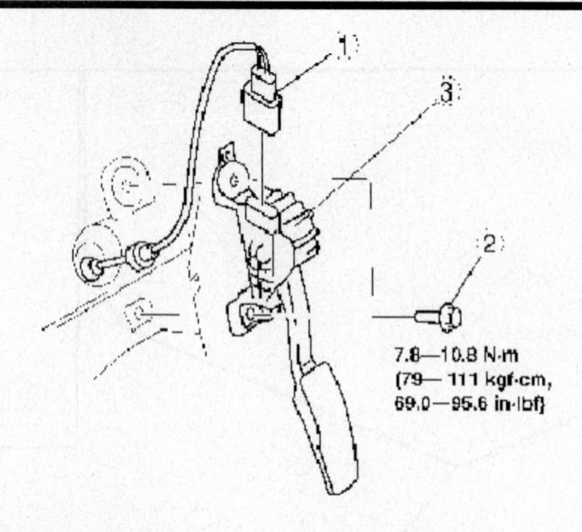

Fig. 1 Accelerator Pedal Position Removal & Installation

   - APP2 %—20.2–21.4% with the accelerator pedal released
   - APP2 %—58.8–70.8% with the accelerator pedal depressed
4. If any check above does not meet the specifications, check connectors and wiring between the sensor and PCM. If ok, replace PCM.
5. If the measured values are outside the standard value range, or if the load % does not change smoothly, replace the sensor.

2004–06 Mazda 3

1. Connect a suitable scan tool to the diagnostic connector. Select APP1 or APP2 percentage.
2. Turn the ignition switch to the ON position.
3. Measure the voltage at the PCM terminal and ground with connectors connected:
   - APP1—8% load or 0.4 volts with the accelerator pedal released
   - APP1—60% load or 3.0 volts with the accelerator pedal depressed
   - APP2 —8% load or 0.4 volts with the accelerator pedal released
   - APP2 —60% load or 3.0 volts with the accelerator pedal depressed
4. If any check above does not meet the specifications, check connectors and wiring between the sensor and PCM. If ok, replace PCM.
5. If the measured values are outside the standard value range, or if the load % and/or voltage does not change smoothly, replace the sensor.

2006 Miata

1. Connect a suitable scan tool to the diagnostic connector. Select APP1 or APP2 percentage.
2. Turn the ignition switch to the ON position.
3. Measure the voltage at the PCM terminal and ground with connectors connected:
   - APP1—32% load or 1.6 volts with the accelerator pedal released
   - APP1—78% load or 3.9 volts with the accelerator pedal depressed
   - APP2 —21% load or 1.0 volts with the accelerator pedal released
   - APP2 —67% load or 3.4 volts with the accelerator pedal depressed
4. If any check above does not meet the specifications, check connectors and wiring between the sensor and PCM. If ok, replace PCM.
5. If the measured values are outside the standard value range, or if the load % and/or voltage does not change smoothly, replace the sensor.

Camshaft Position (CMP) Sensor

LOCATION

Refer to Component Location illustration above.

OPERATION

The CMP sensor provides the camshaft position information, called the CMP signal, which is used by the Powertrain Control Module (PCM) for fuel synchronization.

REMOVAL & INSTALLATION

1995–1998 Models

➡ **The camshaft position sensor requires a scan tool in order to properly test the component.**

1. Remove the distributor from the engine.

2. Detach the 3-wire connector from the distributor.

3. Enable the scan tool to record the camshaft position sensor output.

4. Rotate the distributor drive by hand and verify that there were four 5 volt pulses within one revolution of the distributor drive.

5. If not as specified, replace the distributor assembly.

1999–2003 Models

See Figure 2.

1. Disconnect the battery negative cable.
2. Disconnect the CMP sensor connector.
3. Remove the timing belt cover and timing belt.
4. Remove the cylinder head cover.
5. Hold the camshaft using a wrench on the cast hexagon and loosen the camshaft pulley lock bolt.
6. Remove the camshaft pulley lock bolt.
7. Remove the camshaft pulley.
8. Remove the camshaft position sensor.
9. Install in the reverse order of removal. Tightening torque:
   Timing pulley bolt—91–103 ft lbs
   Sensor bolt— 69.5–95.4 ft lbs

2004–2006 Models except Mazda 3

See Figure 3.

1. Disconnect the negative battery cable.
2. Disconnect the CMP sensor connector.
3. Remove the CMP sensor installation bolt.
4. Remove the CMP sensor.
5. Make sure that the CMP sensor is free of any metallic shavings or particles.

➡ **If metallic shavings or particles are found on the sensor, clean them off.**

6. Install the CMP sensor in the reverse order of removal. Tightening torque is 69.5–95.4 in lbs

2004–06 Mazda 3

See Figure 4.

1. Remove the battery cover.
2. Disconnect the negative battery cable.
3. Remove the plug hole cover.
4. Disconnect the CMP sensor connector.
5. Remove the CMP sensor installation bolt.
6. Remove the CMP sensor from the cylinder head cover.
7. Install in the reverse order of removal. Tightening torque 49–66 in lbs

TESTING

1995–97 Models

1. Connect a scan tool to the diagnostic link.

2. Detach the fuel injector wire harness connectors. This will ensure that the engine does not try to start and run while performing the test.

3. Remove the distributor from the engine.

4. Detach the 3-wire connector from the distributor.

5. Enable the scan tool to record the camshaft position sensor output.

6. Rotate the distributor drive by hand and verify that there were four 5 volt pulses within one revolution of the distributor drive.

7. If not as specified, replace the distributor assembly.

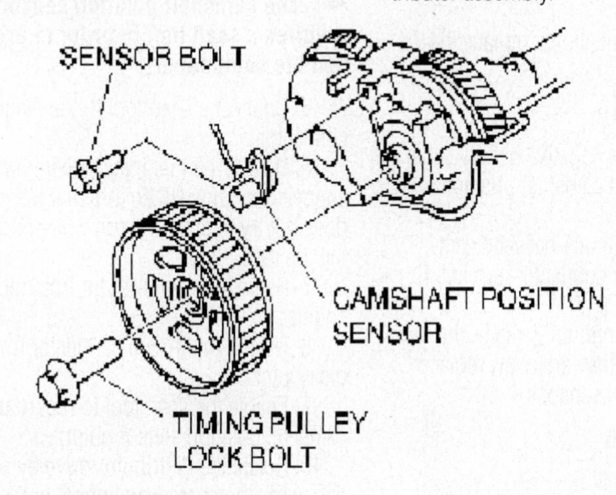

Fig. 2 Camshaft Position Sensor Removal & Installation – 1999-02 626 V6

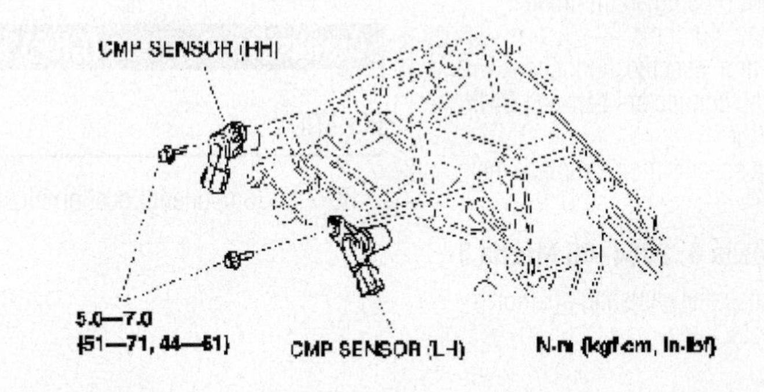

Fig. 3 Camshaft Position Sensor Removal & Installation – 2003-06 Mazda 6 V6

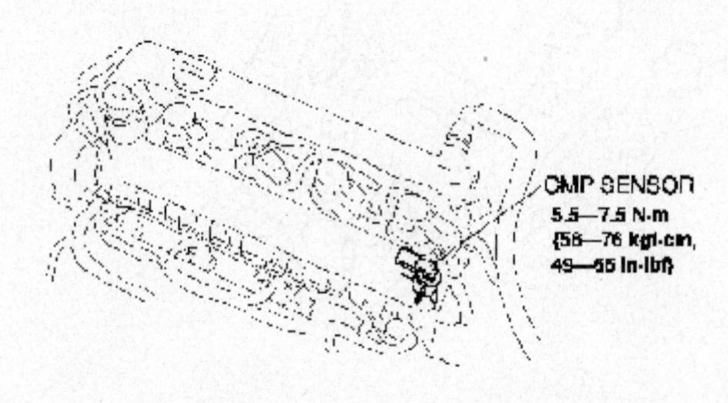

Fig. 4 Camshaft Position Sensor Removal & Installation – 2004-06 Mazda 3

1998–2002 626 L4, 1998–2003 Protege

See Figure 5.

1. Disconnect the camshaft position sensor connector.
2. Measure the resistance between terminals A and B using an ohmmeter. Standard value: 950–1250 ohms
3. If not within specification, check wiring harness and connectors between CMP sensor and PCM.
4. If within specifications, replace the sensor.

1998–2002 Millenia

1. Disconnect the negative battery cable.
2. Disconnect the camshaft position sensor connector.
3. Verify the continuity between camshaft position sensor connector terminal B and C.
4. If there is no continuity, repair the related harnesses. If they are okay, replace the camshaft position sensor.

1998–2002 626 V6

See Figure 6.

1. Disconnect the camshaft position sensor connector.
2. Measure the resistance between terminals A and B using an ohmmeter. Standard value: 550 ohms
3. If not within specification, check wiring harness and connectors between CMP sensor and PCM.
4. If within specifications, replace the sensor.

2003–06 Mazda 6, 2004–06 Mazda 3

1. Disconnect the camshaft position sensor connector.
2. Measure the resistance between terminals A and B using an ohmmeter. Standard value:
   - 4 cylinder—400–550 ohms
   - 6 cylinder—467–571 ohms
3. If not within specification, check wiring harness and connectors between CMP sensor and PCM.
4. If within specifications, replace the sensor.

➡ **The camshaft position sensor requires a scan tool in order to properly test the component.**

5. Connect a scan tool to the diagnostic connector.
6. Detach the fuel injector wire harness connectors. This will ensure that the engine does not try to start and run while performing the test.
7. Remove the distributor from the engine.
8. Detach the 3-wire connector from the distributor.
9. Enable the scan tool to record the camshaft position sensor output.
10. Rotate the distributor drive by hand and verify that there were four 5-volt pulses within one revolution of the distributor drive.
11. If not as specified, replace the distributor assembly.

Crankshaft Position (CKP) Sensor

LOCATION

Refer to Component Location illustration above.

OPERATION

The crankshaft position sensor reads the rotation of the crankshaft, which the computer translates to determine engine-timing sequences. The computer uses this reading, in conjunction with the camshaft position sensor, to determine proper fuel injection and ignition timing.

REMOVAL & INSTALLATION

1995–2002 Models

1. Disconnect the negative battery cable.
2. Remove the necessary accessory drive belts to access the sensor.
3. If necessary, remove the engine oil dipstick tube.
4. Remove the sensor-attaching bolt, and then withdraw the sensor from the engine.
5. Installation is the reverse of the removal procedure.

2003–2006 Models

REMOVAL

See Figures 7, 8 and 9.

1. Remove the right front wheel.
2. Remove the splash shield.
3. Disconnect the CKP sensor connector.
4. Remove the installation bolts to remove the CKP sensor.

INSTALLATION

➡ **Perform the following procedure so that piston No.1 is at the top dead center.**

1. Remove the right side driveshaft.

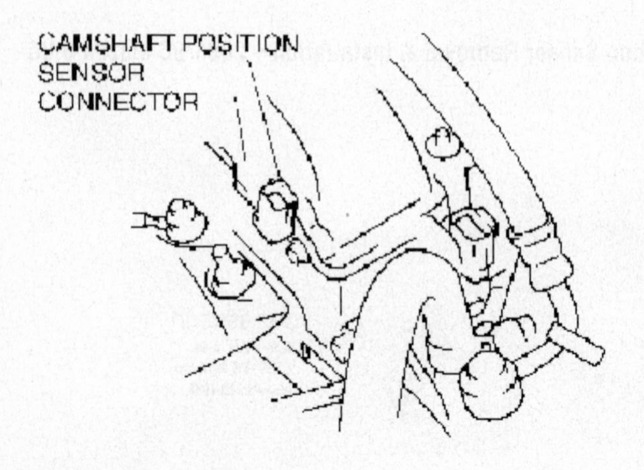

Fig. 5 Camshaft Position Sensor Connector – 1998-2002 626 L4

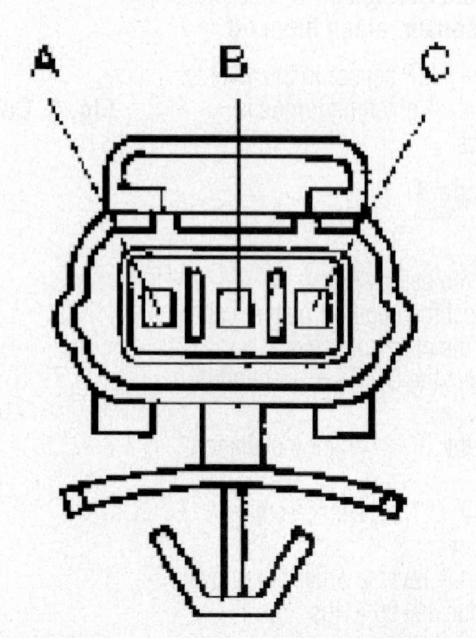

Fig. 6 Camshaft Position Sensor Connector – 1998-02 626 V6

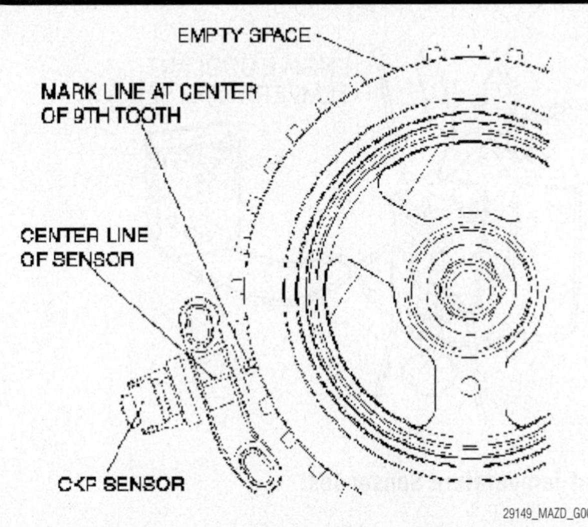

29149_MAZD_G0027

Fig. 7 Crankshaft Position Sensor Removal & Installation – 2003-06 Mazda 6 L4

29149_MAZD_G0028

Fig. 8 Crankshaft Position Sensor TDC Locator – 2003-06 Mazda 6 L4

2. Remove the cylinder block lower blind plug and install the special service tool or equivalent.

3. Turn the crankshaft pulley to the clockwise until it stops.

4. Using a straight edge, draw a straight line directly in the center of the ninth tooth of the crankshaft pulley pulse wheel (counting counterclockwise from the empty space).

➡ **If the line is not accurately drawn, ignition timing, fuel injection and other engine control systems will be adversely affected. Drawn the straight line carefully using a straight edge.**

5. Align the centerline of the crankshaft position sensor and the line drawn, then install the sensor.

6. Install the CKP sensor fitting bolts. Tightening torque 4.1–5.5 ft lbs

7. Remove the special service tool or equivalent then install the cylinder block lower blind plug. Tightening torque 15 ft lbs

8. Install the right side driveshaft.

9. Install the splash shield.

10. Install the right front wheel

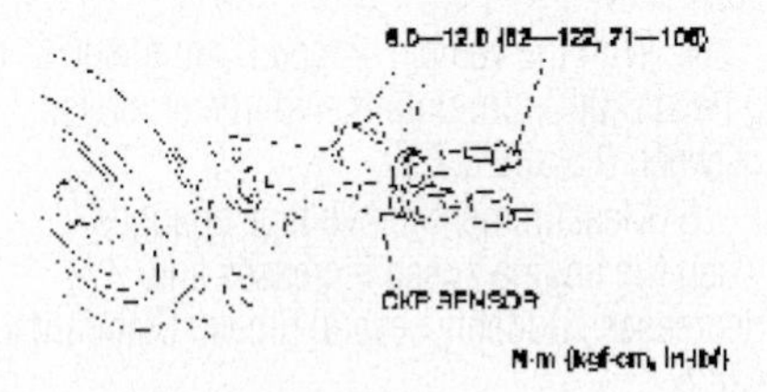

29149_MAZD_G0029

Fig. 9 Crankshaft Position Sensor Removal & Installation – 2003-06 Mazda 6 V6

TESTING

1995–2002 Models except 1995–1998 MPV Models

1. Disconnect the crankshaft position sensor wire harness plug.
2. Connect an ohmmeter to the sensor terminals A and B and measure the resistance.
3. The reading should be 520–580 ohms at 68°F.
4. If not as specified, replace the sensor.
5. Measure the air gap of the sensor between the crankshaft pulley and the sensor.
6. Proper air gap should be 0.040–0.080 in.
7. If not as specified, inspect the crankshaft pulley and/or replace the sensor.

1995 MPV Models

1. Disconnect the negative battery cable.
2. Remove the distributor assembly.
3. Plug in the distributor connector.
4. Unplug the fuel injector connector.
5. Turn the ignition switch to the ON position.
6. Connect the Engine Signal Monitor and Adapter Harness to the engine control module as illustrated.
7. Set the Engine Signal Monitor.
8. Turn the distributor drive by hand and measure the output voltage.
9. If not as specified, replace the distributor assembly.

1996–98 MPV Models

1. Disconnect the negative battery cable.
2. Unplug the crankshaft position sensor wiring harness connector.
3. Using an ohmmeter, measure the resistance between terminals A and B.
4. The measurement should read 950–1250 ohms at 68°F (20°C).
5. If not as specified, replace the crankshaft position sensor.

2003–2006 Models Mazda 6, 2004 Mazda 3

See Figure 10.

1. Disconnect the CKP position sensor connector.
2. Measure the resistance between terminals A and B using an ohmmeter. Standard value:
   - 4 cylinder—400–550 ohms
   - 6 cylinder—315–385 ohms
3. If not within specification, check wiring harness and connectors between CKP sensor and PCM.
4. If within specifications, replace the sensor.

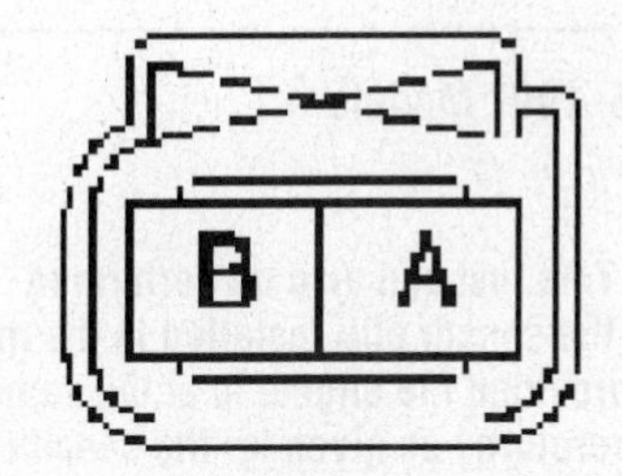

29149_MAZD_G0030

Fig. 10 Crankshaft Position Sensor Connector – 2003-06 Mazda 6 V6

Engine Coolant Temperature (ECT) Sensor

LOCATION

Refer to Component Location illustration above.

OPERATION

The coolant temperature sensor is a variable resistor that is influenced by temperature. Which means, as the temperature changes, so does the resistance inside the sensor. The computer reads this change in resistance to determine the operating temperature of the engine. It uses this reading to determine proper air/fuel mixture and ignition timing settings.

REMOVAL & INSTALLATION

1. Partially drain the engine cooling system until the coolant level is below the ECT sensor-mounting hole.
2. Disconnect the negative battery cable.
3. Detach the wiring harness connector from the ECT sensor.
4. Using an open-end wrench, remove the coolant temperature sensor from the intake manifold or thermostat housing.

To install:

5. Thread the sensor into the intake manifold, or thermostat housing, by hand, then tighten it securely.
6. Connect the negative battery cable.
7. Refill the engine cooling system.
8. Start the engine, check for coolant leaks and top off the cooling system.

TESTING

1995–1997 Models

See Figure 11.

➡ **This test can also be performed with the sensor still installed in the motor, provided the engine is at the same temperatures as given for the specifications.**

1. Remove the ECT.
2. Place the sensor in water with a thermometer and heat the water gradually.
3. Measure the resistance of the sensor and compare to the following values:
   - -4°F (-20°C)—14.6–17.8 kohms
   - 68°F (20°C)—2.2–2.7 kohms
   - 104°F (40°C)—1.0–1.3 kohms
   - 140°F (60°C)—0.50–0.65 kohms
   - 176°F (80°C)—0.29–0.35 kohms

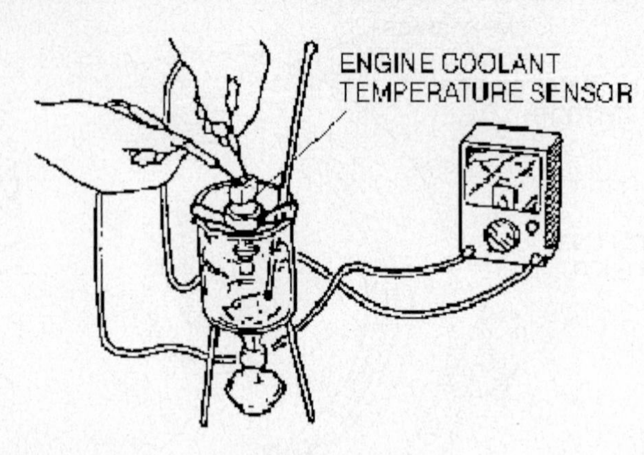

Fig. 11 Engine Coolant Temperature Sensor Test

4. If not as specified, replace the ECT.
5. If within specification, check wiring harness and connectors between ECT and PCM.

1998–2006 Models

1. Remove the ECT.
2. Place the sensor in water with a thermometer and heat the water gradually.
3. Measure the resistance of the sensor and compare to the following values:
   - 68°F (20°C)—35.5–39.2 kohms
   - 140°F (60°C)—7.18–7.92 kohms
4. If not as specified, replace the ECT.
5. If within specification, check wiring harness and connectors between ECT and PCM.

Heated Oxygen (HO2S) Sensor

LOCATION

Refer to Component Location illustration above.

OPERATION

The oxygen sensor supplies the computer with a signal that indicates a rich or lean condition during engine operation. The input information assists the computer in determining the proper air/fuel ratio. A low voltage signal from the sensor indicates too much oxygen in the exhaust (lean condition) and, conversely, a high voltage signal indicates too little oxygen in the exhaust (rich condition).

REMOVAL & INSTALLATION

**** CAUTION**

The temperature of the exhaust system is extremely high after the engine has been run. To prevent personal injury, allow the exhaust system to cool completely before removing sensor from the exhaust system.

1. Disconnect the negative battery cable.
2. Raise and safely support the vehicle on jack stands.
3. Disconnect the HO2S from the engine control sensor wiring.

➡ **If excessive force is needed to remove the sensors lubricate the sensor with penetrating oil prior to removal.**

4. Remove the sensors with a sensor removal tool, such as Ford Tool T94P-9472-A or equivalent.

To install:

5. Install the sensor in the mounting boss, and then tighten it to 27–33 ft. lbs.
6. Connect the sensor electrical wiring connector to the engine wiring harness.
7. Lower the vehicle.
8. Connect the negative battery cable.

TESTING

OXYGEN SENSOR CIRCUIT

See Figures 12 through 17.

1. Warm up the engine and run it at idle.
2. Disconnect the HO2S connector.
3. Connect the voltmeter positive test lead terminal A and negative test lead to terminal B.
4. With the vehicle stopped, run the engine at 3,000 rpm until the voltmeter moves between 0.5 and 0.7 V.
5. Measurement the voltage changes when the engine speed increases and decreases suddenly several times. Standard value:
   a. Acceleration—0.5–1.0 volt
   b. Deceleration—0–0.5 volt
6. If not as specified, replace the HO2S.
7. If within specification, check wiring harness and connectors between HO2S and PCM.

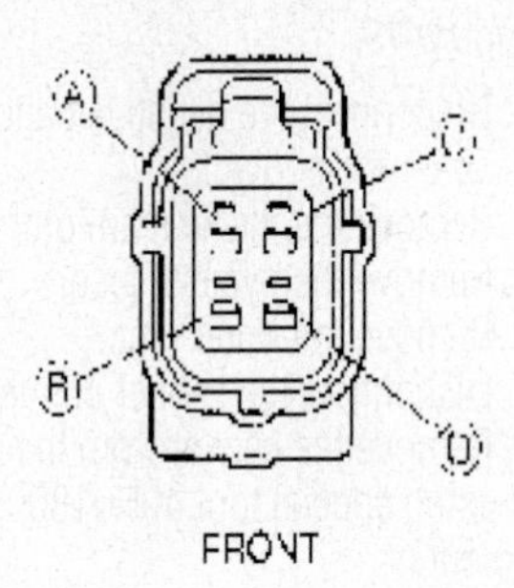

29149_MAZD_G0032

Fig. 12 Front Heated Oxygen Sensor Connector – 1995-98 Millenia

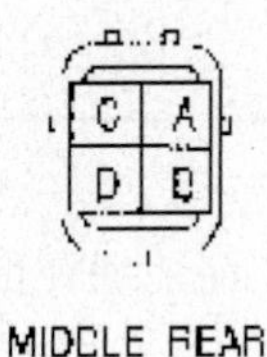

29149_MAZD_G0033

Fig. 13 Rear Heated Oxygen Sensor Connector – 1995-98 Millenia

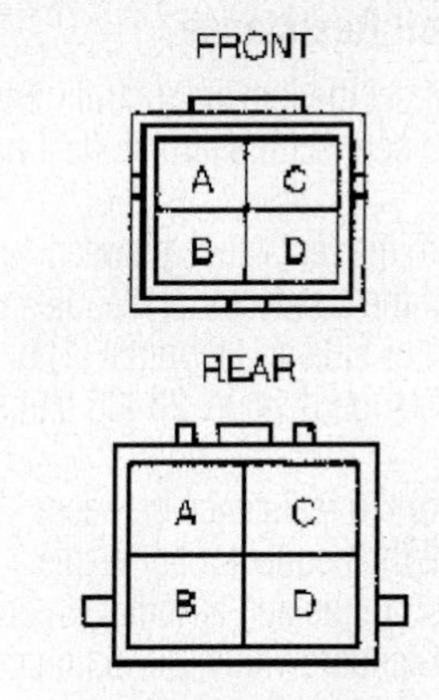

29149_MAZD_G0034

Fig. 14 Front Heated Oxygen Sensor Connector – 1995-2001 626 L4

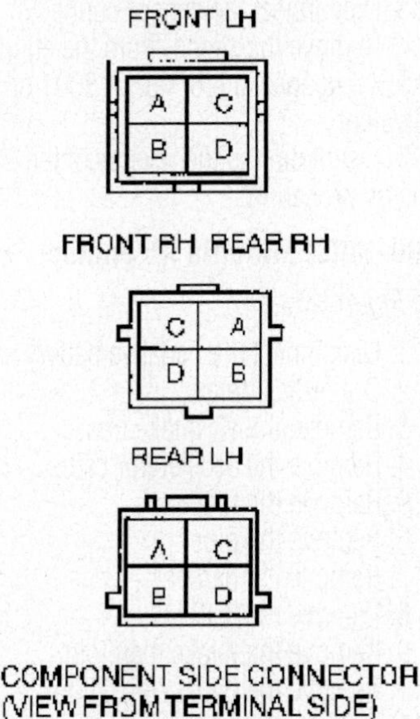

29149_MAZD_G0035

Fig. 15 Rear Heated Oxygen Sensor Connector – 1995-2001 626 V6

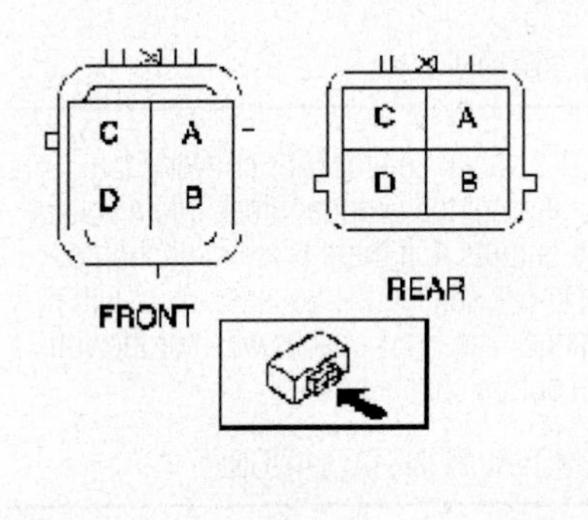

29149_MAZD_G0036

Fig. 16 Front Heated Oxygen Sensor Connector – 2003-06 Mazda 6 L4

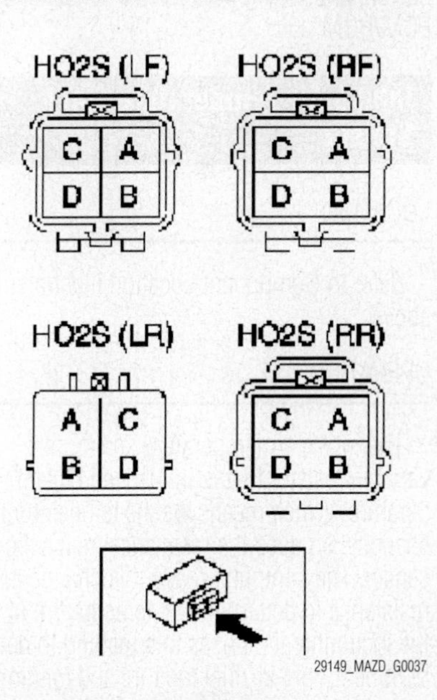

29149_MAZD_G0037

Fig. 17 Rear Heated Oxygen Sensor Connector – 2003-06

HEATER CIRCUIT

1. Disconnect the HO2S connector.
2. Measure the resistance between the HO2S terminals C and D. Standard value:
   a. Front sensor—3.0–3.6 ohms
   b. Rear sensor—5.0–7.0 ohms
3. If not as specified, replace the HO2S.
4. If within specifications, check connectors and wiring between the sensor and ECM/PCM.

Idle Air Control (IAC) Valve

LOCATION

Refer to Component Location illustration above.

OPERATION

The Idle Air Control (IAC) valve controls the engine idle speed and dashpot functions. The valve is located on the side of the throttle body. This valve allows air, determined by the Powertrain Control Module (PCM) and controlled by a duty cycle signal, to bypass the throttle plate in order to maintain the proper idle speed.

REMOVAL & INSTALLATION

1. Disconnect the negative battery cable.
2. Detach the electrical connector at the sensor.
3. Remove the sensor from the throttle body.

To install:

4. Install the sensor in the opening in the throttle body and tighten the sensor.
5. Attach the electrical connector to the sensor.
6. Connect the negative battery cable.

TESTING

Checking Operation Sound

1. Check that the operating sound of the stepper motor can be heard over the idle air control motor when the ignition switch is turned to the ON position (without starting the engine).
2. If no operating sound can be heard, check the stepper motor drive circuit. (If the circuit is good, a defective stepper motor or engine control module is suspected.)

Checking Coil Resistance

1. Disconnect the idle air control motor connector and connect the jumper (test harness).
2. Measure the resistance between terminal (2) of the connector at the idle air control motor side and terminal (1) or terminal (3). Standard value: 28–33 ohms at 68°F
3. Measure the resistance between terminal (5) of the connector at the idle air control motor side and terminal (6) or terminal (4). Standard value: 28–33 ohms at 68°F
4. If not within specifications, replace the IAC Valve.
5. If within specifications, check connectors and wiring between the sensor and ECM/PCM.

Intake Air Temperature (IAT) Sensor

LOCATION

Refer to Component Location illustration above.

OPERATION

The intake air temperature sensor is a variable resistor that is influenced by temperature. Which means, as the temperature changes, so does the resistance inside the sensor. The computer reads this change in resistance to determine the temperature of the incoming air. It uses this reading to determine proper air/fuel mixture and ignition timing settings.

REMOVAL & INSTALLATION

1. Disconnect the negative battery cable.
2. Remove the IAT sensor from the air cleaner housing or intake air pipe.

To install:

3. Install a new sealing washer to the thermosensor.

TESTING

See Figure 18.

1. Disconnect the negative battery cable.
2. Detach the IAT sensor connector.
3. Connect an ohmmeter across the sensor connectors.
4. Measure the resistance of the sensor and compare to the following values:
   - 77°F (25°C)—29.7–36.3 kohms
   - 185°F (85°C)—3.3–3.7 kohms
5. If not as specified, replace the sensor.

Knock Sensor (KS)

LOCATION

Refer to Component Location illustration above.

OPERATION

The Knock Sensor (KS) converts the vibration of the cylinder block into a voltage and outputs it. If there is a malfunction of the knock sensor, the voltage output will not change. The PCM checks whether the voltage output changes.

REMOVAL & INSTALLATION

1995–2002 626 L4

1. Disconnect the negative battery cable.
2. Remove the oil filter element
3. Disconnect the sensor connector.
4. Remove the sensor from the engine block using special tool 49H018001 or equivalent.
5. Installation is the reverse of the removal procedure.
6. Installation new oil filter element.

1995–2002 626 V6

See Figure 19.

1. Disconnect the negative battery cable.
2. Drain the coolant.
3. Remove the intake manifold.
4. Remove the bypass hose.
5. Remove the water pipe.
6. Disconnect the sensor connector.
7. Remove the sensor from the engine block using special tool 49H018001 or equivalent.
8. Installation is the reverse of the removal procedure.

1995–2006 Miata

See Figure 20.

1. Disconnect the negative battery cable.
2. Remove the intake manifold bracket.
3. Disconnect the sensor connector.
4. Remove the sensor from the engine block using special tool 49H018001 or equivalent.
5. Installation is the reverse of the removal procedure.

1999–2002 Millenia KL engine

See Figure 21.

1. Disconnect the negative battery cable.
2. Drain the coolant.
3. Remove the air intake hose.
4. Remove the accelerator cable.
5. Remove the fuel hose.
6. Remove the pipe.
7. Remove the harness.
8. Remove the EGR valve.
9. Remove the intake manifold.
10. Remove the bypass pipe.
11. Remove the water pipe.
12. Disconnect the sensor connector.
13. Remove the sensor from the engine block using special tool 49H018001 or equivalent.
14. Installation is the reverse of the removal procedure.

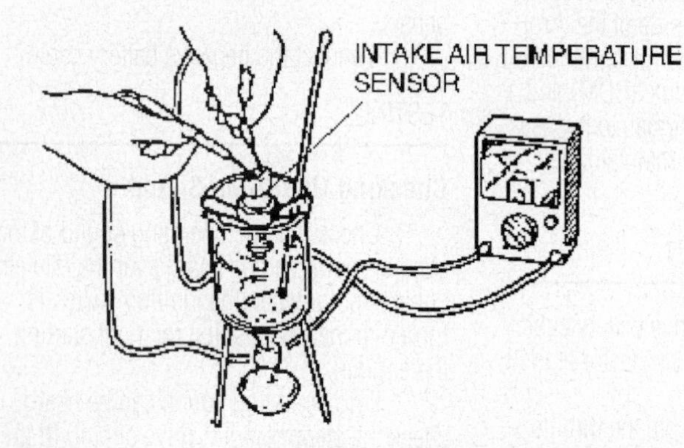

Fig. 18 Intake Air Temperature Testing

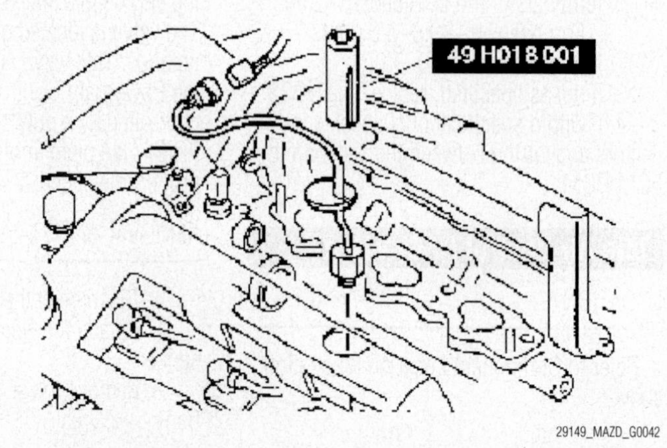

Fig. 19 Knock Sensor Removal & Installation – 1995-99 626 V6

1995–2002 Millenia KJ engine

See Figure 22.

1. Disconnect the negative battery cable.
2. Drain the coolant.
3. Remove the intake air system.
4. Remove the water inlet pipe.
5. Remove the Lysholm compressor bracket.
6. Disconnect the sensor connector.
7. Remove the sensor from the engine block using special tool 49H018001 or equivalent.
8. Installation is the reverse of the removal procedure.

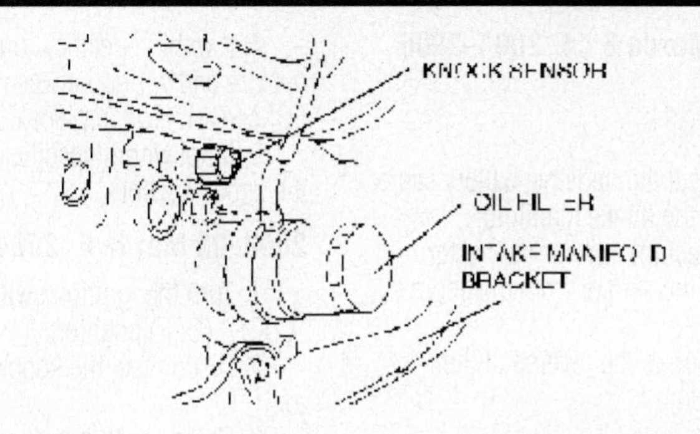

Fig. 20 Knock Sensor Removal & Installation – 2000 Miata

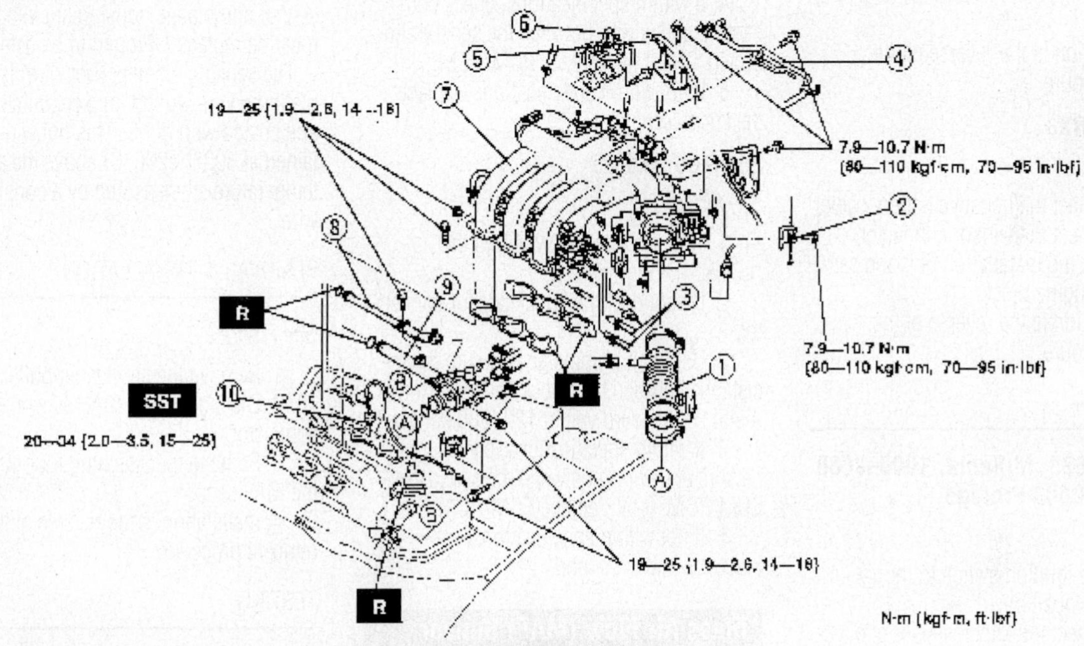

Fig. 21 Knock Sensor Removal & Installation – 1995-99 Millenia V6 (KL)

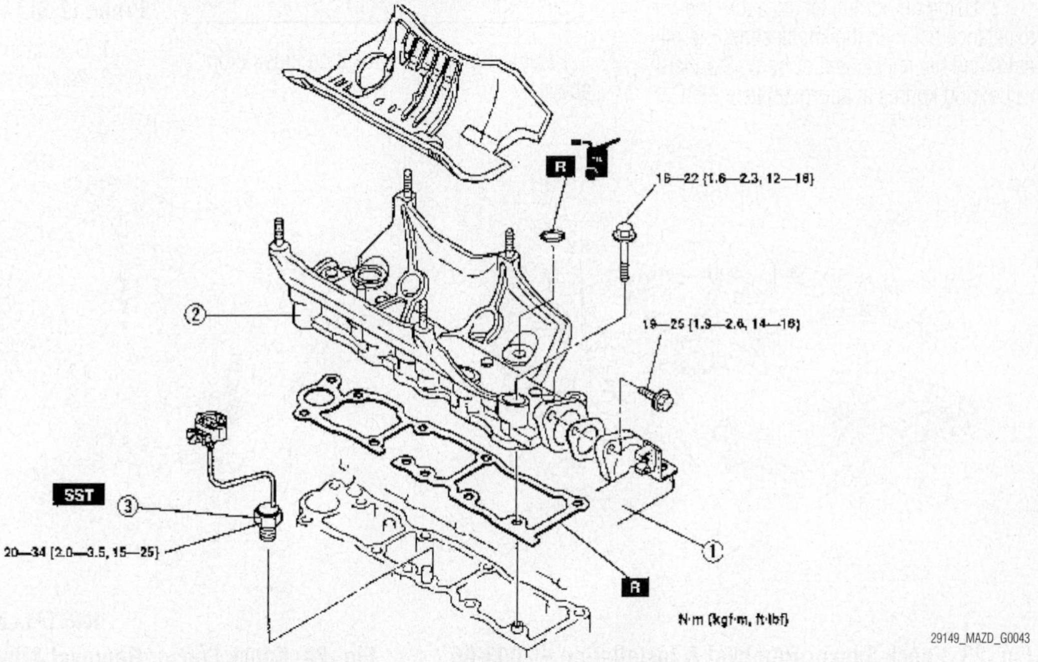

16—22 {1.6—2.3, 12—16}
19—25 {1.9—2.6, 14—18}
SST
20—34 {2.0—3.5, 15—25}
N·m {kgf·m, ft·lbf}
29149_MAZD_G0043

Fig. 22 Knock Sensor Removal & Installation – 1995-99 Millenia V6 (KJ)

2003–2006 Mazda 6 L4, 2004–2006 Mazda 3

See Figure 23.

1. Disconnect the negative battery cable.
2. Remove the intake manifold.
3. Disconnect the sensor connector.
4. Remove the sensor from the engine block.
5. Installation is the reverse of the removal procedure.

2003–2006 Mazda 6 V6

1. Disconnect the negative battery cable.
2. Disconnect the sensor connector.
3. Remove the sensor from the engine block.
4. Installation is the reverse of the removal procedure.

2004–2006 RX8

See Figure 24.

1. Disconnect the negative battery cable.
2. Disconnect the sensor connector.
3. Remove the installation bolt and sensor from the engine block.
4. Installation is the reverse of the removal procedure.

TESTING

1995–2002 626, Millenia, 1995–2006 MPV, 1995–2003 Protege

See Figure 25.

1. Turn the ignition switch to the "LOCK" (OFF) position.
2. Disconnect the knock sensor connector.
3. Using an ohmmeter, measure the resistance between the knock sensor terminal A and the knock sensor body. Standard value: 560 kohms at approximately 68°F.
4. If within specifications, check connectors and wiring between the sensor and ECM/PCM. If ok, replace ECM/PCM.
5. If not within specifications, replace the knock sensor.

2003–06 Mazda 6, 2004–06 Mazda 3

1. Turn the ignition switch to the "LOCK" (OFF) position.
2. Disconnect the knock sensor connector.
3. Using an ohmmeter, measure the resistance between the knock sensor terminal A and B. Standard value: 4870 kohms at approximately 68°F.
4. If within specifications, check connectors and wiring between the sensor and ECM/PCM. If ok, replace ECM/PCM.
5. If not within specifications, replace the knock sensor.

2004–06 RX8

See Figure 26.

1. Turn the ignition switch to the "LOCK" (OFF) position.
2. Disconnect the knock sensor connector.
3. Using an ohmmeter, measure the resistance between the knock sensor terminal A and B. Standard value: 120–280 kohms.
4. If within specifications, check connectors and wiring between the sensor and ECM/PCM. If ok, replace ECM/PCM.
5. If not within specifications, replace the knock sensor.

Mass Air Flow (MAF) Sensor

LOCATION

Refer to Component Location illustration above.

OPERATION

The Mass Air Flow (MAF) sensor directly measures the amount of the air flowing into the engine. The sensor is mounted between the air cleaner assembly and the air cleaner outlet tube.

The sensor utilizes a hot wire-sensing element to measure the amount of air entering the engine. The sensor does this by sending a signal, generated by the sensor when the incoming air cools the hot wire down, to the PCM. The signal is used by the PCM to calculate the injector pulse width, which controls the air/fuel ratio in the engine. The sensor and plastic housing are integral and must be replaced if found to be defective.

The sensing element (hot wire) is a thin platinum wire wound on a ceramic bobbin and coated with glass. This hot wire is maintained at 392°F (200°C) above the ambient temperature as measured by a constant "cold wire".

REMOVAL & INSTALLATION

See Figure 27.

1. Remove the air cleaner outlet tube.
2. Disconnect the MAF sensor electrical connector.
3. Remove the attaching four nuts and the MAF sensor.
4. Installation is the reverse of the removal procedure.

TESTING

MX-3 (1.8L), 1995–98 626/MX-6/Probe (2.5L) with V6 Engines

1. Disconnect the negative battery cable.
2. Remove the airflow meter assembly.

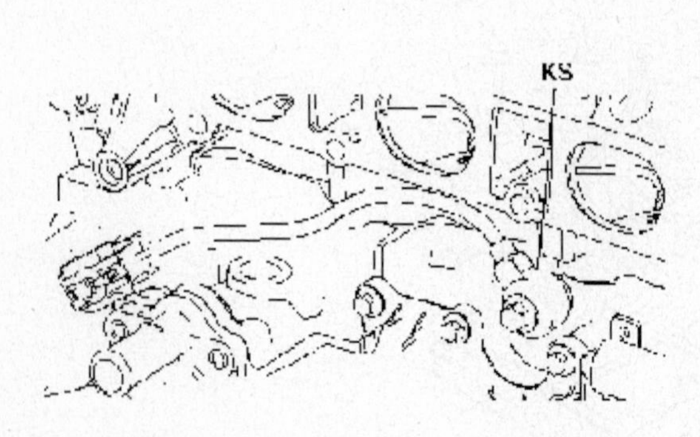

Fig. 23 Knock Sensor Removal & Installation – 2003-06 Mazda 6 L4, 2004-06 Mazda 3

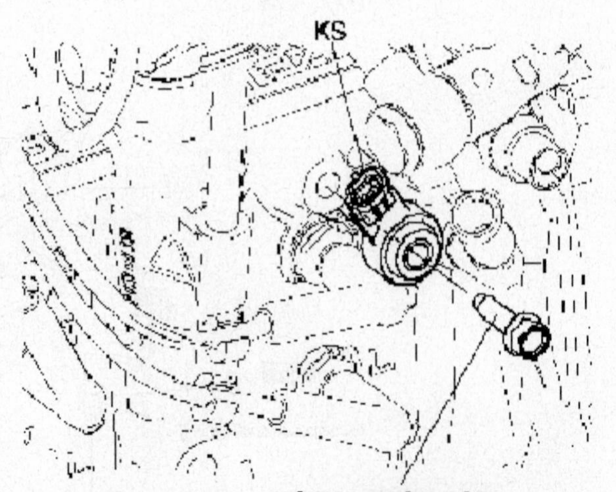

Fig. 24 Knock Sensor Removal & Installation – 2004-06 RX8

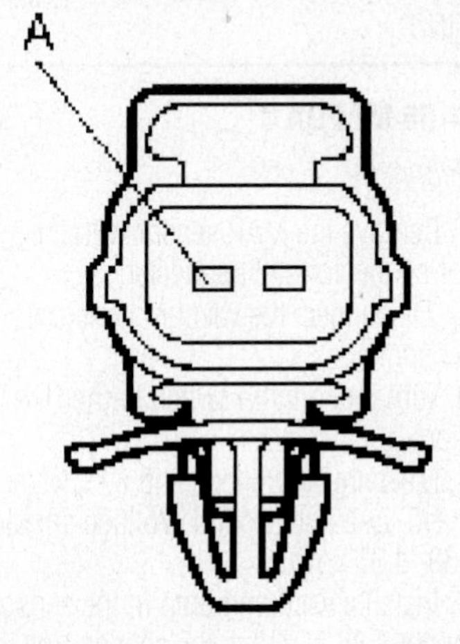

29149_MAZD_G0040

Fig. 25 Knock Sensor Connector – 1995-99 626 V6

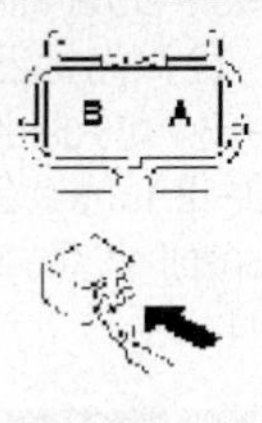

29149_MAZD_G0041

Fig. 26 Knock Sensor Connector – 2004 RX8

3. Inspect the meter body for cracks and for smooth operation of the measuring cone.

4. Using an ohmmeter, check the resistance between the terminals of the airflow meter cone.

5. When measuring the resistance between terminals E2 and VS, the reading should be 200–1000 ohms with the measuring cone fully closed and 20–800 ohms when fully opened.

6. When measuring the resistance between terminals E2 and VC, the reading should be 200–400 ohms with the metering cone in any position.

7. When measuring the resistance between terminals E2 and THA (intake air thermosensor), the reading should be 2000–3000 ohms at 68°F (20°C) and 400–700 ohms at 140°F (60°C) with the metering cone in any position.

8. If not as specified, the airflow meter must be replaced.

626/MX-6/Probe with 2.0L Engine and Millenia

1. Back probe terminal B (center wire) of the air flow sensor electrical connector using a high impedance volt meter.

2. Connect the other terminal of the meter to ground.

3. Turn the ignition switch to the O**N** position.

4. The meter should read 1.0–1.5 volts.

5. Start the engine and allow to idle.

6. The voltmeter should read 1.5–2.5 volts.

7. If the readings are not as specified, replace the sensor.

Navajo and B Series Pick-up Models

1. With the engine running at idle, use a DVOM to verify there is at least 10.5 volts between terminals **A** and **B** of the MAF sensor connector. This indicates the power input to the sensor is correct. Then, measure the voltage between MAF sensor connector terminals **C** and **D**. If the reading is approximately 0.34–1.96 volts, the sensor is functioning properly.

MPV Models

1996–98—3.0L ENGINE

1. Inspect the MAF sensor for damage and cracks.

2. Start the engine and allow it to warm up to operating temperature.

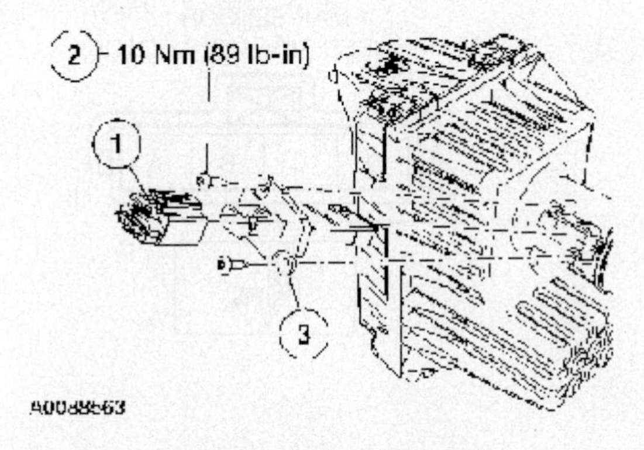

29149_MAZD_G0048

Fig. 27 Mass Air Flow Sensor Removal & Installation

3. Shift the gear selector lever to the **P** (PARK) position.

4. Make sure that all electrical loads are turned OFF. (Ex. headlights, blower motor, etc.)

5. Connect the New Generation Star (NGS) tester, or equivalent scan tool, to the data link connector, located under the driver side dashboard.

6. Select the "PID/DATA MONITOR AND RECORD" function of the tester display.

7. Select "MAF V" on the tester display. The voltage should read as follows:

- Ignition switch ON: 0.5–1.0 Volts
- Engine at idle: 1.0–2.0 Volts

8. If not as specified, inspect the following:

- Harness continuity between PCM terminal 3B and MAF sensor connector terminal B
- Harness continuity between PCM terminal 4A and MAF sensor connector terminal E
- Harness continuity between main relay terminal D and MAF sensor connector terminal A
- Turn the ignition switch to the ON position and measure for battery positive voltage at MAF sensor connector terminal A.

9. Measure the resistance between MAF sensor connector terminals C and D. The resistance should measure 2.21–2.69k ohms at 68°F.

10. If there is incorrect terminal voltage, harness continuity or resistance measurements, replace the MAF sensor.

2004–05 RX8

1. Remove the MAF/IAT sensor without disconnect the MAF/IAT sensor connector.

2. Turn the ignition switch to ON position.

3. As the air gradually approaches the MAF detection part of the MAF/IAT sensor, verify that the voltage at PCM terminal 5N (WDS PID: MAF) varies. Standard value:

- AT—4.4–5.3 g/s at idle
- AT—9.5–12.5 g/s at 2500 rpm
- MT—3.8–4.7 g/s at idle
- MT—8.7–11.7 g/s at 2500 rpm

4. If not as specified, the MAF sensor must be replaced.

2006 RX8

1. Remove the MAF/IAT sensor without disconnect the MAF/IAT sensor connector.
2. Turn the ignition switch to ON position.
3. As the air gradually approaches the MAF detection part of the MAF/IAT sensor, verify that the voltage at PCM terminal 5N (WDS PID: MAF) varies. Standard value:

- 4-sp AT—6.9–8.3 g/s at idle
- 4-sp AT—11.4–15.0 g/s at 2500 rpm
- 6-sp AT—6.3–7.7 g/s at idle
- 6-sp AT—11.9–15.4 g/s at 2500 rpm
- MT—6.0–7.4 g/s at idle
- MT—10.4–14.0 g/s at 2500 rpm

4. If not as specified, the MAF sensor must be replaced.

1999–2002 626 L4

1. Turn the ignition switch to ON position.
2. Using a suitable scan tool, monitor the following.

Standard value:

- AT—2.2–3.10 g/s at idle
- AT—8.7–9.8 g/s at 2500 rpm
- MT—1.9–3.0 g/s at idle
- MT—6.7–9.1 g/s at 2500 rpm

3. If not as specified, the MAF sensor must be replaced.

1999–2002 626 V6

1. Turn the ignition switch to ON position.
2. Using a suitable scan tool, monitor the following.

Standard value:

- AT—2.9–3.7 g/s at idle
- AT—9.9–12.7 g/s at 2500 rpm
- MT—2.9–3.7 g/s at idle
- MT—9.9–12.1 g/s at 2500 rpm

3. If not as specified, the MAF sensor must be replaced.

Manifold Absolute Pressure (MAP) Sensor

LOCATION

See Figure 28.

OPERATION

A 5–volt voltage is supplied to the Manifold Absolute Pressure (MAP) Sensor power terminal from the PCM. The ground terminal is grounded through the PCM. A voltage that is proportional to the intake manifold pressure is sent to the PCM the sensor output terminal.

REMOVAL & INSTALLATION

2004-06 MAZDA 3

1. Remove the battery cover.
2. Disconnect the negative battery cable.
3. Remove the plughole cover.
4. Disconnect the vacuum hose.
5. Disconnect sensor connector.
6. Remove sensor mounting screw.
7. Remove the sensor from intake manifold.
8. Installation is the reverse of the removal procedure. Tightening torque to 32 in. lbs.

TESTING

2004-06 MAZDA 3

See Figure 29.

1. Remove the MAP sensor with the sensor connector still connected
2. Disconnect the vacuum hose from the sensor.
3. Turn the ignition switch to the "ON" position.
4. Measure the voltage at the PCM terminal No. 2AL and ground. Voltage should be 2.69–4.37 volts.
5. Install a vacuum pump to the sensor and apply 8.86 in. Hg to the sensor. Voltage variance should be 1.16–1.27 volts.
6. If not within specification, check connectors and wiring between MAP sensor and PCM. Repair/replace as necessary. If ok, replace MAP sensor
7. If within specifications, replace PCM.

Powertrain Control Module (PCM)

LOCATION

Refer to Component Location illustration above.

OPERATION

The PCM/ECM performs many functions on your car. The module accepts information from various engine sensors and computes the required fuel flow rate necessary to maintain the correct amount of air/fuel ratio throughout the entire engine operational range.

Based on the information that is received and programmed into the PCM's memory, the PCM generates output signals to control relays, actuators and solenoids. The PCM

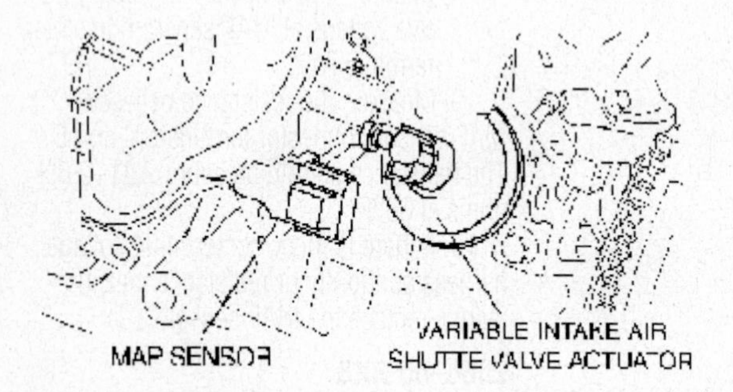

Fig. 28 Manifold Absolute Pressure Sensor Location

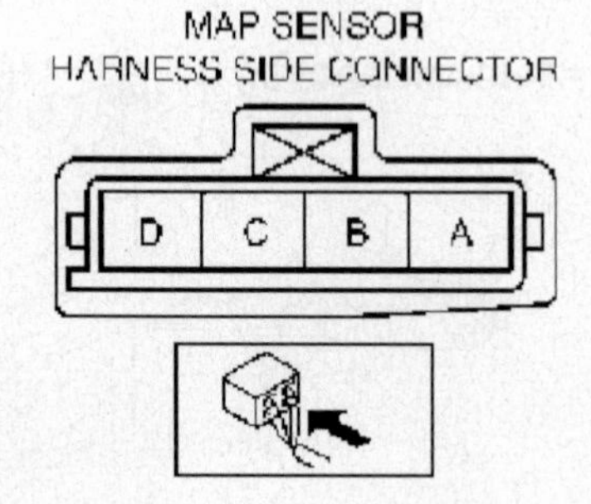

Fig. 29 Manifold Absolute Pressure Sensor Testing

also sends out a command to the fuel injectors that meters the appropriate quantity of fuel. The module automatically senses and compensates for any changes in altitude when driving your vehicle.

REMOVAL & INSTALLATION

1995-98 Miata and Millenia

1. If necessary, remove the driver side under dash cover.
2. Remove the passenger and driver side cover walls from the center console unit.
3. If applicable, detach the rear heater ducts behind the side covers to gain access to the ECU electrical connectors.
4. Installation is the reverse of the removal procedure.

1995-98 Protege

1. If necessary, remove the driver side under dash cover.
2. Remove the passenger and driver side cover walls from the center console unit.
3. Remove the front center console cover as well.
4. If applicable, detach the rear heater ducts behind the side covers to gain access to the ECU electrical connectors.
5. Installation is the reverse of the removal procedure.

1995–98 626/MX-6

1. Remove the center console trim plate around the gear select lever.
2. Remove the ECU mounting bolts/nuts and detach the ECU electrical connectors.
3. Installation is the reverse of the removal procedure.

B Series Pick-up and Navajo

1. Disconnect the negative battery cable.
2. Disengage the wiring harness connector from the PCM by loosening the connector retaining bolt, then pulling the connector from the module.
3. Remove the two nuts and the PCM cover.
4. Remove the PCM from the bracket by pulling the unit outward.

To install:

5. Install the PCM in the mounting bracket.
6. Install the PCM cover and tighten the two nuts.
7. Attach the wiring harness connector to the module, then tighten the connector-retaining bolt.
8. Connect the negative battery cable.

1995–1998 MPV

The PCM is mounted in the vehicle's interior, under the dashboard on the right (passenger) side.

1. Disconnect the negative battery cable.
2. Remove the right side scuff plate and right front side trim.
3. Lift up the front mat.
4. Remove the protector cover.
5. Unplug the wiring harness connector from the control module.
6. Loosen the mounting fasteners and remove the PCM from the vehicle.

To install:

7. Place the PCM into the vehicle in correct position.
8. Install the mounting fasteners and tighten to 70–95 in lbs.
9. Plug the wiring harness connector into the PCM.
10. Install the protector cover and place the mat back in proper position.
11. Install the right front side trim and the right side scuff plate.
12. Connect the negative battery cable.

1995–2002 626

1. Disconnect the negative battery cable.
2. Remove the rear console.
3. Unplug the wiring harness connector from the control module.
4. Loosen the mounting fasteners and remove the PCM from the vehicle.
5. Installation is the reverse of the removal procedure. Tighten mounting fasteners to 70–95 in lbs.

1995–2003 Protege

1. Disconnect the negative battery cable.
2. Remove the front passenger side scuff plate.
3. Remove the front passenger side trim
4. Partially peel off the floor covering from the front of the passenger's side.
5. Remove the PCM mounting plate

**** CAUTION**

The PCM plate edges are sharp. Care should be taken when handling the plate.

6. Unplug the wiring harness connector from the control module.
7. Loosen the mounting fasteners and remove the PCM from the vehicle.
8. Installation is the reverse of the removal procedure. Tighten mounting fasteners to 70–95 in lbs.

1995–2002 Millenia

1. Disconnect the negative battery cable.
2. Remove the rear console box, center panel and under cover.
3. If applicable, detach the rear heater ducts behind the side covers to gain access to the ECU electrical connectors.
4. Unplug the wiring harness connector from the control module.
5. Loosen the mounting fasteners and remove the PCM from the vehicle.
6. Installation is the reverse of the removal procedure. Tighten mounting fasteners to 70–95 in lbs.

2003–2006 Mazda 6

1. Disconnect the negative battery cable.
2. Remove the PCM cover.
3. Remove the PCM harness connectors.
4. Remove the mounting fasteners and remove the PCM from the vehicle.
5. Installation is the reverse of the removal procedure. Tighten mounting fasteners to 70–95 in lbs.

2004–2006 Mazda 3

1. Disconnect the negative battery cable.
2. Remove the battery cover.
3. Remove the battery duct.
4. Remove the battery and battery tray with the PCM.
5. Remove the PCM harness connectors.
6. Remove the PCM from the battery tray.
7. Installation is the reverse of the removal procedure.

2004–2006 RX8

1. Remove the battery cover.
2. Remove the PCM cover.
3. Remove the PCM cooler.
4. Remove the PCM harness connectors.
5. Remove the PCM from the PCM cooler.
6. Installation is the reverse of the removal procedure.

2006 Mazda 5

1. Disconnect the negative battery cable.
2. Remove the battery cover.
3. Remove the battery duct.
4. Remove the battery and battery tray with the PCM.
5. Remove the PCM harness connectors.
6. Remove the PCM from the battery tray.
7. Installation is the reverse of the removal procedure.

Throttle Position Sensor (TPS)

LOCATION

Refer to Component Location illustration above.

OPERATION

The TPS is a variable resistor. As the accelerator pedal is depressed and released, the resistance inside the sensor changes. The computer uses this reading to determine proper air/fuel mixture and ignition timing settings.

REMOVAL & INSTALLATION

1. Disconnect the negative battery cable.
2. Remove the necessary air intake components to access the TPS.
3. Detach the electrical wire harness plug from the sensor.
4. Paint an alignment mark on the sensor housing to the throttle body.
5. Remove the sensor attaching bolts.
6. Remove the sensor from the throttle body.
7. Installation is the reverse of the removal procedure.

TESTING

B Series Pick-up and Navajo Models

1. Disconnect the negative battery cable.
2. Disengage the wiring harness connector from the TP sensor.
3. Using a Digital Volt-Ohmmeter (DVOM) set on ohmmeter function, probe the terminals, which correspond to the Brown/White and the Gray/White connector wires, on the TP sensor. Do not measure the wiring harness connector terminals, rather the terminals on the sensor itself.
4. Slowly rotate the throttle shaft and monitor the ohmmeter for a continuous, steady change in resistance. Any sudden jumps, or irregularities (such as jumping back and forth) in resistance indicates a malfunctioning sensor.
5. Reconnect the negative battery cable.
6. Turn the DVOM to the voltmeter setting.

**** WARNING**

Ensuring the DVOM is on the voltmeter function is vitally important, because if you measure circuit resistance (ohmmeter function) with the battery cable connected, your DVOM will be destroyed.

7. Detach the wiring harness connector from the PCM (located behind the lower right-hand kick panel in the passengers' compartment), then install a breakout box between the wiring harness connector and the PCM connector.
8. Turn the ignition switch **ON** and using the DVOM on voltmeter function, measure the voltage between terminals 89 and 90 of the breakout box. The specification is 0.9 volts.
9. If the voltage is outside the standard value or if it does not change smoothly, inspect the circuit wiring and/or replace the TP sensor.

MPV Models

1. Verify that the throttle valve is at the closed throttle position.
2. Disengage the wiring harness connector from the TP sensor.
3. Using a Digital Volt-Ohmmeter (DVOM) set on ohmmeter function, probe terminals C and D on the TP sensor.
4. Insert a 0.020 inch (0.50mm) feeler gauge between the throttle adjusting screw and the throttle lever. Verify that there is no continuity.
5. If there is no continuity, adjust the TPS as follows:
   a. Loosen the TPS mounting screws.
   b. Insert a feeler gauge between the throttle adjusting screw and the throttle lever.
   c. There should be continuity when inserting a 0.006 inch (0.15mm) feeler gauge and there should be NO continuity when inserting a 0.020 inch (0.50mm) feeler gauge.
   d. Tighten the mounting screws.
6. Replace the TPS if not as specified.

MX-3 with 1.6L SOHC Engine, 323 and Protege

MANUAL TRANSAXLE

1. Detach the connector from the TPS.
2. Connect an ohmmeter between terminals IDL and E.
3. Insert a 0.004 in. (0.1mm) feeler gauge between the throttle stop screw and stop lever.
4. Verify there is continuity between terminals IDL and E.
5. Then replace the feeler gauge with a 0.039 in. (1.0mm) feeler gauge, verify there is no continuity between terminals **IDL** and **E**.
6. Open the throttle wide and verify there is no continuity again between terminals IDL and E.
7. Next, connect the ohmmeter between terminals POW and E.
8. Insert a 0.004 in. (0.1mm) feeler gauge between the throttle stop screw and stop lever.
9. Verify there is no continuity between terminals POW and E.
10. Replace the feeler gauge with 0.039 in. (1.0mm), verify there is no continuity between terminals POW and E.
11. Open the throttle wide and verify there is continuity between terminals POW and E.
12. If not as specified, adjust or replace the throttle sensor.

AUTOMATIC TRANSAXLE

1. Detach the connector from the TPS.
2. Connect an ohmmeter between terminals IDL and E.
3. Insert a 0.004 in. (0.1mm) feeler gauge between the throttle stop screw and stop lever.
4. Verify there is continuity between terminals IDL and E.
5. Insert a 0.024 in. (0.6mm) feeler gauge between the throttle stop screw and stop lever.
6. Verify there is continuity no between terminals IDL and E.
7. Connect an ohmmeter to the throttle sensor terminals Vt and E.
8. Verify that resistance increases as throttle valve opening increase.
9. With throttle valve fully closed the resistance should be below 1 kohm and as throttle valve is fully opened resistance should increase to approximately 5 kohms.
10. If not as specified, adjust or replace the throttle sensor.

626 and MX-6 with 2.2L Engine

1. Remove the air hose from the throttle body.
2. Detach the 3-pin throttle sensor connector.
3. Connect the 49-G018-901 testing harness or equivalent between the throttle sensor and the wiring harness.
4. Turn the ignition switch **ON**.
5. Verify the throttle valve is fully closed.
6. Measure the voltage at the black and the red wires of the testing harness using a precision voltmeter with a scale of 0.01 volts, the voltage at the black wire should be approximately 0 volts and the voltage at the red wire should be 4.5–5.5 volts.
7. If the voltage reading is not as specified, check the battery voltage and wiring harness, if these are okay, replace the engine control unit.
8. Record the red wire voltage.

9. Measure the voltage of the blue wire, verify that the blue wire voltage is within specification according to the red wire voltage. For example; if the red wire voltage reading is 4.50–4.59 volts, then the blue wire voltage reading would have to be within 0.37–0.54 volts.

10. Hold the throttle valve fully open.

11. Measure the blue wire voltage, verify that the blue wire voltage is within specification according to that of the red wire voltage. For example; if the red wire voltage reading is 4.50–4.59 volts, then the blue wire voltage reading would have to be within 3.58–4.23 volts.

12. Check that blue wire voltage increases smoothly when opening the throttle valve from closed to fully open.

13. If the throttle sensor does not perform as specified, adjust or replace the sensor.

14. Turn the ignition **OFF**.

15. Detach the testing harness and reattach the throttle sensor connector.

16. Disconnect the negative battery terminal and depress the brake pedal for at least 5 seconds to eliminate the control unit malfunction memory.

MX-3 with 1.6L DOHC and 626/MX-6/Probe with 2.0L Engine

MANUAL TRANSAXLE

1. Detach the connector from the TPS.

2. Connect an ohmmeter between terminals IDL and E.

3. Insert a 0.004 in. (0.1mm) feeler gauge between the throttle stop screw and stop lever.

4. Verify there is continuity between terminals IDL and E.

5. Then replace the feeler gauge with a 0.027 in. (0.7mm) feeler gauge, verify there is no continuity between terminals IDL and E

6. Then open the throttle wide and verify there is no continuity again between terminals IDL and E.

7. Next, connect the ohmmeter between terminals POW and E.

8. Insert a 0.004 in. (0.1mm) feeler gauge between the throttle stop screw and stop lever.

9. Verify there is no continuity between terminals POW and E.

10. Then replace the feeler gauge with 0.027 in (0.7mm), verify there is no continuity between terminals POW and E.

11. Then open the throttle wide and verify there is continuity between terminals POW and E.

12. If not as specified, adjust or replace the throttle

AUTOMATIC TRANSAXLE

1. Detach the connector from the TPS.

2. Connect an ohmmeter between the terminals IDL and E.

3. Insert a 0.004 in. (0.1mm) feeler gauge between the throttle stop screw and stop lever.

4. Verify there is continuity between terminals IDL and E.

5. Insert a 0.024 in. (0.6mm) feeler gauge between the throttle stop screw and stop lever.

6. Verify there is no continuity between terminals IDL and E.

7. Connect an ohmmeter to the throttle sensor terminals Vt and E.

8. Verify that resistance increases as throttle valve opening increase.

9. With throttle valve fully closed the resistance should be below 1 kohm and as throttle valve is fully opened resistance should increase to approximately 5 kohms.

10. If not as specified, adjust or replace the throttle sensor.

Millenia, MX-3 with 1.8L (K8) and 1995–1998 626/MX-6/Probe with 2.5L Engines

1. Detach the connector from the TPS.

2. Connect an ohmmeter between terminals IDL and GND.

3. Rotate the throttle linkage by hand. With the throttle valve fully closed, the ohmmeter should read 0.1–1.1 volts.

4. With the throttle valve fully open, the ohmmeter should read 3.1–4.4 volts.

5. If not as specified, adjust or replace the TPS.

1999–2002 626 L4, 1999–2003 Protege

See Figure 30.

1. Disconnect the battery negative cable.

2. Disconnect the TPS connector.

3. Using an ohmmeter, measure the resistance between TPS terminals A and C. Should be 4.0–6.0 kohms.

4. If within specification, check wiring and/or connector and repair/replace as necessary.

5. If not as specified, replace the TPS.

1999–2002 626 V6

See Figure 31.

1. Disconnect the battery negative cable.

2. Disconnect the TPS connector.

3. Using an ohmmeter, measure the resistance between TPS terminals A and B. Should be 2.5–5.0 kohms.

4. If within specification, check wiring and/or connector and repair/replace as necessary.

5. If not as specified, replace the TPS.

2003–2006 Mazda 6 L4

1. Turn the ignition switch to ON position (Engine OFF).

2. Measure voltage between the PCM terminal 3M and ground

Standard value:

a. Closed throttle—0.40–0.60 volts.

b. Wide open throttle—4.25–4.75 volts.

3. Measure voltage between the PCM terminal 3J and ground.

Standard value:

a. Closed throttle—4.40–4.60 volts

b. Wide open throttle—0.25–0.75 volts.

4. If within specifications, check wiring and/or connectors and repair/replace as necessary.

5. If not within specifications, replace the throttle body.

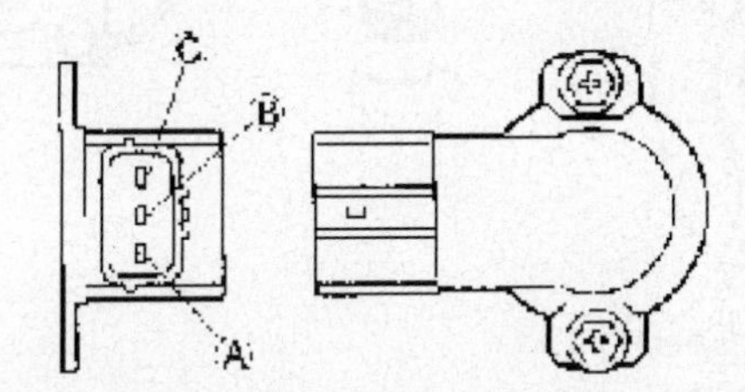

29149_MAZD_G0051

Fig. 30 Throttle Position Sensor Connector 626 L4

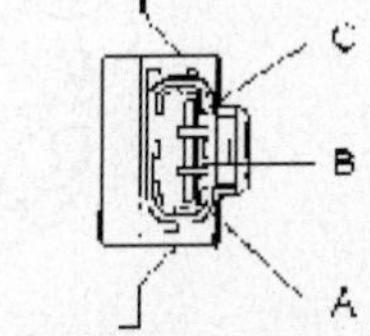

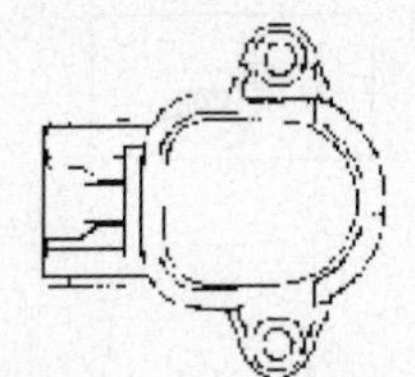

29149_MAZD_G0052

Fig. 31 Throttle Position Sensor Connector 626 V6

2003–2006 Mazda 6 V6

See Figure 32.

1. Connect a suitable scan tool to the diagnostic connector.
2. Turn the ignition switch to ON position.
3. Monitor TP1 and TP2 percentage values. Standard value:
   a. TP1 Closed throttle—83.04–91.54 %.
   b. TP1 Wide open throttle—5.84–25.62 %.
   c. TP2 Closed throttle—6.16–21.26 %.
   d. TP2 Wide open throttle—84.54–92.96 %.
4. If within specifications, check wiring and/or connectors and repair/replace as necessary.
5. If not within specifications, replace the throttle body.

2004–2006 RX8

1. Turn the ignition switch to ON position (Engine OFF).
2. Measure voltage between the PCM terminal 1J and ground
   Standard value:
   a. Closed throttle—0.40–0.80 volts.
   b. Wide open throttle—3.815–4.095 volts.
3. Measure voltage between the PCM terminal 1M and ground.
   Standard value:
   a. Closed throttle—1.18–1.78 volts
   b. Wide open throttle—4.033–4.303 volts.
4. If within specifications, check wiring and/or connectors and repair/replace as necessary.
5. If not within specifications, replace the throttle body.

Vehicle Speed Sensor (VSS)

LOCATION

Refer to Component Location illustration above.

OPERATION

The Vehicle Speed Sensor (VSS) is a magnetic pick-up that sends a signal to the Powertrain Control Module (PCM). The sensor measures the rotation of the transmission and the PCM determines the corresponding vehicle speed.

REMOVAL & INSTALLATION

All Rear Wheel Drive Models

The VSS is located half-way down the right-hand side of the transmission assembly.

1. Apply parking brake, block the rear wheels, then raise and safely support the front of the vehicle on jack stands.
2. From under the right-hand side of the vehicle, disengage the wiring harness connector from the VSS.
3. Loosen the VSS hold-down bolt, then pull the VSS out of the transmission housing.

To install:

4. If a new sensor is being installed, transfer the driven gear retainer and gear to the new sensor.
5. Ensure that the O-ring is properly seated in the VSS housing.
6. For ease of assembly, engage the wiring harness connector to the VSS, then insert the VSS into the transmission assembly.
7. Install and tighten the VSS hold-down bolt to 62–88 in lbs.
8. Lower the vehicle and remove the wheel blocks.

All Front Wheel Drive Models

See Figure 32.

1. Disconnect the negative battery cable.
2. Remove the battery component.
3. Remove the air cleaner component.
4. Disconnect the VSS connector.
5. Remove the VSS.

To install:

6. Apply ATF to a new O-ring and install it on a VSS.
7. Install the VSS.
8. Connect the VSS connector.
9. Install the battery component.
10. Install the air cleaner component.
11. Connect the negative battery cable.

TESTING

All Models

See Figure 33.

1. Remove the VSS
2. Using a low voltage voltmeter, measure the voltage between two terminals of the VSS while rotating the driven gear by hand. Voltmeter needle should move below 5 volts.
3. If the voltmeter needle moves, check wiring and/or connectors and repair/replace as necessary.
4. If the voltmeter needle does not move, replace the VSS.

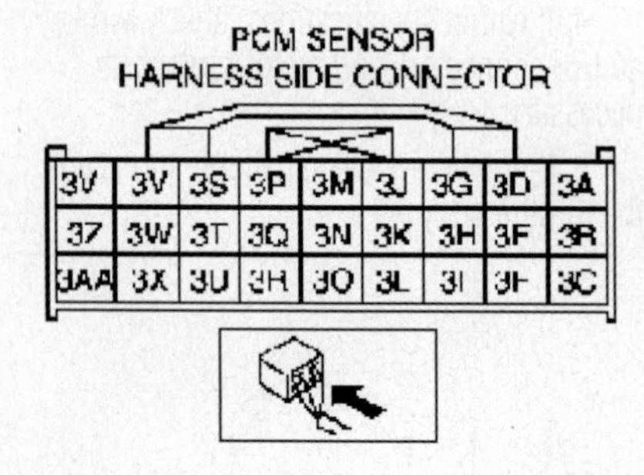

Fig. 32 Throttle Position Sensor Connector Mazda 6

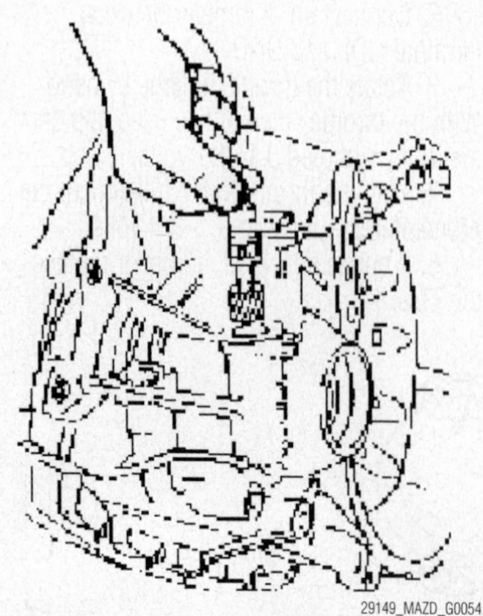

Fig. 33 Vehicle Speed Sensor Removal & Installation

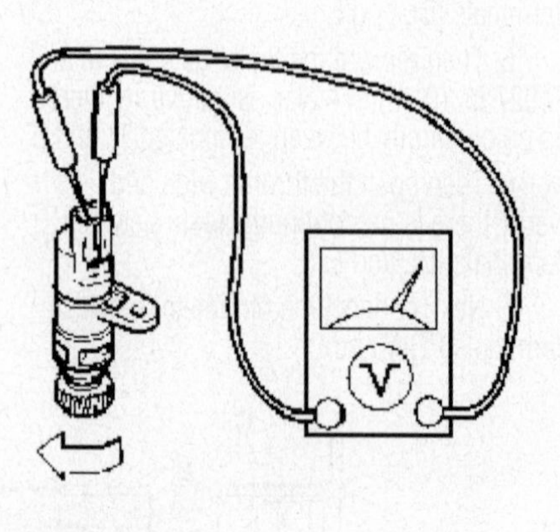

Fig. 34 Vehicle Speed Sensor Testing

MITSUBISHI
DIAGNOSTIC TROUBLE CODES

14

TABLE OF CONTENTS

VEHICLE APPLICATIONS 14-2
P0XXX 14-3
P1XXX 14-49
P2XXX 14-58
U1XXX 14-61

OBD II Vehicle Applications

CONTENTS

3000GT

1995-1999
- 3.0L V6. . . . . VIN H
- 3.0L V6. . . . . VIN J
- 3.0L V6. . . . . VIN K

Diamante

1995-1996
- 3.0L V6. . . . . VIN H
- 3.0L V6. . . . . VIN J

1997-2004
- 3.5L V6. . . . . VIN P

Eclipse

1995-1999
- 2.0L I4 . . . . VIN Y
- 2.0L I4 . . . . VIN F
- 2.4L I4 . . . . VIN G

2000-2005
- 2.4L I4 . . . . VIN G
- 3.0L V6. . . . . VIN H

2006
- 2.4L I4 . . . . VIN F
- 3.8L V6. . . . . VIN T

Eclipse Spyder

1996-1999
- 2.0L I4 . . . . VIN F
- 2.4L I4 . . . . VIN G

2001-2005
- 2.4L I4 . . . . VIN G
- 3.0L V6. . . . . VIN H

Endeavor

2004-2006
- 3.8L V6. . . . . VIN S

Evo

2003-2006
- 2.0L I4 . . . . VIN F

Galant

1995-1998
- 2.4L I4 . . . . VIN G

1999-2003
- 2.4L I4 . . . . VIN G
- 3.0L V6. . . . . VIN H

2004-2006
- 2.4L I4 . . . . VIN F
- 3.8L V6. . . . . VIN S

Lancer

2002-2006
- 2.0L I4 . . . . VIN E
- 2.4L I4 . . . . VIN F

Lancer Sportback

2004-2006
- 2.4L I4 . . . . VIN F

Mirage

1995-2002
- 1.5L I4 . . . . VIN A
- 1.8L I4 . . . . VIN C

Montero

1995-2000
- 3.0L V6. . . . . VIN J
- 3.5L V6. . . . . VIN M

2001-2006
- 3.5L V6. . . . . VIN R
- 3.8L V6. . . . . VIN S

Montero Sport

1997-2004
- 2.4L I4 . . . . VIN G
- 3.0L V6. . . . . VIN P
- 3.5L V6. . . . . VIN R

Outlander

2003-2006
- 2.4L I4 . . . . VIN G
- 2.4L I4 . . . . VIN F

Gas Engine OBD II Trouble Code List (P0XXX Codes)

DTC	Trouble Code Title, Conditions & Possible Causes
DTC: P0011 **2T CCM MIL: YES** **2006** **Models:** Evolution **Engines:** All **Transmissions:** M/T	**Variable Valve Timing System** Engine started and running for greater than 20 seconds, engine speed is 1200 rpm or greater, engine coolant temperature is greater than 169°F, the phase angle of the intake camshaft is 5 degrees or greater for 5 seconds. **Possible Causes:** • Oil feeder control valve failed • Oil passage of variable valve timing control system clogged • Variable valve timing sprocket operation mechanism stuck • ECM failed
DTC: P0031 **2T CCM MIL: YES** **2006** **Models: All** **Engines:** All **Transmissions:** A/T, M/T	**Heated Oxygen Sensor Heater Control Circuit Low (Sensor 1)** Engine started and running for greater than 2 seconds, battery voltage is between 11 and 16.5 volts, with the front heated oxygen sensor on, current continues to be less than 0.17 amps for 2 seconds, and/or with the front heated oxygen sensor is off, the voltage is less than 2 volts for 2 seconds. **Possible Causes:** • Front heated HO2S circuit open or shorted • PCM has failed
DTC: P0032 **2T CCM MIL: YES** **2006** **Models:** All **Engines:** All **Transmissions:** A/T, M/T	**Heated Oxygen Sensor Heater Control Circuit High (Sensor 1)** Engine started and running for greater than 2 seconds, battery voltage is between 11 and 16.5 volts, with the front heated oxygen sensor on, current continues to be greater than 10.5 amps for 2 seconds. **Possible Causes:** • Front heated HO2S circuit open or shorted • PCM has failed
DTC: P0037 **2T CCM MIL: YES** **2006** **Models:** All **Engines:** All **Transmissions:** A/T, M/T	**Heated Oxygen Sensor Heater Control Circuit Low (Sensor 2)** Engine started and running for greater than 2 seconds, battery voltage is between 11 and 16.5 volts, with the rear heated oxygen sensor on, current continues to be less than 0.17 amps for 2 seconds, and/or with the front heated oxygen sensor is off, the voltage is less than 2 volts for 2 seconds. **Possible Causes:** • Rear heated HO2S circuit open or shorted • PCM has failed
DTC: P0038 **2T CCM MIL: YES** **2006** **Models:** All **Engines:** All **Transmissions:** A/T, M/T	**Heated Oxygen Sensor Heater Control Circuit High (Sensor 2)** Engine started and running for greater than 2 seconds, battery voltage is between 11 and 16.5 volts, with the rear heated oxygen sensor on, current continues to be greater than 10.5 amps for 2 seconds. **Possible Causes:** • Rear heated HO2S circuit open or shorted • PCM has failed
DTC: P0051 **2T CCM MIL: YES** **2006** **Models:** Eclipse, Endeavor, Galant, Montero **Engines:** All V6 **Transmissions:** A/T, M/T	**Heated Oxygen Sensor Heater Control Circuit Low (Bank 2, Sensor 1)** Engine started and running for greater than 2 seconds, battery voltage is between 11 and 16.5 volts, with the front heated oxygen sensor on, current continues to be less than 0.17 amps for 2 seconds, and/or with the front heated oxygen sensor is off, the voltage is less than 2 volts for 2 seconds. **Possible Causes:** • Front heated HO2S circuit open or shorted • PCM has failed
DTC: P0052 **2T CCM MIL: YES** **2006** **Models:** Eclipse, Endeavor, Galant, Montero **Engines:** All V6 **Transmissions:** A/T, M/T	**Heated Oxygen Sensor Heater Control Circuit High (Bank 2, Sensor 1)** Engine started and running for greater than 2 seconds, battery voltage is between 11 and 16.5 volts, with the front heated oxygen sensor on, current continues to be greater than 10.5 amps for 2 seconds. **Possible Causes:** • Front heated HO2S circuit open or shorted • PCM has failed
DTC: P0057 **2T CCM MIL: YES** **2006** **Models:** Eclipse, Endeavor, Galant, Montero **Engines:** All V6 **Transmissions:** A/T, M/T	**Heated Oxygen Sensor Heater Control Circuit Low (Bank 2, Sensor 2)** Engine started and running for greater than 2 seconds, battery voltage is between 11 and 16.5 volts, with the rear heated oxygen sensor on, current continues to be less than 0.17 amps for 2 seconds, and/or with the front heated oxygen sensor is off, the voltage is less than 2 volts for 2 seconds. **Possible Causes:** • Rear heated HO2S circuit open or shorted • PCM has failed

DTC	Trouble Code Title, Conditions & Possible Causes
DTC: P0058 **2T CCM MIL: YES** **2006** **Models:** Eclipse, Endeavor, Galant, Montero **Engines:** All V6 **Transmissions:** A/T, M/T	**Heated Oxygen Sensor Heater Control Circuit High (Bank 2, Sensor 2)** Engine started and running for greater than 2 seconds, battery voltage is between 11 and 16.5 volts, with the rear heated oxygen sensor on, current continues to be greater than 10.5 amps for 2 seconds. **Possible Causes:** • Rear heated HO2S circuit open or shorted • PCM has failed
DTC: P0069 **2T CCM MIL: YES** **2005, 2006** **Models:** All **Engines:** All **Transmissions:** A/T, M/T	**Abnormal Correlation Between Manifold Absolute Pressure Sensor And Barometric Pressure Sensor** Engine off, ignition switch is in OFF position, after two seconds pass from the time when the engine is stopped and engine coolant temperature is higher than 0°C (32°F) and the difference between manifold absolute pressure sensor output and barometric pressure sensor output is more than 9 kPa (2.7 in.Hg) for 2 seconds. **Possible Causes:** • MAP sensor failed • Barometric pressure sensor failed • PCM has failed
DTC: P0090 **2T CCM MIL: YES** **2003, 2004, 2005, 2006** **Models:** Evolution **Engines:** All **Transmissions:** M/T	**Fuel Pressure Solenoid Circuit** Engine being cranked, battery positive voltage is between 10 and 16 volts, the fuel pressure solenoid coil surge voltage (battery positive voltage + 2 volts) is not detected for 0.2 seconds. Fuel pressure solenoid is ON, turbocharger wastegate solenoid is OFF for more than 1 second has elapsed after these conditions have been met the fuel pressure solenoid coil surge voltage is not detected for 1 second. **Possible Causes:** • Fuel pressure solenoid failed • Open or shorted fuel pressure circuit, harness damage and/or connector damage • ECM failed
DTC: P0100 **2T CCM MIL: YES** **1995, 1996, 1997, 1998, 1999, 2000, 2001** **Models:** 3000GT, Diamante, Galant, Mirage, Montero, Montero Sport **Engines:** All **Transmissions:** A/T, M/T	**Mass Airflow Sensor Circuit Malfunction** Engine started, engine speed over 500 rpm, and PCM detected the MAF sensor signal was 3.3 Hz or less; or with the engine speed less than 2000 rpm and the TP angle under 2v, it detected the signal was over 800 Hz; or with the engine speed over 2000 rpm and the TP angle over 1.5v, the MAF signal was under 60 Hz for 2 seconds. **Possible Causes:** • Intake air leaks after the MAF sensor location • MAF sensor signal circuit is open or shorted to ground • MAF sensor signal circuit is shorted to VREF or system power • MAF sensor power circuit is open (check power from MFI relay) • MAF sensor is damaged or has failed • PCM has failed
DTC: P0101 **2T CCM MIL: YES** **2001, 2002, 2003, 2004, 2005, 2006** **Models:** Diamante, Eclipse, Endeavor, Evo, Galant, Lancer, Montero, Montero Sport, Outlander **Engines:** All **Transmissions:** A/T, M/T	**Mass Airflow Sensor Range/Performance** Engine started, engine speed over 500 rpm, and PCM detected the MAF sensor signal remained at 3.3 Hz or less for 3 seconds. **Possible Causes:** • Intake air leak between the MAF sensor and the throttle body • MAF sensor signal circuit is open or shorted to ground • MAF sensor is damaged or has failed • PCM has failed
DTC: P0101 **2T CCM MIL: YES** **2001, 2002** **Models:** Mirage **Engines:** 1.8L VIN C **Transmissions:** A/T, M/T	**Mass Airflow Sensor Range/Performance** Engine started, engine speed over 500 rpm, and PCM detected the MAF sensor signal remained at 3.3 Hz or less for 3 seconds. **Possible Causes:** • Intake air leak between the MAF sensor and the throttle body • MAF sensor signal circuit is open or shorted to ground • MAF sensor is damaged or has failed • PCM has failed
DTC: P0102 **2T CCM MIL: YES** **2001, 2002, 2003, 2004, 2005, 2006** **Models:** Diamante, Eclipse, Endeavor, Evo, Galant, Lancer, Montero, Montero Sport, Outlander **Engines:** All **Transmissions:** A/T, M/T	**Mass Airflow Sensor Circuit Low Input** Engine started, engine speed more than 2000 rpm with the TP sensor signal at 1.5v or higher, and the PCM detected the MAF sensor indicated less than 60 Hz for 3 seconds during the CCM test. **Possible Causes:** • MAF sensor signal circuit is open or shorted to ground • MAF sensor is damaged or has failed • PCM has failed

DTC	Trouble Code Title, Conditions & Possible Causes
DTC: P0102 **2T CCM MIL: YES** **2001, 2002** **Models:** Mirage **Engines:** 1.8L VIN C **Transmissions:** A/T, M/T	**Mass Airflow Sensor Circuit Low Input** Engine started, engine speed more than 2000 rpm with the TP sensor signal at 1.5v or higher, and the PCM detected the MAF sensor indicated less than 60 Hz for 3 seconds during the CCM test. **Possible Causes:** • MAF sensor signal circuit is open or shorted to ground • MAF sensor is damaged or has failed • PCM has failed
DTC: P0103 **2T CCM MIL: YES** **2001** **Models:** Eclipse, Galant, Montero Sport **Engines:** All **Transmissions:** A/T, M/T	**Mass Airflow Sensor Circuit High Input** Engine started, engine speed more than 2000 rpm with the TP sensor signal at 1.5v or higher, and the PCM detected the MAF sensor indicate more than 800 Hz for 3 seconds in the CCM test. **Possible Causes:** • MAF sensor signal circuit is shorted to system power • MAF sensor is damaged or has failed • PCM has failed
DTC: P0103 **2T CCM MIL: YES** **2001** **Models:** Mirage **Engines:** 1.8L VIN C **Transmissions:** A/T, M/T	**Mass Airflow Sensor Circuit High Input** Engine started, engine speed more than 2000 rpm with the TP sensor signal at 1.5v or higher, and the PCM detected the MAF sensor indicate more than 800 Hz for 3 seconds in the CCM test. **Possible Causes:** • MAF sensor signal circuit is shorted to system power • MAF sensor is damaged or has failed • PCM has failed
DTC: P0103 **2T CCM MIL: YES** **2004, 2005, 2006** **Models:** Galant, Lancer, Endeavor, Outlander **Engines:** All **Transmissions:** A/T, M/T	**Mass Airflow Sensor Circuit High Input** Engine started for 3 seconds or more and the MAF sensor output voltage is greater than 4.9 volts for 2 seconds. **Possible Causes:** • MAF sensor signal circuit is shorted to system power • MAF sensor circuit open • MAF sensor is damaged or has failed • PCM has failed
DTC: P0105 **2T CCM MIL: YES** **1996, 1997, 1998, 1999, 2000, 2001** **Models:** 3000GT, Diamante, Galant, Mirage, Montero, Montero Sport **Engines:** All **Transmissions:** A/T, M/T	**Barometric Pressure Sensor Circuit Malfunction** Engine started, engine runtime over 2 seconds, system voltage over 8v, and the PCM detected the BARO signal indicated more than 4.45v, or that it indicated less than 0.20v for 4 seconds. **Note: The BARO sensor is located in the Airflow Meter on this engine application.** **Possible Causes:** • BARO sensor signal circuit is open or shorted to ground • BARO sensor signal circuit is shorted to VREF • BARO sensor is damaged or has failed • PCM has failed
DTC: P0105 **2T CCM MIL: YES** **1995, 1996, 1997, 1998, 1999, 2000** **Models:** Eclipse **Engines:** 2.0L VIN F, 2.0L VIN Y, 2.4L VIN G **Transmissions:** A/T, M/T	**Barometric Pressure Sensor Circuit Malfunction** Engine started, engine runtime over 2 seconds, system voltage over 8v, and the PCM detected the BARO signal indicated more than 4.50v (Scan Tool reads 114 kPa), or that it indicated less than 1.95v (Scan Tool reads 50 kPa), either condition met for 10 seconds **Possible Causes:** • BARO sensor signal circuit is open or shorted to ground • BARO sensor signal circuit is shorted to VREF • BARO sensor is damaged or has failed • PCM has failed
DTC: P0105 **2T CCM MIL: YES** **1996, 1997, 1998, 1999, 2000** **Models:** Mirage **Engines:** All **Transmissions:** A/T, M/T	**Barometric Pressure Sensor Circuit Malfunction** Engine started, engine runtime over 2 seconds, system voltage over 8v, and the PCM detected the BARO signal indicated more than 4.50v (Scan Tool reads 114 kPa), or that it indicated less than 1.95v (Scan Tool reads 50 kPa), either condition met for 10 seconds **Possible Causes:** • BARO sensor signal circuit is open or shorted to ground • BARO sensor signal circuit is shorted to VREF • BARO sensor is damaged or has failed • PCM has failed
DTC: P0106 **2T CCM MIL: YES** **1998, 1999** **Models:** Eclipse **Engines:** 2.0L VIN Y **Transmissions:** A/T, M/T	**Manifold Absolute Pressure Sensor Range/Performance** Key on for less than 350 ms with the engine speed below 250 rpm, and the PCM detected the MAP sensor signal was less 2.196v and more than 0.019v for 300 ms during the CCM test. **Possible Causes:** • MAP sensor signal circuit to the PCM interrupted (intermittent) • MAP sensor is damaged or has failed • PCM has failed

DTC	Trouble Code Title, Conditions & Possible Causes
DTC: P0106 **2T CCM MIL: YES** **2001** **Models:** Mirage **Engines:** 1.5L VIN A **Transmissions:** A/T, M/T	**Manifold Absolute Pressure Circuit Range/Performance Problem** Key on, sensor output voltage is less than 4.5 volts for 2 seconds or engine running at idle, sensor voltage is less than 0.2 volts for 2 seconds. **Possible Causes:** • MAP sensor is open or shorted to ground • MAP sensor signal circuit is shorted to VREF • MAP sensor is damaged or has failed • PCM has failed
DTC: P0106 **2T CCM MIL: YES** **2002, 2003, 2004, 2005, 2006** **Models:** Diamante, Eclipse, Galant, Lancer, Mirage, Montero, Montero Sport **Engines:** All **Transmissions:** A/T, M/T	**Barometric Pressure Sensor Range/Performance** Key on for less than 350 ms with the engine speed below 250 rpm, and the PCM detected the MAP sensor signal was less 2.196v and more than 0.019v for 300 ms during the CCM test. **Possible Causes:** • MAP sensor signal circuit to the PCM interrupted (intermittent) • MAP sensor is damaged or has failed • PCM has failed
DTC: P0107 **2T CCM MIL: YES** **1998, 1999** **Models:** Eclipse **Engines:** 2.0L VIN Y **Transmissions:** A/T, M/T	**Manifold Air Pressure Sensor Circuit Low Input** Engine started, engine speed from 400-1500 rpm, TP sensor less than 1.30v, and the PCM detected the MAP signal was less than 0.02v for 2 seconds during the CCM test. **Possible Causes:** • MAP sensor signal circuit is shorted to ground • MAP sensor power circuit is open (VREF signal from the PCM) • MAP sensor is damaged or has failed • PCM has failed
DTC: P0107 **2T CCM MIL: YES** **2001, 2002, 2003** **Models:** Mirage, Lancer **Engines:** 1.5L VIN A, 1.8L VIN B **Transmissions:** A/T, M/T	**Manifold Air Pressure Sensor Circuit Low Input** Engine started, engine speed is 2000 rpm or more, TP sensor greater than 3.5 volts or more, and the PCM detected the MAP output voltage is less than 1.1 volt for 2 seconds. **Possible Causes:** • MAP sensor signal circuit is shorted to ground • MAP sensor power circuit is open (VREF signal from the PCM) • MAP sensor is damaged or has failed • PCM has failed
DTC: P0107 **2T CCM MIL: YES** **2002, 2003, 2004, 2005, 2006** **Models:** Diamante, Galant, Lancer, Mirage, Montero, Montero Sport **Engines:** All **Transmissions:** A/T, M/T	**Manifold Air Pressure Sensor Circuit Low Signal** Engine started, engine speed from 400-1500 rpm, TP sensor less than 1.30v, and the PCM detected the MAP signal was less than 0.02v for 2 seconds during the CCM test. **Possible Causes:** • MAP sensor signal circuit is shorted to ground • MAP sensor power circuit is open (VREF signal from the PCM) • MAP sensor is damaged or has failed • PCM has failed
DTC: P0108 **2T CCM MIL: YES** **1998, 1999** **Models:** Eclipse **Engines:** 2.0L VIN Y **Transmissions:** A/T, M/T	**Manifold Air Pressure Sensor Circuit High Input** Engine started, engine speed from 400-1500 rpm, TP sensor less than 1.30v, and the PCM detected the MAP signal was more than 4.70v for 2 seconds during the CCM test. **Possible Causes:** • MAP sensor signal circuit is open between the sensor and PCM • MAP sensor signal circuit is shorted to VREF or system power • MAP sensor ground circuit is open between sensor and PCM • MAP sensor is damaged or has failed • PCM has failed
DTC: P0108 **2T CCM MIL: YES** **2001, 2002, 2003** **Models:** Mirage, Lancer **Engines:** 1.5L VIN A, 1.8L VIN B **Transmissions:** A/T, M/T	**Manifold Air Pressure Sensor Circuit High Input** Engine started and 8 minutes have passed, manifold absolute pressure is 34.6 in. Hg (117 kPa) or higher for 2 seconds. **Possible Causes:** • MAP sensor signal circuit is open between the sensor and PCM • MAP sensor signal circuit is shorted to VREF or system power • MAP sensor ground circuit is open between sensor and PCM • MAP sensor is damaged or has failed • PCM has failed

DTC	Trouble Code Title, Conditions & Possible Causes
DTC: P0108 **2T CCM MIL: YES** **2002, 2003, 2004, 2005, 2006** **Models:** Diamante, Eclipse, Endeavor, Evo, Galant, Lancer, Mirage, Montero, Montero Sport, Outlander **Engines:** All **Transmissions:** A/T, M/T	**Manifold Air Pressure Sensor Circuit High Input** Engine started and 8 minutes have passed, manifold absolute pressure is 34.6 in. Hg (117 kPa) or higher for 2 seconds. **Possible Causes:** • MAP sensor signal circuit is open between the sensor and PCM • MAP sensor signal circuit is shorted to VREF or system power • MAP sensor ground circuit is open between sensor and PCM • MAP sensor is damaged or has failed • PCM has failed
DTC: P0110 **2T CCM MIL: YES** **1996, 1997, 1998, 1999, 2000, 2001** **Models:** 3000GT, Diamante, Galant, Mirage, Montero, Montero Sport **Engines:** All **Transmissions:** A/T, M/T	**Intake Air Temperature Sensor Circuit Malfunction** Key on for over 60 seconds, or right after engine startup, and the PCM detected the IAT sensor indicated more than 4.60v or less than 0.20v, condition met for 4 seconds during the CCM test. **Possible Causes:** • IAT sensor signal circuit is open or shorted to ground • IAT sensor signal circuit shorted to VREF or to system power • IAT sensor is damaged or has failed (out of calibration) • PCM has failed
DTC: P0110 **2T CCM MIL: YES** **1995, 1996, 1997, 1998, 1999** **Models:** Eclipse **Engines:** 2.0L VIN F, 2.0L VIN Y, 2.4L VIN G **Transmissions:** A/T, M/T	**Intake Air Temperature Sensor Circuit Malfunction** Key on for over 60 seconds, or right after engine startup, and the PCM detected the IAT sensor indicated more than 4.60v or less than 0.20v, condition met for 4 seconds during the CCM test. **Possible Causes:** • IAT sensor signal circuit is open or shorted to ground • IAT sensor signal circuit shorted to VREF or to system power • IAT sensor is damaged or has failed (out of calibration) • PCM has failed
DTC: P0111 **2T CCM MIL: YES** **2001, 2002** **Models:** All **Engines:** All **Transmissions:** A/T, M/T	**Intake Air Temperature Sensor Range/Performance** Engine started, engine runtime over 2 seconds, and the PCM detected the Intake Air Temperature Sensor (IAT) sensor indicated more than 4.6v (Scan Tool reads -49°F); or that it indicated less than 0.2v (Scan Tool reads 257°F) for 2 seconds during the CCM test. **Possible Causes:** • IAT sensor signal circuit is open or shorted to ground • IAT sensor is contaminated, out-of-calibration or has failed • PCM has failed
DTC: P0111 **2T CCM MIL: YES** **2003, 2004, 2005, 2006** **Models:** All **Engines:** All **Transmissions:** A/T, M/T	**Intake Air Temperature Sensor Range/Performance** Engine coolant temperature greater than 169°F, (76°C), start and driven with vehicle speed greater than 31mph (50km/h) for more than 60 seconds, then stop vehicle for greater than 30 seconds with a change in the intake air temperature is less than 1.8°F (1°C). **Possible Causes:** • IAT sensor signal circuit is open or shorted to ground • IAT sensor has failed • PCM has failed
DTC: P0112 **2T CCM MIL: YES** **1995, 1996, 1997, 1998, 1999** **Models:** Eclipse **Engines:** 2.0L VIN Y **Transmissions:** A/T, M/T	**Intake Air Temperature Sensor Circuit Low Signal** Key on or engine running, and the PCM detected the Intake Air Temperature (IAT) sensor signal was less than 0.2v for 3 seconds. **Possible Causes:** • IAT sensor signal circuit is shorted to ground • IAT sensor is damaged or has failed (it may be shorted) • PCM has failed
DTC: P0112 **2T CCM MIL: YES** **2002, 2003, 2004, 2005, 2006** **Models:** Diamante, Eclipse, Endeavor, Evo, Galant, Lancer, Mirage, Montero, Montero Sport, Outlander **Engines:** All **Transmissions:** A/T, M/T	**Intake Air Temperature Sensor Circuit Low Signal** Key on or engine running, and the PCM detected the Intake Air Temperature (IAT) sensor signal was less than 0.2v for 3 seconds. **Possible Causes:** • IAT sensor signal circuit is shorted to ground • IAT sensor is damaged or has failed (it may be shorted) • PCM has failed

DTC	Trouble Code Title, Conditions & Possible Causes
DTC: P0113 **2T CCM MIL: YES** **1995, 1996, 1997, 1998, 1999** **Models:** Eclipse **Engines:** 2.0L VIN Y **Transmissions:** A/T, M/T	**Intake Air Temperature Sensor Circuit High Signal** Key on or engine running, and the PCM detected the Intake Air Temperature (IAT) signal was more than 4.96v for 3 seconds. **Possible Causes:** • IAT sensor signal circuit is open between sensor and the PCM • IAT sensor signal circuit is shorted to VREF or system power • IAT sensor ground circuit is open between sensor and the PCM • IAT sensor is damaged or has failed (an internal short) • PCM has failed
DTC: P0113 **2T CCM MIL: YES** **2002, 2003, 2004, 2005, 2006** **Models:** Diamante, Eclipse, Endeavor, Evo, Galant, Lancer, Mirage, Montero, Montero Sport, Outlander **Engines:** All **Transmissions:** A/T, M/T	**Intake Air Temperature Sensor Circuit High Signal** Key on or engine running, and the PCM detected the Intake Air Temperature (IAT) signal was more than 4.96v for 3 seconds. **Possible Causes:** • IAT sensor signal circuit is open between sensor and the PCM • IAT sensor signal circuit is shorted to VREF or system power • IAT sensor ground circuit is open between sensor and the PCM • IAT sensor is damaged or has failed (an internal short) • PCM has failed
DTC: P0115 **2T CCM MIL: YES** **1996, 1997, 1998, 1999, 2000, 2001** **Models:** 3000GT, Diamante, Galant, Mirage, Montero, Montero Sport **Engines:** All **Transmissions:** A/T, M/T	**Engine Coolant Temperature Sensor Circuit Malfunction** Key on or right after startup, and the PCM detected the ECT signal was over 4.6v, or under 0.1v for 4 seconds; or with the engine speed over 500 rpm, the PCM detected the ECT signal changed less than 1.60v, or with the ECT and IAT signals over 68°F at startup, the PCM detected it took over 5 minutes for the ECT to reach 122°F. **Possible Causes:** • ECT sensor circuit is open, shorted to ground or to VREF • ECT sensor is damaged or has failed (out of calibration) • PCM has failed
DTC: P0115 **2T CCM MIL: YES** **1995, 1996, 1997, 1998, 1999, 2000, 2001** **Models:** Eclipse **Engines:** 2.0L VIN F, 2.0L VIN Y, 2.4L VIN G **Transmissions:** A/T, M/T	**Engine Coolant Temperature Sensor Circuit Malfunction** Key on or right after startup, and the PCM detected the ECT signal was over 4.6v, or under 0.1v for 4 seconds; or with the engine speed over 500 rpm, the PCM detected the ECT signal changed less than 1.60v, or with the ECT and IAT signals over 68°F at startup, the PCM detected it took over 5 minutes for the ECT to reach 122°F. **Possible Causes:** • ECT sensor circuit is open, shorted to ground or VREF • ECT sensor is damaged or has failed (out of calibration) • PCM has failed
DTC: P0116 **2T CCM MIL: YES** **2000, 2001, 2002** **Models:** All **Engines:** All **Transmissions:** A/T, M/T	**Engine Coolant Temperature Sensor Range/Performance** Engine started, engine runtime over 2 seconds, and the PCM detected the Engine Coolant Temperature Sensor (ECT) sensor indicated more than 4.6v (Scan Tool reads -49°F); or that it indicated less than 0.2v (Scan Tool reads 257°F) for 2 seconds in the test. **Possible Causes:** • ECT sensor signal circuit is open or shorted to ground • ECT sensor is contaminated, out-of-calibration or has failed • PCM has failed
DTC: P0116 **2T CCM MIL: YES** **2003, 2004, 2005, 2006** **Models:** All **Engines:** All **Transmissions:** A/T, M/T	**Engine Coolant Temperature Circuit Range/Performance Problem** Engine started and engine coolant temperature is or was more than 45°F (7°C), the engine coolant temperature fluctuates within 1.8°F (1°C) after 5 minutes have passed since starting the engine. The time is not counted if the intake air temperature is 140°F (60°C) or greater, MAF sensor output frequency is 70 Hz or less and/or during fuel shut-off operation. **Possible Causes:** • ECT sensor signal circuit is open or shorted to ground • ECT sensor has failed • PCM has failed
DTC: P0117 **2T CCM MIL: YES** **2001, 2002, 2003, 2004, 2005, 2006** **Models:** All **Engines:** All **Transmissions:** A/T, M/T	**Engine Coolant Temperature Sensor Circuit Low Signal** Key on or engine running, and the PCM detected the Engine Coolant Temperature (ECT) signal indicated less than 0.1v or lower for 2 seconds. **Possible Causes:** • ECT sensor signal circuit is shorted to ground • ECT sensor is damaged or has failed (it may be shorted) • PCM has failed

DTC	Trouble Code Title, Conditions & Possible Causes
DTC: P0117 **2T CCM MIL: YES** **1995, 1996, 1997, 1998, 1999** **Models:** Eclipse **Engines:** 2.0L VIN Y **Transmissions:** A/T, M/T	**Engine Coolant Temperature Sensor Circuit Low Signal** Key on or engine running, and the PCM detected the Engine Coolant Temperature (ECT) signal indicated less than 0.51v for 3 seconds. **Possible Causes:** • ECT sensor signal circuit is shorted to ground • ECT sensor is damaged or has failed (it may be shorted) • PCM has failed
DTC: P0118 **2T CCM MIL: YES** **1995, 1996, 1997, 1998, 1999** **Models:** Eclipse **Engines:** 2.0L VIN Y **Transmissions:** A/T, M/T	**Engine Coolant Temperature Sensor Circuit High Signal** Key on or engine running, and the PCM detected the Engine Coolant Temperature (ECT) signal indicated more than 4.96v for 3 seconds. **Possible Causes:** • ECT sensor signal circuit is open between sensor and the PCM • ECT sensor signal circuit is shorted to VREF or system power • ECT sensor ground circuit is open between sensor and PCM • ECT sensor is damaged or has failed (an internal short) • PCM has failed
DTC: P0118 **2T CCM MIL: YES** **2002, 2003, 2004, 2005, 2006** **Models:** All **Engines:** All **Transmissions:** A/T, M/T	**Engine Coolant Temperature Sensor Circuit High Signal** Key on or engine running, and the PCM detected the Engine Coolant Temperature (ECT) signal indicated more than 4.6v or higher for 2 seconds. **Possible Causes:** • ECT sensor signal circuit is open between sensor and the PCM • ECT sensor signal circuit is shorted to VREF or system power • ECT sensor ground circuit is open between sensor and PCM • ECT sensor is damaged or has failed (an internal short) • PCM has failed
DTC: P0120 **2T CCM MIL: YES** **1996, 1997, 1998, 1999, 2000, 2001** **Models:** 3000GT, Diamante, Galant, Mirage, Montero, Montero Sport **Engines:** All **Transmissions:** A/T, M/T	**TP Sensor Circuit Malfunction** Engine started, engine runtime over 2 seconds, Closed Throttle switch "on", and the PCM detected the TP sensor signal was 2.0v or more, or that it was 0.20v or less for 4 seconds, or with engine at 500-3000 rpm and under low load, the TP signal was over 4.6v for 10 seconds. **Possible Causes:** • TP sensor signal circuit open or shorted to ground • TP sensor ground circuit is open • TP sensor power circuit is open (check VREF circuit at PCM) • TP sensor is damaged or has failed • PCM has failed • TSB 99-13-008 (7/99) contains information related to this code
DTC: P0120 **2T CCM MIL: YES** **1995, 1996, 1997, 1998, 1999** **Models:** Eclipse **Engines:** 2.0L VIN F, 2.0L VIN Y, 2.4L VIN G **Transmissions:** A/T, M/T	**TP Sensor Circuit Malfunction** Engine started, engine runtime over 2 seconds, Closed Throttle switch "on", and the PCM detected the TP sensor signal was 2.0v or more, or that it was 0.20v or less for 4 seconds, or with engine at 500-3000 rpm and under low load, the TP signal was over 4.6v for 10 seconds. **Possible Causes:** • TP sensor signal circuit open or shorted to ground • TP sensor ground circuit is open • TP sensor power circuit is open (check VREF circuit at PCM) • TP sensor is damaged or has failed • PCM has failed
DTC: P0121 **2T CCM MIL: YES** **1995, 1996, 1997, 1998, 1999** **Models:** Eclipse **Engines:** 2.0L VIN Y **Transmissions:** A/T, M/T	**TP Sensor Does Not Agree with MAP Sensor** Engine started, vehicle driven to a speed over 25 mph at an engine speed over 1500 rpm, TP sensor signal from 3.75-4.71v, and the PCM detected the TP sensor signal indicated from 0.16-0.70v, and that it did not correlate to the MAF sensor signal in the CCM test. **Possible Causes:** • TP sensor signal circuit is open to the PCM (intermittent fault) • TP sensor ground circuit is open (an intermittent fault) • MAP sensor is out of calibration • Throttle body is damaged or throttle linkage is bent or binding • TP sensor is damaged or has failed

DTC	Trouble Code Title, Conditions & Possible Causes
DTC: P0121 **2T CCM MIL: YES** **2001, 2002, 2003, 2004, 2005, 2006** **Models:** All **Engines:** All **Transmissions:** A/T, M/T	**Throttle Position Sensor Circuit Range/Performance Problem** Engine started for more than 2 seconds and engine speed is greater than 2000 rpm and volumetric efficiency is greater than 60% and the TPS output voltage has continued to be 0.8 volts or less for 2 seconds and/or engine speed is less than 3000 rpm and volumetric efficiency is less than 30% and the TPS output voltage has continued to be 4.6 volts or greater for 2 seconds. **Possible Causes:** • TP sensor signal circuit is open to the PCM (intermittent fault) • TP sensor ground circuit is open (an intermittent fault) • TP sensor is misadjusted • Throttle body is damaged or throttle linkage is bent or binding • TP sensor is damaged or has failed
DTC: P0122 **2T CCM MIL: YES** **1995, 1996, 1997, 1998, 1999** **Models:** Eclipse **Engines:** 2.0L VIN Y **Transmissions:** A/T, M/T	**TP Sensor Circuit Low Signal** Key on or engine running, and the PCM detected the TP sensor signal was less than 0.20v for 1 second during the CCM test. **Possible Causes:** • TP sensor signal circuit is shorted to ground • TP sensor power circuit is open (check VREF from the PCM) • TP sensor is damaged or failed (it may be shorted internally) • PCM has failed
DTC: P0122 **2T CCM MIL: YES** **2001, 2002, 2003, 2004, 2005, 2006** **Models:** All **Engines:** All **Transmissions:** A/T, M/T	**Throttle Position Sensor Circuit Low Input** Key on or engine running for more than 2 seconds and the PCM detected the sensor signal was less than 0.20v for 2 second. **Possible Causes:** • TP sensor signal circuit is shorted to ground • TP sensor power circuit is open (check VREF from the PCM) • TP sensor has failed or is misadjusted • PCM has failed
DTC: P0123 **2T CCM MIL: YES** **1995, 1996, 1997, 1998, 1999** **Models:** Eclipse **Engines:** 2.0L VIN Y **Transmissions:** A/T, M/T	**TP Sensor Circuit High Signal** Key on or engine running, and the PCM detected the TP sensor signal was more than 4.70v for 1 second during the CCM test. **Possible Causes:** • TP sensor signal circuit is open between sensor and the PCM • TP sensor ground circuit is open between sensor and the PCM • TP sensor is damaged or has failed (it may be open internally) • PCM has failed
DTC: P0123 **2T CCM MIL: YES** **2001, 2002, 2003, 2004, 2005, 2006** **Models:** All **Engines:** All **Transmissions:** A/T, M/T	**Throttle Position Sensor Circuit High Input** Key on or engine running for more than 2 seconds, engine speed is less than 1000 rpm, volumetric efficiency is less than 60% and the PCM detected the sensor signal was greater than 2.0v for 2 second. **Possible Causes:** • TP sensor signal circuit is shorted to ground • TP sensor power circuit is open (check VREF from the PCM) • TP sensor has failed or is misadjusted • PCM has failed
DTC: P0125 **1T ECT MIL: YES** **1996, 1997, 1998, 1999, 2000, 2001** **Models:** 3000GT, Diamante, Galant, Mirage, Montero, Montero Sport **Engines:** All **Transmissions:** A/T, M/T	**Excessive Time To Enter Closed Loop** Engine started, engine runtime over 30 seconds, engine speed 1800-4000 rpm with HO2S signal varying, volumetric efficiency from 16-62%, ECT signal over 178°F, MAP sensor signal from 24-77 kPa, and the PCM the engine did not reach closed loop for 30 seconds. **Possible Causes:** • Check the operation of the thermostat (it may be stuck open) • ECT sensor signal circuit has high resistance • ECT sensor has failed • Inspect for low coolant level or an incorrect coolant mixture
DTC: P0125 **1T ECT MIL: YES** **1995, 1996, 1997, 1998, 1999, 2000, 2001** **Models:** Eclipse **Engines:** All **Transmissions:** A/T, M/T	**Excessive Time To Enter Closed Loop** Engine running for 30 seconds, engine speed 1800-4000 rpm with HO2S signal varying, volumetric efficiency from 16-62%, ECT signal over 178°F, MAP sensor signal from 24-77 kPa, and the PCM the engine did not reach closed loop for 30 seconds. **Possible Causes:** • Check the operation of the thermostat (it may be stuck open) • ECT sensor signal circuit has high resistance • ECT sensor has failed • Inspect for low coolant level or an incorrect coolant mixture

DTC	Trouble Code Title, Conditions & Possible Causes
DTC: P0125 **1T ECT MIL: YES** **2002, 2003, 2004, 2005, 2006** **Models:** All **Engines:** All **Transmissions:** A/T, M/T	**Insufficient Coolant Temperature for Closed Loop Fuel Control** Engine running for approximately 60 to 300 seconds and engine coolant temperature rises to 45°F (7°C) after start and/or engine coolant temperature decreases from greater than 104°F (40°C) to less than 104°F (40°C) for 5 minutes. **Possible Causes:** • Check the operation of the thermostat (it may be stuck open) • ECT sensor circuit shorted or open • ECT sensor has failed • PCM failed
DTC: P0128 **2T ECT MIL: YES** **2000, 2001, 2002** **Models:** Eclipse, Montero, Montero Sport **Engines:** All **Transmissions:** A/T, M/T	**Thermostat Malfunction** Cold engine startup (ECT sensor from 14-180°F and the difference in the ECT and IAT sensor signal less than 9°F), engine running with the MAF sensor signal from 50-100 Hz for 100 seconds, and the PCM detected the ECT sensor signal did not reach 180°F after a 15-20 minute period had elapsed during the CCM Rationality test. **Possible Causes:** • Check the operation of the thermostat (it may be stuck open) • ECT sensor is damaged or out-of-calibration (it is "skewed") • PCM has failed
DTC: P0128 **2T ECT MIL: YES** **2002, 2003, 2004, 2005, 2006** **Models:** All **Engines:** All **Transmissions:** A/T, M/T	**Coolant Thermostat Malfunction** Cold engine startup (ECT sensor from 14-171°F and the difference in the ECT and IAT sensor signal less than 9°F), engine running with the MAF sensor signal from 50-100 Hz for 300 seconds or less, and the PCM detected the ECT sensor signal did not reach 171°F after a 15-20 minute period had elapsed. **Possible Causes:** • Check the operation of the thermostat (it may be stuck open) • ECT sensor is damaged • PCM has failed
DTC: P0130 **2T CCM MIL: YES** **1996, 1997, 1998, 1999, 2000, 2001, 2002** **Models:** 3000GT, Diamante, Galant, Mirage, Montero, Montero Sport **Engines:** All **Transmissions:** A/T, M/T	**HO2S-11 (Bank 1 Sensor 1) Circuit Malfunction** Engine started, engine running in closed loop mode, and after the fuel injector pulse width was increased or decreased with an override, the PCM detected the HO2S-11 response time was too slow/ it detected the HO2S-11 signal circuit was open during the CCM test. **Possible Causes:** • HO2S signal circuit is open between the sensor and the PCM • HO2S signal circuit is shorted to sensor or chassis ground • HO2S signal circuit is shorted to VREF or system power (B+) • HO2S is damaged, contaminated or it has failed • PCM has failed
DTC: P0130 **2T CCM MIL: YES** **1995, 1996, 1997, 1998, 1999, 2000, 2001** **Models:** Eclipse **Engines:** 2.0L VIN F, 2.4L VIN G, 3.0L VIN H **Transmissions:** A/T, M/T	**HO2S-11 (Bank 1 Sensor 1) Circuit Malfunction** Engine started, engine running in closed loop mode, and after the fuel injector pulse width was increased or decreased with an override, the PCM detected the HO2S-11 response time was too slow/ it detected the HO2S-11 signal circuit was open during the CCM test. **Possible Causes:** • HO2S signal circuit is open between the sensor and the PCM • HO2S signal circuit is shorted to sensor or chassis ground • HO2S signal circuit is shorted to VREF or system power (B+) • HO2S is damaged, contaminated or it has failed • PCM has failed
DTC: P0130 **2T CCM MIL: YES** **2003, 2004, 2005** **Models:** Eclipse, Galant **Engines:** 2.4L VIN G **Transmissions:** A/T, M/T	**Heated Oxygen Sensor Circuit (Sensor 1)** Engine started and running for greater than 3 minutes, the front heated oxygen sensor signal voltage has continued to be less than 0.2 volts, engine coolant temperature is greater than 169°F, engine speed is greater than 1200 rpm, volumetric efficiency is greater than 25%, the input voltage supplied to the PCM interface circuit is greater than 4.5 volts when 5 volts is applied to the front heated oxygen sensor output line via a resistor for at least 7 seconds. **Possible Causes:** • HO2S signal circuit is open between the sensor and the PCM • HO2S signal circuit is shorted to sensor or chassis ground • HO2S signal circuit is shorted to VREF or system power (B+) • HO2S is damaged, contaminated or it has failed • PCM has failed

DTC	Trouble Code Title, Conditions & Possible Causes
DTC: P0130 **2T CCM MIL: YES** **2003, 2004, 2005** **Models:** Diamante, Eclipse, Endeavor, Galant, Montero, Montero Sport, Outlander **Engines:** All except 2.4L VIN G **Transmissions:** A/T, M/T	**Heated Oxygen Sensor Circuit (Bank 1, Sensor 1)** Engine started and running for greater than 3 minutes, the right front heated oxygen sensor signal voltage has continued to be less than 0.2 volts, engine coolant temperature is greater than 169°F, engine speed is greater than 1200 rpm, volumetric efficiency is greater than 25%, the input voltage supplied to the PCM interface circuit is greater than 4.5 volts when 5 volts is applied to the right front heated oxygen sensor output line via a resistor for at least 7 seconds. **Possible Causes:** • HO2S signal circuit is open between the sensor and the PCM • HO2S signal circuit is shorted to sensor or chassis ground • HO2S signal circuit is shorted to VREF or system power (B+) • HO2S is damaged, contaminated or it has failed • PCM has failed
DTC: P0131 **2T CCM MIL: YES** **1997, 1998, 1999** **Models:** Eclipse **Engines:** 2.0L VIN Y **Transmissions:** A/T, M/T Number of Trips to Set Code:	**HO2S-11 (Bank 1 Sensor 1) Signal Low Signal** Engine started, ECT sensor signal less than 120°F at engine startup, engine runtime over 3 seconds, and the PCM detected the HO2S-11 signal was less than 0.16v during the CCM test. **Possible Causes:** • HO2S signal circuit is open • HO2S signal circuit is shorted to ground • HO2S is damaged or it has failed • PCM has failed
DTC: P0131 **2T CCM MIL: YES** **2003, 2004, 2005, 2006** **Models:** All **Engines:** All 4 Cylinder Engines **Transmissions:** A/T, M/T	**Heated Oxygen Sensor Circuit Low Voltage (Sensor 1)** Engine started and running for greater than 3 minutes, the front heated oxygen sensor signal voltage has continued to be less than 0.2 volts, engine coolant temperature is greater than 169°F, engine speed is greater than 1200 rpm, volumetric efficiency is greater than 25%, MAF sensor output frequency is greater than 75 Hz, at least 20 seconds have passed since fuel shut-off control was canceled, and the changes in the front heated oxygen sensor is less than 0.078 voltage the front heated oxygen sensor output voltage increase beyond 0.2 volts after making the air/fuel ratio 15% richer for 10 seconds. **Possible Causes:** • HO2S signal circuit is open between the sensor and the PCM • HO2S signal circuit is shorted to sensor or chassis ground • HO2S signal circuit is shorted to VREF or system power (B+) • HO2S is damaged, contaminated or it has failed • PCM has failed
DTC: P0131 **2T CCM MIL: YES** **2003, 2004, 2005, 2006** **Models:** All **Engines:** All 6 Cylinder Engines **Transmissions:** A/T, M/T	**Heated Oxygen Sensor Circuit Low Voltage (Bank 1, Sensor 1)** Engine started and running for greater than 3 minutes, the right front heated oxygen sensor signal voltage has continued to be less than 0.2 volts, engine coolant temperature is greater than 169°F, engine speed is greater than 1200 rpm, volumetric efficiency is greater than 25%, MAF sensor output frequency is greater than 75 Hz, at least 20 seconds have passed since fuel shut-off control was canceled, and the changes in the right front heated oxygen sensor is less than 0.078 voltage the right front heated oxygen sensor output voltage increase beyond 0.2 volts after making the air/fuel ratio 15% richer for 10 seconds. **Possible Causes:** • HO2S signal circuit is open between the sensor and the PCM • HO2S signal circuit is shorted to sensor or chassis ground • HO2S signal circuit is shorted to VREF or system power (B+) • HO2S is damaged, contaminated or it has failed • PCM has failed
DTC: P0132 **2T CCM MIL: YES** **1995, 1996, 1997, 1998, 1999** **Models:** Eclipse **Engines:** 2.0L VIN Y **Transmissions:** A/T, M/T	**HO2S-11 (Bank 1 Sensor 1) Signal High Signal** Engine started, engine runtime over 2 minutes, ECT sensor more than 176°F, and the PCM detected the HO2S-11 signal was more than 1.2v for 3 seconds during the CCM test. **Possible Causes:** • HO2S signal circuit shorted to power (check the heater circuit) • HO2S is damaged or it has failed • PCM has failed
DTC: P0132 **2T CCM MIL: YES** **2002, 2003, 2004, 2005, 2006** **Models:** Diamante, Eclipse, Galant, Lancer, Mirage, Montero, Montero Sport **Engines:** All **Transmissions:** A/T, M/T	**HO2S-11 (Bank 1 Sensor 1) Signal High Signal** Engine started, engine runtime over 2 seconds, and the PCM detected the HO2S-11 signal was more than 1.2v for 2 seconds. **Possible Causes:** • HO2S signal circuit shorted to power (check the heater circuit) • HO2S is damaged or it has failed • PCM has failed

DTC	Trouble Code Title, Conditions & Possible Causes
DTC: P0133 **2T O2S2 MIL: YES** **1995, 1996, 1997, 1998, 1999** **Models:** Eclipse **Engines:** 2.0L VIN Y **Transmissions:** A/T, M/T	**HO2S-11 (Bank 1 Sensor 1) Slow Response** Engine started, engine runtime over 3 minutes, ECT sensor more than 170°F, vehicle driven to a speed over 24 mph for 75 seconds, then at idle speed from 512-864 rpm, PSPS signal "off", A/C clutch not cycling, and the PCM detected the HO2S-11 signal that did not reach 670 mv, or it did not switch between 0.39-6.0v engine times in a 6 second period during the Oxygen Sensor Monitor test. **Possible Causes:** • Exhaust leak present in the exhaust manifold or exhaust pipes • HO2S element fuel contamination • HO2S element has deteriorated • PCM has failed
DTC: P0133 **2T O2S2 MIL: YES** **2002, 2003, 2004, 2005, 2006** **Models:** Diamante, Eclipse, Endeavor, Evo, Galant, Lancer, Mirage, Montero, Montero Sport, Outlander **Engines:** All **Transmissions:** A/T, M/T	**HO2S-11 (Bank 1 Sensor 1) Slow Response** Engine started, engine runtime over 3 minutes, ECT sensor more than 170°F, vehicle driven to a speed over 24 mph for 75 seconds, then at idle speed from 512-864 rpm, PSPS signal "off", A/C clutch not cycling, and the PCM detected the HO2S-11 signal that did not reach 670 mv, or it did not switch between 0.39-6.0v engine times in a 6 second period during the Oxygen Sensor Monitor test. **Possible Causes:** • Exhaust leak present in the exhaust manifold or exhaust pipes • HO2S element fuel contamination • HO2S element has deteriorated • PCM has failed
DTC: P0134 **2T O2S1 MIL: YES** **1995, 1996, 1997, 1998, 1999** **Models:** Eclipse **Engines:** 2.0L VIN Y **Transmissions:** A/T, M/T	**HO2S-11 (Bank 1 Sensor 1) Remains At Center** Engine started, engine runtime over 2 minutes, ECT sensor more than 176°F, and the PCM detected the HO2S-11 signal remained fixed near 0.50v for 1.5 minutes in the Oxygen Sensor Monitor test. **Possible Causes:** • Exhaust leak present in exhaust manifold or exhaust pipes • HO2S element fuel contamination or has deteriorated • HO2S signal circuit or the ground circuit has high resistance • HO2S heater element has failed, or the heater circuit is open • PCM has failed • TSB 2-13-002 (2/02) contains information related to this code
DTC: P0134 **2T O2S1 MIL: YES** **2001, 2002, 2003, 2004, 2005, 2006** **Models:** All **Engines:** All **Transmissions:** A/T, M/T	**HO2S-11 (Bank 1 Sensor 1) No Activity Detected** Engine started, engine runtime over 30 seconds, ECT sensor more than 169°F, engine speed is greater than 1200 rpm, volumetric efficiency is between 30% and 95%, TPS output voltage is less than 4 volts and the PCM detected the HO2S-11 signal remained fixed near 0.50v for 30 seconds in the Oxygen Sensor Monitor test. **Possible Causes:** • Exhaust leak present in exhaust manifold or exhaust pipes • HO2S element fuel contamination or has deteriorated • HO2S signal circuit or the ground circuit has high resistance • HO2S heater element has failed, or the heater circuit is open • PCM has failed
DTC: P0135 **2T O2S HTR2 MIL: YES** **1996, 1997, 1998, 1999, 2000, 2001, 2002** **Models:** 3000GT, Diamante, Galant, Mirage, Montero, Montero Sport **Engines:** All **Transmissions:** A/T, M/T	**HO2S-11 (Bank 1 Sensor 1) Heater Circuit Malfunction** Engine started, ECT sensor more than 68°F, system voltage from 11-16v, and the PCM detected the HO2S-11 heater current was less than 0.2, or that it was more than 3.5 amps for 6 seconds in the test. **Possible Causes:** • HO2S heater control circuit is open or shorted to ground • HO2S heater control circuit is shorted to power • HO2S heater power circuit is open (check power from the relay) • HO2S heater is damaged or has failed • PCM has failed
DTC: P0135 **2T O2S HTR2 MIL: YES** **1995, 1996, 1997, 1998, 1999** **Models:** Eclipse **Engines:** 2.0L VIN F, 2.0L VIN Y, 2.4L VIN G **Transmissions:** A/T, M/T	**HO2S-11 (Bank 1 Sensor 1) Heater Circuit Malfunction** Engine started, ECT sensor more than 68°F, system voltage from 11-16v, and the PCM detected the HO2S-11 heater current was less than 0.2, or that it was more than 3.5 amps for 6 seconds in the test. **Possible Causes:** • HO2S heater control circuit is open or shorted to ground • HO2S heater control circuit is shorted to power • HO2S heater power circuit is open (check power from the relay) • HO2S heater is damaged or has failed • PCM has failed

DTC	Trouble Code Title, Conditions & Possible Causes
DTC: P0135 **2T O2S HTR1 MIL: YES** **1995, 1996, 1997, 1998, 1999** **Models:** Eclipse **Engines:** 2.0L VIN Y **Transmissions:** A/T, M/T	**HO2S-11 (Bank 1 Sensor 1) Heater Circuit Malfunction** Key off for 5 seconds, then after the HO2S signal rose by over 490 mv within 144 seconds after key off period, and the initial rise in HO2S voltage was less than 1.57v, the PCM detected the HO2S-11 did not show a correct amount of voltage decrease during the test. **Possible Causes:** • HO2S heater power circuit is open (check for power from the MFI relay to the heater circuit) • HO2S heater control circuit is open between sensor and PCM • HO2S heater element has high resistance or has failed • PCM has failed
DTC: P0135 **2T O2S HTR2 MIL: YES** **2000, 2001, 2002, 2003, 2004, 2005** **Models:** All **Engines:** All **Transmissions:** A/T, M/T	**HO2S-11 (Bank 1 Sensor 1) Heater Circuit Malfunction** Engine started for more than 60 seconds, ECT sensor more than 68°F, system voltage from 11-16v, and the PCM detected the HO2S-11 heater current was less than 0.16 amps, or it was more than 7.5 amps for 4 seconds during the HO2S Heater Monitor test. **Possible Causes:** • HO2S heater control circuit is open or shorted to ground • HO2S heater control circuit is shorted to power • HO2S heater power circuit is open (check power from the relay) • HO2S heater is damaged or has failed • PCM has failed
DTC: P0136 **2T CCM MIL: YES** **1996, 1997, 1998, 1999, 2000, 2001** **Models:** 3000GT, Diamante, Galant, Mirage, Montero, Montero Sport **Engines:** All **Transmissions:** A/T, M/T	**HO2S-12 (Bank 1 Sensor 2) Circuit Malfunction** Engine started, engine running in closed loop, and after the PCM increased or decreased the fuel injector pulse width with an override, it detected the HO2S-12 response time was slow, or it detected the HO2S-12 signal circuit was open during the CCM test. **Possible Causes:** • HO2S signal circuit is open between the sensor and the PCM • HO2S signal circuit is shorted to sensor or chassis ground • HO2S signal circuit is shorted to VREF or system power (B+) • HO2S is damaged, contaminated or it has failed • PCM has failed
DTC: P0136 **2T CCM MIL: YES** **1995, 1996, 1997, 1998, 1999** **Models:** Eclipse **Engines:** 2.0L VIN F, 2.0L VIN Y, 2.4L VIN G **Transmissions:** A/T, M/T	**HO2S-12 (Bank 1 Sensor 2) Circuit Malfunction** Engine started, engine running in closed loop, and after the PCM increased or decreased the fuel injector pulse width with an override, it detected the HO2S-12 response time was slow, or it detected the HO2S-12 signal circuit was open during the CCM test. **Possible Causes:** • HO2S signal circuit is open between the sensor and the PCM • HO2S signal circuit is shorted to sensor or chassis ground • HO2S signal circuit is shorted to VREF or system power (B+) • HO2S is damaged, contaminated or it has failed • PCM has failed
DTC: P0136 **2T CCM MIL: YES** **2003, 2004, 2005, 2006** **Models:** All **Engines:** All **Transmissions:** A/T, M/T	**HO2S-12 (Bank 1 Sensor 2) Circuit Malfunction** Engine started and running for 3 minutes or more, rear HO2S signal voltage is 0.15 or less, engine coolant temperature is greater than 169°F, engine speed greater than 1200 rpm, volumetric efficiency is greater than 25%, the input voltage supplied to the PCM interface circuit is greater than 4.5 volts when 5 volts is applied to the rear HO2S output line via a resister. **Possible Causes:** • HO2S signal circuit is open between the sensor and the PCM • HO2S signal circuit is shorted to sensor or chassis ground • HO2S signal circuit is shorted to VREF or system power (B+) • HO2S is damaged, contaminated or it has failed • PCM has failed
DTC: P0137 **2T CCM MIL: YES** **1998, 1999, 2000, 2001, 2002** **Models:** Eclipse **Engines:** 2.0L VIN Y **Transmissions:** A/T, M/T	**HO2S-12 (Bank 1 Sensor 2) Signal Low Input** Engine started, ECT sensor signal less than 120°F at engine startup, engine runtime over 3 seconds, and the PCM detected the HO2S-12 signal was less than 0.16v during the CCM test. **Possible Causes:** • HO2S signal circuit is open • HO2S signal circuit is shorted to ground • HO2S is damaged or it has failed • PCM has failed
DTC: P0137 **2T CCM MIL: YES** **2002, 2003, 2004, 2005, 2006** **Models:** Diamante, Endeavor, Evo, Eclipse, Galant, Lancer, Mirage, Montero, Montero Sport, Outlander **Engines:** All **Transmissions:** A/T, M/T	**HO2S-12 (Bank 1 Sensor 2) Signal Low Input** Engine started, ECT sensor signal less than 120°F at engine startup, engine runtime over 3 seconds, and the PCM detected the HO2S-12 signal was less than 0.16v during the CCM test. **Possible Causes:** • HO2S signal circuit is open • HO2S signal circuit is shorted to ground • HO2S is damaged or it has failed • PCM has failed

DTC	Trouble Code Title, Conditions & Possible Causes
DTC: P0138 **2T CCM MIL: YES** **1996, 1997, 1998, 1999** **Models:** Eclipse **Engines:** 2.0L VIN Y **Transmissions:** A/T, M/T	**HO2S-12 (Bank 1 Sensor 2) Signal High Input** Engine started, engine runtime over 2 minutes, ECT sensor more than 176°F, and the PCM detected the HO2S-11 signal was more than 1.2v for 3 seconds during the CCM test. **Possible Causes:** • HO2S signal circuit shorted to power (check the heater circuit) • HO2S is damaged or it has failed • PCM has failed
DTC: P0138 **2T CCM MIL: YES** **2002, 2003, 2004, 2005, 2006** **Models:** Diamante, Eclipse, Galant, Lancer, Mirage, Montero, Montero Sport **Engines:** All **Transmissions:** A/T, M/T	**HO2S-12 (Bank 1 Sensor 2) Signal High Input** Engine started, engine runtime over 2 seconds and the PCM detected the HO2S-12 signal was more than 1.2v for 2 seconds. **Possible Causes:** • HO2S signal circuit shorted to power (check the heater circuit) • HO2S is damaged or it has failed • PCM has failed
DTC: P0139 **2T O2S1 MIL: YES** **1995, 1996** **Models:** Eclipse **Engines:** 2.0L VIN Y **Transmissions:** A/T, M/T	**HO2S-12 (Bank 1 Sensor 2) No Activity Detected** Engine started, engine runtime over 2 minutes, ECT sensor more than 176°F, and the PCM detected the HO2S-11 signal remained fixed near 0.50v for 1.5 minutes in the Oxygen Sensor Monitor test. **Possible Causes:** • HO2S signal circuit is open or it has high resistance • HO2S element fuel contamination or has deteriorated • HO2S is damaged or has failed • PCM has failed
DTC: P0139 **2T O2S1 MIL: YES** **2001, 2002, 2003, 2004, 2005, 2006** **Models:** All **Engines:** All **Transmissions:** A/T, M/T	**HO2S-12 Circuit Slow Response (Bank 1 Sensor 2)** Engine started, engine runtime over 10 seconds, ECT sensor more than 169°F, front heated oxygen sensor is active, MAF sensor output frequency is greater than 4000 Hz, engine speed is greater than 1500 rpm, volumetric efficiency is greater than 40%, vehicle speed is greater than 19 mph and all is repeated 3 or more times, the change in the output voltage of the HO2S-12 is less than 0.312 volts. **Possible Causes:** • HO2S signal circuit is open or it has high resistance • HO2S element fuel contamination or has deteriorated • HO2S is damaged or has failed • PCM has failed
DTC: P0140 **2T O2S1 MIL: YES** **1997, 1998, 1999** **Models:** Eclipse **Engines:** 2.0L VIN F, 2.0L VIN Y, 2.4L VIN G **Transmissions:** A/T, M/T	**HO2S-12 (Bank 1 Sensor 2) Remains At Center** Engine started, engine runtime over 2 minutes, ECT sensor more than 176°F, and the PCM detected the HO2S-11 signal remained fixed near 0.50v for 1.5 minutes in the Oxygen Sensor Monitor test. **Possible Causes:** • HO2S signal circuit is open or it has high resistance • HO2S element fuel contamination or has deteriorated • HO2S is damaged or has failed • PCM has failed
DTC: P0140 **2T CCM MIL: YES** **2006** **Models:** All **Engines:** All V6 **Transmissions:** A/T, M/T	**HO2S-12 (Bank 1 Sensor 2) Circuit No Activity Detected** Engine started, engine runtime over 10 seconds, engine coolant temperature greater than 169°F, front HO2S active, MAF output is greater than 1638 g., engine speed greater than 1500 rpm, volumetric efficiency is greater than 40%, vehicle speed greater than 19 mph, then stop vehicle, repeated 3 or more times, the change in the output voltage in the rear HO2S is less than 0.313 volts. **Possible Causes:** • HO2S-12 failed • HO2S-12 circuit connector damaged • PCM has failed
DTC: P0141 **1T O2S HTR1 MIL: YES** **1996, 1997, 1998, 1999, 2000, 2001, 2002** **Models:** 3000GT, Diamante, Galant, Mirage, Montero, Montero Sport **Engines:** All **Transmissions:** A/T, M/T	**HO2S-12 (Bank 1 Sensor 2) Heater Circuit Malfunction** Engine started, ECT sensor more than 68°F, system voltage from 11-16v, and the PCM detected the HO2S-12 heater current was less than 0.20v, or it was more than 3.5 amps for 6 seconds in the test. **Possible Causes:** • HO2S heater control circuit is open or shorted to ground • HO2S heater control circuit is shorted to power • HO2S heater power circuit is open (check power from the relay) • HO2S heater is damaged or has failed • PCM has failed

DTC	Trouble Code Title, Conditions & Possible Causes
DTC: P0141 **1T O2S HTR1 MIL: YES** **1995, 1996, 1997, 1998, 1999** **Models:** Eclipse **Engines:** 2.0L VIN F, 2.0L VIN Y, 2.4L VIN G **Transmissions:** A/T, M/T	**HO2S-12 (Bank 1 Sensor 2) Heater Circuit Malfunction** Engine started, ECT sensor more than 68°F, system voltage from 11-16v, and the PCM detected the HO2S-12 heater current was less than 0.2, or that it was more than 3.5 amps for 6 seconds in the test. **Possible Causes:** • HO2S heater control circuit is open or shorted to ground • HO2S heater control circuit is shorted to power • HO2S heater power circuit is open (check power from the relay) • HO2S heater is damaged or has failed • PCM has failed
DTC: P0141 **1T O2S HTR1 MIL: YES** **1995, 1996, 1997, 1998, 1999** **Models:** Eclipse **Engines:** 2.0L VIN Y **Transmissions:** A/T, M/T	**HO2S-12 (Bank 1 Sensor 2) Heater Circuit Malfunction** Key off for 5 seconds, then after the HO2S signal rose by over 490 mv within 144 seconds after key off period, and the initial rise in HO2S voltage was less than 1.57v, the PCM detected the HO2S-12 did not show a correct amount of voltage decrease during the test. **Possible Causes:** • HO2S heater power circuit is open (check for power from the MFI relay to the heater circuit) • HO2S heater control circuit is open between sensor and PCM • HO2S heater element has high resistance or has failed • PCM has failed
DTC: P0141 **1T O2S HTR1 MIL: YES** **2002, 2003, 2004, 2005** **Models:** Diamante, Eclipse, Galant, Lancer, Mirage, Montero, Montero Sport **Engines:** All **Transmissions:** A/T, M/T	**HO2S-12 (Bank 1 Sensor 2) Heater Circuit Malfunction** Engine started, greater than 60 seconds from the start of the previous monitoring, ECT sensor more than 68°F, system voltage from 11-16v, and the PCM detected the HO2S-12 heater current is less than 0.16 amps, or that it was more than 5 amps for 4 seconds during the HO2S Heater Monitor test. **Possible Causes:** • HO2S heater control circuit is open or shorted to ground • HO2S heater control circuit is shorted to power • HO2S heater power circuit is open (check power from the relay) • HO2S heater is damaged or has failed • PCM has failed
DTC: P0150 **2T CCM MIL: YES** **1996, 1997, 1998, 1999, 2000, 2001** **Models:** Montero, Montero Sport **Engines:** 3.0L VIN H, 3.0L VIN P, 3.5L VIN M, 3.5L VIN R **Transmissions:** A/T, M/T	**HO2S-21 (Bank 2 Sensor 1) Circuit Malfunction** Engine started, engine running in closed loop, and after the PCM increased or decreased the fuel injector pulse width with an override, it detected the HO2S-21 response time was slow; or it detected the HO2S-21 signal circuit was open during the CCM test. **Possible Causes:** • HO2S signal circuit is open between the sensor and the PCM • HO2S signal circuit is shorted to sensor or chassis ground • HO2S signal circuit is shorted to VREF or system power (B+) • HO2S is damaged, contaminated or it has failed • PCM has failed
DTC: P0150 **2T CCM MIL: YES** **1999, 2000, 2001** **Models:** Eclipse, Galant **Engines:** 3.0L VIN H, 3.0L VIN L **Transmissions:** A/T, M/T	**HO2S-21 (Bank 2 Sensor 1) Circuit Malfunction** Engine started, engine running in closed loop, and after the PCM increased or decreased the fuel injector pulse width with an override, it detected the HO2S-21 response time was slow; or it detected the HO2S-21 signal circuit was open during the CCM test. **Possible Causes:** • HO2S signal circuit is open between the sensor and the PCM • HO2S signal circuit is shorted to sensor or chassis ground • HO2S signal circuit is shorted to VREF or system power (B+) • HO2S is damaged, contaminated or it has failed • PCM has failed
DTC: P0150 **2T CCM MIL: YES** **2003, 2004, 2005, 2006** **Models:** All **Engines:** All V6 **Transmissions:** A/T, M/T	**HO2S-21 (Bank 2 Sensor 1) Circuit Malfunction** Engine started and running for 3 minutes or more, left front HO2S signal voltage is 0.2 or less, engine coolant temperature is greater than 169°F, engine speed greater than 1200 rpm, volumetric efficiency is greater than 25%, the input voltage supplied to the PCM interface circuit is greater than 4.5 volts when 5 volts is applied to the left front HO2S output line via a resister. **Possible Causes:** • HO2S signal circuit is open between the sensor and the PCM • HO2S signal circuit is shorted to sensor or chassis ground • HO2S signal circuit is shorted to VREF or system power (B+) • HO2S is damaged, contaminated or it has failed • PCM has failed

DTC	Trouble Code Title, Conditions & Possible Causes
DTC: P0151 **2T CCM MIL: YES** **2003, 2004, 2005, 2006** **Models:** All **Engines:** All V6 **Transmissions:** A/T, M/T	**HO2S-21 (Bank 2 Sensor 1) Circuit Low Voltage** Engine started and running for greater than 3 minutes, the left front heated oxygen sensor signal voltage has continued to be less than 0.2 volts, engine coolant temperature is greater than 169°F, engine speed is greater than 1200 rpm, volumetric efficiency is greater than 25%, MAF sensor output frequency is greater than 75 Hz, at least 20 seconds have passed since fuel shut-off control was canceled, and the changes in the front heated oxygen sensor is less than 0.078 voltage the front heated oxygen sensor output voltage increase beyond 0.2 volts after making the air/fuel ratio 15% richer for 10 seconds. **Possible Causes:** • HO2S signal circuit is open between the sensor and the PCM • HO2S signal circuit is shorted to sensor or chassis ground • HO2S signal circuit is shorted to VREF or system power (B+) • HO2S is damaged, contaminated or it has failed • PCM has failed
DTC: P0152 **2T CCM MIL: YES** **2002, 2003, 2004, 2005, 2006** **Models:** All **Engines:** All V6 **Transmissions:** A/T, M/T	**HO2S-21 (Bank 2 Sensor 1) Signal High Signal** Engine started, engine runtime over 2 minutes at cruise speed, ECT sensor more than 176°F, and the PCM detected the HO2S-21 signal was not less than 4.5v when 5.0v was applied to the HO2S-21 circuit for 3 seconds during the CCM test. **Possible Causes:** • HO2S signal circuit is open • HO2S signal circuit is shorted to the heater power circuit • HO2S is contaminated, damaged or it has failed • PCM has failed
DTC: P0153 **2T CCM MIL: YES** **2002, 2003, 2004, 2005, 2006** **Models:** All **Engines:** All V6 **Transmissions:** A/T, M/T	**HO2S-21 (Bank 2 Sensor 1) Slow Response** Engine started, engine runtime over 3 minutes, ECT sensor more than 170°F, vehicle driven to a speed over 24 mph for 75 seconds, then at idle speed from 512-864 rpm, PSPS signal "off", A/C clutch not cycling, and the PCM detected the HO2S-11 signal that did not reach 670 mv, or it did not switch between 0.39-6.0v engine times in a 6 second period during the Oxygen Sensor Monitor test. **Possible Causes:** • Exhaust leak present in the exhaust manifold or exhaust pipes • HO2S element fuel contamination • HO2S element has deteriorated • PCM has failed
DTC: P0154 **2T CCM MIL: YES** **2002, 2003, 2004, 2005, 2006** **Models:** All **Engines:** All V6 **Models:** Diamante, Montero, Montero Sport **Engines:** All **Transmissions:** A/T, M/T	**HO2S-11 (Bank 1 Sensor 1) No Activity Detected** Engine started, engine runtime over 2 minutes, ECT sensor more than 176°F, and the PCM detected the HO2S-11 signal remained fixed near 0.50v for 1.5 minutes in the Oxygen Sensor Monitor test. **Possible Causes:** • Exhaust leak present in exhaust manifold or exhaust pipes • HO2S element fuel contamination or has deteriorated • HO2S signal circuit or the ground circuit has high resistance • HO2S heater element has failed, or the heater circuit is open • PCM has failed
DTC: P0155 **2T O2S HTR2 MIL: YES** **2000, 2001, 2002** **Models:** Eclipse **Engines:** 3.0L VIN H **Transmissions:** A/T, M/T	**HO2S-21 (Bank 2 Sensor 1) Heater Circuit Malfunction** Key off for 5 seconds, then after the HO2S signal rose by over 490 mv within 144 seconds after key off period, and the initial rise in HO2S voltage was less than 1.57v, the PCM detected the HO2S-21 did not show a correct amount of voltage decrease during the test. **Possible Causes:** • HO2S heater power circuit is open (check for power from the MFI relay to the heater circuit) • HO2S heater control circuit is open between sensor and PCM • HO2S heater element has high resistance or has failed • PCM has failed
DTC: P0155 **2T O2S HTR1 MIL: YES** **1996, 1997, 1998, 1999, 2000, 2001, 2002, 2003, 2004, 2005** **Models:** All **Engines:** All V6 **Transmissions:** A/T, M/T	**HO2S-21 (Bank 2 Sensor 1) Heater Circuit Malfunction** Engine started, ECT sensor more than 68°F, system voltage from 11-16v, and the PCM detected the HO2S-21 heater current was less than 0.20v; or it was more than 3.5 amps for 6 seconds in the test. **Possible Causes:** • HO2S heater control circuit is open or shorted to ground • HO2S heater control circuit is shorted to power • HO2S heater power circuit is open (check power from the relay) • HO2S heater is damaged or has failed • PCM has failed

DTC	Trouble Code Title, Conditions & Possible Causes
DTC: P0155 **2T O2S HTR2 MIL: YES** **1996, 1997, 1998, 1999, 2000, 2001** **Models:** Montero, Montero Sport **Engines:** 3.0L VIN H, 3.0L VIN P, 3.5L VIN M, 3.5L VIN R **Transmissions:** A/T, M/T	**HO2S-21 (Bank 2 Sensor 1) Heater Circuit Malfunction** Key off for 5 seconds, then after the HO2S signal rose by over 490 mv within 144 seconds after key off period, and the initial rise in HO2S voltage was less than 1.57v, the PCM detected the HO2S-21 did not show a correct amount of voltage decrease during the test. **Possible Causes:** • HO2S heater power circuit is open (check for power from the MFI relay to the heater circuit) • HO2S heater control circuit is open between sensor and PCM • HO2S heater element has high resistance or has failed • PCM has failed
DTC: P0156 **2T CCM MIL: YES** **1996, 1997, 1998, 1999, 2000, 2001, 2002, 2003, 2004, 2005, 2006** **Models:** All **Engines:** All V6 **Transmissions:** A/T, M/T	**HO2S-22 (Bank 2 Sensor 2) Circuit Malfunction** Engine started, engine running in closed loop, and after the PCM increased or decreased the fuel injector pulse width with an override, it detected the HO2S-22 response time was slow, or it detected the HO2S-22 signal circuit was open during the CCM test. **Possible Causes:** • HO2S signal circuit is open between the sensor and the PCM • HO2S signal circuit is shorted to sensor or chassis ground • HO2S signal circuit is shorted to VREF or system power (B+) • HO2S is damaged, contaminated or it has failed • PCM has failed
DTC: P0157 **2T CCM MIL: YES** **2002, 2003, 2004, 2005, 2006** **Models:** All **Engines:** All V6 **Transmissions:** A/T, M/T	**HO2S-22 (Bank 2 Sensor 2) Signal Low Input** Engine started, ECT sensor signal less than 120°F at engine startup, engine runtime over 3 seconds, and the PCM detected the HO2S-22 signal was less than 0.16v during the CCM test. **Possible Causes:** • HO2S signal circuit is open • HO2S signal circuit is shorted to ground • HO2S is damaged or it has failed • PCM has failed
DTC: P0158 **2T CCM MIL: YES** **2002, 2003, 2004, 2005, 2006** **Models:** All **Engines:** All V6 **Transmissions:** A/T, M/T	**HO2S-22 (Bank 2 Sensor 2) Signal High Input** Engine started, engine runtime over 2 minutes, ECT sensor more than 176°F, and the PCM detected the HO2S-22 signal was more than 1.2v for 3 seconds during the CCM test. **Possible Causes:** • HO2S signal circuit shorted to power (check the heater circuit) • HO2S is damaged or it has failed • PCM has failed
DTC: P0159 **2T O2S1 MIL: YES** **2002, 2003, 2004, 2005** **Models:** All **Engines:** All V6 **Transmissions:** A/T, M/T	**HO2S-22 (Bank 2 Sensor 2) No Activity Detected** Engine started, engine runtime over 2 minutes, ECT sensor more than 176°F, and the PCM detected the HO2S-22 signal remained fixed near 0.50v for 1.5 minutes in the Oxygen Sensor Monitor test. **Possible Causes:** • HO2S signal circuit is open or it has high resistance • HO2S element fuel contamination or has deteriorated • HO2S is damaged or has failed • PCM has failed
DTC: P0159 **2T CCM MIL: YES** **2006** **Models:** All **Engines:** All V6 **Transmissions:** A/T, M/T	**HO2S-22 (Bank 2 Sensor 2) Slow Response** Engine started, engine runtime over 3 minutes, ECT sensor more than 170°F, vehicle driven to a speed over 24 mph for 75 seconds, then at idle speed from 512-864 rpm, PSPS signal "off", A/C clutch not cycling, and the PCM detected the HO2S-21 signal that did not reach 670 mv, or it did not switch between 0.39-6.0v engine times in a 6 second period during the Oxygen Sensor Monitor test. **Possible Causes:** • Exhaust leak present in the exhaust manifold or exhaust pipes • HO2S element fuel contamination • HO2S element has deteriorated • PCM has failed
DTC: P0160 **2T CCM MIL: YES** **2006** **Models:** All **Engines:** All V6 **Transmissions:** A/T, M/T	**HO2S-22 (Bank 2 Sensor 2) No Activity Detected** Engine started, engine runtime over 2 minutes, ECT sensor more than 176°F, and the PCM detected the HO2S-22 signal remained fixed near 0.50v for 1.5 minutes in the Oxygen Sensor Monitor test. **Possible Causes:** • Exhaust leak present in exhaust manifold or exhaust pipes • HO2S element fuel contamination or has deteriorated • HO2S signal circuit or the ground circuit has high resistance • HO2S heater element has failed, or the heater circuit is open • PCM has failed

DTC	Trouble Code Title, Conditions & Possible Causes
DTC: P0161 **1T O2S HTR1 MIL: YES** **1996, 1997, 1998, 1999, 2000, 2001, 2002** **Models:** 3000GT, Diamante **Engines:** All **Transmissions:** A/T, M/T	**HO2S-12 (Bank 2 Sensor 2) Heater Circuit Malfunction** Engine started, ECT sensor more than 68°F, system voltage from 11-16v, and the PCM detected the HO2S-22 heater current was less than 0.20v, or it was more than 3.5 amps for 6 seconds in the test. **Possible Causes:** • HO2S heater control circuit is open or shorted to ground • HO2S heater control circuit is shorted to power • HO2S heater power circuit is open (check power from the relay) • HO2S heater is damaged or has failed • PCM has failed
DTC: P0161 **1T O2S HTR1 MIL: YES** **1999, 2000, 2001, 2002** **Models:** Eclipse, Galant **Engines:** 3.0L VIN H, 3.0L VIN L **Transmissions:** A/T, M/T	**HO2S-12 (Bank 2 Sensor 2) Heater Circuit Malfunction** Engine started, ECT sensor more than 68°F, system voltage from 11-16v, and the PCM detected the HO2S-22 heater current was less than 0.20v, or it was more than 3.5 amps for 6 seconds in the test. **Possible Causes:** • HO2S heater control circuit is open or shorted to ground • HO2S heater control circuit is shorted to power • HO2S heater power circuit is open (check power from the relay) • HO2S heater is damaged or has failed • PCM has failed
DTC: P0161 **2T O2S HTR1 MIL: YES 1996, 1997, 1998, 1999, 2000, 2001, 2002** **Models:** Montero, Montero Sport **Engines:** 3.0L VIN H, 3.0L VIN P, 3.5L VIN M, 3.5L VIN R **Transmissions:** A/T, M/T	**HO2S-22 (Bank 2 Sensor 2) Heater Circuit Malfunction** Key off for 5 seconds, then after the HO2S signal rose by over 490 mv within 144 seconds after key off period, and the initial rise in HO2S voltage was less than 1.57v, the PCM detected the HO2S-22 did not show a correct amount of voltage decrease during the test. **Possible Causes:** • HO2S heater power circuit is open (check for power from the MFI relay to the heater circuit) • HO2S heater control circuit is open between sensor and PCM • HO2S heater element has high resistance or has failed • PCM has failed
DTC: P0170 **2T FUEL MIL: YES** **1995, 1996, 1997, 1998, 1999** **Models:** Eclipse **Engines:** 2.0L VIN F, 2.0L VIN Y, 2.4L VIN G **Transmissions:** A/T, M/T	**Fuel Trim Fault (Bank 1)** DTC P0100, P0105, P0110, P0115, P0120, P0130, P0135, P0136, P0141, P0151, P0155, P0156, P0161, P0300, P0301-306, P0440, P0500 and P0505 not set, ECT and IAT sensors signals more than 14°F, MAP sensor over 76 kPa, MAF sensor more than 80 Hz, vehicle driven at a constant speed of less than 65 mph, and the PCM detected the Short Term fuel trim was less than -12.5% or more than +12.5%; or it detected the Long Term fuel trim value was less than -12.5%, or more than +22.4% for 5-10 seconds in the test. **Possible Causes:** • Air leaks after the MAF sensor, or in the EGR or PCV system • Base engine "mechanical" fault affecting one or more cylinders • Exhaust leaks located in front of the A/FS or HO2S location • Fuel control sensor is out of calibration (i.e., ECT, IAT or MAP) • Fuel delivery system supplying too little fuel during cruise or idle periods (e.g., faulty fuel pump or dirty, restricted fuel filter) • Fuel injector (one or more) dirty or pressure regulator has failed • HO2S is contaminated, deteriorated or it has failed • Vehicle driven low on fuel or until it ran out of fuel
DTC: P0170 **2T FUEL MIL: YES** **1996, 1997, 1998, 1999, 2000, 2001** **Models:** 3000GT, Diamante, Galant, Mirage, Montero, Montero Sport **Engines:** All **Transmissions:** A/T, M/T	**Fuel Trim Fault (Bank 1)** DTC P0100, P0105, P0110, P0115, P0120, P0130, P0135, P0136, P0141, P0151, P0155, P0156, P0161, P0300, P0301-306, P0440, P0500 and P0505 not set, ECT and IAT sensors signals more than 14°F, MAP sensor over 76 kPa, MAF sensor more than 80 Hz, vehicle driven at a constant speed of less than 65 mph, and the PCM detected the Short Term fuel trim was less than -12.5% or more than +12.5%; or it detected the Long Term fuel trim value was less than -12.5%, or more than +22.4% for 5-10 seconds in the test. **Possible Causes:** • Air leaks after the MAF sensor, or in the EGR or PCV system • Base engine "mechanical" fault affecting one or more cylinders • Exhaust leaks located in front of the A/FS or HO2S location • Fuel control sensor is out of calibration (i.e., ECT, IAT or MAP) • Fuel delivery system supplying too little fuel during cruise or idle periods (e.g., faulty fuel pump or dirty, restricted fuel filter) • Fuel injector (one or more) dirty or pressure regulator has failed • HO2S is contaminated, deteriorated or it has failed • Vehicle driven low on fuel or until it ran out of fuel

DTC	Trouble Code Title, Conditions & Possible Causes
DTC: P0171 **2T FUEL MIL: YES** **1996, 1997, 1998, 1999** **Models:** Eclipse **Engines:** 2.0L VIN Y **Transmissions:** A/T, M/T	**Fuel System Lean (Bank 1)** DTC P0106, P0107, P0108, P0111, P0112, P0113, P0117, P0118, P0121, P0122, P0123, P0131, P0132, P0133, P0134, P0135, P0138, P0141, P0201-P0204, P0300, P0301-304, P0441, P0500 and P0505 not set, ECT signal over 170°F, vehicle driven at a steady speed of over 62 mph at a speed over 1500 rpm, and the PCM detected the Long Term fuel trim was +25% and the Short Term fuel trim was +12% indicating a lean condition during the test. **Possible Causes:** • Air leaks after the MAF sensor, or in the EGR or PCV system • Base engine "mechanical" fault affecting one or more cylinders • Fuel control sensor is out of calibration (i.e., ECT, IAT or MAP) • Fuel delivery system supplying too little fuel during cruise or idle periods (e.g., faulty fuel pump or dirty, restricted fuel filter) • Fuel injector (one or more) dirty or pressure regulator has failed • HO2S is contaminated, deteriorated or it has failed • Vehicle driven low on fuel or until it ran out of fuel
DTC: P0171 **2T FUEL MIL: YES** **2002, 2003, 2004, 2005, 2006** **Models:** Diamante, Endeavor, Evo, Eclipse, Galant, Lancer, Mirage, Montero, Montero Sport, Outlander **Engines:** All **Transmissions:** A/T, M/T	**Fuel System Lean (Bank 1)** DTC P0101, P0102, P0103, P0106, P0107, P0108, P0111, P0112, P0113, P0117, P0118, P0121, P0122, P0123, P0131, P0132, P0133, P0134, P0135, P0138, P0141, P0151, P0152, P0153, P0154, P0155, P0157, P0158, P0159, P0161, P0201-P0206, P0300, P0301-306, P0441, P0500 and P0505 not set, ECT signal over 170°F, vehicle driven at a steady speed of over 62 mph at a speed over 1500 rpm, and the PCM detected the Long Term fuel trim was +25% and the Short Term fuel trim was +12% indicating a lean condition during the test. **Possible Causes:** • Air leaks after the MAF sensor, or in the EGR or PCV system • Base engine "mechanical" fault affecting one or more cylinders • Fuel control sensor is out of calibration (i.e., ECT, IAT or MAP) • Fuel delivery system supplying too little fuel during cruise or idle periods (e.g., faulty fuel pump or dirty, restricted fuel filter) • Fuel injector (one or more) dirty or pressure regulator has failed • HO2S is contaminated, deteriorated or it has failed • Vehicle driven low on fuel or until it ran out of fuel
DTC: P0172 **2T FUEL MIL: YES** **1996, 1997, 1998, 1999** **Models:** Eclipse **Engines:** 2.0L VIN Y **Transmissions:** A/T, M/T	**Fuel System Rich (Bank 1)** DTC P0106, P0107, P0108, P0111, P0112, P0113, P0117, P0118, P0121, P0122, P0123, P0131, P0132, P0133, P0134, P0135, P0138, P0141, P0201-P0204, P0300, P0301-304, P0441, P0500 and P0505 not set, ECT signal over 170°F, vehicle driven at a steady speed of over 62 mph at a speed over 1500 rpm, and the PCM detected the Long Term fuel trim was -25% and the Short Term fuel trim was -12% indicating a rich condition during the test. **Possible Causes:** • Base engine "mechanical" fault affecting one or more cylinders • EVAP system component has failed or canister fuel saturated • Exhaust leaks located in front of the A/FS or HO2S location • Fuel control sensor is out of calibration (i.e., ECT, IAT or MAF) • Fuel delivery system supplying too much fuel during cruise or idle periods (e.g., faulty fuel pump, or faulty pressure regulator) • Fuel injector(s) is leaking or stuck partially open (one or more) • HO2S is contaminated, deteriorated or it has failed
DTC: P0172 **2T FUEL MIL: YES** **2002, 2003, 2004, 2005, 2006** **Models:** Diamante, Endeavor, Evo, Eclipse, Galant, Lancer, Mirage, Montero, Montero Sport, Outlander **Engines:** All **Transmissions:** A/T, M/T	**Fuel System Rich (Bank 1)** DTC P0101, P0102, P0103, P0106, P0107, P0108, P0111, P0112, P0113, P0117, P0118, P0121, P0122, P0123, P0131, P0132, P0133, P0134, P0135, P0138, P0141, P0151, P0152, P0153, P0154, P0155, P0157, P0158, P0159, P0161, P0201-P0206, P0300, P0301-306, P0441, P0500 and P0505 not set, ECT signal over 170°F, vehicle driven at a steady speed of over 62 mph at a speed over 1500 rpm, and the PCM detected the Long Term fuel trim was -25% and the Short Term fuel trim was -12% indicating a lean condition during the test. **Possible Causes:** • Base engine "mechanical" fault affecting one or more cylinders • EVAP system component has failed or canister fuel saturated • Exhaust leaks located in front of the A/FS or HO2S location • Fuel control sensor is out of calibration (i.e., ECT, IAT or MAF) • Fuel delivery system supplying too much fuel during cruise or idle periods (e.g., faulty fuel pump, or faulty pressure regulator) • Fuel injector(s) is leaking or stuck partially open (one or more) • HO2S is contaminated, deteriorated or it has failed

DTC	Trouble Code Title, Conditions & Possible Causes
DTC: P0173 **2T FUEL MIL: YES** **1996, 1997, 1998, 1999, 2000, 2001** **Models:** 3000GT, Diamante, Montero, Montero Sport **Engines:** All **Transmissions:** A/T, M/T	**Fuel Trim Fault (Bank 2)** DTC P0100, P0105, P0110, P0115, P0120, P0130, P0135, P0136, P0141, P0151, P0155, P0156, P0161, P0300, P0301-306, P0440, P0500 and P0505 not set, ECT and IAT sensors signals more than 14°F, MAP sensor over 76 kPa, MAF sensor more than 80 Hz, vehicle driven at a constant speed of less than 65 mph, and the PCM detected the Short Term fuel trim was less than -12.5% or more than +12.5%; or it detected the Long Term fuel trim value was less than -12.5%, or more than +22.4% for 5-10 seconds in the test. **Possible Causes:** • Air leaks after the MAF sensor, or in the EGR or PCV system • Base engine "mechanical" fault affecting one or more cylinders • Exhaust leaks located in front of the A/FS or HO2S location • Fuel control sensor is out of calibration (i.e., ECT, IAT or MAP) • Fuel delivery system supplying too little fuel during cruise or idle periods (e.g., faulty fuel pump or dirty, restricted fuel filter) • Fuel injector (one or more) dirty or pressure regulator has failed • HO2S is contaminated, deteriorated or it has failed • Vehicle driven low on fuel or until it ran out of fuel
DTC: P0173 **2T FUEL MIL: YES** **1999, 2000, 2001** **Models:** Galant, Eclipse **Engines:** 3.0L VIN L **Transmissions:** A/T, M/T	**Fuel Trim Fault (Bank 2)** DTC P0100, P0105, P0110, P0115, P0120, P0130, P0135, P0136, P0141, P0151, P0155, P0156, P0161, P0300, P0301-306, P0440, P0500 and P0505 not set, ECT and IAT sensors signals more than 14°F, MAP sensor over 76 kPa, MAF sensor more than 80 Hz, vehicle driven at a constant speed of less than 65 mph, and the PCM detected the Short Term fuel trim was less than -12.5% or more than +12.5%; or it detected the Long Term fuel trim value was less than -12.5%, or more than +22.4% for 5-10 seconds in the test. **Possible Causes:** • Air leaks after the MAF sensor, or in the EGR or PCV system • Base engine "mechanical" fault affecting one or more cylinders • Exhaust leaks located in front of the A/FS or HO2S location • Fuel control sensor is out of calibration (i.e., ECT, IAT or MAP) • Fuel delivery system supplying too little fuel during cruise or idle periods (e.g., faulty fuel pump or dirty, restricted fuel filter) • Fuel injector (one or more) dirty or pressure regulator has failed • HO2S is contaminated, deteriorated or it has failed • Vehicle driven low on fuel or until it ran out of fuel
DTC: P0173 **2T FUEL MIL: YES** **2002, 2003, 2004** **Models:** Diamante **Engines:** All **Transmissions:** A/T	**Fuel Trim Fault (Bank 2)** DTC P0100, P0105, P0110, P0115, P0120, P0130, P0135, P0136, P0141, P0151, P0155, P0156, P0161, P0300, P0301-306, P0440, P0500 and P0505 not set, ECT and IAT sensors signals more than 14°F, MAP sensor over 76 kPa, MAF sensor more than 80 Hz, vehicle driven at a constant speed of less than 65 mph, and the PCM detected the Short Term fuel trim was less than -12.5% or more than +12.5%; or it detected the Long Term fuel trim value was less than -12.5%, or more than +22.4% for 5-10 seconds in the test. **Possible Causes:** • Air leaks after the MAF sensor, or in the EGR or PCV system • Base engine "mechanical" fault affecting one or more cylinders • Exhaust leaks located in front of the A/FS or HO2S location • Fuel control sensor is out of calibration (i.e., ECT, IAT or MAP) • Fuel delivery system supplying too little fuel during cruise or idle periods (e.g., faulty fuel pump or dirty, restricted fuel filter) • Fuel injector (one or more) dirty or pressure regulator has failed • HO2S is contaminated, deteriorated or it has failed • Vehicle driven low on fuel or until it ran out of fuel
DTC: P0174 **2T FUEL MIL: YES** **2001, 2002, 2003, 2004, 2005, 2006** **Models:** Diamante, Endeavor, Eclipse, Galant, Montero, Montero Sport **Engines:** All V6 **Transmissions:** A/T, M/T	**Fuel System Lean (Bank 2)** DTC P0101, P0102, P0103, P0106, P0107, P0108, P0111, P0112, P0113, P0117, P0118, P0121, P0122, P0123, P0131, P0132, P0133, P0134, P0135, P0138, P0141, P0151, P0152, P0153, P0154, P0155, P0157, P0158, P0159, P0161, P0201-P0206, P0300, P0301-306, P0441, P0500 and P0505 not set, ECT signal over 170°F, vehicle driven at a steady speed of over 62 mph at a speed over 1500 rpm, and the PCM detected the Long Term fuel trim was +25% and the Short Term fuel trim was +12% indicating a lean condition during the test. **Possible Causes:** • Air leaks after the MAF sensor, or in the EGR or PCV system • Base engine "mechanical" fault affecting one or more cylinders • Fuel control sensor is out of calibration (i.e., ECT, IAT or MAP) • Fuel delivery system supplying too little fuel during cruise or idle periods (e.g., faulty fuel pump or dirty, restricted fuel filter) • Fuel injector (one or more) dirty or pressure regulator has failed • HO2S is contaminated, deteriorated or it has failed • Vehicle driven low on fuel or until it ran out of fuel

DTC	Trouble Code Title, Conditions & Possible Causes
DTC: P0175 **2T FUEL MIL: YES** **2001, 2002, 2003, 2004, 2005, 2006** **Models:** Diamante, Endeavor, Eclipse, Galant, Montero, Montero Sport **Engines:** All V6 **Transmissions:** A/T, M/T	**Fuel System Rich (Bank 2)** DTC P0101, P0102, P0103, P0106, P0107, P0108, P0111, P0112, P0113, P0117, P0118, P0121, P0122, P0123, P0131, P0132, P0133, P0134, P0135, P0138, P0141, P0151, P0152, P0153, P0154, P0155, P0157, P0158, P0159, P0161, P0201-P0206, P0300, P0301-306, P0441, P0500 and P0505 not set, ECT signal over 170°F, vehicle driven at a steady speed of over 62 mph at a speed over 1500 rpm, and the PCM detected the Long Term fuel trim was -25% and the Short Term fuel trim was -12% indicating a lean condition during the test. **Possible Causes:** • Base engine "mechanical" fault affecting one or more cylinders • EVAP system component has failed or canister fuel saturated • Exhaust leaks located in front of the A/FS or HO2S location • Fuel control sensor is out of calibration (i.e., ECT, IAT or MAF) • Fuel delivery system supplying too much fuel during cruise or idle periods (e.g., faulty fuel pump, or faulty pressure regulator) • Fuel injector(s) is leaking or stuck partially open (one or more) • HO2S is contaminated, deteriorated or it has failed
DTC: P0181 **2T CCM MIL: YES** **2001, 2002, 2003, 2004, 2005, 2006** **Models:** Endeavor, Eclipse, Galant, Lancer, Montero, Montero Sport, Outlander **Engines:** All **Transmissions:** A/T, M/T	**Fuel Temperature Sensor Range/Performance** Engine started, ECT sensor from 14-97°F at startup, engine running, IAT sensor more than 9°F, vehicle driven to a speed over 17 mph, ECT sensor more than 140°F, and the PCM detected the Fuel Temperature sensor signal was different than the ECT sensor at startup by more than 27°F during the CCM test. **Possible Causes:** • Fuel temperature sensor signal circuit is open (intermittent) • Fuel temperature sensor signal circuit is shorted to ground • Fuel temperature sensor is damaged or has failed • PCM has failed
DTC: P0182 **2T CCM MIL: YES** **2001, 2002, 2003, 2004, 2005, 2006** **Models:** Endeavor, Eclipse, Galant, Lancer, Montero, Montero Sport, Outlander **Engines:** All **Transmissions:** A/T, M/T	**Fuel Temperature Sensor Circuit Low Input** Key on or engine running, and the PCM detected the Fuel Temperature sensor indicated less than 0.10v for 2 seconds during the CCM test. **Possible Causes:** • Fuel temperature sensor signal circuit is shorted to ground • Fuel temperature sensor is damaged or has failed • PCM has failed
DTC: P0183 **2T CCM MIL: YES** **2001, 2002, 2003, 2004, 2005, 2006** **Models:** Endeavor, Eclipse, Galant, Lancer, Montero, Montero Sport, Outlander **Engines:** All **Transmissions:** A/T, M/T	**Fuel Temperature Sensor Circuit High Input** Key on or engine running, and the PCM detected the Fuel Temperature sensor indicated more than 4.60v for 2 seconds during the CCM test. **Possible Causes:** • Fuel temperature sensor signal circuit is open • Fuel temperature sensor signal circuit shorted to system power • Fuel temperature sensor ground circuit is open • Fuel temperature sensor is damaged or has failed • PCM has failed
DTC: P0201 **2T CCM MIL: YES** **1995, 1996, 1997, 1998, 1999** **Models:** Eclipse **Engines:** All **Transmissions:** A/T, M/T	**Fuel Injector 1 Circuit Malfunction** Engine started, engine speed less than 1000 rpm, TP sensor less than 0.7v, Actuator Tests all "off", and the PCM detected the injector coil surge voltage was too low (i.e., it did not reach system voltage +2v) within 2 seconds after the injector turned "off" during the test. **Possible Causes:** • Injector control circuit is open or shorted to ground • Injector power circuit is open (check power from the MFI relay) • Fuel injector is damaged or has failed
DTC: P0201 **2T CCM MIL: YES** **1996, 1997, 1998, 1999, 2000, 2001, 2002** **Models:** 3000GT, Diamante, Galant, Lancer, Mirage, Montero, Montero Sport **Engines:** All **Transmissions:** A/T, M/T	**Fuel Injector 1 Circuit Malfunction** Engine started, engine speed less than 1000 rpm, TP sensor less than 1.1v, Actuator Tests all "off", and the PCM detected the injector coil surge voltage was too low (i.e., it did not reach system voltage +2v) within 2 seconds after the injector turned "off" during the test. **Possible Causes:** • Injector control circuit is open or shorted to ground • Injector power circuit is open (check power from the MFI relay) • Fuel injector is damaged or has failed

DTC	Trouble Code Title, Conditions & Possible Causes
DTC: P0201 **2T CCM MIL: YES** **2000, 2001, 2002, 2003, 2004, 2005, 2006** **Models:** All **Engines:** All **Transmissions:** A/T, M/T	**Fuel Injector 1 Circuit Malfunction** Engine started, engine speed less than 1000 rpm, TP sensor less than 1.1v, Actuator Tests all "off", and the PCM detected the injector coil surge voltage was too low (i.e., it did not reach system voltage +2v) within 2 seconds after the injector turned "off" during the test. **Possible Causes:** • Injector control circuit is open or shorted to ground • Injector power circuit is open (check power from the MFI relay) • Fuel injector is damaged or has failed
DTC: P0202 **2T CCM MIL: YES** **1995, 1996, 1997, 1998, 1999** **Models:** Eclipse **Engines:** All **Transmissions:** A/T, M/T	**Fuel Injector 2 Circuit Malfunction** Engine started, engine speed less than 1000 rpm, TP sensor less than 0.7v, Actuator Tests all "off", and the PCM detected the injector coil surge voltage was too low (i.e., it did not reach system voltage +2v) within 2 seconds after the injector turned "off" during the test. **Possible Causes:** • Injector control circuit is open or shorted to ground • Injector power circuit is open (check power from the MFI relay) • Fuel injector is damaged or has failed
DTC: P0202 **2T CCM MIL: YES** **1996, 1997, 1998, 1999, 2000, 2001, 2002** **Models:** 3000GT, Diamante, Galant, Lancer Mirage, Montero, Montero Sport **Engines:** All **Transmissions:** A/T, M/T	**Fuel Injector 2 Circuit Malfunction** Engine started, engine speed less than 1000 rpm, TP sensor less than 1.1v, Actuator Tests all "off", and the PCM detected the injector coil surge voltage was too low (i.e., it did not reach system voltage +2v) within 2 seconds after the injector turned "off" during the test. **Possible Causes:** • Injector control circuit is open or shorted to ground • Injector power circuit is open (check power from the MFI relay) • Fuel injector is damaged or has failed
DTC: P0202 **2T CCM MIL: YES** **2000, 2001, 2002, 2003, 2004, 2005, 2006** **Models:** All **Engines:** All **Transmissions:** A/T, M/T	**Fuel Injector 2 Circuit Malfunction** Engine started, engine speed less than 1000 rpm, TP sensor less than 1.1v, Actuator Tests all "off", and the PCM detected the injector coil surge voltage was too low (i.e., it did not reach system voltage +2v) within 2 seconds after the injector turned "off" during the test. **Possible Causes:** • Injector control circuit is open or shorted to ground • Injector power circuit is open (check power from the MFI relay) • Fuel injector is damaged or has failed
DTC: P0203 **2T CCM MIL: YES** **1995, 1996, 1997, 1998, 1999** **Models:** Eclipse **Engines:** All **Transmissions:** A/T, M/T	**Fuel Injector 3 Circuit Malfunction** Engine started, engine speed less than 1000 rpm, TP sensor less than 0.7v, Actuator Tests all "off", and the PCM detected the injector coil surge voltage was too low (i.e., it did not reach system voltage +2v) within 2 seconds after the injector turned "off" during the test. **Possible Causes:** • Injector control circuit is open or shorted to ground • Injector power circuit is open (check power from the MFI relay) • Fuel injector is damaged or has failed
DTC: P0203 **2T CCM MIL: YES** **1996, 1997, 1998, 1999, 2000, 2001, 2002** **Models:** 3000GT, Diamante, Galant, Lancer Mirage, Montero, Montero Sport **Engines:** All **Transmissions:** A/T, M/T	**Fuel Injector 3 Circuit Malfunction** Engine started, engine speed less than 1000 rpm, TP sensor less than 1.1v, Actuator Tests all "off", and the PCM detected the injector coil surge voltage was too low (i.e., it did not reach system voltage +2v) within 2 seconds after the injector turned "off" during the test. **Possible Causes:** • Injector control circuit is open or shorted to ground • Injector power circuit is open (check power from the MFI relay) • Fuel injector is damaged or has failed
DTC: P0203 **2T CCM MIL: YES** **2000, 2001, 2002, 2003, 2004, 2005, 2006** **Models:** All **Engines:** All **Transmissions:** A/T, M/T	**Fuel Injector 3 Circuit Malfunction** Engine started, engine speed less than 1000 rpm, TP sensor less than 1.1v, Actuator Tests all "off", and the PCM detected the injector coil surge voltage was too low (i.e., it did not reach system voltage +2v) within 2 seconds after the injector turned "off" during the test. **Possible Causes:** • Injector control circuit is open or shorted to ground • Injector power circuit is open (check power from the MFI relay) • Fuel injector is damaged or has failed

DTC	Trouble Code Title, Conditions & Possible Causes
DTC: P0204 **2T CCM MIL: YES** **1995, 1996, 1997, 1998, 1999** **Models:** Eclipse **Engines:** All **Transmissions:** A/T, M/T	**Fuel Injector 4 Circuit Malfunction** Engine started, engine speed less than 1000 rpm, TP sensor less than 0.7v, Actuator Tests all "off", and the PCM detected the injector coil surge voltage was too low (i.e., it did not reach system voltage +2v) within 2 seconds after the injector turned "off" during the test. **Possible Causes:** • Injector control circuit is open or shorted to ground • Injector power circuit is open (check power from the MFI relay) • Fuel injector is damaged or has failed
DTC: P0204 **2T CCM MIL: YES** **1996, 1997, 1998, 1999, 2000, 2001, 2002** **Models:** 3000GT, Diamante, Galant, Lancer Mirage, Montero, Montero Sport **Engines:** All **Transmissions:** A/T, M/T	**Fuel Injector 4 Circuit Malfunction** Engine started, engine speed less than 1000 rpm, TP sensor less than 1.1v, Actuator Tests all "off", and the PCM detected the injector coil surge voltage was too low (i.e., it did not reach system voltage +2v) within 2 seconds after the injector turned "off" during the test. **Possible Causes:** • Injector control circuit is open or shorted to ground • Injector power circuit is open (check power from the MFI relay) • Fuel injector is damaged or has failed
DTC: P0204 **2T CCM MIL: YES** **2000, 2001, 2002, 2003, 2004, 2005, 2006** **Models:** All **Engines:** All **Transmissions:** A/T, M/T	**Fuel Injector 4 Circuit Malfunction** Engine started, engine speed less than 1000 rpm, TP sensor less than 1.1v, Actuator Tests all "off", and the PCM detected the injector coil surge voltage was too low (i.e., it did not reach system voltage +2v) within 2 seconds after the injector turned "off" during the test. **Possible Causes:** • Injector control circuit is open or shorted to ground • Injector power circuit is open (check power from the MFI relay) • Fuel injector is damaged or has failed
DTC: P0205 **2T CCM MIL: YES** **1996, 1997, 1998, 1999, 2000, 2001, 2002** **Models:** 3000GT, Diamante **Engines:** All **Transmissions:** A/T, M/T	**Fuel Injector 5 Circuit Malfunction** Engine started, engine speed less than 1000 rpm, TP sensor less than 1.1v, Actuator Tests all "off", and the PCM detected the injector coil surge voltage was too low (i.e., it did not reach system voltage +2v) within 2 seconds after the injector turned "off" during the test. **Possible Causes:** • Injector control circuit is open or shorted to ground • Injector power circuit is open (check power from the MFI relay) • Fuel injector is damaged or has failed
DTC: P0205 **2T CCM MIL: YES** **1996, 1997, 1998, 1999, 2000, 2001, 2002** **Models:** Galant, Montero, Montero Sport **Engines:** 3.0L VIN L, 3.0L VIN H, 3.0L VIN P, 3.5L VIN M, 3.5L VIN R **Transmissions:** A/T, M/T	**Fuel Injector 5 Circuit Malfunction** Engine started, engine speed less than 1000 rpm, TP sensor less than 0.7v, Actuator Tests all "off", and the PCM detected the injector coil surge voltage was too low (i.e., it did not reach system voltage +2v) within 2 seconds after the injector turned "off" during the test. **Possible Causes:** • Injector control circuit is open or shorted to ground • Injector power circuit is open (check power from the MFI relay) • Fuel injector is damaged or has failed
DTC: P0205 **2T CCM MIL: YES** **2000, 2001, 2002, 2003, 2004, 2005, 2006** **Models:** All **Engines:** All V6 **Transmissions:** A/T, M/T	**Fuel Injector 5 Circuit Malfunction** Engine started, engine speed less than 1000 rpm, TP sensor less than 1.1v, Actuator Tests all "off", and the PCM detected the injector coil surge voltage was too low (i.e., it did not reach system voltage +2v) within 2 seconds after the injector turned "off" during the test. **Possible Causes:** • Injector control circuit is open or shorted to ground • Injector power circuit is open (check power from the MFI relay) • Fuel injector is damaged or has failed
DTC: P0206 **2T CCM MIL: YES** **1996, 1997, 1998, 1999, 2000, 2001, 2002** **Models:** 3000GT, Diamante **Engines:** All **Transmissions:** A/T, M/T	**Fuel Injector 6 Circuit Malfunction** Engine started, engine speed less than 1000 rpm, TP sensor less than 1.1v, Actuator Tests all "off", and the PCM detected the injector coil surge voltage was too low (i.e., it did not reach system voltage +2v) within 2 seconds after the injector turned "off" during the test. **Possible Causes:** • Injector control circuit is open or shorted to ground • Injector power circuit is open (check power from the MFI relay) • Fuel injector is damaged or has failed

DTC	Trouble Code Title, Conditions & Possible Causes
DTC: P0206 **2T CCM MIL: YES** **1996, 1997, 1998, 1999, 2000, 2001, 2002** **Models:** Galant, Montero, Montero Sport **Engines:** 3.0L VIN L, 3.0L VIN H, 3.0L VIN P, 3.5L VIN M, 3.5L VIN R **Transmissions:** A/T, M/T	**Fuel Injector 6 Circuit Malfunction** Engine started, engine speed less than 1000 rpm, TP sensor less than 0.7v, Actuator Tests all "off", and the PCM detected the injector coil surge voltage was too low (i.e., it did not reach system voltage +2v) within 2 seconds after the injector turned "off" during the test. **Possible Causes:** • Injector control circuit is open or shorted to ground • Injector power circuit is open (check power from the MFI relay) • Fuel injector is damaged or has failed
DTC: P0206 **2T CCM MIL: YES** **2000, 2001, 2002, 2003, 2004, 2005, 2006** **Models:** All **Engines:** All V6 **Transmissions:** A/T, M/T	**Fuel Injector 6 Circuit Malfunction** Engine started, engine speed less than 1000 rpm, TP sensor less than 1.1v, Actuator Tests all "off", and the PCM detected the injector coil surge voltage was too low (i.e., it did not reach system voltage +2v) within 2 seconds after the injector turned "off" during the test. **Possible Causes:** • Injector control circuit is open or shorted to ground • Injector power circuit is open (check power from the MFI relay) • Fuel injector is damaged or has failed
DTC: P0222 **1T CCM MIL: YES** **2006** **Models:** Except Evo **Engines:** All **Transmissions:** A/T, M/T	Throttle Position Sensor (Sub) Circuit Low Input Key on, TPS (sub) output voltage is 2.5 volts or less for 0.5 seconds. **Possible Causes:** • TPS failed • TPS open or shorted • PCM has failed
DTC: P0223 **1T CCM MIL: YES** **2006** **Models:** Except Evo **Engines:** All **Transmissions:** A/T, M/T	Throttle Position Sensor (Sub) Circuit Low Input Key on, TPS (sub) output voltage is 4.5 volts or more for 0.5 seconds. **Possible Causes:** • TPS failed • TPS open or shorted • PCM has failed
DTC: P0234 **2T CCM MIL: YES** **2006** **Models:** Evo **Engines:** All **Transmissions:** M/T	Turbocharger Wastegate System Malfunction Engine running, volumetric efficiency is greater than 210%–230%. **Possible Causes:** • Turbocharger wastegate actuator failed • Charging pressure control system failed • PCM failed
DTC: P0243 **2T CCM MIL: YES** **2006** **Models:** Evo **Engines:** All **Transmissions:** M/T	**Turbocharger Wastegate Solenoid Circuit Malfunction** Engine started, engine running with the system voltage between 10–16 volts, and the PCM detected the Wastegate solenoid coil surge voltage was not detected with wastegate solenoid is duty cycled between 10% and 90%too low (system voltage +2v) for 1 second. **Possible Causes:** • Turbocharger wastegate solenoid circuit is open or shorted to ground • Turbocharger wastegate solenoid failed • PCM has failed
DTC: P0300 **2T MISFIRE MIL: YES** **1995, 1996, 1997, 1998, 1999, 2000, 2001, 2002, 2003, 2004, 2005** **Models:** Eclipse **Engines:** 2.0L VIN F, 2.0L VIN Y, 2.4L VIN G **Transmissions:** A/T, M/T	**Random Misfire Detected** DTC P0101-P0103, P0106- P0108, P0112, P0113, P0117, P0118, P0120-123, P0125, P0131-P0135, P0137, P0138, P0141, P0441, P0500 and P0505 not set, vehicle driven to a speed over 2 mph at a steady throttle at an engine speed from 440-6500 rpm, ECT and IAT sensors more than 14°F, Crankshaft relearn completed, and the PCM detected multiple cylinders misfiring in over 1.8% of engine cycles in a 200 revolution period (Catalyst Misfire); or it detected multiple cylinders misfiring in over 1.8% of engine cycles within 1000 revolutions (High Emissions Misfire) in the Misfire test. **Note: If the misfire is severe, the MIL will flash on/off on the 1st trip!** **Possible Causes:** • Air leak in the intake manifold, or in the EGR or PCM system • Base engine mechanical fault that affects one or more cylinders • Erratic or interrupted CKP or CMP sensor signals • Fuel delivery component fault that affects one or more cylinders (i.e., a contaminated, dirty or sticking fuel injector) • Ignition system problem (coil or plug) in one or more cylinders

DTC	Trouble Code Title, Conditions & Possible Causes
DTC: P0300 **2T MISFIRE MIL: YES** **1996, 1997, 1998, 1999, 2000, 2001, 2002, 2003, 2004, 2005, 2006** **Models:** 3000GT, Diamante, Endeavor, Evo, Galant, Lancer, Mirage, Montero, Montero Sport, Outlander **Engines:** All **Transmissions:** A/T, M/T	**Random Misfire Detected** DTC P0100, P0105, P0110, P0115, P0120, P0130, P0135, P0136, P0141, P0151, P0155, P0156, P0161, P0440, P0500 and P0505 not set, vehicle driven to a speed of over 2 mph at a steady throttle with the engine speed from 440-6500 rpm, ECT and IAT sensors more than 14°F, CKP sensor "learn" finished, and the PCM detected multiple cylinders misfiring in over 1.8% of engine cycles in 200 revolutions (Catalyst Damaging Misfire); or it detected multiple cylinders misfiring in more than 1.8% of engine cycles within 1000 revolutions (High Emissions Misfire) during the Misfire Monitor test. **Note: If the misfire is severe, the MIL will flash on/off on the 1st trip!** **Possible Causes:** • Air leak in the intake manifold, or in the EGR or PCM system • Base engine mechanical fault that affects one or more cylinders • Erratic or interrupted CKP or CMP sensor signals • Fuel delivery component fault that affects one or more cylinders (i.e., a contaminated, dirty or sticking fuel injector) • Ignition system problem (coil or plug) in one or more cylinders
DTC: P0301 **2T MISFIRE MIL: YES** **1995, 1996, 1997, 1998, 1999, 2000, 2001, 2002, 2003, 2004, 2005** **Models:** Eclipse **Engines:** 2.0L VIN F, 2.0L VIN Y, 2.4L VIN G **Transmissions:** A/T, M/T	**Cylinder 1 Misfire Detected** DTC P0101-P0103, P0106- P0108, P0112, P0113, P0117, P0118, P0120-123, P0125, P0131-P0135, P0137, P0138, P0141, P0441, P0500 and P0505 not set, vehicle driven to a speed over 2 mph at a steady throttle at an engine speed from 440-6500 rpm, ECT and IAT sensors more than 14°F, Crankshaft relearn completed, and the PCM detected a single cylinder misfiring in over 1.8% of engine cycles in a 200 revolution period (Catalyst Misfire); or it detected a single cylinder misfiring in over 1.8% of engine cycles within 1000 revolutions (High Emissions Misfire) in the Misfire test. **Note: If the misfire is severe, the MIL will flash on/off on the 1st trip!** **Possible Causes:** • Air leak in the intake manifold, or in the EGR or PCM system • Base engine mechanical fault that affects only one cylinder • Fuel delivery component fault that affects only one cylinder (i.e., a contaminated, dirty or sticking fuel injector) • Ignition system problem (coil or plug) that affects one cylinder
DTC: P0301 **2T MISFIRE MIL: YES** **1996, 1997, 1998, 1999, 2000, 2001, 2002, 2003, 2004, 2005, 2006** **Models:** 3000GT, Diamante, Endeavor, Evo, Galant, Lancer, Mirage, Montero, Montero Sport, Outlander **Engines:** All **Transmissions:** A/T, M/T	**Cylinder 1 Misfire Detected** DTC P0100, P0105, P0110, P0115, P0120, P0130, P0135, P0136, P0141, P0151, P0155, P0156, P0161, P0440, P0500 and P0505 not set, vehicle driven to a speed of over 2 mph at a steady throttle with the engine speed from 440-6500 rpm, ECT and IAT sensors more than 14°F, CKP sensor "learn" finished, and the PCM detected a misfire in one cylinder in over 1.8% of engine cycles within 200 revolutions (Catalyst Damaging Misfire); or it detected a misfire in one cylinder in more than 1.8% of engine cycles within 1000 revolutions (High Emissions Misfire) in the Misfire Monitor test. **Note: If the misfire is severe, the MIL will flash on/off on the 1st trip!** **Possible Causes:** • Air leak in the intake manifold, or in the EGR or PCM system • Base engine mechanical fault that affects only one cylinder • Fuel delivery component fault that affects only one cylinder (i.e., a contaminated, dirty or sticking fuel injector) • Ignition system problem (coil or plug) that affects one cylinder • TSB 00-11-006 (7/00) has information about this code (V6 only)
DTC: P0302 **2T MISFIRE MIL: YES** **1995, 1996, 1997, 1998, 1999, 2000, 2001, 2002, 2003, 2004, 2005, 2006** **Models:** Eclipse **Engines:** 2.0L VIN F, 2.0L VIN Y, 2.4L VIN G **Transmissions:** A/T, M/T	**Cylinder 2 Misfire Detected** DTC P0101-P0103, P0106- P0108, P0112, P0113, P0117, P0118, P0120-123, P0125, P0131-P0135, P0137, P0138, P0141, P0441, P0500 and P0505 not set, vehicle driven to a speed over 2 mph at a steady throttle at an engine speed from 440-6500 rpm, ECT and IAT sensors more than 14°F, Crankshaft relearn completed, and the PCM detected a single cylinder misfiring in over 1.8% of engine cycles in a 200 revolution period (Catalyst Misfire); or it detected a single cylinder misfiring in over 1.8% of engine cycles within 1000 revolutions (High Emissions Misfire) in the Misfire test. **Note: If the misfire is severe, the MIL will flash on/off on the 1st trip!** **Possible Causes:** • Air leak in the intake manifold, or in the EGR or PCM system • Base engine mechanical fault that affects only one cylinder • Fuel delivery component fault that affects only one cylinder (i.e., a contaminated, dirty or sticking fuel injector) • Ignition system problem (coil or plug) that affects one cylinder
DTC: P0302 **2T MISFIRE MIL: YES** **1996, 1997, 1998, 1999, 2000, 2001, 2002, 2003, 2004, 2005, 2006** **Models:** 3000GT, Diamante, Endeavor, Evo, Galant, Lancer, Mirage, Montero, Montero Sport, Outlander **Engines:** All **Transmissions:** A/T, M/T	**Cylinder 2 Misfire Detected** DTC P0100, P0105, P0110, P0115, P0120, P0130, P0135, P0136, P0141, P0151, P0155, P0156, P0161, P0440, P0500 and P0505 not set, vehicle driven to a speed of over 2 mph at a steady throttle with the engine speed from 440-6500 rpm, ECT and IAT sensors more than 14°F, CKP sensor "learn" finished, and the PCM detected a misfire in one cylinder in over 1.8% of engine cycles within 200 revolutions (Catalyst Damaging Misfire); or it detected a misfire in one cylinder in more than 1.8% of engine cycles within 1000 revolutions (High Emissions Misfire) in the Misfire Monitor test. **Note: If the misfire is severe, the MIL will flash on/off on the 1st trip!** **Possible Causes:** • Air leak in the intake manifold, or in the EGR or PCM system • Base engine mechanical fault that affects only one cylinder • Fuel delivery component fault that affects only one cylinder (i.e., a contaminated, dirty or sticking fuel injector) • Ignition system problem (coil or plug) that affects one cylinder

DTC	Trouble Code Title, Conditions & Possible Causes
DTC: P0303 **2T MISFIRE MIL: YES** **1995, 1996, 1997, 1998, 1999, 2000, 2001, 2002, 2003, 2004, 2005, 2006** **Models:** Eclipse **Engines:** 2.0L VIN F, 2.0L VIN Y, 2.4L VIN G **Transmissions:** A/T, M/T	**Cylinder 3 Misfire Detected** DTC P0101-P0103, P0106- P0108, P0112, P0113, P0117, P0118, P0120-123, P0125, P0131-P0135, P0137, P0138, P0141, P0441, P0500 and P0505 not set, vehicle driven to a speed over 2 mph at a steady throttle at an engine speed from 440-6500 rpm, ECT and IAT sensors more than 14°F, Crankshaft relearn completed, and the PCM detected a single cylinder misfiring in over 1.8% of engine cycles in a 200 revolution period (Catalyst Misfire); or it detected a single cylinder misfiring in over 1.8% of engine cycles within 1000 revolutions (High Emissions Misfire) in the Misfire test. **Note: If the misfire is severe, the MIL will flash on/off on the 1st trip!** **Possible Causes:** • Air leak in the intake manifold, or in the EGR or PCM system • Base engine mechanical fault that affects only one cylinder • Fuel delivery component fault that affects only one cylinder (i.e., a contaminated, dirty or sticking fuel injector) • Ignition system problem (coil or plug) that affects one cylinder
DTC: P0303 **2T MISFIRE MIL: YES** **1996, 1997, 1998, 1999, 2000, 2001, 2002, 2003, 2004, 2005, 2006** **Models:** 3000GT, Diamante, Endeavor, Evo, Galant, Lancer, Mirage, Montero, Montero Sport, Outlander **Engines:** All **Transmissions:** A/T, M/T	**Cylinder 3 Misfire Detected** Engine started, vehicle speed over 1.6 mph at a steady throttle with the engine speed from 440-6500 rpm for 5 seconds, ECT and IAT sensors more than 14°F, CKP sensor "learn" finished, and the PCM detected a misfire in one cylinder in over 1.8% of engine cycles within 200 revolutions (Catalyst Damaging Misfire); or it detected a misfire in one cylinder in more than 1.8% of engine cycles within 1000 revolutions (High Emissions Misfire) in the Misfire Monitor test. **Note: If the misfire is severe, the MIL will flash on/off on the 1st trip!** **Possible Causes:** • Air leak in the intake manifold, or in the EGR or PCM system • Base engine mechanical fault that affects only one cylinder • Fuel delivery component fault that affects only one cylinder (i.e., a contaminated, dirty or sticking fuel injector) • Ignition system problem (coil or plug) that affects one cylinder • TSB 00-11-006 (7/00) has information about this code (V6 only)
DTC: P0304 **2T MISFIRE MIL: YES** **1995, 1996, 1997, 1998, 1999, 2000, 2001, 2002, 2003, 2004, 2005, 2006** **Models:** Eclipse **Engines:** 2.0L VIN F, 2.0L VIN Y, 2.4L VIN G **Transmissions:** A/T, M/T	**Cylinder 4 Misfire Detected** DTC P0101-P0103, P0106- P0108, P0112, P0113, P0117, P0118, P0120-123, P0125, P0131-P0135, P0137, P0138, P0141, P0441, P0500 and P0505 not set, vehicle driven to a speed over 2 mph at a steady throttle at an engine speed from 440-6500 rpm, ECT and IAT sensors more than 14°F, Crankshaft relearn completed, and the PCM detected a single cylinder misfiring in over 1.8% of engine cycles in a 200 revolution period (Catalyst Misfire); or it detected a single cylinder misfiring in over 1.8% of engine cycles within 1000 revolutions (High Emissions Misfire) in the Misfire test. **Note: If the misfire is severe, the MIL will flash on/off on the 1st trip!** **Possible Causes:** • Air leak in the intake manifold, or in the EGR or PCM system • Base engine mechanical fault that affects only one cylinder • Fuel delivery component fault that affects only one cylinder (i.e., a contaminated, dirty or sticking fuel injector) • Ignition system problem (coil or plug) that affects one cylinder
DTC: P0304 **2T MISFIRE MIL: YES** **1996, 1997, 1998, 1999, 2000, 2001, 2002, 2003, 2004, 2005, 2006** **Models:** 3000GT, Diamante, Endeavor, Evo, Galant, Lancer, Mirage, Montero, Montero Sport, Outlander **Engines:** All **Transmissions:** A/T, M/T **MIL: YES**	**Cylinder 4 Misfire Detected** DTC P0100, P0105, P0110, P0115, P0120, P0130, P0135, P0136, P0141, P0151, P0155, P0156, P0161, P0440, P0500 and P0505 not set, vehicle driven to a speed of over 2 mph at a steady throttle with the engine speed from 440-6500 rpm, ECT and IAT sensors more than 14°F, CKP sensor "learn" finished, and the PCM detected a misfire in one cylinder in over 1.8% of engine cycles within 200 revolutions (Catalyst Damaging Misfire); or it detected a misfire in one cylinder in more than 1.8% of engine cycles within 1000 revolutions (High Emissions Misfire) in the Misfire Monitor test. **Note: If the misfire is severe, the MIL will flash on/off on the 1st trip!** **Possible Causes:** • Air leak in the intake manifold, or in the EGR or PCM system • Base engine mechanical fault that affects only one cylinder • Fuel delivery component fault that affects only one cylinder (i.e., a contaminated, dirty or sticking fuel injector) • Ignition system problem (coil or plug) that affects one cylinder
DTC: P0305 **2T MISFIRE MIL: YES** **1996, 1997, 1998, 1999, 2000, 2001, 2002, 2003, 2004, 2005, 2006** **Models:** 3000GT, Diamante, Endeavor, Galant, Montero, Montero Sport **Engines:** All V6 **Transmissions:** A/T, M/T	**Cylinder 5 Misfire Detected** DTC P0100, P0105, P0110, P0115, P0120, P0130, P0135, P0136, P0141, P0151, P0155, P0156, P0161, P0440, P0500 and P0505 not set, vehicle driven to a speed of over 2 mph at a steady throttle with the engine speed from 440-6500 rpm, ECT and IAT sensors more than 14°F, CKP sensor "learn" finished, and the PCM detected a misfire in one cylinder in over 1.8% of engine cycles within 200 revolutions (Catalyst Damaging Misfire); or it detected a misfire in one cylinder in more than 1.8% of engine cycles within 1000 revolutions (High Emissions Misfire) in the Misfire Monitor test. **Note: If the misfire is severe, the MIL will flash on/off on the 1st trip!** **Possible Causes:** • Air leak in the intake manifold, or in the EGR or PCM system • Base engine mechanical fault that affects only one cylinder • Fuel delivery component fault that affects only one cylinder (i.e., a contaminated, dirty or sticking fuel injector) • Ignition system problem (coil or plug) that affects one cylinder • TSB 00-11-006 (7/00) has information about this code (V6 only)

DTC	Trouble Code Title, Conditions & Possible Causes
DTC: P0306 **2T MISFIRE MIL: YES** **1996, 1997, 1998, 1999, 2000, 2001, 2002, 2003, 2004, 2005, 2006** **Models:** 3000GT, Diamante, Endeavor, Galant, Montero, Montero Sport **Engines:** All V6 **Transmissions:** A/T, M/T	**Cylinder 6 Misfire Detected** DTC P0100, P0105, P0110, P0115, P0120, P0130, P0135, P0136, P0141, P0151, P0155, P0156, P0161, P0440, P0500 and P0505 not set, vehicle driven to a speed of over 2 mph at a steady throttle with the engine speed from 440-6500 rpm, ECT and IAT sensors more than 14°F, CKP sensor "learn" finished, and the PCM detected a misfire in one cylinder in over 1.8% of engine cycles within 200 revolutions (Catalyst Damaging Misfire); or it detected a misfire in one cylinder in more than 1.8% of engine cycles within 1000 revolutions (High Emissions Misfire) in the Misfire Monitor test. **Note: If the misfire is severe, the MIL will flash on/off on the 1st trip!** **Possible Causes:** • Air leak in the intake manifold, or in the EGR or PCM system • Base engine mechanical fault that affects only one cylinder • Fuel delivery component fault that affects only one cylinder (i.e., a contaminated, dirty or sticking fuel injector) • Ignition system problem (coil or plug) that affects one cylinder
DTC: P0325 **2T CCM MIL: YES** **1995, 1996, 1997, 1998, 1999, 2000, 2001, 2002, 2003, 2004, 2005, 2006** **Models:** All **Engines:** All **Transmissions:** A/T, M/T	**Knock Sensor 1 Circuit Malfunction** Engine started, engine speed from 2000-3000 rpm, and the PCM detected the change in the Knock sensor 1 signal was less than 0.06v, or the signal was over 5.0v during the last 200 revolutions. **Possible Causes:** • Knock sensor signal circuit is open or shorted to ground • Knock sensor signal circuit is shorted to VREF or system power • Knock sensor is damaged or has failed • Verify the Knock Sensor (KS) is tightened its specification • PCM has failed
DTC: P0335 **1T CCM MIL: YES** **1995, 1996, 1997, 1998, 1999, 2000, 2001, 2002, 2003, 2004, 2005, 2006** **Models:** All **Engines:** All **Transmissions:** A/T, M/T	**Crankshaft Position Sensor Circuit Malfunction** Engine cranking, and the PCM did not detect any Crankshaft Position (CKP) sensor signals; or with the engine running, it detected an abnormal pattern of CKP and CMP signal pulses for 2 seconds during the CCM test. **Possible Causes:** • CKP sensor signal circuit is open or shorted to ground • CKP sensor ground circuit is open • CKP sensor power circuit is open (check power to MFI relay) • CKP sensor is damaged or has failed • PCM has failed (the PCM provides the 5v VREF to the sensor)
DTC: P0340 **2T CCM MIL: YES** **1995, 1996, 1997, 1998, 1999, 2000, 2001, 2002, 2003, 2004, 2005, 2006** **Models:** All **Engines:** All **Transmissions:** A/T, M/T	**Camshaft Position Sensor Circuit Malfunction** Engine cranking, and the PCM did not detect any Crankshaft Position (CMP) sensor signals, or with the engine running, the normal pattern of CKP and CMP signals was not detected for 2 seconds. **Possible Causes:** • CMP sensor signal circuit is open or shorted to ground • CMP sensor ground circuit is open • CMP sensor power circuit is open (check power to MFI relay) • CMP sensor is damaged or has failed • PCM has failed (the PCM provides the 5v VREF to the sensor)
DTC: P0351 **2T CCM MIL: YES** **1995, 1996, 1997, 1998, 1999** **Models:** Eclipse **Engines:** 2.0L VIN Y **Transmissions:** A/T, M/T	**Ignition Coil 1 Primary Circuit Malfunction** Engine started, system voltage over 13.0v, engine speed less than 3000 rpm with the ignition timing stable, and the PCM detected the Coil 1 primary current did not achieve the maximum dwell time for 3 seconds during the CCM test. **Possible Causes:** • Ignition Coil 1 primary circuit is open or shorted to ground • Ignition Coil 1 power circuit is open (test power from the relay) • Ignition Coil 1 is damaged or has failed • PCM has failed
DTC: P0352 **2T CCM MIL: YES** **1995, 1996, 1997, 1998, 1999** **Models:** Eclipse **Engines:** 2.0L VIN Y **Transmissions:** A/T, M/T	**Ignition Coil 2 Primary Circuit Malfunction** Engine started, system voltage over 13.0v, engine speed less than 3000 rpm with the ignition timing stable, and the PCM detected the Coil 2 primary current did not achieve the maximum dwell time for 3 seconds during the CCM test. **Possible Causes:** • Ignition Coil 2 primary circuit is open or shorted to ground • Ignition Coil 2 power circuit is open (test power from the relay) • Ignition Coil 2 is damaged or has failed • PCM has failed

DTC	Trouble Code Title, Conditions & Possible Causes
DTC: P0365 **2T CCM MIL: YES** **2006** **Models:** Evo **Engines: Transmissions:** M/T	**Intake Camshaft Position Sensor Circuit** Engine running with engine speed greater than 50 rpm, the normal signal patter has not been input for the cylinder ID from the crankshaft position sensor signal and intake camshaft position sensor signal for 2 seconds. **Possible Causes:** • Intake camshaft position sensor failed • Intake camshaft position sensor circuit shorted or open • PCM has failed
DTC: P0400 **2T EGR MIL: YES** **1995, 1996, 1997, 1998, 1999** **Models:** Eclipse **Engines:** 2.0L VIN F, 2.0L VIN Y, 2.4L VIN G **Transmissions:** M/T	**Exhaust Gas Recirculation System Flow Malfunction** Engine started, engine runtime over 3 minutes, IAT sensor more than 14°F, ECT sensor more than 176°F, vehicle driven to over 19 mph at a speed of 1500-2000 rpm, MDP sensor from 1.80-2.70v and steady, then with the throttle closed during a deceleration period, the PCM did not detect enough change in the MAP signal after the EGR solenoid was commanded "on" and "off" in the test. **Possible Causes:** • EGR valve source vacuum supply line is open or restricted • EGR exhaust tube is clogged or restricted • EGR valve assembly is damaged or has failed • PCM has failed
DTC: P0400 **2T EGR MIL: YES** **1996, 1997, 1998, 1999, 2000, 2001** **Models:** 3000GT, Diamante, Galant, Mirage, Montero, Montero Sport **Engines:** All **Transmissions:** M/T	**Exhaust Gas Recirculation System Flow Malfunction** Engine started, engine runtime over 3 minutes, IAT sensor more than 14°F, ECT sensor more than 176°F, vehicle driven to over 19 mph at a speed of 940-2000 rpm, MAP signal over 76 kPa with no fluctuation for 1.5 seconds, then with the throttle closed during a deceleration period, the PCM detected too small an amount of change in the MAP signal with the EGR solenoid commanded "on" and then "off" during the EGR Monitor test. **Possible Causes:** • EGR valve source vacuum supply line is open or restricted • EGR exhaust tube is clogged or restricted • EGR valve assembly is damaged or has failed • PCM has failed
DTC: P0401 **2T EGR MIL: YES** **1995, 1996, 1997, 1998, 1999, 2000, 2001, 2002** **Models:** All **Engines:** 2.4L VIN G, 3.0L VIN H **Transmissions:** A/T, M/T	**Insufficient EGR Flow Detected** Engine started, engine runtime over 3 minutes in closed loop, ECT signal over 170°F, vehicle driven to over 3 mph at a speed of 1952-2400 rpm, MAP signal from 1.80-2.70v, TP sensor from 0.6-1.8v, Short Term fuel trim less than +4.4%, and the PCM detected the measured change in the Short Term fuel trim compensation shift was less than 7.4%, or the measured change in Short Term fuel trim compensation was more than 20.5% during the EGR System test. **Possible Causes:** • EGR solenoid control circuit is open or shorted to ground • EGR vacuum hose to source vacuum is loose or disconnected • EGR exhaust tube is clogged or restricted • EGR valve assembly is damaged or has failed • PCM has failed
DTC: P0401 **2T EGR MIL: YES** **1995, 1996, 1997, 1998, 1999** **Models:** Eclipse **Engines:** 2.0L VIN Y **Transmissions:** A/T, M/T	**Insufficient EGR Flow Detected** Engine started, engine runtime over 3 minutes in closed loop, ECT signal over 170°F, vehicle driven to over 3 mph at a speed of 1952-2400 rpm, MAP signal from 1.80-2.70v, TP sensor from 0.6-1.8v, Short Term fuel trim less than +4.4%, and the PCM detected the measured change in the Short Term fuel trim compensation shift was less than 7.4%, or the measured change in Short Term fuel trim compensation was more than 20.5% during the EGR System test. **Possible Causes:** • EGR valve source vacuum supply line is open or restricted • EGR exhaust tube is clogged or restricted • EGR valve assembly is damaged or has failed • PCM has failed
DTC: P0401 **2T EGR MIL: YES** **2002, 2003, 2004, 2005, 2006** **Models:** All **Engines:** All **Transmissions:** A/T, M/T	**Insufficient EGR Flow Detected** Engine started, engine runtime over 20 seconds in closed loop, ECT signal over 169°F, vehicle driven to over 19 mph at a speed of 910-1650 rpm, barometric pressure is greater than 11 psi, at least 90 seconds pass since MAP sensor output voltage fluctuated 1.5v or greater, TP sensor from 0.6-1.8v, volumetric efficiency is less than 28%, and the EGR valve opens to a prescribed opening when the intake manifold pressure fluctuation width is less than 0.37 psi. **Possible Causes:** • EGR valve source vacuum supply line is open or restricted • EGR exhaust tube is clogged or restricted • EGR valve assembly is damaged or has failed • EGR valve circuit shorted or open • MAP sensor failed • PCM has failed

DTC	Trouble Code Title, Conditions & Possible Causes
DTC: P0403 **2T CCM MIL: YES** **1996, 1997, 1998, 1999, 2000, 2001, 2002, 2003, 2004, 2005, 2006** **Models:** 3000GT, Diamante, Endeavor, Evo, Galant, Lancer, Mirage, Montero, Montero Sport, Outlander **Engines:** All **Transmissions:** A/T, M/T	**EGR Control Circuit Malfunction** Engine started, battery positive voltage between 10 and 16 volts, and the PCM detected the EGR Control solenoid coil surge voltage (2 volts) is not detected for 0.2 seconds, EGR solenoid duty cycled ON cycle between 10% and 90%, evaporative emission purge solenoid ON cycle is 0%, evaporative emission ventilation is OFF for more than 1 second, the EGR solenoid coil surge voltage is not detected for 1 second when the EGR solenoid is turned OFF. **Possible Causes:** • EGR solenoid control circuit is open or shorted to ground • EGR solenoid power circuit is open (check power from relay) • EGR solenoid is damaged or has failed • PCM has failed
DTC: P0403 **2T CCM MIL: YES** **1995, 1996, 1997, 1998, 1999, 2000, 2001, 2002** **Models:** Eclipse **Engines:** 2.0L VIN F, 2.0L VIN Y, 2.4L VIN G, 3.0L VIN L **Transmissions:** A/T, M/T	**EGR Control Solenoid Circuit Malfunction** Engine started, then driven at a steady cruise speed in closed loop, and the PCM detected the EGR Control solenoid coil surge voltage was too low (i.e., it did not reach system voltage +2v) after the EGR Control solenoid was turned "on" and then "off" during the CCM test. **Possible Causes:** • EGR solenoid control circuit is open or shorted to ground • EGR solenoid power circuit is open (check power from relay) • EGR solenoid is damaged or has failed • PCM has failed
DTC: P0411 **2T AIR MIL: YES** **1995** **Models:** Eclipse **Engines:** 2.0L VIN Y **Transmissions:** M/T	**Pulsed Secondary AIR System Malfunction** Engine started, engine running at cruise speed in closed loop, ECT sensor less than 140°F and the PCM detected the Short Term fuel trim amount did not change with pulse air was directed downstream. **Possible Causes:** • Pulse Air solenoid hoses and/or pipes are damaged or leaking • Pulse Air solenoid is stuck in closed position or damaged • PCM has failed
DTC: P0412 **2T CCM MIL: YES** **1995** **Models:** Eclipse **Engines:** 2.0L VIN Y **Transmissions:** M/T	**Pulsed Secondary Air Solenoid Circuit Malfunction** Key on or engine running, ECT signal less than 140°F, and the PCM detected an unexpected voltage condition on the Secondary AIR solenoid control circuit during the CCM test. **Possible Causes:** • AIR solenoid control circuit is open or shorted to ground • AIR solenoid power circuit is open (check for power from relay) • AIR solenoid is damaged or has failed • PCM has failed
DTC: P0420 **2T CAT1 MIL: YES** **1995, 1996, 1997, 1998, 1999** **Models:** Eclipse **Engines:** 2.0L VIN F, 2.0L VIN Y **Transmissions:** A/T, M/T	**Catalyst System Efficiency Low (Bank 1)** DTC P0100, P0105, P0110, P0115, P0120, P0130, P0135, P0136, P0141, P0440, P0500 and P0505 not set, engine started, BARO sensor more than 76 kPa, IAT signal more than 14°F, vehicle driven at 45-60 mph in closed loop for 3-5 minutes, MAF signal from 69-169 Hz, and the PCM detected the switch rate of the rear HO2S and the HO2S signals were too similar for 140 seconds during the test. **Possible Causes:** • Air leaks at the exhaust manifold or in the exhaust pipes • Catalytic converter is damaged, contaminated or has failed • Front HO2S or rear HO2S is contaminated with fuel or moisture • Front HO2S and/or the rear HO2S is loose in the mounting hole • Front HO2S older (aged) than the rear HO2S (HO2S is lazy)
DTC: P0420 **2T CAT1 MIL: YES** **1996, 1997, 1998, 1999, 2000, 2001, 2002, 2003, 2004, 2005, 2006** **Models:** Evo, Eclipse, Galant, Outlander, Lancer **Engines:** 2.4L VIN G **Transmissions:** A/T, M/T	**Catalyst System Efficiency Low (Bank 1)** DTC P0100, P0105, P0110, P0115, P0120, P0130, P0135, P0136, P0141, P0440, P0500 and P0505 not set, engine started, BARO sensor more than 76 kPa, IAT signal more than 14°F, vehicle driven to 45-60 mph with the throttle open at a speed of less than 3000 rpm for 3-5 minutes, MAF signal from 63-169 Hz, and the PCM detected the switch rate of the rear HO2S-12 and the front HO2S-11 signals were too similar for 90 seconds during the Catalyst Monitor Test. **Possible Causes:** • Air leaks at the exhaust manifold or in the exhaust pipes • Catalytic converter is damaged, contaminated or has failed • Front HO2S or rear HO2S is contaminated with fuel or moisture • Front HO2S and/or the rear HO2S is loose in the mounting hole • Front HO2S older (aged) than the rear HO2S (HO2S is lazy)

DTC	Trouble Code Title, Conditions & Possible Causes
DTC: P0420 **2T CAT1 MIL: YES** **1996, 1997, 1998, 1999, 2000** **Models:** Mirage, Montero, Montero Sport **Engines:** All **Transmissions:** A/T, M/T	**Catalyst System Efficiency Low (Bank 1)** DTC P0100, P0105, P0110, P0115, P0120, P0130, P0135, P0136, P0141, P0440, P0500 and P0505 not set, BARO signal over 76 kPa, IAT signal more than 14°F, vehicle driven to 45-60 mph with the throttle open at a speed of less than 3000 rpm for 3-5 minutes, MAF signal from 63-169 Hz, and the PCM detected the switch rate of the rear HO2S-12 and the front HO2S-11 signals were too similar for 90 seconds during the Catalyst Efficiency Monitor Test. **Possible Causes:** • Air leaks at the exhaust manifold or in the exhaust pipes • Catalytic converter is damaged, contaminated or has failed • Front HO2S or rear HO2S is contaminated with fuel or moisture • Front HO2S and/or the rear HO2S is loose in the mounting hole • Front HO2S older (aged) than the rear HO2S (HO2S is lazy)
DTC: P0421 **2T CAT2 MIL: YES** **1996, 1997, 1998, 1999, 2000, 2001** **Models:** 3000GT, Diamante, Montero, Montero Sport **Engines:** All **Transmissions:** A/T, M/T	**Warmup Catalyst Efficiency Low (Bank 1)** DTC P0100, P0105, P0110, P0115, P0120, P0130, P0135, P0136, P0141, P0151, P0155, P0156, P0161, P0440, P0500 and P0505 not set, IAT sensor more than 14°F, BARO signal over 76 kPa, vehicle driven at 45-60 mph at a steady throttle in closed loop for 5 minutes, MAF sensor from 69-169 Hz, and the PCM detected switch rate of the rear HO2S and front HO2S signals were similar for 140 seconds. **Possible Causes:** • Air leaks at the exhaust manifold or in the exhaust pipes • Catalytic converter is damaged, contaminated or has failed • Front HO2S or rear HO2S is contaminated with fuel or moisture • Front HO2S and/or the rear HO2S is loose in the mounting hole • Front HO2S older (aged) than the rear HO2S (HO2S is lazy)
DTC: P0421 **2T CAT2 MIL: YES** **2002, 2003, 2004, 2005, 2006** **Models:** Diamante, Evo, Endeavor, Eclipse, Galant, Lancer, Mirage, Montero, Montero Sport, Outlander **Engines:** All **Transmissions:** A/T, M/T	**Warmup Catalyst Efficiency Low (Bank 1)** DTC P0101, P0102, P0103, P0106, P0107, P0108, P0111, P0112, P0113, P0116, P0117, P0118, P0121, P0122, P0123, P0132, P0133, P0134, P0135, P0137, P0139, P0141 and P0500 not set, IAT sensor more than 14°F, BARO signal over 76 kPa, vehicle driven at 45-60 mph in closed loop at a steady throttle for 5 minutes, MAF sensor at 69-169 Hz, and the PCM detected switch rate of the rear HO2S and front HO2S signals were too similar for 140 seconds. **Possible Causes:** • Air leaks at the exhaust manifold or in the exhaust pipes • Catalytic converter is damaged, contaminated or has failed • Front HO2S or rear HO2S is contaminated with fuel or moisture • Front HO2S and/or the rear HO2S is loose in the mounting hole • Front HO2S older (aged) than the rear HO2S (HO2S is lazy)
DTC: P0421 **2T CAT2 MIL: YES** **1996, 1997, 1998, 1999, 2000, 2001** **Models:** Mirage **Engines:** All **Transmissions:** A/T, M/T	**Warmup Catalyst Efficiency Low (Bank 1)** DTC P0100, P0105, P0110, P0115, P0120, P0130, P0135, P0136, P0141, P0151, P0155, P0156, P0161, P0440, P0500 and P0505 not set, IAT sensor more than 14°F, BARO signal over 76 kPa, vehicle driven at 45-60 mph at a steady throttle for 5 minutes in closed loop, MAF sensor at 69-169 Hz, and the PCM detected the switch rate of the rear HO2S and front HO2S signals were similar for 140 seconds. **Possible Causes:** • Air leaks at the exhaust manifold or in the exhaust pipes • Catalytic converter is damaged, contaminated or has failed • Front HO2S or rear HO2S is contaminated with fuel or moisture • Front HO2S and/or the rear HO2S is loose in the mounting hole • Front HO2S older (aged) than the rear HO2S (HO2S is lazy)
DTC: P0421 **2T CAT1 MIL: YES** **1999, 2000, 2001** **Models:** Eclipse, Galant **Engines:** 2.4L VIN G, 3.0L VIN L **Transmissions:** A/T, M/T	**Warmup Catalyst System Efficiency Low (Bank 1)** DTC P0100, P0105, P0110, P0115, P0120, P0130, P0135, P0136, P0141, P0440, P0500 and P0505 not set, BARO signal over 76 kPa, IAT signal more than 14°F, vehicle driven to 45-60 mph with throttle open at a speed of less than 3000 rpm for 3-5 minutes, MAF signal from 63-169 Hz, and the PCM detected the switch rate of the rear HO2S and the front HO2S signals were too similar for 90 seconds. **Possible Causes:** • Air leaks at the exhaust manifold or in the exhaust pipes • Catalytic converter is damaged, contaminated or has failed • Front HO2S or rear HO2S is contaminated with fuel or moisture • Front HO2S and/or the rear HO2S is loose in the mounting hole • Front HO2S older (aged) than the rear HO2S (HO2S is lazy)

DTC	Trouble Code Title, Conditions & Possible Causes
DTC: P0422 **2T CAT1 MIL: YES** **1995, 1996, 1997, 1998, 1999** **Models:** Eclipse **Engines:** 2.0L VIN Y **Transmissions:** A/T, M/T	**Warmup Catalyst Efficiency Low (Bank 1)** DTC P0106, P0107, P0108, P0111, P0112, P0113, P0117, P0118, P0121, P0122, P0123, P0131, P0132, P0133, P0134, P0135, P0138, P0141, P0201-P0204, P0300, P0301-P0304, P0441, P0500 and P0505 not set, ECT signal more than 170°F, vehicle driven to a speed of 45-60 mph at an engine speed of 1248-2400 rpm, MAP at 1.5-2.6v, and the PCM detected the switch rate of the rear HO2S-12 reached 70% of the switch rate of front HO2S-11 during the test. **Possible Causes:** • Air leaks at the exhaust manifold or in the exhaust pipes • Catalytic converter is damaged, contaminated or has failed • Front HO2S or rear HO2S is contaminated with fuel or moisture • Front HO2S and/or the rear HO2S is loose in the mounting hole • Front HO2S older (aged) than the rear HO2S (HO2S is lazy)
DTC: P0431 **2T CAT2 MIL: YES** **1996, 1997, 1998, 1999, 2000, 2001** **Models:** 3000GT, Diamante **Engines:** All **Transmissions:** A/T, M/T	**Warmup Catalyst Efficiency Low (Bank 2)** DTC P0100, P0105, P0110, P0115, P0120, P0130, P0135, P0136, P0141, P0151, P0155, P0156, P0161, P0440, P0500 and P0505 not set, engine started, vehicle driven at 45-60 mph in closed loop at a steady throttle for 5-10 minutes, BARO signal more than 76 kPa, MAF sensor from 69-169 Hz, IAT sensor more than14°F, and the PCM detected the rear HO2S-22 switch rate was similar to the HO2S-21 switch rate for 140 seconds in the Catalyst Monitor test. **Possible Causes:** • Air leaks at the exhaust manifold or in the exhaust pipes • Catalytic converter is damaged, contaminated or has failed • Front HO2S or rear HO2S is contaminated with fuel or moisture • Front HO2S and/or the rear HO2S is loose in the mounting hole • Front HO2S older (aged) than the rear HO2S (HO2S is lazy)
DTC: P0431 **2T CAT2 MIL: YES** **1996, 1997, 1998, 1999, 2000, 2001** **Models:** Galant, Montero, Montero Sport **Engines:** 3.0L VIN L, 3.0L VIN H, 3.0L VIN P, 3.5L VIN M, 3.5L VIN R **Transmissions:** A/T, M/T	**Warmup Catalyst Efficiency Low (Bank 2)** DTC P0100, P0105, P0110, P0115, P0120, P0130, P0135, P0136, P0141, P0151, P0155, P0156, P0161, P0440, P0500 and P0505 not set, engine started, vehicle driven at 45-60 mph in closed loop at a steady throttle for 5-10 minutes, BARO signal more than 76 kPa, MAF sensor from 69-169 Hz, IAT sensor more than14°F, and the PCM detected the rear HO2S-22 switch rate was similar to the HO2S-21 switch rate for 140 seconds in the Catalyst Monitor test. **Possible Causes:** • Air leaks at the exhaust manifold or in the exhaust pipes • Catalytic converter is damaged, contaminated or has failed • Front HO2S or rear HO2S is contaminated with fuel or moisture • Front HO2S and/or the rear HO2S is loose in the mounting hole • Front HO2S older (aged) than the rear HO2S (HO2S is lazy)
DTC: P0431 **2T CAT2 MIL: YES** **2002, 2003, 2004, 2005, 2006** **Models:** Diamante, Endeavor, Eclipse, Galant, Montero, Montero Sport **Engines:** All V6 **Transmissions:** A/T, M/T	**Warmup Catalyst Efficiency Low (Bank 2)** DTC P0101, P0102, P0103, P0106, P0107, P0108, P0111, P0112, P0113, P0116, P0117, P0118, P0121, P0122, P0123, P0132, P0133, P0134, P0135, P0137, P0139, P0141 and P0500 not set, engine started, vehicle driven at 45-60 mph in closed loop at a steady throttle for 5-10 minutes, BARO signal more than 76 kPa, MAF sensor from 69-169 Hz, IAT sensor more than14°F, and the PCM detected the rear HO2S-22 switch rate was similar to the HO2S-21 switch rate for 140 seconds in the Catalyst Monitor test. **Possible Causes:** • Air leaks at the exhaust manifold or in the exhaust pipes • Catalytic converter is damaged, contaminated or has failed • Front HO2S or rear HO2S is contaminated with fuel or moisture • Front HO2S and/or the rear HO2S is loose in the mounting hole • Front HO2S older (aged) than the rear HO2S (HO2S is lazy)
DTC: P0440 **2T EVAP MIL: YES** **1996, 1997, 1998** **Models:** 3000GT, Diamante, Galant, Mirage, Montero, Montero Sport **Engines:** All **Transmissions:** A/T, M/T	**EVAP Control System Malfunction** Cold engine startup (ECT sensor less than 113°F), engine running in closed loop for 3 minutes, ECT sensor more than 176°F, EVAP solenoid commanded "on" for 3 seconds, the PCM detected less than a 3% variation in the Short Term fuel trim value during the test. **Possible Causes:** • Charcoal canister is loaded with fuel or moisture • ECT, IAT, MAF, VSS or TP sensor signal is out-of-calibration • Fuel filler cap loose, cross-threaded, incorrect part or damaged • Fuel tank pressure sensor is damaged or has failed • Fuel tank vapor line(s) blocked, damaged or disconnected • Purge or Vent solenoid control circuit open or shorted to ground • PCM has failed

DTC	Trouble Code Title, Conditions & Possible Causes
DTC: P0440 **2T EVAP MIL: YES** **1995, 1996, 1997, 1998, 1999** **Models:** Eclipse **Engines:** 2.0L VIN F, 2.0L VIN Y, 2.4L VIN G **Transmissions:** A/T, M/T	**EVAP Control System Malfunction** Cold engine startup, engine running in closed loop for 3 minutes at idle speed, ECT sensor more than 176°F, then after the EVAP solenoid was commanded "on" for 3 seconds, the PCM detected less than a 3% variation in the Short Term fuel trim during the test. **Possible Causes:** • Charcoal canister is loaded with fuel or moisture • ECT, IAT, MAF, VSS or TP sensor signal is out-of-calibration • Fuel filler cap loose, cross-threaded, incorrect part or damaged • Fuel tank pressure sensor is damaged or has failed • Fuel tank vapor line(s) blocked, damaged or disconnected • Purge Control solenoid valve circuit open or shorted to ground • PCM has failed
DTC: P0441 **2T EVAP MIL: YES** **1995, 1996, 1997, 1998, 1999** **Models:** Eclipse **Engines:** 2.0L VIN Y **Transmissions:** A/T, M/T	**EVAP Purge Flow System Malfunction** Engine started, MAP sensor at "key on" less than 87 kPa, vehicle driven at a steady speed of 28-48 mph with the engine speed under 2024 rpm, MAP sensor from 1.38-2.0v, the difference between Long Term and Short Term fuel trim less than 5% after the Purge solenoid is cycled "on" to "off", and the PCM did not detect sufficient change in the MAP sensor signal for 2 seconds under these conditions during the CCM Rationality test. **Note: This is a "functionality" test of the EVAP system (flow test).** **Possible Causes:** • Charcoal canister is damaged, clogged or restricted • Purge solenoid circuit is open (fault may be intermittent) • Purge solenoid power circuit is open (check the relay or fuse) • Purge valve vacuum line is clogged, restricted or disconnected • PCM has failed • TSB 1-13-004 (6/01) contains information related to this code
DTC: P0441 **2T EVAP MIL: YES** **2001, 2002, 2003, 2004, 2005, 2006** **Models:** All **Engines:** All **Transmissions:** A/T, M/T	**EVAP Emission Control System Incorrect Purge Flow** During evaporative emission control system monitoring, 20 seconds after the Evap purge solenoid ON duty cycle is 0%, the pressure in the fuel tank is 0.29 psi or less for 0.1 second. **Note: This is a "functionality" test of the EVAP system (flow test).** **Possible Causes:** • Charcoal canister is damaged, clogged or restricted • Purge solenoid circuit is open (fault may be intermittent) • Purge solenoid power circuit is open (check the relay or fuse) • Purge valve vacuum line is clogged, restricted or disconnected • PCM has failed
DTC: P0442 **2T EVAP MIL: YES** **1998, 1999, 2000, 2001, 2002, 2003, 2004, 2005, 2006** **Models:** 3000GT, Diamante, Lancer, Endeavor, Evo, Galant, Mirage, Montero, Montero Sport, Outlander **Engines:** All **Transmissions:** A/T, M/T	**EVAP Control System Small Leak Detected** Cold engine startup (ECT and IAT sensor signals less than 86°F), BARO sensor more than 75 kPa, engine runtime over 16 minutes, volumetric efficiency from 20-80%, ECT sensor more than 140°F, TP sensor from 1-4v, PSP switch "off", vehicle driven to a speed of over 20 mph at an engine speed over 1600 rpm, then with the EVAP purge and vent solenoids both closed and the pressure rise less than 0.065 psi, the PCM detected the change in internal pressure in the fuel tank was more than 0.122 psi (843 kPa) during the leak test. **Possible Causes:** • Canister Purge valve is damaged, leaking or it has failed • Charcoal canister is loaded with fuel or moisture • Fuel filler cap loose, cross-threaded, incorrect part or damaged • Fuel tank is cracked (leaking), or a leak exists in the 'O' ring • Fuel tank pressure sensor is damaged or has failed • Fuel vapor line(s), fuel pipes or hoses damaged or leaking • PCM has failed

DTC	Trouble Code Title, Conditions & Possible Causes
DTC: P0442 **2T EVAP MIL: YES** **1998, 1999** **Models:** Eclipse **Engines:** 2.0L VIN F **Transmissions:** A/T, M/T	**EVAP Control System Small Leak Detected** Engine started (ECT sensor less than 113°F at startup), IAT sensor over 14°F, BARO sensor over 75 kPa, vehicle driven at an engine speed of 1600-3500 rpm, ECT sensor more than 140°F, PSP switch is "off", engine load is 20-80%, then with the EVAP canister vent and Purge solenoids both closed, and the pressure rise in the system less than 490 kPa, the PCM detected the EVAP system pressure fluctuation was less than 667 kPa due to a large leak (0.040"-0.80") in the system for 50-100 seconds during the EVP Monitor Leak Test. **Possible Causes:** • Charcoal canister is loaded with fuel or moisture • Canister Purge solenoid is damaged, leaking or has failed • Canister Vent solenoid is damaged, leaking or it has failed • Fuel control sensor is out of calibration (i.e., ECT, IAT or MDP) • Fuel filler cap loose, cross-threaded, incorrect part or damaged • Fuel tank differential pressure sensor is damaged or has failed • Fuel tank is cracked (leaking), or a leak exists in the 'O' ring • Fuel tank filler tube assembly is damaged or restricted • Fuel vapor line(s), fuel pipes or hoses clogged or restricted • PCM has failed
DTC: P0442 **2T EVAP MIL: YES** **1997, 1998, 1999, 2000, 2001, 2002** **Models:** Eclipse **Engines:** 2.0L VIN Y, 2.4L VIN G, 3.0L VIN H **Transmissions:** A/T, M/T	**EVAP Control System Small Leak Detected** Cold engine startup (ECT and IAT sensor signals less than 86°F), BARO sensor more than 75 kPa, engine runtime over 16 minutes, ECT sensor more than 140°F, TP sensor from 1-4v, vehicle driven to a speed over 20 mph with the engine speed over 1600 rpm, then with the EVAP canister vent solenoid closed, and the Leak Detection pump commanded to a specific duty cycle to raise the pressure in the system, the PCM detected the LDP pump continued cycling "on" and "off" due to a leak (0.040"-0.80") in the system during the test. **Possible Causes:** • Charcoal canister is loaded with fuel or moisture • Canister Vent valve is damaged, leaking or it has failed • Fuel filler cap loose, cross-threaded, incorrect part or damaged • Fuel tank is cracked (leaking), or a leak exists in the 'O' ring • Fuel tank filler tube assembly is damaged or leaking • Fuel vapor line(s), fuel pipes or hoses damaged or leaking • PCM has failed
DTC: P0443 **2T CCM MIL: YES** **1996, 1997, 1998, 1999, 2000, 2001, 2002, 2003, 2004, 2005, 2006** **Models:** 3000GT, Diamante, Endeavor, Evo, Galant, Lancer, Mirage, Montero, Montero Sport **Engines:** All **Transmissions:** A/T, M/T	**EVAP Purge Control Solenoid Circuit Malfunction** Engine started, engine running at cruise speed for 3-5 minutes, system voltage over 10v, and the PCM detected the EVAP Purge Control solenoid coil surge voltage was too low (i.e., it did not reach system voltage +2v) after the EVAP Purge Control solenoid was turned "on" and then "off" during the CCM test. **Possible Causes:** • Purge solenoid circuit control is open or shorted to ground • Purge solenoid power circuit is open (test from the relay) • Purge control solenoid is damaged or has failed • PCM has failed
DTC: P0443 **2T CCM MIL: YES** **1995, 1996, 1997, 1998, 1999** **Models:** Eclipse **Engines:** 2.0L VIN F, 2.0L VIN Y, 2.4L VIN G **Transmissions:** A/T, M/T	**EVAP Purge Control Solenoid Circuit Malfunction** Engine started, engine running at cruise speed for 3-5 minutes, system voltage over 10v, and the PCM detected an unexpected voltage condition on the EVAP Purge Control solenoid control circuit during the CCM test. **Possible Causes:** • EVAP purge solenoid circuit is open or shorted to ground • EVAP purge solenoid power circuit is open (test from the relay) • EVAP purge control solenoid is damaged or has failed • PCM has failed
DTC: P0446 **2T CCM MIL: YES** **1998, 1999, 2000, 2001, 2002, 2003, 2004, 2005, 2006** **Models:** 3000GT, Diamante, Endeavor, Evo, Galant, Lancer, Mirage, Montero, Montero Sport, Outlander **Engines:** All **Transmissions:** A/T, M/T	**EVAP Vent Control Solenoid Circuit Malfunction** Engine started, system voltage over 10v, and the PCM detected the solenoid surge voltage did not reach system voltage (+2v) within 30 ms after the vent solenoid was commanded OFF during the test. **Possible Causes:** • Vent Control solenoid circuit is open or shorted to ground • Vent Control solenoid power circuit is open (test from the relay) • Vent Control solenoid is damaged or has failed • PCM has failed

DTC	Trouble Code Title, Conditions & Possible Causes
DTC: P0446 **2T CCM MIL: YES** **1997, 1998, 1999, 2000, 2001, 2002** **Models:** Eclipse **Engines:** 2.0L VIN F, 2.0L VIN Y, 2.4L VIN G, 3.0L VIN H **Transmissions:** A/T, M/T	**EVAP Vent Control Solenoid Circuit Malfunction** Engine started, system voltage over 10v, and the PCM detected the solenoid surge voltage did not reach system voltage (+2v) within 30 ms after the purge solenoid was commanded "on" during the test. **Possible Causes:** • Vent Control solenoid circuit is open or shorted to ground • Vent Control solenoid power circuit is open (test from the relay) • Vent Control solenoid is damaged or has failed • PCM has failed
DTC: P0450 **2T CCM MIL: YES** **1998, 1999, 2000, 2001** **Models:** 3000GT, Diamante, Galant, Mirage, Montero, Montero Sport **Engines:** All **Transmissions:** A/T, M/T	**Fuel Tank Pressure Sensor Circuit Malfunction** Idle Test Engine running at cruise, followed by a deceleration period to under 1 mph from over 10 mph from an engine speed of 2500 rpm with the volumetric efficiency over 55%, and the PCM detected a sudden pressure change of over 0.20v occurred 20 times in a 5 ms period. Cruise Test Engine started, IAT sensor from 41-113°F, vehicle driven to a speed of over 19 mph at an engine speed over 1600 rpm, volumetric efficiency from 20-80%, and the PCM detected the FTP sensor signal was more than 4.0v with Purge solenoid driven at a 100% duty cycle, or it detected the FTP sensor was less than 1v with the purge solenoid "off". **Possible Causes:** • FTP sensor signal circuit open or shorted to ground • FTP sensor ground circuit is open • FTP sensor power (VREF) circuit is open • FTP sensor is damaged or has failed • PCM has failed
DTC: P0450 **2T CCM MIL: YES** **1997, 1998, 1999** **Models:** Eclipse **Engines:** 2.0L VIN F, 2.0L VIN Y, 2.4L VIN G **Transmissions:** A/T, M/T	**Fuel Tank Pressure Sensor Circuit Malfunction** Idle Test Engine running at cruise, followed by a deceleration period to under 1 mph from over 10 mph from an engine speed of 2500 rpm with the volumetric efficiency over 55%, and the PCM detected a sudden pressure change of over 0.20v occurred 20 times in a 5 ms period. Cruise Test Engine started, IAT sensor from 41-113°F, vehicle driven to a speed over 19 mph at a speed over 1600 rpm, volumetric efficiency from 20-80%, and the PCM detected the FTP sensor indicated more than 4.0v with Purge solenoid duty cycle at 100%, or it detected the FTP sensor indicated less than 1.0v with the purge solenoid turned "off". **Possible Causes:** • FTP sensor signal circuit open or shorted to ground • FTP sensor ground circuit is open • FTP sensor power (VREF) circuit is open • FTP sensor is damaged or has failed • PCM has failed
DTC: P0450 **2T CCM MIL: YES** **2005, 2006** **Models:** All **Engines:** All **Transmissions:** A/T, M/T	**Evaporative Emission System Pressure Sensor Malfunction** Engine started, IAT sensor greater than 41°F, engine speed 1600 rpm or greater, volumetric efficiency is between 20% and 70%, Evap purge solenoid is OFF the fuel differential pressure sensor output voltage remains at 1 volt or less for 10 seconds and/or the Evap purge solenoid is fully operational the fuel differential pressure sensor output voltage remains at 4 volts or greater for 10 seconds. **Possible Causes:** • FTDP sensor signal circuit open or shorted to ground • FTDP sensor ground circuit is open • FTDP sensor power (VREF) circuit is open • FTDP sensor is damaged or has failed • PCM has failed
DTC: P0451 **2T CCM MIL: YES** **1998, 1999, 2000, 2001** **Models:** 3000GT, Diamante **Engines:** All **Transmissions:** A/T, M/T	**Fuel Tank Pressure Sensor Range/Performance** Engine started, IAT sensor from 41-113°F, vehicle driven to a speed over 19 mph at a speed of over 1600 rpm, volumetric efficiency from 20-80%, EVAP purge command at 100%, and the PCM detected the FTP sensor signal was over 4.0v; or less than 1.0v for 10 seconds. **Possible Causes:** • Fuel tank pressure sensor vacuum hoses loose or damaged • Fuel tank pressure sensor is damaged or out-of-calibration • PCM has failed

DTC	Trouble Code Title, Conditions & Possible Causes
DTC: P0451 **2T CCM MIL: YES** **2002, 2003, 1004, 2005, 2006** **Models:** Diamante, Eclipse, Galant, Lancer, Mirage, Montero, Montero Sport, Outlander, Lancer, Evdeaver **Engines:** All **Transmissions:** A/T, M/T	**Emission Control System Pressure Sensor Performance** Engine started, IAT sensor from 41-113°F, vehicle driven to a speed over 19 mph at a speed of over 1600 rpm, volumetric efficiency from 20-80%, EVAP purge command at 100%, and the PCM detected the FTP sensor signal was over 4.0v; or less than 1.0v for 10 seconds. **Possible Causes:** • Fuel tank pressure sensor vacuum hoses loose or damaged • Fuel tank pressure sensor is damaged or out-of-calibration • PCM has failed
DTC: P0452 **2T CCM MIL: YES** **2002** **Models:** Diamante, Endeavor, Outlander, Evo, Eclipse, Galant, Lancer, Mirage, Montero, Montero Sport **Engines:** All **Transmissions:** A/T, M/T	**Emission Control System Pressure Sensor Circuit Low Input** Engine started, IAT sensor from 41-113°F, vehicle driven to an engine speed of over 1600 rpm, volumetric efficiency from 20-80%, EVAP purge commanded "off", and the PCM detected the FTP sensor signal was less than 1.0v for 10 seconds during the test. **Possible Causes:** • Fuel tank pressure sensor signal circuit is shorted to ground • Fuel tank pressure sensor power circuit is open • Fuel tank pressure sensor is damaged or out-of-calibration • PCM has failed
DTC: P0453 **2T CCM MIL: YES** **2002, 2003, 2004, 2005, 2006** **Models:** Diamante, Endeavor, Outlander, Evo, Eclipse, Galant, Lancer, Mirage, Montero, Montero Sport **Transmissions:** A/T, M/T **Engines:** All	**Emission Control System Pressure Sensor Circuit High Input** Engine started, IAT sensor from 41-113°F, vehicle driven to an engine speed of over 1600 rpm, volumetric efficiency from 20-80%, EVAP purge commanded "off", and the PCM detected the FTP sensor signal was less than 1.0v for 10 seconds during the test. **Possible Causes:** • Fuel tank pressure sensor signal circuit is shorted to ground • Fuel tank pressure sensor power circuit is open • Fuel tank pressure sensor is damaged or out-of-calibration • PCM has failed
DTC: P0455 **2T CCM MIL: YES** **2002, 2003, 2004, 2005, 2006** **Models:** Diamante, Endeavor, Outlander, Evo, Eclipse, Galant, Lancer, Mirage, Montero, Montero Sport **Engines:** All **Transmissions:** A/T, M/T	**EVAP Leak Monitor Gross Leak (0.080") Detected** ECT and IAT signals under 86°F at startup, BARO over 75 kPa, ECT signal over 140°F during testing, engine runtime over 16 minutes, TP sensor signal from 1-4v, volumetric efficiency from 20-80%, PSP switch "off", vehicle driven to a speed of over 20 mph at an engine speed over 1600 rpm, then with both the purge and vent solenoids closed and the EVAP pressure rise less than 0.065 psi, the PCM detected the fluctuation of the pressure in fuel tank was less than 324 kPa (0.047 psi) for 20 seconds during the EVAP Leak test. **Possible Causes:** • Canister vent (CV) solenoid may be stuck in open position • EVAP canister tube, EVAP canister purge outlet tube or EVAP return tube disconnected or cracked, or canister is damaged • EVAP canister purge valve stuck closed, or canister damaged • Fuel filler cap missing, loose (not tightened) or the wrong part • Fuel vapor hoses/tubes blocked or restricted, or fuel vapor control valve tube or fuel vapor vent valve assembly blocked • Fuel tank pressure (FTP) sensor has failed (mechanical fault) • Fuel tank control valve is contaminated, damaged or has failed
DTC: P0455 **2T EVAP MIL: YES** **1998, 1999** **Models:** Eclipse **Engines:** 2.0L VIN F, 2.0L VIN Y, 2.4L VIN G **Transmissions:** A/T, M/T	**EVAP Control System Very Large Leak (0.080") Detected** Engine started (ECT sensor less than 113°F at startup), IAT sensor over 14°F, BARO sensor over 75 kPa, vehicle driven at an engine speed of 1600-3500 rpm, ECT sensor more than 140°F, PSP switch is "off", engine load is 20-80%, then with the EVAP canister vent and Purge solenoids both closed, and the pressure rise in the system less than 490 kPa, the PCM detected the EVAP system pressure fluctuation was less than 667 kPa due to a very large leak (over 0.80") in the system for 50-100 seconds during the EVAP Leak Test. **Possible Causes:** • Charcoal canister is loaded with fuel or moisture • Canister Purge solenoid is damaged, leaking or has failed • Canister Vent solenoid is damaged, leaking or it has failed • Fuel control sensor is out of calibration (i.e., ECT, IAT or MDP) • Fuel filler cap loose, cross-threaded, incorrect part or damaged • Fuel tank differential pressure sensor is damaged or has failed • Fuel tank is cracked (leaking), or a leak exists in the 'O' ring • Fuel tank filler tube assembly is damaged or restricted • Fuel vapor line(s), fuel pipes or hoses clogged or restricted • PCM has failed

DTC	Trouble Code Title, Conditions & Possible Causes
DTC: P0455 **2T EVAP MIL: YES** **1996, 1997, 1998, 1999, 2000, 2001, 2002** **Models:** Eclipse **Engines:** 2.0L VIN Y, 2.0L VIN G, 3.0L VIN H **Transmissions:** A/T, M/T	**EVAP Leak Monitor Gross Leak (0.080") Detected** Cold engine startup (ECT and IAT sensor signals less than 86°F), BARO sensor more than 75 kPa, engine runtime over 16 minutes, ECT sensor more than 140°F, TP sensor from 1-4v, vehicle driven to a speed over 20 mph with the engine speed over 1600 rpm, then with the EVAP canister vent solenoid closed, and the Leak Detection pump commanded to a specific duty cycle to raise the pressure in the system, the PCM detected the LDP pump continued to operate on and off due to a gross leak in the EVAP system during the test. **Possible Causes:** • Charcoal canister is loaded with fuel or moisture • Canister Vent valve is damaged, leaking or it has failed • Fuel filler cap loose, cross-threaded, incorrect part or missing • Fuel tank is cracked (leaking), or a leak exists in the 'O' ring • Fuel tank filler tube assembly is damaged or leaking • Fuel vapor line(s), fuel pipes or hoses damaged or leaking • PCM has failed
DTC: P0456 **2T CCM MIL: YES** **2002, 2003, 2004, 2005, 2006** **Models:** Diamante, Endeavor, Outlander, Evo, Eclipse, Galant, Lancer, Mirage, Montero, Montero Sport **Transmissions:** A/T, M/T **Engines:** All	**EVAP Leak Monitor Very Small Leak (0.020") Detected** Cold engine startup (ECT and IAT sensor signals less than 97°F), BARO sensor more than 75 kPa, fuel temperature less than 97°F, volumetric efficiency from 20-70%, vehicle driven and after the fuel tank pressure sensor indicates 1-4v, and the PCM detected the internal fuel tank pressure changed more than 2 kPa in 128 seconds after the fuel tank and vapor line were closed during the test. This small change in pressure indicated a very small leak in the system. **Possible Causes:** • Canister Vent valve is damaged, leaking or it has failed • Fuel filler cap loose, cross-threaded or the incorrect part • Fuel tank is cracked (leaking), or a leak exists in the 'O' ring • Fuel tank filler tube assembly is damaged or leaking • Fuel vapor line(s), fuel pipes or hoses damaged or leaking • PCM has failed
DTC: P0461 **2T CCM MIL: YES** **2002, 2003, 2004, 2005, 2006** **Models:** All **Engines:** All **Transmissions:** A/T, M/T	**Fuel Level Sensor (Main) Circuit Range/Performance** When the fuel consumption calculated from the operation time of the injector amounts to approximately 8 gallons, the diversity of the amount of fuel in the tank calculated from the fuel level sensor is 0.5 gallons or less. **Possible Causes:** • Fuel pump module failed • Fuel level sensor failed • Fuel level sensor circuit shorted or open • PCM has failed
DTC: P0462 **2T CCM MIL: YES** **2005, 2006** **Models:** Eclipse, Galant, Lancer, Montero, Outlander **Engines:** All **Transmissions:** A/T, M/T	**Fuel Level Sensor Circuit Low Input** Engine started and running for 2 seconds or more, battery voltage is between 11 and 16.5 volts, fuel level sensor output voltage has continued to be less than 0.3 volts for 2 seconds. **Possible Causes:** • Fuel level sensor failed • Fuel level sensor circuit shorted • PCM has failed
DTC: P0463 **2T CCM MIL: YES** **2005, 2006** **Models:** Eclipse, Evo, Galant, Lancer, Montero, Outlander **Engines:** All **Transmissions:** A/T, M/T	**Fuel Level Sensor Circuit High Input** Engine started and running for 2 seconds or more, battery voltage is between 11 and 16.5 volts, fuel level sensor output voltage has continued to be greater than 4.6 volts for 2 seconds. **Possible Causes:** • Fuel level sensor failed • Fuel level sensor circuit shorted • PCM has failed
DTC: P0500 **2T CCM MIL: YES** **1996, 1997, 1998, 1999, 2000, 2001, 2002** **Models:** 3000GT, Diamante, Galant, Mirage, Montero, Montero Sport **Engines:** All **Transmissions:** A/T	**Vehicle Speed Sensor Circuit Malfunction** Engine started, the difference between atmospheric and intake manifold pressure more than 34 kPa, vehicle driven at an engine speed under 3000 rpm for 30 seconds with the throttle valve open and the brakes "off", ECT sensor more than 176°F, gear selector not in P/N, and the PCM detected the vehicle speed was below 1 mph for 4 seconds. **Possible Causes:** • VSS signal circuit is open or shorted to ground • VSS power circuit is open (check for power from MFI relay) • VSS ground circuit is open between sensor and chassis ground • VSS is damaged or has failed • PCM has failed (the PCM provides 5v on the VSS signal circuit)

DTC	Trouble Code Title, Conditions & Possible Causes
DTC: P0500 **2T CCM MIL: YES** **2003, 2004, 2005, 2006** **Models:** All **Engines:** All **Transmissions:** M/T	**Vehicle Speed Sensor Circuit Malfunction** Engine started for 2 seconds or more, engine speed is between 2000 and 4000 rpm, volumetric efficiency is between 45% and 100%, vehicle speed sensor output voltage has not changed (no pulse signal) for 2 seconds. **Possible Causes:** • VSS circuit is open or shorted to ground • VSS is damaged or has failed • PCM has failed
DTC: P0500 **2T CCM MIL: YES** **1996, 1997, 1998, 1999, 2000, 2001, 2002** **Models:** 3000GT, Galant, Lancer, Mirage, Montero, Montero Sport **Engines:** All **Transmissions:** M/T	**Vehicle Speed Sensor Circuit Malfunction** Engine started, the difference between atmospheric and intake manifold pressure more than 34 kPa, vehicle driven at an engine speed under 3000 rpm for 30 seconds with the throttle valve open and the brakes "off", ECT sensor more than 176°F, and the PCM detected the vehicle speed was less than 1 mph for 4-10 seconds. **Possible Causes:** • VSS signal circuit is open or shorted to ground • VSS power circuit is open (check for power from MFI relay) • VSS ground circuit is open between sensor and chassis ground • VSS is damaged or has failed • PCM has failed (the PCM provides 5v on the VSS signal circuit)
DTC: P0500 **2T CCM MIL: YES** **1995, 1996, 1997, 1998, 1999, 2000, 2001, 2002** **Models:** Eclipse **Engines:** 2.0L VIN F, VIN Y, 2.4L VIN G, 3.0L VIN H **Transmissions:** A/T	**Vehicle Speed Sensor Circuit Malfunction** Engine started, engine running for over 31 seconds, vehicle driven with the engine speed over 1800 rpm, ECT sensor more than 180°F, throttle valve open with the brakes "off", gear selector not in P/N, the difference between the BARO and MAP sensors is over 34 kPa, and the PCM detected the VSS signal was below 1 mph for 11 seconds. **Possible Causes:** • Speedometer pinion is damaged or has failed • VSS signal circuit is open or shorted to ground • VSS power circuit is open (check for 9v VREF input from PCM) • VSS ground circuit is open between sensor and chassis ground • VSS is damaged or has failed • PCM has failed (the PCM provides 5v on the VSS signal circuit)
DTC: P0500 **2T CCM MIL: YES** **1995, 1997, 1998, 1999** **Models:** Eclipse **Engines:** 2.0L VIN F, VIN Y **Transmissions:** M/T	**Vehicle Speed Sensor Circuit Malfunction** Engine started, engine running for over 31 seconds, vehicle driven with the engine speed over 1800 rpm, ECT sensor more than 180°F, throttle valve open with the brakes "off", the difference between the BARO and MAP sensors is more than 34 kPa, and the PCM detected the VSS signal was below 1 mph for 11 seconds. **Possible Causes:** • Speedometer pinion is damaged or has failed • VSS signal circuit is open or shorted to ground • VSS power circuit is open (check for 9v VREF input from PCM) • VSS ground circuit is open between sensor and chassis ground • VSS is damaged or has failed • PCM has failed (the PCM provides 5v on the VSS signal circuit)
DTC: P0505 **2T CCM MIL: YES** **1996, 1997, 1998, 1999, 2000, 2001** **Models:** 3000GT, Diamante, Galant, Mirage, Montero, Montero Sport **Engines:** All **Transmissions:** A/T, M/T	**Idle Speed Control System Malfunction** Engine started, engine running at cruise speed for 3-5 minutes, then back to idle speed in closed loop, ECT sensor more than 176°F, IAT sensor more than 14°F, system voltage over 10v, and the PCM detected the Actual idle speed was more than 100 rpm higher than the Target idle speed for 10 seconds in the CCM Rationality test. **Possible Causes:** • Stepper motor Coil A1 or A2 circuit is open or shorted to ground • Stepper motor Coil B1 or B2 circuit is open or shorted to ground • Stepper motor coil circuit(s) shorted to system power (B+) • Stepper motor power circuit is open (check power at MFI relay) • Stepper motor is damaged or has failed • PCM has failed
DTC: P0505 **2T CCM MIL: YES** **1995, 1996, 1997, 1998, 1999** **Models:** Eclipse **Engines:** 2.0L VIN F, 2.0L VIN Y, 2.4L VIN G **Transmissions:** A/T, M/T	**Idle Air Control Motor Circuit Malfunction** Engine started, system voltage over 10v, engine running in closed loop, and the PCM detected an unexpected voltage condition on the IAC Stepper Motor circuit for 3 seconds during the CCM test. **Possible Causes:** • Stepper motor Coil A1 or A2 circuit is open or shorted to ground • Stepper motor Coil B1 or B2 circuit is open or shorted to ground • Stepper motor coil circuit(s) shorted to system power (B+) • Stepper motor power circuit is open (check power at MFI relay) • Stepper motor is damaged or has failed • PCM has failed

DTC	Trouble Code Title, Conditions & Possible Causes
DTC: P0506 **2T CCM MIL: YES** **2002, 2003, 2004, 2005, 2006** **Models:** Diamante, Endeavor, Evo, Eclipse, Galant, Lancer, Montero, Montero Sport, Outlander **Engines:** All **Transmissions:** A/T, M/T	**Idle Speed Control System Lower Than Expected** ECT signal over 180°F, system voltage over 10v, BARO signal over 76 kPa, volumetric efficiency less than 40%, IAT signal more than 14°F, and the PCM detected the Actual idle speed was more than 100 rpm lower than the Target idle speed for over 12 seconds. **Possible Causes:** • IAC motor control circuit A1, A2, B1 or B2 is open • IAC motor control circuit A1, A2, B1 or B2 is shorted to ground • IAC motor is damaged or has failed (it may be dirty or sticking) • Throttle plate is carbon fouled (it may need to be cleaned) • PCM has failed
DTC: P0506 **2T CCM MIL: YES** **2000, 2001, 2002** **Models:** Mirage **Engines:** All **Transmissions:** A/T, M/T	**Idle Speed Control System Lower Than Expected** ECT signal over 171°F, system voltage over 10v, BARO signal over 76 kPa, volumetric efficiency less than 40%, IAT signal more than 14°F, and the PCM detected the Actual idle speed was more than 100 rpm lower than the Target idle speed for over 12 seconds. **Possible Causes:** • IAC motor control circuit A1, A2, B1 or B2 is open • IAC motor control circuit A1, A2, B1 or B2 is shorted to ground • IAC motor is damaged or has failed (it may be dirty or sticking) • Throttle plate is carbon fouled (it may need to be cleaned) • PCM has failed
DTC: P0507 **2T CCM MIL: YES** **2000, 2001, 2002** **Models:** Mirage **Engines:** All **Transmissions:** A/T, M/T	**Idle Speed Control System Higher Than Expected** ECT signal over 171°F, system voltage over 10v, BARO signal over 76 kPa, volumetric efficiency less than 40%, IAT signal more than 14°F, and the PCM detected the Actual idle speed was more than 100 rpm higher than the Target idle speed for over 12 seconds. **Possible Causes:** • IAC motor control circuit A1, A2, B1 or B2 is open • IAC motor control circuit A1, A2, B1 or B2 is shorted to ground • IAC motor is damaged or has failed (it may be dirty or sticking) • Throttle plate is carbon fouled (it may need to be cleaned) • PCM has failed
DTC: P0507 **2T CCM MIL: YES** **2002, 2003, 2004, 2005, 2006** **Models:** Diamante, Endeavor, Evo, Eclipse, Galant, Lancer, Montero, Montero Sport, Outlander **Engines:** All **Transmissions:** A/T, M/T	**Idle Speed Control System Higher Than Expected** ECT signal over 171°F, system voltage over 10v, BARO signal over 76 kPa, volumetric efficiency less than 40%, IAT signal more than 14°F, and the PCM detected the Actual idle speed was more than 100 rpm higher than the Target idle speed for over 12 seconds. **Possible Causes:** • IAC motor control circuit A1, A2, B1 or B2 is open • IAC motor control circuit A1, A2, B1 or B2 is shorted to ground • IAC motor is damaged or has failed (it may be dirty or sticking) • Throttle plate is carbon fouled (it may need to be cleaned) • PCM has failed
DTC: P0510 **2T CCM MIL: YES** **1996, 1997, 1998, 1999, 2000, 2001, 2002, 2003, 2004** **Models:** 3000GT, Diamante **Engines:** All **Transmissions:** A/T, M/T	**Closed Throttle Position Switch Circuit Malfunction** Engine started, vehicle driven at over 30 mph and then back to a stop at least 15 times, TP sensor signal over 2.0v at least once, and the PCM detected the CTP switch remained "off" for over 2 seconds. **Possible Causes:** • Closed throttle position switch signal circuit is open or grounded • Closed throttle position switch signal circuit is shorted to power • Closed throttle position switch or TP sensor damaged or failed • PCM has failed
DTC: P0510 **2T CCM MIL: YES** **1995, 1996, 1997, 1998, 1999** **Models:** Eclipse **Engines:** 2.0L VIN F, 2.0L VIN Y, 2.4L VIN G **Transmissions:** A/T, M/T	**Closed Throttle Position Switch Circuit Malfunction** Engine started, vehicle driven at over 30 mph and then back to a stop at least 15 times, TP sensor signal over 2.0v at least once, and the PCM detected the CTP switch remained "off" for over 2 seconds. **Possible Causes:** • Closed throttle position switch signal circuit is open or grounded • Closed throttle position switch signal circuit is shorted to power • Closed throttle position switch or TP sensor damaged or failed • PCM has failed
DTC: P0510 **2T CCM MIL: YES** **1996, 1997, 1998, 1999, 2000** **Models:** Montero, Montero Sport **Engines:** All **Transmissions:** A/T, M/T	**Closed Throttle Position Switch Circuit Malfunction** Engine started, vehicle driven at over 30 mph and then back to a stop at least 15 times, TP sensor signal over 2.0v at least once, and the PCM detected the CTP switch remained "off" for over 2 seconds. **Possible Causes:** • Closed throttle position switch signal circuit is open or grounded • Closed throttle position switch signal circuit is shorted to power • Closed throttle position switch or TP sensor damaged or failed • PCM has failed

DTC	Trouble Code Title, Conditions & Possible Causes
DTC: P0510 **2T CCM MIL: YES** **1996, 1997, 1998, 1999, 2000, 2001, 2002** **Models:** Galant, Mirage **Engines:** 2.4L VIN G, 1.5L VIN A, 1.8L VIN C **Transmissions:** A/T, M/T	**Closed Throttle Position Switch Circuit Malfunction** Engine started, vehicle driven at over 30 mph and then back to a stop at least 15 times, TP sensor signal over 2.0v at least once, and the PCM detected the CTP switch remained "off" for over 2 seconds. **Possible Causes:** • Closed throttle position switch signal circuit is open or grounded • Closed throttle position switch signal circuit is shorted to power • Closed throttle position switch or TP sensor damaged or failed • PCM has failed
DTC: P0510 **2T CCM MIL: YES** **1995, 1996, 1997, 1998, 1999** **Models:** Eclipse **Engines:** 2.0L VIN F, 2.0L VIN Y **Transmissions:** A/T, M/T	**Closed Throttle Position Switch Circuit Malfunction** Engine started, vehicle driven at cruise speed at over 1500 rpm, MAF sensor signal over 100 Hz for 2 seconds, TP sensor over 2v; or with the engine speed less than 800 rpm at least 15 times, the PCM detected the Closed Throttle Position switch indicated "off" under either of these operating conditions during the CCM Rationality test. **Possible Causes:** • Closed throttle position switch signal circuit is open or grounded • Closed throttle position switch signal circuit is shorted to power • Closed throttle position switch or TP sensor damaged or failed • PCM has failed
DTC: P0510 **2T CCM MIL: YES** **2005, 2006** **Models:** Lancer, Montero, Outlander **Engines:** 2.4L VIN F, 3.8L VIN S **Transmissions:** A/T, M/T	**Accelerator Pedal Position Switch Circuit Malfunction** Engine started, vehicle driven at a speed greater than 19 mph and then back to a stop for 2 seconds, 15 or more times, and the accelerator pedal position switch remains OFF. **Possible Causes:** • Accelerator position switch failed • Accelerator position switch circuit is shorted or open • PCM has failed
DTC: P0513 **2T CCM MIL: YES** **2003, 2004, 2005, 2006** **Models:** Eclipse, Evo, Galant, Endeavor, Montero, Montero Sport **Engines:** All **Transmissions:** A/T, M/T	**Immobilizer Malfunction** Ignition switch ON when an error occurs between the PCM and the Immobilizer ECU for 2 seconds or more. **Possible Causes:** • Harness or connector between the PCM and immobilizer ECU • Immobilizer ECU failed • PCM has failed
DTC: P0551 **2T CCM MIL: YES** **1997, 1998, 1999, 2000, 2001, 2002** **Models:** 3000GT, Diamante, Galant, Lancer, Mirage, Montero, Montero Sport **Engines:** All **Transmissions:** A/T, M/T	**Power Steering Pressure Switch Circuit Malfunction** Engine started, ECT sensor more than 50°F, vehicle driven to an engine speed of 2500 rpm, volumetric efficiency over 55% for 2 seconds, then after the vehicle returns to idle speed (800 rpm or less), the PCM detected the Power Steering Pressure switch signal indicated "off" at least 10 times during the CCM Rationality test. **Possible Causes:** • Power steering pressure switch signal circuit is open • Power steering pressure switch signal circuit shorted to ground • Power steering pressure switch is damaged or has failed • PCM has failed
DTC: P0551 **2T CCM MIL: YES** **1995, 1996, 1997, 1998, 1999, 2000, 2001, 2002** **Models:** Eclipse **Engines:** 2.0L VIN Y, 2.4L VIN G, 3.0L VIN H **Transmissions:** A/T, M/T	**Power Steering Pressure Switch Circuit Malfunction** Engine started, BARO signal over 75 kPa, IAT signal more than 14°F, ECT sensor more than 86°F, vehicle driven to a speed over 55 mph, then back to idle speed at least 10 times, and the PCM detected the PSPS signal remained "on" during the CCM test. **Possible Causes:** • Power steering pressure switch signal circuit is open • Power steering pressure switch signal circuit shorted to ground • Power steering pressure switch is damaged or has failed • PCM has failed
DTC: P0638 **1T PCM MIL: YES** **2003, 2004, 2005, 2006** **Models:** Galant, Lancer, Endeavor, Montero, Outlander **Engines:** All **Transmissions:** All	**Throttle Actuator Control Motor Circuit Range/Performance Problem** Key on, battery positive voltage is greater than 8.3 volts, TPS output voltage is between 0.35 and 4.8 volts, drop of TPS output voltage per 100 ms is greater than 0.04 volts and the TPS output voltage has continued to be greater than the target TPS voltage by 0.5 volt or more for 0.5 seconds and/or difference between the TPS output voltage and the target TPS voltage is 1 volt or greater for 4 seconds. **Possible Causes:** • Throttle valve return spring failed • Throttle valve operation failed • Throttle actuator control motor failed • Throttle actuator control motor circuit shorted or open • PCM has failed

DTC	Trouble Code Title, Conditions & Possible Causes
DTC: P0642 **1T PCM MIL: YES** **2003, 2004, 2005, 2006** **Models:** Galant, Lancer, Endeavor, Montero, Outlander **Engines:** All **Transmissions:** All	**Throttle Position Sensor Power Supply** Key on, battery positive voltage is greater than 6.3 volts, TPS power voltage should be 4.1 volts or less for 0.5 seconds. **Possible Causes:** • PCM has failed
DTC: P0657 **1T PCM MIL: YES** **2003, 2004, 2005, 2006** **Models:** Galant, Lancer, Endeavor, Montero, Outlander **Engines:** All **Transmissions:** All	**Throttle Actuator Control Motor Relay Circuit Malfunction** Key on, and the electronic controlled throttle valve system should be 4 volts or less for 1 second. **Possible Causes:** • Throttle actuator control motor failed • Throttle actuator control motor circuit shorted or open • PCM has failed
DTC: P0660 **2T PCM MIL: YES** **2003, 2004, 2005, 2006** **Models:** Eclipse, Montero **Engines:** 3.0L VIN H, 3.8L VIN S **Transmissions:** All	**Intake Manifold Tuning Circuit Malfunction** Engine cranking, battery voltage between 10 and 16 volts, intake manifold tuning solenoid is ON and greater than 1 second has elapsed after conditions have been met, the intake manifold tuning solenoid coil surge voltage (battery voltage 2 volts) is not detected for 0.2 seconds, the PCM monitors for this condition once during the drive cycle and/or intake manifold tuning solenoid coil surge voltage is not detected for 1 second when the intake manifold tuning solenoid is turned off. **Possible Causes:** • Intake manifold tuning solenoid failed • Intake manifold tuning solenoid circuit shorted or open • PCM has failed
DTC: P0700 **2T CCM MIL: YES** **1996, 1997, 1998, 1999, 2000, 2001, 2002** **Models:** Eclipse **Engines:** 2.0L VIN Y **Transmissions:** A/T	**Transmission Control Module Signal** Key on or engine running, and the PCM received a signal from the TCM that indicating an internal problem with the TCM had occurred. **Possible Causes:** • Clear the trouble codes and retest for this trouble code. If the same trouble code resets, the TCM has failed and must be replaced to repair this problem.
DTC: P0703 **2T CCM MIL: YES** **1999, 2000, 2001, 2002** **Models:** Diamante **Engines:** All **Transmissions:** A/T	**Transmission Brake Switch Signal** Key on or engine running, and the PCM detected the Brake Switch signal did not cycle from "high" to "low" as the brake pedal was pressed and then released during the CCM test. **Possible Causes:** • Brake switch signal circuit is open or shorted to ground • Brake switch power circuit is open (check power from the relay) • Brake switch is damaged or has failed • TCM has failed
DTC: P705 **2T CCM MIL: YES** **1996, 1997, 1998, 1999, 2000, 2001, 2002, 2003, 2004, 2005, 2006** **Models:** 3000GT, Diamante, Eclipse, Galant, Lancer, Mirage, Endeavor, Montero, Montero Sport, Outlander **Engines:** All **Transmissions:** A/T	**Transmission Range Sensor Circuit Malfunction** Key on or engine running, and the PCM received a signal from the TCM indicating a problem in the Transmission Range Sensor (P/N Switch) was detected during the CCM test. **Possible Causes:** • P/N switch signal circuit is open or shorted to ground • P/N switch power circuit is open (check for power at MFI relay) • P/N switch is out-of-adjustment • P/N switch is damaged or has failed • PCM has failed
DTC: P0705 **2T CCM MIL: YES** **1995, 1996, 1997, 1998, 1999** **Models:** Eclipse **Engines:** 2.0L VIN F, VIN Y **Transmissions:** A/T	**A/T Check Shifter Signal Circuit Malfunction** Key on or engine running, and the PCM detected an invalid PRNDL switch signal occurred (i.e., a PRNDL switch signal that should never occur was detected) for 100 ms during the CCM test. **Note: This problem must appear 3 times in one trip to set this code.** **Possible Causes:** • Incorrect TCM part number for a given application • Manual Lever (Rooster Comb) is worn out (check the contacts) • TR sensor signal circuit is open, shorted to ground or to power • TR sensor is damaged or has failed • TCM has failed

DTC	Trouble Code Title, Conditions & Possible Causes
DTC: P0705 **2T CCM MIL: YES** **1996, 1997, 1998, 1999, 2000, 2001, 2002** **Models:** Montero, Montero Sport **Engines:** All **Transmissions:** A/T	**Transmission Range Sensor Circuit Malfunction** Key on or engine running, and the PCM received a signal from the TCM indicating a problem in the Transmission Range Sensor (P/N Switch) was detected during the CCM test. **Possible Causes:** • P/N switch signal circuit is open or shorted to ground • P/N switch power circuit is open (check for power at MFI relay) • P/N switch is out-of-adjustment • P/N switch is damaged or has failed • PCM has failed
DTC: P0710 **2T CCM MIL: YES** **1996, 1997, 1998, 1999, 2000, 2001** **Models:** Montero, Montero Sport **Engines:** All **Transmissions:** A/T	**Transmission Fluid Temperature Sensor Circuit Malfunction** Key on or engine running, and the PCM received a signal from the TCM indicating that a problem with the TFT sensor (i.e., an out-of-range value) had been detected during the CCM test. **Possible Causes:** • TFT sensor signal circuit is open or shorted to ground • TFT sensor signal circuit is shorted to VREF or system power • TFT sensor is damaged or has failed • PCM has failed
DTC: P0710 **2T CCM MIL: YES** **1997, 1998, 1999** **Models:** Eclipse **Engines:** 2.0L VIN F, VIN Y **Transmissions:** A/T	**Transmission Fluid Temperature Sensor Circuit Malfunction** Key on or engine running, and the PCM received a signal from the TCM indicating that a problem with the TFT sensor (i.e., an out-of-range sensor value) had been detected during the CCM test. **Possible Causes:** • TFT sensor signal circuit is open or shorted to ground • TFT sensor signal circuit is shorted to VREF or system power • TFT sensor is damaged or has failed • PCM has failed
DTC: P0712 **2T CCM MIL: YES** **2002, 2003, 2004, 2005, 2006** **Models:** Diamante, Eclipse, Endeavor, Galant, Lancer, Mirage, Montero, Montero Sport, Outlander **Engines:** All **Transmissions:** A/T	**Transmission Fluid Temperature Sensor Circuit Low Input** Engine started, engine running, gear selector in any position except for Neutral, and the PCM detected the TFT sensor indicated a "low" signal (Scan Tool reads more than 315°F) during the CCM test. **Possible Causes:** • TFT sensor signal circuit is shorted to ground • TFT sensor is damaged or has failed • PCM has failed
DTC: P0713 **1T CCM MIL: YES** **2002, 2003, 2004, 2005, 2006** **Models:** Diamante, Eclipse, Endeavor, Galant, Lancer, Mirage, Montero, Montero Sport, Outlander **Engines:** All **Transmissions:** A/T	**Transmission Fluid Temperature Sensor Circuit High Input** Engine started, engine running, gear selector in any position except for Neutral, and the PCM detected the TFT sensor indicated a "high" signal (Scan Tool reads less than -40°F) during the CCM test. **Possible Causes:** • TFT sensor signal circuit is open between the sensor and PCM • TFT sensor ground circuit is open between sensor and PCM • TFT sensor is damaged or has failed • PCM has failed
DTC: P0715 **2T CCM MIL: YES** **1997, 1998, 1999, 2000, 2001, 2002** **Models:** Galant, Mirage **Engines:** 3.0L VIN L, 1.5L VIN A, 1.8L VIN C **Transmissions:** A/T	**A/T Input Speed Sensor Circuit Malfunction** Engine started, vehicle driven at cruise speed for 3-5 minutes, and the TCM detected too large a change in the gear/speed ratio from the Input Speed sensor signal during the CCM Rationality test. **Possible Causes:** • Input speed sensor signal circuit is open or shorted to ground • Input speed sensor signal circuit is shorted to VREF or power • Input speed sensor is damaged or has failed • TCM has failed
DTC: P0715 **2T CCM MIL: YES** **1997, 1998, 1999** **Models:** Eclipse **Engines:** 2.0L VIN F **Transmissions:** A/T	**A/T Input Speed Sensor Circuit Malfunction** Engine started, vehicle driven at cruise speed for 3-5 minutes, and the TCM detected too large a change in the gear/speed ratio from the Input Speed sensor signal during the CCM Rationality test. **Possible Causes:** • Input speed sensor signal circuit is open or shorted to ground • Input speed sensor signal circuit is shorted to VREF or power • Input speed sensor is damaged or has failed • TCM has failed

DTC	Trouble Code Title, Conditions & Possible Causes
DTC: P0715 **2T CCM MIL: YES** **2000, 2001, 2002** **Models:** Montero, Montero Sport **Engines:** All **Transmissions:** A/T	**A/T Input Speed Sensor Circuit Malfunction** Engine started, vehicle driven at cruise speed for 3-5 minutes, and the TCM detected too large a change in the gear/speed ratio from the Input Speed sensor signal during the CCM Rationality test. **Possible Causes:** • Input speed sensor signal circuit is open or shorted to ground • Input speed sensor signal circuit is shorted to VREF or power • Input speed sensor is damaged or has failed • TCM has failed
DTC: P0715 **1T CCM MIL: YES** **2002, 2003, 2004, 2005, 2006** **Models:** Diamante, Eclipse, Endeavor, Galant, Lancer, Mirage, Montero, Montero Sport, Outlander **Engines:** All **Transmissions:** A/T	**A/T Input/Turbine Speed Sensor Circuit Malfunction** Engine started, vehicle driven at cruise speed for 3-5 minutes, and the TCM detected too large a change in the gear/speed ratio from the Input Speed sensor signal during the CCM Rationality test. **Possible Causes:** • Input speed sensor signal circuit is open or shorted to ground • Input speed sensor signal circuit is shorted to VREF or power • Input speed sensor is damaged or has failed • TCM has failed
DTC: P0720 **2T CCM MIL: YES** **1997, 1998, 1999, 2000, 2001, 2002** **Models:** Galant, Mirage **Engines:** 3.0L VIN L, 1.5L VIN A, 1.8L VIN C **Transmissions:** A/T	**A/T Output Speed Sensor Circuit Malfunction** Engine started, vehicle driven at cruise speed for 3-5 minutes, and the TCM detected too large a change in the gear/speed ratio from the Output Speed sensor signal during the CCM Rationality test. **Possible Causes:** • Output speed sensor signal circuit is open or shorted to ground • Output speed sensor signal circuit is shorted to VREF or power • Output speed sensor is damaged or has failed • TCM has failed
DTC: P0720 **2T CCM MIL: YES** **1997, 1998, 1999** **Models:** Eclipse **Engines:** 2.0L VIN Y **Transmissions:** A/T	**A/T Output Speed Sensor Circuit Malfunction** Engine started, vehicle driven at cruise speed for 3-5 minutes, and the TCM detected too large a change in the gear/speed ratio from the Output Speed sensor signal during the CCM Rationality test. **Possible Causes:** • Output speed sensor signal circuit is open or shorted to ground • Output speed sensor signal circuit is shorted to VREF or power • Output speed sensor is damaged or has failed • TCM has failed
DTC: P0720 **2T CCM MIL: YES** **2000, 2001, 2002** **Models:** Montero, Montero Sport **Engines:** All **Transmissions:** A/T	**A/T Output Speed Sensor Circuit Malfunction** Engine started, vehicle driven at cruise speed for 3-5 minutes, and the TCM detected too large a change in the gear/speed ratio from the Output Speed sensor signal during the CCM Rationality test. **Possible Causes:** • Output speed sensor signal circuit is open or shorted to ground • Output speed sensor signal circuit is shorted to VREF or power • Output speed sensor is damaged or has failed • TCM has failed
DTC: P0720 **1T CCM MIL: YES** **2002, 2003, 2004, 2005, 2006** **Models:** Diamante, Eclipse, Endeavor, Galant, Lancer, Mirage, Montero, Montero Sport, Outlander **Engines:** All **Transmissions:** A/T	**A/T Output Speed Sensor Circuit Malfunction** Engine started, vehicle driven at cruise speed for 3-5 minutes, and the TCM detected too large a change in the gear/speed ratio from the Output Speed sensor signal during the CCM Rationality test. **Possible Causes:** • Output speed sensor signal circuit is open or shorted to ground • Output speed sensor signal circuit is shorted to VREF or power • Output speed sensor is damaged or has failed • TCM has failed
DTC: P0725 **2T CCM MIL: YES** **1997, 1998, 1999** **Models:** Eclipse **Engines:** 2.0L VIN Y **Transmissions:** A/T	**Engine Speed Sensor Signal to TCM** Engine started, engine runtime over 2 minutes, vehicle driven in one or more forward gears for 2 minutes, then returned to P/N position for 1.5 seconds, then with the OSS signal over 400 rpm and the TSS signal over 600 rpm, the TCM detected the engine speed was not the same as the speed reported over the CCD bus during the test. **Possible Causes:** • CKP sensor signal circuit to TCM is open or shorted to ground • CKP sensor is damaged, erratic or has failed • TCM or the PCM (controller) has failed

DTC	Trouble Code Title, Conditions & Possible Causes
DTC: P0725 **2T CCM MIL: YES** **1997, 1998, 1999, 2000, 2001** **Models:** Galant, Mirage **Engines:** 3.0L VIN L, 1.5L VIN A, 1.8L VIN C **Transmissions:** A/T	**Engine Speed Sensor Signal to TCM** Engine started, engine runtime over 2 minutes, vehicle driven in one or more forward gears for 2 minutes, then returned to P/N position for 1.5 seconds, then with the OSS signal over 400 rpm and the TSS signal over 600 rpm, the TCM detected the engine speed was not the same as the speed reported over the CCD bus during the test. **Possible Causes:** • CKP sensor signal circuit to TCM is open or shorted to ground • CKP sensor is damaged, erratic or has failed • TCM or the PCM (controller) has failed
DTC: P0731 **2T CCM MIL: YES** **1997, 1998, 1999, 2000, 2001, 2002** **Models:** Eclipse **Engines:** 2.0L VIN F, VIN Y, 2.4L VIN G, 3.0L VIN H **Transmissions:** A/T	**A/T Incorrect First Gear Ratio** DTC P0500 not set, engine started, vehicle driven to a speed of over 3 mph with 1st Gear commanded "on", and the PCM detected an incorrect 1st gear/speed ratio during the CCM Rationality test. **Possible Causes:** • Solenoid or related pressure switch is damaged or has failed • Problems related to the Input Speed or Output Speed sensor • LR clutch is damaged, leaking or has failed • Problems related to the transmission valve body
DTC: P0731 **1T CCM MIL: YES** **2002, 2003, 2004, 2005, 2006** **Models:** Diamante, Endeavor, Galant, Lancer, Mirage, Montero, Montero Sport, Outlander **Engines:** All **Transmissions:** A/T	**A/T Incorrect First Gear Ratio** DTC P0500 not set, engine started, vehicle driven to a speed of over 3 mph with 1st Gear commanded "on", and the PCM detected an incorrect 1st gear/speed ratio during the CCM Rationality test. **Possible Causes:** • Solenoid or related pressure switch is damaged or has failed • Problems related to the Input Speed or Output Speed sensor • LR clutch is damaged, leaking or has failed • Problems related to the transmission valve body
DTC: P0732 **2T CCM MIL: YES** **1997, 1998, 1999, 2000, 2001, 2002** **Models:** Eclipse **Engines:** 2.0L VIN F, VIN Y, 2.4L VIN G, 3.0L VIN H **Transmissions:** A/T	**A/T Incorrect Second Gear Ratio** DTC P0500 not set, engine started, vehicle driven to a speed of over 3 mph with 2nd Gear commanded "on", and the PCM detected an incorrect 2nd gear/speed ratio during the CCM Rationality test. **Possible Causes:** • Solenoid or related pressure switch is damaged or has failed • Problems related to the Input Speed or Output Speed sensor • LR clutch is damaged, leaking or has failed • Problems related to the transmission valve body
DTC: P0732 **1T CCM MIL: YES** **2002, 2003, 2004, 2005, 2006** **Models:** Diamante, Endeavor, Galant, Lancer, Mirage, Montero, Montero Sport, Outlander **Engines:** All **Transmissions:** A/T	**A/T Incorrect Second Gear Ratio** DTC P0500 not set, engine started, vehicle driven to a speed of over 3 mph with 2nd Gear commanded "on", and the PCM detected an incorrect 2nd gear/speed ratio during the CCM Rationality test. **Possible Causes:** • Solenoid or related pressure switch is damaged or has failed • Problems related to the Input Speed or Output Speed sensor • LR clutch is damaged, leaking or has failed • Problems related to the transmission valve body
DTC: P0733 **2T CCM MIL: YES** **1997, 1998, 1999, 2000, 2001, 2002** **Models:** Eclipse **Engines:** 2.0L VIN F, VIN Y, 2.4L VIN G, 3.0L VIN H **Transmissions:** A/T	**A/T Incorrect Third Gear Ratio** DTC P0500 not set, engine started, vehicle driven to a speed of over 3 mph with 3rd Gear commanded "on", and the PCM detected an incorrect 3rd Gear ratio during the CCM Rationality test. **Possible Causes:** • Solenoid or related pressure switch is damaged or has failed • Problems related to the Input Speed or Output Speed sensor • LR clutch is damaged, leaking or has failed • Problems related to the transmission valve body
DTC: P0733 **1T CCM MIL: YES** **2002, 2003, 2004, 2005, 2006** **Models:** Diamante, Endeavor, Galant, Lancer, Mirage, Montero, Montero Sport, Outlander **Engines:** All **Transmissions:** A/T	**A/T Incorrect Third Gear Ratio** DTC P0500 not set, engine started, vehicle driven to a speed of over 3 mph with 3rd Gear commanded "on", and the PCM detected an incorrect 3rd Gear ratio during the CCM Rationality test. **Possible Causes:** • Solenoid or related pressure switch is damaged or has failed • Problems related to the Input Speed or Output Speed sensor • LR clutch is damaged, leaking or has failed • Problems related to the transmission valve body

DTC	Trouble Code Title, Conditions & Possible Causes
DTC: P0734 **2T CCM MIL: YES** **1997, 1998, 1999, 2000, 2001, 2002** **Models:** Eclipse **Engines:** 2.0L VIN F, VIN Y, 2.4L VIN G, 3.0L VIN H **Transmissions:** A/T	**A/T Incorrect Fourth Gear Ratio** DTC P0500 not set, engine started, vehicle driven to a speed of over 3 mph with 4th Gear commanded "on", and the PCM detected an incorrect Gear ratio during the CCM test. **Possible Causes:** • Solenoid or related pressure switch is damaged or has failed • LR clutch is damaged, leaking or has failed • Problems related to the Input Speed or Output Speed sensor • Problems related to the transmission valve body
DTC: P0734 **1T CCM MIL: YES** **2002, 2003, 2004, 2005, 2006** **Models:** Diamante, Endeavor, Galant, Lancer, Mirage, Montero, Montero Sport, Outlander **Engines:** All **Transmissions:** A/T	**A/T Incorrect Fourth Gear Ratio** DTC P0500 not set, engine started, vehicle driven to a speed of over 3 mph with 4th Gear commanded "on", and the PCM detected an incorrect Gear ratio during the CCM test. **Possible Causes:** • Solenoid or related pressure switch is damaged or has failed • LR clutch is damaged, leaking or has failed • Problems related to the Input Speed or Output Speed sensor • Problems related to the transmission valve body
DTC: P0735 **2T CCM MIL: YES** **2003, 2004, 2005, 2006** **Models:** Galant, Montero, Montero Sport **Engines:** All V6 **Transmissions:** A/T	**A/T Incorrect Fifth Gear Ratio** DTC P0500 not set, engine started, vehicle driven to a speed of over 3 mph with 5th Gear commanded "on", and the PCM detected an incorrect 5th Gear ratio during the CCM Rationality test. **Possible Causes:** • 5th Gear or pressure switch is damaged or has failed • LR clutch is damaged, leaking or has failed • Problems related to the Input Speed or Output Speed sensor • Problems related to the transmission valve body
DTC: P0736 **2T CCM MIL: YES** **1997, 1998, 1999, 2000, 2001, 2002** **Models:** Eclipse **Engines:** 2.0L VIN F, VIN Y, 2.4L VIN G, 3.0L VIN H **Transmissions:** A/T	**A/T Incorrect Reverse Gear Ratio** DTC P0500 not set, engine started, vehicle driven to a speed of over 3 mph with Reverse Gear commanded "on", and the PCM detected an incorrect Reverse Gear ratio during the CCM Rationality test. **Possible Causes:** • Reverse Gear is damaged or has failed • LR clutch is damaged, leaking or has failed • Problems related to the Input Speed or Output Speed sensor • Problems related to the transmission valve body
DTC: P0736 **1T CCM MIL: YES** **2002, 2003, 2004, 2005, 2006** **Models:** Diamante, Endeavor, Galant, Lancer, Mirage, Montero, Montero Sport, Outlander **Engines:** All **Transmissions:** A/T	**A/T Incorrect Reverse Gear Ratio** DTC P0500 not set, engine started, vehicle driven to a speed of over 3 mph with Reverse Gear commanded "on", and the PCM detected an incorrect Reverse Gear ratio during the CCM Rationality test. **Possible Causes:** • Reverse Gear is damaged or has failed • LR clutch is damaged, leaking or has failed • Problems related to the Input Speed or Output Speed sensor • Problems related to the transmission valve body
DTC: P0740 **1T CCM MIL: YES** **1997, 1998, 1999** **Models:** Eclipse **Engines:** 2.0L VIN Y **Transmissions:** A/T, M/T	**A/T Lockup Control Out-Of-Range Malfunction** Engine started, vehicle driven in 2nd, 3rd or 4th gear at an Input speed over 1750 rpm, throttle angle less than 30 degrees, and the TCM detected the TCC slip value exceeded 100 rpm for 10 seconds at least 3 times in one drive cycle during TCC operation. **Possible Causes:** • Transmission fluid is contaminated, or the fluid level is too low • Worn pump bushing and/or the TCC is damaged or has failed • Valve body lockup accumulator diameter is out of specification
DTC: P0740 **1T CCM MIL: YES** **1997, 1998, 1999, 2000, 2001** **Models:** Galant, Mirage **Engines:** 3.0L VIN L, 1.5L VIN A, 1.8L VIN C **Transmissions:** A/T	**A/T Lockup Control Out-Of-Range Malfunction** Engine started, vehicle driven in 2nd, 3rd or 4th gear at an Input speed over 1750 rpm, throttle angle less than 30 degrees, and the TCM detected the TCC slip value exceeded 100 rpm for 10 seconds at least 3 times in one drive cycle during TCC operation. **Possible Causes:** • Transmission fluid is contaminated, or the fluid level is too low • Worn pump bushing and/or the TCC is damaged or has failed • Valve body lockup accumulator diameter is out of specification
DTC: P0740 **1T CCM MIL: YES** **2000, 2001** **Models:** Montero, Montero Sport **Engines:** All **Transmissions:** A/T	**A/T Lockup Control Out-Of-Range Malfunction** Engine started, vehicle driven in 2nd, 3rd or 4th gear at an Input speed over 1750 rpm, throttle angle less than 30 degrees, and the TCM detected the TCC slip value exceeded 100 rpm for 10 seconds at least 3 times in one drive cycle during TCC operation. **Possible Causes:** • Transmission fluid is contaminated, or the fluid level is too low • Worn pump bushing and/or the TCC is damaged or has failed • Valve body lockup accumulator diameter is out of specification

DTC	Trouble Code Title, Conditions & Possible Causes
DTC: P0741 **2T CCM MIL: YES** **2002, 2003, 2004, 2005, 2006** **Models:** Diamante, Endeavor, Galant, Lancer, Mirage, Montero, Montero Sport, Outlander **Engines:** All **Transmissions:** A/T	**A/T Torque Converter Clutch Performance (Stuck Off)** Engine started, vehicle driven in gear with VSS signals received, and the PCM detected excessive slippage during normal operation. **Possible Causes:** • TCC solenoid has a mechanical failure (solenoid is stuck off) • TCC solenoid has a hydraulic failure • PCM has failed
DTC: P0742 **2T CCM MIL: YES** **2002, 2003, 2004, 2005, 2006** **Models:** Diamante, Endeavor, Galant, Lancer, Mirage, Montero, Montero Sport, Outlander **Engines:** All **Transmissions:** A/T	**A/T Torque Converter Clutch Performance (Stuck On)** Engine started, vehicle driven in gear with VSS signals received, and the PCM detected unusual TCC lockup during normal operation. **Possible Causes:** • TCC solenoid has a mechanical failure (solenoid is stuck on) • TCC solenoid has a hydraulic failure • PCM has failed
DTC: P0743 **1T CCM MIL: YES** **2002, 2003, 2004, 2005, 2006** **Models:** Diamante, Endeavor, Galant, Lancer, Mirage, Montero, Montero Sport, Outlander **Engines:** All **Transmissions:** A/T	**A/T Torque Converter Clutch Performance (Stuck On)** Engine started, vehicle driven in gear with VSS signals received, and the PCM detected an unexpected voltage condition on the TCC solenoid control circuit during the CCM test. **Possible Causes:** • TCC solenoid control circuit open or shorted to ground • TCC solenoid wiring harness connector damaged • TCC solenoid is damaged or has failed • PCM has failed
DTC: P0750 **2T CCM MIL: YES** **1997, 1998, 1999** **Models:** Eclipse **Engines:** 2.0L VIN Y **Transmissions:** A/T	**A/T LR Solenoid Circuit Malfunction** Engine started, engine running with VSS signals received, and the PCM received a message indicating the TCM had detected an unexpected voltage condition (no voltage spike with the solenoid energized) on the Low-Reverse (LR) solenoid circuit during the test. **Possible Causes:** • LR solenoid control circuit is open, shorted to ground or power • LR pressure switch is open, shorted to ground or it has failed • LR solenoid is damaged or has failed • TCM power ground circuit is open, or the TCM has failed
DTC: P0750 **2T CCM MIL: YES** **1997, 1998, 1999, 2000, 2001** **Models:** Galant, Mirage **Engines:** 3.0L VIN L, 1.5L VIN A, 1.8L VIN C **Transmissions:** A/T	**A/T LR Solenoid Circuit Malfunction** Engine started, engine running with VSS signals received, and the PCM received a message indicating the TCM had detected an unexpected voltage condition (no voltage spike with the solenoid energized) on the Low-Reverse (LR) solenoid circuit during the test. **Possible Causes:** • LR solenoid control circuit is open, shorted to ground or power • LR pressure switch is open, shorted to ground or it has failed • LR solenoid is damaged or has failed • TCM power ground circuit is open, or the TCM has failed
DTC: P0750 **2T CCM MIL: YES** **2000, 2001** **Models:** Montero, Montero Sport **Engines:** All **Transmissions:** A/T	**A/T LR Solenoid Circuit Malfunction** Engine started, engine running with VSS signals received, and the PCM received a message indicating the TCM had detected an unexpected voltage condition (no voltage spike with the solenoid energized) on the Low-Reverse (LR) solenoid circuit during the test. **Possible Causes:** • LR solenoid control circuit is open, shorted to ground or power • LR pressure switch is open, shorted to ground or it has failed • LR solenoid is damaged or has failed • TCM power ground circuit is open, or the TCM has failed
DTC: P0753 **1T CCM MIL: YES** **2002, 2003, 2004, 2005, 2006** **Models:** Diamante, Endeavor, Galant, Lancer, Mirage, Montero, Montero Sport, Outlander **Engines:** All **Transmissions:** A/T	**A/T Shift Solenoid 'A' Circuit Malfunction** Engine started, vehicle driven in gear with VSS signals received, and the PCM detected an unexpected voltage condition on the Shift Solenoid 'A' control circuit during the CCM test. **Possible Causes:** • SSA solenoid control circuit open or shorted to ground • SSA solenoid wiring harness connector damaged • SSA solenoid is damaged or has failed • PCM has failed

DTC	Trouble Code Title, Conditions & Possible Causes
DTC: P0755 **2T CCM MIL: YES** **1997, 1998, 1999** **Models:** Eclipse **Engines:** 2.0L VIN Y **Transmissions:** A/T	**A/T 2-3 Solenoid Circuit Malfunction** Engine started, engine running with VSS signals received, and the PCM received a message indicating the TCM had detected an unexpected voltage condition (no voltage spike with the solenoid energized) on the 2-3 Shift Solenoid circuit during the test. **Possible Causes:** • 2-3 solenoid control circuit is open, shorted to ground or power • 2-3 pressure switch is open, shorted to ground or it has failed • 2-3 solenoid is damaged or has failed • TCM power ground circuit is open, or the TCM has failed
DTC: P0755 **2T CCM MIL: YES** **1997, 1998, 1999, 2000, 2001** **Models:** Galant, Mirage **Engines:** 3.0L VIN L, 1.5L VIN A, 1.8L VIN C **Transmissions:** A/T	**A/T 2-3 Solenoid Circuit Malfunction** Engine started, engine running with VSS signals received, and the PCM received a message indicating the TCM had detected an unexpected voltage condition (no voltage spike with the solenoid energized) on the 2-3 Shift Solenoid circuit during the test. **Possible Causes:** • 2-3 solenoid control circuit is open, shorted to ground or power • 2-3 pressure switch is open, shorted to ground or it has failed • 2-3 solenoid is damaged or has failed • TCM power ground circuit is open, or the TCM has failed
DTC: P0758 **1T CCM MIL: YES** **2002, 2003, 2004, 2005, 2006** **Models:** Diamante, Endeavor, Galant, Lancer, Mirage, Montero, Montero Sport, Outlander **Engines:** All **Transmissions:** A/T	**A/T Shift Solenoid 'B' Circuit Malfunction** Engine started, vehicle driven in gear with VSS signals received, and the PCM detected an unexpected voltage condition on the Shift Solenoid 'B' control circuit during the CCM test. **Possible Causes:** • SSB solenoid control circuit open or shorted to ground • SSB solenoid wiring harness connector damaged • SSB solenoid is damaged or has failed • PCM has failed
DTC: P0760 **2T CCM MIL: YES** **1997, 1998, 1999, 2000, 2001, 2002** **Models:** Galant, Mirage **Engines:** 3.0L VIN L, 1.5L VIN A, 1.8L VIN C **Transmissions:** A/T	**A/T OD Solenoid Circuit Malfunction** Engine started, engine running with VSS signals received, and the PCM received a message indicating the TCM had detected an unexpected voltage condition (no voltage spike with the solenoid energized) on the Overdrive (OD) Solenoid circuit during the test. **Possible Causes:** • OD solenoid control circuit is open, shorted to ground or power • OD pressure switch is open, shorted to ground or it has failed • OD solenoid is damaged or has failed • TCM power ground circuit is open, or the TCM has failed
DTC: P0760 **2T CCM MIL: YES** **1997, 1998, 1999** **Models:** Eclipse **Engines:** 2.0L VIN Y **Transmissions:** A/T	**A/T OD Solenoid Circuit Malfunction** Engine started, engine running with VSS signals received, and the PCM received a message indicating the TCM had detected an unexpected voltage condition (no voltage spike with the solenoid energized) on the Overdrive (OD) Solenoid circuit during the test. **Possible Causes:** • OD solenoid control circuit is open, shorted to ground or power • OD pressure switch is open, shorted to ground or it has failed • OD solenoid is damaged or has failed • TCM power ground circuit is open, or the TCM has failed
DTC: P0760 **2T CCM MIL: YES** **2000, 2001** **Models:** Montero, Montero Sport **Engines:** All **Transmissions:** A/T	**A/T OD Solenoid Circuit Malfunction** Engine started, engine running with VSS signals received, and the PCM received a message indicating the TCM had detected an unexpected voltage condition (no voltage spike with the solenoid energized) on the Overdrive (OD) Solenoid circuit during the test. **Possible Causes:** • OD solenoid control circuit is open, shorted to ground or power • OD pressure switch is open, shorted to ground or it has failed • OD solenoid is damaged or has failed • TCM power ground circuit is open, or the TCM has failed
DTC: P0763 **1T CCM MIL: YES** **2002, 2003, 2004, 2005, 2006** **Models:** Diamante, Endeavor, Galant, Lancer, Mirage, Montero, Montero Sport, Outlander **Engines:** All **Transmissions:** A/T	**A/T Shift Solenoid 'C' Circuit Malfunction** Engine started, vehicle driven in gear with VSS signals received, and the PCM detected an unexpected voltage condition on the Shift Solenoid 'C' control circuit during the CCM test. **Possible Causes:** • SSC solenoid control circuit open or shorted to ground • SSC solenoid wiring harness connector damaged • SSC solenoid is damaged or has failed • PCM has failed

DTC	Trouble Code Title, Conditions & Possible Causes
DTC: P0765 **2T CCM MIL: YES** **1997, 1998, 1999** **Models:** Eclipse **Engines:** 2.0L VIN Y **Transmissions:** A/T	**A/T UD Solenoid Circuit Malfunction** Engine started, engine running with VSS signals received, and the PCM received a message indicating the TCM had detected an unexpected voltage condition (no voltage spike with the solenoid energized) on the UD Solenoid circuit during the CCM test. **Possible Causes:** • UD solenoid control circuit is open, shorted to ground or power • UD pressure switch is open, shorted to ground or it has failed • UD solenoid is damaged or has failed • TCM power ground circuit is open, or the TCM has failed
DTC: P0765 **2T CCM MIL: YES** **1997, 1998, 1999, 2000, 2001** **Models:** Galant, Mirage **Engines:** 3.0L VIN L, 1.5L VIN A, 1.8L VIN C **Transmissions:** A/T	**A/T UD Solenoid Circuit Malfunction** Engine started, engine running with VSS signals received, and the PCM received a message indicating the TCM had detected an unexpected voltage condition (no voltage spike with the solenoid energized) on the UD Solenoid circuit during the CCM test. **Possible Causes:** • UD solenoid control circuit is open, shorted to ground or power • UD pressure switch is open, shorted to ground or it has failed • UD solenoid is damaged or has failed • TCM power ground circuit is open, or the TCM has failed
DTC: P0765 **2T CCM MIL: YES** **2000, 2001** **Models:** Montero, Montero Sport **Engines:** All **Transmissions:** A/T	**A/T UD Solenoid Circuit Malfunction** Engine started, engine running with VSS signals received, and the PCM received a message indicating the TCM had detected an unexpected voltage condition (no voltage spike with the solenoid energized) on the UD Solenoid circuit during the CCM test. **Possible Causes:** • UD solenoid control circuit is open, shorted to ground or power • UD pressure switch is open, shorted to ground or it has failed • UD solenoid is damaged or has failed • TCM power ground circuit is open, or the TCM has failed
DTC: P0768 **1T CCM MIL: YES** **2002, 2003, 2004, 2005, 2006** **Models:** Diamante, Endeavor, Galant, Lancer, Mirage, Montero, Montero Sport, Outlander **Engines:** All **Transmissions:** A/T	**A/T Shift Solenoid 'D' Circuit Malfunction** Engine started, vehicle driven in gear with VSS signals received, and the PCM detected an unexpected voltage condition on the Shift Solenoid 'D' control circuit during the CCM test. **Possible Causes:** • SSD solenoid control circuit open or shorted to ground • SSD solenoid wiring harness connector damaged • SSD solenoid is damaged or has failed • PCM has failed
DTC: P0773 **1T CCM MIL: YES** **2002, 2003, 2004, 2005, 2006** **Models:** Montero **Engines:** All **Transmissions:** A/T	**A/T Shift Solenoid 'E' Circuit Malfunction** Engine started, vehicle driven in gear with VSS signals received, and the PCM detected an unexpected voltage condition on the Shift Solenoid 'E' control circuit during the CCM test. **Possible Causes:** • SSE solenoid control circuit open or shorted to ground • SSE solenoid wiring harness connector damaged • SSE solenoid is damaged or has failed • PCM has failed
DTC: P0783 **2T CCM MIL: YES** **1997, 1998, 1999** **Models:** Eclipse **Engines:** 2.0L VIN Y **Transmissions:** A/T	**A/T Solenoid Switch Valve Locked In The LR Position** Engine started, engine running with VSS signals received, and the PCM received a message indicating the TCM had detected an unexpected voltage condition (no voltage spike with the solenoid energized) on the UD Solenoid circuit during the CCM test. **Possible Causes:** • LR pressure switch circuit open, shorted to ground (intermittent) • LR solenoid or pressure switch is damaged or has failed • Problems related to valve body (solenoid stuck in LR position) • TCM power ground circuit is open, or the TCM has failed
DTC: P0830 **2T CCM MIL: YES** **2003, 2004, 2005, 2006** **Models:** Evo **Engines:** All **Transmissions:** M/T	**Clutch Pedal Position Switch Circuit Range/Performance** Engine started and vehicle is driven a minimum speed of 19 mph, the toggling of the signal (high/low) of the clutch pedal position switch is not detected even once. **Possible Causes:** • Clutch pedal position switch failed • Clutch pedal position switch circuit open or shorted • PCM has failed

Gas Engine OBD II Trouble Code List (P1XXX Codes)

DTC	Trouble Code Title, Conditions & Possible Causes
DTC: P1020 **2T CCM MIL: YES** **2004, 2005, 2006** **Models:** Galant, Lancer, Outlander **Engines:** 2.4L VIN F **Transmissions:** A/T, M/T	**Mitsubishi Innovative Valve Timing and Lift Electronic Control System (MIVEC) Performance Problem** Engine started, engine runtime 30 seconds or more, engine coolant temperature 171°F or greater, battery voltage over 10v, engine speed is 3000 rpm or lower and engine oil pressure switch is OFF for 2 seconds, or, engine speed is 4000 rpm or higher and engine oil pressure switch is ON for 2 seconds. **Possible Causes:** • Engine oil pressure switch failed • Engine oil control valve failed • Engine oil pressure switch circuit open or shorted induction control motor circuit is open • PCM has failed
DTC: P01021 **2T CCM MIL: YES** **2004, 2005, 2006** **Models:** Galant, Lancer, Outlander **Engines:** 2.4L VIN F **Transmissions:** A/T, M/T	**Engine Oil Control Valve Circuit** Engine started, battery voltage over 10v, MIVEC operating in the low speed mode the engine control valve circuit to the PCM voltage is less than 1.5 volts for 2 seconds. **Possible Causes:** • Engine oil control valve failed • Engine oil control valve circuit open or shorted • PCM has failed
DTC: P1022 **2T CCM MIL: YES** **2006** **Models:** Galant **Engines:** 3.8L VIN T **Transmissions:** A/T, M/T	**Mitsubishi Innovative Valve Timing Electronic Control System (MIVEC) Performance Problem (bank2)** Engine started, engine runtime 30 seconds or more, engine coolant temperature 171°F or greater, battery voltage over 10v, engine speed is 3500 rpm or lower and engine oil pressure switch is OFF for 5 seconds, or, engine speed is 4500 rpm or higher and engine oil pressure switch is ON for 5 seconds. **Possible Causes:** • Engine oil pressure switch failed • Engine oil control valve failed • Engine oil pressure switch circuit open or shorted induction control motor circuit is open • PCM has failed
DTC: P01023 **2T CCM MIL: YES** **2006** **Models:** Galant **Engines:** 3.8L VIN T **Transmissions:** A/T, M/T	**Engine Oil Control Valve Circuit (bank 2)** Engine started, battery voltage over 10v, MIVEC operating in the low speed mode the engine control valve circuit to the PCM voltage is less than 1.5 volts for 2 seconds. **Possible Causes:** • Engine oil control valve failed • Engine oil control valve circuit open or shorted • PCM has failed
DTC: P1100 **1T CCM MIL: YES** **1996** **Models:** Diamante **Engines:** All **Transmissions:** A/T, M/T	**Induction Control Motor Position Sensor** Engine started, engine runtime 1 minute, system voltage over 10v, engine running at 500 rpm or higher, and the PCM detected the Intake Air Control Valve did not reach the Target position even after the Intake Air Control Valve Drive Motor was commanded to reach its Desired position at least 4 times during the CCM Rationality test. **Possible Causes:** • ICM position sensor signal circuit is open or shorted to ground • ICM position sensor is damaged or has failed • Variable induction control motor circuit is open • Variable induction control motor circuit is shorted to ground • Variable induction control motor is damaged or has failed • PCM has failed
DTC: P1101 **2T CCM MIL: YES** **1996, 1997, 1998, 1999, 2000, 2001, 2002, 2003, 2004** **Models:** Diamante **Engines:** All **Transmissions:** A/T	**Traction Control Vacuum Solenoid Circuit Malfunction** Engine started, engine runtime 1 minute, system voltage over 10v, Scan Tool forced actuation test "off", and the PCM detected the T/C Vacuum solenoid drive or non-drive commands with the solenoid "on" were not the same during the CCM Rationality test. **Possible Causes:** • T/C vacuum solenoid circuit is open or shorted to ground • T/C vacuum solenoid circuit is shorted to system power (B+) • T/C vacuum solenoid circuit is damaged or has failed • PCM has failed • TSB 2-13-002 (2/02) contains information related to this code

DTC	Trouble Code Title, Conditions & Possible Causes
DTC: P1102 **2T CCM MIL: YES** **1996, 1997, 1998, 1999, 2000, 2001, 2002, 2003, 2004** **Models:** Diamante **Engines:** All **Transmissions:** A/T	**Traction Control Ventilation Solenoid Circuit Malfunction** Engine started, engine runtime 1 minute, system voltage over 10v, Scan Tool forced actuation test "off", and the PCM detected the T/C Ventilation solenoid drive or non-drive commands with the solenoid "on" were not the same during the CCM Rationality test. **Possible Causes:** • T/C ventilation solenoid circuit is open or shorted to ground • T/C ventilation solenoid circuit is shorted to system power (B+) • T/C ventilation solenoid circuit is damaged or has failed • PCM has failed • TSB 2-13-002 (2/02) contains information related to this code
DTC: P1103 **2T CCM MIL: YES** **1996, 1997, 1998, 1999** **Models:** 3000GT **Engines:** 3.0L VIN K **Transmissions:** A/T, M/T	**Turbocharger Waste Gate Actuator Circuit Malfunction** Engine started, ECT sensor more than 176°F, TP sensor signal less than 4.2v, and the PCM detected the volumetric efficiency remained at over 200% for 1.5 seconds during the CCM Rationality test. **Note: The PCM continuously monitors the intake air amount in order to determine if the engine is in an Overboost condition. If the PCM judges the engine is in an Overboost condition, it cuts of the fuel supply to the injectors in order to protect the engine from damage.** **Possible Causes:** • Turbocharger Wastegate actuator is damaged or has failed • Boost Pressure Control system has failed • PCM has failed
DTC: P1103 **2T CCM MIL: YES** **1995, 1996, 1997, 1998, 1999** **Models:** Eclipse **Engines:** 2.0L VIN F **Transmissions:** A/T, M/T	**Turbocharger Waste Gate Actuator Circuit Malfunction** Engine started, ECT sensor more than 176°F, TP sensor signal less than 4.2v, and the PCM detected the volumetric efficiency remained at over 200% for 1.5 seconds during the CCM Rationality test. **Note: The PCM continuously monitors the intake air amount in order to determine if the engine is in an Overboost condition. If the PCM judges the engine is in an Overboost condition, it cuts of the fuel supply to the injectors in order to protect the engine from damage.** **Possible Causes:** • Turbocharger Wastegate actuator is damaged or has failed • Boost Pressure Control system has failed • PCM has failed
DTC: P1104 **2T CCM MIL: YES** **1996, 1997, 1998, 1999** **Models:** 3000GT **Engines:** 3.0L VIN K **Transmissions:** A/T, M/T	**Turbocharger Waste Gate Solenoid Circuit Malfunction** Engine started, engine running with the system voltage over 10v, and the PCM detected the Waste Gate solenoid coil surge voltage was too low (less than system voltage +2v) during the CCM test. **Possible Causes:** • Waste Gate solenoid circuit is open or shorted to ground • Waste Gate solenoid power circuit is open (check the relay) • Waste Gate solenoid is damaged or has failed • PCM has failed
DTC: P1104 **2T CCM MIL: YES** **1995, 1996, 1997, 1998, 1999** **Models:** Eclipse **Engines:** 2.0L VIN F **Transmissions:** A/T, M/T	**Turbocharger Waste Gate Solenoid Circuit Malfunction** Engine started, engine running with the system voltage over 10v, and the PCM detected the Waste Gate solenoid coil surge voltage was too low (less than system voltage +2v) during the CCM test. **Possible Causes:** • Waste Gate solenoid control circuit is open • Waste Gate solenoid control circuit is shorted to ground • Waste Gate solenoid power circuit is open (check the relay) • Waste Gate solenoid is damaged or has failed • PCM has failed
DTC: P1105 **2T CCM MIL: YES** **1996, 1997, 1998, 1999** **Models:** 3000GT **Engines:** 3.0L VIN K **Transmissions:** A/T, M/T	**Fuel Pressure Solenoid Circuit Malfunction** Engine started, engine running with the system voltage over 10v, and the PCM detected the Fuel Pressure solenoid coil surge voltage was too low (less than system voltage +2v) during the CCM test. **Possible Causes:** • Fuel Pressure solenoid control circuit is open • Fuel Pressure solenoid control circuit is shorted to ground • Fuel Pressure solenoid power circuit is open (check the relay) • Fuel Pressure solenoid is damaged or has failed • PCM has failed

DTC	Trouble Code Title, Conditions & Possible Causes
DTC: P1105 **2T CCM MIL: YES** **1995, 1996, 1997, 1998, 1999** **Models:** Eclipse **Engines:** 2.0L VIN F **Transmissions:** A/T, M/T	**Fuel Pressure Solenoid Circuit Malfunction** Engine started, engine running with the system voltage over 10v, and the PCM detected the Fuel Pressure solenoid coil surge voltage was too low (less than system voltage +2v) during the CCM test. **Possible Causes:** • Fuel Pressure solenoid control circuit is open • Fuel Pressure solenoid control circuit is shorted to ground • Fuel Pressure solenoid power circuit is open (check the relay) • Fuel Pressure solenoid is damaged or has failed • PCM has failed
DTC: P1294 **2T CCM MIL: YES** **1995, 1996, 1997, 1998, 1999** **Models:** Eclipse **Engines:** 2.0L VIN Y **Transmissions:** A/T, M/T	**Target Idle Speed Not Reached** Engine started, engine at hot idle speed with the brake applied, and the PCM detected the Actual idle speed was more than 200 rpm higher or lower than the Target idle speed for over 12 seconds during the CCM Rationality test. **Possible Causes:** • IAC motor circuit is open or shorted to ground (intermittent) • IAC motor is contaminated, damaged or has failed • PCM has failed
DTC: P1295 **2T CCM MIL: YES** **1997, 1998, 1999** **Models:** Eclipse **Engines:** 2.0L VIN Y **Transmissions:** A/T, M/T	**TP Sensor 5-Volt Supply Circuit Malfunction** Engine started, vehicle driven to a speed of over 20 mph at an engine speed over 1500 rpm, MAP sensor less than 13 kPa, and the PCM detected the TP sensor signal was less than an amount stored in memory during the CCM test. **Possible Causes:** • TP sensor signal circuit is shorted to ground • TP sensor is damaged or has failed • PCM has failed
DTC: P1296 **2T CCM MIL: YES** **1997, 1998, 1999** **Models:** Eclipse **Engines:** 2.0L VIN Y **Transmissions:** A/T, M/T	**MAP Sensor 5-Volt Supply Circuit Malfunction** Immediately after an engine shutdown period and prior to the next engine startup event, the PCM detected the MAP sensor signal indicated less than under a specified value during the CCM test. **Possible Causes:** • MAP sensor 5-volt supply (VREF) circuit is open • MAP sensor signal circuit is shorted to ground (intermittent) • MAP sensor is damaged or has failed • PCM has failed
DTC: P1297 **2T CCM MIL: YES** **1995, 1996, 1997, 1998, 1999** **Models:** Eclipse **Engines:** 2.0L VIN Y **Transmissions:** A/T, M/T	**No Change In MAP From Start to Run** Engine started, engine at hot idle speed with the Actual idle speed close to the Target idle speed, vehicle not moving, and the PCM detected the difference between the MAP sensor signal at key "on", and the MAP sensor signal at idle for 2 seconds was too small. **Possible Causes:** • MAP sensor vacuum line is clogged or it contains moisture/ice • MAP sensor is damaged or has failed • PCM has failed
DTC: P1300 **2T CCM MIL: YES** **1996** **Models:** 3000GT, Diamante, Galant, Mirage, Montero **Engines:** All **Transmissions:** A/T, M/T	**Ignition Timing Adjustment Circuit Malfunction** Engine started, engine running, and the PCM detected an unexpected "low" voltage condition on the Ignition Timing Adjustment circuit during the CCM test. **Note: This code will set during the ignition timing set procedure.** **Possible Causes:** • Ignition timing adjustment circuit is shorted to ground • PCM has failed
DTC: P1300 **2T CCM MIL: YES** **1995, 1996** **Models:** Eclipse **Engines:** 2.0L VIN F **Transmissions:** A/T, M/T	**Ignition Timing Adjustment Circuit Malfunction** Engine started, engine running, and the PCM detected an unexpected "low" voltage condition on the Ignition Timing Adjustment circuit during the CCM test. **Note: This code will set during the ignition timing set procedure.** **Possible Causes:** • Ignition timing adjustment circuit is shorted to ground • PCM has failed

DTC	Trouble Code Title, Conditions & Possible Causes
DTC: P1390 **2T CCM MIL: YES** **1995, 1996, 1997, 1998, 1999** **Models:** Eclipse **Engines:** 2.0L VIN Y **Transmissions:** A/T, M/T	**Timing Belt Skipped One Tooth or More** Engine started, engine running, no "inhibit" conditions exist (i.e., a cold engine, large amount of change in MAP, insufficient start-to-run time, wide open throttle, engine speed outside of a given range), and the PCM detected the CMP sensor signal was offset from the CKP sensor signal by more than one tooth in the CCM Rationality test. **Possible Causes:** • Improperly installed timing belt or crankshaft position sensor • Improperly installed CMP or CKP sensor, or loose connector • Camshaft position has not be "learned" following servicing of the camshaft, camshaft sprocket, timing belt or its tensioner, cylinder head or gasket, crankshaft or crankshaft sprocket • PCM has been replaced
DTC: P1391 **2T CCM MIL: YES** **1995, 1996, 1997, 1998, 1999** **Models:** Eclipse **Engines:** 2.0L VIN Y **Transmissions:** A/T, M/T	**Intermittent Loss of CKP or CMP Sensor Signal** Engine started, engine running, and the PCM detected the CMP and CKP sensor signals were out of synchronization, event detected at least 5 times during the CCM Rationality test. **Possible Causes:** • CKP or CMP sensor improperly installed or sensor is damaged • CMP sensor signal circuit is open (an intermittent fault) • CKP sensor signal circuit is open (an intermittent fault) • Camshaft position sensor connector loose or damaged • Crankshaft position sensor connector loose or damaged
DTC: P1398 **2T CCM MIL: YES** **1998, 1999, 2000, 2001, 2002** **Models:** Eclipse **Engines:** 2.0L VIN Y **Transmissions:** A/T, M/T	**Misfire Adaptive Numerator at Limit** Engine started, ECT sensor more than 75°F higher than the sensor reading at engine startup, A/C clutch is "off", vehicle driven to a speed of over 35 mph in 1st gear, or to a speed over 65 mph in high gear, followed by a closed throttle deceleration period, and the PCM detected at least one of the CMP sensor target windows varied more than 2.86% from the CMP sensor reference window during the test. **Possible Causes:** • Base engine problem in one or more engine cylinders • CKP sensor improperly installed or the sensor has failed • CKP sensor crankshaft target variation is present • Crankshaft bearings are severely worn • CKP sensor wiring harness or connectors loose or open circuit • PCM has failed
DTC: P1400 **2T CCM MIL: YES** **1996, 1997, 1998, 1999, 2000, 2001, 2002, 2003, 2004** **Models:** 3000GT, Diamante, Galant, Evo, Lancer, Montero, Montero Sport **Engines:** All **Transmissions:** A/T, M/T	**Manifold Differential Pressure Sensor Circuit Malfunction** Engine started, ECT sensor more than 65.4°F at startup, engine running at light engine load in closed loop, and the PCM detected the MDP sensor was more than 4.5v; or it detected the MDP sensor signal was less than 0.2v during the CCM test. **Possible Causes:** • MDP sensor signal circuit is open or shorted to ground • MDP sensor signal circuit is shorted to VREF or system power • MDP sensor is damaged or has failed • PCM has failed
DTC: P1400 **2T CCM MIL: YES** **1995, 1996, 1997, 1998, 1999, 2000, 2001, 2002, 2003** **Models:** Eclipse **Engines:** 2.0L VIN F, 2.4L VIN G, 3.0L VIN H **Transmissions:** A/T, M/T	**Manifold Differential Pressure Sensor Circuit Malfunction** Engine started, ECT sensor more than 65.4°F at startup, engine running at light engine load in closed loop, and the PCM detected the MDP sensor was more than 4.5v; or it detected the MDP sensor signal was less than 0.2v during the CCM test. **Possible Causes:** • MDP sensor signal circuit is open or shorted to ground • MDP sensor signal circuit is shorted to VREF or system power • MDP sensor is damaged or has failed • PCM has failed
DTC: P1443 **2T CCM MIL: YES** **1996** **Models:** Montero **Engines:** All **Transmissions:** A/T, M/T	**EVAP Purge Control Solenoid 2 Circuit Malfunction** Engine started, engine running at cruise speed for 3-5 minutes, and the PCM did not detect any EVAP Purge Solenoid 2 coil surge voltage after the solenoid was commanded "on" and "off" in the test. **Possible Causes:** • Purge solenoid control circuit is open or shorted to ground • Purge solenoid power circuit is open (check power at the relay) • Purge solenoid is damaged or has failed • PCM has failed

DTC	Trouble Code Title, Conditions & Possible Causes
DTC: P1486 **2T EVAP MIL: YES** **1996, 1997, 1998, 1999** **Models:** Eclipse **Engines:** 2.0L VIN Y **Transmissions:** A/T, M/T	**EVAP Leak Monitor Pinched Hose Detected** Cold engine startup (ECT and IAT sensor signals less than 86°F), BARO sensor more than 75 kPa, engine runtime over 16 minutes, ECT sensor more than 140°F, TP sensor from 1-4v, vehicle driven to a speed over 20 mph with the engine speed over 1600 rpm, then with the EVAP canister vent solenoid closed, and the Leak Detection pump commanded to a specific duty cycle to raise the pressure in the system, the PCM detected the LDP pump operated for a very short period of time due to a restriction in the system during the test. **Possible Causes:** • Vapor lines are clogged or restricted between the EVAP vent solenoid and the fuel tank • Canister Vent solenoid is damaged or has failed • PCM has failed
DTC: P1487 **2T CCM MIL: YES** **1996, 1997, 1998, 1999** **Models:** Eclipse **Engines:** 2.0L VIN Y **Transmissions:** A/T, M/T	**High Speed Condenser Fan Relay Circuit Malfunction** Engine started, engine running, system voltage over 10v, and the PCM detected an unexpected voltage condition on the High Speed Condenser Fan Control Relay circuit during the CCM test. **Possible Causes:** • Condenser fan relay control circuit is open or shorted to ground • Condenser fan relay power circuit is open (test power at switch) • Condenser fan relay is damaged or has failed • PCM has failed
DTC: P1489 **2T CCM MIL: YES** **1995** **Models:** Eclipse **Engines:** 2.0L VIN Y **Transmissions:** A/T, M/T	**High Speed Condenser Fan Relay Circuit Malfunction** Engine started, engine running, system voltage over 10v, and the PCM detected an unexpected voltage condition on the High Speed Condenser Fan Control Relay circuit during the CCM test. **Possible Causes:** • Condenser fan relay control circuit is open or shorted to ground • Condenser fan relay power circuit is open (test power at switch) • Condenser fan relay is damaged or has failed • PCM has failed
DTC: P1490 **2T CCM MIL: YES** **1995, 1996, 1997, 1998, 1999** **Models:** Eclipse **Engines:** 2.0L VIN Y **Transmissions:** A/T, M/T	**Low Speed Fan Relay Circuit Malfunction** Engine started, engine running, system voltage over 10v, and the PCM detected an unexpected voltage condition on the Low Speed Fan Control Relay circuit during the CCM test. **Possible Causes:** • Low speed fan relay circuit is open or shorted to ground • Low speed fan relay power circuit is open (test power at switch) • Low speed fan relay is damaged or has failed • PCM has failed
DTC: P1492 **2T CCM MIL: YES** **1996, 1997, 1998, 1999** **Models:** Eclipse **Engines:** 2.0L VIN Y **Transmissions:** A/T, M/T	**Battery Temperature Sensor High Signal** Key on or engine running, and the PCM detected the Battery Temperature Sensor (BTS) signal was more than 4.90v for 3 seconds during the CCM test. **Possible Causes:** • BTS signal circuit is open • BTS signal circuit is shorted to VREF • BTS is damaged or has failed • PCM has failed
DTC: P1493 **2T CCM MIL: YES** **1996, 1997, 1998, 1999** **Models:** Eclipse **Engines:** 2.0L VIN Y **Transmissions:** A/T, M/T	**Battery Temperature Sensor Low Signal** Key on or engine running, and the PCM detected the Battery Temperature Sensor (BTS) signal was less than 0.20v for 3 seconds during the CCM test. **Possible Causes:** • BTS signal circuit is shorted to ground • BTS is damaged or has failed • PCM has failed
DTC: P1494 **2T CCM MIL: YES** **1996, 1997, 1998, 1999** **Models:** Eclipse **Engines:** 2.0L VIN Y **Transmissions:** A/T, M/T	**EVAP Ventilation Switch Circuit Malfunction** Engine started, engine runtime over 1 minute, and the PCM detected (with LDP commanded "on" momentarily) that the EVAP Ventilation switch did not cycle from "on" to "off" during the test. **Possible Causes:** • Ventilation switch signal circuit is open or shorted to ground • Ventilation switch ground circuit is open • Ventilation switch is damaged or has failed • PCM has failed

DTC	Trouble Code Title, Conditions & Possible Causes
DTC: P1495 **2T CCM MIL: YES** **1996, 1997, 1998, 1999** **Models:** Eclipse **Engines:** 2.0L VIN Y **Transmissions:** A/T, M/T	**EVAP Ventilation Solenoid Circuit Malfunction** Key on or engine running for 1 minute, and the PCM detected an unexpected voltage condition on the EVAP Ventilation solenoid circuit during the CCM test. **Possible Causes:** • Ventilation solenoid control circuit is open or shorted to ground • Ventilation solenoid control circuit is shorted to system power • Ventilation solenoid is damaged or has failed • PCM has failed
DTC: P1496 **2T CCM MIL: YES** **1996, 1997, 1998, 1999, 2000, 2001, 2002** **Models:** Eclipse **Engines:** 2.0L VIN Y **Transmissions:** A/T, M/T	**5-Volt Supply Output Too Low** Key on or engine running, system voltage over 10v, and the PCM detected the 5-volt supply circuit to the MAP and TP sensors was less than 4v, condition met for more than 4 seconds during the test. **Possible Causes:** • 5-volt supply (VREF) circuit is shorted to ground (intermittent) • PCM has failed
DTC: P1500 **2T CCM MIL: YES** **1996, 1997, 1998, 1999, 2000, 2001, 2002** **Models:** Eclipse, Galant, Mirage **Engines:** All **Transmissions:** A/T, M/T	**Generator 'FR' Terminal Circuit Malfunction** Engine started, engine running, and the PCM detected the Generator 'FR' terminal signal remained above 4.5v for more than 20 seconds during the CCM test. **Possible Causes:** • Generator "FR" circuit is open between Generator and the PCM • Generator is damaged or has failed • PCM has failed
DTC: P1506 **2T CCM MIL: YES** **2006** **Models:** Eclipse, Galant, Lancer, Endeavor **Engines:** All **Transmissions:** A/T, M/T	**Idle Control System RPM Lower Than Expected At Low Temperature** Engine started, engine running under closed loop idle speed control, ECT between 45°F and 106°F, battery voltage greater than 10 volts, power steering pressure switch off, volumetric efficiency is less than 40%, barometric pressure is greater than 22.4 in. Hg, IAT is greater than 14°, throttle actuator control motor position is greater than 255 steps, the actual idle speed is greater than 100 rpm lower than target idle speed for 12 seconds. **Possible Causes:** • Throttle valve area dirty • Throttle body assembly failed is damaged or has failed • PCM has failed
DTC: P1507 **2T CCM MIL: YES** **2006** **Models:** Eclipse, Galant, Lancer, Endeavor **Engines:** All **Transmissions:** A/T, M/T	**Idle Control System RPM Higher Than Expected At Low Temperature** Engine started, engine running under closed loop idle speed control, ECT between 45°F and 106°F, battery voltage greater than 10 volts, power steering pressure switch off, volumetric efficiency is less than 40%, barometric pressure is greater than 22.4 in. Hg, IAT is greater than 14°, throttle actuator control motor position is 0 steps, the actual idle speed is greater than 200 rpm lower than target idle speed for 12 seconds. **Possible Causes:** • Intake system vacuum leak • Throttle body assembly failed is damaged or has failed • PCM has failed
DTC: P1530 **2T CCM MIL: YES** **2006** **Models:** Lancer, Evo, Montero, Outlander **Engines:** All **Transmissions:** A/T, M/T	**A/C1 Switch Circuit Intermittent** Engine started, the repeating ON–OFF switch of the A/C switch 255 times per second. **Possible Causes:** • A/C ECU has failed
DTC: P1600 **1T PCM MIL: NO** **1997, 1998, 1999, 2000, 2001** **Models:** 3000GT, Diamante, Galant, Mirage, Montero, Montero Sport **Engines:** All **Transmissions:** A/T	**PCM/TCM Serial Communication Link Malfunction** Engine started, engine runtime over 2 seconds, and the PCM did not receive any communication signals from the TCM or it detected an abnormality in signals received over the Serial Communication Link. **Possible Causes:** • Serial communication link between the TCM, PCM and the DLC is open, shorted to ground or shorted to system power • TCM and/or PCM has failed
DTC: P1600 **1T PCM MIL: NO** **1996, 1997, 1998, 1999** **Models:** Eclipse **Engines:** 2.0L VIN F, 2.4L VIN G **Transmissions:** A/T	**PCM/TCM Serial Communication Link Malfunction** Engine started, engine runtime over 2 seconds, and the PCM did not receive any communication signals from the TCM or it detected an abnormality in signals received over the Serial Communication Link. **Possible Causes:** • Serial communication link between the TCM, PCM and the DLC is open, shorted to ground or shorted to system power • TCM and/or PCM has failed

DTC	Trouble Code Title, Conditions & Possible Causes
DTC: P1601 **1T PCM MIL: YES** **2003, 2004, 2005** **Models:** Montero, Galant, Endeavor **Engines:** All **Transmissions:** All	**Communication Malfunction (between PCM and Throttle Actuator Control Unit)** Battery voltage is greater than 6.3 volts, the PCM detects an error in communication with the throttle actuator control module for 0.05 seconds and throttle actuator control unit detects an error in communication with the PCM for 0.125 seconds. **Possible Causes:** • PCM has failed
DTC: P1602 **1T PCM MIL: YES** **2004, 2005, 2006** **Models:** Galant, Lancer, Outlander, Endeavor, Montero **Engines:** All **Transmissions:** All	**Communication Malfunction (between PCM Main Processor and System LSI)** Ignition ON, the PCM detects an error in communication with the system LSI 0.07. **Possible Causes:** • PCM has failed
DTC: P1603 **1T CCM MIL: NO** **2002, 2003, 2004, 2005, 2006** **Models:** Eclipse, Galant, Lancer, Mirage, Montero, Montero Sport **Engines:** All **Transmissions:** A/T, M/T	**Battery Backup Circuit Malfunction** Engine started, engine running, and the PCM detected the voltage on the Battery Backup circuit was less than 6.0v for 2 seconds. **Possible Causes:** • Battery backup circuit is open (check the power through the fusible link and Dedicated Fuse No. 7 for an open circuit) • Battery terminals corroded, or loose connections • PCM has failed
DTC: P1610 **2T CCM MIL: NO** **2000, 2001, 2002** **Models:** Diamante, Eclipse, Galant, Lancer, Montero, Montero Sport **Engines:** All **Transmissions:** A/T, M/T	**Immobilizer System Malfunction** Key on or engine running, and the PCM received a signal from the Immobilizer controller that it had detected a problem in one of its related circuits, or it sent a signal to the PCM to prevent the engine from starting during the CCM test. **Possible Causes:** • Immobilizer data circuit to the PCM is open • Immobilizer data circuit to the PCM is shorted to ground • Immobilizer (controller) is damaged or has failed
DTC: P1696 **2T PCM MIL: NO** **1995, 1996, 1997, 1998, 1999** **Models:** Eclipse **Engines:** 2.0L VIN Y **Transmissions:** A/T, M/T	**PCM EEPROM Write Denied** Key on, and the PCM detected that it was unable to write to one or more EEPROM locations during the initial self-test procedure. **Possible Causes:** • Clear the trouble codes and retest for this trouble code. If the same trouble code resets, the PCM has failed and must be replaced to repair this problem.
DTC: P1697 **2T PCM MIL: NO** **1995, 1996, 1997, 1998, 1999** **Models:** Eclipse **Engines:** 2.0L VIN Y **Transmissions:** A/T, M/T	**PCM Malfunction - SRI Miles Not Stored** Key on, and the PCM detected that it was unable to write to the Service Reminder Indicator (SRI or EMR) during the initial self-test. **Possible Causes:** • Clear the trouble codes and retest for this trouble code. If the same trouble code resets, the PCM has failed and must be replaced to repair this problem.
DTC: P1698 **2T PCM MIL: NO** **1995, 1996, 1997, 1998, 1999** **Models:** Eclipse **Engines:** 2.0L VIN F, 2.0L VIN Y **Transmissions:** A/T	**No CCD Messages Received From The TCM** Key on, and the PCM detected that it did not receive any messages from the TCM during the initial communication phase of the self-test. **Possible Causes:** • Clear the trouble codes and retest for this trouble code. If the same trouble code resets, the PCM has failed and must be replaced to repair this problem.
DTC: P1715 **2T CCM MIL: YES** **1996, 1997, 1998, 1999, 2000, 2001** **Models:** 3000GT, Diamante, Galant, Montero, Montero Sport **Engines:** All **Transmissions:** A/T	**TCM Pulse Generator Circuit Malfunction** Engine started, engine running with the gear selector in Drive, and the PCM did not detect any TCM Pulse Generator signals in the test. **Possible Causes:** • A/T Pulse generator (+) circuit is open or shorted to ground • A/T Pulse generator (-) circuit is open or shorted to ground • A/T Pulse generator (+) or (-) circuit is shorted to system power • A/T pulse generator is damaged or has failed • PCM has failed

DTC	Trouble Code Title, Conditions & Possible Causes
DTC: P1715 **2T CCM MIL: YES** **1997, 1998, 1999** **Models:** Eclipse **Engines:** 2.0L VIN F, 2.0L VIN Y, 2.4L VIN G **Transmissions:** A/T	**TCM Pulse Generator Circuit Malfunction** Engine started, engine running with the gear selector in Drive, and the PCM did not detect any TCM Pulse Generator signals in the test. **Possible Causes:** • A/T Pulse generator (+) circuit is open or shorted to ground • A/T Pulse generator (-) circuit is open or shorted to ground • A/T Pulse generator (+) or (-) circuit is shorted to system power • A/T pulse generator is damaged or has failed • PCM has failed
DTC: P1720 **2T CCM MIL: YES** **1996, 1997, 1998, 1999, 2000, 2001** **Models:** Diamante **Engines:** All **Transmissions:** A/T	**Pressure Control, Shift Control, TCC Solenoid Circuit Malfunction** Engine started, engine speed over 500 rpm, transaxle operating while in gear, and the PCM detected an unexpected signal condition in the Pressure solenoid, Shift Control solenoid or Torque Converter Clutch solenoid control circuit during the CCM test. **Possible Causes:** • Pressure control solenoid, Shift Control solenoid or TCC control circuit is open, shorted to ground or shorted to system power • Pressure control solenoid, Shift control solenoid or TCC solenoid is damaged or has failed • PCM has failed
DTC: P1720 **2T CCM MIL: YES** **1996, 1997, 1998, 1999, 2000, 2001** **Models:** Galant, Montero, Montero Sport **Engines:** 3.0L VIN L, 3.0L VIN H, 3.0L VIN P, 3.5L VIN M, 3.5L VIN R **Transmissions:** A/T	**Pressure Control, Shift Control, TCC Solenoid Circuit Malfunction** Engine started, engine speed over 500 rpm, transaxle operating while in gear, and the PCM detected an unexpected signal condition in the Pressure solenoid, Shift Control solenoid or Torque Converter Clutch solenoid control circuit during the CCM test. **Possible Causes:** • Pressure control solenoid, Shift Control solenoid or TCC control circuit is open, shorted to ground or shorted to system power • Pressure control solenoid, Shift control solenoid or TCC solenoid is damaged or has failed • PCM has failed
DTC: P1738 **2T CCM MIL: YES** **1997, 1998, 1999** **Models:** Eclipse **Engines:** 2.0L VIN Y **Transmissions:** A/T	**A/T High Temperature Operations Activated** Engine started, engine runtime from 5-10 minutes, and the PCM detected the Transmission Fluid Temperature (TFT) sensor signal indicated more than 240°F at any time during the CCM test. **Possible Causes:** • Customer driving pattern (towing) requires additional cooling • Engine thermostat is stuck closed, or low engine coolant level • Radiator corroded or clogged, air passages filled with debris • Transmission fluid is overfull • Transmission oil cooler clogged or restricted
DTC: P1750 **2T CCM MIL: YES** **1996, 1997, 1998, 1999, 2000, 2001** **Models:** 3000GT, Galant, Montero, Montero Sport **Engines:** All **Transmissions:** A/T	**Pressure Control, Shift Control, TCC Solenoid Circuit Malfunction** Engine started, engine speed over 500 rpm, transaxle operating while in gear, and the PCM detected an unexpected signal condition in the Pressure solenoid, Shift Control solenoid or Torque Converter Clutch solenoid control circuit during the CCM test. **Possible Causes:** • Pressure control solenoid, Shift Control solenoid or TCC control circuit is open, shorted to ground or shorted to system power • Pressure control solenoid, Shift control solenoid or TCC solenoid is damaged or has failed • PCM has failed
DTC: P1750 **2T CCM MIL: YES** **1997, 1998, 1999** **Models:** Eclipse **Engines:** 2.0L VIN F, 2.4L VIN G **Transmissions:** A/T	**Pressure Control, Shift Control, TCC Solenoid Circuit Malfunction** Engine started, engine speed over 500 rpm, transaxle operating while in gear, and the PCM detected an unexpected signal condition in the Pressure solenoid, Shift Control solenoid or Torque Converter Clutch solenoid control circuit during the CCM test. **Possible Causes:** • Pressure control solenoid, Shift Control solenoid or TCC control circuit is open, shorted to ground or shorted to system power • Pressure control solenoid, Shift control solenoid or TCC solenoid is damaged or has failed • PCM has failed
DTC: P1751 **2T CCM MIL: YES** **1999, 2000, 2001** **Models:** Galant, Mirage **Engines:** 3.0L VIN L, 1.5L VIN A, 1.8L VIN C **Transmissions:** A/T, M/T	**A/T Control Relay Circuit Malfunction** Engine started, vehicle driven to a speed of over 3 mph, and the PCM received as signal from the TCM indicating a problem in the A/T Control Relay circuit during the CCM test. Possible Causes • A/T Control Relay circuit is open or shorted to ground • A/T Control Relay circuit is shorted to system power (B+) • A/T Control Relay is damaged or has failed • PCM has failed

DTC	Trouble Code Title, Conditions & Possible Causes
DTC: P1751 **2T CCM MIL: YES** **2002, 2003, 2004, 2005, 2006** **Models:** Eclipse, Galant, Diamante, Lancer, Montero, Montero Sport, Outlander **Engines:** All **Transmissions:** A/T	**A/T Control Relay Circuit Malfunction** Engine started, vehicle driven to a speed of over 3 mph, and the PCM received as signal from the TCM indicating a problem in the A/T Control Relay circuit during the CCM test. Possible Causes • A/T Control Relay circuit is open or shorted to ground • A/T Control Relay circuit is shorted to system power (B+) • A/T Control Relay is damaged or has failed • PCM has failed
DTC: P1791 **2T CCM MIL: YES** **1997, 1998, 1999, 2000, 2001** **Models:** 3000GT, Montero, Montero Sport **Engines:** All **Transmissions:** A/T	**PCM ECT Level Signal to TCM Circuit Malfunction** Key on or engine running, and the PCM detected a signal from the TCM indicating a problem was detected in the ECT sensor signal. **Possible Causes:** • ECT sensor signal circuit to the TCM is open • ECT sensor signal circuit to the TCM is shorted to ground • ECT sensor is out-of-calibration, damaged or has failed • TCM or PCM has failed
DTC: P1791 **2T CCM MIL: YES** **1997, 1998, 1999** **Models:** Eclipse **Engines:** 2.0L VIN F, 2.4L VIN G **Transmissions:** A/T	**PCM ECT Level Signal to TCM Circuit Malfunction** Key on or engine running, and the PCM detected a signal from the TCM indicating a problem was detected in the ECT sensor signal. **Possible Causes:** • ECT sensor signal circuit to the TCM is open • ECT sensor signal circuit to the TCM is shorted to ground • ECT sensor is out-of-calibration, damaged or has failed • TCM or PCM has failed
DTC: P1795 **2T CCM MIL: YES** **1997, 1998, 1999, 2000, 2001** **Models:** Diamante **Engines:** All **Transmissions:** A/T	**TP Sensor Signal Circuit to TCM** Engine started, vehicle driven to a speed of over 3 mph, and the PCM received as signal from the TCM indicating it had detected a problem in the TP sensor signal circuit during the CCM test. **Possible Causes:** • TP sensor signal circuit to TCM is open or shorted to ground • TP sensor is damaged or it has failed • PCM has failed
DTC: P1795 **2T CCM MIL: YES** **1999, 2000, 2001** **Models:** Galant, Mirage **Engines:** 3.0L VIN L, 1.5L VIN A, 1.8L VIN C **Transmissions:** A/T	**TP Sensor Signal Circuit to TCM** Engine started, vehicle driven to a speed of over 3 mph, and the PCM received as signal from the TCM indicating it had detected a problem in the TP sensor signal circuit during the CCM test. **Possible Causes:** • TP sensor signal circuit to TCM is open or shorted to ground • TP sensor is damaged or it has failed • PCM has failed
DTC: P1899 **2T CCM MIL: YES** **1996, 1997, 1998, 1999, 2000, 2001** **Models:** Eclipse **Engines:** 2.0L VIN Y **Transmissions:** A/T	**P/N Position/Transaxle Range Switch Circuit Malfunction** Engine started, vehicle driven to a speed of over 50 mph at an engine speed of 1984-4480 rpm, TP sensor signal at least 0.49v over the minimum voltage, MAP sensor signal more than 66% of the BARO sensor value, and the PCM detected an unexpected "high" voltage condition on the TR sensor signal circuit during the test. **Possible Causes:** • TR sensor signal circuit is open or shorted to system power • TR sensor is damaged or has failed • PCM has failed

Gas Engine OBD II Trouble Code List (P2XXX Codes)

DTC	Trouble Code Title, Conditions & Possible Causes
DTC: P2066 **2T CCM MIL: YES** **2003, 2004, 2005, 2006** **Models:** Evo, Outlander, Galant, Endeavor **Engines:** All **Transmissions:** All	**Fuel Level Sensor (Sub) Circuit Range/ Performance** When fuel consumption calculated from the operating time of the injector amounts to 8 gallons, the diversity of the amount of the fuel in the tank calculated from the fuel level sensor is 0.5 gallons or less. **Possible Causes:** • Fuel pump module or fuel level sensor (sub) failed • PCM has failed
DTC: P2100 **2T CCM MIL: YES** **2003, 2004, 2005, 2006** **Models:** Galant, Lancer, Endeavor, Montero, Outlander **Engines:** All **Transmissions:** All	**Throttle Actuator Control Motor Circuit Open** Battery voltage is greater than 8.3 volts, the throttle actuator control motor circuit current should be 0.1 amps or less for 0.72 seconds. **Possible Causes:** • Throttle actuator control motor failed • Throttle actuator control motor circuit open • PCM has failed
DTC: P2101 **2T CCM MIL: YES** **2003, 2004, 2005, 2006** **Models:** Galant, Lancer, Endeavor, Montero, Outlander **Engines:** All **Transmissions:** All	**Throttle Actuator Control Motor Magneto Malfunction** Battery voltage is greater than 8.3 volts, the coil temperature of the throttle actuator control motor should be 356°F or greater for 0.8 seconds. **Possible Causes:** • Throttle actuator control motor failed • Throttle actuator control motor circuit shorted • PCM has failed
DTC: P2102 **2T CCM MIL: YES** **2003, 2004, 2005, 2006** **Models:** Galant, Lancer, Endeavor, Montero, Outlander **Engines:** All **Transmissions:** All	**Throttle Actuator Control Motor Circuit Shorted Low** Battery voltage is greater than 8.3 volts, the throttle actuator control motor circuit current should is 12 amps or greater for 0.8 seconds. **Possible Causes:** • Throttle actuator control motor failed • Throttle actuator control motor circuit shorted • PCM has failed
DTC: P2103 **2T CCM MIL: YES** **2003, 2004, 2005, 2006** **Models:** Galant, Lancer, Endeavor, Montero, Outlander **Engines:** All **Transmissions:** All	**Throttle Actuator Control Motor Circuit Shorted High** Battery voltage is greater than 8.3 volts, the throttle actuator control motor circuit current is 8 amps or greater for 0.8 seconds. **Possible Causes:** • Throttle actuator control motor failed • Throttle actuator control motor circuit shorted • PCM has failed
DTC: P2108 **2T CCM MIL: YES** **2003, 2004** **Models:** Galant, Endeavor, Montero **Engines:** All **Transmissions:** All	**Throttle Actuator Control Processor Malfunction** Key on, TPS (main) output voltage is between 0.35 and 4.8 volts, the accelerator pedal position sensor (main) is normal, TPS is normal, the difference between the TPS (main) output voltage and the target TPS (main) voltage should be 1 volt or greater for 1 second. **Possible Causes:** • PCM has failed
DTC: P2121 **2T CCM MIL: YES** **2003, 2004, 2005, 2006** **Models:** Lancer, Endeavor, Montero, Outlander **Engines:** All **Transmissions:** All	**Accelerator Pedal Position Sensor (Main) Circuit Range/Performance Problem** Key on, accelerator pedal position switch ON, accelerator pedal position sensor (sub) output voltage is 1.88 volts or less, the accelerator pedal position sensor (main) output voltage is 1.88 volts or greater for 1 second. **Possible Causes:** • Accelerator pedal position sensor misadjusted or failed • Accelerator pedal position sensor (main) open or shorted • Accelerator pedal position switch misadjusted or failed • Accelerator pedal position switch open or shorted • PCM has failed
DTC: P2122 **2T CCM MIL: YES** **2003, 2004, 2005, 2006** **Models:** Galant, Lancer, Endeavor, Montero, Outlander **Engines:** All **Transmissions:** All	Accelerator Pedal Position Sensor (Main) Circuit Low Input Key on, accelerator pedal position sensor (main) output voltage is 0.2 volts or less for 1 second. **Possible Causes:** • Accelerator pedal position sensor misadjusted or failed • Accelerator pedal position sensor (main) open or shorted • PCM has failed

DTC	Trouble Code Title, Conditions & Possible Causes
DTC: P2123 **2T CCM MIL: YES** **2003, 2004, 2005, 2006** **Models:** Galant, Lancer, Endeavor, Montero, Outlander **Engines:** All **Transmissions:** All	Accelerator Pedal Position Sensor (Main) Circuit High Input Key on, accelerator pedal position sensor (sub) output voltage between 0.2 volts and 2.5, the accelerator pedal position sensor (main) output voltage should be 4.5 volts or greater for 1 second. **Possible Causes:** • Accelerator pedal position sensor misadjusted or failed • Accelerator pedal position sensor (main) open or shorted • PCM has failed
DTC: P2126 **2T CCM MIL: YES** **2003, 2004, 2005, 2006** **Models:** Lancer, Montero, Outlander **Engines:** All **Transmissions:** All	**Accelerator Pedal Position Sensor (Sub) Circuit Range/Performance Problem** Key on, accelerator pedal position switch ON, accelerator pedal position sensor (main) failure detected, the accelerator pedal position sensor (sub) output voltage is 2.5 volts or greater for 1 second. **Possible Causes:** • Accelerator pedal position sensor misadjusted or failed • Accelerator pedal position sensor (sub) open or shorted • Accelerator pedal position switch misadjusted or failed • Accelerator pedal position switch open or shorted • PCM has failed
DTC: P2127 **2T CCM MIL: YES** **2003, 2004, 2005, 2006** **Models:** Galant, Lancer, Endeavor, Montero, Outlander **Engines:** All **Transmissions:** All	Accelerator Pedal Position Sensor (Main) Circuit Low Input Key on, accelerator pedal position sensor (sub) output voltage is 0.2 volts or less for 1 second. **Possible Causes:** • Accelerator pedal position sensor misadjusted or failed • Accelerator pedal position sensor (sub) open or shorted • PCM has failed
DTC: P2128 **2T CCM MIL: YES** **2003, 2004, 2005, 2006** **Models:** Galant, Lancer, Endeavor, Montero, Outlander **Engines:** All **Transmissions:** All	Accelerator Pedal Position Sensor (Main) Circuit High Input Key on, accelerator pedal position sensor (main) output voltage between 0.2 volts and 2.5, the accelerator pedal position sensor (sub) output voltage should be 4.5 volts or greater for 1 second. **Possible Causes:** • Accelerator pedal position sensor misadjusted or failed • Accelerator pedal position sensor (main) open or shorted • PCM has failed
DTC: P2135 **2T CCM MIL: YES** **2003, 2004, 2005, 2006** **Models:** Galant, Lancer, Endeavor, Montero, Outlander **Engines:** All **Transmissions:** All	Throttle Position Sensor (Main and Sub) Range/Performance Problem Key on, TPS (main) output voltage is between 2.5 and 4.8 volts, TPS (sub) output voltage is greater than 2.25 volts, the TPS (sub) output voltage is 4.2 volts or lower. **Possible Causes:** • TPS failed • TPS shorted • PCM has failed
DTC: P2138 **2T CCM MIL: YES** **2003, 2004, 2005, 2006** **Models:** Galant, Lancer, Endeavor, Montero, Outlander **Engines:** All **Transmissions:** All	Accelerator Pedal Position Sensor (Main and Sub) Range/Performance Problem Key on, accelerator pedal position sensor (main) output voltage between 0.2 volts and 4.5, the accelerator pedal position sensor (sub) output voltage between 0.2 and 4.5 volts, change of accelerator pedal position sensor (sub) output voltage per 25 ms is less than 0.1 volt, the accelerator pedal position sensor (sub) output voltage minus the accelerator pedal position sensor (main) output voltage should be 1 volt or greater for 1 second and/or accelerator pedal position sensor (main) output voltage minus accelerator pedal position sensor (sub) output voltage should be 1 volt or greater for 0.2 seconds. **Possible Causes:** • Accelerator pedal position sensor misadjusted or failed • Accelerator pedal position sensor (main) open or shorted • PCM has failed
DTC: P2173 **2T CCM MIL: YES** **2004, 2005, 2006** **Models:** Lancer, Outlander **Engines:** All **Transmissions:** All	**Abnormal Intake Air Amount** Engine running, the actual intake air amount is greater than the allowable intake air amount for 1.5 seconds. **Possible Causes:** • MAF sensor failed • Throttle valve faulty operation • TPS failed • PCM has failed
DTC: P2195 **2T CCM MIL: YES** **2006** **Models:** All **Engines:** All **Transmissions:** All	**HO2S Inactive (Bank 1, Sensor 1)** Engine running for more than 20 seconds, engine coolant temperature greater than 44°F, intake air temperature is greater than 13°F, the front HO2S output voltage is less than 0.5 volts for 128 seconds. **Possible Causes:** • Front HO2S deteriorated or failed • PCM has failed

DTC	Trouble Code Title, Conditions & Possible Causes
DTC: P2197 **2T CCM MIL: YES** **2006** **Models:** All **Engines:** All **Transmissions:** All	**HO2S Inactive (Bank 2, Sensor 1)** Engine running for more than 20 seconds, engine coolant temperature greater than 44°F, intake air temperature is greater than 13°F, the front HO2S output voltage is less than 0.5 volts for 128 seconds. **Possible Causes:** • Front HO2S deteriorated or failed • PCM has failed
DTC: P2227 **2T CCM MIL: YES** **2004** **Models:** Evo, Eclipse, Galant, Lancer, Endeavor, Montero, Montero Sport, Outlander **Engines:** All **Transmissions:** All	**Barometric Pressure Circuit Range/Performance Problem** Engine running, barometric pressure is less than 22.4 in. Hg, during 15 times of driving with the engine coolant temperature 160°F or greater and with each start, engine coolant temperature has increase to 40°F or greater, the changes in the sensor output voltage should be 0.015 volts (equal to 0.12 in. Hg) or less. **Possible Causes:** • PCM has failed
DTC: P2228 **2T CCM MIL: YES** **2004, 2005, 2006** **Models:** Evo, Eclipse, Galant, Lancer, Endeavor, Montero, Montero Sport, Outlander **Engines:** All **Transmissions:** All	**Barometric Pressure Circuit Low Input** Engine running for 2 seconds or more and battery voltage is greater than 8 volts, barometric pressure sensor output has continued to be 14.6 in. Hg or less (about 15,000 ft above sea level) for 10 seconds. **Possible Causes:** • PCM has failed
DTC: P2229 **2T CCM MIL: YES** **2004, 2005, 2006** **Models:** Evo, Eclipse, Galant, Lancer, Endeavor, Montero, Montero Sport, Outlander **Engines:** All **Transmissions:** All	**Barometric Pressure Circuit High Input** Engine running for 2 seconds or more and battery voltage is greater than 8 volts, barometric pressure sensor output has continued to be 33.3 in. Hg or less (about 4,000 ft above sea level) for 10 seconds. **Possible Causes:** • PCM has failed
DTC: P2252 **2T CCM MIL: YES** **2006** **Models:** Eclipse, Galant, Lancer, Endeavor **Engines:** All **Transmissions:** All	**HO2S Offset Circuit Low Voltage** Engine running for 2 seconds or more, the HO2S offset voltage is less than 0.4 volts for 2 seconds. **Possible Causes:** • PCM failed
DTC: P2253 **2T CCM MIL: YES** **2006** **Models:** Eclipse, Galant, Lancer, Endeavor **Engines:** All **Transmissions:** All	**HO2S Offset Circuit High Voltage** Engine running for 2 seconds or more, the HO2S offset voltage is greater than 0.6 volts for 2 seconds. **Possible Causes:** • PCM failed
DTC: P2263 **2T CCM MIL: YES** **2003, 2004, 2005, 2006** **Models:** Evo **Engines:** All **Transmissions:** All	**Intake Charge System Malfunction** Engine running with engine speed between 3000 and 5000 rpm and IAT sensor, barometric pressure sensor, MAF sensor, TPS are normal, TPS output voltage is higher than 2 volts; the volumetric efficiency is lower than 80 %. **Possible Causes:** • Turbocharger wastegate actuator failed • Turbocharger wastegate regulating valve failed • Charging pressure control system failed • Intake charge pressure leak • PCM failed

Gas Engine OBD II Trouble Code List (U1XXX Codes)

DTC	Trouble Code Title, Conditions & Possible Causes
DTC: U1073 **2T CCM MIL: YES** **2004, 2005, 2006** **Models:** Galant, Lancer, Endeavor, Montero, Outlander **Engines:** All **Transmissions:** All	**Bus Off** Bus off error detected **Possible Causes:** • CAN line harness or connectors damage • PCM failed
DTC: U1102 **2T CCM MIL: YES** **2004, 2005, 2006** **Models:** Galant, Lancer, Endeavor, Montero, Outlander **Engines:** All **Transmissions:** All	**ABS-ECU Timeout** Engine not cranking or 3 seconds have passed since the engine was cranked and battery voltage is 10 volts or greater, the ABS-ECU was unable to receive a signal through the CAN bus line. **Possible Causes:** • CAN line harness or connectors damage • ABS-ECU power supply system failed • ABS-ECU failed • PCM failed
DTC: U1108 **2T CCM MIL: YES** **2004, 2005, 2006** **Models:** Galant, Lancer, Endeavor, Montero, Outlander **Engines:** All **Transmissions:** All	**Combination Meter Timeout** Engine not cranking or 3 seconds have passed since the engine was cranked and battery voltage is 10 volts or greater, the combination meter was unable to receive a signal through the CAN bus line. **Possible Causes:** • CAN line harness or connectors damage • Combination meter power supply system failed • Combination meter failed • PCM failed
DTC: U1109 **2T CCM MIL: YES** **2006** **Models:** Galant **Engines:** All **Transmissions:** All	**ETACS–ECU Timeout** Engine not cranking or 3 seconds have passed since the engine was cranked and battery voltage is 10 volts or greater, the ETACS-ECU was unable to receive a signal through the CAN bus line. **Possible Causes:** • CAN line harness or connectors damage • ETACS–ECU power supply system failed • ETACS–ECU failed • PCM failed
DTC: U1110 **2T CCM MIL: YES** **2004, 2005, 2006** **Models:** Galant, Lancer, Endeavor, Montero, Outlander **Engines:** All **Transmissions:** All	**A/C–ECU Timeout** Engine not cranking or 3 seconds have passed since the engine was cranked and battery voltage is 10 volts or greater, the A/C-ECU was unable to receive a signal through the CAN bus line. **Possible Causes:** • CAN line harness or connectors damage • A/C–ECU power supply system failed • A/C–ECU failed • PCM failed
DTC: U1117 **2T CCM MIL: YES** **2006** **Models:** Galant **Engines:** All **Transmissions:** All	**Immobilizer–ECU Timeout** Ignition switch is ON, the ETACS-ECU (immobilizer–ECU) was unable to receive a signal through the CAN bus line. **Possible Causes:** • CAN line harness or connectors damage • ETACS–ECU (immobilizer–ECU) power supply system failed • ETACS–ECU (immobilizer–ECU) failed • PCM failed

MITSUBISHI
COMPONENT TESTING

15

TABLE OF CONTENTS

Component Locations 15-2
- 1995-1999 3000GT 15-2
- 1997-2004 Diamante 15-3
- 1995-1999 Eclipse 3.0L Non-Turbo 15-4
- 1995-1999 Eclipse 3.0L Turbo 15-5
- 1995-1999 Eclipse 2.4L 15-6
- 2000-2005 Eclipse 2.4L 15-7
- 2000-2005 Eclipse 3.0L 15-8
- 2003 Endeavor 15-9
- 2004-2006 Endeavor 15-10
- 20015-2006 Evo 15-11
- 2004-2006 Galant 15-12
- 2002-2006 Lancer 2.0L 15-13
- 2002-2006 Lancer 2.4L 15-14
- 1995-2002 Mirage 1.5L 15-15
- 1995-2002 Mirage 1.8L 15-16
- 1995-2000 Montero 15-17
- 2001-2006 Montero 15-18
- 1997-2004 Montero Sport 15-19
- 20015-2006 Outlander 15-20

Component Testing 15-21
- Accelerator Pedal Position 15-21
- Air Change Temperature Sensor 15-21
- Barometric Pressure Sensor 15-22
- Camshaft Position Sensor 15-22
- Crankshaft Position Sensor 15-24
- Engine Control Unit 15-25
- Engine Coolant Temperature Sensor 15-25
- Heated Oxygen Sensor 15-26
- Idle Air Control Valve 15-26
- Intake Air Temperature Sensor 15-27
- Knock Sensor 15-27
- Mass Air Flow Sensor 15-28
- Manifold Absolute Pressure Sensor 15-29
- Throttle Position Sensor 15-29
- Vehicle Speed Sensor 15-30

COMPONENT TESTING

Component Locations

1995–1999 3000GT

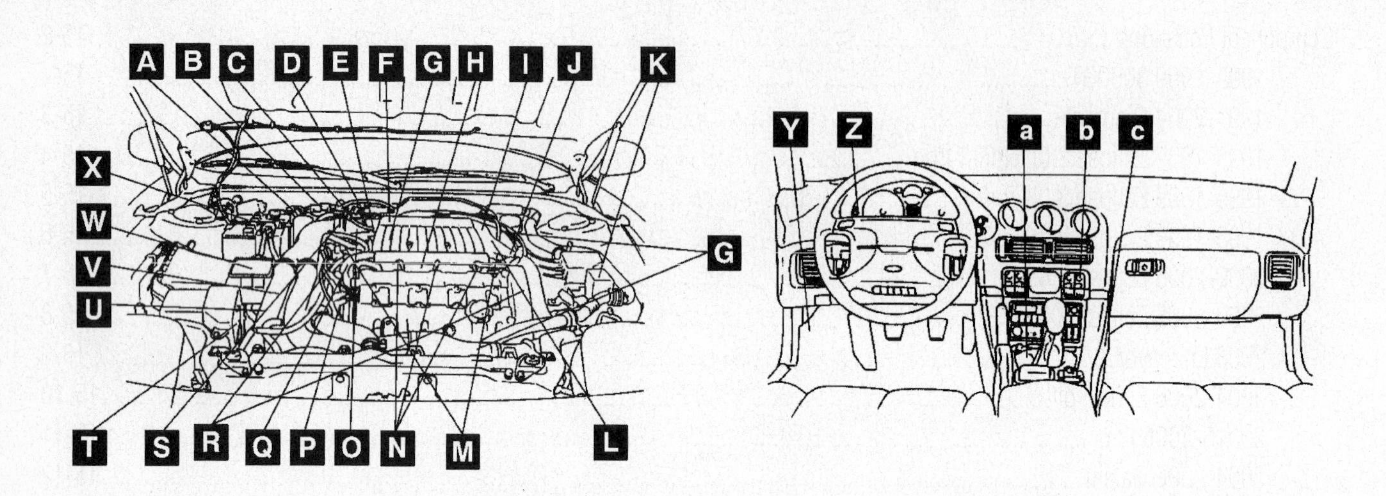

Component Location – 1995-1999 3000GT

Air-conditioning relay
K
Idle air control motor (stepper motor)
S
Air-conditioning switch
b
Ignition coil (ignition power transistor)
P
Camshaft position sensor
M
Ignition timing terminal
X
Check engine/malfunction indicator lamp
Z
Injector
N
Crankshaft position sensor
R
Knock sensor
H
Diagnostic output terminal and diagnostic test mode control terminal
Y

Manifold differential pressure sensor
F
EGR solenoid <Non-Turbo>
L
Multiport fuel injection (MFI) relay
c
EGR solenoid <Turbo>
C
Park/Neutral position switch <A/T>
U
Engine control module
A
Power steering pressure switch
J
Engine coolant temperature sensor
Q
Resistor <Turbo>
A

Evaporative emission purge solenoid <Non-Turbo>
L
Throttle position sensor (with built-in closed throttle position switch)
O
Evaporative emission purge solenoid <Turbo>
E
Evaporative emission ventilation solenoid <Non-Turbo>
T
Turbocharger waste gate solenoid <Turbo>
D
Evaporative emission ventilation solenoid <Turbo>
T
Variable induction control motor (DC motor) (with built-in induction control valve position sensor) <Non-Turbo>
I

Fuel pressure solenoid <Turbo>
B
Fuel tank differential pressure sensor
d
Vehicle speed sensor
V
Heated oxygen sensor <Non-Turbo>
G
Volume air flow sensor (with built-in intake air temperature sensor and barometric pressure sensor)
W
Heated oxygen sensor <Turbo>
G

1997–2004 Diamante

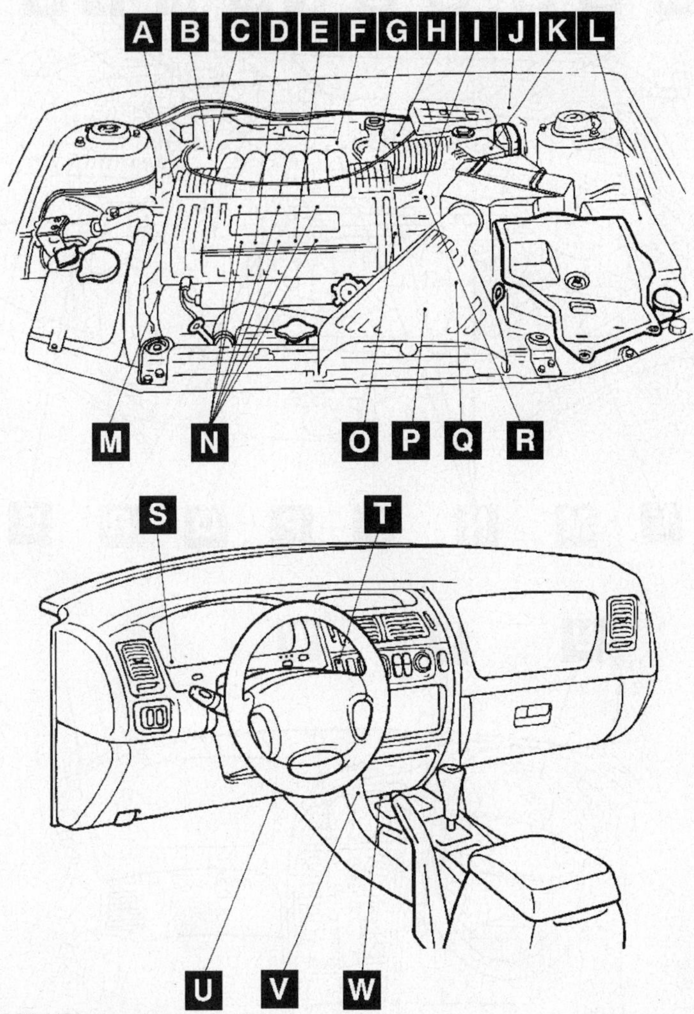

Component Location – 1997-2004 Diamante

Air conditioning compressor clutch relay
L
Heated oxygen sensor (Rear)
C
Air conditioner switch
T
Idle air control motor
H
Camshaft position sensor
I
Ignition coil (Ignition power transistor)
R

Check engine/malfunction indicator lamp
S
Injector
N
Crankshaft position sensor
M
Manifold differential pressure sensor
F
Data link connector
U
Multiport fuel injection (MFI) relay/Fuel pump relay
V

EGR solenoid
E
Park/Neutral position switch
Q
Engine control module
W
Power steering pressure switch
A
Engine coolant temperature sensor
O
Throttle position sensor (with built-in closed throttle position switch)
G

Evaporative emission purge solenoid
D
Vehicle speed sensor
P
Fuel pump check terminal
J
Volume air flow sensor (with built-in intake air temperature sensor and barometric pressure sensor)
K
Heated oxygen sensor (Front)
B

1995–1999 Eclipse 2.0L Non-Turbo

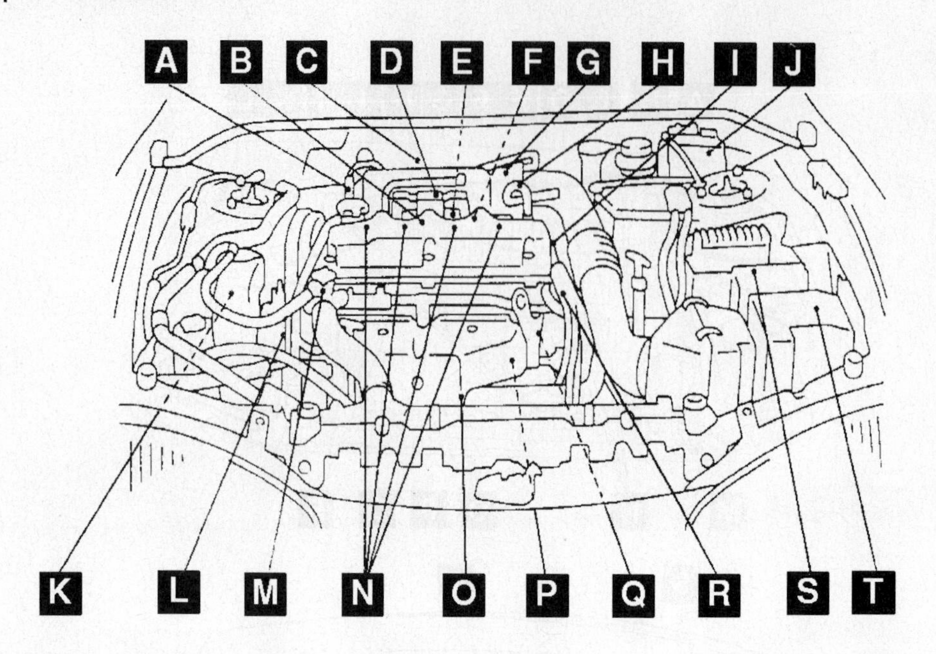

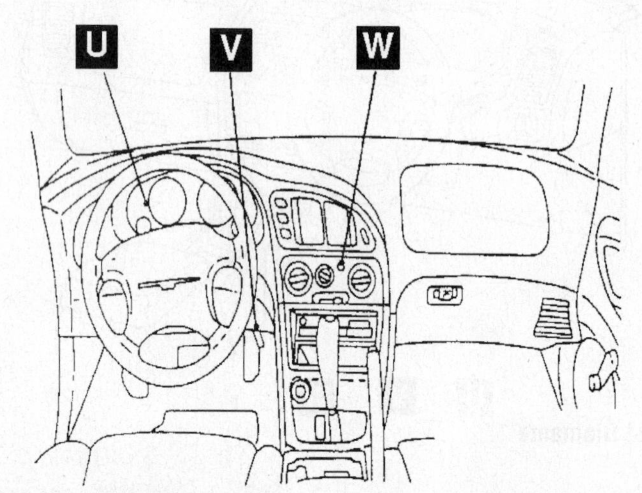

Component Location – 1995-1999 Eclipse 2.0L Non-Turbo

29149_MITS_G0102

Air conditioning compressor clutch relay
T

Idle air control motor
G

Air conditioning switch
W

Ignition coil
D

Camshaft position sensor
I

Injector
N

Check engine/Malfunction indicator lamp
U

Intake air temperature sensor
C

Crankshaft position sensor
B

Knock sensor
F

Data link connector
V

Multiport fuel injection (MFI) relay (Auto shutdown relay)
J

Electric EGR transducer solenoid
R

Engine coolant temperature sensor
M

Manifold absolute pressure (MAP) sensor
A

Evaporative emission purge solenoid
K

Park/Neutral position switch (Transaxle range switch)
Q

Evaporative emission ventilation solenoid
K

Power steering pressure switch
L

Fuel pump relay
J

Powertrain control module (PCM)
S

Heated oxygen sensor (Front)
O

Throttle position sensor
H

Heated oxygen sensor (Rear)
E

Vehicle speed sensor
P

1995–1999 Eclipse 2.0L Turbo

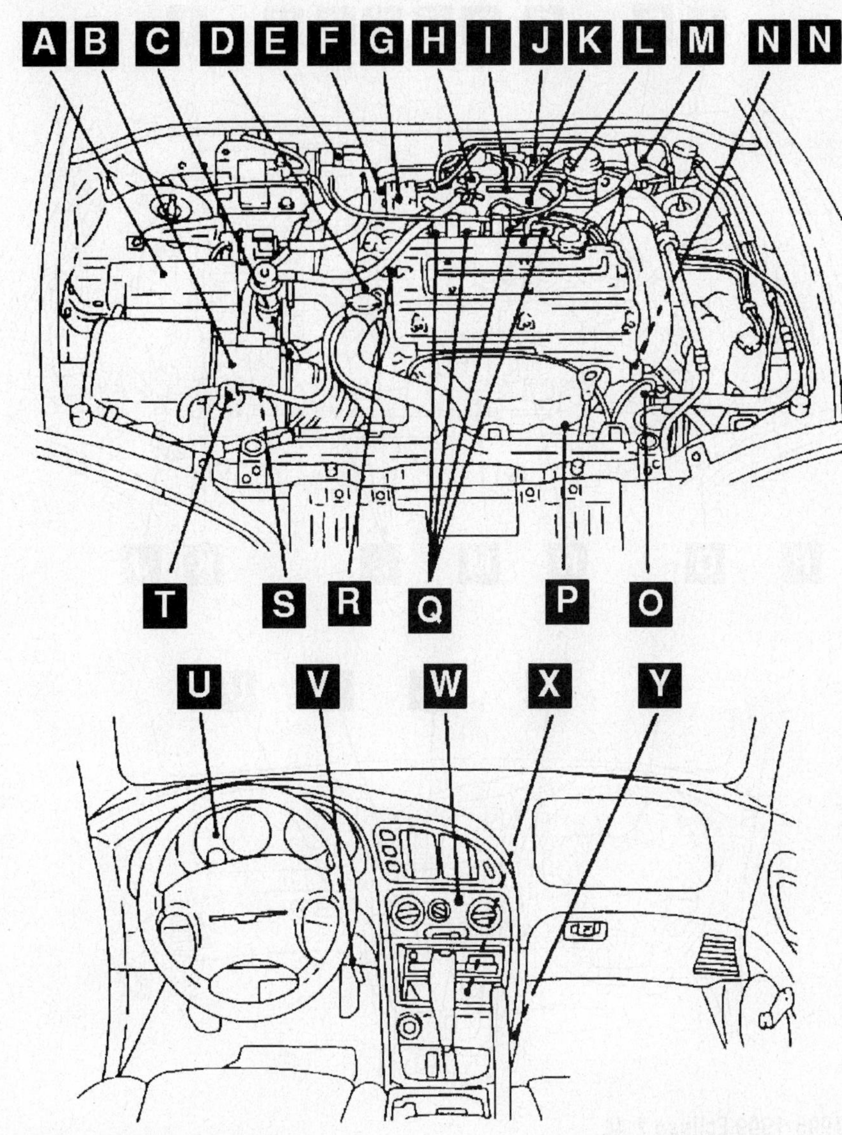

Component Location – 1995-1999 Eclipse 2.0L Turbo

29149_MITS_G0103

Air conditioning compressor clutch relay	A
Injector	Q
Air conditioning switch	W
Camshaft position sensor	R
Knock sensor	L
Check engine/Malfunction indicator lamp	U
Manifold differential pressure (MDP) sensor	H
Crankshaft position sensor	N
Data link connector	V
Multiport fuel injection (MFI) relay	Y
EGR solenoid	J
Park/Neutral position switch	T
Engine control module (ECM)	X
Power steering pressure switch	O
Engine coolant temperature sensor	D
Resistor	I
Evaporative emission purge solenoid	J
Throttle position sensor (with built-in closed throttle position switch)	G
Fuel pressure solenoid	M
Fuel pump check terminal	E
Turbocharger waste gate solenoid	S
Fuel pump relay	Y
Vehicle speed sensor	C
Heated oxygen sensor (Front)	P
Heated oxygen sensor (Rear)	Z
Volume air flow sensor (with built-in intake air temperature sensor and barometric pressure sensor)	B
Idle air control motor	F
Ignition coil (Ignition power transistor)	K

1995–1999 Eclipse 2.4L

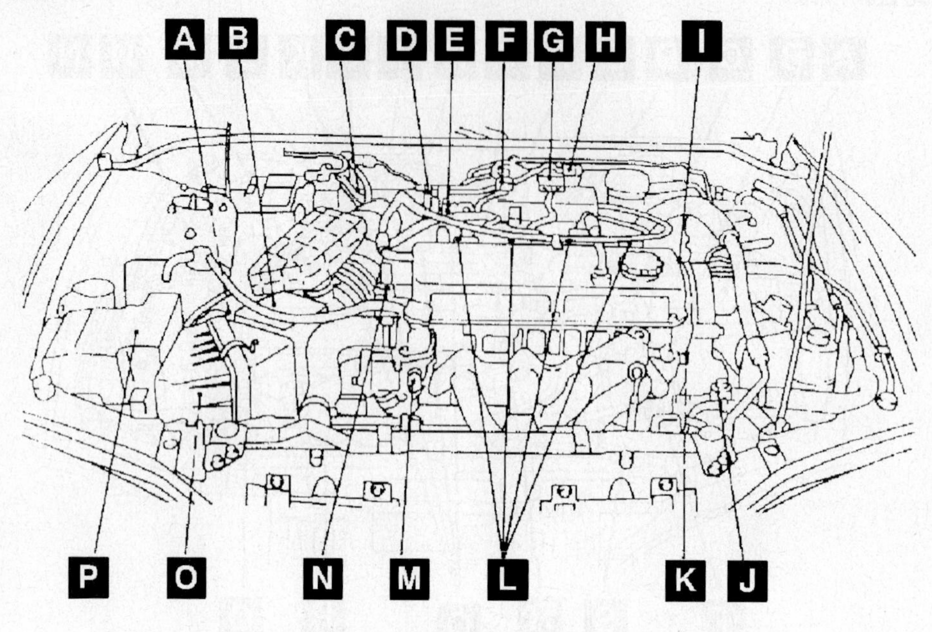

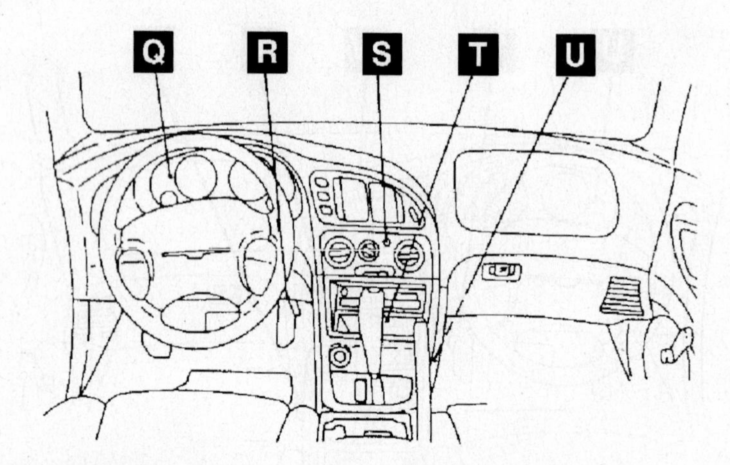

Component Location – 1995-1999 Eclipse 2.4L

29149_MITS_G0104

Air conditioning compressor clutch relay	P
Idle air control motor	D
Air conditioning switch	S
Ignition coil (Ignition power transistor)	I
Camshaft position sensor	N
Ignition power transistor	G
Check engine/Malfunction indicator lamp	Q
Crankshaft position sensor	K
Injector	L
Data link connector	R
Manifold differential pressure (MDP) sensor	F
EGR solenoid	H
Multiport fuel injection (MFI) relay	U
Engine control module (ECM)	T
Park/Neutral position switch	O
Engine coolant temperature sensor	M
Power steering pressure switch	J
Evaporative emission purge solenoid	H
Throttle position sensor (with built-in closed throttle position switch)	E
Fuel pump check terminal	C
Fuel pump relay	U
Vehicle speed sensor	A
Fuel pump relay module	T
Heated oxygen sensor (Front)	V
Vehicle speed sensor	A
Heated oxygen sensor (Rear)	W
Volume air flow sensor (with built-in intake air temperature sensor and barometric pressure sensor)	B

2000–2005 Eclipse 2.4L

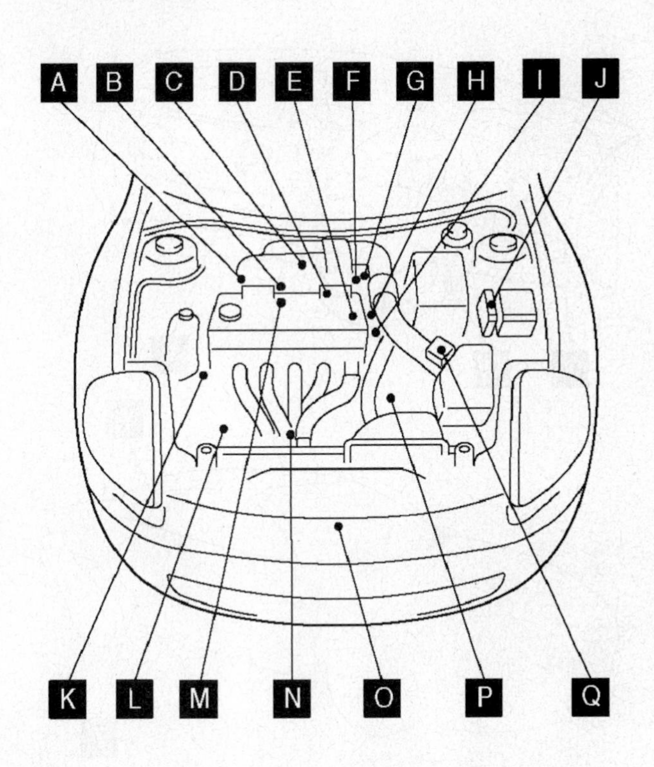

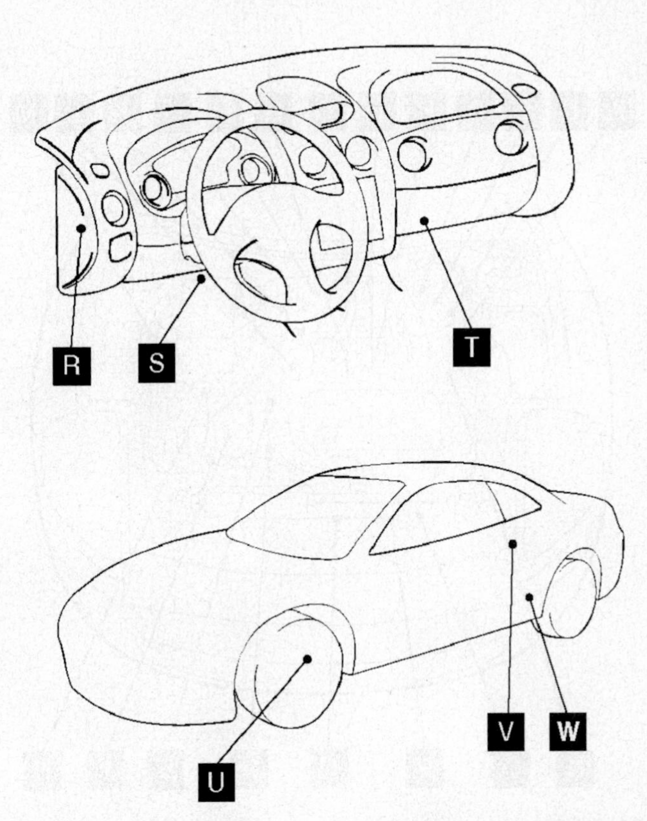

Component Location – 2000-2005 Eclipse 2.4L

29149_MITS_G0111

Component	Location
Air conditioning compressor clutch relay	J
Heated oxygen sensor (front)	N
Camshaft position sensor	I
Heated oxygen sensor (rear)	V
Crankshaft position sensor	K
Idle air control motor	F
Data link connector	T
Ignition coil	M
EGR solenoid	C
Injector	B
Engine control module (ECM) <M/T>	U
Knock sensor	D
Engine coolant temperature sensor	H
Manifold differential pressure sensor	A
Engine speed detection connector	J
Multiport fuel injection (MFI) relay	J
Evaporative emission purge solenoid	C
Park/neutral position switch <A/T>	P
Evaporative emission ventilation solenoid	X
Powertrain control module (PCM) <A/T>	U
Fan controller	O
Power steering pressure switch	L
Fuel tank differential pressure sensor	W
Throttle position sensor	G
Fuel temperature sensor	W
Vehicle speed sensor <M/T>	E
Fuel pump relay 1, 2	S
Volume air flow sensor (with built-in intake air temperature sensor and barometric pressure sensor)	Q

2000–2005 Eclipse 3.0L

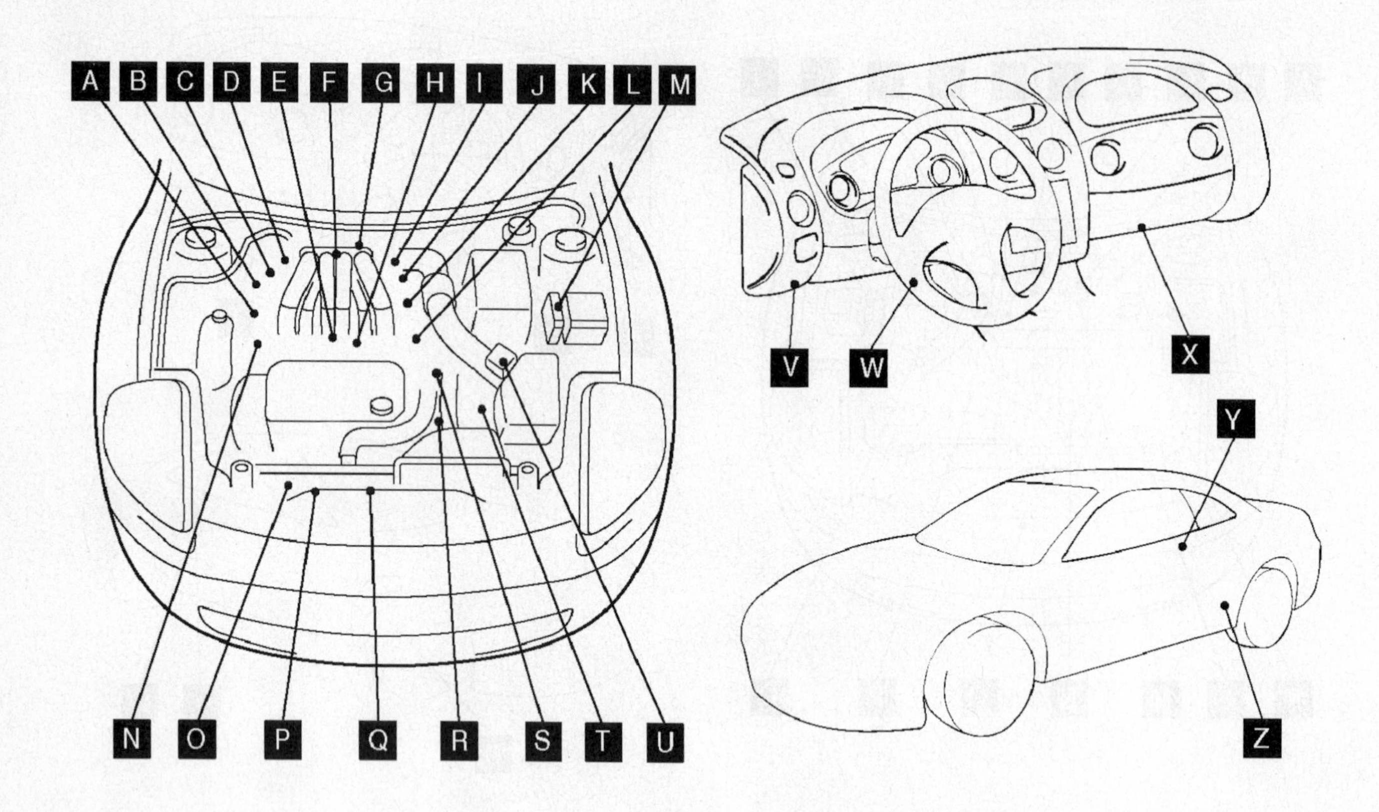

Component Location – 2000-2005 Eclipse 3.0L

29149_MITS_G0112

Air conditioning compressor clutch relay
M

Injector
H

Crankshaft position sensor
N

Knock sensor
E

Data link connector
W

Left bank heated oxygen sensor (front)
P

Distributor (built-in camshaft position sensor and ignition coil)
L

Left bank heated oxygen sensor (rear)
O

EGR solenoid
B

Manifold differential pressure sensor
G

Engine control module (ECM) <M/T>
X

Multiport fuel injection (MFI) relay
M

Engine coolant temperature sensor
S

Park/neutral position switch <A/T>
T

Engine speed detection connector
M

Powertrain control module <A/T>
X

Evaporative emission purge solenoid <without variable induction system>
B

Power steering pressure switch
A

Evaporative emission purge solenoid <with variable induction system>
K

Right bank heated oxygen sensor (front)
D

Evaporative emission ventilation solenoid
Z

Right bank heated oxygen sensor (rear)
F

Fan controller
Q

Throttle position sensor
I

Fuel tank differential pressure sensor
Y

Variable induction control solenoid <with variable induction system>
C

Fuel pump module (Fuel temperature sensor)
Y

Variable induction control solenoid <with variable induction system>
R

Fuel pump relay 1, 2
V

Volume air flow sensor (with built-in intake air temperature sensor and barometric pressure sensor)
U

Idle air control motor
J

2003 Endeavor

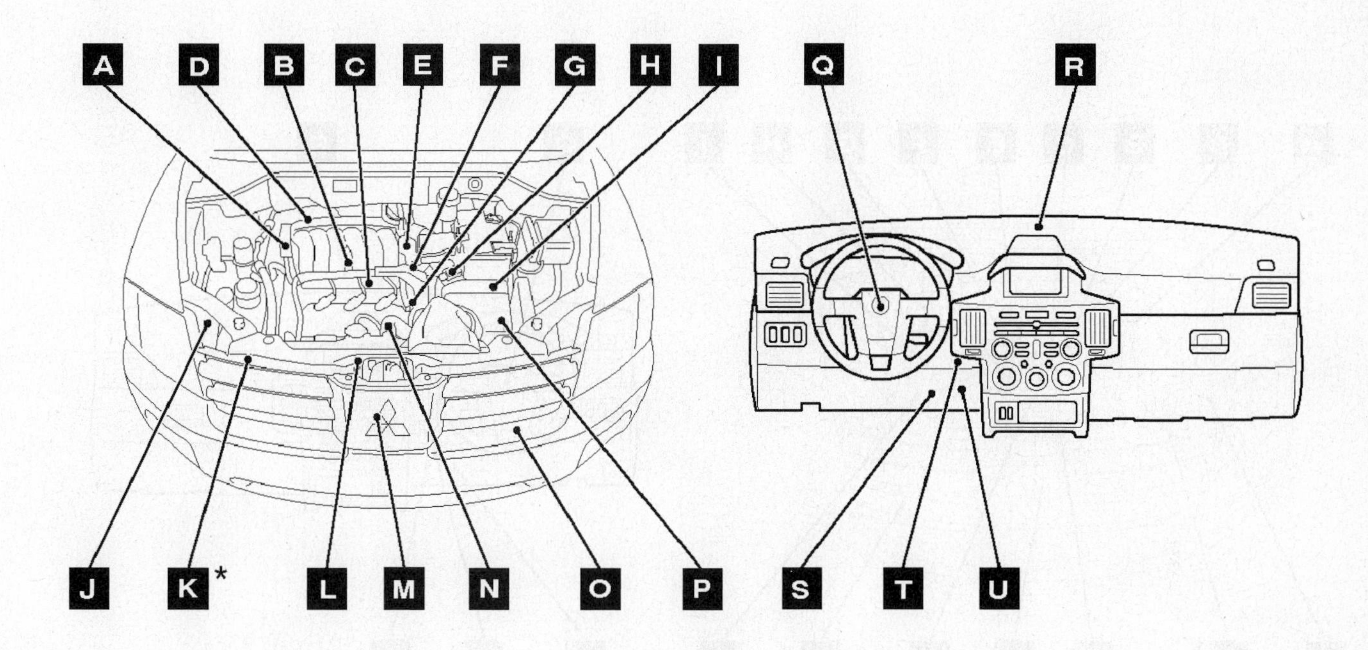

Component Location – 2003 Endeavor

29149_MITS_G0115

A/C pressure sensor
J
Left bank heated oxygen sensor (front)
L
A/T control solenoid valve assembly (incorporating transmission fluid temperature sensor)
P
Left bank heated oxygen sensor (rear)
M
ABS G sensor <ABS-AWD>
a
Manifold absolute pressure sensor
D
Accelerator pedal position sensor
S
Mass airflow sensor (incorporating intake air temperature sensor)
H
Air thermo sensor
T

Occupant classification sensor (LH), (RH)
V
Ambient temperature sensor
O
Output shaft speed sensor
I
Camshaft position sensor
G
Photo sensor <Automatic A/C>
R
Crankshaft position sensor
A
Right bank heated oxygen sensor (front)
C
Engine coolant temperature sensor
F

Right bank heated oxygen sensor (rear)
N
Front impact sensor
K
Seat slide sensor
b
Fuel level sensor (sub)
X
Side impact sensor <Side air bag>
c
Fuel pump module (incorporating fuel level sensor (main) and fuel tank temperature sensor)
X
Steering wheel sensor <ASC>
Q
Fuel tank differential pressure sensor
X
Sunroof motor assembly (incorporating motor sensor unit)
W

G and yaw rate sensor <ASC>
a
Throttle body assembly (incorporating throttle position sensor)
E
Input shaft speed sensor
I
TPMS transmitter (incorporating tire pressure sensor)
Y
Interior temperature sensor <Automatic A/C>
U
Wheel speed sensor (front)
Z
Knock sensor
B
Wheel speed sensor (rear)
d

2004–2006 Endeavor

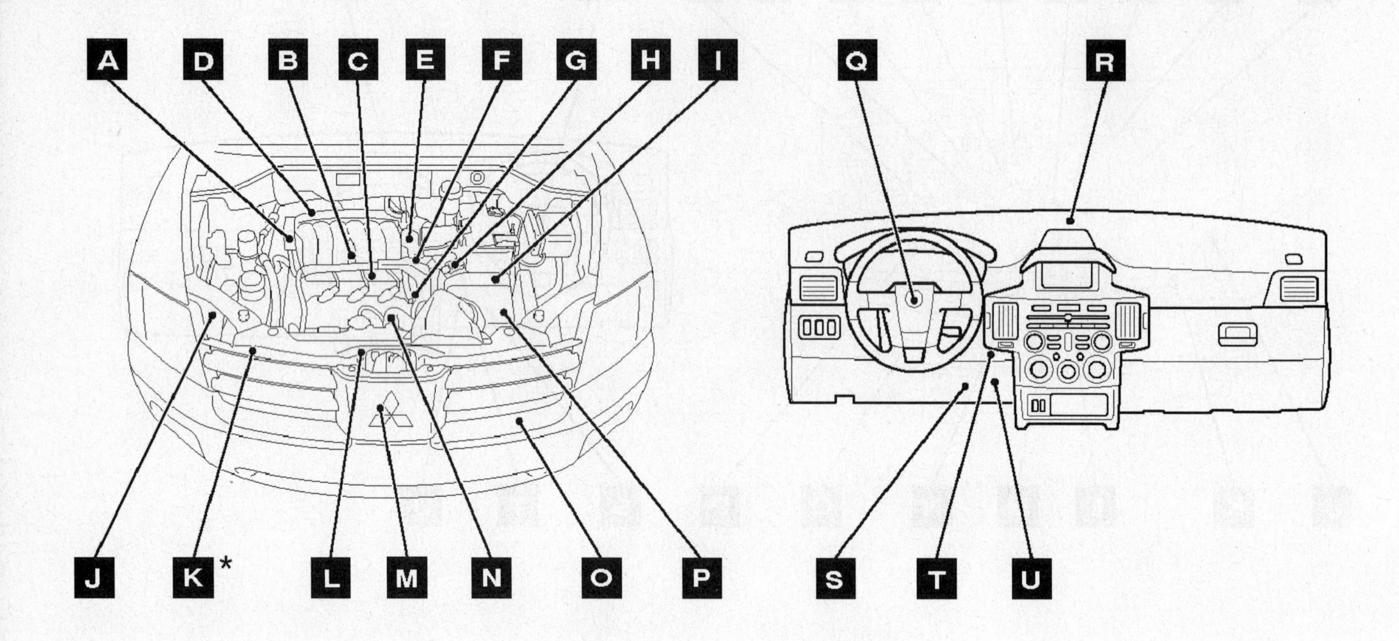

Component Location – 2004-2006 Endeavor

29149_MITS_G0116

A/C pressure sensor
J
Left bank heated oxygen sensor (front)
L
A/T control solenoid valve assembly (incorporating transmission fluid temperature sensor)
P
Left bank heated oxygen sensor (rear)
M
ABS G sensor <ABS-AWD>
a
Manifold absolute pressure sensor
D
Accelerator pedal position sensor
S
Mass airflow sensor (incorporating intake air temperature sensor)
H
Air thermo sensor
T

Occupant classification sensor (LH), (RH)
V
Ambient temperature sensor
O
Output shaft speed sensor
I
Camshaft position sensor
G
Photo sensor <Automatic A/C>
R
Crankshaft position sensor
A
Right bank heated oxygen sensor (front)
C
Engine coolant temperature sensor
F

Right bank heated oxygen sensor (rear)
N
Front impact sensor
K
Seat slide sensor
b
Fuel level sensor (sub)
X
Side impact sensor <Side air bag>
c
Fuel pump module (incorporating fuel level sensor (main) and fuel tank temperature sensor)
X
Steering wheel sensor <ASC>
Q
Fuel tank differential pressure sensor
X
Sunroof motor assembly (incorporating motor sensor unit)
W

G and yaw rate sensor <ASC>
a
Throttle body assembly (incorporating throttle position sensor)
E
Input shaft speed sensor
I
TPMS transmitter (incorporating tire pressure sensor)
Y
Interior temperature sensor <Automatic A/C>
U
Wheel speed sensor (front)
Z
Knock sensor
B
Wheel speed sensor (rear)
d

2003–2006 Evo

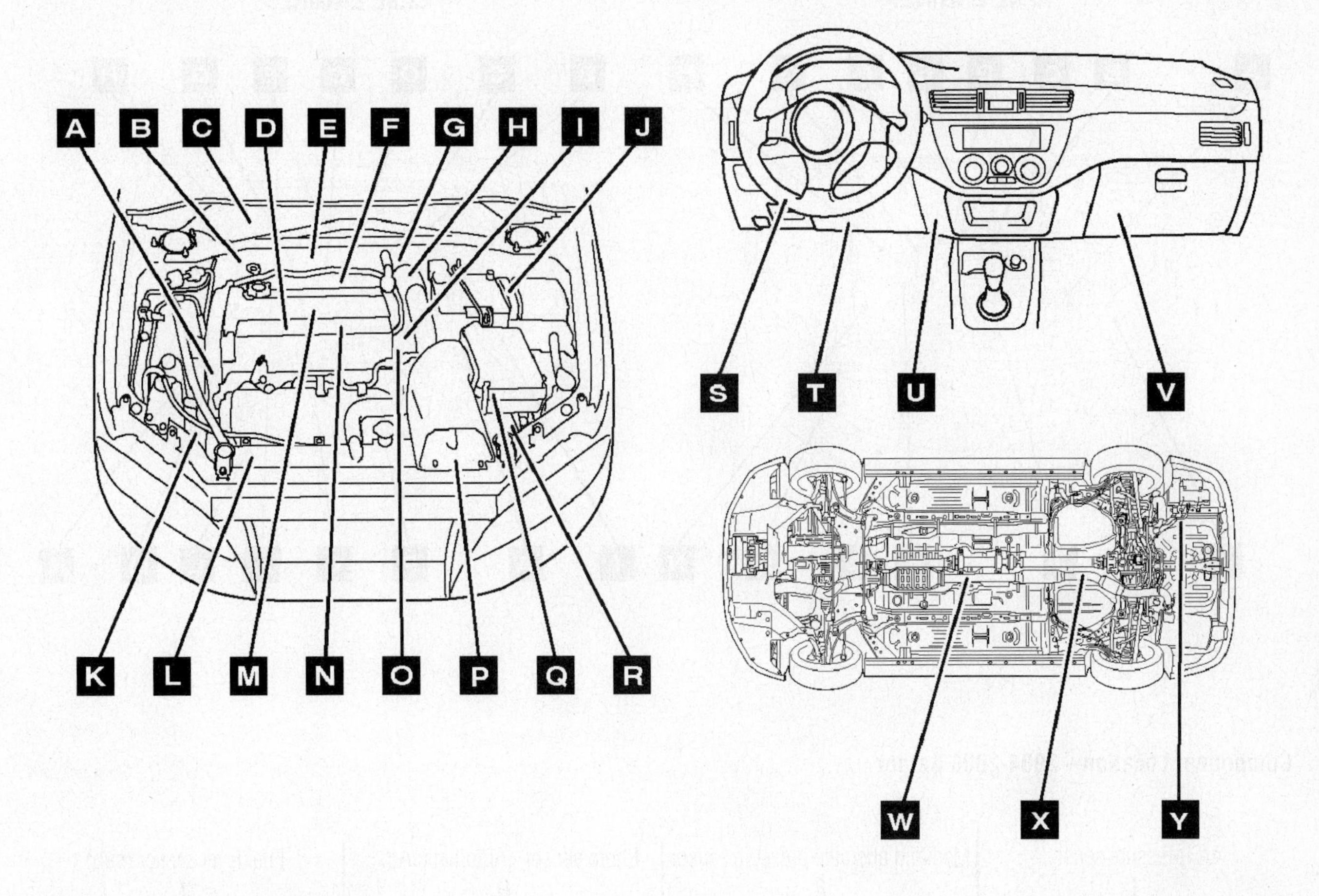

Component Location – 2003-2006 Evo

29149_MITS_G0117

Air conditioning compressor clutch relay **J**	Data link connector **U**	Engine speed detection connector **J**	Fuel pressure solenoid **B**
Fuel pump relay 3 **C**	Idle air control motor **H**	Knock sensor **N**	Throttle position sensor **G**
Camshaft position sensor **O**	EGR vacuum regulator solenoid valve **D**	Evaporative emission purge solenoid **B**	Fuel tank differential pressure sensor **X**
Fuel pump resistor **C**	Ignition coil **M**	Manifold differential pressure sensor **E**	Turbocharger wastegate solenoid **R**
Crankshaft position sensor **A**	Engine control module (ECM) **V**	Evaporative emission ventilation solenoid Y	Fuel tank temperature sensor **X**
Heated oxygen sensor (front) **L**	Injector **F**	Multiport fuel injection (MFI) relay **J**	Vehicle speed sensor **I**
Clutch pedal position switch **T**	Engine coolant temperature sensor **O**	Fan control module **P**	Fuel pump relay 1, 2 **S**
Heated oxygen sensor (rear) **W**	Injector resistor **C**	Power steering pressure switch **K**	Volume airflow sensor (with built-in intake air temperature sensor and barometric pressure sensor) **Q**

2004–2006 Galant

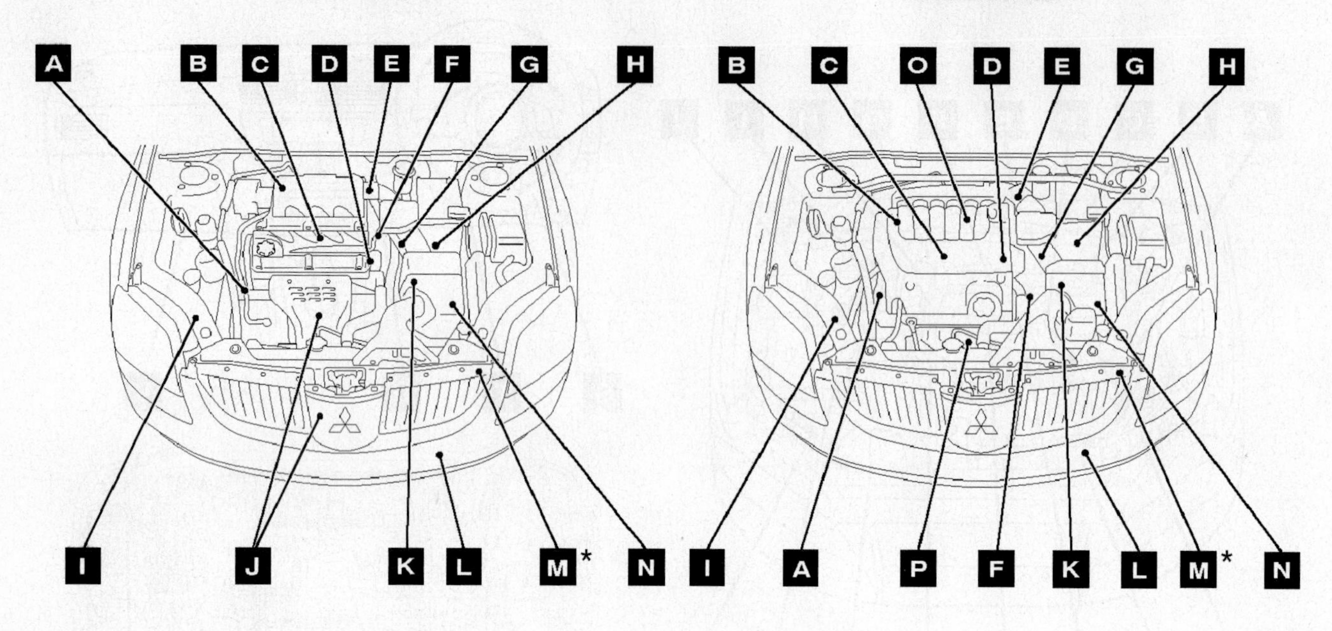

Component Location – 2004-2006 Galant

29149_MITS_G0118

A/C pressure sensor **I** Interior temperature sensor <Automatic A/C> **S** A/T control solenoid valve assembly (Incorporating transmission fluid temperature sensor) **N** Knock sensor **C** Accelerator pedal position sensor **R** Left bank heated oxygen sensor (front) <3.8L Engine> **P** Air thermo sensor **T** Left bank heated oxygen sensor (rear) <3.8L Engine> **P** Ambient temperature sensor **L**	Manifold absolute pressure sensor **B** Camshaft position sensor **D** Mass airflow sensor **G** Crankshaft position sensor **A** Occupant classification sensor (LH), (RH) **X** Cylinder 1, 4 heated oxygen sensor (front) <2.4L Engine> **J** Output shaft speed sensor **H** Cylinder 1, 4 heated oxygen sensor (rear) <2.4L Engine> **J**	Photo sensor <Automatic A/C> **Q** Cylinder 2, 3 heated oxygen sensor (front) <2.4L Engine> **J** Right bank heated oxygen sensor (front) <3.8L Engine> **O** Cylinder 2, 3 heated oxygen sensor (rear) <2.4L Engine> **J** Right bank heated oxygen sensor (rear) <3.8L Engine> **O** Engine coolant temperature sensor **F** Seat slide sensor **Z** Front impact sensor **M** Side impact sensor <Side air bag> **a**	Fuel level sensor (sub) **V** Sunroof assembly (Incorporating pulse sensor) **U** Fuel pump module (Incorporating fuel level sensor (main) and fuel tank temperature sensor) **W** Throttle body assembly (Incorporating throttle position sensor) **E** Fuel tank differential pressure sensor **V** Wheel speed sensor (front) <ABS> **Y** Input shaft speed sensor **K** Wheel speed sensor (rear) <ABS> **b**

2002–2006 Lancer 2.0L

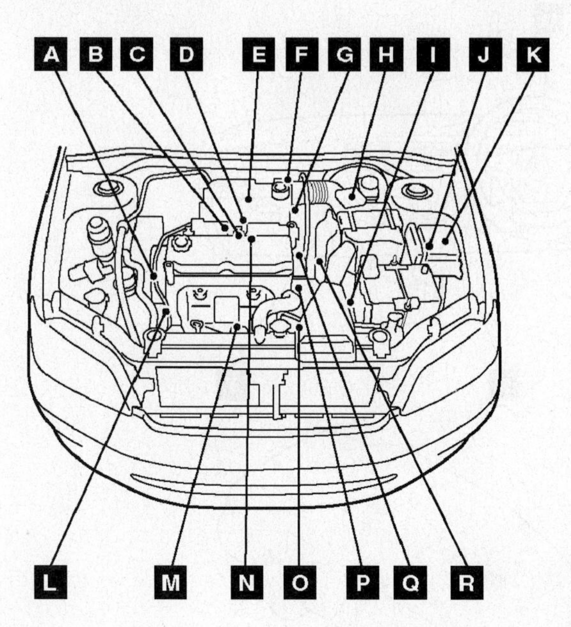

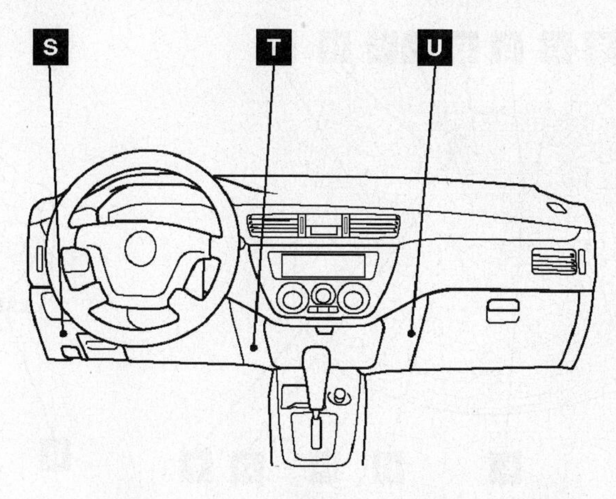

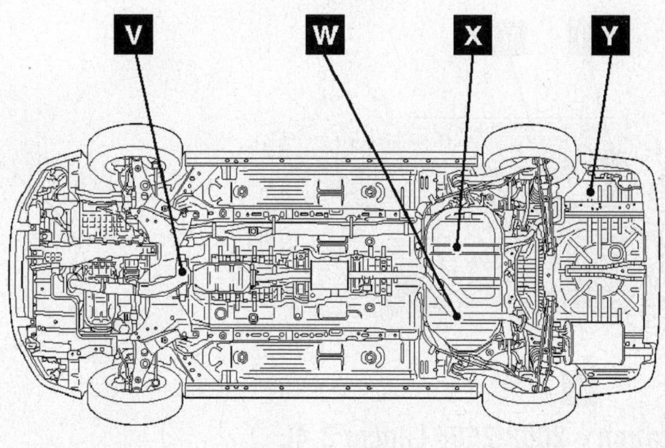

Component Location – 2002-2006 Lancer 2.0L

9149_MITS_G0113

Component	Location
A/C compressor clutch relay	K
Heated oxygen sensor (front)	M
Camshaft position sensor	Q
Heated oxygen sensor (rear)	V
Crankshaft position sensor	A
Idle air control motor	G
Data link connector	T
Ignition coil	B
EGR vacuum regulator solenoid valve	D
Injector	C
Engine control module (ECM) <M/T>	U
Knock sensor	N
Engine coolant temperature sensor	P
Manifold differential pressure sensor	E
Engine speed detection connector	J
Multiport fuel injection (MFI) relay	K
Evaporative emission purge solenoid	D
Transmission range switch <A/T>	I
Evaporative emission ventilation solenoid	Y
Powertrain control module (PCM) <A/T>	U
Fan controller	O
Power steering pressure switch	L
Fuel level sensor (Fuel tank temperature sensor)	W
Throttle position sensor	F
Fuel tank differential pressure sensor	X
Vehicle speed sensor <M/T>	R
Fuel pump relay 1, 2	S
Volume airflow sensor (with built-in intake air temperature sensor and barometric pressure sensor)	H

2002–2006 Lancer 2.4L

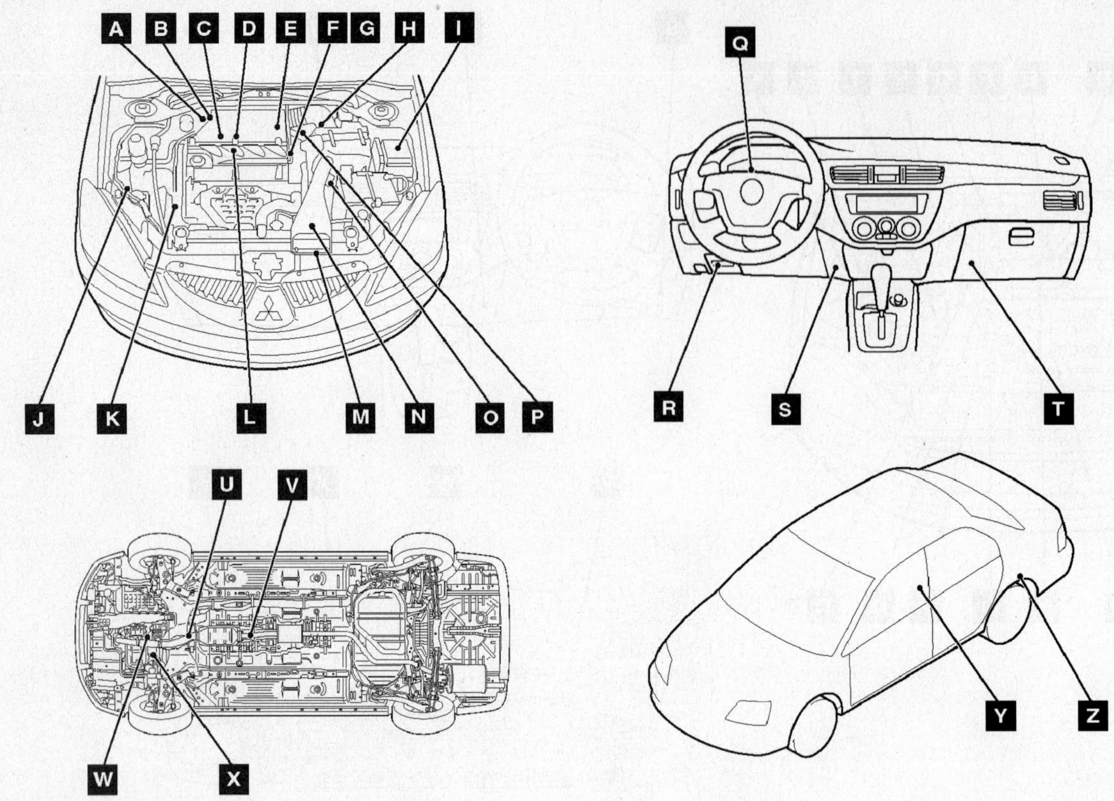

Component Location – 2002-2006 Lancer 2.4L

29149_MITS_G0114

Component	Location
Accelerator pedal position sensor (with built-in accelerator pedal position switch)	J
Heated oxygen sensor (front)	U
Air conditioning compressor clutch relay	I
Heated oxygen sensor (rear)	V
Camshaft position sensor	F
Ignition coil	C
Crankshaft position sensor	K
Injector	D
Data link connector	S
Knock sensor	W
Engine control module (ECM) <M/T>	T
Malfunction indicator lamp (SERVICE ENGINE SOON or Check Engine Lamp)	Q
Engine coolant temperature sensor	F
Manifold absolute pressure sensor	A
Engine oil control valve	G
Mass airflow sensor (with built-in intake air temperature sensor)	H
Engine oil pressure switch	G
Multiport fuel injection (MFI) relay	I
Evaporative emission purge solenoid	B
Output shaft speed sensor	O
Evaporative emission ventilation solenoid	Z
Power steering pressure switch	X
Exhaust gas recirculation valve	L
Powertrain control module (PCM) <A/T>	T
Fan control module	M
Throttle actuator control motor	E
Fuel level sensor	Y
Throttle actuator control motor relay	I
Fuel pump relay	R
Throttle position sensor	E
Fuel tank differential pressure sensor	Y
Transmission range switch <A/T>	N
Fuel tank temperature sensor	Y
Vehicle speed sensor <M/T>	P

1995–2002 Mirage 1.5L

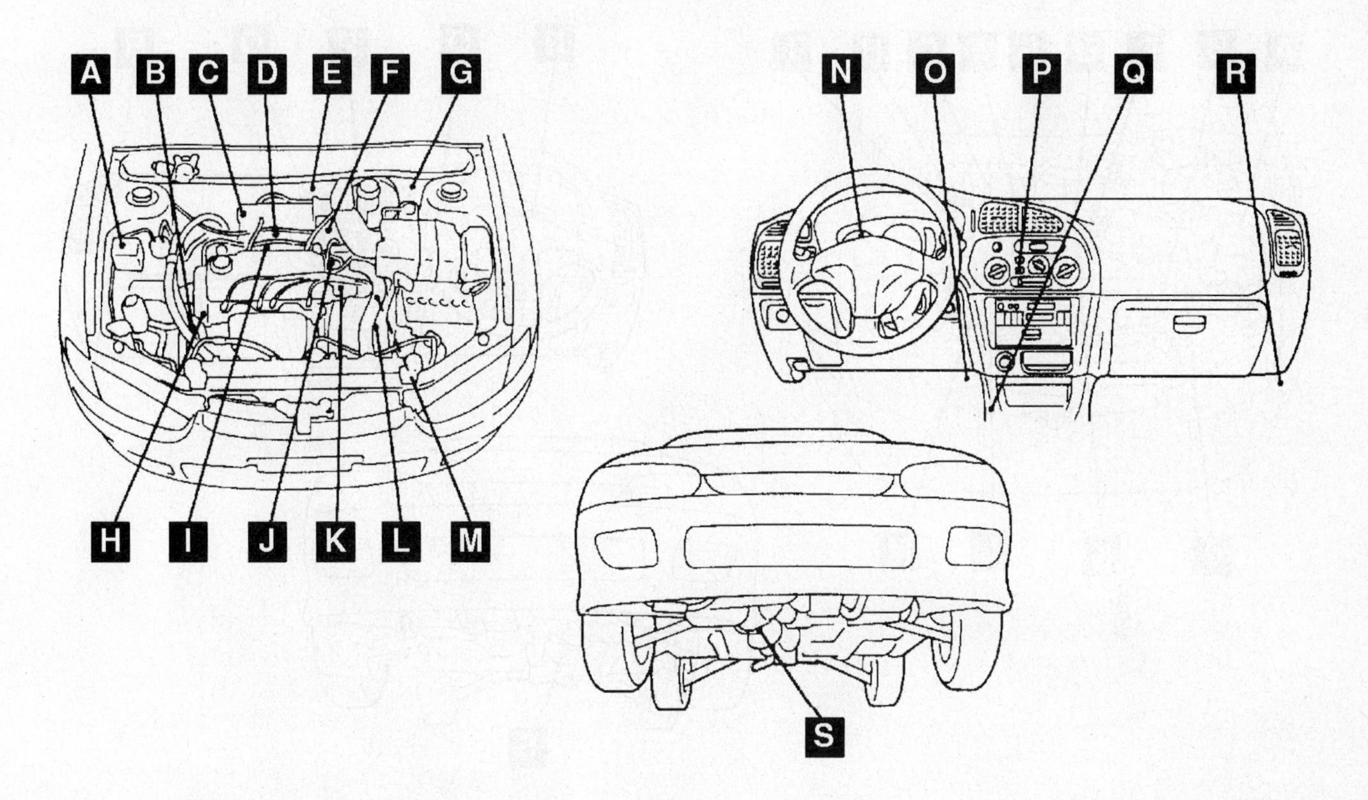

Component Location – 1995-2002 Mirage 1.5L

29149_MITS_G0106

Air conditioning compressor clutch relay	A
Heated oxygen sensor <Federal> Front	S
Air conditioning switch	P
Rear	S
Camshaft position sensor	K
Idle air control motor	F
Check Engine/Malfunction Indicator Lamp	N
Ignition coil/Ignition power transistor	K
Crankshaft position sensor	H
Injector	I
Data link connector	O
Intake air temperature sensor	C
EGR Solenoid	D
Manifold absolute pressure sensor	E
Engine control module	R
Multiport fuel injection (MFI) relay/Fuel pump relay	Q
Engine coolant temperature sensor	J
Park/Neutral position switch	L
Evaporative emission purge solenoid	D
Power steering pressure switch	B
Fuel pump check connector	G
Throttle position sensor (with built-in closed throttle position switch)	F
Heated oxygen sensor <California> Front	S
Vehicle speed sensor	M
Rear	S

1995–2002 Mirage 1.8L

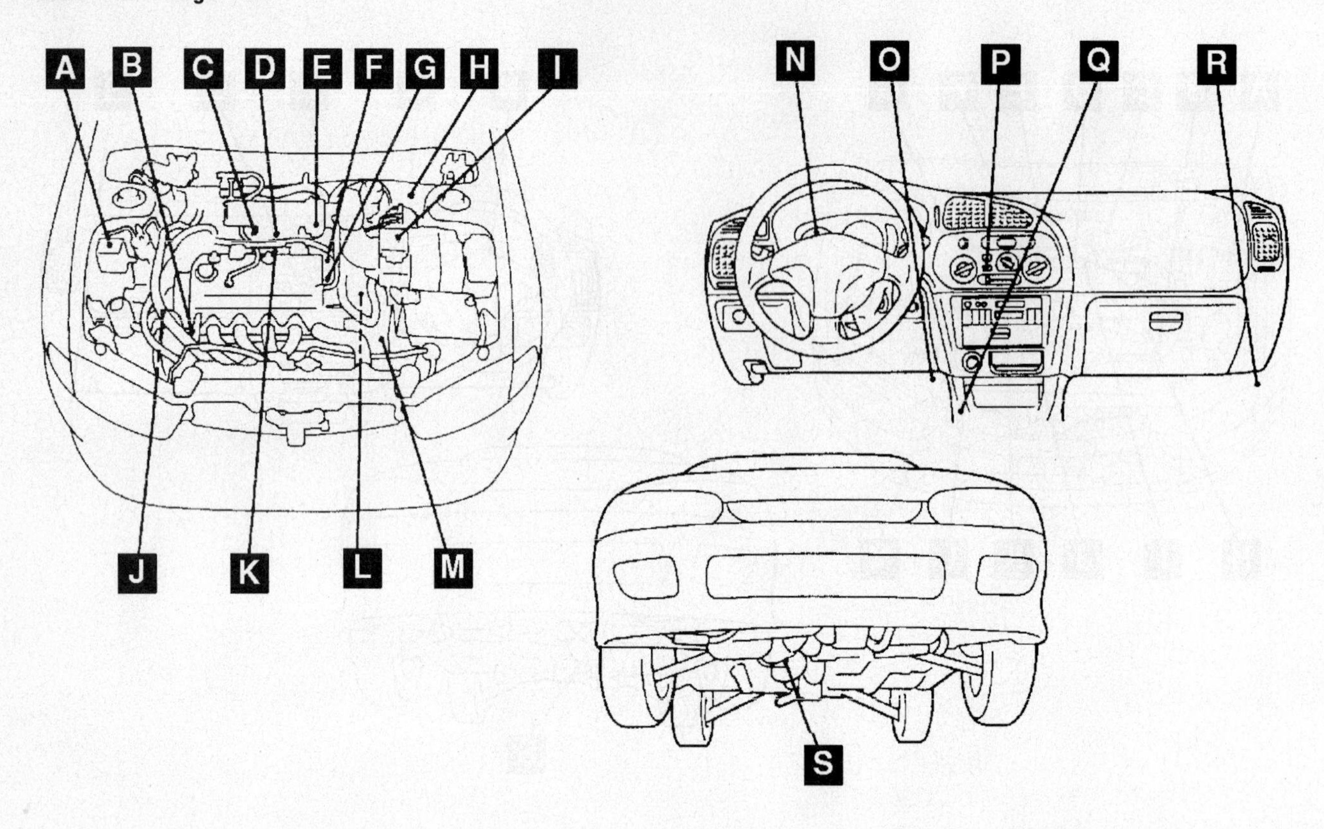

Component Location – 1995-2002 Mirage 1.8L

29149_MITS_G0107

Air conditioning compressor clutch relay
A
Heated oxygen sensor <Federal>
Front
S
Air conditioning switch
P
Rear
S
Camshaft position sensor
F
Idle air control motor
E
Check Engine/Malfunction Indicator Lamp
N

Ignition coil (with built-in ignition power transistor)
K
Crankshaft position sensor
J
Ignition failure sensor
F
Data link connector
O
Injector
K
EGR Solenoid
C
Manifold differential pressure sensor
D

Engine control module
R
Multiport fuel injection (MFI) relay/Fuel pump relay
Q
Engine coolant temperature sensor
G
Park/Neutral position switch
M
Evaporative emission purge solenoid
C
Power steering pressure switch
B

Fuel pump check connector
H
Throttle position sensor (with built-in closed throttle position switch)
E
Heated oxygen sensor <California>
Front
S
Vehicle speed sensor
L
Rear
S
Volume air flow sensor (with built-in intake air temperature sensor and barometric pressure sensor)
I

1995–2000 Montero

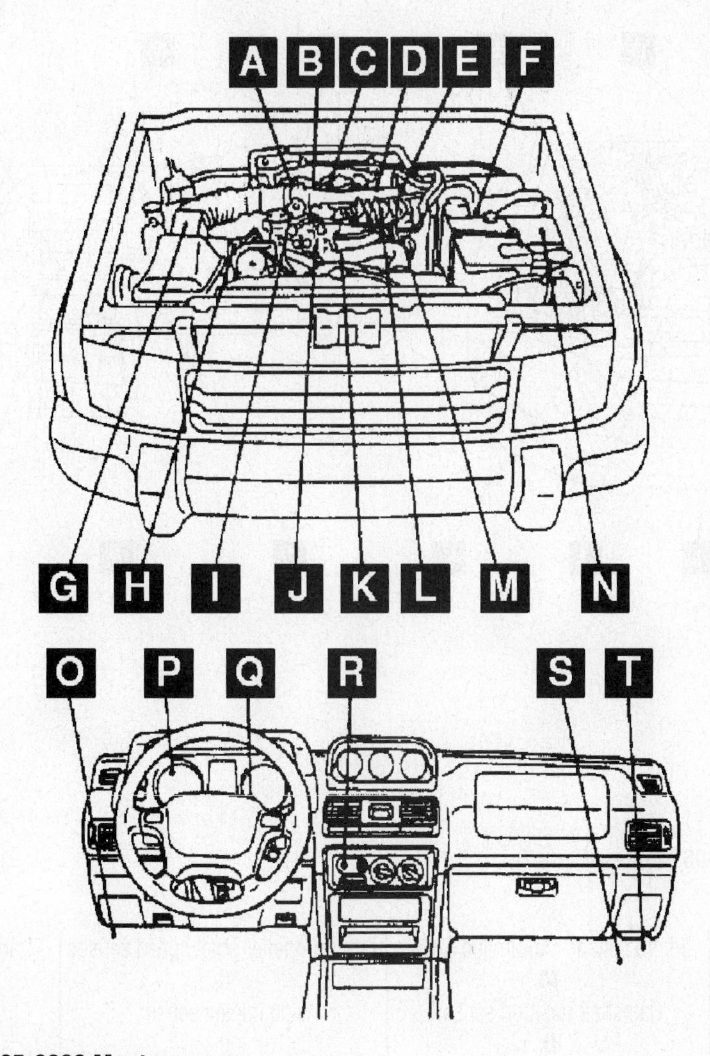

Component Location – 1995-2000 Montero

29149_MITS_G0108

Air conditioning compressor clutch relay	N
Heated oxygen sensor <California>	V
Air conditioner switch	R
Idle air control motor	I
Camshaft position sensor	M
Ignition coil (Ignition power transistor)	L
Crankshaft position sensor	K
Injector	D
Data link connector	O
Manifold differential pressure sensor	E
EGR solenoid	B
Multiport fuel injection (MFI) relay/Fuel pump relay	T
Engine control module	S
Park/Neutral position switch	W
Engine coolant temperature sensor	J
Power steering pressure switch	H
Evaporative emission purge solenoid	C
Service engine soon/malfunction indicator lamp	Q
Evaporative emission ventilation solenoid	F
Throttle position sensor (with built-in closed throttle position switch)	A
Fuel tank differential pressure sensor	X
Vehicle speed sensor	P
Heated oxygen sensor <Federal>	U
Volume air flow sensor (with built-in intake air temperature sensor and barometric pressure sensor)	G

2001–2006 Montero

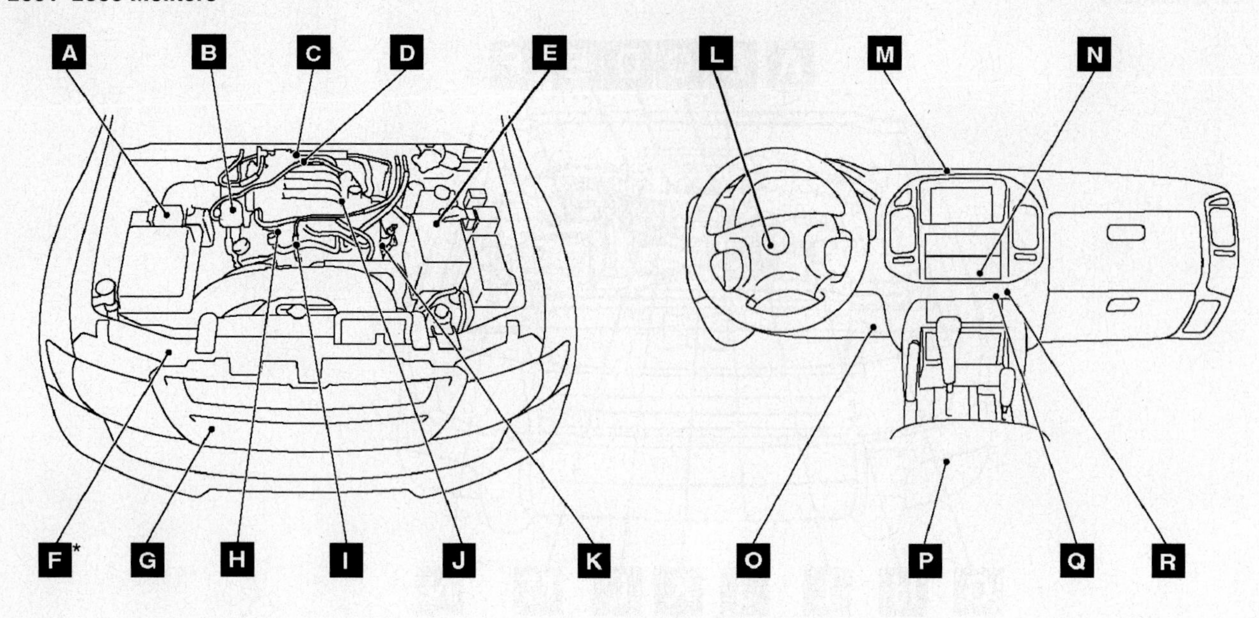

Component Location – 2001-2006 Montero

Accelerator pedal position sensor
O

Left bank heated oxygen sensor (rear)
T

A/C sensor <Automatic A/C>
Q

Lock sensor
K

A/C sensor <Manual A/C>
R

Manifold absolute pressure sensor
C

A/C sensor <Rear cooler, rear heater and rear A/C>
e

Output shaft speed sensor
Y

Ambient air temperature sensor
G

Photo sensor <Automatic A/C>
M

Camshaft position sensor
D

Pressure sensor <M-ASTC>
E

Crankshaft position sensor
I

Rear propeller shaft speed sensor
Z

Engine coolant temperature sensor
H

Right bank heated oxygen sensor (front)
V

Front impact sensor
F

Right bank heated oxygen sensor (rear)
W

Front propeller shaft speed sensor
d

Side impact sensor
g

Fuel level sensor (Incorporating fuel tank temperature sensor)
h

Steering wheel sensor <M-ASTC>
L

Fuel tank differential pressure sensor
h

Throttle position sensor
B

G and yaw rate sensor
P

TPMS transmitter (Incorporating pressure sensor)
f

Heater water temperature sensor <Automatic A/C>
Q

Transmission fluid temperature sensor
b

Input shaft speed sensor
c

Vehicle speed sensor
a

Inside air temperature sensor <Automatic A/C>
N

Volume airflow sensor
A

Knock sensor
J

Wheel speed sensor (front)
U

Left bank heated oxygen sensor (front)
S

Wheel speed sensor (rear)
X

1997–2004 Montero Sport

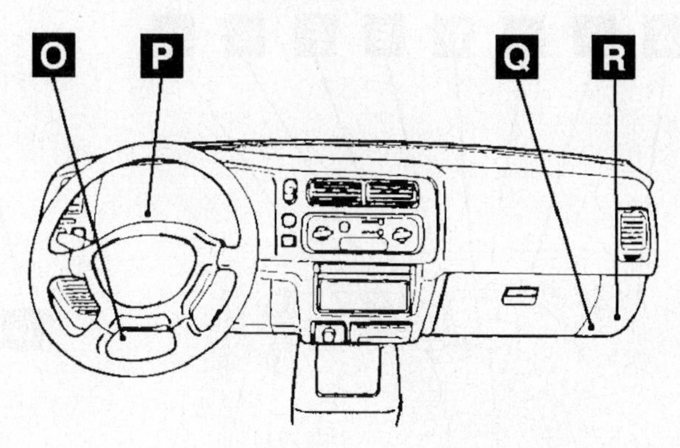

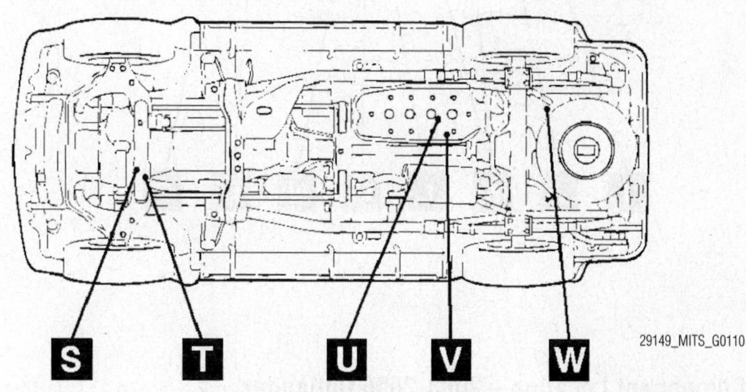

29149_MITS_G0110

Component Location – 1997-2004 Montero Sport

Air conditioning compressor clutch relay
J
Injector
G
Camshaft position sensor
M
Left bank heated oxygen sensor (front)
I
Crankshaft position sensor
L
Left bank heated oxygen sensor (rear)
T

Data link connector
O
Manifold differential pressure sensor
H
EGR solenoid
E
Multiport fuel injection (MFI) relay/fuel pump relay
Q
Engine coolant temperature sensor
F
Power steering pressure switch
K

Evaporative emission purge solenoid
E
Powertrain control module
R
Evaporative emission ventilation solenoid
W
Right bank heated oxygen sensor (front)
D
Fuel tank differential pressure sensor
U
Right bank heated oxygen sensor (rear)
S

Fuel temperature sensor
V
Service engine soon/malfunction indicator lamp
P
Idle air control motor
B
Throttle position sensor (with built-in closed throttle position switch <3.0L Engine>)
C
Ignition coil/ignition power transistor
N
Volume air flow sensor (with built-in intake air temperature sensor and barometric pressure sensor
A

2003–2006 Outlander

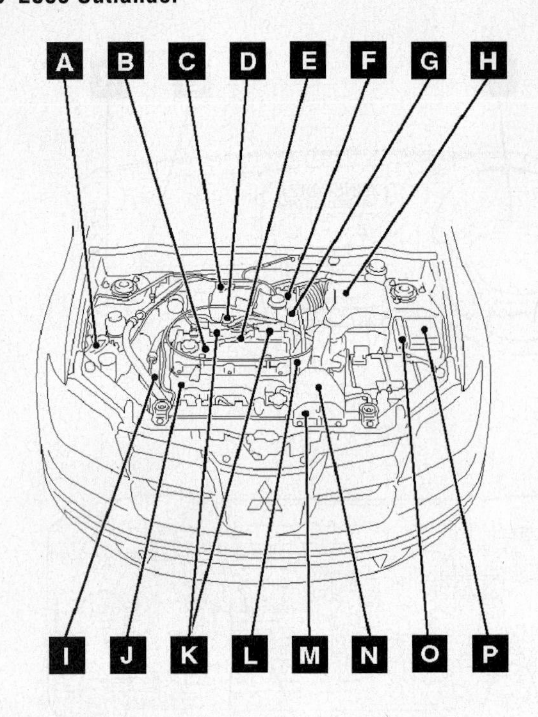

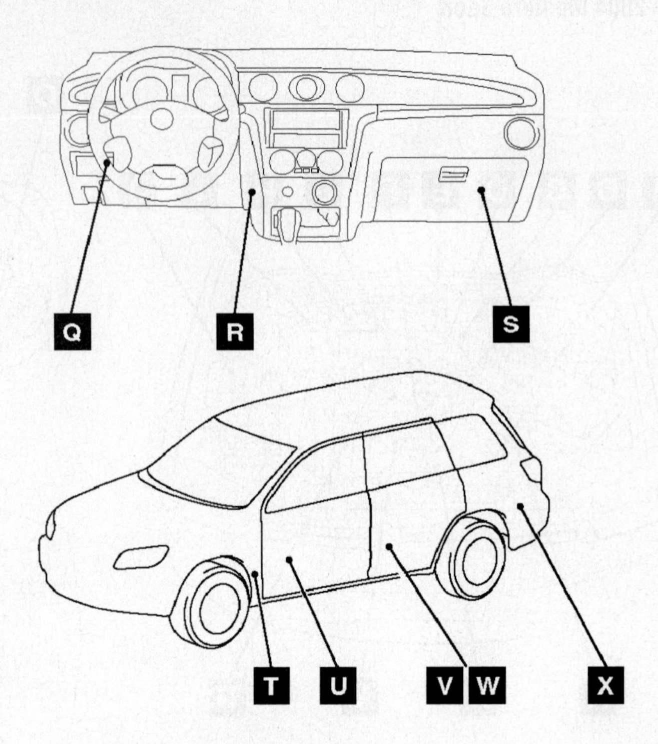

Component Location – 2003-2006 Outlander

Air conditioning compressor clutch relay
P

Fuel pump relay 1, 2
Q

Air conditioning pressure sensor
A

Heated oxygen sensor (front)
T

Camshaft position sensor
L

Heated oxygen sensor (rear)
U

Crankshaft position sensor
J

Idle air control motor
G

Data link connector
R

Ignition coil
K

EGR vacuum regulator solenoid valve
D

Injector
E

Engine coolant temperature sensor
L

Knock sensor
B

Engine speed detection connector
O

Manifold differential pressure sensor
C

Evaporative emission purge solenoid
D

Multiport fuel injection (MFI) relay
P

Evaporative emission ventilation solenoid
X

Transmission range switch
N

Fan controller
M

Powertrain control module (PCM)
S

Fuel tank differential pressure sensor <FWD>
V

Power steering pressure switch
I

Fuel tank differential pressure sensor <4WD>
W

Throttle position sensor
F

Fuel pump module <FWD>
V

Volume airflow sensor (with built-in intake air temperature sensor and barometric pressure sensor)
H

Fuel pump module (with built-in fuel tank temperature sensor) <4WD>
W

Accelerator Pedal Position (APP) Sensor

LOCATION

2003–06 Vehicles

See Figure 1.

OPERATION

2003–06 Vehicles

A 5–volt power supply is applied on the Accelerator Pedal Position (APP) Sensor (main) power terminal from the PCM. The ground terminal is grounded with PCM. When the accelerator pedal is moved from the idle position to the fully opened position, the resistance between the accelerator pedal position sensor (main) output terminal and ground terminal will increase according to the depression.

REMOVAL & INSTALLATION

2003–06 Vehicles

1. Disconnect the negative battery cable.
2. Detach the electrical connector at the sensor.
3. Remove the sensor from the bracket.
4. Installation is the reverse of the removal procedure.

TESTING

2003–06 Vehicles

See Figure 2.

1. Disconnect the sensor connector.
2. Measure the resistance between the sensor connector terminal No. 1 sensor (main) earth and terminal No. 2 sensor (main) power supply, and between terminal No.7 sensor (sub) earth and terminal No. 8 sensor (sub) power supply. Standard value: 3.5–6.5 Kohms
3. Measure the resistance between sensor connector terminal No. 2 sensor (main) power supply and terminal No. 3 sensor (main) output; and between terminal No.8 sensor (sub) power supply and terminal No. 6 sensor (sub) output. When accelerator pedal is gently depressed increases in the resistance are comparatively smooth in proportion to the accelerator pedal depression amount.
4. If the measured values are outside the standard value range, or if the resistance does not change smoothly, replace the sensor.

Air Charge Temperature (ACT) Sensor

LOCATION

See Figure 3.

OPERATION

Approximately 5–volts are applied to the Air Charge Temperature (ACT) Sensor output terminal from the ECM/PCM via the resistor in the ECM/PCM. The ground terminal is grounded with ECM/PCM. The sensor is a negative temperature coefficient type of resistor. When the intake air temperature rises, the resistance decreases. The sensor output voltage increases when the resistance increases and decreases when the resistance decreases.

REMOVAL & INSTALLATION

1. Disconnect the negative battery cable.
2. Replacing the sensor requires disconnecting the electrical connector, then carefully removing the lid of the air filter housing.

➡ **Handle the sensor assembly carefully, protecting it from impact, extremes of temperature and/or exposure to shop chemicals.**

3. Installation is the reverse of the removal procedure.

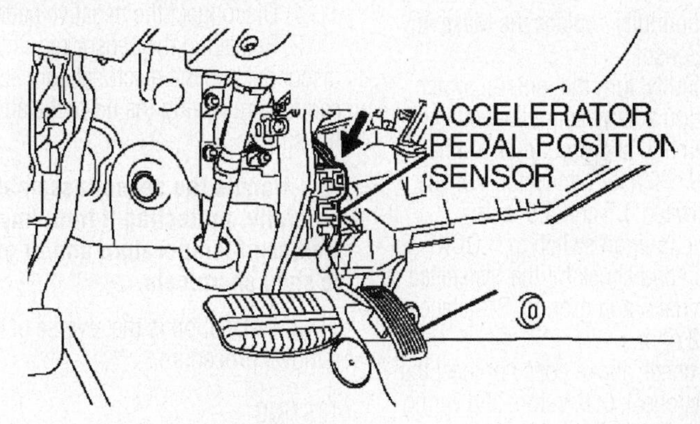

Fig. 1 Accelerator Pedal Position Sensor

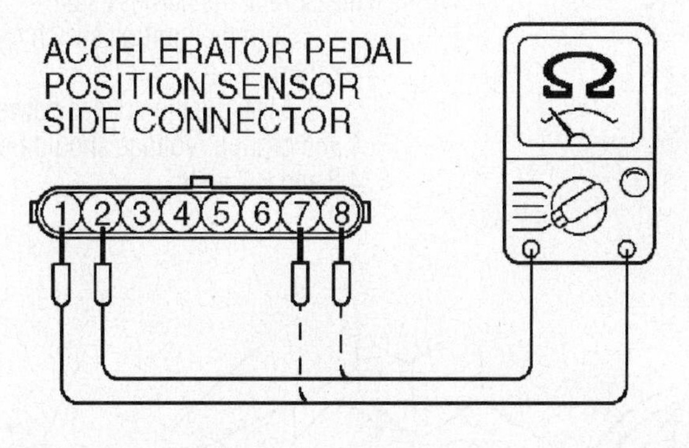

Fig. 2 Accelerator Pedal Position Connector

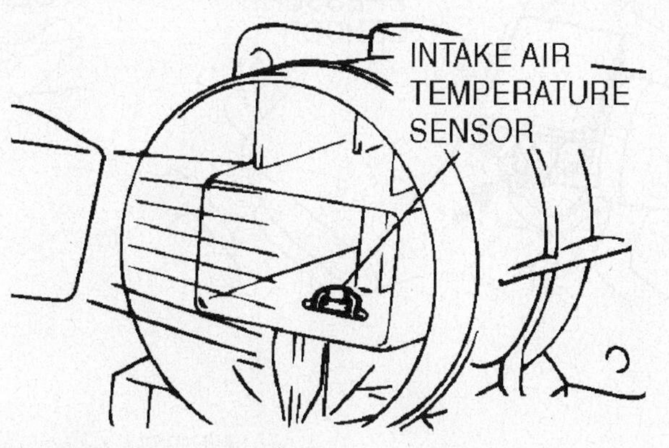

Fig. 3 Intake Air Temperature Sensor

TESTING

1. Disconnect the sensor connector.

2. Measure the resistance between the sensor side connector terminals 5 and 6.

Standard value:
5.3–6.7 Kohms at 32 °F
2.3–3.0 Kohms at 68 °F
1.0–1.5 Kohms at 68 °F
0.30–0.42 Kohms at 176 °F

3. Measure resistance while heating the sensor using warm, dry air. As the temperature rises, the resistance value should drop. If within specifications, replace the ECM or PCM.

➡ **Check that the circuit is not Open circuit.**

4. If no continuity, replace the Mass Air Flow (MAF) Sensor.

5. If continuity, turn the ignition switch to "ON" position and measure the voltage between terminal 6 and ground on the harness side of the sensor connector. Voltage should be between 4.5 and 4.9 volts.

6. Turn the ignition switch to "LOCK" (OFF) position and check for the resistance between terminal 5 and ground. Resistance be less than 2 ohm.

7. If any check above does not meet the specifications, check connectors and wiring between the sensor and ECM/PCM. If ok, replace ECM/PCM.

Barometric Pressure (BARO) Sensor

LOCATION

See Figure 4.

OPERATION

A 5–volt reference and ground is supplied to the Barometric Pressure Sensor from the ECM/PCM. A sensor voltage that is proportional to the atmospheric pressure is sent to the ECM/PCM.

REMOVAL & INSTALLATION

1. Disconnect the negative battery cable.

2. Replacing the sensor requires disconnecting the electrical connector, then carefully removing the lid of the air filter housing.

➡ **Handle the sensor assembly carefully, protecting it from impact, extremes of temperature and/or exposure to shop chemicals.**

3. Installation is the reverse of the removal procedure.

TESTING

1. Disconnect the sensor connector and measure at the harness side.

2. Turn the ignition switch to the "ON" position.

3. Measure the voltage between terminal 1 and ground. Voltage should be between 4.8 and 5.2 volts.

4. Turn the ignition switch to the "LOCK" (OFF) position.

5. Check for the resistance between terminal 5 and ground. Resistance should be less than 2 ohm.

6. If any check above does not meet the specifications, check connectors and wiring between the sensor and ECM/PCM. If ok, replace ECM/PCM.

7. If all checks above meet the specifications, disconnect sensor connector and, using a jumper test harness in between, turn the ignition switch to "ON" position and measure the voltage between terminal 2 and ground.

Standard value:
3.7–4.3 V at sea level.
3.4–4.0 V at 5,906 ft.
3.2–3.8 V at 3,937 ft.
2.9–3.5 V at 1,969 ft.

8. If not within specifications, replace the Mass Air Flow (MAF) Sensor.

Camshaft Position (CMP) Sensor

LOCATION

See Figure 5.

OPERATION

A 5–volt voltage is applied on the Camshaft Position (CMP) Sensor output terminal from the ECM/PCM. The sensor generates a pulse signal when the output terminal is opened and grounded.

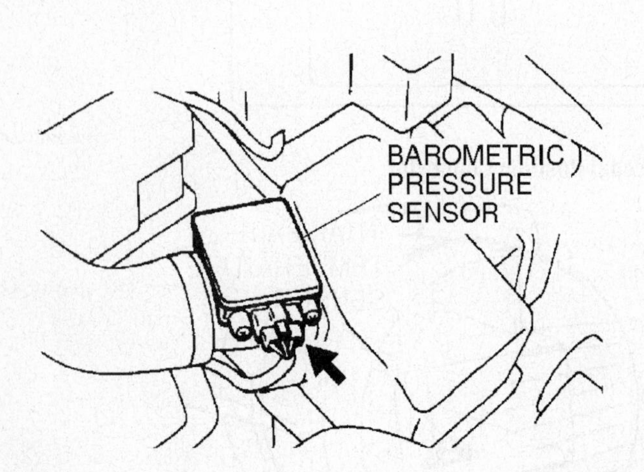

Fig. 4 Barometric Senso

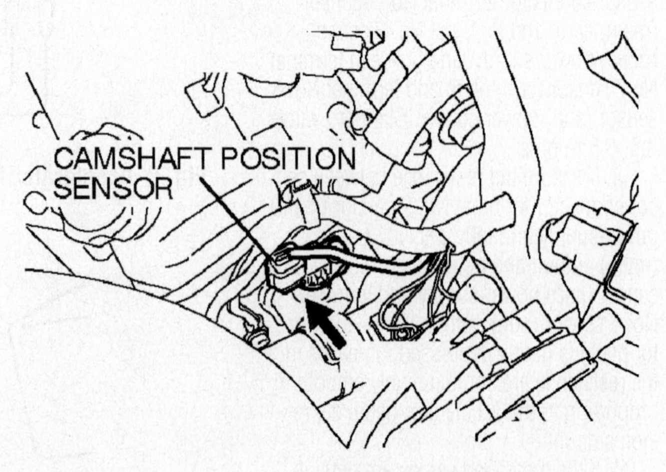

Fig. 5 Camshaft Position Sensor

REMOVAL & INSTALLATION

1995–98 Eclipse

2.0L NON-TURBO ENGINES

1. Disconnect the negative battery cable.
2. Detach the air intake hose.
3. Detach the camshaft position sensor connector.
4. Unfasten the retaining bolts, then remove the camshaft position sensor.
5. If necessary, remove the retaining bolts, then remove the target magnet.

To install:

6. If removed, install the target magnet by aligning the off-set tabs of the magnet with the off-set holes of the camshaft. Tighten the retaining bolts to 2 ft. lbs. (3 Nm).
7. Position the sensor, and secure with the retaining bolts. Tighten to 7 ft. lbs. (9 Nm).
8. Attach the sensor electrical connector.
9. Install the air intake hose, then connect the negative battery cable.

All Except 1995–1998 Eclipse

See Figure 6

1. Disconnect the negative battery cable.
2. Detach the sensor connector.
3. Unfasten the retaining bolt, then remove the sensor and the O-ring.
4. Installation is the reverse of the removal procedure.

TESTING

1. Disconnect the sensor connector and measure at the harness side.
2. Turn the ignition switch to the "ON" position.
3. Measure the voltage between terminal No. 2 and ground. Voltage should be between 4.8 and 5.2 volts.
4. Turn the ignition switch to the "LOCK" (OFF) position.
5. If not within specifications, check connectors and wiring between the sensor and ECM/PCM. Repair/replace as necessary.
6. If within specifications, disconnect the sensor connector and measure at the harness side.
7. Turn the ignition switch to the "ON" position.
8. Measure the voltage between terminal No. 3 and ground. Voltage should be battery positive voltage.
9. Turn the ignition switch to the "LOCK" (OFF) position.
10. If not within specifications, check connectors and wiring between the sensor and MFI relay. Repair/replace as necessary.
11. If within specifications, disconnect the sensor connector and measure at the harness side.
12. Check for the resistance between terminal No. 1 and ground. Resistance should be less than 2 ohms.
13. If not within specifications, check connectors and wiring between the sensor and ECM/PCM. If connectors and wiring between sensor and ECM/PCM are good, replace the ECM/PCM.
14. If within specifications, check the camshaft position sensing cylinder. Repair if necessary.
15. If the camshaft position sensing cylinder in good, replace the CMP sensor.

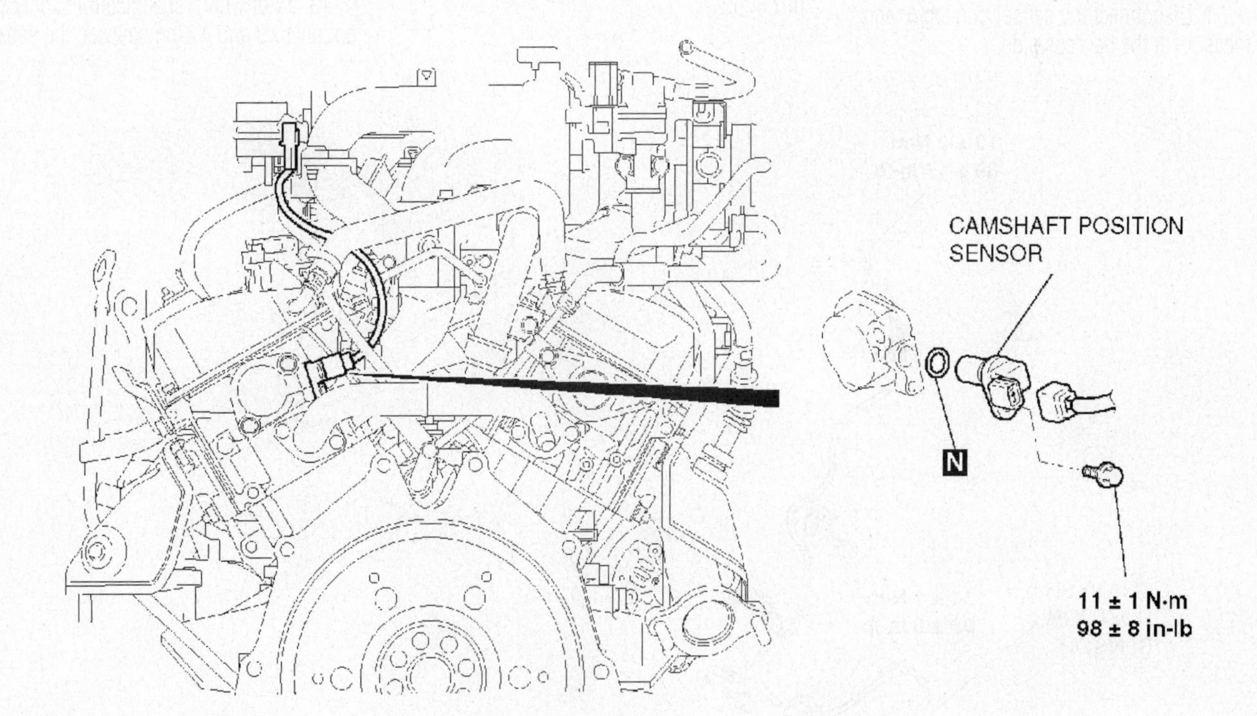

Fig. 6 Camshaft Position Sensor Removal & Installation

Crankshaft Position (CKP) Sensor

LOCATION

See Figure 7.

OPERATION

A 5–volt voltage is applied on the Crankshaft Position (CKP) Sensor output terminal from the ECM/PCM. The sensor generates a pulse signal when the output terminal is opened and grounded.

REMOVAL & INSTALLATION

See Figure 8

1. Disconnect the negative battery cable.
2. Detach the crankshaft position sensor electrical connector.
3. Unfasten the retaining bolt, then remove the sensor.

To install:

4. Position the sensor and install the retaining bolt.
5. Attach the crankshaft position sensor electrical connector.
6. Connect the negative battery cable.

TESTING

1. Disconnect the sensor connector and measure at the harness side.

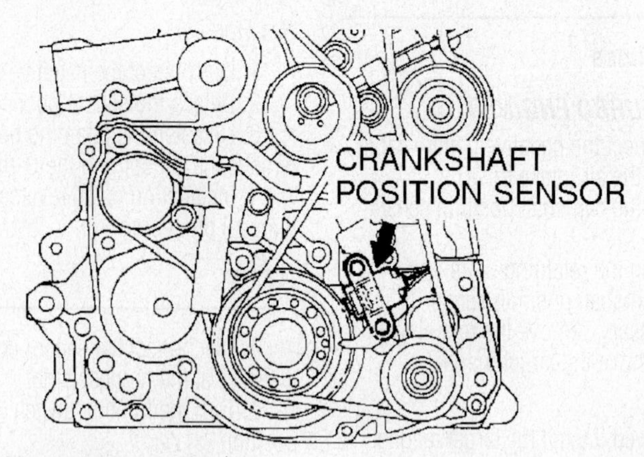

Fig. 7 Crankshaft Position Sensor

2. Turn the ignition switch to the "ON" position.
3. Measure the voltage between terminal No. 2 and ground. Voltage should be between 4.8 and 5.2 volts.
4. Turn the ignition switch to the "LOCK" (OFF) position.
5. If not within specifications, check connectors and wiring between the sensor and ECM/PCM. Repair/replace as necessary.
6. If within specifications, disconnect the sensor connector and measure at the harness side.
7. Turn the ignition switch to the "ON" position.
8. Measure the voltage between terminal No. 3 and ground. Voltage should be battery positive voltage.
9. Turn the ignition switch to the "LOCK" (OFF) position.
10. If not within specifications, check connectors and wiring between the sensor and MFI relay. Repair/replace as necessary.
11. If within specifications, disconnect the sensor connector and measure at the harness side.
12. Check for the resistance between terminal No. 1 and ground. Resistance should be less than 2 ohms.
13. If not within specifications, check connectors and wiring between the sensor

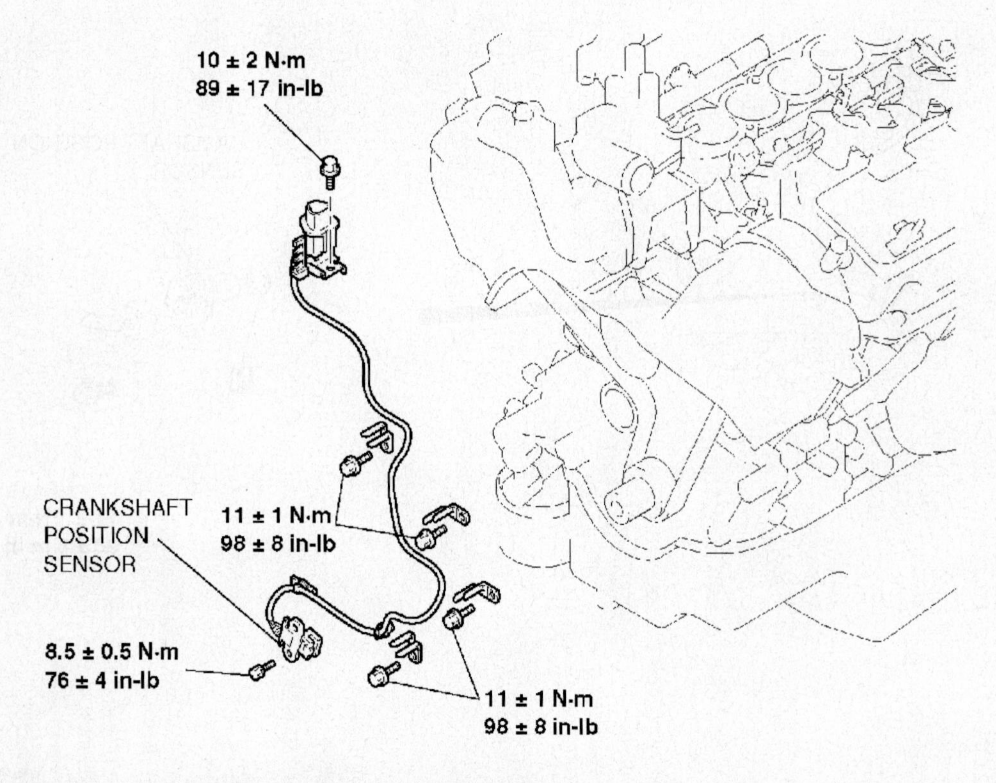

Fig. 8 Crankshaft Position Sensor Removal & Installation

and ECM/PCM. If connectors and wiring between sensor and ECM/PCM are good, replace the ECM/PCM.

14. If within specifications, check the CKP sensor–sensing blade.

15. If the sensor–sensing blade is damaged, replace it.

16. If the sensor–sensing blade is good, replace the CKP sensor.

Engine Control Unit (ECU)

GENERAL INFORMATION

➡ **When the term Electronic Control Unit (ECU) is, it will refer to the engine control computer regardless that it may be a Electronic Control Unit (ECU), Powertrain Control Module (PCM) or Engine Control Module (ECM).**

The heart of the electronic control system, which is found on the vehicles covered by this manual, is a computer control module. The module gathers information from various sensors, then controls fuel supply and engine emission systems. Most early model vehicles are equipped with an Engine Control Module (ECM) which, as its name implies, controls the engine and related emissions systems. Some ECMs may also control the Torque Converter Clutch (TCC) on automatic transaxle vehicles or the manual upshift light on manual transmission vehicles. Later model vehicles may be equipped with a Powertrain Control Module (PCM). This is similar to the original ECMs, but is designed to control additional systems as well. The PCM may control the manual transmission shift lamp or the shift functions of the electronically controlled automatic transmission.

Regardless of the name, all computer control modules are serviced in a similar manner. Care must be taken when handling these expensive components in order to protect them from damage. Carefully follow all instructions included with the replacement part. Avoid touching pins or connectors to prevent damage from static electricity.

**** WARNING**

To prevent the possibility of permanent control module damage, the ignition switch MUST always be OFF when disconnecting power from or reconnecting power to the module. This includes unplugging the module connector, disconnecting the negative battery cable, removing the module fuse or even attempting to jump your dead battery using jumper cables.

LOCATION

See Figure 9.

REMOVAL & INSTALLATION

1. Turn the ignition switch to the **OFF** position.
2. If the ECU is mounted under the dash, remove the left and/or right side panel from the center console, or remove the under dash panel.
3. Remove the bolts holding the ECU to the mounting bracket.
4. Disconnect the wiring harness from the ECU and remove ECU from the vehicle.

To install:

5. Connect the electrical harness to the ECU. Make certain the multi–pin connector is firmly and squarely seated to the ECU.
6. Install the ECU in the mounting bracket and secure in position.
7. If necessary, install the side panels to the center console or dash panel.
8. Connect the negative battery cable.

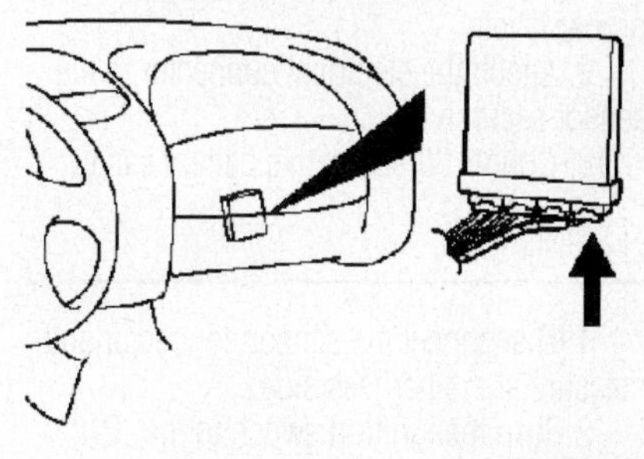

Fig. 9 Powertrain Control Module

Engine Coolant Temperature (ECT) Sensor

LOCATION

See Figure 10.

OPERATION

A 5–volt voltage is applied to the Engine Coolant Temperature (ECT) sensor output terminal from the ECM/PCM via the resistor in the ECM/PCM. The ground terminal is grounded with ECM/PCM. The engine coolant temperature sensor is a negative temperature coefficient type of resistor. It has the characteristic that when the engine coolant temperature rises the resistance decreases. The engine coolant temperature sensor output voltage increases when the resistance increases and decreases when the resistance decreases.

REMOVAL & INSTALLATION

1. Disconnect the negative battery cable.
2. Drain the engine coolant to a level below the intake manifold.
3. Unplug the sensor wiring harness.
4. Unplug the ECT sensor electrical connector
5. Unthread and remove the sensor from the engine.

Use a deep socket and an extension to reach the ECT sensor, remove the ECT sensor from the thermostat housing

To install:

6. Coat the threads of the sensor with a suitable sealant and thread into the housing.
7. Tighten the sensor to 22 ft. lbs. (30 Nm) for all except 1995–98 2.0L non-turbo engines. For 1995–98 2.0L non-turbo engines, tighten the sensor to 5 ft. lbs. (7 Nm).

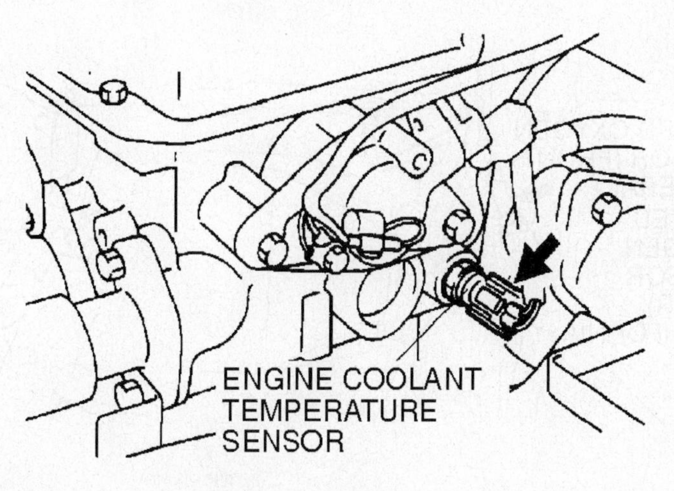

Fig. 10 Engine Coolant Temperature Sensor

8. Refill the cooling system to the proper level.
9. Attach the electrical connector to the sensor securely.
10. Connect the negative battery cable.

TESTING

1. Disconnect the sensor connector and measure at the harness side.
2. Turn the ignition switch to the "ON" position.
3. Measure the voltage between terminal 1 and ground. Voltage should be between 4.5 and 4.9 volts.
4. Turn the ignition switch to the "LOCK" (OFF) position.
5. Check for the resistance between terminal 2 and ground. Resistance should be less than 2 ohm.
6. If any check above does not meet the specifications, check connectors and wiring between the sensor and ECM/PCM. If ok, replace ECM/PCM.
7. If all checks above meet the specifications, measure the resistance between the sensor side connector terminals 1 and 2. There should be resistance (50–72 Kohms).

➡ **Check that the circuit is not open circuit.**

8. If no continuity, replace the sensor.
9. If sensor has continuity, remove the sensor.
10. With the temperature sensing portion of sensor immersed in hot water, check resistance.
Standard value:
5.1–6.5 Kohms at 32 °F
2.1–2.7 Kohms at 68 °F
0.9–1.3 Kohms [at 40 °C (68 °F)
0.26–0.36 Kohms [at 80 °C (176 °F)
11. If not within specifications, replace the sensor.

Heated Oxygen (HO2S) Sensor

LOCATION

See Figure 11.

OPERATION

A voltage corresponding to the Heated Oxygen (HO2S) Sensor concentration in the exhaust gas is sent to the ECM/PCM from the output terminal of the sensor.
The sensor is grounded with the ECM/PCM.

REMOVAL & INSTALLATION

**** CAUTION**

The temperature of the exhaust system is extremely high after the engine has been run. To prevent personal injury, allow the exhaust system to cool completely before removing sensor from the exhaust system.

1. Detach the negative battery cable.
2. Raise and safely support the vehicle.
3. Detach the electrical connector from the oxygen sensor.
4. Using socket MD998770, or equivalent oxygen sensor socket, remove the oxygen sensor.

To install:

5. If installing old oxygen sensor, coat the threads with anti-seize compound. New sensors are already coated. Take care not to contaminate the oxygen sensor probe with the anti-seize compound.
6. Install the oxygen sensor into the exhaust manifold. Tighten the sensor, using the correct tool, to 33 ft. lbs. (45 Nm)
7. Attach the wiring to the sensor.
8. Carefully lower the vehicle.
9. Connect the negative battery cable.

TESTING

1. Disconnect the Heated Oxygen Sensor connector and connect a test harness to the connector on the sensor side.
2. Start the engine and warm up the engine until engine coolant is 80 °C (176 °F) or higher.

➡ **Be very careful when connecting the jumper wires; incorrect connection can damage the sensor.**

3. Use the jumper wires to connect terminal 3 of the sensor connector to the positive battery terminal and terminal 4 to the negative battery terminal.
4. Connect a digital voltmeter between terminal 1 and terminal 2.
5. While repeatedly revving the engine, measure the sensor output voltage. Standard value: 0.6–1.0 V
6. If not within specifications, replace the sensor.
7. If within specifications, check connectors and wiring between the sensor and ECM/PCM.

Idle Air Control (IAC) Valve

LOCATION

See Figure 12.

OPERATION

The amount of air taken in during idling is regulated by the opening and closing of the servo valve located in the air passage that bypasses the throttle valve. The servo valve is opened or closed by the activation of the stepper motor (incorporated within the idle air control motor in the forward or

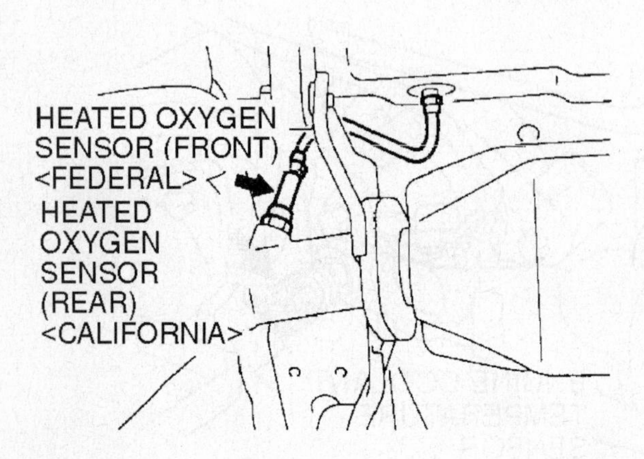

Fig. 11 Heated Oxygen Sensor

Fig. 12 Idle Air Control Valve

reverse direction. Battery positive voltage is supplied, by way of the MFI relay, to the coil of the stepper motor. The engine control module switches ON the power transistors (located within the engine control module) in sequential order, and, when current flows to the stepper motor coil, the stepper motor is activated in the forward or reverse direction.

REMOVAL & INSTALLATION

1. Disconnect the negative battery cable.
2. Detach the electrical connector at the sensor.
3. Remove the sensor from the throttle body.

To install:

4. Install the sensor in the opening in the throttle body and tighten the sensor.
5. Attach the electrical connector to the sensor.
6. Connect the negative battery cable.

TESTING

Checking Operation Sound

1. Check that the operating sound of the stepper motor can be heard over the idle air control motor when the ignition switch is turned to the ON position (without starting the engine).
2. If no operating sound can be heard, check the stepper motor drive circuit. (If the circuit is good, a defective stepper motor or engine control module is suspected.)

Checking Coil Resistance

1. Disconnect the idle air control motor connector and connect the jumper (test harness).
2. Measure the resistance between terminal (2) of the connector at the idle air control motor side and terminal (1) or terminal (3). Standard value: 28–33 ohms at 68 °F
3. Measure the resistance between terminal (5) of the connector at the idle air control motor side and terminal (6) or terminal (4). Standard value: 28–33 ohms at 68 °F
4. If not within specifications, replace the IAC Valve.
5. If within specifications, check connectors and wiring between the sensor and ECM/PCM.

Intake Air Temperature (IAT) Sensor

LOCATION

See Figure 13.

OPERATION

Approximately 5–volts are applied to the Intake Air Temperature (IAT) Sensor output terminal from the PCM via the resistor in the PCM. The ground terminal is grounded with PCM. The sensor is a negative temperature coefficient type of resistor. When the intake air temperature rises, the resistance decreases. The sensor output voltage increases when the resistance increases and decreases when the resistance decreases.

REMOVAL & INSTALLATION

1. Disconnect the negative battery cable.
2. Replacing the sensor requires disconnecting the electrical connector, then carefully removing the lid of the air filter housing.

➡ **Handle the sensor assembly carefully, protecting it from impact, extremes of temperature and/or exposure to shop chemicals.**

3. Installation is the reverse of the removal procedure.

TESTING

1. Disconnect the sensor connector.
2. Measure the resistance between the sensor side connector terminals 1 and 4. Standard value:
Should be 13–17 Kohms at –4 °F
Should be 5.3–6.7 Kohms at 32 °F
Should be 2.3–3.0 Kohms at 68 °F
Should be 1.0–1.5 Kohms at 104 °F
Should be 0.56–0.76 Kohms at 140 °F
Should be 0.30–0.45 Kohms at 176 °F
3. Measure resistance while heating the sensor using warm, dry air. As the temperature rises, the resistance value should drop. If within specifications, replace the ECM or PCM.

➡ **Check that the circuit is not Open circuit.**

4. If no continuity, replace the Mass Air Flow (MAF) Sensor.
5. If continuity, turn the ignition switch to "ON" position and measure the voltage between terminal 1 and ground on the harness side of the sensor connector. Voltage should be between 4.5 and 4.9 volts.
6. Turn the ignition switch to "LOCK" (OFF) position and check for the resistance between terminal 4 and ground. Resistance should be less than 2 ohm.
7. If any check above does not meet the specifications, check connectors and wiring between the sensor and ECM/PCM. If ok, replace ECM/PCM.

Knock Sensor (KS)

LOCATION

See Figure 14.

OPERATION

The Knock Sensor (KS) converts the vibration of the cylinder block into a voltage and outputs it. If there is a malfunction of the knock sensor, the voltage output will not change. The PCM checks whether the voltage output changes.

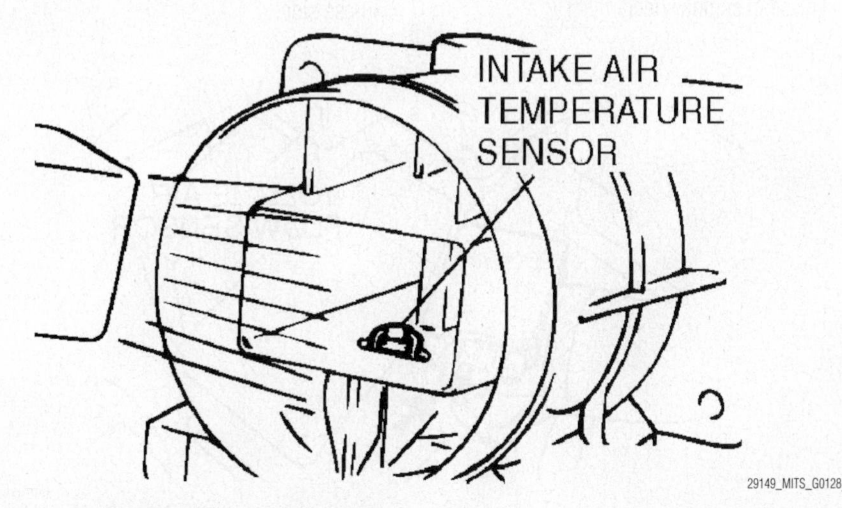

Fig. 13 Intake Air Temperature Sensor

REMOVAL & INSTALLATION

See Figure 15.

1. Disconnect the negative battery cable.
2. Detach the electrical connector at the sensor.
3. Remove the sensor from the engine block.

To install:

4. Install the knock sensor in the opening in the engine block and tighten the sensor retainer.
5. Attach the electrical connector to the sensor.
6. Connect the negative battery cable.

TESTING

1. Disconnect the knock sensor connector.
2. Start the engine and run at idle.
3. Measure the voltage between knock sensor side connector terminal No. 1 (output) and No. 2 (ground).
4. Gradually increase the engine speed. The voltage increases with the increase in the engine speed.
5. Turn the ignition switch to the "LOCK" (OFF) position.
6. If within specifications, check connectors and wiring between the sensor and ECM/PCM. If ok, replace ECM/PCM.
7. If not within specifications, replace the knock sensor.

Mass Air Flow (MAF) Sensor

LOCATION

See Figure 16.

OPERATION

A 5-volt power is applied to the Mass Air Flow (MAF) Sensor output terminal from the ECM/PCM. The sensor generates a pulse signal when the output terminal and ground are opened/closed (opened/short).

REMOVAL & INSTALLATION

1. Disconnect the negative battery cable.
2. Replacing the sensor requires disconnecting the electrical connector, then carefully removing the lid of the air filter housing.

➡ **Handle the sensor assembly carefully, protecting it from impact, extremes of temperature and/or exposure to shop chemicals.**

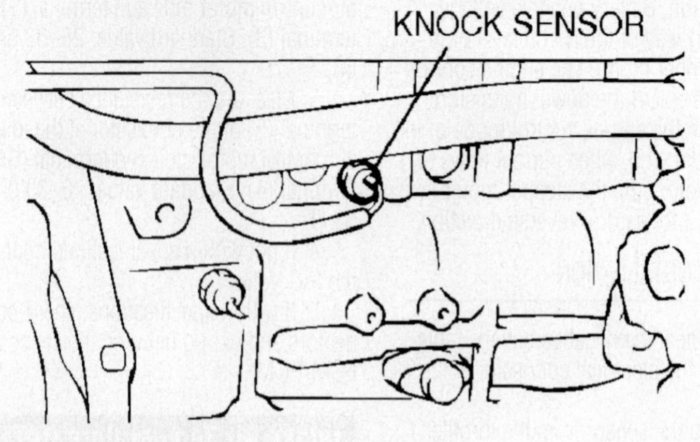

Fig. 14 Knock Sensor

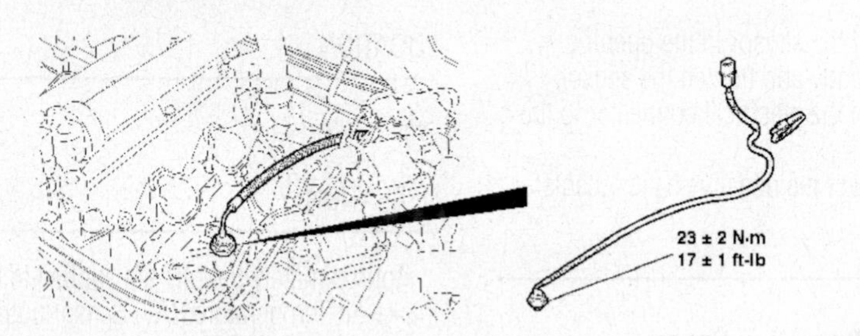

Fig. 15 Knock Sensor Removal & Installation

3. Installation is the reverse of the removal procedure.

TESTING

1. Connect a suitable scan tool to read data.
2. Start the engine and run at idle. Warm up the engine to normal operating temperature: 176 ° to 205 °F.
   a. The standard value during idling should be 10 Hz or more.
   b. When the engine is revved, the frequency should increase according to the increase in engine speed.
3. If the checks above do not meet the specification, disconnect the sensor connector and measure at the harness side.
4. Turn the ignition switch to "ON" position, and measure the voltage between terminal 4 and ground. Voltage should be battery positive voltage.
5. Turn the ignition switch to "LOCK" (OFF) position.
6. If not within specifications, check and repair/replace the connector(s), harness wire between MFI relay and/or the MFI relay.
7. If within specifications, disconnect the sensor connector and measure at the harness side.

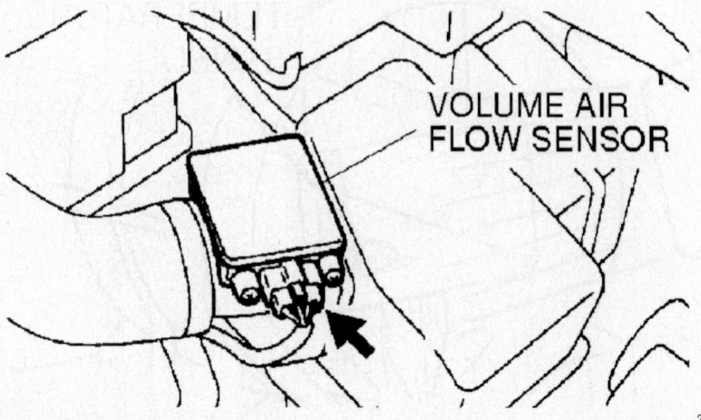

Fig. 16 Mass Air Flow Sensor

8. Turn the ignition switch to "ON" position and measure the voltage between terminal 3 and ground. Voltage should be between 4.8 and 5.2 volts.

9. Turn the ignition switch to "LOCK" (OFF) position and check for the resistance between terminal 5 and ground. Resistance be less than 2 ohm.

10. If all checks above meet the specifications, replace the sensor.

11. If any check above does not meet the specifications, replace the ECM/PCM.

Manifold Absolute Pressure (MAP) Sensor

LOCATION

See Figure 17.

OPERATION

A 5–volt voltage is supplied to the Manifold Absolute Pressure (MAP) Sensor power terminal from the PCM. The ground terminal is grounded through the PCM. A voltage that is proportional to the intake manifold pressure is sent to the PCM the sensor output terminal.

REMOVAL & INSTALLATION

1. Disconnect the negative battery cable.
2. Detach the electrical connector at the sensor.
3. Remove the sensor.
4. Installation is the reverse of the removal procedure.

TESTING

1. Disconnect the MAP sensor connector.
2. Turn the ignition switch to the "ON" position.
3. Measure the voltage between terminal No. 3 of the MAP sensor harness connector and ground. Voltage should be between 4.9 and 5.1 volts.
4. Measure the voltage between terminal No. 2 of the MAP sensor harness connector and ground. Voltage should be 0.5 volt or less.
5. Turn the ignition switch to the "LOCK" (OFF) position.
6. If any check above does not meet the specifications, check connectors and wiring between the sensor and ECM/PCM. Repair/replace as necessary. If ok, replace ECM/PCM.
7. If within specifications, reconnect the MAP sensor connector.
8. Turn the ignition switch to the "ON" position.
9. Measure the voltage between terminal No. 1 and ground by backprobing the MAP sensor connector. Voltage should measure 3.7 and 4.3 volts when altitude is 0 m (0 foot), 3.4 and 4.0 volts when altitude is 600 m (1,969 feet), 3.2 and 3.8 volts when altitude is 1,200 m (3,937 feet), 2.9 and 3.5 volts when altitude is 1,800 m (5,906 feet).
10. Turn the ignition switch to the "LOCK" (OFF) position.
11. If within specifications, check connectors and wiring between the MAP sensor and the ECM/PCM. If ok, replace the ECM/PCM.
12. If not within specifications, replace the MAP sensor.

Throttle Position Sensor (TPS)

LOCATION

See Figure 18.

OPERATION

A 5 volt power supply is applied to the Throttle Position Sensor (TPS) power terminal from the ECM/PCM. The ground terminal is grounded with ECM/PCM. When the throttle valve shaft is turned from the idle position to the fully opened position, the resistance between the TPS output terminal and the ground terminal would increase according to the rotation.

REMOVAL & INSTALLATION

1. Disconnect negative battery cable.
2. Detach the electrical connector from the throttle position sensor.
3. Remove the mounting screws from the sensor, being careful not to round the Phillips screw head.
4. Remove the sensor from the throttle body.

To install:

5. Install the throttle position sensor onto the throttle body and rotate the sensor counterclockwise on the throttle shaft and temporarily tighten the screws.
6. Connect the electrical harness to the sensor.
7. Tighten the retainer screws to 1.8 ft. lbs. (2.5 Nm).
8. Connect the negative battery cable.

TESTING

1. Disconnect the TPS connector.
2. Measure resistance between the TPS side connector terminals 1 and 4. Standard value: 3.5–6.5 Kohms
3. Measure resistance between the TPS side connector terminals 1 and 3. With test probes in place, move the throttle valve from the idle position to the full–open position.

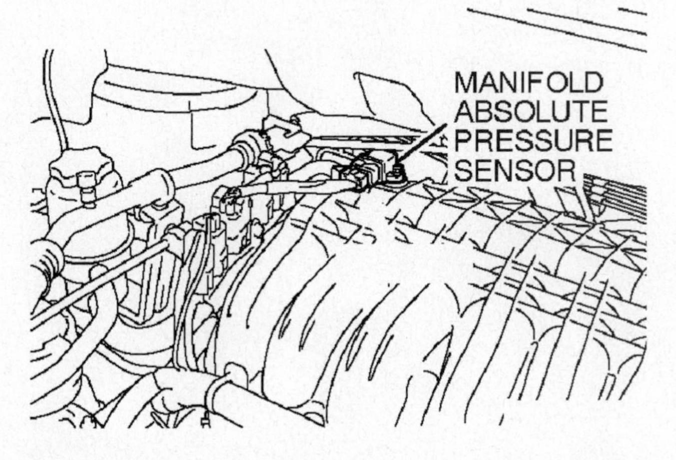

Fig. 17 Manifold Absolute Pressure Sensor

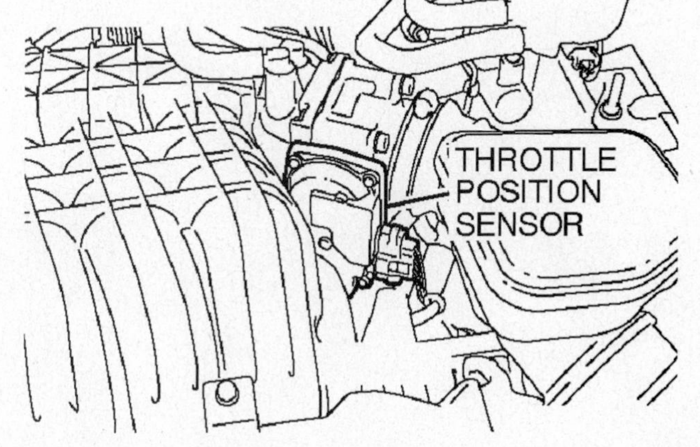

Fig. 18 Throttle Position Sensor

Resistance should change smoothly in proportion to the opening angle of the throttle valve.

4. If any check above does not meet the specifications, replace the TPS.

5. If all checks above meet the specifications, disconnect the sensor connector and measure at the harness side.

6. Turn the ignition switch to "ON" position and measure the voltage between terminal 4 and ground. Voltage should be between 4.8 and 5.2 volts.

7. Turn the ignition switch to "LOCK" (OFF) position and check for the resistance between terminal 1 and ground. Resistance should be less than 2 ohm.

8. If any check above does not meet the specifications, check connectors and wiring between the sensor and ECM/PCM. If ok, replace ECM/PCM.

Vehicle Speed Sensor (VSS)

LOCATION

See Figure 19.

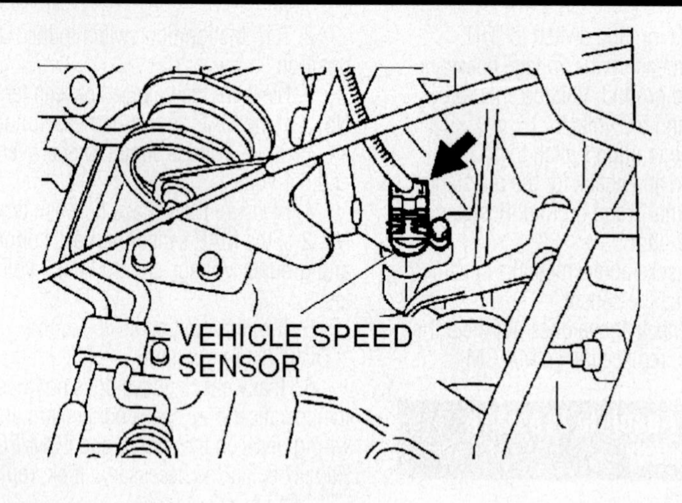

Fig. 19 Vehicle Speed Sensor

OPERATION

1. A 5 volt voltage is applied to the Vehicle Speed Sensor (VSS) output terminal from the ECM/PCM. The sensor generates a pulse signal when the output terminal is opened and grounded.

TESTING

1. Disconnect the sensor connector and measure at the harness side.

2. Turn the ignition switch to "ON" position.

3. Measure the voltage between terminal 86 (M/T) or terminal 80 (A/T) and ground. Voltage should be between 8 and 12 volts.

4. Turn the ignition switch to "LOCK" (OFF) position.

5. If any check above does not meet the specifications, check connectors and wiring between the sensor and ECM/PCM. If ok, replace ECM/PCM.

6. If checks above meet the specifications, replace the VSS.

NISSAN
DIAGNOSTIC TROUBLE CODES

16

TABLE OF CONTENTS

VEHICLE APPLICATIONS 16-2
P0XXX 16-3
P1XXX 16-19

DIAGNOSTIC TROUBLE CODES

OBD II VEHICLE APPLICATIONS

NISSAN

200SX
1996-1998
1.6L I4 .... VIN A
1996-1998
2.0L I4 .... VIN B

240SX
1996-1998
2.4L I4 .... VIN A

300ZX
1996
3.0L V6.......... VIN C
1996
3.0L V6.......... VIN R

350ZX
2003
3.5L V6.......... VIN A

Altima
1996-1997
2.4L I4.......... VIN B
1998-2001
2.4L I4 .... VIN D
2002-2003
2.5L I4 .... VIN A
2002-2003
3.5L V6.......... VIN B

Altra EV
1999-2000
This is an electric vehicle and as such has no VIN code.

Frontier
1998-2003
2.4L I4 .... VIN D
1999-2003
3.3L V6.......... VIN E
2001-2003
3.3L V6.......... VIN M

Maxima
1996-2001
3.0L V6.......... VIN C
2002-2003
3.5L V6.......... VIN D

Murano
2003
3.5L V6.......... VIN D

Pathfinder
1996-2000
3.3L V6.......... VIN A
2001-2003
3.5L V6.......... VIN D

Pickup
1996-1997
2.4L I4 .... VIN S
1996
3.0L V6.......... VIN H

Quest
1996-1998
3.0L V6.......... VIN D
1999-2002
3.3L V6.......... VIN T

Sentra
1996-2000
1.6L I4 .... VIN A
2000-2003
1.8L I4 .... VIN C
1998-2001
2.0L I4 .... VIN B
2002-2003
2.5L I4 .... VIN A

Xterra
2000-2003
2.4L I4 .... VIN D
2000-2003
3.3L V6 .... VIN E
2002-2003
3.3L V6..........

DTC	Trouble Code Title, Conditions & Possible Causes
DTC: P0157 **2T CCM, MIL: Yes** **1996-2002** **Models:** Altima, Frontier, Pathfinder, Xterra **Engines:** 3.3L VIN A, 3.3L VIN E, 3.3L VIN M, 3.5L VIN B, 3.5L VIN D **Transmissions:** All	**HO2S-22 (Bank 2 Sensor 2) Circuit Low Input (Lean Shift)** Engine running in closed loop at a speed of over 20 mph for at least 20 seconds, IAT sensor signal from 14-122°F, fuel level over 25%, and the PCM detected the HO2S signal did not reach a maximum voltage level of 0.60v, or it remained fixed at approximately 300 mv. **Possible Causes:** • Low fuel pressure, fuel filter restricted or fuel injectors plugged • HO2S may be contaminated or it has failed • HO2S heater is damaged or has failed • PCM has failed
DTC: P0158 **2T CCM, MIL: Yes** **1996-2002** **Models: 300ZX, Maxima** **Engines:** All **Transmissions:** All	**HO2S-22 (Bank 2 Sensor 2) Circuit High Input (Rich Shift)** Engine running in closed loop at a speed of over 20 mph for at least 20 seconds, IAT sensor signal from 14-122°F, fuel level over 25%, and the PCM detected the minimum HO2S signal was 600 mv. **Possible Causes:** • Fuel pressure regulator leaking or fuel injectors leaking • HO2S may be contaminated or it has failed • HO2S heater is damaged or has failed • PCM has failed
DTC: P0158 **2T CCM, MIL: Yes** **1996-2002** **Models:** Altima, Frontier, Pathfinder, Xterra **Engines:** 3.3L VIN A, 3.3L VIN E, 3.3L VIN M, 3.5L VIN B, 3.5L VIN D **Transmissions:** All	**HO2S-22 (Bank 2 Sensor 2) Circuit High Input (Rich Shift)** Engine running in closed loop at a speed of over 20 mph for at least 20 seconds, IAT sensor signal from 14-122°F, fuel level over 25%, and the PCM detected the minimum HO2S signal was 600 mv. **Possible Causes:** • Fuel pressure regulator leaking or fuel injectors leaking • HO2S may be contaminated or it has failed • HO2S heater is damaged or has failed • PCM has failed
DTC: P0159 **2T O2S, MIL: Yes** **1996-2002** **Models:** 300ZX, Maxima **Engines:** All **Transmissions:** All	**HO2S-22 (Bank 2 Sensor 2) Slow Response** Engine running in closed loop at over 20 mph for 20 seconds, IAT sensor signal from 14-122°F, fuel level over 25%, and the PCM detected the average HO2S response time was more than 1 second. **Possible Causes:** • HO2S signal circuit is open or shorted to ground • HO2S element is contaminated, or HO2S heater has failed • Intake air leaks, exhaust manifold leaks or PCV system leaks • MAF sensor out of calibration (it may be dirty or contaminated)
DTC: P0159 **2T O2S, MIL: Yes** **1996-2002** **Models:** Altima, Frontier, Pathfinder, Xterra **Engines:** 3.3L VIN A, 3.3L VIN E, 3.3L VIN M, 3.5L VIN B, 3.5L VIN D **Transmissions:** All	**HO2S-22 (Bank 2 Sensor 2) Slow Response** Engine running in closed loop at over 20 mph for 20 seconds, IAT sensor signal from 14-122°F, fuel level over 25%, and the PCM detected the average HO2S response time was more than 1 second. **Possible Causes:** • HO2S signal circuit is open or shorted to ground • HO2S element is contaminated, or HO2S heater has failed • Intake air leaks, exhaust manifold leaks or PCV system leaks • MAF sensor out of calibration (it may be dirty or contaminated)
DTC: P0160 **2T O2S, MIL: Yes** **1996-2002** **Models:** 300ZX, Maxima **Engines:** All **Transmissions:** All	**HO2S-22 (Bank 2 Sensor 2) Insufficient Activity Detected** Engine running in closed loop at a speed of over 20 mph for at least 20 seconds, IAT sensor signal from 14-122°F, fuel level over 25%, and the PCM detected the HO2S signal was fixed under 300 mv, or the HO2S signal was fixed over 600 mv for 10 seconds, or the HO2S signal switched from rich to lean less than 5 times in 10 seconds. **Possible Causes:** • HO2S element is contaminated • HO2S heater is damaged or has failed • HO2S signal circuit is open or shorted to ground • HO2S is damaged or has failed • TSB 01-016 (9/01) contains information related to this code

DTC	Trouble Code Title, Conditions & Possible Causes
DTC: P0160 **2T O2S, MIL: Yes** **1996-2002** **Models:** Altima, Frontier, Pathfinder, Xterra **Engines:** 3.3L VIN A, 3.3L VIN E, 3.3L VIN M, 3.5L VIN B, 3.5L VIN D **Transmissions:** All	**HO2S-22 (Bank 2 Sensor 2) Insufficient Activity Detected** Engine running in closed loop at a speed of over 20 mph for at least 20 seconds, IAT sensor signal from 14-122°F, fuel level over 25%, and the PCM detected the HO2S signal was fixed under 300 mv, or the HO2S signal was fixed over 600 mv for 10 seconds, or the HO2S signal switched from rich to lean less than 5 times in 10 seconds. **Possible Causes:** • HO2S element is contaminated • HO2S heater is damaged or has failed • HO2S signal circuit is open or shorted to ground • HO2S is damaged or has failed
DTC: P0160 **2T O2S, MIL: Yes** **1996-2002** **Models:** 300ZX, Maxima **Engines:** All **Transmissions:** All	**HO2S-22 (Bank 2 Sensor 2) Heater Circuit Malfunction** Engine running in closed loop at less than 3000 rpm, and the PCM detected an unexpected voltage on the HO2S heater circuit. **Note: The current level of the HO2S circuit was too high or too low.** **Possible Causes:** • HO2S heater control circuit is open or shorted to ground • HO2S heater control circuit is shorted to power • HO2S heater is damaged or has failed • PCM has failed
DTC: P0161 **2T O2S, MIL: Yes** **1996-2002** **Models:** Altima, Frontier, Pathfinder, Xterra **Engines:** 3.3L VIN A, 3.3L VIN E, 3.3L VIN M, 3.5L VIN B, 3.5L VIN D **Transmissions:** All	**HO2S-22 (Bank 2 Sensor 2) Heater Circuit Malfunction** Engine running in closed loop at less than 3000 rpm, and the PCM detected an unexpected voltage on the HO2S heater circuit. **Note: The current level of the HO2S circuit was too high or too low.** **Possible Causes:** • HO2S heater control circuit is open or shorted to ground • HO2S heater control circuit is shorted to power • HO2S heater is damaged or has failed • PCM has failed
DTC: P0171 **2T FUEL, MIL: Yes** **1996-2002** **Models:** All **Engines:** All **Transmissions:** All	**Fuel Trim Lean (Bank 1)** DTC P0100, P0105, P0110, P0115, P0120, P0130, P0131- P0141, P0150, P0151-P0161, P0301-P0304 or P0301-P0306, P0400, P0400 and P0600 not set, engine started, vehicle driven at cruise speed in closed loop, and the PCM detected the Bank 1 Fuel Trim had exceeded the lean correction limit during the Fuel System test. **Possible Causes:** • Air leaks after the MAF sensor, or in the EGR or PCV system • Base engine "mechanical" fault affecting one or more cylinders • Exhaust leaks before or near where the front HO2S is mounted • Fuel control sensor is out of calibration (i.e., ECT, IAT or MAP) • Fuel delivery system supplying too little fuel during cruise or idle periods (e.g., faulty fuel pump or dirty, restricted fuel filter) • Fuel injector (one or more) dirty or pressure regulator has failed • HO2S is contaminated, deteriorated or it has failed • Vehicle driven low on fuel or until it ran out of fuel
DTC: P0172 **2T FUEL, MIL: Yes** **1996-2002** **Models:** All **Engines:** All **Transmissions:** All	**Fuel Trim Rich (Bank 1)** DTC P0100, P0105, P0110, P0115, P0120, P0130, P0131- P0141, P0150, P0151-P0161, P0301-P0304 or P0301-P0306, P0400, P0400 and P0600 not set, engine started, vehicle driven at cruise speed in closed loop, and the PCM detected the Bank 1 Fuel Trim had exceeded the rich correction limit during the Fuel System test. **Possible Causes:** • Base engine "mechanical" fault affecting one or more cylinders • EVAP system component has failed or canister fuel saturated • Fuel control sensor is out of calibration (i.e., ECT, IAT or MAP) • Fuel delivery system supplying too much fuel during cruise or idle periods (e.g., faulty fuel pump, or faulty pressure regulator) • Fuel injector(s) is leaking or stuck partially open (one or more) • HO2S is contaminated, deteriorated or it has failed

DTC	Trouble Code Title, Conditions & Possible Causes
DTC: P0174 **2T FUEL, MIL: Yes** **1996-2002** **Models:** 300SZ, Altima, Frontier, Maxima, Pathfinder, Xterra **Engines:** 3.0L VIN C, 3.0L VIN R, 3.3L VIN A, 3.3L VIN E, 3.3L VIN M, 3.5L VIN B, 3.5L VIN D **Transmissions:** All	**Fuel Trim Lean (Bank 2)** DTC P0100, P0105, P0110, P0115, P0120, P0130, P0131- P0141, P0150, P0151-P0161, P0301-P0304 or P0301-P0306, P0400, P0400 and P0600 not set, engine started, vehicle driven at cruise speed in closed loop, and the PCM detected the Bank 1 Fuel Trim had exceeded the lean correction limit during the Fuel System test. **Possible Causes:** • Air leaks after the MAF sensor, or in the EGR or PCV system • Base engine "mechanical" fault affecting one or more cylinders • Exhaust leaks before or near where the front HO2S is mounted • Fuel control sensor is out of calibration (i.e., ECT, IAT or MAP) • Fuel delivery system supplying too little fuel during cruise or idle periods (e.g., faulty fuel pump or dirty, restricted fuel filter) • Fuel injector (one or more) dirty or pressure regulator has failed • HO2S is contaminated, deteriorated or it has failed • Vehicle driven low on fuel or until it ran out of fuel
DTC: P0175 **2T FUEL, MIL: Yes** **1996-2002** **Models:** 300SZ, Altima, Frontier, Maxima, Pathfinder, Xterra **Engines:** 3.0L VIN C, 3.0L VIN R, 3.3L VIN A, 3.3L VIN E, 3.3L VIN M, 3.5L VIN B, 3.5L VIN D **Transmissions:** All	**Fuel Trim Rich (Bank 2)** DTC P0100, P0105, P0110, P0115, P0120, P0130, P0131- P0141, P0150, P0151-P0161, P0301-P0304 or P0301-P0306, P0400, P0400 and P0600 not set, engine started, vehicle driven at cruise speed in closed loop, and the PCM detected the Bank 1 Fuel Trim had exceeded the rich correction limit during the Fuel System test. **Possible Causes:** • Base engine "mechanical" fault affecting one or more cylinders • EVAP system component has failed or canister fuel saturated • Fuel control sensor is out of calibration (i.e., ECT, IAT or MAP) • Fuel delivery system supplying too much fuel during cruise or idle periods (e.g., faulty fuel pump, or faulty pressure regulator) • Fuel injector(s) is leaking or stuck partially open (one or more) • HO2S is contaminated, deteriorated or it has failed
DTC: P0180 **2T CCM, MIL: Yes** **1996-2002** **Models:** All **Engines:** All **Transmissions:** All	**Fuel Tank Temperature Sensor Circuit Malfunction** Key on for over 5 seconds, and the PCM detected the Fuel Tank Temperature signal was too high or tool low, or that the value was not plausible when compared to the ECT and IAT sensor signals. **Note: The fuel tank temperature sensor should read 3.5v at 68°F.** **Possible Causes:** • Fuel tank temperature sensor signal open or shorted to ground • Fuel tank temperature sensor is damaged or has failed • PCM has failed
DTC: P0217 **1T CCM, MIL: Yes** **2000-02** **Models:** All **Engines:** All **Transmissions:** All	**Engine Over-Temperature** Engine running in closed loop, IAT sensor signal from 14-122°F, and the PCM detected the ECT sensor signal indicated that the engine temperature was too high under low engine load conditions. **Possible Causes:** • Cooling fan control circuit open or shorted to ground • Cooling system problems (low coolant, thermostat closed) • Blocked air passages at front of the vehicle (recent damage?) • Blocked or restricted radiator passages
DTC: P0300 **2T MISFIRE, MIL: Yes** **1996-2002** **Models:** All **Engines:** All **Transmissions:** All	**Multiple Misfire Detected** DTC P0100, P0105, P0110, P0120, P0335, P0340 and P0500 not set, Engine speed from 400-3500 rpm, VSS indicating over 3 mph, and the PCM detected irregular CKP sensor signals that indicated a multiple misfire condition present in two or more cylinders during the 200-rpm (Catalyst) or 1000 revolution (High Emissions) Misfire Test. Note: If the misfire is severe, the MIL will flash on/off on the 1st trip! **Possible Causes:** • Air leak in the intake manifold, or in the EGR or PCM system • Base engine problem affecting two or more cylinders • CMP or CKP sensor signals erratic or out of phase • EGR valve stuck open, or EVAP purge system has failed • Fuel delivery component fault that affects more than 1 cylinder (i.e., contaminated, dirty or sticking fuel injectors) • Ignition system problem affecting two or more cylinders • Vehicle driven while quite low or until it ran out of fuel

DTC	Trouble Code Title, Conditions & Possible Causes
DTC: P0301 **2T MISFIRE, MIL: Yes** **1996-2002** **Models:** All **Engines:** All **Transmissions:** All	**Cylinder 1 Misfire Detected** DTC P0100, P0105, P0110, P0120, P0335, P0340 and P0500 not set, Engine speed from 400-3500 rpm, VSS indicating over 3 mph, and the PCM detected irregular CKP sensor signals indicating that a multiple misfire condition existed in Cylinder 1 during the 200-rpm (Catalyst) or 1000 revolution (High Emissions) Misfire Test. Note: If the misfire is severe, the MIL will flash on/off on the 1st trip! **Possible Causes:** • Air leak in the intake manifold, or in the EGR or PCM system • Base engine mechanical fault that affects only one cylinder • Fuel delivery component fault that affects only one cylinder (i.e., a contaminated, dirty or sticking fuel injector) • Ignition system problem (coil or plug) that affects one cylinder • Vehicle driven while quite low or until it ran out of fuel
DTC: P0302 **2T MISFIRE, MIL: Yes** **1996-2002** **Models:** All **Engines:** All **Transmissions:** All	**Cylinder 2 Misfire Detected** DTC P0100, P0105, P0110, P0120, P0335, P0340 and P0500 not set, Engine speed from 400-3500 rpm, VSS indicating over 3 mph, and the PCM detected irregular CKP sensor signals indicating that a multiple misfire condition existed in Cylinder 2 during the 200-rpm (Catalyst) or 1000 revolution (High Emissions) Misfire Test. Note: If the misfire is severe, the MIL will flash on/off on the 1st trip! **Possible Causes:** • Air leak in the intake manifold, or in the EGR or PCM system • Base engine mechanical fault that affects only one cylinder • Fuel delivery component fault that affects only one cylinder (i.e., a contaminated, dirty or sticking fuel injector) • Ignition system problem (coil or plug) that affects one cylinder • Vehicle driven while quite low or until it ran out of fuel
DTC: P0303 **2T MISFIRE, MIL: Yes** **1996-2002** **Models:** All **Engines:** All **Transmissions:** All	**Cylinder 3 Misfire Detected** DTC P0100, P0105, P0110, P0120, P0335, P0340 and P0500 not set, Engine speed from 400-3500 rpm, VSS indicating over 3 mph, and the PCM detected irregular CKP sensor signals indicating that a multiple misfire condition existed in Cylinder 3 during the 200-rpm (Catalyst) or 1000 revolution (High Emissions) Misfire Test. Note: If the misfire is severe, the MIL will flash on/off on the 1st trip! **Possible Causes:** • Air leak in the intake manifold, or in the EGR or PCM system • Base engine mechanical fault that affects only one cylinder • Fuel delivery component fault that affects only one cylinder (i.e., a contaminated, dirty or sticking fuel injector) • Ignition system problem (coil or plug) that affects one cylinder • Vehicle driven while quite low or until it ran out of fuel
DTC: P0304 **2T MISFIRE, MIL: Yes** **1996-2002** **Models:** All **Engines:** All **Transmissions:** All	**Cylinder 4 Misfire Detected** DTC P0100, P0105, P0110, P0120, P0335, P0340 and P0500 not set, Engine speed from 400-3500 rpm, VSS indicating over 3 mph, and the PCM detected irregular CKP sensor signals indicating that a multiple misfire condition existed in Cylinder 4 during the 200-rpm (Catalyst) or 1000 revolution (High Emissions) Misfire Test. Note: If the misfire is severe, the MIL will flash on/off on the 1st trip! **Possible Causes:** • Air leak in the intake manifold, or in the EGR or PCM system • Base engine mechanical fault that affects only one cylinder • Fuel delivery component fault that affects only one cylinder (i.e., a contaminated, dirty or sticking fuel injector) • Ignition system problem (coil or plug) that affects one cylinder • Vehicle driven while quite low or until it ran out of fuel
DTC: P0305 **2T MISFIRE, MIL: Yes** **1996-2002** **Models:** 300SZ, Altima, Frontier, Maxima, Pathfinder, Xterra **Engines:** 3.0L VIN C, 3.0L VIN R, 3.3L VIN A, 3.3L VIN E, 3.3L VIN M, 3.5L VIN B, 3.5L VIN D **Transmissions:** All	**Cylinder 5 Misfire Detected** DTC P0100, P0105, P0110, P0120, P0335, P0340 and P0500 not set, Engine speed from 400-3500 rpm, VSS indicating over 3 mph, and the PCM detected irregular CKP sensor signals indicating that a multiple misfire condition existed in Cylinder 5 during the 200-rpm (Catalyst) or 1000 revolution (High Emissions) Misfire Test. Note: If the misfire is severe, the MIL will flash on/off on the 1st trip! **Possible Causes:** • Air leak in the intake manifold, or in the EGR or PCM system • Base engine mechanical fault that affects only one cylinder • Fuel delivery component fault that affects only one cylinder (i.e., a contaminated, dirty or sticking fuel injector) • Ignition system problem (coil or plug) that affects one cylinder • Vehicle driven while quite low or until it ran out of fuel

DTC	Trouble Code Title, Conditions & Possible Causes
DTC: P0306 **2T MISFIRE, MIL: Yes** **1996-2002** **Models:** 300SZ, Altima, Frontier, Maxima, Pathfinder, Xterra **Engines:** 3.0L VIN C, 3.0L VIN R, 3.3L VIN A, 3.3L VIN E, 3.3L VIN M, 3.5L VIN B, 3.5L VIN D **Transmissions:** All	**Cylinder 6 Misfire Detected** DTC P0100, P0105, P0110, P0120, P0335, P0340 and P0500 not set, Engine speed from 400-3500 rpm, VSS indicating over 3 mph, and the PCM detected irregular CKP sensor signals indicating that a multiple misfire condition existed in Cylinder 6 during the 200-rpm (Catalyst) or 1000 revolution (High Emissions) Misfire Test. Note: If the misfire is severe, the MIL will flash on/off on the 1st trip! **Possible Causes:** • Air leak in the intake manifold, or in the EGR or PCM system • Base engine mechanical fault that affects only one cylinder • Fuel delivery component fault that affects only one cylinder (i.e., a contaminated, dirty or sticking fuel injector) • Ignition system problem (coil or plug) that affects one cylinder • Vehicle driven while quite low or until it ran out of fuel
DTC: P0325 **1T CCM, MIL: No** **1996-2002** **Models:** All **Engines:** All **Transmissions:** All	**Knock Sensor Circuit Malfunction (Bank 1)** Key on or Engine running, system voltage over 10.0v, and the PCM detected the Knock sensor signal was too high or too low in the test. **Note: The Knock sensor signal will read 2.5v at idle if no fault exists.** **Possible Causes:** • Knock sensor signal circuit is open or shorted to ground • Knock sensor signal circuit is shorted to VREF or system power • Knock sensor is damaged or has failed • PCM has failed
DTC: P0335 **2T CCM, MIL: Yes** **1996-2002** **Models:** All **Engines:** All **Transmissions:** All	**Crankshaft Position Sensor Circuit Malfunction** Engine cranking for over 2 seconds, and the PCM did not detect a proper CKP sensor (1°) signal, or with Engine running, it did not detect a normal pattern of CKP sensor signals during the CCM test. **Possible Causes:** • CKP sensor signal circuit is open or shorted to ground • CKP sensor signal is shorted to VREF or system power • CKP sensor is damaged or has failed • PCM has failed
DTC: P0340 **2T CCM, MIL: Yes** **1996-2002** **Models:** All **Engines:** All **Transmissions:** All	**Camshaft Position Sensor Circuit Malfunction** Engine cranking for over 2 seconds, and the PCM did not detect any CMP sensor signals, or with Engine running, the PCM did not detect a normal pattern of CMP signals during the CCM test. **Possible Causes:** • CMP sensor signal circuit is open or shorted to ground • CMP sensor signal is shorted to VREF or system power • CMP sensor is damaged or has failed • PCM has failed
DTC: P0400 **2T EGR, MIL: Yes** **1996-2002** **Models:** All **Engines:** All **Transmissions:** All	**EGR System Recirculation Flow** Engine speed from 1952-2400 rpm in closed loop, ECT sensor signal more than 158°F, vehicle speed over 19-35 mph, and the PCM detected the EGR Temperature sensor signal indicated too little or too much EGR flow with the Dual EGR/EVAP solenoid switched "on" and "off" during the EGR Monitor test. **Possible Causes:** • EGR gas temperature sensor is damaged or has failed • EGR valve is stuck partially open or closed • Dual EGR/EVAP solenoid is damaged or has failed • Exhaust system is damaged or has collapsed • PCM has failed
DTC: P0402 **2T CCM, MIL: Yes** **1996-2002** **Models:** All **Engines:** All **Transmissions:** All	**EGRC-BPT Valve Function Conditions** Engine started, vehicle driven at a speed of 19-35 mph at 1952-2400 rpm in closed loop, ECT sensor more than 158°F, and the PCM detected a problem in the operation of EGRC-BPT valve or circuit. **Possible Causes:** • EGRC-BPT solenoid control circuit is open or shorted to ground • EGRC-BPT valve vacuum hose is clogged or disconnected • Orifice missing in vacuum hose between EGRC-BPT valve and the EGRC solenoid valve • Exhaust system is damaged or has collapsed

DTC	Trouble Code Title, Conditions & Possible Causes
DTC: P0403 **1T CCM, MIL: Yes** **2000-02** **Models:** Sentra **Engines:** All **Transmissions:** All	**EGR Volume Control Valve Circuit Malfunction** Engine running, vehicle driven to a speed of over 35 mph, system voltage over 10v, and the PCM detected an unexpected voltage condition on the EGR Volume Control valve circuit in the CCM test. **Possible Causes:** • EGR volume control valve circuit open or shorted to ground • EGR volume control valve circuit shorted to system power (B+) • EGR volume control valve is damaged or has failed • PCM has failed
DTC: P0420 **1T CAT, MIL: Yes** **1996-2002** **Models:** All **Engines:** All **Transmissions:** All	**Catalyst Efficiency Below Normal (Bank 1)** DTC P0100, P0105, P0110, P0115, P0120, P0130, P0131- P0141, P0301-P0304 or P0301-P0306, P0400, P0400 and P0600 not set, engine started, vehicle driven at 45-60 mph in closed loop for 3-5 minutes, ECT sensor signal over 158°F, and the PCM detected the rear HO2S-12 switch rate and amplitude was similar to the front HO2S-11 switch rate and amplitude for 3 seconds during the test. **Possible Causes:** • Air leaks at the exhaust manifold or in the exhaust pipes • Catalytic converter is damaged, contaminated or has failed • Front HO2S or rear HO2S is contaminated with fuel or moisture • Front HO2S and/or the rear HO2S is loose in the mounting hole • Front HO2S older (aged) than the rear HO2S (HO2S is lazy) • TSB 01-068B (03/01) contains information related to this code
DTC: P0430 **1T CAT, MIL: Yes** **1996-2002** **Models:** 300SZ, Altima, Frontier, Maxima, Pathfinder, Xterra **Engines:** 3.0L VIN C, 3.0L VIN R, 3.3L VIN A, 3.3L VIN E, 3.3L VIN M, 3.5L VIN B, 3.5L VIN D **Transmissions:** All	**Catalyst Efficiency Below Normal (Bank 2)** DTC P0100, P0105, P0110, P0115, P0120, P0130, P0131- P0141, P0301-P0304 or P0301-P0306, P0400, P0400 and P0600 not set, engine started, vehicle driven at 45-60 mph in closed loop for 3-5 minutes, ECT sensor signal over 158°F, and the PCM detected the rear HO2S-22 switch rate and amplitude was similar to the front HO2S-21 switch rate and amplitude for 3 seconds during the test. **Possible Causes:** • Air leaks at the exhaust manifold or in the exhaust pipes • Catalytic converter is damaged, contaminated or has failed • Front HO2S or rear HO2S is contaminated with fuel or moisture • Front HO2S and/or the rear HO2S is loose in the mounting hole • Front HO2S older (aged) than the rear HO2S (HO2S is lazy) • TSB 01-068B (03/01) contains information related to this code
DTC: P0440 **2T EVAP, MIL: Yes** **1996-2002** **Models:** All **Engines:** All **Transmissions:** All	**EVAP System Small Leak (0.040") Detected** DTC P1440 and P1448 not set, IAT sensor from 32-86°F, ECT sensor from 32-158°F, fuel level from 25-75%, and the PCM detected there was a leak in the EVAP system with the EVAP Purge solenoid commanded "on" (closed) during the EVAP leak test. Note: If DTC P1448 is set, repair the cause of this trouble code first. **Possible Causes:** • Fuel tank cap loose, damaged or missing, or vacuum line is off • EVAP purge solenoid is damaged or has failed • EVAP emission canister clogged or restricted • TSB 00-060 (7/00) contains information related to this code
DTC: P0443 **2T CCM, MIL: Yes** **1996-2002** **Models:** All **Engines:** All **Transmissions:** All	**EVAP Canister Purge Solenoid Circuit Malfunction** Engine started, Engine running at cruise speed under light engine load, system voltage from 11-16v, and the PCM detected an unexpected voltage condition on the Purge solenoid circuit, or it detected an invalid EVAP signal present when the Purge solenoid was commanded "on" and "off" during the CCM test. **Possible Causes:** • Purge solenoid control circuit open, shorted to ground or power • Purge solenoid is shorted to system power (B+) • Purge solenoid is damaged or has failed • PCM has failed • TSB 00-060 (7/00) contains information related to this code
DTC: P0446 **2T CCM, MIL: Yes** **1996-2002** **Models:** All **Engines:** All **Transmissions:** All	**EVAP Vent Control Solenoid Circuit Malfunction** Engine started, vehicle driven at cruise speed at light engine load, system voltage from 11-16v, and the PCM detected an unexpected voltage condition on the Vent Control solenoid circuit during the test. **Possible Causes:** • Vent solenoid circuit open, shorted to ground or to power (B+) • Vent control solenoid is damaged or has failed • PCM has failed • TSB 00-060 (7/00) contains information related to this code

DTC	Trouble Code Title, Conditions & Possible Causes
DTC: P0450 **2T CCM, MIL: Yes** **1998-2002** **Models:** All **Engines:** All **Transmissions:** All	**EVAP Pressure Sensor Circuit Malfunction** Engine running at idle speed, vehicle speed indicating 0 mph, EVAP purge commanded "on", and the PCM detected an unexpected condition on the Pressure sensor signal in the purge line in the test. **Possible Causes:** • EVAP pressure sensor circuit is open or shorted to ground • EVAP pressure sensor circuit is shorted to VREF or power (B+) • EVAP pressure sensor is damaged or has failed • PCM is damaged • TSB 00-060 (7/00) contains information related to this code
DTC: P0455 **2T EVAP, MIL: Yes** **2000-02** **Models:** All **Engines:** All **Transmissions:** All	**EVAP System Gross Leak (0.080") Detected** Engine started, vehicle driven at a speed of from 53-60 mph for 6-8 minutes, ECT sensor signal from 32-158°F, IAT sensor signal from 32-86°F, fuel level from 25-75%, and the PCM detected a large leak (larger than 0.080") between the fuel tank and the EVAP canister purge volume control solenoid during the EVAP Monitor leak test. **Possible Causes:** • Fuel tank cap is loose or damaged, or vacuum line is loose • EVAP purge solenoid is damaged or has failed • EVAP emission canister clogged or restricted • TSB 00-060 (7/00) contains information related to this code
DTC: P0460 **2T CCM, MIL: Yes** **2000-02** **Models:** All **Engines:** All **Transmissions:** All	**Fuel Level Sensor Slosh** Engine at idle speed for 30 seconds in closed loop, vehicle speed indicating 0 mph, and the PCM detected too much variation in the Fuel Level sensor input (indicating fuel slosh with the engine at idle). **Possible Causes:** • Fuel level sensor signal open or shorted to ground (intermittent) • Fuel level sensor is damaged or has failed • PCM has failed
DTC: P0461 **2T CCM, MIL: Yes** **2000-02** **Models:** All **Engines:** All **Transmissions:** All	**Fuel Level Sensor Performance** Engine running, then after the vehicle traveled a distance of more than 30 miles, the PCM did not detect any change in the Fuel Level sensor signal (i.e., no change after driving for several miles). **Possible Causes:** • Fuel level sensor signal open or shorted to ground • Fuel level sensor is damaged (stuck) or has failed • PCM has failed
DTC: P0464 **2T CCM, MIL: Yes** **1996-2002** **Models:** All **Engines:** All **Transmissions:** All	**Fuel Level Sensor Circuit Malfunction** Engine running for 5 seconds, and the PCM detected an unexpected voltage condition on the Fuel Level sensor circuit during the test. **Possible Causes:** • Fuel level sensor signal circuit is open or shorted to ground • Fuel level sensor signal circuit shorted to VREF or power (B+) • Fuel level sensor is damaged or has failed • PCM is damaged
DTC: P0500 **2T CCM, MIL: Yes** **1996-2002** **Models:** All **Engines:** All **Transmissions:** All	**Vehicle Speed Sensor Circuit Malfunction** Engine started, vehicle driven at over 30 mph under light engine load condition at an engine speed of over 1500 rpm for over 10 seconds, and the PCM did not detect any VSS signals during the CCM test. **Possible Causes:** • VSS signal circuit is open or shorted to ground • VSS signal circuit is shorted to VREF or system power (B+) • VSS is damaged or has failed • PCM has failed
DTC: P0505 **2T CCM, MIL: Yes** **1996-2002** **Models:** All **Engines:** All **Transmissions:** All	**Idle Air Control, Auxiliary Air Control Valve Circuit Malfunction** Engine at warm idle speed for 30 seconds, and the PCM detected the IAC control volume was incorrect, or after the IAC valve was commanded open and closed, it detected the fluctuation in air volume did not correlate to the engine air volume (determined by comparing the air volume to the volume of the MAF sensor signal). **Possible Causes:** • IACV-AAC valve control circuit is open or shorted to ground • IACV-AAC valve control circuit is shorted to system power (B+) • IACV-AAC valve is damaged or has failed • PCM is damaged or has failed

DTC	Trouble Code Title, Conditions & Possible Causes
DTC: P0510 **2T CCM, MIL: Yes** **1998-2002** **Models:** All **Engines:** All **Transmissions:** All	**Closed Throttle Position Switch Circuit Malfunction** Engine started, vehicle driven at a speed of 5-20 mph, then back to idle speed, and the PCM did not detect any change of status in the Closed Throttle Position switch circuit during the CCM test. **Possible Causes:** • Closed throttle position switch signal circuit is open or grounded • Closed throttle position switch signal circuit is shorted to power • Closed throttle position switch or TP sensor damaged or failed • PCM has failed
DTC: P0600 **2T PCM, MIL: No** **1996-2002** **Models:** All **Engines:** All **Transmissions: A/T**	**TCM (A/T) Communication Line Circuit Malfunction** Key on or Engine running, system voltage over 10v, and the PCM detected continuous incorrect communication data from the TCM. **Possible Causes:** • TCM communication line circuit is open or shorted to ground • TCM communication line circuit shorted to VREF or power (B+) • TCM or PCM has failed
DTC; P0605 **2T PCM, MIL: Yes** **1996-2002** **Models:** All **Engines:** All **Transmissions:** All	**PCM Internal Error** Engine started, engine runtime over 30 seconds, and the PCM detected an internal calculation function error. **Possible Causes:** • Clear the code and retest for the same code. If the same code reset, the PCM will have to be replaced.
DTC: P0705 **2T CCM, MIL: Yes** **1996-2002** **Models:** All **Engines:** All **Transmissions: A/T**	**Park Neutral Position Switch Circuit Malfunction** Engine started, vehicle driven to a speed of over 3 mph, and the PCM detected multiple P/N switch inputs, or it did not detect a change in the P/N switch status with the vehicle moving. **Possible Causes:** • P/N switch signal circuit is open or shorted to ground • P/N switch signal circuit is shorted to VREF or system power • P/N switch is damaged or has failed • PCM has failed
DTC: P0710 **2T CCM, MIL: Yes** **1996-2002** **Models:** All **Engines:** All **Transmissions: A/T**	**Transmission Fluid Temperature Sensor Circuit Malfunction** Vehicle drive a speed of over 6 mph in Drive with the TP signal over 1.2v, and the PCM detected the TFT sensor signal from the TCM was either too high or low during the CCM test. **Possible Causes:** • TFT sensor signal circuit is open or shorted to ground • TFT sensor signal circuit is shorted to VREF or system power • TFT sensor is damaged or has failed • PCM has failed
DTC: P0720 **2T CCM, MIL: Yes** **1996-2002** **Models:** All **Engines:** All **Transmissions: A/T**	**TCM Revolution Sensor Circuit Malfunction** Vehicle driven at a speed of 19-35 mph in Drive with the TP signal over 1.2v for 30 seconds, and the TCM did not detect any signals from the Revolution sensor during the CCM test. **Possible Causes:** • Revolution sensor signal circuit is open or shorted to ground • Revolution sensor signal circuit shorted to VREF or power (B+) • Revolution sensor is damaged or has failed • PCM has failed
DTC: P0725 **2T CCM, MIL: Yes** **1996-2002** **Models:** All **Engines:** All **Transmissions: A/T**	**TCM Engine Speed Signal** Vehicle driven at a speed of 19-35 mph in Drive with the TP signal over 1.2v for 30 seconds, and the TCM did not detect any Engine Speed signals during the CCM test. **Possible Causes:** • Engine speed signal is open or shorted to ground • Engine speed sensor signal shorted to VREF or power (B+) • Engine speed sensor is damaged or has failed • PCM has failed

DTC	Trouble Code Title, Conditions & Possible Causes
DTC: P0731 **2T CCM, MIL: Yes** **2000-02** **Models:** All **Engines:** All **Transmissions: A/T**	**A/T First Gear Circuit Malfunction** Engine running at cruise speed, and the TCM detected an incorrect voltage condition on the 1st Gear circuit during the CCM test. **Note: During this test, the CCM monitors the Actual gear position by checking the torque converter slip ratio. The slip ratio is calculated as this equation: A X C/B (where 'A' is the output shaft revolution signal, 'B' is the engine speed signal from the PCM, and 'C' is the gear ratio inferred TCM from other inputs. If the Actual gear ratio is higher than the gear position (1st2nd) inferred by the TCM, the slip ratio will be too high. If the slip ratio exceeds a certain value, the TCM determines that a fault exists and signals the PCM.** **Possible Causes:** • Shift Solenoid 'A' is stuck in open position • Shift Solenoid 'B' is stuck in open position
DTC: P0732 **2T CCM, MIL: Yes** **2000-02** **Models:** All **Engines:** All **Transmissions: A/T**	**A/T Second Gear Circuit Malfunction** Engine running at cruise speed, and the TCM detected an incorrect voltage condition on the 2nd Gear circuit during the CCM test. **Note: During this test, the CCM monitors the Actual gear position by checking the torque converter slip ratio. The slip ratio is calculated as this equation: A X C/B (where 'A' is the output shaft revolution signal, 'B' is the engine speed signal from the PCM, and 'C' is the gear ratio inferred TCM from other inputs. If the Actual gear ratio is higher than the gear position (2nd) inferred by the TCM, the slip ratio will be too high. If the slip ratio exceeds a certain value, the TCM determines that a fault exists and signals the PCM.** **Possible Causes:** • Control valve sticking or binding • Solenoid valve is damaged or not operating
DTC: P0733 **2T CCM, MIL: Yes** **2000-02** **Models:** All **Engines:** All **Transmissions: A/T**	**A/T Third Gear Circuit Malfunction** Engine running at cruise speed, and the TCM detected an incorrect voltage condition on the 3rd Gear circuit during the CCM test. **Note: During this test, the CCM monitors the Actual gear position by checking the torque converter slip ratio. The slip ratio is calculated as this equation: A X C/B (where 'A' is the output shaft revolution signal, 'B' is the engine speed signal from the PCM, and 'C' is the gear ratio inferred TCM from other inputs. If the Actual gear ratio is higher than the gear position (3rd) inferred by the TCM, the slip ratio will be too high. If the slip ratio exceeds a certain value, the TCM determines that a fault exists and signals the PCM.** **Possible Causes:** • Control valve sticking, solenoid valve damaged or not operating • Servo piston or brake band is damaged or is not operating
DTC: P0734 **2T CCM, MIL: Yes** **2000-02** **Models:** All **Engines:** All **Transmissions: A/T**	**A/T Fourth Gear Circuit Malfunction** Engine running at cruise speed, and the TCM detected an incorrect voltage condition on the 4th Gear circuit during the CCM test. **Note: During this test, the CCM monitors the Actual gear position by checking the torque converter slip ratio. The slip ratio is calculated as this equation: A X C/B (where 'A' is the output shaft revolution signal, 'B' is the engine speed signal from the PCM, and 'C' is the gear ratio inferred TCM from other inputs. If the Actual gear ratio is higher than the gear position (4th) inferred by the TCM, the slip ratio will be too high. If the slip ratio exceeds a certain value, the TCM determines that a fault exists and signals the PCM.** **Possible Causes:** • Oil pump is damaged, or the TCC is not operating • Shift Solenoid 'B' may be stuck closed
DTC: P0740 **2T CCM, MIL: Yes** **2000-02** **Models:** All **Engines:** All **Transmissions: A/T**	**A/T TCC Solenoid Circuit Malfunction** Engine running at cruise speed, and the TCM detected an unexpected voltage condition on the TCC solenoid control circuit. **Possible Causes:** • TCC solenoid control circuit open or shorted to ground • TCC solenoid control circuit shorted to system power (B+) • TCC solenoid is damaged or has failed • PCM has failed
DTC: P0744 **2T CCM, MIL: Yes** **2000-02** **Models:** All **Engines:** All **Transmissions: A/T**	**A/T Servo Valve Solenoid Circuit Malfunction** Engine running at cruise speed, and the TCM detected an open or shorted condition in Servo Valve solenoid circuit. **Possible Causes:** • Servo valve solenoid control circuit open or shorted to ground • Servo valve solenoid control circuit shorted to system power • Servo valve solenoid is damaged or has failed • PCM has failed

DTC	Trouble Code Title, Conditions & Possible Causes
DTC: P0745 **2T CCM, MIL: Yes** **2000-02** **Models:** All **Engines:** All **Transmissions: A/T**	**A/T Low Pressure Solenoid Circuit Malfunction** Engine running at cruise speed, and the TCM detected an open or shorted condition in Low Pressure solenoid circuit. **Possible Causes:** • Low pressure solenoid control circuit open or shorted to ground • Low pressure solenoid control circuit shorted to system power • Low pressure solenoid is damaged or has failed • PCM has failed
DTC: P0750 **2T CCM, MIL: No** **2000-02** **Models:** All **Engines:** All **Transmissions: A/T**	**A/T Shift Solenoid 'A' Circuit Malfunction** Engine running at cruise speed, and the TCM detected an open or shorted condition in Shift Solenoid 'A' circuit. **Possible Causes:** • Shift Solenoid 'A' control circuit open or shorted to ground • Shift Solenoid 'A' control circuit shorted to VREF or power (B+) • Shift Solenoid 'A' is damaged or has failed • PCM has failed
DTC: P0755 **2T CCM, MIL: No** **2000-02** **Models:** All **Engines:** All **Transmissions: A/T**	**A/T Shift Solenoid 'B' Circuit Malfunction** Engine running at cruise speed, and the TCM detected an open or shorted condition in Shift Solenoid 'B' circuit. **Possible Causes:** • Shift Solenoid 'B' control circuit open or shorted to ground • Shift Solenoid 'B' control circuit shorted to VREF or power (B+) • Shift Solenoid 'B' is damaged or has failed • PCM has failed

OBD II Trouble Code List (P1xxx Codes)

DTC	Trouble Code Title, Conditions & Possible Causes
DTC: P1108 **2T CCM, MIL: Yes** **2000-02** **Models:** Sentra **Engines:** 1.8L VIN C **Transmissions:** All	**Manifold Absolute Pressure Sensor Circuit Malfunction** Key on or Engine running, and the PCM detected an unexpected voltage condition on the MAP sensor signal during the CCM test. **Possible Causes:** • MAP sensor signal circuit is open or shorted to ground • MAP sensor signal circuit is shorted to VREF or system power • MAP sensor is damaged or has failed • PCM has failed
DTC: P1110 **2T CCM, MIL: Yes** **2000-02** **Models:** Sentra **Engines:** 1.8L VIN C **Transmissions:** All	**Intake Valve Timing Control Performance** Engine started, vehicle driven to over 4 mph under high engine load condition with the engine speed from 1100-4600 rpm, ECT sensor from 59-230°F, and the PCM detected an incorrect correlation when it compared the Intake Valve Timing solenoid "on" position to the correlate to the solenoid "off" position. **Possible Causes:** • Intake valve timing control position sensor or its circuit is open • Intake valve timing position control sensor is damaged or failed • Signal pickup portion of camshaft is contaminated with debris
DTC: P1111 **2T CCM, MIL: Yes** **2000-02** **Models:** Sentra **Engines:** 1.8L VIN C **Transmissions:** All	**Intake Valve Timing Control Circuit Malfunction** Engine running, and the PCM detected an unexpected voltage condition on the Intake Valve Timing Control solenoid circuit during the CCM test. **Possible Causes:** • Intake valve timing control solenoid circuit is open • Intake valve timing control solenoid circuit is shorted to ground • Intake valve timing position control solenoid is damaged/failed • PCM has failed
DTC: P1126 **2T ECT, MIL: Yes** **1996-2002** **Models:** All **Engines:** All **Transmissions:** All	**Thermostat Malfunction (Stuck Open)** DTC P0115 not set, engine started, IAT sensor more than 14°F and ECT sensor from 14-149°F at engine startup, engine runtime over 10 minutes, and the PCM detected that the engine temperature did not reach at least 176°F under these conditions. **Possible Causes:** • Inspect for low coolant level • Inspect for incorrect coolant mixture • Check the operation of the thermostat (it may be stuck open)
DTC: P1132 **2T CCM, MIL: Yes** **2000-02** **Models:** Sentra **Engines:** 1.8L VIN C **Transmissions:** All	**Swirl Control Valve Circuit Malfunction** Engine started, Engine running at cruise speed, then back to idle speed, and the PCM detected an unexpected "high" or "low" voltage signal on the SCV Control Position sensor circuit during the test. **Note: The SWL CV (B1) PID on the Scan Tool should read from 0-5 counts with the engine at idle speed and the ECT sensor less than 111°F; and it should read from 115-120 steps with the ECT sensor signal indicating a temperature of more than 113°F.** **Possible Causes:** • Swirl Control valve control circuit is open or shorted • Swirl Control valve control circuit is shorted to system power • Swirl Control valve is damaged or has failed • PCM has failed
DTC: P1137 **2T CCM, MIL: Yes** **2000-02** **Models:** Sentra **Engines:** 1.8L VIN C **Transmissions:** All	**Swirl Control Valve Control Position Sensor Circuit Malfunction** Engine started, Engine running at idle speed, and the PCM detected an unexpected "high" or "low" signal on the SCV Control Position sensor circuit during the CCM test. **Note: The SWL/C POSI SE PID on the Scan Tool should read close to 0 deg. with the engine at idle speed and the ECT sensor less than 111°F; and it should read close 80 deg. with the ECT sensor signal indicating a temperature of more than 113°F.** **Possible Causes:** • Swirl Control valve control position sensor open or shorted • Swirl Control valve control position sensor is damaged • Swirl Control valve control position sensor has failed • PCM has failed

DTC	Trouble Code Title, Conditions & Possible Causes
DTC: P1138 **2T CCM, MIL: Yes** **2000-02** **Models:** Sentra **Engines:** 1.8L VIN C **Transmissions:** All	**Swirl Control Valve Control Performance** Engine started, Engine running at cruise speed and then back to idle speed, and the PCM detected the Actual Swirl Valve Control Position sensor value did not agree with the Target opening angle of the Swirl Control valve (as controlled by the PCM). **Note: The SWL CV (B1) PID on the Scan Tool should read from 0-5 counts with the engine at idle speed and the ECT sensor less than 111°F; and it should read from 115-120 steps with the ECT sensor signal indicating a temperature of more than 113°F.** **Possible Causes:** • Swirl Control valve circuit is open or shorted (intermittent fault) • Swirl Control valve is contaminated, damaged or has failed • PCM has failed
DTC: P1140 **2T CCM, MIL: Yes** **2000-02** **Models:** Sentra **Engines:** 1.8L VIN C **Transmissions:** All	**Intake Valve Timing Control Position Sensor Circuit Malfunction (Bank 1)** Engine started, Engine running at idle speed, and the PCM detected an incorrect signal from the Bank 1 Intake Valve Timing Control Position sensor circuit during the CCM test. **Possible Causes:** • Intake valve timing control position sensor open or shorted • Intake valve timing control position sensor is damaged • Intake valve timing control position sensor has failed • PCM has failed
DTC: P1148 **2T CCM, MIL: Yes** **1998-2002** **Models:** All **Engines:** All **Transmissions:** All	**Closed Loop Malfunction Detected (Bank 1)** Engine running in closed loop for over 2 minutes, and the PCM detected that the engine was not operating in closed loop mode. **Possible Causes:** • A/FS or HO2S signal circuit is open or shorted to ground • A/FS or HO2S heater is damaged or has failed • A/FS or HO2S is damaged, contaminated or has failed • PCM has failed
DTC: P1168 **2T CCM, MIL: Yes** **1998-2002** **Models:** All **Engines:** All **Transmissions:** All	**Closed Loop Malfunction Detected (Bank 2)** Engine started, engine runtime from 2-5 minutes, and the PCM detected the engine was not operating in closed loop during the test. **Possible Causes:** • A/FS or HO2S signal circuit is open or shorted to ground • A/FS or HO2S heater is damaged or has failed • A/FS or HO2S is damaged, contaminated or has failed • PCM has failed
DTC: P1217 **2T CCM, MIL: No** **2000-02** **Models:** All **Engines:** All **Transmissions:** All	**Engine Over-Temperature Condition** Engine started, Engine running in closed loop for 3-5 minutes, and the PCM detected an engine overheated (engine over temperature) condition for too long a period during the CCM Rationality test. **Possible Causes:** • Engine coolant low or an incorrect coolant mixture exists • Engine cooling fan circuit(s) open or shorted to ground • Engine cooling fan is damaged or has failed • Check cooling system components (radiator hose, cap, etc.) • Check the thermostat operation (it may be stuck partly closed)
DTC: P1271 **2T CCM, MIL: Yes** **2000-02** **Models:** Sentra **Engines:** 1.8L VIN C **Transmissions:** All	**A/F Sensor-11 (Bank 1 Sensor 1) Circuit Malfunction** Engine started, Engine running under closed loop conditions, and the PCM detected the Air Fuel Sensor (AFS1) signal indicated close to 0.00v during the CCM test. **Possible Causes:** • Air Fuel Sensor (AFS) signal circuit is open • Air Fuel Sensor (AFS) signal circuit is shorted to ground • Air Fuel Sensor (AFS) is damaged or has failed • PCM had failed
DTC: P1272 **2T CCM, MIL: Yes** **2000-02** **Models:** Sentra **Engines:** 1.8L VIN C **Transmissions:** All	**Air Fuel Sensor (Bank 1 Sensor 1) High Input** Engine started, Engine running under closed loop conditions, and the PCM detected the computed Air Fuel Sensor (AFS1) signal indicated 4.5v during the CCM test. **Possible Causes:** • Air Fuel Sensor (AFS) signal circuit is open • Air Fuel Sensor (AFS) signal circuit is shorted to VREF • Air Fuel Sensor (AFS) signal circuit is shorted to power (B+) • Air Fuel Sensor (AFS) is damaged or has failed

DTC	Trouble Code Title, Conditions & Possible Causes
DTC: P1493 **2T EVAP, MIL: Yes** **1998-2002** **Models:** All **Engines:** All **Transmissions:** All	**EVAP Canister Purge Control Solenoid Performance (Open)** Engine running in closed loop under light engine load conditions, and the PCM detected that the EVAP Canister Purge Control Valve did not operate correctly (it may be stuck in open position). **Possible Causes:** • EVAP canister purge control valve is damaged or has failed • EVAP canister purge control solenoid valve is damaged/failed • EVAP vacuum hoses are clogged or disconnected • EVAP canister vent control valve is damaged or has failed • EVAP canister is full of water, or the vapor separator has failed • TSB 00-060 (7/00) contains information related to this code
DTC: P1605 **1T CCM, MIL: Yes** **1996-2002** **Models:** All **Engines:** All **Transmissions: A/T**	**TCM A/T Diagnosis Communication Line Malfunction** Engine runtime over 30 seconds, system voltage over 10.5v, and the PCM detected an unexpected voltage condition on the A/T Diagnosis Communication Line during the test. **Possible Causes:** • TCM communication line circuit is open or shorted to ground • TCM communication line circuit shorted to VREF or power (B+) • TCM or PCM has failed • TSB 01-004 (01/01) contains information related to this code
DTC: P1705 **1T CCM, MIL: Yes** **1996-2002** **Models:** All **Engines:** All **Transmissions: A/T**	**Throttle Position Sensor Circuit Malfunction** Engine started, engine runtime over 30 seconds, and the TCM detected an unexpected "high" or "low" voltage on the TP Switch circuit, regardless of the accelerator position (i.e., the signal did not correlate properly when the TP Idle Switch was either "on" or "off"). **Possible Causes:** • TP sensor signal circuit is open between the sensor and TCM • TP sensor signal circuit is shorted between sensor and TCM • TP sensor is damaged or has failed • TCM has failed
DTC: P1706 **2T CCM, MIL: Yes** **1998-2002** **Models:** All **Engines:** All **Transmissions: A/T**	**Park Neutral Position Switch Circuit Malfunction** Engine started, Engine running for over 30 seconds, vehicle speed over 5 mph, and the TCM detected the PNP switch signal was not plausible with the vehicle moving in a forward gear, or it detected the PNP switch signal did not change during the start to run transition. **Possible Causes:** • PNP switch signal circuit is open or shorted to ground • PNP switch is shorted to VREF or system power (B+) • PNP switch or the PNP relay is damaged or has failed • PCM has failed
DTC: P1760 **2T CCM, MIL: Yes** **1996-2002** **Models:** All **Engines:** All **Transmissions: A/T**	**Overrun Clutch Solenoid Valve Circuit Malfunction** Engine runtime over 30 seconds, and the TCM detected an incorrect voltage reading while operating the Overrun Clutch Solenoid valve. **Possible Causes:** • Overrun clutch solenoid control circuit is open • Overrun clutch solenoid control circuit is shorted to ground • Overrun clutch solenoid is damaged or has failed • TCM has failed
DTC: P1900 **2T CCM, MIL: Yes** **1996, 1997** **Models:** All **Engines:** All **Transmissions:** All	**Cooling Fan Control Circuit Malfunction** DTC P0115 not set, engine run time over 5 seconds, hot engine condition present, and the PCM detected the Cooling Fan did not operate, or the Cooling Fan system did not operate. **Note: This trouble code can be set even with the coolant level okay.** **Possible Causes:** • Cooling fan control circuit is open or shorted to ground • Cooling fan is damaged or has failed • Cooling system component failure (i.e., radiator hose, radiator cap, water pump or thermostat may be stuck partially closed

NISSAN
COMPONENT TESTING

17

TABLE OF CONTENTS

Component Locations 17-2
- 240SX 17-2
- 300ZX 17-3
- 350Z 17-4
- Altima 17-9
- Armada 17-21
- Frontier 17-27
- Maxima 17-45
- Murano 17-56
- Pathfinder 17-62
- Pickup 17-74
- Quest 17-75
- Sentra 17-86
- Titan 17-116
- Xterra 17-122

Component Testing 17-134
- Camshaft Position Sensor 17-134
- Crankshaft Position Sensor 17-140
- Engine Coolant Temperature Sensor 17-143
- EGR Temperature Sensor 17-145
- Intake Air Temperature Sensor 17-147
- Knock Sensor 17-148
- Mass Air Flow Sensor 17-149
- Throttle Position Sensor 17-152

COMPONENT TESTING

Component Locations

240SX

1996–1998
KA24DE I4 ENGINE

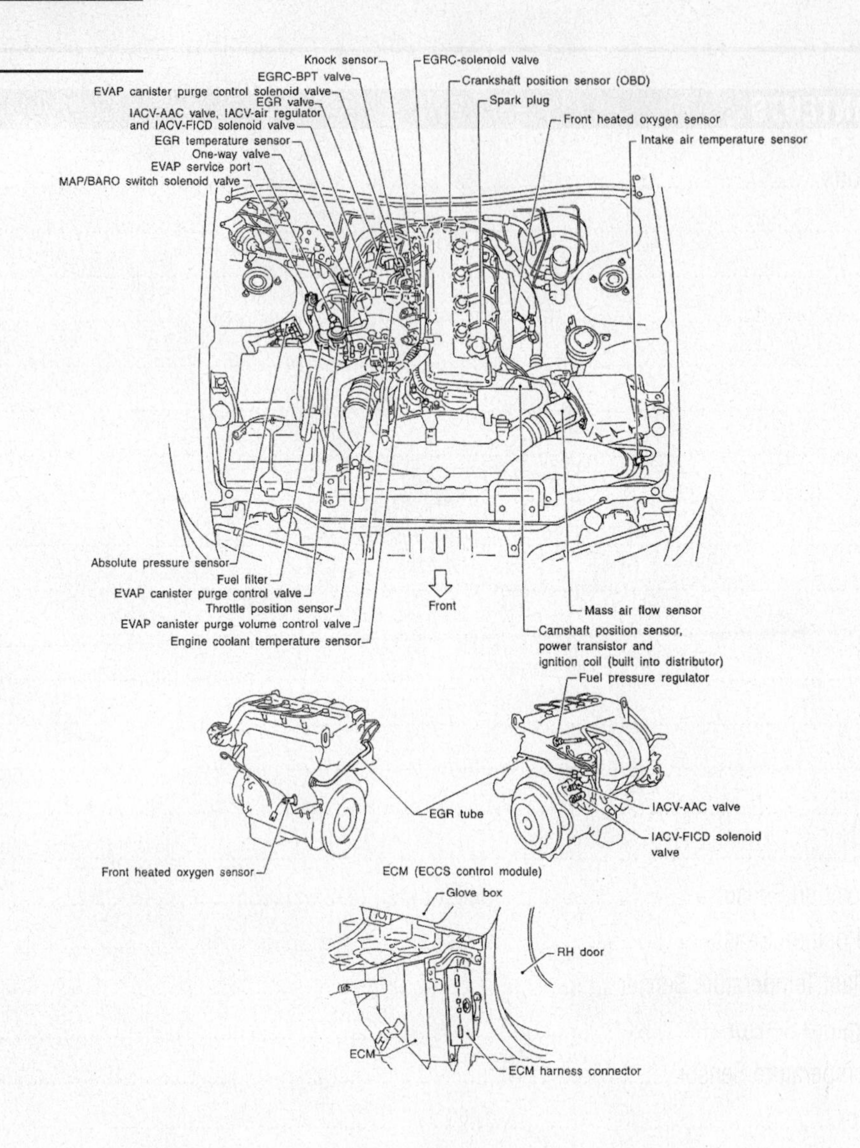

29149_NISS_G0001

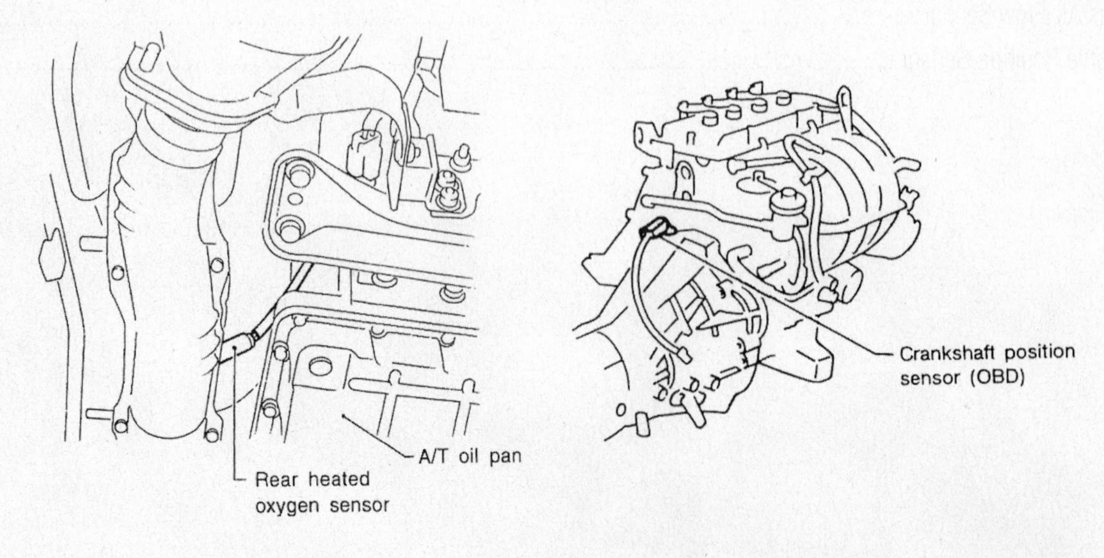

29149_NISS_G0002

300ZX

1996

VG30DE, VG30DETT V6 ENGINES

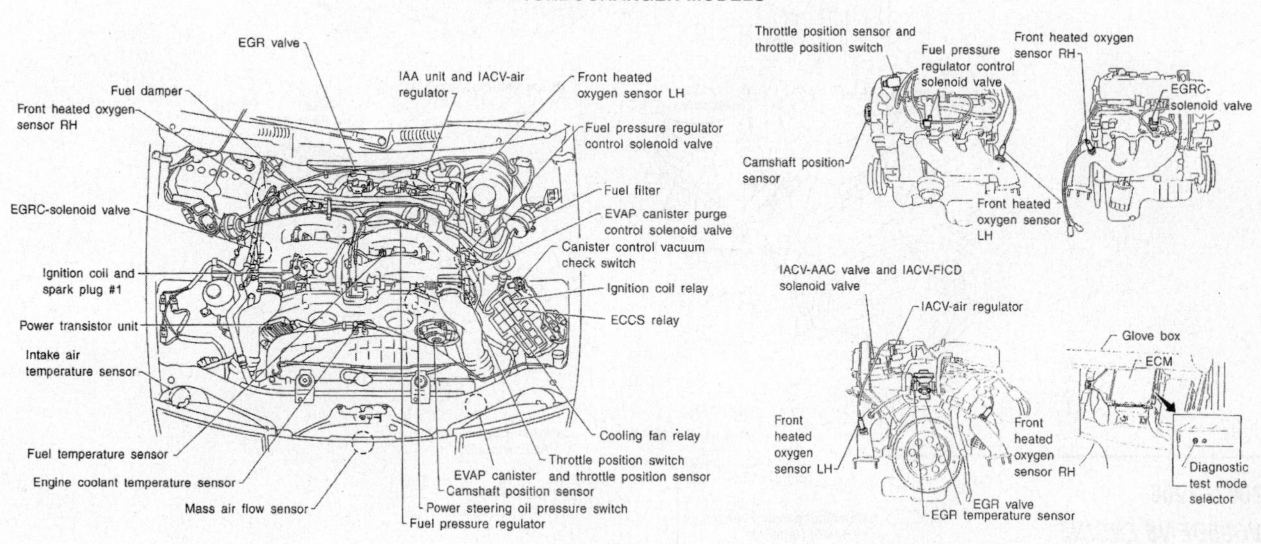

29149_NISS_G0003

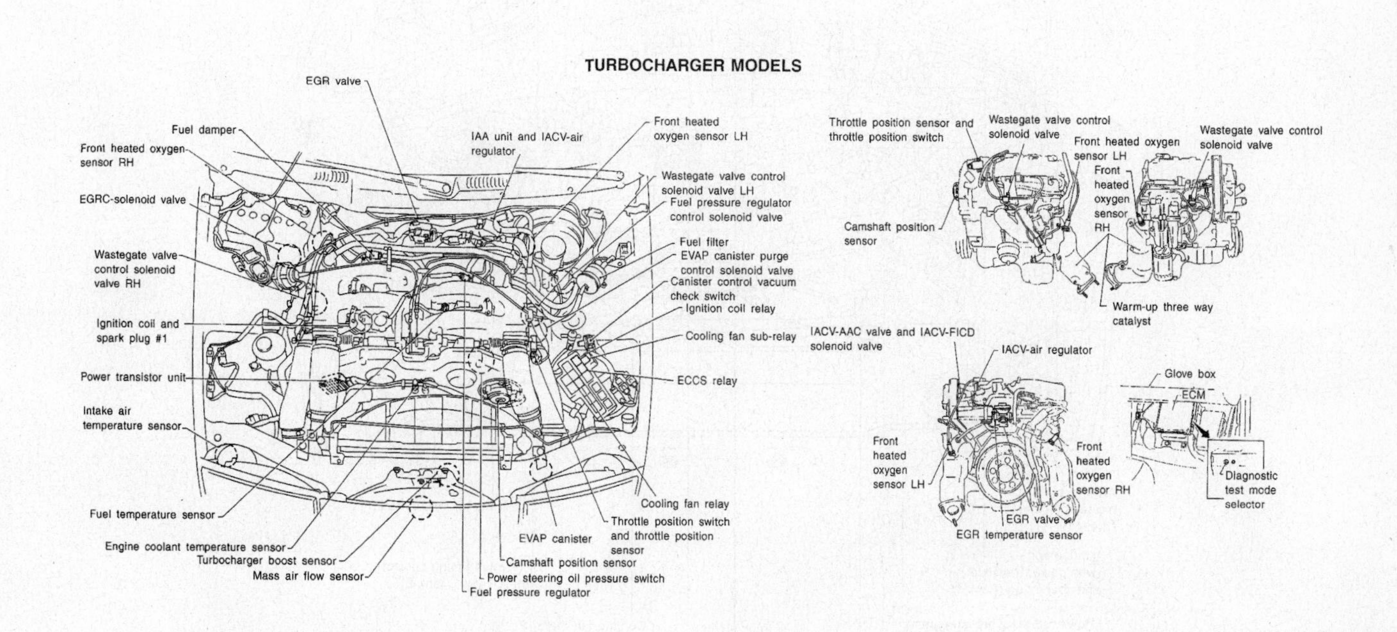

29149_NISS_G0004

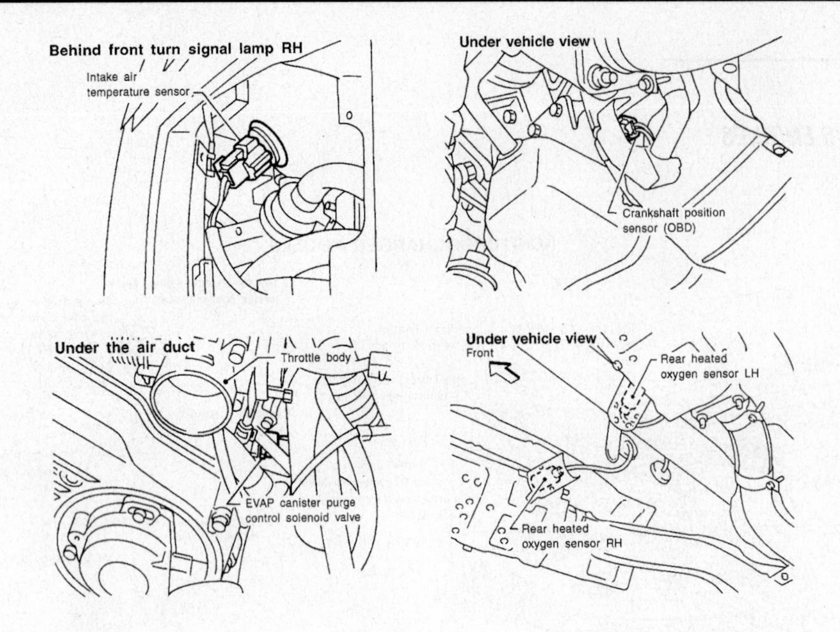
Behind front turn signal lamp RH
Intake air temperature sensor
Under vehicle view
Crankshaft position sensor (OBD)
Under the air duct
Throttle body
EVAP canister purge control solenoid valve
Under vehicle view
Front
Rear heated oxygen sensor LH
Rear heated oxygen sensor RH

29149_NISS_G0005

350Z

2003–2006

VQ35DE V6 ENGINE

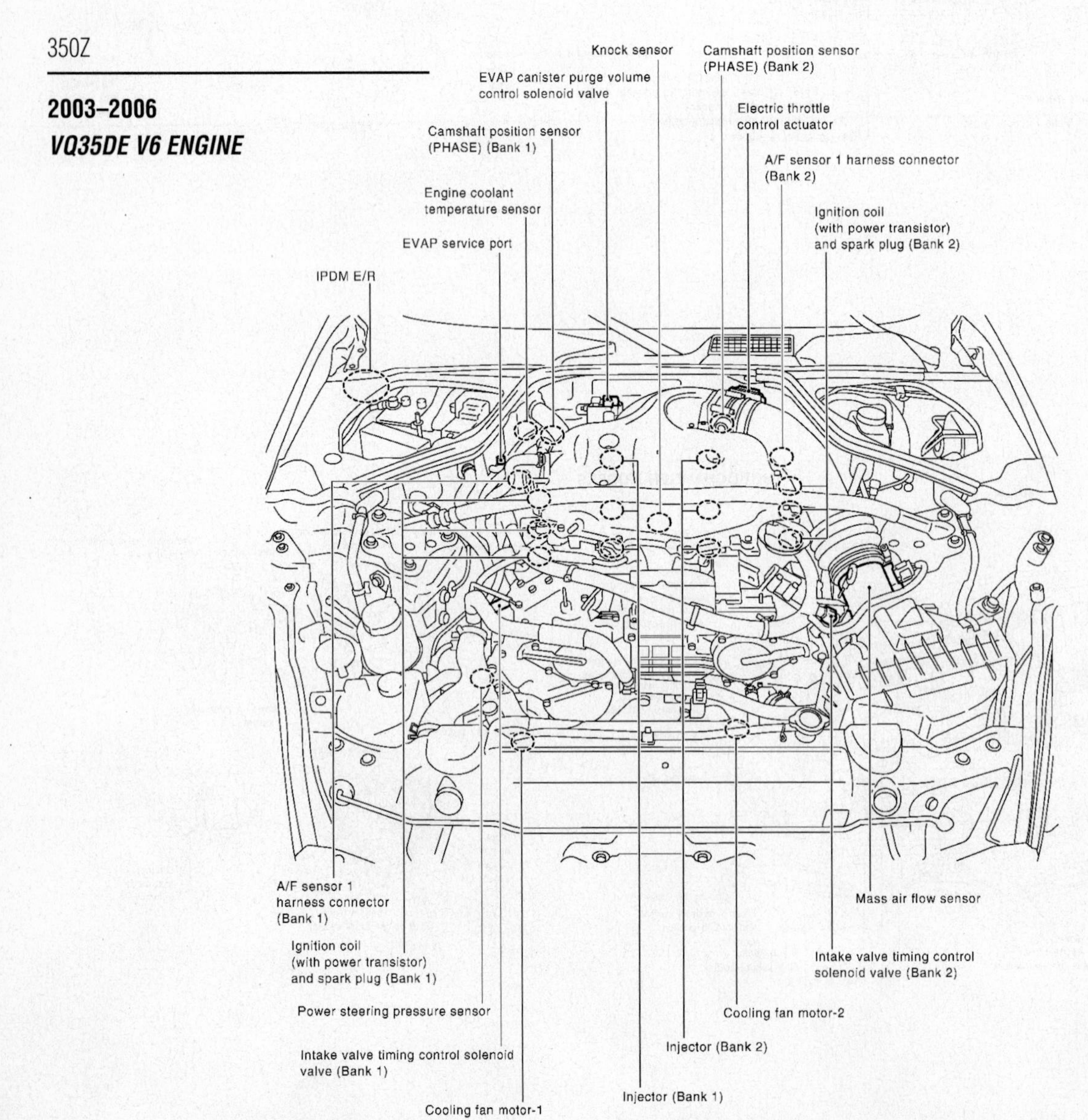
Knock sensor
Camshaft position sensor (PHASE) (Bank 2)
EVAP canister purge volume control solenoid valve
Electric throttle control actuator
Camshaft position sensor (PHASE) (Bank 1)
A/F sensor 1 harness connector (Bank 2)
Engine coolant temperature sensor
Ignition coil (with power transistor) and spark plug (Bank 2)
EVAP service port
IPDM E/R
A/F sensor 1 harness connector (Bank 1)
Mass air flow sensor
Ignition coil (with power transistor) and spark plug (Bank 1)
Intake valve timing control solenoid valve (Bank 2)
Power steering pressure sensor
Cooling fan motor-2
Injector (Bank 2)
Intake valve timing control solenoid valve (Bank 1)
Injector (Bank 1)
Cooling fan motor-1

29149_NISS_G0074

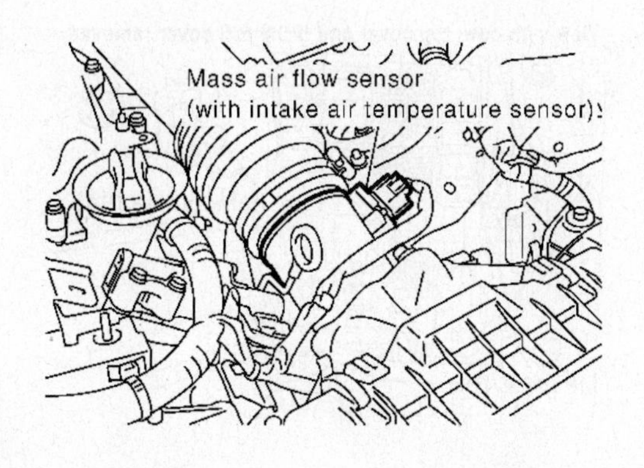
Mass air flow sensor
(with intake air temperature sensor)

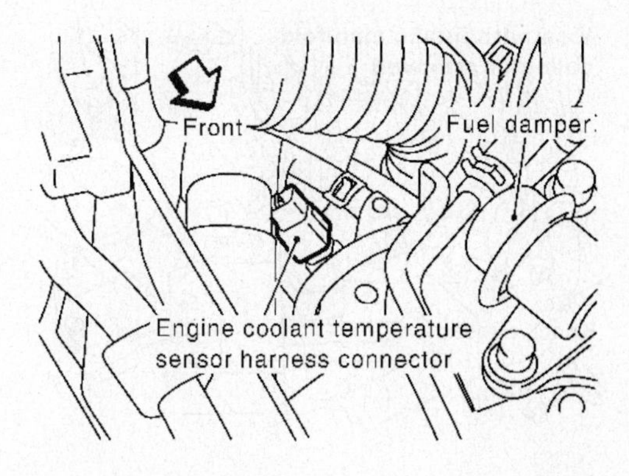
Front
Fuel damper
Engine coolant temperature
sensor harness connector

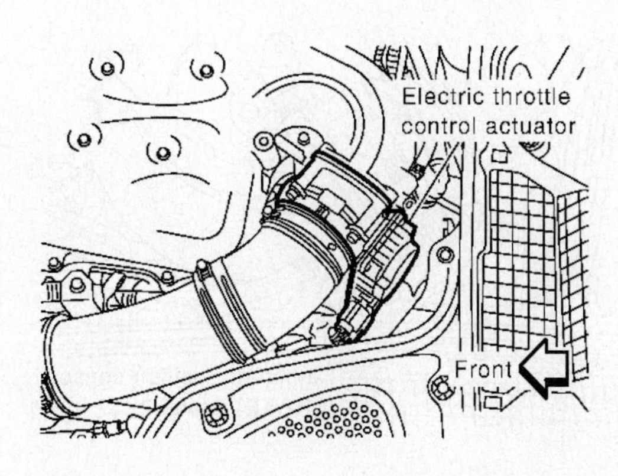
Electric throttle
control actuator
Front

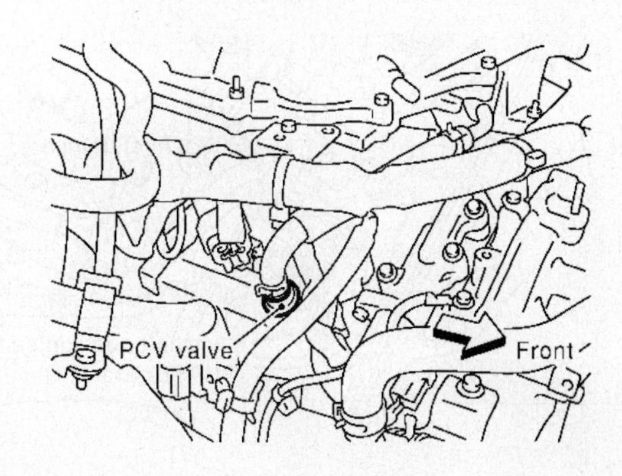
PCV valve
Front

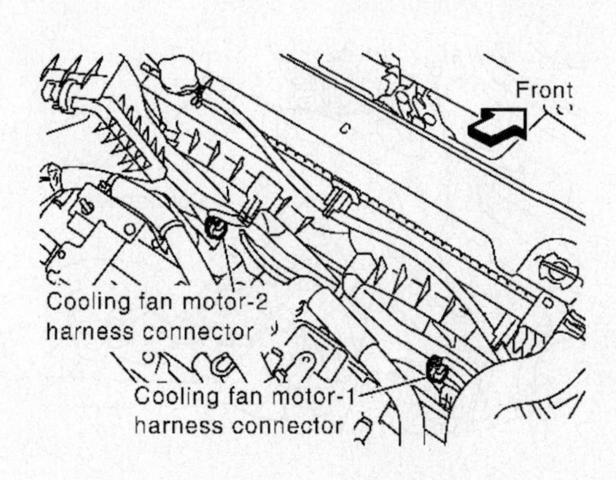
Front
Cooling fan motor-2
harness connector
Cooling fan motor-1
harness connector

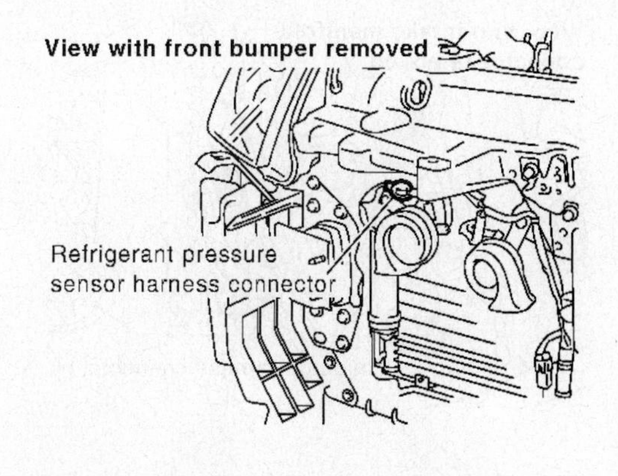
View with front bumper removed
Refrigerant pressure
sensor harness connector

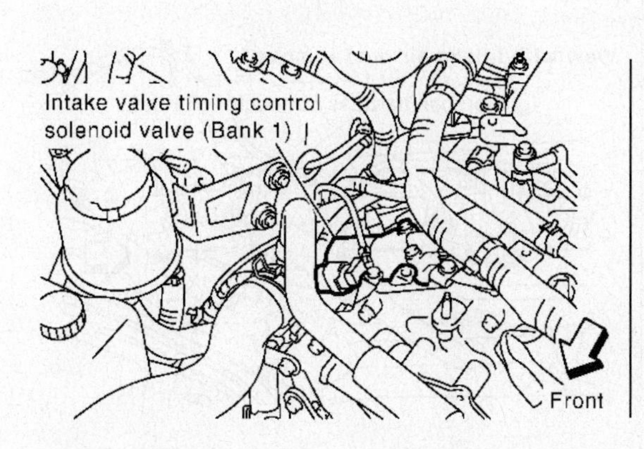
Intake valve timing control
solenoid valve (Bank 1)
Front

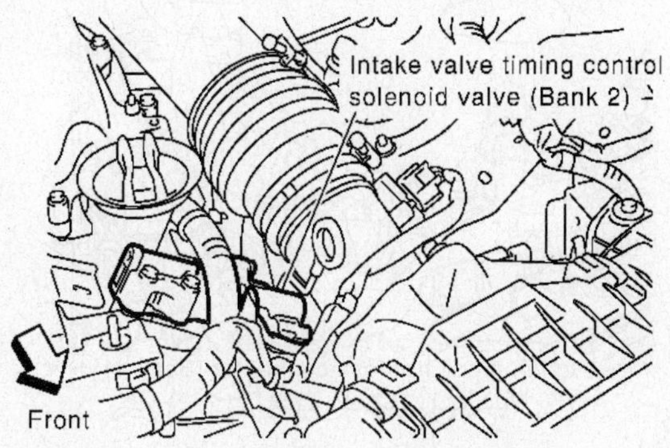
Intake valve timing control
solenoid valve (Bank 2)
Front

29149_NISS_G0075

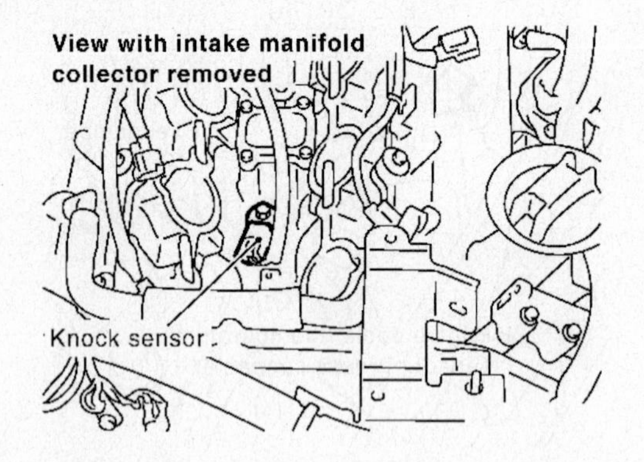
View with intake manifold collector removed
Knock sensor

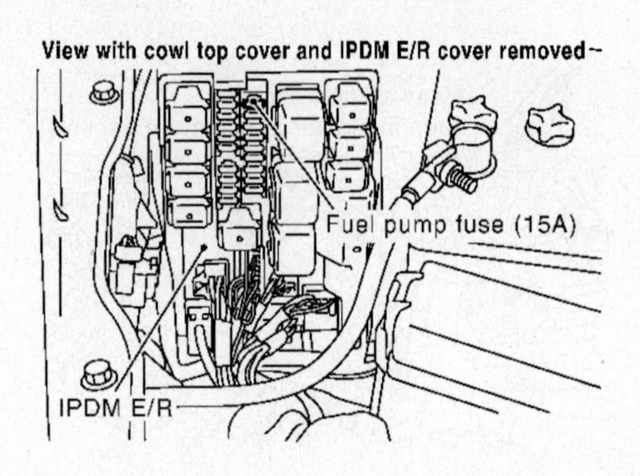
View with cowl top cover and IPDM E/R cover removed
Fuel pump fuse (15A)
IPDM E/R

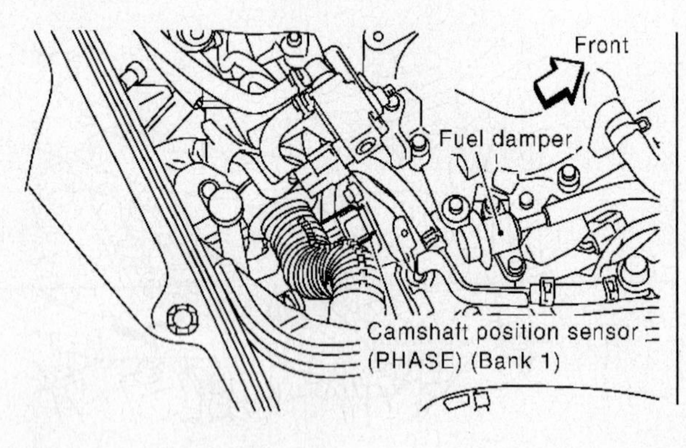
Front
Fuel damper
Camshaft position sensor (PHASE) (Bank 1)

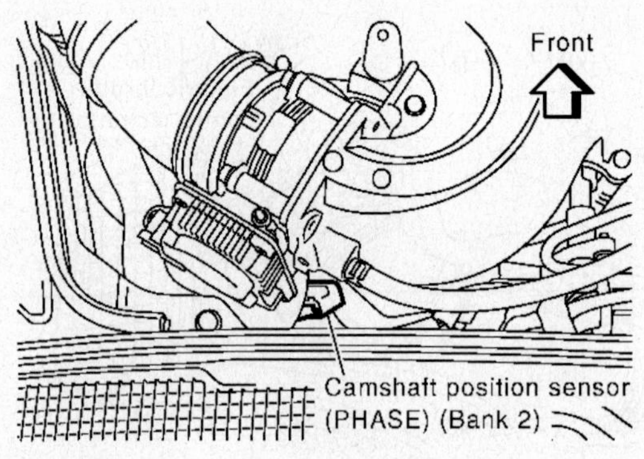
Front
Camshaft position sensor (PHASE) (Bank 2)

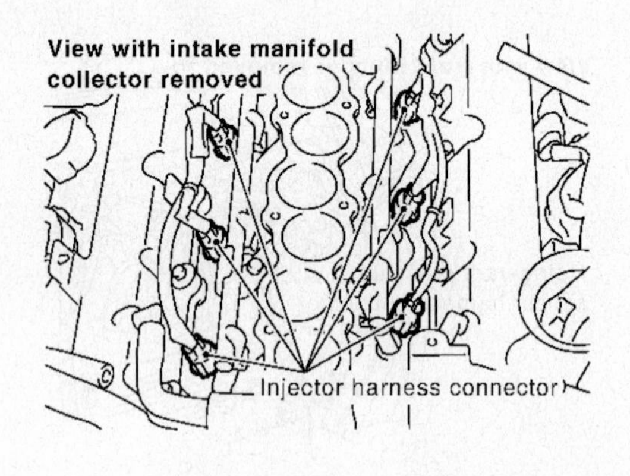
View with intake manifold collector removed
Injector harness connector

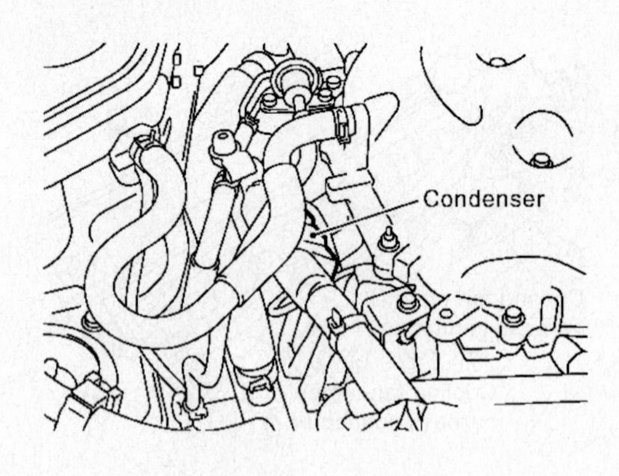
Condenser

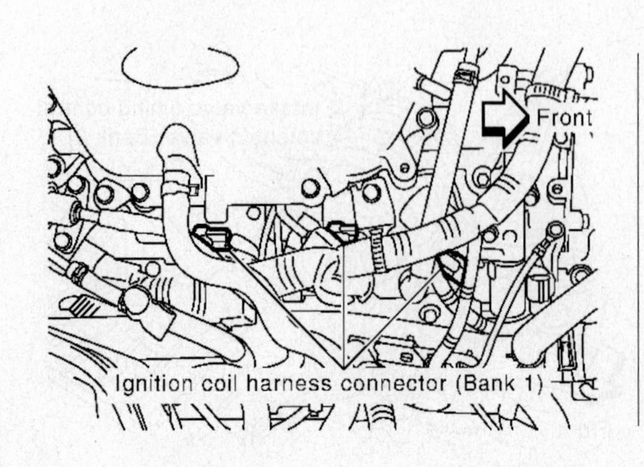
Front
Ignition coil harness connector (Bank 1)

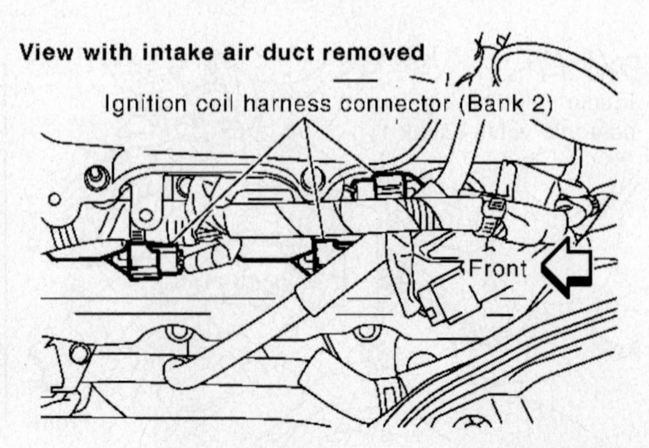
View with intake air duct removed
Ignition coil harness connector (Bank 2)
Front

29149_NISS_G0076

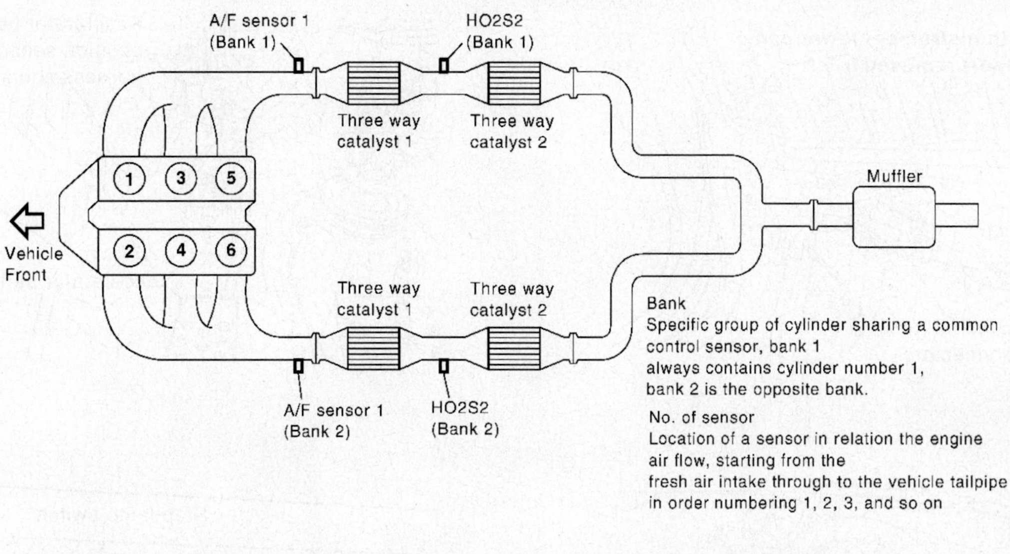

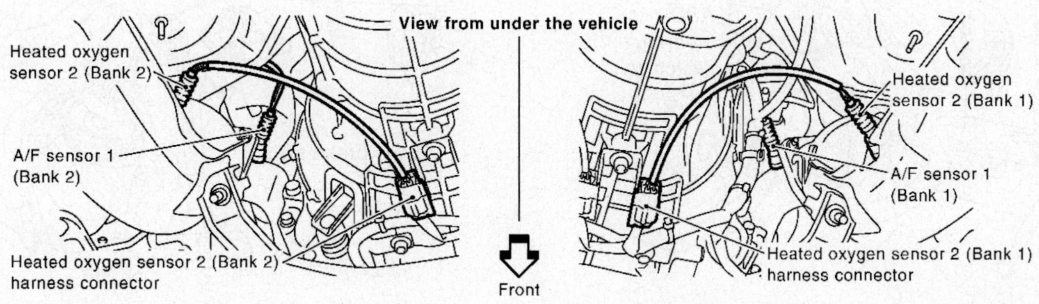

29149_NISS_G0077

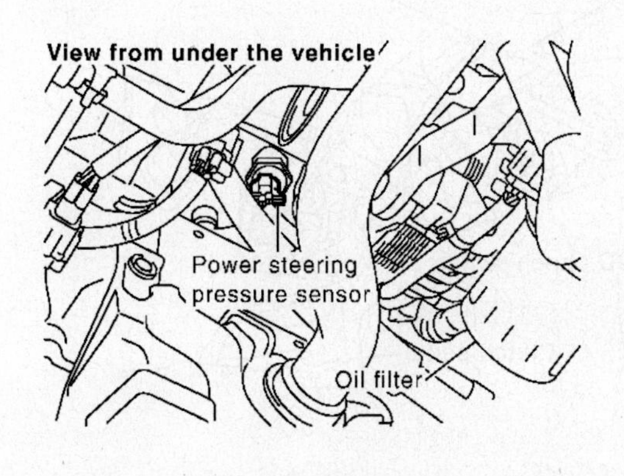

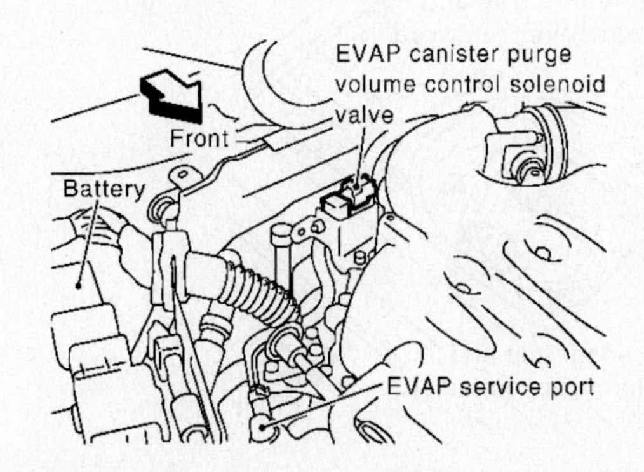

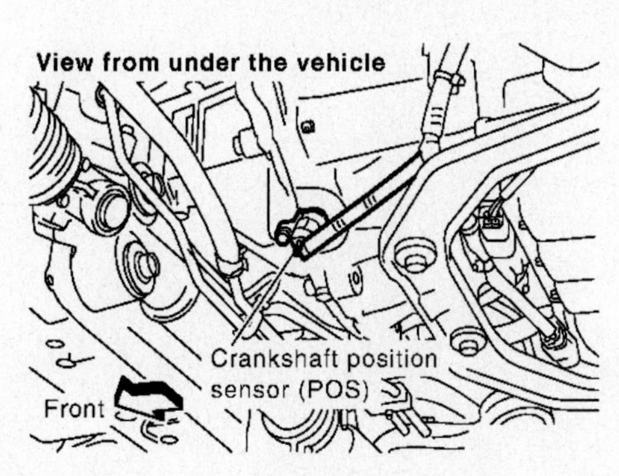

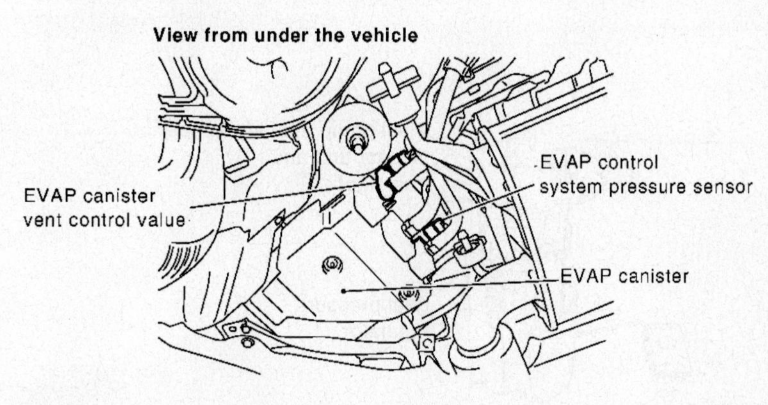

29149_NISS_G0078

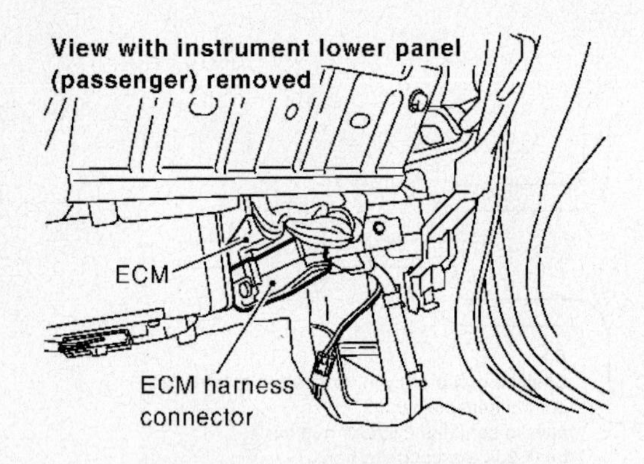

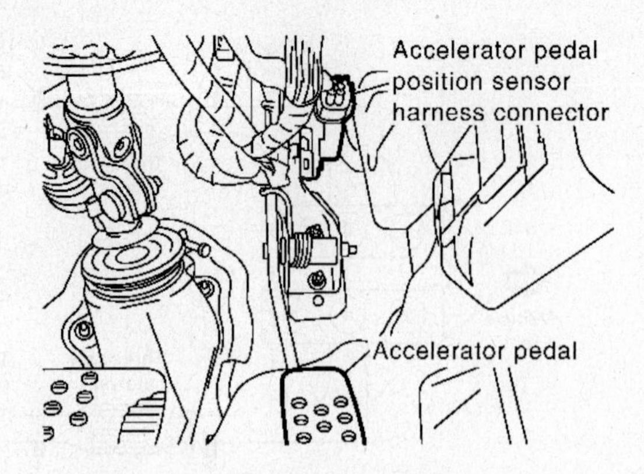

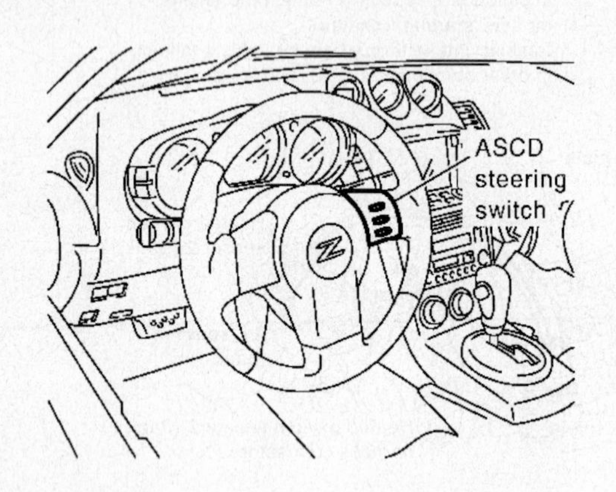

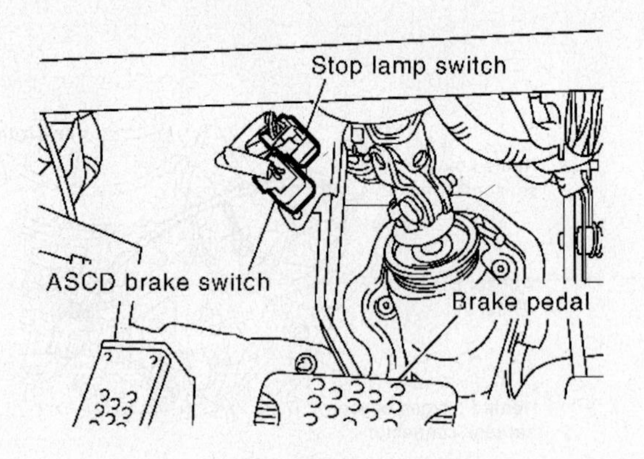

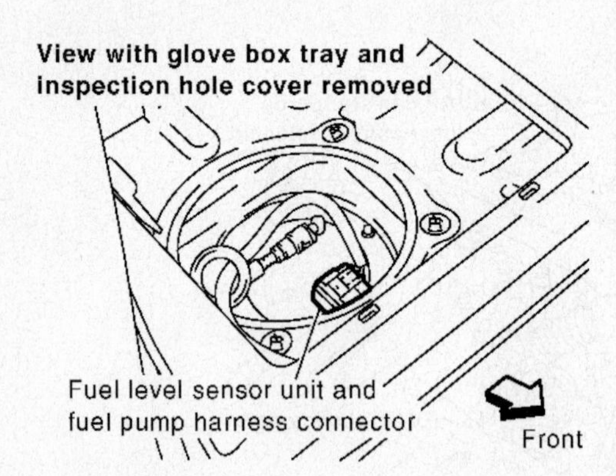

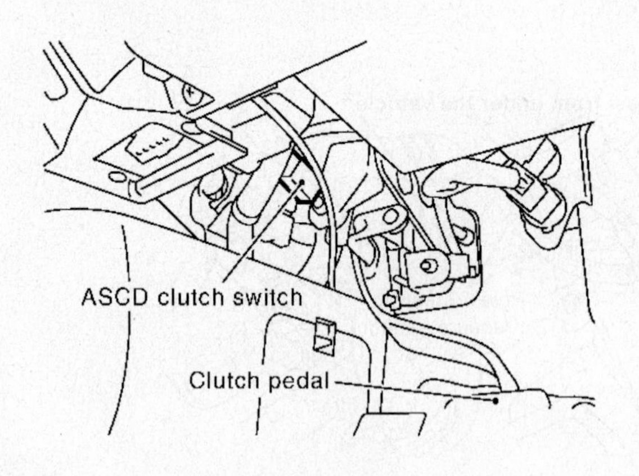

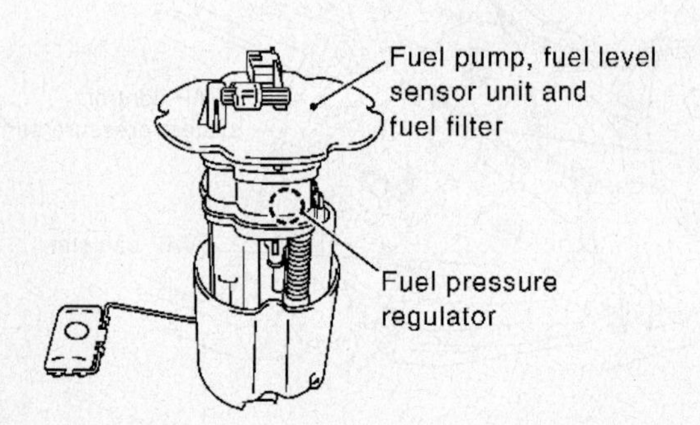

29149_NISS_G0079

ALTIMA

1996–2001

KA24DE I4 ENGINE

NON-CALIFORNIA MODELS

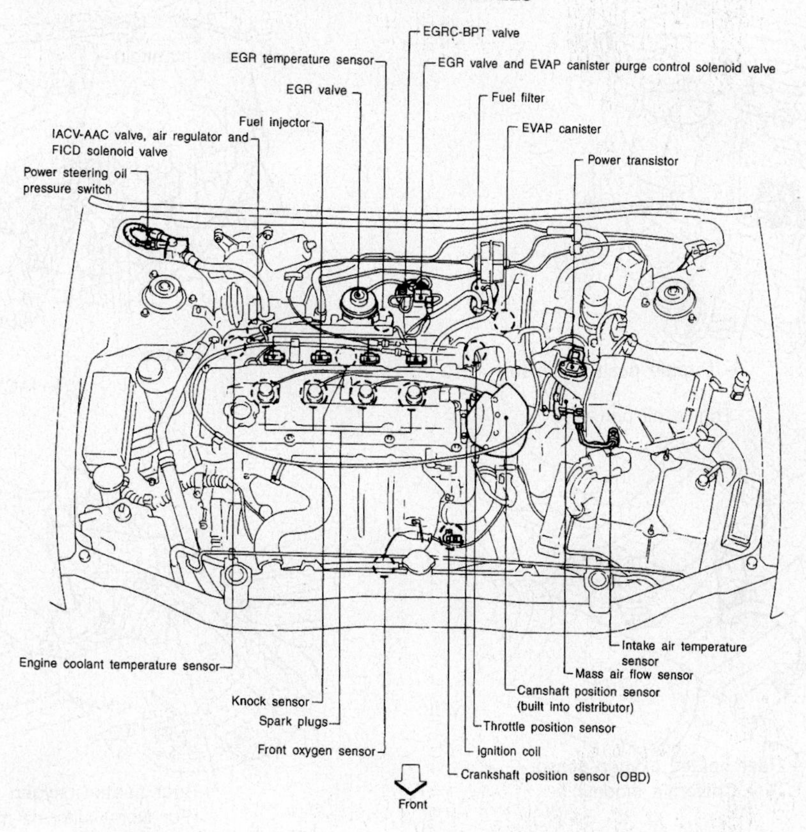

29149_NISS_G0006

CALIFORNIA MODELS

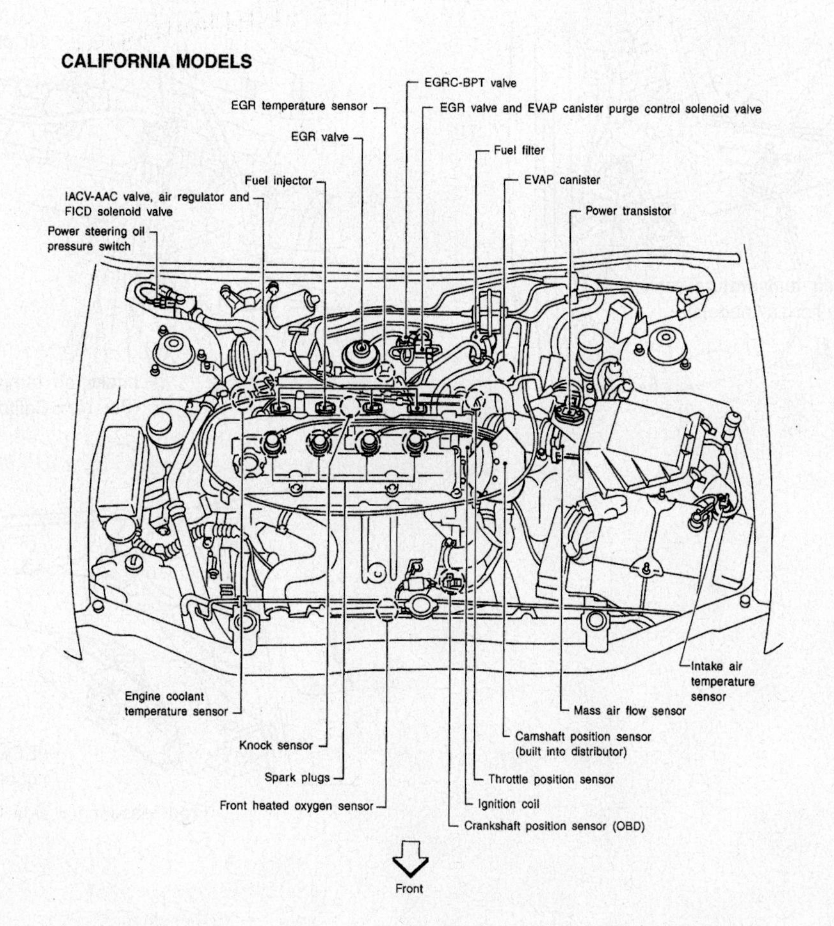

29149_NISS_G0007

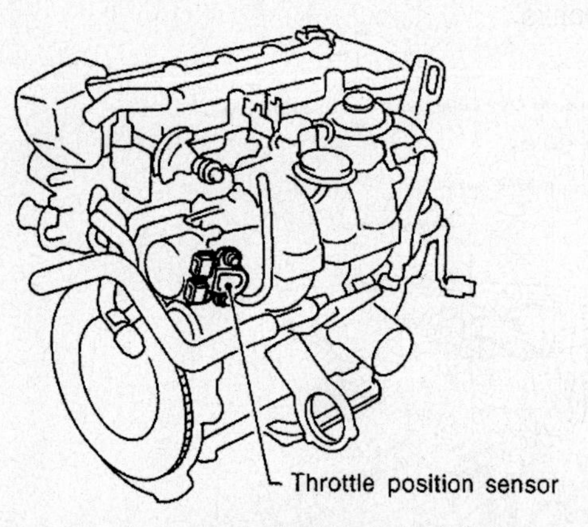
Throttle position sensor

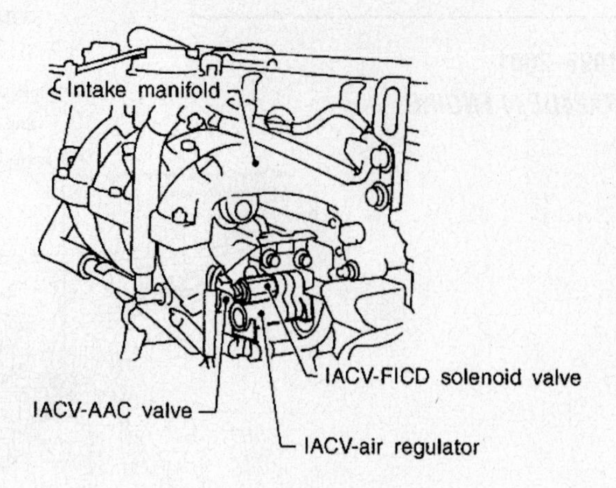
Intake manifold
IACV-FICD solenoid valve
IACV-AAC valve
IACV-air regulator

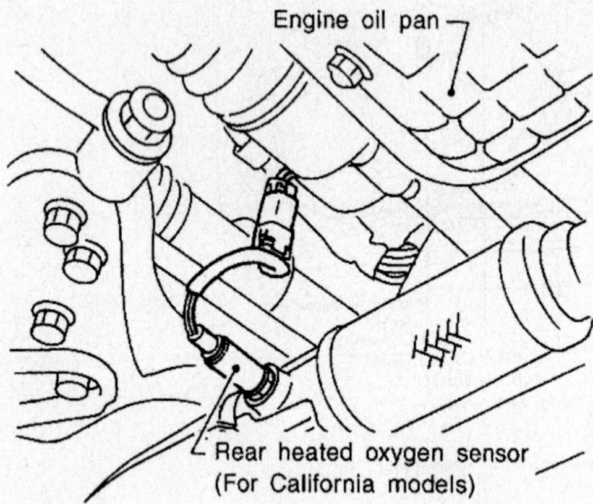
Engine oil pan
Rear heated oxygen sensor
(For California models)

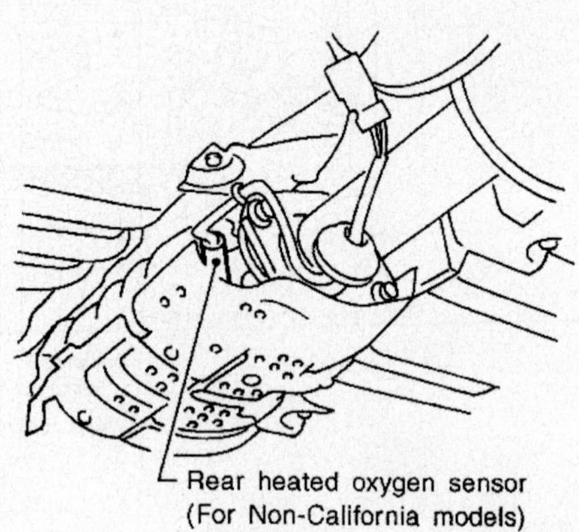
Rear heated oxygen sensor
(For Non-California models)

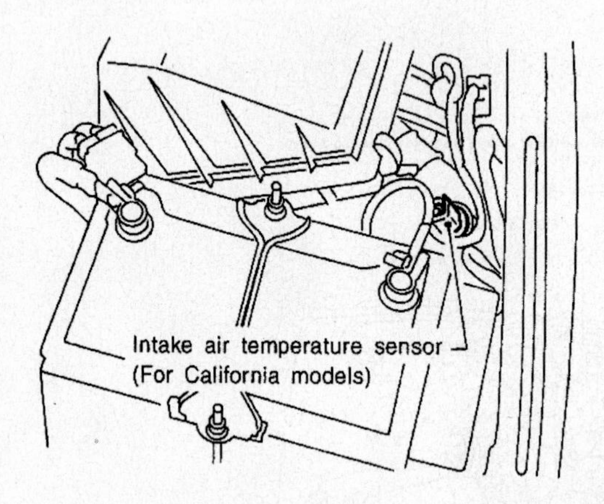
Intake air temperature sensor
(For California models)

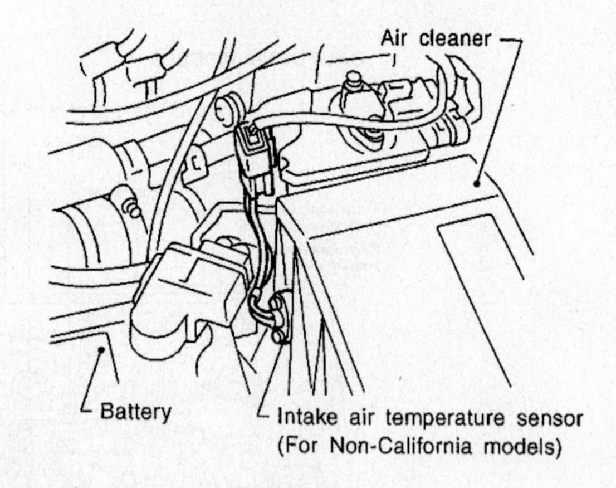
Air cleaner
Battery
Intake air temperature sensor
(For Non-California models)

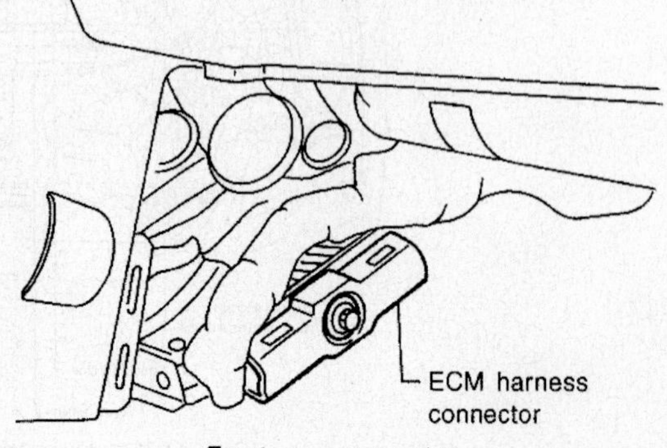
ECM harness
connector
Front passenger side

29149_NISS_G0008

2002–2006

QR25DE I4 ENGINE

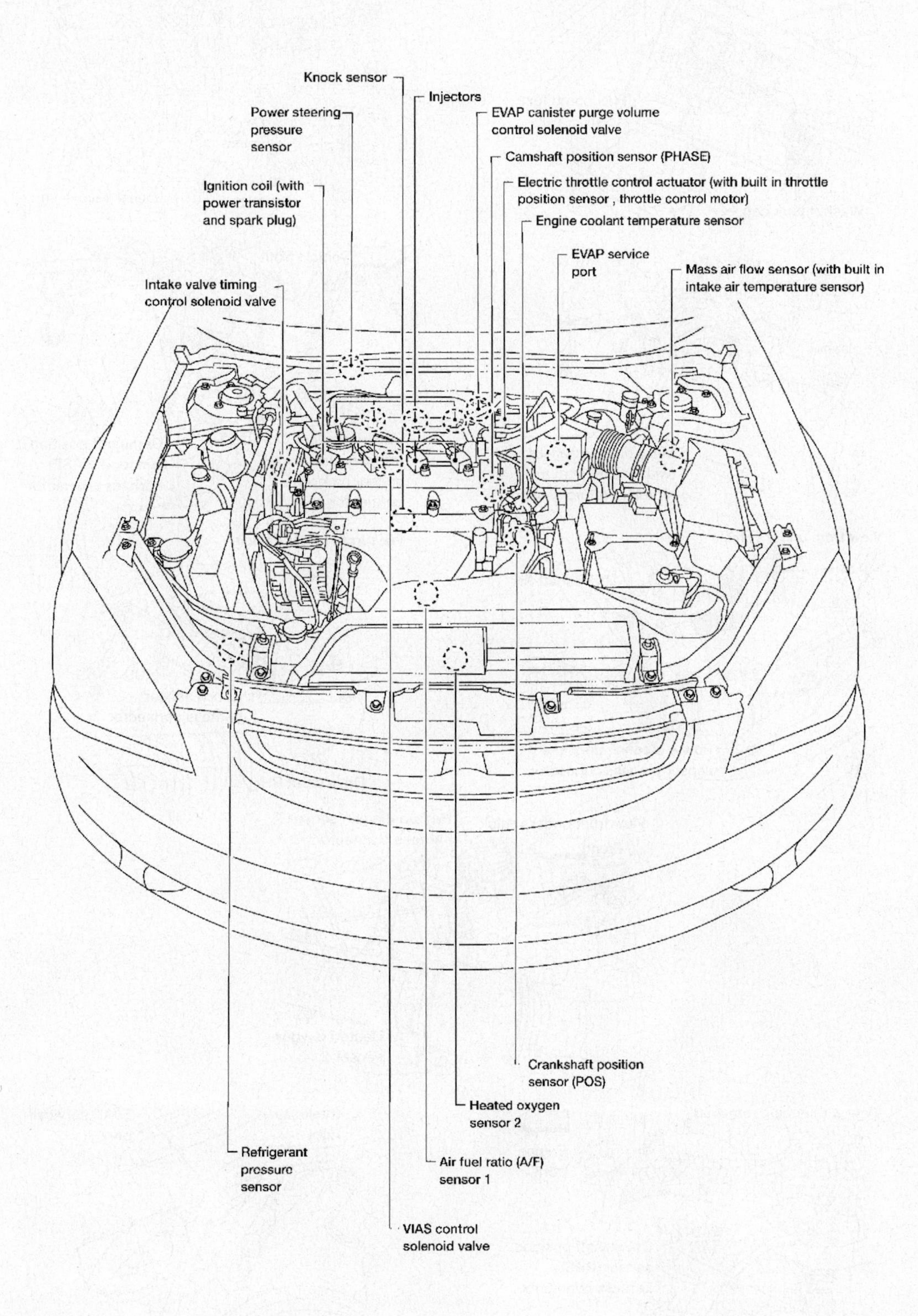
Knock sensor
Injectors
Power steering pressure sensor
EVAP canister purge volume control solenoid valve
Camshaft position sensor (PHASE)
Ignition coil (with power transistor and spark plug)
Electric throttle control actuator (with built in throttle position sensor , throttle control motor)
Engine coolant temperature sensor
EVAP service port
Mass air flow sensor (with built in intake air temperature sensor)
Intake valve timing control solenoid valve
Crankshaft position sensor (POS)
Heated oxygen sensor 2
Refrigerant pressure sensor
Air fuel ratio (A/F) sensor 1
VIAS control solenoid valve

29149_NISS_G0080

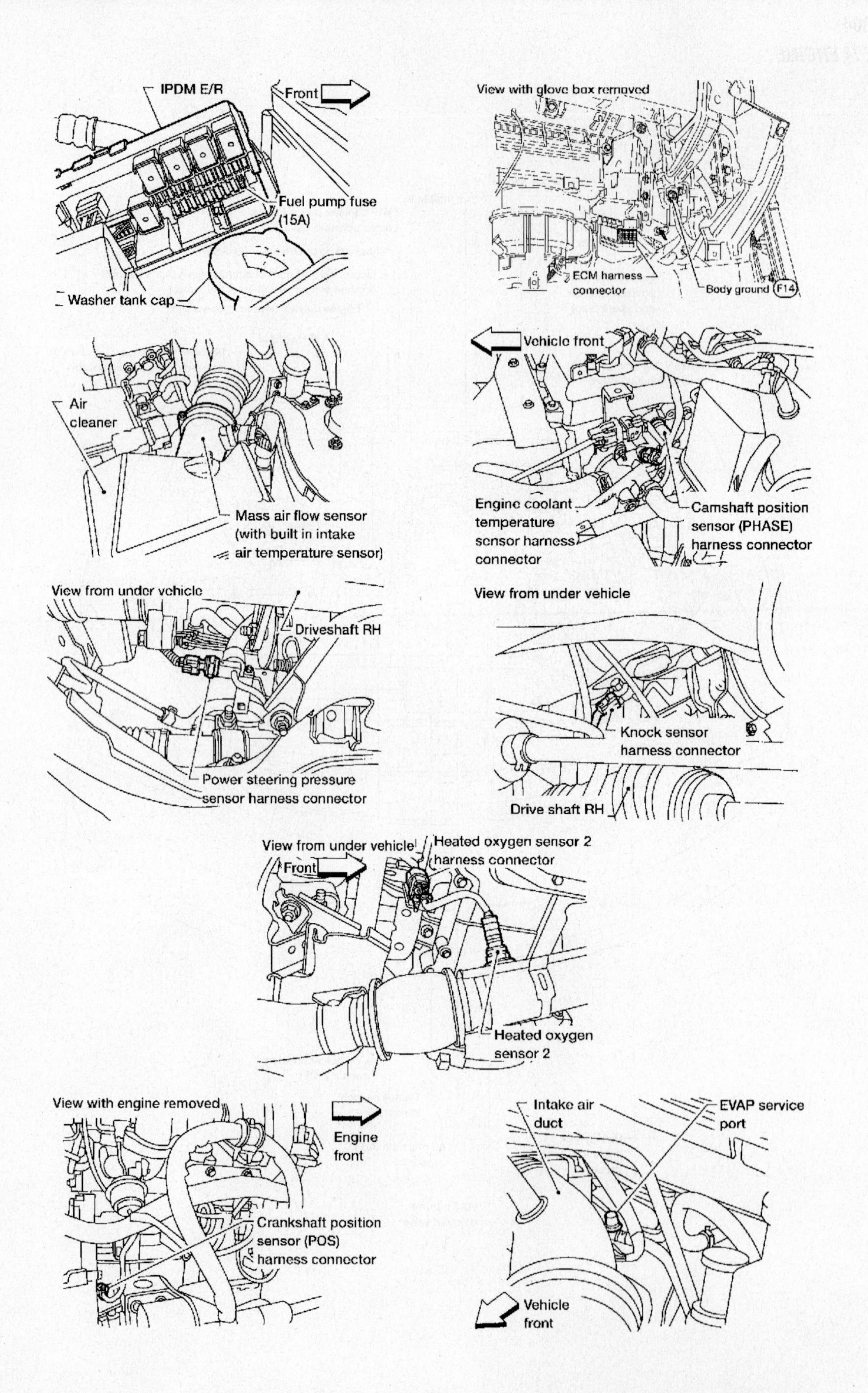
IPDM E/R
Front
Fuel pump fuse
(15A)
Washer tank cap
View with glove box removed
ECM harness
connector
Body ground
F14
Air
cleaner
Mass air flow sensor
(with built in intake
air temperature sensor)
Vehicle front
Engine coolant
temperature
sensor harness
connector
Camshaft position
sensor (PHASE)
harness connector
View from under vehicle
Driveshaft RH
Power steering pressure
sensor harness connector
View from under vehicle
Knock sensor
harness connector
Drive shaft RH
View from under vehicle
Front
Heated oxygen sensor 2
harness connector
Heated oxygen
sensor 2
View with engine removed
Engine
front
Crankshaft position
sensor (POS)
harness connector
Intake air
duct
EVAP service
port
Vehicle
front

29149_NISS_G0081

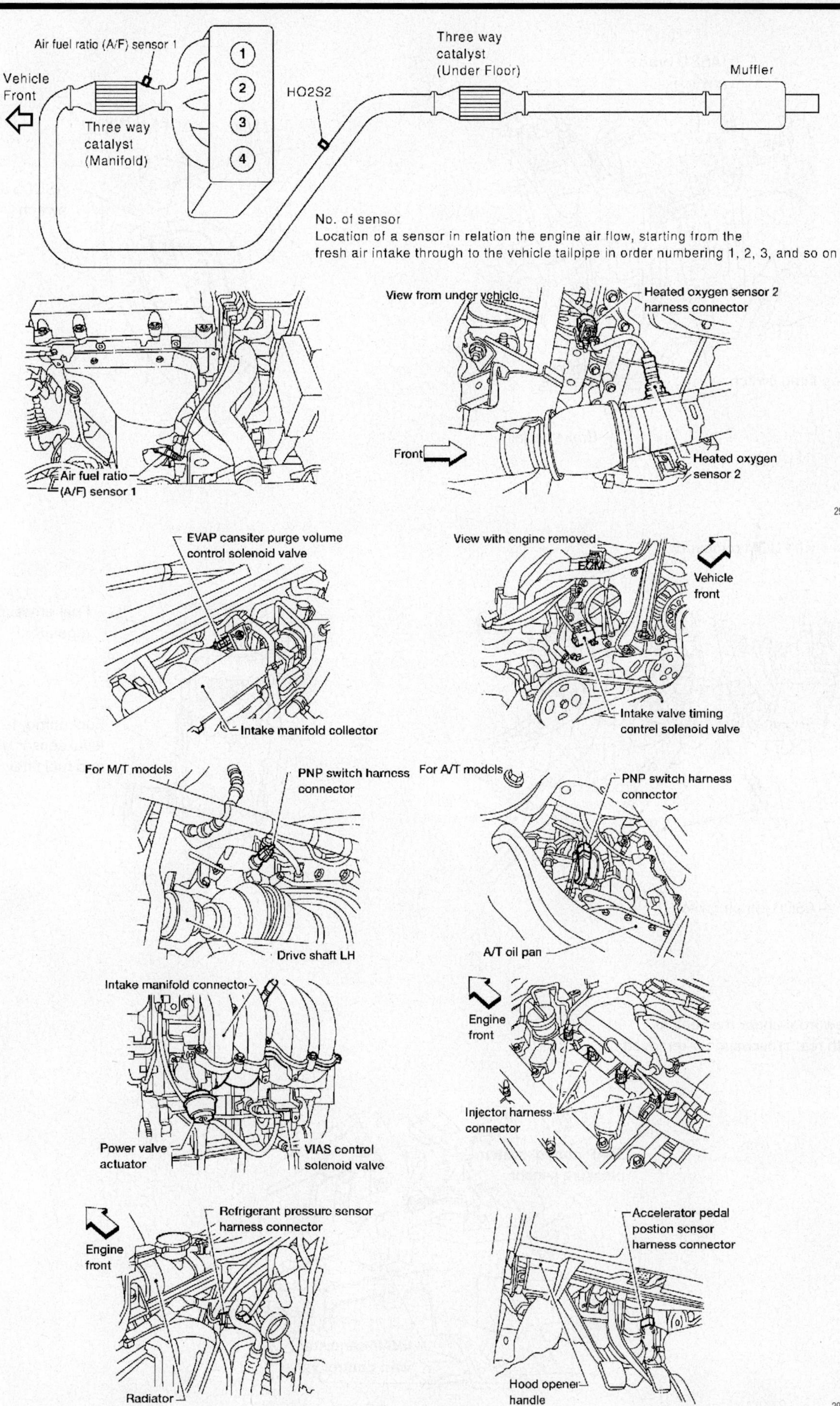
Air fuel ratio (A/F) sensor 1
Vehicle Front
Three way catalyst (Manifold)
1
2
3
4
HO2S2
Three way catalyst (Under Floor)
Muffler
No. of sensor
Location of a sensor in relation the engine air flow, starting from the fresh air intake through to the vehicle tailpipe in order numbering 1, 2, 3, and so on
Air fuel ratio (A/F) sensor 1
View from under vehicle
Heated oxygen sensor 2 harness connector
Front
Heated oxygen sensor 2
29149_NISS_G0082
EVAP cansiter purge volume control solenoid valve
Intake manifold collector
View with engine removed
ECM
Vehicle front
Intake valve timing contrel solenoid valve
For M/T models
PNP switch harness connector
Drive shaft LH
For A/T models
PNP switch harness connector
A/T oil pan
Intake manifold connector
Power valve actuator
VIAS control solenoid valve
Engine front
Injector harness connector
Engine front
Refrigerant pressure sensor harness connector
Radiator
Accelerator pedal postion sensor harness connector
Hood opener handle
29149_NISS_G0083

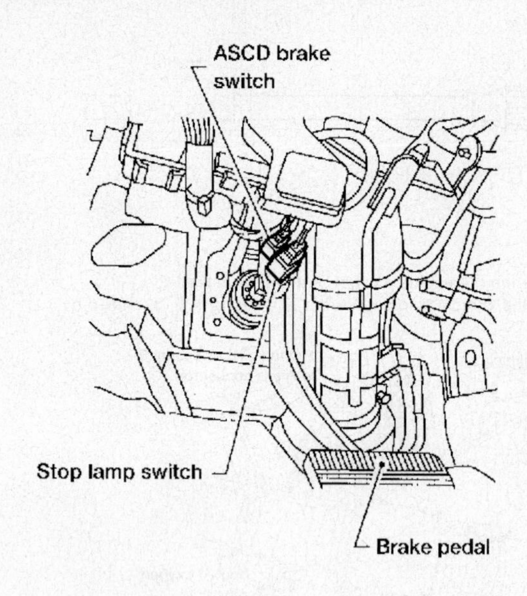

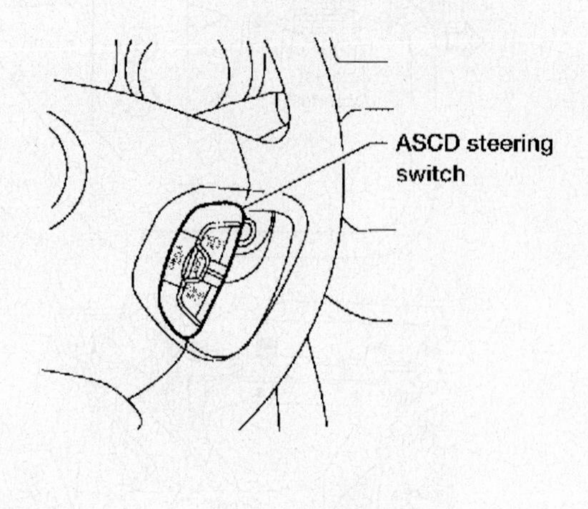

View with BCM removed

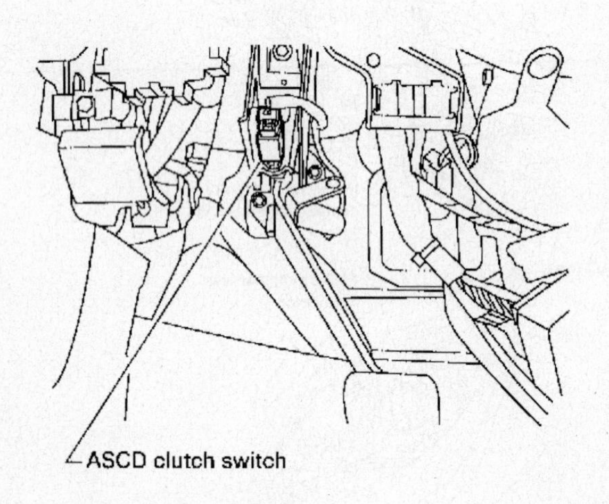

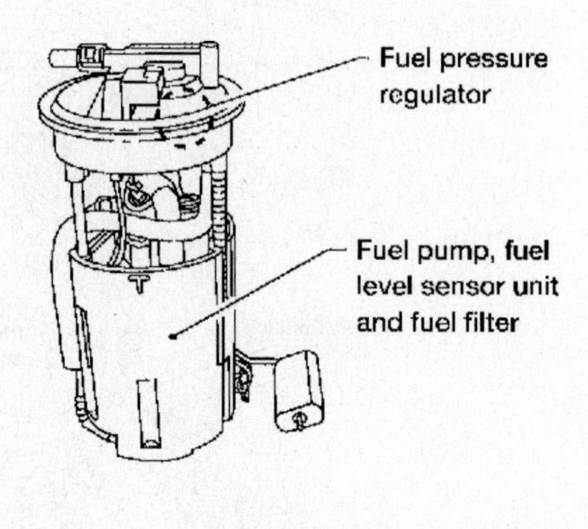

View from under the vehicle with rear crossmember removed

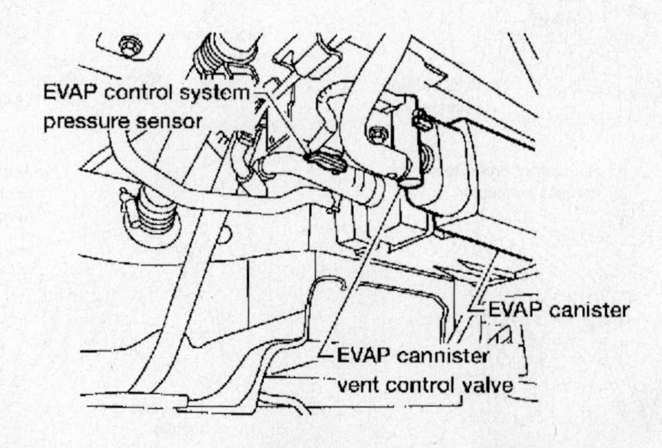

29149_NISS_G0084

2003–2006

VQ35DE V6 ENGINE

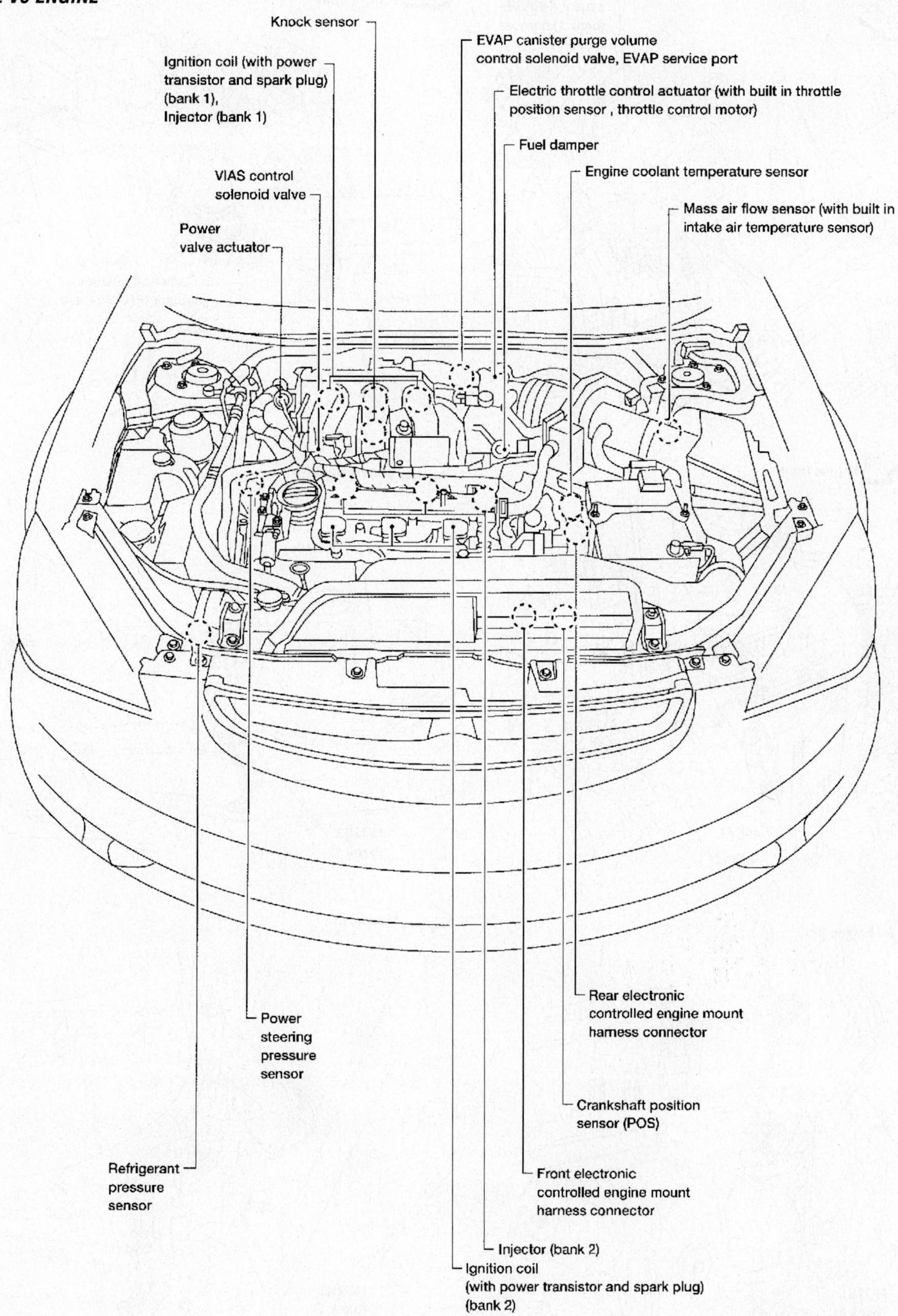
Knock sensor
Ignition coil (with power transistor and spark plug) (bank 1), Injector (bank 1)
VIAS control solenoid valve
Power valve actuator
EVAP canister purge volume control solenoid valve, EVAP service port
Electric throttle control actuator (with built in throttle position sensor , throttle control motor)
Fuel damper
Engine coolant temperature sensor
Mass air flow sensor (with built in intake air temperature sensor)
Rear electronic controlled engine mount harness connector
Power steering pressure sensor
Crankshaft position sensor (POS)
Refrigerant pressure sensor
Front electronic controlled engine mount harness connector
Injector (bank 2)
Ignition coil (with power transistor and spark plug) (bank 2)

29149_NISS_G0085

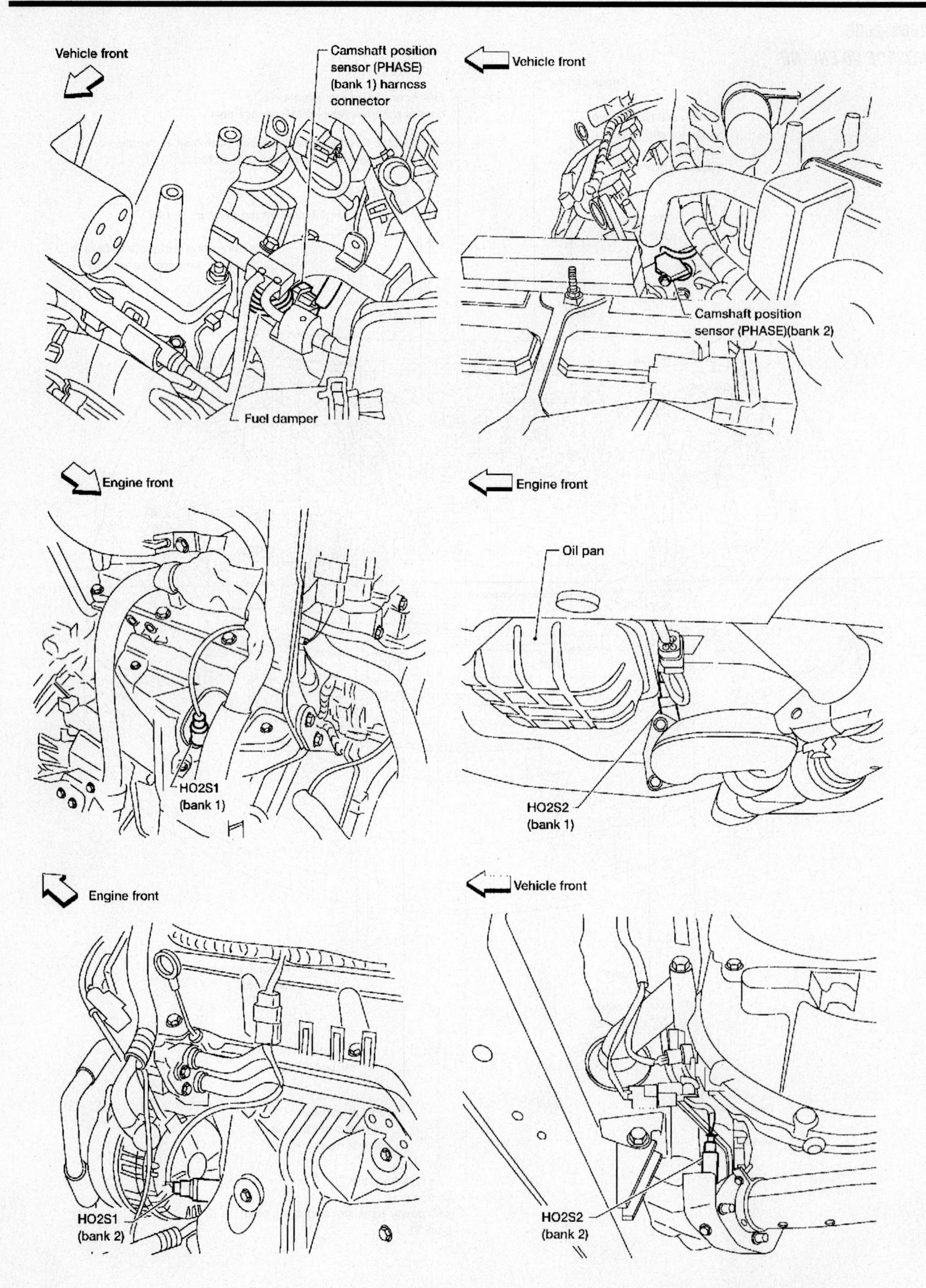
Vehicle front
Camshaft position
sensor (PHASE)
(bank 1) harness
connector
Fuel damper
Vehicle front
Camshaft position
sensor (PHASE)(bank 2)
Engine front
HO2S1
(bank 1)
Engine front
Oil pan
HO2S2
(bank 1)
Engine front
HO2S1
(bank 2)
Vehicle front
HO2S2
(bank 2)

29149_NISS_G0086

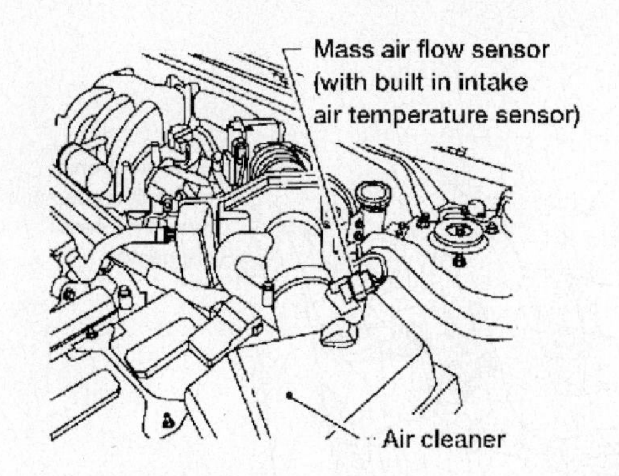
Mass air flow sensor
(with built in intake
air temperature sensor)
Air cleaner

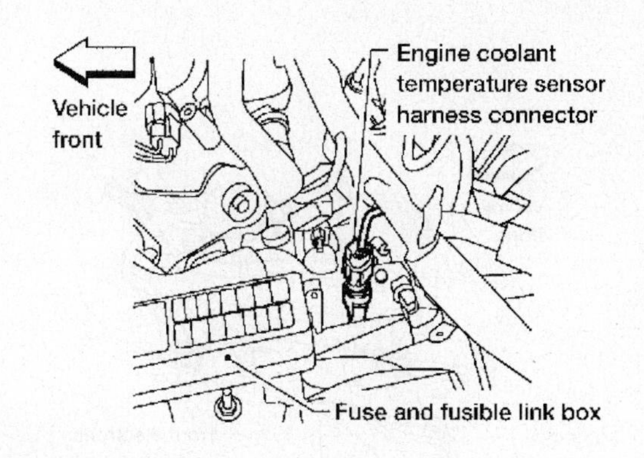
Vehicle
front
Engine coolant
temperature sensor
harness connector
Fuse and fusible link box

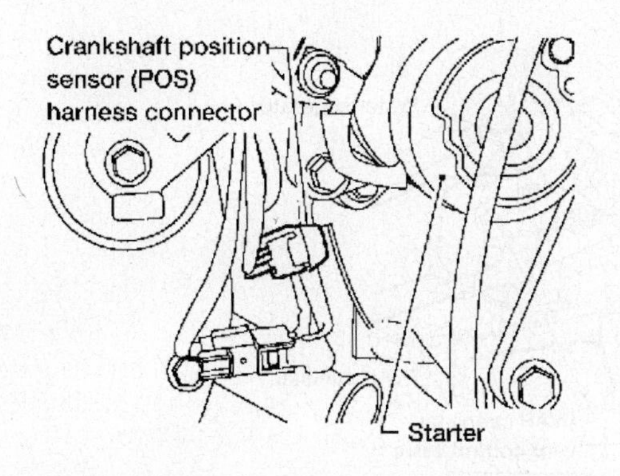
Crankshaft position
sensor (POS)
harness connector
Starter

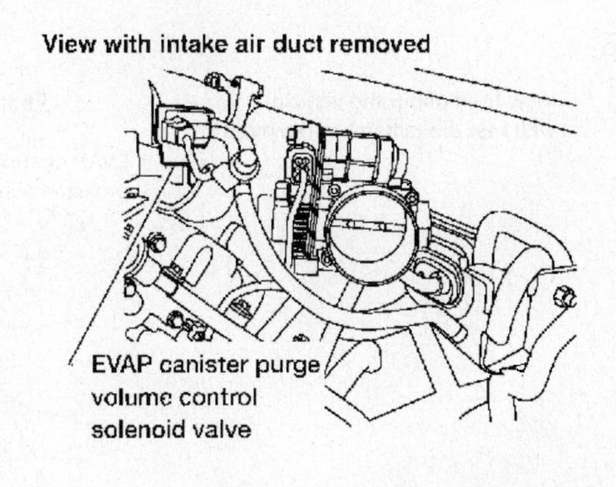
View with intake air duct removed
EVAP canister purge
volume control
solenoid valve

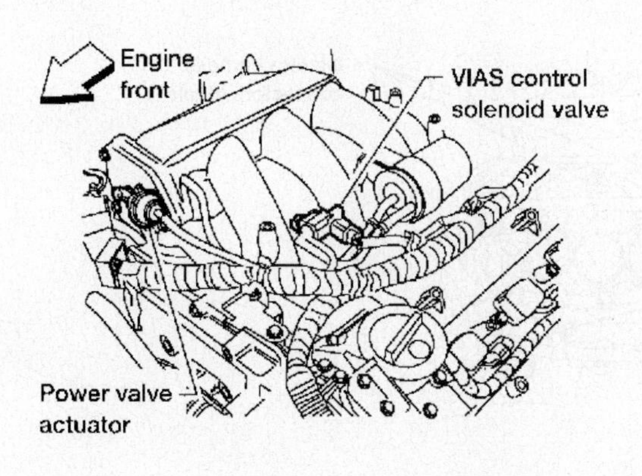
Engine
front
VIAS control
solenoid valve
Power valve
actuator

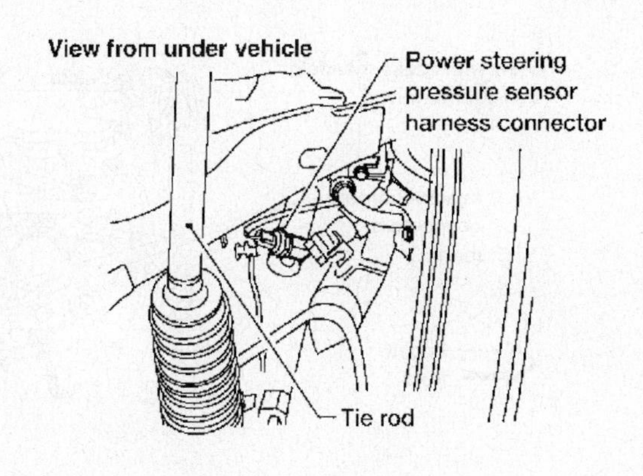
View from under vehicle
Power steering
pressure sensor
harness connector
Tie rod

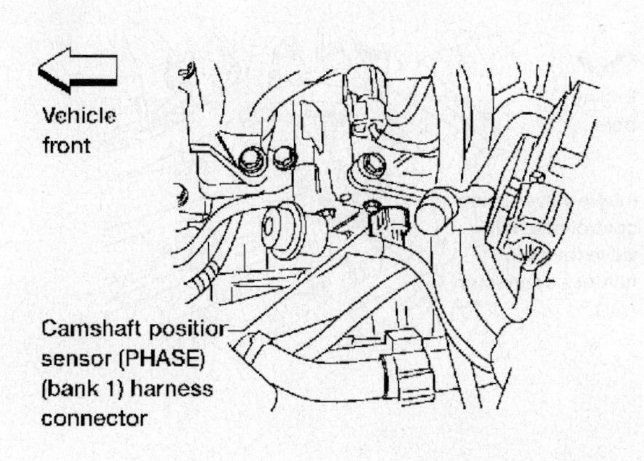
Vehicle
front
Camshaft position
sensor (PHASE)
(bank 1) harness
connector

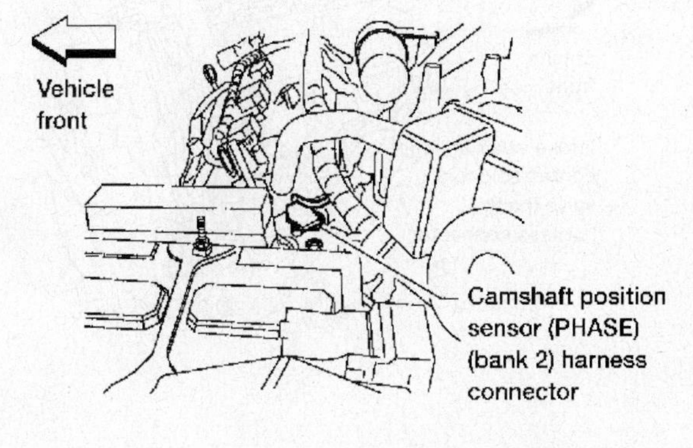
Vehicle
front
Camshaft position
sensor (PHASE)
(bank 2) harness
connector

29149_NISS_G0087

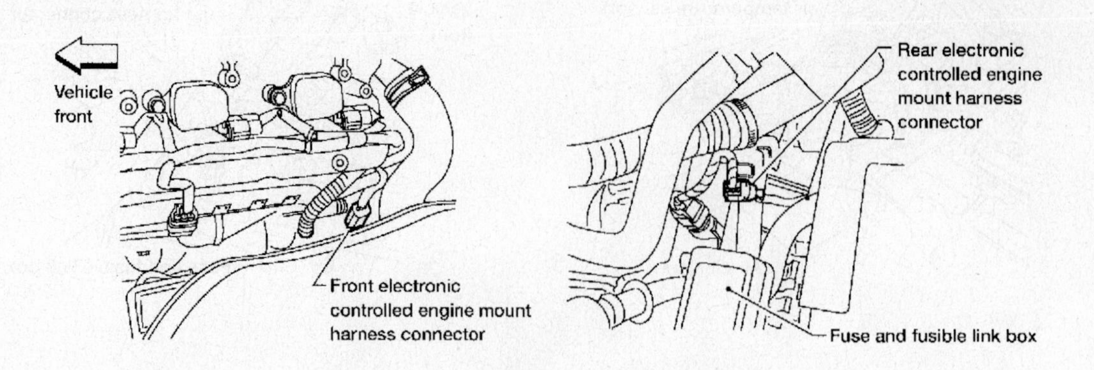
Vehicle
front
Front electronic
controlled engine mount
harness connector
Rear electronic
controlled engine
mount harness
connector
Fuse and fusible link box

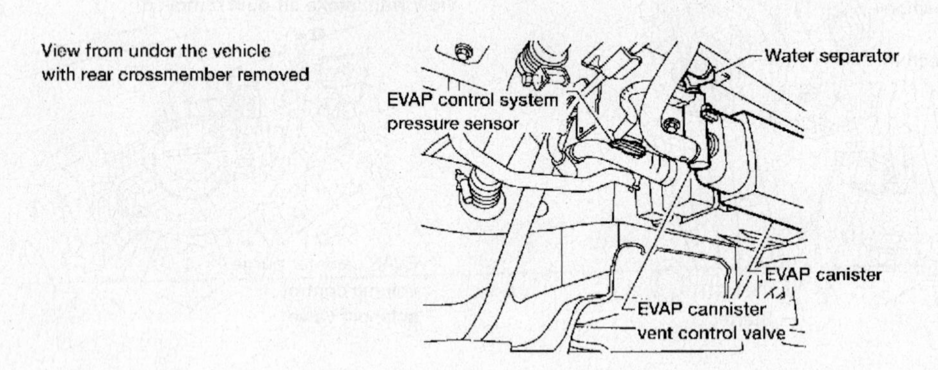
View from under the vehicle
with rear crossmember removed
EVAP control system
pressure sensor
Water separator
EVAP canister
EVAP cannister
vent control valve

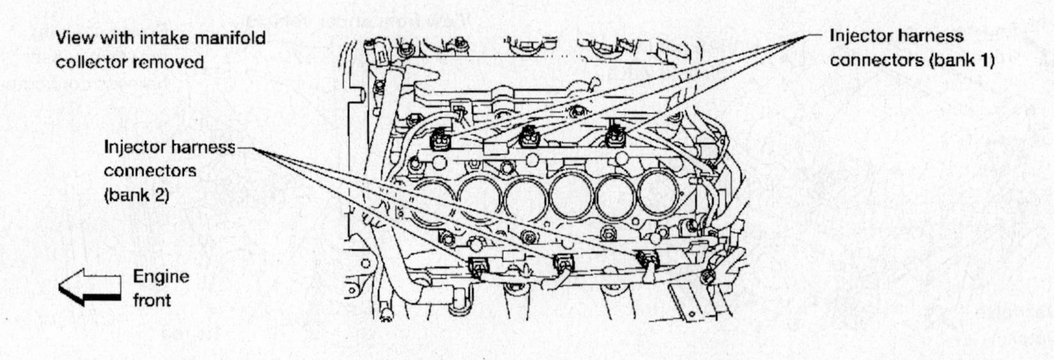
View with intake manifold
collector removed
Injector harness
connectors (bank 1)
Injector harness
connectors
(bank 2)
Engine
front

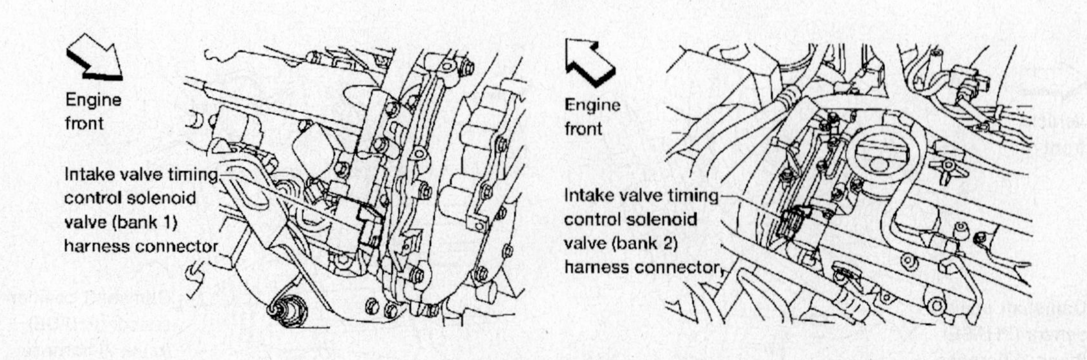
Engine
front
Intake valve timing
control solenoid
valve (bank 1)
harness connector
Engine
front
Intake valve timing
control solenoid
valve (bank 2)
harness connector

29149_NISS_G0088

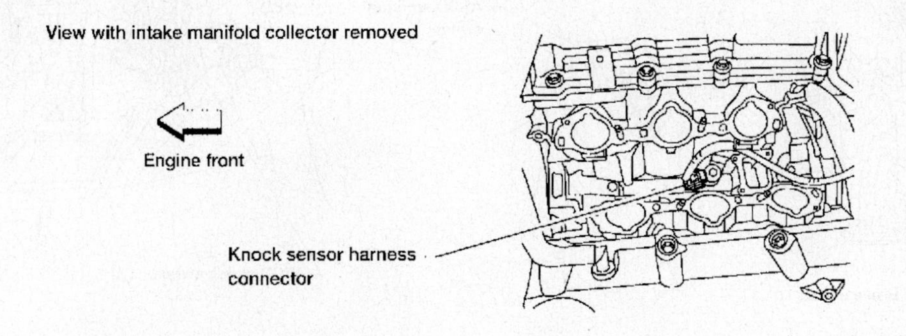
View with intake manifold collector removed
Engine front
Knock sensor harness connector

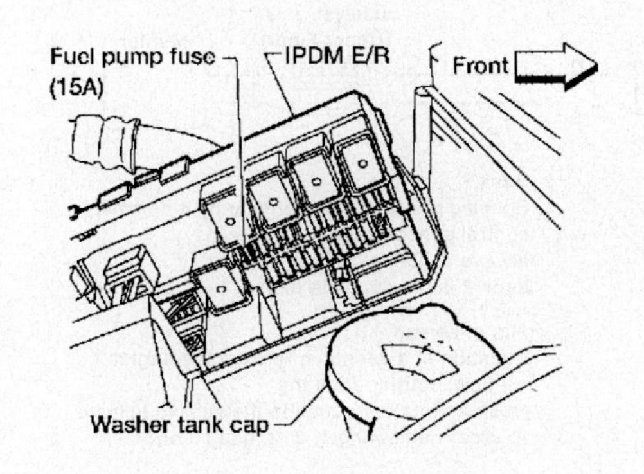
Fuel pump fuse (15A)
IPDM E/R
Front
Washer tank cap

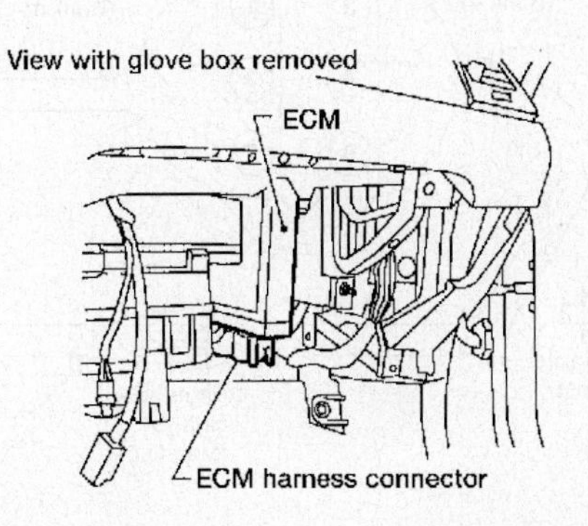
View with glove box removed
ECM
ECM harness connector

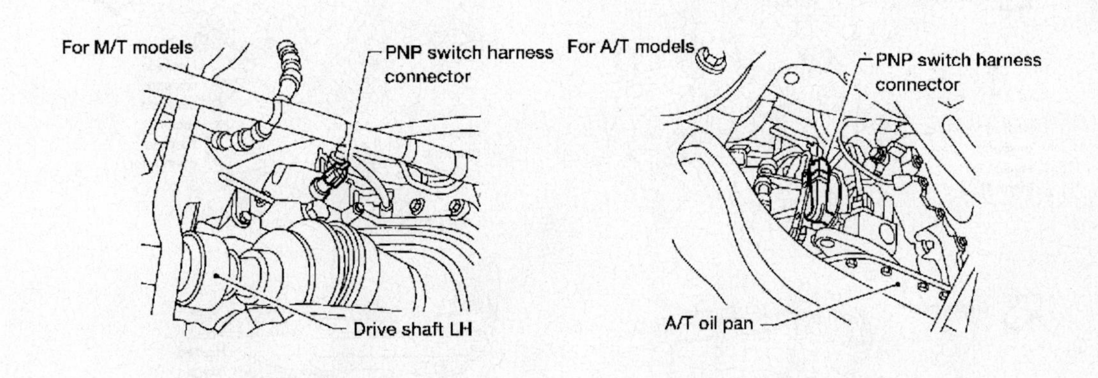
For M/T models
PNP switch harness connector
Drive shaft LH
For A/T models
PNP switch harness connector
A/T oil pan

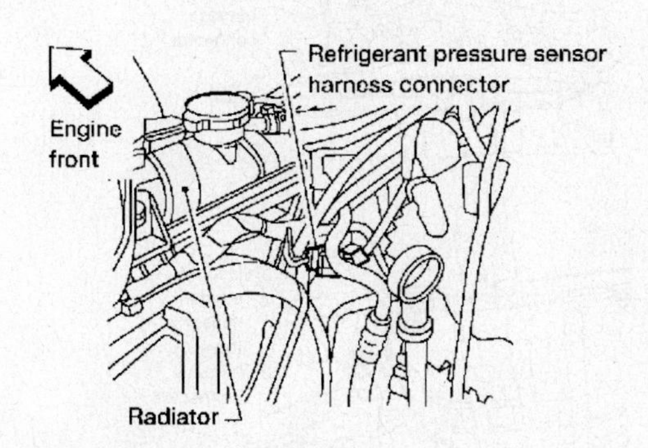
Engine front
Refrigerant pressure sensor harness connector
Radiator

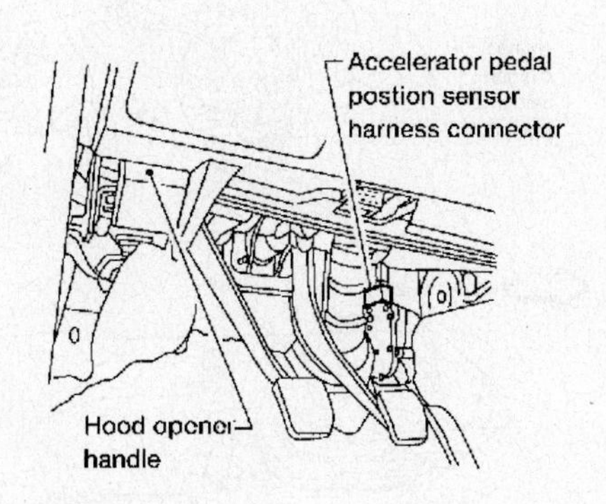
Accelerator pedal postion sensor harness connector
Hood opener handle

29149_NISS_G0089

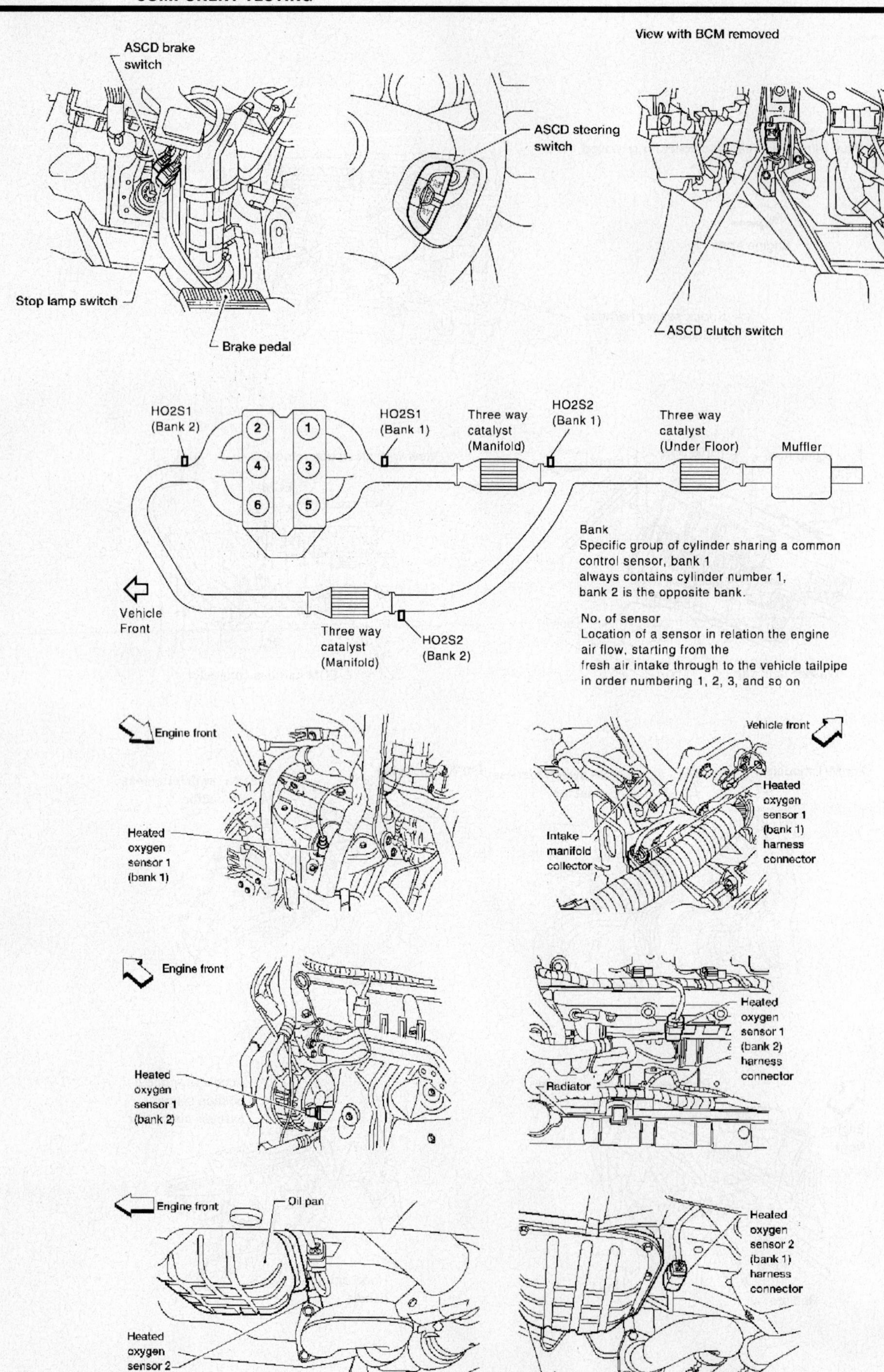

29149_NISS_G0091

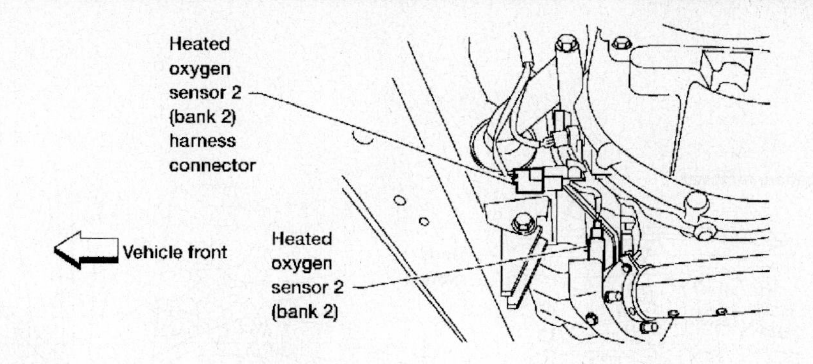
Heated
oxygen
sensor 2
(bank 2)
harness
connector
Vehicle front
Heated
oxygen
sensor 2
(bank 2)

29149_NISS_G0092

ARMADA

2005–2006

VK56DE V8 ENGINE

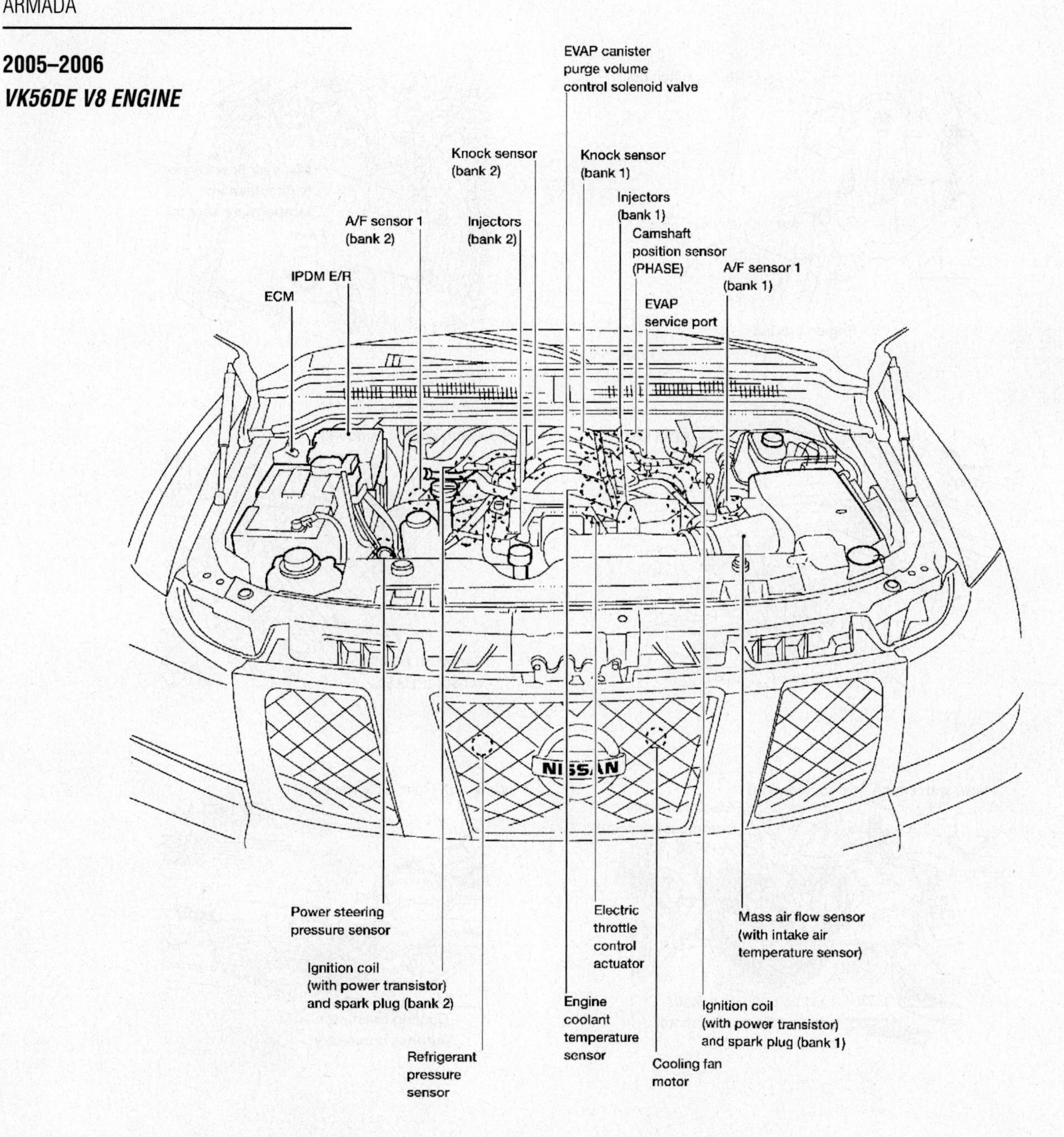
EVAP canister
purge volume
control solenoid valve
Knock sensor
(bank 2)
Knock sensor
(bank 1)
Injectors
(bank 1)
A/F sensor 1
(bank 2)
Injectors
(bank 2)
Camshaft
position sensor
(PHASE)
A/F sensor 1
(bank 1)
IPDM E/R
ECM
EVAP
service port
Power steering
pressure sensor
Ignition coil
(with power transistor)
and spark plug (bank 2)
Refrigerant
pressure
sensor
Electric
throttle
control
actuator
Engine
coolant
temperature
sensor
Cooling fan
motor
Mass air flow sensor
(with intake air
temperature sensor)
Ignition coil
(with power transistor)
and spark plug (bank 1)

29149_NISS_G0135

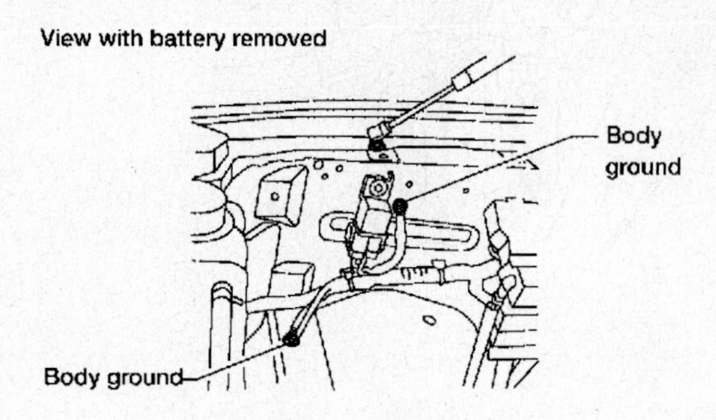
View with battery removed
Body
ground
Body ground

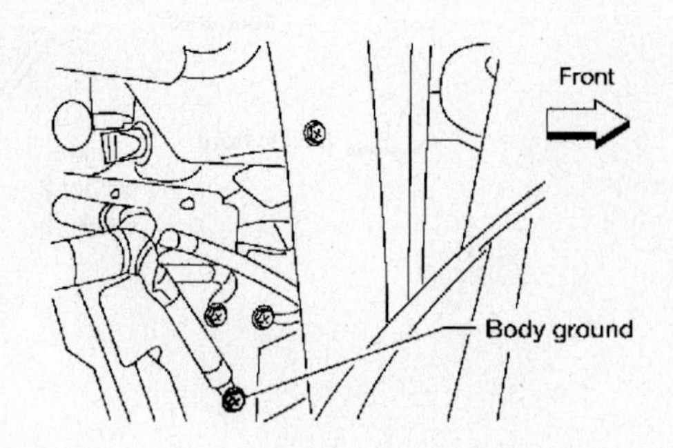
Front
Body ground

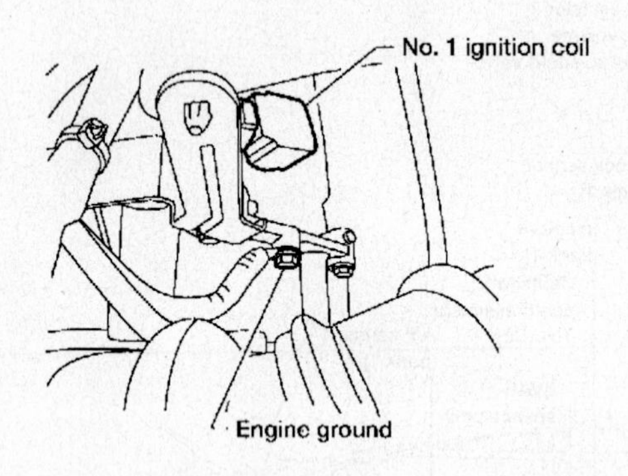
No. 1 ignition coil
Engine ground

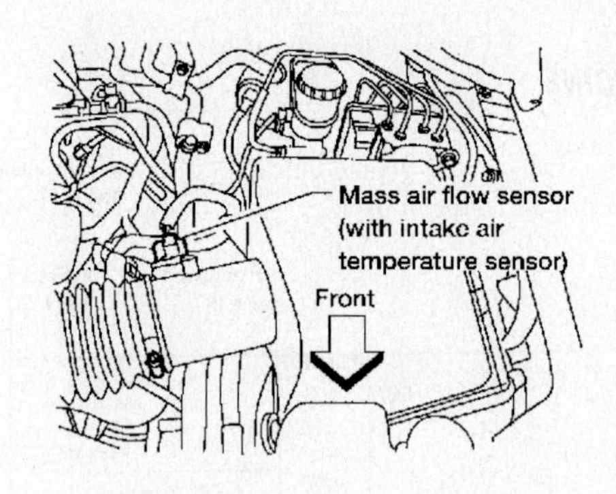
Mass air flow sensor
(with intake air
temperature sensor)
Front

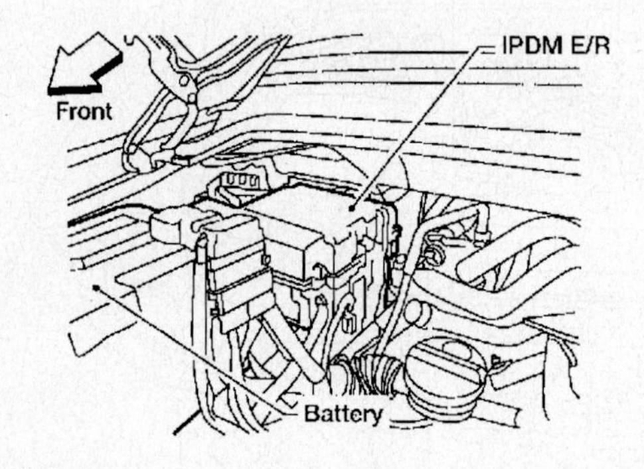
IPDM E/R
Front
Battery

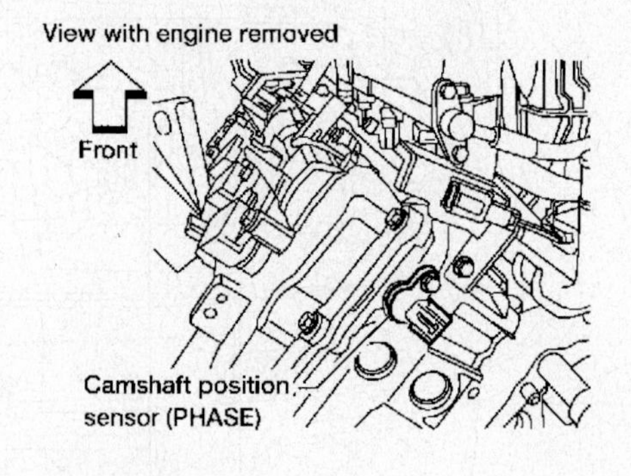
View with engine removed
Front
Camshaft position
sensor (PHASE)

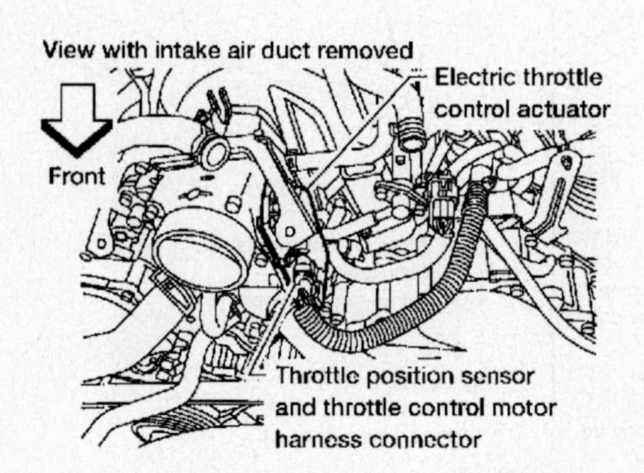
View with intake air duct removed
Electric throttle
control actuator
Front
Throttle position sensor
and throttle control motor
harness connector

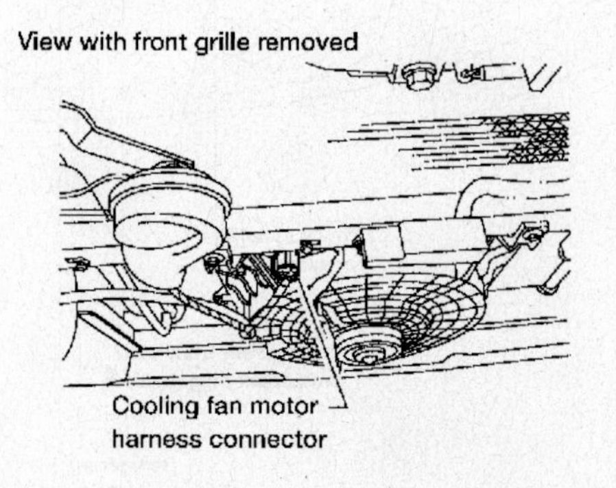
View with front grille removed
Cooling fan motor
harness connector

29149_NISS_G0136

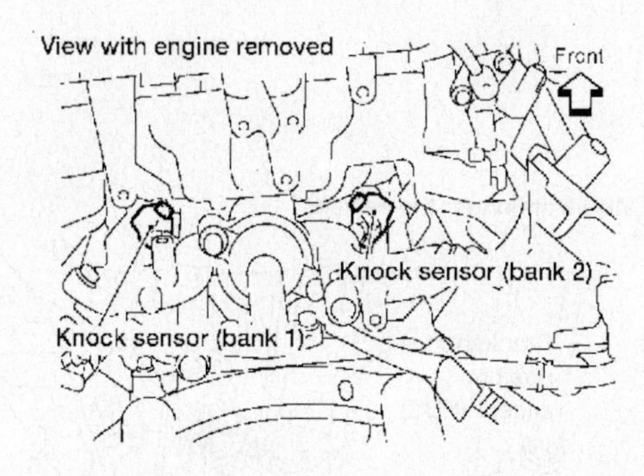

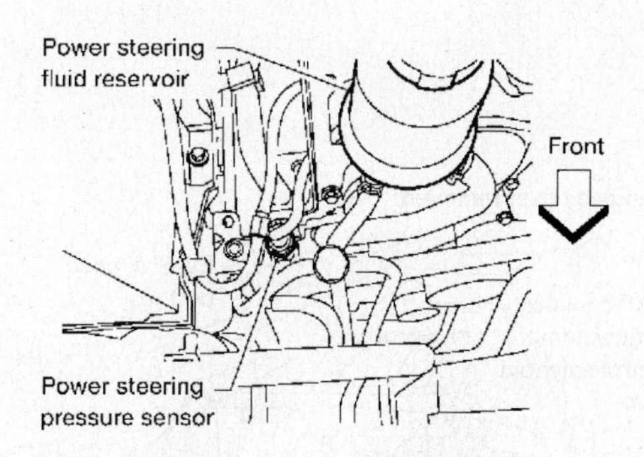

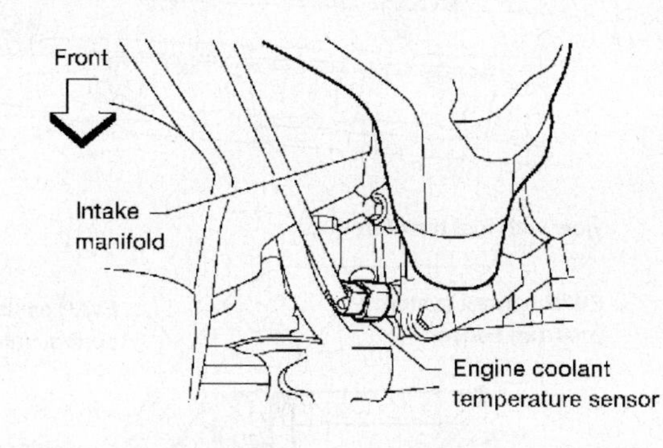

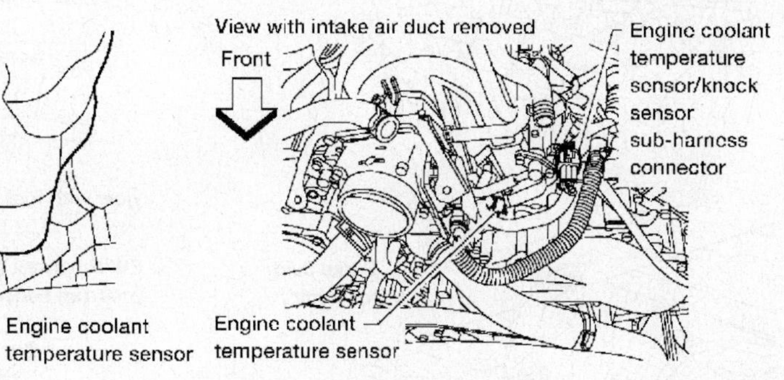

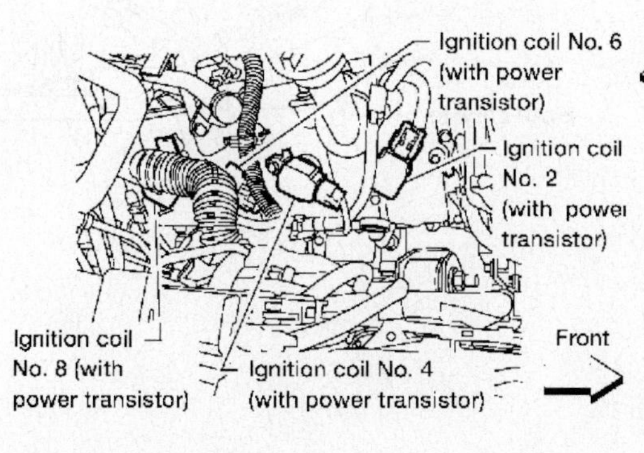

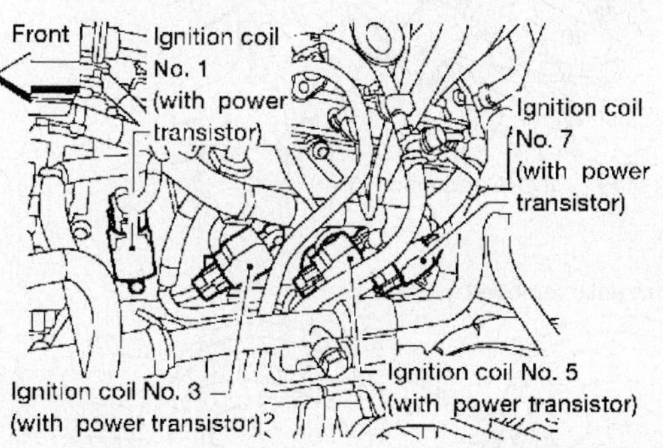

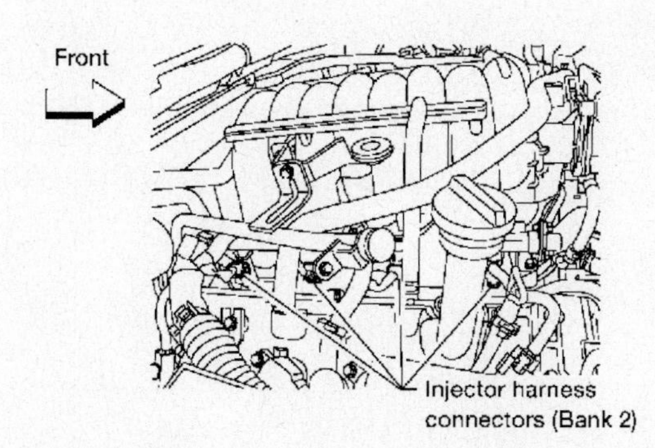

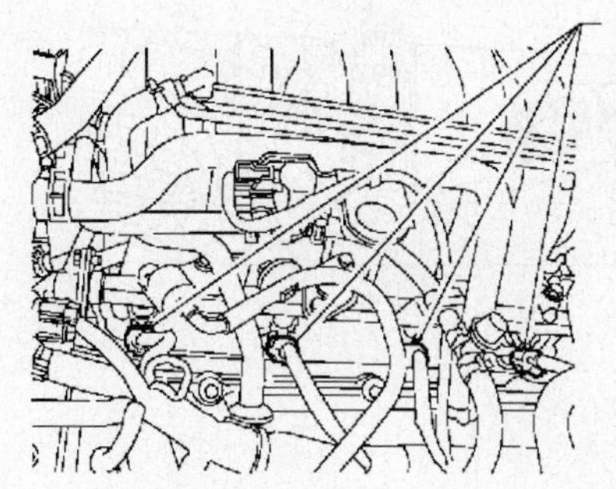

29149_NISS_G0137

View with engine cover removed

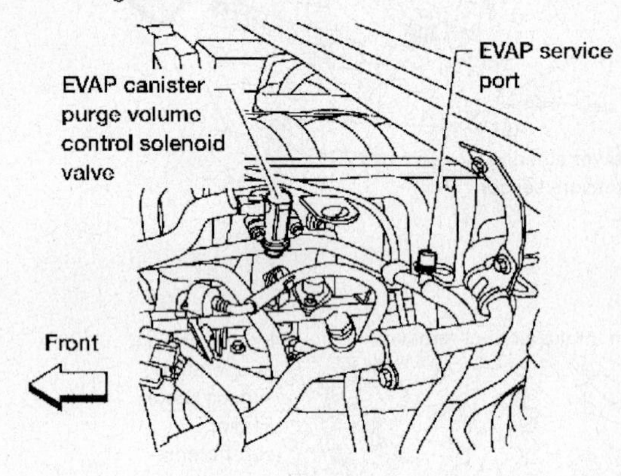

View from under the vehicle

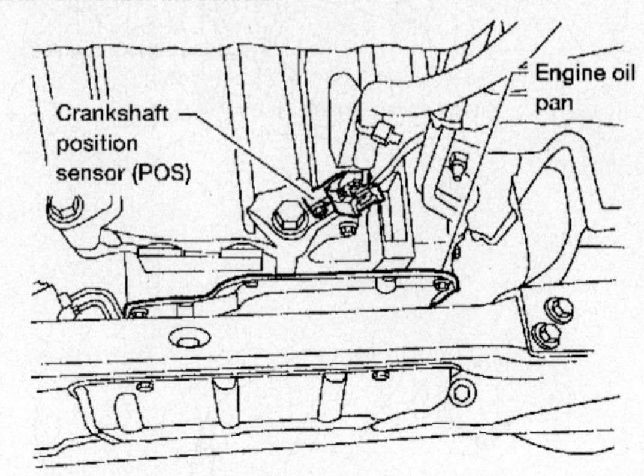

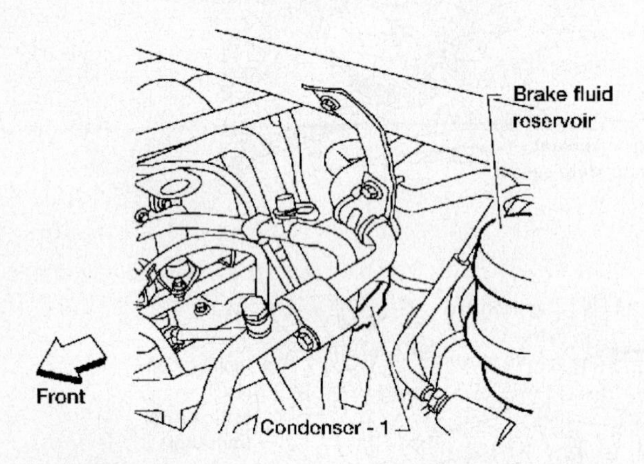

View with fuel tank removed

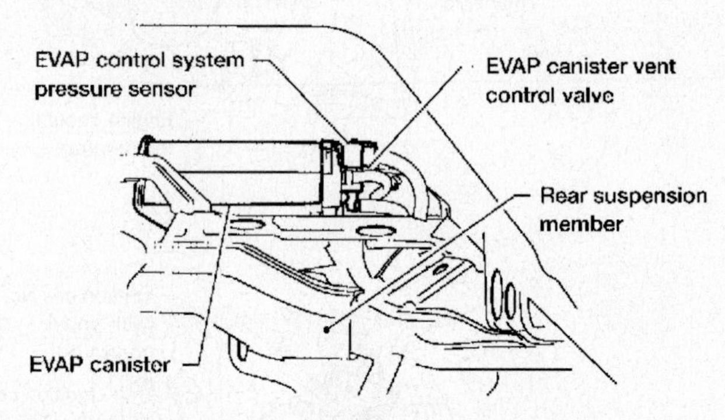

View with front grille removed

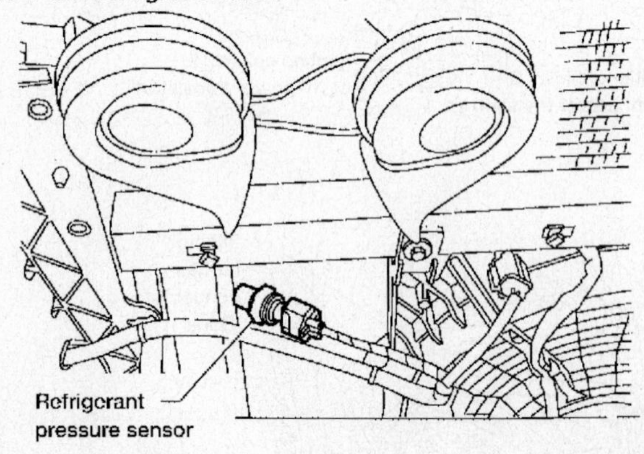

29149_NISS_G0138

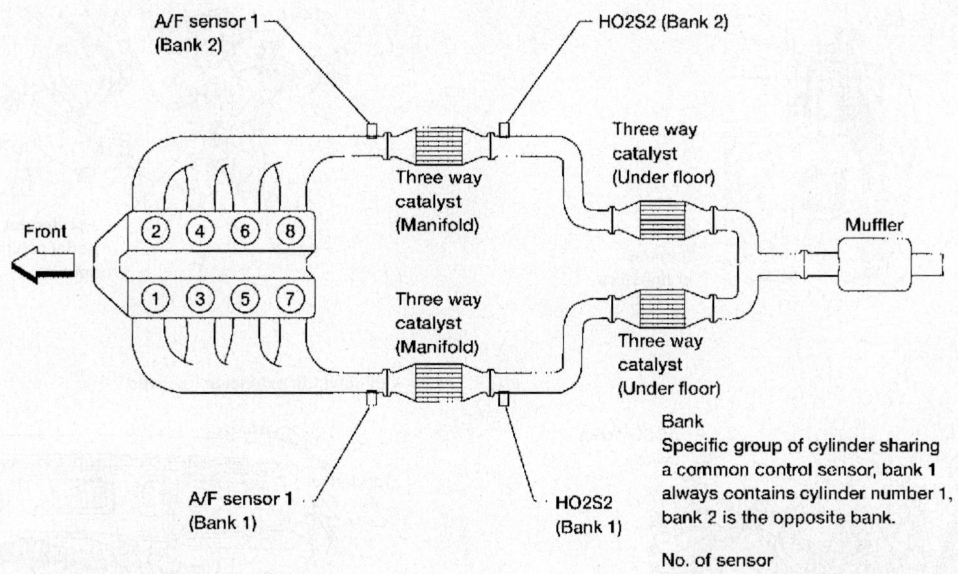

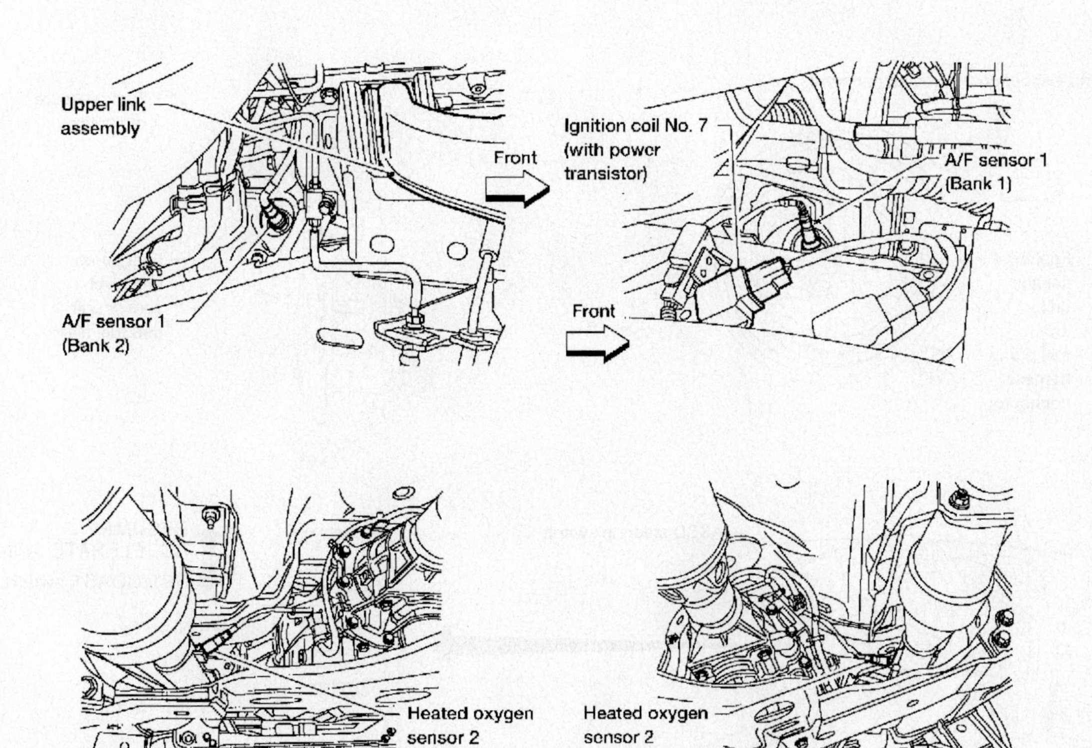

29149_NISS_G0139

View with battery removed

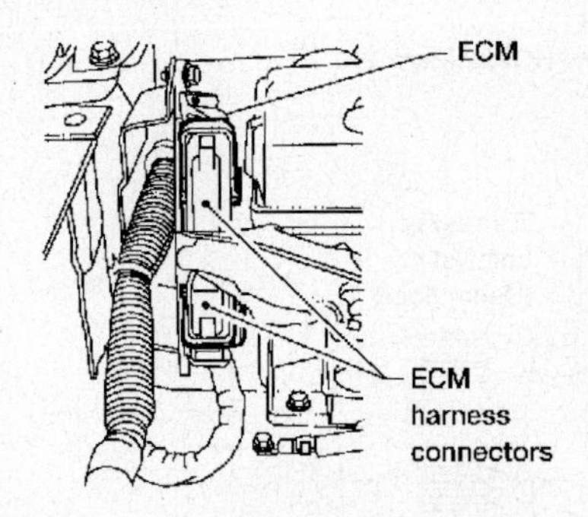

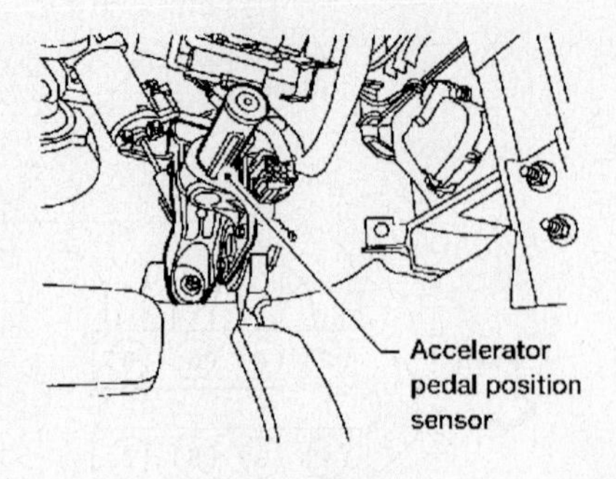

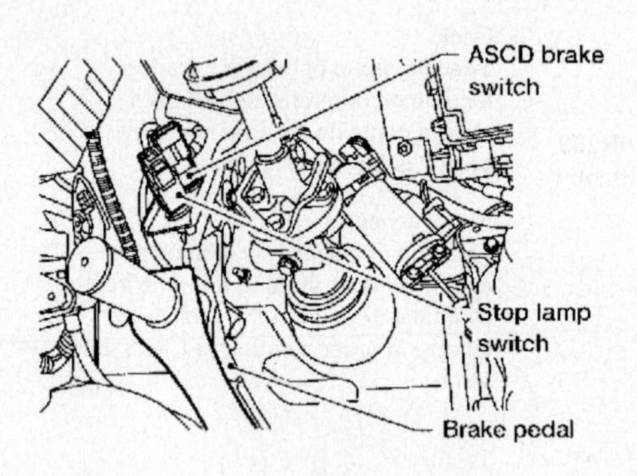

View with cowl top extension removed

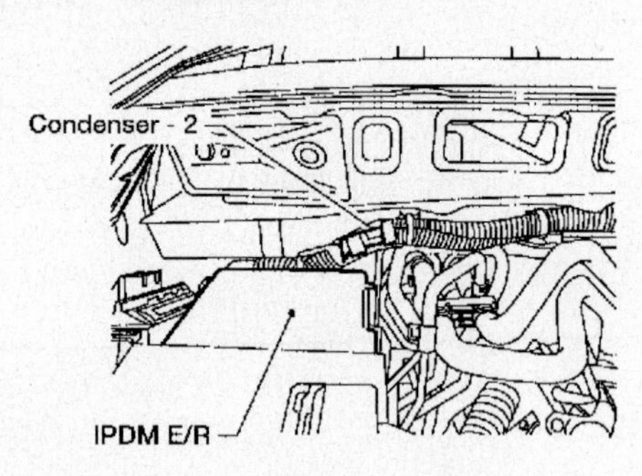

View with inspection hole cover removed

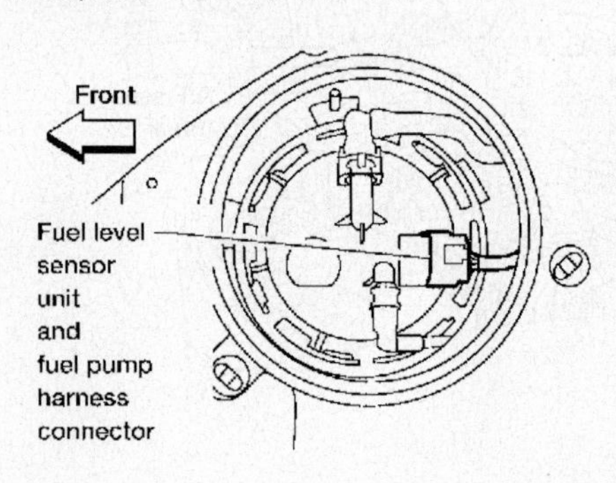

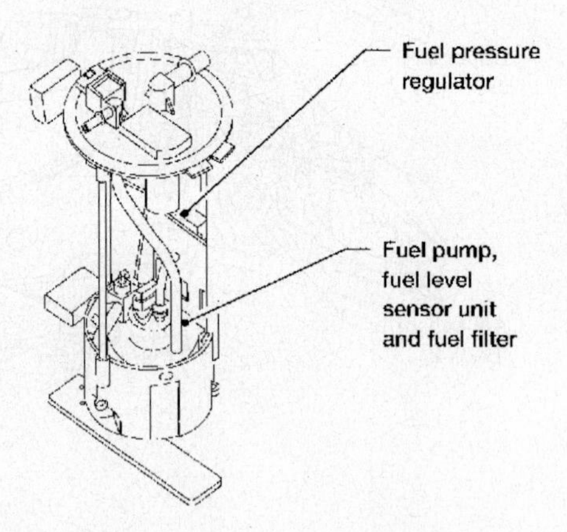

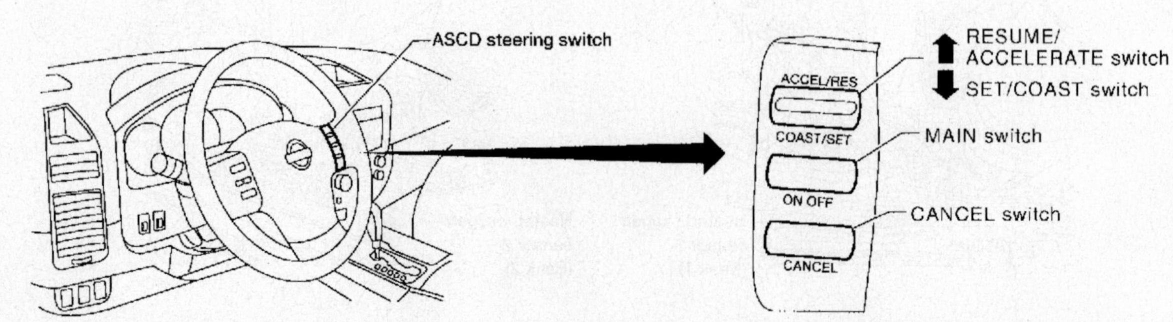

29149_NISS_G0140

FRONTIER

1998–2004

KA24DE I4 ENGINE

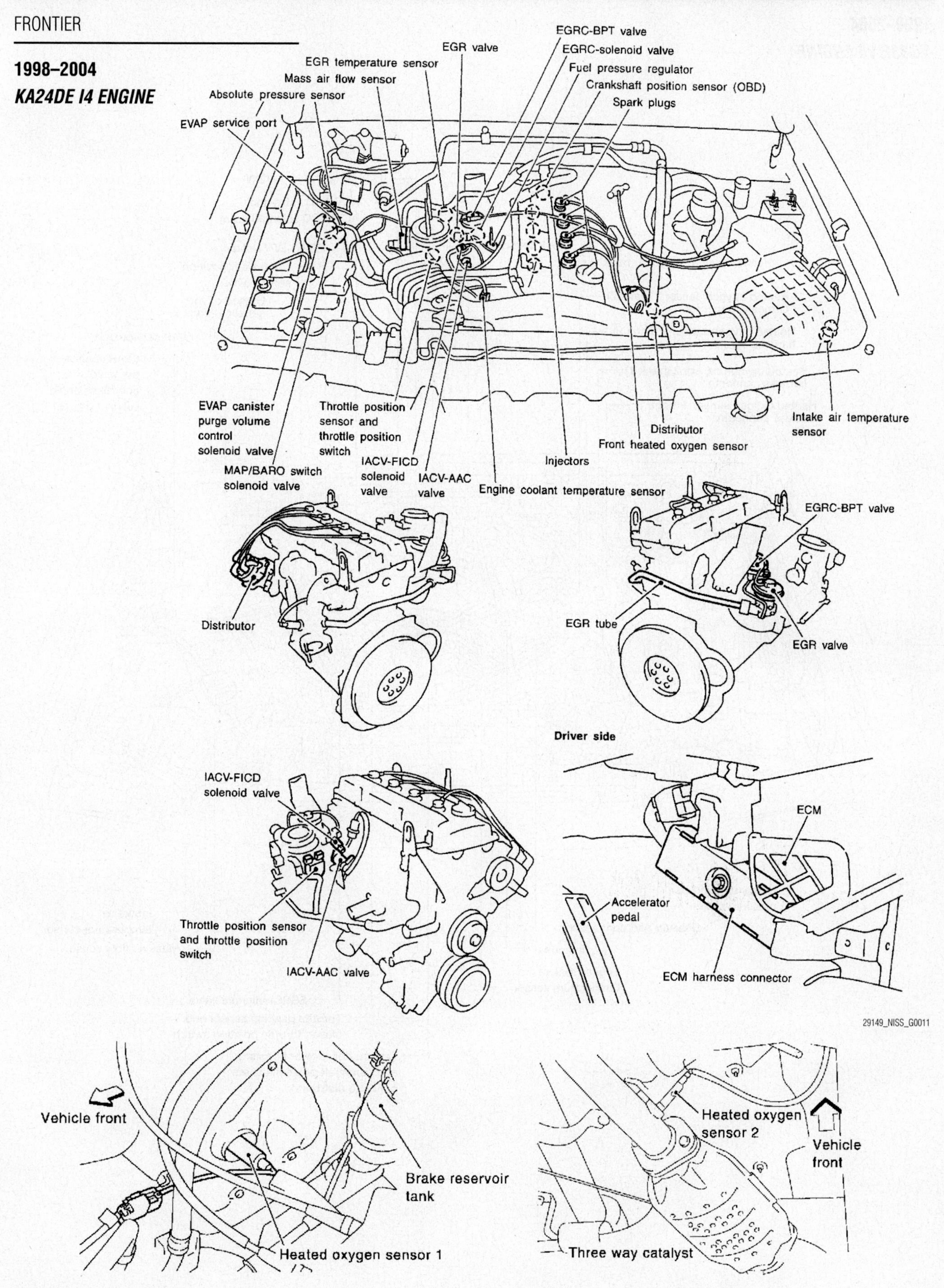

1999–2004

VG33E V6 ENGINE

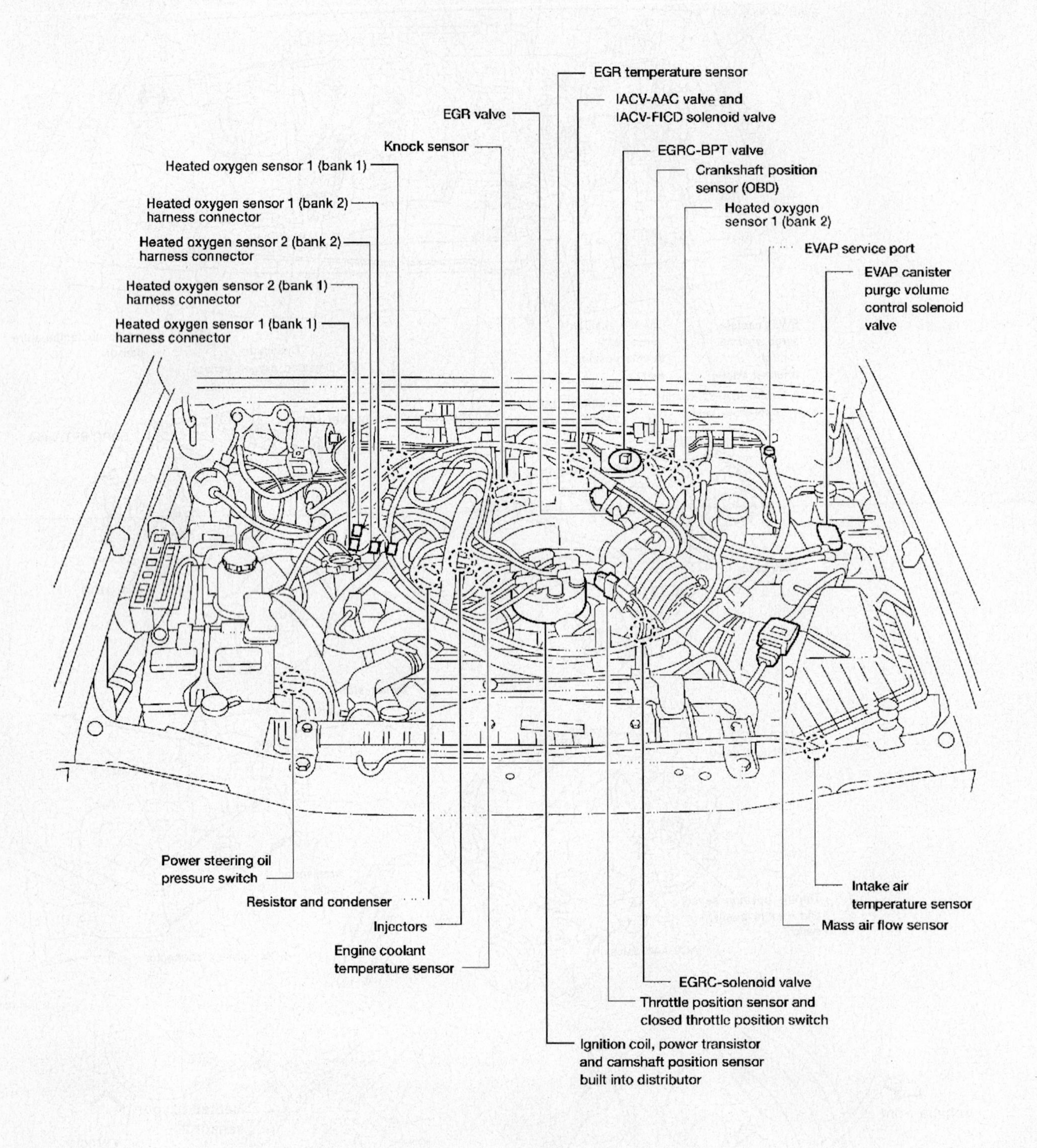
EGR temperature sensor
IACV-AAC valve and
IACV-FICD solenoid valve
EGR valve
Knock sensor
EGRC-BPT valve
Crankshaft position
sensor (OBD)
Heated oxygen sensor 1 (bank 1)
Heated oxygen sensor 1 (bank 2)
harness connector
Heated oxygen
sensor 1 (bank 2)
Heated oxygen sensor 2 (bank 2)
harness connector
EVAP service port
EVAP canister
purge volume
control solenoid
valve
Heated oxygen sensor 2 (bank 1)
harness connector
Heated oxygen sensor 1 (bank 1)
harness connector
Power steering oil
pressure switch
Resistor and condenser
Injectors
Engine coolant
temperature sensor
Intake air
temperature sensor
Mass air flow sensor
EGRC-solenoid valve
Throttle position sensor and
closed throttle position switch
Ignition coil, power transistor
and camshaft position sensor
built into distributor

29149_NISS_G0013

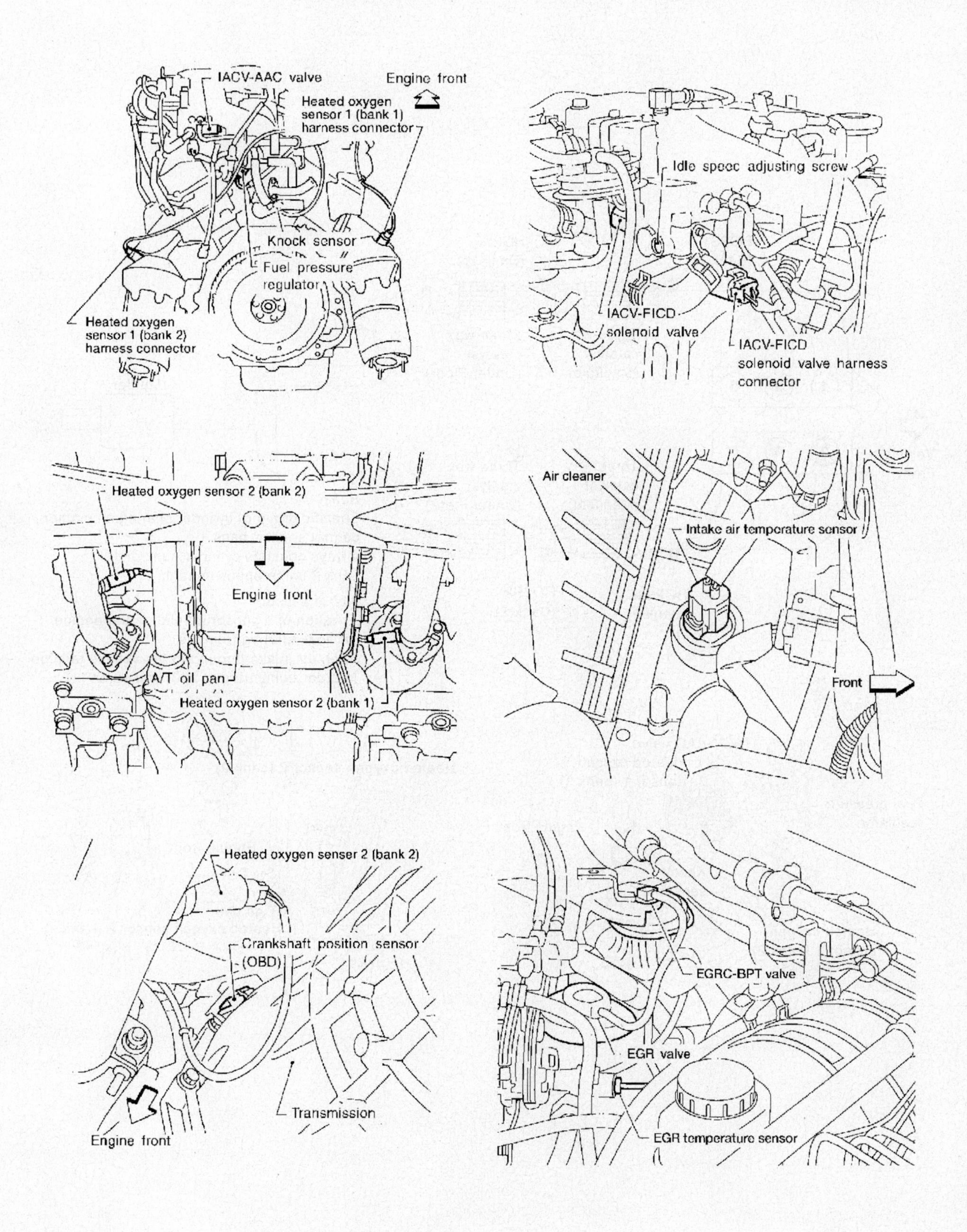
IACV-AAC valve
Engine front
Heated oxygen sensor 1 (bank 1) harness connector
Knock sensor
Fuel pressure regulator
Heated oxygen sensor 1 (bank 2) harness connector
Idle speed adjusting screw
IACV-FICD solenoid valve
IACV-FICD solenoid valve harness connector
Heated oxygen sensor 2 (bank 2)
Engine front
A/T oil pan
Heated oxygen sensor 2 (bank 1)
Air cleaner
Intake air temperature sensor
Front
Heated oxygen senser 2 (bank 2)
Crankshaft position sensor (OBD)
Transmission
Engine front
EGRC-BPT valve
EGR valve
EGR temperature sensor

29149_NISS_G0014

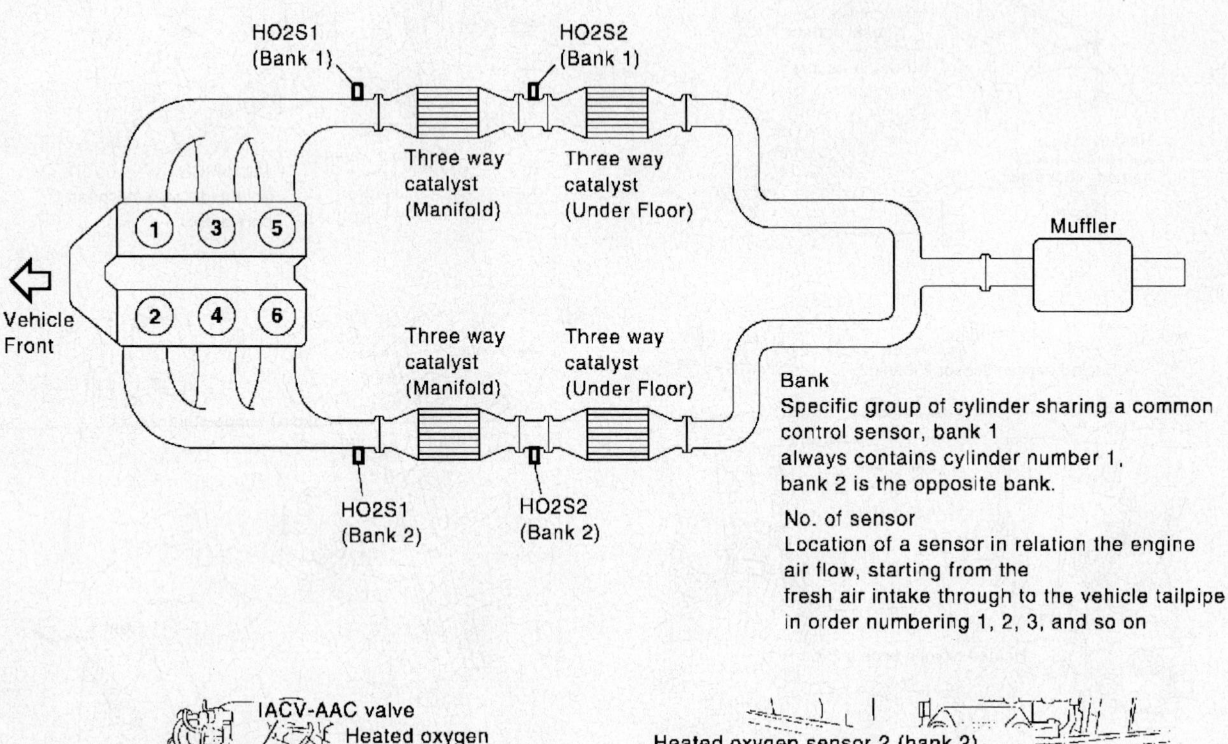

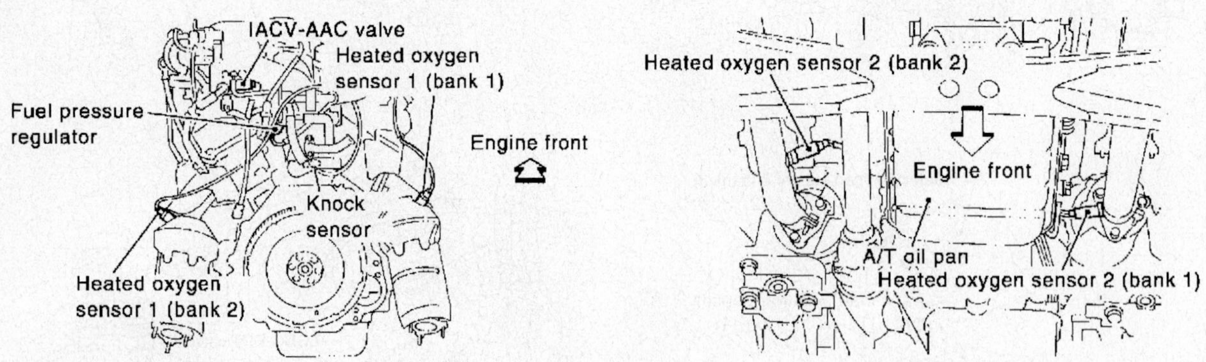

29149_NISS_G0015

2001–2004

VG33ER V6 ENGINE

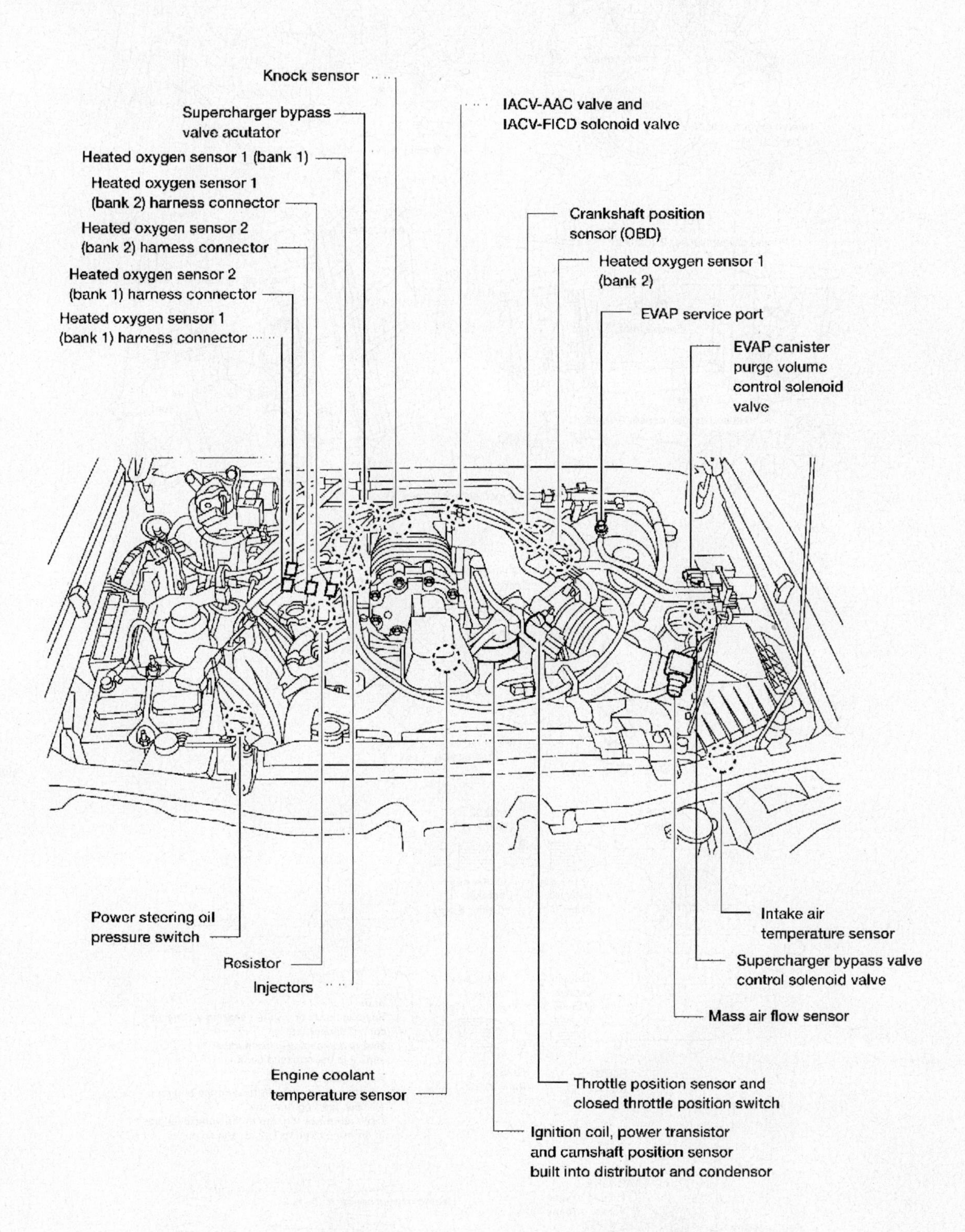

29149_NISS_G0093

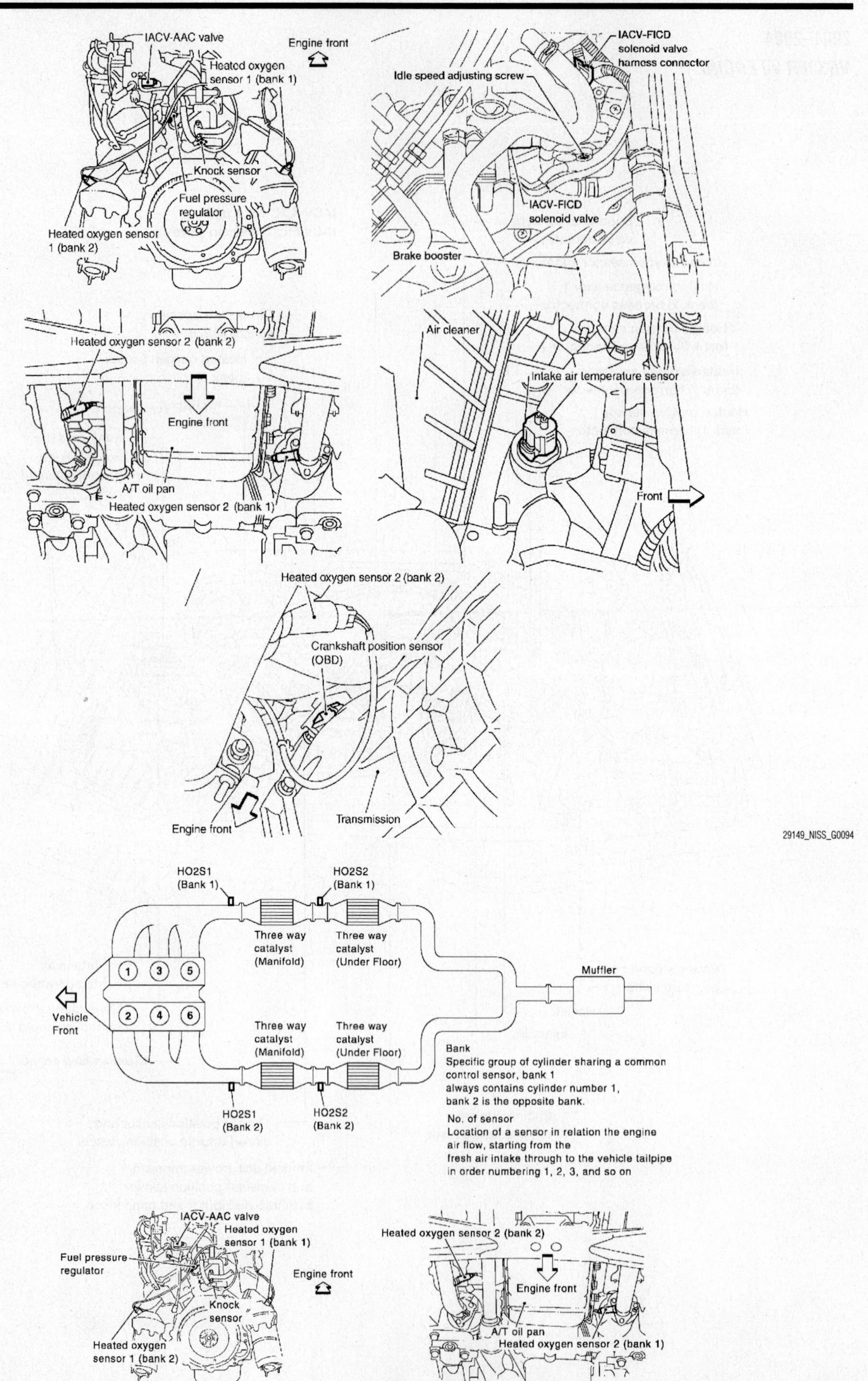
IACV-AAC valve
Engine front
Heated oxygen sensor 1 (bank 1)
Knock sensor
Fuel pressure regulator
Heated oxygen sensor 1 (bank 2)
IACV-FICD solenoid valve harness connector
Idle speed adjusting screw
IACV-FICD solenoid valve
Brake booster
Heated oxygen sensor 2 (bank 2)
Engine front
A/T oil pan
Heated oxygen sensor 2 (bank 1)
Air cleaner
Intake air temperature sensor
Front
Heated oxygen sensor 2 (bank 2)
Crankshaft position sensor (OBD)
Engine front
Transmission
29149_NISS_G0094
HO2S1 (Bank 1)
HO2S2 (Bank 1)
Three way catalyst (Manifold)
Three way catalyst (Under Floor)
Muffler
Vehicle Front
Three way catalyst (Manifold)
Three way catalyst (Under Floor)
Bank
Specific group of cylinder sharing a common control sensor, bank 1 always contains cylinder number 1, bank 2 is the opposite bank.
No. of sensor
Location of a sensor in relation the engine air flow, starting from the fresh air intake through to the vehicle tailpipe in order numbering 1, 2, 3, and so on
HO2S1 (Bank 2)
HO2S2 (Bank 2)
IACV-AAC valve
Heated oxygen sensor 1 (bank 1)
Fuel pressure regulator
Engine front
Knock sensor
Heated oxygen sensor 1 (bank 2)
Heated oxygen sensor 2 (bank 2)
Engine front
A/T oil pan
Heated oxygen sensor 2 (bank 1)
29149_NISS_G0095

2005–2006

QR25DE I4 ENGINE

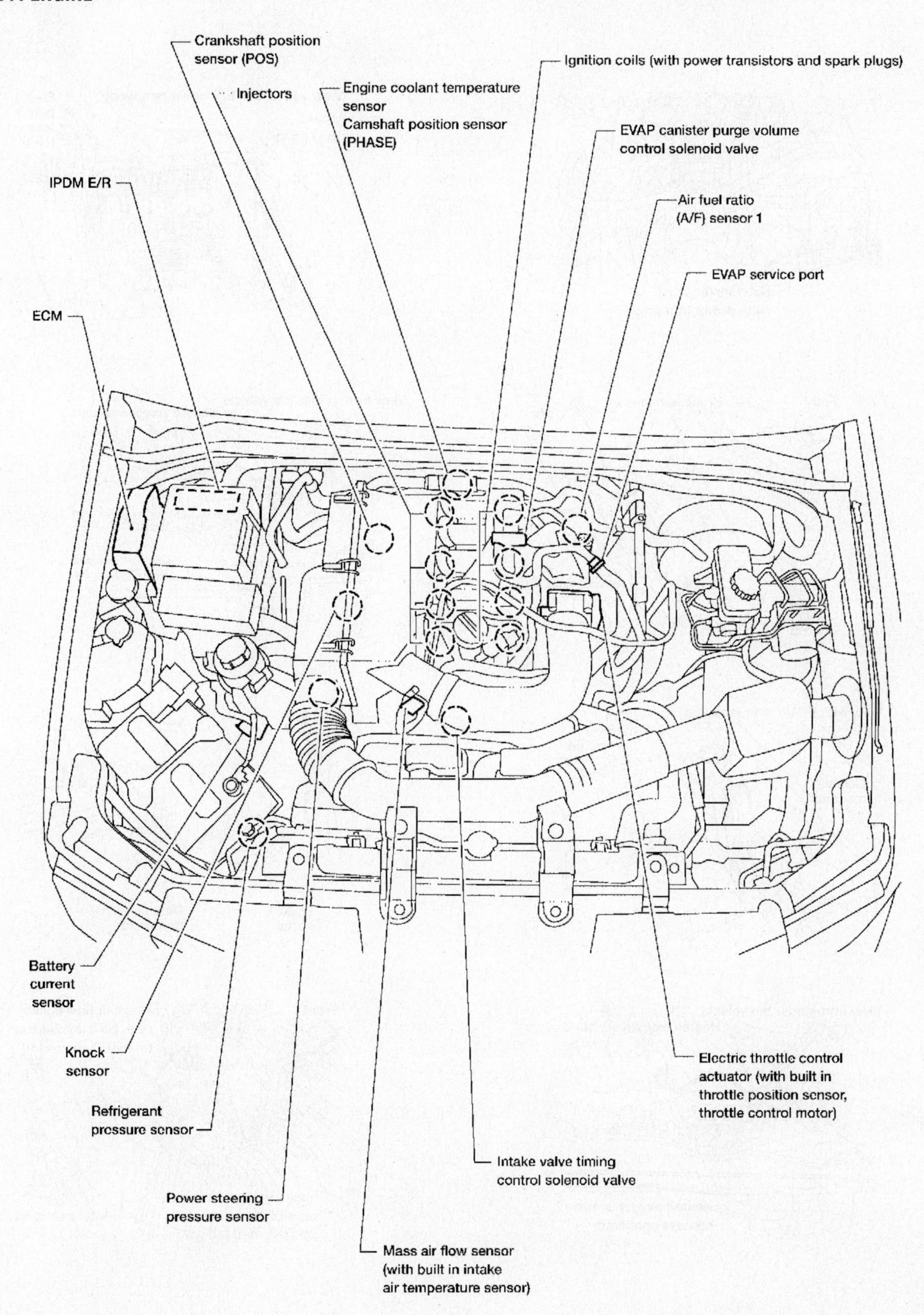
Crankshaft position sensor (POS)
Injectors
Engine coolant temperature sensor
Camshaft position sensor (PHASE)
Ignition coils (with power transistors and spark plugs)
EVAP canister purge volume control solenoid valve
IPDM E/R
Air fuel ratio (A/F) sensor 1
EVAP service port
ECM
Battery current sensor
Knock sensor
Refrigerant pressure sensor
Power steering pressure sensor
Mass air flow sensor (with built in intake air temperature sensor)
Intake valve timing control solenoid valve
Electric throttle control actuator (with built in throttle position sensor, throttle control motor)

29149_NISS_G0096

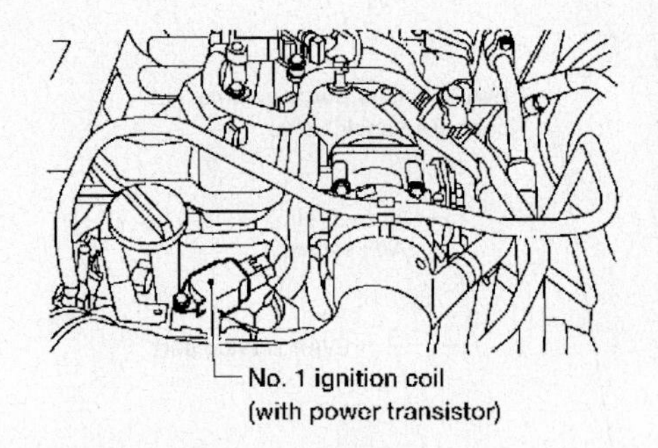
No. 1 ignition coil
(with power transistor)

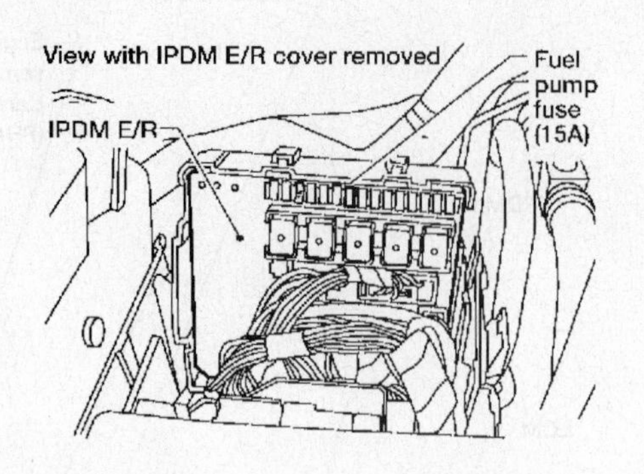
View with IPDM E/R cover removed
IPDM E/R
Fuel
pump
fuse
(15A)

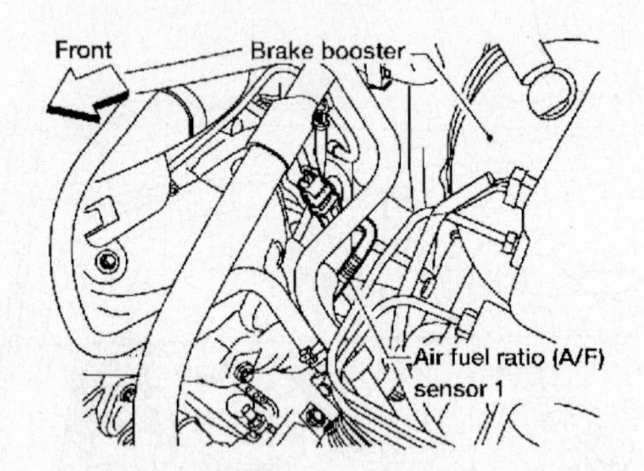
Front
Brake booster
Air fuel ratio (A/F)
sensor 1

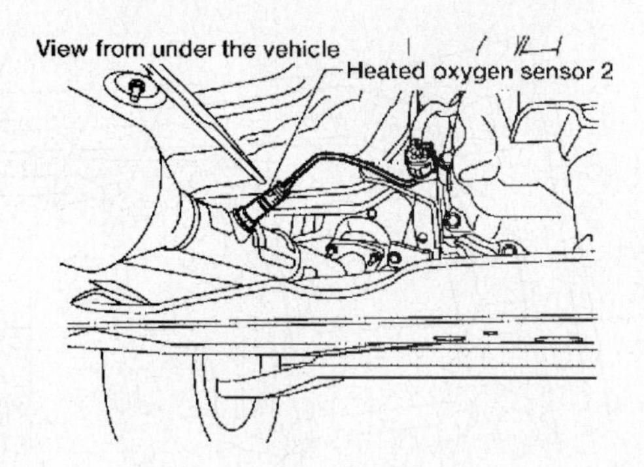
View from under the vehicle
Heated oxygen sensor 2

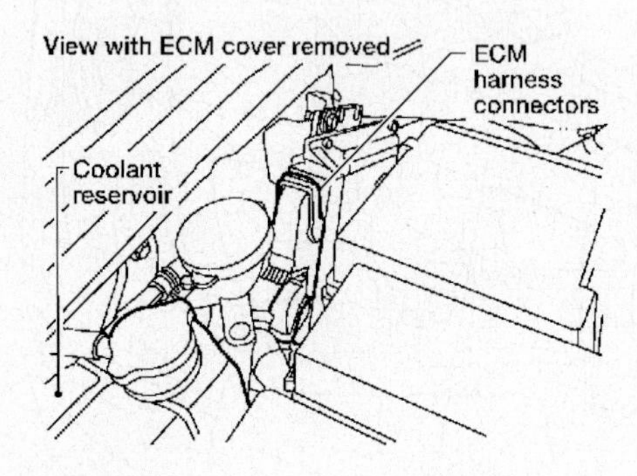
View with ECM cover removed
ECM
harness
connectors
Coolant
reservoir

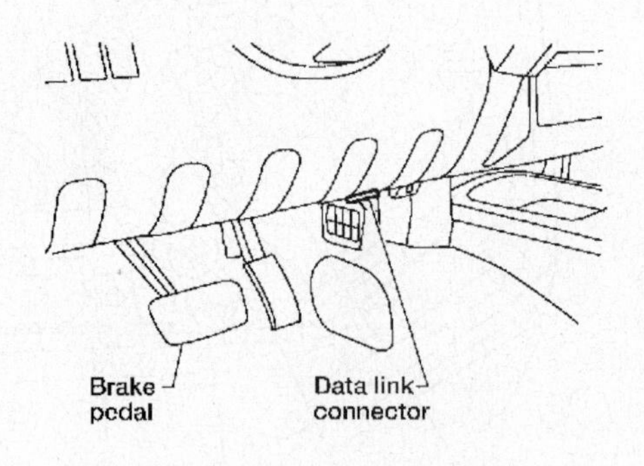
Brake
pedal
Data link
connector

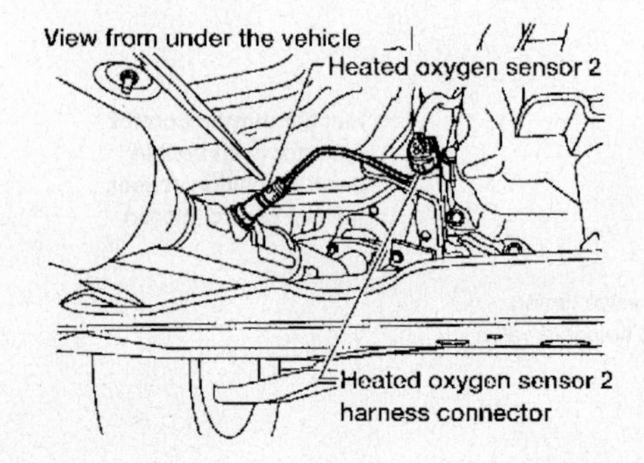
View from under the vehicle
Heated oxygen sensor 2
Heated oxygen sensor 2
harness connector

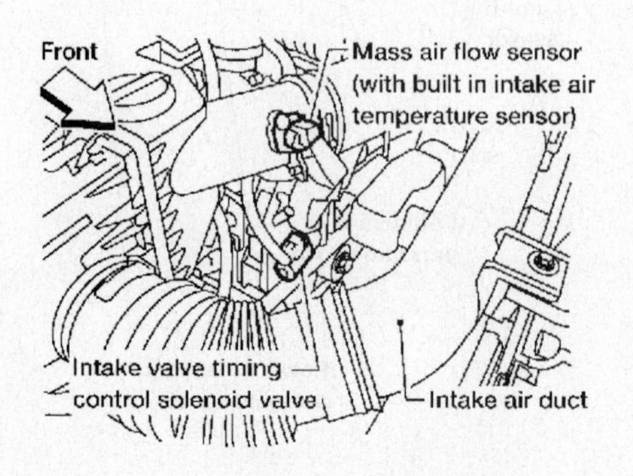
Front
Mass air flow sensor
(with built in intake air
temperature sensor)
Intake valve timing
control solenoid valve
Intake air duct

29149_NISS_G0097

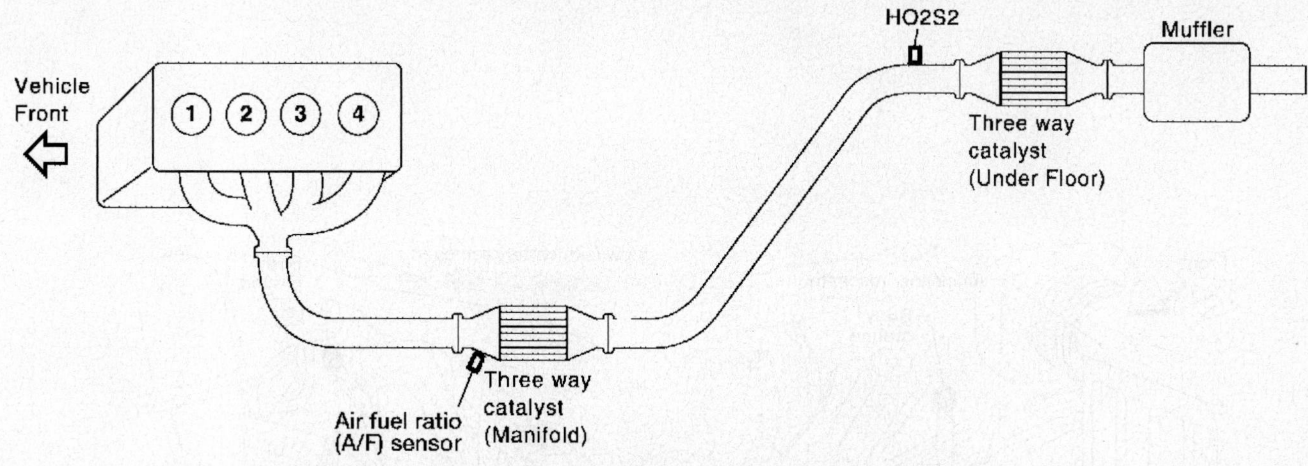

No. of sensor
Location of a sensor in relation to the engine air flow, starting from the fresh air intake through to the vehicle tailpipe in order numbering 1, 2, 3, and so on

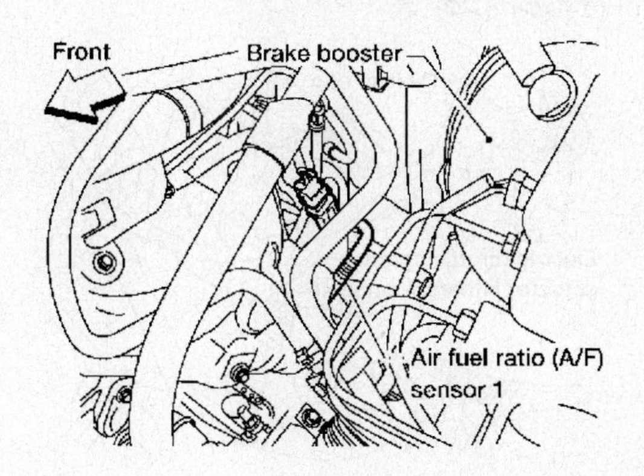

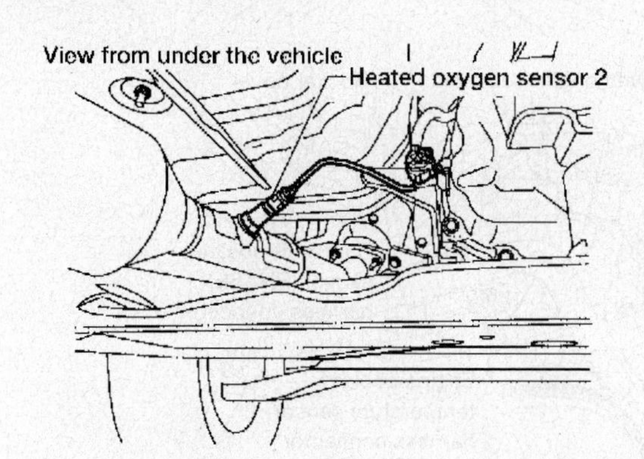

29149_NISS_G0098

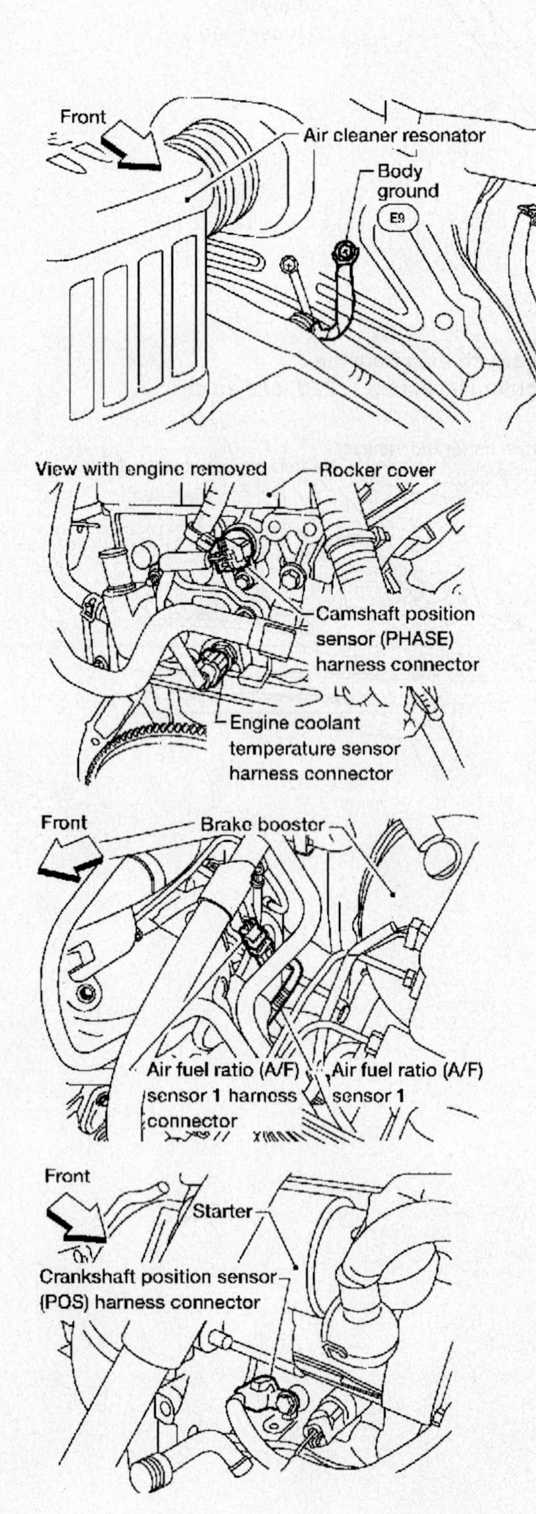
Front
Air cleaner resonator
Body
ground
E9
View with engine removed
Rocker cover
Camshaft position
sensor (PHASE)
harness connector
Engine coolant
temperature sensor
harness connector
Front
Brake booster
Air fuel ratio (A/F)
sensor 1 harness
connector
Air fuel ratio (A/F)
sensor 1
Front
Starter
Crankshaft position sensor
(POS) harness connector

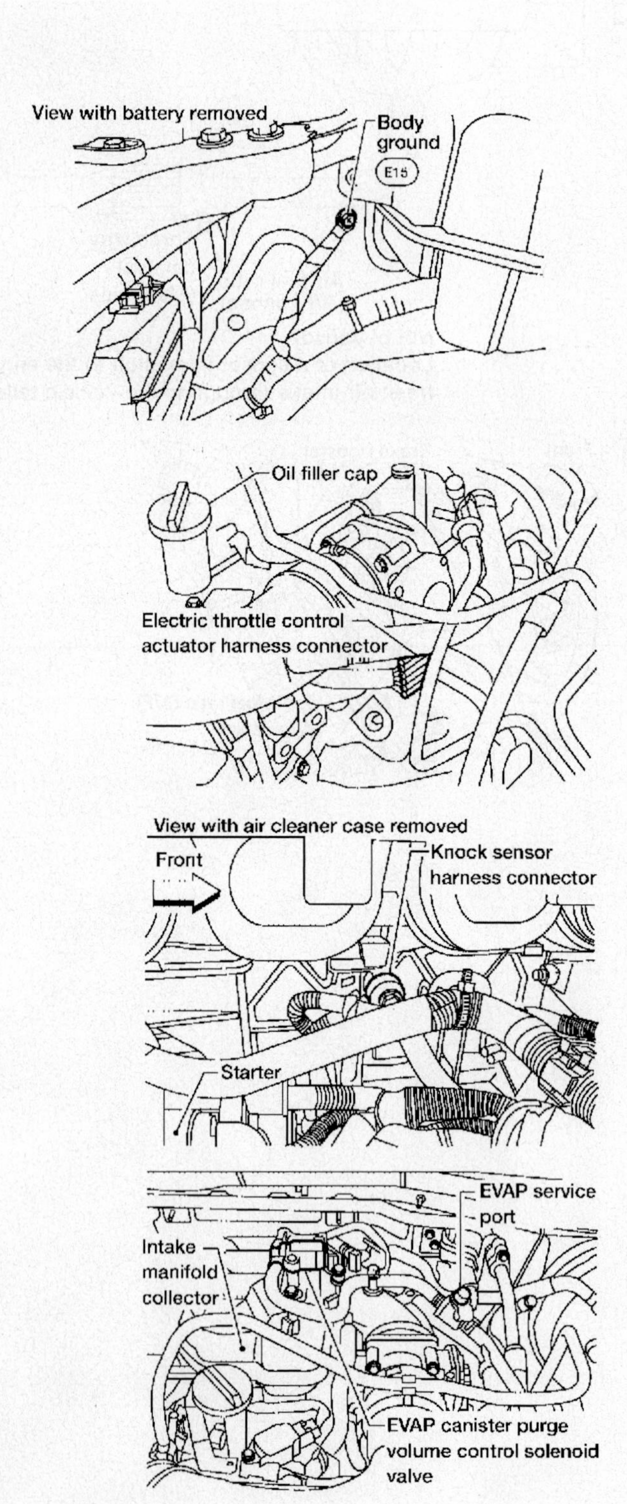
View with battery removed
Body
ground
E15
Oil filler cap
Electric throttle control
actuator harness connector
View with air cleaner case removed
Front
Knock sensor
harness connector
Starter
EVAP service
port
Intake
manifold
collector
EVAP canister purge
volume control solenoid
valve

29149_NISS_G0099

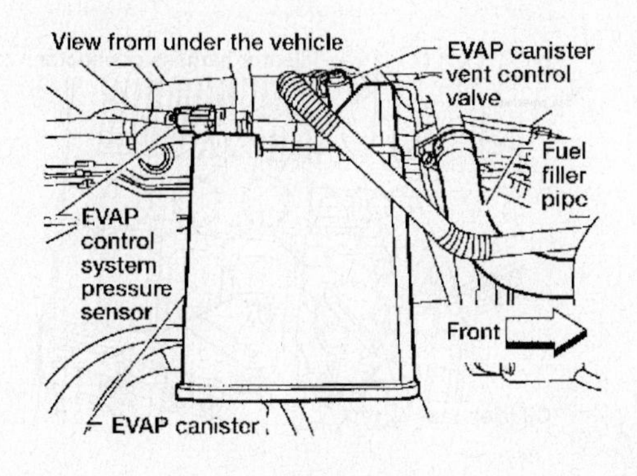
View from under the vehicle
EVAP canister vent control valve
Fuel filler pipe
EVAP control system pressure sensor
Front
EVAP canister

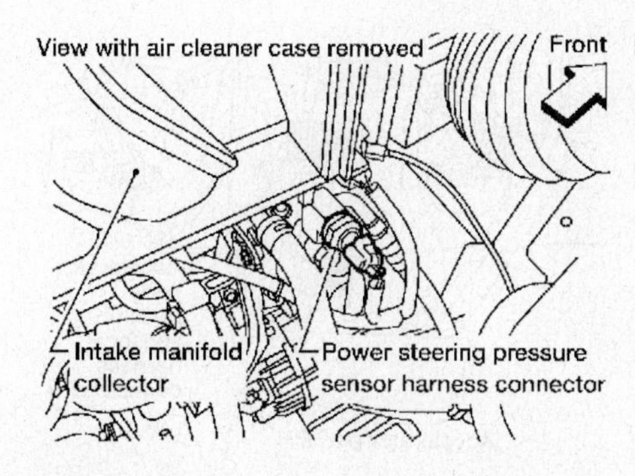
View with air cleaner case removed
Front
Intake manifold collector
Power steering pressure sensor harness connector

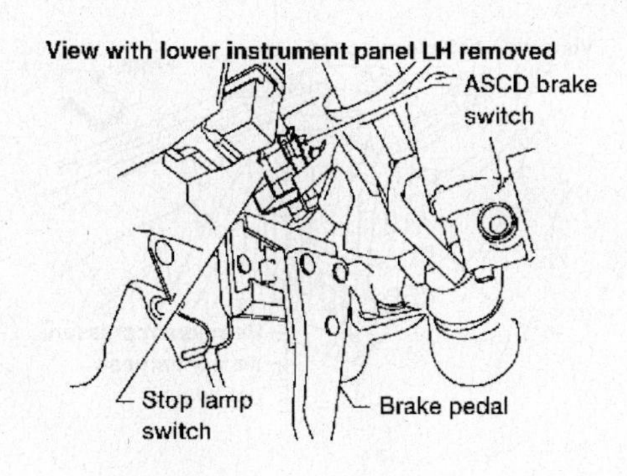
View with lower instrument panel LH removed
ASCD brake switch
Stop lamp switch
Brake pedal

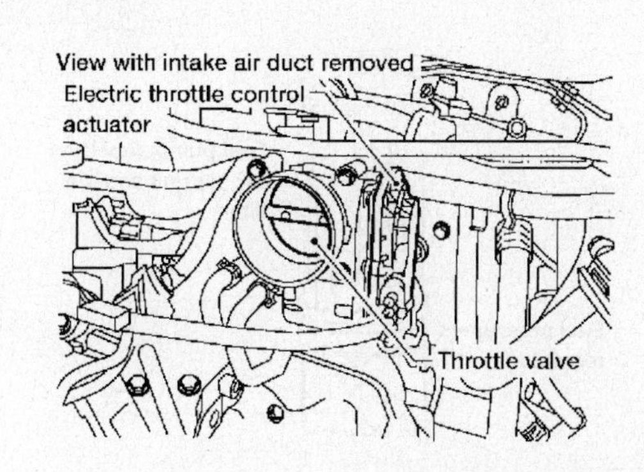
View with intake air duct removed
Electric throttle control actuator
Throttle valve

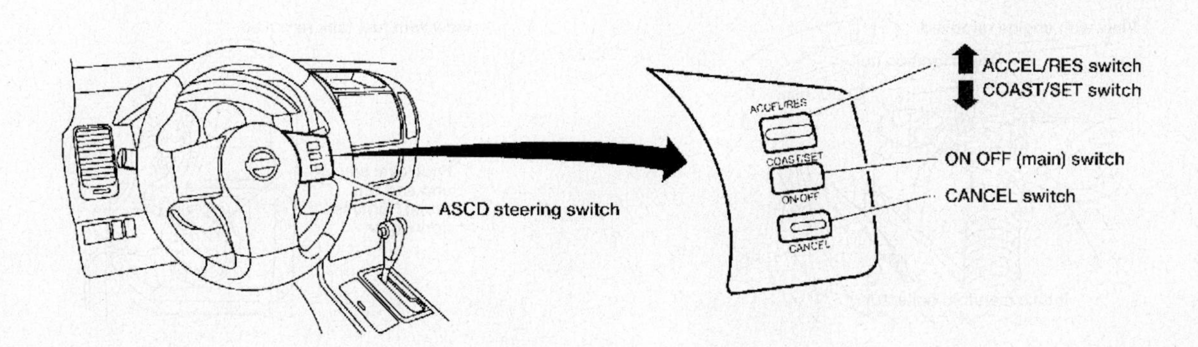
ASCD steering switch
ACCEL/RES switch
COAST/SET switch
ON OFF (main) switch
CANCEL switch

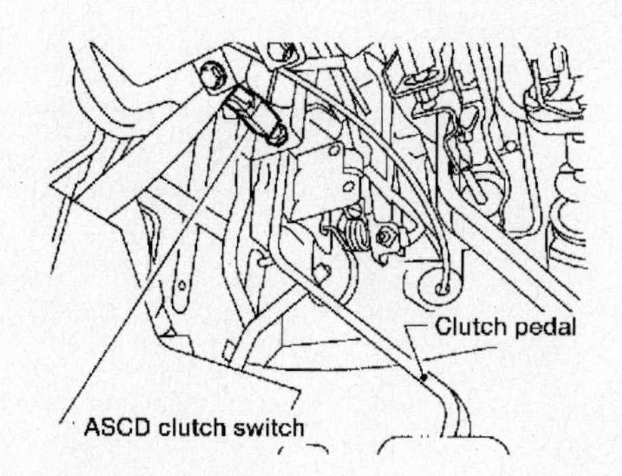
Clutch pedal
ASCD clutch switch

29149_NISS_G0100

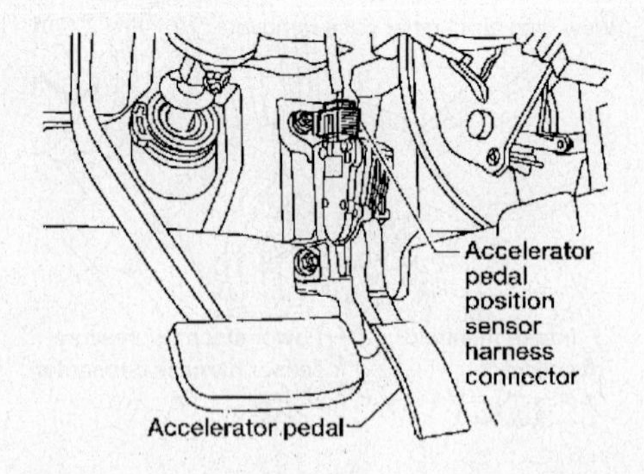

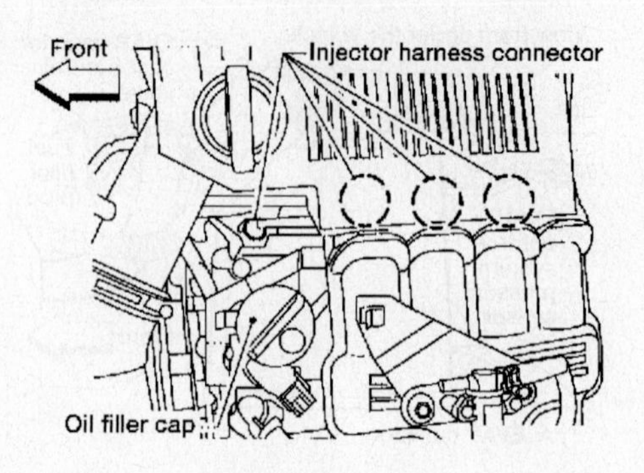

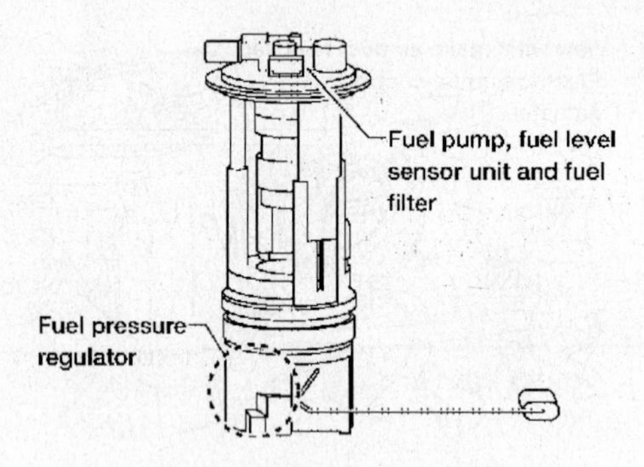

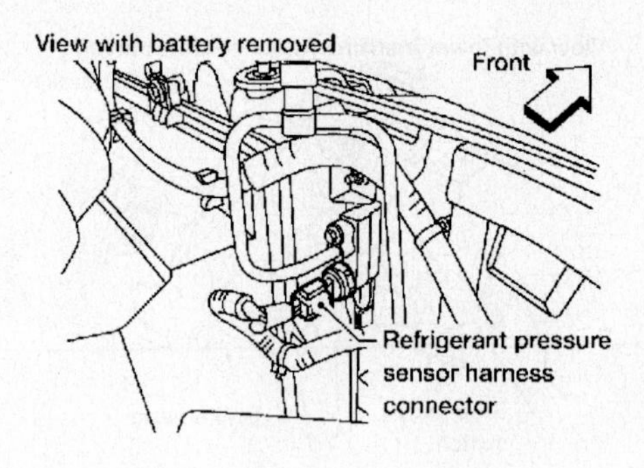

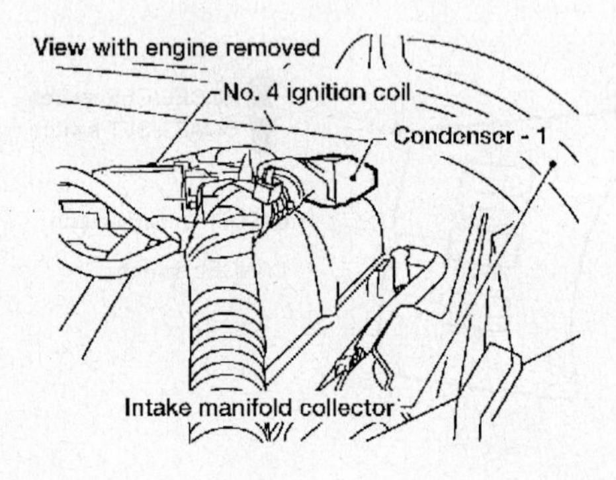

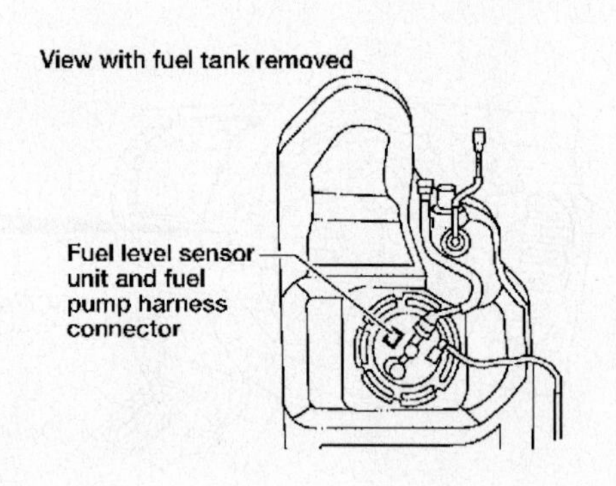

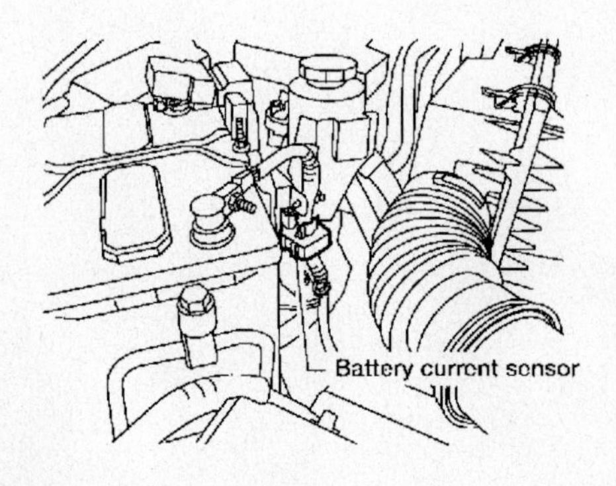

29149_NISS_G0101

2005–2006

VQ40DE V6 ENGINE

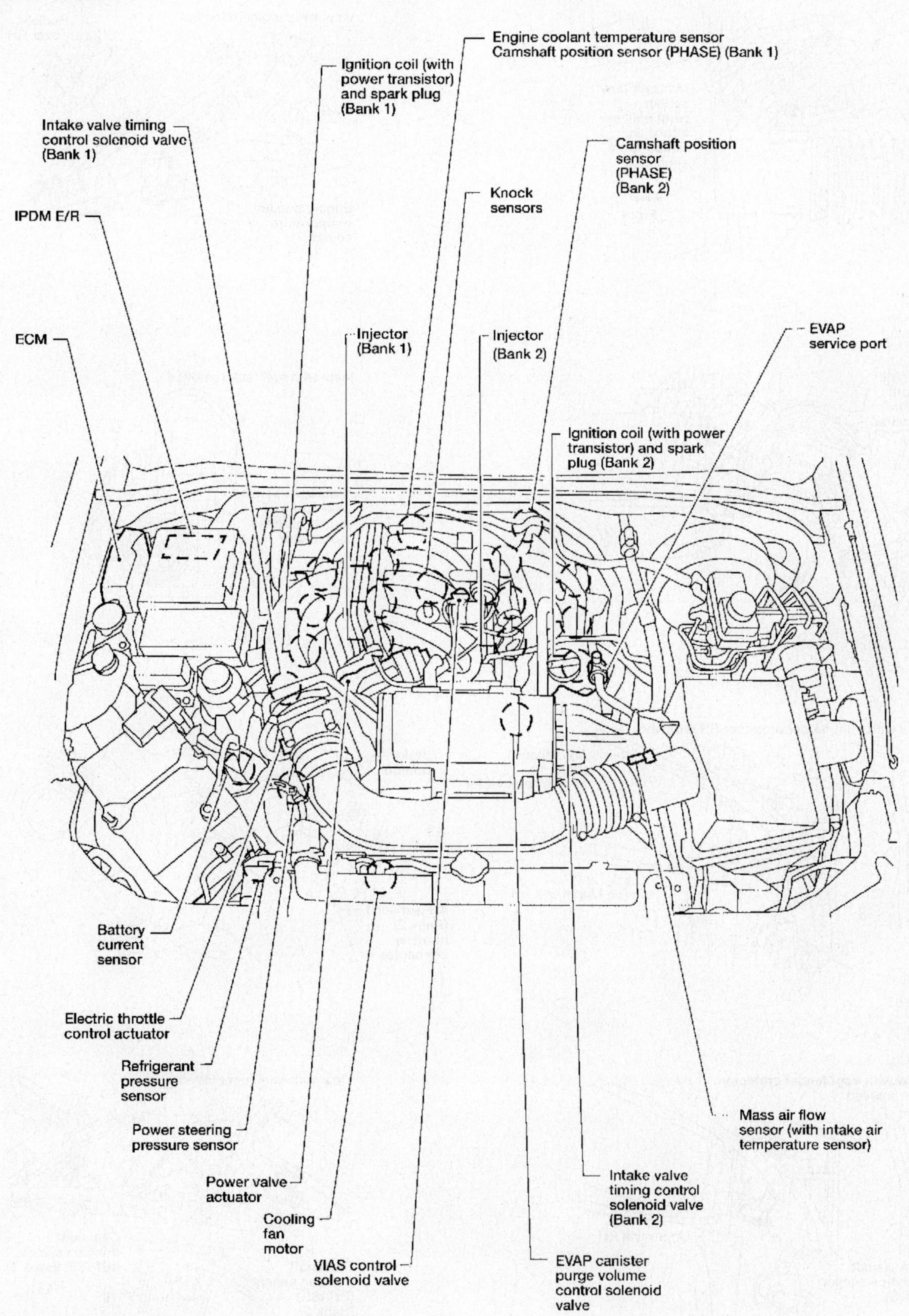
Engine coolant temperature sensor
Camshaft position sensor (PHASE) (Bank 1)
Ignition coil (with power transistor) and spark plug (Bank 1)
Intake valve timing control solenoid valve (Bank 1)
Camshaft position sensor (PHASE) (Bank 2)
Knock sensors
IPDM E/R
ECM
Injector (Bank 1)
Injector (Bank 2)
EVAP service port
Ignition coil (with power transistor) and spark plug (Bank 2)
Battery current sensor
Electric throttle control actuator
Refrigerant pressure sensor
Power steering pressure sensor
Power valve actuator
Cooling fan motor
VIAS control solenoid valve
Mass air flow sensor (with intake air temperature sensor)
Intake valve timing control solenoid valve (Bank 2)
EVAP canister purge volume control solenoid valve

29149_NISS_G0102

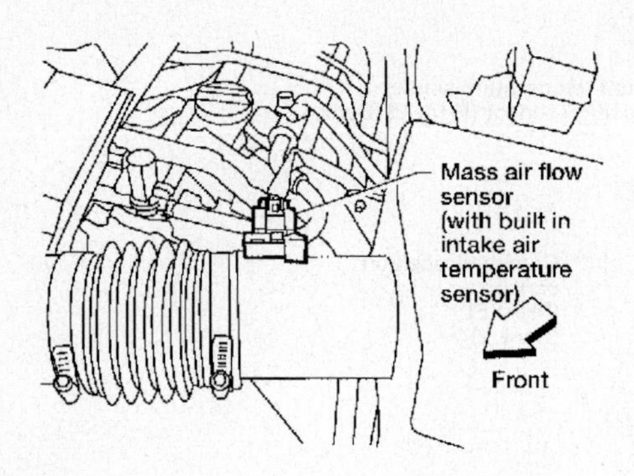
Mass air flow sensor (with built in intake air temperature sensor)
Front

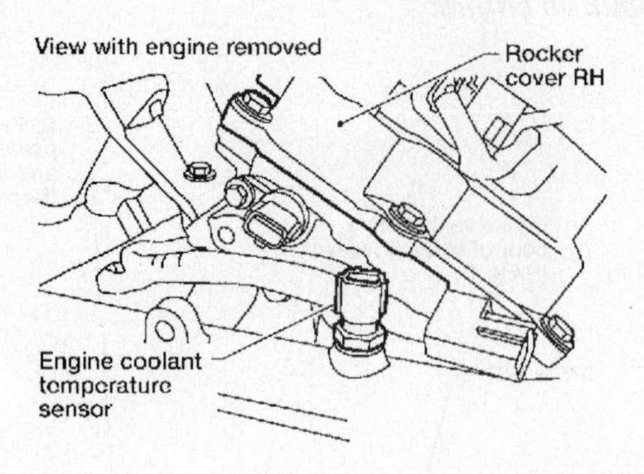
View with engine removed
Rocker cover RH
Engine coolant temperature sensor

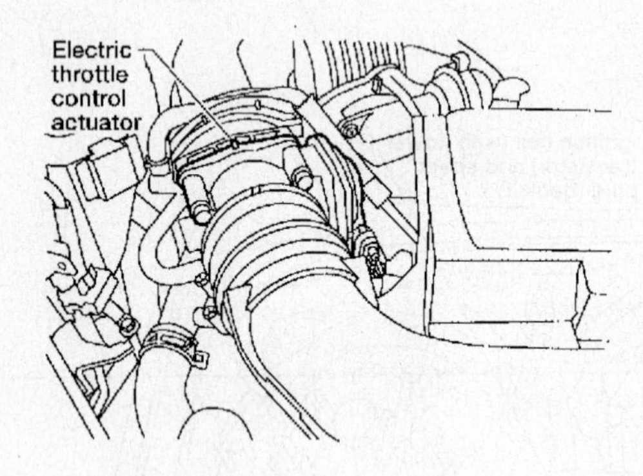
Electric throttle control actuator

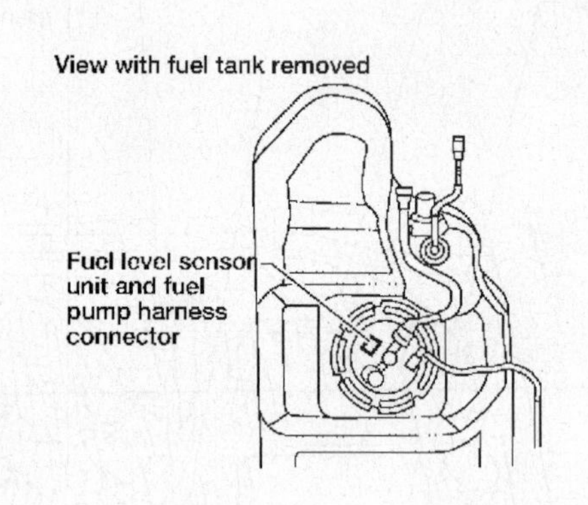
View with fuel tank removed
Fuel level sensor unit and fuel pump harness connector

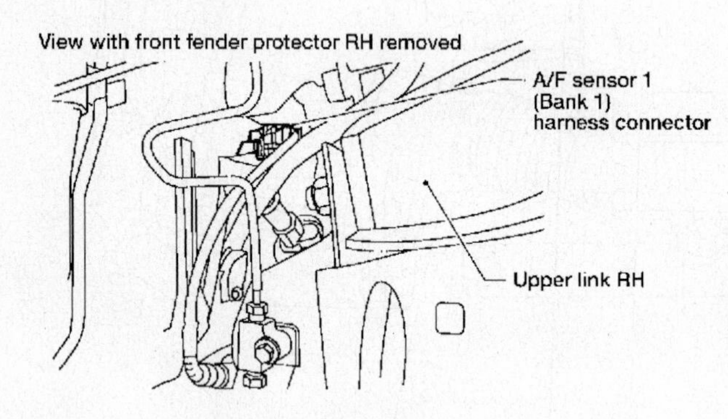
View with front fender protector RH removed
A/F sensor 1 (Bank 1) harness connector
Upper link RH

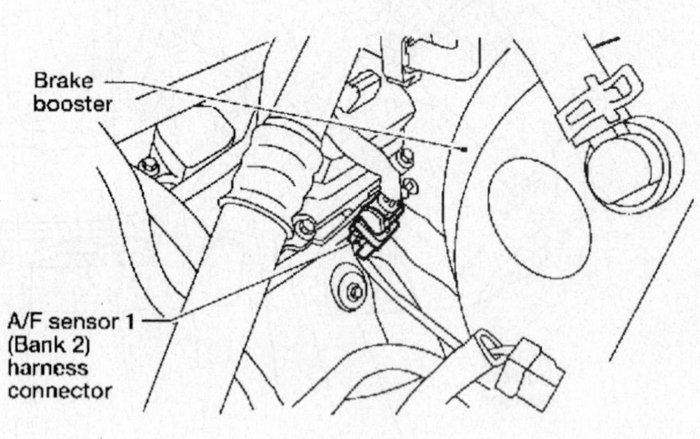
Brake booster
A/F sensor 1 (Bank 2) harness connector

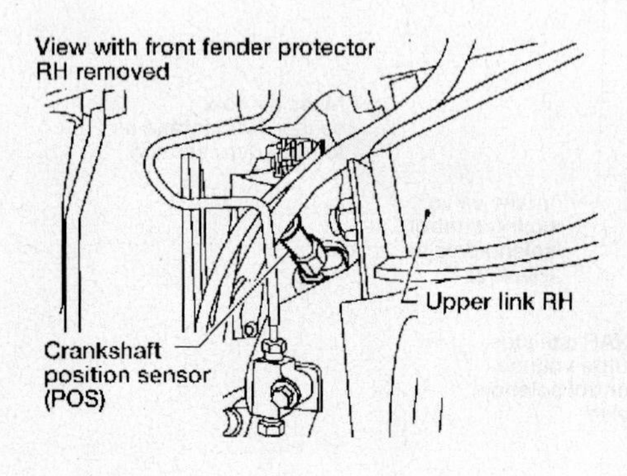
View with front fender protector RH removed
Upper link RH
Crankshaft position sensor (POS)

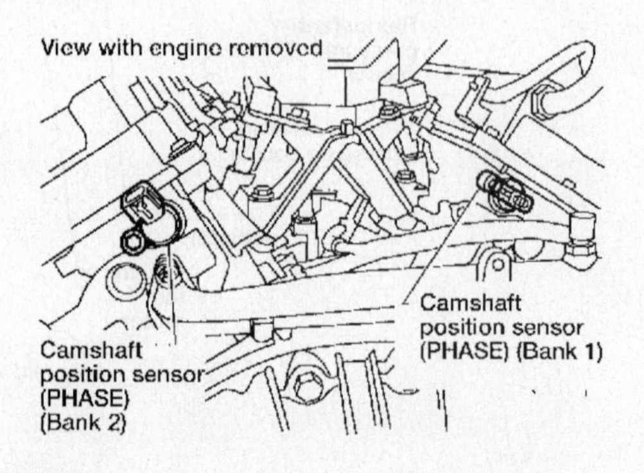
View with engine removed
Camshaft position sensor (PHASE) (Bank 1)
Camshaft position sensor (PHASE) (Bank 2)

29149_NISS_G0103

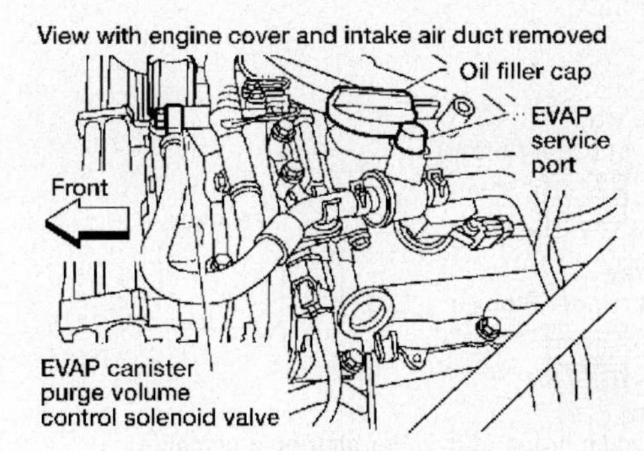
View with engine cover and intake air duct removed
Oil filler cap
EVAP
service
port
Front
EVAP canister
purge volume
control solenoid valve

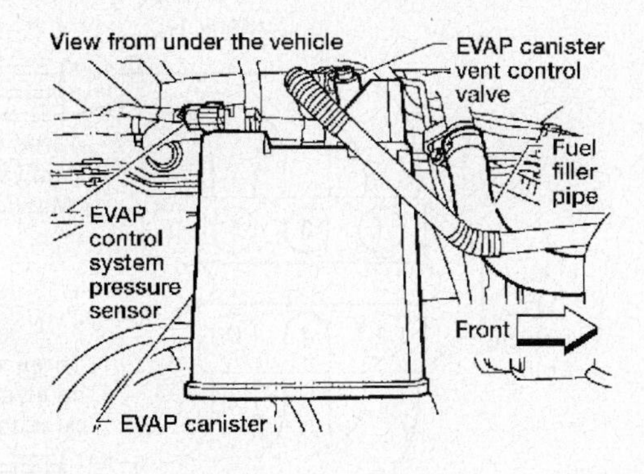
View from under the vehicle
EVAP canister
vent control
valve
Fuel
filler
pipe
EVAP
control
system
pressure
sensor
Front
EVAP canister

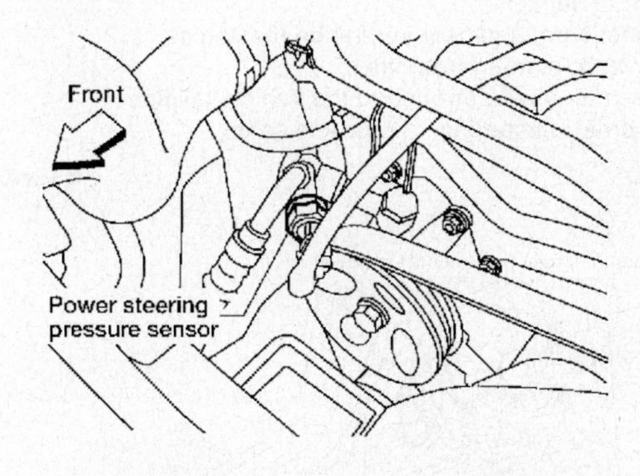
Front
Power steering
pressure sensor

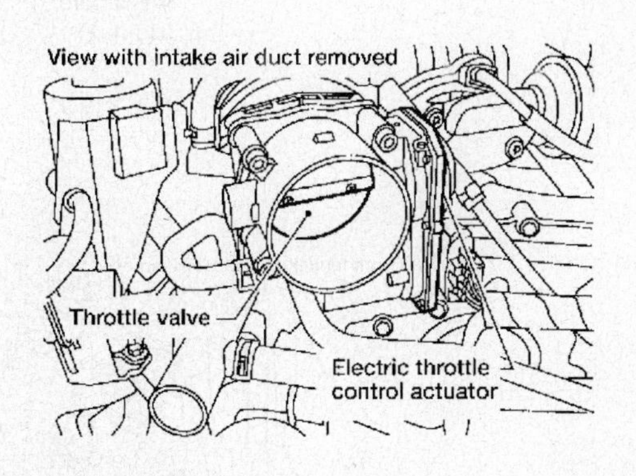
View with intake air duct removed
Throttle valve
Electric throttle
control actuator

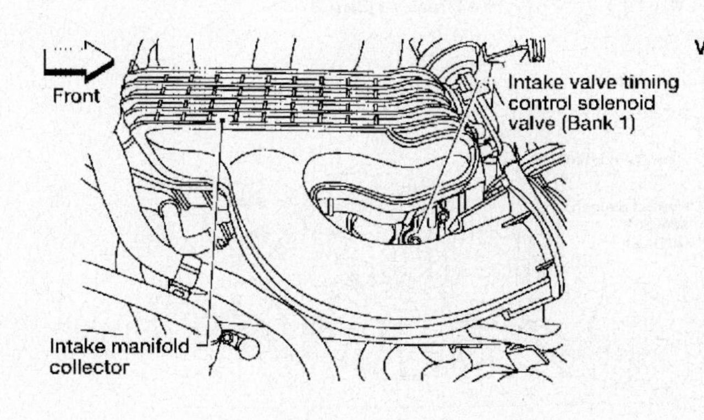
Front
Intake valve timing
control solenoid
valve (Bank 1)
Intake manifold
collector

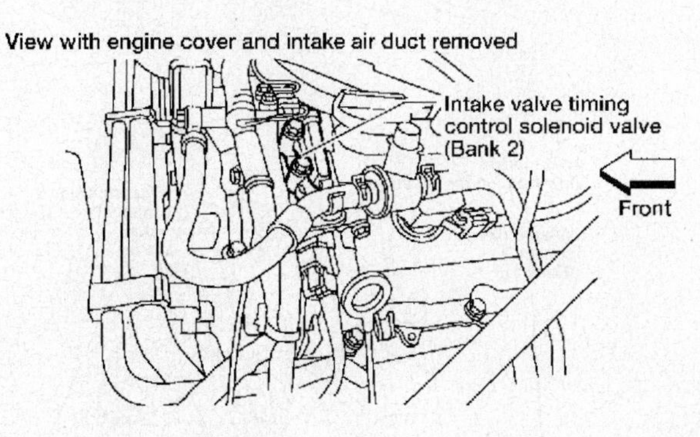
View with engine cover and intake air duct removed
Intake valve timing
control solenoid valve
(Bank 2)
Front

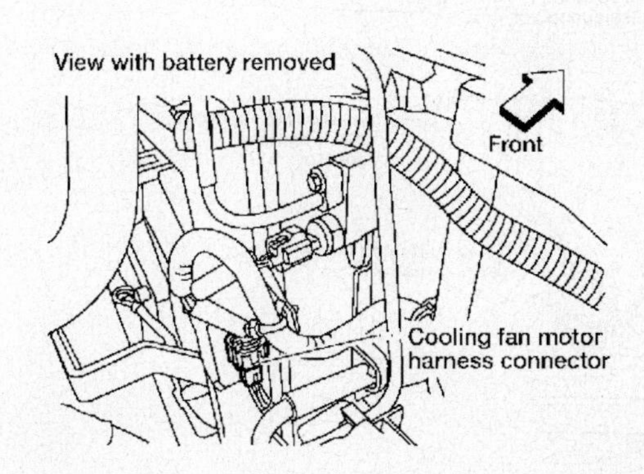
View with battery removed
Front
Cooling fan motor
harness connector

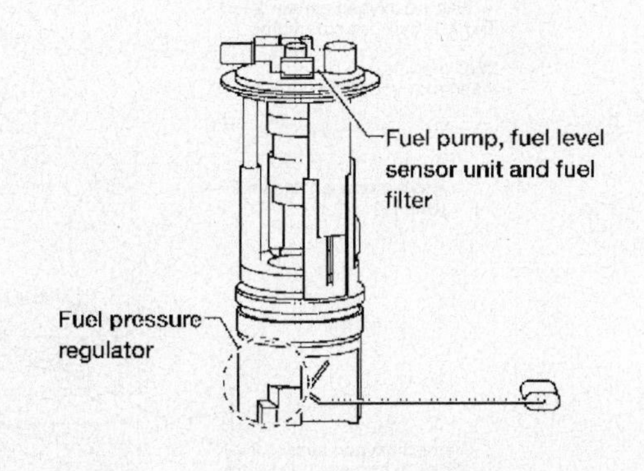
Fuel pump, fuel level
sensor unit and fuel
filter
Fuel pressure
regulator

29149_NISS_G0104

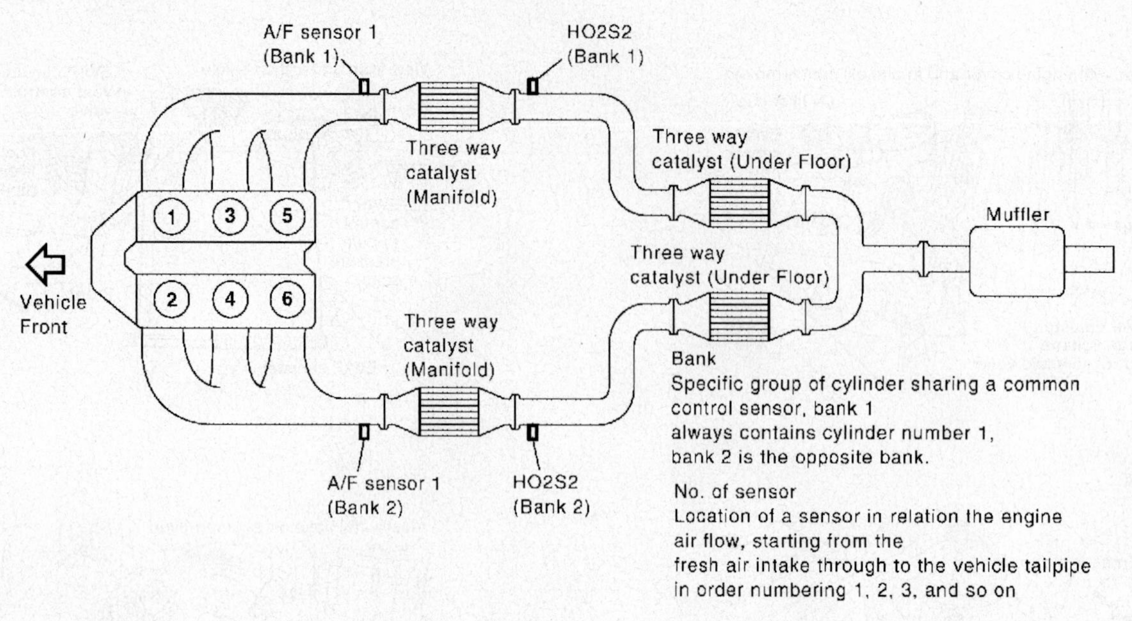

29149_NISS_G0105

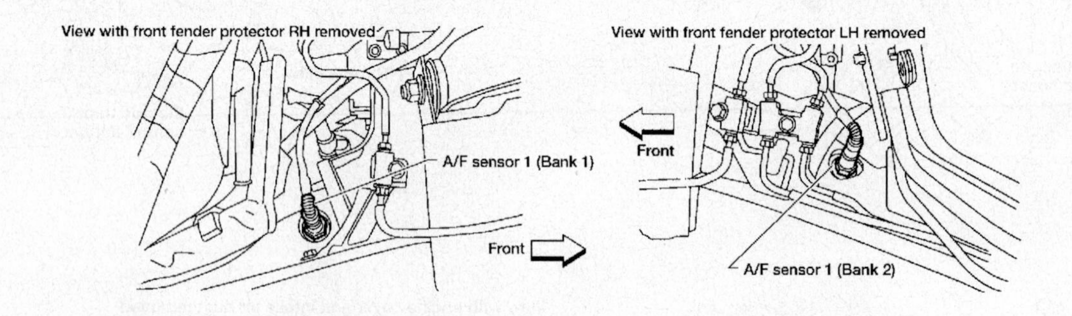

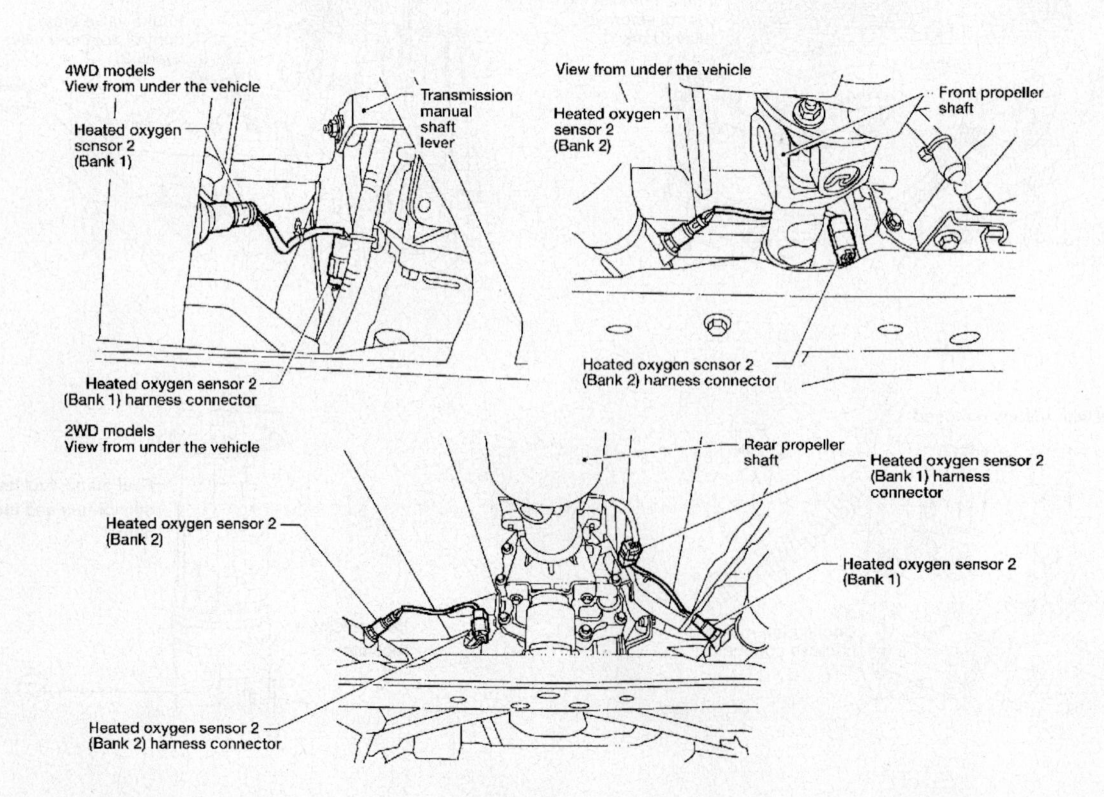

29149_NISS_G0106

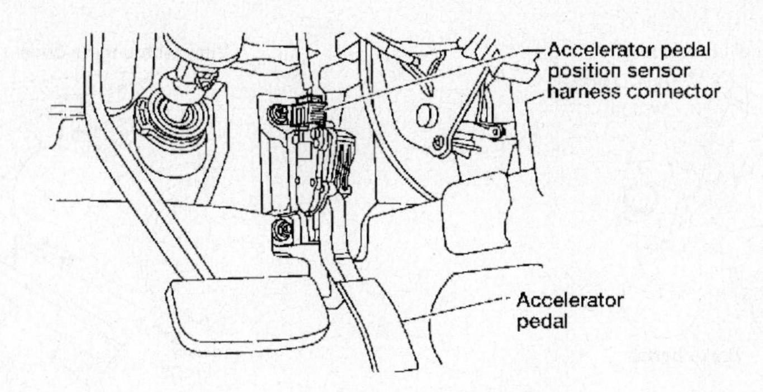
Accelerator pedal position sensor harness connector
Accelerator pedal

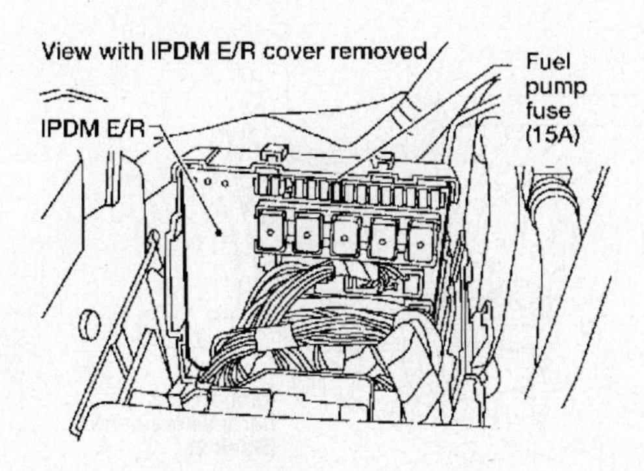
View with IPDM E/R cover removed
IPDM E/R
Fuel pump fuse (15A)

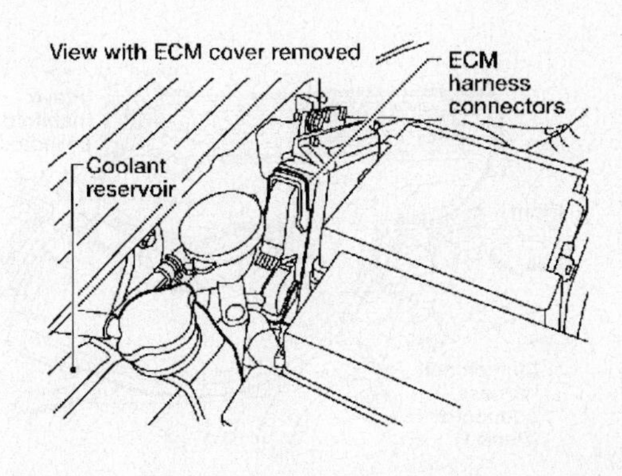
View with ECM cover removed
ECM harness connectors
Coolant reservoir

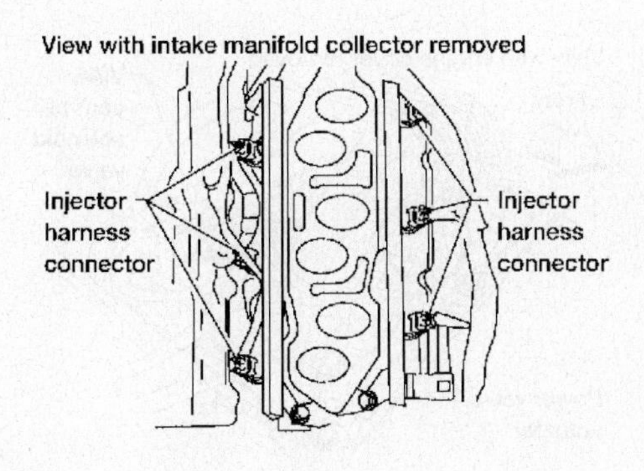
View with intake manifold collector removed
Injector harness connector
Injector harness connector

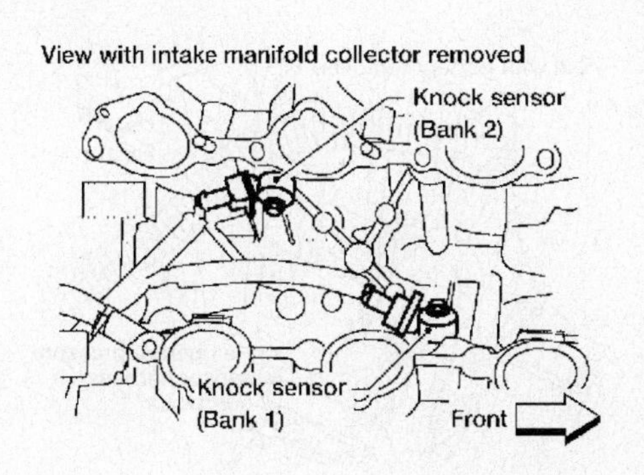
View with intake manifold collector removed
Knock sensor (Bank 2)
Knock sensor (Bank 1)
Front

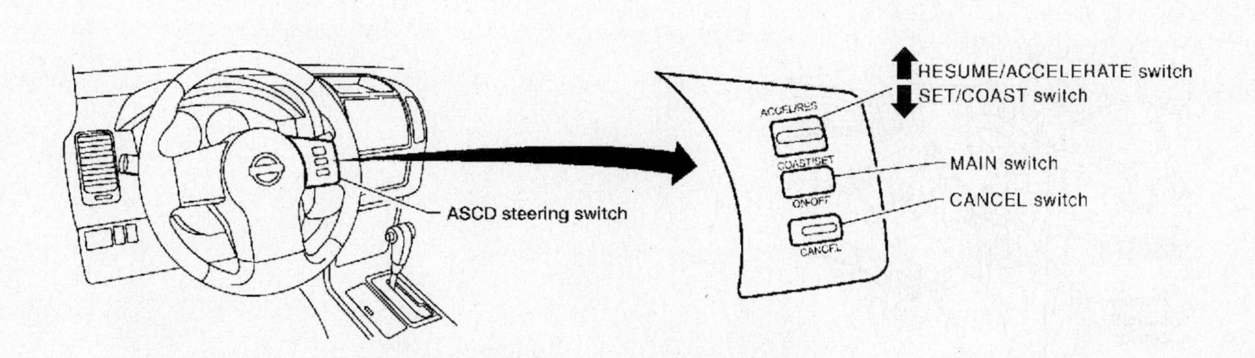
ASCD steering switch
RESUME/ACCELERATE switch
SET/COAST switch
MAIN switch
CANCEL switch

29149_NISS_G0107

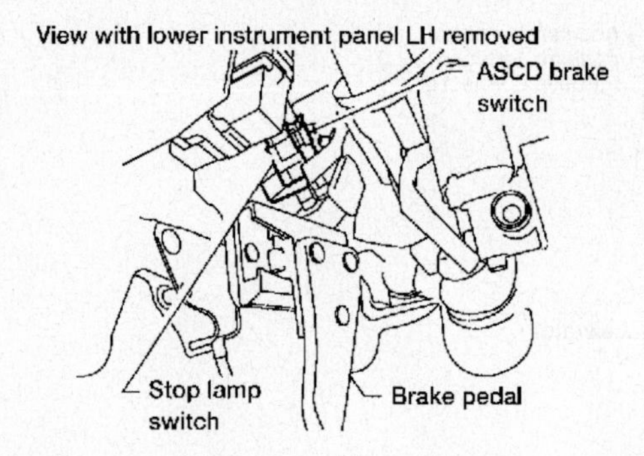
View with lower instrument panel LH removed
ASCD brake switch
Stop lamp switch
Brake pedal

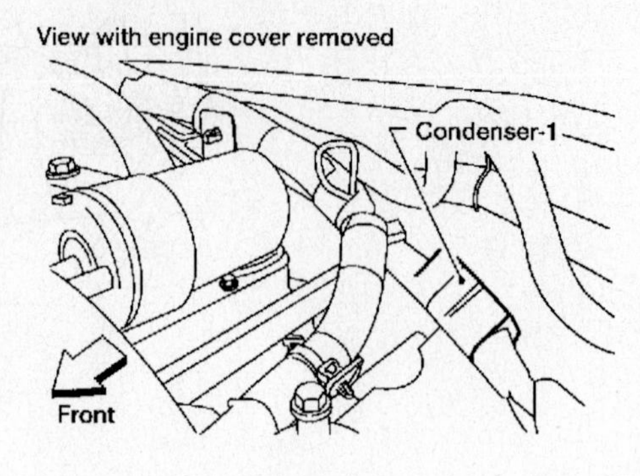
View with engine cover removed
Condenser-1
Front

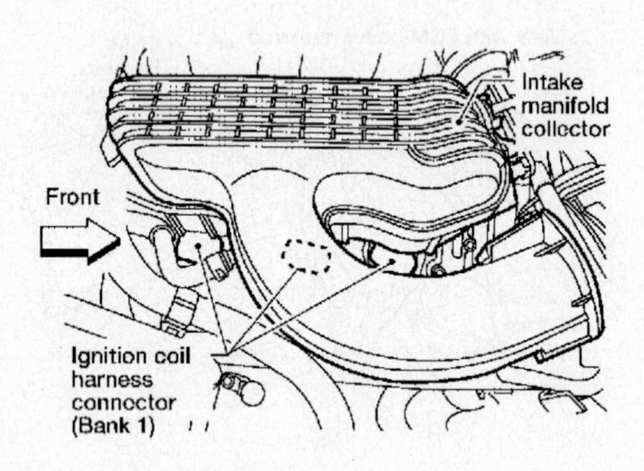
Intake manifold collector
Front
Ignition coil harness connector (Bank 1)

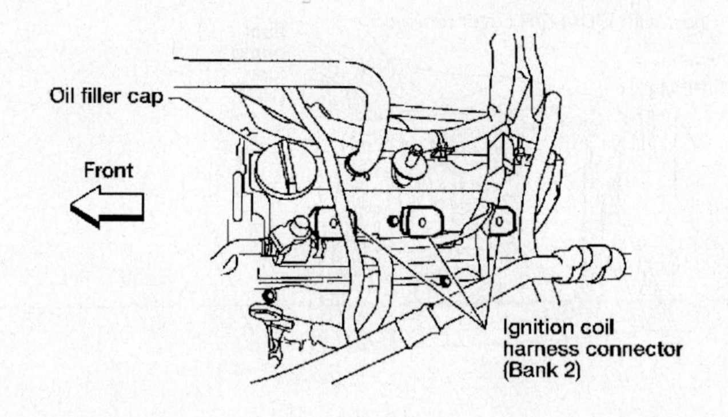
Oil filler cap
Front
Ignition coil harness connector (Bank 2)

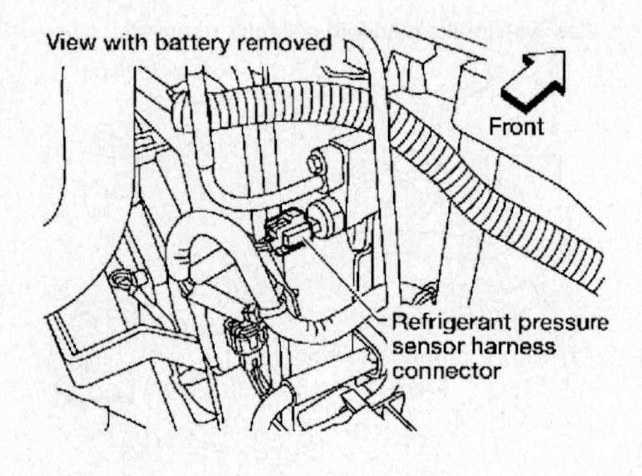
View with battery removed
Front
Refrigerant pressure sensor harness connector

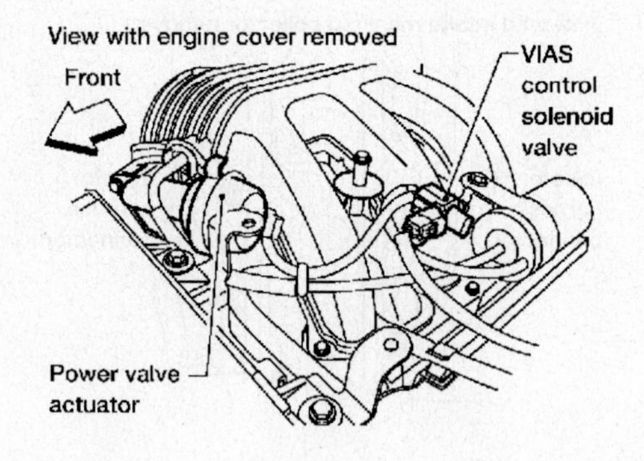
View with engine cover removed
Front
VIAS control solenoid valve
Power valve actuator

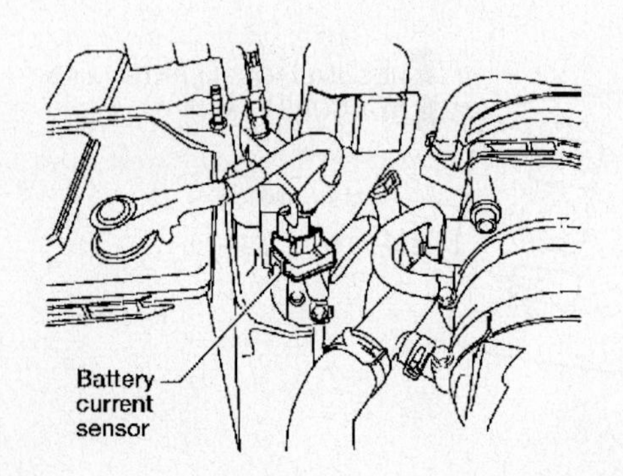
Battery current sensor

29149_NISS_G0108

MAXIMA

1996–2001

VQ30DE V6 ENGINE

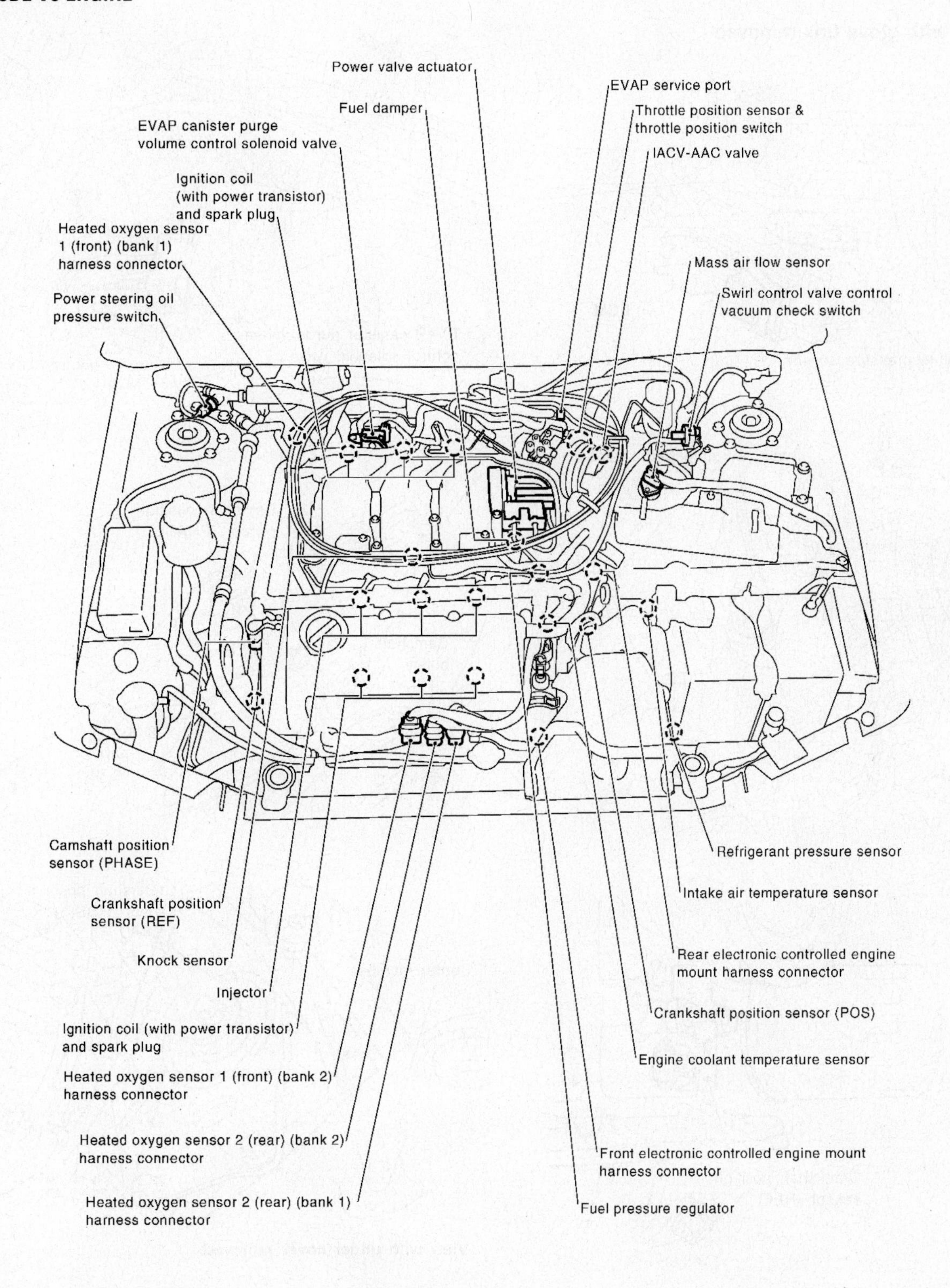
Power valve actuator
Fuel damper
EVAP canister purge volume control solenoid valve
Ignition coil (with power transistor) and spark plug
Heated oxygen sensor 1 (front) (bank 1) harness connector
Power steering oil pressure switch
EVAP service port
Throttle position sensor & throttle position switch
IACV-AAC valve
Mass air flow sensor
Swirl control valve control vacuum check switch
Camshaft position sensor (PHASE)
Crankshaft position sensor (REF)
Knock sensor
Injector
Ignition coil (with power transistor) and spark plug
Heated oxygen sensor 1 (front) (bank 2) harness connector
Heated oxygen sensor 2 (rear) (bank 2) harness connector
Heated oxygen sensor 2 (rear) (bank 1) harness connector
Refrigerant pressure sensor
Intake air temperature sensor
Rear electronic controlled engine mount harness connector
Crankshaft position sensor (POS)
Engine coolant temperature sensor
Front electronic controlled engine mount harness connector
Fuel pressure regulator

29149_NISS_G0016

View with glove box removed

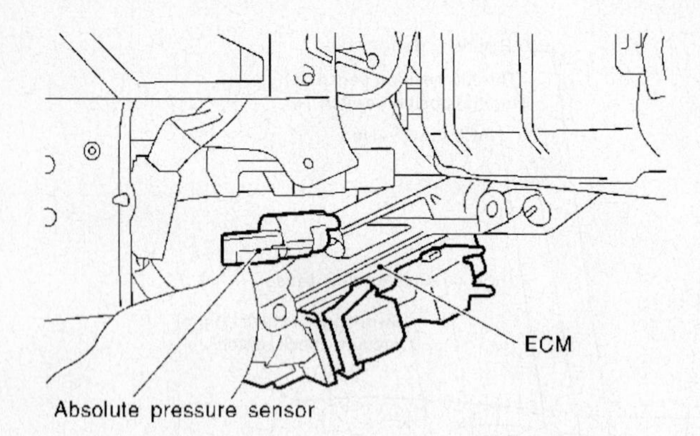

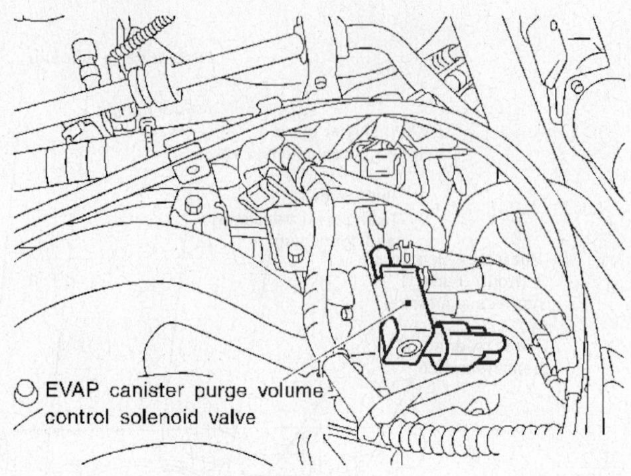

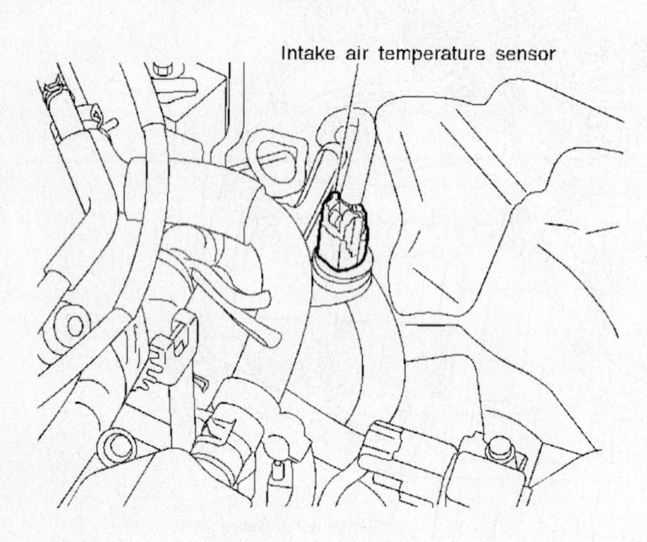

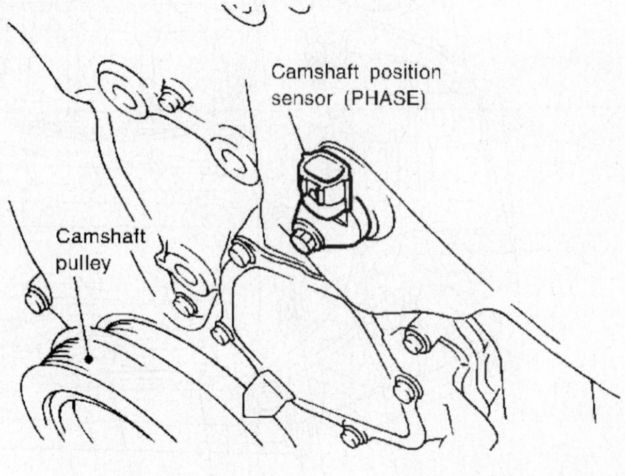

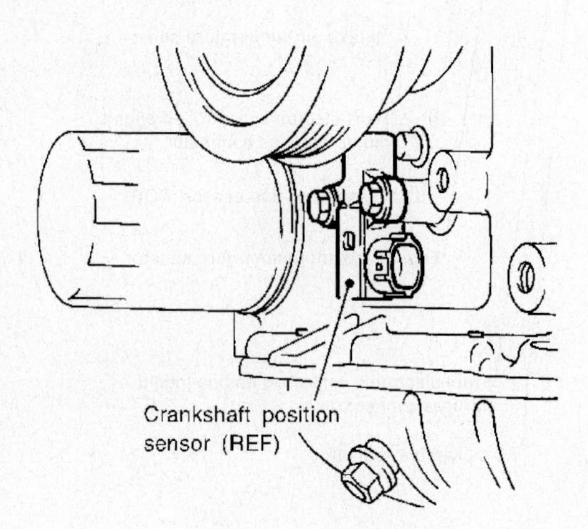

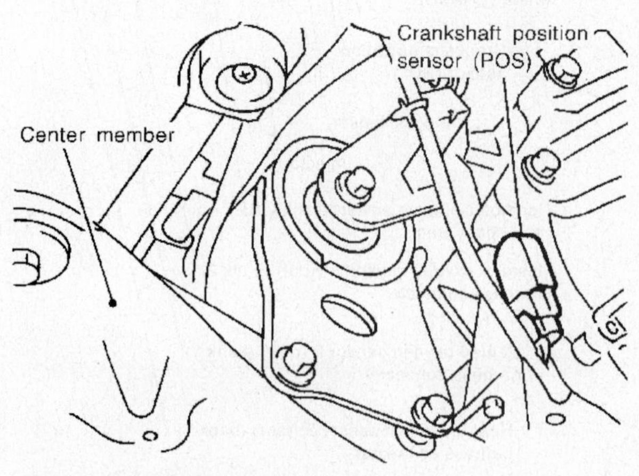

View with under cover removed

29149_NISS_G0017

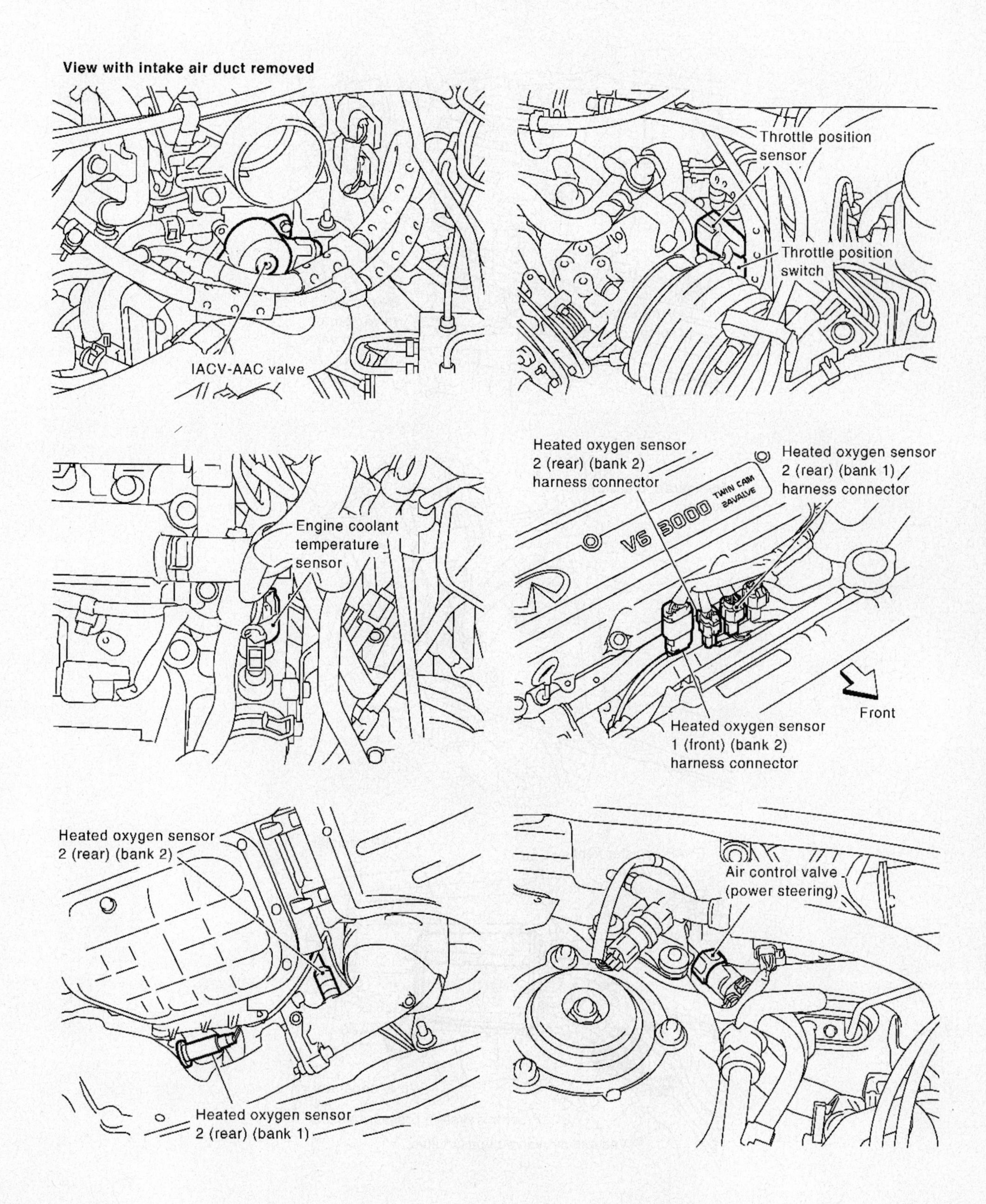
View with intake air duct removed
IACV-AAC valve
Throttle position sensor
Throttle position switch
Engine coolant temperature sensor
Heated oxygen sensor 2 (rear) (bank 2) harness connector
Heated oxygen sensor 2 (rear) (bank 1) harness connector
V6 3000
TWIN CAM 24VALVE
Front
Heated oxygen sensor 1 (front) (bank 2) harness connector
Heated oxygen sensor 2 (rear) (bank 2)
Heated oxygen sensor 2 (rear) (bank 1)
Air control valve (power steering)

29149_NISS_G0018

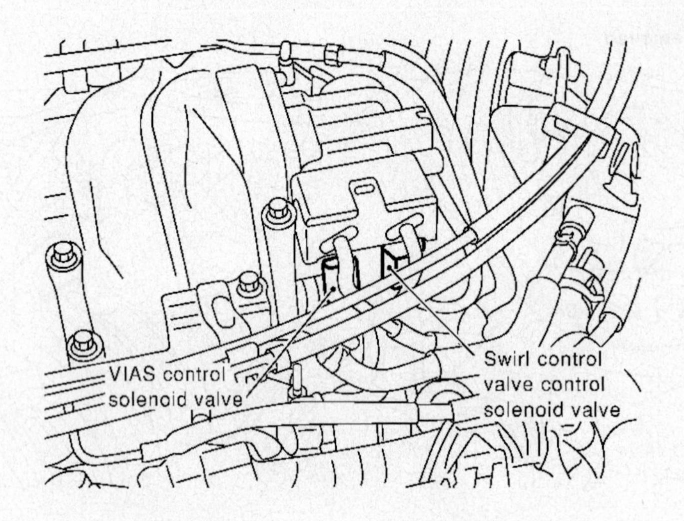
VIAS control
solenoid valve
Swirl control
valve control
solenoid valve

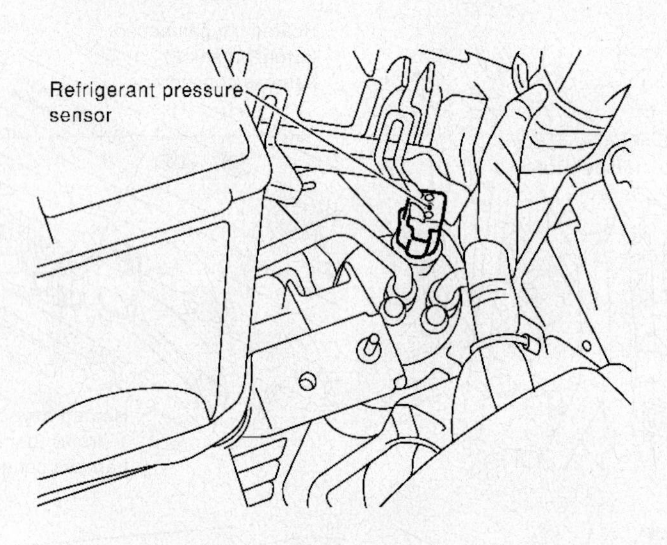
Refrigerant pressure
sensor

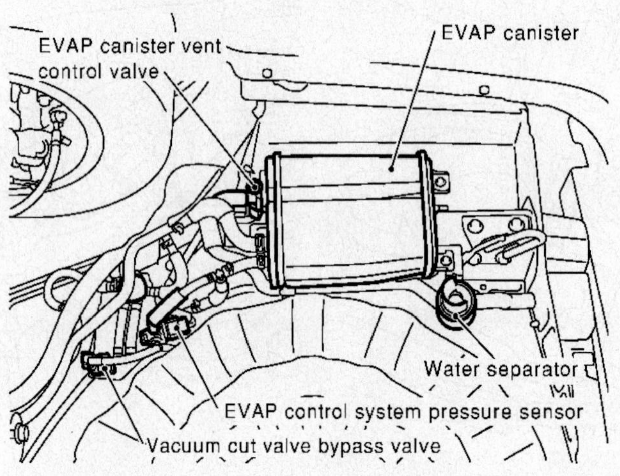
EVAP canister vent
control valve
EVAP canister
Water separator
EVAP control system pressure sensor
Vacuum cut valve bypass valve

29149_NISS_G0019

2002–2006
VQ35DE V6 ENGINE

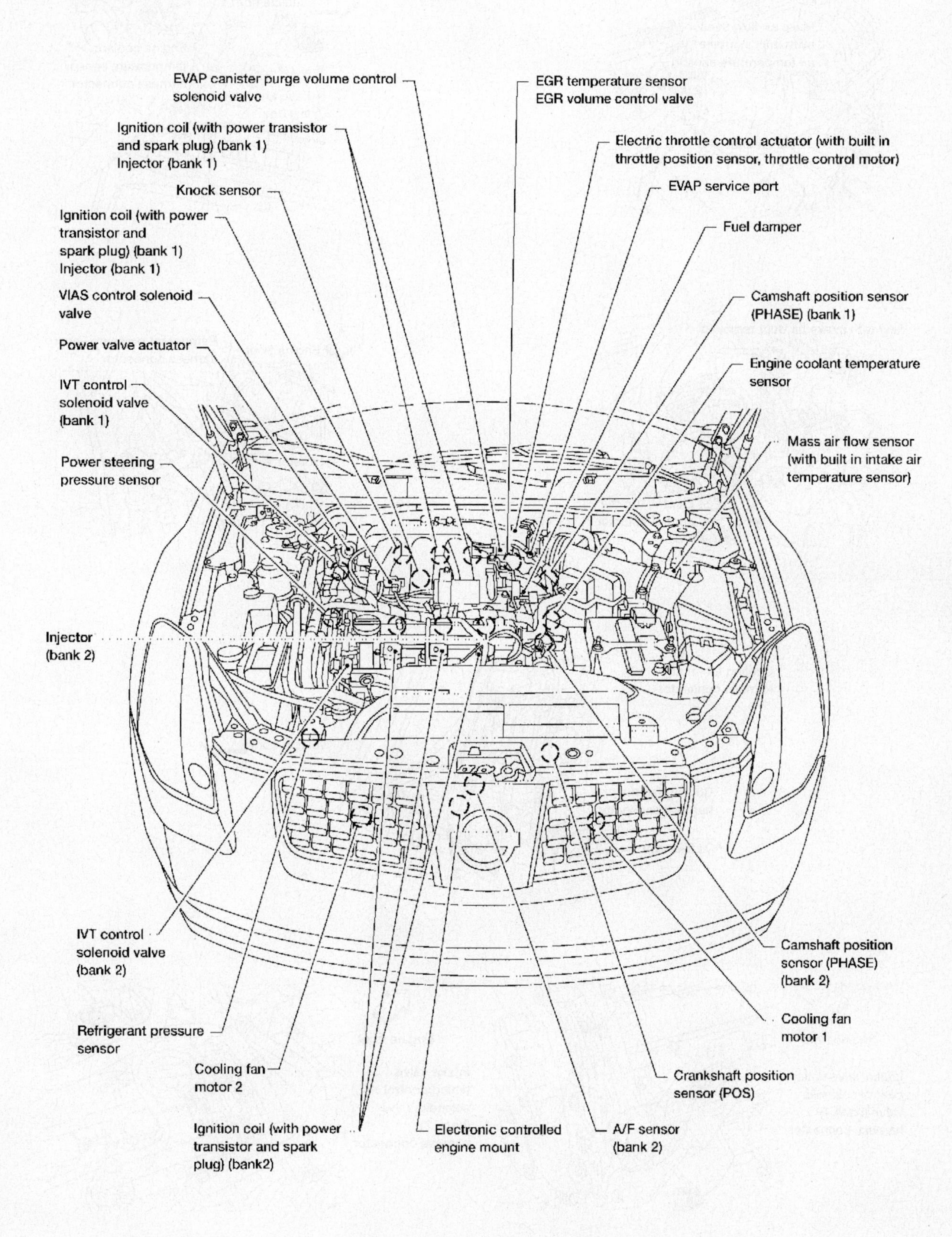
EVAP canister purge volume control solenoid valve
EGR temperature sensor
EGR volume control valve
Ignition coil (with power transistor and spark plug) (bank 1)
Injector (bank 1)
Electric throttle control actuator (with built in throttle position sensor, throttle control motor)
Knock sensor
EVAP service port
Ignition coil (with power transistor and spark plug) (bank 1)
Injector (bank 1)
Fuel damper
VIAS control solenoid valve
Camshaft position sensor (PHASE) (bank 1)
Power valve actuator
Engine coolant temperature sensor
IVT control solenoid valve (bank 1)
Mass air flow sensor (with built in intake air temperature sensor)
Power steering pressure sensor
Injector (bank 2)
IVT control solenoid valve (bank 2)
Camshaft position sensor (PHASE) (bank 2)
Cooling fan motor 1
Refrigerant pressure sensor
Cooling fan motor 2
Crankshaft position sensor (POS)
Ignition coil (with power transistor and spark plug) (bank2)
Electronic controlled engine mount
A/F sensor (bank 2)

29149_NISS_G0109

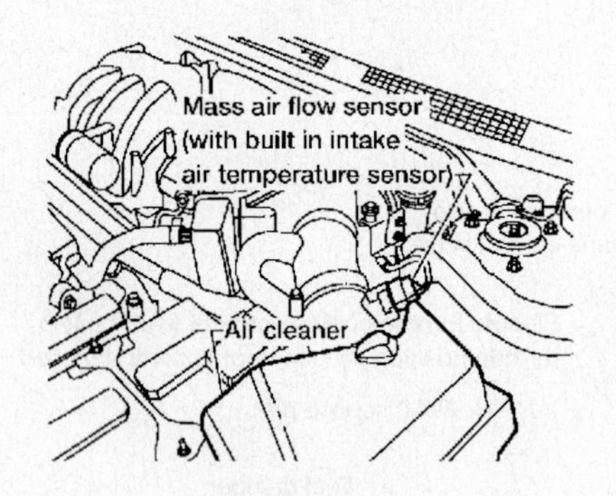
Mass air flow sensor
(with built in intake
air temperature sensor)
Air cleaner

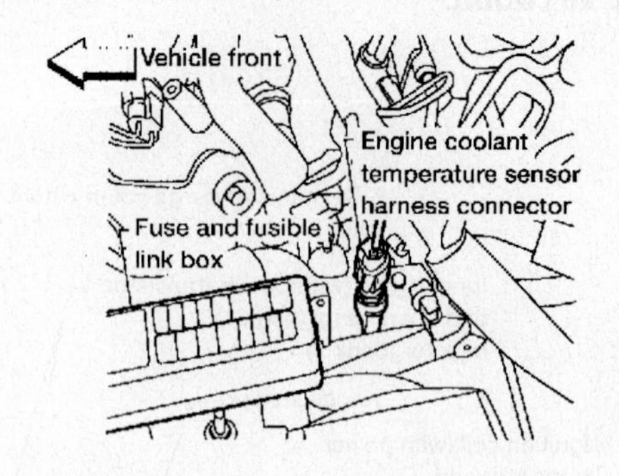
Vehicle front
Engine coolant
temperature sensor
harness connector
Fuse and fusible
link box

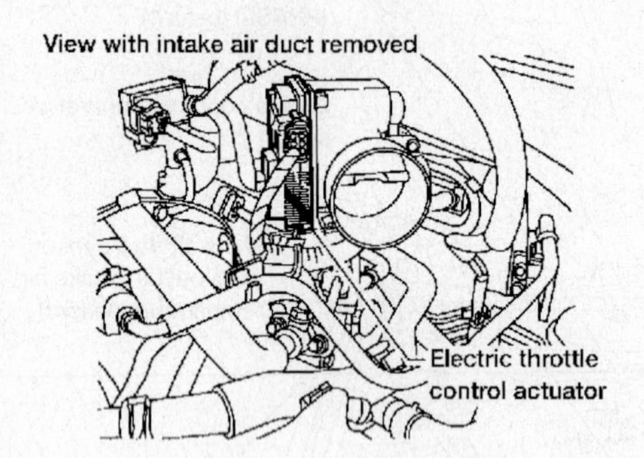
View with intake air duct removed
Electric throttle
control actuator

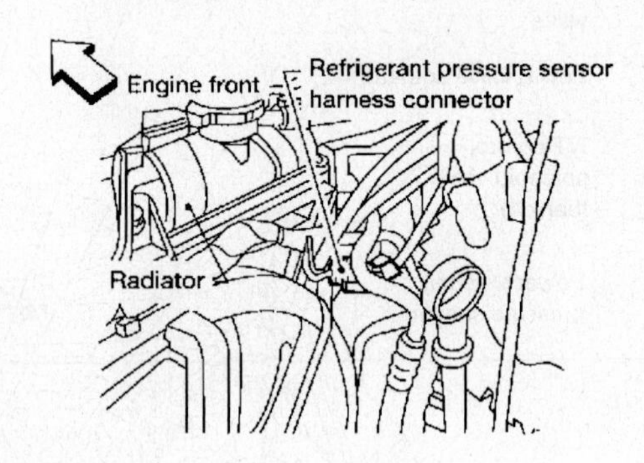
Engine front
Refrigerant pressure sensor
harness connector
Radiator

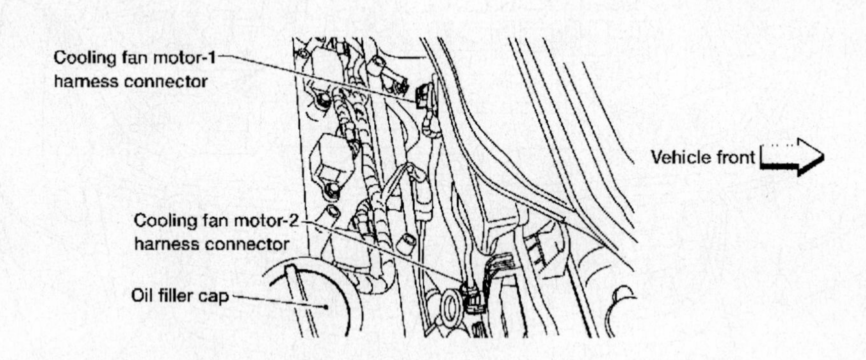
Cooling fan motor-1
harness connector
Vehicle front
Cooling fan motor-2
harness connector
Oil filler cap

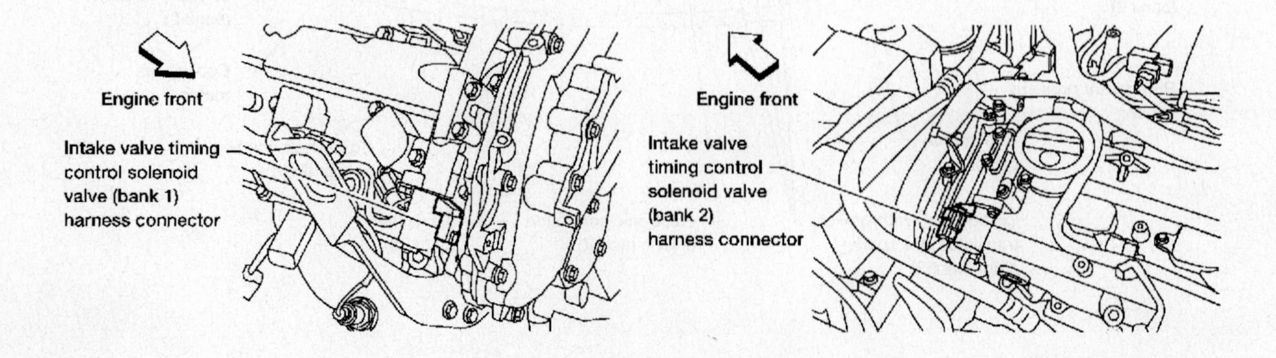
Engine front
Intake valve timing
control solenoid
valve (bank 1)
harness connector
Engine front
Intake valve
timing control
solenoid valve
(bank 2)
harness connector

29149_NISS_G0110

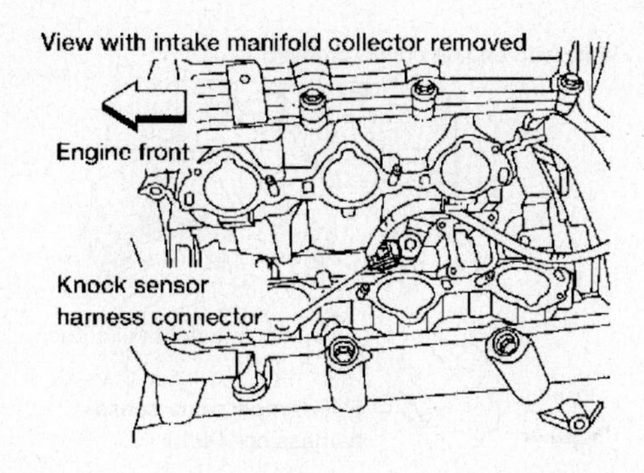
View with intake manifold collector removed
Engine front
Knock sensor
harness connector

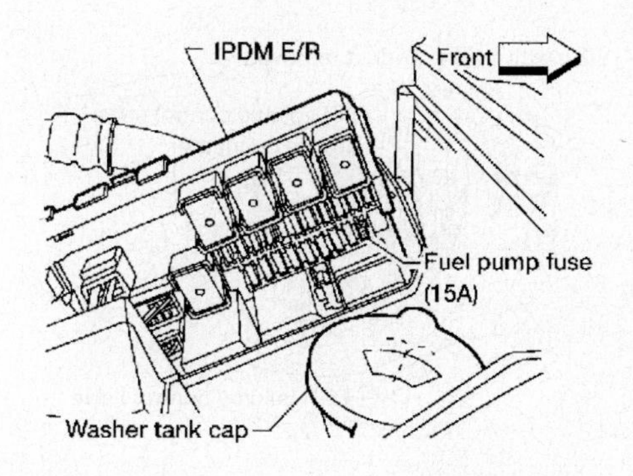
IPDM E/R
Front
Fuel pump fuse
(15A)
Washer tank cap

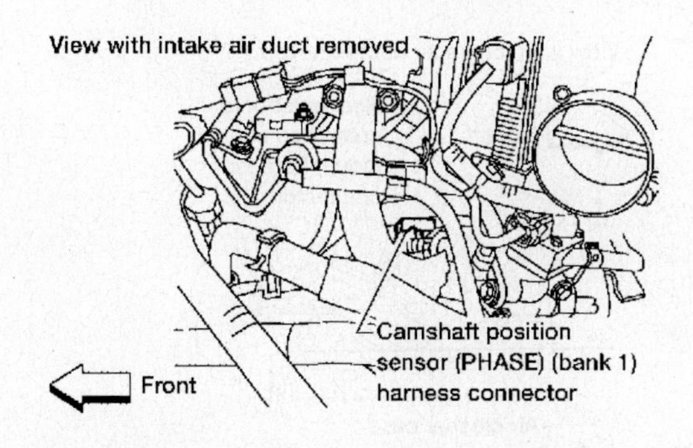
View with intake air duct removed
Camshaft position
sensor (PHASE) (bank 1)
harness connector
Front

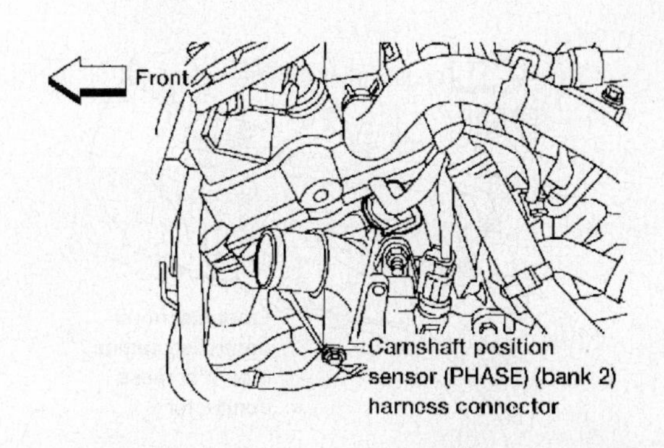
Front
Camshaft position
sensor (PHASE) (bank 2)
harness connector

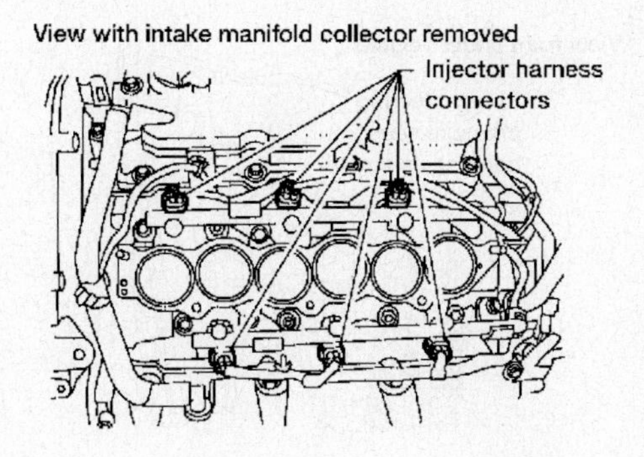
View with intake manifold collector removed
Injector harness
connectors

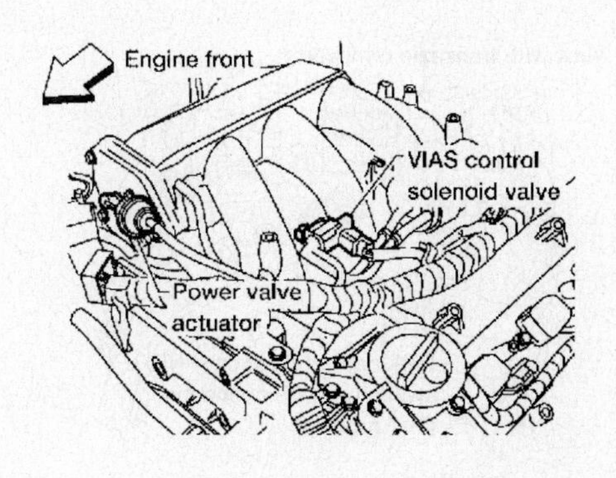
Engine front
VIAS control
solenoid valve
Power valve
actuator

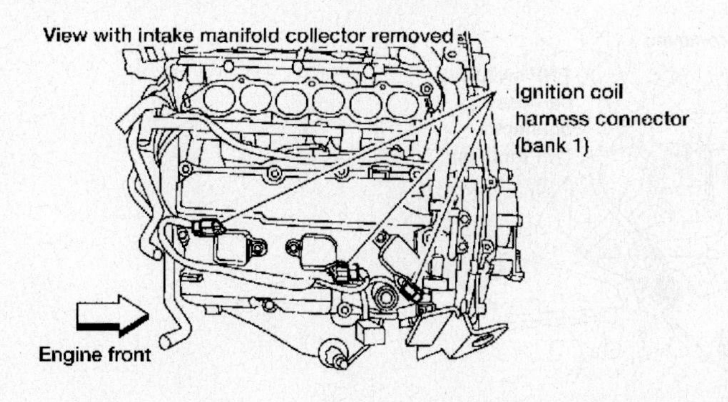
View with intake manifold collector removed
Ignition coil
harness connector
(bank 1)
Engine front

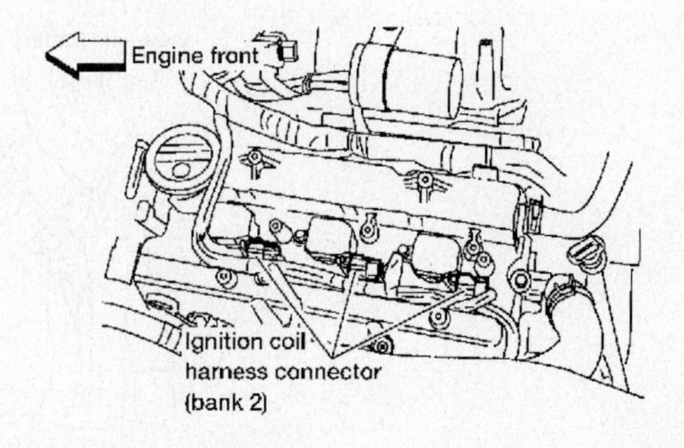
Engine front
Ignition coil
harness connector
(bank 2)

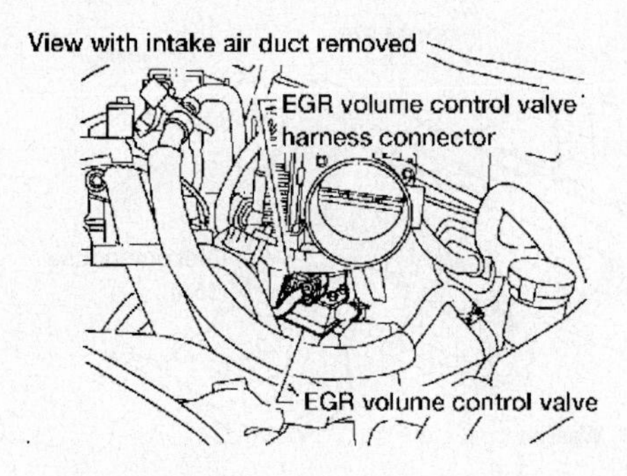
View with intake air duct removed
EGR volume control valve harness connector
EGR volume control valve

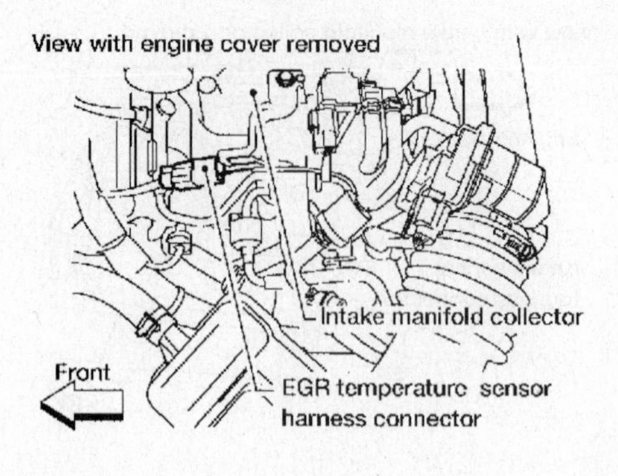
View with engine cover removed
Intake manifold collector
Front
EGR temperature sensor harness connector

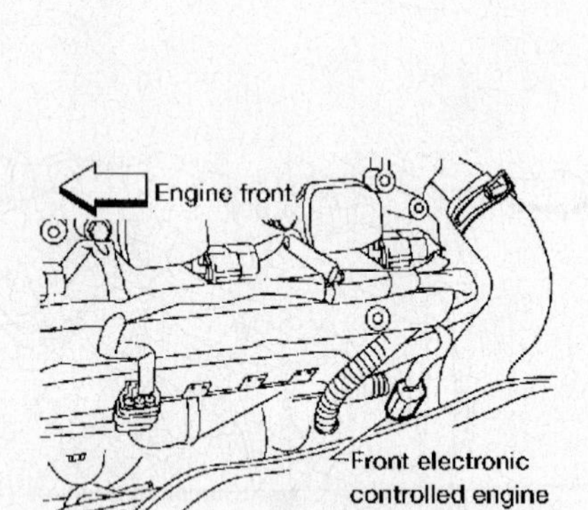
Engine front
Front electronic controlled engine mount harness connector

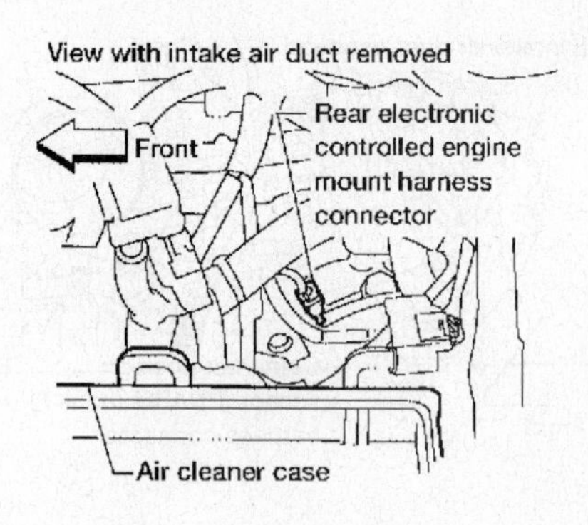
View with intake air duct removed
Front
Rear electronic controlled engine mount harness connector
Air cleaner case

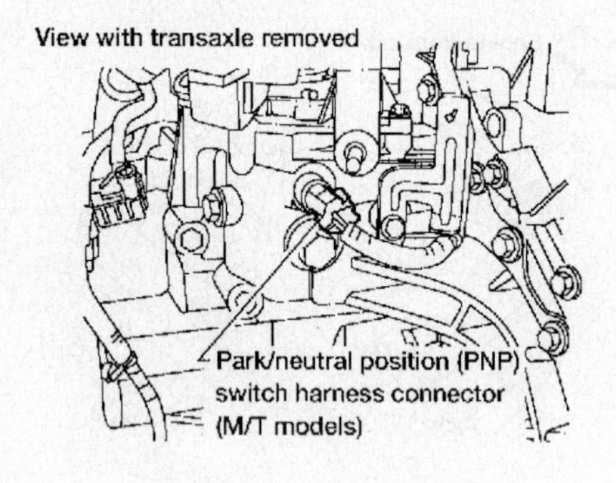
View with transaxle removed
Park/neutral position (PNP) switch harness connector (M/T models)

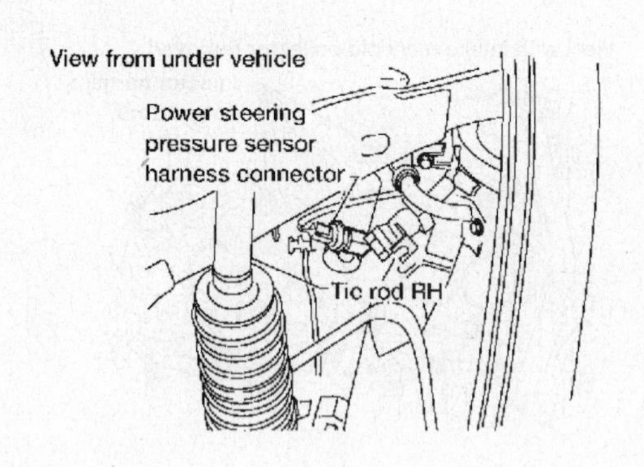
View from under vehicle
Power steering pressure sensor harness connector
Tie rod RH

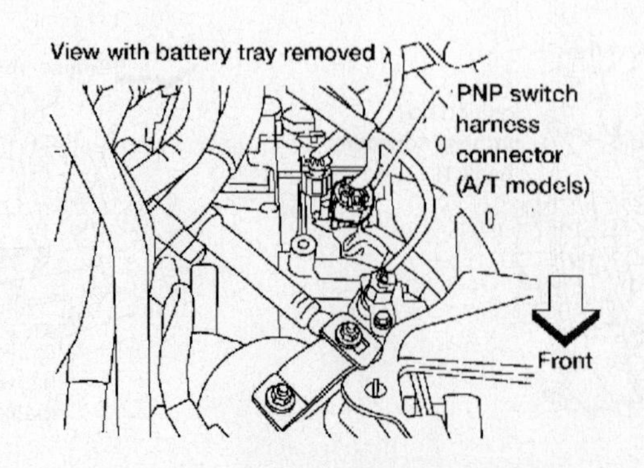
View with battery tray removed
PNP switch harness connector (A/T models)
Front

29149_NISS_G0112

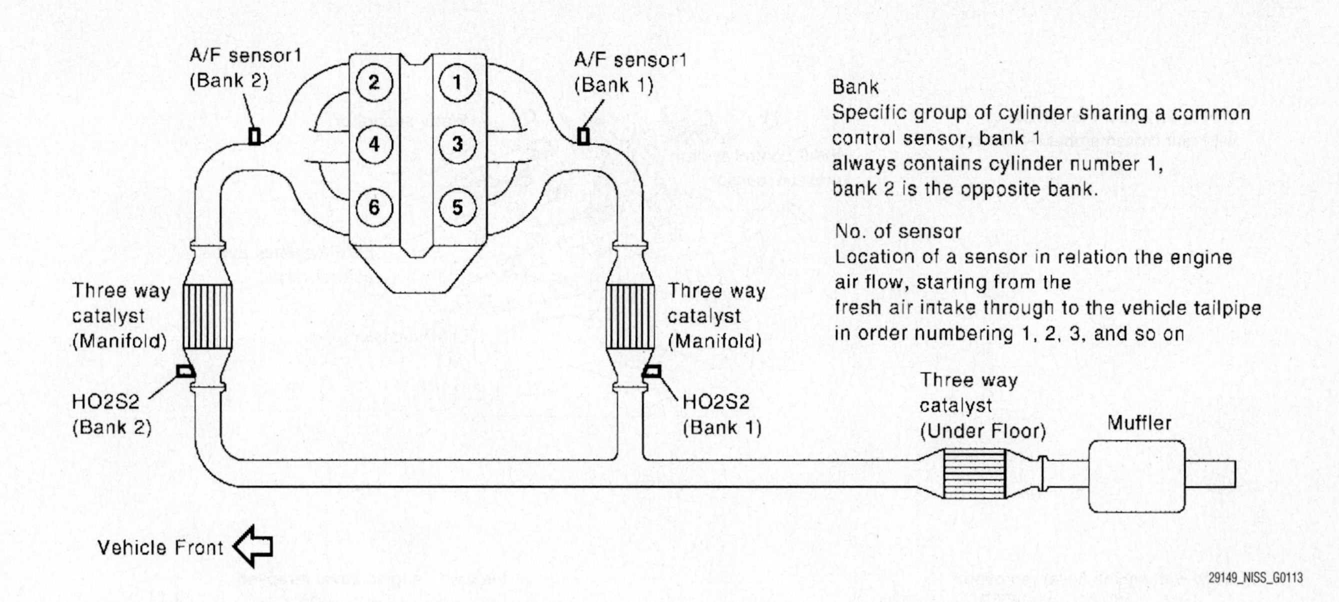

29149_NISS_G0113

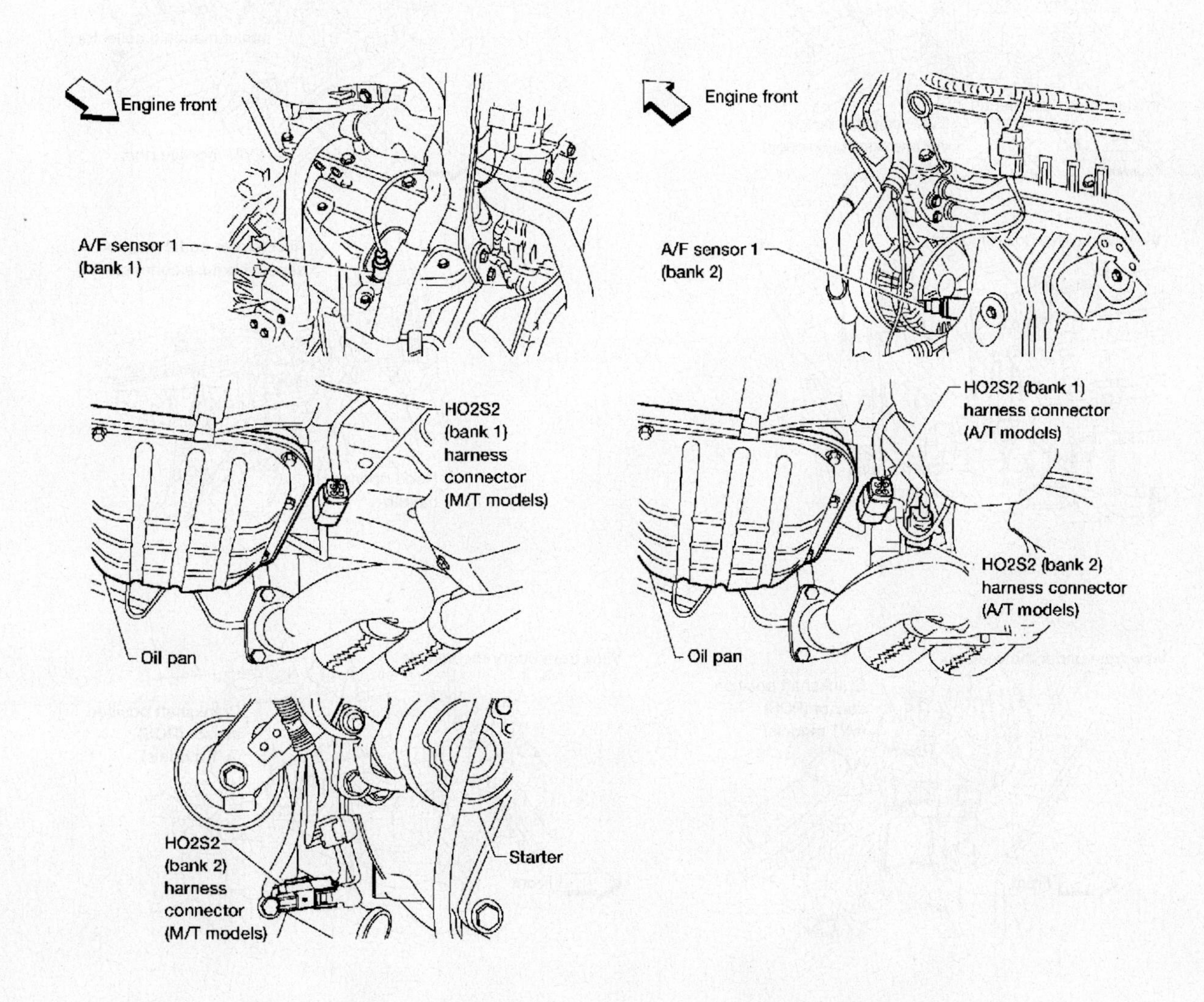

29149_NISS_G0114

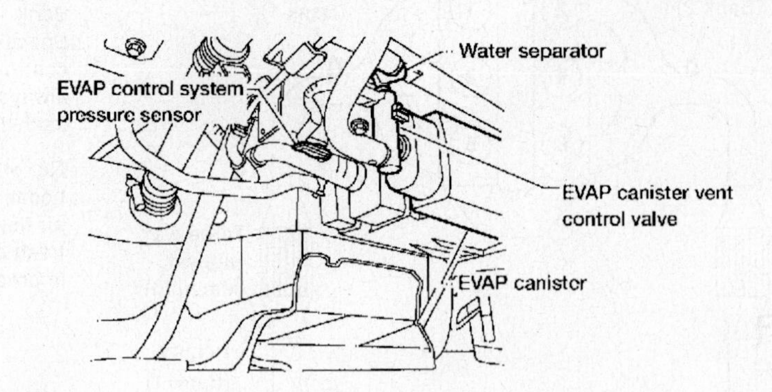
View from under the vehicle
with rear crossmember removed
Water separator
EVAP control system
pressure sensor
EVAP canister vent
control valve
EVAP canister

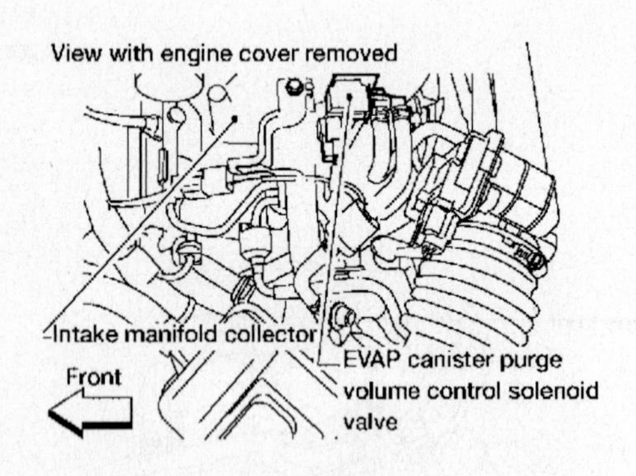
View with engine cover removed
Intake manifold collector
EVAP canister purge
volume control solenoid
valve
Front

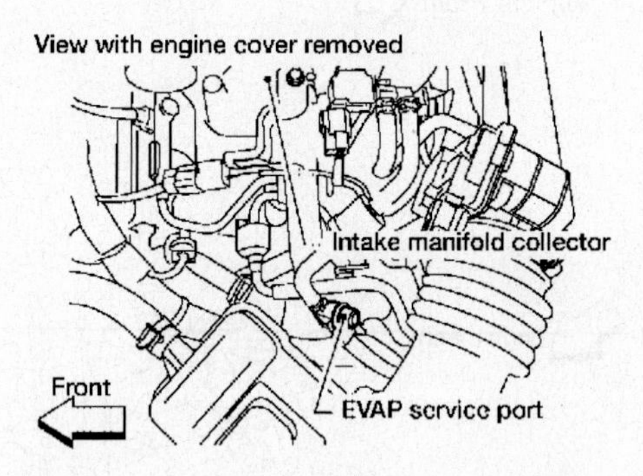
View with engine cover removed
Intake manifold collector
Front
EVAP service port

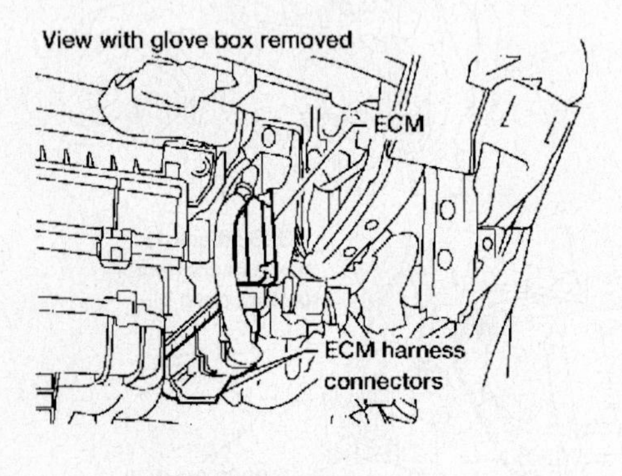
View with glove box removed
ECM
ECM harness
connectors

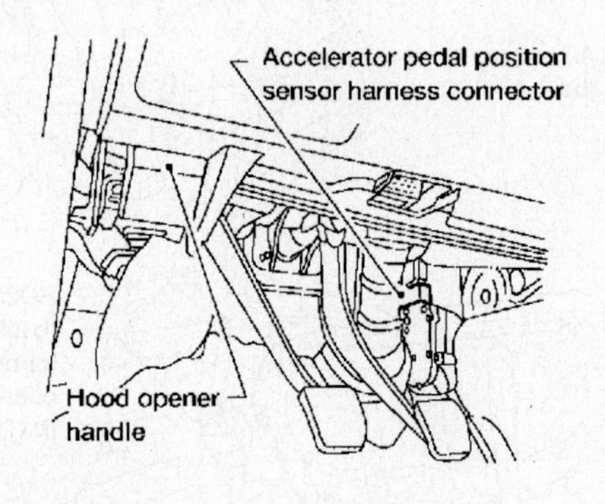
Accelerator pedal position
sensor harness connector
Hood opener
handle

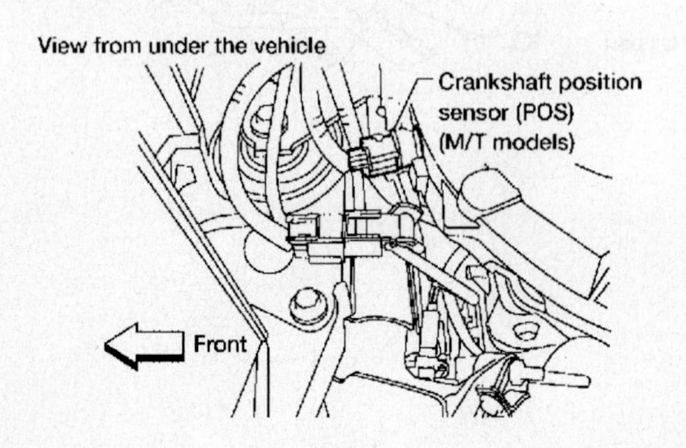
View from under the vehicle
Crankshaft position
sensor (POS)
(M/T models)
Front

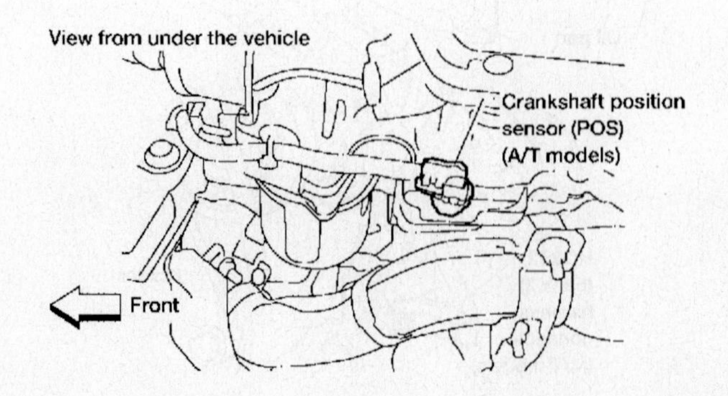
View from under the vehicle
Crankshaft position
sensor (POS)
(A/T models)
Front

29149_NISS_G0115

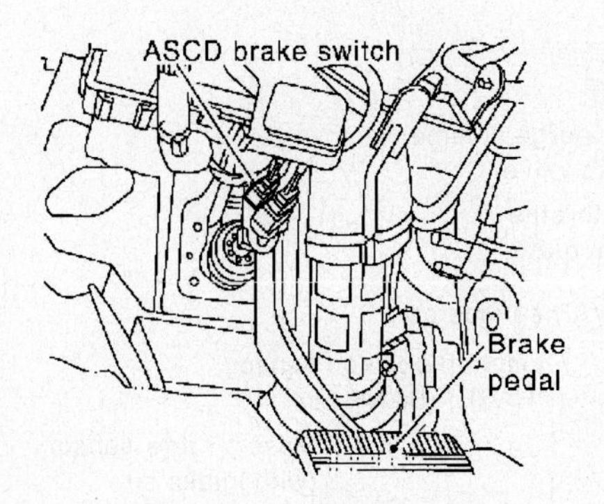
ASCD brake switch
Brake
pedal

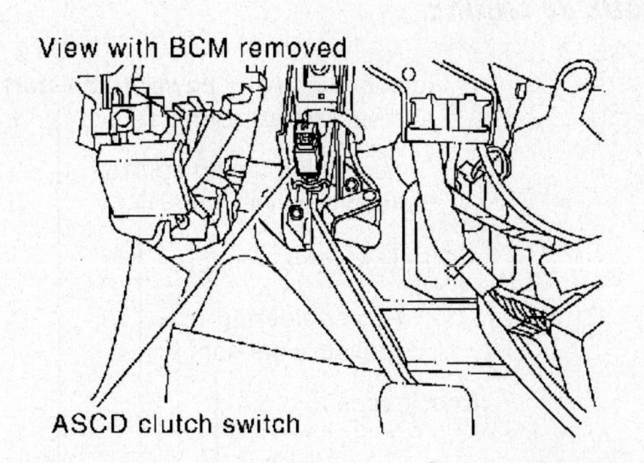
View with BCM removed
ASCD clutch switch

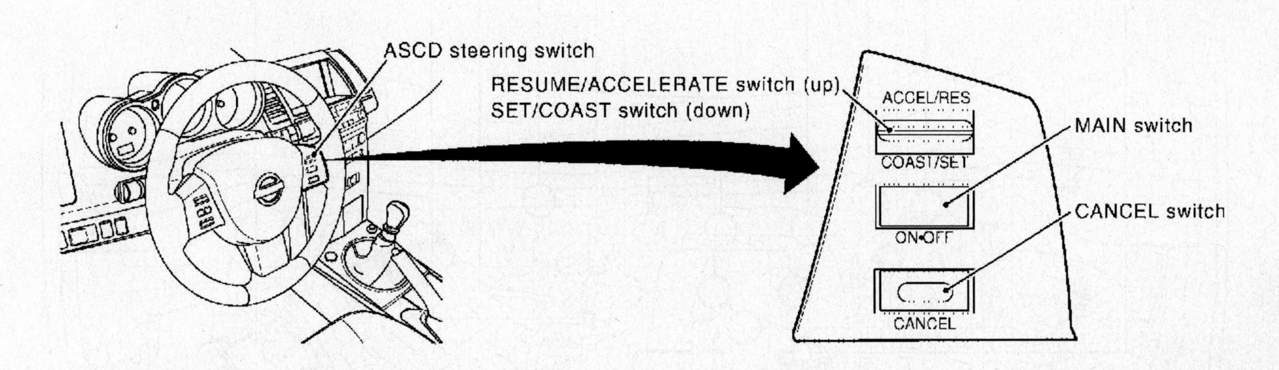
ASCD steering switch
RESUME/ACCELERATE switch (up)
SET/COAST switch (down)
ACCEL/RES
COAST/SET
MAIN switch
ON•OFF
CANCEL switch
CANCEL

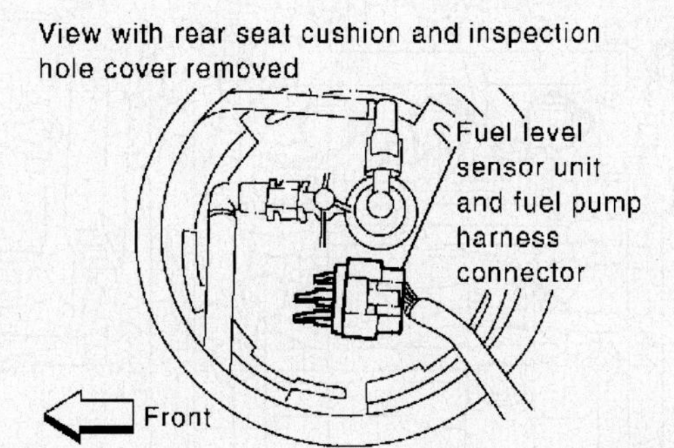
View with rear seat cushion and inspection
hole cover removed
Fuel level
sensor unit
and fuel pump
harness
connector
Front

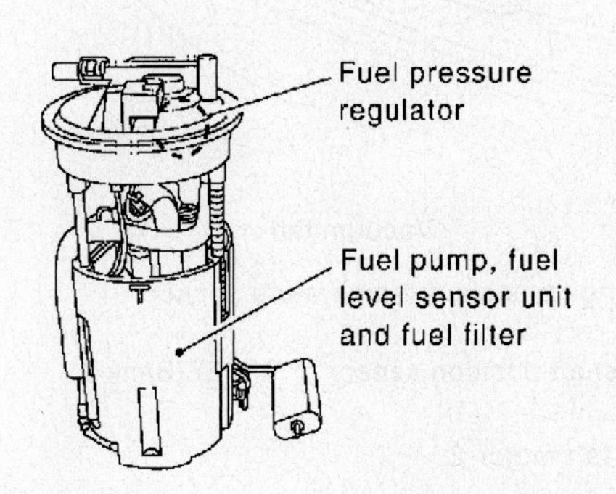
Fuel pressure
regulator
Fuel pump, fuel
level sensor unit
and fuel filter

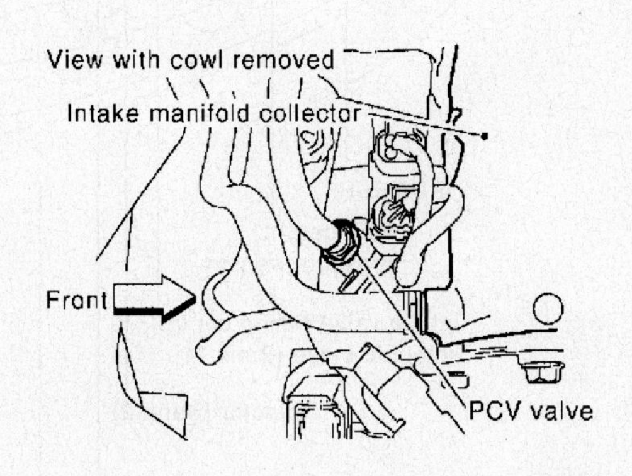
View with cowl removed
Intake manifold collector
Front
PCV valve

29149_NISS_G0116

MURANO

2003–2006

VQ35DE V6 ENGINE

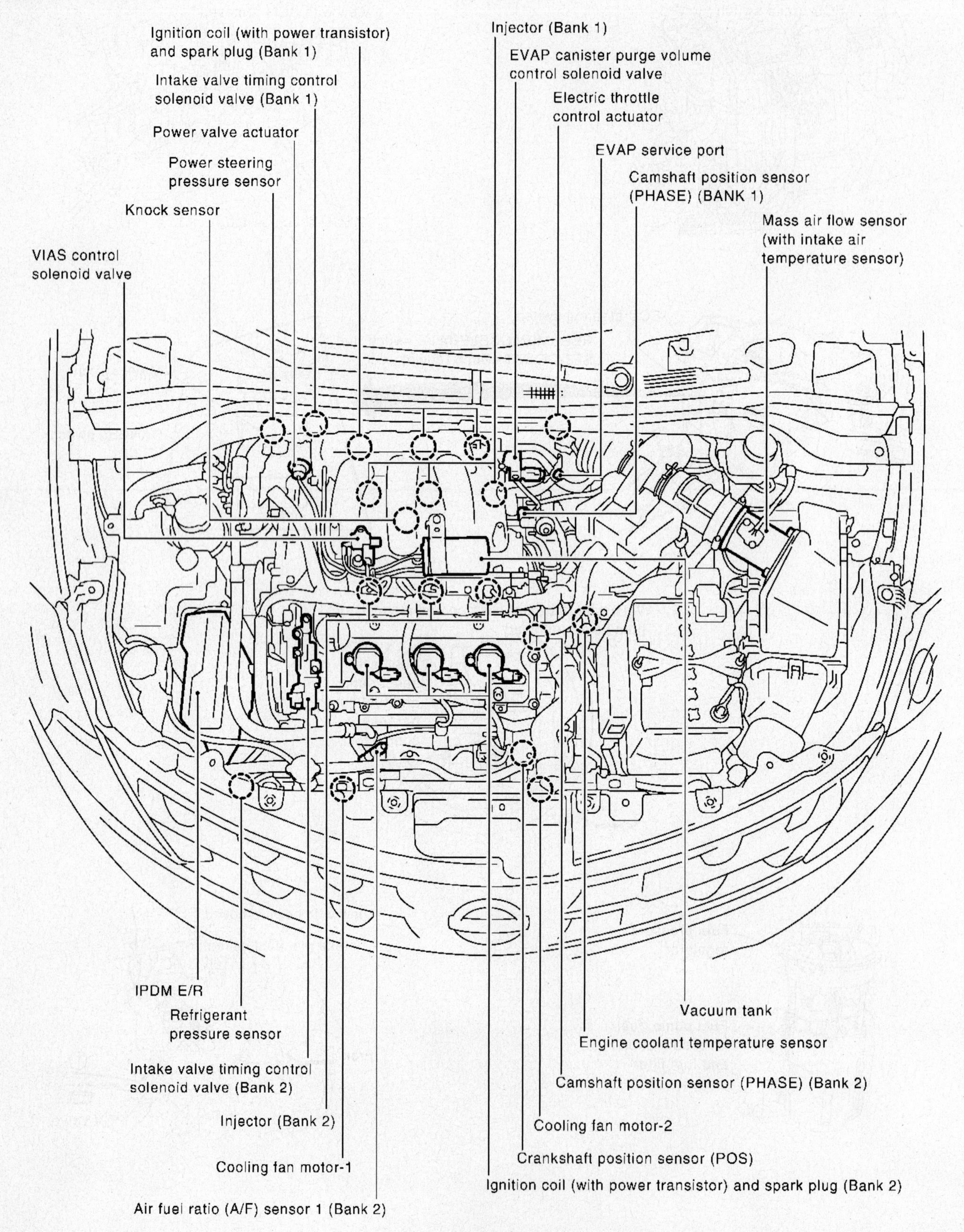
Ignition coil (with power transistor) and spark plug (Bank 1)
Intake valve timing control solenoid valve (Bank 1)
Power valve actuator
Power steering pressure sensor
Knock sensor
VIAS control solenoid valve
Injector (Bank 1)
EVAP canister purge volume control solenoid valve
Electric throttle control actuator
EVAP service port
Camshaft position sensor (PHASE) (BANK 1)
Mass air flow sensor (with intake air temperature sensor)
IPDM E/R
Refrigerant pressure sensor
Intake valve timing control solenoid valve (Bank 2)
Injector (Bank 2)
Cooling fan motor-1
Air fuel ratio (A/F) sensor 1 (Bank 2)
Vacuum tank
Engine coolant temperature sensor
Camshaft position sensor (PHASE) (Bank 2)
Cooling fan motor-2
Crankshaft position sensor (POS)
Ignition coil (with power transistor) and spark plug (Bank 2)

29149_NISS_G0117

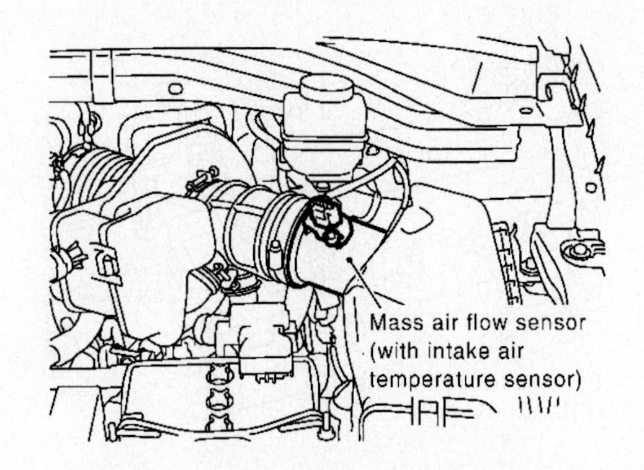

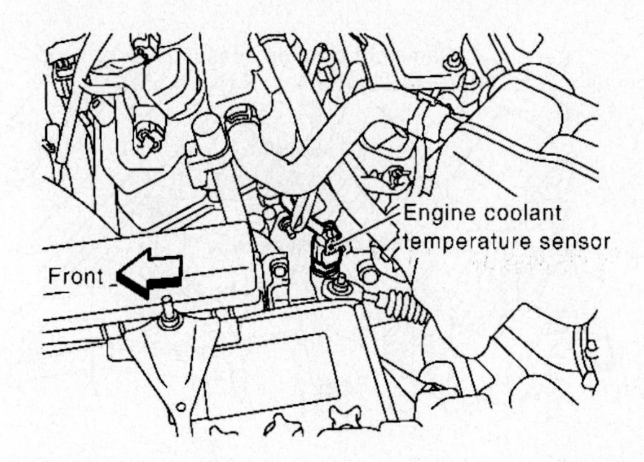

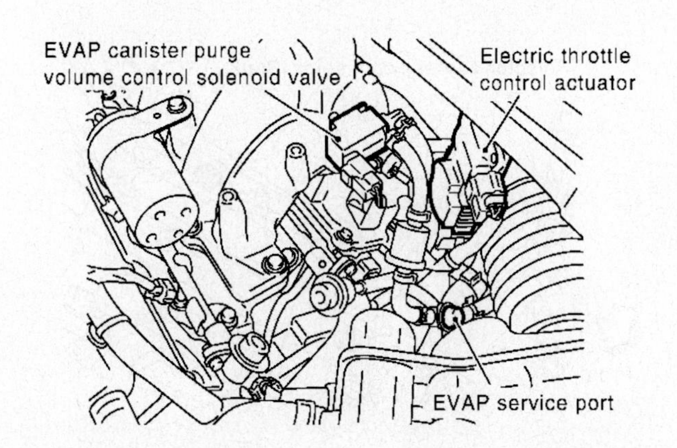

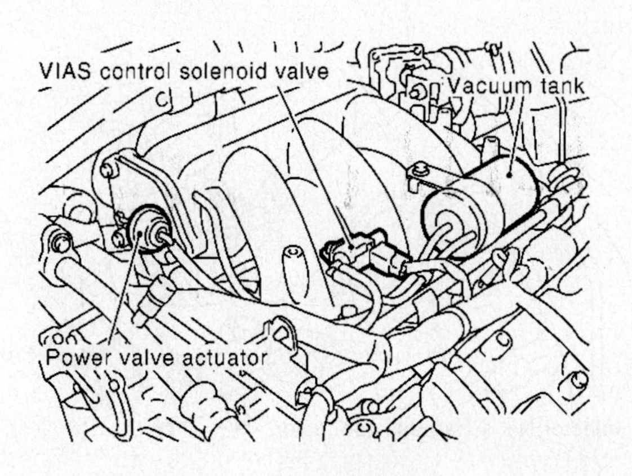

View with cowl top cover and cowl box removed

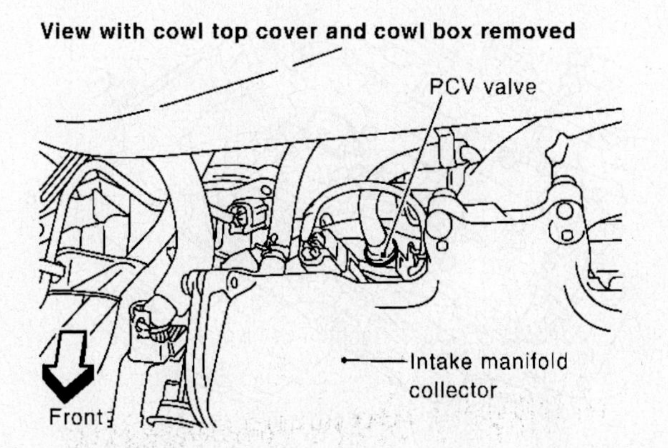

View with IPDM E/R cover removed
and IPDM E/R lifted

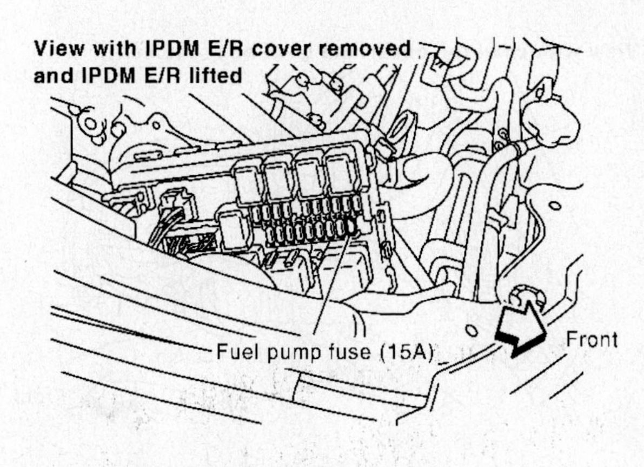

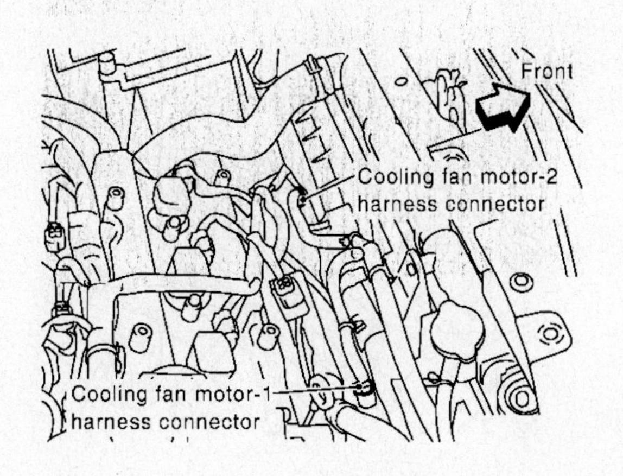

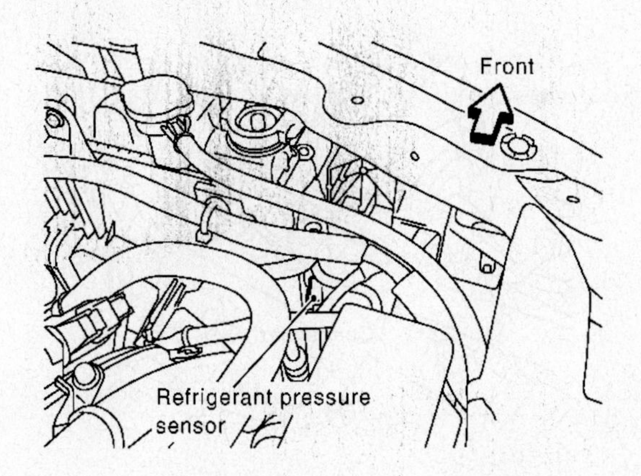

29149_NISS_G0118

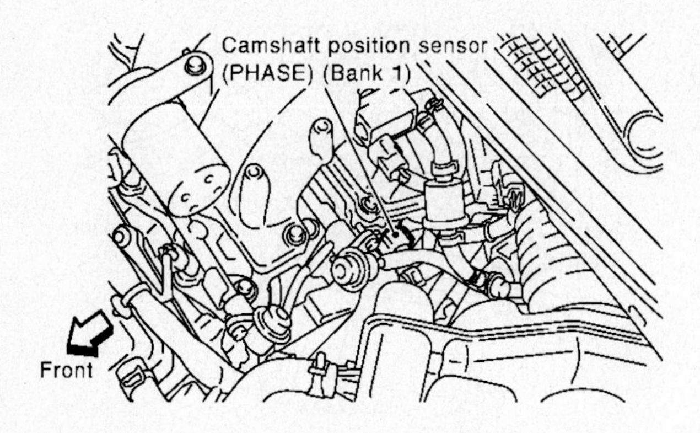
Camshaft position sensor
(PHASE) (Bank 1)
Front

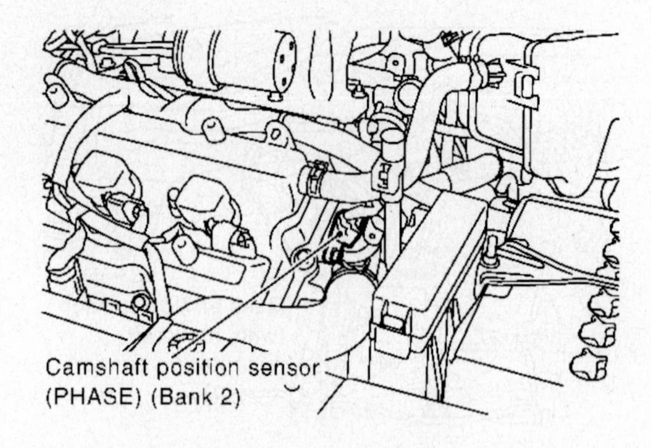
Camshaft position sensor
(PHASE) (Bank 2)

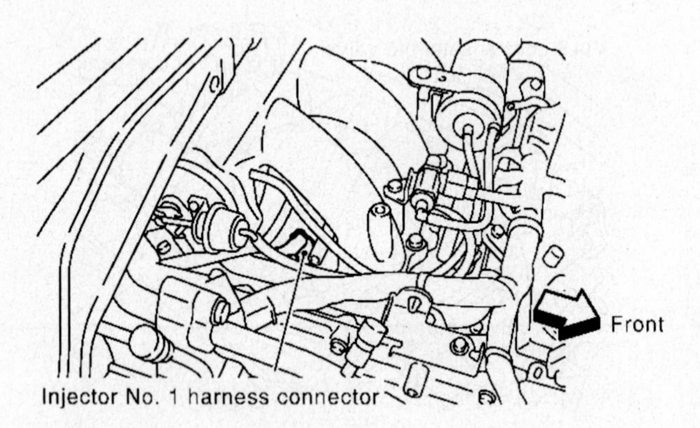
Front
Injector No. 1 harness connector

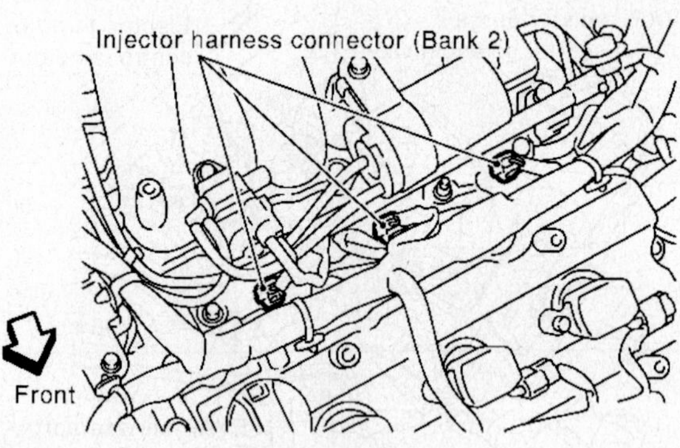
Injector harness connector (Bank 2)
Front

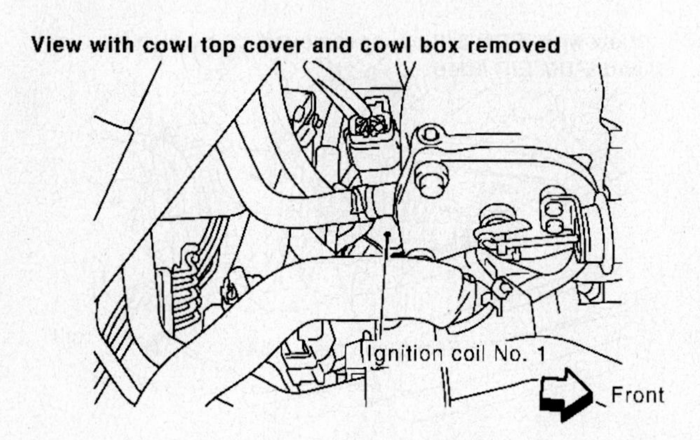
View with cowl top cover and cowl box removed
Ignition coil No. 1
Front

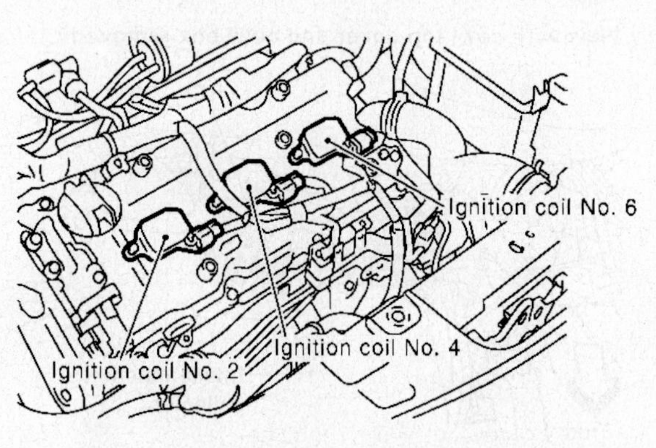
Ignition coil No. 6
Ignition coil No. 4
Ignition coil No. 2

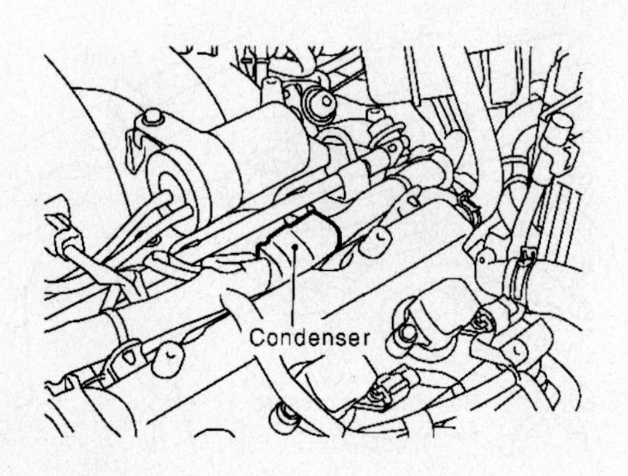
Condenser

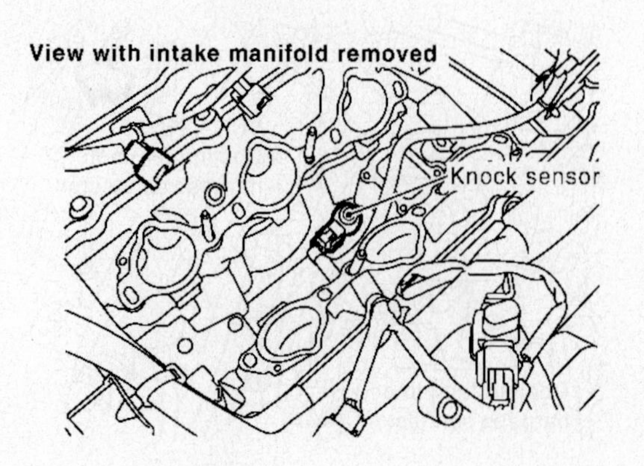
View with intake manifold removed
Knock sensor

29149_NISS_G0119

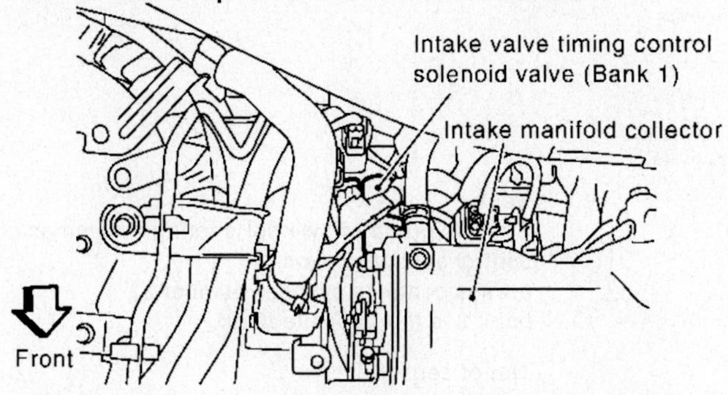
View with cowl top cover and cowl box removed
Intake valve timing control
solenoid valve (Bank 1)
Intake manifold collector
Front

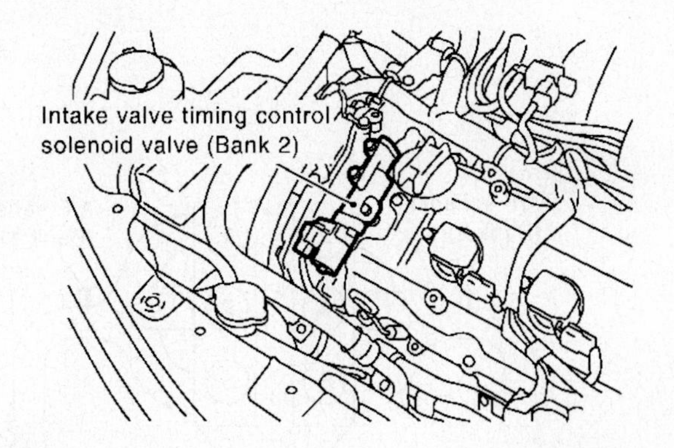
Intake valve timing control
solenoid valve (Bank 2)

View from under the vehicle

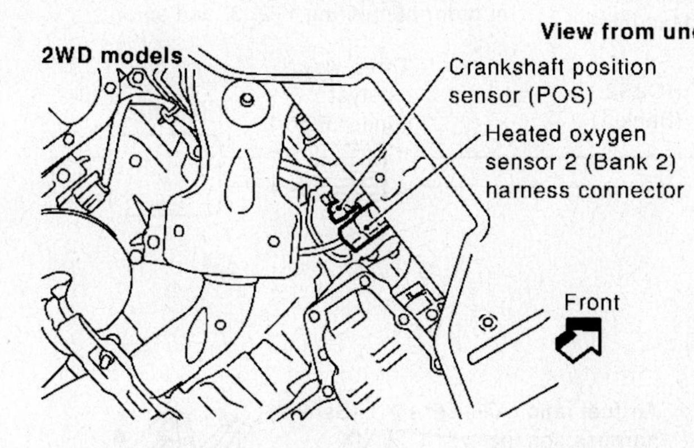
2WD models
Crankshaft position
sensor (POS)
Heated oxygen
sensor 2 (Bank 2)
harness connector
Front

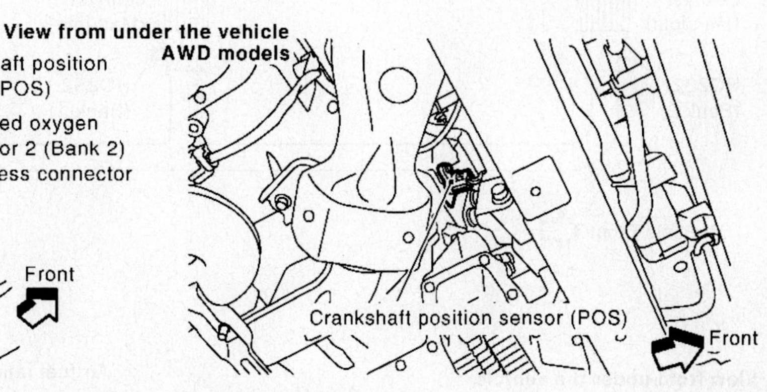
AWD models
Crankshaft position sensor (POS)
Front

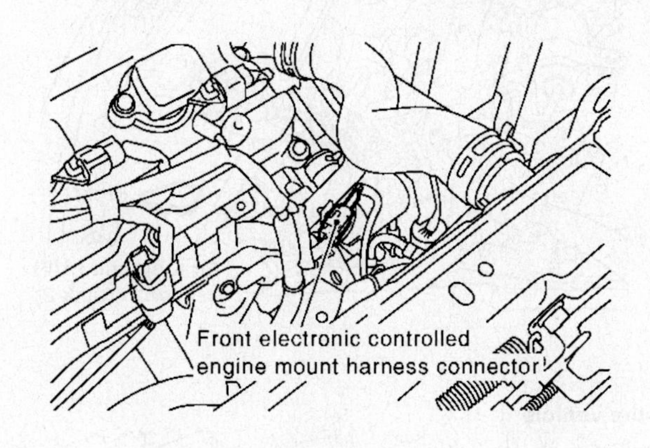
Front electronic controlled
engine mount harness connector

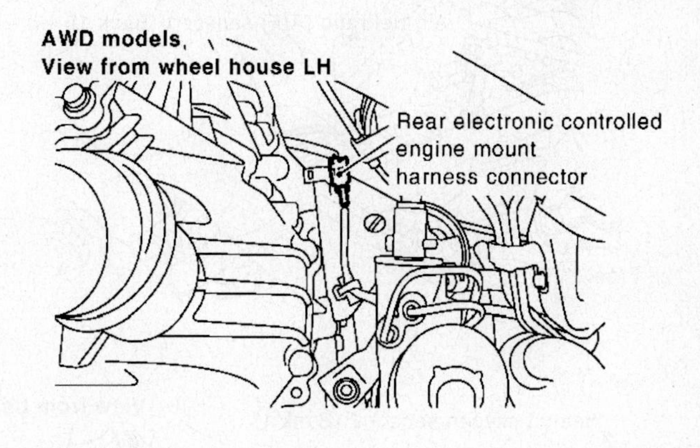
AWD models,
View from wheel house LH
Rear electronic controlled
engine mount
harness connector

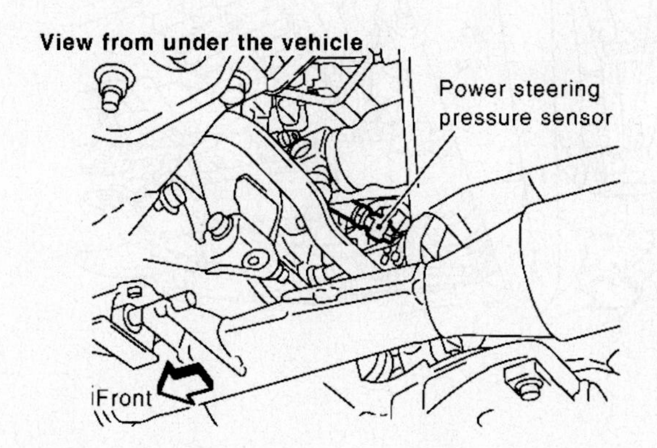
View from under the vehicle
Power steering
pressure sensor
Front

29149_NISS_G0120

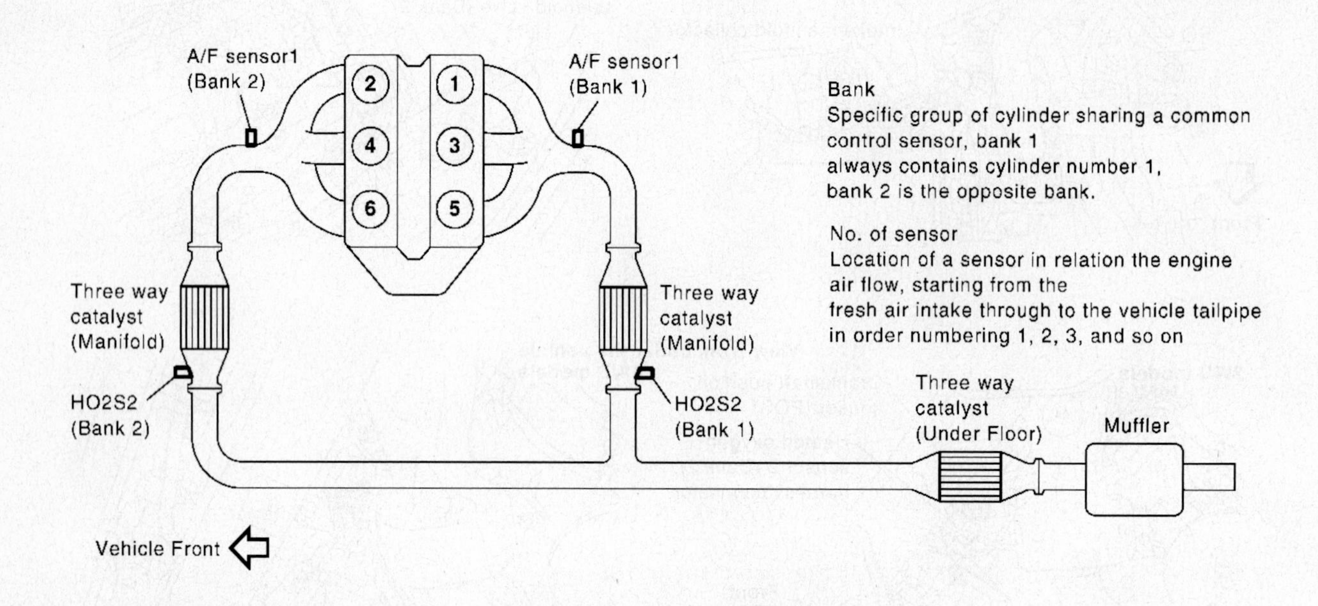

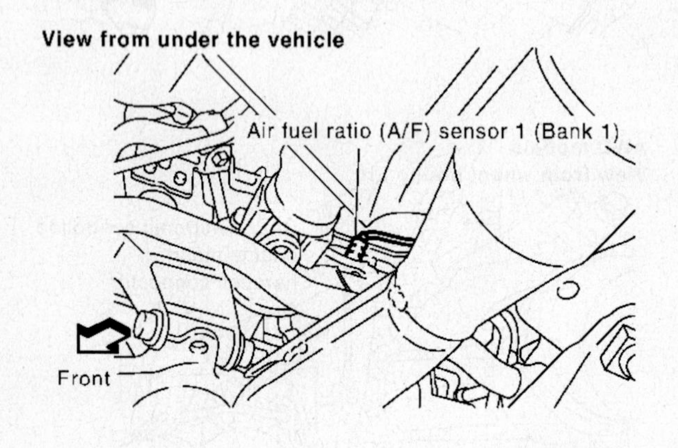

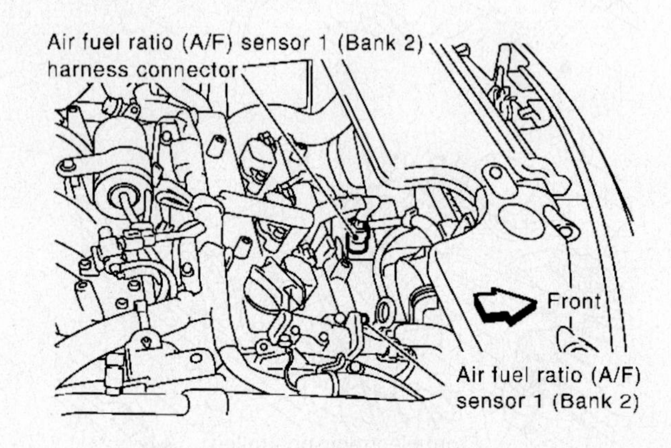

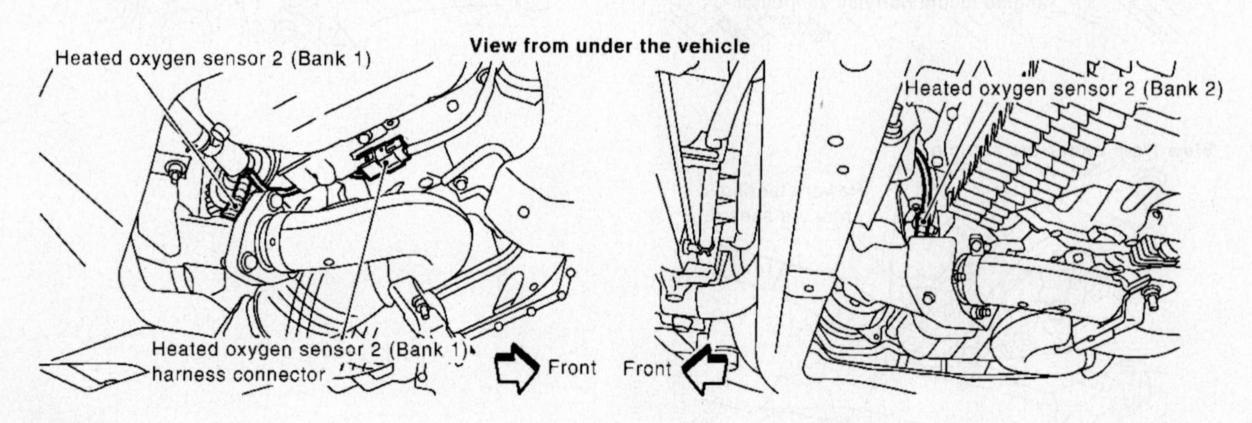

29149_NISS_G0121

View from under the vehicle

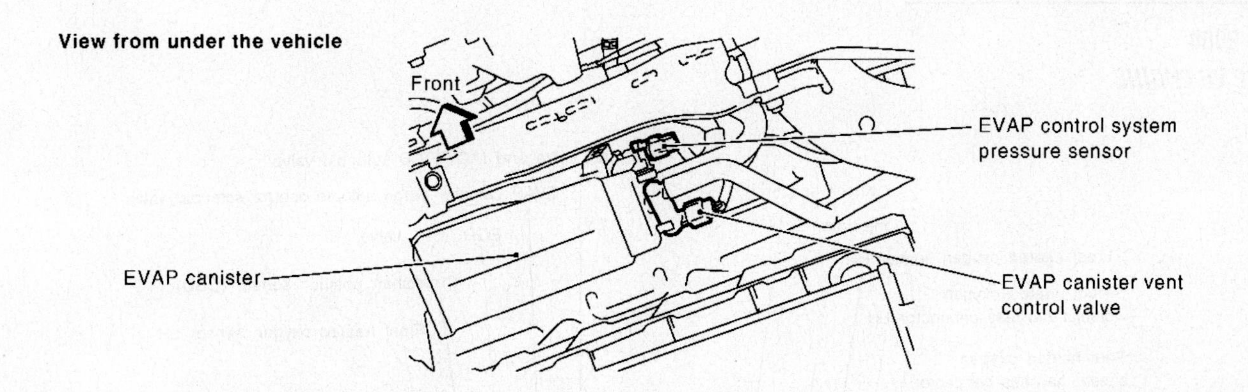

View with glove box removed

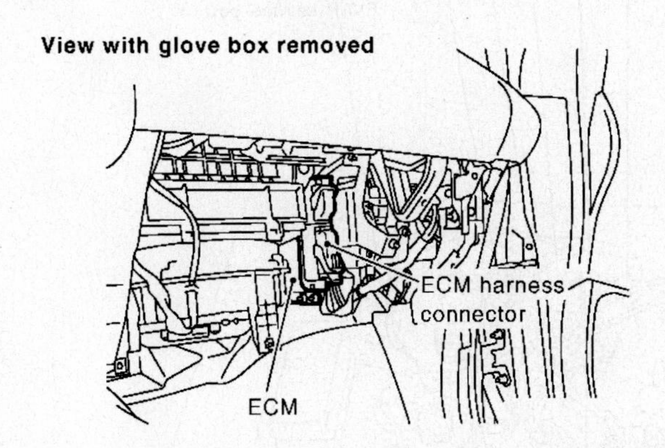

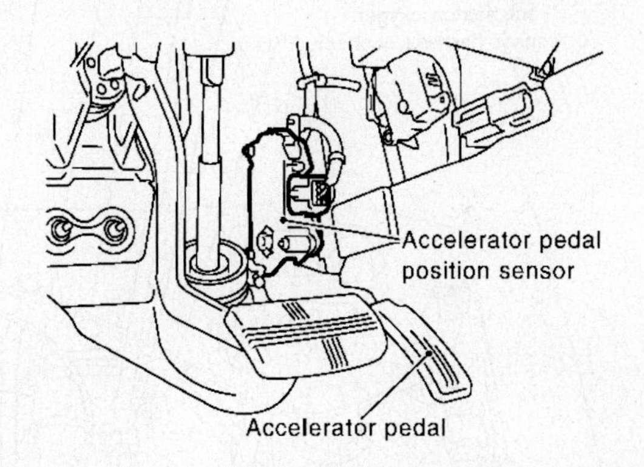

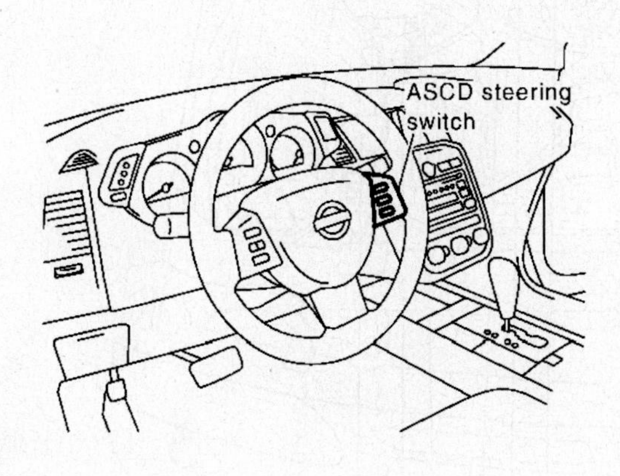

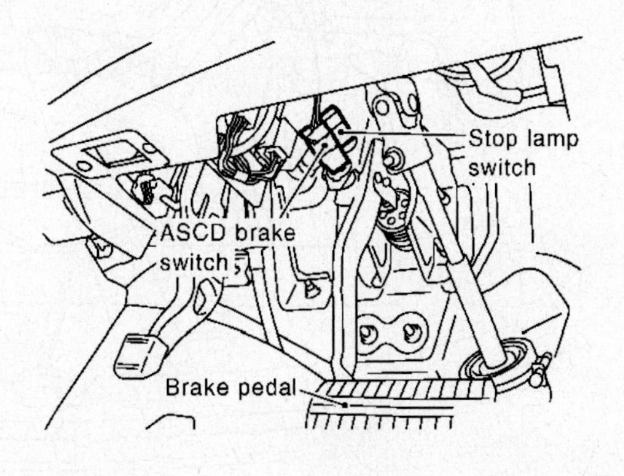

View with rear seat cushion removed

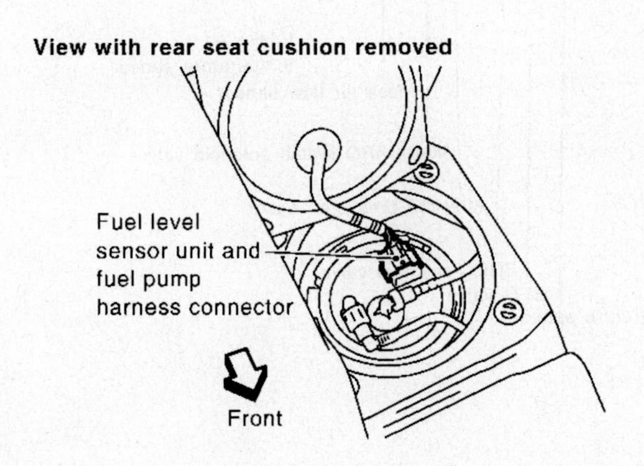

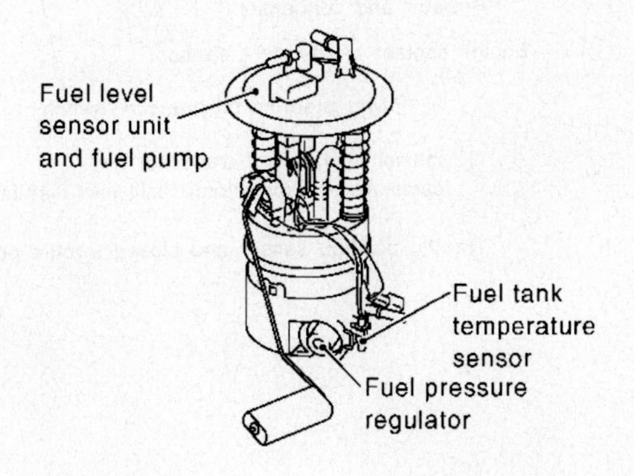

29149_NISS_G0122

PATHFINDER

1996–2000
VG33E V6 ENGINE

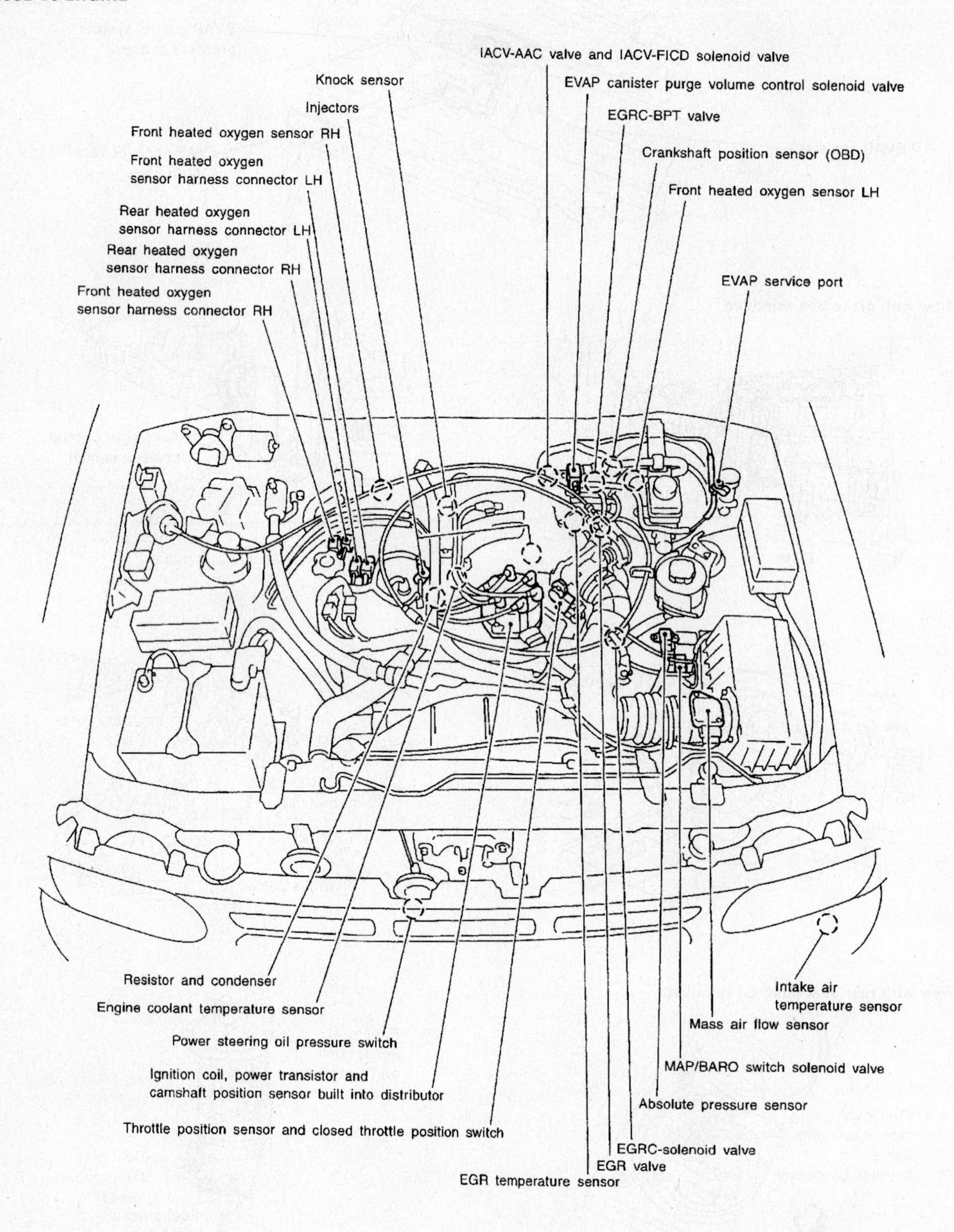
IACV-AAC valve and IACV-FICD solenoid valve
Knock sensor
EVAP canister purge volume control solenoid valve
Injectors
EGRC-BPT valve
Front heated oxygen sensor RH
Front heated oxygen
sensor harness connector LH
Crankshaft position sensor (OBD)
Front heated oxygen sensor LH
Rear heated oxygen
sensor harness connector LH
Rear heated oxygen
sensor harness connector RH
EVAP service port
Front heated oxygen
sensor harness connector RH
Resistor and condenser
Intake air
temperature sensor
Engine coolant temperature sensor
Mass air flow sensor
Power steering oil pressure switch
MAP/BARO switch solenoid valve
Ignition coil, power transistor and
camshaft position sensor built into distributor
Absolute pressure sensor
Throttle position sensor and closed throttle position switch
EGRC-solenoid valve
EGR valve
EGR temperature sensor

29149_NISS_G0020

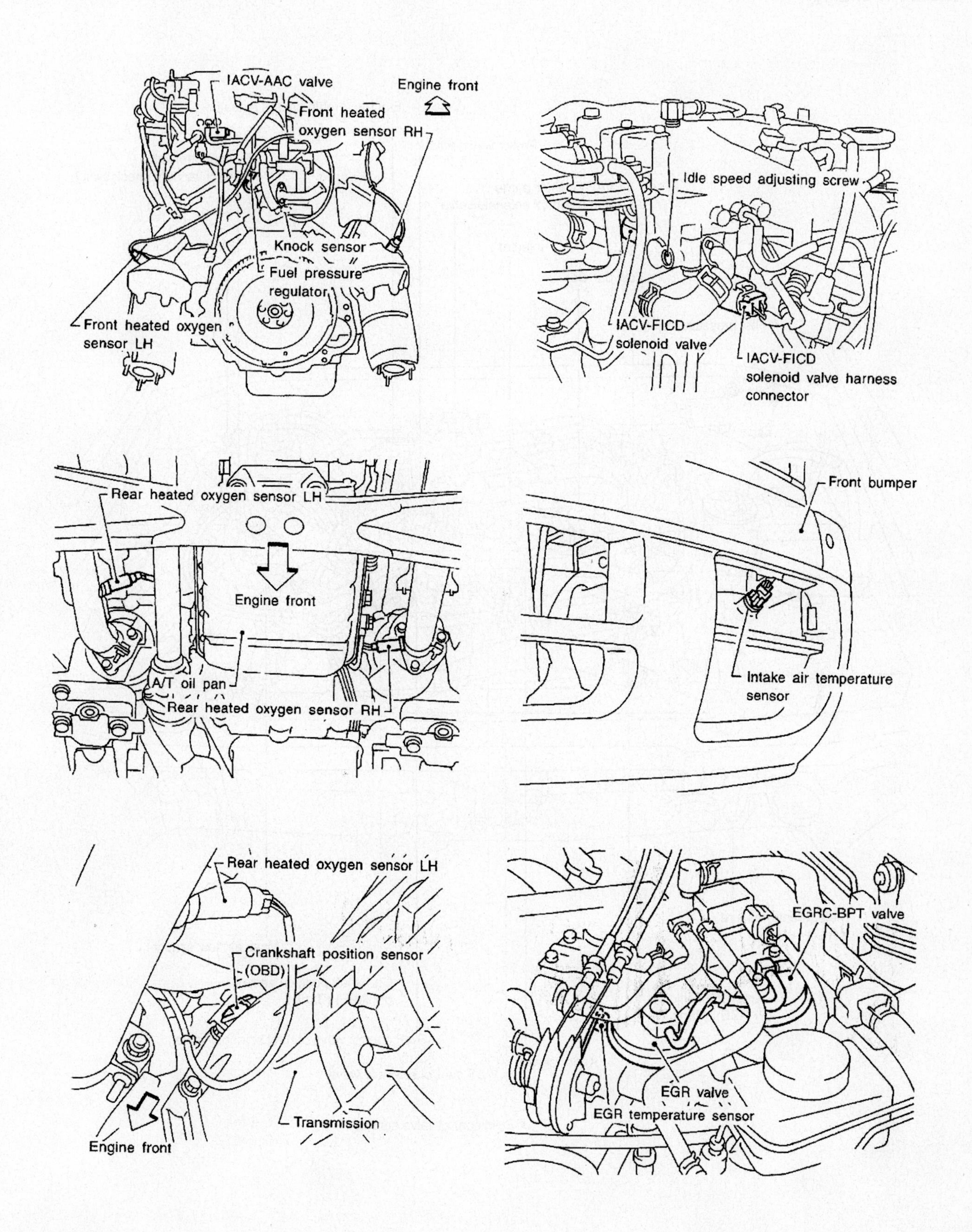
IACV-AAC valve
Engine front
Front heated oxygen sensor RH
Knock sensor
Fuel pressure regulator
Front heated oxygen sensor LH
Idle speed adjusting screw
IACV-FICD solenoid valve
IACV-FICD solenoid valve harness connector
Rear heated oxygen sensor LH
Engine front
A/T oil pan
Rear heated oxygen sensor RH
Front bumper
Intake air temperature sensor
Rear heated oxygen sensor LH
Crankshaft position sensor (OBD)
Transmission
Engine front
EGRC-BPT valve
EGR valve
EGR temperature sensor

2001–2004

VQ35DE V6 ENGINE

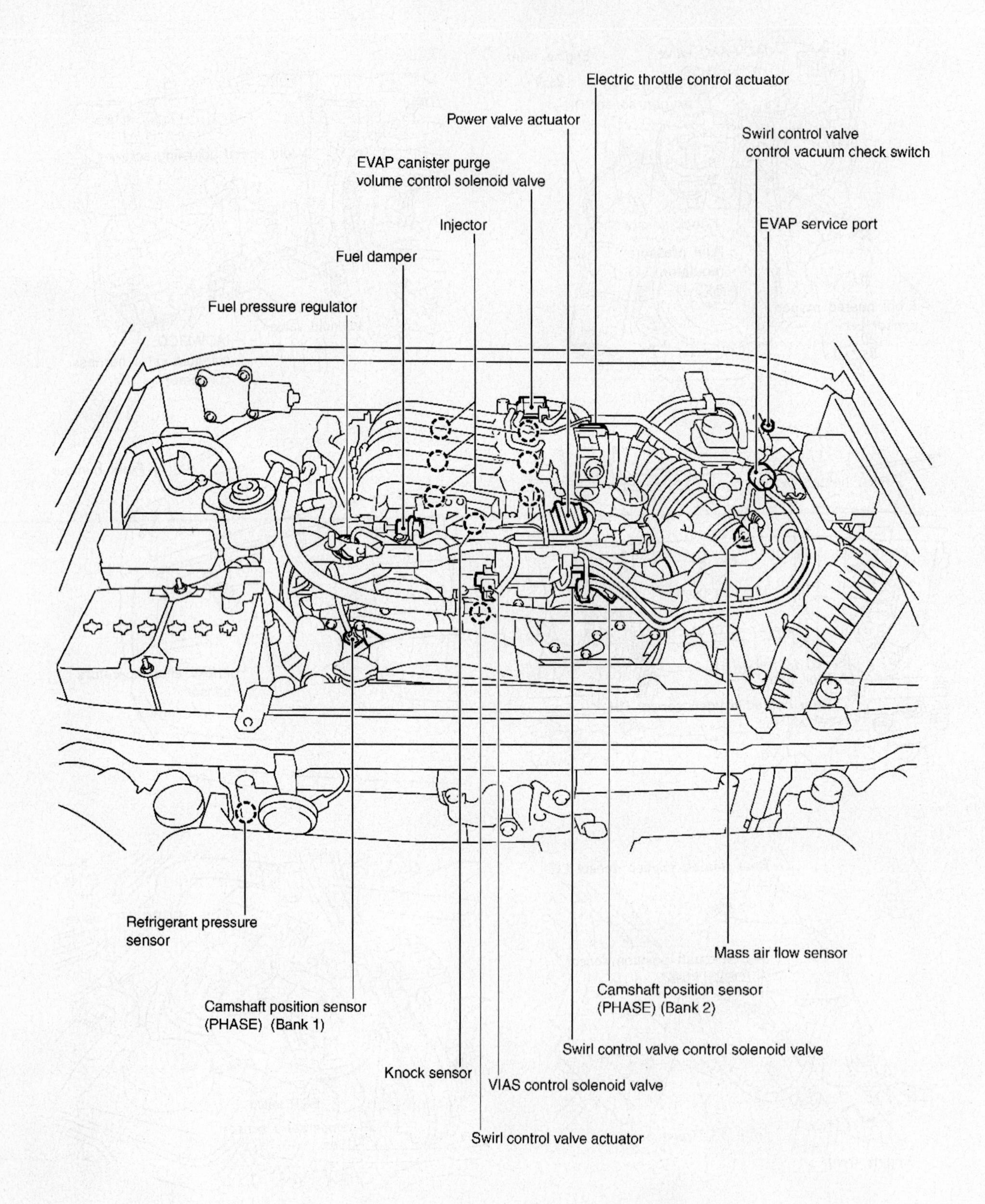
Electric throttle control actuator
Power valve actuator
Swirl control valve control vacuum check switch
EVAP canister purge volume control solenoid valve
Injector
EVAP service port
Fuel damper
Fuel pressure regulator
Refrigerant pressure sensor
Mass air flow sensor
Camshaft position sensor (PHASE) (Bank 2)
Camshaft position sensor (PHASE) (Bank 1)
Swirl control valve control solenoid valve
Knock sensor
VIAS control solenoid valve
Swirl control valve actuator

29149_NISS_G0123

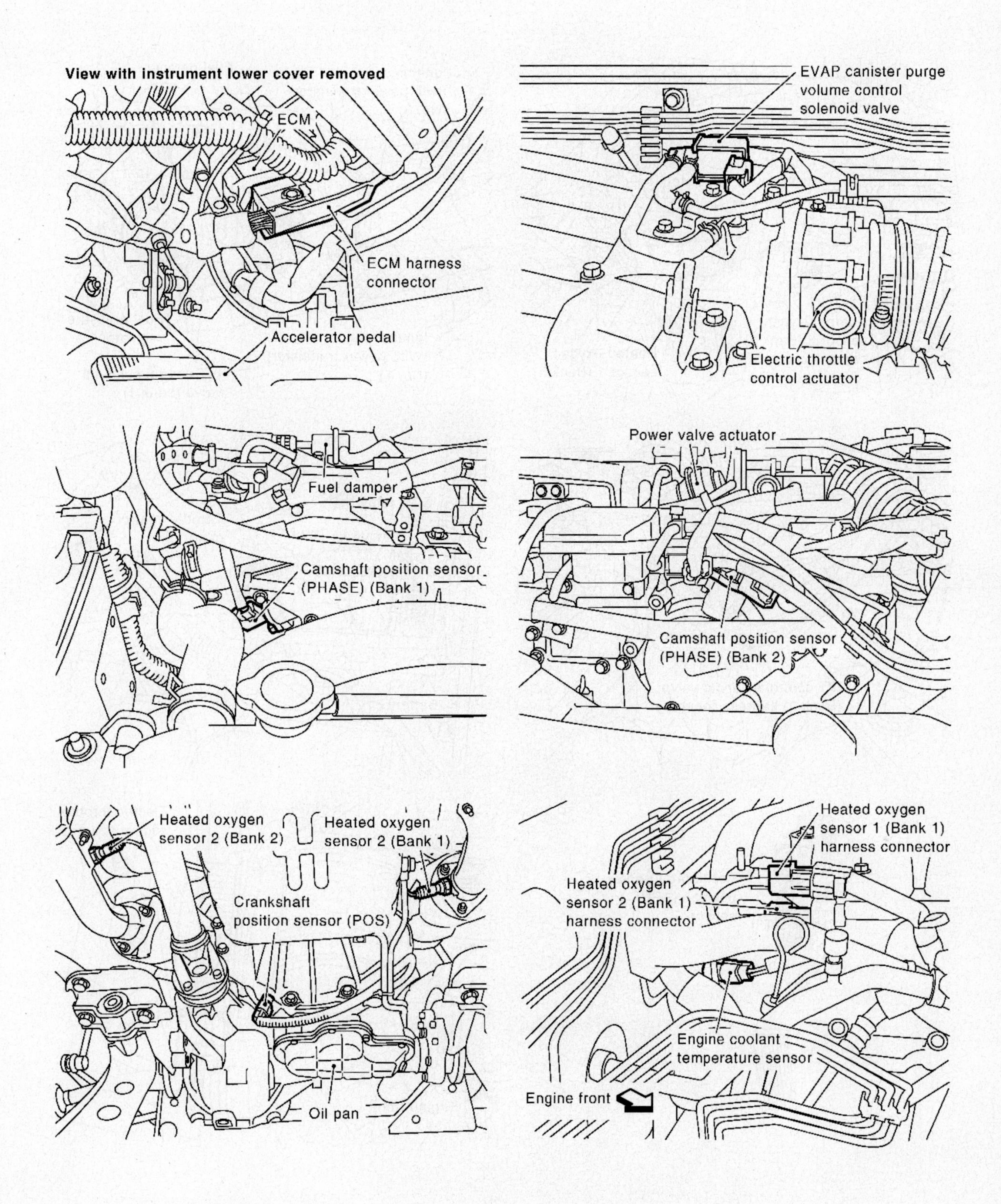
View with instrument lower cover removed
ECM
ECM harness connector
Accelerator pedal
EVAP canister purge volume control solenoid valve
Electric throttle control actuator
Fuel damper
Camshaft position sensor (PHASE) (Bank 1)
Power valve actuator
Camshaft position sensor (PHASE) (Bank 2)
Heated oxygen sensor 2 (Bank 2)
Heated oxygen sensor 2 (Bank 1)
Crankshaft position sensor (POS)
Oil pan
Heated oxygen sensor 1 (Bank 1) harness connector
Heated oxygen sensor 2 (Bank 1) harness connector
Engine coolant temperature sensor
Engine front

29149_NISS_G0124

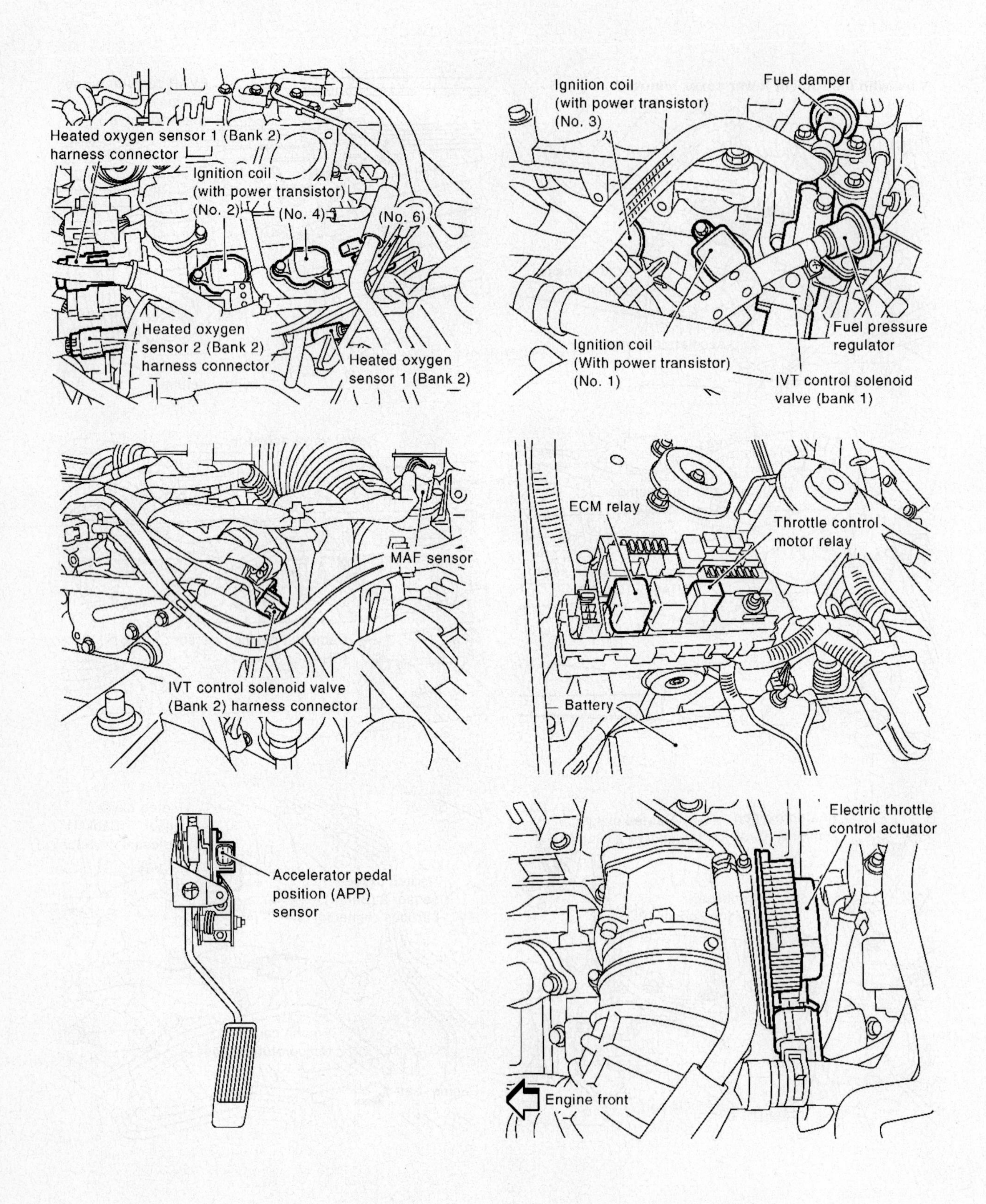
Heated oxygen sensor 1 (Bank 2)
harness connector
Ignition coil
(with power transistor)
(No. 2)
(No. 4)
(No. 6)
Heated oxygen
sensor 2 (Bank 2)
harness connector
Heated oxygen
sensor 1 (Bank 2)
Ignition coil
(with power transistor)
(No. 3)
Fuel damper
Ignition coil
(With power transistor)
(No. 1)
Fuel pressure
regulator
IVT control solenoid
valve (bank 1)
MAF sensor
IVT control solenoid valve
(Bank 2) harness connector
ECM relay
Throttle control
motor relay
Battery
Accelerator pedal
position (APP)
sensor
Electric throttle
control actuator
Engine front

29149_NISS_G0125

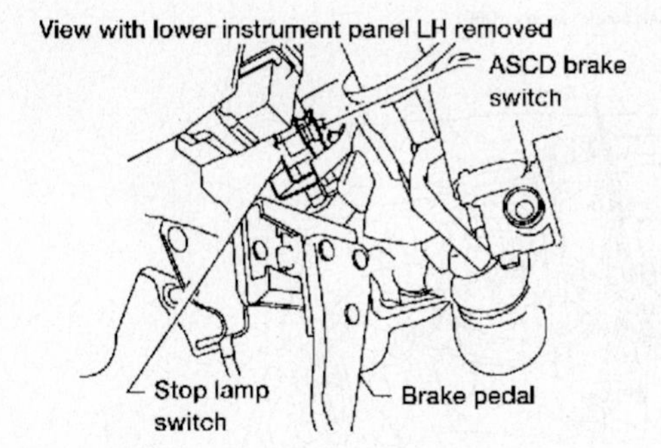

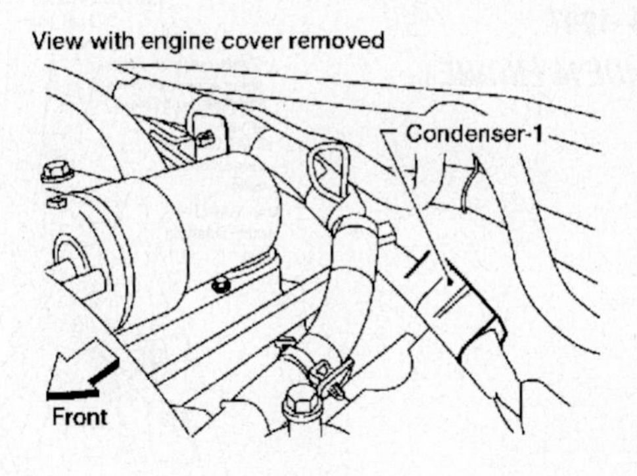

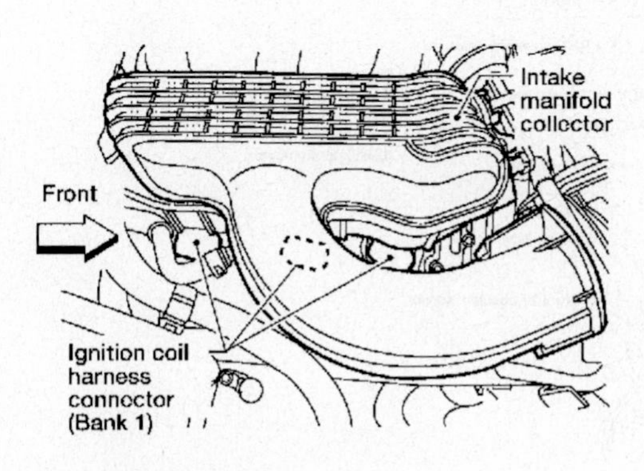

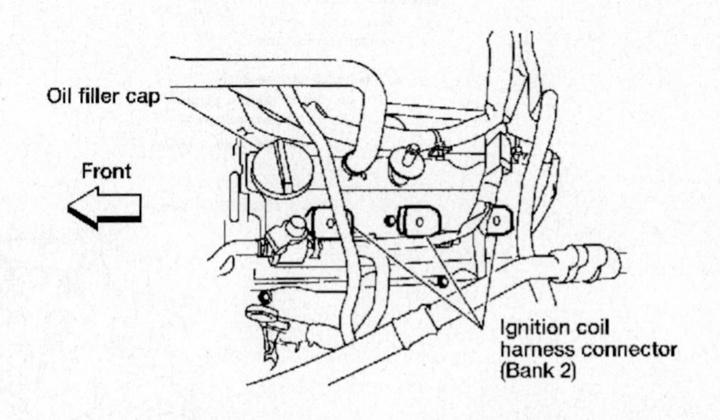

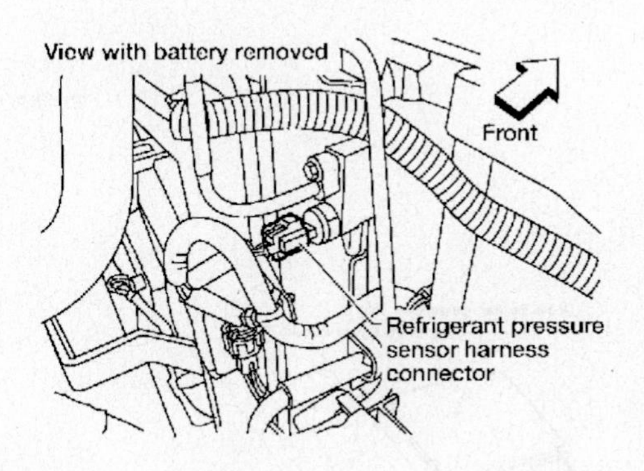

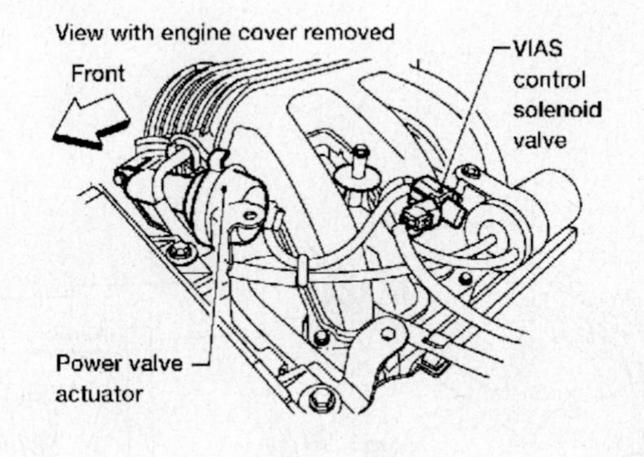

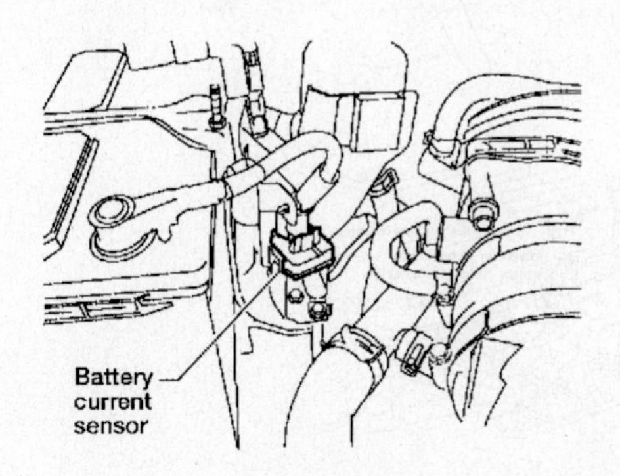

29149_NISS_G0134

PICKUP

1996–1997

KA24DE I4 ENGINE

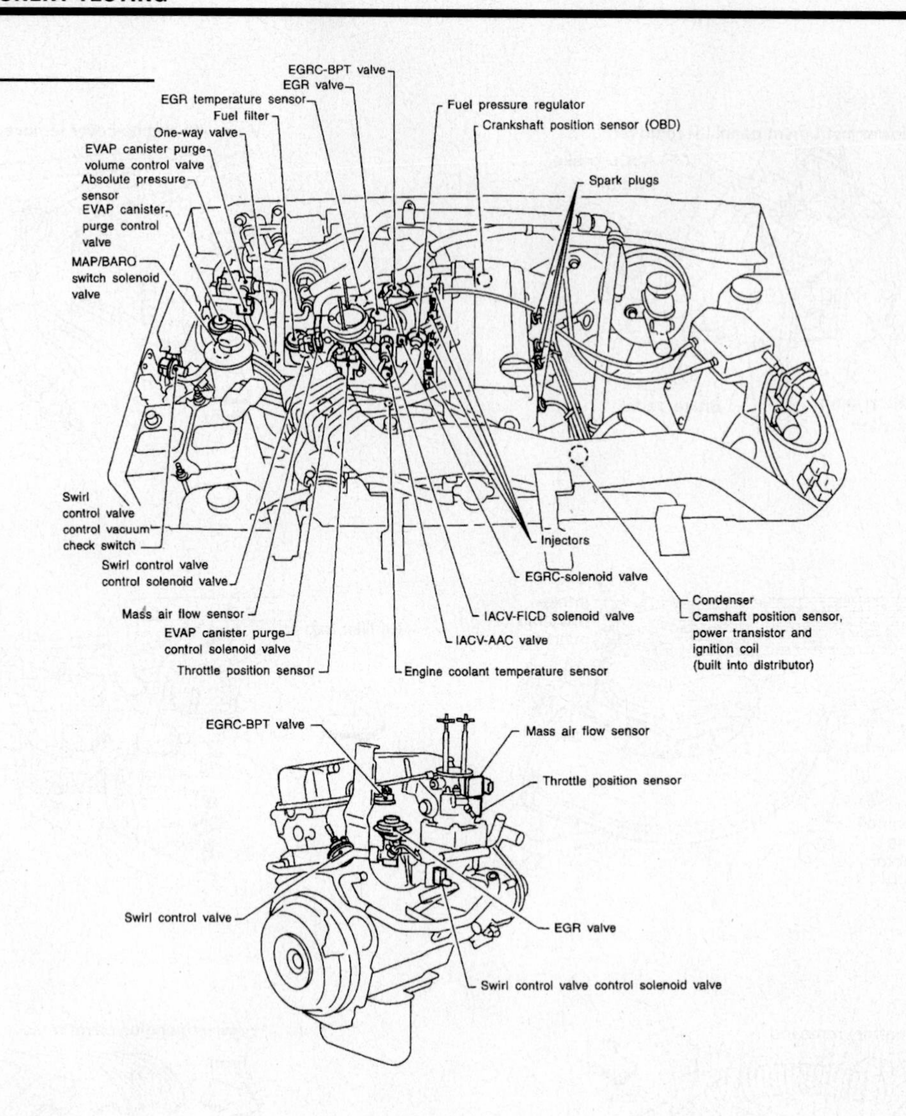
EGRC-BPT valve
EGR valve
EGR temperature sensor
Fuel filter
One-way valve
EVAP canister purge volume control valve
Absolute pressure sensor
EVAP canister purge control valve
MAP/BARO switch solenoid valve
Fuel pressure regulator
Crankshaft position sensor (OBD)
Spark plugs
Swirl control valve control vacuum check switch
Swirl control valve control solenoid valve
Mass air flow sensor
EVAP canister purge control solenoid valve
Throttle position sensor
Injectors
EGRC-solenoid valve
IACV-FICD solenoid valve
IACV-AAC valve
Engine coolant temperature sensor
Condenser
Camshaft position sensor, power transistor and ignition coil (built into distributor)
EGRC-BPT valve
Mass air flow sensor
Throttle position sensor
Swirl control valve
EGR valve
Swirl control valve control solenoid valve

29149_NISS_G0009

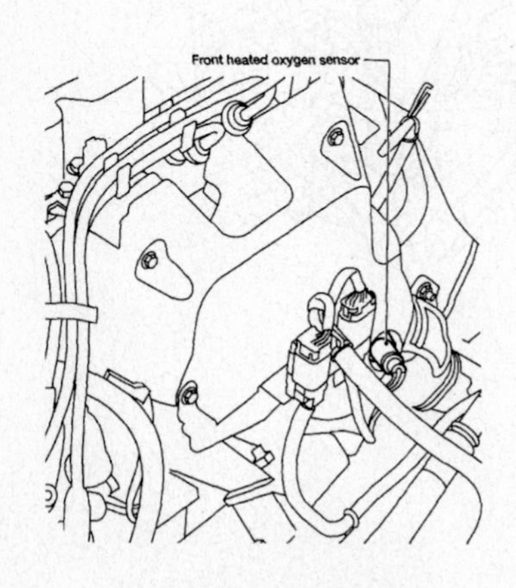
Front heated oxygen sensor

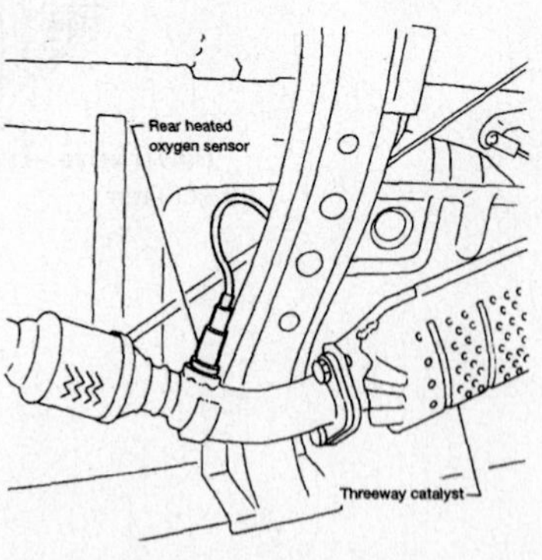
Rear heated oxygen sensor
Threeway catalyst

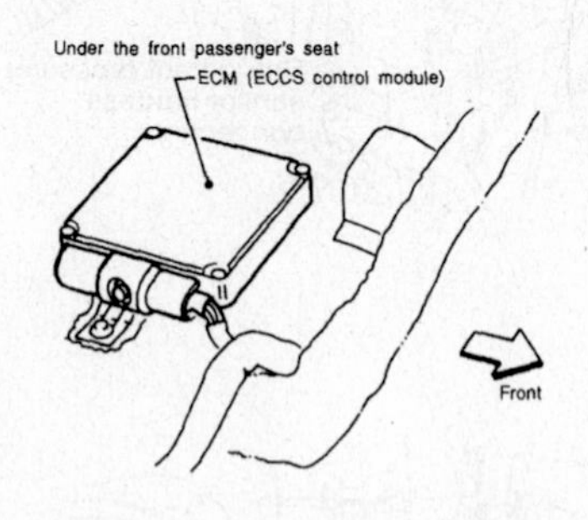
Under the front passenger's seat
ECM (ECCS control module)
Front

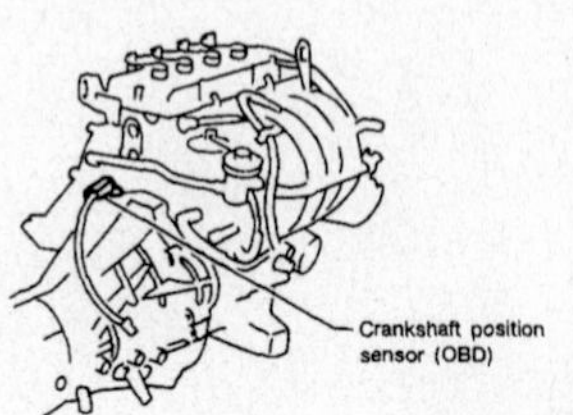
Crankshaft position sensor (OBD)

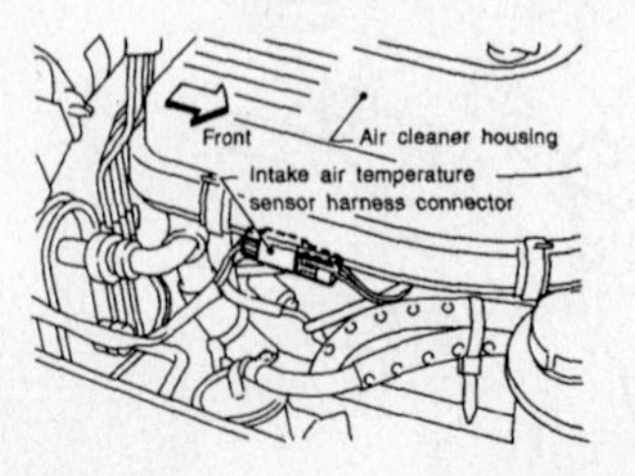
Front
Air cleaner housing
Intake air temperature sensor harness connector

29149_NISS_G0010

QUEST

1996–1998

VG30E V6 ENGINE

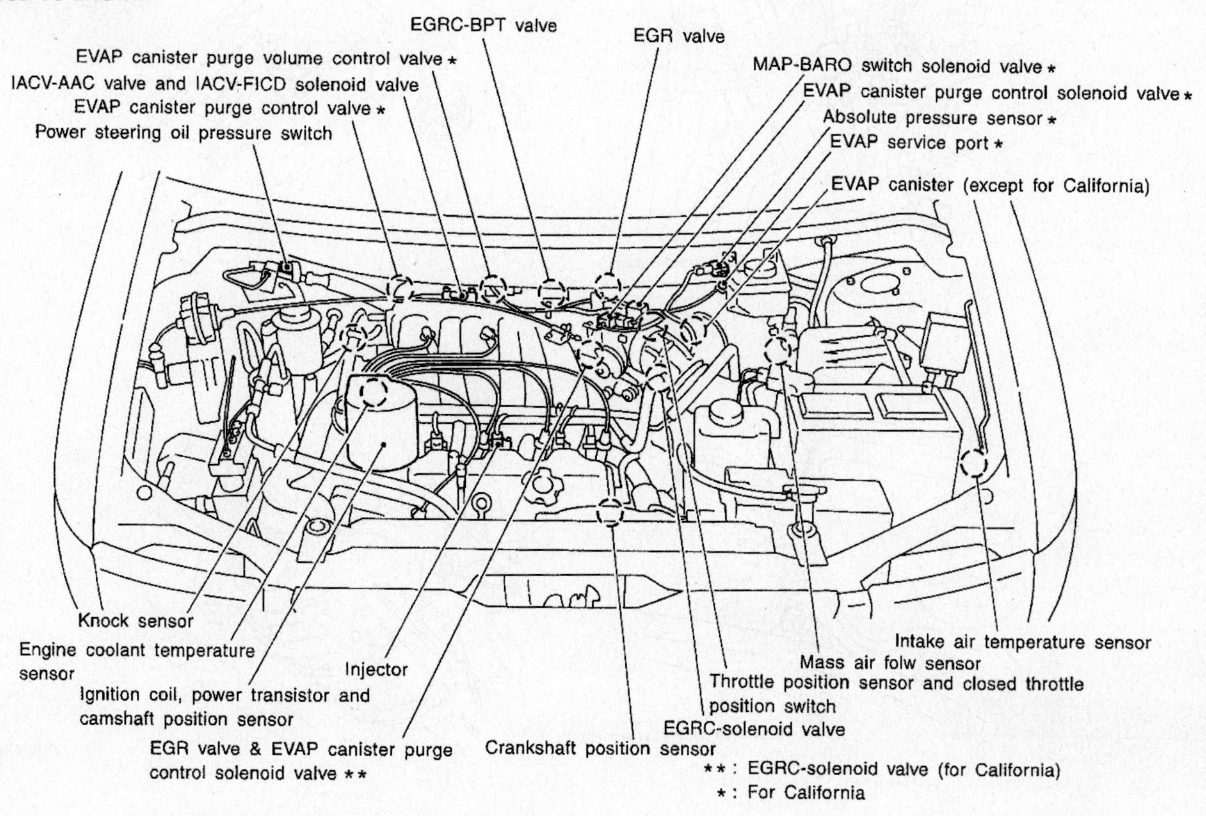
EGRC-BPT valve
EGR valve
EVAP canister purge volume control valve *
IACV-AAC valve and IACV-FICD solenoid valve
EVAP canister purge control valve *
Power steering oil pressure switch
MAP-BARO switch solenoid valve *
EVAP canister purge control solenoid valve *
Absolute pressure sensor *
EVAP service port *
EVAP canister (except for California)
Knock sensor
Engine coolant temperature sensor
Injector
Ignition coil, power transistor and camshaft position sensor
EGR valve & EVAP canister purge control solenoid valve **
Crankshaft position sensor
Intake air temperature sensor
Mass air folw sensor
Throttle position sensor and closed throttle position switch
EGRC-solenoid valve
** : EGRC-solenoid valve (for California)
* : For California

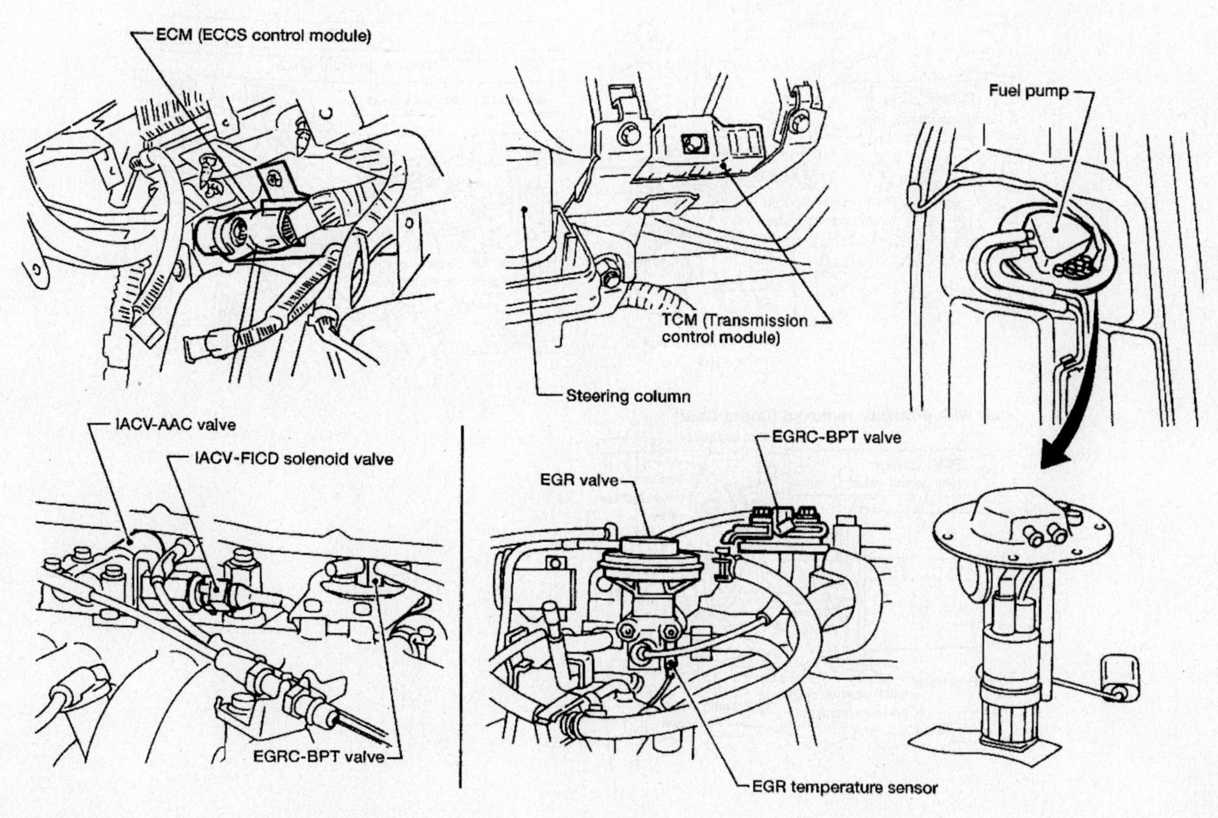
ECM (ECCS control module)
TCM (Transmission control module)
Steering column
Fuel pump
IACV-AAC valve
IACV-FICD solenoid valve
EGRC-BPT valve
EGR valve
EGRC-BPT valve
EGR temperature sensor

29149_NISS_G0022

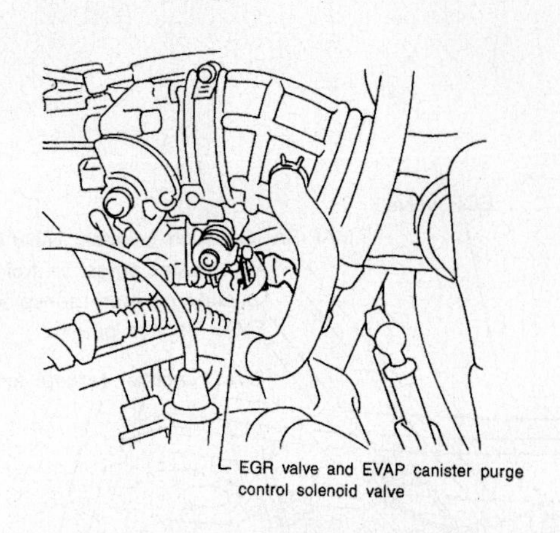
EGR valve and EVAP canister purge
control solenoid valve

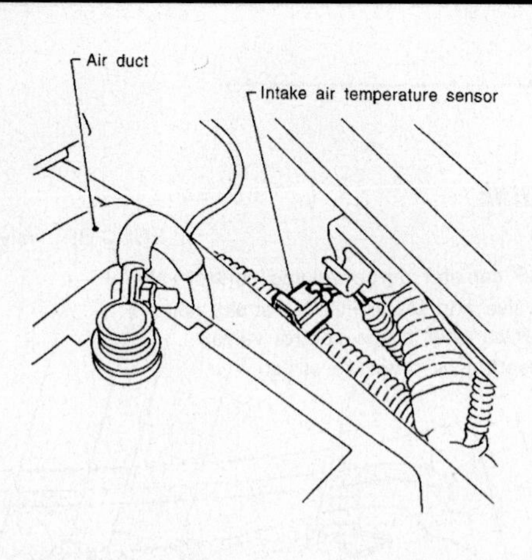
Air duct
Intake air temperature sensor

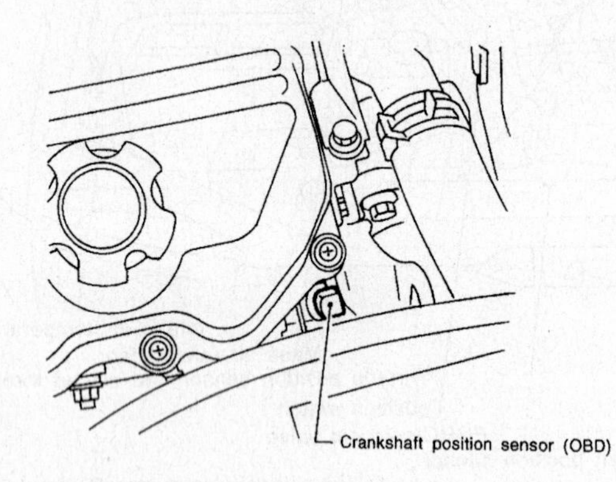
Crankshaft position sensor (OBD)

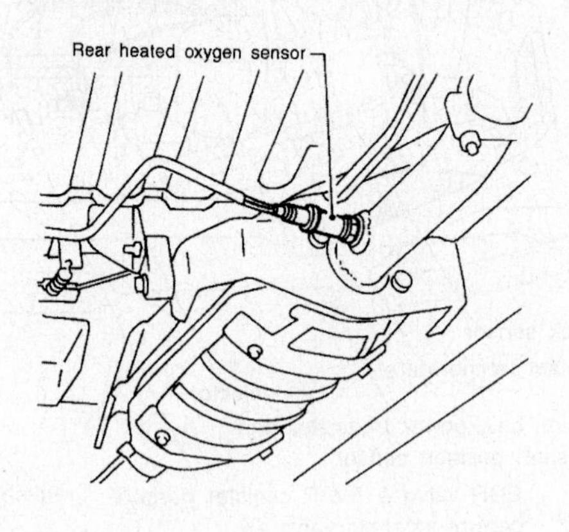
Rear heated oxygen sensor

29149_NISS_G0023

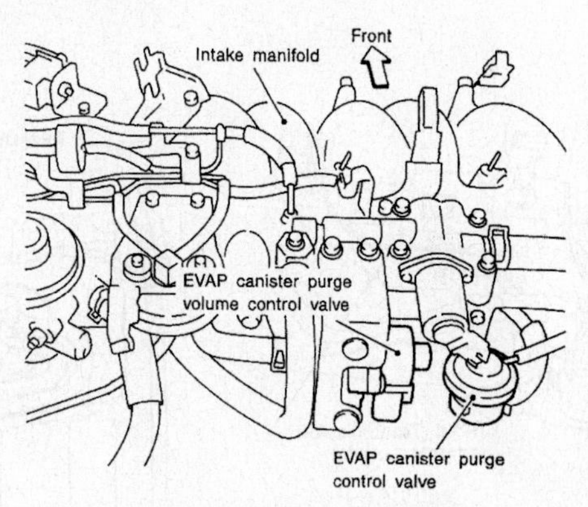
Front
Intake manifold
EVAP canister purge
volume control valve
EVAP canister purge
control valve

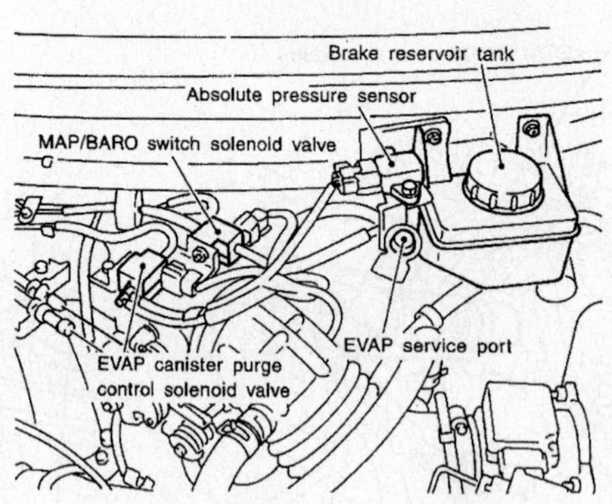
Brake reservoir tank
Absolute pressure sensor
MAP/BARO switch solenoid valve
EVAP service port
EVAP canister purge
control solenoid valve

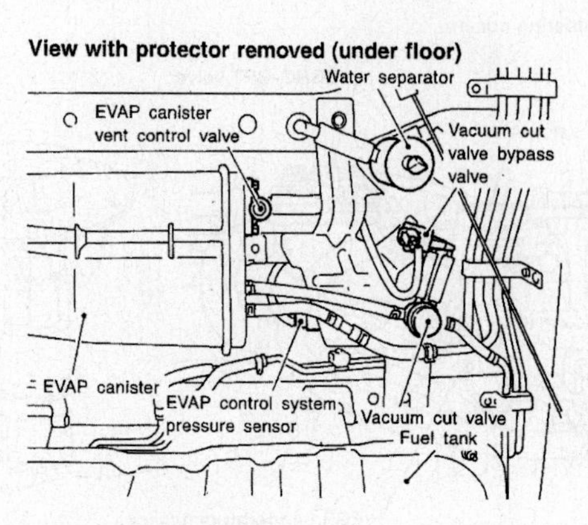
View with protector removed (under floor)
Water separator
EVAP canister
vent control valve
Vacuum cut
valve bypass
valve
EVAP canister
EVAP control system
pressure sensor
Vacuum cut valve
Fuel tank

29149_NISS_G0024

1999–2002
VG33E V6 ENGINE

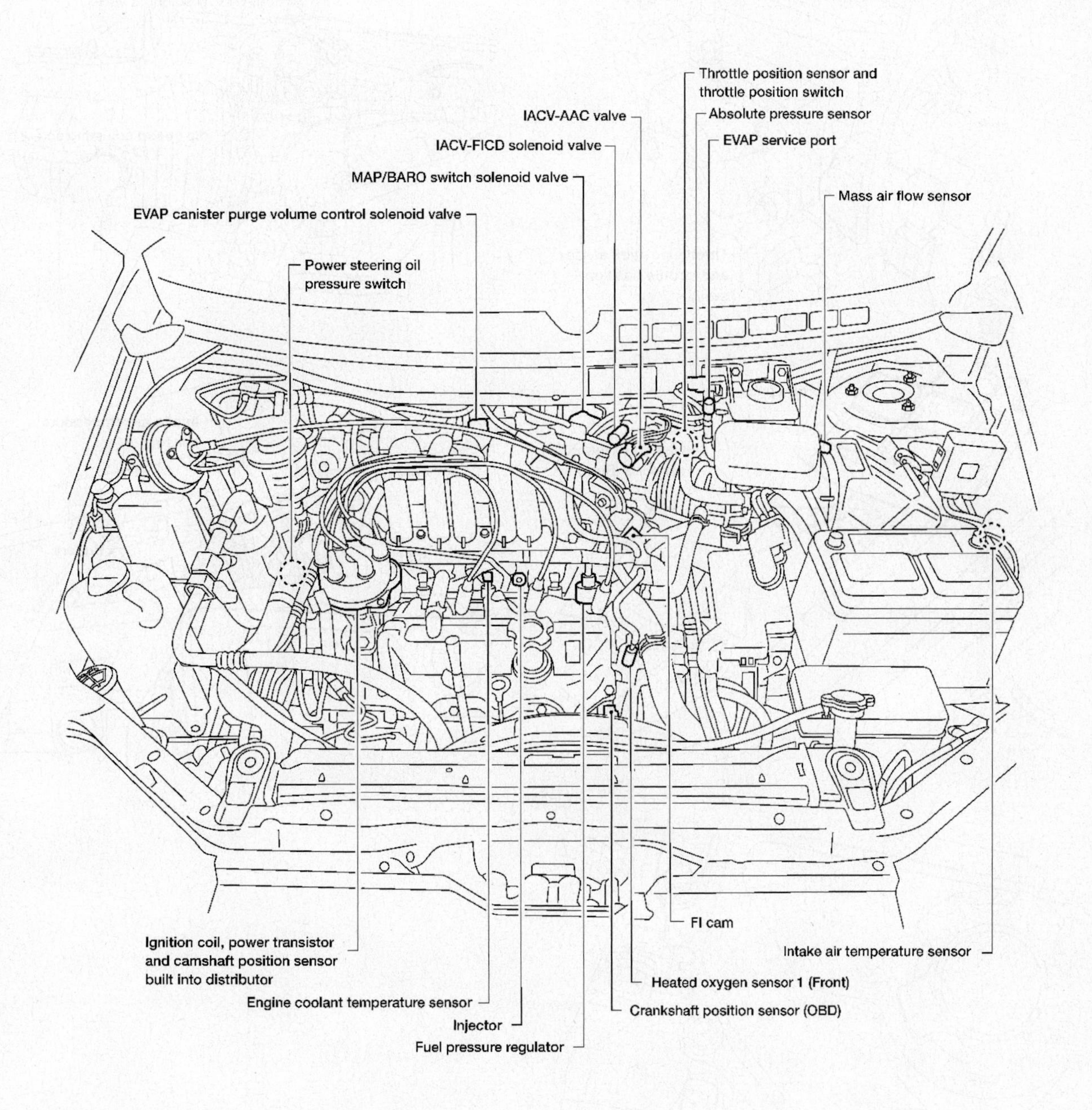
Throttle position sensor and
throttle position switch
IACV-AAC valve
Absolute pressure sensor
IACV-FICD solenoid valve
EVAP service port
MAP/BARO switch solenoid valve
Mass air flow sensor
EVAP canister purge volume control solenoid valve
Power steering oil
pressure switch
FI cam
Ignition coil, power transistor
and camshaft position sensor
built into distributor
Intake air temperature sensor
Heated oxygen sensor 1 (Front)
Engine coolant temperature sensor
Crankshaft position sensor (OBD)
Injector
Fuel pressure regulator

29149_NISS_G0025

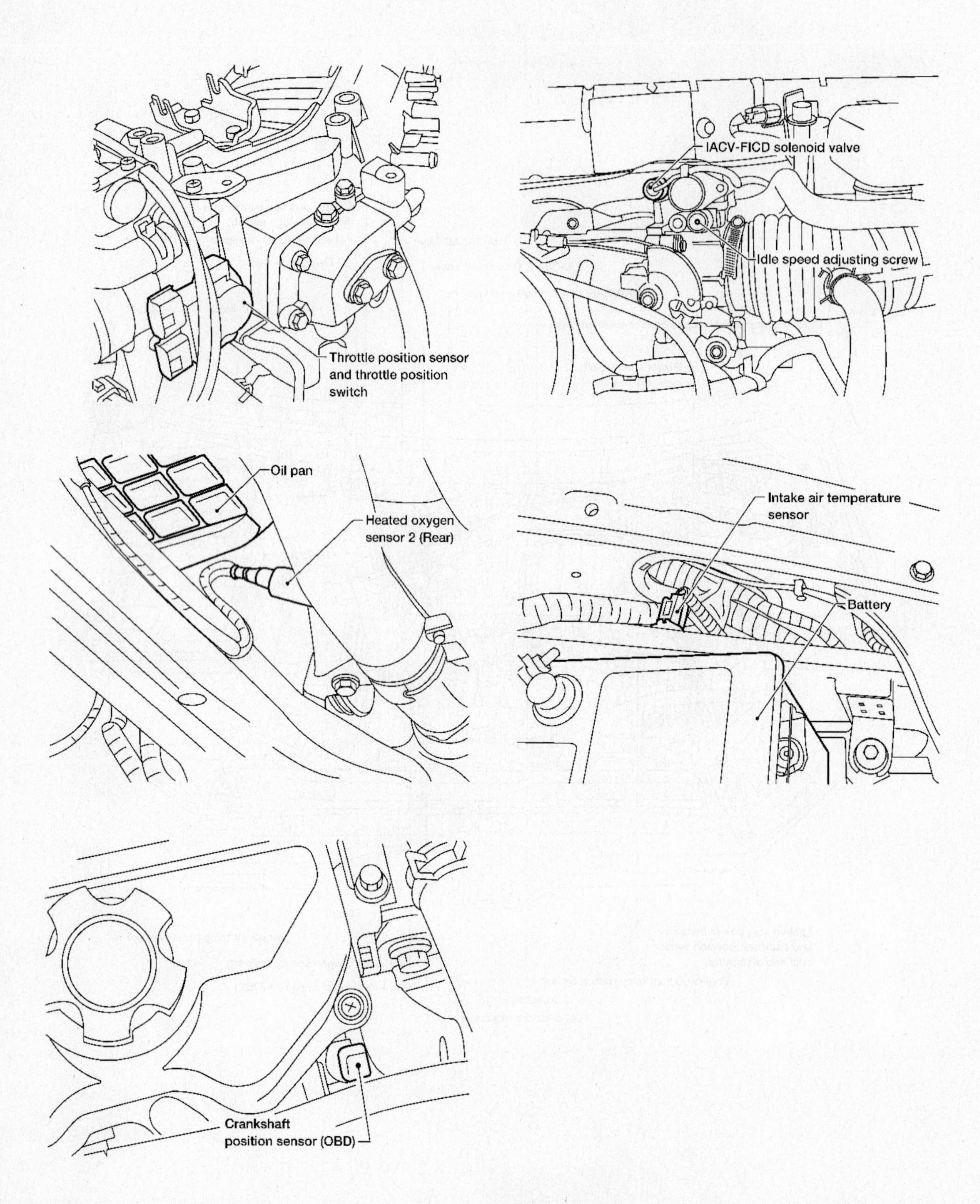
Throttle position sensor
and throttle position
switch
IACV-FICD solenoid valve
Idle speed adjusting screw
Oil pan
Heated oxygen
sensor 2 (Rear)
Intake air temperature
sensor
Battery
Crankshaft
position sensor (OBD)

29149_NISS_G0026

2004–2006

VQ35DE V6 ENGINE

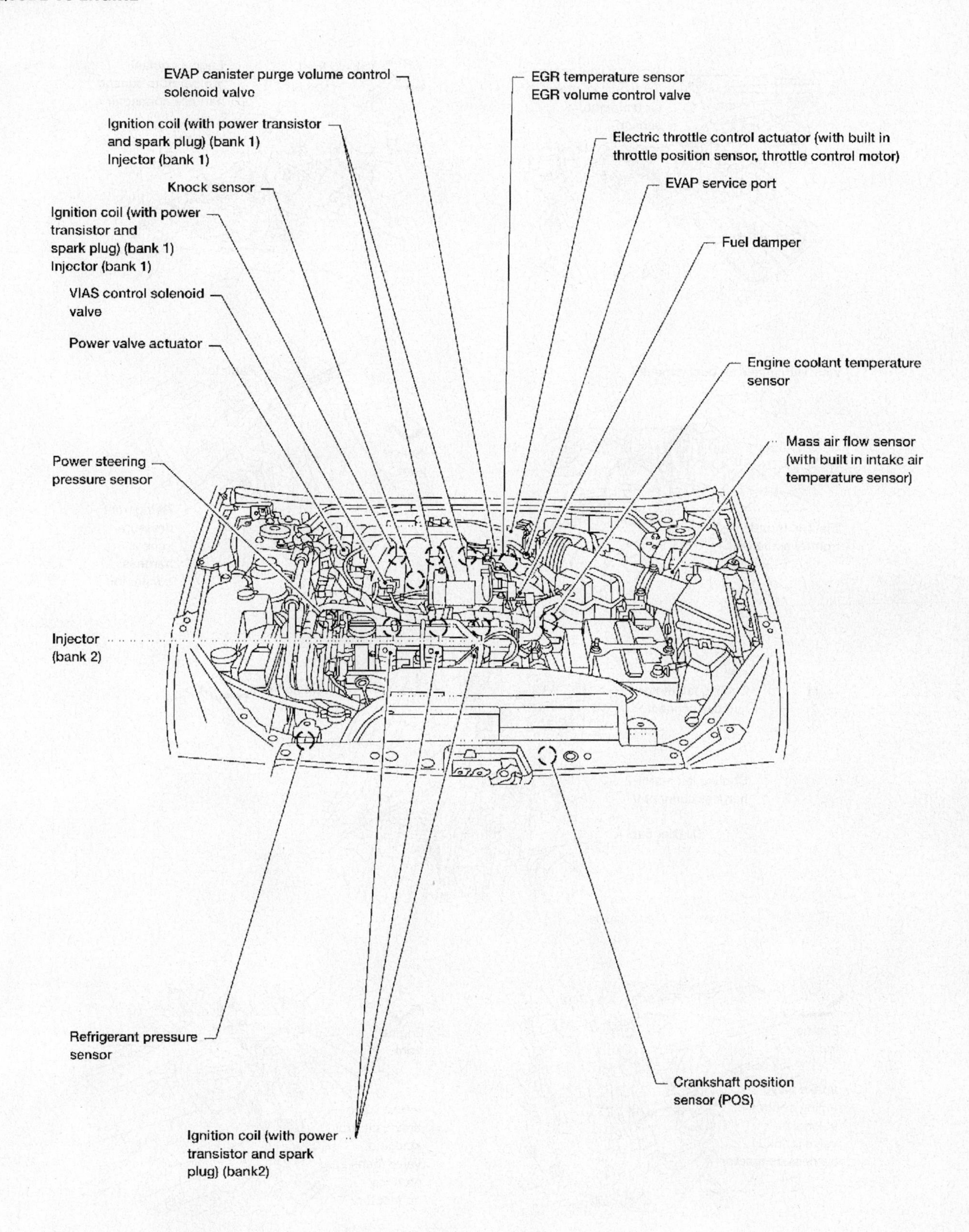

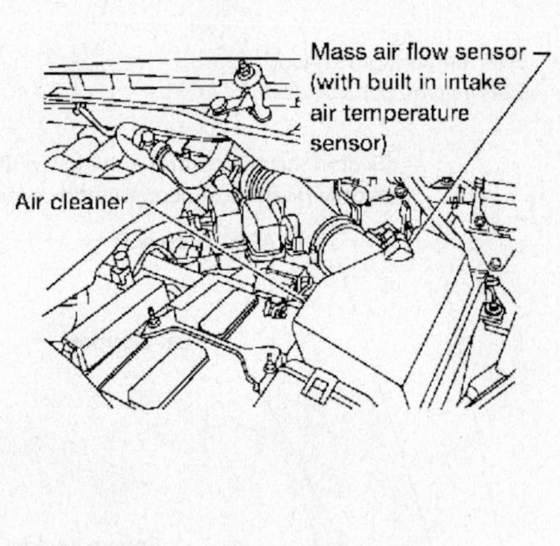

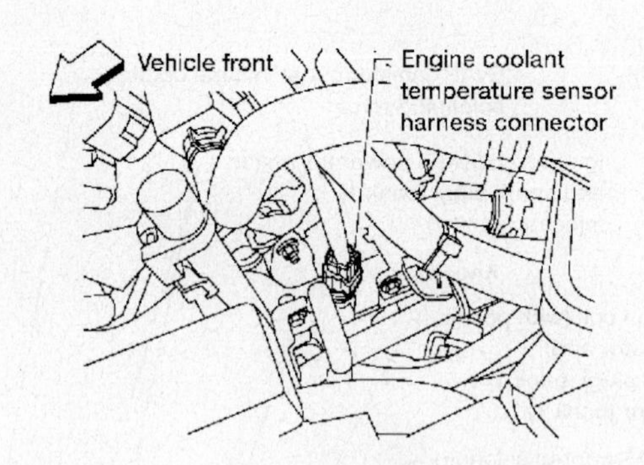

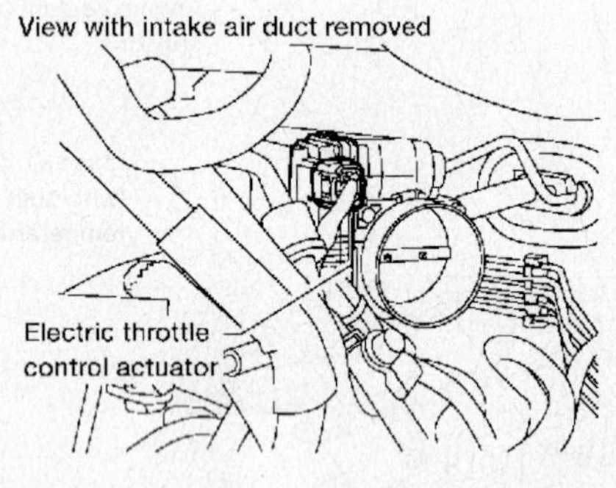

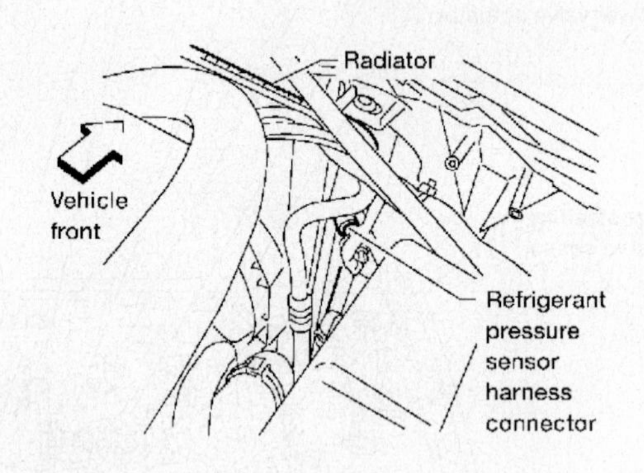

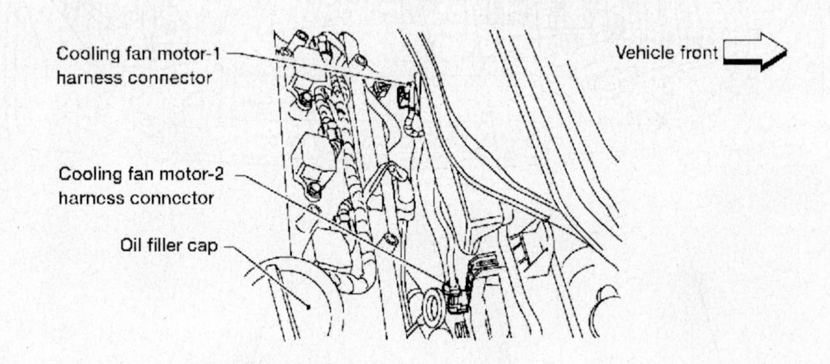

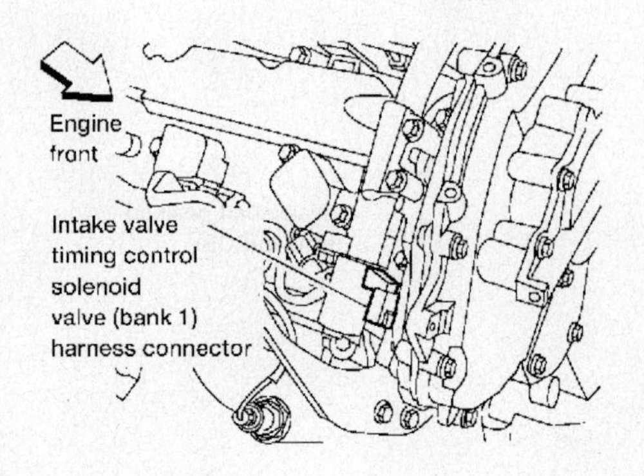

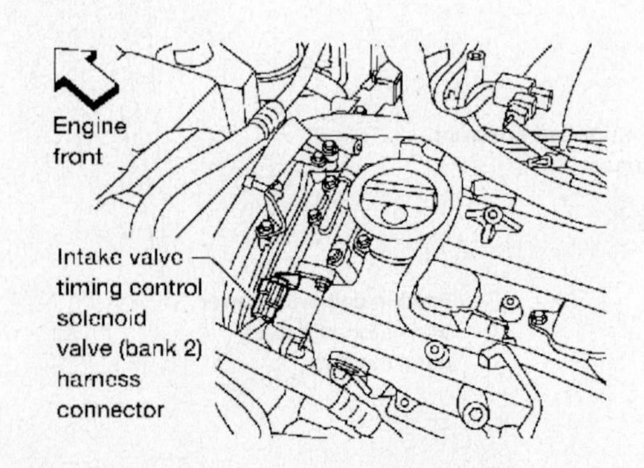

29149_NISS_G0142

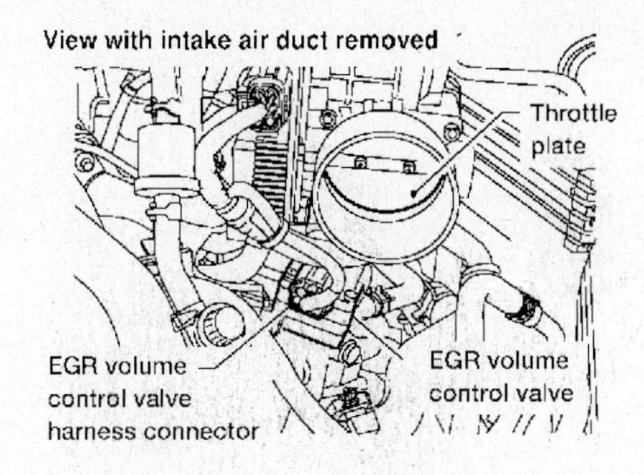
View with intake air duct removed
Throttle plate
EGR volume control valve harness connector
EGR volume control valve

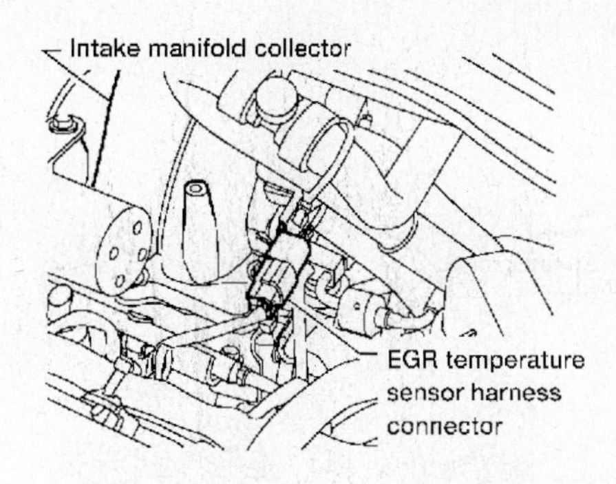
Intake manifold collector
EGR temperature sensor harness connector

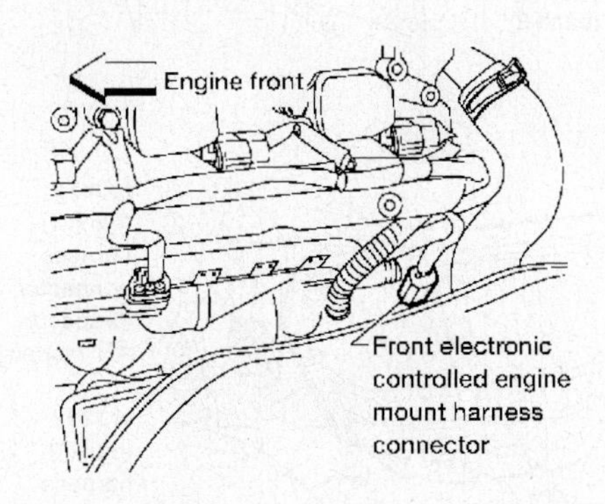
Engine front
Front electronic controlled engine mount harness connector

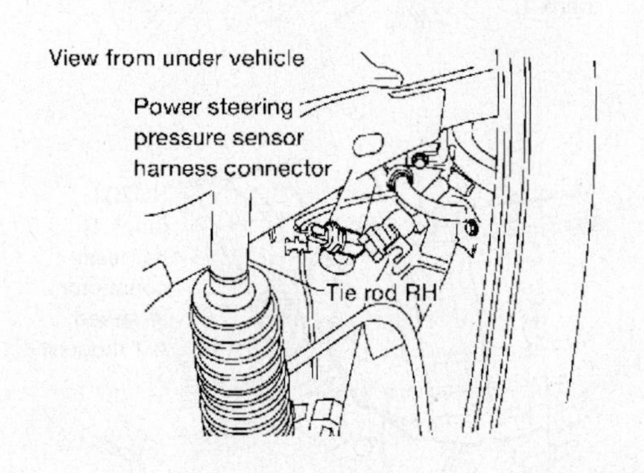
View from under vehicle
Power steering pressure sensor harness connector
Tie rod RH

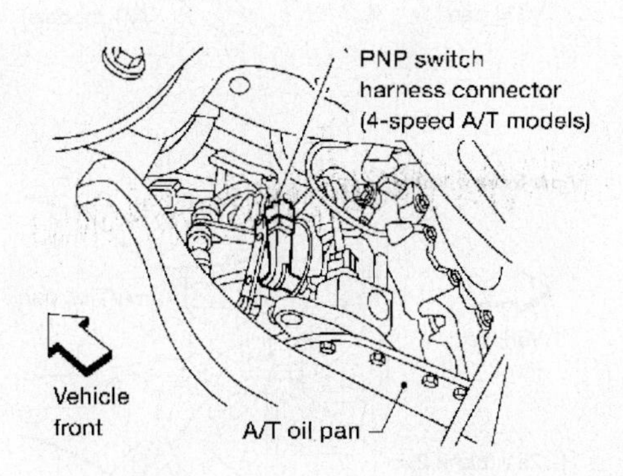
PNP switch harness connector (4-speed A/T models)
Vehicle front
A/T oil pan

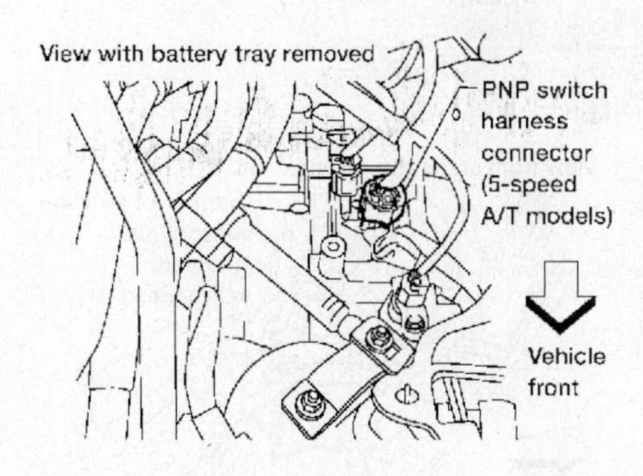
View with battery tray removed
PNP switch harness connector (5-speed A/T models)
Vehicle front

29149_NISS_G0143

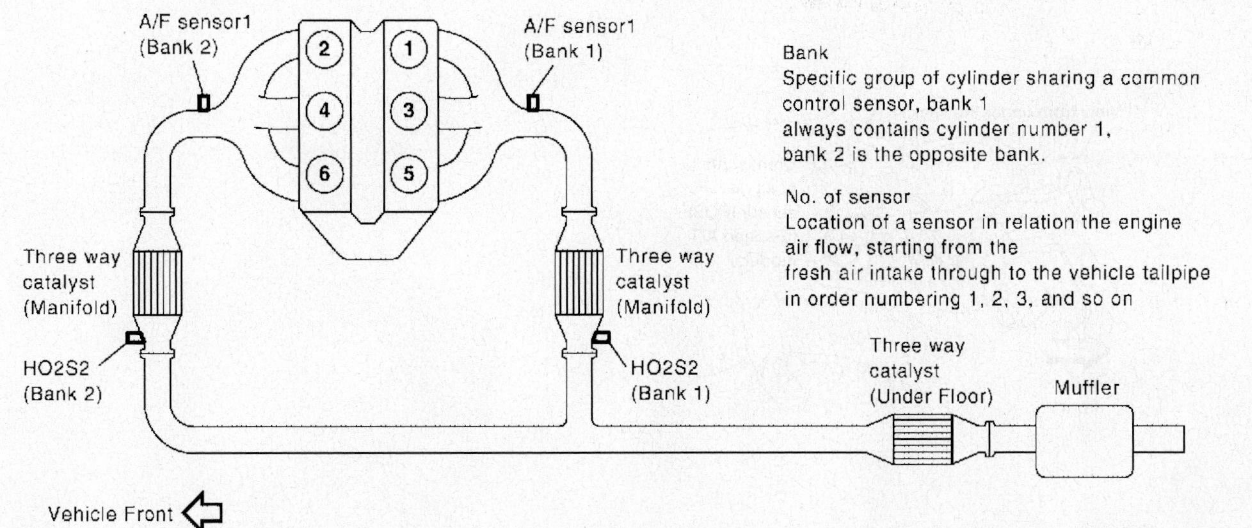
A/F sensor1 (Bank 2)
A/F sensor1 (Bank 1)
2
1
4
3
6
5
Three way catalyst (Manifold)
Three way catalyst (Manifold)
HO2S2 (Bank 2)
HO2S2 (Bank 1)
Three way catalyst (Under Floor)
Muffler
Vehicle Front
Bank
Specific group of cylinder sharing a common control sensor, bank 1 always contains cylinder number 1, bank 2 is the opposite bank.
No. of sensor
Location of a sensor in relation the engine air flow, starting from the fresh air intake through to the vehicle tailpipe in order numbering 1, 2, 3, and so on

29149_NISS_G0144

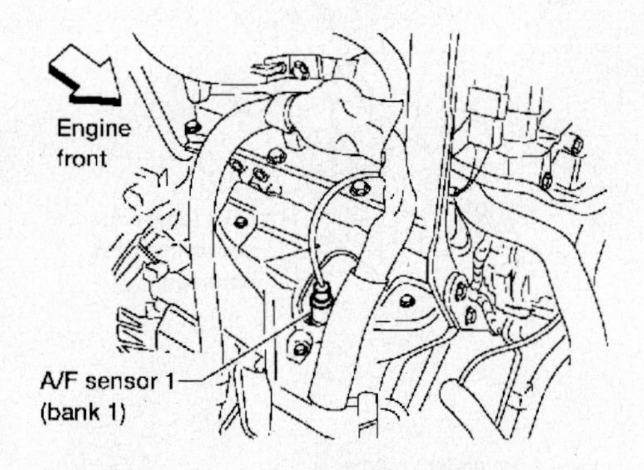

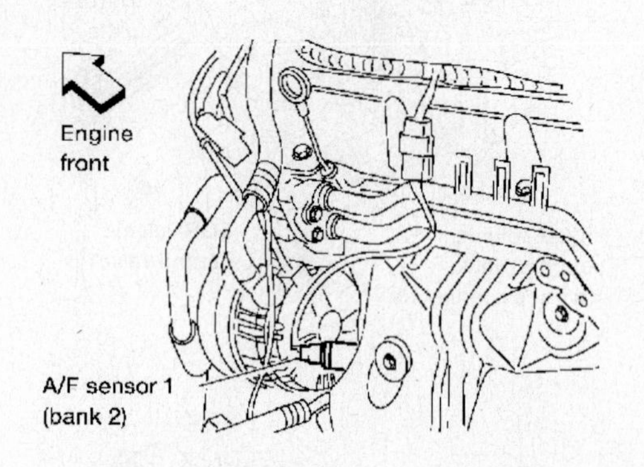

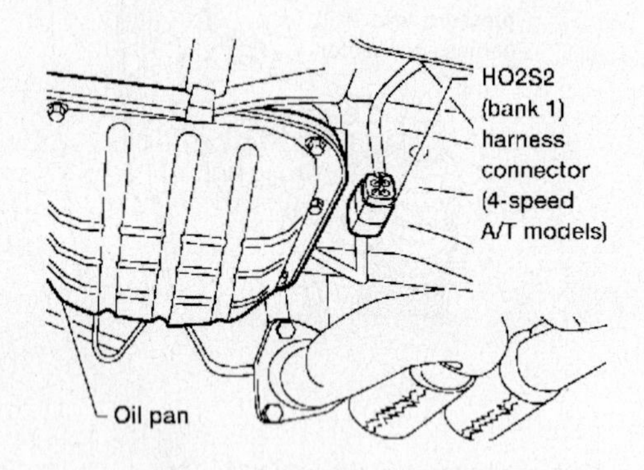

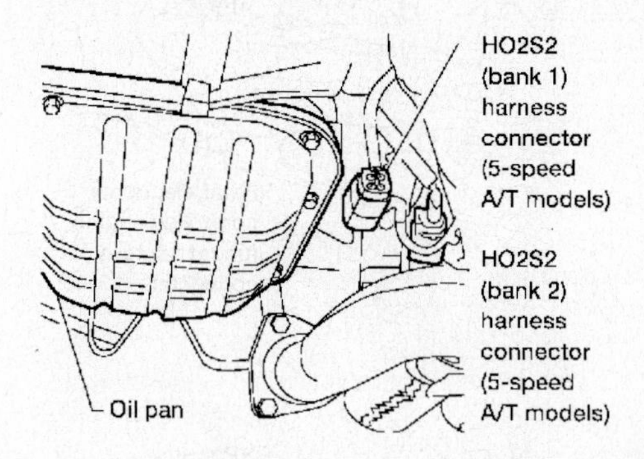

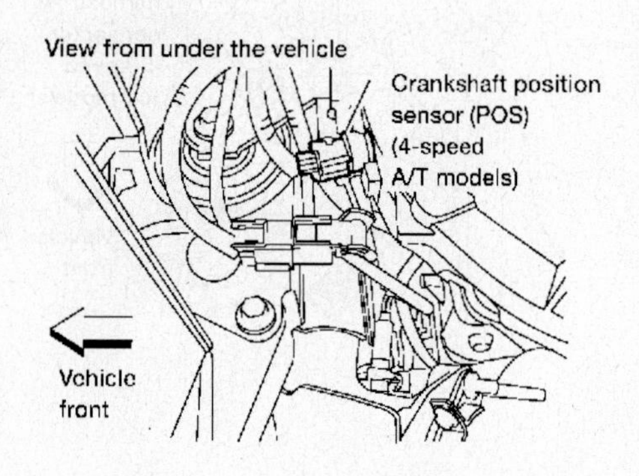

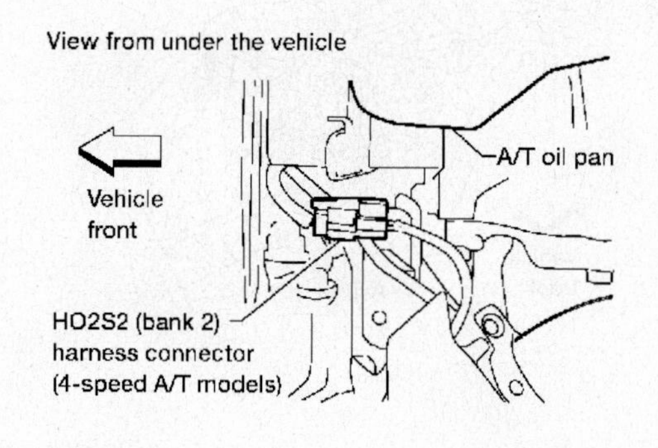

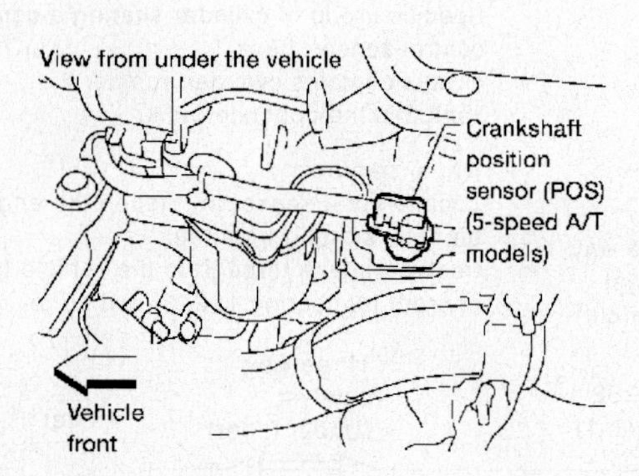

29149_NISS_G0145

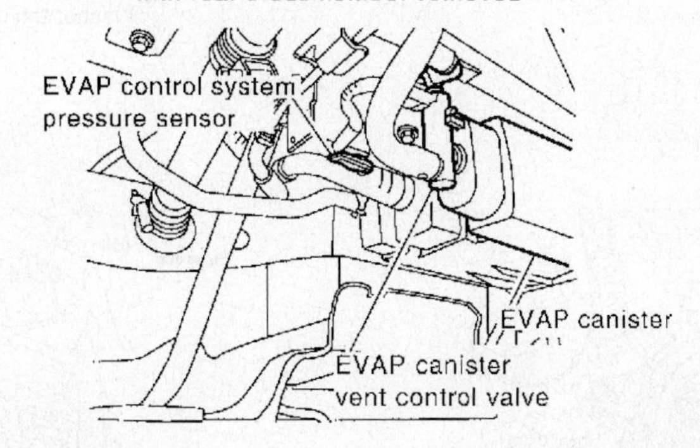

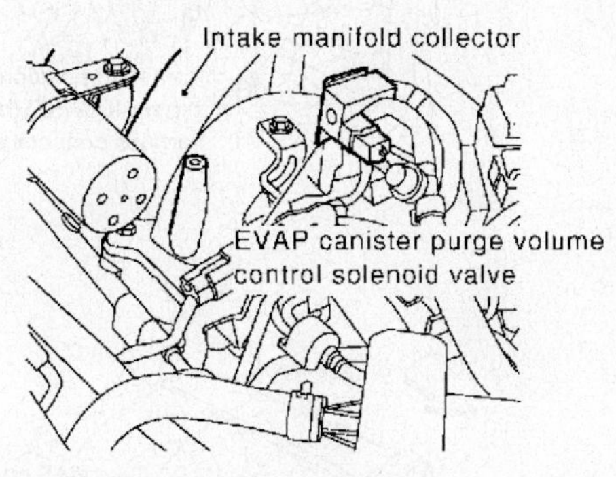

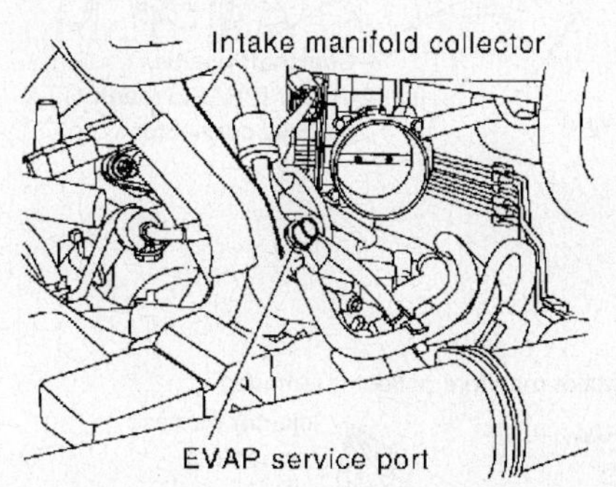

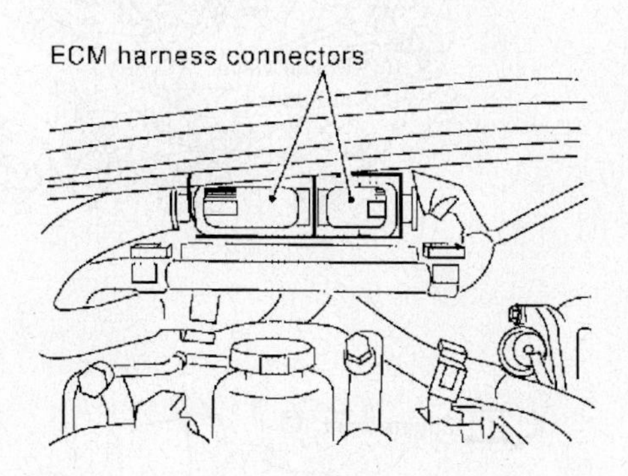

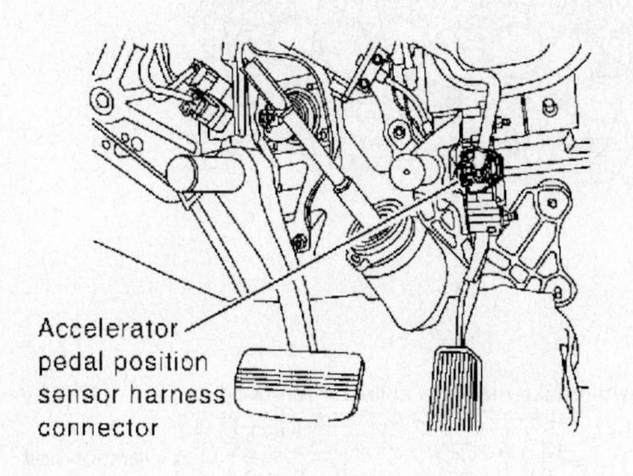

29149_NISS_G0146

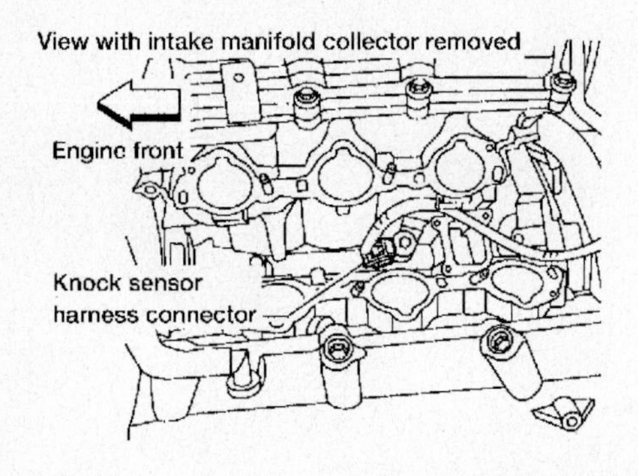

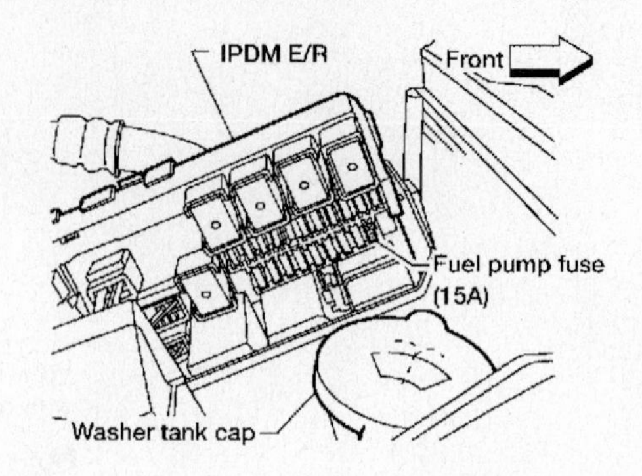

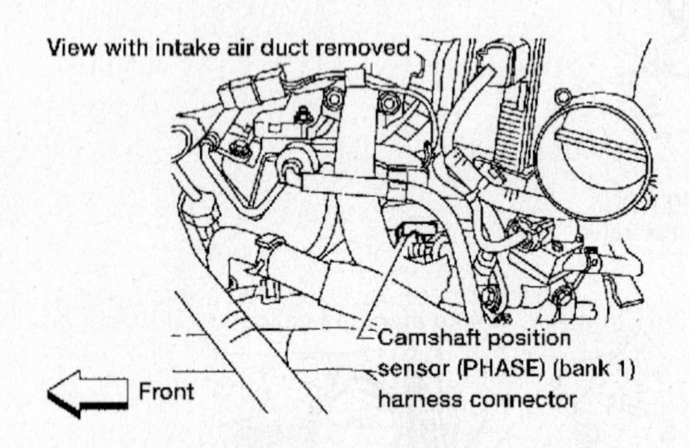

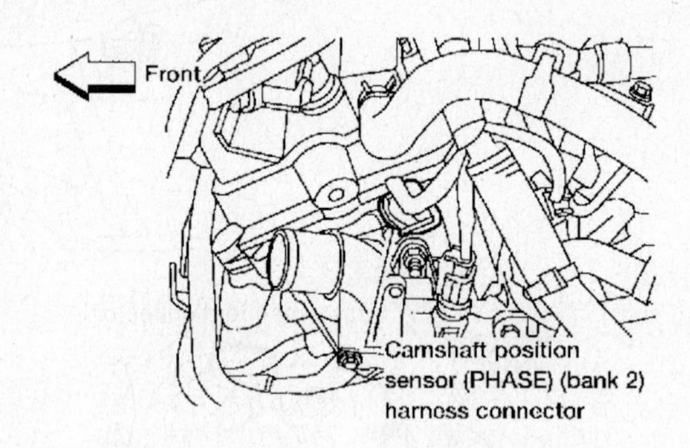

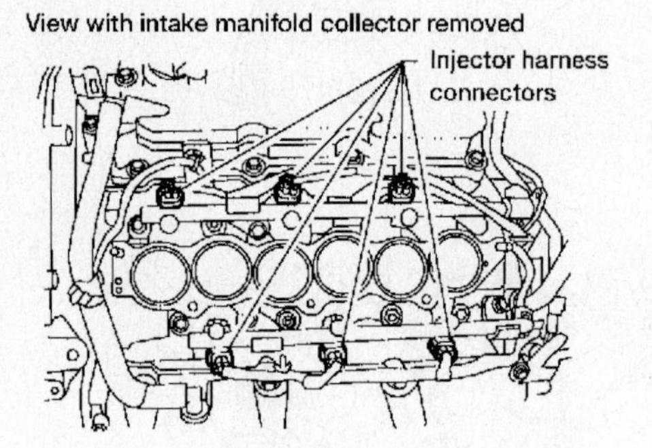

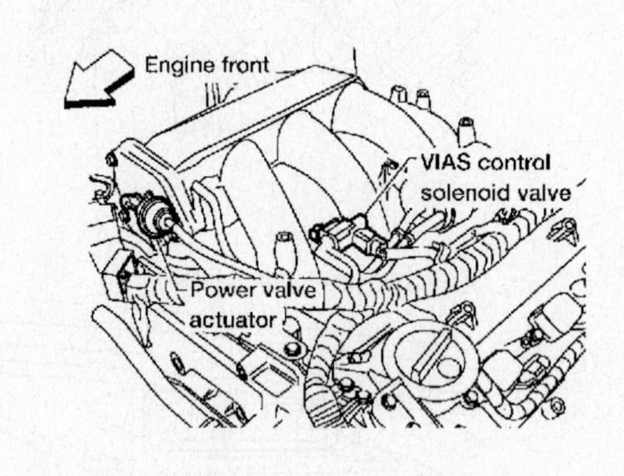

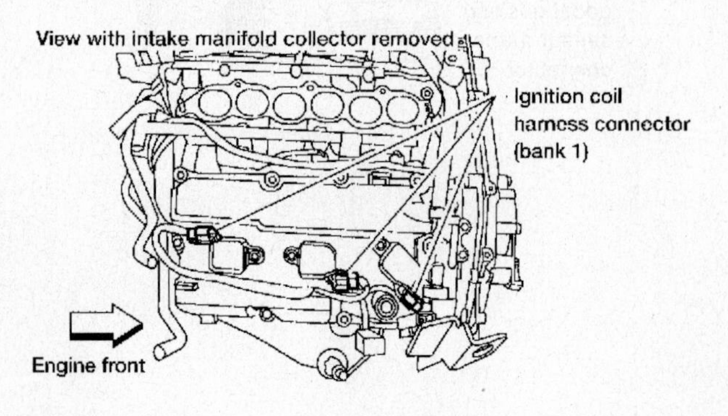

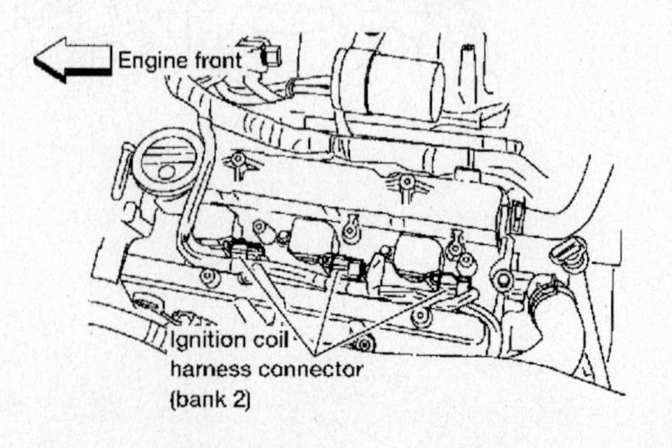

29149_NISS_G0147

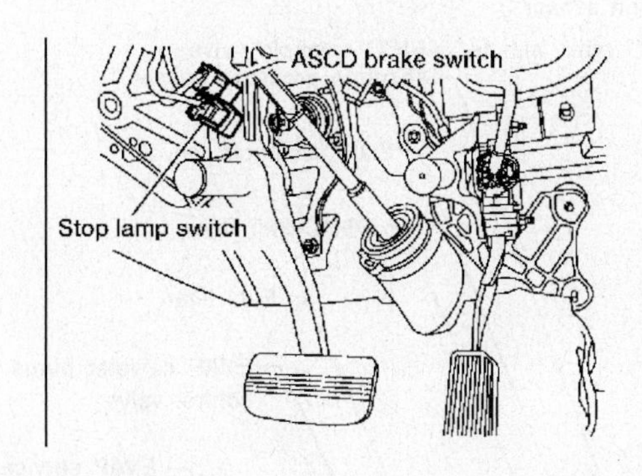
ASCD brake switch
Stop lamp switch

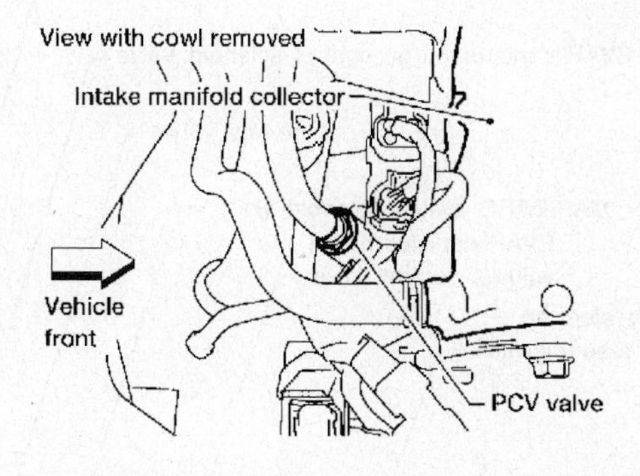
View with cowl removed
Intake manifold collector
Vehicle front
PCV valve

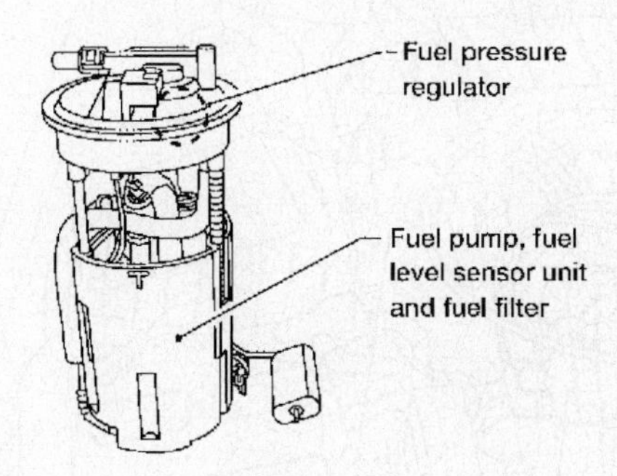
Fuel pressure regulator
Fuel pump, fuel level sensor unit and fuel filter

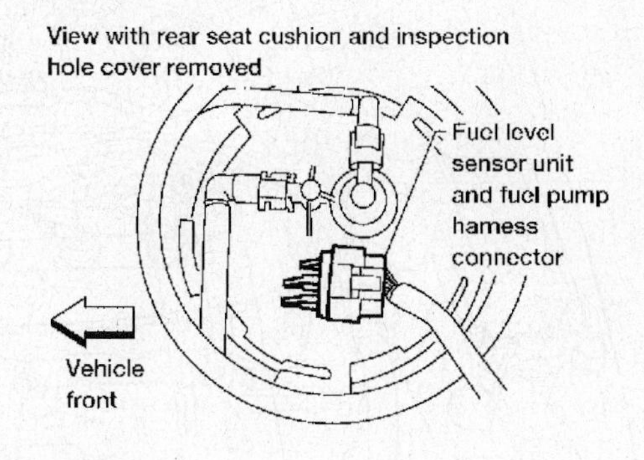
View with rear seat cushion and inspection hole cover removed
Fuel level sensor unit and fuel pump harness connector
Vehicle front

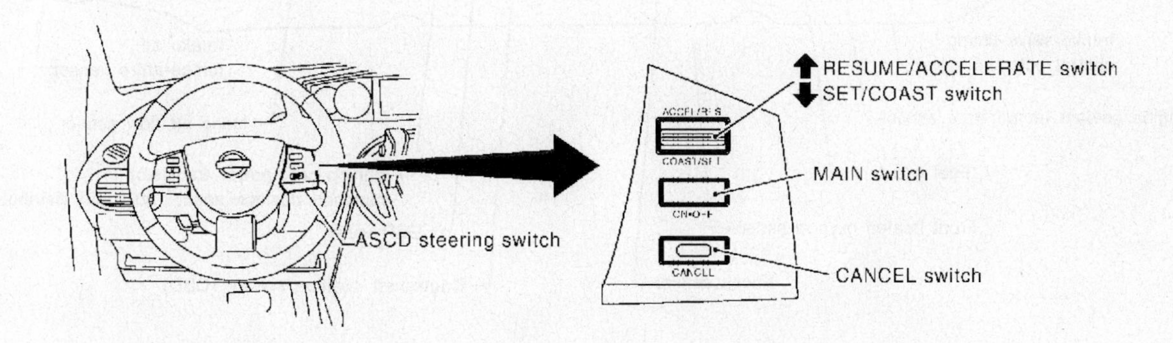
ASCD steering switch
RESUME/ACCELERATE switch
SET/COAST switch
MAIN switch
CANCEL switch

SENTRA

1996–1999

GA16DE I4 ENGINE

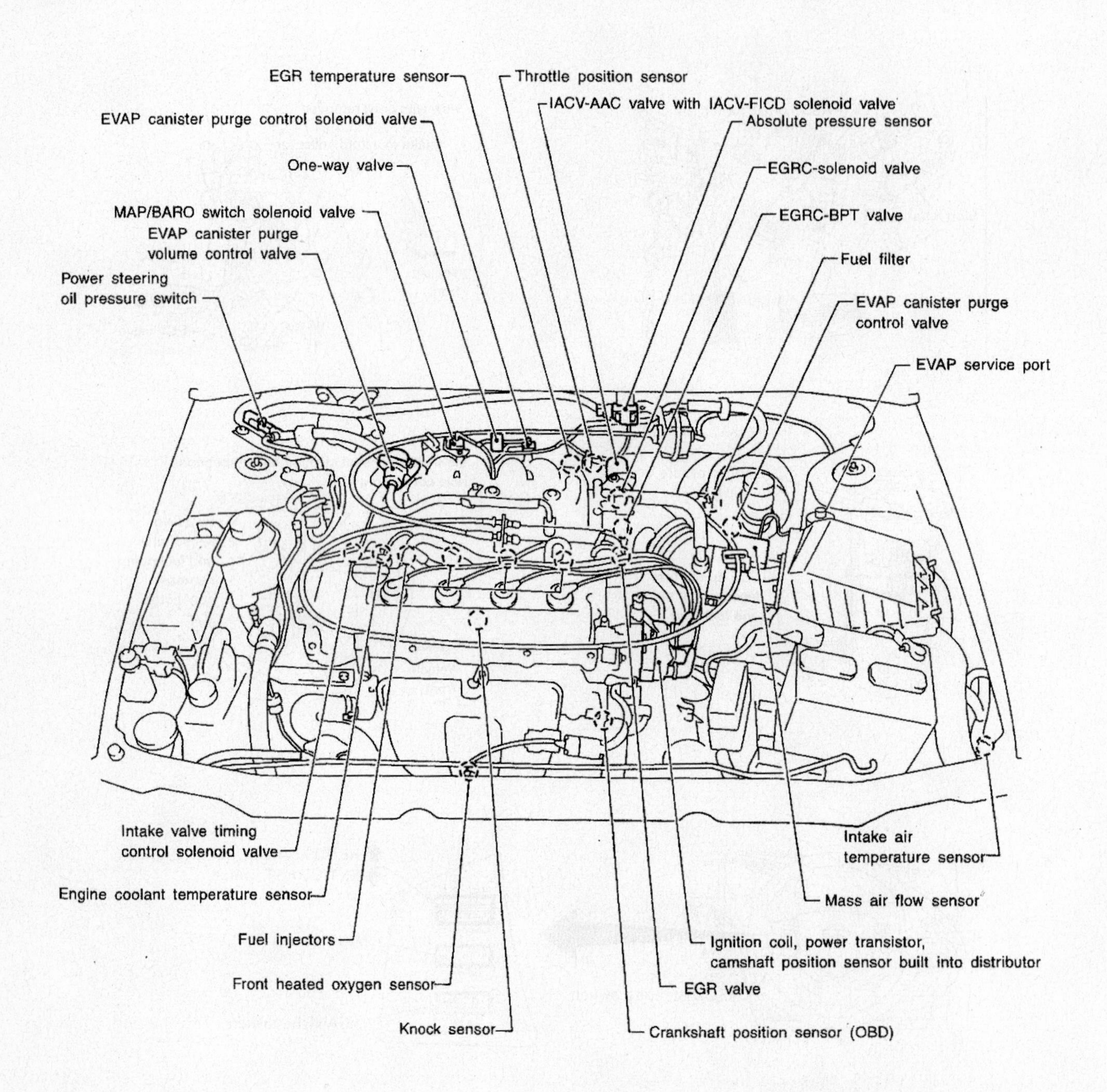
EGR temperature sensor
Throttle position sensor
IACV-AAC valve with IACV-FICD solenoid valve
EVAP canister purge control solenoid valve
Absolute pressure sensor
One-way valve
EGRC-solenoid valve
MAP/BARO switch solenoid valve
EGRC-BPT valve
EVAP canister purge
volume control valve
Fuel filter
Power steering
oil pressure switch
EVAP canister purge
control valve
EVAP service port
Intake valve timing
control solenoid valve
Intake air
temperature sensor
Engine coolant temperature sensor
Mass air flow sensor
Fuel injectors
Ignition coil, power transistor,
camshaft position sensor built into distributor
Front heated oxygen sensor
EGR valve
Knock sensor
Crankshaft position sensor (OBD)

29149_NISS_G0027

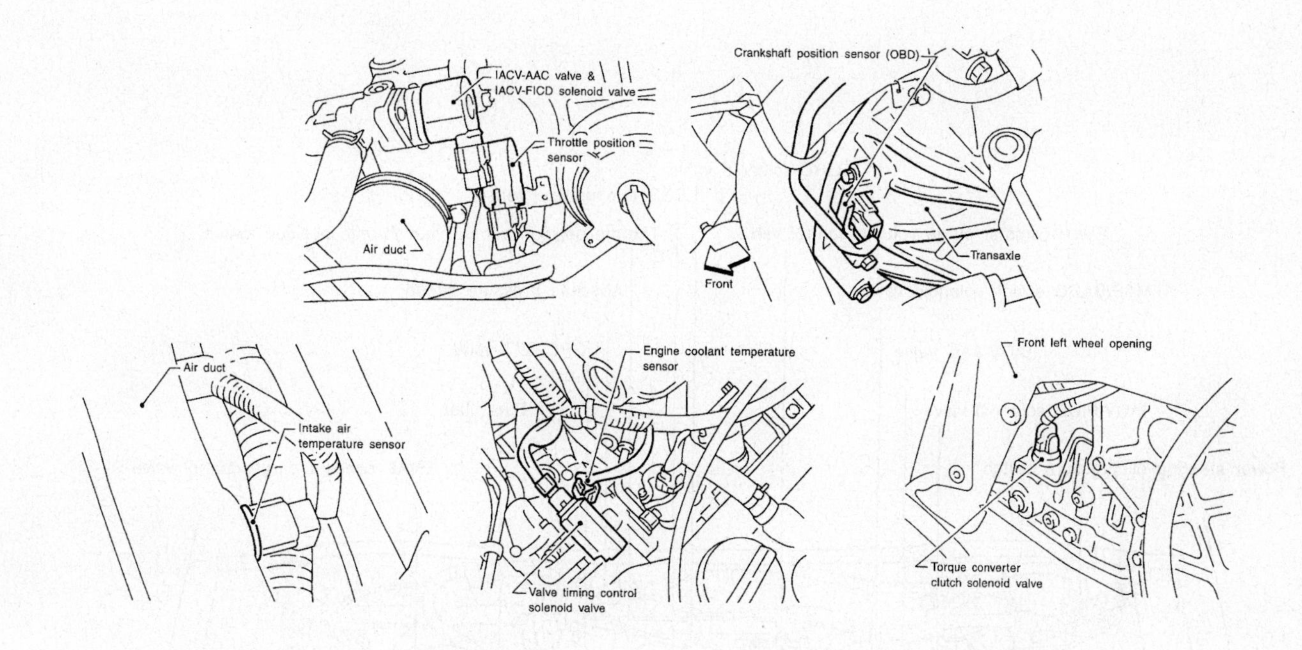
IACV-AAC valve &
IACV-FICD solenoid valve
Throttle position sensor
Air duct
Crankshaft position sensor (OBD)
Transaxle
Front
Air duct
Intake air temperature sensor
Engine coolant temperature sensor
Valve timing control solenoid valve
Front left wheel opening
Torque converter clutch solenoid valve

29149_NISS_G0028

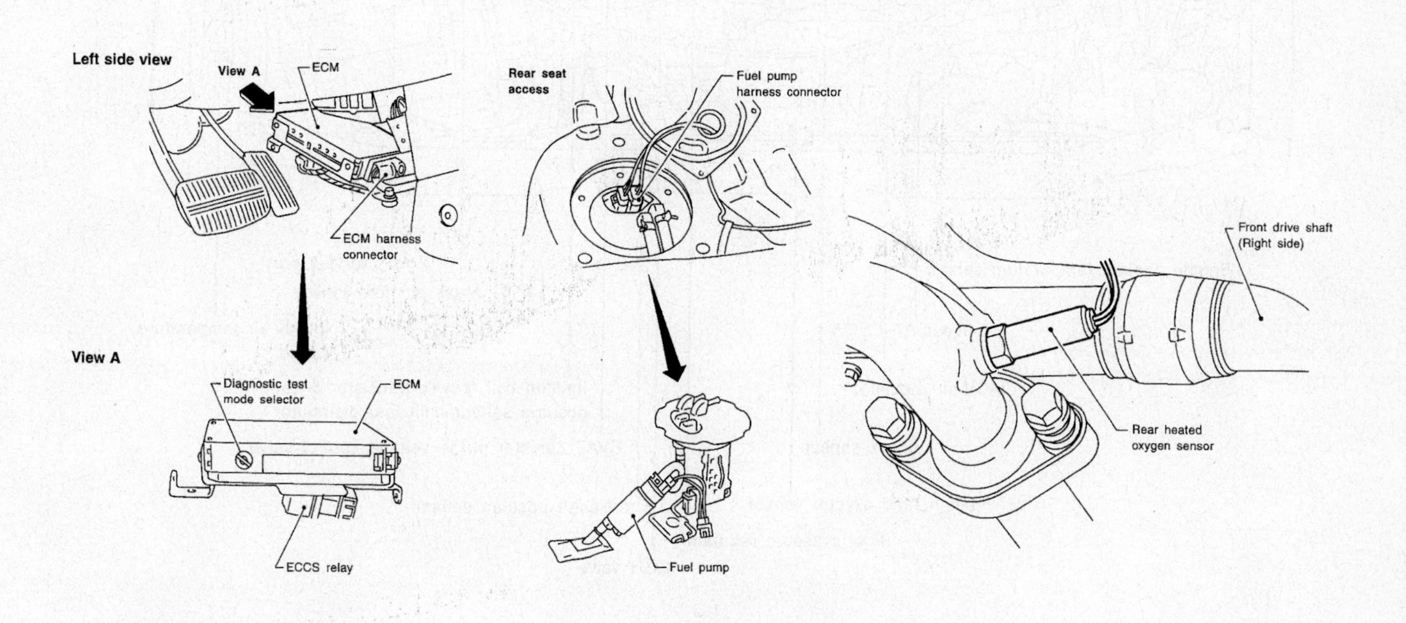
Left side view
View A
ECM
ECM harness connector
View A
Diagnostic test mode selector
ECM
ECCS relay
Rear seat access
Fuel pump harness connector
Fuel pump
Front drive shaft (Right side)
Rear heated oxygen sensor

29149_NISS_G0029

1996–2001
SR20DE I4 ENGINE

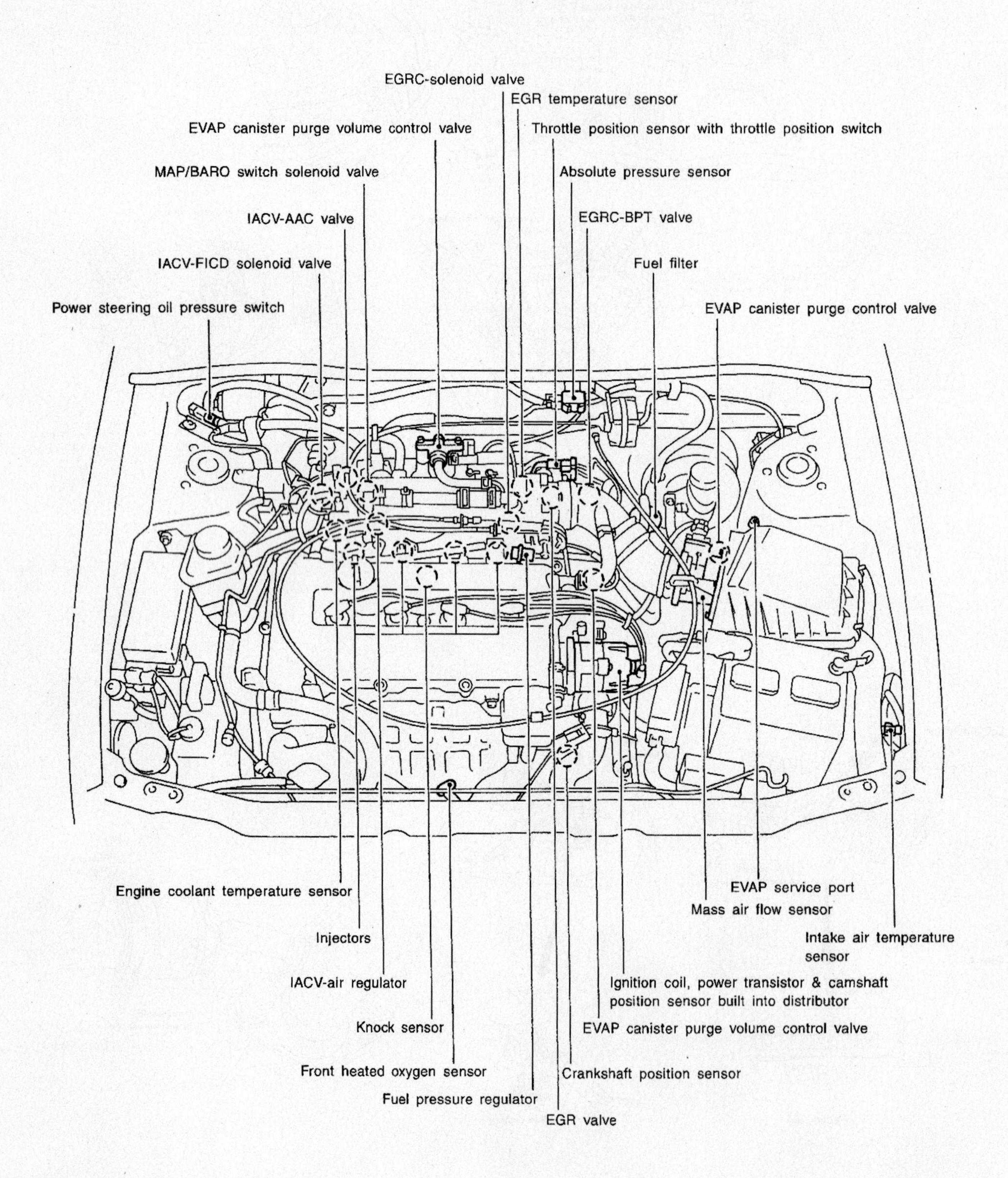
EGRC-solenoid valve
EGR temperature sensor
EVAP canister purge volume control valve
Throttle position sensor with throttle position switch
MAP/BARO switch solenoid valve
Absolute pressure sensor
IACV-AAC valve
EGRC-BPT valve
IACV-FICD solenoid valve
Fuel filter
Power steering oil pressure switch
EVAP canister purge control valve
Engine coolant temperature sensor
EVAP service port
Mass air flow sensor
Injectors
Intake air temperature sensor
IACV-air regulator
Ignition coil, power transistor & camshaft position sensor built into distributor
Knock sensor
EVAP canister purge volume control valve
Front heated oxygen sensor
Crankshaft position sensor
Fuel pressure regulator
EGR valve

29149_NISS_G0030

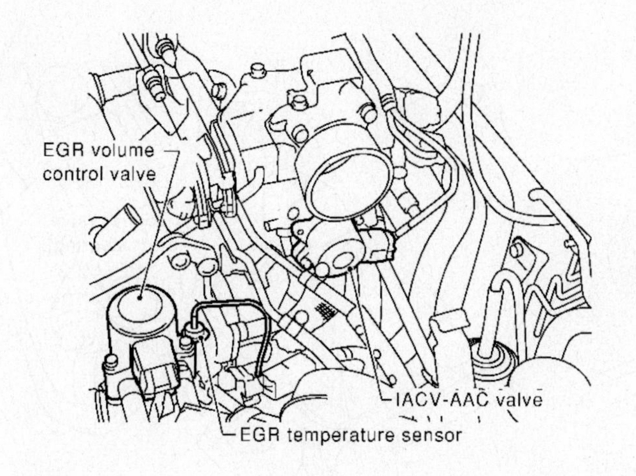
EGR volume
control valve
IACV-AAC valve
EGR temperature sensor

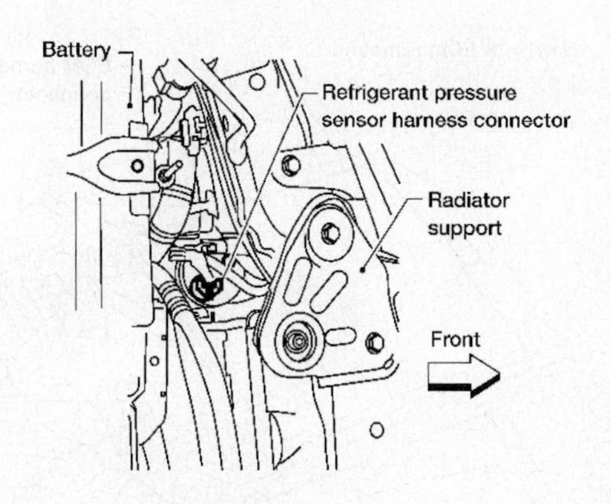
Battery
Refrigerant pressure
sensor harness connector
Radiator
support
Front

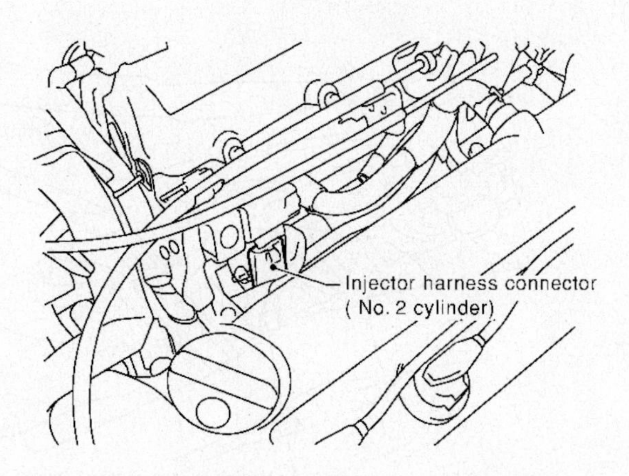
Injector harness connector
(No. 2 cylinder)

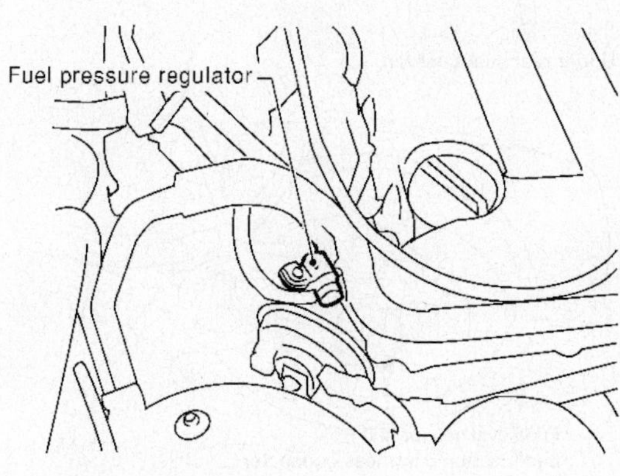
Fuel pressure regulator

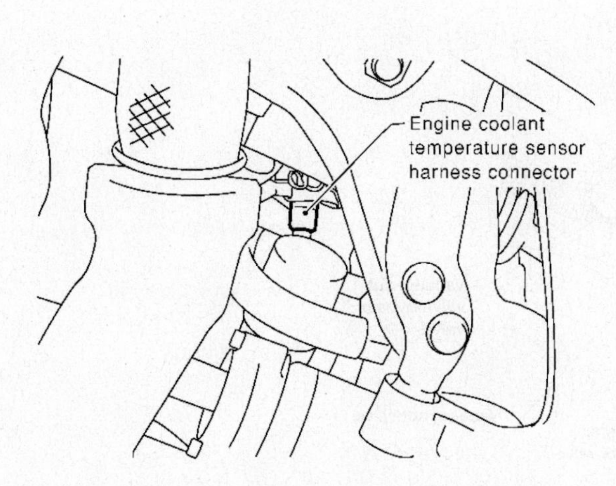
Engine coolant
temperature sensor
harness connector

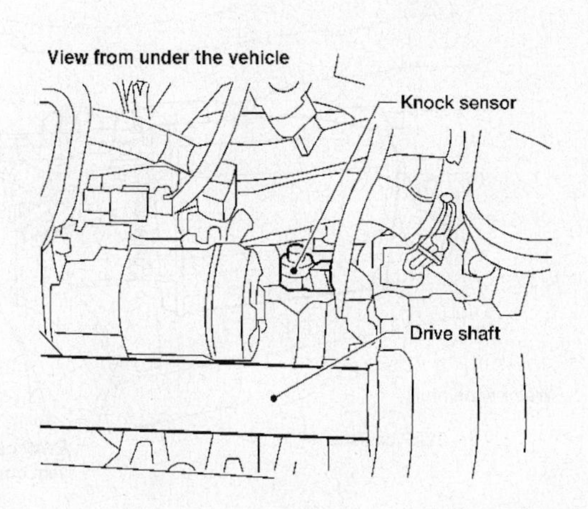
View from under the vehicle
Knock sensor
Drive shaft

29149_NISS_G0031

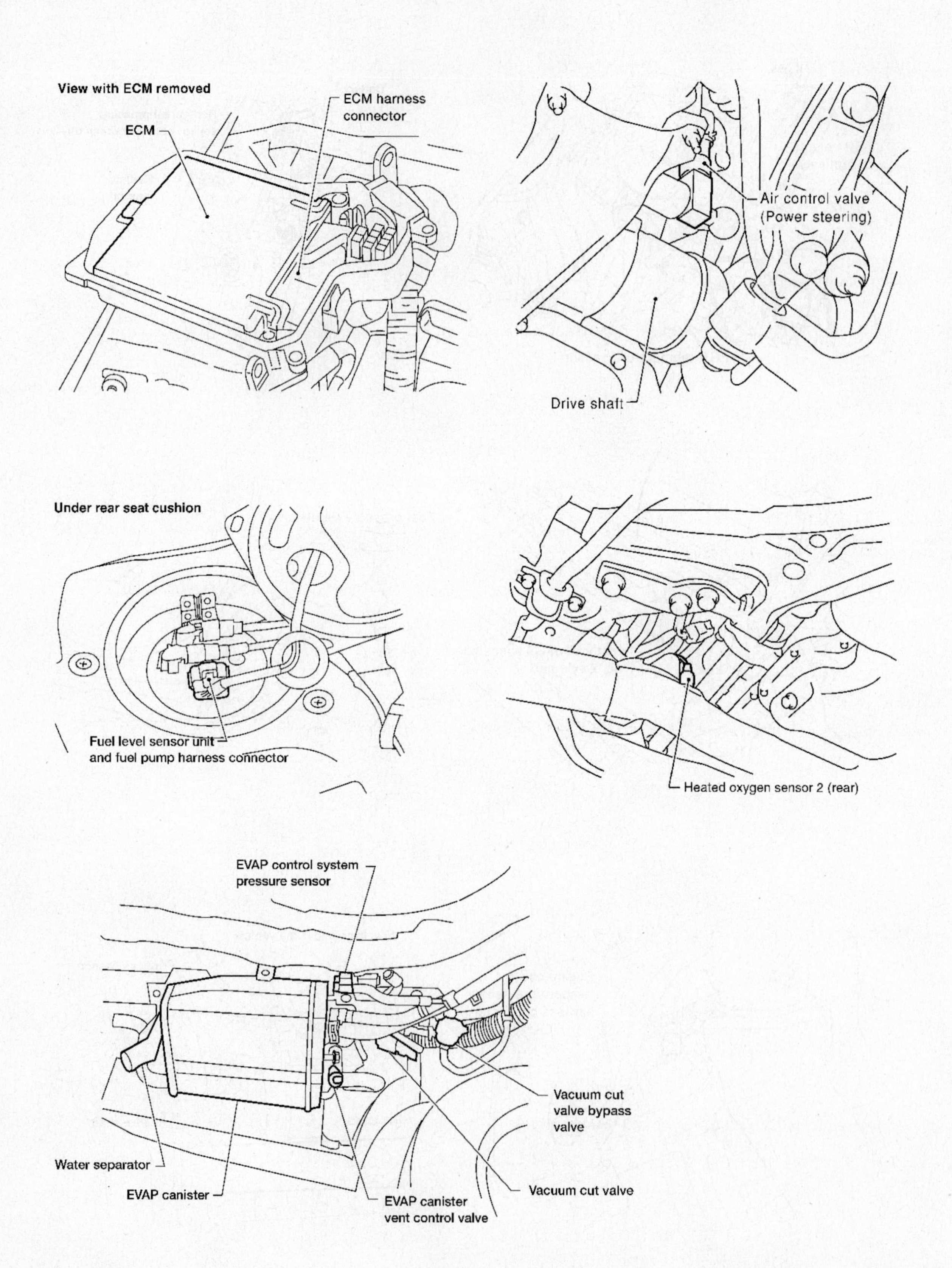
View with ECM removed
ECM harness connector
ECM
Air control valve (Power steering)
Drive shaft
Under rear seat cushion
Fuel level sensor unit and fuel pump harness connector
Heated oxygen sensor 2 (rear)
EVAP control system pressure sensor
Vacuum cut valve bypass valve
Water separator
EVAP canister
Vacuum cut valve
EVAP canister vent control valve

29149_NISS_G0032

2000–2006

QG18DE I4 SULEV ENGINE

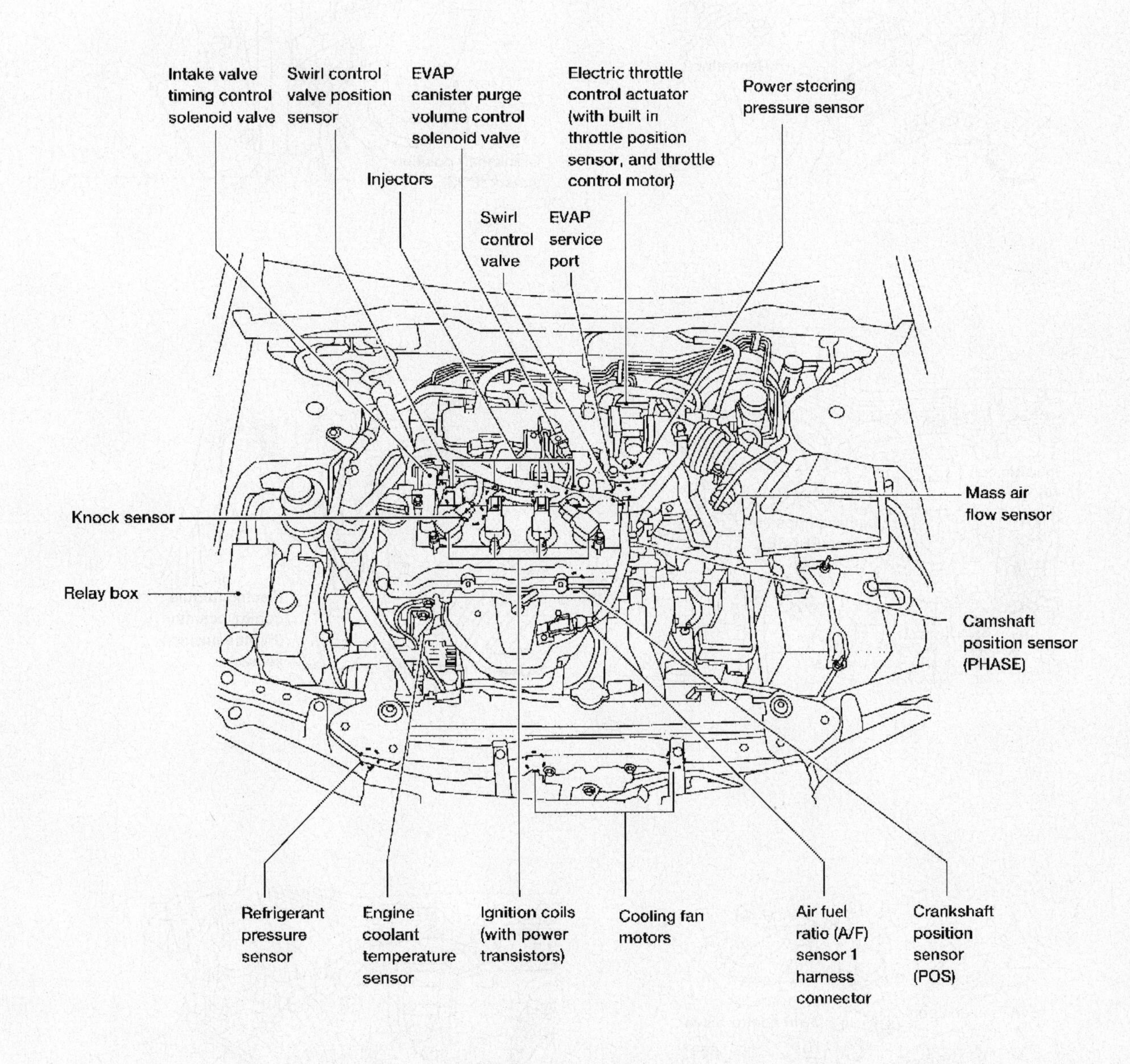
Intake valve timing control solenoid valve
Swirl control valve position sensor
EVAP canister purge volume control solenoid valve
Injectors
Electric throttle control actuator (with built in throttle position sensor, and throttle control motor)
Power steering pressure sensor
Swirl control valve
EVAP service port
Knock sensor
Relay box
Mass air flow sensor
Camshaft position sensor (PHASE)
Refrigerant pressure sensor
Engine coolant temperature sensor
Ignition coils (with power transistors)
Cooling fan motors
Air fuel ratio (A/F) sensor 1 harness connector
Crankshaft position sensor (POS)

29149_NISS_G0149

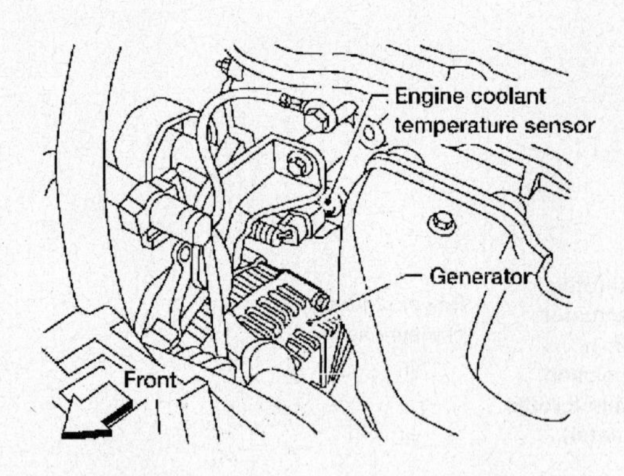
Engine coolant temperature sensor
Generator
Front

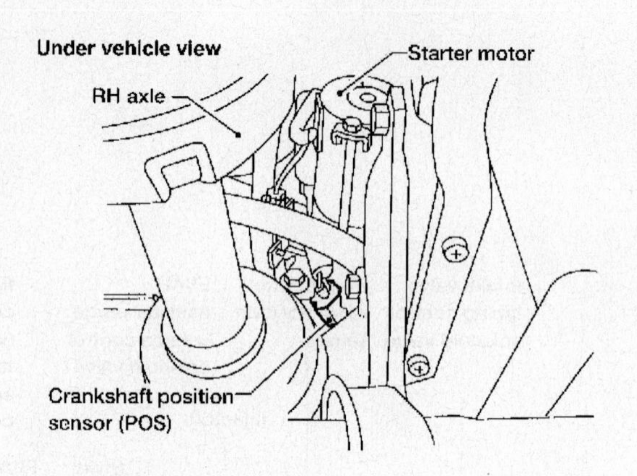
Under vehicle view
Starter motor
RH axle
Crankshaft position sensor (POS)

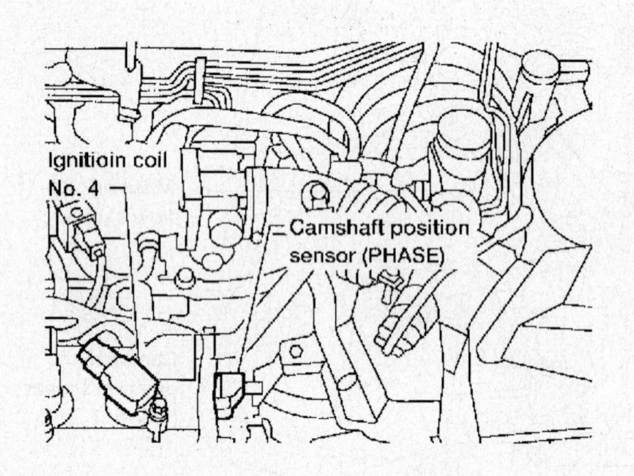
Ignitioin coil No. 4
Camshaft position sensor (PHASE)

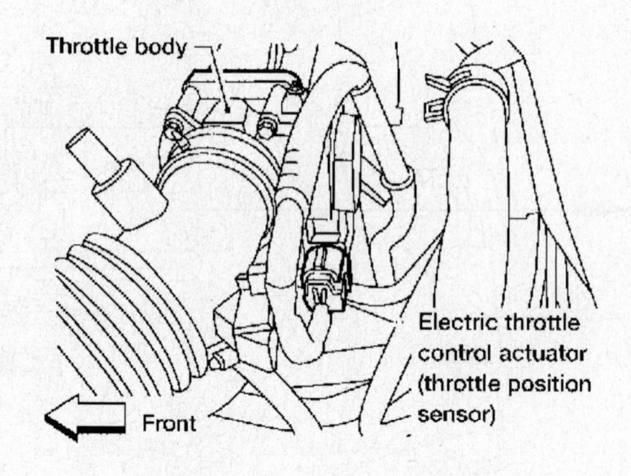
Throttle body
Electric throttle control actuator (throttle position sensor)
Front

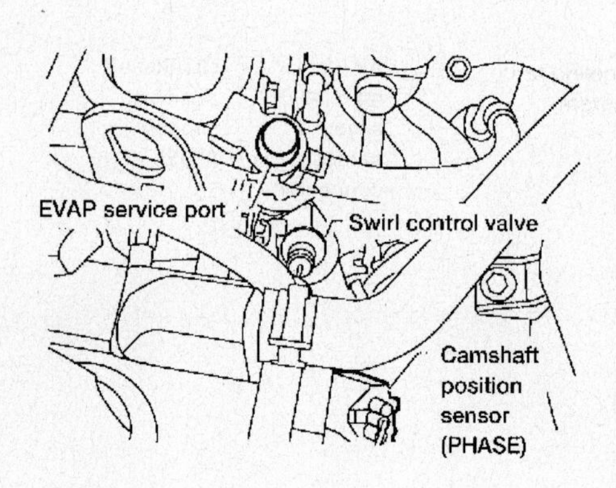
EVAP service port
Swirl control valve
Camshaft position sensor (PHASE)

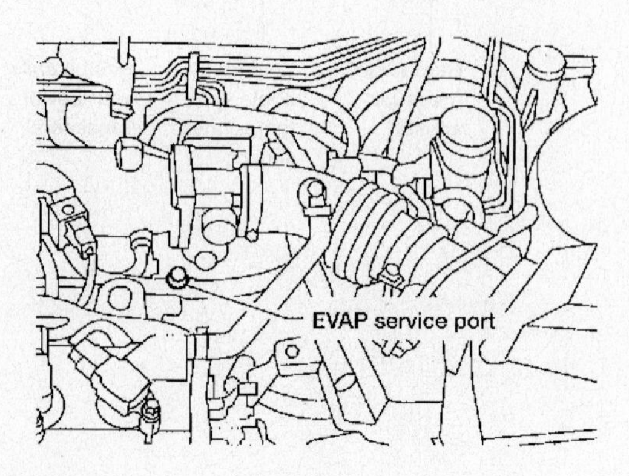
EVAP service port

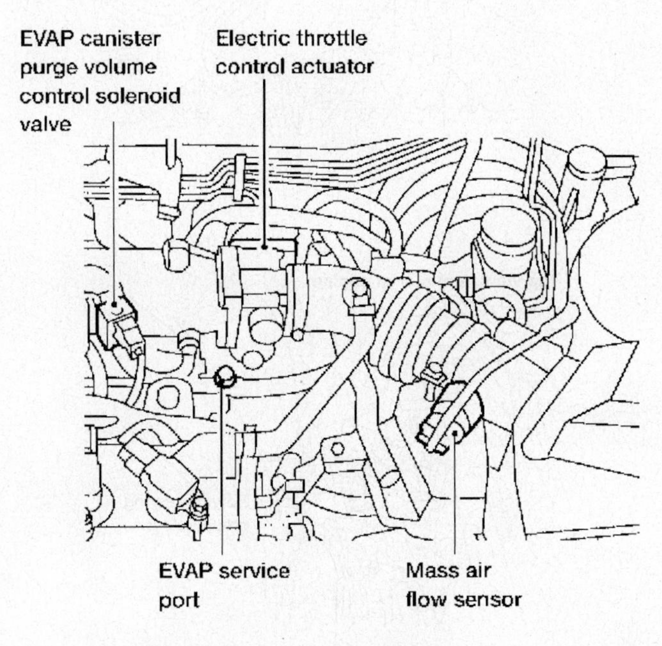
EVAP canister purge volume control solenoid valve
Electric throttle control actuator
EVAP service port
Mass air flow sensor

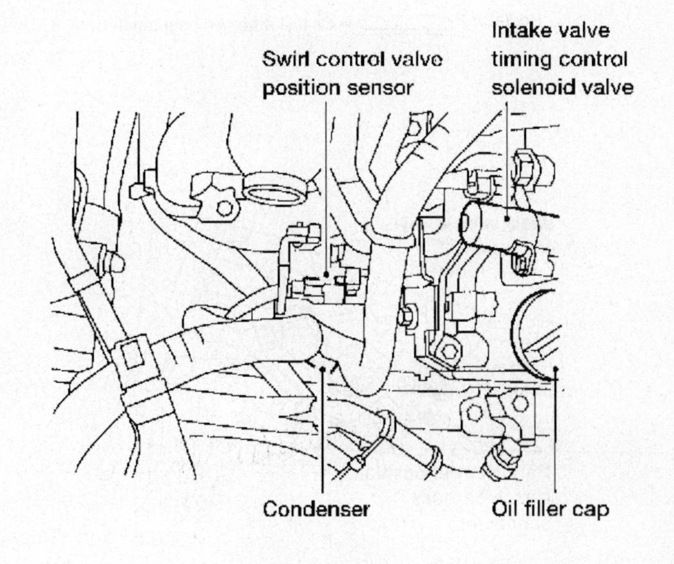
Swirl control valve position sensor
Intake valve timing control solenoid valve
Condenser
Oil filler cap

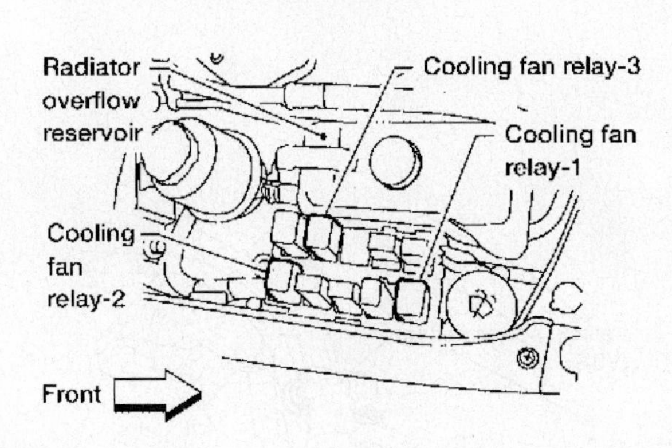
View with relay box cover removed
Radiator overflow reservoir
Cooling fan relay-3
Cooling fan relay-1
Cooling fan relay-2
Front

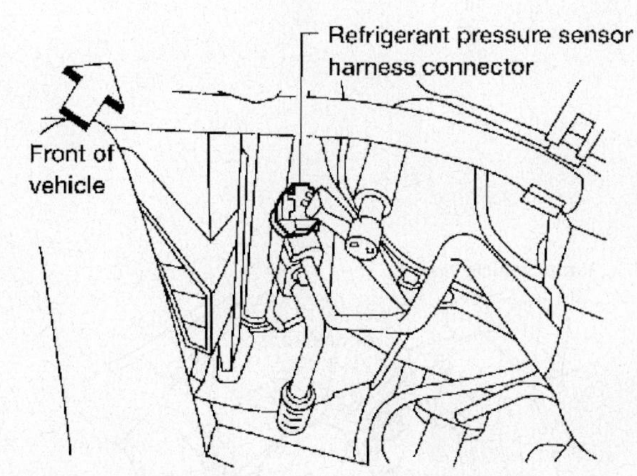
Refrigerant pressure sensor harness connector
Front of vehicle

29149_NISS_G0151

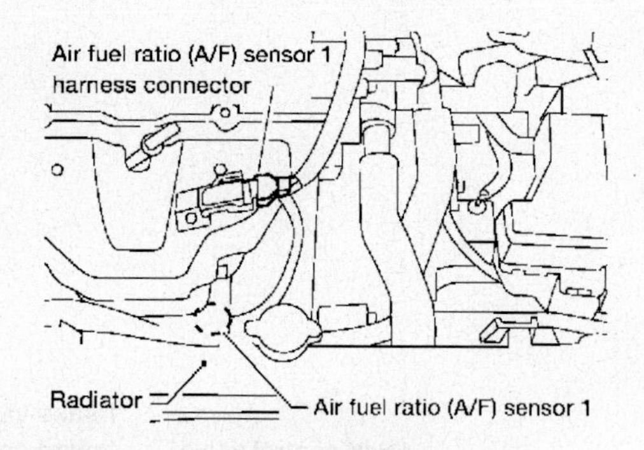
Air fuel ratio (A/F) sensor 1
harness connector
Radiator
Air fuel ratio (A/F) sensor 1

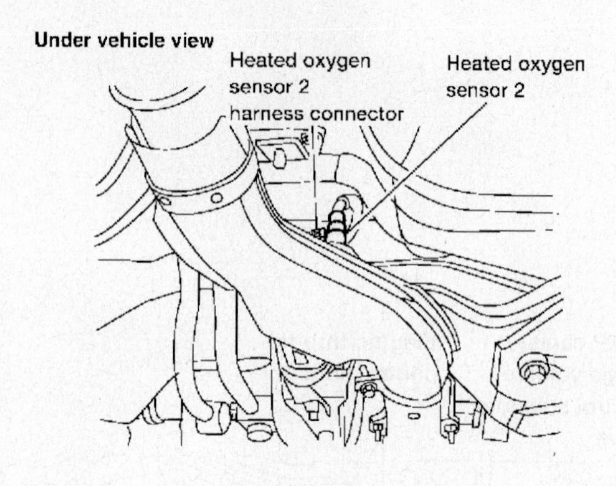
Under vehicle view
Heated oxygen
sensor 2
harness connector
Heated oxygen
sensor 2

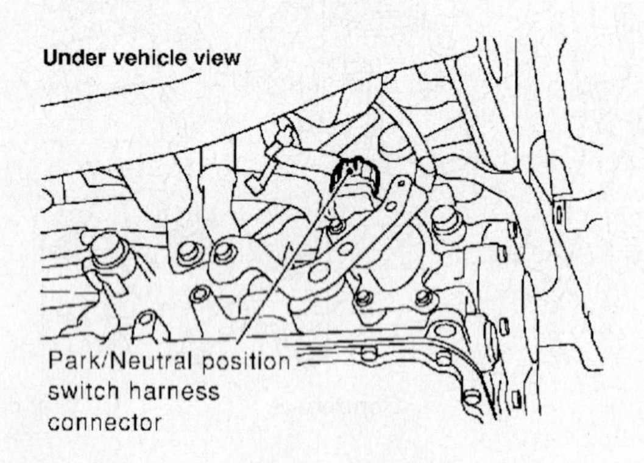
Under vehicle view
Park/Neutral position
switch harness
connector

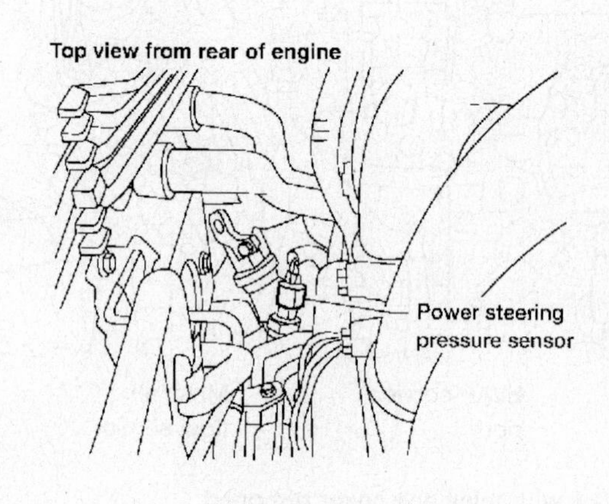
Top view from rear of engine
Power steering
pressure sensor

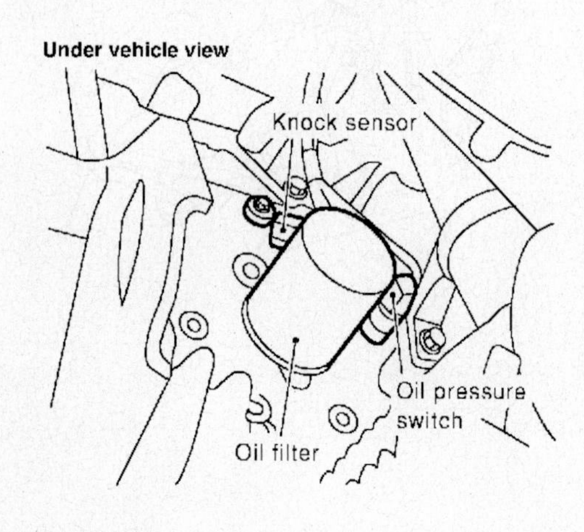
Under vehicle view
Knock sensor
Oil pressure
switch
Oil filter

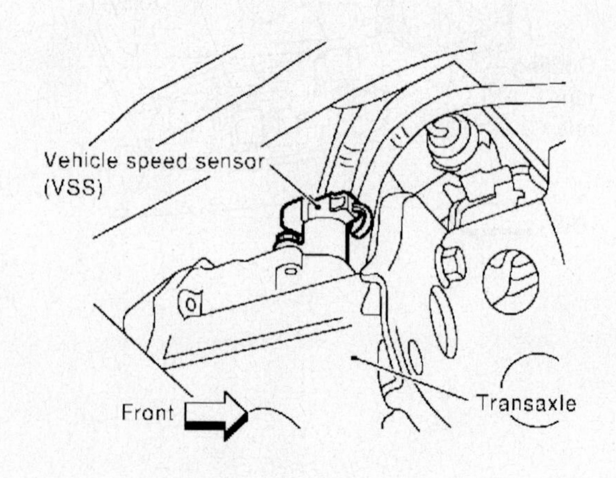
Vehicle speed sensor
(VSS)
Front
Transaxle

29149_NISS_G0152

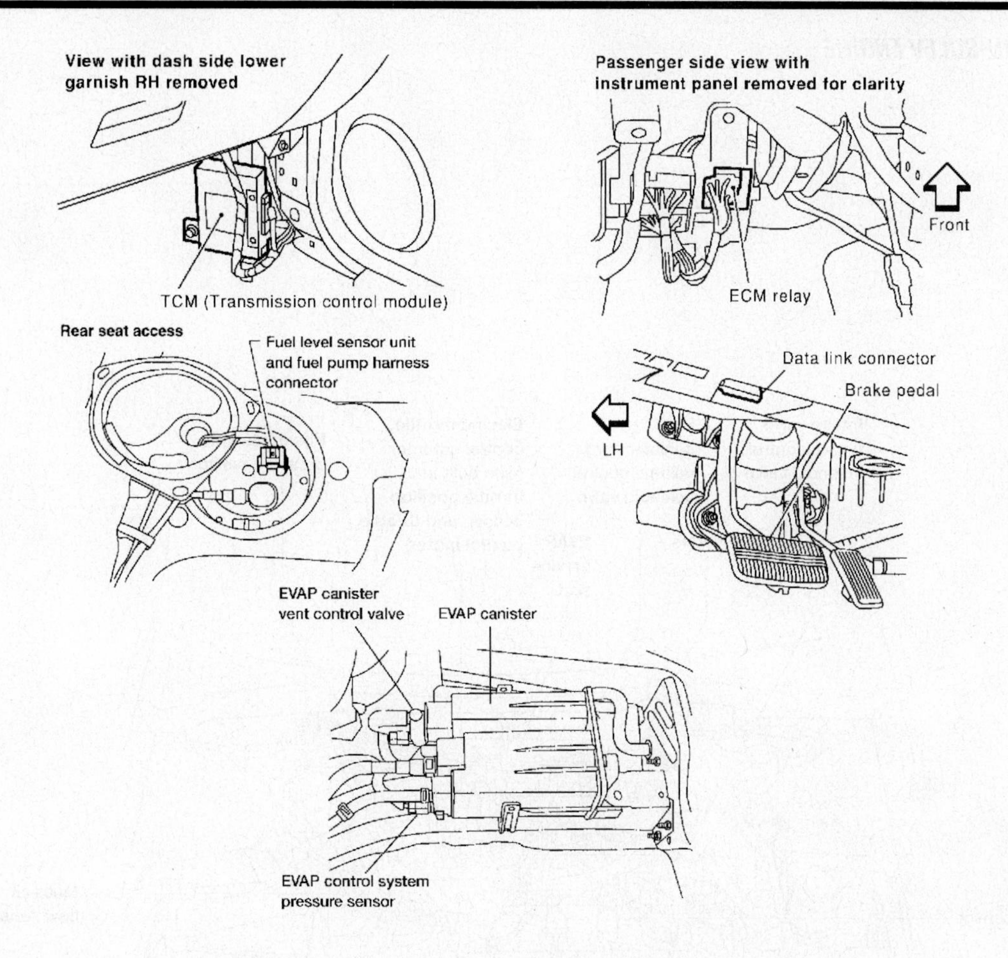

29149_NISS_G0153

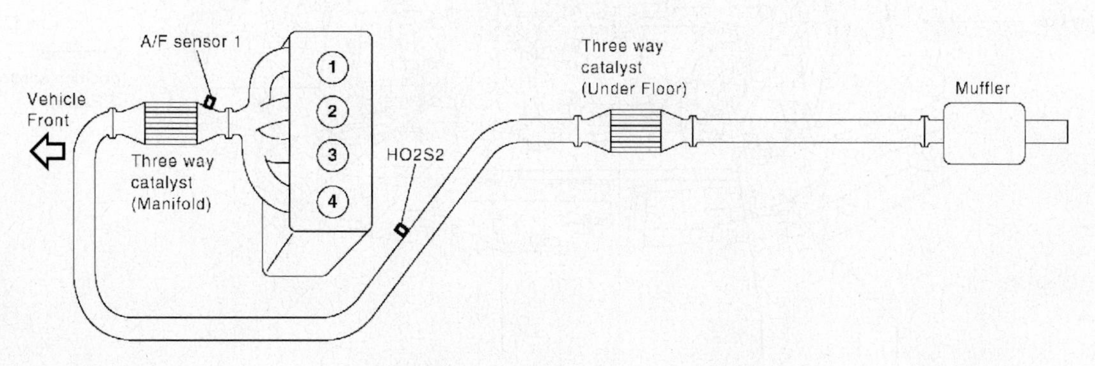

No. of sensor
Location of a sensor in relation the engine air flow, starting from the fresh air intake through to the vehicle tailpipe in order numbering 1, 2, 3, and so on

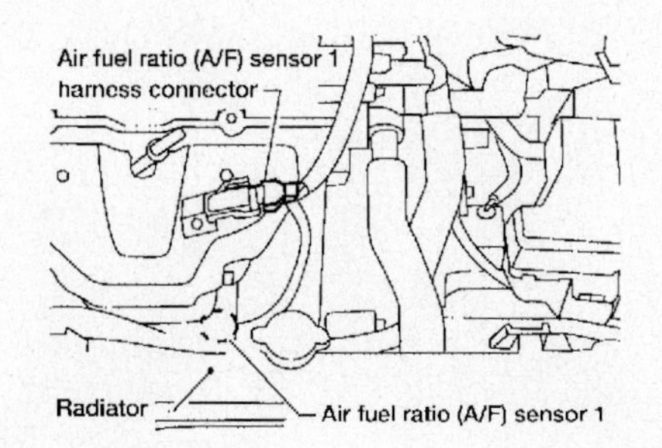

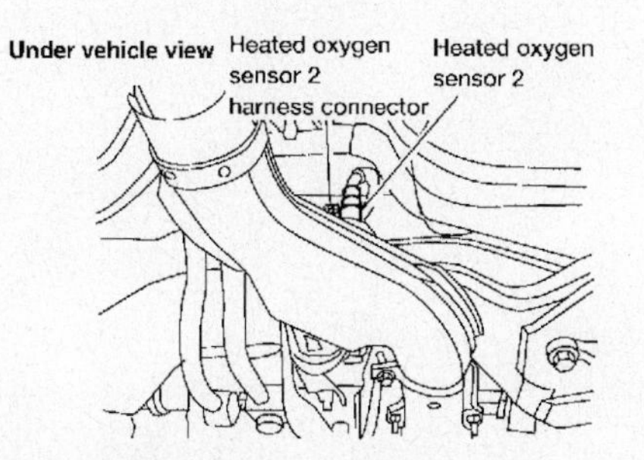

29149_NISS_G0154

QG18DE I4 NON-SULEV ENGINE

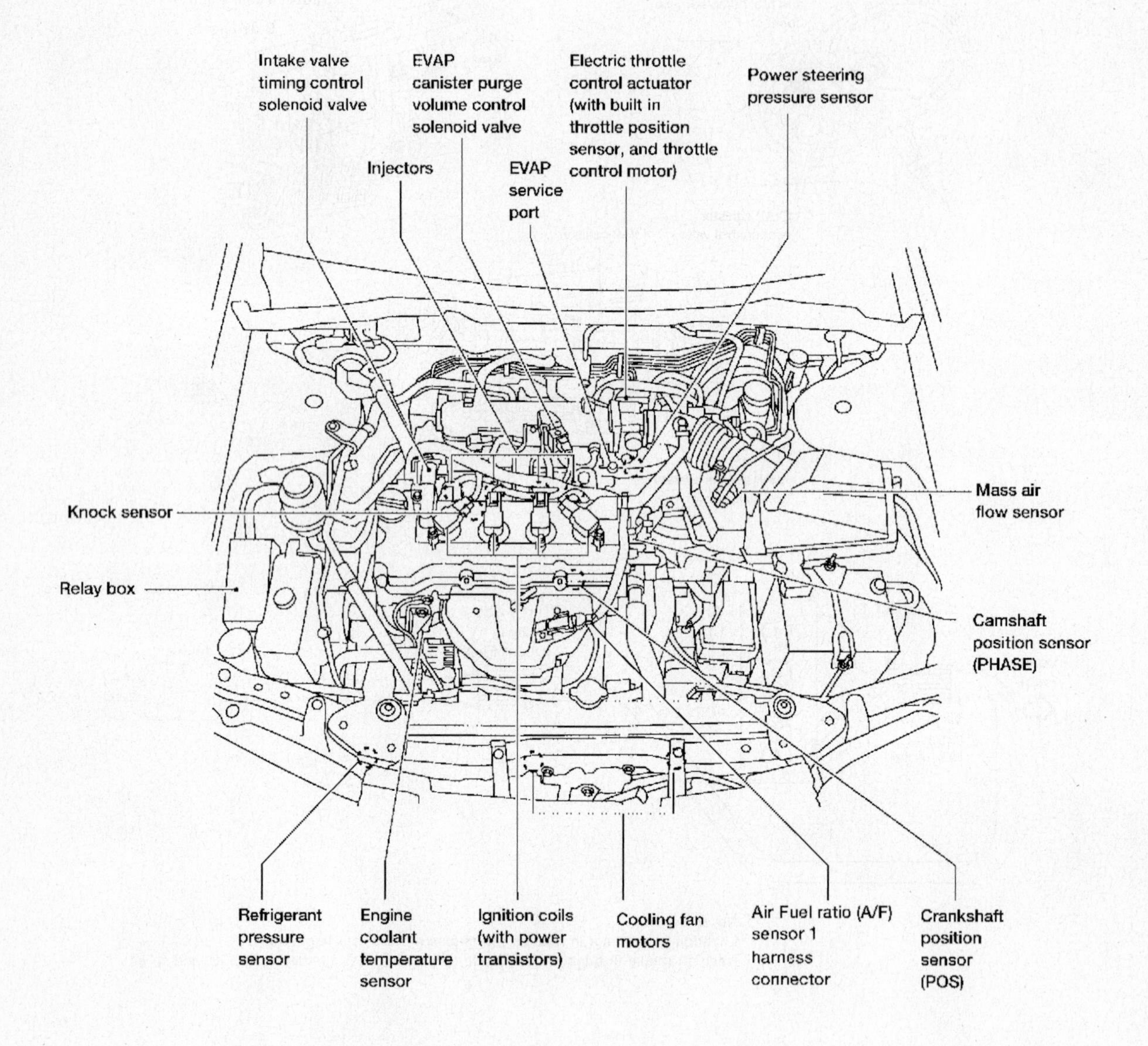

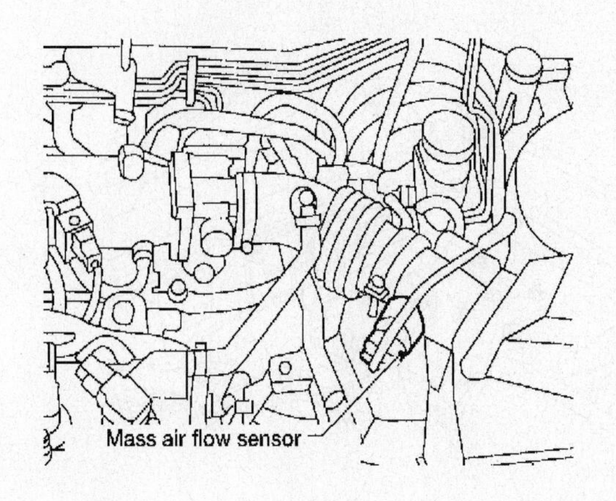
Mass air flow sensor

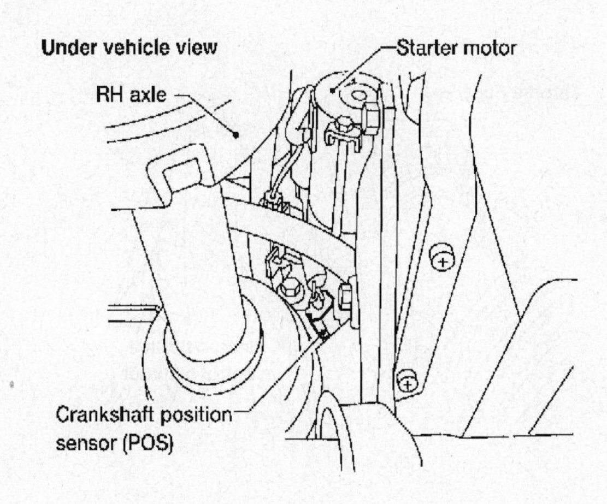
Under vehicle view
Starter motor
RH axle
Crankshaft position
sensor (POS)

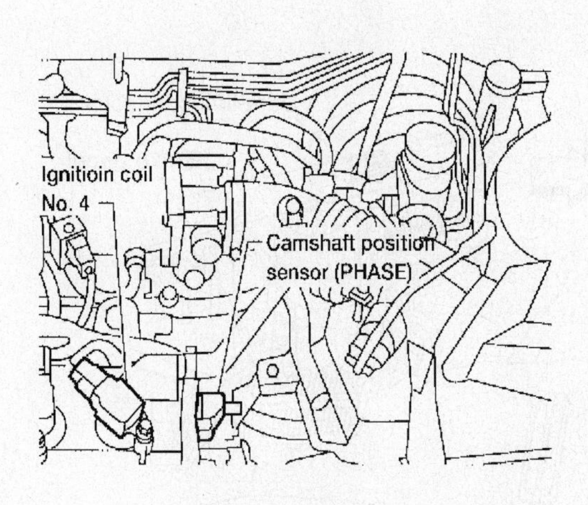
Ignitioin coil
No. 4
Camshaft position
sensor (PHASE)

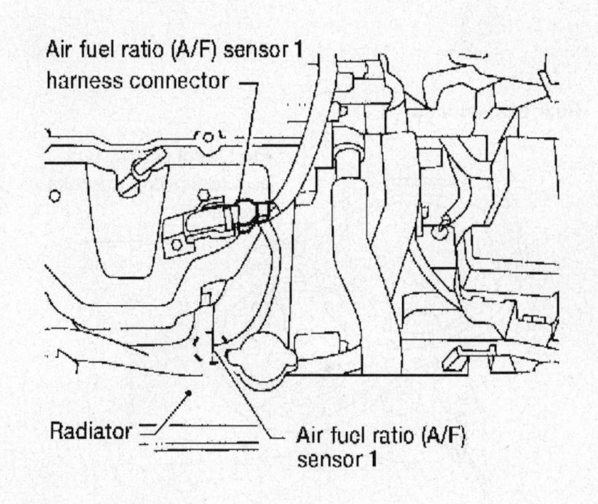
Air fuel ratio (A/F) sensor 1
harness connector
Radiator
Air fuel ratio (A/F)
sensor 1

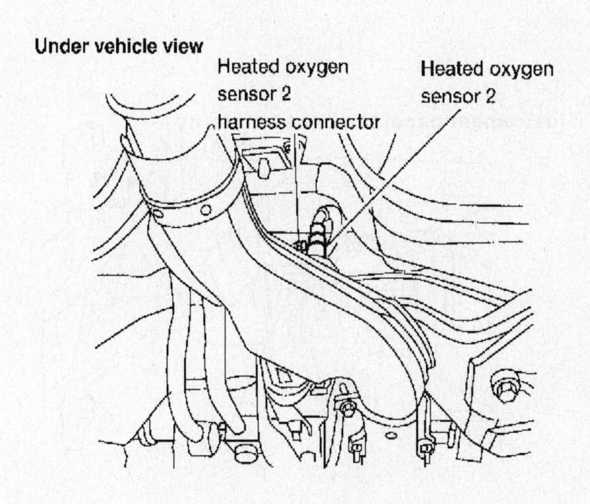
Under vehicle view
Heated oxygen
sensor 2
harness connector
Heated oxygen
sensor 2

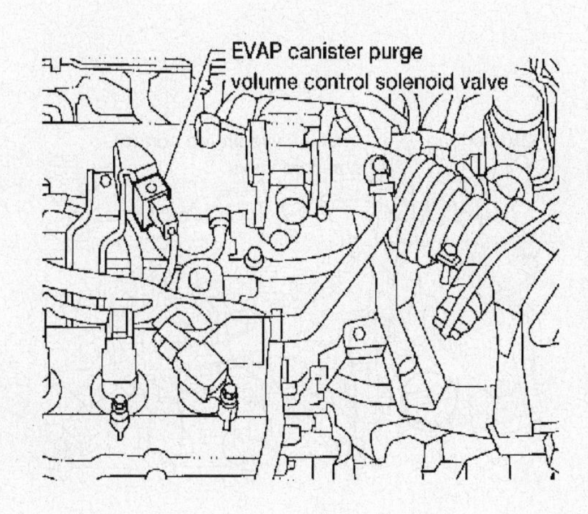
EVAP canister purge
volume control solenoid valve

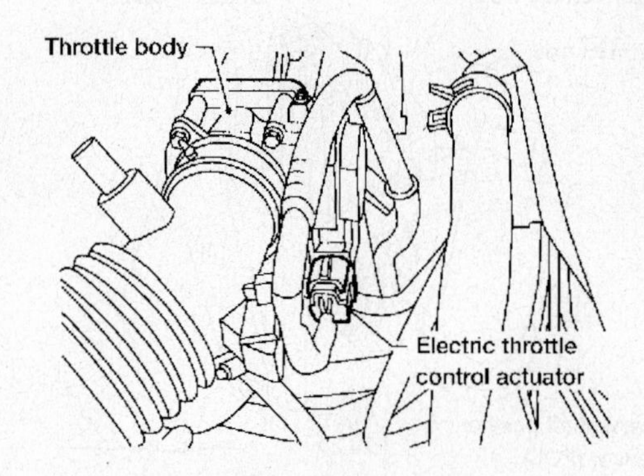

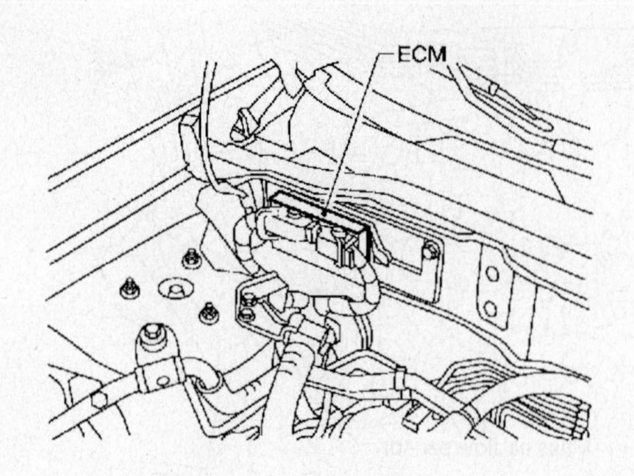

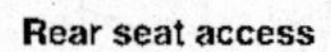

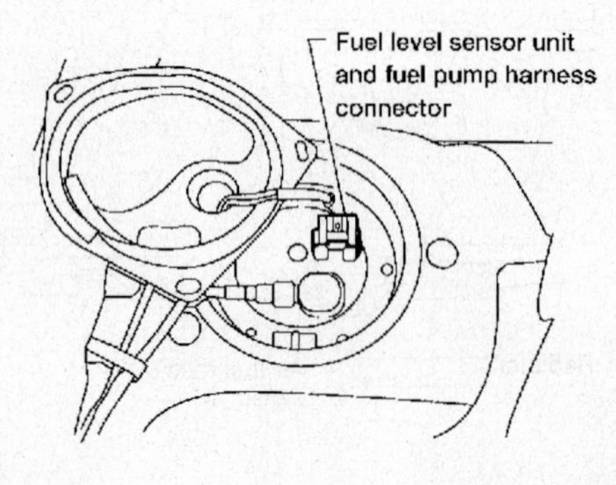

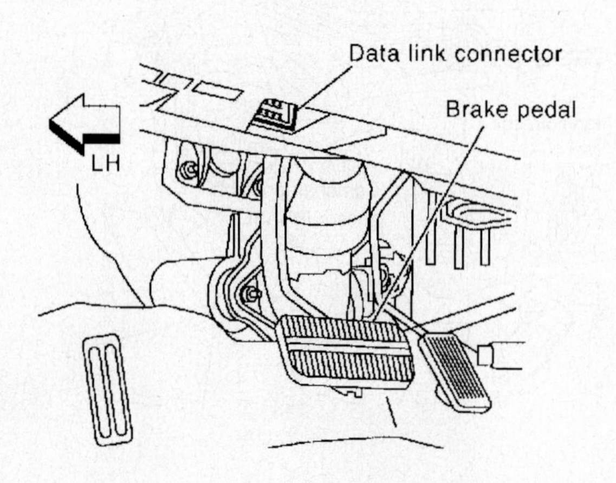

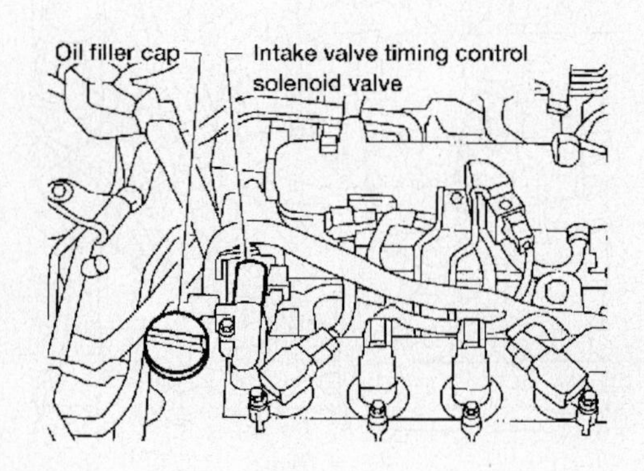

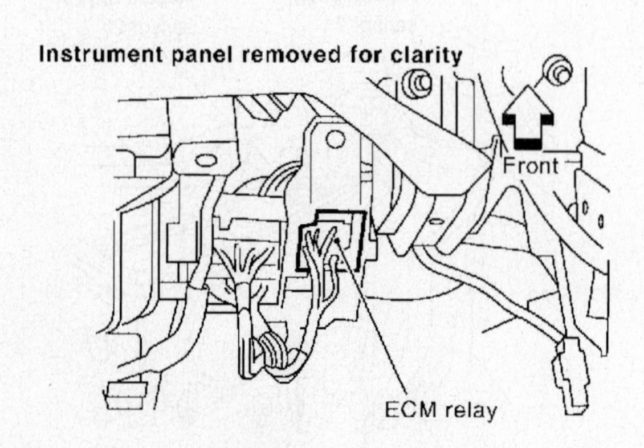

29149_NISS_G0157

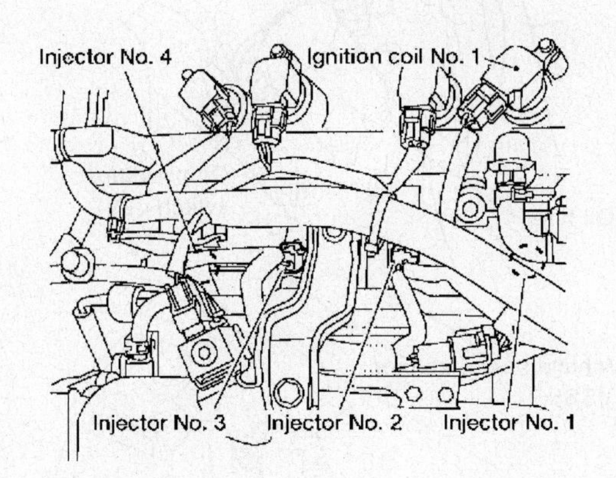
Injector No. 4
Ignition coil No. 1
Injector No. 3
Injector No. 2
Injector No. 1

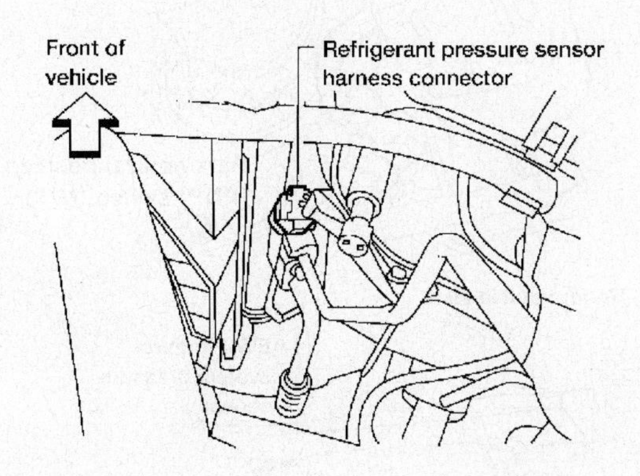
Front of
vehicle
Refrigerant pressure sensor
harness connector

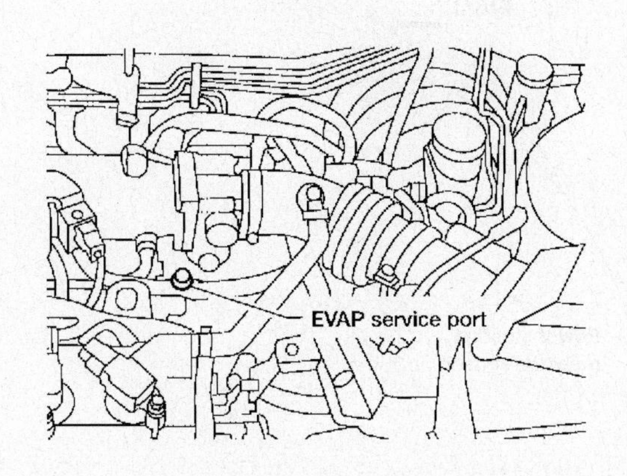
EVAP service port

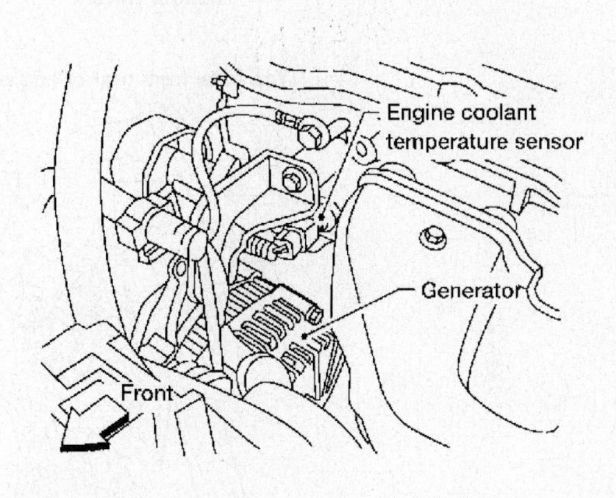
Engine coolant
temperature sensor
Generator
Front

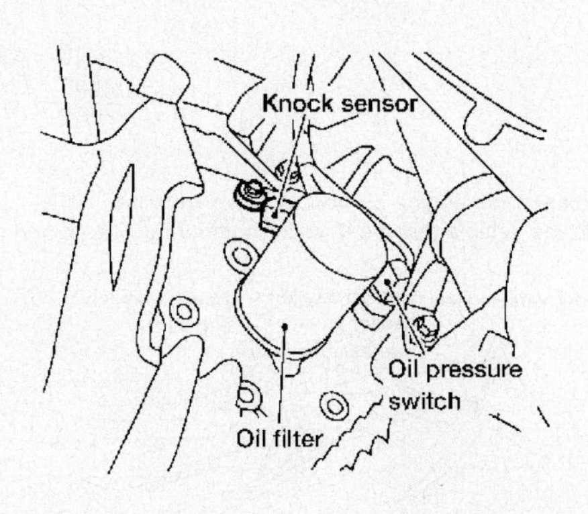
Knock sensor
Oil pressure
switch
Oil filter

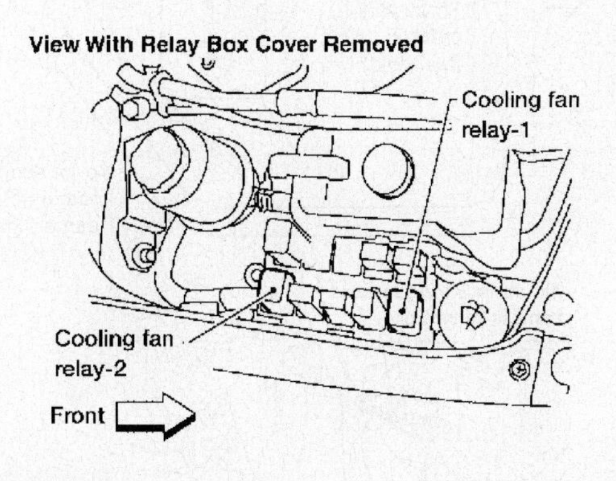
View With Relay Box Cover Removed
Cooling fan
relay-1
Cooling fan
relay-2
Front

29149_NISS_G0158

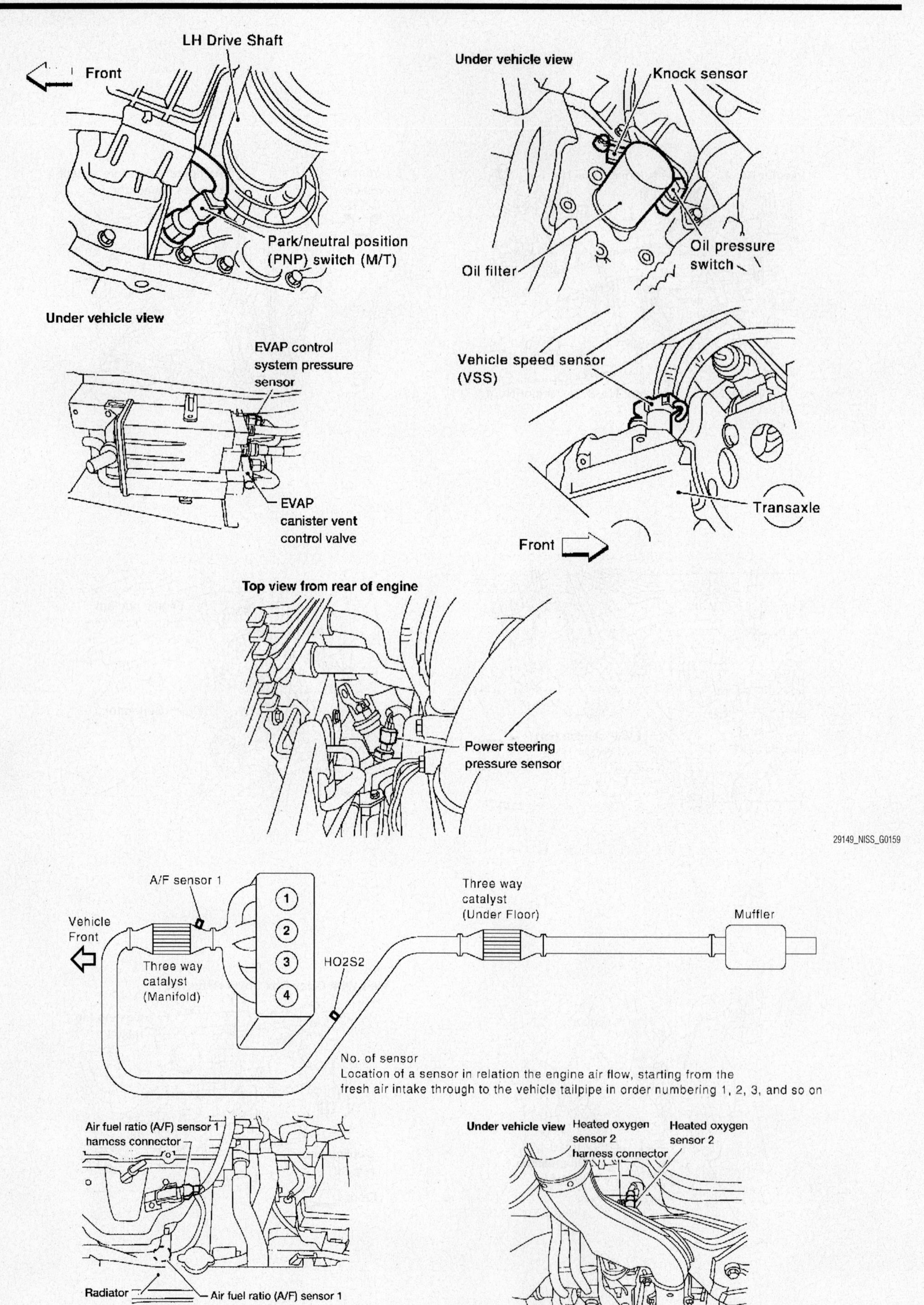
LH Drive Shaft
Front
Park/neutral position
(PNP) switch (M/T)
Under vehicle view
Knock sensor
Oil filter
Oil pressure
switch
Under vehicle view
EVAP control
system pressure
sensor
EVAP
canister vent
control valve
Vehicle speed sensor
(VSS)
Transaxle
Front
Top view from rear of engine
Power steering
pressure sensor
29149_NISS_G0159
A/F sensor 1
Vehicle
Front
Three way
catalyst
(Manifold)
1
2
3
4
HO2S2
Three way
catalyst
(Under Floor)
Muffler
No. of sensor
Location of a sensor in relation the engine air flow, starting from the
fresh air intake through to the vehicle tailpipe in order numbering 1, 2, 3, and so on
Air fuel ratio (A/F) sensor 1
harness connector
Radiator
Air fuel ratio (A/F) sensor 1
Under vehicle view
Heated oxygen
sensor 2
harness connector
Heated oxygen
sensor 2
29149_NISS_G0160

2002–2006

QR25DE I4 ENGINE W/ MANUAL TRANSMISSION

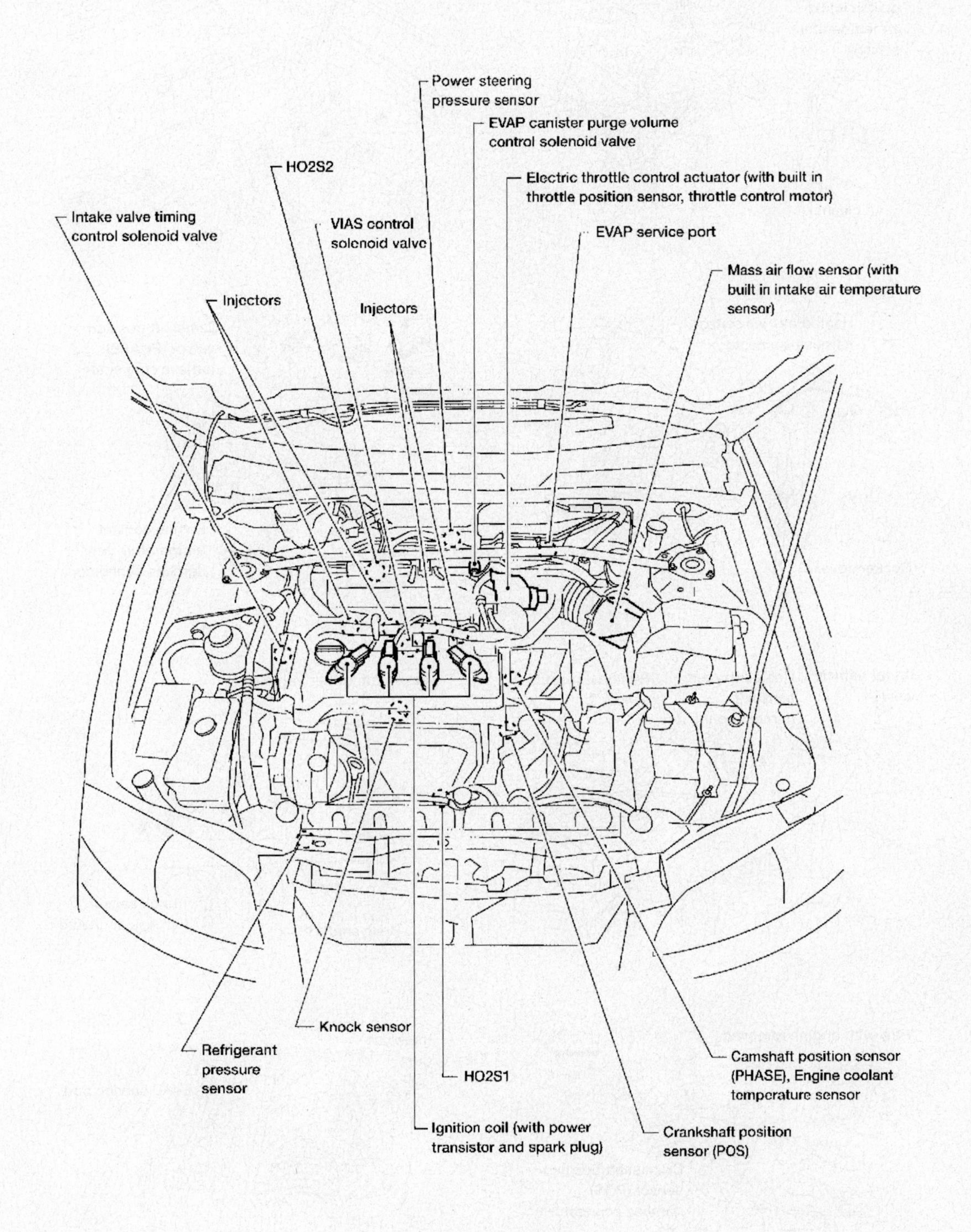

29149_NISS_G0161

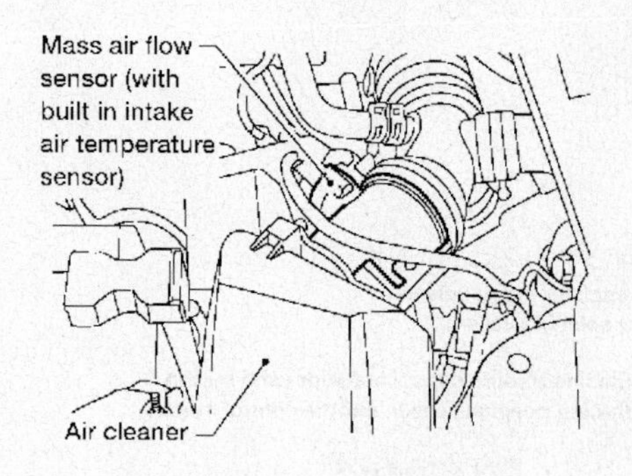
Mass air flow sensor (with built in intake air temperature sensor)
Air cleaner

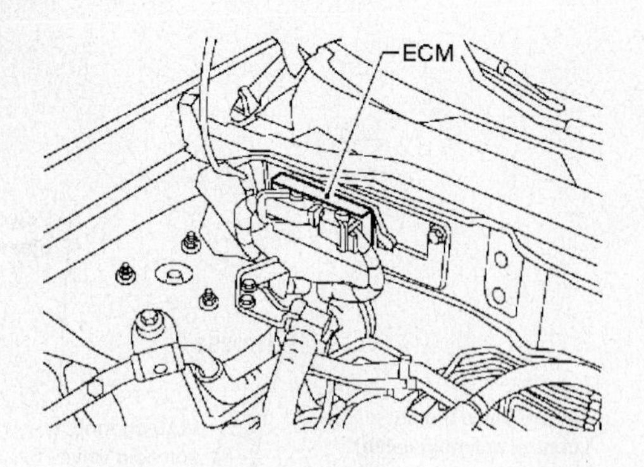
ECM

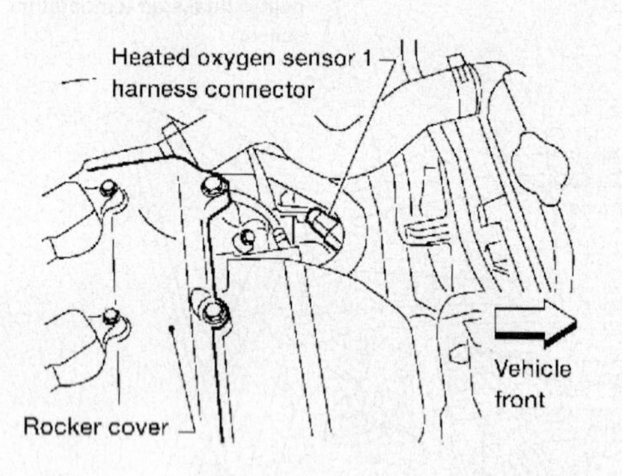
Heated oxygen sensor 1 harness connector
Vehicle front
Rocker cover

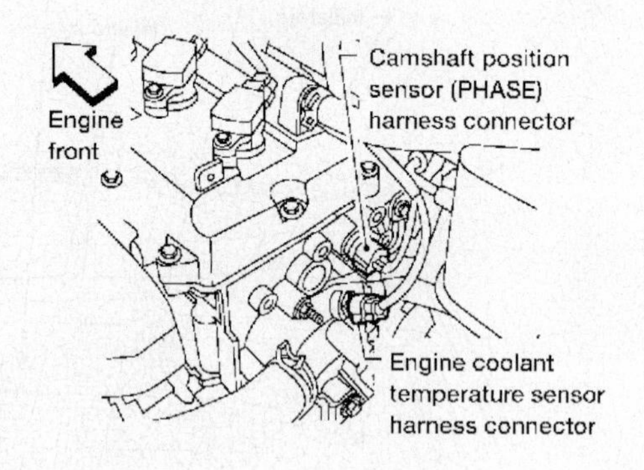
Engine front
Camshaft position sensor (PHASE) harness connector
Engine coolant temperature sensor harness connector

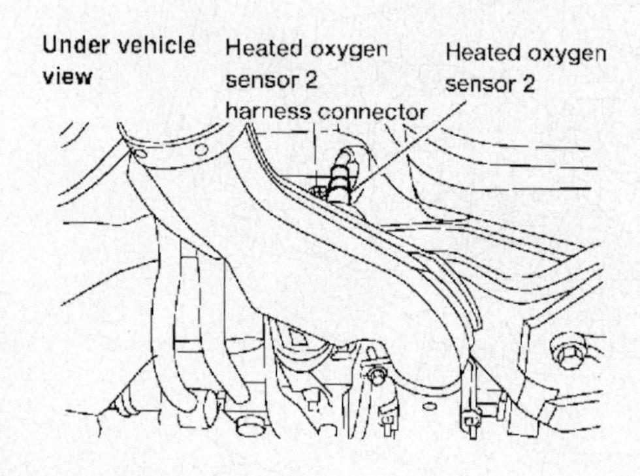
Under vehicle view
Heated oxygen sensor 2 harness connector
Heated oxygen sensor 2

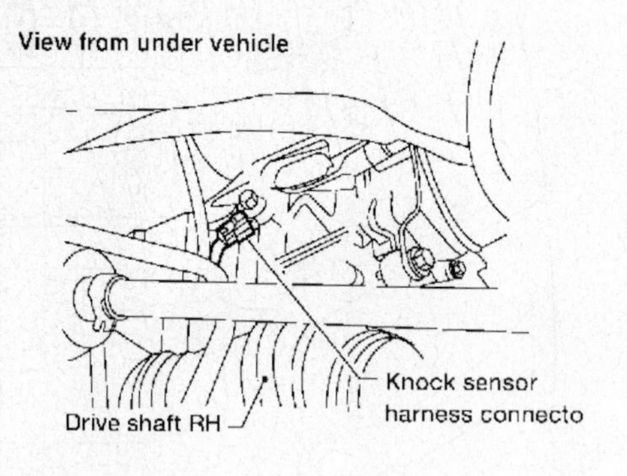
View from under vehicle
Knock sensor harness connecto
Drive shaft RH

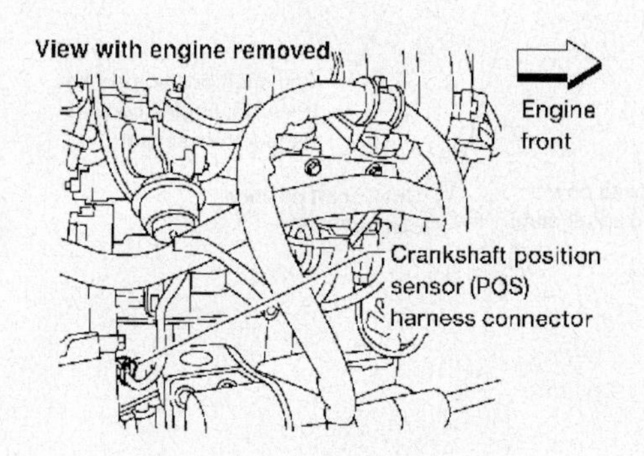
View with engine removed
Engine front
Crankshaft position sensor (POS) harness connector

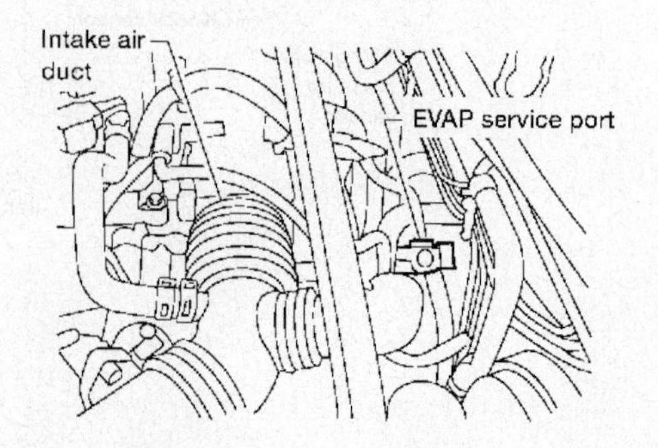
Intake air duct
EVAP service port

29149_NISS_G0162

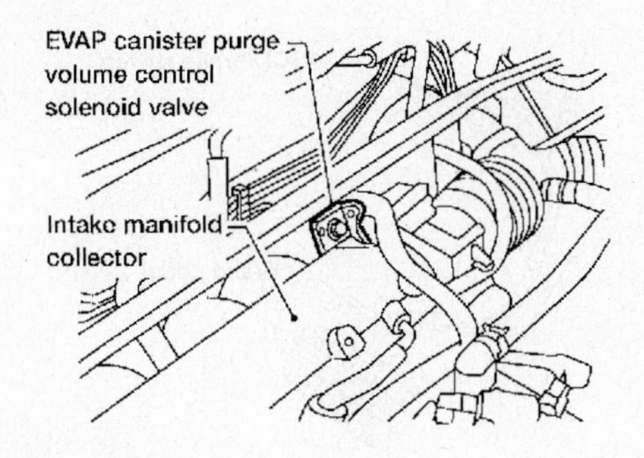
EVAP canister purge volume control solenoid valve
Intake manifold collector

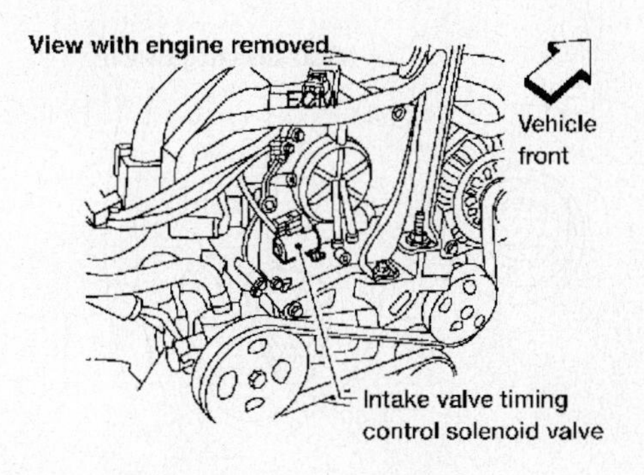
View with engine removed
ECM
Vehicle front
Intake valve timing control solenoid valve

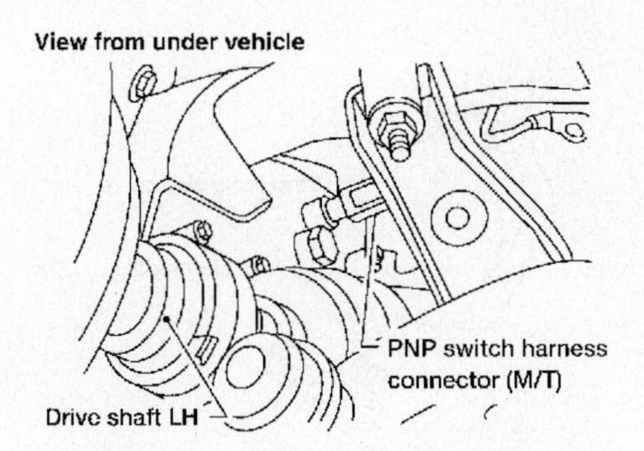
View from under vehicle
PNP switch harness connector (M/T)
Drive shaft LH

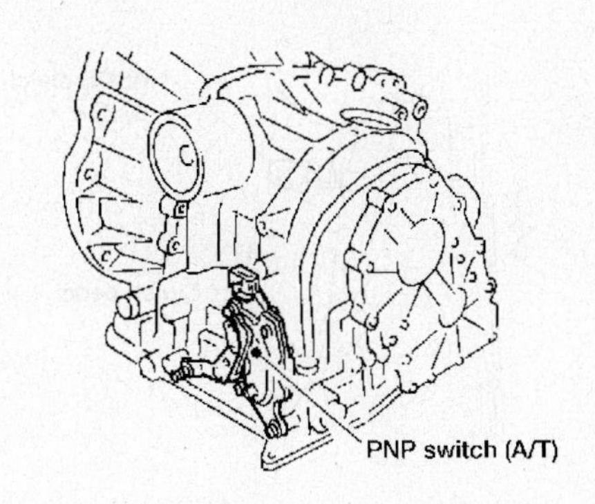
PNP switch (A/T)

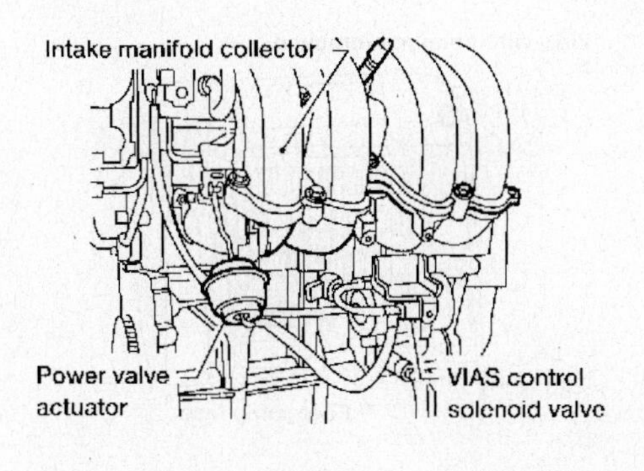
Intake manifold collector
Power valve actuator
VIAS control solenoid valve

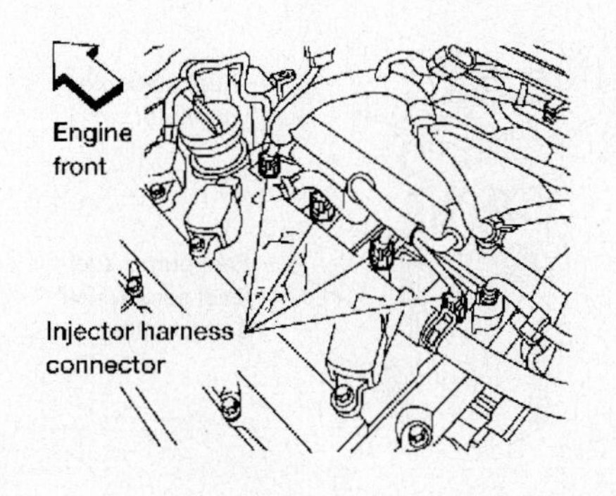
Engine front
Injector harness connector

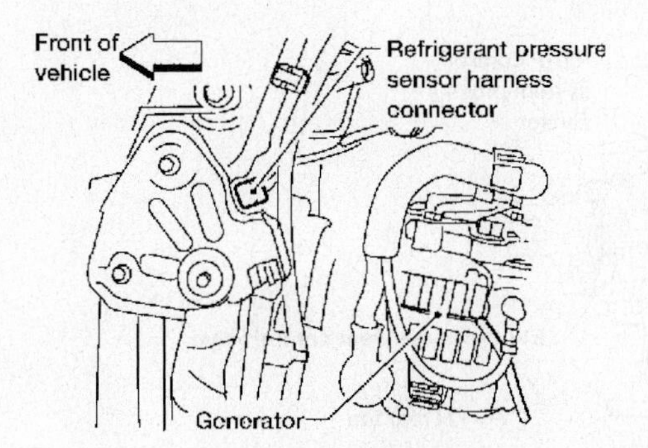
Front of vehicle
Refrigerant pressure sensor harness connector
Generator

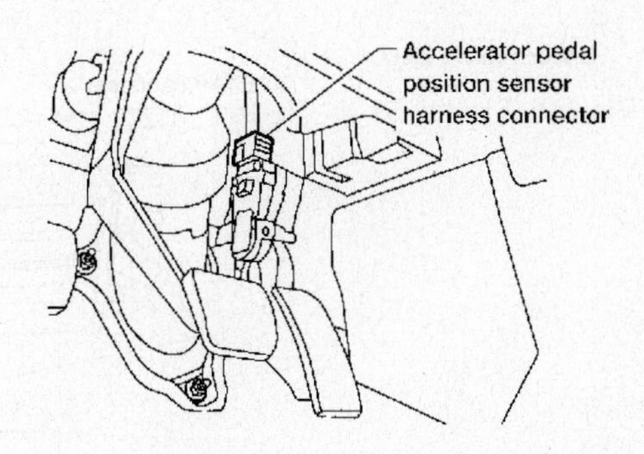
Accelerator pedal position sensor harness connector

29149_NISS_G0163

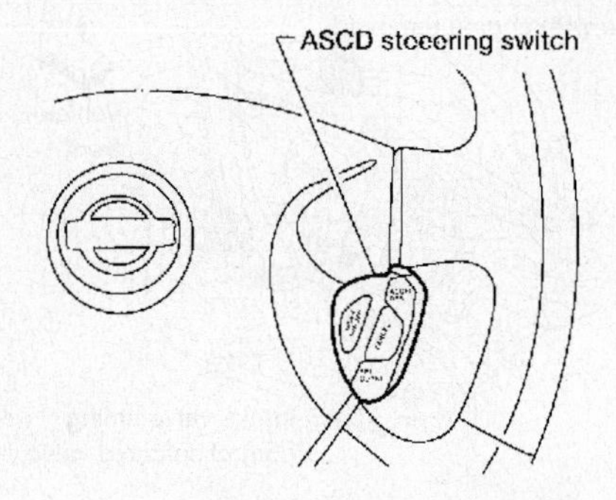

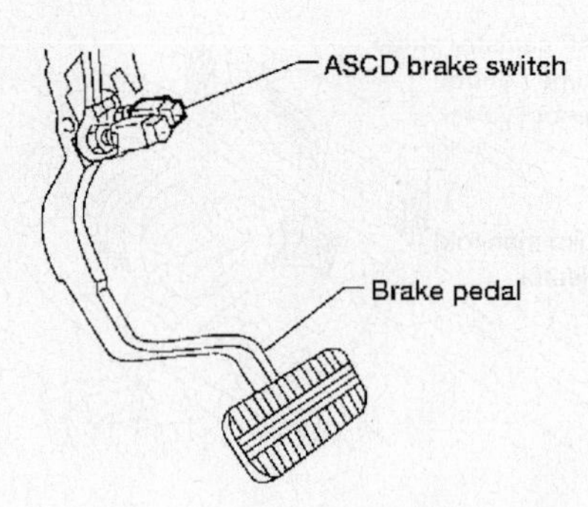

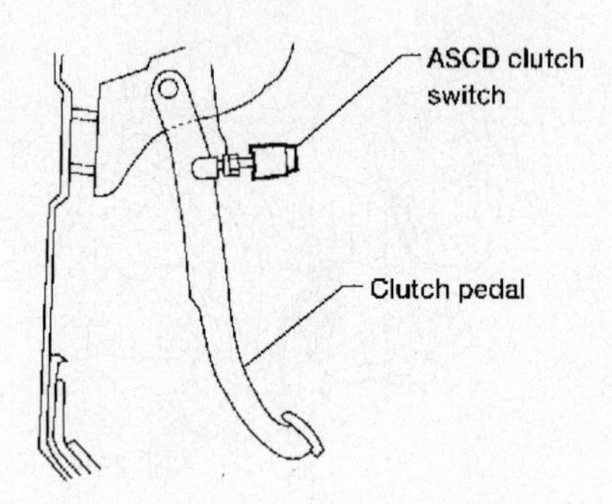

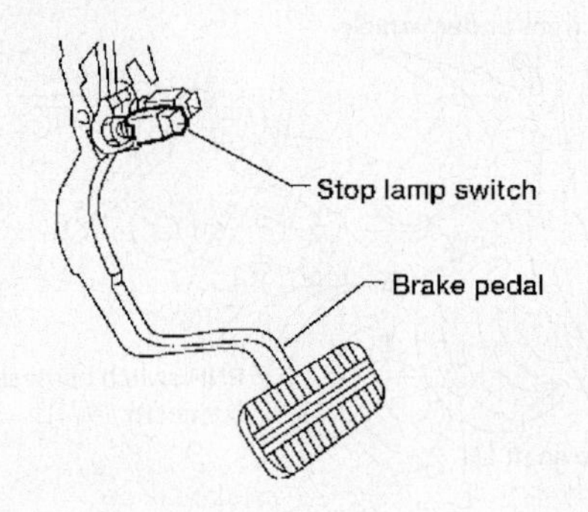

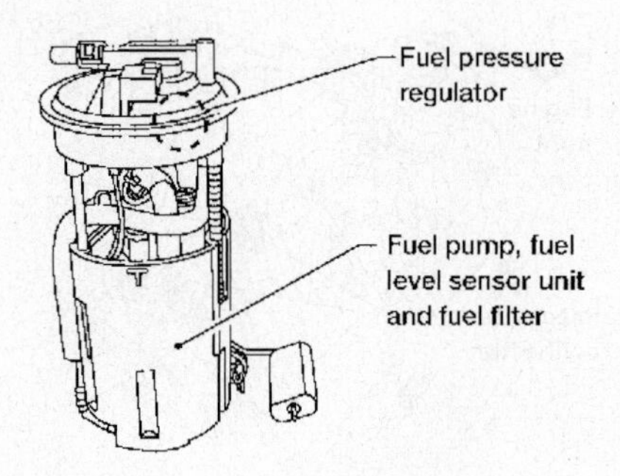

View with coin box removed

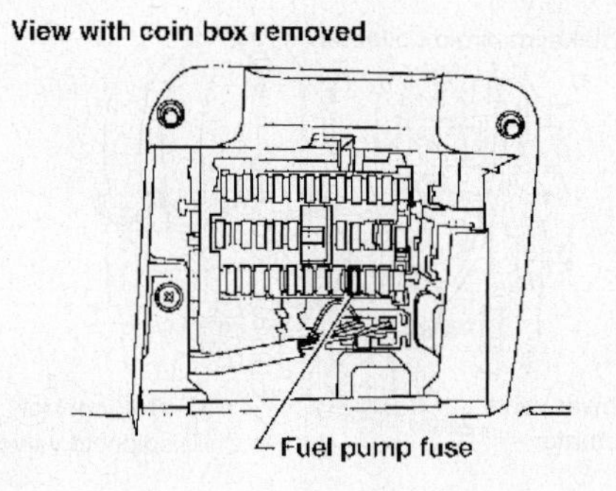

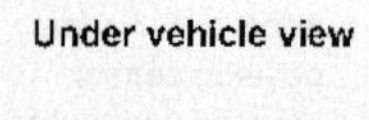

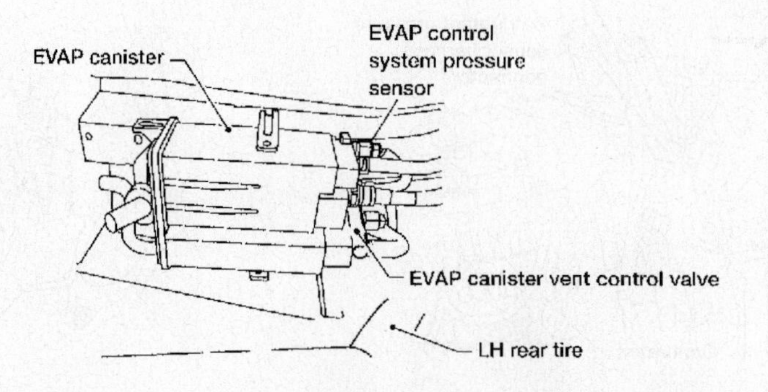

29149_NISS_G0164

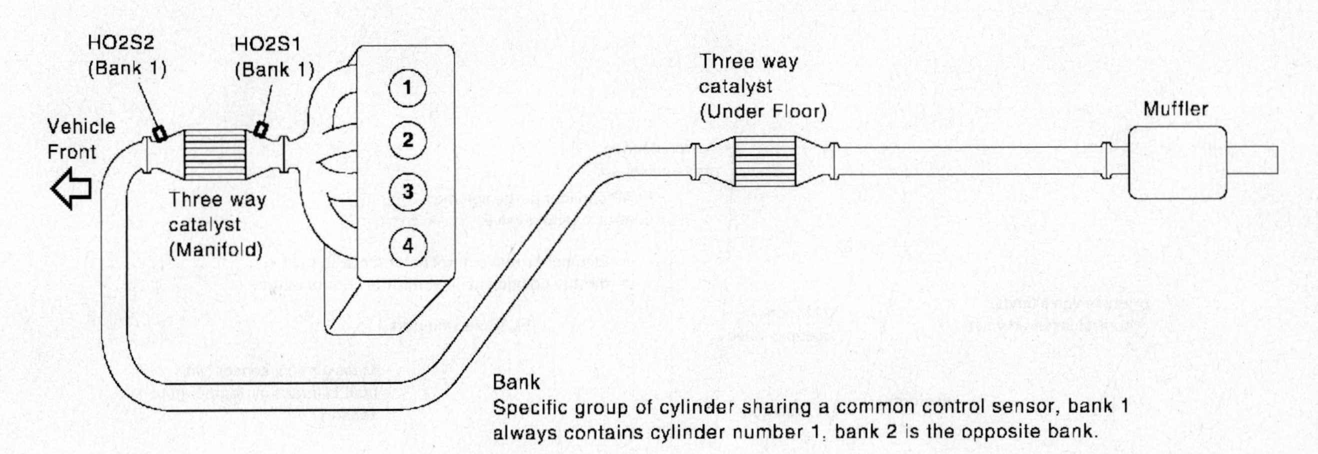
HO2S2
(Bank 1)
HO2S1
(Bank 1)
Vehicle
Front
Three way
catalyst
(Manifold)
1
2
3
4
Three way
catalyst
(Under Floor)
Muffler
Bank
Specific group of cylinder sharing a common control sensor, bank 1
always contains cylinder number 1, bank 2 is the opposite bank.
No. of sensor
Location of a sensor in relation the engine air flow, starting from the
fresh air intake through to the vehicle tailpipe in order numbering 1, 2, 3, and so on

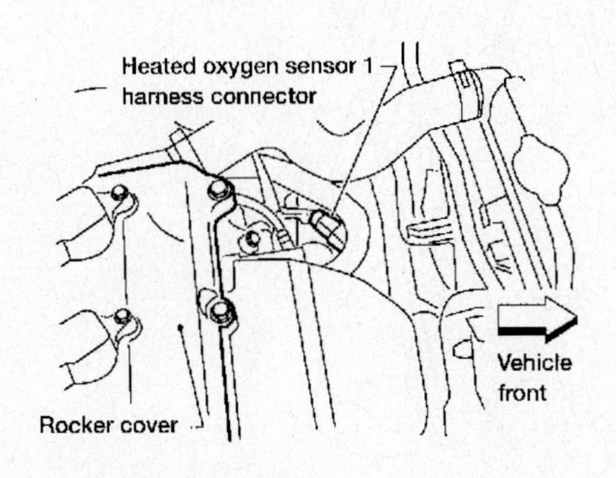
Heated oxygen sensor 1
harness connector
Vehicle
front
Rocker cover

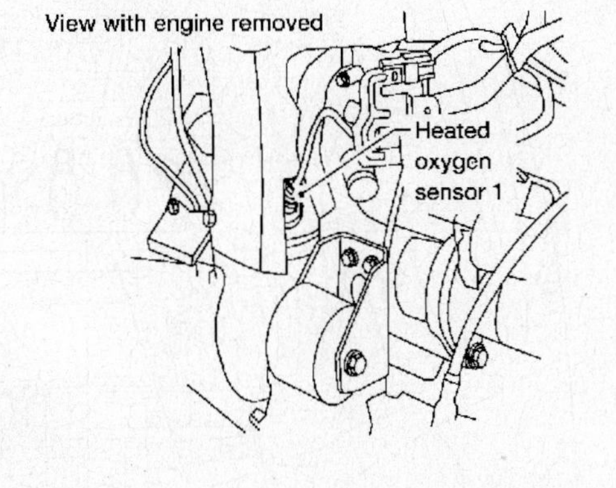
View with engine removed
Heated
oxygen
sensor 1

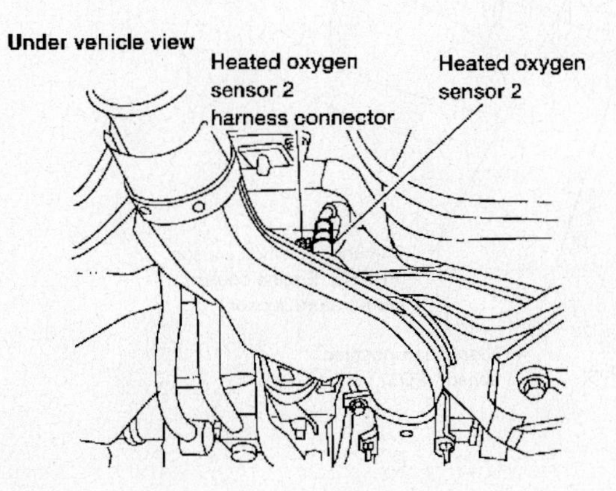
Under vehicle view
Heated oxygen
sensor 2
harness connector
Heated oxygen
sensor 2

QR25DE I4 ULEV ENGINE W/AUTOMATIC TRANSMISSION

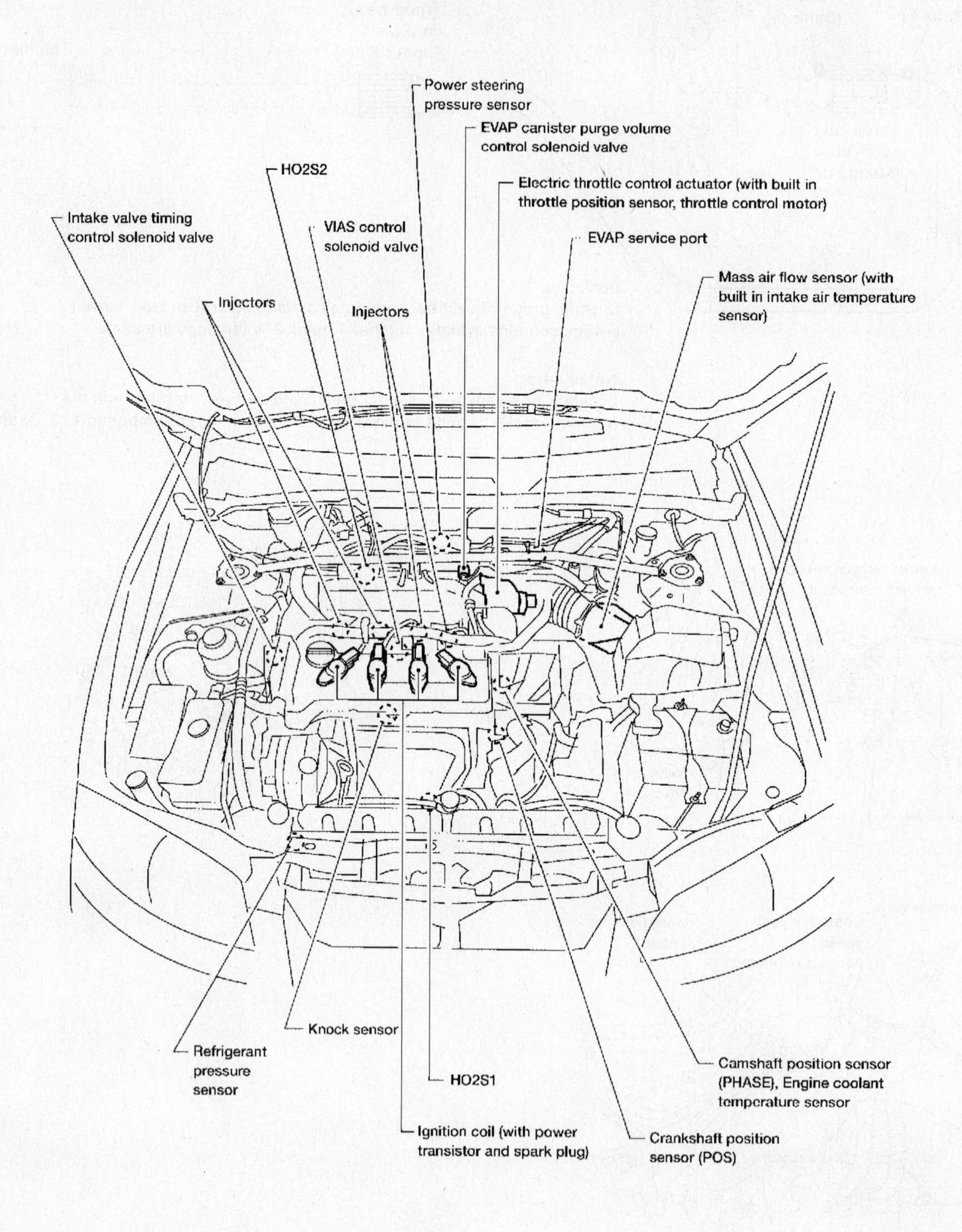

29149_NISS_G0161

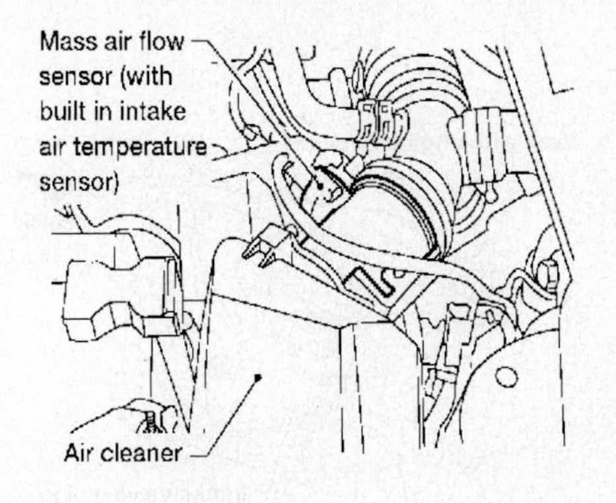

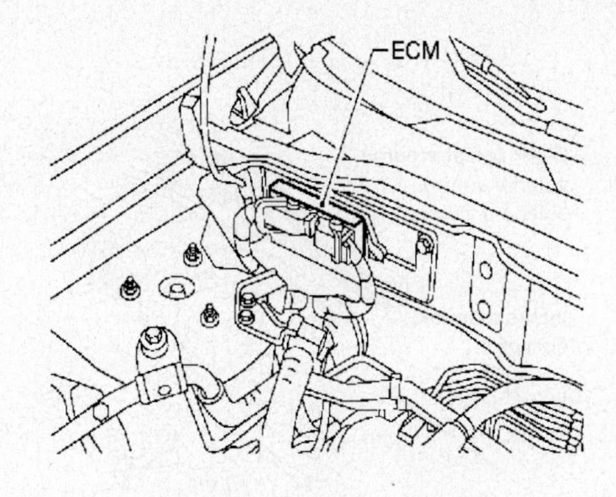

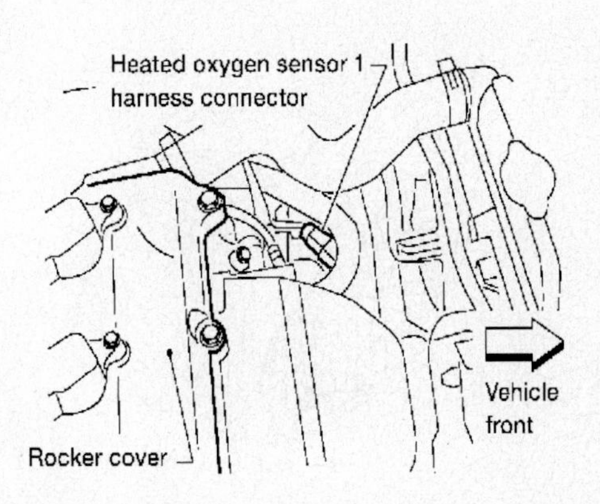

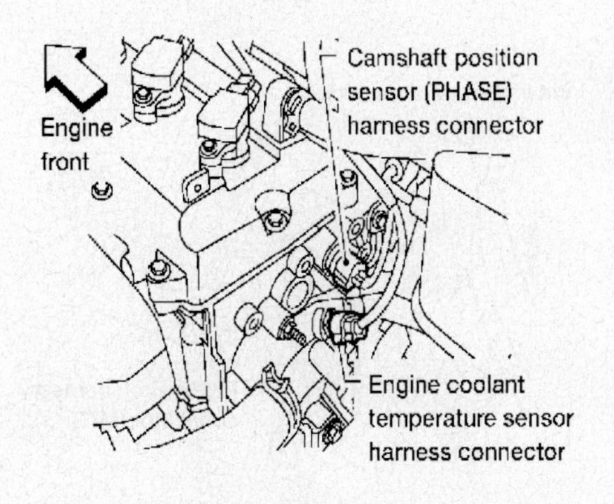

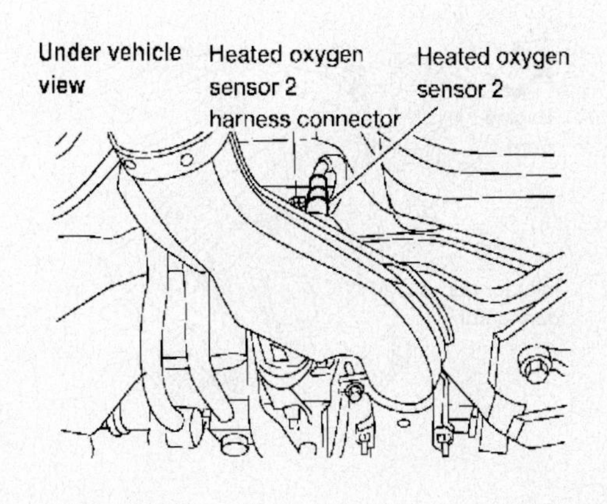

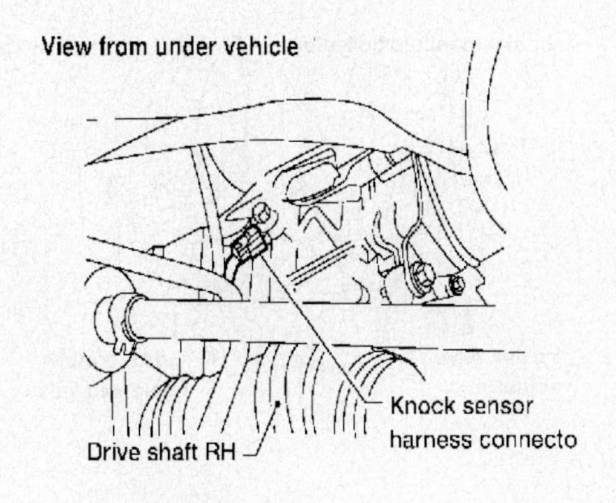

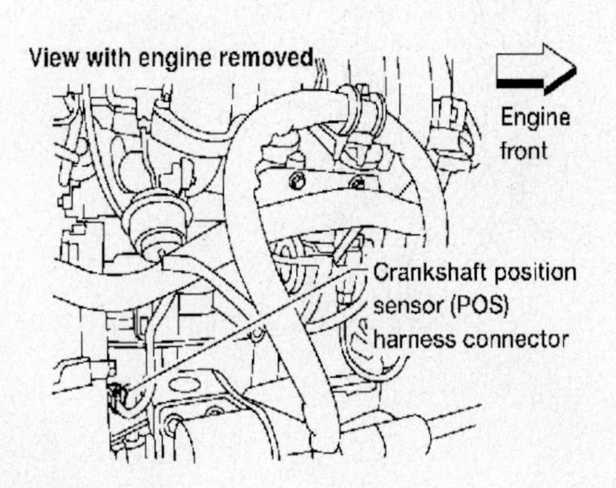

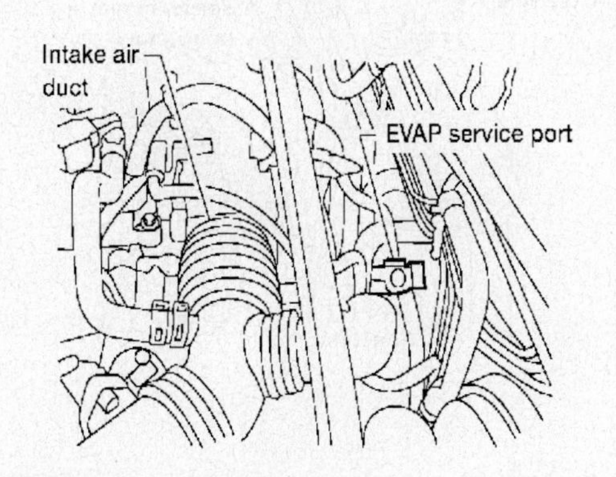

29149_NISS_G0162

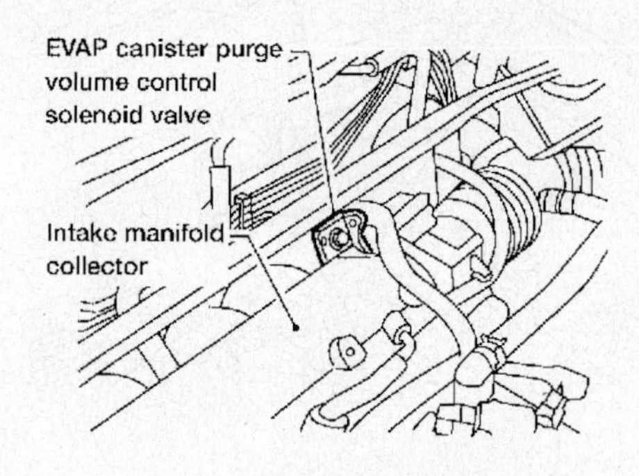

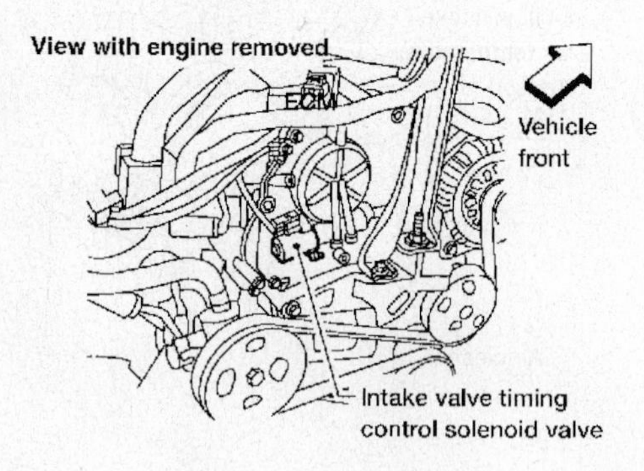

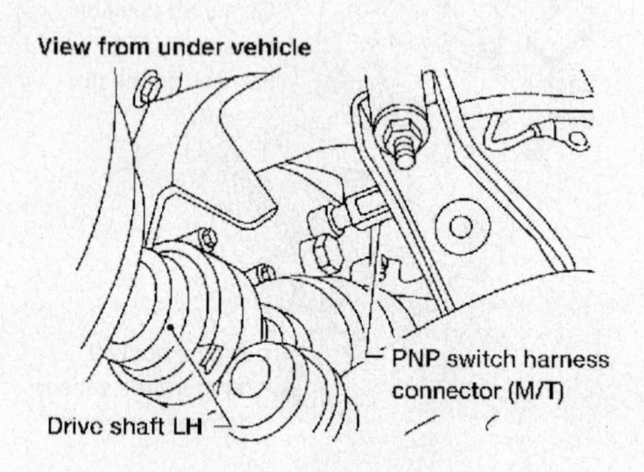

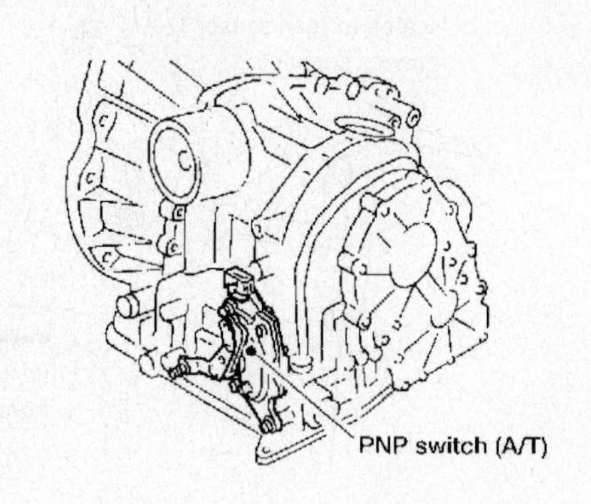

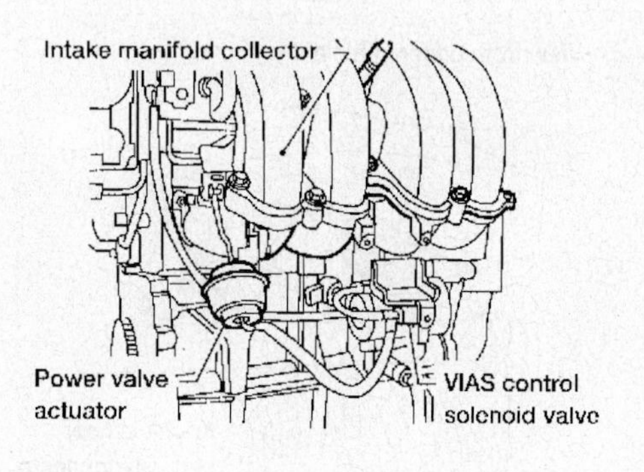

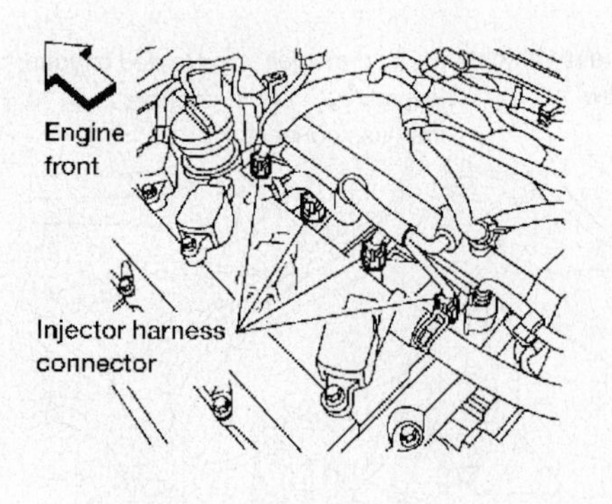

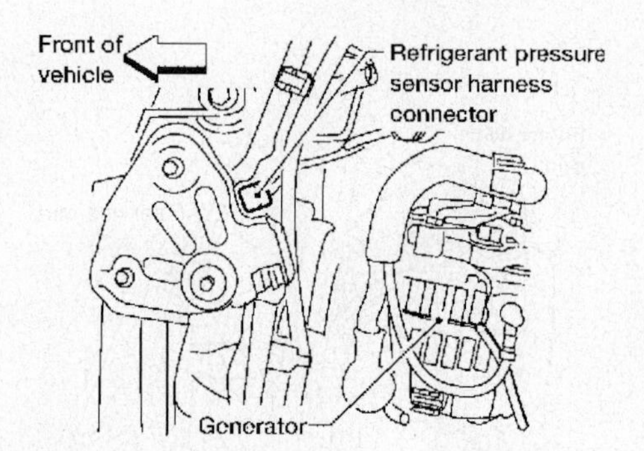

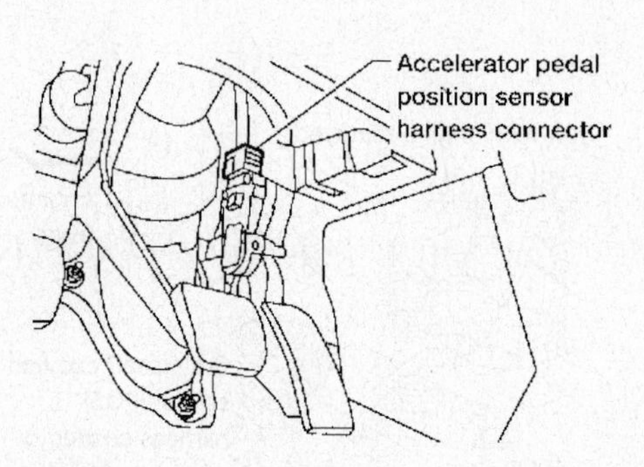

29149_NISS_G0163

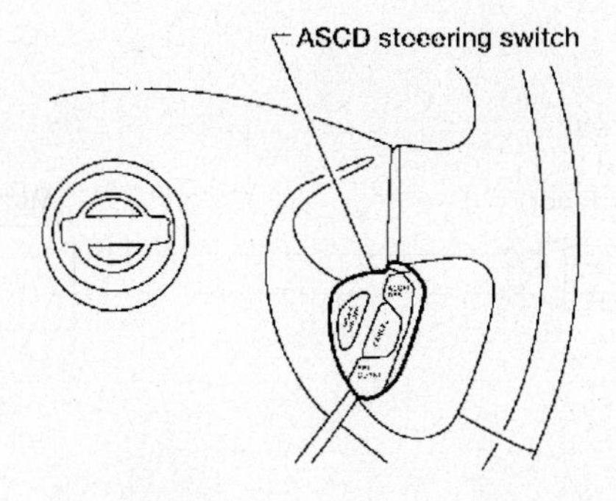

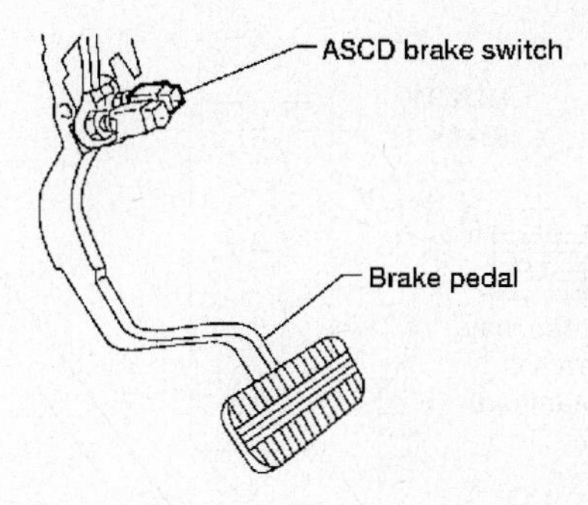

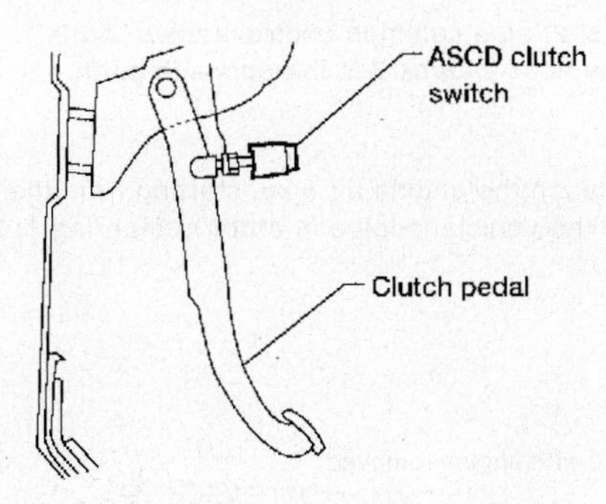

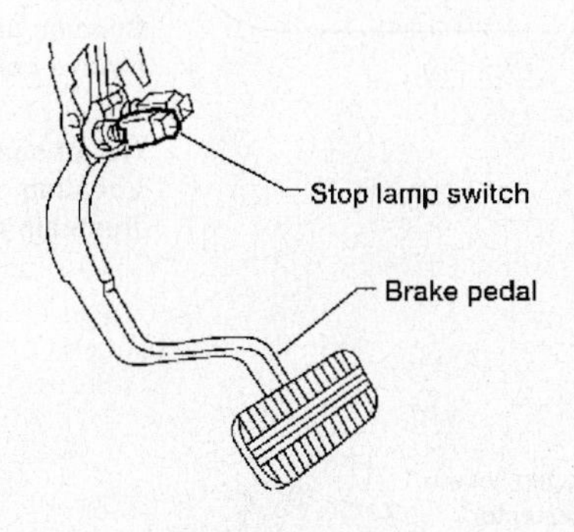

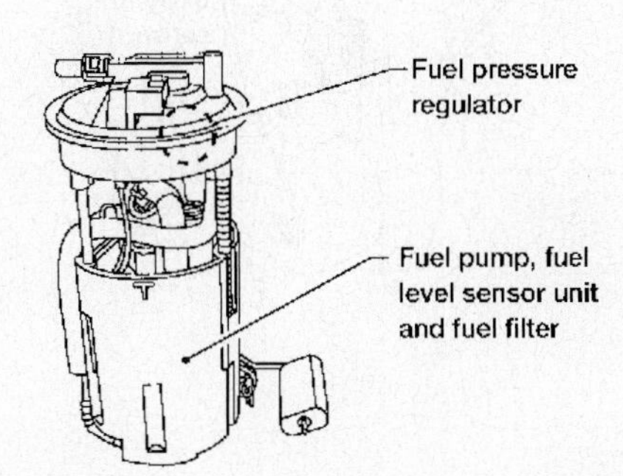

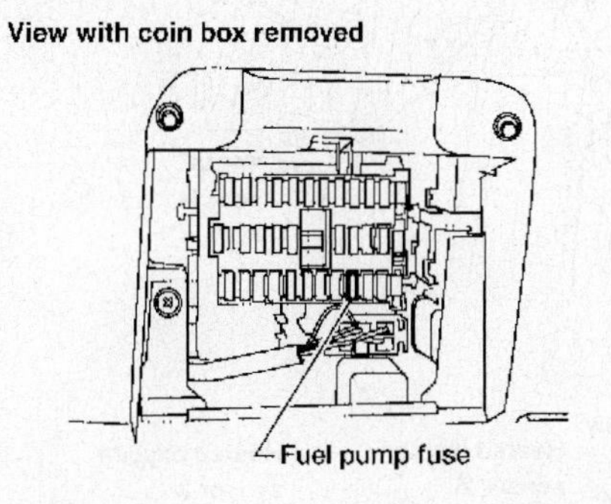

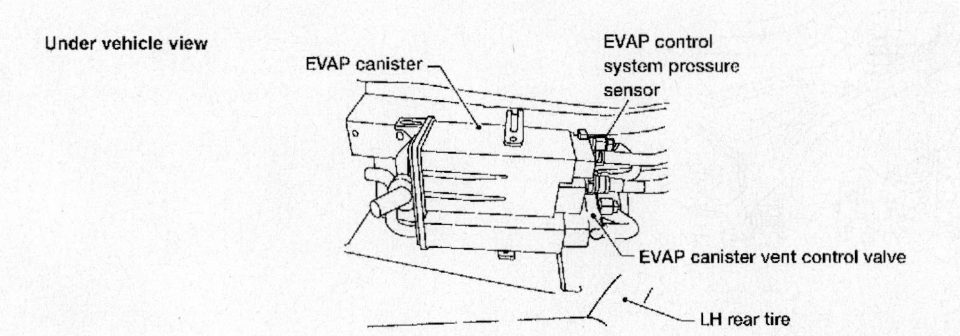

29149_NISS_G0164

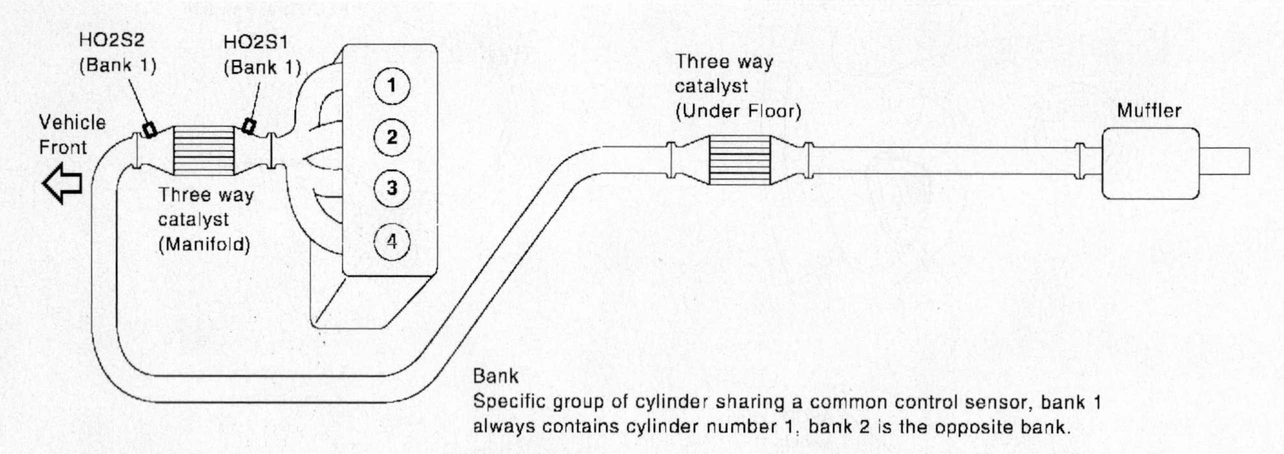

Bank
Specific group of cylinder sharing a common control sensor, bank 1 always contains cylinder number 1, bank 2 is the opposite bank.

No. of sensor
Location of a sensor in relation the engine air flow, starting from the fresh air intake through to the vehicle tailpipe in order numbering 1, 2, 3, and so on

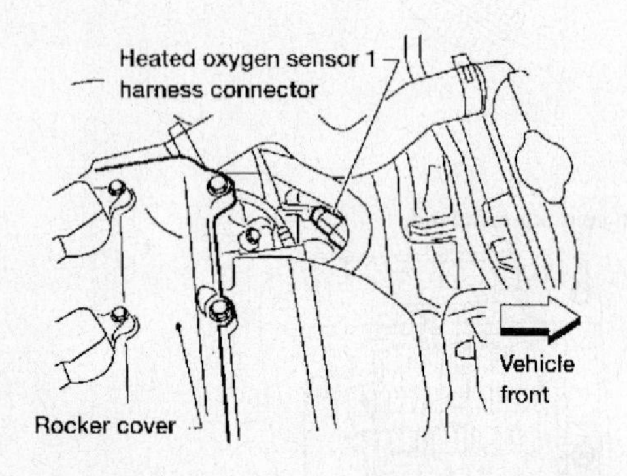

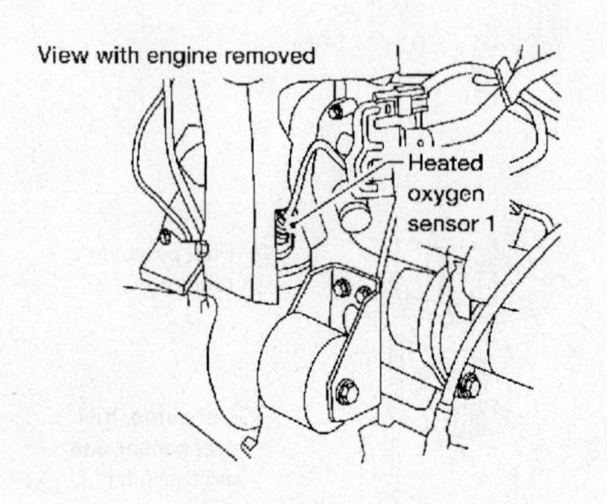

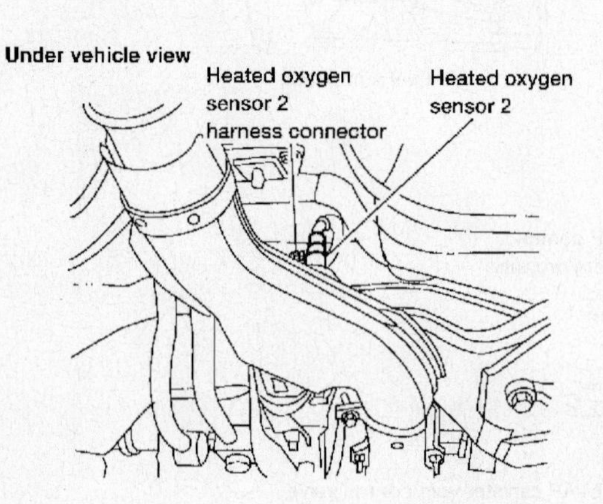

29149_NISS_G0165

QR25DE I4 NON-ULEV ENGINE W/AUTOMATIC TRANSMISSION

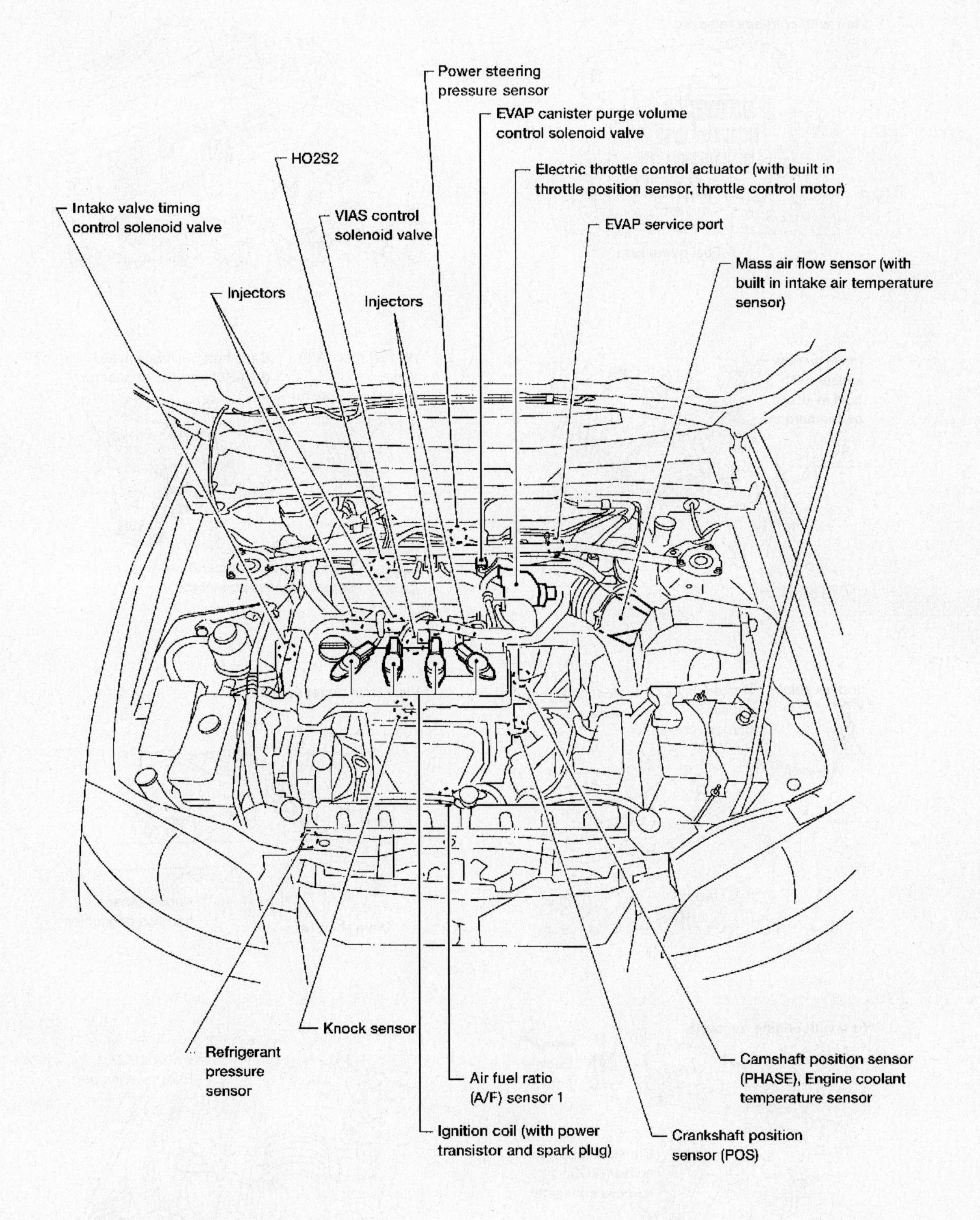

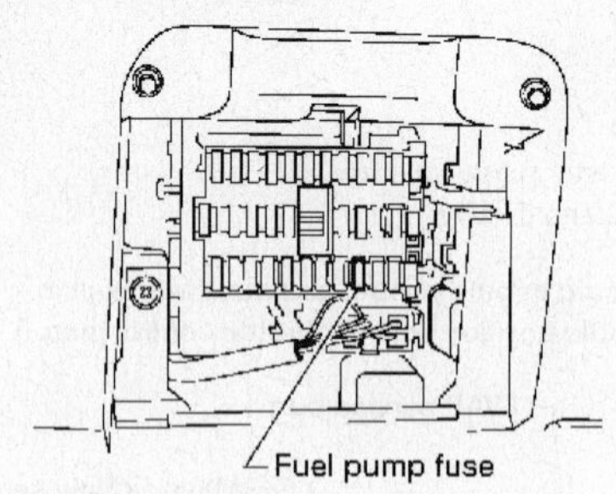

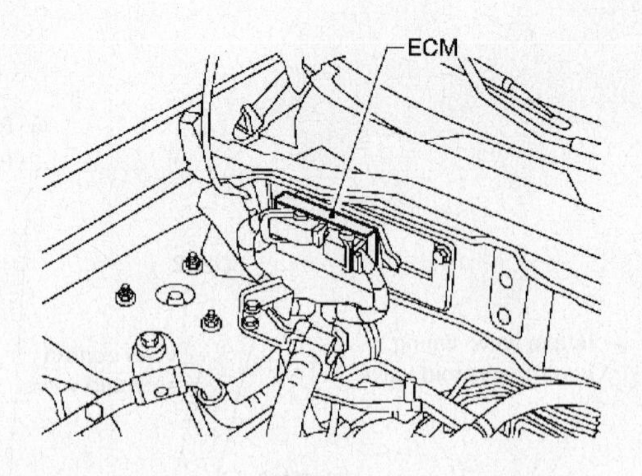

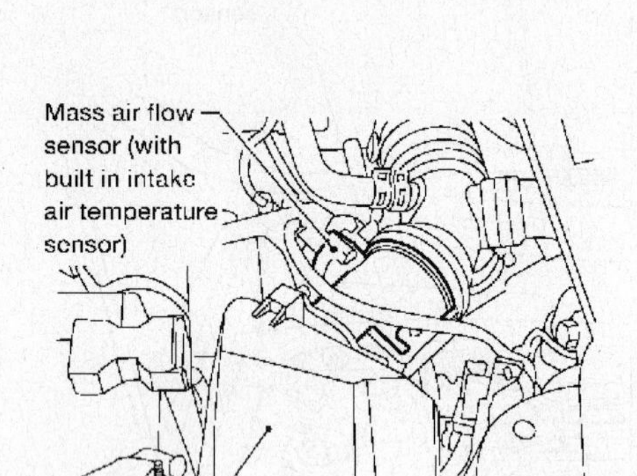

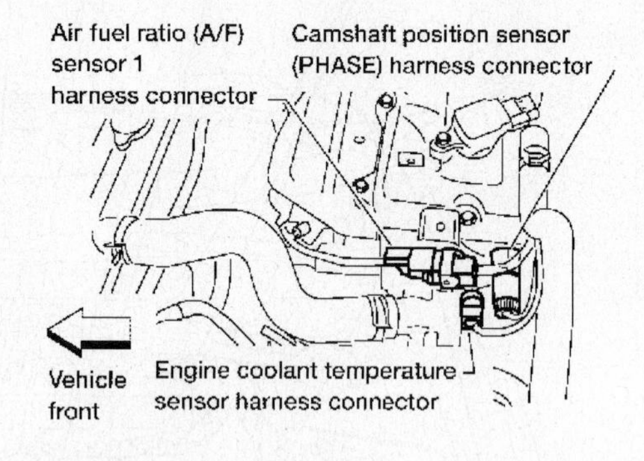

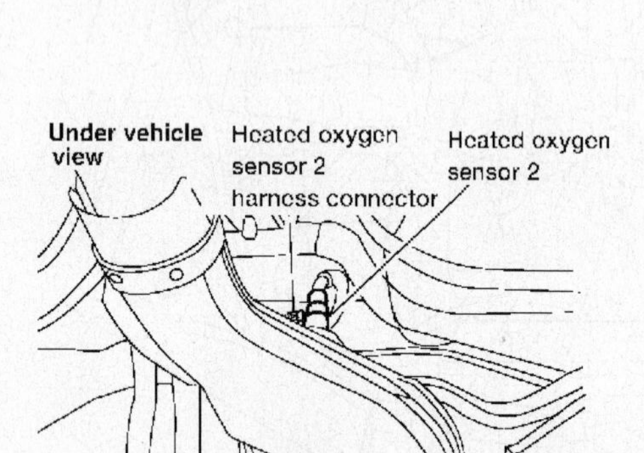

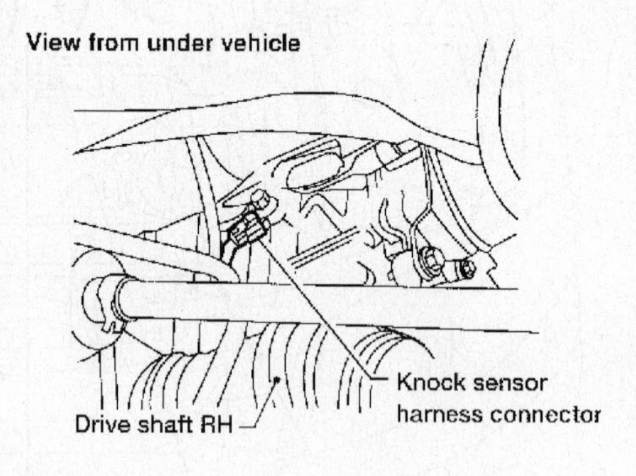

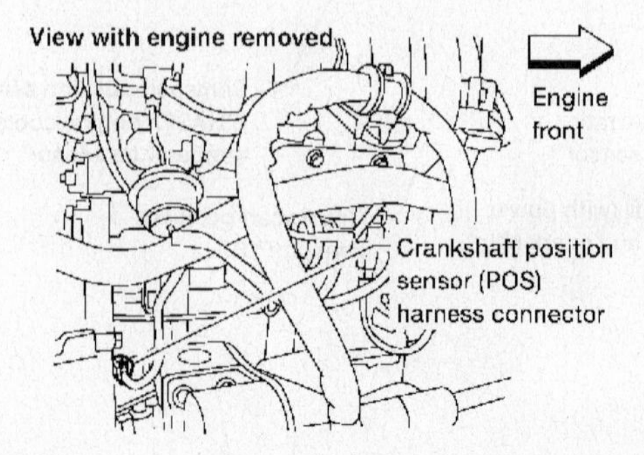

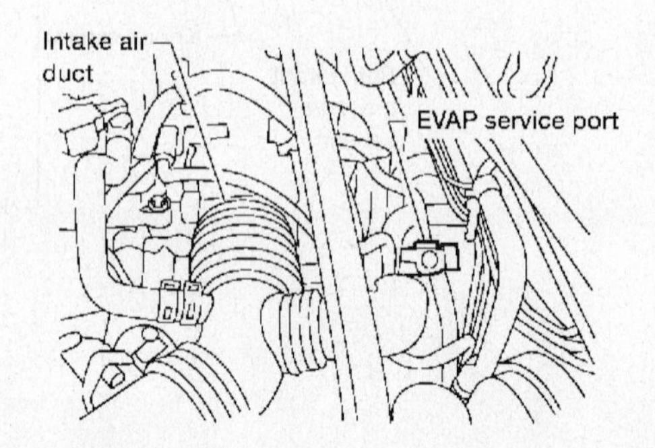

29149_NISS_G0167

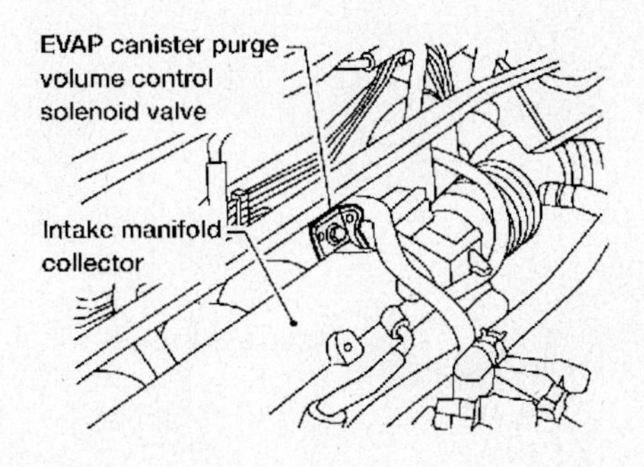

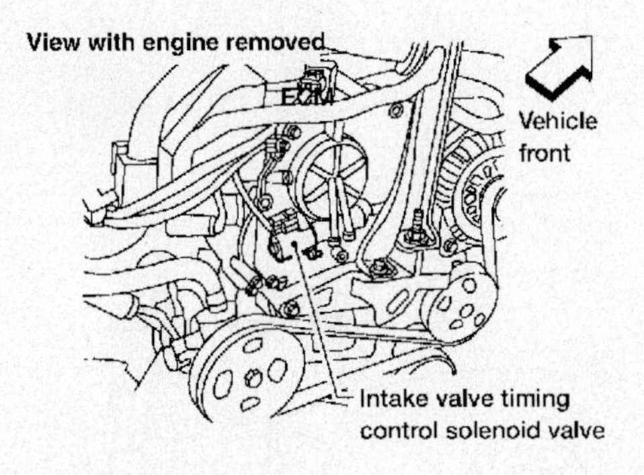

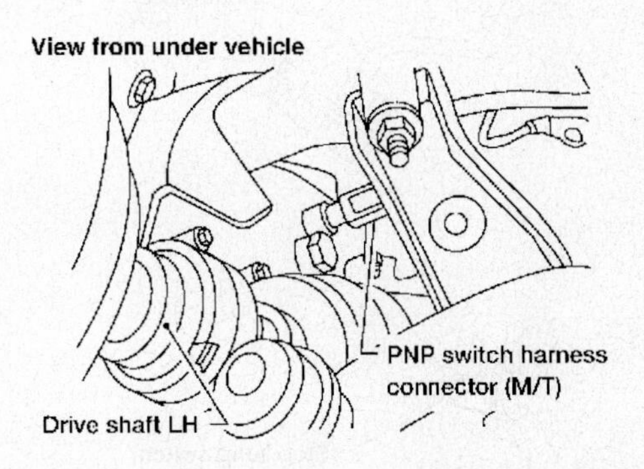

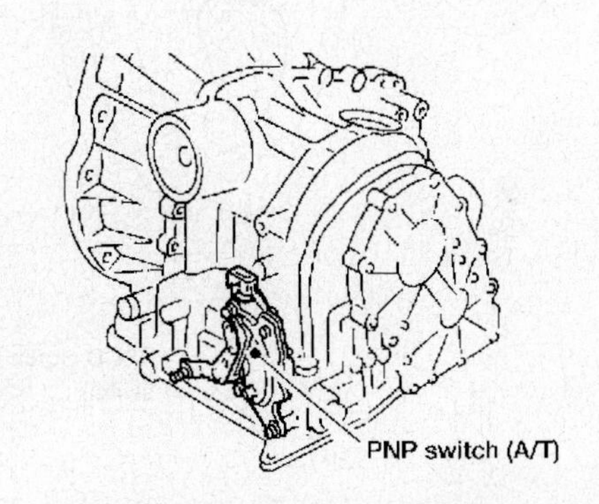

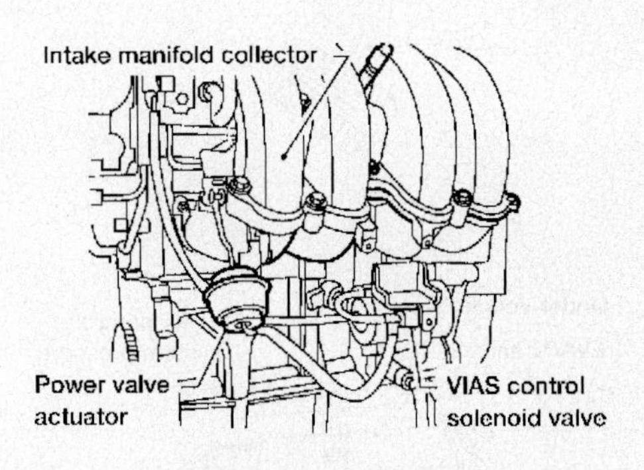

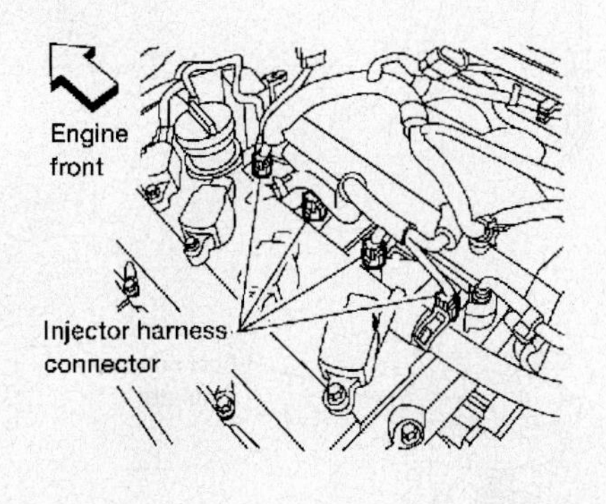

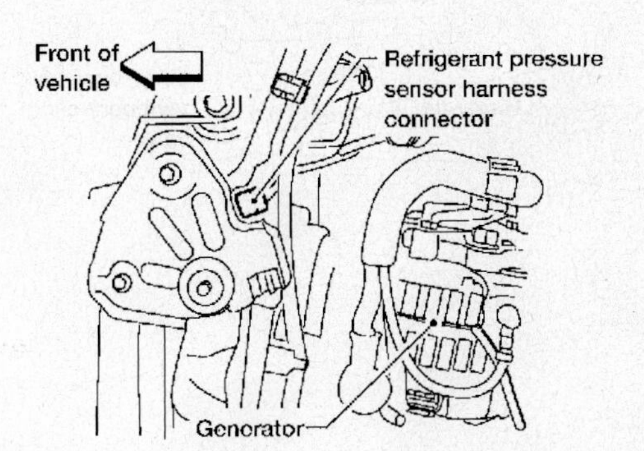

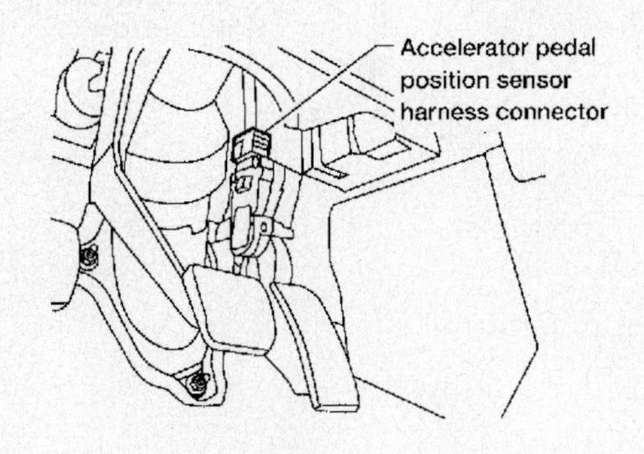

29149_NISS_G0168

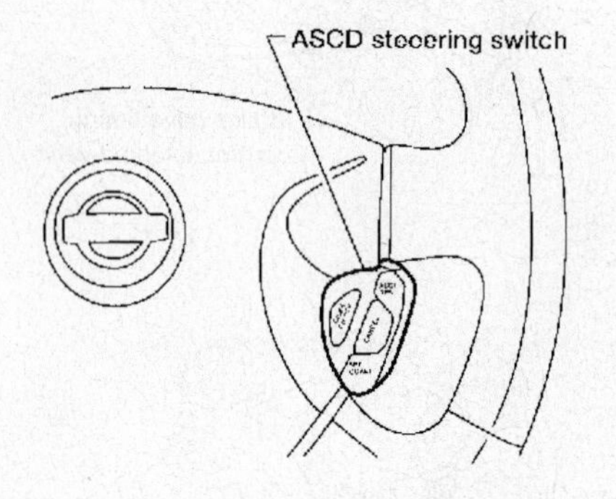

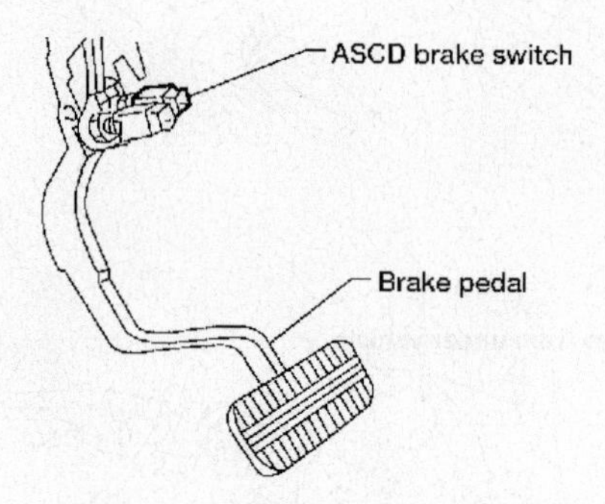

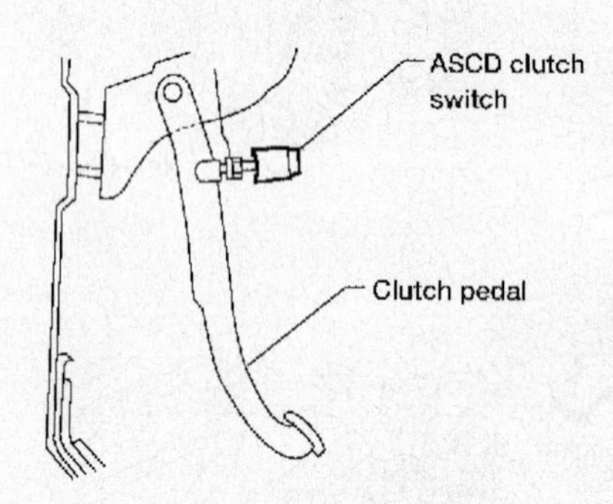

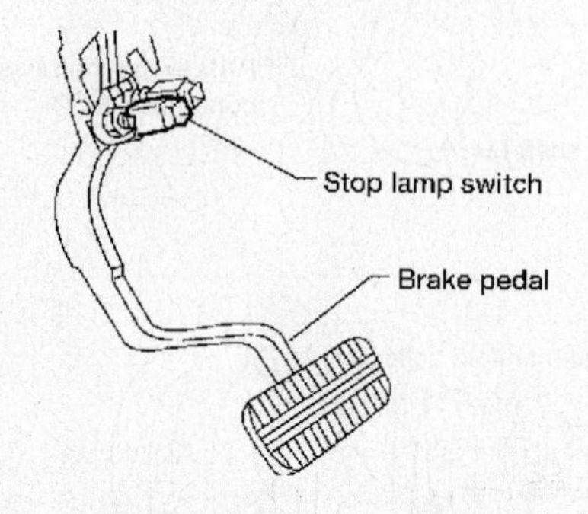

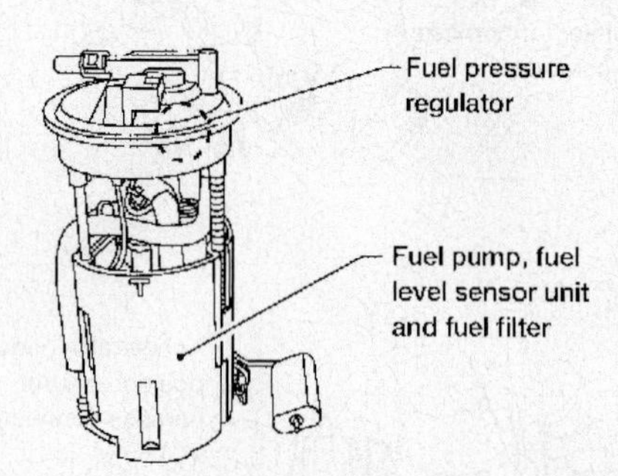

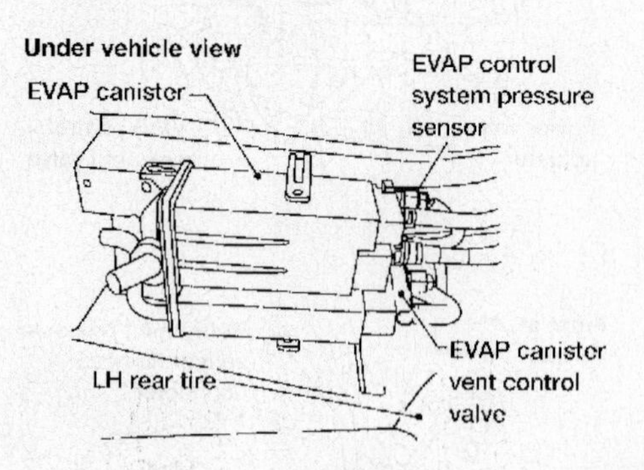

29149_NISS_G0169

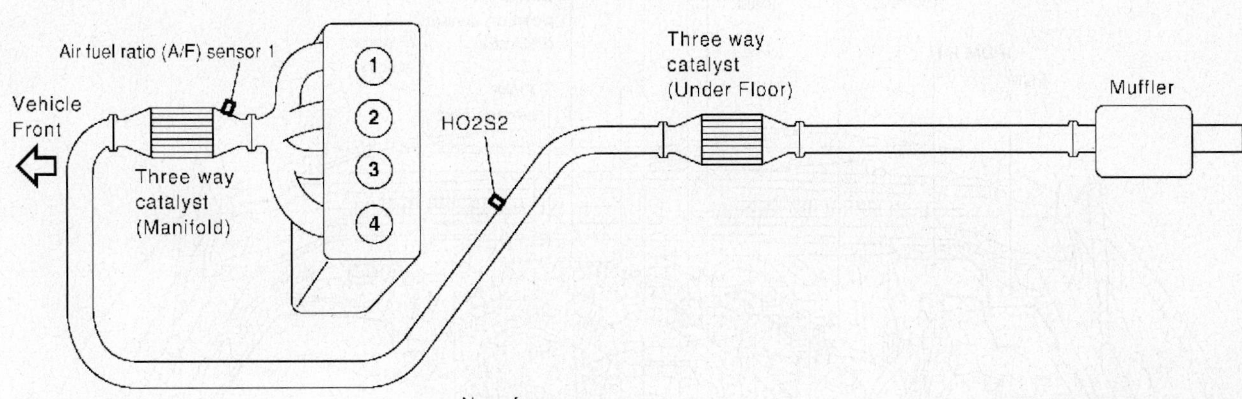

No. of sensor
Location of a sensor in relation the engine air flow, starting from the fresh air intake through to the vehicle tailpipe in order numbering 1, 2, 3, and so on

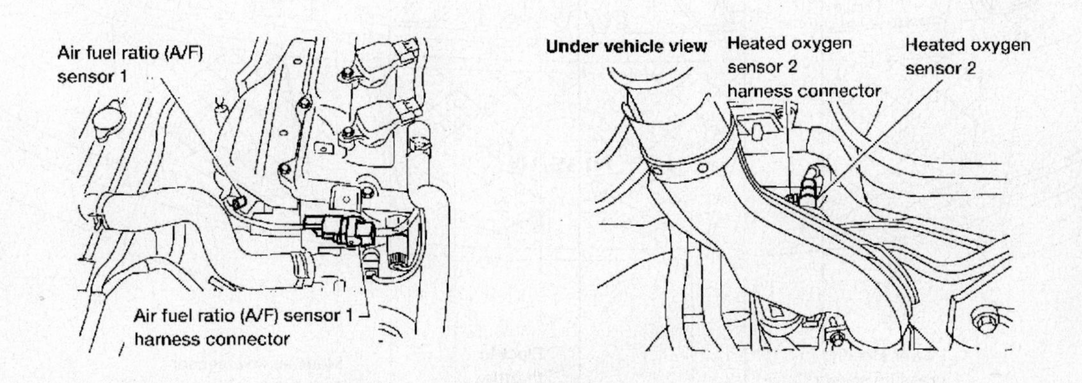

29149_NISS_G0170

TITAN

2004–2006

VK56DE V8 ENGINE

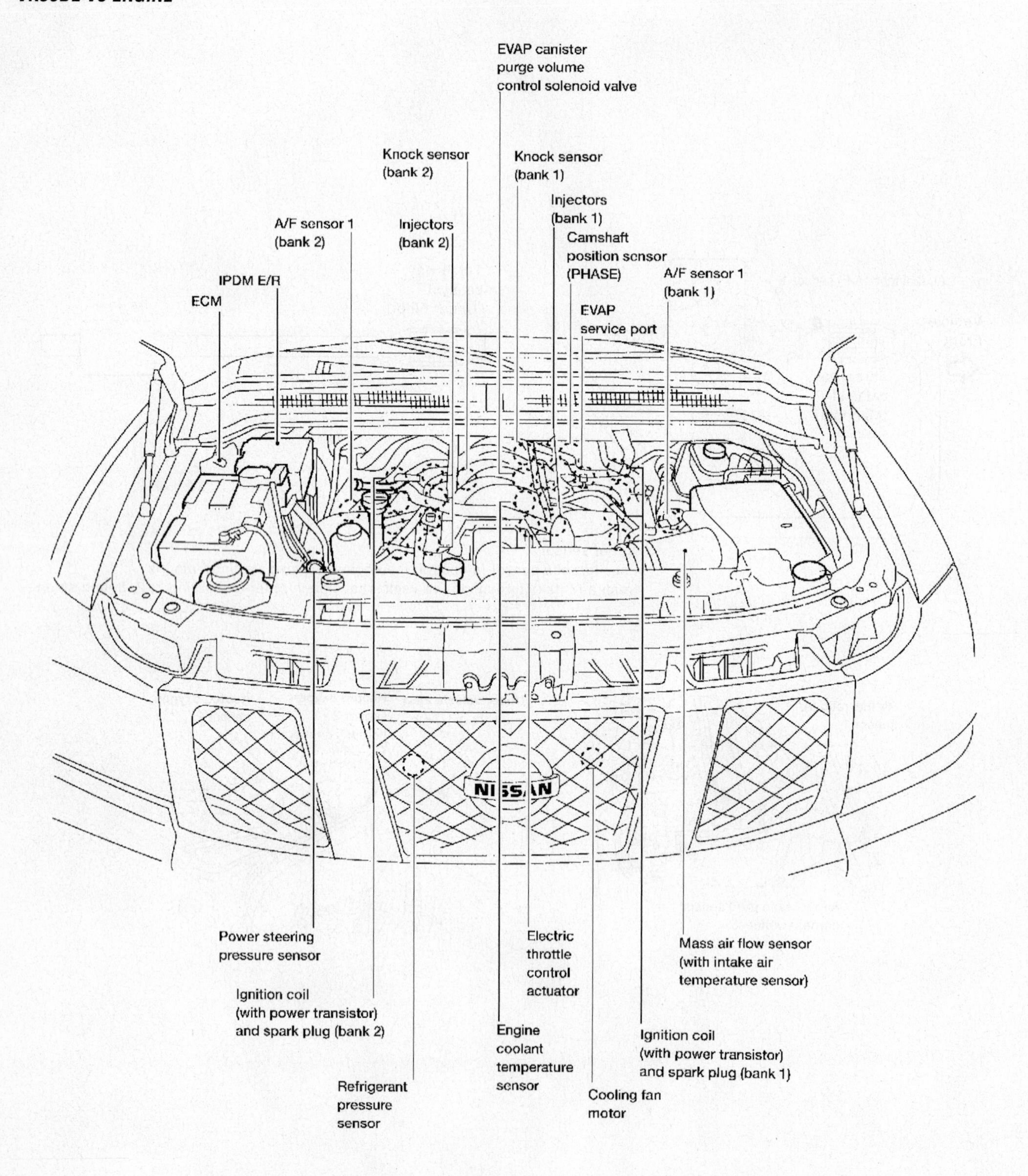
EVAP canister
purge volume
control solenoid valve
Knock sensor
(bank 2)
Knock sensor
(bank 1)
Injectors
(bank 1)
A/F sensor 1
(bank 2)
Injectors
(bank 2)
Camshaft
position sensor
(PHASE)
A/F sensor 1
(bank 1)
IPDM E/R
ECM
EVAP
service port
NISSAN
Power steering
pressure sensor
Electric
throttle
control
actuator
Mass air flow sensor
(with intake air
temperature sensor)
Ignition coil
(with power transistor)
and spark plug (bank 2)
Engine
coolant
temperature
sensor
Ignition coil
(with power transistor)
and spark plug (bank 1)
Refrigerant
pressure
sensor
Cooling fan
motor

29149_NISS_G0171

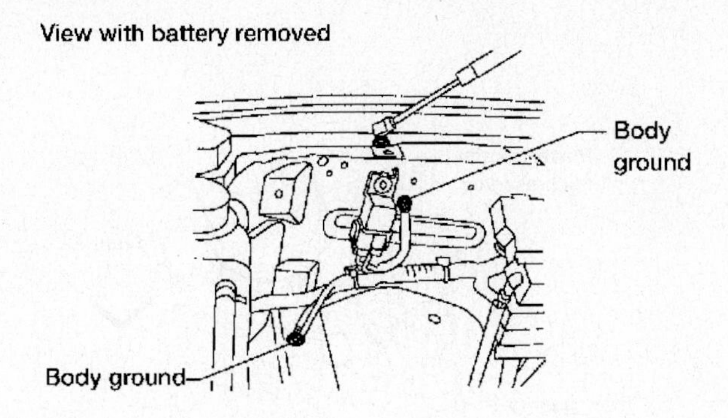
View with battery removed
Body
ground
Body ground

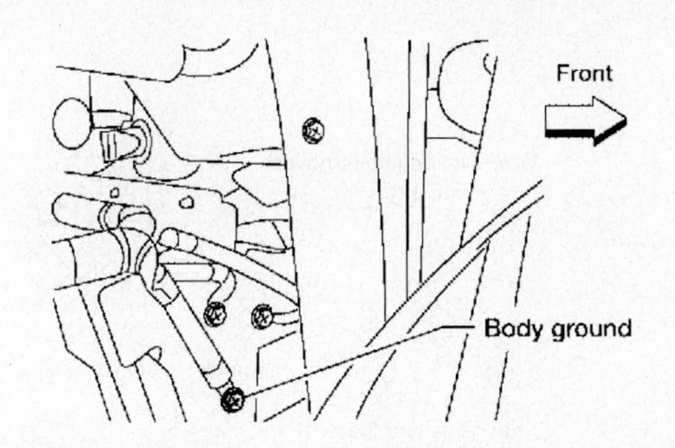
Front
Body ground

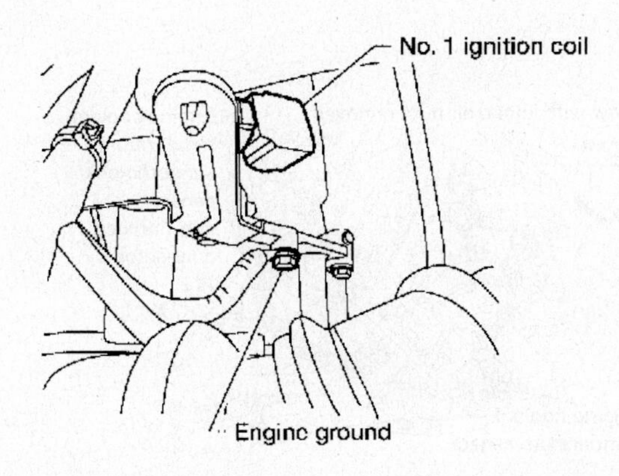
No. 1 ignition coil
Engine ground

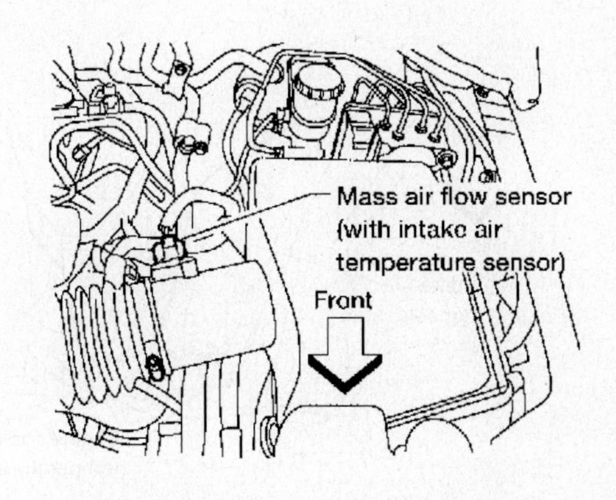
Mass air flow sensor
(with intake air
temperature sensor)
Front

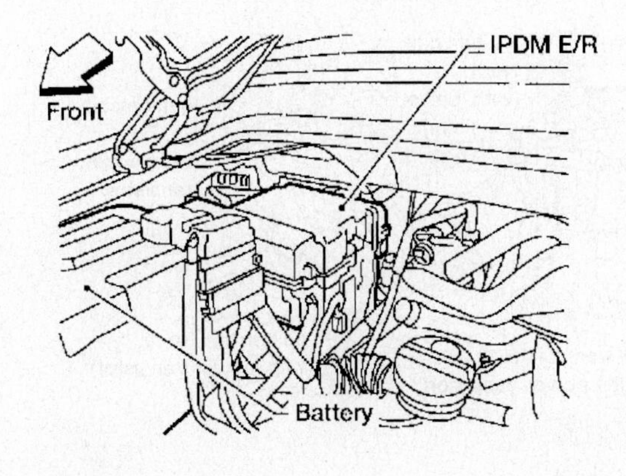
IPDM E/R
Front
Battery

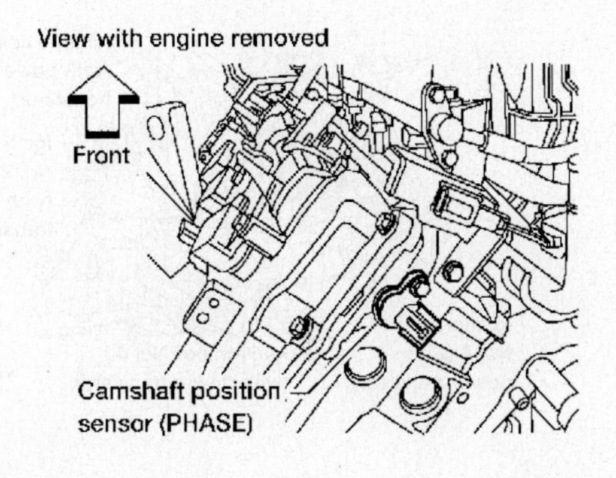
View with engine removed
Front
Camshaft position
sensor (PHASE)

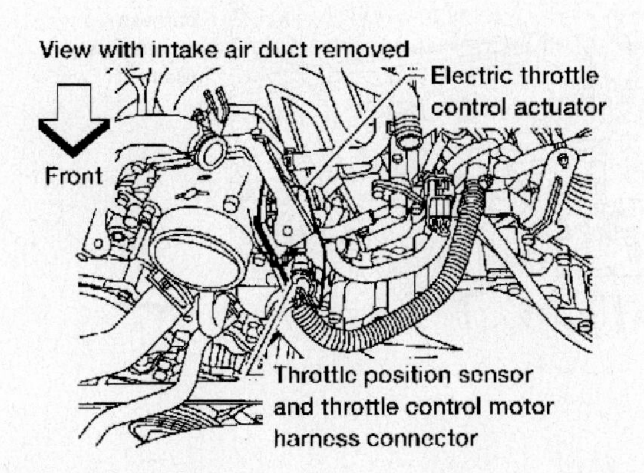
View with intake air duct removed
Electric throttle
control actuator
Front
Throttle position sensor
and throttle control motor
harness connector

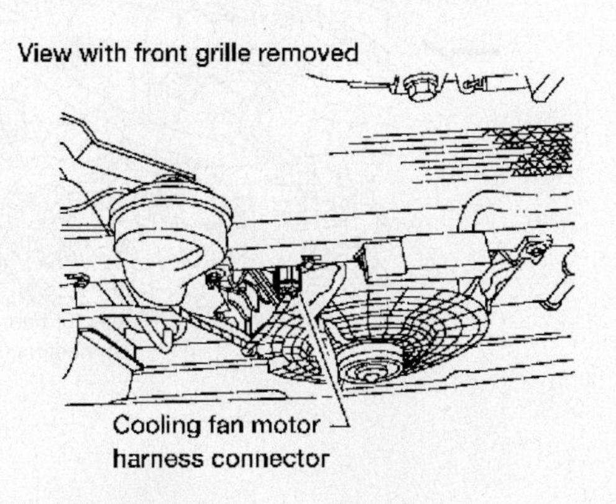
View with front grille removed
Cooling fan motor
harness connector

29149_NISS_G0172

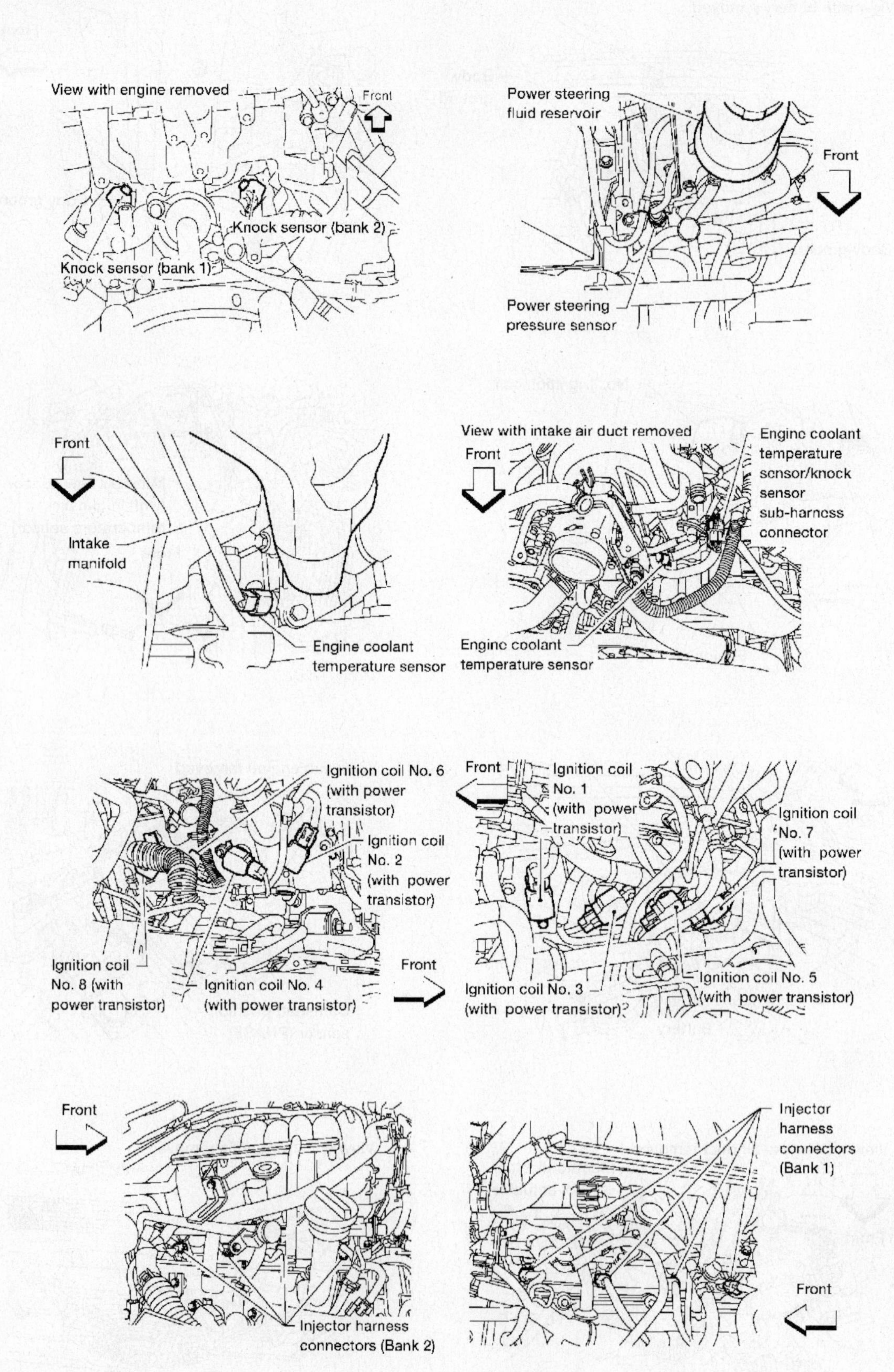

29149_NISS_G0173

View with engine cover removed

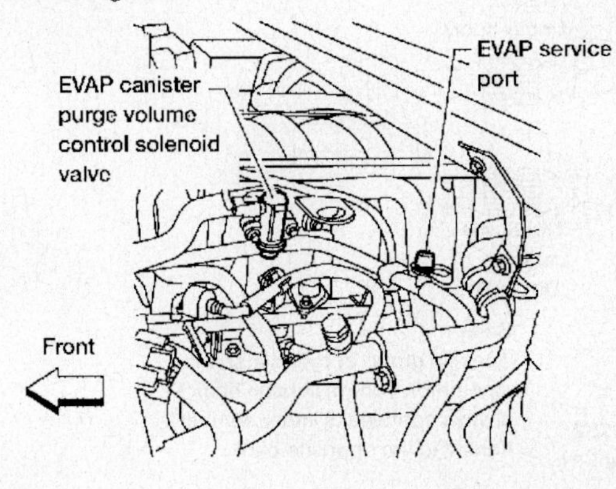

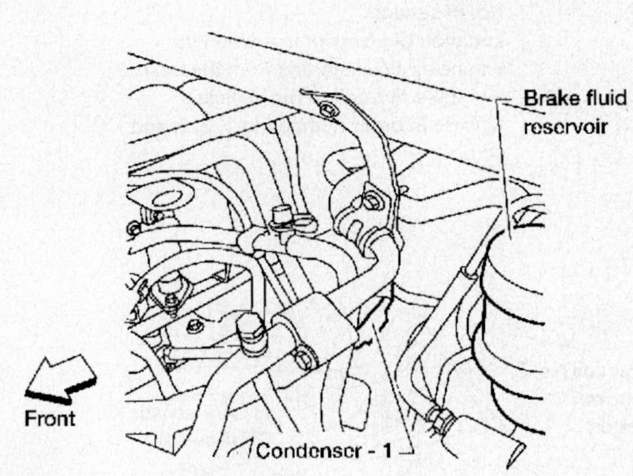

View with front grille removed

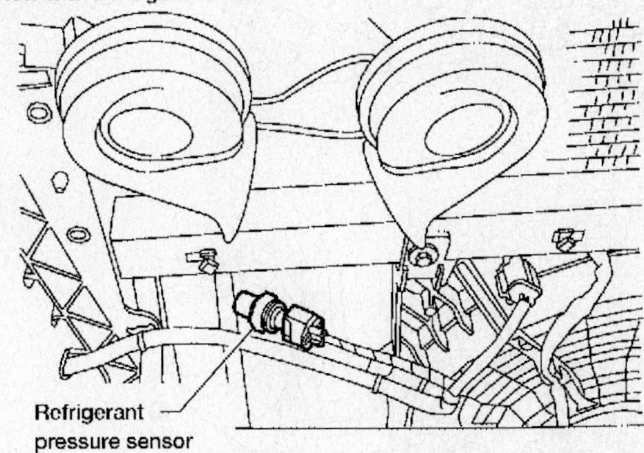

View from under the vehicle

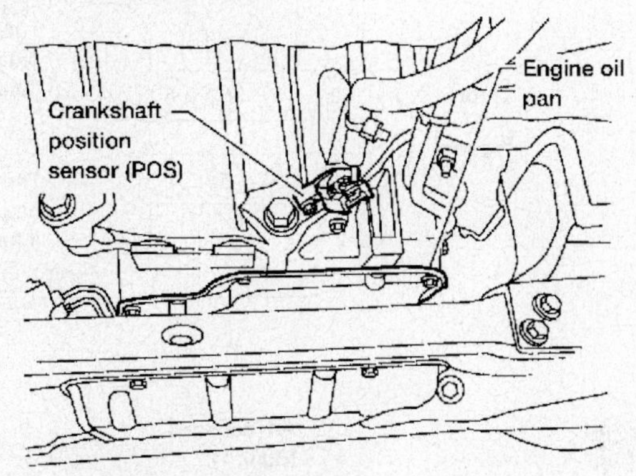

View with fuel tank removed

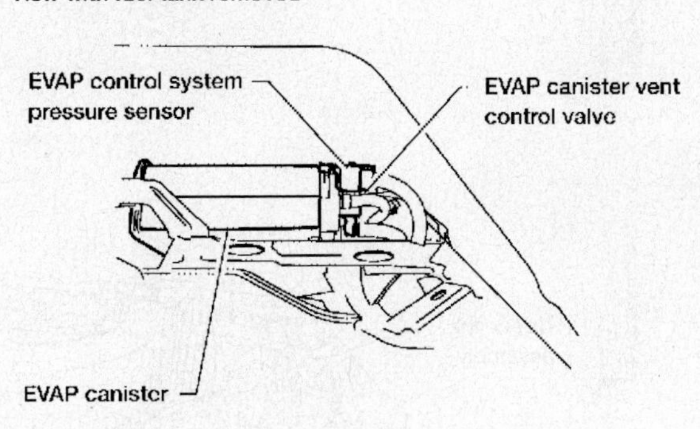

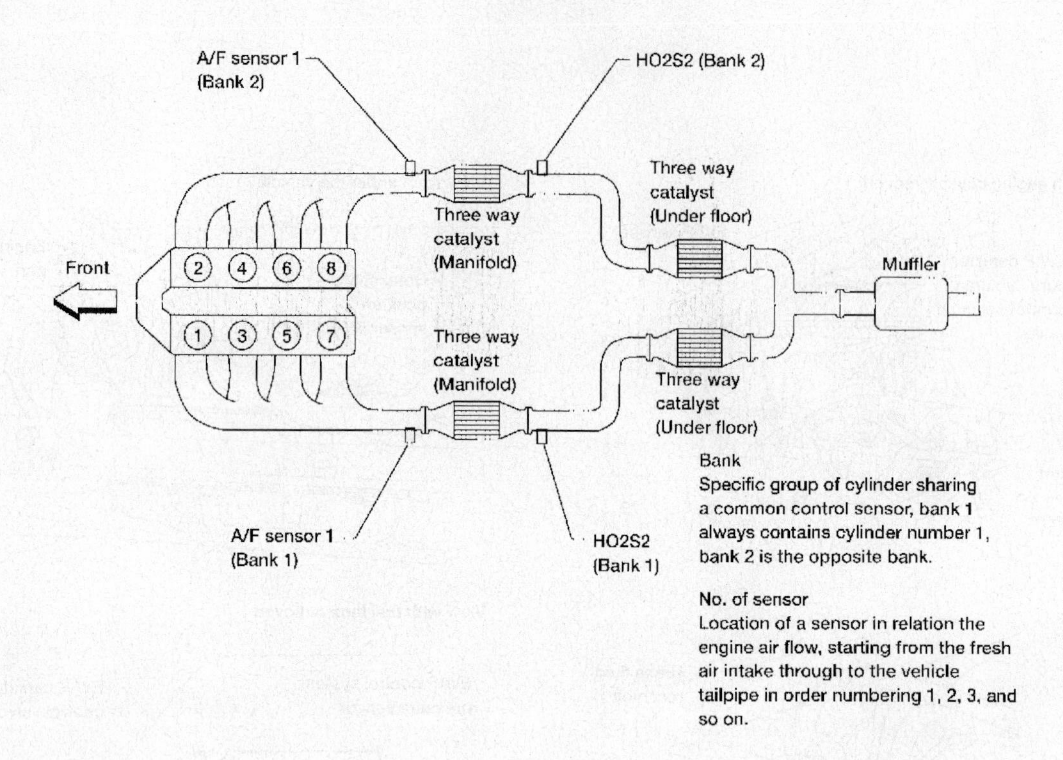
A/F sensor 1
(Bank 2)
HO2S2 (Bank 2)
Three way
catalyst
(Under floor)
Three way
catalyst
(Manifold)
Front
Muffler
Three way
catalyst
(Manifold)
Three way
catalyst
(Under floor)
A/F sensor 1
(Bank 1)
HO2S2
(Bank 1)
Bank
Specific group of cylinder sharing a common control sensor, bank 1 always contains cylinder number 1, bank 2 is the opposite bank.
No. of sensor
Location of a sensor in relation the engine air flow, starting from the fresh air intake through to the vehicle tailpipe in order numbering 1, 2, 3, and so on.

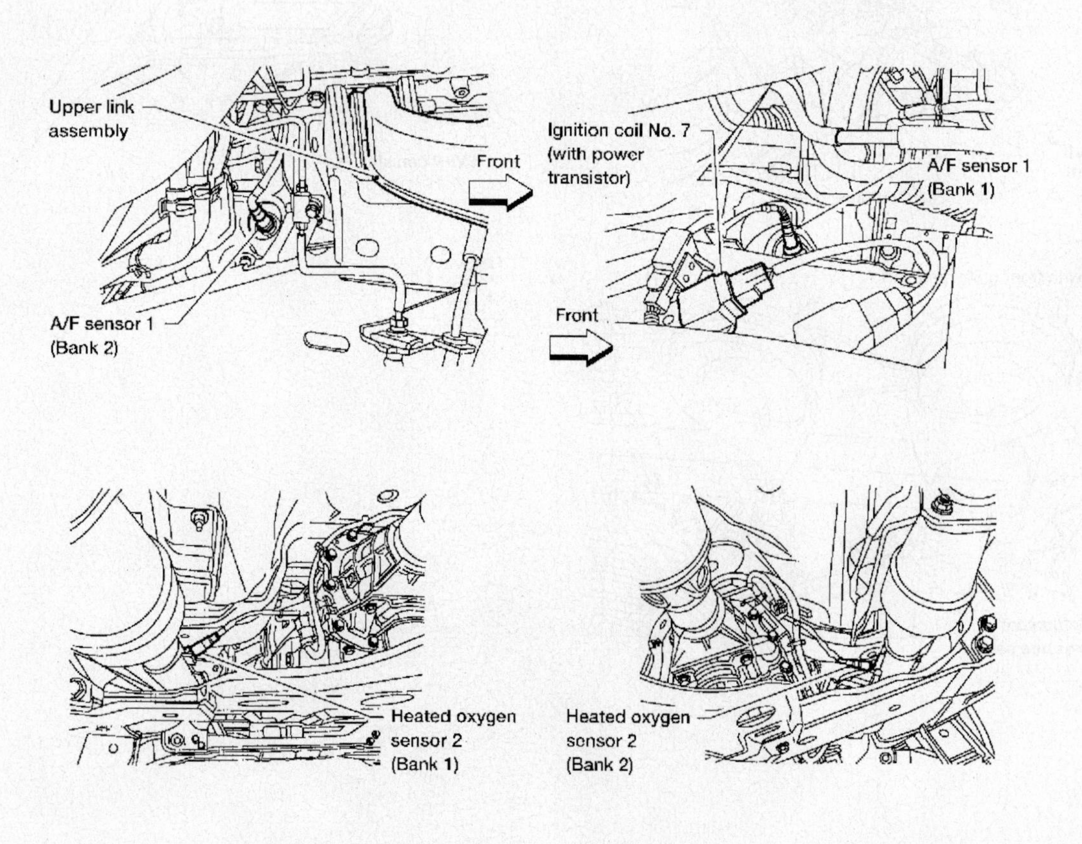
Upper link
assembly
Front
A/F sensor 1
(Bank 2)
Ignition coil No. 7
(with power
transistor)
A/F sensor 1
(Bank 1)
Front
Heated oxygen
sensor 2
(Bank 1)
Heated oxygen
sensor 2
(Bank 2)

View with battery removed

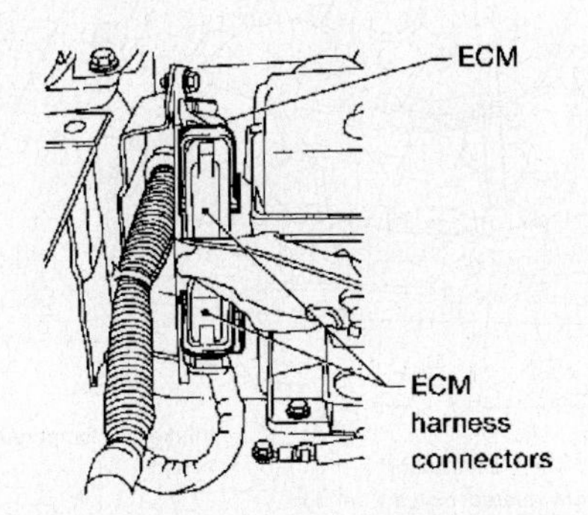

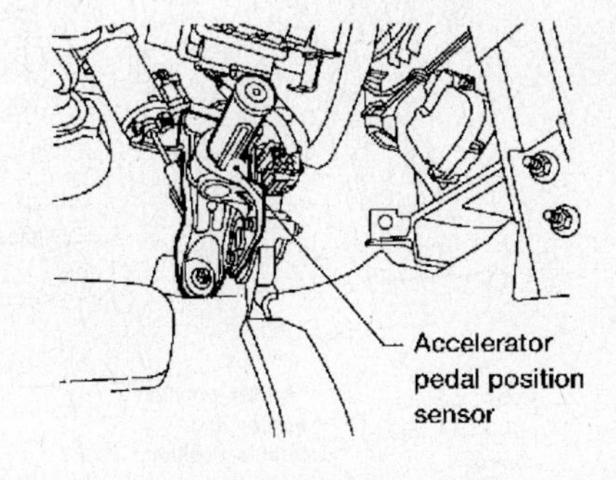

View with cowl top extension removed

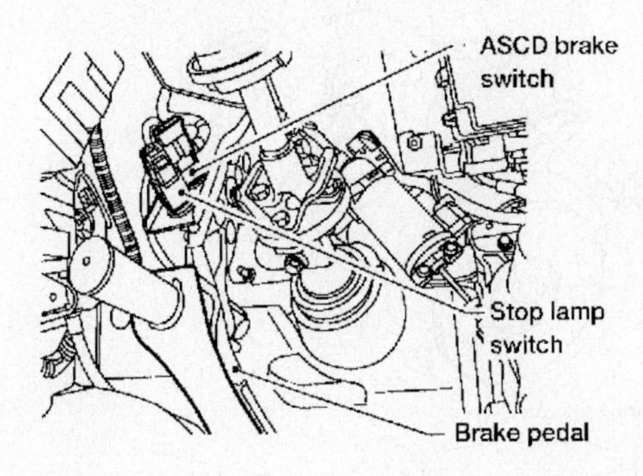

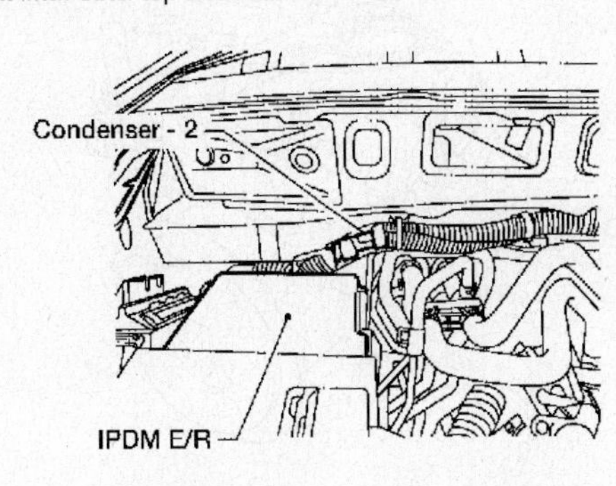

View with fuel tank removed

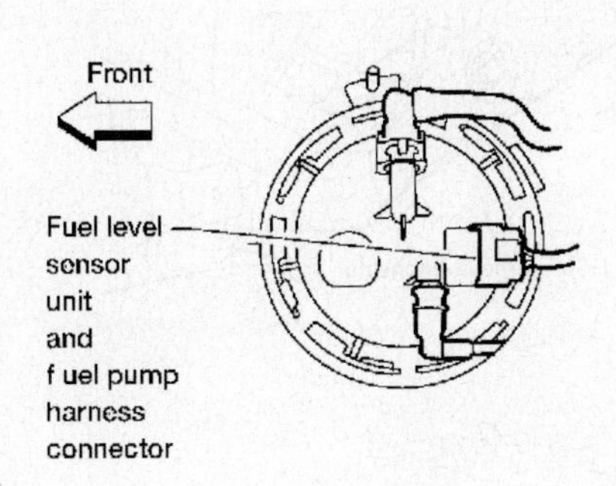

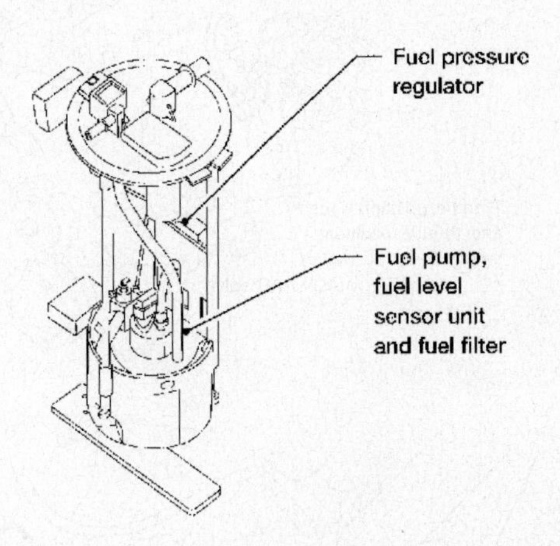

29149_NISS_G0176

XTERRA

2000–2004

KA24DE I4 ENGINE

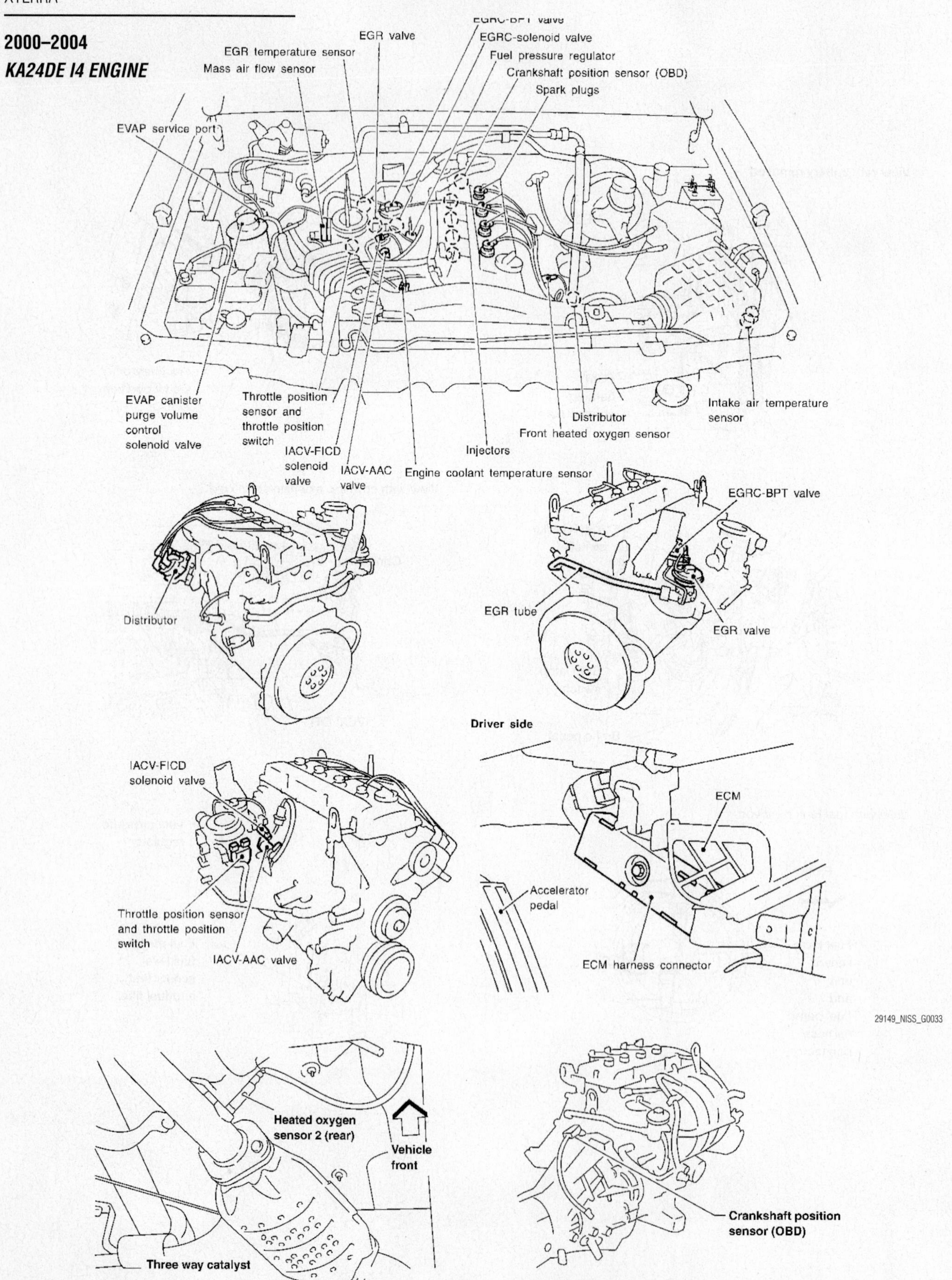
EGRC-BPT valve
EGR valve
EGRC-solenoid valve
EGR temperature sensor
Fuel pressure regulator
Mass air flow sensor
Crankshaft position sensor (OBD)
Spark plugs
EVAP service port
EVAP canister purge volume control solenoid valve
Throttle position sensor and throttle position switch
IACV-FICD solenoid valve
IACV-AAC valve
Engine coolant temperature sensor
Injectors
Distributor
Front heated oxygen sensor
Intake air temperature sensor
Distributor
EGRC-BPT valve
EGR tube
EGR valve
Driver side
IACV-FICD solenoid valve
ECM
Throttle position sensor and throttle position switch
Accelerator pedal
IACV-AAC valve
ECM harness connector
Heated oxygen sensor 2 (rear)
Vehicle front
Three way catalyst
Crankshaft position sensor (OBD)

29149_NISS_G0033

29149_NISS_G0034

2000–2004

VG33E V6 ENGINE

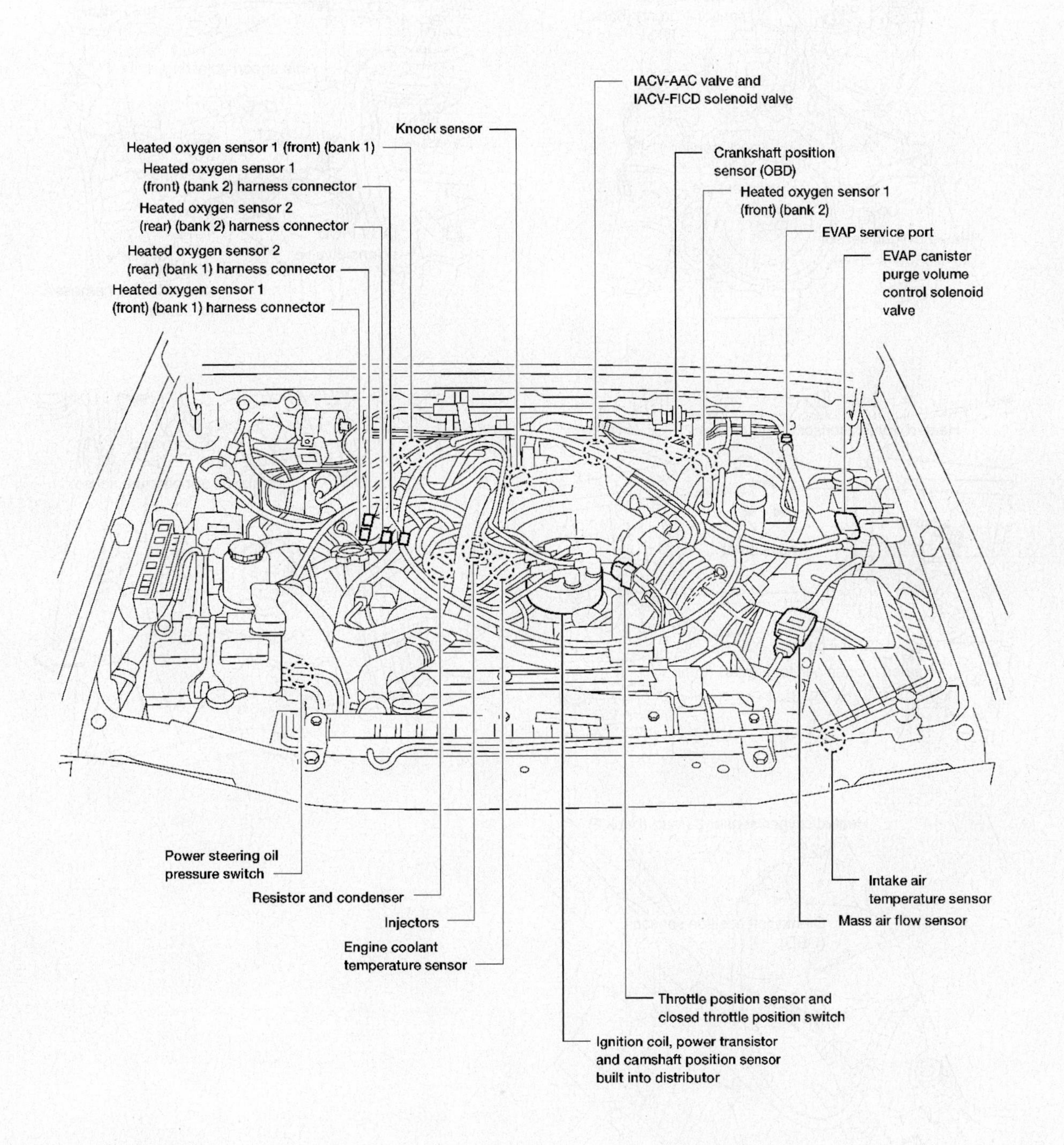
IACV-AAC valve and
IACV-FICD solenoid valve
Knock sensor
Heated oxygen sensor 1 (front) (bank 1)
Heated oxygen sensor 1
(front) (bank 2) harness connector
Heated oxygen sensor 2
(rear) (bank 2) harness connector
Heated oxygen sensor 2
(rear) (bank 1) harness connector
Heated oxygen sensor 1
(front) (bank 1) harness connector
Crankshaft position
sensor (OBD)
Heated oxygen sensor 1
(front) (bank 2)
EVAP service port
EVAP canister
purge volume
control solenoid
valve
Power steering oil
pressure switch
Resistor and condenser
Injectors
Engine coolant
temperature sensor
Intake air
temperature sensor
Mass air flow sensor
Throttle position sensor and
closed throttle position switch
Ignition coil, power transistor
and camshaft position sensor
built into distributor

29149_NISS_G0035

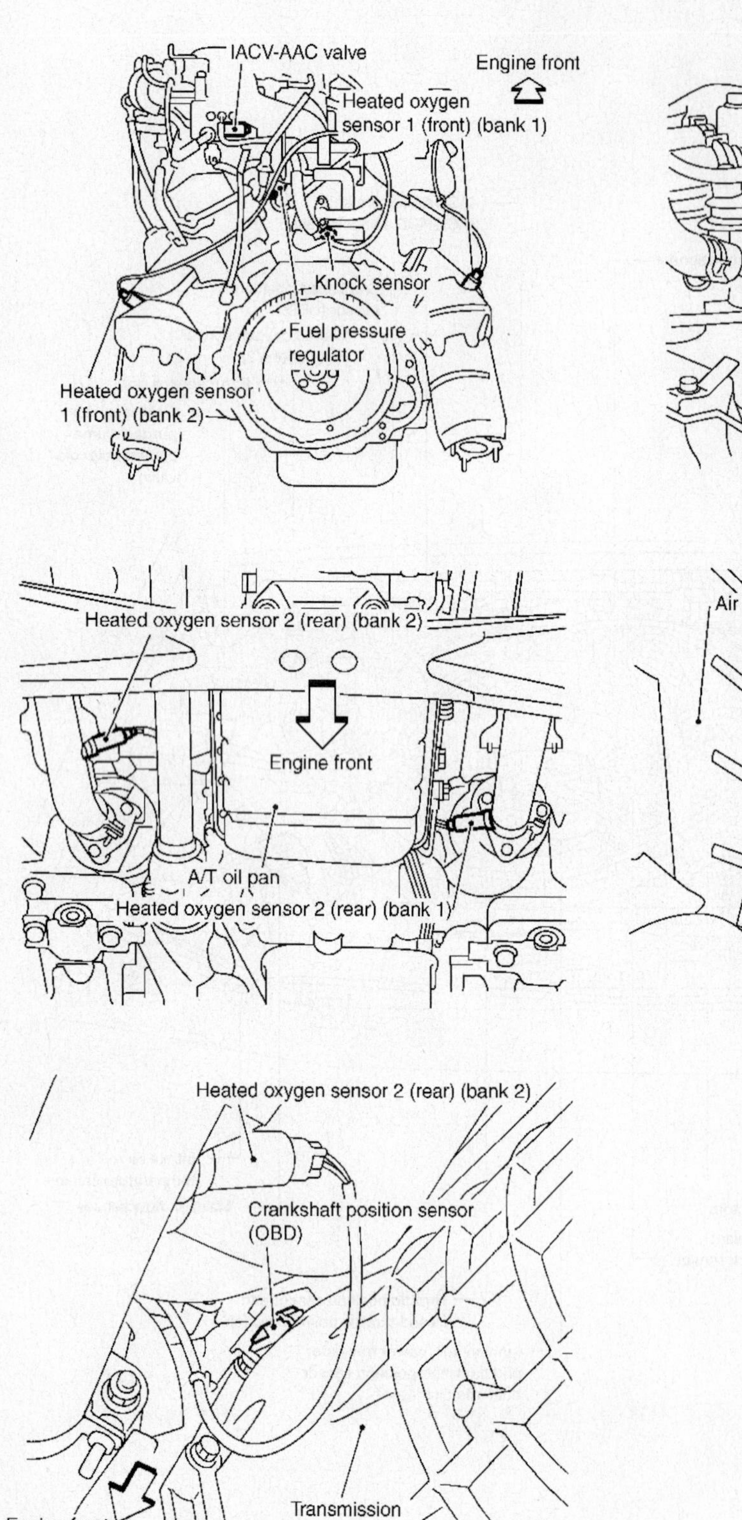
IACV-AAC valve
Engine front
Heated oxygen sensor 1 (front) (bank 1)
Knock sensor
Fuel pressure regulator
Heated oxygen sensor 1 (front) (bank 2)
Heated oxygen sensor 2 (rear) (bank 2)
Engine front
A/T oil pan
Heated oxygen sensor 2 (rear) (bank 1)
Heated oxygen sensor 2 (rear) (bank 2)
Crankshaft position sensor (OBD)
Engine front
Transmission

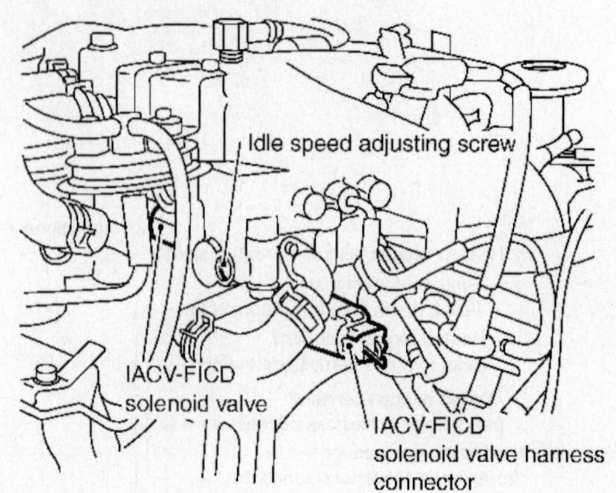
Idle speed adjusting screw
IACV-FICD solenoid valve
IACV-FICD solenoid valve harness connector

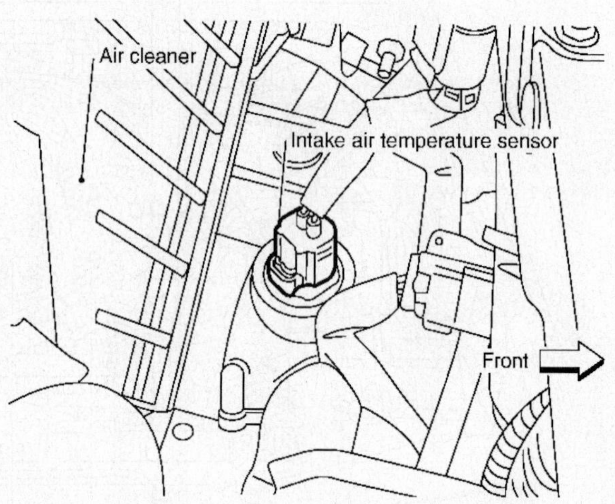
Air cleaner
Intake air temperature sensor
Front

29149_NISS_G0036

2002–2004

VG33ER V6 ENGINE

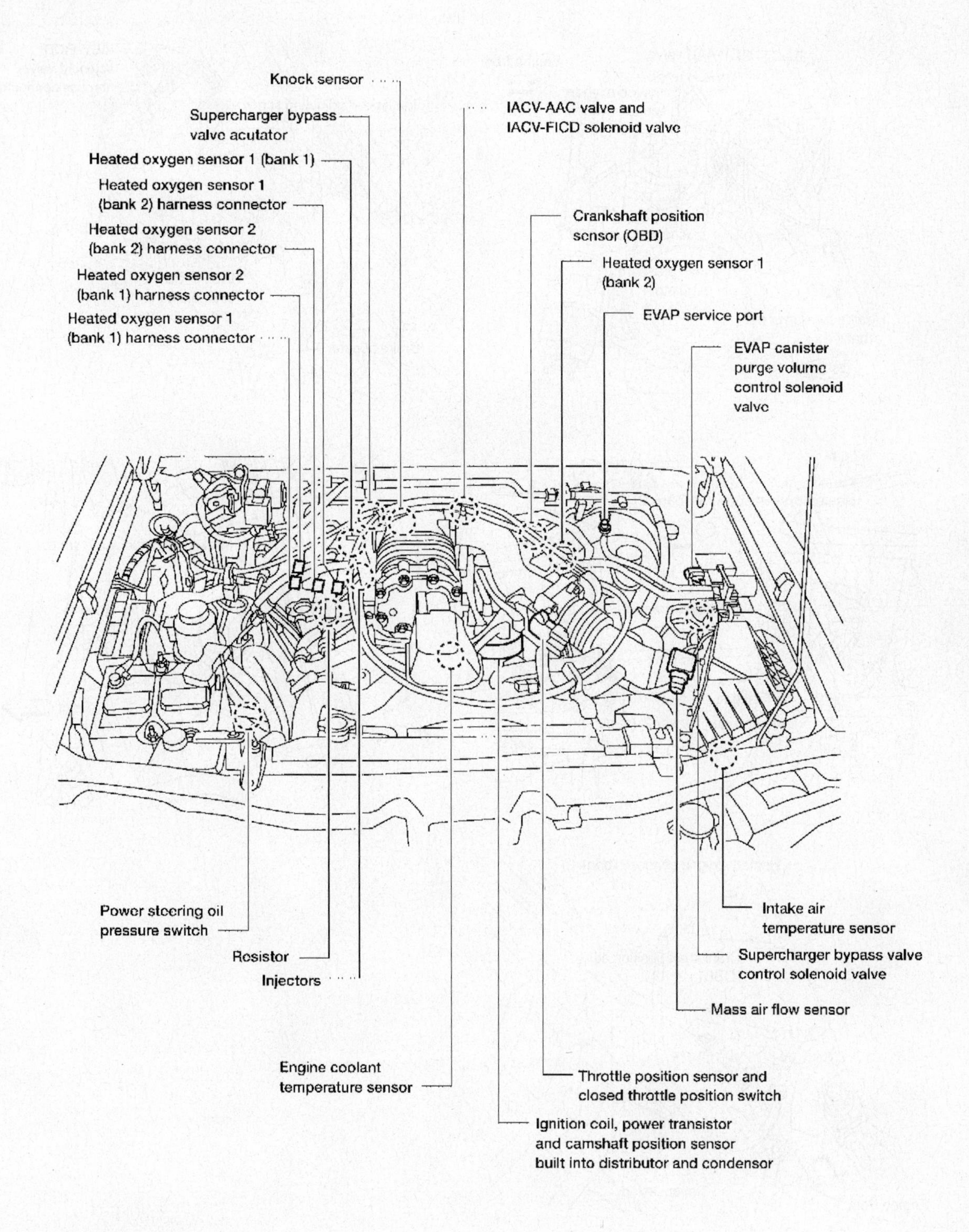

29149_NISS_G0177

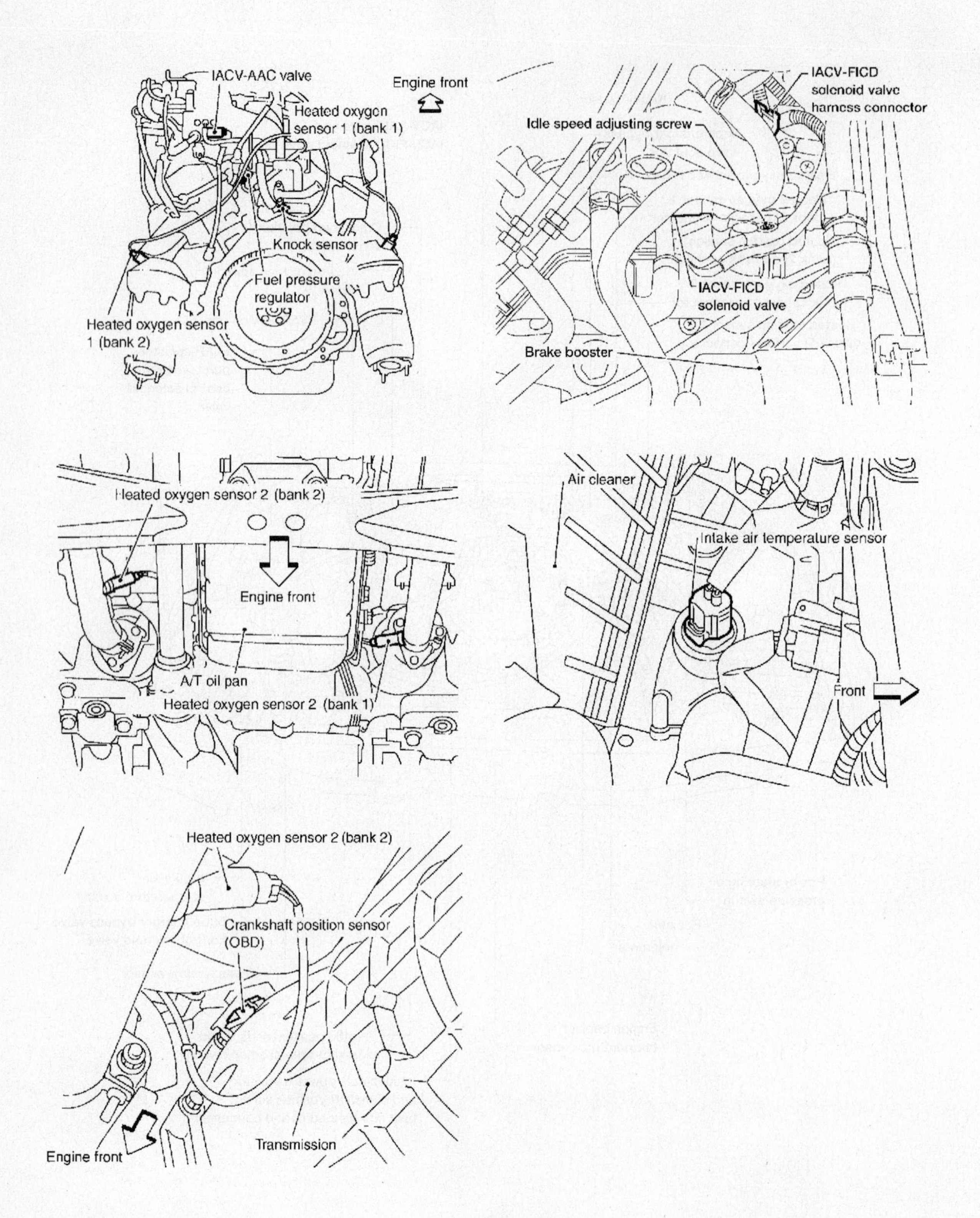
IACV-AAC valve
Engine front
Heated oxygen sensor 1 (bank 1)
Knock sensor
Fuel pressure regulator
Heated oxygen sensor 1 (bank 2)
IACV-FICD solenoid valve harness connector
Idle speed adjusting screw
IACV-FICD solenoid valve
Brake booster
Heated oxygen sensor 2 (bank 2)
Engine front
A/T oil pan
Heated oxygen sensor 2 (bank 1)
Air cleaner
Intake air temperature sensor
Front
Heated oxygen sensor 2 (bank 2)
Crankshaft position sensor (OBD)
Engine front
Transmission

29149_NISS_G0178

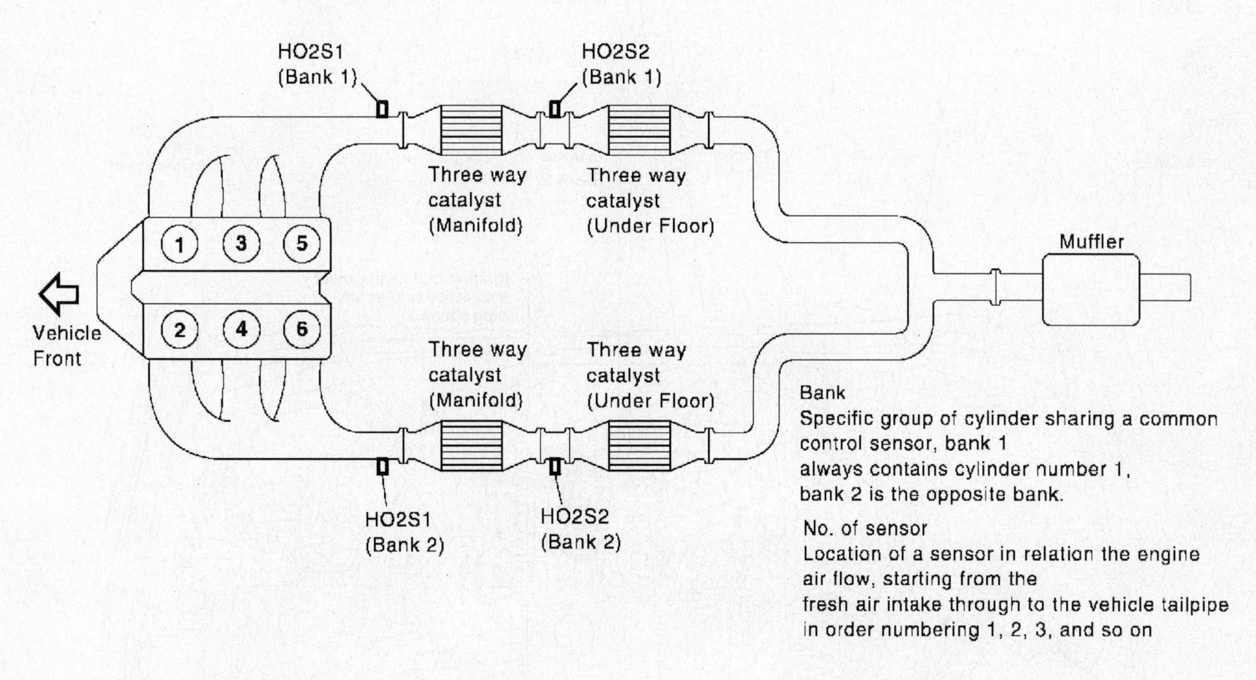
HO2S1
(Bank 1)
HO2S2
(Bank 1)
Three way
catalyst
(Manifold)
Three way
catalyst
(Under Floor)
Muffler
1
3
5
2
4
6
Vehicle
Front
Three way
catalyst
(Manifold)
Three way
catalyst
(Under Floor)
HO2S1
(Bank 2)
HO2S2
(Bank 2)
Bank
Specific group of cylinder sharing a common
control sensor, bank 1
always contains cylinder number 1,
bank 2 is the opposite bank.
No. of sensor
Location of a sensor in relation the engine
air flow, starting from the
fresh air intake through to the vehicle tailpipe
in order numbering 1, 2, 3, and so on

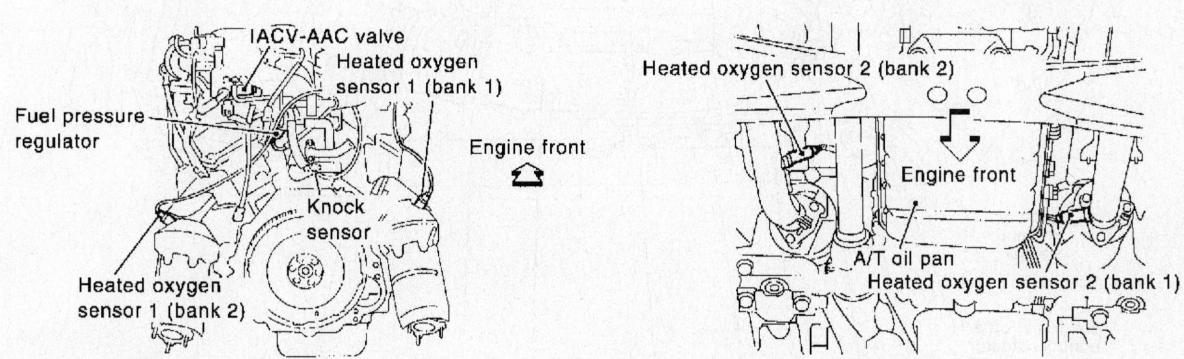
IACV-AAC valve
Heated oxygen
sensor 1 (bank 1)
Fuel pressure
regulator
Engine front
Knock
sensor
Heated oxygen
sensor 1 (bank 2)
Heated oxygen sensor 2 (bank 2)
Engine front
A/T oil pan
Heated oxygen sensor 2 (bank 1)

2005–2006

VQ40DE V6 ENGINE

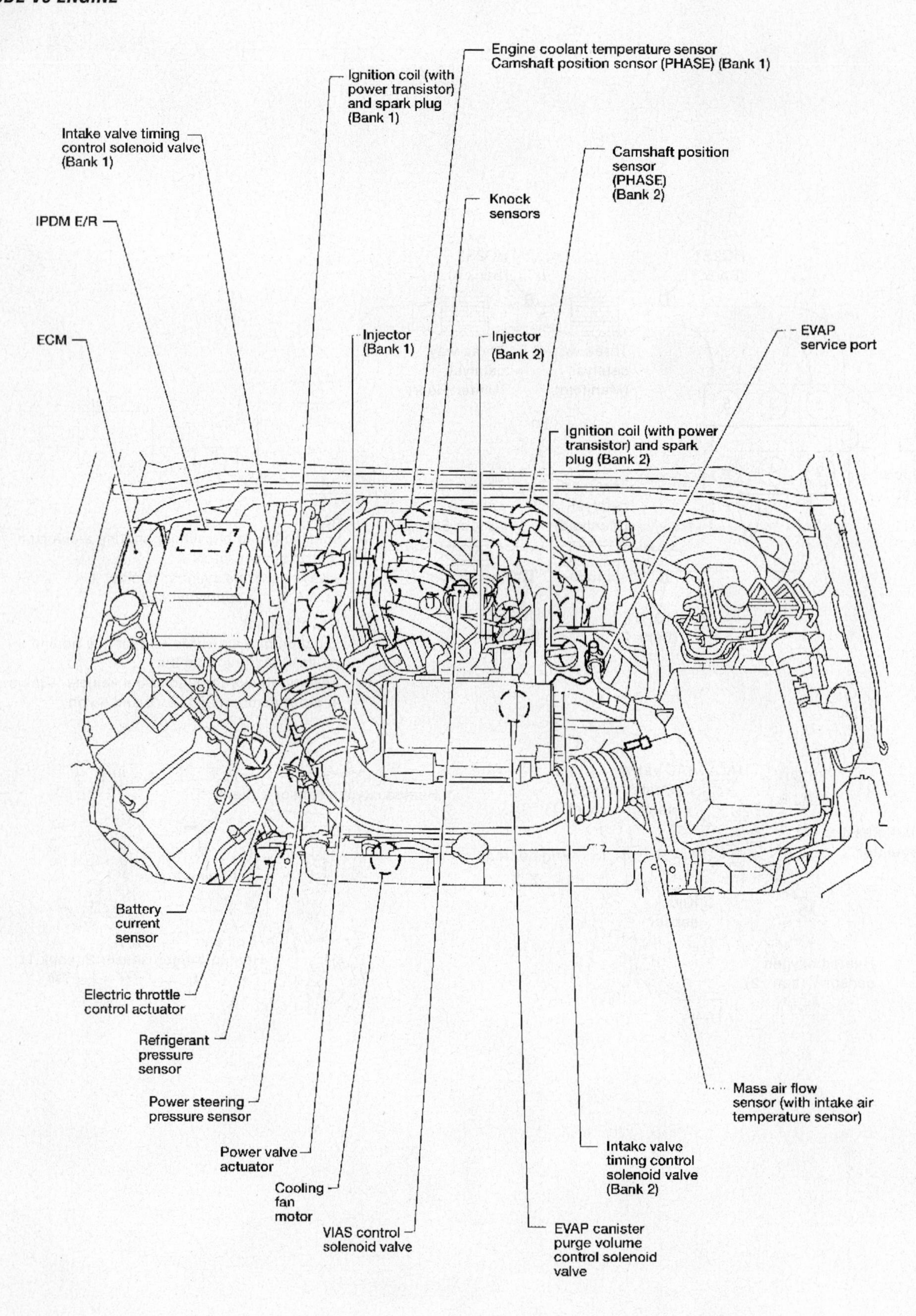
Engine coolant temperature sensor
Camshaft position sensor (PHASE) (Bank 1)
Ignition coil (with power transistor) and spark plug (Bank 1)
Intake valve timing control solenoid valve (Bank 1)
Camshaft position sensor (PHASE) (Bank 2)
Knock sensors
IPDM E/R
ECM
Injector (Bank 1)
Injector (Bank 2)
EVAP service port
Ignition coil (with power transistor) and spark plug (Bank 2)
Battery current sensor
Electric throttle control actuator
Refrigerant pressure sensor
Power steering pressure sensor
Mass air flow sensor (with intake air temperature sensor)
Power valve actuator
Intake valve timing control solenoid valve (Bank 2)
Cooling fan motor
VIAS control solenoid valve
EVAP canister purge volume control solenoid valve

29149_NISS_G0180

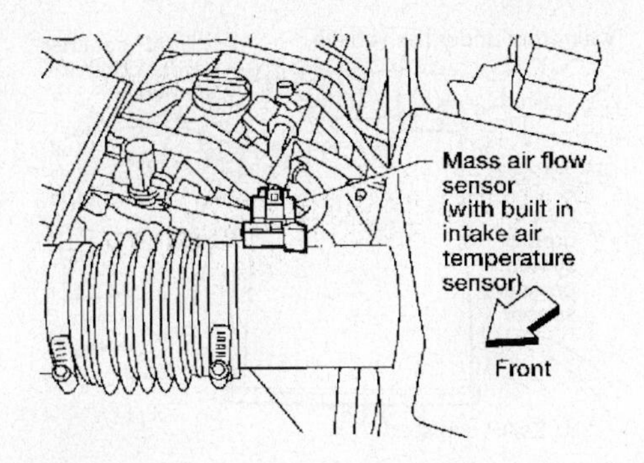
Mass air flow sensor (with built in intake air temperature sensor)
Front

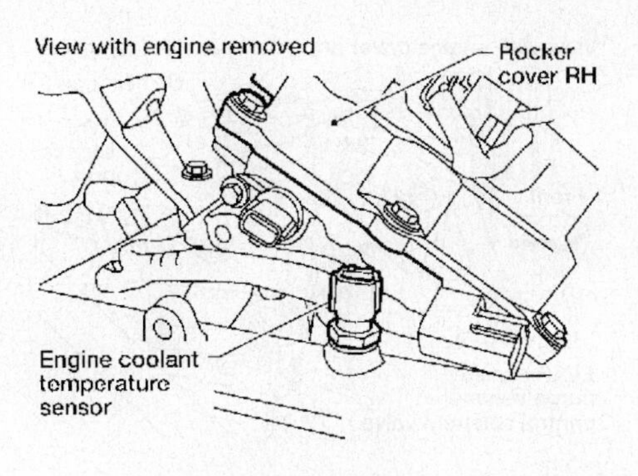
View with engine removed
Rocker cover RH
Engine coolant temperature sensor

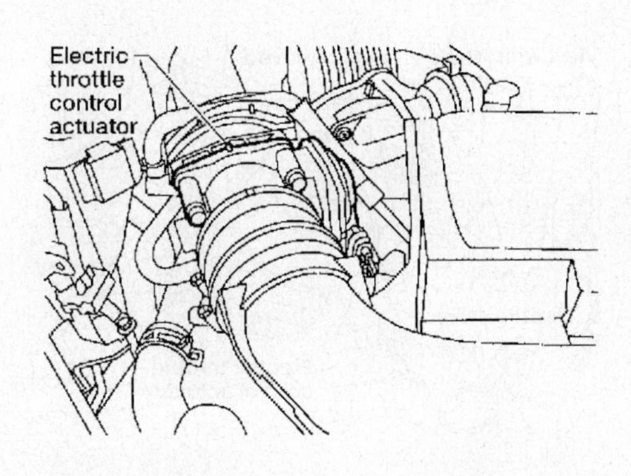
Electric throttle control actuator

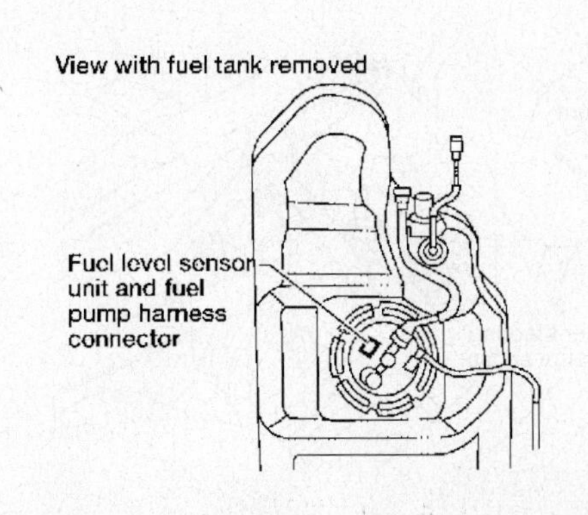
View with fuel tank removed
Fuel level sensor unit and fuel pump harness connector

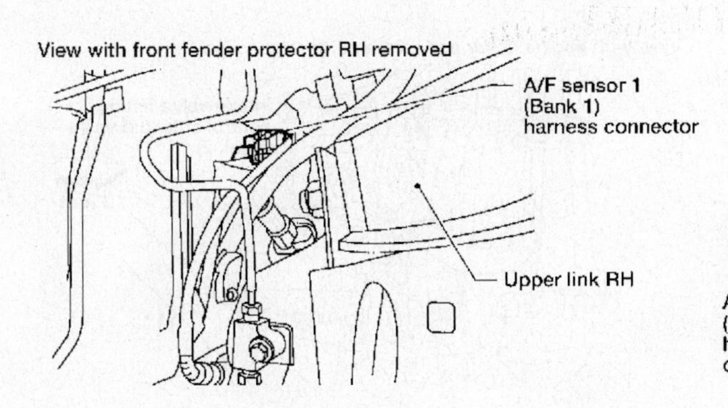
View with front fender protector RH removed
A/F sensor 1 (Bank 1) harness connector
Upper link RH

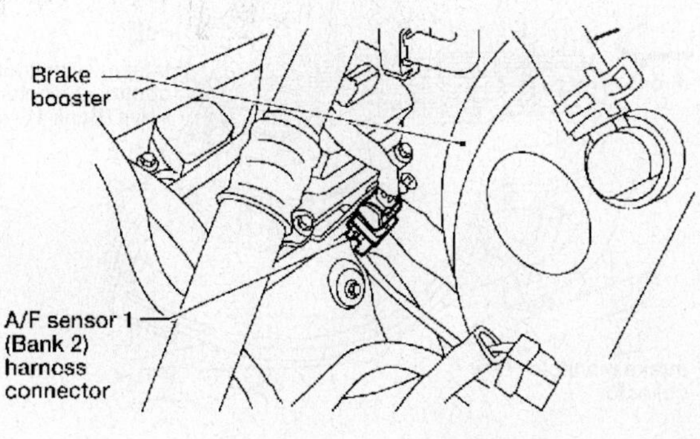
Brake booster
A/F sensor 1 (Bank 2) harness connector

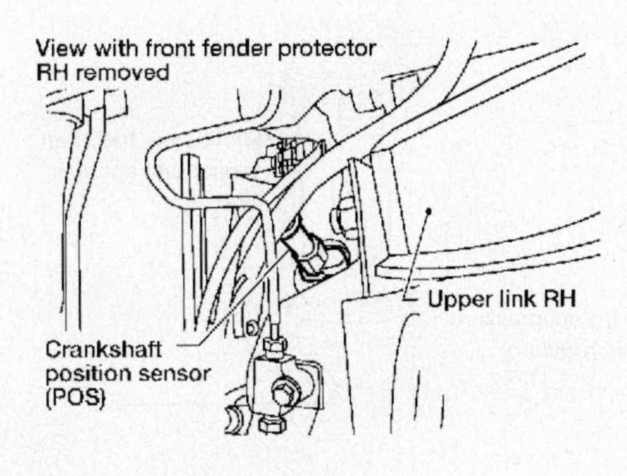
View with front fender protector RH removed
Upper link RH
Crankshaft position sensor (POS)

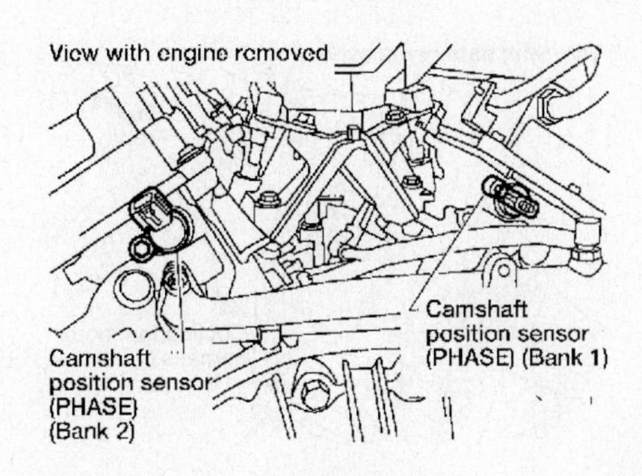
View with engine removed
Camshaft position sensor (PHASE) (Bank 1)
Camshaft position sensor (PHASE) (Bank 2)

29149_NISS_G0181

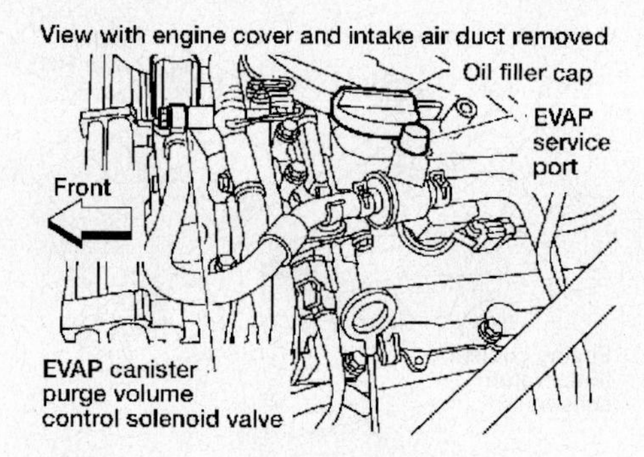
View with engine cover and intake air duct removed
Oil filler cap
EVAP
service
port
Front
EVAP canister
purge volume
control solenoid valve

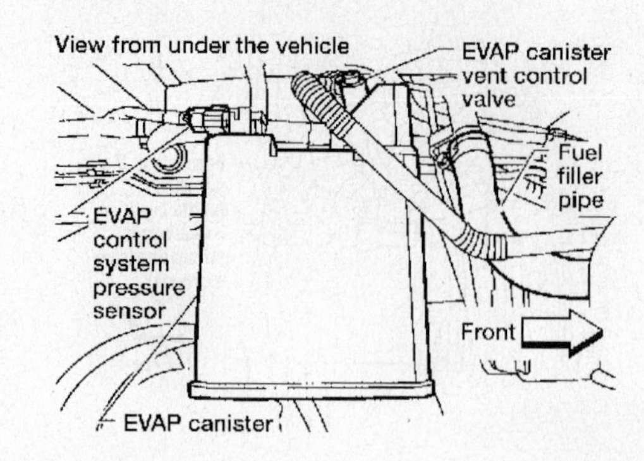
View from under the vehicle
EVAP canister
vent control
valve
Fuel
filler
pipe
EVAP
control
system
pressure
sensor
Front
EVAP canister

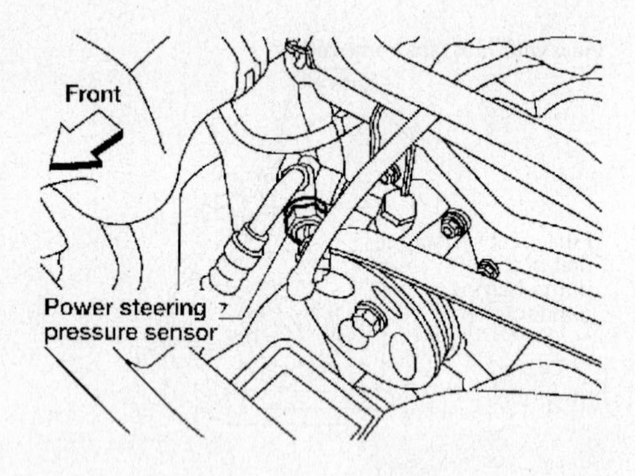
Front
Power steering
pressure sensor

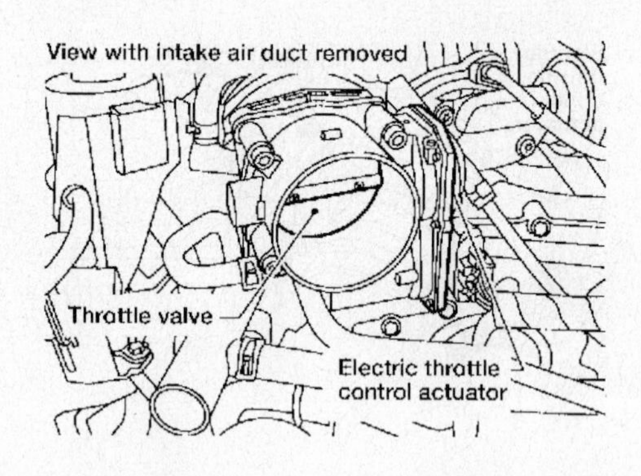
View with intake air duct removed
Throttle valve
Electric throttle
control actuator

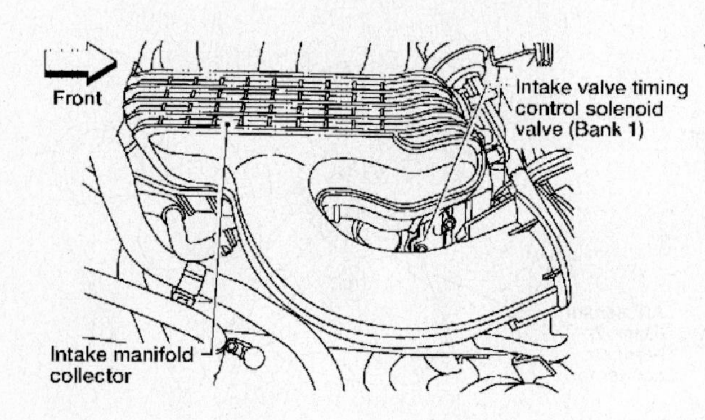
Front
Intake valve timing
control solenoid
valve (Bank 1)
Intake manifold
collector

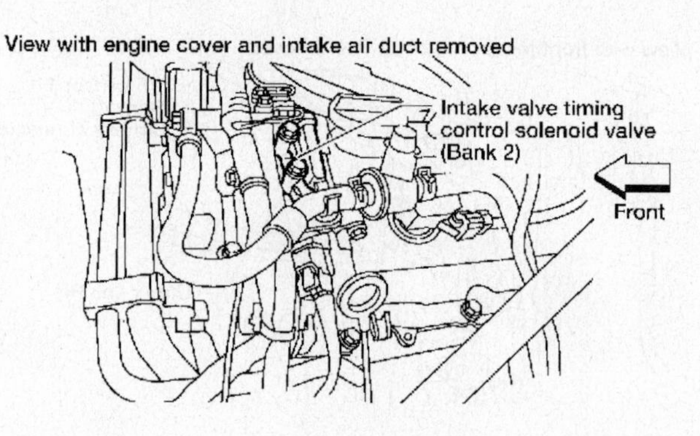
View with engine cover and intake air duct removed
Intake valve timing
control solenoid valve
(Bank 2)
Front

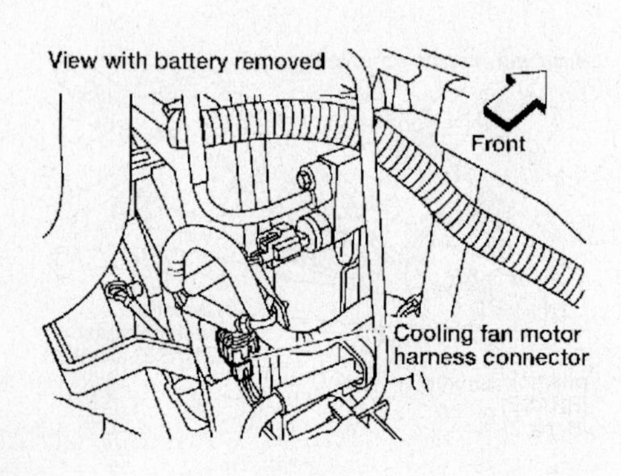
View with battery removed
Front
Cooling fan motor
harness connector

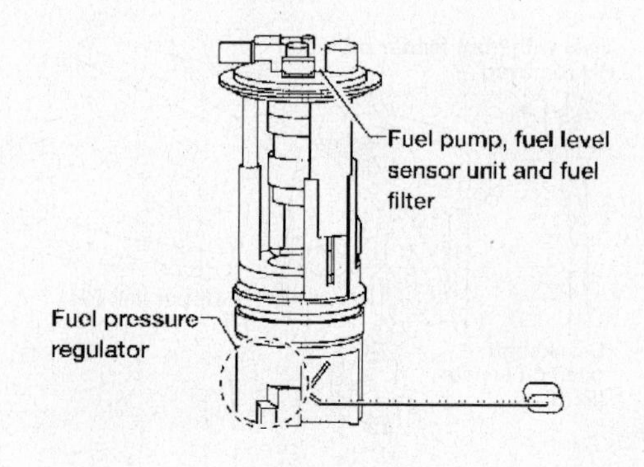
Fuel pump, fuel level
sensor unit and fuel
filter
Fuel pressure
regulator

29149_NISS_G0182

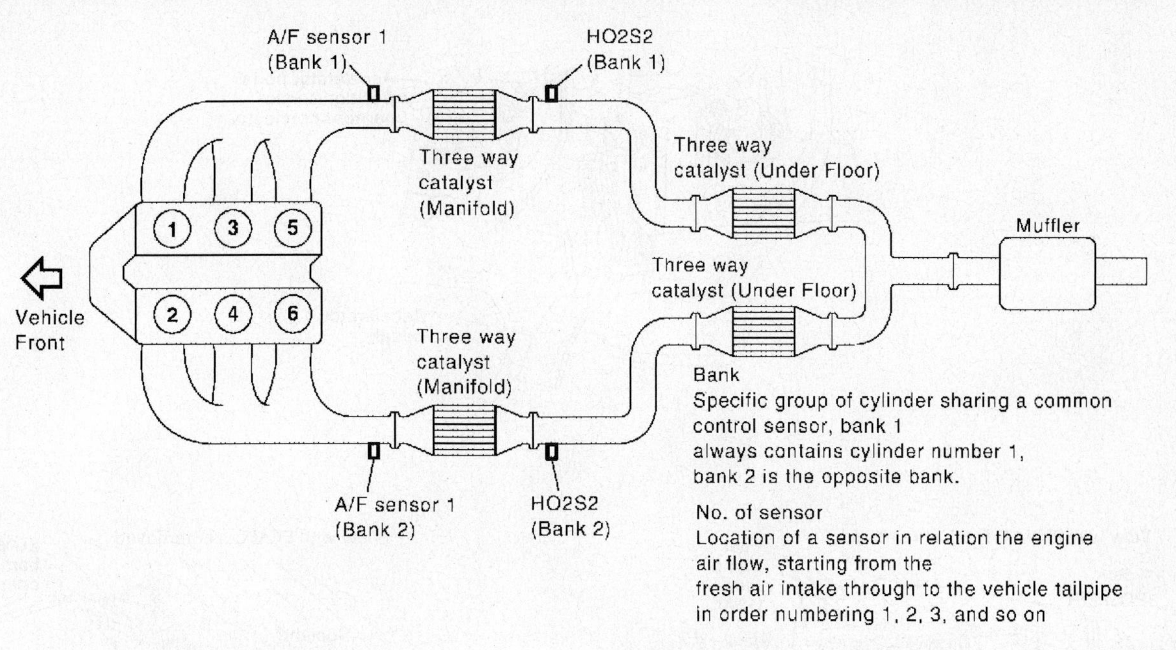

29149_NISS_G0183

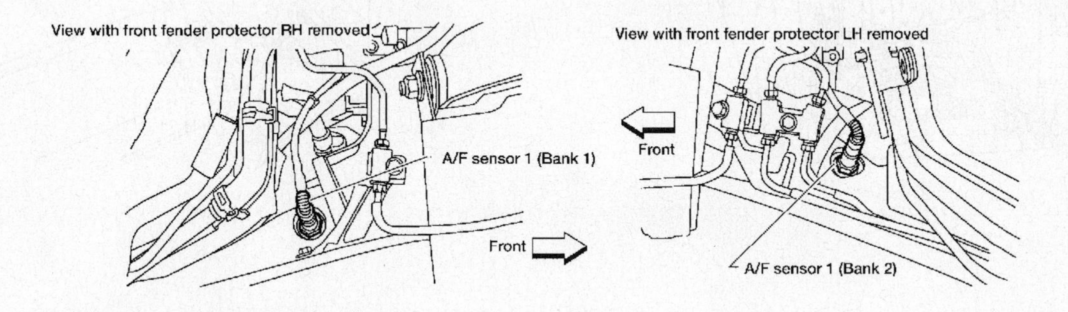

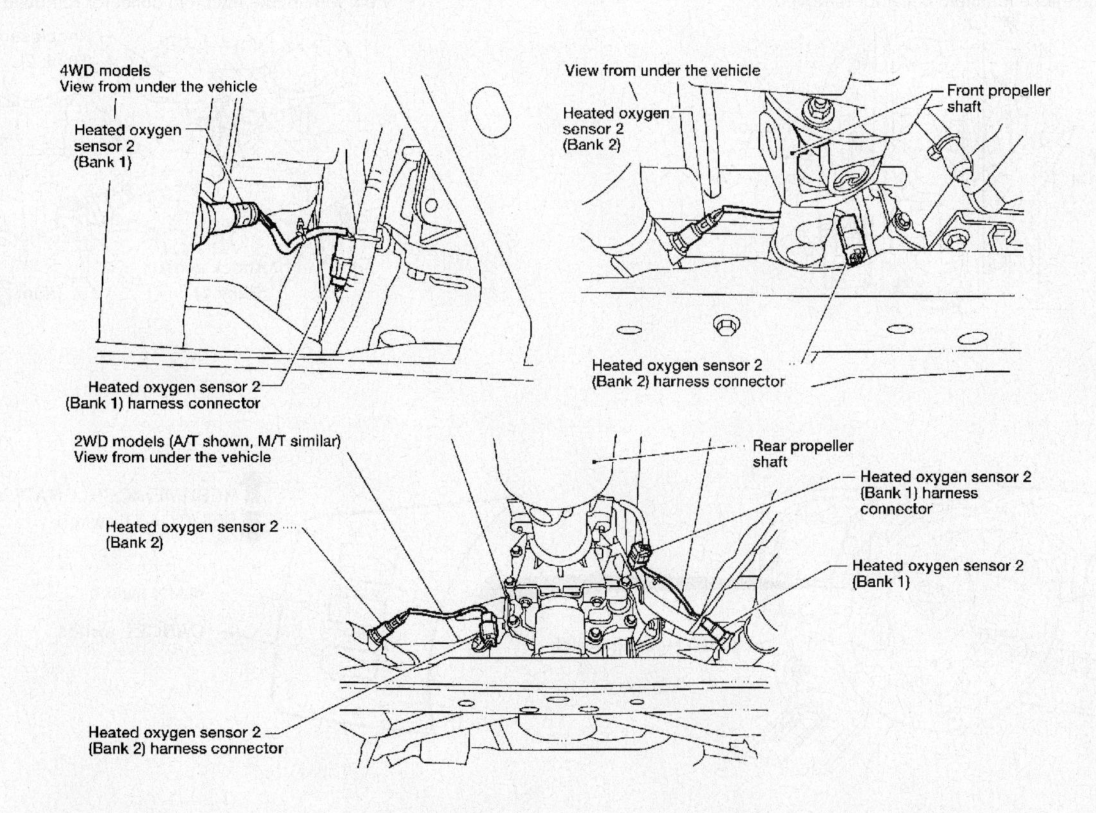

29149_NISS_G0184

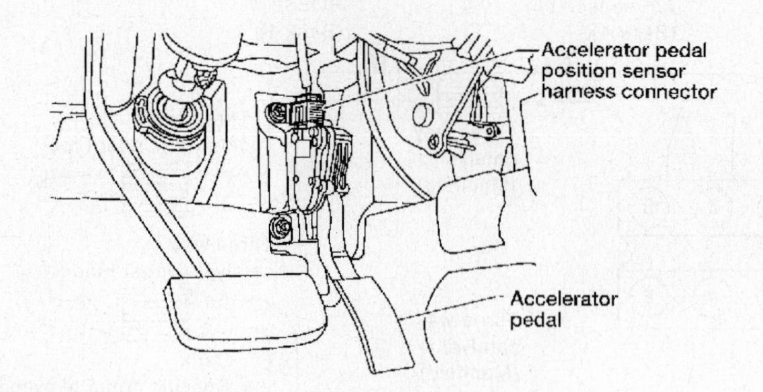

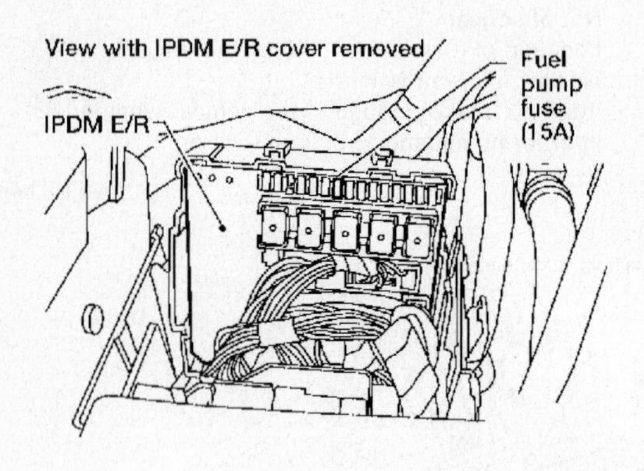

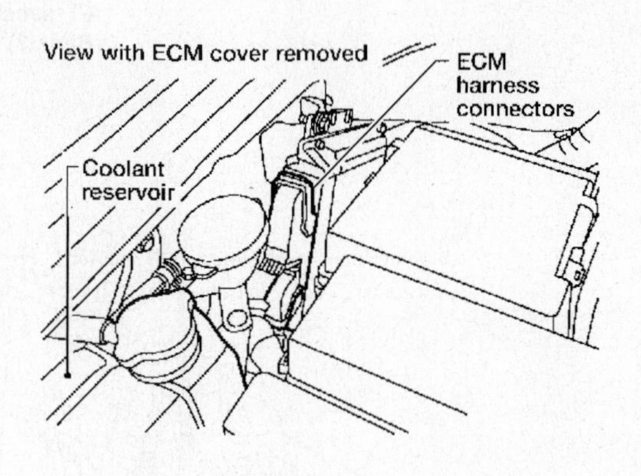

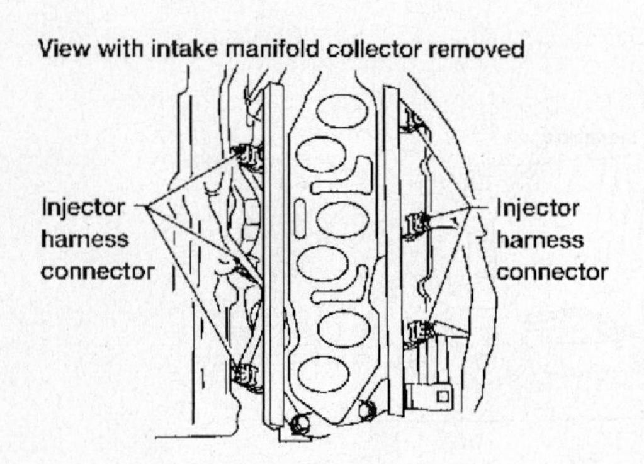

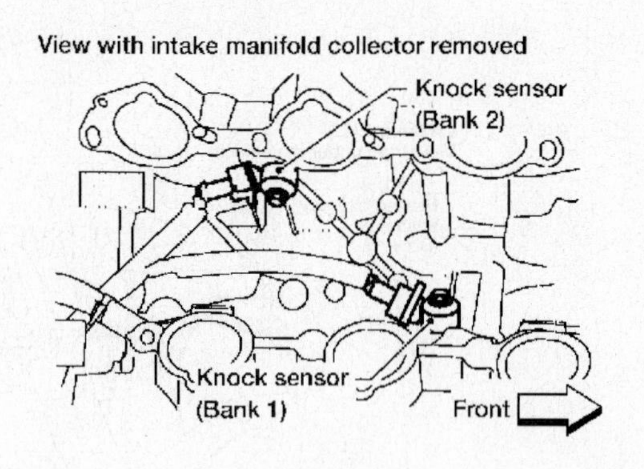

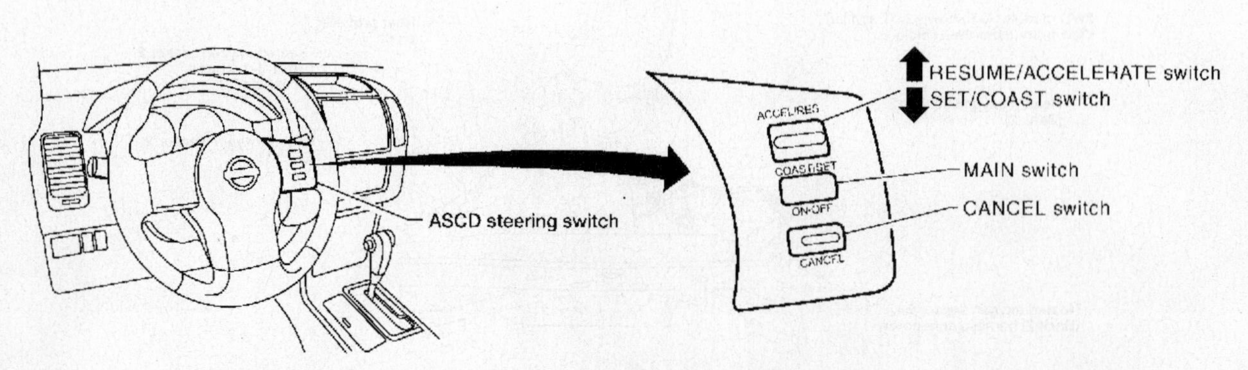

29149_NISS_G0185

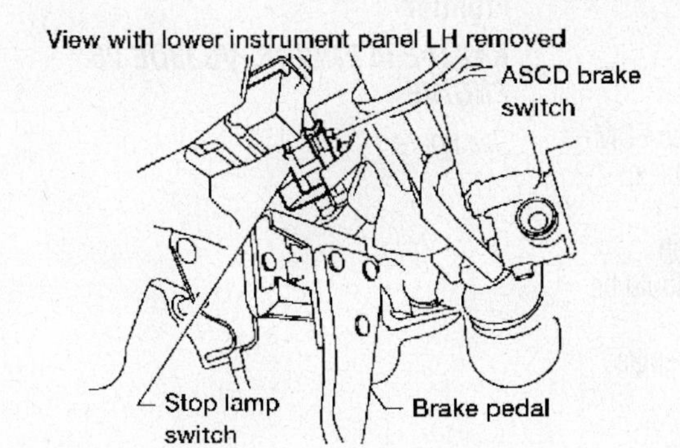

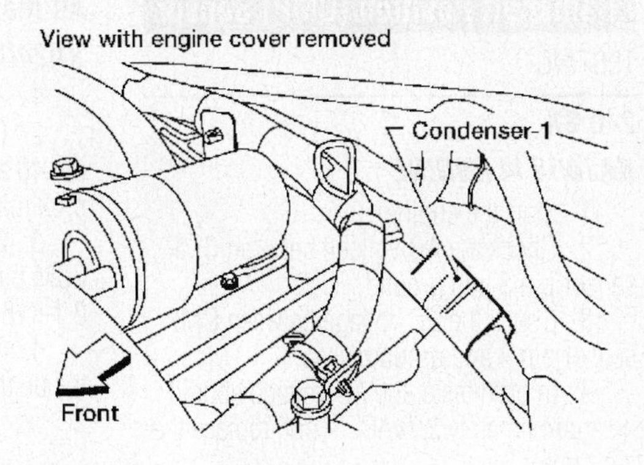

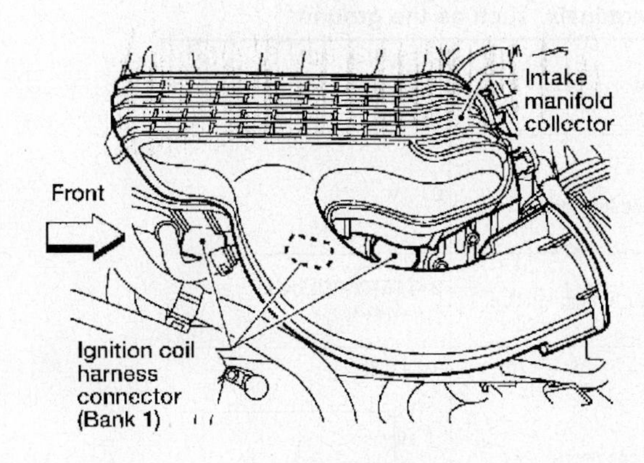

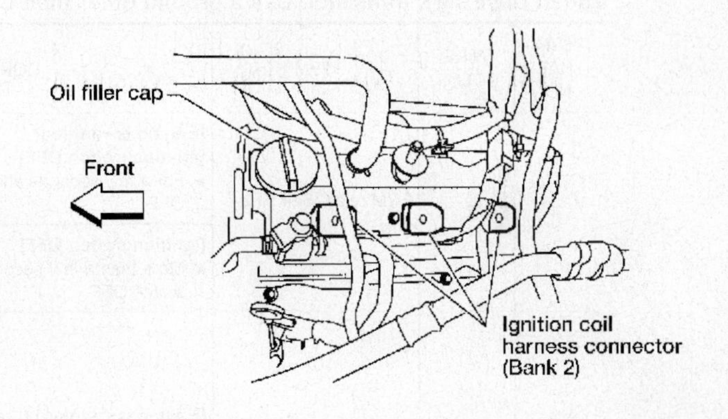

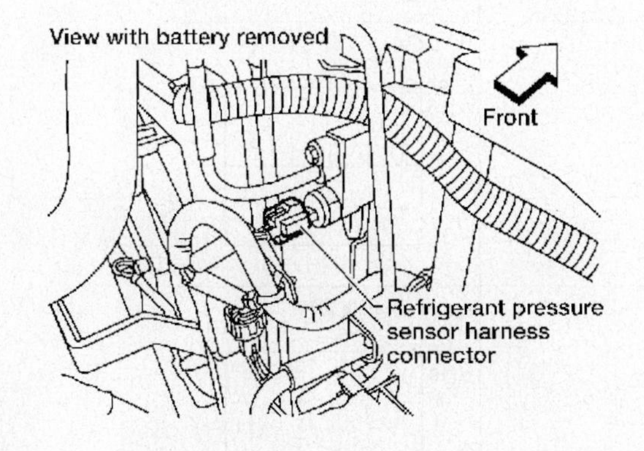

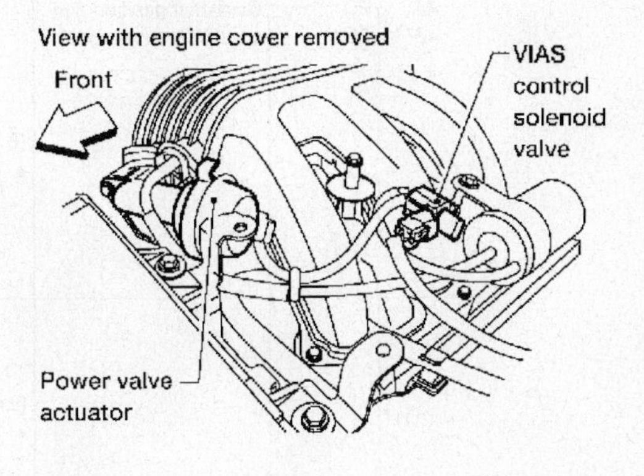

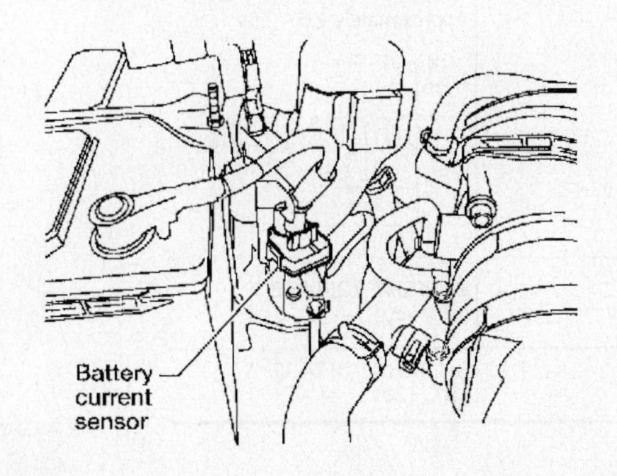

29149_NISS_G0186

Camshaft Position (CMP) Sensor

TESTING

240 SX

KA24DE I4 ENGINE

1. Start the engine.
2. Check the AC voltage between CMP sensor pin 3 and ground.
3. Check the AC voltage between CMP sensor pin 4 and ground.
4. In both tests, the AC voltage should be approximately 2.7vAC. If not, replace the distributor.

Altima

KA24DE I4 ENGINE

1. Start engine.
2. Check the DC voltage between ECM pin 40 and ground. Voltage should be approximately 2.5vDC.
3. Check the DC voltage between ECM pin 41 and ground. Voltage should be 0.1–0.5 vDC.
4. If voltages are not as listed, replace the distributor.

Frontier

KA24DE I4 ENGINE, VG33DE V6 ENGINE

See figure 1.

Do not use ECM ground terminals when measuring input/output voltage. Doing so may result in damage to the ECM's transistor. Use a ground other than ECM terminals, such as the ground.

TERMINAL NO.	WIRE COLOR	ITEM	CONDITION	DATA (DC Voltage)
4	LG/R	ECM relay (Self shut-off)	**[Engine is running]** **[Ignition switch OFF]** • For a few seconds after turning ignition switch OFF	0 - 1V
			[Ignition switch OFF] • More than a few seconds after turning ignition switch OFF	BATTERY VOLTAGE (11 - 14V)
44 48	PU PU	Camshaft position sensor (Reference signal)	**[Engine is running]** (Warm-up condition) • Idle speed	0.2 - 0.5V (waveform: (V) 10, 5, 0; 10 ms)
			[Engine is running] • Engine speed is 2,000 rpm	0 - 0.5V (waveform: (V) 10, 5, 0; 10ms)
49	LG	Camshaft position sensor (Position signal)	**[Engine is running]** • **Warm-up condition** • Idle speed	Approximately 2.6V (waveform: (V) 10, 5, 0; 0.2ms)
			[Engine is running] • Engine speed is 2,000 rpm	Approximately 2.5 - 2.6V (waveform: (V) 10, 5, 0; 0.2ms)
67 72	B/P B/P	Power supply for ECM	**[Ignition switch ON]**	BATTERY VOLTAGE (11 - 14V)
117	B/P	Current return	**[Engine is running]** • Idle speed	BATTERY VOLTAGE (11 - 14V)

29149_NISS_G0052

Fig. 1 Camshaft Position Sensor testing—Frontier

Maxima

VQ30DE V6 ENGINE

See figure 2.

1. With engine running, measure DC voltage between ECM pins 44 and 25.

2. Measure DC voltage between ECM pins 48 and 25.

3. Compare measured results with the illustration.

Pathfinder

VG33E V6 ENGINE

See figure 3.

ECM TERMINALS AND REFERENCE VALUE

Specification data are reference values, and are measured between each terminal and ㉕ (ECM ground).

TERMINAL NO.	WIRE COLOR	ITEM	CONDITION	DATA (DC voltage)
44 48	W W	Camshaft position sensor (PHASE)	Engine is running. (Warm-up condition) Idle speed	Approximately 4.2V★ (AC voltage) 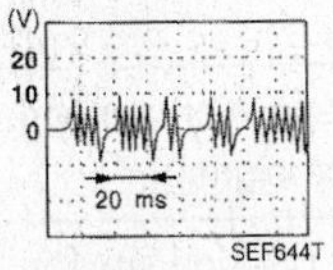 (V) 20 10 0, 20 ms SEF644T

★: Average voltage for pulse signal (Actual pulse signal can be confirmed by oscilloscope.)

29149_NISS_G0056

Fig. 2 Camshaft Position Sensor testing—Maxima

CAUTION:
Do not use ECM ground terminals when measuring input/output voltage. Doing so, may result in damage to the ECM's transistor. Use a ground other than ECM terminals, such as the ground.

TERMINAL NO.	WIRE COLOR	ITEM	CONDITION	DATA (DC Voltage)
4	L/B	ECCS relay (Self-shut-off)	**[Engine is running]** **[Ignition switch "OFF"]** For a few seconds after turning ignition switch "OFF"	0 - 1.5V
			[Ignition switch "OFF"] A few seconds passed after turning ignition switch "OFF"	BATTERY VOLTAGE (11 - 14V)
44	B/W	Camshaft position sensor (Position signal)	**[Engine is running]** **Warm-up condition** Idle speed	Approximately 2.5V 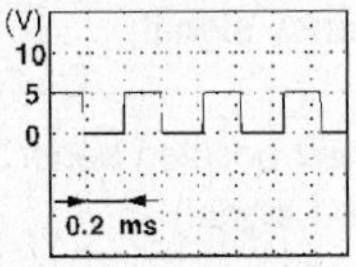 (V) 10 5 0, 0.2 ms
			[Engine is running] Engine speed is 2,000 rpm	Approximately 2.5V 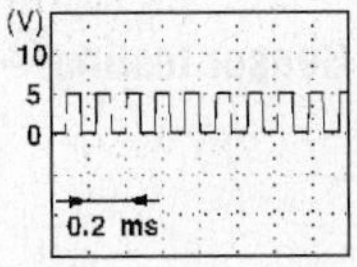 (V) 10 5 0, 0.2 ms
49	L	Camshaft position sensor (Reference signal)	**[Engine is running]** Idle speed	0.3 - 0.5V 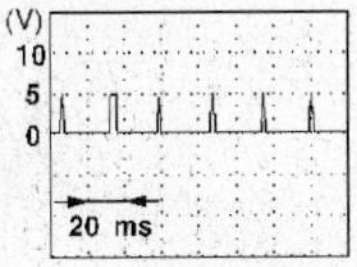 (V) 10 5 0, 20 ms
53	L		**[Engine is running]** Engine speed is 2,000 rpm	0.3 - 0.5V 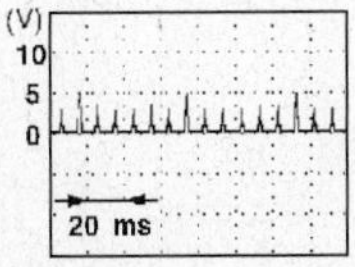 (V) 10 5 0, 20 ms
67	B/W	Power supply for ECM	**[Ignition switch "ON"]**	BATTERY VOLTAGE (11 - 14V)
72	B/W			
117	B/W	Current return	**[Engine is running]** Idle speed	BATTERY VOLTAGE (11 - 14V)

29149_NISS_G0058

Fig. 3 Camshaft Position Sensor testing—Pathfinder

Pickup

KA24DE I4 ENGINE

See figure 4.

1. With the engine running, measure DC voltage between ECM pins 43 and 40.
2. Measure the DC voltage between ECM pins 43 and 44.
3. Measure the DC voltage between ECM pins 43 and 41.
4. Compare measured results with the illustration.

Quest

VG30E V6 ENGINE

See figure 5.

ECM TERMINALS AND REFERENCE VALUE

Specification data are reference values and are measured between each terminal and (43) (ECCS ground).

TER-MINAL NO.	WIRE COLOR	ITEM	CONDITION	DATA (DC voltage)
40 44	GY R	Camshaft position sensor (Reference signal)	Engine is running.	Approximately 0.3V*
41	G	Camshaft position sensor (Position signal)	Engine is running.	Approximately 2.5V*

*: Average voltage for pulse signal (Actual pulse signal can be confirmed by oscilloscope.)

Fig. 4 Camshaft Position Sensor Voltage Specifications—Pickup

29149_NISS_G0047

ECM TERMINALS AND REFERENCE VALUE

Specification data are reference values and are measured between each terminal and (43) (ECCS ground).

TER-MINAL NO.	WIRE COLOR	ITEM	CONDITION	DATA (DC Voltage)
40	G/B	Camshaft position sensor (Reference signal)	Engine is running. └ Idle speed	0.2 - 0.5V*
44	G/B			
41	G/Y	Camshaft position sensor (Position signal)	Engine is running. └ Idle speed	2.0 - 3.0V*
45	G/Y			

*: Average voltage for pulse signal (Actual pulse signal can be confirmed by oscilloscope.)

Fig. 5 Camshaft Position Sensor testing—Quest with VG30E Engine

29149_NISS_G0064

Xterra

KA24DE I4 ENGINE, VG33E V6 ENGINE

See figure 20.

1. With engine running, measure AC voltage between ECM pin 47 and ground.
2. Compare the measured results with the illustration.

Engine Coolant Temperature (ECT) Sensor

TESTING

240 SX

KA24DE I4 ENGINE

See figure 21.

1. Measure resistance of the ETC sensor and compare values with the illustration.

300ZX

VG30DE, VG30DETT V6 ENGINES

See figure 22.

1. Measure resistance of the ETC sensor and compare values with the illustration.

Do not use ECM ground terminals when measuring input/output voltage. Doing so may result in damage to the ECM's transistor. Use a ground other than ECM terminals, such as the ground.

TERMINAL NO.	WIRE COLOR	ITEM	CONDITION	DATA (AC Voltage)
47	L	Crankshaft position sensor (OBD)	**[Engine is running]** • **Warm-up condition** • Idle speed	(V) 10 5 0 0.2 ms
			[Engine is running] • Engine speed is 2,000 rpm	(V) 10 5 0 0.2 ms

29149_NISS_G0050

Fig. 20 Crankshaft Position Sensor voltage specifications—Xterra

Engine Coolant Temperature Sensor

Temperature °C (°F)	Resistance kΩ
20 (68)	2.1 - 2.9
50 (122)	0.68 - 1.00
90 (194)	0.236 - 0.260

29149_NISS_G0037

Fig. 21 Engine Coolant Temperature Sensor specifications—240 SX

Engine coolant temperature °C (°F)	Voltage (V)	Resistance (kΩ)
−10 (14)	4.4	7.0 - 11.4
20 (68)	3.5	2.1 - 2.9
50 (122)	2.2	0.68 - 1.00
90 (194)	0.98	0.236 - 0.260
110 (230)	0.64	0.143 - 0.153

29149_NISS_G0043

Fig. 22 Engine Coolant Temperature Sensor specifications—300 ZX

Altima

KA24DE I4 ENGINE

See figure 23.

1. Measure resistance of the ETC sensor and compare values with the illustration.

Frontier

KA24DE I4 ENGINE, VG33DE V6 ENGINE

See figure 24.

1. Measure resistance of the ETC sensor and compare values with the illustration.

Maxima

VQ30DE V6 ENGINE

See figure 25.

1. Measure resistance of the ETC sensor and compare values with the illustration.

Pathfinder

VG33E V6 ENGINE

See figure 26.

1. Measure resistance of the ETC sensor and compare values with the illustration.

Pickup

KA24DE I4 ENGINE

See figure 27.

1. Measure resistance of the ETC sensor and compare values with the illustration.

Quest

VG30E V6 ENGINE

See figure 28.

1. Measure resistance of the ETC sensor and compare values with the illustration.

VG33E V6 ENGINE

See figure 29.

1. Measure resistance of the ETC sensor and compare values with the illustration.

Sentra

GA16DE I4 ENGINE

See figure 30.

1. Measure resistance of the ETC sensor and compare values with the illustration.

SR20DE I4 ENGINE

See figure 31.

1. Measure resistance of the ETC sensor and compare values with the illustration.

Engine Coolant Temperature Sensor

Temperature °C (°F)	Resistance kΩ
20 (68)	2.1 - 2.9
50 (122)	0.68 - 1.00
90 (194)	0.236 - 0.260

29149_NISS_G0037

Fig. 23 Engine Coolant Temperature Sensor specifications—Altima

Engine coolant temperature °C (°F)	Voltage* V	Resistance kΩ
-10 (14)	4.4	7.0 - 11.4
20 (68)	3.5	2.1 - 2.9
50 (122)	2.2	0.68 - 1.00
90 (194)	0.9	0.236 - 0.260

*: These data are reference values and are measured between ECM terminal 59 (Engine coolant temperature sensor) and ground.

CAUTION:

Do not use ECM ground terminals when measuring input/output voltage. Doing so, may result in damage to the ECM's transistor. Use a ground other than ECM terminals, such as the ground.

29149_NISS_G0049

Fig. 24 Engine Coolant Temperature Sensor specifications—Frontier

Engine coolant temperature °C (°F)	Voltage* V	Resistance kΩ
-10 (14)	4.4	7.0 - 11.4
20 (68)	3.5	2.1 - 2.9
50 (122)	2.2	0.68 - 1.00
90 (194)	0.9	0.236 - 0.260

*: These data are reference values and are measured between ECM terminal 59 (Engine coolant temperature sensor) and ground.

CAUTION:

Do not use ECM ground terminals when measuring input/output voltage. Doing so, may result in damage to the ECM's transistor. Use a ground other than ECM terminals, such as the ground.

29149_NISS_G0049

Fig. 25 Engine Coolant Temperature Sensor specifications—Maxima

Engine coolant temperature °C (°F)	Voltage* V	Resistance kΩ
-10 (14)	4.4	7.0 - 11.4
20 (68)	3.5	2.1 - 2.9
50 (122)	2.2	0.68 - 1.00
90 (194)	0.9	0.236 - 0.260

*: These data are reference values and are measured between ECM terminal 59 (Engine coolant temperature sensor) and ground.

CAUTION:

Do not use ECM ground terminals when measuring input/output voltage. Doing so, may result in damage to the ECM's transistor. Use a ground other than ECM terminals, such as the ground.

29149_NISS_G0049

Fig. 26 Engine Coolant Temperature Sensor specifications—Pathfinder

Engine Coolant Temperature Sensor

Temperature °C (°F)	Resistance kΩ
20 (68)	2.1 - 2.9
50 (122)	0.68 - 1.00
90 (194)	0.236 - 0.260

29149_NISS_G0037

Fig. 27 Engine Coolant Temperature Sensor specifications—Pickup

Engine coolant temperature °C (°F)	Voltage* (V)	Resistance (kΩ)
-10 (14)	4.4	7.0 - 11.4
20 (68)	3.5	2.1 - 2.9
50 (122)	2.2	0.68 - 1.00
90 (194)	0.9	0.236 - 0.260

*: These data are reference values and are measured between ECM terminal ⑤① (Engine coolant temperature sensor) and ECM terminal ④③ (ECCS ground).

29149_NISS__G0061

Fig. 28 Engine Coolant Temperature Sensor specifications—Quest with VG30E Engine

Engine coolant temperature °C (°F)	Voltage* V	Resistance kΩ
-10 (14)	4.4	7.0 - 11.4
20 (68)	3.5	2.1 - 2.9
50 (122)	2.2	0.68 - 1.00
90 (194)	0.9	0.236 - 0.260

*: These data are reference values and are measured between ECM terminal 59 (Engine coolant temperature sensor) and ground.

CAUTION:
Do not use ECM ground terminals when measuring input/output voltage. Doing so, may result in damage to the ECM's transistor. Use a ground other than ECM terminals, such as the ground.

29149_NISS__G0049

Fig. 29 Engine Coolant Temperature Sensor specifications—Quest with VG33E Engine

Engine coolant temperature °C (°F)	Voltage* (V)	Resistance (kΩ)
-10 (14)	4.4	7.0 - 11.4
20 (68)	3.5	2.1 - 2.9
50 (122)	2.2	0.68 - 1.00
90 (194)	0.9	0.236 - 0.260

*: These data are reference values and are measured between ECM terminal ⑤① (Engine coolant temperature sensor) and ECM terminal ④③ (ECCS ground).

29149_NISS__G0061

Fig. 30 Engine Coolant Temperature Sensor Specifications—Sentra with GA16DE Engine

Engine coolant temperature °C (°F)	Voltage* V	Resistance kΩ
-10 (14)	4.4	7.0 - 11.4
20 (68)	3.5	2.1 - 2.9
50 (122)	2.2	0.68 - 1.00
90 (194)	0.9	0.236 - 0.260

*: These data are reference values and are measured between ECM terminal ⑤① (Engine coolant temperature sensor) and engine ground.

29149_NISS__G0072

Fig. 31 Engine Coolant Temperature Sensor Specifications—Sentra with SR20DE Engine

EGR Temperature Sensor

240 SX

KA24DE I4 ENGINE

See figure 32.

1. Measure the resistance of the EGR temperature sensor and compare with the specifications in the illustration.

300 ZX

VG30DE, VG30DETT V6 ENGINES

See figure 33.

1. Measure the resistance of the EGR temperature sensor and compare with the specifications in the illustration.

Altima

KA24DE I4 ENGINE

See figure 34.

1. Measure the resistance of the EGR temperature sensor and compare with the specifications in the illustration.

EGR temperature °C (°F)	Voltage (V)	Resistance (MΩ)
0 (32)	4.81	7.9 - 9.7
50 (122)	2.82	0.57 - 0.70
100 (212)	0.8	0.08 - 0.10

29149_NISS__G0041

Fig. 32 EGR Temperature Sensor specifications—240 SX

EGR temperature °C (°F)	Voltage (V)	Resistance (MΩ)
0 (32)	4.81	7.9 - 9.7
50 (122)	2.82	0.57 - 0.70
100 (212)	0.8	0.08 - 0.10

29149_NISS__G0041

Fig. 33 EGR Temperature Sensor specifications—300 ZX

EGR temperature °C (°F)	Voltage (V)	Resistance (MΩ)
0 (32)	4.81	7.9 - 9.7
50 (122)	2.82	0.57 - 0.70
100 (212)	0.8	0.08 - 0.10

29149_NISS__G0041

Fig. 34 EGR Temperature Sensor specifications—Altima

Frontier

KA24DE I4 ENGINE

See figure 35.

1. Measure the resistance of the EGR temperature sensor and compare with the specifications in the illustration.

Maxima

VQ30DE V6 ENGINE

See figure 36.

1. With the engine **OFF**, disconnect the ECM wiring connector.
2. Measure resistance between pins 63 and 43 of the ECM connector.
3. Compare the results with the specifications in the illustration.

Pathfinder

VG33E V6 ENGINE

See figure 37.

1. With the engine **OFF**, disconnect the ECM wiring connector.
2. Measure resistance between ECM pin 63 and ground.
3. Compare the results with the specifications in the illustration.

Pickup

KA24DE I4 ENGINE

See figure 38.

1. Measure the resistance of the EGR temperature sensor and compare with the specifications in the illustration.

Quest

VG30E V6 ENGINE

See figure 39.

1. With the engine **OFF**, disconnect the ECM wiring connector.
2. Measure resistance between ECM pin 63 and ground.
3. Compare the results with the specifications in the illustration.

Sentra

GA16DE I4 ENGINE, SR20DE I4 ENGINE

See figure 40.

1. With the engine **OFF**, disconnect the ECM wiring connector.
2. Measure resistance between ECM pin 62 and ground.
3. Compare the results with the specifications in the illustration.

Xterra

KA24DE I4 ENGINE & VG33E V6 ENGINE

See figure 41.

1. Measure resistance of the ETC sensor and compare values with the illustration.

EGR temperature °C (°F)	Voltage* V	Resistance MΩ
0 (32)	4.56	0.62 - 1.05
50 (122)	2.25	0.065 - 0.094
100 (212)	0.59	0.011 - 0.015

*: These data are reference values and are measured between ECM terminal 63 (EGR temperature sensor) and ground.
When EGR system is operating.
Voltage: 0 - 1.5V

29149_NISS__G0053

Fig. 35 EGR Temperature Sensor specifications–Frontier

EGR temperature °C (°F)	Voltage* V	Resistance MΩ
0 (32)	4.56	0.62 - 1.05
50 (122)	2.25	0.065 - 0.094
100 (212)	0.59	0.011 - 0.015

*: These data are reference values and are measured between ECM terminal 63 (EGR temperature sensor) and ground.
When EGR system is operating.
Voltage: 0 - 1.5V

29149_NISS__G0053

Fig. 36 EGR Temperature Sensor specifications—Maxima

EGR temperature °C (°F)	Voltage* V	Resistance MΩ
0 (32)	4.81	7.9 - 9.7
50 (122)	2.82	0.57 - 0.70
100 (212)	0.8	0.08 - 0.10

*: These data are reference values and are measured between ECM terminal 63 (EGR temperature sensor) and ground.
When EGR system is operating.
Voltage: 0 - 1.5V

CAUTION:
Do not use ECM ground terminals when measuring input/output voltage. Doing so, may result in damage to the ECM's transistor. Use a ground other than ECM terminals, such as the ground.

29149_NISS__G0059

Fig. 37 EGR Temperature Sensor specifications—Pathfinder

EGR temperature °C (°F)	Voltage* V	Resistance MΩ
0 (32)	4.61	0.68 - 1.11
50 (122)	2.53	0.09 - 0.12
100 (212)	0.87	0.017 - 0.024

*: These data are reference values and are measured between ECM terminal ⑥② (EGR temperature sensor) and ground.
When EGR system is operating.
Voltage: 0 - 1.5V

29149_NISS__G0045

Fig. 38 EGR Temperature Sensor specifications–Pickup

EGR temperature °C (°F)	Voltage* V	Resistance MΩ
0 (32)	4.81	7.9 - 9.7
50 (122)	2.82	0.57 - 0.70
100 (212)	0.8	0.08 - 0.10

*: These data are reference values and are measured between ECM terminal 63 (EGR temperature sensor) and ground.
When EGR system is operating.
Voltage: 0 - 1.5V

CAUTION:
Do not use ECM ground terminals when measuring input/output voltage. Doing so, may result in damage to the ECM's transistor. Use a ground other than ECM terminals, such as the ground.

29149_NISS__G0059

Fig. 39 EGR Temperature Sensor specifications—Quest

EGR temperature °C (°F)	Voltage* V	Resistance MΩ
0 (32)	4.81	7.9 - 9.7
50 (122)	2.82	0.57 - 0.70
100 (212)	0.8	0.08 - 0.10

*: These data are reference values and are measured between ECM terminal (62) (EGR temperature sensor) and ECM terminal (43) (ECCS ground).
When EGR system is operating.
Voltage: 0 - 1.5V

29149_NISS__G0070

Fig. 40 EGR Temperature Sensor specifications—Sentra

Engine coolant temperature °C (°F)	Voltage* V	Resistance kΩ
−10 (14)	4.4	7.0 - 11.4
20 (68)	3.5	2.1 - 2.9
50 (122)	2.2	0.68 - 1.00
90 (194)	0.9	0.236 - 0.260

*: These data are reference values and are measured between ECM terminal 59 (Engine coolant temperature sensor) and ground.

CAUTION:
Do not use ECM ground terminals when measuring input/output voltage. Doing so, may result in damage to the ECM's transistor. Use a ground other than ECM terminals, such as the ground.

29149_NISS__G0049

Fig. 41 Engine Coolant Temperature Sensor specifications—Xterra

Intake Air Temperature (IAT) Sensor

TESTING

See figure 42.

Intake air temperature °C (°F)	Voltage V	Resistance kΩ
20 (68)	3.5	2.1 - 2.9
80 (176)	1.23	0.27 - 0.38

CAUTION:
Do not use ECM ground terminals when measuring input/output voltage. Doing so, may result in damage to the ECM's transistor. Use a ground other than ECM terminals, such as the ground.

29149_NISS__G0042

Fig. 42 Intake Air Temperature Sensor specifications

240 SX

KA24DE I4 ENGINE

1. Measure the resistance of the IAT sensor.
2. The sensor resistance should be 2.1–2.9 kohms at 68°F and decrease as air temperature rises.

300 ZX

VG30DE, VG30DETT V6 ENGINES

1. Measure the resistance of the IAT sensor.
2. The sensor resistance should be 2.1–2.9 kohms at 68°F and decrease as air temperature rises.

Altima

KA24DE I4 ENGINE

1. Measure the resistance of the IAT sensor.
2. The sensor resistance should be 2.1–2.9 kohms at 68°F and decrease as air temperature rises.

Frontier

KA24DE I4 ENGINE, VG33DE V6 ENGINE

1. Measure the resistance of the IAT sensor.
2. The sensor resistance should be 2.1–2.9 kohms at 68°F and decrease as air temperature rises.

Maxima

VQ30DE V6 ENGINE

1. Measure the resistance of the IAT sensor.
2. The sensor resistance should be 2.1–2.9 kohms at 68°F and decrease as air temperature rises.

Pathfinder

VG33E V6 ENGINE

1. Measure the resistance of the IAT sensor.
2. The sensor resistance should be 2.1–2.9 kohms at 68°F and decrease as air temperature rises.

Pickup

KA24DE I4 ENGINE

1. Measure the resistance of the IAT sensor.
2. The sensor resistance should be 2.1–2.9 kohms at 68°F and decrease as air temperature rises.

Quest

VG30E V6 ENGINE, VG33E V6 ENGINE

1. Measure the resistance of the IAT sensor.
2. The sensor resistance should be 2.1–2.9 kohms at 68°F and decrease as air temperature rises.

Sentra

GA16DE I4 ENGINE, SR20DE I4 ENGINE

1. Measure the resistance of the IAT sensor.
2. The sensor resistance should be 2.1–2.9 kohms at 68°F and decrease as air temperature rises.

Xterra

KA24DE I4 ENGINE, VG33E V6 ENGINE

1. Measure the resistance of the IAT sensor.
2. The sensor resistance should be 2.1–2.9 kohms at 68°F and decrease as air temperature rises.

Knock Sensor (KS)

TESTING

240 SX

KA24DE I4 ENGINE

See figure 43.

1. Start engine and run at idle.
2. Measure voltage at ECM pin 54.
3. Voltage should be approximately 2.5 vDC.

Altima

KA24DE I4 ENGINE

See figure 44.

1. Start engine and run at idle.
2. Measure voltage at ECM pin 54.
3. Voltage should be approximately 2.5 vDC.

Frontier

KA24DE I4 ENGINE

See figure 45.

1. Start engine and run at idle.
2. Measure voltage at ECM pin 64.
3. Voltage should be approximately 2.4 vDC.

Maxima

VQ30DE V6 ENGINE

See figure 46.

1. Start engine and run at idle.
2. Measure voltage at ECM pin 64.
3. Voltage should be approximately 2.4 vDC.

Pathfinder

VG33E V6 ENGINE

See figure 47.

1. Start engine and run at idle.
2. Measure voltage at ECM pin 64.
3. Voltage should be approximately 2.4 vDC.

Quest

VG30E V6 ENGINE

See figure 48.

1. Start engine and run at idle.
2. Measure voltage at ECM pin 54.
3. Voltage should be approximately 2.5 vDC.

VG33E V6 ENGINE

See figure 49.

1. Start engine and run at idle.
2. Measure voltage at ECM pin 64.
3. Voltage should be approximately 2.4 vDC.

ECM TERMINALS AND REFERENCE VALUE

Specification data are reference values and are measured between each terminal and ㊸ (ECCS ground).

TER-MINAL NO.	WIRE COLOR	ITEM	CONDITION	DATA (DC Voltage)
54	W	Knock sensor	Engine is running. └ Idle speed	2.0 - 3.0V

29149_NISS__G0063

Fig. 43 Knock Sensor specification—240 SX

ECM TERMINALS AND REFERENCE VALUE

Specification data are reference values and are measured between each terminal and ㊸ (ECCS ground).

TER-MINAL NO.	WIRE COLOR	ITEM	CONDITION	DATA (DC Voltage)
54	W	Knock sensor	Engine is running. └ Idle speed	2.0 - 3.0V

29149_NISS__G0063

Fig. 44 Knock Sensor specification—Altima

Do not use ECM ground terminals when measuring input/output voltage. Doing so may result in damage to the ECM's transistor. Use a ground other than ECM terminals, such as the ground.

TERMINAL NO.	WIRE COLOR	ITEM	CONDITION	DATA (DC Voltage)
64	W	Knock sensor	**[Engine is running]** • Idle speed	Approximately 2.4V

29149_NISS__G0051

Fig. 45 Knock Sensor specification—Frontier

Do not use ECM ground terminals when measuring input/output voltage. Doing so may result in damage to the ECM's transistor. Use a ground other than ECM terminals, such as the ground.

TERMINAL NO.	WIRE COLOR	ITEM	CONDITION	DATA (DC Voltage)
64	W	Knock sensor	**[Engine is running]** • Idle speed	Approximately 2.4V

29149_NISS__G0051

Fig. 46 Knock Sensor specification—Maxima

Do not use ECM ground terminals when measuring input/output voltage. Doing so may result in damage to the ECM's transistor. Use a ground other than ECM terminals, such as the ground.

TERMINAL NO.	WIRE COLOR	ITEM	CONDITION	DATA (DC Voltage)
64	W	Knock sensor	**[Engine is running]** • Idle speed	Approximately 2.4V

29149_NISS__G0051

Fig. 47 Knock Sensor specification—Pathfinder

ECM TERMINALS AND REFERENCE VALUE

Specification data are reference values and are measured between each terminal and ㊸ (ECCS ground).

TER-MINAL NO.	WIRE COLOR	ITEM	CONDITION	DATA (DC Voltage)
54	W	Knock sensor	Engine is running. └ Idle speed	2.0 - 3.0V

29149_NISS__G0063

Fig. 48 Knock Sensor specification—Quest with VG30E Engine

Do not use ECM ground terminals when measuring input/output voltage. Doing so may result in damage to the ECM's transistor. Use a ground other than ECM terminals, such as the ground.

TERMINAL NO.	WIRE COLOR	ITEM	CONDITION	DATA (DC Voltage)
64	W	Knock sensor	**[Engine is running]** • Idle speed	Approximately 2.4V

29149_NISS__G0051

Fig. 49 Knock Sensor specification— Quest with VG33E Engine

Sentra

GA16DE I4 ENGINE, SR20DE I4 ENGINE

See figure 50.

1. Start engine and run at idle.
2. Measure voltage at ECM pin 54.
3. Voltage should be 2.0–3.0 vDC.

Xterra

KA24DE I4 ENGINE, VG33E V6 ENGINE

See figure 51.

1. Start engine and run at idle.
2. Measure voltage at ECM pin 64.
3. Voltage should be approximately 2.4 vDC.

Mass Air Flow (MAF) Sensor

TESTING

240SX

KA24DE I4 ENGINE

See figure 52.

1. Turn ignition switch **ON**.
2. Start engine and warm it up sufficiently.
3. Check the voltage between ECM pin 47 and ground.
4. Check for linear voltage rise in response to increases in engine RPM.

300ZX

VG30DE, VG30DETT V6 ENGINES

See figure 53.

1. Turn ignition switch **ON**.
2. Start engine and warm it up sufficiently.
3. Check the voltage between ECM pin 54 and ground.
4. Check for linear voltage rise in response to increases in engine RPM.

Altima

KA24DE I4 ENGINE

See figure 54.

1. Turn ignition switch **ON**.
2. Start engine and warm it up sufficiently.
3. Check the voltage between ECM pin 47 and ground.
4. Check for linear voltage rise in response to increases in engine RPM.

ECM TERMINALS AND REFERENCE VALUE

Specification data are reference values and are measured between each terminal and ㊸ (ECCS ground).

TERMINAL NO.	WIRE COLOR	ITEM	CONDITION	DATA (DC voltage)
54	W	Knock sensor	Engine is running. └ Idle speed	2.0 - 3.0V

29149_NISS__G0068

Fig. 50 Knock Sensor specification—Sentra

Do not use ECM ground terminals when measuring input/output voltage. Doing so may result in damage to the ECM's transistor. Use a ground other than ECM terminals, such as the ground.

TERMINAL NO.	WIRE COLOR	ITEM	CONDITION	DATA (DC Voltage)
64	W	Knock sensor	**[Engine is running]** • Idle speed	Approximately 2.4V

29149_NISS__G0051

Fig. 51 Knock Sensor specifications—Xterra

ECM TERMINALS AND REFERENCE VALUE

Specification data are reference values and are measured between each terminal and ㊸ (ECCS ground).

TERMINAL NO.	WIRE COLOR	ITEM	CONDITION	DATA (DC Voltage)
47	W	Mass air flow sensor	Engine is running. (Warm-up condition) └ Idle speed	1.3 - 1.7V
			Engine is running. (Warm-up condition) └ Engine speed is 2,500 rpm.	1.7 - 2.1V
48	B	Mass air flow sensor ground	Engine is running. (Warm-up condition) └ Idle speed	Approximately 0V

29149_NISS__G0060

Fig. 52 Mass Air Flow Sensor testing—240SX

Do not use ECM ground terminals when measuring input/output voltage. Doing so may result in damage to the ECM's transistor. Use a ground other than ECM terminals, such as the ground.

TERMINAL NO.	WIRE COLOR	ITEM	CONDITION	DATA (DC Voltage)
54	R	Mass air flow sensor	**[Engine is running]** • **Warm-up condition** • Idle speed	0.9 - 1.8V
			[Engine is running] • **Warm-up condition** • Engine speed is 2,500 rpm	1.9 - 2.3V
55	G	Mass air flow sensor ground	**[Engine is running]** • **Warm-up condition** • Idle speed	Approximately 0V

Fig. 53 Mass Air Flow Sensor testing—300ZX

29149_NISS__G0048

ECM TERMINALS AND REFERENCE VALUE

Specification data are reference values and are measured between each terminal and ㊸ (ECCS ground).

TERMINAL NO.	WIRE COLOR	ITEM	CONDITION	DATA (DC Voltage)
47	W	Mass air flow sensor	Engine is running. (Warm-up condition) └ Idle speed	1.3 - 1.7V
			Engine is running. (Warm-up condition) └ Engine speed is 2,500 rpm.	1.7 - 2.1V
48	B	Mass air flow sensor ground	Engine is running. (Warm-up condition) └ Idle speed	Approximately 0V

29149_NISS__G0060

Fig. 54 Mass Air Flow Sensor testing—Altima

Frontier

KA24DE I4 ENGINE, VG33DE V6 ENGINE

See figure 55.

1. Turn ignition switch **ON**.
2. Start engine and warm it up sufficiently.
3. Check the voltage between ECM pin 54 and ground.
4. Check for linear voltage rise in response to increases in engine RPM.

Maxima

VQ30DE V6 ENGINE

See figure 56.

Pathfinder

VG33E V6 ENGINE

See figure 57.

1. Turn ignition switch **ON**.
2. Start engine and warm it up sufficiently.
3. Check the voltage between ECM pin 54 and ground.
4. Check for linear voltage rise in response to increases in engine RPM.

Pickup

KA24DE I4 ENGINE

See figure 58.

1. Turn ignition switch **ON**.
2. Start engine and warm it up sufficiently.
3. Check the voltage between ECM pin 47 and ground.
4. Check for linear voltage rise in response to increases in engine RPM.

Do not use ECM ground terminals when measuring input/output voltage. Doing so may result in damage to the ECM's transistor. Use a ground other than ECM terminals, such as the ground.

TERMINAL NO.	WIRE COLOR	ITEM	CONDITION	DATA (DC Voltage)
54	R	Mass air flow sensor	**[Engine is running]** • **Warm-up condition** • Idle speed	0.9 - 1.8V
			[Engine is running] • **Warm-up condition** • Engine speed is 2,500 rpm	1.9 - 2.3V
55	G	Mass air flow sensor ground	**[Engine is running]** • **Warm-up condition** • Idle speed	Approximately 0V

29149_NISS__G0048

Fig. 55 Mass Air Flow Sensor testing—Frontier

ECM TERMINALS AND REFERENCE VALUE

Specification data are reference values, and are measured between each terminal and ㉕ (ECM ground).

TERMINAL NO.	WIRE COLOR	ITEM	CONDITION	DATA (DC voltage)
54	W	Mass air flow sensor	Engine is running. (Warm-up condition) └ Idle speed	1.0 - 1.7V
			Engine is running. (Warm-up condition) └ Engine speed is 2,500 rpm.	1.5 - 2.1V
55	B	Mass air flow sensor ground	Engine is running. (Warm-up condition) └ Idle speed	Approximately 0V
4	W/B	ECCS relay (Self-shutoff)	Engine is running. Ignition switch "OFF" └ For a few seconds after turning ignition switch "OFF"	0 - 1V
			Ignition switch "OFF" └ A few seconds passed after turning ignition switch "OFF"	BATTERY VOLTAGE (11 - 14V)
67 72	R R	Power supply for ECM	Ignition switch "ON"	BATTERY VOLTAGE (11 - 14V)

29149_NISS__G0054

Fig. 56 Mass Air Flow Sensor Testing—Maxima

Do not use ECM ground terminals when measuring input/output voltage. Doing so may result in damage to the ECM's transistor. Use a ground other than ECM terminals, such as the ground.

TERMINAL NO.	WIRE COLOR	ITEM	CONDITION	DATA (DC Voltage)
54	R	Mass air flow sensor	**[Engine is running]** • **Warm-up condition** • Idle speed	0.9 - 1.8V
			[Engine is running] • **Warm-up condition** • Engine speed is 2,500 rpm	1.9 - 2.3V
55	G	Mass air flow sensor ground	**[Engine is running]** • **Warm-up condition** • Idle speed	Approximately 0V

29149_NISS__G0048

Fig. 57 Mass Air Flow Sensor testing—Pathfinder

ECM TERMINALS AND REFERENCE VALUE

Specification data are reference values and are measured between each terminal and ㊸ (ECCS ground).

TERMINAL NO.	WIRE COLOR	ITEM	CONDITION	DATA (DC Voltage)
47	W	Mass air flow sensor	Engine is running. (Warm-up condition) └ Idle speed	1.3 - 1.7V
			Engine is running. (Warm-up condition) └ Engine speed is 2,500 rpm.	1.7 - 2.1V
48	B	Mass air flow sensor ground	Engine is running. (Warm-up condition) └ Idle speed	Approximately 0V

29149_NISS__G0060

Fig. 58 Mass Air Flow Sensor testing—Pickup

Quest

VG30E V6 ENGINE

See figure 59.

1. Turn ignition switch **ON**.
2. Start engine and warm it up sufficiently.
3. Check the voltage between ECM pin 47 and ground.
4. Check for linear voltage rise in response to increases in engine RPM.

VG33E V6 ENGINE

See figure 60.

1. Turn ignition switch **ON**.
2. Start engine and warm it up sufficiently.
3. Check the voltage between ECM pin 54 and ground.
4. Check for linear voltage rise in response to increases in engine RPM.

Sentra

GA16DE I4 ENGINE

See figure 61.

1. Turn ignition switch **ON**.
2. Start engine and warm it up sufficiently.
3. Check the voltage between ECM pin 47 and ground.
4. Check for linear voltage rise in response to increases in engine RPM.

SR20DE I4 ENGINE

See figure 62.

1. Turn ignition switch **ON**.
2. Start engine and warm it up sufficiently.
3. Check the voltage between ECM pin 47 and ground.
4. Check for linear voltage rise in response to increases in engine RPM.

ECM TERMINALS AND REFERENCE VALUE

Specification data are reference values and are measured between each terminal and ㊸ (ECCS ground).

TERMINAL NO.	WIRE COLOR	ITEM	CONDITION	DATA (DC Voltage)
47	W	Mass air flow sensor	Engine is running. (Warm-up condition) Idle speed	1.3 - 1.7V
			Engine is running. (Warm-up condition) Engine speed is 2,500 rpm.	1.7 - 2.1V
48	B	Mass air flow sensor ground	Engine is running. (Warm-up condition) Idle speed	Approximately 0V

29149_NISS__G0060

Fig. 59 Mass Air Flow Sensor testing—Quest with VG30E Engine

Do not use ECM ground terminals when measuring input/output voltage. Doing so may result in damage to the ECM's transistor. Use a ground other than ECM terminals, such as the ground.

TERMINAL NO.	WIRE COLOR	ITEM	CONDITION	DATA (DC Voltage)
54	R	Mass air flow sensor	**[Engine is running]** • **Warm-up condition** • Idle speed	0.9 - 1.8V
			[Engine is running] • **Warm-up condition** • Engine speed is 2,500 rpm	1.9 - 2.3V
55	G	Mass air flow sensor ground	**[Engine is running]** • **Warm-up condition** • Idle speed	Approximately 0V

29149_NISS__G0048

Fig. 60 Mass Air Flow Sensor testing—Quest with VG33E Engine

ECM TERMINALS AND REFERENCE VALUE

Specification data are reference values and are measured between each terminal and ㊸ (ECCS ground).

TERMINAL NO.	WIRE COLOR	ITEM	CONDITION	DATA (DC Voltage)
47	G	Mass air flow sensor	Engine is running. (Warm-up condition) Idle speed	1.0 - 1.7V
			Engine is running. (Warm-up condition) Engine speed is 2,500 rpm.	1.5 - 2.1V
48	R	Mass air flow sensor ground	Engine is running. (Warm-up condition) Idle speed	0.005 - 0.02V

29149_NISS__G0066

Fig. 61 Mass Air Flow Sensor testing—Sentra with GA16DE Engine

ECM TERMINALS AND REFERENCE VALUE

Specification data are reference values and are measured between each terminal and engine ground.

TERMINAL NO.	WIRE COLOR	ITEM	CONDITION	DATA (DC voltage)
47	OR	Mass air flow sensor	Engine is running. (Warm-up condition) Idle speed	1.3 - 1.7V
			Engine is running. (Warm-up condition) Engine speed is 2,500 rpm.	1.8 - 2.4V
48	W	Mass air flow sensor ground	Engine is running. (Warm-up condition) Idle speed	Approximately 0V

29149_NISS__G0071

Fig. 62 Mass Air Flow Sensor testing—Sentra with SR20DE Engine

Xterra

KA24DE I4 ENGINE, VG33E V6 ENGINE

See figure 63.

1. Turn ignition switch **ON**.
2. Start engine and warm it up sufficiently.
3. Check the voltage between ECM pin 54 and ground.
4. Check for linear voltage rise in response to increases in engine RPM.

Do not use ECM ground terminals when measuring input/output voltage. Doing so may result in damage to the ECM's transistor. Use a ground other than ECM terminals, such as the ground.

TERMINAL NO.	WIRE COLOR	ITEM	CONDITION	DATA (DC Voltage)
54	R	Mass air flow sensor	**[Engine is running]** • **Warm-up condition** • Idle speed	0.9 - 1.8V
			[Engine is running] • **Warm-up condition** • Engine speed is 2,500 rpm	1.9 - 2.3V
55	G	Mass air flow sensor ground	**[Engine is running]** • **Warm-up condition** • Idle speed	Approximately 0V

29149_NISS__G0048

Fig. 63 Mass Air Flow Sensor testing—Xterra

Throttle Position Sensor (TPS)

TESTING

See figures 64 and 65.

240SX

KA24DE I4 ENGINE

1. Check for 5 volt supply voltage between pins 1 and 3.
2. Check output voltage between pins 2 and 3.
3. Compare measured values with the specifications in the illustration.

300ZX

VG30DE, VG30DETT V6 ENGINES

1. Check for 5 volt supply voltage between pins 1 and 3.
2. Check output voltage between pins 2 and 3.
3. Compare measured values with the specifications in the illustration.

Altima

KA24DE I4 ENGINE

1. Check for 5 volt supply voltage between pins 1 and 3.
2. Check output voltage between pins 2 and 3.
3. Compare measured values with the specifications in the illustration.

Frontier

KA24DE I4 ENGINE, VG33DE V6 ENGINE

1. Check for 5 volt supply voltage between pins 1 and 3.
2. Check output voltage between pins 2 and 3.
3. Compare measured values with the specifications in the illustration.

Maxima

VQ30DE V6 ENGINE

1. Check for 5 volt supply voltage between pins 1 and 3.
2. Check output voltage between pins 2 and 3.
3. Compare measured values with the specifications in the illustration.

Pathfinder

VG33E V6 ENGINE

1. Check for 5 volt supply voltage between pins 1 and 3.
2. Check output voltage between pins 2 and 3.
3. Compare measured values with the specifications in the illustration.

Pickup

KA24DE I4 ENGINE

1. Check for 5 volt supply voltage between pins 1 and 3.
2. Check output voltage between pins 2 and 3.
3. Compare measured values with the specifications in the illustration.

Quest

VG30E V6 ENGINE, VG33E V6 ENGINE

1. Check for 5 volt supply voltage between pins 1 and 3.
2. Check output voltage between pins 2 and 3.
3. Compare measured values with the specifications in the illustration.

Sentra

GA16DE I4 ENGINE, SR20DE I4 ENGINE

1. Check for 5 volt supply voltage between pins 1 and 3.
2. Check output voltage between pins 2 and 3.
3. Compare measured values with the specifications in the illustration.

Xterra

KA24DE I4 ENGINE, VG33E V6 ENGINE

1. Check for 5 volt supply voltage between pins 1 and 3.
2. Check output voltage between pins 2 and 3.
3. Compare measured values with the specifications in the illustration.

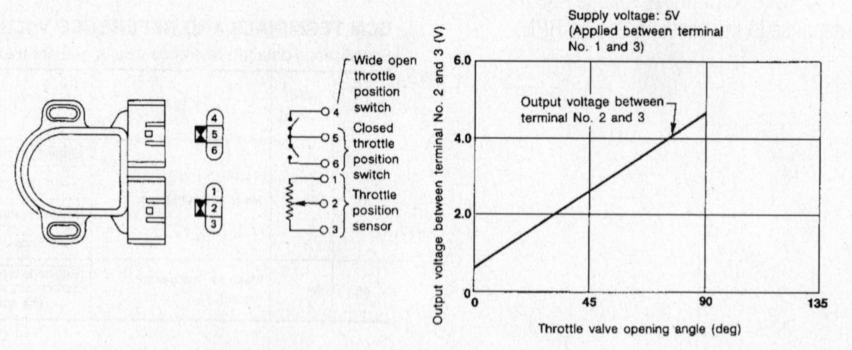

29149_NISS__G0038

Fig. 64 Throttle Position Sensor Testing

Throttle Position Sensor

Throttle valve conditions	Voltage (at normal operating temperature, engine off, ignition switch ON, throttle opener disengaged)
Completely closed (a)	0.15 - 0.85V
Partially open	Between (a) and (b)
Completely open (b)	3.5 - 4.7V

Fig. 65 Throttle Position Sensor specifications

29149_NISS__G0039

THOMSON
DELMAR LEARNING

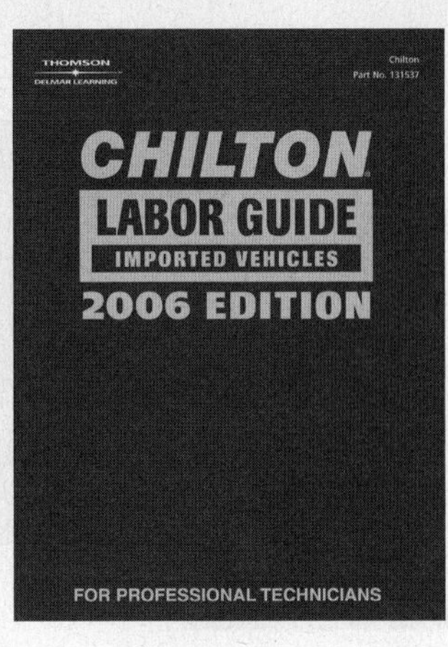

Chilton 2006 Labor Guide Manuals ISBN 1-4180-1688-8/Part No. 131688

We've added hundreds of new labor operations-including maintenance services and electronic system diagnosis to the Chilton 2006 Labor Guide Manuals. All labor times for the 1981 through 2005 (and available 2006) domestic and imported vehicles consider the real world environment in which technicians work. The parts terminology is more standardized across different OEMs to simplify reference. You can reference any of three labor times for many models: Chilton's Standard and Severe Service times, plus OEM warranty time. Each OEM is arranged alphabetically by section for easy reference, and improved indexing means easier access to today's repair industry standards. Chilton labor times are accepted by most insurance and extended warranty companies. Vehicle makes and models conform to current Automotive Aftermarket Industry Association standards. Make sure you have the latest edition of these manuals because our experts have updated hundreds of labor times for earlier models as well!

Labor Guide Manual Benefits:

- a total of 2,500 pages of updated Chilton labor times split into two volumes (Domestic and Imported vehicles)
- enjoy quicker referencing than ever by using separate, easier-to-handle, domestic and imported vehicle manuals
- find it fast by using tabs that display contents by manufacturer and model, two indexes — labor operations and systems — in each model group, and page numbering that includes manufacturer code so you know where you are in the book
- manufacturers are arranged alphabetically within each volume

Hardcover Manuals are 8 1/2" x 11", ©2006

Labor Guide CD-ROM Benefits:

- automatically calculates labor charges, taxes, and your parts as total job estimates
- creates professional estimates for your customers and worksheets for your technicians, allowing them to print them whenever needed
- keeps track of customers, prior estimates, and your own parts or package jobs

CD ISBN 1-4180-0605-X/Part No. 130605
©2006

Chilton 2006 Mechanical Service Manuals

The *Chilton® 2006 Mechanical Service Manuals* provide updated coverage through 2005 models and even many 2006 models, as made available from original equipment manufacturers (OEMs). Chilton is still your reliable source for fast, accurate repairs and reassembly and it still provides the lowest-priced professional repair manuals on the market! These manuals are organized by make, model, and system so information gathering is easier. Now with even more illustrations and a streamlined index, it's no wonder more automotive professionals turn to Chilton Professional Manuals for their mechanical service and repair information.

Mechanical Service Manual Benefits:

- access up-to-date service and repair information covering model years 2002-2006, all logically arranged by manufacturer
- follow clear, step-by-step procedures—from drive train to chassis—to yield fast, accurate results
- service more mechanical systems, including brakes, engines, suspensions, steering, and related components
- know what special tools are required for specific jobs, as Chilton editors describe and illustrate them to make repair work go more smoothly

2006 Editions

Chilton 2006 DaimlerChrysler Mechanical Service Manual—ISBN 1-4180-0600-9/Part No. 130600
Chilton 2006 Ford Mechanical Service Manual—ISBN 1-4180-0601-7/Part No. 130601
Chilton 2006 General Motors Mechanical Service Manual—ISBN 1-4180-0602-5/Part No. 130602
Chilton 2006 Asian Mechanical Service Manual—Volume I—ISBN 1-4180-0947-4/Part No. 130947
Chilton 2006 Asian Mechanical Service Manual—Volume II—ISBN 1-4180-0948-2/Part No. 130948
Chilton 2006 Asian Mechanical Service Manual—Volume III—ISBN 1-4180-0949-0/Part No. 130949
Chilton 2006 Asian Mechanical Service Manual—3 Volume Set—ISBN 1-4180-0603-3/Part No. 130603
Chilton 2006 European Mechanical Service Manual—ISBN 1-4180-0604-1/Part No. 130604

Manuals are 8 1/2" x 11"

2005 Editions

Chilton 2005 General Motors Mechanical Service Manual—ISBN 1-4018-7146-1/Part No. 27146
Chilton 2005 Chrysler Mechanical Service Manual—ISBN 1-4018-6718-9/Part No. 26718
Chilton 2005 Ford Mechanical Service Manual—ISBN 1-4018-6719-7/Part No. 26719
Chilton 2005 European Mechanical Service Manual—ISBN 1-4018-6720-0/Part No. 126720
Chilton 2005 Asian Mechanical Service Manual – Volume I—(Acura-Mazda) ISBN 1-4018-6716-2/Part No. 26716
Chilton 2005 Asian Mechanical Service Manual – Volume II—(Mitsubishi-Toyota)
ISBN 1-4018-6717-0/Part No. 26717
Chilton 2005 Asian Mechanical Service Manual – 2 Volume Set—ISBN 1-4018-7180-1/Part No. 27180

Manuals are 8 1/2" x 11", ©2005

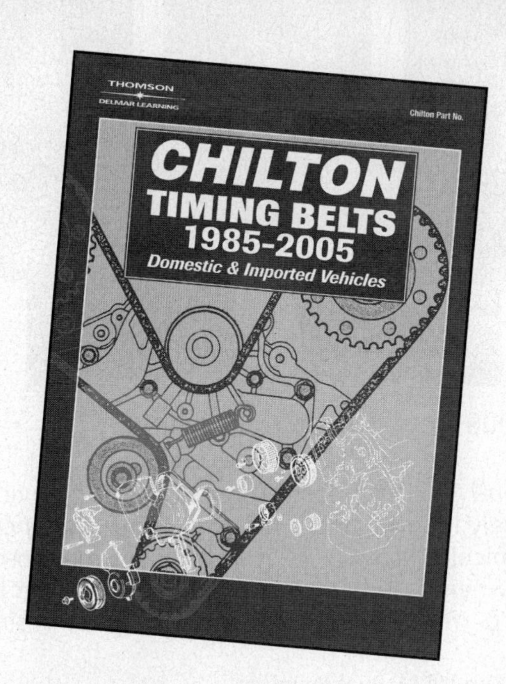

Chilton Timing Belts, 1985-2005

Chilton

ISBN 1-4018-9880-7/Part No. 129880

Timing belt procedures can represent increased profits for automotive repair shops and service stations, and this manual contains all the information automotive technicians need to properly service timing belts on domestic and imported cars, vans, and light trucks through 2005 models. Clear, straightforward procedures, illustrations, and specifications help to communicate 20 years of vehicle applications for fast, accurate inspection, replacement, and tensioning of timing belts. Readers will learn step-by-step how to perform key procedures both quickly and safely, while learning the correct labor time to charge for the service. OEM-recommended replacement intervals for proper maintenance of customer's vehicles are also featured. Professional technicians, trainers, industry professionals, and automotive enthusiasts will all benefit from this manual.

Benefits

- simplify your work - use illustrations showing camshaft and crankshaft timing alignment marks, exploded views of timing belt components, belt routings, special tools, and tensioning adjustments
- take advantage of manufacturer tips on performing correct procedures and adjustments
- prevent engine damage - identify interference or free-wheeling engines
- be confident - use updated manufacturer-recommended timing belt replacement procedures, intervals, and torque specifications
- provide accurate estimates - use trusted Chilton labor times for standard or severe service, including additional time for air conditioning, power steering, balance shaft belt, and water pump removal
- save time - quickly identify year, make, and model in the contents, turn to the start of each engine section, and confirm engine application or VIN code, labor times, and replacement intervals all in one table

544 pp, 8 1/2" x 11", softcover, ©2006

Chilton 2006 Diagnostic Service Manuals

The *Chilton® 2006 Diagnostic Service Manuals* provide technicians with the critical diagnostic information they need to accurately identify and solve engine performance problems. Clear explanations, specifications, and illustrations help technicians diagnose second generation on-board diagnostic (OBD-II) systems. *Chilton Diagnostic Service Manuals*, when used with an engine analyzer, scan tool, or lab scope, allow diagnosticians to understand functions of engine performance components and systems, simplify testing procedures, and diagnose trouble codes.

Diagnostic Service Manual Benefits:
- provide training information in addition to reference material
- explain engine performance components and system operation
- function as exceptional diagnostic companions when analyzing automotive drive-train performance problems
- provide a comprehensive list of trouble code titles, conditions, and possible causes
- reduce diagnostic and repair time using expert testing procedures and troubleshooting hints

2006 Editions

Chilton 2006 DaimlerChrysler Diagnostic Service Manual
ISBN 1-4180-2118-0/Part No. 132118
Chilton 2006 Ford Diagnostic Service Manual
ISBN 1-4180-2119-9/Part No. 132119
Chilton 2006 General Motors Diagnostic Service Manual
ISBN 1-4180-2120-2/Part No. 132120
Chilton 2006 Asian Diagnostic Service Manual, Volume I
ISBN 1-4180-2913-0/Part No. 132913
Chilton 2006 Asian Diagnostic Service Manual, Volume II
ISBN 1-4180-2914-9/Part No. 132914
Chilton 2006 Asian Diagnostic Service Manual, Volume III
ISBN 1-4180-2915-7/Part No. 132915
Chilton 2006 Asian Diagnostic Service Manual, 3 Volume Set
ISBN 1-4180-2986-6/Part No. 132986
Chilton 2006 European Diagnostic Service Manual
ISBN 1-4180-2924-6/Part No. 132924

Manuals are 8 1/2" x 11"

2005 Editions

Chilton 2005 General Motors Diagnostic Service Manual
ISBN 1-4180-0552-5/Part No. 130552
Chilton 2005 Chrysler Diagnostic Service Manual
ISBN 1-4180-0550-9/Part No. 130550
Chilton 2005 Ford Diagnostic Service Manual
ISBN 1-4180-0551-7/Part No. 130551
Chilton 2005 Asian Diagnostic Service Manual
ISBN 1-4180-0553-3/Part No. 130553

Manuals are 8 1/2" x 11", ©2005

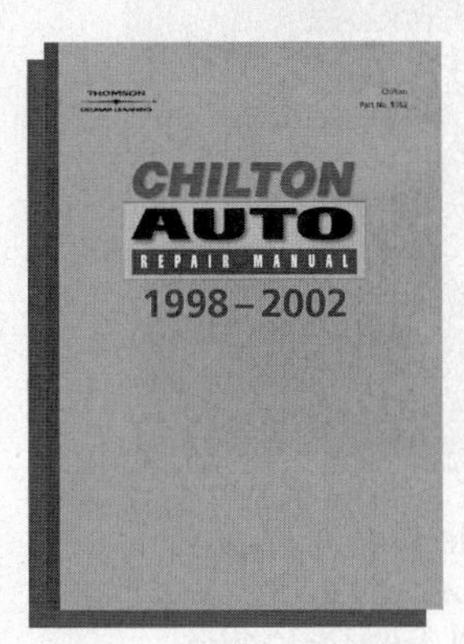

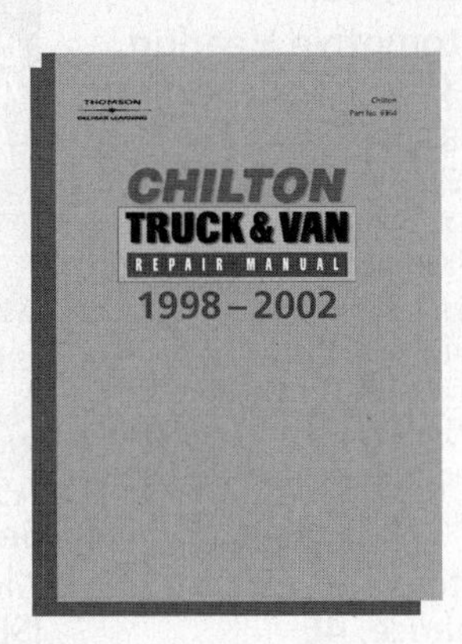

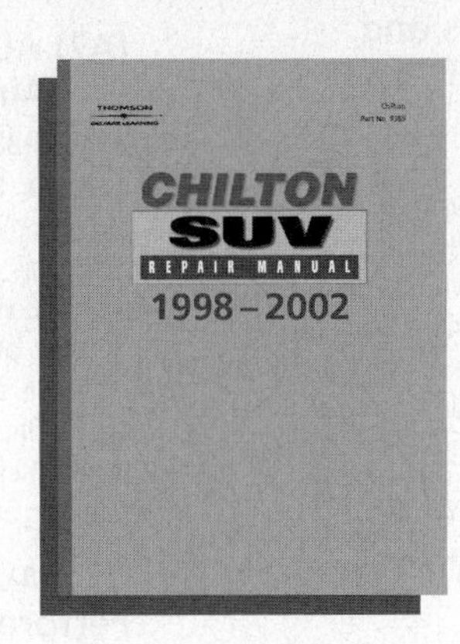

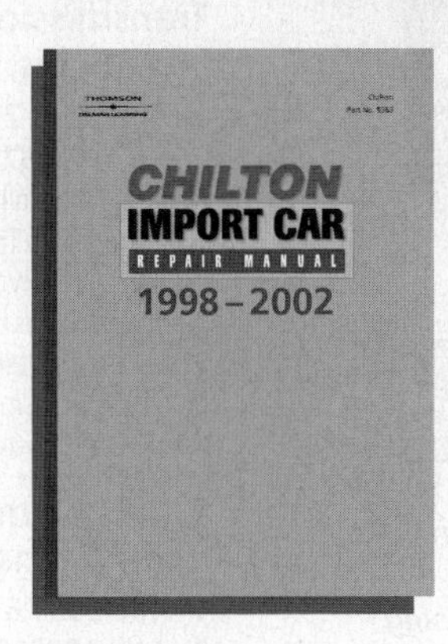

Chilton Service Manuals - Perennial Editions

The *Chilton® Perennial Editions* contain repair and maintenance information for popular mechanical systems that may not be available elsewhere. They offer a wide range of repair information on cars, trucks, vans, and SUVs dating back to the early 1960s, and as current as 2002. Information for 1993 and later model years includes scheduled maintenance interval charts.

Benefits:

- covers the most common vehicle models found in the repair aftermarket today
- gain quick understanding of systems using exploded-view illustrations, diagrams, and charts
- simplify tough jobs with easy-to-follow removal and installation instructions for heater core and other components
- obtain complete coverage of repair procedures from drive train to chassis and associated components

Auto Repair Manual, 1998-2002, 1,426 pages
ISBN 0-8019-9362-8/Part No. 9362
Auto Repair Manual, 1993-1997, 2,064 pages
ISBN 0-8019-7919-6/Part No. 7919
Auto Repair Manual, 1988-1992, 1,284 pages
ISBN 0-8019-7906-4/Part No. 7906
Auto Repair Manual, 1980-1987, 1,344 pages
ISBN 0-8019-7670-7/Part No. 7670

Import Car Repair Manual, 1998-2002, 1,792 pps
ISBN 0-8019-9363-6/Part No. 9363
Import Car Repair Manual, 1993-1997, 2,080 pps
ISBN 0-8019-7920-X/Part No. 7920
Import Car Repair Manual, 1988-1992, 1,632 pages
ISBN 0-8019-7907-2/Part No. 7907
Import Car Repair Manual, 1980-1987, 1,488 pages
ISBN 0-8019-7672-3/Part No. 7672

Truck & Van Repair Manual, 1998-2002, 1,408 pages
ISBN 0-8019-9364-4/Part No. 9364
Truck & Van Repair Manual, 1993-1997, 2,096 pages
ISBN 0-8019-7921-8/Part No. 7921
Truck & Van Repair Manual, 1991-1995, 1,664 pages
ISBN 0-8019-7911-0/Part No. 7911
Truck & Van Repair Manual, 1986-1990, 1,536 pages
ISBN 0-8019-7902-1/Part No. 7902
Truck & Van Repair Manual, 1979-1986, 1,440 pages
ISBN 0-8019-7655-3/Part No. 7655

SUV Repair Manual, 1998-2002, 1,292 pages
ISBN 0-8019-9365-2/Part No. 9365

Hardcover manuals are 8 1/2" x 11"

Chilton Collector's Editions—*Reference Manuals for Vintage Vehicles*
Auto Repair Manual, 1964-1971, ISBN 0-8019-5974-8/Part No. 5974,
Truck & Van Repair Manual, 1971-1978, ISBN 0-8019-7012-1/Part No. 7012

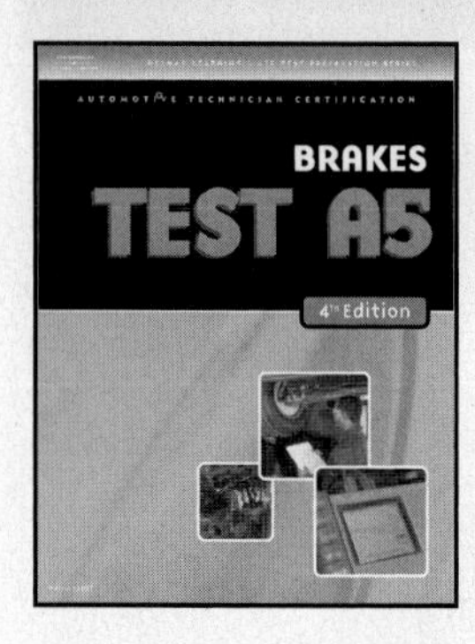

ASE Test Preparation Series

Thomson Delmar Learning

ISBN 1-4180-3954-3
Part No. 133954
(Complete Set: A1-A8, L1, P2 X1, C1)

Thomson Delmar Learning has developed comprehensive ASE Test Preparation Manuals to help automotive technicians increase their success on these certification programs. The material covers the topics one might find during the test process. The booklets include many review questions and answers, as well as detailed descriptions of the repairs involved. Designed to look like the actual test, participants will feel more comfortable with practice, which will translate into greater success in taking the actual tests. The design of the Thomson Delmar Learning product also includes helpful test taking hints and student preparation ideas designed to enhance success.

Benefits

- history of the ASE
- test-taking strategies
- tasks lists and overview
- sample test questions
- ASE-style exams
- explanations to the answers (right and wrong)
- glossary of terms

(A1) Automotive Engine Repair, 4E

1-4180-3878-4
Part No. 133878
Includes the following topics: General Engine Diagnosis, Cylinder Head and Valve Train Diagnosis and Repair, Engine Block Diagnosis and Repair, Lubrication and Cooling Systems Diagnosis and Repair, and Fuel, Electrical, Ignition and Exhaust Systems Inspection and Service.

(A2) Automotive Transmissions and Transaxles, 4E

1-4180-3879-2
Part No. 133879
Includes the following topics: General Transmission/Transaxle Diagnosis (Mechanical/Hydraulic Systems and Electronic Systems), Transmission/Transaxle Maintenance and Adjustment, In-Vehicle Transmission/Transaxle Repair, Off-Vehicle Transmission/Transaxle Repair.

(A3) Automotive Manual Drive Trains and Axles, 4E

1-4180-3880-6
Part No.133880
Includes the following topics: Clutch Diagnosis and Repair, Transmission Diagnosis and Repair, Transaxle Diagnosis and Repair, Drive Shaft/Half Shaft and Universal Joint/Constant Velocity (CV) Joint Diagnosis and Repair (Front and Rear Wheel Drive), Rear Axle Diagnosis and Repair, Four Wheel Drive/All Wheel Drive Component Diagnosis and Repair.

(A4) Automotive Suspension and Steering, 4E

1-4180-3881-4
Part No. 133881
Includes the following topics: Steering Systems Diagnosis and Repair (Steering Columns and Manual Steering Gears, Power Assisted Steering Units, Steering Linkage), Suspension Systems Diagnosis and Repair (Front Suspensions, Rear Suspensions, Miscellaneous Services), Wheel Alignment Diagnosis, Adjustment and Repair, and Wheel and Tire Diagnosis and Repair.

(A5) Automotive Brakes, 4E

1-4180-3882-2
Part No. 133882
Iincludes the following topics: Hydraulic System Diagnosis and Repair, Drum Brake Diagnosis and Repair, Disc Brake Diagnosis and Repair, Power Assist Units Diagnosis and Repair, Miscellaneous Systems Diagnosis and Repair, Antilock Brake Systems (ABS) Diagnosis and Repair.

(A6) Automotive Electrical-Electronic Systems, 4E

1-4180-3883-0
Part No.133883
Includes the following topics: General Electrical/Electronic Systems Diagnosis, Battery Diagnosis and Service, Starting Systems Diagnosis and Repair, Charging Systems Diagnosis and Repair, Lighting Systems Diagnosis and Repair, Gauges, Warning Devices and Driver Information Systems Diagnosis and Repair, Horn and Wiper/Washer Diagnosis and Repair.

(A7) Automotive Heating and Air Conditioning, 4E

1-4180-3884-9
Part No. 133884
Includes the following topics: A/C System Diagnosis and Repair, Refrigeration System Component Diagnosis and Repair, Heating and Engine Cooling Systems Diagnosis and Repair, Operating Systems and Related Controls Diagnosis and Repair, Refrigerant Recovery, Recycling, Handling and Retrofit.

(A8) Automotive Engine Performance, 4E

1-4180-3885-7
Part No. 133885
Includes the following topics: General Engine Diagnosis, Ignition System Diagnosis and Repair, Fuel, Air Induction and Exhaust Systems Diagnosis and Repair, Emissions Control Systems Diagnosis and Repair (Including OBDII), Computerized Engine Controls Diagnosis and Repair (Including OBDII), Engine Electrical Systems Diagnosis and Repair.

(L1) Automotive Advanced Engine Performance, 4E

1-4180-3888-1
Part No. 133888
Includes the following topics: General Powertrain Diagnosis, Computerized Powertrain Controls Diagnosis (Including OBDII), Ignition System Diagnosis, Fuel Systems and Air Induction Systems Diagnosis, Emission Control Systems Diagnosis, I/M Failure Diagnosis.

(P2) Automobile Parts Specialist, 4E

1-4180-3887-3
Part No. 133887
Includes the following topics: General Operations, Customer Relations and Sales Skills, Vehicle Systems Knowledge, Vehicle Identification, Cataloging Skills, Inventory Management, Merchandising.

(X1) Exhaust Systems, 4E

1-4180-3886-5
Part No. 133886
Includes the following topics: Exhaust Systems Inspection and Repair, Emissions Systems Diagnosis, Exhaust System Fabrication, Exhaust System Installation, Exhaust System Repair Regulations.

(C1) Automotive Service Consultant, 4E

See next page for details

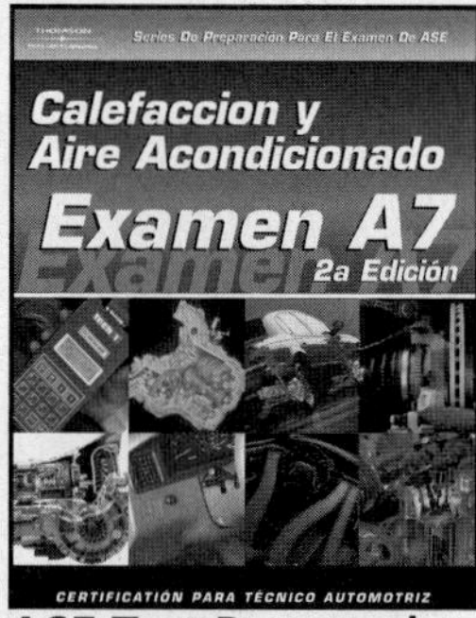

ASE Test Preparation Series in Español!

Thomson Delmar Learning

ISBN 1-4018-1530-8
(Complete Set: A1-A8, L1, P2, X1)

Now available in Español – the first of its kind for Spanish-speaking technicians! This comprehensive package of ASE test preparation booklets are intended for any Spanish-speaking automotive technician who is preparing to take an ASE examination. The series includes questions that relate to each competency required for certification by ASE. In addition to a multitude of questions, the reason why each answer is right or wrong is explained, along with task lists and overview, test-taking strategies, and more.

(A1) Reparación de Motores, 2A Edición
1-4018-1014-4/Part No. 21014

(A2) Transmision Automática/ Eje de Transmision Automática, 2A Edición
1-4018-1015-2/Part No. 21015

(A3) Tren de y Mando Ejes Manuales, 2A Edición
1-4018-1016-0/Part No. 21016

(A4) Suspensión y Dirección, 2A Edición
1-4018-1017-9/Part No. 21017

(A5) Frenos, 2A Edición
1-4018-1018-7/Part No. 21018

(A6) Sistemas Eléctricos/ Electrónicos, 2A Edición
1-4018-1019-5/Part No. 21019

(A7) Calefacción y Aire Acondicionado, 2A Edición
1-4018-1020-9/Part No. 21020

(A8) Funcionamiento de Motores, 2A Edición
1-4018-1021-7/Part No. 21021

(L1) Especialista en el Funciommiato Avansado de Motores, 2A Edición
1-4018-1022-5/Part No. 21022

(P2) Especialista en Partes de Automovil, 2A Edición
1-4018-1023-3/Part No. 21023

(X1) Sistemas de Escape, 2A Edición
1-4018-1024-1/Part No. 21024

ASE Test Preparation Manual—C1 Automotive Service Consultant, 4E

Thomson Delmar Learning

ISBN 1-4180-3889-X

Part No.133889

Prepare to pass the Service Consultant ASE Exam with help from this new test preparation booklet. The new C1 Exam is designed to measure systems knowledge and people skills of those who come in contact with the customer. It will contain questions on Communications, Product Knowledge, Sales Skills, and Shop Operations.

Benefits:

- the ASE task list is fully up-to-date, while current test prep questions reflect the most recent ASE task changes for the broadest knowledge possible
- hundreds of ASE-style exam questions adequately prepare readers to successfully pass the ASE exam
- readers are given multiple opportunities to check their understanding of critical concepts through sample problems, refresher materials, and competency-specific test questions
- overviews of each task provide a great reference point to help answer difficult ASE questions
- explanations for each answer help the user understand why the response is correct or incorrect

Softcover manual is 8 1/2" x 11", ©2006

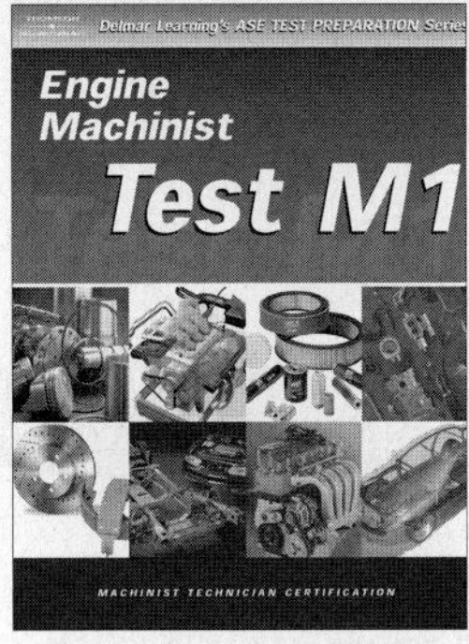

ASE Test Preparation Manuals—M1 Engine Machinist

Thomson Delmar Learning

ISBN 0-7668-6283-6

Part No. 16283

(Complete Set: M1-M3)

With an abundance of quality content, Thomson Delmar Learning's ASE Test Preparation M-Series contains detailed information designed to help you pass the ASE exams. Each manual combines refresher materials with an abundance of sample test questions, as well as a wealth of information regarding test-taking strategies and the types of questions found in an ASE exam. In addition to the questions, thorough explanations are provided as to why each answer is correct or incorrect.

Benefits:

- the History section explains why the exams are important to the industry
- test-taking strategies help prepare technicians for the environment they will encounter during the actual exam

(M1) Cylinder Head Specialist
0-7668-6280-1/Part No. 16280

(M2) Cylinder Block Specialist
0-7668-6281-X/
Part No. 16281

(M3) Assembly Specialist
0-7668-6282-8/
Part No. 16282

Softcover manuals are 8 1/2" x 11", ©2002

ASE Test Preparation Manuals—B2-B6 Collison Repair and Refinishing

Thomson Delmar Learning

ISBN 1-4018-5120-7/Part No. 25120

(Complete Set: B2-B6)

This fully expanded third edition has been contains high quality ASE test preparation material designed to increase the test taking success of collision repair and refinish technicians. Each book in the series provides valuable preparation for technicians seeking certification in one or more of the ASE collision repair areas. Readers are afforded scores of opportunities to ascertain their knowledge of critical concepts through the extensive array of sample problems, ASE-style exams, and competency-specific test questions required for certification by ASE.

Benefits:

- current, job-related ASE-style exam questions reflecting the most recent ASE task changes test the skills that technicians need to know on the job
- each book contains a general knowledge pretest, a sample test, and additional practice learning that add up to the most real-test practice time available

(B2) Painting and Refinishing, 3E
1-4018-3664-X/Part No. 23664

(B3) Non-Structural Analysis and Damage Repair, 3E
1-4018-3665-8/Part No. 23665

(B4) Structural Analysis and Damage Repair, 3E
1-4018-3666-6/Part No. 23666

(B5) Mechanical and Electrical Components, 3E
1-4018-3667-4/Part No. 23667

(B6) Damage Analysis and Estimation, 3E
1-4018-3668-2/Part No. 23668

Softcover manuals are 8 1/2" x 11", ©2006

Technician Test Preparation—Automotive Bilingual Series

Thomson Delmar Learning

Now both English and Spanish speaking technicians seeking ASE certification can access online test preparation material with ease! The *TTP-Automotive (Bilingual)* series for automotive training and certification provides up-to-date technology and content for tests A1-A8, L1, P2, X1, and C1. An easy-to-use format combined with helpful remediation addresses the unique needs of technicians by clearly demonstrating text-based theory for enhanced learning and retention. Not only is *TTP-Automotive (Bilingual)* the ultimate in test preparation, but it is also an excellent learning tool!

Technician Test Preparation Benefits:

- maps to the latest ASE task lists to familiarize users with the actual work they should be able to do as technicians when taking the ASE tests
- well-illustrated remediation offered via digitized video clips, animations, and high impact graphics further explains key concepts for a more effective learning process
- practice questions provide helpful hints, insight into right and wrong answers, and links to further study specific task areas
- detailed reports provide accurate test results and instant feedback for selected test types so that users can pinpoint the task areas needing improvement
- switch between Spanish and English versions at the click of a button

Call Your Thomson Delmar Learning Sales Rep for Part Numbers & Pricing

Visit **www.techniciantestprep.com** to see the latest modules and a free demo!

The ASE "Passing Lane" Package

Thomson Delmar Learning

ISBN 0-7668-4338-6

(Complete Set: A1-A8, L1, P2)

The most comprehensive test preparation for Automotive Tests A1-A8, L1, and P2. Combining the most thorough ASE Test Preparation books with the latest in ASE videos, this package provides a program of self-study for the automotive ASE Tests.

Book Benefits:

- test-taking strategies
- tasks lists and overview
- sample test questions
- ASE-style exams
- explanations to the answers
- glossary of terms

Video Benefits:

- lively, easy to follow videos emphasize safety throughout
- covers major task areas and topics for each of the ASE exams
- accompanying Activity Sheets help users comprehend and retain information

(A1) Automotive Engine Repair Book/Video,
0-7668-4181-2
(A2) Automotive Transmissions and Transaxles Book/Video,
0-7668-4182-0
(A3) Automotive Manual Drive Trains and Axles Book/Video,
0-7668-4183-9
(A4) Automotive Suspension and Steering Book/Video,
0-7668-4184-7
(A5) Automotive Brakes Book/Video,
0-7668-4185-5
(A6) Automotive Electrical-Electronics Systems Book/Video,
0-7668-4186-3
(A7) Automotive Heating and Air Conditioning Book/Video,
0-7668-4187-1
(A8) Automotive Engine Performance Book/Video,
0-7668-4188-X
(L1) Automotive Advanced Engine Performance Book/Video,
0-7668-4189-8
(P2) Automobile Parts Specialist Book/Video,
0-7668-4190-1

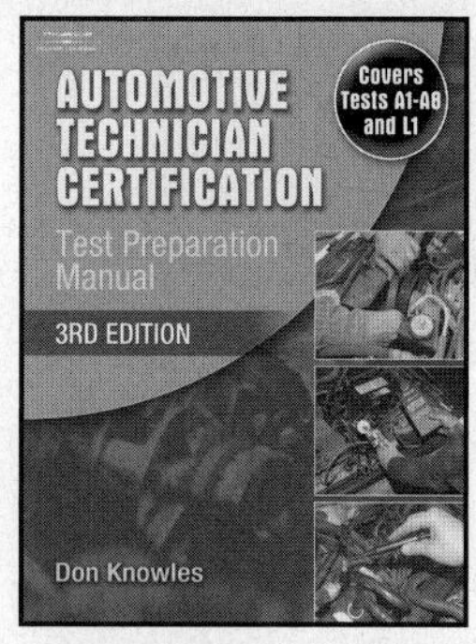

Automotive Technician Certification Test Preparation Manual, 3E

Don Knowles

ISBN 1-4180-4926-3/ Part No. 134926

Filled with updated task list theory, practice tests, and abundant, demonstrative graphics, this revised edition provides all the latest information required to sufficiently prepare technicians to pass each of the A1-A8 and L1 ASE certification exams. Each chapter begins with a pretest that indicates the depth of preparation required to become familiar with the information in the chapter, followed by a description of each ASE task and the must-have information related to the task. ASE-type questions at the end of each chapter appear in the same format as on actual ASE tests to further prepare users to pass each exam.

Benefits

- current information provides practice questions which match the latest ASE task list
- answers to pretest questions and helpful analyses at the end of each chapter provide learners with faster access to accurate information
- supportive "Hints" throughout each chapter help users work through the process of determining the correct answers to the questions

CONTENTS

Chapter 1 - Engine Repair. Chapter 2 - Automatic Transmission/Transaxle. Chapter 3 - Manual Drive Train and Axles. Chapter 4 - Suspension and Steering. Chapter 5 - Brakes. Chapter 6 - Electrical/Electronic Systems. Chapter 7 - Heating, Ventilation, and Air Conditioning Systems. Chapter 8 - Engine Performance. Chapter 9 - Advanced Engine Performance.

656 pp, 8½" x 11", softcover, ©2007

ATC Challenge 3.0 CD-ROM

Thomson Delmar Learning

ISBN 0-7668-2982-0

These exciting interactive CD-ROMs have been designed to prepare technicians for successful completion of the Automotive ASE task areas (A1-A8, L1, P2, and F1). This multimedia software assesses strengths and weaknesses by identifying topics needing further study while allowing users to review ASE task areas at their own pace. Explanations, hints, notes, and a glossary aid the user in comprehension, critical thinking and retention. These CD-ROMs offer hundreds of ASE-style questions, a test taking strategy section and LAN compatibility.

CD-ROM, ©2001

Site License Available for Multiple Unit Purchases or Multiple Workstations for ATC Challenge 3.0:
User 1: Full Price (List or Net)
Users 2-5:
$80/workstation + Full Price
Users 6-10:
$70/workstation + Full Price
Users 11-20:
$60/workstation + Full Price
Users 21+:
$50/workstation + Full Price

ATC Challenge for P2

Thomson Delmar Learning

ISBN 0-7668-1827-6

This interactive CD-ROM contains material that will help prepare technicians for the Automotive Parts Specialist (P2) ASE certification exam.
CD-ROM, ©2000

NATEF Standards Job Sheets, 2E

Thomson Delmar Learning

ISBN 1-4180-2082-6

(Complete Set: A1-A8)

New from today's leading automotive education publisher, each of our eight NATEF (National Automotive Technicians Education Foundation) Standards Job Sheets workbooks has been thoughtfully designed to assist users in gaining valuable job preparedness skills and mastering specific technical competencies required for success as a professional automotive technician. Ideal for use as a stand-alone item, or with any comprehensive or topic-specific automotive text, the entire series is based on the 2005 NATEF tasks and consists of individual books for each of the following areas: Engine Repair, Automatic Transmissions/Transaxles, Manual Drive Trains and Axles, Suspension and Steering, Brakes, Electricity/Electronics, Heating and Air Conditioning, and Engine Performance.

Key Features

- manuals are not keyed to a specific text, making it easy to use one or more of them in any automotive training program in which NATEF coverage is desired
- NATEF tasks are addressed in each manual, providing students with a first-quality, comprehensive learning experience

JOB SHEETS AVAILABLE FOR:

(A1) Automotive Engine Repair, 1-4180-2074-5
(A2) Automatic Transmissions and Transaxles, 1-4180-2075-3
(A3) Manual Drive Trains and Axles, 1-4180-2076-1
(A4) Automotive Suspension and Steering, 1-4180-2077-X
(A5) Automotive Brakes, 1-4180-2078-8
(A6) Automotive Electrical and Electronic Systems, 1-4180-2079-6
(A7) Automotive Heating and Air Conditioning, 1-4180-2080-X
(A8) Automotive Engine Performance, 1-4180-2081-8

Softcover manuals are 8½" x 11", ©2006

The Service Consultant: Working in an Automotive Facility

Ronald A. Garner & C. William Garner

ISBN 1-4018-7990-X
Part No. 127990

This book examines the multi-faceted responsibilities of an automotive service consultant. It outlines task-oriented procedures for day-to day operations and provides an understanding of how service techniques are used to maximize customer satisfaction and profitability. Content follows the tasks identified by ASE for Automotive Service Consultant (C1). ASE terminology is used throughout to describe the people and businesses servicing the driving public. Coverage examines communications specific to customer relations and sales as well as internal communications, relations, and supervision. Customer delivery and follow up round out this thorough exploration of the functions of a successful automotive service consultant.

Key Features

- content correlating to the ASE tasks for Automotive Service Consultant (C1) helps readers better prepare for the ASE certification exam
- activities that require readers to interview local service facility managers provide opportunities for establishing important industry contacts for the future
- careful attention to the sequence of job duties for the service consultant gives readers a clear picture of the types of work and responsibilities that will be expected of them when they enter the workforce

224 pp, 7 3/8" x 9", hardcover, ©2005

Managing Automotive Businesses: Strategic Planning, Personnel and Finances

Ronald A. Garner & C. William Garner

ISBN 1-4018-9896-3
Part No. 129896

The success of any organization most often depends on the execution and management of such strategic issues as business development, personnel, and fiscal operations. This new book introduces readers to the duties and practices assigned to service managers in the successful operation of an automotive service facility. Coverage begins with a general discussion of the management structure and the service manager's role in facility operations. Consideration is then given to navigation of the personnel process from the recruitment of workers to supervision of their performance. The financial business practices of a service manager familiarizes readers with the importance of fiscal responsibility in the operation of a lucrative automotive service business.

Key Features

- fosters a thorough understanding of strategic planning from the owner's perspective to the service manager's level in the establishment of a productive service facility
- information on the recruitment and retention of employees and the establishment of rules and performance measures gives readers valuable insight into the "real world" of the automobile business
- the analysis of financial statements exposes readers to the basic activities and duties that are typical of an automotive service manager's position

272 pp, 7 3/8" x 9 1/4", hardcover, ©2006

AUTOMOTIVE SERVICE MANAGEMENT SERIES

This pioneering eight-book series offers automotive service shop owners and those wanting to be shop owners the necessary business and customer service skills to run a successful automotive service facility.

The series covers three main topic areas: personnel management, business management, and sales and marketing. Each book provides a framework to help technicians make consistent, high-quality, and productive service a part of every day shop operations. According to the author, "Great performance coupled with increased customer loyalty, trust, and operational excellence will almost always result in increased profits."

Automotive Service Management Series Benefits:

- real-world approach reflects author's experience as a fourth generation technician, a repair & service company owner, and an automotive industry trainer
- all-inclusive coverage spans from designing an automotive repair facility floor plan through financial management techniques, customer/staff relations, and more
- length of each book makes it easy to incorporate this series into workshops, seminars, and training/education courses
- information is available "as is" or for customization

Total Customer Relationship Management
ISBN 1-4018-2657-1/Part No. 22657
From Intent to Implementation
ISBN 1-4018-2658-X/Part No. 22658
Operational Excellence
ISBN 1-4018-2659-8/Part No. 22659
Building a Team
ISBN 1-4018-2660-1/Part No. 22660
The High Performance Shop
ISBN 1-4018-2661-X/Part No. 22661
Safety Communications
ISBN 1-4018-2662-8/Part No. 22662
Managing Dollars with Sense
ISBN 1-4018-2663-6/Part No. 22663
Operations Management
ISBN 1-4018-2665-2/Part No. 22665
Entire Set of 8 Books
ISBN 1-4018-2499-4/Part No. 22499

Softcover manuals are 8 1/2" x 11", ©2003

ABOUT THE AUTHOR

Mitch Schneider is a fourth generation mechanic/technician and is a frequent speaker at major conventions and meetings of automotive industry trade organizations. Schneider is also an award-winning journalist and is a regular contributor and senior contributing editor for Motor Age magazine. He provides commentary on the evolving relationship between service dealers, jobbers, warehouse directors and manufacturers.

Schneider has also appeared on the TNN cable show "Truckin' USA" where he hosted the "Tech Tips" segment. In addition to operating the award-winning Schneider's Automotive for 22 years in Simi Valley, CA, he is also the president and founder of Schneider's Future-Tech, a service company specializing in conducting management seminars for automotive service dealers, jobbers, warehouse distribution companies, and manufacturers.

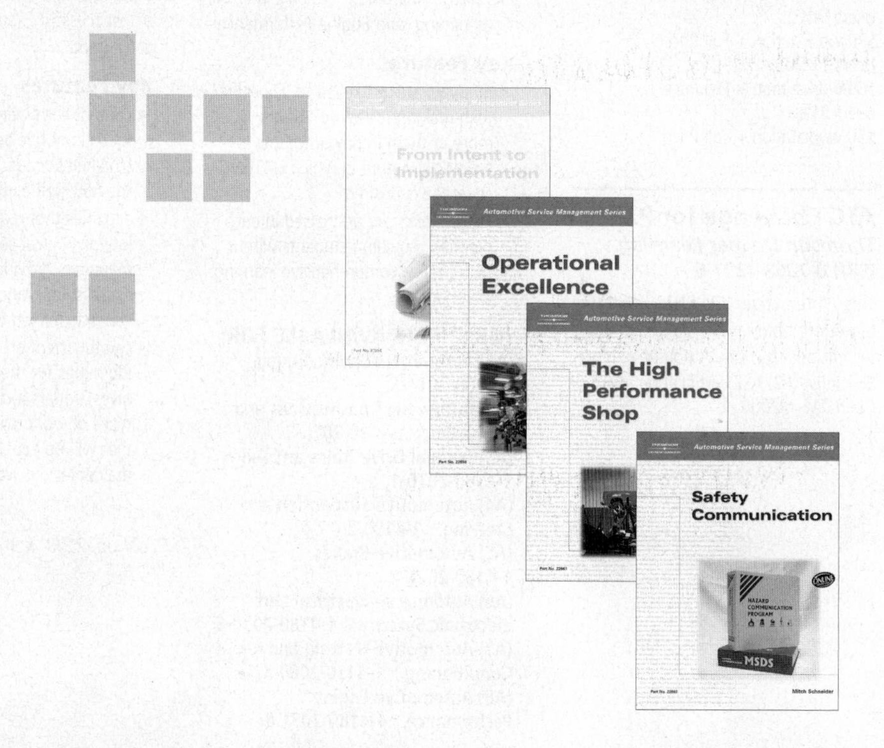

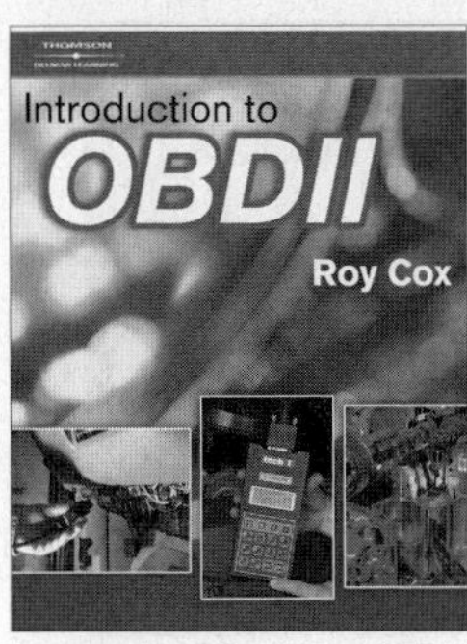

Introduction to OBDII
Roy Cox
ISBN 1-4180-1220-3
Part No. 131220

Here's an easy-to-understand, logical guide to the diagnosis and repair of today's complex and sophisticated automotive control systems! Introduction to On-Board Diagnostics (OBD II) readers will learn the fundamentals of how to perform diagnostic procedures, and be provided with valuable reference material for diagnosing and troubleshooting components and circuits. This book provides a simple, logical approach to explain the operation of the OBD II process and will teach the reader how to quickly spot problems and identify components that are not functioning correctly.

Benefits

- "quick hit" troubleshooting tricks teach readers how to diagnose problems when there is no stored OBDII trouble code, as well as how to handle situations where the trouble code is actually set by a basic mechanical problem rather than a failure of the indicated component
- information is useful for those who wish to expand their capabilities from more basic, mechanical repairs to complex electronics and drivability diagnosis and repair
- focuses on logical troubleshooting that can be done without expensive, complicated test equipment and special tools

CONTENTS

Chapter 1- Introduction, Chapter 2- Evolution of OBD, Chapter 3- OBDII Terminology, Chapter 4- System Operating Protocols, Chapter 5- System Monitors, Chapter 6- Drive Cycles, Chapter 7- Diagnostic Trouble Codes (DTC's), Chapter 8- Diagnostic Routines

256 pp, 8 1/2" x 11", softcover, ©2006

SUPPLEMENTS
Diagnostic Tool CD-ROM
1-4180-1221-1/Part No. 131221
Instructor's Guide 1-4180-1222-X

Professional Automotive Technician Training Series
Thomson Delmar Learning

This self-paced, interactive learning series contains must-have training for today's professional technicians. Each course in the series contains the most up-to-date content, reinforced using engaging graphics, animations, and user interactions. Section review questions, as well as the end of course review, are designed to reinforce user learning and progression.

These thought-provoking products combine theory, diagnosis, and repair information into one easy-to-use training tool! More than 8.5 hours of state-of-the-art instruction per course is provided. Available in both CD-ROM(CBT) and Web-based (WBT) formats, these training tools are ideal for all automotive technicians, and they have been developed to comply with both AICC and SCORM compliance standards.

Benefits:

- regular use of highly engaging animations and interactivity keeps users engaged throughout all the material
- bookmarking technology enables users to track their progress from beginning to end
- periodic progress checks and end-of-section reviews are integrated throughout to ensure the highest level of retention
- certificates of completion can be printed by users achieving a score of 80% or higher on the final course review
- all material is up-to-date to the latest ASE standards

Basic Automotive Service and Maintenance
CBT: ISBN 1-4180-4100-9/Part No. 134100
WBT: ISBN 1-4180-4101-7/Part No. 134101

Engine Performance
CBT: ISBN 1-4180-4239-0/Part No. 134239
WBT: ISBN 1-4180-4240-4/Part No. 134240

Brakes
CBT: ISBN 1-4180-4235-8/Part No. 134235
WBT: ISBN 1-4180-4236-6/Part No. 134236

Suspension and Steering
CBT: ISBN 1-4180-4237-4/Part No. 134237
WBT: ISBN 1-4180-4238-2/Part No. 134238

Electricity and Electronics
CBT: ISBN 1-4180-4241-2/Part No. 134241
WBT: ISBN 1-4180-4242-0/Part No. 134242

©2007

CSAT AUTO Comprehensive Skill Assessment Tool

Comprehensive Skill Assessment Tool (CSAT)—Automotive
Thomson Delmar Learning

The Comprehensive Skill Assessment Tool for Automotive is an online skill gap analysis product, designed to help instructors and trainers implement targeted training programs. Strategic learning areas are measured to account for a technicians knowledge of theory, hands-on/application, and diagnostic knowledge across key automotive skill areas. While the pre-assessment will assist with the identification of areas needing improvement, the combined phases of education and training, and post-assessment allow trainers to track skill level growth and prove out return on investment against their training dollars invested. This is a must have tool for any training prganization.

Benefits

- a low-cost solution benefiting trainers and students
- individual users can take tests online to identify areas of strength and areas needing improvement
- account set-up that enables instructors to assess and track the results of individual users

Call Your Thomson Delmar Learning Sales Rep for Part Numbers & Pricing

Visit **www.skillanalysis.com** to see the latest modules and a free demo!

Improving Fuel Economy

Roy Cox

ISBN 1-4018-8367-2/Part No. 28367

With gasoline prices on the rise, every driver will appreciate this complete, concise guide to fuel economy and how to improve it in today's automobiles! The only publication of its kind, *Fuel Economy* thoroughly discusses ways in which modern automotive systems work together and the interrelationships that affect fuel economy. Author Roy Cox uses simple, straightforward language to raise such important topics as conditions and components that cause specific problems, actions that can be taken to make your car use less fuel, and when to seek professional repairs.

"Drive slower," says Cox "Driving at 50 to 55 mph is measurably more economical. Drivers should also try to avoid braking and drive off-hours to as not to get caught in rush hour traffic jams." Fuel Economy details the beneficial effects of fuel economy maintenance. Cox includes step-by-step instructions for common do-it-yourself procedures. "Vehicle owners should drive less and drive better." Says Cox. "Regular maintenance and unclogging of air filters is important, too." This clever combination of "how it works" and "how to do the maintenance" approach fosters new insight into the key factors affecting fuel economy in cars, and what drivers can do to keep them performing at peak power and economy.

Benefits

- useful advice, such as what fuel choice to make, what motor oil and filter to buy, and what factors to consider when buying replacement parts, makes this a handy resource for new motorists and auto enthusiasts alike
- well illustrated charts provide valuable quick-reference information that can be stored in the glove box and consulted again and again
- useful guidance regarding what maintenance is appropriate to do oneself and what is best left to the professionals helps prevent consumers from getting in "over their heads"
- a beneficial glossary of common automotive acronyms and abbreviations helps readers sort out technical jargon and understand the reasons for recommended repairs and maintenance

CONTENTS

1. Factors that affect fuel economy, 2. Things you can do to make your car use less fuel, 3. When to seek professional repairs, 4. Keep the performance your car was born with: recipe for a long and happy engine life, Glossary of commonly used automotive abbreviations and acronyms

52 pp, 4 1/2" x 7 1/2", softcover, ©2005

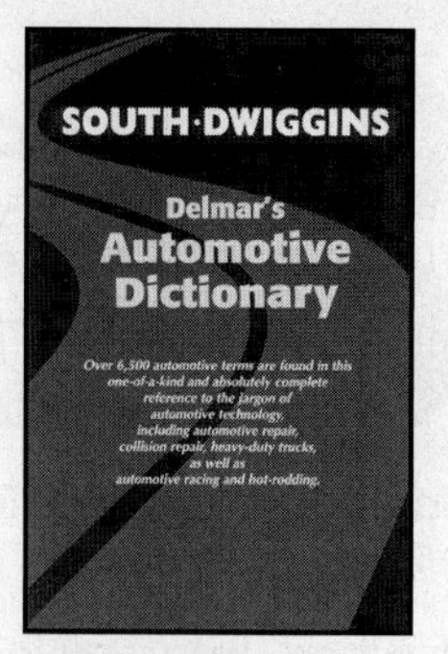

Delmar's Automotive Dictionary

David W. South & Boyce Dwiggins

ISBN 0-8273-7405-4

This handy, ready-reference dictionary provides the automotive engineer, technician, mechanic, student, enthusiast or layperson with a single source for the most up-to-date definitions available of technical, professional and informal terminology used in today's automotive world. It is descriptive and covers the wide scope of terms pertinent to the automotive field. With multiple definitions and aids, and proper pronunciation of terms, this dictionary is a must for all!

Benefits

- over 3000 terms comprehensively covering more than 100 subject areas
- enhanced by a list of acronyms and abbreviations
- up-to-date definitions of today's automotive terminology
- aids for proper pronunciation
- each term has multiple definitions

281 pp, 6" x 9", softcover, ©1997

Math for the Automotive Trade, 3E

John C. Peterson & William deKryger

ISBN 0-8273-6712-0

Math for Automotive Trades, 3E provides excellent examples and problems that reflect technological requirements of workers in automotive technology. The text has three parts: review of basic mathematics skills, math applications to specific automotive situations, and an examination of measurement aspects beginning with angle and linear measurements and ending with an extensive look at measurement tools used in the automotive trade.

345 pp, 8½" x 11", softcover, ©1995

Instructor's Manual **0-8273-6713-9**

Practical Problems in Mathematics for Automotive Technicians, 6E Text:

Larry Sformo, Todd Sformo, and George Moore

ISBN 1-4018-3999-1

Comprehensive and easy-to use, this updated edition covers every type of practical math problem that automotive technicians will face on the job. The subject matter is organized in a knowledge-building format that progresses from the basics of whole number operations into percentages, linear measurements, ratios, and the use of more complex formulas. Complete coverage of fundamentals, as well as more advanced computations make this book suitable for both beginning and advanced technicians. With a special section on graphs, scales, test meters, estimation, and invoices used in the workplace, this book is tailor-made for any automotive course of study!

Benefits

- new section on conversion of measurements makes English-to-metric conversions quick and easy
- proficiency using fractions is encouraged to assist with real world math applications
- step-by-step instructions, diagrams, charts, and examples cultivate problem-solving skills that are crucial to success on the job
- problems proceed from the simple to the more complex so that readers have the ample opportunities to develop their skills and confidence
- solid understanding of the practical applications for each mathematical process fosters a strong foundation in fundamental principles for beginners and advanced technicians

CONTENTS

Whole Numbers. Common Fractions. Decimal Fractions. Percent and Percentage. Measurement. Ratio and Proportion. Powers and Roots. Formulas. Graphs. Invoices.

288 pp, 7 7/8" x 9 1/4", softcover, ©2005

Instructor's Manual **1-4018-4000-0**